## CONNECT . . .

McGraw-Hill *Connect Biology* is a web-based assignment and assessment platform that gives you the means to better connect with your coursework, with your instructors, and with the important concepts that you will need to know for success now and in the future. With *Connect Biology*, instructors can deliver assignments, quizzes, and tests easily online, and you can practice important skills at your own pace and on your own schedule. With *Connect Biology Plus*, you also get 24/7 online access to an eBook—an online edition of the text—to aid you in successfully completing your work, wherever and whenever you choose. Be sure to request *Connect Biology* from your instructor today!

## GO GREEN!

***Go Green. Save Green. McGraw-Hill eBooks.***

Green…it's on everybody's minds these days. It's not only about saving trees; it's also about saving money. Available for a greatly reduced price, McGraw-Hill eBooks are an eco-friendly and cost-saving alternative to the traditional print textbook. So, you do some good for the environment…and you do some good for your wallet.

Visit **www.mhhe.com/ebooks** for details.

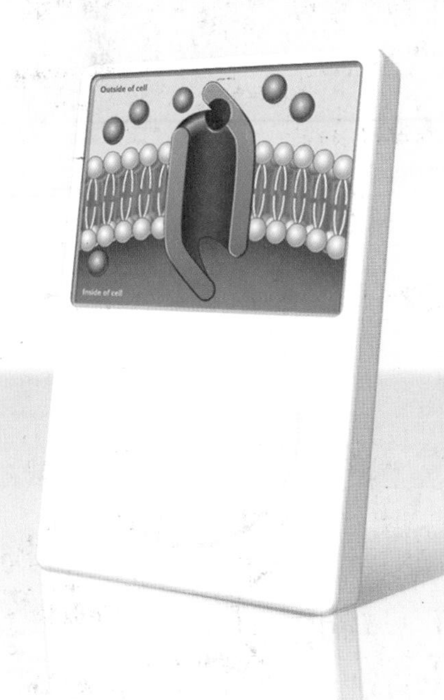

## ON THE GO . . .

Study anywhere, anytime with McGraw-Hill Study on the Fly! You can download video files to play on your computer or portable media players. Study on the Fly makes studying easy and convenient, so you can review content anytime and many times, to get that A!!

# Biology
tenth edition


**Sylvia S. Mader**

*with significant contributions by*

**Andrew Baldwin**
*Mesa Community College*

**Rebecca Roush**
*Sandhills Community College*

**Stephanie Songer**
*North Georgia College and State University*

**Michael Thompson**
*Middle Tennessee State University*



McGraw Hill Higher Education

Boston Burr Ridge, IL Dubuque, IA New York San Francisco St. Louis
Bangkok Bogotá Caracas Kuala Lampur Lisbon London Madrid Mexico City
Milan Montreal New Delhi Santiago Seoul Singapore Sydney Taipai Toronto

The McGraw·Hill Companies

BIOLOGY, TENTH EDITION

Published by McGraw-Hill, a business unit of The McGraw-Hill Companies, Inc., 1221 Avenue of the Americas, New York, NY 10020. 

 This book is printed on recycled, acid-free paper containing 10% postconsumer waste.

3 4 5 6 7 8 9 0 WDQ/WDQ 10

ISBN 978–0–07–352543–3
MHID 0–07–352543–X

Publisher: *Janice Roerig-Blong*
Executive Editor: *Michael S. Hackett*
Director of Development: *Kristine Tibbetts*
Senior Developmental Editor: *Lisa A. Bruflodt*
Marketing Manager: *Tamara Maury*
Senior Project Manager: *Jayne L. Klein*
Lead Production Supervisor: *Sandy Ludovissy*
Senior Media Project Manager: *Jodi K. Banowetz*
Senior Designer: *David W. Hash*
Cover/Interior Designer: *Christopher Reese*
(USE) Cover Image: *Blue-Footed Booby, Galápagos Islands, ©Michael Melford/Getty Images*
Senior Photo Research Coordinator: *Lori Hancock*
Photo Research: *Evelyn Jo Johnson*
Supplement Producer: *Mary Jane Lampe*
Compositor: *Electronic Publishing Services Inc., NYC*
Typeface: *10/12 Palatino*
Printer: *Worldcolor Dubuque, IA*

The credits section for this book begins on page C-1 and is considered an extension of the copyright page.

**Library of Congress Cataloging-in-Publication Data**

Mader, Sylvia S.
Biology / Sylvia S. Mader. -- 10th ed.
p. cm.
Includes index.
ISBN 978-0-07-352543-3 --- ISBN 0-07-352543-X (hard copy : alk. paper) 1. Biology. I. Title.
QH308.2.M23 2010
570--dc22

2008034142

www.mhhe.com

# BRIEF CONTENTS

# PREFACE

The mission of my text, *Biology*, has always been to give students an understanding of biological concepts and a working knowledge of the scientific process. If one understands the concepts of biology and the methodology of science, they can be used to understand the particulars of new ideas or a system on any scale from the cell to the biosphere. By now, we are well into the twenty-first century, and the field of biology has been flooded with exciting new discoveries and insights way beyond our predictions even a few short years ago. It is our task, as instructors, to make these findings available to our students so they will have the background to keep up with the many discoveries still to come. At the same time, we must provide students with a firm foundation in those core principles on which biology is founded. This means that the tenth edition of *Biology* is both new and old at the same time. With this edition, instructors will be confident that they are "up to date," while still teaching the fundamental concepts of biology in a way that allows students to apply them in new and different ways. In this edition you will find:

- *Increased Evolutionary Coverage*
- *Currency of Coverage*
- *Media Integration*

## Birth of *Biology*

I am an instructor of biology as are the contributors that have lent their several talents to this edition of *Biology*. Collectively, we have taught students for many years from the community college to the university level. We are all dedicated to the desire that students develop a particular view of the world—a biological view. When I wrote the first edition of *Biology,* it seemed to me that a thorough grounding in biological principles would lead to an appreciation of the structure and function of individual organisms, how they evolved, and how they interact in the biosphere. This caused me to use the levels of biological organization as my guide—thus, this edition, like the previous editions, begins with chemistry and ends with the biosphere.

Students need to be aware that our knowledge of biology is built on theories that have survived the rigors of scientific testing. The first chapter explains the process of science and thoroughly reviews examples of how scientists come to conclusions. Throughout the text, biologists are introduced, and their experiments are explained. An appreciation of how science progresses should lead to the perception that, without the scientific process, biology could not exist.

## Evolution of *Biology*

While I have always guided the development of each new edition of *Biology*, many instructors have lent their talents to ensuring its increasing success. I give my utmost thanks to all the reviewers and contributors that have been so generous with their time and expertise. This edition, I want to particularly thank Andrew Baldwin, of Mesa Community College, who revised the ecology chapters; Rebecca Roush, of Sandhills Community College, for her work on Part VI; Michael Thompson, of Middle Tennessee State University, who did the first chapter and the genetics chapters; and Stephanie Songer, of North Georgia College and State University, who revised Part IV and many chapters in Part V. My involvement ensured that each of these chapters, along with the chapters I revised, are written and illustrated in the familiar Mader style.

The brilliance of the illustrations and the eye-catching paging of *Biology* are due to the talented staff of EPS (Electronic Publishing Services Inc.), who took my first attempts and altered them to produce the most detailed, refined, and pedagogically sound presentations ever developed for an introductory biology book.

## The Learning System

Mader books excel in pedagogy, and *Biology* is consistent with the usual high standard. Pages xii–xv of this preface review "The Learning System" of *Biology*. As explained, each part opening page introduces that part in a new engaging way that explains the rationale of that part. The chapter opening page lists the key concepts under the major sections for that chapter. In this way, students are given an overview of the chapter and its concepts. The opening vignette captures student interest and encourages them to begin their study of the chapter. New to this edition, major sections end with "Check Your Progress" questions designed to foster confidence as they proceed through the chapter. "Connecting the Concepts" at the end of the chapter ties the concepts of this chapter to those in other chapters. The end matter gives students an opportunity to review the chapter and test themselves on how well they understand the concepts.

The Mader writing style is well known for its clarity and a simplicity of style that appeals to students because it meets them where they are and assists them in achieving mastery of the concept. Concepts are only grasped if a student comes away with "take-home messages." Once students have internalized the fundamental concepts of biology, they will have developed a biological view of the world that is essential in the twenty-first century.

## Changes in *Biology*, Tenth Edition

The tenth edition builds on the visual appeal of the previous edition. New illustrations have been developed that are just as stunning as those prepared for the ninth edition, and many new photographs and micrographs have been added.

*Biology* has a new table of contents that consolidates chapters so that the book is shorter by some forty pages compared to the last edition. No individual chapter is overly long, however. In Part II, certain material from Chapter 12 was moved into Chapter 10, *Meiosis and Sexual Reproduction* and Chapter 11, *Mendelian Patterns of Inheritance*. In Part III, *Speciation and Macroevolution* is a much needed new chapter. In Part VI, the two invertebrate evolution chapters from the previous edition have become Chapter 28, *Invertebrates*. In Part VIII, Chapter 45, *Community and Ecosystem Ecology* is a consolidation of two chapters from the previous edition.

I believe you will be interested in knowing about these chapters that demonstrate the quality of *Biology*, Tenth Edition:

- Chapter 1, *A View of Life*, was revised to have a new section: "Evolution, the Unifying Concept of *Biology*." This section presents basic evolutionary principles and contains a depiction of the Tree of Life, which introduces the three domains of life and the various types of eukaryotes. Prokaryotes and eukaryotes are also pictorially displayed.

### *Part I The Cell*

- Chapter 5, *Membrane Structure and Function*, introduces the concept of cell signaling. New to this edition, the plasma membrane art now depicts the extracellular matrix (see Fig. 5.1), which has a role in cell signaling—a topic that is further explored in the Science Focus, "How Cells Talk to One Another."
- Chapter 8, *Cellular Respiration*, begins with a new section that now emphasizes that cellular respiration is the reason we eat and breathe (see Figure 8.1). The fermentation section in this edition precedes the events that occur in mitochondria and is enhanced by a new Science Focus box, "Fermentation Helps Produce Numerous Food Products." The chapter now ends with a comparison of photosynthesis to cellular respiration (see Fig. 8.12).

## Overview of Changes to *Biology*, Tenth Edition

### VISUALS

The brilliant visuals program of the previous edition is enhanced even more by the addition of many new micrographs and innovative page layouts.

### CELLULAR BIOLOGY

Cell signaling receives expanded coverage as a mechanism of cellular metabolism and cell division control.

### GENETICS

Reorganization of the genetics chapters results in increased genome coverage, including the role of small RNA molecules in regulation.

### SYSTEMATICS

Cladistics is better explained, and new evolutionary trees are presented for protists, plants, and animals.

### EVOLUTION

A new chapter, *Speciation and Macroevolution*, points to the possible role of Hox genes in punctuated evolution.

### PLANT EVOLUTION

A reorganization of Chapter 23 better describes the evolution of plants from an aquatic green algal ancestor.

### ANIMAL EVOLUTION

Reorganization of Part VI results in two new animal diversity chapters: the invertebrates and the vertebrates.

## Part II Genetic Basis of Life

- Chapter 9, *The Cell Cycle and Cellular Reproduction*, builds on the topic of cell signaling that was introduced in Chapter 5. Cell signaling is the means by which the cell cycle, and, therefore, cell division is regulated. A new Science Focus box shows how the $G_1$ checkpoint is highly regulated by cell signaling, and Figure 9.8 dramatically illustrates how a breakdown in cell cycle regulation may contribute to cancer.
- Chapter 13, *Regulation of Gene Activity*, is an excellent chapter that instructors will not want to overlook because it explains how humans can make do with far fewer protein-coding genes than have been discovered by DNA sequencing of our genome. The chapter is updated by continued emphasis on chromatin structure, many references to the regulatory role of RNA molecules including a new Science Focus box, "Alternative mRNA Splicing in Disease."
- Chapter 14, *Biotechnology and Genomics*, has an expanded section on genomics. Much of chromatin consists of introns and intergenic sequences which may have important functions still to be discovered (see Fig. 14.8). Molecular geneticists are seeking a new definition of a gene that can apply to both protein-coding and non-protein-coding sequences. The chapter also discusses genomic diversity. The new Science Focus box, "DNA Microarray Technology," explains how this technique is now being applied to identify genes involved in health and disease. Another new Science Focus box, "Copy Number Variations," gives another example of genetic diversity within the population and its relationship to health and disease.

## Part III Evolution

- Chapter 16, *How Populations Evolve*, is an exciting new chapter that begins with an introduction based on community acquired MRSA. This chapter is also enhanced by new figures: an example of genetic diversity (see Fig. 16.1), the gene pool (see Fig. 16.2), microevolution (see Fig. 16.3), and a natural selection experiment (see Fig. 16.10) are included. Also, sexual selection is now included in this chapter.
- Chapter 17, *Speciation and Macroevolution*, is new to this edition. This chapter begins by describing species concepts, and examples of both allopatric and sympatric speciation are given. The concepts of gradualistic and punctuated equilibrium are discussed with reference to the Burgess Shale as an example of rapid evolution to produce many species, and Hox genes are offered as a possible mechanism to bring it about.

## Part IV Microbiology and Evolution

- Chapter 21, *Protist Evolution and Diversity*, has been revised because protist classification has undergone dramatic changes in recent years. This chapter is reorganized accordingly, but the biological and ecological relevance of each type of protist is still discussed.

## Part V Plant Evolution and Biology

- Chapter 23, *Plant Evolution and Diversity*, employs a new evolutionary tree based in part on molecular data. Land plants and stoneworts, which are charophytes, share a common green algal ancestor. All land plants protect the embryo, and thereafter each of five innovations can be associated with a particular group of land plants.

### *Part VI Animal Evolution and Diversity*

- Chapter 28, *Invertebrates*, has been thoroughly updated and revised in this edition. The chapter better defines an animal and explains the colonial flagellate hypothesis on the origin of animals. The organization of this chapter follows a new evolutionary tree based on molecular and developmental data; the biology of each group is discussed as before.
- Chapter 29, *Vertebrates*, has been reorganized, and each vertebrate group is now a major section. In keeping with modern findings, birds are considered reptiles. Each section begins with a listing of characteristics for that group and is followed by a discussion of the evolution and then the diversity of that group.

### *Part VII Comparative Animal Biology*

- Chapter 33, *Lymph Transport and Immunity*, has been reorganized and revised so that both nonspecific defense (innate immunity) and specific defense (acquired immunity) have their own major section. All concepts regarding antibodies have been brought together in the specific defense section. Immunity side effects has new illustrations; Cytokines and Cancer Therapy is a new subsection.
- Chapter 35, *Respiratory Systems*, is much improved in this edition from an increased emphasis on diversity to a better description of the human respiratory tract and transport of gases (see Figs. 35.3, 35.6, and 35.12). This chapter now ends with a dramatic photo of emphysema and lung cancer (see Fig. 35.15). "Connecting the Concepts" emphasizes the contribution of the respiratory system to homeostasis by description and art.
- Chapter 41, *Reproductive Systems*, now begins with a revised comparative section that includes more photos. An illustration depicting contraceptives replaces a table, and there is a new Health Focus, "Preimplantation Genetic Diagnosis." Sexually transmitted diseases have been updated to reflect current statistics. A new bioethical issue concerns the use of fertility drugs.

### *Part VIII Behavior and Ecology*

- Chapter 43, *Behavioral Ecology*, has an evolutionary emphasis culminating in a new section entitled "Behaviors that Increase Fitness" in which several types of societal interactions are explored as a means to increase representation of genes in the next generation. Orientation and migratory behavior and cognitive learning are ways of learning not discussed previously.
- Chapter 45, *Community and Ecosystem Ecology*, is a combined chapter that allows instructors to cover the basics of ecology in one chapter. A discussion of symbiotic relationships and ecological succession precede the concepts of chemical cycling and energy flow in ecosystems.

## About the Author

Dr. Sylvia S. Mader has authored several nationally recognized biology texts published by McGraw-Hill. Educated at Bryn Mawr College, Harvard University, Tufts University, and Nova Southeastern University, she holds degrees in both Biology and Education. Over the years, she has taught at the University of Massachusetts–Lowell, Massachusetts Bay Community College, Suffolk University, and Nathan Matthew Seminars. Her ability to reach out to science-shy students led to the writing of her first text, *Inquiry into Life*, which is now in its twelfth edition. Highly acclaimed for her crisp and entertaining writing style, her books have become models for others who write in the field of biology.

Although her writing schedule is always quite demanding, Dr. Mader enjoys taking time to visit and explore the various ecosystems of the biosphere. Her several trips to the Florida Everglades and Caribbean coral reefs resulted in talks she has given to various groups around the country. She has visited the tundra in Alaska, the taiga in the Canadian Rockies, the Sonoran Desert in Arizona, and tropical rain forests in South America and Australia. A photo safari to the Serengeti in Kenya resulted in a number of photographs for her texts. She was thrilled to think of walking in Darwin's steps when she journeyed to the Galápagos Islands with a group of biology educators. Dr. Mader was also a member of a group of biology educators who traveled to China to meet with their Chinese counterparts and exchange ideas about the teaching of modern-day biology.

*For My Children*

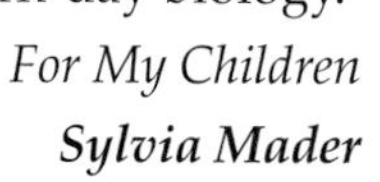

***Sylvia Mader***

# Guided Tour

## Increased Evolutionary Coverage

**NEW CHAPTERS**

16 (*How Populations Evolve*) and 17 (*Speciation and Macroevolution*) highlight new evolutionary coverage.

**FIGURE 16.3 Microevolution.**

Microevolution has occurred when there is a change in gene pool frequencies—in this case, due to natural selection. On the left, birds cannot see light-colored peppered moths, *Biston betularia*, against light-colored vegetation—and, therefore, light-colored moths are more frequent in the population. On the right, after vegetation has been darkened due to pollution, birds are less likely to see dark-colored moths against dark vegetation, and dark moths are more frequent in the population.

The **Hardy-Weinberg principle** states that an equilibrium of gene pool frequencies, calculated by using the bi-
ression, will remain in effect in each succeeding
of a sexually reproducing population, as long as
ons are met:

utations: Allele changes do not occur, or changes
direction are balanced by changes in the
ite direction.
ne flow: Migration of alleles into or out of the
ation does not occur.
m mating: Individuals pair by chance, not
ing to their genotypes or phenotypes.
netic drift: The population is very large, and
es in allele frequencies due to chance alone are
ificant.
ection: Selective forces do not favor one
pe over another.

l life, these conditions are rarely, if ever, met,
frequencies in the gene pool of a population
from one generation to the next. Therefore,
has occurred. The significance of the Hardy-
principle is that it tells us what factors cause
-those that violate the conditions listed. Micro-
an be detected by noting any deviation from a
nberg equilibrium of allele frequencies in the
f a population.
nge in allele frequencies may result in a change
e frequencies. Our calculation of gene pool frequencies in Figure 16.3 assumes that industrial melanism may have started but was not fully in force yet. **Industrial melanism** refers to a darkening of moths once industrialization has begun in a country. Prior to the Industrial Revolution in Great Britain, light-colored peppered moths living on the light-colored, unpolluted vegetation, were more common than dark-colored peppered moths. When dark-colored moths landed on light vegetation, they were seen and eaten by predators. In Figure 16.3, *left*, we suppose that only 36% of the population were dark-colored, while 64% were light-colored. With the advent of industry and an increase in pollution, the vegetation was stained darker. Now, light-colored moths were easy prey for predators. Figure 16.3, *right*, assumes that the gene pool frequencies switched, and now the dark-colored moths are 64% of the population. Can you calculate the change in gene pool frequencies using Figure 16.2 as a guide?

Just before the Clean Air legislation in the mid-1950s, the numbers of dark-colored moths exceeded a frequency of 80% in some populations. After the legislation, a dramatic reversal in the ratio of light-colored moths to dark-colored moths occurred once again as light-colored moths became more and more frequent. Aside from showing that natural selection can occur within a short period of time, our example shows that a change in gene pool frequencies does occur as microevolution occurs. Recall that microevolution occurs below the species level.

### Causes of Microevolution

The list of conditions for a Hardy-Weinberg equilibrium implies that the opposite conditions can cause evolutionary

# 16 How Populations Evolve

*concepts*

**16.1 POPULATION GENETICS**

- Genetic diversity is a necessity for microevolution to occur, and today investigators are interested in DNA sequence differences between individuals. It might be possible to associate particular variations with illnesses. 284
- The Hardy-Weinberg principle provides a way to know if a population has evolved. Allele frequency changes in the next generation signify that microevolution has occurred. 285–86
- Microevolution will occur unless five conditions are met: no mutations, no gene flow, mating is random, no genetic drift, and no selection of a particular trait. 286–88

**16.2 NATURAL SELECTION**

- A change in phenotype frequencies occurs if a population has undergone stabilizing selection, directional selection, or disruptive selection. 289–90
- Sexual selection fostered by male competition and female choice is also a type of natural selection because it influences reproductive success. 291–92

**16.3 MAINTENANCE OF DIVERSITY**

- Genetic diversity is maintained within a population; for example, by the diploid genotype and also when the heterozygote is the most adaptive genotype. 294–95

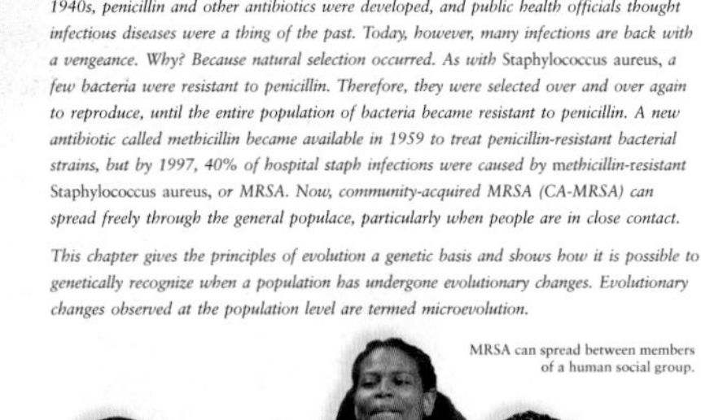

*When your grandparents were young, infectious diseases, such as tuberculosis, pneumonia, and syphilis, killed thousands of people every year. Then in the 1940s, penicillin and other antibiotics were developed, and public health officials thought infectious diseases were a thing of the past. Today, however, many infections are back with a vengeance. Why? Because natural selection occurred. As with* Staphylococcus aureus, *a few bacteria were resistant to penicillin. Therefore, they were selected over and over again to reproduce, until the entire population of bacteria became resistant to penicillin. A new antibiotic called methicillin became available in 1959 to treat penicillin-resistant bacterial strains, but by 1997, 40% of hospital staph infections were caused by methicillin-resistant* Staphylococcus aureus, *or MRSA. Now, community-acquired MRSA (CA-MRSA) can spread freely through the general populace, particularly when people are in close contact.*

*This chapter gives the principles of evolution a genetic basis and shows how it is possible to genetically recognize when a population has undergone evolutionary changes. Evolutionary changes observed at the population level are termed microevolution.*

MRSA can spread between members of a human social group.

# 17 Speciation and Macroevolution

*concepts*

**17.1 SEPARATION OF THE SPECIES**

- Species can be recognized by their traits, by reproductive isolation, and by DNA differences. 300–301
- Mechanisms that prevent reproduction between species are divided into those that prevent attempts at reproduction and those that prevent development of an offspring or cause the offspring to be infertile. 302–3

**17.2 MODES OF SPECIATION**

- Allopatric speciation occurs when a new species evolves in geographic isolation from an ancestral species. 304–5
- Adaptive radiation, during which a single species gives rise to a number of different species, is an example of allopatric speciation. 306
- Sympatric speciation occurs when a new species evolves without geographic isolation. 307
- The Burgess Shale gives us a glimpse of marine life some 540 million years ago. 308–9

**17.3 PRINCIPLES OF MACROEVOLUTION**

- Macroevolution is phenotypic changes at the species and higher levels of taxonomy up to a domain. 310
- The tempo of speciation can be rapid or slow. Developmental genes provide a mechanism for rapid speciation. 311–12
- Macroevolution involves speciation, diversification, and extinction, as observed in the evolution of the horse. Macroevolution is not goal directed and, instead, represents adaptation to varied environments through time. 313–14

*The immense liger featured here is an offspring of a lion and a tiger, two normally reproductively isolated animal species. Ligers are the largest of all known cats, measuring up to 12 feet tall when standing on their hind legs and weighing as much as 1,000 lbs. Their coat color is usually tan with tiger stripes on the back and hindquarters and lion cub spots on the abdomen. A liger can produce both the "chuff" sound of a tiger and the roar of a lion. Male ligers may have a modest lion mane or no mane at all. Most ligers like to be near water and love to swim. Generally, ligers have a gentle disposition; however, considering their size and heritage, handlers should be extremely careful. By what criteria could a liger be considered a new species? Only if they, in turn, were reproductively isolated and only mated with ligers. In this chapter, we will explore the definition of a species and how species arise. In so doing, we will begin our discussion of macroevolution, which we continue in the next chapter.*

This liger is a hybrid because it has a lion father and a tiger mother.

## 1.2 Evolution, the Unifiying Concept of Biology

Despite diversity in form, function, and lifestyle, organisms share the same basic characteristics. As mentioned, they are all composed of cells organized in a similar manner. Their genes are composed of DNA, and they carry out the same metabolic reactions to acquire energy and maintain their organization. The unity of living things suggests that they are descended from a common ancestor—the first cell or cells.

An evolutionary tree is like a family tree (Fig. 1.5). Just as a family tree shows how a group of people have descended from one couple, an evolutionary tree traces the ancestry of life on Earth to a common ancestor. One couple can have diverse children, and likewise a population can be a common ancestor to several other groups, each adapted to a particular set of environmental conditions. In this way, over time, diverse life-forms have arisen. Evolution may be considered the unifying concept of biology because it explains so many aspects of biology, including how living organisms arose from a single ancestor.

### Organizing Diversity

Because life is so diverse, it is helpful to group organisms into categories. **Taxonomy** [Gk. *tasso*, arrange, and *nomos*, usage] is the discipline of identifying and grouping organisms according to certain rules. Taxonomy makes sense out of the bewildering variety of life on Earth and is meant to provide valuable insight into evolution. As more is learned about living things, including the evolutionary relationships between species, taxonomy changes. DNA technology is now being used to revise current information and to discover previously unknown relationships between organisms.

Several of the basic classification categories, or *taxa*, going from least inclusive to most inclusive, are **species, genus, family, order, class, phylum, kingdom,** and **domain** (Table 1.1). The least inclusive category, species [L. *species*, model, kind], is defined as a group of interbreeding individuals. Each successive classification category above species contains more types of organisms than the preceding one. Species placed within one genus share many specific characteristics and are the most closely related, while species placed in the same kingdom share only general characteristics with one another. For example, all species in the genus *Pisum* look pretty much the same—that is, like pea plants—but species in the plant kingdom can be quite varied, as is evident when we compare grasses to trees. Species placed in different domains are the most distantly related.

**TABLE 1.1**

**Levels of Classification**

| *Category* | *Human* | *Corn* |
|---|---|---|
| Domain | Eukarya | Eukarya |
| Kingdom | Animalia | Plantae |
| Phylum | Chordata | Anthophyta |
| Class | Mammalia | Monocotyledones |
| Order | Primates | Commelinales |
| Family | Hominidae | Poaceae |
| Genus | *Homo* | *Zea* |
| Species* | *H. sapiens* | *Z. mays* |

*To specify an organism, you must use the full binomial name, such as *Homo sapiens*.

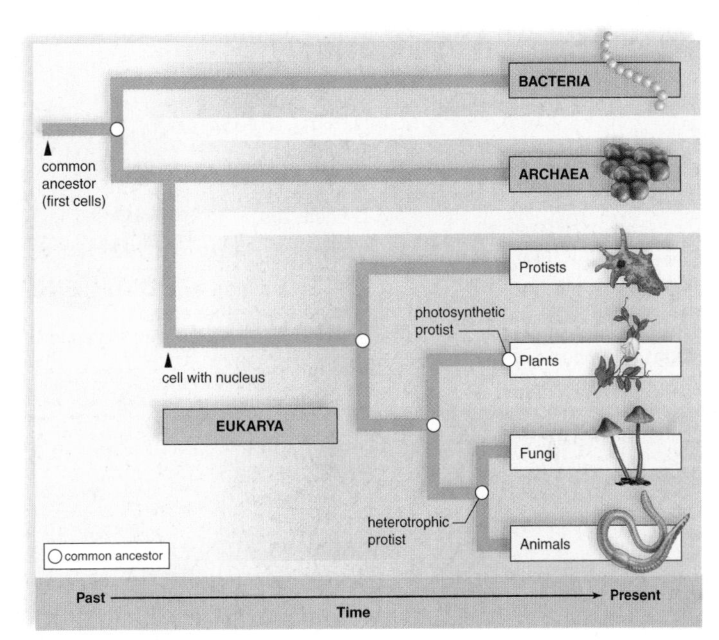


**FIGURE 1.5 Evolutionary tree of life.**

As existing organisms change over time, they give rise to new species. Evolutionary studies show that all living organisms arose from a common ancestor about 4 billion years ago. Domain Archaea includes prokaryotes capable of surviving in extreme environments, such as those with high salinity and temperature and low pH. Domain Bacteria includes metabolically diverse prokaryotes widely distributed in various environments. The domain Eukarya includes both unicellular and multicellular organisms that possess a membrane-bounded nucleus.

**NEW SECTION**

Chapter 1 includes a new section that covers basic evolutionary principles and a new depiction of the Tree of Life which introduces the three domains of life.

# A Stunning Visuals Program

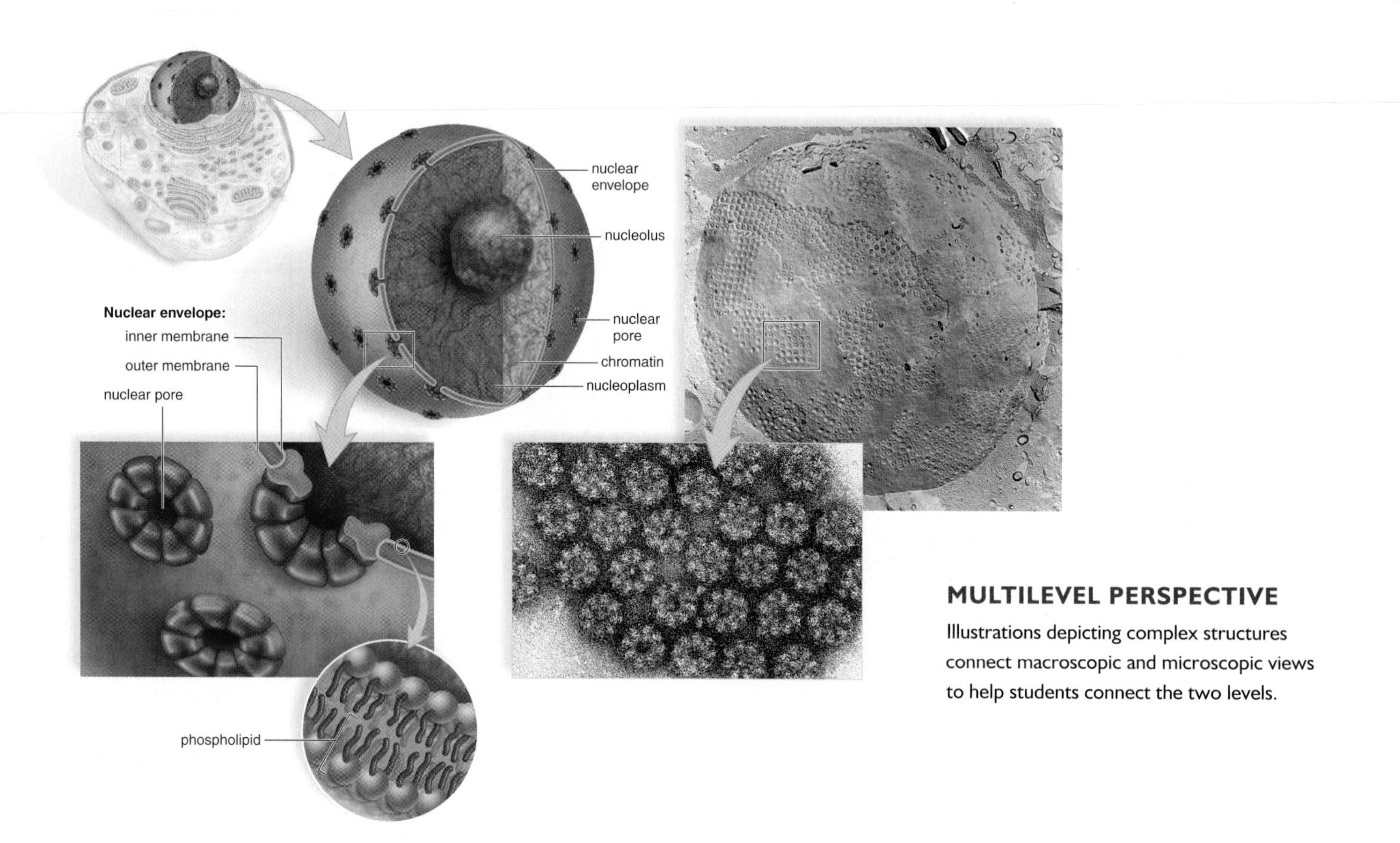


## MULTILEVEL PERSPECTIVE

Illustrations depicting complex structures connect macroscopic and microscopic views to help students connect the two levels.

## COMBINATION ART

Drawings of structures are often paired with micrographs to enhance visualization.

**Loose fibrous connective tissue**
- has space between components.
- occurs beneath skin and most epithelial layers.
- functions in support and binds organs.

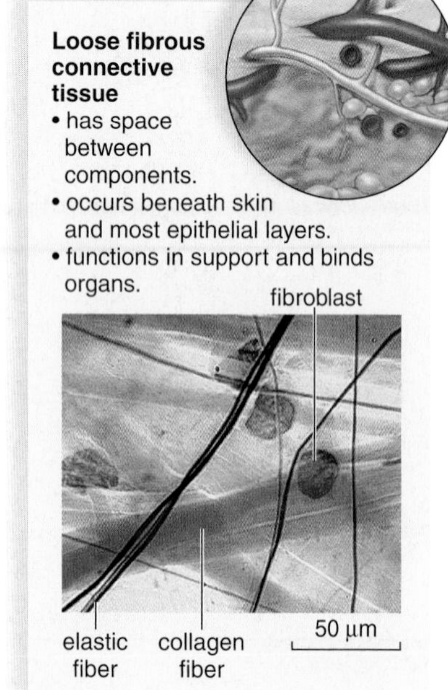


a.

**Adipose tissue**
- cells are filled with fat.
- occurs beneath skin, around heart and other organs.
- functions in insulation, stores fat.

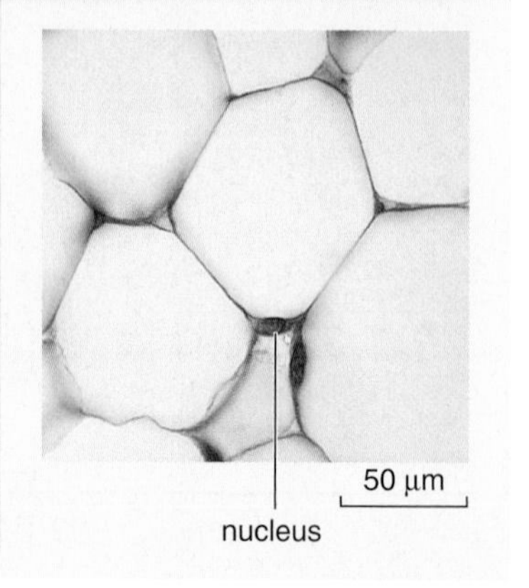


b.

**Dense fibrous connective tissue**
- has collagenous fibers closely packed.
- in dermis of skin, tendons, ligaments.
- functions in support.

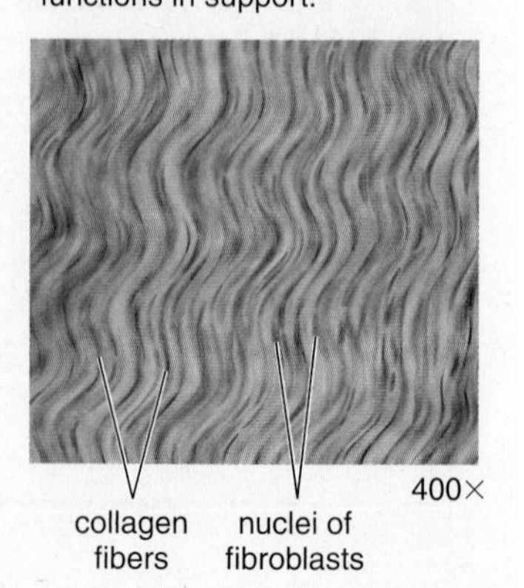


c.

## PROCESS FIGURES

These figures break down processes into a series of smaller steps and organize them in an easy-to-follow format.

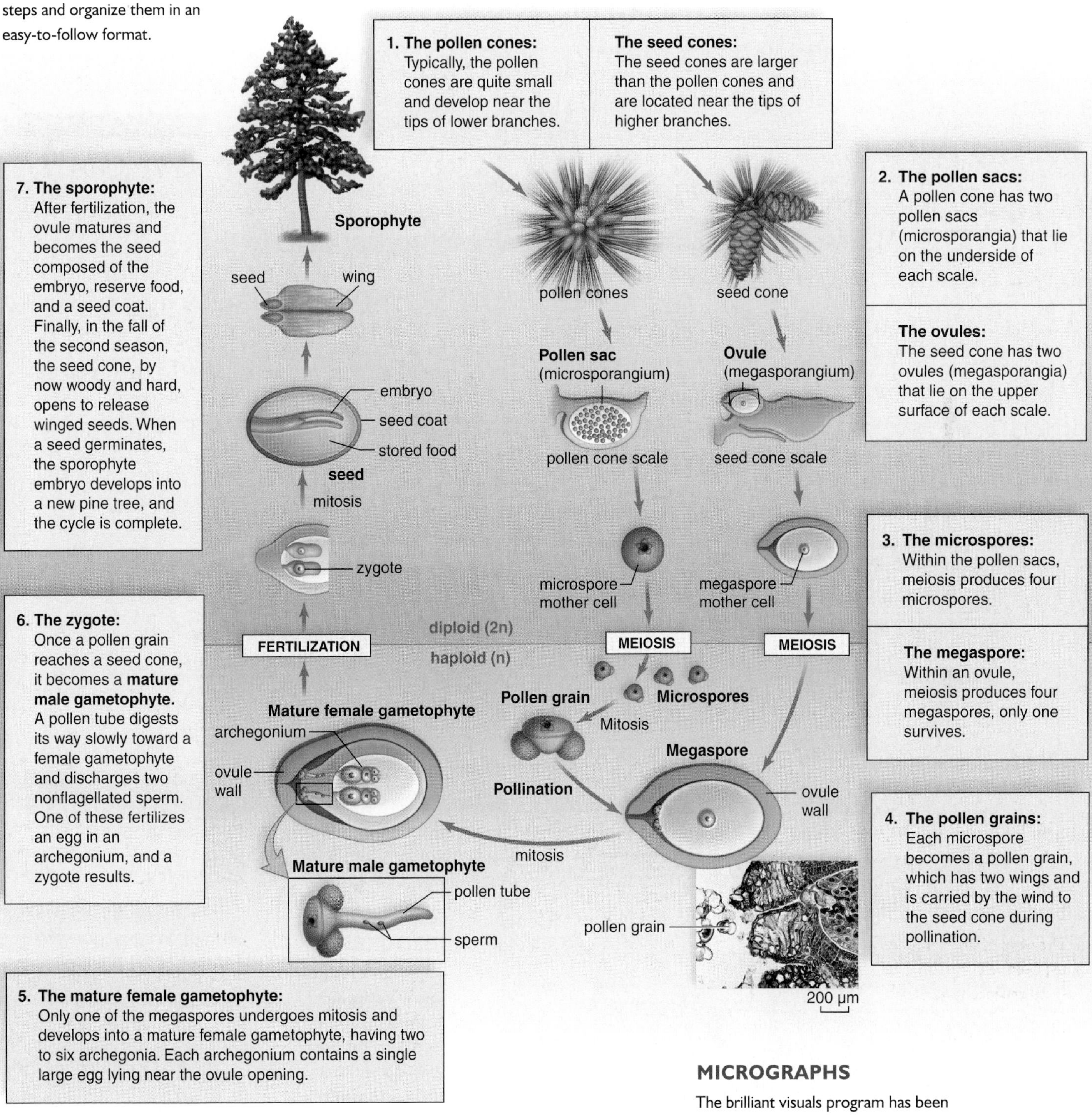

**1. The pollen cones:** Typically, the pollen cones are quite small and develop near the tips of lower branches.

**The seed cones:** The seed cones are larger than the pollen cones and are located near the tips of higher branches.

**2. The pollen sacs:** A pollen cone has two pollen sacs (microsporangia) that lie on the underside of each scale.

**The ovules:** The seed cone has two ovules (megasporangia) that lie on the upper surface of each scale.

**3. The microspores:** Within the pollen sacs, meiosis produces four microspores.

**The megaspore:** Within an ovule, meiosis produces four megaspores, only one survives.

**4. The pollen grains:** Each microspore becomes a pollen grain, which has two wings and is carried by the wind to the seed cone during pollination.

**5. The mature female gametophyte:** Only one of the megaspores undergoes mitosis and develops into a mature female gametophyte, having two to six archegonia. Each archegonium contains a single large egg lying near the ovule opening.

**6. The zygote:** Once a pollen grain reaches a seed cone, it becomes a **mature male gametophyte.** A pollen tube digests its way slowly toward a female gametophyte and discharges two nonflagellated sperm. One of these fertilizes an egg in an archegonium, and a zygote results.

**7. The sporophyte:** After fertilization, the ovule matures and becomes the seed composed of the embryo, reserve food, and a seed coat. Finally, in the fall of the second season, the seed cone, by now woody and hard, opens to release winged seeds. When a seed germinates, the sporophyte embryo develops into a new pine tree, and the cycle is complete.

Sporophyte
seed
wing
embryo
seed coat
stored food
seed
mitosis
zygote
pollen cones
seed cone
Pollen sac (microsporangium)
Ovule (megasporangium)
pollen cone scale
seed cone scale
microspore mother cell
megaspore mother cell
diploid (2n)
haploid (n)
FERTILIZATION
MEIOSIS
MEIOSIS
Pollen grain
Microspores
Mitosis
Megaspore
Mature female gametophyte
archegonium
ovule wall
Pollination
ovule wall
mitosis
Mature male gametophyte
pollen tube
sperm
pollen grain
200 μm

## MICROGRAPHS

The brilliant visuals program has been enhanced by many new micrographs.

# The Learning System

## Proven Pedagogical Features That Will Facilitate Your Understanding of Biology

### CHAPTER CONCEPTS

The chapter begins with an integrated outline that numbers the major topics of the chapter and lists the concepts for each topic.

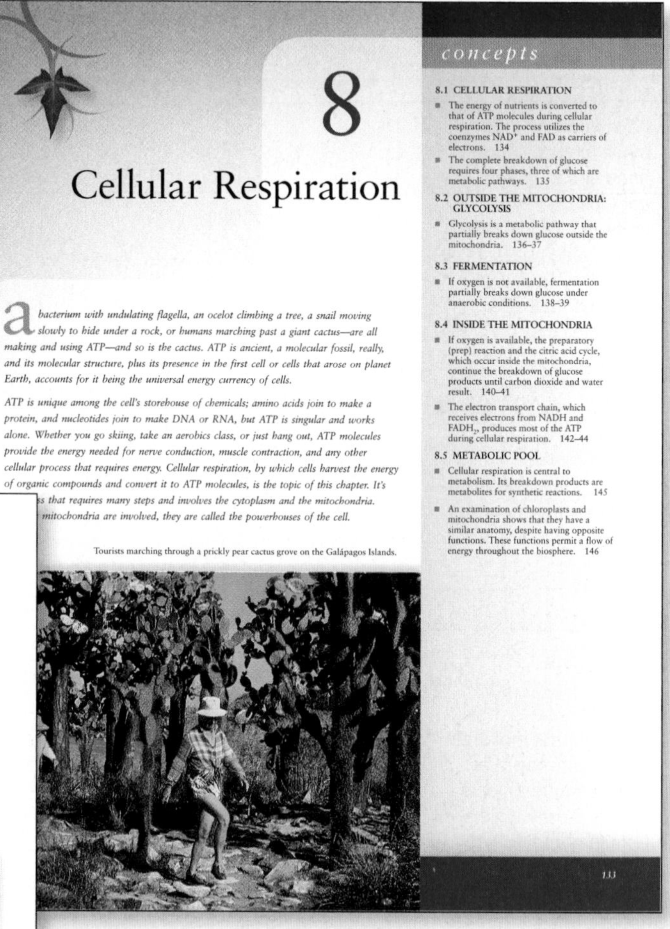

8

Cellular Respiration

a bacterium with undulating flagella, an ocelot climbing a tree, a snail moving slowly to hide under a rock, or humans marching past a giant cactus—are all making and using ATP—and so is the cactus. ATP is ancient, a molecular fossil, really, and its molecular structure, plus its presence in the first cell or cells that arose on planet Earth, accounts for it being the universal energy currency of cells.

ATP is unique among the cell's storehouse of chemicals; amino acids join to make a protein, and nucleotides join to make DNA or RNA, but ATP is singular and works alone. Whether you go skiing, take an aerobics class, or just hang out, ATP molecules provide the energy needed for nerve conduction, muscle contraction, and any other cellular process that requires energy. Cellular respiration, by which cells harvest the energy of organic compounds and convert it to ATP molecules, is the topic of this chapter. It's ...s that requires many steps and involves the cytoplasm and the mitochondria. ... mitochondria are involved, they are called the powerhouses of the cell.

Tourists marching through a prickly pear cactus grove on the Galápagos Islands.

concepts

8.1 CELLULAR RESPIRATION
- The energy of nutrients is converted to that of ATP molecules during cellular respiration. The process utilizes the coenzymes $NAD^+$ and FAD as carriers of electrons. 134
- The complete breakdown of glucose requires four phases, three of which are metabolic pathways. 135

8.2 OUTSIDE THE MITOCHONDRIA: GLYCOLYSIS
- Glycolysis is a metabolic pathway that partially breaks down glucose outside the mitochondria. 136–37

8.3 FERMENTATION
- If oxygen is not available, fermentation partially breaks down glucose under anaerobic conditions. 138–39

8.4 INSIDE THE MITOCHONDRIA
- If oxygen is available, the preparatory (prep) reaction and the citric acid cycle, which occur inside the mitochondria, continue the breakdown of glucose products until carbon dioxide and water result. 140–41
- The electron transport chain, which receives electrons from NADH and $FADH_2$, produces most of the ATP during cellular respiration. 142–44

8.5 METABOLIC POOL
- Cellular respiration is central to metabolism. Its breakdown products are metabolites for synthetic reactions. 145
- An examination of chloroplasts and mitochondria shows that they have a similar anatomy, despite having opposite functions. These functions permit a flow of energy throughout the biosphere. 146

133

CHAPTER 8 Cellular Respiration 135

**Phases of Cellular Respiration**

Cellular respiration involves four phases: glycolysis, the preparatory reaction, the citric acid cycle, and the electron transport chain (Fig. 8.2). Glycolysis takes place outside the mitochondria and does not require the presence of oxygen. Therefore, glycolysis is **anaerobic.** The other phases of cellular respiration take place inside the mitochondria, where oxygen is the final acceptor of electrons. Because they require oxygen, these phases are called **aerobic.**

During these phases, notice where $CO_2$ and $H_2O$, the end products of cellular respiration, are produced.

- **Glycolysis** [Gk. *glycos*, sugar, and *lysis*, splitting] is the breakdown of glucose to two molecules of pyruvate. Oxidation results in NADH and provides enough energy for the net gain of two ATP molecules.
- The **preparatory (prep) reaction** takes place in the matrix of the mitochondrion. Pyruvate is broken down to a 2-carbon acetyl group, and $CO_2$ is released. Since glycolysis ends with two molecules of pyruvate, the prep reaction occurs twice per glucose molecule.
- The **citric acid cycle** also takes place in the matrix of the mitochondrion. As oxidation occurs, NADH and $FADH_2$ results, and more $CO_2$ is released. The citric acid cycle is able to produce one ATP per turn. Because two acetyl groups enter the cycle per glucose molecule, the cycle turns twice.
- The **electron transport chain (ETC)** is a series of carriers on the cristae of the mitochondria. NADH and $FADH_2$ give up electrons to the chain. Energy is released and captured as the electrons move from a higher-energy to a lower-energy state. Later, this energy will be used for the production of ATP by chemiosmosis. After oxygen receives electrons, it combines with hydrogen ions ($H^+$) and becomes water ($H_2O$).

**Pyruvate,** the end product of glycolysis, is a pivotal metabolite; its further treatment is dependent on whether oxygen is available. If oxygen is available, pyruvate enters a mitochondrion and is broken down completely to $CO_2$ and $H_2O$. If oxygen is not available, pyruvate is further metabolized in the cytoplasm by an anaerobic process called **fermentation.** Fermentation results in a net gain of only two ATP per glucose molecule.

**Check Your Progress 8.1**

1. Explain why glucose is broken down slowly, rather than quickly, during cellular respiration.
2. List the four phases of complete glucose breakdown. Tell which ones release $CO_2$ and which produces $H_2O$.

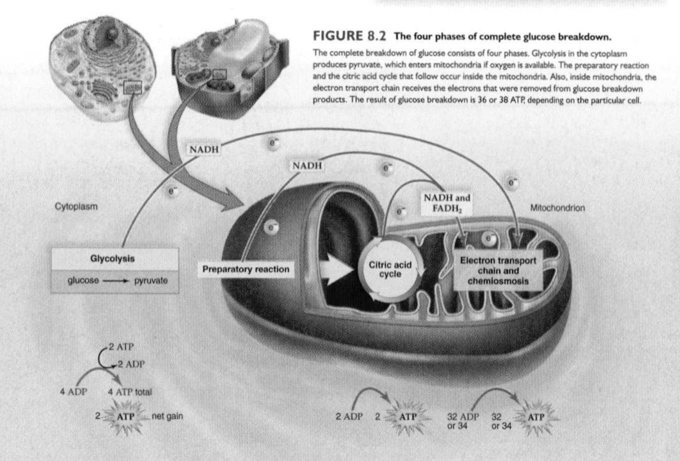


**FIGURE 8.2 The four phases of complete glucose breakdown.** The complete breakdown of glucose consists of four phases. Glycolysis in the cytoplasm produces pyruvate, which enters mitochondria if oxygen is available. The preparatory reaction and the citric acid cycle that follow occur inside the mitochondria. Also, inside mitochondria, the electron transport chain receives the electrons that were removed from glucose breakdown products. The result of glucose breakdown is 36 or 38 ATP, depending on the particular cell.

### CHECK YOUR PROGRESS

Check Your Progress questions appear at the end of each major section of the chapter to help students focus on the key concepts.

# Three Types of Boxed Readings

**Science Focus** readings describe how experimentation and observations have contributed to our knowledge about the living world.



*science focus*

## Fermentation Helps Produce Numerous Food Products

At the grocery store, you will find such items as bread, yogurt, soy sauce, pickles, and maybe even wine (Fig. 8A). These are just a few of the many foods that are produced when microorganisms ferment (break down sugar in the absence of oxygen). Foods produced by fermentation last longer because the fermenting organisms have removed many of the nutrients that would attract other organisms. The products of fermentation can even be dangerous to the very organisms that produced them, as when yeasts are killed by the alcohol they produce.

### Yeast Fermentation

Baker's yeast, *Saccharomyces cerevisiae*, is added to bread for the purpose of leavening—the dough rises when the yeasts give off $CO_2$. The ethyl alcohol produced by the fermenting yeast evaporates during baking. The many different varieties of sourdough breads obtain their leavening from a starter composed of fermenting yeasts along with bacteria from the environment. Depending on the community of microorganisms in the starter, the flavor of the bread may range from sour and tangy, as in San Francisco–style sourdough, to a milder taste, such as that produced by most Amish friendship bread recipes.

Ethyl alcohol is desired when yeasts are used to produce wine and beer. When yeasts ferment the carbohydrates of fruits, the end result is wine. If they ferment grain, beer results. A few specialized varieties of beer, such as traditional wheat beers, have a distinctive sour taste because they are produced with the assistance of lactic acid–producing bacteria, such as those of the genus *Lactobacillus*. Stronger alcoholic drinks (e.g., whiskey and vodka) require distillation to concentrate the alcohol content.

The acetic acid bacteria, including *Acetobacter aceti*, spoil wine. These bacteria convert the alcohol in wine or cider to acetic acid (vinegar). Until the renowned nineteenth-century scientist Louis Pasteur invented the process of pasteurization, acetic acid bacteria commonly caused wine to spoil. Although today we generally associate the process of pasteurization with making milk safe to drink, it was originally developed to reduce bacterial contamination in wine so that limited acetic acid would be produced.

### Bacterial Fermentation

Yogurt, sour cream, and cheese are produced through the action of various lactic acid bacteria that cause milk to sour. Milk contains lactose, which these bacteria use as a substrate for fermentation. Yogurt, for example, is made by adding lactic acid bacteria, such as *Streptococcus thermophilus* and *Lactobacillus bulgaricus*, to milk and then incubating it to encourage the bacteria to act on lactose. During the production of cheese, an enzyme called rennin must also be added to the milk to cause it to coagulate and become solid.

Old-fashioned brine cucumber pickles, sauerkraut, and kimchi are pickled vegetables produced by the action of acid-producing, fermenting bacteria that can survive in high-salt environments. Salt is used to draw liquid out of the vegetables and aid in their preservation. The bacteria need not be added to the vegetables, because they are already present on the surfaces of the plants.

### Soy Sauce Production

Soy sauce is traditionally made by adding a mold, *Aspergillus*, and a combination of yeasts and fermenting bacteria to soybeans and wheat. The mold breaks down starch, supplying the fermenting microorganisms with sugar they can use to produce alcohol and organic acids.

**FIGURE 8A Products from fermentation.** *Fermentation helps make the products shown on this page.*

**Ecology Focus** readings show how the concepts of the chapter can be applied to ecological concerns.



*ecology focus*

## Carboniferous Forests

Our industrial society runs on fossil fuels such as coal. The term *fossil fuel* might seem odd at first until one realizes that it refers to the remains of organic material from ancient times. During the Carboniferous period more than 300 million years ago, a great swamp forest (Fig. 23A) encompassed what is now northern Europe, the Ukraine, and the Appalachian Mountains in the United States. The weather was warm and humid, and the trees grew very tall. These are not the trees we know today; instead, they are related to today's seedless vascular plants: the lycophytes, horsetails, and ferns! Lycophytes today may stand as high as 30 cm, but their ancient relatives were 35 m tall and 1 m wide. The stroboli were up to 30 cm long, and some had leaves more than 1 m long. Horsetails too—at 18 m tall—were giants compared to today's specimens. Tree ferns were also taller than tree ferns found in the tropics today. The progymnosperms, including "seed ferns," were significant plants of a Carboniferous swamp. Seed ferns are misnamed because they were actually progymnosperms.

The amount of biomass in a Carboniferous swamp forest was enormous, and occasionally the swampy water rose and the trees fell. Trees under water do not decompose well, and their partially decayed remains became covered by sediment that sometimes changed to sedimentary rock. Exposed to pressure from sedimentary rock, the organic material then became coal, a fossil fuel. This process continued for millions of years, resulting in immense deposits of coal. Geological upheavals raised the deposits to the level where they can be mined today.

With a change of climate, the trees of the Carboniferous period became extinct, and only their herbaceous relatives survived to our time. Without these ancient forests, our life today would be far different because they helped bring about our industrialized society.

Fossil seed ferns

lycophytes, horsetail, seed fern, progymnosperm, fern

**FIGURE 23A Swamp** ... *Nonvascular plants and ... their leaves looked like f...*

**Health Focus** readings review procedures and technology that can contribute to our well-being.





*health focus*

## Prevention of Cardiovascular Disease

All of us can take steps to prevent cardiovascular disease, the most frequent cause of death in the United States. Certain genetic factors predispose an individual to cardiovascular disease, such as family history of heart attack under age 55, male gender, and ethnicity (African Americans are at greater risk). People with one or more of these risk factors need not despair, however. It means only that they should pay particular attention to the following guidelines for a heart-healthy lifestyle.

### The Don'ts

*Smoking*

Hypertension is well recognized as a major contributor to cardiovascular disease. When a person smokes, the drug nicotine, present in cigarette smoke, enters the bloodstream. Nicotine causes the arterioles to constrict and the blood pressure to rise. Restricted blood flow and cold hands are associated with smoking in most people. More serious is the need for the heart to pump harder to propel the blood through the lungs at a time when the oxygen-carrying capacity of the blood is reduced.

*Drug Abuse*

Stimulants, such as cocaine and amphetamines, can cause an irregular heartbeat and lead to heart attacks and strokes in people who are using drugs even for the first time. Intravenous drug use may result in a cerebral embolism.

Too much alcohol can destroy just about every organ in the body, the heart included. But investigators have discovered that people who take an occasional drink have a 20% lower risk of heart disease than do teetotalers. Two to four drinks a week is the recommended limit for men; one to three drinks for women.

*Weight Gain*

Hypertension is prevalent in persons who are more than 20% above the recommended weight for their height. In those who are overweight, more tissues require servicing, and the heart sends the extra blood out under greater pressure. It may be harder to lose weight once it is gained, and therefore it is recommended that weight control be a lifelong endeavor. Even a slight decrease in weight can bring with it a reduction in hypertension. A 4.5-kg weight (about 10 lbs) loss doubles the chance that blood pressure can be normalized without drugs.

### The Dos

*Healthy Diet*

Diet influences the amount of cholesterol in the blood. Cholesterol is ferried by two types of plasma proteins, called LDL (low-density lipoprotein) and HDL (high-density lipoprotein). LDL (called "bad" lipoprotein) takes cholesterol from the liver to the tissues, and HDL (called "good" lipoprotein) transports cholesterol out of the tissues to the liver. When the LDL level in blood is high or the HDL level is abnormally low, plaque, which interferes with circulation, accumulates on arterial walls (Fig. 32A).

Eating foods high in saturated fat (red meat, cream, and butter) and foods containing so-called trans-fats (most margarines, commercially baked goods, and deep-fried foods) raises the LDL-cholesterol level. Replacement of these harmful fats with healthier ones, such as monounsaturated fats (olive and canola oils) and polyunsaturated fats (corn, safflower, and soybean oils), is recommended. Cold water fish (e.g., halibut, sardines, tuna, and salmon) contain polyunsaturated fatty acids and especially omega-3 polyunsaturated fatty acids, which can reduce plaque.

Evidence is mounting to suggest a role for antioxidant vitamins (A, E, and C) in preventing cardiovascular disease. Antioxidants protect the body from free radicals that oxidize cholesterol and damage the lining of an artery, leading to a blood clot that can block blood vessels. Nutritionists believe that consuming at least five servings of fruits and vegetables a day may protect against cardiovascular disease.

*Cholesterol Profile*

Starting at age 20, all adults are advised to have their cholesterol levels tested at least every five years. Even in healthy individuals, an LDL level above 160 mg/100 ml and an HDL level below 40 mg/100 ml are matters of concern. If a person has heart disease or is at risk for heart disease, an LDL level below 100 mg/100 ml is now recommended. Medications will most likely be prescribed for individuals who do not meet these minimum guidelines.

*Exercise*

People who exercise are less apt to have cardiovascular disease. One study found that moderately active men who spent an average of 48 minutes a day on a leisure-time activity such as gardening, bowling, or dancing had one-third fewer heart attacks than peers who spent an average of only 16 minutes each day being active. Exercise helps keep weight under control, may help minimize stress, and reduces hypertension. The heart beats faster when exercising, bu... slowly increases its capacity. This ... the heart can beat slower when ... rest and still do the same amour... One physician recommends that ... cular patients walk for one hour, t... a week, and, in addition, practice ... and yogalike stretching and brea... cises to reduce stress.

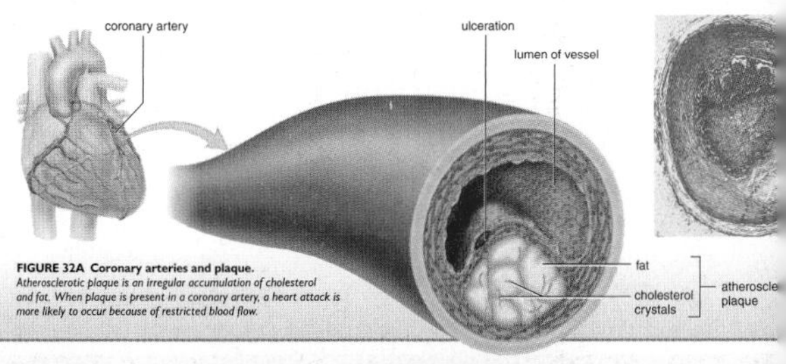


**FIGURE 32A Coronary arteries and plaque.** *Atherosclerotic plaque is an irregular accumulation of cholesterol and fat. When plaque is present in a coronary artery, a heart attack is more likely to occur because of restricted blood flow.*

# End of Chapter Study Tools

## CONNECTING THE CONCEPTS

These appear at the close of the text portion of the chapter, and they stimulate critical thinking by showing how the concepts of the chapter are related to other concepts in the text.

## CHAPTER SUMMARY

The summary is organized according to the major sections in the chapter and helps students review the important topics and concepts.



### Connecting the Concepts

food in
$O_2$ in
Digestive System
Respiratory System
excretion of $CO_2$
liver
nutrients
$O_2$ $CO_2$
Cardiovascular System
wastes
nutrients and water
Urinary System
$O_2$ and nutrients
$CO_2$ and waste
excretion of metabolic wastes
tissue fluid
cells

In mammals, the respiratory system consists of the respiratory tract with the nasal passages (or mouth) at one end and the lungs at the other end. Inspired air is 20% $O_2$ and 0.04% $CO_2$, while expired air is about 14% $O_2$ and 6% $CO_2$. Gas exchange in the lungs accounts for the difference in composition of inspired and expired air.

In the lungs, oxygen is absorbed into the bloodstream and from there it is transported by red blood cells to the capillaries, where it exits and enters tissue fluid. On the other hand, carbon dioxide enters capillaries at the tissues and is transported largely as the bicarbonate ion to the lungs, where it is converted to carbon dioxide and exits the body. Diffusion alone accounts for gas exchange in the lungs, called external respiration, and gas exchange in the tissues, called internal respiration. Energy is not needed, as gases follow their concentration gradients according to their partial pressures.

Internal gas exchange is extremely critical because cells use oxygen and release carbon dioxide as a result of cellular respiration, the process that generates ATP in cells. External gas exchange has the benefit of helping to keep the pH of the blood constant as required for homeostasis. When carbon dioxide exits, the blood pH returns to normal. In Chapter 36, we consider the contribution of the kidneys to homeostasis.

### summary

**35.1 Gas Exchange Surfaces**
Some aquatic animals, such as hydras and planarians, use their entire body surface for gas exchange. Most animals have a specialized gas-exchange area. Large aquatic animals usually pass water through gills. In bony fishes, blood in the capillaries flows in the direction opposite that of the water. Blood takes up almost all of the oxygen in the water as a result of this countercurrent flow. On land, insects use tracheal systems, and vertebrates have lungs. In insects, air enters the tracheae at openings called spiracles. From there, the air moves to ever smaller tracheoles until gas exchange takes place at the cells themselves. Lungs are found inside the body, where water loss is reduced. To ventilate the lungs, some vertebrates use positive pressure, but most inhale, using muscular contraction to produce a negative pressure that causes air to rush into the lungs. When the breathing muscles relax, air is exhaled.

Birds have a series of air sacs attached to the lungs. When a bird inhales, air enters the posterior air sacs, and when a bird exhales, air moves through the lungs to the anterior air sacs before exiting the respiratory tract. The one-way flow of air through the lungs allows more fresh air to be present in the lungs with each breath, and this leads to greater uptake of oxygen from one breath of air.

**35.2 Breathing and Transport of Gases**
During inspiration, air enters the body at nasal cavities and then passes from the pharynx through the glottis, larynx, trachea, bronchi, and bronchioles to the alveoli of the lungs, where exchange occurs, and during expiration air passes in the opposite direction. Humans breathe by negative pressure, as do other mammals. During inspiration, the rib cage goes up and out, and the diaphragm lowers. The lungs expand ushing in. During expiration, the rib cage goes down diaphragm rises. Therefore, air rushes out.
f breathing increases when the amount of $H^+$ and in the blood rises, as detected by chemoreceptors tic and carotid bodies.
ge in the lungs and tissues is brought about by diffusion. nsports oxygen in the blood; carbon dioxide is mainly lasma as the bicarbonate ion. Excess hydrogen ions are transported by hemoglobin. The enzyme carbonic anhydrase found in red blood cells speeds the formation of the bicarbonate ion.

**35.3 Respiration and Health**
The respiratory tract is subject to infections such as pneumonia and pulmonary tuberculosis. New strains of tuberculosis are resistant to the usual antibiotic therapy.

Major lung disorders are usually due to cigarette smoking. In chronic bronchitis the air passages are inflamed, mucus is common, and the cilia that line the respiratory tract are gone. Emphysema and lung cancer are two of the most serious consequences of smoking cigarettes. When the lungs of these patients are removed upon death, they are blackened by smoke.

### understanding the terms

alveolus (pl., alveoli) 654
aortic body 657
bicarbonate ion 659
bronchiole 655
bronchus (pl., bronchi) 655
carbaminohemoglobin 659
carbonic anhydrase 659
carotid body 657
countercurrent exchange 652
diaphragm 656
epiglottis 654
expiration 656
external respiration 650
gills 651
glottis 654
heme 659
hemoglobin (Hb) 659
inspiration 656
internal respiration 650
larynx 654
lungs 651
oxyhemoglobin 659
partial pressure 658
pharynx 654
respiration 650
respiratory center 657
trachea (pl., tracheae) 653, 654
ventilation 650
vocal cord 654

Match the terms to these definitions:

a. ______________ In terrestrial vertebrates, the mechanical act of moving air in and out of the lungs; breathing.
b. ______________ Dome-shaped muscularized sheet separating the thoracic cavity from the abdominal cavity in mammals.
c. ______________ Fold of tissue within the larynx; creates vocal sounds when it vibrates.



d. ______________ Respiratory organ in most aquatic animals; in fish, an outward extension of the pharynx.
e. ______________ Stage during breathing when air is pushed out of the lungs.

### reviewing this chapter

1. Compare the respiratory organs of aquatic animals to those of terrestrial animals. 650–54
2. How does the countercurrent flow of blood within gill capillaries and water passing across the gills assist respiration in fishes? 652
3. Why is it beneficial for the body wall of earthworms to be moist? Why don't insects require circulatory system involvement in air transport? 653
4. Name the parts of the human respiratory system, and list a function for each part. How is the air reaching the lungs cleansed? 654
5. Explain the phrase "breathing by using negative pressure." 656
6. Contrast the tidal ventilation mechanism in humans with the one-way ventilation mechanism in birds, and explain the benefits of the ventilation mechanism in birds. 656–57
7. The concentration of what substances in blood controls the breathing rate in humans? Explain. 658
8. How are oxygen and carbon dioxide transported in blood? What does carbonic anhydrase do? 659
9. Which conditions depicted in Figure 35.14 are due to infection? Which are due to behavioral or environmental factors? Explain. 660–61

### testing yourself

Choose the best answer for each question.

1. Label the following diagram depicting respiration.

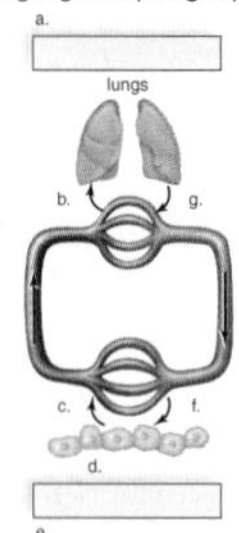


2. One problem faced by terrestrial animals with lungs, but not by aquatic animals with gills, is that
   a. gas exchange involves water loss.
   b. breathing requires considerable energy.
   c. oxygen diffuses very slowly in air.
   d. the concentration of oxygen in water is greater than that in air.
   e. All of these are correct.
3. In which animal is the circulatory system not involved in gas transport?
   a. mouse
   b. dragonfly
   c. trout
   d. sparrow
   e. human
4. Birds have more efficient lungs than humans because the flow of air
   a. is the same during both inspiration and expiration.
   b. travels in only one direction through the lungs.
   c. never backs up as it does in human lungs.
   d. is not hindered by a larynx.
   e. enters their bones.
5. Which animal breathes by positive pressure?
   a. fish
   b. human
   c. bird
   d. frog
   e. planarian
6. Which of these is a true statement?
   a. In lung capillaries, carbon dioxide combines with water to produce carbonic acid.
   b. In tissue capillaries, carbonic acid breaks down to carbon dioxide and water.
   c. In lung capillaries, carbonic acid breaks down to carbon dioxide and water.
   d. In tissue capillaries, carbonic acid combines with hydrogen ions to form the carbonate ion.
   e. All of these statements are true.
7. Air enters the human lungs because
   a. atmospheric pressure is less than the pressure inside the lungs.
   b. atmospheric pressure is greater than the pressure inside the lungs.
   c. although the pressures are the same inside and outside, the partial pressure of oxygen is lower within the lungs.
   d. the residual air in the lungs causes the partial pressure of oxygen to be less than it is outside.
   e. the process of breathing pushes air into the lungs.
8. If the digestive and respiratory tracts were completely separate in humans, there would be no need for
   a. swallowing.
   b. a nose.
   c. an epiglottis.
   d. a diaphragm.
   e. All of these are correct.
9. In tracing the path of air in humans, you would list the trachea
   a. directly after the nose.
   b. directly before the bronchi.
   c. before the pharynx.
   d. directly before the lungs.
   e. Both a and c are correct.
10. In humans, the respiratory control center
    a. is stimulated by carbon dioxide.
    b. is located in the medulla oblongata.
    c. controls the rate of breathing.
    d. is stimulated by hydrogen ion concentration.
    e. All of these are correct.
11. Carbon dioxide is carried in the plasma
    a. in combination with hemoglobin.
    b. as the bicarbonate ion.
    c. combined with carbonic anhydrase.
    d. only as a part of tissue fluid.
    e. All of these are correct.
12. Which of these is anatomically incorrect?
    a. The nose has two nasal cavities.
    b. The pharynx connects the nasal and oral cavities to the larynx.
    c. The larynx contains the vocal cords.
    d. The trachea enters the lungs.
    e. The lungs contain many alveoli.

## UNDERSTANDING THE TERMS

The boldface terms in the chapter are page referenced, and a matching exercise allows you to test your knowledge of the terms.

## REVIEWING THIS CHAPTER

These page-referenced study questions follow the sequence of the chapter.

## TESTING YOURSELF

These objective questions allow you to test your ability to answer recall-based questions. Answers to Testing Yourself questions are provided in Appendix A.

13. How is inhaled air modified before it reaches the lungs?
    a. It must be humidified.
    b. It must be warmed.
    c. It must be filtered and cleansed.
    d. All of these are correct.
14. Internal respiration refers to
    a. the exchange of gases between the air and the blood in the lungs.
    b. the movement of air into the lungs.
    c. the exchange of gases between the blood and tissue fluid.
    d. cellular respiration, resulting in the production of ATP.
15. The chemical reaction that converts carbon dioxide to a bicarbonate ion takes place in
    a. the blood plasma.
    b. red blood cells.
    c. the alveolus.
    d. the hemoglobin molecule.
16. Which of these would affect hemoglobin's $O_2$-binding capacity?
    a. pH
    b. partial pressure of oxygen
    c. blood pressure
    d. temperature
    e. All of these except c are correct.
17. The enzyme carbonic anhydrase
    a. causes the blood to be more basic in the tissues.
    b. speeds the conversion of carbonic acid to carbon dioxide and water.
    c. actively transports carbon dioxide out of capillaries.
    d. is active only at high altitudes.
    e. All of these are correct.
18. Which of these is incorrect concerning inspiration?
    a. Rib cage moves up and out.
    b. Diaphragm contracts and moves down.
    c. Pressure in lungs decreases, and air comes rushing in.
    d. The lungs expand because air comes rushing in.
19. Label this diagram of the human respiratory system.

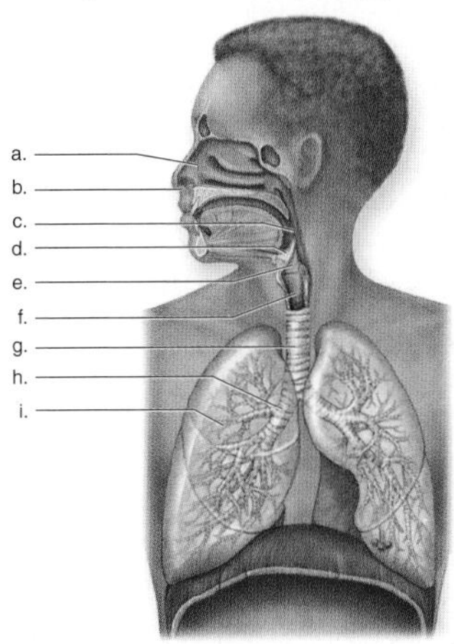


## thinking scientifically

1. You are a physician who witnessed Christopher Reeve's riding accident. Why might you immediately use mouth to mouth resuscitation until mechanical ventilation becomes available?
2. Fetal hemoglobin picks up oxygen from the maternal blood. If the oxygen-binding characteristics of hemoglobin in the fetus were identical to the hemoglobin of the mother, oxygen could never be transferred at the placenta to fetal circulation. What hypothesis about the oxygen-binding characteristics of fetal hemoglobin would explain how fetuses get the oxygen they need?

## bioethical issue

### Antibiotic Therapy

Antibiotics cure respiratory infections, but there are problems associated with antibiotic therapy. Aside from a possible allergic reaction, antibiotics not only kill off disease-causing bacteria, but they also reduce the number of beneficial bacteria in the intestinal tract and other locations. These beneficial bacteria hold in check the growth of other pathogens that now begin to flourish. Diarrhea can result, as can a vaginal yeast infection. The use of antibiotics can also prevent natural immunity from occurring, leading to the need for recurring antibiotic therapy. Especially alarming at this time is the occurrence of resistance. Resistance takes place when vulnerable bacteria are killed off by an antibiotic, and this allows resistant bacteria to become prevalent. The bacteria that cause ear, nose, and throat infections as well as scarlet fever and pneumonia are becoming widely resistant because we have not been using antibiotics properly. Tuberculosis is on the rise, and the new strains are resistant to the usual combined antibiotic therapy.

Every citizen needs to be aware of our present crisis situation. Stuart Levy, a Tufts University School of Medicine microbiologist, says that we should do what is ethical for society and ourselves. What is needed? Antibiotics kill bacteria, not viruses—therefore, we shouldn't take antibiotics unless we know for sure we have a bacterial infection. And we shouldn't take them prophylactically—that is, just in case we might need one. If antibiotics are taken in low dosages and intermittently, resistant strains are bound to take over. Animal and agricultural use should be pared down, and household disinfectants should no longer be spiked with antibacterial agents. Perhaps then, Levy says, vulnerable bacteria will begin to supplant the resistant ones in the population. Are you doing all you can to prevent bacteria from becoming resistant?

## *Biology* website

The companion website for *Biology* provides a wealth of information organized and integrated by chapter. You will find practice tests, animations, videos, and much more that will complement your learning and understanding of general biology.

**http://www.mhhe.com/maderbiology10**

### THINKING SCIENTIFICALLY

Critical thinking questions give you an opportunity to reason as a scientist. Detailed answers to these questions are found on ARIS, the Biology, Tenth Edition website. Answers to these questions are found in Appendix A.

### BIOETHICAL ISSUE

A Bioethical Issue is found at the end of most chapters. These short readings discuss a variety of controversial topics that confront our society. Each reading ends with appropriate questions to help you fully consider the issue and arrive at an opinion.

### WEBSITE REMINDER

Located at the end of the chapter is this reminder that additional study questions and other learning activities are on the *Biology*, Tenth Edition website.

# ACKNOWLEDGMENTS

The hard work of many dedicated and talented individuals helped to vastly improve this edition of *Biology*. Let me begin by thanking the people who guided this revision at McGraw-Hill. I am very grateful for the help of so many professionals who were involved in bringing this book to fruition. In particular, let me thank Janice Roerig-Blong, who guided us as we shaped the content and pedagogy of the book. Lisa Bruflodt, the developmental editor, who kept everyone on target as the book was developed. The biology editor was Michael Hackett, who became a member of the team this past year. The project manager, Jayne Klein, faithfully and carefully steered the book through the publication process. Tamara Maury, the marketing manager, tirelessly promoted the text and educated the sales reps on its message.

The design of the book is the result of the creative talents of David Hash and many others who assisted in deciding the appearance of each element in the text. EPS followed their guidelines as they created and reworked each illustration, emphasizing pedagogy and beauty to arrive at the best presentation on the page. Lori Hancock and Jo Johnson did a superb job of finding just the right photographs and micrographs.

My assistant, Beth Butler, worked faithfully to do a preliminary paging of the book, helped proof the chapters, and made sure all was well before the book went to press. As always, my family was extremely patient with me as I remained determined to make every deadline on the road to publication. My husband, Arthur Cohen, is also a teacher of biology. The many discussions we have about the minutest detail to the gravest concept are invaluable to me.

As stated previously, the content of the tenth edition of *Biology* is not due to my efforts alone. I want to thank the many specialists who were willing to share their knowledge to improve *Biology*. Also, this edition was enriched by four contributors: Michael Thompson revised the genetics chapters, Stephanie Songer reworked the microbiology chapters and several animal biology chapters, Rebecca Roush contributed to the animal diversity chapters, and Andy Baldwin oversaw the ecology chapters. The tenth edition of *Biology* would not have the same excellent quality without the input of these contributors and those of the many reviewers who are listed on page xvii.

## 360 Development

McGraw-Hill's 360° Development Process is an ongoing, never-ending, market-oriented approach to building accurate and innovative print and digital products. It is dedicated to continual large-scale and incremental improvement driven by multiple customer feedback loops and checkpoints. This is initiated during the early planning stages of our new products, and intensifies during the development and production stages, then begins again upon publication in anticipation of the next edition.

This process is designed to provide a broad, comprehensive spectrum of feedback for refinement and innovation of our learning tools, for both student and instructor. The 360° Development Process includes market research, content reviews, course- and product-specific symposia, accuracy checks, and art reviews. We appreciate the expertise of the many individuals involved in this process.

## Ancillary Authors

Instructor's Manual – Andrea Thomason, MassBay Community College
Practice Tests – Raymond Burton, Germanna Community College
Media Asset Correlations – Eric Rabitoy, Citrus College
Test Bank – Deborah Dardis, Southeastern Louisiana University
Lecture Outlines/Image PowerPoints – Isaac Barjis, New York City College of Technology
BioInteractive Questions – Eileen Preston, Tarrant County College NW

## Tenth Edition Reviewers

**Mike Aaron**
*Shelton State Community College*

**John Aliff**
*Georgia Perimeter College*

**Michael Bell**
*Richland College*

**Danita Bradshaw-Ward**
*Eastfield College*

**Eric Buckers**
*Dillard University*

**Frank Campo**
*Southeastern Louisiana University*

**Pam Cole**
*Shelton State Community College*

**Denise Conover**
*University of Cincinnati*

**Janice Cooley**
*Mississippi Gulf Coast Community College*

**David Corey**
*Midlands Technical College*

**David Cox**
*Lincoln Land Community College*

**Jason Curtis**
*Purdue University–Westville*

**Philip Darby**
*University of West Florida*

**Lewis Deaton**
*University of Louisiana–Lafayette*

**Domenica Devine**
*Armstrong Atlantic State University*

**Waneene Dorsey**
*Grambling State University*

**Salman Elawad**
*Chattahoochee Valley Community College*

**Eugene Fenster**
*Longview Community College – Lees Summit*

**Jennifer Fernandes-Miller**
*Eastfield College*

**Julie Fischer**
*Wallace Community College–Dothan*

**Paul Florence**
*Jefferson Community College*

**Theresa Fulcher**
*Pellissippi State Technical College*

**Michelle Geary**
*West Valley College*

**Ann Gathers**
*University of Tennessee–Martin*

**Melanie Glasscock**
*Wallace State Community College–Hanceville*

**Lonnie Guralnick**
*Western Oregon University*

**Leslie Hendricks**
*Southeastern Louisiana University*

**Jennifer Herzog**
*Herkimer County Community College*

**Dagne Hill**
*Grambling State University*

**Harold Horn**
*Lincoln Land Community College*

**Walter Judd**
*University of Florida–Gainesville*

**Laina Karthikeyan**
*New York City College of Technology*

**Jenny Knoth**
*Armstrong Atlantic State University*

**Yaser Maksoud**
*Olive Harvey College*

**Dan Matusiak**
*St. Charles Community College*

**Scott Murdoch**
*Moraine Valley Community College*

**Necia Nicholas**
*Calhoun Community College*

**Joshua Parker**
*Community College of Southern Nevada*

**Theresa Poole**
*Georgia State University*

**Eric Rabitoy**
*Citrus College*

**Jose Ramirez-Domenech**
*Dillard University*

**Kathleen Richardson**
*Portland Community College–Sylvania*

**David Saunders**
*Augusta State University*

**Tanita Shannon-Gragg**
*Southeast Arkansas College*

**Linda Smith-Staton**
*Pellissippi State Technical College*

**Staria Vanderpool**
*Arkansas State University*

**Wendy Vermillion**
*Columbus State Community College*

**Heather Wilkins**
*University of Tennessee–Martin*

**Albert Wilson**
*Spokane Falls Community College*

**Bryant Wright**
*Jefferson College*

**Ted Zerucha**
*Appalachian State University*

# SUPPLEMENTS

Dedicated to providing high-quality and effective supplements for instructors and students, the following supplements were developed for *Biology*.

## For Instructors

### *Laboratory Manual*

The *Biology Laboratory Manual*, Tenth Edition, is written by Dr. Sylvia Mader. With few exceptions, each chapter in the text has an accompanying laboratory exercise in the manual. Every laboratory has been written to help students learn the fundamental concepts of biology and the specific content of the chapter to which the lab relates, and to gain a better understanding of the scientific method.

ISBN (13) 978–0–07–722617–6
ISBN (10) 0 07–722617–8

### *Connect Biology*

McGraw-Hill Connect Biology is a web-based assignment and assessment platform that gives students the means to better connect with their coursework, with their instructors, and with the important concepts that they will need to know for success now and in the future. With Connect Biology, instructors can deliver assignments, quizzes and tests easily online. Students can practice important skills at their own pace and on their own schedule. With Connect Biology Plus, students also get 24/7 online access to an eBook—an online edition of the text—to aid them in successfully completing their work, wherever and whenever they choose.

### *Companion Website*

The companion website contains the following resources for instructors:

- **Presentation Tools** Everything you need for outstanding presentation in one place! This easy-to-use table of assets includes
  - Enhanced image PowerPoints—including every piece of art that has been sized and cropped specifically for superior presentations as well as labels that you can edit. Also included are tables, photographs, and unlabeled art pieces.
  - Animation PowerPoints—Numerous full-color animations illustrating important processes are also provided. Harness the visual impact of concepts in motion by importing these files into classroom presentations or online course materials.
  - Lecture PowerPoints with animations fully embedded.
  - Labeled and unlabeled JPEG images—Full-color digital files of all illustrations that can be readily incorporated into presentations, exams, or custom-made classroom materials.
- **Presentation Center** In addition to the images from your book, this online digital library contains photos, artwork, animations, and other media from an array of McGraw-Hill textbooks that can be used to create customized lectures, visually enhanced tests and quizzes, compelling course websites, or attractive printed support materials. All assets are copyrighted by McGraw-Hill Higher Education, but can be used by instructors for classroom purposes.
- **Instructor's Manual** The instructor's manual contains chapter outlines, lecture enrichment ideas, and critical thinking questions.
- **Computerized Test Bank** A comprehensive bank of test questions is provided within a computerized test bank powered by McGraw-Hill's flexible electronic testing program EZ Test Online. EZ Test Online allows you to create paper and online tests or quizzes in this easy to use program! A new tagging scheme allows you to sort questions by difficulty level, topic, and section. Imagine being able to create and access your test or quiz anywhere, at any time, without installing the testing software. Now, with EZ Test Online, instructors can select questions from multiple McGraw-Hill test banks or author their own, and then either print the test for paper distribution or give it online.

#### *Test Creation*

- Author/edit questions online using the 14 different question type templates
- Create question pools to offer multiple versions online—great for practice
- Export your tests for use in WebCT, Blackboard, PageOut, and Apple's iQuiz
- Sharing tests with colleagues, adjuncts, TAs is easy

### *Online Test Management*

- Set availability dates and time limits for your quiz or test
- Assign points by question or question type with dropdown menu
- Provide immediate feedback to students or delay feedback until all finish the test
- Create practice tests online to enable student mastery
- Your roster can be uploaded to enable student self-registration

### *Online Scoring and Reporting*

- Automated scoring for most of EZ Test's numerous question types
- Allows manual scoring for essay and other open-response questions
- Manual rescoring and feedback are also available
- EZ Test's grade book is designed to easily export to your grade book
- View basic statistical reports

### *Support and Help*

- Flash tutorials for getting started on the support site
- Support Website: **www.mhhe.com/eztest**
- Product specialist available at **1–800–331–5094**
- Online Training: **http://auth.mhhe.com/mpss/workshops**

Go to **www.mhhe.com/maderbiology10e** to learn more.

### *McGraw-Hill: Biology Digitized Video Clips*

ISBN (13) 978–0–312155–0
ISBN (10) 0–07–312155–X

McGraw-Hill is pleased to offer an outstanding presentation tool to text adopting instructors—digitized biology video clips on DVD! Licensed from some of the highest-quality science video producers in the world, these brief segments range from about five seconds to just under three minutes in length and cover all areas of general biology from cells to ecosystems. Engaging and informative, McGraw-Hill's digitized videos will help capture students' interest while illustrating key biological concepts and processes such as mitosis, how cilia and flagella work, and how some plants have evolved into carnivores.

### *Student Response System*

Wireless technology brings interactivity into the classroom or lecture hall. Instructors and students receive immediate feedback through wireless response pads that are easy to use and engage students. This system can be used by instructors to take attendance, administer quizzes and tests, create a lecture with intermittent questions, manage lectures and student comprehension through the use of the grade book, and integrate interactivity into their PowerPoint presentations.

## For Students

### *Companion Website*

The Mader: *Biology* website is an electronic study system that offers students a digital portal of knowledge. Students can readily access a variety of digital learning objects that include:

- Chapter-level quizzing with pretest and posttest
- Bio Tutorial Animations with quizzing
- Vocabulary flashcards
- Virtual Labs
- Vocabulary flashcards
- Biology Prep, also available on the companion website, helps students to prepare for their upcoming coursework in biology. This website enables students to perform self assessments, conduct self-study sessions with tutorials, and perform a post-assessment of their knowledge in the following areas: introductory biology skills, basic math, metric system, chemistry, and lab reports.
  - Introductory Biology Skills
  - Basic Math Review I and II
  - Chemistry
  - Metric System
  - Lab Reports and Referencing

### *Electronic Books*

If you or your students are ready for an alternative version of the traditional textbook, McGraw-Hill eBooks offer a cheaper and eco-friendly alternative to traditional textbooks. By purchasing eBooks from McGraw-Hill students can save as much as 50% on selected titles delivered on the most advanced E-book platform available. Contact your McGraw-Hill sales representative to discuss E-book packaging options.

### *How to Study Science*

ISBN (13) 978–0–07–234693–0
ISBN (10) 0–07–234693–0

This workbook offers students helpful suggestions for meeting the considerable challenges of a science course. It gives practical advice on such topics as how to take notes, how to get the most out of laboratories, and how to overcome science anxiety.

### *Photo Atlas for General Biology*

ISBN (13) 978–0–07–284610–2
ISBN (10) 0–07–284610–0

Atlas was developed to support our numerous general biology titles. It can be used as a supplement for a general biology lecture or laboratory course.

# CONTENTS

## *part* VII: Comparative Animal Biology 576

## *part* VIII: Behavior and Ecology 798

# READINGS

## Ecology Focus



## Health Focus



## Science Focus

# 1 A View of Life

## concepts

**1.1 HOW TO DEFINE LIFE**

- The living organisms on Earth share many common characteristics: they are organized, acquire materials and energy, respond to the environment, and reproduce and develop. 2–5
- Still, living things are diverse because they are adapted to their different environments. 5

**1.2 EVOLUTION, THE UNIFYING CONCEPT OF BIOLOGY**

- The theory of evolution states that all types of living organisms share a common ancestor and change over time. Therefore, evolution explains both the unity and diversity on Earth. 6–8

**1.3 HOW THE BIOSPHERE IS ORGANIZED**

- Living things interact with each other and with the physical environment to form ecosystems. 9–10
- Due to human activities, many diverse ecosystems are currently endangered. Biologists are concerned about the current rate of extinctions, and believe that we should take steps to preserve biodiversity. 10

**1.4 THE PROCESS OF SCIENCE**

- The scientific process is used by biologists to gather information and come to conclusions about the natural world before reporting it to other scientists and to the public. 11
- The process includes observation, proposing a hypothesis, performing an experiment, analyzing the results, and making conclusions. Conclusions usually lead to a new hypothesis. 11–16

*At a height of nearly 3 m (10 ft), the titan arum slowly unfurls its enormous flower, which heats up, turns red, and emits an overpowering stench reminiscent of rotting meat. Its home is the forests of Sumatra, and the smell attracts the beetles and flies that ordinarily pollinate the flower. Now the plant is cultivated in botanical gardens around the world to the delight of curious onlookers.*

*The Earth hosts a wide variety of ecosystems, from which spring a mind-boggling diversity of life, including the titan arum. Even so, all Earth's organisms, regardless of form, are united by a number of common characteristics, such as the need to acquire nutrients, the ability to respond to a changing environment, and to reproduce their own kind. Incredibly, even organisms as diverse as the titan arum and a human being share similar characteristics, including a common chemistry and genetic code. As you read this chapter, reflect on the staggering diversity of life on Earth and on the many ties that bind even the most diverse organisms, from bacteria to the titan arum to humans. It is through these ties that our fates are linked together in the web of life.*

The titan arum (*Amorphophallus titanum*).

## 1.1 How to Define Life

Life on Earth takes on a staggering variety of forms, often functioning and behaving in ways strange to humans. For example, gastric-brooding frogs swallow their embryos and give birth to them later by throwing them up! Some species of puffballs, a type of fungus, are capable of producing trillions of spores when they reproduce. Fetal sand sharks kill and eat their siblings while still inside their mother. Some *Ophrys* orchids look so much like female bees that male bees try to mate with them. Octopuses and squid have remarkable problem-solving abilities despite a small brain. Some bacteria live their entire life in 15 minutes, while bristlecone pine trees outlive ten generations of humans. Simply put, from the deepest oceanic trenches to the upper reaches of the atmosphere, life is plentiful and diverse.

Figure 1.1 illustrates the major groups of living things, also called **organisms.** From left to right, bacteria are widely distributed, tiny, microscopic organisms with a very simple structure. A *Paramecium* is an example of a microscopic protist. Protists are larger in size and more complex than bacteria. The other organisms in Figure 1.1 are easily seen with the naked eye. They can be distinguished by how they get their food. A morel is a fungus that digests its food externally. A sunflower is a photosynthetic plant that makes its own food, and a snow goose is an animal that ingests its food.

Because life is so diverse, it seems reasonable that it cannot be defined in a straightforward manner. Instead, life is best defined by several basic characteristics shared by all organisms. Like nonliving things, organisms are composed of chemical elements. Also, organisms obey the same laws of chemistry and physics that govern everything within the universe. The characteristics of life, however, will provide great insight into the unique nature of organisms and will help us distinguish living things from nonliving things.

### Living Things Are Organized

The levels of organization depicted in Figure 1.2 begin with atoms, which are the basic units of matter. Atoms combine with other atoms of the same or different elements to form molecules. The **cell,** which is composed of a variety of molecules working together, is the basic unit of structure and function of all living things. Some cells, such as **unicellular** paramecia, live independently. Other cells, for example, the colonial alga *Volvox,* cluster together in microscopic colonies.

Many living things are **multicellular,** meaning they contain more than one cell. In multicellular organisms, similar cells combine to form a tissue; nerve tissue is a common tissue in animals. Tissues make up organs, as when various tissues combine to form the brain. Organs work together in systems; for example, the brain works with the spinal cord and a network of nerves to form the nervous system. Organ systems are joined together to form a complete living thing, or organism, such as an elephant.

The levels of biological organization extend beyond the individual organism. All the members of one species in a particular area belong to a population. A nearby forest may have a population of gray squirrels and a population of white oaks, for example. The populations of various animals and plants in the forest make up a community. The community of populations interacts with the physical environment and forms an ecosystem. Finally, all the Earth's ecosystems make up the biosphere.

#### Emergent Properties

Each level of biological organization builds upon the previous level, and is more complex. Moving up the hierarchy, each level acquires new **emergent properties** that are determined by the interactions between the individual parts. When cells are broken down into bits of membrane and liquids, these parts themselves cannot carry out the business of living. For example, you can take apart a lump of coal, rearrange the pieces in any order, and still have a lump of coal with the same function as the original one. But, if you slice apart a living plant and rearrange the pieces, the plant is no longer functional as a complete plant, because it depends on the exact order of those pieces.

In the living world, the whole is indeed more than the sum of its parts. The emergent properties created by the interactions between levels of biological organization are new, unique characteristics. These properties are governed by the laws of chemistry and physics.

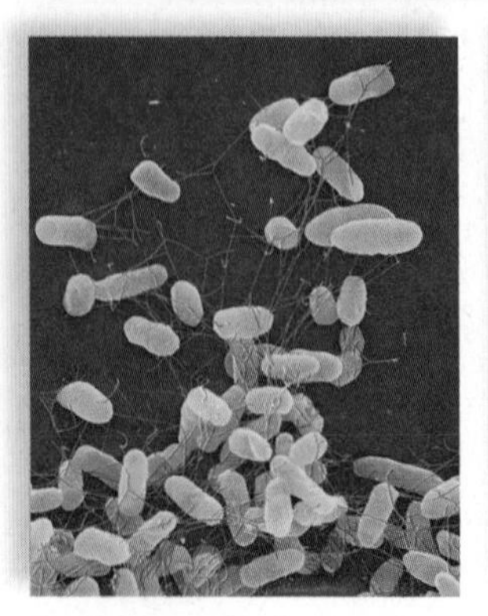

Bacteria

*Paramecium*

Morel

Sunflower

Snow goose

**FIGURE 1.1 Diversity of life.**

Biology is the scientific study of life. Many diverse forms of life are found on planet Earth.

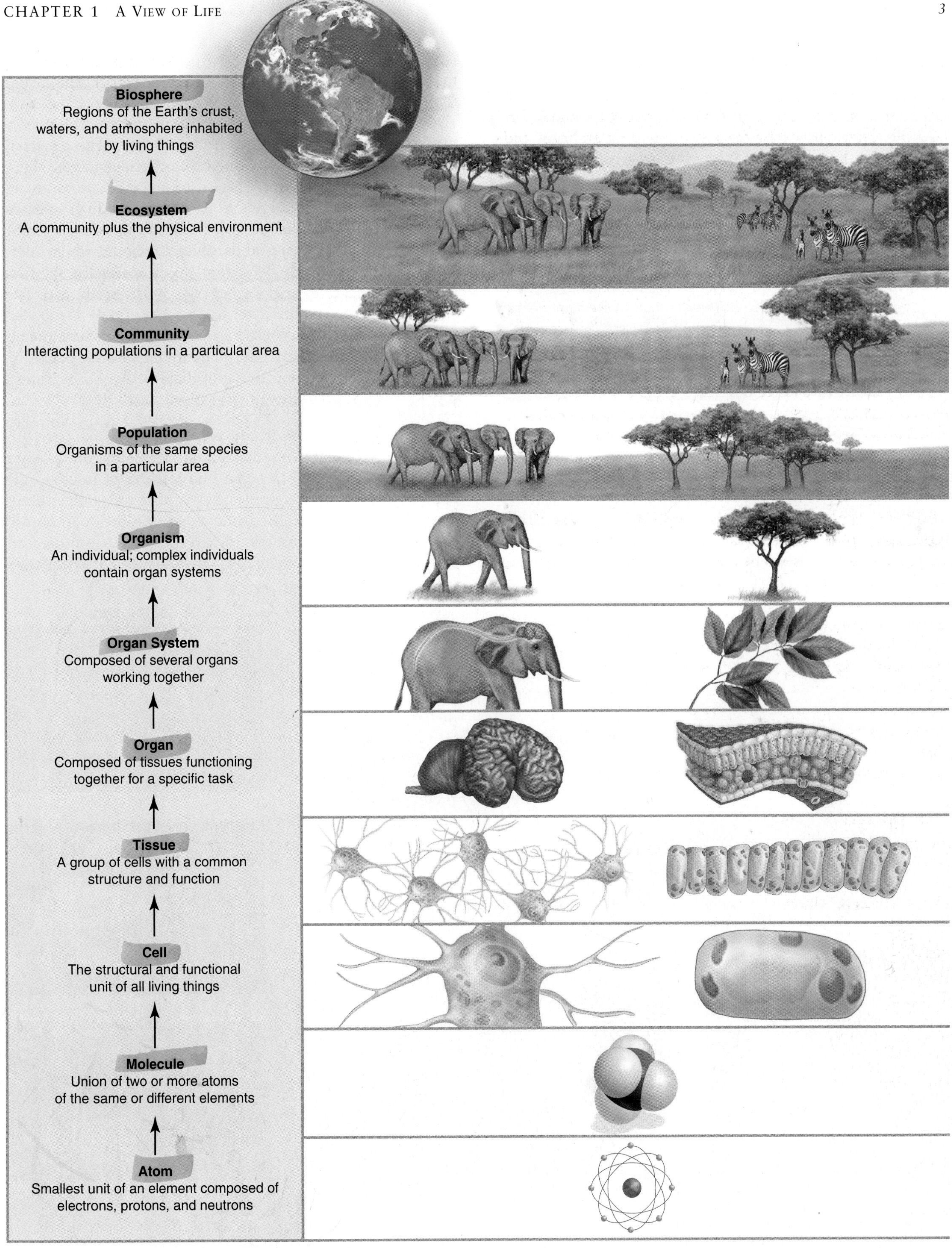

**Biosphere**
Regions of the Earth's crust, waters, and atmosphere inhabited by living things

**Ecosystem**
A community plus the physical environment

**Community**
Interacting populations in a particular area

**Population**
Organisms of the same species in a particular area

**Organism**
An individual; complex individuals contain organ systems

**Organ System**
Composed of several organs working together

**Organ**
Composed of tissues functioning together for a specific task

**Tissue**
A group of cells with a common structure and function

**Cell**
The structural and functional unit of all living things

**Molecule**
Union of two or more atoms of the same or different elements

**Atom**
Smallest unit of an element composed of electrons, protons, and neutrons

**FIGURE 1.2** **Levels of biological organization.**

## Living Things Acquire Materials and Energy

Living things cannot maintain their organization or carry on life's activities without an outside source of nutrients and energy (Fig. 1.3). Food provides nutrients, which are used as building blocks or for energy. **Energy** is the capacity to do work, and it takes work to maintain the organization of the cell and the organism. When cells use nutrient molecules to make their parts and products, they carry out a sequence of chemical reactions. The term **metabolism** [Gk. *meta*, change] encompasses all the chemical reactions that occur in a cell.

The ultimate source of energy for nearly all life on Earth is the sun. Plants and certain other organisms are able to capture solar energy and carry on **photosynthesis,** a process that transforms solar energy into the chemical energy of organic nutrient molecules. All life on Earth acquires energy by metabolizing nutrient molecules made by photosynthesizers. This applies even to plants.

### *Remaining Homeostatic*

To survive, it is imperative that an organism maintain a state of biological balance or **homeostasis** [Gk. *homoios*, like, and *stasis*, the same]. For life to continue, temperature, moisture level, acidity, and other physiological factors must remain within the tolerance range of the organism. Homeostasis is maintained by systems that monitor internal conditions and make routine and necessary adjustments.

Organisms have intricate feedback and control mechanisms that do not require any conscious activity. These mechanisms may be controlled by one or more tissues themselves, or by the nervous system. When a student is so engrossed in her textbook that she forgets to eat lunch, her liver releases stored sugar to keep blood sugar levels within normal limits. Many organisms depend on behavior to regulate their internal environment. These behaviors are controlled by the nervous system, and are usually not consciously controlled. The same student may realize that she is hungry and decide to visit the local diner. A lizard may raise its internal temperature by basking in the sun or cool down by moving into the shade.

## Living Things Respond

Living things interact with the environment as well as with other living things. Even unicellular organisms can respond to their environment. In some, the beating of microscopic hairs or, in others, the snapping of whiplike tails moves them toward or away from light or chemicals. Multicellular organisms can manage more complex responses. A vulture can detect a carcass a kilometer away and soar toward dinner. A

a.

b.

c.

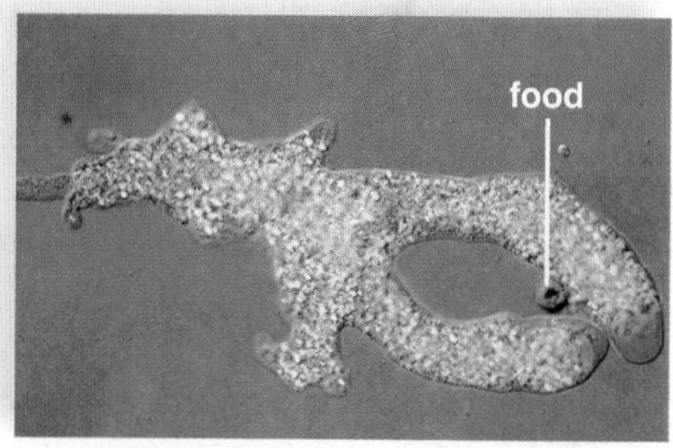

d.

e.

f.

**FIGURE 1.3 Acquiring nutrients and energy.**

**a.** An eagle ingesting fish. **b.** A human eating an apple. **c.** A cypress tree capturing sunlight. **d.** An amoeba engulfing food. **e.** A fungus feeding on a tree. **f.** A bison eating grass.

monarch butterfly can sense the approach of fall and begin its flight south where resources are still abundant.

The ability to respond often results in movement: the leaves of a land plant turn toward the sun, and animals dart toward safety. Appropriate responses help ensure survival of the organism and allow it to carry on its daily activities. All together, these activities are termed the behavior of the organism. Organisms display a variety of behaviors as they maintain homeostasis and search and compete for energy, nutrients, shelter, and mates. Many organisms display complex communication, hunting, and defense behaviors.

## Living Things Reproduce and Develop

Life comes only from life. Every type of living thing can **reproduce,** or make another organism like itself (Fig. 1.4). Bacteria, protists, and other unicellular organisms simply split in two. In most multicellular organisms, the reproductive process begins with the pairing of a sperm from one partner and an egg from the other partner. The union of sperm and egg, followed by many cell divisions, results in an immature stage, which grows and develops through various stages to become the adult.

An embryo develops into a humpback whale or a purple iris because of a blueprint inherited from its parents. The instructions, or blueprint, for an organism's metabolism and organization are encoded in genes. The **genes,** which contain specific information for how the organism is to be ordered, are made of long molecules of DNA (deoxyribonucleic acid). DNA has a shape resembling a spiral staircase with millions of steps. Housed within this spiral staircase is the genetic code that is shared by all living things.

When living things reproduce, their genes are passed on to the next generation. Random combinations of sperm and egg, each of which contains a unique collection of genes, ensure that the new individual has new and different characteristics. The DNA of organisms, over time, also undergoes mutations (changes) that may be passed on to the next generation. These events help to create a staggering diversity of life, even within a group of otherwise identical organisms. Sometimes, organisms inherit characteristics that allow them to be more suited to their way of life.

## Living Things Have Adaptations

**Adaptations** [L. *ad*, toward, and *aptus*, suitable] are modifications that make organisms better able to function in a particular environment. For example, penguins are adapted to an aquatic existence in the Antarctic. An extra layer of downy feathers is covered by short, thick feathers that form a waterproof coat. Layers of blubber also keep the birds warm in cold water. Most birds have forelimbs proportioned for flying, but penguins have stubby, flattened wings suitable for swimming. Their feet and tails serve as rudders

**FIGURE 1.4 Rockhopper penguins with their offspring.**
Rockhopper penguins, which are named for their skill in leaping from rock to rock, produce one or two offspring at a time. Both male and female have a brood patch, a feather-free area of skin containing many blood vessels, which keeps the egg(s) warm when either parent sits on the nest.

in the water, but the flat feet also allow them to walk on land. Rockhopper penguins have a bill adapted to eating small shellfish.

Penguins also have many behavioral adaptations to living in the Antarctic. Penguins often slide on their bellies across the snow in order to conserve energy when moving quickly. Their eggs—one or at most two—are carried on the feet, where they are protected by a pouch of skin. This also allows the birds to huddle together for warmth while standing erect and incubating eggs.

From penguins to fire ants, life on Earth is very diverse because over long periods of time, organisms respond to ever-changing environments by developing new adaptations. **Evolution** [L. *evolutio*, an unrolling] includes the way in which populations of organisms change over the course of many generations to become more suited to their environments. Evolution constantly reshapes the species, providing a way for organisms to persist, despite a changing environment.

### Check Your Progress 1.1

1. What are common characteristics of living organisms?
2. In what ways do viruses (p. 356) not specifically meet all of the above characteristics?
3. What adaptations would suit an organism, such as a cactus, to life in a desert?

## 1.2 Evolution, the Unifiying Concept of Biology

Despite diversity in form, function, and lifestyle, organisms share the same basic characteristics. As mentioned, they are all composed of cells organized in a similar manner. Their genes are composed of DNA, and they carry out the same metabolic reactions to acquire energy and maintain their organization. The unity of living things suggests that they are descended from a common ancestor—the first cell or cells.

An evolutionary tree is like a family tree (Fig. 1.5). Just as a family tree shows how a group of people have descended from one couple, an evolutionary tree traces the ancestry of life on Earth to a common ancestor. One couple can have diverse children, and likewise a population can be a common ancestor to several other groups, each adapted to a particular set of environmental conditions. In this way, over time, diverse life-forms have arisen. Evolution may be considered the unifying concept of biology because it explains so many aspects of biology, including how living organisms arose from a single ancestor.

### Organizing Diversity

Because life is so diverse, it is helpful to group organisms into categories. **Taxonomy** [Gk. *tasso*, arrange, and *nomos*, usage] is the discipline of identifying and grouping organisms according to certain rules. Taxonomy makes sense out of the bewildering variety of life on Earth and is meant to provide valuable insight into evolution. As more is learned about living things, including the evolutionary relationships between species, taxonomy changes. DNA technology is now being used to revise current information and to discover previously unknown relationships between organisms.

Several of the basic classification categories, or *taxa*, going from least inclusive to most inclusive, are **species, genus, family, order, class, phylum, kingdom,** and **domain** (Table 1.1). The least inclusive category, species [L. *species*, model, kind], is defined as a group of interbreeding individuals. Each successive classification category above species contains more types of organisms than the preceding one. Species placed within one genus share many specific characteristics and are the most closely related, while species placed in the same kingdom share only general characteristics with one another. For example, all species in the genus *Pisum* look pretty much the same—that is, like pea plants—but species in the plant kingdom can be quite varied, as is evident when we compare grasses to trees. Species placed in different domains are the most distantly related.

**TABLE 1.1**

**Levels of Classification**

| *Category* | *Human* | *Corn* |
|---|---|---|
| Domain | Eukarya | Eukarya |
| Kingdom | Animalia | Plantae |
| Phylum | Chordata | Anthophyta |
| Class | Mammalia | Monocotyledones |
| Order | Primates | Commelinales |
| Family | Hominidae | Poaceae |
| Genus | *Homo* | *Zea* |
| Species* | *H. sapiens* | *Z. mays* |

*To specify an organism, you must use the full binomial name, such as *Homo sapiens*.

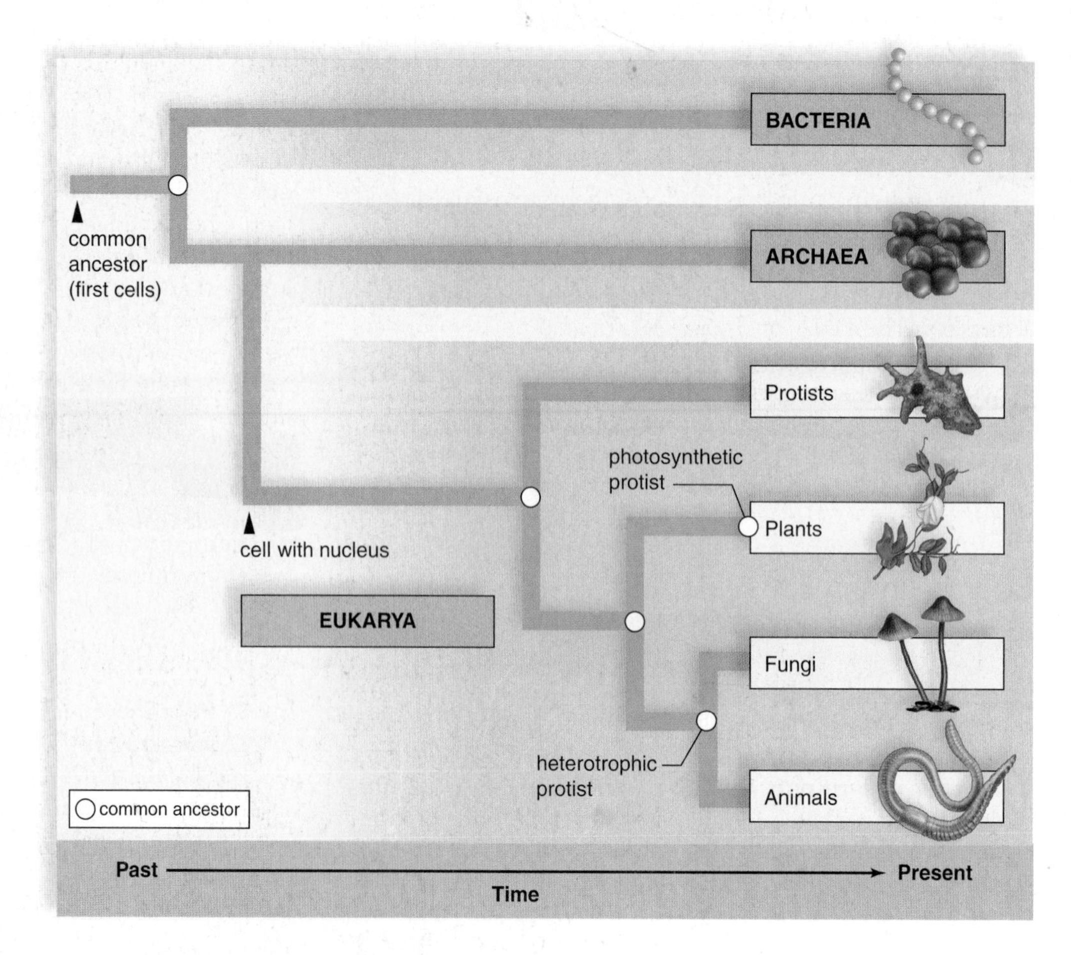


**FIGURE 1.5 Evolutionary tree of life.**

As existing organisms change over time, they give rise to new species. Evolutionary studies show that all living organisms arose from a common ancestor about 4 billion years ago. Domain Archaea includes prokaryotes capable of surviving in extreme environments, such as those with high salinity and temperature and low pH. Domain Bacteria includes metabolically diverse prokaryotes widely distributed in various environments. The domain Eukarya includes both unicellular and multicellular organisms that possess a membrane-bounded nucleus.

## *Domains*

Biochemical evidence suggests that there are only three domains: **domain Bacteria, domain Archaea,** and **domain Eukarya.** Figure 1.5 shows how the domains are believed to be related. Both domain Bacteria and domain Archaea may have evolved from the first common ancestor soon after life began. These two domains contain the **prokaryotes,** which lack the membrane-bounded nucleus found in the **eukaryotes** of domain Eukarya. However, archaea organize their DNA differently than bacteria, and their cell walls and membranes are chemically more similar to eukaryotes than to bacteria. So, the conclusion is that eukarya split off from the archaeal line of descent.

Prokaryotes are structurally simple but metabolically complex. Archaea (Fig. 1.6) can live in aquatic environments that lack oxygen or are too salty, too hot, or too acidic for most other organisms. Perhaps these environments are similar to those of the primitive Earth, and archaea (Gk. *archae,* ancient) are the least evolved forms of life, as their name implies. Bacteria (Fig. 1.7) are variously adapted to living almost anywhere—in the water, soil, and atmosphere, as well as on our skin and in our mouths and large intestines.

Taxonomists are in the process of deciding how to categorize archaea and bacteria into kingdoms. Domain Eukarya, on the other hand, contains four major groups of organisms (Fig. 1.8). **Protists,** which now comprise a number of kingdoms, range from unicellular forms to a few multicellular ones. Some are photosynthesizers, and some must acquire their food. Common protists include algae, the protozoans, and the water molds. Figure 1.5 shows that plants, fungi, and animals most likely evolved from protists. **Plants** (kingdom Plantae) are multicellular photosynthetic organisms. Example plants include azaleas, zinnias, and pines. Among the **fungi**

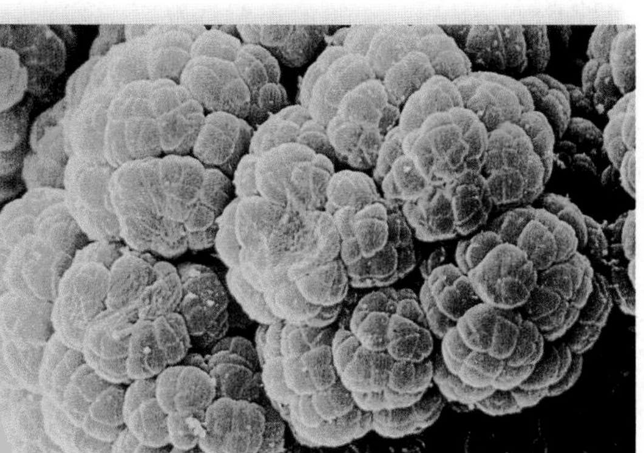

*Methanosarcina mazei,* an archaeon 1.6 μm

- Prokaryotic cells of various shapes
- Adaptations to extreme environments
- Absorb or chemosynthesize food
- Unique chemical characteristics

**FIGURE 1.6** **Domain Archaea.**

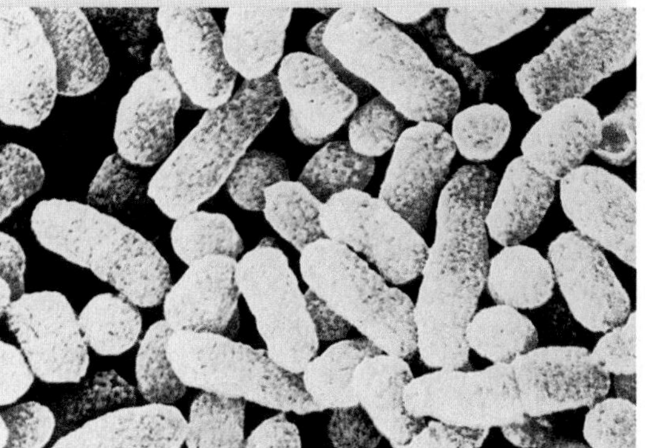

*Escherichia coli,* a bacterium 1.5 μm

- Prokaryotic cells of various shapes
- Adaptations to all environments
- Absorb, photosynthesize, or chemosynthesize food
- Unique chemical characteristics

**FIGURE 1.7** **Domain Bacteria.**

**Protists**

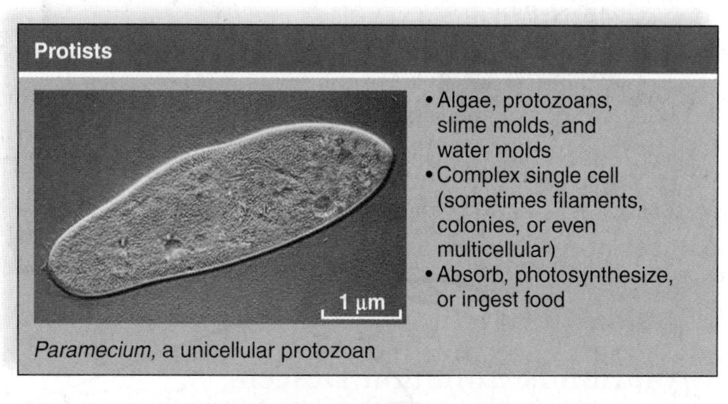


- Algae, protozoans, slime molds, and water molds
- Complex single cell (sometimes filaments, colonies, or even multicellular)
- Absorb, photosynthesize, or ingest food

*Paramecium,* a unicellular protozoan

**KINGDOM: Fungi**

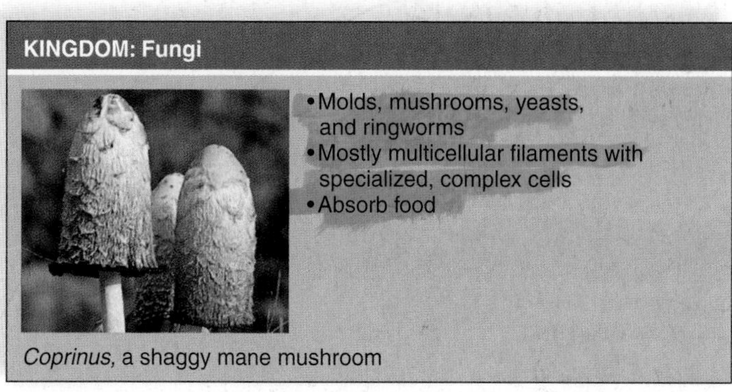

- Molds, mushrooms, yeasts, and ringworms
- Mostly multicellular filaments with specialized, complex cells
- Absorb food

*Coprinus,* a shaggy mane mushroom

**KINGDOM: Plants**

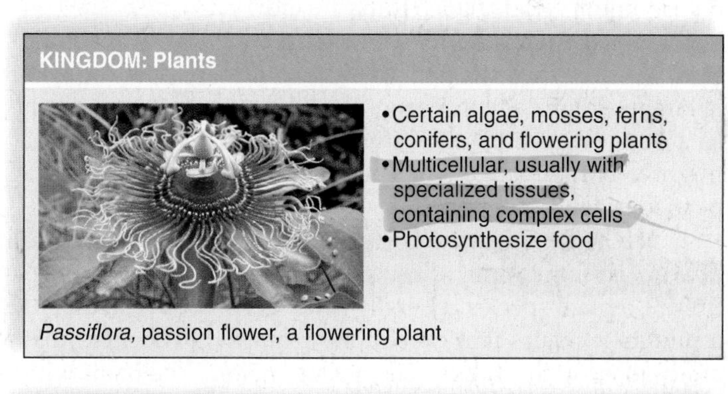

- Certain algae, mosses, ferns, conifers, and flowering plants
- Multicellular, usually with specialized tissues, containing complex cells
- Photosynthesize food

*Passiflora,* passion flower, a flowering plant

**KINGDOM: Animals**

- Sponges, worms, insects, fishes, frogs, turtles, birds, and mammals
- Multicellular with specialized tissues containing complex cells
- Ingest food

*Vulpes,* a red fox

**FIGURE 1.8** **Domain Eukarya.**

(kingdom Fungi) are the familiar molds and mushrooms that, along with bacteria, help decompose dead organisms. **Animals** (kingdom Animalia) are multicellular organisms that must ingest and process their food. Aardvarks, jellyfish, and zebras are representative animals.

### *Scientific Name*

Biologists use **binomial nomenclature** [L. *bi*, two, and *nomen*, name] to assign each living thing a two-part name called a scientific name. For example, the scientific name for mistletoe is *Phoradendron tomentosum*. The first word is the genus, and the second word is the specific epithet of a species within a genus. The genus may be abbreviated (e.g., *P. tomentosum*) and the species may simply be indicated if it is unknown (e.g., *Phoradendron* sp.). Scientific names are universally used by biologists to avoid confusion. Common names tend to overlap and often are in the language of a particular country. But scientific names are based on Latin, a universal language that not too long ago was well known by most scholars.

## Evolution Is Common Descent with Modification

The phrase "common descent with modification" sums up the process of evolution because it means that, as descent occurs from common ancestors, so do modifications that cause organisms to be adapted to the environment. Through many observations and experiments, Charles Darwin came to the conclusion that **natural selection** was the process that made modification—that is, adaptation—possible.

### *Natural Selection*

During the process of natural selection, some aspect of the environment selects which traits are more apt to be passed on to the next generation. The selective agent can be an abiotic agent (part of the physical environment, such as altitude) or it can be a biotic agent (part of the living environment, such as a deer). Figure 1.9 shows how the dietary habits of deer might eventually affect the characteristics of the leaves of a particular land plant.

Mutations fuel natural selection because mutation introduces variations among the members of a population. In Figure 1.9, a plant species generally produces smooth leaves, but a mutation occurs that causes one plant to have leaves that are covered with small extensions or "hairs." The plant with hairy leaves has an advantage because the deer (the selective agent) prefer to eat smooth leaves and not hairy leaves. Therefore, the plant with hairy leaves survives best and produces more seeds than most of its neighbors. As a result, generations later most plants of this species produce hairy leaves.

As with this example, Darwin realized that although all individuals within a population have the ability to reproduce, not all do so with the same success. Prevention of reproduction can run the gamut from an inability to capture resources, as when long-neck, but not short-neck, giraffes can reach their food source, to an inability to escape being eaten because long legs but not short legs can carry an animal to safety. Whatever the example, it can be seen that living things with advantageous traits can produce more offspring than those that lack them. In this way, living things change over time, and these changes are passed on from one generation to the next. Over long periods of time, the introduction of newer, more advantageous traits into a population may drastically reshape a species. Natural selection tends to sculpt a species to fit its environment and lifestyle and can create new species from existing ones. The end result is the diversity of life classified into the three domains of life (see Fig. 1.5).

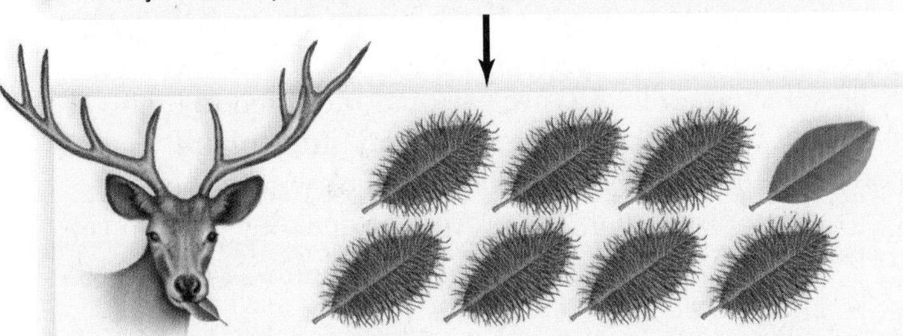


**FIGURE 1.9 Natural selection.**

Natural selection selects for or against new traits introduced into a population by mutations. Over many generations, selective forces such as competition, predation, and the physical environment alter the makeup of a population to more suit its environment and lifestyle.

**Check Your Progress 1.2**

1. List the levels of taxonomic classification from most inclusive to least inclusive.
2. What differences might be used to distinguish the various kingdoms of domain Eukarya?
3. Explain how natural selection results in new adaptations within a species.

# 1.3 How the Biosphere Is Organized

The organization of life extends beyond the individual organism to the **biosphere,** the zone of air, land, and water at the surface of the Earth where organisms exist (see Fig. 1.2). Individual organisms belong to a **population,** which is all the members of a species within a particular area. The populations of a **community** interact among themselves and with the physical environment (e.g., soil, atmosphere, and chemicals), thereby forming an **ecosystem.**

Figure 1.10 depicts a grassland inhabited by populations of rabbits, mice, snakes, hawks, and various types of land plants. These populations exchange gases with and give off heat to the atmosphere. They also take in water from and give off water to the physical environment. In addition, populations interact by forming food chains in which one population feeds on another. Mice feed on plants and seeds, snakes feed on mice, and hawks feed on rabbits and snakes, for example. Interactions between the various food chains make up a food web.

Ecosystems are characterized by chemical cycling and energy flow, both of which begin when photosynthetic plants, aquatic algae, and some bacteria take in solar energy and inorganic nutrients to produce food in the form of organic nutrients. The gray arrows in Figure 1.10 represent chemical cycling—chemicals move from one population to another in a food chain, until with death and decomposition, inorganic nutrients are returned to living plants once again. The yellow to red arrows represent energy flow. Energy flows from the sun through plants and other members of the food chain as one population feeds on another. With each transfer some energy is lost as heat. Eventually, all the energy taken in by photosynthesizers has dissipated into the atmosphere. Because energy flows and does not cycle, ecosystems could not stay in existence without a constant input of solar energy and the ability of photosynthesizers to absorb it.

## The Human Population

Humans possess the unique ability to modify existing ecosystems, which can greatly upset their natural nutrient cycles. When an ecosystem's natural energy flow has been disrupted by eliminating food sources for other animal populations even the human population can eventually suffer harm. Humans clear forests or grasslands to grow crops; later, they build houses on what was once farmland; and finally, they convert small towns into cities. Coastal ecosystems are most vulnerable. As they are developed, humans send sediments, sewage, and other pollutants into the sea. Human activities destroy valuable coastal wetlands, which serve as protection against storms and as nurseries for a myriad of invertebrates and vertebrates.

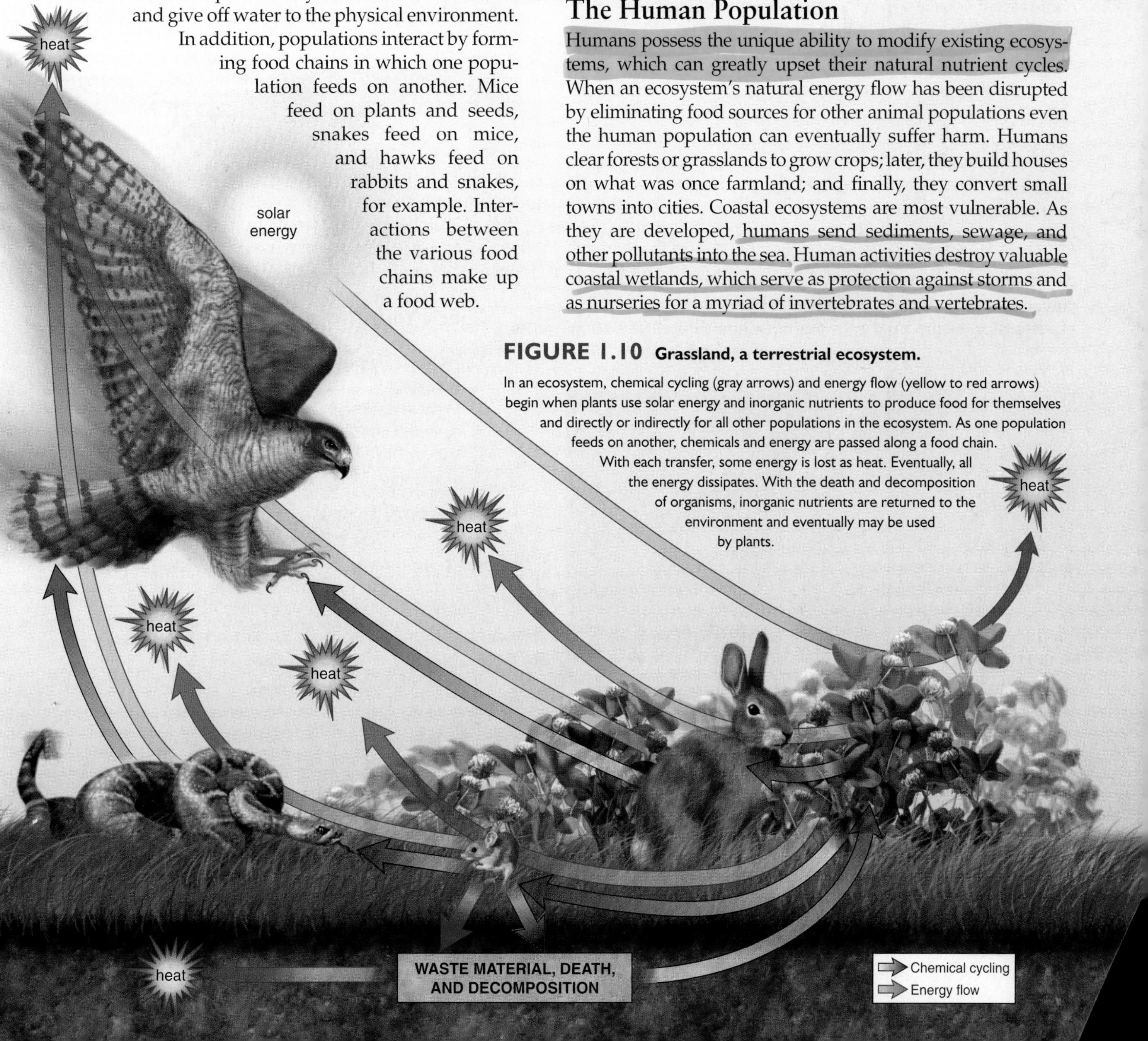

**FIGURE 1.10 Grassland, a terrestrial ecosystem.**

In an ecosystem, chemical cycling (gray arrows) and energy flow (yellow to red arrows) begin when plants use solar energy and inorganic nutrients to produce food for themselves and directly or indirectly for all other populations in the ecosystem. As one population feeds on another, chemicals and energy are passed along a food chain. With each transfer, some energy is lost as heat. Eventually, all the energy dissipates. With the death and decomposition of organisms, inorganic nutrients are returned to the environment and eventually may be used by plants.

## Biodiversity

The two most biologically diverse ecosystems—tropical rain forests and coral reefs—are home to many organisms. These ecosystems are also threatened by human activities. The canopy of the tropical rain forest alone supports a variety of organisms including orchids, insects, and monkeys. Coral reefs, which are found just offshore of the continents and islands of the Southern Hemisphere, are built up from calcium carbonate skeletons of sea animals called corals. Reefs provide a habitat for many animals, including jellyfish, sponges, snails, crabs, lobsters, sea turtles, moray eels, and some of the world's most colorful fishes (Fig. 1.11*a*). Like tropical rain forests, coral reefs are severely threatened as the human population increases in size. Some reefs are 50 million years old, and yet in just a few decades, human activities have destroyed 10% of all coral reefs and seriously degraded another 30% (Fig. 1.11*b*). At this rate, nearly three-quarters could be destroyed within 50 years. Similar statistics are available for tropical rain forests.

Destruction of healthy ecosystems has many unintended effects. For example, we depend on them for food, medicines, and various raw materials. Draining the natural wetlands of the Mississippi and Ohio rivers and the construction of levees has worsened flooding problems, making once fertile farmland undesirable. The destruction of South American rain forests has killed many species that may have yielded the next miracle drug and has also decreased the availability of many types of lumber.

We are only now beginning to realize that we depend on ecosystems even more for the services they provide. Just as chemical cycling occurs within a single ecosystem, so all ecosystems keep chemicals cycling throughout the entire biosphere. The workings of ecosystems ensure that the environmental conditions of the biosphere are suitable for the continued existence of humans. And several studies show that ecosystems cannot function properly unless they remain biologically diverse.

**Biodiversity** is the total number and relative abundance of species, the variability of their genes, and the different ecosystems in which they live. The present biodiversity of our planet has been estimated to be as high as 15 million species, and so far, less than 2 million have been identified and named. **Extinction** is the death of a species or larger classification category. It is estimated that presently we are losing as many as 400 species per day due to human activities. For example, several species of fishes have all but disappeared from the coral reefs of Indonesia and along the African coast because of overfishing. Many biologists are alarmed about the present rate of extinction and hypothesize it may eventually rival the rates of the five mass extinctions that have occurred during our planet's history. The last mass extinction, about 65 million years ago, caused many plant and animal species, including the dinosaurs, to become extinct.

It would seem that the primary bioethical issue of our time is preservation of ecosystems. Just as a native fisherman who assists in overfishing a reef is doing away with his own food source, so are we as a society contributing to the destruction of our home, the biosphere. If instead we adopt a conservation ethic that preserves the biosphere, we would help ensure the continued existence of our own species.

### Check Your Progress 1.3

1. How do various communities interact to form an ecosystem?
2. What are some unintentional ways in which human activities affect ecosystems?
3. Why might ecosystems with high biodiversity be more vulnerable to destruction by human activities?

a. Healthy coral reef

1975 Minimal coral death

1985 Some coral death with no fish present

1995 Coral bleaching with limited chance of recovery

2004 Coral is black from sedimentation; bleaching still evident

b.

**FIGURE 1.11 Coral reef, a marine ecosystem.**

**a.** Coral reefs, a type of ecosystem found in tropical seas, contain many diverse forms of life, a few of which are shown here. **b.** Various human activities have caused catastrophic damage to this coral reef off the coast of Florida, as shown over the course of 19 years. Preserving biodiversity is a modern-day challenge of great proportions.

# 1.4 The Process of Science

The process of science pertains to **biology,** the scientific study of life. Biology consists of many disciplines and areas of specialty because life has numerous aspects. Some biological disciplines are cytology, the study of cells; anatomy, the study of structure; physiology, the study of function; botany, the study of plants; zoology, the study of animals; genetics, the study of heredity; and ecology, the study of the interrelationships between organisms and their environment.

Religion, aesthetics, ethics, and science are all ways in which human beings seek order in the natural world. Science differs from these other ways of knowing and learning because the scientific process uses the **scientific method,** a standard series of steps used in gaining new knowledge that is widely accepted among scientists. The steps of the scientific method are often applicable to other situations, and begin with observation (Fig. 1.12).

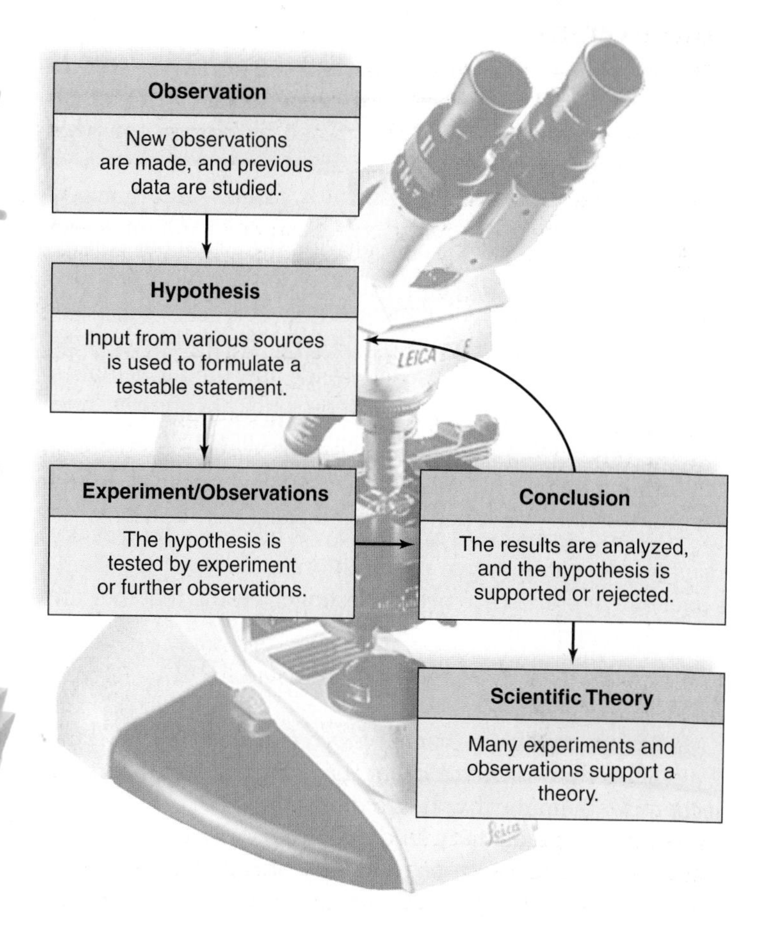


**FIGURE 1.12 Flow diagram for the scientific method.**
On the basis of new and/or previous observations, a scientist formulates a hypothesis. The hypothesis is tested by further observations and/or experiments, and new data either support or do not support the hypothesis. The return arrow indicates that a scientist often chooses to retest the same hypothesis or to test a related hypothesis. Conclusions from many different but related experiments may lead to the development of a scientific theory. For example, studies pertaining to development, anatomy, and fossil remains all support the theory of evolution.

## Observation

Scientists believe that nature is orderly and measurable—that natural laws, such as the law of gravity, do not change with time, and that a natural event, or **phenomenon,** can be understood more fully through observation. Scientists use all of their senses in making **observations.** The behavior of chimpanzees can be observed through visual means, the disposition of a skunk can be observed through olfactory means, and the warning rattles of a rattlesnake provide auditory information of imminent danger. Scientists also extend the ability of their senses by using instruments; for example, the microscope enables us to see objects that could never be seen by the naked eye. Finally, scientists may expand their understanding even further by taking advantage of the knowledge and experiences of other scientists. For instance, they may look up past studies at the library or on the Internet, or they may write or speak to others who are researching similar topics.

Nevertheless, chance alone can help a scientist get an idea. The most famous case pertains to penicillin. When examining a petri dish, Alexander Fleming observed an area around a mold that was free of bacteria. Upon investigating, Fleming found that the mold, a *Penicillium* species, produced an antibacterial substance he called penicillin, and he thought that perhaps penicillin would be useful in humans. This discovery changed medicine and has saved countless lives.

## Hypothesis

After making observations and gathering knowledge about a phenomenon, a scientist uses inductive reasoning. **Inductive reasoning** occurs whenever a person uses creative thinking to combine isolated facts into a cohesive whole. In this way, a scientist comes up with a **hypothesis,** a possible explanation for a natural event. The hypothesis is a statement that can be tested in a manner suited to the process of science.

All of a scientist's past experiences, no matter what they might be, will most likely influence the formation of a hypothesis. But a scientist only considers hypotheses that can be tested. Moral and religious beliefs, while very important to the lives of many people, differ between cultures and through time and may not be testable.

## Experiments/Further Observations

Testing a hypothesis involves either conducting an **experiment** or making further observations. To determine how to test a hypothesis, a scientist uses deductive reasoning. **Deductive reasoning** involves "if, then" logic. For example, a scientist might reason, if organisms are composed of cells, then microscopic examination of any part of an organism should reveal cells. We can also say that the scientist has made a **prediction** that the hypothesis can be supported by doing microscopic studies. Making a prediction helps a scientist know what to do next.

The manner in which a scientist intends to conduct an experiment is called the **experimental design.** A good experimental design ensures that scientists are testing what they want to test and that their results will be meaningful. It is always best for an experiment to include a control group. Often, a control group, or simply the **control,** goes through all the steps of an experiment but lacks the factor (is not exposed to the factor) being tested.

In some cases, scientists may use a **model** as a representation of the actual object because altering the actual object may be physically impossible, very expensive, or morally questionable. Later in this section, a scientist uses bluebird models because it would have been impossible to get live birds to cooperate. Computer models are used to decide how human activities will affect climate, because of expense, ethical concerns, and physical limitations. Scientists often use mice instead of humans for medical research because of ethical concerns. Bacteria are used in much genetic research because they are inexpensive to grow and reproduce very quickly. While these models are usually relevant and give useful information, they are themselves still hypotheses in need of testing to ensure that they are valid representations.

### *Data*

The results of an experiment are referred to as the **data.** Data should be observable and objective, rather than subjective. Mathematical data are often displayed in the form of a graph or table. Many studies, such as the one discussed in the Science Focus on page 13, rely on statistical data. As a hypothetical example, let's say an investigator wants to know if eating onions can prevent women from getting osteoporosis (weak bones). The scientist conducts a survey asking women about their onion-eating habits and then correlates this data with the condition of their bones. Other scientists critiquing this study would want to know: How many women were surveyed? How old were the women? What were their exercise habits? What proportion of the diet consisted of onions? And what criteria were used to determine the condition of their bones? Should the investigators conclude that eating onions does protect a woman from osteoporosis, other scientists might want to know the statistical probability of error. The probability of error is a mathematical calculation based on the conditions and methods of the experiment. If the results are significant at a 0.30 level, then the probability that the correlation is incorrect is 30% or less. (This would be considered a high probability of error.) The greater the variance in the data, the greater the probability of error. Even if this study had a low probability of error, it would be considered hypothetical until we learn of some ingredient in onions that has a direct biochemical or physiological effect on bones. Therefore, scientists must be skeptics who always pressure one another to continue investigating a particular topic.

### Conclusion

Scientists must analyze the data in order to reach a **conclusion** as to whether the hypothesis is supported or not (see Fig. 1.12). Because science progresses, the conclusion of one experiment can lead to the hypothesis for another experiment, as represented by the return arrow in Figure 1.12. Results that do not support one hypothesis can often help a scientist formulate another hypothesis to be tested. Scientists report their findings in scientific journals so that their methodology and data are available to other scientists for critique. Experiments and observations must be repeatable—that is, the reporting scientist and any scientist who repeats the experiment must get the same results, or else the data are suspect.

## Scientific Theory

The ultimate goal of science is to understand the natural world in terms of **scientific theories,** which are concepts that join together well-supported and related hypotheses. In ordinary speech, the word *theory* refers to a speculative idea. In contrast, a scientific theory is supported by a broad range of observations, experiments, and data often from a variety of disciplines. Some of the basic theories of biology are:

| *Theory* | *Concept* |
|---|---|
| Cell | All organisms are composed of cells, and new cells only come from preexisting cells. |
| Homeostasis | The internal environment of an organism stays relatively constant—within a range that is protective of life. |
| Gene | Organisms contain coded information that dictates their form, function, and behavior. |
| Ecosystem | Organisms are members of populations, which interact with each other and the physical environment within a particular locale. |
| Evolution | All living things have a common ancestor, but each is adapted to a particular way of life. |

As stated earlier, the theory of evolution is the unifying concept of biology because it pertains to many different aspects of living things. For example, the theory of evolution enables scientists to understand the history of living things, and the anatomy, physiology, and embryological development of organisms. Even behavior can be described through evolution, as we shall see in a study discussed later in this chapter.

The theory of evolution has been a fruitful scientific theory, meaning that it has helped scientists generate new hypotheses. Because this theory has been supported by so many observations and experiments for over 100 years, some biologists refer to the **principle** of evolution, a term sometimes used for theories that are generally accepted by an overwhelming number of scientists. The term **law** instead of principle is preferred by some. For instance, in a subsequent chapter concerning energy relationships, we will examine the laws of thermodynamics.

*science focus*

# The Benefits and Limitations of Statistical Studies

Many of the studies published in scientific journals and reported in the news are statistical studies, so it behooves us to be aware of their benefits and limitations. At the start, you should know that a statistical study will gather numerical information from various sources and then try to make sense out of it, for the purpose of coming to a conclusion.

## Example of a Statistical Study

Let's take a look at a study that allows us to conclude that babies conceived 18 months to five years after a previous birth are healthier than those conceived at shorter or longer intervals. In other words, spacing children about two to five years apart is a good idea (Fig. 1A). Here is how the authors collected their data and the results they published in the *Journal of the American Medical Association*.*

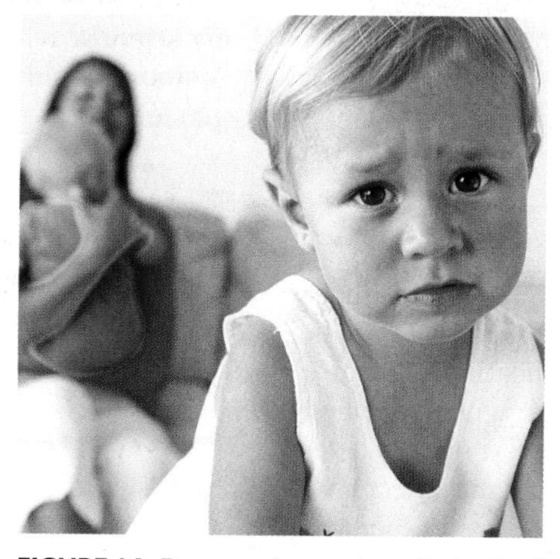

**FIGURE 1A Does spacing pregnancies lead to healthier children?** A recent statistical study suggests that it does. If so, which mother, left or right, may have a healthier younger child?

**Objective.** To determine if there is an association between birth spacing and a healthy baby when data are corrected for maternal characteristics or socioeconomic status.

**Data.** The authors collected data from studies performed around the world in 1966 through January 2006. The studies were published in various journals, reported on at professional meetings, or were known to the authors by personal contact. The authors gathered a very large pool of data that included over 11 million pregnancies from 67 individual studies. Twenty of the studies were from the United States, with the remaining 47 coming from 61 different countries. The authors attempted to adjust the data (by elimination of certain data) for factors such as mother's age, wealth, access to prenatal care, and breastfeeding. These adjustments allow the findings to be applied to both developed and developing countries.

**Conclusion.**

1. A pregnancy that begins less than six months after a previous birth has a 77% higher chance of being preterm and a 39% higher chance of lower birth weight.
2. For up to 18 months between pregnancies, the chance of a preterm birth decreases by 2% per month, and the chance of a low-weight birth decreases by 3% per month as the 18-month time period is approached.
3. Babies conceived after 59 months have the same risk as those conceived in the less-than-six-months group.
4. The optimum spacing between pregnancies appears to be 18 months to five years after a previous birth.

The study leader, Agustin Conde-Agudelo, said, "Health officials should counsel women who have just given birth to delay their next conception by 18 to 59 months."

## Limitations of Experimental Studies

The expression "statistical study" is a bit of a misnomer because most scientists collect quantitative data and use them to come to a conclusion. However, if we compare this study to experimental studies, we can see that the experimental studies include both a control group and test groups. The groups are treated the same except for the experimental variable. Obviously, you wouldn't be able to divide women of the same childbearing age into various groups and tell each group when they will conceive their children for the purpose of deciding the best interval between pregnancies for the health of the newborn. So, what is the next best thing? Do a statistical study utilizing data already available about women who became pregnant at different intervals.

A statistical study is really a correlation study. In our example, the authors studied the correlation between birth spacing and the health of a newborn. The more data collected from more varied sources make a correlation study more reliable. The study by Conde-Agudelo has a very large sample size, which goes a long way to validating the results. Even so, a correlation does not necessarily translate to causation. So, it is not surprising that Dr. Mark A. Klebanoff, director of the National Institute of Child Health and Human Development, commented that many factors will affect birth spacing and that the study is not detailed enough to take all factors into consideration. Is any statistical study detailed enough? Most likely not.

## Benefits of Statistical Studies

Before we give up on statistical studies, let's consider that they do provide us with information not attainable otherwise. Regardless of whether we understand the intricacies of statistical analysis, statistical studies do allow scientists to gain information and insights into many problems. True, further study is needed to find out if the observed correlation does mean causation, but science is always a work in progress, with additional findings being published every day.

* Agustin Conde-Agudelo, MD, MPH, Anyeli Rosas-Bermudez, MPH, Ana Cecilia Kafury-Goeta, MD. "Birth Spacing and Risk of Adverse Perinatal Outcomes." *JAMA*, 2006;295:1809–23. Abstract.

## Using the Scientific Method

Scientists using the scientific method often do controlled studies to ensure that the outcome is due to the **experimental variable** or independent variable, the component or factor being tested. The result is called the **responding variable** or dependent variable because it is due to the experimental variable:

| **Experimental Variable** (Independent Variable) | **Responding Variable** (Dependent Variable) |
|---|---|
| Factor of the experiment being tested | Result or change that occurs due to the experimental variable |

### *Observation*

Researchers doing this study knew that nitrogen fertilizer in the short run enhances yield and increases food supplies. However, excessive nitrogen fertilizer application can cause pollution by adding toxic levels of nitrates to water supplies. Also, applying nitrogen fertilizer year after year may alter soil properties to the point that crop yields may decrease, instead of increase. Then the only solution is to let the land remain unplanted for several years until the soil recovers naturally.

An alternative to the use of nitrogen fertilizers is the use of legumes, plants such as peas and beans, that increase soil nitrogen. Legumes provide a home for bacteria that convert atmospheric nitrogen to a form usable by the plant. The bacteria live in nodules on the roots (Fig. 1.13). The bacteria supply the plant with nitrogen compounds, and in turn, the plant passes the product of photosyntheis to the nodules.

Numerous legume crops can be rotated (planted every other season) with any number of cereal crops. The nitrogen added to the soil by the legume crop is a natural fertilizer that increases the yield of cereal crops. The particular rotation used by farmers tends to depend on the location, climate, and market demand. In this study, researchers perform an experiment in which method of fertilization is the experimental variable and enhanced yield is the responding variable.

**FIGURE 1.13 Root nodules.**

Bacteria that live in nodules on the roots of legumes, such as pea plants, convert nitrogen in the air to a form that land plants can use to make proteins and other nitrogen-containing molecules.

### *Hypothesis*

Researchers doing this study knew that the pigeon pea plant is a legume with a high rate of atmospheric nitrogen conversion. This plant is widely grown as a food crop in India, Kenya, Uganda, Pakistan, and other subtropical countries. Researchers formulated the hypothesis that a pigeon pea/winter wheat rotation would be a reasonable alternative to the use of nitrogen fertilizer to increase the yield of winter wheat.

HYPOTHESIS: A pigeon pea/winter wheat rotation will cause winter wheat production to increase as well as or better than the use of nitrogen fertilizer.

PREDICTION: Wheat biomass following the growth of pigeon peas will surpass wheat biomass following nitrogen fertilizer treatment.

### *Experiment*

In this study, the investigators decided on the following experimental design (Fig. 1.14*a*):

CONTROL POTS

- Winter wheat was planted in pots of soil that received no fertilization treatment—that is, no nitrogen fertilizer and no preplanting of pigeon peas.

TEST POTS

- Winter wheat was grown in clay pots in soil treated with nitrogen fertilizer equivalent to 45 kilograms (kg)/hectare (ha).
- Winter wheat was grown in clay pots in soil treated with nitrogen fertilizer equivalent to 90 kg/ha.
- Pigeon pea plants were grown in clay pots in the summer. The pigeon pea plants were then tilled into the soil and winter wheat was planted in the same pots.

To ensure a controlled experiment, the conditions for the control pots and the test pots were identical; the plants were exposed to the same environmental conditions and watered equally. During the following spring, the wheat plants were dried and weighed to determine wheat biomass production in each of the pots.

### *Data*

After the first year, wheat biomass was higher in certain test pots than in the control pots (Fig. 1.14*b*). Specifically, test pots with 45 kg/ha of nitrogen fertilizer (orange) had only slightly more wheat biomass production than the control pots, but test pots that received 90 kg/ha treatment (green) demonstrated nearly twice the biomass production of the control pots. To the surprise of investigators, wheat production following summer planting of pigeon peas did not demonstrate as high a biomass production as the control pots.

### *Conclusion and Further Investigation*

Wheat biomass following the growth of pigeon peas is not as great as that obtained with nitrogen fertilizer treatments, meaning that the data from the experiment did not support the investigators' hypothesis. This is not an

**FIGURE 1.14 Pigeon pea/winter wheat rotation study.**

**a.** Experiment involves control pots and test pots of three types: test pots that received 45 kg/ha of nitrogen; test pots that received 90 kg/ha of nitrogen; and test pots in which pigeon peas rotated with winter wheat. **b.** The graph compares wheat biomass for each of three years. Wheat biomass in test pots that received the most nitrogen fertilizer (green) declined while wheat biomass in test pots with pigeon pea/winter wheat rotation (brown) increased dramatically.

uncommon event in scientific investigations. However, the investigators decided to continue the experiment using the same design and the same pots as before, to see if the buildup of residual soil nitrogen from pigeon peas would eventually increase wheat biomass. So they proposed a new hypothesis.

> HYPOTHESIS: A sustained pigeon pea/winter wheat rotation will eventually cause an increase in winter wheat production.
>
> PREDICTION: Wheat biomass following two years of pigeon pea/winter wheat rotation will surpass wheat biomass following nitrogen fertilizer treatment.

After two years, the yield following 90 kg/ha nitrogen treatment (green) was not as much as it was the first year (Fig. 1.14*b*). Indeed, wheat biomass following summer planting of pigeon peas (brown) was the highest of all treatments, suggesting that buildup of residual nitrogen from pigeon peas had the potential to provide fertilization for winter wheat growth.

> CONCLUSION: The hypothesis is supported. At the end of two years, the yield of winter wheat following a pigeon pea/winter wheat rotation was better than for the other type pots.

The researchers continued their experiment for still another year. After three years, winter wheat biomass production had decreased in the control pots and in the pots treated with nitrogen fertilizer. Pots treated with nitrogen fertilizer still had increased wheat biomass production compared with the control pots but not nearly as much as pots following summer planting of pigeon peas. Compared to the first year, wheat biomass increased almost fourfold in pots having a pigeon pea/winter wheat rotation (brown, Fig. 1.14*b*). The researchers suggested that the soil was improved by the organic matter as well as the addition of nitrogen from the pigeon peas. The researchers published their results in a scientific journal.[1]

[1] Bidlack, J. E., Rao, S. C., and Demezas, D. H. 2001. Nodulation, nitrogenase activity, and dry weight of chickpea and pigeon pea cultivars using different *Bradyrhizobium* strains. *Journal of Plant Nutrition* 24:549–60.

## A Field Study

A scientist, David Barash, while observing the mating behavior of mountain bluebirds (Fig. 1.15*a*, *b*), formulated the hypothesis that aggression of the male varies during the reproductive cycle. To test this hypothesis, he reasoned that he should evaluate the intensity of male aggression at three stages: after the nest is built, after the first egg is laid, and after the eggs hatch.

HYPOTHESIS: Male bluebird aggression varies during the reproductive cycle.

PREDICTION: Aggression intensity will change after the nest is built, after the first egg is laid, and after hatching.

### *Testing the Hypothesis*

For his experiment, Barash decided to measure aggression intensity by recording the "number of approaches per minute" a male made toward a rival male and his own female mate. To provide a rival, Barash posted a male bluebird model near the nests while resident males were out foraging. The aggressive behavior (approaches) of the resident male was noted during the first 10 minutes of the male's return (Fig. 1.15*c*). To give his results validity, Barash included a control group. For his control, Barash posted a male robin model instead of a male bluebird near certain nests.

Resident males of the control group did not exhibit any aggressive behavior, but resident males of the experimental groups did exhibit aggressive behavior. Barash graphed his mathematical data (Fig. 1.15*d*). By examining the graph, you can see that the resident male was more aggressive toward the rival male model than toward his female mate, and that he was most aggressive while the nest was under construction, less aggressive after the first egg was laid, and least aggressive after the eggs hatched.

### *The Conclusion*

The results allowed Barash to conclude that aggression in male bluebirds is related to their reproductive cycle. Therefore, his hypothesis was supported. If male bluebirds were always aggressive, even toward male robin models, his hypothesis would not have been supported.

CONCLUSION: The hypothesis is supported. Male bluebird aggression does vary during the reproductive cycle.

Barash reported his experiment in *The American Naturalist*.[2] In this article, Barash gave an evolutionary interpretation to his results. It was adaptive, he said, for male bluebirds to be less aggressive after the first egg is laid because by then the male bird is "sure the offspring is his own." It was maladaptive for the male bird to waste energy being aggressive after hatching because his offspring are already present.

**Check Your Progress 1.4**

1. What is the benefit of an experimental control?
2. How might using a model affect the conclusions drawn from an experiment?
3. What are the possible disadvantages of the peer review system?

[2] Barash, D.P. 1976. Male response to apparent female adultery in the mountain bluebird (*Sialia currucoides*): an evolutionary interpretation. *The American Naturalist* 110:1097–1101.

**FIGURE 1.15** **A field study.**

Observation of normal male bluebird behavior (**a** and **b**) allowed David Barash to formulate a testable hypothesis. He (**c**) collected data, which was (**d**) displayed in a graph. Then, he came to a conclusion.

## Connecting the Concepts

The diversity of life on Earth is staggering, but organisms are united by a number of common features that define them as living. Among these features is the ability to adapt, and descent with modification occurs when these adaptations are passed from one generation to the next over long periods of time. Evolution is a unifying theory in biology that accounts for the differences that divide and the unity that joins all living things. All living things are organized and function similarly because they share a common evolution extending back through time to the first cells on Earth.

What we know about biology and what we'll learn in the future result from objective observation and testing of the natural world through the scientific method. The ultimate goal of science is to understand the natural world in terms of theories—conceptual schemes supported by abundant research. Scientists should provide the public with as much information as possible, especially when such issues as recombinant DNA technology or human impacts on the biosphere are being debated. Then they, along with other citizens, can help make intelligent decisions about what is most likely best for society. Everyone has a responsibility to decide how to use scientific knowledge so that it benefits all living things, including the human species.

This textbook was written to help you understand the scientific process and learn the basic concepts of general biology so that you will be better informed. This chapter has introduced you to the levels of biological organization, from the cell to the biosphere. The cell, the simplest of living things, is composed of nonliving molecules. Therefore, we must begin our study of biology with a brief look at cellular chemistry. In the next two chapters, you will study some important inorganic and organic molecules as they relate to cells. Then, you will learn how the cell makes use of energy and materials to maintain itself and to reproduce.

## summary

### 1.1 How to Define Life

Although living things are diverse, they have certain characteristics in common. Living things (a) are organized, and their levels of organization extend from the cell to ecosystems; (b) need an outside source of materials and energy; (c) respond to external stimuli; (d) reproduce and develop, passing on genes to their offspring; and (e) have adaptations suitable to their way of life in a particular environment. Together, these characteristics unify life on Earth.

### 1.2 Evolution, the Unifying Concept of Biology

Life on Earth is diverse, but the theory of evolution unifies life and describes how all living organisms evolved from a common ancestor. Taxonomists assign each living thing an italicized binomial name that consists of the genus and the specific epithet. From the least inclusive to the most inclusive category, each species belongs to a genus, family, order, class, phylum, kingdom, and finally domain.

The three domains of life are Archaea, Bacteria, and Eukarya. The first two domains contain prokaryotic organisms that are structurally simple but metabolically complex. Domain Eukarya contains the protists, fungi, plants, and animals. Protists range from unicellular to multicellular organisms and include the protozoans and most algae. Among the fungi are the familiar molds and mushrooms. Plants are well known as the multicellular photosynthesizers of the world, while animals are multicellular and ingest their food. An evolutionary tree shows how the domains are related by way of common ancestors.

Natural selection describes the process by which living organisms are descended from a common ancestor. Mutations occur within a population, creating new traits. The agents of natural selection, present in both biological and physical environments, shape species over time and may create new species from existing ones.

### 1.3 How the Biosphere Is Organized

Within an ecosystem, populations interact with one another and with the physical environment. Nutrients cycle within and between ecosystems, but energy flows unidirectionally and is eventually lost as unusable forms. Adaptations of organisms allow them to play particular roles within an ecosystem.

### 1.4 The Process of Science

When studying the natural world, scientists use the scientific process. Observations, along with previous data, are used to formulate a hypothesis. New observations and/or experiments are carried out in order to test the hypothesis. A good experimental design includes an experimental variable and a control group. The experimental and observational results are analyzed, and the scientist comes to a conclusion as to whether the results support the hypothesis or do not support the hypothesis.

Several conclusions in a particular area may allow scientists to arrive at a theory, such as the cell theory, the gene theory, or the theory of evolution. The theory of evolution is a unifying concept of biology.

## understanding the terms

protist 7
reproduce 5
responding variable 14
scientific method 11
scientific theory 12
species 6
taxonomy 6
unicellular 2

Match the terms to these definitions:

a. _______________ All of the chemical reactions that occur in a cell during growth and repair.
b. _______________ Changes that occur among members of a species with the passage of time, often resulting in increased adaptation to the prevailing environment.
c. _______________ Component in an experiment that is manipulated as a means of testing it.
d. _______________ Process by which plants use solar energy to make their own organic food.
e. _______________ Sample that goes through all the steps of an experiment but lacks the factor being tested.

## reviewing this chapter

1. What are the common characteristics of life listed in the chapter? 2–5
2. Describe the levels of biological organization. 2
3. Why do living things require an outside source of nutrients and energy? Describe these sources. 4
4. What is passed from generation to generation when organisms reproduce? What has to happen to the hereditary material DNA for evolution to occur? 5
5. How does evolution explain both the unity and the diversity of life? 5–6
6. What are the categories of classification? How does the domain Eukarya differ from domain Bacteria and domain Archaea? 6
7. Explain the scientific name of an organism. 6
8. How does natural selection result in adaptation to the environment? 8
9. What is an ecosystem, and why should human beings preserve ecosystems? 9–10
10. Describe the series of steps involved in the scientific method. 11–12
11. What is the ultimate goal of science? Give an example that supports your answer. 12
12. Give an example of a controlled study. Name the experimental variable and the responding variable. 14–15
13. What is a field study, and how does it differ from a controlled study? How are they similar? 16

## testing yourself

Choose the best answer for each question.

1. Which of these is not a property of all living organisms?
   a. organization
   b. acquisition of materials and energy
   c. care for their offspring
   d. reproduction
   e. responding to the environment
2. Describe an emergent property that might arise when moving from a single neuron (nerve cell) to nervous tissue.
3. The level of organization that includes cells of similar structure and function would be
   a. an organ.
   b. a tissue.
   c. an organ system.
   d. an organism.
4. The color, temperature, and foul odor of the flowers of the titan arum are examples of
   a. obtaining materials
   b. adaptations
   c. organizations
   d. homeostasis
5. Which of the following is an example of adaptation?
   a. In a very wet year, some plants grow unusually tall stalks and large leaves.
   b. Over millions of years, the eyes of cave salamanders lose their function.
   c. An escaped dog joins a pack of wild dogs and begins interbreeding with them.
   d. A harsh winter kills many birds within a population, especially the smallest ones.
6. Energy is brought into ecosystems by which of the following?
   a. fungi and other decomposers
   b. cows and other organisms that graze on grass
   c. meat-eating animals
   d. organisms that photosynthesize, such as plants
   e. All of these are correct.
7. We use the scientific method every day. Suppose one morning that your car does not start. Which of the following is a testable hypothesis stemming from this observation?
   a. I'm going to be late.
   b. My battery is dead.
   c. Check to see if I left the lights on.
   d. Kick the tires.
   e. I will add a quart of oil.
8. Which of the following statements is a hypothesis?
   a. Will increasing my cat's food increase her weight?
   b. Increasing my cat's food consumption will result in a 25% increase in her weight.
   c. I will feed my cat more food.
   d. My cat has gained weight; therefore, she is eating more food.
9. After formulating a hypothesis, a scientist
   a. proves the hypothesis true or false.
   b. tests the hypothesis.
   c. decides how to best avoid having a control.
   d. makes sure environmental conditions are just right.
   e. formulates a scientific theory.
10. The experimental variable in the bluebird experiment was the
    a. use of a model male bluebird.
    b. observations of the experimenter.
    c. various behavior of the males.
    d. identification of what bluebirds to study.
    e. All of these are correct.
11. The control group in the pigeon pea/winter wheat experiment was the pots that were
    a. planted with pigeon peas.
    b. treated with nitrogen fertilizer.
    c. not treated.
    d. not watered.
    e. Both c and d are correct.
12. Which of the following are agents of natural selection?
    a. changes in the environment
    b. competition among individuals for food and water
    c. predation by another species
    d. competition among members of a population for prime nesting sites
    e. All of these are correct.

13. Which of the following is an example of natural selection?
    a. In a very wet year, some plants grow unusually tall stalks and large leaves.
    b. After several unusually cold winters, squirrels with an extra layer of fat have more offspring.
    c. Squirrels may have long or short tails.
    d. Dogs with longer legs are able to run faster than dogs with shorter legs.
14. Which of the following statements regarding evolution is false?
    a. Adaptations may be physical or behavioral.
    b. Natural selection always results in organisms becoming more adapted to the environment.
    c. A trait selected for may suddenly become selected against when the environment changes.
    d. Some traits are neither selected for nor against.

For questions 15–17, write a brief answer.

15. Why is it said that all energy used by living organisms originates from the sun?
16. Carbon dioxide emissions have been blamed for climate change by many scientists. How might excessive amounts of carbon dioxide affect nutrient cycling?
17. Would the accidental introduction of a new species to an ecosystem necessarily have a negative effect on biodiversity? Why or why not?

## thinking scientifically

1. An investigator spills dye on a culture plate and notices that the bacteria live despite exposure to sunlight. He decides to test if the dye is protective against ultraviolet (UV) light. He exposes one group of culture plates containing bacteria and dye and another group containing only bacteria to UV light. The bacteria on all plates die. Complete the following diagram.

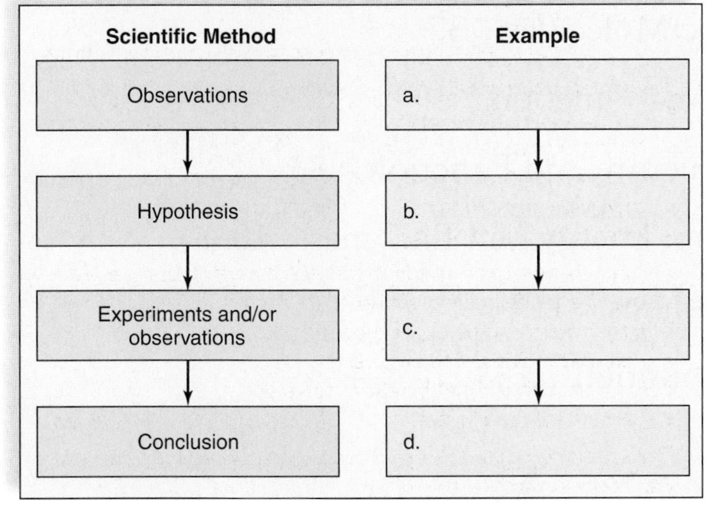


2. You want to grow large tomatoes and notice that a name-brand fertilizer claims to produce larger produce than a generic brand. How would you test this claim?
3. A scientist wishes to test her hypothesis that a commonly used drug causes heart attacks in some individuals. What kind of study should she initiate? What would you expect her experimental and responding variables to be?

## bioethical issue

### Oil Drilling in the Arctic

Established by an act of Congress in 1980, the Arctic National Wildlife Refuge (ANWR) covers a total of 19 million acres of northernmost Alaska far above the Arctic Circle. ANWR is home to a variety of wildlife, such as caribou, migratory birds, grizzly and polar bears, wolves, and musk oxen. But it is also home to substantial oil reserves, which has led to an ongoing contentious debate over its future: Should Congress allow development of ANWR for oil exploration and drilling?

Those who favor oil drilling in ANWR insist that first and foremost, the impact on the land would be minimal. The affected area would be roughly the size of an airport in a total area roughly the size of the state of South Carolina. They contend that the effect would mainly be underground because new techniques allow us to go lower and spread out beneath the surface to get the oil. Waste treatment and disposal methods have also improved. Acquiring the oil, advocates say, would also protect jobs and national security in the United States by lessening dependence on often hostile foreign countries for oil, and would have the added benefit of insulating the U.S. economy from oil price spikes and supply shocks.

Those who do not favor oil drilling in ANWR are eager to point out that at current levels of consumption, the oil coming from ANWR would hardly have a noticeable impact on prices and supply levels. Furthermore, they believe that the best solution to the current energy crunch would be for U.S. citizens to adopt simple energy conservation measures and invest in research on alternative fuels. They suggest that this would save many times the oil that could come from drilling in the Arctic refuge and that, by using a renewable energy resource, the environment in the lower 48 states would be protected, in addition to protecting the wildlife in the Arctic National Wildlife Refuge. Using renewable energy would lessen the need for foreign oil, and would also protect our national security.

Should Congress approve oil drilling in ANWR? Or should Congress invest in alternative and renewable energy forms, and insist that citizens adopt energy conservation measures? Should public tax monies be made available to Congress for oil exploration or for investment in alternative energy sources?

## *Biology* website

The companion website for *Biology* provides a wealth of information organized and integrated by chapter. You will find practice tests, animations, videos, and much more that will complement your learning and understanding of general biology.

**http://www.mhhe.com/maderbiology10**

part I

# The Cell

*We're going to take a fairly long journey through the various levels of biological organization from atoms to ecosystems, as shown in Figure 1.2. Whenever you get ready to go on a trip, you think about what you should bring with you and how to pack your suitcase. Similarly, you can think of the chapters in Part I as the necessities you are going to bring with you as we take our biological journey.*

*The chapters in Part I will teach you certain principles of biology that will apply to every chapter in the book. Chapters 2 and 3 introduce you to chemistry because all organisms are composed of chemicals, some of them quite unique to living things. In Chapters 4 and 5, we will see how these chemicals are arranged to form the structure of a cell, the basic unit of life. Some organisms are single cells and some are multicellular, but all are made up of cells. Chapters 6, 7, and 8 are about the physiology of cells—how they stay alive.*

*If these chapters are well understood, you will happily launch forth to study the other parts of the book, secure in the knowledge that you have left nothing behind.*

# 2

# Basic Chemistry

Can we understand a bottle-nosed dolphin, without a fundamental knowledge and respect for its chemistry? After all, a dolphin has a certain salinity tolerance, can only stay underwater for so long, and must have a particular diet to keep its complex organ systems functioning. Chemistry also plays a role in the behavior of the dolphin, whether it is playing in the Gulf of Mexico or performing at Sea World. A dolphin cannot jump unless its nervous system is prepared to chemically direct its muscles to contract. In fact, all aspects of a dolphin's biology involve molecular chemistry.*

*At one time, it was believed that organisms contained a vital force, and this force accounted for their "vitality." Such a hypothesis has never been supported, and instead, today we know that living things are composed of the same elements as inanimate objects. It is true, though, that they differ as to which elements are most common, as we shall see. This chapter reviews inorganic chemistry, which largely pertains to nonliving things. It also explores the composition and chemistry of water, an inorganic substance that is so intimately connected to the life of organisms.*

Bottle-nosed dolphin, *Tursiops truncatus*.

## concepts

# 2.1 Chemical Elements

Turn the page, throw a ball, pat your dog, rake leaves; everything we touch—from the water we drink to the air we breathe—is composed of matter. **Matter** refers to anything that takes up space and has mass. Although matter has many diverse forms—anything from molten lava to kidney stones—it only exists in three distinct states: solid, liquid, and gas.

## Elements

All matter, both nonliving and living, is composed of certain basic substances called **elements.** An element is a substance that cannot be broken down to simpler substances with different properties (a property is a physical or chemical characteristic, such as density, solubility, melting point, and reactivity) by ordinary chemical means. It is quite remarkable that there are only 92 naturally occurring elements that serve as the building blocks of matter. Other elements have been "human-made" and are not biologically important.

Both the Earth's crust and all organisms are composed of elements, but they differ as to which ones are common. Only six elements—carbon, hydrogen, nitrogen, oxygen, phosphorus, and sulfur—are basic to life and make up about 95% of the body weight of organisms. The acronym CHNOPS helps us remember these six elements. The properties of these elements are essential to the uniqueness of cells and organisms, such as the macaws in Figure 2.1. The macaws have gathered on a salt lick in South America. Salt contains the elements sodium and chlorine and is commonly sought after by many forms of life. Potassium, calcium, iron, and magnesium are still other elements found in living things.

## Atoms

In the early 1800s, the English scientist John Dalton championed the atomic theory, which says that elements consist of tiny particles called **atoms** [Gk. *atomos*, uncut, indivisible]. An atom is the smallest part of an element that displays the properties of the element. An element and its atoms share the same name. One or two letters create the **atomic symbol,** which stands for this name. For example, the symbol H means a hydrogen atom, the symbol Rn stands for radon, and the symbol Na (for *natrium* in Latin) is used for a sodium atom.

Physicists have identified a number of subatomic particles that make up atoms. The three best known subatomic particles include positively charged **protons,** uncharged **neutrons,** and negatively charged **electrons** [Gk. *elektron*, electricity]. Protons and neutrons are located within the nucleus of an atom, and electrons move about the nucleus. Figure 2.2 shows the arrangement of the subatomic particles in a helium atom, which has only two electrons. In Figure 2.2*a*, the stippling shows the probable location of electrons, and in Figure 2.2*b*, the circle represents an **electron shell,** the average location of electrons.

The concept of an atom has changed greatly since Dalton's day. If an atom could be drawn the size of a football field, the nucleus would be like a gumball in the center of the field, and the electrons would be tiny specks whirling about in the upper stands. Most of an atom is empty space. We should also realize that we can only indicate where the electrons are expected to be most of the time. In our analogy, the electrons might very well stray outside the stadium at times.

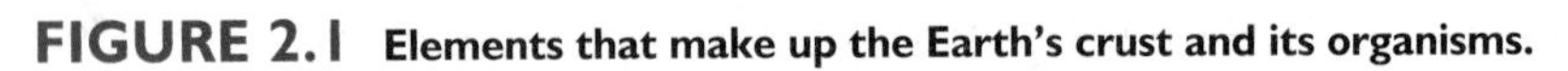

**FIGURE 2.1 Elements that make up the Earth's crust and its organisms.**

Scarlet macaws gather on a salt lick in South America. The graph inset shows the Earth's crust primarily contains the elements silicon (Si), aluminum (Al), and oxygen (O). Organisms primarily contain the elements oxygen (O), nitrogen (N), carbon (C), and hydrogen (H). Along with sulfur (S) and phosphorus (P), these elements make up biological molecules.

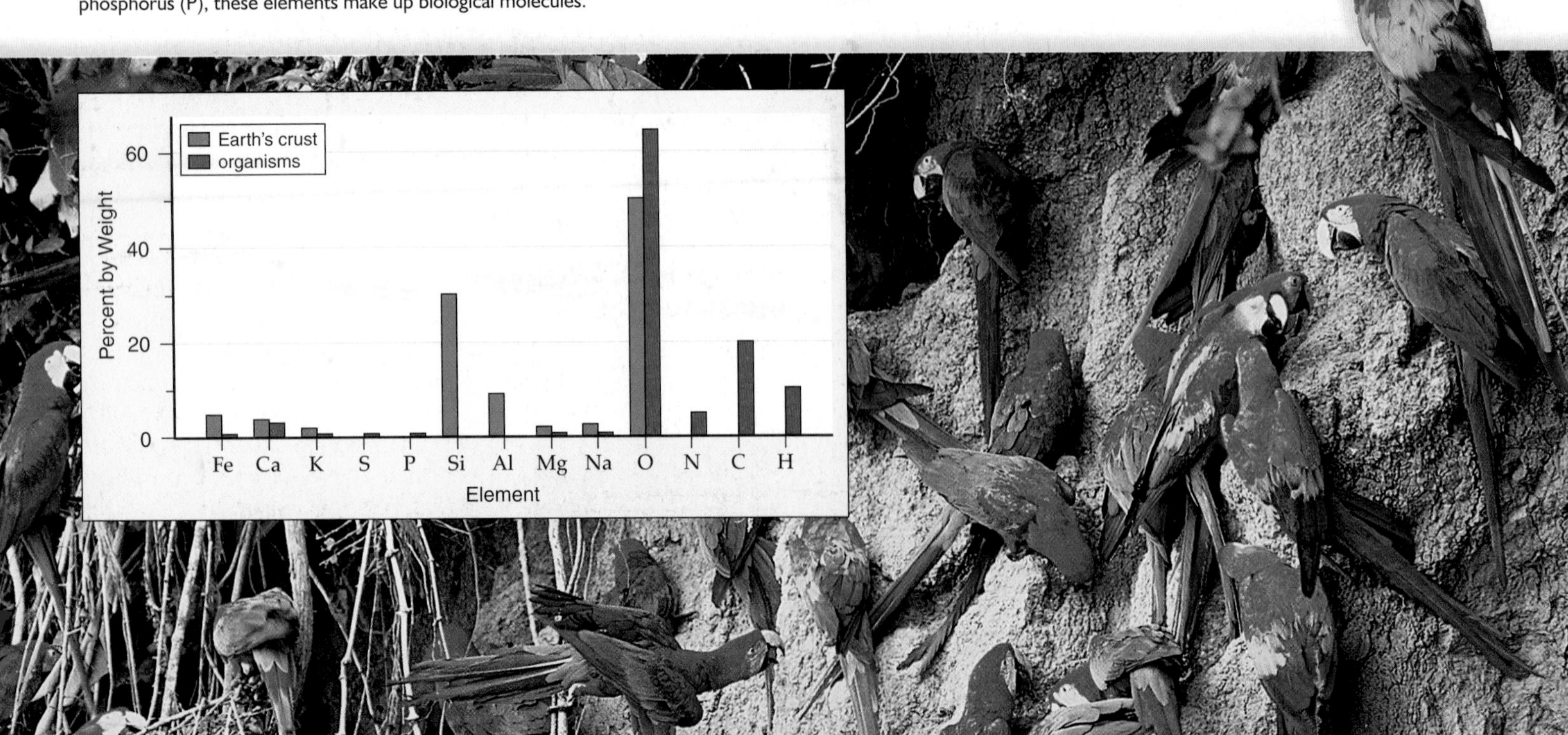

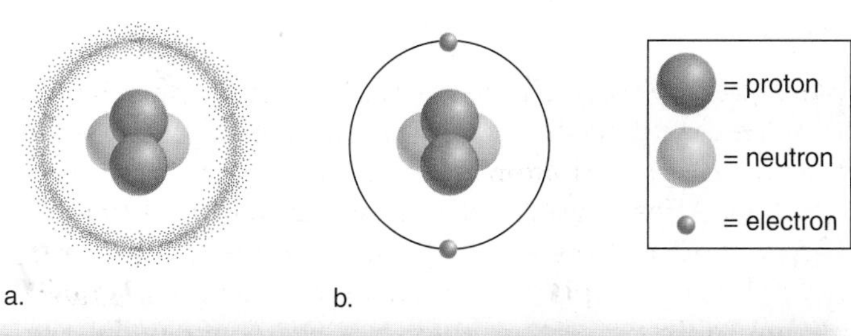


| Subatomic Particles | | | |
|---|---|---|---|
| **Particle** | **Electric Charge** | **Atomic Mass Unit (AMU)** | **Location** |
| Proton | +1 | 1 | Nucleus |
| Neutron | 0 | 1 | Nucleus |
| Electron | −1 | 0 | Electron shell |

c.

**FIGURE 2.2 Model of helium (He).**

Atoms contain subatomic particles, which are located as shown. Protons and neutrons are found within the nucleus, and electrons are outside the nucleus. **a.** The stippling shows the probable location of the electrons in the helium atom. **b.** The average location of an electron is sometimes represented by a circle termed an electron shell. **c.** The electric charge and the atomic mass units (AMU) of the subatomic particles vary as shown.

Periods / Groups (1 = atomic number; H = atomic symbol; 4.003 = atomic mass)

| I | II | III | IV | V | VI | VII | VIII |
|---|---|---|---|---|---|---|---|
| 1 **H** 1.008 | | | | | | | 2 **He** 4.003 |
| 3 **Li** 6.941 | 4 **Be** 9.012 | 5 **B** 10.81 | 6 **C** 12.01 | 7 **N** 14.01 | 8 **O** 16.00 | 9 **F** 19.00 | 10 **Ne** 20.18 |
| 11 **Na** 22.99 | 12 **Mg** 24.31 | 13 **Al** 26.98 | 14 **Si** 28.09 | 15 **P** 30.97 | 16 **S** 32.07 | 17 **Cl** 35.45 | 18 **Ar** 39.95 |
| 19 **K** 39.10 | 20 **Ca** 40.08 | 31 **Ga** 69.72 | 32 **Ge** 72.59 | 33 **As** 74.92 | 34 **Se** 78.96 | 35 **Br** 79.90 | 36 **Kr** 83.60 |

**FIGURE 2.3 A portion of the periodic table.**

In the periodic table, the elements, and therefore atoms, are in the order of their atomic numbers but arranged so that they are placed in groups (vertical columns) and periods (horizontal rows). All the atoms in a particular group have chemical characteristics in common. These four periods contain the elements that are most important in biology; the complete periodic table is in Appendix D.

# Atomic Number and Mass Number

Atoms not only have an atomic symbol, they also have an atomic number and mass number. All atoms of an element have the same number of protons housed in the nucleus. This is called the **atomic number,** which accounts for the unique properties of this type atom.

Each atom also has its own mass number dependent on the number of subatomic particles in that atom. Protons and neutrons are assigned one atomic mass unit (AMU) each. Electrons are so small that their AMU is considered zero in most calculations (Fig. 2.2*c*). Therefore, the **mass number** of an atom is the sum of protons and neutrons in the nucleus.

The term mass is used, and not *weight,* because mass is constant, while weight changes according to the gravitational force of a body. The gravitational force of the Earth is greater than that of the moon; therefore, substances weigh less on the moon, even though their mass has not changed.

By convention, when an atom stands alone (and not in the periodic table, discussed next), the atomic number is written as a subscript to the lower left of the atomic symbol. The mass number is written as a superscript to the upper left of the atomic symbol. Regardless of position, the smaller number is always the atomic number, as shown here for carbon.

mass number — $^{12}_{6}C$ — atomic symbol
atomic number

# The Periodic Table

Once chemists discovered a number of the elements, they began to realize that even though each element consists of a different atom, certain chemical and physical characteristics recur. The periodic table, developed by the Russian chemist Dmitri Mendeleev (1834–1907), was constructed as a way to group the elements, and therefore atoms, according to these characteristics.

Figure 2.3 is a portion of the periodic table, which is shown in total in Appendix D. The atoms shown in the periodic table are assumed to be electrically neutral. Therefore, the atomic number not only tells you the number of protons, it also tells you the number of electrons. The **atomic mass** is the average of the AMU for all the isotopes (discussed next) of that atom. To determine the number of neutrons, subtract the number of protons from the atomic mass, and take the closest whole number.

In the periodic table, every atom is in a particular period (the horizontal rows) and in a particular group (the vertical columns). The atomic number of every atom in a period increases by one if you read from left to right. All the atoms in a group share the same binding characteristics. For example, all the atoms in group VII react with one atom at a time, for reasons we will soon explore. The atoms in group VIII are called the noble gases because they are inert and rarely react with another atom. Notice that helium and krypton are noble gases.

## Isotopes

**Isotopes** [Gk. *isos*, equal, and *topos*, place] are atoms of the same element that differ in the number of neutrons. Isotopes have the same number of protons, but they have different atomic masses. For example, the element carbon has three common isotopes:

$$^{12}_{6}C \qquad ^{13}_{6}C \qquad ^{14}_{6}C^*$$

*radioactive

Carbon 12 has six neutrons, carbon 13 has seven neutrons, and carbon 14 has eight neutrons. Unlike the other two isotopes of carbon, carbon 14 is unstable; it changes over time into nitrogen 14, which is a stable isotope of the element nitrogen. As carbon 14 decays, it releases various types of energy in the form of rays and subatomic particles, and therefore it is a radioactive isotope. The radiation given off by radioactive isotopes can be detected in various ways. The Geiger counter is an instrument that is commonly used to detect radiation. In 1860, the French physicist Antoine-Henri Becquerel discovered that a sample of uranium would produce a bright image on a photographic plate because it was radioactive. A similar method of detecting radiation is still in use today. Marie Curie, who worked with Becquerel, contributed much to the study of radioactivity, as she named it. Today, radiation is used by biologists to date objects, create images, and trace the movement of substances.

### *Low Levels of Radiation*

The chemical behavior of a radioactive isotope is essentially the same as that of the stable isotopes of an element. This means that you can put a small amount of radioactive isotope in a sample and it becomes a **tracer** by which to detect molecular changes. Melvin Calvin and his co-workers used carbon 14 to detect all the various reactions that occur during the process of photosynthesis.

The importance of chemistry to medicine is nowhere more evident than in the many medical uses of radioactive isotopes. Specific tracers are used in imaging the body's organs and tissues. For example, after a patient drinks a solution containing a minute amount of $^{131}I$, it becomes concentrated in the thyroid—the only organ to take it up. A subsequent image of the thyroid indicates whether it is healthy in structure and function (Fig. 2.4*a*). Positron-emission tomography (PET) is a way to determine the comparative activity of tissues. Radioactively labeled glucose, which emits a subatomic particle known as a positron, is injected into the body. The radiation given off is detected by sensors and analyzed by a computer. The result is a color image that shows which tissues took up glucose and are metabolically active. The red areas surrounded by green in Figure 2.4*b* indicate which areas of the brain are most active. PET scans of the brain are used to evaluate patients who have memory disorders of an undetermined cause or suspected brain tumors or seizure disorders that could possibly benefit from surgery. PET scans, utilizing radioactive thallium, can detect signs of coronary artery disease and low blood flow to the heart.

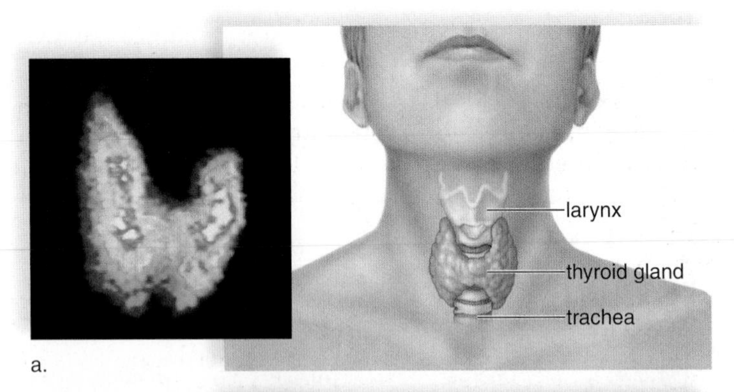


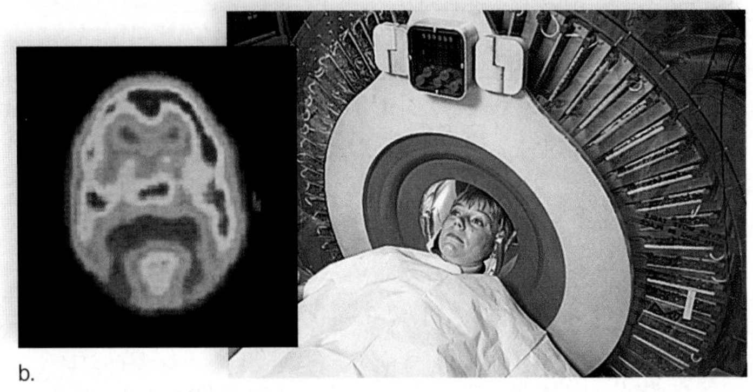


**FIGURE 2.4 Low levels of radiation.**

**a.** Incomplete scan of the thyroid gland on the left indicates the presence of a tumor that does not take up the radioactive iodine. **b.** A PET (positron-emission tomography) scan reveals which portions of the brain are most active (green and red colors).

### *High Levels of Radiation*

Radioactive substances in the environment can harm cells, damage DNA, and cause cancer. When Marie Curie was studying radiation, its harmful effects were not known, and she and many of her co-workers developed cancer. The release of radioactive particles following a nuclear power plant accident can have far-reaching and long-lasting effects on human health. The harmful effects of radiation can be put to good use, however (Fig. 2.5). Radiation from radioactive isotopes has been used for many years to sterilize medical

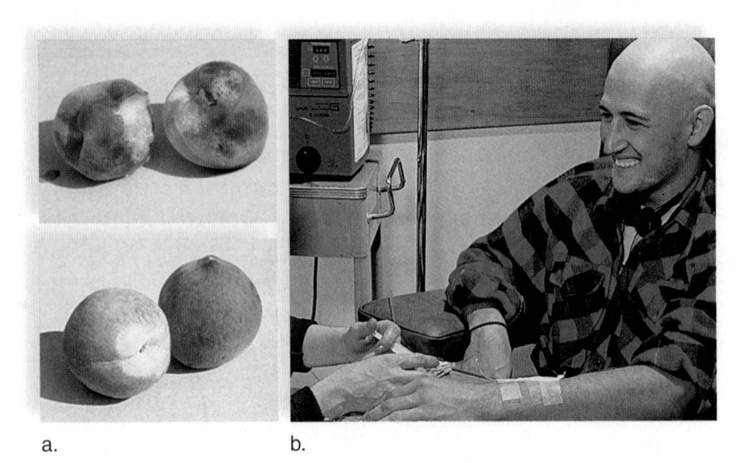


**FIGURE 2.5 High levels of radiation.**

**a.** Radiation kills bacteria and fungi. After irradiation, peaches spoil less quickly and can be kept for a longer length of time. **b.** Physicians use targeted radiation therapy to kill cancer cells.

and dental products. Now it can be used to sterilize the U.S. mail and other packages to free them of possible pathogens, such as anthrax spores. The ability of radiation to kill cells is often applied to cancer cells. Targeted radioisotopes can be introduced into the body so that the subatomic particles emitted destroy only cancer cells, with little risk to the rest of the body.

## Electrons and Energy

In an electrically neutral atom, the positive charges of the protons in the nucleus are balanced by the negative charges of electrons moving about the nucleus. Various models in years past have attempted to illustrate the precise location of electrons. Figure 2.6 uses the Bohr model, which is named after the physicist Niels Bohr. The Bohr model is useful, but we need to realize that today's physicists tell us it is not possible to determine the precise location of any individual electron at any given moment.

In the Bohr model, the electron shells about the nucleus also represent energy levels. It seems reasonable to suggest that negatively charged electrons are attracted to the positively charged nucleus, and that it takes energy to push them away and keep them in their own shell. Further, the more distant the shell, the more energy it takes. Therefore, it is proper to speak of electrons as being at particular energy levels in relation to the nucleus. When you study photosynthesis, you will learn that when atoms absorb the energy of the sun, electrons are boosted to a higher energy level. Later, as the electrons return to their original energy level, energy is released and transformed into chemical energy. This chemical energy supports all life on Earth and therefore our very existence is dependent on the energy of electrons.

You will want to learn to draw a Bohr model for each of the elements that occurs in the periodic table shown in Figure 2.3. Let's begin by examining the models depicted in Figure 2.6. Notice that the first shell (closest to the nucleus) can contain two electrons; thereafter, each additional shell can contain eight electrons. Also, each lower level is filled with electrons before the next higher level contains any electrons.

The sulfur atom, with an atomic number of 16, has two electrons in the first shell, eight electrons in the second shell, and six electrons in the third, or outer, shell. Revisit the periodic table (see Fig. 2.3), and note that sulfur is in the third period. In other words, the period tells you how many shells an atom has. Also note that sulfur is in group VI. The group tells you how many electrons an atom has in its outer shell.

If an atom has only one shell, the outer shell is complete when it has two electrons. Otherwise, the **octet rule,** which states that the outer shell is most stable when it has eight electrons, holds. As mentioned previously, atoms in group VIII of the periodic table are called the noble gases because they do not ordinarily react. Stability exists because an outer shell with eight electrons has less energy. In general, lower energy states represent stability, as we will have an opportunity to point out again in Chapter 6.

Just as you sometimes communicate with and react to other people by using your hands, so atoms use the electrons in their outer shells to undergo reactions. Atoms with fewer than eight electrons in the outer shell react with other atoms in such a way that after the reaction, each has a stable outer shell. As we shall see, the number of electrons in an atom's outer shell, called the **valence shell,** determines whether it gives up, accepts, or shares electrons to acquire eight electrons in the outer shell.

### Check Your Progress 2.1

1. Contrast atomic number and mass number.
2. **a.** How do group III elements differ in the periodic table? **b.** How do period III elements differ?
3. List some uses of radioactive isotopes in biology and medicine.

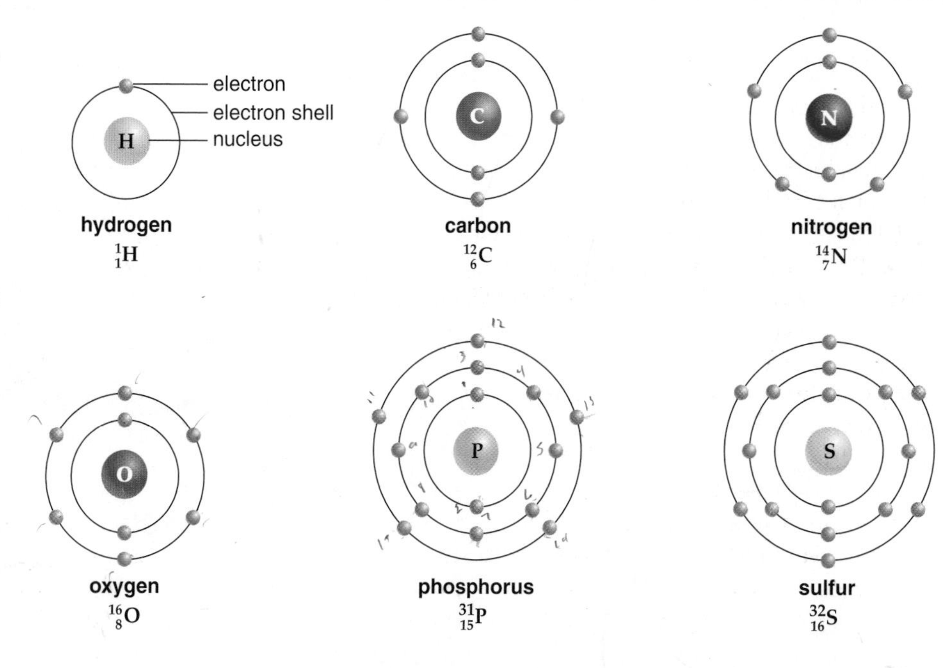


**FIGURE 2.6 Bohr models of atoms.**

Electrons orbit the nucleus at particular energy levels (electron shells): The first shell contains up to two electrons, and each shell thereafter can contain up to eight electrons as long as we consider only atoms with an atomic number of 20 or below. Each shell is filled before electrons are placed in the next shell. Why does carbon have only two shells while phosphorus and sulfur have three shells?

## 2.2 Compounds and Molecules

A **compound** exists when two or more elements have bonded together. A **molecule** [L. *moles*, mass] is the smallest part of a compound that still has the properties of the particular compound. In practice, these two terms are used interchangeably, but in biology, we usually speak of molecules. Water ($H_2O$) is a molecule that contains atoms of hydrogen and oxygen. A **formula** tells you the number of each kind of atom in a molecule. For example, in glucose:

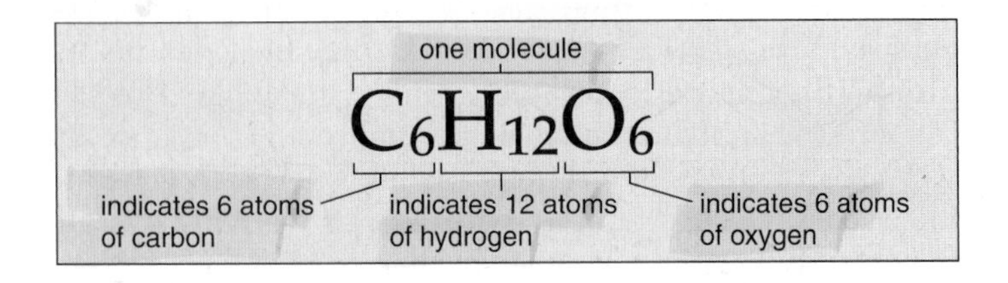


Electrons possess energy, and the bonds that exist between atoms also contain energy. Organisms are directly dependent on chemical-bond energy to maintain their organization. As you may know, organisms routinely break down glucose, the sugar shown above, to obtain energy. When a chemical reaction occurs, as when glucose is broken down, electrons shift in their relationship to one another, and energy is released. Spontaneous reactions, which are ones that occur freely, always release energy.

### Ionic Bonding

Sodium (Na), with only one electron in its third shell, tends to be an electron donor (Fig. 2.7*a*). Once it gives up this electron, the second shell, with eight electrons, becomes its outer shell. Chlorine (Cl), on the other hand, tends to be an electron acceptor. Its outer shell has seven electrons, so if it acquires only one more electron it has a completed outer shell. When a sodium atom and a chlorine atom come together, an electron is transferred from the sodium atom to the chlorine atom. Now both atoms have eight electrons in their outer shells.

This electron transfer, however, causes a charge imbalance in each atom. The sodium atom has one more proton than it has electrons; therefore, it has a net charge of +1 (symbolized by $Na^+$). The chlorine atom has one more electron than it has protons; therefore, it has a net charge of −1 (symbolized by $Cl^-$). Such charged particles are called **ions.** Sodium ($Na^+$) and chloride ($Cl^-$) are not the only biologically important ions. Some, such as potassium ($K^+$), are formed by the transfer of a single electron to another atom; others, such as calcium ($Ca^{2+}$) and magnesium ($Mg^{2+}$), are formed by the transfer of two electrons.

Ionic compounds are held together by an attraction between negatively and positively charged ions called an **ionic bond.** When sodium reacts with chlorine, an ionic compound called sodium chloride (NaCl) results. Sodium chloride is a salt, commonly known as table salt, because it is used to season our food (Fig. 2.7*b*). **Salts** are solid substances that usually separate and exist as individual ions in water, as discussed on page 30.

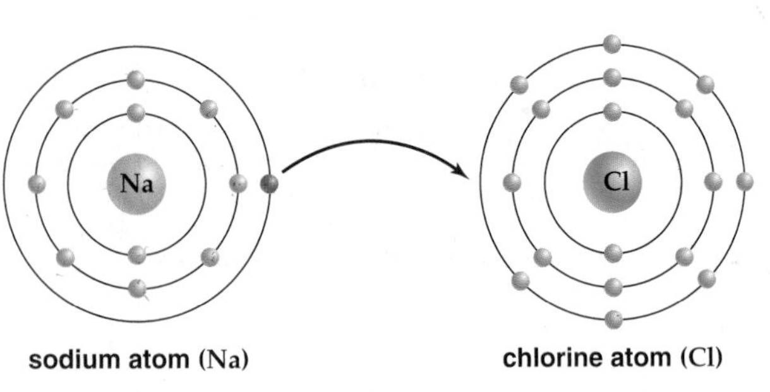


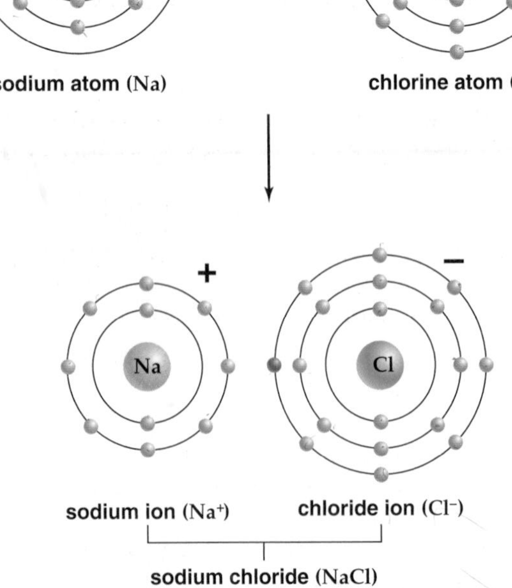

**FIGURE 2.7 Formation of sodium chloride (table salt).**

**a.** During the formation of sodium chloride, an electron is transferred from the sodium atom to the chlorine atom. At the completion of the reaction, each atom has eight electrons in the outer shell, but each also carries a charge as shown. **b.** In a sodium chloride crystal, ionic bonding between $Na^+$ and $Cl^-$ causes the atoms to assume a three-dimensional lattice in which each sodium ion is surrounded by six chloride ions, and each chloride ion is surrounded by six sodium ions. The result is crystals of salt as in table salt.

## Covalent Bonding

A **covalent bond** [L. *co*, together, with, and *valens*, strength] results when two atoms share electrons in such a way that each atom has an octet of electrons in the outer shell (or two electrons, in the case of hydrogen). In a hydrogen atom, the outer shell is complete when it contains two electrons. If hydrogen is in the presence of a strong electron acceptor, it gives up its electron to become a hydrogen ion ($H^+$). But if this is not possible, hydrogen can share with another atom and thereby have a completed outer shell. For example, one hydrogen atom will share with another hydrogen atom. Their two electron shells overlap and the electrons are shared between them (Fig. 2.8*a*). Because they share the electron pair, each atom has a completed outer shell.

A more common way to symbolize that atoms are sharing electrons is to draw a line between the two atoms, as in the structural formula H—H. Just as a handshake requires two hands, one from each person, a covalent bond between two atoms requires two electrons, one from each atom. In a molecular formula, the line is omitted and the molecule is simply written as $H_2$.

Sometimes, atoms share more than one pair of electrons to complete their octets. A double covalent bond occurs when two atoms share two pairs of electrons (Fig. 2.8*b*). To show that oxygen gas ($O_2$) contains a double bond, the molecule can be written as O=O. It is also possible for atoms to form triple covalent bonds, as in nitrogen gas ($N_2$), which can be written as N≡N. Single covalent bonds between atoms are quite strong, but double and triple bonds are even stronger.

| Electron Model | Structural Formula | Molecular Formula |
|---|---|---|
| H H | H—H | $H_2$ |

a. Hydrogen gas

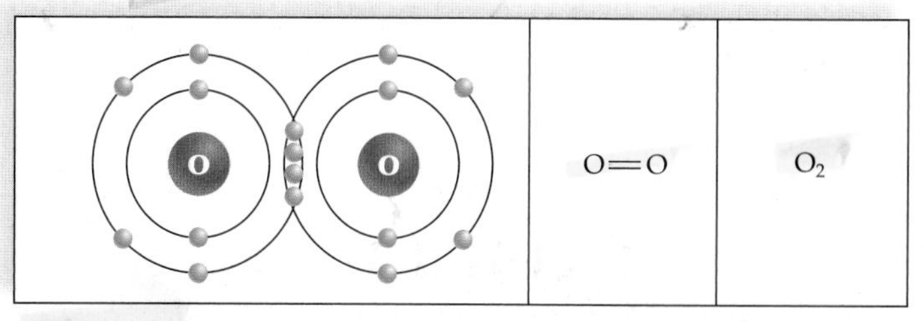


b. Oxygen gas

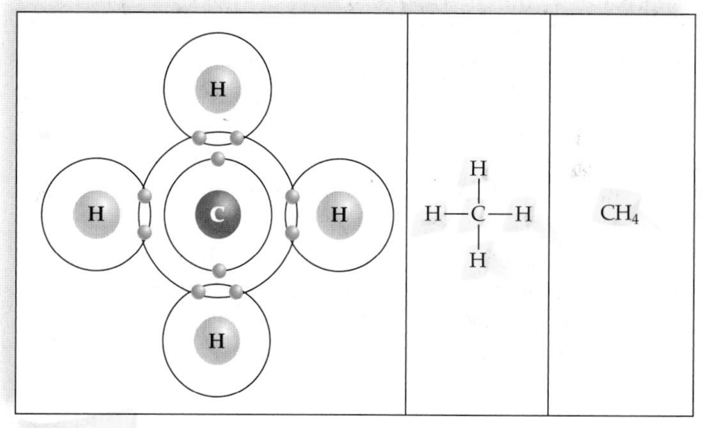


c. Methane

**FIGURE 2.8 Covalently bonded molecules.**

In a covalent bond, atoms share electrons, allowing each atom to have a completed outer shell. **a.** A molecule of hydrogen ($H_2$) contains two hydrogen atoms sharing a pair of electrons. This single covalent bond can be represented in any of the three ways shown. **b.** A molecule of oxygen ($O_2$) contains two oxygen atoms sharing two pairs of electrons. This results in a double covalent bond. **c.** A molecule of methane ($CH_4$) contains one carbon atom bonded to four hydrogen atoms.

### *Nonpolar and Polar Covalent Bonds*

When the sharing of electrons between two atoms is equal, the covalent bond is said to be a **nonpolar covalent bond.** If one atom is able to attract electrons to a greater degree than the other atom, it is the more electronegative atom. **Electronegativity** is dependent on the number of protons—the greater the number of protons, the greater the electronegativity. When electrons are not shared equally, the covalent bond is a **polar covalent bond.**

You can readily see that the bonds in methane (Fig. 2.8*c*) must be polar because carbon has more protons than a hydrogen atom. However, methane is a symmetrical molecule and the polarities cancel each other out—methane is a nonpolar molecule. Not so in water, which has this shape:

Oxygen is partially negative ($\delta^-$)

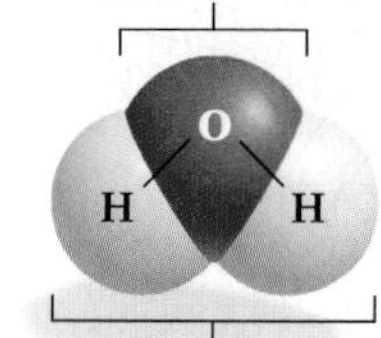


Hydrogens are partially positive ($\delta^+$)

In water, the oxygen atom is more electronegative than the hydrogen atoms and the bonds are polar. Moreover, because of its nonsymmetrical shape, the polar bonds cannot cancel each other and water is a polar molecule. The more electronegative end of the molecule is designated slightly negative ($\delta^-$), and the hydrogens are designated slightly positive ($\delta^+$).

Water is not the only polar molecule in living things. For example, the amine group ($—NH_2$) is polar, and this causes amino acids and nucleic acids to exhibit polarity, as we shall see in the next chapter. The polarity of molecules affects how they interact with other molecules.

### Check Your Progress 2.2

1. Contrast an ionic bond with a covalent bond.
2. Why would you expect calcium to become an ion that carries two plus charges?
3. Explain how it is that all the atoms in methane ($CH_4$) have a complete outer shell.

## 2.3 Chemistry of Water

Figure 2.9*a* recaps what we know about the water molecule. The structural formula on the far left shows that when water forms, an oxygen atom is sharing electrons with two hydrogen atoms. The ball-and-stick model in the center shows that the covalent bonds between oxygen and each of the hydrogens are at an angle of 104.5°. Finally, the space-filling molecule gives us the three-dimensional shape of the molecule and indicates its polarity.

The shape of water and of all organic molecules is necessary to the structural and functional roles they play in living things. For example, hormones have specific shapes that allow them to be recognized by the cells in the body. We can stay well only when antibodies combine with disease-causing agents, like a key fits a lock. Similarly, homeostasis is only maintained when enzymes have the proper shape to carry out their particular reactions in cells. The shape of a water molecule and its polarity makes hydrogen bonding possible. A **hydrogen bond** is the attraction of a slightly positive hydrogen to a slightly negative atom in the vicinity. In carbon dioxide, O=C=O, there is also a slight difference in polarity between carbon and the oxygens but because carbon dioxide is symmetrical, the opposing charges cancel one another and hydrogen bonding does not occur.

### Hydrogen Bonding

The dotted lines in Figure 2.9*b* indicate that the hydrogen atoms in one water molecule are attracted to the oxygen atoms in other water molecules. This attraction, which is weaker than an ionic or covalent bond, is called a hydrogen bond. The dotted lines indicate that hydrogen bonds are more easily broken than covalent bonds. Hydrogen bonding is not unique to water. Other biological molecules, such as DNA, have polar covalent bonds involving an electropositive hydrogen and usually an electronegative oxygen or nitrogen. In these instances, a hydrogen bond can occur within the same molecule or between nearby molecules.

Although a hydrogen bond is more easily broken than a covalent bond, many hydrogen bonds taken together are quite strong. Hydrogen bonds between cellular molecules help maintain their proper structure and function. For example, hydrogen bonds hold the two strands of DNA together. When DNA makes a copy of itself, hydrogen bonds easily break, allowing DNA to unzip. But normally, the hydrogen bonds add stability to the DNA molecule. Similarly, the shape of protein molecules is often maintained by hydrogen bonding between parts of the same molecule. As we shall see, many of the important properties of water are the result of hydrogen bonding.

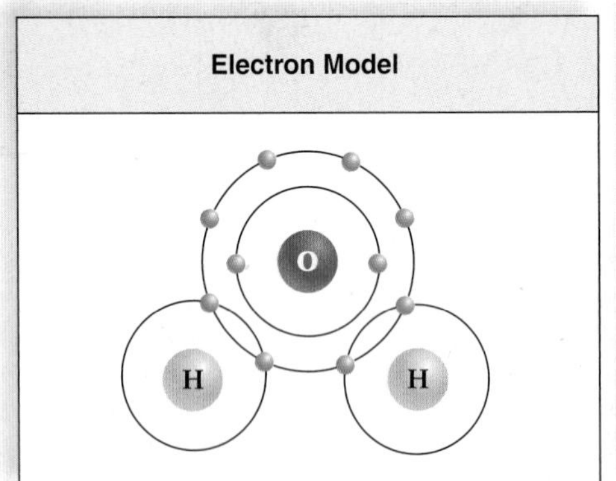


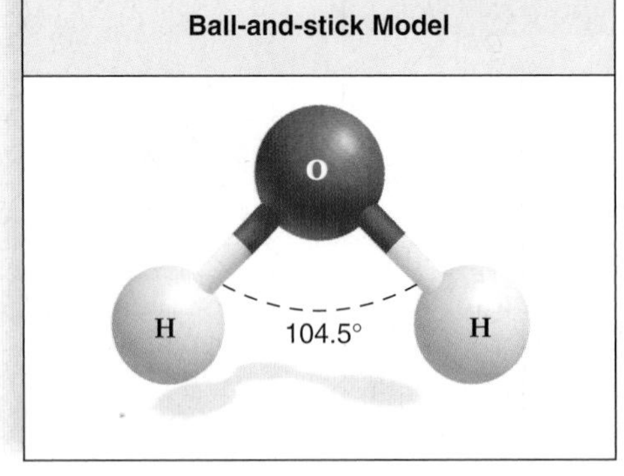


Space-filling Model

Oxygen attracts the shared electrons and is partially negative.

δ⁻ O H H δ⁺ δ⁺

Hydrogens are partially positive.

a. Water ($H_2O$)

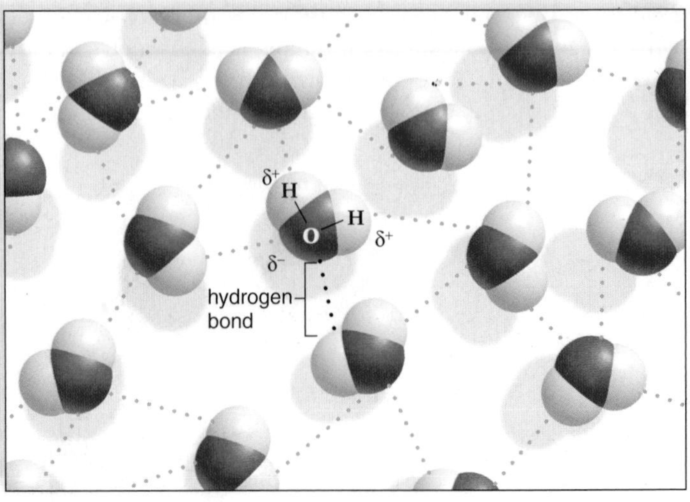


b. Hydrogen bonding between water molecules

**FIGURE 2.9 Water molecule.**

**a.** Three models for the structure of water. The electron model does not indicate the shape of the molecule. The ball-and-stick model shows that the two bonds in a water molecule are angled at 104.5°. The space-filling model also shows the V shape of a water molecule. **b.** Hydrogen bonding between water molecules. Each water molecule can hydrogen-bond to four other molecules. When water is in its liquid state, some hydrogen bonds are forming and others are breaking at all times.

## Properties of Water

The first cell(s) evolved in water, and all living things are 70–90% water. Due to hydrogen bonding, water molecules cling together. Without hydrogen bonding between molecules, water would melt at –100°C and boil at –91°C, making most of the water on Earth steam, and life unlikely. But because of hydrogen bonding, water is a liquid at temperatures typically found on the Earth's surface. It melts at 0°C and boils at 100°C. These and other unique properties of water make it essential to the existence of life as we know it. When scientists examine the other planets with the hope of finding life, they first look for signs of water.

**Water Has a High Heat Capacity.** A **calorie** is the amount of heat energy needed to raise the temperature of 1 g of water 1°C. In comparison, other covalently bonded liquids require input of only about half this amount of energy to rise in temperature 1°C. The many hydrogen bonds that link water molecules together help water absorb heat without a great change in temperature. Converting 1 g of the coldest liquid water to ice requires the loss of 80 calories of heat energy (Fig. 2.10*a*). Water holds onto its heat, and its temperature falls more slowly than that of other liquids. This property of water is important not only for aquatic organisms but also for all living things.

Because the temperature of water rises and falls slowly, organisms are better able to maintain their normal internal temperatures and are protected from rapid temperature changes.

**Water Has a High Heat of Evaporation.** When water boils, it evaporates—that is, vaporizes into the environment. Converting 1 g of the hottest water to a gas requires an input of 540 calories of energy. Water has a high heat of evaporation because hydrogen bonds must be broken before water boils.

Water's high heat of vaporization gives animals in a hot environment an efficient way to release excess body heat. When an animal sweats, or gets splashed, body heat is used to vaporize water, thus cooling the animal (Fig. 2.10*b*). Because of water's high heat of vaporization and ability to hold onto its heat, temperatures along the coasts are moderate. During the summer, the ocean absorbs and stores solar heat, and during the winter, the ocean releases it slowly. In contrast, the interior regions of continents experience abrupt changes in temperatures.

**Water Is a Solvent.** Due to its polarity, water facilitates chemical reactions, both outside and within living systems. It dissolves a great number of substances. A **solution** contains dissolved substances, which are then called **solutes.**

Calories of Heat Energy / g
800
600
400
200
0
Gas
540 calories
Liquid
80 calories
Solid
freezing occurs
evaporation occurs
0 20 40 60 80 100 120
Temperature (°C)

a. Calories lost when 1 g of liquid water freezes and calories required when 1 g of liquid water evaporates.

b. Bodies of organisms cool when their heat is used to evaporate water.

**FIGURE 2.10 Temperature and water.**

**a.** Water can be a solid, a liquid, or a gas at naturally occurring environmental temperatures. At room temperature and pressure, water is a liquid. When water freezes and becomes a solid (ice), it gives off heat, and this heat can help keep the environmental temperature higher than expected. On the other hand, when water evaporates, it takes up a large amount of heat as it changes from a liquid to a gas. **b.** This means that splashing water on the body will help keep body temperature within a normal range. Can you also see why water's properties help keep the coasts moderate in both winter and summer?

When ionic salts—for example, sodium chloride (NaCl)—are put into water, the negative ends of the water molecules are attracted to the sodium ions, and the positive ends of the water molecules are attracted to the chloride ions. This causes the sodium ions and the chloride ions to separate, or dissociate, in water.

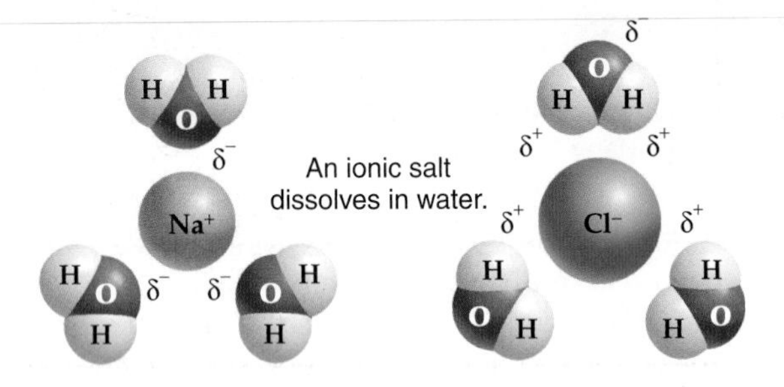


Water is also a solvent for larger polar molecules, such as ammonia ($NH_3$).

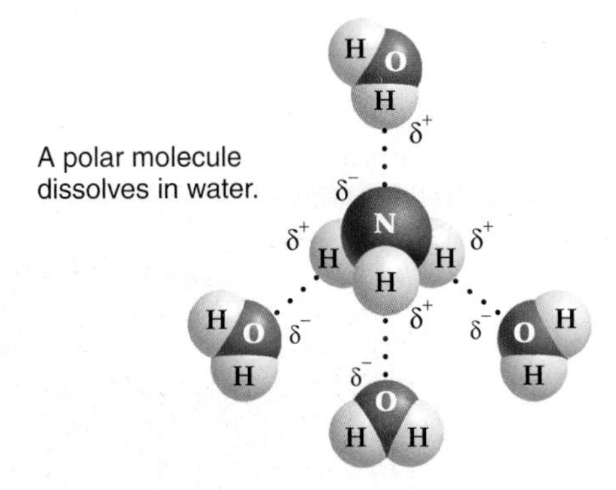


Those molecules that can attract water are said to be **hydrophilic** [Gk. *hydrias,* of water, and *phileo,* love]. When ions and molecules disperse in water, they move about and collide, allowing reactions to occur. Nonionized and nonpolar molecules that cannot attract water are said to be **hydrophobic** [Gk. *hydrias,* of water, and *phobos,* fear]. Gasoline contains nonpolar molecules, and therefore it does not mix with water and is hydrophobic.

**Water Molecules Are Cohesive and Adhesive.** Cohesion refers to the ability of water molecules to cling to each other due to hydrogen bonding. Because of cohesion, water exists as a liquid under ordinary conditions of temperature and pressure. The strong cohesion of water molecules is apparent because water flows freely, yet water molecules do not separate from each other. Adhesion refers to the ability of water molecules to cling to other polar surfaces. This is because of water's polarity. Multicellular animals often contain internal vessels in which water assists the transport of nutrients and wastes because the cohesion and adhesion of water allows blood to fill the tubular vessels of the cardiovascular system. For example, the liquid portion of our blood, which transports dissolved and suspended substances about the body, is 90% water.

Cohesion and adhesion also contribute to the transport of water in plants. Plants have their roots anchored in the soil, where they absorb water, but the leaves are uplifted and exposed to solar energy. Water evaporating from the leaves is immediately replaced with water molecules from transport vessels that extend from the roots to the leaves (Fig. 2.11). Because water molecules are cohesive, a tension is created that pulls the water column up from the roots. Adhesion of water to the walls of the vessels also helps prevent the water column from breaking apart.

Because water molecules are strongly attracted to each other, they cling together at a surface exposed to air. The stronger the force between molecules in a liquid, the greater the **surface tension.** Water's high surface tension makes it possible for humans to skip rocks on water. Water striders, a common insect, can even walk on the surface of a pond without breaking the surface.

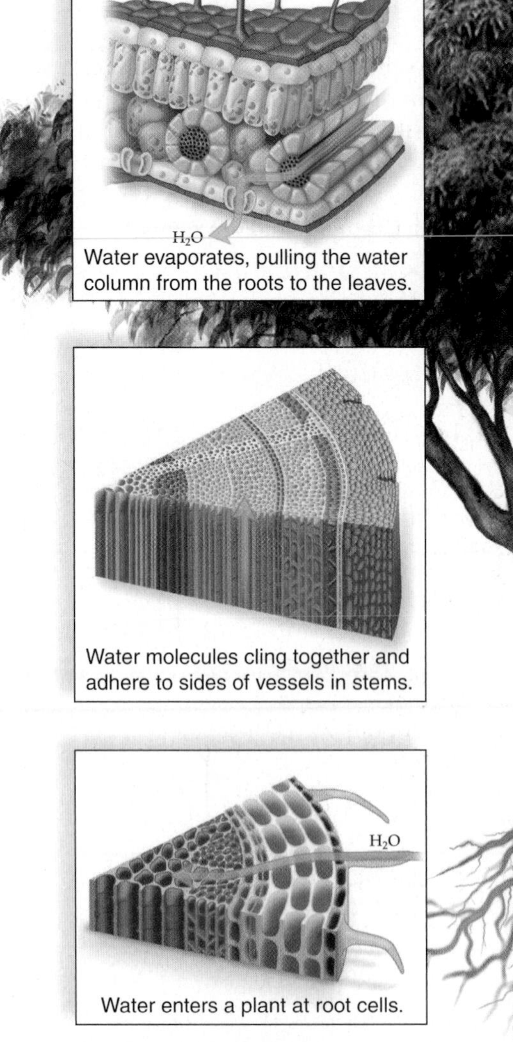


**FIGURE 2.11** **Water as a transport medium.**

How does water rise to the top of tall trees? Vessels are water-filled pipelines from the roots to the leaves. When water evaporates from the leaves, the water column is pulled upward due to the cohesion of water molecules with one another and the adhesion of water molecules to the sides of the vessels.

**Frozen water (ice) is less dense than liquid water.** As liquid water cools, the molecules come closer together. Water is most dense at 4°C, but the water molecules are still moving about (Fig. 2.12). At temperatures below 4°C, there is only vibrational movement, and hydrogen bonding becomes more rigid but also more open. This means that water expands as it freezes, which is why cans of soda burst when placed in a freezer or why frost heaves make northern roads bumpy in the winter. It also means that ice is less dense than liquid water, and therefore ice floats on liquid water.

If ice did not float on water, it would sink, and ponds, lakes, and perhaps even the ocean would freeze solid, making life impossible in the water and also on land. Instead, bodies of water always freeze from the top down. When a body of water freezes on the surface, the ice acts as an insulator to prevent the water below it from freezing. This protects aquatic organisms so that they can survive the winter. As ice melts in the spring, it draws heat from the environment, helping to prevent a sudden change in temperature that might be harmful to life.

### Check Your Progress 2.3

1. Explain why water has a high heat of vaporization.
2. Explain why children in summer can cool off by playing in a sprinkler.
3. Explain why ice skating is possible in the winter.

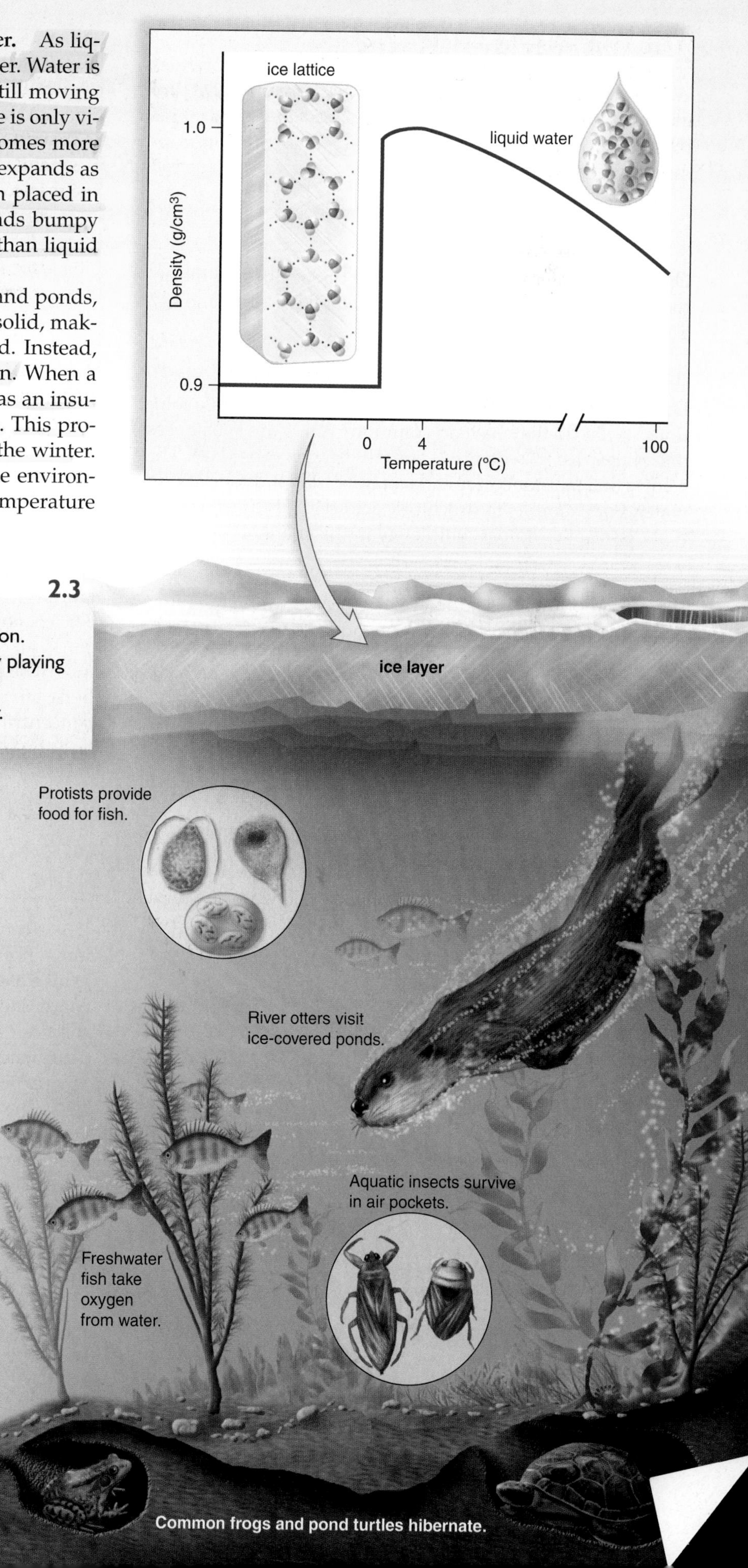


**FIGURE 2.12 A pond in winter.**

*Above:* Remarkably, water is more dense at 4°C than at 0°C. Most substances contract when they solidify, but water expands when it freezes because in ice, water molecules form a lattice in which the hydrogen bonds are farther apart than in liquid water. *Below:* The layer of ice that forms at the top of a pond shields the water and protects the protists, plants, and animals so that they can survive the winter. These animals, except for the otter, are ectothermic, which means that they take on the temperature of the outside environment. This might seem disadvantageous until you realize that water remains relatively warm because of its high heat capacity. During the winter, frogs and turtles hibernate and in this way, lower their oxygen needs. Insects survive in air pockets. Fish, as you will learn later in this text, have an efficient means of extracting oxygen from the water and they need less oxygen than the endothermic otter, which depends on muscle activity to warm its body.

# 2.4 Acids and Bases

When water ionizes, it releases an equal number of **hydrogen ions ($H^+$)** (also called a proton[1]) and **hydroxide ions ($OH^-$)**:

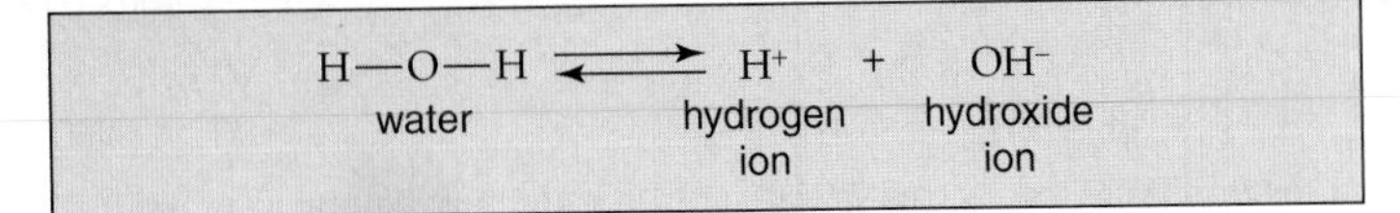


Only a few water molecules at a time dissociate, and the actual number of $H^+$ and $OH^-$ is very small ($1 \times 10^{-7}$ moles/liter).[2]

## *Acidic Solutions (High $H^+$ Concentrations)*

Lemon juice, vinegar, tomatoes, and coffee are all acidic solutions. What do they have in common? **Acids** are substances that dissociate in water, releasing hydrogen ions ($H^+$). The acidity of a substance depends on how fully it dissociates in water. For example, hydrochloric acid (HCl) is a strong acid that dissociates almost completely in this manner:

$$HCl \longrightarrow H^+ + Cl^-$$

If hydrochloric acid is added to a beaker of water, the number of hydrogen ions ($H^+$) increases greatly.

## *Basic Solutions (Low $H^+$ Concentration)*

Milk of magnesia and ammonia are common basic solutions familiar to most people. **Bases** are substances that either take up hydrogen ions ($H^+$) or release hydroxide ions ($OH^-$). For example, sodium hydroxide (NaOH) is a strong base that dissociates almost completely in this manner:

$$NaOH \longrightarrow Na^+ + OH^-$$

If sodium hydroxide is added to a beaker of water, the number of hydroxide ions increases.

## *pH Scale*

The **pH scale** is used to indicate the acidity or basicity (alkalinity) of a solution.[3] The pH scale (Fig. 2.13) ranges from 0 to 14. A pH of 7 represents a neutral state in which the hydrogen ion and hydroxide ion concentrations are equal. A pH below 7 is an acidic solution because the hydrogen ion concentration is greater than the hydroxide concentration. A pH above 7 is basic because the $[OH^-]$ is greater than the $[H^+]$. Further, as we move down the pH scale from pH 14 to pH 0, each unit is 10 times more acidic than the previous unit. As we move up the scale from 0 to 14, each unit is 10 times more basic than the previous unit. Therefore pH 5 is 100 times more acidic than is pH 7 and a 100 times more basic than pH 3.

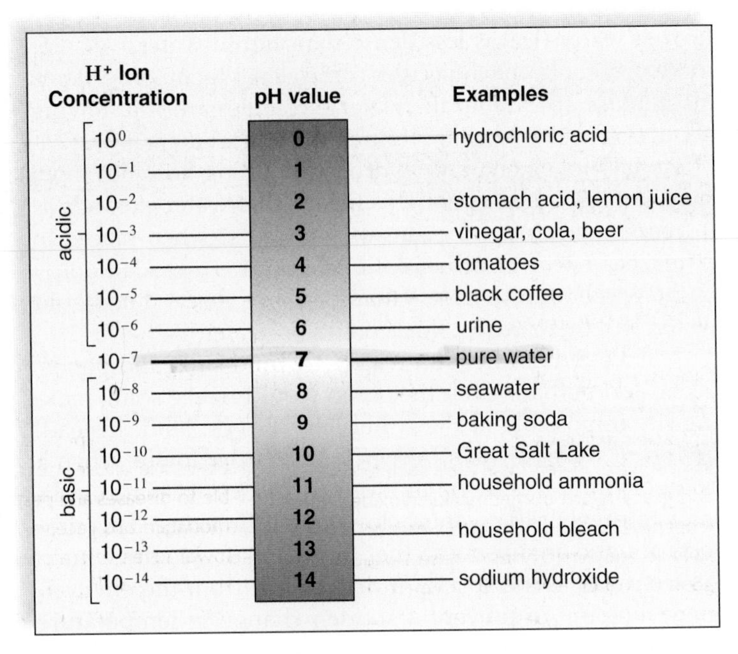


**FIGURE 2.13 The pH scale.**

The pH scale ranges from 0 to 14 with 0 being the most acidic and 14 being the most basic. pH 7 (neutral pH) has equal amounts of hydrogen ions ($H^+$) and hydroxide ions ($OH^-$). An acidic pH has more $H^+$ than $OH^-$ and a basic pH has more $OH^-$ than $H^+$.

The pH scale was devised to eliminate the use of cumbersome numbers. For example, the possible hydrogen ion concentrations of a solution are on the left of this listing and the pH is on the right:

| $[H^+]$ (moles per liter) | | pH |
|---|---|---|
| 0.000001 | $= 1 \times 10^{-6}$ | 6 |
| 0.0000001 | $= 1 \times 10^{-7}$ | 7 |
| 0.00000001 | $= 1 \times 10^{-8}$ | 8 |

To further illustrate the relationship between hydrogen ion concentration and pH, consider the following question. Which of the pH values listed indicates a higher hydrogen ion concentration $[H^+]$ than pH 7, and therefore would be an acidic solution? A number with a smaller negative exponent indicates a greater quantity of hydrogen ions than one with a larger negative exponent. Therefore, pH 6 is an acidic solution.

The Ecology Focus on page 33 describes detrimental environmental consequences to nonliving and living things as rain and snow have become more acidic. In humans, pH needs to be maintained with a narrow range or there are health consequences. The pH of blood is around 7.4, and blood is buffered in the manner described next to keep the pH within a normal range.

## *Buffers and pH*

A **buffer** is a chemical or a combination of chemicals that keeps pH within normal limits. Many commercial products such as Bufferin, shampoos, or deodorants are buffered as an added incentive for us to buy them.

In living things, the pH of body fluids is maintained within a narrow range, or else health suffers. The pH of our blood when we are healthy is always about 7.4—that is, just

[1] A hydrogen atom contains one electron and one proton. A hydrogen ion has only one proton, so it is often simply called a proton.

[2] In chemistry, a mole is defined as the amount of matter that contains as many objects (atoms, molecules, ions) as the number of atoms in exactly 12 g of $^{12}C$.

[3] pH is defined as the negative log of the hydrogen ion concentration $[H^+]$. A log is the power to which 10 must be raised to produce a given number.

*ecology focus*

## The Harm Done by Acid Deposition

Normally, rainwater has a pH of about 5.6 because the carbon dioxide in the air combines with water to give a weak solution of carbonic acid. Acid deposition includes rain or snow that has a pH of less than 5, as well as dry acidic particles that fall to Earth from the atmosphere. When fossil fuels such as coal, oil, and gasoline are burned, sulfur dioxide and nitrogen oxides combine with water to produce sulfuric and nitric acids. These pollutants are generally found eastward of where they originated because of wind patterns. The use of very tall smokestacks causes them to be carried even hundreds of miles away. For example, acid rain in southeastern Canada results from the burning of fossil fuels in factories and power plants in the midwestern United States.

### Impact on Lakes

Acid rain adversely affects lakes, particularly in areas where the soil is thin and lacks limestone (calcium carbonate, or $CaCO_3$), a buffer to acid deposition. Acid deposition leaches toxic aluminum from the soil and converts mercury deposits in lake bottom sediments to toxic methyl mercury, which accumulates in fish. People are now advised against eating fish from the Great Lakes because of high mercury levels. Hundreds of lakes are devoid of fish in Canada and New England, and thousands have suffered the same fate in the Scandinavian countries. Some of these lakes have no signs of life at all.

### Impact on Forests

The leaves of plants damaged by acid rain can no longer carry on photosynthesis as before. When plants are under stress, they become susceptible to diseases and pests of all types. Forests on mountaintops receive more rain than those at lower levels; therefore, they are more affected by acid rain (Fig. 2A*a*). Forests are also damaged when toxic chemicals such as aluminum are leached from the soil. These kill soil fungi that assist roots in acquiring the nutrients trees need. In New England, 1.3 million acres of high-elevation forests have been devastated.

### Impact on Humans and Structures

Humans may be affected by acid rain. Inhaling dry sulfate and nitrate particles appears to increase the occurrence of respiratory illnesses, such as asthma. Buildings and monuments made of limestone and marble break down when exposed to acid rain (Fig. 2A*b*). The paint on homes and automobiles is likewise degraded.

a.

b.

**FIGURE 2A Effects of acid deposition.**
*The burning of gasoline derived from oil, a fossil fuel, leads to acid deposition, which causes (**a**) trees to die and (**b**) statues to deteriorate.*

slightly basic (alkaline). If the blood pH drops to about 7, acidosis results. If the blood pH rises to about 7.8, alkalosis results. Both conditions can be life threatening; the blood pH must be kept around 7.4. Normally, pH stability is possible because the body has built-in mechanisms to prevent pH changes. Buffers are one of these important mechanisms.

Buffers help keep the pH within normal limits because they are chemicals or combinations of chemicals that take up excess hydrogen ions ($H^+$) or hydroxide ions ($OH^-$). For example, carbonic acid ($H_2CO_3$) is a weak acid that minimally dissociates and then re-forms in the following manner:

$$\underset{\text{carbonic acid}}{H_2CO_3} \underset{\text{re-forms}}{\overset{\text{dissociates}}{\rightleftharpoons}} H^+ + \underset{\text{bicarbonate ion}}{HCO_3^-}$$

Blood always contains a combination of some carbonic acid and some bicarbonate ions. When hydrogen ions ($H^+$) are added to blood, the following reaction reduces acidity:

$$H^+ + HCO_3^- \longrightarrow H_2CO_3$$

When hydroxide ions ($OH^-$) are added to blood, this reaction reduces basidity:

$$OH^- + H_2CO_3 \longrightarrow HCO_3^- + H_2O$$

These reactions prevent any significant change in blood pH.

**Check Your Progress** **2.4**

1. Contrast an acid with a base.
2. Give an example to substantiate that acid rain is detrimental to both plants and animals.
3. A substance that absorbs hydrogen ions makes the pH rise. Explain.

## Connecting the Concepts

All matter consists of various combinations of the same 92 elements. Living things consist primarily of just six of these elements—carbon, hydrogen, nitrogen, oxygen, phosphorus, and sulfur (CHNOPS for short). These elements combine to form the unique types of molecules found in living cells. In organisms, many other elements exist in smaller amounts as ions, and their functions are dependent on their charged nature. Cells consist largely of water, a molecule that contains only hydrogen and oxygen. Polar covalent bonding between the atoms and hydrogen bonding between the molecules give water the properties that make life possible. Presently, we are aware of no other planet that has liquid water.

In the next chapter, we will learn that a carbon atom combines covalently with CHNOPS to form the organic molecules of cells. It is these unique molecules that set living forms apart from nonliving objects. Carbon-containing molecules can be modified in numerous ways, and this accounts for life's diversity, such as differences between a bottle-nosed dolphin and a black shoulder peacock. Varying molecular compositions in plants can also tell us, for example, why some trees have leaves that change color in the fall.

It is difficult for us to visualize that a bottle-nosed dolphin, a kangaroo, or a pine tree is a combination of molecules and ions, but later in this text we will learn that even our thoughts about these organisms are simply the result of molecules flowing from one brain cell to another. An atomic, ionic, and molecular understanding of the variety of processes unique to life provides a deeper understanding of the definition of life and offers tools for the improvement of its quality, preservation of its diversity, and appreciation of its beauty.

## summary

### 2.1 Chemical Elements

Both living and nonliving things are composed of matter consisting of elements. The acronym CHNOPS stands for the most significant elements (atoms) found in living things: carbon, hydrogen, nitrogen, oxygen, phosphorus, and sulfur. Elements contain atoms, and atoms contain subatomic particles. Protons and neutrons in the nucleus determine the mass number of an atom. The atomic number indicates the number of protons and the number of electrons in electrically neutral atoms. Protons have positive charges, neutrons are uncharged, and electrons have negative charges. Isotopes are atoms of a single element that differ in their numbers of neutrons. Radioactive isotopes have many uses, including serving as tracers in biological experiments and medical procedures.

Electrons occupy energy levels (electron shells) at discrete distances from the nucleus. The number of electrons in the outer shell determines the reactivity of an atom. The first shell is complete when it is occupied by two electrons. In atoms up through calcium, number 20, every shell beyond the first shell is complete with eight electrons. The octet rule states that atoms react with one another in order to have a completed outer shell. Most atoms, including those common to living things, do not have filled outer shells and this causes them to react with one another to form compounds and/or molecules. Following the reaction, the atoms have completed outer shells.

### 2.2 Compounds and Molecules

Ions form when atoms lose or gain one or more electrons to achieve a completed outer shell. An ionic bond is an attraction between oppositely charged ions. When covalent compounds form, atoms share electrons. A covalent bond is one or more shared pairs of electrons. There are single, double, and triple covalent bonds.

In polar covalent bonds, the sharing of electrons is not equal. If the molecule is polar, the more electronegative atom carries a slightly negative charge and the other atom carries a slightly positive charge.

### 2.3 Chemistry of Water

Water is a polar molecule. The polarity of water molecules allows hydrogen bonding to occur between water molecules. A hydrogen bond is a weak attraction between a slightly positive hydrogen atom and a slightly negative oxygen or nitrogen atom within the same or a different molecule. Hydrogen bonds help maintain the structure and function of cellular molecules.

Water's polarity and hydrogen bonding account for its unique properties. These features allow living things to exist and carry on cellular activities.

### 2.4 Acids and Bases

A small fraction of water molecules dissociate to produce an equal number of hydrogen ions and hydroxide ions. Solutions with equal numbers of $H^+$ and $OH^-$ are termed neutral. In acidic solutions, there are more hydrogen ions than hydroxide ions; these solutions have a pH less than 7. In basic solutions, there are more hydroxide ions than hydrogen ions; these solutions have a pH greater than 7. Cells are sensitive to pH changes. Biological systems often contain buffers that help keep the pH within a normal range.

## understanding the terms

Match the terms to these definitions:

a. _______________ Bond in which the sharing of electrons between atoms is unequal.

b. _______________ Charged particle that carries a negative or positive charge(s).

c. _______________ Molecules tending to raise the hydrogen ion concentration in a solution and to lower its pH numerically.

d. _______________ The smallest part of a compound that still has the properties of that compound.

e. _______________ A chemical or a combination of chemicals that maintains a constant pH upon the addition of small amounts of acid or base.

## reviewing this chapter

1. Name the kinds of subatomic particles studied. What is their atomic mass unit, charge, and location in an atom? 21–23
2. What is an isotope? A radioactive isotope? Radioactivity is always considered dangerous. Why? 24–26
3. Using the Bohr model, draw an atomic structure for a carbon that has six protons and six neutrons. 26
4. Draw an atomic representation for $MgCl_2$. Using the octet rule, explain the structure of the compound. 27
5. Explain whether $CO_2$ (O=C=O) is an ionic or a covalent compound. Why does this arrangement satisfy all atoms involved? 27
6. Of what significance is the shape of molecules in organisms? 28
7. Explain why water is a polar molecule. What does the polarity and shape of water have to do with its ability to form hydrogen bonds? 28
8. Name five properties of water, and relate them to the structure of water, including its polarity and hydrogen bonding between molecules. 28–31
9. On the pH scale, which numbers indicate a solution is acidic? Basic? Neutral? 32
10. What are buffers, and why are they important to life? 32–33

## testing yourself

Choose the best answer for each question.

1. Which of the subatomic particles contributes almost no weight to an atom?
   a. protons in the electron shells
   b. electrons in the nucleus
   c. neutrons in the nucleus
   d. electrons at various energy levels
2. The atomic number tells you the
   a. number of neutrons in the nucleus.
   b. number of protons in the atom.
   c. atomic mass of the atom.
   d. number of its electrons if the atom is neutral.
   e. Both b and d are correct.
3. An atom that has two electrons in the outer shell, such as magnesium, would most likely
   a. share to acquire a completed outer shell.
   b. lose these two electrons and become a negatively charged ion.
   c. lose these two electrons and become a positively charged ion.
   d. bind with carbon by way of hydrogen bonds.
   e. bind with another calcium atom to satisfy its energy needs.
4. Isotopes differ in their
   a. number of protons.
   b. atomic number.
   c. number of neutrons.
   d. number of electrons.
5. When an atom gains electrons, it
   a. forms a negatively charged ion.
   b. forms a positively charged ion.
   c. forms covalent bonds.
   d. forms ionic bonds.
   e. gains atomic mass.
6. A covalent bond is indicated by
   a. plus and minus charges attached to atoms.
   b. dotted lines between hydrogen atoms.
   c. concentric circles about a nucleus.
   d. overlapping electron shells or a straight line between atomic symbols.
   e. the touching of atomic nuclei.
7. The shape of a molecule
   a. is dependent in part on the angle of bonds between its atoms.
   b. influences its biological function.
   c. is dependent on its electronegativity.
   d. is dependent on its place in the periodic table.
   e. Both a and b are correct.
8. In which of these are the electrons always shared unequally?
   a. double covalent bond
   b. triple covalent bond
   c. hydrogen bond
   d. polar covalent bond
   e. ionic and covalent bonds
9. In the molecule

```
     H
     |
  H—C—H
     |
     H
```

   a. all atoms have eight electrons in the outer shell.
   b. all atoms are sharing electrons.
   c. carbon could accept more hydrogen atoms.
   d. the bonds point to the corners of a square.
   e. All of these are correct.
10. Which of these properties of water cannot be attributed to hydrogen bonding between water molecules?
   a. Water stabilizes temperature inside and outside the cell.
   b. Water molecules are cohesive.
   c. Water is a solvent for many molecules.
   d. Ice floats on liquid water.
   e. Both b and c are correct.
11. Complete this diagram by placing an O for oxygen or an H for hydrogen on the appropriate atoms. Place partial charges where they belong.

12. $H_2CO_3/NaHCO_3$ is a buffer system in the body. What effect will the addition of an acid have on the pH of a solution that is buffered?
    a. The pH will rise.
    b. The pH will lower.
    c. The pH will not change.
    d. All of these are correct.
13. Rainwater has a pH of about 5.6; therefore, rainwater is
    a. a neutral solution.
    b. an acidic solution.
    c. a basic solution.
    d. It depends if it is buffered.
14. Acids
    a. release hydrogen ions in solution.
    b. cause the pH of a solution to rise above 7.
    c. take up hydroxide ions and become neutral.
    d. increase the number of water molecules.
    e. Both a and b are correct.
15. Which type of bond results from the sharing of electrons between atoms?
    a. covalent
    b. ionic
    c. hydrogen
    d. neutral
16. Complete this diagram of a nitrogen atom by placing the correct number of protons and neutrons in the nucleus and electrons in the shells. Explain why the correct formula for ammonia is $NH_3$, not $NH_4$.

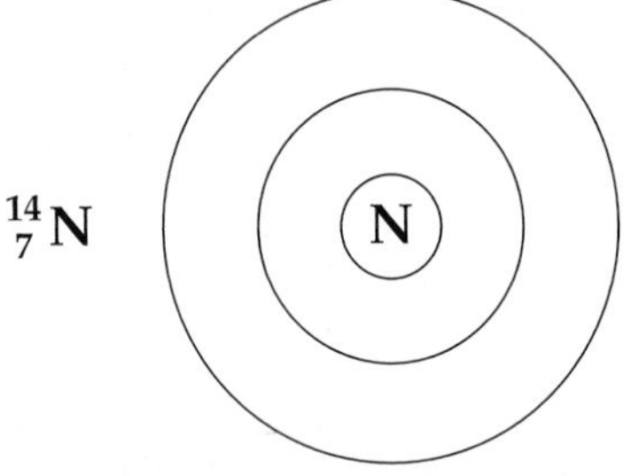


17. Why is $—NH_2$ a polar group?
    a. Nitrogen is more electronegative than hydrogen.
    b. The bonds are not symmetrical.
    c. Because hydrogen bonding takes place.
    d. Both a and b are correct.
18. If a chemical accepted $H^+$ from the surrounding solution, the chemical could be
    a. a base.
    b. an acid.
    c. a buffer.
    d. None of the above are correct.
    e. Both a and c are correct.
19. The periodic table tells us
    a. the atomic number, symbol, and mass.
    b. how many shells an atom has.
    c. how many electrons are in the outer shell.
    d. whether the atom will react or not.
    e. All of these are correct.
20. Which of these best describes the changes that occur when a solution goes from pH 5 to pH 7?
    a. The solution is now 100 times more acidic.
    b. The solution is now 100 times more basic.
    c. The hydrogen ion concentration decreases by only a factor of 20, as the solution goes from basic to acidic.
    d. The hydrogen ion concentration changes by only a factor of 20, as the solution goes from acidic to basic.
21. A hydrogen bond is not
    a. involved in maintaining the shape of certain molecules.
    b. necessary to the properties of water.
    c. as strong as a covalent bond.
    d. represented by a dotted line.
    e. More than one of these is correct.

For questions 22–25, match the statements with a property of water in the key.

**KEY:**

a. Water flows because it is cohesive.
b. Water holds its heat.
c. Water has neutral pH.
d. Water has a high heat of vaporization.

22. Sweating helps cool us off.
23. Our blood is composed mostly of water and cells.
24. Our blood is just about pH 7.
25. We usually maintain a normal body temperature.

## thinking scientifically

1. Natural phenomena often require an explanation. Based on how sodium chloride dissociates in water (see pages 29–30) and Figure 2.12, explain why the oceans don't freeze.
2. Melvin Calvin used radioactive carbon (as a tracer) to discover a series of molecules that form during photosynthesis. Explain why carbon behaves chemically the same, even when radioactive.

## bioethical issue

### The Right to Refuse an IV

When a person gets sick or endures physical stress—as, for example, during childbirth—pH levels may dip or rise too far, endangering that person's life. In most U.S. hospitals, doctors routinely administer IVs, or intravenous infusions, of certain fluids to maintain a patient's pH level. Some people who oppose IVs for philosophical reasons may refuse an IV. That's relatively safe, as long as the person is healthy.

Problems arise when hospital policy dictates an IV, even though a patient does not want one. Should a patient be allowed to refuse an IV? Or does a hospital have the right to insist, for health reasons, that patients accept IV fluids? And what role should doctors play—patient advocates or hospital representatives?

## *Biology* website

The companion website for *Biology* provides a wealth of information organized and integrated by chapter. You will find practice tests, animations, videos, and much more that will complement your learning and understanding of general biology.

**http://www.mhhe.com/maderbiology10**

# 3

# The Chemistry of Organic Molecules

*We might have trouble thinking of ways that plants and animals are similar, but we all know that vegetarians have no trouble sustaining themselves by eating plants, as long as they include a variety of plants in their diet. That's because plants and humans generally have the same molecules in their cells—namely, carbohydrates, lipids, proteins, and nucleic acids. When we feed on plants, we digest their macromolecules to smaller molecules, and then we use these smaller molecules to build our own types of carbohydrates, lipids, proteins, and nucleic acids.*

*A similarity in chemistry between plants and humans is especially evident when we acquire vitamins from plants and use them exactly as plants do, because vitamins assist the same enzymes found in all organisms. The differences between plants and humans are due to their genes. But, then, all genes are made of DNA, and the way genes function in cells is the same in all organisms. In this chapter, we continue our look at basic chemistry by considering the molecules found in all living things. These are the types of molecules that account for the structure and function of all cells in any type of organism.*

Granddaughter Sylvia with azalea plant.

## concepts

# 3.1 Organic Molecules

Because chemists of the nineteenth century thought that the molecules of cells must contain a vital force, they divided chemistry into **organic chemistry,** the chemistry of organisms, and **inorganic chemistry,** the chemistry of the nonliving world. This terminology is still with us even though many types of organic molecules can now be synthesized in the laboratory. Today, we simply define **organic molecules** as molecules that contain both carbon and hydrogen atoms (Table 3.1).

**TABLE 3.1**

**Inorganic Versus Organic Molecules**

| *Inorganic Molecules* | *Organic Molecules* |
|---|---|
| Usually contain positive and negative ions | Always contain carbon and hydrogen |
| Usually ionic bonding | Always covalent bonding |
| Always contain a small number of atoms | Often quite large, with many atoms |
| Often associated with nonliving matter | Usually associated with living organisms |

There are only four classes of organic compounds in any living thing: carbohydrates, lipids, proteins, and nucleic acids. Despite the limited number of classes, the so-called **biomolecules** in cells are quite diverse. A bacterial cell contains some 5,000 different organic molecules, and a plant or animal cell has twice that number. This diversity of organic molecules makes the diversity of life possible (Fig. 3.1). It is quite remarkable that the variety of organic molecules can be traced to the unique chemical properties of the carbon atom.

## The Carbon Atom

What is there about carbon that makes organic molecules the same and also different? Carbon is quite small, with only a total of six electrons: two electrons in the first shell and four electrons in the outer shell. In order to acquire four electrons to complete its outer shell, a carbon atom almost always shares electrons with—you guessed it—CHNOPS, the elements that make up most of the weight of living things (see Fig. 2.1).

Because carbon needs four electrons to complete its outer shell, it can share with as many as four other elements, and this spells diversity. But even more significant to the shape, and therefore the function, of biomolecules, carbon often shares electrons with another carbon atom. The C—C bond is quite stable, and the result is carbon chains that can be quite long. Hydrocarbons are chains of carbon atoms bonded exclusively to hydrogen atoms.

```
    H   H   H   H   H   H   H   H
    |   |   |   |   |   |   |   |
H—C—C—C—C—C—C—C—C—H
    |   |   |   |   |   |   |   |
    H   H   H   H   H   H   H   H
```

octane

Branching at any carbon atom is possible, and also a hydrocarbon can turn back on itself to form a ring compound when placed in water:

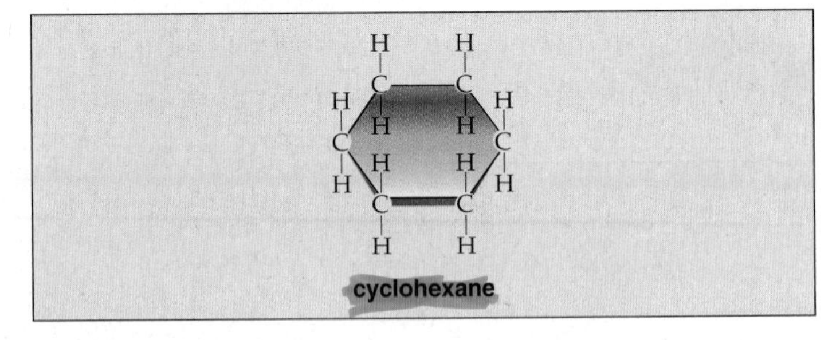


Carbon can form double bonds with itself and other atoms. Double bonds restrict the movement of bonded atoms, and in that way contribute to the shape of the molecule. As in acetylene, H—C≡C—H, carbon is also capable of forming a triple bond with itself.

The diversity of organic molecules is further enhanced by the presence of particular functional groups, as discussed next. Contrast the structure of cyclohexane, above, with the structure of glucose in Figure 3.6. The difference in structure can be attributed to the functional groups added to the same number of carbons.

a. Cell walls contain cellulose. b. Shell contains chitin. c. Cell walls contain peptidoglycan.

**FIGURE 3.1 Carbohydrates as structural materials.**

**a.** Plants, such as cacti, have the carbohydrate cellulose in their cell walls. **b.** The shell of a crab contains chitin, a different carbohydrate. **c.** The cell walls of bacteria contain another type of carbohydrate known as peptidoglycan.

## The Carbon Skeleton and Functional Groups

The carbon chain of an organic molecule is called its skeleton or backbone. The terminology is appropriate because just as a skeleton accounts for your shape, so does the carbon skeleton of an organic molecule account for its shape. Vertebrates look very different, even though they all have a backbone of vertebrae. We recognize them by their shape and also by their appendages, whether they have fins, wings, or limbs, for example. So, the diversity of organic molecules comes about when different functional groups are added to the carbon skeleton. A **functional group** is a specific combination of bonded atoms that always reacts in the same way, regardless of the particular carbon skeleton. As in Figure 3.2, it is even acceptable to use an *R* to stand for the remainder of the molecule, which is the carbon skeleton, because only the functional group is involved in a reaction.

Notice that when a particular functional group is added to a carbon skeleton, the molecule becomes a certain type of compound. For example, the addition of an —OH (hydroxyl group) to a carbon skeleton turns that molecule into an alcohol. When an —OH replaces one of the hydrogens in ethane, a 2-carbon hydrocarbon, it becomes ethanol, a type of alcohol that is familiar because it is consumable by humans. Whereas ethane, like other hydrocarbons, is **hydrophobic** (not soluble in water), ethanol is **hydrophilic** (soluble in water) because the —OH functional group is polar. Since cells are 70–90% water, the ability to interact with and be soluble in water profoundly affects the function of organic molecules in cells.

Organic molecules containing carboxyl (acidic) groups (—COOH) are highly polar. They tend to ionize and release hydrogen ions in solution:

$$-\mathrm{COOH} \longrightarrow -\mathrm{COO}^- + \mathrm{H}^+$$

The attached functional groups determine the polarity of an organic molecule and also the types of reactions it will undergo. We will see that alcohols react with carboxyl groups when a fat forms, and that carboxyl groups react with amino groups during protein formation.

| Functional Groups | | | |
|---|---|---|---|
| **Group** | **Structure** | **Compound** | **Significance** |
| Hydroxyl | *R*—OH | Alcohol as in ethanol | Polar, forms hydrogen bond, present in sugars and some amino acids |
| Carbonyl | *R*—C(=O)—H | Aldehyde as in formaldehyde | Polar, present in sugars |
| | *R*—C(=O)—*R* | Ketone as in acetone | Polar, present in sugars |
| Carboxyl (acidic) | *R*—C(=O)—OH | Carboxylic acid as in acetic acid | Polar, acidic, present in fatty acids and amino acids |
| Amino | *R*—N(—H)—H | Amine as in tryptophan | Polar, basic, forms hydrogen bonds, present in amino acids |
| Sulfhydryl | *R*—SH | Thiol as in ethanethiol | Forms disulfide bonds, present in some amino acids |
| Phosphate | *R*—O—P(=O)(—OH)—OH | Organic phosphate as in phosphorylated molecules | Polar, acidic, present in nucleotides and phospholipids |

*R* = remainder of molecule

**FIGURE 3.2 Functional groups.**

Molecules with the same carbon skeleton can still differ according to the type of functional group attached to the carbon skeleton. Many of these functional groups are polar, helping to make the molecule soluble in water. In this illustration, the remainder of the molecule (does not include the functional group) is represented by an *R*.

### *Isomers*

**Isomers** [Gk. *isos*, equal, and *meros*, part, portion] are organic molecules that have identical molecular formulas but a different arrangement of atoms. In essence, isomers are variations in the molecular architecture of a molecule. Isomers are another example of how the chemistry of carbon leads to variations in organic molecules.

The two molecules in Figure 3.3 are isomers of one another; they have the same molecular formula but different functional groups. Therefore, we would expect them to react differently in chemical reactions.

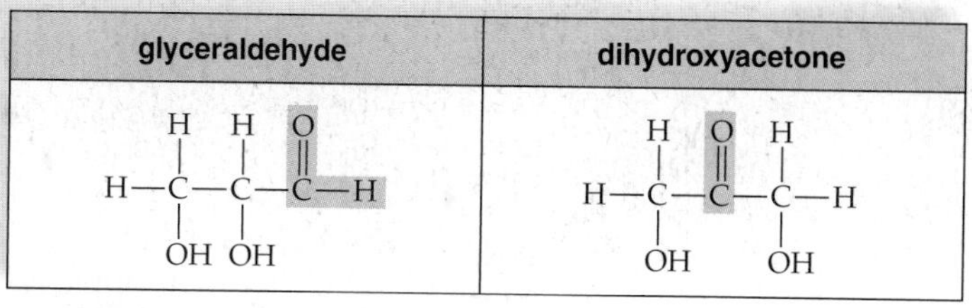

| glyceraldehyde | dihydroxyacetone |
|---|---|
| H—C(H)(OH)—C(H)(OH)—C(=O)—H | H—C(H)(OH)—C(=O)—C(H)(OH)—H |

**FIGURE 3.3 Isomers.**

Isomers have the same molecular formula but different atomic configurations. Both of these compounds have the formula $C_3H_6O_3$. In glyceraldehyde, oxygen is double-bonded to an end carbon. In dihydroxyacetone, oxygen is double-bonded to the middle carbon.

| Biomolecules | | |
|---|---|---|
| **Category** | **Example** | **Subunit(s)** |
| Carbohydrates* | Polysaccharide | Monosaccharide |
| Lipids | Fat | Glycerol and fatty acids |
| Proteins* | Polypeptide | Amino acids |
| Nucleic acids* | DNA, RNA | Nucleotide |

*Polymers

**FIGURE 3.4 Common foods.**

Carbohydrates in bread and pasta are digested to sugars; lipids such as oils are digested to glycerol and fatty acids; and proteins in meat are digested to amino acids. Cells use these subunit molecules to build their own biomolecules and as a source of energy.

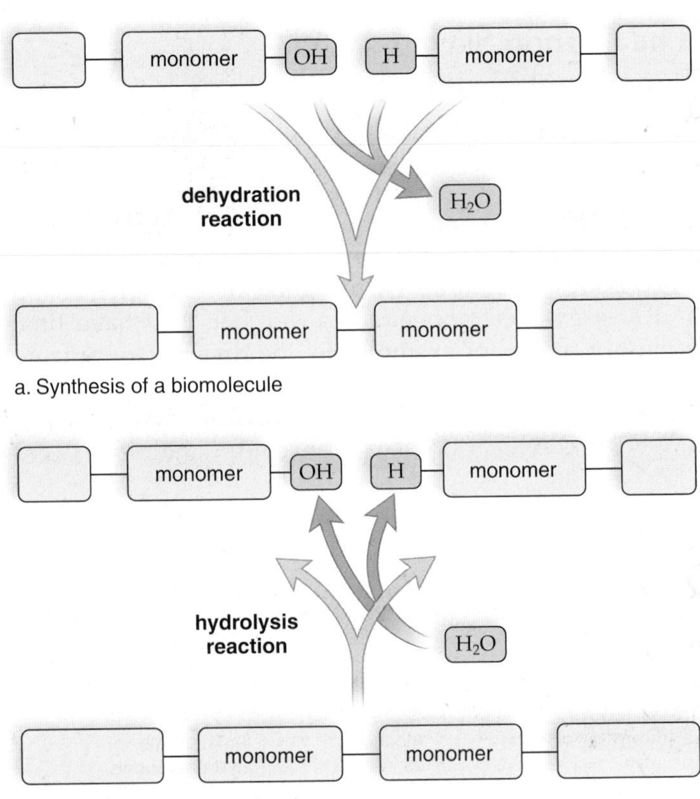


**FIGURE 3.5 Synthesis and degradation of biomolecules.**

**a.** In cells, synthesis often occurs when subunits bond during a dehydration reaction (removal of $H_2O$). **b.** Degradation occurs when the subunits separate during a hydrolysis reaction (the addition of $H_2O$).

## The Biomolecules of Cells

You are very familiar with the names of biomolecules—carbohydrates, lipids, proteins, and nucleic acids—because certain foods are known to be rich in them, as illustrated in Figure 3.4. For example, bread is rich in carbohydrate, and meat is rich in protein. When you digest food, it gets broken down into smaller molecules that are subunits for biomolecules. Digestion of bread releases glucose molecules, digestion of meat releases amino acids. Your body then takes these subunits and builds from them the particular carbohydrates and proteins that make up your cells (Fig. 3.4, *below*).

### *Synthesis and Degradation*

A cell uses a condensation reaction to synthesize (build up) any type of biomolecule. It's called a **dehydration reaction** because the equivalent of a water molecule—that is, an —OH (hydroxyl group) and an —H (hydrogen atom), is removed as subunits are joined. Therefore, water molecules result as biomolecules are synthesized (Fig. 3.5*a*).

To break down biomolecules, a cell uses an opposite type of reaction. During a **hydrolysis** [Gk. *hydro*, water, and *lyse*, break] **reaction,** an —OH group from water attaches to one subunit, and an —H from water attaches to the other subunit. In other words, biomolecules are broken down by adding water to them (Fig. 3.5*b*).

Enzymes are required for cells to carry out dehydration and hydrolysis reactions. An **enzyme** is a molecule that speeds a reaction by bringing reactants together, and the enzyme may even participate in the reaction but it is unchanged by it.

**Polymers.** The largest of the biomolecules are called **polymers** and like all biomolecules, polymers are constructed by linking together a large number of the same type of subunit. However, in the case of polymers, the subunits are called **monomers.** A polysaccharide, a protein and a nucleic acid, is a polymer that contains innumerable monomers. Just as a train increases in length when boxcars are hitched together one by one, so a polymer gets longer as monomers bond to one another.

**Check Your Progress 3.1**

1. Describe the properties of a carbon atom that make it ideally suited to produce varied carbon skeletons.
2. **a.** How could two pearl necklaces be both the same and different? **b.** How could two protein polymers be both the same and different?

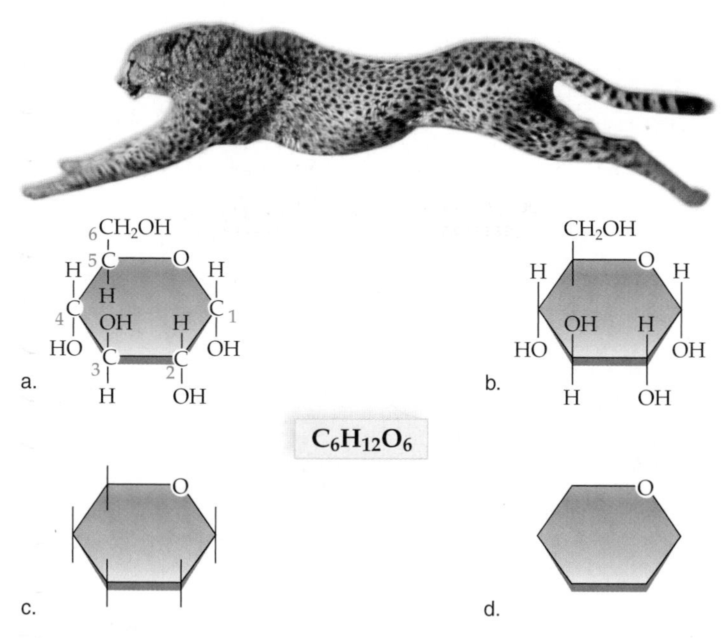


**FIGURE 3.6 Glucose.**

Glucose provides energy for organisms, such as this cheetah. Each of these structural formulas is glucose. **a.** The carbon skeleton and all attached groups are shown. **b.** The carbon skeleton is omitted. **c.** The carbon skeleton and attached groups are omitted. **d.** Only the ring shape, which includes one oxygen atom, remains.

# 3.2 Carbohydrates

**Carbohydrates** are almost universally used as an immediate energy source in living things, but they also play structural roles in a variety of organisms (see Fig. 3.1). The majority of carbohydrates have a carbon to hydrogen to oxygen ratio of 1:2:1. The term *carbohydrate* includes single sugar molecules and also chains of sugars. Chain length varies from a few sugars to hundreds of sugars. The long chains are thus polymers. The monomers of carbohydrates are monosaccharides.

## Monosaccharides: Ready Energy

**Monosaccharides** [Gk. *monos,* single, and *sacchar,* sugar], consisting of only a single sugar molecule, are called simple sugars. A simple sugar can have a carbon backbone of three to seven carbons. The molecular formula for a simple sugar is some multiple of $CH_2O$, suggesting that every carbon atom is bonded to an —H and an —OH. This is not strictly correct, as you can see by examining the structural formula for glucose (Fig. 3.6). Still, sugars do have many hydroxyl groups, and this polar functional group makes them soluble in water.

**Glucose,** with six carbon atoms, is a **hexose** [Gk. *hex,* six] and has a molecular formula of $C_6H_{12}O_6$. Despite the fact that glucose has several isomers, such as fructose and galactose, we usually think of $C_6H_{12}O_6$ as glucose. This signifies that glucose has a special place in the chemistry of organisms. This simple sugar is the major source of cellular fuel for all living things. Glucose is transported in the blood of animals, and it is the molecule that is broken down in nearly all types of organisms during cellular respiration, with the resulting buildup of ATP molecules.

**Ribose** and **deoxyribose,** with five carbon atoms, are **pentoses** [Gk. *pent,* five] of significance because they are found respectively in the nucleic acids RNA and DNA. RNA and DNA are discussed later in the chapter.

## Disaccharides: Varied Uses

A **disaccharide** contains two monosaccharides that have joined during a dehydration reaction. Figure 3.7 shows how the disaccharide maltose (an ingredient used in brewing) arises when two glucose molecules bond together. Note the position of the bond that results when the —OH groups participating in the reaction project below the ring. When our hydrolytic digestive juices break this bond, the result is two glucose molecules.

Sucrose (the structure shown at right) is another disaccharide of special interest because it is sugar we use at home to sweeten our food. Sucrose is also the form in which sugar is transported in plants. We acquire sucrose from plants such as sugarcane and sugar beets. You may also have heard of lactose, a disaccharide found in milk. Lactose is glucose combined with galactose. Individuals that are lactose intolerant cannot break this disaccharide down and have subsequent medical problems. To prevent problems they can buy foods in which lactose has been broken down into its subunits.

$CH_2OH$ $CH_2OH$
glucose O fructose
$CH_2OH$
sucrose

$CH_2OH$ H + H $CH_2OH$ OH HO

dehydration reaction

hydrolysis reaction

$CH_2OH$ $CH_2OH$ O + $H_2O$

glucose $C_6H_{12}O_6$ glucose $C_6H_{12}O_6$ maltose $C_{12}H_{22}O_{11}$ water

monosaccharide + monosaccharide ⇄ disaccharide + water

**FIGURE 3.7 Synthesis and degradation of maltose, a disaccharide.**

Synthesis of maltose occurs following a dehydration reaction when a bond forms between two glucose molecules, and water is removed. Degradation of maltose occurs following a hydrolysis reaction when this bond is broken by the addition of water.

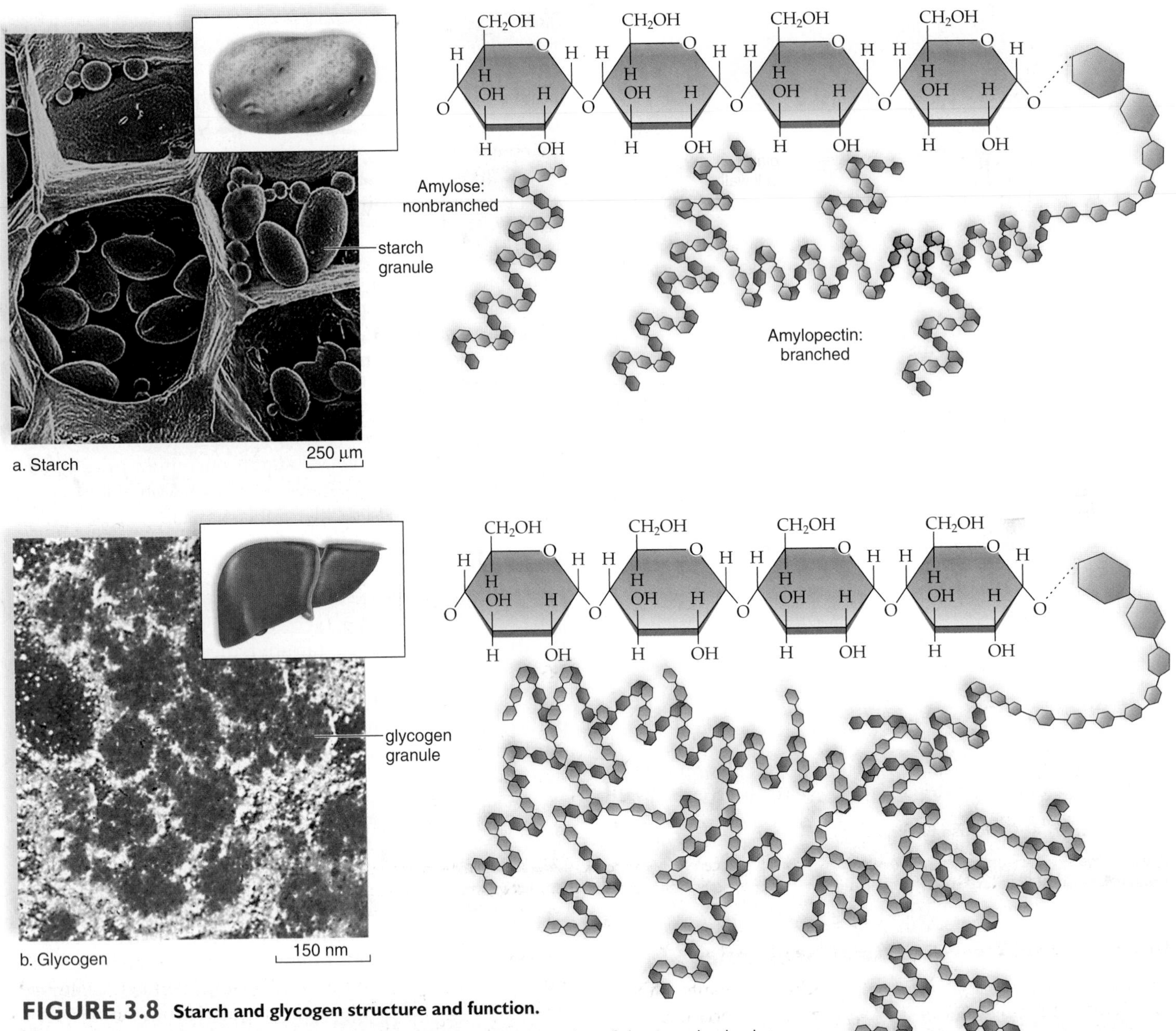


**FIGURE 3.8 Starch and glycogen structure and function.**

**a.** The electron micrograph shows the location of starch in plant cells. Starch is a chain of glucose molecules that can be nonbranched or branched. **b.** The electron micrograph shows glycogen deposits in a portion of a liver cell. Glycogen is a highly branched polymer of glucose molecules.

## Polysaccharides: Energy Storage Molecules

**Polysaccharides** are polymers of monosaccharides. Some types of polysaccharides function as short-term energy storage molecules. When an organism requires energy, the polysaccharide is broken down to release sugar molecules. The helical shape of the polysaccharides in Figure 3.8 exposes the sugar linkages to the hydrolytic enzymes that can break them down.

Plants store glucose as **starch.** The cells of a potato contain granules where starch resides during winter until energy is needed for growth in the spring. Notice in Figure 3.8*a* that starch exists in two forms: One form (amylose) is nonbranched and the other (amylopectin) is branched. When a polysaccharide is branched, there is no main carbon chain because new chains occur at regular intervals, always at the sixth carbon of the monomer.

Animals store glucose as **glycogen.** In our bodies and those of other vertebrates, liver cells contain granules where glycogen is being stored until needed. The storage and release of glucose from liver cells is under the control of hormones. After we eat, the release of the hormone insulin from the pancreas promotes the storage of glucose as glycogen. Notice in Figure 3.8*b* that glycogen is even more branched than starch.

Polysaccharides serve as storage molecules because they are not as soluble in water, and are much larger than a sugar. Therefore, polysaccharides cannot easily pass through the plasma membrane, a sheetlike structure that encloses cells.

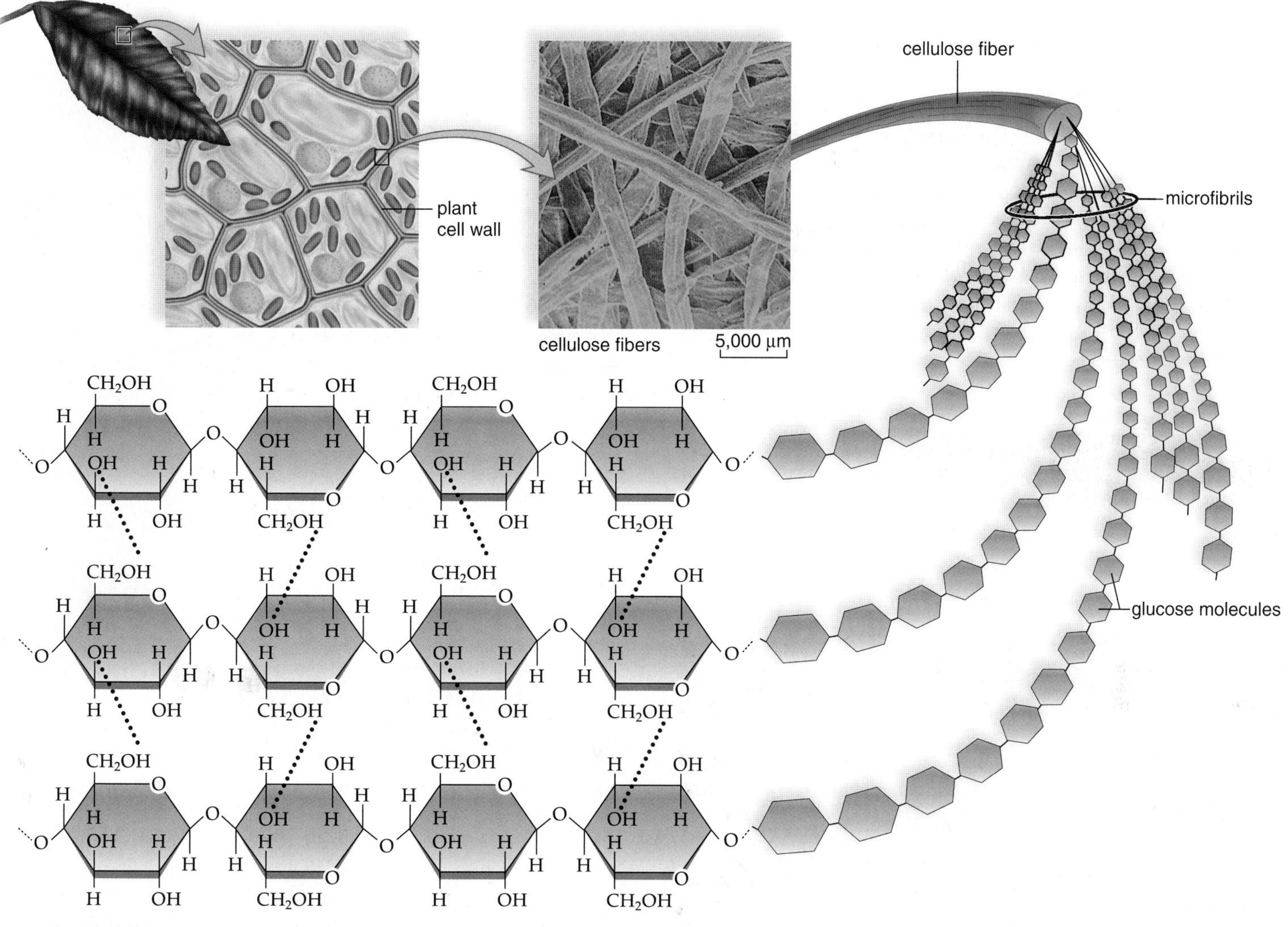


**FIGURE 3.9 Cellulose fibrils.**

Cellulose fibers criss-cross in plant cell walls for added strength. A cellulose fiber contains several microfibrils, each a polymer of glucose molecules—notice that the linkage bonds differ from those of starch. Every other glucose is flipped, permitting hydrogen bonding between the microfibrils.

## Polysaccharides: Structural Molecules

Structural polysaccharides include **cellulose** in plants, **chitin** in animals and fungi, and **peptidoglycan** in bacteria (see Fig. 3.1). In all three, monomers are joined by the type of bond shown for cellulose in Figure 3.9. The cellulose monomer is simply glucose, but in chitin, the monomer has an attached amino group. The structure of peptidoglycan is even more complex because each monomer also has an amino acid chain.

Cellulose is the most abundant carbohydrate and, indeed, the most abundant organic molecule on Earth—over 100 billion tons of cellulose is produced by plants each year. Wood, a cellulose plant product, is used for construction, and cotton is used for cloth. Microorganisms, but not animals, are able to digest the bond between glucose monomers in cellulose. The protozoans in the gut of termites allows termites to digest wood. In cows and other ruminants, microorganisms break down cellulose in a special pouch before the "cud" is returned to the mouth for more chewing and reswallowing. In rabbits, microorganisms digest cellulose in a pouch where it is packaged into pellets. In order to make use of these nutrient pellets, rabbits have to reswallow them as soon as they pass out at the anus. For animals, such as humans, that have no means of digesting cellulose, cellulose is dietary fiber, which maintains regularity of elimination.

Chitin [Gk. *chiton*, tunic] is found in fungal cell walls and in the exoskeletons of crabs and related animals, such as lobsters, scorpions, and insects. Chitin, like cellulose, cannot be digested by animals; however, humans have found many other good uses for chitin. Seeds are coated with chitin, and this protects them from attack by soil fungi. Because chitin also has antibacterial and antiviral properties, it is processed and used in medicine as a wound dressing and suture material. Chitin is even useful during the production of cosmetics and various foods.

### Check Your Progress 3.2

1. Explain why humans cannot utilize the glucose in cellulose as a nutrient source.
2. Compare and contrast the structure and function of cellulose with chitin.

# 3.3 Lipids

A variety of organic compounds are classified as **lipids** [Gk. *lipos*, fat] (Table 3.2). These compounds are insoluble in water due to their hydrocarbon chains. Hydrogens bonded only to carbon have no tendency to form hydrogen bonds with water molecules. Fat, a well-known lipid, is used for both insulation and long-term energy storage by animals. Fat below the skin of marine mammals is called blubber (Fig. 3.10); in humans, it is given slang expressions such as "spare tire" and "love handles." Plants use oil instead of fat for long-term energy storage. We are familiar with fats and oils because we use them as foods and for cooking.

Phospholipids and steroids are also important lipids found in living things. They serve as major components of the plasma membrane in cells. Waxes, which are sticky, not greasy like fats and oils, tend to have a protective function in living things.

## Triglycerides: Long-Term Energy Storage

**Fats** and **oils** contain two types of subunit molecules: fatty acids and glycerol. Each **fatty acid** consists of a long hydrocarbon chain with a —COOH (carboxyl) group at one end. Most of the fatty acids in cells contain 16 or 18 carbon atoms per molecule, although smaller ones are also found. Fatty acids are either saturated or unsaturated. **Saturated fatty acids** have no double bonds between the carbon atoms. The carbon chain is saturated, so to speak, with all the hydrogens that can be held. **Unsaturated fatty acids** have double bonds in the carbon chain wherever the number of hydrogens is less than two per carbon atom.

**Glycerol** is a compound with three —OH groups. The —OH groups are polar; therefore, glycerol is soluble in water. When a fat or oil forms, the acid portions of three fatty acids react with the —OH groups of glycerol during a dehydration reaction (Fig. 3.11*a*). In addition to a fat molecule, three molecules of water result. Fats and oils are degraded following a hydrolysis reaction. Because there are three fatty acids attached to each glycerol molecule, fats and oils are sometimes called **triglycerides.** Notice that triglycerides have many C—H bonds; therefore, they do not mix with water. Despite the liquid nature of both cooking oils and water, cooking oils separate out of water even after shaking.

**FIGURE 3.10** **Blubber.**

The fat (blubber) beneath the skin of marine mammals protects them well from the cold. Blubber accounts for about 25% of their body weight.

Triglycerides containing fatty acids with unsaturated bonds melt at a lower temperature than those containing only saturated fatty acids. This is because a double bond creates a kink in the fatty acid chain that prevents close packing between the hydrocarbon chains (Fig. 3.11*a*). We can reason that butter, a fat that is solid at room temperature, must contain primarily saturated fatty acids, while corn oil, which is a liquid even when placed in the refrigerator, must contain primarily unsaturated fatty acids (Fig. 3.11*b*). This difference is useful to living things. For example, the feet of reindeer and penguins contain unsaturated triglycerides, and this helps protect those exposed parts from freezing.

In general, however, fats, which are most often of animal origin, are solid at room temperature, and oils, which are liquid at room temperature, are of plant origin. Diets high in animal fat have been associated with circulatory disorders because fatty material accumulates inside the lining of blood vessels and blocks blood flow. Replacement of fat whenever possible with oils such as olive oil and canola oil has been suggested.

Nearly all animals use fat in preference to glycogen for long-term energy storage. Gram per gram, fat stores more energy than glycogen. The C—H bonds of fatty acids make them a richer source of chemical energy than glycogen, because glycogen has many C—OH bonds. Also, fat droplets, being nonpolar, do not contain water. Small birds, like the broad-tailed hummingbird, store a great deal of fat before they start their long spring and fall migratory flights. About 0.15 g of fat per gram of body weight is accumulated each day. If the same amount of energy were stored as glycogen, a bird would be so heavy it would not be able to fly.

**TABLE 3.2**

**Lipids**

| *Type* | *Functions* | *Human Uses* |
|---|---|---|
| Fats | Long-term energy storage and insulation in animals | Butter, lard |
| Oils | Long-term energy storage in plants and their seeds | Cooking oils |
| Phospholipids | Component of plasma membrane | — |
| Steroids | Component of plasma membrane (cholesterol), sex hormones | Medicines |
| Waxes | Protection, prevent water loss (cuticle of plant surfaces), beeswax, earwax | Candles, polishes |

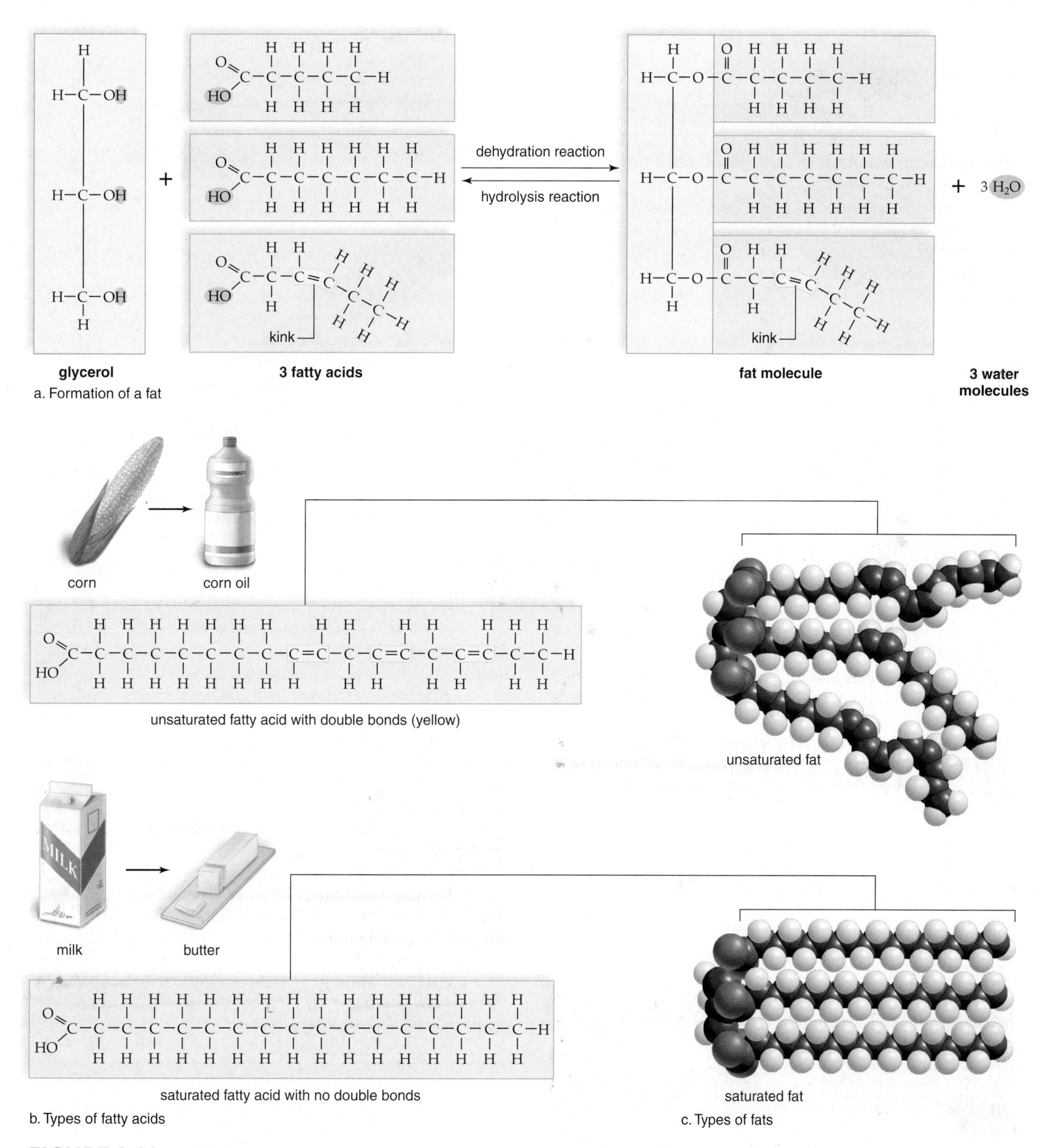


## FIGURE 3.11 Fat and fatty acids.

**a.** Following a dehydration reaction, glycerol is bonded to three fatty acid molecules as fat forms and water is given off. Following a hydrolysis reaction, the bonds are broken due to the addition of water. **b.** A fatty acid has a carboxyl group attached to a long hydrocarbon chain. If there are double bonds between some of the carbons in the chain, the fatty acid is unsaturated and a kink occurs in the chain. If there are no double bonds, the fatty acid is saturated. **c.** Space-filling models of an unsaturated fat and a saturated fat.

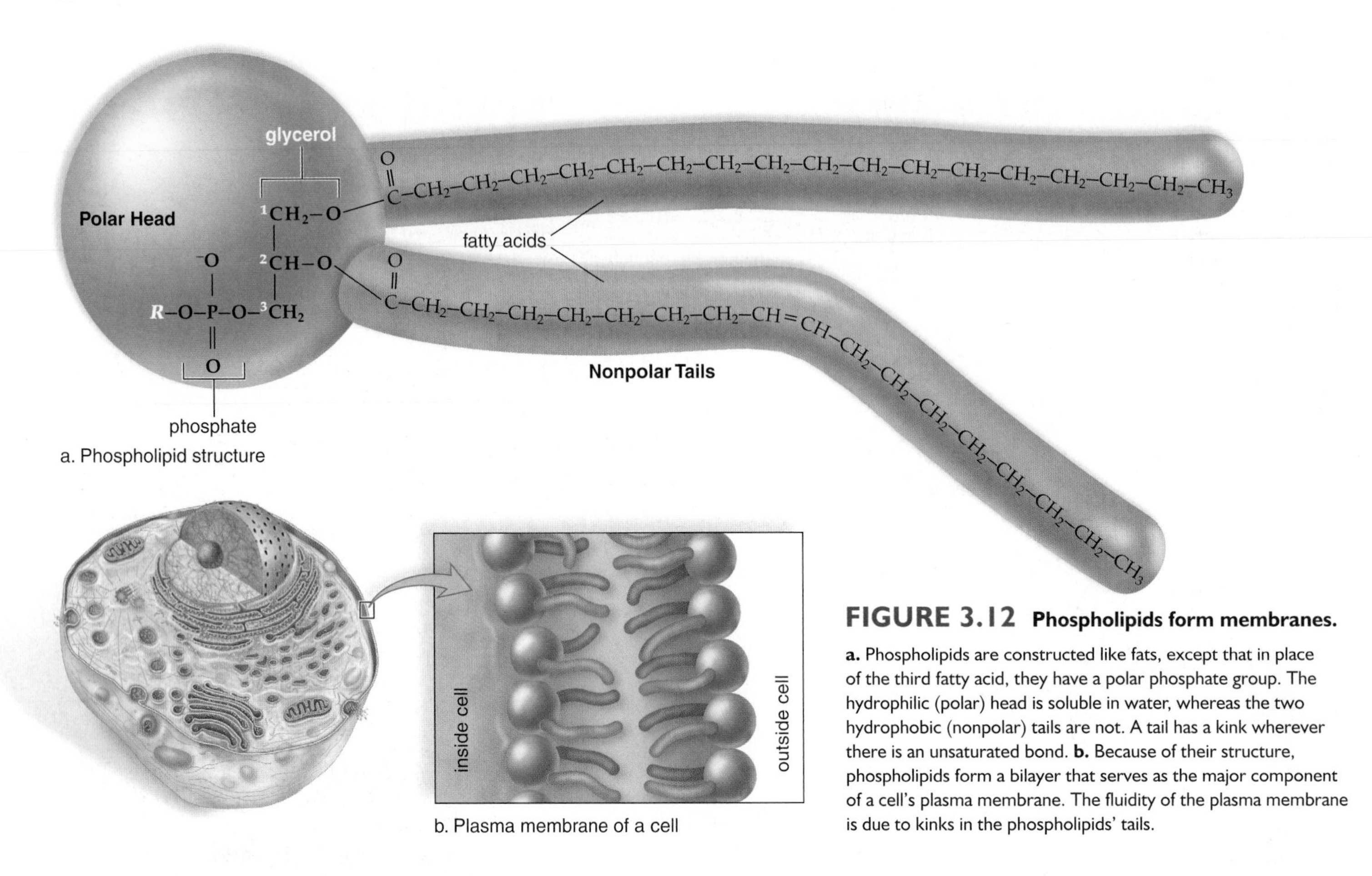


a. Phospholipid structure

b. Plasma membrane of a cell

**FIGURE 3.12** **Phospholipids form membranes.**

**a.** Phospholipids are constructed like fats, except that in place of the third fatty acid, they have a polar phosphate group. The hydrophilic (polar) head is soluble in water, whereas the two hydrophobic (nonpolar) tails are not. A tail has a kink wherever there is an unsaturated bond. **b.** Because of their structure, phospholipids form a bilayer that serves as the major component of a cell's plasma membrane. The fluidity of the plasma membrane is due to kinks in the phospholipids' tails.

## Phospholipids: Membrane Components

**Phospholipids** [Gk. *phos,* light, and *lipos,* fat], as implied by their name, contain a phosphate group. Essentially, a phospholipid is constructed like a fat, except that in place of the third fatty acid attached to glycerol, there is a polar phosphate group. The phosphate group is usually bonded to another organic group, indicated by *R* in Figure 3.12*a*. This portion of the molecule becomes the polar head, while the hydrocarbon chains of the fatty acids become the nonpolar tails. Note that a double bond causes a tail to kink.

Because phospholipids have hydrophilic heads and hydrophobic tails, they tend to arrange themselves so that only the polar heads are adjacent to a watery medium. Therefore, when surrounded by water, phospholipids become a bilayer (double layer) in which the hydrophilic heads project outward and the hydrophobic tails project inward.

The plasma membrane that surrounds cells consists primarily of a phospholipid bilayer (Fig. 3.12*b*). The presence of kinks in the tail cause the plasma membrane to be fluid in nature. A plasma membrane is absolutely essential to the structure and function of a cell, and this signifies the importance of phospholipids to living things.

## Steroids: Four Fused Rings

**Steroids** are lipids that have entirely different structures from those of fats. Steroid molecules have skeletons of four fused carbon rings (Fig. 3.13*a*). Each type of steroid differs primarily by the types of functional groups attached to the carbon skeleton.

Cholesterol is an essential component of an animal cell's plasma membrane, where it provides physical stability. Cholesterol is the precursor of several other steroids, such as the sex hormones testosterone and estrogen (Fig. 3.13*b, c*). The male sex hormone, testosterone, is formed primarily in the testes, and the female sex hormone, estrogen, is formed primarily in the ovaries. Testosterone and estrogen differ only by the functional groups attached to the same carbon skeleton, and yet they each have their own profound effect on the body and the sexuality of an animal. Human and plant estrogen are similar in structure and, if estrogen therapy is recommended, some women prefer taking soy products in preference to estrogen from animals.

Not only saturated fats, but also cholesterol can contribute to circulatory disorders. The presence of cholesterol encourages the accumulation of fatty material inside the lining of blood vessels and, therefore, high blood pressure. Cholesterol-lowering medication is available.

**FIGURE 3.13**
**Steroid diversity.**
**a.** Built like cholesterol, (**b**) testosterone and (**c**) estrogen have different effects on the body due to different functional groups attached to the same carbon skeleton. Testosterone is the male sex hormone active in peacocks (*left*), and estrogen is the female sex hormone active in peahens (*right*).

## Waxes

In **waxes,** long-chain fatty acids bond with long-chain alcohols:

**long-chain fatty acid**

$$O{=}C{-}CH_2{-}CH_2{-}CH_2{-}CH_2{-}CH_2{-}CH_2{-}CH_2{-}CH_2{-}CH_2{-}CH_3$$
$$O{-}\overset{H}{\underset{H}{C}}{-}CH_2{-}CH_2{-}CH_2{-}CH_2{-}CH_2{-}CH_2{-}CH_2{-}CH_2{-}CH_3$$

**long-chain alcohol**

Waxes are solid at normal temperatures because they have a high melting point. Being hydrophobic, they are also waterproof and resistant to degradation. In many plants, waxes, along with other molecules, form a protective cuticle (covering) that retards the loss of water for all exposed parts (Fig. 3.14*a*). In many animals, waxes are involved in skin and fur maintenance. In humans, wax is produced by glands in the outer ear canal. Earwax contains cerumin, an organic compound that at the very least repels insects, and in some cases even kills them. It also traps dust and dirt, preventing them from reaching the eardrum.

A honeybee produces beeswax in glands on the underside of its abdomen. Beeswax is used to make the six-sided cells of the comb where honey is stored (Fig. 3.14*b*). Honey contains the sugars fructose and glucose, breakdown products of the sugar sucrose.

Humans have found a myriad of uses for waxes, from making candles to polishing cars, furniture, and floors.

### Check Your Progress 3.3

1. **a.** Compare and contrast a saturated fatty acid with an unsaturated fatty acid. **b.** Which of these is preferred in the diet and why?
2. Explain why phospholipids form a bilayer in a watery medium.

**FIGURE 3.14 Waxes.**

Waxes are a type of lipid. **a.** Fruits are protected by a waxy coating that is visible on these plums. **b.** Bees secrete the wax that allows them to build a comb where they store honey. This bee has collected pollen (yellow) to feed growing larvae.

a.

b.

## 3.4 Proteins

**Proteins** [Gk. *proteios*, first place], as their Greek derivation implies, are of primary importance to the structure and function of cells. As much as 50% of the dry weight of most cells consists of proteins. Presently, over 100,000 proteins have been identified. Here are some of their many functions in animals:

**Metabolism** Enzymes bring reactants together and thereby speed chemical reactions in cells. They are specific for one particular type of reaction and can function at body temperature.

**Support** Some proteins have a structural function. For example, keratin makes up hair and nails, while collagen lends support to ligaments, tendons, and skin.

**Transport** Channel and carrier proteins in the plasma membrane allow substances to enter and exit cells. Some other proteins transport molecules in the blood of animals; **hemoglobin** is a complex protein that transports oxygen.

**Defense** Antibodies are proteins. They combine with foreign substances, called antigens. In this way, they prevent antigens from destroying cells and upsetting homeostasis.

**Regulation** Hormones are regulatory proteins. They serve as intercellular messengers that influence the metabolism of cells. The hormone insulin regulates the content of glucose in the blood and in cells; the presence of growth hormone determines the height of an individual.

**Motion** The contractile proteins actin and myosin allow parts of cells to move and cause muscles to contract. Muscle contraction accounts for the movement of animals from place to place. All cells contain proteins that allow cell components to move from place to place. Without such proteins, cells would not be able to function.

Proteins are such a major part of living organisms that tissues and cells of the body can sometimes be characterized by the proteins they contain or produce. For example, muscle cells contain large amounts of actin and myosin for contraction; red blood cells are filled with hemoglobin for oxygen transport; and support tissues, such as ligaments and tendons, contain the protein collagen, which is composed of tough fibers.

### Peptides

Proteins are polymers with amino acid monomers. Figure 3.15 shows how two amino acids join by a dehydration reaction between the carboxyl group of one and the amino group of another. The resulting covalent bond between two amino acids is called a **peptide bond.** The atoms associated with the peptide bond share the electrons unevenly because oxygen is more electronegative than nitrogen. Therefore, the hydrogen attached to the nitrogen has a slightly positive charge, while the oxygen has a slightly negative charge:

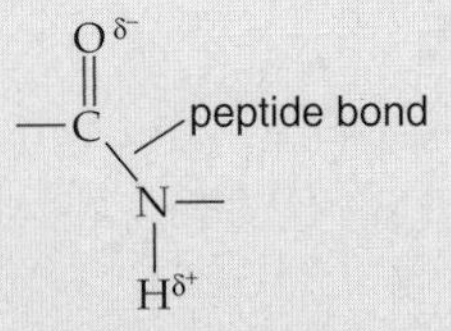

The polarity of the peptide bond means that hydrogen bonding is possible between the —CO of one amino acid and the —NH of another amino acid in a polypeptide.

A **peptide** is two or more amino acids bonded together, and a **polypeptide** is a chain of many amino acids joined by peptide bonds. A protein may contain more than one polypeptide chain; therefore, you can see why a protein could have a very large number of amino acids. In 1953, Frederick Sanger developed a method to determine the sequence of amino acids in a polypeptide. Now that we know the sequences of thousands of polypeptides, it is clear that each polypeptide has its own normal sequence. This sequence influences the final three-dimensional shape of the protein. Proteins that have an abnormal sequence have the wrong shape and cannot function properly.

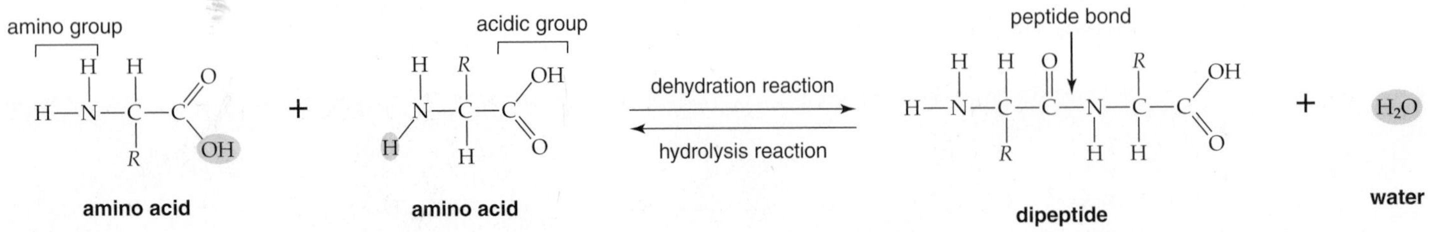

**FIGURE 3.15 Synthesis and degradation of a peptide.**
Following a dehydration reaction, a peptide bond joins two amino acids and a water molecule is released. Following a hydrolysis reaction, the bond is broken due to the addition of water.

## Amino Acids: Protein Monomers

The name **amino acid** is appropriate because one of these groups is an $—NH_2$ (amino group) and another is a $—COOH$ (an acid group). The third group is called an *R* group for an amino acid:

amino group | acid group

$$\begin{array}{c} H \\ | \\ H_2N—C—COOH \\ | \\ R \end{array}$$

*R* = rest of molecule

Note that the central carbon atom in an amino acid bonds to a hydrogen atom and also to three other groups of atoms, one of which is the *R* group (Fig. 3.15). Amino acids differ according to their particular *R* group, shaded in blue in Figure 3.16. The *R* groups range in complexity from a single hydrogen atom to a complicated ring compound. Some *R* groups are polar and some are not. Also, the amino acid cysteine has an *R* group that ends with an —SH group, which often serves to connect one chain of amino acids to another by a disulfide bond, —S—S—. Several other amino acids commonly found in cells are shown in Figure 3.16. Each protein has a definite sequence of amino acids, and this leads to levels of structure and a particular shape per protein.

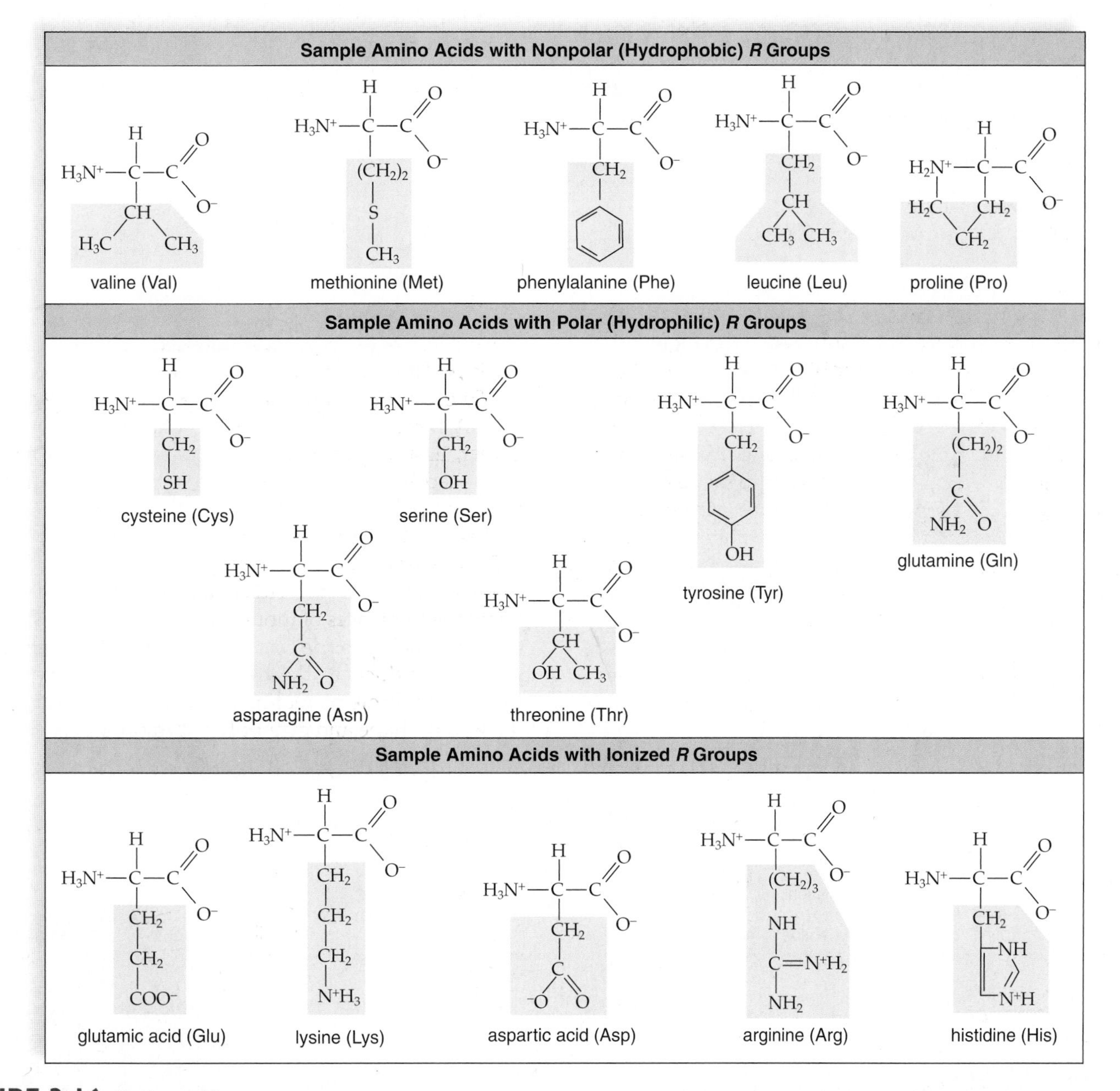


**FIGURE 3.16 Amino acids.**

Polypeptides contain 20 different kinds of amino acids, some of which are shown here. Amino acids differ by the particular *R* group (blue) attached to the central carbon. Some *R* groups are nonpolar and hydrophobic, some are polar and hydrophilic, and some are ionized and hydrophilic. The amino acids are shown in ionized form.

## Shape of Proteins

A protein can have up to four levels of structure, but not all proteins have all four levels.

### *Primary Structure*

The primary structure of one protein is its own particular sequence of amino acids. The following analogy can help you see that hundreds of thousands of different polypeptides can be built from just 20 amino acids: The English alphabet contains only 26 letters, but an almost infinite number of words can be constructed by varying the number and sequence of these few letters. In the same way, many different proteins can result by varying the number and sequence of just 20 amino acids.

### *Secondary Structure*

The secondary structure of a protein occurs when the polypeptide coils or folds in a particular way (Fig. 3.17).

Linus Pauling and Robert Corey, who began studying the structure of amino acids in the late 1930s, concluded that a coiling they called an α (alpha) helix and a pleated sheet they called the β (beta) sheet were two basic patterns of amino acids within a polypeptide. The names came from the fact that the α helix was the first pattern they discovered, and the β sheet was the second pattern they discovered.

Hydrogen bonding often holds the secondary structure of a polypeptide in place. Hydrogen bonding between every fourth amino acid accounts for the spiral shape of the helix. In a β sheet, the polypeptide turns back upon itself,

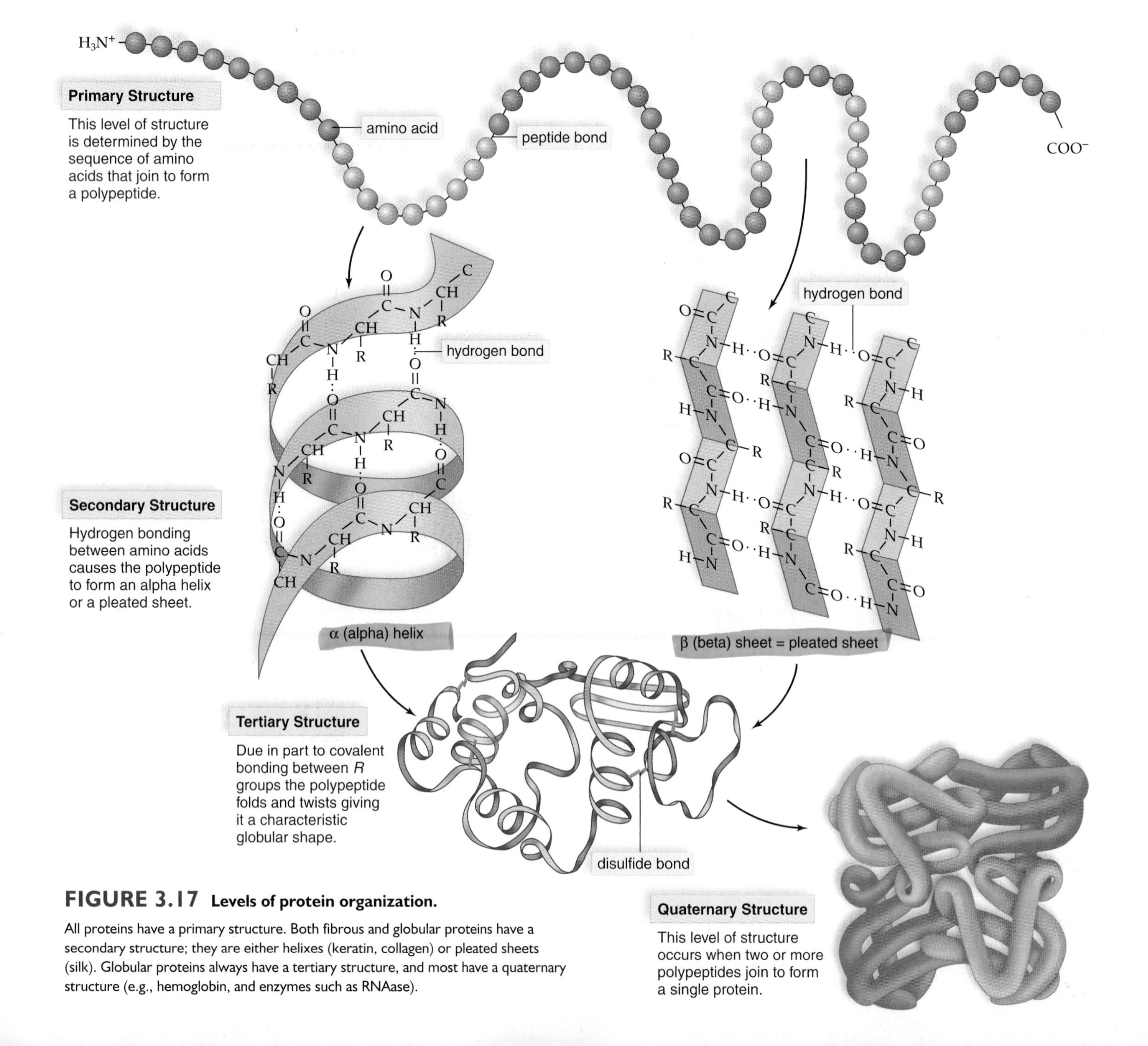


**FIGURE 3.17 Levels of protein organization.**

All proteins have a primary structure. Both fibrous and globular proteins have a secondary structure; they are either helixes (keratin, collagen) or pleated sheets (silk). Globular proteins always have a tertiary structure, and most have a quaternary structure (e.g., hemoglobin, and enzymes such as RNAase).

and hydrogen bonding occurs between extended lengths of the polypeptide. **Fibrous proteins,** which are structural proteins, exist as helices or pleated sheets that hydrogen-bond to each other. Examples are keratin, a protein in hair and silk, a protein that forms spider webs. Both of these proteins have only a secondary structure (Fig. 3.18).

### *Tertiary Structure*

A tertiary structure is the folding that results in the final three-dimensional shape of a polypeptide. So-called **globular proteins,** which tend to ball up into rounded shapes, have a tertiary structure. Hydrogen bonds, ionic bonds, and covalent bonds between *R* groups all contribute to the tertiary structure of a polypeptide. Strong disulfide linkages in particular help maintain the tertiary shape. Hydrophobic *R* groups do not bond with other *R* groups, and they tend to collect in a common region where they are not exposed to water and can interact. Although hydrophobic interactions are not as strong as hydrogen bonds, they are important in creating and stabilizing the tertiary structure.

Enzymes are globular proteins. Enzymes work best at body temperature, and each one also has an optimal pH at which the rate of the reaction is highest. At this temperature and pH, the enzyme has its normal shape. A high temperature and change in pH can disrupt the interactions that maintain the shape of the enzyme. When a protein loses its natural shape, it is said to be **denatured.**

### *Quaternary Structure*

Some proteins have a quaternary structure because they consist of more than one polypeptide. Hemoglobin is a much-studied globular protein that consists of four polypeptides, and therefore it has a quaternary structure. Each polypeptide in hemoglobin has a primary, secondary, and tertiary structure.

## Protein-Folding Diseases

Proteins cannot function properly unless they fold into their correct shape. In recent years it has been shown that the cell contains **chaperone proteins,** which help new proteins fold into their normal shape. At first it seemed as if chaperone proteins ensured that proteins folded properly, but now it seems that they might correct any misfolding of a new protein. In any case, without fully functioning chaperone proteins, a cell's proteins may not be functional because they have misfolded. Several diseases in humans, such as cystic fibrosis and Alzheimer disease, are associated with misshapen proteins. The possibility exists that the diseases are due to missing or malfunctioning chaperone proteins.

Other diseases in humans are due to misfolded proteins, but the cause may be different. For years, investigators have been studying fatal brain diseases, known as TSEs,[1] that have no cure because no infective agent can be found. Mad cow disease is a well-known example of a TSE disease. Now it appears that TSE diseases could be due to misfolded proteins, called **prions,** that cause other proteins of the same type to fold the wrong way too. A possible relationship between prions and the functioning of chaperone proteins is now under investigation.

**Check Your Progress 3.4**

1. Which of the protein functions in animals is shared by plants (see page 48)?
2. What is the primary structure of a protein?
3. **a.** What does the peptide bond have to do with the secondary structure of a protein? **b.** What type of bonding maintains the tertiary structure of a protein?

[1] TSEs (transmissible spongiform encephalopathies)

a.

b.

c.

**FIGURE 3.18 Fibrous proteins.**

Fibrous proteins are structural proteins. **a.** Keratin—found, for example, in hair, horns, and hoofs—exemplifies fibrous proteins that are helical for most of their length. Keratin is a hydrogen-bonded triple helix. In this photo, Drew Barrymore has straight hair. **b.** In order to give her curly hair, water was used to disrupt the hydrogen bonds, and when the hair dried, new hydrogen bonding allowed it to take on the shape of a curler. A permanent-wave lotion induces new covalent bonds within the helix. **c.** Silk made by spiders and silkworms exemplifies fibrous proteins that are pleated sheets for most of their length. Hydrogen bonding between parts of the molecule occurs as the pleated sheet doubles back on itself.

# 3.5 Nucleic Acids

**Nucleic acids** are polymers of nucleotides with very specific functions in cells. **DNA (deoxyribonucleic acid)** is the genetic material that stores information regarding its own replication and the order in which amino acids are to be joined to make a protein. **RNA (ribonucleic acid)** is another type of nucleic acid. One type of RNA molecule called messenger RNA (mRNA) is an intermediary in the process of protein synthesis, conveying information from DNA regarding the amino acid sequence in a protein.

Some nucleotides have independent metabolic functions in cells. For example, some are components of **coenzymes,** nonprotein organic molecules that facilitate enzymatic reactions. **ATP (adenosine triphosphate)** is a nucleotide that supplies energy for synthetic reactions and for various other energy-requiring processes in cells.

## Structure of DNA and RNA

Every **nucleotide** is a molecular complex of three types of molecules: phosphate (phosphoric acid), a pentose sugar, and a nitrogen-containing base (Fig. 3.19*a*). In DNA, the pentose sugar is deoxyribose, and in RNA the pentose sugar is ribose. A difference in the structure of these 5-carbon sugars accounts for their respective names because deoxyribose lacks an oxygen atom found in ribose (Fig. 3.19*b*).

There are four types of nucleotides in DNA and four types of nucleotides in RNA (Fig. 3.19*c*). The base of a nucleotide can be a pyrimidine with a single ring or a purine with a double ring. In DNA, the pyrimidine bases are cytosine and thymine; in RNA, the pyrimidine bases are cytosine and uracil. In both DNA and RNA, the purine bases are adenine or guanine. These molecules are called bases because their presence raises the pH of a solution.

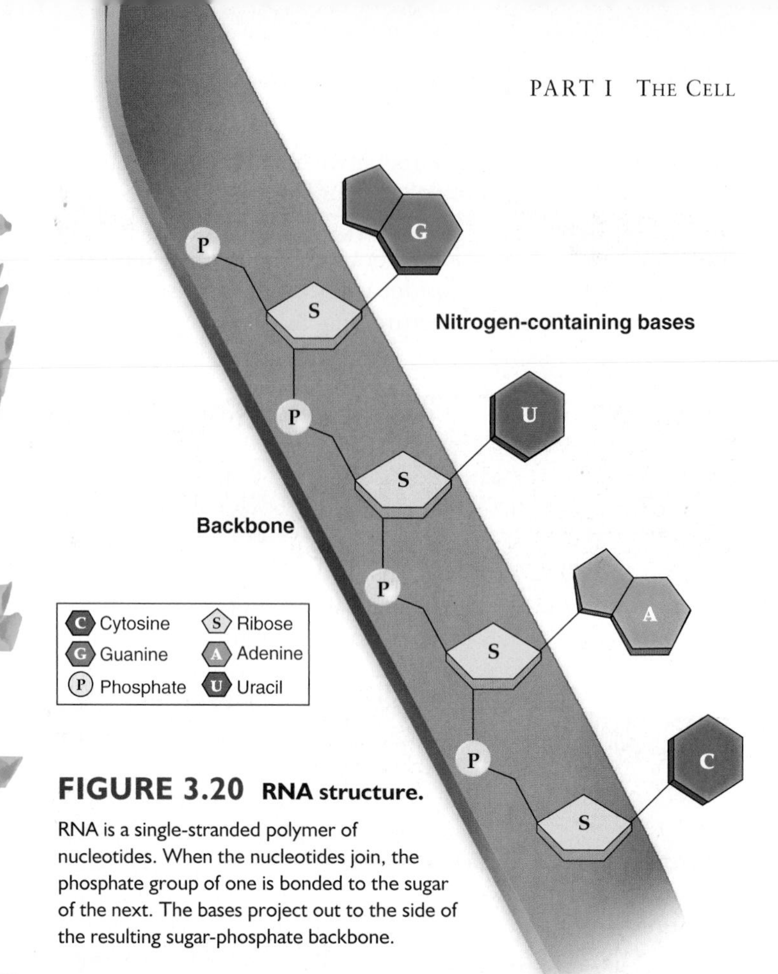


**FIGURE 3.20 RNA structure.**

RNA is a single-stranded polymer of nucleotides. When the nucleotides join, the phosphate group of one is bonded to the sugar of the next. The bases project out to the side of the resulting sugar-phosphate backbone.

Nucleotides join in a definite sequence by a series of dehydration reactions when DNA and RNA form. The polynucleotide is a linear molecule called a strand in which the backbone is made up of a series of sugar-phosphate-sugar-phosphate molecules. The bases project to one side of the backbone. Since the nucleotides occur in a definite order, so do the bases. RNA is single stranded (Fig. 3.20).

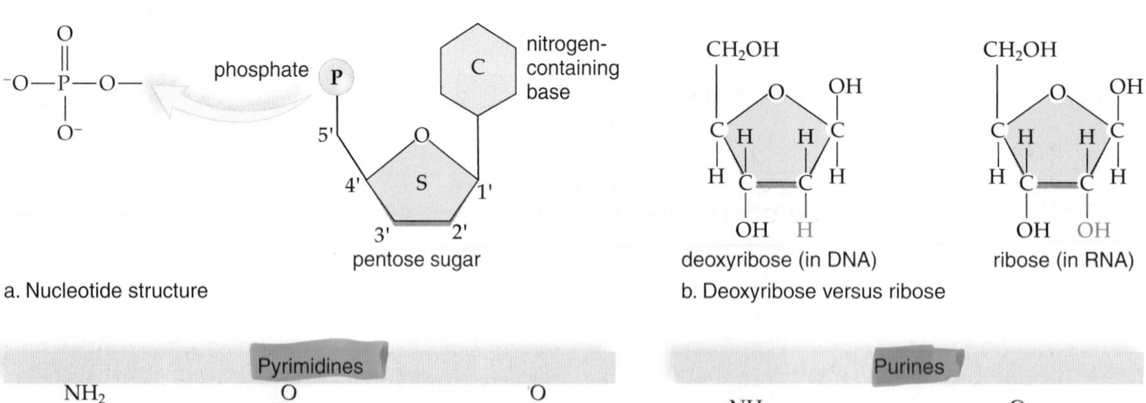


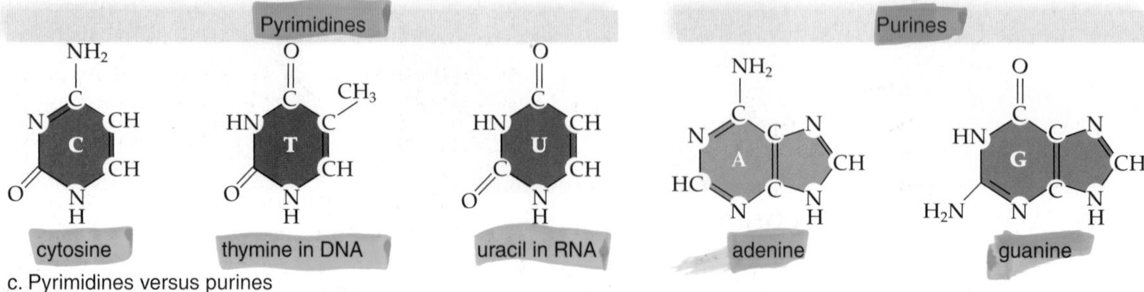


**FIGURE 3.19 Nucleotides.**

**a.** A nucleotide consists of a phosphate molecule, a pentose sugar, and a nitrogen-containing base. **b.** DNA contains the sugar deoxyribose, and RNA contains the sugar ribose. **c.** DNA contains the pyrimidines C and T and the purines A and G. RNA contains the pyrimidines C and U and the purines A and G.

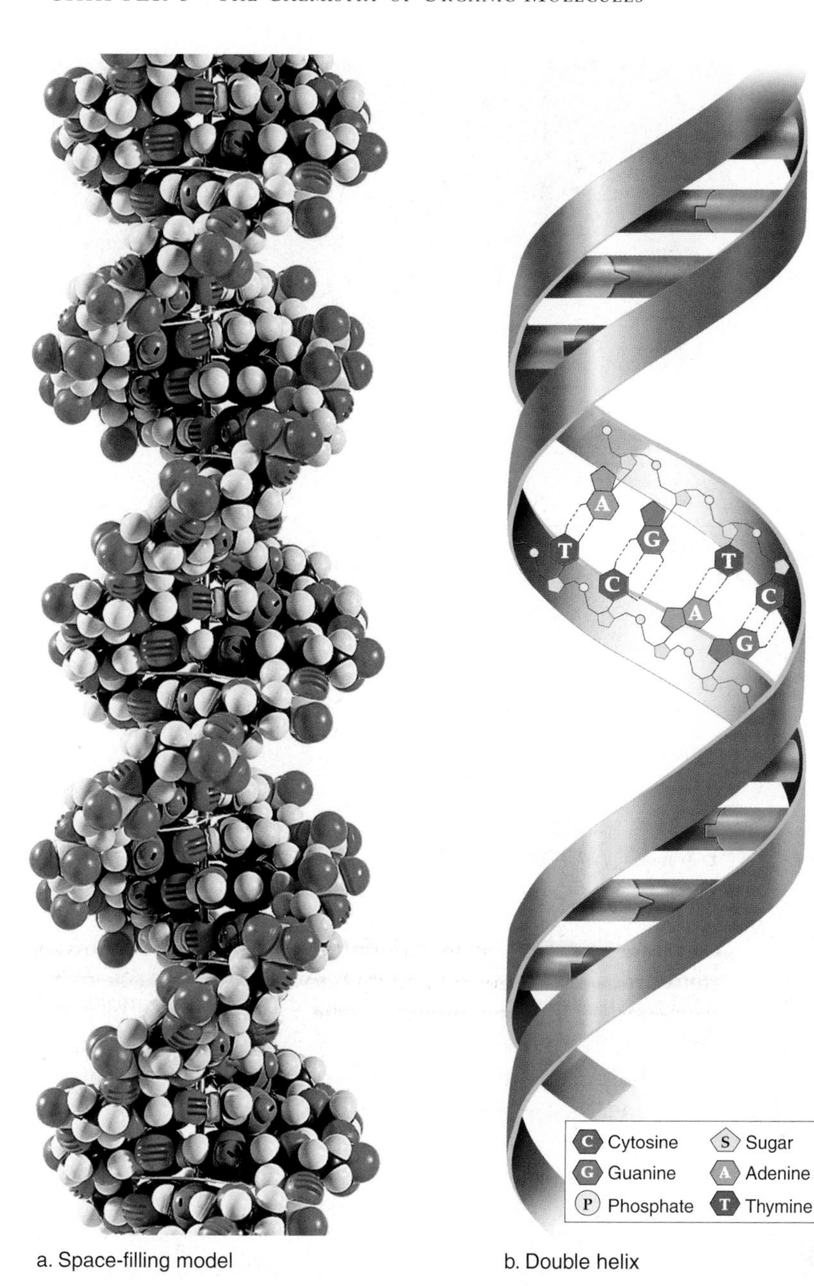


a. Space-filling model

b. Double helix

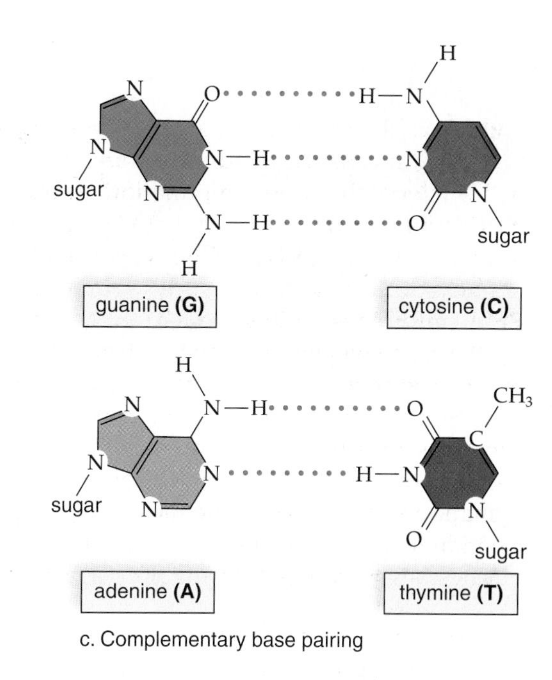


c. Complementary base pairing

**TABLE 3.3**

**DNA Structure Compared to RNA Structure**

| | *DNA* | *RNA* |
|---|---|---|
| Sugar | Deoxyribose | Ribose |
| Bases | Adenine, guanine, thymine, cytosine | Adenine, guanine, uracil, cytosine |
| Strands | Double stranded with base pairing | Single stranded |
| Helix | Yes | No |

**FIGURE 3.21 DNA structure.**

**a.** Space-filling model of DNA. **b.** DNA is a double helix in which the two polynucleotide strands twist about each other. **c.** Hydrogen bonds (dotted lines) occur between the complementarily paired bases: C is always paired with G, and A is always paired with T.

DNA is double stranded, with the two strands usually twisted about each other in the form of a double helix (Fig. 3.21*a, b*). The two strands are held together by hydrogen bonds between pyrimidine and purine bases. The bases can be in any order within a strand, but between strands, thymine (T) is always paired with adenine (A), and guanine (G) is always paired with cytosine (C). This is called **complementary base pairing.** Therefore, regardless of the order or the quantity of any particular base pair, the number of purine bases (A + G) always equals the number of pyrimidine bases (T + C) (Fig. 3.21*c*).

Table 3.3 summarizes the differences between DNA and RNA.

## ATP (Adenosine Triphosphate)

ATP is a nucleotide in which **adenosine** is composed of adenine and ribose. Triphosphate stands for the three phosphate groups that are attached together and to ribose, the pentose sugar (Fig. 3.22).

ATP is a high-energy molecule because the last two phosphate bonds are unstable and are easily broken. In cells, the terminal phosphate bond is usually hydrolyzed to give the molecule **ADP (adenosine diphosphate)** and a phosphate molecule Ⓟ.

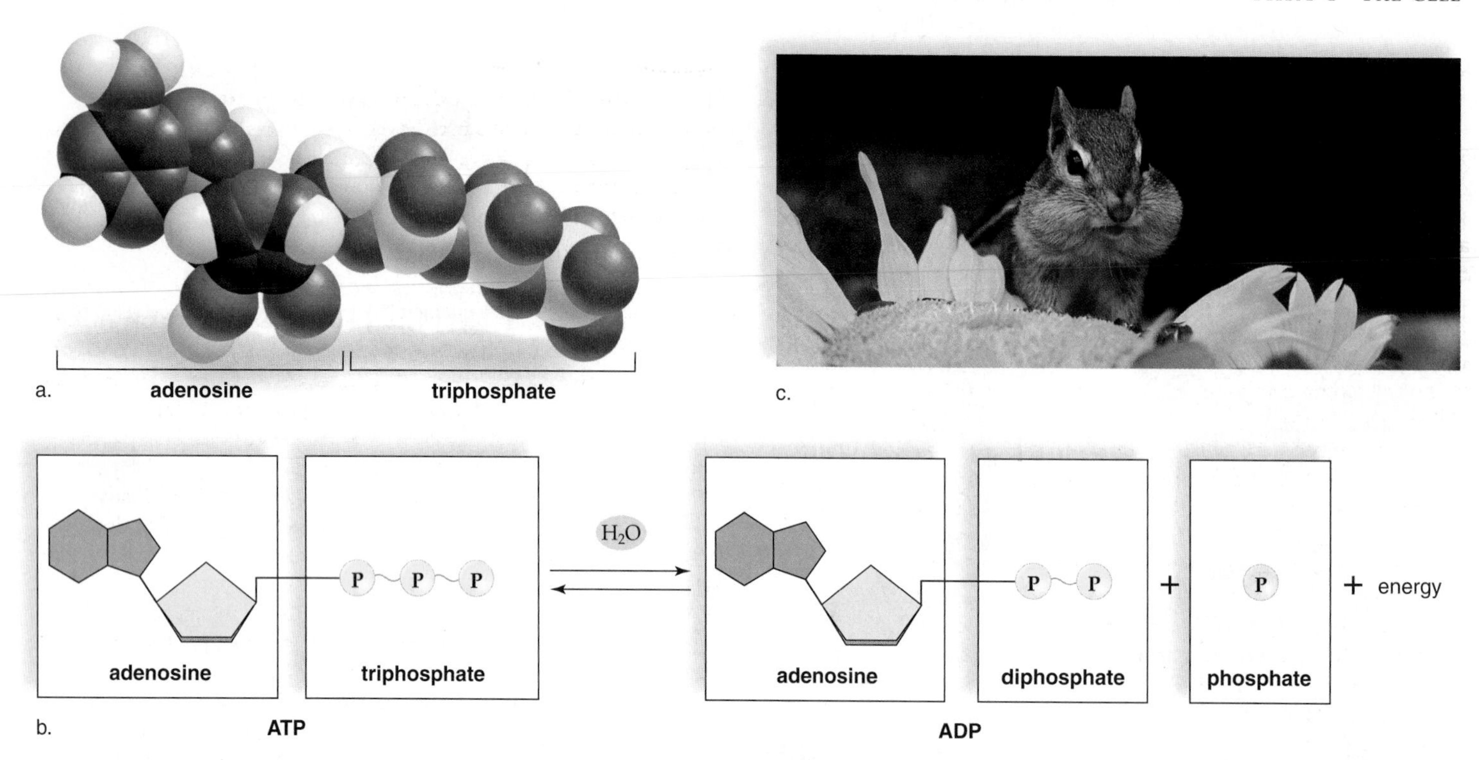


**FIGURE 3.22** **ATP.**

ATP, the universal energy currency of cells, is composed of adenosine and three phosphate groups. **a.** Space-filling model of ATP. **b.** When cells require energy, ATP becomes ADP + Ⓟ, and energy is released. **c.** The breakdown of ATP provides the energy that an animal, such as a chipmunk, uses to acquire food and make more ATP.

The energy that is released by ATP breakdown is coupled to energy-requiring processes in cells. For example, the energy required for the synthesis of macromolecules, such as carbohydrates and proteins, is derived from ATP breakdown. ATP also supplies the energy for muscle contraction and nerve impulse conduction. Just as you spend money when you pay for a product or a service, cells "spend" ATP when they need something. Therefore, ATP is called the energy currency of cells.

Because energy is released when the last phosphate bond of ATP is hydrolyzed, it is sometimes called a high-energy bond, symbolized by a wavy line. But this terminology is misleading—the breakdown of ATP releases energy because the products of hydrolysis (ADP and Ⓟ) are more stable than the original reactant ATP. It is the entire molecule that releases energy, not a particular bond.

### Check Your Progress 3.5

1. List the three components of a nucleotide.
2. What is complementary base pairing in and between nucleic acids?
3. What property of ATP makes it a carrier of energy?

## Connecting the Concepts

What does the term *organic* mean? For some, organic means that food products have been grown without the use of chemicals or have been minimally processed. Biochemically speaking, organic refers to molecules containing carbon and hydrogen. In biology, organic also refers to living things or anything that has been alive in the past. Therefore, the food we eat and the wood we burn are organic substances. Fossil fuels (coal and oil) formed over 300 million years ago from plant and animal life that, by chance, did not fully decompose are also organic. When burned, they release carbon dioxide into the atmosphere just as we do when we breathe!

Although living things are very complex, certain biomolecules are simply polymers of small organic molecules. Simple sugars are the monomers of complex carbohydrates; amino acids are the monomers of proteins; nucleotides are the monomers of nucleic acids. Fats are composed of fatty acids and glycerol.

This system of forming macromolecules still allows for diversity. Monomers exist in modified forms and can combine in slightly different ways; therefore, a variety of macromolecules can come about. In cellulose, a plant product, glucose monomers are linked in a slightly different way than glucose monomers in glycogen, an animal product. One protein differs from another by the number and/or sequence of the same 20 amino acids.

There is no doubt that the chemistry of carbon is the chemistry of life. The groups of molecules discussed in this chapter, as well as other small molecules and ions, are assembled into structures that make up cells. As discussed in Chapter 4, each structure has a specific function necessary to the life of a cell.

## summary

### 3.1 Organic Molecules

The chemistry of carbon accounts for the diversity of organic molecules found in living things. Carbon can bond with as many as four other atoms. It can also bond with itself to form both chains and rings. Differences in the carbon skeleton and attached functional groups cause organic molecules to have different chemical properties. The chemical properties of a molecule determine how it interacts with other molecules and the role the molecule plays in the cell. Some functional groups are hydrophobic and others are hydrophilic.

There are four classes of biomolecules in cells: carbohydrates, lipids, proteins, and nucleic acids (Table 3.4). Polysaccharides, the largest of the carbohydrates, are polymers of simple sugars called monosaccharides. The polypeptides of proteins are polymers of amino acids, and nucleic acids are polymers of nucleotides. Polymers are formed by the joining together of monomers. For each bond formed during a dehydration reaction, a molecule of water is removed, and for each bond broken during a hydrolysis reaction, a molecule of water is added.

### 3.2 Carbohydrates

Monosaccharides, disaccharides, and polysaccharides are all carbohydrates. Therefore, the term *carbohydrate* includes both the monomers (e.g., glucose) and the polymers (e.g., starch, glycogen, and cellulose). Glucose is the immediate energy source of cells. Polysaccharides such as starch, glycogen, and cellulose are polymers of glucose. Starch in plants and glycogen in animals are energy storage compounds, but cellulose in plants and chitin in crabs and related animals, as well as fungi, have structural roles. Chitin's monomer is glucose with an attached amino group.

### 3.3 Lipids

Lipids include a wide variety of compounds that are insoluble in water. Fats and oils, which allow long-term energy storage, contain one glycerol and three fatty acids. Both glycerol and fatty acids have polar groups, but fats and oils are nonpolar, and this accounts for their insolubility in water. Fats tend to contain saturated fatty acids, and oils tend to contain unsaturated fatty acids. Saturated fatty acids do not have carbon–carbon double bonds, but unsaturated fatty acids do have double bonds in their hydrocarbon chain. The double bond causes a kink in the molecule that accounts for the liquid nature of oils.

In a phospholipid, one of the fatty acids is replaced by a phosphate group. In the presence of water, phospholipids form a bilayer because the head of each molecule is hydrophilic and the tails are hydrophobic. Steroids have the same four-ring structure as cholesterol, but each differs by the groups attached to these rings. Waxes are composed of a fatty acid with a long hydrocarbon chain bonded to an alcohol, also with a long hydrocarbon chain.

**TABLE 3.4**

**Organic Compounds in Cells**

| | *Categories* | *Elements* | *Examples* | *Functions* |
|---|---|---|---|---|
| Carbohydrates | Monosaccharides | C, H, O | | |
| | 6-carbon sugar | | Glucose | Immediate energy source |
| | 5-carbon sugar | | Deoxyribose, ribose | Found in DNA, RNA |
| | Disaccharides<br>12-carbon sugar | C, H, O | Sucrose | Transport sugar in plants |
| | Polysaccharides<br>Polymer of glucose | C, H, O | Starch, glycogen,<br>Cellulose | Energy storage in plants, animals<br>Plant cell wall structure |
| Lipids | Triglycerides<br>1 glycerol + 3 fatty acids | C, H, O | Fats, oils | Long-term energy storage |
| | Phospholipids<br>Like triglyceride except the head group contains phosphate | C, H, O, P | Lecithin | Plasma membrane phospholipid bilayer |
| | Steroids<br>Backbone of 4 fused rings | C, H, O | Cholesterol<br>Testosterone, estrogen | Plasma membrane component<br>Sex hormones |
| | Waxes<br>Fatty acid + alcohol | C, H, O | Cuticle<br>Earwax | Protective covering in plants<br>Protective wax in ears |
| Proteins | Polypeptides<br>Polymer of amino acids | C, H, O, N, S | Enzymes<br>Myosinand actin<br>Insulin<br>Hemoglobin<br>Collagen | Speed cellular reactions<br>Movement of muscle cells<br>Hormonal control of blood sugar<br>Transport of oxygen in blood<br>Fibrous support of body parts |
| Nucleic Acids | Nucleic acids<br>Polymer of nucleotides | C, H, O, N, P | DNA<br>RNA | Genetic material<br>Protein synthesis |
| | Nucleotides | | ATP<br>Coenzymes | Energy carrier<br>Assist enzymes |

### 3.4 Proteins

Proteins carry out many diverse functions in cells and organisms, including support, metabolism, transport, defense, regulation, and motion. Proteins contain polymers of amino acids.

A polypeptide is a long chain of amino acids joined by peptide bonds. There are 20 different amino acids in cells, and they differ only by their *R* groups. Presence or absence of polarity is an important aspect of the *R* groups. A polypeptide has up to four levels of structure: The primary level is the sequence of the amino acids, which varies between polypeptides; the secondary level contains $\alpha$ helices and $\beta$ (pleated) sheets held in place by hydrogen bonding between amino acids along the polypeptide chain; and the tertiary level is the final folding of the polypeptide, which is held in place by bonding and hydrophobic interactions between *R* groups. Proteins that contain more than one polypeptide have a quaternary level of structure as well.

The shape of an enzyme is important to its function. Both high temperatures and a change in pH can cause proteins to denature and lose their shape.

### 3.5 Nucleic Acids

The nucleic acids DNA and RNA are polymers of nucleotides. Variety is possible because the nucleotides can be in any order. Each nucleotide has three components: a phosphate (phosphoric acid), a 5-carbon sugar, and a nitrogen-containing base.

DNA, which contains the sugar deoxyribose, is the genetic material that stores information for its own replication and for the order in which amino acids are to be sequenced in proteins. DNA, with the help of mRNA, specifies protein synthesis. DNA, which contains phosphate, the sugar deoxyribose, and nitrogen-containing bases, is a double-stranded helix in which A pairs with T and C pairs with G through hydrogen bonding. RNA, containing phosphate, the sugar ribose, and the bases A, U, C, and G, is single stranded.

ATP, with its unstable phosphate bonds, is the energy currency of cells. Hydrolysis of ATP to ADP + Ⓟ releases energy, which is used by the cell to make a product or do any other type of metabolic work.

## understanding the terms

adenosine 53
ADP (adenosine diphosphate) 53
amino acid 49
ATP (adenosine triphosphate) 52
biomolecule 38
carbohydrate 41
cellulose 43
chaperone protein 51
chitin 43
coenzyme 52
complementary base pairing 53
dehydration reaction 40
denatured 51
deoxyribose 41
disaccharide 41
DNA (deoxyribonucleic acid) 52
fat 44
fatty acid 44
fibrous protein 51
functional group 39
globular protein 51
glucose 41
glycerol 44
glycogen 42
hemoglobin 48
hexose 41
hydrolysis reaction 40
hydrophilic 39
hydrophobic 39
inorganic chemistry 38
isomer 39
lipid 44
monomer 40
monosaccharide 41
nucleic acid 52
nucleotide 52
oil 44
organic chemistry 38
organic molecule 38
pentose 41
peptide 48
peptide bond 48
peptidoglycan 43
phospholipid 46
polymer 40
polypeptide 48
polysaccharide 42
prion 51
protein 48
ribose 41
RNA (ribonucleic acid) 52
saturated fatty acid 44
starch 42
steroid 46
triglyceride 44
unsaturated fatty acid 44
wax 47

Match the terms to these definitions:

a. ______________ Class of organic compounds that includes monosaccharides, disaccharides, and polysaccharides.
b. ______________ Class of organic compounds that tend to be soluble in nonpolar solvents such as alcohol but insoluble in water.
c. ______________ Biomolecule consisting of covalently bonded monomers.
d. ______________ Molecules that have the same molecular formula but a different structure and, therefore, shape.
e. ______________ Two or more amino acids joined together by covalent bonding.

## reviewing this chapter

1. How do the chemical characteristics of carbon affect the structure of organic molecules? 38–39
2. Give examples of functional groups, and discuss the importance of these groups being hydrophobic or hydrophilic. 39
3. What biomolecules are monomers of the polymers studied in this chapter? How do monomers join to produce polymers, and how are polymers broken down to monomers? 40
4. Name several monosaccharides, disaccharides, and polysaccharides, and give a function of each. How are these molecules structurally distinguishable? 41–42
5. What is the difference between a saturated and an unsaturated fatty acid? Explain the structure of a fat molecule by stating its components and how they join together. 44–45
6. How does the structure of a phospholipid differ from that of a fat? How do phospholipids form a bilayer in the presence of water? 46
7. Describe the structure of a generalized steroid. How does one steroid differ from another? 46–47
8. Draw the structure of an amino acid and a peptide, pointing out the peptide bond. 48
9. Discuss the four possible levels of protein structure, and relate each level to particular bonding patterns. 50–51
10. How do nucleotides bond to form nucleic acids? State and explain several differences between the structure of DNA and that of RNA. 52–53
11. Discuss the structure and function of ATP. 53–54

## testing yourself

Choose the best answer for each question.

1. Which of these is not a characteristic of carbon?
   a. forms four covalent bonds
   b. bonds with other carbon atoms
   c. is sometimes ionic
   d. can form long chains
   e. sometimes shares two pairs of electrons with another atom
2. The functional group —COOH is
   a. acidic.
   b. basic.
   c. never ionized.
   d. found only in nucleotides.
   e. All of these are correct.

3. A hydrophilic group is
   a. attracted to water.
   b. a polar and/or ionized group.
   c. found at the end of fatty acids.
   d. the opposite of a hydrophobic group.
   e. All of these are correct.
4. Which of these is an example of a hydrolysis reaction?
   a. amino acid + amino acid ⟶ dipeptide + $H_2O$
   b. dipeptide + $H_2O$ ⟶ amino acid + amino acid
   c. denaturation of a polypeptide
   d. Both a and b are correct.
   e. Both b and c are correct.
5. Which of these makes cellulose nondigestible in humans?
   a. a polymer of glucose subunits
   b. a fibrous protein
   c. the linkage between the glucose molecules
   d. the peptide linkage between the amino acid molecules
   e. The carboxyl groups ionize.
6. A fatty acid is unsaturated if it
   a. contains hydrogen.
   b. contains carbon–carbon double bonds.
   c. contains a carboxyl (acidic) group.
   d. bonds to glycogen.
   e. bonds to a nucleotide.
7. Which of these is not a lipid?
   a. steroid
   b. fat
   c. polysaccharide
   d. wax
   e. phospholipid
8. The difference between one amino acid and another is found in the
   a. amino group.
   b. carboxyl group.
   c. *R* group.
   d. peptide bond.
   e. carbon atoms.
9. The shape of a polypeptide is
   a. maintained by bonding between parts of the polypeptide.
   b. ultimately dependent on the primary structure.
   c. necessary to its function.
   d. All of these are correct.
10. Which of these illustrates a peptide bond?

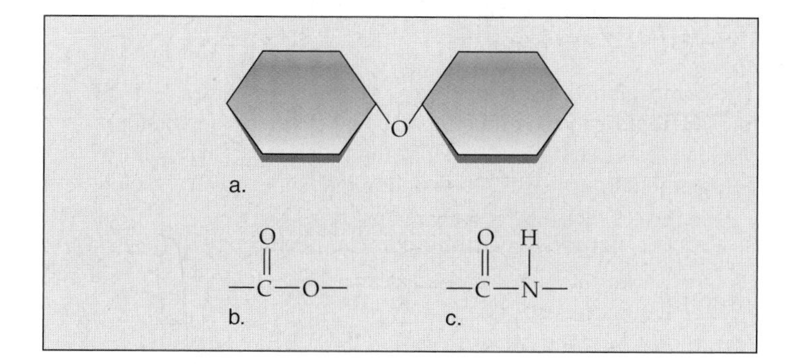


11. Nucleotides
    a. contain a sugar, a nitrogen-containing base, and a phosphate group.
    b. are the monomers of fats and polysaccharides.
    c. join together by covalent bonding between the bases.
    d. are present in both DNA and RNA.
    e. Both a and d are correct.
12. ATP
    a. is an amino acid.
    b. has a helical structure.
    c. is a high-energy molecule that can break down to ADP and phosphate.
    d. provides enzymes for metabolism.
    e. is most energetic when in the ADP state.
13. Label the following diagram using the terms $H_2O$, monomer, hydrolysis reaction, dehydration reaction, and polymer. Terms can be used more than once and a term need not be used.

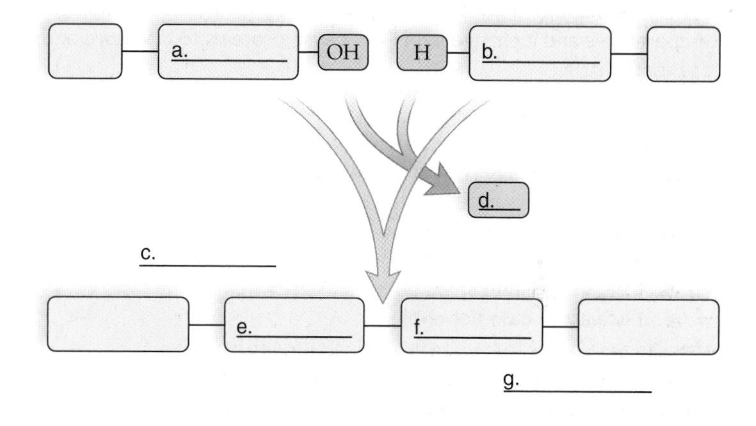


14. The monomer of a carbohydrate is
    a. an amino acid.
    b. a nucleic acid.
    c. a monosaccharide.
    d. a fatty acid.
15. The joining of two adjacent amino acids is called
    a. a peptide bond.
    b. a dehydration reaction.
    c. a covalent bond.
    d. All of these are correct.
16. The characteristic globular shape of a polypeptide is the
    a. primary structure.
    b. secondary structure.
    c. tertiary structure.
    d. quaternary structure.
17. The shape of a polypeptide
    a. is maintained by bonding between parts of the polypeptide.
    b. is ultimately dependent on the primary structure.
    c. involves hydrogen bonding.
    d. All of these are correct.
18. Which of the following pertains to an RNA nucleotide and not to a DNA nucleotide?
    a. contains the sugar ribose
    b. contains a nitrogen-containing base
    c. contains a phosphate molecule
    d. becomes bonded to other nucleotides following a dehydration reaction
19. Which is a carbohydrate?
    a. disaccharide
    b. amino acid
    c. dipeptide
    d. Both a and c are correct.

For questions 20–27, match the items to those in the key. Some answers are used more than once.

**KEY:**

a. carbohydrate
b. fats and oils
c. protein
d. nucleic acid

20. contains the bases adenine, guanine, cytosine, and thymine
21. the 6-carbon sugar, glucose
22. polymer of amino acids
23. glycerol and fatty acids
24. enzymes
25. long-term energy storage
26. genes
27. plant cell walls
28. muscle cells
29. butter
30. potato
31. Which of these does not apply to DNA?
    a. sequence of nucleotides
    b. sugar-phosphate backbone
    c. A-T and C-G
    d. sequence of amino acids
    e. Both a and c do not apply.
32. Which is a correct statement about carbohydrates?
    a. All polysaccharides serve as energy storage molecules.
    b. Glucose is broken down for immediate energy.
    c. Glucose is not a carbohydrate, only polysaccharides are.
    d. Starch, glycogen, and cellulose have different monomers.
    e. Both a and c are correct.
33. In phospholipids,
    a. heads are polar.
    b. tails are nonpolar.
    c. heads contain phosphate.
    d. All of these are correct.

For questions 34–38, match the items to those in the key.

**KEY:**

a. Most enzymes are globular.
b. DNA is a double helix.
c. Steroids differ by their attached groups.
d. The tails of a phospholipid can contain nonsaturated fatty acids.
e. Hydrogen bonding occurs between microfibrils of cellulose.

34. Strands held together by hydrogen bonding between strands.
35. Four fused rings plus functional groups.
36. Tertiary level of organization of a protein.
37. Provides added strength for plant cell wall.
38. Makes plasma membrane a fluid bilayer.

## thinking scientifically

1. The seeds of temperate plants tend to contain unsaturated fatty acids, while the seeds of tropical plants tend to have saturated fatty acids. **a.** How would you test your hypothesis. **b.** Assuming your hypothesis is supported, give an explanation.
2. Chemical analysis reveals that an abnormal form of an enzyme contains a polar amino acid at the location where the normal form has a nonpolar amino acid. Formulate a testable hypothesis concerning the abnormal enzyme.

## bioethical issue

### Organic Pollutants

Organic compounds include the carbohydrates, proteins, lipids, and nucleic acids that make up our bodies. Modern industry also uses all sorts of organic compounds that are synthetically produced. Indeed, our modern way of life wouldn't be possible without synthetic organic compounds.

Pesticides, herbicides, disinfectants, plastics, and textiles contain organic substances that are termed pollutants when they enter the natural environment and cause harm to living things. Global use of pesticides has increased dramatically since the 1950s, and modern pesticides are ten times more toxic than those of the 1950s. The Centers for Disease Control and Prevention in Atlanta reports that 40% of children working in agricultural fields now show signs of pesticide poisoning. The U.S. Geological Survey estimates that 32 million people in urban areas and 10 million people in rural areas are using groundwater that contains organic pollutants. J. Charles Fox, an official of the Environmental Protection Agency, says that "over the life of a person, ingestion of these chemicals has been shown to have adverse health effects such as cancer, reproductive problems, and developmental effects."

At one time, people failed to realize that everything in the environment is connected to everything else. In other words, they didn't know that an organic chemical can wander far from the site of its entry into the environment and that eventually these chemicals can enter our own bodies and cause harm. Now that we are aware of this outcome, we have to decide as a society how to proceed. We might decide to do nothing if the percentage of people dying from exposure to organic pollutants is small. Or we might decide to regulate the use of industrial compounds more strictly than has been done in the past. We could also decide that we need better ways of purifying public and private water supplies so that they do not contain organic pollutants.

## *Biology* website

The companion website for *Biology* provides a wealth of information organized and integrated by chapter. You will find practice tests, animations, videos, and much more that will complement your learning and understanding of general biology.

**http://www.mhhe.com/maderbiology10**

# 4

# Cell Structure and Function

the Dutch shopkeeper Antoni van Leeuwenhoek (1632–1723) may have been the first person to see living cells. Using a microscope he built himself, he looked at everything possible, including his own stool. He wrote, "I have usually of a morning a well-formed stool. But, hitherto, I have had sometimes a looseness of bowels, so I went to stool some twice, thrice, or four times a day. My excrement being so thin, I was at diverse times constrained to examine it. Wherein I have sometimes seen animalcules a moving prettily. Their bodies were somewhat longer than broad, and the belly, which was flat-lie, furnished with sundry little paws. . . ."
*November 9, 1681.*

*In this way, Antoni van Leeuwenhoek reported seeing the parasite* Giardia lamblia *in his feces. Giardia are unicellular organisms, while humans are multicellular organisms. In this chapter, we will see that cells are the fundamental building blocks of organisms organized to carry out basic metabolic functions. We will concentrate on the generalized bacterial, animal, and plant cell, while still realizing that all cells are specialized in particular ways.*

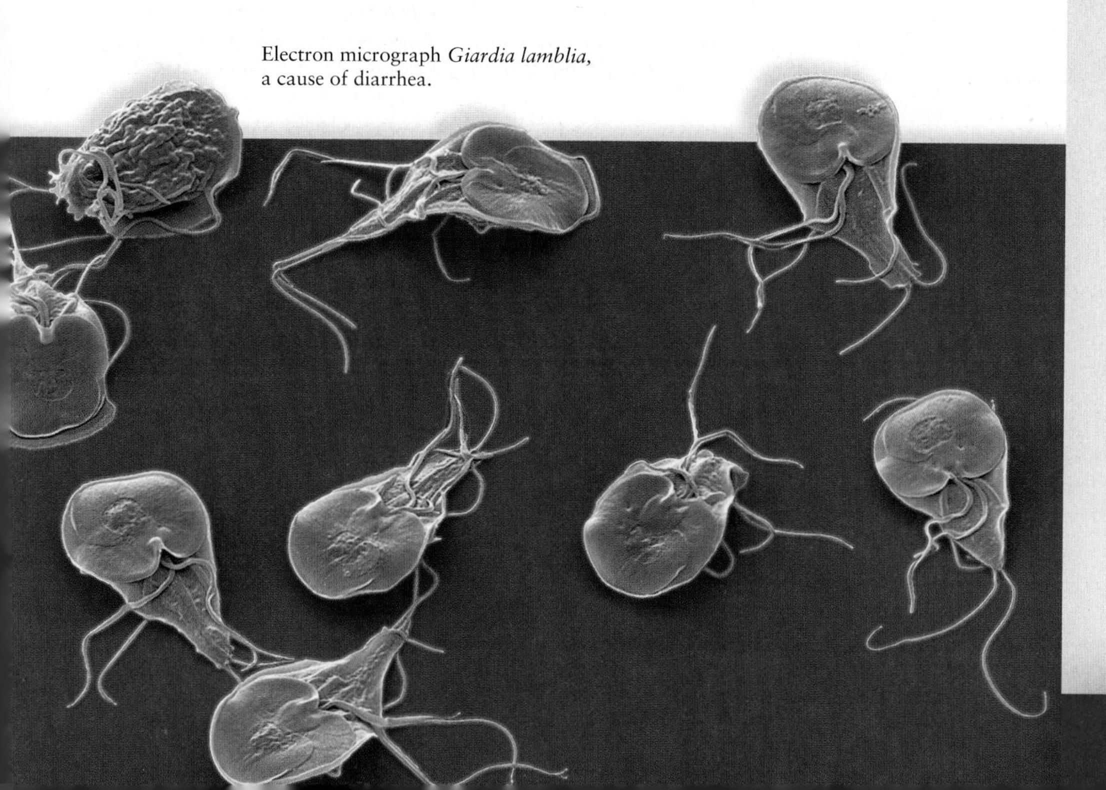

Electron micrograph *Giardia lamblia*, a cause of diarrhea.

## concepts

**4.1 CELLULAR LEVEL OF ORGANIZATION**

- All organisms are composed of cells, which arise from preexisting cells. 60
- A microscope is needed to see cells; their small size results in a favorable surface-area-to-volume relationship. 61–63

**4.2 PROKARYOTIC CELLS**

- Prokaryotic cells do have a plasma membrane but do not have a membrane-bounded nucleus, nor the various membranous organelles of eukaryotic cells. 64–65

**4.3 INTRODUCING EUKARYOTIC CELLS**

- Eukaryotic cells have a plasma membrane, a membrane-bounded nucleus, and a cytoplasm that contains a cytoskeleton and various organelles. 66–69

**4.4 THE NUCLEUS AND RIBOSOMES**

- The nucleus houses the chromosomes, and therefore the genes that work with the ribosomes to bring about protein synthesis. 70–71

**4.5 THE ENDOMEMBRANE SYSTEM**

- The endomembrane system consists of several organelles that communicate with one another, often resulting in the secretion of proteins. 72–74

**4.6 OTHER VESICLES AND VACUOLES**

- The cell has numerous and varied vesicles and vacuoles with varied functions. 75

**4.7 THE ENERGY-RELATED ORGANELLES**

- Chloroplasts and mitochondria are organelles that process energy. Chloroplasts use solar energy to produce carbohydrates, and mitochondria break down these molecules to produce ATP. 76–77

**4.8 THE CYTOSKELETON**

- The cytoskeleton, a complex system of filaments and tubules and associated motor proteins, gives the cell its shape and accounts for the movement of the cell and its organelles. 78–81

## 4.1 Cellular Level of Organization

Figure 4.1 illustrates that in our daily lives we observe whole organisms, but if it were possible to view them internally with a microscope, we would see their cellular nature. This became clear to microscopists during the 1830s.

In 1831, the English botanist Robert Brown described the nucleus of cells. In 1838, the German botanist Matthais Schleiden stated that all plants are composed of cells. A year later, the German zoologist Theodor Schwann declared that all animals are composed of cells. As a result of their work, the field of cytology (study of cells) began, and we can conclude that a **cell** is the smallest unit of living matter.

In the 1850s, the German physician Rudolph Virchow viewed the human body as a state in which each cell was a citizen. Today, we know that various illnesses of the body, such as diabetes and prostate cancer, are due to a malfunctioning of cells, rather than the organ itself. It also means that a cell is the basic unit of function as well as structure in organisms.

Virchow was the first to tell us that cells reproduce and "every cell comes from a preexisting cell." When unicellular organisms reproduce, a single cell divides, and when multicellular organisms grow, many cells divide. Cells are also involved in the sexual reproduction of multicellular organisms. In reality, there is a continuity of cells from generation to generation, even back to the very first cell (or cells) in the history of life. Due to countless investigations, which began with the work of Virchow, it is evident that cells are capable of self-reproduction.

The **cell theory** is based upon the work of Schleiden, Schwann, and Virchow. It states that:

1. all organisms are composed of cells,
2. cells are the basic units of structure and function in organisms, and
3. cells come only from preexisting cells because cells are self-reproducing.

**FIGURE 4.1 Organisms and cells.**

All organisms, including plants and animals, are composed of cells. This is not readily apparent because a microscope is usually needed to see the cells. **a.** Lilac plant. **b.** Light micrograph of a cross section of a lilac leaf showing many individual cells. **c.** Rabbit. **d.** Light micrograph of a rabbit's intestinal lining showing that it, too, is composed of cells. The dark-staining bodies are nuclei.

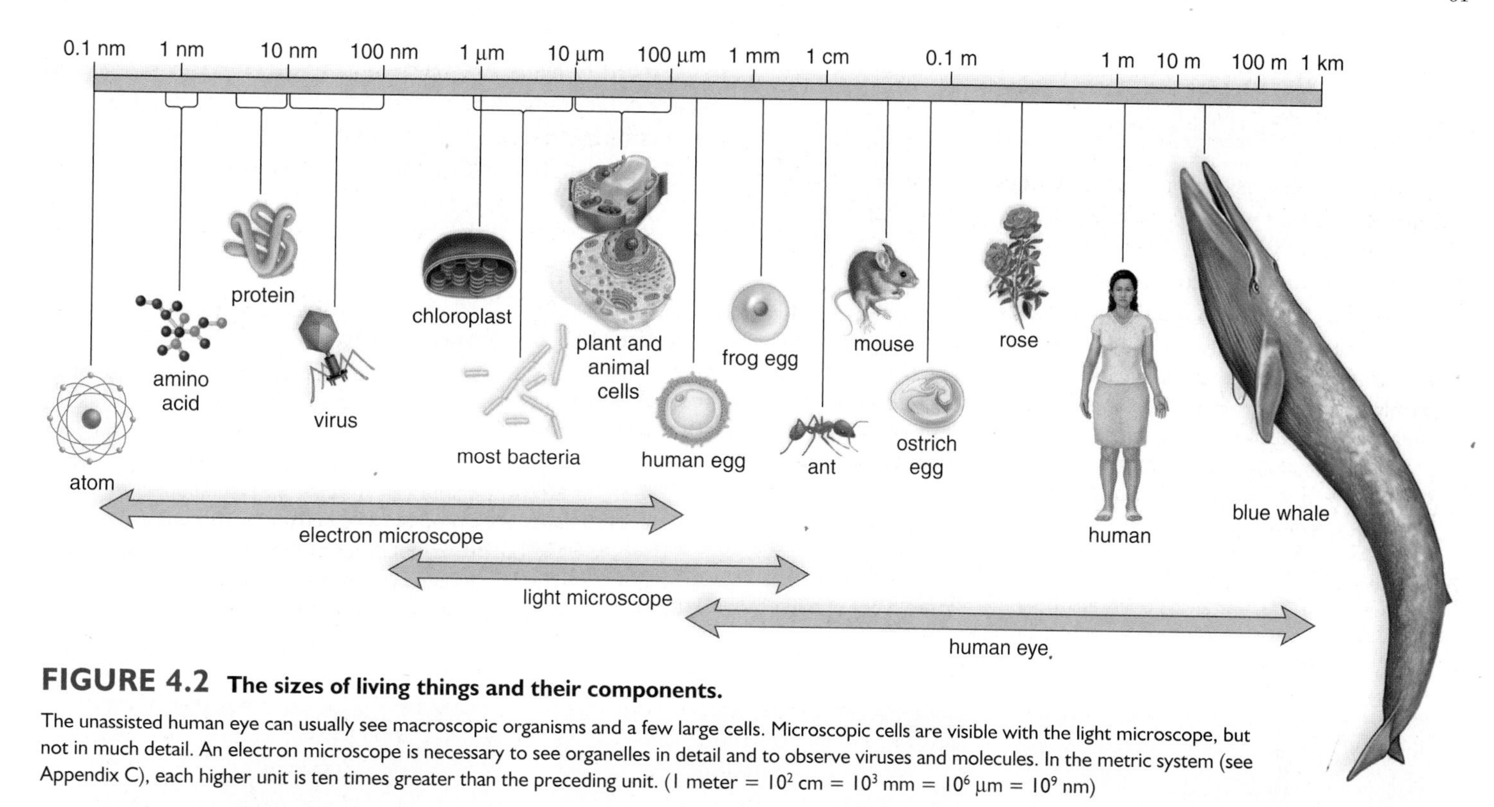


**FIGURE 4.2 The sizes of living things and their components.**

The unassisted human eye can usually see macroscopic organisms and a few large cells. Microscopic cells are visible with the light microscope, but not in much detail. An electron microscope is necessary to see organelles in detail and to observe viruses and molecules. In the metric system (see Appendix C), each higher unit is ten times greater than the preceding unit. (1 meter = $10^2$ cm = $10^3$ mm = $10^6$ µm = $10^9$ nm)

## Cell Size

Cells are quite small. A frog's egg, at about 1 millimeter (mm) in diameter, is large enough to be seen by the human eye. But most cells are far smaller than 1 mm; some are even as small as 1 micrometer (µm)—one thousandth of a millimeter. Cell inclusions and macromolecules are smaller than a micrometer and are measured in terms of nanometers (nm). Figure 4.2 outlines the visual range of the eye, light microscope, and electron microscope, and the discussion of microscopy in the Science Focus on pages 62 and 63 explains why the electron microscope allows us to see so much more detail than the light microscope does.

Why are cells so small? To answer this question, consider that a cell needs a surface area large enough to allow adequate nutrients to enter and to rid itself of wastes. Small cells, not large cells, are likely to have an adequate surface area for exchanging wastes for nutrients. For example, Figure 4.3 visually demonstrates that cutting a large cube into smaller cubes provides a lot more surface area per volume. The calculations show that a 4-cm cube has a **surface-area-to-volume ratio** of only 1.5:1, whereas a 1-cm cube has a surface-area-to-volume ratio of 6:1.

A mental image might help you realize that a smaller cell has more surface area per volume than a large cell. Imagine a small room and a large room packed with people. The small room has only two doors and the large room has four doors. But in case of fire, the smaller room has the more favorable ratio of doors to people. Similarly, the small cell is more advantageous for the exchange of molecules because of its greater surface-area-to-volume ratio.

We would expect then that actively metabolizing cells would have to remain small. A chicken's egg is several centimeters in diameter, but the egg is not actively metabolizing. Once the egg is incubated and metabolic activity begins, the egg divides repeatedly without growth. Cell division restores the amount of surface area needed for adequate exchange of materials.

| | **One 4-cm cube** | **Eight 2-cm cubes** | **Sixty-four 1-cm cubes** |
|---|---|---|---|
| **Total surface area** (height × width × number of sides × number of cubes) | 96 cm$^2$ | 192 cm$^2$ | 384 cm$^2$ |
| **Total volume** (height × width × length × number of cubes) | 64 cm$^3$ | 64 cm$^3$ | 64 cm$^3$ |
| **Surface area: Volume per cube** (surface area ÷ volume) | 1.5:1 | 3:1 | 6:1 |

**FIGURE 4.3 Surface-area-to-volume relationships.**

As cell size decreases from 4 cm$^3$ to 1 cm$^3$, the surface-area-to-volume ratio increases.

### Check Your Progress 4.1

1. Explain why a large surface-area-to-volume ratio is needed for the proper functioning of cells.

*science focus*

## Microscopy Today

Cells were not discovered until the seventeenth century (when the microscope was invented). Since that time, various types of microscopes have been developed for the study of cells and their components.

In the *compound light microscope,* light rays passing through a specimen are brought into focus by a set of glass lenses, and the resulting image is then viewed by the human eye. In the *transmission electron microscope (TEM),* electrons passing through a specimen are brought into focus by a set of electromagnetic lenses, and the resulting image is projected onto a fluorescent screen or photographic film. In the *scanning electron microscope (SEM),* a narrow beam of electrons is scanned over the surface of the specimen, which is coated with a thin metal layer. The metal gives off secondary electrons that are collected by a detector to produce an image on a television screen. The SEM permits the development of three-dimensional images. Figure 4A shows these three types of microscopic images.

### Magnification, Resolution, and Contrast

**Magnification** is the ratio between size of an image and its actual size. The electron microscope magnifies to a greater extent than does the compound light microscope. A light microscope can magnify objects about a thousand times, but an electron microscope can magnify them hundreds of thousands of times. The difference lies in the means of illumination. The path of light rays and electrons moving through

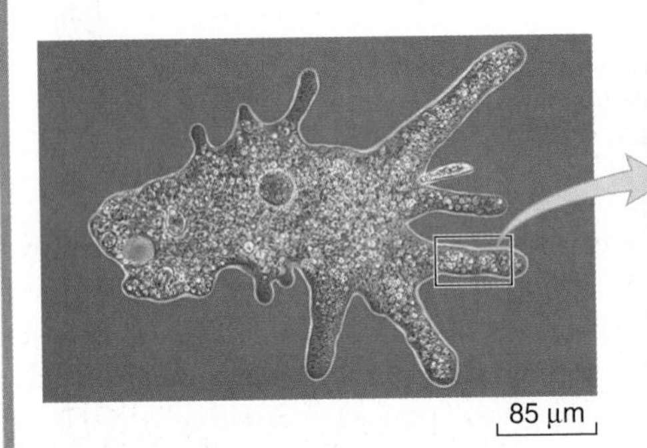

amoeba, light micrograph

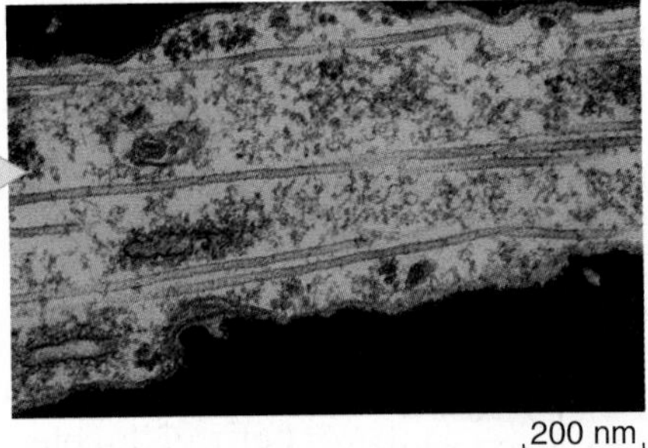

pseudopod segment, transmission electron micrograph

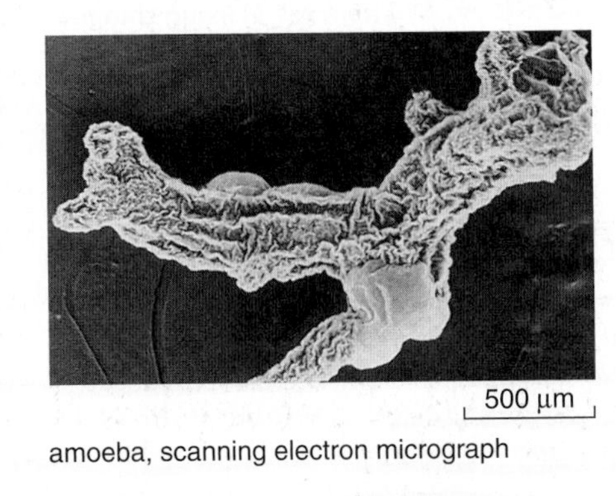

amoeba, scanning electron micrograph

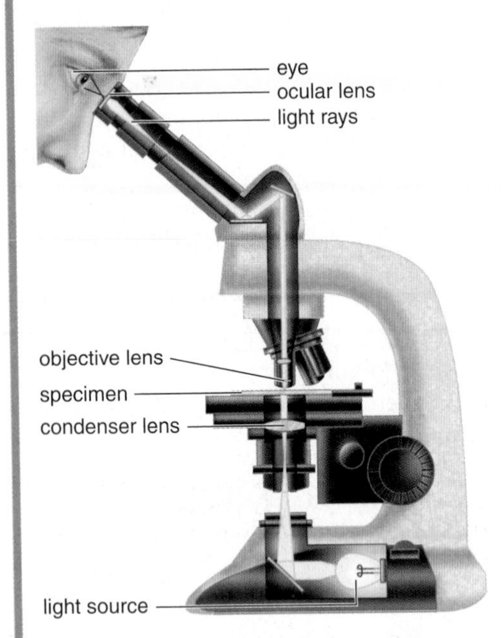

a. Compound light microscope

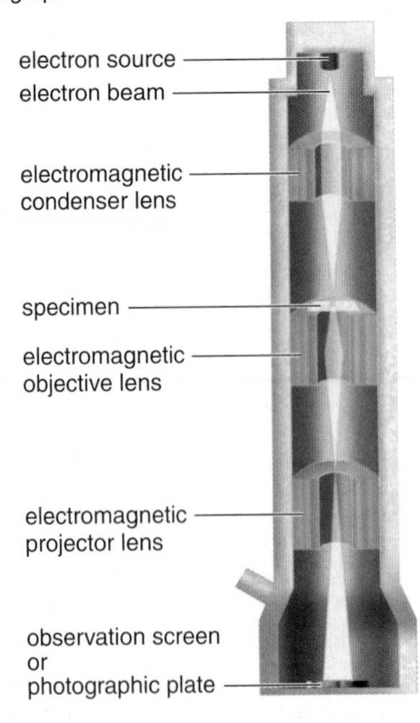

b. Transmission electron microscope

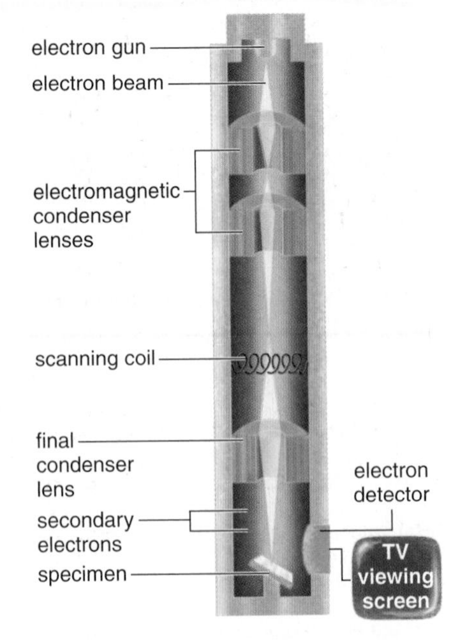

c. Scanning electron microscope

**FIGURE 4A Diagram of microscopes with accompanying micrographs of *Amoeba proteus.***
***a.** The compound light microscope and **(b)** the transmission electron microscope provide an internal view of an organism. **c.** The scanning electron microscope provides an external view of an organism.*

space is wavelike, but the wavelength of electrons is much shorter than the wavelength of light. This difference in wavelength accounts for the electron microscope's greater magnifying capability and its greater resolving power.

**Resolution** is the minimum distance between two objects that allows them to be seen as two separate objects. A microscope with poor resolution might enable a student to see only one cellular granule, while the microscope with the better resolution would show two granules next to each other. The greater the resolving power, the greater the detail eventually seen. If oil is placed between the sample and the objective lens of the compound light microscope, the resolving power is increased, and if ultraviolet light is used instead of visible light, it is also increased. But typically, a light microscope can resolve down to 0.2 μm, while the transmission electron microscope can resolve down to 0.0002 μm. If the resolving power of the average human eye is set at 1.0, then that of the typical compound light microscope is about 500, and that of the transmission electron microscope is 100,000 (Fig. 4A*b*).

The ability to make out a particular object can depend on **contrast,** a difference in the shading of an object compared to its background. Contrast is often achieved by staining cells with colored dyes (light microscopy) or with electron-dense metals (electron microscopy). Another way to increase contrast is to use optical methods such as phase contrast and differential interference contrast (Fig. 4B). Then, too, fluorescent antibodies will attach to particular proteins in a cell to reveal their presence (see Fig. 4.18).

## Illumination, Viewing, and Recording

Light rays can be bent (refracted) and brought to focus as they pass through glass lenses, but electrons do not pass through glass. Electrons have a charge that allows them to be brought into focus by electromagnetic lenses. The human eye uses light to see an object but cannot use electrons for the same purpose. Therefore, electrons leaving the specimen in the electron microscope are directed toward a screen or a photographic plate that is sensitive to their presence. Humans can view the image on the screen or photograph.

A major advancement in illumination has been the introduction of *confocal microscopy,* which uses a laser beam scanned across the specimen to focus on a single shallow plane within the cell. The microscopist can "optically section" the specimen by focusing up and down, and a series of optical sections can be combined in a computer to create a three-dimensional image, which can be displayed and rotated on the computer screen.

An image from a microscope may be recorded by replacing the human eye with a television camera. The television camera converts the light image into an electronic image, which can be entered into a computer. In *video-enhanced contrast microscopy,* the computer makes the darkest areas of the original image much darker and the lightest areas of the original much lighter. The result is a high-contrast image with deep blacks and bright whites. Even more contrast can be introduced by the computer if shades of gray are replaced by colors.

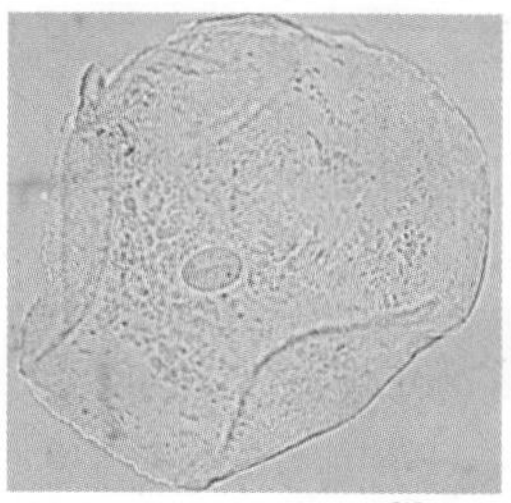


**Bright-field.** Light passing through the specimen is brought directly into focus. Usually, the low level of contrast within the specimen interferes with viewing all but its largest components.

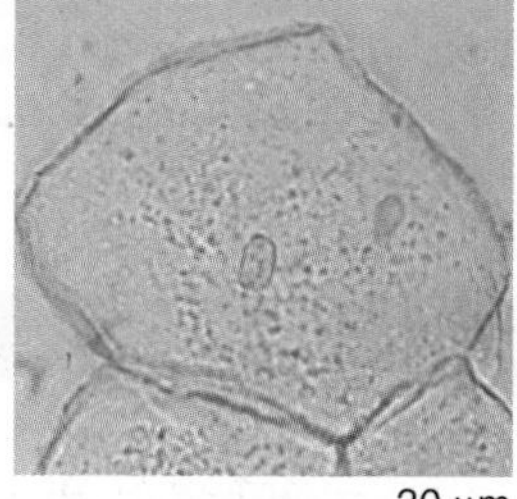


**Bright-field (stained).** Dyes are used to stain the specimen. Certain components take up the dye more than other components, and therefore contrast is enhanced.

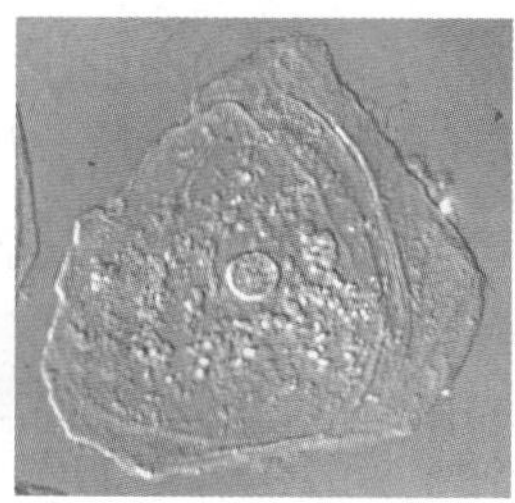


**Differential interference contrast.** Optical methods are used to enhance density differences within the specimen so that certain regions appear brighter than others. This technique is used to view living cells, chromosomes, and organelle masses.

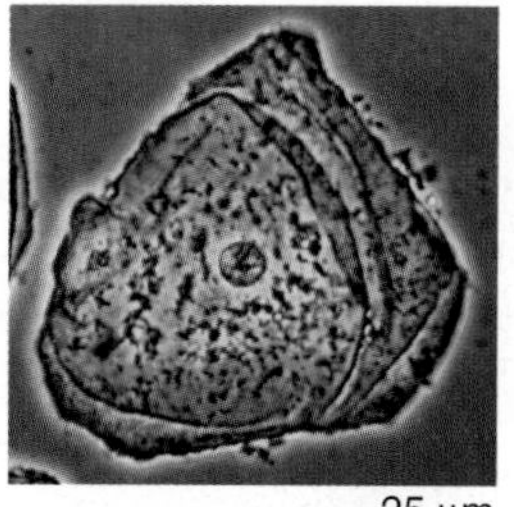


**Phase contrast.** Density differences in the specimen cause light rays to come out of "phase." The microscope enhances these phase differences so that some regions of the specimen appear brighter or darker than others. The technique is widely used to observe living cells and organelles.

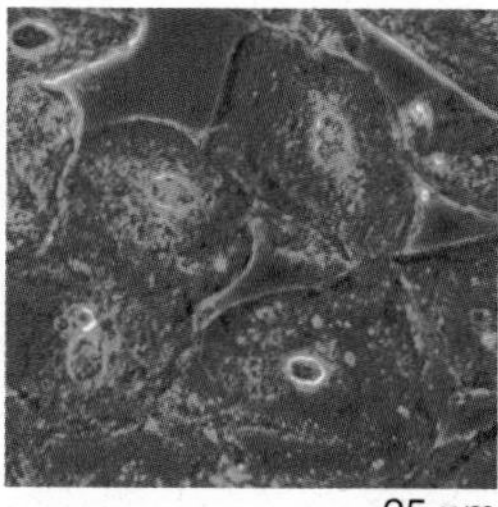


**Dark-field.** Light is passed through the specimen at an oblique angle so that the objective lens receives only light diffracted and scattered by the object. This technique is used to view organelles, which appear quite bright against a dark field.

**FIGURE 4B Photomicrographs of cheek cells.**
*Bright-field microscopy is the most common form used with a compound light microscope. Other types of microscopy include differential interference contrast, phase contrast, and dark-field.*

# 4.2 Prokaryotic Cells

Fundamentally, two different types of cells exist. **Prokaryotic cells** [Gk. *pro,* before, and *karyon,* kernel, nucleus] are so named because they lack a membrane-bounded nucleus. The other type of cell, called a **eukaryotic cell** [Gk. *eu,* true, and *karyon,* kernel, nucleus], has a nucleus (see Figs. 4.6 and 4.7). Prokaryotes are present in great numbers in the air, in bodies of water, in the soil, and they also live in and on other organisms. Although they are structurally less complicated than eukaryotes, their metabolic capabilities as a group far exceed those of eukaryotes. Prokaryotes are an extremely successful group of organisms whose evolutionary history dates back to the first cells on Earth.

Prokaryotic cells are divided into two types, largely based on DNA and RNA base sequence differences. The two groups of prokaryotes are so biochemically different that they have been placed in separate domains, called domain Bacteria and domain Archaea. Bacteria are well known because they cause some serious diseases, such as tuberculosis, anthrax, tetanus, throat infections, and gonorrhea. But many species of bacteria are important to the environment because they decompose the remains of dead organisms and contribute to ecological cycles. Bacteria also assist humans in still another way—we use them to manufacture all sorts of products, from industrial chemicals to foodstuffs and drugs. For example, today we know how to place human genes in bacteria so that they will produce human insulin, a necessary hormone for the treatment of diabetes.

## The Structure of Prokaryotes

Prokaryotes are quite small; an average size is 1.1–1.5 µm wide and 2.0–6.0 µm long. While other prokaryote shapes have been identified, three basic shapes are most common:

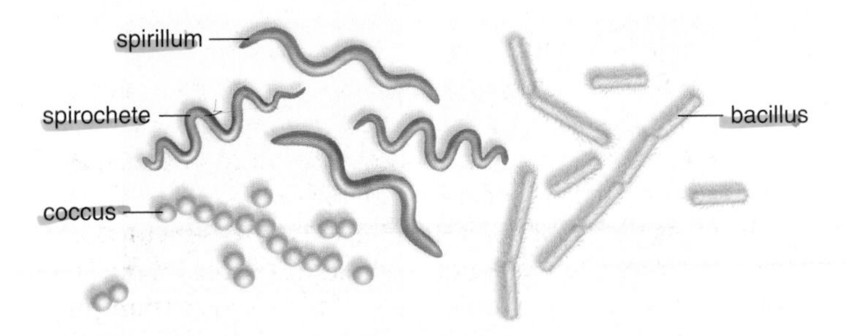


A rod-shaped bacterium is called a **bacillus,** while a spherical-shaped bacterium is a **coccus.** Both of these can occur as pairs or chains, and in addition, cocci can occur as clusters. Some long rods are twisted into spirals, in which case they are **spirilla** if they are rigid or **spirochetes** if they are flexible.

Figure 4.4 shows the generalized structure of a bacterium. This means that not all bacteria have all the structures depicted, and some have more than one of each. Also, for the sake of discussion, we will divide the organization of bacteria into the cell envelope, the cytoplasm, and the appendages.

### *Cell Envelope*

In bacteria, the **cell envelope** includes the plasma membrane, the cell wall, and the glycocalyx. The **plasma membrane** is a phospholipid bilayer with embedded proteins:

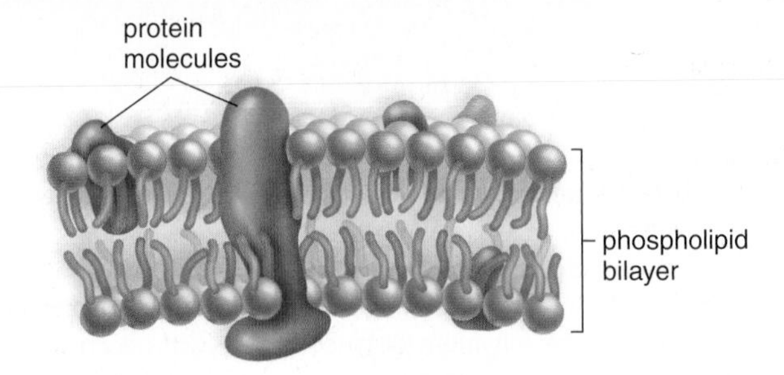


The plasma membrane has the important function of regulating the entrance and exit of substances into and out of the cytoplasm. After all, the cytoplasm has a normal composition that needs to be maintained.

In prokaryotes, the plasma membrane can form internal pouches called mesosomes. **Mesosomes** most likely increase the internal surface area for the attachment of enzymes that are carrying on metabolic activities.

The **cell wall,** when present, maintains the shape of the cell, even if the cytoplasm should happen to take up an abundance of water. You may recall that the cell wall of a plant cell is strengthened by the presence of cellulose, while the cell wall of a bacterium contains peptidoglycan, a complex molecule containing a unique amino disaccharide and peptide fragments.

The **glycocalyx** is a layer of polysaccharides lying outside the cell wall in some bacteria. When the layer is well organized and not easily washed off, it is called a **capsule.** A slime layer, on the other hand, is not well organized and is easily removed. The glycocalyx aids against drying out and helps bacteria resist a host's immune system. It also helps bacteria attach to almost any surface.

### *Cytoplasm*

The **cytoplasm** is a semifluid solution composed of water and inorganic and organic molecules encased by a plasma membrane. Among the organic molecules are a variety of enzymes, which speed the many types of chemical reactions involved in metabolism.

The DNA of a prokaryote is found in a chromosome that coils up and is located in a region called the **nucleoid.** Many bacteria also have an extrachromosomal piece of circular DNA called a **plasmid.** Plasmids are routinely used in biotechnology laboratories as vectors to transport DNA into a bacterium—even human DNA can be put into a bacterium by using a plasmid as a vector. This technology is important in the production of new medicines.

The many proteins specified for by bacterial DNA are synthesized on tiny particles called **ribosomes.** A bacterial cell contains thousands of ribosomes that are smaller than eukaryotic ribosomes. However, bacterial ribosomes still

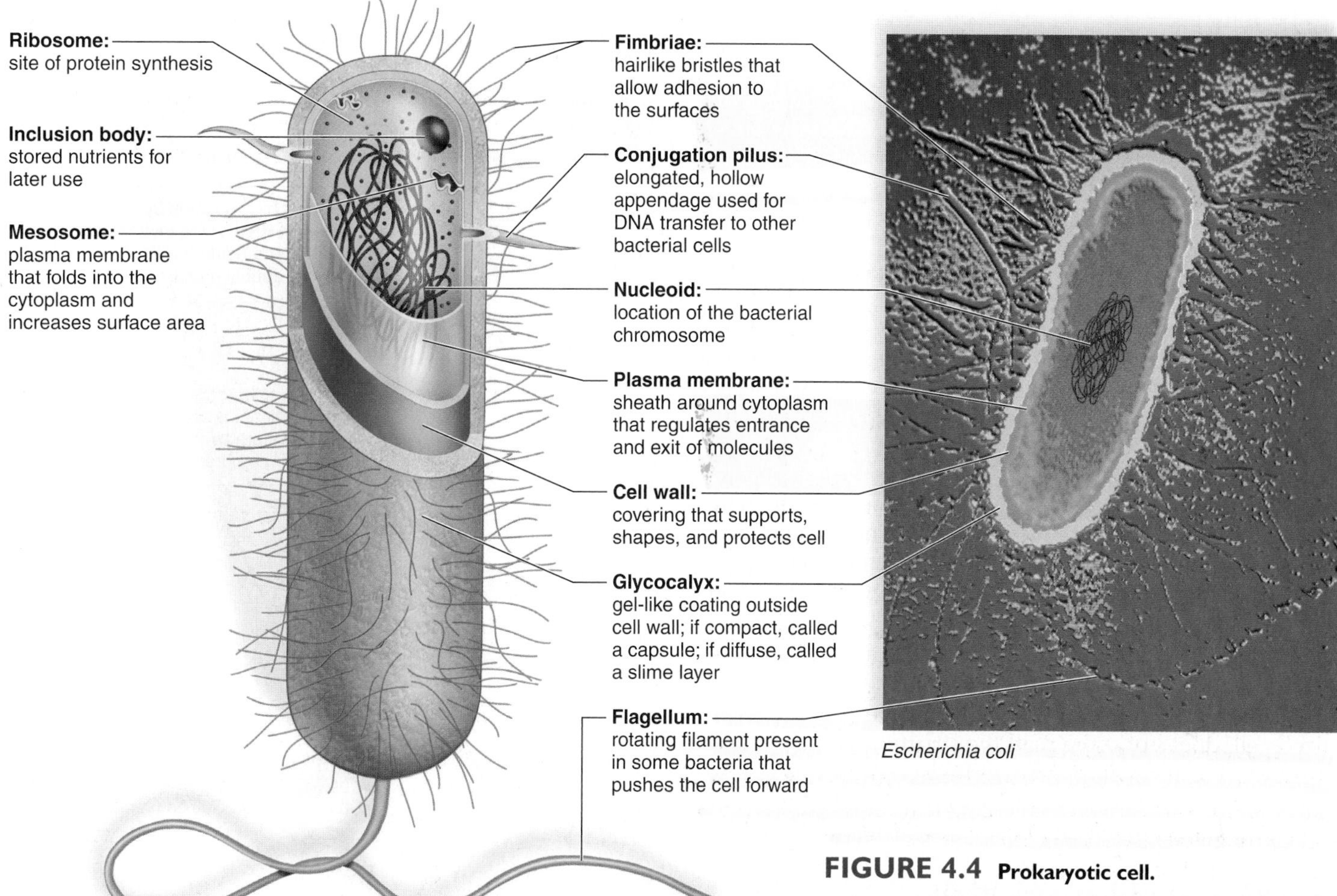


**FIGURE 4.4 Prokaryotic cell.**

Prokaryotic cells lack membrane-bounded organelles, as well as a nucleus. Their DNA is located in a region called a nucleoid.

contain RNA and protein in two subunits, as do eukaryotic ribosomes. The **inclusion bodies** found in the cytoplasm are stored granules of various substances. Some are nutrients that can be broken down when needed.

Most bacteria metabolize in the same manner as animals, but the **cyanobacteria** are bacteria that photosynthesize in the same manner as plants. These organisms live in water, in ditches, on buildings, and on the bark of trees. Their cytoplasm contains extensive internal membranes called **thylakoids** [Gk. *thylakon,* small sac], where chlorophyll and other pigments absorb solar energy for the production of carbohydrates. Cyanobacteria are called the blue-green bacteria because some have a pigment that adds a shade of blue to the cell, in addition to the green color of chlorophyll. The cyanobacteria release oxygen as a side product of photosynthesis, and perhaps ancestral cyanobacteria were the first types of organisms on Earth to do so. The addition of oxygen changed the composition of the Earth's atmosphere.

## *Appendages*

The appendages of a bacterium, namely the flagella, fimbriae, and conjugation pili, are made of protein. Motile bacteria can propel themselves in water by the means of appendages called **flagella** (usually 20 nm in diameter and 1–70 nm long). The bacterial flagellum has a filament, a hook, and a basal body. The basal body is a series of rings anchored in the cell wall and membrane. The hook rotates 360° within the basal body, and this motion propels bacteria—the bacterial flagellum does not move back and forth like a whip. Sometimes flagella occur only at the two ends of a cell, and sometimes they are dispersed randomly over the surface. The number and location of flagella can be used to help distinguish different types of bacteria.

**Fimbriae** are small, bristlelike fibers that sprout from the cell surface. They are not involved in locomotion; instead, fimbriae attach bacteria to a surface. **Conjugation pili** are rigid tubular structures used by bacteria to pass DNA from cell to cell. Bacteria reproduce asexually by binary fission, but they can exchange DNA by way of the conjugation pili. They can also take up DNA from the external medium or by way of viruses.

### Check Your Progress 4.2

1. What is the major distinction between a prokaryotic cell and a eukaryotic cell?
2. Which of the structures shown in Figure 4.4 pertain to the cell envelope, the cytoplasm, and the appendages?

# 4.3 Introducing Eukaryotic Cells

Eukaryotic cells, like prokaryotic cells, have a plasma membrane that separates the contents of the cell from the environment and regulates the passage of molecules into and out of the cytoplasm. The plasma membrane is a phospholipid bilayer with embedded proteins. It has been suggested by some scientists that the nucleus evolved as the result of the invagination of the plasma membrane (Fig. 4.5).

## Origin of the Eukaryotic Cell

While Figure 4.5 suggests that the nucleus evolved as a result of plasma membrane invagination, the **endosymbiotic theory** says that mitochondria and chloroplasts, the two energy-related organelles, arose when a large eukaryotic cell engulfed independent prokaryotes. This explains why they are bounded by a double membrane and contain their own genetic material separate from that of the nucleus. We will be mentioning this theory again when the structure and function of mitochondria and chloroplasts are discussed in more detail later in the chapter. Figures 4.6 and 4.7 can represent the fully evolved animal and plant cell, but they are generalized cells. A specialized cell, as opposed to a generalized cell, does not contain all the structures depicted and may have more copies of any particular organelle. A generalized cell is useful for study purposes, but the body of a plant or animal is made up of specialized cells.

## Structure of Eukaryotic Cell

Some eukaryotic cells, notably plant cells and those of fungi and many protists, have a cell wall in addition to a plasma membrane. A plant cell wall contains cellulose fibrils and, therefore, has a different composition from the bacterial cell wall.

As shown in Figures 4.6 and 4.7, eukaryotic cells are compartmentalized—they have compartments. The compartments of a eukaryotic cell, typically called **organelles,** are membranous. The nucleus is a compartment that houses the genetic material within eukaryotic chromosomes. The nucleus communicates with ribosomes in the cytoplasm, and the organelles of the endomembrane system—notably the endoplasmic reticulum and the Golgi apparatus—communicate with one another. Because each organelle has its own particular set of enzymes it produces its own products, and the products move from one organelle to the other. The products are carried between organelles by little transport **vesicles,** membranous sacs that enclose the molecules and keep them separate from the cytoplasm. For example, the endoplasmic reticulum communicates with the Golgi apparatus by means of transport vesicles. Communication with the energy-related organelles—mitochondria and chloroplasts—is less obvious but it does occur because they are capable of importing particular molecules from the cytoplasm. An animal cell has only mitochondria, while a plant cell has both mitochondria and chloroplasts.

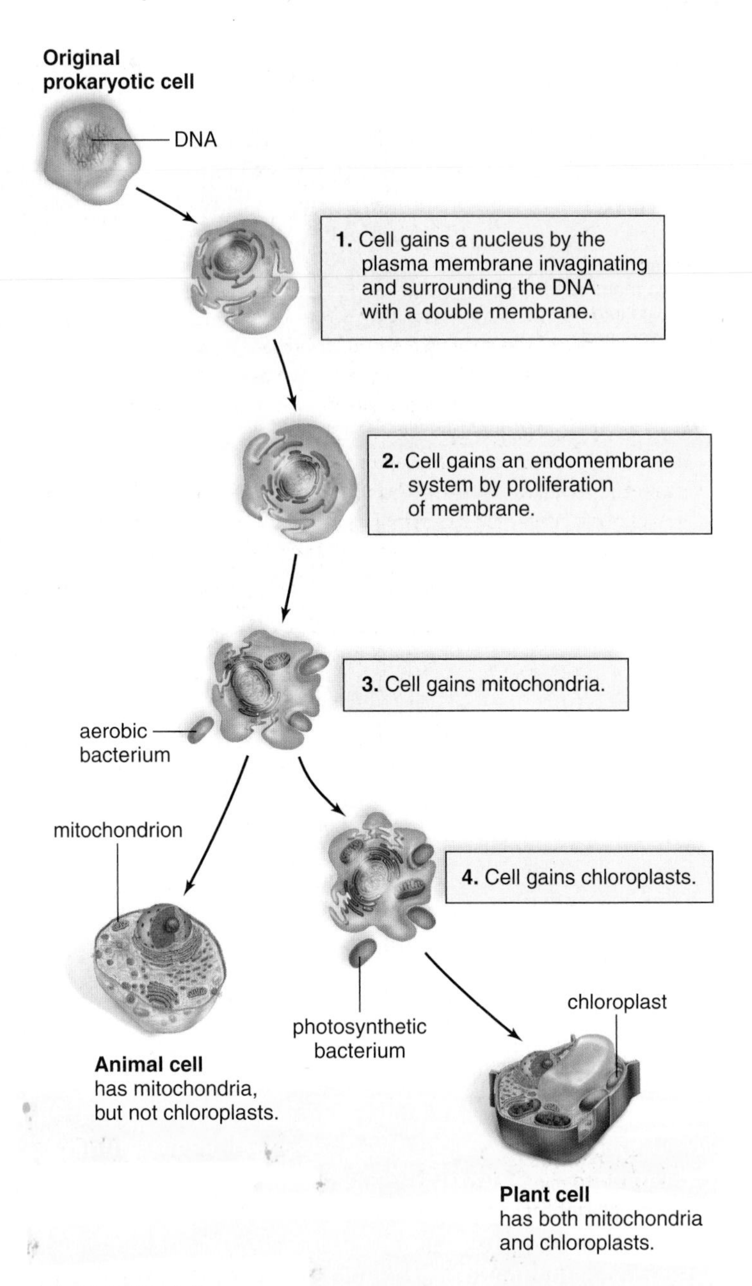


**FIGURE 4.5 Origin of organelles.**

Invagination of the plasma membrane could have created the nuclear envelope and an endomembrane system that involves several organelles. The endosymbiotic theory states that mitochondria and chloroplasts were independent prokaryotes that took up residence in a eukaryotic cell. Mitochondria carry on cellular respiration, and chloroplasts carry on photosynthesis. Endosymbiosis was a first step toward the origin of the plant and animal cell during the evolutionary history of life.

Each membranous organelle has a specific structure and function. This is possible because all the molecules necessary to specificity can be concentrated inside an organelle. The internal membrane of organelles provides a large surface for the attachment of enzymes. Having organelles also means that cells can become specialized by the presence or absence of particular organelles. The final result has been the complexity we associate with an organism that has

## science focus

### Separating the Contents of Cells

Modern microscopy can be counted on to reveal the structure and distribution of organelles in a cell. But how do researchers separate the different types of organelles from a cell so that they can determine their function? Suppose, for example, you wanted to study the function of ribosomes. How would you acquire some ribosomes? First, researchers remove cells from an organism or cell culture and place them in a sugar or salt solution. Then they fractionate (break open) the cells in a tube.

A process called *differential centrifugation* allows researchers to separate the parts of a cell by size and density. A centrifuge works like the spin cycle of a washing machine. Only when the centrifuge spins do cell components come out of suspension and form a sediment. The faster the centrifuge spins, the smaller the components that settle out.

Figure 4C shows that the slowest spin cycle separates out the nuclei, and then progressively faster cycles separate out ever smaller components. In between spins, the fluid portion of the previous cycle must be poured into another test tube. Why? If you didn't start with a fresh tube, all the different cell parts would pile up in the sediment of one tube.

By using different concentrations of salt solutions and different centrifuge speeds, researchers can obtain essentially pure preparations of almost any cell component. Biochemical analysis and manipulation then allow them to determine the functions of that cell component.

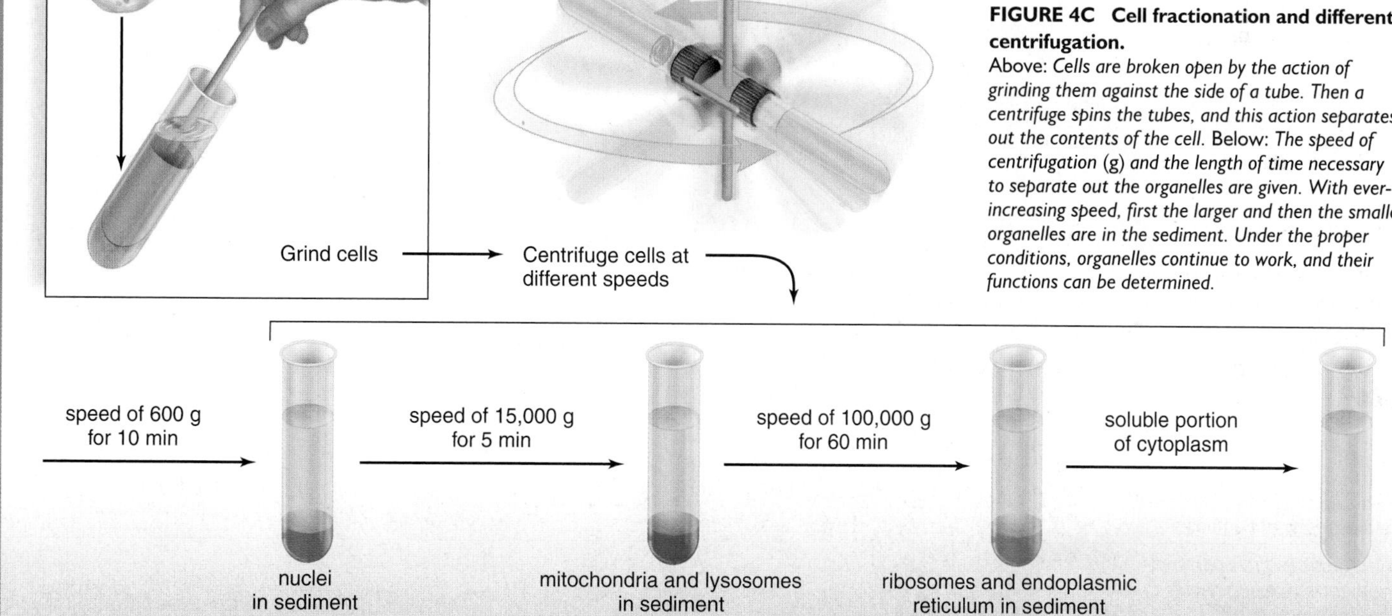


**FIGURE 4C Cell fractionation and differential centrifugation.**
Above: *Cells are broken open by the action of grinding them against the side of a tube. Then a centrifuge spins the tubes, and this action separates out the contents of the cell.* Below: *The speed of centrifugation (g) and the length of time necessary to separate out the organelles are given. With ever-increasing speed, first the larger and then the smaller organelles are in the sediment. Under the proper conditions, organelles continue to work, and their functions can be determined.*

different tissues arranged in organs, each with a particular structure and function. The Science Focus above describes the process by which investigators were able to discover the structure and function of various organelles.

The **cytoskeleton** is a lattice of protein fibers that maintains the shape of the cell and assists in the movement of organelles. The protein fibers serve as tracks for the transport vesicles that are taking molecules from one organelle to another. In other words, the tracks direct and speed them on their way. The manner in which vesicles and other types of organelles move along these tracks will be discussed in more detail later in the chapter. Without a cytoskeleton, a eukaryotic cell would not have an efficient means of moving organelles and their products within the cell and possibly could not exist.

### Check Your Progress 4.3

1. Name three benefits of compartmentalization found in cells.
2. How did the energy-related organelles arise?

## FIGURE 4.6

**Animal cell anatomy.**

Micrograph of an insect cell (*above*) and drawing of a generalized animal cell (*below*).

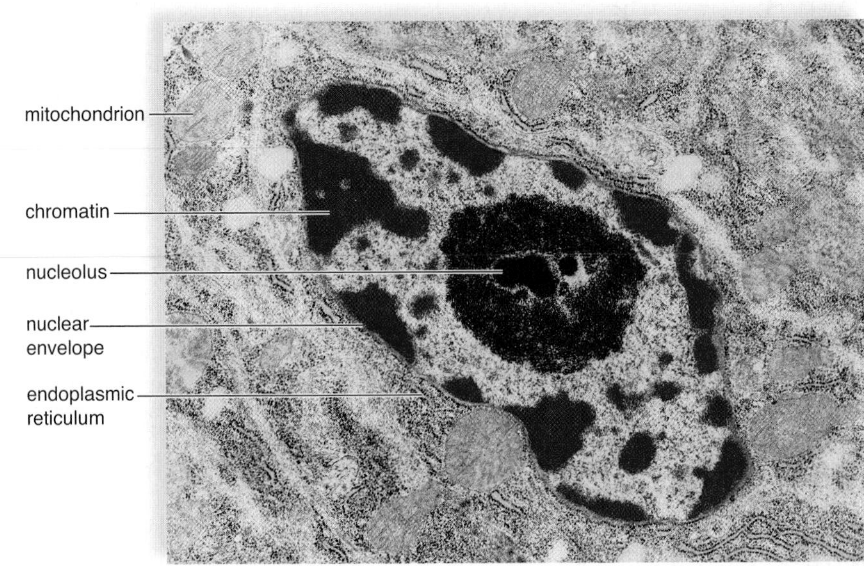


**Plasma membrane:** outer surface that regulates entrance and exit of molecules

protein

phospholipid

**Cytoskeleton:** maintains cell shape and assists movement of cell parts:

**Microtubules:** protein cylinders that move organelles

**Intermediate filaments:** protein fibers that provide stability of shape

**Actin filaments:** protein fibers that play a role in change of shape

**Centrioles*:** short cylinders of microtubules of unknown function

**Centrosome:** microtubule organizing center that contains a pair of centrioles

**Lysosome*:** vesicle that digests macromolecules and even cell parts

**Vesicle:** small membrane-bounded sac that stores and transports substances

**Cytoplasm:** semifluid matrix outside nucleus that contains organelles

**Nucleus:** command center of cell

**Nuclear envelope:** double membrane with nuclear pores that encloses nucleus

**Chromatin:** diffuse threads containing DNA and protein

**Nucleolus:** region that produces subunits of ribosomes

**Endoplasmic reticulum:** protein and lipid metabolism

**Rough ER:** studded with ribosomes that synthesize proteins

**Smooth ER:** lacks ribosomes, synthesizes lipid molecules

**Peroxisome:** vesicle that is involved in fatty acid metabolism

**Ribosomes:** particles that carry out protein synthesis

**Polyribosome:** string of ribosomes simultaneously synthesizing same protein

**Mitochondrion:** organelle that carries out cellular respiration, producing ATP molecules

**Golgi apparatus:** processes, packages, and secretes modified proteins

*not in plant cells

peroxisome
mitochondrion
nucleus
ribosomes
central vacuole
plasma membrane
cell wall
chloroplast
1 μm

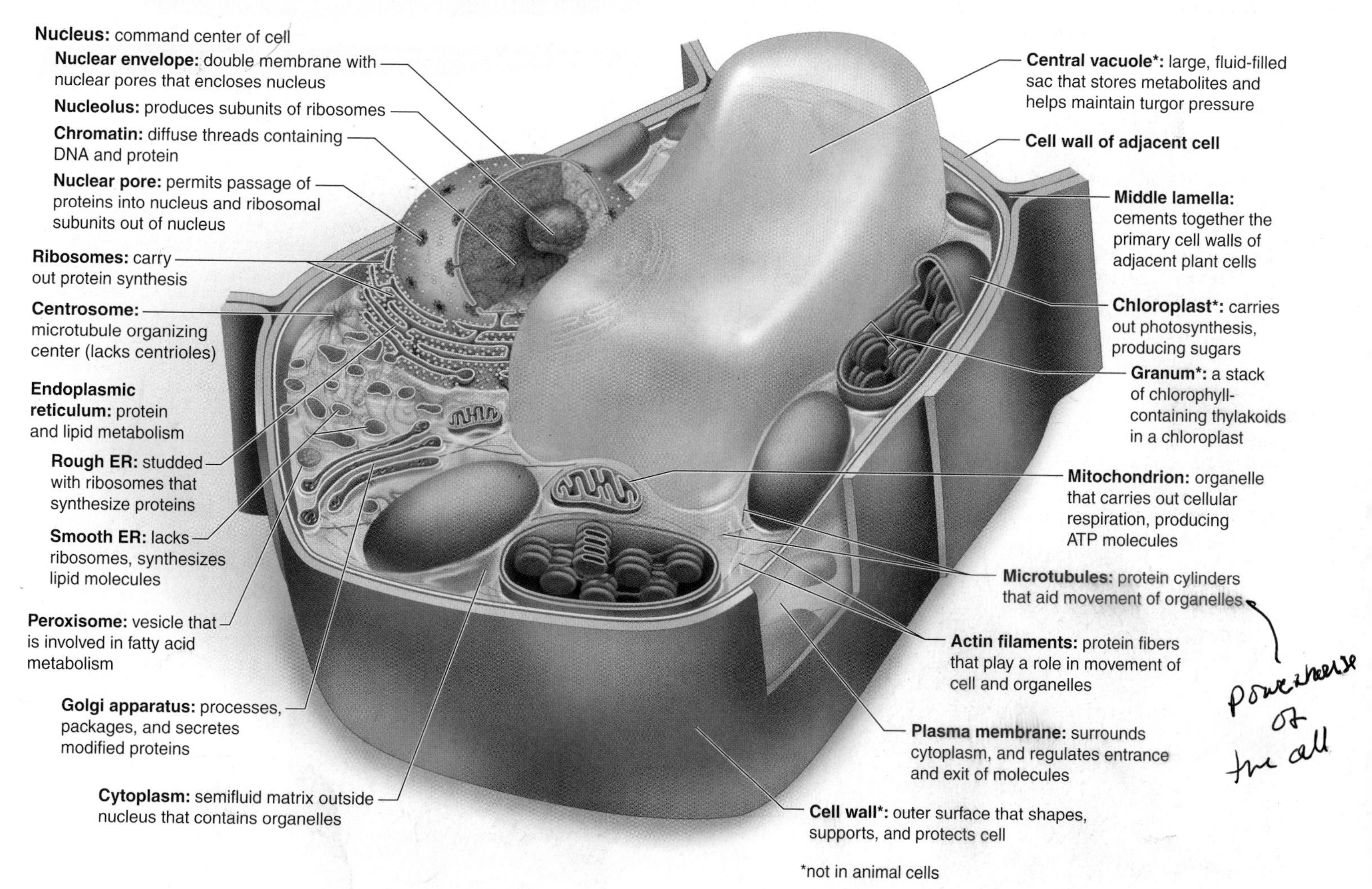


## FIGURE 4.7

**Plant cell anatomy.**

False-colored micrograph of a young plant cell (*above*) and drawing of a generalized plant cell (*below*).

## 4.4 The Nucleus and Ribosomes

The nucleus is essential to the life of a cell. It contains the genetic information that is passed on from cell to cell and from generation to generation. The ribosomes use this information to carry out protein synthesis.

### The Nucleus

The nucleus, which has a diameter of about 5 μm, is a prominent structure in the eukaryotic cell (Fig. 4.8). It generally appears as an oval structure located near the center of most cells. Some cells, such as skeletal muscle cells, can have more than one nucleus. The nucleus contains **chromatin** [Gk. *chroma*, color, and *teino*, stretch] in a semifluid matrix called the **nucleoplasm.** Chromatin looks grainy, but actually it is a network of strands that condenses and undergoes coiling into rodlike structures called **chromosomes** [Gk. *chroma*, color, and *soma*, body], just before the cell divides. All the cells of an individual contain the same number of chromosomes, and the mechanics of nuclear division ensure that each daughter cell receives the normal number of chromosomes, except for the egg and sperm, which usually have half this number. This alone suggested to early investigators that the chromosomes are the carriers of genetic information and that the nucleus is the command center of the cell.

Chromatin, and therefore chromosomes, contains DNA, protein, and some RNA (ribonucleic acid). **Genes,** composed of DNA, are units of heredity located on the chromosomes.

Three types of RNA are produced in the nucleus: *ribosomal RNA (rRNA), messenger RNA (mRNA),* and *transfer RNA (tRNA).* Ribosomal RNA is produced in the **nucleolus,** a dark region of chromatin where rRNA joins with proteins to form the subunits of ribosomes. Ribosomes are small bodies in the cytoplasm where protein synthesis occurs. Messenger RNA acts as an intermediary for DNA, which specifies the sequence of amino acids in a protein. Transfer RNA participates in the assembly of amino acids during protein synthesis. The proteins of a cell determine its structure and functions.

The nucleus is separated from the cytoplasm by a double membrane known as the **nuclear envelope.** Even so, the nucleus communicates with the cytoplasm. The nuclear envelope has **nuclear pores** of sufficient size (100 nm) to per-

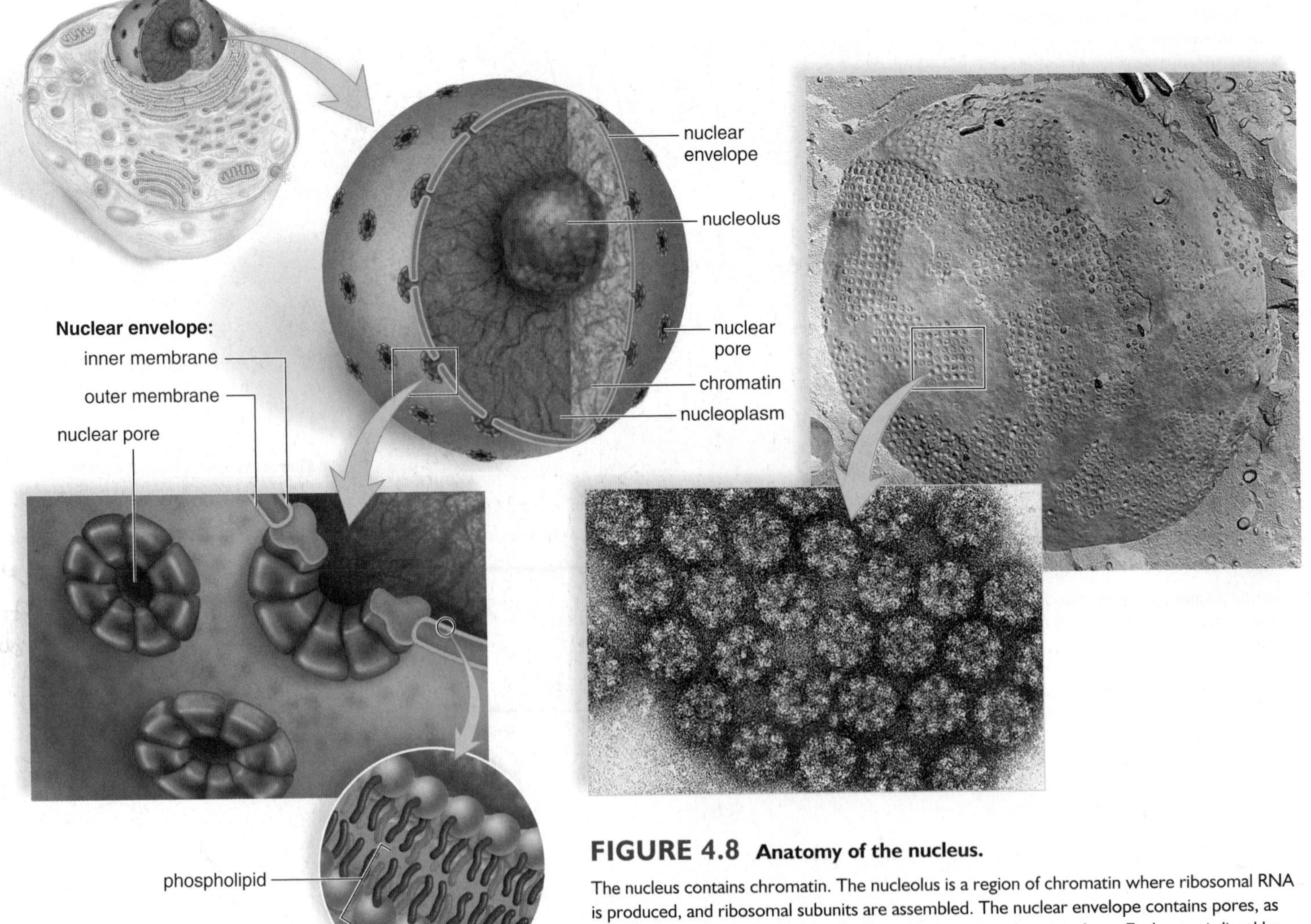


**FIGURE 4.8 Anatomy of the nucleus.**

The nucleus contains chromatin. The nucleolus is a region of chromatin where ribosomal RNA is produced, and ribosomal subunits are assembled. The nuclear envelope contains pores, as shown in the larger micrograph of a freeze-fractured nuclear envelope. Each pore is lined by a complex of eight proteins, as shown in the smaller micrograph and drawing. Nuclear pores serve as passageways for substances to pass into and out of the nucleus.

mit the passage of ribosomal subunits and mRNA out of the nucleus into the cytoplasm and the passage of proteins from the cytoplasm into the nucleus. High-power electron micrographs show that nonmembranous components associated with the pores form a nuclear pore complex.

## Ribosomes

**Ribosomes** are particles where protein synthesis occurs. In eukaryotes, ribosomes are 20 nm by 30 nm, and in prokaryotes they are slightly smaller. In both types of cells, ribosomes are composed of two subunits, one large and one small. Each subunit has its own mix of proteins and rRNA. The number of ribosomes in a cell varies depending on its functions. For example, pancreatic cells and those of other glands have many ribosomes because they produce secretions that contain proteins.

In eukaryotic cells, some ribosomes occur freely within the cytoplasm, either singly or in groups called **polyribosomes,** and others are attached to the endoplasmic reticulum (ER), a membranous system of flattened saccules (small sacs) and tubules, which is discussed more fully on the next page. Ribosomes receive mRNA from the nucleus, and this mRNA carries a coded message from DNA indicating the correct sequence of amino acids in a particular protein. Proteins synthesized by cytoplasmic ribosomes are used in the cytoplasm, and those synthesized by attached ribosomes end up in the ER.

What causes a ribosome to bind to the endoplasmic reticulum? Binding occurs only if the protein being synthesized by a ribosome begins with a sequence of amino acids called a **signal peptide.** The signal peptide binds a particle (signal recognition particle, SRP), which then binds to a receptor on the ER. Once the protein enters the ER, an enzyme cleaves off the signal peptide, and the protein ends up within the lumen (interior) of the ER, where it folds into its final shape (Fig. 4.9).

### Check Your Progress 4.4

1. List the components of the nucleus and give a function for each.
2. Where are ribosomes found in the cell, and what do they do?

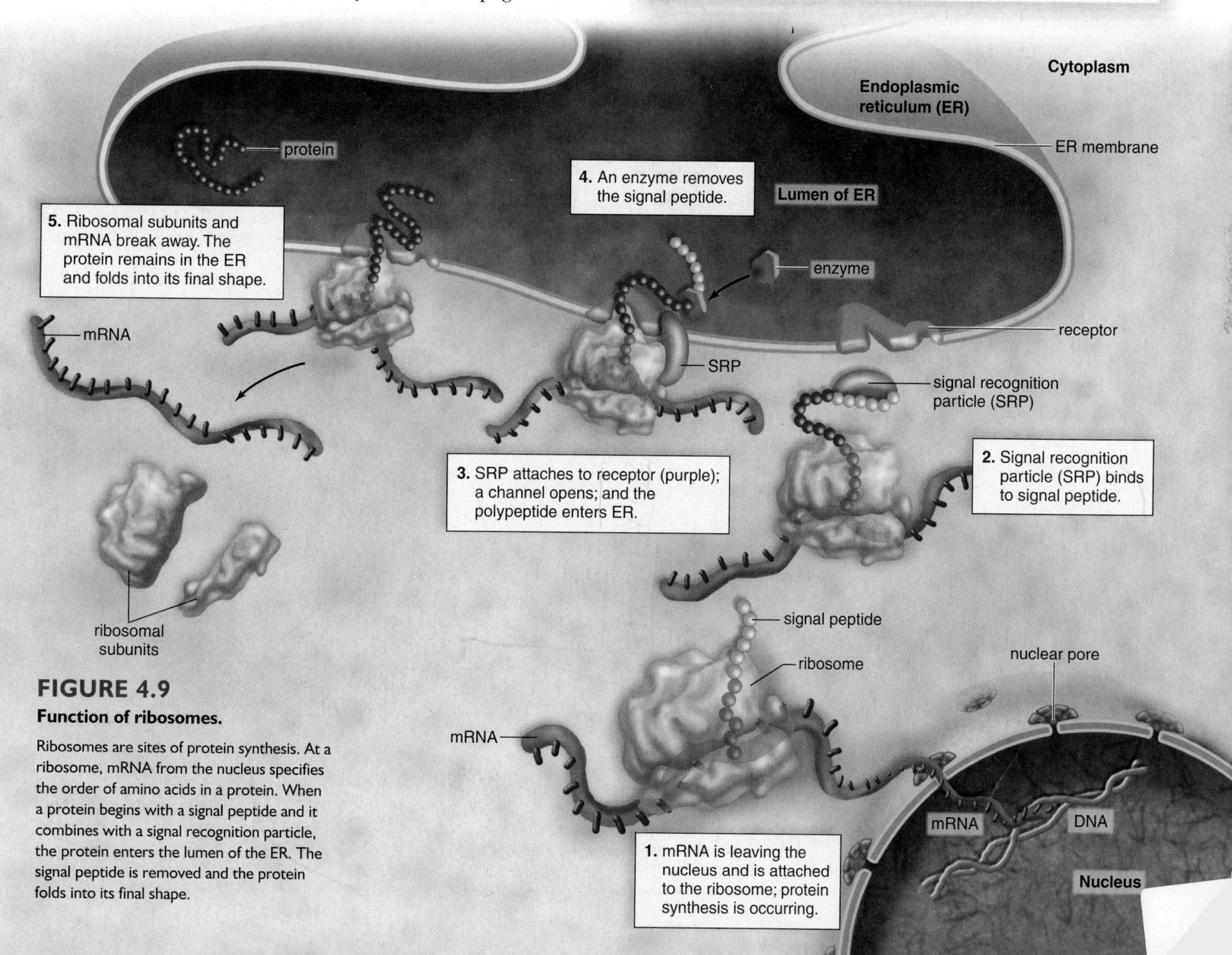


**FIGURE 4.9**
**Function of ribosomes.**

Ribosomes are sites of protein synthesis. At a ribosome, mRNA from the nucleus specifies the order of amino acids in a protein. When a protein begins with a signal peptide and it combines with a signal recognition particle, the protein enters the lumen of the ER. The signal peptide is removed and the protein folds into its final shape.

# 4.5 The Endomembrane System

The **endomembrane system** consists of the nuclear envelope, the membranes of the endoplasmic reticulum, the Golgi apparatus, and several types of vesicles. This system compartmentalizes the cell so that particular enzymatic reactions are restricted to specific regions. The vesicles transport molecules from one part of the system to another.

## Endoplasmic Reticulum

The **endoplasmic reticulum (ER)** [Gk. *endon*, within; *plasma*, something molded; L. *reticulum*, net], consisting of a complicated system of membranous channels and saccules (flattened vesicles), is physically continuous with the nuclear envelope (Fig. 4.10). The ER consists of rough ER and smooth ER, which have a different structure and functions. Only **rough ER** is studded with ribosomes on the side of the membrane that faces the cytoplasm, and because of this, rough ER has the capacity to produce proteins. Inside its lumen, rough ER contains enzymes that can add carbohydrate (sugar) chains to proteins, and then these proteins are called glycoproteins. While in the ER, proteins fold and take on their final three-dimensional shape.

**Smooth ER,** which is continuous with rough ER, does not have attached ribosomes. Certain organs contain an abundance of smooth ER and its function depends on the organ. In some organs, smooth ER is associated with the production of lipids. For example, in the testes, smooth ER produces testosterone, a steroid hormone. In the liver, smooth ER helps detoxify drugs. The smooth ER of the liver increases in quantity when a person consumes alcohol or takes barbiturates on a regular basis. Regardless of a difference in their functions, both rough and smooth ER form vesicles that transport molecules to other parts of the cell, notably the Golgi apparatus.

## The Golgi Apparatus

The **Golgi apparatus** is named for Camillo Golgi, who discovered its presence in cells in 1898. The Golgi apparatus typically consists of a stack of three to twenty slightly curved, flattened saccules whose appearance can be compared to a stack of pancakes (Fig. 4.11). In animal cells, one side of the stack (the cis or inner face) is directed toward the ER, and the other side of the stack (the trans or outer face) is directed toward the plasma membrane. Vesicles can frequently be seen at the edges of the saccules.

Protein-filled vesicles that bud from the rough ER and lipid-filled vesicles that bud from the smooth ER are received by the Golgi apparatus at its inner face. Thereafter, the apparatus alters these substances as they move through its saccules. For example, the Golgi apparatus contains enzymes that modify the carbohydrate chains first attached to proteins in the rough ER. It can change one sugar for another sugar. In some cases, the modified carbohydrate chain serves as a signal molecule that determines the protein's final destination in the cell.

The Golgi apparatus sorts the modified molecules and packages them into vesicles that depart from the outer face. In animal cells, some of these vesicles are lysosomes, which are discussed next. Other vesicles may return to the ER or proceed to the plasma membrane, where they become part of the membrane as they discharge their contents during **secretion.** Secretion is termed exocytosis because the substance exits the cytoplasm.

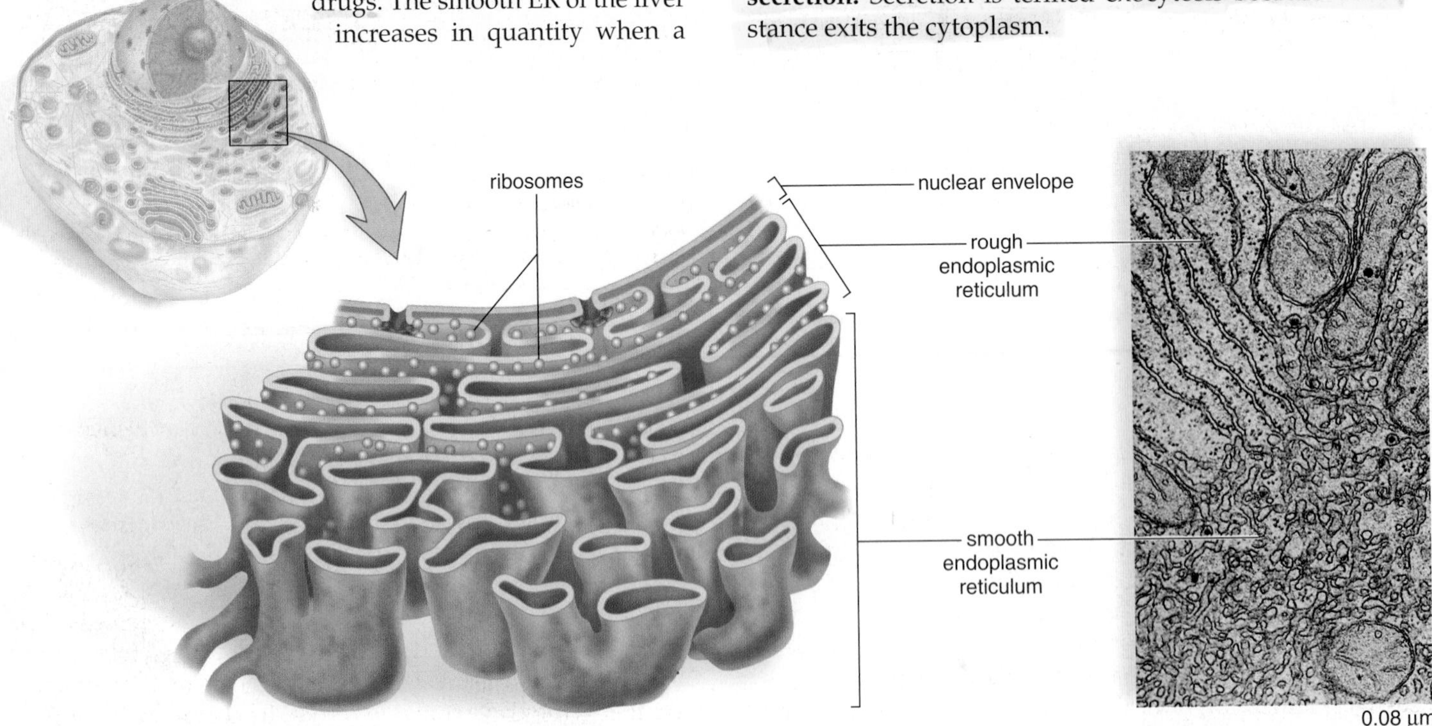


**FIGURE 4.10 Endoplasmic reticulum (ER).**

Ribosomes are present on rough ER, which consists of flattened saccules, but not on smooth ER, which is more tubular. Proteins are synthesized by rough ER, which can also attach carbohydrate chains to proteins after they enter its lumen, as described in Figure 4.9. Smooth ER is involved in lipid synthesis, detoxification reactions, and several other possible functions.

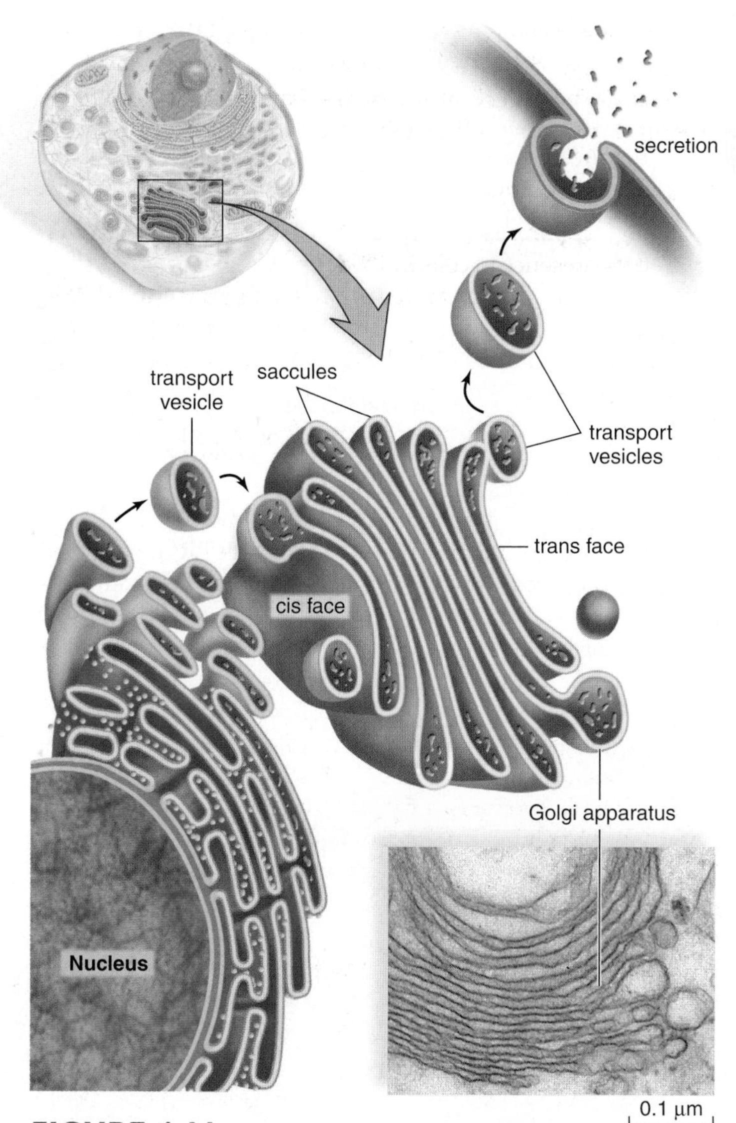


**FIGURE 4.11 Golgi apparatus.**

The Golgi apparatus is a stack of flattened, curved saccules. It processes proteins and lipids and packages them in transport vesicles that distribute these molecules to various locations.

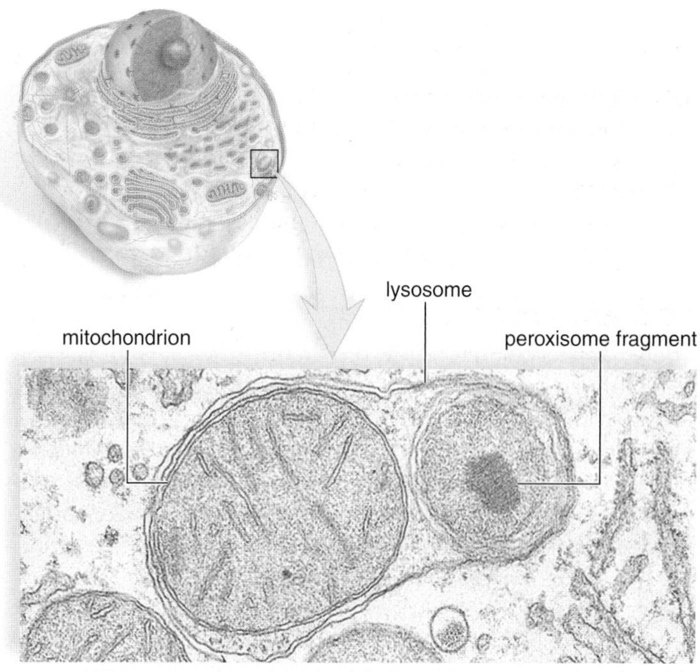


a. Mitochondrion and a peroxisome in a lysosome

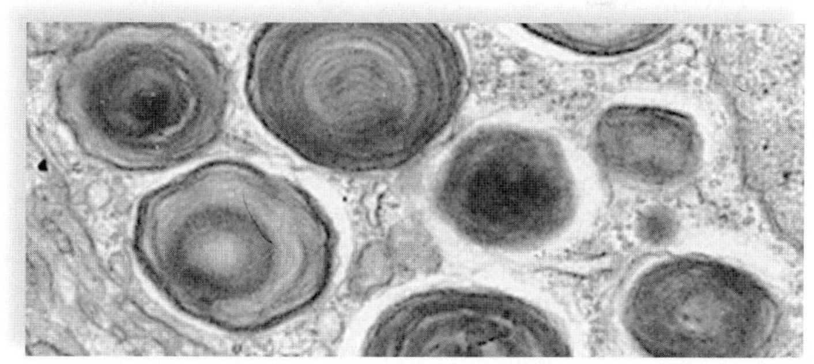

b. Storage bodies in a cell with defective lysosomes

**FIGURE 4.12 Lysosomes.**

**a.** Lysosomes, which bud off the Golgi apparatus in cells, are filled with hydrolytic enzymes that digest molecules and parts of the cell. Here a lysosome digests a worn mitochondrion and a peroxisome. **b.** The nerve cells of a person with Tay-Sachs disease are filled with membranous cytoplasmic bodies storing a fat that lysosomes are unable to digest.

## Lysosomes

**Lysosomes** [Gk. *lyo,* loose, and *soma,* body] are membrane-bounded vesicles produced by the Golgi apparatus. They have a very low pH and store powerful hydrolytic-digestive enzymes in an inactive state. Lysosomes assist in digesting material taken into the cell, and they destroy nonfunctional organelles and portions of cytoplasm (Fig. 4.12).

Materials can be brought into a cell by vesicle or vacuole formation at the plasma membrane. When a lysosome fuses with either, the lysosomal enzymes are activated and digest the material into simpler subunits that then enter the cytoplasm. Some white blood cells defend the body by engulfing bacteria that are then enclosed within vacuoles. When lysosomes fuse with these vacuoles, the bacteria are digested. Sometimes a small amount of residue is left and then it is ejected from the cell at the plasma membrane

A number of human lysosomal storage diseases are due to a missing lysosomal enzyme. In Tay-Sachs disease, the missing enzyme digests a fatty substance that helps insulate nerve cells and increases their efficiency. The fatty substance accumulates in so many storage bodies that nerve cells die off. Affected individuals appear normal at birth but begin to develop neurological problems at four to six months of age. Eventually, the child suffers cerebral degeneration, slow paralysis, blindness, and loss of motor function. Children with Tay-Sachs disease live only about three to four years. In the future, it may be possible to provide the missing enzyme and, in that way, prevent lysosomal storage diseases.

## Endomembrane System Summary

We have seen that the endomembrane system is a series of membranous organelles that work together and communicate by means of transport vesicles. The endoplasmic reticulum (ER) and the Golgi apparatus are essentially flattened saccules, and lysosomes are specialized vesicles.

Figure 4.13 shows how the components of the endomembrane system work together. Proteins produced in rough ER and lipids produced in smooth ER are carried in transport vesicles to the Golgi apparatus, where they are further modified before being packaged in vesicles that leave the Golgi. Using signaling sequences, the Golgi apparatus sorts proteins and packages them into vesicles that transport them to various cellular destinations. Secretory vesicles take the proteins to the plasma membrane, where they exit the cell when the vesicles fuse with the membrane. This is called secretion by exocytosis. For example, secretion into ducts occurs when the mammary glands produce milk or the pancreas produces digestive enzymes.

In animal cells, the Golgi apparatus also produces specialized vesicles called lysosomes that contain stored hydrolytic enzymes. Lysosomes fuse with incoming vesicles from the plasma membrane and digest macromolecules and/or even debris brought into a certain cell. White blood cells are well known for engulfing pathogens (e.g., disease-causing viruses and bacteria) that are then broken down in lysosomes.

### Check Your Progress 4.5

1. Contrast the structure and functions of rough endoplasmic reticulum with those of smooth endoplasmic reticulum.
2. Describe the relationship between the components of the endomembrane system.

**FIGURE 4.13 Endomembrane system.**

The organelles in the endomembrane system work together to carry out the functions noted. Plant cells do not have lysosomes, nor do they have incoming and outgoing (secretory) vesicles.

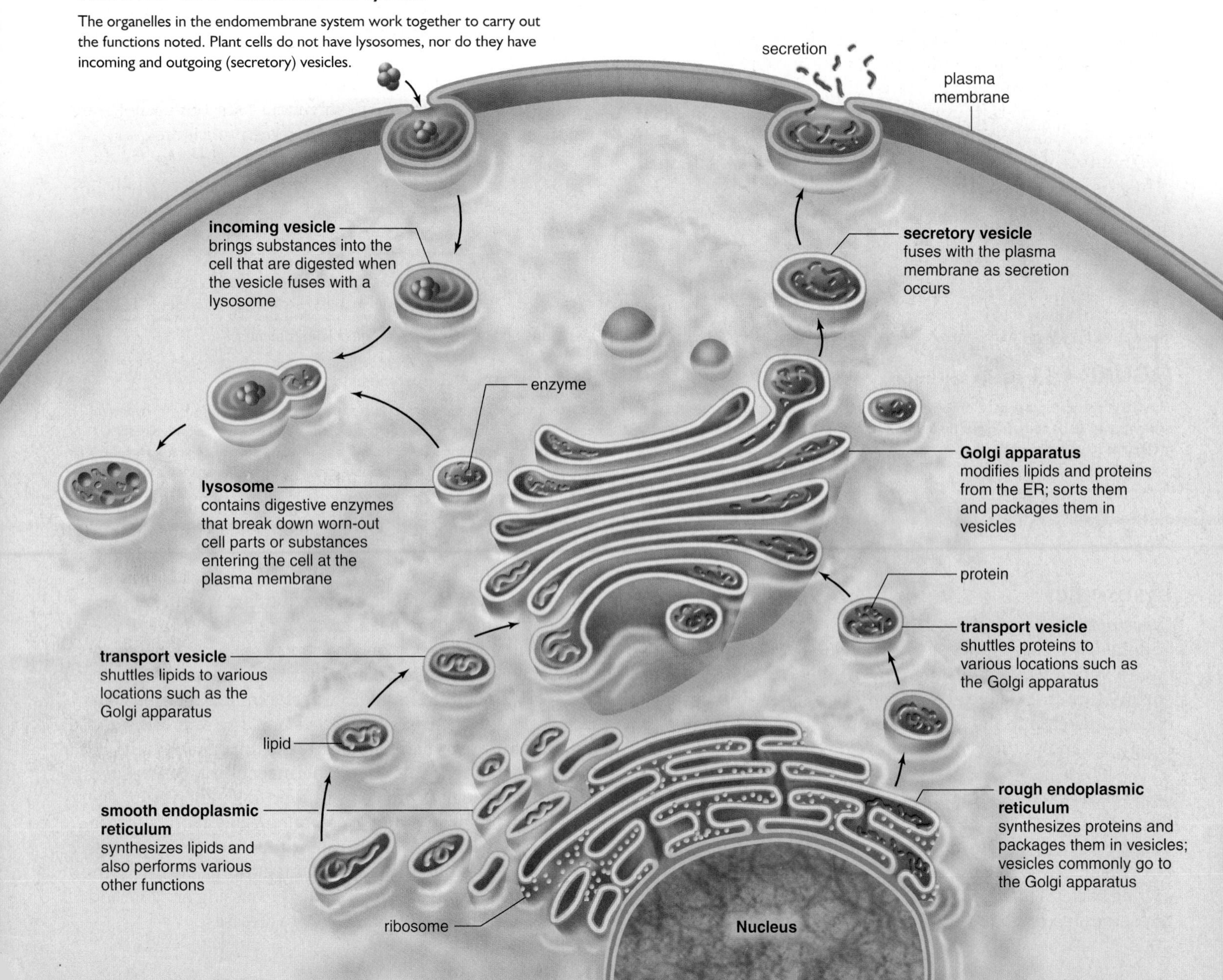

# 4.6 Other Vesicles and Vacuoles

Peroxisomes and the vacuoles of cells do not communicate with the organelles of the endomembrane system, and therefore are not part of it.

## Peroxisomes

**Peroxisomes,** similar to lysosomes, are membrane-bounded vesicles that enclose enzymes. However, the enzymes in peroxisomes are synthesized by free ribosomes and transported into a peroxisome from the cytoplasm. All peroxisomes contain enzymes whose actions result in hydrogen peroxide ($H_2O_2$):

$$RH_2 + O_2 \longrightarrow R + H_2O_2$$

Hydrogen peroxide, a toxic molecule, is immediately broken down to water and oxygen by another peroxisomal enzyme called catalase. When hydrogen peroxide is applied to a wound, bubbling occurs as catalase breaks it down.

Peroxisomes are metabolic assistants to the other organelles. They have varied functions but are especially prevalent in cells that are synthesizing and breaking down lipids. In the liver, some peroxisomes produce bile salts from cholesterol, and others break down fats. In a 1992 movie, *Lorenzo's Oil,* the peroxisomes in a boy's cells lack a membrane protein needed to import a specific enzyme and/or long chain fatty acids from the cytoplasm. As a result, long chain fatty acids accumulate in his brain, and he suffers neurological damage. This disorder is known as adrenoleukodystrophy.

Plant cells also have peroxisomes (Fig. 4.14). In germinating seeds, they oxidize fatty acids into molecules that can be converted to sugars needed by the growing plant. In leaves, peroxisomes can carry out a reaction that is opposite to photosynthesis—the reaction uses up oxygen and releases carbon dioxide.

**FIGURE 4.14 Peroxisomes.**

Peroxisomes contain one or more enzymes that can oxidize various organic substances. Peroxisomes also contain the enzyme catalase, which breaks down hydrogen peroxide ($H_2O_2$), which builds up after organic substances are oxidized.

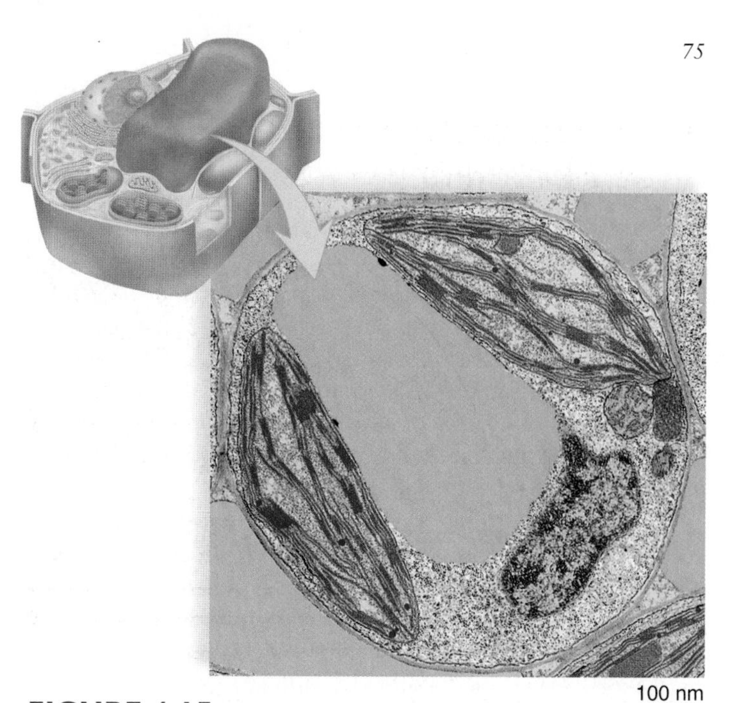


**FIGURE 4.15 Plant cell central vacuole.**

The large central vacuole of plant cells has numerous functions, from storing molecules to helping the cell increase in size.

## Vacuoles

Like vesicles, **vacuoles** are membranous sacs, but vacuoles are larger than vesicles. The vacuoles of some protists are quite specialized including contractile vacuoles for ridding the cell of excess water and digestive vacuoles for breaking down nutrients. Vacuoles usually store substances. Few animal cells contain vacuoles, but fat cells contain a very large lipid-engorged vacuole that takes up nearly two-thirds of the volume of the cell!

Plant vacuoles contain not only water, sugars, and salts but also water-soluble pigments and toxic molecules. The pigments are responsible for many of the red, blue, or purple colors of flowers and some leaves. The toxic substances help protect a land plant from herbivorous animals.

### *Plant Cell Central Vacuole*

Typically, plant cells have a large **central vacuole** that may take up to 90% of the volume of the cell. The vacuole is filled with a watery fluid called cell sap that gives added support to the cell (Fig. 4.15). The central vacuole maintains hydrostatic pressure or turgor pressure in plant cells. A plant cell can rapidly increase in size by enlarging its vacuole. Eventually, a plant cell also produces more cytoplasm.

The central vacuole functions in storage of both nutrients and waste products. Metabolic waste products are pumped across the vacuole membrane and stored permanently in the central vacuole. As organelles age and become nonfunctional, they fuse with the vacuole, where digestive enzymes break them down. This is a function carried out by lysosomes in animal cells.

**Check Your Progress 4.6**

1. How is a peroxisome like, and how is it different from, a lysosome?
2. How is the plant cell central vacuole like, and how is it different from, a lysosome?

# 4.7 The Energy-Related Organelles

Life is possible only because a constant input of energy maintains the structure of cells. Chloroplasts and mitochondria are the two eukaryotic membranous organelles that specialize in converting energy to a form that can be used by the cell.

During photosynthesis, **chloroplasts** [Gk. *chloros*, green, and *plastos*, formed, molded], use solar energy to synthesize carbohydrates, which serve as organic nutrient molecules for plants and all living things on Earth. *Photosynthesis* can be represented by this equation:

solar energy + carbon dioxide + water → carbohydrate + oxygen

Plants, algae, and cyanobacteria are capable of carrying on photosynthesis in this manner, but only plants and algae have chloroplasts because they are eukaryotes.

Cellular respiration is the process by which carbohydrate-derived products are broken down in **mitochondria** (sing., mitochondrion) to produce ATP (adenosine triphosphate). *Cellular respiration* can be represented by this equation:

carbohydrate + oxygen → carbon dioxide + water + energy

Here the word *energy* stands for ATP molecules. When a cell needs energy, ATP supplies it. The energy of ATP is used for synthetic reactions, active transport, and all energy-requiring processes in cells.

## Chloroplasts

Some algal cells have only one chloroplast, while some plant cells have as many as a hundred. Chloroplasts can be quite large, being twice as wide and as much as five times the length of a mitochondrion. Chloroplasts have a three-membrane system (Fig. 4.16). They are bounded by a double membrane, which includes an outer membrane and an inner membrane. The double membrane encloses the semifluid **stroma,** which contains enzymes and **thylakoids,** disklike sacs formed from a third chloroplast membrane. A stack of thylakoids is a **granum.** The lumens of the thylakoids are believed to form a large internal compartment called the thylakoid space. Chlorophyll and the other pigments that capture solar energy are located in the thylakoid membrane, and the enzymes that synthesize carbohydrates are located outside the thylakoid in the fluid of the stroma.

The endosymbiotic theory says that chloroplasts are derived from a photosynthetic bacterium that was engulfed by a eukaryotic cell. This certainly explains why a chloroplast is bounded by a double membrane—one membrane is derived from the vesicle that brought the prokaryote into the cell, while the inner membrane is derived from the prokaryote. The endosymbiotic theory is also supported by the finding that chloroplasts have their own prokaryotic-type chromosome and ribosomes, and they produce some of their own enzymes even today!

### Other Types of Plastids

A chloroplast is a type of plastid. **Plastids** are plant organelles that are surrounded by a double membrane and having varied functions. **Chromoplasts** contain pigments that result in a yellow, orange, or red color. Chromoplasts are responsible for the color of autumn leaves, fruits, carrots, and some flowers. **Leucoplasts** are generally colorless plastids that synthesize and store starches and oils. A microscopic examination of potato tissue yields a number of leucoplasts.

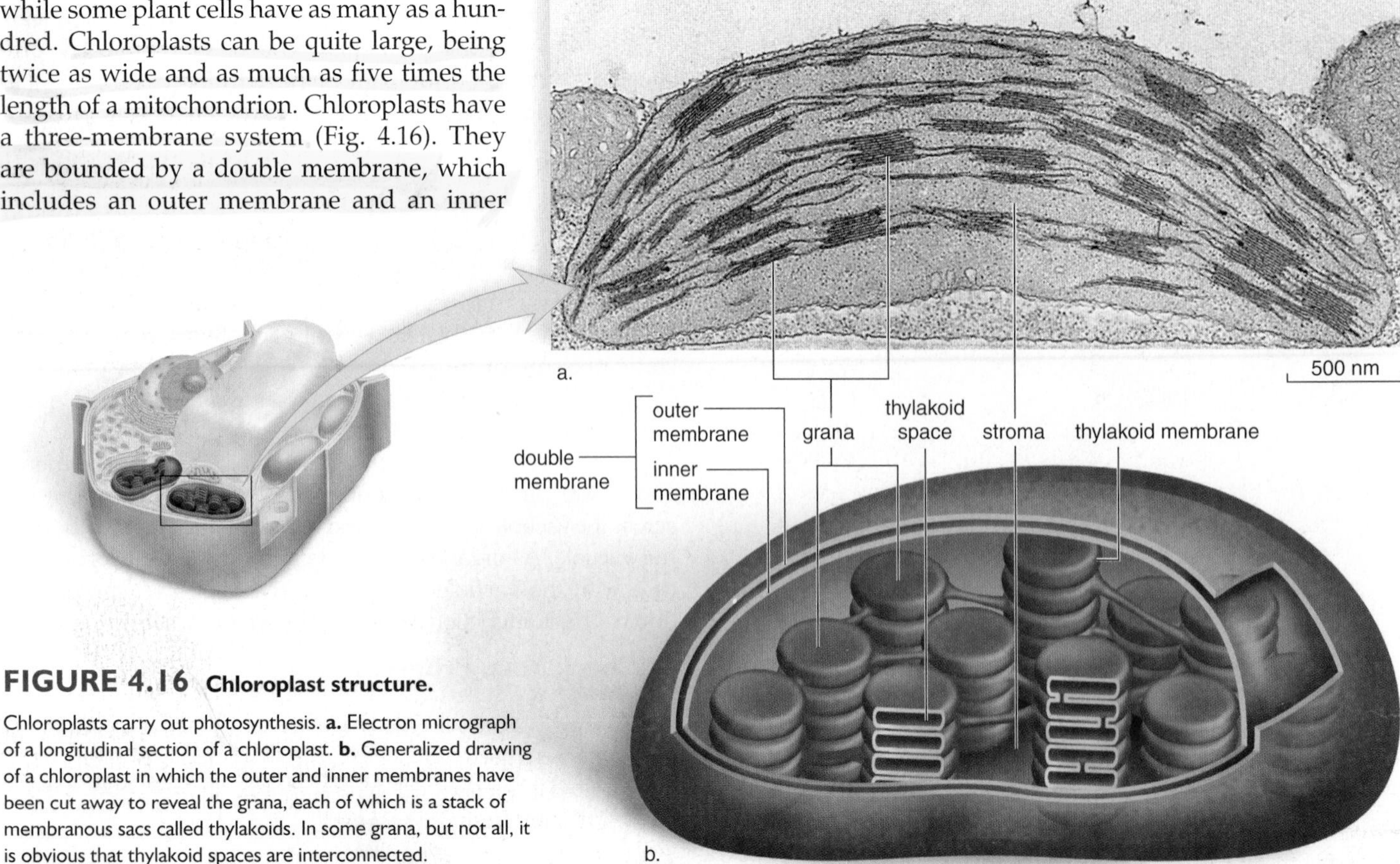


**FIGURE 4.16 Chloroplast structure.**

Chloroplasts carry out photosynthesis. **a.** Electron micrograph of a longitudinal section of a chloroplast. **b.** Generalized drawing of a chloroplast in which the outer and inner membranes have been cut away to reveal the grana, each of which is a stack of membranous sacs called thylakoids. In some grana, but not all, it is obvious that thylakoid spaces are interconnected.

## Mitochondria

Nearly all eukaryotic cells, and certainly all plant and algal cells in addition to animal cells, contain mitochondria. Even though mitochondria are smaller than chloroplasts, they can usually be seen when using a light microscope. The number of mitochondria can vary in cells depending on their activities. Some cells, such as liver cells, may have as many as 1,000 mitochondria. We think of mitochondria as having a shape like that shown in Figure 4.17, but actually they often change shape to be longer and thinner or shorter and broader. Mitochondria can form long, moving chains, or they can remain fixed in one location—typically where energy is most needed. For example, they are packed between the contractile elements of cardiac cells and wrapped around the interior of a sperm's flagellum. Fat cells contain few mitochondria—they function in fat storage, which does not require energy.

Mitochondria have two membranes, the outer membrane and the inner membrane. The inner membrane is highly convoluted into **cristae** that project into the matrix. These cristae increase the surface area of the inner membrane so much that in a liver cell they account for about one-third the total membrane in the cell. The inner membrane encloses a semifluid **matrix,** which contains mitochondrial DNA and ribosomes. Again, the presence of a double membrane and mitochondrial genes is consistent with the endosymbiotic theory regarding the origin of mitochondria, which was illustrated in Figure 4.5.

Mitochondria are often called the powerhouses of the cell because they produce most of the ATP utilized by the cell. The procedure described in the Science Focus on page 67 allowed investigators to separate the inner membrane, the outer membrane, and the matrix from each other. Then they discovered that the matrix is a highly concentrated mixture of enzymes that break down carbohydrates and other nutrient molecules. These reactions supply the chemical energy that permits a chain of proteins on the inner membrane to create the conditions that allow ATP synthesis to take place. The entire process, which also involves the cytoplasm, is called cellular respiration because oxygen is used and carbon dioxide is given off, as shown on the previous page.

### *Mitochondrial Diseases*

So far, more than 40 different mitochondrial diseases that affect the brain, muscles, kidneys, heart, liver, eyes, ears, or pancreas have been identified. The common factor among these genetic diseases is that the patient's mitochondria are unable to completely metabolize organic molecules to produce ATP. As a result, toxins accumulate inside the mitochondria and the body. The toxins can be free radicals (substances that readily form harmful compounds when they react with other molecules), and these compounds damage mitochondria over time. In the United States, between 1,000 and 4,000 children per year are born with a mitochondrial disease. In addition, it is possible that many diseases of aging are due to malfunctioning mitochondria.

**Check Your Progress 4.7**

1. Compare and contrast the structure and function of chloroplasts with those of mitochondria.

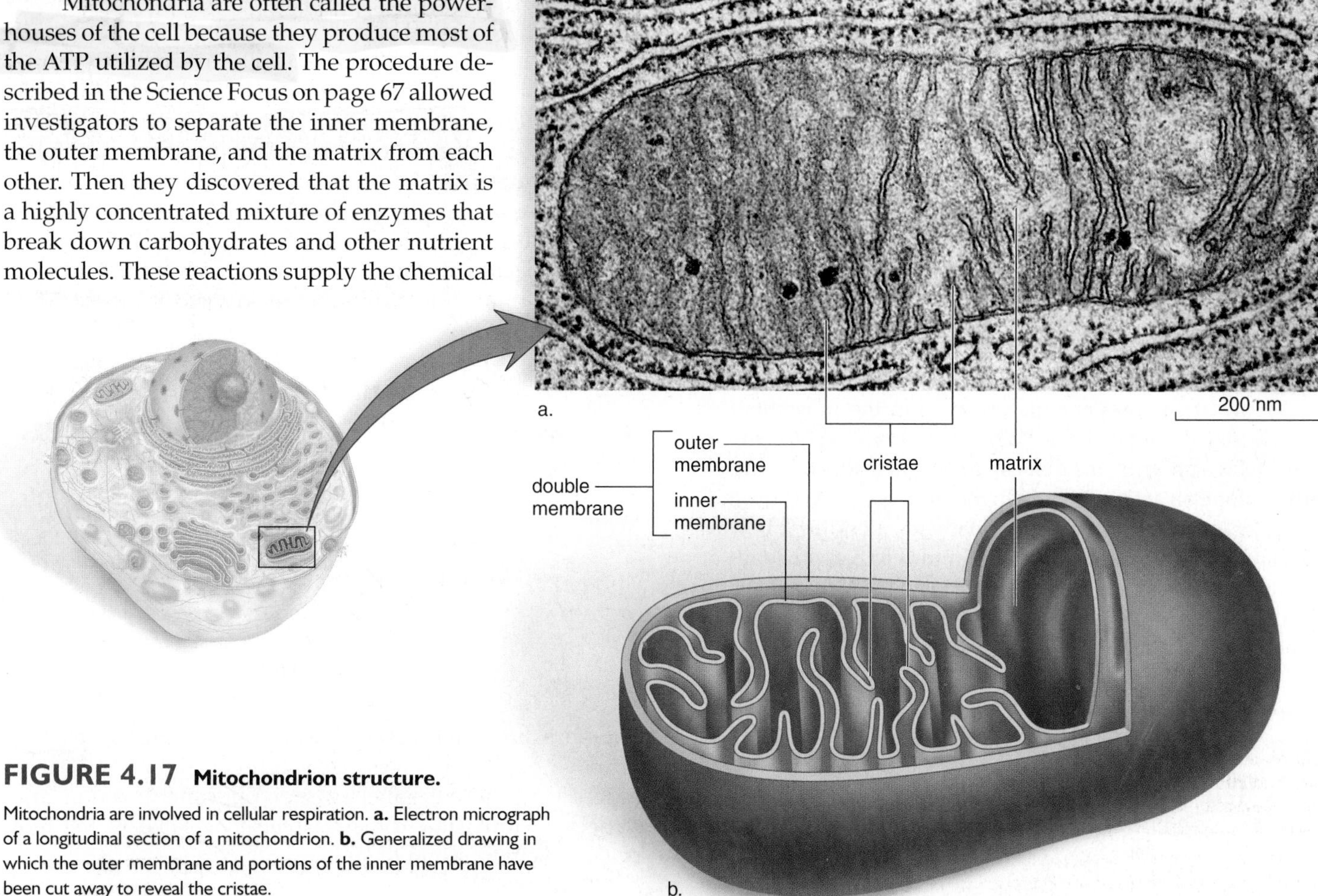


**FIGURE 4.17 Mitochondrion structure.**

Mitochondria are involved in cellular respiration. **a.** Electron micrograph of a longitudinal section of a mitochondrion. **b.** Generalized drawing in which the outer membrane and portions of the inner membrane have been cut away to reveal the cristae.

# 4.8 The Cytoskeleton

The protein components of the cytoskeleton [Gk. *kytos*, cell, and *skeleton*, dried body] interconnect and extend from the nucleus to the plasma membrane in eukaryotic cells. Prior to the 1970s, it was believed that the cytoplasm was an unorganized mixture of organic molecules. Then, high-voltage electron microscopes, which can penetrate thicker specimens, showed instead that the cytoplasm was highly organized. The technique of immunofluorescence microscopy identified the makeup of the protein components within the cytoskeletal network (Fig. 4.18).

The cytoskeleton contains actin filaments, intermediate filaments, and microtubules, which maintain cell shape and allow the cell and its organelles to move. Therefore, the cytoskeleton is often compared to the bones and muscles of an animal. However, the cytoskeleton is dynamic, especially because its protein components can assemble and disassemble as appropriate. Apparently a number of different mechanisms regulate this process, including protein kinases that phosphorylate proteins. Phosphorylation leads to disassembly, and dephosphorylation causes assembly.

## Actin Filaments

**Actin filaments** (formerly called microfilaments) are long, extremely thin, flexible fibers (about 7 nm in diameter) that occur in bundles or meshlike networks. Each actin filament contains two chains of globular actin monomers twisted about one another in a helical manner.

Actin filaments play a structural role when they form a dense, complex web just under the plasma membrane, to which they are anchored by special proteins. They are also seen in the microvilli that project from intestinal cells, and their presence most likely accounts for the ability of microvilli to alternately shorten and extend into the intestine. In plant cells, actin filaments apparently form the tracks along which chloroplasts circulate in a particular direction; doing so is called cytoplasmic streaming. Also, the presence of a network of actin filaments lying beneath the plasma membrane accounts for the formation of **pseudopods** [L. *pseudo*, false, and *pod*, feet], extensions that allow certain cells to move in an amoeboid fashion.

How are actin filaments involved in the movement of the cell and its organelles? They interact with **motor molecules,** which are proteins that can attach, detach, and reattach farther along an actin filament. In the presence of ATP, the motor molecule myosin pulls actin filaments along in this way. Myosin has both a head and a tail. In muscle cells, the tails of several muscle myosin molecules are joined to form a thick filament. In nonmuscle cells, cytoplasmic myosin tails are bound to membranes, but the heads still interact with actin:

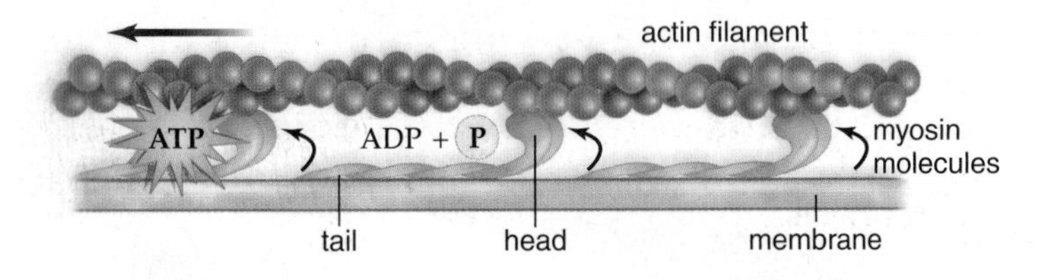


During animal cell division, the two new cells form when actin, in conjunction with myosin, pinches off the cells from one another.

## Intermediate Filaments

**Intermediate filaments** (8–11 nm in diameter) are intermediate in size between actin filaments and microtubules. They are a ropelike assembly of fibrous polypeptides, but the specific filament type varies according to the tissue. Some intermediate filaments support the nuclear envelope, whereas others support the plasma membrane and take part in the formation of cell-to-cell junctions. In the skin, intermediate filaments, made of the protein keratin, give great mechanical strength to skin cells. We now know that intermediate filaments are also highly dynamic and will disassemble when phosphate is added to them by a kinase.

## Microtubules

**Microtubules** [Gk. *mikros*, small, little; L. *tubus*, tube] are small, hollow cylinders about 25 nm in diameter and from 0.2–25 μm in length.

Microtubules are made of a globular protein called tubulin, which is of two types called $\alpha$ and $\beta$. There is a slightly different amino acid sequence in $\alpha$ tubulin compared to $\beta$ tubulin. When assembly occurs, $\alpha$ and $\beta$ tubulin molecules come together as dimers, and the dimers arrange themselves in rows. Microtubules have 13 rows of tubulin dimers, surrounding what appears in electron micrographs to be an empty central core.

The regulation of microtubule assembly is under the control of a microtubule organizing center (MTOC). In most eukaryotic cells, the main MTOC is in the **centrosome** [Gk. *centrum*, center, and *soma*, body], which lies near the nucleus. Microtubules radiate from the centrosome, helping to maintain the shape of the cell and acting as tracks along which organelles can move. Whereas the motor molecule myosin is associated with actin filaments, the motor molecules kinesin and dynein are associated with microtubules:

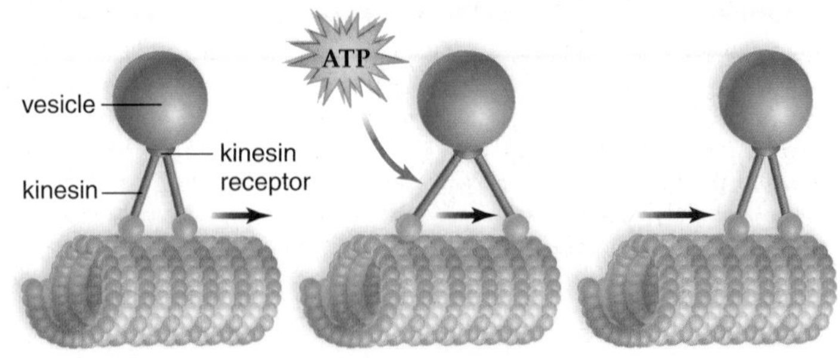


There are different types of kinesin proteins, each specialized to move one kind of vesicle or cellular organelle. Kinesin moves vesicles or organelles in an opposite direction from dynein. Cytoplasmic dynein is closely related to the molecule dynein found in flagella.

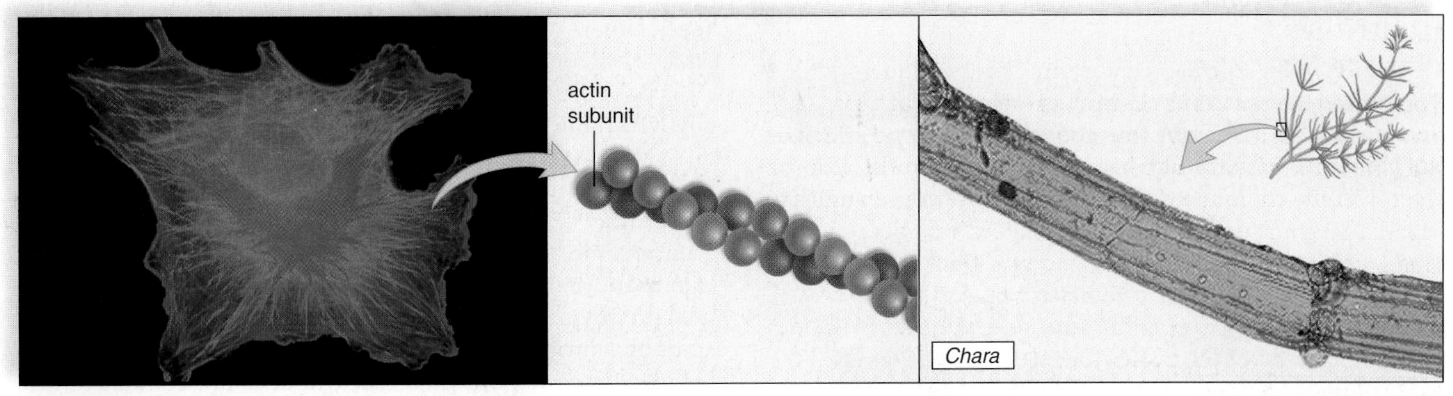


a. Actin filaments

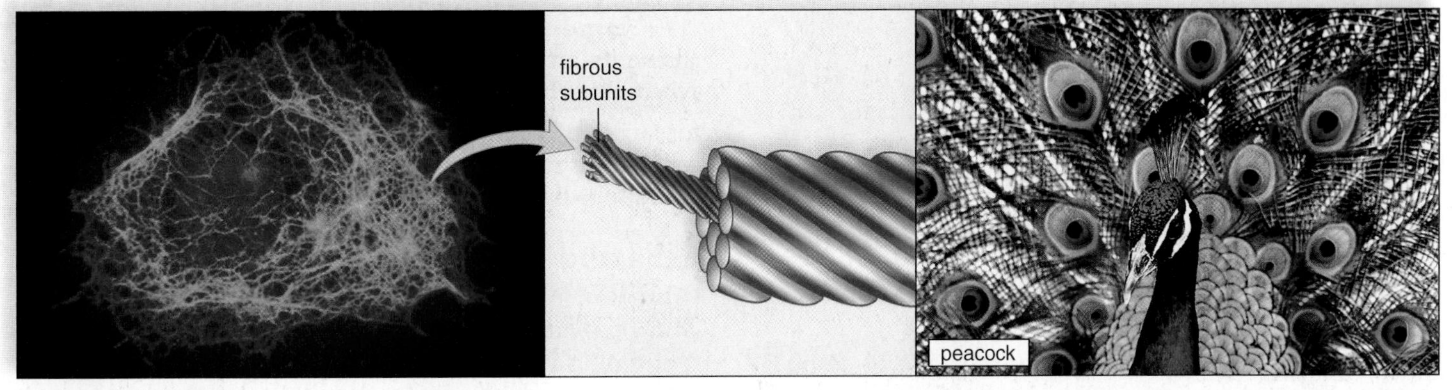


b. Intermediate filaments

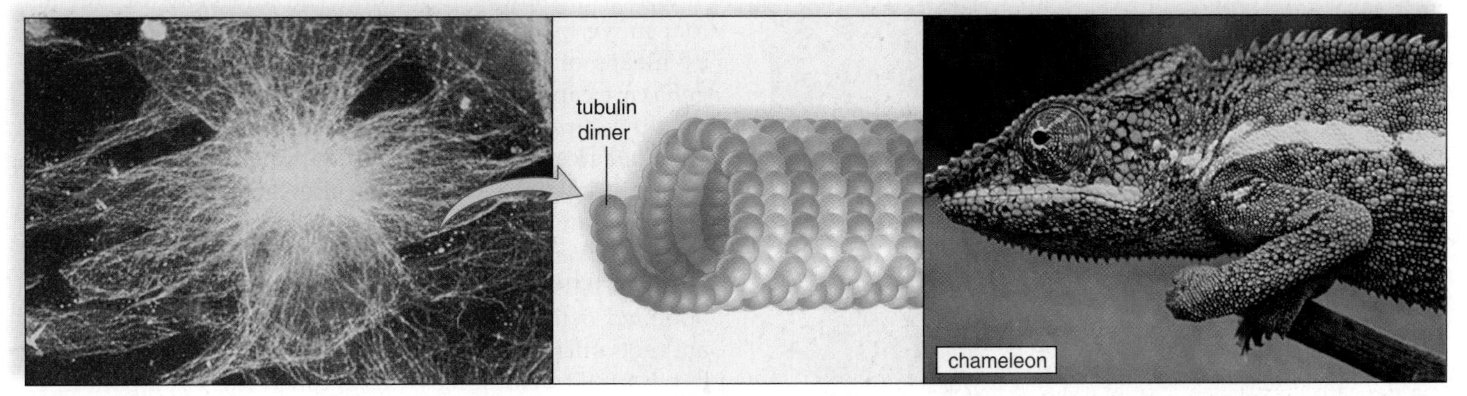


c. Microtubules

**FIGURE 4.18 The cytoskeleton.**

The cytoskeleton maintains the shape of the cell and allows its parts to move. Three types of protein components make up the cytoskeleton. They can be detected in cells by using a special fluorescent technique that reveals only one type of component at a time. **a.** (*left* to *right*) Animal cells are treated so that actin filaments can be microscopically detected; the drawing shows that actin filaments are composed of a twisted double chain of actin subunits. The giant cells of the green alga *Chara* rely on actin filaments to move organelles from one end of the cell to another. **b.** (*left* to *right*) Animal cells are treated so that intermediate filaments can be microscopically detected; the drawing shows that fibrous proteins account for the ropelike structure of intermediate filaments. A peacock's colorful feathers are strengthened by the presence of intermediate filaments. **c.** (*left* to *right*) Animal cells are treated so that microtubules can be microscopically detected; the drawing shows that microtubules are hollow tubes composed of tubulin dimers. The skin cells of a chameleon rely on microtubules to move pigment granules around so that they take on the color of their environment.

Before a cell divides, microtubules disassemble and then reassemble into a structure called a spindle that distributes chromosomes in an orderly manner. At the end of cell division, the spindle disassembles, and microtubules reassemble once again into their former array. In the arms race between plants and herbivores, plants have evolved various types of poisons that prevent them from being eaten. Colchicine is a plant poison that binds tubulin and blocks the assembly of microtubules.

## Centrioles

**Centrioles** [Gk. *centrum,* center] are short cylinders with a 9 + 0 pattern of microtubule triplets—nine sets of triplets are arranged in an outer ring, but the center of a centriole does not contain a microtubule. In animal cells and most protists, a centrosome contains two centrioles lying at right angles to each other. A centrosome, as mentioned previously, is the major microtubule-organizing center for the cell. Therefore, it is possible that centrioles are also involved in the process by which microtubules assemble and disassemble.

Before an animal cell divides, the centrioles replicate, and the members of each pair are at right angles to one another (Fig. 4.19). Then each pair becomes part of a separate centrosome. During cell division, the centrosomes move apart and most likely function to organize the mitotic spindle. In any case, each new cell has its own centrosome and pair of centrioles. Plant and fungal cells have the equivalent of a centrosome, but this structure does not contain centrioles, suggesting that centrioles are not necessary to the assembly of cytoplasmic microtubules.

A **basal body** is an organelle that lies at the base of cilia and flagella and may direct the organization of microtubules within these structures. In other words, a basal body may do for a cilium or flagellum what the centrosome does for the cell. In cells with cilia and flagella, centrioles are believed to give rise to basal bodies.

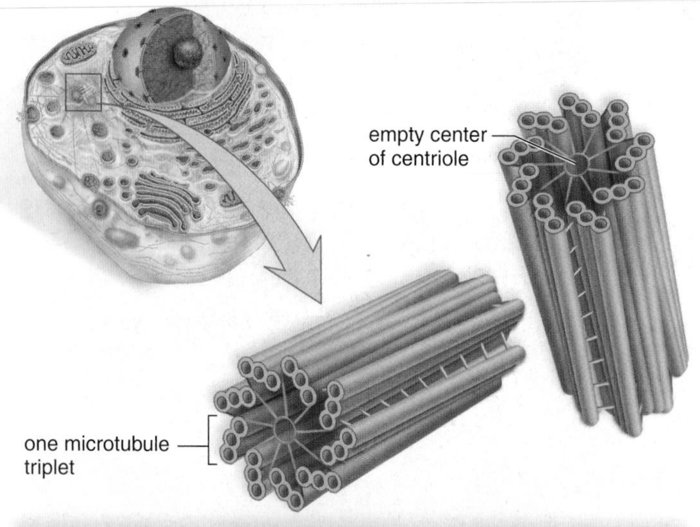


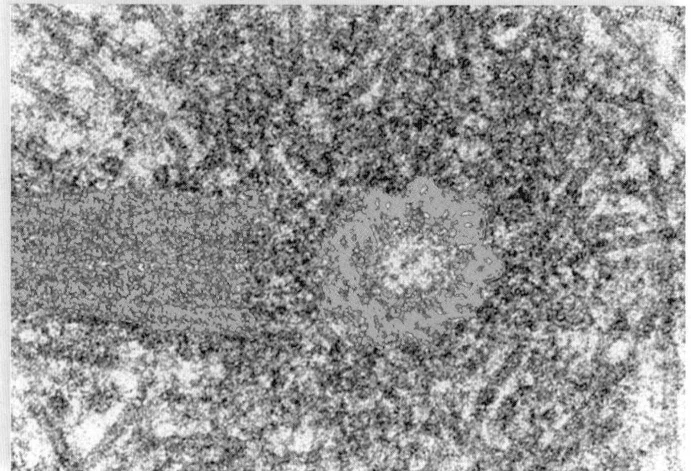


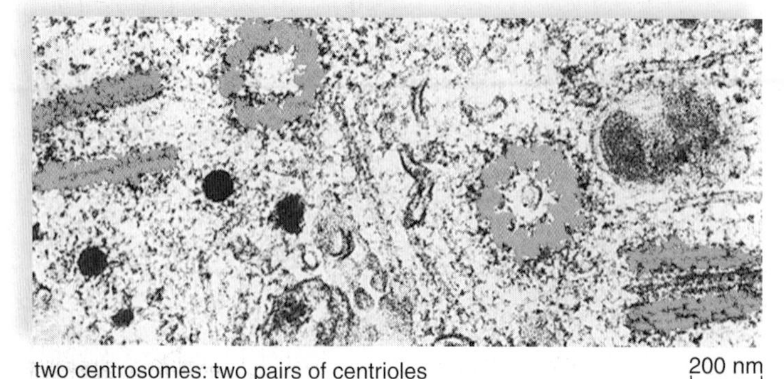


**FIGURE 4.19 Centrioles.**

In a nondividing animal cell, there is a single pair of centrioles in the centrosome located just outside the nucleus. Just before a cell divides, the centrioles replicate, producing two centrosomes. During cell division, the centrosomes separate so that each new cell has one centrosome containing one pair of centrioles.

## Cilia and Flagella

**Cilia** [L. *cilium,* eyelash, hair] and **flagella** [L. *flagello,* whip] are hairlike projections that can move either in an undulating fashion, like a whip, or stiffly, like an oar. If a cell is not attached, cilia (or flagella) move the cell through liquid. For example, unicellular paramecia are organisms that move by means of cilia, whereas sperm cells move by means of flagella. If the cell is attached, cilia (or flagella) are capable of moving liquid over the cell. The cells that line our upper respiratory tract are attached and have cilia that sweep debris trapped within mucus back up into the throat, where it can be swallowed. This action helps keep the lungs clean.

In eukaryotic cells, cilia are much shorter than flagella, but they have a similar construction. Both are membrane-bounded cylinders enclosing a matrix area. In the matrix are nine microtubule doublets arranged in a circle around two central microtubules; this is called the 9 + 2 pattern of microtubules (Fig. 4.20). Cilia and flagella move when the microtubule doublets slide past one another.

As mentioned, each cilium and flagellum has a basal body lying in the cytoplasm at its base. Basal bodies have the same circular arrangement of microtubule triplets as centrioles and are believed to be derived from them. It is possible that basal bodies organize the microtubules within cilia and flagella, but this is not supported by the observation that cilia and flagella grow by the addition of tubulin dimers to their tips.

### Check Your Progress 4.8

1. List the components of the cytoskeleton.
2. Explain the structure of cilia and flagella.
3. Give an example of a cell that has cilia and one that has flagella. Describe the functions of these cells.

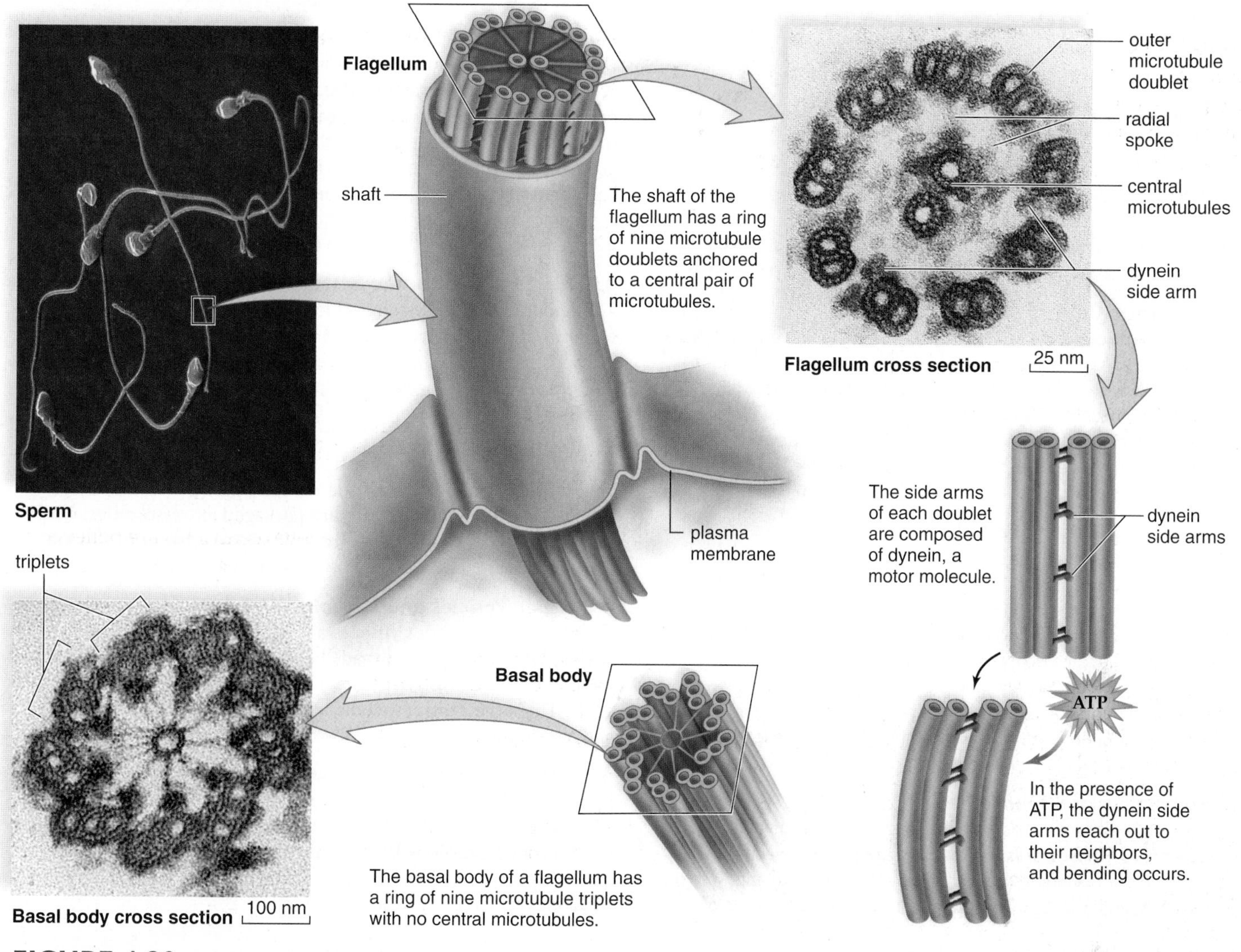


**FIGURE 4.20 Structure of a flagellum.**

(*below, left*) The basal body of a flagellum has a 9 + 0 pattern of microtubule triplets. Notice the ring of nine triplets, with no central microtubules. (*above, left*) In sperm, the shaft of the flagellum has a 9 + 2 pattern (a ring of nine microtubule doublets surrounds a central pair of microtubules). (*middle, right*) In place of the triplets seen in a basal body, a flagellum's outer doublets have side arms of dynein, a motor molecule. (*below, right*) In the presence of ATP, the dynein side arms reach out and attempt to move along their neighboring doublet. Because of the radial spokes connecting the doublets to the central microtubules, bending occurs.

## Connecting the Concepts

Our knowledge of cell anatomy has been gathered by studying micrographs of cells. This has allowed cytologists (biologists who study cells) to arrive at a picture of generalized cells, such as those depicted for the animal and plant cells in Figures 4.6 and 4.7. The Science Focus on page 67 describes the methodology for studying the function of organelles.

Eukaryotic cells, taken as a whole, contain several types of organelles, and the chapter concepts for the chapter suggest that you should know the structure and function of each one. A concept to keep in mind is that "structure suits function." For example, ribosome subunits move from the nucleus to the cytoplasm; therefore, it seems reasonable that the nuclear envelope has pores. Finding relationships between structure and function will give you a deeper understanding of the cell, which will boost your memory capabilities.

Not all eukaryotic cells contain every type of organelle depicted. Cells actually have many specializations of structure that are consistent with their particular functions. Because red blood cells lack a nucleus, more room is made available for molecules of hemoglobin, the molecule that transports oxygen in the blood. Muscle cells are quite large and contain many specialized contractile organelles not discussed in this chapter. They also contain many mitochondria that supply the ATP needed for muscle contraction. Therefore, it can be seen that eukaryotic cells are specialized according to the organelles they contain or do not contain. This leads to the specialization of tissues and organs found in complex multicellular organisms.

In Chapter 5, we continue our study of the generalized cell by considering some functions that are common to all cells. We will see that all cells exchange substances across the plasma membrane and maintain a saltwater balance within certain limits. This is an example of homeostasis, or the relative constancy of the internal environment. Another such example was mentioned in Chapter 2, when we considered that organisms contain buffers that help maintain the pH of body fluids within limits suitable to life.

# summary

## 4.1 Cellular Level of Organization

All organisms are composed of cells, the smallest units of living matter. Cells are capable of self-reproduction, and existing cells come only from preexisting cells. Cells are very small and are measured in micrometers. The plasma membrane regulates exchange of materials between the cell and the external environment. Cells must remain small in order to have an adequate amount of surface area to volume.

## 4.2 Prokaryotic Cells

There are two major groups of prokaryotic cells: the bacteria and the archaea. Prokaryotic cells lack the nucleus of eukaryotic cells. The cell envelope of bacteria includes a plasma membrane, a cell wall, and an outer glycocalyx. The cytoplasm contains ribosomes, inclusion bodies, and a nucleoid that is not bounded by a nuclear envelope. The cytoplasm of cyanobacteria also includes thylakoids. The appendages of a bacterium are the flagella, the fimbriae, and the conjugation pili.

## 4.3 Introducing Eukaryotic Cells

Eukaryotic cells are much larger than prokaryotic cells, but they are compartmentalized by the presence of organelles, each with a specific structure and function (Table 4.1). The nuclear envelope most likely evolved through invagination of the plasma membrane, but mitochondria and chloroplasts may have arisen when a eukaryotic cell took up bacteria and algae in separate events. Perhaps this accounts for why the mitochondria and chloroplasts function fairly independently. Other membranous organelles are in constant communication by way of transport vesicles.

## 4.4 The Nucleus and Ribosomes

The nucleus of eukaryotic cells is bounded by a nuclear envelope containing pores. These pores serve as passageways between the cytoplasm and the nucleoplasm. Within the nucleus, chromatin, which contains DNA, undergoes coiling into chromosomes at the time of cell division. The nucleolus is a special region of the chromatin where rRNA is produced and ribosomal subunits are formed.

Ribosomes are organelles that function in protein synthesis. When protein synthesis occurs, mRNA leaves the nucleus with a coded message from DNA that specifies the sequence of amino acids in that protein. After mRNA attaches to a ribosome, it binds to the ER if it has a signal peptide. (Specifically, the signal peptide attaches to a signal recognition particle (SRP) that, in turn, binds to an SRP receptor on the ER.) When completed, the protein remains in the lumen of the ER.

## 4.5 The Endomembrane System

The endomembrane system includes the ER (both rough and smooth), the Golgi apparatus, the lysosomes (in animal cells), and transport vesicles. Newly produced proteins are modified in the ER before they are packaged in transport vesicles, many of which go to the Golgi apparatus. The smooth ER has various metabolic functions, depending on the cell type, but it also forms vesicles that carry lipids to different locations, particularly to the Golgi apparatus. The Golgi apparatus modifies, sorts, and repackages proteins and also processes lipids. Some proteins are packaged into lysosomes, which carry out intracellular digestion, or into vesicles that fuse with the plasma membrane. Following fusion, secretion occurs.

## 4.6 Other Vesicles and Vacuoles

Cells contain numerous vesicles and vacuoles, some of which, such as lysosomes, have already been discussed. Peroxisomes are vesicles that are involved in the metabolism of long chain fatty acids. The large central vacuole in plant cells functions in storage and also in the breakdown of molecules and cell parts.

## 4.7 The Energy-Related Organelles

Cells require a constant input of energy to maintain their structure. Chloroplasts capture the energy of the sun and carry on photosynthesis, which produces carbohydrates. Carbohydrate-derived products are broken down in mitochondria as ATP is produced. This is an oxygen-requiring process called cellular respiration.

**TABLE 4.1**

**Comparison of Prokaryotic Cells and Eukaryotic Cells**

| | Prokaryotic Cells (1–20 μm in diameter) | Eukaryotic Cells (10–100 μm in diameter) Animal | Eukaryotic Cells (10–100 μm in diameter) Plant |
|---|---|---|---|
| Cell wall | Usually (peptidoglycan) | No | Yes (cellulose) |
| Plasma membrane | Yes | Yes | Yes |
| Nucleus | No | Yes | Yes |
| Nucleolus | No | Yes | Yes |
| Ribosomes | Yes (smaller) | Yes | Yes |
| Endoplasmic reticulum | No | Yes | Yes |
| Golgi apparatus | No | Yes | Yes |
| Lysosomes | No | Yes | No |
| Mitochondria | No | Yes | Yes |
| Chloroplasts | No | No | Yes |
| Peroxisomes | No | Usually | Usually |
| Cytoskeleton | No | Yes | Yes |
| Centrioles | No | Yes | No |
| 9 + 2 cilia or flagella | No | Often | No (in flowering plants)<br>Yes (sperm of bryophytes, ferns, and cycads) |

### 4.8 The Cytoskeleton

The cytoskeleton contains actin filaments, intermediate filaments, and microtubules. These maintain cell shape and allow it and the organelles to move. Actin filaments, the thinnest filaments, interact with the motor molecule myosin in muscle cells to bring about contraction; in other cells, they pinch off daughter cells and have other dynamic functions. Intermediate filaments support the nuclear envelope and the plasma membrane and probably participate in cell-to-cell junctions. Microtubules radiate out from the centrosome and are present in centrioles, cilia, and flagella. They serve as tracks along which vesicles and other organelles move, due to the action of specific motor molecules.

## understanding the terms

actin filament 78
bacillus 64
basal body 80
capsule 64
cell 60
cell envelope (of prokaryotes) 64
cell theory 60
cell wall 64
central vacuole (of plant cell) 75
centriole 80
centrosome 78
chloroplast 76
chromatin 70
chromoplast 76
chromosome 70
cilium 80
coccus 64
conjugation pili 65
contrast 63
cristae 77
cyanobacteria 65
cytoplasm 64
cytoskeleton 67
endomembrane system 72
endoplasmic reticulum (ER) 72
endosymbiotic theory 66
eukaryotic cell 64
fimbriae 65
flagellum (pl., flagella) 65, 80
gene 70
glycocalyx 64
Golgi apparatus 72
granum 76
inclusion body 65
intermediate filament 78
leucoplast 76
lysosome 73
magnification 62
matrix 77
mesosome 64
microtubule 78
mitochondrion 76
motor molecule 78
nuclear envelope 70
nuclear pore 70
nucleoid 64
nucleolus 70
nucleoplasm 70
organelle 66
peroxisome 75
plasma membrane 64
plasmid 64
plastid 76
polyribosome 71
prokaryotic cell 64
pseudopod 78
resolution 63
ribosome 64, 71
rough ER 72
secretion 72
signal peptide 71
smooth ER 72
spirillum 64
spirochete 64
stroma 76
surface-area-to-volume ratio 61
thylakoid 65, 76
vacuole 75
vesicle 66

Match the terms to these definitions:

a. ______________ Organelle, consisting of saccules and vesicles, that processes, packages, and distributes molecules about or from the cell.
b. ______________ Especially active in lipid metabolism; always produces $H_2O_2$.
c. ______________ Dark-staining, spherical body in the cell nucleus that produces ribosomal subunits.
d. ______________ Internal framework of the cell, consisting of microtubules, actin filaments, and intermediate filaments.
e. ______________ Allows prokaryotic cells to attach to other cells.

## reviewing this chapter

1. What are the three basic principles of the cell theory? 60
2. Why is it advantageous for cells to be small? 61
3. Roughly sketch a bacterial (prokaryotic) cell, label its parts, and state a function for each of these. 65
4. How do eukaryotic and prokaryotic cells differ? 66
5. Describe how the nucleus, the chloroplast, and the mitochondrion may have become a part of the eukaryotic cell. 66
6. What does it mean to say that the eukaryotic cell is compartmentalized? 66–67
7. Describe the structure and the function of the nuclear envelope and the nuclear pores. 70–71
8. Distinguish between the nucleolus, rRNA, and ribosomes. 70–71
9. Name organelles that are a part of the endomembrane system and explain the term. 72
10. Trace the path of a protein from rough ER to the plasma membrane. 74
11. Give the overall equations for photosynthesis and cellular respiration, contrast the two, and tell how they are related. 76
12. Describe the structure and function of chloroplasts and mitochondria. How are these two organelles related to one another? 76–77
13. What are the three components of the cytoskeleton? What are their structures and functions? 78–79
14. Relate the structure of flagella (and cilia) to centrioles, and discuss the function of both. 80

## testing yourself

Choose the best answer for each question.

1. The small size of cells best correlates with
   a. the fact that they are self-reproducing.
   b. their prokaryotic versus eukaryotic nature.
   c. an adequate surface area for exchange of materials.
   d. the fact that they come in multiple sizes.
   e. All of these are correct.
2. Which of these is not a true comparison of the compound light microscope and the transmission electron microscope?

| | LIGHT | ELECTRON |
|---|---|---|
| a. | Uses light to "view" object | Uses electrons to "view" object |
| b. | Uses glass lenses for focusing | Uses magnetic lenses for focusing |
| c. | Specimen must be killed and stained | Specimen may be alive and nonstained |
| d. | Magnification is not as great | Magnification is greater |
| e. | Resolution is not as great | Resolution is greater |

3. Which of these best distinguishes a prokaryotic cell from a eukaryotic cell?
   a. Prokaryotic cells have a cell wall, but eukaryotic cells never do.
   b. Prokaryotic cells are much larger than eukaryotic cells.
   c. Prokaryotic cells have flagella, but eukaryotic cells do not.
   d. Prokaryotic cells do not have a membrane-bounded nucleus, but eukaryotic cells do have such a nucleus.
   e. Prokaryotic cells have ribosomes, but eukaryotic cells do not have ribosomes.
4. Which of these is not found in the nucleus?
   a. functioning ribosomes
   b. chromatin that condenses to chromosomes
   c. nucleolus that produces rRNA
   d. nucleoplasm instead of cytoplasm
   e. all forms of RNA

5. Vesicles from the ER most likely are on their way to
   a. the rough ER.
   b. the lysosomes.
   c. the Golgi apparatus.
   d. the plant cell vacuole only.
   e. the location suitable to their size.
6. Lysosomes function in
   a. protein synthesis.
   b. processing and packaging.
   c. intracellular digestion.
   d. lipid synthesis.
   e. production of hydrogen peroxide.
7. Mitochondria
   a. are involved in cellular respiration.
   b. break down ATP to release energy for cells.
   c. contain grana and cristae.
   d. are present in animal cells but not plant cells.
   e. All of these are correct.
8. Which organelle releases oxygen?
   a. ribosome
   b. Golgi apparatus
   c. chloroplast
   d. smooth ER
9. Label only the parts of the cell that are involved in protein synthesis and modification. Give a function for each structure.

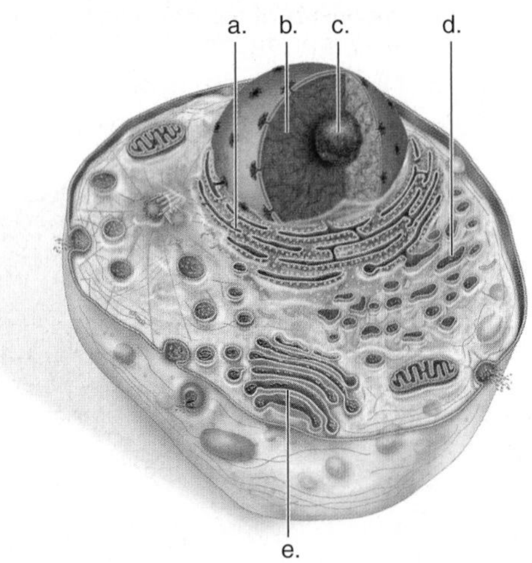


10. Which of these is not true?
    a. Actin filaments are found in muscle cells.
    b. Microtubules radiate out from the ER.
    c. Intermediate filaments sometimes contain keratin.
    d. Motor molecules use microtubules as tracks.
11. Cilia and flagella
    a. have a 9 + 0 pattern of microtubules, same as basal bodies.
    b. contain myosin that pulls on actin filaments.
    c. are organized by basal bodies derived from centrioles.
    d. are constructed similarly in prokaryotes and eukaryotes.
    e. Both a and c are correct.
12. Which of the following organelles contains its (their) own DNA, suggesting they were once independent prokaryotes?
    a. Golgi apparatus
    b. mitochondria
    c. chloroplasts
    d. ribosomes
    e. Both b and c are correct.
13. Which organelle most likely originated by invagination of the plasma membrane?
    a. mitochondria
    b. flagella
    c. nucleus
    d. chloroplasts
    e. All of these are correct.
14. Which structures are found in a prokaryotic cell?
    a. cell wall, ribosomes, thylakoids, chromosome
    b. cell wall, plasma membrane, nucleus, flagellum
    c. nucleoid, ribosomes, chloroplasts, capsule
    d. plasmid, ribosomes, enzymes, DNA, mitochondria
    e. chlorophyll, enzymes, Golgi apparatus, plasmids
15. Study the example given in (a) below. Then for each other organelle listed, state another that is structurally and functionally related. Tell why you paired these two organelles.
    a. The nucleus can be paired with nucleoli because nucleoli are found in the nucleus. Nucleoli occur where chromatin is producing rRNA.
    b. mitochondria
    c. centrioles
    d. ER

## thinking scientifically

1. The protists that cause malaria contribute to infections associated with AIDS. Scientists have discovered that an antibiotic that inhibits prokaryotic enzymes will kill the parasite because it is effective against the plastids in the protist. What can you conclude about the origin of the plastids?
2. For your cytology study, you have decided to label and, thereby, detect the presence of the base uracil in an animal cell. In what parts of the cell do you expect to find your radioactive tracer?

## bioethical issue

### Stem Cells

A stem cell is an immature cell that is capable of producing cells that will differentiate into mature cells. Stem cells exist in the various organs of the human body; however, they are difficult to obtain, except for those that reside in red bone marrow and produce all types of blood cells. One method of obtaining stem cells is to take a 2n adult nucleus from, say, skin, manipulate it genetically, and put it in an enucleated egg cell. If all goes well, development will begin, and the cells that result can be pried apart and used to make neurological tissues that could possibly cure Alzheimer or Parkinson disease or any other type of neurological disorder. However, if development were to continue, a clone of the human that donated the 2n nucleus could possibly result.

Is it bioethical to continue investigating such research? Especially when you consider that the "embryo" that provided the stem cells was not produced by the normal method of having a sperm fertilize an egg? Or, is it wrong to produce an embryo only to serve as a source of stem cells?

## *Biology* website

The companion website for *Biology* provides a wealth of information organized and integrated by chapter. You will find practice tests, animations, videos, and much more that will complement your learning and understanding of general biology.

**http://www.mhhe.com/maderbiology10**

# 5

# Membrane Structure and Function

*An African pygmy, an overweight diabetic, and a young child with cystic fibrosis suffer from a defect in their cells' plasma membrane. Growth hormone does not bind to the pygmy's plasma membrane, the diabetic's does not respond properly to insulin, and the membrane does not transport chloride from the cells of a child who has cystic fibrosis.*

*A plasma membrane encloses every cell, whether the cell is a unicellular amoeba or one of many from the body of a squid, carnation, mushroom, or human. Universally, a plasma membrane protects a cell by acting as a barrier between its living contents and the surrounding environment. It regulates what goes into and out of the cell and serves as a means of communication between cells. Inside eukaryotic cells, membrane compartmentalizes the cell so that specific enzymes for particular functions are isolated from one another. This chapter describes the plasma membrane and its numerous functions. It also discusses various ways cells communicate so that the activities of tissues and organs are coordinated.*

A eukaryotic cell is surrounded by a plasma membrane, and membrane also compartmentalizes the cell into various organelles.

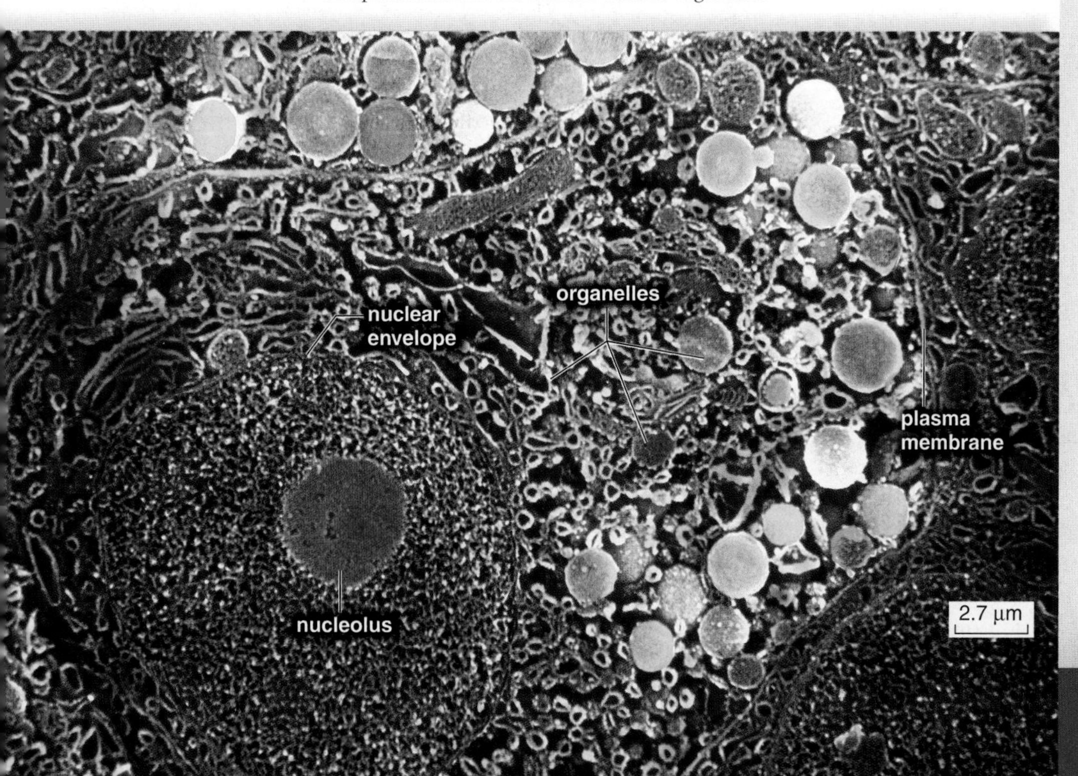


## concepts

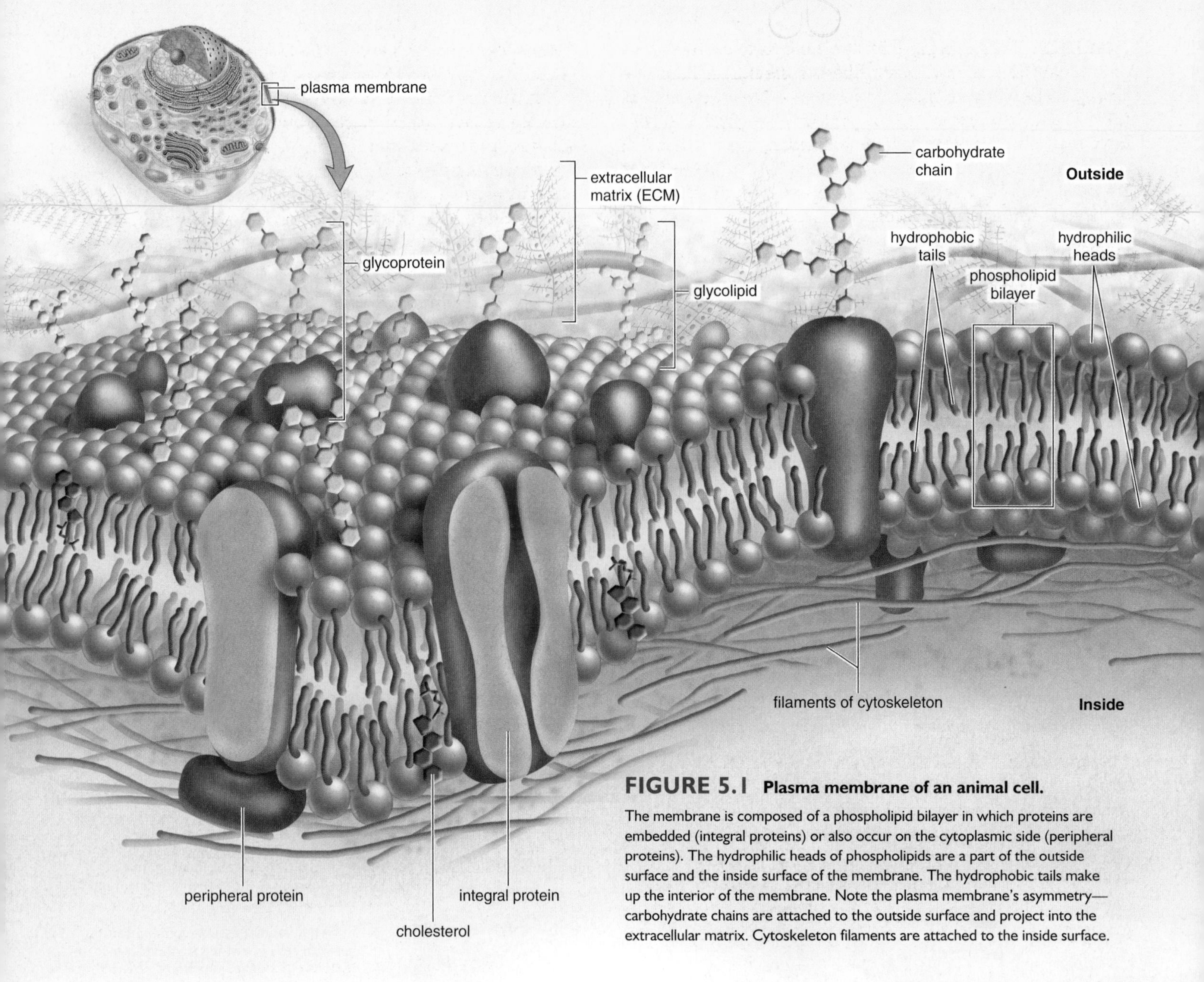


**FIGURE 5.1 Plasma membrane of an animal cell.**

The membrane is composed of a phospholipid bilayer in which proteins are embedded (integral proteins) or also occur on the cytoplasmic side (peripheral proteins). The hydrophilic heads of phospholipids are a part of the outside surface and the inside surface of the membrane. The hydrophobic tails make up the interior of the membrane. Note the plasma membrane's asymmetry—carbohydrate chains are attached to the outside surface and project into the extracellular matrix. Cytoskeleton filaments are attached to the inside surface.

# 5.1 Plasma Membrane Structure and Function

The structure of an animal cell's plasma membrane is depicted in Figure 5.1. The drawing shows that the membrane is a phospholipid bilayer in which protein molecules are either partially or wholly embedded. A phospholipid is an *amphipathic molecule,* meaning that it has both a hydrophilic (water-loving) region and a hydrophobic (water-fearing) region. The amphipathic nature of phospholipids largely explains why the membrane is a bilayer—has two layers of phospholipids. The hydrophilic polar heads of the phospholipid molecules naturally face the outside and inside of the cell, where water is found. The hydrophobic nonpolar tails face each other. **Cholesterol** is another lipid found in the animal plasma membrane; related steroids are found in the plasma membrane of plants. Cholesterol helps modify the fluidity of the membrane, as discussed later.

As shown in Figure 5.1, the proteins are scattered throughout the membrane in an irregular pattern, and this pattern can vary from membrane to membrane. Electron micrographs verify that many of the proteins are embedded within the membrane. During freeze-fracture, the membrane is first frozen and then split so that the upper layer is separated from the lower layer. The proteins remain intact and go with one layer or the other. The embedded proteins are termed integral proteins, and other proteins that occur only on the cytoplasmic side of the membrane are termed peripheral proteins. Some integral proteins protrude from only one surface of the bilayer but most span the membrane, with a hydrophobic region within the membrane, while their hydrophilic heads protrude from both surfaces of the bilayer. These proteins can be held in place by attachments to protein fibers of the cytoskeleton (inside) and fibers of the extracellular matrix (outside). Only animal cells have an **extracellular matrix (ECM),** which contains various protein fibers and also very large and complex carbohydrate molecules. The ECM, which is discussed in greater detail at the end of the chapter, has various functions, from lending support to the plasma membrane to assisting communication between cells.

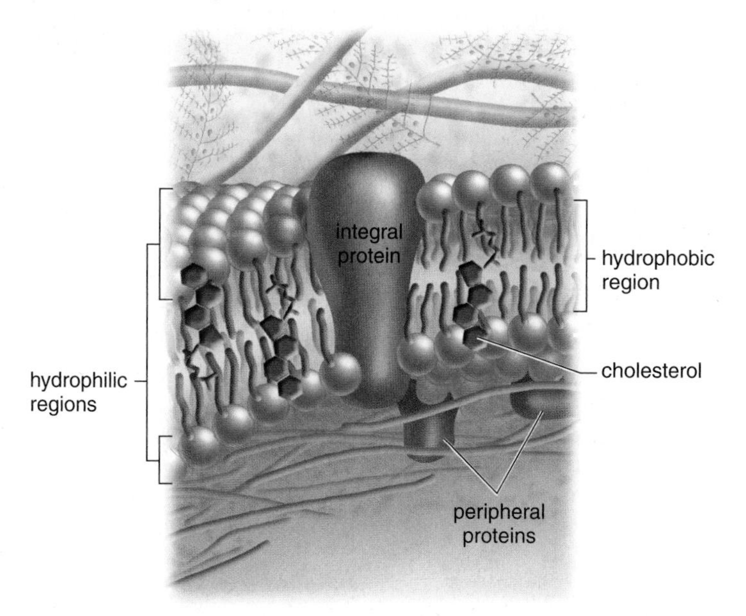


## Fluid-Mosaic Model

The model currently in use to describe the plasma membrane is called the **fluid-mosaic model.** The fluidity of the membrane is due to its lipid component. At body temperature, the phospholipid bilayer of the plasma membrane has the consistency of olive oil. The greater the concentration of unsaturated fatty acid residues, the more fluid is the bilayer. In each monolayer, the hydrocarbon tails wiggle, and the entire phospholipid molecule can move sideways at a rate averaging about 2 μm—the length of a prokaryotic cell—per second. (Phospholipid molecules rarely flip-flop from one layer to the other, because this would require the hydrophilic head to move through the hydrophobic center of the membrane.) The fluidity of a phospholipid bilayer means that cells are pliable. Imagine if they were not—the long nerve fibers in your neck would crack whenever you nodded your head! The fluidity of the membrane also prevents it from solidifying as external temperatures drop.

The presence of cholesterol molecules in the plasma membrane affects its fluidity. At higher temperatures, cholesterol stiffens the membrane and makes it less fluid than it would otherwise be. At lower temperatures, cholesterol helps prevent the membrane from freezing by not allowing contact between certain phospholipid tails.

The mosaic nature of the plasma membrane is due to its protein content. The number and kinds of proteins can vary in the plasma membrane and in the membrane of the various organelles. The presence of various proteins that seem to have no set positions is consistent with the idea that they form a mosaic pattern. Further, it was once thought that the proteins could freely move sideways within the fluid bilayer. Figure 5.2 describes an experiment in which the proteins were tagged prior to allowing mouse and human cells to fuse. An hour after fusion, the proteins were completely mixed. Such an experiment suggests that at least some proteins are able to move sideways in the membrane. Today, however, we know that proteins are often bond to either or both the ECM and the cytoskeleton. These connections hold a protein in place and prevent it from moving in the fluid phospholipid bilayer.

## Carbohydrate Chains

Both phospholipids and proteins can have attached carbohydrate (sugar) chains. If so, these molecules are called **glycolipids** and **glycoproteins,** respectively. Since the carbohydrate chains occur only on the outside surface and peripheral proteins occur asymmetrically on one surface or the other, the two sides of the membrane are not identical.

In animal cells, the carbohydrate chains of proteins give the cell a "sugar coat," more properly called the glycocalyx. The glycocalyx protects the cell and has various other functions. For example, it facilitates adhesion between cells, reception of signaling molecules, and cell-to-cell recognition.

The possible diversity of the carbohydrate (sugar) chains is enormous. The chains can vary by the number (15 is usual, but there can be several hundred) and sequence of sugars and by whether the chain is branched. Each cell within the individual has its own particular "fingerprint" because of these chains. As you probably know, transplanted tissues are often rejected by the recipient. This is because the

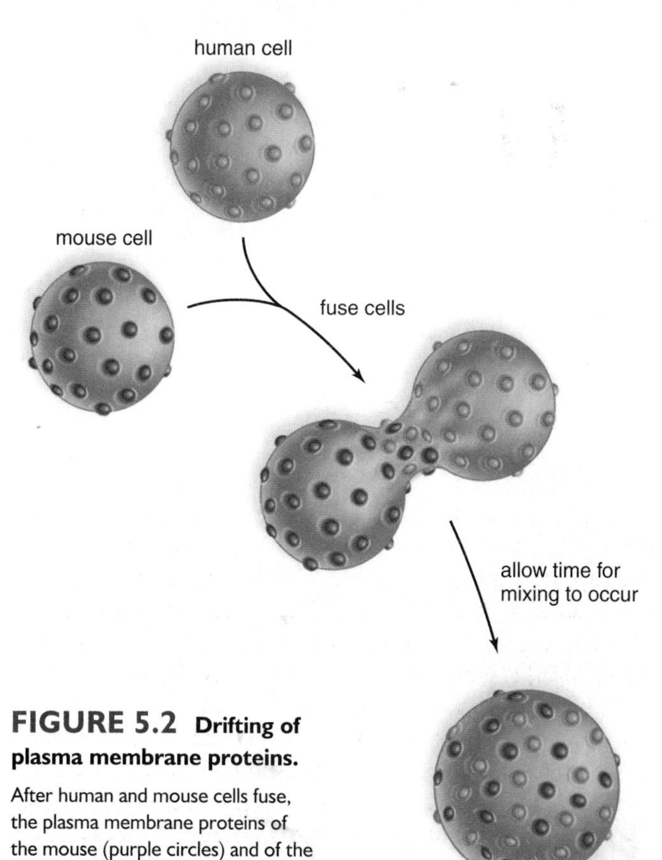


**FIGURE 5.2 Drifting of plasma membrane proteins.**

After human and mouse cells fuse, the plasma membrane proteins of the mouse (purple circles) and of the human cell (orange circles) mix within a short time.

immune system is able to recognize that the foreign tissue's cells do not have the appropriate carbohydrate chains. In humans, carbohydrate chains are also the basis for the A, B, and O blood groups.

## The Functions of the Proteins

While the plasma membranes of various cells and the membranes of various organelles can contain various proteins at different times, these types of proteins are apt to be present:

**Channel proteins** Channel proteins are involved in the passage of molecules through the membrane. They have a channel that allows a substance to simply move across the membrane (Fig. 5.3*a*). For example, a channel protein allows hydrogen ions to flow across the inner mitochondrial membrane. Without this movement of hydrogen ions, ATP would never be produced.

**Carrier proteins** Carrier proteins are also involved in the passage of molecules through the membrane. They combine with a substance and help it move across the membrane (Fig. 5.3*b*). A carrier protein transports sodium and potassium ions across the plasma membrane of a nerve cell. Without this carrier protein, nerve conduction would be impossible.

**Cell recognition proteins** Cell recognition proteins are glycoproteins (Fig. 5.3*c*). Among other functions, these proteins help the body recognize when it is being invaded by pathogens so that an immune response can occur. Without this recognition, pathogens would be able to freely invade the body.

**Receptor proteins** Receptor proteins have a shape that allows a specific molecule to bind to it (Fig. 5.3*d*). The binding of this molecule causes the protein to change its shape and thereby bring about a cellular response. The coordination of the body's organs is totally dependent on such signaling molecules. For example, the liver stores glucose after it is signaled to do so by insulin.

**Enzymatic proteins** Some plasma membrane proteins are enzymatic proteins that carry out metabolic reactions directly (Fig. 5.3*e*). Without the presence of enzymes, some of which are attached to the various membranes of the cell, a cell would never be able to perform the metabolic reactions necessary to its proper function.

**Junction proteins** As discussed on page 98, proteins are involved in forming various types of junctions between animal cells (Fig. 5.3*f*). Signaling molecules that pass through gap junctions allow the cilia of cells that line your respiratory tract to beat in unison.

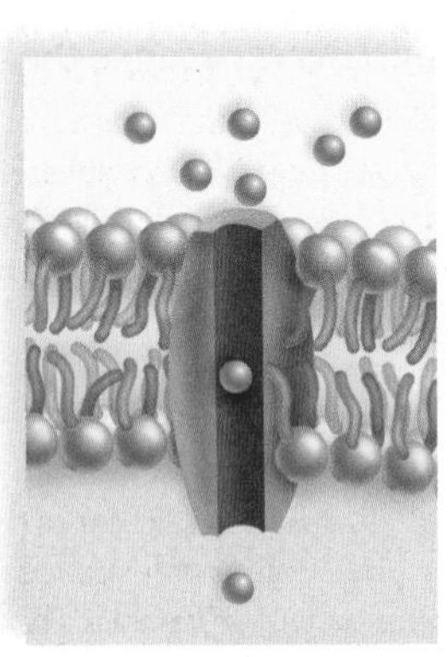

**Channel Protein:** Allows a particular molecule or ion to cross the plasma membrane freely. Cystic fibrosis, an inherited disorder, is caused by a faulty chloride ($Cl^-$) channel; a thick mucus collects in airways and in pancreatic and liver ducts.

a.

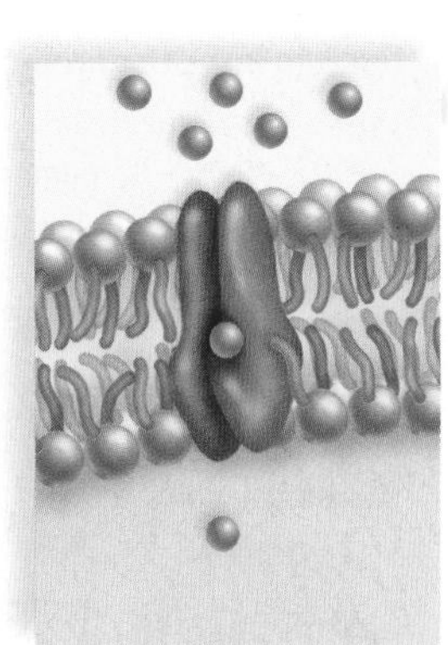

**Carrier Protein:** Selectively interacts with a specific molecule or ion so that it can cross the plasma membrane. The inability of some persons to use energy for sodium-potassium ($Na^+$–$K^+$) transport has been suggested as the cause of their obesity.

b.

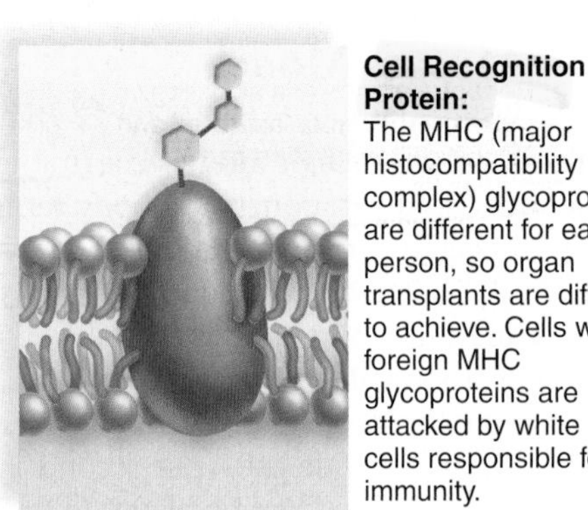

**Cell Recognition Protein:** The MHC (major histocompatibility complex) glycoproteins are different for each person, so organ transplants are difficult to achieve. Cells with foreign MHC glycoproteins are attacked by white blood cells responsible for immunity.

c.

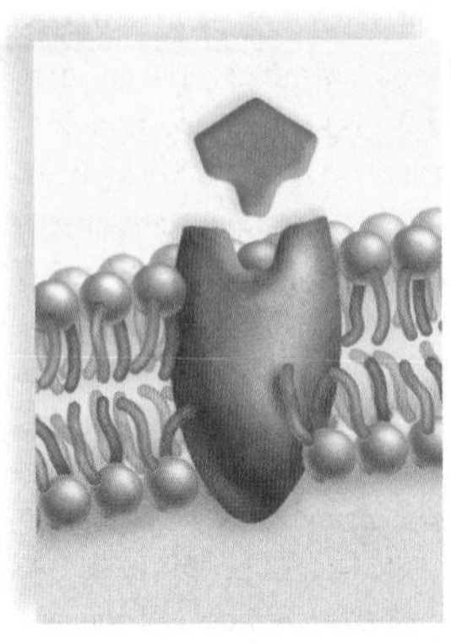

**Receptor Protein:** Is shaped in such a way that a specific molecule can bind to it. Pygmies are short, not because they do not produce enough growth hormone, but because their plasma membrane growth hormone receptors are faulty and cannot interact with growth hormone.

d.

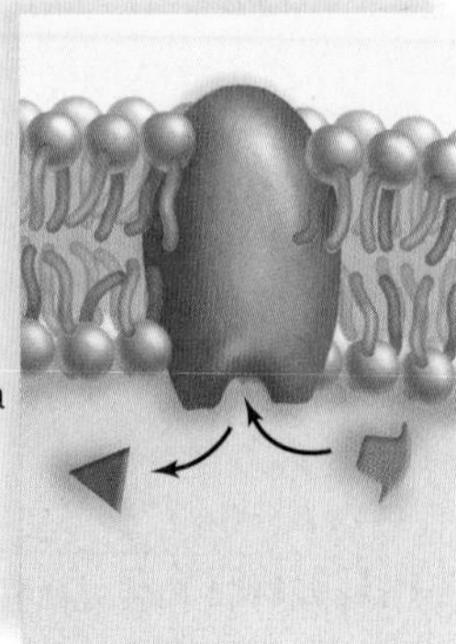

**Enzymatic Protein:** Catalyzes a specific reaction. The membrane protein, adenylate cyclase, is involved in ATP metabolism. Cholera bacteria release a toxin that interferes with the proper functioning of adenylate cyclase; sodium ($Na^+$) and water leave intestinal cells, and the individual may die from severe diarrhea.

e.

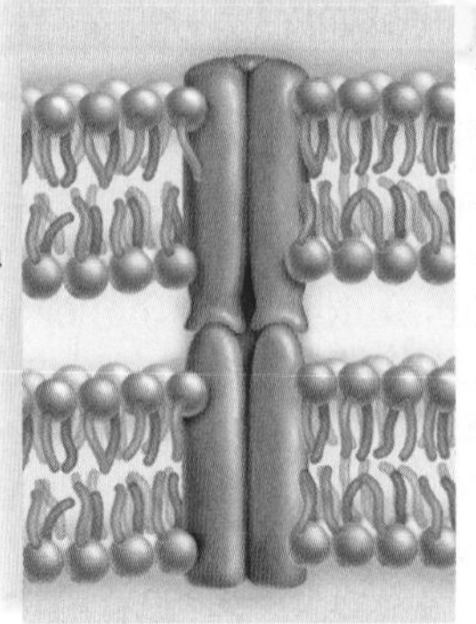

**Junction Proteins:** Tight junctions join cells so that a tissue can fulfill a function, as when a tissue pinches off the neural tube during development. Without this cooperation between cells, an animal embryo would have no nervous system.

f.

**FIGURE 5.3 Membrane protein diversity.**

These are some of the functions performed by proteins found in the plasma membrane.

*science focus*

# How Cells Talk to One Another

All organisms are able to sense and respond to specific signals in their environment. A bacterium that has taken up residence in your body is responding to signaling molecules when it finds food and escapes immune cells in order to stay alive. Signaling helps bread mold on stale bread in your refrigerator detect the presence of an opposite mating strain and begin its sexual life cycle. Similarly, the cells of an embryo are responding to signaling molecules when they move to specific locations and assume the shape and perform the functions of specific tissues (Fig. 5A*a*). In the newborn, signaling is still required because the functions of a specific tissue may be necessary only on occasion, or one tissue may need to perform one of its various functions only at particular times. In plants, external signals, such as a change in the amount of light, tells them when it is time to resume growth or flower. Internal signaling molecules enable plants to coordinate the activities of roots, stems, and leaves.

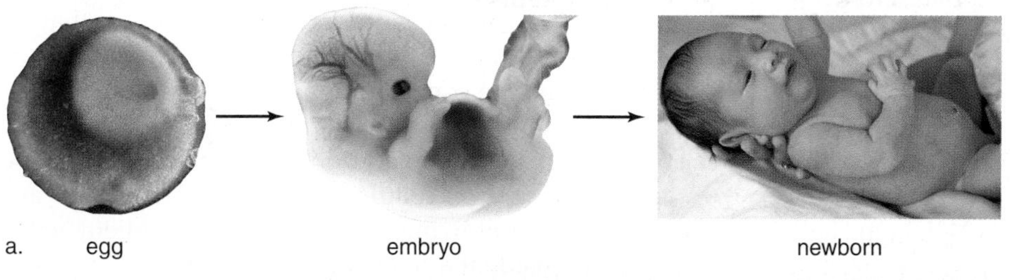


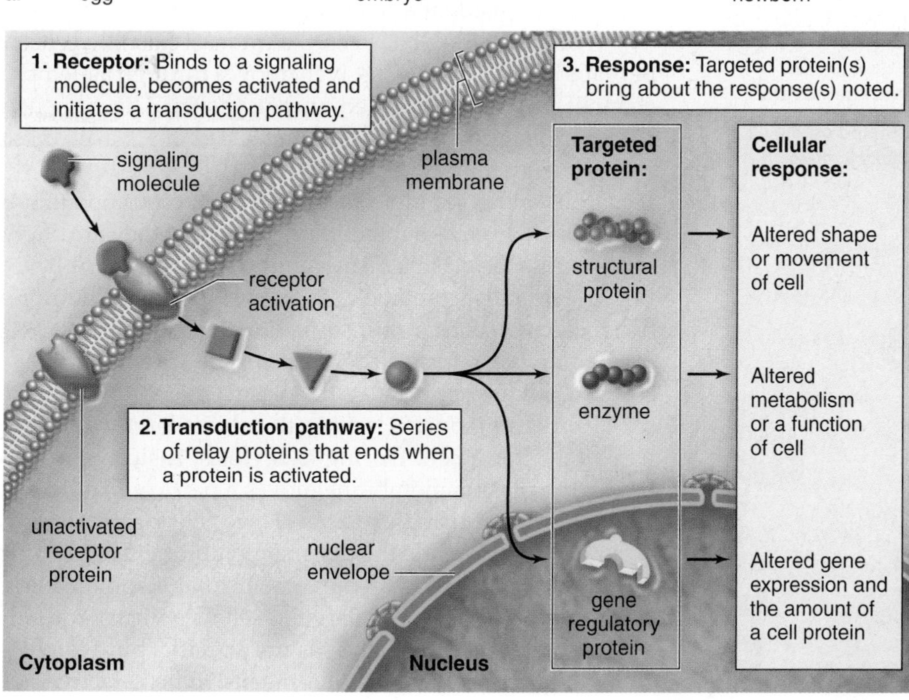


**FIGURE 5A Cell signaling.**
***a.*** *The process of signaling helps account for the transformation of an egg into an embryo and then an embryo into a newborn.* ***b.*** *The process of signaling involves three steps: binding of the signaling molecule, transduction of the signal, and response of the cell depending on what type protein is targeted.*

## Cell Signaling

The cells of a multicellular organism "talk" to one another by using signaling molecules, sometimes called chemical messengers. Some messengers are produced at a distance from a target tissue and, in animals, are carried by the circulatory system to various sites around the body. For example, the pancreas releases a hormone called insulin, which is transported in blood vessels to the liver, and thereafter, the liver stores glucose as glycogen. Failure of the liver to respond appropriately results in a medical condition called diabetes. In Chapter 9, we are particularly interested in growth factors, which act locally as signaling molecules and cause cells to divide. Overreacting to growth factors can result in a tumor characterized by unlimited cell division. The importance of cell signaling causes much research to be directed toward understanding the intricacies of the process.

We have learned that cells respond to only certain signaling molecules. Why? Because they must bind to a receptor protein, and cells have receptors for only certain signaling molecules. Each cell has receptors for numerous signaling molecules and often the final response is due to a summing up of all the various signals received. These molecules tell a cell what it should be doing at the moment, and without any signals, the cell dies.

Signaling not only involves a receptor protein, it also involves a pathway called a transduction pathway and a response. To understand the process, consider an analogy. When a TV camera (the receptor) is shooting a scene, the picture is converted to electrical signals (transduction pathway) that are understood by the TV in your house and are converted to a picture on your screen (the response). The process in cells is more complicated because each member of the pathway can turn on the activity of a number of other proteins. As shown in Figure 5A*b*, the cell response to a transduction pathway can be a change in the shape or movement of a cell, the activation of a particular enzyme, or the activation of a specific gene. We will be mentioning and giving examples of cell signaling between cells throughout the text.

**TABLE 5.1**

**Passage of Molecules into and out of the Cell**

| | *Name* | *Direction* | *Requirement* | *Examples* |
|---|---|---|---|---|
| Energy Not Required | Diffusion | Toward lower concentration | Concentration gradient | Lipid-soluble molecules, and gases |
| | Facilitated transport | Toward lower concentration | Channels or carrier and concentration gradient | Some sugars, and amino acids |
| Energy Required | Active transport | Toward higher concentration | Carrier plus energy | Sugars, amino acids, and ions |
| | Bulk transport | Toward outside or inside | Vesicle utilization | Macromolecules |

## Permeability of the Plasma Membrane

The plasma membrane regulates the passage of molecules into and out of the cell. This function is critical because the life of the cell depends on maintenance of its normal composition. The plasma membrane can carry out this function because it is **differentially** (selectively) **permeable,** meaning that certain substances can move across the membrane while others cannot.

Table 5.1 lists, and Figure 5.4 illustrates, which types of molecules can passively (no energy required) cross a membrane and which may require transport by a carrier protein and/or an expenditure of energy. In general, small, noncharged molecules, such as carbon dioxide, oxygen, glycerol, and alcohol, can freely cross the membrane. They are able to slip between the hydrophilic heads of the phospholipids and pass through the hydrophobic tails of the membrane. These molecules are said to follow their **concentration gradient** as they move from an area where their concentration is high to an area where their concentration is low. Consider that a cell is always using oxygen when it carries on cellular respiration. Therefore, the concentration of oxygen is always lower inside a cell than outside a cell, and so oxygen has a tendency to enter a cell. Carbon dioxide, on the other hand, is produced when a cell carries on cellular respiration. Therefore, carbon dioxide is also following a concentration gradient when it moves from inside the cell to outside the cell.

A new finding has been that at least in some cells, and perhaps all cells, water passively moves through a membrane channel protein now called an **aquaporin.** The presence of aquaporins accounts for why water can cross a membrane more quickly than expected.

Ions and polar molecules, such as glucose and amino acids, can slowly cross a membrane. Therefore, they are often assisted across the plasma membrane by carrier proteins. The carrier protein must combine with an ion, such as sodium ($Na^+$), or a molecule, such as glucose, before transporting it across the membrane. Therefore, carrier proteins are specific for the substances they transport across the plasma membrane.

**Bulk transport** is a way that large particles can exit a cell or enter a cell. During exocytosis, fusion of a vesicle with the plasma membrane moves a particle to outside the membrane. During endocytosis, vesicle formation moves a particle to inside the plasma membrane. Vesicle formation is reserved for movement of macromolecules or even for something larger, such as a virus. You might think that endocytosis is not specific, but we will see that a cell does have a means to be selective about what enters by endocytosis.



**FIGURE 5.4** **How molecules cross the plasma membrane.**

The curved arrows indicate that these substances cannot passively cross the plasma membrane, and the long back-and-forth arrows indicate that these substances can diffuse across the plasma membrane.

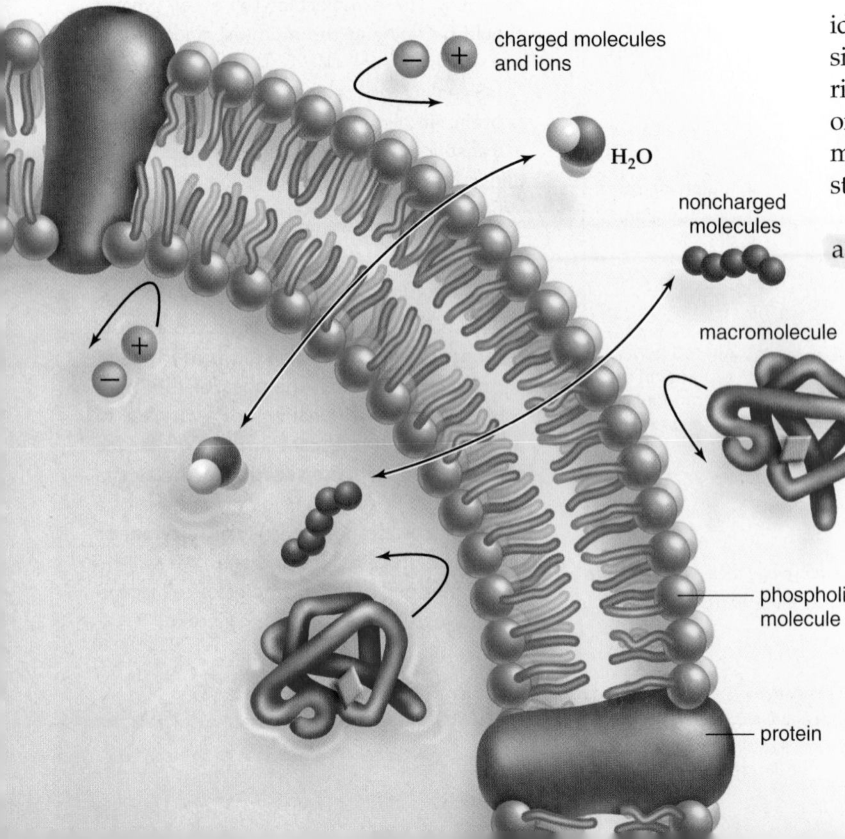

### Check Your Progress 5.1

1. Briefly describe the structure of the plasma membrane.
2. List six types of proteins found in the plasma membrane.

## 5.2 Passive Transport Across a Membrane

**Diffusion** is the movement of molecules from a higher to a lower concentration—that is, down their concentration gradient—until equilibrium is achieved and they are distributed equally. Diffusion is a physical process due to random molecular motion that can be observed with any type of molecule. For example, when a crystal of dye is placed in water (Fig. 5.5), the dye and water molecules move in various directions, but their net movement, which is the sum of their motion, is toward the region of lower concentration. Eventually, the dye is dissolved in the water, resulting in equilibrium and a colored solution.

A **solution** contains both a solute, usually a solid, and a solvent, usually a liquid. In this case, the **solute** is the dye and the **solvent** is the water molecules. Once the solute and solvent are evenly distributed, they continue to move about, but there is no net movement of either one in any direction.

The chemical and physical properties of the plasma membrane allow only a few types of molecules to enter and exit a cell simply by diffusion. Gases can diffuse through the lipid bilayer; this is the mechanism by which oxygen enters cells and carbon dioxide exits cells. Also, consider the movement of oxygen from the alveoli (air sacs) of the lungs to the blood in the lung capillaries (Fig. 5.6). After inhalation (breathing in), the concentration of oxygen in the alveoli is higher than that in the blood; therefore, oxygen diffuses into the blood.

Several factors influence the rate of diffusion. Among these factors are temperature, pressure, electrical currents, and molecular size. For example, as temperature increases, the rate of diffusion increases. The movement of fishes in the tank would certainly speed the rate of diffusion (Fig. 5.5).

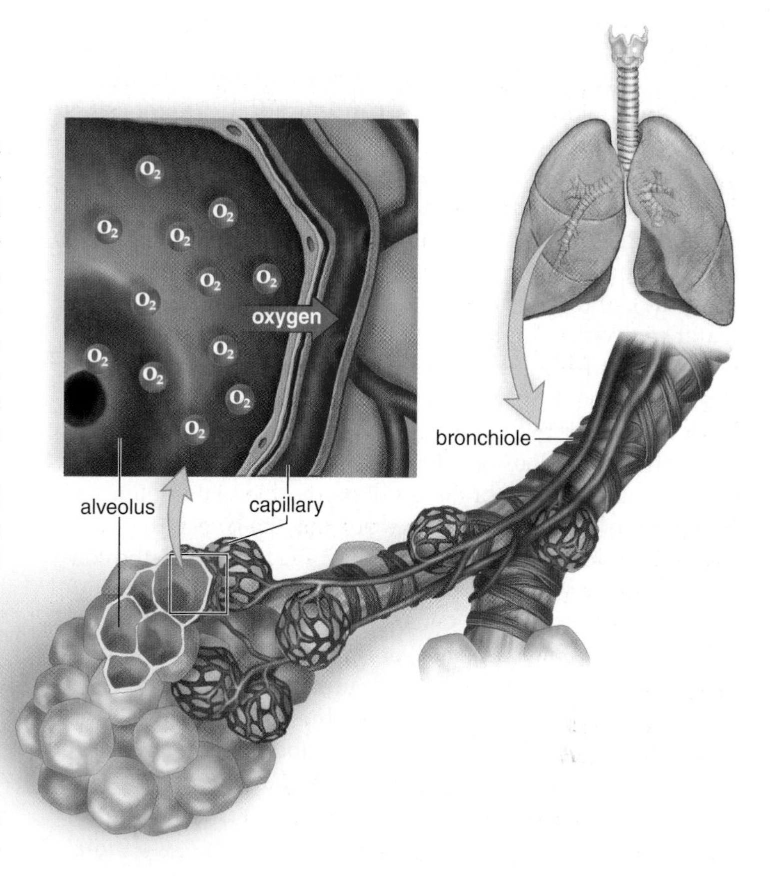


**FIGURE 5.6 Gas exchange in lungs.**

Oxygen ($O_2$) diffuses into the capillaries of the lungs because there is a higher concentration of oxygen in the alveoli (air sacs) than in the capillaries.

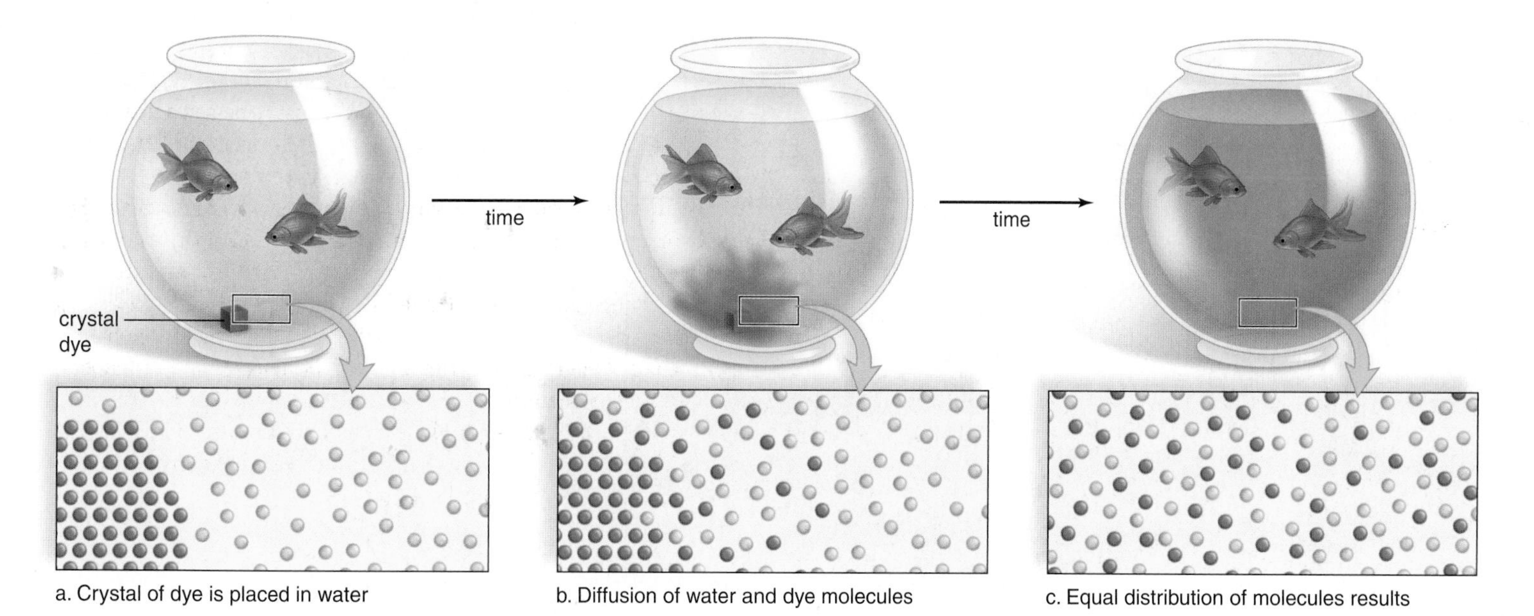


**FIGURE 5.5 Process of diffusion.**

Diffusion is spontaneous, and no chemical energy is required to bring it about. **a.** When a dye crystal is placed in water, it is concentrated in one area. **b.** The dye dissolves in the water, and there is a net movement of dye molecules from a higher to a lower concentration. There is also a net movement of water molecules from a higher to a lower concentration. **c.** Eventually, the water and the dye molecules are equally distributed throughout the container.

## Osmosis

The diffusion of water across a differentially (selectively) permeable membrane due to concentration differences is called **osmosis.** To illustrate osmosis, a thistle tube containing a 10% solute solution[1] is covered at one end by a differentially permeable membrane and then placed in a beaker containing a 5% solute solution (Fig. 5.7*a*). The beaker has a higher concentration of water molecules (lower percentage of solute), and the thistle tube has a lower concentration of water molecules (higher percentage of solute). Diffusion always occurs from higher to lower concentration. Therefore, a net movement of water takes place across the membrane from the beaker to the inside of the thistle tube (Fig. 5.7*b*).

The solute does not diffuse out of the thistle tube. Why not? Because the membrane is not permeable to the solute. As water enters and the solute does not exit, the level of the solution within the thistle tube rises (Fig. 5.7*c*). In the end, the concentration of solute in the thistle tube is less than 10%. Why? Because there is now less solute per unit volume. And the concentration of solute in the beaker is greater than 5%. Why? Because there is now more solute per unit volume.

Water enters the thistle tube due to the osmotic pressure of the solution within the thistle tube. **Osmotic pressure** is the pressure that develops in a system due to osmosis.[2] In other words, the greater the possible osmotic pressure, the more likely it is that water will diffuse in that direction. Due to osmotic pressure, water is absorbed by the kidneys and taken up by capillaries in the tissues. Osmosis also occurs across the plasma membrane, as we shall now see (Fig. 5.8).

[1] Percent solutions are grams of solute per 100 mL of solvent. Therefore, a 10% solution is 10 g of sugar with water added to make 100 mL of solution.

[2] Osmotic pressure is measured by placing a solution in an osmometer and then immersing the osmometer in pure water. The pressure that develops is the osmotic pressure of a solution.

### *Isotonic Solution*

In the laboratory, cells are normally placed in **isotonic solutions**—that is, the solute concentration and the water concentration both inside and outside the cell are equal, and therefore there is no net gain or loss of water. The prefix *iso* means "the same as," and the term **tonicity** refers to the strength of the solution. A 0.9% solution of the salt sodium chloride (NaCl) is known to be isotonic to red blood cells. Therefore, intravenous solutions medically administered usually have this tonicity. Terrestrial animals can usually take in either water or salt as needed to maintain the tonicity of their internal environment. Many animals living in an estuary, such as oysters, blue crabs, and some fishes, are able to cope with changes in the salinity (salt concentrations) of their environment. Their kidneys, gills, and other structures help them do this.

### *Hypotonic Solution*

Solutions that cause cells to swell, or even to burst, due to an intake of water are said to be **hypotonic solutions.** The prefix *hypo* means "less than" and refers to a solution with a lower concentration of solute (higher concentration of water) than inside the cell. If a cell is placed in a hypotonic solution, water enters the cell; the net movement of water is from the outside to the inside of the cell.

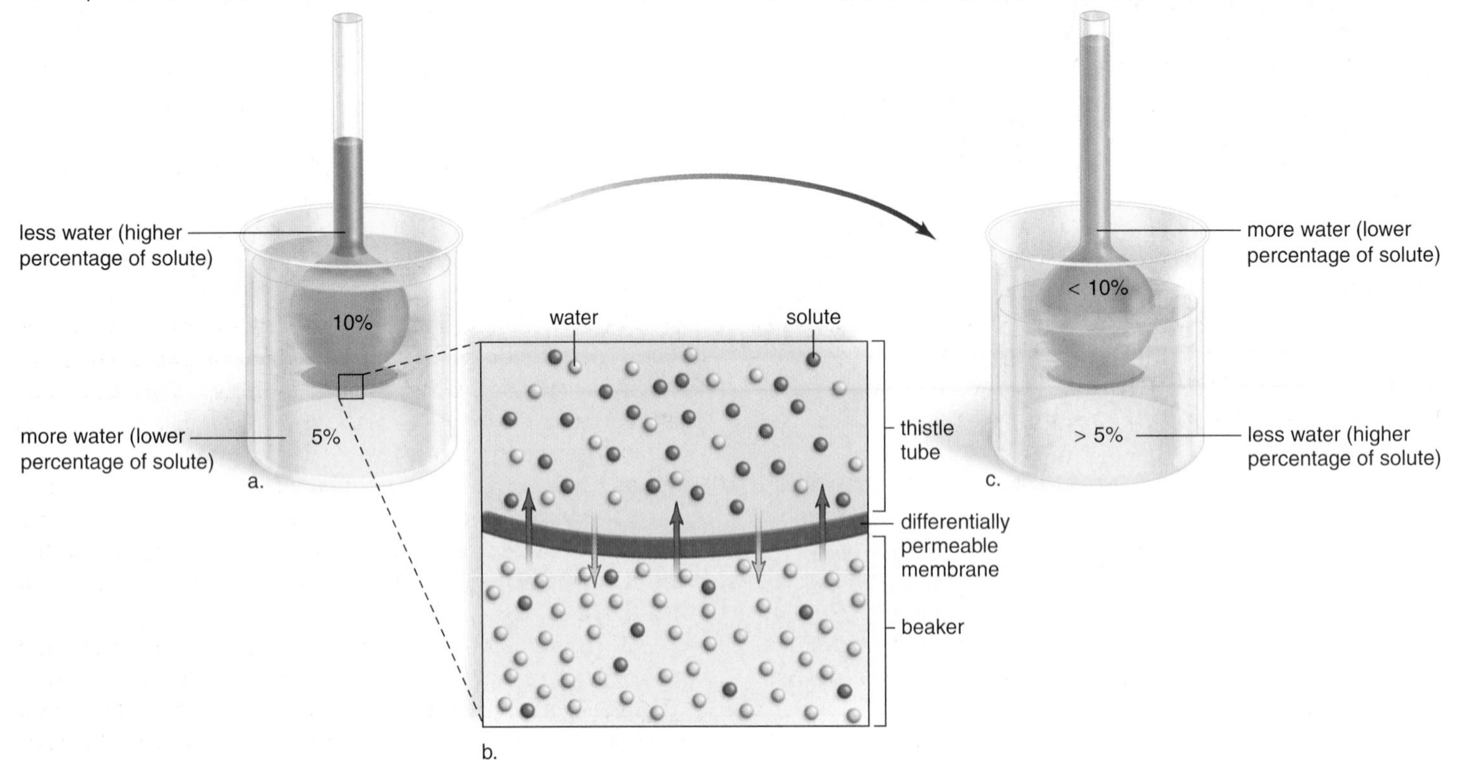


**FIGURE 5.7 Osmosis demonstration.**

**a.** A thistle tube, covered at the broad end by a differentially permeable membrane, contains a 10% solute solution. The beaker contains a 5% solute solution. **b.** The solute (green circles) is unable to pass through the membrane, but the water (blue circles) passes through in both directions. There is a net movement of water toward the inside of the thistle tube, where there is a lower percentage of water molecules. **c.** Due to the incoming water molecules, the level of the solution rises in the thistle tube.

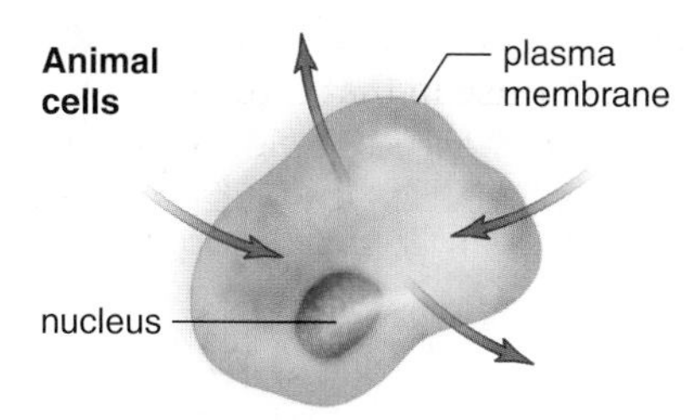


In an isotonic solution, there is no net movement of water.

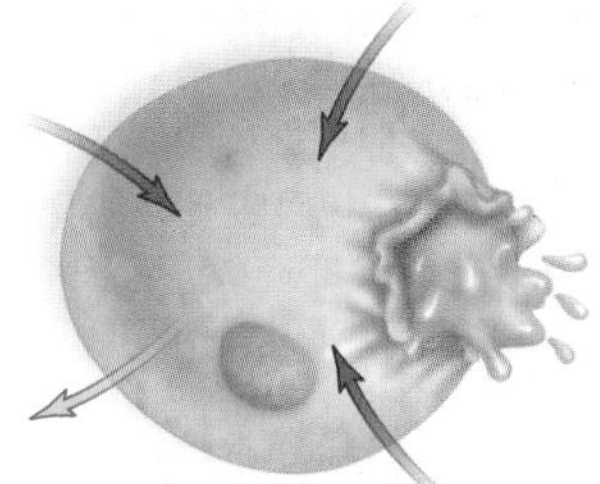

In a hypotonic solution, water mainly enters the cell, which may burst (lysis).

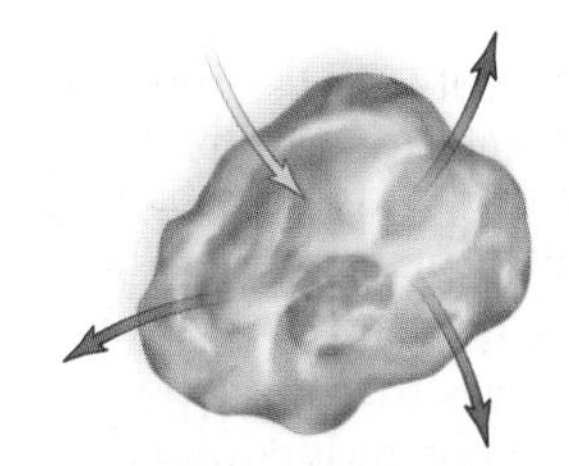

In a hypertonic solution, water mainly leaves the cell, which shrivels (crenation).

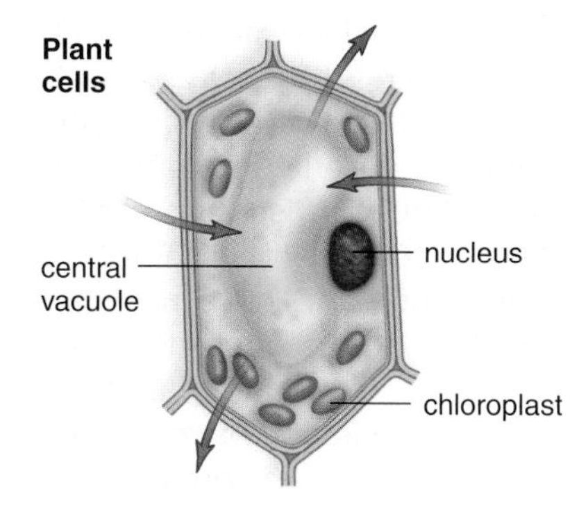


In an isotonic solution, there is no net movement of water.

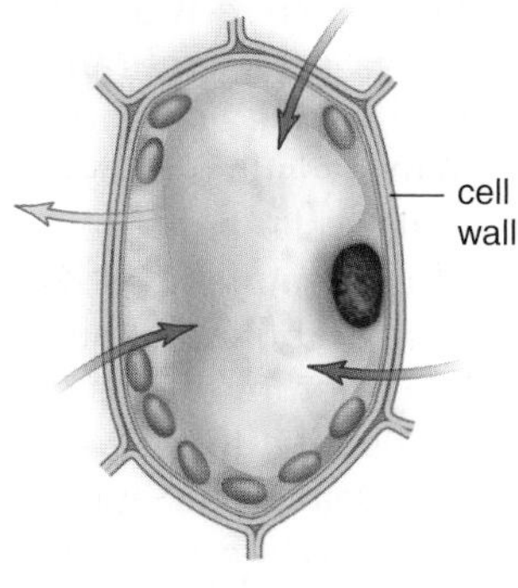


In a hypotonic solution, vacuoles fill with water, turgor pressure develops, and chloroplasts are seen next to the cell wall.

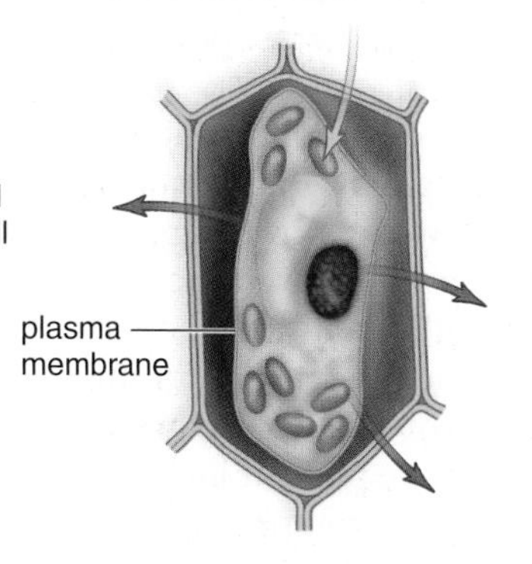


In a hypertonic solution, vacuoles lose water, the cytoplasm shrinks (plasmolysis), and chloroplasts are seen in the center of the cell.

**FIGURE 5.8 Osmosis in animal and plant cells.**

The arrows indicate the movement of water molecules. To determine the net movement of water, compare the number of dark blue arrows that are taking water molecules into the cell versus the number of light blue arrows that are taking water out of the cell. In an isotonic solution, a cell neither gains nor loses water; in a hypotonic solution, a cell gains water; and in a hypertonic solution, a cell loses water.

Any concentration of a salt solution lower than 0.9% is hypotonic to red blood cells. Animal cells placed in such a solution expand and sometimes burst due to the buildup of pressure. The term *cytolysis* is used to refer to disrupted cells; hemolysis, then, is disrupted red blood cells.

The swelling of a plant cell in a hypotonic solution creates **turgor pressure.** When a plant cell is placed in a hypotonic solution, we observe expansion of the cytoplasm because the large central vacuole gains water and the plasma membrane pushes against the rigid cell wall. The plant cell does not burst because the cell wall does not give way. Turgor pressure in plant cells is extremely important to the maintenance of the plant's erect position. If you forget to water your plants, they wilt due to decreased turgor pressure.

Organisms that live in fresh water have to prevent the uptake of too much water. Many protozoans, such as paramecia, have contractile vacuoles that rid the body of excess water. Freshwater fishes have well-developed kidneys that excrete a large volume of dilute urine. Even so, they have to take in salts at their gills. Even though freshwater fishes are good osmoregulators, they would not be able to survive in either distilled water or a marine environment.

## *Hypertonic Solution*

Solutions that cause cells to shrink or shrivel due to loss of water are said to be **hypertonic solutions.** The prefix *hyper* means "more than" and refers to a solution with a higher percentage of solute (lower concentration of water) than the cell. If a cell is placed in a hypertonic solution, water leaves the cell; the net movement of water is from the inside to the outside of the cell.

Any concentration of a salt solution higher than 0.9% is hypertonic to red blood cells. If animal cells are placed in this solution, they shrink. The term **crenation** refers to red blood cells in this condition. Meats are sometimes preserved by salting them. The bacteria are not killed by the salt but by the lack of water in the meat.

When a plant cell is placed in a hypertonic solution, the plasma membrane pulls away from the cell wall as the large central vacuole loses water. This is an example of **plasmolysis,** a shrinking of the cytoplasm due to osmosis. The dead plants you may see along a salted roadside died because they were exposed to a hypertonic solution during the winter. Also, when salt water invades coastal marshes due to storms and human activities, coastal plants die. Without roots to hold the soil, it washes into the sea, doing away with many acres of valuable wetlands.

Marine animals cope with their hypertonic environment in various ways that prevent them from losing water to the environment. Sharks increase or decrease urea in their blood until their blood is isotonic with the environment and in this way do not lose excessive water. Marine fishes and other types of animals drink no water but excrete salts across their gills. Have you ever seen a marine turtle cry? It is ridding its body of salt by means of glands near the eye.

## Facilitated Transport

The plasma membrane impedes the passage of all but a few substances. Yet, biologically useful molecules are able to enter and exit the cell at a rapid rate either by way of a channel protein or because of carrier proteins in the membrane. These transport proteins are specific; each can transport with only a certain type of molecule or ion, which is then transported through the membrane. It is not completely understood how carrier proteins function, but after a carrier combines with a molecule, the carrier is believed to undergo a change in shape that moves the molecule across the membrane. Carrier proteins are utilized for both facilitated transport and active transport (see Table 5.1).

**Facilitated transport** explains the rapid passage of water and also such molecules as glucose and amino acids across the plasma membrane. Whereas water moves through a channel protein, the passage of glucose and amino acids is facilitated by their reversible combination with carrier proteins, which transport them through the plasma membrane. These carrier proteins are specific. For example, various sugar molecules of identical size might be present inside or outside the cell, but glucose can cross the membrane hundreds of times faster than the other sugars. As stated earlier, this is the reason the membrane can be called differentially permeable.

A model for facilitated transport (Fig. 5.9) shows that after a carrier has assisted the movement of a molecule to the other side of the membrane, it is free to assist the passage of other similar molecules. Neither diffusion nor facilitated transport requires an expenditure of energy because the molecules are moving down their concentration gradient in the same direction they tend to move anyway.

**Check Your Progress 5.2**

1. Use the terms solute and solvent to describe a hypotonic and hypertonic solution.
2. Compare and contrast diffusion with facilitated transport.

# 5.3 Active Transport Across a Membrane

During **active transport,** molecules or ions move through the plasma membrane, accumulating either inside or outside the cell. For example, iodine collects in the cells of the thyroid gland; glucose is completely absorbed from the gut by the cells lining the digestive tract; and sodium can be almost completely withdrawn from urine by cells lining the kidney tubules. In these instances, molecules have moved to the region of higher concentration, exactly opposite to the process of diffusion.

Both carrier proteins and an expenditure of energy are needed to transport molecules against their concentration gradient. In this case, chemical energy (ATP molecules usually) is required for the carrier to combine with the substance to be transported. Therefore, it is not surprising that cells involved primarily in active transport, such as kidney cells, have a large number of mitochondria near membranes where active transport is occurring.

Proteins involved in active transport often are called pumps because, just as a water pump uses energy to move water against the force of gravity, proteins use energy to move a substance against its concentration gradient. One

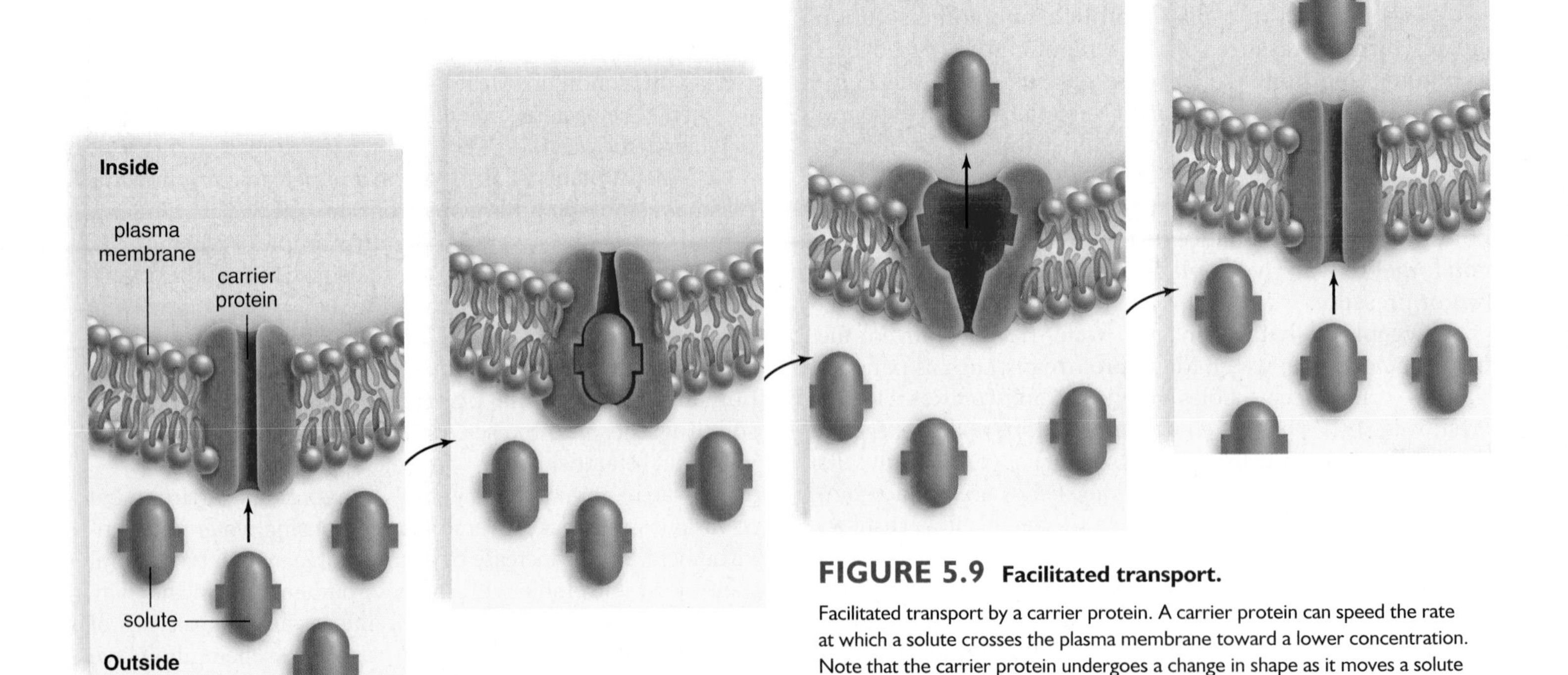


**FIGURE 5.9 Facilitated transport.**

Facilitated transport by a carrier protein. A carrier protein can speed the rate at which a solute crosses the plasma membrane toward a lower concentration. Note that the carrier protein undergoes a change in shape as it moves a solute across the membrane.

type of pump that is active in all animal cells, but is especially associated with nerve and muscle cells, moves sodium ions ($Na^+$) to the outside of the cell and potassium ions ($K^+$) to the inside of the cell. These two events are linked, and the carrier protein is called a **sodium-potassium pump.** A change in carrier shape after the attachment and again after the detachment of a phosphate group allows it to combine alternately with sodium ions and potassium ions (Fig. 5.10). The phosphate group is donated by ATP when it is broken down enzymatically by the carrier. The sodium-potassium pump results in both a solute concentration gradient and an electrical gradient for these ions across the plasma membrane.

The passage of salt (NaCl) across a plasma membrane is of primary importance to most cells. The chloride ion ($Cl^-$) usually crosses the plasma membrane because it is attracted by positively charged sodium ions ($Na^+$). First sodium ions are pumped across a membrane, and then chloride ions simply diffuse through channels that allow their passage.

As noted in Figure 5.3*a*, the genetic disorder cystic fibrosis results from a faulty chloride channel. When chloride is unable to exit a cell, water stays behind. The lack of water causes abnormally thick mucus in the bronchial tubes and pancreatic ducts, thus interfering with the function of the lungs and pancreas.

**FIGURE 5.10 The sodium-potassium pump.**

The same carrier protein transports sodium ions ($Na^+$) to the outside of the cell and potassium ions ($K^+$) to the inside of the cell because it undergoes an ATP-dependent change in shape. Three sodium ions are carried outward for every two potassium ions carried inward; therefore, the inside of the cell is negatively charged compared to the outside.

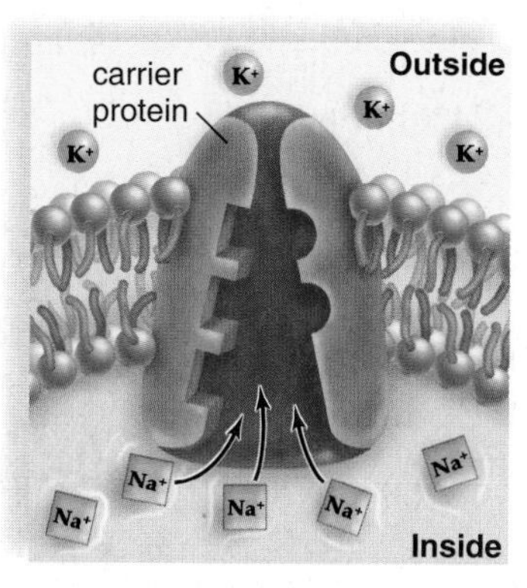


**1.** Carrier has a shape that allows it to take up 3 $Na^+$.

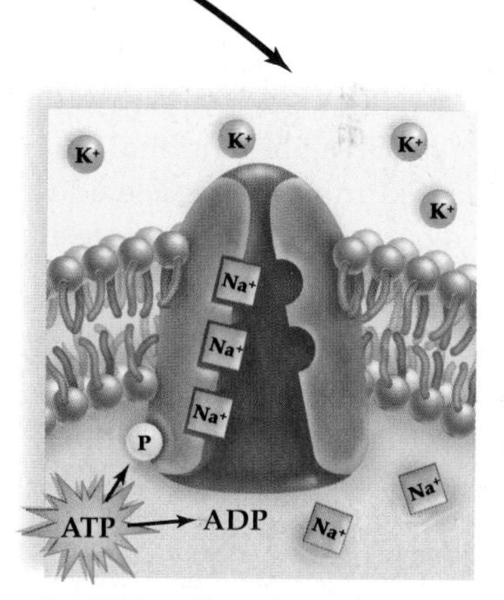


**2.** ATP is split, and phosphate group attaches to carrier.

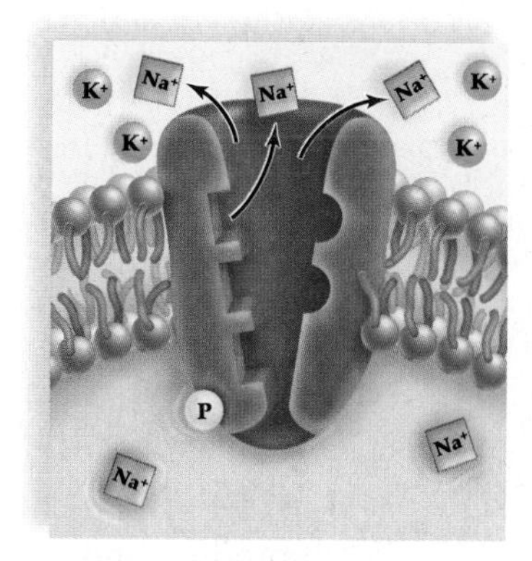


**3.** Change in shape results and causes carrier to release 3 $Na^+$ outside the cell.

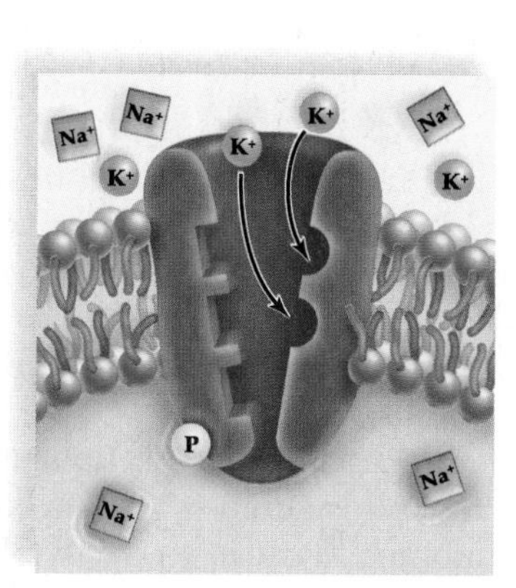


**4.** Carrier has a shape that allows it to take up 2 $K^+$.

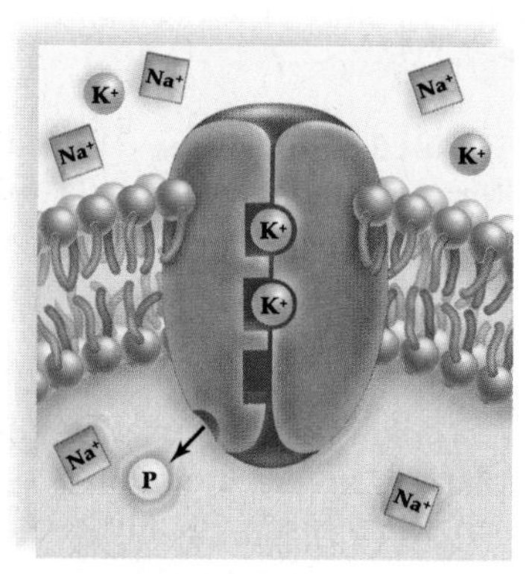


**5.** Phosphate group is released from carrier.

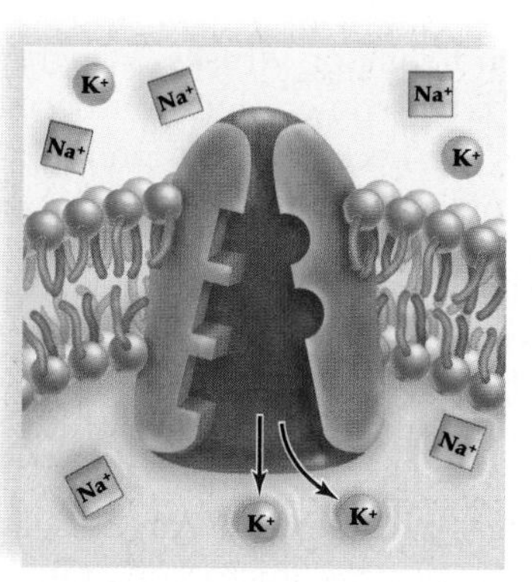


**6.** Change in shape results and causes carrier to release 2 $K^+$ inside the cell.

## Bulk Transport

How do macromolecules such as polypeptides, polysaccharides, or polynucleotides enter and exit a cell? Because they are too large to be transported by carrier proteins, macromolecules are transported into and out of the cell by vesicle formation. Vesicle formation is membrane-assisted transport because membrane is needed to form the vesicle. Vesicle formation requires an expenditure of cellular energy, but vesicle formation has the added benefit that the vesicle membrane keeps the contained macromolecules from mixing with molecules within the cytoplasm. Exocytosis is a way substances can exit a cell, and endocytosis is a way substances can enter a cell.

### *Exocytosis*

During **exocytosis,** a vesicle fuses with the plasma membrane as secretion occurs (Fig. 5.11). Hormones, neurotransmitters, and digestive enzymes are secreted from cells in this manner. The Golgi body often produces the vesicles that carry these cell products to the membrane. During exocytosis, the membrane of the vesicle becomes a part of the plasma membrane, which is thereby enlarged. For this reason, exocytosis can be a normal part of cell growth. The proteins released from the vesicle adhere to the cell surface or become incorporated in an extracellular matrix.

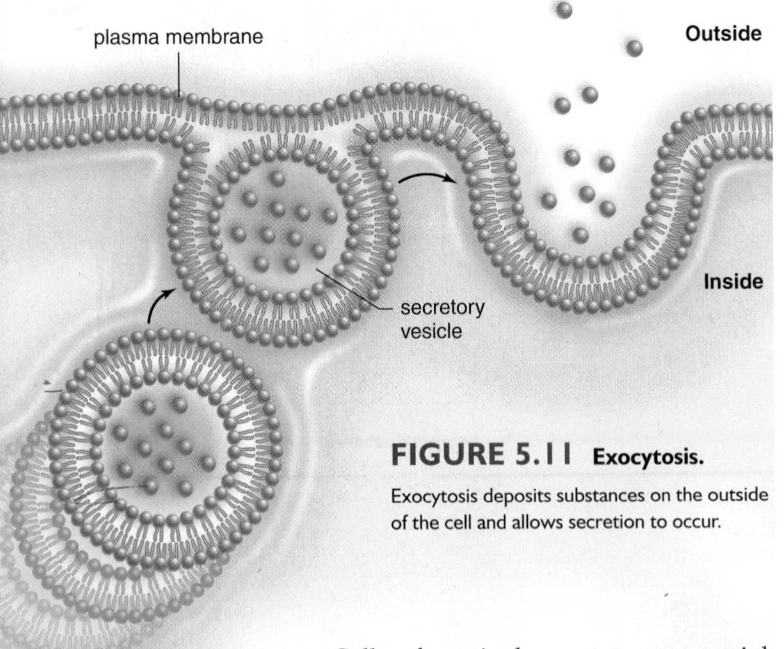


**FIGURE 5.11 Exocytosis.**

Exocytosis deposits substances on the outside of the cell and allows secretion to occur.

Cells of particular organs are specialized to produce and export molecules. For example, pancreatic cells produce digestive enzymes or insulin, and anterior pituitary cells produce growth hormone, among other hormones. In these cells, secretory vesicles accumulate near the plasma membrane, and the vesicles release their contents only when the cell is stimulated by a signal received at the plasma membrane. A rise in blood sugar, for example, signals pancreatic cells to release the hormone insulin. This is called regulated secretion, because vesicles fuse with the plasma membrane only when it is appropriate to the needs of the body.

### *Endocytosis*

During **endocytosis,** cells take in substances by vesicle formation. A portion of the plasma membrane invaginates to envelop the substance, and then the membrane pinches off to form an intracellular vesicle. Endocytosis occurs in one of three ways, as illustrated in Figure 5.12. Phagocytosis transports large substances, such as a virus, and pinocytosis transports small substances, such as a macromolecule, into a cell. Receptor-mediated endocytosis is a special form of pinocytosis.

**Phagocytosis.** When the material taken in by endocytosis is large, such as a food particle or another cell, the process is called **phagocytosis** [Gk. *phagein,* to eat]. Phagocytosis is common in unicellular organisms such as amoebas (Fig. 5.12*a*). It also occurs in humans. Certain types of human white blood cells are amoeboid—that is, they are mobile like an amoeba, and they are able to engulf debris such as worn-out red blood cells or viruses. When an endocytic vesicle fuses with a lysosome, digestion occurs. We will see that this process is a necessary and preliminary step toward the development of immunity to bacterial diseases.

**Pinocytosis. Pinocytosis** [Gk. *pinein,* to drink] occurs when vesicles form around a liquid or around very small particles (Fig. 5.12*b*). Blood cells, cells that line the kidney tubules or the intestinal wall, and plant root cells all use pinocytosis to ingest substances.

Whereas phagocytosis can be seen with the light microscope, the electron microscope must be used to observe pinocytic vesicles, which are no larger than 0.1–0.2 μm. Still, pinocytosis involves a significant amount of the plasma membrane because it occurs continuously. The loss of plasma membrane due to pinocytosis is balanced by the occurrence of exocytosis, however.

**Receptor-Mediated Endocytosis. Receptor-mediated endocytosis** is a form of pinocytosis that is quite specific because it uses a receptor protein shaped in such a way that a specific molecule such as a vitamin, peptide hormone, or lipoprotein can bind to it (Fig. 5.12*c*). The receptors for these substances are found at one location in the plasma membrane. This location is called a coated pit because there is a layer of protein on the cytoplasmic side of the pit. Once formed, the vesicle is uncoated and may fuse with a lysosome. When an empty, used vesicle fuses with the plasma membrane, the receptors return to their former location.

Receptor-mediated endocytosis is selective and much more efficient than ordinary pinocytosis. It is involved in uptake and also in the transfer and exchange of substances between cells. Such exchanges take place when substances move from maternal blood into fetal blood at the placenta, for example.

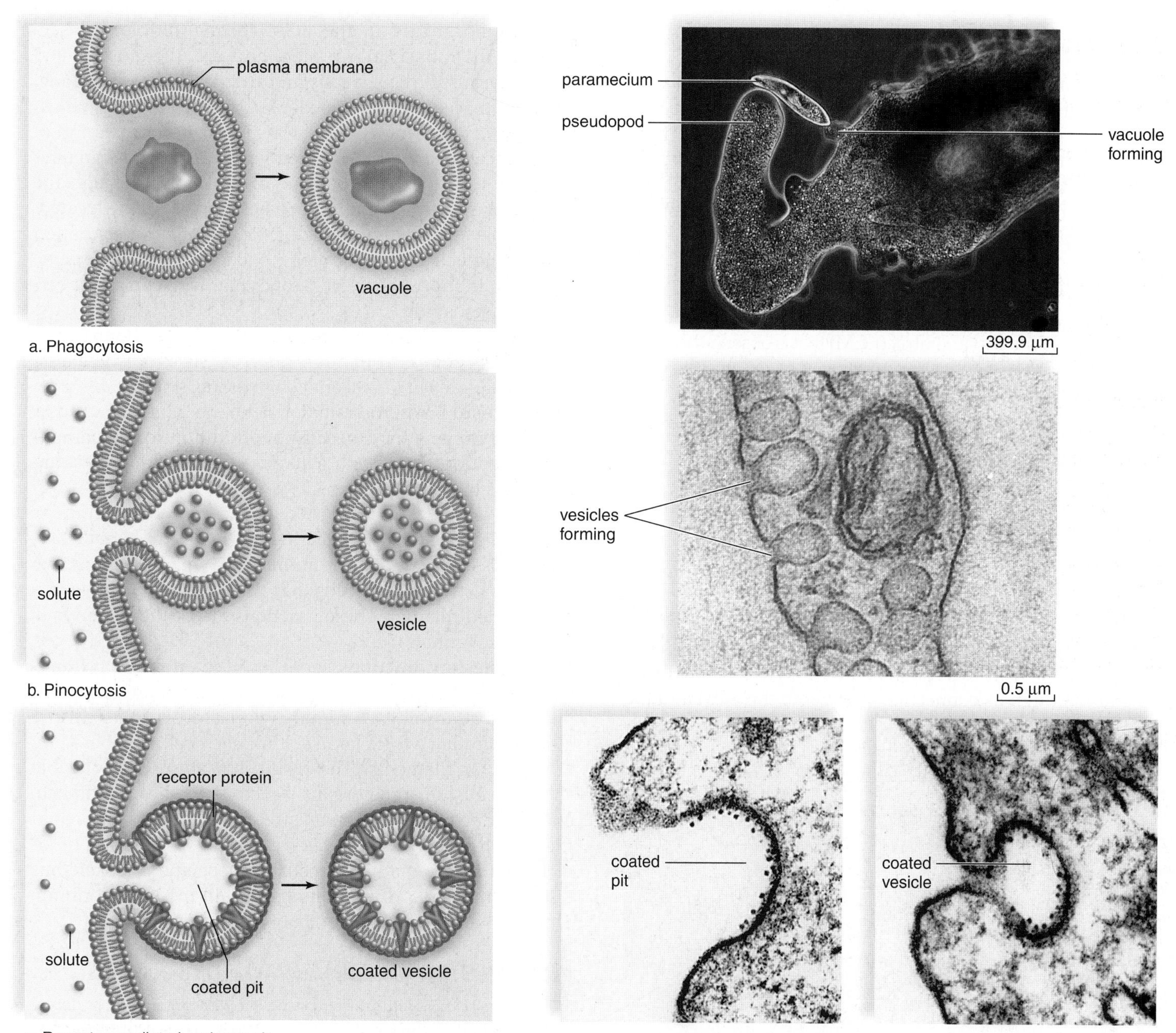


**FIGURE 5.12 Three methods of endocytosis.**

**a.** Phagocytosis occurs when the substance to be transported into the cell is large; amoebas ingest by phagocytosis. Digestion occurs when the resulting vacuole fuses with a lysosome. **b.** Pinocytosis occurs when a macromolecule such as a polypeptide is transported into the cell. The result is a vesicle (small vacuole). **c.** Receptor-mediated endocytosis is a form of pinocytosis. Molecules first bind to specific receptor proteins, which migrate to or are already in a coated pit. The vesicle that forms contains the molecules and their receptors.

The importance of receptor-mediated endocytosis is demonstrated by a genetic disorder called familial hypercholesterolemia. Cholesterol is transported in blood by a complex of lipids and proteins called low-density lipoprotein (LDL). Ordinarily, body cells take up LDL when LDL receptors gather in a coated pit. But in some individuals, the LDL receptor is unable to properly bind to the coated pit, and the cells are unable to take up cholesterol. Instead, cholesterol accumulates in the walls of arterial blood vessels, leading to high blood pressure, occluded (blocked) arteries, and heart attacks.

### Check Your Progress 5.3

1. Compare facilitated transport with active transport.
2. Compare and contrast exocytosis and endocytosis.

## 5.4 Modification of Cell Surfaces

Extracellular structures take shape from materials the cell produces and secretes across its plasma membrane. In plants, prokaryotes, fungi, and most algae, the extracellular component of the cell is a fairly rigid cell wall. A cell wall occurs in organisms that have a rather inactive lifestyle. Animals that have an active way of life have a more varied extracellular anatomy appropriate to the particular tissue type.

### Cell Surfaces in Animals

We will consider two different types of animal cell surface features: (1) the extracellular matrix (ECM) that is observed outside cells and (2) junctions that occur between some types of cells. Both of these can connect to the cytoskeleton and contribute to communication between cells and, therefore, tissue formation.

#### *Extracellular Matrix*

A protective extracellular matrix is a meshwork of proteins and polysaccharides in close association with the cell that produced them (Fig. 5.13). Collagen and elastin fibers are two well-known structural proteins in the ECM; collagen resists stretching and elastin gives the ECM resilience. Fibronectin is an adhesive protein, colored green in Figure 5.13, that binds to a protein in the plasma membrane called integrin. Notice that integrin also makes contact with the cytoskeleton inside the cell. Through its connections with both the ECM and the cytoskeleton, integrin plays a role in cell signaling, permitting the ECM to influence the activities of the cytoskeleton and, therefore, the shape and activities of the cell.

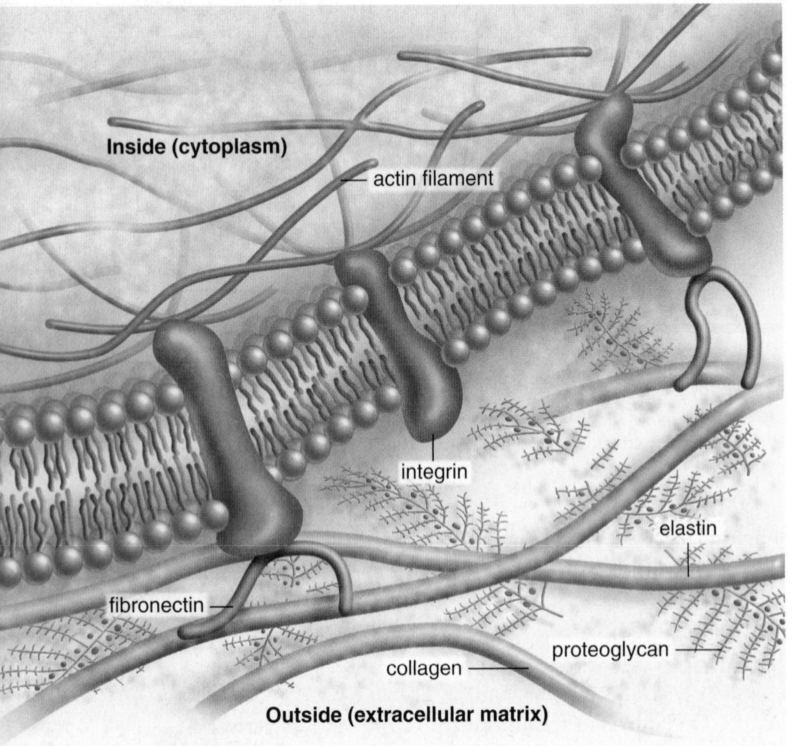


**FIGURE 5.13 Animal cell extracellular matrix.**

In the extracellular matrix, collagen and elastin have a support function, while fibronectins bind to integrin, and in this way, assist communication between ECM and the cytoskeleton.

Amino sugars in the ECM form multiple polysaccharides that attach to a protein and are, therefore, called proteoglycans. Proteoglycans, in turn, attach to a very long, centrally placed polysaccharide. The entire structure, which looks like an enormous bottle brush, resists compression of the extracellular matrix. Proteoglycans assist cell signaling when they regulate the passage of molecules through the ECM to the plasma membrane, where receptors are located. During development, they help bring about differentiation by guiding cell migration along collagen fibers to specific locations. In short, the ECM has a dynamic role in all aspects of a cell's behavior.

When we study tissues, we will see that the extracellular matrix varies in quantity and in consistency from being quite flexible, as in loose connective tissue, semiflexible as in cartilage, and being rock solid, as in bone. The extracellular matrix of bone is hard because, in addition to the components mentioned, mineral salts, notably calcium salts, are deposited outside the cell.

#### *Junctions Between Cells*

Certain tissues of vertebrate animals are well known to have junctions between their cells that allow them to behave in a coordinated manner. These junctions are of the three types shown in Figure 5.14.

**Adhesion junctions** serve to mechanically attach adjacent cells. Two types of adhesion junctions are described here. In **desmosomes,** internal cytoplasmic plaques, firmly attached to the cytoskeleton within each cell, are joined by intercellular filaments. The result is a sturdy but flexible sheet of cells. In some organs—such as the heart, stomach, and bladder, where tissues get stretched—desmosomes hold the cells together. At a *hemidesmosome,* a single point of attachment between adjacent cells connects the cytoskeletons of adjacent cells. Adhesion junctions are the most common type of intercellular junction between skin cells.

**FIGURE 5.14 Junctions between cells of the intestinal wall.**

**a.** In adhesion junctions such as a desmosome, intercellular filaments run between two cells. **b.** Tight junctions between cells form an impermeable barrier because their adjacent plasma membranes are joined. **c.** Gap junctions allow communication between two cells because adjacent plasma membrane channels are joined.

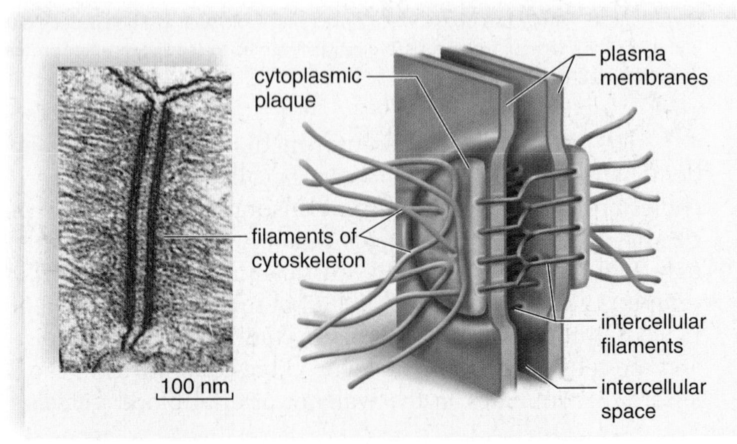


a. Adhesion junction

Adjacent cells are even more closely joined by **tight junctions,** in which plasma membrane proteins actually attach to each other, producing a zipperlike fastening. The cells of tissues that serve as barriers are held together by tight junctions; in the intestine, the digestive juices stay out of the body, and in the kidneys the urine stays within kidney tubules, because the cells are joined by tight junctions.

A **gap junction** allows cells to communicate. A gap junction is formed when two identical plasma membrane channels join. The channel of each cell is lined by six plasma membrane proteins. A gap junction lends strength to the cells, but it also allows small molecules and ions to pass between them. Gap junctions are important in heart muscle and smooth muscle because they permit a flow of ions that is required for the cells to contract as a unit.

## Plant Cell Walls

In addition to a plasma membrane, plant cells are surrounded by a porous **cell wall** that varies in thickness, depending on the function of the cell. All plant cells have a primary cell wall. The primary cell wall contains cellulose fibrils in which microfibrils are held together by noncellulose substances. Pectins allow the wall to stretch when the cell is growing, and noncellulose polysaccharides harden the wall when the cell is mature. Pectins are especially abundant in the middle lamella, which is a layer of adhesive substances that holds the cells together. Some cells in woody plants have a secondary wall that forms inside the primary cell wall. The secondary wall has a greater quantity of cellulose fibrils than the primary wall, and layers of cellulose fibrils are laid down at right angles to one another. Lignin, a substance that adds strength, is a common ingredient of secondary cell walls in woody plants.

In a plant, the cytoplasm of living cells is connected by **plasmodesmata** (sing., plasmodesma), numerous narrow, membrane-lined channels that pass through the cell wall (Fig. 5.15). Cytoplasmic strands within these channels allow direct exchange of some materials between adjacent plant cells and eventually all the cells of a plant. The plasmodesmata are large enough to allow only water and small solutes to pass freely from cell to cell. This limitation means that plant cells can maintain their own concentrations of larger substances and differentiate into particular cell types.

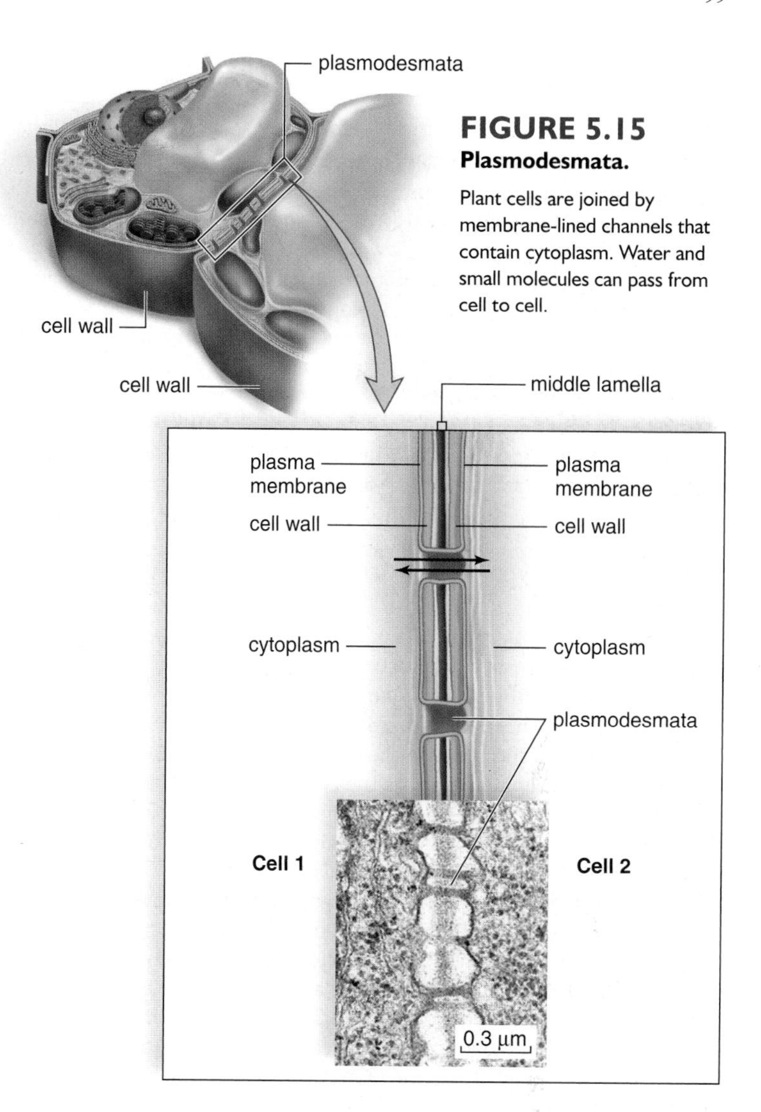


**FIGURE 5.15**
**Plasmodesmata.**
Plant cells are joined by membrane-lined channels that contain cytoplasm. Water and small molecules can pass from cell to cell.

### Check Your Progress 5.4

1. Describe the chemical composition of the extracellular matrix of an animal cell.
2. Give a function for an adhesion junction, tight junction, and gap junction.
3. Contrast a plant's primary cell wall with its secondary cell wall.

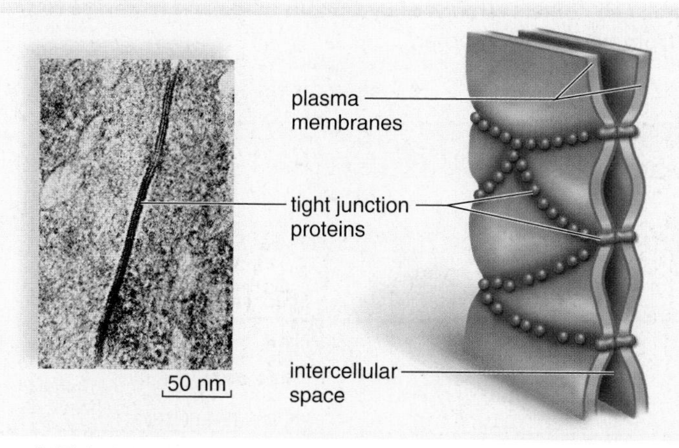


b. Tight junction

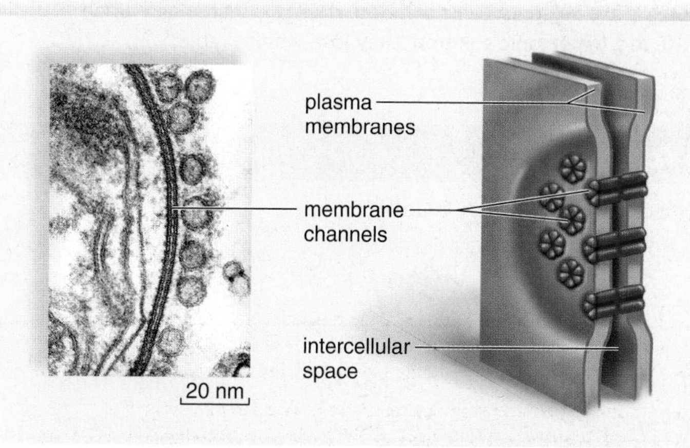


c. Gap junction

## Connecting the Concepts

The plasma membrane is quite appropriately called the gatekeeper of the cell because it maintains the integrity of the cell and stands guard over what enters and leaves. But we have seen that the plasma membrane also does much more than this. Its glycoproteins and glycolipids mark the cell as belonging to the organism. Its numerous proteins allow communication between cells and, thereby, enable tissues to function as a whole. A new endeavor is to understand how the extracellular matrix in animal cells assists the plasma membrane in its varied functions.

The progression in our knowledge about the plasma membrane illustrates how science works. The concepts and techniques of science evolve and change, and the knowledge we have today will be amended and expanded by new investigative work. Also, basic science has applications that promote the health of human beings. To know that the plasma membrane is malfunctioning in a person who has diabetes or cystic fibrosis or in someone who has a high cholesterol count is a first step toward curing these conditions. Even cancer is sometimes due to receptor proteins that signal the cell to divide even when no growth factor is present.

Our ability to understand the functioning of the plasma membrane is dependent on a working knowledge of the molecules that make up the cell. We continue this theme as we discuss metabolism in the next three chapters.

## summary

### 5.1 Plasma Membrane Structure and Function

Two components of the plasma membrane are lipids and proteins. In the lipid bilayer, phospholipids are arranged with their hydrophilic (polar) heads at the surfaces and their hydrophobic (nonpolar) tails in the interior. The lipid bilayer has the consistency of oil but acts as a barrier to the entrance and exit of most biological molecules. Membrane glycolipids and glycoproteins are involved in marking the cell as belonging to a particular individual and tissue.

The hydrophobic portion of an integral protein lies in the lipid bilayer of the plasma membrane, and the hydrophilic portion lies at the surfaces. Proteins act as receptors, carry on enzymatic reactions, join cells together, form channels, or act as carriers to move substances across the membrane. Some of these proteins make contact with the extracellular matrix (ECM) outside and with the cytoskeleton inside. Thus, the ECM can influence the happenings inside the cell.

### 5.2 Passive Transport Across a Membrane

The plasma membrane is differentially permeable. Some molecules (lipid-soluble compounds, water, and gases) simply diffuse across the membrane from the area of higher concentration to the area of lower concentration. No metabolic energy is required for diffusion to occur.

The diffusion of water across a differentially permeable membrane is called osmosis. Water moves across the membrane into the area of higher solute (less water) content per volume. When cells are in an isotonic solution, they neither gain nor lose water. When cells are in a hypotonic solution, they gain water, and when they are in a hypertonic solution, they lose water (Table 5.2).

Other molecules are transported across the membrane either by a channel protein or by carrier proteins that span the membrane. During facilitated transport, a substance moves down its concentration gradient. No energy is required.

### 5.3 Active Transport Across a Membrane

During active transport, a carrier protein acts as a pump that causes a substance to move against its concentration gradient. The sodium-potassium pump carries $Na^+$ to the outside of the cell and $K^+$ to the inside of the cell. Energy in the form of ATP molecules is required for active transport to occur.

Larger substances can enter and exit a membrane by exocytosis and endocytosis. Exocytosis involves secretion. Endocytosis includes phagocytosis, pinocytosis, and receptor-mediated endocytosis. Receptor-mediated endocytosis makes use of receptor proteins in the plasma membrane. Once a specific solute binds to receptors, a coated pit becomes a coated vesicle. After losing the coat, the vesicle can join with the lysosome, or after discharging the substance, the receptor-containing vesicle can fuse with the plasma membrane.

### 5.4 Modification of Cell Surfaces

Animal cells have an extracellular matrix (ECM) that influences their shape and behavior. Tissues vary as to the amount and character of the ECM. Some animal cells have junction proteins that join them to other cells of the same tissue. Adhesion junctions and tight junctions help hold cells together; gap junctions allow passage of small molecules between cells.

Plant cells have a freely permeable cell wall, with cellulose as its main component. Also, plant cells are joined by narrow, membrane-lined channels called plasmodesmata that span the cell wall and contain strands of cytoplasm that allow materials to pass from one cell to another.

**TABLE 5.2**

**Effect of Osmosis on a Cell**

| | *Concentrations* | | | |
|---|---|---|---|---|
| *Tonicity of Solution* | *Solute* | *Water* | *Net Movement of Water* | *Effect on Cell* |
| Isotonic | Same as cell | Same as cell | None | None |
| Hypotonic | Less than cell | More than cell | Cell gains water | Swells, turgor pressure |
| Hypertonic | More than cell | Less than cell | Cell loses water | Shrinks, plasmolysis |

## understanding the terms

active transport 94
adhesion junction 98
aquaporin 90
bulk transport 90
carrier protein 88
cell recognition protein 88
cell wall 99
channel protein 88
cholesterol 86
concentration gradient 90
crenation 93
desmosome 98
differentially permeable 90
diffusion 91
endocytosis 96
enzymatic protein 88
exocytosis 96
extracellular matrix (ECM) 87
facilitated transport 94
fluid-mosaic model 87
gap junction 99
glycolipid 87
glycoprotein 87
hypertonic solution 93
hypotonic solution 92
isotonic solution 92
junction protein 88
osmosis 92
osmotic pressure 92
phagocytosis 96
pinocytosis 96
plasmodesmata 99
plasmolysis 93
receptor-mediated endocytosis 96
receptor protein 88
sodium-potassium pump 94
solute 91
solution 91
solvent 91
tight junction 99
tonicity 92
turgor pressure 93

Match the terms to these definitions:

a. _______________ Characteristic of the plasma membrane due to its ability to allow certain molecules but not others to pass through.
b. _______________ Diffusion of water through the plasma membrane of cells.
c. _______________ Higher solute concentration (less water) than the cytoplasm of a cell; causes cell to lose water by osmosis.
d. _______________ Protein in plasma membrane that bears a carbohydrate chain.
e. _______________ Process by which a cell engulfs a substance, forming an intracellular vacuole.

## reviewing this chapter

1. Describe the fluid-mosaic model of membrane structure. 86–87
2. Tell how the phospholipids are arranged in the plasma membrane. What other lipid is present in the membrane, and what functions does it serve? 87–88
3. Describe the possible functions of proteins in the plasma membrane. 88
4. What is cell signaling and how does it occur? 89
5. Define diffusion. What factors can influence the rate of diffusion? What substances can diffuse through a differentially permeable membrane? 90–91
6. Define osmosis. Describe verbally and with drawings what happens to an animal cell and a plant cell when placed in isotonic, hypotonic, and hypertonic solutions. 92–93
7. Why do most substances have to be assisted through the plasma membrane? Contrast movement by facilitated transport with movement by active transport. 94–95
8. Draw and explain a diagram that shows how the sodium-potassium pump works. 94–95
9. Describe and contrast three methods of endocytosis. 96–97
10. Describe the structure and function of animal and plant cell modifications. 98–99

## testing yourself

Choose the best answer for each question.

1. Write hypotonic solution or hypertonic solution beneath each cell. Justify your conclusions.

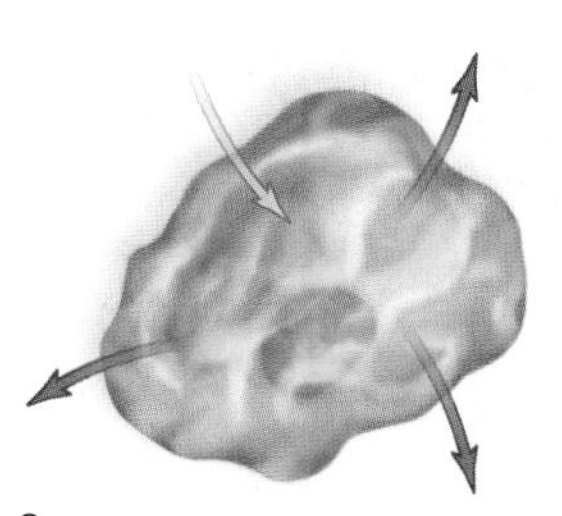

a. _______________

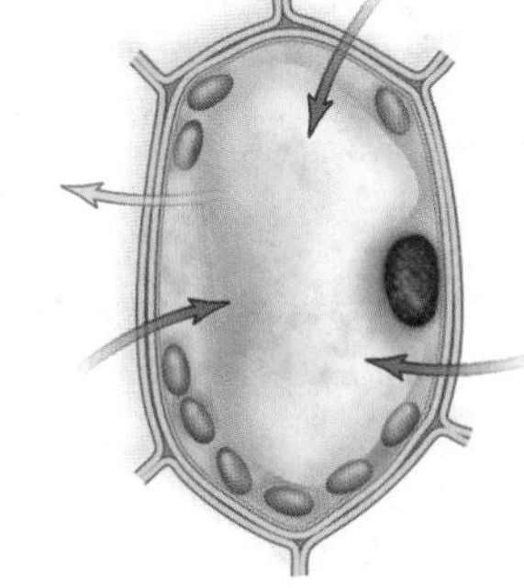

b. _______________

2. Electron micrographs following freeze-fracture of the plasma membrane indicate that
   a. the membrane is a phospholipid bilayer.
   b. some proteins span the membrane.
   c. protein is found only on the surfaces of the membrane.
   d. glycolipids and glycoproteins are antigenic.
   e. there are receptors in the membrane.
3. A phospholipid molecule has a head and two tails. The tails are found
   a. at the surfaces of the membrane.
   b. in the interior of the membrane.
   c. spanning the membrane.
   d. where the environment is hydrophilic.
   e. Both a and b are correct.
4. During diffusion,
   a. solvents move from the area of higher to lower concentration, but solutes do not.
   b. there is a net movement of molecules from the area of higher to lower concentration.
   c. a cell must be present for any movement of molecules to occur.
   d. molecules move against their concentration gradient if they are small and charged.
   e. All of these are correct.
5. When a cell is placed in a hypotonic solution,
   a. solute exits the cell to equalize the concentration on both sides of the membrane.
   b. water exits the cell toward the area of lower solute concentration.
   c. water enters the cell toward the area of higher solute concentration.
   d. solute exits and water enters the cell.
   e. Both c and d are correct.
6. When a cell is placed in a hypertonic solution,
   a. solute exits the cell to equalize the concentration on both sides of the membrane.
   b. water exits the cell toward the area of lower solute concentration.
   c. water exits the cell toward the area of higher solute concentration.
   d. solute exits and water enters the cell.
   e. Both a and c are correct.

7. Active transport
   a. requires a carrier protein.
   b. moves a molecule against its concentration gradient.
   c. requires a supply of chemical energy.
   d. does not occur during facilitated transport.
   e. All of these are correct.
8. The sodium-potassium pump
   a. helps establish an electrochemical gradient across the membrane.
   b. concentrates sodium on the outside of the membrane.
   c. uses a carrier protein and chemical energy.
   d. is present in the plasma membrane.
   e. All of these are correct.
9. Receptor-mediated endocytosis
   a. is no different from phagocytosis.
   b. brings specific solutes into the cell.
   c. helps concentrate proteins in vesicles.
   d. results in high osmotic pressure.
   e. All of these are correct.
10. Plant cells
    a. always have a secondary cell wall, even though the primary one may disappear.
    b. have channels between cells that allow strands of cytoplasm to pass from cell to cell.
    c. develop turgor pressure when water enters the nucleus.
    d. do not have cell-to-cell junctions like animal cells.
    e. All of these are correct.
11. Label this diagram of the plasma membrane.

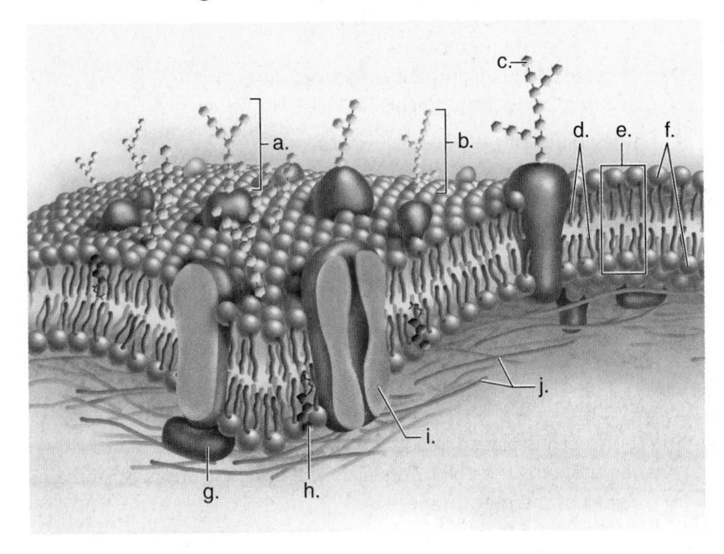


12. The fluid-mosaic model of membrane structure refers to
    a. the fluidity of proteins and the pattern of phospholipids in the membrane.
    b. the fluidity of phospholipids and the pattern of proteins in the membrane.
    c. the fluidity of cholesterol and the pattern of carbohydrate chains outside the membrane.
    d. the lack of fluidity of internal membranes compared to the plasma membrane, and the ability of the proteins to move laterally in the membrane.
    e. the fluidity of hydrophobic regions, proteins, and the mosaic pattern of hydrophilic regions.
13. Which of the following is not a function of proteins present in the plasma membrane? Proteins
    a. assist the passage of materials into the cell.
    b. interact and recognize other cells.
    c. bind with specific hormones.
    d. carry out specific metabolic reactions.
    e. produce lipid molecules.
14. The carbohydrate chains projecting from the plasma membrane are involved in
    a. adhesion between cells.
    b. reception of molecules.
    c. cell-to-cell recognition.
    d. All of these are correct.
15. Plants wilt on a hot summer day because of a decrease in
    a. turgor pressure.
    b. evaporation.
    c. condensation.
    d. diffusion.
16. The extracellular matrix
    a. assists in the movement of substances across the plasma membrane.
    b. prevents the loss of water when cells are placed in a hypertonic solution.
    c. has numerous functions that affect the shape and activities of the cell that produced it.
    d. contains the junctions that sometimes occur between cells.
    e. All of these are correct.

## thinking scientifically

1. The mucus in bronchial tubes must be thin enough for cilia to move bacteria and viruses up into the throat away from the lungs. Which way would $Cl^-$ normally cross the plasma membrane of bronchial tube cells in order for mucus to be thin (see Fig. 5.3*a*)? Use the concept of osmosis to explain your answer.
2. Winter wheat is planted in the early fall, grows over the winter when the weather is colder, and is harvested in the spring. As the temperature drops, the makeup of the plasma membrane of winter wheat changes. Unsaturated fatty acids replace saturated fatty acids in the phospholipids of the membrane. Why is this a suitable adaptation?

## *Biology* website

The companion website for *Biology* provides a wealth of information organized and integrated by chapter. You will find practice tests, animations, videos, and much more that will complement your learning and understanding of general biology.

**http://www.mhhe.com/maderbiology10**

# 6

# Metabolism: Energy and Enzymes

*Photosynthesizing grasses on an African plain provide impalas with building blocks and the energy they need to evade being caught by a cheetah. Eating impalas provides cheetahs with food and the energy they need to be quick enough to catch impalas!*

*All life on Earth is dependent on the flow of energy coming from the sun. You, like the cheetah, are dependent on energy from the sun. Even as you digest your food, be it veggies or meat, energy escapes into the environment as heat. This heat is no longer usable by photosynthesizers; it is too diffuse. Solar energy is concentrated enough to allow plants to keep on photosynthesizing and, in that way, provide a continual supply of food for you and the biosphere.*

*Energy, so important to metabolism and enzymatic reactions, is the first topic we consider in this chapter. Without enzymes, you and the cheetah would not be able to make use of energy to maintain your bodies, nor to carry on any type of activity.*

The cheetah, and more directly the impala, is dependent on solar energy captured by photosynthesizers.

## concepts

# 6.1 Cells and the Flow of Energy

In order to maintain their organization and carry out metabolic activities, cells—as well as organisms—need a constant supply of energy. **Energy**, defined as the ability to do work or bring about a change, allows living things to carry on the processes of life, including growth, development, metabolism, and reproduction.

Organic nutrients, produced by photosynthesizers (algae, plants, and some bacteria), directly provide organisms with energy. But, consider that photosynthesizers use solar energy to produce organic nutrients; therefore, life on Earth is ultimately dependent on solar energy.

## Forms of Energy

Energy occurs in two forms: kinetic and potential energy. **Kinetic energy** is the energy of motion, as when a ball rolls down a hill or a moose walks through grass. **Potential energy** is stored energy—its capacity to accomplish work is not being used at the moment. The food we eat has potential energy because it can be converted into various types of kinetic energy. Food is specifically called **chemical energy** because it is composed of organic molecules such as carbohydrates, proteins, and fat. When a moose walks, it has converted chemical energy into a type of kinetic energy called **mechanical energy** (Fig. 6.1).

## Two Laws of Thermodynamics

Figure 6.1 illustrates the flow of energy in a terrestrial ecosystem. Plants capture only a small portion of solar energy, and much of it dissipates as **heat.** When plants photosynthesize and then make use of the food they produce, more heat results. Still, there is enough remaining to sustain a moose and the other organisms in an ecosystem. As they metabolize nutrient molecules, all the captured solar energy eventually dissipates as heat. Therefore, energy flows and does not cycle. Two **laws of thermodynamics** explain why energy flows through ecosystems and through cells. These laws were formulated by early researchers who studied energy relationships and exchanges:

> The first law of thermodynamics—the law of conservation of energy—states energy cannot be created or destroyed, but it can be changed from one form to another.

When leaf cells photosynthesize, they use solar energy to form carbohydrate molecules from carbon dioxide and water. (Carbohydrates are energy-rich molecules, while carbon dioxide and water are energy-poor molecules.) Not all of the captured solar energy becomes carbohydrates; some becomes heat:

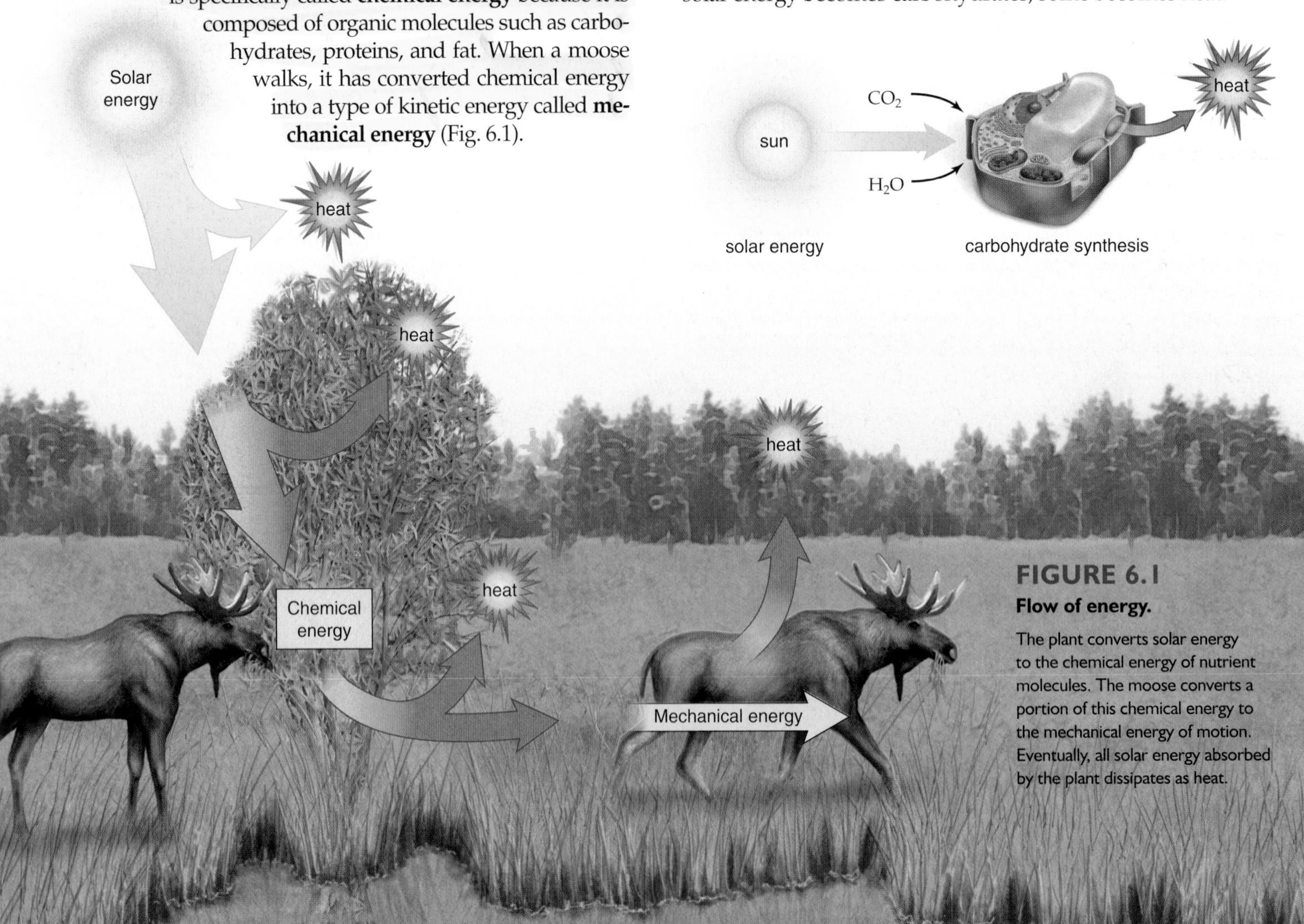


**FIGURE 6.1**
**Flow of energy.**
The plant converts solar energy to the chemical energy of nutrient molecules. The moose converts a portion of this chemical energy to the mechanical energy of motion. Eventually, all solar energy absorbed by the plant dissipates as heat.

Obviously, plant cells do not create the energy they use to produce carbohydrate molecules; that energy comes from the sun. Is any energy destroyed? No, because the heat they give off is also a form of energy. Similarly, a moose uses the energy derived from carbohydrates to power its muscles. And as its cells use this energy, none is destroyed, but some becomes heat, which dissipates into the environment:

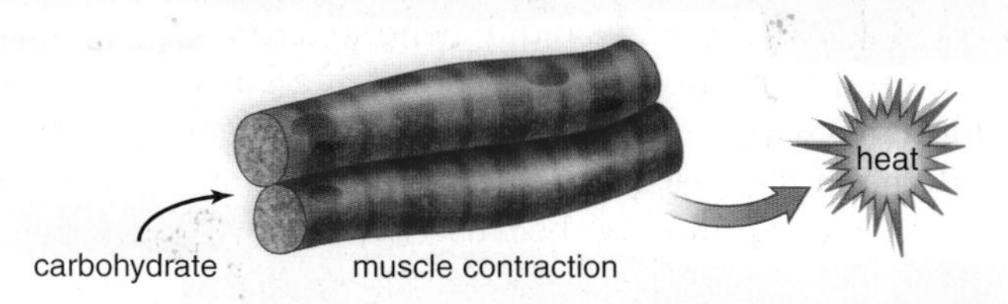


The second law of thermodynamics therefore applies to living systems:

The second law of thermodynamics states energy cannot be changed from one form to another without a loss of usable energy.

In our example, this law is upheld because some of the solar energy taken in by the plant and some of the chemical energy within the nutrient molecules taken in by the moose become heat. When heat dissipates into the environment, it is no longer usable—that is, it is not available to do work. With transformation upon transformation, eventually all usable forms of energy become heat that is lost to the environment. Heat that dissipates into the environment cannot be captured and converted to one of the other forms of energy.

As a result of the second law of thermodynamics, no process requiring a conversion of energy is ever 100% efficient. Much of the energy is lost in the form of heat. In automobiles, the gasoline engine is between 20% and 30% efficient in converting chemical energy into mechanical energy. The majority of energy is obviously lost as heat. Cells are capable of about 40% efficiency, with the remaining energy being given off to the surroundings as heat.

## Cells and Entropy

The second law of thermodynamics can be stated another way: Every energy transformation makes the universe less organized and more disordered. The term **entropy** [Gk. *entrope*, a turning inward] is used to indicate the relative amount of disorganization. Since the processes that occur in cells are energy transformations, the second law means that every process that occurs in cells always does so in a way that increases the total entropy of the universe. Then, too, any one of these processes makes less energy available to do useful work in the future.

Figure 6.2 shows two processes that occur in cells. The second law of thermodynamics tells us that glucose tends to break apart into carbon dioxide and water. Why? Because glucose is more organized, and therefore less stable, than its breakdown products. Also, hydrogen ions on one side of a membrane tend to move to the other side unless they are prevented from doing so. Why? Because when they are distributed randomly, entropy has increased. As an analogy, you know from experience that a neat room is more organized but less stable than a messy room, which is disorganized but more stable. How do you know a neat room is less stable than a messy room? Consider that a neat room always tends to become more messy.

On the other hand, you know that some cells can make glucose out of carbon dioxide and water, and all cells can actively move ions to one side of the membrane. How do they do it? These cellular processes obviously require an input of energy from an outside source. This energy ultimately comes from the sun. Living things depend on a constant supply of energy from the sun because the ultimate fate of all solar energy in the biosphere is to become randomized in the universe as heat. A living cell is a temporary repository of order purchased at the cost of a constant flow of energy.

**Check Your Progress 6.1**

1. Contrast potential energy with kinetic energy.
2. Explain how the second energy law is related to entropy.

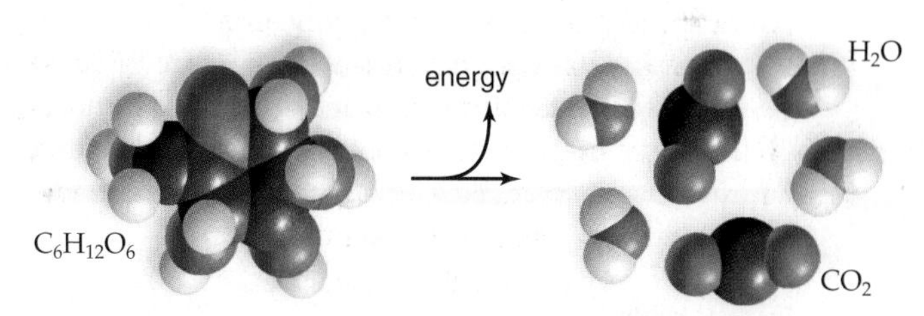


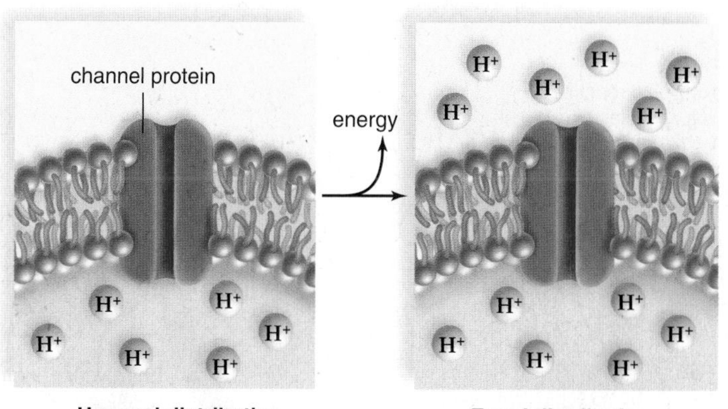


**FIGURE 6.2 Cells and entropy.**

The second law of thermodynamics tells us that (**a**) glucose, which is more organized, tends to break down to carbon dioxide and water, which are less organized. **b.** Similarly, hydrogen ions ($H^+$) on one side of a membrane tend to move to the other side so that the ions are randomly distributed. Both processes result in an increase in entropy.

## 6.2 Metabolic Reactions and Energy Transformations

**Metabolism** is the sum of all the chemical reactions that occur in a cell. **Reactants** are substances that participate in a reaction, while **products** are substances that form as a result of a reaction. In the reaction A + B $\longrightarrow$ C + D, A and B are the reactants while C and D are the products. How would you know that this reaction will occur spontaneously—that is, without an input of energy? Using the concept of entropy, it is possible to state that a reaction will occur spontaneously if it increases the entropy of the universe. But in cell biology, we do not usually wish to consider the entire universe. We simply want to consider this reaction. In such instances, cell biologists use the concept of free energy instead of entropy. **Free energy** is the amount of energy available—that is, energy that is still "free" to do work—after a chemical reaction has occurred. The change in free energy after a reaction occurs is calculated by subtracting the free energy content of the reactants from that of the products. A negative result means that the products have less free energy than the reactants, and the reaction will occur spontaneously. In our reaction, if C and D have less free energy than A and B, then the reaction will "go."

**Exergonic reactions** are spontaneous and release energy, while **endergonic reactions** require an input of energy to occur. In the body, many reactions, such as protein synthesis, nerve conduction, or muscle contraction, are endergonic, and they occur because exergonic reactions, which release energy, can be used to drive endergonic reactions, which require energy. ATP is a carrier of energy between exergonic and endergonic reactions.

### ATP: Energy for Cells

**ATP (adenosine triphosphate)** is the common energy currency of cells; when cells require energy, they "spend" ATP. A sedentary oak tree as well as a flying bat requires vast amounts of ATP. The more active the organism, the greater the demand for ATP. However, the amount on hand at any one moment is minimal because ATP is constantly being generated from **ADP (adenosine diphosphate)** and a molecule of inorganic phosphate Ⓟ (Fig. 6.3). A cell is assured of a supply of ATP, because glucose breakdown during cellular respiration provides the energy for the buildup of ATP in mitochondria. Only 39% of the free energy of glucose is transformed to ATP; the rest is lost as heat.

There are many biological advantages to the use of ATP as an energy carrier in living systems. ATP provides a common and universal energy currency because it can be used in many different types of reactions. Also, when ATP is converted to energy, ADP, and Ⓟ, the amount of energy released is sufficient for a particular biological function, and there is little waste of energy. In addition, ATP breakdown can be coupled to endergonic reactions in such a way that it minimizes energy loss.

#### *Structure of ATP*

ATP is a nucleotide composed of the nitrogen-containing base adenine and the 5-carbon sugar ribose (together called adenosine) and three phosphate groups. ATP is called a "high-energy" compound because a phosphate group can be easily removed. Under cellular conditions, the amount of energy released when ATP is hydrolyzed to ADP + Ⓟ is about 7.3 kcal per mole.[1]

[1] A mole is the number of molecules present in the molecular weight of a substance (in grams).

adenosine triphosphate

P — P — P

Energy from exergonic reactions (e.g., cellular respiration)

ATP

Energy for endergonic reactions (e.g., protein synthesis, nerve conduction, muscle contraction)

ADP + P

P — P + P

adenosine diphosphate + phosphate

a.

b. 2.25×

**FIGURE 6.3 The ATP cycle.**

**a.** In cells, ATP carries energy between exergonic reactions and endergonic reactions. When a phosphate group is removed by hydrolysis, ATP releases the appropriate amount of energy for most metabolic reactions. **b.** In order to produce light, a firefly breaks down ATP.

## Coupled Reactions

How can the energy released by ATP hydrolysis be transferred to a reaction that requires energy, and therefore would not ordinarily occur? In other words, how does ATP act as a carrier of chemical energy? The answer is that ATP breakdown is coupled to the energy-requiring reaction. **Coupled reactions** are reactions that occur in the same place, at the same time, and in such a way that an energy-releasing (exergonic) reaction drives an energy requiring (endergonic) reaction. Usually the energy-releasing reaction is the hydrolysis of ATP. Because the cleavage of ATP's phosphate groups releases more energy than the amount consumed by the energy-requiring reaction, entropy will increase, and both reactions will proceed. The simplest way to represent a coupled reaction is like this:

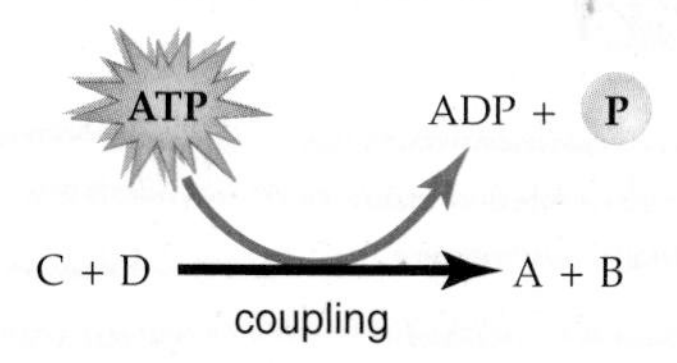


This reaction tells you that coupling occurs, but it does not show how coupling is achieved. A cell has two main ways to couple ATP hydrolysis to an energy-requiring reaction: ATP is used to energize a reactant, or ATP is used to change the shape of a reactant. Both can be achieved by transferring a phosphate group to the reactant so that the product is *phosphorylated.*

For example, when an ion moves across the plasma membrane of a cell, ATP is hydrolyzed and, instead of the last phosphate group floating away, an enzyme attaches it to a carrier protein. This causes the protein to undergo a change in shape that allows it to move the ion into or out of the cell. As a contrasting example, when a polypeptide is synthesized at a ribosome, an enzyme transfers a phosphate group from ATP to each amino acid in turn, and this transfer supplies the energy that allows an amino acid to bond with another amino acid.

Figure 6.4 shows how ATP hydrolysis provides the necessary energy for muscle contraction. During muscle contraction, myosin filaments pull actin filaments to the center of the cell, and the muscle shortens. **1** Myosin head combines with ATP (three connected green triangles) and takes on its resting shape. **2** ATP breaks down to ADP (two connected green triangles) plus Ⓟ (one green triangle). Now a change in shape allows myosin to attach to actin. **3** The release of ADP and Ⓟ from myosin head causes it to change its shape again and pull on the actin filament. The cycle begins again at **1**, when myosin head combines with ATP and takes on its resting shape. During this cycle, chemical energy has been transformed to mechanical energy, and entropy has increased.

Through coupled reactions, ATP drives forward energetically unfavorable processes that must occur to create the high degree of order essential for life. Macromolecules must be made and organized to form cells and tissues; the internal composition of the cell and the organism must be maintained; and movement of cellular organelles and the organism must occur if life is to continue.

### Check Your Progress 6.2

1. Explain why ATP is a good short-term energy storage molecule.
2. Briefly explain the function of ATP in coupled reactions.

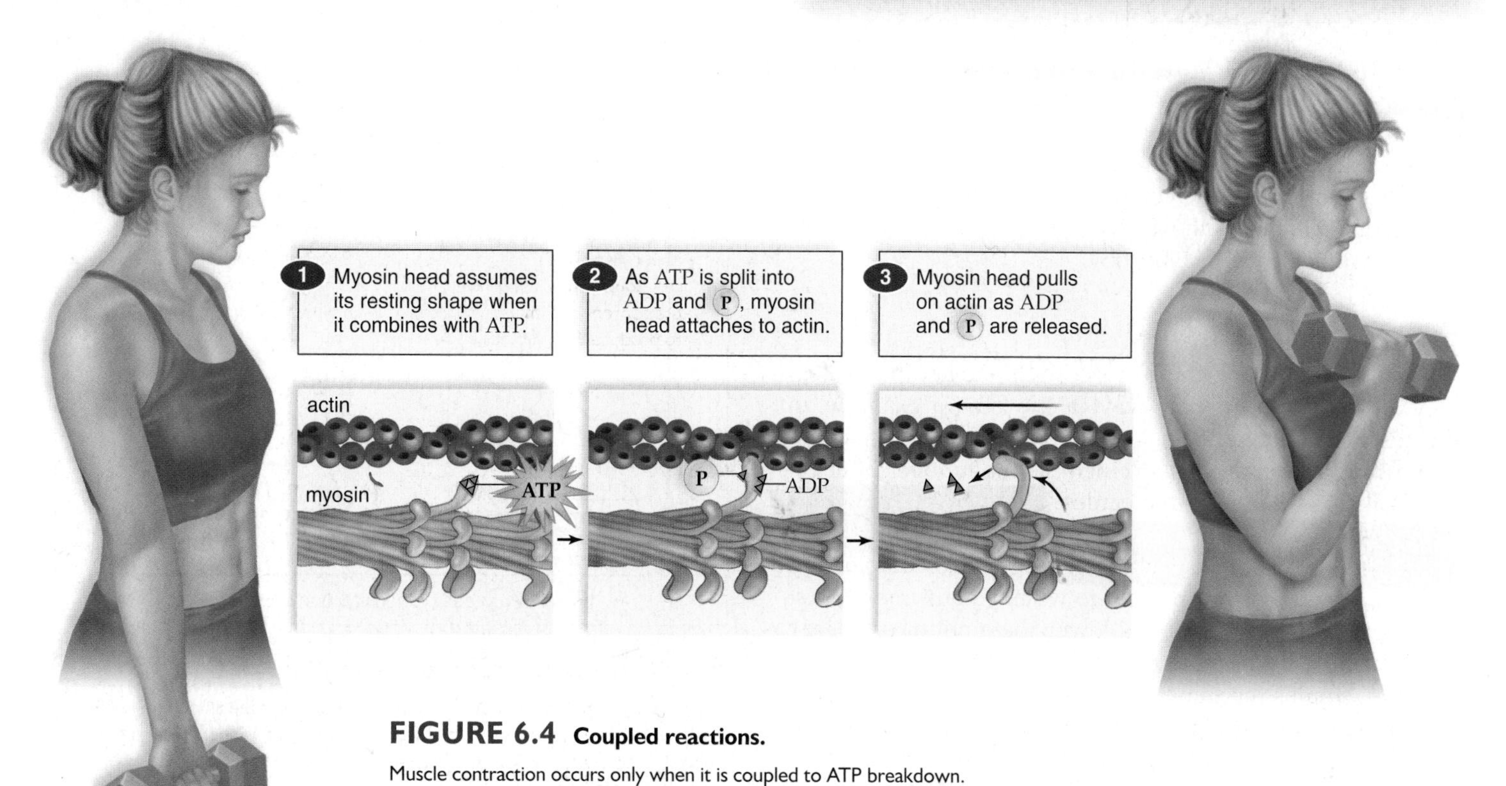


**FIGURE 6.4 Coupled reactions.**

Muscle contraction occurs only when it is coupled to ATP breakdown.

# 6.3 Metabolic Pathways and Enzymes

An **enzyme** is a protein molecule that functions as an organic catalyst to speed a chemical reaction without itself being affected by the reaction. In this part of the chapter, we will also see that enzymes allow reactions to occur under mild conditions, and that they regulate metabolism, including the elimination of side reactions.

First, we will mention that not all enzymes are proteins. **Ribozymes,** which are made of RNA instead of proteins, can also serve as biological catalysts. Ribozymes are involved in the synthesis of RNA and the synthesis of proteins at the ribosomes.

Let's also recognize that reactions do not occur haphazardly in cells; they are usually part of a **metabolic pathway,** a series of linked reactions. Metabolic pathways begin with a particular reactant and terminate with an end product. While it is possible to write an overall equation for a pathway as if the beginning reactant went to the end product in one step, actually many specific steps occur in between. In the pathway, one reaction leads to the next reaction, which leads to the next reaction, and so forth, in an organized, highly structured manner. This arrangement makes it possible for one pathway to lead to several others, because various pathways have several molecules in common. Also, metabolic energy is captured and used more easily if it is released in small increments rather than all at once.

A metabolic pathway can be represented by the following diagram:

$$A \xrightarrow{E_1} B \xrightarrow{E_2} C \xrightarrow{E_3} D \xrightarrow{E_4} E \xrightarrow{E_5} F \xrightarrow{E_6} G$$

In this diagram, the letters A–F are reactants and the letters B–G are products in the various reactions. In other words, the products from the previous reaction become the reactants of the next reaction. The letters $E_1$–$E_6$ are enzymes. Any one of the molecules (A–G) in this linear pathway could also be a reactant in another pathway. A diagram showing all the possibilities would be highly branched.

## Energy of Activation

Molecules frequently do not react with one another unless they are activated in some way. In the lab, for example, in the absence of an enzyme, activation is very often achieved by heating a reaction flask to increase the number of effective collisions between molecules. The energy that must be added to cause molecules to react with one another is called the **energy of activation** ($E_a$). Figure 6.5 compares $E_a$ when an enzyme is not present to when an enzyme is present, illustrating that enzymes lower the amount of energy required for activation to occur. The energy content of the product remains the same, however.

Enzymes allow reactions to occur under mild conditions by bringing reactants into contact with one another and even by participating in the reaction at times.

## Enzyme-Substrate Complex

The reactants in an enzymatic reaction are called the **substrates** for that enzyme. Considering the metabolic pathway shown previously, A is the substrate for $E_1$, and B is the product. Now B becomes the substrate for $E_2$, and C is the product. This process continues until the final product G forms.

The following equation, which is pictorially shown in Figure 6.6, is often used to indicate that an enzyme forms a complex with its substrate:

| E | + | S | → | ES | → | E | + | P |
|---|---|---|---|---|---|---|---|---|
| enzyme | | substrate | | enzyme-substrate complex | | enzyme | | product |

In most instances, only one small part of the enzyme, called the **active site,** binds with the substrate(s). It is here that the enzyme and substrate fit together, seemingly like a key fits a lock; however, it is now known that the active site undergoes a slight change in shape to accommodate the substrate(s). This is called the **induced fit model** because the enzyme is induced to undergo a slight alteration to achieve optimum fit.

The change in shape of the active site facilitates the reaction that now occurs. After the reaction has been completed, the product(s) is released, and the active site returns to its original state, ready to bind to another substrate molecule. Only a small amount of enzyme is actually needed in a cell because enzymes are not used up by the reaction.

Some enzymes do more than simply complex with their substrate(s); they participate in the reaction. Trypsin digests protein by breaking peptide bonds. The active site of trypsin contains three amino acids with *R* groups that

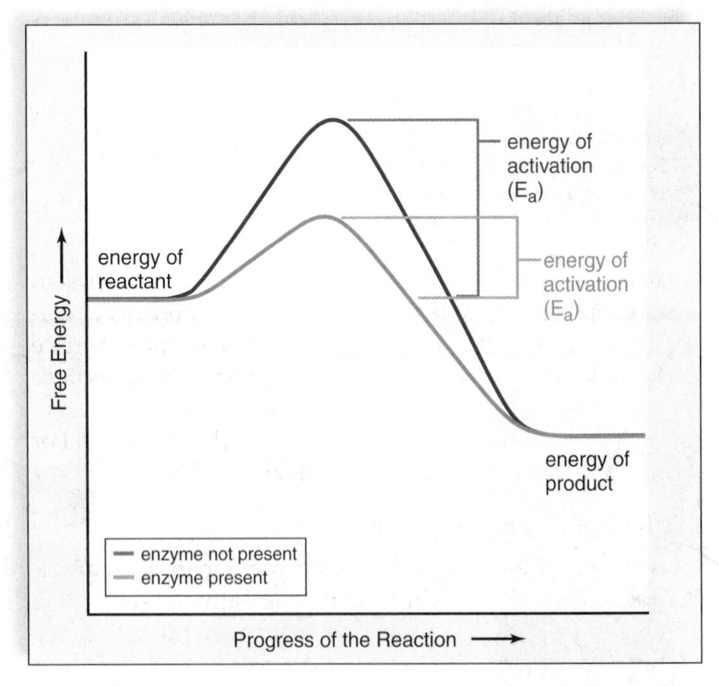


**FIGURE 6.5 Energy of activation ($E_a$).**

Enzymes speed the rate of reactions because they lower the amount of energy required for the reactants to react. Even reactions like this one, in which the energy of the product is less than the energy of the reactant, speed up when an enzyme is present.

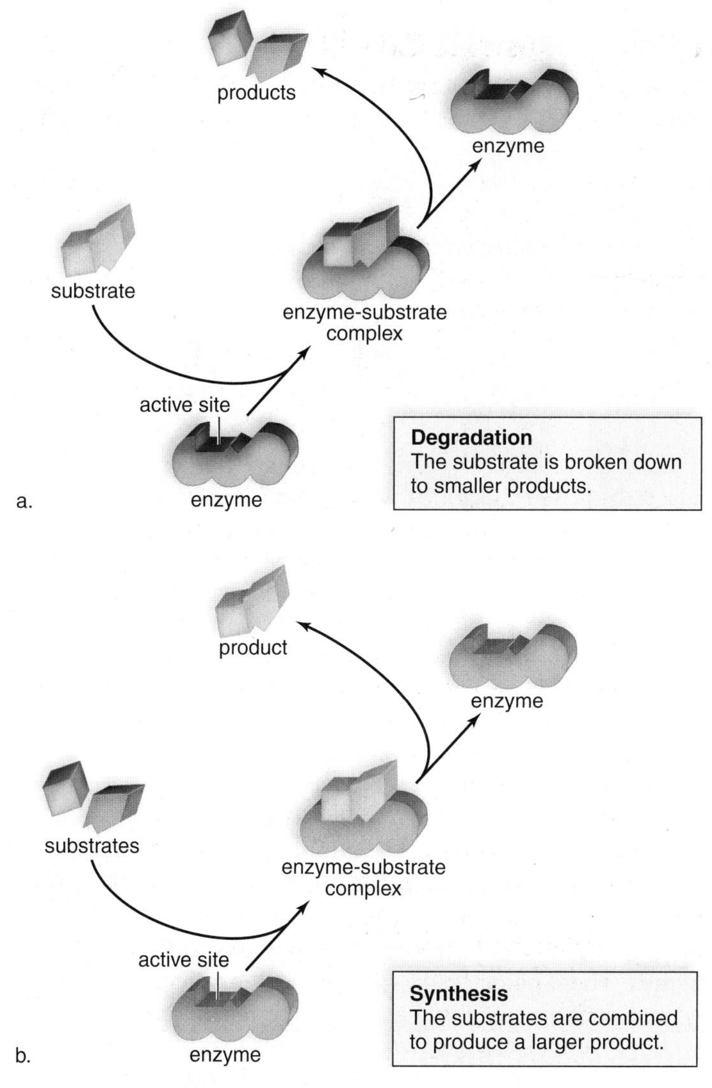


**FIGURE 6.6 Enzymatic actions.**

Enzymes have an active site where the substrate(s) and the enzyme fit together so the reaction will occur. Following the reaction, the product(s) is released, and the enzyme is free to act again. Certain enzymes carry out (**a**) degradation and others carry out (**b**) synthesis.

actually interact with members of the peptide bond—first to break the bond and then to introduce the components of water. This illustrates that the formation of the enzyme-substrate complex is very important in speeding the reaction. Because enzymes only complex with their substrates, they are sometimes named for their substrates, and usually end in *ase*. For example, lipase is involved in hydrolyzing lipids.

### *Regulation of Metabolism*

Because enzymes are specific, they participate in regulating metabolism. First, which metabolic pathways are being utilized is dependent on which enzymes are present. Since metabolic pathways can intersect, the presence of particular enzymes can determine the direction of metabolism. Then, too, some particular reactants can produce more than one type of product. Therefore, which enzyme is present determines which product is produced and the direction of metabolism without several side pathways being activated.

## Factors Affecting Enzymatic Speed

Generally, enzymes work quickly, and in some instances they can increase the reaction rate more than 10 million times. The rate of a reaction is the amount of product produced per unit time. The amount of product per unit time depends on how much substrate is at the active sites of enzymes. Therefore, increasing the amount of substrate and also the amount of enzyme can increase the rate of the reaction. Any factor that alters the shape of the active site, such as pH or temperature or an inhibitor, can decrease the rate of a reaction. Finally, some enzymes require cofactors that help speed the rate of the reaction because they help bind the substrate to the active site or they participate in the reaction at the active site.

### *Substrate Concentration*

Molecules must collide to react. Generally, enzyme activity increases as substrate concentration increases because there are more collisions between substrate molecules and the enzyme. As more substrate molecules fill active sites, more product results per unit time. But when the active sites are filled almost continuously with substrate, the rate of the reaction cannot increase any more. Maximum rate has been reached.

Just as the amount of substrate can increase or limit the rate of an enzymatic reaction, so the amount of active enzyme can also increase or limit the rate of an enzymatic reaction.

### *Optimal pH*

Each enzyme also has an optimal pH at which the rate of the reaction is highest. Figure 6.7 shows the optimal pH for the enzymes pepsin and trypsin. At this pH value, these enzymes have their normal configurations. The globular structure of an enzyme is dependent on interactions, such as hydrogen bonding, between *R* groups. A change in pH can alter the ionization of these side chains and disrupt normal interactions, and under extreme conditions of pH, the enzyme becomes inactive. Inactivity occurs because the enzyme has an altered shape and is then unable to combine efficiently with its substrate.

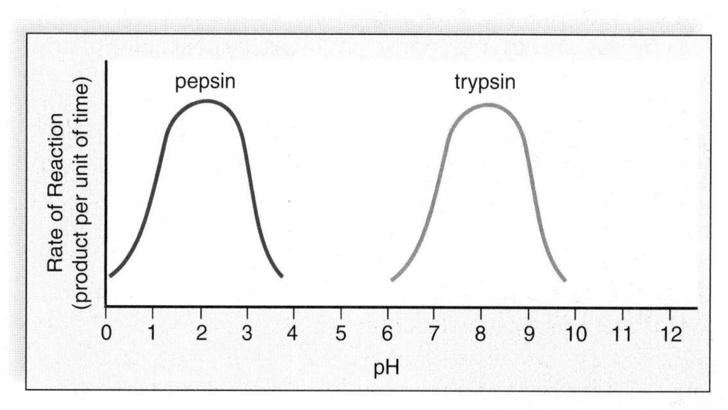


**FIGURE 6.7 The effect of pH on rate of reaction.**

The optimal pH for pepsin, an enzyme that acts in the stomach, is about 2, while the optimal pH for trypsin, an enzyme that acts in the small intestine, is about 8. The optimal pH of an enzyme maintains its shape so that it can bind with its substrates.

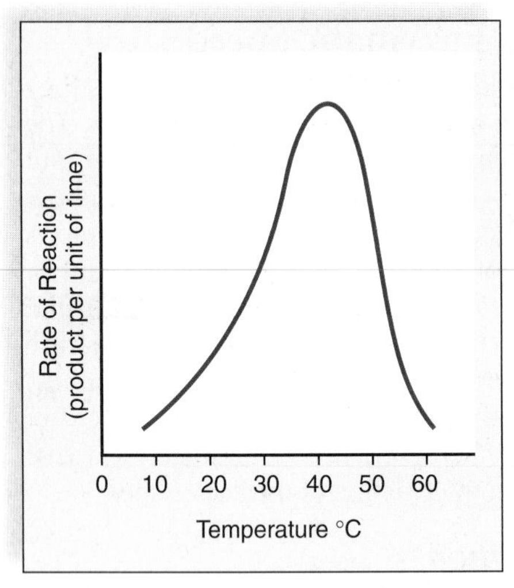


a. Rate of reaction as a function of temperature

b. Body temperature of ectothermic animals often limits rates of reactions.

c. Body temperature of endothermic animals promotes rates of reactions.

**FIGURE 6.8 The effect of temperature on rate of reaction.**

**a.** Usually, the rate of an enzymatic reaction doubles with every 10°C rise in temperature. This enzymatic reaction is maximum at about 40°C; then it decreases until the reaction stops altogether, because the enzyme has become denatured. **b.** The body temperature of ectothermic animals, such as an iguana, which take on the temperature of their environment, often limits rates of reactions. **c.** The body temperature of endothermic animals, such as a polar bear, promotes rates of reaction.

## *Temperature*

Typically, as temperature rises, enzyme activity increases (Fig. 6.8*a*). This occurs because warmer temperatures cause more effective collisions between enzyme and substrate. The body temperature of an animal seems to affect whether it is normally active or inactive (Fig. 6.8*b, c*). It has been suggested that mammals are more prevalent today than reptiles because they maintain a warm internal temperature that allows their enzymes to work at a rapid rate.

In the laboratory, if the temperature rises beyond a certain point, enzyme activity eventually levels out and then declines rapidly because the enzyme is **denatured.** An enzyme's shape changes during denaturation, and then it can no longer bind its substrate(s) efficiently. Exceptions to this generalization do occur. For example, some prokaryotes can live in hot springs because their enzymes do not denature. These organisms are responsible for the brilliant colors of the hot springs. Another exception involves the coat color of animals. Siamese cats have inherited a mutation that causes an enzyme to be active only at cooler body temperatures! Their activity causes the cooler regions of the body—the face, ears, legs, and tail—to be dark in color (Fig. 6.9). The coat color pattern in several other animals can be explained similarly.

**FIGURE 6.9 The effect of temperature on enzymes.**

Siamese cats have inherited a mutation that causes an enzyme to be active only at cooler body temperatures. Therefore, only certain regions of the body are dark in color.

## *Enzyme Cofactors*

Many enzymes require the presence of an inorganic ion or nonprotein organic molecule at the active site in order to be active; these necessary ions or molecules are called **cofactors** (Fig. 6.10). The inorganic ions are metals such as copper, zinc, or iron. The nonprotein organic molecules are called **coenzymes.** These cofactors participate in the reaction and may even accept or contribute atoms to the reactions. In the next section, we will discuss two coenzymes that play significant roles in photosynthesis and cellular respiration, respectively.

**Vitamins** are relatively small organic molecules that are required in trace amounts in our diet and in the diets of other animals for synthesis of coenzymes. The vitamin becomes part of a coenzyme's molecular structure. If a vitamin is not available, enzymatic activity will decrease, and the result will be a vitamin-deficiency disorder: Niacin deficiency results in a skin disease called pellagra, and riboflavin deficiency results in cracks at the corners of the mouth.

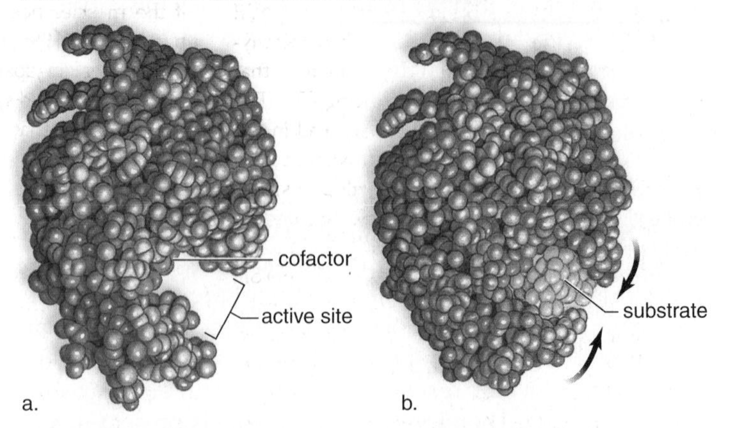


**FIGURE 6.10 Cofactors at active site.**

**a.** Cofactors, including inorganic ions and organic coenzymes, may participate in the reaction at the active site (**b**).

### *Enzyme Inhibition*

**Enzyme inhibition** occurs when a molecule (the inhibitor) binds to an enzyme and decreases its activity. (1) In Figure 6.11, F is the end product of a metabolic pathway that can act as an inhibitor. This is beneficial because once sufficient end product of a metabolic pathway is present, it is best to inhibit further production to conserve raw materials and energy.

(2) Figure 6.11 also illustrates **noncompetitive inhibition** because the inhibitor (F, the end product) binds to the enzyme $E_1$ at a location other than the active site. The site is called an **allosteric site.** When an inhibitor is at the allosteric site, the active site of the enzyme changes shape.

(3) The enzyme $E_1$ is inhibited because it is unable to bind to A, its substrate. The inhibition of $E_1$ means that the metabolic pathway is inhibited and no more end product will be produced.

In contrast to noncompetitive inhibition, **competitive inhibition** occurs when an inhibitor and the substrate compete for the active site of an enzyme. Product will form only when the substrate, not the inhibitor, is at the active site. In this way, the amount of product is regulated.

Normally, enzyme inhibition is reversible, and the enzyme is not damaged by being inhibited. When enzyme inhibition is irreversible, the inhibitor permanently inactivates or destroys an enzyme.

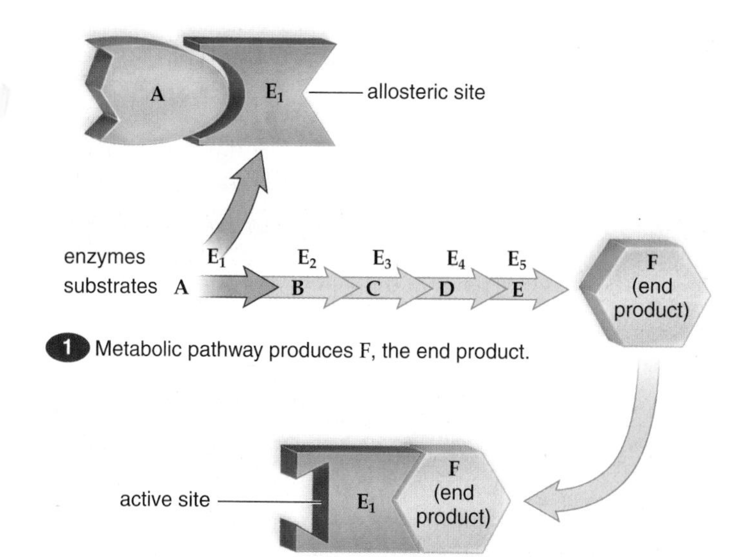

(2) F binds to allosteric site and the active site of $E_1$ changes shape.

(3) A cannot bind to $E_1$; the enzyme has been inhibited by F.

**FIGURE 6.11 Noncompetitive inhibition of an enzyme.**

In the pathway, A–E are substrates, $E_1$–$E_5$ are enzymes, and F is the end product of the pathway that inhibits the enzyme $E_1$.

#### Check Your Progress 6.3

1. How do enzymes lower the energy of activation?
2. What factors can affect the speed of an enzymatic reaction?

## *health focus*

## Enzyme Inhibitors Can Spell Death

Cyanide gas was formerly used to execute people. How did it work? Cyanide can be fatal because it binds to a mitochondrial enzyme necessary for the production of ATP. MPTP (1-methyl-4-phenyl-1,2,3.6-tetrahydropyridine) is another enzyme inhibitor that stops mitochondria from producing ATP. The toxic nature of MPTP was discovered in the early 1980s, when a group of intravenous drug users in California suddenly developed symptoms of Parkinson disease, including uncontrollable tremors and rigidity. All of the drug users had injected a synthetic form of heroin that was contaminated with MPTP. Parkinson disease is characterized by the death of brain cells, the very ones that are also destroyed by MPTP.

Sarin is a chemical that inhibits an enzyme at neuromuscular junctions, where nerves stimulate muscles. When the enzyme is inhibited, the signal for muscle contraction cannot be turned off, so the muscles are unable to relax and become paralyzed. Sarin can be fatal if the muscles needed for breathing become paralyzed. In 1995, terrorists released sarin gas on a subway in Japan (Fig. 6A). Although many people developed symptoms, only 17 died.

A fungus that contaminates and causes spoilage of sweet clover produces a chemical called warfarin. Cattle that eat the spoiled feed die from internal bleeding because warfarin inhibits a crucial enzyme for blood clotting. Today, warfarin is widely used as a rat poison. Unfortunately, it is not uncommon for warfarin to be mistakenly eaten by pets and even very small children, with tragic results.

Many people are prescribed a medicine called Coumadin to prevent inappropriate blood clotting. For example, those who have received an artificial heart valve need such a medication. Coumadin contains a nonlethal dose of warfarin.

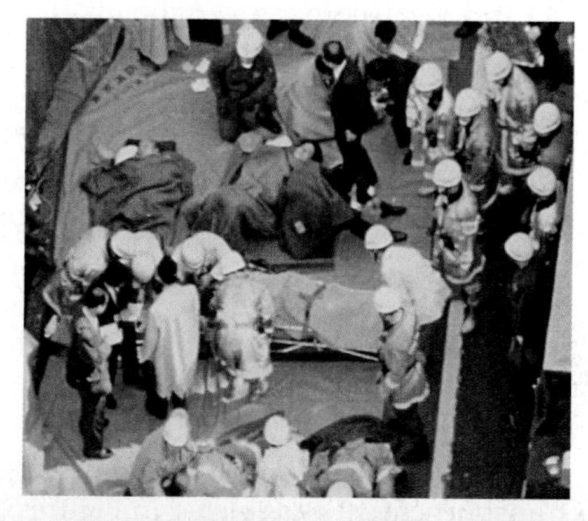

**Figure 6A Sarin gas.**
The aftermath when sarin, a nerve gas that results in the inability to breathe, was released by terrorists in a Japanese subway in 1995.

# 6.4 Organelles and the Flow of Energy

Two organelles are particularly involved in the flow of energy from the sun through all living things. Photosynthesis, a process that captures solar energy to produce carbohydrates, takes place in chloroplasts. Cellular respiration, which breaks down carbohydrates, takes place in mitochondria.

## Photosynthesis

The overall reaction for photosynthesis can be written like this:

$$\underset{\text{carbon dioxide}}{6\,CO_2} + \underset{\text{water}}{6\,H_2O} + \text{energy} \longrightarrow \underset{\text{glucose}}{C_6H_{12}O_6} + \underset{\text{oxygen}}{6\,O_2}$$

This equation shows that hydrogen atoms ($H^+ + e^-$) are transferred from water to carbon dioxide, when glucose is formed. **Oxidation** is defined as the loss of electrons and **reduction** is the gain of electrons. Therefore, water has been oxidized and carbon dioxide has been reduced.

Since glucose is a high-energy molecule, an input of energy is needed to make the reaction go. Chloroplasts are able to capture solar energy and convert it by way of an electron transport chain (discussed next) to the chemical energy of ATP molecules. ATP is then used along with hydrogen atoms to reduce carbon dioxide to glucose.

A coenzyme of oxidation-reduction called **$NADP^+$ (nicotinamide adenine dinucleotide phosphate)** is active during photosynthesis. This molecule carries a positive charge, and therefore is written as $NADP^+$. During photosynthesis, $NADP^+$ accepts electrons and a hydrogen ion derived from water and later passes them by way of a metabolic pathway to carbon dioxide, forming glucose. The reaction that reduces $NADP^+$ is:

$$NADP^+ + 2\,e^- + H^+ \longrightarrow NADPH$$

## Cellular Respiration

The overall equation for cellular respiration is opposite to that for photosynthesis:

$$\underset{\text{glucose}}{C_6H_{12}O_6} + \underset{\text{oxygen}}{6\,O_2} \longrightarrow \underset{\text{carbon dioxide}}{6\,CO_2} + \underset{\text{water}}{6\,H_2O} + \text{energy}$$

In this reaction, glucose has lost hydrogen atoms (been oxidized), and oxygen has gained hydrogen atoms (been reduced). The hydrogen atoms that were formerly bonded to carbon are now bonded to oxygen. Glucose is a high-energy molecule, while its breakdown products, carbon dioxide and water, are low-energy molecules; therefore, energy is released. Mitochondria use the energy released from glucose breakdown to build ATP molecules by way of an electron transport chain, as depicted in Figure 6.12.

In metabolic pathways, most oxidations such as those that occur during cellular respiration involve a coenzyme called **$NAD^+$ (nicotinamide adenine dinucleotide).** This molecule carries a positive charge, and therefore it is represented as $NAD^+$. During oxidation reactions, $NAD^+$ accepts two electrons but only one hydrogen ion. The reaction that reduces $NAD^+$ is:

$$NAD^+ + 2\,e^- + H^+ \longrightarrow NADH$$

## Electron Transport Chain

As previously mentioned, chloroplasts use solar energy to generate ATP, and mitochondria use glucose energy to generate ATP by way of an electron transport chain. An **electron transport chain (ETC)** is a series of membrane-bound carriers that pass electrons from one carrier to another. High-energy electrons are delivered to the chain, and low-energy electrons leave it. If a hot potato were passed from one person to another, it would lose heat with each transfer. In the same manner, every time electrons are transferred to a new carrier, energy is released. However, unlike the hot potato transfer example, the cell is able to capture the released energy and use it to produce ATP molecules (Fig. 6.12).

In certain redox reactions, the result is release of energy, and in others, energy is required. In an ETC, each carrier is reduced and then oxidized in turn. The overall effect of oxidation-reduction as electrons are passed from carrier to carrier of the electron transport chain is the release of energy for ATP production.

## ATP Production

For many years, it was known that ATP synthesis was somehow coupled to the ETC, but the exact mechanism could not be determined. Peter Mitchell, a British biochemist, received a Nobel Prize in 1978 for his theory of ATP production in both mitochondria and chloroplasts.

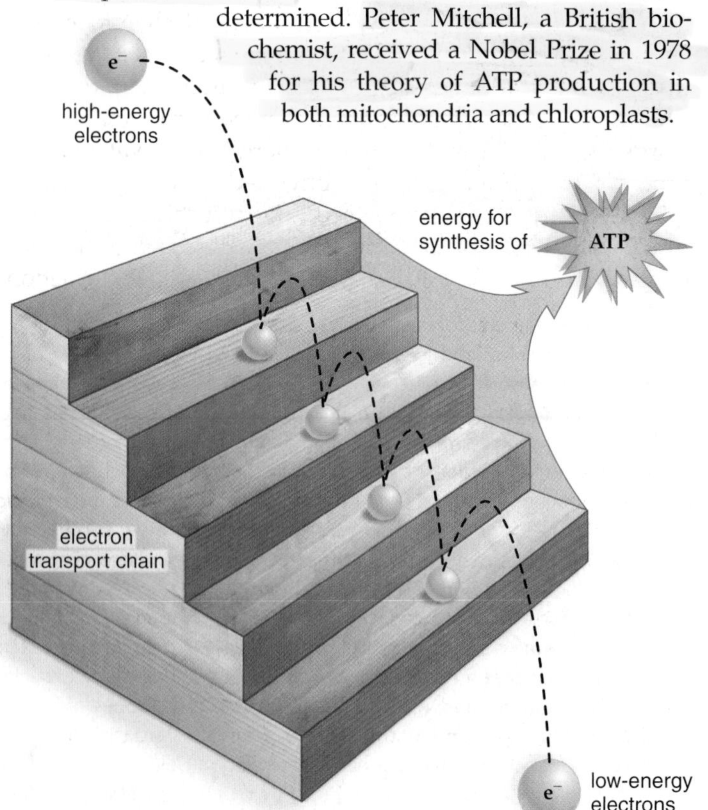


**FIGURE 6.12 Electron transport chain.**

High-energy electrons are delivered to the chain and, with each step as they pass from carrier to carrier, energy is released and used for ATP production.

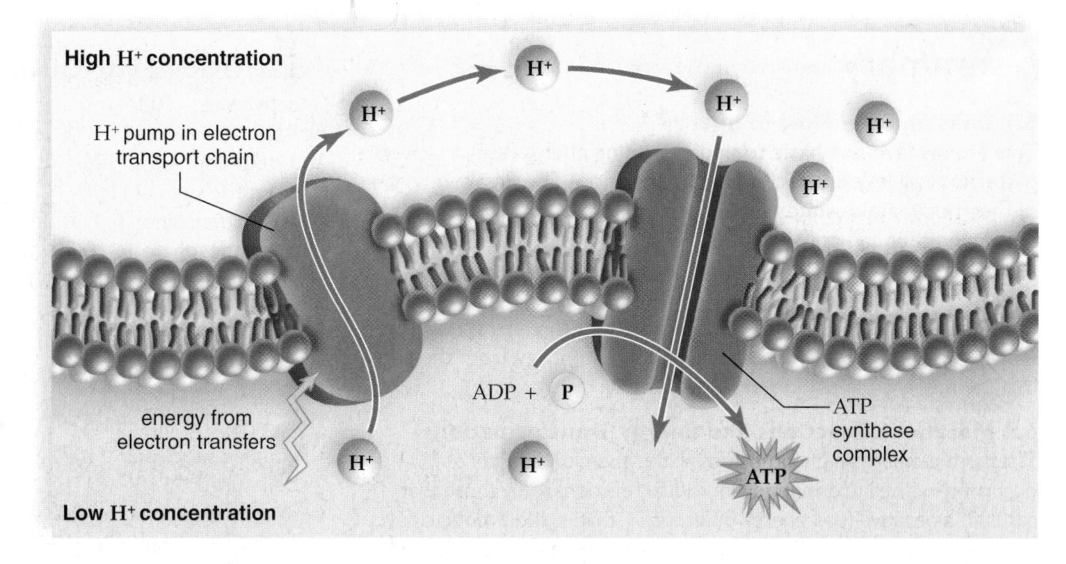


**FIGURE 6.13 Chemiosmosis.** Carriers in the electron transport chain pump hydrogen ions ($H^+$) across a membrane. When the hydrogen ions flow back across the membrane through an ATP synthase complex, ATP is synthesized by an enzyme called ATP synthase. Chemiosmosis occurs in chloroplasts and mitochondria.

In chloroplasts and mitochondria, the carriers of the ETC are located within a membrane: thylakoid membranes in chloroplasts and cristae in mitochondria. Hydrogen ions ($H^+$), which are often referred to as protons in this context, tend to collect on one side of the membrane because they are pumped there by certain carriers of the electron transport chain. This establishes an electrochemical gradient across the membrane that can be used to provide energy for ATP production. Enzymes and their carrier proteins, called **ATP synthase complexes,** span the membrane. Each complex contains a channel that allows hydrogen ions to flow down their electrochemical gradient. The flow of hydrogen ions through the channel provides the energy for the ATP synthase enzyme to produce ATP from ADP + Ⓟ (Fig. 6.13). The production of ATP due to a hydrogen ion gradient across a membrane is called **chemiosmosis** [Gk. *osmos*, push].

Consider this analogy to understand chemiosmosis. The sun's rays evaporate water from the seas and help create the winds that blow clouds to the mountains, where water falls in the form of rain and snow. The water in a mountain reservoir has a higher potential energy than water in the ocean. The potential energy is converted to electrical energy when water is released and used to turn turbines in an electrochemical dam before it makes its way to the ocean. The continual release of water results in a continual production of electricity.

Similarly, during photosynthesis, solar energy collected by chloroplasts continually leads to ATP production. Energized electrons lead to the pumping of hydrogen ions across a thylakoid membrane, which acts like a dam to retain them. The hydrogen ions flow through the channel of an ATP synthase complex. This complex couples the flow of hydrogen ions to the formation of ATP, just as the turbines in a hydroelectric dam system couple the flow of water to the formation of electricity.

Similarly, during cellular respiration, glucose breakdown provides the energy to establish a hydrogen ion gradient across the cristae of mitochondria. And again, hydrogen ions flow through the channel within an ATP synthase complex that couples the flow of hydrogen ions to the formation of ATP.

### Check Your Progress 6.4

1. **a.** In particular, what molecule does the grass make available to the impala as a source of energy? **b.** What happens to this molecule during cellular respiration?
2. Carbon dioxide is **(a)** ______ to produce glucose during photosynthesis and glucose is **(b)** ______ to produce ATP molecules during cellular respiration.

## Connecting the Concepts

All cells use energy. Energy is the ability to do work, to bring about change, and to make things happen, whether it's a leaf growing or a human running. The metabolic pathways inside cells use the chemical energy of ATP to synthesize molecules, cause muscle contraction, and even allow you to read these words.

A metabolic pathway consists of a series of individual chemical reactions, each with its own enzyme. The cell can regulate the activity of the many hundreds of different enzymes taking part in cellular metabolism. Enzymes are proteins, and as such they are sensitive to environmental conditions, including pH, temperature, and even certain pollutants, as will be discussed in later chapters.

ATP is called the universal energy "currency" of life. This is an apt analogy—before we can spend currency (e.g., money), we must first make some money. Similarly, before the cell can spend ATP molecules, it must make them. Cellular respiration in mitochondria transforms the chemical energy of carbohydrates into that of ATP molecules. ATP is spent when it is hydrolyzed, and the resulting energy is coupled to an endergonic reaction. All cells are continually making and breaking down ATP. If ATP is lacking, the organism dies.

What is the ultimate source of energy for ATP production? In Chapter 7, we will see that, except for a few deep ocean vents and certain cave communities, the answer is the sun. Photosynthesis inside chloroplasts transforms solar energy into the chemical energy of carbohydrates. And then in Chapter 8 we will discuss how carbohydrate products are broken down in mitochondria as ATP is built up. Chloroplasts and mitochondria are the cellular organelles that permit a flow of energy from the sun through all living things.

## summary


### 6.1 Cells and the Flow of Energy

Two energy laws are basic to understanding energy-use patterns at all levels of biological organization. The first law of thermodynamics states that energy cannot be created or destroyed, but can only be transferred or transformed. The second law of thermodynamics states that one usable form of energy cannot be completely converted into another usable form. As a result of these laws, we know that the entropy of the universe is increasing and that only a flow of energy from the sun maintains the organization of living things.

### 6.2 Metabolic Reactions and Energy Transformations

The term *metabolism* encompasses all the chemical reactions occurring in a cell. Considering individual reactions, only those that result in a negative free-energy difference—that is, the products have less usable energy than the reactants—occur spontaneously. Such reactions, called exergonic reactions, release energy. Endergonic reactions, which require an input of energy, occur only in cells because it is possible to couple an exergonic process with an endergonic process. For example, glucose breakdown is an exergonic metabolic pathway that drives the buildup of many ATP molecules. ATP goes through a cycle in which it is constantly being built up from, and then broken down to, ADP + Ⓟ. When ATP breaks down, energy is released that can drive forward energy requiring metabolic reactions, if the two reactions are coupled. In general, ATP is used to energize a reactant or change the shape of a reactant so the reaction occurs.

### 6.3 Metabolic Pathways and Enzymes

A metabolic pathway is a series of reactions that proceed in an orderly, step-by-step manner. Enzymes speed reactions by lowering the energy of activation when they form a complex with their substrates. Enzymes regulate metabolism because, in general, no reaction occurs unless its enzyme is present. Which enzymes are present determine which metabolic pathways will be utilized.

Generally, enzyme activity increases as substrate concentration increases; once all active sites are filled, maximum rate has been achieved. Any environmental factor, such as temperature or pH, affects the shape of a protein and, therefore, also affects the ability of an enzyme to do its job. Many enzymes need cofactors or coenzymes to carry out their reactions. The activity of most metabolic pathways is regulated by feedback inhibition.

### 6.4 Organelles and the Flow of Energy

A flow of energy occurs through organisms because (1) photosynthesis in chloroplasts captures solar energy and produces carbohydrates, and (2) cellular respiration in mitochondria breaks down this carbohydrate to produce ATP molecules, which (3) are used to provide energy for metabolic reactions. The overall equation for photosynthesis is the opposite of that for cellular respiration. During photosynthesis, the coenzyme NADPH reduces substrates, while during cellular respiration, the coenzyme $NAD^+$ oxidizes substrates.

Both processes make use of an electron transport chain in which electrons are transferred from one carrier to the next with the release of energy that is ultimately used to produce ATP molecules. Chemiosmosis explains how the electron transport chain produces ATP. The carriers of this system deposit hydrogen ions ($H^+$) on one side of a membrane. When hydrogen ions flow down an electrochemical gradient through an ATP synthase complex, an enzyme uses the release of energy to make ATP from ADP and Ⓟ.


## understanding the terms

active site 108
ADP (adenosine diphosphate) 106
allosteric site 111
ATP (adenosine triphosphate) 106
ATP synthase complex 113
chemical energy 104
chemiosmosis 113
coenzyme 110
cofactor 110
competitive inhibition 111
coupled reactions 107
denatured 110
electron transport chain (ETC) 112
endergonic reaction 106
energy 104
energy of activation 108
entropy 105
enzyme 108
enzyme inhibition 111
exergonic reaction 106
free energy 106
heat 104
induced fit model 108
kinetic energy 104
laws of thermodynamics 104
mechanical energy 104
metabolic pathway 108
metabolism 106
$NAD^+$ (nicotinamide adenine dinucleotide) 112
$NADP^+$ (nicotinamide adenine dinucleotide phosphate) 112
noncompetitive inhibition 111
oxidation 112
potential energy 104
product 106
reactant 106
reduction 112
ribozyme 108
substrate 108
vitamin 110

Match the terms to these definitions:

a. ______________ All of the chemical reactions that occur in a cell during growth and repair.
b. ______________ Stored energy as a result of location or spatial arrangement.
c. ______________ Essential requirement in the diet, needed in small amounts. They are often part of coenzymes.
d ______________ Measure of disorder or randomness.
e ______________ Nonprotein organic molecule that aids the action of the enzyme to which it is loosely bound.
f. ______________ Loss of one or more electrons from an atom or molecule; in biological systems, generally the loss of hydrogen atoms.

## reviewing this chapter

1. State the first law of thermodynamics, and give an example. 104
2. State the second law of thermodynamics, and give an example. 104–5
3. Explain why the entropy of the universe is always increasing and why an organized system such as an organism requires a constant input of useful energy. 105
4. What is the difference between exergonic reactions and endergonic reactions? Why can exergonic but not endergonic reactions occur spontaneously? 106
5. Why is ATP called the energy currency of cells? What is the ATP cycle? 106
6. Define coupling, and write an equation that shows an endergonic reaction being coupled to ATP breakdown. 107
7. Diagram a metabolic pathway. Label the reactants, products, and enzymes. Explain how enzymes regulate metabolism. 108–9
8. Why is less energy needed for a reaction to occur when an enzyme is present? 108
9. Why are enzymes specific, and why can't each one speed many different reactions? 108–9
10. Name and explain the manner in which at least three environmental factors can influence the speed of an enzymatic reaction. How do cells regulate the activity of enzymes? 109–11

11. What are cofactors and coenzymes? 110
12. Compare and contrast competitive and noncompetitive inhibition. 111
13. How do chloroplasts and mitochondria permit a flow of energy through all organisms. What role is played by oxidation and reduction? 112
14. Describe an electron transport chain. 112
15. Tell how cells form ATP during chemiosmosis. 112–13

## testing yourself

Choose the best answer for each question.

1. A form of potential energy is
   a. a boulder at the top of a hill.
   b. the bonds of a glucose molecule.
   c. a starch molecule.
   d. stored fat tissue.
   e. All of these are correct.
2. A lit lightbulb can be used to explain the
   a. creation of heat energy.
   b. second law of thermodynamics.
   c. conversion of electrical energy into heat energy.
   d. first law of thermodynamics.
   e. All of the above except a are correct.
3. Consider this reaction: A + B ⟶ C + D + energy.
   a. This reaction is exergonic.
   b. An enzyme could still speed the reaction.
   c. ATP is not needed to make the reaction go.
   d. A and B are reactants; C and D are products.
   e. All of these are correct.
4. The active site of an enzyme
   a. is similar to that of any other enzyme.
   b. is the part of the enzyme where its substrate can fit.
   c. can be used over and over again.
   d. is not affected by environmental factors, such as pH and temperature.
   e. Both b and c are correct.
5. If you want to increase the amount of product per unit time of an enzymatic reaction, do not increase the
   a. amount of substrate.
   b. amount of enzyme.
   c. temperature somewhat.
   d. pH.
   e. All of these are correct.
6. An allosteric site on an enzyme is
   a. the same as the active site.
   b. nonprotein in nature.
   c. where ATP attaches and gives up its energy.
   d. often involved in feedback inhibition.
   e. All of these are correct.
7. During photosynthesis, carbon dioxide
   a. is oxidized to oxygen.
   b. is reduced to glucose.
   c. gives up water to the environment.
   d. is a coenzyme of oxidation-reduction.
   e. All of these are correct.
8. Use these terms to label the following diagram: substrates, enzyme (used twice), active site, product, and enzyme-substrate complex. Explain the importance of an enzyme's shape to its activity.

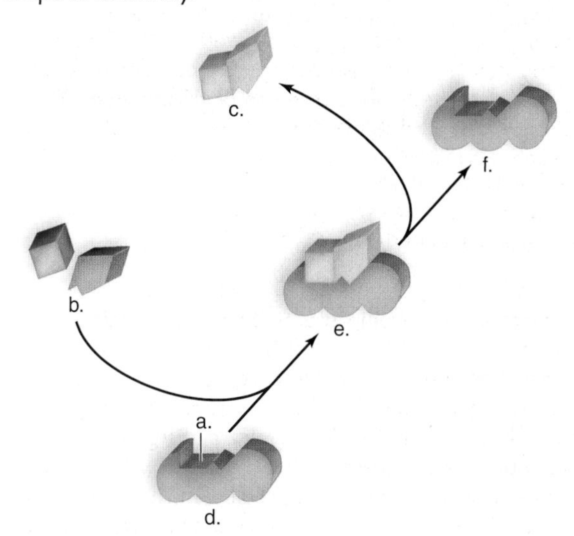


9. Coenzymes
   a. have specific functions in reactions.
   b. have an active site just as enzymes do.
   c. can be carriers for proteins.
   d. always have a phosphate group.
   e. are used in photosynthesis, but not in cellular respiration.

For questions 10–16, match each description to a process in the key.

**KEY:**

a. photosynthesis
b. cellular respiration
c. Both
d. Neither

10. captures solar energy
11. requires enzymes and coenzymes
12. releases $CO_2$ and $H_2O$
13. utilizes an electron transport chain
14. performed by plants
15. transforms one form of energy into another form with the release of heat
16. creates energy for the living world

For questions 17–22, match each pair to a description in the key. Choose more than one answer if correct.

**KEY:**

a. first includes the other
b. first breaks down to the other
c. have nothing to do with each other
d. work together

17. metabolic pathway, enzyme
18. allosteric site, reduction
19. kinetic energy, mechanical energy
20. ATP, ADP + Ⓟ
21. enzyme, coenzyme
22. chemiosmosis, electron transport chain

23. Oxidation
    a. is the opposite of reduction.
    b. sometimes uses $NAD^+$.
    c. is involved in cellular respiration.
    d. occurs when ATP goes to ADP + Ⓟ.
    e. All of these but d are correct.
24. $NAD^+$ is the ______________ form, and when it later becomes NADH, it is said to be ______________.
    a. reduced, oxidized
    b. neutral, a coenzyme
    c. oxidized, reduced
    d. active, denatured
25. Electron transport chains
    a. are found in both mitochondria and chloroplasts.
    b. release energy as electrons are transferred.
    c. are involved in the production of ATP.
    d. are located in a membrane.
    e. All of these are correct.
26. Chemiosmosis is dependent on
    a. the diffusion of water across a differentially permeable membrane.
    b. an outside supply of phosphate and other chemicals.
    c. the establishment of an electrochemical hydrogen ion ($H^+$) gradient.
    d. the ability of ADP to join with Ⓟ even in the absence of a supply of energy.
    e. All of these are correct.
27. The difference between $NAD^+$ and $NADP^+$ is that
    a. only $NAD^+$ production requires niacin in the diet.
    b. one is an organic molecule, and the other is inorganic because it contains phosphate.
    c. one carries electrons to the electron transport chain, and the other carries them to synthetic reactions.
    d. one is involved in cellular respiration, and the other is involved in photosynthesis.
    e. Both c and d are correct.
28. Label this diagram describing chemiosmosis.

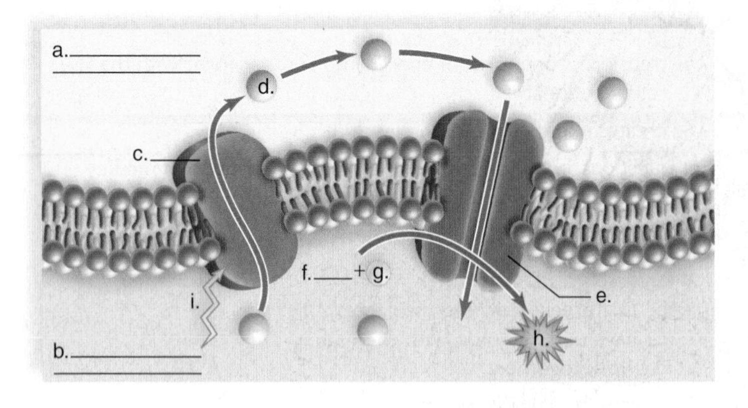


## thinking scientifically

1. A flower generates heat in order to attract pollinating insects. Why might the flower break down a sugar and not ATP to produce heat?
2. You decide to calculate how much energy is released when sucrose is broken down by a flower and run into complications because you have to first heat the sucrose before it breaks down. Explain why this complication is not a problem for the flower.

## bioethical issue

### Global Warming and Emerging Diseases

In this chapter, we learned that a rise in temperature fosters enzymatic reactions. Could a rise in environmental temperatures due to global warming cause an increase in the number of pathogens? For example, a 2006 outbreak of diarrhea in Washington state was due to eating raw or partly cooked shellfish infected with *Vibrio* bacteria. Warmer-than-usual ocean waters may have caused the extensive growth of *Vibrio* bacteria that infected the shellfish and led to the outbreak. The connection between global warming and emerging diseases can be more subtle. In 1993, the hantavirus strain emerged from the common deer mouse and killed about 60 young people in the Southwest. In this instance, we know that climate was involved. An unusually mild winter and wet spring caused piñon trees to bloom well and provide pine nuts to the mice. The increasing deer mouse population came into contact with humans, and the hantavirus leaped easily from mice to humans.

Evidence suggests that global warming, caused in part by the burning of fossil fuels, as explained on page 125, contributes to outbreaks of hantavirus as well as malaria, dengue and yellow fevers, filariasis, encephalitis, schistosomiasis, and cholera. Clearly, any connection between global warming and emerging diseases offers another reason why fossil fuel consumption should be curtailed. Would you as a homeowner or a CEO of a company be willing to switch to renewable energy supplies because a warming of the environment may increase the incidence of human illnesses? Instead, would you approve of giving companies monetary incentives to use renewable energy supplies that do not contribute to global warming? Or, do you think we should wait for more confirmation that global warming is due to human activities and leads to an increase in diseases that could affect us and our families? What type of confirmation would you be looking for?

## *Biology* website

The companion website for *Biology* provides a wealth of information organized and integrated by chapter. You will find practice tests, animations, videos, and much more that will complement your learning and understanding of general biology.

**http://www.mhhe.com/maderbiology10**

# 7 Photosynthesis

**W**hite light, the kind that shines down on us everyday, contains different colors of light, from violet to green, yellow, orange, and red. Plants use all the colors of light, except green, when they photosynthesize—and that's why we see them as green! Does this mean that if plants weren't so wasteful and used green light, in addition to other colors, they would appear black to us? Yes, natural areas like the one pictured below would be black, as shown on the right.

*How did it happen that plants do not use green light for photosynthesis? When the ancestors of plants arose in the ocean, green light was already being absorbed by other photosynthesizers, so natural selection favored the evolution of a pigment such as chlorophyll, which does not absorb green light. On land, there is plentiful sunlight, and a more efficient pigment has no advantage. As discussed in this chapter, two interconnected pathways allow chloroplasts to produce carbohydrate while releasing oxygen. Such a remarkable process deserves our close attention.*

Plants appear green because chlorophyll reflects green light (*left*). Otherwise, plants would be black (*right*).

## concepts

## 7.1 Photosynthetic Organisms

**Photosynthesis** converts solar energy into the chemical energy of a carbohydrate. Photosynthetic organisms, including land plants, algae, and cyanobacteria, are called **autotrophs** because they produce their own food (Fig. 7.1). Photosynthesis produces an enormous amount of carbohydrate. So much that, if it were instantly converted to coal and the coal were loaded into standard railroad cars (each car holding about 50 tons), the photosynthesizers of the biosphere would fill more than 100 cars per second with coal.

No wonder photosynthetic organisms are able to sustain themselves and all other living things on Earth. With few exceptions, it is possible to trace any food chain back to plants and algae. In other words, producers, which have the ability to synthesize carbohydrates, feed not only themselves but also consumers, which must take in preformed organic molecules. Collectively, consumers are called **heterotrophs.** Both autotrophs and heterotrophs use organic molecules produced by photosynthesis as a source of building blocks for growth and repair and as a source of chemical energy for cellular work.

Photosynthesizers also produce copious amounts of oxygen as a by-product. Oxygen, which is required by organisms when they carry on cellular respiration, rises high into the atmosphere, where it forms an ozone shield that filters out ultraviolet radiation and makes terrestrial life possible.

Our analogy about photosynthetic products becoming coal is apt because the bodies of many ancient plants did become the coal we burn today, usually to produce electricity. Coal formation happened several hundred million years ago, and that is why coal is called a fossil fuel. Today's trees are also commonly used as fuel. Then, too, the fermentation of plant materials produces ethanol, which can be used directly to fuel automobiles or as a gasoline additive.

The products of photosynthesis are critical to humankind in a number of other ways. They serve as a source of building materials, fabrics, paper, and pharmaceuticals. And while we are thanking green plants for their services, let's not forget the simple beauty of a magnolia blossom or the majesty of the Earth's forests.

**FIGURE 7.1 Photosynthetic organisms.**

Photosynthetic organisms include plants, such as trees, garden plants, and mosses, which typically live on land; photosynthetic protists, such as *Euglena,* diatoms, and kelp, which typically live in water; and cyanobacteria, a type of bacterium that lives in water, damp soil, and rocks.

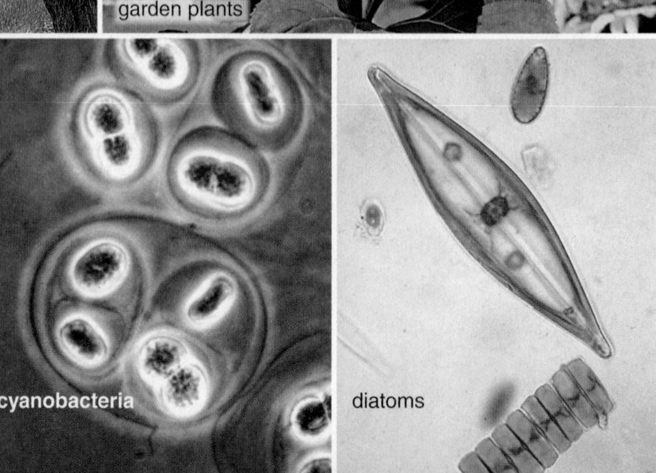

## Flowering Plants as Photosynthesizers

Photosynthesis takes place in the green portions of plants. The leaves of a flowering plant contain mesophyll tissue in which cells are specialized for photosynthesis (Fig. 7.2). The raw materials for photosynthesis are water and carbon dioxide. The roots of a plant absorb water, which then moves in vascular tissue up the stem to a leaf by way of the leaf veins. Carbon dioxide in the air enters a leaf through small openings called **stomata** (sing., stoma). After entering a leaf, carbon dioxide and water diffuse into **chloroplasts** [Gk. *chloros*, green, and *plastos*, formed, molded], the organelles that carry on photosynthesis.

A double membrane surrounds a chloroplast, and its semifluid interior called the **stroma** [Gk. *stroma*, bed, mattress]. A different membrane system within the stroma forms flattened sacs called **thylakoids** [Gk. *thylakos*, sack, and *eides*, like, resembling], which in some places are stacked to form **grana** (sing., granum), so called because they looked like piles of seeds to early microscopists. The space of each thylakoid is thought to be connected to the space of every other thylakoid within a chloroplast, thereby forming an inner compartment within chloroplasts called the thylakoid space.

The thylakoid membrane contains **chlorophyll** and other pigments that are capable of absorbing solar energy, the type of energy that drives photosynthesis. The stroma contains an enzyme-rich solution where carbon dioxide is first attached to an organic compound and is then reduced to a carbohydrate.

Therefore, it is proper to associate the absorption of solar energy with the thylakoid membranes making up the grana and to associate the reduction of carbon dioxide to a carbohydrate with the stroma of a chloroplast.

Human beings, and indeed nearly all organisms, release carbon dioxide into the air. This is some of the same carbon dioxide that enters a leaf through the stoma and is converted to carbohydrate. Carbohydrate, in the form of glucose, is the chief source of chemical energy for most organisms.

### Check Your Progress 7.1

1. List three major groups of photosynthetic organisms.
2. Which part of a chloroplast absorbs solar energy, and which part forms a carbohydrate?

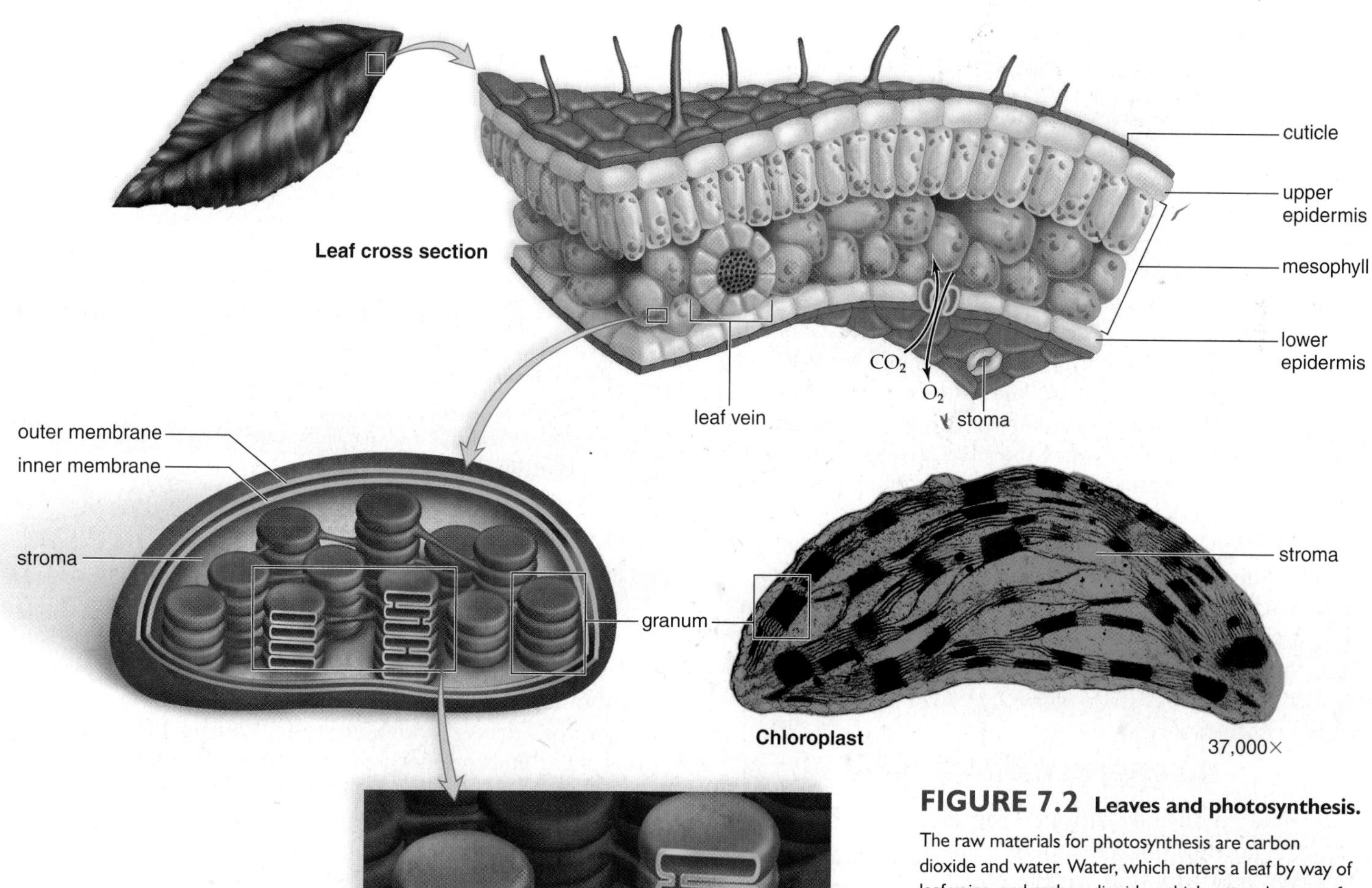


**FIGURE 7.2 Leaves and photosynthesis.** The raw materials for photosynthesis are carbon dioxide and water. Water, which enters a leaf by way of leaf veins, and carbon dioxide, which enters by way of the stomata, diffuse into chloroplasts. Chloroplasts have two major parts. The grana are made up of thylakoids, which are membranous disks. Their membrane contains photosynthetic pigments such as chlorophylls *a* and *b*. These pigments absorb solar energy. The stroma is a semifluid interior where carbon dioxide is enzymatically reduced to a carbohydrate.

# 7.2 The Process of Photosynthesis

This overall equation can be used to represent the process of photosynthesis:

$$CO_2 + 2H_2O \xrightarrow{\text{solar energy}} (CH_2O) + H_2O + O_2$$

In this equation, $(CH_2O)$ represents carbohydrate. If the equation were multiplied by six, the carbohydrate would be $C_6H_{12}O_6$, or glucose.

The overall equation shows that photosynthesis involves oxidation-reduction (redox) and the movement of electrons from one molecule to another. Recall that oxidation is the loss of electrons, and reduction is the gain of electrons. In living things, as discussed on page 112, the electrons are very often accompanied by hydrogen ions so that oxidation is the loss of hydrogen atoms $(H^+ + e^-)$, and reduction is the gain of hydrogen atoms. This simplified rewrite of the above equation makes it clear that carbon dioxide has been reduced, and water has been oxidized:

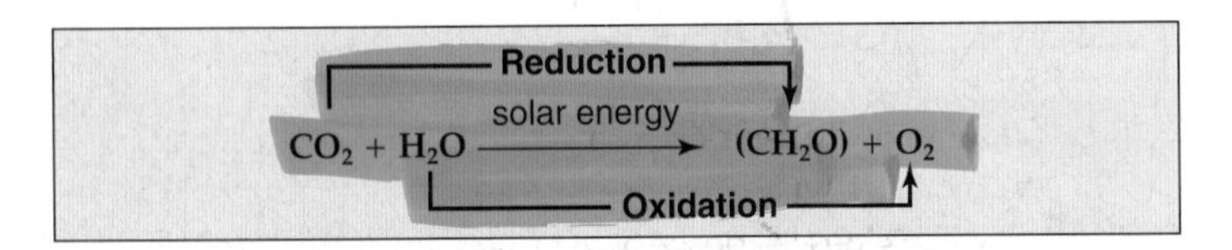


It takes hydrogen atoms and also energy to reduce carbon dioxide. From our study of energy and enzymes in Chapter 6, you expect that solar energy will not be used directly during photosynthesis, and instead it will be converted to ATP molecules. ATP is the energy currency of cells and, when cells need something, they spend ATP. In this case, solar energy will be used to generate the ATP needed to reduce carbon dioxide to a carbohydrate. Of course, we always want to keep in mind that this carbohydrate represents the food produced by land plants, algae, and cyanobacteria that feeds the biosphere.

A review of page 112 will also lead you to suspect that the electrons needed to reduce carbon dioxide will be carried by a coenzyme. $NADP^+$ is the coenzyme of oxidation-reduction (redox coenzyme) active during photosynthesis. When $NADP^+$ is reduced, it has accepted two electrons and one hydrogen atom, and when it is oxidized, it gives up its electrons:

$$NADP^+ + 2\,e^- + H^+ \longrightarrow NADPH$$

What molecule supplies the electrons that reduce NADPH during photosynthesis? Put a sprig of *Elodea* in a beaker, and supply it with light, and you will observe a bubbling (Fig. 7.3). The bubbling occurs because the plant is releasing oxygen as it photosynthesizes. A very famous experiment performed by C. B. van Niel of Stanford University found that the oxygen given off by photosynthesizers comes from water. This was the first step toward discovering that water splits during photosynthesis. When water splits, oxygen is released and the hydrogen atoms $(H^+ + e^-)$ are taken up by NADPH. Later, NADH reduces carbon dioxide to a carbohydrate.

**FIGURE 7.3**
**Photosynthesis releases oxygen.**
Bubbling indicates that the aquatic plant *Elodea* releases $O_2$ gas when it photosynthesizes.

Van Niel performed two separate experiments. When an isotope of oxygen, $^{18}O$, was a part of water, the $O_2$ given off by the plant contain $^{18}O$. When $^{18}O$ was a part of carbon dioxide supplied to a plant, the $O_2$ given off by a plant did not contain the $^{18}O$. Why not?

## Two Sets of Reactions

Many investigators have contributed to our understanding of the overall equation of photosynthesis and to our current realization that photosynthesis consists of two separate sets of reactions. F. F. Blackman was the first to suggest, in 1905, that enzymes must be involved in the reduction of carbon dioxide to a carbohydrate and that, therefore, the process must consist of two separate sets of reactions. We will call the two sets of reactions the light reactions and the Calvin cycle reactions.

### *Light Reactions*

The **light reactions** are so named because they only occur when solar energy is available (during daylight hours). The overall equation for photosynthesis gives no hint that the green pigment chlorophyll, present in thylakoid membranes, is largely responsible for absorbing the solar energy that drives photosynthesis. During the light reactions, solar energy energizes electrons that move down an electron transport chain (see Figure 6.12). As the electrons move down the chain, energy is released and captured for the production of ATP molecules. Energized electrons are also taken up by $NADP^+$, which becomes NADPH. This equation can be used to summarize the light reactions because, during the light reactions, solar energy is converted to chemical energy:

$$\text{solar energy} \longrightarrow \text{chemical energy (ATP, NADPH)}$$

## *Calvin Cycle Reactions*

The **Calvin cycle reactions** are named for Melvin Calvin, who received a Nobel Prize for discovering the enzymatic reactions that reduce carbon dioxide to a carbohydrate in the stroma of chloroplast (Fig. 7.5). The enzymes that are able to speed the reduction of carbon dioxide during both day and night are located in the semifluid substance of the stroma.

During the Calvin cycle reactions, $CO_2$ is taken up and then reduced to a carbohydrate that can later be converted to glucose. This equation can be used to summarize the Calvin cycle reactions because, during these reactions, the ATP and NADPH formed during the light reactions are used to reduce carbon dioxide:

## *Summary*

Figure 7.4 can be used to summarize our discussion so far. This figure shows that during the light reactions, (1) solar energy is absorbed, (2) water is split so that oxygen is released, and (3) ATP and NADPH are produced.

During the Calvin cycle reactions, (1) $CO_2$ is absorbed and (2) reduced to a carbohydrate ($CH_2O$) by utilizing ATP and NADPH from the light reactions (see bottom set of red arrows). The top set of red arrows takes ADP + Ⓟ and $NADP^+$ back to light reactions, where they become ATP and NADPH once more so that carbohydrate production can continue.

**FIGURE 7.5 Melvin Calvin in the laboratory.**

Melvin Calvin used tracers to discover the cycle of reactions that reduce $CO_2$ to a carbohydrate.

### Check Your Progress 7.2

1. Show that the overall equation for photosynthesis is a redox reaction.
2. In general terms, describe the light reactions and the Calvin cycle reactions.

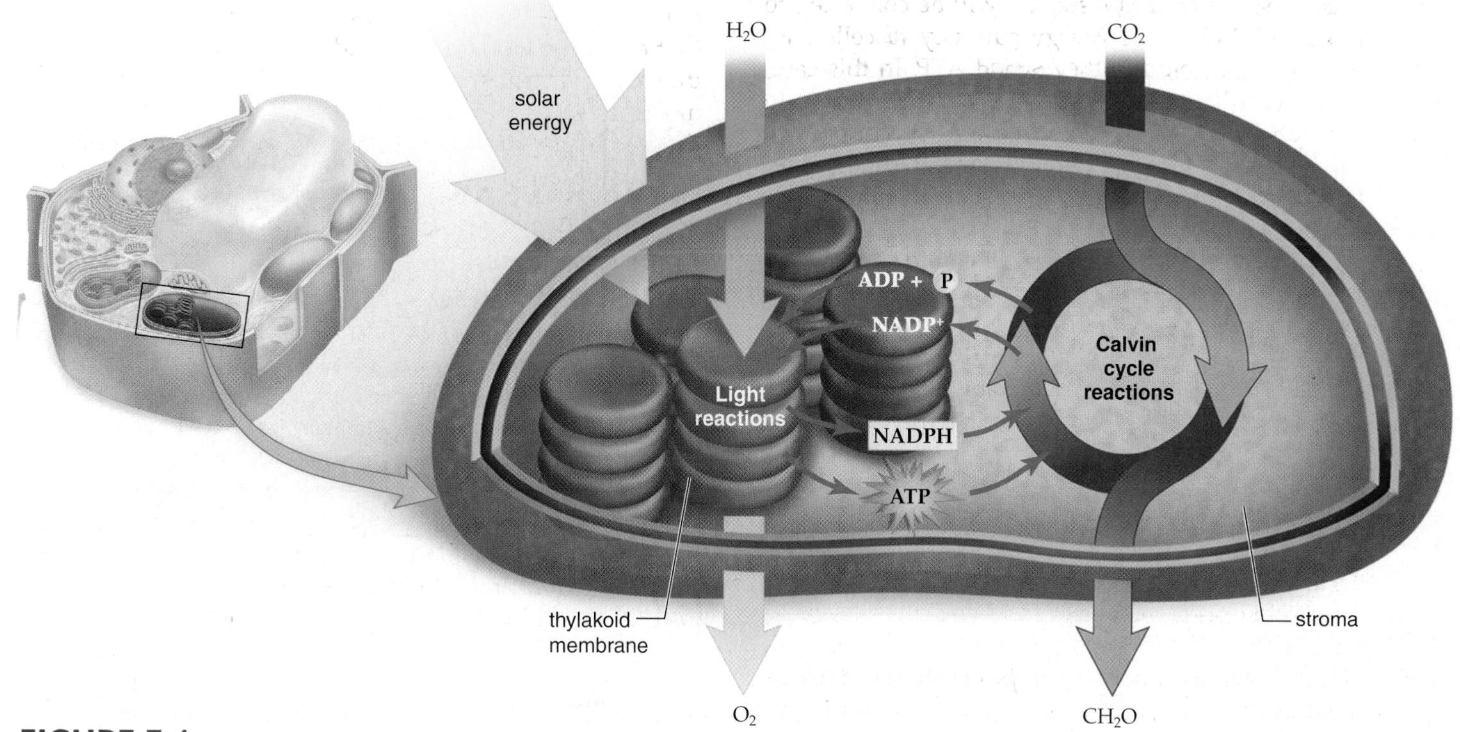


**FIGURE 7.4 Overview of photosynthesis.**

The process of photosynthesis consists of the light reactions and the Calvin cycle reactions. The light reactions, which produce ATP and NADPH, occur in the thylakoid membrane. These molecules are used in the Calvin cycle reactions which take place in the stroma. The Calvin cycle reactions reduce carbon dioxide to a carbohydrate.

## 7.3 Plants as Solar Energy Converters

Solar energy can be described in terms of its wavelength and its energy content. Figure 7.6*a* lists the different types of radiant energy from the shortest wavelength, gamma rays, to the longest, radio waves. Most of the radiation reaching the Earth is within the visible-light range. Higher-energy wavelengths are screened out by the ozone layer in the atmosphere, and lower-energy wavelengths are screened out by water vapor and carbon dioxide before they reach the Earth's surface. The conclusion is, then, that organic molecules and processes within organisms, such as vision and photosynthesis, are chemically adapted to the radiation that is most prevalent in the environment—**visible light** (Fig. 7.6*a*).

Pigment molecules absorb wavelengths of light. Most pigments absorb only some wavelengths; they reflect or transmit the other wavelengths. The pigments found in chloroplasts are capable of absorbing various portions of visible light. This is called their **absorption spectrum.** Photosynthetic organisms differ by the type of chlorophyll they contain. In plants, chlorophyll *a* and chlorophyll *b* play prominent roles in photosynthesis. **Carotenoids** play an accessory role. Both chlorophylls *a* and *b* absorb violet, blue, and red light better than the light of other colors. Because green light is transmitted and reflected by chlorophyll, plant leaves appear green to us. The carotenoids, which are shades of yellow and orange, are able to absorb light in the violet-blue-green range. These pigments become noticeable in the fall when chlorophyll breaks down.

How do you determine the absorption spectrum of pigments? To identify the absorption spectrum of a particular pigment, a purified sample is exposed to different wavelengths of light inside an instrument called a spectrophotometer. A spectrophotometer measures the amount of light that passes through the sample, and from this it is possible to calculate how much was absorbed. The amount of light absorbed at each wavelength is plotted on a graph, and the result is a record of the pigment's absorption spectrum (Fig. 7.6*b*).

### Light Reactions

The light reactions utilize two photosystems, called photosystem I (PS I) and photosystem II (PS II). The photosystems are named for the order in which they were discovered, not for the order in which they occur in the thylakoid membrane or participate in the photosynthetic process. A **photosystem** consists of a pigment complex (molecules of chlorophyll *a*, chlorophyll *b*, and the carotenoids) and electron acceptor molecules within the thylakoid membrane. The pigment complex serves as an "antenna" for gathering solar energy.

#### *Noncyclic Pathway*

During the light reactions, electrons usually follow a **noncyclic pathway** that begins with photosystem II (Fig. 7.7). The pigment complex absorbs solar energy, which is then passed from one pigment to the other until it is concentrated in a particular pair of chlorophyll *a* molecules, called the *reaction center.* Electrons ($e^-$) in the reaction center become so energized that they escape from the reaction center and move to nearby electron acceptor molecules.

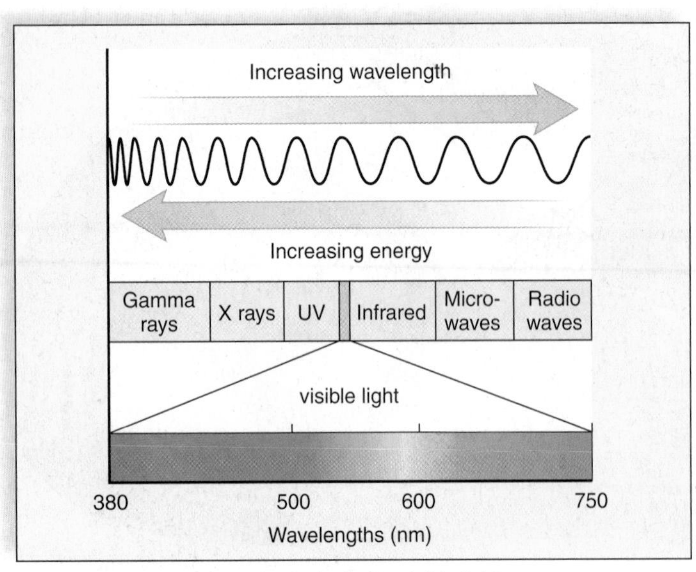


a. The electromagnetic spectrum includes visible light.

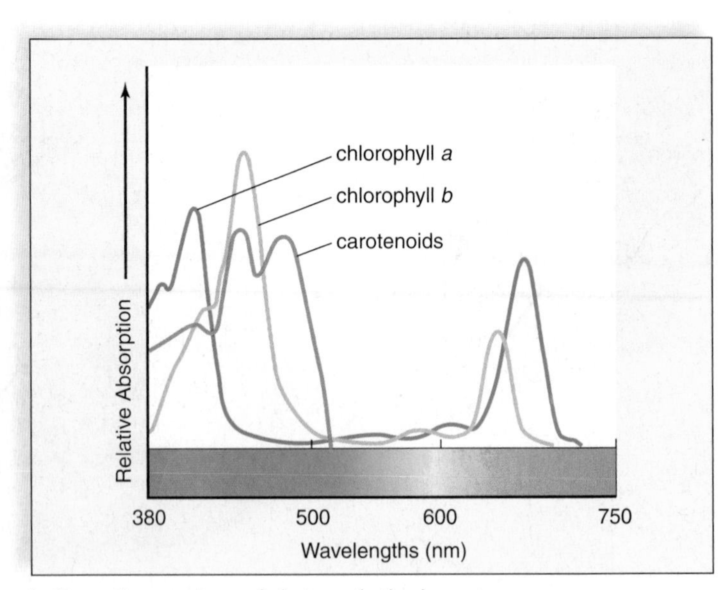


b. Absorption spectrum of photosynthetic pigments.

**FIGURE 7.6 Photosynthetic pigments and photosynthesis.**

**a.** The wavelengths in visible light differ according to energy content and color. **b.** The photosynthetic pigments in chlorophylls *a* and *b* and the carotenoids absorb certain wavelengths within visible light. This is their absorption spectrum.

PS II would disintegrate without replacement electrons, and these are removed from water, which splits, releasing oxygen to the atmosphere. Notice that with the loss of electrons, water has been oxidized, and that indeed, the oxygen released during photosynthesis does come from water. Many organisms, including plants and even ourselves, use this oxygen within their mitochondria. The hydrogen ions ($H^+$) stay in the thylakoid space and contribute to the formation of a hydrogen ion gradient.

An electron acceptor sends energized electrons, received from the reaction center, down an electron transport chain (ETC), a series of carriers that pass electrons from one to the other (see Fig. 6.13). As the electrons pass from one carrier to the next, energy is captured and stored in the form of a hydrogen ion ($H^+$) gradient. When these hydrogen ions flow down their electrochemical gradient through ATP synthase complexes, ATP production occurs (see page 124). Notice that this ATP will be used by the Calvin cycle reactions in the stroma to reduce carbon dioxide to a carbohydrate.

When the PS I pigment complex absorbs solar energy, energized electrons leave its reaction center and are captured by electron acceptors. (Low-energy electrons from the *electron transport chain* adjacent to PS II replace those lost by PS I.) The electron acceptors in PS I pass their electrons to $NADP^+$ molecules. Each one accepts two electrons and an $H^+$ to become a reduced form of the molecule, that is, NADPH. This NADPH will be used by the Calvin cycle reactions in the stroma to reduce carbon dioxide to a carbohydrate.

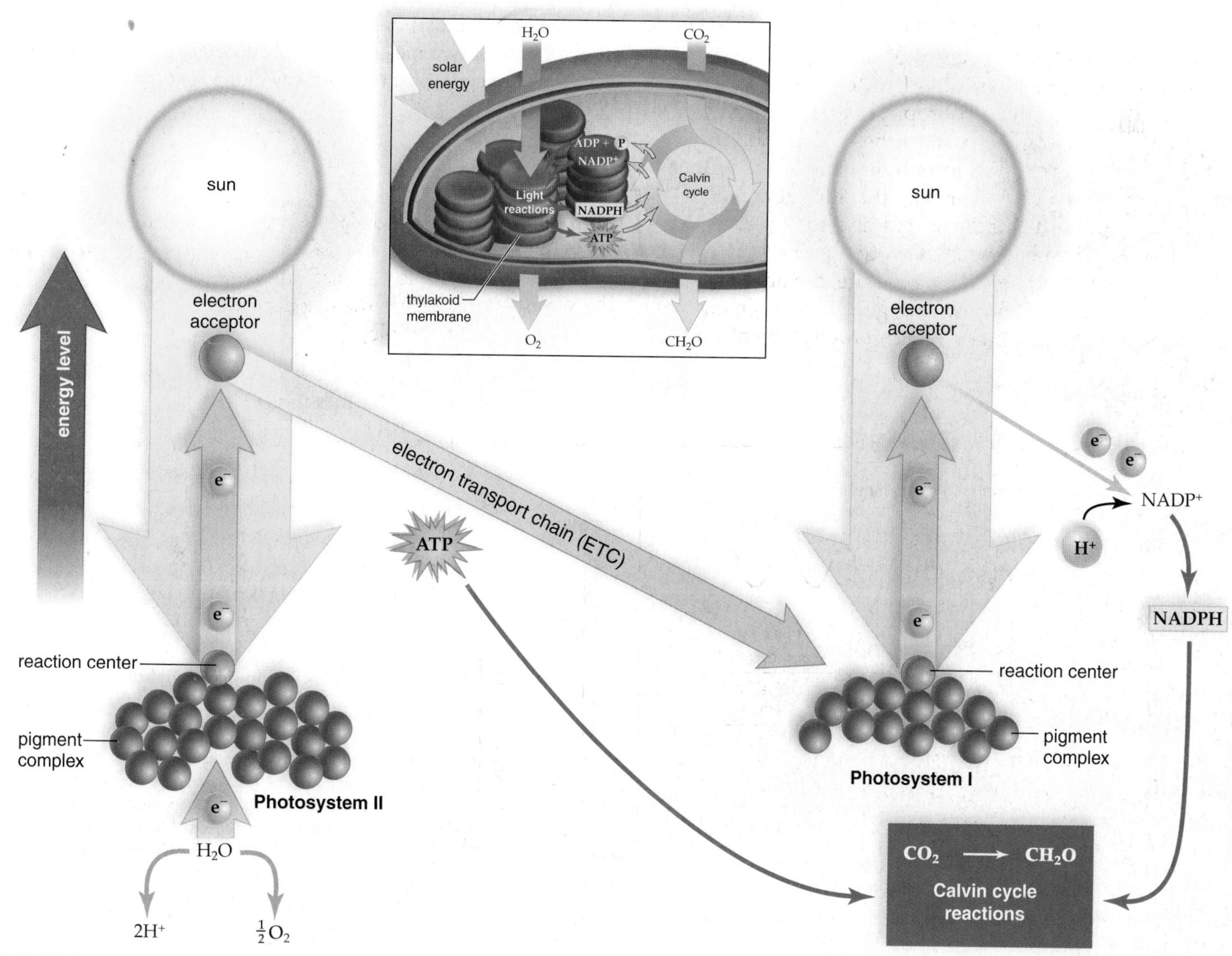


**FIGURE 7.7 Noncyclic pathway: Electrons move from water to $NADP^+$.**

Energized electrons (replaced from water, which splits, releasing oxygen) leave photosystem II and pass down an electron transport chain, leading to the formation of ATP. Energized electrons (replaced by photosystem II by way of the ETC) leave photosystem I and pass to $NADP^+$, which then combines with $H^+$, becoming NADPH.

## The Organization of the Thylakoid Membrane

As we have discussed, the following molecular complexes are in the thylakoid membrane (Fig. 7.8):

PS II, which consists of a pigment complex and electron-acceptor molecules, receives electrons from water as water splits, releasing oxygen.

The electron transport chain (ETC), consisting of Pq (plastoquinone) and cytochrome complexes, carries electrons from PS II to PS I. Pq also pumps $H^+$ from the stroma into the thylakoid space.

PS I, which also consists of a pigment complex and electron-acceptor molecules, is adjacent to NADP reductase, which reduces $NADP^+$ to NADPH.

The ATP synthase complex has a channel and a protruding ATP synthase, an enzyme that joins ADP + Ⓟ.

## ATP Production

The thylakoid space acts as a reservoir for hydrogen ions ($H^+$). First, each time water is oxidized, two $H^+$ remain in the thylakoid space. Second, as the electrons move from carrier to carrier along the electron transport chain, the electrons give up energy, which is used to pump $H^+$ from the stroma into the thylakoid space. Therefore, there are more $H^+$ in the thylakoid space than in the stroma. The flow of $H^+$ (often referred to as protons in this context) from high to low concentration provides kinetic energy that allows an **ATP synthase complex** enzyme to enzymatically produce ATP from ADP + Ⓟ. This method of producing ATP is called **chemiosmosis** because ATP production is tied to the establishment of an $H^+$ gradient (see Fig. 6.13).

### Check Your Progress 7.3

1. What part of the electromagnetic spectrum is utilized for photosynthesis?
2. What two molecules are produced as a result of the noncyclic electron pathway of the light reactions?

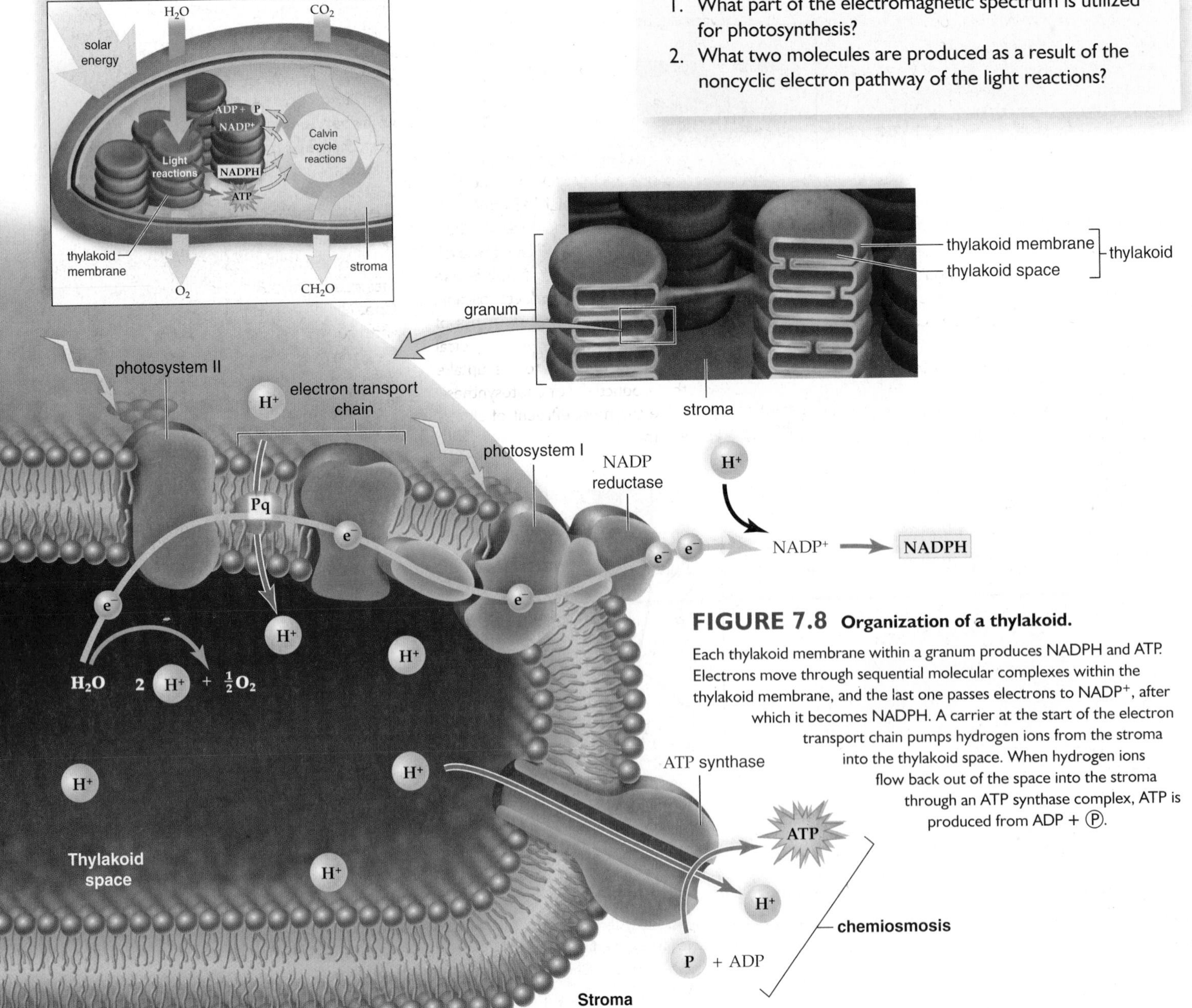


**FIGURE 7.8 Organization of a thylakoid.** Each thylakoid membrane within a granum produces NADPH and ATP. Electrons move through sequential molecular complexes within the thylakoid membrane, and the last one passes electrons to $NADP^+$, after which it becomes NADPH. A carrier at the start of the electron transport chain pumps hydrogen ions from the stroma into the thylakoid space. When hydrogen ions flow back out of the space into the stroma through an ATP synthase complex, ATP is produced from ADP + Ⓟ.

*ecology focus*

# Tropical Rain Forest Destruction and Global Warming

Al Gore, former presidential candidate, won the 2007 Nobel Peace Prize for raising public awareness concerning global warming. The Nobel Committee said that "global warming could induce large-scale migrations and lead to greater competition for the Earth's resources. As such, it may increase the danger of violent conflicts and wars, within and between countries." **Global warming** refers to an expected rise in the average global temperature during the twenty-first century due to the introduction of certain gases into the atmosphere. For at least a thousand years prior to 1850, atmospheric carbon dioxide ($CO_2$) levels remained fairly constant at 0.028%. Since the 1850s, when industrialization began, the amount of $CO_2$ in the atmosphere has increased to 0.038% (Fig. 7A*a*).

## Role of Carbon Dioxide

In much the same way as the panes of a greenhouse, $CO_2$ and other gases in our atmosphere trap radiant heat from the sun. Therefore, these gases are called greenhouse gases. Without any greenhouse gases, the Earth's temperature would be about 33°C cooler than it is now. Likewise, increasing the concentration of greenhouse gases is predicted to cause global warming.

Certainly, the burning of fossil fuels adds $CO_2$ to the atmosphere. But another factor that contributes to an increase in atmospheric $CO_2$ is tropical rain forest destruction.

## Role of Tropical Rain Forests

Between 10 and 30 million hectares of rain forests are lost every year to ranching, logging, mining, and otherwise developing areas of the forest for human needs. The clearing of forests often involves burning them (Fig. 7A*b*). Each year, deforestation in tropical rain forests accounts for 20–30% of all $CO_2$ in the atmosphere. The consequence of burning forests is greater trouble for global warming because burning a forest adds $CO_2$ to the atmosphere and, at the same time, removes trees that would ordinarily absorb $CO_2$.

## The Argument for Preserving Forests

The process of photosynthesis and also the oceans act as a sink for $CO_2$. Despite their reduction in size from an original 14% to 6% of land surface today, tropical rain forests make a substantial contribution to global $CO_2$ removal. Taking into account all ecosystems, marine and terrestrial, photosynthesis produces organic matter that is 300 to 600 times the mass of people currently on Earth this year. Tropical rain forests contribute greatly to the uptake of $CO_2$ and the productivity of photosynthesis because they are the most efficient of all terrestrial ecosystems.

Tropical rain forests occur near the equator. They can exist wherever temperatures are above 26°C and rainfall is heavy (from 100–200 cm) and regular. Huge trees with buttressed trunks and broad, undivided, dark-green leaves predominate. Nearly all land plants in a tropical rain forest are woody, and woody vines are also abundant.

It might be hypothesized that an increased amount of $CO_2$ in the atmosphere will cause photosynthesis to increase in the remaining portion of the forest. To study this possibility, investigators measured atmospheric $CO_2$ levels, daily temperature levels, and tree girth in La Selva, Costa Rica, for 16 years. The data collected demonstrated relatively lower forest productivity at higher temperatures. These findings suggest that, as temperatures rise, tropical rain forests may add to ongoing atmospheric $CO_2$ accumulation and accelerated global warming rather than the reverse. All the more reason to slow global warming and preserve forests.

Some countries have programs to combat the problem of deforestation. In the mid-1970s, Costa Rica established a system of national parks and reserves to protect 12% of the country's land area from degradation. The current Costa Rican government wants to expand the goal by increasing protected areas to 25% in the near future. Similar efforts in other countries may help slow the ever-increasing threat of global warming.

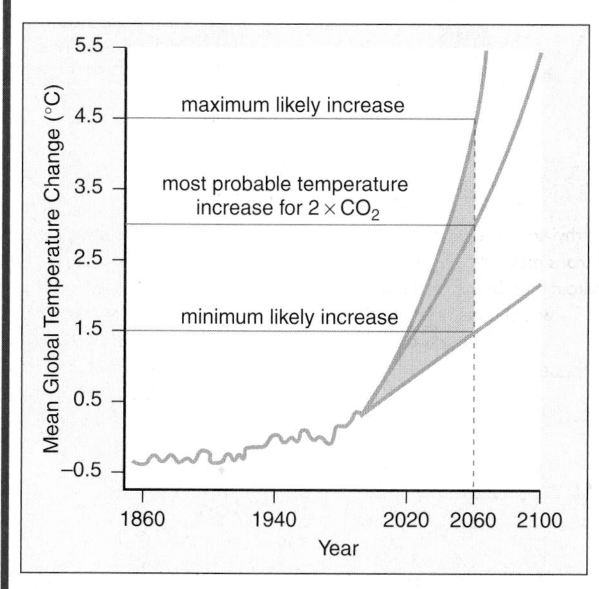


a.

b.

**FIGURE 7A Global warming.**
***a.*** *Mean global temperature change is expected to rise due to the introduction of greenhouse gases into the atmosphere.* ***b.*** *The burning of tropical rain forests adds $CO_2$ to the atmosphere and at the same time removes a sink for $CO_2$.*

# 7.4 Calvin Cycle Reactions

The Calvin cycle reactions occur after the light reactions. The Calvin cycle is a series of reactions that produce carbohydrate before returning to the starting point once more (Fig. 7.9). The cycle is named for Melvin Calvin, who, with colleagues, used the radioactive isotope $^{14}C$ as a tracer to discover the reactions making up the cycle.

This series of reactions uses carbon dioxide from the atmosphere to produce carbohydrate. How does carbon dioxide get into the atmosphere? We and most other organisms take in oxygen from the atmosphere and release carbon dioxide to the atmosphere. The Calvin cycle includes (1) carbon dioxide fixation, (2) carbon dioxide reduction, and (3) regeneration of RuBP (ribulose-1,5-bisphosphate).

## Fixation of Carbon Dioxide

**Carbon dioxide ($CO_2$) fixation** is the first step of the Calvin cycle. During this reaction, carbon dioxide from the atmosphere is attached to RuBP, a 5-carbon molecule. The result is one 6-carbon molecule, which splits into two 3-carbon molecules.

The enzyme that speeds this reaction, called **RuBP carboxylase,** is a protein that makes up about 20–50% of the protein content in chloroplasts. The reason for its abundance may be that it is unusually slow (it processes only a few molecules of substrate per second compared to thousands per second for a typical enzyme), and so there has to be a lot of it to keep the Calvin cycle going.

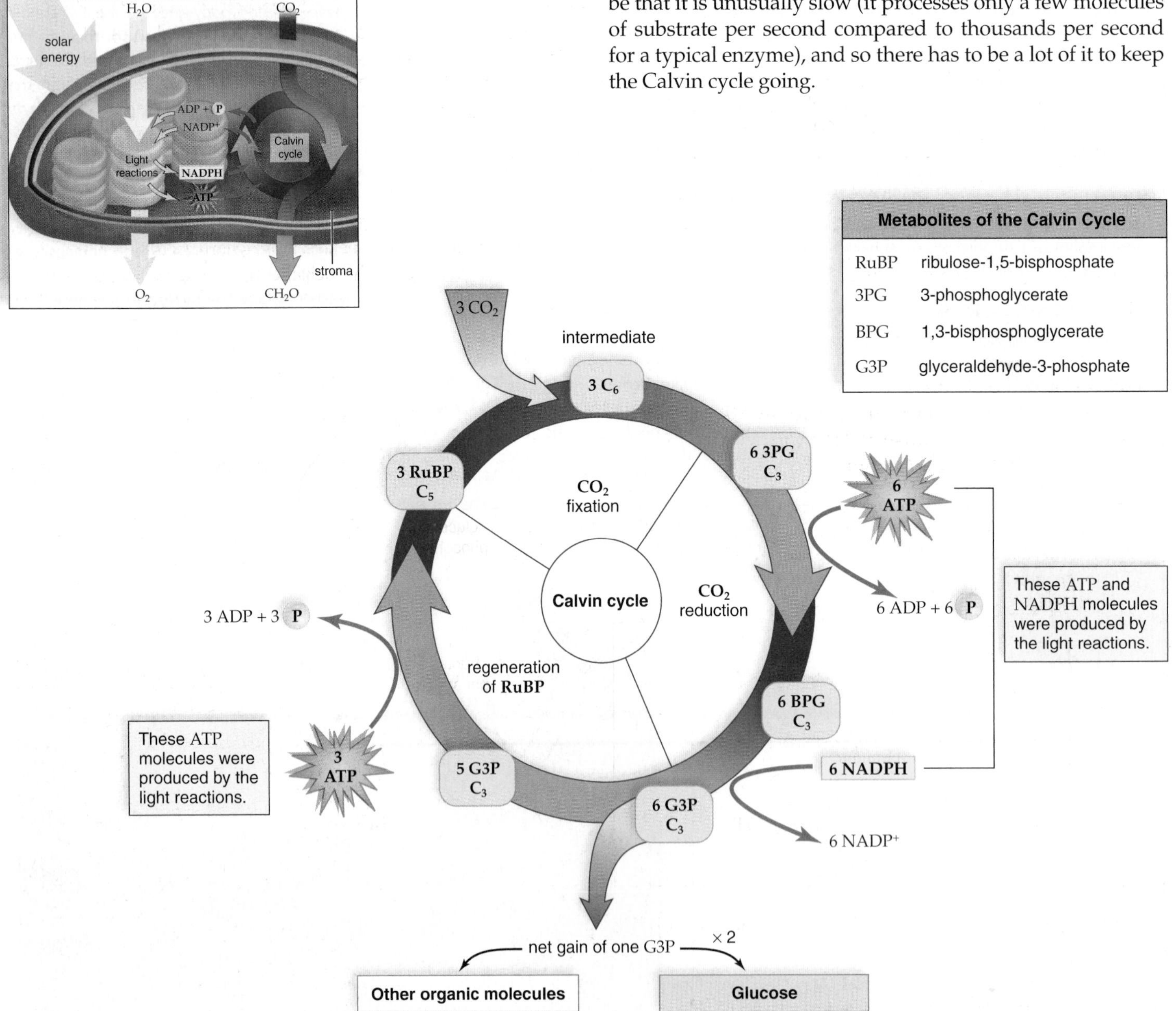


**FIGURE 7.9 The Calvin cycle reactions.**

The Calvin cycle is divided into three portions: $CO_2$ fixation, $CO_2$ reduction, and regeneration of RuBP. Because five G3P are needed to re-form three RuBP, it takes three turns of the cycle to have a net gain of one G3P. Two G3P molecules are needed to form glucose.

## Reduction of Carbon Dioxide

The first 3-carbon molecule in the Calvin cycle is called 3PG (3-phosphoglycerate). Each of two 3PG molecules undergoes reduction to G3P in two steps:

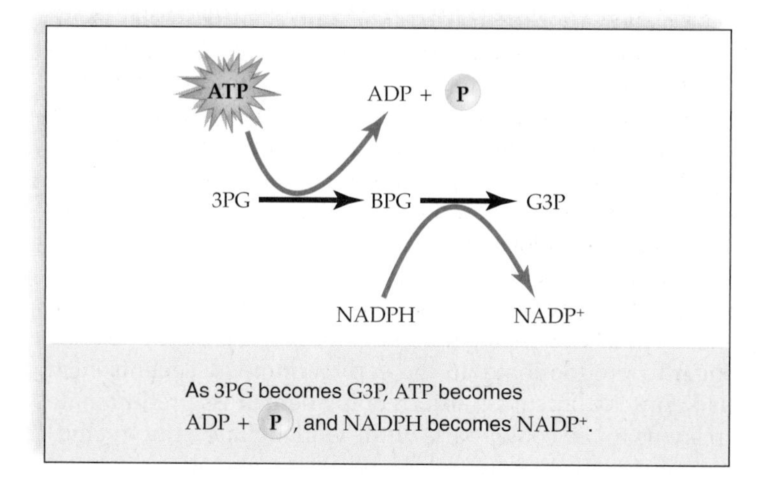


This is the sequence of reactions that uses some ATP and NADPH from the light reactions. This sequence signifies the reduction of carbon dioxide to a carbohydrate because $R$—$CO_2$ has become $R$—$CH_2O$. Energy and electrons are needed for this reduction reaction, and these are supplied by ATP and NADPH.

## Regeneration of RuBP

Notice that the Calvin cycle reactions in Figure 7.9 are multiplied by three because it takes three turns of the Calvin cycle to allow one G3P to exit. Why? Because, for every three turns of the Calvin cycle, five molecules of G3P are used to re-form three molecules of RuBP and the cycle continues. Notice that $5 \times 3$ (carbons in G3P) = $3 \times 5$ (carbons in RuBP):

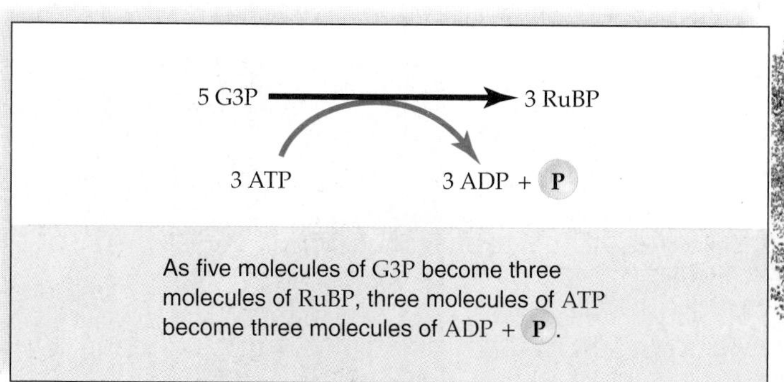


This reaction also uses some of the ATP produced by the light reactions.

## The Importance of the Calvin Cycle

G3P (glyceraldehyde-3-phosphate) is the product of the Calvin cycle that can be converted to other molecules a plant needs. Notice that glucose phosphate is among the organic molecules that result from G3P metabolism (Fig. 7.10). This is of interest to us because glucose is the molecule that plants and animals most often metabolize to produce the ATP molecules they require for their energy needs.

Glucose phosphate can be combined with fructose (and the phosphate removed) to form sucrose, the molecule that plants use to transport carbohydrates from one part of the plant to the other.

Glucose phosphate is also the starting point for the synthesis of starch and cellulose. Starch is the storage form of glucose. Some starch is stored in chloroplasts, but most starch is stored in amyloplasts in roots. Cellulose is a structural component of plant cell walls and becomes fiber in our diet because we are unable to digest it.

A plant can use the hydrocarbon skeleton of G3P to form fatty acids and glycerol, which are combined in plant oils. We are all familiar with corn oil, sunflower oil, or olive oil used in cooking. Also, when nitrogen is added to the hydrocarbon skeleton derived from G3P, amino acids are formed.

**Check Your Progress 7.4**

1. What are three major steps of the Calvin cycle?
2. List the substances that a plant cell can make from G3P, the product of the Calvin cycle.

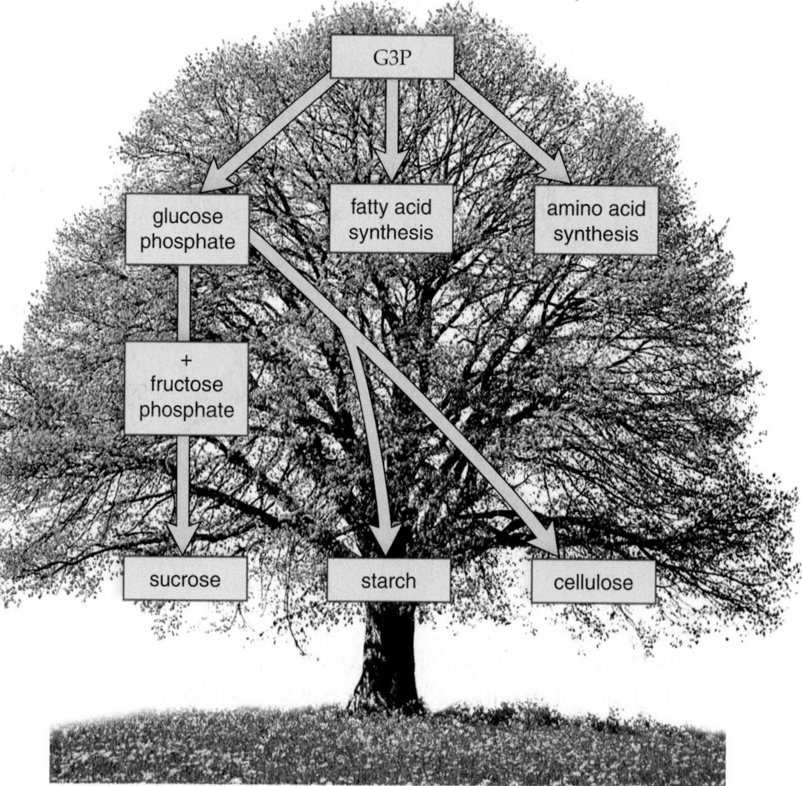


**FIGURE 7.10 Fate of G3P.**

G3P is the first reactant in a number of plant cell metabolic pathways. Two G3Ps are needed to form glucose phosphate; glucose is often considered the end product of photosynthesis. Sucrose is the transport sugar in plants; starch is the storage form of glucose; and cellulose is a major constituent of plant cell walls.

# 7.5 Other Types of Photosynthesis

The majority of land plants such as azaleas, maples, and tulips carry on photosynthesis as described and are called **$C_3$ plants** (Fig. 7.11*a*). $C_3$ plants use the enzyme RuBP carboxylase to fix $CO_2$ to RuBP in mesophyll cells. The first detected molecule following fixation is the 3-carbon molecule 3PG:

$$RuBP + CO_2 \xrightarrow{\text{RuBP carboxylase}} 2\ 3PG$$

As shown in Figure 7.2, leaves have small openings called stomata through which water can leave and carbon dioxide ($CO_2$) can enter. If the weather is hot and dry, the stomata close, conserving water. (Water loss might cause the plant to wilt and die.) Now the concentration of $CO_2$ decreases in leaves, while $O_2$, a by-product of photosynthesis, increases. When $O_2$ rises in $C_3$ plants, RuBP carboxylase combines it with RuBP instead of $CO_2$. The result is one molecule of 3PG and the eventual release of $CO_2$. This is called **photorespiration** because in the presence of light *(photo)*, oxygen is taken up and $CO_2$ is released *(respiration)*.

An adaptation called $C_4$ photosynthesis enables some plants to avoid photorespiration.

## $C_4$ Photosynthesis

In a $C_3$ plant, the mesophyll cells contain well-formed chloroplasts and are arranged in parallel layers. In a $C_4$ leaf, the bundle sheath cells, as well as the mesophyll cells, contain chloroplasts. Further, the mesophyll cells are arranged concentrically around the bundle sheath cells:

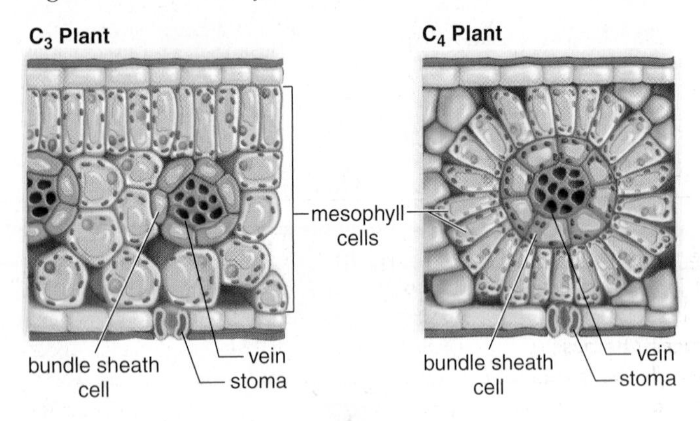


**$C_4$ plants** use the enzyme PEP carboxylase (PEPCase) to fix $CO_2$ to PEP (phosphoenolpyruvate, a $C_3$ molecule). The result is oxaloacetate, a $C_4$ molecule:

$$PEP + CO_2 \xrightarrow{\text{PEPCase}} \text{oxaloacetate}$$

In a $C_4$ plant, $CO_2$ is taken up in mesophyll cells, and then malate, a reduced form of oxaloacetate, is pumped into the bundle sheath cells (Fig. 7.11*b*). Here, and only here, does $CO_2$ enter the Calvin cycle. It takes energy to pump molecules, and you would think that the $C_4$ pathway would be disadvantageous. Yet in hot, dry climates, the net photosynthetic rate of $C_4$ plants such as sugarcane, corn, and Bermuda grass is about two to three times that of $C_3$ plants such as wheat, rice, and oats. Why do $C_4$ plants enjoy such an advantage? The answer is that they can avoid photorespiration, discussed previously. Photorespiration is wasteful because it is not part of the Calvin cycle. Photorespiration does not occur in $C_4$ leaves because PEPCase, unlike RuBP carboxylase, does not combine with $O_2$. Even when stomata are closed, $CO_2$ is delivered to the Calvin cycle in the bundle sheath cells.

When the weather is moderate, $C_3$ plants ordinarily have the advantage, but when the weather becomes hot and dry, $C_4$ plants have the advantage, and we can expect them to predominate. In the early summer, $C_3$ plants such as Kentucky bluegrass and creeping bent grass predominate in lawns in the cooler parts of the United States, but by midsummer, crabgrass, a $C_4$ plant, begins to take over.

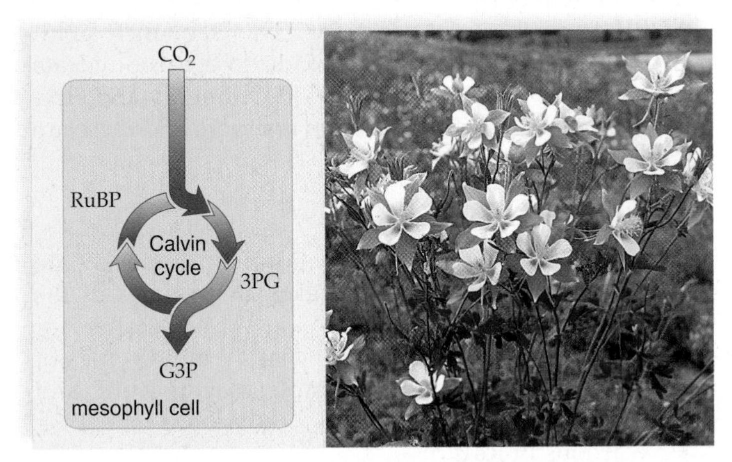


a. $CO_2$ fixation in a $C_3$ plant, blue columbine, *Aquilegia caerulea*

b. $CO_2$ fixation in a $C_4$ plant, corn, *Zea mays*

**FIGURE 7.11 Carbon dioxide fixation in $C_3$ and $C_4$ plants.**

**a.** In $C_3$ plants, $CO_2$ is taken up by the Calvin cycle directly in mesophyll cells. **b.** $C_4$ plants form a $C_4$ molecule in mesophyll cells prior to releasing $CO_2$ to the Calvin cycle in bundle sheath cells.

## CAM Photosynthesis

**CAM** stands for crassulacean-acid metabolism; the Crassulaceae is a family of flowering succulent (water-containing) plants that live in warm, dry regions of the world. CAM was first discovered in these plants, but now it is known to be prevalent among other groups of plants.

Whereas a $C_4$ plant represents partitioning in space—carbon dioxide fixation occurs in mesophyll cells and the Calvin cycle occurs in bundle sheath cells—CAM is partitioning by the use of time. During the night, CAM plants use PEPCase to fix some $CO_2$, forming $C_4$ molecules, which are stored in large vacuoles in mesophyll cells. During the day, $C_4$ molecules (malate) release $CO_2$ to the Calvin cycle when NADPH and ATP are available from the light reactions (Fig. 7.12). The primary advantage for this partitioning again has to do with the conservation of water. CAM plants open their stomata only at night, and therefore only at that time does atmospheric $CO_2$ enter the plant. During the day, the stomata close; this conserves water, but $CO_2$ cannot enter the plant.

Photosynthesis in a CAM plant is minimal because a limited amount of $CO_2$ is fixed at night, but it does allow CAM plants to live under stressful conditions.

$CO_2$ fixation in a CAM plant, pineapple, *Ananas comosus*

**FIGURE 7.12 Carbon dioxide fixation in a CAM plant.** CAM plants, such as pineapple, fix $CO_2$ at night, forming a $C_4$ molecule that is released to the Calvin cycle during the day.

## Photosynthesis and Adaptation to the Environment

The different types of photosynthesis give us an opportunity to consider that organisms are metabolically adapted to their environment. Each method of photosynthesis has its advantages and disadvantages, depending on the climate.

$C_4$ plants most likely evolved in, and are adapted to, areas of high light intensities, high temperatures, and limited rainfall. $C_4$ plants, however, are more sensitive to cold, and $C_3$ plants do better than $C_4$ plants below 25°C. CAM plants, on the other hand, compete well with either type of plant when the environment is extremely arid. Surprisingly, CAM is quite widespread and has evolved in 23 families of flowering plants, including some lilies and orchids! And it is found among nonflowering plants, including some ferns and cone-bearing trees.

### Check Your Progress 7.5

1. Name some plants that use a method of photosynthesis other than $C_3$ photosynthesis.
2. Explain why $C_4$ photosynthesis is advantageous in hot, dry conditions.

## Connecting the Concepts

"Have You Thanked a Green Plant Today?" is a bumper sticker that you may have puzzled over until now. Plants, you now know, capture solar energy and store it in carbon-based organic nutrients that are passed to other organisms when they feed on plants and/or on other organisms. In this context, plants are called autotrophs because they make their own organic food. Heterotrophs are organisms that take in preformed organic food.

The next chapter considers cellular respiration, the process that produces ATP molecules. We have to keep in mind that cells cannot create energy, and therefore when we say that the cell produces ATP, we mean that it converts the energy within glucose molecules to that found in ATP molecules, with a loss of heat, of course. Why do cells carry out this wasteful process? Because ATP is the energy currency of cells and is able to contribute energy to many different cellular processes and reactions.

In the carbon cycle, living and dead organisms contain organic carbon and serve as a reservoir of carbon. Some 300 million years ago, a host of plants died and did not decompose. These plants were compressed to form the coal that we mine and burn today. (Oil has a similar origin, but it most likely formed in marine sedimentary rocks that included animal remains.)

The amount of carbon dioxide in the atmosphere is increasing steadily, in part because we humans burn fossil fuels to run our modern industrial society. This buildup of carbon dioxide will contribute to global warming. Autotrophs such as plants take in carbon dioxide when they photosynthesize. Carbon dioxide is returned to the atmosphere when autotrophs and heterotrophs carry on cellular respiration. In this way, the very same carbon atoms cycle from the atmosphere to autotrophs, then to heterotrophs, and then back to autotrophs again. Energy does not cycle, and therefore all life is dependent on the ability of plants to capture solar energy and produce carbohydrate molecules.

## summary

### 7.1 Photosynthetic Organisms

Photosynthesis produces carbohydrates and releases oxygen, both of which are used by the majority of living things. Cyanobacteria, algae, and land plants carry on photosynthesis. In plants, photosynthesis takes place in chloroplasts. A chloroplast is bounded by a double membrane and contains two main components: the semifluid stroma and the membranous grana made up of thylakoids.

### 7.2 The Process of Photosynthesis

The overall equation for photosynthesis shows that it is a redox reaction. Carbon dioxide is reduced, and water is oxidized. During photosynthesis, the light reactions take place in the thylakoid membranes, and the Calvin cycle reactions take place in the stroma.

### 7.3 Plants as Solar Energy Converters

Photosynthesis uses solar energy in the visible-light range. Specifically, chlorophylls *a* and *b* absorb violet, blue, and red wavelengths best. This causes chlorophyll to appear green to us. The carotenoids absorb light in the violet-blue-green range and are yellow to orange pigments.

The noncyclic electron pathway of the light reactions begins when solar energy enters PS II. In PS II, energized electrons are picked up by electron acceptors. The oxidation (splitting) of water replaces these electrons in the reaction-center chlorophyll *a* molecules. Oxygen is released to the atmosphere, and hydrogen ions ($H^+$) remain in the thylakoid space. An electron acceptor molecule passes electrons to PS I by way of an electron transport chain. When solar energy is absorbed by PS I, energized electrons leave and are ultimately received by $NADP^+$, which also combines with $H^+$ from the stroma to become NADPH.

Chemiosmosis requires an organized membrane. The thylakoid membrane is highly organized: PS II is associated with an enzyme that oxidizes (splits) water, the cytochrome complexes transport electrons and pump $H^+$; PS I is associated with an enzyme that reduces $NADP^+$, and ATP synthase produces ATP.

The energy made available by the passage of electrons down the electron transport chain allows carriers to pump $H^+$ into the thylakoid space. The buildup of $H^+$ establishes an electrochemical gradient. When $H^+$ flows down this gradient through the channel present in ATP synthase complexes, ATP is synthesized from ADP and Ⓟ by ATP synthase. This method of producing ATP is called chemiosmosis.

### 7.4 Calvin Cycle Reactions

The energy yield of the light reactions is stored in ATP and NADPH. These molecules are used by the Calvin cycle reactions to reduce $CO_2$ to carbohydrate, namely G3P, which is then converted to all the organic molecules a plant needs.

During the first stage of the Calvin cycle, the enzyme RuBP carboxylase fixes $CO_2$ to RuBP, producing a 6-carbon molecule that immediately breaks down to two $C_3$ molecules. During the second stage, $CO_2$ (incorporated into an organic molecule) is reduced to carbohydrate ($CH_2O$). This step requires the NADPH and some of the ATP from the light reactions. For every three turns of the Calvin cycle, the net gain is one G3P molecule; the other five G3P molecules are used to re-form three molecules of RuBP. This step also requires ATP for energy. It takes two G3P molecules to make one glucose molecule.

### 7.5 Other Types of Photosynthesis

In $C_4$ plants, as opposed to the $C_3$ plants just described, the enzyme PEPCase fixes carbon dioxide to PEP to form a 4-carbon molecule, oxaloacetate, within mesophyll cells. A reduced form of this molecule is pumped into bundle sheath cells, where $CO_2$ is released to the Calvin cycle. PEPCase has an advantage over RuBP carboxylase because RuBP carboxylase, but not PEPCase, combines $O_2$ with RuBP instead of $CO_2$ when the stomata close and the concentration of $O_2$ rises. $C_4$ plants avoid this complication by a partitioning of pathways in space. Carbon dioxide fixation occurs utilizing PEPCase in mesophyll cells, and the Calvin cycle occurs in bundle sheath cells.

In CAM plants, the stomata are open only at night, conserving water. PEPCase fixes $CO_2$ to PEP only at night, and the next day, $CO_2$ is released and enters the Calvin cycle within the same cells. This represents a partitioning of pathways in time: Carbon dioxide fixation occurs at night, and the Calvin cycle occurs during the day. CAM was discovered in desert plants, but since then it has been discovered in many different types of plants.

## understanding the terms



Match the terms to these definitions:

a. ______________ Energy-capturing portion of photosynthesis that takes place in thylakoid membranes of chloroplasts and cannot proceed without solar energy; it produces ATP and NADPH.

b. ______________ Photosynthetic unit where solar energy is absorbed and high-energy electrons are generated; contains an antenna complex and an electron acceptor.

c. ______________ Process usually occurring within chloroplasts, whereby chlorophyll traps solar energy and carbon dioxide is reduced to a carbohydrate.

d. ______________ Series of photosynthetic reactions in which carbon dioxide is fixed and reduced to G3P.

## reviewing this chapter

1. Why is it proper to say that almost all living things are dependent on solar energy? 118
2. Name the two major components of chloroplasts, and associate each with one of two sets of reactions that occur during photosynthesis. How are the two sets of reactions related? 119–21
3. Write the overall equation of photosynthesis and associate each participant with either the light reactions or the Calvin cycle reactions. 120–21
4. Discuss the electromagnetic spectrum and the combined absorption spectrum of chlorophylls *a* and *b* and the carotenoids. Why is chlorophyll a green pigment, and the carotenoids a yellow-orange pigment? 122
5. Trace the noncyclic electron pathway, naming and explaining all the events that occur as the electrons move from water to $NADP^+$. 122–23

6. How is the thylakoid membrane organized? Name the main complexes in the membrane. Give a function for each. 124
7. Explain what is meant by chemiosmosis, and relate this process to the electron transport chain present in the thylakoid membrane. 124
8. Describe the three stages of the Calvin cycle. Which stage uses the ATP and NADPH from the light reactions? 126–27
9. Compare $C_3$ and $C_4$ photosynthesis, contrasting the actions of RuBP carboxylase and PEPCase. 128–29
10. Explain CAM photosynthesis, contrasting it to $C_4$ photosynthesis in terms of partitioning a pathway. 129

## testing yourself

Choose the best answer for each question.

1. The absorption spectrum of chlorophyll
   a. is not the same as that of carotenoids.
   b. approximates the action spectrum of photosynthesis.
   c. explains why chlorophyll is a green pigment.
   d. shows that some colors of light are absorbed more than others.
   e. All of these are correct.
2. The final acceptor of electrons during the noncyclic electron pathway is
   a. PS I.
   b. PS II.
   c. ATP.
   d. NADP⁺.
   e. water.
3. A photosystem contains
   a. pigments, a reaction center, and electron acceptors.
   b. ADP, Ⓟ, and hydrogen ions ($H^+$).
   c. protons, photons, and pigments.
   d. cytochromes only.
   e. Both b and c are correct.

For questions 4–8, match each item to those in the key. Use an answer more than once, if possible.

**KEY:**

a. solar energy
b. chlorophyll
c. chemiosmosis
d. Calvin cycle

4. light energy
5. ATP synthase
6. thylakoid membrane
7. green pigment
8. RuBP

For questions 9–11, indicate whether the statement is true (T) or false (F).

9. RuBP carboxylase is the enzyme that fixes carbon dioxide to RuBP in the Calvin cycle. ________
10. When 3PG becomes G3P during the light reactions, carbon dioxide is reduced to carbohydrate. ________
11. NADPH and ATP cycle between the Calvin cycle and the light reactions constantly. ________
12. The NADPH and ATP from the light reactions are used to
    a. split water.
    b. cause RuBP carboxylase to fix $CO_2$.
    c. re-form the photosystems.
    d. cause electrons to move along their pathways.
    e. convert 3PG to G3P.
13. Chemiosmosis
    a. depends on complexes in the thylakoid membrane.
    b. depends on an electrochemical gradient.
    c. depends on a difference in $H^+$ concentration between the thylakoid space and the stroma.
    d. results in ATP formation.
    e. All of these are correct.
14. The function of the light reactions is to
    a. obtain $CO_2$.
    b. make carbohydrate.
    c. convert light energy into a usable form of chemical energy.
    d. regenerate RuBP.
15. Label the following diagram of a chloroplast:

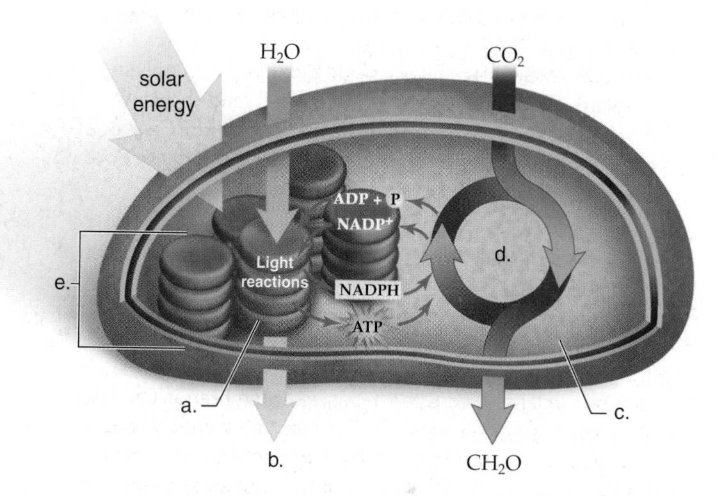


    f. The light reactions occur in which part of a chloroplast?
    g. The Calvin cycle reactions occur in which part of a chloroplast?
16. The oxygen given off by photosynthesis comes from
    a. $H_2O$.
    b. $CO_2$.
    c. glucose.
    d. RuBP.
17. Label the following diagram using these labels: water, carbohydrate, carbon dioxide, oxygen, ATP, ADP + Ⓟ, NADPH, and NADP⁺.

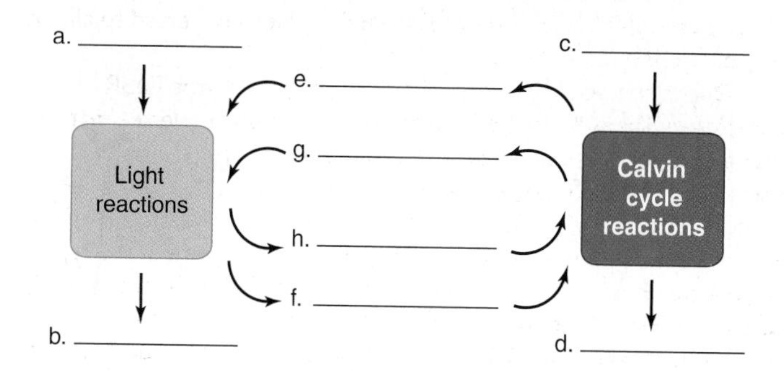


18. The glucose formed by photosynthesis can be used by plants to make
    a. starch.
    b. cellulose.
    c. lipids and oils.
    d. proteins.
    e. All of these are correct.

19. The Calvin cycle reactions
    a. produce carbohydrate.
    b. convert one form of chemical energy into a different form of chemical energy.
    c. regenerate more RuBP.
    d. use the products of the light reactions.
    e. All of these are correct.
20. CAM photosynthesis
    a. is the same as $C_4$ photosynthesis.
    b. is an adaptation to cold environments in the Southern Hemisphere.
    c. is prevalent in desert plants that close their stomata during the day.
    d. occurs in plants that live in marshy areas.
    e. stands for chloroplasts and mitochondria.
21. Compared to RuBP carboxylase, PEPCase has the advantage that
    a. PEPCase is present in both mesophyll and bundle sheath cells, but RuBP carboxylase is not.
    b. RuBP carboxylase fixes carbon dioxide ($CO_2$) only in $C_4$ plants, but PEPCase does it in both $C_3$ and $C_4$ plants.
    c. RuBP carboxylase combines with $O_2$, but PEPCase does not.
    d. PEPCase conserves energy, but RuBP carboxylase does not.
    e. Both b and c are correct.
22. $C_4$ photosynthesis
    a. is the same as $C_3$ photosynthesis because it takes place in chloroplasts.
    b. occurs in plants whose bundle sheath cells contain chloroplasts.
    c. takes place in plants such as wheat, rice, and oats.
    d. is an advantage when the weather is hot and dry.
    e. Both b and d are correct.

## thinking scientifically

1. In 1882, T. W. Engelmann carried out an ingenious experiment to demostrate that chlorophyll absorbs light in the blue and red portions of the spectrum. He placed a single filament of a green alga in a drop of water on a microscope slide. Then he passed light through a prism and onto the string of algal cells. The slide also contained aerobic bacterial cells. After some time, he peered into the microscope and saw the bacteria clustered around the regions of the algal filament that were receiving blue light and red light, as shown in the illustration. Why do you suppose the bacterial cells were clustered in this manner?

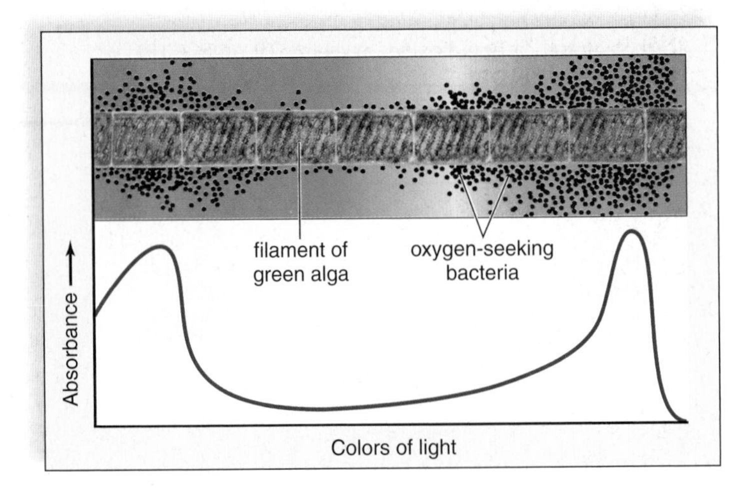


2. In the fall of the year, the leaves of many trees change from green to red or yellow. Two hypotheses can explain this color change: (**a**) In the fall, chlorophyll degenerates, and red or yellow pigments that were earlier masked by chlorophyll become apparent. (**b**) In the fall, red or yellow pigments are synthesized, and they mask the color of chlorophyll. How could you test these two hypotheses?

## bioethical issue

### The World's Food Supply

The Food and Agriculture Organization of the United Nations warns that the world's food supply is dwindling rapidly and food prices are soaring to historic levels. Their records show that the reserves of cereals are severely depleted, and presently only 12 weeks of the world's total consumption is stored, which is much less than the average of 18 weeks' consumption in storage during the years 2000–2005. Only 8 weeks of corn are in storage compared to 11 weeks during this same time period.

Various reasons are offered for a possible calamitous shortfall in the world's grain supplies in the near future. Possible causes are an ever larger world population, water shortages, climate change, and the growing costs of fertilizer. Also of concern is the converting of corn into ethanol because of possible huge profits.

There are apparently no quick fixes to boost supplies. In years past, newly-developed hybrid crops led to enormous increases in yield per acre, but they also caused pollution problems that degrade the environment. Even if the promised biotech advances in drought-, cold-, and disease-resistant crops are made, they will not immediately boost food supplies. Possible solutions have been offered. Rather than exporting food to needy countries, it may be better to improve their ability to grow their own food, especially when you consider that transportation costs are soaring. Also, it would be beneficial to achieve zero population growth as quickly as possible and use renewable energy supplies other than converting corn to ethanol. The use of ethanol only contributes to global warming, which is expected to be a contributing factor to producing less grain. What do you think should be done to solve the expected shortage in the world's food supply, and how should your solution be brought about?

## *Biology* website

The companion website for *Biology* provides a wealth of information organized and integrated by chapter. You will find practice tests, animations, videos, and much more that will complement your learning and understanding of general biology.

**http://www.mhhe.com/maderbiology10**

# 8

# Cellular Respiration

*a bacterium with undulating flagella, an ocelot climbing a tree, a snail moving slowly to hide under a rock, or humans marching past giant cacti—are all making and using ATP—and so are the cacti. ATP is ancient, a molecular fossil, really. Its molecular structure, plus its presence in the first cell or cells that arose on planet Earth, accounts for it being the universal energy currency of cells.*

*ATP is unique among the cell's storehouse of chemicals; amino acids join to make a protein, and nucleotides join to make DNA or RNA, but ATP is singular and works alone. Whether you go skiing, take an aerobics class, or just hang out, ATP molecules provide the energy needed for nerve conduction, muscle contraction, and any other cellular process that requires energy. Cellular respiration, by which cells harvest the energy of organic compounds and convert it to ATP molecules, is the topic of this chapter. It's a process that requires many steps and involves the cytoplasm and the mitochondria. Because mitochondria are involved, they are called the powerhouses of the cell.*

Tourists marching through a prickly pear cactus grove on the Galápagos Islands.

## concepts

## 8.1 Cellular Respiration

**Cellular respiration** is the process by which cells acquire energy by breaking down nutrient molecules produced by photosynthesizers. Its very name implies that cellular respiration requires oxygen ($O_2$) and gives off carbon dioxide ($CO_2$). In fact, it is the reason any animal, such as an ocelot or human, breathes (Fig. 8.1) and why plants also require a supply of oxygen. Most often, cellular respiration involves the complete breakdown of glucose to carbon dioxide and water ($H_2O$):

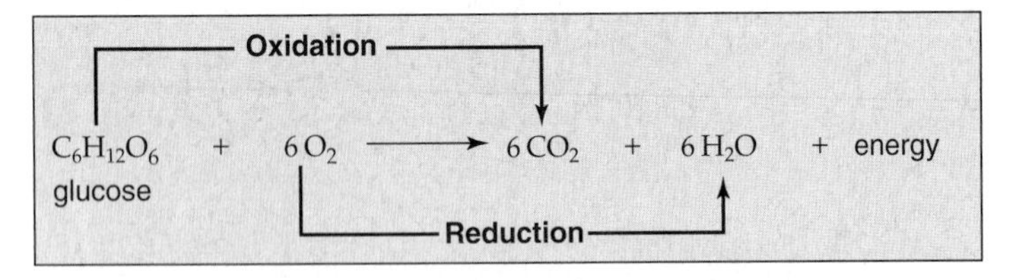


This equation points out that cellular respiration is an oxidation-reduction reaction. Recall that oxidation is the loss of electrons, and reduction is the gain of electrons; therefore, glucose has been oxidized and $O_2$ has been reduced. But, remember that a hydrogen atom consists of a hydrogen ion plus an electron ($H^+ + e^-$). Therefore, when hydrogen atoms are removed from glucose, so are electrons, and similarly, when hydrogen atoms are added to oxygen, so are electrons.

Glucose is a high-energy molecule, and its breakdown products, $CO_2$ and $H_2O$, are low-energy molecules. Therefore, as the equation shows, energy is released. This is the energy that will be used to produce ATP molecules. The cell carries out cellular respiration in order to build up ATP molecules!

The pathways of cellular respiration allow the energy within a glucose molecule to be released slowly so that ATP can be produced gradually. Cells would lose a tremendous amount of energy if glucose breakdown occurred all at once—much energy would become nonusable heat. The step-by-step breakdown of glucose to $CO_2$ and $H_2O$ usually realizes a maximum yield of 36 or 38 ATP molecules, dependent on conditions to be discussed later. The energy in these ATP molecules is equivalent to about 39% of the energy that was available in glucose. This conversion is more efficient than many others; for example, only between 20% and 30% of the energy within gasoline is converted to the motion of a car.

### $NAD^+$ and FAD

Cellular respiration involves many individual metabolic reactions, each one catalyzed by its own enzyme. Enzymes of particular significance are those that use **$NAD^+$**, a coenzyme of oxidation-reduction sometimes called a redox coenzyme. When a metabolite is oxidized, $NAD^+$ accepts two electrons plus a hydrogen ion ($H^+$), and NADH results. The electrons received by $NAD^+$ are high-energy electrons that are usually carried to the electron transport chain (see Fig. 6.12):

$$NAD^+ + 2e^- + H^+ \longrightarrow NADH$$

$NAD^+$ can oxidize a metabolite by accepting electrons and can reduce a metabolite by giving up electrons. Only a small amount of $NAD^+$ need be present in a cell, because each $NAD^+$ molecule is used over and over again. **FAD,** another coenzyme of oxidation-reduction, is sometimes used instead of $NAD^+$. FAD accepts two electrons and two hydrogen ions ($H^+$) to become $FADH_2$.

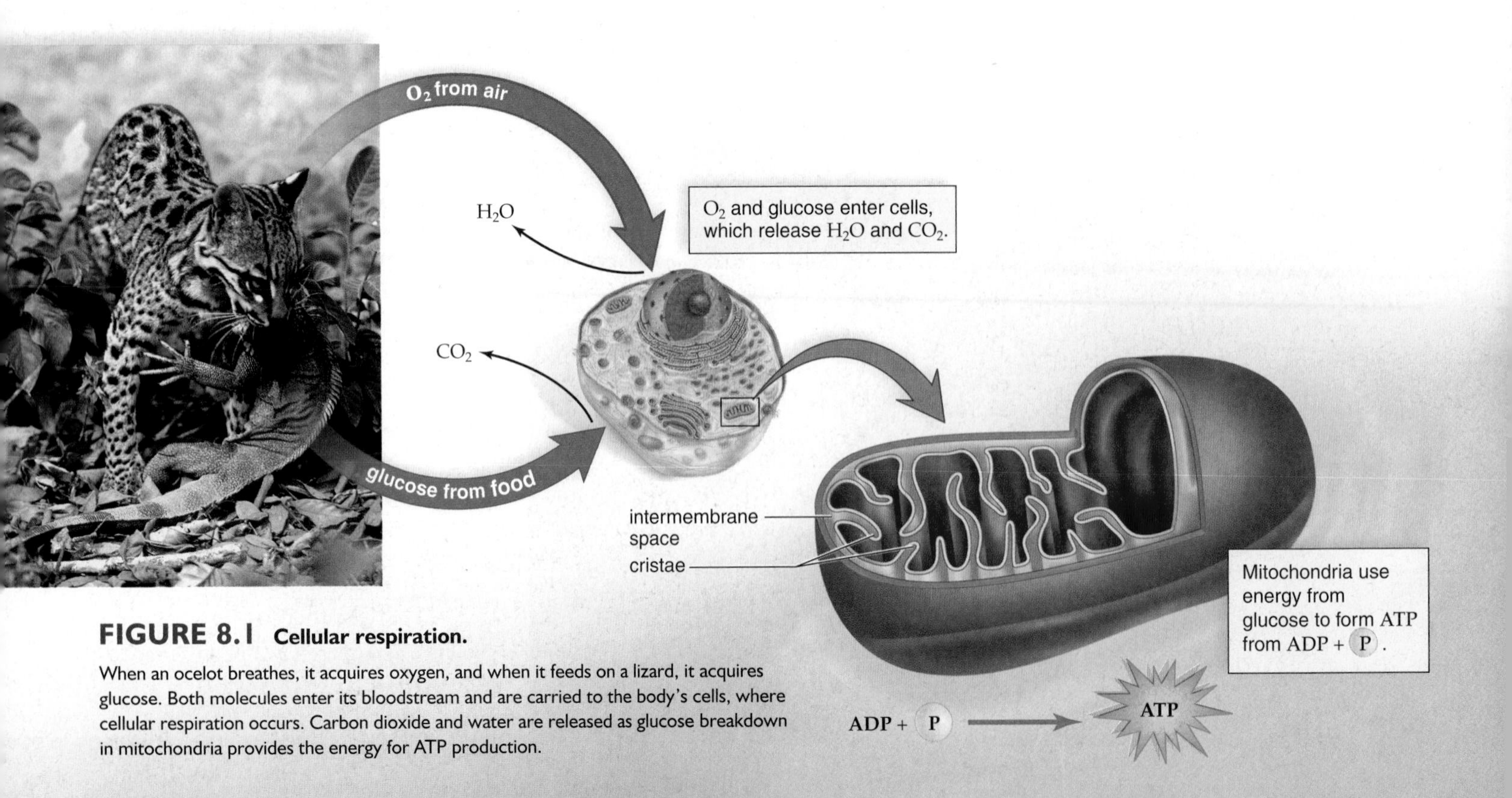


**FIGURE 8.1 Cellular respiration.**

When an ocelot breathes, it acquires oxygen, and when it feeds on a lizard, it acquires glucose. Both molecules enter its bloodstream and are carried to the body's cells, where cellular respiration occurs. Carbon dioxide and water are released as glucose breakdown in mitochondria provides the energy for ATP production.

## Phases of Cellular Respiration

Cellular respiration involves four phases: glycolysis, the preparatory reaction, the citric acid cycle, and the electron transport chain (Fig. 8.2). Glycolysis takes place outside the mitochondria and does not require the presence of oxygen. Therefore, glycolysis is **anaerobic.** The other phases of cellular respiration take place inside the mitochondria, where oxygen is the final acceptor of electrons. Because they require oxygen, these phases are called **aerobic.**

During these phases, notice where $CO_2$ and $H_2O$, the end products of cellular respiration, and ATP are produced.

- **Glycolysis** [Gk. *glycos*, sugar, and *lysis*, splitting] is the breakdown of glucose to two molecules of pyruvate. Oxidation results in NADH and provides enough energy for the net gain of two ATP molecules.
- The **preparatory (prep) reaction** takes place in the matrix of the mitochondrion. Pyruvate is broken down to a 2-carbon ($C_2$) acetyl group, and $CO_2$ is released. Since glycolysis ends with two molecules of pyruvate, the prep reaction occurs twice per glucose molecule.
- The **citric acid cycle** also takes place in the matrix of the mitochondrion. As oxidation occurs, NADH and $FADH_2$ results, and more $CO_2$ is released. The citric acid cycle is able to produce one ATP per turn. Because two acetyl groups enter the cycle per glucose molecule, the cycle turns twice.
- The **electron transport chain (ETC)** is a series of carriers on the cristae of the mitochondria. NADH and $FADH_2$ give up electrons to the chain. Energy is released and captured as the electrons move from a higher-energy to a lower-energy state. Later, this energy will be used for the production of ATP by chemiosmosis. After oxygen receives electrons, it combines with hydrogen ions ($H^+$) and becomes water ($H_2O$).

**Pyruvate,** the end product of glycolysis, is a pivotal metabolite; its further treatment is dependent on whether oxygen is available. If oxygen is available, pyruvate enters a mitochondrion and is broken down completely to $CO_2$ and $H_2O$. If oxygen is not available, pyruvate is further metabolized in the cytoplasm by an anaerobic process called **fermentation.** Fermentation results in a net gain of only two ATP per glucose molecule.

### Check Your Progress 8.1

1. Explain the benefit of slow glucose breakdown rather than rapid breakdown during cellular respiration.
2. List the four phases of complete glucose breakdown. Tell which ones release $CO_2$ and which produces $H_2O$.

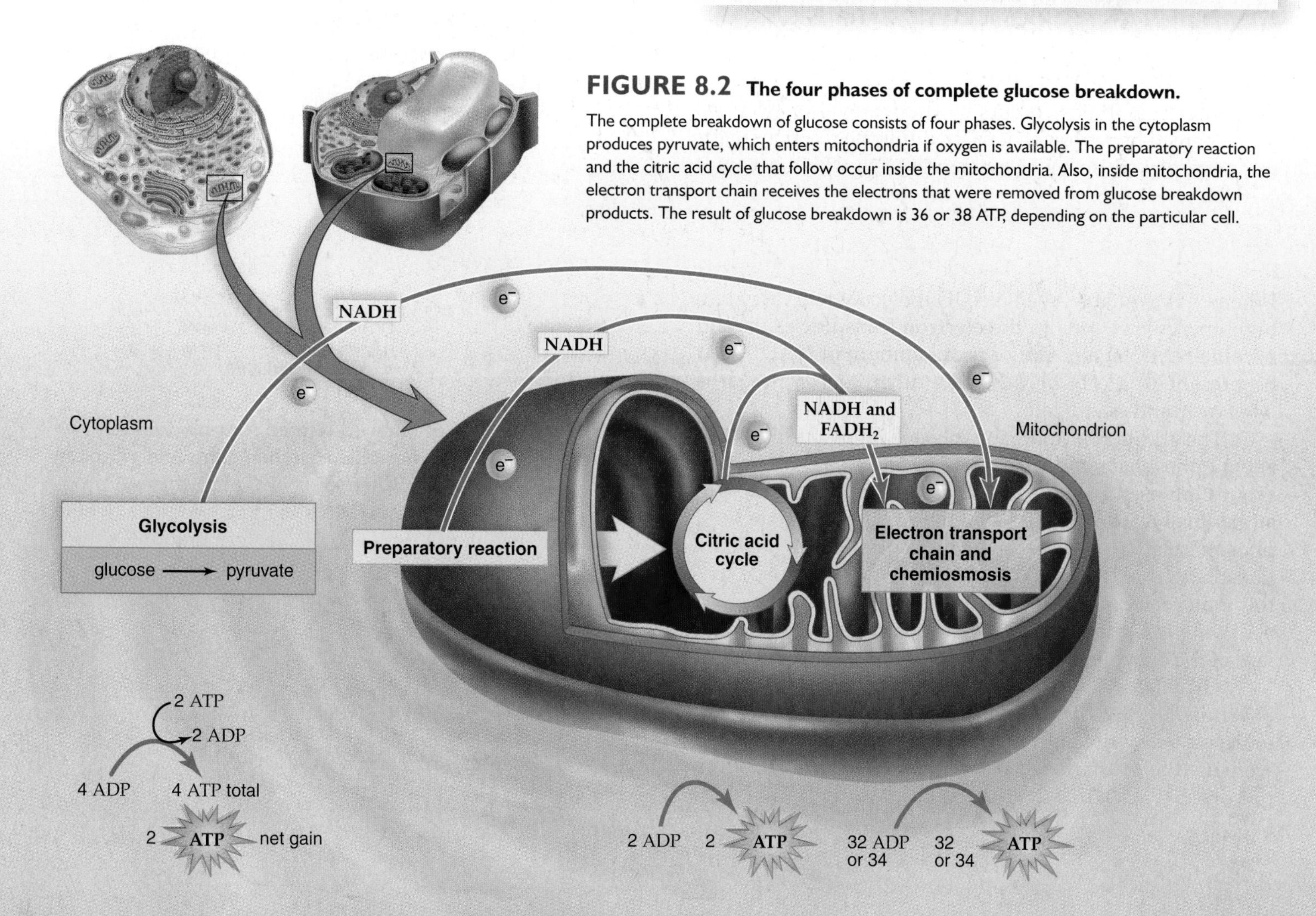


**FIGURE 8.2 The four phases of complete glucose breakdown.**
The complete breakdown of glucose consists of four phases. Glycolysis in the cytoplasm produces pyruvate, which enters mitochondria if oxygen is available. The preparatory reaction and the citric acid cycle that follow occur inside the mitochondria. Also, inside mitochondria, the electron transport chain receives the electrons that were removed from glucose breakdown products. The result of glucose breakdown is 36 or 38 ATP, depending on the particular cell.

# 8.2 Outside the Mitochondria: Glycolysis

**Glycolysis,** which takes place within the cytoplasm outside the mitochondria, is the breakdown of glucose to two pyruvate molecules. Since glycolysis occurs universally in organisms, it most likely evolved before the citric acid cycle and the electron transport chain. This may be why glycolysis occurs in the cytoplasm and does not require oxygen. There was no free oxygen in the early atmosphere of the Earth.

Glycolysis is a long series of reactions, and just as you would expect for a metabolic pathway, each step has its own enzyme. The pathway can be conveniently divided into the energy-investment step and the energy-harvesting steps. During the energy-investment step, ATP is used to "jump-start" glycolysis. During the energy-harvesting steps, more ATP is made than was used to get started.

## Energy-Investment Step

As glycolysis begins, two ATP are used to activate glucose, a $C_6$ (6-carbon) molecule that splits into two $C_3$ molecules known as G3P. Each G3P has a phosphate group. From this point on, each $C_3$ molecule undergoes the same series of reactions.

## Energy-Harvesting Step

Oxidation of G3P now occurs by the removal of electrons accompanied by hydrogen ions. In duplicate reactions, electrons are picked up by coenzyme $NAD^+$, which becomes NADH:

$$2\,NAD^+ + 4\,e^- + 2\,H^+ \longrightarrow 2\,NADH$$

When $O_2$ is available, each NADH molecule will carry two high-energy electrons to the electron transport chain and become $NAD^+$ again. Only a small amount of $NAD^+$ need be present in a cell, because like other coenzymes, it is used over and over again.

The addition of inorganic phosphate result in a high-energy phosphate group per $C_3$ molecule. These phosphate groups are used to synthesize two ATP. This is called **substrate-level ATP synthesis** (sometimes called substrate-level phosphorylation) because an enzyme passes a high-energy phosphate to ADP, and ATP results (Fig. 8.3). Notice that this is an example of coupling: An energy-releasing reaction is driving forward an energy-requiring reaction on the surface of the enzyme.

Oxidation occurs again but by the removal of $H_2O$. Substrate-level ATP synthesis occurs again per $C_3$, and two molecules of pyruvate result. Subtracting the two ATP that were used to get started, there is a net gain of two ATP from glycolysis (Fig. 8.4).

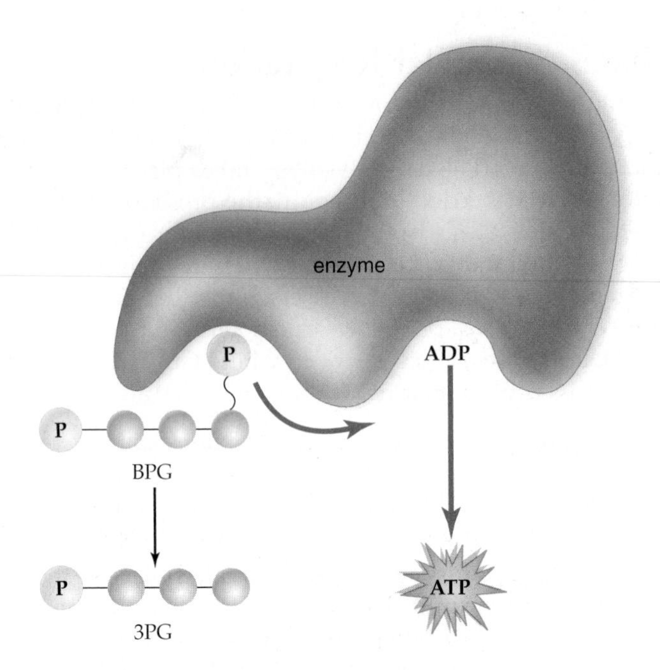


**FIGURE 8.3 Substrate-level ATP synthesis.**

Substrates participating in the reaction are oriented on the enzyme. A phosphate group is transferred to ADP, producing one ATP molecule. During glycolysis (see Fig. 8.4), BPG is a $C_3$ substrate (each gray ball is a carbon atom) that gives up a phosphate group to ADP. This reaction occurs twice per glucose molecule.

### *Inputs and Outputs of Glycolysis*

Altogether, the inputs and outputs of glycolysis are as follows:

**Glycolysis**

| inputs | outputs |
|---|---|
| glucose | 2 pyruvate |
| 2 $NAD^+$ | 2 NADH |
| 2 ATP | 2 ADP |
| 4 ADP + 4 P | 4 ATP total |

2 ATP net gain

Notice that, so far, we have accounted for only two out of a possible 36 or 38 ATP per glucose when completely broken down to $CO_2$ and $H_2O$. When $O_2$ is available, the end product of glycolysis, pyruvate, enters the mitochondria, where it is metabolized. If $O_2$ is not available, fermentation, which is discussed next, will occur.

**Check Your Progress 8.2**

1. Contrast the energy-investment step of glycolysis with the energy-harvesting steps.
2. What happens to pyruvate when oxygen is not available in a cell? When it is available?

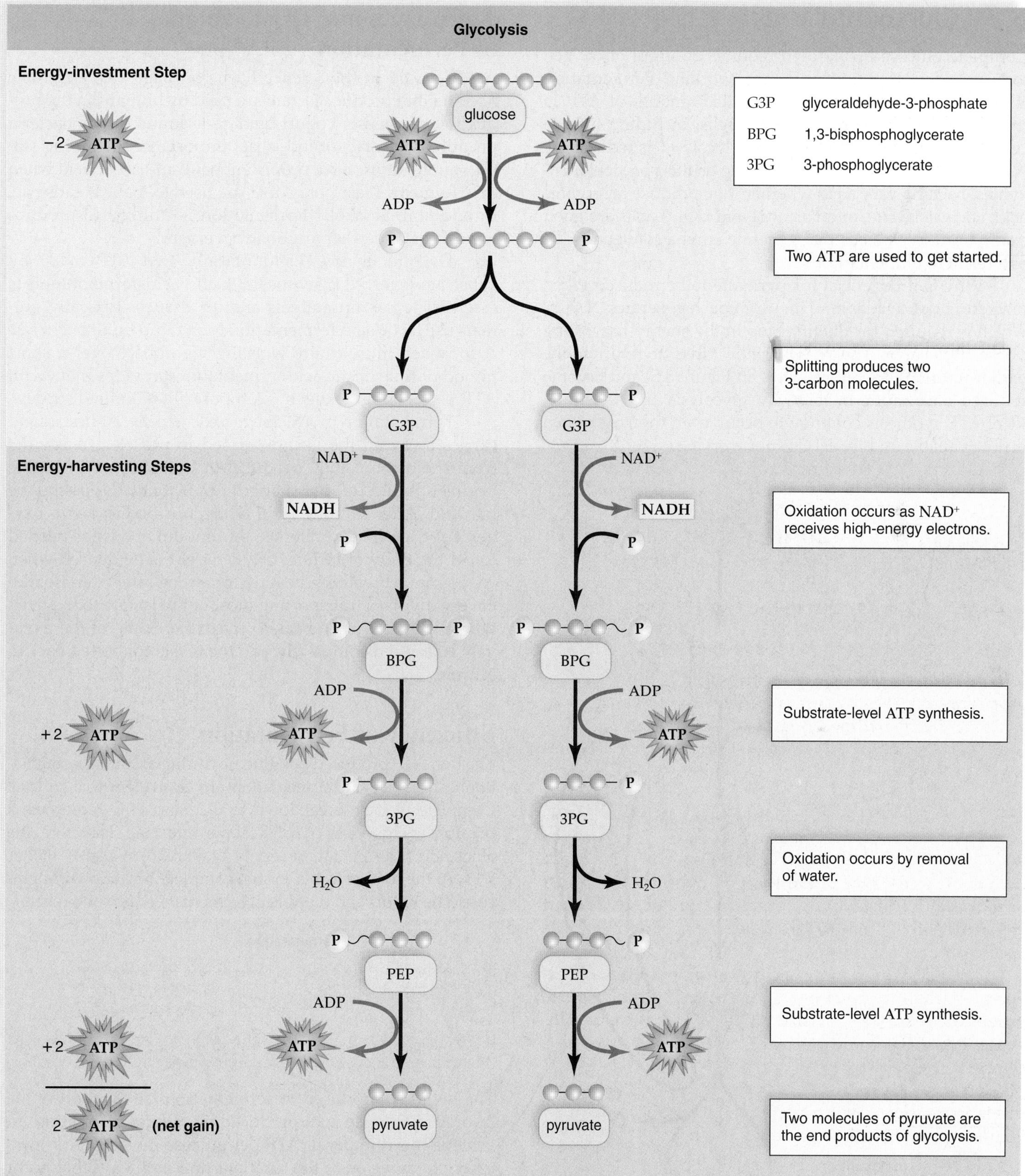


**FIGURE 8.4** **Glycolysis.**

This metabolic pathway begins with $C_6$ glucose (each gray ball is a carbon atom) and ends with two $C_3$ pyruvate molecules. Net gain of two ATP molecules can be calculated by subtracting those expended during the energy-investment step from those produced during the energy-harvesting steps.

# 8.3 Fermentation

Complete glucose breakdown requires an input of oxygen to keep the electron transport chain working. **Fermentation** is anaerobic, and it produces a limited amount of ATP in the absence of oxygen. In animal cells, including human cells, pyruvate, the end product of glycolysis, is reduced by NADH to lactate (Fig. 8.5). Depending on their particular enzymes, bacteria vary as to whether they produce an organic acid, such as lactate, or an alcohol and $CO_2$. Yeasts are good examples of organisms that generate ethyl alcohol and $CO_2$ as a result of fermentation.

Why is it beneficial for pyruvate to be reduced when oxygen is not available? This reaction regenerates $NAD^+$, which is required for the first step in the energy-harvesting phase of glycolyis. This $NAD^+$ is now "free" to return to the earlier reaction (see return arrow in Figure 8.5) and become reduced once more. In this way, glycolysis and substrate-level ATP synthesis continue to occur, even though oxygen is not available and the ETC is not working.

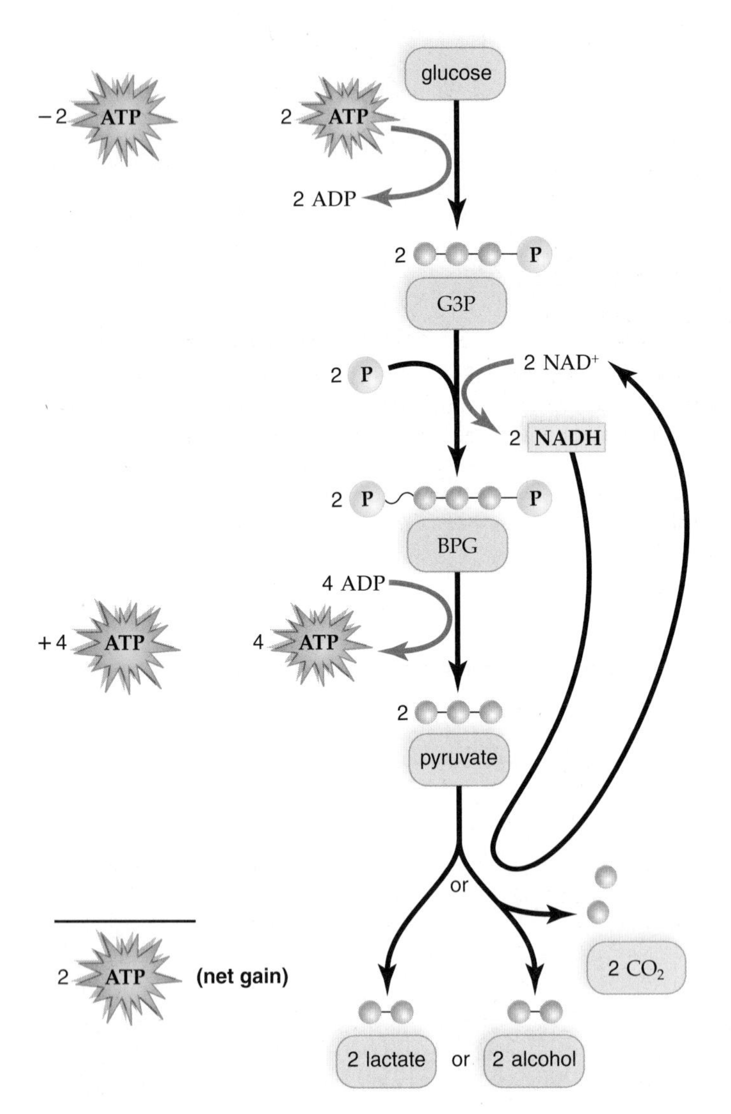


**FIGURE 8.5 Fermentation.**

Fermentation consists of glycolysis followed by a reduction of pyruvate. This "frees" $NAD^+$ and it returns to the glycolytic pathway to pick up more electrons.

## Advantages and Disadvantages of Fermentation

As discussed in the Science Focus on page 139, anaerobic bacteria that produce lactate are used by humans in the production of cheese, yogurt, and sauerkraut. Other bacteria produce chemicals of industrial importance, including isopropanol, butyric acid, proprionic acid, and acetic acid when they ferment. Yeasts, of course, are used to make breads rise. In addition, alcoholic fermentation is utilized to produce wine, beer, and other alcoholic beverages.

Despite its low yield of only two ATP made by substrate-level ATP synthesis, lactic acid fermentation is essential to certain animals and/or tissues. Typically, animals use lactic acid fermentation for a rapid burst of energy. Also, when muscles are working vigorously over a short period of time, lactic acid fermentation provides them with ATP, even though oxygen is temporarily in limited supply.

Fermentation products are toxic to cells. At first, blood carries away all the lactate formed in muscles. Yeasts die from the alcohol they produce. In humans, when lactate begins to build up, pH changes occur that can possibly be harmful. After running for a while, our bodies are in **oxygen debt,** a term that refers to the amount of oxygen needed to rid the body of lactate. Oxygen debt is evidenced when we continue breathing heavily for a time after exercise. Recovery involves transporting most of the lactate to the liver, where it is converted back to pyruvate. Some of the pyruvate is respired completely, and the rest is converted back to glucose.

## Efficiency of Fermentation

The two ATP produced per glucose during alcoholic fermentation and lactic acid fermentation are equivalent to 14.6 kcal. Complete glucose breakdown to $CO_2$ and $H_2O$ represents a possible energy yield of 686 kcal per molecule. Therefore, the efficiency of fermentation is only 14.6 kcal/686 kcal $\times$ 100, or 2.1% of the total possible for the complete breakdown of glucose. The inputs and outputs of fermentation are shown here:

**Fermentation**

| inputs | outputs |
|---|---|
| glucose | 2 lactate or<br>2 alcohol and 2 $CO_2$ |
| 2 ADP + 2 P | 2 ATP net gain |

The two ATP produced by fermentation fall far short of the 36 or 38 ATP molecules produced by cellular respiration. To achieve this number of ATP per glucose molecule, it is necessary to move on to the reactions and pathways that occur in the mitochondria.

### Check Your Progress 8.3

1. What are the drawbacks and benefits of fermentation?

*science focus*

## Fermentation Helps Produce Numerous Food Products

At the grocery store, you will find such items as bread, yogurt, soy sauce, pickles, and maybe even wine (Fig. 8A). These are just a few of the many foods that are produced when microorganisms ferment (break down sugar in the absence of oxygen). Foods produced by fermentation last longer because the fermenting organisms have removed many of the nutrients that would attract other organisms. The products of fermentation can even be dangerous to the very organisms that produced them, as when yeasts are killed by the alcohol they produce.

### Yeast Fermentation

Baker's yeast, *Saccharomyces cerevisiae,* is added to bread for the purpose of leavening—the dough rises when the yeasts give off $CO_2$. The ethyl alcohol produced by the fermenting yeast evaporates during baking. The many different varieties of sourdough breads obtain their leavening from a starter composed of fermenting yeasts along with bacteria from the environment. Depending on the community of microorganisms in the starter, the flavor of the bread may range from sour and tangy, as in San Francisco–style sourdough, to a milder taste, such as that produced by most Amish friendship bread recipes.

Ethyl alcohol is desired when yeasts are used to produce wine and beer. When yeasts ferment the carbohydrates of fruits, the end result is wine. If they ferment grain, beer results. A few specialized varieties of beer, such as traditional wheat beers, have a distinctive sour taste because they are produced with the assistance of lactic acid–producing bacteria, such as those of the genus *Lactobacillus.* Stronger alcoholic drinks (e.g., whiskey and vodka) require distillation to concentrate the alcohol content.

The acetic acid bacteria, including *Acetobacter aceti,* spoil wine. These bacteria convert the alcohol in wine or cider to acetic acid (vinegar). Until the renowned nineteenth-century scientist Louis Pasteur invented the process of pasteurization, acetic acid bacteria commonly caused wine to spoil. Although today we generally associate the process of pasteurization with making milk safe to drink, it was originally developed to reduce bacterial contamination in wine so that limited acetic acid would be produced.

### Bacterial Fermentation

Yogurt, sour cream, and cheese are produced through the action of various lactic acid bacteria that cause milk to sour. Milk contains lactose, which these bacteria use as a substrate for fermentation. Yogurt, for example, is made by adding lactic acid bacteria, such as *Streptococcus thermophilus* and *Lactobacillus bulgaricus,* to milk and then incubating it to encourage the bacteria to act on lactose. During the production of cheese, an enzyme called rennin must also be added to the milk to cause it to coagulate and become solid.

Old-fashioned brine cucumber pickles, sauerkraut, and kimchi are pickled vegetables produced by the action of acid-producing, fermenting bacteria that can survive in high-salt environments. Salt is used to draw liquid out of the vegetables and aid in their preservation. The bacteria need not be added to the vegetables, because they are already present on the surfaces of the plants.

### Soy Sauce Production

Soy sauce is traditionally made by adding a mold, *Aspergillus,* and a combination of yeasts and fermenting bacteria to soybeans and wheat. The mold breaks down starch, supplying the fermenting microorganisms with sugar they can use to produce alcohol and organic acids.

**FIGURE 8A Products from fermentation.**
*Fermentation helps make the products shown on this page.*

# 8.4 Inside the Mitochondria

The preparatory (prep) reaction, the citric acid cycle, and the electron transport chain, which are needed for the complete breakdown of glucose, take place within the mitochondria. A **mitochondrion** has a double membrane with an intermembrane space (between the outer and inner membrane). Cristae are folds of inner membrane that jut out into the matrix, the innermost compartment, which is filled with a gel-like fluid (Fig. 8.6). Just like a chloroplast, a mitochondrion is highly structured, and we would expect reactions to be located in particular parts of this organelle.

The enzymes that speed the prep reaction and the citric acid cycle are arranged in the matrix, and the electron transport chain is located in the cristae in a very organized manner. Most of the ATP from cellular respiration is produced in mitochondria; therefore, mitochondria are often called the powerhouses of the cell.

## The Preparatory Reaction

The **preparatory (prep) reaction** is so called because it occurs before the citric acid cycle. In this reaction, the $C_3$ pyruvate is converted to a $C_2$ acetyl group and $CO_2$ is given off. This is an oxidation reaction in which electrons are removed from pyruvate by $NAD^+$, and NADH is formed. One prep reaction occurs per pyruvate, so altogether, the prep reaction occurs twice per glucose molecule:

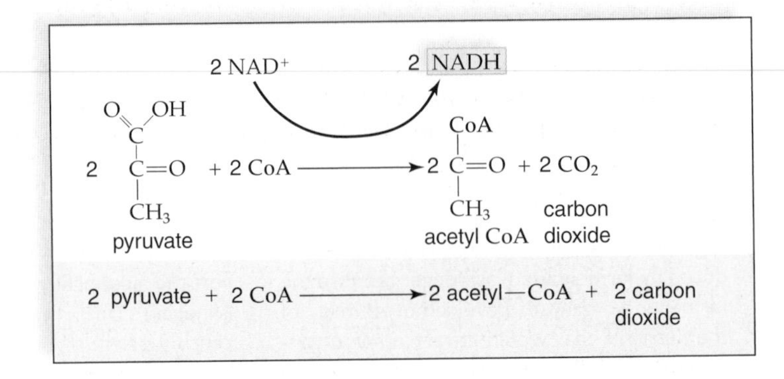


The $C_2$ acetyl group is combined with a molecule known as CoA. CoA will carry the acetyl group to the citric acid cycle. The two NADH carry electrons to the electron transport chain. What about the $CO_2$? In vertebrates, such as ourselves, $CO_2$ freely diffuses out of cells into the blood, which transports it to the lungs where it is exhaled.

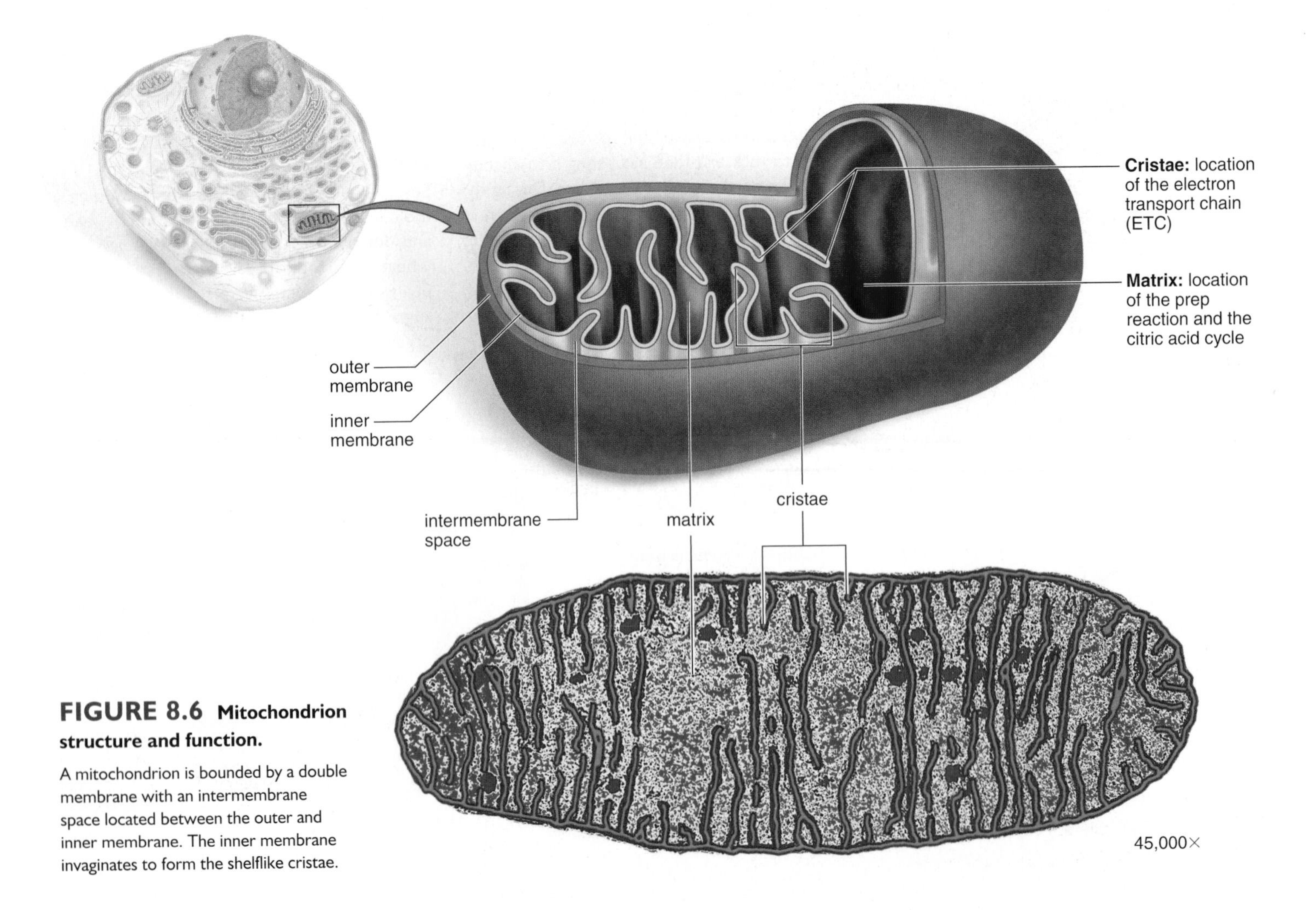


**FIGURE 8.6 Mitochondrion structure and function.**

A mitochondrion is bounded by a double membrane with an intermembrane space located between the outer and inner membrane. The inner membrane invaginates to form the shelflike cristae.

## Citric Acid Cycle

The **citric acid cycle** is a cyclical metabolic pathway located in the matrix of mitochondria (Fig. 8.7). The citric acid cycle is also known as the Krebs cycle, after Hans Krebs, the chemist who worked out the fundamentals of the process in the 1930s.

At the start of the citric acid cycle, the ($C_2$) acetyl group carried by CoA joins with a $C_4$ molecule, and a $C_6$ citrate molecule results. During the cycle, oxidation occurs when electrons are accepted by $NAD^+$ in three instances and by FAD in one instance. Therefore, three NADH and one $FADH_2$ are formed as a result of the citric acid cycle. Also, the acetyl group received from the prep reaction is oxidized to two $CO_2$ molecules. Substrate-level ATP synthesis is also an important event of the citric acid cycle. In substrate-level ATP synthesis, you will recall, an enzyme passes a high-energy phosphate to ADP, and ATP results.

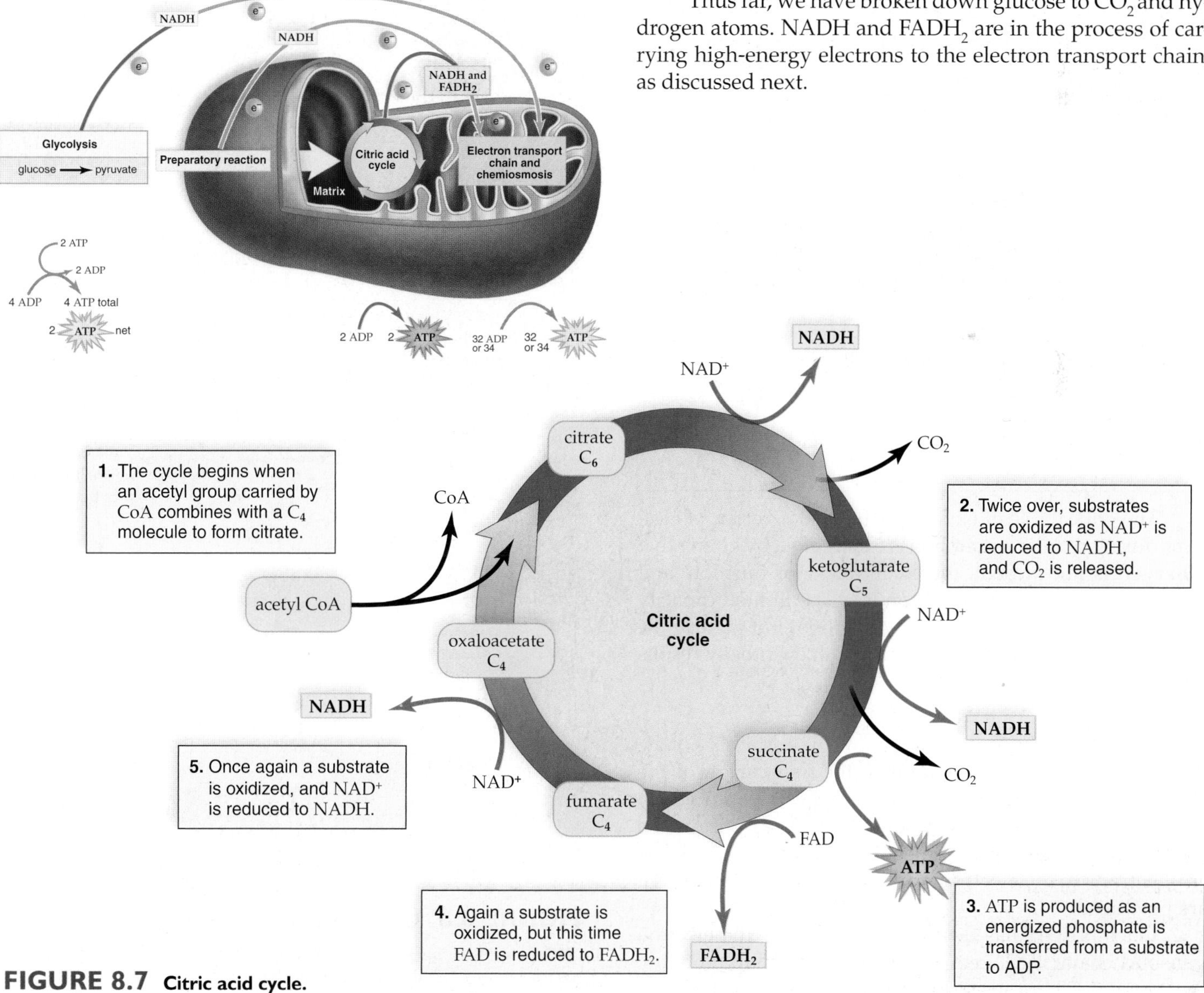


**FIGURE 8.7 Citric acid cycle.**

Citric acid cycle turns twice per glucose molecule.

Because the citric acid cycle turns twice for each original glucose molecule, the inputs and outputs of the citric acid cycle per glucose molecule are as follows:

**Citric acid cycle**

| inputs | outputs |
|---|---|
| 2 acetyl groups | 4 $CO_2$ |
| 6 $NAD^+$ | 6 NADH |
| 2 FAD | 2 $FADH_2$ |
| 2 ADP + 2 P | 2 ATP |

### *Production of $CO_2$*

The six carbon atoms originally located in a glucose molecule have now become $CO_2$. The prep reaction produces two $CO_2$, and the citric acid cycle produces four $CO_2$ per glucose molecule. We have already mentioned that this is the $CO_2$ humans and other animals breathe out.

Thus far, we have broken down glucose to $CO_2$ and hydrogen atoms. NADH and $FADH_2$ are in the process of carrying high-energy electrons to the electron transport chain, as discussed next.

## Electron Transport Chain

The **electron transport chain (ETC)** located in the cristae of the mitochondria and the plasma membrane of aerobic prokaryotes is a series of carriers that pass electrons from one to the other. The electrons that enter the electron transport chain are carried by NADH and $FADH_2$. Figure 8.8 is arranged to show that high-energy electrons enter the chain, and low-energy electrons leave the chain.

### *Members of the Chain*

When NADH gives up its electrons, it becomes $NAD^+$, and when $FADH_2$ gives up its electrons, it becomes FAD. The next carrier gains the electrons and is reduced. This oxidation-reduction reaction starts the process, and each of the carriers, in turn, becomes reduced and then oxidized as the electrons move down the chain.

Many of the carriers are cytochrome molecules. A **cytochrome** is a protein that has a tightly bound heme group with a central atom of iron, the same as hemoglobin does. When the iron accepts electrons, it becomes reduced, and when iron gives them up, it becomes oxidized. A number of poisons, such as cyanide, cause death by binding to and blocking the function of cytochromes. As the pair of electrons is passed from carrier to carrier, energy is captured and eventually used to form ATP molecules.

What is the role of oxygen in cellular respiration and the reason we breathe to take in oxygen? Oxygen is the final acceptor of electrons from the electron transport chain. Oxygen receives the energy-spent electrons from the last of the carriers (i.e., cytochrome oxidase). After receiving electrons, oxygen combines with hydrogen ions, and water forms:

$$\frac{1}{2}O_2 + 2e^- + 2H^+ \longrightarrow H_2O$$

The critical role of oxygen as the final acceptor of electrons during cellular respiration is exemplified by noting that if oxygen is not present, the chain does not function, and no ATP is produced by mitochondria. The limited capacity of the body to form ATP in a way that does not involve the electron transport chain means that death eventually results if oxygen is not available.

### *Cycling of Carriers*

When NADH delivers electrons to the first carrier of the electron transport chain, enough energy is captured by the time the electrons are received by $O_2$ to permit the production of three ATP molecules. When $FADH_2$ delivers electrons to the electron transport chain, only two ATP are produced.

Once NADH has delivered electrons to the electron transport chain, it is "free" to return and pick up more hydrogen atoms. The reuse of coenzymes increases cellular efficiency since it does away with the necessity to synthesize them anew.

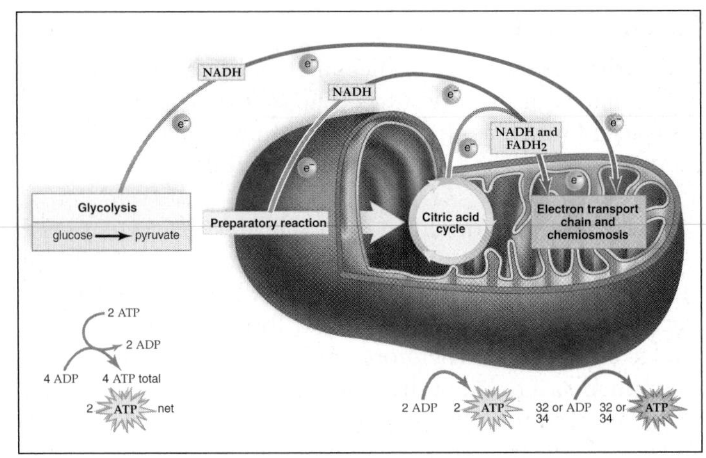


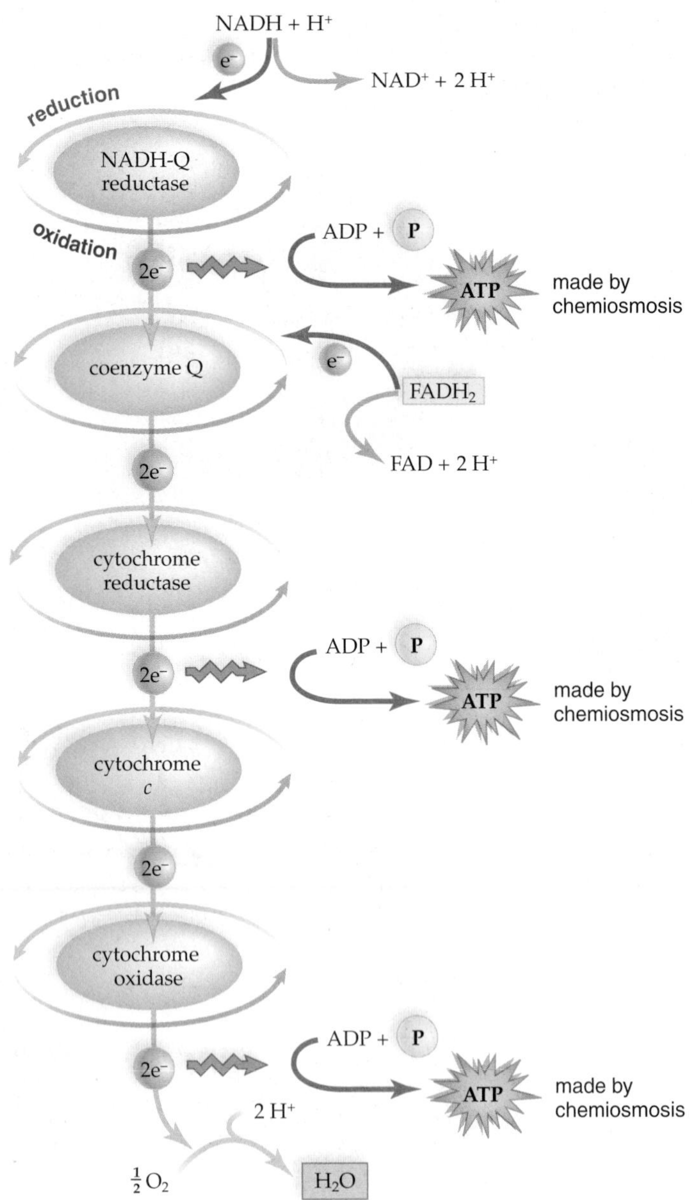


**FIGURE 8.8 The electron transport chain (ETC).**

NADH and $FADH_2$ bring electrons to the electron transport chain. As the electrons move down the chain, energy is captured and used to form ATP. For every pair of electrons that enters by way of NADH, three ATP result. For every pair of electrons that enters by way of $FADH_2$, two ATP result. Oxygen, the final acceptor of the electrons, becomes a part of water.

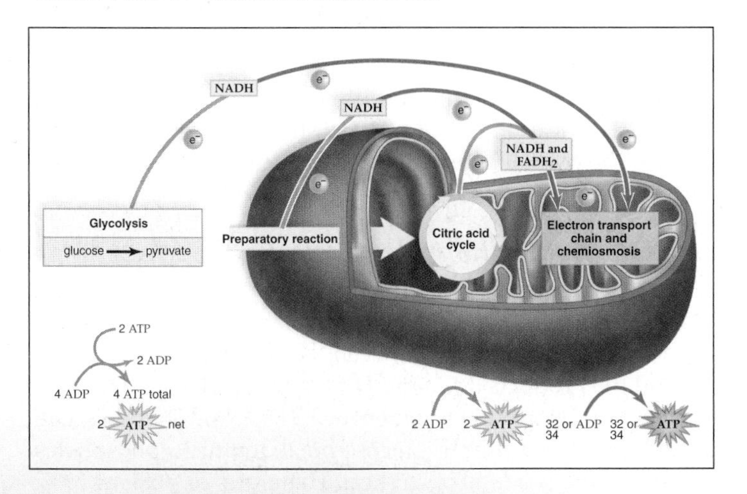


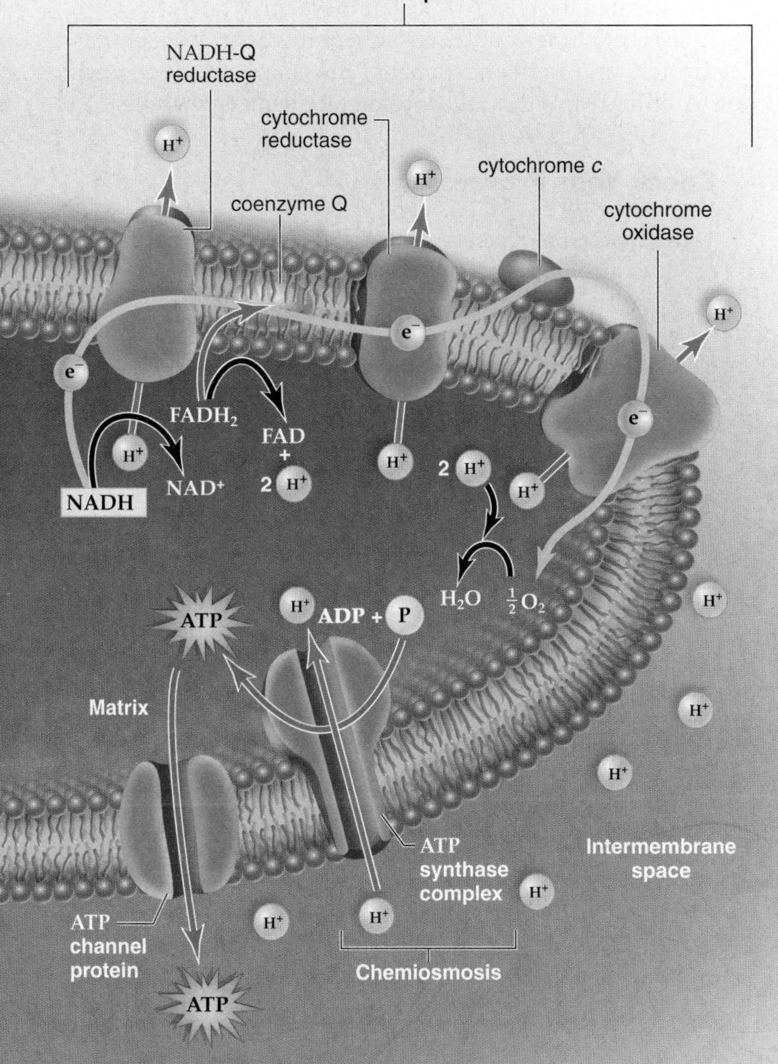


**FIGURE 8.9 Organization and function of cristae.**

The electron transport chain is located in the cristae. As electrons move from one protein complex to the other, hydrogen ions ($H^+$) are pumped from the matrix into the intermembrane space. As hydrogen ions flow down a concentration gradient from the intermembrane space into the mitochondrial matrix, ATP is synthesized by the enzyme ATP synthase. ATP leaves the matrix by way of a channel protein.

## *The Cristae of a Mitochondrion and Chemiosmosis*

The carriers of the electron transport chain and the proteins concerned with ATP synthesis are spatially arranged on the cristae of mitochondria. Their arrangement on the cristae allows the production of ATP to occur.

**The ETC Pumps Hydrogen Ions** Essentially, the electron transport chain consists of three protein complexes and two carriers. The three protein complexes include NADH-Q reductase complex, the cytochrome reductase complex, and cytochrome oxidase complex. The two other carriers that transport electrons between the complexes are coenzyme Q and cytochrome *c* (Fig. 8.9).

We have already seen that the members of the electron transport chain accept electrons, which they pass from one to the other. What happens to the hydrogen ions ($H^+$) carried by NADH and $FADH_2$? The complexes of the electron transport chain use the released energy to pump these hydrogen ions from the matrix into the intermembrane space of a mitochondrion. The vertical arrows in Figure 8.9 show that all the protein complexes of the electron transport chain all pump $H^+$ into the intermembrane space. Just as the walls of a dam hold back water, allowing it to collect, so do cristae hold back hydrogen ions. Eventually, a strong electrochemical gradient develops; there are about ten times as many $H^+$ in the intermembrane space as there are in the matrix.

**The ATP Synthase Complex Produces ATP** The ATP synthase complex can be likened to the gates of a dam. When the gates of a hydroelectric dam are opened, water rushes through, and electricity (energy) is produced. Similarly, when $H^+$ flows down a gradient from the intermembrane space into the matrix, the enzyme ATP synthase synthesizes ATP from ADP + Ⓟ. This process is called **chemiosmosis** because ATP production is tied to the establishment of an $H^+$ gradient.

Once formed, ATP moves out of mitochondria and is used to perform cellular work, during which it breaks down to ADP and Ⓟ. Then these molecules are returned to mitochondria for recycling. At any given time, the amount of ATP in a human would sustain life for only about a minute; therefore, ATP synthase must constantly produce ATP. It is estimated that mitochondria produce our body weight in ATP every day.

**Active Tissues Contain Mitochondria** Active tissues, such as muscles, require greater amounts of ATP and have more mitochondria than less active cells. When a burst of energy is required, however, muscles still ferment.

As an example of the relative amounts of ATP, consider that the dark meat of chickens, the legs, contains more mitochondria than the white meat of the breast. This suggests that chickens mainly walk or run, rather than fly about the barnyard.

## Energy Yield from Glucose Metabolism

Figure 8.10 calculates the ATP yield for the complete breakdown of glucose to $CO_2$ and $H_2O$ during cellular respiration. Notice that the diagram includes the number of ATP produced directly by glycolysis and the citric acid cycle (to the left), as well as the number produced as a result of electrons passing down the electron transport chain (to the right). Thirty-two or 34 ATP molecules are produced by the electron transport chain.

### *Substrate-Level ATP Synthesis*

Per glucose molecule, there is a net gain of two ATP from glycolysis, which takes place in the cytoplasm. The citric acid cycle, which occurs in the matrix of mitochondria, accounts for two ATP per glucose molecule. This means that a total of four ATP are formed by substrate-level ATP synthesis outside the electron transport chain.

### *ETC and Chemiosmosis*

Most ATP is produced by the electron transport chain and chemiosmosis. Per glucose molecule, ten NADH and two $FADH_2$ take electrons to the electron transport chain. For each NADH formed *inside* the mitochondria by the citric acid cycle, three ATP result, but for each $FADH_2$, only two ATP are produced. Figure 8.8 explains the reason for this difference: $FADH_2$ delivers its electrons to the transport chain after NADH, and therefore these electrons cannot account for as much ATP production.

What about the ATP yield per NADH generated *outside* the mitochondria by the glycolytic pathway? In some cells, NADH cannot cross mitochondrial membranes, but a "shuttle" mechanism allows its electrons to be delivered to the electron transport chain inside the mitochondria. The cost to the cell is one ATP for each NADH that is shuttled to the ETC. This reduces the overall count of ATP produced as a result of glycolysis, in some cells, to four instead of six ATP.

### *Efficiency of Cellular Respiration*

It is interesting to calculate how much of the energy in a glucose molecule eventually becomes available to the cell. The difference in energy content between the reactants (glucose and $O_2$) and the products ($CO_2$ and $H_2O$) is 686 kcal. An ATP phosphate bond has an energy content of 7.3 kcal, and 36 of these are usually produced during glucose breakdown; 36 phosphates are equivalent to a total of 263 kcal. Therefore, 263/686, or 39%, of the available energy is usually transferred from glucose to ATP. The rest of the energy is lost in the form of heat.

This concludes our discussion of the phases of cellular respiration, and in the next part of the chapter, we consider how cellular respiration fits into metabolism as a whole.

**Check Your Progress 8.4**

1. A $C_2$ acetyl group enters the citric acid cycle. Where does it come from?
2. What are the products of the citric acid cycle per glucose breakdown?
3. Compare the function of the mitochondrial inner molecule to a hydroelectric dam.

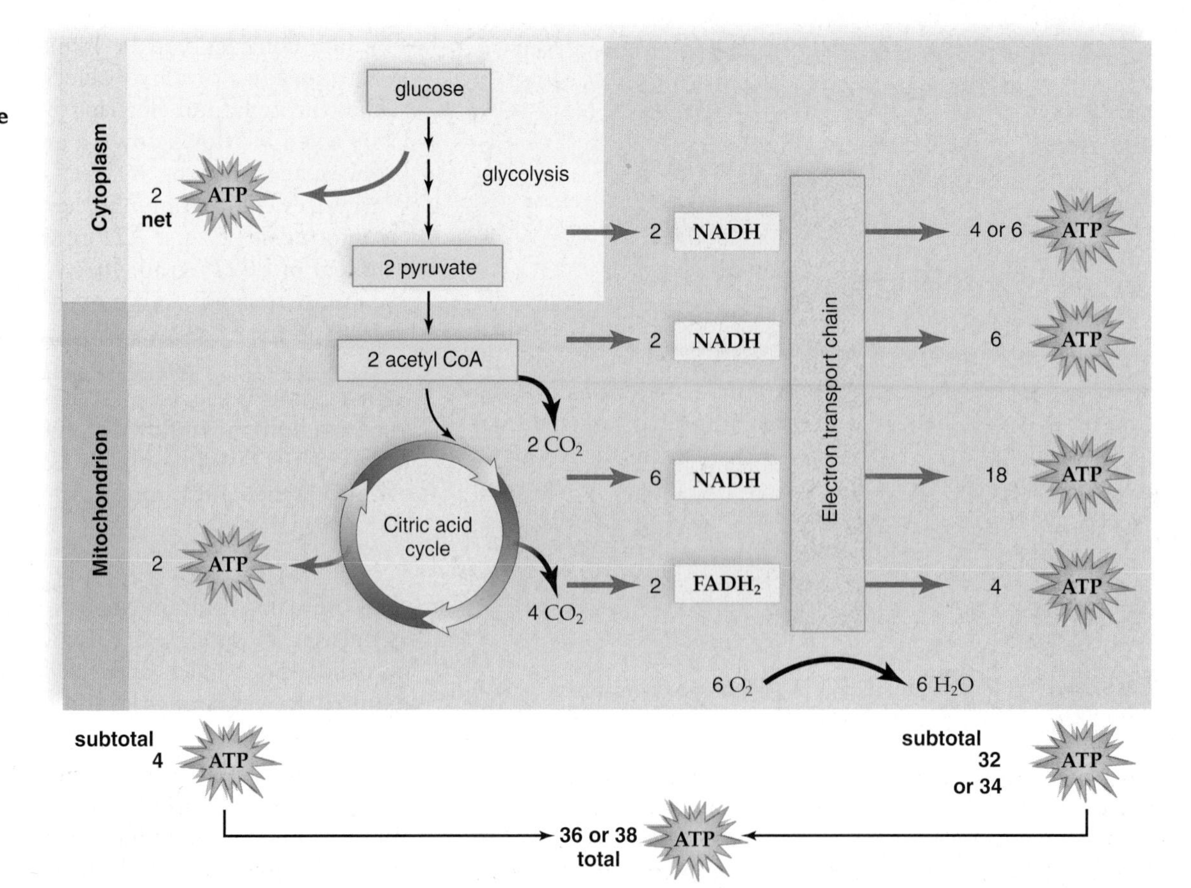


**FIGURE 8.10 Accounting of energy yield per glucose molecule breakdown.**

Substrate-level ATP synthesis during glycolysis and the citric acid cycle accounts for 4 ATP. The electron transport chain accounts for 32 or 34 ATP, and the grand total of ATP is therefore 36 or 38 ATP. Cells differ as to the delivery of the electrons from NADH generated outside the mitochondria. If they are delivered by a shuttle mechanism to the start of the electron transport chain, 6 ATP result; otherwise, 4 ATP result.

# 8.5 Metabolic Pool

Certain substrates recur in various key metabolic pathways, and therefore they form a **metabolic pool.** In the metabolic pool, these substrates serve as entry points for the degradation or synthesis of larger molecules (Fig. 8.11). Degradative reactions break down molecules and collectively participate in **catabolism.** The cellular respiration pathways make a significant contribution to catabolism. Synthetic reactions are to be contrasted with catabolic reactions because they build up molecules and collectively participate in anabolism.

## Catabolism

We already know that glucose is broken down during cellular respiration. However, other molecules can also undergo catabolism. When a fat is used as an energy source, it breaks down to glycerol and three fatty acids. As Figure 8.11 indicates, glycerol can enter glycolysis. The fatty acids are converted to acetyl CoA, and the acetyl group enters the citric acid cycle. An 18-carbon fatty acid results in nine acetyl CoA molecules. Calculation shows that respiration of these can produce a total of 108 ATP molecules. For this reason, fats are an efficient form of stored energy—there are three long fatty acid chains per fat molecule.

The carbon skeleton of amino acids can enter glycolysis, be converted to acetyl group, or enter the citric acid cycle at some other juncture. The carbon skeleton is produced in the liver when an amino acid undergoes **deamination,** or the removal of the amino group. The amino group becomes ammonia ($NH_3$), which enters the urea cycle and becomes part of urea, the primary excretory product of humans. Just where the carbon skeleton begins degradation depends on the length of the *R* group, since this determines the number of carbons left after deamination.

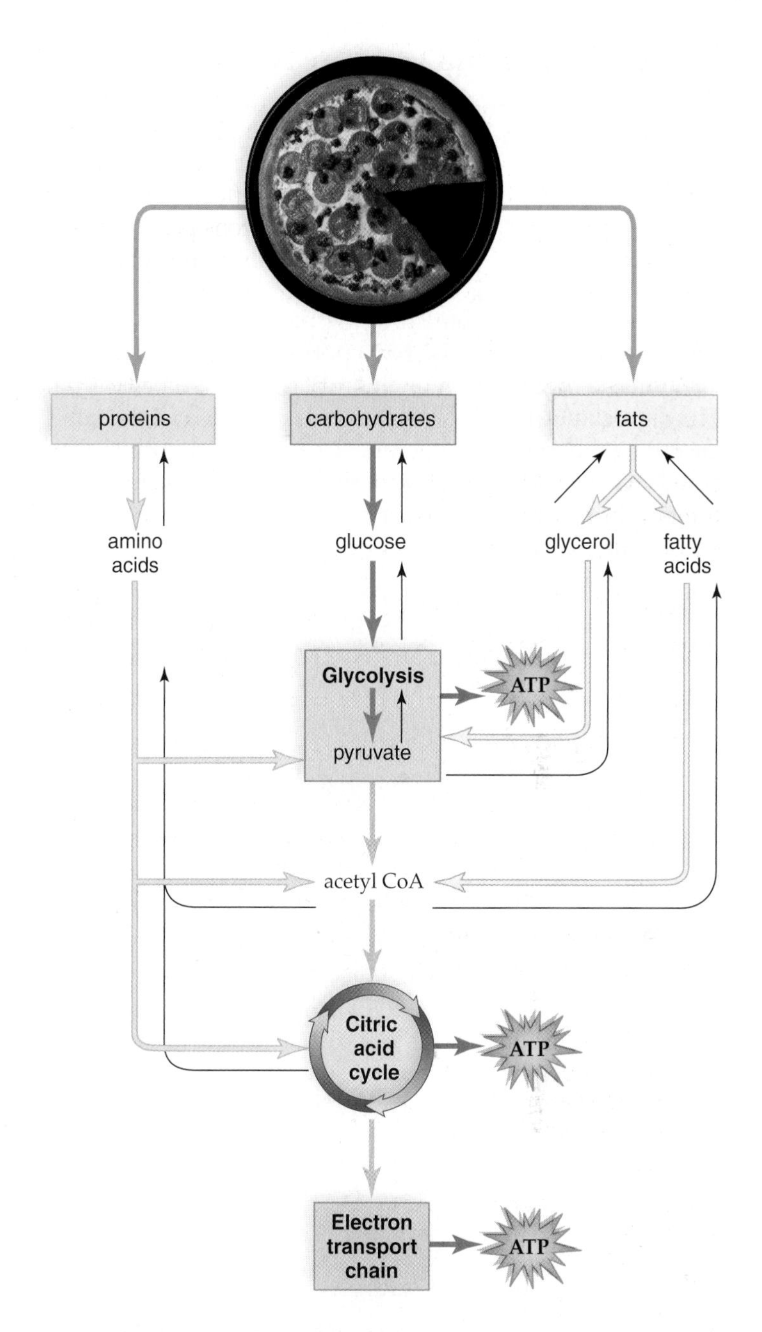


**FIGURE 8.11 The metabolic pool concept.**

Carbohydrates, fats, and proteins can be used as energy sources, and their monomers (carbohydrates and proteins) or subunits (fats) enter degradative pathways at specific points. Catabolism produces molecules that can also be used for anabolism of other compounds.

## Anabolism

We have already mentioned that the ATP produced during catabolism drives anabolism. But catabolism is also related to anabolism in another way. The substrates making up the pathways in Figure 8.11 can be used as starting materials for synthetic reactions. In other words, compounds that enter the pathways are oxidized to substrates that can be used for biosynthesis. This is the cell's metabolic pool, in which one type of molecule can be converted to another. In this way, carbohydrate intake can result in the formation of fat. G3P from glycolysis can be converted to glycerol, and acetyl groups from glycolysis can be joined to form fatty acids. Fat synthesis follows. This explains why you gain weight from eating too much candy, ice cream, or cake.

Some substrates of the citric acid cycle can be converted to amino acids through transamination—the transfer of an amino group to an organic acid, forming a different amino acid. Plants are able to synthesize all of the amino acids they need. Animals, however, lack some of the enzymes necessary for synthesis of all amino acids. Adult humans, for example, can synthesize 11 of the common amino acids, but they cannot synthesize the other 9. The amino acids that cannot be synthesized must be supplied by the diet; they are called the essential amino acids. (The amino acids that can be synthesized are called nonessential.) It is quite possible for animals to suffer from protein deficiency if their diets do not contain adequate quantities of all the essential amino acids.

## The Energy Organelles Revisited

The equation for photosynthesis in a chloroplast is opposite to that of cellular respiration in a mitochondrion (Fig. 8.12):

$$\text{energy} + 6\,CO_2 + 6\,H_2O \underset{\text{cellular respiration}}{\overset{\text{photosynthesis}}{\rightleftharpoons}} C_6H_{12}O_6 + 6\,O_2$$

Even so, while you were studying photosynthesis and cellular respiration, you may have noticed a remarkable similarity in the organization of chloroplasts and mitochondria. Through evolution, all organisms are related, and the similar organization of these organelles suggests that they may be related also. This list summarizes the likeness of the two organelles as they carry out opposite processes:

1. Use of membrane. In a chloroplast, an inner membrane forms the thylakoids of the grana. In a mitochondrion, an inner membrane forms the convoluted cristae.
2. Electron transport chain (ETC). An ETC is located on the thylakoid membrane of chloroplasts and the cristae of mitochondria. In chloroplasts, the electrons passed down the ETC have been energized by the sun; in mitochondria, energized electrons have been removed from glucose and glucose products. In both, the ETC establishes an electrochemical gradient of $H^+$ with subsequence ATP production by chemiosmosis.
3. Enzymes. In a chloroplast, the stroma contains the enzymes of the Calvin cycle and in mitochondria, the matrix contains the enzymes of the citric acid cycle. In the Calvin cycle, NADPH and ATP are used to reduce carbon dioxide to a carbohydrate. In the citric acid cycle, the oxidation of glucose products occurs as NADH and ATP are produced.

### *Flow of Energy*

The ultimate source of energy for producing a carbohydrate in chloroplasts is the sun; the ultimate goal of cellular respiration in a mitochondrion is the conversion of carbohydrate energy into that of ATP molecules. Therefore, there is a flow of energy through chloroplasts to carbohydrates and then through mitochondria to ATP molecules.

This flow of energy maintains biological organization at all levels from molecules, organisms, and the biosphere. In keeping with the energy laws, useful energy is lost with each chemical transformation: as carbohydrate is made and as food is captured and used by all members of food chains. Eventually, the solar energy captured by plants is lost in the form of heat. Therefore, living things are dependent on a continual input of solar energy.

Although energy flows through organisms, chemicals cycle. All living things utilize the carbohydrate and oxygen produced by chloroplasts, but they return to chloroplasts, and the carbon dioxide produced by mitochondria return to chloroplasts. Therefore, chloroplasts and mitochondria are instrumental in not only allowing a flow of energy through living things, they also permit a cycling of chemicals.

#### Check Your Progress 8.5

1. In Chapter 3, you learned the terms dehydration reaction and hydrolytic reaction. **a.** Which type of reaction is catabolic? Anabolic? **b.** Which term could be associated with ATP breakdown?
2. Compare the structure and function of chloroplasts and mitochondria.

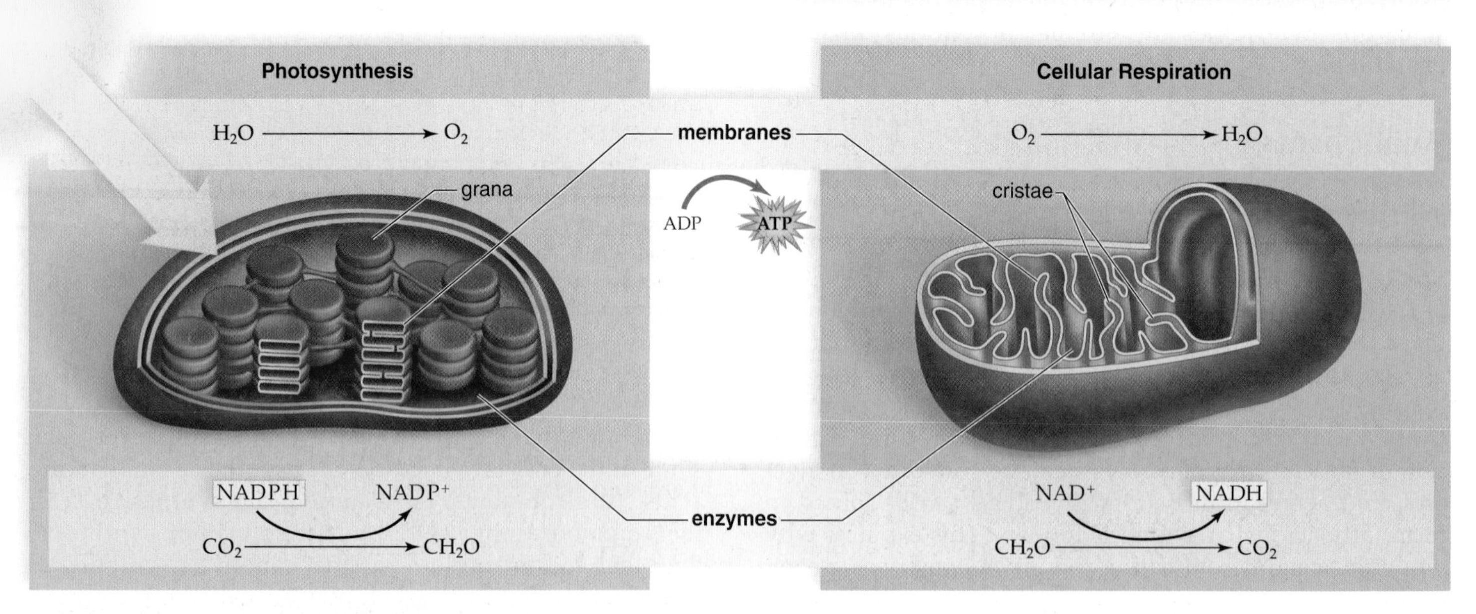


**FIGURE 8.12 Photosynthesis versus cellular respiration.**

*Above:* In photosynthesis, water is oxidized and oxygen is released; in cellular respiration, oxygen is reduced to water. *Middle:* Both processes have an electron transport chain located within membranes (the grana of chloroplasts and the cristae of mitochondria), where ATP is produced by chemiosmosis. *Below:* Both have enzyme-catalyzed reactions within the semifluid interior. In photosynthesis, $CO_2$ is reduced to a carbohydrate; in cellular respiration, a carbohydrate is oxidized to $CO_2$.

# Connecting the Concepts

Chloroplasts and mitochondria play a significant role in metabolism and their enzyme-requiring pathways permit a flow of energy through all living things. The energy transformations that take place in these organelles results in a loss of energy in the form of heat. Therefore, all organisms are in need of a constant supply of energy, which they get from their food. Food is ultimately produced by plants, which have the ability to capture solar energy.

"Structure suits function" is a concept well exemplified by chloroplasts and mitochondria. Their membranous structure is well suited to the isolation of enzymatic reactions in the interior from complexes located on the membrane. As high-energy electrons make energy available, these complexes pump $H^+$ ions into the thylakoid space of chloroplasts and the intermembrane space of mitochondria. When $H^+$ flows down its concentration gradient through ATP synthase complexes, ATP synthesis results. ATP production in mitochondria is traceable back to the ability of chloroplasts to capture solar energy.

In the next part of the text, we depart from metabolism briefly to learn certain laws of genetics. Mendelian genetics will allow you to predict chances an offspring will have particular traits, including genetic disorders. Such disorders stem from a faulty genetic code that results in malformed proteins, including enzymes. The twentieth century was the age of genetics, during which scientists discovered that the genes are on the chromosomes housed in the nucleus. These scientists defined genes as a sequence of nitrogen-containing bases that code for a sequence of amino acids in a protein. When we study the details of protein synthesis, we are once again studying metabolism. Scientists of the twenty-first century are in the process of redefining genes as a result of analyzing the data gathered from the Human Genome Project.

## summary

### 8.1 Cellular Respiration

Cellular respiration, during which glucose is completely broken down to $CO_2$ and $H_2O$, consists of four phases: glycolysis, the prep reaction, the citric acid cycle, and the passage of electrons along the electron transport chain. Oxidation of substrates involves the removal of hydrogen atoms ($H^+ + e^-$), usually by redox coenzymes. $NAD^+$ becomes NADH, and FAD becomes $FADH_2$.

### 8.2 Outside the Mitochondria: Glycolysis

Glycolysis, the breakdown of glucose to two molecules of pyruvate, is a series of enzymatic reactions that occur in the cytoplasm and is anaerobic. Breakdown releases enough energy to immediately give a net gain of two ATP by substrate-level ATP synthesis and the production of 2 NADH.

### 8.3 Fermentation

Fermentation involves glycolysis followed by the reduction of pyruvate by NADH either to lactate (animals) or to alcohol and carbon dioxide ($CO_2$) (yeast). The reduction process "frees" $NAD^+$ so that it can accept more hydrogen atoms from glycolysis.

Although fermentation results in only two ATP molecules, it still serves a purpose. Many of the products of fermentation are used in the baking and brewing industries. In vertebrates, it provides a quick burst of ATP energy for short-term, strenuous muscular activity. The accumulation of lactate puts the individual in oxygen debt because oxygen is needed when lactate is completely metabolized to $CO_2$ and $H_2O$.

### 8.4 Inside the Mitochondria

When oxygen is available, pyruvate from glycolysis enters the mitochondrion, where the prep reaction takes place. During this reaction, oxidation occurs as $CO_2$ is removed from pyruvate. $NAD^+$ is reduced, and CoA receives the $C_2$ acetyl group that remains. Since the reaction must take place twice per glucose molecule, two NADH result.

The acetyl group enters the citric acid cycle, a cyclical series of reactions located in the mitochondrial matrix. Complete oxidation follows, as two $CO_2$ molecules, three NADH molecules, and one $FADH_2$ molecule are formed. The cycle also produces one ATP molecule. The entire cycle must turn twice per glucose molecule.

The final stage of glucose breakdown involves the electron transport chain located in the cristae of the mitochondria. The electrons received from NADH and $FADH_2$ are passed down a chain of carriers until they are finally received by oxygen, which combines with $H^+$ to produce water. As the electrons pass down the chain, energy is captured and stored for ATP production.

The cristae of mitochondria contain complexes of the electron transport chain that not only pass electrons from one to the other but also pump $H^+$ into the intermembrane space, setting up an electrochemical gradient. When $H^+$ flows down this gradient through an ATP synthase complex, energy is captured and used to form ATP molecules from ADP and Ⓟ. This is ATP synthesis by chemiosmosis.

Of the 36 or 38 ATP formed by complete glucose breakdown, four are the result of substrate-level ATP synthesis and the rest are produced as a result of the electron transport chain. For most NADH molecules that donate electrons to the electron transport chain, three ATP molecules are produced. However, in some cells, each NADH formed in the cytoplasm results in only two ATP molecules because a shuttle, rather than NADH, takes electrons through the mitochondrial membrane. $FADH_2$ results in the formation of only two ATP because its electrons enter the electron transport chain at a lower energy level.

### 8.5 Metabolic Pool

Carbohydrate, protein, and fat can be metabolized by entering the degradative pathways at different locations. These pathways also provide metabolites needed for the anabolism of various important substances. Therefore, catabolism and anabolism both use the same pools of metabolites.

Similar to the metabolic pool concept, photosynthesis and cellular respiration can be compared. For example, both utilize an ETC and chemiosmosis. As a result of the ETC in chloroplasts, water is split, while in mitochondria, water is formed. The enzymatic reactions in chloroplasts reduce $CO_2$ to a carbohydrate, while the enzymatic reactions in mitochondria oxidize carbohydrate with the release of $CO_2$.

## understanding the terms



Match the terms to these definitions:

a. _______________ A metabolic pathway that begins with glucose and ends with two molecules of pyruvate.
b. _______________ Occurs due to the accumulation of lactate following vigorous exercise.
c. _______________ Metabolic process that degrades molecules and tends to be exergonic.
d. _______________ Metabolites that are the products of and/or the substrates for key reactions in cells.
e. _______________ Uses a hydrogen ion gradient to drive ATP formation.

## reviewing this chapter

1. What is the overall chemical equation for the complete breakdown of glucose to $CO_2$ and $H_2O$? Explain how this is an oxidation-reduction reaction. 134
2. What are $NAD^+$ and FAD? What are their functions? 134
3. Briefly describe the four phases of cellular respiration. 135
4. What are the main events of glycolysis? How is ATP formed? 136–37
5. What is fermentation, and how does it differ from glycolysis? Mention the benefit of pyruvate reduction during fermentation. What types of organisms carry out lactic acid fermentation, and what types carry out alcoholic fermentation? 138–39
6. Give the substrates and products of the prep reaction. Where does it take place? 140
7. What are the main events of the citric acid cycle? 141
8. What is the electron transport chain, and what are its functions? 142
9. Describe the organization of protein complexes within the cristae. Explain how the complexes are involved in ATP production. 143
10. Calculate the energy yield of glycolysis and complete glucose breakdown. Compare the yields from substrate-level ATP synthesis and from the electron transport chain. 144
11. Give examples to support the concept of the metabolic pool. 145
12. Compare the structure and function of chloroplasts and mitochondria. Explain the flow of energy concept. 146

## testing yourself

Choose the best answer for each question.
For questions 1–8, identify the pathway involved by matching each description to the terms in the key.

**KEY:**

a. glycolysis
b. citric acid cycle
c. electron transport chain

1. carbon dioxide ($CO_2$) given off
2. water ($H_2O$) formed
3. G3P
4. NADH becomes $NAD^+$
5. pump $H^+$
6. cytochrome carriers
7. pyruvate
8. FAD becomes $FADH_2$
9. The prep reaction
   a. connects glycolysis to the citric acid cycle.
   b. gives off $CO_2$.
   c. uses $NAD^+$.
   d. results in an acetyl group.
   e. All of these are correct.
10. The greatest contributor of electrons to the electron transport chain is
    a. oxygen.
    b. glycolysis.
    c. the citric acid cycle.
    d. the prep reaction.
    e. fermentation.
11. Substrate-level ATP synthesis takes place in
    a. glycolysis and the citric acid cycle.
    b. the electron transport chain and the prep reaction.
    c. glycolysis and the electron transport chain.
    d. the citric acid cycle and the prep reaction.
    e. Both b and d are correct.
12. Which of these is not true of fermentation?
    a. net gain of only two ATP
    b. occurs in cytoplasm
    c. NADH donates electrons to electron transport chain
    d. begins with glucose
    e. carried on by yeast
13. Fatty acids are broken down to
    a. pyruvate molecules, which take electrons to the electron transport chain.
    b. acetyl groups, which enter the citric acid cycle.
    c. amino acids, which excrete ammonia.
    d. glycerol, which is found in fats.
    e. All of these are correct.
14. How many ATP molecules are usually produced per NADH?
    a. 1
    b. 3
    c. 36
    d. 10
15. How many NADH molecules are produced during the complete breakdown of one molecule of glucose?
    a. 5
    b. 30
    c. 10
    d. 6

16. What is the name of the process that adds the third phosphate to an ADP molecule using the flow of hydrogen ions?
    a. substrate-level ATP synthesis
    b. fermentation
    c. reduction
    d. chemiosmosis
17. Which are possible products of fermentation?
    a. lactic acid
    b. alcohol
    c. $CO_2$
    d. All of these are possible.
18. The metabolic process that produces the most ATP molecules is
    a. glycolysis.
    b. citric acid cycle.
    c. electron transport chain.
    d. fermentation.
19. Which of these is not true of the citric acid cycle? The citric acid cycle
    a. includes the prep reaction.
    b. produces ATP by substrate-level ATP synthesis.
    c. occurs in the mitochondria.
    d. is a metabolic pathway, as is glycolysis.
20. Which of these is not true of the electron transport chain? The electron transport chain
    a. is located on the cristae.
    b. produces more NADH than any metabolic pathway.
    c. contains cytochrome molecules.
    d. ends when oxygen accepts electrons.
21. Which of these is not true of the prep reaction? The prep reaction
    a. begins with pyruvate and ends with acetyl CoA.
    b. produces more NADH than does glycolysis.
    c. occurs in the mitochondria.
    d. occurs after glycolysis and before the citric acid cycle.
22. The oxygen required by cellular respiration is reduced and becomes part of which molecule?
    a. ATP
    b. $H_2O$
    c. pyruvate
    d. $CO_2$

For questions 23–26, match each pathway to metabolite in the key. Choose more than one if correct.

**KEY:**
   a. pyruvate
   b. acetyl CoA
   c. G3P
   d. NADH
   e. None of these are correct.

23. electron transport chain
24. glycolysis
25. citric acid cycle
26. prep reaction
27. Which of these is not true of glycolysis? Glycolysis
    a. is anaerobic.
    b. occurs in the cytoplasm.
    c. is a part of fermentation.
    d. evolved after the citric acid cycle.
28. Label this diagram of a mitochondrion, and state a function for each portion indicated.

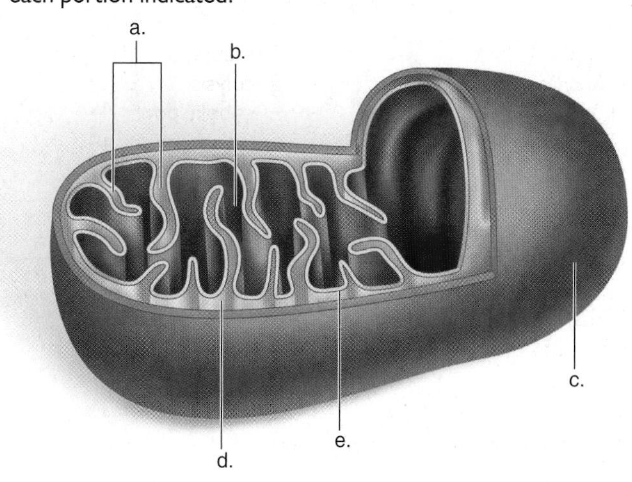


## thinking scientifically

1. You are able to extract mitochondria from the cell and remove the outer membrane. You want to show that the mitochondria can still produce ATP if placed in the right solution. The solution should be isotonic, but at what pH? Why?
2. You are working with acetyl CoA molecules that contain only radioactive carbon. They are incubated with all the components of the citric acid cycle long enough for one turn of the cycle. Examine Figure 8.7 and explain why the carbon dioxide given off is radioactive.

## bioethical issue

### Alternative Medicine

Feeling tired and run-down? Want to jump-start your mitochondria? If you seem to have no specific ailment, you might be tempted to turn to what is now called alternative medicine. Alternative medicine includes such nonconventional therapies as herbal supplements, acupuncture, chiropractic therapy, homeopathy, osteopathy, and therapeutic touch (e.g., laying on of hands).

Advocates of alternative medicine have made some headway in having alternative medicine practices accepted by almost anyone. For example, Congress has established the National Center for Complementary and Alternative Medicine. It has also passed the Dietary Supplement Health and Education Act, which allows vitamins, minerals, and herbs to be marketed without first being approved by the Food and Drug Administration (FDA).

But is this a mistake? Many physicians believe control studies are needed to test the efficacy of alternative medications and practices. Do you agree, or is word of mouth good enough?

## *Biology* website

The companion website for *Biology* provides a wealth of information organized and integrated by chapter. You will find practice tests, animations, videos, and much more that will complement your learning and understanding of general biology.

**http://www.mhhe.com/maderbiology10**

part II

# Genetic Basis of Life

*This part gives you a wonderful opportunity to become acquainted with the basics of Mendelian and molecular genetics. Mendelian genetics can help you predict your chances of having a child with a genetic disorder, and DNA technology can suggest possible procedures to prevent such an occurrence. The potential to cure genetic diseases has expanded greatly now that the human genome has been sequenced, and we know the order of the base pairs in our DNA. Yet, genetic advances are fraught with ethical decisions, such as whether the cloning of humans should be permissible or how far to go in shaping the traits of our children.*

*The field of genetics is making progress in other areas too. We are beginning to understand how cell division is regulated by various genes. Improper regulation of cell division leads to cancer, and therefore we need to know as much as possible about proper regulation if cancer is to be prevented. At every turn, it is clear that you can't be a part of the happenings of the twenty-first century without a knowledge of genetics, and this is your chance to become a part of the action.*

# 9

# The Cell Cycle and Cellular Reproduction

*Consider the development of a human being. A new life begins as one cell—an egg fertilized by a sperm. Yet in nine short months, a human becomes a complex organism consisting of trillions of cells. How is such a feat possible? Cell division enables a single cell to eventually produce many cells, allowing an organism to grow and develop. Cell division also occurs when repair is needed and worn-out tissues have to be replaced. Adult humans have over 200 different types of specialized cells working together in harmony.*

*Genes code for signaling molecules that turn on and off the process of cell division. During the first part of an organism's life, all cells divide. When adulthood is reached, however, only specific cells—human blood and skin cells, for example—continue to divide daily. Other cells, such as nerve cells, no longer routinely divide and produce new cells. Cancer results when the genes that code for signaling proteins mutate and cell division occurs nonstop. The following chapter describes the process of cell division, how it is regulated, and how cancer may develop when regulatory mechanisms malfunction.*

Cancer cell dividing.

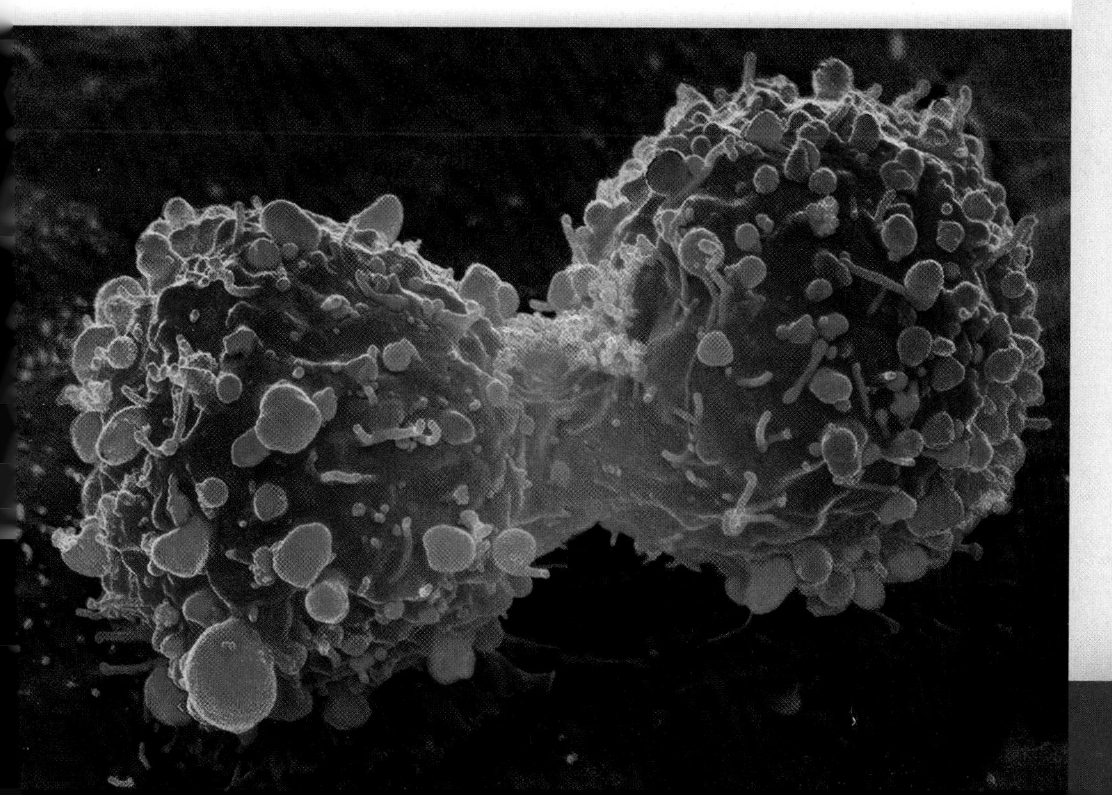

## *concepts*

# 9.1 The Cell Cycle

The **cell cycle** is an orderly set of stages that take place between the time a eukaryotic cell divides and the time the resulting daughter cells also divide. When a cell is going to divide, it grows larger, the number of organelles doubles, and the amount of DNA doubles as DNA replication occurs. The two portions of the cell cycle are interphase, which includes a number of stages, and the mitotic stage when mitosis and cytokinesis occur.

## Interphase

As Figure 9.1 shows, most of the cell cycle is spent in **interphase.** This is the time when a cell performs its usual functions, depending on its location in the body. The amount of time the cell takes for interphase varies widely. Embryonic cells complete the entire cell cycle in just a few hours. For adult mammalian cells, interphase lasts for about 20 hours, which is 90% of the cell cycle. In the past, interphase was known as the resting stage. However, today it is known that interphase is very busy, and that preparations are being made for mitosis. Interphase consists of three stages, referred to as $G_1$, S, and $G_2$.

### $G_1$ Stage

Cell biologists named the stage before DNA replication $G_1$, and they named the stage after DNA replication $G_2$. G stood for "gap," but now that we know how metabolically active the cell is, it is better to think of G as standing for "growth." During $G_1$, the cell recovers from the previous division. Then, the cell increases in size, doubles its organelles (such as mitochondria and ribosomes), and accumulates materials that will be used for DNA synthesis. Otherwise, during $G_1$, cells are constantly performing their normal daily functions, including communicating with other cells, secreting substances, and carrying out cellular respiration.

Some cells, such as nerve and muscle cells, typically do not complete the cell cycle and are permanently arrested. These cells are said to have entered a $G_0$ stage. While the cells continue to perform normal everyday processes, there are no preparations being made for cell division, and cells may not leave $G_0$ stage without proper signals from other cells and other parts of the body. Thus, completion of the cell cycle is very tightly controlled.

### S Stage

Following $G_1$, the cell enters the S stage, when DNA synthesis or replication occurs. At the beginning of the S stage, each chromosome is composed of one DNA double helix. Following DNA replication, each chromosome is composed of two identical DNA double helix molecules. Each double helix is called a **chromatid.** Another way of expressing these events is to say that DNA replication has resulted in duplicated chromosomes, and the two chromatids will remain attached until they are separated during mitosis.

### $G_2$ Stage

Following the S stage, $G_2$ is the stage from the completion of DNA replication to the onset of mitosis. During this stage, the cell synthesizes proteins that will assist cell division. For example, it makes the proteins that form microtubules. Microtubules are used during the mitotic stage to form the mitotic spindle.

## M (Mitotic) Stage

Following interphase, the cell enters the M (for mitotic) stage. This cell division stage includes **mitosis** (nuclear division) and **cytokinesis** (division of the cytoplasm). During mitosis, daughter chromosomes are distributed by the **mitotic spindle** to two daughter nuclei. When division of the cytoplasm is complete, two daughter cells are present.

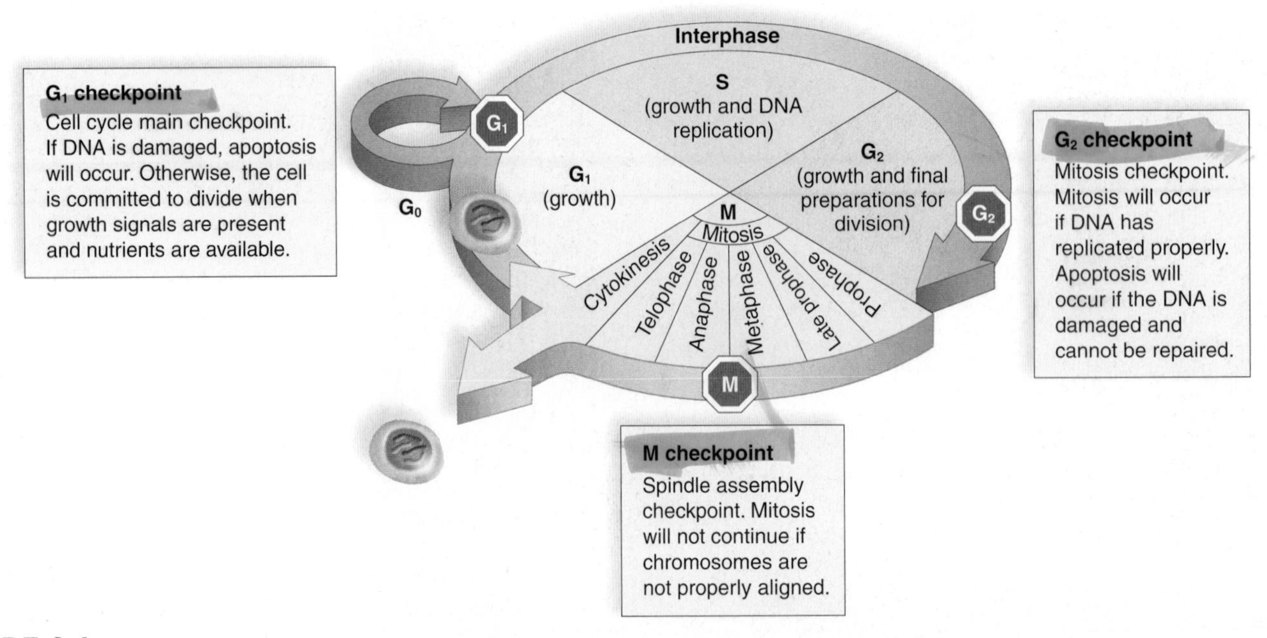


**FIGURE 9.1 The cell cycle.**

Cells go through a cycle that consists of four stages: $G_1$, S, $G_2$, and M. The major activities and checkpoints for each stage are given.

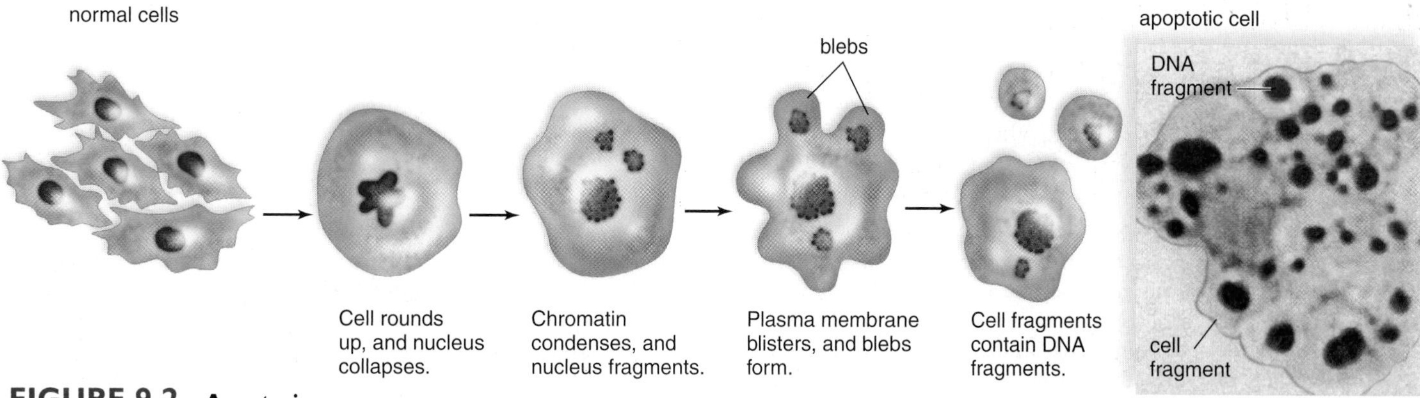


**FIGURE 9.2 Apoptosis.**

Apoptosis is a sequence of events that results in a fragmented cell. The fragments are phagocytized (engulfed) by white blood cells and neighboring tissue cells.

## Control of the Cell Cycle

A **signal** is an agent that influences the activities of a cell. **Growth factors** are signaling proteins received at the plasma membrane. Even cells arrested in $G_0$ will finish the cell cycle if stimulated to do so by growth factors. In general, signals ensure that the cell cycle stages follow one another in the normal sequence.

### *Cell Cycle Checkpoints*

The red barriers in Figure 9.1 represent three checkpoints when the cell cycle either stops or continues on, depending on the internal signal it receives. Researchers have identified a family of internal signaling proteins called **cyclins** that increase and decrease as the cell cycle continues. Specific cyclins must be present for the cell to proceed from the $G_1$ stage to the S stage and for the cell to proceed from the $G_2$ stage to the M stage.

As discussed in the Science Focus on the next page, the primary checkpoint of the cell cycle is the $G_1$ checkpoint. In mammalian cells, the signaling protein **p53** (p stands for *pro*tein and 53 stands for a molecular weight of 53,000 g) stops the cycle at the $G_1$ checkpoint when DNA is damaged. First, p53 attempts to initiate DNA repair, but rising levels bring about **apoptosis,** which is programmed cell death (Fig. 9.2). Another protein, called **RB,** is responsible for interpreting growth signals and also nutrient availability signals. RB stands for *r*etino*b*lastoma, a cancer of the retina that occurs when the *RB* gene undergoes a mutation.

The cell cycle may also stop at the $G_2$ checkpoint if DNA has not finished replicating. This prevents the initiation of the M stage before completion of the S stage. If DNA is physically damaged, such as from exposure to solar radiation or X-rays, the $G_2$ checkpoint also offers the opportunity for DNA to be repaired.

Another cell cycle checkpoint occurs during the mitotic stage. The cycle stops if the chromosomes are not properly attached to the mitotic spindle. Normally, the mitotic spindle ensures that the chromosomes are distributed accurately to the daughter cells.

### *Apoptosis*

Apoptosis is often defined as programmed cell death because the cell progresses through a usual series of events that bring about its destruction (Fig. 9.2). The cell rounds up, causing it to lose contact with its neighbors. The nucleus fragments, and the plasma membrane develops blisters. Finally, the cell fragments are engulfed by white blood cells and/or neighboring cells. A remarkable finding of the past few years is that the enzymes that bring about apoptosis, called **caspases,** are always present in the cell. The enzymes are ordinarily held in check by inhibitors but can be unleashed by either internal or external signals.

**Apoptosis and Cell Division.** In living systems, opposing events keep the body in balance and maintain homeostasis. For now, consider that some carrier proteins transport molecules into the cell, and others transport molecules out of the cell. Some hormones increase the level of blood glucose, and others decrease the level. Similarly, cell division and apoptosis are two opposing processes that keep the number of cells in the body at an appropriate level. **Cell division increases and apoptosis decreases the number of somatic (body) cells.** Both mitosis and apoptosis are normal parts of growth and development. An organism begins as a single cell that repeatedly divides to produce many cells, but eventually some cells must die for the organism to take shape. For example, when a tadpole becomes a frog, the tail disappears as apoptosis occurs. The fingers and toes of a human embryo are at first webbed, but then they are usually freed from one another as a result of apoptosis.

Cell division occurs during your entire life. Even now, your body is producing thousands of new red blood cells, skin cells, and cells that line your respiratory and digestive tracts. Also, if you suffer a cut, cell division repairs the injury. Apoptosis occurs all the time too, particularly if an abnormal cell that could become cancerous appears, or a cell becomes infected with a virus. Death through apoptosis prevents a tumor from developing and helps to limit the spread of viruses.

**Check Your Progress 9.1**

1. What are the four stages of the cell cycle? During which of these stages is the DNA replicated, and when does cell division occur?
2. What conditions might cause a cell to halt the cell cycle?

*science focus*

# The $G_1$ Checkpoint

Cell division is very tightly regulated so that only certain cells in an adult body are actively dividing. After cell division occurs, cells enter the $G_1$ stage. Upon completing $G_1$, they will divide again, but before this happens they have to pass through the $G_1$ checkpoint. The $G_1$ checkpoint ensures that conditions are right for making the commitment to divide by evaluating the meaning of growth signals, determining the availability of nutrients, and assessing the integrity of DNA. Failure to meet any one of these criteria results in a cell halting the cell cycle and entering $G_0$ stage, or undergoing apoptosis if the problems are severe.

## Evaluating Growth Signals

Multicellular organisms tightly control cell division so that it occurs only when needed. Signaling molecules, such as hormones, may be sent from nearby cells or distant tissues to encourage or discourage cells from entering the cell cycle. Such signals may cause a cell to enter a $G_0$ stage, or complete $G_1$ and enter the S stage. Growth signals that promote cell division cause a cyclin-dependent-kinase (CDK) to add a phosphate group to RB, a major regulator of the $G_1$ checkpoint.

Ordinarily, a protein called E2F is bound to RB, but when RB is phosphorylated, its shape changes and it releases E2F. Now, E2F binds to DNA, activating certain genes whose products are needed to complete the cell cycle (Fig. 9A*a*). Likewise, growth signals prompt cells that are in $G_0$ stage to reenter the $G_1$ stage, complete it, and enter the S stage. If growth signals are sufficient, a cell passes through the $G_1$ checkpoint and cell division occurs.

## Determining Nutrient Availability

Much as an experienced hiker would ensure that she has sufficient food for her journey, a cell ensures that nutrient levels to support cell division are adequate before committing to it. For example, scientists know that starving cells in culture enter $G_0$. At that time, phosphate groups are removed from RB (see reverse arrows in Figure 9Aa); RB does not release E2F; and the proteins needed to complete the cell cycle are not produced. When, nutrients *become* available, CDKs bring about the phosphorylation of RB, which then releases E2F (see forward arrows in Figure 9A*a*). After E2F binds to DNA, proteins needed to complete the cell cycle are produced. As mentioned, growth signals prompt a cell that is in the $G_0$ stage to reenter the $G_1$ stage, complete it, and enter the S stage. Again, we can note that cells do not commit to divide until conditions are conducive for them to do so.

## Assessing DNA Integrity

For cell division to occur, DNA must be free of errors and damage. The p53 protein is involved in this quality control function. Ordinarily, p53 is broken down because it has no job to do. In response to DNA damage, CDK phosphorylates p53 (Fig. 9A*b*). Now, the molecule will not be broken down as usual, and instead its level in the nucleus begins to rise. Phosphorylated p53 binds to DNA; certain genes are activated; and DNA repair proteins are produced. If the DNA damage cannot be repaired, p53 levels continue to rise, and apoptosis is triggered. If the damage is successfully repaired, p53 levels fall, and the cell is allowed to complete $G_1$ stage as long as growth signals and nutrients are present, for example. Actually, there are many criteria that must be met in order for a cell to commit to cell division, and the failure to meet any one of them may cause the cell cycle to be halted and/or apoptosis to be initiated. The $G_1$ checkpoint is currently an area of intense research because understanding it holds the key to possibly curing cancer and for unleashing the power of normal, healthy cells to regenerate tissues that could be used to cure many other human conditions.

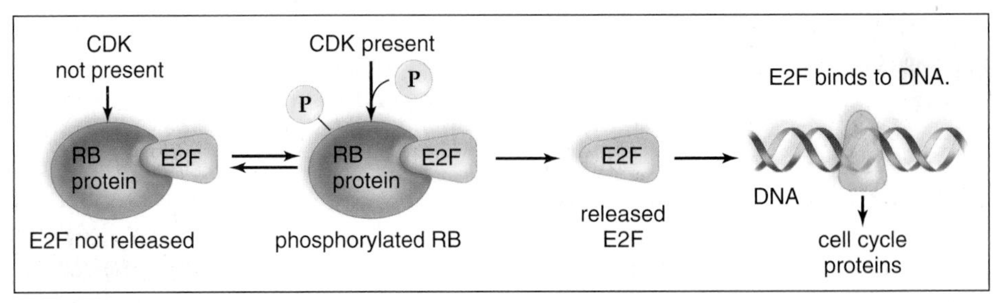


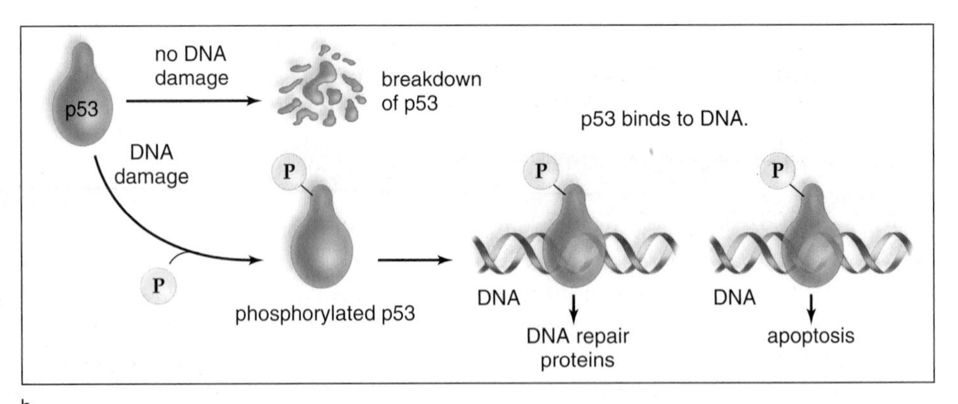


**FIGURE 9A Regulation of the G1 checkpoint.**
***a.*** *When CDK (cyclin-dependent-kinase) is not present, RB retains E2F. When CDK is present, a phosphorylated RB releases E2F, and after it binds to DNA, proteins necessary to completing cell division are produced.* ***b.*** *If DNA is damaged, p53 is not broken down, and instead is involved in the production of DNA repair enzymes and in triggering apoptosis when repair is impossible.*

# 9.2 Mitosis and Cytokinesis

As mentioned, cell division in eukaryotes involves mitosis, which is nuclear division, and cytokinesis, which is division of the cytoplasm. During mitosis, chromosomes are distributed to two daughter cells.

## Eukaryotic Chromosomes

The DNA in the chromosomes of eukaryotes is associated with various proteins, including **histones** that are especially involved in organizing chromosomes. When a eukaryotic cell is not undergoing division, the DNA (and associated proteins) are located within **chromatin** which has the appearance of a tangled mass of thin threads. Before mitosis begins, chromatin becomes highly coiled and condensed, and it is easy to see the individual chromosomes.

When the chromosomes are visible, it is possible to photograph and count them. Each species has a characteristic chromosome number (Table 9.1). This is the full or **diploid (2n) number** [Gk. *diplos,* twofold, and *-eides,* like] of chromosomes that is found in all cells of the individual. The diploid number includes two chromosomes of each kind. Half the diploid number, called the **haploid (n) number** [Gk. *haplos,* single, and *-eides,* like] of chromosomes, contains only one chromosome of each kind. Typically, only sperm and eggs have the haploid number of chromosomes in the life cycle of animals.

## Preparations for Mitosis

During interphase, a cell must make preparations for cell division. These arrangements include replicating the chromosomes and duplicating most cellular organelles, including the centrosome, which will organize the spindle apparatus necessary for movement of chromosomes.

### *Chromosome Duplication*

During mitosis, a 2n nucleus divides to produce daughter nuclei that are also 2n. The dividing cell is called the *parent cell,* and the resulting cells are called the *daughter cells.* Before nuclear division takes place, DNA replicates, duplicating the chromosomes in the parent cell. This occurs during S stage of interphase. Now each chromosome has two identical double helical molecules. Each double helix is a chromatid, and the two identical chromatids are called **sister chromatids** (Fig. 9.3). Sister chromatids are constricted and attached to each other at a region called the **centromere.** Protein complexes called **kinetochores** develop on either side of the centromere during cell division.

**TABLE 9.1**

**Diploid Chromosome Numbers of Some Eukaryotes**

| *Type of Organism* | *Name of Chromosome* | *Chromosome Number* |
|---|---|---|
| Fungi | *Saccharomyces cerevisiae* (yeast) | 32 |
| Plants | *Pisum sativum* (garden pea) | 14 |
| | *Solanum tuberosum* (potato) | 48 |
| | *Ophioglossum vulgatum* (Southern adder's tongue fern) | 1,320 |
| Animals | *Drosophila melanogaster* (fruit fly) | 8 |
| | *Homo sapiens* (human) | 46 |
| | *Carassius auratus* (goldfish) | 94 |

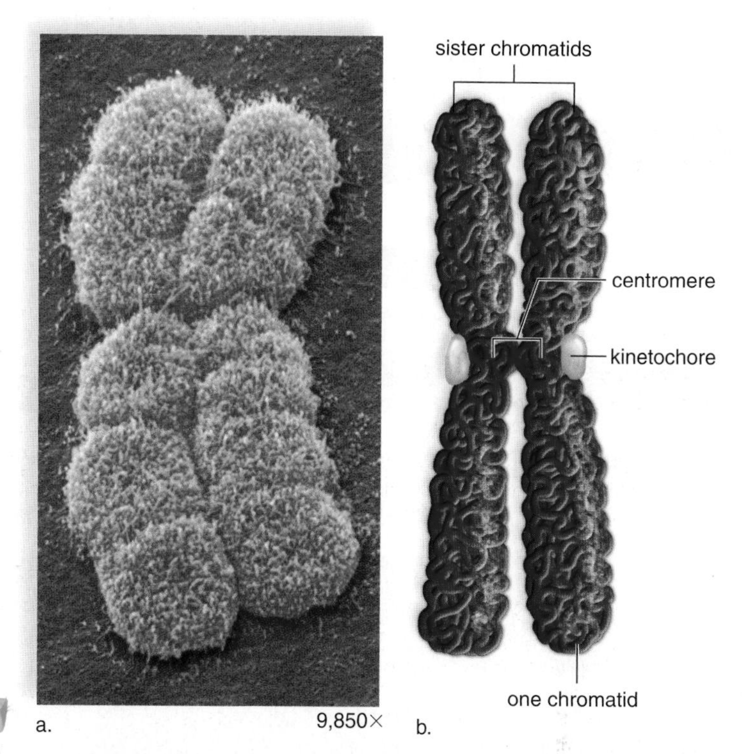


**FIGURE 9.3 Duplicated chromosomes.**

A duplicated chromosome contains two sister chromatids, each with a copy of the same genes. **a.** Electron micrograph of a highly coiled and condensed chromosome, typical of a nucleus about to divide. **b.** Diagrammatic drawing of a condensed chromosome. The chromatids are held together at a region called the centromere.

During nuclear division, the two sister chromatids separate at the centromere, and in this way each duplicated chromosome gives rise to two daughter chromosomes. Each daughter chromosome has only one double helix molecule. The daughter chromosomes are distributed equally to the daughter cells. In this way, each daughter nucleus gets a copy of each chromosome that was in the parent cell.

### *Division of the Centrosome*

The **centrosome** [Gk. *centrum,* center, and *soma,* body], the main microtubule-organizing center of the cell, also divides before mitosis begins. Each centrosome in an animal cell—but not a plant cell—contains a pair of barrel-shaped organelles called **centrioles.**

The centrosomes organize the mitotic spindle, which contains many fibers, each composed of a bundle of microtubules. Microtubules are hollow cylinders made up of the protein tubulin. They assemble when tubulin subunits join, and when they disassemble, tubulin subunits become free once more. The microtubules of the cytoskeleton disassemble when spindle fibers begin forming. Most likely, this provides tubulin for the formation of the spindle fibers, or may allow the cell to change shape as needed for cell division.

## Phases of Mitosis

Mitosis is a continuous process that is arbitrarily divided into five phases for convenience of description: prophase, prometaphase, metaphase, anaphase, and telophase (Fig. 9.4).

### *Prophase*

It is apparent during **prophase** that nuclear division is about to occur because chromatin has condensed and the chromosomes are visible. Recall that DNA replication occurred during interphase, and therefore the *parental chromosomes are already duplicated and composed of two sister chromatids held together at a centromere.* Counting the number of centromeres in diagrammatic drawings gives the number of chromosomes for the cell depicted.

During prophase, the nucleolus disappears and the nuclear envelope fragments. The spindle begins to assemble as the two centrosomes migrate away from one another. In animal cells, an array of microtubules radiates toward the plasma membrane from the centrosomes. These structures are called **asters.** It is thought that asters serve to brace the centrioles during later stages of cell division. Notice that the chromosomes have no particular orientation because the spindle has not yet formed.

**FIGURE 9.4 Phases of mitosis in animal and plant cells.**

The blue chromosomes were inherited from one parent and the red from the other parent.

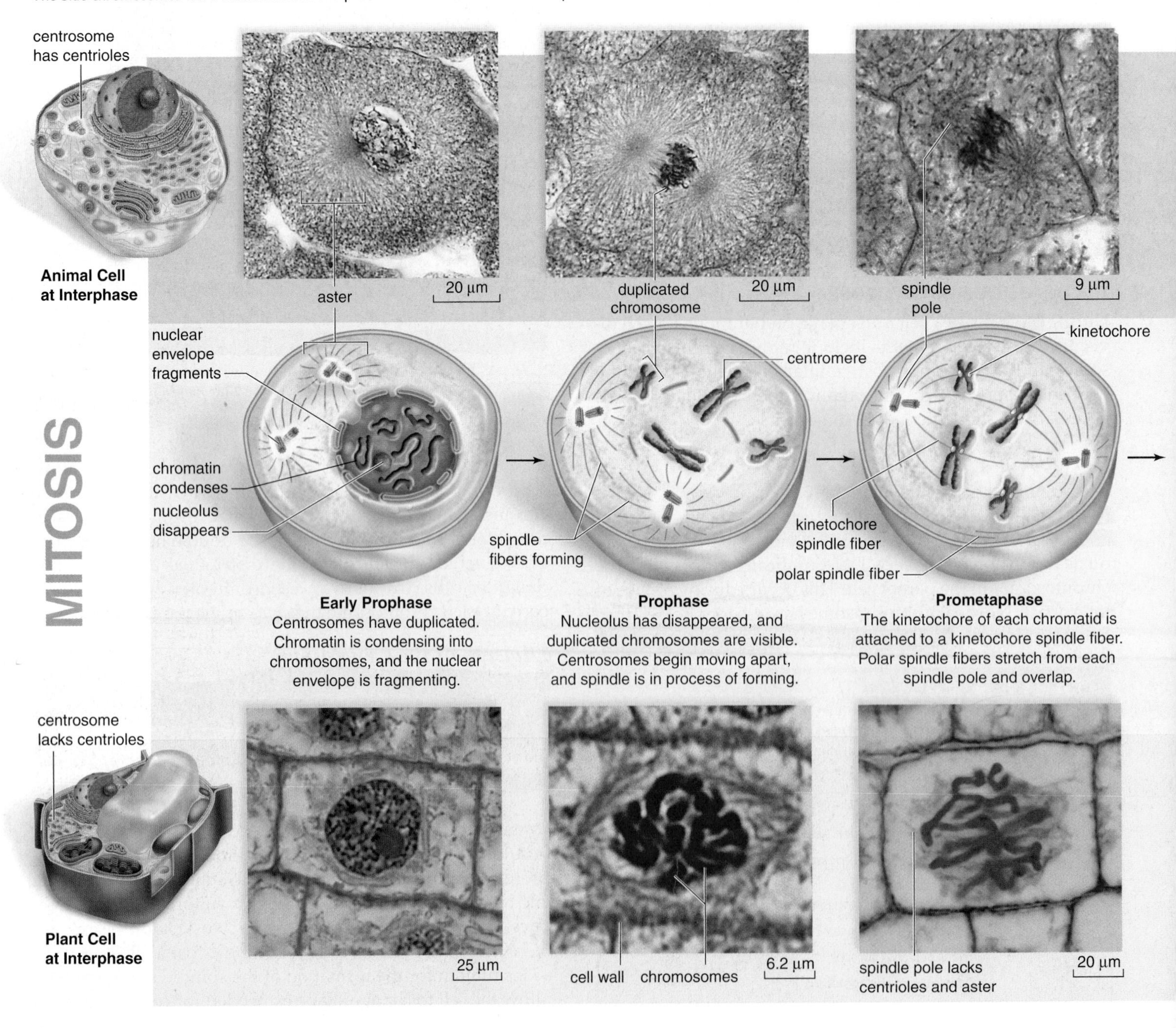

### *Prometaphase (Late Prophase)*

During **prometaphase**, preparations for sister chromatid separation are evident. Kinetochores appear on each side of the centromere, and these attach sister chromatids to the so-called kinetochore spindle fibers. These fibers extend from the poles to the chromosomes, which will soon be located at the center of the spindle.

The kinetochore fibers attach the sister chromatids to opposite poles of the spindle, and the chromosomes are pulled first toward one pole and then toward the other before the chromosomes come into alignment. Notice that even though the chromosomes are attached to the spindle fibers in prometaphase, they are still not in alignment.

### *Metaphase*

During **metaphase,** the centromeres of chromosomes are now in alignment on a single plane at the center of the cell. The chromosomes usually appear as a straight line across the middle of the cell when viewed under a light microscope. An imaginary plane that is perpendicular and passes through this circle is called the **metaphase plate.** It indicates the future axis of cell division. Several nonattached spindle fibers called *polar spindle fibers* reach beyond the metaphase plate and overlap. A cell cycle checkpoint, the M checkpoint, delays the start of anaphase until the kinetochores of each chromosome are attached properly to spindle fibers and the chromosomes are properly aligned along the metaphase plate.

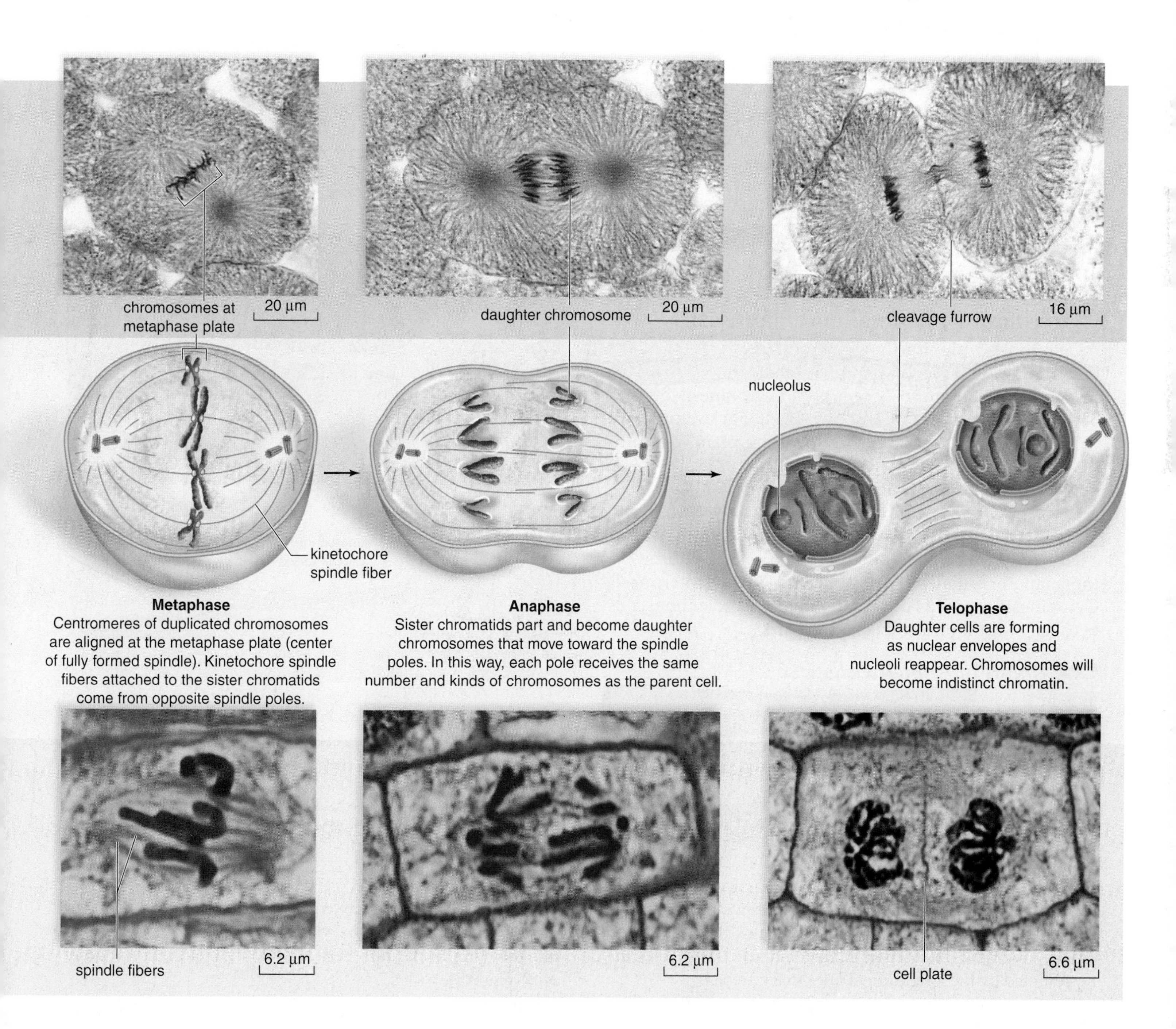


**Metaphase**
Centromeres of duplicated chromosomes are aligned at the metaphase plate (center of fully formed spindle). Kinetochore spindle fibers attached to the sister chromatids come from opposite spindle poles.

**Anaphase**
Sister chromatids part and become daughter chromosomes that move toward the spindle poles. In this way, each pole receives the same number and kinds of chromosomes as the parent cell.

**Telophase**
Daughter cells are forming as nuclear envelopes and nucleoli reappear. Chromosomes will become indistinct chromatin.

### *Anaphase*

At the start of **anaphase,** the two sister chromatids of each duplicated chromosome separate at the centromere, giving rise to two daughter chromosomes. Daughter chromosomes, each with a centromere and single chromatid composed of a single double helix, appear to move toward opposite poles. Actually, the daughter chromosomes are being pulled to the opposite poles as the kinetochore spindle fibers disassemble at the region of the kinetochores. Even as the daughter chromosomes move toward the spindle poles, the poles themselves are moving farther apart because the polar spindle fibers are sliding past one another. Microtubule-associated proteins such as the motor molecules kinesin and dynein are involved in the sliding process. Anaphase is the shortest phase of mitosis.

### *Telophase*

During **telophase,** the spindle disappears as new nuclear envelopes form around the daughter chromosomes. Each daughter nucleus contains the same number and kinds of chromosomes as the original parent cell. Remnants of the polar spindle fibers are still visible between the two nuclei.

The chromosomes become more diffuse chromatin once again, and a nucleolus appears in each daughter nucleus. Division of the cytoplasm requires cytokinesis, which is discussed in the next section.

## Cytokinesis in Animal and Plant Cells

As mentioned previously, cytokinesis is division of the cytoplasm. Cytokinesis accompanies mitosis in most cells but not all. When mitosis occurs but cytokinesis doesn't occur, the result is a multinucleated cell. For example, we will see that the embryo sac in flowering plants is multinucleated.

Division of the cytoplasm begins in anaphase, continues in telophase, but does not reach completion until the following interphase begins. By the end of mitosis each newly forming cell has received a share of the cytoplasmic organelles that duplicated during interphase. Cytokinesis proceeds differently in plant and animal cells because of differences in cell structure.

### *Cytokinesis in Animal Cells*

In animal cells, a cleavage furrow, which is an indentation of the membrane between the two daughter nuclei, forms just as anaphase draws to a close. By that time, the newly forming cells have received a share of the cytoplasmic organelles that duplicated during the previous interphase.

The cleavage furrow deepens when a band of actin filaments, called the contractile ring, slowly forms a circular constriction between the two daughter cells. The action of the contractile ring can be likened to pulling a drawstring ever tighter about the middle of a balloon. As the drawstring is pulled tight, the balloon constricts in the middle as the material on either side of the constriction gathers in folds. These folds are represented by the longitudinal lines in Figure 9.5.

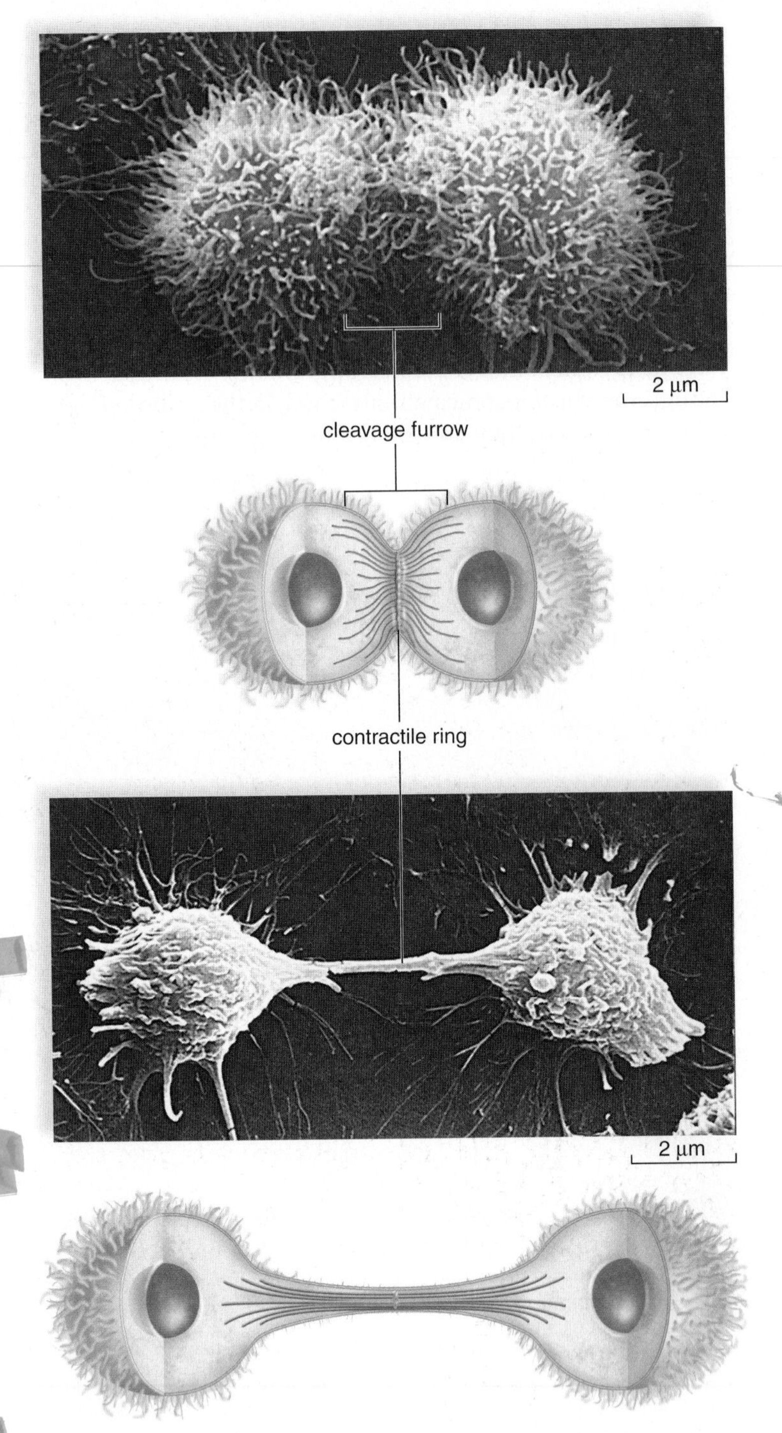


**FIGURE 9.5 Cytokinesis in animal cells.**

A single cell becomes two cells by a furrowing process. A contractile ring composed of actin filaments gradually gets smaller, and the cleavage furrow pinches the cell into two cells.



A narrow bridge between the two cells can be seen during telophase, and then the contractile ring continues to separate the cytoplasm until there are two independent daughter cells (Fig. 9.5).

### *Cytokinesis in Plant Cells*

Cytokinesis in plant cells occurs by a process different from that seen in animal cells (Fig. 9.6). The rigid cell wall that surrounds plant cells does not permit cytokinesis by furrowing. Instead, cytokinesis in plant cells involves the building of new cell walls between the daughter cells.

Cytokinesis is apparent when a small, flattened disk appears between the two daughter plant cells near the site where the metaphase plate once was. In electron micrographs, it is possible to see that the disk is at right angles to a set of microtubules that radiate outward from the forming nuclei. The Golgi apparatus produces vesicles, which move along the microtubules to the region of the disk. As more vesicles arrive and fuse, a cell plate can be seen. The **cell plate** is simply newly formed plasma membrane that expands outward until it reaches the old plasma membrane and fuses with this membrane. The new membrane releases molecules that form the new plant cell walls. These cell walls, known as primary cell walls, are later strengthened by the addition of cellulose fibrils. The space between the daughter cells becomes filled with middle lamella, which cements the primary cell walls together.

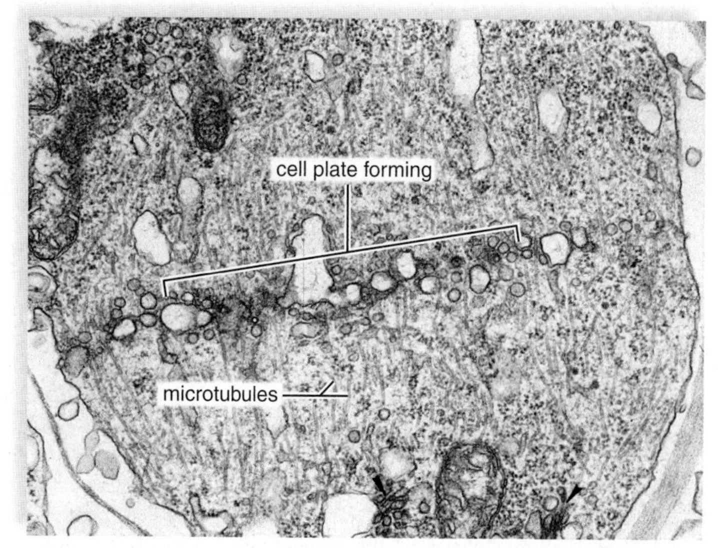

Vesicles containing cell wall components fusing to form cell plate

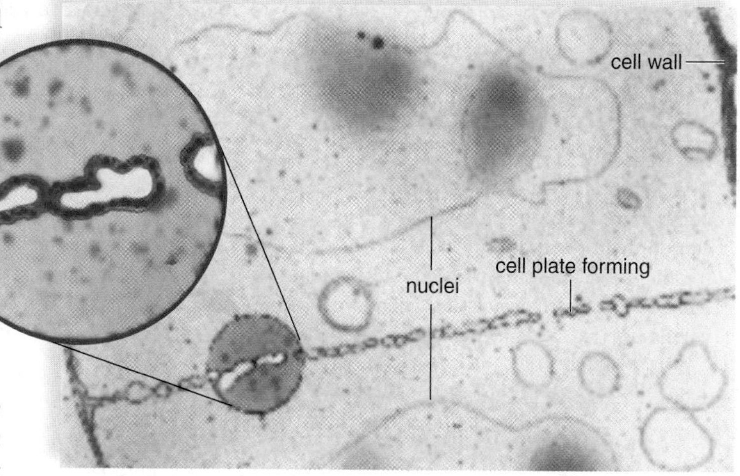


**FIGURE 9.6 Cytokinesis in plant cells.**

During cytokinesis in a plant cell, a cell plate forms midway between two daughter nuclei and extends to the plasma membrane.

## The Functions of Mitosis

Mitosis permits growth and repair. In both plants and animals, mitosis is required during development as a single cell develops into an individual. In plants, the individual could be a fern or daisy, while in animals, the individual could be a grasshopper or a human being.

In flowering plants, meristematic tissue retains the ability to divide throughout the life of a plant. Meristematic tissue at the shoot tip accounts for an increase in the height of a plant for as long as it lives. Then, too, lateral meristem accounts for the ability of trees to increase their girth each growing season.

In human beings and other mammals, mitosis is necessary as a fertilized egg becomes an embryo and as the embryo becomes a fetus. Mitosis also occurs after birth as a child becomes an adult. Throughout life, mitosis allows a cut to heal or a broken bone to mend.

## Stem Cells

Earlier, you learned that the cell cycle is tightly controlled, and that most cells of the body at adulthood are permanently arrested in $G_0$ stage. However, mitosis is needed to repair injuries, such as a cut or a broken bone. Many mammalian organs contain stem cells (often called adult stem cells) that retain the ability to divide. In the body, red bone marrow stem cells repeatedly divide to produce millions of cells that go on to become various types of blood cells. The possibility exists that researchers can learn to manipulate the production of various types of tissues from red bone marrow stem cells in the laboratory. If so, these tissues could be used to cure illnesses. As discussed in the Science Focus on page 160, **therapeutic cloning,** which is used to produce human tissues, can begin with either adult stem cells or embryonic stem cells. Embryonic stem cells can also be used for **reproductive cloning,** the production of a new individual.

### Check Your Progress 9.2

1. What are the major events that occur during prophase, and why are these events important to the process of cell division?
2. How does cytokinesis differ between animal and plant cells? Why is this difference necessary?

science focus

# Reproductive and Therapeutic Cloning

Our knowledge of how the cell cycle is controlled has yielded major technological breakthroughs, including reproductive cloning—the ability to clone an adult animal from a normal body cell, and therapeutic cloning, which allows the rapid production of mature cells of a specific type. Both types of cloning are a direct result of recent discoveries about how the cell cycle is controlled.

Reproductive cloning, or the cloning of adult animals, was once thought to be impossible because investigators found it difficult to have the nucleus of an adult cell "start over" with the cell cycle, even when it was placed in an egg cell that had its own nucleus removed.

In 1997, Dolly the sheep demonstrated that reproductive cloning is indeed possible. The donor cells were starved before the cell's nucleus was placed in an enucleated egg. This caused them to stop dividing and go into a $G_0$ (resting) stage, and this made the nuclei amenable to cytoplasmic signals for initiation of development (Fig. 9B*a*). This advance has made it possible to clone all sorts of farm animals that have desirable traits and even to clone rare animals that might otherwise become extinct. Despite the encouraging results, however, there are still obstacles to be overcome, and a ban on the use of federal funds in experiments to clone human beings remains firmly in place.

In therapeutic cloning, however, the objective is to produce mature cells of various cell types rather than an individual organism. The purpose of therapeutic cloning is (1) to learn more about how specialization of cells occurs and (2) to provide cells and tissues that could be used to treat human illnesses, such as diabetes, or major injuries like strokes or spinal cord injuries. There are two possible ways to carry out therapeutic cloning. The first way is to use the exact same procedure as reproductive cloning, except embryonic cells, called *embryonic stem cells,* are separated and each is subjected to a treatment that causes it to develop into a particular type of cell, such as red blood cells, muscle cells, or nerve cells (Fig. 9B*b*). Some have ethical concerns about this type of therapeutic cloning, which is still experimental, because if the embryo were allowed to continue development, it would become an individual.

The second way to carry out therapeutic cloning is to use *adult stem cells.* Stem cells are found in many organs of the adult's body; for example, the bone marrow has stem cells that produce new blood cells. However, adult stem cells are limited in the possible number of cell types that they may become. Nevertheless, a recent advance shows promise in overcoming this obstacle. By adding just four genes to adult skin stem cells, Japanese scientists were able to coax the cells, called fibroblasts, into becoming very similar to embryonic stem cells. The researchers were then able to create heart and brain cells from the adult stem cells. Ultimately, this technique may provide a way to make tissues and organs for transplantation that carry no risk of rejection. In the future, this new technology promises to overcome current limitations and alleviate ethical concerns.

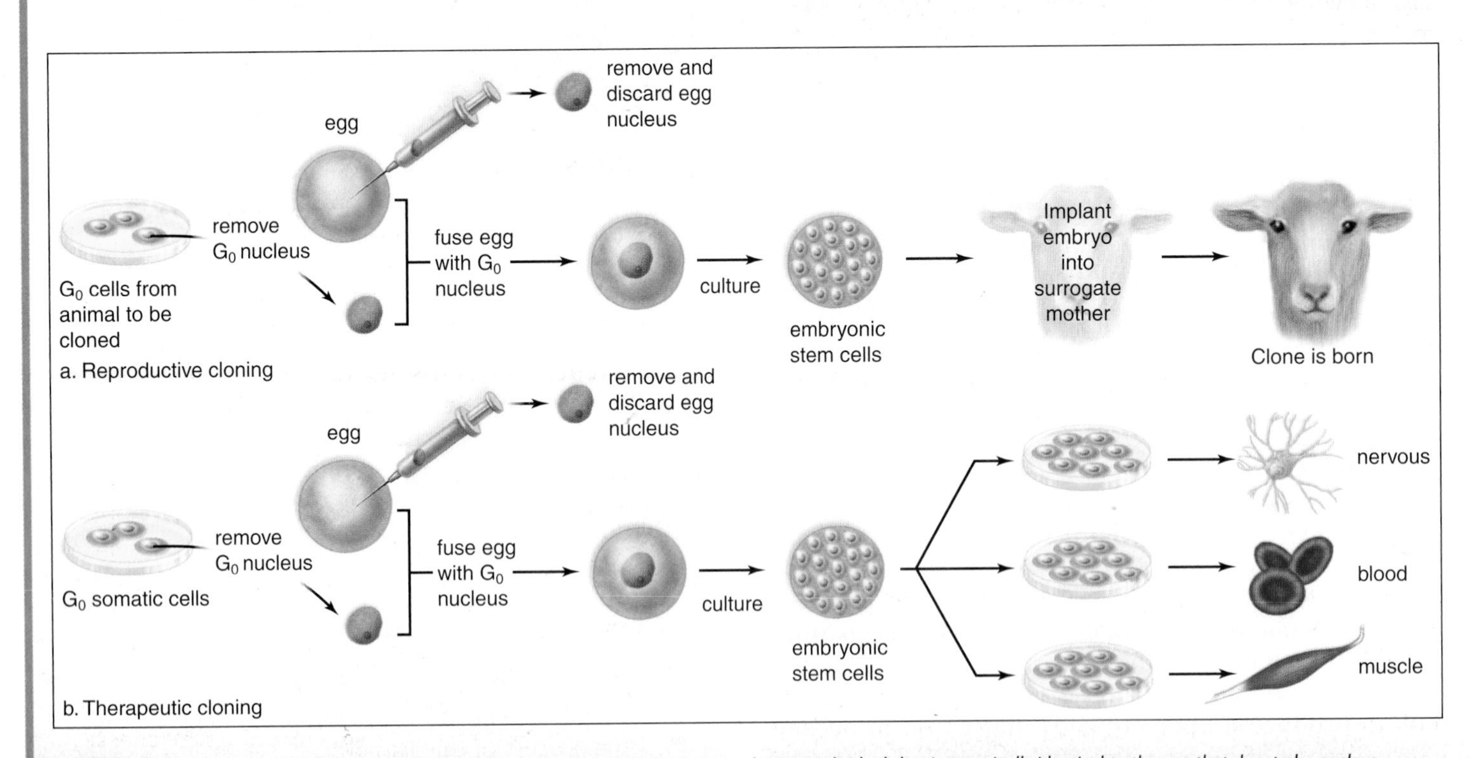


**FIGURE 9B Two types of cloning.** ***a.*** *The purpose of somatic cell cloning is to produce an individual that is genetically identical to the one that donated a nucleus. The nucleus is placed in an enucleated egg, and, after several mitotic divisions, the embryo is implanted into a surrogate mother for further development.* ***b.*** *The purpose of therapeutic cloning is to produce specialized tissue cells. A nucleus is placed in an enucleated egg, and, after several mitotic divisions, the embryonic cells (called embryonic stem cells) are separated and treated to become specialized cells.*

# 9.3 The Cell Cycle and Cancer

**Cancer** is a cellular growth disorder that occurs when cells divide uncontrollably. Although causes widely differ, most cancers are the result of accumulating mutations that ultimately cause a loss of control of the cell cycle.

Although cancers vary greatly, they usually follow a common multistep progression (Fig. 9.7). Most cancers begin as an abnormal cell growth that is **benign,** or not cancerous, and usually does not grow larger. However, additional mutations may occur, causing the abnormal cells to fail to respond to inhibiting signals that control the cell cycle. When this occurs, the growth becomes **malignant,** meaning that it is cancerous and possesses the ability to spread.

## Characteristics of Cancer Cells

The development of cancer is gradual. A mutation in a cell may cause it to become precancerous, but many other regulatory processes within the body prevent it from becoming cancerous. In fact, it may be decades before a cell possesses most or all of the characteristics of a cancer cell (Table 9.2 and Fig. 9.7). Although cancers vary greatly, cells that possess the following characteristics are generally recognized as cancerous:

*Cancer cells lack differentiation.* Cancer cells are not specialized and do not contribute to the functioning of a tissue. Although cancer cells may still possess many of the characteristics of surrounding normal cells, they usually look distinctly abnormal. Normal cells can enter the cell cycle about 50 times before they are incapable of dividing again. Cancer cells can enter the cell cycle repeatedly, and in this way seem immortal.

*Cancer cells have abnormal nuclei.* The nuclei of cancer cells are enlarged and may contain an abnormal number of chromosomes. Often, extra copies of one or more chromosomes may be present. Often, there are also duplicated portions of some chromosomes present, which causes gene amplification, or extra copies of specific genes. Some chromosomes may also possess deleted portions.

*Cancer cells do not undergo apoptosis.* Ordinarily, cells with damaged DNA undergo apoptosis, or programmed cell death. The immune system can also recognize abnormal cells and trigger apoptosis, which normally prevents tumors from developing. Cancer cells fail to undergo apoptosis even though they are abnormal cells.

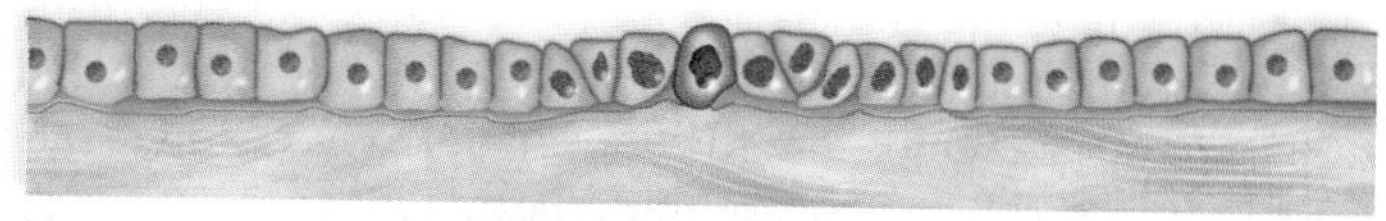

New mutations arise, and one cell (brown) has the ability to start a tumor.

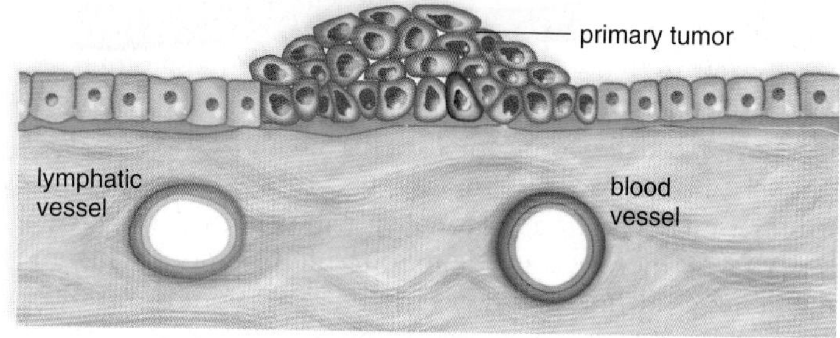


Cancer in situ. The tumor is at its place of origin. One cell (purple) mutates further.

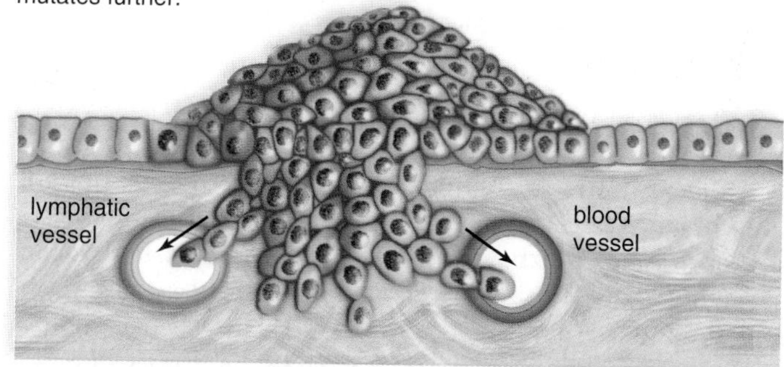


Cancer cells now have the ability to invade lymphatic and blood vessels and travel throughout the body.

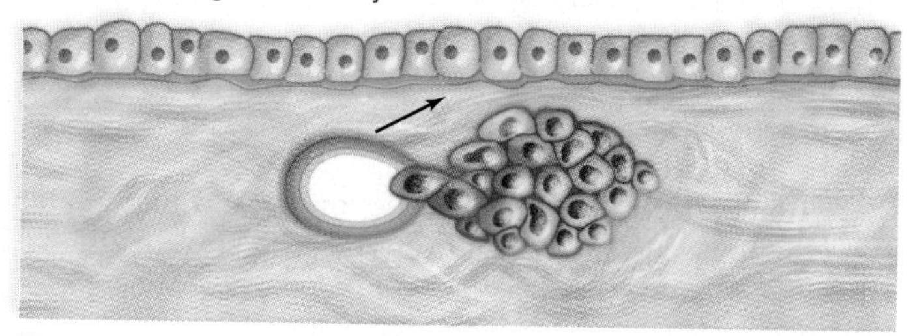

New metastatic tumors are found some distance from the primary tumor.

**FIGURE 9.7 Progression of cancer.**

The development of cancer requires a series of mutations leading first to a localized tumor and then to metastatic tumors. With each successive step toward cancer, the most genetically altered and aggressive cell becomes the dominant type of tumor. The cells take on characteristics of embryonic cells; they are not differentiated, they can divide uncontrollably; and they are able to metastasize and spread to other tissues.

**TABLE 9.2**

**Cancer Cells Versus Normal Cells**

| *Cancer Cells* | *Normal Cells* |
|---|---|
| Nondifferentiated cells | Differentiated cells |
| Abnormal nuclei | Normal nuclei |
| Do not undergo apoptosis | Undergo apoptosis |
| No contact inhibition | Contact inhibition |
| Disorganized, multilayered | One organized layer |
| Undergo metastasis and angiogenesis | |

*Cancer cells form tumors.* Normal cells anchor themselves to a substratum and/or adhere to their neighbors. They exhibit contact inhibition—in other words, when they come in contact with a neighbor, they stop dividing. Cancer cells have lost all restraint and do not exhibit contact inhibition. The abnormal cancer cells pile on top of one another and grow in multiple layers, forming a **tumor.** During carcinogenesis, the most aggressive cell becomes the dominant cell of the tumor.

*Cancer cells undergo metastasis and angiogenesis.* Additional mutations may cause a benign tumor, which is usually contained within a capsule and cannot invade adjacent tissue, to become malignant, and spread throughout the body, forming new tumors distant from the primary tumor. These cells now produce enzymes that they normally do not express, allowing tumor cells to invade underlying tissues. Then, they travel through the blood and lymph, to start tumors elsewhere in the body. This process is known as **metastasis.**

Tumors that are actively growing soon encounter another obstacle—the blood vessels supplying nutrients to the tumor cells become insufficient to support the sudden growth of the tumor. In order to grow further, the cells of the tumor must receive additional nutrition. Thus, the formation of new blood vessels is required to bring nutrients and oxygen to support further growth. Additional mutations occurring in tumor cells allow them to direct the growth of new blood vessels into the tumor in a process called **angiogenesis.** Some modes of cancer treatment are aimed at preventing angiogenesis from occurring.

## Origin of Cancer

Normal growth and maintenance of body tissues depend on a balance between signals that promote and inhibit cell division. When this balance is upset, conditions such as cancer may occur. Thus, cancer is usually caused by mutations affecting genes that directly or indirectly affect this balance, such as those shown in Figure 9.8. These two types of genes are usually affected:

1. **Proto-oncogenes** code for proteins that promote the cell cycle and prevent apoptosis. They are often likened to the gas pedal of a car because they cause the cell cycle to go or speed up.
2. **Tumor suppressor genes** code for proteins that inhibit the cell cycle and promote apoptosis. They are often likened to the brakes of a car because they cause the cell cycle to go more slowly or even stop.

### *Proto-oncogenes Become Oncogenes*

Proto-oncogenes are normal genes that promote progression through the cell cycle. They are often at the end of a *stimulatory pathway* extending from the plasma membrane to the nucleus. A stimulus, such as an injury, results in the release of a growth factor that binds to a receptor protein in the plasma membrane. This sets in motion a whole series of enzymatic reactions leading to the activation of genes that promote the cell cycle, both directly and indirectly. Proto-oncogenes include the receptors and signal molecules that make up these pathways.

When mutations occur in proto-oncogenes, they become **oncogenes,** or cancer-causing genes. Oncogenes are under constant stimulation and keep on promoting the cell cycle regardless of circumstances. For example, an oncogene may code for a faulty receptor in the stimulatory pathway that stimulates the cell cycle even when no growth factor is present! Or, an oncogene may specify either an abnormal protein product or produce abnormally high levels of a normal product that stimulates the cell cycle to begin or to go to completion. As a result, uncontrolled cell division may occur.

Researchers have identified perhaps 100 oncogenes that can cause increased growth and lead to tumors. The oncogenes most frequently involved in human cancers belong to the *ras* gene family. Mutant forms of the *BRCA1* oncogene (breast cancer predisposition gene 1) are associated with certain hereditary forms of breast and ovarian cancer.

### *Tumor Suppressor Genes Become Inactive*

Tumor suppressor genes, on the other hand, directly or indirectly inhibit the cell cycle and prevent cells from dividing uncontrollably. Some tumor suppressor genes prevent progression of the cell cycle when DNA is damaged. Other tumor suppressor genes may promote apoptosis as a last resort.

A mutation in a tumor suppressor gene is much like brake failure in a car; when the mechanism that slows down and stops cell division does not function, the cell cycle accelerates and does not halt. Researchers have identified about a half-dozen tumor suppressor genes. Among these are *RB* and *p53* genes that code for RB and p53. The Science Focus on page 154 discussed the function of these proteins in controlling the cell cycle. The *RB* tumor suppressor gene was discovered when the inherited condition retinoblastoma was being studied, but malfunctions of this gene have now been identified in many other cancers as well, including breast, prostate, and bladder cancers. Another major tumor suppressor gene is *p53,* a gene that turns on the expression of other genes that inhibit the cell cycle. The p53 protein can also stimulate apoptosis, programmed cell death. It is estimated that over half of human cancers involve an abnormal or deleted *p53* gene.

### *Other Causes of Cancer*

As mentioned previously, cancer develops when the delicate balance between promotion and inhibition of cell division is tilted towards uncontrolled cell division. Thus, other mutations may occur within a cell that affect this balance. For example, while a mutation affecting the cell's DNA repair system will not immediately cause cancer, it leads to a much greater chance of a mutation occurring within a proto-oncogene or tumor suppressor gene. And in some cancer cells, mutation of an enzyme that regulates the length of **telomeres,** or the ends of chromosomes, causes telomeres to remain at a constant length. Since cells with shortened telomeres normally stop dividing, keeping the telomeres at a constant length allows the cancer cells to continue dividing over and over again.

#### Check Your Progress 9.3

1. What are the major characteristics of cancer cells that distinguish them from normal cells?
2. What are the usual steps in development of a malignant tumor from a benign tumor?
3. Compare and contrast the effect on the cell cycle of **(a)** a mutation in a proto-oncogene to **(b)** a mutation in a tumor suppressor gene.

**FIGURE 9.8 Causes of cancer.**

**a.** Mutated genes that cause cancer can be due to the influences noted. **b.** A growth factor that binds to a receptor protein initiates a reaction that triggers a stimulatory pathway. **c.** A stimulatory pathway that begins at the plasma membrane turns on proto-oncogenes. The products of these genes promote the cell cycle and double back to become part of the stimulatory pathway. When proto-oncogenes become oncogenes, they are turned on all the time. An inhibitory pathway begins with tumor suppressor genes whose products inhibit the cell cycle. When tumor suppressor genes mutate, the cell cycle is no longer inhibited. **d.** Cancerous skin cell.

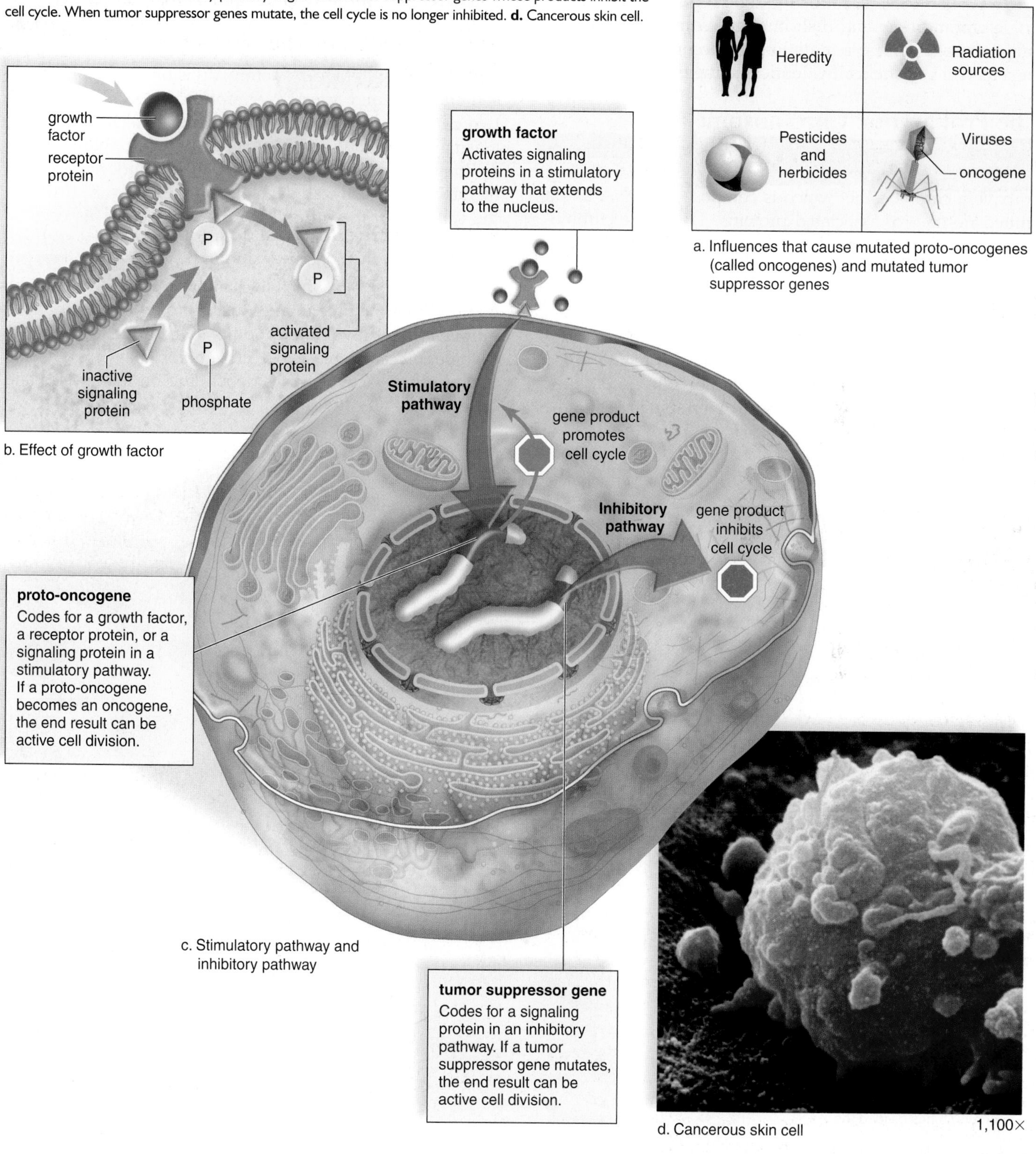

Heredity
Radiation sources
Pesticides and herbicides
Viruses
oncogene

a. Influences that cause mutated proto-oncogenes (called oncogenes) and mutated tumor suppressor genes

growth factor
receptor protein
P
activated signaling protein
inactive signaling protein
phosphate

b. Effect of growth factor

**growth factor**
Activates signaling proteins in a stimulatory pathway that extends to the nucleus.

**Stimulatory pathway**
gene product promotes cell cycle
**Inhibitory pathway**
gene product inhibits cell cycle

**proto-oncogene**
Codes for a growth factor, a receptor protein, or a signaling protein in a stimulatory pathway. If a proto-oncogene becomes an oncogene, the end result can be active cell division.

**tumor suppressor gene**
Codes for a signaling protein in an inhibitory pathway. If a tumor suppressor gene mutates, the end result can be active cell division.

c. Stimulatory pathway and inhibitory pathway

d. Cancerous skin cell 1,100×

## 9.4 Prokaryotic Cell Division

Cell division in unicellular organisms, such as prokaryotes, produces two new individuals. This is **asexual reproduction** in which the offspring are genetically identical to the parent. In prokaryotes, reproduction consists of duplicating the single chromosome and distributing a copy to each of the daughter cells. Unless a mutation has occurred, the daughter cells will be genetically identical to the parent cell.

### The Prokaryotic Chromosome

Prokaryotes (bacteria and archaea) lack a nucleus and other membranous organelles found in eukaryotic cells. Still, they do have a chromosome, which is composed of DNA and a limited number of associated proteins. The single chromosome of prokaryotes contains just a few proteins and is organized differently from eukaryotic chromosomes. A eukaryotic chromosome has many more proteins than a prokaryotic chromosome.

In electron micrographs, the bacterial chromosome appears as an electron-dense, irregularly shaped region called the **nucleoid** [L. *nucleus,* nucleus, kernel; Gk. *-eides,* like], which is not enclosed by membrane. When stretched out, the chromosome is seen to be a circular loop with a length that is up to about a thousand times the length of the cell. No wonder it is folded when inside the cell.

### Binary Fission

Prokaryotes reproduce asexually by binary fission. The process is termed **binary fission** because division (fission) produces two (binary) daughter cells that are identical to the original parent cell. Before division takes place, the cell enlarges, and after DNA replication occurs, there are two chromosomes. These chromosomes attach to a special plasma membrane site and separate by an elongation of the cell that pulls them apart. During this period, new plasma membrane and cell wall develop and grow inward to divide the cell. When the cell is approximately twice its original length, the new cell wall and plasma membrane for each cell are complete (Fig. 9.9).

*Escherichia coli,* which lives in our intestines, has a generation time (the time it takes the cell to divide) of about 20 minutes under favorable conditions. In about seven hours, a single cell can increase to over 1 million cells! The division rate of other bacteria varies depending on the species and conditions.

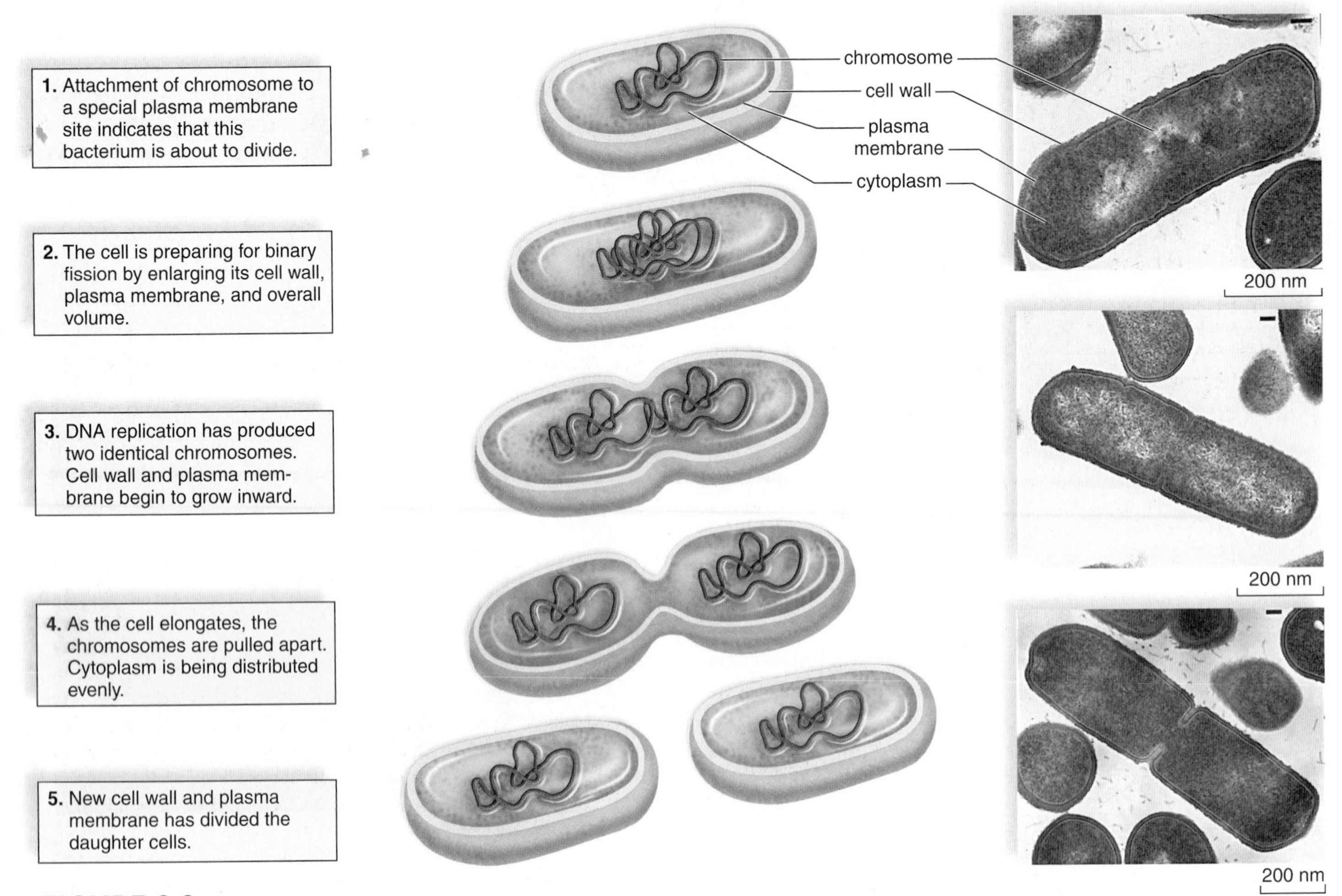


**FIGURE 9.9 Binary fission.**

First, DNA replicates, and as the cell lengthens, the two chromosomes separate, and the cells become divided. The two resulting bacteria are identical.

## Comparing Prokaryotes and Eukaryotes

Both binary fission and mitosis ensure that each daughter cell is genetically identical to the parent cell. The genes are portions of DNA found in the chromosomes.

Prokaryotes (bacteria and archaea), protists (many algae and protozoans), and some fungi (yeasts) are unicellular. Cell division in unicellular organisms produces two new individuals:

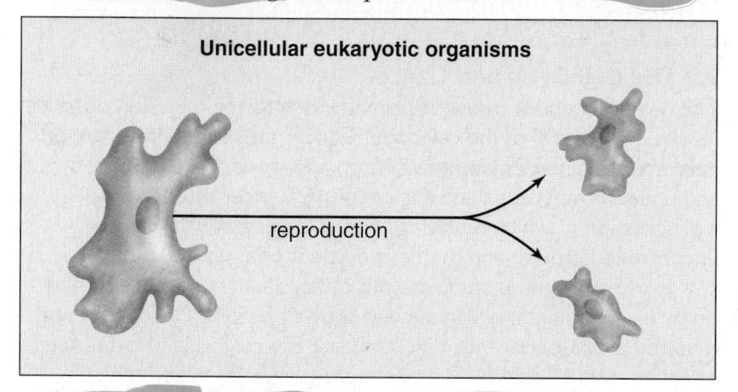


This is a form of asexual reproduction because one parent has produced identical offspring (Table 9.3).

In multicellular fungi (molds and mushrooms), plants, and animals, cell division is part of the growth process. It produces the multicellular form we recognize as the mature organism. Cell division is also important in multicellular forms for renewal and repair:

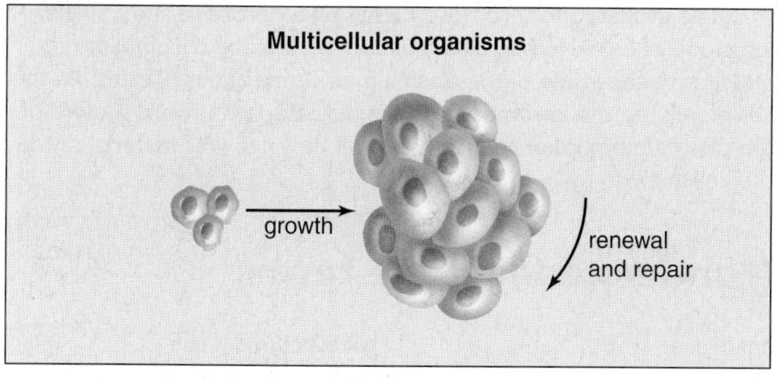


The chromosomes of eukaryotic cells are composed of DNA and many associated proteins. The histone proteins organize a chromosome, allowing it to extend as chromatin during interphase and to coil and condense just prior to mitosis. Each species of multicellular eukaryotes has a characteristic number of chromosomes in the nuclei. As a result of mitosis, each daughter cell receives the same number and kinds of chromosomes as the parent cell. The spindle, which appears during mitosis, is involved in distributing the daughter chromosomes to the daughter nuclei. Cytokinesis, either by the formation of a cell plate (plant cells) or by furrowing (animal cells), is division of the cytoplasm.

In prokaryotes, the single chromosome consists largely of DNA with some associated proteins. During binary fission, this chromosome duplicates, and each daughter cell receives one copy as the parent cell elongates, and a new cell wall and plasma membrane form between the daughter cells. No spindle is involved in binary fission.

**TABLE 9.3**

**Functions of Cell Division**

| *Type of Organism* | *Cell Division* | *Function* |
|---|---|---|
| **Prokaryotes** Bacteria and archaea | Binary fission | Asexual reproduction |
| **Eukaryotes** | | |
| Protists, and some fungi (yeast) | Mitosis and cytokinesis | Asexual reproduction |
| Other fungi, plants, and animals | Mitosis and cytokinesis | Development, growth, and repair |

### Check Your Progress 9.4

1. How does binary fission in prokaryotes differ from mitosis and cytokinesis in eukaryotes?
2. How are prokaryotic and eukaryotic chromosomes different?

## Connecting the Concepts

Cell division is a remarkable, complex process that is only a small part of the cell cycle, the life cycle of a cell. The cell cycle is heavily regulated to ensure that conditions are favorable and that it is permissible for the cell to divide, because there may be serious consequences if control of cell division breaks down. For example, in humans, overproduction of skin cells due to an overstimulated cell cycle produces a chronic inflammatory condition known as psoriasis. On the other hand, aggressive inhibition of the cell cycle that destroys the reproductive capacity of all the body's cells leads to a condition called progeria, which causes young people to grow old and die at an early age. Many different types of cancer can result when the signals that keep the cell cycle in check are not transmitted or received properly. Learning how the cell cycle is regulated and how to control it may lead to many important scientific advances, such as the possibility of therapeutic cloning and tissue engineering which forms organs in the laboratory.

Mitosis involves division of the nucleus and the distribution of its contents, the chromosomes, into the daughter cells. Before this occurs, the chromosomes must be duplicated so that each daughter cell can receive one of each kind of chromosome. However, as we will soon see, a special type of cell division, meiosis, reduces the chromosome number in order to produce gametes.

## summary

### 9.1 The Cell Cycle

Eukaryotic cells go through a cell cycle that includes (1) interphase and (2) a mitotic stage that consists of mitosis and cytokinesis. Interphase, in turn, is composed of three stages: $G_1$ (growth as certain organelles double), S (the synthesis stage, where the chromosomes are duplicated), and $G_2$ (growth as the cell prepares to divide). Most cells of the body are no longer dividing and are said to be arrested in a $G_0$ state, from which cells must receive signals to return to $G_1$ stage and complete the cell cycle. During the mitotic stage (M), the chromosomes are sorted into two daughter cells so that each receives a full complement of chromosomes.

The cell cycle is regulated by three well-known checkpoints—the restriction point, or $G_1$ checkpoint, the $G_2$ checkpoint prior to the M stage, and the M stage checkpoint, or spindle assembly checkpoint, immediately before anaphase. The $G_1$ checkpoint ensures that conditions are favorable and that the proper signals are present, and also checks the DNA for damage. If the DNA is damaged beyond repair, the p53 protein may initiate apoptosis. During apoptosis, enzymes called caspases bring about destruction of the nucleus and the rest of the cell. Cell division and apoptosis are two opposing processes that keep the number of healthy cells in balance.

### 9.2 Mitosis and Cytokinesis

Interphase represents the portion of the cell cycle between nuclear divisions, and during this time, preparations are made for cell division. These preparations include duplication of most cellular contents, including the centrosome, which organizes the mitotic spindle. The DNA is duplicated during S stage, at which time the chromosomes, which consisted of a single chromatid each, are duplicated. The $G_2$ checkpoint ensures that DNA has replicated properly. This results in a nucleus containing the same number of chromosomes, with each now consisting of two chromatids attached at the centromere. During interphase, the chromosomes are not distinct and are collectively called chromatin. Each eukaryotic species has a characteristic number of chromosomes. The total number is called the diploid number, and half this number is the haploid number.

Among eukaryotes, cell division involves both mitosis (nuclear division) and division of the cytoplasm (cytokinesis). As a result of mitosis, the chromosome number stays constant because each chromosome is duplicated and gives rise to two daughter chromosomes that consist of a single chromatid each.

Mitosis consists of five phases:

Prophase—The nucleolus disappears, the nuclear envelope fragments, and the spindle forms between centrosomes. The chromosomes condense and become visible under a light microscope. In animal cells, asters radiate from the centrioles within the centrosomes. Plant cells lack centrioles and, therefore, asters. Even so, the mitotic spindle forms.

Prometaphase (late prophase)—The kinetochores of sister chromatids attach to kinetochore spindle fibers extending from opposite poles. The chromosomes move back and forth until they are aligned at the metaphase plate.

Metaphase—The spindle is fully formed, and the duplicated chromosomes are aligned at the metaphase plate. The spindle consists of polar spindle fibers that overlap at the metaphase plate and kinetochore spindle fibers that are attached to chromosomes. The M stage checkpoint, or spindle assembly checkpoint, must be satisfied before progressing to the next phase.

Anaphase—Sister chromatids separate, becoming daughter chromosomes that move toward the poles. The polar spindle fibers slide past one another, and the kinetochore spindle fibers disassemble. Cytokinesis by furrowing begins.

Telophase—Nuclear envelopes re-form, chromosomes begin changing back to chromatin, the nucleoli reappear, and the spindle disappears. Cytokinesis continues, and is complete by the end of telophase.

Cytokinesis in animal cells is a furrowing process that divides the cytoplasm. Cytokinesis in plant cells involves the formation of a cell plate from which the plasma membrane and cell wall are completed.

### 9.3 The Cell Cycle and Cancer

The development of cancer is primarily due to the mutation of genes involved in control of the cell cycle. Cancer cells lack differentiation, have abnormal nuclei, do not undergo apoptosis, form tumors, and undergo metastasis and angiogenesis. Cancer often follows a progression in which mutations accumulate, gradually causing uncontrolled growth and the development of a tumor.

Proto-oncogenes stimulate the cell cycle after they are turned on by environmental signals such as growth factors. Oncogenes are mutated proto-oncogenes that stimulate the cell cycle without need of environmental signals. Tumor suppressor genes inhibit the cell cycle. Mutated tumor suppressor genes no longer inhibit the cell cycle, allowing unchecked cell division.

### 9.4 Prokaryotic Cell Division

Binary fission (in prokaryotes) and mitosis (in unicellular eukaryotic protists and fungi) allow organisms to reproduce asexually. Mitosis in multicellular eukaryotes is primarily for the purpose of development, growth, and repair of tissues.

The prokaryotic chromosome has a few proteins and a single, long loop of DNA. When binary fission occurs, the chromosome attaches to the inside of the plasma membrane and replicates. As the cell elongates, the chromosomes are pulled apart. Inward growth of the plasma membrane and formation of new cell wall material divide the cell in two.

## understanding the terms

Match the terms to these definitions:

a ______________ Central microtubule organizing center of cells, consisting of granular material. In animal cells, it contains two centrioles.
b. ______________ Constriction where sister chromatids of a chromosome are held together.
c. ______________ Microtubule structure that brings about chromosome movement during nuclear division.
d. ______________ One of two genetically identical chromosome units that are the result of DNA replication.
e. ______________ Programmed cell death that is carried out by enzymes routinely present in the cell

## reviewing this chapter

1. Describe the cell cycle, including its different stages. 152
2. Describe three checkpoints of the cell cycle. 153
3. What is apoptosis, and what are its functions? 153
4. Distinguish between chromosome, chromatin, chromatid, centriole, cytokinesis, centromere, and kinetochore. 152–55
5. Describe the preparations for mitosis. 155
6. Describe the events that occur during the phases of mitosis. 156–58
7. How does plant cell mitosis differ from animal cell mitosis? 156–59
8. Contrast cytokinesis in animal cells and plant cells. 158–59
9. List and discuss characteristics of cancer cells that distinguish them from normal cells. 161–62
10. Compare and contrast the functions of proto-oncogenes and tumor suppressor genes in controlling the cell cycle. 162–63
11. Describe the prokaryotic chromosome and the process of binary fission. 164
12. Contrast the function of cell division in prokaryotic and eukaryotic cells. 165

## testing yourself

Choose the best answer for each question.

1. In contrast to a eukaryotic chromosome, a prokaryotic chromosome
   a. is shorter and fatter.
   b. has a single loop of DNA.
   c. never replicates.
   d. contains many histones.
   e. All of these are correct.
2. The diploid number of chromosomes
   a. is the 2n number.
   b. is in a parent cell and therefore in the two daughter cells following mitosis.
   c. varies according to the particular organism.
   d. is in every somatic cell.
   e. All of these are correct.

For questions 3–5, match the descriptions that follow to the terms in the key.

**KEY:**

a. centrosome
b. chromosome
c. centromere
d. cyclin

3. Point of attachment for sister chromatids
4. Found at a spindle pole in the center of an aster
5. Coiled and condensed chromatin
6. If a parent cell has 14 chromosomes prior to mitosis, how many chromosomes will each daughter cell have?
   a. 28 because each chromatid is a chromosome
   b. 14 because the chromatids separate
   c. only 7 after mitosis is finished
   d. any number between 7 and 28
   e. 7 in the nucleus and 7 in the cytoplasm, for a total of 14
7. In which phase of mitosis are the kinetochores of the chromosomes being attached to spindle fibers?
   a. prophase
   b prometaphase
   c. metaphase
   d. anaphase
   e. telophase
8. Interphase
   a. is the same as prophase, metaphase, anaphase, and telophase.
   b. is composed of $G_1$, S, and $G_2$ stages.
   c. requires the use of polar spindle fibers and kinetochore spindle fibers.
   d. is the majority of the cell cycle.
   e. Both b and d are correct.
9. At the metaphase plate during metaphase of mitosis, there are
   a. single chromosomes.
   b. duplicated chromosomes.
   c. $G_1$ stage chromosomes.
   d. always 23 chromosomes.
10. During which mitotic phases are duplicated chromosomes present?
    a. all but telophase
    b. prophase and anaphase
    c. all but anaphase and telophase
    d. only during metaphase at the metaphase plate
    e. Both a and b are correct.
11. Which of these is paired incorrectly?
    a. prometaphase—the kinetochores become attached to spindle fibers
    b. anaphase—daughter chromosomes are located at the spindle poles
    c. prophase—the nucleolus disappears and the nuclear envelope disintegrates
    d. metaphase—the chromosomes are aligned in the metaphase plate
    e. telophase—a resting phase between cell division cycles
12. When cancer occurs,
    a. cells cannot pass the $G_1$ checkpoint.
    b. control of the cell cycle is impaired.
    c. apoptosis has occurred.
    d. the cells can no longer enter the cell cycle.
    e. All of these are correct.
13. Which of the following is not characteristic of cancer cells?
    a. Cancer cells often undergo angiogenesis.
    b. Cancer cells tend to be nonspecialized.
    c. Cancer cells undergo apoptosis.
    d. Cancer cells often have abnormal nuclei.
    e. Cancer cells can metastasize.
14. Which of the following statements is true?
    a. Proto-oncogenes cause a loss of control of the cell cycle.
    b. The products of oncogenes may inhibit the cell cycle.
    c. Tumor-suppressor-gene products inhibit the cell cycle.
    d. A mutation in a tumor suppressor gene may inhibit the cell cycle.
    e. A mutation in a proto-oncogene may convert it into a tumor suppressor gene.

For questions 15–18, match the descriptions to a stage in the key.

**KEY:**

a. $G_1$ stage
b. S stage
c. $G_2$ stage
d. M (mitotic) stage

15. At the end of this stage, each chromosome consists of two attached chromatids.
16. During this stage, daughter chromosomes are distributed to two daughter nuclei.
17. The cell doubles its organelles and accumulates the materials needed for DNA synthesis.
18. The cell synthesizes the proteins needed for cell division.
19. Which is not true of the cell cycle?
    a. The cell cycle is controlled by internal/external signals.
    b. Cyclin is a signaling molcule that increases and decreases as the cycle continues.
    c. DNA damage can stop the cell cycle at the $G_1$ checkpoint.
    d. Apoptosis occurs frequently during the cell cycle.
20. In human beings, mitosis is necessary for
    a. growth and repair of tissues.
    b. formation of the gametes.
    c. maintaining the chromosome number in all body cells.
    d. the death of unnecessary cells
    e. Both a and c are correct.
21. Label this diagram. What phase of mitosis does it represent?

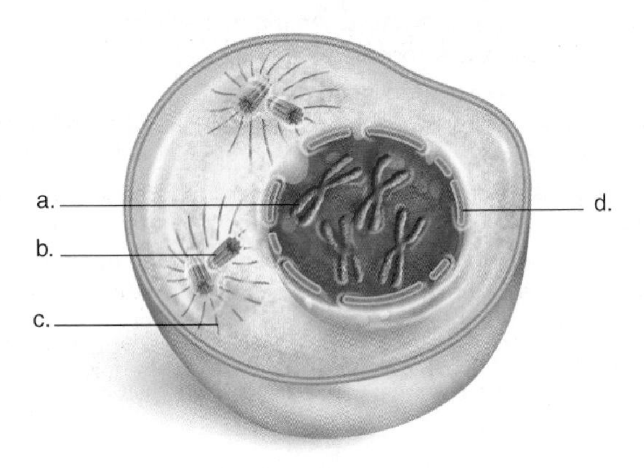


## thinking scientifically

1. After DNA is duplicated in eukaryotes, it must be bound to histones. This requires the synthesis of hundreds of millions of new protein molecules. With reference to Figure 9.1, when in the cell cycle would histones be made?
2. The survivors of the atomic bombs that were dropped on Hiroshima and Nagasaki have been the subjects of long-term studies of the effects of ionizing radiation on cancer incidence. The frequencies of different types of cancer in these individuals varied across the decades. In the 1950s, high levels of leukemia and cancers of the lung and thyroid gland were observed. The 1960s and 1970s brought high levels of breast and salivary gland cancers. In the 1980s, rates of colon cancer were especially high. Why do you suppose the rates of different types of cancer varied across time?

## bioethical issue

### Paying for Cancer Treatment

The risk factors for developing cancer are generally well known. Many lifestyle factors, such as smoking, poor dietary habits, obesity, physical inactivity, risky sexual behavior, and alcohol abuse, among others, have all been linked to higher risks of developing cancer. The greatly increasing rates of cancer over the past few decades have been decried as a public health epidemic. But aside from the cost in human life, the rising tide of cancer is causing a major crisis in today's society—how to pay for it all.

Despite increasing cure rates, effective new drugs, and novel treatments for various types of cancer, the costs of treatment continue to skyrocket. Nowhere is this more apparent than in the pharmaceutical industry. For example, new cancer drugs, while effective, are extremely expensive. Drug companies claim that it costs them between $500 million and $1 billion to bring a single new medicine to market. This cost may seem overblown, especially when you consider that the National Cancer Institute funds basic research into cancer biology and that drug companies often benefit indirectly from the findings. But the drug companies tell us that they need one successful drug to pay for the many drugs they try to develop that do not pay off. Still, it does seem as if successful drug companies try to keep lower-cost competitors out of the market. The question of how much drug companies can charge for drugs and who should pay for them is a thorny one. If drug companies don't show a profit, they may go out of business and there will be no new drugs. The same is true for insurance companies if they can't raise the cost of insurance to pay for expensive drugs. If the government buys drugs for Medicare patients, taxes may go up dramatically.

But how should the cost of treatment be met? Cancer is an illness that can be the direct result of poor lifestyle choices, but it can also occur in otherwise healthy individuals who make proper choices. And with increasing life spans, the incidence of cancer can only be expected to increase in future years. Should people who develop cancer due to poor lifestyle choices be held fully or partly responsible for paying for treatment? And if so, how? And how should the cost of developing new drugs and treatments be borne? There are no easy answers for any of these questions, but as cancer continues to extract a high toll in both human life and financial resources, future generations may face some difficult choices.

## *Biology* website

The companion website for *Biology* provides a wealth of information organized and integrated by chapter. You will find practice tests, animations, videos, and much more that will complement your learning and understanding of general biology.

**http://www.mhhe.com/maderbiology10**

# 10

# Meiosis and Sexual Reproduction

*Nanu Ram Jogi, at 90 years old, recently became the world's oldest new father. As he hoisted his newborn daughter into the air amid a throng of cameras, microphones, and reporters, he boasted that he plans to continue fathering children with his wife, Saburi, now 50, until he is 100. He cannot even recall how many children he has fathered over the many years of his life, but it is estimated that he has at least twelve sons, nine daughters, and twenty grandchildren. Extreme cases such as this remind us of the huge reproductive potential of most species.*

*This chapter discusses meiosis, the process that occurs during sexual reproduction and ensures that offspring will have a different combination of genes compared to their parents. Occasionally, offspring inherit a detrimental number of genes and chromosomes. Such events do not detract from the principle that genetic variations are essential to the survival of species, because they allow them to evolve and become adapted to an ever-changing environment.*

Nanu Ram Jogi, the world's oldest father.

## concepts

# 10.1 Halving the Chromosome Number

In sexually reproducing organisms, **meiosis** [Gk. *mio,* less, and *-sis,* act or process of] is the type of nuclear division that reduces the chromosome number from the diploid (2n) number [Gk. *diplos,* twofold, and *-eides,* like] to the haploid (n) number [Gk. *haplos,* single, and *-eides,* like]. The **diploid (2n) number** refers to the total number of chromosomes. The **haploid (n) number** of chromosomes is half the diploid number. In humans, the diploid number of 46 is reduced to the haploid number of 23. **Gametes** (reproductive cells, often the sperm and egg) usually have the haploid number of chromosomes. Gamete formation and then fusion of gametes to form a cell called a zygote are integral parts of **sexual reproduction.** A **zygote** always has the full or diploid (2n) number of chromosomes. In plants and animals, the zygote undergoes development to become an adult organism.

Obviously, if the gametes contained the same number of chromosomes as the body cells, the number of chromosomes would double with each new generation. Within a few generations, the cells of an animal would be nothing but chromosomes! For example, in humans with a diploid number of 46 chromosomes, in five generations the chromosome number would increase to 1,472 chromosomes ($46 \times 2^5$). In 10 generations this number would increase to a staggering 47,104 chromosomes ($46 \times 2^{10}$). The early cytologists (biologists who study cells) realized this, and Pierre-Joseph van Beneden, a Belgian, was gratified to find in 1883 that the sperm and the egg of the roundworm *Ascaris* each contain only two chromosomes, while the zygote and subsequent embryonic cells always have four chromosomes.

## Homologous Pairs of Chromosomes

In diploid body cells, the chromosomes occur in pairs. Figure 10.1*a,* a pictorial display of human chromosomes, shows the chromosomes arranged according to pairs. The members of each pair are called homologous chromosomes. **Homologous chromosomes** or **homologues** [Gk. *homologos,* agreeing, corresponding] look alike; they have the same length and centromere position. When stained, homologues have a similar banding pattern because they contain genes for the same traits in the same order. But while homologous chromosomes have genes for the same traits, such as finger length, the gene on one homologue may code for short fingers and the gene at the same location on the other homologue may code for long fingers. Alternate forms of a gene (as for long fingers and short fingers) are called **alleles.**

The chromosomes in Figure 10.1*a* are duplicated as they would be just before nuclear division. Recall that during the S stage of the cell cycle, DNA replicates and the chromosomes become duplicated. The results of the duplication process are depicted in Figure 10.1*b.* When duplicated, a chromosome is composed of two identical parts called sister chromatids, each containing one DNA double helix molecule. The sister chromatids are held together at a region called the centromere.

Why does the zygote have two chromosomes of each kind? One member of a homologous pair was inherited from the male parent, and the other was inherited from the female parent by way of the gametes. In Figure 10.1*b* and throughout the chapter, the paternal chromosome is colored blue, and the maternal chromosome is colored red. *Therefore, you should use length and centromere location, not color, to recognize homologues.* We will see how meiosis reduces the chromosome number. Whereas the zygote and body cells have homologous pairs of chromosomes, the gametes have only one chromosome of each kind—derived from either the paternal or maternal homologue.

## Overview of Meiosis

Meiosis requires two nuclear divisions and produces four haploid daughter cells, each having one of each kind of chromosome. Replication occurs only once and the daughter cells have half the total number of chromosomes as were in the diploid parent nucleus. The daughter cells receive one of each kind of parental chromosome, but in different combinations. Therefore, the daughter cells are not genetically identical to the parent cell or to each other.

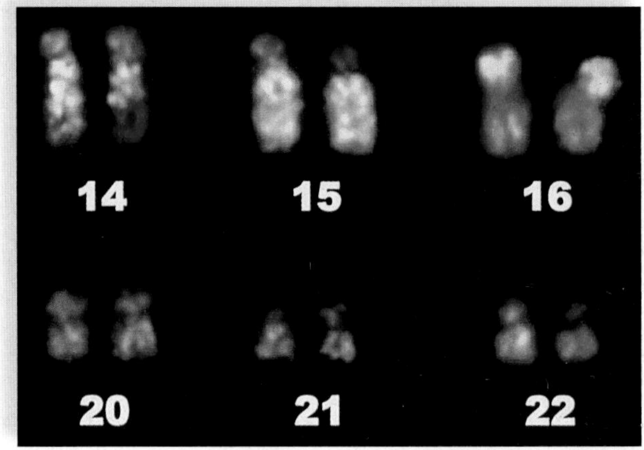


a.

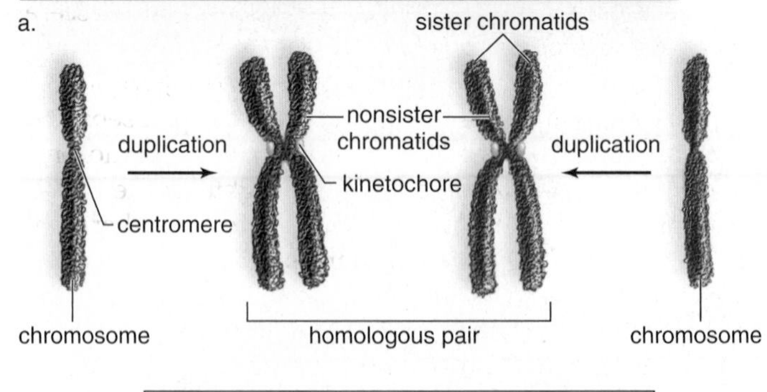


b.

**FIGURE 10.1 Homologous chromosomes.**

In diploid body cells, the chromosomes occur in pairs called homologous chromosomes. **a.** In this micrograph of stained chromosomes from a human cell, the pairs have been numbered. **b.** These chromosomes are duplicated, and each one is composed of two chromatids. The sister chromatids contain the exact same genes; the nonsister chromatids contain genes for the same traits (e.g., type of hair, color of eyes), but they may differ in that one could "call for" dark hair and eyes and the other for light hair and eyes.

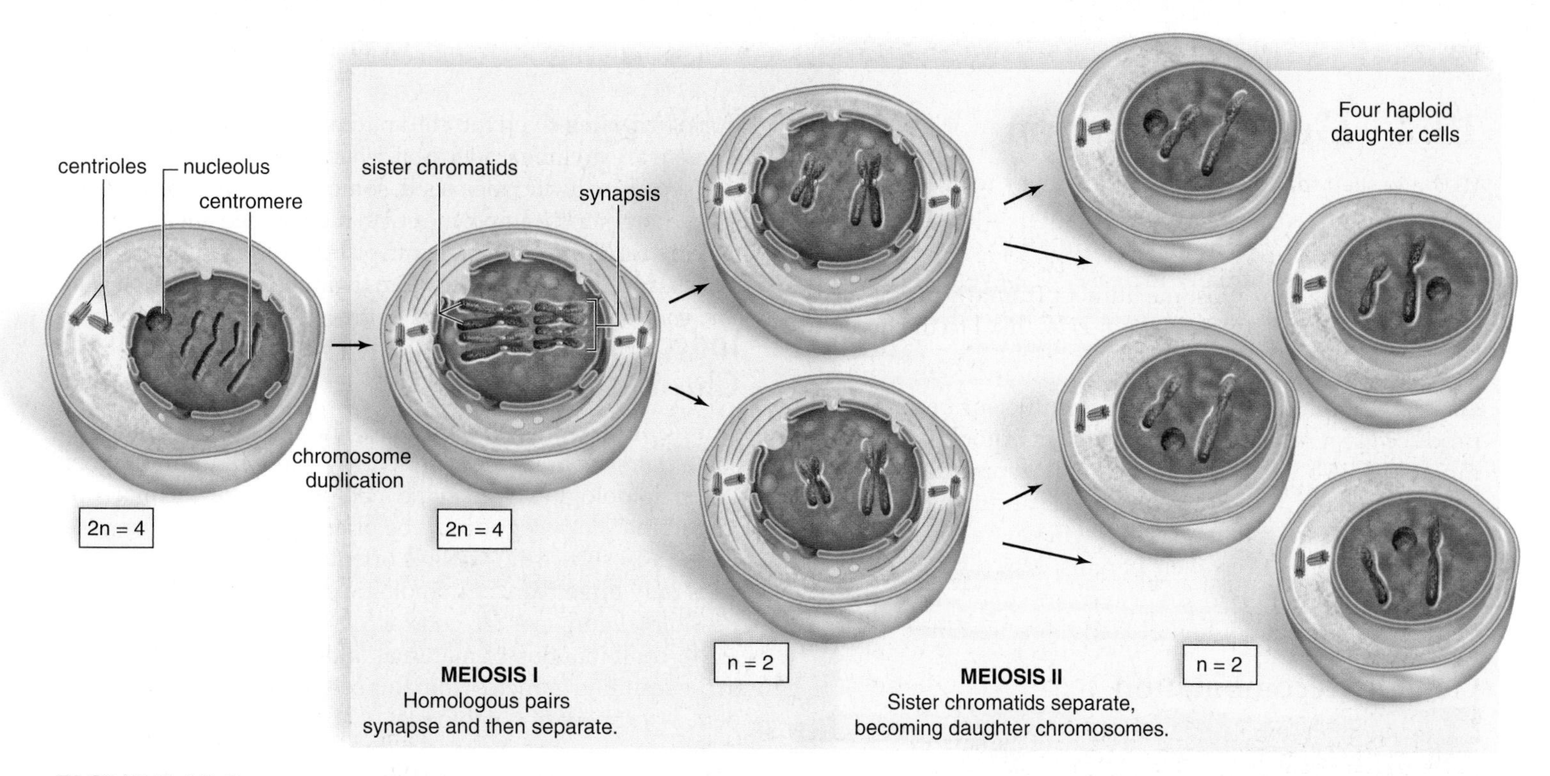


**FIGURE 10.2 Overview of meiosis.**

Following DNA replication, each chromosome is duplicated and consists of two chromatids. During meiosis I, homologous chromosomes pair and separate. During meiosis II, the sister chromatids of each duplicated chromosome separate. At the completion of meiosis, there are four haploid daughter cells. Each daughter cell has one of each kind of chromosome.

Figure 10.2 presents an overview of meiosis, indicating the two cell divisions, meiosis I and meiosis II. Prior to meiosis I, DNA (deoxyribonucleic acid) replication has occurred; therefore, each chromosome has two sister chromatids. During meiosis I, something new happens that does not occur in mitosis. The homologous chromosomes come together and line up side by side due to a means of attraction still unknown. This process is called **synapsis** [Gk. *synaptos*, united, joined together] and results in a **bivalent** [L. *bis*, two, and *valens*, strength]—that is, two homologous chromosomes that stay in close association during the first two phases of meiosis I. Sometimes the term tetrad [Gk. *tetra*, four] is used instead of bivalent because, as you can see, a bivalent contains four chromatids.

Following synapsis, homologous pairs align at the metaphase plate, and then the members of each pair separate. This separation means that only one duplicated chromosome from each homologous pair reaches a daughter nucleus. It is important for each daughter nucleus to have a member from each pair of homologous chromosomes because only in that way can there be a copy of each kind of chromosome in the daughter nuclei. Notice in Figure 10.2 that two possible combinations of chromosomes in the daughter cells are shown: short red with long blue and short blue with long red. Knowing that all daughter cells have to have one short chromosome and one long chromosome, what are the other two possible combinations of chromosomes for these particular cells?

Notice that replication occurs only once during meiosis; no replication of DNA is needed between meiosis I and meiosis II because the chromosomes are still duplicated; they already have two sister chromatids. During meiosis II, the sister chromatids separate, becoming daughter chromosomes that move to opposite poles. The chromosomes in the four daughter cells contain only one DNA double helix molecule because they are not duplicated.

The number of centromeres can be counted to verify that the parent cell has the diploid number of chromosomes. At the end of meiosis I, the chromosome number has been reduced because there are half as many centromeres present, even though each chromosome still consists of two chromatids each. Each daughter cell that forms has the haploid number of chromosomes. At the end of meiosis II, sister chromatids separate, and each daughter cell that forms still contains the haploid number of chromosomes, each consisting of a single chromatid.

## Fate of Daughter Cells

In the plant life cycle, the daughter cells become haploid spores that germinate to become a haploid generation. This generation produces the gametes by mitosis. The plant life cycle is studied in Chapter 24. In the animal life cycle, the daughter cells become the gametes, either sperm or eggs. The body cells of an animal normally contain the diploid number of chromosomes due to the fusion of sperm and egg during fertilization. If meiotic events go wrong, the gametes can contain the wrong number of chromosomes or altered chromosomes. This possibility and its consequences are discussed on pages 180–85.

### Check Your Progress 10.1

1. Define what is meant by a homologous pair of chromosomes.
2. How does chromosome sorting in meiosis I differ from mitosis?

# 10.2 Genetic Variation

We have seen that meiosis provides a way to keep the chromosome number constant generation after generation. Without meiosis, the chromosome number of the next generation would continually increase. The events of meiosis also help ensure that genetic variation occurs with each generation. Asexually reproducing organisms, such as the prokaryotes, depend primarily on mutations to generate variation among offspring. This is sufficient because they produce great numbers of offspring very quickly. Although mutations also occur among sexually reproducing organisms, the reshuffling of genetic material during sexual reproduction ensures that offspring will have a different combination of genes from their parents. Meiosis brings about genetic variation in two key ways: crossing-over and independent assortment of homologous chromosomes.

## Genetic Recombination

**Crossing-over** is an exchange of genetic material between nonsister chromatids of a bivalent during meiosis I. It is estimated that an average of two or three crossovers occur per human chromosome. At synapsis, homologues line up side by side, and a nucleoprotein lattice appears between them (Fig. 10.3). This lattice holds the bivalent together in such a way that the DNA of the duplicated chromosomes of each homologue pair is aligned. This ensures that the genes contained on the nonsister chromatids are directly aligned. Now crossing-over may occur. As the lattice breaks down, homologues are temporarily held together by *chiasmata* (sing., chiasma), regions where the nonsister chromatids are attached due to crossing-over. Then homologues separate and are distributed to different daughter cells. Crossing-over has been shown to be essential for the normal segregation of chromosomes during meiosis. For example, reduced levels of crossing-over have been linked to Down syndrome, which is caused by an extra copy of chromosome 21.

To appreciate the significance of crossing-over, it is necessary to remember that the members of a homologous pair can carry slightly different instructions for the same genetic traits. In the end, due to a swapping of genetic material during crossing-over, the chromatids held together by a centromere are no longer identical. Therefore, when the chromatids separate during meiosis II, some of the daughter cells receive daughter chromosomes with recombined alleles. Due to **genetic recombination,** the offspring have a different set of alleles, and therefore genes, than their parents.

## Independent Assortment of Homologous Chromosomes

During **independent assortment,** the homologous chromosome pairs separate independently, or in a random manner. When homologues align at the metaphase plate, the maternal or paternal homologue may be oriented toward either pole. Figure 10.4 shows the possible orientations for a cell that contains only three pairs of homologous chromosomes. Once all possible alignments are considered, the result will be $2^3$, or eight, combinations of maternal and paternal chromosomes in the resulting gametes from this cell, simply due to independent assortment of homologues.

## Significance of Genetic Variation

In humans, who have 23 pairs of chromosomes, the possible chromosomal combinations in the gametes is a staggering $2^{23}$, or 8,388,608. The variation that results from meiosis is enhanced by **fertilization,** the union of the male and female gametes. The chromosomes donated by the parents are combined, and in humans, this means that there are $(2^{23})^2$, or 70,368,744,000,000, chromosomally different zygotes possible, even assuming no crossing-over. If crossing-over occurs once, then $(4^{23})^2$, or 4,951,760,200,000,000,000,000,000,000, genetically different zygotes are possible for every couple. Keep in mind that crossing-over can occur several times in each chromosome!

The staggering amount of genetic variation achieved through meiosis is particularly important to the long-term survival of a species because it increases genetic variation within a population. (While asexual reproduction passes on exactly the same combination of chromosomes and genes.) The process of sexual reproduction brings about genetic re-

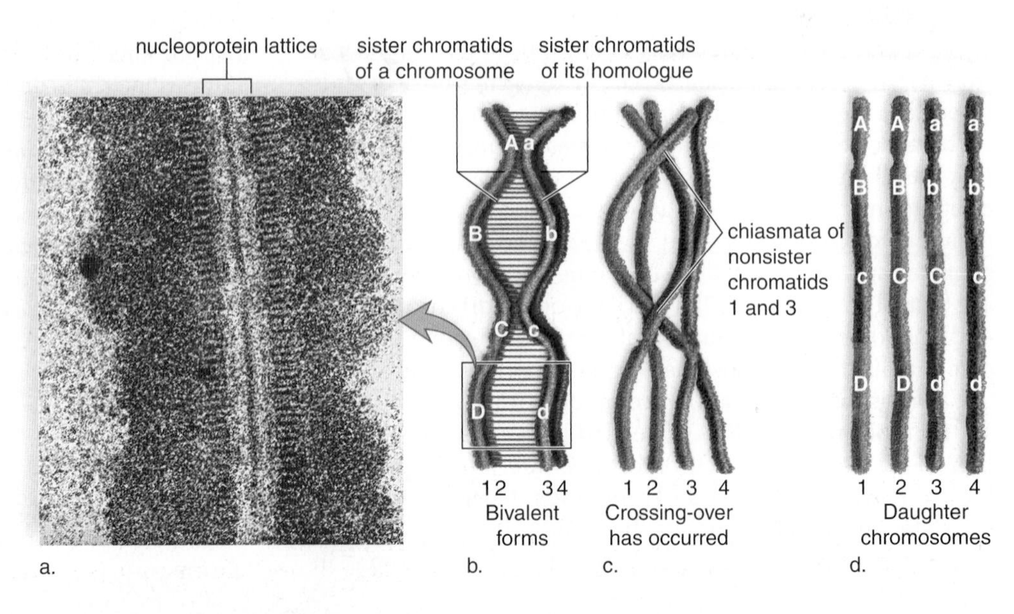


**FIGURE 10.3 Crossing-over occurs during meiosis I.**

**a.** The homologous chromosomes pair up, and a nucleoprotein lattice develops between them. This is an electron micrograph of the lattice. It zippers the members of the bivalent together so that corresponding genes are in alignment. **b.** This diagrammatic representation shows only two places where nonsister chromatids 1 and 3 have come into contact. Actually, the other two nonsister chromatids most likely are also crossing-over. **c.** Chiasmata indicate where crossing-over has occurred. The exchange of color represents the exchange of genetic material. **d.** Following meiosis II, daughter chromosomes have a new combination of genetic material due to crossing-over, which occurred between nonsister chromatids during meiosis I.

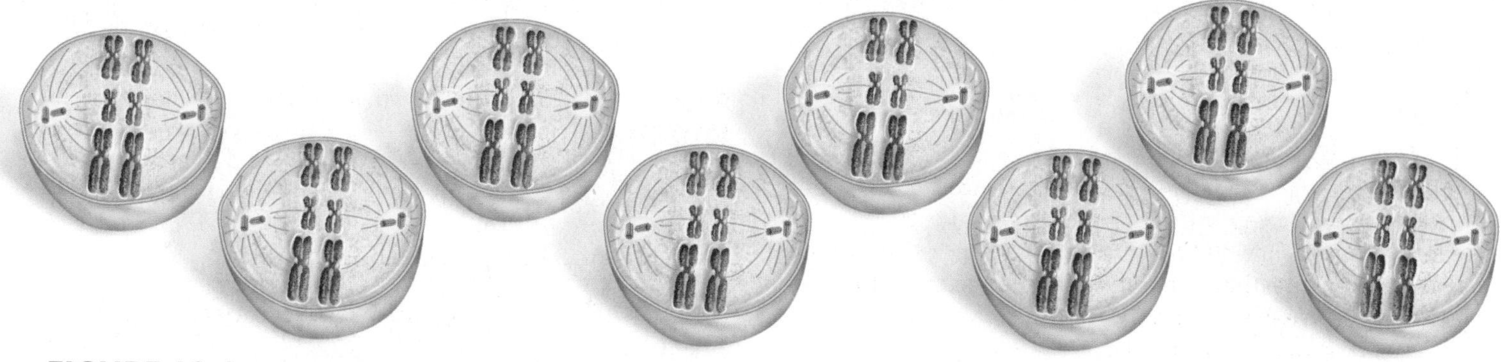

**FIGURE 10.4 Independent assortment.**

When a parent cell has three pairs of homologous chromosomes, there are $2^3$, or 8, possible chromosome alignments at the metaphase plate due to independent assortment. Among the 16 daughter nuclei resulting from these alignments, there are 8 different combinations of chromosomes.

combinations among members of a population (Fig. 10.5). If a parent is already successful in a particular environment, is asexual reproduction advantageous? It would seem so as long as the environment remains unchanged. However, if the environment changes, genetic variability among offspring introduced by sexual reproduction may be advantageous. Under the new conditions, some offspring may have a better chance of survival and reproductive success than others in a population. For example, suppose the ambient temperature were to rise due to global warming. Perhaps a dog with genes for the least amount of fur may have an advantage over other dogs of its generation.

In a changing environment, sexual reproduction, with its reshuffling of genes due to meiosis and fertilization, might give a few offspring a better chance of survival.

**Check Your Progress 10.2**

1. Briefly describe the two main ways in which meiosis contributes to genetic variation.
2. In a cell with four pairs of homologous chromosomes, how many combinations of chromosomes are possible in the gametes?
3. Why are meiosis and sexual reproduction important in responding to the changing environment?

**FIGURE 10.5 Genetic variation.**

Why do the puppies in this litter have a different appearance even though they have the same two parents? Because crossing-over and independent assortment occurred during meiosis, and fertilization brought different gametes together.

# 10.3 The Phases of Meiosis

Meiosis consists of two unique cell divisions, meiosis I and meiosis II. The phases of both meiosis I and meiosis II—prophase, metaphase, anaphase, and telophase—are described.

## Prophase I

It is apparent during prophase I that nuclear division is about to occur because a spindle forms as the centrosomes migrate away from one another. The nuclear envelope fragments, and the nucleolus disappears.

The homologous chromosomes, each having two sister chromatids, undergo synapsis to form bivalents. As depicted in Figure 10.3 by the exchange of color, crossing-over between the nonsister chromatids may occur at this time. After crossing-over, the sister chromatids of a duplicated chromosome are no longer identical.

Throughout prophase I, the chromosomes have been condensing so that by now they have the appearance of metaphase chromosomes.

## Metaphase I

During metaphase I, the bivalents held together by chiasmata (see Fig. 10.3) have moved toward the metaphase plate (equator of the spindle). Metaphase I is characterized by a fully formed spindle and alignment of the bivalents at the metaphase plate. As in mitosis, kinetochores are seen, but the two kinetochores of a duplicated chromosome are attached to the same kinetochore spindle fiber.

Bivalents independently align themselves at the metaphase plate of the spindle. The maternal homologue of each bivalent may be oriented toward either pole, and the paternal homologue of each bivalent may be oriented toward either pole. The orientation of one bivalent is not dependent on the orientation of the other bivalents. This contributes to the genetic variability of the daughter cells because all possible combinations of chromosomes can occur in the daughter cells.

## Anaphase I

During anaphase I, the homologues of each bivalent separate and move to opposite poles, but sister chromatids do not separate. Therefore, each chromosome still has two chromatids (see Fig. 10.6, page 174).

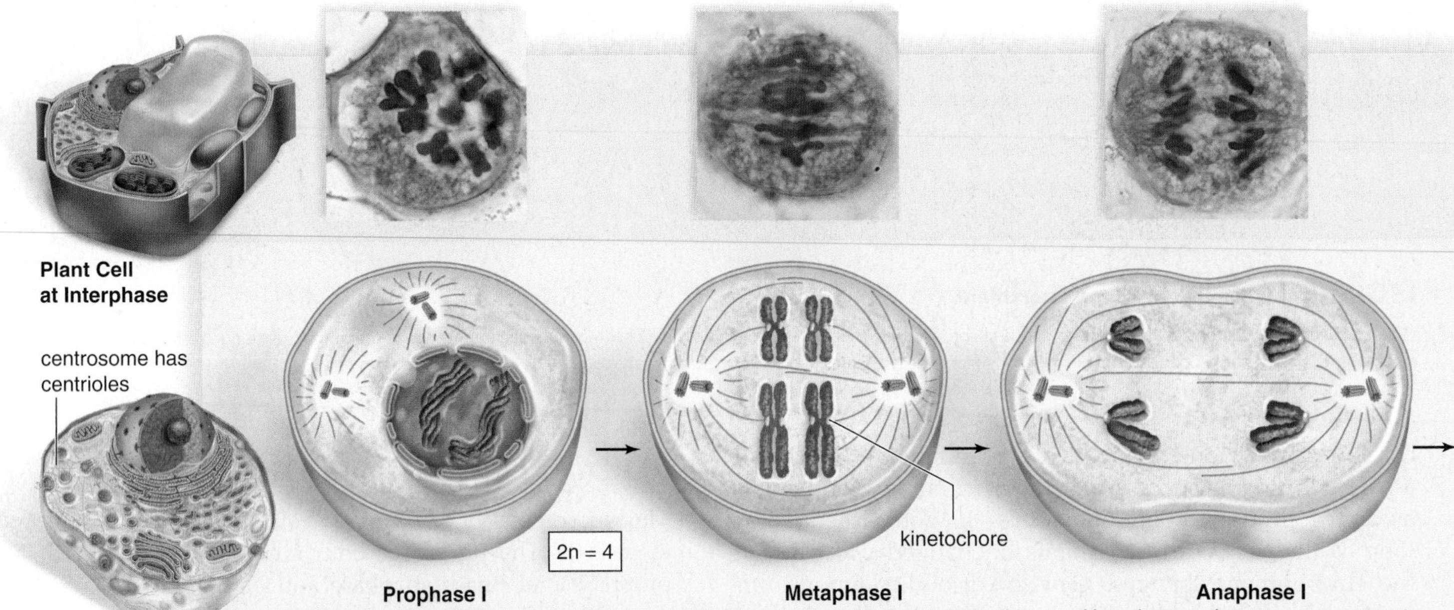


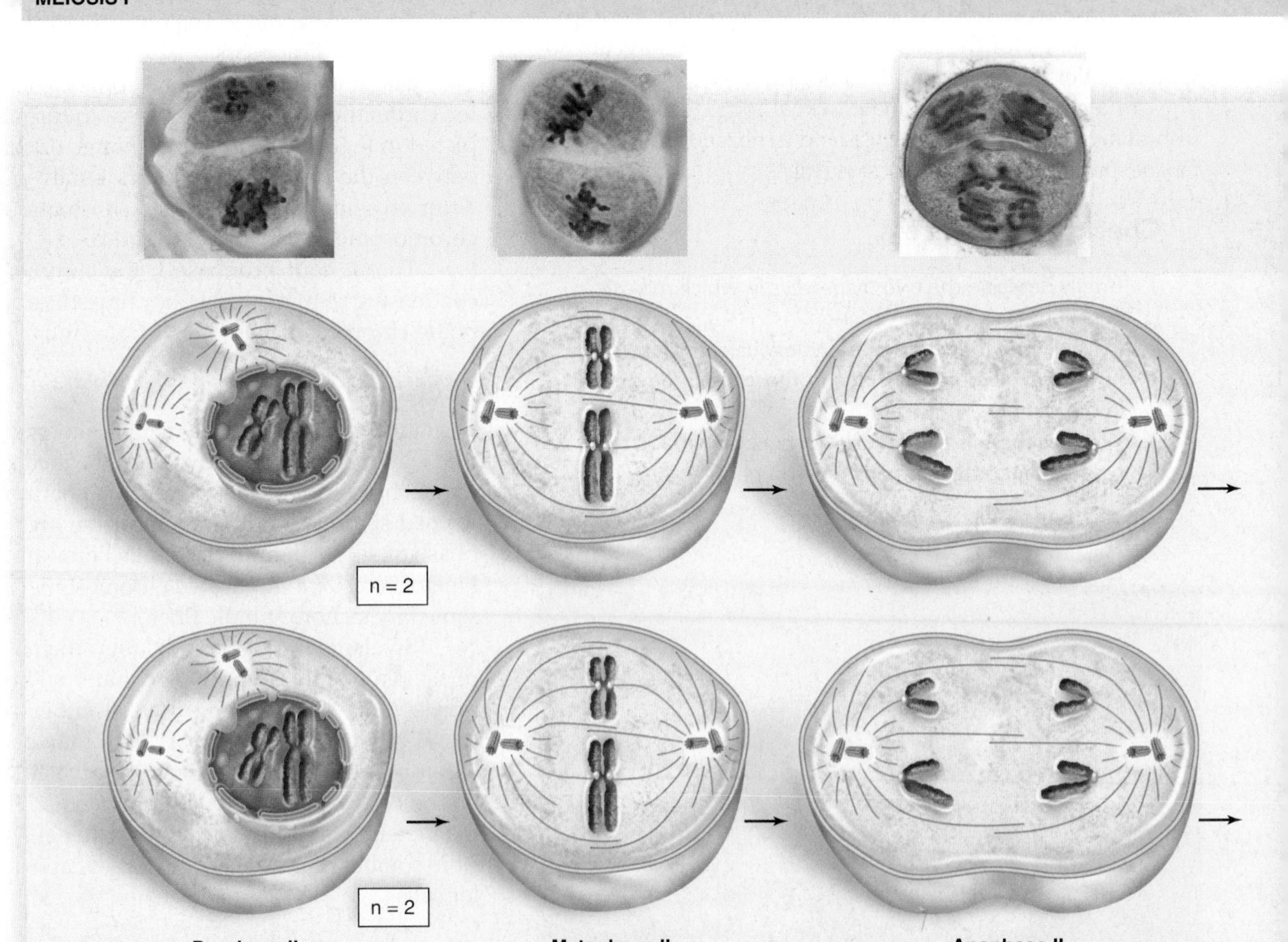


**FIGURE 10.6 Meiosis I and II in plant cell micrographs and animal cell drawings.**

When homologous chromosomes pair during meiosis I, crossing-over occurs as represented by the exchange of color. Pairs of homologous chromosomes separate during meiosis I, and chromatids separate, becoming daughter chromosomes during meiosis II. Following meiosis II, there are four haploid daughter cells.

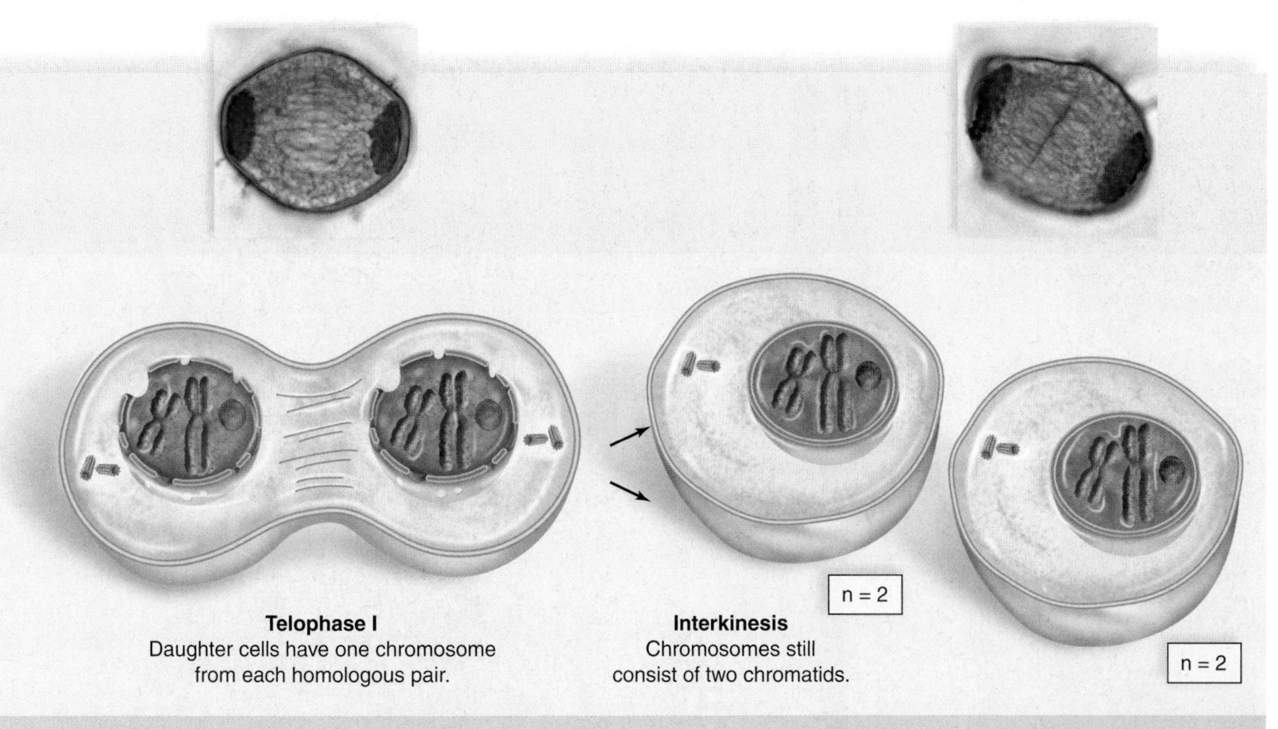
n = 2
Telophase I
Daughter cells have one chromosome from each homologous pair.
Interkinesis
Chromosomes still consist of two chromatids.
n = 2
MEIOSIS I cont'd

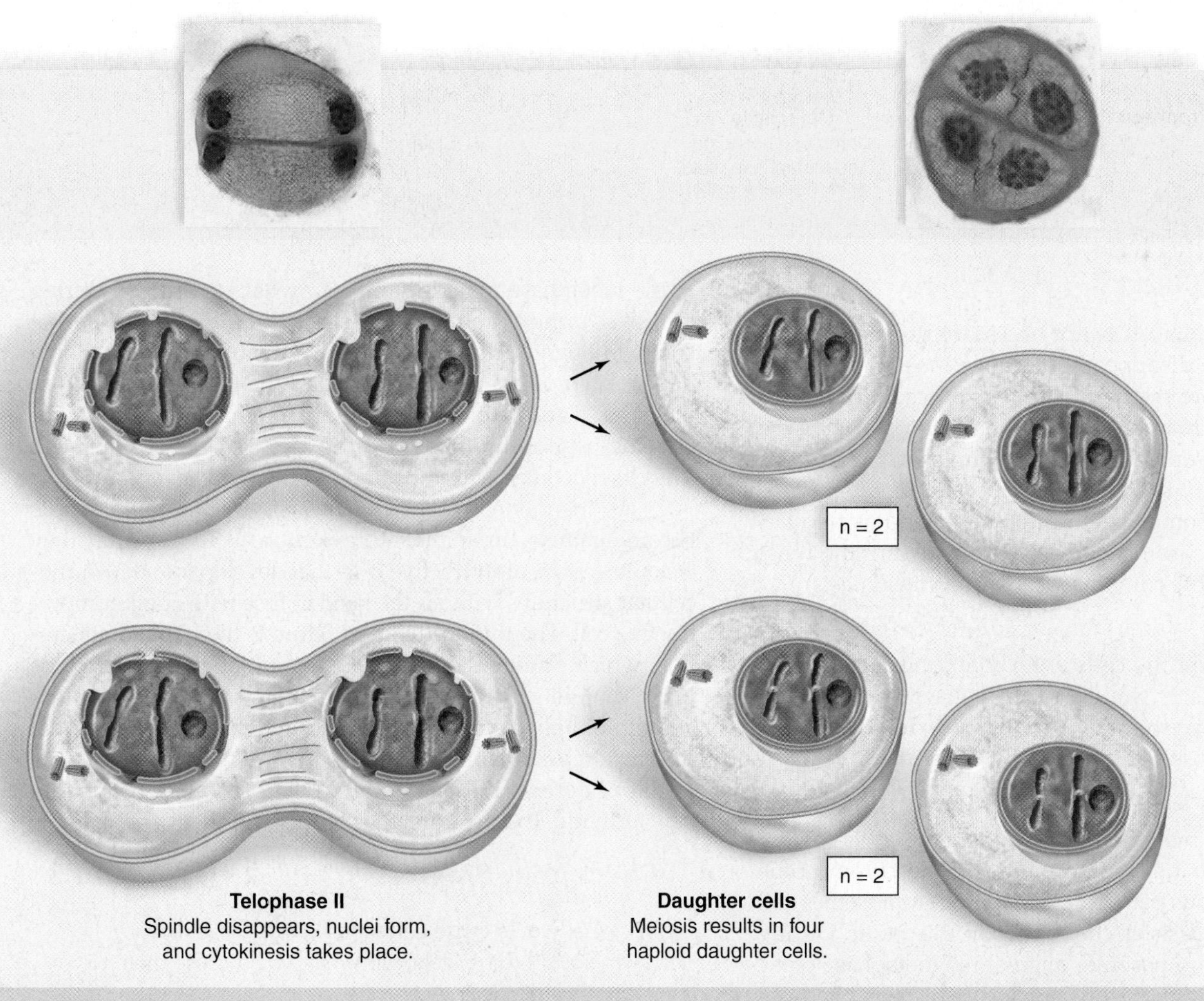
n = 2
n = 2
Telophase II
Spindle disappears, nuclei form, and cytokinesis takes place.
Daughter cells
Meiosis results in four haploid daughter cells.
MEIOSIS II cont'd

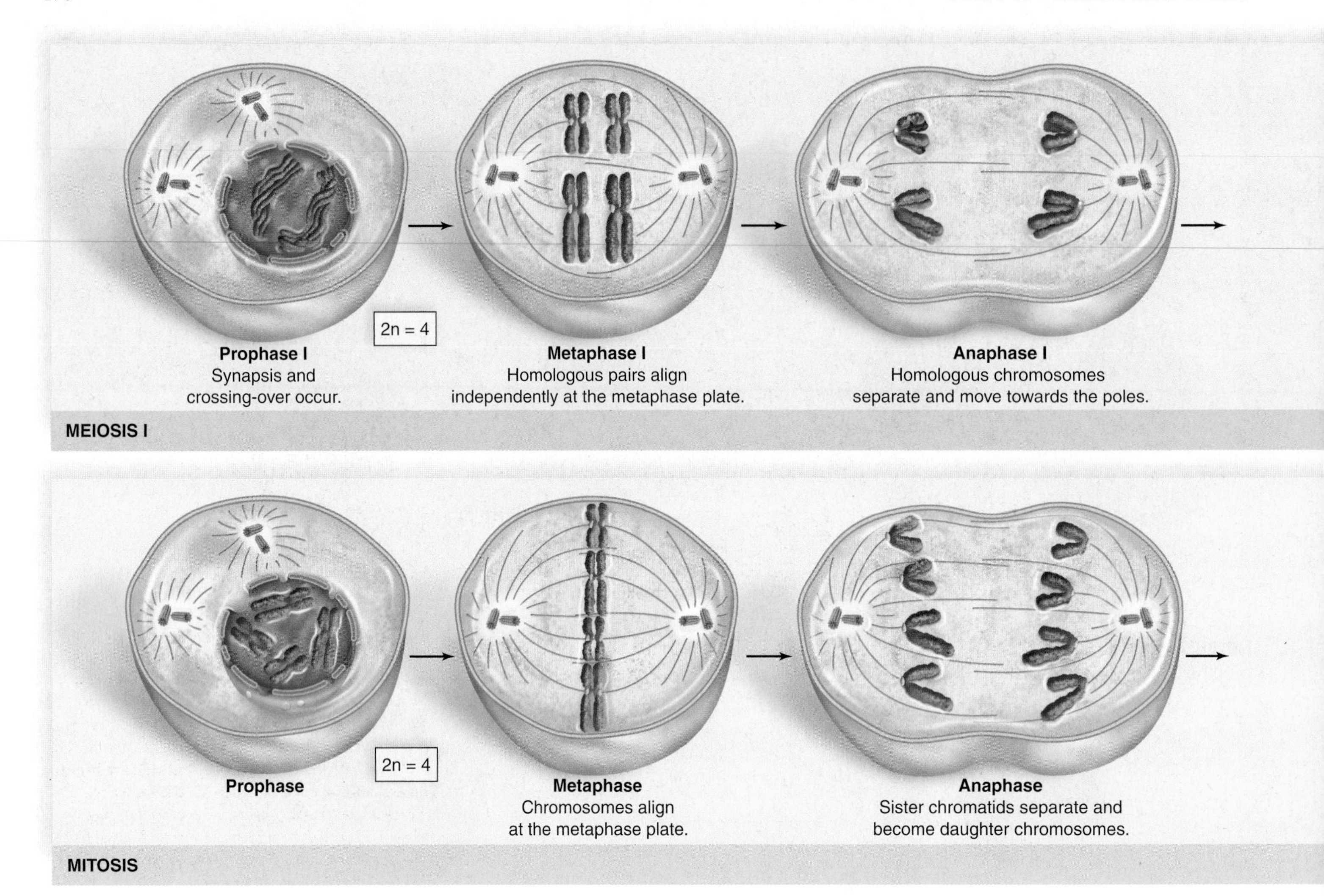


## Telophase I

Completion of telophase I is not necessary during meiosis. That is, the spindle disappears, but new nuclear envelopes need not form before the daughter cells proceed to meiosis II. Also, this phase may or may not be accompanied by cytokinesis, which is separation of the cytoplasm. Figure 10.6, page 174, shows only two of the four possible combinations of haploid chromosomes when the parent cell has two homologous pairs of chromosomes. Can you determine what the other two possible combinations of chromosomes are?

## Interkinesis

Following telophase, the cells enter interkinesis. The process of **interkinesis** is similar to interphase between mitotic divisions except that DNA replication does not occur because the chromosomes are already duplicated.

## Meiosis II and Gamete Formation

At the beginning of meiosis II, the two daughter cells contain the haploid number of chromosomes, or one chromosome from each homologous pair. Note that these chromosomes still consist of duplicated sister chromatids at this point. During metaphase II, the chromosomes align at the metaphase plate, but do not align in homologous pairs as in meiosis I because only one chromosome of each homologous pair is present (see Fig. 10.6, page 174). Thus, the alignment of the chromosomes at the metaphase plate is similar to what is observed during mitosis. During anaphase II, the sister chromatids separate, becoming daughter chromosomes that are not duplicated. These daughter chromosomes move toward the poles. At the end of telophase II and cytokinesis, there are four haploid cells. Due to crossing-over of chromatids during meiosis I, each gamete will most likely contain chromosomes with varied genes.

As mentioned, following meiosis II, the haploid cells become gametes in animals (see Section 10.5). In plants, they become **spores,** reproductive cells that develop into new multicellular structures without the need to fuse with another reproductive cell. The multicellular structure is the haploid generation, which produces gametes. The resulting zygote develops into a diploid generation. Therefore, plants have both haploid and diploid phases in their life cycle, and plants are said to exhibit an **alternation of generations.** In most fungi and algae, the zygote undergoes meiosis, and the daughter cells develop into new individuals. Therefore, the organism is always haploid.

### Check Your Progress 10.3

1. What would cause certain daughter cells following meiosis II to be identical? What would cause them to not be identical?
2. How does interkinesis differ from interphase?

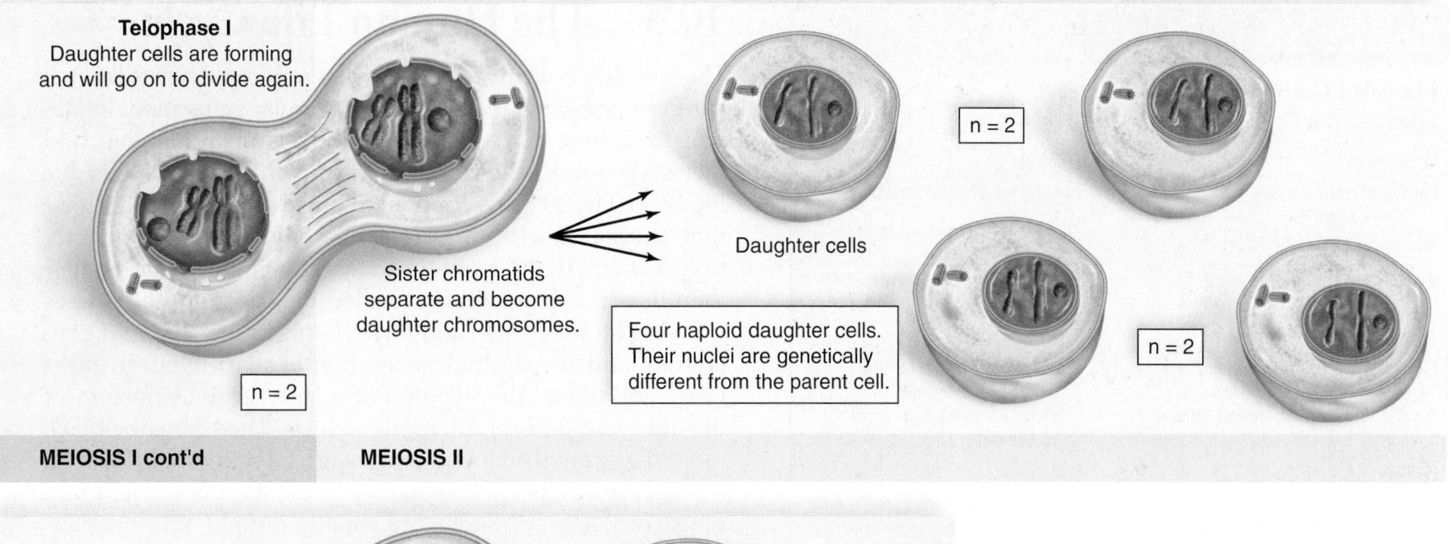


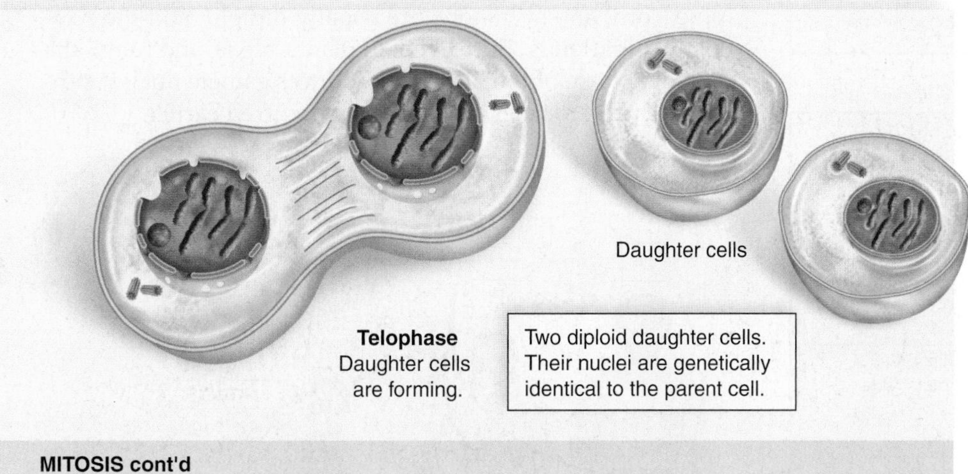


**FIGURE 10.7 Meiosis compared to mitosis.**

Why does meiosis produce daughter cells with half the number while mitosis produces daughter cells with the same number of chromosomes as the parent cell? Compare metaphase I of meiosis to metaphase of mitosis. Only in metaphase I are the homologous chromosomes paired at the metaphase plate. Members of homologous chromosome pairs separate during anaphase I, and therefore the daughter cells are haploid. The exchange of color between nonsister chromatids represents the crossing-over that occurred during meiosis I. The blue chromosomes were inherited from the paternal parent, and the red chromosomes were inherited from the maternal parent.

## 10.4 Meiosis Compared to Mitosis

Figure 10.7 graphically compares meiosis and mitosis. Several of the fundamental differences between the two processes include:

- Meiosis requires two nuclear divisions, but mitosis requires only one nuclear division.
- Meiosis produces four daughter nuclei. Following cytokinesis there are four daughter cells. Mitosis followed by cytokinesis results in two daughter cells.
- Following meiosis, the four daughter cells are haploid and have half the chromosome number as the diploid parent cell. Following mitosis, the daughter cells have the same chromosome number as the parent cell.
- Following meiosis, the daughter cells are neither genetically identical to each other or to the parent cell. Following mitosis, the daughter cells are genetically identical to each other and to the parent cell.

In addition to the fundamental differences between meiosis and mitosis, two specific differences between the two types of nuclear divisions can be categorized. These differences involve occurrence and process.

### Occurrence

Meiosis occurs only at certain times in the life cycle of sexually reproducing organisms. In humans, meiosis occurs only in the reproductive organs and produces the gametes. Mitosis is more common because it occurs in all tissues during growth and repair.

### Process

We will compare both meiosis I and meiosis II to mitosis.

#### *Meiosis I Compared to Mitosis*

Notice that these events distinguish meiosis I from mitosis:

- During prophase I, bivalents form and crossing-over occurs. These events do not occur during mitosis.
- During metaphase I of meiosis, bivalents independently align at the metaphase plate. The paired chromosomes have a total of four chromatids each. During metaphase in mitosis, individual chromosomes align at the metaphase plate. They each have two chromatids.
- During anaphase I of meiosis, homologues of each bivalent separate and duplicated chromosomes (with centromeres intact) move to opposite poles. During anaphase of mitosis, sister chromatids separate, becoming daughter chromosomes that move to opposite poles.

**TABLE 10.1**

**Meiosis I Compared to Mitosis**

| *Meiosis I* | *Mitosis* |
|---|---|
| *Prophase I* | *Prophase* |
| Pairing of homologous chromosomes | No pairing of chromosomes |
| *Metaphase I* | *Metaphase* |
| Bivalents at metaphase plate | Duplicated chromosomes at metaphase plate |
| *Anaphase I* | *Anaphase* |
| Homologues of each bivalent separate and duplicated chromosomes move to poles | Sister chromatids separate, becoming daughter chromosomes that move to the poles |
| *Telophase I* | *Telophase* |
| Two haploid daughter cells, not identical to the parent cell | Two diploid daughter cells, identical to the parent cell |

**TABLE 10.2**

**Meiosis II Compared to Mitosis**

| *Meiosis II* | *Mitosis* |
|---|---|
| *Prophase II* | *Prophase* |
| No pairing of chromosomes | No pairing of chromosomes |
| *Metaphase II* | *Metaphase* |
| Haploid number of duplicated chromosomes at metaphase plate | Diploid number of duplicated chromosomes at metaphase plate |
| *Anaphase II* | *Anaphase* |
| Sister chromatids separate, becoming daughter chromosomes that move to the poles | Sister chromatids separate, becoming daughter chromosomes that move to the poles |
| *Telophase II* | *Telophase* |
| Four haploid daughter cells, not genetically identical | Two diploid daughter cells, identical to the parent cell |

### *Meiosis II Compared to Mitosis*

The events of meiosis II are similar to those of mitosis except in meiosis II, the nuclei contain the haploid number of chromosomes. In mitosis, the original number of chromosomes is maintained. Meiosis II produces two daughter cells from each parent cell that completes meiosis I, for a total of four daughter cells. These daughter cells contain the same number of chromosomes as they did at the end of meiosis I. Tables 10.1 and 10.2 compare meiosis I and II to mitosis.

**Check Your Progress 10.4**

1. How does the alignment of chromosomes in metaphase I differ from the alignment of chromosomes in metaphase of mitosis?
2. How is meiosis II more similar to mitosis than to meiosis I? How does it differ?

## 10.5 The Human Life Cycle

The term **life cycle** refers to all the reproductive events that occur from one generation to the next similar generation. In animals, including humans, the individual is always diploid, and meiosis produces the gametes, the only haploid phase of the life cycle (Fig. 10.8). In contrast, plants have a haploid phase that alternates with a diploid phase. The haploid generation, known as the **gametophyte,** may be larger or smaller than the diploid generation, called the **sporophyte.** Mosses growing on bare rocks and forest floors are the haploid generation, and the diploid generation is short-lived. In most fungi and algae, the zygote is the only diploid portion of the life cycle, and it undergoes meiosis. Therefore, the black mold that grows on bread and the green scum that floats on a pond are haploid. The majority of plant species, including pine, corn, and sycamore, are usually diploid, and the haploid generation is short-lived. In plants, algae, and fungi, the haploid phase of the life cycle produces gamete nuclei without the need for meiosis because it occurred earlier.

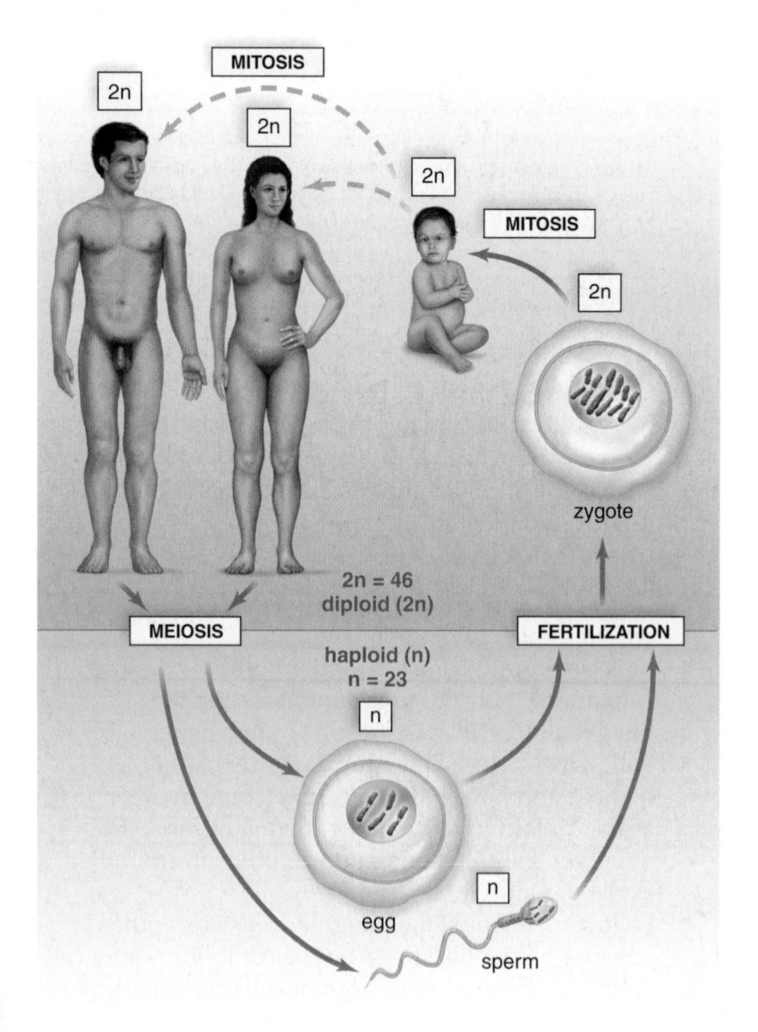


**FIGURE 10.8 Life cycle of humans.**

Meiosis in males is a part of sperm production, and meiosis in females is a part of egg production. When a haploid sperm fertilizes a haploid egg, the zygote is diploid. The zygote undergoes mitosis as it develops into a newborn child. Mitosis continues throughout life during growth and repair.

Animals are diploid and meiosis occurs during the production of gametes **(gametogenesis).** In males, meiosis is a part of **spermatogenesis** [Gk. *sperma*, seed; L. *genitus*, producing], which occurs in the testes and produces sperm. In females, meiosis is a part of **oogenesis** [Gk. *oon*, egg; L. *genitus*, producing], which occurs in the ovaries and produces eggs. A sperm and egg join at fertilization, restoring the diploid chromosome number. The resulting zygote undergoes mitosis during development of the fetus. After birth, mitosis is involved in the continued growth of the child and repair of tissues at any time.

## Spermatogenesis and Oogenesis in Humans

In human males, spermatogenesis occurs within the testes, and in females, oogenesis occurs within the ovaries. The testes contain stem cells called spermatogonia, and these cells keep the testes supplied with primary spermatocytes that undergo spermatogenesis as described in Figure 10.9, *top*. Primary spermatocytes with 46 chromosomes undergo meiosis I to form two secondary spermatocytes, each with 23 duplicated chromosomes. Secondary spermatocytes undergo meiosis II to produce four spermatids with 23 daughter chromosomes. Spermatids then differentiate into viable sperm (spermatozoa). Upon sexual arousal, the sperm enter ducts and exit the penis upon ejaculation.

The ovaries contain stem cells called oogonia that produce many primary oocytes with 46 chromosomes during fetal development. They even begin oogenesis, but only a few continue when a female is sexually mature. The result of meiosis I is two haploid cells with 23 chromosomes each (Fig. 10.9, *bottom*). One of these cells, termed the **secondary oocyte** [Gk, *oon*, egg, and *kytos*, cell], receives almost all the cytoplasm. The other is a **polar body** that may either disintegrate or divide again. The secondary oocyte begins meiosis II but stops at metaphase II. Then the secondary oocyte leaves the ovary and enters an oviduct, where sperm may be present. If no sperm are in the oviduct or one does not enter the secondary oocyte, it eventually disintegrates without completing meiosis. If a sperm does enter the oocyte, some of its contents trigger the completion of meiosis II in the secondary oocyte, and another polar body forms. At the completion of oogenesis, following entrance of a sperm, there is one egg and two to three polar bodies. The polar bodies are a way to dispose of chromosomes while retaining much of the cytoplasm in the egg. Cytoplasmic molecules are needed by a developing embryo following fertilization. Some zygote components, such as the centrosome, are contributed by the sperm.

The mature egg has 23 chromosomes, but the zygote formed when the sperm and egg nuclei fuse has 46 chromosomes. Therefore, fertilization restores the diploid number of chromosomes.

### Check Your Progress 10.5

1. Which cells in humans are capable of meiosis?
2. What is the benefit for one egg per oogenesis?

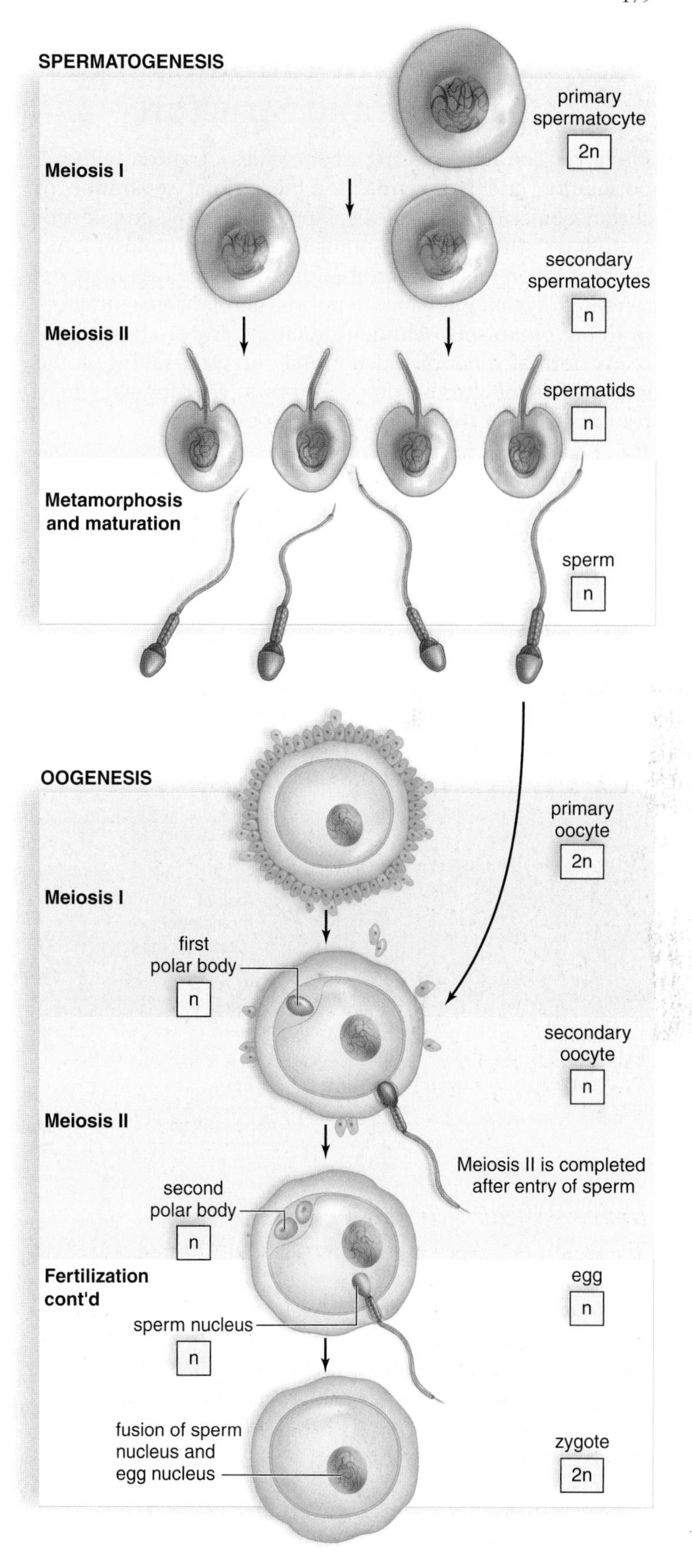


**FIGURE 10.9 Spermatogenesis and oogenesis in mammals.**

Spermatogenesis produces four viable sperm, whereas oogenesis produces one egg and at least two polar bodies. In humans, both sperm and egg have 23 chromosomes each; therefore, following fertilization, the zygote has 46 chromosomes.

# 10.6 Changes in Chromosome Number and Structure

We have seen that crossing-over creates variation within a population and is essential for the normal separation of chromosomes during meiosis. Furthermore, the proper separation of homologous chromosomes during meiosis I and the separation of sister chromatids during meiosis II are essential for the maintenance of normal chromosome numbers in living organisms. Although meiosis almost always proceeds normally, nondisjunction may occur, resulting in the gain or loss of chromosomes. Errors in crossing-over may result in extra or missing parts of chromosomes.

## Aneuploidy

The correct number of chromosomes in a species is known as **euploidy.** A change in the chromosome number resulting from nondisjunction during meiosis is called **aneuploidy.** Aneuploidy is seen in both plants and animals. Monosomy and trisomy are two aneuploid states.

**Monosomy** (2n − 1) occurs when an individual has only one of a particular type of chromosome, and **trisomy** (2n + 1) occurs when an individual has three of a particular type of chromosome. Both monosomy and trisomy are the result of nondisjunction, or the failure of chromosomes to separate normally during mitosis or meiosis. *Primary nondisjunction* occurs during meiosis I when both members of a homologous pair go into the same daughter cell (Fig. 10.10*a*). *Secondary nondisjunction* occurs during meiosis II when the sister chromatids fail to separate and both daughter chromosomes go into the same gamete (Fig. 10.10*b*). Notice that when secondary nondisjunction occurs, there are two normal gametes and two aneuploid gametes. However, when primary nondisjunction occurs, there are no normal gametes produced.

In animals, autosomal monosomies and trisomies are generally lethal, but a trisomic individual is more likely to survive than a monosomic one. In humans, only three autosomal trisomic conditions are known to be viable beyond birth: trisomy 13, 18, and 21. Only trisomy 21 is viable beyond early childhood, and is characterized by a distinctive set of physical and mental abnormalities. On the other hand, sex chromosome aneuploids are better tolerated in animals and have a better chance of producing survivors.

### *Trisomy 21*

The most common autosomal trisomy seen among humans is trisomy 21, also called Down syndrome. This syndrome is easily recognized by these characteristics: short stature; an eyelid fold; a flat face; stubby fingers; a wide gap between

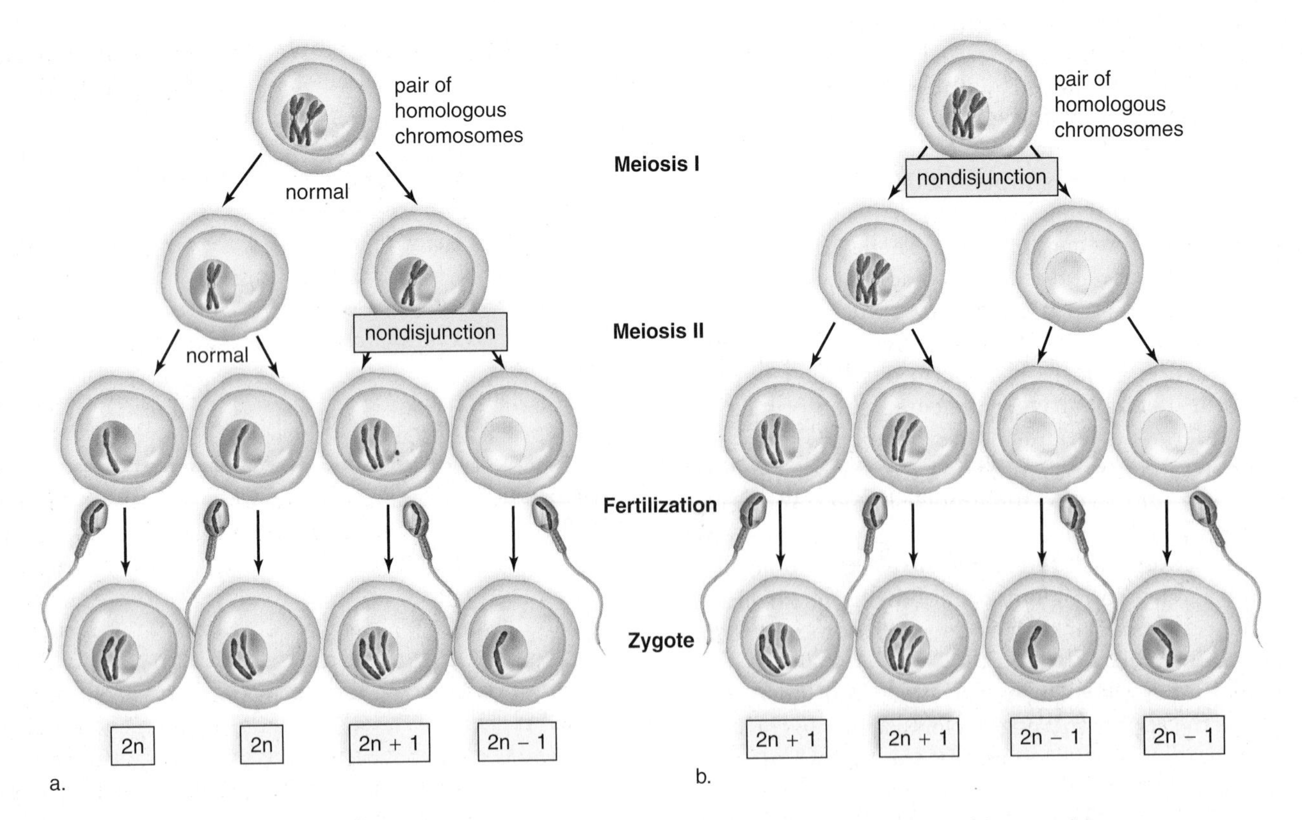


**FIGURE 10.10 Nondisjunction of chromosomes during oogenesis, followed by fertilization with normal sperm.**

**a.** Nondisjunction can occur during meiosis II if the sister chromatids separate but the resulting chromosomes go into the same daughter cell. Then the egg will have one more or one less than the usual number of chromosomes. Fertilization of these abnormal eggs with normal sperm produces an abnormal zygote with abnormal chromosome numbers. **b.** Nondisjunction can also occur during meiosis I and result in abnormal eggs that also have one more or one less than the normal number of chromosomes. Fertilization of these abnormal eggs with normal sperm results in a zygote with abnormal chromosome numbers. 2n = diploid number of chromosomes.

a.

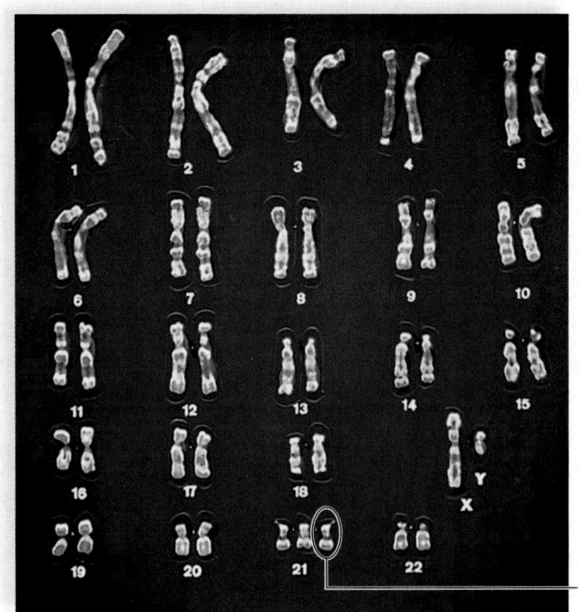


b.

**FIGURE 10.11 Trisomy 21.**

Persons with Down syndrome, or trisomy 21, have an extra chromosome 21. **a.** Common characteristics of the syndrome include a wide, rounded face and a fold on the upper eyelids. Mental disabilities, along with an enlarged tongue, make it difficult for a person with Down syndrome to speak distinctly. **b.** The karyotype of an individual with Down syndrome shows three copies of chromosome 21. Therefore, the individual has three copies instead of two copies of each gene on chromosome 21. Researchers are using new techniques to discover which genes on chromosome 21 are causing the syndrome's disabilities.

the first and second toes; a large, fissured tongue; a round head; a distinctive palm crease; heart problems; and, unfortunately, mental retardation, which can sometimes be severe. Individuals with Down syndrome also have a greatly increased risk of developing leukemia and tend to age rapidly, resulting in a shortened life expectancy. In addition, these individuals have an increased chance of developing Alzheimer disease later in life.

Many scientists agree that the symptoms of Down syndrome are caused by gene dosage effects resulting from the presence of the extra chromosome. However, recent studies indicate that not all of the genes on the chromosome are expressed at a level of 150%, challenging this theory. However, scientists have identified several genes that have been linked to increased risk of leukemia, cataracts, aging, and mental retardation.

The chances of a woman having a child with Down syndrome increase rapidly with age. In women age 20 to 30, 1 in 1,400 births have Down syndrome, and in women 30 to 35, about 1 in 750 births have Down syndrome. It is thought that the longer the oocytes are stored in the female, the greater the chances of nondisjunction occurring. However, even though an older woman is more likely to have a Down syndrome child, most babies with Down syndrome are born to women younger than age 40 because this is the age group having the most babies. Furthermore, some recent research also indicate that in 23% of the cases studied, the sperm contributed an extra chromosome. A **karyotype,** a visual display of the chromosomes arranged by size, shape, and banding pattern, may be performed to identify babies with Down syndrome and other aneuploid conditions (Fig. 10.11).

### *Changes in Sex Chromosome Number*

An abnormal sex chromosome number is the result of inheriting too many or too few X or Y chromosomes. Nondisjunction during oogenesis or spermatogenesis can result in gametes with an abnormal number of sex chromosomes. However, extra copies of the sex chromosomes are much more easily tolerated in humans than are extra copies of autosomes.

A person with Turner syndrome (XO) is a female, and a person with Klinefelter syndrome (XXY) is a male. However, deletion of the *SRY* gene on the short arm of the Y chromosome results in Swyer syndrome, or an "XY female." Individuals with Swyer syndrome lack a hormone called testis-determining factor, which plays a critical role in the development of male genitals. Furthermore, movement of this same gene onto the X chromosome may result in de la Chapelle syndrome, or an "XX male." Men with de la Chapelle syndrome exhibit undersized testes, sterility, and rudimentary breast development. Together, these observations suggest that in humans, the presence of the *SRY* gene, not the number of X chromosomes, determines maleness. In its absence, a person develops as a female.

Why are newborns with an abnormal sex chromosome number more likely to survive than those with an abnormal autosome number? Since females have two X chromosomes and males have only one, one would expect females to produce twice the amount of each gene from this chromosome, but both males and females produce roughly the same amount. In reality, both males and females only have one functioning X chromosome. In females, and in males with extra X chromosomes, the others become an inactive mass called a **Barr body,** named after Murray Barr, the person who discovered it. This provides a natural method for gene dosage compensation of the sex chromosomes and explains why extra sex chromosomes are more easily tolerated than extra autosomes.

**Turner Syndrome.** From birth, an XO individual with Turner syndrome has only one sex chromosome, an X; the O signifies the absence of a second sex chromosome (Fig. 10.12*a*). Therefore, the nucleus does not contain a Barr body. The approximate incidence is 1 in 10,000 females.

Turner females are short, with a broad chest and widely spaced nipples. These individuals also have a low posterior hairline and neck webbing. The ovaries, oviducts, and uterus are very small and underdeveloped. Turner females do not undergo puberty or menstruate, and their breasts do not develop. However, some have given birth following in vitro fertilization using donor eggs. They usually are of normal intelligence and can lead fairly normal lives if they receive hormone supplements.

**Klinefelter Syndrome.** A male with Klinefelter syndrome has two or more X chromosomes in addition to a Y chromosome (Fig. 10.12*b*). The extra X chromosomes become Barr bodies. The approximate incidence for Klinefelter syndrome is 1 in 500 to 1,000 males.

In Klinefelter males, the testes and prostate gland are underdeveloped and there is no facial hair. But there may be some breast development. Affected individuals have large hands and feet and very long arms and legs. They are usually slow to learn but not mentally retarded unless they inherit more than two X chromosomes. No matter how many X chromosomes there are, an individual with a Y chromosome is a male.

While males with Klinefelter syndrome exhibit no other major health abnormalities, there is an increased risk of some disorders, including breast cancer, osteoporosis, and lupus,

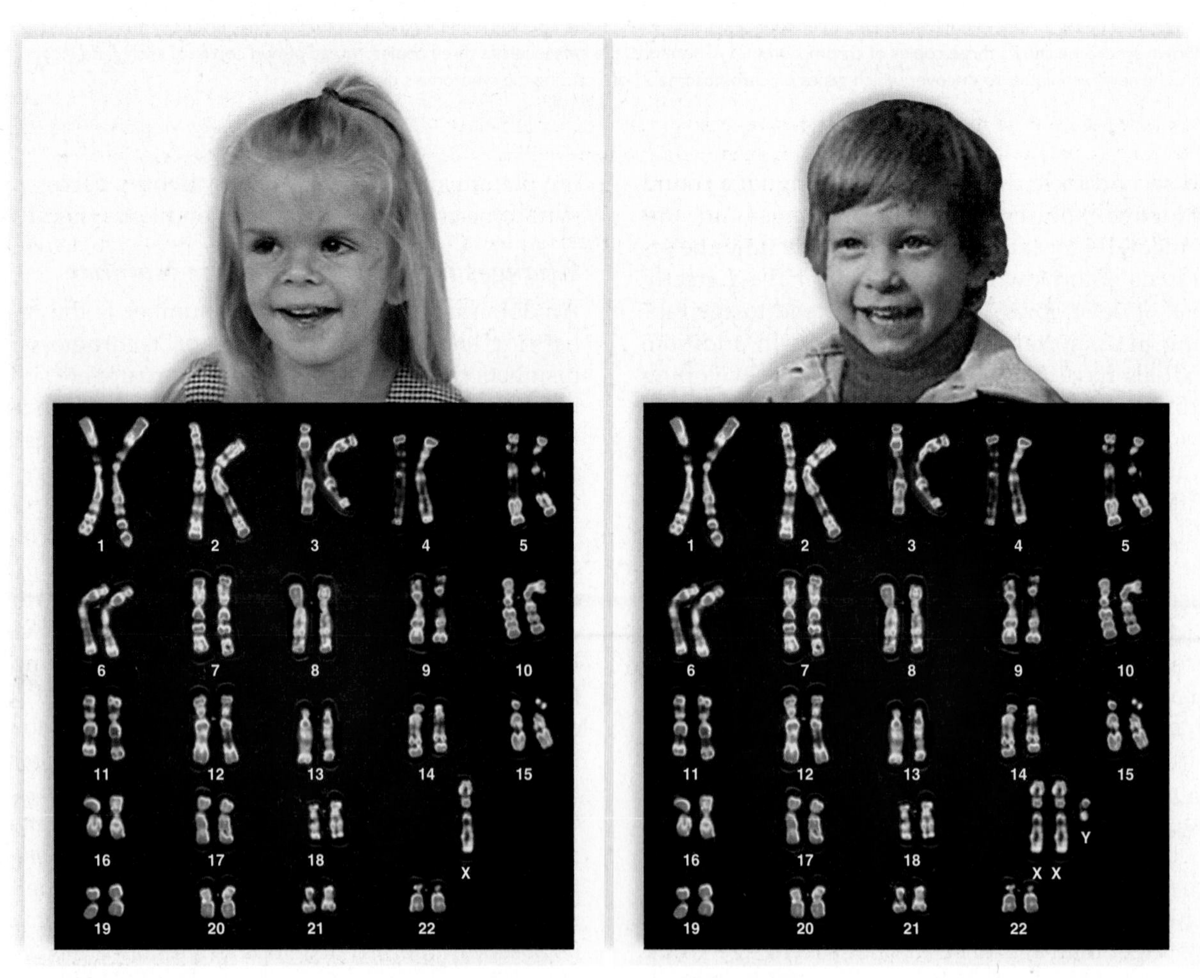


a. Turner syndrome b. Klinefelter syndrome

**FIGURE 10.12 Abnormal sex chromosome number.**

People with (**a**) Turner syndrome, who have only one sex chromosome, an X, as shown, and (**b**) Klinefelter syndrome, who have more than one X chromosome plus a Y chromosome, as shown, can look relatively normal (especially as children) and can lead relatively normal lives.

*health focus*

## Living with Klinefelter Syndrome

In 1996, at the age of 25, I was diagnosed with Klinefelter syndrome (KS). Being diagnosed has changed my life for the better.

I was a happy baby, but when I was still very young, my parents began to believe that there was something wrong with me. I knew something was different about me, too, as early on as five years old. I was very shy and had trouble making friends. One minute I'd be well behaved, and the next I'd be picking fights and flying into a rage. Many psychologists, therapists, and doctors tested me because of school and social problems and severe mood changes. Their only diagnosis was "learning disabilities" in such areas as reading comprehension, abstract thinking, word retrieval, and auditory processing. No one could figure out what the real problem was, and I hated the tutoring sessions I had. In the seventh grade, a psychologist told me that I was stupid and lazy, I would probably live at home for the rest of my life, and I would never amount to anything. For the next five years, he was basically right, and I barely graduated from high school.

I believe, though, that I have succeeded because I was told that I would fail. I quit the tutoring sessions when I enrolled at a community college; I decided I could figure things out on my own. I received an associate degree there, then transferred to a small liberal arts college. I never told anyone about my learning disabilities and never sought special help. However, I never had a semester below a 3.0, and I graduated with two B.S. degrees. I was accepted into a graduate program but decided instead to accept a job as a software engineer even though I did not have an educational background in this field. As I later learned, many KS'ers excel in computer skills. I had been using a computer for many years and had learned everything I needed to know on my own, through trial and error.

Around the time I started the computer job, I went to my physician for a physical. He sent me for blood tests because he noticed that my testes were smaller than usual. The results were conclusive: Klinefelter syndrome with sex chromosomes XXY. I initially felt denial, depression, and anger, even though I now had an explanation for many of the problems I had experienced all my life. But then I decided to learn as much as I could about the condition and treatments available. I now give myself a testosterone injection once every two weeks, and it has made me a different person, with improved learning abilities and stronger thought processes in addition to a more outgoing personality.

I found, though, that the best possible path I could take was to help others live with the condition. I attended my first support group meeting four months after I was diagnosed. By spring 1997, I had developed an interest in KS that was more than just a part-time hobby. I wanted to be able to work with this condition and help people forever. I have been very involved in KS conferences and have helped to start support groups in the United States, Spain, and Australia.

Since my diagnosis, it has been my dream to have a son with KS, although when I was diagnosed, I found out it was unlikely that I could have biological children. Through my work with KS, I had the opportunity to meet my fiancée Chris. She has two wonderful children: a daughter, and a son who has the same condition that I do. There are a lot of similarities between my stepson and me, and I am happy I will be able to help him get the head start in coping with KS that I never had. I also look forward to many more years of helping other people seek diagnosis and live a good life with Klinefelter syndrome.

*Stefan Schwarz*

which disproportionately affect females. Although men with Klinefelter syndrome typically do not need medical treatment, some have found that testosterone therapy may help increase muscle strength, sex drive, and concentration ability. Testosterone treatment, however, will not reverse the sterility associated with Klinefelter syndrome due to the incomplete testicle development associated with it.

The Health Focus on this page tells of the experiences of a person with Klinefelter syndrome. He suggests that it is best for parents to know right away that they have a child with this abnormality because much can be done to help the child lead a normal life.

**Poly-X Females.** A poly-X female, sometimes called a superfemale, has more than two X chromosomes and, therefore, extra Barr bodies in the nucleus. Females with three X chromosomes have no distinctive phenotype aside from a tendency to be tall and thin. Although some have delayed motor and language development, as well as learning problems, most poly-X females are not mentally retarded. Some may have menstrual difficulties, but many menstruate regularly and are fertile. Children usually have a normal karyotype. The incidence for poly-X females is about 1 in 1,500 females.

Females with more than three X chromosomes occur rarely. Unlike XXX females, XXXX females are usually tall and severely retarded. Various physical abnormalities are seen, but they may menstruate normally.

**Jacobs Syndrome.** XYY males with Jacobs syndrome can only result from nondisjunction during spermatogenesis. These individuals are sometimes called supermales. Among all live male births, the frequency of the XYY karyotype is about 1 in 1,000. Affected males are usually taller than average, suffer from persistent acne, and tend to have speech and reading problems, but are fertile and may have children. Based upon the number of XYY individuals in prisons and mental facilities, at one time it was suggested that these men were likely to be criminally aggressive, but it has since been shown that the incidence of such behavior among them may be no greater than among XY males.

## Changes in Chromosome Structure

Changes in chromosome structure are another type of chromosomal mutation. Some, but not all, changes in chromosome structure can be detected microscopically. Various agents in the environment, such as radiation, certain organic chemicals, or even viruses, can cause chromosomes to break. Ordinarily, when breaks occur in chromosomes, the two broken ends reunite to give the same sequence of genes. Sometimes, however, the broken ends of one or more chromosomes do not rejoin in the same pattern as before, and the result is various types of chromosomal mutations.

Changes in chromosome structure include deletions, duplications, translocations, and inversions of chromosome segments. A **deletion** occurs when an end of a chromosome breaks off or when two simultaneous breaks lead to the loss of an internal segment (Fig. 10.13*a*). Even when only one member of a pair of chromosomes is affected, a deletion often causes abnormalities.

A **duplication** is the presence of a chromosomal segment more than once in the same chromosome (Fig. 10.13*b*). Duplications may or may not cause visible abnormalities, depending on the size of the duplicated region. An **inversion** has occurred when a segment of a chromosome is turned around 180° (Fig. 10.13*c*). Most individuals with inversions exhibit no abnormalities, but this reversed sequence of genes can result in duplications or deletions being passed on to their children, as described in Figure 10.14.

A **translocation** is the movement of a chromosome segment from one chromosome to another, nonhomologous chromosome. The translocation shown in Figure 10.13*d* is *balanced*, meaning that there is a reciprocal swap of one piece of the chromosome for the other. Often, there are no visible effects of the swap, but if the individual has children, they will receive one normal copy of the chromosome from the normal parent and one of the abnormal chromosomes. The translocation is now *unbalanced*, and there is extra material from one chromosome and missing material from another chromosome. Unbalanced translocations usually miscarry, but those that do not often have severe symptoms.

Some Down syndrome cases are caused by an unbalanced translocation between chromosomes 21 and 14. In other words, because a portion of chromosome 21 is now attached to a portion of chromosome 14, the individual has three copies of the genes that bring about Down syndrome when they are present in triplet copy. In these cases, Down syndrome is not caused by nondisjunction during meiosis, but is passed on normally like any other genetic trait as described in Chapter 11.

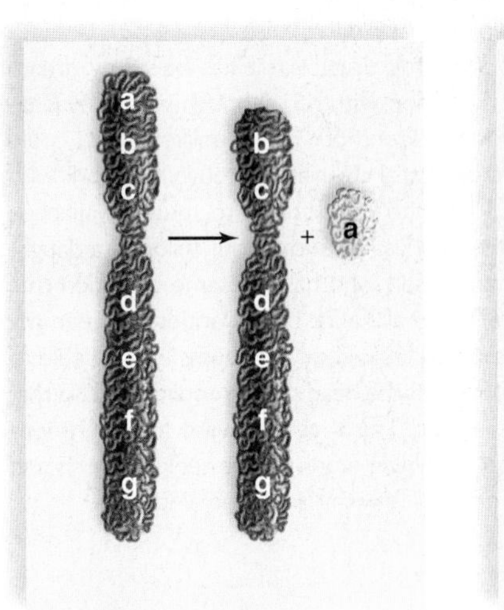


a. Deletion

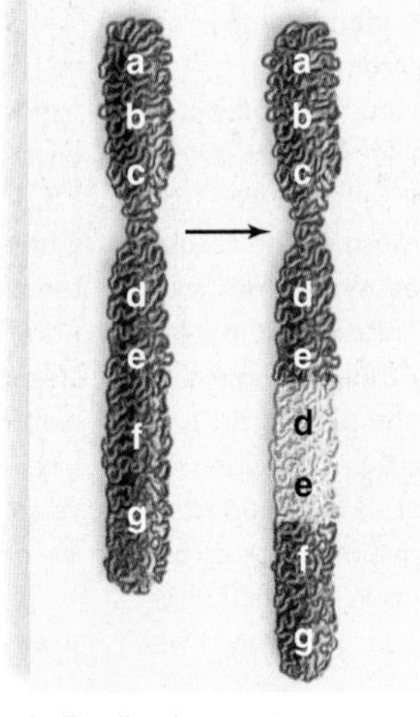


b. Duplication

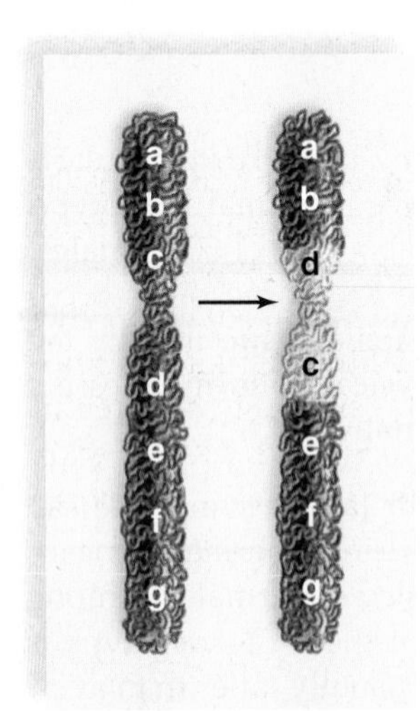


c. Inversion

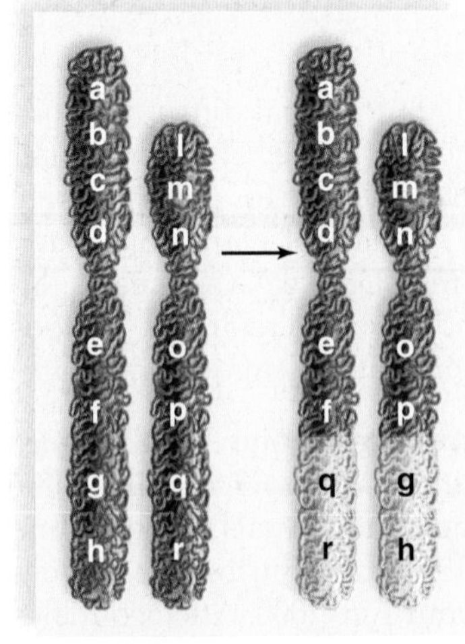


d. Translocation

**FIGURE 10.13 Types of chromosomal mutations.**

**a.** Deletion is the loss of a chromosome piece. **b.** Duplication occurs when the same piece is repeated within the chromosome. **c.** Inversion occurs when a piece of chromosome breaks loose and then rejoins in the reversed direction. **d.** Translocation is the exchange of chromosome pieces between nonhomologous pairs.

### *Human Syndromes*

Changes in chromosome structure occur in humans and lead to various syndromes, many of which are just now being discovered. Sometimes changes in chromosome structure can be detected in humans by doing a karyotype. They may also be discovered by studying the inheritance pattern of a disorder in a particular family.

**Deletion Syndromes.** Williams syndrome occurs when chromosome 7 loses a tiny end piece (Fig. 10.14). Children who have this syndrome look like pixies, with turned-up noses, wide mouths, a small chin, and large ears. Although their academic skills are poor, they exhibit excellent verbal and musical abilities. The gene that governs the production of the protein elastin is missing, and this affects the health of the cardiovascular system and causes their skin to age prematurely. Such individuals are very friendly but need an ordered life, perhaps because of the loss of a gene for a protein that is normally active in the brain.

Cri du chat (cat's cry) syndrome is seen when chromosome 5 is missing an end piece. The affected individual has a small head, is mentally retarded, and has facial abnormalities. Abnormal development of the glottis and larynx results in the most characteristic symptom—the infant's cry resembles that of a cat.

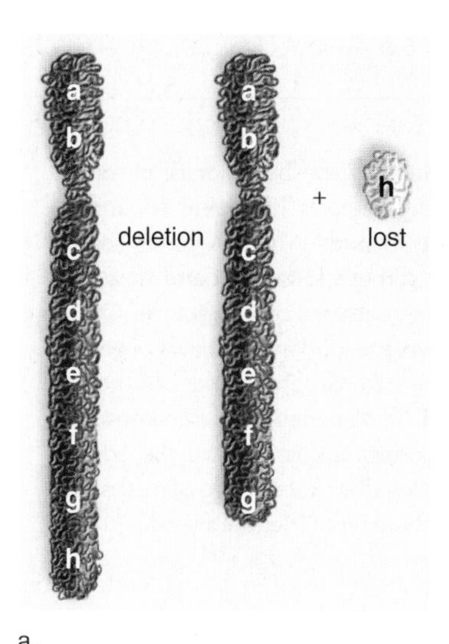

**FIGURE 10.14**
**Deletion.**

**a.** When chromosome 7 loses an end piece, the result is Williams syndrome. **b.** These children, although unrelated, have the same appearance, health, and behavioral problems.

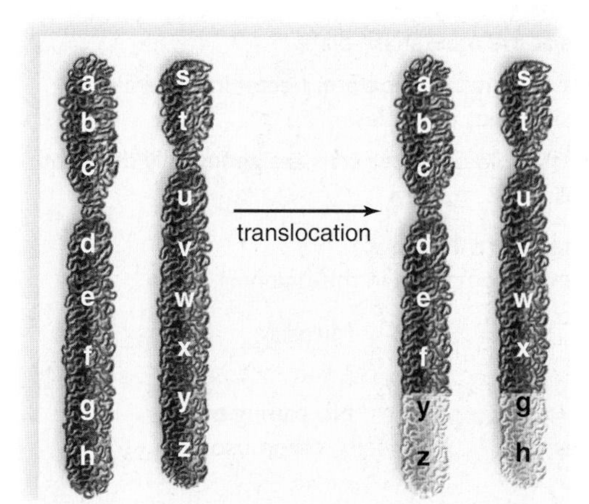


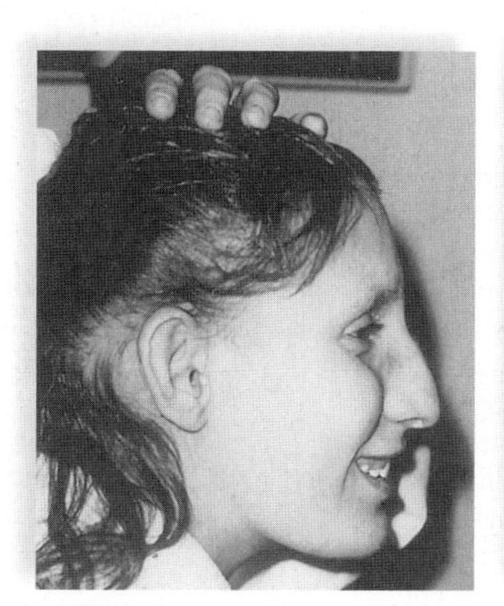

**FIGURE 10.15**
**Translocation.**

**a.** When chromosomes 2 and 20 exchange segments, (**b**) Alagille syndrome, with distinctive facial features, sometimes results because the translocation disrupts an allele on chromosome 20.

**Translocation Syndromes.** A person who has both of the chromosomes involved in a translocation has the normal amount of genetic material and is healthy, unless the chromosome exchange breaks an allele into two pieces. The person who inherits only one of the translocated chromosomes will no doubt have only one copy of certain alleles and three copies of certain other alleles. A genetic counselor begins to suspect a translocation has occurred when spontaneous abortions are commonplace and family members suffer from various syndromes. A special microscopic technique allows a technician to determine that a translocation has occurred.

Figure 10.15 shows a daughter and her father who have a translocation between chromosomes 2 and 20. Although they have the normal amount of genetic material, they have the distinctive face, abnormalities of the eyes and internal organs, and severe itching characteristic of Alagille syndrome. People with this syndrome ordinarily have a deletion on chromosome 20; therefore, it can be deduced that the translocation disrupted an allele on chromosome 20 in the father. The symptoms of Alagille syndrome range from mild to severe, so some people may not be aware they have the syndrome. This father did not realize it until he had a child with the syndrome.

Translocations can also be responsible for a variety of other disorders including certain types of cancer. In the 1970s, new staining techniques identified that a translocation from a portion of chromosome 22 to chromosome 9 was responsible for many cases of chronic myelogenous leukemia. This translocated chromosome was called Philadelphia chromosome. In Burkitt lymphoma, a cancer common in children in equatorial Africa, a large tumor develops from lymph glands in the region of the jaw. This disorder involves a translocation from a portion of chromosome 8 to chromosome 14.

### Check Your Progress 10.6

1. What kind of changes in chromosome number may be caused by nondisjunction in meiosis?
2. Why is sex chromosome aneuploidy more common than autosome aneuploidy?
3. Describe the difference between an inversion and a translocation.

## Connecting the Concepts

Meiosis is similar to mitosis except that meiosis is a more elaborate process. Like the cell cycle and mitosis, meiosis is tightly controlled. Regulatory mechanisms exist to ensure that homologous chromosomes first pair and then separate during the first division and that sister chromatids do not separate until the second division. In addition, meiosis only occurs in certain types of cells during a restricted period of an organism's life span.

Meiosis facilitates sexual reproduction, and there are both evolutionary costs and benefits involved. Although the increased number of genes controlling the process can lead to an increased chance of mutations and chromosomal abnormalities, and therefore the possibility of faulty gametes, there is the advantage in that sexually reproducing species have a greater likelihood of survival than asexually reproducing species because of the greater genetic diversity that sexual reproduction introduces.

Understanding the behavior of chromosomes during meiosis is critical to understanding the manner in which genes segregate during gamete formation and how this contributes to patterns of inheritance. Chapter 11 reviews the fundamental laws of genetics established by Gregor Mendel. Although Mendel had no knowledge of chromosome behavior, modern students have the advantage of applying their knowledge of meiosis to their understanding of Mendel's laws.

## summary

### 10.1 Halving the Chromosome Number

Meiosis ensures that the chromosome number in offspring stays constant generation after generation. The nucleus contains pairs of chromosomes, called homologous chromosomes (homologues).

Meiosis requires two cell divisions and results in four daughter cells. Replication of DNA takes place before meiosis begins. During meiosis I, the homologues undergo synapsis (resulting in a bivalent) and align independently at the metaphase plate. The daughter cells receive one member of each pair of homologous chromosomes. There is no replication of DNA during interkinesis. During meiosis II, the sister chromatids separate, becoming daughter chromosomes that move to opposite poles as they do in mitosis. The four daughter cells contain the haploid number of chromosomes and only one of each kind.

### 10.2 Genetic Variation

Sexual reproduction ensures that the offspring have a different genetic makeup than the parents. Meiosis contributes to genetic variability in two ways: crossing-over and independent assortment of the homologous chromosomes. When homologous chromosomes lie side by side during synapsis, nonsister chromatids may exchange genetic material. Due to crossing-over, the chromatids that separate during meiosis II have a different combination of genes.

When the homologous chromosomes align at the metaphase plate during metaphase I, either the maternal or the paternal chromosome can be facing either pole. Therefore, there will be all possible combinations of chromosomes in the gametes.

### 10.3 The Phases of Meiosis

Meiosis I is divided into four phases:

Prophase I—Bivalents form, and crossing-over occurs as chromosomes condense; the nuclear envelope fragments.

Metaphase I—Bivalents independently align at the metaphase plate.

Anaphase I—Homologous chromosomes separate, and duplicated chromosomes move to poles.

Telophase I—Nuclei become haploid, having received one duplicated chromosome from each homologous pair.

Meiosis II is divided into four phases:

Prophase II—Chromosomes condense, and the nuclear envelope fragments.

Metaphase II—The haploid number of still duplicated chromosomes align at the metaphase plate.

Anaphase II—Sister chromatids separate, becoming daughter chromosomes that move to the poles.

Telophase II—Four haploid daughter cells are genetically different from the parent cell.

### 10.4 Meiosis Compared to Mitosis

Mitosis and meiosis can be compared in this manner:

| Meiosis I | Mitosis |
|---|---|
| *Prophase* | |
| Pairing of homologous chromosomes | No pairing of chromosomes |
| *Metaphase* | |
| Bivalents at metaphase plate | Duplicated chromosomes at metaphase plate |
| *Anaphase* | |
| Homologous chromosomes separate and move to poles | Sister chromatids separate, becoming daughter chromosomes that move to the poles |
| *Telophase* | |
| Daughter nuclei have the haploid number of chromosomes | Daughter nuclei have the parent cell chromosome number |

Meiosis II is like mitosis except the nuclei are haploid.

### 10.5 The Human Life Cycle

Meiosis occurs in any life cycle that involves sexual reproduction. In the animal life cycle, only the gametes are haploid; in plants, meiosis produces spores that develop into a multicellular haploid adult that produces the gametes. In unicellular protists and fungi, the zygote undergoes meiosis, and spores become a haploid adult that gives rise to gametes.

During the life cycle of humans and other animals, meiosis is involved in spermatogenesis and oogenesis. Whereas spermatogenesis produces four sperm per meiosis, oogenesis produces one egg and two to three nonfunctional polar bodies. Spermatogenesis occurs in males, and oogenesis occurs in females. When a sperm fertilizes an egg, the zygote has the diploid number of chromosomes. Mitosis, which is involved in growth and repair, also occurs during the life cycle of all animals.

**10.6 Changes in Chromosome Number and Structure**
Nondisjunction during meiosis I or meiosis II may result in aneuploidy (extra or missing copies of chromosomes). Monosomy occurs when an individual has only one of a particular type of chromosome (2n − 1); trisomy occurs when an individual has three of a particular type of chromosome (2n + 1). Down syndrome is a well-known trisomy in human beings resulting from an extra copy of chromosome 21.

Aneuploidy of the sex chromosomes is tolerated more easily than aneuploidy of the autosomes. Turner syndrome, Klinefelter syndrome, poly-X females, and Jacobs syndrome are examples of sex chromosome aneuploidy.

Abnormalities in crossing-over may result in deletions, duplications, inversions, and translocations within chromosomes. Many human syndromes, including Williams syndrome, cri du chat syndrome, and Alagille syndrome, result from changes in chromosome structure.

## understanding the terms

allele 170
alternation of generations 176
aneuploidy 180
Barr body 182
bivalent 171
crossing-over 172
deletion 184
diploid (2n) number 170
duplication 184
euploidy 180
fertilization 172
gamete 170
gametogenesis 179
gametophyte 178
genetic recombination 172
haploid (n) number 170
homologous chromosome 170
homologue 170
independent assortment 172
interkinesis 176
inversion 184
karyotype 181
life cycle 178
meiosis 170
monosomy 180
oogenesis 179
polar body 179
secondary oocyte 179
sexual reproduction 170
spermatogenesis 179
spore 176
sporophyte 178
synapsis 171
translocation 184
trisomy 180
zygote 170

Match the terms to these definitions:

a. ______________ Production of sperm in males by the process of meiosis and maturation.
b. ______________ Pair of homologous chromosomes at the metaphase plate during meiosis I.
c. ______________ A nonfunctional product of oogenesis.
d. ______________ The functional product of meiosis I in oogenesis becomes the egg.
e. ______________ Member of a pair of chromosomes in which both members carry genes for the same traits.

## reviewing this chapter

1. Why did early investigators predict that there must be a reduction division in the sexual reproduction process? 170
2. What are homologous chromosomes? Contrast the genetic makeup of sister chromatids with that of nonsister chromatids. 170–71
3. Draw and explain a diagram that illustrates crossing-over and another that shows all possible results from independent assortment of homologous pairs. How do these events ensure genetic variation among the gametes? 172–73
4. Draw and explain a series of diagrams that illustrate the stages of meiosis I and meiosis II. 173–76
5. What accounts for **(a)** the genetic similarity between daughter cells and the parent cell following mitosis, and **(b)** the genetic dissimilarity between daughter cells and the parent cell following meiosis? 176–78
6. Explain the human (animal) life cycle and the roles of meiosis and mitosis. 178–79
7. Compare spermatogenesis in males to oogenesis in females. 179
8. How does aneuploidy occur? Why is sex chromosome aneuploidy more common than autosomal aneuploidy? What are some human syndromes associated with aneuploidy? 180–83
9. Name and explain four types of changes in chromosome structure. 184
10. Name some syndromes that occur in humans due to changes in chromosome structure. 184–85

## testing yourself

Choose the best answer for each question.

1. A bivalent is
   a. a homologous chromosome.
   b. the paired homologous chromosomes.
   c. a duplicated chromosome composed of sister chromatids.
   d. the two daughter cells after meiosis I.
   e. the two centrioles in a centrosome.
2. If a parent cell has 16 chromosomes, then each of the daughter cells following meiosis will have
   a. 48 chromosomes.
   b. 32 chromosomes.
   c. 16 chromosomes.
   d. 8 chromosomes.
3. At the metaphase plate during metaphase I of meiosis, there are
   a. chromosomes consisting of one chromatid.
   b. unpaired duplicated chromosomes.
   c. bivalents.
   d. homologous pairs of chromosomes.
   e. Both c and d are correct.
4. At the metaphase plate during metaphase II of meiosis, there are
   a. chromosomes consisting of one chromatid.
   b. unpaired duplicated chromosomes.
   c. bivalents.
   d. homologous pairs of chromosomes.
   e. Both c and d are correct.
5. Gametes contain one of each kind of chromosome because
   a. the homologous chromosomes separate during meiosis.
   b. the chromatids separate during meiosis.
   c. only one replication of DNA occurs during meiosis.
   d. crossing-over occurs during prophase I.
   e. the parental cell contains only one of each kind of chromosome.
6. Crossing-over occurs between
   a. sister chromatids of the same chromosome.
   b. two different kinds of bivalents.
   c. two different kinds of chromosomes.
   d. nonsister chromatids of a bivalent.
   e. two daughter nuclei.
7. During which phase of meiosis do homologous chromosomes separate?
   a. prophase I
   b. telophase I
   c. anaphase I
   d. anaphase II

8. During which phase of meiosis does crossing-over occur?
   a. prophase I
   b. interkinesis
   c. metaphase II
   d. anaphase I
9. Which of the following statements is false?
   a. Oogenesis occurs in females, and spermatogenesis occurs in males.
   b. Spermatogenesis produces four viable gametes, while oogenesis only produces one.
   c. Daughter cells produced from oogenesis are diploid, while daughter cells produced by spermatogenesis are haploid.
   d. Spermatogenesis goes to completion, while oogenesis does not always go to completion.
10. Nondisjunction during meiosis I of oogenesis will result in eggs that have
    a. the normal number of chromosomes.
    b. one too many chromosomes.
    c. one less than the normal number of chromosomes.
    d. Both b and c are correct.
11. Which two of these chromosomal mutations are most likely to occur when an inverted chromosome is undergoing synapsis?
    a. deletion and translocation
    b. deletion and duplication
    c. duplication and translocation
    d. inversion and duplication
12. A male with underdeveloped testes and some breast development most likely has
    a. Down syndrome.
    b. Jacobs syndrome.
    c. Turner syndrome.
    d. Klinefelter syndrome.

For questions 13–17, fill in the blanks.

13. If the parent cell has 24 chromosomes, the daughter cells following mitosis will have ____________ chromosomes and following meiosis will have ____________ chromosomes.
14. Meiosis in males is a part of ____________, and meiosis in females is a part of ____________.
15. Oogenesis will not go to completion unless ____________ occurs.
16. In humans, meiosis produces ____________, and in plants, meiosis produces ____________.
17. During oogenesis, the primary oocyte has the ____________ and the secondary oocyte has the ____________ number of chromosomes.

For questions 18–24, match the statements that follow to the items in the key. Answers may be used more than once, and more than one answer may be used.

**KEY:**

a. mitosis
b. meiosis I
c. meiosis II
d. Both meiosis I and meiosis II are correct.
e. All of these are correct.

18. Spindle fibers are attached to kinetochores.
19. A parent cell with ten duplicated chromosomes will produce daughter cells with five duplicated chromosomes each.
20. Involves pairing of duplicated homologous chromosomes.
21. A parent cell with five duplicated chromosomes will produce daughter cells with five chromosomes consisting of one chromatid each.
22. Nondisjunction may occur, causing abnormal gametes to form.
23. A parent cell with ten duplicated chromosomes will produce daughter cells with ten chromosomes consisting of one chromatid each.
24. Involved in growth and repair of tissues.

## thinking scientifically

1. Why is the first meiotic division considered to be the reduction division for chromosome number?
2. Recall that during interphase, the $G_2$ checkpoint ensures that the DNA has been faithfully replicated before the cell is allowed to divide by mitosis. Would you expect this checkpoint to be active during interkinesis? How might you set up an experiment to test your hypothesis?
3. A man has a balanced translocation between chromosome 2 and 6. If he reproduces with a normal woman could the child have the same translocation? Why or why not?

## bioethical issue

### The Risks of Advanced Maternal Age

In today's society, it is commonplace for women to embark on careers and pursue higher education, delaying marriage and childbirth until later years. Between 1991 and 2001, the birthrate among women aged 35 to 39 increased over 30%, while the birthrate among women aged 40 to 44 leaped by almost 70%. These increases have occurred as society has changed, spurred by the elimination of the social stigmas, better prenatal care, and new medical technologies that can overcome the decline in fertility associated with age and treat at-risk children.

The decision to delay childbirth does carry risks. Although the reasons are not well understood, the risk of many disorders associated with meiotic nondisjunction, such as Down syndrome, increase greatly with age, rising from nearly 1 in 900 at age 30 to 1 in 109 by age 40. The risk of complications to the mother, such as gestational diabetes, are also much higher in women over 30. Thus, the medical community has embarked on a campaign to ensure that women who are pregnant and over age 35 are offered more intensive prenatal care. Many people are concerned about the ultimate cost to society, through increased insurance premiums and increased costs to governments to pay for it.

While there are definitely risks associated with advanced maternal age, others contend that having children later in life provides many advantages. Women over age 35 are usually at a later stage in their careers and have higher salaries, lessening the need for many social welfare programs. Furthermore, women over 35 are much less likely to divorce or give birth out of wedlock and are often able to devote more time to the child than younger women. Therefore, while older mothers require more medical attention, the overall costs to society are lower.

Considering both the benefits and the disadvantages, are we as a society obligated to fund intense screening and prenatal care for women of advanced maternal age, and to pay for treating the maladies associated with it? As birthrates among women over age 30 continue to soar, the debate over advanced maternal age is not likely to abate any time soon.

## *Biology* website

The companion website for *Biology* provides a wealth of information organized and integrated by chapter. You will find practice tests, animations, videos, and much more that will complement your learning and understanding of general biology.

**http://www.mhhe.com/maderbiology10**

# 11 Mendelian Patterns of Inheritance

Camille was painfully aware of her foul body odor because the children teased her relentlessly, calling her "Miss Fishy" and other nasty names. Little did she know, however, that she suffers from trimethylaminuria, or "fish odor syndrome," an extremely rare genetic disorder she shares with possibly 1 in 10,000 people. People with this syndrome all have a defective gene whose product is unable to break down the smelly chemical trimethylamine, and it ends up in their urine, sweat, and at times even in their breath.*

*Rare genetic disorders like Camille's constantly pique our curiosity about how traits are inherited from one generation to the next. In the following chapter, you will learn that the process of meiosis can be used to predict the inheritance of a trait. You will also learn how Mendel discovered that certain traits, such as trimethylaminuria, are recessive and it takes two copies of a gene before you are affected. In this chapter, you will be introduced to other human genetic disorders that can be definitely linked to specific genes on the chromosomes.*

The other kids teased Camille because she had a fishy smell.

## concepts

## 11.1 Gregor Mendel

The science of genetics explains the stability of inheritance (why you are human as are your parents) and also variations between offspring from one generation to the next (why you have a different combination of traits than your parents). Virtually every culture in history has attempted to explain observed inheritance patterns. An understanding of these patterns has always been important to agriculture, animal husbandry (the science of breeding animals), and medicine.

### The Blending Concept of Inheritance

Until the late nineteenth century, most plant and animal breeders believed that traits were inherited by the blending concept of inheritance, which stated that an offspring's genetic makeup was intermediate to that of its parents. While they acknowledged that both sexes contribute equally to a new individual, they believed that parents of contrasting appearance always produce offspring of intermediate appearance. Therefore, according to this concept, a cross between plants with red flowers and plants with white flowers would yield only plants with pink flowers. However, this theory did not always explain observed inheritance patterns. For example, red and white flowers reappeared in future generations even though the parents had pink flowers. The breeders mistakenly attributed this to instability of the genetic material. The blending concept of inheritance offered little help to Charles Darwin, the father of evolution, whose treatise on natural selection lacked a strong genetic basis. If populations contained only intermediate individuals and normally lacked variations, how could diverse forms evolve?

**FIGURE 11.1 Gregor Mendel, 1822–84.**

Mendel grew and tended the pea plants he used for his experiments. For each experiment, he observed as many offspring as possible. For a cross that required him to count the number of round seeds to wrinkled seeds, he observed and counted a total of 7,324 peas!

### Mendel's Particulate Theory of Inheritance

Gregor Mendel was an Austrian monk who developed a particulate theory of inheritance after performing a series of ingenious experiments in the 1860s (Fig. 11.1). Mendel studied science and mathematics at the University of Vienna, and at the time of his genetic research, he was a substitute natural science teacher at a local high school. Mendel was a successful scientist for several reasons. First, he was one of the first scientists to apply mathematics to biology. Most likely his background in mathematics prompted him to apply statistical methods and the laws of probability to his breeding experiments. He was also a careful, deliberate scientist who followed the scientific method very closely and kept very detailed, accurate records. He prepared for his experiments carefully and conducted many preliminary studies with various animals and plants.

Mendel's theory of inheritance is called a particulate theory because it is based on the existence of minute particles or hereditary units we now call genes. Inheritance involves the reshuffling of the same genes from generation to generation. His laws of segregation and the law of independent assortment describe the behavior of these particulate units of heredity as they are passed from one generation to the next. Much of modern genetics is based upon Mendel's theories, which have withstood the test of time and have been supported by innumerable experiments.

### Mendel Worked with the Garden Pea

Mendel's preliminary experiments prompted him to choose the garden pea, *Pisum sativum* (Fig. 11.2*a*), as his experimental material. The garden pea was a good choice for many reasons. The plants were easy to cultivate and had a short generation time. Although peas normally self-pollinate (pollen only goes to the same flower), they could be cross-pollinated by hand by transferring pollen from the anther to the stigma. Many varieties of peas were available, and Mendel chose 22 for his experiments. When these varieties self-pollinated, they were *true-breeding*—meaning that the offspring were like the parent plants and like each other. In contrast to his predecessors, Mendel studied the inheritance of relatively simple and discrete traits that were not subjective and were easy to observe, such as seed shape, seed color, and flower color. In his crosses, Mendel observed either dominant or recessive characteristics but no intermediate ones (Fig. 11.2*b*).

**Check Your Progress 11.1**

1. What made Gregor Mendel's experiments successful?
2. Why was the garden pea a good choice for Mendel's experiments?

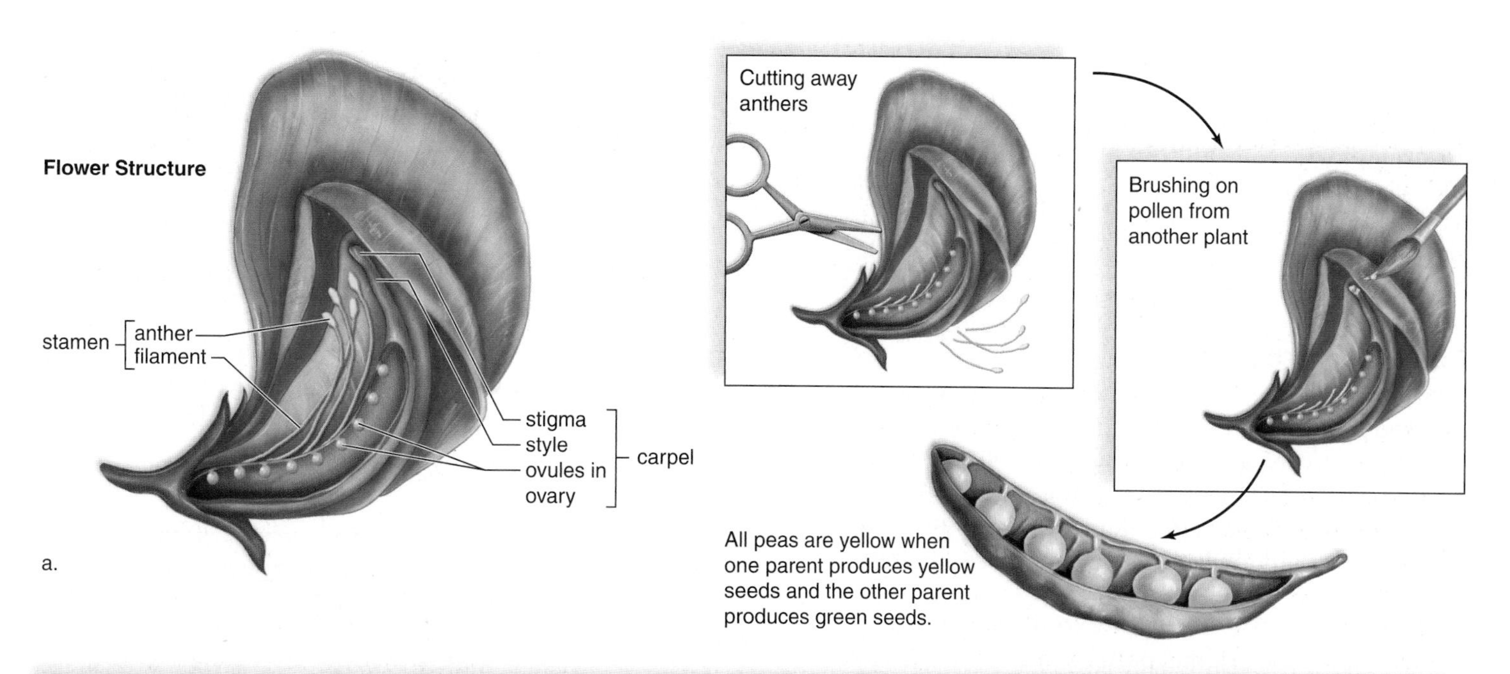


| Trait | Characteristics *Dominant | Characteristics *Recessive | F2 Results* Dominant | F2 Results* Recessive |
|---|---|---|---|---|
| Stem length | Tall | Short | 787 | 277 |
| Pod shape | Inflated | Constricted | 882 | 299 |
| Seed shape | Round | Wrinkled | 5,474 | 1,850 |
| Seed color | Yellow | Green | 6,022 | 2,001 |
| Flower position | Axial | Terminal | 651 | 207 |
| Flower color | Purple | White | 705 | 224 |
| Pod color | Green | Yellow | 428 | 152 |

*All of these produce approximately a 3:1 ratio. For example, $\frac{787}{277} = \frac{3}{1}$.

b.

**FIGURE 11.2 Garden pea anatomy and a few traits.**

**a.** In the garden pea, *Pisum sativum,* pollen grains produced in the anther contain sperm, and ovules in the ovary contain eggs. When Mendel performed crosses, he brushed pollen from one plant onto the stigma of another plant. After sperm fertilized eggs, the ovules developed into seeds (peas). The open pod shows the results of a cross between plants with yellow seeds and plants with green seeds. **b.** Mendel selected traits like these for study. He made sure his parent (P generation) plants bred true, and then he cross-pollinated the plants. The offspring called $F_1$ (first filial) generation always resembled the parent with the dominant characteristic (*left*). Mendel then allowed the $F_1$ plants to self-pollinate. In the $F_2$ (second filial) generation, he always achieved a 3:1 (dominant to recessive) ratio. The text explains how Mendel went on to interpret these results.

# 11.2 Mendel's Laws

After ensuring that his pea plants were true-breeding—for example, that his tall plants always had tall offspring and his short plants always had short offspring—Mendel was ready to perform cross-pollination experiments (see Fig. 11.2*a*). These crosses allowed Mendel to formulate his law of segregation.

## Law of Segregation

For these initial experiments, Mendel chose varieties that differed in only one trait. If the blending theory of inheritance were correct, the cross should yield offspring with an intermediate appearance compared to the parents. For example, the offspring of a cross between a tall plant and a short plant should be intermediate in height.

### Mendel's Experimental Design and Results

Mendel called the original parents the *P generation* and the first generation the $F_1$, or filial [L. *filius*, sons and daughters], *generation* (Fig. 11.3). He performed *reciprocal crosses:* First he dusted the pollen of tall plants onto the stigmas of short plants, and then he dusted the pollen of short plants onto the stigmas of tall plants. In both cases, all $F_1$ offspring resembled the tall parent.

Certainly, these results were contrary to those predicted by the blending theory of inheritance. Rather than being intermediate, the $F_1$ plants were tall and resembled only one parent. Did these results mean that the other characteristic (i.e., shortness) had disappeared permanently? Apparently not, because when Mendel allowed the $F_1$ plants to self-pollinate, ¾ of the $F_2$ *generation* were tall and ¼ were short, a 3:1 ratio (Fig. 11.3). Therefore, the $F_1$ plants were able to pass on a factor for shortness—it didn't disappear, it just skipped a generation. Perhaps the $F_1$ plants were tall because tallness was dominant to shortness?

Mendel counted many plants. For this particular cross, called a **monohybrid cross** because the parents are hybrids in one way, he counted a total of 1,064 plants, of which 787 were tall and 277 were short. In all crosses that he performed, he found a 3:1 ratio in the $F_2$ generation (see Fig. 11.2*b*). The characteristic that had disappeared in the $F_1$ generation reappeared in ¼ of the $F_2$ offspring. *Today, we know that the expected phenotypic results of a monohybrid cross are always 3:1.*

### Mendel's Conclusion

His mathematical approach led Mendel to interpret his results differently from previous breeders. He knew that the same ratio was obtained among the $F_2$ generation time and time again when he did a monohybrid cross involving the seven traits he was studying. Eventually Mendel arrived at this explanation: A 3:1 ratio among the $F_2$ offspring was possible if (1) the $F_1$ parents contained two separate copies of each hereditary factor, one of these being dominant and the other recessive; (2) the factors separated when the gametes were formed, and each gamete carried only one copy of each factor; and (3) random fusion of all possible gametes occurred upon fertilization. Only in this way could shortness reoccur in the $F_2$ generation. Thinking this, Mendel arrived at the first of his laws of inheritance—the law of segregation. The law of segregation is a cornerstone of his particulate theory of inheritance.

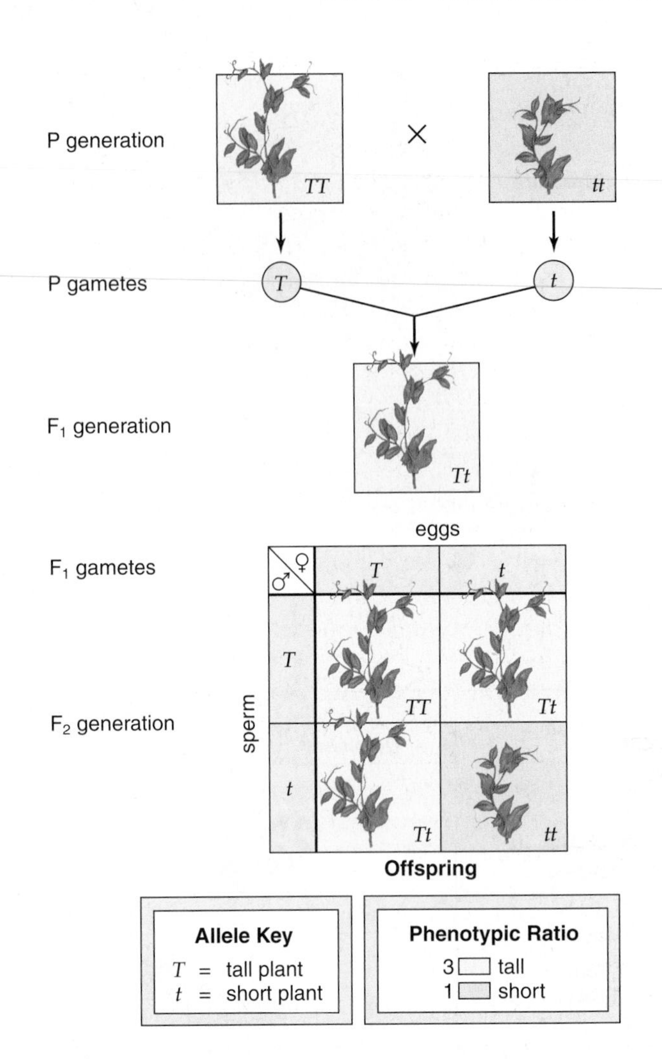


**FIGURE 11.3 Monohybrid cross done by Mendel.**
The P generation plants differ in one regard—length of the stem. The $F_1$ generation plants are all tall, but the factor for short has not disappeared because ¼ of the $F_2$ generation plants are short. The 3:1 ratio allowed Mendel to deduce that individuals have two discrete and separate genetic factors for each trait.

The law of segregation states the following:

- Each individual has two factors for each trait.
- The factors segregate (separate) during the formation of the gametes.
- Each gamete contains only one factor from each pair of factors.
- Fertilization gives each new individual two factors for each trait.

## Mendel's Cross as Viewed by Classical Genetics

Figure 11.3 also shows how classical scientists interpreted the results of Mendel's experiments on inheritance of stem length in peas. Stem length in peas is controlled by a single gene. This gene occurs on a homologous pair of chromosomes at a particular location that is called the **gene locus** (Fig. 11.4). Alternative versions of a gene are called **alleles** [Gk. *allelon*, reciprocal, parallel]. The **dominant allele** is so named because of its ability to mask the expression of the other allele, called the **recessive allele.** The dominant allele is identified by a capital letter and the recessive allele by the same but lowercase letter. Usually, the first letter designating a trait is chosen to identify the allele. With reference to the cross being discussed, there is an allele for tallness (*T*) and an allele for shortness (*t*).

Meiosis is the type of cell division that reduces the chromosome number. During meiosis I, the members of bivalents (homologous chromosomes each having sister chromatids) separate. This means that the two alleles for each gene separate from each other during meiosis (see Fig. 11A). Therefore, the process of meiosis gives an explanation for Mendel's law of segregation, and why only one allele for each trait is in a gamete.

In Mendel's cross, the original parents (P generation) were true-breeding; therefore, the tall plants had two alleles for tallness (*TT*), and the short plants had two alleles for shortness (*tt*). When an organism has two identical alleles, as these had, we say it is **homozygous** [Gk. *homo*, same, and *zygos*, balance, yoke]. Because the parents were homozygous, all gametes produced by the tall plant contained the allele for tallness (*T*), and all gametes produced by the short plant contained an allele for shortness (*t*).

After cross-pollination, all the individuals of the resulting $F_1$ generation had one allele for tallness and one for shortness (*Tt*). When an organism has two different alleles at a gene locus, we say that it is **heterozygous** [Gk. *hetero*, different, and *zygos*, balance, yoke]. Although the plants of the $F_1$ generation had one of each type of allele, they were all tall. The allele that is expressed in a heterozygous individual is the dominant allele. The allele that is not expressed in a heterozygote is the recessive allele. This explains why shortness, the recessive trait, skipped a generation in Mendel's experiment.

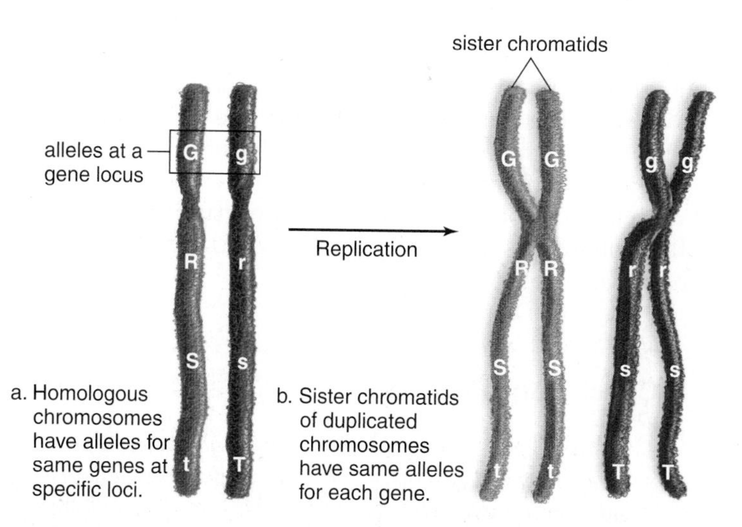


**FIGURE 11.4 Classical view of homologous chromosomes.**

**a.** The letters represent alleles; that is, alternate forms of a gene. Each allelic pair, such as *Gg* or *Tt*, is located on homologous chromosomes at a particular gene locus. **b.** Sister chromatids carry the same alleles in the same order.

**TABLE 11.1**

**Genotype Versus Phenotype**

| *Genotype* | *Genotype* | *Phenotype* |
|---|---|---|
| *TT* | Homozygous dominant | Tall plant |
| *Tt* | Heterozygous | Tall plant |
| *tt* | Homozygous recessive | Short plant |

Continuing with the discussion of Mendel's cross (see Fig. 11.3), the $F_1$ plants produce gametes in which 50% have the dominant allele *T* and 50% have the recessive allele *t*. During the process of fertilization, we assume that all types of sperm (i.e., *T* or *t*) have an equal chance to fertilize all types of eggs (i.e., *T* or *t*). When this occurs, such a monohybrid cross will always produce a 3:1 (dominant to recessive) ratio among the offspring. Figure 11.2*b* gives Mendel's results for several monohybrid crosses, and you can see that the results were always close to 3:1.

### *Genotype Versus Phenotype*

It is obvious from our discussion that two organisms with different allelic combinations for a trait can have the same outward appearance. (*TT* and *Tt* pea plants are both tall.) For this reason, it is necessary to distinguish between the alleles present in an organism and the appearance of that organism.

The word **genotype** [Gk. *genos*, birth, origin, race, and *typos*, image, shape] refers to the alleles an individual receives at fertilization. Genotype may be indicated by letters or by short, descriptive phrases. Genotype *TT* is called homozygous dominant, and genotype *tt* is called homozygous recessive. Genotype *Tt* is called heterozygous.

The word **phenotype** [Gk. *phaino*, appear, and *typos*, image, shape] refers to the physical appearance of the individual. The homozygous dominant (*TT*) individual and the heterozygous (*Tt*) individual both show the dominant phenotype and are tall, while the homozygous recessive individual shows the recessive phenotype and is short (Table 11.1). The phenotype is dependent upon the genotype of the individual.

**Check Your Progress 11.2A**

1. For each of the following genotypes, list all possible gametes, noting the proportion of each for the individual. **a.** *WW*; **b.** *Ww*; **c.** *Tt*; **d.** *TT*
2. In rabbits, if *B* = black and *b* = white, which of these genotypes (*Bb*, *BB*, *bb*) could a white rabbit have?
3. If a heterozygous rabbit reproduces with one of its own kind, what phenotypic ratio do you expect among the offspring? If there are 120 rabbits, how many are expected to be white?

## Mendel's Law of Independent Assortment

Mendel performed a second series of crosses in which true-breeding plants differed in two traits. For example, he crossed tall plants having green pods with short plants having yellow pods (Fig. 11.5). The $F_1$ plants showed both dominant characteristics. As before, Mendel then allowed the $F_1$ plants to self-pollinate. This $F_1$ cross is known as a **dihybrid cross** because the plants are hybrid in two ways. Two possible results could occur in the $F_2$ generation:

1. If the dominant factors (*TG*) always segregate into the $F_1$ gametes together, and the recessive factors (*tg*) always stay together, then there would be two phenotypes among the $F_2$ plants—tall plants with green pods and short plants with yellow pods.
2. If the four factors segregate into the $F_1$ gametes independently, then there would be four phenotypes among the $F_2$ plants—tall plants with green pods, tall plants with yellow pods, short plants with green pods, and short plants with yellow pods.

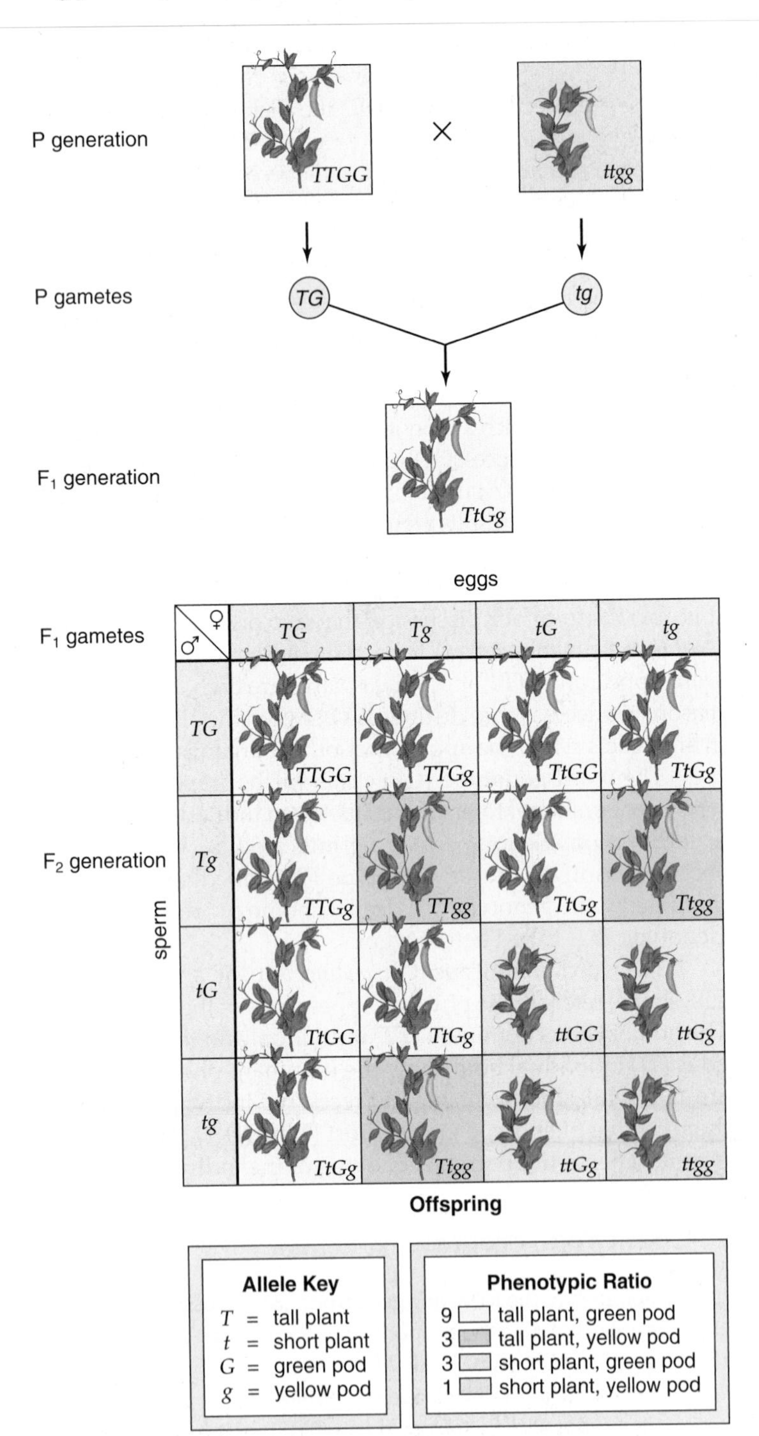


**FIGURE 11.5 Dihybrid cross done by Mendel.**

P generation plants differ in two regards—length of the stem and color of the pod. The $F_1$ generation shows only the dominant traits, but all possible phenotypes appear among the $F_2$ generation. The 9:3:3:1 ratio allowed Mendel to deduce that factors segregate into gametes independently of other factors.

Figure 11.5 shows that Mendel observed four phenotypes among the $F_2$ plants, supporting the second hypothesis. Therefore, Mendel formulated his second law of heredity—the law of independent assortment.

The law of independent assortment states the following:

- Each pair of factors segregates (assorts) independently of the other pairs.
- All possible combinations of factors can occur in the gametes.

The law of independent assortment applies only to alleles on different chromosomes. Each chromosome carries a large number of alleles.

Again, we know that the process of meiosis explains why the $F_1$ plants produced every possible type of gamete and, therefore, four phenotypes appear among the $F_2$ generation of plants. As was explained in the Science Focus on page 195, there are no rules regarding the alignment of homologues at the metaphase plate—the daughter cells produced have all possible combinations of alleles. The possible gametes are the two dominants (such as *TG*), the two recessives (such as *tg*), and the ones that have a dominant and recessive (such as *Tg* and *tG*). When all possible sperm have an opportunity to fertilize all possible eggs, *the expected phenotypic ratio of a dihybrid cross is always 9:3:3:1.*

### Check Your Progress 11.2B

1. In fruit flies, $L$ = long wings and $l$ = short wings; $G$ = gray body and $g$ = black body. List all possible gametes for a heterozygote.
2. What phenotypic ratio is expected when two dihybrids reproduce?

*science focus*

## Mendel's Laws and Meiosis

Today, we realize that the genes are on the chromosomes and that Mendel's laws hold because of the events of meiosis. Figure 11A assumes a parent cell that has two homologous pairs of chromosomes and that the alleles *A,a* are on one pair and the alleles *B,b* are on the other pair. Following duplication of the chromosomes, the parent cell undergoes meiosis as a first step toward the production of gametes. At metaphase I, the homologous pairs line up independently and, therefore, all alignments of homologous chromosomes can occur at the metaphase plate. Then the pairs of homologous chromosomes separate.

In keeping with Mendel's law of independent assortment and law of segregation, each pair of chromosomes and alleles segregates independently of the other pairs. It matters not which member of a homologous pair faces which spindle pole. Therefore, the daughter cells from meiosis I have all possible combinations of alleles. One daughter cell has both dominant alleles, namely *A* and *B*. Another daughter cell has both recessive alleles, namely *a* and *b*. The other two are mixed: *A* with *b* and *a* with *B*. Therefore, all possible combinations of alleles occur in the gametes.

When you form the gametes for any genetic cross, you are following the dictates of Mendel's laws but also mentally taking the chromosomes and alleles through the process of meiosis. We can also note that fertilization restores both the diploid chromosome number and the paired condition of alleles in the zygote.

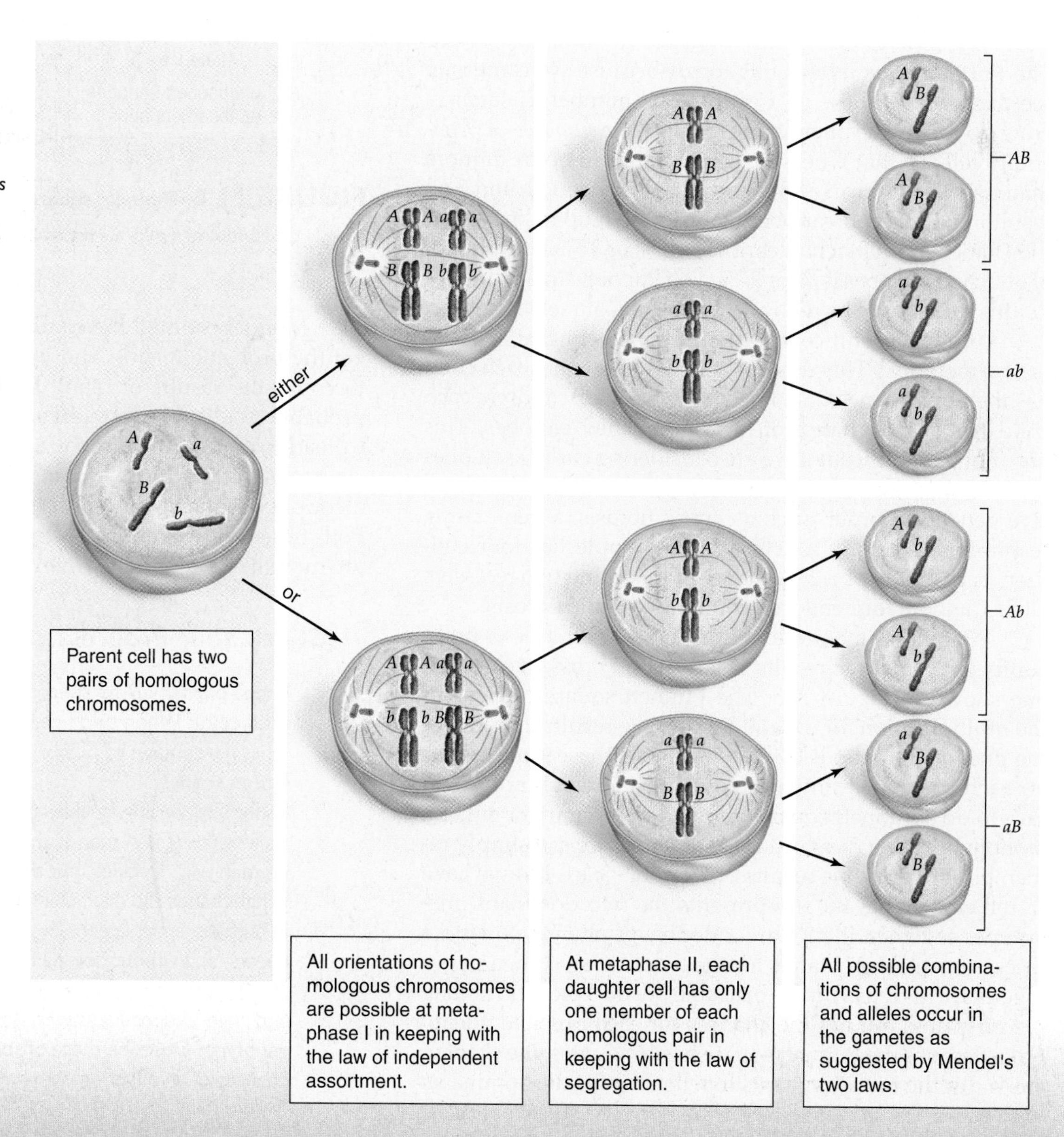


**FIGURE 11A Independent assortment and segregation during meiosis.**

*Mendel's laws hold because of the events of meiosis. The homologous pairs of chromosomes line up randomly at the metaphase plate during meiosis I. Therefore, the homologous chromosomes, and alleles they carry, segregate independently during gamete formation. All possible combinations of chromosomes and alleles occur in the gametes.*

## Mendel's Laws of Probability

The diagram we have been using to calculate the results of a cross is called a Punnett square. The **Punnett square** allows us to easily calculate the chances, or the probability, of genotypes and phenotypes among the offspring. Like flipping a coin, an offspring of the cross illustrated in the Punnett square in Figure 11.6 has a 50% (or $\frac{1}{2}$) chance of receiving an *E* for unattached earlobe or an *e* for attached earlobe from each parent:

The chance of $E = \frac{1}{2}$
The chance of $e = \frac{1}{2}$

How likely is it that an offspring will inherit a specific set of two alleles, one from each parent? The *product rule* of probability tells us that we have to multiply the chances of independent events to get the answer:

1. The chance of *EE* $= \frac{1}{2} \times \frac{1}{2} = \frac{1}{4}$
2. The chance of *Ee* $= \frac{1}{2} \times \frac{1}{2} = \frac{1}{4}$
3. The chance of *eE* $= \frac{1}{2} \times \frac{1}{2} = \frac{1}{4}$
4. The chance of *ee* $= \frac{1}{2} \times \frac{1}{2} = \frac{1}{4}$

The Punnett square does this for us because we can easily see that each of these is $\frac{1}{4}$ of the total number of squares. How do we get the phenotypic results? The *sum rule* of probability tells us that when the same event can occur in more than one way, we can add the results. Because 1, 2, and 3 all result in unattached earlobes, we add them up to know that the chance of unattached earlobes is $\frac{3}{4}$, or 75%. The chance of attached earlobes is $\frac{1}{4}$, or 25%. The Punnett square doesn't do this for us—we have to add the results ourselves.

Another useful concept is the statement that "chance has no memory." This concept helps us know that each child has the same chances. So, if a couple has four children, each child has a 25% chance of having attached earlobes. This may not be significant if we are considering earlobes. It does become significant, however, if we are considering a recessive genetic disorder, such as cystic fibrosis, a debilitating respiratory illness. If a heterozygous couple has four children, each child has a 25% chance of inheriting two recessive alleles, and all four children could have cystic fibrosis.

We can use the product rule and the sum rule of probability to predict the results of a dihybrid cross, such as the one shown in Figure 11.5. The Punnett square carries out the multiplication for us, and we add the results to find that the phenotypic ratio is 9:3:3:1. We expect these same results for each and every dihybrid cross. Therefore, it is not necessary to do a Punnett square over and over again for either a monohybrid or a dihybrid cross. Instead, we can simply remember the probable results of 3:1 and 9:3:3:1. But we have to remember that the 9 represents the two dominant phenotypes together, the 3's are a dominant phenotype with a recessive, and the 1 stands for the two recessive phenotypes together. This tells you the probable phenotypic ratio among the offspring, but not the chances for each possible phenotype. Because the Punnett square has 16 squares, the chances are $\frac{9}{16}$ for the two dominants together, $\frac{3}{16}$ for the dominants with each recessive, and $\frac{1}{16}$ for the two recessives together.

Parents

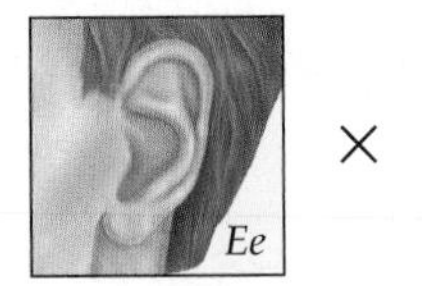


×

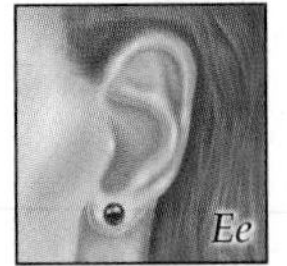


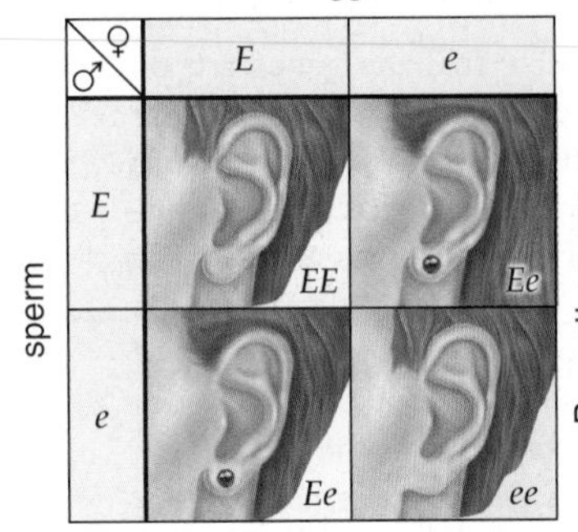


| Allele Key |
|---|
| *E* = unattached earlobes |
| *e* = attached earlobes |

| Phenotypic Ratio |
|---|
| 3 unattached earlobes |
| 1 attached earlobes |

**FIGURE 11.6 Punnett square.**

Use of Punnett square to calculate probable results in this case a 3 : 1 phenotypic ratio.

Mendel counted the results of many similar crosses to get the probable results, and in the laboratory, we too have to count the results of many individual crosses to get the probable results for a monohybrid or a dihybrid cross. Why? Consider that each time you toss a coin, you have a 50% chance of getting heads or tails. If you tossed the coin only a couple of times, you might very well have heads or tails both times. However, if you toss the coin many times, you are more likely to finally achieve 50% heads and 50% tails.

### Check Your Progress 11.2C

1. In pea plants, yellow seed color is dominant over green seed color. When two heterozygous plants are crossed, what percentage of plants would have yellow seeds? Green seeds?
2. In humans, having freckles (*F*) is dominant over having no freckles (*f*). A man with freckles reproduces with a woman with freckles, but the children have no freckles. What chance did each child have for having freckles?
3. In humans, short fingers (*S*) are dominant over long fingers (*s*). Without doing a Punnett square, what phenotypic ratio is probable when a dihybrid for freckles and fingers reproduces with another having the same genotype? Describe these offspring. What are the chances of an offspring with no freckles and long fingers?

## Testcrosses

To confirm that the $F_1$ plants of his one-trait crosses were heterozygous, Mendel crossed his $F_1$ generation plants with true-breeding, short (homozygous recessive) plants. Mendel performed these so-called **testcrosses** because they allowed him to support the law of segregation. For the cross in Figure 11.7, he reasoned that half the offspring should be tall and half should be short, producing a 1:1 phenotypic ratio. His results supported the hypothesis that alleles segregate when gametes are formed. In Figure 11.7*a*, the homozygous recessive parent can produce only one type of gamete—*t*—and so the Punnett square has only one column. The use of one column signifies that all the gametes carry a *t*. *The expected phenotypic ratio for this type of one-trait cross (heterozygous × recessive) is always 1:1.*

### *One-Trait Testcross*

Today, a one-trait testcross is used to determine if an individual with the dominant phenotype is homozygous dominant (e.g., *TT*) or heterozygous (e.g., *Tt*). Since both of these genotypes produce the dominant phenotype, it is not possible to determine the genotype by observation. Figure 11.7*b* shows that if the individual is homozygous dominant, all the offspring will be tall. Each parent has only one type of gamete and, therefore, a Punnett square is not required to determine the results.

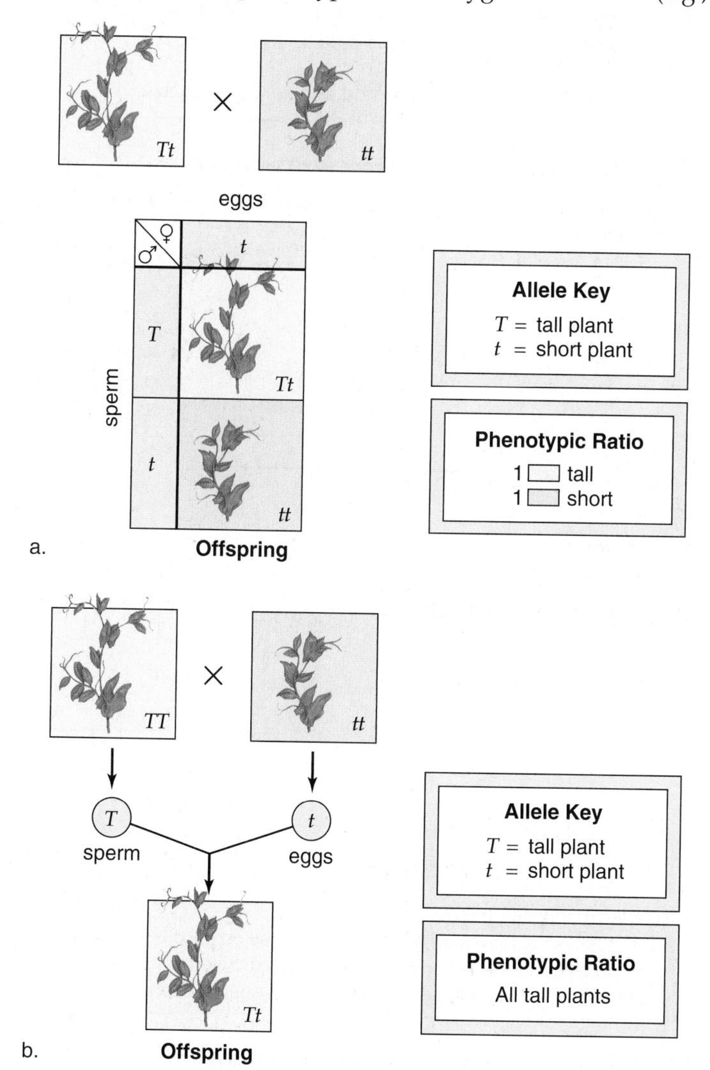


**FIGURE 11.7 One-trait testcrosses.**

**a.** One-trait testcross when the individual with the dominant phenotype is heterozygous. **b.** One-trait testcross when the individual with the dominant phenotype is homozygous.

### *Two-Trait Testcross*

When doing a two-trait testcross, an individual with the dominant phenotype is crossed with one having the recessive phenotype. Suppose you are working with fruit flies in which:

| | |
|---|---|
| *L* = long wings | *G* = gray bodies |
| *l* = vestigial (short) wings | *g* = black bodies |

You wouldn't know by examination whether the fly on the left was homozygous or heterozygous for wing and body color. In order to find out the genotype of the test fly, you cross it with the one on the right. You know by examination that this vestigial-winged and black-bodied fly is homozygous recessive for both traits.

If the test fly is homozygous dominant for both traits with the genotype *LLGG*, it will form only one gamete: *LG*. Therefore, all the offspring from the proposed cross will have long wings and a gray body.

However, if the test fly is heterozygous for both traits with the genotype *LlGg*, it will form four different types of gametes:

Gametes: *LG* *Lg* *lG* *lg*

and could have four different offspring:

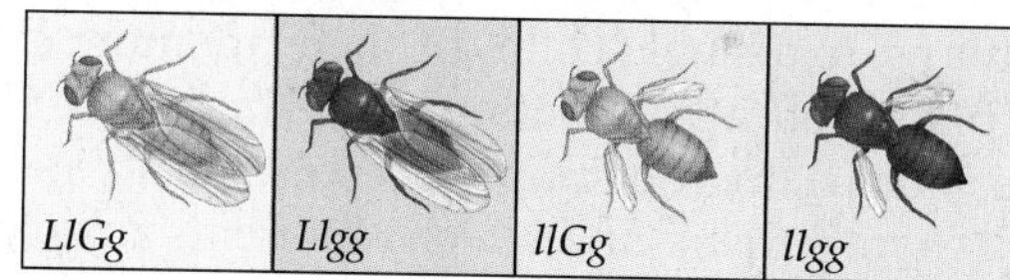


The presence of the offspring with vestigial wings and a black body shows that the test fly is heterozygous for both traits and has the genotype *LlGg*. Otherwise, it could not have this offspring. In general, you will want to remember that *the expected phenotypic ratio for this type of two-trait cross (heterozygous for two traits × recessive for both traits) is always 1:1:1:1.*

**Check Your Progress** **11.2D**

1. A heterozygous fruit fly (*LlGg*) is crossed with a homozygous recessive (*llgg*). What are the chances of offspring with long wings and a black body?
2. Using the key above for fruit flies, what are the most likely genotypes of the parents if a student gets the following phenotypic results? **a.** 1:1:1:1 **b.** 9:3:3:1
3. In horses, trotter (*T*) is dominant over pacer (*t*). A trotter is mated to a pacer, and the offspring is a pacer. Give the genotype of all the horses.

## Mendel's Laws and Human Genetic Disorders

Many traits and disorders in humans, and other organisms also, are genetic in origin and follow Mendel's laws. These traits are controlled by a single pair of alleles on the autosomal chromosomes. An **autosome** is any chromosome other than a sex (X or Y) chromosome.

### *Autosomal Patterns of Inheritance*

When a genetic disorder is autosomal dominant, the normal allele (*a*) is recessive, and an individual with the alleles *AA* or *Aa* has the disorder. When a genetic disorder is autosomal recessive, the normal allele (*A*) is dominant, and only individuals with the alleles *aa* have the disorder. A pedigree shows the pattern of inheritance for a particular condition and can be used by genetic counselors to determine whether a condition is dominant or recessive. Consider these two possible patterns of inheritance:

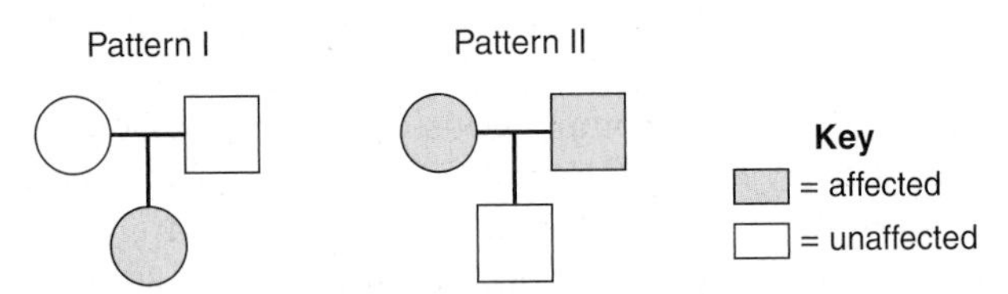


In both patterns, males are designated by squares and females by circles. Shaded circles and squares are affected individuals. The shaded boxes do not indicate whether the condition is dominant or recessive, only that the individual exhibits the trait. A line between a square and a circle represents a union. A vertical line going downward leads, in these patterns, to a single child. (If there are more children, they are placed off a horizontal line.) Which pattern of inheritance (I or II) do you suppose represents an autosomal dominant characteristic, and which represents an autosomal recessive characteristic?

In pattern I, the child is affected, but neither parent is; this can happen if the condition is recessive and both parents are *Aa*. Notice that the parents are **carriers** because they appear normal (do not express the trait) but are capable of having a child with the genetic disorder. In pattern II, the child is unaffected, but the parents are affected. This can happen if the condition is dominant and the parents are *Aa*.

Figure 11.8 shows other ways to recognize an autosomal recessive pattern of inheritance, and Figure 11.9 shows other ways to recognize an autosomal dominant pattern of inheritance. In these pedigrees, generations are indicated by Roman numerals placed on the left side. Notice in the third generation of Figure 11.8 that two closely related individuals have produced three children, two of which have the affected phenotype. In this case, a double line denotes consanguineous reproduction, or inbreeding, which is reproduction between two closely related individuals. This illustrates that inbreeding significantly increases the chances of children inheriting two copies of a potentially harmful recessive allele.

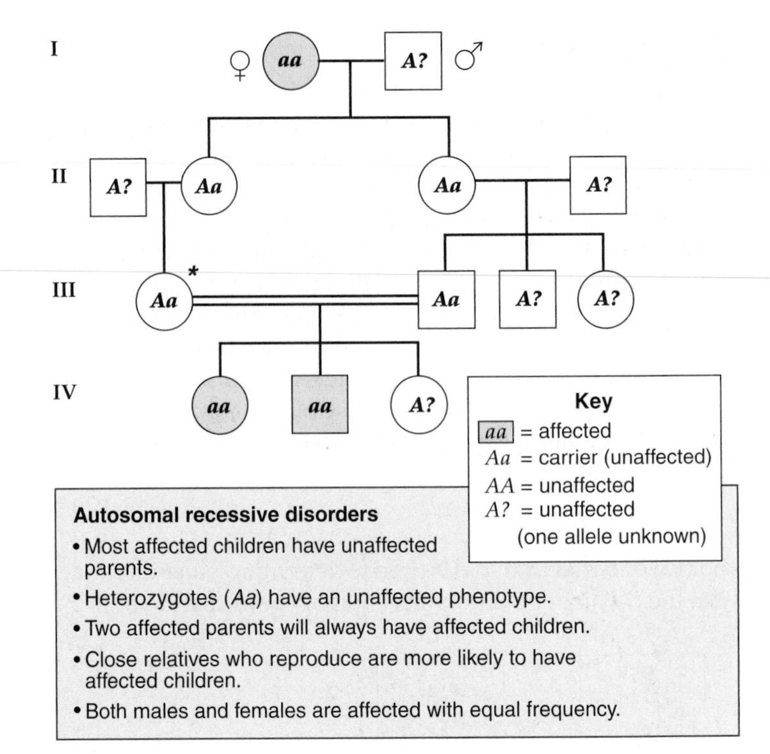


**FIGURE 11.8 Autosomal recessive pedigree.**

The list gives ways to recognize an autosomal recessive disorder. How would you know the individual at the asterisk is heterozygous?[1]

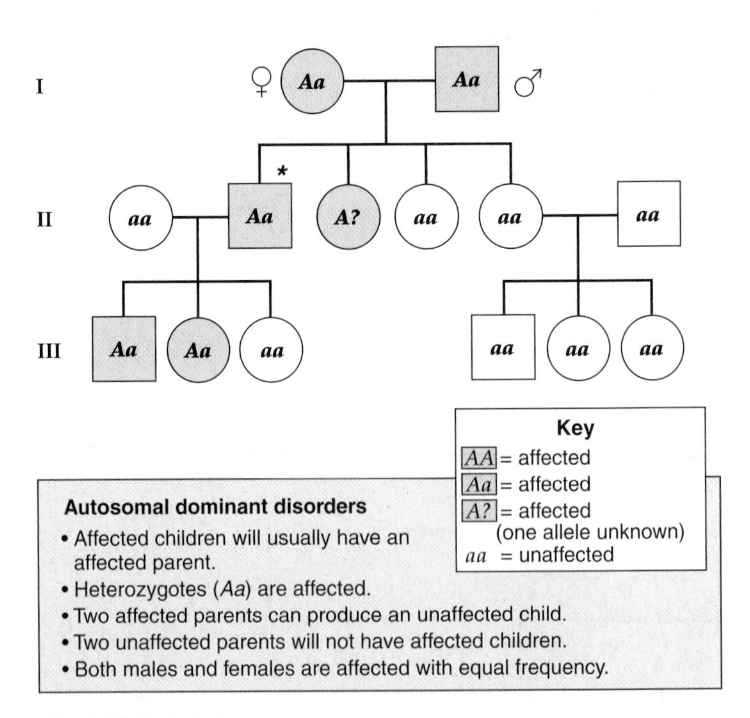


**FIGURE 11.9 Autosomal dominant pedigree.**

The list gives ways to recognize an autosomal dominant disorder. How would you know the individual at the asterisk is heterozygous?[1]

1. See Appendix A for answers.

## Autosomal Recessive Disorders

In humans, a number of autosomal recessive disorders have been identified. Here, we discuss methemoglobinemia, cystic fibrosis, and Niemann-Pick disease.

### *Methemoglobinemia*

Methemoglobinemia is a relatively harmless disorder that results from an accumulation of methemoglobin in the blood. While this disorder has been documented for centuries, the exact cause and genetic link remained mysterious. Although rarely mentioned, hemoglobin, the main oxygen-carrying protein in the blood, is usually converted at a slow rate to an alternate form called methemoglobin. Unlike hemoglobin, which is bright red when carrying oxygen, methemoglobin has a bluish color, similar to that of oxygen-poor blood. Although this process is harmless, individuals with methemoglobinemia are unable to clear the abnormal blue protein from their blood, causing their skin to appear bluish-purple in color (Fig. 11.10)!

A persistent and determined physician finally solved the age-old mystery of what causes methemoglobinemia by doing blood tests and pedigree analysis involving a family known as the blue Fugates of Troublesome Creek. Enzyme tests indicated that the blue Fugates lacked the enzyme diaphorase, coded for by a gene on chromosome 22. The enzyme normally converts methemoglobin back to hemoglobin. The physician treated the disorder in a simple, but rather unconventional manner. He injected the Fugates with a dye called methylene blue! This unusual dye can donate electrons to other compounds, successfully converting the excess methemoglobin back into normal hemoglobin. The results were striking but immediate—the patient's skin quickly turned pink after treatment.

A pedigree analysis of the Fugate family indicated that the trait is common in the family because so many carried the recessive allele.

### *Cystic Fibrosis*

Cystic fibrosis (CF) is the most common lethal genetic disease among Caucasians in the United States (Fig. 11.11). About 1 in 20 Caucasians is a carrier, and about 1 in 2,000 newborns has the disorder. CF patients exhibit a number of characteristic symptoms, the most obvious being extremely salty sweat. In children with CF, the mucus in the bronchial tubes and pancreatic ducts is particularly thick and viscous, interfering with the function of the lungs and pancreas. To ease breathing, the thick mucus in the lungs has to be loosened periodically, but still the lungs frequently become infected. The clogged pancreatic ducts prevent digestive enzymes from reaching the small intestine, and to improve digestion, patients take digestive enzymes mixed with applesauce before every meal.

Cystic fibrosis is caused by a defective chloride ion channel that is encoded by the *CFTR* allele on chromosome 7. Research has demonstrated that chloride ions ($Cl^-$) fail to pass through the defective version of the CFTR chloride ion channel, which is located on the plasma membrane. Ordinarily, after chloride ions have passed through the channel to the other side of the membrane, sodium ions ($Na^+$) and water follow. It is believed that lack of water is the cause of the abnormally thick mucus in the bronchial tubes and pancreatic ducts.

In the past few years, new treatments have raised the average life expectancy for CF patients to as much as 35 years of age. It is hoped that other novel treatments, such as gene therapy, may be able to correct the defect by placing a normal copy of the gene in patients to replace the faulty ones. To explain the persistence of the mutated *CFTR* allele in a population, it has been suggested that those heterozygous for CF are less likely to die from potentially fatal diseases, such as cholera.

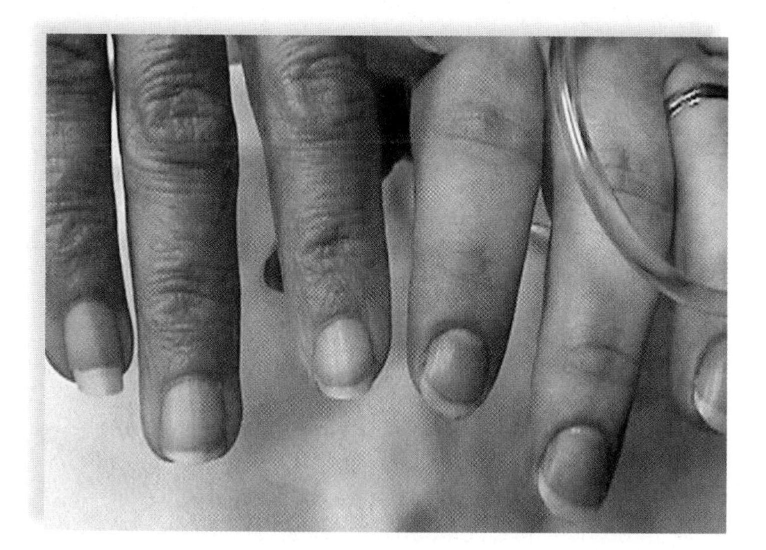

**FIGURE 11.10 Methemoglobinemia.**

The hands of the woman on the right appear blue due to chemically induced methemoglobinemia.

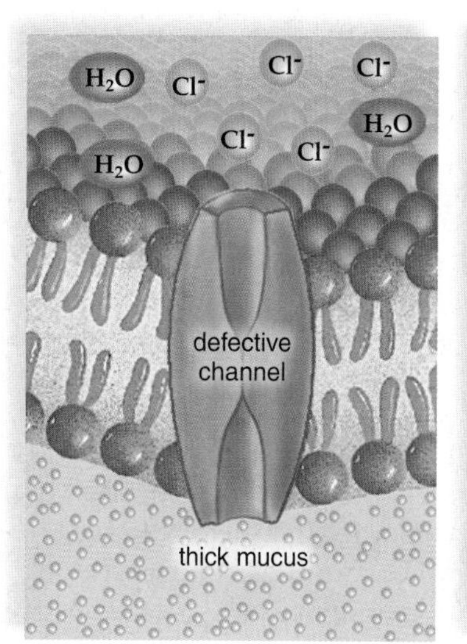


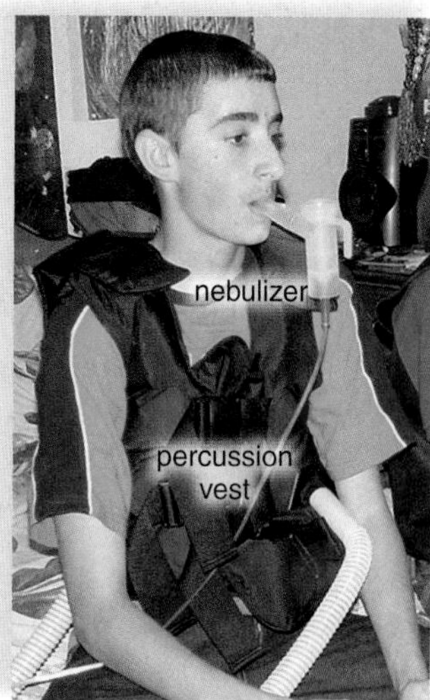


**FIGURE 11.11 Cystic fibrosis.**

Cystic fibrosis is due to a faulty protein that is supposed to regulate the flow of chloride ions into and out of cells through a channel protein.

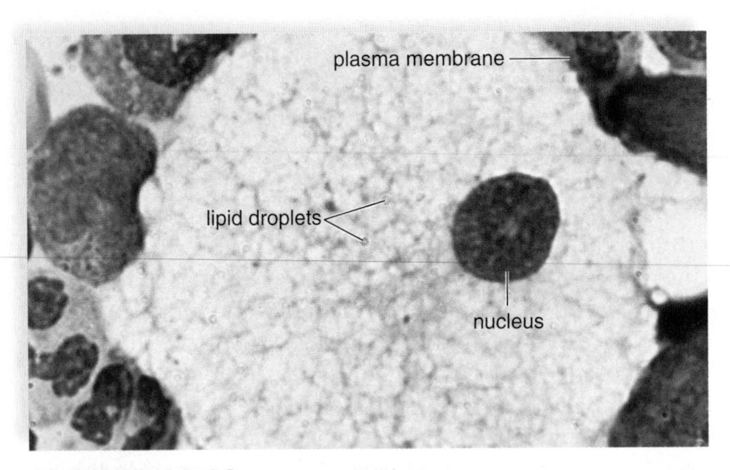


**FIGURE 11.12 Niemann-Pick disease.**

Persons with Niemann-Pick disease accumulate lipid droplets containing sphingomyelin within the cytoplasm of cells in the liver, spleen, and lymph nodes.

### *Niemann-Pick Disease*

In infants, a persistent jaundice, feeding difficulties, an enlarged abdomen, and pronounced mental retardation may signal to a medical professional that the child has Niemann-Pick disease.

Type A and B forms of Niemann-Pick disease are caused by defective versions of the same gene located on chromosome 11. This gene codes for acid sphingomyelinase, an enzyme that normally breaks down a lipid called sphingomyelin. Affected individuals accumulate lipid droplets within cells of the liver, lymph nodes, and spleen (Fig. 11.12). The abnormal accumulation of lipids causes enlargement of the abdomen, one of the hallmarks of the disease. In more severe cases, the lipids build up within the brain as well, causing the severe neurological problems characteristic of type A. Although both A and B forms of Niemann-Pick disease are caused by defective versions of the same gene, type B is the milder form because the protein product of its allele has some activity, while the protein product of the type A allele is totally inactive.

## Autosomal Dominant Disorders

A number of autosomal dominant disorders have been identified in humans. Two relatively well-known autosomal dominant disorders include osteogenesis imperfecta and hereditary spherocytosis.

### *Osteogenesis Imperfecta*

Osteogenesis [L. *os*, bone, *genesis*, origin] imperfecta is an autosomal dominant genetic disorder that results in weakened, brittle bones. Although there are at least nine types of the disorder, most are linked to mutations in two genes necessary to the synthesis of a type I collagen—one of the most abundant proteins in the human body. Collagen has many roles, including providing strength and rigidity to bone and forming the framework for most of the body's tissues. Osteogenesis imperfecta leads to a defective collagen I that causes the bones to be brittle and weak. Because the mutant collagen can cause structural defects even when combined with normal collagen I, osteogenesis imperfecta is generally considered to be dominant.

Osteogenesis imperfecta, which has an incidence of approximately 1 in 5,000 live births, affects all racial groups similarly, and has been documented as long as 300 years ago. Because he was often carried into battle on a shield and was known as Ivar, the Boneless, some historians suspect that the Viking chieftain, Ivar Ragnarsson had the condition. In most cases, the diagnosis is made in young children who visit the emergency room frequently due to broken bones. Some children with the disorder have an unusual blue tint in the sclera, the white portion of the eye, reduced skin elasticity, weakened teeth, and occasionally heart valve abnormalities. Currently, the disorder is treatable with a number of drugs that help to increase bone mass, but these drugs must be taken long-term.

### *Hereditary Spherocytosis*

Hereditary spherocytosis is an autosomal dominant genetic blood disorder that results from a defective copy of the ankyrin-1 gene found on chromosome 8. The protein encoded by this gene serves as a structural component of red blood cells, and is responsible for maintaining their disk-like shape. The abnormal spherocytosis protein is unable to perform its usual function, causing the affected person's red blood cells to adopt a spherical shape. As a result, the abnormal cells are fragile and burst easily, especially under osmotic stress. Enlargement of the spleen is also commonly seen in people with the disorder.

With an incidence of approximately 1 in 5,000, hereditary spherocytosis is one of the most common hereditary blood disorders. Roughly one-fourth of these cases result from new mutations and are not inherited from either parent. Hereditary spherocytosis exhibits incomplete penetrance, so not all individuals who inherit the mutant allele will exhibit the trait. The cause of incomplete penetrance in these cases and others remains poorly understood.

**Check Your Progress 11.2E**

1. What is the genotype of the child in Figure 11.11? What are the genotypes of his parents if neither parent has cystic fibrosis? (Use this key: *C* = normal; *c* = cystic fibrosis)
2. What is the chance that the parents in the above problem will have a child with cystic fibrosis?
3. What is the genotype of the woman in Figure 11B*a* if she is heterozygous? What is the genotype of a husband who is homozygous recessive? (Use this key: *H* = Huntington; *h* = unaffected)
4. What is the probability that the parents in the above problem will have a child with Huntington disease?

*science focus*

## Testing for Genetic Disorders

Many human genetic disorders such as Huntington disease and cystic fibrosis are the result of inheriting faulty genes. Huntington disease (Fig. 11B*a*) is a devastating neurological disease caused by the inheritance of a single dominant allele, while cystic fibrosis, being a recessive disorder, requires the inheritance of two recessive alleles. Many adults want to be tested to see if they have a particular genetic disease or if they are a carrier for a disease. A carrier appears to be normal but is capable of passing on the recessive allele for the disorder. When you are tested for a genetic disorder, what does the technician test? Your DNA, of course! Tests have been developed that can detect a particular sequence of bases, and this sequence tells whether you have the genetic disorder.

When researchers set out to develop a test for Huntington disease, they first obtained multiple **family pedigrees,** such as the one shown in Figure 11B*b*. This pedigree meets the requirements for a dominant allele: Every individual who is affected (shaded box or circle) has a parent who is also affected, heterozygotes are affected, and both males and females are affected in equal numbers. Each offspring of an affected individual has a 50% chance of getting the faulty gene and having Huntington disease, which doesn't appear until later in life.

The letters under the square or circle mean the individual has undergone a blood test that resulted in an analysis of their DNA. A computer was employed to search the DNA of all these individuals for similar base sequences. The computer found that a large number of individuals either had a sequence designated as J, K, or L. Only the sequence of bases designated as L appears in all the individuals with Huntington disease. Is this sequence a part of the gene for Huntington or is it in a gene that is linked to Huntington? Apparently, it is not in the gene for Huntington because at least one individual has the sequence but does not have Huntington disease. Several alleles can occur on the same chromosome, and these alleles are said to be linked. Linked alleles tend to go into the same gamete together, and this is the reason that alleles must be on separate chromosomes for the law of independent assortment to hold. Still, even genes that are closely linked can undergo crossing-over and become unlinked on occasion. Testable sequences that are closely linked to that of the faulty gene are called genetic markers, and genetic markers can be used as tests for genetic disorders, such as Huntington disease.

Association studies are another way for researchers to find possible sequences that indicate someone has a genetic disorder. During an association study, the DNA of a diverse sample of the general population is tested to find similar DNA sequences. If, for example, it turns out that many people who have type 2 diabetes have a particular sequence, this sequence might be used as a genetic marker for type 2 diabetes. With the advent of the human genome project, which resulted in the sequencing of all the bases in human DNA, it has been possible to successfully identify many genes that were formerly only tied to a particular chromosome by use of markers.

The mapping of disorders to genes within the human genome, while often painstaking and difficult to accomplish, has yielded much valuable information to the scientific community. The information can be used in prenatal genetic testing, for diagnosis of the disorder in individuals before symptoms occur, and for carrier testing in the case of disorders that are recessive. This information can be used to further understand the origin, progression, and pathology of the disorder, which may also lead to novel treatment methods. New techniques and technologies have greatly accelerated this process, but the tried and true methods of family pedigrees and association studies are still the primary techniques used by geneticists in pursuing the cure for many human genetic ailments.

a.

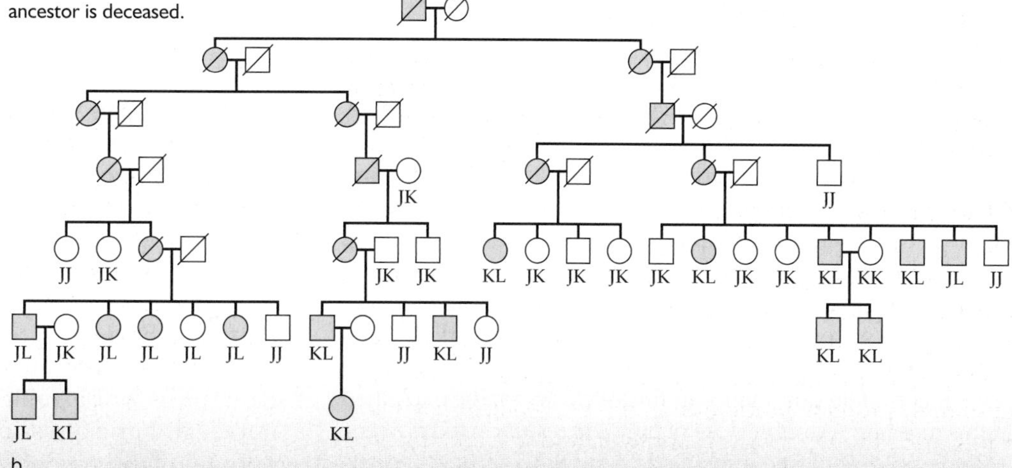


b.

**FIGURE 11B Blood sample testing.**
**a.** Huntington disease is a devastating neurological condition. **b.** In order to develop a test for Huntingon disease, researchers used white blood cells to discover that a particular sequence of DNA bases (L) is always present when a person has Huntington. The pedigree chart shows that (L) is not present unless an ancestor had Huntington disease. The backslash indicates that the ancestor is deceased.

# 11.3 Extending the Range of Mendelian Genetics

Mendelian genetics can also be applied to complex patterns of inheritance, such as multiple alleles, incomplete dominance, pleiotropy, and polygenic inheritance.

## Multiple Allelic Traits

When a trait is controlled by **multiple alleles,** the gene exists in several allelic forms. For example, while a person's ABO blood type is controlled by a single gene pair, there are three possible alleles that determine the blood type. These alleles determine the presence or absence of antigens on red blood cells.

$I^A$ = A antigen on red blood cells
$I^B$ = B antigen on red blood cells
$i$ = Neither A nor B antigen on red blood cells

The possible phenotypes and genotypes for blood type are as follows:

| *Phenotype* | *Genotype* |
|---|---|
| A | $I^AI^A$, $I^Ai$ |
| B | $I^BI^B$, $I^Bi$ |
| AB | $I^AI^B$ |
| O | $ii$ |

The inheritance of the ABO blood group in humans is also an example of **codominance** because both $I^A$ and $I^B$ are fully expressed in the presence of the other. Therefore, a person inheriting one of each of these alleles will have type AB blood. On the other hand, both $I^A$ and $I^B$ are dominant over $i$. Therefore, there are two possible genotypes for type A blood, and two possible genotypes for type B blood. Use a Punnett square to confirm that reproduction between a heterozygote with type A blood and a heterozygote with type B blood can result in any one of the four blood types. Such a cross makes it clear that an offspring can have a different blood type from either parent, and for this reason, DNA fingerprinting is now used to identify the parents of an individual instead of blood type.

## Incomplete Dominance and Incomplete Penetrance

**Incomplete dominance** is exhibited when the heterozygote has an intermediate phenotype between that of either homozygote. In a cross between a true-breeding, red-flowered four-o'clock strain and a true-breeding, white-flowered strain, the offspring have pink flowers. But this is not an example of the blending inheritance. When the pink plants self-pollinate, the offspring have a phenotypic ratio of 1 red-flowered : 2 pink-flowered : 1 white-flowered plant. The reappearance of the three phenotypes in this generation makes it clear that we are still dealing with a single pair of alleles (Fig. 11.13).

Incomplete dominance in four-o'clocks can be explained in this manner: A double dose of pigment results in red flowers; a single dose of pigment results in pink flowers; and because white flowers produce no pigment, the flowers are white.

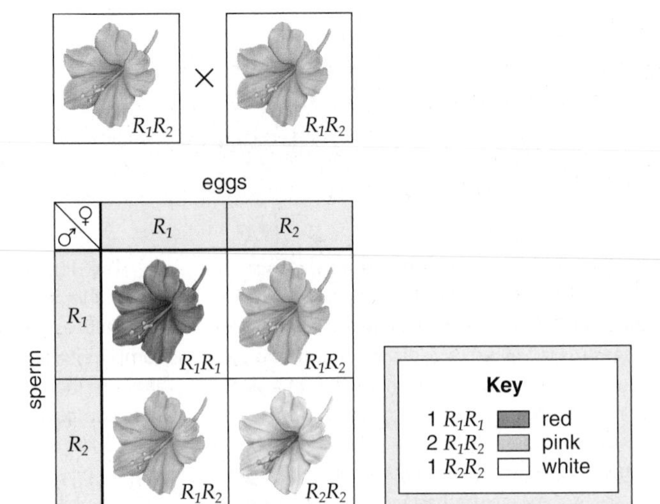


**FIGURE 11.13 Incomplete dominance.**

When pink four-o'clocks self-pollinate, the results show three phenotypes. This is only possible if the pink parents had an allele for red pigment ($R_1$) and an allele for no pigment ($R_2$). Note that alleles involved in incomplete dominance are both given a capital letter.

### *Human Examples of Incomplete Dominance*

In humans, familial hypercholesterolemia (FH) is an example of incomplete dominance. An individual with two alleles for this disorder develops fatty deposits in the skin and tendons and may have a heart attack as a child. An individual with one normal allele and one *FH* allele may suffer a heart attack as a young adult, and an individual with two normal alleles does not have the disorder.

Perhaps the inheritance pattern of other human disorders should be considered one of incomplete dominance. To detect the carriers of cystic fibrosis, for example, it is customary to determine the amount of cellular activity of the gene. When the activity is one-half that of the dominant homozygote, the individual is a carrier, even though the individual does not exhibit the genetic disease. In other words, at the level of gene expression, the homozygotes and heterozygotes do differ in the same manner as four-o'clock plants.

In some cases, a dominant allele may not always lead to the dominant phenotype in a heterozygote, even when the alleles show a true dominant/recessive relationship. The dominant allele in this case does not always determine the phenotype of the individual, so we describe these traits as showing **incomplete penetrance.** Many dominant alleles exhibit varying degrees of penetrance.

The best-known example is polydactyly, the presence of one or more extra digits on hands, feet, or both. Polydactyly is inherited in an autosomal dominant manner; however, not all individuals who inherit the dominant allele will exhibit the trait. The reasons for this are not clear, but expression of polydactyly may require additional environmental factors or be influenced by other genes, as discussed again later.

## Pleiotropic Effects

**Pleiotropy** occurs when a single mutant gene affects two or more distinct and seemingly unrelated traits. For example, persons with Marfan syndrome have disproportionately long arms, legs, hands, and feet; a weakened aorta; poor eyesight; and other characteristics (Fig. 11.14). All of these characteristics are due to the production of abnormal connective tissue. Marfan syndrome has been linked to a mutated gene ($FBN_1$) on chromosome 15 that ordinarily specifies a functional protein called fibrillin. Fibrillin is essential for the formation of elastic fibers in connective tissue. Without the structural support of normal connective tissue, the aorta can burst, particularly if the person is engaged in a strenuous sport, such as volleyball or basketball. Flo Hyman may have been the best American woman volleyball player ever, but she fell to the floor and died at the age of only 31 because her aorta gave way during a game. Now that coaches are aware of Marfan syndrome, they are on the lookout for it among very tall basketball players. Chris Weisheit, whose career was cut short after he was diagnosed with Marfan syndrome, said, "I don't want to die playing basketball."

Many other disorders, including porphyria and sickle-cell disease, are examples of pleiotropic traits. Porphyria is caused by a chemical insufficiency in the production of hemoglobin, the pigment that makes red blood cells red. The symptoms of porphyria are photosensitivity, strong abdominal pain, port-wine-colored urine, and paralysis in the arms and legs. Many members of the British royal family in the late 1700s and early 1800s suffered from this disorder, which can lead to epileptic convulsions, bizarre behavior, and coma.

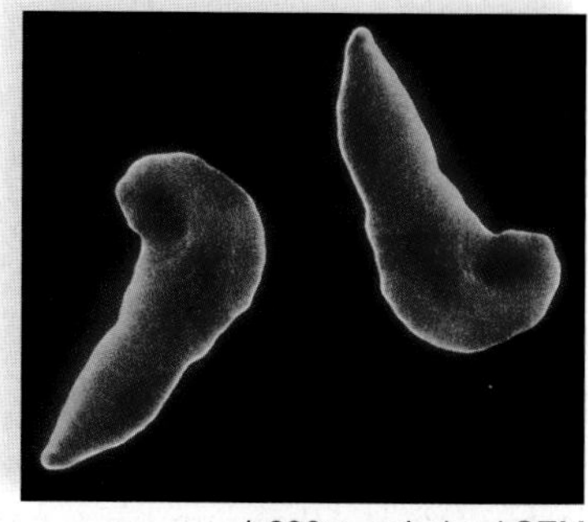

1,600×, colorized SEM
Sickled red blood cell

In a person suffering from sickle-cell disease ($Hb^SHb^S$), the cells are sickle-shaped. The underlying mutation is in a gene that codes for a type of polypeptide chain in hemoglobin. Of 146 amino acids, the mutation changes only one amino acid, but the result is a less soluble polypeptide chain that stacks up and causes red blood cells to be sickle-shaped. The abnormally shaped sickle cells slow down blood flow and clog small blood vessels. In addition, sickled red blood cells have a shorter life span than normal red blood cells. Affected individuals may exhibit a number of symptoms, including severe anemia, physical weakness, poor circulation, impaired mental function, pain and high fever, rheumatism, paralysis, spleen damage, low resistance to disease, and kidney and heart failure. All of thse effects are due to the tendency of sickled red blood cells to break down and to the resulting decreased oxygen-carrying capacity of the blood and the damage the body suffers as a result of the condition. Although sickle-cell disease is a devastating disorder, it provides heterozygous individuals with a survival advantage. People who have sickle-cell trait are resistant to the protozoan parasite that causes malaria. The parasite spends part of its life cycle in red blood cells feeding on hemoglobin, but it cannot complete its life cycle when sickle-shaped cells form and break down earlier than usual.

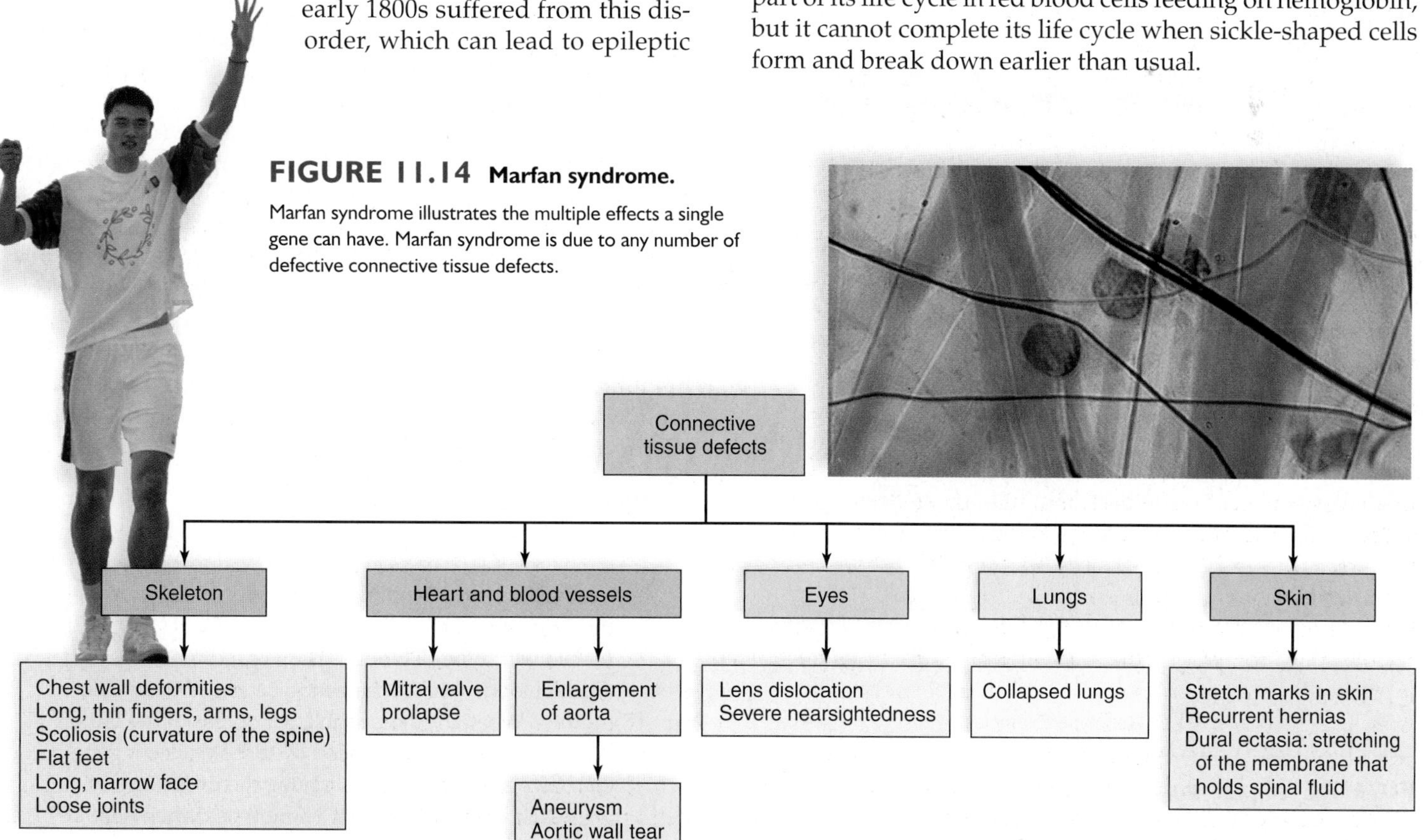


**FIGURE 11.14 Marfan syndrome.**

Marfan syndrome illustrates the multiple effects a single gene can have. Marfan syndrome is due to any number of defective connective tissue defects.

## Polygenic Inheritance

**Polygenic inheritance** [Gk. *poly*, many; L. *genitus*, producing] occurs when a trait is governed by two or more sets of alleles. The individual has a copy of all allelic pairs, possibly located on many different pairs of chromosomes. Each dominant allele has a quantitative effect on the phenotype, and these effects are additive. Therefore, a population is expected to exhibit continuous phenotypic variations. In Figure 11.15, a cross between genotypes *AABBCC* and *aabbcc* yields $F_1$ hybrids with the genotype *AaBbCc*. A range of genotypes and phenotypes results in the $F_2$ generation that can be depicted as a bell-shaped curve (Fig. 11.15). **Multifactorial traits** are controlled by polygenes subject to environmental influences. We observed previously (see Fig. 6.9) that the coat color of a Siamese cat is darker in color at the ears, nose, paws, and tails because an enzyme involved in the production of melanin is active only at a low temperature. Similarly, multifactoral traits are controlled by polygenes subject to environmental affects.

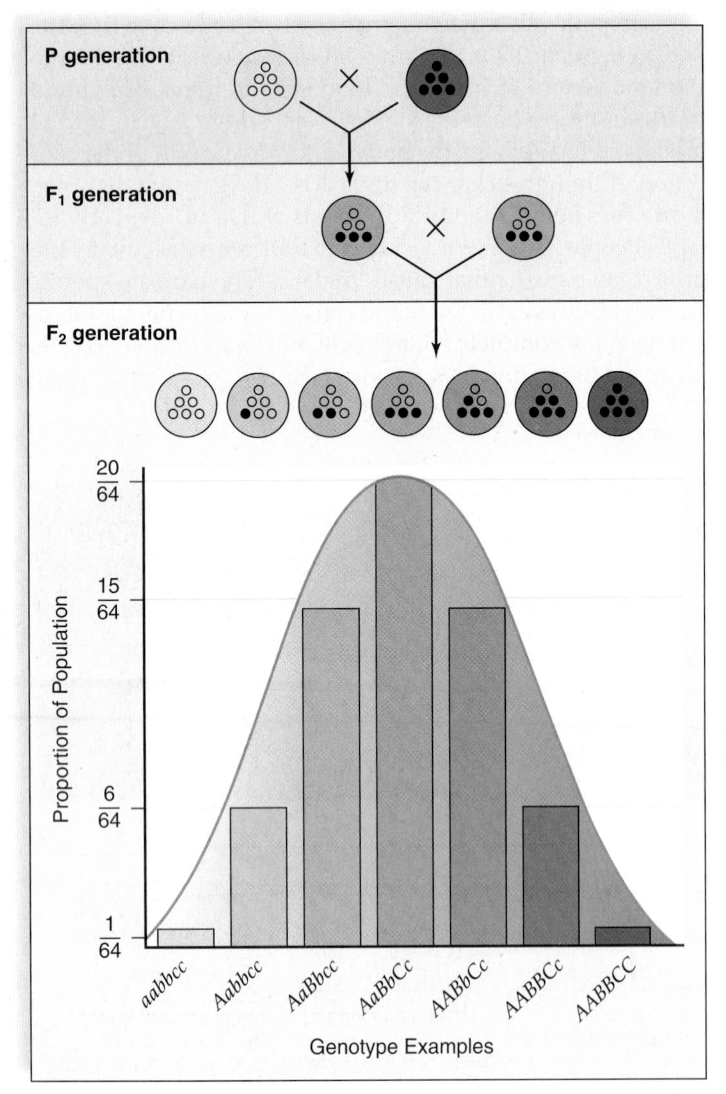


**FIGURE 11.15 Polygenic inheritance.**

In polygenic inheritance, a number of pairs of genes control the trait. *Above:* Black dots and intensity of blue shading stand for the number of dominant alleles. *Below:* Orange shading shows the degree of environmental influences.

### *Human Examples of Multifactorial Inheritance*

Human skin color and height are examples of polygenic traits affected by the environment. For example, exposure to the sun can affect skin color and nutrition can affect human height. Just how many pairs of alleles control skin color is not known, but a range in colors can be explained on the basis of just two pairs when *each capital letter contributes equally to the pigment in the skin.*

| *Genotypes* | *Phenotypes* |
|---|---|
| *AABB* | Very dark |
| *AABb* or *AaBB* | Dark |
| *AaBb* or *AAbb* or *aaBB* | Medium brown |
| *Aabb* or *aaBb* | Light |
| *aabb* | Very light |

Eye color is also a polygenic trait. The amount of melanin deposited in the iris increases the darker color of the eye. Different eye colors from the brightest of blue to nearly black eyes are thought to be the result of two genes with alleles each interacting in an additive manner.

Many human disorders, such as cleft lip and/or palate, clubfoot, congenital dislocations of the hip, hypertension, diabetes, schizophrenia, and even allergies and cancers, are most likely due to the combined action of many genes plus environmental influences. In recent years, reports have surfaced that all sorts of behavioral traits, such as alcoholism, phobias, and even suicide, can be associated with particular genes. The relative importance of genetic and environmental influences on the phenotype can vary, but in some instances the role of the environment is clear. For example, cardiovascular disease is more prevalent among those whose biological or adoptive parents have cardiovascular disease. Can you suggest environmental reasons for this correlation, based on your study of Chapter 3?

Many investigators are trying to determine what percentage of various traits is due to nature (inheritance) and what percentage is due to nurture (the environment). Some studies use twins separated since birth, because if identical twins in different environments share the same trait, the trait is most likely inherited. Identical twins are more similar in their intellectual talents, personality traits, and levels of lifelong happiness than are fraternal twins separated at birth. Biologists conclude that all behavioral traits are partly heritable, and that genes exert their effects by acting together in complex combinations susceptible to environmental influences.

### Check Your Progress 11.3A

1. If the inheritance pattern for a genetic disorder was exemplified by incomplete dominance, what would be the genotype of the heterozygote (see Fig. 11.13)?
2. A child with type O blood is born to a mother with type A blood. What is the genotype of the child? The mother? What are the possible genotypes of the father?
3. A polygenic trait is controlled by three different gene loci. Give seven genotypes among the offspring that will result in seven different phenotypes when *AaBbCc* is crossed with *AaBbCc*.

## X-Linked Inheritance

The X and Y chromosomes in mammals determine the gender of the individual. Females are XX and males are XY. These chromosomes carry genes that control development and, in particular, if the Y chromosome contains an *SRY* gene, the embryo becomes a male. The term **X-linked** is used for genes that have nothing to do with gender, and yet they are carried on the X chromosome. The Y chromosome does not carry these genes and indeed carries very few genes. This type of inheritance was discovered in the early 1900s by a group at Columbia University, headed by Thomas Hunt Morgan. Morgan performed experiments with fruit flies, whose scientific name is *Drosophila melanogaster.* Fruit flies are even better subjects for genetic studies than garden peas. They can be easily and inexpensively raised in simple laboratory glassware: Females mate and then lay hundreds of eggs during their lifetimes; the generation time is short, taking only about ten days from egg to adult. Fruit flies have the same sex chromosome pattern as humans, and therefore Morgan's experiments with X-linked genes apply directly to humans.

### *Morgan's Experiment*

Morgan took a newly discovered mutant male with white eyes and crossed it with a red-eyed female:

| | ♀ | | ♂ |
|---|---|---|---|
| P | red-eyed | × | white-eyed |
| $F_1$ | red-eyed | | red-eyed |

From these results, he knew that red eyes are the dominant characteristic and white eyes are the recessive characteristic. He then crossed the $F_1$ flies. In the $F_2$ generation, there was the expected 3 red-eyed : 1 white-eyed ratio, but it struck him as odd that all of the white-eyed flies were males:

| | ♀ | | ♂ |
|---|---|---|---|
| $F_1 \times F_1$ | red-eyed | × | red-eyed |
| $F_2$ | red-eyed | | 1 red-eyed : 1 white-eyed |

Obviously, a major difference between the male flies and the female flies was their sex chromosomes. Could it be possible that an allele for eye color was on the Y chromosome but not on the X? This idea could be quickly discarded because usually females have red eyes, and they have no Y chromosome. Perhaps an allele for eye color was on the X, but not on the Y, chromosome. Figure 11.16 indicates that this explanation would match the results obtained in the experiment. These results support the chromosome theory of inheritance by showing that the behavior of a specific allele corresponds exactly with that of a specific chromosome—the X chromosome in *Drosophila*.

Notice that X-linked alleles have a different pattern of inheritance than alleles that are on the autosomes because the Y chromosome is lacking for these alleles, and the inheritance of a Y chromosome cannot offset the inheritance of an X-linked recessive allele. For the same reason, males always receive an X-linked recessive mutant allele from the female parent—they receive the Y chromosome from the male parent, and therefore sex-linked recessive traits appear much more frequently in males than in females.

### *Solving X-Linked Genetics Problems*

Recall that when solving autosomal genetics problems, the allele key and genotypes can be represented as follows:

| Allele key | Genotypes |
|---|---|
| $L$ = long wings | $LL, Ll, ll$ |
| $l$ = short wings | |

When predicting inheritance of sex-linked traits, however, it is necessary to indicate the sex chromosomes of

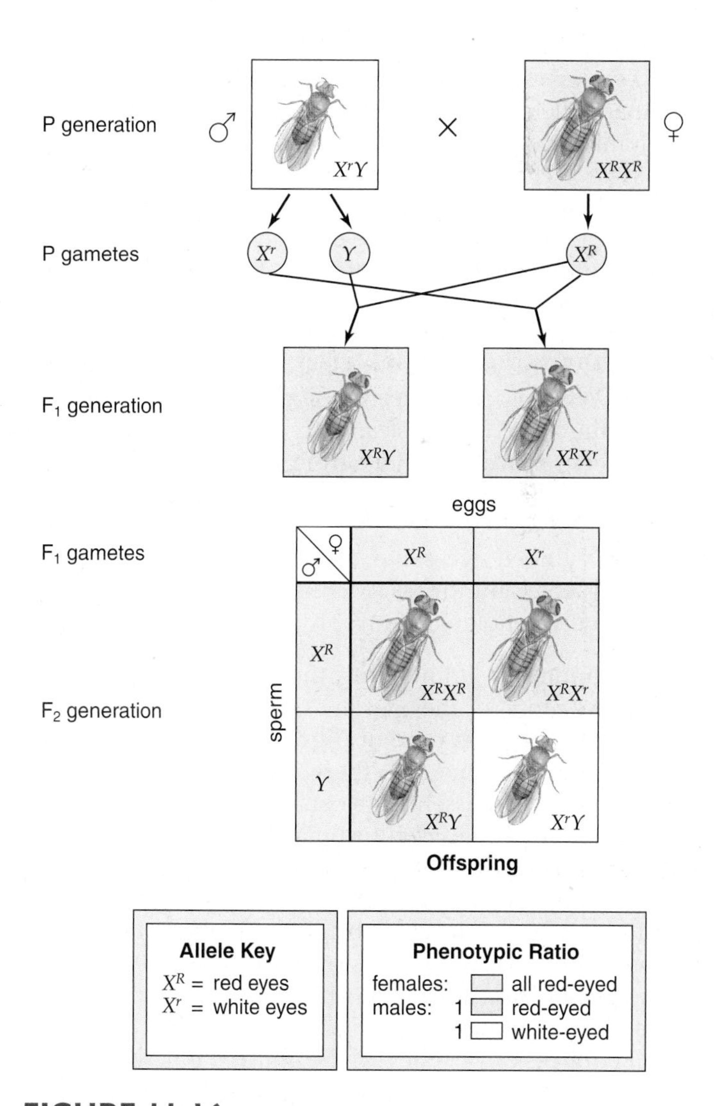


**FIGURE 11.16 X-linked inheritance.**

Once researchers deduced that the alleles for red/white eye color are on the X chromosome in *Drosophila*, they were able to explain their experimental results. Males with white eyes in the $F_2$ generation inherit the recessive allele only from the female parent; they receive a Y chromosome lacking the allele for eye color from the male parent.

each individual. As noted in Figure 11.16, however, the allele key for an X-linked gene shows an allele attached to the X:

Allele key
$X^R$ = red eyes
$X^r$ = white eyes

The possible genotypes in both males and females are as follows:

$X^RX^R$ = red-eyed female
$X^RX^r$ = red-eyed female
$X^rX^r$ = white-eyed female
$X^RY$ = red-eyed male
$X^rY$ = white-eyed male

Notice that there are three possible genotypes for females but only two for males. Females can be heterozygous $X^RX^r$, in which case they are carriers. Carriers usually do not show a recessive abnormality, but they are capable of passing on a recessive allele for an abnormality. But unlike autosomal traits, males cannot be carriers for X-linked traits; if the dominant allele is on the single X chromosome, they show the dominant phenotype, and if the recessive allele is on the single X chromosome, they show the recessive phenotype. For this reason, males are considered **hemizygous** for X-linked traits, because a male only possesses one allele for the trait and, therefore, expresses whatever allele is present on the X chromosome.

We know that male fruit flies have white eyes when they receive the mutant recessive allele from the female parent. What is the inheritance pattern when females have white eyes? Females can only have white eyes when they receive a recessive allele from both parents.

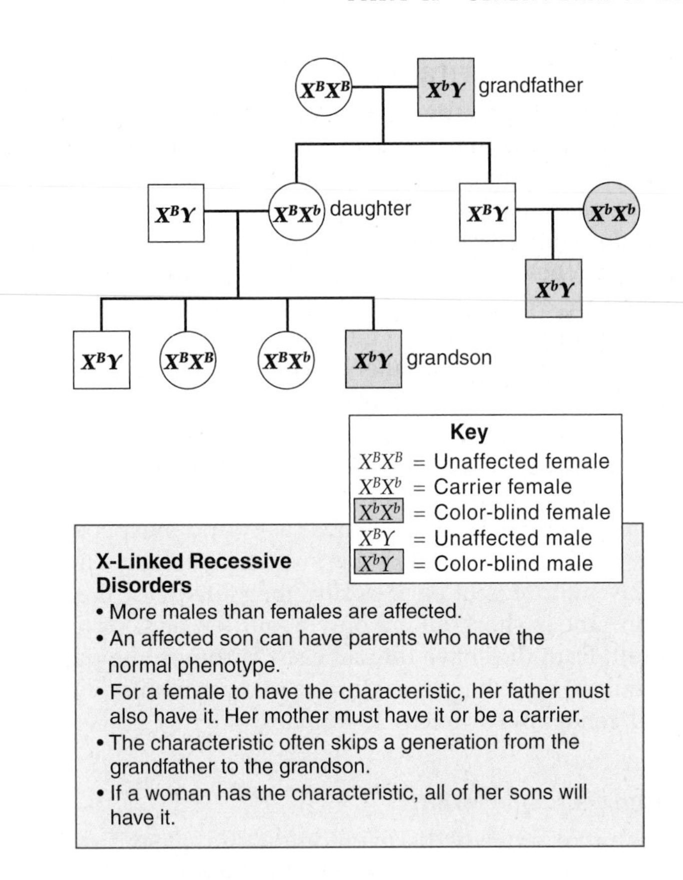


**FIGURE 11.17 X-linked recessive pedigree.**

This pedigree for color blindness exemplifies the inheritance pattern of an X-linked recessive disorder. The list gives various ways of recognizing the X-linked recessive pattern of inheritance.

## *Human X-Linked Disorders*

Several X-linked recessive disorders occur in humans including color blindness, Menkes syndrome, muscular dystrophy, adrenoleukodystrophy, and hemophilia.

**Color Blindness.** In humans, the receptors for color vision in the retina of the eyes are three different classes of cone cells. Only one type of pigment protein is present in each class of cone cell; there are blue-sensitive, red-sensitive, and green-sensitive cone cells. The allele for the blue-sensitive protein is autosomal, but the alleles for the red- and green-sensitive pigments are on the X chromosome. About 8% of Caucasian men have red-green color blindness. Most of these see brighter greens as tans, olive greens as browns, and reds as reddish browns. A few cannot tell reds from greens at all. They see only yellows, blues, blacks, whites, and grays.

Pedigrees can also reveal the unusual inheritance pattern seen in sex-linked traits. For example, the pedigree in Figure 11.17 shows the usual pattern of inheritance for color blindness. More males than females have the trait because recessive alleles on the X chromosome are expressed in males. The disorder often passes from grandfather to grandson through a carrier daughter.

**Menkes Syndrome.** Menkes syndrome, or kinky hair syndrome, is caused by a defective allele on the X chromosome. Normally, the gene product controls the movement of the metal copper in and out of cells. The symptoms of Menkes syndrome are due to accumulation of copper in some parts of the body, and the lack of the metal in other parts.

Symptoms of Menkes syndrome include poor muscle tone, seizures, abnormally low body temperature, skeletal anomalies, and the characteristic brittle, steely hair associated with the disorder. Although the condition is relatively rare, affecting approximately 1 in 100,000, mostly males, the prognosis for people with Menkes syndrome is poor, and most individuals die within the first few years of life. In recent years, some people with Menkes syndrome have been treated with injections of copper directly underneath the skin, but with mixed results, and treatment must begin very early in life to be effective.

**Muscular Dystrophy.** Muscular dystrophy, as the name implies, is characterized by a wasting away of the muscles. The most common form, Duchenne muscular dystrophy, is X-linked and occurs in about 1 out of every 3,600 male births (Fig. 11.18). Symptoms, such as waddling gait, toe walking, frequent falls, and difficulty in rising, may appear as soon as the child starts to walk. Muscle weakness intensifies until the individual is confined to a wheelchair. Death usually occurs by age 20; therefore, affected males are rarely fathers. The recessive allele remains in the population through passage from carrier mother to carrier daughter.

The allele for Duchenne muscular dystrophy has been isolated, and it was discovered that the absence of a protein called dystrophin causes the disorder. Much investigative work determined that dystrophin is involved in the release of calcium from the sarcoplasmic reticulum in muscle fibers. The lack of dystrophin causes calcium to leak into the cell, which promotes the action of an enzyme that dissolves muscle fibers. When the body attempts to repair the tissue, fibrous tissue forms, and this cuts off the blood supply so that more and more cells die.

A test is now available to detect carriers of Duchenne muscular dystrophy. Also, various treatments have been tried. Immature muscle cells can be injected into muscles, and for every 100,000 cells injected, dystrophin production occurs in 30–40% of muscle fibers. The allele for dystrophin has been inserted into thigh muscle cells, and about 1% of these cells then produced dystrophin.

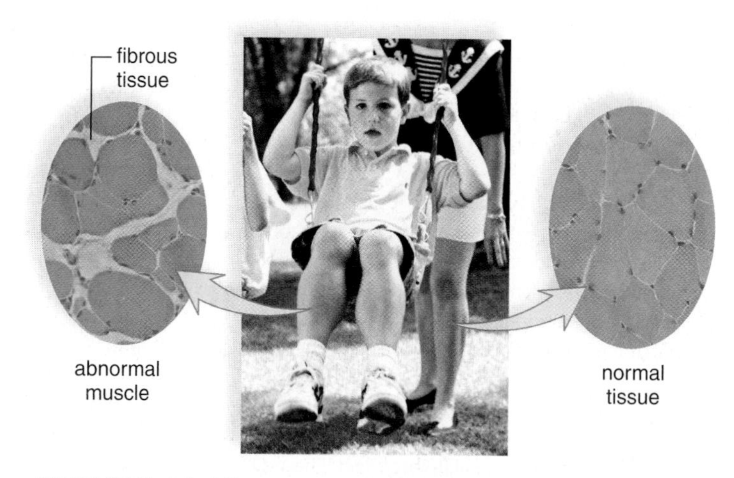


**FIGURE 11.18 Muscular dystrophy.**

In muscular dystrophy, an X-linked recessive disorder, calves enlarge because fibrous tissue develops as muscles waste away, due to lack of the protein dystrophin.

**Adrenoleukodystrophy.** Adrenoleukodystrophy, or ALD, is an X-linked recessive disorder due to the failure of a carrier protein to move either an enzyme or very long chain fatty acid (24–30 carbon atoms) into peroxisomes. As a result, these fatty acids are not broken down, and they accumulate inside the cell and the result is severe nervous system damage.

Children with ALD fail to develop properly after age 5, lose adrenal gland function, exhibit very poor coordination, and show a progressive loss of hearing, speech, and vision. The condition is usually fatal, with no known cure, but the onset and severity of symptoms in patients not yet showing symptoms may be mitigated by treatment with a mixture of lipids derived from olive oil. The disease was made famous by the 1992 movie *Lorenzo's Oil*, detailing a mother's and father's determination to devise a treatment for their son who was suffering from ALD.

**Hemophilia.** About 1 in 10,000 males is a hemophiliac. There are two common types of hemophilia: Hemophilia A is due to the absence or minimal presence of a clotting factor known as factor VIII, and hemophilia B is due to the absence of clotting factor IX. Hemophilia is called the bleeder's disease because the affected person's blood either does not clot or clots very slowly. Although hemophiliacs bleed externally after an injury, they also bleed internally, particularly around joints. Hemorrhages can be stopped with transfusions of fresh blood (or plasma) or concentrates of the clotting protein. Also, clotting factors are now available as biotechnology products.

At the turn of the century, hemophilia was prevalent among the royal families of Europe, and all of the affected males could trace their ancestry to Queen Victoria of England. Of Queen Victoria's 26 grandchildren, four grandsons had hemophilia and four granddaughters were carriers. Because none of Queen Victoria's relatives were affected, it seems that the faulty allele she carried arose by mutation either in Victoria or in one of her parents. Her carrier daughters Alice and Beatrice introduced the allele into the ruling houses of Russia and Spain, respectively. Alexis, the last heir to the Russian throne before the Russian Revolution, was a hemophiliac. There are no hemophiliacs in the present British royal family because Victoria's eldest son, King Edward VII, did not receive the allele.

### Check Your Progress 11.3B

1. In *Drosophila*, if a homozygous red-eyed female is crossed with a red-eyed male, what would be the possible genotypes of their offspring?
2. A woman is color-blind. **a.** What are the chances that her sons will be color-blind? **b.** If she is married to a man with normal vision, what are the chances that her daughters will be color-blind? **c.** Will be carriers?
3. In a cross between a brown-haired female and a black-haired male, all male offspring have brown hair and all female offspring have black hair. What is the genotype of all individuals involved, assuming X-linkage?

# Connecting the Concepts

A good experimental design and a bit of luck allowed Mendel to discover his laws of inheritance.

Although humans do not usually produce a large number of offspring, it has been possible to conclude that Mendel's laws do apply to humans in many instances. Good historical records of inheritance in large families, such as Mormon families, have allowed researchers to show that a number of human genetic disorders are indeed controlled by a single allelic pair. Such disorders include methemoglobinemia, cystic fibrosis, osteogenesis imperfecta, hereditary spherocytosis, and Marfan syndrome.

Mendel was lucky in that he chose to study an organism, namely the garden pea, whose observable traits are often determined by a single allelic pair. In most cases, however, traits are often determined by several genes or are affected by additional factors, such as the environment. These other types of inheritance patterns that differ from simple Mendelian inheritance are also discussed in this chapter.

The work of Morgan and others showed that the sex chromosomes contain genes unrelated to gender. Geneticists later discovered that some human genetic diseases, such as hemophilia, are caused by faulty genes on the X chromosome. With the help of Mendelian genetics, pedigree analysis, and statistics, scientists have been able to link many human diseases to specific genes on certain chromosomes. This knowledge later fueled an intense interest in deciphering exactly how these faulty genes could lead to such devastating diseases. As you will learn in the next chapter, genes on chromosomes direct the production of proteins in the cytoplasm of a cell through an RNA intermediate. It is the activity, or inactivity, of these proteins that leads to the observed phenotypes.

## summary

### 11.1 Gregor Mendel

Gregor Mendel used the garden pea as the subject in his genetic studies. In contrast to preceding plant breeders, his study involved nonblending traits of the garden pea. Mendel applied mathematics, followed the scientific method very closely, and kept careful records. Therefore, he arrived at a particulate theory of inheritance, effectively disproving the blending theory of inheritance.

### 11.2 Mendel's Laws

When Mendel crossed heterozygous plants with other heterozygous plants, he found that the recessive phenotype reappeared in about 1/4 of the $F_2$ plants; there was a 3:1 phenotypic ratio. This allowed Mendel to propose his law of segregation, which states that the individual has two factors for each trait, and the factors segregate into the gametes.

Mendel conducted two-trait crosses, in which the $F_1$ individuals showed both dominant characteristics, but there were four phenotypes among the $F_2$ offspring. (The actual phenotypic ratio was 9:3:3:1.) This allowed Mendel to deduce the law of independent assortment, which states that the members of one pair of factors separate independently of those of another pair. Therefore, all possible combinations of parental factors can occur in the gametes.

The laws of probability can be used to calculate the expected phenotypic ratio of a cross. A large number of offspring must be counted in order to observe the expected results, and to ensure that all possible types of sperm have fertilized all possible types of eggs, as is done in a Punnett square. The Punnett square uses the product law of probability to arrive at possible genotypes among the offspring, and then the sum law can be used to arrive at the phenotypic ratio.

Mendel also crossed the $F_1$ plants having the dominant phenotype with homozygous recessive plants. The 1:1 results indicated that the recessive factor was present in these $F_1$ plants (i.e., that they were heterozygous). Today, we call this a testcross, because it is used to test whether an individual showing the dominant characteristic is homozygous dominant or heterozygous. The two-trait testcross allows an investigator to test whether an individual showing two dominant characteristics is homozygous dominant for both traits or for one trait only, or is heterozygous for both traits.

Studies have shown that many human traits and genetic disorders can be explained on the basis of simple Mendelian inheritance. When studying human genetic disorders, biologists often construct pedigrees to show the pattern of inheritance of a characteristic within a family. The particular pattern indicates the manner in which a characteristic is inherited. Sample pedigrees for autosomal recessive and autosomal dominant patterns appear in Figures 11.8 and 11.9.

### 11.3 Extending the Range of Mendelian Genetics

Other patterns of inheritance have been discovered since Mendel's original contribution. For example, some genes have multiple alleles, although each individual organism has only two alleles, as in the inheritance of blood type in human beings. Inheritance of blood type also illustrates codominance. With incomplete dominance, the $F_1$ individuals are intermediate between the parent phenotypes; this does not support the blending theory because the parent phenotypes reappear in $F_2$. With incomplete penetrance, some traits that are dominant may not be expressed due to unknown reasons.

In pleiotropy, one gene has multiple effects as with Marfan syndrome and sickle-cell disease. Polygenic traits are controlled by several genes that have an additive effect on the phenotype, resulting in quantitative variations. A bell-shaped curve is seen because environmental influences bring about many intervening phenotypes, as in the inheritance of height in human beings. Skin color and eye color are also examples of multifactorial inheritance (polygenes plus the environment).

In *Drosophila*, as in humans, the sex chromosomes determined the sex of the individual, with XX being female and XY being male. Experimental support for the chromosome theory of inheritance came when Morgan and his group were able to determine that the gene for a trait unrelated to sex determination, the white-eyed allele in *Drosophila*, is on the X chromosome.

Alleles on the X chromosome are called X-linked alleles. Therefore, when doing X-linked genetics problems, it is the custom to indicate the sexes by using sex chromosomes and to indicate the alleles by superscripts attached to the X. The Y is blank because it does not carry these genes. Color blindness, Menkes syndrome, adrenoleukodystrophy, and hemophilia are X-linked recessive disorders in humans.

## understanding the terms

allele 193
autosome 198
carrier 198
codominance 202
dihybrid cross 194
dominant allele 193
family pedigree 201
gene locus 193
genotype 193
hemizygous 206
heterozygous 193
homozygous 193
incomplete dominance 202
incomplete penetrance 202
monohybrid cross 192
multifactorial trait 204
multiple alleles 202
phenotype 193
pleiotropy 203
polygenic inheritance 204
Punnett square 196
recessive allele 193
testcross 197
X-linked 205

Match the terms to these definitions:

a. ______________ Allele that exerts its phenotypic effect only in the homozygote; its expression is masked by a dominant allele.
b. ______________ Alternative form of a gene that occurs at the same locus on homologous chromosomes.
c. ______________ Polygenic trait that is subject to environmental affects.
d. ______________ Cross between an individual with the dominant phenotype and an individual with the recessive phenotype to see if the individual with the dominant phenotype is homozygous or heterozygous.
e. ______________ Genes of an organism for a particular trait or traits; for example, *BB* or *Aa*.

## reviewing this chapter

1. How did Mendel's procedure differ from that of his predecessors? What is his theory of inheritance called? 190
2. How does the $F_2$ of Mendel's one-trait cross refute the blending concept of inheritance? Using Mendel's one-trait cross as an example, trace his reasoning to arrive at the law of segregation. 190–92
3. Using Mendel's two-trait cross as an example, trace his reasoning to arrive at the law of independent assortment. 194
4. What are the two laws of probability, and how do they apply to a Punnett square? 196
5. What is a testcross, and when is it used? 197
6. How might you distinguish an autosomal dominant trait from an autosomal recessive trait when viewing a pedigree? 198
7. For autosomal recessive disorders, what are the chances of two carriers having an affected child? 199–200
8. For most autosomal dominant disorders, what are the chances of a heterozygote and a normal individual having an affected child? 200
9. Explain inheritance by multiple alleles. List the human blood types, and give the possible genotypes for each. 202
10. Explain the inheritance of incompletely dominant alleles and why this is not an example of blending inheritance. 202
11. Explain why traits controlled by polygenes show continuous variation and produce a distribution in the $F_2$ generation that follows a bell-shaped curve. 204
12. How do you recognize a pedigree for an X-linked recessive allele in human beings? 205–6

## testing yourself

Choose the best answer for each question. For questions 1–4, match each item to those in the key.

**KEY:**

a. 3:1
b. 9:3:3:1
c. 1:1
d. 1:1:1:1
e. 3:1:3:1

1. *TtYy* × *TtYy*
2. *Tt* × *Tt*
3. *Tt* × *tt*
4. *TtYy* × *ttyy*
5. Which of these could be a normal gamete?
   a. *GgRr*
   b. *GRr*
   c. *Gr*
   d. *GgR*
   e. None of these are correct.
6. Which of these properly describes a cross between an individual who is homozygous dominant for hairline but heterozygous for finger length and an individual who is recessive for both characteristics? (*W* = widow's peak, *w* = straight hairline, *S* = short fingers, *s* = long fingers)
   a. *WwSs* × *WwSs*
   b. *WWSs* × *wwSs*
   c. *Ws* × *ws*
   d. *WWSs* × *wwss*
7. In peas, yellow seed (*Y*) is dominant over green seed (*y*). In the $F_2$ generation of a monohybrid cross that begins when a dominant homozygote is crossed with a recessive homozygote, you would expect
   a. three plants with yellow seeds to every plant with green seeds.
   b. plants with one yellow seed for every green seed.
   c. only plants with the genotype *Yy*.
   d. only plants that produce yellow seeds.
   e. Both c and d are correct.
8. In humans, pointed eyebrows (*B*) are dominant over smooth eyebrows (*b*). Mary's father has pointed eyebrows, but she and her mother have smooth. What is the genotype of the father?
   a. *BB*
   b. *Bb*
   c. *bb*
   d. *BbBb*
   e. Any one of these is correct.
9. In guinea pigs, smooth coat (*S*) is dominant over rough coat (*s*), and black coat (*B*) is dominant over white coat (*b*). In the cross *SsBb* × *SsBb*, how many of the offspring will have a smooth black coat on average?
   a. 9 only
   b. about $^9/_{16}$
   c. $^1/_{16}$
   d. $^6/_{16}$
   e. $^2/_6$
10. In horses, *B* = black coat, *b* = brown coat, *T* = trotter, and *t* = pacer. A black trotter that has a brown pacer offspring would have which of the following genotypes?
    a. *BT*
    b. *BbTt*
    c. *bbtt*
    d. *BBtt*
    e. *BBTT*

11. In tomatoes, red fruit (*R*) is dominant over yellow fruit (*r*), and tallness (*T*) is dominant over shortness (*t*). A plant that is *RrTT* is crossed with a plant that is *rrTt*. What are the chances of an offspring possessing both recessive traits?
   a. none
   b. 1/2
   c. 1/4
   d. 3/4
12. In the cross *RrTt* × *rrtt*,
   a. all the offspring will be tall with red fruit.
   b. 75% (3/4) will be tall with red fruit.
   c. 50% (1/2) will be tall with red fruit.
   d. 25% (1/4) will be tall with red fruit.
13. A boy is color-blind (X-linked recessive) and has a straight hairline (autosomal recessive). Which could be the genotype of his mother?
   a. *bbww*
   b. $X^bYWw$
   c. $bbX^wX^w$
   d. $X^BX^bWw$
   e. $X^wX^wBb$
14. Which of the following would you *not* find in a pedigree when a male has an X-linked recessive disorder?
   a. Neither parent has the disorder.
   b. Only males in the pedigree have the disorder.
   c. Only females in the pedigree have the disorder.
   d. The sons of a female with the disorder will all have the disorder.
   e. Both a and c would not be seen.

For questions 15–17, match the statements to the items in the key.

**KEY:**

a. multiple alleles
b. polygenes
c. pleiotropic gene

15. People with sickle cell disease have many cardiovascular complications.
16. Although most people have an IQ of about 100, IQ generally ranges from about 50 to 150.
17. In humans, there are three possible alleles at the chromosomal locus that determine blood type.
18. Alice and Henry are at the opposite extremes for a polygenic trait. Their children will
   a. be bell-shaped.
   b. be a phenotype typical of a 9:3:3:1 ratio.
   c. have the middle phenotype between their two parents.
   d. look like one parent or the other.
19. Determine if the characteristic possessed by the shaded squares (males) and circles (females) is an autosomal dominant, autosomal recessive, or X-linked recessive.

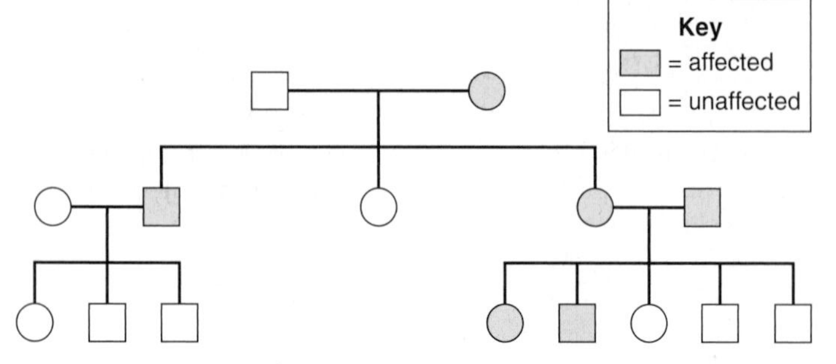


## additional genetics problems*

1. If a man homozygous for widow's peak (dominant) reproduces with a woman homozygous for straight hairline (recessive), what are the chances of their children having a widow's peak? A straight hairline?
2. A son with cystic fibrosis (autosomal recessive) is born to a couple who appear to be normal. What are the chances that any child born to this couple will have cystic fibrosis?
3. In horses, $B$ = black coat and $b$ = brown coat. What type of cross should be done to best determine whether a black-coated horse is homozygous dominant or heterozygous?
4. In a fruit fly experiment (see key on page 197), two gray-bodied fruit flies produce mostly gray-bodied offspring, but some offspring have black bodies. If there are 280 offspring, how many do you predict will have gray bodies and how many will have black bodies? How many of the 280 offspring do you predict will be heterozygous?
5. In humans, the allele for short fingers is dominant over that for long fingers. If a person with short fingers who had one parent with long fingers reproduces with a person having long fingers, what are the chances of each child having short fingers?
6. In humans, short fingers and widow's peak are dominant over long fingers and straight hairline. A heterozygote in both regards produces with a similar heterozygote. What is the chance of any one child having the same phenotype as the parents?
7. A man has type AB blood. What is his genotype? Could this man be the father of a child with type B blood? If so, what blood types could the child's mother have?
8. Is it possible for a woman who is homozygous dominant for normal color vision and a color-blind man to have a son who is color-blind? Why or why not?
9. Both the mother and father of a male hemophiliac appear normal. From whom did the son inherit the allele for hemophilia? What are the genotypes of the mother, the father, and the son?

*Answers to Additional Genetics Problems appear in Appendix A.

## thinking scientifically

1. You want to determine whether a newly found *Drosophila* characteristic is dominant or recessive. Would you wait to cross this male fly with another of its own kind or cross it now with a fly that lacks the characteristic?
2. You want to test if the leaf pattern of a plant is influenced by the amount of fertilizer in the environment. What would you do?

## *Biology* website

The companion website for *Biology* provides a wealth of information organized and integrated by chapter. You will find practice tests, animations, videos, and much more that will complement your learning and understanding of general biology.

**http://www.mhhe.com/maderbiology10**

# 12

# Molecular Biology of the Gene

*nearly 1.5 million different species of organisms have been discovered and named. This number represents a small portion of the total number of species on Earth. It certainly represents a small fraction of the total number of species that have ever lived. Yet one gene differs from another only by the sequence of the nucleotide bases in DNA. How does a difference in base sequence determine the uniqueness of a species—for example, whether an individual is a daffodil or a gorilla? Or, for that matter, whether a human has blue, brown, or hazel eyes? By studying the activity of genes in cells, geneticists have confirmed that proteins are the link between the genotype and the phenotype. Mendel's peas are smooth or wrinkled according to the presence or absence of a starch-forming enzyme. The allele* S *in peas dictates the presence of the starch-forming enzyme, whereas the allel* s *does not.*

*Through its ability to specify proteins, DNA brings about the development of the unique structures that make up a particular type of organism. When studying gene expression in this chapter, keep in mind this flow diagram: DNA's sequence of nucleotides → sequences of amino acids → specific enzymes → structures in organism.*

The diversity of life is dependent on gene activity.

## *concepts*

# 12.1 The Genetic Material

The middle of the twentieth century was an exciting period of scientific discovery. On one hand, geneticists were busy determining that *DNA (deoxyribonucleic acid)* is the genetic material of living things. On the other hand, biochemists were in a frantic race to describe the structure of DNA. The classic experiments performed during this era set the stage for an explosion in our knowledge of modern molecular biology.

When researchers began their work, they knew that the genetic material must be

1. able to *store information* that pertains to the development, structure, and metabolic activities of the cell or organism;
2. stable so that it *can be replicated* with high fidelity during cell division and be transmitted from generation to generation;
3. able to *undergo rare changes* called mutations [L. *muta*, change] that provide the genetic variability required for evolution to occur.

This chapter will show, as the researchers of the twentieth century did, that DNA can fulfill these functions.

## Transformation of Bacteria

During the late 1920s, the bacteriologist Frederick Griffith was attempting to develop a vaccine against *Streptococcus pneumoniae* (pneumococcus), which causes pneumonia in mammals. In 1931, he performed a classic experiment with the bacterium. He noticed that when these bacteria are grown on culture plates, some, called S strain bacteria, produce shiny, smooth colonies, and others, called R strain bacteria, produce colonies that have a rough appearance. Under the microscope, S strain bacteria have a capsule (mucous coat) but R strain bacteria do not. When Griffith injected mice with the S strain of bacteria, the mice died, and when he injected mice with the R strain, the mice did not die (Fig. 12.1). In an effort to determine if the capsule alone was responsible for the virulence (ability to kill) of the S strain bacteria, he injected mice with heat-killed S strain bacteria. The mice did not die.

Finally, Griffith injected the mice with a mixture of heat-killed S strain and live R strain bacteria. Most unexpectedly, the mice died and living S strain bacteria were recovered from the bodies! Griffith concluded that some substance necessary for the bacteria to produce a capsule and be virulent must have passed from the dead S strain bacteria to the living R strain bacteria so that the R strain bacteria were *transformed* (Fig. 12.1*d*). This change in the phenotype of the R strain bacteria must be due to a change in their genotype. Indeed, couldn't the transforming substance that passed from S strain to R strain be genetic material? Reasoning such as this prompted investigators at the time to begin looking for the transforming substance to determine the chemical nature of the genetic material.

### *DNA: The Transforming Substance*

By the time the next group of investigators, led by Oswald Avery, began their work, it was known that the genes are on the chromosomes and that the chromosomes contain both proteins and nucleic acids. Investigators were having a much heated debate about whether protein or DNA was the genetic material. Many thought that the protein component of chromosomes must be the genetic material because proteins contain 20 different amino acids that can be sequenced in any particular way. On the other hand, nucleic acids—DNA and RNA—contain only four types of **nucleotides.** Perhaps DNA did not have enough variability to be able to store information and be the genetic material!

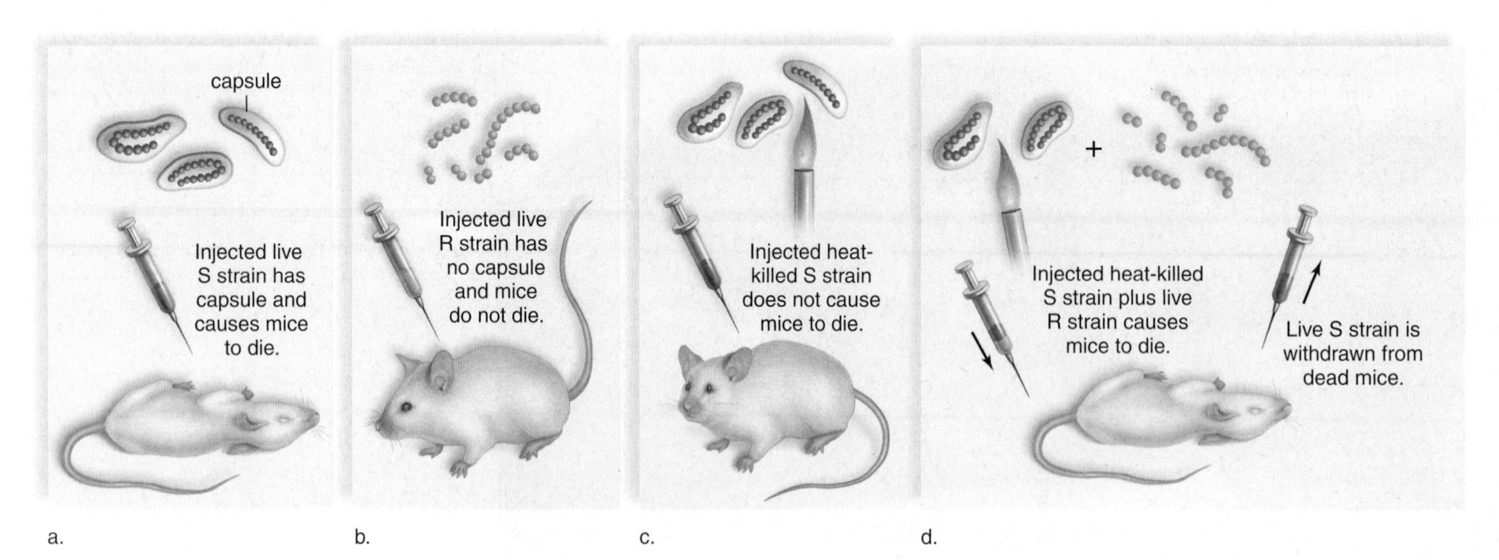


**FIGURE 12.1 Griffith's transformation experiment.**

**a.** Encapsulated S strain is virulent and kills mice. **b.** Nonencapsulated R strain is not virulent and does not kill mice. **c.** Heat-killed S strain bacteria do not kill mice. **d.** If heat-killed S strain and R strain are both injected into mice, they die because the R strain bacteria have been transformed into the virulent S strain.

Not surprising, Oswald Avery did not work with mice—it's inconvenient to be looking for the substance that transforms bacteria in mice. Avery and his group did in vitro experiments (in laboratory glassware). In 1944, after 16 years of research, Avery and his coinvestigators, Colin MacLeod and Maclyn McCarty, published a paper demonstrating that the transforming substance that allows *Streptococcus* to produce a capsule and be virulent is DNA. This meant that DNA is the genetic material. Their evidence included the following data:

1. DNA from S strain bacteria causes R strain bacteria to be transformed so that they can produce a capsule and be virulent. We know today that DNA codes for the enzymes that allow bacteria to make a capsule.
2. The addition of DNase, an enzyme that digests DNA, prevents transformation from occurring. This supports the hypothesis that DNA is the genetic material.
3. The molecular weight of the transforming substance is so great that it must contain about 1,600 nucleotides! Certainly this suggests the possibility of genetic variability.
4. The addition of enzymes that degrade proteins have no affect on the transforming substance nor does RNase, an enzyme that digests RNA. This shows that neither protein nor RNA is the genetic material.

These experiments certainly showed that DNA is the transforming substance and, therefore, the genetic material. Although some remained skeptical, many felt that the evidence for DNA being the genetic material was overwhelming.

## *Transformation of Organisms Today*

Transformation of organisms, resulting in so-called genetically modified organisms (GMOs), is an invaluable tool in modern biotechnology today. As discussed further in the next chapter, transformation of bacteria and other organisms has resulted in commercial products that are currently much used. Early biotechnologists seeking a dramatic way to show that it was possible to transfer a gene from one type of organism to another decided to make use of a jellyfish gene that codes for a green fluorescent protein (GFP).

When this gene is transferred to another organism, the organism glows in the dark (Fig. 12.2)! The basic technique is relatively simple. First, isolate the jellyfish gene and then transfer it to a bacterium, or the embryo of a plant, pig, or mouse. The result is a bioluminescent organism. Genes have no difficulty crossing the species barrier. Mammalian genes work just as well in bacteria, and an invertebrate gene, such as the GFP gene, has no trouble functioning in a bacterium, plant, or animal.

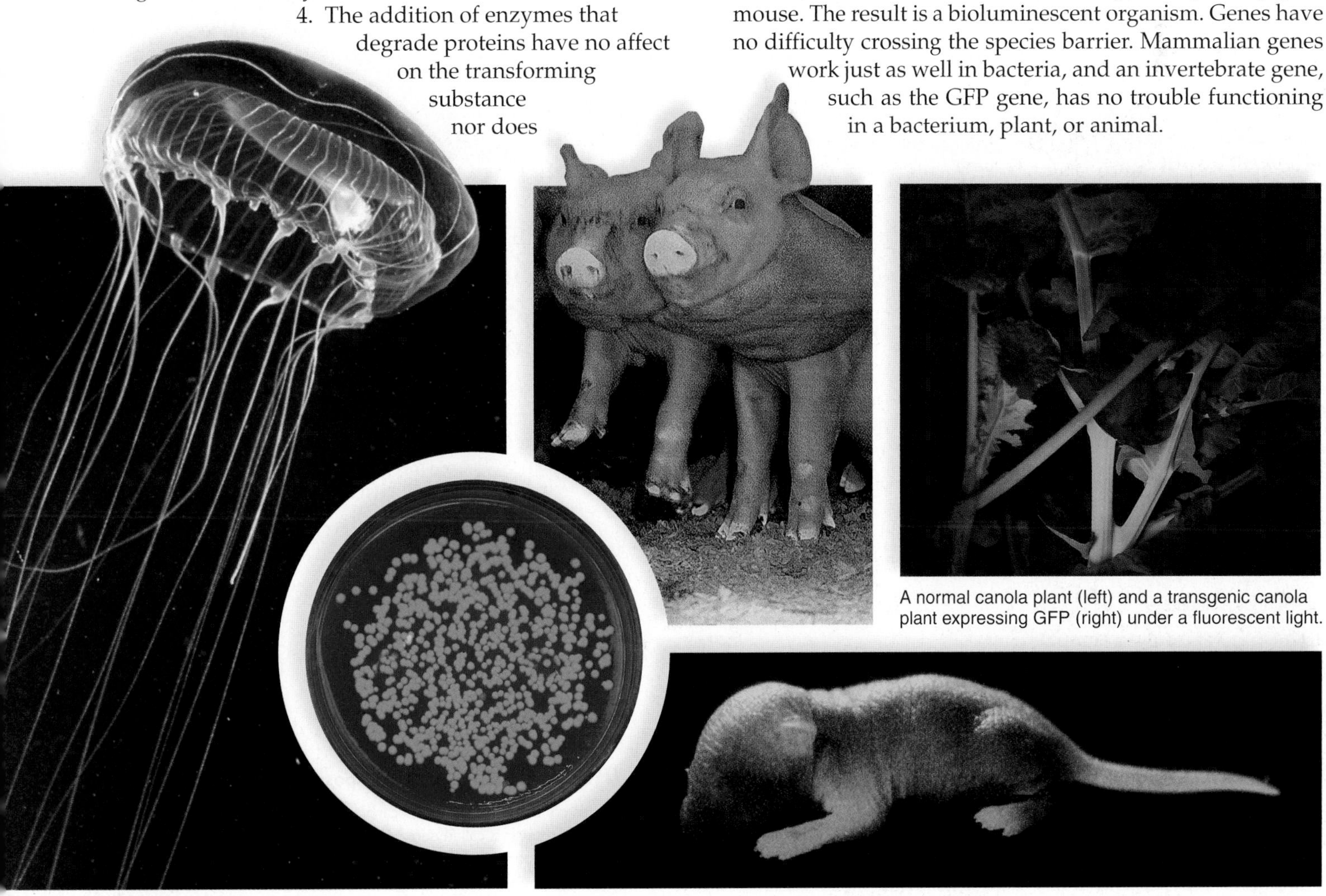

A normal canola plant (left) and a transgenic canola plant expressing GFP (right) under a fluorescent light.

**FIGURE 12.2 Transformation of organisms.**

When bacteria, plants, pigs, and mice are given a jellyfish gene for green fluorescent protein (GFP), these organisms glow in the dark.

## The Structure of DNA

By the 1950s, DNA was widely accepted as the genetic material of living things. But another fundamental question remained—what exactly is the structure of DNA, and how can a molecule with only four different nucleotides produce the great diversity of life on Earth?

One obstacle in describing the structure of DNA is understanding the base composition of DNA. To accomplish this, it is possible to turn to the work of Erwin Chargaff, who used new chemical techniques developed in the 1940s to analyze in detail the base content of DNA. It was known that DNA contains four different types of nucleotides: two with *purine* bases, **adenine (A)** and **guanine (G),** which have a double ring, and two with *pyrimidine* bases, **thymine (T)** and **cytosine (C),** which have a single ring (Fig. 12.3*a, b*). At first, chemists hypothesized that DNA has repeating units, each unit having four nucleotides—one for each of the four bases. If so, the DNA of every species would contain 25% of each kind of nucleotide.

A sample of Chargaff's data is seen in Figure 12.3*c*. You can see that while some species—*E. coli* and *Zea mays* (corn), for example—do have approximately 25% of each type of nucleotide, most do not. Further, the percentage of each type of nucleotide differs from species to species. Therefore, the nucleotide content of DNA is not fixed, and DNA does have the *variability* between species required of the genetic material.

Within each species, however, DNA was found to have the *constancy* required of the genetic material—that is, all members of a species have the same base composition. Also, the percentage of A always equals the percentage of T, and the percentage of G equals the percentage of C. The percentage of A + G equals 50%, and the percentage of T + C equals 50%. These relationships are called Chargaff's rules.

Chargaff's rules:

1. The amount of A, T, G, and C in DNA varies from species to species.
2. In each species, the amount of A = T and the amount of G = C.

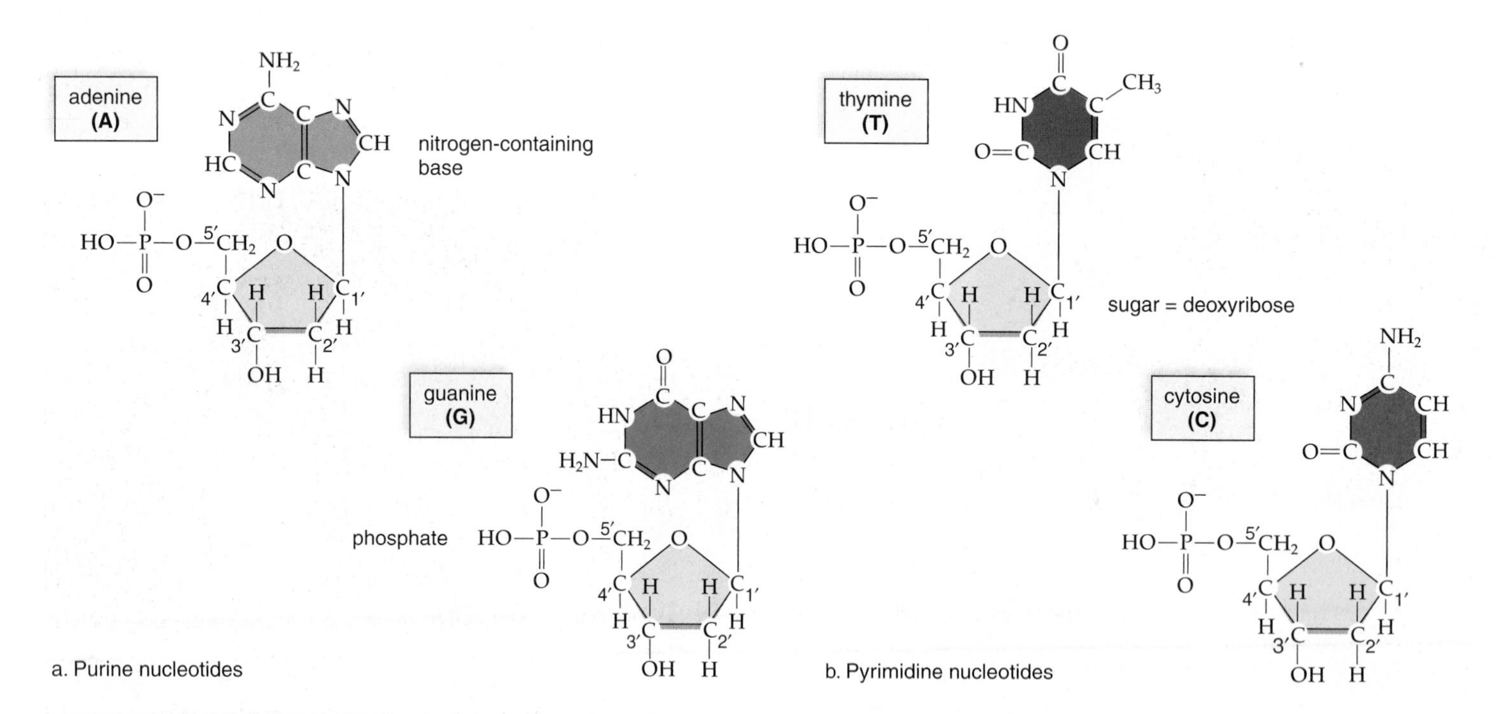


| DNA Composition in Various Species (%) | | | | |
|---|---|---|---|---|
| **Species** | **A** | **T** | **G** | **C** |
| *Homo sapiens* (human) | 31.0 | 31.5 | 19.1 | 18.4 |
| *Drosophila melanogaster* (fruit fly) | 27.3 | 27.6 | 22.5 | 22.5 |
| *Zea mays* (corn) | 25.6 | 25.3 | 24.5 | 24.6 |
| *Neurospora crassa* (fungus) | 23.0 | 23.3 | 27.1 | 26.6 |
| *Escherichia coli* (bacterium) | 24.6 | 24.3 | 25.5 | 25.6 |
| *Bacillus subtilis* (bacterium) | 28.4 | 29.0 | 21.0 | 21.6 |

c. Chargaff's data

**FIGURE 12.3 Nucleotide composition of DNA.**

All nucleotides contain phosphate, a 5-carbon sugar, and a nitrogen-containing base. In DNA, the sugar is called deoxyribose because it lacks an oxygen atom in the 2′ position, compared to ribose. The nitrogen-containing bases are (**a**) the purines adenine and guanine, which have a double ring, and (**b**) the pyrimidines thymine and cytosine, which have a single ring. **c.** Chargaff's data show that the DNA of various species differs. For example, in humans the A and T percentages are about 31%, but in fruit flies these percentages are about 27%.

While there are only four possible bases in each nucleotide position in DNA, the sheer length of most DNA molecules is more than sufficient to provide for variability. For example, it has been calculated that each human chromosome usually contains about 140 million base pairs. This provides for a staggering number of possible sequences of nucleotides. Because any of the four possible nucleotides can be present at each nucleotide position, the total number of possible nucleotide sequences is $4^{140 \times 10^6}$ or $4^{140,000,000}$. No wonder each species has its own base percentages!

## X-Ray Diffraction of DNA

Rosalind Franklin (Fig. 12.4*a*), a researcher in the laboratory of Maurice H. F. Wilkins at King's College in London, studied the structure of DNA using X-rays. She found that if a concentrated, viscous solution of DNA is made, it can be separated into fibers. Under the right conditions, the fibers are enough like a crystal (a solid substance whose atoms are arranged in a definite manner) that when X-rayed, an X-ray diffraction pattern results (Fig. 12.4*b*). The X-ray diffraction pattern of DNA shows that DNA is a double helix. The helical shape is indicated by the crossed (X) pattern in the center of the photograph in Figure 12.4*c*. The dark portions at the top and bottom of the photograph indicate that some portion of the helix is repeated.

## The Watson and Crick Model

James Watson, an American, was on a postdoctoral fellowship at Cavendish Laboratories in Cambridge, England, and while there he began to work with the biophysicist Francis H. C. Crick. Using the data provided from X-ray diffraction and other sources, they constructed a model of DNA for which they received a Nobel Prize in 1962.

Watson and Crick knew, of course, that DNA is a polymer of nucleotides, but they did not know how the nucleotides were arranged within the molecule. However, they deduced that DNA is a **double helix** with sugar-phosphate backbones on the outside and paired bases on the inside. This arrangement fits the mathematical measurements provided by Franklin's X-ray diffraction data for the spacing between the base pairs (0.34 nm) and for a complete turn of the double helix (3.4 nm).

According to Watson and Crick's model, the two DNA strands of the double helix are antiparallel, meaning that the sugar-phosphate groups of each strand are oriented in opposite directions. This means that the 5′ end of one strand is paired to the 3′ end of the other strand, and vice versa.

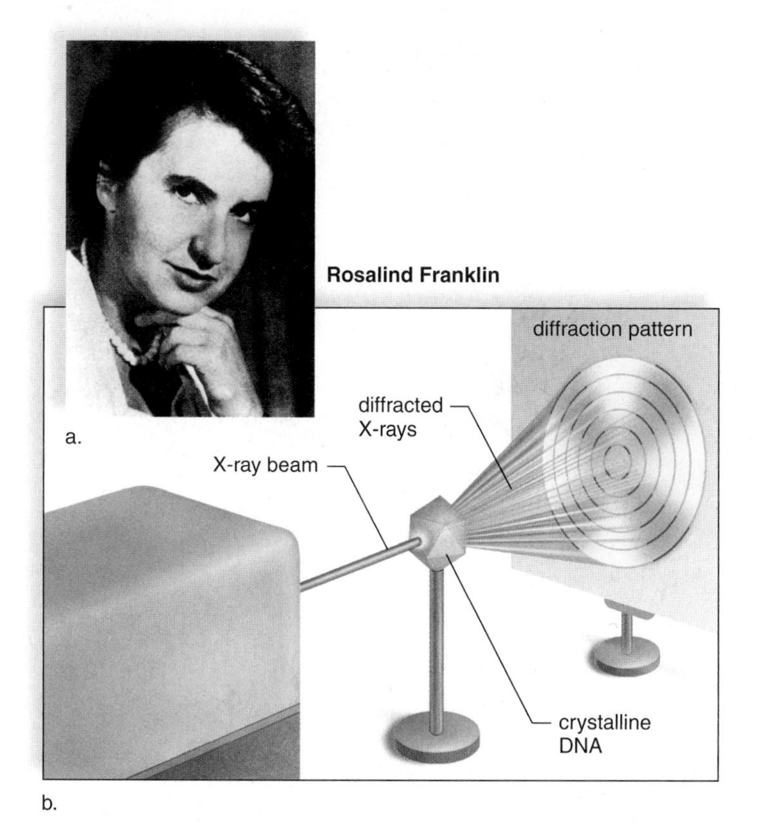


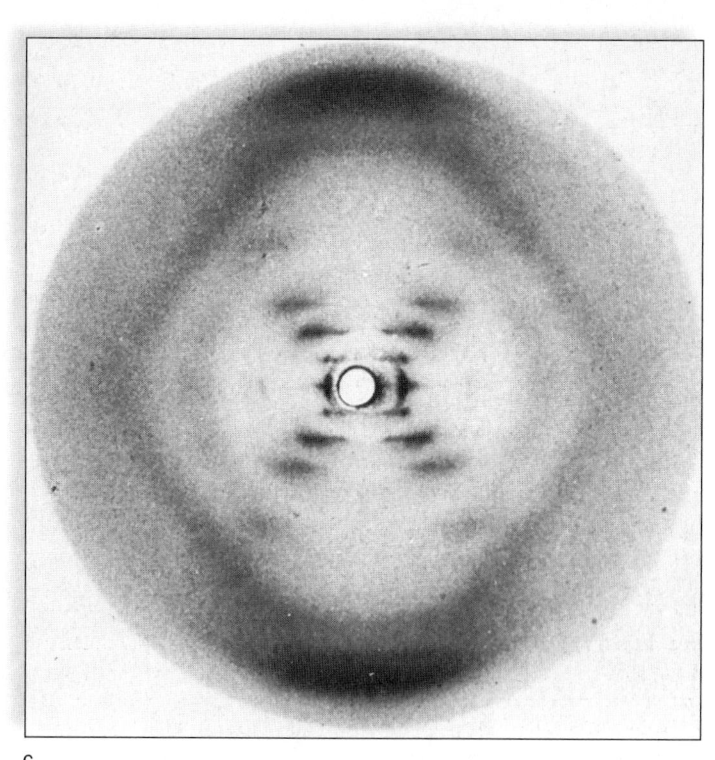


**FIGURE 12.4 X-ray diffraction of DNA.**

**a.** Rosalind Franklin, 1920–1958. **b.** When a crystal is X-rayed, the way in which the beam is diffracted reflects the pattern of the molecules in the crystal. The closer together two repeating structures are in the crystal, the farther from the center the beam is diffracted. **c.** The diffraction pattern of DNA produced by Rosalind Franklin. The crossed (X) pattern in the center told investigators that DNA is a helix, and the dark portions at the top and the bottom told them that some feature is repeated over and over. Watson and Crick determined that this feature was the hydrogen-bonded bases.

This model also agreed with Chargaff's rules, which said that A = T and G = C. Figure 12.5 shows that A is hydrogen-bonded to T, and G is hydrogen-bonded to C. This so-called **complementary base pairing** means that a purine is always bonded to a pyrimidine. The antiparallel arrangement of the two strands ensures that the bases are oriented properly so that they can interact. Only in this way will the molecule have the width revealed by Franklin's X-ray diffraction pattern, since two pyrimidines together are too narrow, and two purines together are too wide (Fig. 12.5).

The information stored within DNA must always be read in the correct order. As explained on page 218, each nucleotide possesses a phosphate group located at the 5′ position of the sugar. Nucleotides are joined together by linking the 5′ phosphate of one nucleotide to a free hydroxyl (—OH) located at the 3′ position on the sugar of the preceding nucleotide, giving the molecule directionality. Thus, a DNA strand is usually made in a 5′ to 3′ direction.

## Check Your Progress 12.1

1. What are the requirements for DNA to be the genetic material?
2. What are the major features of DNA structure?

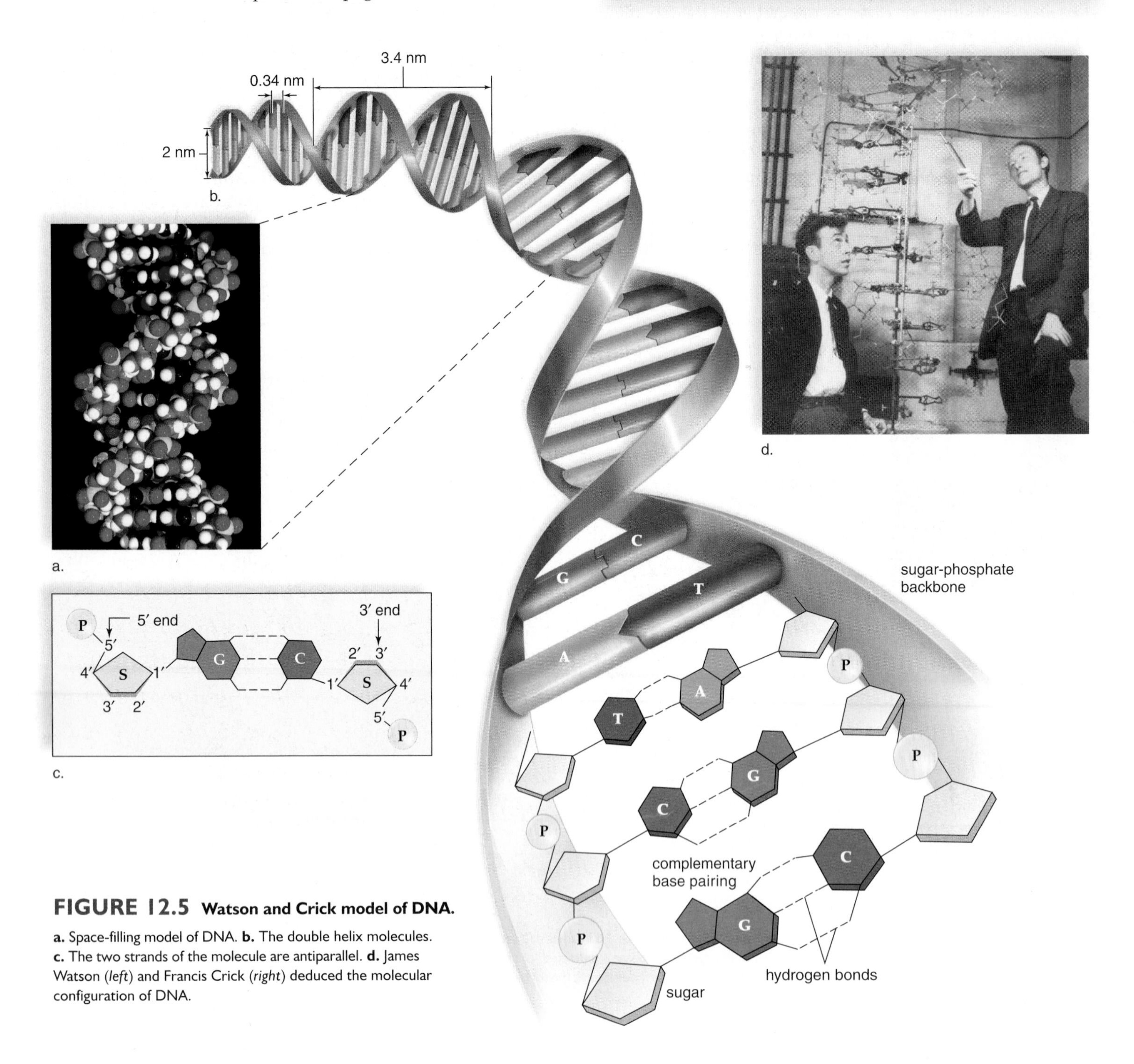


**FIGURE 12.5 Watson and Crick model of DNA.**

**a.** Space-filling model of DNA. **b.** The double helix molecules. **c.** The two strands of the molecule are antiparallel. **d.** James Watson (*left*) and Francis Crick (*right*) deduced the molecular configuration of DNA.

## 12.2 Replication of DNA

The term **DNA replication** refers to the process of copying a DNA molecule. Following replication, there is usually an exact copy of the parental DNA double helix. As soon as Watson and Crick developed their double-helix model, they commented, "It has not escaped our notice that the specific pairing we have postulated immediately suggests a possible copying mechanism for the genetic material."

A **template** is most often a mold used to produce a shape complementary to itself. During DNA replication, each DNA strand of the parental double helix serves as a template for a new strand in a daughter molecule (Fig. 12.6). DNA replication is termed **semiconservative replication** because each daughter DNA double helix contains an old strand from the parental DNA double helix and a new strand.

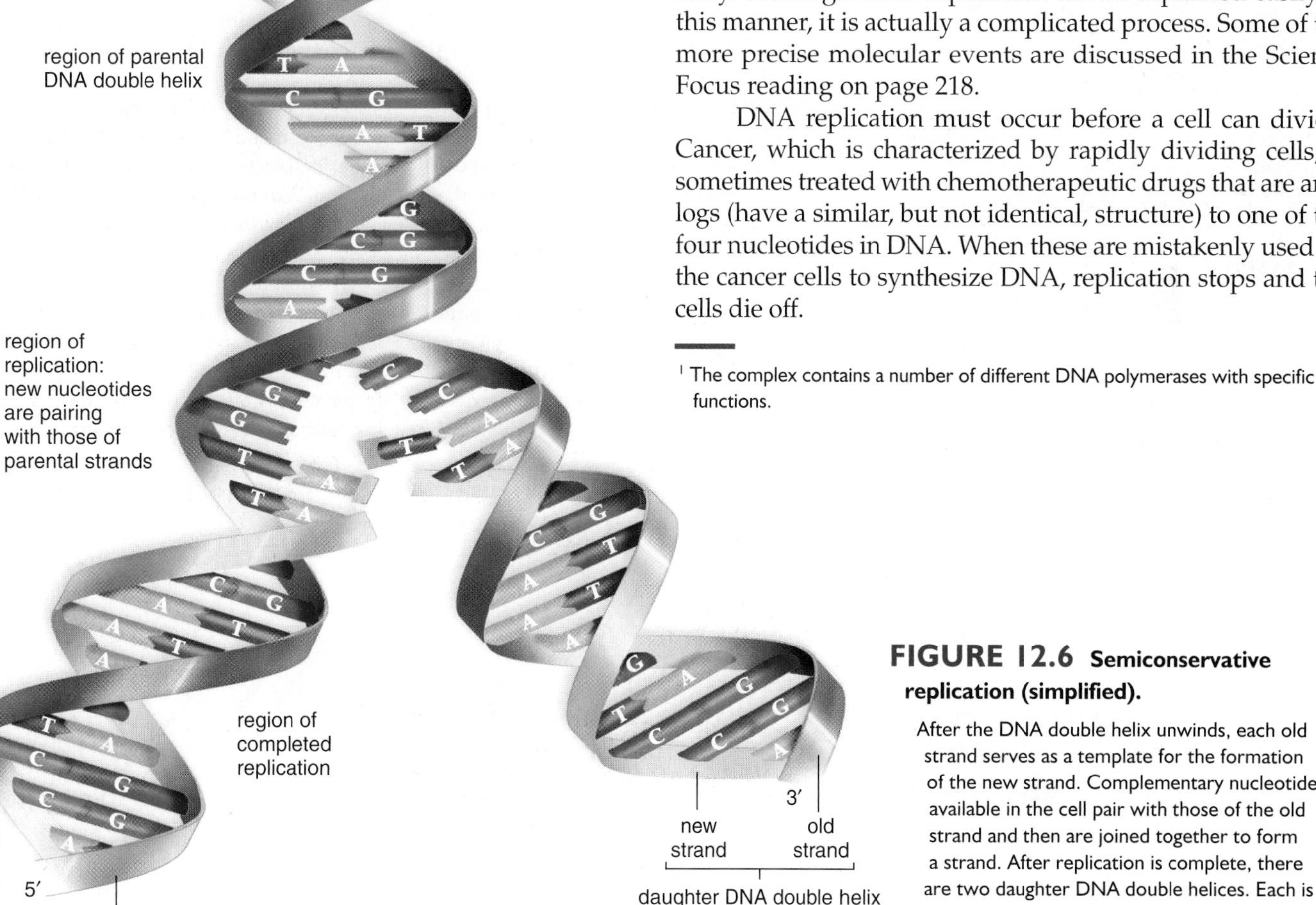


**FIGURE 12.6 Semiconservative replication (simplified).** After the DNA double helix unwinds, each old strand serves as a template for the formation of the new strand. Complementary nucleotides available in the cell pair with those of the old strand and then are joined together to form a strand. After replication is complete, there are two daughter DNA double helices. Each is composed of an old strand and a new strand. Each daughter double helix has the same sequence of base pairs as the parental double helix had before unwinding occurred.

Replication requires the following steps:

1. *Unwinding.* The old strands that make up the parental DNA molecule are unwound and "unzipped" (i.e., the weak hydrogen bonds between the paired bases are broken). A special enzyme called helicase unwinds the molecule.
2. *Complementary base pairing.* New complementary nucleotides, always present in the nucleus, are positioned by the process of complementary base pairing.
3. *Joining.* The complementary nucleotides join to form new strands. Each daughter DNA molecule contains an old strand and a new strand.

Steps 2 and 3 are carried out by an enzyme complex called **DNA polymerase.**[1] DNA polymerase works in the test tube as well as in cells.

In Figure 12.6, the backbones of the parental DNA molecule are bluish, and each base is given a particular color. Following replication, the daughter molecules each have a greenish backbone (new strand) and a bluish backbone (old strand). A daughter DNA double helix has the same sequence of bases as the parental DNA double helix had originally. Although DNA replication can be explained easily in this manner, it is actually a complicated process. Some of the more precise molecular events are discussed in the Science Focus reading on page 218.

DNA replication must occur before a cell can divide. Cancer, which is characterized by rapidly dividing cells, is sometimes treated with chemotherapeutic drugs that are analogs (have a similar, but not identical, structure) to one of the four nucleotides in DNA. When these are mistakenly used by the cancer cells to synthesize DNA, replication stops and the cells die off.

[1] The complex contains a number of different DNA polymerases with specific functions.

*science focus*

## Aspects of DNA Replication

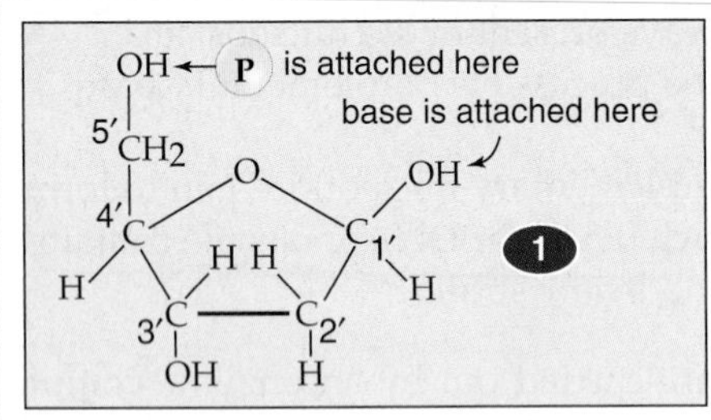


Deoxyribose molecule

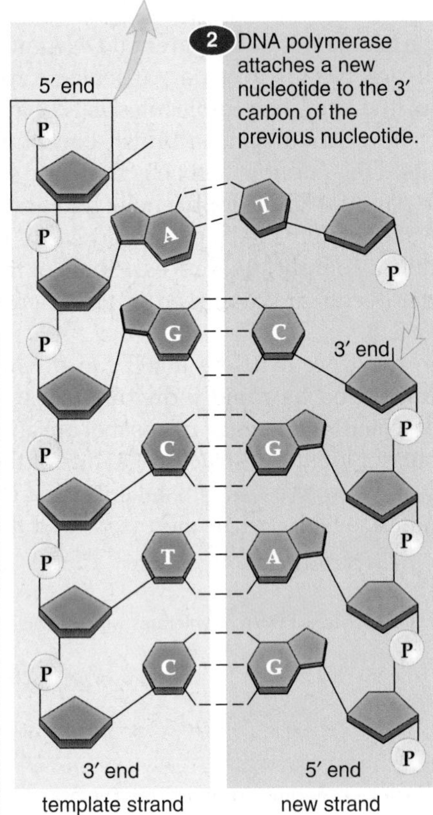


Direction of replication

Watson and Crick realized that the strands in DNA had to be antiparallel to allow for complementary base pairing. This opposite polarity of the strands introduces complications for DNA replication, as we will now see. In Figure 12A, ❶ take a look at a deoxyribose molecule, in which the carbon atoms are numbered. Use the structure to see that ❷ the DNA strand in the blue box runs opposite from the DNA strand in the green box. In other words, the strand in the blue box has a 5′ end at the top, and the strand in the green box has a 3′ end at the top. During replication, DNA polymerase has to join the nucleotides of the new strand so that the 3′ end is uppermost. Why? Because DNA polymerase can only join a nucleotide to the free 3′ end of the previous nucleotide, as shown. Also, *DNA polymerase cannot start the synthesis of a DNA chain.* Therefore, an RNA polymerase (see page 222) lays down a short amount of RNA, called an RNA primer, that is complementary to the template strand being replicated. After that, DNA polymerase can join DNA nucleotides to the 3′ end of the growing new strand.

❸ As a helicase enzyme unwinds DNA, one template strand can be copied in the direction of the replication fork. (Binding proteins serve to stabilize the newly formed, single-stranded regions.) ❹ This strand is called the leading new strand. The other template strand has to be copied in the direction away from the fork. Therefore, replication must begin over and over again as the DNA molecule unwinds. ❺ Replication of this so-called lagging new strand is, therefore, discontinuous, and it results in segments called ❻ Okazaki fragments, after the Japanese scientist Reiji Okazaki, who discovered them.

Replication is only complete when the RNA primers are removed. This works out well for the lagging new strand. While proofreading, DNA polymerase removes the RNA primers and replaces them with complementary DNA nucleotides. ❼ Another enzyme, called DNA ligase, joins the fragments. However, there is no way for DNA polymerase to replicate the 5′ ends of both new strands after RNA primers are removed. This means that DNA molecules get shorter as one replication follows another. The ends of eukaryotic DNA molecules have a special nucleotide sequence called a telomere. **Telomeres** do not code for proteins and, instead, are repeats of a short nucleotide sequence, such as TTAGGG.

Mammalian cells grown in a culture divide about 50 times and then stop. After this number of divisions, the loss of telomeres apparently signals the cell to stop dividing. Ordinarily, telomeres are only added to chromosomes in stem cells by an enzyme called telomerase. This enzyme, unfortunately, is often mistakenly turned on in cancer cells, an event that contributes to the ability of cancer cells to keep on dividing without limit.

5′
3′
template strand
4 leading new strand
DNA polymerase
3′
3 helicase at replication fork
RNA primer
6 Okazaki fragment
template strand
5 lagging strand
5′
3′
5′
parental DNA helix
7 DNA ligase
DNA polymerase
3′

Replication fork introduces complications

**FIGURE 12A DNA replication (in depth).**

## Prokaryotic Versus Eukaryotic Replication

The process of DNA replication is distinctly different in prokaryotic and eukaryotic cells (Fig. 12.7).

### *Prokaryotic DNA Replication*

Bacteria have a single circular loop of DNA that must be replicated before the cell divides. In some circular DNA molecules, replication moves around the DNA molecule in one direction only. In others, as shown in Figure 12.7*a*, replication occurs in two directions. The process always occurs in the 5′ to 3′ direction.

The process begins at the *origin of replication,* a specific site on the bacterial chromosome. The strands are separated and unwound, and a DNA polymerase binds to each side of the opening and begins the copying process. When the two DNA polymerases meet at a termination region, replication is halted, and the two copies of the chromosome are separated.

Bacterial cells require about 40 minutes to replicate the complete chromosome. Because bacterial cells are able to divide as often as once every 20 minutes, it is possible for a new round of DNA replication to begin even before the previous round is completed!

### *Eukaryotic DNA Replication*

In eukaryotes, DNA replication begins at numerous origins of replication along the length of the chromosome, and the so-called replication bubbles spread bidirectionally until they meet. Notice in Figure 12.7*b* that there is a V shape wherever DNA is being replicated. This is called a **replication fork.**

The chromosomes of eukaryotes are long and linear, making replication a more time-consuming process. Eukaryotes replicate their DNA at a slower rate—500–5,000 base pairs per minute—but there are many individual origins of replication to accelerate the process. Therefore, eukaryotic cells complete the replication of the diploid amount of DNA (in humans, over 6 billion base pairs) in a matter of hours!

The linear chromosomes of eukaryotes also pose another problem—DNA polymerase is unable to replicate the ends of the chromosomes. The ends of eukaryotic chromosomes are composed of telomeres, which are short DNA sequences that are repeated over and over. Telomeres are not copied by DNA polymerase; rather, they are added by an enzyme called telomerase, which adds the repeats after the chromosome is replicated. In stem cells, this process preserves the ends of the chromosomes and prevents the loss of DNA after successive rounds of replication.

### *Accuracy of Replication*

A DNA polymerase is very accurate and makes a mistake approximately once per 100,000 base pairs at most. This error rate, however, would result in many errors accumulating over the course of several cell divisions. DNA polymerase is also capable of proofreading the daughter strand it is making. It can recognize a mismatched nucleotide and remove it from a daughter strand by reversing direction and removing several nucleotides. Once it has removed the mismatched nucleotide, it changes direction again and resumes making DNA. Overall, the error rate for the bacterial DNA polymerase is only one in 100 million base pairs!

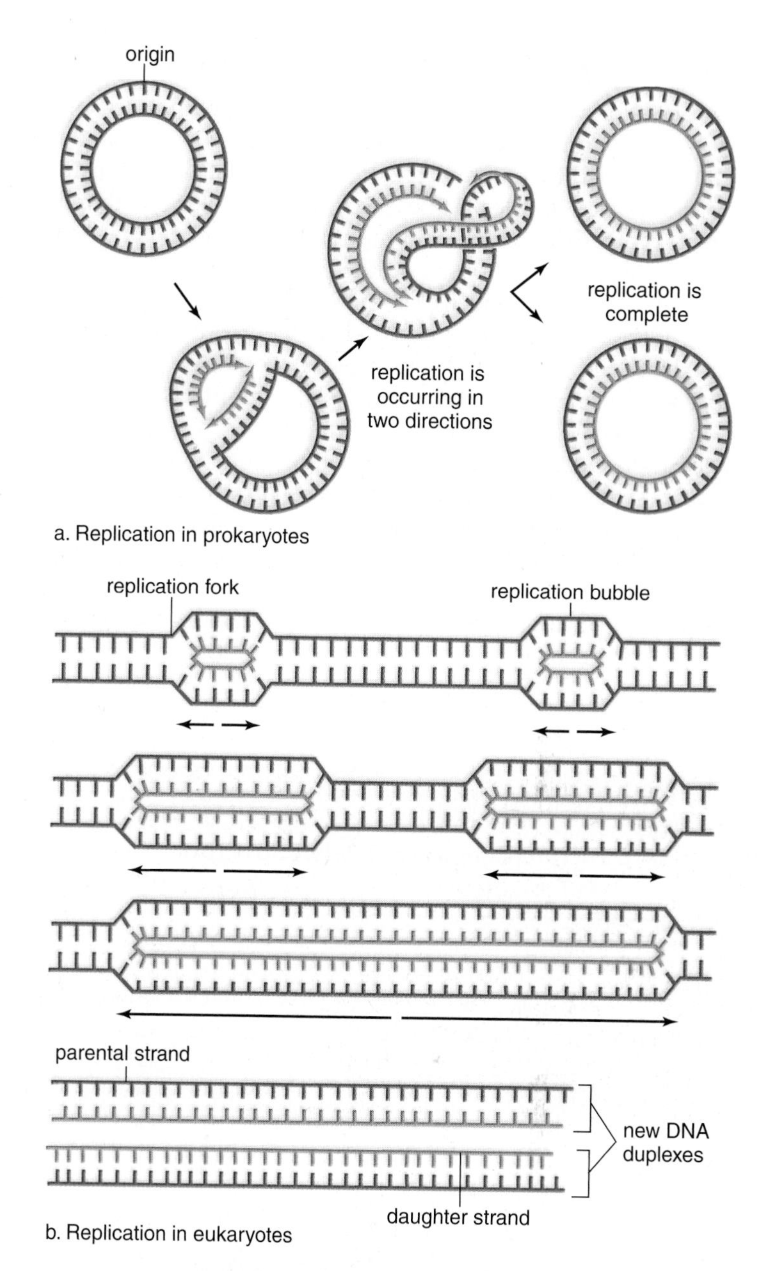


**FIGURE 12.7 Prokaryotic versus eukaryotic replication.**
**a.** In prokaryotes, replication can occur in two directions at once because the DNA molecule is circular. **b.** In eukaryotes, replication occurs at numerous replication forks. The bubbles thereby created spread out until they meet.

**Check Your Progress 12.2**

1. Describe the three major steps in DNA replication.
2. Why is DNA replication referred to as semiconservative?
3. How does DNA replication in eukaryotes differ from prokaryotic DNA replication?

# 12.3 The Genetic Code of Life

Evidence began to mount in the 1900s that metabolic disorders can be inherited. An English physician, Sir Archibald Garrod, called them "inborn errors of metabolism." Investigators George Beadle and Edward Tatum, working with red bread mold, discovered what they called the "one gene, one enzyme hypothesis," based on the observation that a defective gene caused a defective enzyme.

Other investigators decided to see if the protein hemoglobin in persons with the inherited condition sickle-cell disease (see page 203) has a structure different from normal hemoglobin. They found that the amino acid glutamate had been replaced by the amino acid valine in one location. This causes sickle cell hemoglobin to stack up into long, semirigid rods distorting red blood cells into the sickle shape.

These examples support the hypothesis that DNA, the genetic material, specifies proteins. It has been shown many times over by now that DNA specifies proteins with the help of RNA molecules.

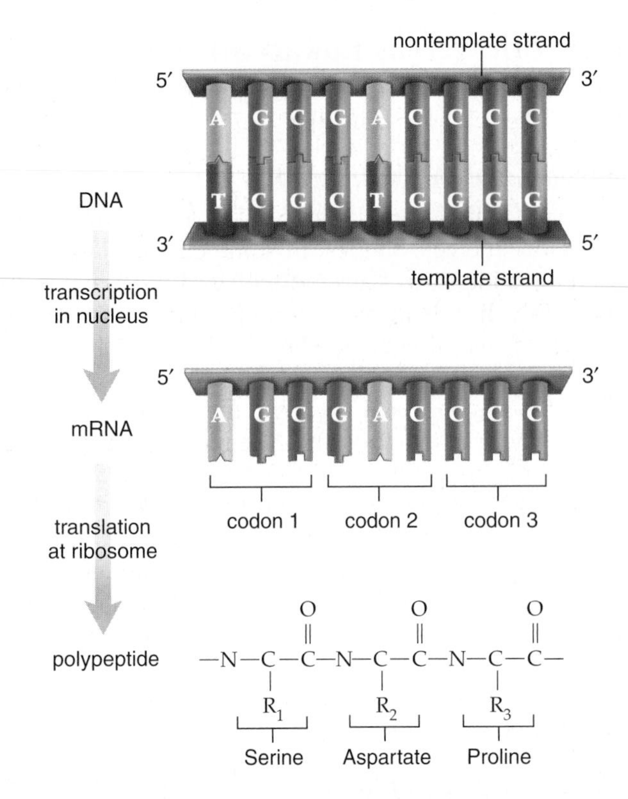


**FIGURE 12.9 The central dogma of molecular biology.**

One strand of DNA acts as a template for mRNA synthesis, and the sequence of bases in mRNA determines the sequence of amino acids in a polypeptide.

**TABLE 12.1**

**RNA Structure Compared to DNA Structure**

| | RNA | DNA |
|---|---|---|
| Sugar | Ribose | Deoxyribose |
| Bases | Adenine, guanine, uracil, cytosine | Adenine, guanine, thymine, cytosine |
| Strands | Single stranded | Double stranded with base pairing |
| Helix | No | Yes |

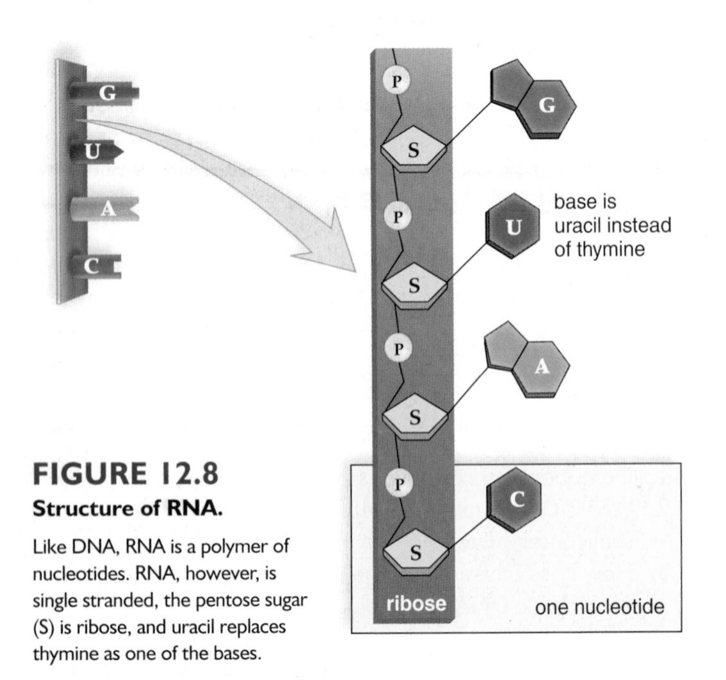


**FIGURE 12.8**
**Structure of RNA.**

Like DNA, RNA is a polymer of nucleotides. RNA, however, is single stranded, the pentose sugar (S) is ribose, and uracil replaces thymine as one of the bases.

## RNA Carries the Information

Like DNA, *RNA (ribonucleic acid)* is a polymer composed of nucleotides. The nucleotides in RNA, however, contain the sugar ribose and the bases adenine (A), cytosine (C), guanine (G), and **uracil (U).** In RNA, the base uracil replaces the thymine found in DNA. Finally, RNA is single stranded and does not form a double helix in the same manner as DNA (Table 12.1 and Fig. 12.8).

There are three major classes of RNA. Each class of RNA has its own unique size, shape, and function in protein synthesis.

**Messenger RNA (mRNA)** takes a message from DNA in the nucleus to the ribosomes in the cytoplasm.

**Transfer RNA (tRNA)** transfers amino acids to the ribosomes.

**Ribosomal RNA (rRNA),** along with ribosomal proteins, makes up the ribosomes, where polypeptides are synthesized.

## The Genetic Code

There are two major steps in synthesizing a protein based on the information stored in DNA (Fig. 12.9). First, during **transcription** [L. *trans*, across, and *scriptio*, a writing], DNA serves as a template for RNA formation. DNA is transcribed monomer by monomer into another type of polynucleotide (RNA). Second, during **translation** [L. *trans*, across, and *latus*, carry or bear], the mRNA transcript directs the sequence of amino acids in a polypeptide. Like a translator who understands two languages, the cell changes a nucleotide sequence into an amino acid sequence. With the help of the three types of RNA, a gene (a segment of DNA) specifies the sequence of amino acids in a polypeptide. Together, the flow of information from DNA to protein is known as the central dogma of molecular biology.

Therefore, it is obvious that the sequence of nucleotides in DNA and mRNA specify the order of amino acids in a polypeptide. It would seem then that there must be a **genetic code** for each of the 20 amino acids found in proteins. But can four nucleotides provide enough combinations to code for 20 amino acids? If each code word, called a **codon,** were made up of two bases, such as AG, there could be only 16 codons. But if each codon were made up of three bases, such as AGC, there would be 64 codons—more than enough to code for 20 amino acids:

| number of bases in genetic code | | | number of different amino acids specified |
|---|---|---|---|
| | 1 | 4 | |
| | 2 | 16 | |
| | 3 | 64 | |

The genetic code is a **triplet code.** Each codon consists of three nucleotide bases, such as AUC.

## *Finding the Genetic Code*

In 1961, Marshall Nirenberg and J. Heinrich Matthei performed an experiment that laid the groundwork for cracking the genetic code. First, they found that a cellular enzyme could be used to construct a synthetic RNA (one that does not occur in cells), and then they found that the synthetic RNA polymer could be translated in a test tube that contains the cytoplasmic contents of a cell. Their first synthetic RNA was composed only of uracil, and the protein that resulted was composed only of the amino acid phenylalanine. Therefore, the mRNA codon for phenylalanine was known to be UUU. Later, they were able to translate just three nucleotides at a time; in that way, it was possible to assign an amino acid to each of the mRNA codons (Fig. 12.10).

A number of important properties of the genetic code can be seen by careful inspection of Figure 12.10.

1. The genetic code is degenerate. This means that most amino acids have more than one codon; leucine, serine, and arginine have six different codons, for example. The degeneracy of the code protects against potentially harmful effects of mutations.
2. The genetic code is unambiguous. Each triplet codon has only one meaning.
3. The code has start and stop signals. There is only one start signal, but there are three stop signals.

## *The Code Is Universal*

With a few exceptions, the genetic code (Fig. 12.10) is universal to all living things. In 1979, however, researchers discovered that the genetic code used by mammalian mitochondria and chloroplasts differs slightly from the more familiar genetic code.

| First Base | Second Base U | Second Base C | Second Base A | Second Base G | Third Base |
|---|---|---|---|---|---|
| U | UUU phenylalanine | UCU serine | UAU tyrosine | UGU cysteine | U |
| U | UUC phenylalanine | UCC serine | UAC tyrosine | UGC cysteine | C |
| U | UUA leucine | UCA serine | UAA *stop* | UGA *stop* | A |
| U | UUG leucine | UCG serine | UAG *stop* | UGG tryptophan | G |
| C | CUU leucine | CCU proline | CAU histidine | CGU arginine | U |
| C | CUC leucine | CCC proline | CAC histidine | CGC arginine | C |
| C | CUA leucine | CCA proline | CAA glutamine | CGA arginine | A |
| C | CUG leucine | CCG proline | CAG glutamine | CGG arginine | G |
| A | AUU isoleucine | ACU threonine | AAU asparagine | AGU serine | U |
| A | AUC isoleucine | ACC threonine | AAC asparagine | AGC serine | C |
| A | AUA isoleucine | ACA threonine | AAA lysine | AGA arginine | A |
| A | **AUG *(start)* methionine** | ACG threonine | AAG lysine | AGG arginine | G |
| G | GUU valine | GCU alanine | GAU aspartate | GGU glycine | U |
| G | GUC valine | GCC alanine | GAC aspartate | GGC glycine | C |
| G | GUA valine | GCA alanine | GAA glutamate | GGA glycine | A |
| G | GUG valine | GCG alanine | GAG glutamate | GGG glycine | G |

**FIGURE 12.10 Messenger RNA codons.**

Notice that in this chart, each of the codons (in boxes) is composed of three letters representing the first base, second base, and third base. For example, find the box where C for the first base and A for the second base intersect. You will see that U, C, A, or G can be the third base. The bases CAU and CAC are codons for histidine; the bases CAA and CAG are codons for glutamine.

The universal nature of the genetic code provides strong evidence that all living things share a common evolutionary heritage. Since the same genetic code is used by all living things, it is possible to transfer genes from one organism to another. Many commercial and medicinal products such as insulin can be produced in this manner. Genetic engineering has also produced some unusual organisms such as mice that literally glow in the dark. In this case, the gene responsible for bioluminescence in jellyfish was placed into mouse embryos (see Fig. 12.2).

### Check Your Progress 12.3

1. What are the three major classes of RNA, and what are their functions?
2. What does it mean to say that the genetic code is degenerate?

# 12.4 First Step: Transcription

During *transcription*, a segment of the DNA serves as a template for the production of an RNA molecule. Although all three classes of RNA are formed by transcription, we will focus right now on transcription to form mRNA, the type of RNA that eventually leads to building a polypeptide as a gene product.

## Messenger RNA Is Formed

An mRNA molecule has a sequence of bases complementary to a portion of one DNA strand; wherever A, T, G, or C is present in the DNA template, U, A, C, or G, respectively, is incorporated into the mRNA molecule (Fig. 12.11). When a gene is transcribed, a segment of the DNA helix unwinds and unzips, and complementary RNA nucleotides pair with DNA nucleotides of the strand opposite the gene. This strand is known as the *template strand*; the other strand is the nontemplate strand. An RNA polymerase joins the nucleotides together in the 5′ ⟶ 3′ direction. In other words, an **RNA polymerase** only adds a nucleotide to the 3′ end of the polymer under construction.

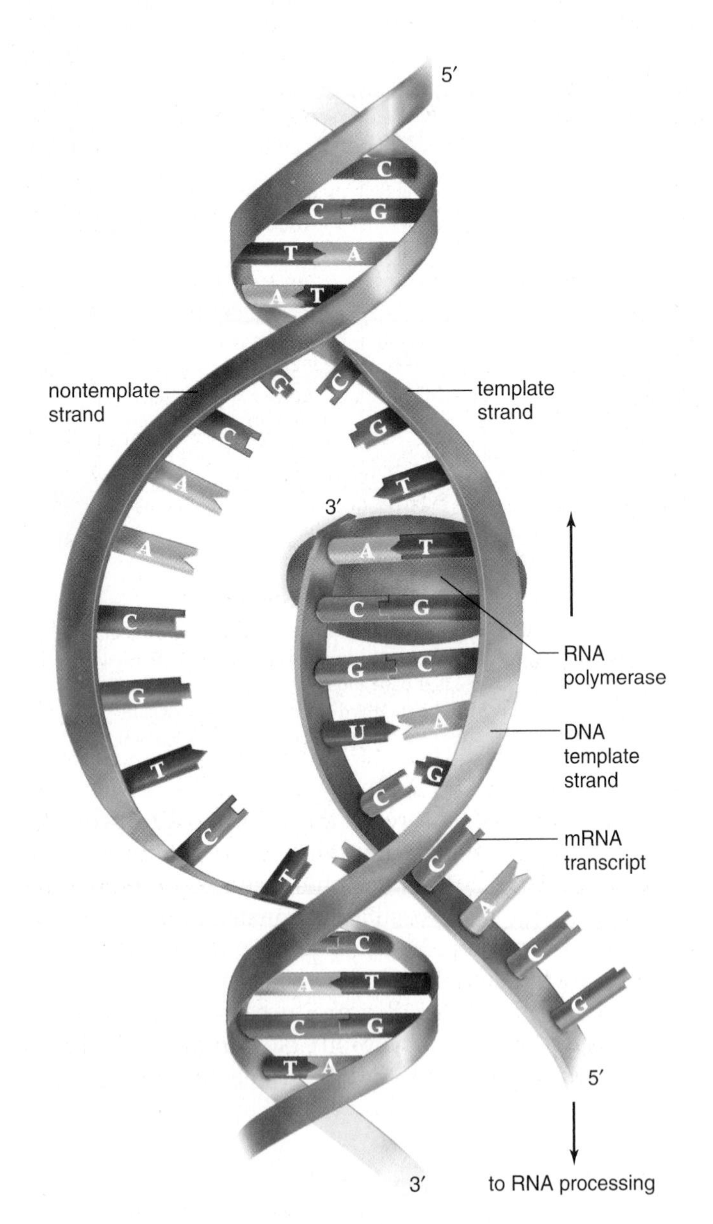


**FIGURE 12.11 Transcription.**

During transcription, complementary RNA is made from a DNA template. At the point of attachment of RNA polymerase, the DNA helix unwinds and unzips, and complementary RNA nucleotides are joined together. After RNA polymerase has passed by, the DNA strands rejoin and the mRNA transcript dangles to the side.

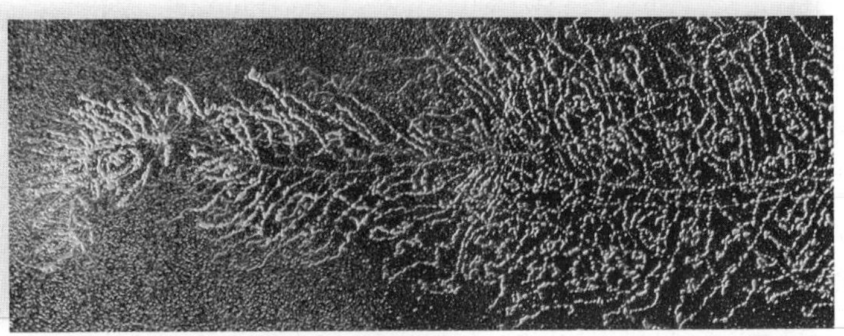


a.

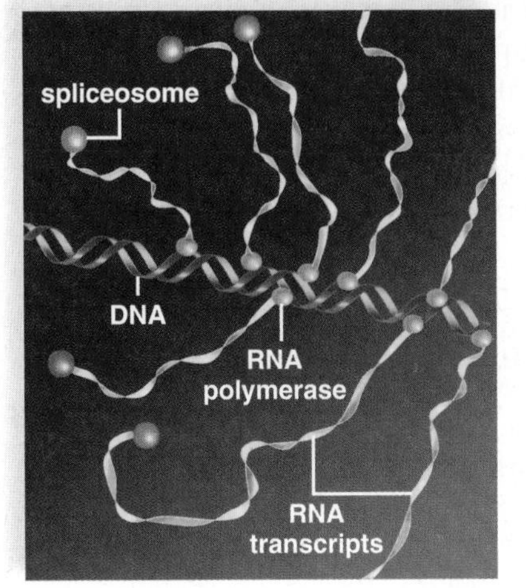


b.

**FIGURE 12.12**
**RNA polymerase.**

**a.** Numerous RNA transcripts extend from a horizontal gene in an amphibian egg cell. **b.** The strands get progressively longer because transcription begins to the left. The dots along the DNA are RNA polymerase molecules. The dots at the end of the strands are spliceosomes involved in RNA processing (see Fig. 12.13).

Transcription begins when RNA polymerase attaches to a region of DNA called a promoter. A **promoter** defines the start of transcription, the direction of transcription, and the strand to be transcribed. The binding of RNA polymerase to the promoter is the *initiation* of transcription. The RNA-DNA association is not as stable as the DNA helix. Therefore, only the newest portion of an RNA molecule that is associated with RNA polymerase is bound to the DNA, and the rest dangles off to the side. *Elongation* of the mRNA molecule continues until RNA polymerase comes to a DNA stop sequence. The stop sequence causes RNA polymerase to stop transcribing the DNA and to release the mRNA molecule, now called an **mRNA transcript.**

Many RNA polymerase molecules can be working to produce mRNA transcripts at the same time (Fig. 12.12). This allows the cell to produce many thousands of copies of the same mRNA molecule, and eventually many copies of the same protein, within a shorter period of time than if the single copy of DNA were used to direct protein synthesis.

It is of interest that either strand of DNA can be a template strand. In other words, each strand of DNA can be a template strand but for a different gene.

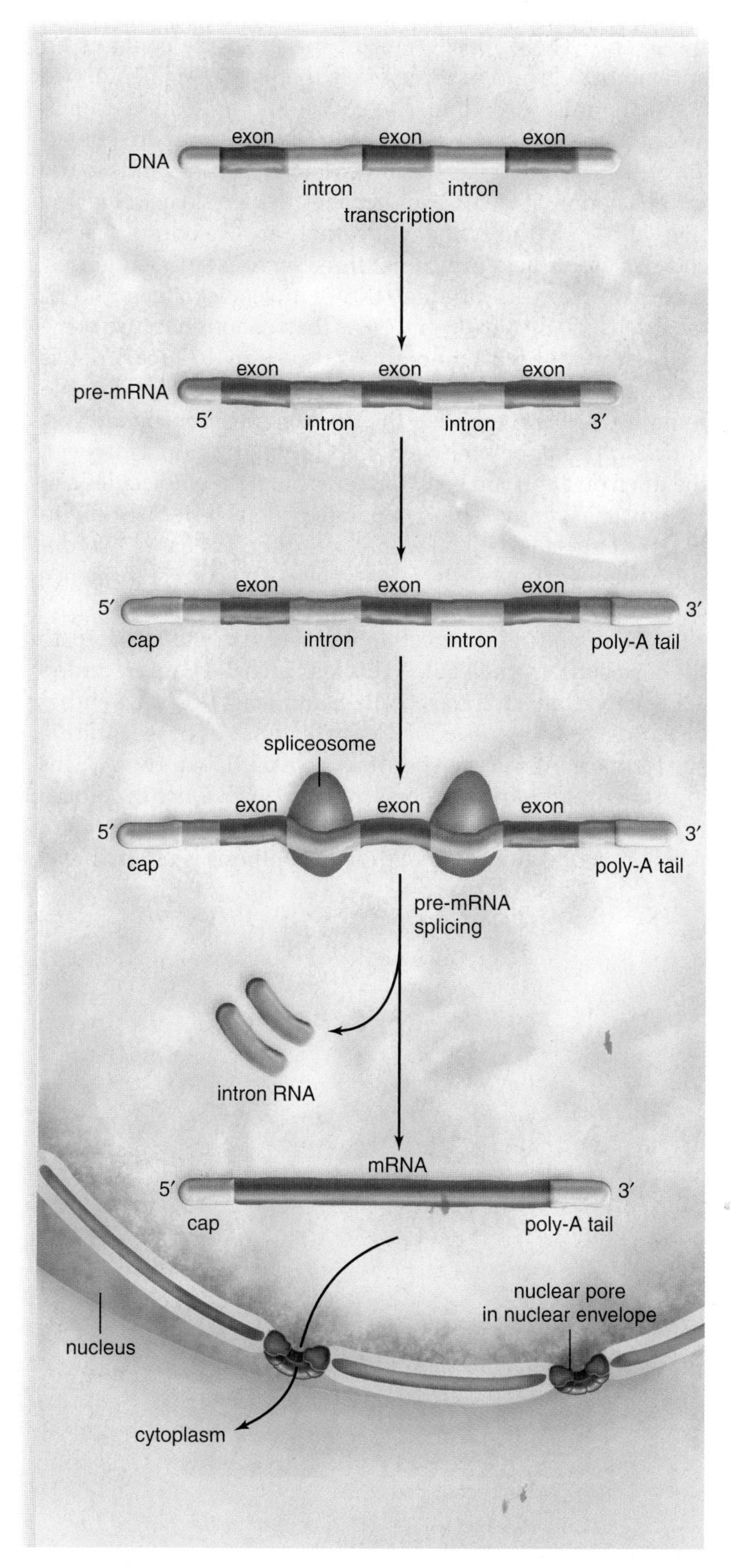


**FIGURE 12.13 Messenger RNA (mRNA) processing in eukaryotes.**

DNA contains both exons (protein-coding sequences) and introns (non-protein-coding sequences). Both of these are transcribed and are present in pre-mRNA. During processing, a cap and a poly-A tail (a series of adenine nucleotides) are added to the molecule. Also, there is excision of the introns and a splicing together of the exons. This is accomplished by complexes called spliceosomes. Then the mRNA molecule is ready to leave the nucleus.

## RNA Molecules Are Processed

A newly formed RNA transcript, called a pre-mRNA, is modified before leaving the eukaryotic nucleus. For example, the molecule receives a cap at the 5′ end and a tail at the 3′ end (Fig. 12.13). The *cap* is a modified guanine (G) nucleotide that helps tell a ribosome where to attach when translation begins. The tail consists of a chain of 150–200 adenine (A) nucleotides. This so-called *poly-A tail* facilitates the transport of mRNA out of the nucleus and also inhibits degradation of mRNA by hydrolytic enzymes.

Also, the pre-mRNA, particularly in multicellular eukaryotes, is composed of exons and introns. The exons of the pre-mRNA molecule will be *ex*pressed, but not the **introns,** which occur *in* between the **exons.** During pre-mRNA splicing, the introns are removed. In prokaryotes, introns are removed by "self-splicing"—that is, the intron itself has the capability of enzymatically splicing itself out of a pre-mRNA. In eukaryotes, the RNA splicing is done by spliceosomes, which contain *small nuclear RNAs (snRNAs).* By means of complementary base pairing, snRNAs are capable of identifying the introns to be removed. A spliceosome utilizes a ribozyme when it removes the introns. **Ribozymes,** also found in prokaryotes, are RNA molecules that possess catalytic activity in the same manner as enzymes composed of protein. Following splicing, an mRNA is ready to leave the nucleus.

Another type of RNA called small nucleolar RNA (snoRNA) is present in the nucleolus, where it helps process rRNA and tRNA molecules.

### *Function of Introns*

The presence of introns allows a cell to pick and choose which exons will go into a particular mRNA (see pages 242–43). That is, it has been discovered that an mRNA can contain only some of the possible exons available from a DNA sequence. Therefore, what is an exon in one mRNA could be an intron in another mRNA. This is called *alternative mRNA splicing*. Because the snRNAs play a role in determining what is an exon or intron for a particular mRNA, they take on greater significance in eukaryotes. Some introns give rise to *microRNAs (miRNAs),* which are involved in regulating the translation of mRNAs. These molecules bond with the mRNA through complementary base pairing and, in that way, prevent translation from occurring.

It is also possible that the presence of introns encourages crossing-over during meiosis, and this permits so-called *exon shuffling,* which can play a role in the evolution of new genes.

**Check Your Progress 12.4**

1. In which direction along the template DNA strand does transcription proceed, and in which direction is the mRNA molecule built?
2. What are the three major modifications that occur during the processing of an mRNA?

# 12.5 Second Step: Translation

*Translation,* which takes place in the cytoplasm of eukaryotic cells, is the second step by which gene expression leads to protein synthesis. During translation, the sequence of codons in the mRNA at a ribosome directs the sequence of amino acids in a polypeptide. In other words, one language (nucleic acids) gets translated into another language (protein).

## The Role of Transfer RNA

Transfer RNA (tRNA) molecules transfer amino acids to the ribosomes. A tRNA molecule is a single-stranded nucleic acid that doubles back on itself to create regions where complementary bases are hydrogen-bonded to one another. The structure of a tRNA molecule is generally drawn as a flat cloverleaf, but a space-filling model shows the molecule's three-dimensional shape (Fig. 12.14).

There is at least one tRNA molecule for each of the 20 amino acids found in proteins. The amino acid binds to the 3′ end. The opposite end of the molecule contains an **anticodon,** a group of three bases that is complementary to a specific mRNA codon. The codon and anticodon pair in an antiparallel fashion, just as two DNA strands do. For example, a tRNA that has the anticodon 5′ AAG 3′ binds to the mRNA codon 5′ CUU 3′ and carries the amino acid leucine (Fig. 12.14*a*). In the genetic code, there are 61 codons that encode for amino acids; the other three serve as stop sequences. Approximately 40 different tRNA molecules are found in most cells. There are fewer tRNAs than codons because some tRNAs can pair with more than one codon. In 1966, Francis Crick observed this phenomenon and called it the **wobble hypothesis.** He stated that the first two positions in a tRNA anticodon pair obey the A–U/G–C configuration. However, the third position can be variable. Some tRNA molecules can recognize as many as four separate codons differing only in the third nucleotide. The wobble effect helps ensure that despite changes in DNA base sequences, the correct sequence of amino acids will result in a protein.

How does the correct amino acid become attached to the correct tRNA molecule? This task is carried out by amino acid–activating enzymes, called aminoacyl-tRNA synthetases. Just as a key fits a lock, each enzyme has a recognition site for the amino acid to be joined to a particular tRNA. This is an energy-requiring process that uses ATP. Once the amino acid–tRNA complex is formed, it travels through the cytoplasm to a ribosome, where protein synthesis is occurring.

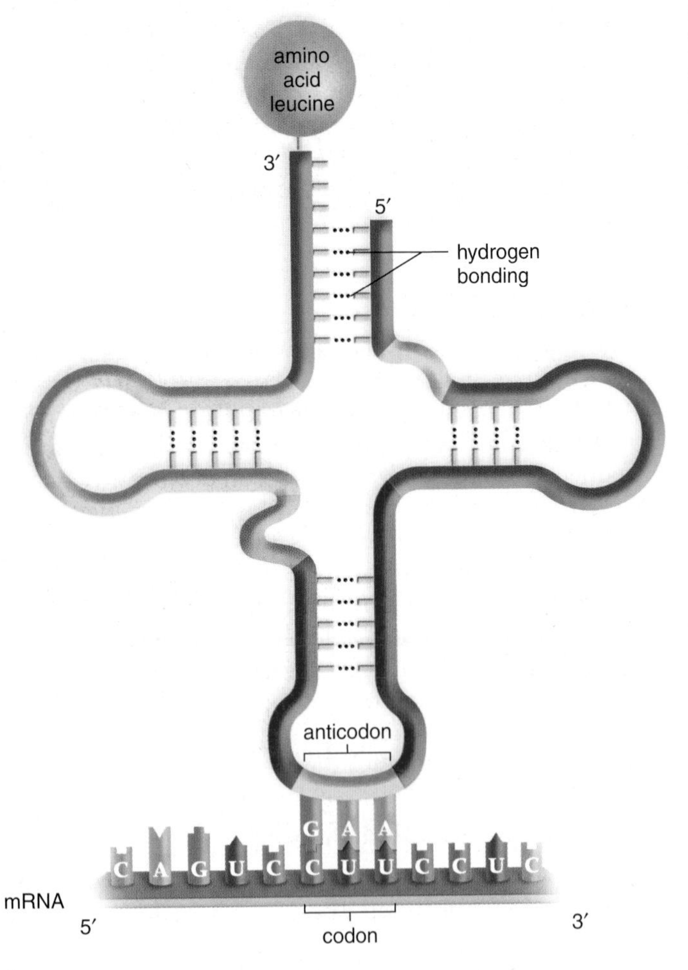


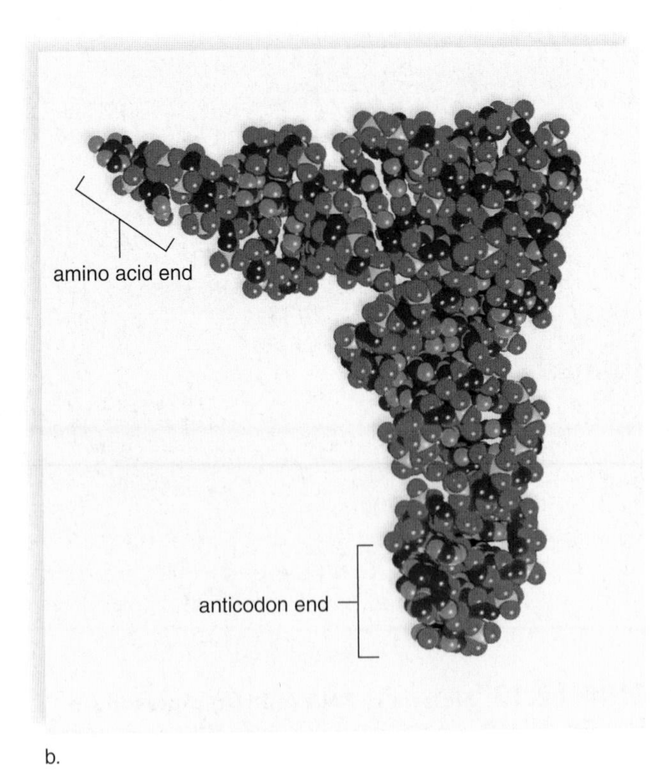


**FIGURE 12.14 Structure of a transfer RNA (tRNA) molecule.**

**a.** Complementary base pairing indicated by hydrogen bonding occurs between nucleotides of the molecule, and this causes it to form its characteristic loops. The anticodon that base-pairs with a particular messenger RNA (mRNA) codon occurs at one end of the folded molecule; the other two loops help hold the molecule at the ribosome. An appropriate amino acid is attached at the 3′ end of the molecule. For this mRNA codon and tRNA anticodon, the specific amino acid is leucine. **b.** Space-filling model of tRNA molecule.

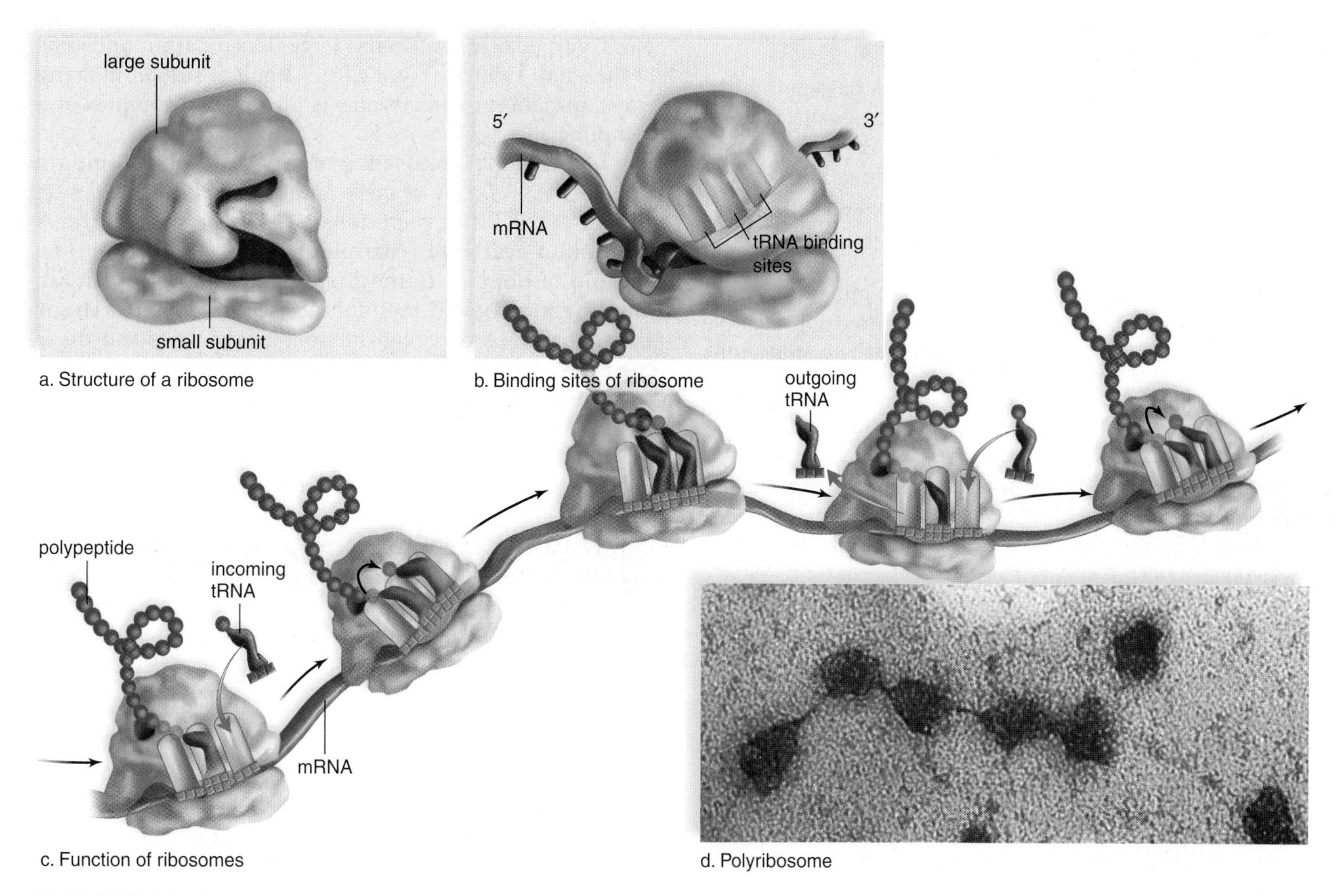


**FIGURE 12.15 Ribosome structure and function.**

**a.** Side view of a ribosome shows that it is composed of two subunits: a small subunit and a large subunit. **b.** Frontal view of a ribosome shows its binding sites. mRNA is bound to the small subunit, and the large subunit has three binding sites for tRNAs. **c.** Overview of protein synthesis. A polypeptide increases by one amino acid at a time because a peptide-bearing tRNA passes the peptide to an amino acid-bearing tRNA at a ribosome. Freed of its burden, the "empty" tRNA exits, and the peptide-bearing tRNA moves over one binding site. The polypeptide is formed as this process is repeated. **d.** Electron micrograph of a polyribosome, a number of ribosomes all translating the same mRNA molecule.

# The Role of Ribosomal RNA

The structure of a ribosome is suitable to its function.

## *Structure of a Ribosome*

In eukaryotes, ribosomal RNA (rRNA) is produced from a DNA template in the nucleolus of a nucleus. The rRNA is packaged with a variety of proteins into two ribosomal subunits, one of which is larger than the other. Then the subunits move separately through nuclear envelope pores into the cytoplasm, where they combine when translation begins (Fig. 12.15*a*). Ribosomes can remain in the cytoplasm, or they can become attached to endoplasmic reticulum.

## *Function of a Ribosome*

Both prokaryotic and eukaryotic cells contain thousands of ribosomes per cell because they play a significant role in protein synthesis. Ribosomes have a binding site for mRNA and three binding sites for transfer RNA (tRNA) molecules (Fig. 12.15*b*). The tRNA binding sites facilitate complementary base pairing between tRNA anticodons and mRNA codons. A ribosomal RNA (i.e., a ribozyme) is now known to join one amino acid to another amino acid as a polypeptide is synthesized by the ribosome.

When a ribosome moves down an mRNA molecule, the polypeptide increases by one amino acid at a time (Fig. 12.15*c*). Translation terminates at a stop codon. Once transcription is complete, the polypeptide dissociates from the translation complex and adopts its normal shape. In Chapter 3 we observed that a polypeptide twists and bends into a definite shape. This so-called folding process begins as soon as the polypeptide emerges from a ribosome, and chaperone molecules are often present in the cytoplasm and in the ER to make sure that all goes well. Some proteins contain more than one polypeptide, and if so they join to produce the final three-dimensional structure of a functional protein.

Several ribosomes are often attached to and translating the same mRNA. As soon as the initial portion of mRNA has been translated by one ribosome, and the ribosome has begun to move down the mRNA, another ribosome attaches to the mRNA. The entire complex is called a **polyribosome** (Fig. 12.15*d*) and greatly increases the efficiency of translation.

## Translation Requires Three Steps

During translation, the codons of an mRNA base pair with the anticodons of tRNA molecules carrying specific amino acids. The order of the codons determines the order of the tRNA molecules at a ribosome and the sequence of amino acids in a polypeptide. The process of translation must be extremely orderly so that the amino acids of a polypeptide are sequenced correctly.

Protein synthesis involves three steps: initiation, elongation, and termination. Enzymes are required for each of the three steps to function properly. The first two steps, initiation and elongation, require energy.

### *Initiation*

**Initiation** is the step that brings all the translation components together. Proteins called initiation factors are required to assemble the small ribosomal subunit, mRNA, initiator tRNA, and the large ribosomal subunit for the start of protein synthesis.

Initiation is shown in Figure 12.16. In prokaryotes, a small ribosomal subunit attaches to the mRNA in the vicinity of the *start codon* (AUG). The first or initiator tRNA pairs with this codon. Then, a large ribosomal subunit joins to the small subunit (Fig. 12.16). Although similar in many ways, initiation in eukaryotes is much more complex and complicated.

As already discussed, a ribosome has three binding sites for tRNAs. One of these is called the E (for exit) site, second is the P (for peptide) site, and the third is the A (for amino acid) site. The initiator tRNA happens to be capable of binding to the P site, even though it carries only the amino acid methionine (see Fig. 12.10). The A site is for tRNA carrying the next amino acid, and the E site is for any tRNAs that are leaving a ribosome. Following initiation, translation continues with elongation and then termination.

### *Elongation*

**Elongation** is the protein synthesis step in which a polypeptide increases in length one amino acid at a time. In addition to the participation of tRNAs, elongation requires elongation factors, which facilitate the binding of tRNA anticodons to mRNA codons at a ribosome.

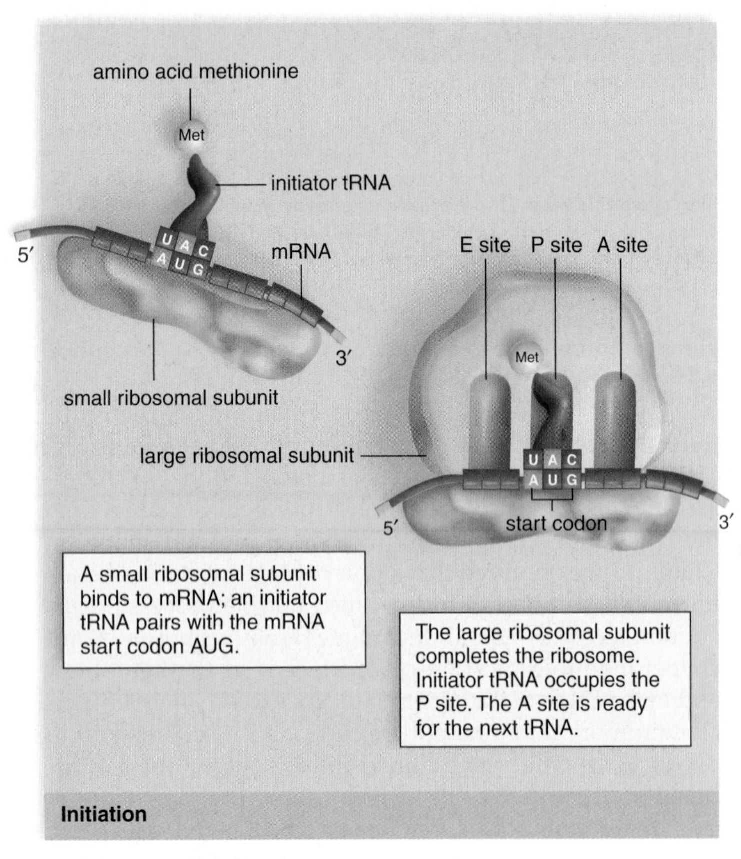


**FIGURE 12.16 Initiation.**

In prokaryotes, participants in the translation process assemble as shown. The first amino acid is typically methionine.

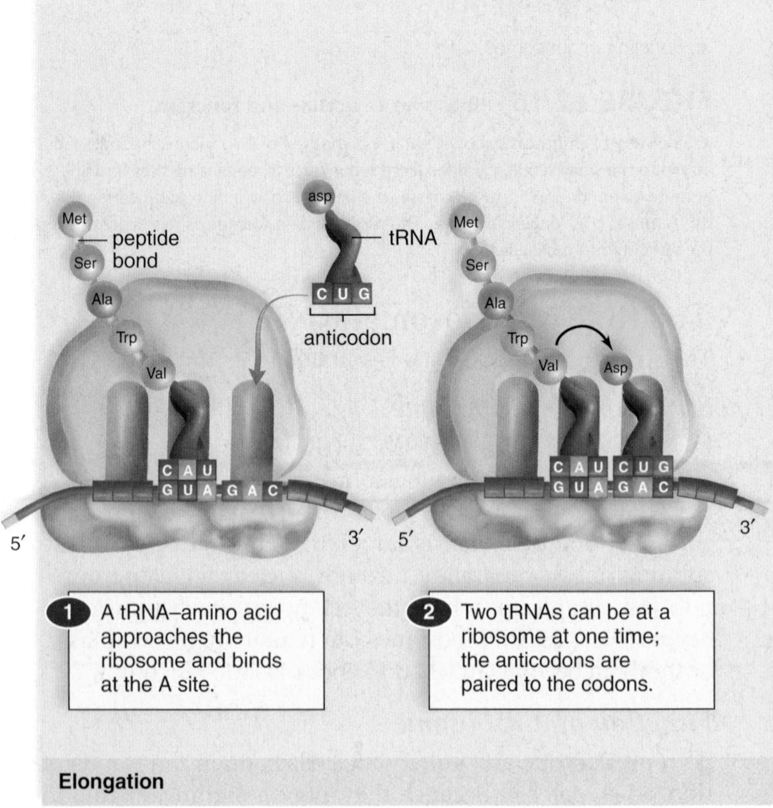


**FIGURE 12.17 Elongation.**

Note that a polypeptide is already at the P site. During elongation, polypeptide synthesis occurs as amino acids are added one at a time to the growing chain.

❶ Elongation is shown in Figure 12.17, where a tRNA with an attached peptide is already at the P site, and a tRNA carrying its appropriate amino acid is just arriving at the A site. ❷ Once a ribosome has verified that the incoming tRNA matches the codon and is firmly in place at the A site, the peptide will be transferred to this tRNA. A ribozyme, which is a part of the larger ribosomal subunit, and energy are needed to bring about this transfer. ❸ Following peptide bond formation the peptide is one amino acid longer than it was before. ❹ Next, **translocation** occurs: The ribosome moves forward, and the peptide-bearing tRNA is now at the P site of the ribosome. The spent tRNA is now at the E site, and it exits. A new codon is at the A site and is ready to receive another tRNA.

The complete cycle—complementary base pairing of new tRNA, transfer of peptide chain, and translocation—is repeated at a rapid rate (about 15 times each second in the bacterium *Escherichia coli*).

Eventually, the ribosome reaches a stop codon, and termination occurs, during which the polypeptide is released.

## *Termination*

**Termination** is the final step in protein synthesis. During termination, as shown in Figure 12.18, the polypeptide and the assembled components that carried out protein synthesis are separated from one another.

Termination of polypeptide synthesis occurs at a *stop codon*—that is, a codon that does not code for an amino acid. Termination requires a protein called a release factor, which can bind to a stop codon and also cleave the polypeptide from the last tRNA. After this occurs, the polypeptide is set free and begins to take on its three-dimensional shape. The ribosome dissociates into its two subunits.

The next section reviews the entire process of protein synthesis (recall that a protein contains one or more polypeptides) and the role of the rough endoplasmic reticulum in the production of a polypeptide. Proteins do the work of the cell, whether they reside in a cellular membrane or free in the cytoplasm. A whole new field of biology called **proteomics** is now dedicated to understanding the structure of proteins and how they function in metabolic pathways. One of the important goals of proteomics is to understand how proteins are modified in the endoplasmic reticulum and the Golgi apparatus.

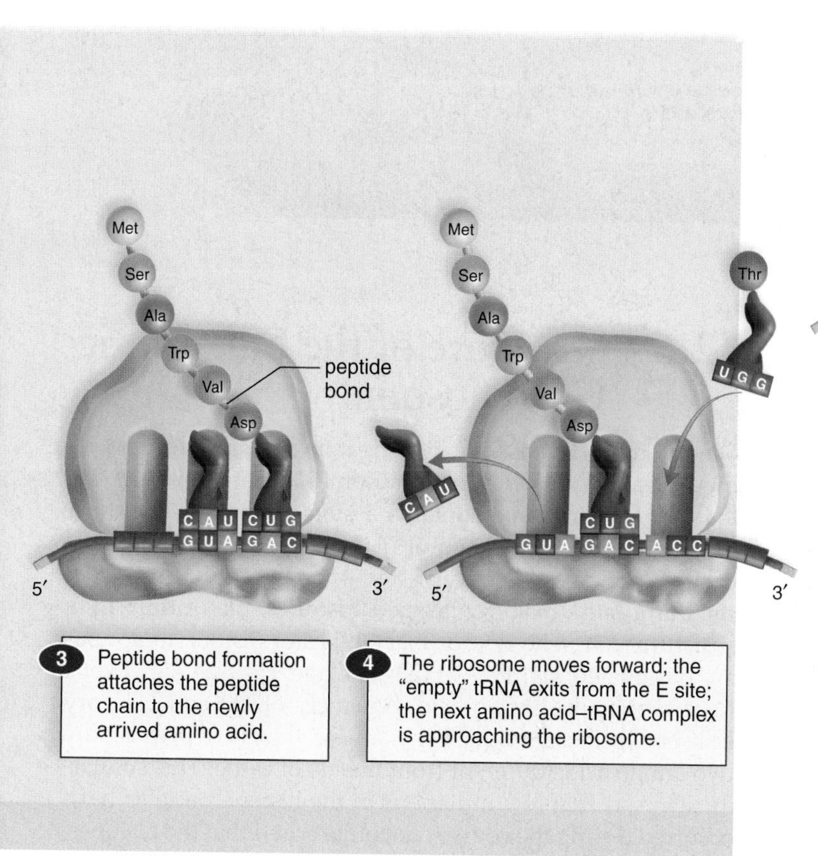


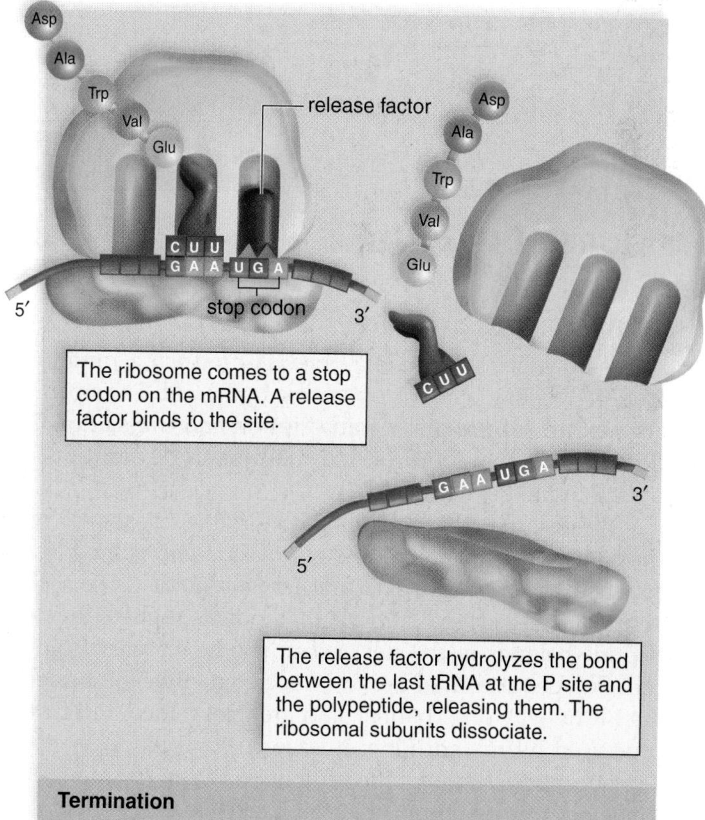


**FIGURE 12.18 Termination.**

During termination, the finished polypeptide is released, as is the mRNA and the last tRNA.

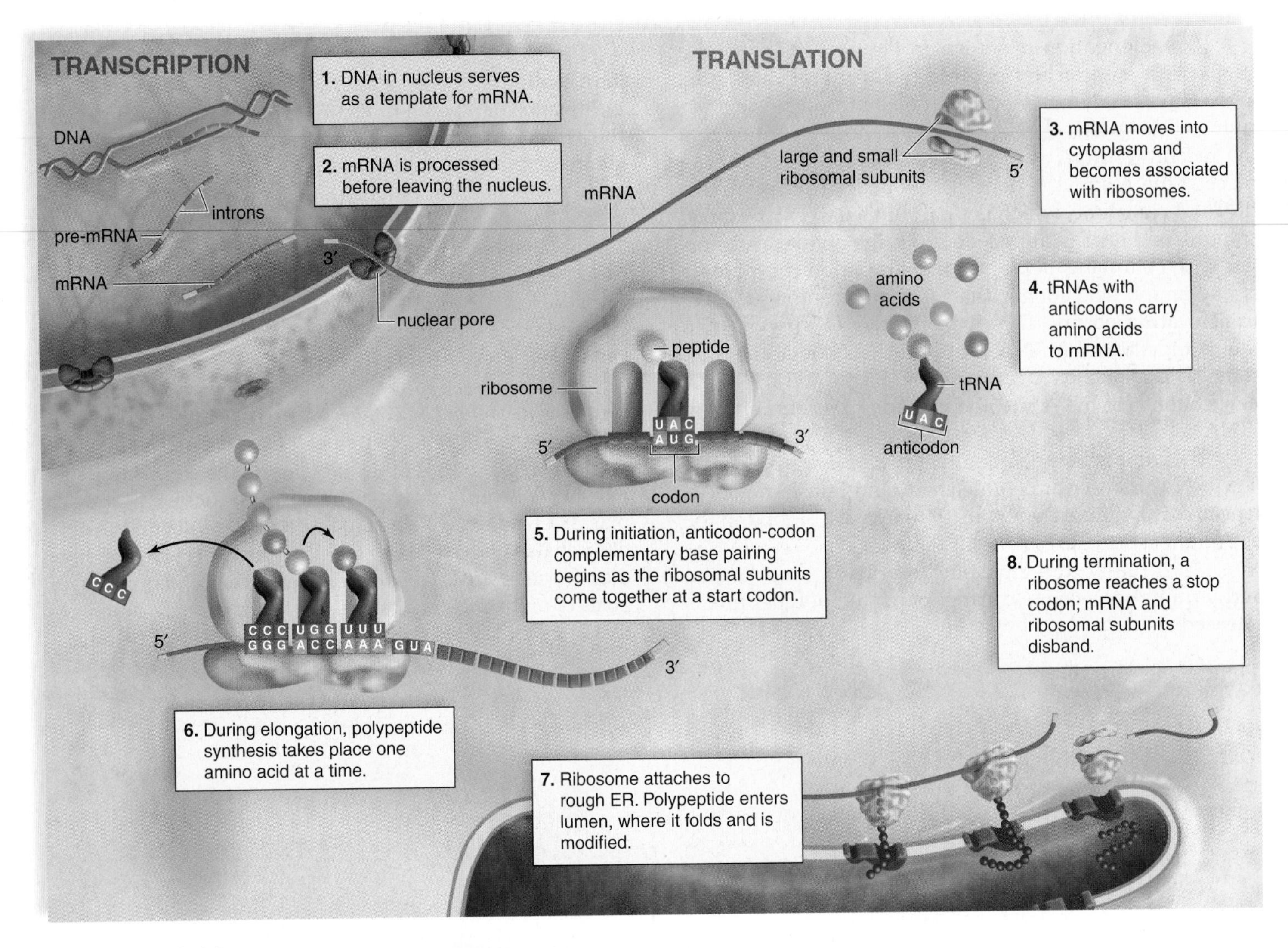


**FIGURE 12.19** **Summary of protein synthesis in eukaryotes.**

## Gene Expression

A gene has been expressed once its product, a protein (or an RNA), is made and is operating in the cell. For a protein, gene expression requires transcription and translation (Fig. 12.19) and it also requires that the protein be active as discussed in the next chapter.

Translation occurs at ribosomes. Some ribosomes (polyribosomes) remain free in the cytoplasm, and some become attached to rough ER. The first few amino acids of a polypeptide act as a signal peptide that indicates where the polypeptide belongs in the cell or if it is to be secreted from the cell. Polypeptides that are to be secreted enter the lumen of the ER by way of a channel, and are then folded and further processed by the addition of sugars, phosphates, or lipids. Transport vesicles carry the proteins between organelles and to the plasma membrane as appropriate for that protein.

### Check Your Progress 12.5

1. What is the role of transfer RNA in translation?
2. Briefly describe the structure of a ribosome.
3. Describe the three major steps of translation.

# 12.6 Structure of the Eukaryotic Chromosome

Only in recent years have investigators been able to produce models suggesting how chromosomes are organized. A eukaryotic chromosome contains a single double helix DNA molecule, but is composed of more than 50% protein. Some of these proteins are concerned with DNA and RNA synthesis, but a large majority, termed **histones,** play primarily a structural role. The five primary types of histone molecules are designated H1, H2A, H2B, H3, and H4 (see Fig. 13.5*b*). Remarkably, the amino acid sequences of H3 and H4 vary little between organisms. For example, the H4 of peas is only two amino acids different from the H4 of cattle. This similarity suggests that few mutations in the histone proteins have occurred during the course of evolution and that the histones, therefore, have important functions.

A human cell contains at least 2 m of DNA. Yet, all of this DNA is packed into a nucleus that is about 5 μm in diameter. The histones are responsible for packaging the DNA so that it can fit into such a small space. First, the DNA double helix is wound at intervals around a core of eight histone molecules (two copies each of H2A, H2B, H3, and H4), giving the appearance of a string of beads (Fig. 12.20*a*). Each bead is called a **nucleosome,** and the nucleosomes are said to be joined by "linker" DNA. This string is compacted by folding into a zigzag structure, further shortening the DNA strand (Fig. 12.20*b*). Histone H1 appears to mediate this coiling process. The fiber then loops back and forth into radial loops (Fig. 12.20*c*). This loosely coiled **euchromatin** represents the active chromatin containing genes that are being transcribed. The DNA of euchromatin may be accessed by RNA polymerase and other factors that are needed to promote transcription.

Under a microscope, one often observes dark-stained fibers within the nucleus of the cell. These areas within the nucleus represent a more highly compacted form of the chromosome called **heterochromatin** (Fig. 12.20*d*). Most chromosomes exhibit both levels of compaction in a living cell, depending on which portions of the chromosome are being used more frequently. Heterochromatin is considered inactive chromatin because the genes contained on it are infrequently transcribed, if at all. Prior to cell division, a protein scaffold helps to further condense the chromosome into a form that is characteristic of metaphase chromosomes (Fig. 12.20*e*). No doubt, compact chromosomes are easier to move about than extended chromatin.

### Check Your Progress 12.6

1. What is the typical compaction state of euchromatin, and how does this differ from heterochromatin?

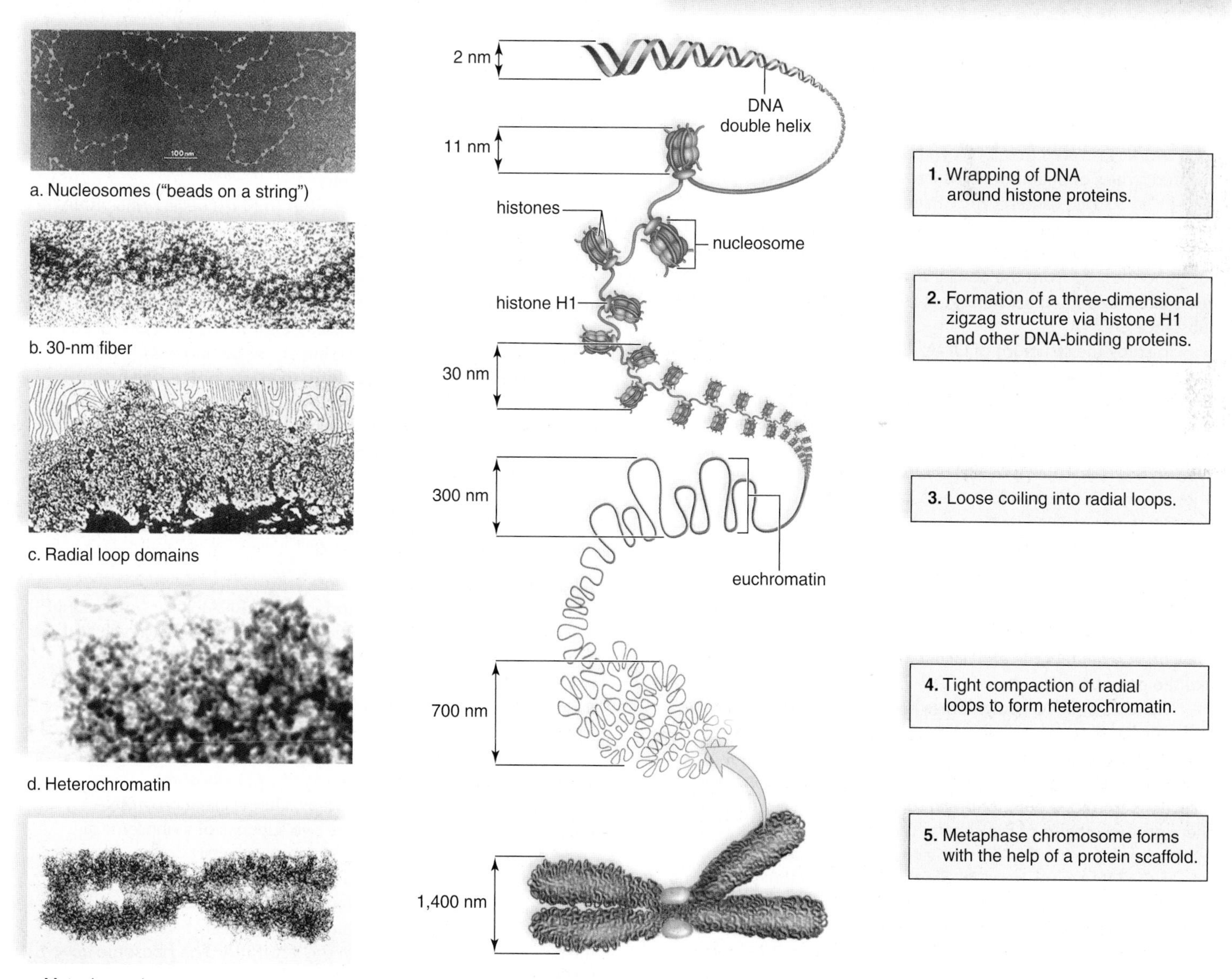


**FIGURE 12.20 Structure of eukaryotic chromosomes.**

The DNA molecule of a chromosome is compacted at several levels. **a.** The DNA strand is wound around histones to form nucleosomes. **b.** The strand is further shortened by folding it into a zigzag structure. **c.** The fiber loops back and forth into radial loops. **d.** In heterochromatin, additional proteins further compact the radial loops. **e.** A metaphase chromosome forms.

## Connecting the Concepts

Early investigators who did their work between 1950 and 1990 came to the conclusion that DNA is organized into discrete units called genes. Genes specify proteins through the steps of transcription and translation. During transcription, a strand of DNA is used as a template for the production of an mRNA molecule. The actions of RNA polymerase, the enzyme that carries out transcription, and DNA polymerase, which is required for DNA replication, are similar enough to suggest that both enzymes evolved from a common ancestral enzyme.

Scientists are now discovering that the rest of the DNA that does not specify proteins may also have valuable functions. Specifically, it now appears that RNA may play a prominent role in the regulation of the genome. Some believe this is evidence that RNA may have preceded DNA in the evolutionary history of cells. Many biologists believe that we need a new definition of a gene that recognizes that much of our DNA results in RNA molecules rather than protein products. Nevertheless, both protein-coding and non-protein-coding DNA provides the blueprint for building and developing an entire organism. But just as a blueprint is useless without a team of engineers, architects, and construction workers to execute it, the expression of genes requires a large cadre of proteins and other factors to control it. As you will see in the following chapter, regulatory proteins may turn genes on or off, and genes can be combined in many different ways to alter the proteins that are made. Together, these mechanisms contribute to the great complexity and diversity of living organisms.

## summary

### 12.1 The Genetic Material

Early work illustrated that DNA was the hereditary material. Griffith injected strains of pneumococcus into mice and observed that when heat-killed S strain bacteria were injected along with live R strain bacteria, virulent S strain bacteria were recovered from the dead mice. Griffith said that the R strain had been transformed by some substance passing from the dead S strain to the live R strain. Twenty years later, Avery and his colleagues reported that the transforming substance is DNA.

To study the structure of DNA, Chargaff performed a chemical analysis of DNA and found that A = T and G = C, and that the amount of purine equals the amount of pyrimidine. Franklin prepared an X-ray photograph of DNA that showed it is helical, has repeating structural features, and has certain dimensions. Watson and Crick built a model of DNA in which the sugar-phosphate molecules made up the sides of a twisted ladder, and the complementary-paired bases were the rungs of the ladder.

### 12.2 Replication of DNA

The Watson and Crick model immediately suggested a method by which DNA could be replicated. Basically, the two strands unwind and unzip, and each parental strand acts as a template for a new (daughter) strand. In the end, each new helix is like the other and like the parental helix.

The enzyme DNA polymerase joins the nucleotides together and proofreads them to make sure the bases have been paired correctly. Incorrect base pairs that survive the process are a mutation. Replication in prokaryotes typically proceeds in both directions from one point of origin to a termination region until there are two copies of the circular chromosome. Replication in eukaryotes has many points of origin and many bubbles (places where the DNA strands are separating and replication is occurring). Replication occurs at the ends of the bubbles—at replication forks. Since eukaryotes have linear chromosomes, they cannot replicate the very ends of them. Therefore, the ends (telomeres) get shorter with each replication.

### 12.3 The Genetic Code of Life

The central dogma of molecular biology says that (1) DNA is a template for its own replication and also for RNA formation during transcription, and (2) the sequence of nucleotides in mRNA directs the correct sequence of amino acids of a polypeptide during translation.

The genetic code is a triplet code, and each codon (code word) consists of three bases. The code is degenerate—that is, more than one codon exists for most amino acids. There are also one start and three stop codons. The genetic code is considered universal, but there are a few exceptions.

### 12.4 First Step: Transcription

Transcription to form messenger RNA (mRNA) begins when RNA polymerase attaches to the promoter of a gene. Elongation occurs until RNA polymerase reaches a stop sequence. The mRNA is processed following transcription. A cap is put onto the 5′ end, and a poly-A tail is put onto the 3′ end, and introns are removed in eukaryotes by spliceosomes. Small nuclear RNAs (snRNAs) present in spliceosomes help identify the introns to be removed. Small nucleolar RNAs (snoRNAs) perform the same processing function in the nucleolus. These snRNAs play a role in alternative mRNA splicing, which allows a single eukaryotic gene to code for different proteins, depending on which segments of the gene serve as introns and which serve as exons. Some introns serve as microRNAs (mRNAs), which help regulate the translation of mRNAs. Research is now directed to discovering the many ways small RNAs influence the production of proteins in a cell.

### 12.5 Second Step: Translation

Translation requires mRNA, transfer RNA (tRNA), and ribosomal RNA (rRNA). Each tRNA has an anticodon at one end and an amino acid at the other; amino acid–activating enzymes ensure that the correct amino acid is attached to the correct tRNA. When tRNAs bind with their codon at a ribosome, the amino acids are correctly sequenced in a polypeptide according to the order predetermined by DNA.

In the cytoplasm, many ribosomes move along the same mRNA at a time. Collectively, these are called a polyribosome.

Translation requires these steps: During initiation, mRNA, the first (initiator) tRNA, and the two subunits of a ribosome all come together in the proper orientation at a start codon. During elongation, as the tRNA anticodons bind to their codons, the growing peptide chain is transferred by peptide bonding to the next amino acid in a polypeptide. During termination at a stop codon, the polypeptide is cleaved from the last tRNA. The ribosome now dissociates.

### 12.6 Structure of the Eukaryotic Chromosome

Eukaryotic cells contain nearly 2 m of DNA, yet must pack it all into a nucleus no more than 20 μm in diameter. Thus, the DNA is compacted by winding it around DNA-binding proteins called histones to make nucleosomes. The nucleosomes are further compacted into a zigzag structure, which is then folded upon itself many times to form radial loops, which is the usual compaction state of euchromatin. Heterochromatin is further compacted by scaffold proteins, and further compaction can be achieved prior to mitosis and meiosis.

## understanding the terms

adenine (A) 214
anticodon 224
codon 221
complementary base pairing 216
cytosine (C) 214
DNA polymerase 217
DNA replication 217
double helix 215
elongation 226
euchromatin 229
exon 223
genetic code 221
guanine (G) 214
heterochromatin 229
histone 228
initiation 226
intron 223
messenger RNA (mRNA) 220
mRNA transcript 222
nucleosome 229
nucleotide 212
polyribosome 225
promoter 222
proteomics 227
replication fork 219
ribosomal RNA (rRNA) 220
ribozyme 223
RNA polymerase 222
semiconservative replication 217
telomere 218
template 217
termination 227
thymine (T) 214
transcription 220
transfer RNA (tRNA) 220
translation 220
translocation 227
triplet code 221
uracil (U) 220
wobble hypothesis 224

Match the terms to these definitions:

a. ______________ Noncoding segment of DNA that is transcribed but is removed from the transcript before leaving the nucleus.
b. ______________ During replication, an enzyme that joins the nucleotides complementary to a DNA template.
c. ______________ A type of repetitive DNA element that may be distributed across multiple chromosomes.
d. ______________ Events by which the sequence of codons in mRNA determines the sequence of amino acids in a polypeptide.

## reviewing this chapter

1. List and discuss the requirements for genetic material. 212
2. How did Avery and his colleagues demonstrate that the transforming substance is DNA? 212–13
3. Describe the Watson and Crick model of DNA structure. How did it fit the data provided by Chargaff and the X-ray diffraction patterns of Franklin? 214–16
4. Explain how DNA replicates semiconservatively. What role does DNA polymerase play? What role does helicase play? 217
5. List and discuss differences between prokaryotic and eukaryotic replication of DNA. 219
6. How did investigators reason that the code must be a triplet code, and in what manner was the code cracked? Why is it said that the code is degenerate, unambiguous, and almost universal? 220–21
7. What two steps are required for the expression of a gene? 222, 224
8. What specific steps occur during transcription of RNA off a DNA template? 222
9. How is messenger RNA (mRNA) processed before leaving the eukaryotic nucleus? 222–23
10. What is the role of snRNAs in the nucleus and the role of snoRNAs in the nucleolus? 223
11. Compare the functions of mRNA, transfer RNA (tRNA), and ribosomal RNA (rRNA) during protein synthesis. What are the specific events of translation? 224–27
12. What are the various levels of chromosome structure? 228–29

## testing yourself

Choose the best answer for each question.

1. If 30% of an organism's DNA is thymine, then
   a. 70% is purine.
   b. 20% is guanine.
   c. 30% is adenine.
   d. 70% is pyrimidine.
   e. Both c and d are correct.
2. The double-helix model of DNA resembles a twisted ladder in which the rungs of the ladder are
   a. a purine paired with a pyrimidine.
   b. A paired with G and C paired with T.
   c. sugar-phosphate paired with sugar-phosphate.
   d. a 5′ end paired with a 3′ end.
   e. Both a and b are correct.
3. In a DNA molecule,
   a. the bases are covalently bonded to the sugars.
   b. the sugars are covalently bonded to the phosphates.
   c. the bases are hydrogen-bonded to one another.
   d. the nucleotides are covalently bonded to one another.
   e. All of these are correct.
4. DNA replication is said to be semiconservative because
   a. one of the new molecules conserves both of the original DNA strands.
   b. the new DNA molecule contains two new DNA strands.
   c. both of the new molecules contain one new strand and one old strand.
   d. DNA polymerase conserves both of the old strands.
5. If the sequence of bases in one strand of DNA is 5′ TAGCCT 3′, then the sequence of bases in the other strand will be
   a. 3′ TCCGAT 5′.
   b. 3′ ATCGGA 5′.
   c. 3′ TAGCCT 5′.
   d. 3′ AACGGUA 5′.
6. Transformation occurs when
   a. DNA is transformed into RNA.
   b. DNA is transformed into protein.
   c. bacteria cannot grow on penicillin.
   d. organisms receive foreign DNA and thereby acquire a new characteristic.
7. Pyrimidines
   a. are always paired with a purine.
   b. are thymine and cytosine.
   c. keep DNA from replicating too often.
   d. are adenine and guanine.
   e. Both a and b are correct.
8. Watson and Crick incorporated which of the following into their model of DNA structure?
   a. Franklin's diffraction data
   b. Chargaff's rules
   c. complementary base pairing
   d. alternating sugar-phosphate backbone
   e. All of these are correct.

9. A nucleotide
   a. is smaller than a base.
   b. is a subunit of nucleic acids.
   c. has a lot of variable parts.
   d. has at least four phosphates.
   e. always joins with other nucleotides.
10. This is a segment of a DNA molecule. What are (a) the RNA codons, (b) the tRNA anticodons, and (c) the sequence of amino acids in a protein?

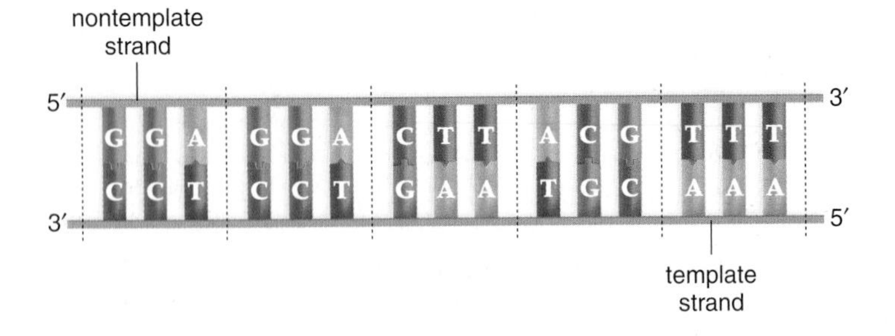


11. During replication, separation of DNA strands requires
   a. backbones to split.
   b. nucleotides to join together.
   c. hydrolysis and synthesis to occur.
   d. hydrogen bonds to unzip.
   e. All of these are correct.
12. In prokaryotes,
   a. replication can occur in two directions at once because their DNA molecule is circular.
   b. bubbles thereby created spread out until they meet.
   c. replication occurs at numerous replication forks.
   d. a new round of DNA replication cannot begin before the previous round is complete.
   e. Both a and b are correct.
13. The central dogma of molecular biology
   a. states that DNA is a template for all RNA production.
   b. states that DNA is a template only for DNA replication.
   c. states that translation precedes transcription.
   d. states that RNA is a template for DNA replication.
   e. All of these are correct.
14. Transcription of a gene results in the production of
   a. an mRNA.
   b. proteins.
   c. an rRNA.
   d. ribozymes.
15. Which of these does not characterize the process of transcription? Choose more than one answer if correct.
   a. RNA is made with one strand of the DNA serving as a template.
   b. In making RNA, the base uracil of RNA pairs with the base thymine of DNA.
   c. The enzyme RNA polymerase synthesizes RNA.
   d. RNA is made in the cytoplasm of eukaryotic cells.
16. Because there are more codons than amino acids,
   a. some amino acids are specified by more than one codon.
   b. some codons specify more than one amino acid.
   c. some codons do not specify any amino acid.
   d. some amino acids do not have codons.
17. If the sequence of bases in the coding strand of a DNA is TAGC, then the sequence of bases in the mRNA will be
   a. AUCG.
   b. TAGC.
   c. UAGC.
   d. CGAU.
18. During protein synthesis, an anticodon on transfer RNA (tRNA) pairs with
   a. DNA nucleotide bases.
   b. ribosomal RNA (rRNA) nucleotide bases.
   c. messenger RNA (mRNA) nucleotide bases.
   d. other tRNA nucleotide bases.
   e. Any one of these can occur.
19. If the sequence of DNA on the template strand of a gene is AAA, the mRNA codon produced by transcription will be ______________ and will specify the amino acid ______________.
   a. AAA, lysine
   b. AAA, phenylalanine
   c. TTT, arginine
   d. UUU, phenylalanine
   e. TTT, lysine
20. Euchromatin
   a. is organized into radial loops.
   b. is less condensed than heterochromatin.
   c. contains nucleosomes.
   d. All of these are correct.
21. Which of the following statements about the organization of eukaryotic genes is true? The protein-coding region of a eukaryotic gene
   a. is divided into exons.
   b. is always the same in every cell.
   c. is determined in part by small nuclear RNAs that influence which segments of a gene will be introns.
   d. undergoes both transcription and translation.
   e. Both a and c are true.
22. Which of these can influence the final product of a eukaryotic protein-coding gene?
   a. The introns and exons of a gene,
   b. The snRNAs present in the spliceosome.
   c. The work of microRNAs, which are derived from introns.
   d. The nuclear envelope.
   e. All but d are correct.

## thinking scientifically

1. How would you test a hypothesis that a genetic condition, such as neurofibromatosis, is due to a transposon?
2. Knowing that a plant will grow from a single cell in tissue culture, how could you transform a plant so that it glows in the dark?

## *Biology* website

The companion website for *Biology* provides a wealth of information organized and integrated by chapter. You will find practice tests, animations, videos, and much more that will complement your learning and understanding of general biology.

**http://www.mhhe.com/maderbiology10**

# 13

# Regulation of Gene Activity

*The human genome project revealed that humans have about 20,500 genes, scant more than a nematode, which contains nearly 20,000. So how do humans and other complex eukaryotes get by with so few genes? The answer, surprisingly, may lie in the regulation of pre-mRNA splicing to allow the production of a myriad of proteins from a single gene. DSCAM is a gene associated with Down Syndrome that is present in the brain of many animals. In fruit flies DSCAM has four regions of alternative exons. The result is over 38,000 different possible combinations of functional mRNAs and, therefore, the DSCAM gene is able to specify 38,000 different proteins. Such a huge number of proteins is sufficient to provide each nerve cell with a unique identity as it communicates with others within a brain.*

*Complex alternative splicing, and other regulatory mechanisms, could very well account for how humans generate so many different proteins from so few genes. This chapter introduces you to regulatory mechanisms in both prokaryotes and eukaryotes, allowing you to see how these mechanisms influence the processes of transcription and translation that you learned about in the previous chapter.*

Individual neurons may use unique forms of the Dscam proteins in their plasma membranes to identify themselves.

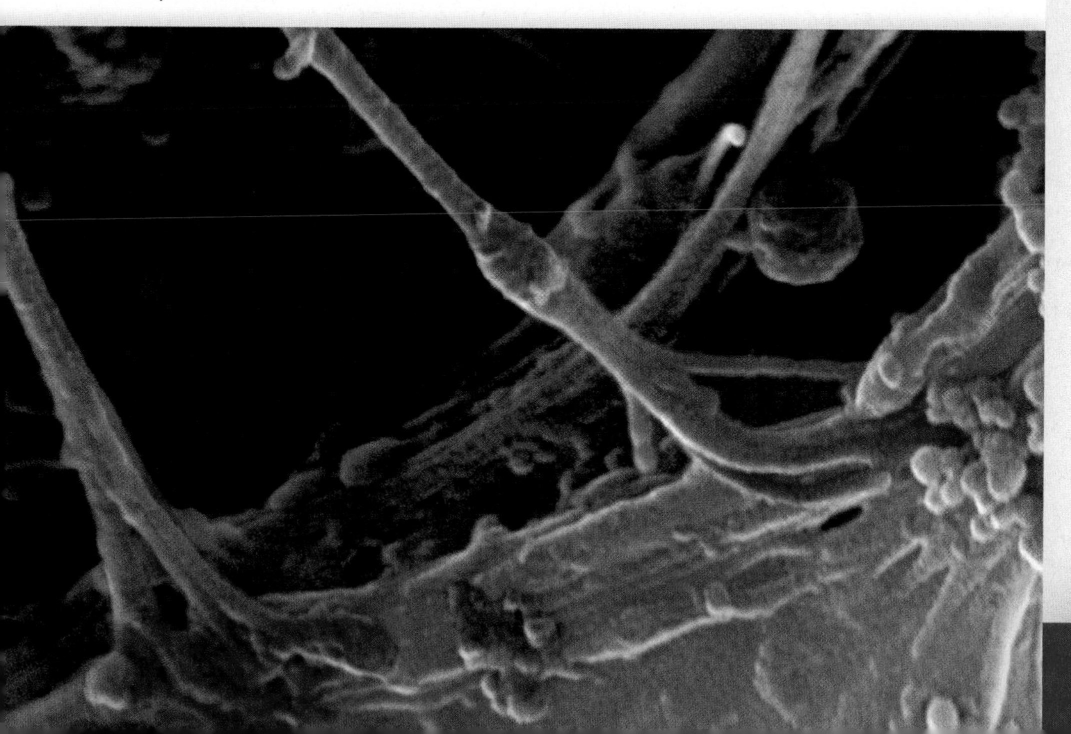

## concepts

## 13.1 Prokaryotic Regulation

Because their environment is ever changing, bacteria do not need the same enzymes and possibly other proteins all the time. In 1961, French microbiologists François Jacob and Jacques Monod showed that *Escherichia coli* is capable of regulating the expression of its genes. They observed that the genes for a metabolic pathway, called structural genes, are grouped on a chromosome and subsequently are transcribed at the same time. Jacob and Monod, therefore, proposed the **operon** [L. *opera*, works] model to explain gene regulation in prokaryotes and later received a Nobel Prize for their investigations. An operon typically includes the following elements:

**Promoter**—A short sequence of DNA where RNA polymerase first attaches to begin transcription of the grouped genes. Basically, a promoter signals where transcription is to begin.

**Operator**—A short portion of DNA where an active repressor binds. When an active **repressor** binds to the operator, RNA polymerase cannot attach to the promoter, and transcription cannot occur. In this way, the operator controls transcription of structural genes.

**Structural genes**—One to several genes coding for the primary structure of enzymes in a metabolic pathway transcribed as a unit.

A **regulator gene,** normally located outside the operon and controlled by its own promoter, codes for a repressor that controls whether the operon is active or not.

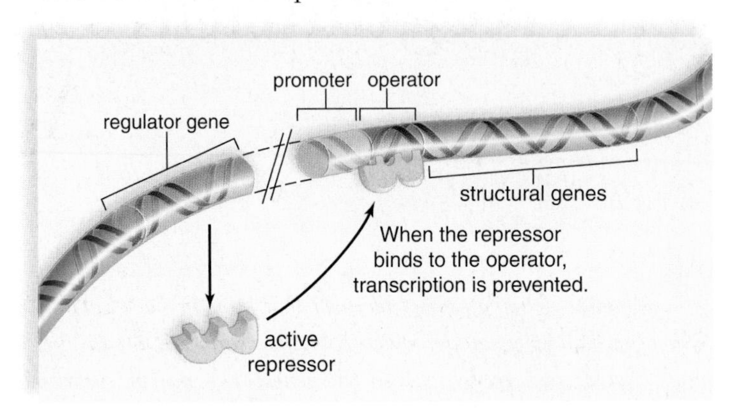


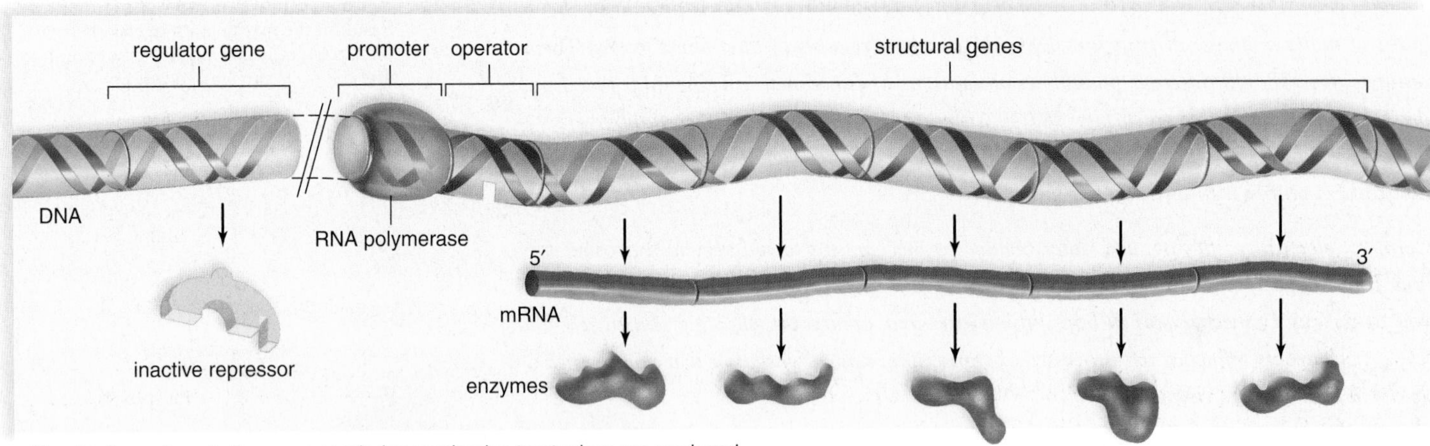


a. **Tryptophan absent.** Enzymes needed to synthesize tryptophan are produced.

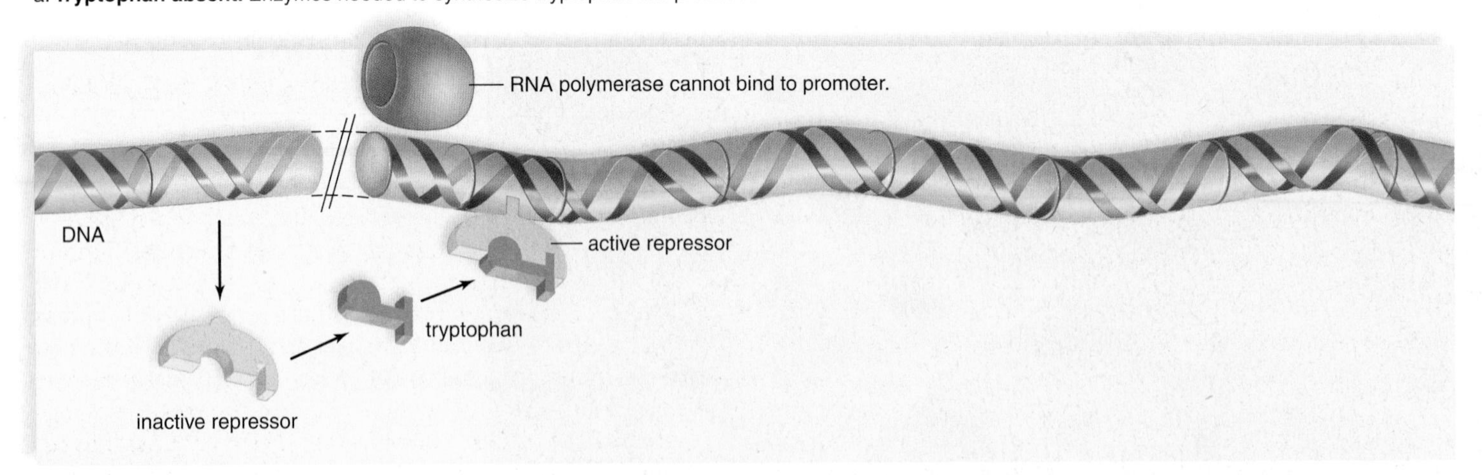


b. **Tryptophan present.** Presence of tryptophan prevents production of enzymes used to synthesize tryptophan.

**FIGURE 13.1 The *trp* operon.**

**a.** The regulator gene codes for a repressor protein that is normally inactive. RNA polymerase attaches to the promoter, and the structural genes are expressed. **b.** When the nutrient tryptophan is present, it binds to the repressor, changing its shape. Now the repressor is active and can bind to the operator. RNA polymerase cannot attach to the promoter, and the structural genes are not expressed.

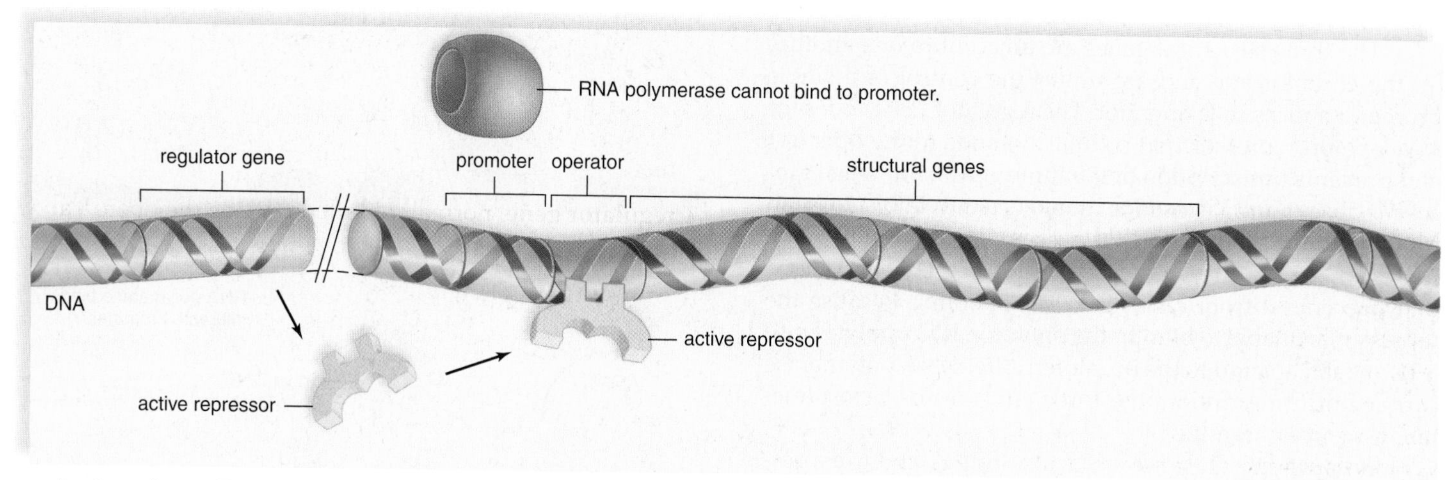


a. **Lactose absent.** Enzymes needed to take up and use lactose are not produced.

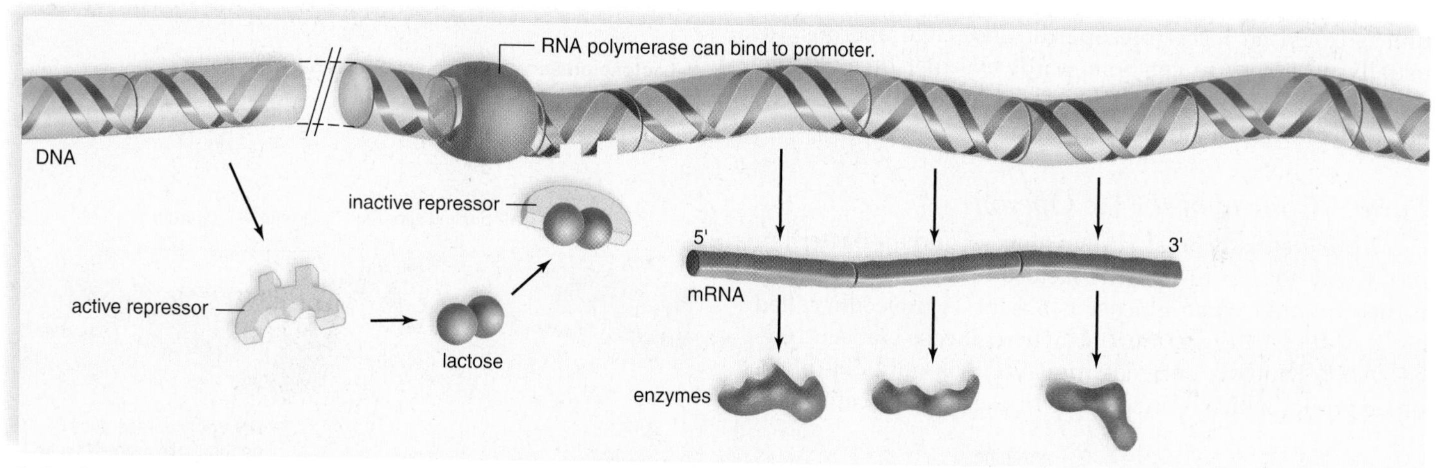


b. **Lactose present.** Enzymes needed to take up and use lactose are produced only when lactose is present.

**FIGURE 13.2 The *lac* operon.**

**a.** The regulator gene codes for a repressor that is normally active. When it binds to the operator, RNA polymerase cannot attach to the promoter, and structural genes are not expressed. **b.** When lactose is present, it binds to the repressor, changing its shape so that it is inactive and cannot bind to the operator. Now, RNA polymerase binds to the promoter, and the structural genes are expressed.

## The *trp* Operon

Investigators, including Jacob and Monod, found that some operons in *E. coli* usually exist in the "on" rather than "off" condition. For example, in the *trp* operon, the regulator codes for a repressor that ordinarily is unable to attach to the operator. Therefore, RNA polymerase is able to bind to the promoter, and the structural genes of the operon are ordinarily expressed (Fig. 13.1). Their products, five different enzymes, are part of an anabolic pathway for the synthesis of the amino acid tryptophan.

If tryptophan happens to be already present in the medium, these enzymes are not needed by the cell, and the operon is turned off by the following method. Tryptophan binds to the repressor. A change in shape now allows the repressor to bind to the operator, and the structural genes are not expressed. The enzymes are said to be repressible, and the entire unit is called a **repressible operon.** Tryptophan is called the **corepressor.** Repressible operons are usually involved in anabolic pathways that synthesize a substance needed by the cell.

## The *lac* Operon

Bacteria metabolism is remarkably efficient; when there is no need for certain proteins or enzymes, the genes that are used to make them are usually inactive. For example, if the milk sugar lactose is not present, there is no need to express genes for enzymes involved in lactose catabolism. But when *E. coli* is denied glucose and is given the milk sugar lactose instead, it immediately begins to make the three enzymes needed for the metabolism of lactose. These enzymes are encoded by three genes (Fig. 13.2): One gene is for an enzyme called β-galactosidase, which breaks down the disaccharide lactose to glucose and galactose; a second gene codes for a permease that facilitates the entry of lactose into the cell; and a third gene codes for an enzyme called transacetylase, which has an accessory function in lactose metabolism.

The three structural genes are adjacent to one another on the chromosome and are under the control of a single promoter and a single operator. The regulator gene codes for a *lac* operon repressor that ordinarily binds to the operator and prevents transcription of the three genes. But when glucose is absent and lactose (or more correctly, allolactose, an isomer formed from lactose) is present, lactose binds to the repressor, and the repressor undergoes a change in shape that prevents it from binding to the operator. Because the repressor is unable to bind to the operator, RNA polymerase is better able to bind to the promoter. After RNA polymerase carries out transcription, the three enzymes of lactose metabolism are synthesized.

Because the presence of lactose brings about expression of genes, it is called an **inducer** of the *lac* operon: The enzymes are said to be inducible enzymes, and the entire unit is called an **inducible operon.** Inducible operons are usually necessary to catabolic pathways that break down a nutrient. Why is that beneficial? Because these enzymes need only be active when the nutrient is present.

## *Further Control of the* lac *Operon*

*E. coli* preferentially breaks down glucose, and the bacterium has a way to ensure that the lactose operon is maximally turned on only when glucose is absent. A molecule called *cyclic AMP (cAMP)* accumulates when glucose is absent. Cyclic AMP, which is derived from ATP, has only one phosphate group, which is attached to ribose at two locations:

adenine
5′ $CH_2$ O
P
3′
OH

**cyclic AMP**
(cAMP)

Cyclic AMP binds to a molecule called a *catabolite activator protein (CAP),* and the complex attaches to a CAP binding site next to the *lac* promoter. When CAP binds to DNA, DNA bends, exposing the promoter to RNA polymerase. RNA polymerase is now better able to bind to the promoter so that the *lac* operon structural genes are transcribed, leading to their expression (Fig. 13.3).

When glucose is present, there is little cAMP in the cell; CAP is inactive, and the lactose operon does not function maximally. CAP affects other operons as well and takes its name for activating the catabolism of various other metabolites when glucose is absent. A cell's ability to encourage the metabolism of lactose and other metabolites when glucose is absent provides a backup system for survival when the preferred energy source glucose is absent.

The CAP protein's regulation of the *lac* operon is an example of positive control. Why? Because when this molecule is active, it promotes the activity of an operon. The use of repressors, on the other hand, is an example of negative control because when active they shut down an operon. A positive control mechanism allows the cell to fine-tune its response. In the case of the *lac* operon, the operon is only maximally active when glucose is absent and lactose is present. If both glucose and lactose are present, the cell preferentially metabolizes glucose.

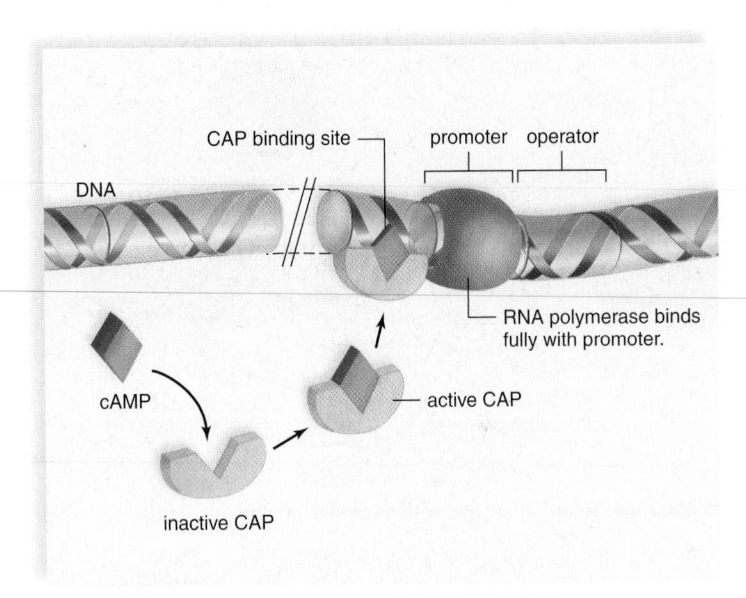


a. **Lactose present, glucose absent (cAMP level high)**

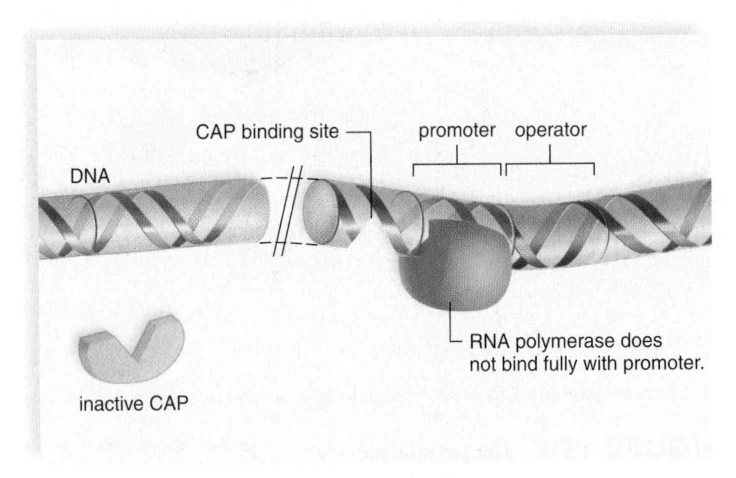


b. **Lactose present, glucose present (cAMP level low)**

**FIGURE 13.3 Action of CAP.**

When active CAP binds to its site on DNA, the RNA polymerase is better able to bind to the promoter so that the structural genes of the *lac* operon are expressed. **a.** CAP becomes active in the presence of cAMP, a molecule that is prevalent when glucose is absent. Therefore, transcription of lactose enzymes increases, and lactose is metabolized. **b.** If glucose is present, CAP is inactive, and RNA polymerase does not completely bind to the promoter. Therefore, transcription of lactose enzymes decreases, and less metabolism of lactose occurs.

### Check Your Progress 13.1

1. What is an operon? Why is it advantageous for prokaryotes to organize their genes in this manner?
2. Explain the difference between positive control and negative control of gene expression.

## 13.2 Eukaryotic Regulation

Each cell in multicellular eukaryotes, including humans, has a copy of all genes; however, different genes are actively expressed in different cells. Muscle cells, for example, have a different set of genes that are turned on in the nucleus and a different set of proteins that are active in the cytoplasm than do nerve cells.

Like prokaryotic cells, a variety of mechanisms regulate gene expression in eukaryotic cells. These mechanisms can be grouped under five primary levels of control; three of them pertain to the nucleus, and two pertain to the cytoplasm (Fig. 13.4). In other words, control of gene activity in eukaryotes extends from transcription to protein activity. These are the types of control in eukaryotic cells that can modify the amount of the gene product:

1. *Chromatin structure:* Chromatin packing is used as a way to keep genes turned off. If genes are not accessible to RNA polymerase, they cannot be transcribed.

   In the nucleus, highly condensed chromatin is not available for transcription, while more loosely condensed chromatin is available for transcription. Chromatin structure is a part of **epigenetic** [Gk. *epi*, besides] **inheritance,** the transmission of genetic information outside the coding sequences of a gene.
2. *Transcriptional control:* The degree to which a gene is transcribed into mRNA determines the amount of gene product. In the nucleus, transcription factors may promote or repress transcription, the first step in gene expression.
3. *Posttranscriptional control:* Posttranscriptional control involves mRNA processing and how fast mRNA leaves the nucleus. We now know that mRNA processing differences can determine the type of protein product made by a cell.

   Also, mRNA processing differences can affect how fast mRNA leaves the nucleus and the amount of gene product within a given amount of time.
4. *Translational control:* Translational control occurs in the cytoplasm and affects when translation begins and how long it continues. Any influence that can cause the persistence of the 5′ cap and 3′ poly-A tail can affect the length of translation. Excised introns are now believed to be involved in a regulatory system that directly affects the life span of mRNA.

   Some mRNAs may need further processing before they are translated. The possibility of an RNA-based regulatory system is being investigated.
5. *Posttranslational control:* Posttranslational control, which also takes place in the cytoplasm, occurs after protein synthesis. Only a functional protein is an active gene product.

   The polypeptide product may have to undergo additional changes before it is biologically functional. Also, a functional enzyme is subject to feedback control so that it is no longer able to speed its reaction.

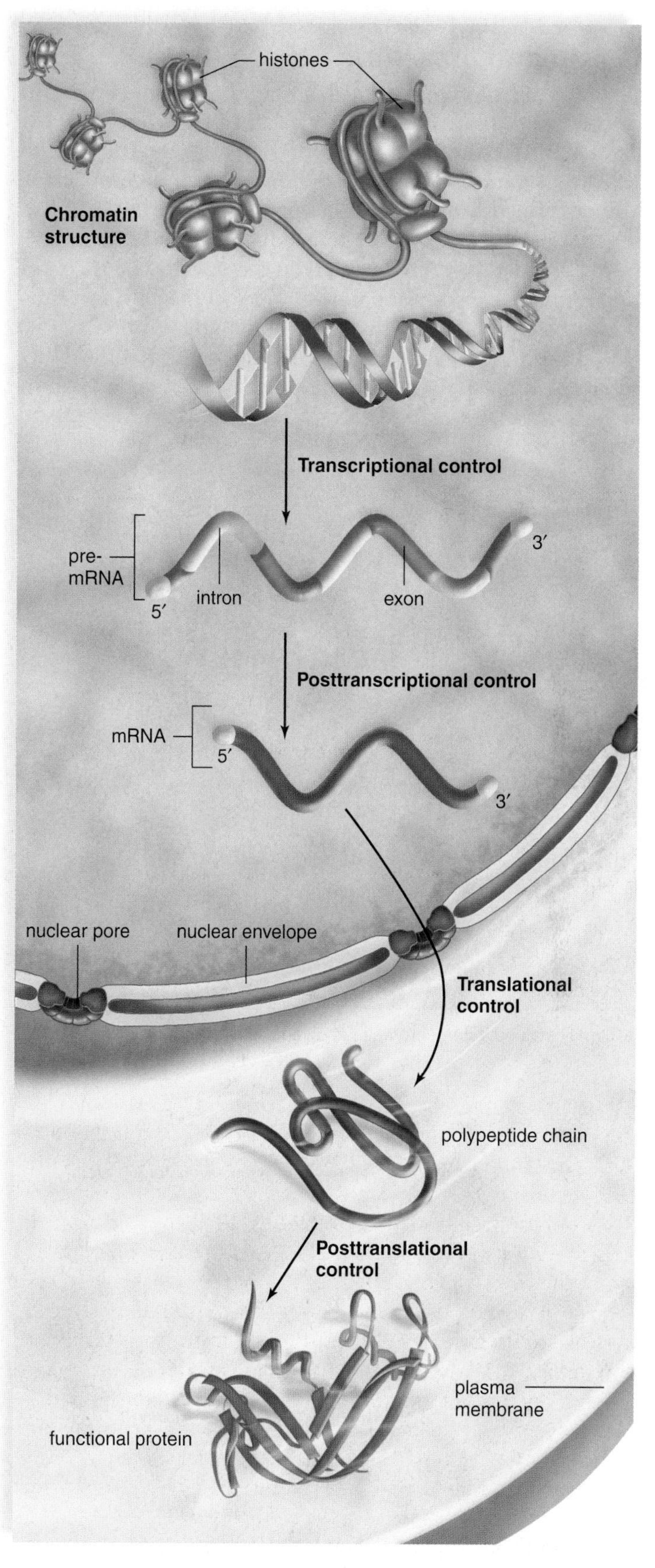


**FIGURE 13.4 Levels at which control of gene expression occurs in eukaryotic cells.**

The five levels of control are (1) chromatin structure, (2) transcriptional control, and (3) posttranscriptional control, which occur in the nucleus; and (4) translational and (5) posttranslational control, which occur in the cytoplasm.

## Chromatin Structure

Genetic control in eukaryotes is bound to be more complicated than prokaryotes if only because they have a great deal more DNA than prokaryotes. We learned in Figure 12.20 that various levels of condensation and compaction are necessary in order to fit a very large amount of DNA into a much smaller nucleus. The degree to which chromatin is compacted greatly affects the accessibility of the chromatin to the transcriptional machinery of the cell, and thus the expression levels of the genes contained within.

In general, highly condensed heterochromatin is inaccessible to RNA polymerase, and the genes contained within are seldom or never transcribed. **Heterochromatin** appears as darkly stained portions within the nucleus in electron micrographs (Fig. 13.5*a*). A dramatic example of heterochromatin is the Barr body in mammalian females. Females have a small, darkly staining mass of condensed chromatin adhering to the inner edge of the nuclear membrane. This structure, called a **Barr body** after its discoverer, is an inactive X chromosome. One of the X chromosomes undergoes inactivation in the cells of female embryos. The inactive X chromosome does not produce gene products, and therefore female cells have a reduced amount of product from genes on the X chromosome.

How do we know that Barr bodies are inactive X chromosomes that are not producing gene product? Suppose 50% of the cells have one X chromosome active and 50% have

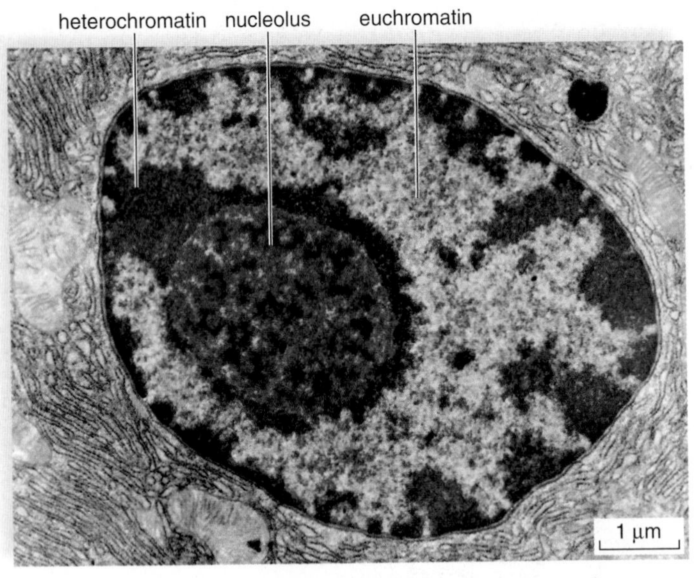


a. Darkly stained heterochromatin and lightly stained euchromatin

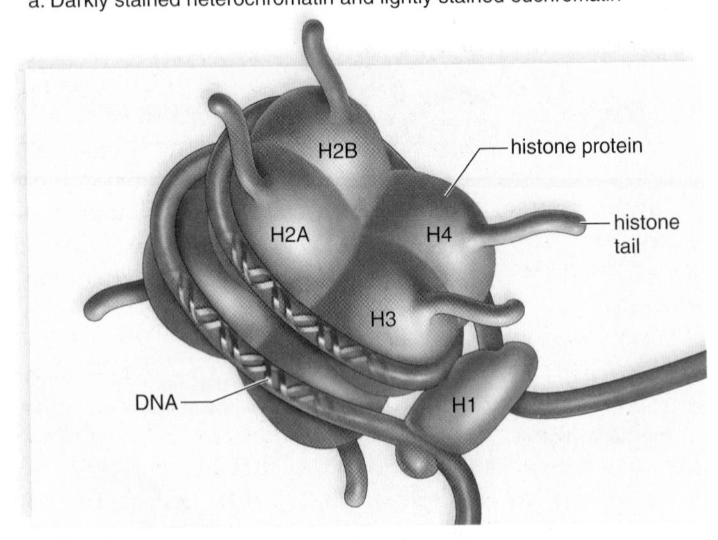


b. A nucleosome

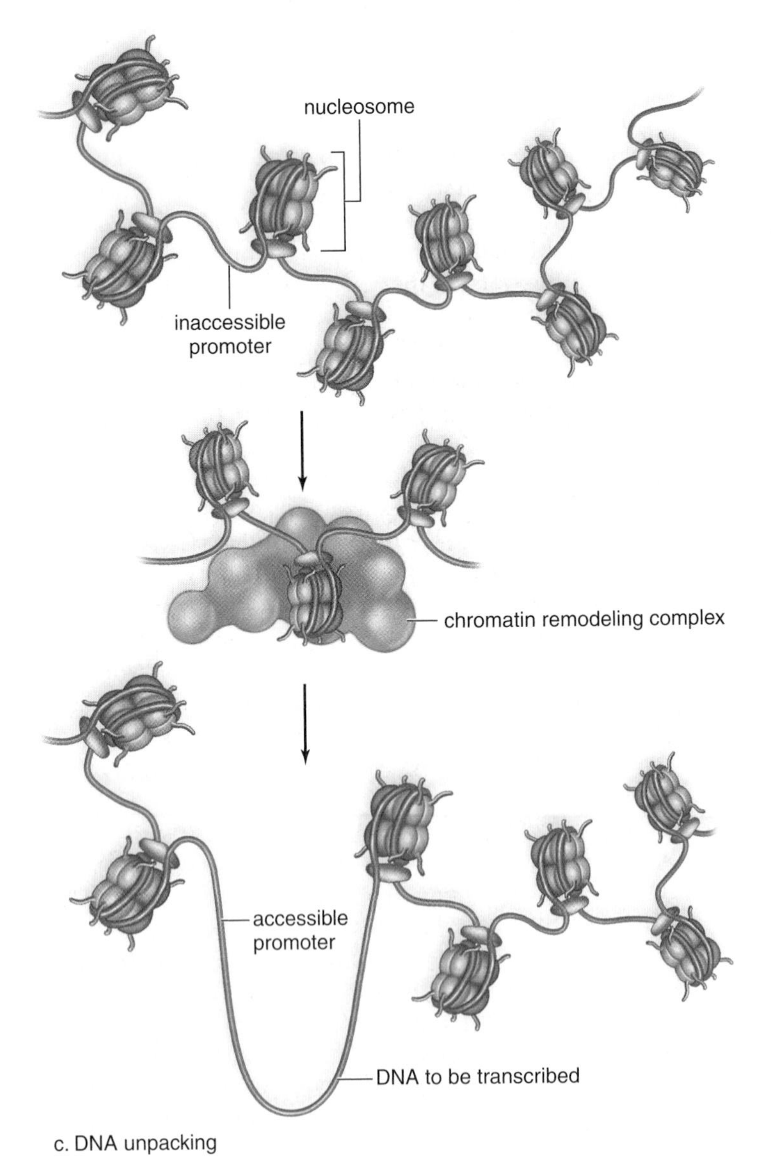


c. DNA unpacking

**FIGURE 13.5 Chromatin structure regulates gene expression.**

**a.** A eukaryotic nucleus contains heterochromatin (darkly stained, highly condensed chromatin) and euchromatin, which is not as condensed. **b.** Nucleosomes ordinarily prevent access to DNA so that transcription cannot take place. If histone tails are acetylated, access can be achieved; if the tails are methylated, access is more difficult. **c.** A chromatin remodeling complex works on euchromatin to make a promoter accessible for transcription.

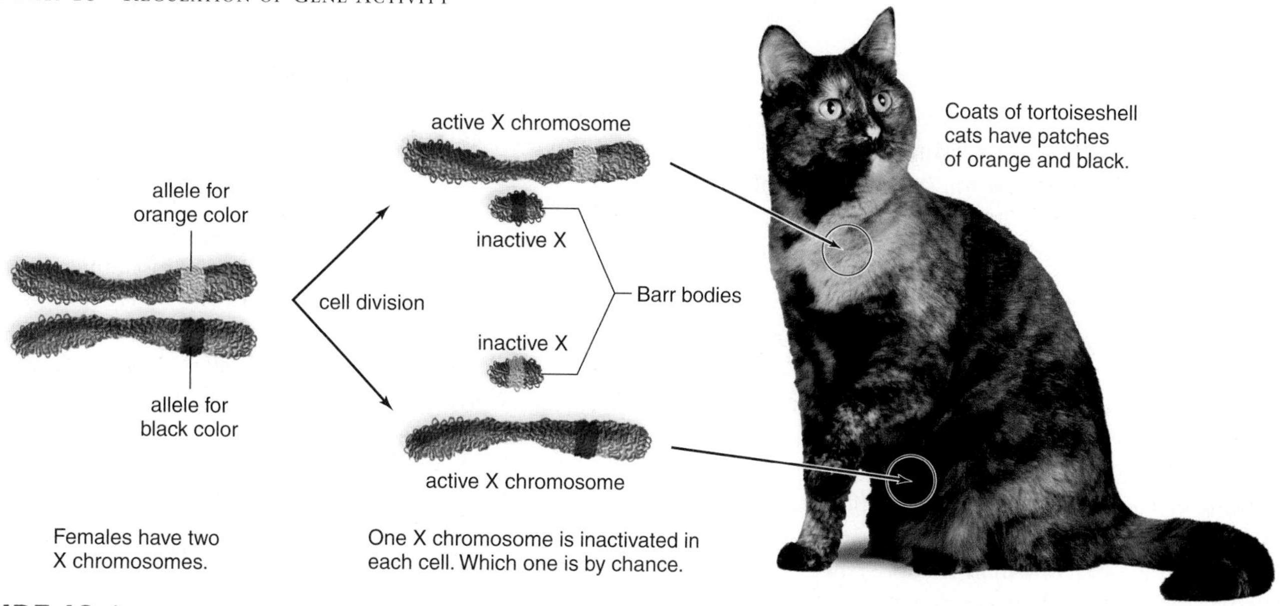


**FIGURE 13.6 X-inactivation in mammalian females.**

In cats, the alleles for black or orange coat color are carried on the X chromosomes. Random X-inactivation occurs in females. Therefore, in heterozygous females, 50% of the cells have an allele for black coat color and 50% of cells have an allele for orange coat color. The result is tortoiseshell cats that have coats with patches of both black and orange.

the other X chromosome active. Wouldn't the body of a heterozygous female be a mosaic, with "patches" of genetically different cells? This is exactly what investigators have discovered. Human females who are heterozygous for an X-linked recessive form of ocular albinism have patches of pigmented and nonpigmented cells at the back of the eye. Women heterozygous for Duchenne muscular dystrophy have patches of normal muscle tissue and degenerative muscle tissue (the normal tissue increases in size and strength to make up for the defective tissue). And women who are heterozygous for X-linked hereditary absence of sweat glands have patches of skin lacking sweat glands. The female tortoiseshell cat also provides dramatic support for a difference in X-inactivation in its cells. In these cats, an allele for black coat color is on one X chromosome, and a corresponding allele for orange coat color is on the other X chromosome. The patches of black and orange in the coat can be related to which X chromosome is in the Barr bodies of the cells found in the patches (Fig. 13.6).

## *DNA Unpacking*

Active genes in eukaryotic cells are associated with more loosely packed chromatin called **euchromatin** (see Fig. 12.20). What regulates whether chromatin exists as heterochromatin or euchromatin? You learned in Figure 12.20 that in a 30 nm fiber a *nucleosome* is a portion of DNA wrapped around a group of histone molecules. Histone molecules have *tails,* strings of amino acids that extend beyond the main portion of a nucleosome (Fig. 13.5*b*). In heterochromatin, the histone tails tend to bear methyl groups (—$CH_3$); in euchromatin, the histone tails tend to be acetylated and have attached acetyl groups (—$COCH_3$).

Histones regulate accessibility to DNA, and euchromatin becomes genetically active when histones no longer bar access to DNA. When DNA in euchromatin is transcribed, a so-called *chromatin remodeling complex* pushes aside the histone portion of a nucleosome so that access to DNA is not barred and transcription can begin (Fig. 13.5*c*). After *unpacking* occurs, many decondensed loops radiate from the central axis of the chromosome. These chromosomes have been named lampbrush chromosomes because their feathery appearance resembles the brushes that were once used to clean kerosene lamps.

In addition to physically moving nucleosomes aside to expose promoters, chromatin remodeling complexes may also affect gene expression by adding acetyl or methyl groups to histone tails.

## *Epigenetic Inheritance*

Histone modification is sometimes linked to a phenomenon termed *epigenetic inheritance,* in which the pattern of inheritance does not depend on the genes themselves. When histones are methylated, sometimes the DNA itself becomes methylated as well. Some genes undergo a phenomenon called *genomic imprinting,* and either the mother's or father's allele is methylated during gamete formation. If an inherited allele is highly methylated, the gene is not expressed, even if it is a normal gene in every other way. For traits that exhibit genomic imprinting, the expression of the gene depends on whether it was inherited from the mother or the father.

The expression epigenetic inheritance is now used broadly for other inheritance patterns that do not depend on the genes themselves. In this instance, epigenetic inheritance depends on whether acetyl and methyl groups surround and adhere to DNA. Epigenetic inheritance explains unusual inheritance patterns and also may play an important role in growth, aging, and cancer. Researchers are even hopeful that it will be easier to develop drugs to modify this level of inheritance rather than trying to change the DNA itself.

## Transcriptional Control

Although eukaryotes have various levels of genetic control (see Fig. 13.4), **transcriptional control** remains the most critical of these levels. The first step toward transcription is availability of DNA for transcription, which involves DNA unpacking (see page 239). Transcriptional control also involves the participation of transcription factors, activators, and repressors.

### *Transcription Factors, Activators, and Repressors*

Although no operons like those of prokaryotic cells have been found in eukaryotic cells, transcription is still controlled by DNA-binding proteins. Every cell contains many different types of **transcription factors,** proteins that help regulate transcription. The same transcription factors, but not the same mix, are used over again at other promoters, so it is easy to imagine that if one malfunctions, the result could be disastrous to the cell. A group of transcription factors binds to a promoter adjacent to a gene, and then the complex attracts and binds RNA polymerase, but transcription may still not begin.

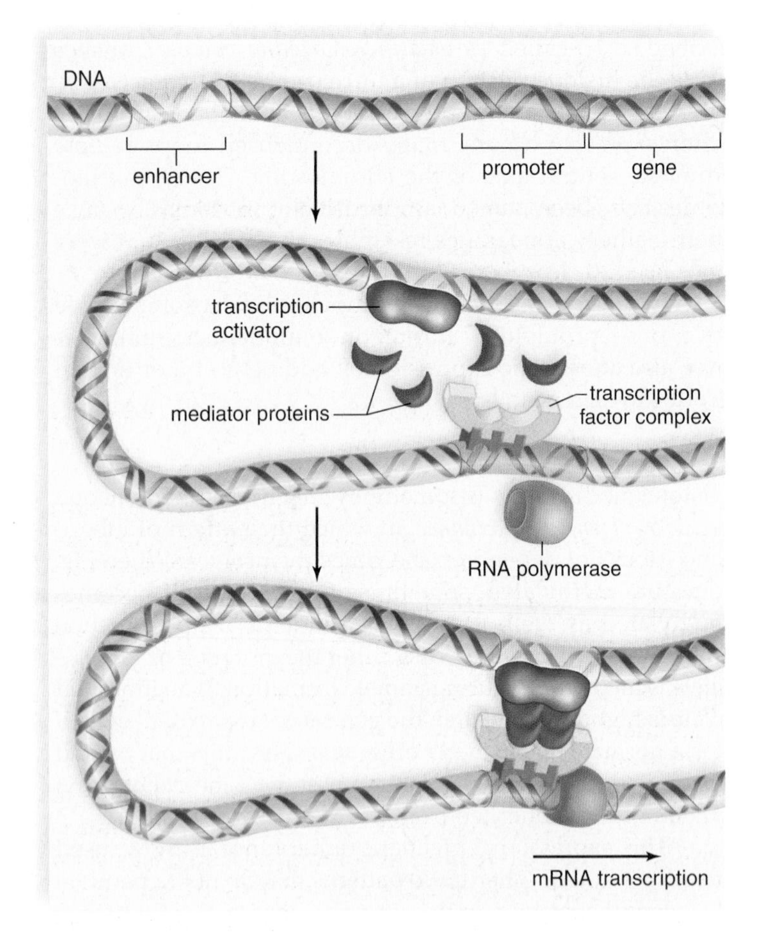


**FIGURE 13.7 Eukaryotic transcription factors.**

Transcription in eukaryotic cells requires that transcription factors bind to the promoter and transcription activators bind to an enhancer. The enhancer is far from the promoter, but the DNA loops and mediator proteins act as a bridge joining activators to factors. Only then does transcription begin.

In eukaryotes, **transcription activators** are DNA-binding proteins that speed transcription dramatically. They bind to a region of DNA called an **enhancer** that can be quite a distance from the promoter. A hairpin loop in the DNA brings the transcription activators attached to the enhancer into contact with the transcription factor complex. Likewise, the binding of repressors to silencers within the promoter may prohibit the transcription of certain genes. Most genes are subject to regulation by both activators and repressors also.

The promoter structure of eukaryotic genes is often very complex, and there is a large variety of regulatory proteins that may interact with each other and with transcription factors to affect a gene's transcription level. Mediator proteins act as a bridge between transcription factors and transcription activators at the promoter. Now RNA polymerase begins the transcription process (Fig. 13.7). Such protein-to-protein interactions are a hallmark of eukaryotic gene regulation. Together, these mechanisms can fine-tune a gene's transcription level in response to a large variety of conditions.

Transcription factors, activators, and repressors are always present in the nucleus of a cell, but they most likely have to be activated in some way before they will bind to DNA. Activation often occurs when they are phosphorylated by a kinase. Kinases, which add a phosphate group to molecules, and phosphatases, which remove a phosphate group, are known to be signaling proteins involved in a growth regulatory network that reaches from receptors in the plasma membrane to the genes in the nucleus.

## Posttranscriptional Control

**Posttranscriptional control** of gene expression occurs in the nucleus and includes alternative mRNA splicing and controlling the speed with which mRNA leaves the nucleus.

During pre-mRNA splicing, introns (noncoding regions) are excised, and exons (expressed regions) are joined together to form an mRNA (see Fig. 12.13). When introns are removed from pre-mRNA, differential splicing of exons can occur, and this affects gene expression. For example, an exon that is normally included in an mRNA transcript may be skipped, and it is excised along with the flanking introns (Fig. 13.8). The resulting mature mRNA has an altered sequence, and the protein encoded differs. Sometimes introns remain in an mRNA transcript; when this occurs, the protein coding sequence will also change.

Examples of alternative pre-mRNA splicing abound. Both the hypothalamus and the thyroid gland produce a protein hormone called calcitonin, but the mRNA that leaves the nucleus is not the same in both types of cells. This results in the thyroid releasing a slightly different version of calcitonin than the hypothalamus. Evidence of alternative mRNA splicing is found in other cells, such as those that produce neurotransmitters, muscle regulatory proteins, and antibodies. This process allows humans and other complex organisms to recombine their genes in many new and novel ways to create the great variety of proteins found in these organisms. Researchers are busy determining how small nuclear

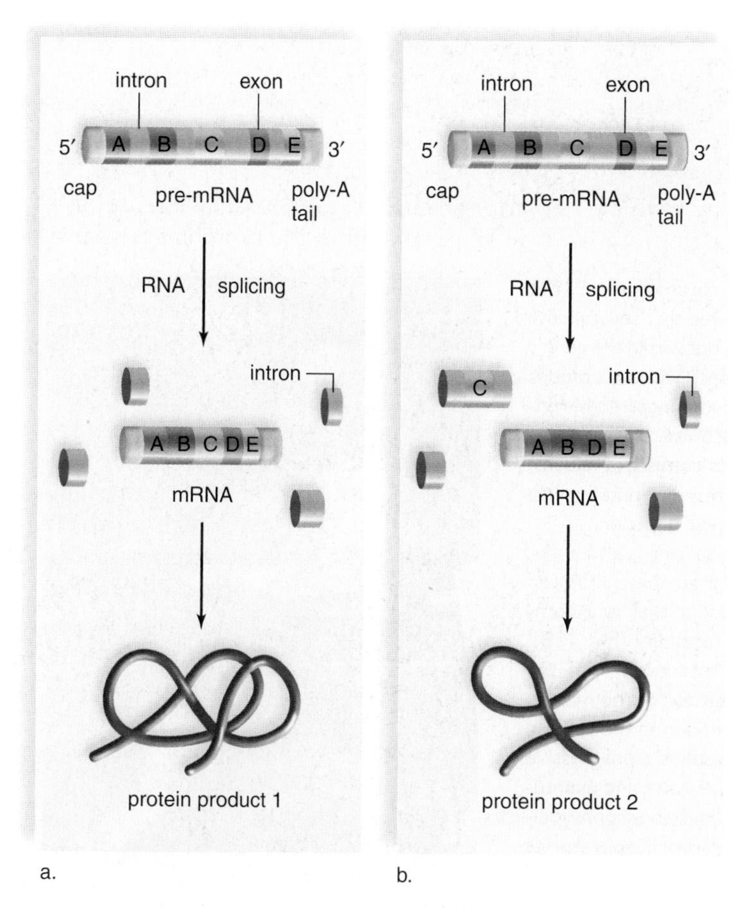


**FIGURE 13.8 Alternative processing of pre-mRNA.**
Because the pre-mRNAs are processed differently in these two cells (**a** and **b**), distinct proteins result. This is a form of posttranscriptional control of gene expression.

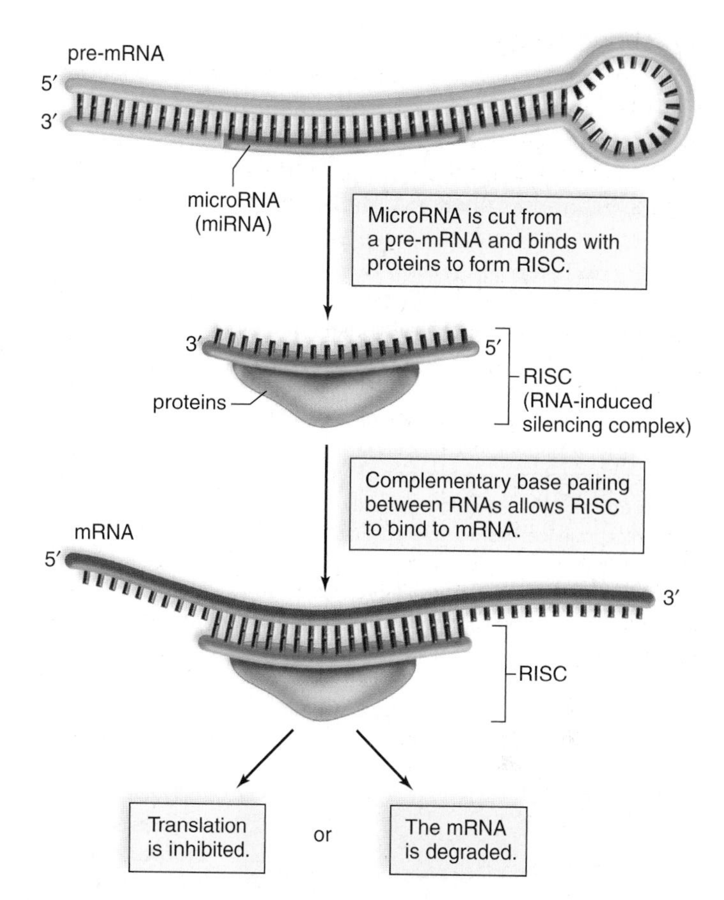


**FIGURE 13.9 Function of microRNAs (miRNAs).**
MicroRNAs are one of several types of RNAs now known to function in the nucleus, nucleolus, or cytoplasm to regulate protein-coding gene expression.

RNAs (snRNAs) affect the splicing of pre-mRNA. They also know that, sometimes, alternative mRNA splicing can result in the inclusion of an intron that results in destruction of the mRNA before it leaves the nucleus.

Further posttranscriptional control of gene expression is achieved by modifying the speed of transport of mRNA from the nucleus into the cytoplasm. Evidence indicates there is a difference in the length of time it takes various mRNA molecules to pass through a nuclear pore, affecting the amount of gene product realized per unit time following transcription.

## Translational Control

**Translational control** begins when the processed mRNA molecule reaches the cytoplasm and before there is a protein product. Translational control involves the activity of mRNA for translation at the ribosome.

Presence or absence of the 5′ cap and the length of the poly-A (adenine nucleotide) tail at the 3′ end of a mature mRNA transcript can determine whether translation takes place and how long the mRNA is active. The long life of mRNAs that code for hemoglobin in mammalian red blood cells is attributed to the persistence of their 5′ end caps and their long 3′ poly-A tails. On the other hand, any influence that affects the length of the poly-A tail or leads to removal of the cap may trigger the destruction of an mRNA.

Scientists studying **microRNAs** (miRNAs) have shown that these mysterious non-protein-coding RNAs can regulate translation by causing the destruction of mRNAs before they can be translated. Cut directly from a pre-mRNA transcript, miRNAs regulate gene activity by interfering with translation of a target mRNA or RNAs (Fig 13.9). Before the miRNA leaves the nucleus, it is enzymatically processed and bound to proteins to form an RNA-induced silencing complex (RISC). An active miRNA complex is complementary to a specific target mRNA, with which it base-pairs to form a double-stranded RNA complex. The end result is inhibition of translation or, in some cases, the destruction of the mRNA itself. Much like a dimmer switch on a light, miRNAs can fine-tune the expression of genes. Scientists have since learned how to use this pathway to study the function of genes by turning them off with artificial miRNAs. Andrew Fire and Craig Mello received the 2006 Nobel Prize in Physiology or Medicine for developing this new technique.

*science focus*

## Alternative mRNA Splicing in Disease

As you have read, the ability to combine the exons and introns of genes into new and novel combinations through alternative mRNA splicing is one of the mechanisms that allows humans to achieve a higher degree of complexity than simpler organisms without a huge increase in the number of genes. In more advanced organisms, the number of alternatively spliced mRNAs increases greatly. Recently, medical science has discovered that when this process goes awry, disease may result.

It has been known for many years that Gorlin syndrome, an autosomal dominant syndrome that includes aggressive skin cancer (Fig. 13A), multiple other tumors, either benign or cancerous, and cysts in various organs. Gorlin syndrome is linked to the tumor suppressor gene called *patched* located on chromosome 9. But a recent finding demonstrated that several new mutations in *patched* were not within the gene's exons, but within its introns! These mutations caused the mRNA to be spliced incorrectly, rendering the protein nonfunctional. Since at least 95% of most human genes consists of introns, there are likely many more such mutations to be discovered in other genetic disorders.

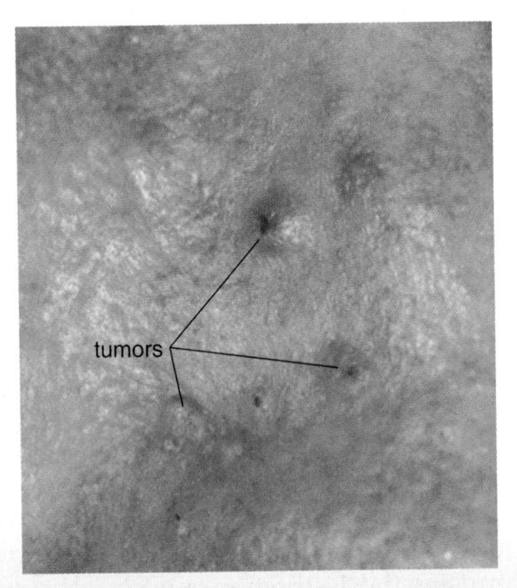


**FIGURE 13A Skin tumors in patient with Gorlin syndrome.**

Defective pre-mRNA splicing is also a major cause of spinal muscular atrophy (SMA), an autosomal recessive disorder that is a common cause of childhood mortality (Fig. 13B). Recent research shows that exon 7 of the *SMN2* gene tends to be left out of the mature mRNA in SMA patients, rendering the protein nonfunctional. The end result is a progressive loss of spinal cord motor neurons, and eventually paralysis and skeletal muscle atrophy. Scientists at Cold Spring Harbor Laboratories turned to antisense oligonucleotide technology, a relatively new technique, in an attempt to reverse the defect. The results were stunning—several of the oligonucleotides tested were able to promote the inclusion of exon 7 in the mature mRNA both in vitro and in cultured cells. These results raise the possibility that targeting aberrant pre-mRNA splicing may ultimately be a viable treatment for many disorders.

Alternative pre-mRNA splicing is also causing scientists to rethink strategies in disease treatment. For example, recent research indicates that the common drug acetaminophen actually targets an alternative version of the COX-1 protein in neurons. This protein variant arises from alternative splicing of the mRNA encoding COX-1 that only occurs in certain neurons. Ultimately, such new findings may allow investigators to design more powerful pain relievers with fewer and less severe side effects.

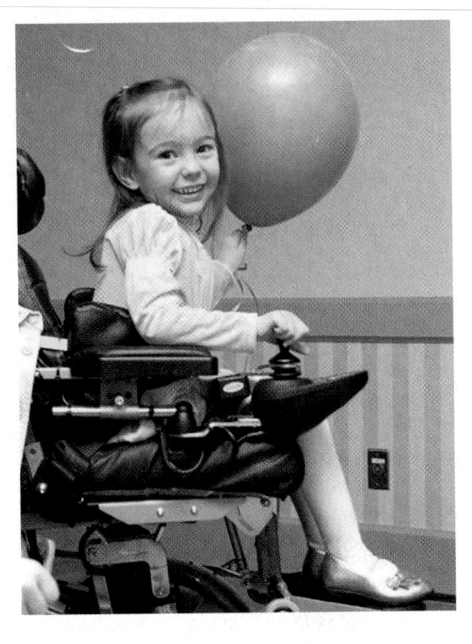

**FIGURE 13B Spinal muscle atrophy.**

Geneticists estimate that 80% or more of human genes undergo alternative mRNA splicing, and the estimate is constantly being revised upward. It is perhaps not surprising that this new frontier in gene regulation is redefining the standard approach to identifying the causes of illness and presenting new targets for the development of therapeutics.

## Posttranslational Control

**Posttranslational control** begins once a protein has been synthesized and has become active. Posttranslational control represents the last chance a cell has for influencing gene expression.

Some proteins are not immediately active after synthesis. For example, at first bovine proinsulin is a single, long polypeptide that folds into a three-dimensional structure. Cleavage results in two smaller chains that are bonded together by disulfide (S—S) bonds. Only then is active insulin present. This ensures that some proteins only become active when it is appropriate for them to do so.

Some proteins are short-lived in cells because they are degraded or destroyed. For example, the cyclin proteins that control the cell cycle are only temporarily present. The cell has giant protein complexes called proteosomes that carry out the task of destroying proteins.

### Check Your Progress 13.2

1. What are the five levels of genetic control in eukaryotes?
2. How does chromatin structure affect gene expression?
3. How does alternative processing of mRNA lead to genetic and phenotypic diversity?

# 13.3 Regulation Through Gene Mutations

A **gene mutation** is a permanent change in the sequence of bases in DNA. The effect of a DNA base sequence change on protein activity can range from no effect to complete inactivity. Germ-line mutations are those that occur in sex cells and can be passed to subsequent generations. Somatic mutations occur in body cells and, therefore, they may affect only a small number of cells in a tissue. Somatic mutations are not passed on to future generations, but they can lead to the development of cancer.

## Causes of Mutations

Some mutations are spontaneous—they happen for no apparent reason—while others are induced by environmental influences. In most cases, **spontaneous mutations** arise as a result of abnormalities in normal biological processes. **Induced mutations** may result from exposure to toxic chemicals or radiation, which induce changes in the base sequence of DNA.

### *Spontaneous Mutations*

Spontaneous mutations can be associated with any number of normal processes. For example, the movement of transposons from one chromosomal location to another can disrupt a gene and lead to an abnormal product. On rare occasions, a base in DNA can undergo a chemical change that leads to a miss pairing during replication. A subsequent base pair change may be carried forth in future generations. Spontaneous mutations due to DNA replication errors, however, are rare. DNA polymerase, the enzyme that carries out replication, proofreads the new strand against the old strand and detects any mismatched nucleotides, and each is usually replaced with a correct nucleotide. In the end, only about one mistake occurs for every 1 billion nucleotide pairs replicated.

### *Induced Mutations*

Induced mutations are caused by **mutagens,** environmental factors that can alter the base composition of DNA. Among the best-known mutagens are radiation and organic chemicals. Many mutagens are also **carcinogens** (cancer-causing).

Chemical mutagens are present in many sources, including some of the food we eat and many industrial chemicals. The mutagenic potency of AF-2, a food additive once widely used in Japan, and of safrole, a flavoring agent once used to flavor root beer, caused them to be banned. Surprisingly, many naturally occurring substances like aflatoxin, produced in moldy grain and peanuts (and present in peanut butter at an average level of 2 parts per billion), and acrylamide, a natural product found in French fries, are also suspected mutagens.

Tobacco smoke contains a number of organic chemicals that are known carcinogens, and it is estimated that one-third of all cancer deaths can be attributed to smoking. Lung cancer is the most frequent lethal cancer in the United States, and smoking is also implicated in the development of cancers of the mouth, larynx, bladder, kidney, and pancreas. The greater the number of cigarettes smoked per day, the earlier the habit starts, and the higher the tar content, the greater the possibility of these cancers. When smoking is combined with drinking alcohol, the risk of these cancers increases even more.

Scientists use the Ames test for mutagenicity to hypothesize that a chemical can be carcinogenic (Fig. 13.10). In the Ames test, a histidine-requiring strain of bacteria is exposed to a chemical. If the chemical is mutagenic, the bacteria can grow without histidine. A large number of chemicals used in agriculture and industry give a positive Ames test. Examples are ethylene dibromide (EDB), which is added to leaded gasoline (to vaporize lead deposits in the engine and send them out the exhaust), and ziram, which is used to prevent fungus disease on crops. Some drugs, such as isoniazed (used to prevent tuberculosis), are mutagenic according to the Ames test.

Aside from chemicals, certain forms of radiation, such as X-rays and gamma rays, are called ionizing radiation

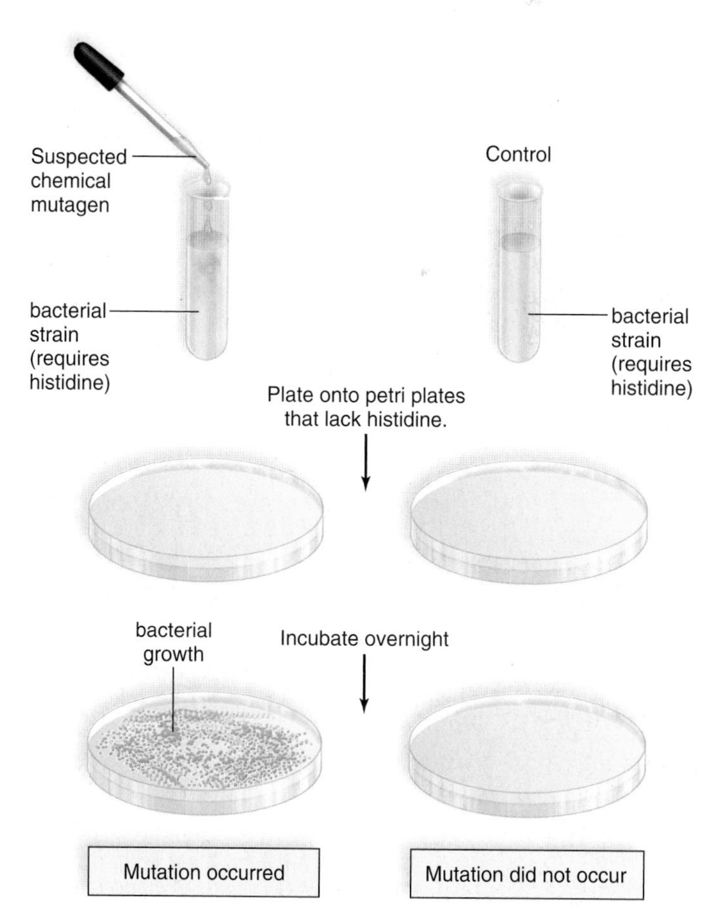


**FIGURE 13.10 The Ames test for mutagenicity.**

A bacterial strain that requires histidine as a nutrient is exposed to a suspected chemical mutagen, but a control is not exposed. The bacteria are plated on a medium that lacks histidine and only the bacteria exposed to the chemical show growth. A mutation allowed the bacteria to grow; therefore, the chemical can be carcinogenic.

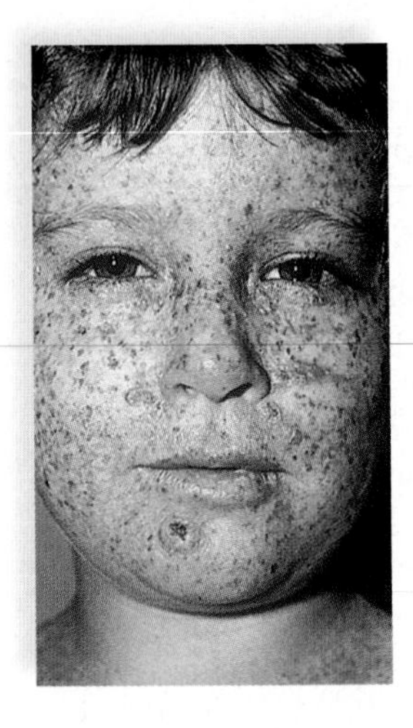

**FIGURE 13.11**
**Xeroderma pigmentosum.**
In xeroderma pigmentosum, deficient DNA repair enzymes leave the skin cells vulnerable to the mutagenic effects of ultraviolet light, allowing many induced mutations to accumulate. Hundreds of skin cancers (small dark spots) appear on the skin exposed to the sun. This individual also has a tumor on the bridge of the nose.

because they create free radicals, ionized atoms with unpaired electrons. Free radicals react with and alter the structure of other molecules, including DNA. Ultraviolet (UV) radiation is easily absorbed by the pyrimidines in DNA. Wherever there are two thymine molecules next to one another, ultraviolet radiation may cause them to bond together, forming *thymine dimers.* A kink results in the DNA. Usually, these dimers are removed by **DNA repair enzymes,** which constantly monitor DNA and fix any irregularities. One enzyme excises a portion of DNA that contains the dimer; another makes a new section by using the other strand as a template; and still another seals the new section in place. The importance of these repair enzymes is exemplified by individuals with the condition known as xeroderma pigmentosum. They lack some of the repair enzymes, and as a consequence, these individuals have a high incidence of skin cancer because of the large number of mutations that accumulate over time (Fig. 13.11). Also, repair enzymes can fail as when skin cancer develops because of excessive sunbathing or prolonged exposure to X-rays.

## Effect of Mutations on Protein Activity

**Point mutations** involve a change in a single DNA nucleotide and, therefore, a possible change in a specific amino acid. The base change in the second row of Figure 13.12*a* has no effect on the resulting amino acid in hemoglobin; the change in the third row, however, codes for the amino acid glutamic acid instead of valine. This base change accounts for the genetic disorder sickle-cell disease because the incorporation of valine, instead of glutamic acid, causes hemoglobin molecules to form semirigid rods, and the red blood cells become sickle shaped. (Compare Figure 13.12*b* to Figure 13.12*c*.) Sickle-shaped cells clog blood vessels and die off more quickly than normal-shaped cells. The base change in the fourth row of Figure 13.12*a* may also have drastic results because the DNA now codes for a stop codon.

**Frameshift mutations** occur most often because one or more nucleotides are either inserted or deleted from DNA. The result of a frameshift mutation can be a completely new sequence of codons and nonfunctional protein. Here is how this occurs: The sequence of codons is read from a specific starting point, as in this sentence, THE CAT ATE THE RAT. If the letter C is deleted from this sentence and the reading frame is shifted, we read THE ATA TET HER AT—something that doesn't make sense.

### *Nonfunctional Proteins*

A single nonfunctioning protein can have a dramatic effect on the phenotype, because enzymes are often a part of metabolic pathways. One particular metabolic pathway in cells is as follows:

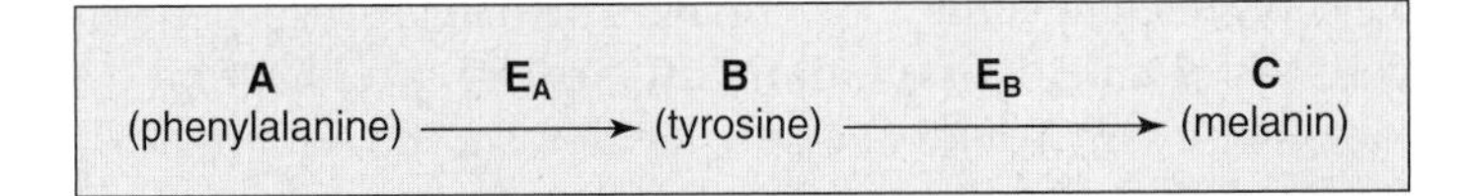


If a faulty code for enzyme $E_A$ is inherited, a person is unable to convert the molecule A to B. Phenylalanine builds

No mutation
3′ 5′
CACGTGGAGTGAGGTCTCCTC
Val His Leu Thr Pro Glu Glu

His → His
(normal protein)
CACGTAGAGTGAGGTCTCCTC
Val His Leu Thr Pro Glu Glu

Glu → Val
(abnormal protein)
CACGTGGAGTGAGGTCACCTC
Val His Leu Thr Pro Val Glu

Glu → Stop
(incomplete protein)
CACGTGGAGTGAGGTATCCTC
Val His Leu Thr Pro Stop

a.

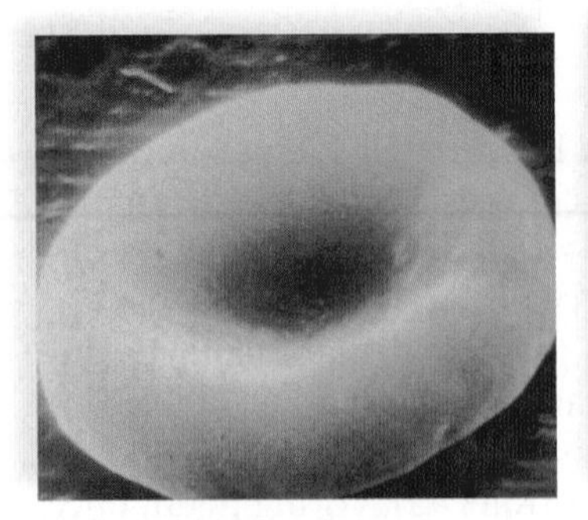

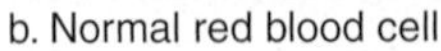

b. Normal red blood cell

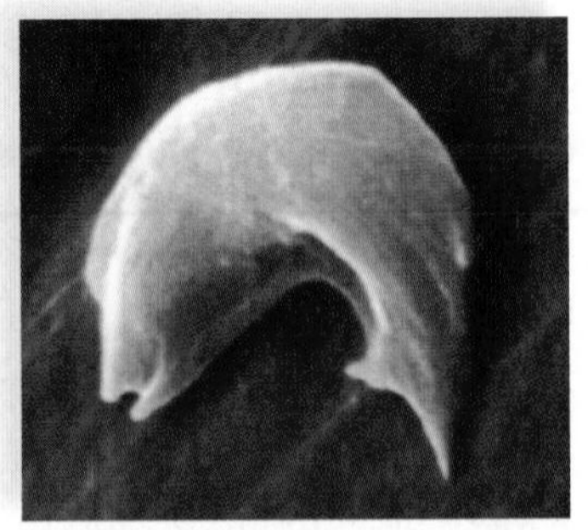

c. Sickled red blood cell

**FIGURE 13.12** **Point mutations in hemoglobin.**
The effect of a point mutation can vary. **a.** Starting at the *top:* Normal sequence of bases in hemoglobin; next, the base change has no effect; next, due to base change, DNA now codes for valine instead of glutamic acid, and the result is that normal red blood cells (**b**) become sickle shaped **c**; next, base change will cause DNA to code for termination and the protein will be incomplete.

up in the system, and the excess causes mental retardation and the other symptoms of the genetic disorder phenylketonuria (PKU). In the same pathway, if a person inherits a faulty code for enzyme $E_B$, then B cannot be converted to C, and the individual is an albino.

A rare condition called androgen insensitivity is due to a faulty receptor for androgens, which are male sex hormones such as testosterone. Although there is plenty of testosterone in the blood, the cells are unable to respond to it. Female instead of male external genitals form, and female instead of male secondary sex characteristics occur. The individual, who appears to be a normal female, may be prompted to seek medical advice when menstruation never occurs. The karyotype is that of a male rather than a female, and the individual does not have the internal sexual organs of a female.

## Mutations Can Cause Cancer

It is estimated that one-third of the children born in 1999 will develop cancer at some time in their lives. Of these affected individuals, one-third of the females and one-fourth of the males will die due to cancer. In the United States, the three deadliest forms of cancer are lung cancer, colon and rectal cancer, and breast cancer.

The development of cancer involves a series of accumulating mutations that can be different for each type of cancer. As discussed in Chapter 9, tumor suppressor genes ordinarily act as brakes on cell division, especially when it begins to occur abnormally. Proto-oncogenes stimulate cell division but are usually turned off in fully differentiated nondividing cells. When proto-oncogenes mutate, they become oncogenes that are active all the time. Carcinogenesis begins with the loss of tumor suppressor gene activity and/or the gain of oncogene activity. When tumor suppressor genes are inactive and oncogenes are active, cell division occurs uncontrollably because a cell signaling pathway that reaches from the plasma membrane to the nucleus no longer functions as it should (Fig. 13.13 and Fig. 13.14).

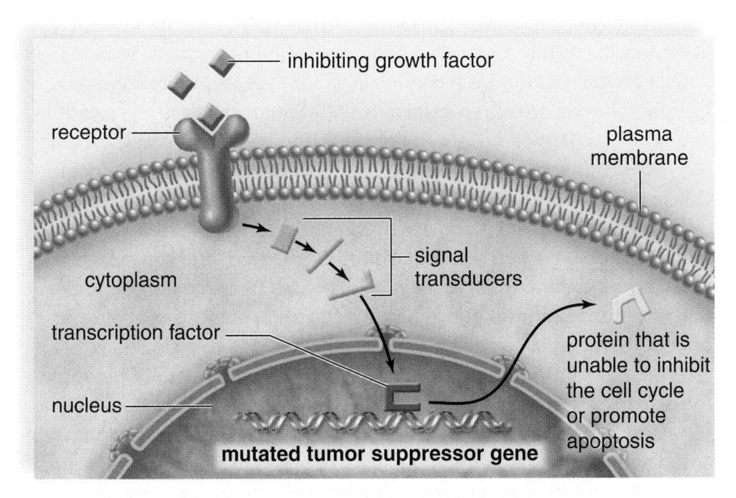


**FIGURE 13.13 Cell signaling pathway that stimulates a mutated tumor suppressor gene.**

A mutated tumor suppressor gene codes for a product that directly or indirectly stimulates the cell cycle.

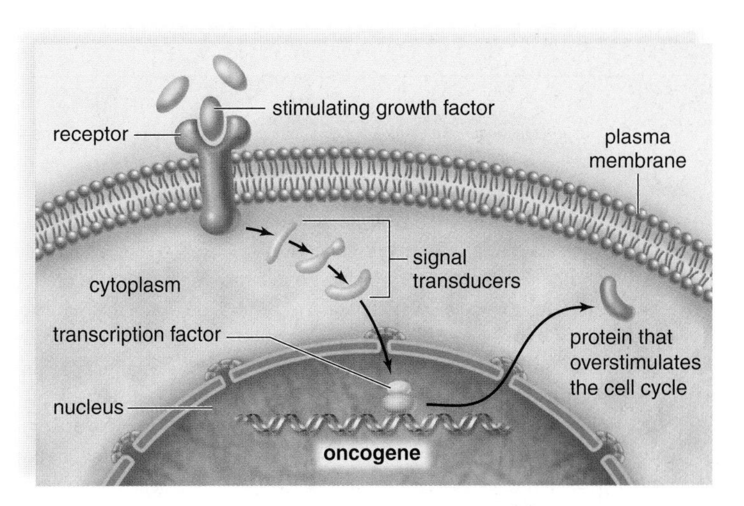


**FIGURE 13.14 Cell signaling pathway that stimulates an oncogene.**

An oncogene codes for a product that either directly or indirectly overstimulates the cell cycle.

It often happens that tumor suppressor genes and proto-oncogenes code for transcription factors or proteins that control transcription factors. As we have seen, transcription factors are a part of the rich and diverse types of mechanisms that control gene expression in cells. They are of fundamental importance to DNA replication and repair, cell growth and division, control of apoptosis, and cellular differentiation. Therefore, it is not surprising that inherited or acquired defects in transcription factor structure and function contribute to the development of cancer.

To take an example, a major tumor suppressor gene called *p53* is more frequently mutated in human cancers than any other known gene. It has been found that the p53 protein acts as a transcription factor, and as such is involved in turning on the expression of genes whose products are cell cycle inhibitors (see page 153). *p53* also promotes apoptosis (programmed cell death) when it is needed. The retinoblastoma protein (RB) controls the activity of a transcription factor for cyclin D and other genes whose products promote entry into the S stage of the cell cycle. When the tumor suppressor gene *p16* mutates, the RB protein is always available, and the result is too much active cyclin D in the cell.

Mutations in many other genes also contribute to the development of cancer. Several proto-oncogenes code for ras proteins, which are needed for cells to grow, to make new DNA, and to not grow out of control. A point mutation is sufficient to turn a normally functioning *ras* proto-oncogene into an oncogene. Abnormal growth results.

### Check Your Progress 13.3

1. What are some common causes of spontaneous and induced mutations?
2. Explain how a frameshift mutation may disrupt a gene's function.

## Connecting the Concepts

The characteristics of specialized cells, such as nerve cells, muscle fibers, and reproductive cells, are maintained by the differences in genes that each cell type expresses. Gene regulation determines whether a gene is expressed and/or the degree to which a gene is expressed. The collective work of many researchers was required to determine that the regulation of genes can occur at most stages during the processes of transcription and translation and that even the chromatin structure of the chromosome containing the gene can influence whether or not a gene is active.

Further control of gene expression, and thus the type of proteins produced by a cell, is achieved through mechanisms such as alternative pre-mRNA splicing, by influencing the rate at which an mRNA leaves the nucleus, and through regulation by miRNAs. Together, these mechanisms alter gene expression to suit an organism's needs.

All living things are subject to genetic mutations or changes in the base sequence of DNA. The effects of mutations vary greatly, but in some cases, mutations may cause genes to be abnormally turned off or on, or expressed in abnormal quantity. Such mutations can seriously affect a developing embryo or lead to development of cancer. Cancers can arise when proto-oncogenes, which are normally not expressed, become oncogenes that are expressed. On the other hand, cancers can also arise when tumor suppressor genes fail to be adequately expressed.

Although mutations in regulatory genes may represent a major evolutionary force, the mechanisms by which gene expression is regulated also make a major contribution. The proteins in chimpanzees and humans are strikingly similar in amino acid sequence. It's possible that the presence or absence of particular proteins in particular cells may cause this, that new and novel combinations of exons of certain genes may create new proteins that differentiate the two species, or that other changes in the DNA, such as changes in the chromatin structure around certain genes, may account for their phenotypic differences. Throughout the past half century, our knowledge of the mechanisms that regulate gene expression and our understanding of how mutations drive the evolutionary process have blossomed. In Chapter 14, you will see how this knowledge is being harnessed to develop cutting-edge technologies that promise to revolutionize agriculture and medicine.

## summary

### 13.1 Prokaryotic Regulation

Prokaryotes often organize genes that are involved in a common process or pathway into operons in which the genes are coordinately regulated. Gene expression in prokaryotes is usually regulated at the level of transcription. The operon model developed by Jacob and Monod says that a regulator gene codes for a repressor, which sometimes binds to the operator. When it does, RNA polymerase is unable to bind to the promoter, and transcription of the structural genes of the operon cannot take place. However, we now know that operons may be regulated by both activators and repressors.

The *trp* operon is an example of a repressible operon because when tryptophan, the corepressor, is present, it binds to the repressor. The repressor is then able to bind to the operator, and transcription of structural genes does not take place.

The *lac* operon is an example of an inducible operon because when lactose, the inducer, is present, it binds to the repressor. The repressor is unable to bind to the operator, and transcription of structural genes takes place if glucose is absent.

Both the *lac* and *trp* operons exhibit negative control, because a repressor is involved. However, positive control also occurs. There are also examples of positive control. The structural genes in the *lac* operon are not maximally expressed unless glucose is absent and lactose is present. At that time, cAMP attaches to a molecule called CAP, and then CAP binds to a site next to the promoter. Now RNA polymerase is better able to bind to the promoter, and transcription occurs.

### 13.2 Eukaryotic Regulation

The following levels of control of gene expression are possible in eukaryotes: chromatin structure, transcriptional control, posttranscriptional control, translational control, and posttranslational control.

Chromatin structure helps regulate transcription. Highly condensed heterochromatin is genetically inactive, as exemplified by Barr bodies. Less-condensed euchromatin is genetically active, as exemplified by lampbrush chromosomes in vertebrates.

Regulatory proteins called transcription factors, as well as DNA sequences called enhancers and silencers, play a role in controlling transcription in eukaryotes. Transcription factors bind to the promoter, and transcription activators bind to an enhancer.

Posttranscriptional control is achieved by creating variations in messenger RNA (mRNA) splicing, which may yield multiple mRNA messages from the same gene, and by altering the speed with which a particular mRNA molecule leaves the nucleus.

Translational control affects mRNA translation and the length of time it is translated, primarily by altering the stability of an mRNA. MicroRNAs are a unique example of translational control. Posttranslational control affects whether or not an enzyme is active and how long it is active.

### 13.3 Regulation Through Gene Mutations

In molecular terms, a gene is a sequence of DNA nucleotide bases, and a genetic mutation is a change in this sequence. Mutations can be spontaneous or due to environmental mutagens such as radiation and organic chemicals. Carcinogens are mutagens that cause cancer.

Point mutations can range in effect, depending on the particular codon change. Sickle cell disease is an example of a point mutation that greatly changes the activity of the affected gene. Frameshift mutations result when a base is added or deleted, and the result is usually a nonfunctional protein. Most cases of cystic fibrosis are due to a frameshift mutation. Nonfunctional proteins can affect the phenotype drastically, as in albinism, which is due to a single faulty enzyme, and androgen insensitivity, which is due to a faulty receptor for testosterone.

Cancer is often due to an accumulation of genetic mutations among genes that code for regulatory proteins. The cell cycle occurs inappropriately when proto-oncogenes become oncogenes and tumor suppressor genes are no longer effective. Mutations that affect transcription factors and other regulators of gene expression are frequent causes of cancer.

## understanding the terms

Barr body 238
carcinogen 243
corepressor 235
DNA repair enzyme 244
enhancer 240
epigenetic inheritance 237
euchromatin 239
frameshift mutation 244
gene mutation 243
heterochromatin 238
induced mutation 243
inducer 236
inducible operon 236
microRNA 241
mutagen 243
operator 234
operon 234
point mutation 244
posttranscriptional control 240
posttranslational control 242
promoter 234
regulator gene 235
repressible operon 235
repressor 234
spontaneous mutation 243
structural gene 234
transcription activator 240
transcriptional control 240
transcription factor 240
translational control 241

Match the terms to these definitions:

a. ______________ A regulation of gene expression that begins once there is an mRNA transcript.
b. ______________ Genes involved in a common function that are located and regulated together.
c. ______________ Dark-staining body in the nuclei of female mammals that contains a condensed, inactive X chromosome.
d. ______________ Changes in the base sequence of DNA that arise as a result of environmental influences.
e. ______________ Environmental agent that causes mutations leading to the development of cancer.

## reviewing this chapter

1. Name and state the function of the three components of operons. 234
2. Explain the operation of the *trp* operon, and note why it is considered a repressible operon. 235
3. Explain the operation of the *lac* operon, and note why it is considered an inducible operon. 235–36
4. What are the five levels of genetic regulatory control in eukaryotes? 237
5. Relate heterochromatin and euchromatin to levels of chromatin organization. 238–39
6. With regard to transcriptional control in eukaryotes, explain how Barr bodies show that heterochromatin is genetically inactive. 238–39
7. Explain how lampbrush chromosomes in vertebrates show that euchromatin is genetically active. 239
8. What do transcription factors do in eukaryotic cells? What are enhancers? 240
9. Explain how alternative mRNA processing may create multiple mRNAs from a single gene. 240–41
10. Give examples of translational and posttranslational control in eukaryotes. 241–42
11. Name some causes of mutations. 243
12. What are two major types of mutations, and what effect can they have on protein activity? 243–44
13. Mutations in what types of genes, in particular, can cause cancer? 245

## testing yourself

Choose the best answer for each question.

1. Which of the following illustrates negative control?
   a. A repressor that becomes active when bound to a corepressor and inhibits transcription.
   b. A gene that binds a repressor and becomes active.
   c. An activator that becomes active when bound to a coactivator and activates transcription.
   d. A repressor that binds a gene and becomes inactive.
2. In regulation of the *lac* operon, when lactose is present and glucose is absent,
   a. the repressor is able to bind to the operator.
   b. the repressor is unable to bind to the operator.
   c. transcription of structural genes occurs.
   d. transcription of lactose occurs.
   e. Both b and c are correct.
3. In regulation of the *trp* operon, when tryptophan is present,
   a. the repressor is able to bind to the operator.
   b. the repressor is unable to bind to the operator.
   c. transcription of the repressor in inhibited.
   d. transcription of the structural genes, operator, and promoter occurs.
4. In operon models, the function of the promoter is to
   a. code for the repressor protein.
   b. bind with RNA polymerase.
   c. bind to the repressor.
   d. code for the regulator gene.
5. Which of the following statements is/are true regarding operons?
   a. The regulator gene is transcribed with the structural genes.
   b. The structural genes are always transcribed.
   c. All genes are always transcribed.
   d. The regulator gene has its own promoter.
6. Which of the following regulate gene expression in the eukaryotic nucleus?
   a. posttranslational control
   b. transcriptional control
   c. translational control
   d. posttranscriptional control
   e. Both b and d are correct.
7. Which of the following mechanisms may create multiple mRNAs from the same gene?
   a. posttranslational control
   b. alternative mRNA splicing
   c. binding of a transcription factor
   d. chromatin remodeling
   e. miRNAs
8. Translational control of gene expression occurs within the
   a. nucleus.
   b. cytoplasm.
   c. nucleolus.
   d. mitochondria.
9. Alternative mRNA splicing is an example of which type of regulation of gene expression?
   a. transcriptional
   b. posttranscriptional
   c. translational
   d. posttranslational
10. A scientist adds radioactive uridine (label for RNA) to a culture of cells and examines an autoradiograph. Which type of chromatin is apt to show the label?
    a. heterochromatin
    b. euchromatin
    c. the histones, not the DNA
    d. the DNA, not the histones
    e. Both a and d are correct.

11. Barr bodies are
    a. genetically active X chromosomes in males.
    b. genetically inactive X chromosomes in females.
    c. genetically active Y chromosomes in males.
    d. genetically inactive Y chromosomes in females.
12. Which of these might cause a proto-oncogene to become an oncogene?
    a. exposure of the cell to radiation
    b. exposure of the cell to certain chemicals
    c. viral infection of the cell
    d. exposure of the cell to pollutants
    e. All of these are correct.
13. A cell is cancerous. You might find an abnormality in
    a. a proto-oncogene.
    b. a tumor suppressor gene.
    c. regulation of the cell cycle.
    d. tumor cells.
    e. All of these are correct.
14. A tumor suppressor gene
    a. inhibits cell division.
    b. opposes oncogenes.
    c. prevents cancer.
    d. is subject to mutations.
    e. All of these are correct.
15. Label this diagram of an operon.

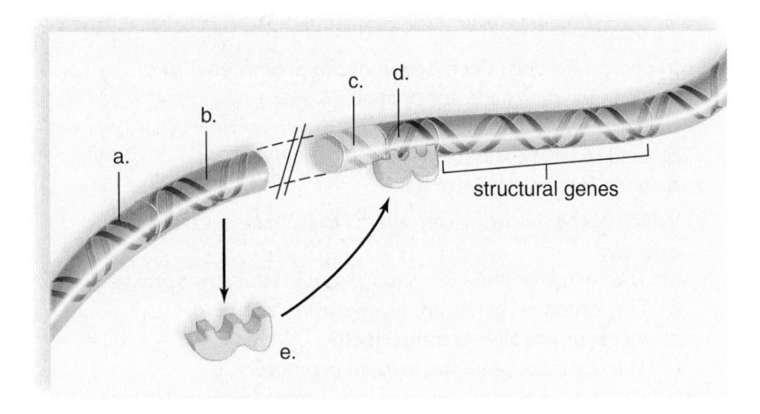


16. If the DNA codons are CAT CAT CAT, and a guanine base is added at the beginning, which would result?
    a. CAT CAT CAT G
    b. G CAT CAT CAT
    c. GCA TCA TCA T
    d. GC ATC ATC AT
17. A mutation in a DNA molecule involving the replacement of one nucleotide base pair with another is called a(n)
    a. frameshift mutation.
    b. transposon.
    c. deletion mutation.
    d. point mutation.
    e. insertion mutation.
18. Which of these is characteristic of cancer?
    a. It may involve a lack of mutations over a length of time.
    b. It cannot be tied to particular environmental factors.
    c. Apoptosis is one of the first developmental effects.
    d. Mutations in certain types of genes.
    e. It typically develops within a short period of time.
19. Which is not evidence that eukaryotes control transcription?
    a. euchromatin/heterochromatin
    b. existence of transcription factors
    c. lampbrush chromosomes
    d. occurrence of mutations
    e. All of these are correct.

## thinking scientifically

1. In patients with chronic myelogenous leukemia, an odd chromosome is seen in all the cancerous cells. A small piece of chromosome 9 is connected to chromosome 22. This 9:22 translocation has been termed the Philadelphia chromosome. How could a translocation cause genetic changes that result in cancer?
2. New findings indicate that mutations outside of genes may cause disease, such as in some cases of Hirschsprung disease and multiple endocrine neoplasia. Explain how such a mutation might alter the expression of a gene.

## bioethical issue

### Environmental Estrogens and Mutation

You have learned from this chapter that many types of carcinogens, such as those found in cigarette smoke, may alter the base sequence of DNA. However, environmental estrogens are a recently identified type of carcinogen that is generating much attention and concern in recent years. Environmental estrogens are estrogen-like compounds that can disrupt normal endocrine system function in animals by competing with normal sex hormones for receptors, inadvertently activating and inactivating transcription factors and greatly affecting gene expression. They have been linked to increased mutation rates, to deformed genitals in alligators and fish, to promotion of cell division in cultured breast cancer cells, and to inhibition of sperm development in humans.

Environmental estrogens are sometimes found naturally at low concentrations in foods such as soybeans and flax seeds. However, many of these compounds are artificial, originating from chemicals such as polychlorinated biphenyls (PCBs), phthalates (found extensively in many plastics), and atrazine, a compound found in many commercial weed killers. Many people, including scientists at the EPA, contend that these artificial compounds, even at very low doses, are a major threat to the environment, to many animal species, and to human health.

However, some critics contend that the concentrations of these compounds in the soil, air, and water are far below concentrations necessary to cause problems in most animal species, including humans. They also tout studies showing high concentrations of environmental estrogens in many grains, fruits, and vegetables, and that many of these compounds are rendered harmless by the body before they have a chance to cause mutations.

Should known environmental estrogens, such as those found in plastics, herbicides, and insecticides, be closely monitored by the government, and maximal permissible levels set for their emission into the environment? And where should money to fund these regulations be derived? Or, as some critics insist, are we worried about a problem that simply does not exist?

## *Biology* website

The companion website for *Biology* provides a wealth of information organized and integrated by chapter. You will find practice tests, animations, videos, and much more that will complement your learning and understanding of general biology.

**http://www.mhhe.com/maderbiology10**

# 14

# Biotechnology and Genomics

*biotechnology is used in a myriad of ways today, even to make ice cream more smooth and creamy. An eel-like fish, the ocean pout, produces a natural antifreeze protein that is now made by genetically modified bacteria. The product is readily available to all, even ice-cream manufacturers who want their product to be free of ice crystals. Modern biotechnology has also made it possible for farmers, bioengineers, and medical scientists to alter the genotype and subsequently the phenotype of plants, animals, and humans. Genetically modified crops are resistant to disease and able to grow under stressful conditions. Farm animals grow larger than usual, and humans are supplied with normal genes to make up for ones that do not function as they should.*

*But others worry that genetically-modified bacteria and plants might harm the environment and that the products produced by altered organisms might not be healthy for humans. Other ethical concerns abound. Is it ethical to give a cat a gene that makes it glow? To what extent would it be proper to improve the human genome? Everyone should be knowledgeable about modern genetics so they can participate in deciding these issues.*

A biotechnology product derived from the ocean pout (*Zoarces americanus*) improves the texture of ice cream.

## *concepts*

## 14.1 DNA Cloning

In biology, **cloning** is the production of genetically identical copies of DNA, cells, or organisms through some asexual means. When an underground stem or root sends up new shoots, the resulting plants are clones of one another. The members of a bacterial colony on a petri dish are clones because they all came from the division of the same original cell. Human identical twins are also considered clones. The first two cells of the embryo separate, and each becomes a complete individual.

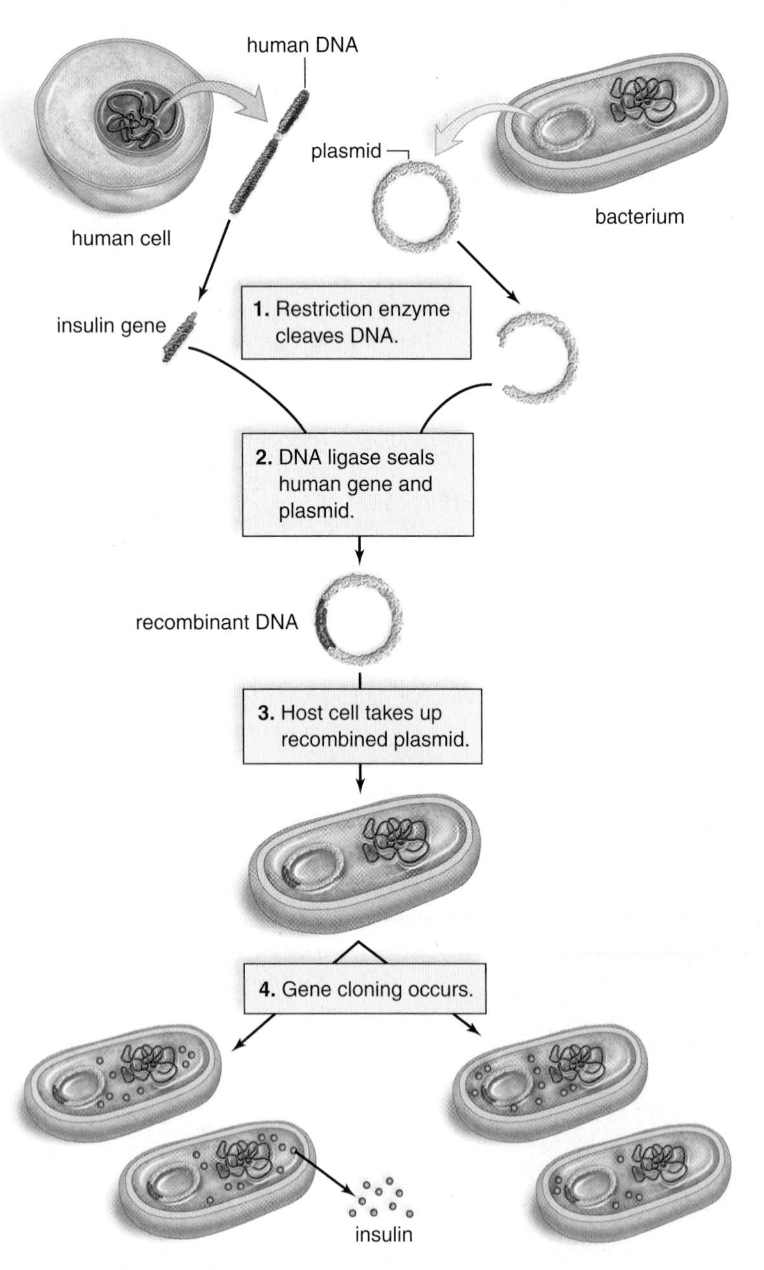


**FIGURE 14.1 Cloning a human gene.**

Human DNA and plasmid DNA are cleaved by a specific type of restriction enzyme. Then the human DNA, perhaps containing the insulin gene, is spliced into a plasmid by the enzyme DNA ligase. Gene cloning is achieved after a bacterium takes up the plasmid. If the gene functions normally as expected, the product (e.g., insulin) may also be retrieved.

DNA cloning can be done to produce many identical copies of the same gene; that is, for the purpose of **gene cloning.** Scientists clone genes for a number of reasons. They might want to determine the difference in base sequence between a normal gene and a mutated gene. Or they might use the genes to genetically modify organisms in a beneficial way. When cloned genes are used to modify a human, the process is called **gene therapy.** Otherwise, the organisms are called **transgenic organisms** [L. *trans,* across, through; Gk. *genic,* producing]. Transgenic organisms are frequently used today to produce a product desired by humans.

**Recombinant DNA (rDNA)** technology and the **polymerase chain reaction (PCR)** are two procedures that scientists can use to clone DNA.

### Recombinant DNA Technology

Recombinant DNA (rDNA) contains DNA from two or more different sources, such as a human cell and a bacterial cell, as shown in Figure 14.1. To make rDNA, a technician needs a **vector** [L. *vehere,* to carry], by which rDNA will be introduced into a host cell. One common vector is a plasmid. **Plasmids** are small accessory rings of DNA found in bacteria. The ring is not part of the bacterial chromosome and replicates on its own. Plasmids were discovered by investigators studying the bacterium *Escherichia coli (E. coli).*

Two enzymes are needed to introduce foreign DNA into vector DNA: (1) a **restriction enzyme,** which cleaves DNA, and (2) an enzyme called **DNA ligase** [L. *ligo,* bind], which seals DNA into an opening created by the restriction enzyme. Hundreds of restriction enzymes occur naturally in bacteria, where they cut up any viral DNA that enters the cell. They are called restriction enzymes because they *restrict* the growth of viruses, but they also act as molecular scissors to cleave any piece of DNA at a specific site. For example, the restriction enzyme called *Eco*RI always cuts double-stranded DNA at this sequence of bases and in this manner:

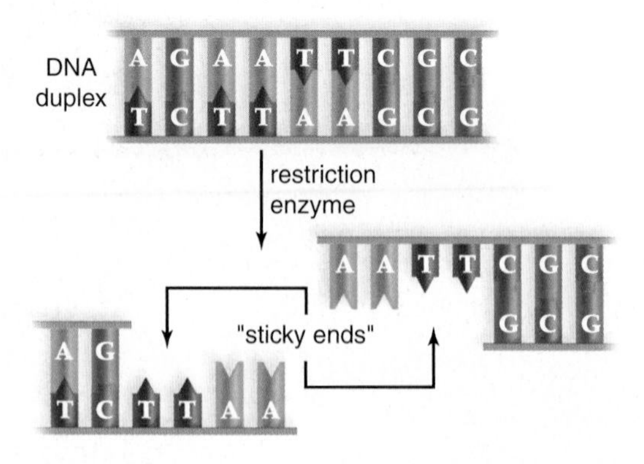


Notice that there is now a gap into which a piece of foreign DNA can be placed if it ends in bases complementary to those exposed by the restriction enzyme. The single-stranded, but complementary, ends of the two DNA molecules are called "sticky ends" because they can bind a piece of foreign DNA by complementary base-pairing. Sticky ends facilitate the insertion of foreign DNA into vector DNA as long as both are cleansed by the same restriction enzyme.

Next, genetic engineers use the enzyme DNA ligase to seal the foreign piece of DNA into the vector. DNA splicing is now complete; an rDNA molecule has been prepared (Fig. 14.1). Bacterial cells take up recombinant plasmids, especially if they are treated to make their cell membranes more permeable. Thereafter, as the plasmid replicates, DNA is cloned.

In order for bacteria to express a human gene, the cloned gene has to be accompanied by regulatory regions unique to bacteria. Also, the gene should not contain introns because bacteria don't have introns. However, it is possible to make a human gene that lacks introns. The enzyme called reverse transcriptase can be used to make a DNA copy of human mRNA. The DNA molecule, called **complementary DNA (cDNA),** does not contain introns. Bacteria may then transcribe and translate the cloned cDNA to produce a human protein because the genetic code is the same in humans and bacteria.

## The Polymerase Chain Reaction

The polymerase chain reaction (PCR), developed by Kary Mullis in 1985, can create copies of a segment of DNA quickly in a test tube. PCR is very specific—it amplifies (makes copies of) a targeted DNA sequence. The targeted sequence can be less than one part in a million of the total DNA sample!

PCR requires the use of DNA polymerase, the enzyme that carries out DNA replication, and a supply of nucleotides for the new DNA strands. PCR is a chain reaction because the targeted DNA is repeatedly replicated as long as the process continues. The colors in Figure 14.2 distinguish the old strand from the new DNA strand. Notice that the amount of DNA doubles with each replication cycle.

PCR has been in use for years, and now almost every laboratory has automated PCR machines to carry out the procedure. Automation became possible after a temperature-insensitive (thermostable) DNA polymerase was extracted from the bacterium *Thermus aquaticus,* which lives in hot springs. The enzyme can withstand the high temperature used to separate double-stranded DNA; therefore, replication does not have to be interrupted by the need to add more enzyme.

### *Analyzing DNA*

DNA amplified by PCR can be analyzed for various purposes. For example, mitochondrial DNA taken from modern living populations was used to decipher the evolutionary history of human populations. For identification purposes, DNA taken from a corpse burned beyond recognition can be matched to that on the bristles of their toothbrush!

Analysis of DNA following PCR has undergone improvements over the years. At first, the entire genome was treated with restriction enzymes, resulting in a unique collection of different-sized fragments because each person has their own restriction enzyme sites. During a process called gel electrophoresis, the fragments were separated according to their size, and the result was a pattern of distinctive bands that identified the person. Now, **short tandem repeat (STR) profiling** is the method of choice. STRs are the same short sequence of DNA bases that recur several times, as in TCGTCGTCG. STR profiling is advantageous because it doesn't require the use of restriction enzymes. The chromosomal locations for STRs are known and, therefore, it is possible to subject only these locations to PCR and use gel electrophoresis to arrive at a band pattern that is different for each person. The band patterns are different because each person has their own number of repeats at the different locations. The greater the number of STRs at a location, the longer the DNA fragment amplified by PCR. The more STR

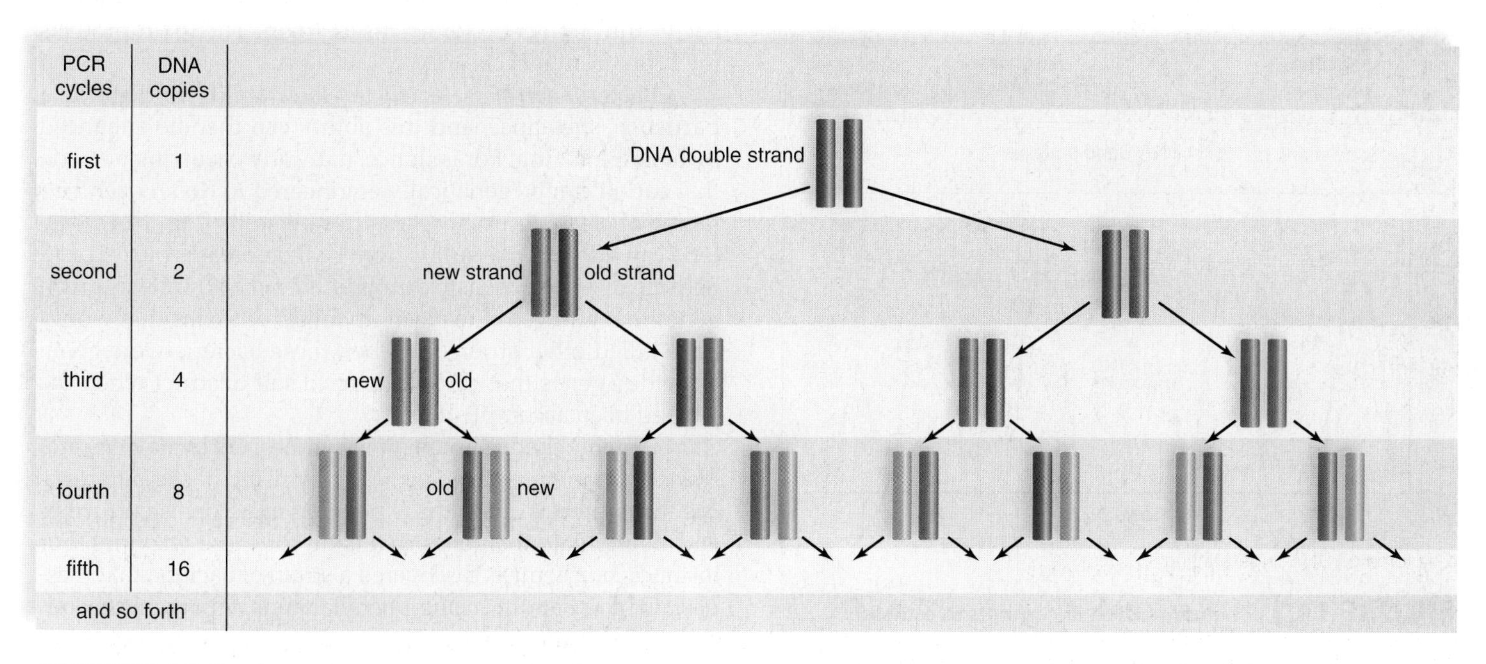


**FIGURE 14.2 Polymerase chain reaction (PCR).**

PCR allows the production of many identical copies of DNA in a laboratory setting.

locations employed, the more confident scientists can be of distinctive results for each person (Fig. 14.3*a*). The newest method of doing fingerprints today does away with the need to use gel electrophoresis: The DNA fragments are fluorescently labeled. Using a particular laboratory instrument, a laser excites the fluorescent STRs, and a detector records the amount of emission for each DNA fragment in terms of peaks and valleys. Therefore, the greater the fluorescence, the greater the number of repeats at a location. The printout, such as the one shown in Figure 14.3*b*, is the DNA fingerprint, and each person has their own particular printout.

Applications of PCR are limited only by our imaginations. When the DNA matches that of a virus or mutated gene, it is known that a viral infection, genetic disorder, or cancer is present. DNA fingerprinted from blood or tissues at a crime scene has been successfully used in convicting criminals. DNA fingerprinting through STR profiling was extensively used to identify the victims of the September 11, 2001 terrorist attacks in the United States. Relatives can be found, paternity suits can be settled (Fig. 14.3*a*), genetic disorders can be detected, and illegally poached ivory and whale meat can be recognized using this technology. PCR has also shed new light on evolutionary studies by comparing extracted DNA from certain fossils with that of living organisms.

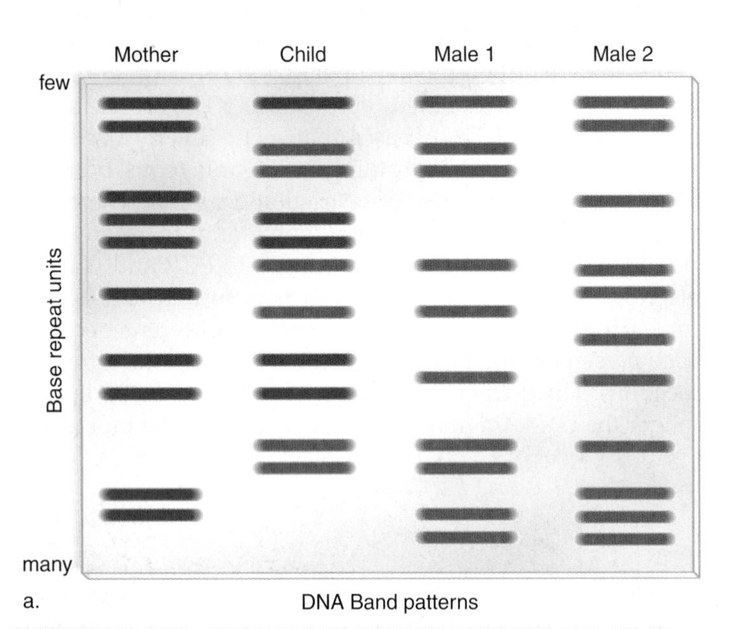


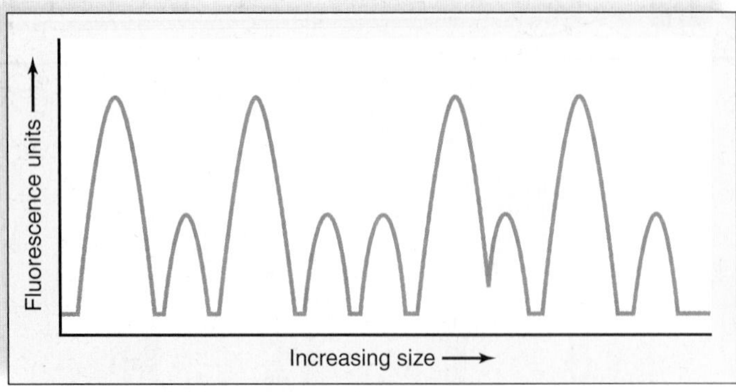


b. Automated DNA fingerprinting

**FIGURE 14.3 The use of DNA fingerprints to establish paternity.**

**a.** In this method, DNA fragments containing STRs are separated by gel electrophoresis. Male 1 is the father. **b.** Each person's fingerprint pattern (only one is shown) can also be printed out by a machine that detects fluorescence.

**Check Your Progress 14.1**

1. Describe the process of creating an rDNA molecule.
2. How can DNA fingerprinting be used to analyze DNA molecules?

# 14.2 Biotechnology Products

Today, transgenic bacteria, plants, and animals are often called **genetically modified organisms (GMOs),** and the products they produce are called **biotechnology products.**

## Genetically Modified Bacteria

Recombinant DNA technology is used to produce transgenic bacteria, which are grown in huge vats called bioreactors. The gene product is usually collected from the medium. Biotechnology products produced by bacteria include insulin, clotting factor VIII, human growth hormone, t-PA (tissue plasminogen activator), and hepatitis B vaccine. Transgenic bacteria have many other uses as well. Some have been produced to promote the health of plants. For example, bacteria that normally live on plants and encourage the formation of ice crystals have been changed from frost-plus to frost-minus bacteria. As a result, new crops such as frost-resistant strawberries are being developed. Also, a bacterium that normally colonizes the roots of corn plants has now been endowed with genes (from another bacterium) that code for an insect toxin. The toxin protects the roots from insects.

Bacteria can be selected for their ability to degrade a particular substance, and this ability can then be enhanced by bioengineering. For instance, naturally occurring bacteria that eat oil can be genetically engineered to do an even better job of cleaning up beaches after oil spills (Fig. 14.4). Bacteria can also remove sulfur from coal before it is burned and help clean up toxic waste dumps. One such strain was given genes that allowed it to clean up levels of toxins that would have killed other strains. Further, these bacteria were given "suicide" genes that caused them to self-destruct when the job had been accomplished.

Organic chemicals are often synthesized by having catalysts act on precursor molecules or by using bacteria to carry out the synthesis. Today, it is possible to go one step further and manipulate the genes that code for these enzymes. For instance, biochemists discovered a strain of bacteria that is especially good at producing phenylalanine, an organic chemical needed to make aspartame, the dipeptide sweetener better known as NutraSweet. They isolated, altered, and formed a vector for the appropriate genes so that various bacteria could be genetically engineered to produce phenylalanine.

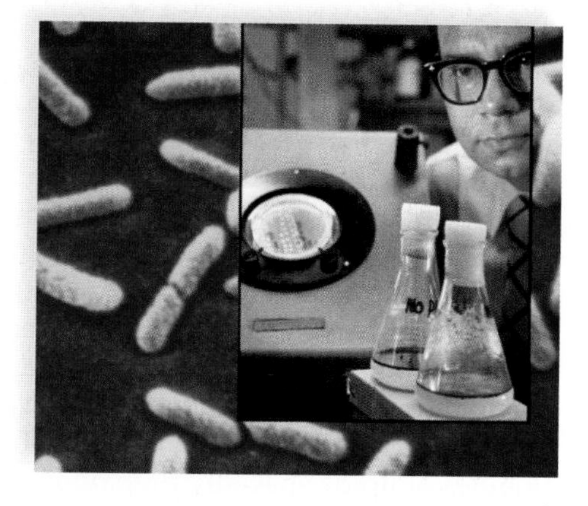

**FIGURE 14.4 Genetically modified bacteria.**

Bacteria capable of decomposing oil have been engineered and patented. In the inset, the flask toward the rear contains oil and no bacteria; the flask toward the front contains the bacteria and is almost clear of oil.

## Genetically Modified Plants

Techniques have been developed to introduce foreign genes into immature plant embryos or into plant cells called *protoplasts* that have had the cell wall removed. It is possible to treat protoplasts with an electric current while they are suspended in a liquid containing foreign DNA. The electric current makes tiny, self-sealing holes in the plasma membrane through which genetic material can enter. Protoplasts go on to develop into mature plants. One altered plant known as the pomato is the result of these technologies. This plant produces potatoes belowground and tomatoes aboveground.

Foreign genes transferred to cotton, corn, and potato strains have made these plants resistant to pests because their cells now produce an insect toxin. Similarly, soybeans have been made resistant to a common herbicide. Some corn and cotton plants are both pest- and herbicide-resistant. These and other genetically modified crops that are expected to have increased yield are now sold commercially.

Like bacteria, plants are also being engineered to produce human proteins, such as hormones, clotting factors, and antibodies, in their seeds. One type of antibody made by corn can deliver radioisotopes to tumor cells, and another made by soybeans can be used to treat genital herpes.

## Genetically Modified Animals

Techniques have been developed to insert genes into the eggs of animals. It is possible to microinject foreign genes into eggs by hand, but another method uses vortex mixing. The eggs are placed in an agitator with DNA and silicon-carbide needles, and the needles make tiny holes through which the DNA can enter. When these eggs are fertilized, the resulting offspring are transgenic animals. Using this technique, many types of animal eggs have taken up the gene for bovine growth hormone (bGH). The procedure has been used to produce larger fishes, cows, pigs, rabbits, and sheep.

**Gene pharming,** the use of transgenic farm animals to produce pharmaceuticals, is being pursued by a number of firms. Genes that code for therapeutic and diagnostic proteins are incorporated into an animal's DNA, and the proteins appear in the animal's milk. Plans are under way to produce drugs for the treatment of cystic fibrosis, cancer, blood diseases, and other disorders by this method. Figure 14.5*a* outlines the procedure for producing transgenic mammals: DNA containing the gene of interest is injected into donor eggs. Following in vitro fertilization, the zygotes are placed in host females, where they develop. After female offspring mature, the product is secreted in their milk.

### *Cloning Transgenic Animals*

For many years, it was believed that adult vertebrate animals could not be cloned because cloning requires that all the genes of an adult cell be turned on if development is to proceed normally. This had long been thought impossible.

In 1997, however, Scottish scientists announced that they had produced a cloned sheep called Dolly. Since then, calves and goats have also been cloned, as described in Figure 14.5*b*. After enucleated eggs from a donor are microinjected with 2n

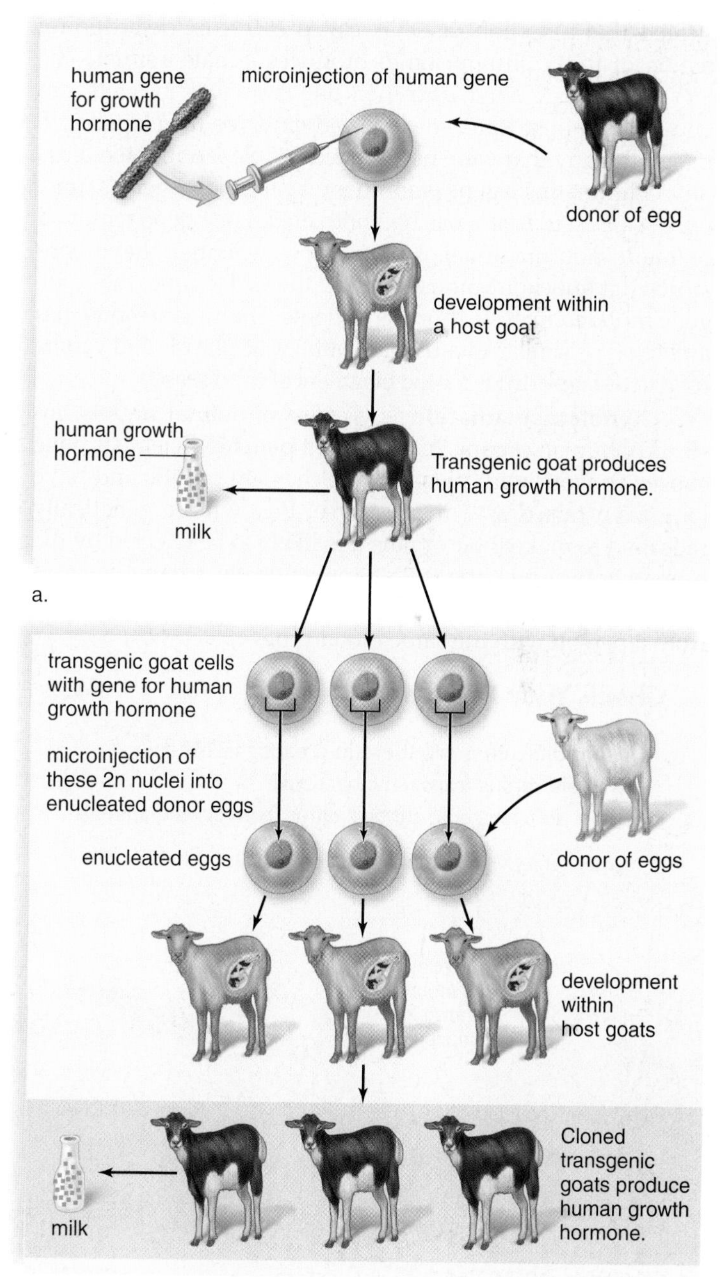


**FIGURE 14.5 Transgenic mammals produce a product.**

**a.** A bioengineered egg develops in a host to create a transgenic goat that produces a biotechnology product in its milk. **b.** Nuclei from the transgenic goat are transferred into donor eggs, which develop into cloned transgenic goats.

nuclei from the same transgenic animal, they are coaxed to begin development in vitro. Development continues in host females until the clones are born. The female clones have the same product in their milk as the donor of the eggs. Now that scientists have a way to clone animals, this procedure will undoubtedly be used routinely to procure biotechnology products. However, animal cloning is a difficult process with a low success rate. The vast majority of cloning attempts are unsuccessful, resulting in the early death of the clone.

### *Applications of Transgenic Animals*

Researchers are using transgenic mice for various research projects. Figure 14.6 shows how this technology has demonstrated that a section of DNA called *SRY* (sex determining region of the Y chromosome) produces a male animal. The *SRY* gene was cloned, and then one copy was injected into one-celled mouse embryos. Injected embryos developed into males, but any that were not injected developed into females.

Eliminating a gene is another way to study a gene's function. A *knockout mouse* has had both alleles of a gene removed or made nonfunctional. For example, scientists have constructed a knockout mouse lacking the *CFTR* gene, the same gene mutated in cystic fibrosis patients. The mutant mouse has a phenotype similar to a human with cystic fibrosis and can be used to test new drugs for the treatment of the disease.

**Xenotransplantation** is the use of animal organs, instead of human organs, in transplant patients. Scientists have chosen to work with pigs because they are prolific and have long been raised as a meat source. Pigs will be genetically modified to make their organs less likely to be rejected by the human body. The hope is that one day a pig organ will be as easily accepted by the human body as a blood transfusion from a person with the same blood type.

**Check Your Progress 14.2**

1. What difficulties are there in creating transgenic animals versus transgenic bacteria?
2. How do transgenic animals differ from cloned animals?

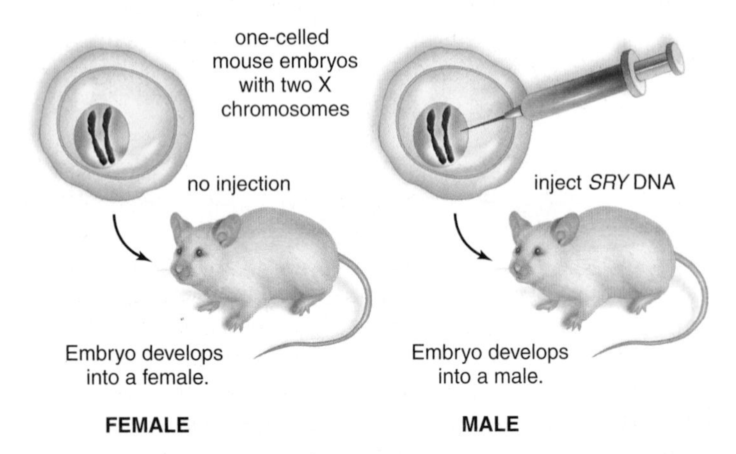


**FIGURE 14.6 Experimental use of mice.**
Bioengineered mice showed that maleness is due to *SRY* DNA.

## 14.3 Gene Therapy

The manipulation of an organism's genes can be extended to humans in a process called gene therapy. Gene therapy is an accepted therapy for the treatment of a disorder. Gene therapy has been used to cure inborn errors of metabolism, as well as to treat more generalized disorders such as cardiovascular disease and cancer. Figure 14.7 shows regions of the body that have received copies of normal genes by various methods of gene transfer. Viruses genetically modified to be safe can be used to ferry a normal gene into the body, and so can liposomes, which are microscopic globules of lipids specially prepared to enclose the normal gene. On the other hand, sometimes the gene is injected directly into a particular region of the body. Below we give examples of **ex vivo gene therapy** (the gene is inserted into cells that have been removed and then returned to the body) and **in vivo gene therapy** (the gene is delivered directly into the body).

### Ex Vivo Gene Therapy

Children who have SCID (severe combined immunodeficiency) lack the enzyme ADA (adenosine deaminase), which is involved in the maturation of immune cells. Therefore, these children are prone to constant infections and may die without treatment. To carry out gene therapy, bone marrow stem cells are removed from the bone marrow of the patient and infected with a virus that carries a normal gene for the enzyme into their DNA. Then the cells are returned to the patient, where it is hoped they will divide to produce more blood cells with the same genes.

Familial hypercholesterolemia is a condition that develops when liver cells lack a receptor protein for removing cholesterol from the blood. The high levels of blood cholesterol make the patient subject to fatal heart attacks at a young age. A small portion of the liver is surgically excised and then infected with a virus containing a normal gene for the receptor before being returned to the patient. Patients are expected to experience lowered serum cholesterol levels following this procedure.

### In Vivo Gene Therapy

Cystic fibrosis patients lack a gene that codes for a transmembrane carrier of the chloride ion. They often suffer from numerous and potentially deadly infections of the respiratory tract. In gene therapy trials, the gene needed to cure cystic fibrosis is sprayed into the nose or delivered to the lower respiratory tract by adenoviruses or by the use of a liposome. So far, this treatment has resulted in limited success.

It has been known for some time that VEGF (vascular endothelial growth factor) can cause the growth of new blood vessels. The gene that codes for this growth factor can be injected alone or within a virus into the heart to stimulate branching of coronary blood vessels to increase circulation to the heart. Patients report that they have less chest pain and can run longer on a treadmill.

In cancer patients, genes are being used to make healthy cells more tolerant of chemotherapy and to make tumors more vulnerable to chemotherapy. The gene *p53* brings about apoptosis, and there is much interest in introducing it into cancer cells that no longer have the gene and in that way killing them off.

**Check Your Progress 14.3**

1. What methods are being used to introduce genes into human beings for gene therapy?
2. Give an example of ex vivo and of in vivo gene therapy.

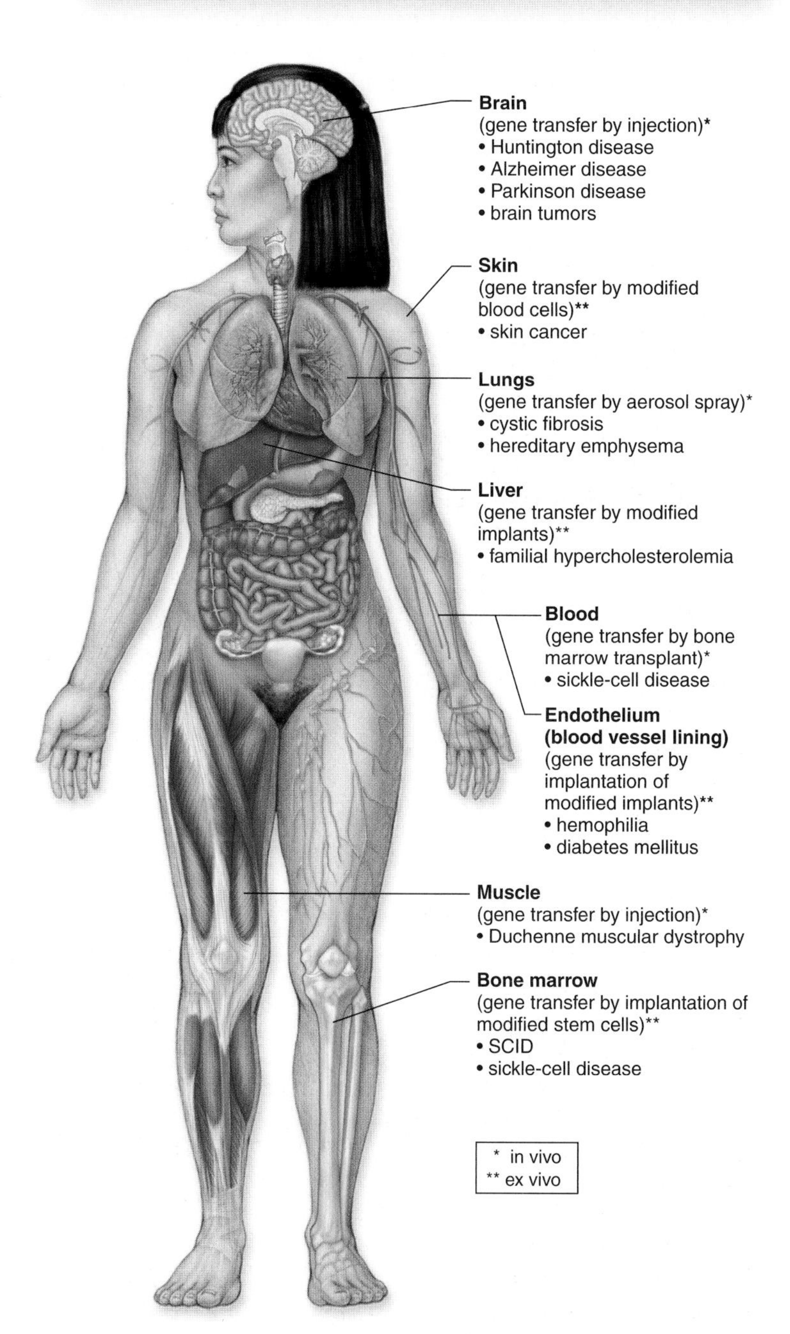


**FIGURE 14.7 Gene therapy.**

Sites of ex vivo and in vivo gene therapy to cure the conditions noted.

# 14.4 Genomics

In the previous century, researchers discovered the structure of DNA, how DNA replicates, and how DNA and RNA are involved in the process of protein synthesis. Genetics in the twenty-first century concerns **genomics,** the study of genomes—our complete genetic makeup and also that of other organisms. Knowing the sequence of bases in genomes is the first step, and thereafter we want to understand the function of our genes and their introns and the intergenic sequences (see Figure 14.8). The enormity of the task can be appreciated by knowing that there are approximately 6 billion base pairs in the 2n human genome. Many other organisms have a larger number of protein-coding genes but less noncoding regions compared to the human genome.

## Sequencing the Genome

We now know the order of the base pairs in our genome. This feat, which has been likened to arriving at the periodic table of the elements in chemistry, was accomplished by the **Human Genome Project (HGP),** a 13-year effort that involved both university and private laboratories around the world. How did they do it? First, investigators developed a laboratory procedure that would allow them to decipher a short sequence of base pairs, and then instruments became available that could carry out sequencing automatically. Over the 13-year span, DNA sequencers were constantly improved, and now modern instruments can automatically analyze up to 2 million base pairs of DNA in a 24-hour period. Where did all this DNA come from? Sperm DNA was the material of choice because it has a much higher ratio of DNA to protein than other types of cells. (Recall that sperm do provide both X and Y chromosomes.) However, white cells from the blood of female donors were also used in order to include female-originated samples. The male and female donors were of European, African, American (both North and South), and Asian ancestry.

Many small regions of DNA that vary among individuals (polymorphisms) were identified during the HGP. Most of these are *single nucleotide polymorphisms (SNPs)* (a difference of only one nucleotide). Many SNPs have no effect; others may contribute to enzymatic differences affecting the phenotype. It's possible that certain SNP patterns change an individual's susceptibility to disease and alter their response to medical treatments (see p. 284).

Determining that humans have 20,500 genes required a number of techniques, many of which relied on identifying RNAs in cells and then working backward to find the DNA that can pair with that RNA. **Structural genomics**—knowing the sequence of the bases and how many genes we have—is now being followed by functional genomics. Most of the known 20,500 human genes are expected to code for proteins. However, much of the human genome was formerly described as "junk" because it does not specify the order of amino acids in a polypeptide. However, it is possible for RNA molecules to have a regulatory effect in cells, as discussed in the next section.

## Eukaryotic Gene Structure

Historically, genes were defined as discrete units of heredity that corresponded to a locus on a chromosome (see Fig. 11.4). While prokaryotes typically possess a single circular chromosome with genes that are packed together very closely, eukaryotic chromosomes are much more complex. The genes are seemingly randomly distributed along the length of a chromosome and are fragmented into exons, with intervening sequences called introns scattered throughout the length of the gene (Fig. 14.8). In general, more complex organisms have more complex genes with more and larger introns. In humans, 95% or more of the average protein-coding gene is introns. Once a gene is transcribed, the introns must be removed and the exons joined together to form a functional mRNA transcript (see Fig. 12.13).

Once regarded as merely intervening sequences, introns are now attracting attention as regulators of gene expression. The presence of introns allows exons to be put together in various sequences so that different mRNAs and proteins can result from a single gene. It could also be that introns function to regulate gene expression and help determine which genes are to be expressed and how they are to be spliced. In fact, entire genes have been found embedded within the introns of other genes.

## Intergenic Sequences

DNA sequences occur between genes and are referred to as **intergenic sequences.** In general, as the complexity of an organism increases, so does the proportion of its non-protein-coding DNA sequences. Intergenic sequences are now known to comprise the vast majority of human chromosomes, and protein-coding genes represent only about 1.5% of our total DNA. The remainder of this DNA, once dismissed as "junk DNA," is now thought to serve many important functions, and has piqued the curiosity of many investigators. There are several basic types of intergenic sequences found in the human genome, including (1) repetitive elements, (2) transposons, and (3) unknown sequences. The majority of intergenic sequences are the unknown sequences.

### *Repetitive DNA Elements*

**Repetitive DNA elements** occur when the same sequence of two or more nucleotides are repeated many times along the length of one or more chromosomes. Repetitive elements are very common—comprising nearly half of the human genome—and, therefore, many scientists believe that their true significance has yet to be discovered. Although many scientists still dismiss them as having no function, others point out that the centromeres and telomeres of chromosomes are composed of repetitive elements and, therefore, repetitive DNA elements may not be as useless as once thought. For one thing, those of the centromere could possibly help with segregating the chromosomes during cell division.

Repetitive DNA elements occur as tandem repeats and interspersed repeats. **Tandem repeat** means that the repeated sequences are next to each other on the chromosome.

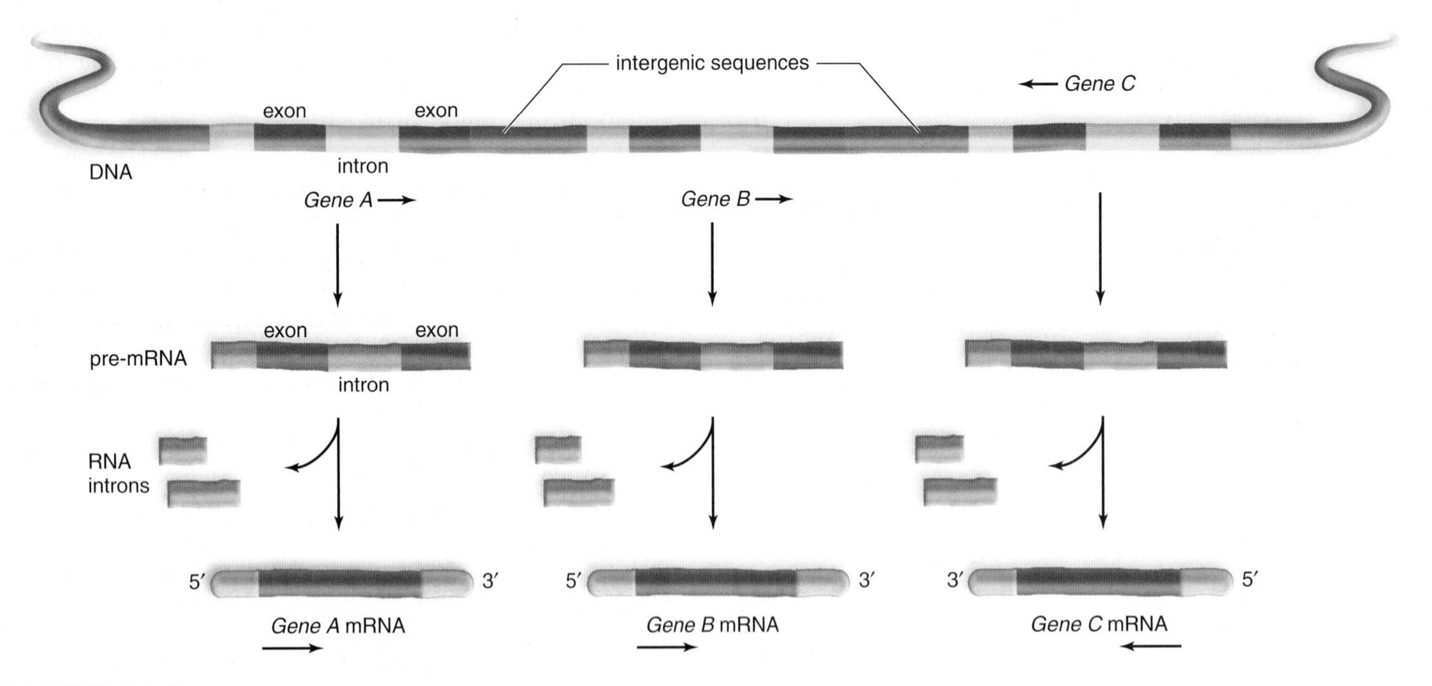


**FIGURE 14.8 Chromosomal DNA.**

Protein-coding DNA sequences are spread throughout eukaryotic chromosomes, often with very long intergenic sequences between them. Which strand of DNA is used as a template strand can vary, and protein-coding sequences may sometimes even be found within the introns of other genes. The functions of non-protein-coding DNA present in introns and intergenic sequences are being investigated extensively today.

Tandem repeats are often referred to as satellite DNA, because they have a different density than the rest of the DNA within the chromosome. The number and types of tandem repeats may vary significantly from one individual to another, making them invaluable as indicators of heritage. One type of tandem repeat sequence, referred to as *short tandem repeats,* or STRs, has become a standard method in forensic science for identifying one individual from another and for determining familial relationships (see page 251).

The second type of repetitive DNA element is called an **interspersed repeat,** meaning that the repetitions may be placed intermittently along a single chromosome, or across multiple chromosomes. For example, a repetitive DNA element, known as the *Alu* sequence, is interspersed every 5,000 base pairs in human DNA and comprises nearly 5–6% of total human DNA. Because of their common occurrence, interspersed repeats are thought to play a role in the evolution of new genes.

### Transposons

**Transposons** are specific DNA sequences that have the remarkable ability to move within and between chromosomes. Their movement to a new location sometimes alters neighboring genes, particularly decreasing their expression. In other words, a transposon sometimes acts like a regulator gene. The movement of transposons throughout the genome is thought to be a driving force in the evolution of living things. In fact, many scientists now think that many repetitive DNA elements were originally derived from transposons.

Although Barbara McClintock first described these "movable elements" in corn over 50 years ago, it took time for the scientific community to fully appreciate this revolutionary idea. In fact, their significance was only realized within the past few decades. So-called jumping genes now have been discovered in bacteria, fruit flies, humans, and many other organisms. McClintock received a Nobel Prize in 1983 for her discovery of transposons and for her pioneering work in genetics.

### Unknown Sequences

While genes comprise a scant 1.5% of the human genome and repetitive DNA elements make up about 44%, the function of the remaining half, called unknown sequences, remains a mystery. Even though this DNA does not appear to contain any protein-coding genes, it has been highly conserved through evolution. In the many millions of years that separates humans from mice, large tracts of this mysterious DNA have remained almost unchanged. But if it has no relevant function, then why has it been so meticulously maintained?

Recently, scientists observed that between 74% and 93% of the genome is transcribed into RNA, including many of these unknown sequences. Thus, what was once thought to be a vast junk DNA wasteland may be much more important than once thought and may play active roles in the cell. Small-sized RNAs may be able to carry out regulatory functions more easily than proteins at times. Therefore, a heretofore-overlooked RNA signaling network may be what allows humans, for example, to achieve structural complexity far beyond anything seen in the unicellular world. Together, these findings have revealed a much more complex, dynamic genome than was envisioned merely a few decades ago.

## What Is a Gene?

Perhaps the modern definition of a gene should take the emphasis away from the chromosome and place it on the results of transcription. Previously, molecular genetics considered a gene to be a nucleic acid sequence that codes for the sequence of amino acids in a protein. In contrast to this definition, we have known for some time all three types of RNA are transcribed from DNA and that these RNAs are useful products. We also know that protein-coding regions can be interrupted by regions that do not code for a protein but do produce RNAs with various functions. In recognition of these new findings, what about using this definition suggested by Mark Gerstein and associates in 2007: "A gene is a genomic sequence (either DNA or RNA) directly encoding functional products, either RNA or protein."[1]

This definition merely expands on the central dogma of genetics and recognizes that a gene product need not be a protein, and a gene need not be a particular locus on a chromosome. The DNA sequence that results in a gene product can be split and be present on one or several chromosomes. Also, any DNA sequence can result in one or more products.

This definition recognizes that some prokaryotes have RNA genes. In other words, the genetic material need not be DNA. Again, we can view this as a simple expansion of the central dogma of genetics.

The definition does not spell out what is meant by functional product. It would seem, then, that sequences of DNA resulting in regulatory RNAs or proteins could be considered genes.

The definition does expand on what is meant by "coding." Coding does not necessarily mean a DNA sequence that codes for a sequence of amino acids. Coding simply means a sequence of DNA bases that are transcribed. The gene product can be RNA molecules, or it can be a protein.

**Check Your Progress** **14.4A**

1. How does a tandem repeat differ from an interspersed repeat?
2. Why is a new definition of a gene required?

[1] Mark B. Gerstein et al., *What is a gene, post-encode? History and Updated Definition*, Cold Spring Harbor Laboratory Press, 2007.

**TABLE 14.1**

**Comparison of Sequenced Genomes**

| Organism | *Homo sapiens* (human) | *Mus musculus* (mouse) | *Drosophila melanogaster* (fruit fly) | *Arabidopsis thaliana* (flowering plant) | *Caenorhabditis elegans* (roundworm) | *Saccharomyces cerevisiae* (yeast) |
|---|---|---|---|---|---|---|
| **Estimated Size** | 2,900 million bases | 2,500 million bases | 180 million bases | 125 million bases | 97 million bases | 12 million bases |
| **Estimated Number of Genes** | ~20,500 | ~30,000 | 13,600 | 25,500 | 19,100 | 6,300 |
| **Chromosome Number** | 46 | 40 | 8 | 10 | 12 | 32 |

## Functional and Comparative Genomics

Since we now know the structure of our genome, the emphasis today is on functional genomics and also on comparative genomics. The aim of **functional genomics** is to understand the exact role of the genome in cells or organisms. To that end, a new technology called **DNA microarrays** can be used to monitor the expression of thousands of genes simultaneously. In other words, the use of a microarray can tell you what genes are turned on in a specific cell or organism at a particular time and under what particular environmental circumstances. The Science Focus on page 259 discusses the importance of DNA microarrays, which are also known as DNA chips, or genome chips. DNA microarrays contain microscopic amounts of known DNA sequences fixed onto a small glass slide or silicon chip in known locations (see Fig. 14A). When mRNA molecules of a cell or organism bind through complementary base pairing with the various DNA sequences, then that gene is active in the cell. As also discussed on page 259, DNA microarrays are available that rapidly identify all the various mutations in the genome of an individual. This is called the person's **genetic profile.** The genetic profile can indicate if any genetic illnesses are likely and what type of drug therapy for an illness might be most appropriate for that individual.

Aside from the protein-coding regions, researchers also want to know how SNPs and non-protein-coding regions, including repeats, affect which proteins are active in cells. As already discussed at length in Chapter 12, much research is now devoted to knowing the function of DNA regions that do not code for proteins.

The aim of **comparative genomics** is to compare the human genome to the genome of other organisms, such as those listed in Table 14.1. Surprisingly, perhaps, functional genomics has also been advanced by sequencing the genome of these organisms called model organisms (Table 14.1). Model organisms are used in genetic analysis because they have many genetic mechanisms and cellular pathways in common with each other and with humans. As described on page 254, much has been learned by genetically modifying mice. However, other model organisms can sometimes be used instead. Scientists inserted a human gene associated with early-onset Parkinson's disease into *Drosophila melanogaster,* and the flies showed symptoms similar to those seen in humans with the disorder. This suggested that we might be able to use these organisms instead of mice to test therapies for Parkinsons.

Comparative genomics also offers a way to study changes in a genome over time because the model organisms have a shorter generation time than humans. Comparing genomes will also help us understand the evolutionary relationships between organisms. One surprising discovery is that the genomes of all vertebrates are similar. Researchers were not surprised to find that the genes of chimpanzees and humans are 98% alike, but they did not expect to find that our sequence is also 85% like that of a mouse. Genomic comparisons will likely reveal evolutionary relationships between organisms never previously considered.

## Proteomics

The entire collection of a species' proteins is the **proteome.** At first, it may be surprising to learn that the proteome is larger than the genome until we consider all the many regulatory mechanisms, such as alternative pre-mRNA splicing, that increase the number of possible proteins in an organism.

**Proteomics** is the study of the structure, function, and interaction of cellular proteins. Specific regulatory mechanisms differ between cells, and these differences account for the specialization of cells. One goal of proteomics is to identify and determine the function of the proteins within a particular cell type. Each cell produces thousands of different proteins that can vary between cells and within the same cell, depending on circulations. Therefore, the goal of proteomics is an overwhelming endeavor. Microarray

*science focus*

## DNA Microarray Technology

With advances in robotic technology, it is now possible to spot all 20,500 known human genes, or even the entire human genome, onto a single microarray (Fig. 14A). The mRNA from the organism or the cell to be tested is labeled with a fluorescent dye and added to the chip. When the mRNAs bind to the microarray, a fluorescent pattern results that is recorded by a computer. Now the investigator knows what DNA is active in that cell or organism. A researcher can use this method to determine the difference in gene expression between two different cell types, such as between liver cells and muscle cells.

A mutation microarray, the most common type, can be used to generate a person's genetic profile. The microarray contains hundreds to thousands of known disease-associated mutant gene alleles. Genomic DNA from the individual to be tested is labeled with a fluorescent dye, and then added to the microarray. The spots on the microarray fluoresce if the individual's DNA binds to the mutant genes on the chip, indicating that the individual may have a particular disorder or is at risk for developing it later in life. This technique can generate a genetic profile much more quickly and inexpensively than older methods involving DNA sequencing.

### Diseased Tissues

DNA microarrays also promise to hasten the identification of genes associated with diseased tissues. In the first instance, mRNA derived from diseased tissue and normal tissue is labeled with different fluorescent dyes. The normal tissue serves as a control. The investigator applies the mRNA from both normal and abnormal tissue to the microarray. The relative intensities of fluorescence from a spot on the microarray indicate the amount of mRNA originating from that gene in the diseased tissue relative to the normal tissue. If a gene is activated in the disease, more copies of mRNA will bind to the microarray than from the control tissue, and the spot will appear more red than green.

Genomic microarrays are also used to identify links between disease and chromosomal variations. In this instance, the chip contains genomic DNA that is cut into fragments. Each spot on the microarray corresponds to a known chromosomal location. Labeled genomic DNA from diseased and control tissues bind to the DNA on the chip, and the relative fluorescence from both dyes is determined. If the number of copies of any particular target DNA has increased, more sample DNA will bind to that spot on the microarray relative to the control DNA, and a difference in fluorescence of the two dyes will be detected. Researchers are currently using this technique to identify disease-associated copy number variations, such as those discussed in the Science Focus on page 260.

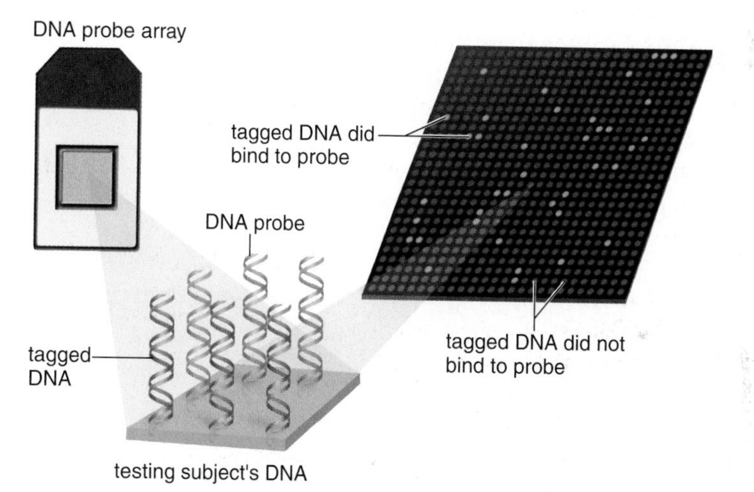


**FIGURE 14A DNA microarray technology.**
*A DNA microarray contains many microscopic samples of DNA bound to known locations on a silicon chip. A fluorescently labeled mRNA from a tissue or organism binds to the DNA on the chip by complementary base-pairing. The fluorescent spots indicate that binding has occurred and that the gene functions in that cell.*

technology can assist with this project and so can today's supercomputers.

Computer modeling of the three-dimensional shape of cellular proteins is also an important part of proteomics. If the primary structure of a protein is known, it should be possible to predict its final three-dimensional shape, and even the effects of DNA mutations on the protein's shape and function.

The study of protein function is viewed as essential to the discovery of new and better drugs. Also, it may be possible one day, to correlate drug treatment to the particular proteome of the individual to increase efficiency and decrease side effects. Proteomics will be a critical field of endeavor for many years to come.

## Bioinformatics

**Bioinformatics** is the application of computer technologies, specially developed software, and statistical techniques to the study of biological information, particularly databases that contain much genomic and proteomic information (Fig. 14.9). The new data produced by structural genomics and proteomics have produced raw data that is stored in databases that are readily available to research scientists. It is called raw data because, as yet, it has little meaning. Functional genomics and proteomics are dependent on computer analysis to find significant patterns in the raw data. For example, BLAST, which stands for *b*asic *l*ocal *a*lignment *s*earch *t*ool, is a computer program that can identify homologous genes among the genomic sequences of model organisms. **Homologous genes** are genes that code for the same proteins, although the base sequence may be slightly different. Finding these differences can help trace the history of evolution among a group of organisms.

Bioinformatics also has various applications in human genetics. For example, researchers found the function of the protein that causes cystic fibrosis by using the computer to search for genes in model organisms that have the same sequence. Because they knew the function of this same gene in model organisms, they could deduce the function in humans. This was a necessary step toward possibly developing specific treatments for cystic fibrosis. The human genome has 3 billion known base pairs, and without the computer it would be almost impossible to make sense of these data. For example, it is now known that an individual's genome often contains multiple copies of a gene. But individuals may differ as to the number of copies—called copy number variations, as discussed in the Science Focus below. Now it seems that the number of copies in a genome can be associated with specific diseases. The computer can help make correlations between genomic differences among large numbers of people and disease.

It is safe to say that without bioinformatics, our progress in determining the function of DNA sequences; in comparing our genome to model organisms; in knowing how genes and proteins interact in cells; and so forth, would be extremely slow. Instead, with the help of bioinformatics, progress should proceed rapidly in these and other areas.

**FIGURE 14.9 Bioinformatics.**

New computer programs are being developed to make sense out of the raw data generated by genomics and proteomics. Bioinformatics allows researchers to study both functional and comparative genomics in a meaningful way.

### Check Your Progress 14.4B

1. How is the information learned through the Human Genome Project being used to improve human health?
2. What kind of information can be learned through proteomics?
3. How is comparative genomics being used to divulge evolutionary relationships between organisms?

*science focus*

## Copy Number Variations

Geneticists have long been aware of large chromosomal duplications, deletions, and rearrangements detectable microscopically (see Fig. 10.13*a*,*b*). However, scientists have recently become aware of small duplications and deletions referred to as copy number variations (CNVs). CNVs occur when genes have changed their number.

The change may arise from so-called fork stalling and template switching. DNA damage or some other difficulty may cause the replication fork to stall. In order to continue, the replication machinery can switch to nearby chromosomal material of the same sequence. The replication fork is soon transferred back to the normal template, but the end result is extra or missing copies of small DNA segments. The fact that repetitive elements facilitate template switching suggests a new function for such sequences in our genome.

Some CNVs have known links to disease. Research shows that individuals with fewer copies of the *CCL3L1* gene are more susceptible to HIV infection than those with more copies. Lupus is much more common among people with fewer copies of the complement component *C4* gene. But more surprising was a recent study that suggested at least some cases of autism can be linked to CNVs. The scientists who published the study examined the total chromosomal content of 1,441 autistic children and compared their DNA to more than 2,800 normal individuals. They found that in autistic children, a 25-gene region of chromosome 16 was missing. Furthermore, analysis of other DNA databases revealed the same result: Approximately 1% of autism cases could be directly linked to the same deletion.

CNVs are also emerging as a possible driving force in evolution. A recent study utilized DNA microarrays to examine the chromosomal structure of 47 individuals from many ethnic backgrounds, and found 119 regions where copy number variations existed. More surprising, none of the CNVs were found exclusively in one ethnic group, suggesting that these variants existed well before the human population spread across the Earth. Perhaps they contributed to the phenotypic variations that developed thereafter. Furthermore, many scientists are suggesting that it be advantageous for a species to have multiple copies of genes—if one or both normal copies of an allele fail to function properly, having a third allele available might be advantageous because it could restore normal function. Conversely, an organism's two normal alleles would free this extra gene copy from having to maintain normal function. This would allow the gene to accumulate mutations without major consequence, which could ultimately lead to the formation of a new, unique gene. Copy number variations may contribute to evolution because they are yet another mechanism for organisms to achieve genetic innovation.

## Connecting the Concepts

Basic research into the nature and organization of genes in various organisms allowed geneticists to produce recombinant DNA molecules. An understanding of transcription and translation made it possible for scientists to manipulate the expression of genes in organisms. These breakthroughs have ushered in a biotechnology revolution. Because the genetic code is almost universal, bacteria and eukaryotic cells are now used to produce vaccines, hormones, and growth factors for use in humans. Plants and animals are also engineered to make a product or to possess characteristics desired by humans. Biotechnology also offers the promise of treating and even some day curing human genetic disorders such as muscular dystrophy, cystic fibrosis, hemophilia, and many others. It also shows promise in creating hardier crops that could help alleviate food shortages in many parts of the world.

The growing field of genomics also shows promise. Now that the entire human genome has been sequenced, scientists can use this information to determine which genes function in particular cells and also to determine people's genetic profiles for the purpose of prescribing medications, diagnosing illness, and preventing future problems. Geneticists have also sequenced the genomes of many other species, and comparisons between them is yielding valuable new insights into the relationships between species, impacting taxonomy and evolutionary biology.

## summary

### 14.1 DNA Cloning

DNA cloning can isolate a gene and produce many copies of it. The gene can be studied in the laboratory or inserted into a bacterium, plant, or animal. Then, this gene may be transcribed and translated to produce a protein, which can become a commercial product or used as a medicine.

Two methods are currently available for making copies of DNA: recombinant DNA technology and the polymerase chain reaction (PCR). Recombinant DNA contains DNA from two different sources. A restriction enzyme is used to cleave plasmid DNA and to cleave foreign DNA. The resulting "sticky ends" facilitate the insertion of foreign DNA into vector DNA. The foreign gene is sealed into the vector DNA by DNA ligase. Both bacterial plasmids and viruses can be used as vectors to carry foreign genes into bacterial host cells.

PCR uses the enzyme DNA polymerase to quickly make multiple copies of a specific piece (target) of DNA. PCR is a chain reaction because the targeted DNA is replicated over and over again. Analysis of DNA segments following PCR has all sorts of uses from assisting genomic research to doing DNA fingerprinting for the purpose of identifying individuals and their paternity.

### 14.2 Biotechnology Products

Transgenic organisms have had a foreign gene inserted into them. Genetically modified bacteria, agricultural plants, and farm animals now produce commercial products of interest to humans, such as hormones and vaccines. Bacteria usually secrete the product. The seeds of plants and the milk of animals contain the product.

Transgenic bacteria have also been engineered to promote the health of plants, perform bioremediation, extract minerals, and produce chemicals. Transgenic crops, engineered to resist herbicides and pests, are commercially available. Transgenic animals have been given various genes, in particular the one for bovine growth hormone (bGH). Cloning of animals is now possible.

### 14.3 Gene Therapy

Gene therapy, by either ex vivo or in vivo methods, is used to correct the genotype of humans and to cure various human ills. Ex vivo gene therapy has apparently helped children with SCID lead normal lives. In vivo treatment for cystic fibrosis has been less successful. A number of in vivo therapies are being employed in the war against cancer and other human illnesses, such as cardiovascular disease.

### 14.4 Genomics

Researchers now know the sequence of all the base pairs along the length of the human chromosomes. So far, researchers have found only 20,500 human genes that code for proteins; the rest of our DNA consists of regions that do not code for a protein. Currently, researchers are placing an emphasis on functional and comparative genomics.

Genes only comprise 1.5% of the human genome. The rest of this DNA is surprisingly more active than once thought. About half of this DNA consists of repetitive DNA elements, which may be in tandem or interspersed throughout several chromosomes. Some of this DNA is made up of mobile DNA sequences called transposons, which are a driving evolutionary force within the genome. The remaining half of the genome remains unclassified, but even these unknown DNA sequences may play an important role in regulation of gene expression, and challenging the classical definition of the gene. Functional genomics aims to understand the function of protein coding regions and noncoding regions of our genome. To that end, researchers are utilizing new tools such as DNA microarrays. Microarrays can also be used to create an individual's genetic profile, which can be helpful in predicting illnesses and how a person will react to particular medications.

Comparative genomics has revealed that there is little difference between the DNA sequence of our bases and those of many other organisms. Genome comparisons have revolutionized our understanding of evolutionary relations by revealing previously unknown relationships between organisms.

Proteomics is the study of which genes are active in producing proteins in which cells under which circumstances. Bioinformatics is the use of the computer to assist proteomics and functional and comparative genomics.

## understanding the terms

Human Genome Project (HGP) 255
intergenic sequence 256
interspersed repeat 257
in vivo gene therapy 254
plasmid 250
polymerase chain reaction (PCR) 250
proteome 258
proteomics 258
recombinant DNA (rDNA) 250
repetitive DNA element 256
restriction enzyme 250
short tandem repeat (STR) profiling 251
structural genomics 255
tandem repeat 256
transgenic organism 250
transposon 257
vector 250
xenotransplantation 254

Match the terms to these definitions:

a. _____________ Bacterial agent that stops viral reproduction by cleaving viral DNA; used to cut DNA at specific points during production of recombinant DNA.
b. _____________ Free-living organism in the environment that has had a foreign gene inserted into it.
c. _____________ Use of animal organs, instead of human organs, in human transplant patients.
d. _____________ Production of identical copies; in genetic engineering, the production of many identical copies of a gene.
e. _____________ Biotechnology method that can quickly produce many duplicate copies of a piece of DNA.

## reviewing this chapter

1. What is the methodology for producing recombinant DNA so useful for gene cloning? 250
2. What is the polymerase chain reaction (PCR), and how is it carried out to produce multiple copies of a DNA segment? 251–52
3. How does STR profiling produce a DNA fingerprint? 251–52
4. What are some practical applications of DNA segment analysis following PCR? 252
5. For what purposes have bacteria, plants, and animals been genetically altered? 252–54
6. Explain and give examples of ex vivo and in vivo gene therapies in humans. 254–55
7. What was the purpose of the Human Genome Project? What is the goal of functional genomics? 255–58
8. What insights into evolutionary relationships between organisms are arising from comparative genomics? 258
9. Describe the various types of intergenic DNA sequences found within the genome. 256–57
10. What are the goals of proteomics and bioinformatics? 258–60

## testing yourself

Choose the best answer for each question.

1. Using this key, put the phrases in the correct order to form a plasmid-carrying recombinant DNA.

**KEY:**

(1) use restriction enzymes
(2) use DNA ligase
(3) remove plasmid from parent bacterium
(4) introduce plasmid into new host bacterium

a. 1, 2, 3, 4
b. 4, 3, 2, 1
c. 3, 1, 2, 4
d. 2, 3, 1, 4

2. Restriction enzymes found in bacterial cells are ordinarily used
   a. during DNA replication.
   b. to degrade the bacterial cell's DNA.
   c. to degrade viral DNA that enters the cell.
   d. to attach pieces of DNA together.

3. A genetic profile can
   a. assist in maintaining good health.
   b. be accomplished utilizing bioinformatics.
   c. show how many genes are normal.
   d. be accomplished utilizing a microarray.
   e. Both a and d are correct.

4. Bacteria are able to successfully transcribe and translate human genes because
   a. both bacteria and humans contain plasmid vectors.
   b. bacteria can replicate their DNA, but humans cannot.
   c. human and bacterial ribosomes are vastly different.
   d. the genetic code is nearly universal.

5. Bioinformatics can
   a. assist genomics and proteomics.
   b. compare our genome to that of a monkey.
   c. depend on computer technology.
   d. match up genes with proteins.
   e. All of these are correct.

6. The polymerase chain reaction
   a. uses RNA polymerase.
   b. takes place in huge bioreactors.
   c. uses a temperature-insensitive enzyme.
   d. makes lots of nonidentical copies of DNA.
   e. All of these are correct.

7. Which is a true statement?
   a. Genomics would be slow going without bioinformatics.
   b. Genomics is related to the field of proteomics.
   c. Genomics has now moved on to functional and comparative genomics.
   d. Genomics shows that we are related to all other organisms tested so far.
   e. All of these are correct.

8. DNA amplified by PCR and then used for fingerprinting could come from
   a. any diploid or haploid cell.
   b. only white blood cells that have been karyotyped.
   c. only skin cells after they are dead.
   d. only purified animal cells.
   e. Both b and d are correct.

9. Which was used to find the function of the cystic fibrosis gene?
   a. microarray
   b. proteomics
   c. comparative genomics and bioinformatics
   d. sequencing the gene

10. Which of these pairs is incorrectly matched?
    a. DNA ligase—mapping human chromosomes
    b. protoplast—plant cell engineering
    c. DNA fragments—DNA fingerprinting
    d. DNA polymerase—PCR

11. Which matches best to proteomics?
 a. Start with known gene sequences and build proteins.
 b. Use a microarray to discover what proteins are active in particular cells.
 c. Use bioinformatics to discover the proteins in the cells of other organisms.
 d. Match up known proteins with known genes.
12. Which is not a correct association with regard to bioengineering?
 a. plasmid as a vector—bacteria
 b. protoplast as a vector—plants
 c. RNA virus as a vector—human stem cells
 d. All of these are correct.
13. Proteomics is used to discover
 a. what genes are active in what cells.
 b. what proteins are active in what cells.
 c. the structure and function of proteins.
 d. how proteins interact.
 e. All but a are correct.
14. Which of these is an incorrect statement?
 a. Bacteria usually secrete the biotechnology product into the medium.
 b. Plants are being engineered to have human proteins in their seeds.
 c. Animals are engineered to have a human protein in their milk.
 d. Animals can be cloned, but plants and bacteria cannot.
15. Repetitive DNA elements
 a. may be tandem or spread across several chromosomes.
 b. are found in centromeres and telomeres.
 c. make up nearly half of human chromosomes.
 d. may be present just a few to many thousands of copies.
 e. All of these are correct.
16. Because of the Human Genome Project, we now know
 a. the sequence of the base pairs of our DNA.
 b. the sequence of all genes along the human chromosomes.
 c. all the mutations that lead to genetic disorders.
 d. All of these are correct.
 e. Only a and c are correct.
17. Which of the following delivery methods is not used in gene therapy?
 a. virus
 b. nasal sprays
 c. liposomes
 d. electric currents
18. The restriction enzyme called *Eco*RI has cut double-stranded DNA in the following manner. The piece of foreign DNA to be inserted has what bases from the left and from the right?

19. Which of these is a true statement?
 a. Plasmids can serve as vectors.
 b. Plasmids can carry recombinant DNA, but viruses cannot.
 c. Vectors carry only the foreign gene into the host cell.
 d. Only gene therapy uses vectors.
 e. Both a and d are correct.
20. Gene therapy
 a. is sometimes used in medicine today.
 b. is always successful.
 c. is only used to cure genetic disorders, such as SCID and cystic fibrosis.
 d. makes use of viruses to carry foreign genes into human cells.
 e. Both a and d are correct.

## thinking scientifically

1. Before the human genome was sequenced, gene discovery was accomplished through the use of DNA libraries. A genomic library is a set of cloned DNA segments that altogether are representative of the genome of an organism, whereas a cDNA library contains only expressed DNA sequences for a particular cell. How might these libraries be used to map the introns and exons of a gene within the genome?
2. The Science Focus on page 260 describes copy number variations within the genome. Copy number variations do not always contain genes. How might having extra or missing copies of intergenic DNA sequences be beneficial? How might it be harmful?

## bioethical issue

### Transgenic Crops

Transgenic plants can possibly allow crop yields to keep up with the ever-increasing worldwide demand for food. And some of these plants have the added benefit of requiring less fertilizer and/or pesticides, which are harmful to human health and the environment.

But some scientists believe transgenic crops pose their own threat to the environment, and many activists believe transgenic plants are themselves dangerous to our health. Studies have shown that wind-carried pollen can cause transgenic crops to hybridize with nearby weedy relatives. Although it has not happened yet, some fear that characteristics acquired in this way might cause weeds to become uncontrollable pests. Or perhaps a toxin produced by transgenic crops could possibly hurt other organisms in the field. Many researchers are conducting tests to see if this might occur. And although transgenic crops have not caused any illnesses in humans so far, some scientists concede the possibility that people could be allergic to the transgene's protein product. After unapproved genetically modified corn was detected in Taco Bell taco shells several years ago, a massive recall pulled about 2.8 million boxes of the product from grocery stores.

Already, transgenic plants must be approved by the Food and Drug Administration before they are considered safe for human consumption, and they must meet certain Environmental Protection Administration standards. Some people believe safety standards for transgenic crops should be further strengthened, while others fear stricter standards will result in less food produced. Another possibility is to retain the current standards but require all biotech foods to be clearly labeled so the buyer can choose whether or not to eat them. Which approach do you prefer?

## *Biology* website

The companion website for *Biology* provides a wealth of information organized and integrated by chapter. You will find practice tests, animations, videos, and much more that will complement your learning and understanding of general biology.

**http://www.mhhe.com/maderbiology10**

part III

# Evolution

*Evolution may seem like a foreign topic to you, but it need not be. Evolution simply refers to the changes that occur as one generation begets the next generation. Just as you can trace your ancestry to your great-grandparents and beyond, so all of life can trace its ancestry to the first living cell or cells. A remarkable finding in recent times has been that some of our genes are even the same as those of prokaryotes. This couldn't be if we were not related to prokaryotes.*

*The changes that occur as a population reproduces assist its members in finding food, mates, a place to live, and even in avoiding dangers. Consider, for example, that resistant bacteria are able to avoid the danger of being killed by an antibiotic. When an antibiotic is taken for a staph infection, a few staph bacteria may be able to survive, and they are the ones that will produce the next generation of staph bacteria. Soon, most members of future generations are resistant. This is the manner in which all adaptations to the environment occur. From carnivorous plants to bats and whales, all life is adapted to its particular environment.*

*Evolution, the topic of this part, explains the world of living things; how it came to be, and why it is so diverse. The next time you are in a natural area, observe the diversity and say, "Evolution!"*

# 15 Darwin and Evolution

**i**n 2006, *a fossil snake was discovered with a pelvic girdle and legs. Charles Darwin, sometimes called the father of evolution, would have been pleased because such a fossil shows us how evolution occurred. Living pythons have leg remnants, so now we have both living and fossil evidence that snakes had legged ancestors.*

*Darwin's most noted contribution was to discover what caused the tree of life to have so many branches. Much data allowed him to conclude that nature (the environment) selects which members of a population will reproduce and pass on their adaptive traits to their offspring. Life is diverse because environments are diverse. This chapter is about Darwin's contribution to the field of evolution. First, we take a look at evolutionary thought when Darwin began his work. Then, we retrace Darwin's trip around the world and present still other data that allowed Darwin to develop his theory of evolution by natural selection. Finally, we take a look at evidence that supports Darwin's statement that evolution is "common descent with modification."*

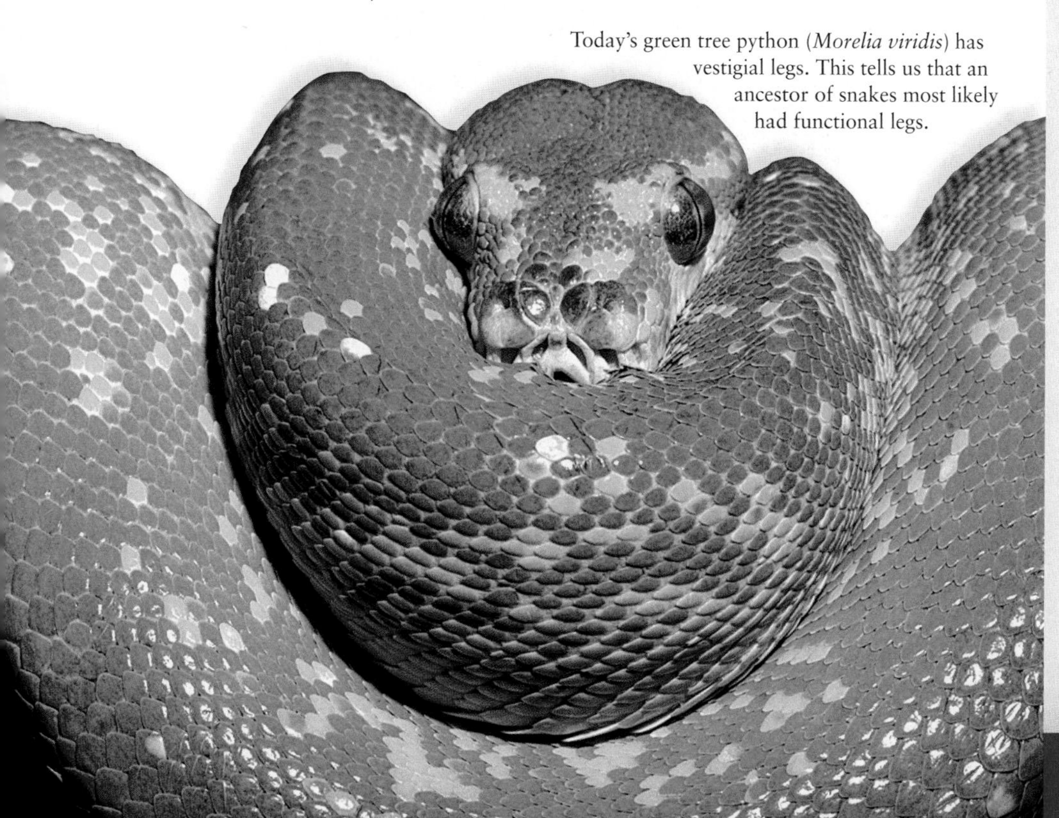

Today's green tree python (*Morelia viridis*) has vestigial legs. This tells us that an ancestor of snakes most likely had functional legs.

## concepts

**15.1 HISTORY OF EVOLUTIONARY THOUGHT**

- In the mid-eighteenth century, scientists became especially interested in classifying and understanding the relationships among the many forms of present and past life. Gradually, in the late-eighteenth century, evidence began to convince scientists that life-forms change over time. Their explanations varied as to how such changes occurred. 266–68

**15.2 DARWIN'S THEORY OF EVOLUTION**

- Charles Darwin's trip around the Southern Hemisphere aboard the HMS *Beagle* provided him with evidence that the Earth is very old and that evolution does occur. 269–73
- Both Darwin and Alfred Russel Wallace proposed natural selection as the mechanism by which adaptation to the environment takes place. This mechanism is consistent with our present-day knowledge of genetics. 274–75

**15.3 EVIDENCE FOR EVOLUTION**

- Darwin told us that evolution has two components: descent from a common ancestor and adaptation to the environment. The fossil record, biogeographical evidence, anatomical evidence, and biochemical evidence support the hypothesis of common descent. 276–79

# 15.1 History of Evolutionary Thought

In December 1831, a new chapter in the history of biology had its humble origins. A 22-year-old naturalist, Charles Darwin (1809–82), set sail on a journey of a lifetime aboard the British naval vessel the HMS *Beagle* (Fig. 15.1). Darwin's primary mission on his journey around the world was to expand the navy's knowledge of potential natural resources, such as water and food in foreign lands. Prior to Darwin, the worldview was forged by deep-seated beliefs that were held to be intractable truths and not by experimentation and observation of the natural word. In contrast, Darwin used a variety of data to come to the conclusion that the Earth is very old, not young, and that biological evolution is the method by which species arise and change. The acceptance of the Darwinian view of the world was fostered by a scientific and intellectual revolution that occurred in both the scientific and social realms of the mid-1800s.

Although it is often believed that Darwin (Fig. 15.2) forged this change in worldview by himself, several biologists during the preceding century and some of Darwin's con-

**FIGURE 15.1**
**Voyage of the HMS *Beagle*.**

**a.** Map shows the journey of the HMS *Beagle* around the world. Notice that the map's encircled colors are keyed to the encircled colors in the photographs, which show us what Charles Darwin may have observed in or near South America. **b.** As Darwin traveled along the east coast of South America, he noted that a bird called a rhea looked like the African ostrich. **c.** The sparse vegetation of the Patagonian Desert is in the southern part of the continent. **d.** The Andes Mountains of the west coast have strata containing fossilized organisms. **e.** The lush tropical rain forest contains a high and unique diversity of life. **f.** On the Galápagos Islands, marine iguanas have large claws to help them cling to rocks and blunt snouts for eating algae growing on rocks. **g.** Galápagos finches are specialized to feed in various ways. This finch is using a cactus spine to probe for insects.

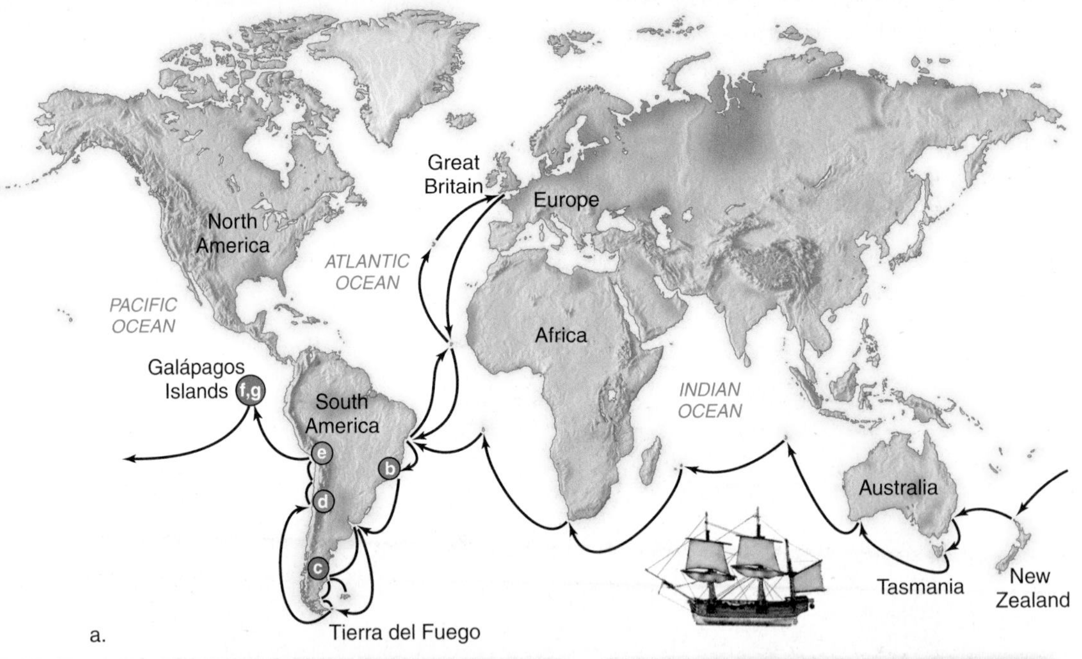

temporaries slowly began to accept the idea that species change over time. This concept would eventually be known as **evolution** [L. *evolutio,* an unrolling] and is now considered the unifying principle of all the biological sciences. Evolution explains both the unity and diversity of life on Earth. First, evolution illustrates that living things share like characteristics because they have a common ancestry. Evolution also explains how species adapt to various habitats and ways of life with the result that life is very diverse.

The history of evolutionary thought is a history of ideas about descent and adaptation. Darwin used the expression "descent with modification," by which he meant that as descent through generations occurs over time, so does diversification. Darwin brilliantly saw the process of adaptation as a means by which the diversity of species arises.

**FIGURE 15.2 Charles Darwin (1809–82).**

The theory of evolution is usually identified with Charles Darwin, who, along with Alfred Wallace, proposed a mechanism for evolution. This portrait of Darwin dates from 1831; he did not publish his authoritative book, *On the Origin of Species*, until 1859.

## Mid-Eighteenth-Century Contributions

Taxonomy, the science of classifying organisms, was an important endeavor during the mid-eighteenth century. Chief among the taxonomists was Carolus Linnaeus (1707–78), who developed the binomial system of nomenclature (a two-part name for species, such as *Homo sapiens*) and a system of classification for living things. In addition to taxonomy, comparative anatomy, the evaluation of similar structures across a variety of species, was of interest to biologists prior to Darwin. By the late eighteenth century, scientists had discovered fossils and knew that they were the remains of plants and animals from the past. Explorers traveled the world bringing back not only **extant** (living species) but also fossils to be compared to living species. At first, they believed that each type fossil had a living descendant, but eventually some fossils did not seem to match up well with known species. Baron Georges Cuvier was the first to suggest that some species known only from the fossil record had become extinct.

Linnaeus, like other taxonomists of his time, believed in the fixity of species. Each species had an "ideal" structure and function and also a place in the *scala naturae,* a sequential ladder of life. The simplest and most material being was on the lowest rung of the ladder, and the most complex and spiritual being was on the highest rung. In this view, human beings occupied the highest rung of the ladder. These ideas, which were consistent with Judeo-Christian teachings about special creation, can be traced to the works of the famous Greek philosophers Plato (427–347 BC) and Aristotle (384–322 BC). Plato said that every object on Earth was an imperfect copy of an ideal form, which can be deduced upon reflection and study. To Plato, individual variations were imperfections that only distract the observer. Aristotle saw that organisms were diverse and some were more complex than others. His belief that all organisms could be arranged in order of increasing complexity became the *scala naturae* just described.

Linnaeus and other taxonomists wanted to determine the ideal characteristics of each species and also wanted to discover the proper rank for each species in the *scala naturae.* Therefore, for most of his working life, Linnaeus did not even consider the possibility of evolutionary change. There is evidence, however, that he did eventually perform hybridization experiments, which made him think that a species might change with time.

Georges-Louis Leclerc, better known by his title, Count Buffon (1707–88), was a French naturalist who devoted many years of his life to writing a 44-volume natural history series of texts that described all known plants and animals. He provided evidence of descent with modification, and he even speculated on various causative mechanisms such as environmental influences, migration, geographic isolation, and the struggle for existence. Buffon seemed to waver, however, as to whether or not he recognized evolutionary descent, and often he professed to believe in special creation and the fixity of species.

Erasmus Darwin (1731–1802), Charles Darwin's grandfather, was a physician and a naturalist. His writings on both botany and zoology, although they were mostly in footnotes and asides, contained many comments that suggested the possibility of common descent. He based his conclusions on changes undergone by animals during development, artificial selection by humans, and the presence of **vestigial structures** (structures or organs that are thought to have been functional in an ancestor but are reduced and nonfunctional in a descendant). Like Buffon, Erasmus Darwin offered no conclusive mechanism by which evolutionary descent might occur.

## Late-Eighteenth-/Early-Nineteenth-Century Contributions

Even though Linnaeus was never an evolutionist, his hierarchical method of classifying organisms is consistent with modern evolutionary thinking. This is the reason that Linnaeus' basic method of classification has been modified even until today. Whether Linnaean classification will always be flexible enough for continual modification is a question that is even now being asked.

### *Cuvier and Catastrophism*

Baron Georges Cuvier (1769–1832), a distinguished zoologist, used comparative anatomy to develop a system of classifying animals. He also founded the science of

**paleontology** [Gk. *palaios*, old; *ontos*, having existed; *-logy*, study of], the study of fossils, and was quite skilled at using fossil bones to deduce the structure of an animal (Fig. 15.3*a*).

Because Cuvier was a staunch advocate of special creation and the fixity of species, he faced a real problem when a particular region showed a succession of life-forms in the Earth's strata (layers). To explain these observations, he hypothesized that a series of local catastrophes or mass extinctions had occurred whenever a new stratum of that region showed a new mix of fossils. After each catastrophe, the region was repopulated by species from surrounding areas, and this accounted for the appearance of new fossils in the new stratum. The result of all these catastrophes was change appearing over time. Some of Cuvier's followers even suggested that there had been worldwide catastrophes and that after each of these events, God created new sets of species. This explanation of the history of life came to be known as **catastrophism** [Gk. *katastrophe*, calamity, misfortune].

## *Lamarck and Acquired Characteristics*

Jean-Baptiste de Lamarck (1744–1829) was the first biologist to offer a mechanism for how evolution occurs and to link diversity with adaptation to the environment. Lamarck's ideas about descent were entirely different from those of Cuvier, perhaps because Lamarck specialized in the study of invertebrates (animals without backbones), while Cuvier was a vertebrate zoologist, who studied animals with backbones. Lamarck concluded, after studying the succession of life-forms in strata, that more complex organisms are descended from less complex organisms. He mistakenly said, however, that increasing complexity is the result of a natural force—a desire for perfection—that is inherent in all living things.

To explain the process of adaptation to the environment, Lamarck supported the idea of **inheritance of acquired characteristics**—that the environment can bring about inherited change. One example that he gave—and for which he is most famous—is that the long neck of a giraffe developed over time because animals stretched their necks to reach food in tall trees and then passed on a long neck to their offspring (Fig. 15.3*b*). His hypothesis of the inheritance of acquired characteristics has never been substantiated by experimentation. The molecular mechanism of inheritance explains why. Phenotypic changes acquired during an organism's lifetime do not result in genetic changes that can be passed to subsequent generations.

### Check Your Progress 15.1

1. According to Lamarck, if someone dyed their hair dark to make their blonde less visible at night, what color hair would be passed on to their offspring?
2. Based on Cuvier's ideas, if an asteroid were to impact the Earth and cause 99% of living species to go extinct, where would replacement species come from?

a.

b.

**FIGURE 15.3 Evolutionary thought before Darwin.**

**a.** Cuvier reconstructed animals such as extinct mastodons and said that catastrophes followed by repopulations could explain why species change over time. **b.** Lamarck explained the long neck of a giraffe according to his ideas about the inheritance of acquired characteristics.

# 15.2 Darwin's Theory of Evolution

When Darwin signed on as the naturalist aboard the HMS *Beagle,* he possessed a suitable background for the position. Since childhood, he was a devoted student of nature and a collector of insects. At age 16, Darwin was sent to medical school to follow in the footsteps of his grandfather and father. However, his sensitive nature prevented him from studying medicine, and he enrolled in the school of divinity at Christ College at Cambridge, intent on becoming a clergyman.

While at Christ College, he attended many lectures in biology and geology to satisfy his interest in natural science. During this time, he became the protégé of his teacher and friend, Reverend John Henslow. Darwin gained valuable experience in geology in the summer of 1831, conducting fieldwork with Adam Sedgewick. Shortly after Darwin was awarded his BA, Henslow recommended him to serve, without pay, as ship's naturalist on the HMS *Beagle*. The trip was to take two years—but ended up taking five years—and the ship was to traverse the Southern Hemisphere (see Fig. 15.1), where life is most abundant and varied. Along the way, Darwin encountered forms of life very different from his native England.

## Occurrence of Descent

Although it was not his original intent, Darwin began to gather evidence that organisms are related through common descent and that adaptation to various environments results in diversity. Darwin also began contemplating the "mystery of mysteries," the origin of new species.

### *Geology and Fossils*

Darwin took Charles Lyell's *Principles of Geology* on the voyage, which presented arguments to support a theory of geological change proposed by James Hutton. In contrast to the catastrophists, Hutton explained that the Earth was subject to slow but continuous cycles of rock formation through erosion and uplift. Weather causes erosion; thereafter, dirt and rock debris are washed into the rivers and transported to the oceans. These loose sediments are deposited in thick layers, which are converted eventually into sedimentary rocks. These sedimentary rocks, which often contain fossils, are then uplifted from below sea level to form land. Hutton concluded that extreme geological changes can be accounted for by slow, natural processes, given enough time. Lyell went on to propose the theory of **uniformitarianism,** which stated that these slow changes occurred at a uniform rate and that the natural processes witnessed today are the same processes that occurred in the past. Hutton's general ideas about slow and continual geological change are still accepted today, although modern geologists realize that rates of change have not always been uniform through history. Darwin was not taken by the idea of uniform change, but he was convinced, as was Lyell, that the Earth's massive geological changes are the result of extremely slow processes and that the Earth, therefore, must be very old.

On his trip, Darwin observed massive geological changes firsthand. When he explored what is now Argentina, he saw raised beaches for great distances along the coast. In Chile, he witnessed the effects of an earthquake that caused the land to rise several feet, leaving marine shells inland, well above sea level. This observation, along with marine shells high in the cliffs of the impressive Andes Mountains, supported Lyell's theory of slow geologic changes of a very old planet. While Darwin was making geological observations, he also collected fossil specimens. For example, on the east coast of South America, he found the fossil remains of an armadillo-like animal (*Glyptodon*), the size of a small modern-day car, and giant ground sloths, the smallest of which stood nearly 3 m tall (Fig. 15.4). Once Darwin accepted the supposition that the Earth must be very old, he began to think that there would have been enough time for descent with modification to occur. Therefore, living forms could be descended from extinct forms known only from the fossil record. It would seem that species were not fixed; instead, they changed over time.

### *Biogeography*

**Biogeography** [Gk. *bios,* life, *geo,* earth, and *grapho,* writing] is the study of the range and geographic distribution of life-forms on Earth. Darwin could not help but compare the animals of South America to those with which he was familiar. For example, instead of rabbits, he found the Patagonian hare in the grasslands of South America. The Patagonian hare has long legs and ears but the face of a guinea pig, a

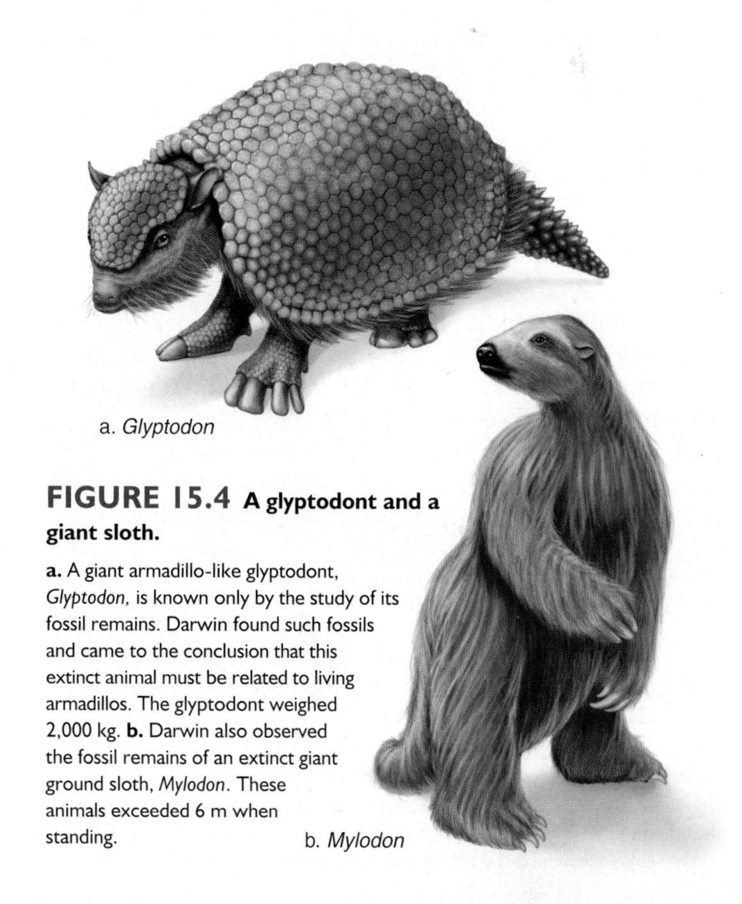


**FIGURE 15.4 A glyptodont and a giant sloth.**

**a.** A giant armadillo-like glyptodont, *Glyptodon*, is known only by the study of its fossil remains. Darwin found such fossils and came to the conclusion that this extinct animal must be related to living armadillos. The glyptodont weighed 2,000 kg. **b.** Darwin also observed the fossil remains of an extinct giant ground sloth, *Mylodon*. These animals exceeded 6 m when standing.

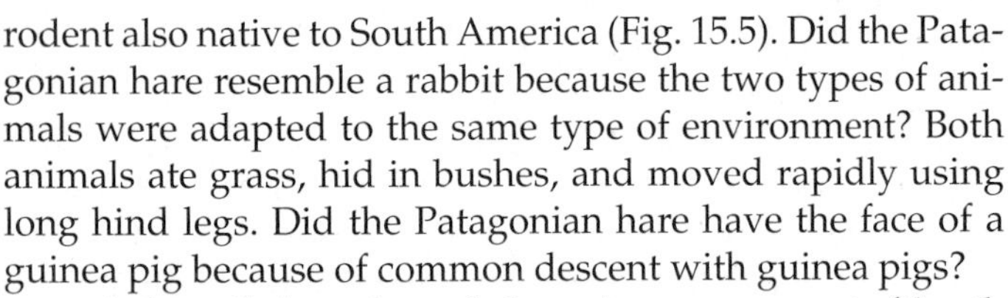


**FIGURE 15.5**
**The European hare, (head only), and the Patagonian hare.**

rodent also native to South America (Fig. 15.5). Did the Patagonian hare resemble a rabbit because the two types of animals were adapted to the same type of environment? Both animals ate grass, hid in bushes, and moved rapidly using long hind legs. Did the Patagonian hare have the face of a guinea pig because of common descent with guinea pigs?

As he sailed southward along the eastern coast of South America, Darwin saw how similar species replaced one another. For example, the greater rhea (an ostrichlike bird) found in the north was replaced by the lesser rhea in the south. Therefore, Darwin reasoned that related species could be modified according to environmental differences (i.e., Northern vs. Southern latitudes). When he explored the Galápagos Islands, he found further evidence of this phenomenon. The Galápagos Islands are a small group of volcanic islands formed 965 km off the western coast of South America when underwater volcanoes emerged from the ocean. These islands are too far from the mainland for most terrestrial animals and plants to colonize, yet life was present there. The types of plants and animals found there were slightly different from species Darwin had observed on the mainland and, even more important, they also varied from island to island. Where did animals and plants inhabiting these islands come from, and why were different species on each island?

**Tortoises.** Each of the Galápagos Islands seemed to have its own type of tortoise, and Darwin began to wonder if this could be correlated with a difference in vegetation among the islands (Fig. 15.6). Long-necked tortoises seemed to inhabit only dry areas, where food was scarce, most likely because the longer neck was helpful in reaching tall-growing cacti. In moist regions with relatively abundant ground foliage, short-necked tortoises were found. Had an ancestral tortoise from the mainland of South America given rise to these different types, each adapted to a different environment?

**Finches.** Darwin almost overlooked the finches because of their unassuming nature compared with many of the other animals in the Galápagos. However, these birds would eventually play a major role in his thoughts about geographic isolation. The finches of the Galápagos Islands seemed to Darwin like mainland finches, but they exhibited significant variety with regard to their beaks (see Fig. 15.10). Today, there are ground-dwelling finches with beaks sized appropriate to the seeds they feed on, tree-dwelling finches with beaks sized according to their insect prey, and a cactus-eating finch with a more pointed beak. The most unusual of the finches is a woodpecker-type finch. This bird has a sharp beak to chisel through tree bark but lacks the long tongue characteristic of a true woodpecker, which probes for insects. To compensate for this, the bird carries a twig or cactus spine in its beak and uses it to poke into crevices (see Fig. 15.1*g*). Once an insect emerges, the finch drops this tool and seizes the insect with its beak.

Later, Darwin speculated as to whether these different species of finches could have descended from a type of mainland finch. In other words, he wondered if a finch from South America was the common ancestor to all the types on the Galápagos Islands. Had speciation occurred because the islands allowed isolated populations of birds to evolve independently? Could the present-day species have resulted from accumulated changes occurring within each of these isolated populations?

a.

b.

**FIGURE 15.6 Galápagos tortoises.**

Darwin wondered if all of the Galápagos tortoises, *Geochelone nigra*, of the various islands were descended from a common ancestor. **a.** The tortoises with dome shells and short necks feed at ground level and occur on well-watered islands where grass is available. **b.** Those with shells that flare up in front have long necks and are able to feed on tall, treelike cacti. They live on arid islands where prickly pear cacti are the main food source. Only on these islands are the cacti treelike.

## Natural Selection and Adaptation

Upon returning to England, Darwin began to reflect on the voyage of the HMS *Beagle* and to collect evidence supporting his ideas about how organisms adapt to the environment. Darwin decided that adaptations develop over time (instead of being the instant work of a creator), and he began to think about a mechanism that would allow this to happen. In the late 1850s, Darwin proposed **natural selection** as a mechanism for evolutionary change. Meanwhile, Alfred Russel Wallace, who is discussed in the Science Focus on page 274, proposed the same concept based on similar observations from the other side of the globe. Natural selection is a process consisting of these components:

- The members of a population have inheritable variations. Variation within a population of a species occurs for a multitude of traits, many of which are inheritable.
- A population produces more offspring in each generation than the environment can support. In any given environment, there is a limited amount of food, water, physical space, and other resources for which individuals must compete.
- Some individuals have favorable traits that enable them to better compete for limited resources. The individuals with favorable traits acquire more resources than the individuals with less favorable traits and can devote more energy to reproduction. Darwin called this ability to have more offspring *differential reproductive success.*
- Natural selection can result in a population adapted to the local environment. An increasing proportion of individuals in each succeeding generation will have the favorable characteristics—characteristics suited to surviving and reproducing in that environment. In this way, evolution brings about adaptation to the environment.

**FIGURE 15.7 Variation in a population.**

For Darwin, variations, such as those seen in a human population, are highly significant and are required for natural selection to result in adaptation to the environment.

Each of these steps leading to adaptation is now examined in more detail.

### *Organisms Have Inheritable Variations*

Darwin emphasized that the members of a population vary in their functional, physical, and behavioral characteristics (Fig. 15.7). Before Darwin, variations were viewed as imperfections that should be ignored since they were not important to the description of a species. Darwin emphasized that variations were essential to the natural selection process. He suspected—but did not have the evidence we have today—that the occurrence of variations is completely random; they arise by accident and for no particular purpose. New variations are just as likely to be harmful as helpful to the organism.

The variations that make adaptation to the environment possible are those that are passed on from generation to generation. The science of genetics was not yet well established, so Darwin was never able to determine the cause of variations or how they are passed on. Today, we realize that genes, along with the environment, determine the phenotype of an organism, and that mutations and recombination of alleles during sexual reproduction can cause new variations to arise.

Natural selection can only operate on variations that are already available in the population's gene pool; it lacks any directedness towards "improvement" or anticipation of future environmental changes—and the environment of living things is constantly changing.

### *Organisms Compete for Resources*

In Darwin's time, a socioeconomist, Thomas Malthus, stressed the reproductive potential of human beings. He proposed that death and famine were inevitable because the human population tends to increase faster than the supply of food. Darwin applied this concept to all organisms and saw that the available resources were not sufficient for all members of a population to survive. He calculated the reproductive potential of elephants, assuming an average life span of 100 years and a breeding span from 30–90 years. Given these assumptions, a single female probably bears no fewer than six young, and if all these young survive and continue to reproduce at the same rate, after only 750 years, the descendants of a single pair of elephants will number about 19 million! Obviously, no environment has the resources to support an elephant population of this magnitude, and no such elephant population has ever existed. This overproduction potential of a species is often referred to as Darwin's geometric ratio of increase.

### *Organisms Differ in Reproductive Success*

**Fitness** is the reproductive success of an individual relative to other members of a population. The most-fit individuals are the ones that capture a disproportionate amount of resources, and convert these resources into a larger number of viable offspring. Since organisms vary anatomically and physiologically and the challenges of local environments vary, what determines fitness varies for different populations. For example, among western diamondback rattlesnakes (*Crotalus atrox*)

**FIGURE 15.8 Artificial selection of animals.**

All dogs, *Canis lupus familiaris*, are descended from the gray wolf, *Canis lupus*, which began to be domesticated about 14,000 years ago. The process of diversification has led to extreme phenotypic differences. Several factors may have contributed: (1) The wolves under domestication were separated from other wolves because human settlements were separate, and (2) humans in each settlement selected for whatever traits appealed to them. Artificial selection of dogs continues even today.

living on lava flows, the most fit are those that are black in color. But among those living on desert soil, the most fit are those with the typical light coloring with brown blotching. Background matching helps an animal both capture prey and avoid being captured; therefore, it is expected to lead to survival and increased reproduction.

In nature, interactions with the environment determine which members of a population reproduce to a greater degree than other members. In contrast to artificial selection, the result of natural selection is not predesired. Natural selection occurs because certain members of a population happen to have a variation that allows them to survive and reproduce to a greater extent than other members. For example, a variation that reduces water loss is beneficial to a desert plant; and one that increases the sense of smell helps a wild dog find its prey. Therefore, we expect organisms with these traits to have increased fitness.

**Artificial Selection.** Darwin noted that when humans help carry out **artificial selection,** the process by which a breeder chooses which traits to perpetuate, they select the animals and plants that will reproduce. For example, prehistoric humans probably noticed desirable variations among wolves and selected particular individuals for breeding. Therefore, the desired traits increased in frequency in the next generation. This same process was repeated many times over. The result today is the existence of many varieties of dogs, all descended from the wolf (Fig. 15.8). To take a modern example, foxes are very shy and normally shun the company of people, but in forty years time, Russian scientists have produced silver foxes that now allow themselves to be petted and even seek attention. They did this by selecting the most docile animals to reproduce. The scientists noted that some physical characteristics changed as well. The legs and tails became shorter, the ears become floppier, and the coat color patterns changed. Artificial selection is only possible because the original population exhibits a range of characteristics, allowing humans to select which traits they prefer to perpetuate. Therefore, several varieties of vegetables can be traced to a single ancestor. Chinese cabbage, brussel sprouts, and kohlrabi are all derived from a single species, *Brassica oleracea* (Fig. 15.9).

**FIGURE 15.9 Artificial selection of plants.**

The vegetables Chinese cabbage, brussel sprouts, and kohlrabi are derived from wild mustard, *Brassica oleracea*. Darwin described artificial selection as a model by which to understand natural selection. With natural selection, however, the environment and not human selection, provides the selective force.

## *Organisms Become Adapted*

An **adaptation** [L. *ad*, toward, and *aptus*, fit, suitable] is a trait that helps an organism be more suited to its environment. Adaptations are especially recognizable when unrelated organisms, living in a particular environment, display similar characteristics. For example, manatees, penguins, and sea turtles all have flippers, which help them move through the water. In Chapter 1, we described other ways in which rockhopper penguins are adapted to their environment. Similarly, it can be noted that a Venus flytrap, a plant that lives in the nitrogen-poor soil of a bog, is able to catch and digest flies because it has leaves adapted to catching them.

Such adaptations to their specific environments result from natural selection. Differential reproduction generation after generation can cause adaptive traits to be increasingly represented in each succeeding generation. There are other processes of evolution aside from natural selection (see pages 286–93), but natural selection is the only process that results in adaptation to the environment.

# *On the Origin of Species* by Darwin

After the HMS *Beagle* returned to England, Darwin waited more than 20 years to publish his ideas. During the intervening years, he used the scientific process to support his hypothesis that today's diverse life-forms are descended from a common ancestor and that natural selection is a mechanism by which species can change and new species can arise. In other words, Darwin hypothesized that new species could arise by gradual changes over time. Thus, when Darwin first published data supporting evolution by natural selection, he called the book *On the Origin of Species.*

Darwin was prompted to publish after receiving a letter from Alfred Russel Wallace in which Wallace also proposed natural selection as a mechanism for evolutionary change. The Science Focus on page 274 tells of this and the many other accomplishments of Wallace. Darwin was stunned to see his own theory being proposed by another and immediately went into print. Many feel that Wallace has not received the credit he deserves and that he should be given equal billing to Darwin for also discovering the mechanism by which evolution occurs. However, Darwin worked for many years to gather data to support his theory of natural selection. He later expanded *On the Origin of Species* and published over fifteen additional treatises with examples over the next two decades. Natural selection is so well supported that today we speak of the theory of evolution by natural selection.

# science focus

## Alfred Russel Wallace

Alfred Russel Wallace (1823–1913) is best known as the English naturalist who independently and simultaneously proposed natural selection as a mechanism for evolution (Fig. 15A). While working as a schoolteacher in his early twenties, he met the entomologist Henry Walter Bates, who interested him in insects. Together, they took a collecting trip to the Amazon, which lasted for several years. Wallace's knowledge of the world's diversity was further expanded by a tour of the Malay Archipelago from 1854–62. After studying the animals of every important island in the region, he divided the islands into a western group, with organisms like those found in Asia, and an eastern group, with organisms like those of Australia. The sharp line dividing these two island groups within the archipelago is now known as Wallace's Line (Fig. 15B). A narrow, but deep, strait occurs at Wallace's Line. At times during the past 50 million years, sea levels have lowered and land bridges have appeared between some of the islands. The strait, however, has always remained as a way to prevent animal dispersal between the western and eastern groups of islands.

In 1855, Wallace wrote an essay entitled "On the Law Which Has Regulated the Introduction of New Species," which stated "every species has come into existence coincident both in time and space with a preexisting closely allied species." It is clear, then, that by this date Wallace saw that species share a common ancestry and change over time. It was not until 1858, while suffering an attack of malaria, that he concluded changes in species are due to changes in the environment through natural selection. These conclusions were written in an essay to Darwin for comment. Darwin was stunned upon its receipt. Here before him was the hypothesis he had formulated as early as 1844. For 14 years, he collected copious data supporting natural selection as a mechanism for evolutionary change. Although a draft of a book was in hand by 1856, Darwin had never dared to publish it because he feared the criticism it would most likely receive. He told his friend and colleague Charles Lyell that Wallace's ideas were so similar to his own that even Wallace's "terms now stand as heads of my chapters."

Darwin suggested that Wallace's paper be published immediately, even though Darwin himself as yet had nothing in print. However, Lyell and others who knew of Darwin's detailed work substantiating the process of natural selection suggested that a joint paper be read to the Linnean Society. The title of Wallace's section was "On the Tendency of Varieties to Depart Indefinitely from the Original Type." Darwin presented an abstract of a paper he had written in 1844 and an abstract of his book *On the Origin of Species*. On July 1, 1858, two authors announced to the world that species evolve via natural selection. One year later, Darwin overshadowed Wallace, who was still in the field at the time, by publishing his famous book. However, many still referred to Wallace as "England's Greatest Living Naturalist" because of his diverse contributions well past the age of 90.

**FIGURE 15A Alfred Russel Wallace.**

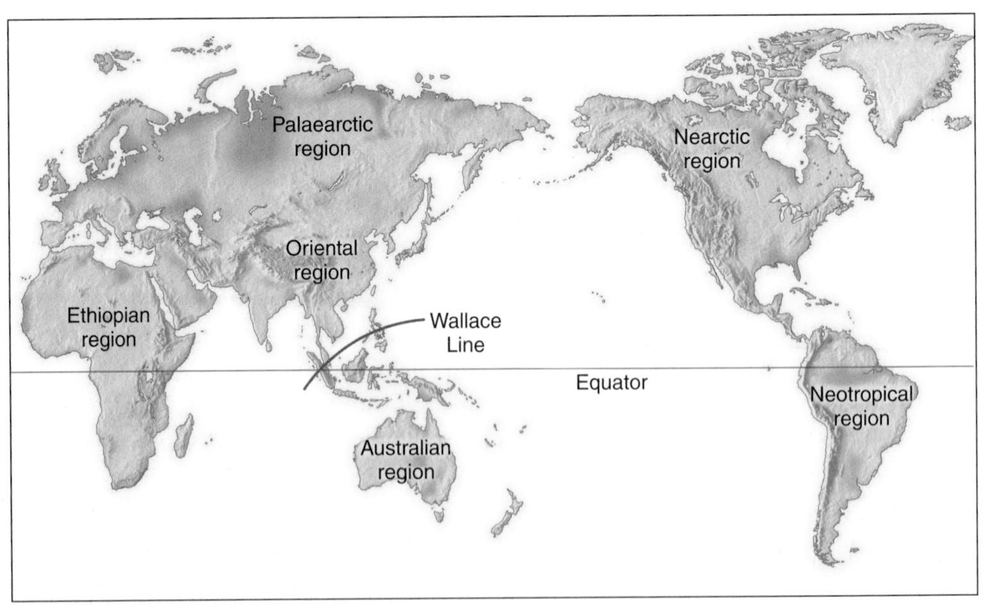


**FIGURE 15B Wallace's Line.**

*Aside from presenting a hypothesis that natural selection explains the origin of new species, Wallace is well known for recognizing the sharp change in animal species inhabiting the islands on either side of a narrow strait bisecting the Malay Archipelago. This deep channel between the Oriental and Australian regions is called Wallace's Line, and serves as an impassable barrier to animal dispersal. By linking geography (the study of maps) to zoology (the study of animals), Wallace is often considered the "Father of Zoogeography."*

a. Large, ground-dwelling finch

b. Warbler finch

c. Cactus-finch

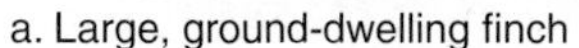

**FIGURE 15.10 Galápagos finches.**

Each of the present-day 13 species of finches has a beak adapted to a particular way of life. For example, (**a**) the heavy beak of the large ground-dwelling finch (*Geospiza magnirostris*) is suited to a diet of large seeds; (**b**) the beak of the warbler-finch (*Certhidea olivacea*)is suited to feeding on insects found among ground vegetation or caught in the air; and (**c**) the longer beak, somewhat decurved, and the split tongue of the cactus-finch (*Cactornis scandens*) are suited to probing a cactus for seeds.

## Natural Selection Can Be Witnessed

Darwin had formed his natural selection hypothesis by observing the distribution of tortoises and finches on the Galápagos Islands. Tortoises with domed shells and short necks live on well-watered islands, where grass is available. Those with shells that flare up in front have long necks and are able to feed on cacti. They live on arid islands, where treelike prickly-pear cactus is the main food source. Similarly, the islands are home to many different types of finches. The heavy beak of the large, ground-dwelling finch is suited to a diet of seeds. The thin, sharp beak of a warbler-finch is able to probe vegetation for and spear insects, while the decurved beak of a cactus-finch can find and feed on cactus seeds (Fig. 15.10).

Today, investigators, such as Peter and Rosemary Grant of Princeton University, are actually watching natural selection as it occurs. In 1973, the Grants began a study of the various finches on Daphne Major, near the center of the Galápagos Islands. The weather swung widely back and forth from wet years to dry years, and they found that the beak size of the medium ground finch, *Geospiza fortis*, adapted to each weather swing, generation after generation (Fig. 15.11). These finches like to eat small, tender seeds that require a smaller beak, but when the weather turns dry, they have to eat larger, drier seeds, which are harder to crush. The birds that have a larger beak depth have an advantage and have more offspring. Therefore, among the next generation of *G. fortis* birds, the beak size has more depth than the previous generation.

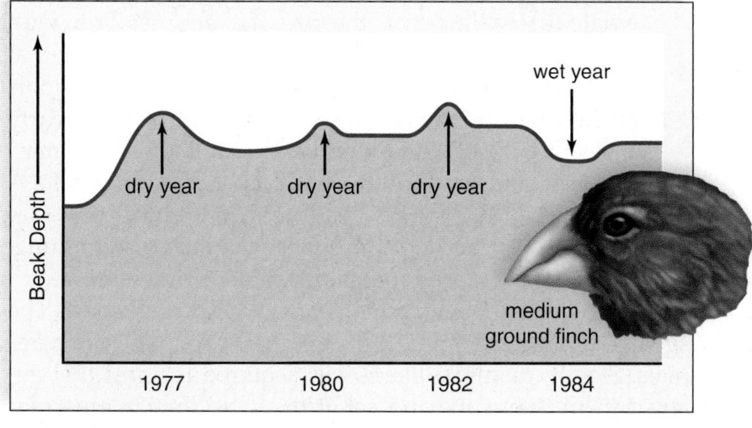


**FIGURE 15.11 Beak depth.**

The beak depth of a ground finch varies from generation to generation, according to the weather.

Among other examples, the shell of the marine snail (*Littorina obtusata*) has changed over time, probably due to being heavily hunted by crabs. Also, the beak length of the scarlet honeycreeper (*Vestiaria coccinea*) was reduced when the bird switched to a new source of nectar because its favorite flowering plants, the lobelloids, were disappearing.

A much-used example of natural selection is industrial melanism. Prior to the Industrial Revolution in Great Britain, light-colored peppered moths, *Biston betularia*, were more common than dark-colored peppered moths. It was estimated that only 10% of the moth population was dark at this time. With the advent of industry and an increase in pollution, the number of dark-colored moths exceeded 80% of the moth population. After legislation to reduce pollution, a dramatic reversal in the ratio of light-colored moths to dark-colored moths occurred. In 1994, one collecting site recorded a drop in the frequency of dark-colored moths to 19%, from a high of 94% in 1960.

The rise in bacterial resistance to antibiotics has occurred within the past 30 years or so. Resistance is an expected way of life now, not only in medicine, but also in agriculture. New chemotherapeutic and HIV drugs are required because of the resistance of cancer cells and HIV, respectively. Also, pesticides and herbicides have created resistant insects and weeds.

### Check Your Progress 15.2

1. What characteristic must a population first have for the process of natural selection to occur?
2. Hypothetically, the members of a rabbit population vary as to length of fur. Some members of the population migrate up a mountain, where it is much colder. After many generations, what would the fur length of rabbits likely be at the bottom and top of the mountain? Explain.

# 15.3 Evidence for Evolution

Many different lines of evidence support the hypothesis that organisms are related through common descent. This is significant, because the more varied the evidence supporting a hypothesis, the more certain it becomes.

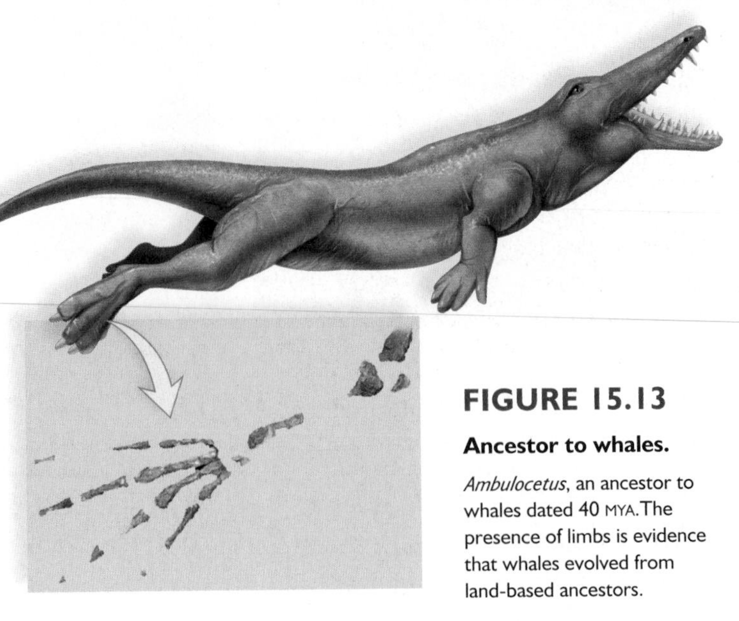

**FIGURE 15.13**

**Ancestor to whales.**

*Ambulocetus*, an ancestor to whales dated 40 MYA. The presence of limbs is evidence that whales evolved from land-based ancestors.

## Fossil Evidence

**Fossils** (L. *fossilis*, dug up) are the remains and traces of past life or any other direct evidence of past life. Traces include trails, footprints, burrows, worm casts, or even preserved droppings. Usually when an organism dies, the soft parts are either consumed by scavengers or decomposed by bacteria. Occasionally, the organism is buried quickly and in such a way that decomposition is never completed or is completed so slowly that the soft parts leave an imprint of their structure. Most fossils, however, consist only of hard parts, such as shells, bones, or teeth, because these are usually not consumed or destroyed.

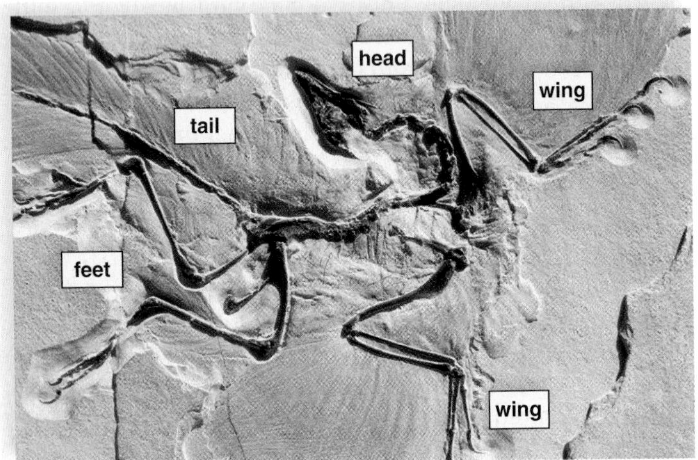


a. *Archaeopteryx* fossil

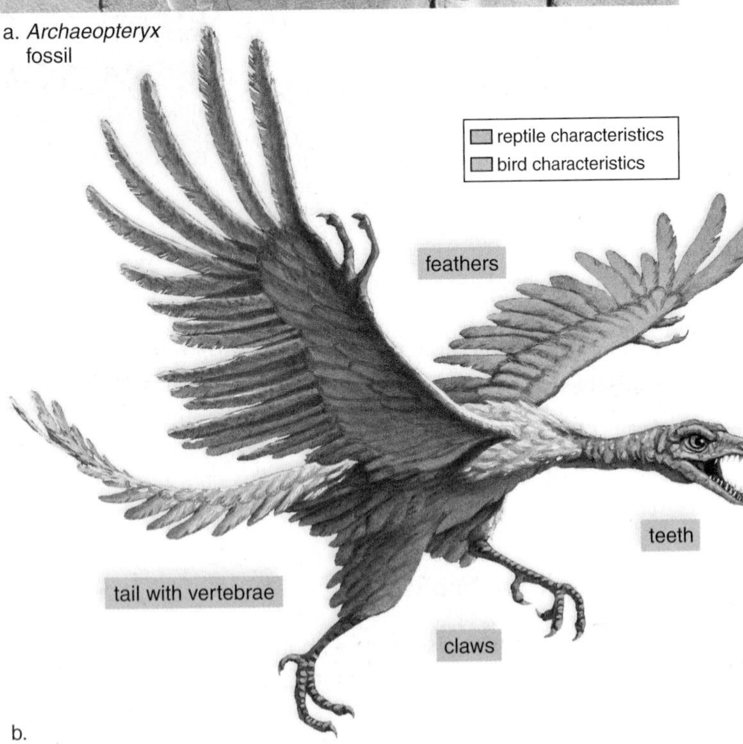


b.

**FIGURE 15.12 Transitional fossils.**

**a.** *Archaeopteryx* was a transitional link between dinosaurs and birds. Fossils indicate it had feathers and wings with claws, and teeth. Most likely, it was a poor flier. **b.** *Archaeopteryx* also had a feather-covered, bony reptilian-type tail that shows up well in this artist's representation. (Orange labels = reptilian characteristics; green label = bird characteristic.)

**Transitional fossils** are a common ancestor for two different groups of organisms, or they are closely related to the common ancestor for these groups. Transitional fossils allow us to trace the descent of organisms. Even in Darwin's day, scientists knew of *Archaeopteryx*, which is an intermediate between dinosaurs and birds (Fig. 15.12). Progressively younger fossils than *Archaeopteryx* have been found: The skeletal remains of *Sinornis* suggest it had wings that could fold against its body like those of modern birds, and its grasping feet had an opposable toe, but it still had a tail. Another fossil, *Confuciusornis*, had the first toothless beak. A third fossil, called *Iberomesornis*, had a breastbone to which powerful flight muscles could attach. Such fossils show how the species of today evolved.

It had always been thought that whales had terrestrial ancestors. Now, fossils have been discovered that support this hypothesis. *Ambulocetus natans* (meaning the walking whale that swims) was the size of a large sea lion, with broad, webbed feet on its forelimbs and hindlimbs that enabled it to both walk and swim. It also had tiny hoofs on its toes and the primitive skull and teeth of early whales (Fig. 15.13). An older fossil, *Pakicetus*, was primarily terrestrial, and yet had the dentition of an early whale. A younger fossil, *Rodhocetus*, had reduced hindlimbs that would have been no help for either walking or swimming, but may have been used for stabilization during mating.

The origin of mammals is also well documented. The synapsids, an early amniote group, gave rise to the premammals. Slowly, mammal-like fossils acquired features that enabled them to breathe and eat at the same time, a muscular diaphragm and rib cage that helped them breathe efficiently, and so forth. The earliest true mammals were shrew-sized creatures that have been unearthed in fossil beds about 200 million years old.

## Biogeographical Evidence

Biogeography is the study of the range and distribution of plants and animals in different places throughout the world. Such distributions are consistent with the hypothesis that, when forms are related, they evolved in one locale and then spread to accessible regions. Therefore, a different mix of plants and animals would be expected whenever geography separates continents, islands, seas, and so on. As previously mentioned, Darwin noted that South America lacked rabbits, even though the environment was quite suitable to them. He concluded there are no rabbits in South America because rabbits evolved somewhere else and had no means of reaching South America.

To take another example, both cacti and spurges (*Euphorbia*) are plants adapted to a hot, dry environment—both are succulent, spiny, flowering plants. Why do cacti grow in the American deserts and most *Euphorbia* grow in African deserts when each would do well on the other continent? It seems obvious that they just happened to evolve on their respective continents.

The islands of the world have many unique species of animals and plants that are found no place else, even when the soil and climate are the same. Why do so many species of finches live on the Galápagos Islands when these same species are not on the mainland? The reasonable explanation is that an ancestral finch originally inhabited the different islands. Geographic isolation allowed the ancestral finch to adapt and evolve into a different species on each island.

Also, in the history of the Earth, South America, Antarctica, and Australia were originally connected (see Fig. 18.15). Marsupials (pouched mammals) had evolved from their egg-laying mammalian ancestors at this time and today are found in both South America and Australia. But when Australia separated and drifted away, the marsupials diversified into many different forms suited to various environments of Australia (Fig. 15.14). They were free to do so because there were few, if any, placental mammals with which to compete in Australia. In South America, where there are placental mammals, marsupials are not as diverse. This supports the hypothesis that evolution is influenced by the mix of plants and animals in a particular continent—that is, by biogeography, not by design.

**FIGURE 15.14 Biogeography.**

Each type of marsupial in Australia is adapted to a different way of life. All of the marsupials in Australia presumably evolved from a common ancestor that entered Australia some 60 million years ago.

Sugar glider, *Petaurus breviceps*, is a tree-dweller and resembles the placental flying squirrel.

The Australian wombat, *Vombatus*, is nocturnal and lives in burrows. It resembles the placental woodchuck.

Kangaroo, *Macropus*, is an herbivore that inhabits plains and forests. It resembles the placental Patagonian cavy of South America.

Tasmanian wolf, *Thylacinus*, now extinct, was a nocturnal carnivore that inhabited deserts and plains. It resembles the placental grey wolf.

The Australian native cat, *Dasyurus*, is a carnivore and inhabits forests. It resembles the placental wild cat.

## Anatomical Evidence

Darwin was able to show that a common descent hypothesis offers a plausible explanation for anatomical similarities among organisms. Vertebrate forelimbs are used for flight (birds and bats), orientation during swimming (whales and seals), running (horses), climbing (arboreal lizards), or swinging from tree branches (monkeys). Yet all vertebrate forelimbs contain the same sets of bones organized in similar ways, despite their dissimilar functions (Fig. 15.15). The most plausible explanation for this unity is that this basic forelimb plan belonged to a common ancestor, and then the plan was modified independently in all of its descendents as each continued along its own evolutionary pathway. Structures that are anatomically similar because they are inherited from a common ancestor are called **homologous structures** [Gk. *homologos,* agreeing, corresponding]. In contrast, **analogous structures** serve the same function, but they are not constructed similarly, nor do they share a common ancestry. The wings of birds and the adaptive streamline shape of a fish and squid are analogous to each other. The presence of homology, not analogy, is evidence that organisms are related.

**Vestigial structures** [L. *vestigium,* trace, footprint] are anatomical features that are fully developed in one group of organisms but are reduced and may have no function in similar groups. Most birds, for example, have well-developed wings used for flight. Some bird species (e.g., ostrich), however, have greatly reduced wings and do not fly. Similarly, snakes and whales, too, have no use for hindlimbs, and yet some have remnants of a pelvic girdle and legs. Humans have a tailbone but no tail. The presence of vestigial structures can be explained by the common descent hypothesis. Vestigial structures occur because organisms inherit their anatomy from their ancestors; they are traces of an organism's evolutionary history.

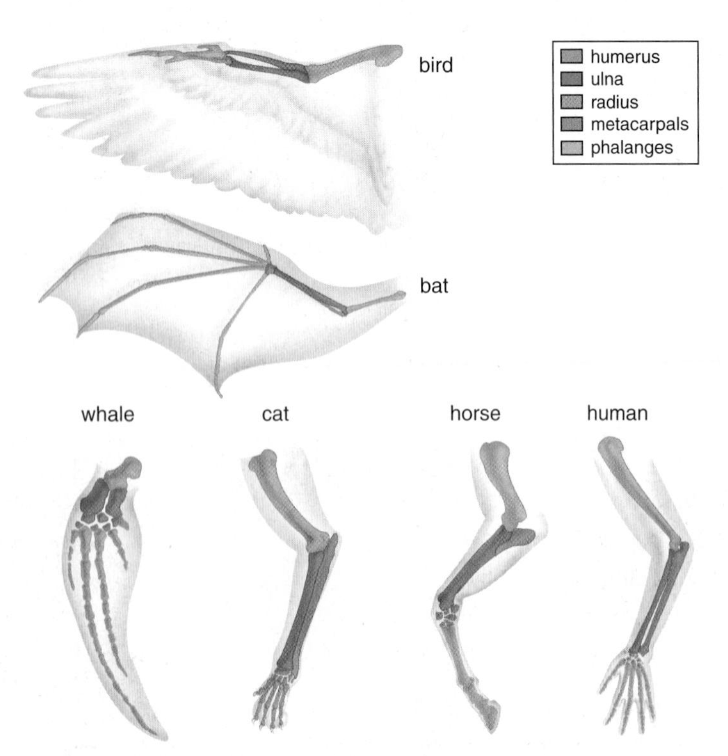


**FIGURE 15.15 Significance of homologous structures.**

Although the specific design details of vertebrate forelimbs are different, the same bones are present (they are color-coded). Homologous structures provide evidence of a common ancestor.

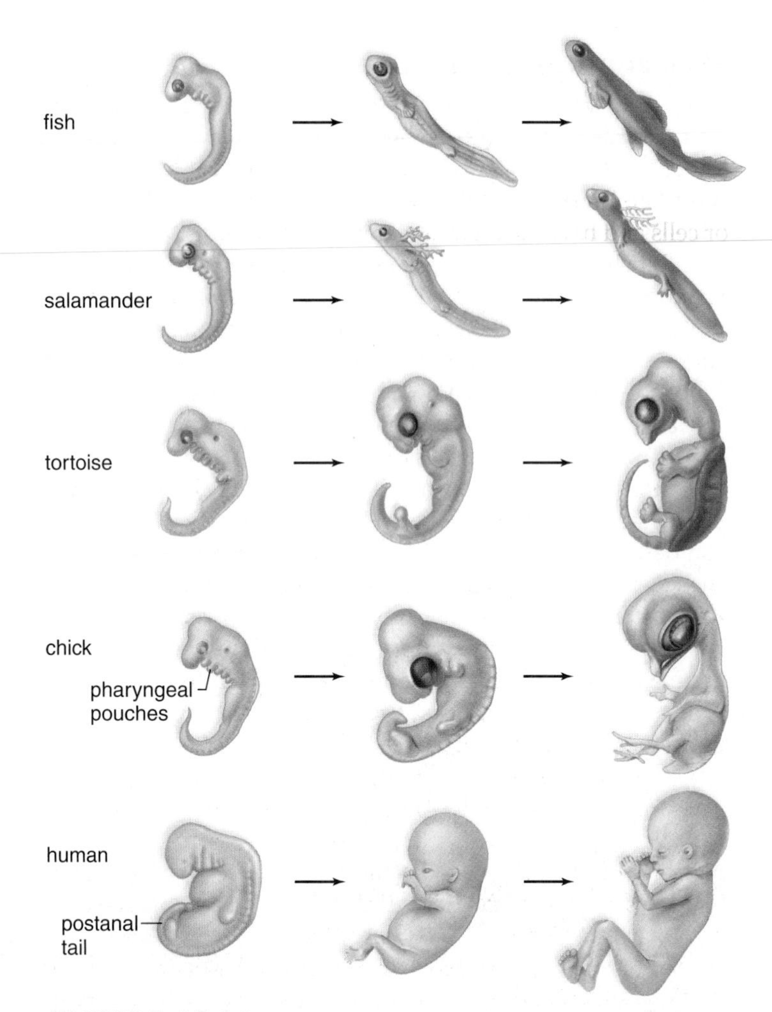


**FIGURE 15.16 Significance of developmental similarities.**

At these comparable developmental stages, vertebrate embryos have many features in common, which suggests they evolved from a common ancestor. (These embryos are not drawn to scale.)

The homology shared by vertebrates extends to their embryological development (Fig. 15.16). At some time during development, all vertebrates have a postanal tail and exhibit paired pharyngeal pouches supported by cartilaginous arches. In fishes and amphibian larvae, these pouches develop into functioning gills. In humans, the first pair of pouches and arches becomes the jawbones, cavity of the middle ear, and the auditory tube. The second pair of pouches becomes the tonsils and facial muscle and nerve, while the third and fourth pairs become the thymus and parathyroid glands. Why should terrestrial vertebrates develop and then modify structures like pharyngeal pouches that have lost their original function? New structures (or structures with novel functions) can only originate by "modifying" the preexisting structures on one's ancestors. All vertebrates inherited the same developmental pattern from their original common ancestor. Each vertebrate group now has a specific set of modifications of this original ancestral pattern.

## Biochemical Evidence

All living organisms use the same basic biochemical molecules, including DNA (deoxyribonucleic acid), RNA (ribonucleic acid), and ATP (adenosine triphosphate). The conclusion is that these molecules were present in the first living cell or cells and have been passed on as life began.

Further, organisms use the same DNA triplet code and the same 20 amino acids in their proteins. Since the sequences of DNA bases in the genomes of many organisms are now known, it has become clear that humans share a large number of genes with much simpler organisms, even prokaryotes. This means that humans and prokaryotes also share many of the same enzymes since genes code for enzymes. Today it is possible to put a human gene in a bacterium, and it will produce a product as if it were in a human cell.

Also of interest, evolutionists who study development have found that many developmental genes are shared in animals ranging from worms to humans. It appears that life's vast diversity has come about by slight differences in genes—perhaps regulatory genes that affect the activity of other genes. The result has been widely divergent types of bodies. For example, a similar gene in arthropods and vertebrates determines the dorsal-ventral axis. Although the base sequences are similar, the genes have differing effects. Therefore, in arthropods, such as fruit flies and crayfish, the nerve cord is ventral, whereas in vertebrates, such as chickens and humans, the nerve cord is dorsal. The nerve cord eventually gives rise to the spinal cord and brain.

When the degree of similarity in DNA base sequences, or the degree of similarity in amino acid sequences, of proteins is examined, the data are as expected, assuming common descent. For example, investigators made a comparative study of the sequence of amino acids in cytochrome *c*, a molecule that is used in the electron transport chain, an important metabolic process. The sequence is expected to stay nearly the same in all organisms, because if it should change greatly, so might the function of cytochrome *c*. Figure 15.17 shows the results of their study. The sequence in a human differs from that in a monkey by only one amino acid, from that in a duck by 11 amino acids, and from that in yeast by 51 amino acids. These data are consistent with other data regarding the anatomical similarities of these organisms and, therefore, their relatedness.

Evolution is no longer considered a hypothesis. It is the great unifying theory of biology. In science, the word *theory* is reserved for those conceptual schemes that are supported by a large number of observations and have not yet been found lacking. The theory of evolution has the same status in biology that the germ theory of disease has in medicine.

### Check Your Progress 15.3

1. Why is fossil evidence the best evidence for evolution? Of what significance is it that fossils can be dated?
2. Why are there so many marsupials in Australia but not in South America?
3. If American cacti and African spurges were closely related, could they be used to show that biogeography supports evolution?

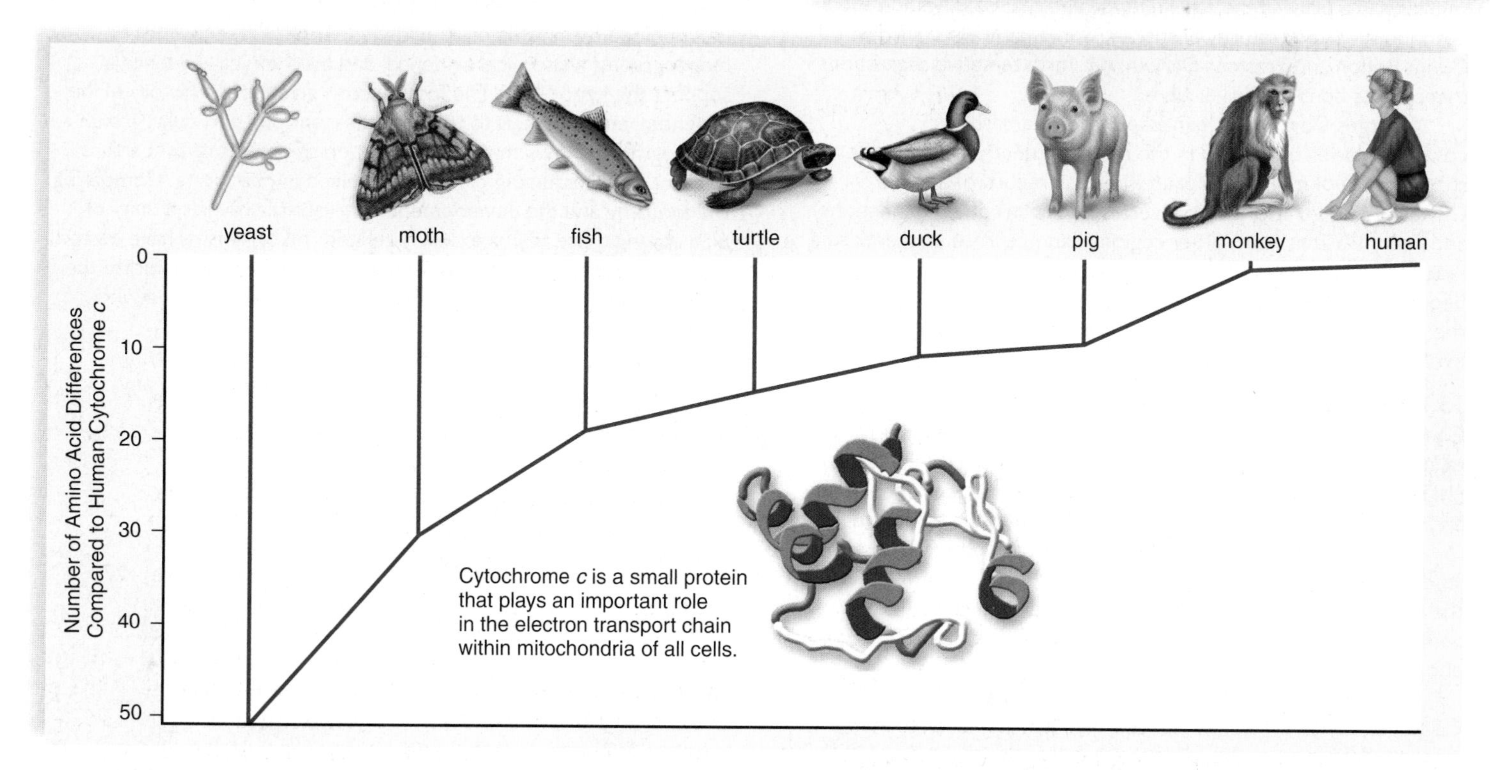


**FIGURE 15.17 Significance of biochemical differences.**

The branch points in this diagram indicate the number of amino acids that differ between human cytochrome *c* and the organisms depicted. These biochemical data are consistent with those provided by a study of the fossil record and comparative anatomy.

# Connecting the Concepts

Before the 1800s, most people believed that each species had been created at the beginning of the world, and that modern organisms were essentially unchanged descendants of their ancestors.

At the time Charles Darwin boarded the HMS *Beagle*, he had studied the writings of his grandfather Erasmus Darwin, James Hutton, Charles Lyell, Jean-Baptiste de Lamarck, Thomas Malthus, Carolus Linnaeus, and other original thinkers. He had ample time during his five-year voyage to reflect on the ideas of these authors, and from them collectively, he built a framework that helped support his theory of descent with modification.

One aspect of scientific genius is the power of astute observation—to see what others miss or fail to appreciate. In this area, Darwin excelled. By the time he reached the Galápagos Islands, Darwin had already begun to hypothesize that species could be modified according to the environment. In other words, like other scientists, Darwin used a testable hypothesis to explain his observations. His observation of the finches on the isolated Galápagos Islands supported his hypothesis. He concluded that the finches on each island varied from one another and from mainland finches because each species had become adapted to a different habitat; therefore, one species had given rise to many. The fact that Alfred Russel Wallace simultaneously proposed natural selection as an evolutionary mechanism suggests that the scientific community was ready for a new conceptual understanding of life's diversity on Earth.

The theory of evolution has quite rightly been called the grand unifying theory (GUT) of biology. Fossils, comparative anatomy and development, biogeography, and biochemical data all indicate that living things share common ancestors. Evolutionary principles help us understand why organisms are both different and alike, and why some species flourish and others die out. As the Earth's habitats change over millions of years, those individuals with the traits best adapted to new environments survive and reproduce; thus, populations change over time.

## summary

### 15.1 History of Evolutionary Thought

In general, the pre-Darwinian worldview was different from the post-Darwinian worldview. The scientific community, however, was ready for a new worldview, and it received widespread acceptance.

A century before Darwin's trip, the classification of organisms had been a main concern of biology. Linnaeus thought that each species had a place in the *scala naturae* and that classification should describe the fixed features of species. Some naturalists, such as Count Buffon and Erasmus Darwin, put forth tentative suggestions that species do change over time.

Georges Cuvier and Jean-Baptiste de Lamarck, contemporaries of Darwin in the late-eighteenth century, differed sharply on evolution. To explain the fossil record of a region, Cuvier proposed that a whole series of catastrophes (extinctions) and repopulations from other regions had occurred. Lamarck said that descent with modification does occur and that organisms do become adapted to their environments; however, he suggested the inheritance of acquired characteristics as a mechanism for evolutionary change.

### 15.2 Darwin's Theory of Evolution

Charles Darwin formulated hypotheses concerning evolution after taking a trip around the world as a naturalist aboard the HMS *Beagle* (1831–36). His hypotheses were that common descent does occur and that natural selection results in adaptation to the environment.

Darwin's trip involved two primary types of observations. His study of geology and fossils caused him to concur with Lyell that the observed massive geological changes were caused by slow, continuous changes. Therefore, he concluded that the Earth is old enough for descent with modification to occur.

Darwin's study of biogeography, including the animals of the Galápagos Islands, allowed him to conclude that adaptation to the environment can cause diversification, including the origin of new species.

Natural selection is the mechanism Darwin proposed for how adaptation comes about. Members of a population exhibit random, but inherited, variations. (In contrast to the previous worldview, variations are highly significant.) Relying on Malthus's ideas regarding overpopulation, Darwin stressed that there was a struggle for existence. The most-fit organisms are those possessing characteristics that allow them to acquire more resources and to survive and reproduce more than the less fit. In this way, natural selection can result in adaptation to an environment.

### 15.3 Evidence for Evolution

The hypothesis that organisms share a common descent is supported by many lines of evidence. The fossil record, biogeography, anatomical evidence, and biochemical evidence all support the hypothesis. The fossil record gives us the history of life in general and allows us to trace the descent of a particular group. Biogeography shows that the distribution of organisms on Earth is explainable by assuming organisms evolved in one locale. Comparing the anatomy and the development of organisms reveals a unity of plan among those that are closely related. All organisms have certain biochemical molecules in common, and any differences indicate the degree of relatedness. A hypothesis is greatly strengthened when many different lines of evidence support it.

Today, the theory of evolution is one of the great unifying theories of biology because it has been supported by so many different lines of evidence.

## understanding the terms

Match the terms to these definitions:

a. _____________ Study of the global distribution of organisms.
b. _____________ Study of fossils that results in knowledge about the history of life.
c. _____________ Poorly developed body part that was complete and functional in an ancestor but is no longer functional in a descendant.
d. _____________ Organism's modification in structure, function, or behavior suitable to the environment.
e. _____________ Lamarckian explanation that organisms become adapted to their environment during their lifetime and pass on these adaptations to their offspring.

## reviewing this chapter

1. In general, contrast the pre-Darwinian worldview with the post-Darwinian worldview. 266–68
2. Cite naturalists who made contributions to biology in the mid-eighteenth century, and state their understandings about evolutionary descent. 267
3. How did Cuvier explain the succession of life-forms in the Earth's strata? 268
4. What is meant by the inheritance of acquired characteristics, a hypothesis that Lamarck used to explain adaptation to the environment? 267–68
5. What were Darwin's views on geology, and what observations did he make regarding geology? 269
6. What observations did Darwin make regarding biogeography? How did these influence his conclusions about the origin of new species? 269–70
7. What are the steps of the natural selection process as proposed by Darwin? 271–73
8. Distinguish between the concepts of fitness and adaptation to the environment. 271–73
9. How do transitional fossils support common descent with modifications? Give an example. 276
10. How does biogeography support the concept of common descent? Explain why a diverse assemblage of marsupials evolved in Australia. 277
11. How does anatomical evidence support the concept of common descent? Explain why vertebrate forelimbs are similar despite different functions. 278
12. How does biochemical evidence support the concept of common descent? Explain why the sequence of amino acids in cytochrome *c* differs between two organisms. 279

## testing yourself

Choose the best answer for each question.

1. Which of these pairs is mismatched?
   a. Charles Darwin—natural selection
   b. Linnaeus—classified organisms according to the *scala naturae*
   c. Cuvier—series of catastrophes explains the fossil record
   d. Lamarck—uniformitarianism
   e. All of these are correct.
2. According to the theory of inheritance of acquired characteristics,
   a. if a man loses his hand, then his children will also be missing a hand.
   b. changes in phenotype are passed on by way of the genotype to the next generation.
   c. organisms are able to bring about a change in their phenotype.
   d. evolution is striving toward improving particular traits.
   e. All of these are correct.
3. Why was it helpful to Darwin to learn that Lyell thought the Earth was very old?
   a. An old Earth has more fossils than a new Earth.
   b. It meant there was enough time for evolution to have occurred slowly.
   c. There was enough time for the same species to spread out into all continents.
   d. Darwin said that artificial selection occurs slowly.
   e. All of these are correct.
4. All the finches on the Galápagos Islands
   a. are unrelated but descended from a common ancestor.
   b. are descended from a common ancestor, and therefore related.
   c. rarely compete for the same food source.
   d. Both a and c are correct.
   e. Both b and c are correct.
5. Organisms
   a. compete with other members of their species.
   b. differ in fitness.
   c. are adapted to their environment.
   d. are related by descent from common ancestors.
   e. All of these are correct.
6. DNA nucleotide similarities between organisms
   a. indicate the degree of relatedness among organisms.
   b. may reflect phenotypic (morphological) similarities.
   c. explain why there are phenotypic similarities.
   d. are to be expected if the organisms are related due to common ancestry.
   e. All of these are correct.
7. If evolution occurs, we would expect different biogeographical regions with similar environments to
   a. all contain the same mix of plants and animals.
   b. each have its own specific mixes of plants and animals.
   c. have plants and animals with similar adaptations.
   d. have plants and animals with different adaptations.
   e. Both b and c are correct.
8. The fossil record offers direct evidence for common descent because you can
   a. see that the types of fossils change over time.
   b. sometimes find evidence of the transitional link between lineages.
   c. sometimes trace the ancestry of a particular group.
   d. trace the biological history of living things.
   e. All of these are correct.
9. Organisms such as birds and insects that are adapted to an aerial way of life
   a. will probably have homologous structures.
   b. will have similar adaptations.
   c. may very well have analogous structures.
   d. will have the same degree of fitness.
   e. Both b and c are correct.

For questions 10–17, match the evolutionary evidence in the key to the description. Choose more than one answer if correct.

**KEY:**

a. biogeographical evidence
b. fossil evidence
c. biochemical evidence
d. anatomical evidence
e. developmental evidence

10. It's possible to trace the evolutionary ancestry of a species.
11. Rabbits are not found in Patagonia.
12. A group of related species have homologous structures.
13. The same types of molecules are found in all living things.
14. Islands have many unique species not found elsewhere.
15. All vertebrate embryos have pharyngeal pouches.
16. More distantly related species have more amino acid differences in cytochrome *c*.
17. Transitional fossils have been found between some major groups of organisms.
18. Which of these is/are necessary to natural selection?
    a. variations
    b. differential reproduction
    c. an environmental catastrophe
    d. Both a and b are correct.
    e. All of these are correct.
19. Which of these is explained incorrectly?
    a. Organisms have variations—mutations and recombination of alleles occur.
    b. Organisms struggle to exist—the environment will support only so many of the same type of species.
    c. Organisms differ in fitness—adaptations enable some members of a species to reproduce more than other members.
    d. Individuals become adapted—individuals ever improve because of differential fitness.
    e. All of these are correct.

For questions 20–24, offer an explanation for each of these observations based on information in the section indicated. Write out your answer.

20. Transitional fossils serve as links between groups of organisms. See Fossil Evidence (page 276).
21. Cacti and spurges (*Euphorbia*) exist on different continents, but both have spiny, water-storing stems. See Biogeographical Evidence (page 277).
22. The forelimbs of vertebrates possess homologous structures (i.e., bones in the forelimbs). See Anatomical Evidence (page 278).
23. Amphibians, reptiles, birds, and mammals all have pharyngeal pouches at some time during development. See Anatomical Evidence (page 278).
24. The base sequence of DNA differs among species but yet maintains some degree of similarity. See Biochemical Evidence (page 279).

## thinking scientifically

1. Mutations occur at random and increase the variation within a population for no particular purpose. Our immune system is capable of detecting and killing certain viruses. Would a virus that has a frequent rate of mutation be less or more able to avoid the immune system? Explain.
2. A cotton farmer applies a new insecticide against the boll weevil to his crop for several years. At first, the treatment was successful, but then the insecticide became ineffective and the boll weevil rebounded. Did evolution occur? Explain.

## bioethical issue

### Theory of Evolution

People are often confused by the terminology "theory of evolution." They believe that the word "theory" is being used in an everyday sense, such as "I have a theory about the win-loss record of the Boston Red Sox." But after studying this text, you realize that the word "theory" in science refers to a major scientific concept that has been so supported by observation and experiments that it is widely accepted by scientists as explaining many phenomena in the natural word. The theory of evolution is often referred to as the unifying theory of biology because it explains so many different aspects of living things.

Laypeople, in particular, who may misunderstand the term "theory," have been known to suggest that other theories, aside from the theory of evolution, should be taught in school to explain the diversity of life. Do you think that the curriculum of a science course should be restricted to content that is traditionally considered scientific, or do you believe that other "theories" that may not have been tested by experimentation should also be presented in a science course? If so, how and by whom? If not, why not?

## *Biology* website

The companion website for *Biology* provides a wealth of information organized and integrated by chapter. You will find practice tests, animations, videos, and much more that will complement your learning and understanding of general biology.

**http://www.mhhe.com/maderbiology10**

# 16

# How Populations Evolve

When your grandparents were young, infectious diseases, such as tuberculosis, pneumonia, and syphilis, killed thousands of people every year. Then in the 1940s, penicillin and other antibiotics were developed, and public health officials thought infectious diseases were a thing of the past. Today, however, many infections are back with a vengeance. Why? Because natural selection occurred. As with Staphylococcus aureus, a few bacteria were resistant to penicillin. Therefore, they were selected over and over again to reproduce, until the entire population of bacteria became resistant to penicillin. A new antibiotic called methicillin became available in 1959 to treat penicillin-resistant bacterial strains, but by 1997, 40% of hospital staph infections were caused by methicillin-resistant Staphylococcus aureus, or MRSA. Now, community-acquired MRSA (CA-MRSA) can spread freely through the general populace, particularly when people are in close contact.

*This chapter gives the principles of evolution a genetic basis and shows how it is possible to genetically recognize when a population has undergone evolutionary changes. Evolutionary changes observed at the population level are termed microevolution.*

MRSA can spread between members of a human social group.

## concepts

# 16.1 Population Genetics

Darwin stressed that diversity exists among the members of a population. A **population** is all the members of a single species occupying a particular area at the same time. **Population genetics,** as its name implies, studies this diversity in terms of allele differences. Since Darwin was unaware of Mendel's work, he never had the opportunity to study the genetics of a population, as we will do in this chapter.

## Genetic Diversity

When we consider that a population can have many subpopulations, such as those illustrated for a human population in Figure 16.1, we begin to realize that a population can have many phenotypic, and therefore genotypic, differences. Many traits in a population are controlled by polygenes, and if these multiple gene loci were to have multiple alleles, genetic diversity becomes plentiful, indeed.

Studies have been done to determine enzyme variations among members of a population. Extracted enzymes are subjected to electrophoresis, a process that separates proteins according to size and shape. The result in *Drosophila* suggests that a fly population has multiple alleles at no less than 30% of its gene loci. Similar results are the rule in all populations.

Increasingly today, instead of studying proteins, investigators go right to sequencing DNA to discover the amount of genetic diversity in a population, as in copy number variation s (see p. 260). This has also allowed them to discover various loci that exhibit **single nucleotide polymorphisms,** or **SNPs** (pronounced snips). These are DNA sequences in a species' genome that differ by a single nucleotide. To take an example of an SNP, compare ACGTACGTA to ACGTACCTA and notice that there is only a single base difference between the two sequences. Investigators would say that the SNP has two alleles, in this case, G and C. SNPs generally have two alleles.

SNPs that occur within a protein-coding DNA sequence can result in a change sequence of amino acids, but not necessarily, due to redundancy of the genetic code (see page 221). SNPs that do not result in a changed amino acid sequence may still cause regulatory differences. Therefore, SNPs are now thought to be an important source of genetic diversity in the populations of all species, including humans (Fig. 16.1).

Another interesting finding is that humans inherit patterns of base-pair differences now called haplotypes (from the terms haploid and genotype). To take an example, if a chromosome has a G rather than a C at a particular location, this change is most likely accompanied by other particular base differences near the G. Researchers are in the process of discovering the most common haplotypes among African, Asian, and European populations. They want to link haplotypes to the risk of specific illnesses, in the hope it will lead to new methods of preventing, diagnosing, and treating disease. Also, certain haplotypes may respond better than others to particular medicines, vaccines, and other treatment strategies.

**FIGURE 16.1 The HapMap project.**

The HapMap project compares DNA base-pair sequences among African, Asian, and European populations to discover unique base pair differences.

## Microevolution

It wasn't until the 1930s that population geneticists worked out a way to describe the diversity in a population in terms of alleles, and to, thereby, develop a way to recognize when evolution had occurred. **Microevolution** pertains to evolutionary changes within a population.

In population genetics, the various alleles at all the gene loci in all individuals make up the **gene pool** of the population. It is customary to describe the gene pool of a population in terms of genotype and allele frequencies. Let's take an example based on the peppered moths we discussed in Chapter 15 (page 275). Suppose you research the literature and find that the color of peppered moths is controlled by a single set of alleles, and you decide to use the following key:

$D$ = dark color

$d$ = light color

Further, in one Great Britain population before pollution fully darkened the trees, only 4% (0.04) of moths were homozygous dominant; 32% (0.32) were heterozygous, and 64% (0.64) were homozygous recessive. From these genotype frequencies, you can calculate the allele and gamete frequencies in the populations:

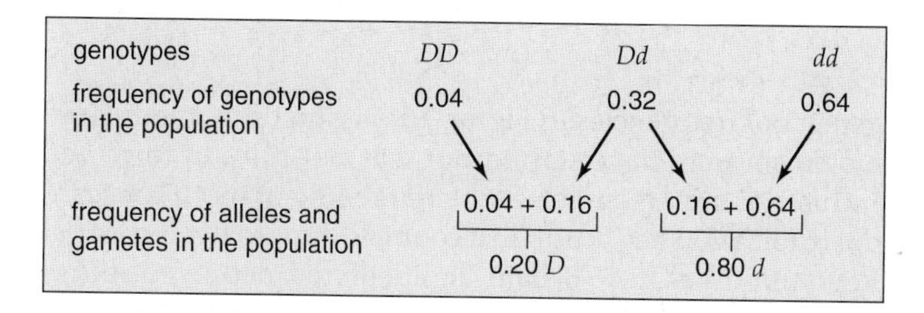


The frequency of the gametes (sperm and egg) produced by this population will necessarily be the same as the allele frequencies. Assuming random mating (all possible gametes have an equal chance to combine with any other), we can use these gamete frequencies to calculate the ratio of genotypes in the next generation by using a Punnett square (Fig. 16.2).

There is an important difference between a Punnett square used for a cross between individuals, as we have done previously, and the one shown in Figure 16.2. In Figure 16.2, we are using the gamete frequencies in the population to determine the genotype frequencies in the next generation. As you can see, the genotype frequencies (and therefore the allele frequencies) in the next generation are the same as they were in the previous generation. In other words, we will find that the homozygous dominant moths are still 0.04; the heterozygous moths are still 0.32; and the homozygous recessive moths are still 0.64. This is an amazing finding, and it tells us that: *Sexual reproduction alone cannot bring about a change in genotype and allele frequencies of a population.* By the way, what percentage of moths are dark-colored, and what percentage of moths are light-colored? Adding the homozygous dominant and the heterozygous moths = 36% are dark-colored, and 64% are light-colored.

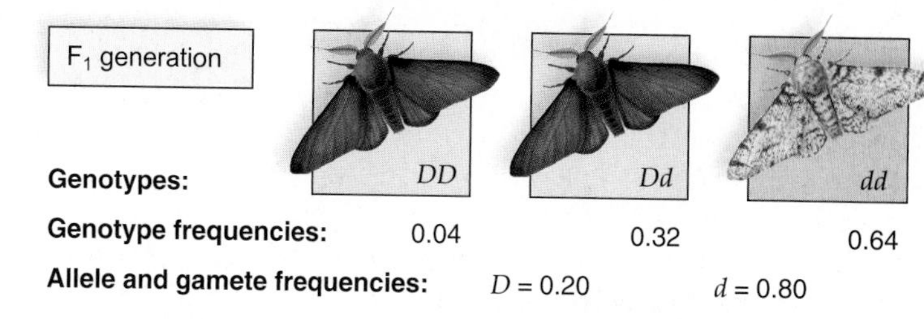


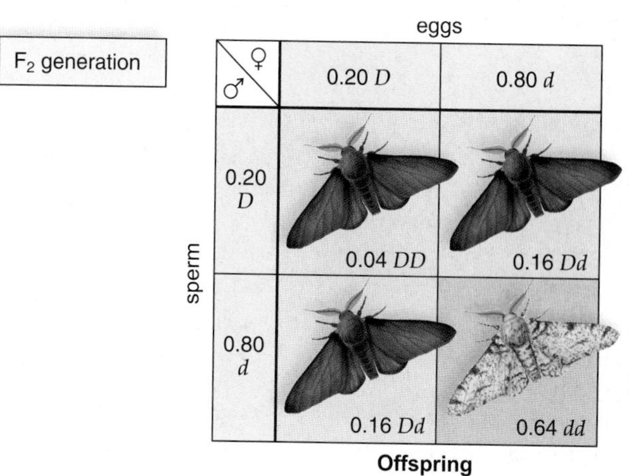


**Genotype frequencies:** $0.04\ DD + 0.32\ Dd + 0.64\ dd = 1$

$p^2 + 2pq + q^2 = 1$

$p^2$ = frequency of $DD$ genotype (dark-colored) = $(0.20)^2$ = 0.04

$2pq$ = frequency of $Dd$ genotype (dark-colored) = $2(0.20)(0.80)$ = 0.32

$q^2$ = frequency of $dd$ genotype (light-colored) = $(0.80)^2$ = 0.64

1.00

**FIGURE 16.2 Hardy-Weinberg equilibrium.**

Using the gamete frequencies in a population, it is possible to use a Punnett square to calculate the genotype frequencies of the next generation. When this is done, it can be shown that sexual reproduction alone does not alter the Hardy-Weinberg equilibrium: the genotype, and therefore allele frequencies, remain the same. Notice the binomial expression is used to calculate the genotype frequencies of a population.

Also, the dominant allele need not increase from one generation to the next. Dominance does not cause an allele to become a common allele. The potential constancy, or equilibrium state, of gene pool frequencies was independently recognized in 1908 by G. H. Hardy, an English mathematician, and W. Weinberg, a German physician. They used the binomial expression to calculate the genotype and allele frequencies of a population, as illustrated beneath the Punnett square in Figure 16.2.

In the Hardy-Weinberg equation: $(p^2 + 2pq + q^2)$

$p^2$ = frequency of the homozygous dominant

$p$ = frequency of the dominant allele

$2pq$ = frequency of the heterozygous genotype

$q^2$ = frequency of the homozygous recessive

$q$ = frequency of the recessive allele

**FIGURE 16.3 Microevolution.**

Microevolution has occurred when there is a change in gene pool frequencies—in this case, due to natural selection. (*Left*) Light-colored moths are more frequent in the population because birds that eat moths are less likely to see light-colored peppered moths against light vegetation. (*Right*) Dark-colored moths are more frequent in the population because birds are less likely to see dark-colored moths against dark vegetation..

The **Hardy-Weinberg principle** states that an equilibrium of gene pool frequencies, calculated by using the binomial expression, will remain in effect in each succeeding generation of a sexually reproducing population, as long as five conditions are met:

1. No mutations: Allele changes do not occur, or changes in one direction are balanced by changes in the opposite direction.
2. No gene flow: Migration of alleles into or out of the population does not occur.
3. Random mating: Individuals pair by chance, not according to their genotypes or phenotypes.
4. No genetic drift: The population is very large, and changes in allele frequencies due to chance alone are insignificant.
5. No selection: Selective forces do not favor one genotype over another.

In real life, these conditions are rarely, if ever, met, and allele frequencies in the gene pool of a population do change from one generation to the next. Therefore, evolution has occurred. The significance of the Hardy-Weinberg principle is that it tells us what factors cause evolution—those that violate the conditions listed. Microevolution can be detected by noting any deviation from a Hardy-Weinberg equilibrium of allele frequencies in the gene pool of a population.

A change of allele frequencies is expected to result in a change of phenotype frequencies. Our calculation of gene pool frequencies in Figure 16.3 assumes that industrial melanism may have started but was not fully in force yet. **Industrial melanism** refers to a darkening of moths once industrialization has begun in a country. Prior to the Industrial Revolution in Great Britain, light-colored peppered moths living on the light-colored, unpolluted vegetation, were more common than dark-colored peppered moths. When dark-colored moths landed on light vegetation, they were seen and eaten by predators. In Figure 16.3, *left*, we suppose that only 36% of the population were dark-colored, while 64% were light-colored. With the advent of industry and an increase in pollution, the vegetation was stained darker. Now, light-colored moths were easy prey for birds that eat moths. Figure 16.3, *right*, shows that the gene pool frequencies switched, and now the dark-colored moths are 64% of the population. Can you calculate the change in gene pool frequencies using Figure 16.2 as a guide?

Just before the Clean Air legislation in the mid-1950s, the numbers of dark-colored moths exceeded a frequency of 80% in some populations. After the legislation, a dramatic reversal in the ratio of light-colored moths to dark-colored moths occurred once again as light-colored moths became more and more frequent. Aside from showing that natural selection can occur within a short period of time, our example shows that a change in gene pool frequencies does occur as microevolution occurs. Recall that microevolution occurs below the species level.

## Causes of Microevolution

The list of conditions for a Hardy-Weinberg equilibrium implies that the opposite conditions can cause evolutionary change. The conditions are mutation, nonrandom mating, gene flow, genetic drift, and natural selection. Only natural selection results in adaptation to the environment.

### *Mutations*

The Hardy-Weinberg principle recognizes new mutations as a force that can cause the allele frequencies to change in a gene pool and cause microevolution to occur. **Mutations,** which are permanent genetic changes, are the raw material for evolutionary change because without mutations, there could be no inheritable phenotypic diversity among members of a population. The rate of mutations is generally very low—on the order of one per 100,000 cell divisions. Also, it is important to realize that evolution is not directed, meaning that no mutation arises because the organism "needs" one. For example, the mutation that causes bacteria to be resistant was already present before antibiotics appeared in the environment.

Mutations are the primary source of genetic differences among prokaryotes that reproduce asexually. Generation time is so short that many mutations can occur quickly, even though the rate is low, and since these organisms are haploid, any mutation that results in a phenotypic change is immediately tested by the environment. In diploid organisms, a recessive mutation can remain hidden and become significant only when a homozygous recessive genotype arises. The importance of recessive alleles increases if the environment is changing; it's possible that the homozygous recessive genotype could be helpful in a new environment, if not the present one. It's even possible that natural selection will maintain a recessive allele if the heterozygote has advantages. As noted on page 284, investigators now know that even SNPs can be a significant source of diversity in a population.

In sexually reproducing organisms, the process of reproduction, consisting of meiosis and fertilization, is just as important as mutation in generating phenotypic differences, because the process can bring together a new and different combination of alleles. This new combination might produce a more successful phenotype. Success, of course, is judged by the environment and evaluated by the relative number of healthy offspring an organism produces.

### *Nonrandom Mating and Gene Flow*

Random mating occurs when individuals pair by chance. You make sure random mating occurs when you do a genetic cross on paper or in the lab, and cross all possible types of sperm with all possible types of eggs. **Nonrandom mating** occurs when certain genotypes or phenotypes mate with one another. **Assortative mating** is a type of nonrandom mating that occurs when individuals tend to mate with those having the *same* phenotype with respect to a certain characteristic. For example, flowers such as the garden pea usually self-pollinate—therefore, the same phenotype has mated with the same phenotype (Fig. 16.4). Assortative mating can also be observed in human society. Men and women tend to marry individuals with characteristics such as intelligence and height that are similar to their own. Assortative mating causes homozygotes for certain gene loci to increase in frequency and heterozygotes for these loci to decrease in frequency.

**Gene flow,** also called gene migration, is the movement of alleles between populations. When animals move between populations, or when pollen is distributed between species (Fig. 16.5), gene flow has occurred. When gene flow brings a new or rare allele into the population, the allele frequency in the next generation changes. When gene flow between adjacent populations is constant, allele frequencies continue to change until an equilibrium is reached. Therefore, continued gene flow tends to make the gene pools similar and reduce the possibility of allele frequency differences between populations.

### *Genetic Drift*

**Genetic drift** refers to changes in the allele frequencies of a gene pool due to chance rather than selection by the environment. Therefore, genetic drift does not necessarily result in

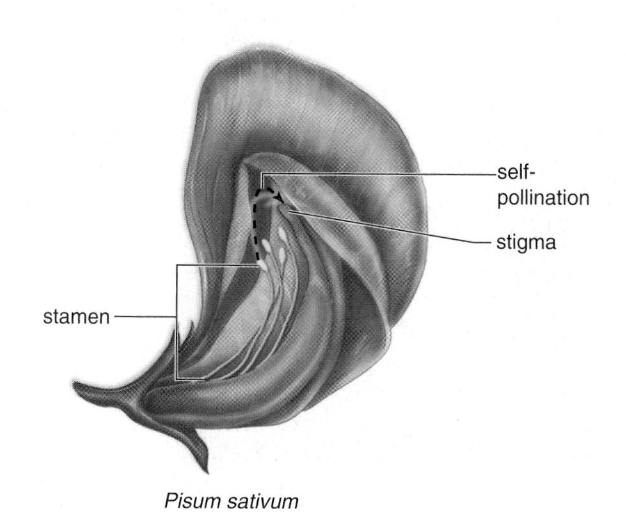


**FIGURE 16.4 Anatomy of the garden pea.**

The anatomy of the garden pea, *Pisum sativum*, ensures self-pollination, an example of nonrandom mating.

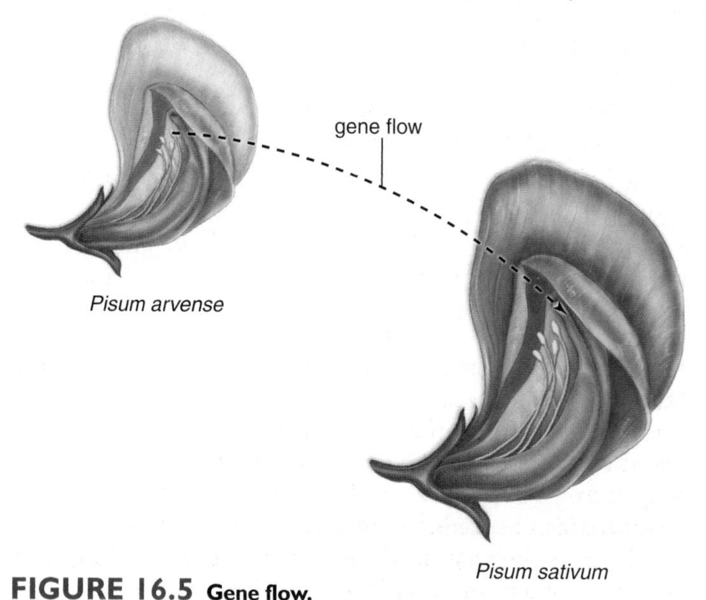


**FIGURE 16.5 Gene flow.**

Occasional cross-pollination between a population of *Pisum sativum* and a population of *Pisum arvense* is an example of gene flow.

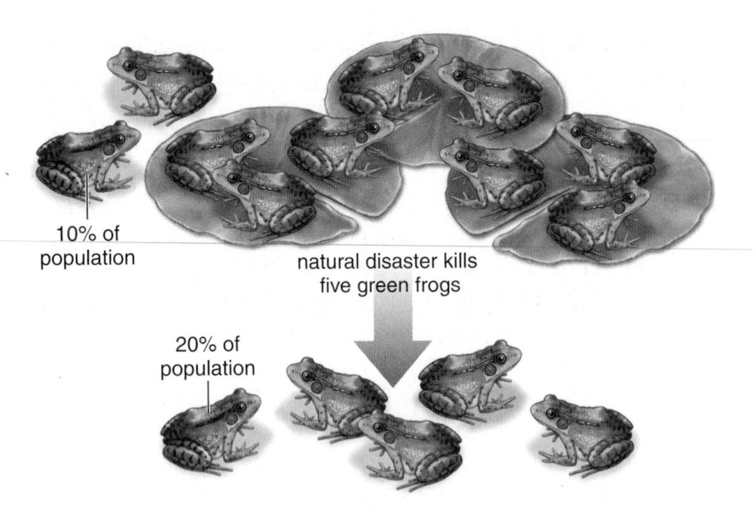


**FIGURE 16.6 Genetic drift.**

Genetic drift occurs when, by chance, only certain members of a population (in this case, green frogs) reproduce and pass on their alleles to the next generation. A natural disaster can cause the allele frequencies of the next generation's gene pool to be markedly different from those of the previous generation.

adaptation to the environment, as does natural selection. For example, in California, there are a number of cypress groves, each a separate population. The phenotypes within each grove are more similar to one another than they are to the phenotypes in the other groves. Some groves have conical-shaped trees, and others have pyramidally-shaped trees. The bark is rough in some colonies and smooth in others. The leaves are gray to bright green or bluish, and the cones are small or large. The environmental conditions are similar for all the groves, and no correlation has been found between phenotype and the environment across groves. Therefore, scientists hypothesize that diversity among the groves are due to genetic drift.

Although genetic drift occurs in populations of all sizes, a smaller population is more likely to show the effects of drift. Suppose the allele *B* (for brown) occurs in 10% of the members in a population of frogs. In a population of 50,000 frogs, 5,000 will have the allele *B*. If a hurricane kills off half the frogs, the frequency of allele *B* may very well remain the same among the survivors. On the other hand, 10% of a population with ten frogs means that only one frog has the allele *B*. Under these circumstances, a natural disaster could very well do away with that one frog, should half the population perish. Or, let's suppose that five green frogs only out of a ten-member population die. Now, the frequency of allele *B* will increase from 10% to 20% (Fig. 16.6).

**Bottleneck and Founder Effects.** When a species is subjected to near extinction because of a natural disaster (e.g., hurricane, earthquake, or fire) or because of overhunting, overharvesting, and habitat loss, it is as if most of the population has stayed behind and only a few survivors have passed through the neck of a bottle. This so-called **bottleneck effect** prevents the majority of genotypes from participating in the production of the next generation. The extreme genetic similarity found in cheetahs is believed to be due to a bottleneck effect. In a study of 47 different enzymes, each of which can come in several different forms, the sequence of amino acids in the enzymes was exactly the same in all the cheetahs. What caused the cheetah bottleneck is not known, but today they suffer from relative infertility because of the intense inbreeding that occurred after the bottleneck. Even if humans were to intervene and the population were to increase in size, without genetic variation, the cheetah could still become extinct. Other organisms pushed to the brink of extinction suffer a similar plight as the cheetah.

The **founder effect** is an example of genetic drift in which rare alleles, or combinations of alleles, occur at a higher frequency in a population isolated from the general population. Founding individuals could contain only a fraction of the total genetic diversity of the original gene pool. Which alleles the founders carry is dictated by chance alone. The Amish of Lancaster County, Pennsylvania, are an isolated group that was begun by German founders. Today, as many as 1 in 14 individuals carries a recessive allele that causes an unusual form of dwarfism (affecting only the lower arms and legs) and polydactylism (extra fingers) (Fig. 16.7). In the general population, only 1 in 1,000 individuals has this allele.

### Check Your Progress 16.1

1. If two genetically different subpopulations of the same species come into contact and gene flow begins, in general, how will the genetic makeup of the merged populations change?
2. Many zoological parks send the offspring of a single breeding pair of animals to zoos around the country. How is this an example of the founder effect?

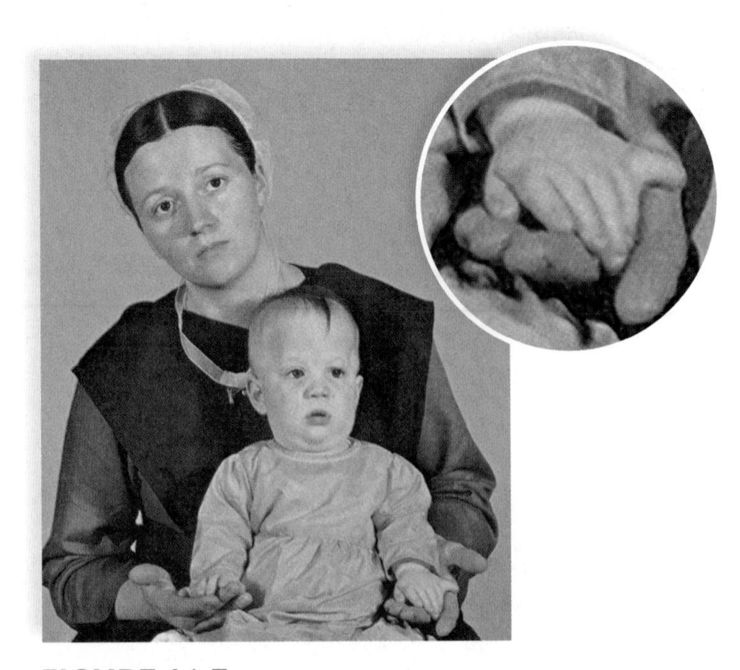

**FIGURE 16.7 Founder effect.**

A member of the founding population of Amish in Pennsylvania had a recessive allele for a rare kind of dwarfism linked with polydactylism. The percentage of the Amish population with this phenotype is much higher compared to that of the general population.

## 16.2 Natural Selection

In this chapter, we wish to consider natural selection in a genetic context. Many traits are polygenic (controlled by many genes), and the continuous variation in phenotypes results in a bell-shaped curve. When a range of phenotypes is exposed to the environment, natural selection favors the one that is most adaptive under the present environmental circumstances. Natural selection acts much the same way as a governing board that decides which applying students will be admitted to a college. Some students will be favored and allowed to enter, while others will be rejected and not allowed to enter. Of course, in the case of natural selection, the chance to reproduce is the prize awarded. In this context, natural selection can be stabilizing, directional, or disruptive (Fig. 16.8).

**Stabilizing selection** occurs when an intermediate phenotype can improve the adaptation of the population to those aspects of the environment that remain constant. With stabilizing selection, extreme phenotypes are selected against, and the intermediate phenotype is favored. As an example, consider that when Swiss starlings lay four to five eggs, more young survive than when the female lays more or less than this number. Genes determining physiological characteristics, such as the production of yolk, and behavioral characteristics, such as how long the female will mate, are involved in determining clutch size.

Human birth weight is another example of stabilizing selection. Through the years, hospital data have shown that human infants born with an intermediate birth weight (3–4 kg) have a better chance of survival than those at either extreme (either much less or much greater than usual). When a baby is small, its systems may not be fully functional, and when a baby is large, it may have experienced a difficult delivery. Stabilizing selection reduces the variability in birth weight in human populations (Fig. 16.9).

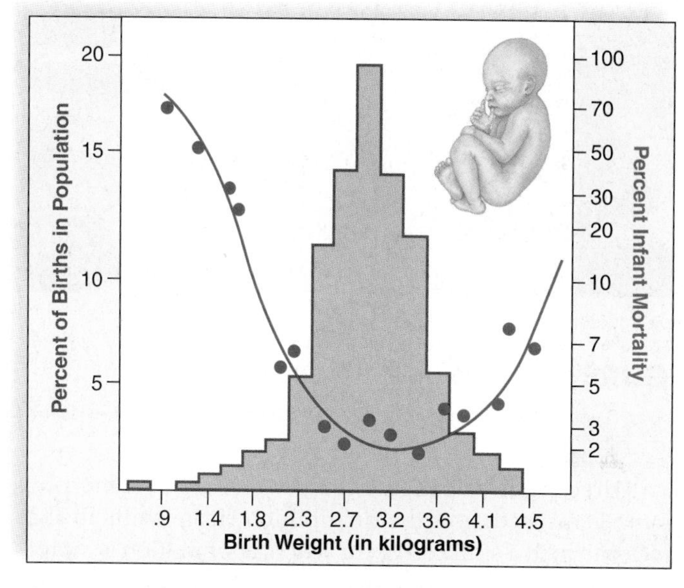


**FIGURE 16.9 Human birth weight.**

The birth weight (blue) is influenced by the chance of death (red).

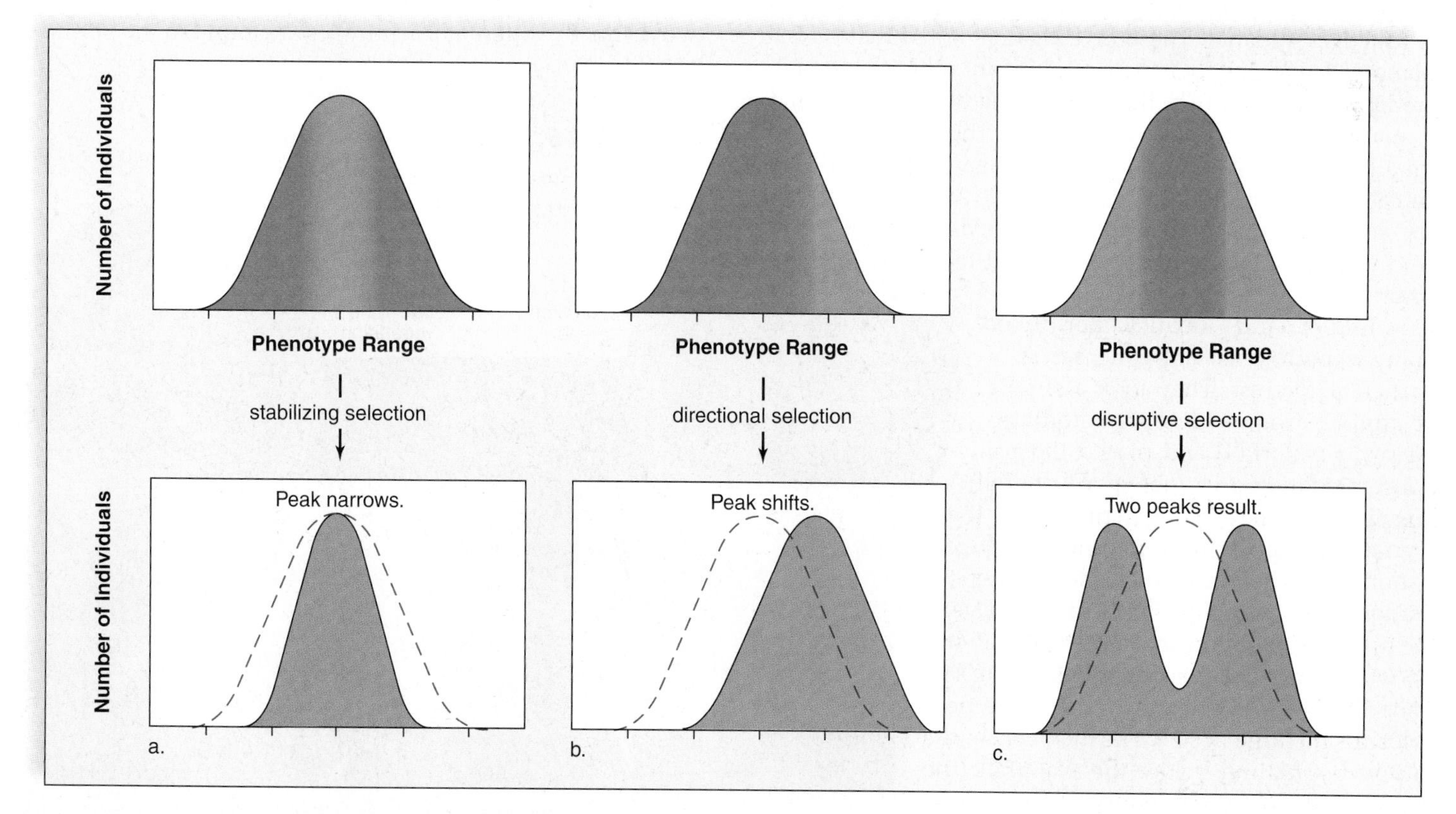


**FIGURE 16.8 Three types of natural selection.**

**a.** During stabilizing selection, the intermediate phenotype is favored; (**b**) during directional selection, an extreme phenotype is favored; and (**c**) during disruptive selection, two extreme phenotypes are favored.

Experimental site

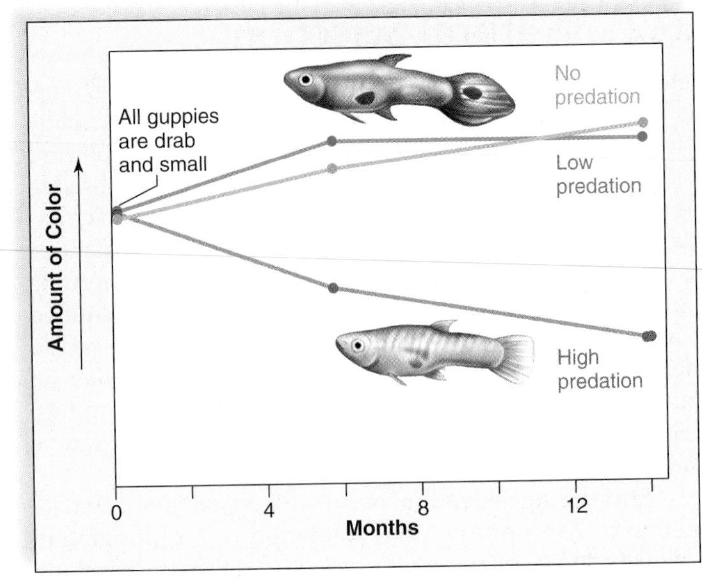


Result

**FIGURE 16.10 Directional selection.**

Guppies, *Poecilia reticulata*, become more colorful (blue, yellow graph lines) in the absence of predation and less colorful (red graph line) when in the presence of predation.

**Directional selection** occurs when an extreme phenotype is favored, and the distribution curve shifts in that direction. Such a shift can occur when a population is adapting to a changing environment.

Two investigators, John Endler and David Reznick, both at the University of California, conducted a study of guppies, which are known for their bright colors and reproductive potential. These investigators noted that on the island of Trinidad, when male guppies are subjected to high predation by other fish, they tend to be drab in color and to mature early and at a smaller size. The drab color and small size are most likely protective against being found and eaten. On the other hand, when male guppies are exposed to minimal or no predation, they tend to be colorful, to mature later, and to attain a larger size.

Endler and Reznick performed many experiments, and one set is of particular interest. They took a supply of guppies from a high-predation area (below a waterfall) and placed them in a low-predation area (above a waterfall) (Fig. 16.10). The waterfall prevented the predator fish (pike) from entering the low-predation area. They monitored the guppy population for 12 months, and during that year, the guppy population above the waterfall underwent directional selection (Fig. 16.10). The male members of the population were now colorful and large in size. The members of the guppy population below the waterfall (the control population) were still drab and small.

In **disruptive selection,** two or more extreme phenotypes are favored over any intermediate phenotype. For example, British land snails have a wide habitat range that includes low-vegetation areas (grass fields and hedgerows) and forests. In forested areas, thrushes feed mainly on light-banded snails, and the snails with dark shells become more prevalent. In low-vegetation areas, thrushes feed mainly on snails with dark shells, and light-banded snails become more prevalent. Therefore, these two distinctly different phenotypes are found in the population (Fig. 16.11).

**FIGURE 16.11 Disruptive selection.**

Due to exposure to two different environments, British land snails have two shell patterns.

## Sexual Selection

**Sexual selection** refers to adaptive changes in males and females that lead to an increased ability to secure a mate. Sexual selection in males may result in an increased ability to compete with other males for a mate, while females may select a male with the best **fitness** (ability to produce surviving offspring). In that way, the female increases her own fitness. Many consider sexual selection a form of natural selection because it affects fitness.

### *Female Choice*

Females produce few eggs, so the choice of a mate becomes a serious consideration. In a study of satin bowerbirds, two opposing hypotheses regarding female choice were tested:

1. *Good genes hypothesis:* Females choose mates on the basis of traits that improve the chance of survival.
2. *Runaway hypothesis:* Females choose mates on the basis of traits that improve male appearance. The term *runaway* pertains to the possibility that the trait will be exaggerated in the male until its mating benefit is checked by the trait's unfavorable survival cost.

**FIGURE 16.12 Dimorphism.**

In the Raggiana Bird of Paradise, *Paradisaea raggiana*, males have brilliantly colored plumage brought about by sexual selection. The drab females tend to choose flamboyant males as mates.

As investigators observed the behavior of satin bowerbirds, they discovered that aggressive males were usually chosen as mates by females. It could be that inherited aggressiveness does improve the chance of survival, or it could be aggressive males are good at stealing blue feathers from other males. Females prefer blue feathers as bower decorations. Therefore, the data did not clearly support either hypothesis.

The Raggiana Bird of Paradise is remarkably *dimorphic*, meaning that males and females differ in size and other traits. The males are larger than the females and have beautiful orange flank plumes. In contrast, the females are drab (Fig. 16.12). Female choice can explain why male birds are more ornate than females. Consistent with the two hypotheses, it is possible that the remarkable plumes of the male signify health and vigor to the female. Or, it's possible that females choose the flamboyant males on the basis that their sons will have an increased chance of being selected by females. Some investigators have hypothesized that extravagant male features could indicate that they are relatively parasite-free. In barn swallows, females also choose those with the longest tails, and investigators have shown that males that are relatively free of parasites have longer tails than otherwise.

### *Male Competition*

Males can father many offspring because they continuously produce sperm in great quantity. We expect males to compete in order to inseminate as many females as possible. **Cost-benefit analyses** have been done to determine if the *benefit* of access to mating is worth the *cost* of competition among males.

Baboons, a type of Old World monkey, live together in a troop. Males and females have separate **dominance hierarchies** in which a higher-ranking animal has greater access to resources than a lower-ranking animal. Dominance is decided by confrontations, resulting in one animal giving way to the other.

Baboons are dimorphic; the males are larger than the females, and they can threaten other members of the troop with their long, sharp canines. One or more males become dominant by frightening the other males. However, the male baboon pays a cost for his dominant position. Being larger means that he needs more food, and being willing and able to fight predators means that he may get hurt, and so forth. Is there a reproductive benefit to his behavior? Yes, in that dominant males do indeed monopolize females when they are most fertile. Nevertheless, there may be other ways to father offspring. A male may act as a helper to a female and her offspring; then, the next time she is in estrus, she may mate preferentially with him instead of a dominant male. Or subordinate males may form a friendship group that opposes a dominant male, making him give up a receptive female.

A **territory** is an area that is defended against competitors. Scientists are able to track an animal in the wild in order

**FIGURE 16.13 A male olive baboon displaying full threat.**

In olive baboons, *Papio anubis*, males are larger than females and have enlarged canines. Competition between males establishes a dominance hierarchy for the distribution of resources.

to determine its home range or territory. **Territoriality** includes the type of defensive behavior needed to defend a territory. Baboons travel within a home range, foraging for food each day and sleeping in trees at night. Dominant males decide where and when the troop will move. If the troop is threatened, dominant males protect the troop as it retreats and attack intruders when necessary. Vocalization and displays, rather than outright fighting, may be sufficient to defend a territory (Fig. 16.13). In songbirds, for example, males use singing to announce their willingness to defend a territory. Other males of the species become reluctant to make use of the same area.

Red deer stags (males) on the Scottish island of Rhum compete to be the harem master of a group of hinds (females) that mate only with them. The reproductive group occupies a territory that the harem master defends against other stags. Harem masters first attempt to repel challengers by roaring. If the challenger remains, the two lock antlers and push against one another (Fig. 16.14). If the challenger then withdraws, the master pursues him for a short distance, roaring the whole time. If the challenger wins, he becomes the harem master.

A harem master can father two dozen offspring at most, because he is at the peak of his fighting ability for only a short time. And there is a cost to being a harem master. Stags must be large and powerful in order to fight; therefore, they grow faster and have less body fat. During bad times, they are more likely to die of starvation, and in general, they have shorter lives. Harem master behavior will persist in the population only if its cost (reduction in the potential number of offspring because of a shorter life) is less than its benefit (increased number of offspring due to harem access).

### Check Your Progress 16.2

1. The evolution of the horse from an animal adapted to living in a forest to one adapted to living on a plain is an example of what types of selection? Explain.
2. Why is sexual selection a form of natural selection?

a.

b.

**FIGURE 16.14 Competition between male red deer.**

Male red deer, *Cervus elaphus*, compete for a harem within a particular territory. **a.** Roaring alone may frighten off a challenger, but (**b**) outright fighting may be necessary, and the victor is most likely the stronger of the two animals.

*science focus*

## Sexual Selection in Humans

A study of sexual selection among humans shows that the concepts of female choice and male competition apply to humans as well as to the animals we have been discussing. Increased fitness (ability to produce surviving offspring) again seems to be the result of sexual selection.

### Human Males Compete

Consider that women, by nature, must invest more in having a child than men. After all, it takes nine months to have a child, and pregnancy is followed by lactation, when a woman may nurse her infant. Men, on the other hand, need only contribute sperm during a sex act that may require only a few minutes. The result is that men are generally more available for mating than are women. Because more men are available, they necessarily have to compete with others for the privilege of mating.

Like many other animals, humans are dimorphic. Males tend to be larger and more aggressive than females, perhaps as a result of past sexual selection by females. As in other animals, males pay a price for their physical attractiveness to females. Male humans live on the average seven years less than females do.

### Females Choose

A study in modern Quebec sampled a large number of respondents on how often they had copulated with different sexual partners in the preceding year. Male mating success correlated best with income—those males who had both wealth and status were much more successful in acquiring mates than those who lacked these attributes. In this study, it would appear that females prefer to mate with a male who is wealthy and has a successful career because these men are more likely to be able to provide them with the resources they need to raise their children (Fig. 16A).

The desire of women for just certain types of men has led to the practice of polygamy in many primitive human societies and even in some modern societies. Women would rather share a husband who can provide resources than to have a one-on-one relationship with a poor man, because the resources provided by the wealthy man make it all the more certain that their children will live to reproduce. On the other hand, polygamy works for wealthy men because having more than one wife will undoubtedly increase their fitness as well. As an alternative to polygamy, modern societies stress monogamy in which the male plays a prominent role in helping to raise the children. This is another way males can raise their fitness.

### Men Also Have a Choice

Just as women choose men who can provide resources, men prefer women who are most likely to present them with children. It has been shown that the "hourglass figure" so touted by men actually correlates with the best distribution of body fat for reproductive purposes! Men responding to questionnaires about their preferences in women list attributes that biologists associate with a strong immune system, good health, high estrogen levels, and especially with youthfulness. Young males prefer partners who are their own age, give or take five years, but as men age, they prefer women who are many years younger than themselves. Men can reproduce for many more years than women can. Therefore, by choosing younger women, older men increase their fitness as judged by the number of children they have (Fig. 16A).

Men, unlike women, do not have the same assurance that a child is their own. Therefore, men put a strong emphasis on having a wife who is faithful to them. Both men and women respondents to questionnaires view adultery in women as more offensive than adultery in men.

**FIGURE 16A King Hussein and family.**
*The tendency of men to mate with fertile younger women is exemplified by King Hussein of Jordan, who was about 16 years older than his wife, Queen Noor. This photo shows some of their children, one of whom is now King Abdullah of Jordan.*

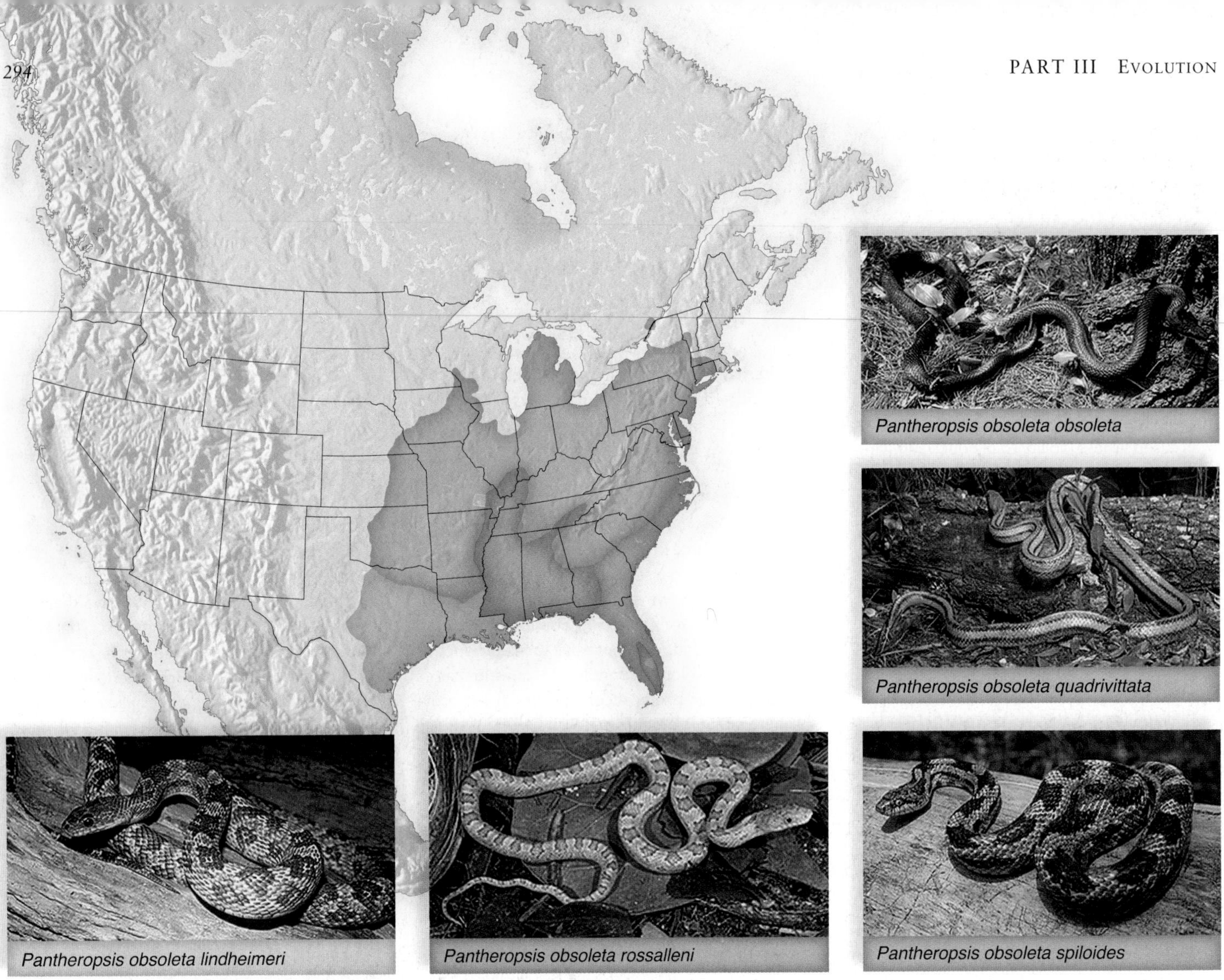

Pantheropsis obsoleta obsoleta

Pantheropsis obsoleta quadrivittata

Pantheropsis obsoleta lindheimeri

Pantheropsis obsoleta rossalleni

Pantheropsis obsoleta spiloides

**FIGURE 16.15 Subspecies help maintain diversity.**

Each subspecies of rat snakes represents a separate population of snakes. Each subspecies has a reservoir of alleles different from another subspecies. Because the populations are adjacent to one another, there is interbreeding, and, therefore, gene flow among the populations. This introduces alleles that may keep each subspecies from fully adapting to their particular environment.

# 16.3 Maintenance of Diversity

Diversity is maintained in a population for any number of reasons. Mutations still create new alleles, and sexual reproduction still recombines alleles due to meiosis and fertilization. Genetic drift also occurs, particularly in small populations, and the end result may be contrary to adaptation to the environment.

## Natural Selection

The process of natural selection itself causes imperfect adaptation to the environment. First, it is important to realize that evolution doesn't start from scratch. Just as you can only bake a cake with the ingredients available to you, evolution is constrained by the available diversity. Lightweight titanium bones might benefit birds, but their bones contain calcium and other minerals the same as other reptiles. When you mix the ingredients for a cake, you probably follow the same steps taught to you by your elders. Similarly, the processes of development prevent the emergence of novel features, and therefore the wing of a bird has the same bones as those of other vertebrate forelimbs.

Imperfections are common because of necessary compromises. The success of humans is attributable to their dexterous hands, but the spine is subject to injury because the vertebrate spine was not originally designed to stand erect. A feature that evolves has a benefit that is worth the cost. For example, the benefit of freeing the hands must have been worth the cost of spinal injuries from assuming an erect posture. We should also consider that sexual selection has a reproductive benefit but not necessarily an adaptive benefit.

Second, we want to realize that the environment plays a role in maintaining diversity. It's easy to see that disruptive selection, dependent on an environment that differs widely, promotes polymorphisms in a population (see Fig. 16.11). Then, too, if a population occupies a wide range, as shown in Figure 16.15, it may have several subpopulations designated as subspecies because of recognizable differences. (Subspecies are given a third name in addition to the usual binomial name.) Each subspecies is partially adapted to its own environment and can serve as a reservoir for a different combination of alleles that flow from one group to the next when adjacent subspecies interbreed.

The environment also includes specific selecting agents that help maintain diversity. We have already seen how insectivorous birds can help maintain the frequencies of both the light-colored and dark-colored moths, depending on the color of background vegetation. Some predators have a search image that causes them to select the most common phenotype among its prey. This promotes the survival of the rare form and helps maintain variation. Or, a herbivore can oscillate in its preference for food. In Figure 15.11, we observed that the medium ground finch on the Galápagos Islands had a different-sized beak dependent on the available food supply. In times of drought, when only large seeds were available, birds with larger beaks were favored. In this case, we can clearly see that maintenance of variation among a population has survival value for the species.

## Heterozygote Advantage

**Heterozygote advantage** occurs when the heterozygote is favored over the two homozygotes. In this way, heterozygote advantage assists the maintenance of genetic, and therefore phenotypic, diversity in future generations.

### *Sickle-Cell Disease*

Sickle-cell disease can be a devastating condition. Patients can have severe anemia, physical weakness, poor circulation, impaired mental function, pain and high fever, rheumatism, paralysis, spleen damage, low resistance to disease, and kidney and heart failure. In these individuals, the red blood cells are sickle-shaped and tend to pile up and block flow through tiny capillaries. The condition is due to an abnormal form of hemoglobin (*Hb*), the molecule that carries oxygen in red blood cells. People with sickle-cell disease ($Hb^SHb^S$) tend to die early and leave few offspring, due to hemorrhaging and organ destruction. Interestingly, however, geneticists studying the distribution of sickle-cell disease in Africa have found that the recessive allele ($Hb^S$) has a higher frequency in regions (blue color) where the disease malaria is also prevalent (Fig. 16.16). Malaria is caused by a protozoan parasite that lives in and destroys the red blood cells of the normal homozygote ($Hb^AHb^A$). Individuals with this genotype also have fewer offspring, due to an early death or to debilitation caused by malaria.

Heterozygous individuals ($Hb^AHb^S$) have an advantage because they don't die from sickle-cell disease, and they don't die from malaria. The parasite causes any red blood cell it infects to become sickle-shaped. Sickle-shaped red blood cells lose potassium, and this causes the parasite to die. Heterozygote advantage causes all three alleles to be maintained in the population. It's as if natural selection were a store owner balancing the advantages and disadvantages of maintaining the recessive allele $Hb^S$ in the warehouse. As long as the protozoan that causes malaria is present in the environment, it is advantageous to maintain the recessive allele.

Heterozygote advantage is also an example of stabilizing selection because the genotype $Hb^AHb^S$ is favored over the two extreme genotypes, $Hb^AHb^A$ and $Hb^SHb^S$. In the parts of Africa where malaria is common, one in five individuals is heterozygous (has sickle-cell trait) and survives malaria, while only 1 in 100 is homozygous and dies of sickle-cell disease. What happens in the United States where malaria is not prevalent? As you would expect, the frequency of the $Hb^S$ allele is declining among African Americans because the heterozygote has no particular advantage in this country.

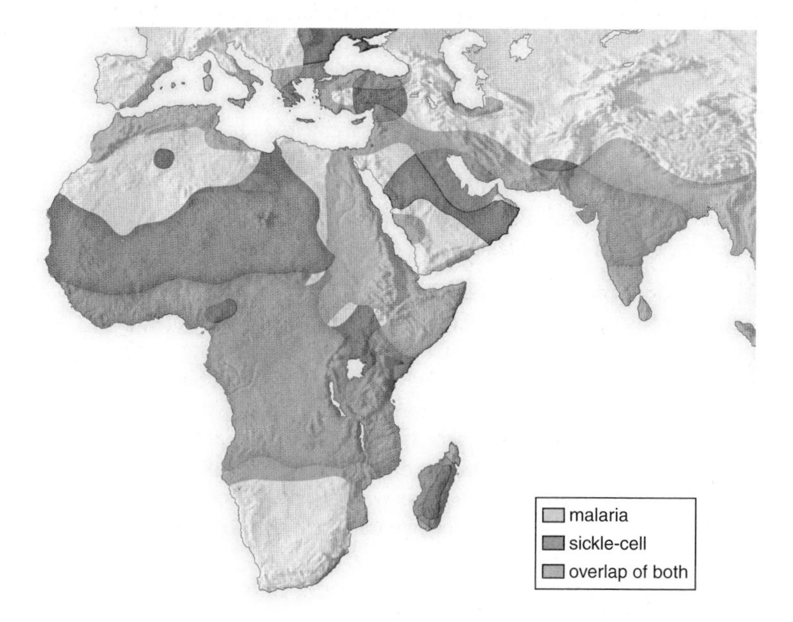


**FIGURE 16.16 Sickle-cell disease.**

Sickle-cell disease is more prevalent in areas of Africa where malaria is more common.

### *Cystic Fibrosis*

Stabilizing selection is also thought to have influenced the frequency of other alleles. Cystic fibrosis is a debilitating condition that leads to lung infections and digestive difficulties. In this instance, the recessive allele, common among individuals of northwestern European descent, causes the person to have a defective plasma membrane protein. The agent that causes typhoid fever can use the normal version of this protein, but not the defective one, to enter cells. Here again, heterozygote superiority caused the recessive allele to be maintained in the population.

**Check Your Progress 16.3**

1. Natural selection cannot do away with diversity in a population. Explain.
2. Use a Hardy-Weinberg equilibrium to explain why heterozygote advantage maintains diversity.

## Connecting the Concepts

We have seen that there are variations among the individuals in any population, whether the population is tuberculosis-causing bacteria in a city, the dandelions on a hill, or the squirrels in your neighborhood. Individuals vary because of the presence of mutations and, in sexually reproducing species, because of the recombination of alleles and chromosomes due to the processes of meiosis and fertilization.

This chapter is about microevolution—that is, gene frequency changes within a population below the level of speciation. The field of population genetics, which uses the Hardy-Weinberg principle, shows us how the study of microevolution is objective rather than subjective. A change in gene pool allele frequencies defines and signifies that microevolution has occurred. There are various agents of microevolutionary change, but only natural selection results in adaptation to the environment. Recent observations and experiments show that natural selection can occur rapidly. The emergence of MRSA, methicillin-resistant *Staphylococcus aureus* bacteria, only took about sixty years. Investigators who noted that guppies have a different appearance according to the presence of predators can also observe this change because it takes only a few months in the wild or even in the laboratory. So, we have to change our perception of natural selection as a process that cannot be observed to one that anyone can observe if they are so inclined.

Sexual selection should be considered a form of natural selection because it affects the reproductive capacity of the individual. Males that are selected to reproduce by females have more offspring than those that are not selected. Sexual selection teaches us that any trait that evolves through natural selection can be subjected to a cost-benefit analysis. It is obvious that some male traits arising through sexual selection bear a cost and may actually decrease the chance of survival. Still, if a trait helps a male leave more fertile offspring than other members of a population, it is beneficial in an evolutionary context.

We have seen that natural selection does not normally reduce variations to the point that no further evolutionary change is possible. Retention of variations is always desirable because a changing environment requires further evolutionary adjustments.

## summary

### 16.1 Population Genetics

Microevolution requires diversity, and this chapter is interested in allele and genotype differences within a population. Diversity can even extend to SNPs, which are single nucleotide polymorphisms. Investigators are beginning to think that SNPs have significance because they may help regulate the expression of genes.

The Hardy-Weinberg equilibrium is a constancy of gene pool allele frequencies that remains from generation to generation if certain conditions are met. The conditions are no mutations, no gene flow, random mating, no genetic drift, and no selection. Since these conditions are rarely met, a change in gene pool frequencies is likely. When gene pool frequencies change, microevolution has occurred. Deviations from a Hardy-Weinberg equilibrium allow us to determine when evolution has taken place.

Mutations, gene flow, nonrandom mating, genetic drift, and natural selection all cause deviations from a Hardy-Weinberg equilibrium. Mutations are the raw material for evolutionary change. Recombination of alleles during sexual reproduction help bring about adaptive genotypes. Gene flow occurs when a breeding individual (in animals) migrates to another population or when gametes and seeds (in plants) are carried into another population. Constant gene flow between two populations causes their gene pools to become similar. Nonrandom mating occurs when relatives mate (inbreeding) or assortative mating takes place. Both of these cause an increase in homozygotes. Genetic drift occurs when allele frequencies are altered only because some individuals, by chance, contribute more alleles to the next generation. Genetic drift can cause the gene pools of two isolated populations to become dissimilar as some alleles are lost and others are fixed. Genetic drift is particularly evident after a bottleneck, when severe inbreeding occurs, or when founders start a new population.

### 16.2 Natural Selection

Most of the traits of evolutionary significance are polygenic; the diversity in a population results in a bell-shaped curve. Three types of selection occur: (1) directional—the curve shifts in one direction, as when bacteria become resistant to antibiotics. (New observations and experiments of late have shown how quickly directional selection can occur. Within months, guppies originally similar in appearance become more colorful if predation is absent and less colorful if predation is present.); (2) stabilizing—the peak of the curve increases, as when there is an optimum clutch size for survival of Swiss starling young; and (3) disruptive—the curve has two peaks, as when *Cepaea* snails vary because a wide geographic range causes selection to vary.

Traits that promote reproductive success are expected to be advantageous overall, despite any possible disadvantage. Males produce many sperm and are expected to compete to inseminate females. Females produce few eggs and are expected to be selective about their mates. Studies of satin bowerbirds and birds of paradise have been done to test hypotheses regarding female choice. A cost-benefit analysis can be applied to competition between males for mates, in reference to a dominance hierarchy (e.g., baboons) and territoriality (e.g., red deer).

It is possible that male competition and female choice also occur among humans. Biological differences between the sexes may promote certain mating behaviors because they increase fitness.

### 16.3 Maintenance of Diversity

Despite constant natural selection, genetic diversity is maintained. Mutations and recombination still occur; gene flow among small populations can introduce new alleles; and natural selection itself sometimes results in variation. In sexually reproducing diploid organisms, the heterozygote acts as a repository for recessive alleles whose frequency is low. In regard to sickle-cell disease, the heterozygote is more fit in areas where malaria occurs, and therefore both homozygotes are maintained in the population.

## understanding the terms

gene flow 287
gene pool 285
genetic drift 287
Hardy-Weinberg principle 286
heterozygote advantage 295
industrial melanism 286
microevolution 285
mutation 287
nonrandom mating 287
population 284
population genetics 284
sexual selection 291
single nucleotide polymorphism (SNP) 284
stabilizing selection 289
territoriality 292
territory 291

Match the terms to these definitions:

a. ______________ Outcome of natural selection in which extreme phenotypes are eliminated and the average phenotype is conserved.

b. ______________ Marking and/or defending a particular area against invasion by another species member.

c. ______________ Change in the genetic makeup of a population due to chance (random) events; important in small populations or when only a few individuals mate.

d. ______________ Total of all the genes of all the individuals in a population.

e. ______________ Sharing of genes between two populations through interbreeding.

## reviewing this chapter

1. The discovery of SNPs is of what significance? 284
2. What is the Hardy-Weinberg principle? 285–86
3. Name and discuss the five conditions of evolutionary change. 286–88
4. What is a population bottleneck, and what is the founder effect? Give examples of each. 288
5. Distinguish among directional, stabilizing, and disruptive selection by giving examples. 289–90
6. What is sexual selection, and why does it foster female choice and male competition during mating? 291–93
7. What is a cost-benefit analysis, and how does it apply to a dominance hierarchy and territoriality? Give examples. 291–92
8. State ways in which diversity is maintained in a population. 294–95

## testing yourself

Choose the best answer for each question.

1. Assuming a Hardy-Weinberg equilibrium, 21% of a population is homozygous dominant, 50% is heterozygous, and 29% is homozygous recessive. What percentage of the next generation is predicted to be homozygous recessive?
   a. 21%
   b. 50%
   c. 29%
   d. 42%
   e. 58%
2. A human population has a higher-than-usual percentage of individuals with a genetic disorder. The most likely explanation is
   a. mutations and gene flow.
   b. mutations and natural selection.
   c. nonrandom mating and founder effect.
   d. nonrandom mating and gene flow.
   e. All of these are correct.
3. The offspring of better-adapted individuals are expected to make up a larger proportion of the next generation. The most likely explanation is
   a. mutations and nonrandom mating.
   b. gene flow and genetic drift.
   c. mutations and natural selection.
   d. mutations and genetic drift.
4. The continued occurrence of sickle-cell disease with malaria in parts of Africa is due to
   a. continual mutation.
   b. gene flow between populations.
   c. relative fitness of the heterozygote.
   d. disruptive selection.
   e. protozoan resistance to DDT.
5. Which of these is necessary to natural selection?
   a. diversity
   b. differential reproduction
   c. inheritance of differences
   d. differential adaptiveness
   e. All of these are correct.
6. When a population is small, there is a greater chance of
   a. gene flow.
   b. genetic drift.
   c. natural selection.
   d. mutations occurring.
   e. sexual selection.
7. Which of these is an example of stabilizing selection?
   a. Over time, *Equus* developed strength, intelligence, speed, and durable grinding teeth.
   b. British land snails mainly have two different phenotypes.
   c. Swiss starlings usually lay four or five eggs, thereby increasing their chances of more offspring.
   d. Drug resistance increases with each generation; the resistant bacteria survive, and the nonresistant bacteria get killed off.
   e. All of these are correct.
8. Which of these cannot occur if a population is to maintain an equilibrium of allele frequencies?
   a. People leave one country and relocate in another.
   b. A disease wipes out the majority of a herd of deer.
   c. Members of an Indian tribe only allow the two tallest people in a tribe to marry each spring.
   d. Large black rats are the preferred males in a population of rats.
   e. All of these are correct.
9. The homozygote $Hb^SHb^S$ persists because
   a. it offers protection against malaria.
   b. the heterozygote offers protection against malaria.
   c. the genotype $Hb^AHb^A$ offers protection against malaria.
   d. sickle-cell disease is worse than sickle-cell trait.
   e. Both b and d are correct.
10. The diagrams represent a distribution of phenotypes in a population. Superimpose another diagram on (a) to show that directional selection has occurred, on (b) to show that stabilizing selection has occurred, and on (c) to show that disruptive selection has occurred.

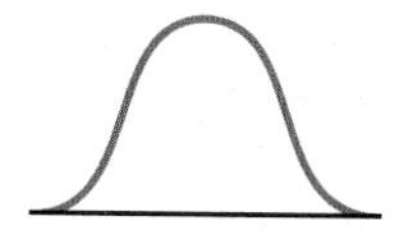
a. Directional selection

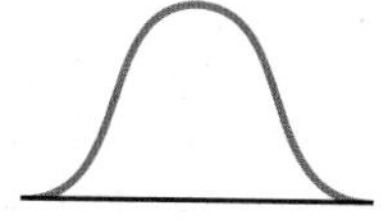
b. Stabilizing selection

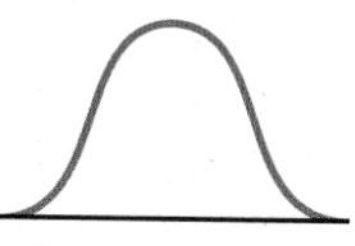
c. Disruptive selection

11. The observation that the most fit male bowerbirds are the ones that can keep their nests intact supports which hypothesis?
   a. good genes hypothesis—females choose mates based on their improved chances of survival
   b. runaway hypothesis—females choose mates based on their appearance
   c. Either hypothesis could be true.
   d. Neither hypothesis is true.

12. In some bird species, the female chooses a mate that is most similar to her in size. This supports
    a. the good genes hypothesis.
    b. the runaway hypothesis.
    c. Either hypothesis could be true.
    d. Neither hypothesis is true.
13. Which of the following are costs that a dominant male baboon must pay in order to gain a reproductive benefit?
    a. He requires more food and must travel larger distances.
    b. He requires more food and must care for his young.
    c. He is more prone to injury and requires more food.
    d. He is more prone to injury and must care for his young.
    e. He must care for his young and travel larger distances.
14. A red deer harem master typically dies earlier than other males because he is
    a. likely to get expelled from the herd and cannot survive alone.
    b. more prone to disease because he interacts with so many animals.
    c. in need of more food than other males.
    d. apt to place himself between a predator and the herd to protect the herd.
15. Which one of the following statements would *not* pertain to a Punnett square that involves the alleles of a gene pool?
    a. The results tell you the chances that an offspring can have a particular condition.
    b. The results tell you the genotype frequencies of the next generation.
    c. The eggs and sperm are the gamete frequencies of the previous generation.
    d. All of these are correct.
16. Which of the following applies to the Hardy-Weinberg expression: $p^2 + 2pq + q^2$?
    a. Knowing either $p^2$ or $q^2$, you can calculate all the other frequencies.
    b. applies to Mendelian traits that are controlled by one pair of alleles
    c. $2pq$ = heterozygous individuals
    d. can be used to determine the genotype and allele frequencies of the previous and the next generations
    e. All of these are correct.
17. Following genetic drift, the
    a. genotype and allele frequencies would not change.
    b. genotype and allele frequencies would change.
    c. adaptation would occur.
    d. population would have more phenotypic variation but less genotypic variation.
18. The high frequency of Huntington disease in a population could be due to
    a. mutation plus nonrandom mating.
    b. the founder effect.
    c. natural selection because Huntington disease has a benefit.
    d. pollution in the environment.
    e. Both a and b are correct.
19. For disruptive selection to occur,
    a. the population has to contain diversity.
    b. the environment has to contain diversity.
    c. pollution must be present.
    d. natural selection must occur.
    e. All but c are correct.
20. Which of these is mismatched?
    a. male competition—males produce many sperm
    b. female choice—females produce few eggs
    c. male choice—males with exaggerated traits get to choose
    d. male competition—dominance hierarchy
21. All vertebrate forelimbs contain the same bones because
    a. they form the best structures for all sorts of adaptations.
    b. they are pliable and able to adapt.
    c. their common ancestor had these bones.
    d. vertebrates have vertebrates in their spine.
    e. All of these are correct.

## additional genetics problems*

1. If $p^2 = 0.36$, what percentage of the population has the recessive phenotype, assuming a Hardy-Weinberg equilibrium?
2. If 1% of a human population has the recessive phenotype, what percentage has the dominant phenotype, assuming a Hardy-Weinberg equilibrium?
3. In a population of snails, ten had no antennae (*aa*); 180 were heterozygous with antennae (*Aa*); and 810 were homozygous with antennae (*AA*). What is the frequency of the *a* allele in the population?

*Answers to Additional Genetics Problems appear in Appendix A.

## thinking scientifically

1. A farmer uses a new pesticide. He applies the pesticide as directed by the manufacturer and loses about 15% of his crop to insects. A farmer in the next state learns of these results, uses three times as much pesticide, and loses only 3% of her crop to insects. Each farmer follows this pattern for five years. At the end of five years, the first farmer is still losing about 15% of his crop to insects, but the second farmer is losing 40% of her crop to insects. How could these observations be interpreted on the basis of natural selection?
2. You are observing a grouse population in which two feather phenotypes are present in males. One is relatively dark and blends into shadows well, and the other is relatively bright and so is more obvious to predators. The females are uniformly dark-feathered. Observing the frequency of mating between females and the two types of males, you have recorded the following:
   matings with dark-feathered males: 13
   matings with bright-feathered males: 32
   Propose a hypothesis to explain why females apparently prefer bright-feathered males. What selective advantage might there be in choosing a male with alleles that make it more susceptible to predation? What data would help test your hypothesis?

## *Biology* website

The companion website for *Biology* provides a wealth of information organized and integrated by chapter. You will find practice tests, animations, videos, and much more that will complement your learning and understanding of general biology.

**http://www.mhhe.com/maderbiology10**

# 17

# Speciation and Macroevolution

*The immense liger featured here is an offspring of a lion and a tiger, two normally reproductively isolated animal species. Ligers are the largest of all known cats, measuring up to 12 feet tall when standing on their hind legs and weighing as much as 1,000 lbs. Their coat color is usually tan with tiger stripes on the back and hindquarters and lion cub spots on the abdomen. A liger can produce both the "chuff" sound of a tiger and the roar of a lion. Male ligers may have a modest lion mane or no mane at all. Most ligers like to be near water and love to swim. Generally, ligers have a gentle disposition; however, considering their size and heritage, handlers should be extremely careful. By what criteria could a liger be considered a new species? Only if they, in turn, were reproductively isolated and only mated with ligers. In this chapter, we will explore the definition of a species and how species arise. In so doing, we will begin our discussion of macroevolution, which we continue in the next chapter.*

This liger is a hybrid because it has a lion father and a tiger mother.

## concepts

# 17.1 Separation of the Species

In Chapter 16, we defined microevolution as any allele frequency change within the gene pool of a population. Macroevolution, which is observed best within the fossil record, requires the origin of species, also called speciation. Speciation is the final result of changes in gene pool allelic and genotypic frequencies. The diversity of life we see about us is absolutely dependent on speciation, so it is important to be able to define a species and to know when speciation has occurred.

## What Is a Species?

Up until now, we have defined a species as a type of living thing, but now we want to characterize a species in more depth. The **evolutionary species concept** recognizes that every species has its own evolutionary history, at least part of which is in the fossil record. As an example, consider that the species depicted in Figure 17.1 are a part of the evolutionary history of toothed whales. Binomial nomenclature, discussed on pages 338–39, was used to name these ancestors of whales, as well as the various species of toothed whales today. The two-part scientific name, when translated from the Latin, often tells you something about the organism. For example, the scientific name of the dinosaur, *Tyrannosaurus rex*, means "tyrant-lizard king."

The evolutionary species concept relies on identification of certain morphological (structural) traits, called diagnostic traits, to distinguish one species from another. As long as these traits are the same, fossils are considered members of the same species. Abrupt changes in these traits indicate the evolution of a new species in the fossil record. In summary, the evolutionary species concept states that members of a species share the same distinct evolutionary pathway and that species can be recognized by morphological trait differences.

One advantage of the evolutionary species concept is that it applies to both sexually and asexually reproducing organisms. A major disadvantage occurs because morphological traits are being used to distinguish species. For example, it's possible that the presence of variations, such as size differences in males and females, might make you think you are dealing with two species instead of one, and the lack of diagnostic differences could cause you to conclude that two fossils are the same species when they are not.

The evolutionary species concept necessarily assumes that the members of a species are reproductively isolated. If members of different species were to reproduce with one another—that is, hybridize—their evolutionary history would be mingled, not separate. By contrast, the **biological species concept** relies primarily on reproductive isolation rather than trait differences to define a species. In other words, although traits can help us distinguish species, the most important criterion, according to the biological species concept, is reproductive isolation—the members of a species have a single gene pool. While useful, the biological species concept cannot be applied to asexually reproducing organisms, organisms known only by the fossil record, or species that could possibly

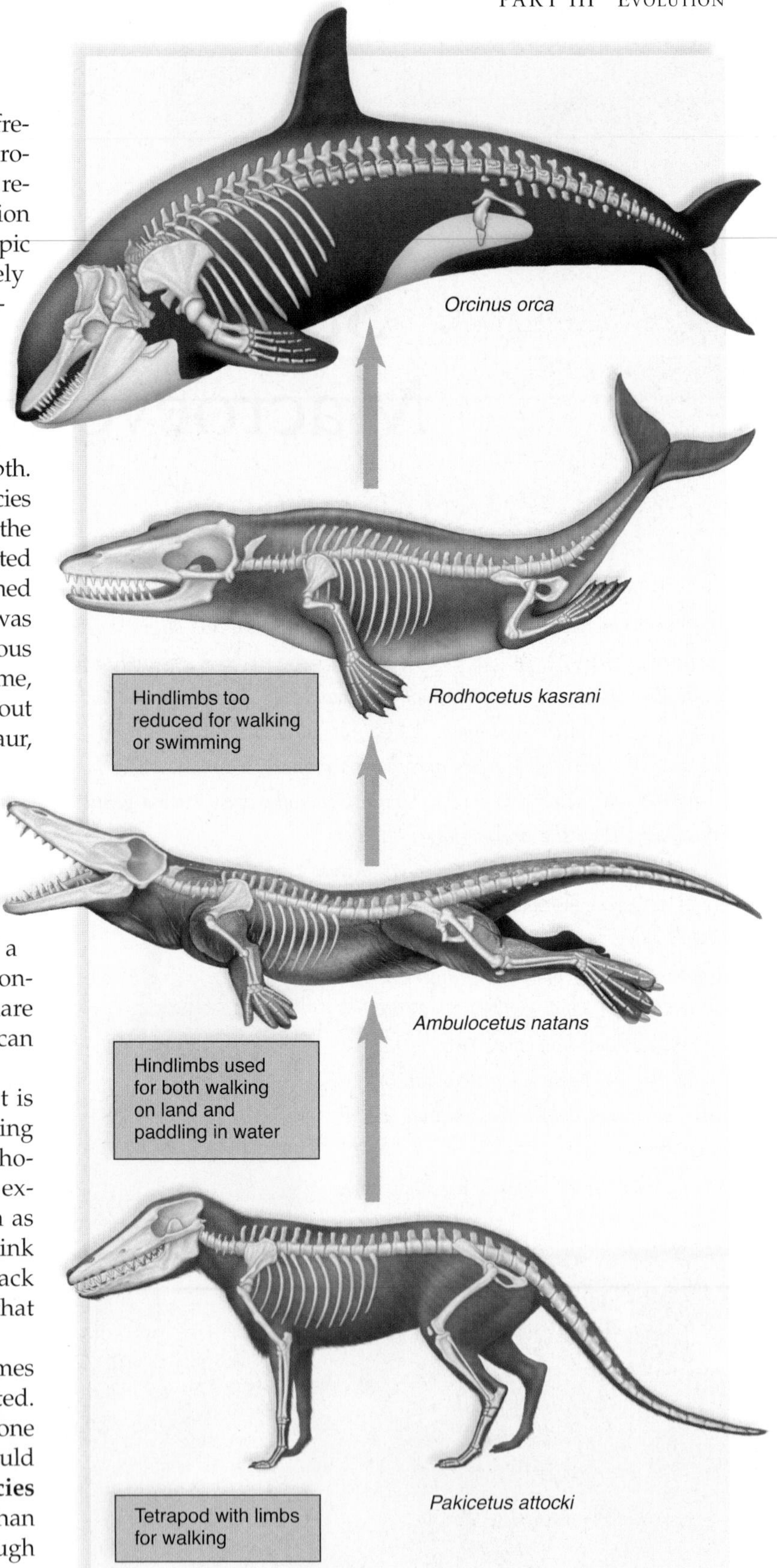


**FIGURE 17.1 Evolutionary species concept.**

Diagnostic traits can be used to distinguish these species known only from the fossil record. Such traits no doubt would include the anatomy of the limbs.

Acadian flycatcher, *Empidonax virescens* Willow flycatcher, *Empidonax traillii* Least flycatcher, *Empidonax minimus*

**FIGURE 17.2 Biological species concept.**

We know these flycatchers are separate species because they are reproductively isolated—the members of each species reproduce only with each other. Each species has a characteristic song and its own particular habitat during the mating season as well.

interbreed if they lived near one another. The benefit of the concept is that it can designate species even when trait differences may be difficult to find. Therefore, the flycatchers in Figure 17.2 are very similar, but they do not reproduce with one another; therefore, they are separate species. They live in different habitats. The Acadian flycatcher inhabits deciduous woods and wooded swamps, especially beeches; the willow flycatcher inhabits thickets, bushy pastures, old orchards, and willows; and the least flycatcher inhabits open woods, orchards, and farms. They also have different calls. Conversely, when anatomic differences are apparent, but reproduction is not deterred, only one species is present. Despite noticeable variations, humans from all over the world can reproduce with one another and belong to one species. The Massai of East Africa and the Eskimos of Alaska are kept apart by geography, but we know that, should they meet, reproduction between them would be possible (Fig. 17.3).

The biological species concept gives us a way to know when speciation has occurred, without regard to anatomic differences. As soon as descendants of a group of organisms are able to reproduce only among themselves, speciation has occurred.

In recent years, the biological species concept has been supplemented by our knowledge of molecular genetics. DNA base sequence data and differences in proteins can indicate the relatedness of groups of organisms. For example, it has recently been proposed that differences in the DNA sequence of the mitochondrial cytochrome oxidase gene could be used to identify many diverse species of animals. A study of 111 specimens of Indian mosquitoes, belonging to 15 genera and identified by morphological traits to be 63 species, was undertaken. It was found that DNA sequence differences for this gene allowed the investigators to identify 62 of the mosquito species. Two closely related species could not be identified as separate because they had negligible genetic differences.

**FIGURE 17.3 Human populations.**

The Massai of East Africa (*left*) and the Eskimos of Alaska (*right*) belong to the same species.

## Reproductive Isolating Mechanisms

For two species to remain separate, they must be reproductively isolated—that is, gene flow must not occur between them. Reproduction, and indeed isolation, is successful when fertile offspring are produced. Only fertile offspring can pass on genes to the next generation. Reproductive barriers that prevent successful reproduction from occurring are called isolating mechanisms (Fig. 17.4) Prezygotic (before the formation of a zygote) isolating mechanisms are considered before postzygotic (after formation of a zygote) isolating mechanisms. A zygote is the first cell that results when a sperm fertilizes an egg.

**Prezygotic isolating mechanisms** prevent reproductive attempts or make it unlikely that fertilization will be successful if mating is attempted. These isolating mechanisms make it highly unlikely hybridization will occur.

*Habitat isolation* When two species occupy different habitats, even within the same geographic range, they are less likely to meet and attempt to reproduce. This is one of the reasons that the flycatchers in Figure 17.2 do not mate, and that red maple and sugar maple trees do not exchange pollen. In tropical rain forests, many animal species are restricted to a particular level of the forest canopy, and in this way they are isolated from similar species.

*Temporal isolation* Several related species can live in the same locale, but if each reproduces at a different time of year, they do not attempt to mate. Five species of frogs of the genus *Rana* are all found at Ithaca, New York (Fig. 17.5). The species remain separate because the period of most active mating differs and so do the breeding sites. For example, wood frogs breed in woodland ponds or shallow water, leopard frogs in lowland swamps, and pickerel frogs in streams and ponds on high ground.

*Behavioral isolation* Many animal species have courtship patterns that allow males and females to recognize one another. The male blue-footed boobie in Figure 17.6 does a dance. Male fireflies are recognized by females of their species by the pattern of their flashings; similarly, female crickets recognize male crickets by their chirping. Many males recognize females of their species by sensing chemical signals called pheromones. For example, female gypsy moths release pheromones that are detected miles away by receptors on the antennae of males.

*Mechanical isolation* When animal genitalia or plant floral structures are incompatible, reproduction cannot occur. Inaccessibility of pollen to certain pollinators can prevent cross-fertilization in plants, and the sexes of many insect species have genitalia that do not match, or other characteristics that make mating impossible. For example, male dragonflies have claspers that are suitable for holding only the females of their own species.

*Gamete isolation* Even if the gametes of two different species meet, they may not fuse to become a zygote. In animals, the sperm of one species may not be able to survive in the reproductive tract of another species, or the egg may have receptors only for sperm of its species. In plants, only certain types of pollen grains can germinate so that sperm successfully reach the egg.

**FIGURE 17.4 Reproductive barriers.**

Prezygotic isolating mechanisms prevent mating attempts or a successful outcome should mating take place. No zygote ever forms. Postzygotic isolating mechanisms prevent the zygote from developing—or should an offspring result, it is not fertile.

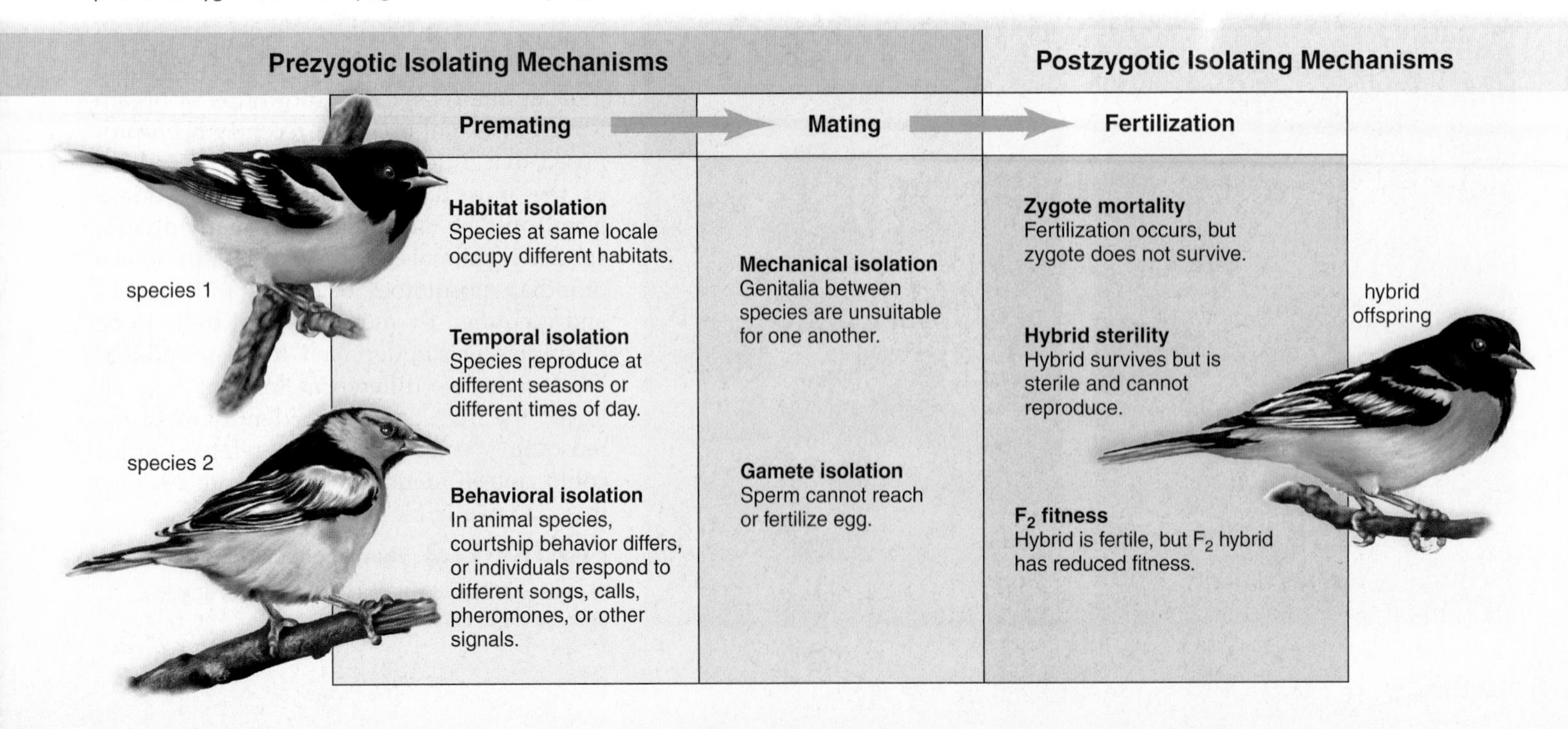

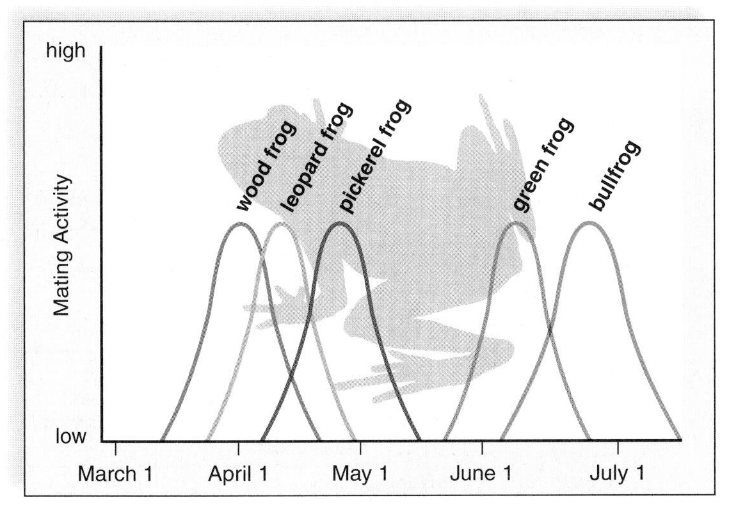


**FIGURE 17.5 Temporal isolation.**

Five species of frogs of the genus *Rana* are all found at Ithaca, New York. The species remain separate due to breeding peaks at different times of the year, as indicated by this graph.

**FIGURE 17.6 Prezygotic isolating mechanism.**

An elaborate courtship display allows the blue-footed boobies of the Galápagos Islands to select a mate. The male lifts up his feet in a ritualized manner that shows off their bright blue color.

**Postzygotic** (after the formation of a zygote) **isolating mechanisms** prevent hybrid offspring from developing, even if reproduction attempts have been successful. Or, if a hybrid is born, it is infertile and cannot reproduce. Either way, the genes of the parents are unable to be passed on.

***Zygote mortality*** A hybrid zygote may not be viable, and so it dies. A zygote with two different chromosome sets may fail to go through mitosis properly, or the developing embryo may receive incompatible instructions from the maternal and paternal genes so that it cannot continue to exist.

***Hybrid sterility*** The hybrid zygote may develop into a sterile adult. As is well known, a cross between a male horse and a female donkey produces a mule, which is usually sterile—it cannot reproduce (Fig. 17.7). Sterility of hybrids generally results from complications in meiosis that lead to an inability to produce viable gametes. Similarly, a cross between a cabbage and a radish produces offspring that cannot form gametes, most likely because the cabbage chromosomes and the radish chromosomes could not align during meiosis.

***$F_2$ fitness*** Even if hybrids can reproduce, their offspring may be unable to reproduce. In some cases, mules are fertile, but their offspring (the $F_2$ generation) are not fertile.

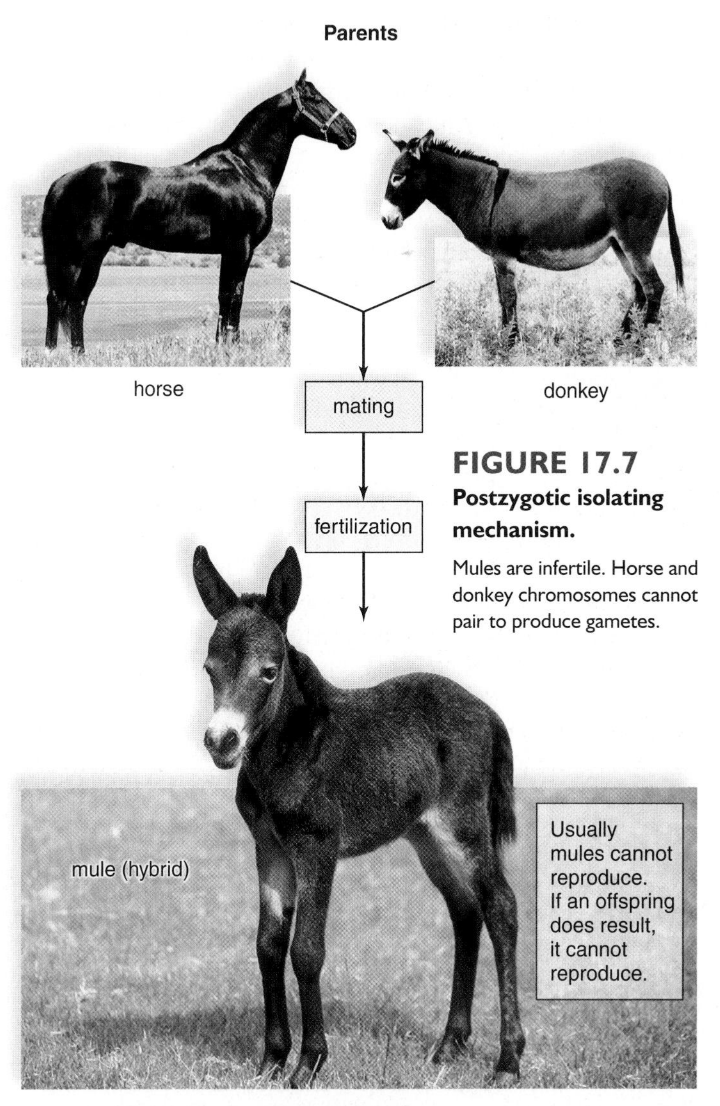


**FIGURE 17.7 Postzygotic isolating mechanism.**

Mules are infertile. Horse and donkey chromosomes cannot pair to produce gametes.

## Check Your Progress 17.1

1. On the basis of the evolutionary species concept, should ligers be considered a species in their own right?
2. Lions and tigers do not meet in the wild, and ligers born in captivity are sterile. What (**a**) pre- and (**b**) postzygotic isolating mechanisms are working?

# 17.2 Modes of Speciation

Researchers recognize two modes of **speciation,** splitting of one species into two or more species or the transformation of one species into a new species over time. One requires geographic isolation and the other one does not. Geographic isolation is helpful because it allows populations to continue on their own evolutionary path, which eventually causes them to be reproductively isolated from other species. Once reproductive isolation has begun, it can be reinforced by the evolution of more traits that prevent breeding with related species. Geographic isolation can repeatedly occur, so one ancestral species can give rise to several other species.

## Allopatric Speciation

In 1942, Ernst Mayr, an evolutionary biologist, published the book *Systematics and the Origin of Species,* in which he proposed the biological species concept and a process by which speciation could occur. He said that when members of a species become isolated, the new populations will start to differ because of genetic drift and natural selection over a period of time. Eventually, the two groups will be unable to mate with one another. At that time, they have evolved into new species. Mayr's hypothesis, termed **allopatric speciation** [Gk. *allo,* other, and *patri,* fatherland], requires that populations be separated by a geographic barrier for speciation to occur.

### *Examples of Allopatric Speciation*

Figure 17.8 features an example of allopatric speciation that has been extensively studied in California. An ancestral population of *Ensatina* salamanders lives in the Pacific Northwest. (1) Members of this ancestral population migrated southward, establishing a series of populations. Each population was exposed to its own selective pressures along the coastal mountains and the Sierra Nevada mountains. (2) Due to the presence of the Central Valley of California, gene flow rarely occurs between the eastern populations and the western populations. (3) Genetic differences increased from north to south, resulting in two distinct forms of *Ensatina* salamanders in Southern California that differ dramatically in color and rarely interbreed.

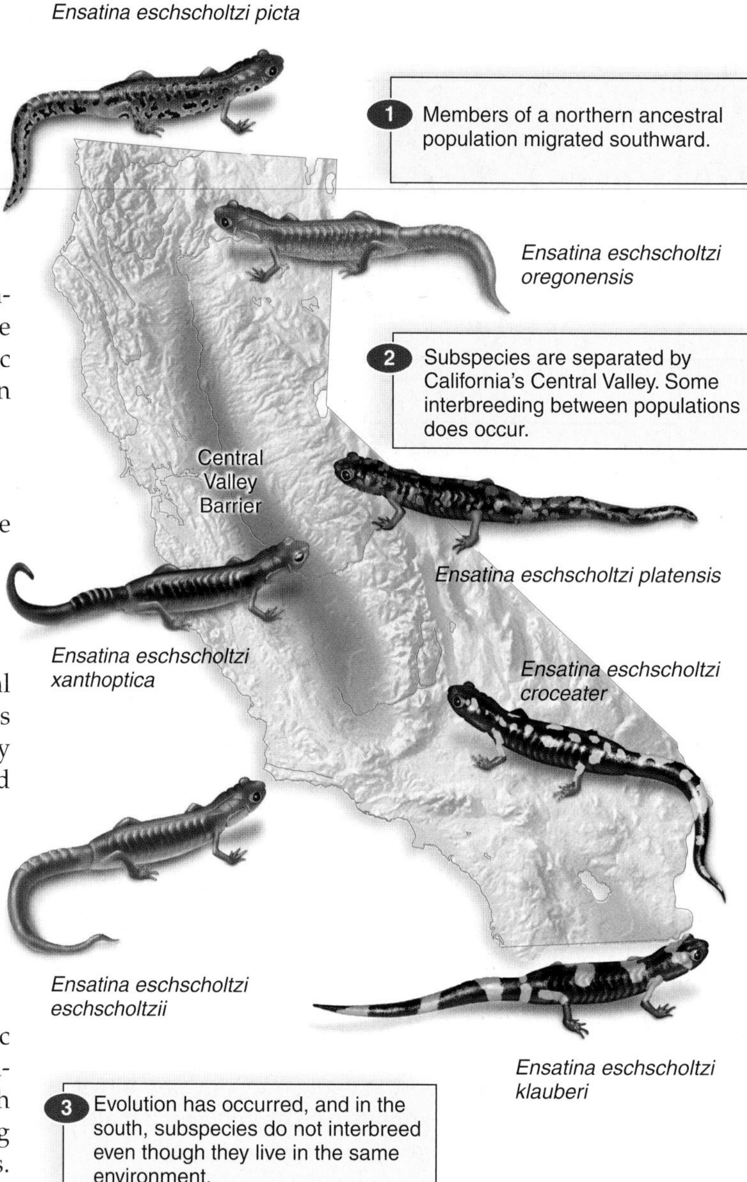


**FIGURE 17.8 Allopatric speciation among *Ensatina* salamanders.**

The Central Valley of California is separating a range of populations descended from the same northern ancestral species. The limited contact between the populations on the west and those on the east allows genetic changes to build up so that two populations, both living in the south, do not reproduce with one another, and therefore can be designated as separate species.

Geographic isolation is even more obvious in other examples. The green iguana of South America is hypothesized to be the common ancestor for both the marine iguana on the Galápagos Islands (to the west) and the rhinoceros iguana on Hispaniola, an island to the north. If so, how could it happen? Green iguanas are strong swimmers, so by chance, a few could have migrated to these islands, where they formed populations separate from each other and from the parent population back in South America. Each population continued on its own evolutionary path as new mutations, genetic drift, and different selection pressures occurred. Eventually, reproductive isolation developed, and the result was three species of iguanas that are reproductively isolated from each other.

A more detailed example of allopatric speciation involves sockeye salmon in Washington State. In the 1930s and 1940s, hundreds of thousands of sockeye salmon were introduced into Lake Washington. Some colonized an area of the lake near Pleasure Point Beach (Fig. 17.9*a*). Others migrated into the Cedar River (Fig. 17.9*b*). Andrew Hendry, a biologist at McGill University, is able to tell a Pleasure Point Beach salmon from a Cedar River salmon because they differ in shape and size due to the demands of reproducing. In the river, where the waters are fast-moving, males tend to be more slender than those at the beach. A slender body is better able to turn sideways in a strong current, and the courtship ritual of a sockeye salmon requires this maneuver. On

a. Sockeye salmon at Pleasure Point Beach, Lake Washington

b. Sockeye salmon in Cedar River. The river connects with Lake Washington.

**FIGURE 17.9 Allopatric speciation among sockeye salmon.**

In Lake Washington, salmon that matured (**a**) at Pleasure Point Beach do not reproduce with those that matured in (**b**) the Cedar River. The females from Cedar River are noticeably larger and the males are more slender than those from Pleasure Point Beach, and these shapes help them reproduce in the river.

the other hand, the females tend to be larger than those at the beach. This larger body helps them dig slightly deeper nests in the gravel beds on the river bottom. Deeper nests are not disturbed by river currents and remain warm enough for egg viability.

Hendry has an independent way of telling beach salmon from river salmon. Ear stones called otoliths reflect variations in water temperature while a fish embryo is developing. Water temperatures at the beach are relatively constant compared to the river temperatures. By checking otoliths in adults, Hendry found that a third of the sockeye males at Pleasure Point Beach had grown up in the river. Yet the distinction between male and female shape and size in the two locations remains. Therefore, these males are not successful breeders along the beach. In other words, reproductive isolation has occurred.

### *Reinforcement of Reproductive Isolation*

As seen in sockeye salmon and other animals, a side effect to adaptive changes involving mating is reproductive isolation. Another example is seen among *Anolis* lizards in which males court females by extending a colorful flap of skin, called a "dewlap." The dewlap must be seen in order to attract mates. Populations of *Anolis* in a dim forest tend to evolve a light-colored dewlap, while populations in open habitats tend to evolve dark-colored ones. This change in dewlap color causes the populations to be reproductively isolated, because females distinguish males of their species by their dewlaps.

As populations become reproductively isolated, postzygotic isolating mechanisms may arise before prezygotic isolating mechanisms. As we have seen, when a horse and a mule reproduce, the hybrid or the offspring of a hybrid is not fertile. Therefore, natural selection would favor any variation in populations that prevents the occurrence of hybrids when they do not have offspring. Indeed, natural selection would favor the continual improvement of prezygotic isolating mechanisms until the two populations are completely reproductively isolated. The term reinforcement is given to the process of natural selection favoring variations that lead to reproductive isolation. An example of reinforcement has been seen in the pied and collared flycatchers of the Czech Republic and Slovakia, where both species occur in close proximity. Only here have the pied flycatchers evolved a different coat color than the collared flycatchers. The difference in color helps the two species recognize and mate with their own species.

## Adaptive Radiation

**Adaptive radiation** is a type of allopatric speciation and occurs when a single ancestral species gives rise to a variety of species, each adapted to a specific environment. An *ecological niche* is where a species lives and how it interacts with other species.

### *Examples of Adaptive Radiation*

When an ancestral finch arrived on the Galápagos Islands, its descendants spread out to occupy various niches. Geographic isolation of the various finch populations caused their gene pools to become isolated. Because of natural selection, each population adapted to a particular habitat on its island. In time, the many populations became so genotypically different that now, when by chance they reside on the same island, they do not interbreed, and are therefore separate species. During mating, finches use beak shape to recognize members of the same species, and suitors with the wrong type of beak are rejected.

Similarly, on the Hawaiian Islands, a wide variety of honeycreepers are descended from a common goldfinchlike ancestor that arrived from Asia or North America about 5 million years ago. Today, honeycreepers have a range of beak sizes and shapes for feeding on various food sources, including seeds, fruits, flowers, and insects (Fig. 17.10). Adaptive radiation also occurs among plants; a good example is the silversword alliance, which includes plants adapted to moist and dry environments and even lava fields.

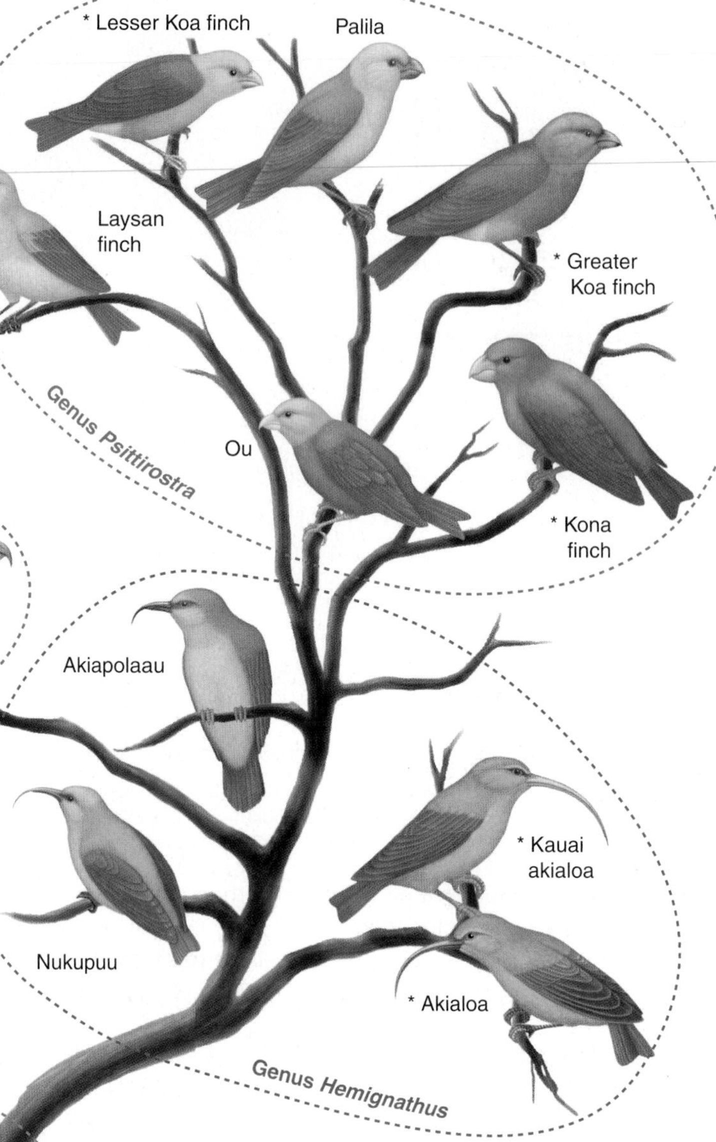


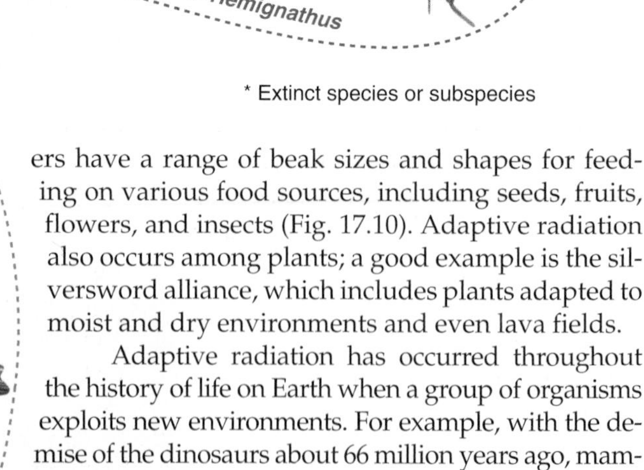


**FIGURE 17.10 Adaptive radiation in Hawaiian honeycreepers.**

More than 20 species, now classified in separate genera, evolved from a single species of a goldfinchlike bird that colonized the Hawaiian Islands.

Adaptive radiation has occurred throughout the history of life on Earth when a group of organisms exploits new environments. For example, with the demise of the dinosaurs about 66 million years ago, mammals underwent adaptive radiation as they exploited environments previously occupied by the dinosaurs. Mammals diversified in just 10 million years to include the early representatives of all the mammalian orders, including hoofed mammals (e.g., horses and pigs), aquatic mammals (e.g., whales and seals), primates (e.g., lemurs and monkeys), flying mammals (e.g., bats), and rodents (e.g., mice and squirrels). A changing world presented new environmental habitats and new food sources also. Insects fed on flowering plants and, in turn, became food for mammals. Primates lived in trees where fruits were available.

## Sympatric Speciation

Speciation without the presence of a geographic barrier is termed **sympatric speciation** [Gk. *sym*, together, and *patri*, fatherland]. Sympatric speciation has been difficult to substantiate in animals. For example, two populations of the Meadow Brown butterfly, *Maniola jurtina*, have different distributions of wing spots. The two populations are both in Cornwall, England, and they maintain the difference in wing spots, even though there is no geographic boundary between them. But, as yet, no reproductive isolating mechanism has been found.

In contrast, sympatric speciation involving **polyploidy** (a chromosome number beyond the diploid [2n] number) is well documented in plants. A polyploid plant can reproduce with itself, but cannot reproduce with the 2n population because not all the chromosomes would be able to pair during meiosis. Two types of polyploidy are known: autoploidy and alloploidy.

**Autoploidy** occurs when a diploid plant produces diploid gametes due to nondisjunction during meiosis (see Fig. 10.10). If this diploid gamete fuses with a haploid gamete, a triploid plant results. A triploid (3n) plant is sterile and cannot produce offspring because the chromosomes cannot pair during meiosis. Humans have found a use for sterile plants because they produce fruits without seeds. Figure 17.11 contrasts a diploid banana with seeds to today's polyploid banana that produces no seeds. If two diploid gametes fuse, the plant is a tetraploid (4n) and the plant is fertile, so long as it reproduces with another of its own kind. The fruits of polyploid plants are much larger than those of diploid plants. The huge strawberries of today are produced by octaploid (8n) plants.

**Alloploidy** [Gk. *allo*, other, and *ploidy*, uncountable] requires a more complicated process than autoploidy because it requires that two different but related species of plants hybridize (Fig. 17.12). Hybridization is followed by doubling of the chromosomes. For example, the Western wildflower, *Clarkia concinna*, is a diploid plant with fourteen chromosomes (seven pairs). The related species, *C. virgata*, is a diploid plant with ten chromosomes (five pairs). A hybrid of these two species is not fertile because seven chromosomes from one plant cannot pair evenly with five chromosomes from the other plant. However, meiosis occurs normally in the hybrid, *C. pulchella*, due to doubling of the chromosome number, which allows the chromosomes to pair during meiosis.

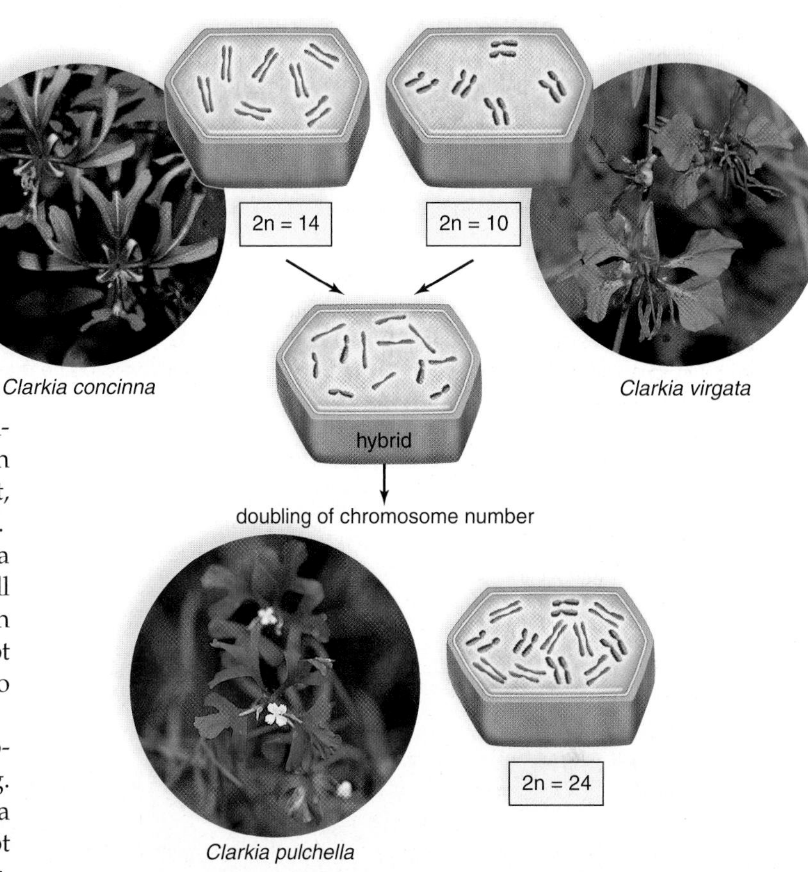


**FIGURE 17.12 Alloploidy produces a new species.**

Reproduction between two species of *Clarkia* is a sterile hybrid. Doubling of the chromosome number results in a fertile third *Clarkia* species that can reproduce with itself only.

Alloploidy also occurred during the evolution of the wheat plant, which is commonly used today to produce bread. The parents of our present-day bread wheat had 28 and 14 chromosomes, respectively. The hybrid with 21 chromosomes is sterile, but bread wheat with 42 chromosomes is fertile because the chromosomes can pair during meiosis. Recent molecular data tell us that polyploidy is common in plants and makes a significant contribution to the evolution of new plants.

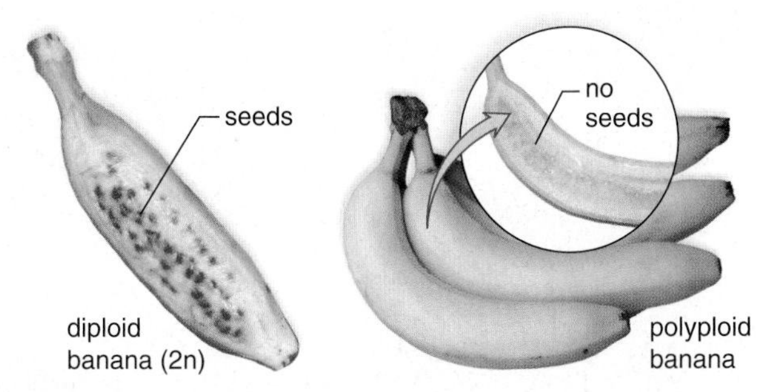


**FIGURE 17.11 Autoploidy produces a new species.**

The small, diploid-seeded banana is contrasted with the large, polyploid banana that produces no seeds.

### Check Your Progress 17.2

1. Five species of big cats are classified in a single genus: *Panthera leo* (lion), *P. tigris* (tiger), *P. pardus* (leopard), *P. onca* (jaguar), and *P. uncia* (snow leopard). What evidence would you need to show that this is a case of adaptive radiation?
2. What fossil evidence might support the hypothesis that the different species of cats arose sympatrically?

science focus

## The Burgess Shale Hosts a Diversity of Life

Finding the Burgess Shale, a rock outcropping in Yoho National Park, British Columbia, was a chance happening. In 1909, Charles Doolittle Walcott of the Smithsonian Institution was out riding when his horse stopped in front of a rock made of shale. He cracked the rock open and saw the now-famous fossils of the animals depicted in Figure 17A. Walcott and his team began working the site and continued on their own for quite a few years. Around 1960, other paleontologists became interested in studying the Burgess Shale fossils.

As a result of uplifting and erosion, the intriguing fossils of the Burgess Shale are relatively common in that particular area. However, the highly delicate impressions and films found in the rocks are very difficult to remove from their matrix. Early attempts to remove the fossils involved splitting the rocks along their sedimentary plane and using rock saws. Unfortunately, these methods were literally "shots in the dark," and many valuable fossils were destroyed in the process. New methods, involving ultraviolet light to see the fossils and diluted acetic acid solutions to remove the matrix, have been more successful in freeing the fossils.

The fossils tell a remarkable story of marine life some 540 MYA (million years ago), during the Precambrian era. In addition to fossils of organisms that had external skeletons, many of the fossils are remains of soft-bodied invertebrates; these are a great find because soft-bodied animals rarely fossilize. During this time, all organisms lived in the sea, and it is believed the barren land was subject to mudslides, which entered the ocean and buried the animals, killing them. Later, the mud turned into shale, and later still, an upheaval raised the shale. Be-

**FIGURE 17A Burgess Shale.**

Burgess Shale quarry (*above*), where many ancient fossils (shown in Fig. 17B) have been found.

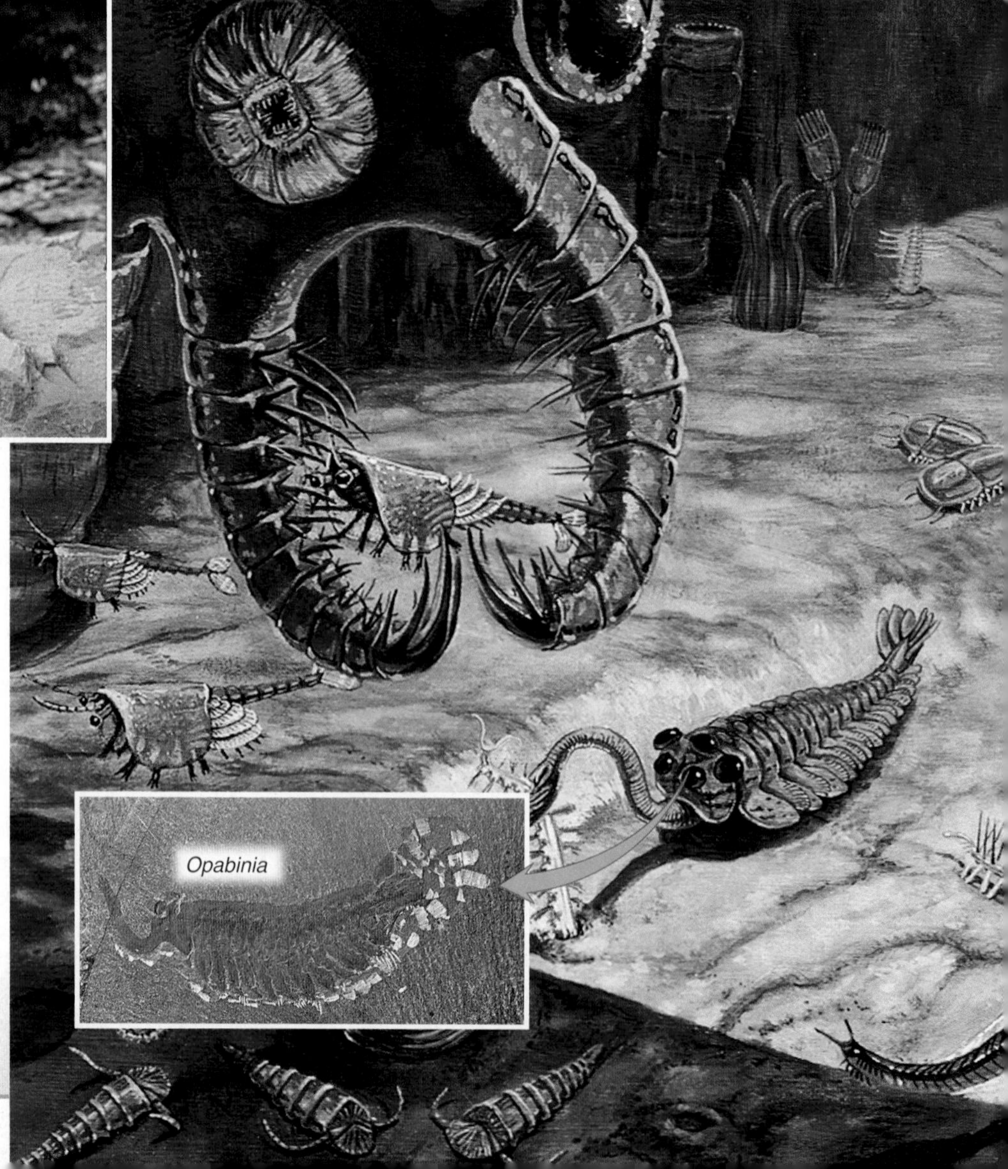


**FIGURE 17B Fossils found at the Burgess Shale.**

Variety of fossils alongside drawings of the animals based on their fossilized remains.

fore the shale formed, fine mud particles filled the spaces in and around the organisms so that the soft tissues were preserved and the fossils became somewhat three-dimensional.

The fossils tell us that the ancient seas were teaming with weird-looking, mostly invertebrate animals (Fig. 17B). All of today's groups of animals can trace their ancestry to one of these strange-looking forms, which include sponges, arthropods, worms, and trilobites, as well as spiked creatures and oversized predators. The vertebrates, like ourselves, are descended from *Pikaia,* the only one of the fossils that has a supporting rod called a notochord. (In vertebrates, the notochord is replaced by the vertebral column during development.)

Unicellular organisms have also been preserved at the Burgess Shale site. They appear to be bacteria, cyanobacteria, dinoflagellates, and other protists. Fragments of algae are preserved in thin, shiny carbon films. A technique has been perfected that allows the films to be peeled off the rocks.

Anyone can travel to Yoho National Park, look at the fossils, and get an idea of the types of animals that dominated the world's oceans for nearly 300 million years. Some of the animals had external skeletons, but many were soft-bodied. Interpretations of the fossils vary. Some authorities hypothesize that the great variety of animals in the Burgess Shale evolved within 20–50 million years, and therefore the site supports the hypothesis of punctuated equilibrium. Others believe that the animals started evolving much earlier and that we are looking at the end result of an adaptive radiation requiring many more millions of years to accomplish. Some investigators present evidence that all the animals are related to today's animals and should be classified as such. Others believe that several of them are unique creatures unrelated to the animals of today. Regardless of the controversies, the fossils tell us that speciation, diversification, and eventual extinction are part of the history of life.

## 17.3 Principles of Macroevolution

Many evolutionists hypothesize, as Darwin did, that **macroevolution,** which is evolution at the species or higher level of classification, occurs gradually. After all, natural selection can only do so much to bring about change in each generation. Therefore, these evolutionists support a *gradualistic model,* which proposes that speciation occurs after populations become isolated, with each group continuing slowly on its own evolutionary pathway. These evolutionists often show the history of groups of organisms by drawing the type of diagram shown in Figure 17.13*a*. Note that in this diagram, an ancestral species has given rise to two separate species, represented by a slow change in plumage color. The gradualistic model suggests that it is difficult to indicate when speciation occurred because there would be so many transitional links (see Fig. 15.12).

After studying the fossil record, some paleontologists tell us that species can appear quite suddenly, and then they remain essentially unchanged phenotypically during a period of stasis (sameness) until they undergo extinction. Based on these findings, they developed a *punctuated equilibrium model* to explain the pace of evolution. This model says that periods of equilibrium (no change) are punctuated (interrupted) by speciation. Figure 17.13*b* shows this way of representing the history of evolution over time. This model suggests that transitional links are less likely to become fossils and less likely to be found. Moreover, speciation is more likely to involve only an isolated population at one locale, because a favorable genotype could spread more rapidly within such a population. Only when this population expands and replaces other species is it apt to show up in the fossil record.

A strong argument can be made that it is not necessary to choose between these two models of evolution and that both could very well assist us in interpreting the fossil record. In other words, some fossil species may fit one model, and some may fit the other model. In a stable environment, a species may be kept in equilibrium by stabilizing selection for a long period. On the other hand, if the environment changes slowly, a species may be able to adapt gradually. If environmental change is rapid, a new species may arise suddenly before the parent species goes on to extinction. The differences between all possible patterns of evolutionary change are rather subtle, especially when we consider that, because geologic time is measured in millions of years, the "sudden" appearance of a new species in the fossil record could actually represent many thousands of years. Using only a small rate of change (.0008/year), two investigators calculated that the brain size in the human lineage could have increased from 900 $cm^3$ to 1,400 $cm^3$ in only 135,000 years. This would appear to be a very rapid change in the fossil record, and it actually took much longer (about 500,000 years), indicating that the real pace was slower than it might have been.

The difficulty of deciding the tempo of evolution from examining the fossil record is exemplified by a review of the forms of life fossilized in the Burgess Shale (see the Science Focus on pages 308–9). It is difficult to tell how rapidly these animals evolved.

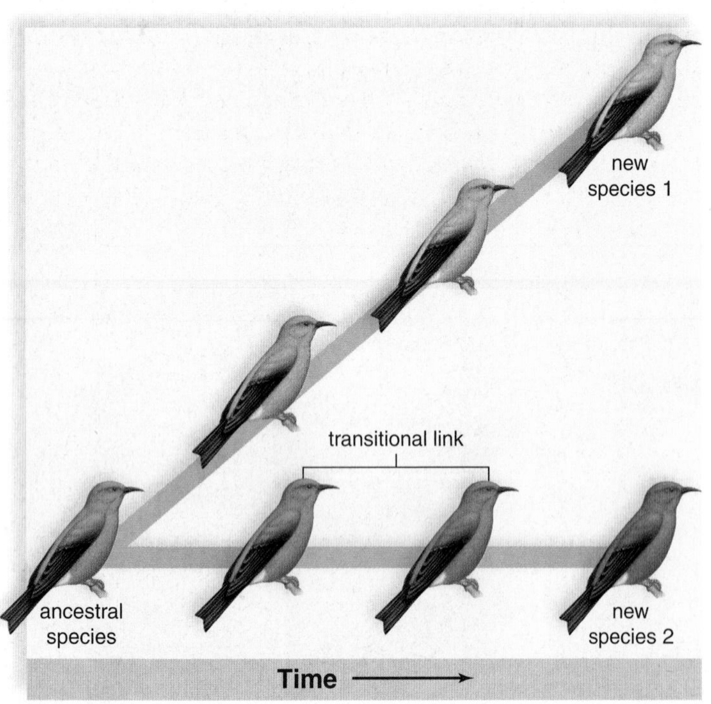


a. Gradualistic model

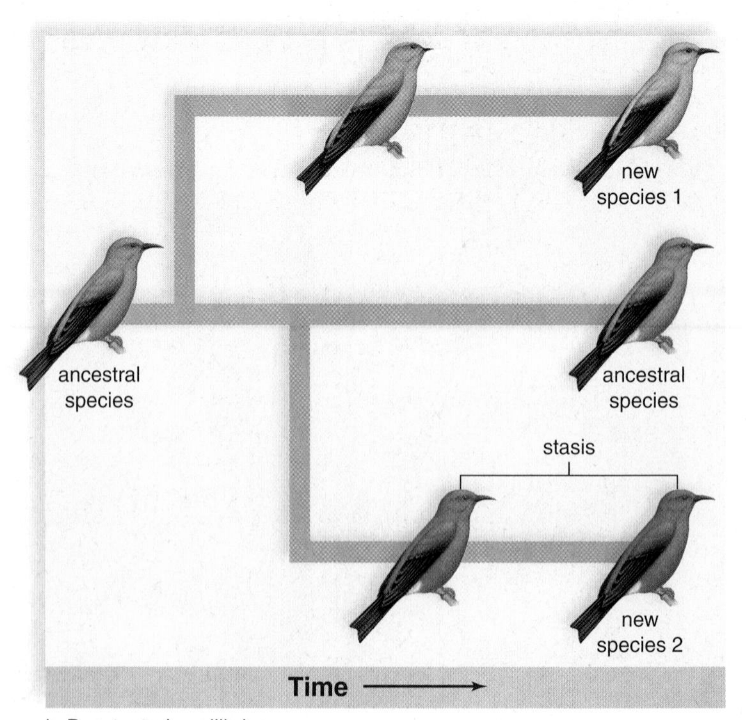


b. Punctuated equilibrium

**FIGURE 17.13 Gradualistic and punctuated equilibrium models.**

**a.** Speciation occurs gradually and many transitional links occur. Therefore, apparent stasis (sameness) is not real. **b.** Speciation occurs rapidly, transitional links do not occur, and stasis is real.

## Developmental Genes and Macroevolution

Investigators have discovered genes that can bring about radical changes in body shapes and organs. For example, it is now known that the *Pax6* gene is involved in eye formation in all animals, and that homeotic (*Hox*) genes determine the location of repeated structures in all vertebrates

## Gene Expression Can Influence Development

Whether slow or fast, how could evolution have produced the myriad of animals in the Burgess Shale and, indeed, in the history of life? Or, to ask the question in a genetic context, how can genetic changes bring about such major differences in form? It has been suggested since the time of Darwin that the answer must involve development processes. In 1917, D'Arcy Thompson asked us to imagine an ancestor in which all parts are developing at a particular rate. A change in gene expression could stop a developmental process or continue it beyond its normal time. For instance, if the growth of limb bones were stopped early, the result would be shorter limbs, and if it were extended, the result would be longer limbs compared to those of an ancestor. Or, if the whole period of growth were extended, a larger animal would result, accounting for why some species of horses are so large today.

Using new kinds of microscopes and the modern techniques of cloning and manipulating genes, investigators have indeed discovered genes whose differential expression can bring about changes in body shapes and organs. This result suggests that these genes must date back to a common ancestor that lived more than 600 MYA (before the Burgess Shale animals), and that despite millions of years of divergent evolution, all animals share the same control switches for development.

### *Development of the Eye*

The animal kingdom contains many different types of eyes, and it was long thought that each type would require its own set of genes. Flies, crabs, and other arthropods have compound eyes that have hundreds of individual visual units. Humans and all other vertebrates have a camera-type eye with a single lens. So do squids and octopuses. Humans are not closely related to either flies or squids, so wouldn't it seem as if all three types of animals evolved "eye" genes separately? Not so. In 1994, Walter Gehring and his colleagues at the University of Basel, Switzerland, discovered that a gene called *Pax6* is required for eye formation in all animals tested (Fig. 17.14). Mutations in the *Pax6* gene lead to failure of eye development in both people and mice, and remarkably, the mouse *Pax6* gene can cause an eye to develop on the leg of a fruit fly (Fig. 17.15).

### *Development of Limbs*

Wings and arms are very different, but both humans and birds express the *Tbx5* gene in developing limb buds. *Tbx5* codes for a transcription factor that turns on the genes needed to make a limb. What seems to have changed as birds and humans evolved are the genes that *Tbx5* turns on. Perhaps in an ancestral tetrapod, the *Tbx5* protein triggered the transcription of only one gene. In humans and birds, a few genes are expressed in response to *Tbx5* protein, but the particular genes are different. There is also the question of timing. Changing the timing of gene expression, as well as which genes are expressed, can result in dramatic changes in shape.

### *Development of Overall Shape*

Vertebrates have repeating segments, as exemplified by the vertebral column. Changes in the number of segments can lead to changes in overall shape. In general, *Hox* genes control the number and appearance of repeated structures along the main body axes of vertebrates. Shifts in when *Hox* genes are expressed in embryos are responsible for why a snake has hundreds of rib-bearing vertebrae and essentially no neck in contrast to other vertebrates, such as a

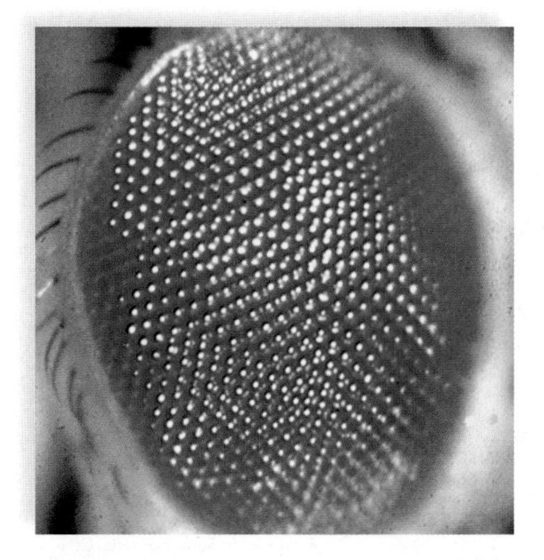
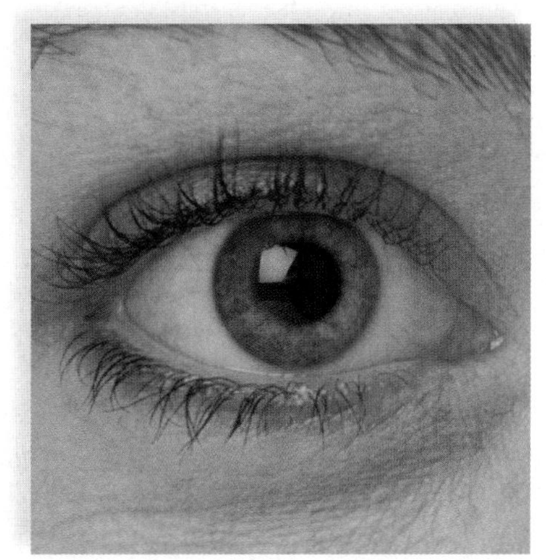
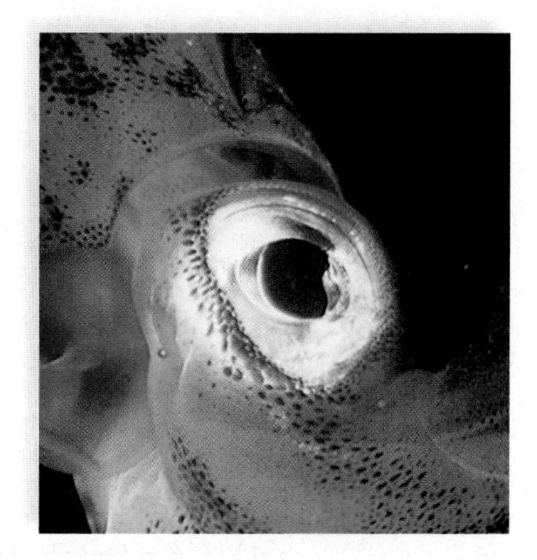

**FIGURE 17.14** ***Pax6* gene and eye development.**
*Pax6* is involved in eye development in a fly, a human, and a squid.

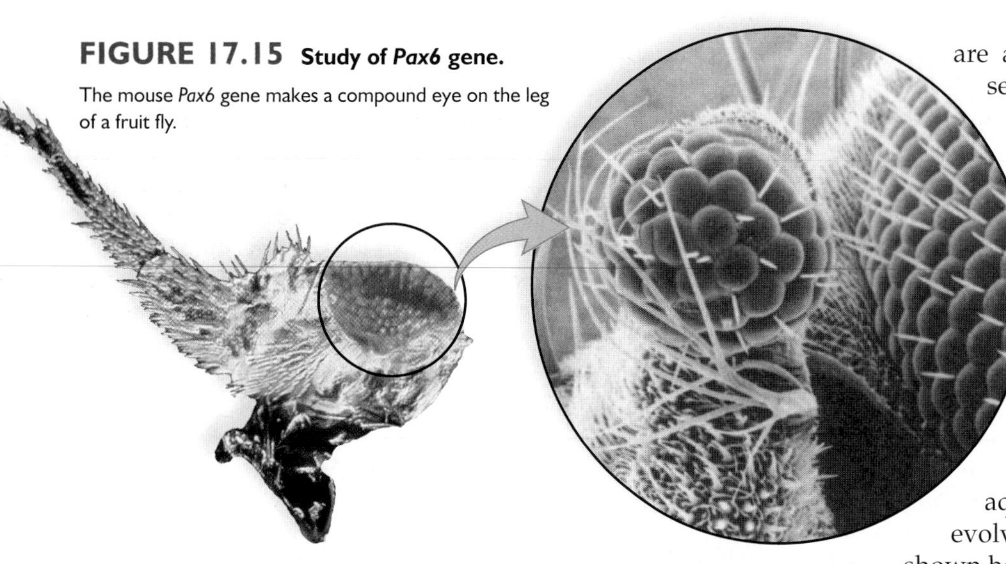

**FIGURE 17.15** **Study of *Pax6* gene.**

The mouse *Pax6* gene makes a compound eye on the leg of a fruit fly.

chick (Fig. 17.16). *Hox* genes have been found in all animals, and other shifts in the expression of these genes can explain why insects have just six legs and other arthropods, such as crayfish, have ten legs. In general, the study of *Hox* genes has shown how animal diversity is due to variations in the expression of ancient genes rather than to wholly new and different genes.

## *Pelvic-Fin Genes*

The three-spined stickleback fish occurs in two forms in North American lakes. In the open waters of a lake, long pelvic spines help protect the stickleback from being eaten by large predators. But on the lake bottom, long pelvic spines are a disadvantage because dragonfly larvae seize and feed on young sticklebacks by grabbing them by their spines.

The presence of short spines in bottom-dwelling fish can be traced to a reduction in the development of the pelvic-fin bud in the embryo, and this reduction is due to the altered expression of a particular gene. Hindlimb reduction has occurred during the evolution of other vertebrates. The hindlimbs became greatly reduced in size as whales and manatees evolved from land-dwelling ancestors into fully aquatic forms. Similarly, legless lizards have evolved many times. The stickleback study has shown how natural selection can lead to major skeletal changes in a relatively short time.

## *Human Evolution*

The sequencing of genomes has shown us that our DNA base sequence is very similar to that of chimpanzees, mice, and, indeed, all vertebrates. Based on this knowledge and the work just described, investigators no longer expect to find new genes to account for the evolution of humans. Instead, they predict that differential gene expression and/or new functions for "old" genes will explain how humans evolved.

Mutations of developmental genes occur by chance, and in the next section, we observe that evolution is not directed toward any particular end.

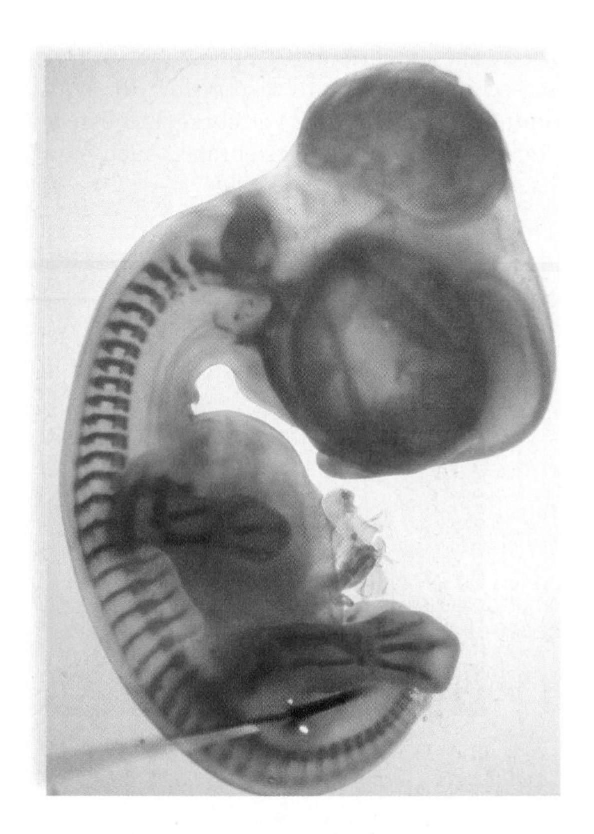

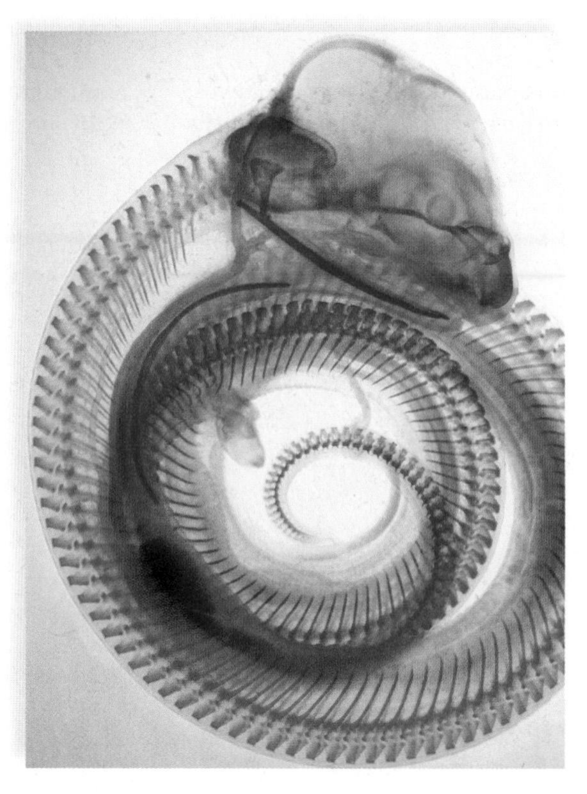

**FIGURE 17.16**
***Hox6* genes.**

Differential expression of *Hox6* genes causes a chick to have seven vertebrae (*purple*) and a snake to have many more vertebrae (*purple*).

Burke, A. C. 2000, *Hox* genes and the global patterning of the somitic mesoderm. In Somitogenesis. C. Ordahl (ed.), *Current Topics in Developmental Biology*, Vol. 47. Academic Press.

## Macroevolution Is Not Goal-Oriented

The evolution of the horse, *Equus*, has been studied since the 1870s, and at first the ancestry of this genus seemed to represent a model for gradual, straight-line evolution until its goal, the modern horse, had been achieved. Three trends were particularly evident during the evolution of the horse: increase in overall size, toe reduction, and change in tooth size and shape.

By now, however, many more fossils have been found, making it easier to tell that the lineage of a horse is complicated by the presence of many ancestors with varied traits. Which fossils represent the direct ancestors to *Equus* is not known. After all, humans, not nature, drew the oversimplified tree in Figure 17.17. The tree is an oversimplification because each of the names is a genus that contains several species, and not all past genera in the horse family are included. It is apparent, then, that the ancestors of *Equus* form a thick bush of many equine species and that straight-line evolution did not occur. Because *Equus* alone remains and the other genera have become extinct, it might seem as if evolution was directed toward producing *Equus*, but this is not the case. Instead, each of these ancestral species was adapted to its environment. Adaptation occurs only because the members of a population with an advantage are able to have more offspring than other members. Natural selection is opportunistic, not goal-directed.

Fossils named *Hyracotherium* have been designated as the first probable members of the horse family, living about 57 MYA. These animals had a wooded habitat, ate leaves and fruit, and were about the size of a dog. Their short legs and broad feet with several toes would have allowed them to scamper from thicket to thicket to avoid predators. *Hyracotherium* was obviously well adapted to its environment because this genus survived for 20 million years.

The family tree of *Equus* does tell us once more that speciation, diversification, and extinction are common occurrences in the fossil record. The first adaptive radiation of horses occurred about 35 MYA. The weather was becoming drier, and grasses were evolving. Eating grass requires tougher teeth, and an increase in size and longer legs would have permitted greater speed to escape enemies. The second adaptive radiation of horses occurred about 15 MYA and included *Merychippus* as a representative of those groups that were speedy grazers who lived on the open plain. By 10 MYA, the horse family was quite diversified. Some species were large forest browsers, some were small forest browsers, and others

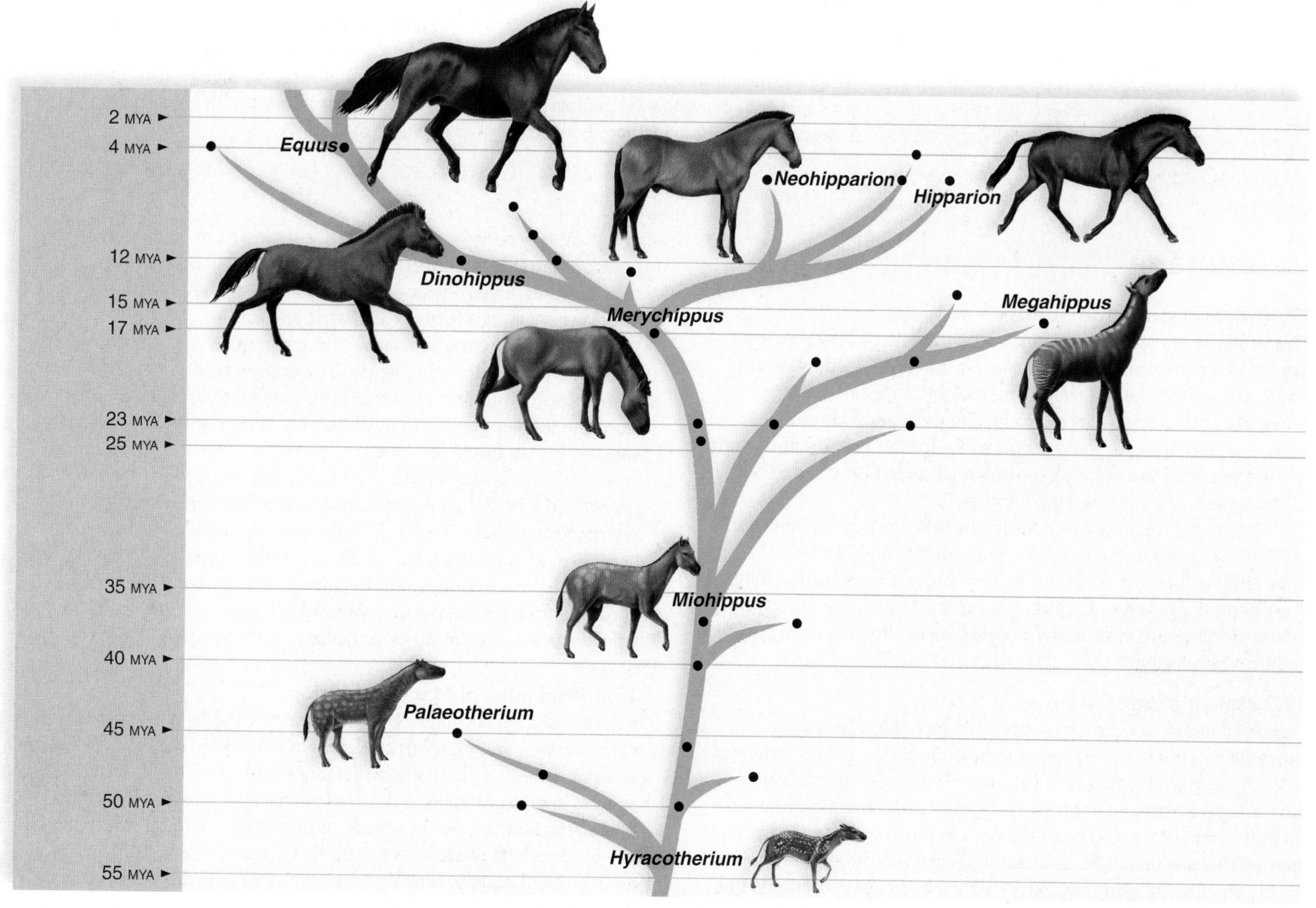


**FIGURE 17.17 Simplified family tree of *Equus*.**

Every dot represents a genus.

were large plains grazers. Many species had three toes, but some had one strong toe. (The hoof of the modern horse includes only one toe.)

Modern horses evolved about 4 MYA from ancestors who had features that are adaptive for living on an open plain, such as large size, long legs, hoofed feet, and strong teeth. The other groups of horses prevalent at the time became extinct no doubt for complex reasons. Humans have corralled modern horses for various purposes, and this makes it difficult to realize that the traits of a modern horse are adaptive for living in a grassland environment.

**Check Your Progress** **17.3**

1. Would the presence of ligers in the fossil record be used as evidence for a gradualistic or a punctuated equilibrium model of evolution? Explain.
2. How does a study of developmental genes support the possibility of rapid speciation in the fossil record? Explain.
3. Why does it seem that differential expression must occur during the development of ligers?
4. There are only five species of cats in the genus *Panthera*. Does this represent a goal of evolution?

## Connecting the Concepts

Macroevolution, the study of the origin and history of the species on Earth, is the subject of this chapter and the next. The biological species concept states that the members of a species have an isolated gene pool and can only reproduce with one another.

This chapter concerns speciation. Speciation usually occurs after two populations have been separated geographically. If two populations of salamanders were suddenly separated by a barrier, each population would become adapted to its particular environment over time. Eventually, the two populations might become so genetically different that even if members of each population came in contact, they would not be able to produce fertile offspring. Because gene flow between the two populations would no longer be possible, the salamanders would be considered separate species. Aided by geographic separation, multiple species can repeatedly arise from an ancestral species, as when a common ancestor from the mainland led to 13 species of Galápagos finches, each adapted to its own particular environment.

Does speciation occur gradually, as Darwin supposed, or rapidly (in geologic time), as described by the punctuated equilibrium model? The fossils of the Burgess Shale support the punctuated equilibrium model. How can genetic changes bring about such major changes in form, whether fast or slow? Investigators have now discovered ancient genes whose differential expression can bring about changes in body shapes and organs.

Evolution is not directed toward any particular end, and the traits of the species alive today arose through common descent with adaptations to a local environment. The subject of Chapter 18 is the origin and history of life.

## summary

### 17.1 Separation of the Species

The evolutionary species concept recognizes that every species has its own evolutionary history and can be recognized by certain diagnostic morphological traits. The biological species concept recognizes that a species is reproductively isolated from other species and, therefore, the members of a species breed only among themselves. The use of DNA sequence data can also be used today to distinguish one species from another.

Prezygotic isolating mechanisms (habitat, temporal, behavior, mechanical, and gamete isolation) prevent mating from being attempted or prevent fertilization from being successful if mating is attempted. Postzygotic isolating mechanisms (zygote mortality, hybrid sterility, and $F_2$ fitness) prevent hybrid offspring from surviving and/or reproducing.

### 17.2 Modes of Speciation

During allopatric speciation, geographic separation precedes reproductive isolation. Geographic isolation allows genetic changes to occur so that the ancestral species and the new species can no longer breed with one another. A series of salamander subspecies on either side of the Central Valley of California has resulted in two species that are unable to successfully reproduce when they come in contact. During sympatric speciation, a geographic barrier is not required, and speciation is simply a change in genotype that prevents successful reproduction. The best example of sympatric speciation is occurrence of polyploidy in plants.

Adaptive radiation is multiple speciation from an ancestral population because varied habitats permit varied adaptations to occur. Adaptive radiation, as exemplified by the Hawaiian honeycreepers, is a form of allopatric speciation.

During sympatric speciation, no geographic separation precedes reproductive isolation. In plants, the occurrence of polyploidy (chromosome number above 2n) reproductively isolates an offspring from the former generation. Autoploidy occurs within the same species. For example, if—due to nondisjunction—a diploid gamete fuses with a haploid gamete, a triploid plant results that cannot reproduce with 2n plants because some of the chromosomes would not pair during meiosis. Alloploidy occurs when two species hybridize. If an odd number of chromosomes results, the hybrid is sterile unless a doubling of the chromosomes occur and the chromosomes can pair during meiosis. The sex chromosomes in animals makes speciation by polyploidy highly unlikely.

### 17.3 Principles of Macroevolution

Macroevolution is evolution of new species and higher levels of classification. The fossil record, such as is found in the Burgess Shale, gives us a view of life many millions of years ago. The hypothesis that species evolve gradually is now being challenged by the hypothesis that speciation can occur rapidly. In that case, the fossil record could show periods of stasis interrupted by spurts of change, for example, a punctuated equilibrium. Transitional fossils would be expected with gradual change but not with punctuated equilibrium.

It could be that both models are seen in the fossil record, but rapid change can occur by differential expression of regulatory genes. The same regulatory gene (*Pax6*) controls the development of both the camera-type and the compound-type eye. *Tbx5* gene controls development of limbs, whether the wing of a bird or the leg of a tetrapod. *Hox* genes control the number and appearance of a repeated structure along the main body axes of vertebrates. The same pelvic-fin genes control the development of a pelvic girdle. Changing the timing of gene expression, as well as which genes are expressed, can result in dramatic changes in shape.

Speciation, diversification, and extinction are seen during the evolution of *Equus*. These three processes are commonplace in the fossil record and illustrate that macroevolution is not goal-directed. The life we see about us represents adaptations to particular environments. Such adaptations have changed in the past and will change in the future.

## understanding the terms

adaptive radiation 306
allopatric speciation 304
alloploidy 307
autoploidy 307
biological species concept 300
evolutionary species concept 300
macroevolution 310
polyploidy 307
postzygotic isolating mechanism 303
prezygotic isolating mechanism 302
speciation 304
sympatric speciation 307

Match the terms to these definitions:

a. ______________ Anatomic or physiological difference between two species that prevents successful reproduction after mating has taken place.

b. ______________ Evolution of a large number of species from a common ancestor.

c. ______________ Origin of new species due to the evolutionary process of descent with modification.

d. ______________ Origin of new species after populations have been separated geographically.

## reviewing this chapter

1. Give the pros and cons of the evolutionary species concept and the biological species concept. Give an example to show that DNA sequence data can distinguish species. 300–301
2. List and discuss five prezygotic isolating mechanisms and three postzygotic isolating mechanisms. 302–3
3. Use the *Ensatina* salamander example to explain allopatric speciation. 304
4. Use the honeycreepers of Hawaii and the Galápagos finches to explain adaptive radiation. 306
5. How does sympatric speciation differ from allopatric speciation, and why is sympatric speciation common in plants but rare in animals. 307
6. Does the Burgess Shale help us determine whether speciation occurs quickly or slowly? Explain. 308–9
7. With regard to the speed of speciation, how do the gradualistic model and the punctuated equilibrium model differ? Which model predicts the occurrence of many transitional fossils? Explain. 310
8. What types of genes are pertinent to the discussion of the speed of speciation? Explain. 311–12
9. Use the evolution of the horse to show that speciation is not goal-oriented. 313–14

## testing yourself

Choose the best answer for each question.

1. A biological species
   a. always looks different from other species.
   b. always has a different chromosome number from that of other species.
   c. is reproductively isolated from other species.
   d. never occupies the same niche as other species.

For questions 2–7, indicate the type of isolating mechanism described in each scenario.

**KEY:**

a. habitat isolation
b. temporal isolation
c. behavioral isolation
d. mechanical isolation
e gamete isolation
f. zygote mortality
g. hybrid sterility
h. low $F_2$ fitness

2. Males of one species do not recognize the courtship behaviors of females of another species.
3. One species reproduces at a different time than another species.
4. A cross between two species produces a zygote that always dies.
5. Two species do not interbreed because they occupy different areas.
6. The sperm of one species cannot survive in the reproductive tract of another species.
7. The offspring of two hybrid individuals exhibit poor vigor.
8. Which of these is a prezygotic isolating mechanism?
   a. habitat isolation
   b. temporal isolation
   c. hybrid sterility
   d. zygote mortality
   e. Both a and b are correct.
9. Male moths recognize females of their species by sensing chemical signals called pheromones. This is an example of
   a. gamete isolation.
   b. habitat isolation.
   c. behavorial isolation.
   d. mechanical isolation.
   e. temporal isolation.
10. Which of these is mechanical isolation?
    a. Sperm cannot reach or fertilize an egg.
    b. Courtship pattern differs.
    c. The organisms live in different locales.
    d. The organisms reproduce at different times of the year.
    e. Genitalia are unsuitable to each other.
11. Complete the following diagram illustrating allopatric speciation by using these phrases: genetic changes (used twice), geographic barrier, species 1, species 2, species 3.

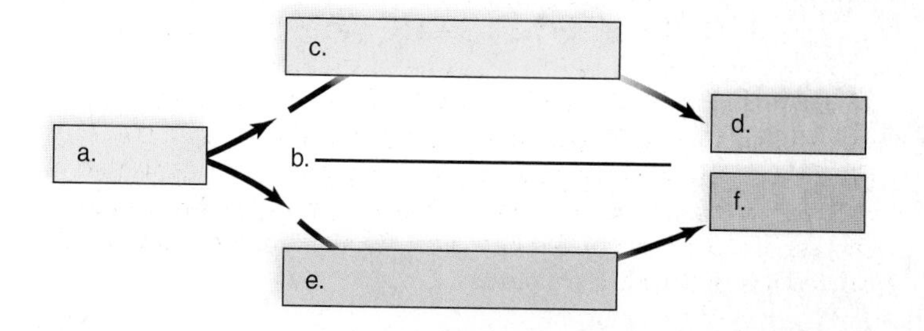

12. The creation of new species due to geographic barriers is called
    a. isolation speciation.
    b. allopatric speciation.
    c. allelomorphic speciation.
    d. sympatric speciation.
    e. symbiotic speciation.
13. The many species of Galápagos finches are each adapted to eating different foods. This is the result of
    a. gene flow.
    b. adaptive radiation.
    c. sympatric speciation.
    d. genetic drift.
    e. All of these are correct.
14. Allopatric, but not sympatric, speciation requires
    a. reproductive isolation.
    b. geographic isolation.
    c. spontaneous differences in males and females.
    d. prior hybridization.
    e. rapid rate of mutation.
15. Which of the following is not a characteristic of plant alloploidy?
    a. hybridization
    b. chromosome doubling
    c. related species mating
    d. All of these are characteristics of plant alloploidy.
16. Corn is an allotetraploid, which means that its
    a. chromosome number is 4n.
    b. occurrence resulted from hybridization.
    c. occurrence required a geographic barrier.
    d. Both a and b are correct.
17. Transitional links are least likely to be found if evolution proceeds according to the
    a. gradualistic model.
    b. punctuated equilibrium model.
18. Adaptive radiation is only possible if evolution is punctuated.
    a. true
    b. false
19. Why are there no fish fossils in the Burgess Shale?
    a. The habitat was not aquatic.
    b. Fish do not fossilize easily because they do not have shells.
    c. The fossils of the Burgess Shale predate vertebrate animals.
    d. There are fish fossils in the Burgess Shale.
20. Which of the following can influence the rapid development of new types of animals?
    a. The influence of molecular clocks.
    b. A change in the expression of regulating genes.
    c. The sequential expression of genes.
    d. All of these are correct.
21. Which gene is incorrectly matched to its function?
    a. *Hox*—body shape
    b. *Pax6*—body segmentation
    c. *Tbx5*—limb development
    d. All of these choices are correctly matched.
22. In the evolution of the modern horse, which was the goal of the evolutionary process?
    a. large size
    b. single toe
    c. Both a and b are correct.
    d. Neither a nor b is correct.
23. Which of the following was not a characteristic of *Hyracotherium*, an ancestral horse genus?
    a. small size
    b. single toe
    c. wooded habitat
    d. All of these are characteristics of *Hyracotherium*.
24. Which statement about speciation is not true?
    a. Speciation can occur rapidly or slowly.
    b. Developmental genes can account for rapid speciation.
    c. The fossil record gives no evidence that speciation can occur rapidly.
    d. Speciation always requires genetic changes, such as mutations, genetic drift, and natural selection.
25. Which statement concerning allopatric speciation would come first?
    a. Genetic and phenotypic changes occur.
    b. Subspecies have a three-part name.
    c. Two subpopulations are separated by a barrier.

## thinking scientifically

1. You want to decide what definition of a species to use in your study. What are the advantages and disadvantages of one based on DNA sequences as opposed to the evolutionary and biological species concept?
2. You decide to create a hybrid by crossing two species of plants. If the hybrid is a fertile plant that produces normal size fruit, what conclusion is possible?

## *Biology* website

The companion website for *Biology* provides a wealth of information organized and integrated by chapter. You will find practice tests, animations, videos, and much more that will complement your learning and understanding of general biology.

**http://www.mhhe.com/maderbiology10**

# 18 Origin and History of Life

*today, paleontologists are setting the record straight about dinosaurs. It now appears that dinosaurs nested in the same manner as birds! Bowl-shaped nests containing dinosaur eggs have been found in Mongolia, Argentina, and Montana. The nests contain fossilized eggs and bones along with eggshell fragments. From this evidence, it seems that baby dinosaurs stayed in the nest after hatching until they were big enough to walk around and fragment the eggshells. The spacing between the nests suggests that this dinosaur,* Maisaura *(meaning good mother lizard in Greek), fed its young. The remains of an enormous herd of about 10,000 found in Montana is further evidence that* Maisaura *was indeed social in their behavior.*

Maisaura *and* Microraptor, *the winged gliding dinosaur featured below, provide us with structural and behavioral evidence of the link between dinosaurs and birds. In this chapter, we trace the origin of life before considering the history of life.*

Many transitional forms, such as *Microraptor* shown here in reconstruction, indicate a link between dinosaurs and birds.

## concepts

## 18.1 Origin of Life

We have no data that suggests life arises spontaneously from nonlife, and we say that "life comes only from life." But if this is so, how did the first form of life come about? Since it was the very first living thing, it had to come from nonliving chemicals. Could there have been an increase in the complexity of the chemicals—could a **chemical evolution** have produced the first cell(s) on early Earth?

### The Early Earth

The sun and the planets, including Earth, probably formed over a 10-billion-year period from aggregates of dust particles and debris. At 4.6 billion years ago (BYA), the solar system was in place. Intense heat produced by gravitational energy and radioactivity produced several stratified layers. Heavier atoms of iron and nickel became the molten liquid core, and dense silicate minerals became the semiliquid mantle. Upwellings of volcanic lava produced the first crust.

The Earth's mass is such that the gravitational field is strong enough to have an atmosphere. Less mass and atmospheric gases would escape into outer space. The early Earth's atmosphere was not the same as today's atmosphere; it was produced primarily by outgassing from the interior, exemplified by volcanic eruptions. The early atmosphere most likely consisted mainly of these inorganic chemicals: water vapor ($H_2O$), nitrogen ($N_2$), and carbon dioxide ($CO_2$), with only small amounts of hydrogen ($H_2$), methane ($CH_4$), ammonia ($NH_3$), hydrogen sulfide ($H_2S$), and carbon monoxide (CO). The early atmosphere may have been a reducing atmosphere, with little free oxygen. If so, that would have been fortuitous because oxygen ($O_2$) attaches to organic molecules, preventing them from joining to form larger molecules.

At first it was so hot that water was present only as a vapor that formed dense, thick clouds. Then, as the early Earth cooled, water vapor condensed to liquid water, and rain began to fall. It rained in such enormous quantity over hundreds of millions of years that the oceans of the world were produced. Our planet is an appropriate distance from the sun: any closer, water would have evaporated; any farther away, water would have frozen.

It is also possible that the oceans were fed by celestial comets that entered the Earth's gravitational field. In 1999, physicist Louis Frank presented images taken by cameras on NASA's *Polar* satellite to substantiate his claim that the Earth is bombarded with 5–30 icy comets the size of a house every minute. The ice becomes water vapor that later comes down as rain, enough rain, says Frank, to raise the oceans' level by an inch in just 10,000 years.

### Monomers Evolve

Several hypotheses suggest how organic monomers could have evolved, and these are that (1) monomers came from outer space, (2) monomers came from reactions in the atmosphere, or (3) monomers came from reactions at hydrothermal vents.

#### *Alternative Hypotheses*

In support of monomers coming from outer space, we know that comets and meteorites have constantly pelted the Earth throughout history. In recent years, scientists have confirmed the presence of organic molecules in some meteorites. Many scientists, championed by Chandra Wickramsinghe, feel that these organic molecules could have seeded the chemical origin of life on early Earth. Others even hypothesize that bacterium-like cells evolved first on another planet and then were carried to Earth. A meteorite from Mars labeled ALH84001 landed on Earth some 13,000 years ago. When examined, experts found tiny rods similar in shape to fossilized bacteria. Therefore this hypothesis is being investigated.

In support of monomers coming from reactions in the atmosphere, Stanley Miller performed a famous experiment in 1953 (Fig. 18.1). He was following up on the suggestion made by A. I. Oparin (a Russian biochemist) and J. B. S. Haldane (a Scottish physiologist and geneticist) who independently suggested in the early 1900s that monomers could have been produced from early atmospheric gases in the presence of strong energy sources. The energy sources on early Earth included heat from volcanoes and meteorites, radioactivity from isotopes, powerful electric discharges in lightning, and solar radiation, especially ultraviolet radiation. Oparin's idea is called abiotic synthesis, the formation of simple monomers (sugars, amino acids, nucleotide bases) from inorganic molecules.

For his experiment, Miller placed a mixture resembling a strongly reducing atmosphere—methane ($CH_4$), ammonia

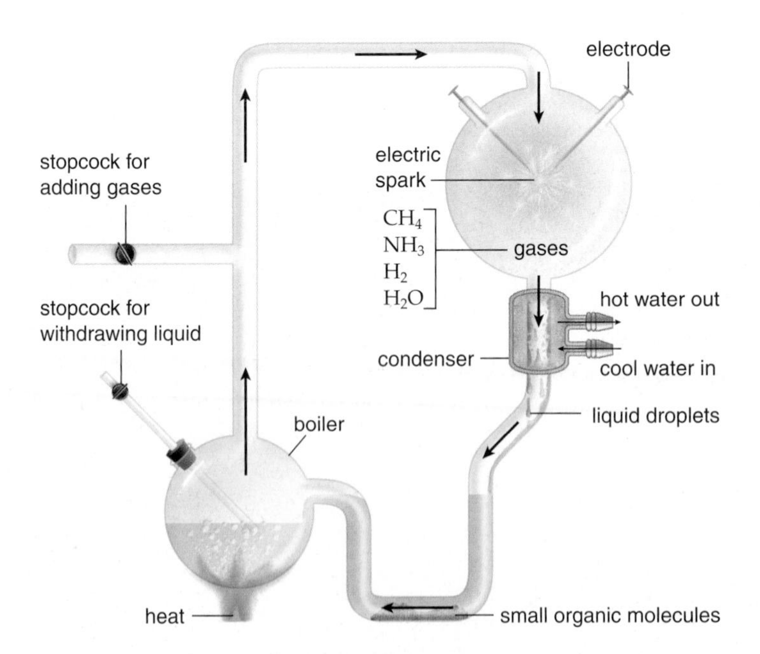


**FIGURE 18.1 Stanley Miller's apparatus and experiment.**

Gases that were thought to be present in the early Earth's atmosphere were admitted to the apparatus, circulated past an energy source (electric spark), and cooled to produce a liquid that could be withdrawn. Upon chemical analysis, the liquid was found to contain various small organic molecules, which could serve as monomers for large cellular polymers.

**FIGURE 18.2 Chemical evolution at hydrothermal vents.**

Minerals that form at deep-sea hydrothermal vents like this one can catalyze the formation of ammonia and even monomers of larger organic molecules in cells.

($NH_3$), hydrogen ($H_2$), and water ($H_2O$)—in a closed system, heated the mixture, and circulated it past an electric spark (simulating lightning). After a week's run, Miller discovered that a variety of amino acids and organic acids had been produced. Since that time, other investigators have achieved similar results by using other, less-reducing combinations of gases dissolved in water.

If atmospheric gases did react with one another to produce small organic compounds, neither oxidation (there was no free oxygen) nor decay (there were no bacteria) would have destroyed these molecules, and rainfall would have washed them into the ocean, where they accumulated for hundreds of millions of years. Therefore, the oceans would have been a thick, warm organic soup.

In support of monomers coming from reactions at hydrothermal vents, a team of investigators at the Carnegie Institution in Washington, D.C. first point out that Miller used ammonia as one of the atmospheric gases. Whereas inert nitrogen gas ($N_2$) would have been abundant in the primitive atmosphere, ammonia ($NH_3$) would have been scarce. However, ammonia would have been plentiful at hydrothermal vents on the ocean floor. These vents line huge **ocean ridges,** where molten magma wells up and adds crust to the ocean floor in each direction. Cool water seeping through the vents is heated to a temperature as high as 350°C, and when it spews back out, it contains various mixed iron-nickel sulfides that can act as catalysts to change $N_2$ to $NH_3$ (Fig. 18.2). A laboratory test of this hypothesis worked perfectly. Under ventlike conditions, 70% of various nitrogen sources were converted to ammonia within 15 minutes. German organic chemists Gunter Wachtershaüser and Claudia Huber have gone one more step. They have shown that organic molecules will react and amino acids will form peptides in the presence of iron-nickel sulfides under ventlike conditions.



## Polymers Evolve

In cells, organic monomers join to form polymers in the presence of enzymes, which of course are proteins. How did the first organic polymers form if there were no proteins yet? As just mentioned, Wachtershaüser and Huber have managed to achieve the formation of peptides using iron-nickel sulfides as inorganic catalysts under ventlike conditions of high temperature and pressure. These minerals have a charged surface that attracts amino acids and provides electrons so they can bond together.

### *Alternative Hypotheses*

Sidney Fox has shown that amino acids polymerize abiotically when exposed to dry heat. He suggests that once amino acids were present in the oceans, they could have collected in shallow puddles along the rocky shore. Then the heat of the sun could have caused them to form **proteinoids,** small polypeptides that have some catalytic properties. When he simulates this scenario in the lab and returns proteinoids to water, they form **microspheres** [Gk. *mikros,* small, little, and *sphaera,* ball], structures composed only of protein that have many properties of a cell (Fig. 18.3*a*). It's possible that even newly formed polypeptides had enzymatic properties, and some proved to be more capable than others. Those that led to the first cell or cells had a selective advantage. Fox's **protein-first hypothesis** assumes that DNA genes came after protein enzymes arose. After all, it is protein enzymes that are needed for DNA replication.

Another hypothesis is put forth by Graham Cairns-Smith. He hypothesizes that clay was especially helpful in causing polymerization of monomers to produce both proteins and nucleic acids at the same time. Clay also attracts small organic molecules and contains iron and zinc, which may have served as inorganic catalysts for polypeptide formation. In addition, clay has a tendency to collect energy from radioactive decay and to discharge it when the temperature and/or humidity changes. This could have been a source of energy for polymerization to take place. Cairns-Smith suggests that RNA nucleotides and amino acids became associated in such a way that polypeptides were ordered by and helped synthesize RNA. This hypothesis suggests that both polypeptides and RNA arose at the same time.

There is still another hypothesis concerning this stage in the origin of life. The **RNA-first hypothesis** suggests that only the macromolecule RNA (ribonucleic acid) was needed to progress toward formation of the first cell or cells. Thomas Cech and Sidney Altman shared a Nobel Prize in 1989 because they discovered that RNA can be both a substrate and an enzyme. Some viruses today have RNA genes; therefore, the first genes could have been RNA. It would seem, then, that RNA could have carried out the processes of life commonly associated today with DNA (deoxyribonucleic acid, the genetic material) and proteins (enzymes). Those who support this hypothesis say that it was an "RNA world" some 4 BYA.

## A Protocell Evolves

Before the first true cell arose, there would have been a **protocell** or **protobiont** [Gk. *protos,* first], a structure that first and foremost has an outer membrane. After all, life requires chemical reactions that take place within a boundary.

### *The Plasma Membrane*

The plasma membrane separates the living interior from the nonliving exterior. There are several hypotheses about the origin of the first membrane. Sidney Fox has shown that if lipids are made available to microspheres, lipids tend to become associated with microspheres (Fig. 18.3*a*), producing a lipid-protein membrane. Microspheres have many interesting properties: They resemble bacteria, they have an electrical potential difference, and they divide and perhaps are subject to selection. Although Fox believes they are a cell, others disagree.

Oparin, who was mentioned previously, showed that under appropriate conditions of temperature, ionic composition, and pH, concentrated mixtures of macromolecules tend to give rise to complex units called **coacervate droplets.** Coacervate droplets have a tendency to absorb and incorporate various substances from the surrounding solution. Eventually, a semipermeable-type boundary may form about the droplet.

In the early 1960s, biophysicist Alec Bangham of the Animal Physiology Institute in Cambridge, England, discovered that when he extracted lipids from egg yolks and placed them in water, the lipids would naturally organize themselves into double-layered bubbles roughly the size of a cell. Bangham's bubbles soon became known as **liposomes** [Gk. *lipos,* fat, and *soma,* body] (Fig. 18.3*b*). Later, biophysicist David Deamer of the University of California and Bangham realized that liposomes might have provided life's first boundary. Perhaps liposomes with a phospholipid membrane engulfed early molecules that had enzymatic, even replicative abilities. The liposomes would have protected the molecules from their surroundings and concentrated them so they could react (and evolve) quickly and efficiently. These investigators called this the "membrane-first" hypothesis, meaning that the first cell had to have a plasma membrane before any of its other parts. Perhaps the first membrane formed in this manner, and the protocell contained only RNA, which functioned as both genetic material and enzymes.

### *Nutrition*

The protocell would have had to carry on nutrition so that it could grow. If organic molecules formed in the atmosphere and were carried by rain into the ocean, nutrition would have been no problem because simple organic molecules could have served as food. This hypothesis suggests that the protocell was a heterotroph [Gk. *hetero,* different, and *trophe,* food], an organism that takes in preformed food. On the other hand, if the protocell evolved at hydrothermal vents, it may have carried out chemosynthesis. Chemoautotrophic bacteria obtain energy for synthesizing organic molecules by oxidizing inorganic compounds, such as hydrogen sulfide ($H_2S$), a molecule that is abundant at the vents. When hydrothermal vents were first discovered in the 1970s, investigators were surprised to discover complex vent ecosystems supported by organic molecules formed by chemosynthesis, a process that does not require the energy of the sun.

At first, the protocell may have used preformed ATP (adenosine triphosphate), but as this supply dwindled,

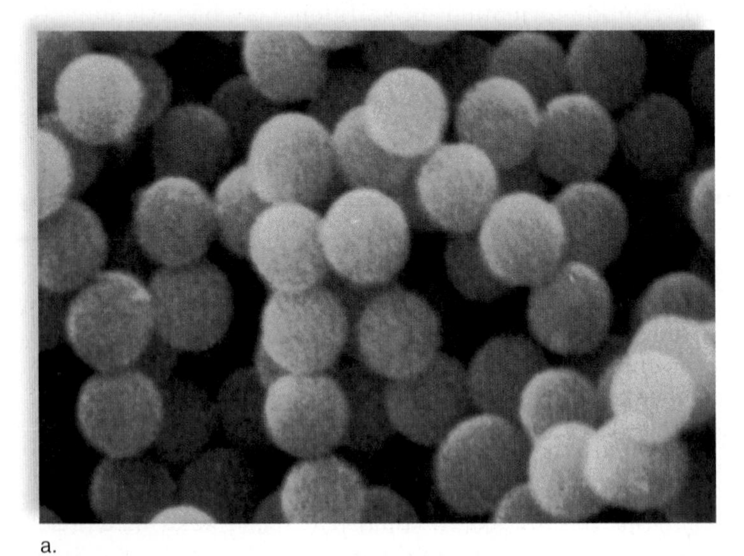

a.

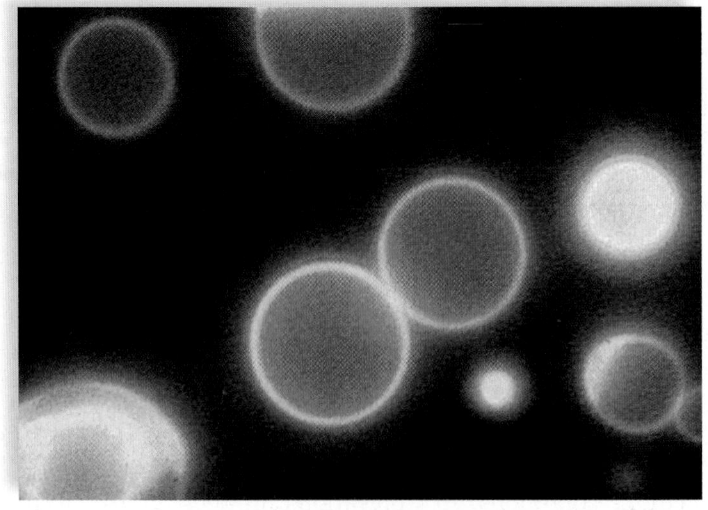

b.

**FIGURE 18.3 Origin of plasma membrane.**

**a.** Microspheres have a number of cellular characteristics and could have evolved into the protocell. For example, microspheres are composed only of protein, but if lipids are added, the proteins combine with them. Perhaps this was the step that led to the first plasma membrane. **b.** Liposomes form automatically when lipids are put into water. Phospholipids also have a tendency to form a circle. Perhaps the first plasma membrane was simply a phospholipid bilayer without any proteins being present.

natural selection favored any cells that could extract energy from carbohydrates in order to transform ADP (adenosine diphosphate) to ATP. Glycolysis is a common metabolic pathway in living things, and this testifies to its early evolution in the history of life. Since there was no free oxygen, we can assume that the protocell carried on a form of fermentation. At first the protocell must have had limited ability to break down organic molecules, and scientists speculate that it took millions of years for glycolysis to evolve completely. Interestingly, Fox has shown that microspheres from which protocells may have evolved have some catalytic ability, and Oparin found that coacervates do incorporate enzymes if they are available in the medium.

## A Self-Replication System Evolves

Today's cell is able to carry on protein synthesis in order to produce the enzymes that allow DNA to replicate. The central dogma of genetics states that DNA directs protein synthesis and that information flows from DNA to RNA to protein. It is possible that this sequence developed in stages.

### *Alternative Hypotheses*

According to the RNA-first hypothesis, RNA would have been the first to evolve, and the first true cell would have had RNA genes. These genes would have directed and enzymatically carried out protein synthesis. Today, ribozymes are enzymatic RNA molecules. Also, today we know there are viruses that have RNA genes. These viruses have a protein enzyme called reverse transcriptase that uses RNA as a template to form DNA. Perhaps with time, reverse transcription occurred within the protocell, and this is how DNA genes arose. If so, RNA was responsible for both DNA and protein formation. Once there were DNA genes, protein synthesis would have been carried out in the manner dictated by the central dogma of genetics.

According to the protein-first hypothesis, proteins, or at least polypeptides, were the first of the three (i.e., DNA, RNA, and protein) to arise. Only after the protocell developed a plasma membrane and sophisticated enzymes did it have the ability to synthesize DNA and RNA from small molecules provided by the ocean. Because a nucleic acid is a complicated molecule, the likelihood RNA arose *de novo* (on its own) is minimal. It seems more likely that enzymes were needed to guide the synthesis of nucleotides and then nucleic acids. Again, once there were DNA genes, protein synthesis would have been carried out in the manner dictated by the central dogma of genetics.

Cairns-Smith proposes that polypeptides and RNA evolved simultaneously. Therefore, the first true cell would have contained RNA genes that could have replicated because of the presence of proteins. This eliminates the baffling chicken-and-egg paradox: Assuming a plasma membrane, which came first, proteins or RNA? It means, however, that two unlikely events would have had to happen at the same time.

After DNA formed, the genetic code had to evolve before DNA could store genetic information. The present genetic code is subject to fewer errors than a million other possible codes. Also, the present code is among the best at minimizing the effect of mutations. A single-base change in a present codon is likely to result in the substitution of a chemically similar amino acid and, therefore, minimal changes in the final protein. This evidence suggests that the genetic code did undergo a natural selection process before finalizing into today's code.

## A Recap of the Steps

Figure 18.4 reviews how most biologists hypothesize that life evolved on early Earth.

1. An abiotic synthesis process created small organic molecules such as amino acids and nucleotides, perhaps in the atmosphere or at hydrothermal vents.
2. These monomers joined together to form polymers along the shoreline (warm seaside rocks or clay) or at the vents. The first polymers could have been proteins or RNA, or they could have evolved together.

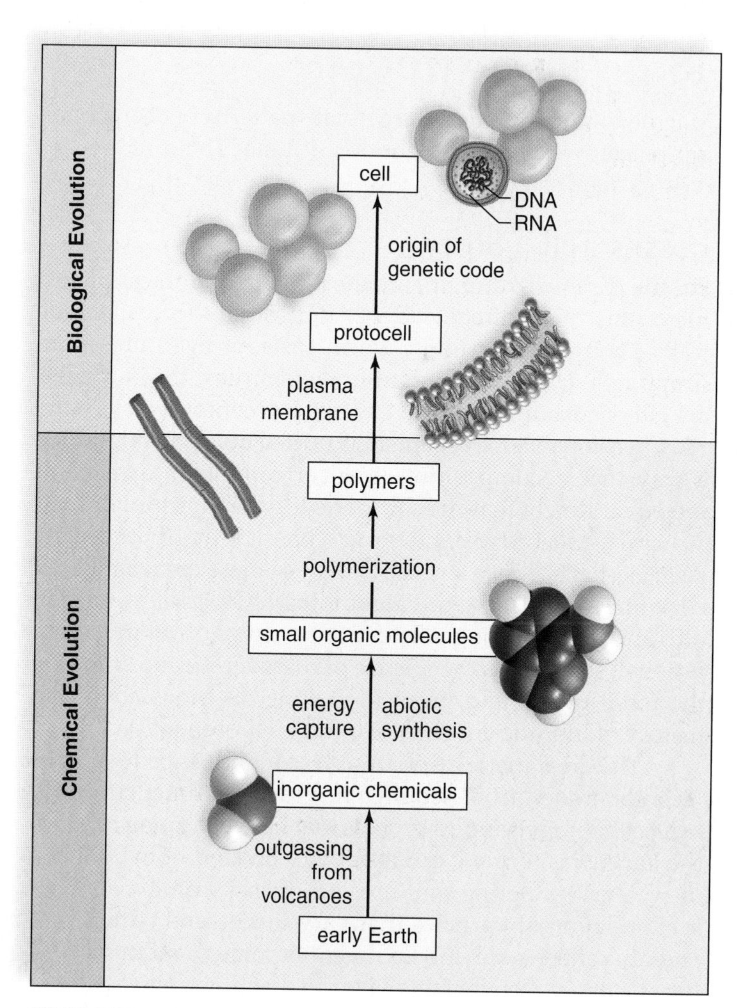


**FIGURE 18.4 Origin of the first cell(s).**

There was an increase in the complexity of macromolecules, leading to a self-replicating system (DNA ⟶ RNA ⟶ protein) enclosed by a plasma membrane. The protocell, a heterotrophic fermenter, underwent biological evolution, becoming a true cell, which then diversified.

3. The aggregation of polymers inside a plasma membrane produced a protocell, which had enzymatic properties such that it could grow. If the protocell developed in the ocean, it was a heterotroph; if it developed at hydrothermal vents, it was a chemoautotroph.
4. Once the protocell contained DNA genes, a true cell had evolved. The first genes may have been RNA molecules, but later DNA became the information storage molecule of heredity. Biological evolution—and the history of life—had begun!

**Check Your Progress 18.1**

1. What function can RNA perform that a protein cannot perform in cells?
2. Without mitochondria, what metabolic pathway would have allowed the first cell or cells to produce ATP molecules?

# 18.2 History of Life

Macroevolution includes large-scale patterns of change taking place over very long periods of time. The fossil record records such changes.

## Fossils Tell a Story

**Fossils** [L. *fossilis,* dug up] are the remains and traces of past life or any other direct evidence of past life. Traces include trails, footprints, burrows, worm casts, or even preserved droppings. Usually when an organism dies, the soft parts are either consumed by scavengers or decomposed by bacteria. Occasionally, the organism is buried quickly and in such a way that decomposition is never completed or is completed so slowly that the soft parts leave an imprint of their structure. Most fossils, however, consist only of hard parts such as shells, bones, or teeth, because these are usually not consumed or destroyed. **Paleontology** [Gk. *palaios,* ancient, old, and *ontos,* having existed; *-logy,* study of, from *logikos,* rational, sensible] is the science of discovering and studying the fossil record and, from it, making decisions about the history of life, ancient climates, and environments.

The great majority of fossils are found embedded in or recently eroded from sedimentary rock. **Sedimentation** [L. *sedimentum,* a settling], a process that has been going on since the Earth was formed, can take place on land or in bodies of water. Weathering and erosion of rocks produce an accumulation of particles that vary in size and nature and are called sediment. Sediment becomes a **stratum** (pl., strata), a recognizable layer in a stratigraphic sequence (Fig. 18.5*a*). Any given stratum is older than the one above it and younger than the one immediately below it. Figure 18.5*b* shows the history of the Earth as if it had occurred during a 24-hour time span that starts at midnight. (The actual years are shown on an inner ring of the diagram.) This figure illustrates dramatically that only unicellular organisms were present during most (about 80%) of the history of the Earth.

If the Earth formed at midnight, prokaryotes do not appear until about 5 A.M., eukaryotes are present at approximately 4 P.M., and multicellular forms do not appear until around 8 P.M. Invasion of the land doesn't occur until about 10 P.M., and humans don't appear until 30 seconds before the end of the day. This timetable has been worked out by studying the fossil record. In addition to sedimentary fossils, more recent fossils can be found in tar, ice, bogs, and amber. Petrified wood, shells, and bones are also relatively common (Fig. 18.6).

### *Relative Dating of Fossils*

In the early nineteenth century, even before the theory of evolution was formulated, geologists sought to correlate the strata worldwide. The problem was that strata change their character over great distances, and therefore a stratum in England might contain different sediments than one of the same age in Russia. Geologists discovered that each stratum of the same age contained certain **index fossils** that serve to identify deposits made at apparently the same time in different parts of the world. These index fossils are used in **relative dating** methods. For example, a particular species of fossil ammonite (an animal related to the chambered nautilus) has been found over a wide range and for a limited time period. Therefore, all strata around the world that contain this fossil must be of the same age.

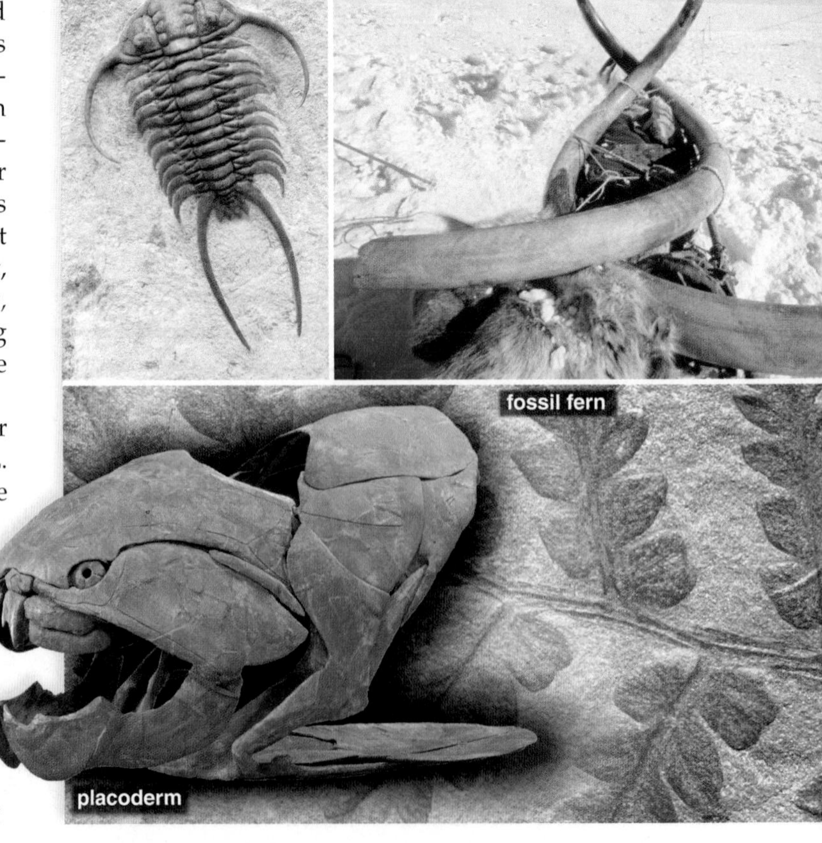

a.

b.

## FIGURE 18.5 The history of life.

**a.** Strata, layers seen in sedimentary rock as exposed by road cuts, are the source of fossils that tell us about the history of life. **b.** The blue ring of this diagram shows the history of life as it would be measured on a 24-hour timescale starting at midnight. (The red ring shows the actual years going back in time to 4.6 BYA.) The fossil record suggests that a very large portion of life's history was devoted to the evolution of unicellular organisms. The first multicellular organisms do not appear in the fossil record until just before 8 P.M., and humans are not on the scene until less than a minute before midnight. BYA = billion years ago

## FIGURE 18.6 Fossils.

Fossils are the remains of past life. They can be impressions left in rocks, footprints, mineralized bones, shells, or any other evidences of life-forms that lived in the past.

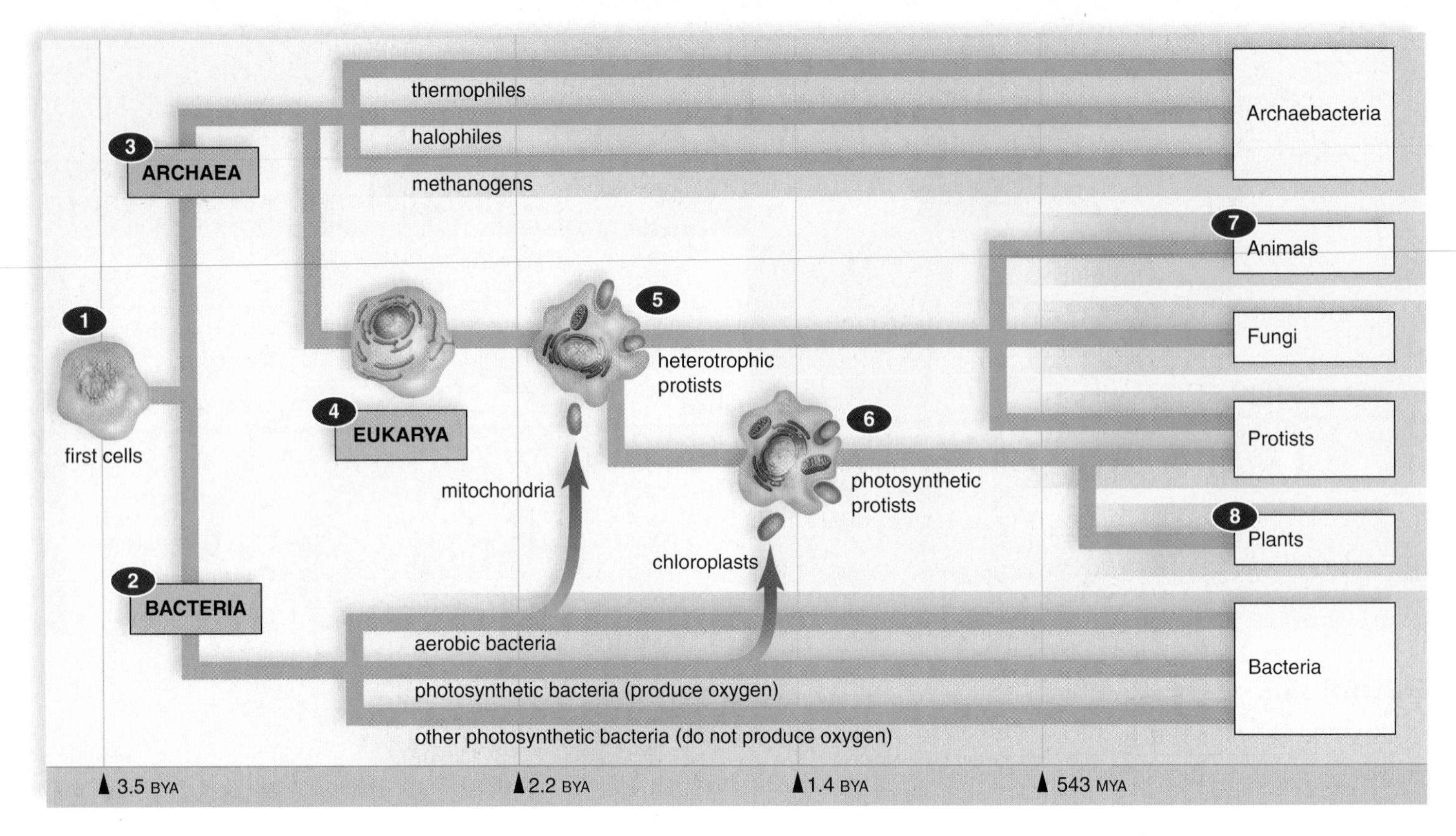


**FIGURE 18.7 The tree of life.**

During the Precambrian time, 1 the first cell or cells give rise to 2 bacteria and 3 archaea; 4 the first eukaryotic cell evolves from archaea. 5 Heterotrophic protists arise when eukaryotic cells gain mitochondria by engulfing aerobic bacteria, and 6 photosynthetic protists arise when these cells gain chloroplasts by engulfing photosynthetic bacteria. 7 Animals (and fungi) evolve from heterotrophic protists, and 8 plants evolve from photosynthetic protists. BYA = billion years ago

### *Absolute Dating of Fossils*

An **absolute dating** method that relies on radioactive dating techniques assigns an actual date to a fossil. All radioactive isotopes have a particular half-life, the length of time it takes for half of the radioactive isotope to change into another stable element. If the fossil has organic matter, half of the carbon 14 ($^{14}C$) will have changed to nitrogen 14 ($^{14}N$) in 5,730 years. To know how much $^{14}C$ was in the organism to begin with, it is reasoned that organic matter always begins with the same amount of $^{14}C$. (In reality, it is known that the $^{14}C$ levels in the air—and therefore the amount in organisms—can vary from time to time.) Now we need only compare the $^{14}C$ radioactivity of the fossil to that of a modern sample of organic matter. The amount of radiation left can be converted to the age of the fossil. After 50,000 years, however, the amount of $^{14}C$ radioactivity is so low that it cannot be used to measure the age of a fossil accurately.

$^{14}C$ is the only radioactive isotope contained within organic matter, but it is possible to use others to date rocks, and from that to infer the age of a fossil contained in the rock. For instance, the ratio of potassium 40 ($^{40}K$) to argon 40 trapped in rock is often used to date rocks.

## The Precambrian Time

As a result of their study of fossils in strata, geologists have devised the **geologic timescale,** which divides the history of the Earth into eras and then periods and epochs (Table 18.1). We will follow the biologist's tradition of first discussing Precambrian time. The Precambrian is a very long period of time, comprising about 87% of the geologic timescale. During this time, life arose and the first cells came into existence (Fig. 18.7). The first cells were probably prokaryotes. Prokaryotes do not have a nucleus or membrane-bounded organelles. Prokaryotes, the archaea and bacteria, can live in the most inhospitable of environments, such as hot springs, salty lakes, and airless swamps—all of which may typify habitats on early Earth. The cell wall, plasma membrane, RNA polymerase, and ribosomes of archaea are more like those of eukaryotes than those of bacteria.

The first identifiable fossils are those of complex prokaryotes. Chemical fingerprints of complex cells are found in sedimentary rocks from southwestern Greenland, dated at 3.8 BYA (billion years ago). But paleobiologist J. William Schopf found the oldest prokaryotic fossils in western Australia. These 3.46-billion-year-old microfossils resemble today's cyanobacteria, prokaryotes that carry on photosynthesis in the same manner as plants. At this time, only volcanic rocks jutted above the waves, and there were as yet no continents. Strange-looking boulders, called **stromatolites,** littered beaches and shallow waters (Fig. 18.8*a*). Living stromatolites can still be found today along Australia's western coast. The outer surface of a stromatolite is alive with cyanobacteria that secrete a mucus. Grains of sand get caught in the mucus and bind with calcium carbonate from the water to form rock. To gain access to sunlight, the photosynthetic organisms move outward

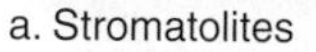

a. Stromatolites

b. *Primaevifilum*

**FIGURE 18.8**
**Prokaryote fossils of the Precambrian.**

**a.** Stromatolites date back to 3.46 BYA. Living stromatolites are located in shallow waters off the shores of western Australia and also in other tropical seas. **b.** The prokaryotic microorganism *Primaevifilum* (with interpretive drawing) was found in stromatolites.

toward the surface before they are cemented in. They leave behind a menagerie of aerobic and then anaerobic bacteria caught in the layers of the rock.

The cyanobacteria in ancient stromatolites added oxygen to the atmosphere (Fig. 18.8*b*). By 2.0 BYA, the presence of oxygen was such that most environments were no longer suitable for anaerobic prokaryotes, and they began to decline in importance. Photosynthetic cyanobacteria and aerobic bacteria proliferated as new metabolic pathways evolved. Due to the presence of oxygen, the atmosphere became an oxidizing one instead of a reducing one. Oxygen in the upper atmosphere forms ozone ($O_3$), which filters out the ultraviolet (UV) rays of the sun. Before the formation of the **ozone shield,** the amount of ultraviolet radiation reaching the Earth could have helped create organic molecules, but it would have destroyed any land-dwelling organisms. Once the ozone shield was in place, living things were sufficiently protected and able to live on land (see page 880).

## *Eukaryotic Cells Arise*

The eukaryotic cell, which originated around 2.1 BYA, is nearly always aerobic and contains a nucleus as well as other membranous organelles. Most likely, the eukaryotic cell acquired its organelles gradually. The nucleus may have developed by an invagination of the plasma membrane. The mitochondria of the eukaryotic cell were once free-living aerobic bacteria, and the chloroplasts were free-living photosynthetic prokaryotes. The **endosymbiotic theory** states that a nucleated cell engulfed these prokaryotes, which then became organelles (see Fig. 18.7). The evidence for the theory is the following:

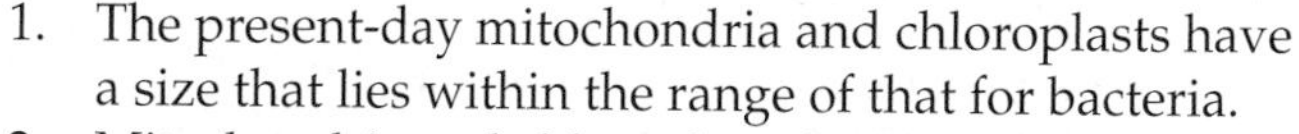

1. The present-day mitochondria and chloroplasts have a size that lies within the range of that for bacteria.
2. Mitochondria and chloroplasts have their own DNA and make some of their own proteins. (The DNA of the nucleus also codes for some of the mitochondrial proteins.)
3. The mitochondria and chloroplasts divide by binary fission, the same as bacteria do.
4. The outer membrane of mitochondria and chloroplasts differ—the outer membrane resembles that of a eukaryotic cell and the inner membrane resembles that of a bacterial cell.

It's been suggested that flagella (and cilia) also arose by endosymbiosis. First, slender undulating prokaryotes could have attached themselves to a host cell to take advantage of food leaking from the host's plasma membrane. Eventually, these prokaryotes adhered to the host cell and became the flagella and cilia we know today. The first eukaryotes were unicellular, as are prokaryotes.

## *Multicellularity Arises*

Fossils identified as multicellular protists and dated 1.4 BYA have been found in arctic Canada. It's possible that the first multicellular organisms practiced sexual reproduction. Among today's protists we find colonial forms in which some cells are specialized to produce gametes needed for sexual reproduction. Separation of germ cells, which produce gametes from somatic cells, may have been an important first step toward the development of the Ediacaran invertebrates discussed next, which appeared about 630 MYA (million years ago) and died out about 545 MYA.

## TABLE 18.1

**The Geologic Timescale: Major Divisions of Geologic Time and Some of the Major Evolutionary Events of Each Time Period**

| *Era* | *Period* | *Epoch* | *Million Years Ago (MYA)* | *Plant Life* | *Animal Life* |
|---|---|---|---|---|---|
| Cenozoic | Quaternary | Holocene | (0.01–0) | Human influence on plant life | Age of *Homo sapiens* |
| | | | | **Significant Mammalian Extinction** | |
| | | Pleistocene | (1.80–0.01) | Herbaceous plants spread and diversify. | Presence of Ice Age mammals. Modern humans appear. |
| | Tertiary | Pliocene | (5.33–1.80) | Herbaceous angiosperms flourish. | First hominids appear. |
| | | Miocene | (23.03–5.33) | Grasslands spread as forests contract. | Apelike mammals and grazing mammals flourish; insects flourish. |
| | | Oligocene | (33.9–23.03) | Many modern families of flowering plants evolve. | Browsing mammals and monkeylike primates appear. |
| | | Eocene | (55.8–33.9) | Subtropical forests with heavy rainfall thrive. | All modern orders of mammals are represented. |
| | | Paleocene | (65.5–55.8) | Flowering plants continue to diversify. | Primitive primates, herbivores, carnivores, and insectivores appear. |
| | | | | **Mass Extinction: Dinosaurs and Most Reptiles** | |
| Mesozoic | Cretaceous | | (145.5–65.5) | Flowering plants spread; conifers persist. | Placental mammals appear; modern insect groups appear. |
| | Jurassic | | (199.6–145.5) | Flowering plants appear. | Dinosaurs flourish; birds appear. |
| | | | | **Mass Extinction** | |
| | Triassic | | (251–199.6) | Forests of conifers and cycads dominate. | First mammals appear; first dinosaurs appear; corals and molluscs dominate seas. |
| | | | | **Mass Extinction** | |
| Paleozoic | Permian | | (299–251) | Gymnosperms diversify. | Reptiles diversify; amphibians decline. |
| | Carboniferous | | (359.2–299) | Age of great coal-forming forests; ferns, club mosses, and horsetails flourish. | Amphibians diversify; first reptiles appear; first great radiation of insects. |
| | | | | **Mass Extinction** | |
| | Devonian | | (416–359.2) | First seed plants appear. Seedless vascular plants diversify. | First insects and first amphibians appear on land. |
| | Silurian | | (443.7–416) | Seedless vascular plants appear. | Jawed fishes diversify and dominate the seas. |
| | | | | **Mass Extinction** | |
| | Ordovician | | (488.3–443.7) | Nonvascular land plants appear on land. | First jawless and then jawed fishes appear. |
| | Cambrian | | (542–488.3) | Marine algae flourish. | All invertebrate phyla present; first chordates appear. |
| Precambrian Time | | | 630 | Soft-bodied invertebrates | |
| | | | 1,000 | Protists diversify. | |
| | | | 2,100 | First eukaryotic cells | |
| | | | 2,700 | $O_2$ accumulates in atmosphere. | |
| | | | 3,500 | First prokaryotic cells | |
| | | | 4,570 | Earth forms. | |

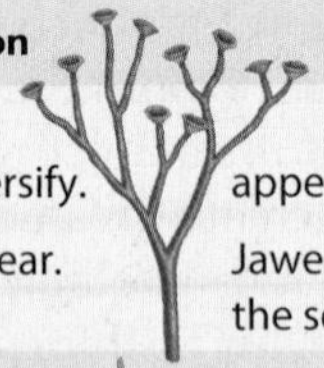

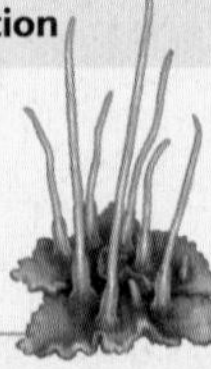

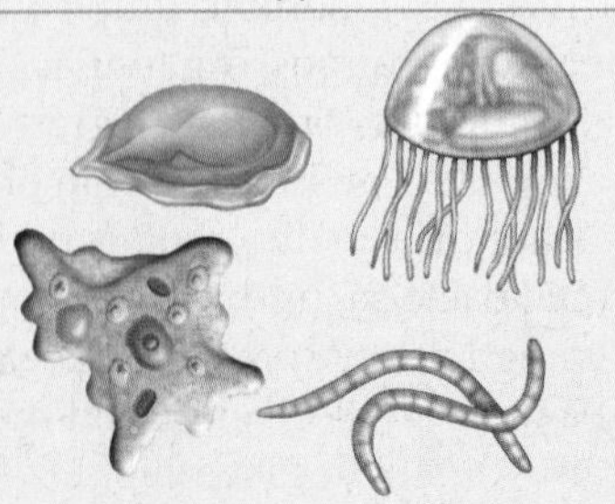

a.

b.

**FIGURE 18.9 Ediacaran fossils.**

The Ediacaran invertebrates lived from about 600–545 MYA. They were all flat, soft-bodied invertebrates. **a.** Classified as *Spriggina*, this bilateral organism had a crescent-shaped head and numerous segments tapering to a posterior end. **b.** Classified as *Dickinsonia*, these fossils are often interpreted to be segmented worms. However, in the opinion of some, they may be cnidarian polyps.

In 1946, R. C. Sprigg, a government geologist assessing abandoned lead mines in southern Australia, discovered the first remains of a remarkable biota that has taken its name from the region, the Ediacara Hills. Since then, similar fossils have been discovered on a number of other continents. Many of the fossils, dated 630–545 MYA, are thought to be of soft-bodied invertebrates (animals without a vertebral column) that most likely lived on mudflats in shallow marine waters. Some may have been mobile, but others were large, immobile, bizarre creatures resembling spoked wheels, corrugated ribbons, and lettucelike fronds. All were flat and probably had two tissue layers; few had any type of skeleton (Fig. 18.9). They apparently had no mouths; perhaps they absorbed nutrients from the sea or else had photosynthetic organisms living on their tissues. These soft-bodied animals could flourish because there were no predators to eat them. Their fossils are like footprints—impressions made in the sandy seafloor before their bodies decayed away. Whether the Ediacaran animals were simply a failed evolutionary experiment or whether they are related to animals of the Cambrian period is not known. With few exceptions they disappear from the fossil record at 545 MYA, but even so some may have given rise to modern cnidarians and related animals. What caused their demise is not known, but they very well could have been eaten by the myriad of animals with mouths that suddenly appear in the Cambrian.

### Check Your Progress 18.2A

1. What sequence of events in Precambrian time led to heterotrophic protists and photosynthetic protists?
2. What additional event is needed before animals, fungi, and plants could arise?

## The Paleozoic Era

The Paleozoic era lasted about 300 million years. Even though the era was quite short compared to the length of the Precambrian, many events occurred during this era, including three major mass extinctions (see Table 18.1). An **extinction** is the total disappearance of all the members of a species or higher taxonomic group. **Mass extinctions,** which are the disappearance of a large number of species or a higher taxonomic group within an interval of just a few million years, are discussed on page 333.

### *Cambrian Animals*

The seas of the Cambrian period, which began at about 542 MYA, teemed with invertebrate life as illustrated in Figure 17B, pages 308–9. Life became so abundant that scientists refer to this period in Earth's history as the Cambrian explosion. All of today's groups of animals can trace their ancestry to this time, and perhaps earlier, according to new molecular clock data. A **molecular clock** is based on the principle that mutations in certain parts of the genome occur at a fixed rate and are not tied to natural selection. Therefore, the number of DNA base-pair differences tells how long two species have been evolving separately.

Even if certain of the animals in Figure 18.10 had evolved earlier, no fossil evidence of them occurs until the Cambrian period, perhaps because they lacked a skeleton. Animals that lived during the Cambrian possessed protective outer skeletons known as exoskeletons. These structures are capable of surviving the forces that are apt to destroy soft-bodied organisms. For example, Cambrian seafloors were dominated by now-extinct trilobites, which had thick, jointed armor covering them from head to tail. Trilobites are classified as arthropods, a major phylum of animals today. (Some Cambrian species, with most unusual eating and locomotion appendages, have been classified in phyla that no longer exist today.)

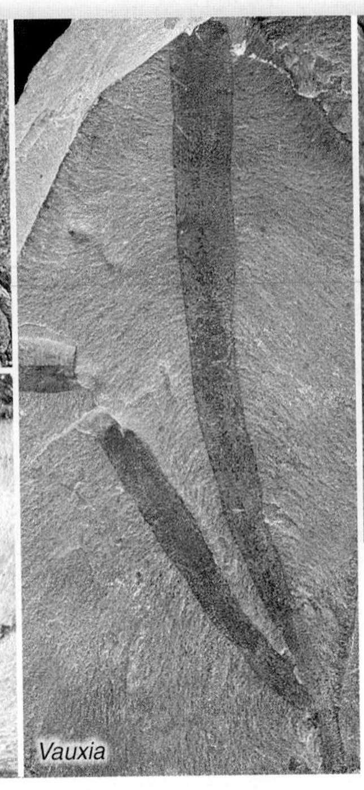

**FIGURE 18.10 Sea life of the Cambrian period.**

The animals depicted here are found as fossils in the Burgess Shale, a formation of the Rocky Mountains of British Columbia, Canada. Some lineages represented by these animals are still evolving today. *Opabinia* has been designated a crustacean, *Thaumaptilon* a sea pen, *Vauxia* a sponge, and *Wiwaxia* a segmented worm.

Paleontologists seek an explanation for why animals of the Cambrian period possessed exoskeletons and why this development did not occur during the Precambrian. By this time, not only cyanobacteria but also various algae, which are floating photosynthetic organisms, were pumping oxygen into the atmosphere at a rapid rate. Perhaps the oxygen supply became great enough to permit aquatic animals to acquire oxygen even though they had outer skeletons. The presence of a skeleton reduces possible access to oxygen in seawater. Steven Stanley of Johns Hopkins University suggests that predation may have played a role. Skeletons may have evolved during the Cambrian period because skeletons help protect animals from predators. If so, the evolutionary arms race came of age in the Cambrian seas.

## *Invasion of Land*

**Plants.** During the Ordovician period, algae, which were abundant in the seas, most likely began to take up residence in bodies of fresh water. Eventually, algae invaded damp areas on land. The first land plants were nonvascular (did not possess water-conducting tissues) similar to the mosses and liverworts that survive today. The lack of water-conducting tissues limited the height of these plants to a few centimeters. Although the Ordovician evidence is scarce, spore fossils from this time support this hypothesis.

Fossils of seedless vascular plants (those having tissue for water and organic nutrient transport) date back to the Silurian period. They later flourished in the warm swamps of the Carboniferous period. Club mosses, horsetails, and seed ferns were the trees of that time, and they grew to enormous size. A wide variety of smaller ferns and fernlike plants formed an underbrush (Fig. 18.11).

**Invertebrates.** The jointed appendages and exoskeleton of arthropods are adaptive for living on land. Various arthropods—spiders, centipedes, mites, and millipedes—all preceded the appearance of insects on land. Insects enter the fossil record in the Carboniferous period. One fossil dragonfly from this period had a wingspan of nearly a meter. The evolution of wings provided advantages that allowed insects to radiate into the most diverse and abundant group of animals today. Flying provides a way to escape enemies, find food, and disperse to new territories.

**Vertebrates.** Vertebrates are animals with a vertebral column. The vertebrate line of descent began in the early Ordovician period with the evolution of jawless fishes. Jawed fishes appeared later in the Silurian period. Fishes are ectothermic (cold-blooded) aquatic vertebrates that have gills, scales, and fins. The cartilaginous and ray-finned fishes made their appearance in the Devonian period, which is called the Age of Fishes.

At this time, the seas were filled with giant predatory fish covered with protective armor made of external bone. Sharks cruised up deep, wide rivers, and smaller lobe-finned fishes lived at the river's edge in waters too shallow for large predators. Fleshy fins helped the small fishes push aside debris or hold their place in strong currents, and the fins may also have allowed these fishes to venture onto land and lay their eggs safely in inland pools. Much data tells us that lobe-finned fishes were ancestral to the amphibians and to modern-day lobe-finned fishes.

Amphibians are thin-skinned vertebrates that are not fully adapted to life on land, particularly because they must return to water to reproduce. The Carboniferous swamp forests provided the water they needed, and amphibians adaptively radiated into many different sizes and shapes. Some superficially resembled alligators and were covered with protective scales; others were small and snakelike; and a few were larger plant eaters. The largest measured 6 m

a.

b.

c.

**FIGURE 18.11 Swamp forests of the Carboniferous period.**

**a.** Vast swamp forests of treelike club mosses and horsetails dominated the land during the Carboniferous period (see Table 18.1). The air contained insects with wide wingspans, such as the predecessors to dragonflies shown here, and amphibians lumbered from pool to pool. **b.** Dragonfly fossil from the Carboniferous period. **c.** Modern-day dragonfly.

from snout to tail. The Carboniferous period is called the Age of Amphibians.

The process that turned the great Carboniferous forests into the coal we use today to fuel our modern society started during the Carboniferous period. The weather turned cold and dry, and this brought an end to the Age of Amphibians. A major mass extinction event occurred at the end of the Permian period, bringing an end to the Paleozoic era and setting the stage for the Mesozoic era.

**Check Your Progress 18.2B**

1. Why is the Carboniferous a significant period?

**FIGURE 18.12 Dinosaurs of the late Cretaceous period.**

*Parasaurolophus walkeri*, although not as large as other dinosaurs, was one of the largest plant-eaters of the late Cretaceous period. Its crest atop the head was about 2 m long and was used to make booming calls. Also living at this time were the rhinolike dinosaurs represented here by *Triceratops* (*left*), another herbivore.

## The Mesozoic Era

Although a severe mass extinction occurred at the end of the Paleozoic era, the evolution of certain types of plants and animals continued into the Triassic, the first period of the Mesozoic era. Nonflowering seed plants (collectively called gymnosperms), which had evolved and then spread during the Paleozoic, became dominant. Cycads are short and stout with palmlike leaves, and they produce large cones. Cycads and related plants were so prevalent during the Triassic and Jurassic periods that these periods are sometimes called the Age of Cycads. Reptiles can be traced back to the Permian period of the Paleozoic era. Unlike amphibians, reptiles can thrive in a dry climate because they have scaly skin and lay a shelled egg that hatches on land. Reptiles underwent an adaptive radiation during the Mesozoic era to produce forms that lived in the air, in the sea, and on the land. One group of reptiles, the therapsids, had several mammalian skeletal traits.

During the Jurassic period, large flying reptiles called pterosaurs ruled the air, and giant marine reptiles with paddlelike limbs ate fishes in the sea. But on land, it was dinosaurs that prevented the evolving mammals from taking center stage.

Although the average size of the dinosaurs was about that of a crow, many giant species developed. The gargantuan *Apatosaurus* and the armored, tractor-sized *Stegosaurus* fed on cycad seeds and conifer trees. The size of a dinosaur such as *Apatosaurus* is hard for us to imagine. It was 4.5 m tall at the hips and 27 m long in length and weighed about 40 tons. How might dinosaurs have benefited from being so large? One hypothesis is that, being ectothermic (cold-blooded), the surface-area-to-volume-ratio was favorable for retaining heat. There is also data that suggests dinosaurs were endothermic (warm-blooded).

During the Cretaceous period, great herds of rhinolike dinosaurs, *Triceratops*, roamed the plains, as did the infamous *Tyrannosaurus rex*, which may have been a carnivore, filling the same ecological role as lions do today. *Parasaurolophus* was a unique-looking, long-crested, duck-billed dinosaur (Fig. 18.12). The long, hollow crest was bigger than the rest of its skull and functioned as a resonating chamber for making booming calls, perhaps used during mating or to help members of a herd locate each other. In comparison to *Apatosaurus, Parasaurolophus* was small. It was less than 3 m tall at the hips and weighed only about 3 tons. Still, it was one of the largest plant-eaters of the late Cretaceous period and fed on pine needles, leaves, and twigs. *Parasaurolophus* was easy prey for large predators; its main defense would have been running away in large herds.

At the end of the Cretaceous period, the dinosaurs became victims of a mass extinction, which will be discussed on page 333.

One group of dinosaurs, called theropods, were bipedal and had an elongate, mobile, S-shaped neck. They most likely gave rise to the birds, whose fossil record includes the famous *Archaeopteryx* (see Fig. 15.12, page 276). Up until 1999, Mesozoic mammal fossils largely consisted of teeth. This changed when a fossil found in China was dated at 120 MYA and named *Jeholodens*. The animal, identified as a mammal, apparently looked like a long-snouted rat. Surprisingly, *Jeholodens* had sprawling hindlimbs as do reptiles, but its forelimbs were under the belly, as in today's mammals.

**FIGURE 18.13 Mammals of the Oligocene epoch.**

The artist's representation of these mammals and their habitat vegetation is based on fossil remains.

**FIGURE 18.14 Woolly mammoth of the Pleistocene epoch.**

Woolly mammoths were animals that lived along the borders of continental glaciers.

## The Cenozoic Era

Classically, the Cenozoic era is divided into two periods, the Tertiary period and the Quaternary period. Another scheme, dividing the Cenozoic into the Paleogene and the Neogene periods is gaining popularity. This new system divides the epochs differently. In any case, we are living in the Holocene epoch.

### *Mammalian Diversification*

At the end of the Mesozoic era, mammals began an adaptive radiation into the many habitats now left vacant by the demise of the dinosaurs. Mammals are endotherms, and they have hair, which helps keep body heat from escaping. Their name refers to the presence of mammary glands, which produce milk to feed their young. At the start of the Paleocene epoch, mammals were small and resembled rats. By the end of the Eocene epoch, mammals had diversified to the point that all of the modern orders were in existence. Mammals adaptively radiated into a number of environments. Several species of mammals, including the bats, conquered the air. Whales, dolphins, manatees, and other mammals returned to the sea from land ancestry. On land, herbivorous hoofed mammals populated the forests and grasslands and were preyed upon by carnivorous mammals. Many of the types of herbivores and carnivores of the Oligocene epoch are extinct today (Fig. 18.13).

### *Evolution of Primates*

Flowering plants (collectively called angiosperms) were already diverse and plentiful by the Cenozoic era. Primates are a type of mammal adapted to living in flowering trees, where there is protection from predators and where food in the form of fruit is plentiful. The ancestors of modern primates appeared during the Eocene epoch about 55 MYA. The first primates were small, squirrel-like animals. Ancestral apes appeared during the Oligocene epoch. These primates were adapted to living in the open grasslands and savannas. Apes diversified during the Miocene and Pliocene epochs and gave rise to the first hominids, a group that includes humans. Many of the skeletal differences between apes and humans relate to the fact that humans walk upright. Exactly what caused humans to adopt bipedalism is still being debated.

The world's climate became progressively colder during the Tertiary period. The Quaternary period begins with the Pleistocene epoch, which is known for multiple ice ages in the Northern Hemisphere. During periods of glaciation, snow and ice covered about one-third of the land surface of the Earth. The Pleistocene epoch was an age of not only humans, but also giant ground sloths, beavers, wolves, bison, woolly rhinoceroses, mastodons, and mammoths (Fig. 18.14). Humans have survived, but what happened to the oversized mammals just mentioned? Some think humans became such skilled hunters that they are at least partially responsible for the extinction of these awe-inspiring animals.

**Check Your Progress** **18.2C**

1. If the geologic timescale was proportional to Figure 18.5*b*, which era would be allowed the least amount of space in the timescale?
2. What significant type of plant and animal was abundant during the Mesozoic era?

# 18.3 Factors That Influence Evolution

In the past, it was thought that the Earth's crust was immobile, that the continents had always been in their present positions, and that the ocean floors were only a catch basin for the debris that washed off the land. But in 1920, Alfred Wegener, a German meteorologist, presented data from a number of disciplines to support his hypothesis of continental drift.

## Continental Drift

**Continental drift** was finally confirmed in the 1960s, establishing that the continents are not fixed; instead, their positions and the positions of the oceans have changed over time (Fig. 18.15). During the Paleozoic era, the continents joined to form one supercontinent that Wegener called Pangaea [Gk. *pangea,* all lands]. First, Pangaea divided into two large subcontinents, called Gondwana and Laurasia, and then these also split to form the continents of today. Presently, the continents are still drifting in relation to one another.

Continental drift explains why the coastlines of several continents are mirror images of each other—for example, the outline of the west coast of Africa matches that of the east coast of South America. The same geological structures are also found in many of the areas where the continents touched. A single mountain range runs through South America, Antarctica, and Australia. Continental drift also explains the unique distribution patterns of several fossils. Fossils of the same species of seed fern *(Glossopteris)* have been found on all the southern continents. No suitable explanation was possible previously, but now it seems plausible that the plant evolved on one continent and spread to the others while they were still joined as one. Similarly, the fossil reptile *Cynognathus* is found in Africa and South America, and *Lystrosaurus,* a mammal-like reptile, has now been discovered in Antarctica, far from Africa and southeast Asia, where it also occurs. With mammalian fossils, the situation is different: Australia, South America, and Africa all have their own distinctive mammals because mammals evolved after the continents separated. The mammalian biological diversity of today's world is the result of isolated evolution on separate continents. For example, why are marsupials prevalent in Australia but no place else? Most likely marsupials started evolving in the Americas and were able to reach Australia when the southern continents were still joined. Once Australia separated off, marsupials were able to diversify because placental mammals on that continent offered little competition. On the other hand, placental mammals are prevalent in the Americas and few marsupials can be found.

### *Plate Tectonics*

Why do the continents drift? An answer has been suggested through a branch of geology known as **plate tectonics** [Gk. *tektos,* fluid, molten, able to flow], which says that the Earth's crust is fragmented into slablike plates that float on a lower hot mantle layer. The continents and the ocean basins are a part of these rigid plates, which move like conveyor belts. At ocean ridges, seafloor spreading occurs as molten mantle rock rises and material is added to the ocean floor. Seafloor spreading causes the continents to move a few centimeters a year on the average. At *subduction zones,* the forward edge of a moving plate sinks into the mantle and is destroyed, forming deep ocean trenches bordered by volcanoes or volcanic island chains. The Earth isn't getting bigger or smaller, so the amount of oceanic crust being formed is as much as that being destroyed. When two continents collide, the result is often a mountain range; for example, the Himalayas resulted when India collided with Eurasia. The place where two plates meet and scrape past one another is called a *transform boundary*. The San Andreas fault in Southern California is at a transform boundary, and the movement of the two plates is responsible for the many earthquakes in that region. No one can see the continents moving. The only visible evidence of movement is an earthquake at transform boundaries.

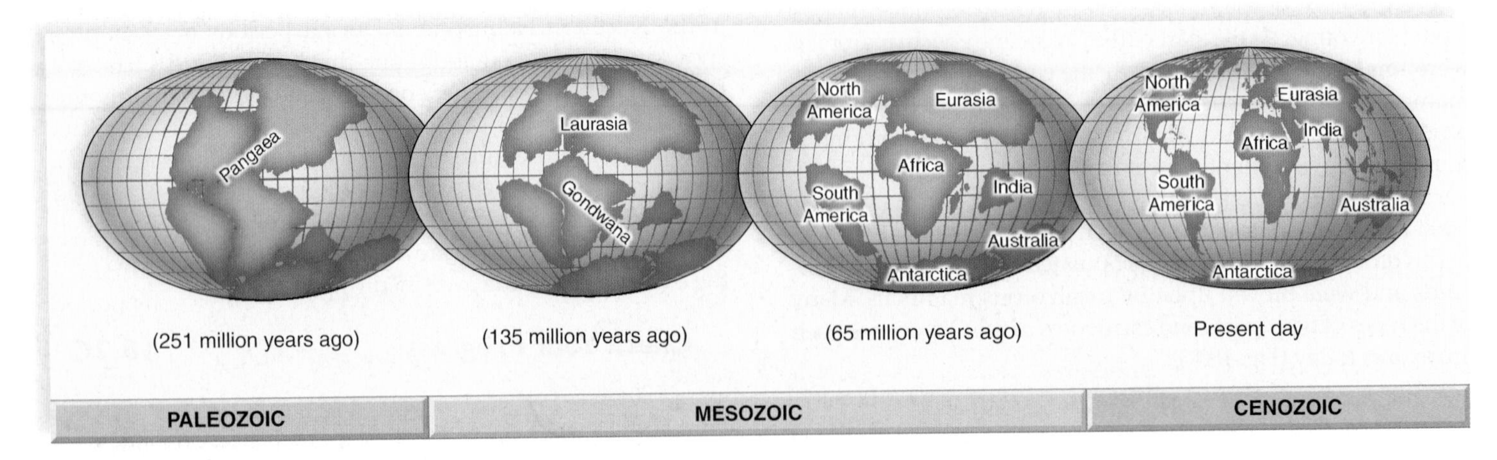


**FIGURE 18.15 Continental drift.**

About 251 MYA, all the continents were joined into a supercontinent called Pangaea. During the Mesozoic era, the joined continents of Pangaea began moving apart, forming two large continents called Laurasia and Gondwana. Then all the continents began to separate. Presently, North America and Europe are drifting apart at a rate of about 2 cm per year.

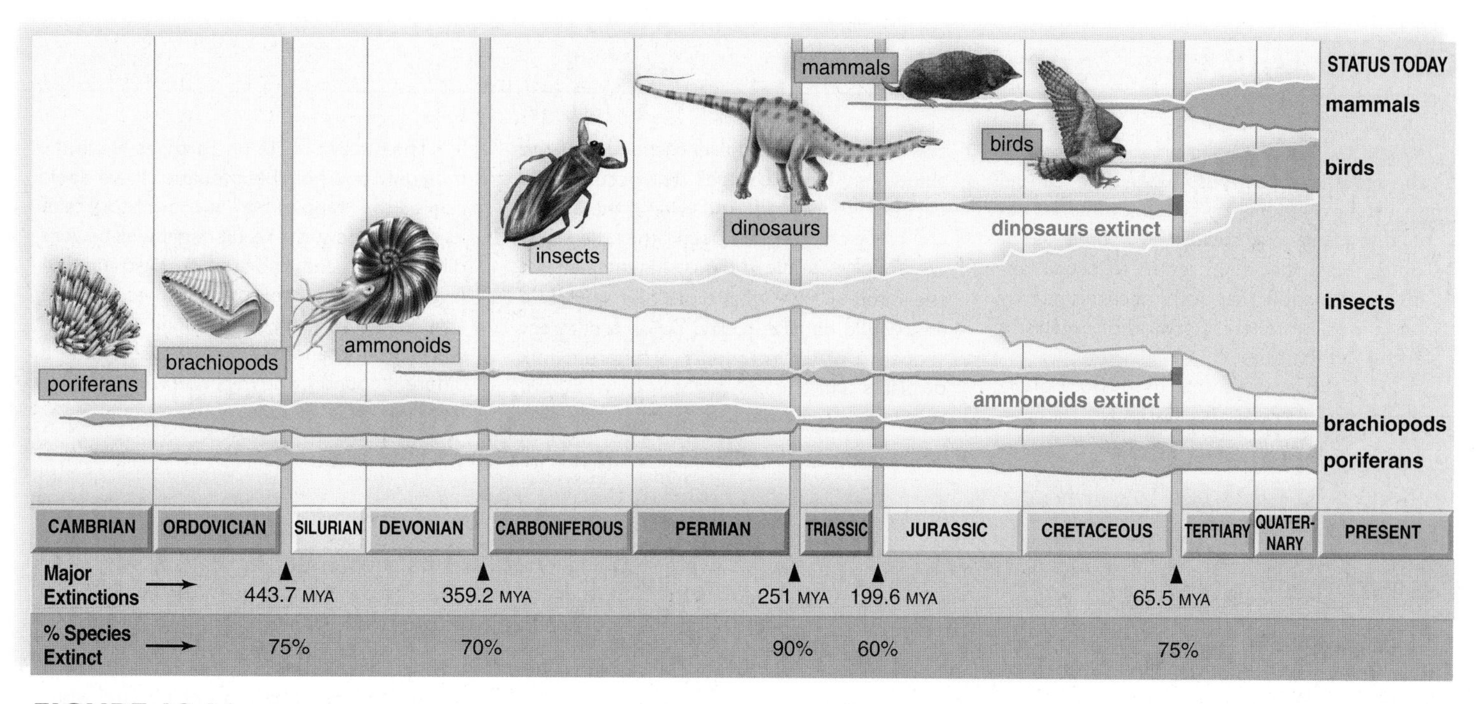


**FIGURE 18.16 Mass extinctions.**

Five significant mass extinctions and their effects on the abundance of certain forms of marine and terrestrial life. The width of the horizontal bars indicates the varying abundance of each life-form considered. MYA=million years ago

# Mass Extinctions

At least five mass extinctions have occurred throughout history: at the ends of the Ordovician, Devonian, Permian, Triassic, and Cretaceous periods (Fig. 18.16; see Table 18.1). Is a mass extinction due to some cataclysmic event, or is it a more gradual process brought on by environmental changes, including tectonic, oceanic, and climatic fluctuations? This question was brought to the fore when Walter and Luis Alvarez proposed in 1977 that the Cretaceous extinction when the dinosaurs died out was due to a bolide. A bolide is an asteroid (minor planet) that explodes, producing meteorites that fall to Earth. They found that Cretaceous clay contains an abnormally high level of iridium, an element that is rare in the Earth's crust but more common in asteroids and meteorites. The result of a large meteorite striking Earth could have been similar to that of a worldwide atomic bomb explosion: A cloud of dust would have mushroomed into the atmosphere, blocking out the sun and causing plants to freeze and die. A layer of soot has been identified in the strata alongside the iridium, and a huge crater that could have been caused by a meteorite was found in the Caribbean–Gulf of Mexico region on the Yucatán peninsula.

Certainly, continental drift contributed to the Ordovician extinction. This extinction occurred after Gondwana arrived at the South Pole. Immense glaciers, which drew water from the oceans, chilled even once-tropical land. Marine invertebrates and coral reefs, which were especially hard hit, didn't recover until Gondwana drifted away from the pole and warmth returned. The mass extinction at the end of the Devonian period saw an end to 70% of marine invertebrates. Helmont Geldsetzer of Canada's Geological Survey notes that iridium has also been found in Devonian rocks in Australia, suggesting it's possible that a bolide event was involved because iridium has been found in Devonian rocks in Australia. Some scientists believe that this mass extinction could have been due to movement of Gondwana back to the South Pole. The extinction at the end of the Permian period was quite severe; 90% of species disappeared. The latest hypothesis attributes the Permian extinction to excess carbon dioxide. When Pangaea formed, there were no polar ice caps to initiate ocean currents. The lack of ocean currents caused organic matter to stagnate at the bottom of the ocean. Then, as the continents drifted into a new configuration, ocean circulation switched back on. Now, the extra carbon on the seafloor was swept up to the surface where it became carbon dioxide, a deadly gas for sea life. The trilobites became extinct, and the crinoids (sea lilies) barely survived. Excess carbon dioxide on land led to a global warming that altered the pattern of vegetation. Areas that were wet and rainy became dry and warm, and vice versa. Burrowing animals that could escape land surface changes seemed to have the best chance of survival.

The extinction at the end of the Triassic period is another that has been attributed to the environmental effects of a meteorite collision with Earth. Central Quebec has a crater half the size of Connecticut that some believe is the impact site. The dinosaurs may have benefited from this event because this is when the first of the gigantic dinosaurs took charge of the land. A second wave occurred in the Cretaceous period but it ended in dinosaur extinction as discussed previously.

### Check Your Progress 18.3

1. Climate permitting, would you expect to find dinosaur bones all over the globe? Explain.
2. Humans did not become extinct during any of the mass extinctions discussed. Explain.

# Connecting the Concepts

Would the history of life on Earth always be the same? In a previous chapter, we learned that the evolutionary process sometimes occurs gradually and steadily over time and at other times speciation seems to occur rapidly. Is it possible that both mechanisms may be at work in different groups of organisms and at different times?

Is it also possible that the history of life could have turned out differently? The species alive today are the end product of the abiotic and biotic changes that occurred on Earth as life evolved. And what if the abiotic and biotic changes had been other than they were? For example, if the continents had not separated 65 MYA, what types of mammals, if any, would be alive today? Given a different sequence of environments, a different mix of plants and animals might very well have resulted.

The history of life on Earth, as we know it, is only one possible scenario. If we could rewind the "tape of life" and let history take its course anew, the result might well be very different, depending on the geologic and biologic events that took place the second time around. As an analogy, consider that if you were born in another time period and in a different country, you might be very different from the "you" of today.

## summary

### 18.1 Origin of Life

The unique conditions of the early Earth allowed a chemical evolution to occur. An abiotic synthesis of small organic molecules such as amino acids and nucleotides occurred, possibly either in the atmosphere or at hydrothermal vents. These monomers joined together to form polymers either on land (warm seaside rocks or clay) or at the vents. The first polymers could have been proteins or RNA, or they could have evolved together. The aggregation of polymers inside a plasma membrane produced a protocell having some enzymatic properties such that it could grow. If the protocell developed in the ocean, it was a heterotroph; if it developed at hydrothermal vents, it was a chemoautotroph. A true cell had evolved once the protocell contained DNA genes. The first genes may have been RNA molecules, but later DNA became the information storage molecule of heredity. Biological evolution now began.

### 18.2 History of Life

The fossil record allows us to trace the history of life. The oldest prokaryotic fossils are cyanobacteria, dated about 3.5 BYA, and they were the first organisms to add oxygen to the atmosphere. The eukaryotic cell evolved about 2.2 BYA, but multicellular animals (the Ediacaran animals) do not occur until 600 MYA.

A rich animal fossil record starts at the Cambrian period of the Paleozoic era. The occurrence of external skeletons, which seems to explain the increased number of fossils at this time, may have been due to the presence of plentiful oxygen in the atmosphere, or perhaps it was due to predation. The fishes were the first vertebrates to diversify and become dominant. Amphibians are descended from lobe-finned fishes.

Plants also invaded land during the Ordovician period. The swamp forests of the Carboniferous period contained seedless vascular plants, insects, and amphibians. This period is sometimes called the Age of Amphibians.

The Mesozoic era was the Age of Cycads and Reptiles. First mammals and then birds evolved from reptilian ancestors. During this era, dinosaurs of enormous size were present. By the end of the Cretaceous period, the dinosaurs were extinct.

The Cenozoic era is divided into the Tertiary period and the Quaternary period. The Tertiary is associated with the adaptive radiation of mammals and flowering plants that formed vast tropical forests. The Quaternary is associated with the evolution of primates; first monkeys appeared, then apes, and then humans. Grasslands were replacing forests, and this put pressure on primates, who were adapted to living in trees. The result may have been the evolution of humans—primates who left the trees.

### 18.3 Factors That Influence Evolution

The continents are on massive plates that move, carrying the land with them. Plate tectonics is the study of the movement of the plates. Continental drift helps explain the distribution pattern of today's land organisms.

Mass extinctions have played a dramatic role in the history of life. It has been suggested that the extinction at the end of the Cretaceous period was caused by the impact of a large meteorite, and evidence indicates that other extinctions have a similar cause as well. It has also been suggested that tectonic, oceanic, and climatic fluctuations, particularly due to continental drift, can bring about mass extinctions.

## understanding the terms

absolute dating (of fossils) 324
chemical evolution 318
coacervate droplet 320
continental drift 332
endosymbiotic theory 325
extinction 327
fossil 322
geologic timescale 324
index fossil 322
liposome 320
mass extinction 327
microsphere 319
molecular clock 327
ocean ridge 319
ozone shield 325
paleontology 322
plate tectonics 332
protein-first hypothesis 319
proteinoid 319
protobiont 320
protocell 320
relative dating (of fossils) 322
RNA-first hypothesis 319
sedimentation 322
stratum 322
stromatolite 324

Match the terms to these definitions:

a. ______________ Concept that rates mutational in certain types of genes is constant over time.

b. ______________ Cell forerunner that possibly developed from cell-like microspheres.

c. ______________ Droplet of lipid molecules formed in a liquid environment.

d. ______________ A region where crust forms and from which it moves laterally in each direction.

e. ______________ Formed from oxygen in the upper atmosphere, it protects the Earth from ultraviolet radiation.

## reviewing this chapter

1. List and describe the various hypotheses concerning the chemical evolution that produced polymers. 318–19
2. Trace in general the steps by which the protocell may have evolved from polymers. 319–20
3. List and describe the various hypotheses concerning the origin of a self-replication system. 320–21
4. Explain how the fossil record develops and how fossils are dated relatively and absolutely. 322, 324
5. When did prokaryotes arise, and what are stromatolites? 324–25
6. When and how might the eukaryotic cell have arisen? 325
7. Describe the first multicellular animals found in the Ediacara Hills in southern Australia. 326
8. Why might there be so many fossils from the Cambrian period? 327
9. Which plants, invertebrates, and vertebrates were present on land during the Carboniferous period? 328–29
10. Which type vertebrate was dominant during the Mesozoic era? Which types began evolving at this time? 330
11. Which type of vertebrate underwent an adaptive radiation in the Cenozoic era? 331
12. What is continental drift, and how is it related to plate tectonics? Give examples to show how biogeography supports the occurrence of continental drift. 332
13. Identify five significant mass extinctions during the history of the Earth. What may have caused mass extinctions? 333

## testing yourself

Choose the best answer for each question.

For questions 1–6, match the statements with events in the key. Answers may be used more than once.

**KEY:**

a. early Earth
b. monomers evolve
c. polymers evolve
d. protocell evolves
e. self-replication system evolves

1. The heat of the sun could have caused amino acids to form proteinoids.
2. In a liquid environment, phospholipid molecules automatically form a membrane.
3. As the Earth cooled, water vapor condensed, and subsequent rain produced the oceans.
4. Miller's experiment shows that under the right conditions, inorganic chemicals can react to form small organic molecules.
5. Some investigators believe that RNA was the first nucleic acid to evolve.
6. An abiotic synthesis may have occurred at hydrothermal vents.
7. Which of these did Stanley Miller place in the experimental system to show that organic monomers could have arisen from inorganic molecules on the early Earth?
   a. microspheres
   b. purines and pyrimidines
   c. early atmospheric gases
   d. only RNA
   e. All of these are correct.
8. Which of these is not a place where polymers found in today's cells may have arisen?
   a. at hydrothermal vents
   b. on rocks beside the sea
   c. in clay
   d. in the atmosphere
   e. Both b and c are correct.
9. Which of these is the chief reason the protocell was probably a fermenter?
   a. The protocell didn't have any enzymes.
   b. The atmosphere didn't have any oxygen.
   c. Fermentation provides the most energy.
   d. There was no ATP yet.
   e. All of these are correct.
10. Liposomes (lipid droplets) are significant because they show that
    a. the first plasma membrane contained protein.
    b. a plasma membrane could have easily evolved.
    c. a biological evolution produced the first cell.
    d. there was water on the early Earth.
    e. the protocell had organelles.
11. Evolution of the DNA ⟶ RNA ⟶ protein system was a milestone because the protocell could now
    a. be a heterotrophic fermenter.
    b. pass on genetic information.
    c. use energy to grow.
    d. take in preformed molecules.
    e. All of these are correct.
12. Fossils
    a. are the remains and traces of past life.
    b. can be dated absolutely according to their location in strata.
    c. are usually found embedded in sedimentary rock.
    d. have been found for all types of animals except humans.
    e. Both a and c are correct.
13. Which of these events did not occur during the Precambrian?
    a. evolution of the prokaryotic cell
    b. evolution of the eukaryotic cell
    c. evolution of multicellularity
    d. evolution of the first animals
    e. All of these occurred during the Precambrian.
14. The organisms with the longest evolutionary history are
    a. prokaryotes that left no fossil record.
    b. eukaryotes that left a fossil record.
    c. prokaryotes that are still evolving today.
    d. animals that had a shell.

For questions 15–19, match the phrases with divisions of geologic time in the key. Answers may be used more than once.

**KEY:**

a. Cenozoic era
b. Mesozoic era
c. Paleozoic era
d. Precambrian time

15. dinosaur diversity, evolution of birds and mammals
16. contains the Carboniferous period
17. prokaryotes abound; eukaryotes evolve and become multicellular
18. mammalian diversification
19. invasion of land

20. Which of these occurred during the Carboniferous period?
    a. Dinosaurs evolved twice and became huge.
    b. Human evolution began.
    c. The great swamp forests contained insects and amphibians.
    d. Prokaryotes evolved.
    e. All of these are correct.
21. Continental drift helps explain
    a. mass extinctions.
    b. the distribution of fossils on the Earth.
    c. geological upheavals such as earthquakes.
    d. climatic changes.
    e. All of these are correct.
22. Which of these pairs is mismatched?
    a. Mesozoic—cycads and dinosaurs
    b. Cenozoic—grasses and humans
    c. Paleozoic—rise of prokaryotes and unicellular eukaryotes
    d. Cambrian—marine organisms with external skeletons
    e. Precambrian—origin of the cell at hydrothermal vents
23. Complete the following listings using these phrases: *$O_2$ accumulates in atmosphere, Ediacaran animals, oldest known fossils, Cambrian animals, protists diversify, oldest eukaryotic fossils*

| | | | |
|---|---|---|---|
| 2.1 BYA | a. ______________ | 1.0 BYA | d. ______________ |
| 2.7 BYA | b. ______________ | 630 MYA | e. ______________ |
| 3.5 BYA | c. ______________ | 542 MYA | f. ______________ |
| 4.6 BYA | formation of the Earth | | |

24. The protocell is hypothesized to have had a membrane boundary as do ______________ and ______________.
25. Once there was a flow of information from DNA to RNA to protein, the protocell became a ______________ cell, and biological evolution began.
26. The evolution of ______________ prokaryotes caused oxygen to enter the atmosphere.
27. Primitive vascular plants and amphibians were large and abundant during the ______________ period.
28. The mammals diversified and human evolution began during the ______________ era.
29. Mass extinctions seem to be due to climatic changes that occur after a ______________ bombards the Earth, or after the continents ______________ into a new configuration.
30. Which statement is not correct?
    a. The geologic timescale divides the history of Earth into eras, then periods, and then epochs.
    b. Eras span the least amount of time and epochs have the longest time frames.
    c. Only the periods of the Cenozoic era are divided into epochs, meaning that more attention is given to the evolution of primates and flowering plants than to the earlier evolving organisms.
    d. Modern civilization is given its own epoch, despite the fact that humans have only been around about 0.4% of the history of life.
    e. All of these are correct.
31. Which statement is not correct?
    a. The geologic timescale shows that evolution has been a series of events leading from the first cells to humans.
    b. The geologic timescale shows all the facets, twists, and turns of the history of life.
    c. Timewise, the events at the bottom of the timescale would be in a lower strata than the events that occur at the top of the timescale.
    d. Humans were present and, therefore, our ancestors had first-hand knowledge about the events that occurred during the history of the Earth.
    e. Both a and d are incorrect.
32. Which statement is not correct? The tree of life shows that
    a. endosymbiotic events can account for at least some of the organelles in a eukaryotic cell.
    b. evolution proceeds from the simple to the complex.
    c. both plants and animals can trace their ancestry to the protists.
    d. humans hold a special place in the evolution of animals.
    e. the prokaryotic cell preceded the eukaryotic cell.
    f. All of these are correct.

## thinking scientifically

1. You were asked to supply an evolutionary tree of life and decided to use Figure 18.7. How is this tree consistent with evolutionary principles?
2. Explain the occurrence of living fossils, such as horseshoe crabs, that closely resemble their ancestors known from the fossil record.

## *Biology* website

The companion website for *Biology* provides a wealth of information organized and integrated by chapter. You will find practice tests, animations, videos, and much more that will complement your learning and understanding of general biology.

**http://www.mhhe.com/maderbiology10**

# 19 Systematics and Phylogeny

*Molecular technology offers powerful new tools to tell who is related to whom. Take the field of orchid biology, for example. With the help of DNA sequencing, it became possible to confidently reconstruct the evolutionary history of orchids and to even trace new origins within groups of orchids.*

*DNA sequencing has predictive value. Suppose, for example, you have discovered a group of plants capable of producing antiviral compounds and want to find other groups capable of producing the same compounds. DNA sequencing can lead you to them! Or, if you are working with orchids and want to win a prize in the next flower show, DNA sequencing will tell you which species might hybridize well to produce bigger and more showy flowers. DNA sequencing can also assist the formulation of conservation strategies because it can single out the rare species existing only in isolated populations. Now you can use conservation dollars to save the rare and endangered species, even among orchids.*

*This chapter introduces you to systematics, which involves reconstructing evolutionary history and then classifying or grouping organisms according to evolutionary findings.*

Orchids are quite varied, and DNA sequencing can be relied on to indicate evolutionary relationships.

## concepts

# 19.1 Systematics

All fields of biology, but especially **systematics** [Gk. *systema*, an orderly arrangement], are dedicated to understanding the evolutionary history of life on Earth, including those on an African plain (Fig. 19.1). Systematics is very analytical and relies on a combination of data from the fossil record and comparative anatomy and development, with an emphasis today on molecular data, to determine evolutionary relationships. **Taxonomy** [Gk. *tasso*, arrange, classify, and *nomos*, usage, law], the branch of biology concerned with identifying, naming, and classifying organisms, is a part of systematics.

## Linnean Taxonomy

Suppose you went to Africa on a photo safari and wanted to classify the organisms shown in Figure 19.1 according to your own system. Most likely, you would begin by making a list, and naturally this would require you to give each organism a name. Then you would start assigning the organisms on your list to particular groups. But what criteria would you use—color, shape, size, how the organisms relate to you? Deciding on the number, types, and arrangement of the groups would not be easy, and periodically you might change your mind or even have to start over. Biologists, too, have not had an easy time deciding how living things should be classified and have made changes in their methods throughout history. These changes are often brought about by an increase in fossil, anatomical, or molecular data. Ideally, classification is based on our understanding of how organisms are related to one another through evolution. A natural system of classification, as opposed to an artificial system, reflects the evolutionary history of organisms.

Taxonomy began with the ancient Greeks and Romans. The famous Greek philosopher Aristotle was interested in taxonomy, and he identified organisms as belonging to a particular group, such as horses, birds, and oaks. In the Middle Ages, these names were translated into Latin, the language still used for scientific names today. Much later, John Ray (1627–1705), a British naturalist of the seventeenth century, believed that each organism should have a set name. He said, "When men do not know the name and properties of natural objects—they cannot see and record accurately."

## The Binomial System

The number of known types of organisms expanded greatly in the mid-eighteenth century as Europeans traveled to distant parts of the world. During this time, Car-

**FIGURE 19.1 Classifying organisms.**

How would you name and classify these organisms? After naming them, how would you assign each to a particular group? Based on what criteria? An artificial system would not take into account how they might be related through evolution, as would a natural system.

a.

b. *Lilium canadense*

c. *Lilium bulbiferum*

**FIGURE 19.2 Carolus Linnaeus.**

**a.** Linnaeus was the father of taxonomy and gave us the binomial system of naming and classifying organisms. His original name was Karl von Linne, but he later latinized it because of his fascination with scientific names. Linnaeus was particularly interested in classifying plants. **b, c.** Each of these two lilies are species in the same genus, *Lilium*.

olus Linnaeus (1707–78) developed **binomial nomenclature,** by which each species receives a two-part name (Fig. 19.2). For example, *Lilium bulbiferum* and *Lilium canadense* are two different species of lily. The first word, *Lilium,* is the genus (pl., genera), a classification category that can contain many species. The second word, the **specific epithet,** refers to one species within that genus. The specific epithet sometimes tells us something descriptive about the organism. Notice that the scientific name is in italics; the genus is capitalized, while the specific epithet is not. Both names are separately underlined when handwritten. The species is designated by the full binomial name—in this case, either *Lilium bulbiferum* or *Lilium canadense.* The specific epithet alone gives no clue as to species—just as the house number alone without the street name gives no clue as to which house is specified. The genus name can be used alone, however, to refer to a group of related species. Also, the genus can be abbreviated to a single letter if used with the specific epithet (e.g., *L. bulbiferum*) and if the full name has been given previously.

Scientific names are derived in a number of ways. Some scientific names are descriptive in nature, for example, *Acer rubrum* for the red maple. Other scientific names may include geographic descriptions such as *Alligator mississippiensis* for the American alligator. Scientific names can also include eponyms (named after someone), such as the owl mite *Strigophilus garylarsonii* (named after the cartoonist). Many scientific names are derived from mythical characters, such as *Iris versicolor,* named for Iris, the goddess of the rainbow. Some scientific names reflect a humorous slant, such as *Ba humbugi* for a species of snail.

Why do organisms need scientific names? And why do scientists use Latin, rather than common names, to describe organisms? There are several reasons. First, a common name will vary from country to country because different countries use different languages. Second, even people who speak the same language sometimes use different common names to describe the same organism. For example, bowfin, grindle, choupique, and cypress trout describe the same common fish, *Amia calva.* Furthermore, between countries, the same common name is sometimes given to different organisms. A "robin" in England is very different from a "robin" in the United States, for example. Latin, on the other hand, is a universal language that not too long ago was well known by most scholars, many of whom were physicians or clerics. When scientists throughout the world use the same scientific binomial name, they know they are speaking of the same organism.

The Linnean Society rules on the appropriateness of the binomial name for each species in the world. Of the estimated 3–30 million species now living on Earth, a million species of animals and a half million species of plants and microorganisms have been named. We are further along on some groups than others; we may have finished the birds, but there may be hundreds of thousands of unnamed insects. The task of identifying and naming the species of the world is a daunting one. A new fast and efficient way of identifying species that is based on their DNA is described in the Science Focus on page 347. This method has been called into question by those who feel that nucleotide base differences in a single gene may not yield enough data to distinguish two closely related species or to recognize when hybridization has occurred. But the method was found to be satisfactory for the identification of mosquito species in India.

## Linnaean Classification Categories

Classification, which begins when an organism is named, includes taxonomy, since genus and species are two classification categories. In the context of classification, a species is a taxonomic category below the rank of genus. A

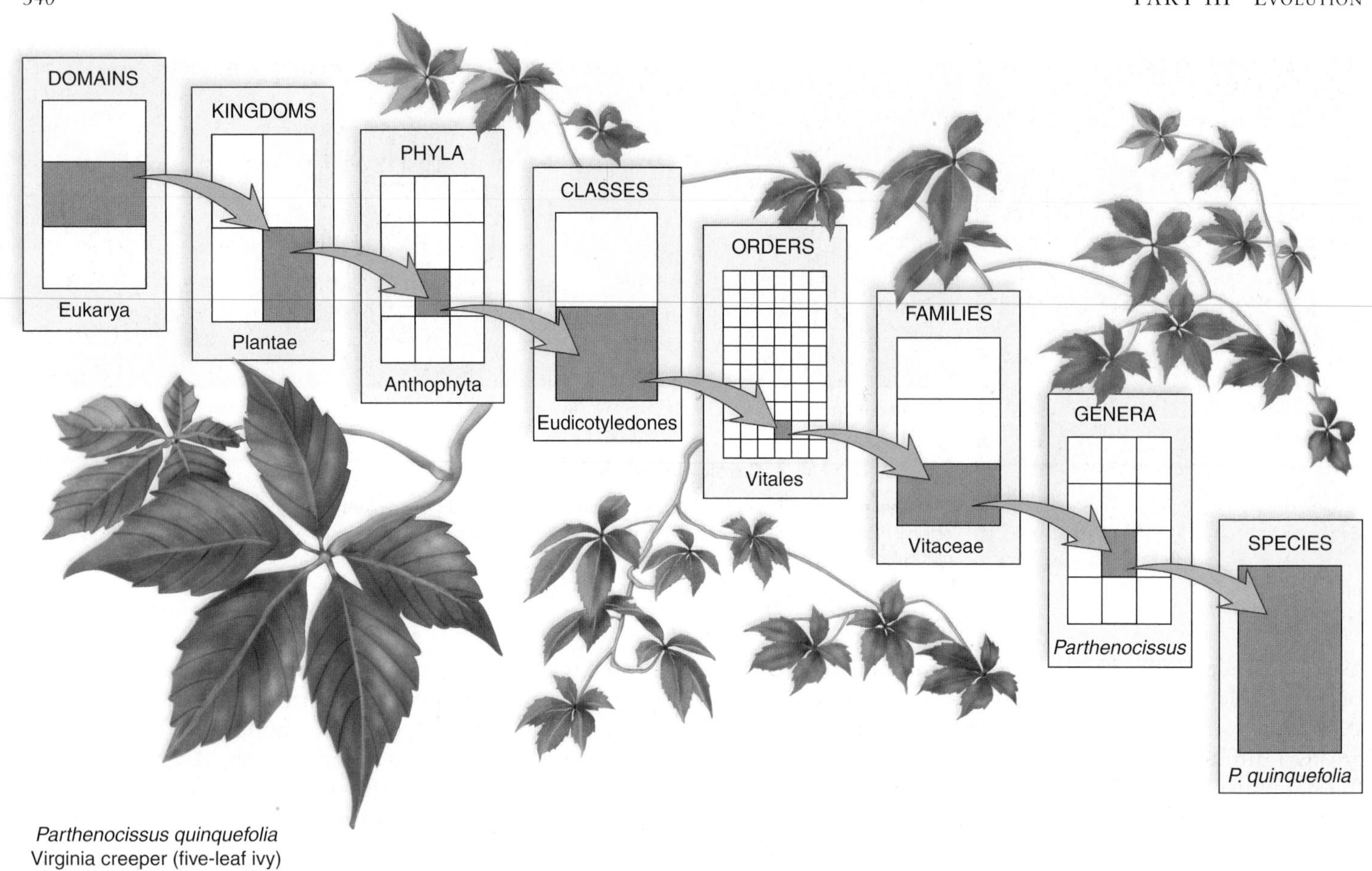


**FIGURE 19.3 Hierarchy of taxa for *Parthenocissus quinquefolia*.**

A domain is the most inclusive of the classification categories. Kingdom Plantae is in the domain Eukarya. Kingdom Plantae contains several phyla, one of which is Anthophyta. The phylum Anthophyta has two classes (the monocots and eudicots). In the class Eudicotyledones, there are many orders, including Vitales. One of the genera in the family Vitaceae is the genus *Parthenocissus*. This genus includes the species *Parthenocissus quinquefolia*. The specific epithet is due to the plant's whorl of five leaves. This illustration is diagrammatic and is not necessarily representative of the correct number of subcategories.

**taxon** (pl., taxa) is a group of organisms that fills a particular category of classification; *Rosa* and *Felis* are taxa at the genus level.

The taxonomists mentioned in this chapter contributed to classification. Aristotle divided living things into 14 groups—mammals, birds, fish, and so on. Then he subdivided the groups according to the size of the organisms. Ray used a more natural system, grouping animals and plants according to how he thought they were related. Linnaeus simply used flower part differences to assign plants to the categories species, genus, order, and class. His studies were published in a book called *Systema Naturae* in 1735.

Today, taxonomists use the following major categories of classification: **species, genus, family, order, class, phylum,** and **kingdom.** Recently, a higher taxonomic category, the **domain,** has been added to this list. There can be several species within a genus, several genera within a family, and so forth—the higher the category, the more inclusive it is (Fig. 19.3). Therefore, there is a hierarchy of categories. You can also say that the categories are **nested** because one group exists inside another group. For example, domain contains many kingdoms, and one kingdom contains many classes and so forth.

The organisms that fill a particular classification category are distinguishable from other organisms by sharing a set of traits, sometimes called characters. Organisms in the same domain have general traits in common; those in the same species have quite specific traits in common. In most cases, categories of classification can be subdivided into three additional categories, as in superorder, order, suborder, and infraorder. Considering these, there are more than 30 categories of classification.

### Check Your Progress 19.1

1. Humans are in the order Primates. What more inclusive categories and what less inclusive categories would be needed to classify humans?
2. The scientific name for modern humans is *Homo sapiens*. Would you suspect that the genus *Homo* contains more than one *Homo*? Explain.

## 19.2 Phylogenetic Trees

One goal of systematics is to determine **phylogeny** [Gk. *phyle*, tribe; L. *genitus*, producing], a depiction of evolutionary history called a **phylogenetic tree** or an evolutionary tree. The tree shows the **common ancestors** (an ancestor to two or more lines of descent) and branches coming off from the common ancestor. Each branch in a tree is a divergence that gives rise to two or more new groups. For example, this portion of an evolutionary tree says that monkeys and apes share a common primate ancestor:

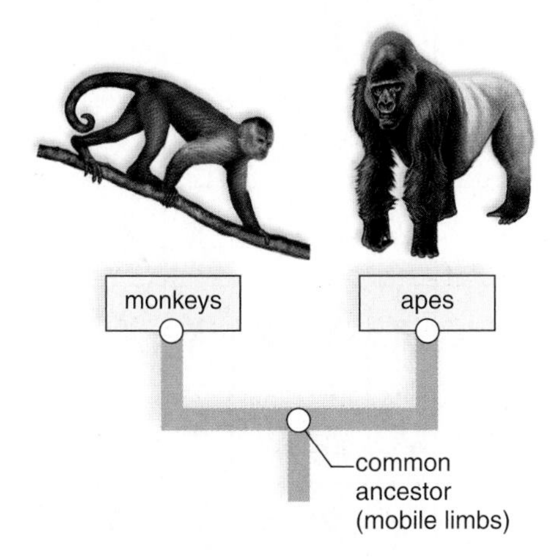


Divergence is presumed because monkeys and apes have their own **derived traits** (traits not seen previously). For example, skeletal differences allow an ape to swing from limb to limb of a tree while monkeys run along the tops of tree branches. The common primate ancestor to both monkeys and apes has traits that are shared by the ancestor and also monkeys and apes. For example, the common primate ancestor had mobile limbs.

A phylogenetic tree has many branch points, and they show that it is possible to trace the ancestry of a group of organisms back farther and farther in the past. For example, reindeer, monkeys, and apes all give birth to live young because they all have a common ancestor that was a placental mammal. This ancestor was also a quadruped, as they all are:

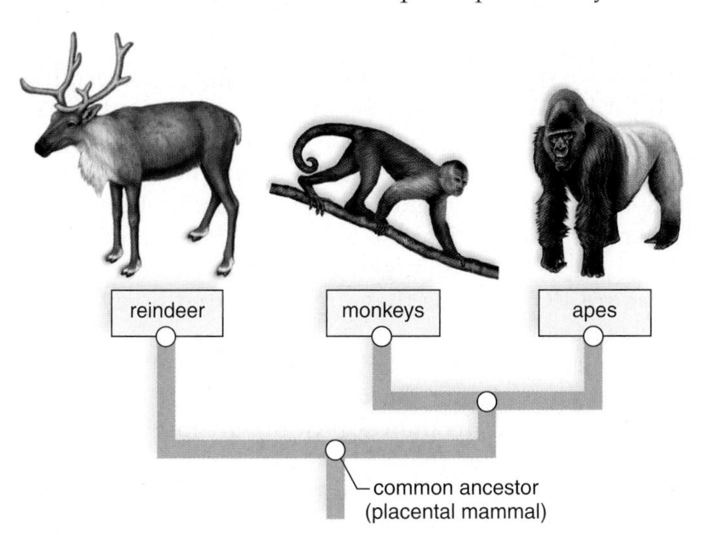


Because classification is hierarchical, it is possible to use classification categories to construct a phylogenetic tree. A species is most closely related to other species in the same genus and then is related, but less so, to genera in the same family, and so forth, from order to class to phylum to kingdom. When we say that two species (or genera, families, etc.) are closely related, we mean that they share a recent common ancestor. For example, all the animals in Figure 19.4 are related because we can trace their ancestry back to the same order. The animals in the order Artiodactyla all have even-toed hoofs. Animals in the family Cervidae have solid horns, called antlers, but they are highly branched in red deer (genus *Cervus*) and palmate (having the shape of a hand) in reindeer (genus *Rangifer*). In contrast, animals in the family Bovidae have hollow horns and, unlike the Cervidae, both males and females have horns, although they are smaller in females.

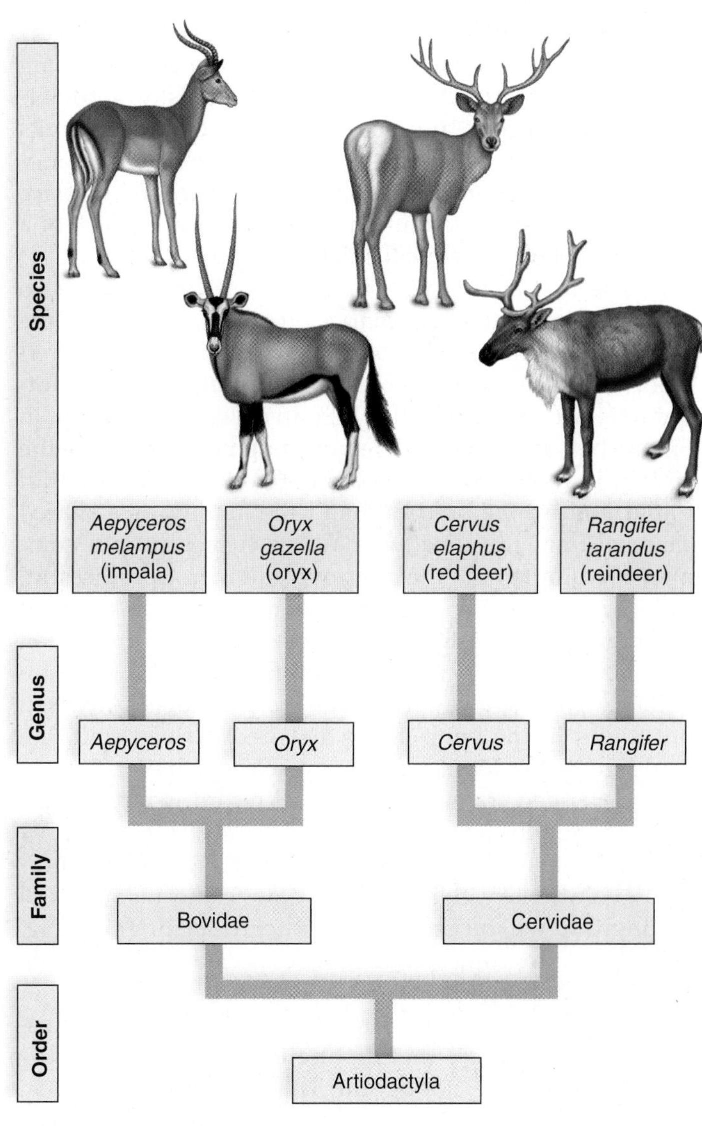


**FIGURE 19.4 Classification and phylogeny.**

The classification and phylogenetic tree for a group of organisms are ideally constructed to reflect their evolutionary history. A species is most closely related to other species in the same genus, more distantly related to species in other genera of the same family, and so forth, on through order, class, phylum, kingdom, and domain.

## Cladistic Phylogenetic Trees

Biologists are always seeking new and improved ways to discover the evolutionary history of life on Earth. Tracing evolutionary history would be easy if similarities alone could be used to trace phylogeny, but this is not the case because evolution is quite variable, sometimes even reversing to a former state. For example, some vertebrates have teeth and some do not, and therefore we need a methodology that will tell us which is the ancestral state—teeth or no teeth. In this instance, the fossil record tells us that possession of teeth is an early characteristic of vertebrates. But if the fossil is unavailable, we need some other method. The most commonly used method to determine evolutionary relationships when a complete fossil record is not available is called cladistics.

### *Methodology of Cladistics*

**Cladistics,** which is based on the work of Willi Hennig, is a way to trace evolutionary history of a group by using shared traits, derived from a common ancestor, to determine which species are most closely related. These traits are then used to construct phylogenetic trees called cladograms. A **cladogram** [Gk. *klados*, branch, stem, and *gramma*, picture] depicts the evolutionary history (phylogeny) of a group based on the available data.

The first step when constructing a cladogram is to draw up a table that summarizes the derived traits of the species being compared (Fig. 19.5). At least one, but preferably several species, is considered an **outgroup.** The outgroup is not part of the study group called the **ingroup.** In Figure 19.5, lancelets are the outgroup because unlike the species in the ingroup they are not vertebrates. Any trait found both in the outgroup and the ingroup is a shared *ancestral* trait, presumed to have been present in a common ancestor to both the outgroup and ingroup. Ancestral traits are not used in the cladogram, so why do we need an outgroup? An outgroup tells us which traits are shared derived traits, also called **synapomorphies** [Gk. *syn*, together with, *apo*, away from, and *morph*, shape]. Any trait not found in the outgroup is a shared derived trait. We could go to the fossil record to discover which traits are shared derived traits, but the fossil record is rarely complete enough to use exclusively.

All the synapomorphies listed in Figure 19.5 indicate evolutionary relationships among the members of the ingroup will be used to construct the cladogram (Fig. 19.6). A cladogram contains several clades; each **clade** includes a common ancestor and all its descendants that share one or more synapomorphies. These synapomorphies are differences that distinguish the clade from the other clades in the cladogram. The common ancestors in our cladogram are indicated by white circles. Because the outgroup does not have vertebrae and all the species in the ingroup do have vertebrae, we know that the first common ancestor for the ingroup was a vertebrate—it had vertebrae. All the ingroup species are in the first clade because they all have vertebrae. The next clade includes all the species that have four limbs, and the members of the next clade all had an amniotic egg. The common ancestor for only the lizard, crocodile, and bird had epidermal scales. The common ancestor for crocodiles and birds had a gizzard. Feathers are not listed in Figure 19.5 because only birds have feathers. The last clade in our cladogram includes the dog and human because they have a common ancestor, which had both hair and mammary glands. Similar to feathers, the canine teeth of the dog and the enlarged brain of the human are not in Figure 19.5 because they are not shared by any other species in our ingroup.

A cladogram is objective because it lists the data that are used to construct the cladogram. Cladists typically use many more traits than appear in our simplified cladogram. They also feel that a cladogram is a hypothesis that can be tested and either corroborated or refuted on the basis of additional data. The terms you need to learn to understand cladistics are given in Table 19.1.

| | Species | | | | | | | |
|---|---|---|---|---|---|---|---|---|
| | ingroup | | | | | | | lancelet (outgroup) |
| Traits | chimpanzee | dog | finch | crocodile | lizard | frog | tuna | |
| mammary glands | X | X | | | | | | |
| hair | X | X | | | | | | |
| gizzard | | | X | X | | | | |
| epidermal scales | | | X | X | X | | | |
| amniotic egg | X | X | X | X | X | | | |
| four limbs | X | X | X | X | X | X | | |
| vertebrae | X | X | X | X | X | X | X | |
| notochord in embryo | X | X | X | X | X | X | X | X |

**FIGURE 19.5 Constructing a cladogram: the data.**

This lancelet is in the outgroup, and all the other species listed are in an ingroup (study group). The species in the ingroup have shared derived traits (synapomorphies), derived because a lancelet does not have the trait, and shared because certain species in the study group do have them. All the species in the ingroup have vertebrae, all but a fish have four limbs, and so forth. The shared derived traits indicate which species are distantly related and which are closely related. For example, a human is more distantly related to a fish, with which it shares only one trait (vertebrae), than an iguana, with which it shares three traits (vertebrae, four limbs, amniotic egg).

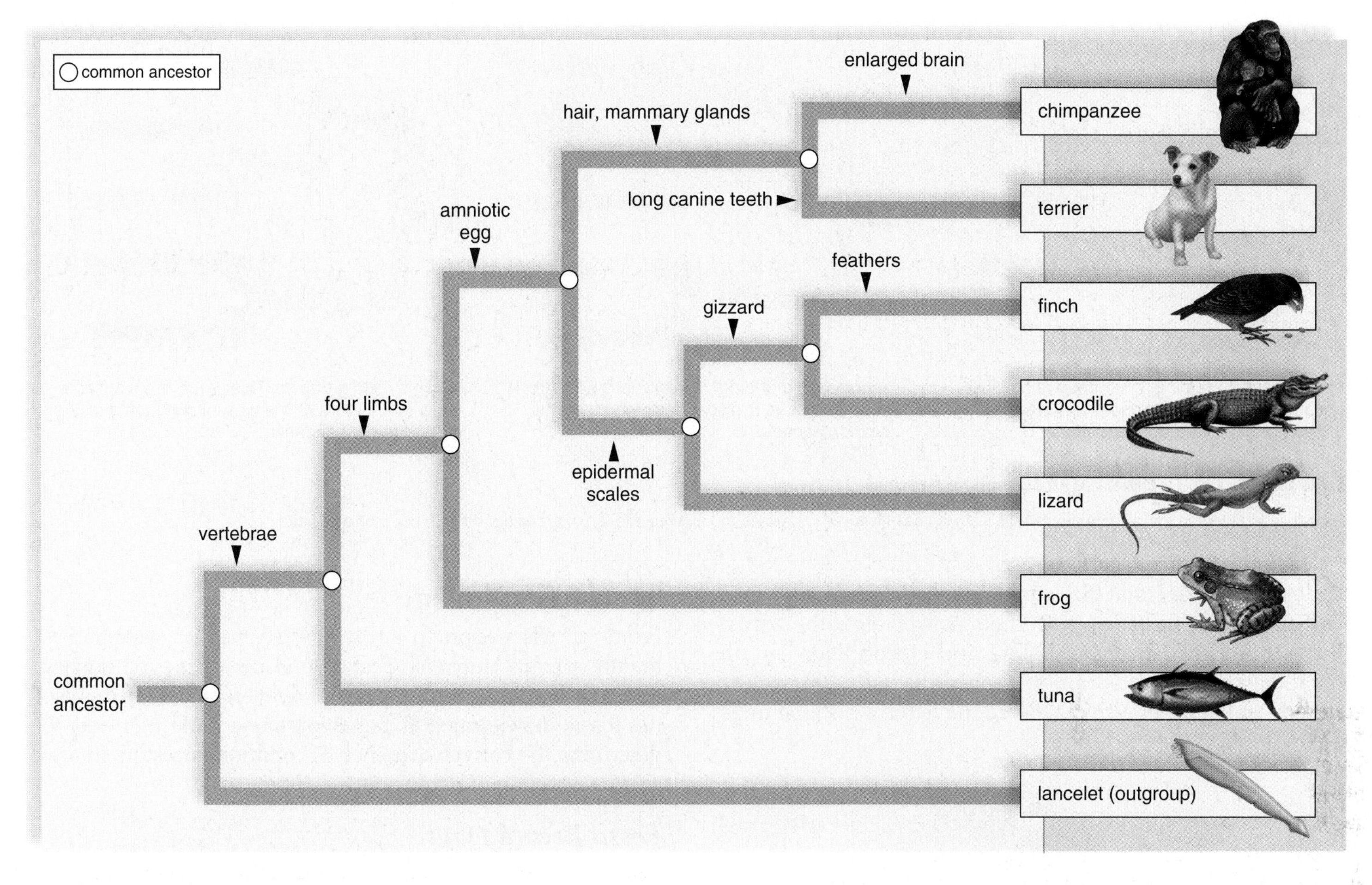


**FIGURE 19.6 Constructing a cladogram: the phylogenetic tree.**

Based on the data shown in Figure 19.5, the ingroup in this phylogenetic tree has six clades. Each clade contains a common ancestor with derived traits that are shared by the members of the clade.

## *How to Judge a Cladogram*

In order to tell if a cladogram has produced the best hypothesis, cladists are often guided by the principle of parsimony, which states that the minimum number of assumptions is the most logical. That is, they construct the cladogram that leaves the fewest number of shared derived characters unexplained or that minimizes the number of evolutionary changes. The rule of parsimony works best for traits that evolve at a slower rate than the frequency of speciation events. A problem with parsimony can arise when DNA sequencing is used to help construct cladograms. Mutations, especially in noncoding DNA, can be quite high, and, if so, base charges are not reliable data to distinguish clades. For this reason, some systematists have begun using statistical tools and not parsimony to help construct phylogenetic trees. This new branch of systematics is called **statistical phylogenetics.** In any case, the reliability of a cladogram is dependent on the knowledge and skill of the particular investigator gathering the data and doing the character analysis.

**TABLE 19.1**

**Terms Used in Cladistics**

|  |  |
|---|---|
| Outgroup | Species that define(s) which study group trait is oldest |
| Ingroup | Species that will be placed into clades in a cladogram |
| Ancestral trait | Traits present in both the outgroup and the ingroup |
| Clade | Evolutionary branch of a cladogram that contains a common ancestor and all its descendant species |
| Shared derived traits | Traits that distinguish a particular clade |
| Monophyletic grouping | Contains a single common ancestor and all its descendant species |
| Parsimony | Results in the simplest cladogram possible |

## *How to Judge a Clade*

Just like Linnean taxonomic categories, clades are often nested inside other clades. For example, all the clades in our cladogram are inside the first one because all the species have vertebrae. Notice also that crocodiles and birds have a common ancestor with a gizzard and, along with the lizard, are in a clade whose common ancestor had epidermal

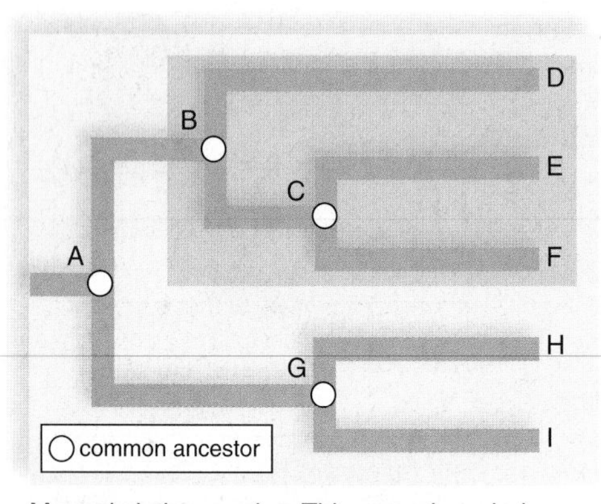

Monophyletic grouping: This group is a clade because it contains B, closest common ancestor, and all the descendants of B.

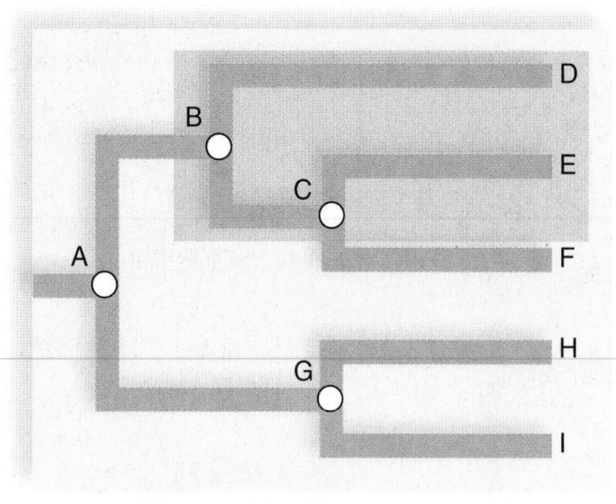

Paraphyletic grouping: This group is not a clade because it lacks F, also a descendant of the common ancestor B.

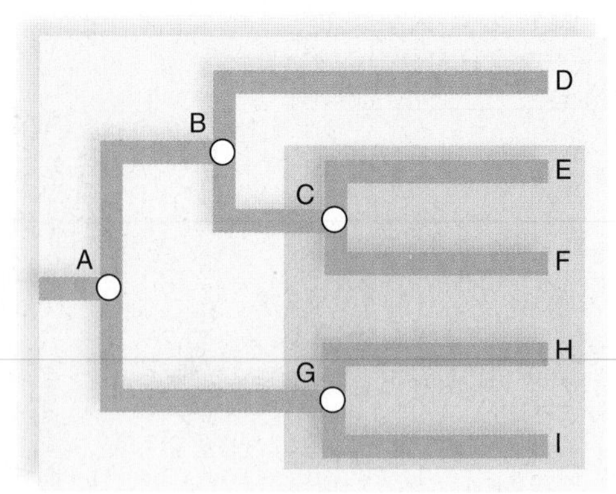

Polyphyletic grouping: This group is not a clade because it lacks A, the closest common ancestor to all the descendants.

**FIGURE 19.7 Different groupings of species.**

A clade in a cladogram must be monophyletic. Linnean classification is criticized for allowing the use of groupings that are not monophyletic.

scales. This means that birds are closely related to crocodiles and should be classified with them as well as with lizards! Birds, dinosaurs, lizards, snakes, and crocodilians can all trace their ancestry to amniotes called diapsids. Dinosaurs, as well as the other animals listed, have the skull openings of a diapsid:

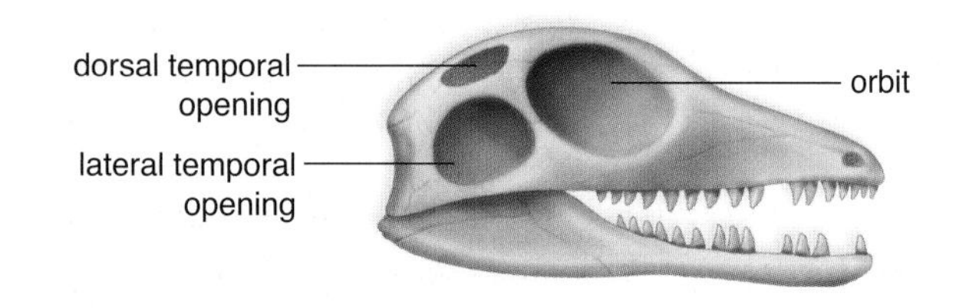


Linnean classification does not group birds with crocodiles nor with reptiles and, in doing so, has broken one of the rules of cladistics. Among the groupings shown in Figure 19.7, cladistics only allows monophyletic groupings. A **monophyletic group** includes a common ancestor and all the descendants of that ancestor. A paraphyletic group contains a common ancestor and does not include all the descendants. A polyphyletic group contains some of the descendants of more than one common ancestor and not all the common ancestors.

Because the Linnean classification system allows groupings other than those that are monophyletic, it is now being severely criticized by some biologists who show that Linnean classification does not truly reflect the evolutionary history of life on Earth. Biologists are presently trying to determine how Linnean classification could be changed to reflect our current understanding of phylogeny.

**Check Your Progress 19.2A**

1. How do you know how many clades your cladogram will have?
2. Why do cladists tell us that birds should be classified as reptiles?

## Tracing Phylogeny

From our discussion so far, it may seem as if systematists mainly rely on morphological data to discover evolutionary relationships between species. However, systematists also use fossil, developmental, behavioral, and molecular data to determine the correct sequence of common ancestors in any particular group of organisms.

### *Fossil Record Data*

One of the advantages of fossils is that they can be dated, but unfortunately it is not always possible to tell to which group, living or extinct, a fossil is related. For example, at present, paleontologists are discussing whether fossil turtles indicate that turtles are distantly or closely related to crocodiles. On the basis of his interpretation of fossil turtles, Olivier C. Rieppel of the Field Museum of Natural History in Chicago is challenging the conventional interpretation that turtles are ancestral (have traits seen in a common ancestor to all reptiles) and are not closely related to crocodiles, which evolved later. His interpretation is being supported by molecular data that show turtles and crocodiles are closely related.

If the fossil record was more complete, there might be fewer controversies about the interpretation of fossils. One reason the fossil record is incomplete is that most fossils exist as only harder body parts, such as bones and teeth. Soft parts are usually eaten or decayed before they have a chance to be buried. This may be one reason it has been difficult to discover when angiosperms (flowering plants) first evolved. A Jurassic fossil recently found, if accepted as an angiosperm by most botanists, may help pin down the date (Fig. 19.8). As paleontologists continue to explore the world, the sometimes stingy fossil record will reveal some of its secrets.

### *Morphological Data*

**Homology** [Gk. *homologos*, agreeing, corresponding] is structural similarity that stems from having a common ancestor. Comparative anatomy, including developmental evidence such as that

**FIGURE 19.8**
**Ancestral angiosperm.**
The fossil *Archaefructus liaoningensis,* dated from the Jurassic period, may be the earliest angiosperm to be discovered. Without knowing the anatomy of the first flowering plant, it has been difficult to determine the ancestry of angiosperms.

shown in Figure 19.9, provides information regarding homology. **Homologous structures** are similar to each other because of common descent. The forelimbs of vertebrates contain the same bones organized just as they were in a common ancestor, despite adaptations to different environments. As Figure 15.15 shows, even though a horse has but a single digit and toe (the hoof), while a bat has four lengthened digits that support its wing, a horse's forelimb and a bat's forelimb contain the same bones.

Deciphering homology is sometimes difficult because of convergent evolution. **Convergent evolution** has occurred when distantly related species have a structure that looks the same only because of adaptation to the same type of environment. Similarity due to convergence is termed **analogy.** The wings of an insect and the wings of a bat are analogous. **Analogous structures** have the same function in different groups but do not have a common ancestry. Both cacti and spurges are adapted similarly to a hot, dry environment, and both are succulent (thick, fleshy) with spiny leaves. However, the details of their flower structure indicate that these plants are not closely related. The construction of phylogenetic trees is dependent on discovering homologous structures and avoiding the use of analogous structures to uncover ancestry.

## *Behavioral Data*

The opening story for Chapter 18 presents the evidence that dinosaurs cared for their young in a manner similar to crocodilians (includes alligators) and birds. These data substantiate the morphological data that dinosaurs, crocodilians, and birds are related through evolution.

## *Molecular Data*

Speciation occurs when mutations bring about changes in the base-pair sequences of DNA. Systematists, therefore, assume that the more closely species are related, the fewer changes there will be in DNA base-pair sequences. Since DNA codes for amino acid sequences in proteins, it also follows that the more closely species are related, the fewer differences there will be in the amino acid sequences within their proteins.

Because molecular data are straightforward and numerical, they can sometimes sort out relationships obscured by inconsequential anatomical variations or convergence. Software breakthroughs have made it possible to analyze nucleotide sequences or amino acid sequences quickly and accurately using a computer. Also, these analyses are available to anyone doing comparative studies through the Internet, so each investigator doesn't have to start from scratch. The combination of accuracy and availability of past data has made molecular systematics a standard way to study the relatedness of groups of organisms today.

**Protein Comparisons.** Before amino acid sequencing became routine, immunological techniques were used to roughly judge the similarity of plasma membrane proteins. In one procedure, antibodies are produced by transfusing a rabbit with the cells of one species. Cells of the second species are exposed to these antibodies, and the degree of the reaction is observed. The stronger the reaction, the more similar the cells from the two species.

Later, it became customary to use amino acid sequencing to determine the number of amino acid differences in a particular protein. Cytochrome *c* is a protein that is found in all aerobic organisms, so its sequence has been determined for a number of different organisms. The amino acid difference in cytochrome *c* between chickens and ducks is only 3, but between chickens and humans there are 13 amino acid differences. From this data you can conclude that, as expected, chickens and ducks are more closely related than are chickens and humans. Since the number of proteins available for study in all living things at all times is limited, most new studies today study differences in RNA and DNA.

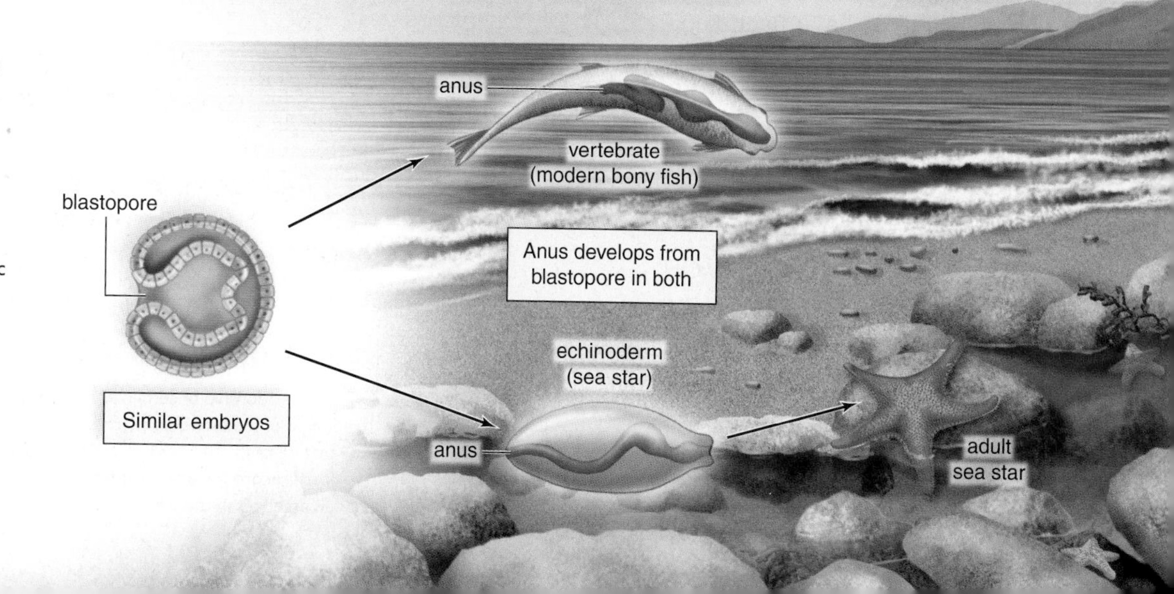


**FIGURE 19.9 Development reveals homologies.**
Among invertebrates, echinoderms, such as a sea star, are most closely related to vertebrates, even though echinoderms have radial symmetry. A study of their embryos shows that they develop similarly—in both, the embryonic blastopore becomes the anus. Later, the echinoderm, but not the vertebrate, becomes radially symmetrical.

**DNA and RNA Comparisons.** In the next part of this chapter, we will mention that a study of RNA differences between prokaryotes and eukaryotes resulted in an acceptance of the three-domain system of classification. In the opening story for this chapter, we discuss how a study of DNA differences has helped make sense of the structural data regarding the evolutionary history of orchids. The same RNA and DNA comparisons can be used in other ways, aside from deciphering evolutionary relationships. For example, they can be used by conservationists to determine that a species is rare and endangered.

DNA differences can substantiate data, help trace the course of macroevolution, and fill in the gaps of the fossil record. The phylogenetic tree of primates shown in Figure 19.10 is based on the fossil record and on DNA differences. No doubt we share a recent common ancestor with chimpanzees, and therefore according to the rules of cladistics, should be classified with them as we now are to the level of subfamily (see Fig. 30.4).

Mitochondrial DNA (mtDNA) mutates ten times faster than nuclear DNA. Therefore, when determining the phylogeny of closely related species, investigators often choose to sequence mtDNA instead of nuclear DNA. One such study concerned North American songbirds. It had long been suggested that these birds diverged into eastern and western subspecies due to retreating glaciers some 250,000–100,000 years ago. Sequencing of mtDNA allowed investigators to conclude that groups of North American songbirds diverged from one another an average of 2.5 million years ago (MYA). Since the old hypothesis based on glaciation is apparently flawed, a new hypothesis is required to explain why eastern and western subspecies arose among these songbirds.

**Molecular Clocks.** When nucleic acid changes are neutral (not tied to adaptation) and accumulate at a fairly constant rate, these changes can be used as a kind of **molecular clock** to indicate relatedness and evolutionary time. The researchers doing comparative mtDNA sequencing used their data as a molecular clock when they equated a 5.1% nucleic acid difference among songbird subspecies to 2.5 mya. In Figure 19.10, the researchers used their DNA sequence data to suggest how long the different types of primates have been separate. The fossil record was used to calibrate the clock: When the fossil record for one divergence is known, it indicates how long it probably takes for each nucleotide pair difference to occur. When the fossil record and molecular clock data agree, researchers have more confidence that the proposed phylogenetic tree is correct.

**FIGURE 19.10 Molecular data.**

The relationship of certain primate species based on a study of their genomes. The length of the branches indicates the relative number of nucleotide pair differences that were found between groups. These data, along with knowledge of the fossil record for one divergence, make it possible to suggest a date for the other divergences in the tree.

## Check Your Progress 19.2B

1. What type of evidence could you use to determine that the wing of an insect and the wing of a bat are analogous or are homologous?
2. What would you expect to find if you compared the DNA differences of a snake, bird, and monkey?

*science focus*

## DNA Bar Coding of Life

Traditionally, taxonomists have often relied on anatomical data to tell species apart. For example, differences in the type of spinning apparatus and the type of web have played a large role in distinguishing one spider from another (Fig. 19A). We can well imagine that if a mother wanted to know if certain spiders in the backyard were dangerous to her children, she might want a faster answer than could be provided by a traditional taxonomist at a university some distance away.

Enter the Consortium for the Barcode of Life (CBOL), which proposes that any scientist, not just taxonomists, will be able to identify a species with the flick of a handheld scanner. Just like the 11-digit Universal Product Code (UPC) used to identify products sold in a supermarket, the consortium believes that a sample of DNA should be able to identify any organism on Earth. The proposed scanner would tap into a bar-code database that contains the bar codes for all species so far identified on planet Earth. Also, a handheld DNA–bar-coding device is expected to provide a fast and inexpensive way for a wide range of researchers, including biology students, to catalog any and all of the world's species that do not yet have a bar code. So far scientists have identified only about 1.5 million species out of a potential 30 million. And there is no central database that keeps track of the known species.

The idea of using bar codes to identify species is not new, but Paul Hebert and his colleagues at the University of Guelph in Canada are the first to suggest it would be possible to use the base sequence in DNA to develop a bar code for each living thing. The order of DNA's nucleotides—A, T, C, and G—within a particular gene common to the organisms in each kingdom would fill the role taken by numbers in the UPC used in warehouses and stores. Hebert believes that the gene

- should contain no more than 650 nucleotides so that sequencing can be accomplished speedily with few mistakes;
- should be easy to extract from an organism's complete genome;
- should have mutated to the degree that each species has its own sequence of bases but not so fast that the sequence differs greatly among individuals within the same species.

Hebert's team decided that a mitochondrial gene known as cytochrome *c* oxidase subunit I, or COI, would be a suitable target gene in animals. (This gene codes for one of the carriers in the electron transport chain; see page 112.) Another researcher, John Kress, a plant taxonomist at the Smithsonian Institution in Washington, D.C., has developed a potential method for bar coding plant species. The Consortium for the Barcode of Life is growing by leaps and bounds and now includes various biotech companies, various museums and universities, the U.S. Food and Drug Administration, and also the U.S. Department of Homeland Security. Hebert has received a $3 million grant from the Gordon and Betty Moore Foundation to start the Biodiversity Institute of Ontario, which will be housed on the University of Guelph campus, where he teaches.

Speedy DNA bar coding would not only be a boon to ordinary citizens and taxonomists, but it would also benefit farmers who need to identify a pest attacking their crops, doctors who need to know the correct antivenin for snakebite victims, and college students who are expected to identify the plants, animals, and protists on an ecological field trip.

a.

b.

**FIGURE 19A Identifying spiders.**

*Identification of spiders at present depends in part on their type of spinning apparatus and the type of web they weave. The orb web of the garden spider* Araneus diadematus *(**a**) differs somewhat from (**b**) the orb web of the New Zealand spider* Waitkera waitkerensis.

## 19.3 The Three-Domain System

From Aristotle's time to the middle of the twentieth century, biologists recognized only two kingdoms: kingdom Plantae (plants) and kingdom Animalia (animals). Plants were literally organisms that were planted and immobile, while animals were animated and moved about. In the 1880s, a German scientist, Ernst Haeckel, proposed adding a third kingdom. The kingdom Protista (protists) included unicellular microscopic organisms but not multicellular, largely macroscopic ones.

In 1969, R. H. Whittaker expanded the classification system to five kingdoms: Monera, Protista, Fungi, Plantae, and Animalia. Organisms were placed in these kingdoms based on the type of cell (prokaryotic or eukaryotic), complexity (unicellular or multicellular), and type of nutrition. Kingdom Monera contained all the prokaryotes, which are organisms that lack a membrane-bounded nucleus. These unicellular organisms were collectively called the bacteria. The other four kingdoms contain types of eukaryotes that we will describe later. We can note, however, that Whittaker gave fungi their own kingdom. He did so because fungi are generally multicellular, yet they are heterotrophic by absorption. Plants, of course, are photosynthesizers while animals are heterotrophic by ingestion.

### The Domains

In the late 1970s, Carl Woese and his colleagues at the University of Illinois were studying relationships among the prokaryotes using rRNA sequences. As mentioned previously, rRNA probably changes only slowly during evolution, and indeed it may change only when there is a major evolutionary event. Woese found that the rRNA sequence of prokaryotes that lived at high temperatures or produced methane was quite different from that of all the other types of prokaryotes and from the eukaryotes. Therefore, he proposed that there are two groups of prokaryotes (rather than one group as in the **five-kingdom system**). Further, Woese said that the rRNA sequences of these two groups, called the bacteria and archaea, are so fundamentally different from each other that they should be assigned to separate domains, a category of classification that is higher than the kingdom category. The two designated domains are **domain Bacteria** and **domain Archaea.** The eukaryotes are in the **domain Eukarya.** The phylogenetic tree shown in Figure 19.11 is based on his rRNA sequencing data. The data suggested that both bacteria and archaea evolved in the history of life from the first common ancestor. Later, the eukarya diverged from the archaea line of descent.

### *Domain Bacteria*

Bacteria are so diversified and plentiful they are found in large numbers nearly everywhere on Earth. In large part, Chapter 20 will be devoted to discussing the bacteria, which differ from the archaea not structurally but biochemically (Table 19.2).

In the meantime, we can note that the cyanobacteria are large photosynthetic prokaryotes. They carry on photosynthesis in the same manner as plants in that they use solar energy to convert carbon dioxide and water to a carbohydrate and in the process give off oxygen. Indeed the cyanobacteria may have been the first organisms to contribute oxygen to early Earth's atmosphere, making it hospitable to the evolution of oxygen-using organisms, including animals.

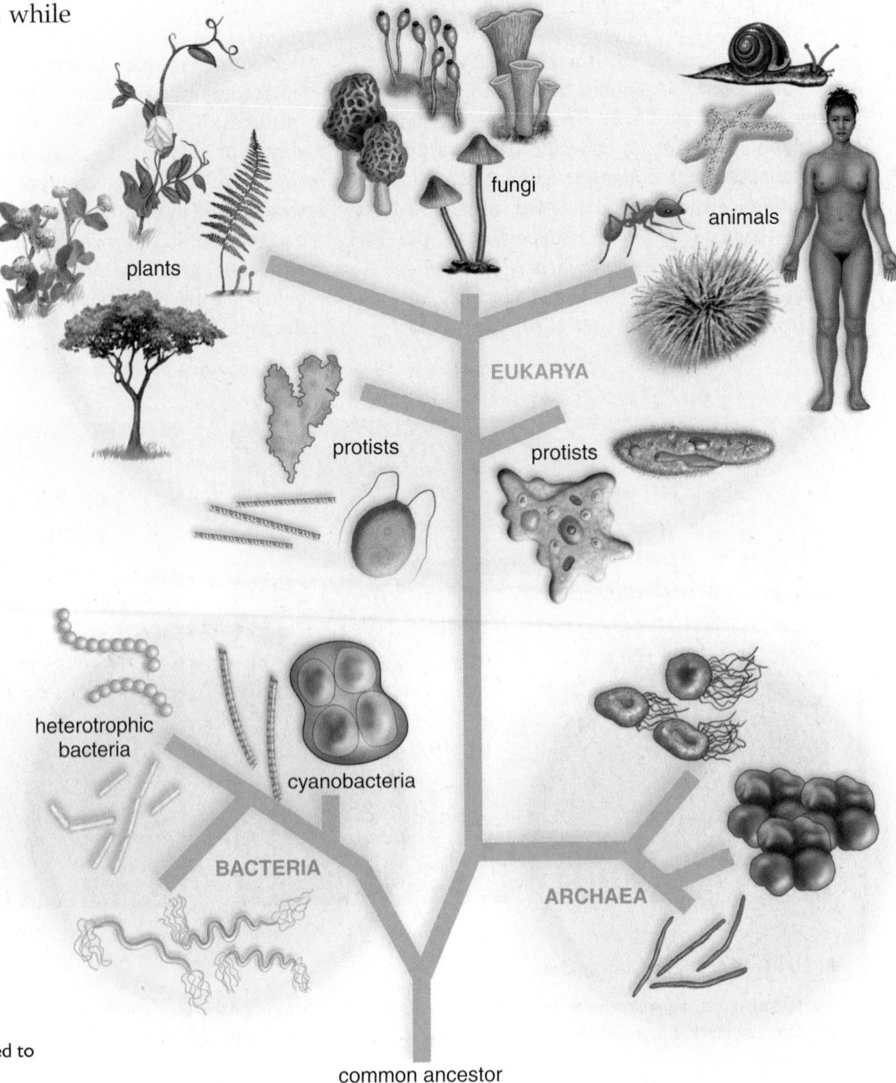


**FIGURE 19.11 A tree of life showing the three domains.**

Representatives of each domain are depicted in the ovals, and the phylogenetic tree shows that domain Archaea is more closely related to domain Eukarya than either is to domain Bacteria.

**TABLE 19.2**

**Major Distinctions Among the Three Domains of Life**

| | *Bacteria* | *Archaea* | *Eukarya* |
|---|---|---|---|
| Unicellularity | Yes | Yes | Some, many multicellular |
| Membrane lipids | Phospholipids, unbranched | Varied branched lipids | Phospholipids, unbranched |
| Cell wall | Yes (contains peptidoglycan) | Yes (no peptidoglycan) | Some yes, some no |
| Nuclear envelope | No | No | Yes |
| Membrane-bounded organelles | No | No | Yes |
| Ribosomes | Yes | Yes | Yes |
| Introns | No | Some | Yes |

All forms of nutrition are found among the bacteria, but most are heterotrophic. *Escherichia coli,* which lives in the human intestine, is heterotrophic as are parasitic forms that cause human disease. *Clostridium tetani* (cause of tetanus), *Bacillus anthracis* (cause of anthrax), and *Vibrio cholerae* (cause of cholera) are disease-causing species of bacteria. Heterotrophic bacteria are beneficial in ecosystems because they are organisms of decay that break down organic remains. Along with fungi, they keep chemical cycling going so that plants always have a source of inorganic nutrients.

### Domain Archaea

Like bacteria, archaea are prokaryotic unicellular organisms that reproduce asexually. Archaea don't look that different from bacteria under the microscope, and the extreme conditions under which many species live has made it difficult to culture them. This may have been the reason that their unique place among the living organisms long went unrecognized.

The archaea are distinguishable from bacteria by a difference in their rRNA base sequences and also by their unique plasma membrane and cell wall chemistry (Table 19.3). The chemical nature of the archaeal cell wall is diverse and never the same as that of the bacterial cell. The branched nature of diverse lipids in the archaeal plasma membrane, for example, could possibly help them live in extreme conditions.

The archaea live in all sorts of environments, but they are known for thriving in extreme environments thought to be similar to those of the early Earth. For example, the methanogens live in anaerobic environments, such as swamps and marshes and the guts of animals; the halophiles are salt lovers living in bodies of water such as the Great Salt Lake in Utah; and the thermoacidophiles are both high temperature and acid loving. These archaea live in extremely hot acidic environments, such as hot springs and geysers.

### Domain Eukarya

Eukaryotes are unicellular to multicellular organisms whose cells have a membrane-bounded nucleus. They also have various organelles that arose through endosymbiosis (see page 325). Sexual reproduction is common, and various types of life cycles are seen. Later in this text, we will be studying the individual kingdoms that occur within the domain Eukarya (Fig. 19.12 and Table 19.3). In the meantime, we can note that protists are a diverse group of organisms that are hard to classify and define. They are eukaryotes and mainly unicellular, but some are filaments, colonies, or multicellular sheets. Even so, protists do not have true tissues. Nutrition is diverse and some are heterotrophic by ingestion or absorption and some are photosynthetic. Green algae, paramecia, and slime molds are representative protists. There has been considerable debate over the classification of protists, and presently they are placed in five supergroups (Fig. 21.2) and not in a kingdom.

Fungi are eukaryotes that form spores, lack flagella, and have cell walls containing chitin. They are multicellular with a few exceptions. Fungi are heterotrophic by absorption—they secrete digestive enzymes and then absorb nutrients from decaying organic matter. Mushrooms, molds, and yeasts are representative fungi.

Despite appearances, molecular data suggest that fungi and animals are more closely related to each other than either are to plants.

Plants are photosynthetic organisms that have become adapted to a land environment. They share a common ancestor, which is an aquatic photosynthetic protist. Land plants possess true tissues and have the organ system level of organization. Examples include cacti, ferns, and cypress trees.

Animals are motile eukaryotic multicellular organisms that evolved from a heterotrophic protist. Like land plants, animals have true tissues and the organ system level of organization. Animals ingest their food; examples include worms, whales, and insects.

**Check Your Progress 19.3**

1. Explain the introduction of the domain level of classification.
2. What type of evidence suggests that fungi are related to animals rather than plants? Why is a close relationship between animals and fungi so unexpected?

**FIGURE 19.12 The three domains of life.**

This pictorial representation of the domains Bacteria, Archaea, and Eukarya includes an example for each of the four types of eukaryotes: protists, fungi, plants, and animals.

## TABLE 19.3

### Classification Criteria for the Three Domains

| | *Domains Bacteria and Archaea* | *Domain Eukarya* | | | |
|---|---|---|---|---|---|
| | | ***Protists*** | ***Kingdom Fungi*** | ***Kingdom Plantae*** | ***Kingdom Animalia*** |
| Type of cell | Prokaryotic | Eukaryotic | Eukaryotic | Eukaryotic | Eukaryotic |
| Complexity | Unicellular | Unicellular usual | Multicellular usual | Multicellular | Multicellular |
| Type of nutrition | Autotrophic or heterotrophic | Photosynthetic or heterotrophic by various means | Heterotrophic by absorption | Autotrophic by photosynthesis | Heterotrophic by ingestion |
| Motility | Sometimes by flagella | Sometimes by flagella (or cilia) | Nonmotile | Nonmotile | Motile by contractile fibers |
| Life cycle | Asexual usual | Various life cycles | Haploid | Alternation of generations | Diploid |
| Internal protection of zygote | No | No | No | Yes | Yes |

# Connecting the Concepts

We have seen in this chapter that identifying, naming, and classifying living organisms is an ongoing process. Carolus Linnaeus's system of binomial nomenclature is still accepted by virtually all biologists, but many species remain to be found and named (most in the rain forests). For years, Whittaker's five-kingdom concept of life on Earth was widely used. Now, new findings suggest that there are three domains of life: Bacteria, Archaea, and Eukarya. The archaea are structurally similar to bacteria, but their rRNA differs from that of bacteria and is instead similar to that of eukaryotes. Also, some archaeal genes are unique only to the archaea. Kingdoms fungi, plants, and animals are still recognized among eukarya, but protists are now placed in several supergroups

Most of today's systematists use evolutionary relationships among organisms for classification purposes. Cladistics offers us mechanisms for determining such relationships and indicates how Linnean classification should be revised. The cladist believes that only shared derived differences in chiefly anatomical traits and DNA sequences should be used to classify organisms. The sequencing of DNA has emerged as a powerful new tool to assist in determining evolutionary relationships and how species should be classified.

## summary

### 19.1 Systematics

Systematics is dedicated to understanding the evolutionary history of life on Earth. Taxonomy, a part of systematics, deals with the naming of organisms; each species is given a binomial name consisting of the genus and specific epithet.

Classification involves the assignment of species to categories. When an organism is named, a species has been assigned to a particular genus. Eight obligatory categories of classification are species, genus, family, order, class, phylum, kingdom, and domain. Each higher category is more inclusive; species in the same kingdom share general characters, and species in the same genus share quite specific characters.

### 19.2 Phylogenetic Trees

Phylogenetic trees are diagrams that should be considered a hypothesis concerning the evolutionary relationships between designated species. The tree shows a sequence of common ancestors by which evolution occurred. Because of the hierarchical nature of Linnean classification, it is possible to draw up a phylogenetic tree based on classification categories.

Rather than using similarities to construct phylogenetic trees, cladistics offers a way to use shared derived traits to distinguish different groups of species from one another. When no original common ancestor can be found in the fossil record, the use of an outgroup allows us to determine with what trait to begin the tree. Based on the rest of the available data, it is possible to determine the sequence of clades in the tree. Clades are monophyletic; they contain the most recent common ancestor along with all its descendants. All the members of a clade share the same derived traits. Linnean classification permits the use of groupings other than those that are monophyletic, and therefore has come under severe criticism.

The fossil record, homology, and molecular data, in particular, are used to help decipher phylogenies. Because fossils can be dated, available fossils can establish the antiquity of a species. If the fossil record is complete enough, we can sometimes trace a lineage through time. Homology helps indicate when species belong to a monophyletic taxon (share a common ancestor); however, convergent evolution sometimes makes it difficult to distinguish homologous structures from analogous structures. DNA base sequence data are commonly used to help determine evolutionary relationships.

### 19.3 The Three-Domain System

On the basis of molecular data, three evolutionary domains have been established: Bacteria, Archaea, and Eukarya. The first two domains contain prokaryotes; the domain Eukarya contains the protists, fungi, plants, and animals.

## understanding the terms



Match the terms to these definitions:

a. ______________ Branch of biology concerned with identifying, describing, and naming organisms.
b. ______________ Diagram that indicates common ancestors and lines of descent.
c. ______________ Group of organisms that fills a particular classification category.
d. ______________ School of systematics that determines the degree of relatedness by analyzing shared derived characters.
e. ______________ Similarity in structure due to having a common ancestor.

## reviewing this chapter

1. Explain the binomial system of naming organisms. Why must species be designated by a complete name? 338–39
2. Why is it necessary to give organisms scientific names? 339
3. What are the eight obligatory classification categories? In what way are they a hierarchy? 340
4. Discuss the principles of cladistics, and explain how to construct a cladogram. 342
5. Explain the difference between monophyletic, paraphyletic, and polyphyletic groupings. 343–44
6. With reference to the phylogenetic tree shown in Figure 19.6, why are birds in a clade with crocodiles? In a clade with other reptiles? 344
7. What types of data help systematists construct phylogenetic trees? 345–46
8. Compare the five-kingdom system of classification to the three-domain system. 348–49
9. Contrast the characteristics of the bacteria, the archaea, and the eukarya. 349–50
10. Contrast the eukaryotic protists, fungi, plants, and animals. 349–50

## testing yourself

Choose the best answer for each question.

1. Which is the scientific name of an organism?
   a. *Rosa rugosa*
   b. *Rosa*
   c. *rugosa*
   d. *Rugosa rugosa*
   e. Both a and d are correct.
2. Which of these describes systematics?
   a. studies evolutionary relationships
   b. includes taxonomy and classification
   c. includes phylogenetic trees
   d. utilizes fossil, morphological, and molecular data
   e. All of these are correct.
3. The classification category below the level of family is
   a. class.
   b. species.
   c. phylum.
   d. genus.
   e. order.
4. Which of these are domains? Choose more than one answer if correct.
   a. Bacteria
   b. Archaea
   c. Eukarya
   d. Animals
   e. Plants
5. Which of these are eukaryotes? Choose more than one answer if correct.
   a. bacteria
   b. archaea
   c. eukarya
   d. animals
   e. plants
6. Which of these characteristics is shared by bacteria and archaea? Choose more than one answer if correct.
   a. presence of a nucleus
   b. absence of a nucleus
   c. presence of ribosomes
   d. absence of membrane-bounded organelles
   e. presence of a cell wall
7. Which is mismatched?
   a. Fungi—prokaryotic single cells
   b. Plants—nucleated
   c. Plants—flowers and mosses
   d. Animals—arthropods and humans
   e. Protists—unicellular eukaryotes
8. Which is mismatched?
   a. Fungi—heterotrophic by absorption
   b. Plants—usually photosynthetic
   c. Animals—rarely ingestive
   d. Protists—various modes of nutrition
   e. Both c and d are mismatched.
9. Concerning a phylogenetic tree, which is incorrect?
   a. Dates of divergence are always given.
   b. Common ancestors give rise to descendants.
   c. The more recently evolved are always at the top of the tree.
   d. Ancestors have primitive characters.
10. Which pair is mismatched?
   a. homology—character similarity due to a common ancestor
   b. molecular data—DNA strands match
   c. fossil record—bones and teeth
   d. homology—functions always differ
   e. molecular data—molecular clock
11. One benefit of the fossil record is
   a. that hard parts are more likely to fossilize.
   b. fossils can be dated.
   c. its completeness.
   d. fossils congregate in one place.
   e. All of these are correct.
12. The discovery of common ancestors in the fossil record, the presence of homologies, and nucleic acid similarities help scientists decide
   a. how to classify organisms.
   b. the proper cladogram.
   c. how to construct phylogenetic trees.
   d. how evolution occurred.
   e. All of these are correct.
13. Molecular clock data are based on
   a. common adaptations among animals.
   b. DNA dissimilarities in living species.
   c. DNA fingerprinting of fossils.
   d. finding homologies among plants.
   e. All of these are correct.
14. In cladistics,
   a. a clade must contain the common ancestor plus all its descendants.
   b. shared derived traits help construct cladograms.
   c. data for the cladogram are presented.
   d. the species in a clade share homologous structures.
   e. All of these are correct.

15. Linnean classification is being criticized because
    a. it doesn't always use monophyletic groupings.
    b. it cannot be reconciled in general with the principles of systematics.
    c. some scientists are behind the times.
    d. it doesn't lend itself to the construction of phylogenetic trees.
    e. All of these are correct.
16. Which of these pairs is mismatched?
    a. cladogram—shows shared derived characters
    b. any phylogenetic tree—must use names of classification categories
    c. cladogram—based on monophyletic groups
    d. any phylogenetic tree—shows common ancestors
    e. All of these are properly matched.
17. Lancelets are the outgroup. Lancelets
    a. are related to all the species in the ingroup.
    b. have traits shared by both lancelets and the species in the ingroup.
    c. cannot be a member of any clade in the phylogenetic tree.
    d. don't have any of the traits listed in the tree.
    e. All of these are correct.
18. Birds have a gizzard that mammals don't have and mammals have hair that birds don't have.
    a. Even so, both birds and mammals can be in a clade together because they both descended from an ancestor that had an amniotic egg.
    b. Regardless, birds and mammals cannot be in the same clade because they have other traits that are different.
    c. If they and their common ancestors are grouped together, the grouping would be polyphyletic.
    d. Both a and b are correct.
    e. Both b and c are correct.
19. Which of these statements is correct?
    a. All the species in the ingroup have all the traits in common.
    b. All the species in the ingroup have only the first trait(s) of the phylogenetic tree in common.
    c. Only the last species to evolve necessarily has all the traits mentioned in the tree.
    d. Every clade has only unique derived traits not shared by any other clade.
    e. All of these are correct.
20. This clade is paraphyletic because it does

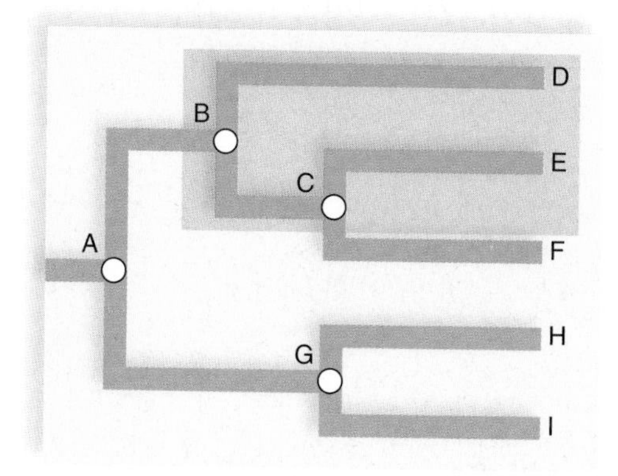


    a. not include the common ancestor for the species in the clade.
    b. not include all the descendants of a common ancestor.
    c. include species from two different common ancestors.
    d. include species from a number of different clades.
    e. All of these are correct.
21. How is Linnean systematics like that of cladistic systematics? They both
    a. allow monophyletic and paraphyletic groups.
    b. have an outgroup to decide which traits are ancestral.
    c. are nested—have groups within groups.
    d. are tied to a strict classification system.
    e. coordinate well with the geological timescale.

## thinking scientifically

1. Recent DNA evidence suggests to some plant taxonomists that the traditional way of classifying flowering plants is not correct, and that flowering plants need to be completely reclassified. Other botanists disagree, saying it would be chaotic and unwise to disregard the historical classification groups. Argue for and against keeping traditional classification schemes.
2. What data might make you conclude that the eukaryotes should be in more than one domain? What domains do you hypothesize might be required?

## bioethical issue

### Classifying Chimpanzees

Because the genomes of chimpanzees and humans are almost identical, and the differences between them are no greater than between any two human beings, their classification has been changed. Chimpanzees and humans are placed in the same family and subfamily. They are in different "tribes," which is a rarely used classification category between subfamily and genus.

The former classification of chimpanzees and humans placed the two animals in different families. Do you believe the chimpanzees should be classified in the same family and subfamily as humans, or do you prefer the classification used formerly? Which way seems prejudicial? Give your reasons for preferring one method over the other.

## *Biology* website

The companion website for *Biology* provides a wealth of information organized and integrated by chapter. You will find practice tests, animations, videos, and much more that will complement your learning and understanding of general biology.

**http://www.mhhe.com/maderbiology10**

part IV

# Microbiology and Evolution

*Microbes occupy a world unseen by the naked eye. It's richly populated because it includes the viruses, the prokaryotic bacteria and archaea, and the eukaryotic protists and fungi. These organisms occur everywhere from the highest mountain peaks to the deepest ocean trenches and in every type of environment, even those that are extremely hot and acidic.*

*At the outset, let's acknowledge that microbes cause serious diseases in plants and animals, including ourselves. But we make use of microbes in innumerable ways. For example, bacteria help us accomplish gene cloning and genetic engineering; make food stuffs and antibiotics; and help dispose of sewage and environmental pollutants. The biosphere is totally dependent on the services of microorganisms. While we often mention how much we rely on land plants, we may fail to acknowledge that without microorganisms, land plants could not exist. Decomposing fungi and bacteria make inorganic nutrients available to plants, which they can absorb all the better because their roots are coated with friendly fungi. Photosynthetic bacteria first put oxygen in the atmosphere, and they, along with certain protists, are the producers of food in the oceans.*

*Microbes are our ancestors. They alone were on Earth for about 2.5 billion years, and unicellular protists gave rise to animals and land plants that populate the macroscopic world. This part discusses microbes, organisms that contribute so much to our world, even though we cannot see them without the use of a microscope.*

# 20

# Viruses, Bacteria, and Archaea

*Viruses are noncellular particles that parasitize all forms of life, even bacteria. As the micrograph below shows, many viruses can attack a single bacterium where they will take up residence in order to reproduce themselves. Bacteria also cause diseases, such as gonorrhea and tuberculosis, in humans. Still, because they are cellular, researchers can use bacteria to commercially carry out protein synthesis; manufacture foods, industrial chemicals, and medicines; mine minerals, clean up oil spills, and treat sewage. In ecosystems, bacteria decompose dead organisms and recycle their nutrients. Some bacteria carry on photosynthesis and produce much of the oxygen we breathe. Archaea live in extreme environments such as the hot springs of Yellowstone National Park and the anaerobic soils of swamps. But they live in all kinds of moderate environments, as well. Surprisingly, molecular biologists tell us we are more closely related to archaea than to bacteria.*

*Microbes are extremely numerous and diverse. Our skin is home to about 182 different bacterial species. For every human cell, ten microbial cells are present in the body. In the following chapter, we will further examine these amazing microbes.*

Micrograph of viruses attached to bacterium, *Escherichia coli*.

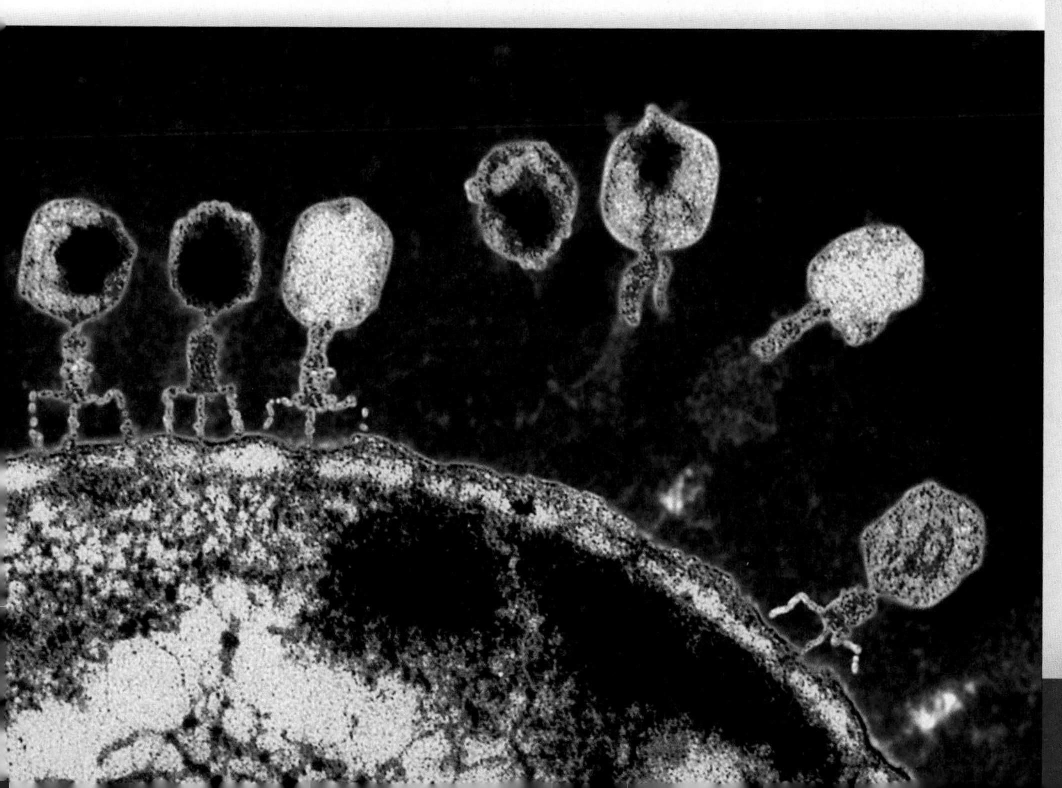

## concepts

# 20.1 Viruses, Viroids, and Prions

The term **virus** [L. *virus*, poison] is associated with a number of plant, animal, and human diseases (see Table 20.1). The mere mention of the term brings to mind serious illnesses such as polio, rabies, and AIDS (acquired immunodeficiency syndrome), as well as common childhood maladies such as measles, chickenpox, and mumps. Viral diseases are of concern to everyone; it is estimated that the average person catches a cold two or three times a year.

**TABLE 20.1**

**Viral Diseases in Humans**

| *Category* | *Disease* |
|---|---|
| Sexually transmitted diseases | AIDS (HIV), genital warts, genital herpes |
| Childhood diseases | Mumps, measles, chickenpox, German measles |
| Respiratory diseases | Common cold, influenza, severe acute respiratory infection (SARS) |
| Skin diseases | Warts, fever blisters, shingles |
| Digestive tract diseases | Gastroenteritis, diarrhea |
| Nervous system diseases | Poliomyelitis, rabies, encephalitis |
| Other diseases | Smallpox, hemorrhagic fevers, cancer, hepatitis, mononucleosis, yellow fever, dengue fever, conjunctivitis, hepatitis C |

The viruses are a biological enigma. They have a DNA or RNA genome, but they can reproduce only by using the metabolic machinery of a host cell. Viruses are noncellular, and therefore cannot be assigned a two-part binomial name, as are organisms.

Our knowledge of viruses began in 1884 when the French chemist Louis Pasteur (1822–95) suggested that something smaller than a bacterium was the cause of rabies, and it was he who chose the word *virus* from a Latin word meaning poison. In 1892, Dimitri Ivanowsky (1864–1920), a Russian microbiologist, was studying a disease of tobacco leaves, called

Leaf infected with tobacco mosaic virus.

**FIGURE 20.1 Viruses.**

Despite their diversity, all viruses have an outer capsid composed of protein subunits and a nucleic acid core—composed of either DNA or RNA, but not both. Some types of viruses also have a membranous envelope.

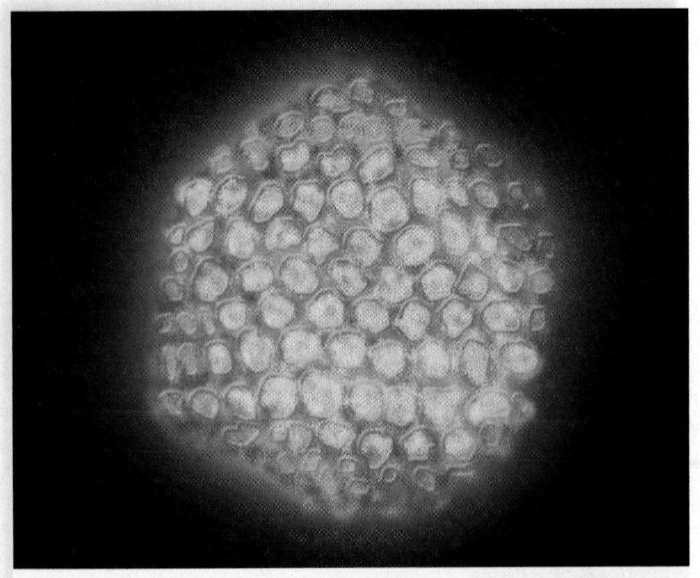

TEM 80,000×

Adenovirus: DNA virus with a polyhedral capsid and a fiber at each corner.

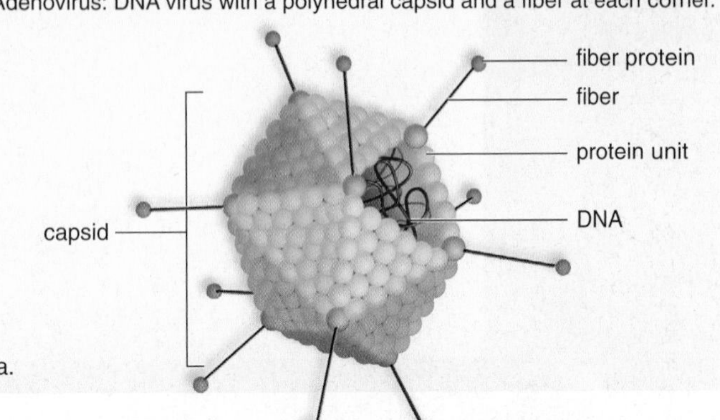



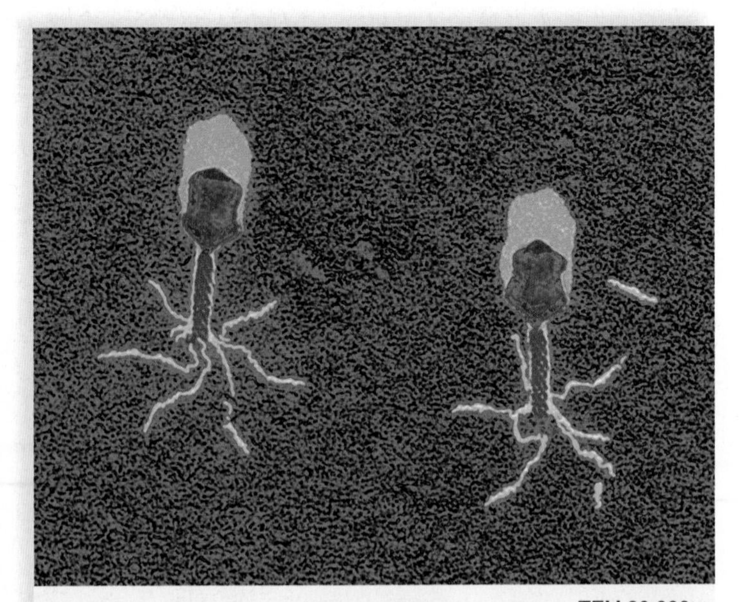

TEM 90,000×

T-even bacteriophage: DNA virus with a polyhedral head and a helical tail.

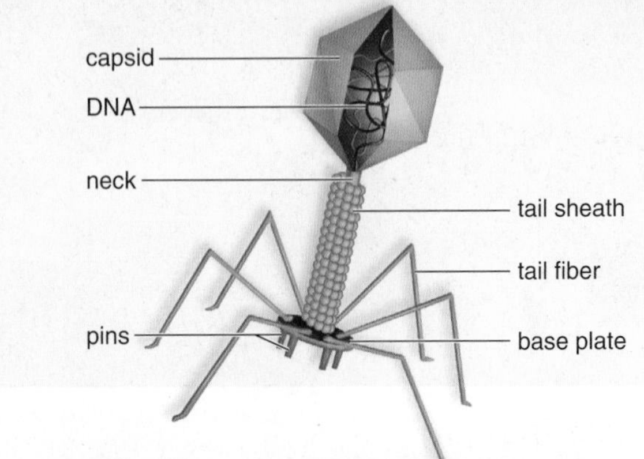

tobacco mosaic disease because of the leaves' mottled appearance. He noticed that even when an infective extract was filtered through a fine-pore porcelain filter that retains bacteria, it still caused disease. This substantiated Pasteur's belief because it meant that the disease-causing agent was smaller than any known bacterium. In the next century, electron microscopy was born, and viruses were seen for the first time. By the 1950s, virology was an active field of research; the study of viruses, and now also viroids and prions, has contributed much to our understanding of disease, genetics, and even the characteristics of living things.

## Viral Structure

The size of a virus is comparable to that of a large protein macromolecule, and ranges in size from 10–400 nm. Viruses are best studied through electron microscopy. Many viruses can be purified and crystallized, and the crystals can be stored just as chemicals are stored. Still, viral crystals will become infectious when the viral particles they contain are given the opportunity to invade a host cell.

Viruses are categorized by (1) their size and shape; (2) their type of nucleic acid, including whether it is single stranded or double stranded; and (3) the presence or absence of an outer envelope.

Viruses vary in shape from threadlike to polyhedral (Fig. 20.1). However, all viruses possess the same basic anatomy: an outer **capsid** composed of protein subunits and an inner core of nucleic acid—either DNA (deoxyribonucleic acid) or RNA (ribonucleic acid), but not both. A viral genome has as little as three and as many as 100 genes; a human cell contains tens of thousands of genes. The viral capsid may be surrounded by an outer membranous envelope; if not, the virus is said to be naked. Figure 20.1*a, b, c* gives examples of naked viruses, while Figure 20.1*d* is an example of an enveloped virus. The envelope is actually a piece of the host's plasma membrane that also contains viral glycoprotein spikes. Aside from its genome, a viral particle may also contain various proteins, especially enzymes such as the polymerases, needed to produce viral DNA and/or RNA.

The following diagram summarizes viral structure:

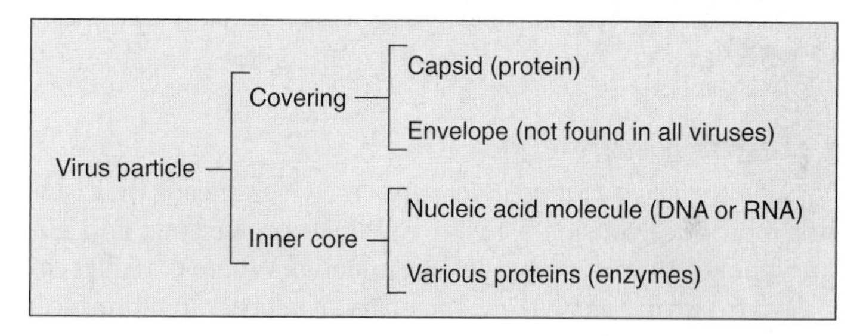


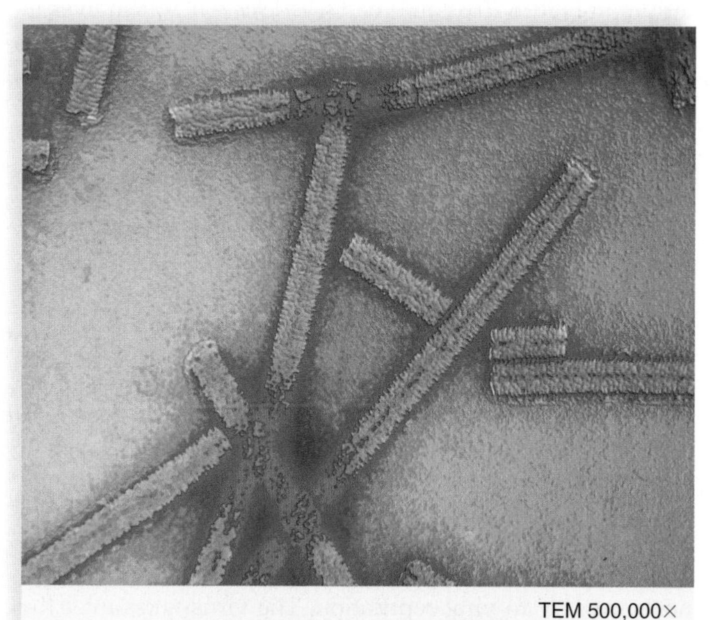


Tobacco mosaic virus: RNA virus with a helical capsid.

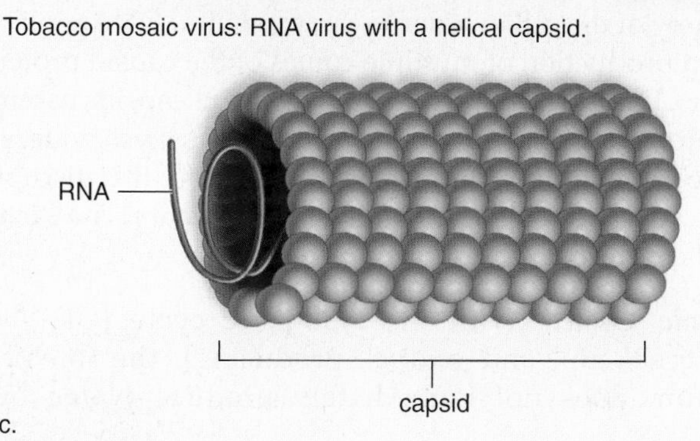


c.

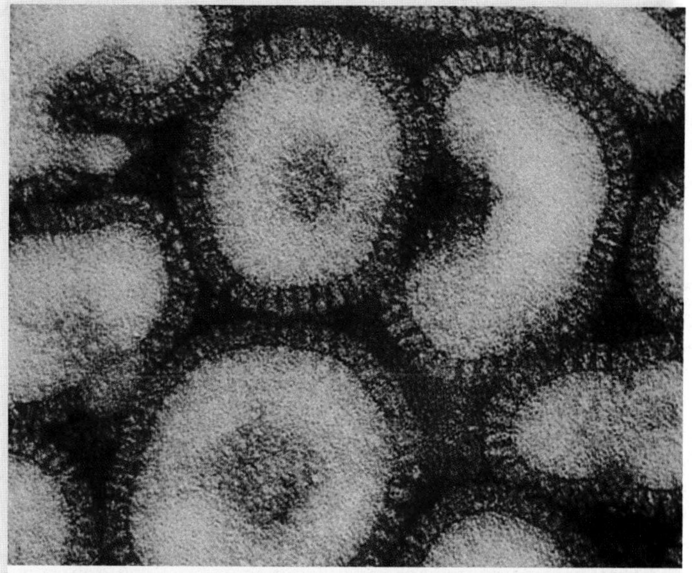


Influenza virus: RNA virus with a helical capsid surrounded by an envelope with spikes.

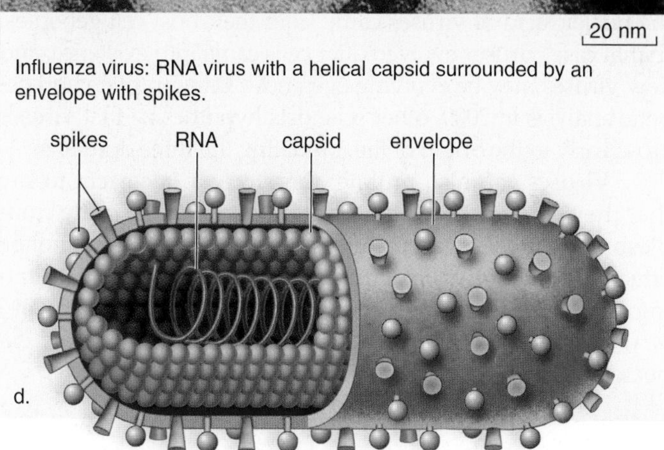


d.

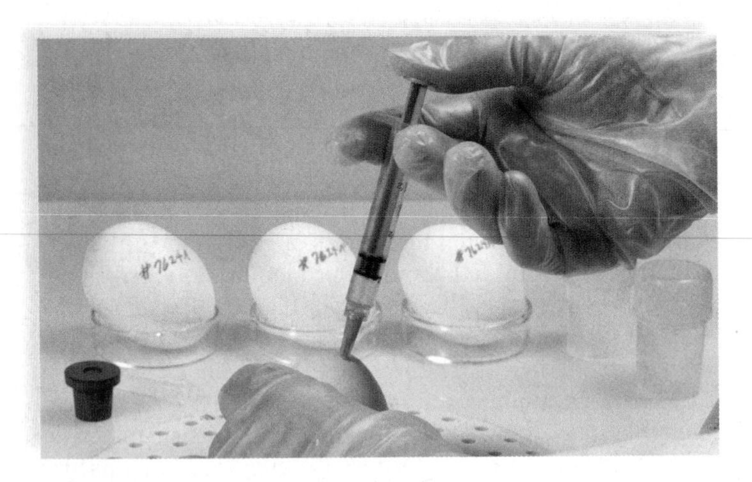

**FIGURE 20.2 Culturing viruses.**

To culture a virus, scientists can inoculate live chicken eggs with viral particles. A virus reproduces only inside a living cell because it takes over the machinery of the cell.

### *Parasitic Nature*

Viruses are *obligate intracellular parasites,* which means they cannot reproduce outside a living cell. Like prokaryotic and eukaryotic cells, viruses have genetic material. Whereas a cell is capable of copying its own genetic material in order to reproduce, a virus cannot duplicate its genetic material or any of its other components. For a virus to reproduce, it must infect a living cell. The infected cell duplicates the nucleic acid and other parts of the virus, including the capsid, viral enzymes, and for some viruses, the envelope.

To maintain animal viruses in the laboratory, they are sometimes injected into live chicken embryos (Fig. 20.2). Today, host cells are often maintained in tissue (cell) culture by simply placing a few cells in a glass or plastic container with appropriate nutrients to sustain the cells. The cells can then be infected with the animal virus to be studied. Viruses infect a variety of cells, but they are **host specific.** Bacteriophages infect only bacteria, the tobacco mosaic virus infects only certain species of plants, and the rabies virus infects only mammals, for example. Some human viruses even specialize in a particular tissue. Human immunodeficiency virus (HIV) enters only certain blood cells, the polio virus reproduces in spinal nerve cells, and the hepatitis viruses infect only liver cells. What could cause this remarkable parasite–host cell correlation? Some scientists hypothesize that viruses are derived from the very cell they infect; the nucleic acid of viruses came from their host cell genomes! In that case, viruses evolved after cells came into existence, and new viruses may be evolving even now. Using protein and genetic analysis in 2000, other scientists hypothesize that viruses arose early in the origin of life, predating the three domains.

Viruses can also mutate; therefore, it is correct to say that they evolve. Those that mutate often can be quite troublesome because a vaccine that is effective today may not be effective tomorrow. Flu viruses are well known for mutating, and this is why it is necessary to have a flu shot every year—antibodies generated from last year's shot are not expected to be effective this year.

## Viral Reproduction

Viruses are microscopic pirates, commandeering the metabolic machinery of a host cell. Viruses gain entry into a host cell because portions of a naked capsid (or one of the types of envelope spikes) attach in a lock-and-key manner with a receptor on the host cell's outer surface. The attachment of the capsid or spikes of a virus to particular host cell receptors is responsible for the remarkable specificity between viruses and their host cells. A virus cannot infect a host cell to which it is unable to attach. For example, the tobacco mosaic virus cannot infect an exposed human because its capsid cannot attach to the receptors on the surfaces of human cells.

After a virus has become attached to a suitable host cell, the viral nucleic acid enters the cell. Once inside, the nucleic acid codes for the protein units in the capsid. In addition, the virus may have genes for special enzymes needed for the virus to reproduce and exit from the host cell. In large measure, however, a virus relies on the host's enzymes, ribosomes, transfer RNA (tRNA), and ATP (adenosine triphosphate) for its own reproduction. Because the host cell's metabolism is diverted from meeting the needs of the cell, infected cells may have an abnormal appearance.

### *Reproduction of Bacteriophages*

**Bacteriophages** [Gk. *bacterion,* rod, and *phagein,* to eat], or simply phages, are viruses that parasitize bacteria; the bacterium in Figure 20.3 could be *Escherichia coli,* which lives in our intestines, for example. As the figure shows, there are two types of bacteriophage life cycles, termed the lytic cycle and the lysogenic cycle. In the lytic cycle, viral reproduction occurs, and the host cell undergoes *lysis,* a breaking open of the cell to release viral particles. In the lysogenic cycle, viral reproduction does not immediately occur, but reproduction may take place sometime in the future. The following discussion is based on the DNA bacteriophage lambda, which undergoes both lytic and lysogenic cycles.

**Lytic Cycle.** The **lytic cycle** [Gk. *lyo,* loose] may be divided into five stages: attachment, penetration, biosynthesis, maturation, and release. During *attachment,* portions of the capsid combine with a receptor on the rigid bacterial cell wall in a lock-and-key manner. During *penetration,* a viral enzyme digests away part of the cell wall, and viral DNA is injected into the bacterial cell. *Biosynthesis* of viral components begins after the virus brings about inactivation of host genes not necessary to viral replication. The virus takes over the machinery of the cell in order to carry out viral DNA replication and production of multiple copies of the capsid protein subunits. During *maturation,* viral DNA and capsids assemble to produce several hundred viral particles. Lysozyme, an enzyme coded for by a viral gene, is produced; this disrupts the cell wall, and the *release* of new viruses occurs. The bacterial cell dies as a result.

**Lysogenic Cycle.** With the **lysogenic cycle** [Gk. *lyo,* loose, break up, and *genitus,* producing], the infected bacterium does not immediately produce phage but

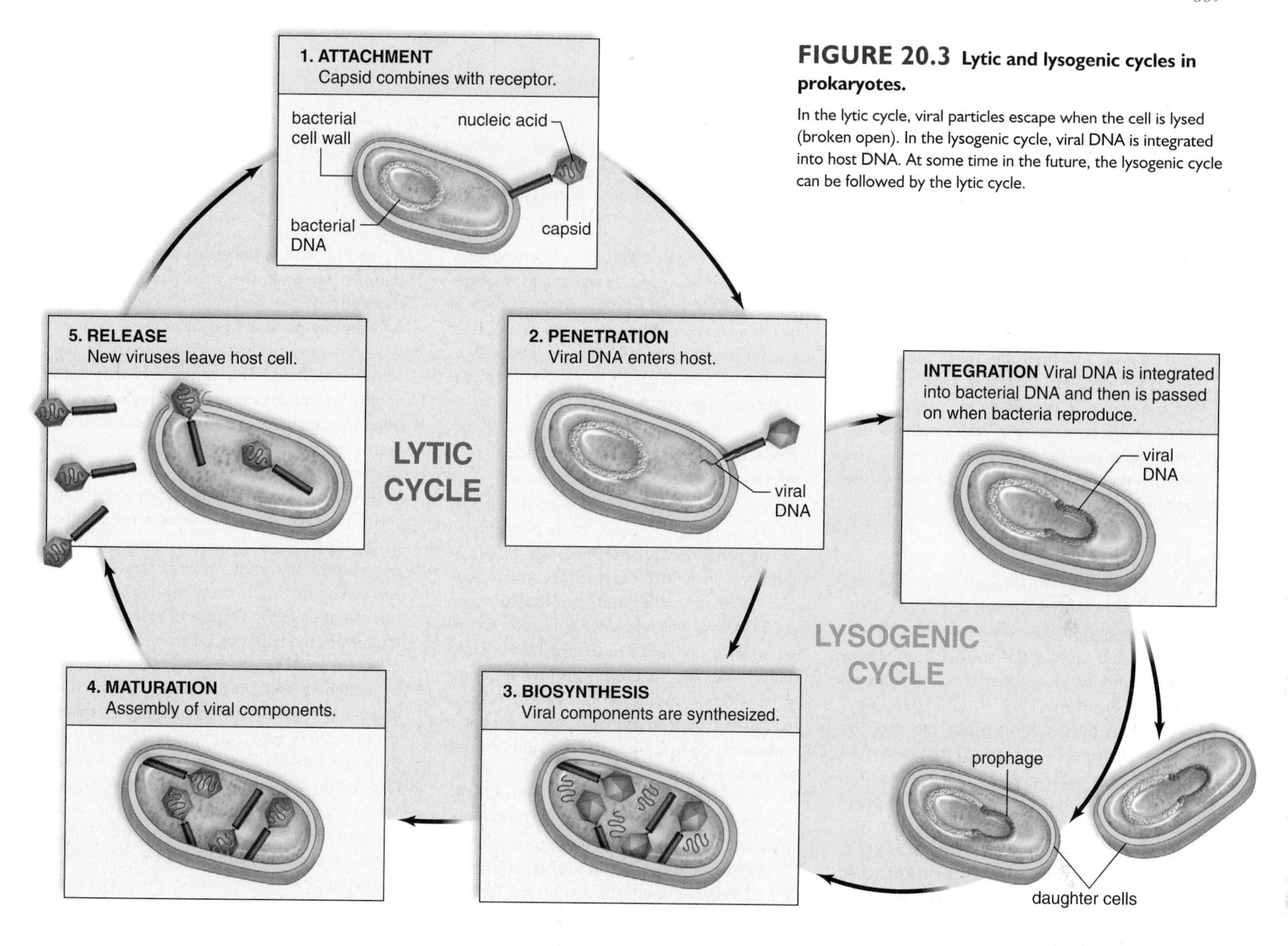


**FIGURE 20.3 Lytic and lysogenic cycles in prokaryotes.**

In the lytic cycle, viral particles escape when the cell is lysed (broken open). In the lysogenic cycle, viral DNA is integrated into host DNA. At some time in the future, the lysogenic cycle can be followed by the lytic cycle.

may do so sometime in the future. In the meantime, the phage is *latent*—not actively replicating. Following attachment and penetration, *integration* occurs: Viral DNA becomes incorporated into bacterial DNA with no destruction of host DNA. While latent, the viral DNA is called a *prophage*. The prophage is replicated along with the host DNA, and all subsequent cells, called **lysogenic cells,** carry a copy of the prophage. Lysogenic bacterial cells may have distinctive properties due to the prophage genes they carry. The presence of a prophage may cause a bacterial cell to produce a toxin. For example, if the same bacterium that causes strep throat happens to carry a certain prophage, then it will cause scarlet fever, so-named because the toxin causes a widespread red skin rash as it spreads through the body. Likewise, diphtheria is caused by a bacterium carrying a prophage. The diphtheria toxin damages the lining of the upper respiratory tract, resulting in the formation of a thick membrane that restricts breathing. Certain environmental factors, such as ultraviolet radiation, can induce the prophage to enter the lytic stage of biosynthesis, followed by maturation and release.

### *Reproduction of Animal Viruses*

Animal viruses reproduce in a manner similar to that of bacteriophages, but there are modifications. Various animal viruses have different ways of introducing their genetic material into their host cells. For some enveloped viruses, the process is as simple as attachment and fusion of the spike-studded envelope with the host cell's plasma membrane. Many naked and some enveloped viruses are taken into host cells by endocytosis. Once inside, the virus is uncoated—that is, the capsid and, if necessary, the envelope are removed. The viral genome, either DNA or RNA, is now free of its covering, and biosynthesis plus the other steps then proceed. Viral release is just as variable as penetration for animal viruses. Some mature viruses are released by budding. During budding, the virus picks up its envelope consisting of lipids, proteins, and carbohydrates from the host cell. Most enveloped animal viruses acquire their envelope from the plasma membrane of the host cell, but some take envelopes from other membranes, such as the nuclear envelope or Golgi apparatus. Envelope markers, such as the glycoprotein spikes that allow the virus to enter a host cell, are coded for by viral genes. Naked animal viruses are usually released by host cell lysis.

*health focus*

## Flu Pandemic

If you've ever had seasonal flu (influenza), you know how miserable it can be. The flu is a viral infection that causes runny nose, cough, chills, fever, head and body aches, and nausea. You catch the flu by inhaling virus-laden droplets that have been coughed or sneezed into the air by an infected person, or by contact with contaminated objects, such as door handles or bedding. The viruses then attach to and infect cells of the respiratory tract.

### Flu Viruses

A flu virus has an H (hemagglutinin) spike and an N (neuraminidase) spike (Fig. 20A*a*, *left*). Its H spike allows the virus to bind to its receptor, and its N spike attacks host plasma membranes in a way that allows mature viruses to exit the cell.

Just as purses and wallets can each be shaped differently, so can H spikes and N spikes: 16 types of H and 9 types of N spikes are known. Worst yet, just as any shape purse or wallet can be a different color, so each type of spike can occur in different varieties called subtypes. Many of the flu viruses are assigned specific codes based on the type of spike. For example, H5N1 virus gets its name from its variety of H5 spikes and its variety of N1 spikes. Our immune system only recognizes the particular variety of H spikes and N spikes it has been exposed to in the past by infection or immunization. When a new flu virus arises, one for which there is little or no immunity in the human population, a flu pandemic (global outbreak) may occur.

### Possible Bird Flu Pandemic of the Future

Currently, the H5N1 subtype of flu virus is of great concern because of its potential to reach pandemic proportions. An H5N1 is common in wild birds such as waterfowl, and can readily infect domestic poultry such as chickens, which is why it is referred to as an avian influenza or a bird flu virus. An H5N1 virus has infected waterfowl for some time without causing serious illness. A more pathogenic version of H5N1 appeared about a decade ago in China, and promptly started to cause widespread and severe illness in domestic chickens. Scientists are still trying to determine what made H5N1 become so lethal, first to chickens, and then to humans.

Why can the bird flu H5N1 infect humans? Because the virus can attach to both a bird flu receptor and to a human flu receptor. Close contact between domestic poultry and humans is necessary for this to happen. At this time, the virus has rarely been transmitted from one human to another and only among people who have close contact with one another, such as members of the same household. The concern is that with additional mutations, the H5N1 virus could become capable of sustained human-to-human transmission, and then spread around the world. How could H5N1 become better at spreading within the human population? At this time, bird flu H5N1 infects mostly the lungs. Most human flu viruses infect the upper respiratory tract, trachea, and bronchi and can be spread by coughing. If a spontaneous mutation in the H spike of H5N1 enabled it to attack the upper respiratory tract, then it could be easily spread from human to human by coughing and sneezing (Fig. 20A*a*). Or, another possibility is a combining of spikes could occur in a person who is infected with both the bird flu and the human flu viruses (Fig. 20A*b*). According to the CDC (Centers for Disease Control and Prevention), over the past decade an increasing number of humans infected with an H5N1 virus have been reported in Asia, the Pacific, the Near East, Africa, and Europe. Over half of these people have died. Currently, there are no available vaccines for an H5N1 virus.

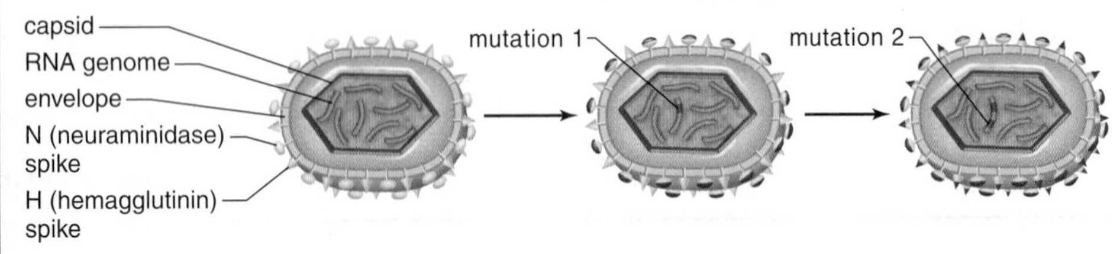


a. Viral genetic mutations occur in a bird host

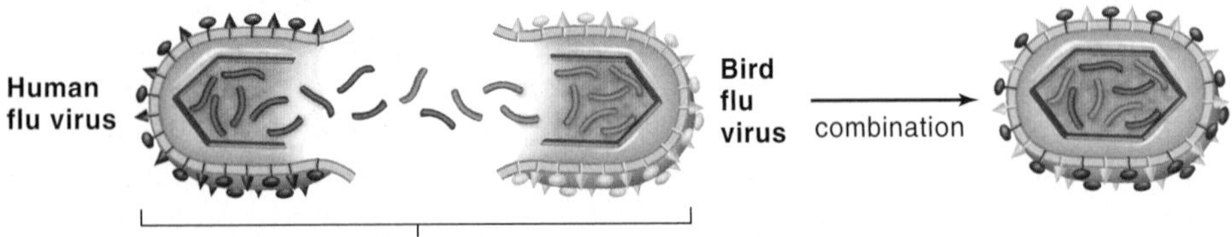


b. Combination of viral genes occurs in human host

**FIGURE 20A Spikes of bird flu virus.**
***a.*** *Genetic mutations in bird flu viral spikes could allow the virus to infect the human upper respiratory tract.*
***b.*** *Alternatively, combination of bird flu and human spikes could allow the virus to infect the human upper respiratory tract.*

### How to Be Prepared

A flu pandemic presents many challenges, including rapid spread of the virus, the overload of our health-care systems, inadequate medical supplies, and economic and social disruption. Vaccines and antiviral medications will become very short in supply. Prevention can be surprisingly simple. One of the easiest practices to prevent the spread of a flu virus is cleaning your hands thoroughly and often. Soap and water or an alcohol-based sanitizer is best. Keeping your hands away from your eyes, nose, and mouth also helps prevent the virus from entering your body. Education and outreach will be the keys to preparing for a pandemic. Knowing what a pandemic is, what needs to be done to prepare for one, and what could happen during a pandemic will help us as individuals and citizens to make wise decisions. For more information on flu pandemics, visit www.pandemicflu.gov.

**Retroviruses.** **Retroviruses** [L. *retro*, backward, and *virus*, poison] are RNA animal viruses that have a DNA stage. Figure 20.4 illustrates the reproduction of a retrovirus—namely, HIV (human immunodeficiency virus), the cause of AIDS. A retrovirus contains a special enzyme called **reverse transcriptase,** which carries out RNA ⟶ DNA transcription. First, the enzyme synthesizes a DNA strand called cDNA because it is complementary to viral RNA, and then the enzyme uses cDNA as a template to form double-stranded DNA. Using host enzymes, double-stranded DNA is integrated into the host genome. The viral DNA remains in the host genome and is replicated when host DNA is replicated. When and if this DNA is transcribed, new viruses are produced by the steps we have already cited: biosynthesis, maturation, and release; in this instance, by plasma membrane budding.

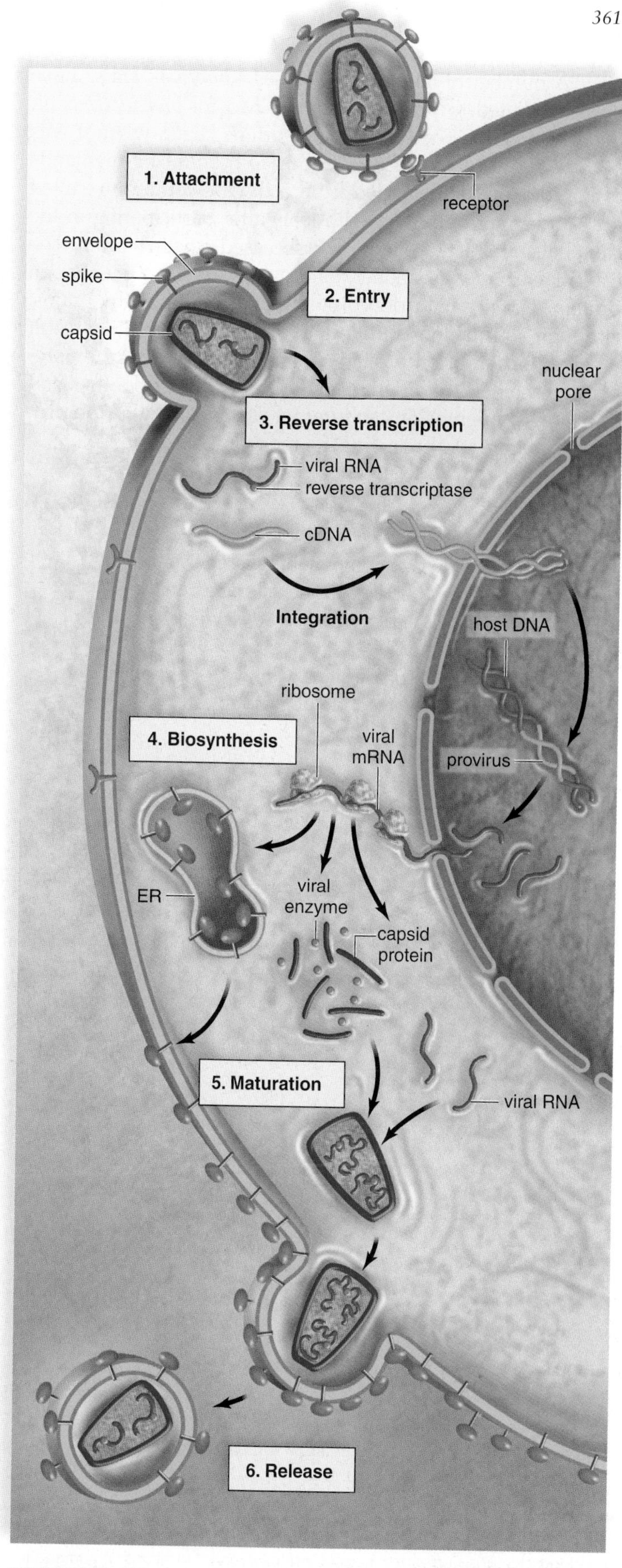


**FIGURE 20.4** **Reproduction of the retrovirus HIV.**

HIV uses reverse transcription to produce a DNA copy (cDNA) of RNA genes; double-stranded DNA integrates into the cell's chromosomes before the virus reproduces and buds from the cell. The steps in color are unique to retroviruses.

## Viral Infections of Special Concern

All viral infections of humans and other organisms are challenging to treat, especially because antibiotics designed to interfere with bacterial metabolism have no effect on viral illnesses. By learning how viruses replicate in cells, scientists have been able to design drugs to help treat certain viral infections. For example, the reverse transcriptase inhibitor AZT is used to block the replication of HIV. Acyclovir, a drug used to treat herpes, inhibits the replication of viral DNA. Certain viral infections are especially serious because they lead to even more severe diseases. In humans, papillomaviruses, herpesviruses, hepatitis viruses, adenoviruses, and retroviruses are associated with specific types of cancer. Emerging viruses are of recent concern.

### *Emerging Viruses*

Some emerging diseases—new or previously uncommon illnesses—are caused by viruses that are now able to infect large numbers of humans. These viruses are known as **emerging viruses.** Examples of emerging viral diseases are AIDS, West Nile encephalitis, hantavirus pulmonary syndrome (HPS), severe acute respiratory syndrome (SARS), Ebola hemorrhagic fever, and avian influenza (bird flu).

Several different types of events can cause a viral disease to suddenly "emerge" and start causing a widespread human illness. A virus can extend its range when it is transported from one part of the world to another. West Nile encephalitis is a virus that extended its range after being transported into the United States, where it took hold in bird and mosquito populations. SARS was transported from Southeast Asia to Toronto, Canada.

Viruses are well known for their high mutation rates. Sometimes viruses that formerly infected animals other than humans can "jump" species and start infecting humans due to a change in their capsids or spikes that enables them to attach to human cell receptors, as discussed in the Health Focus on page 360.

## Viroids and Prions

At least a thousand different viruses cause diseases in plants. About a dozen diseases of crops, including potatoes, coconuts, and citrus, have been attributed not to viruses but to **viroids,** which are naked strands of RNA (not covered by a capsid). Like viruses, though, viroids direct the cell to produce more viroids.

A number of fatal brain diseases, known as TSEs,[1] have been attributed to **prions,** a term coined for *proteinaceous infectious* particles. The discovery of prions began when it was observed that members of a primitive tribe in the highlands of Papua New Guinea died from a disease commonly called kuru (meaning trembling with fear) after participating in the cannibalistic practice of eating a deceased person's brain. The causative agent was smaller than a virus—it was a misshapen protein. The normal prion protein is found in healthy brains, although its function is unknown. It appears that TSEs result when a normal prion protein changes shape so that the polypeptide chain is in a different configuration. It is hypothesized that a misshapen prion can interact with a normal prion protein to change its shape, but the mechanism is unclear.

**Check Your Progress 20.1**

1. What are the two components shared by all viruses?
2. Viruses are generally considered to be nonliving. Should viroids and prions also be viewed as nonliving? Why or why not?
3. From an evolutionary standpoint, why is it beneficial to a virus if its host lives and does not die?

[1]TSEs (transmissible spongiform encephalopathies)

# 20.2 The Prokaryotes

As previously mentioned, the **prokaryotes** include bacteria and archaea, which are fully functioning cells. Because they are microscopic, the prokaryotes were not discovered until the Dutch microscopist Antonie van Leeuwenhoek (1632–1723), better known as the father of the microscope, first described them along with many other microorganisms. Leeuwenhoek and others after him believed that the "little animals" that he observed could arise spontaneously from inanimate matter. For about 200 years, scientists carried out various experiments to determine the origin of microorganisms in laboratory cultures. Finally, in about 1850, Louis Pasteur devised an experiment for the French Academy of Sciences that is described in Figure 20.5. It showed that a previously sterilized broth cannot become cloudy with growth unless it is exposed directly to the air, where bacteria are abundant. Today we know that bacteria are plentiful in air, water, and soil and on most objects. In the pages following, the general characteristics of prokaryotes are discussed before those specific to the bacteria (domain Bacteria) and then the archaea (domain Archaea) are considered in more detail.

## Structure of Prokaryotes

Prokaryotes generally range in size from 1 to 10 μm in length and from 0.7 to 1.5 μm in width. The term *prokaryote* means "before a nucleus," and these organisms lack a eukaryotic

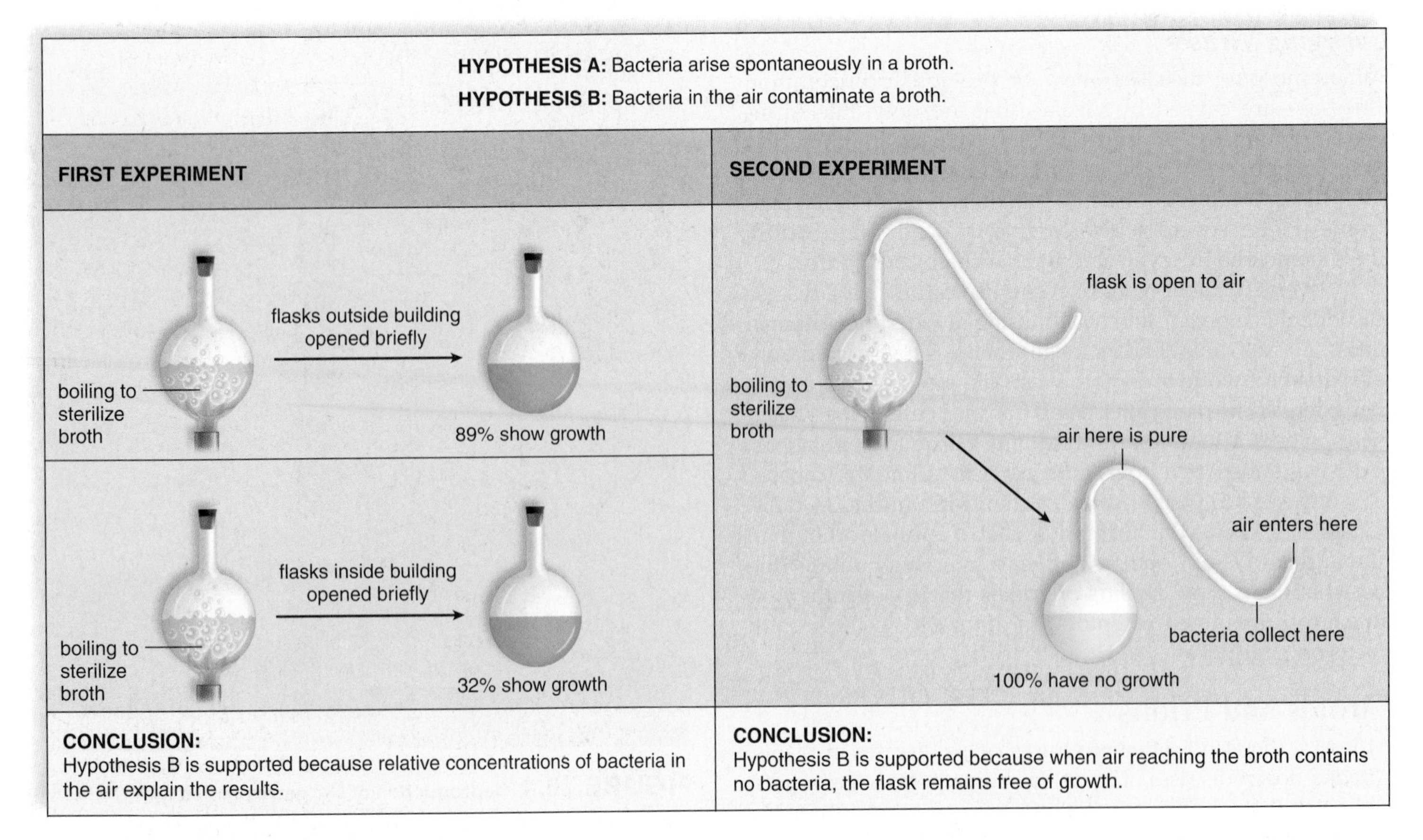


**FIGURE 20.5 Pasteur's experiment.**

Pasteur disproved the theory of spontaneous generation of microbes by performing these types of experiments.

nucleus. There are prokaryotic fossils dated as long ago as 3.5 billion years, and the fossil record indicates that the prokaryotes were alone on Earth for at least 1.3 billion years. During that time, they became extremely diverse in structure and especially diverse in metabolic capabilities. Prokaryotes are adapted to living in most environments because the various types differ in the ways they acquire and use energy.

A typical prokaryotic cell has a cell wall situated outside the plasma membrane. The cell wall prevents a prokaryote from bursting or collapsing due to osmotic changes. Yet another layer may exist outside the cell wall. The structure and composition of this layer vary among the different kinds of prokaryotes. In many bacteria, the cell wall is surrounded by a layer of polysaccharides called a glycocalyx. A well-organized glycocalyx is called a capsule, while a loosely organized one is called a slime layer. Many bacteria and archaea have a layer comprised of protein, or glycoprotein, instead of a glycocalyx; such a layer is called an S-layer. In parasitic forms of bacteria, these outer coverings help protect the cell from host defenses.

Some prokaryotes move by means of **flagella** (Fig. 20.6). A bacterial flagellum has a filament composed of strands of the protein flagellin wound in a helix. The filament is inserted into a hook that is anchored by a basal body. The 360° rotation of the flagellum causes the cell to spin and move forward. The archaeal flagellum is similar, but more slender and apparently lacking a basal body. Many prokaryotes adhere to surfaces by means of **fimbriae,** short bristlelike fibers extending from the surface (Fig. 20.7). The fimbriae of the bacterium *Neisseria gonorrhoeae* allow it to attach to host cells and cause gonorrhea.

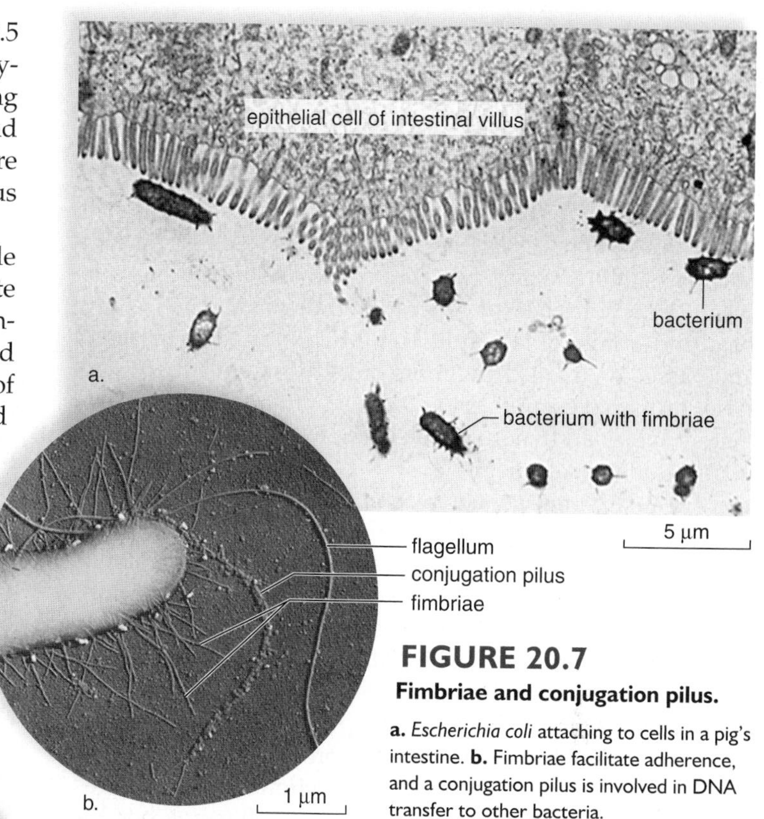


**FIGURE 20.7**
**Fimbriae and conjugation pilus.**

**a.** *Escherichia coli* attaching to cells in a pig's intestine. **b.** Fimbriae facilitate adherence, and a conjugation pilus is involved in DNA transfer to other bacteria.

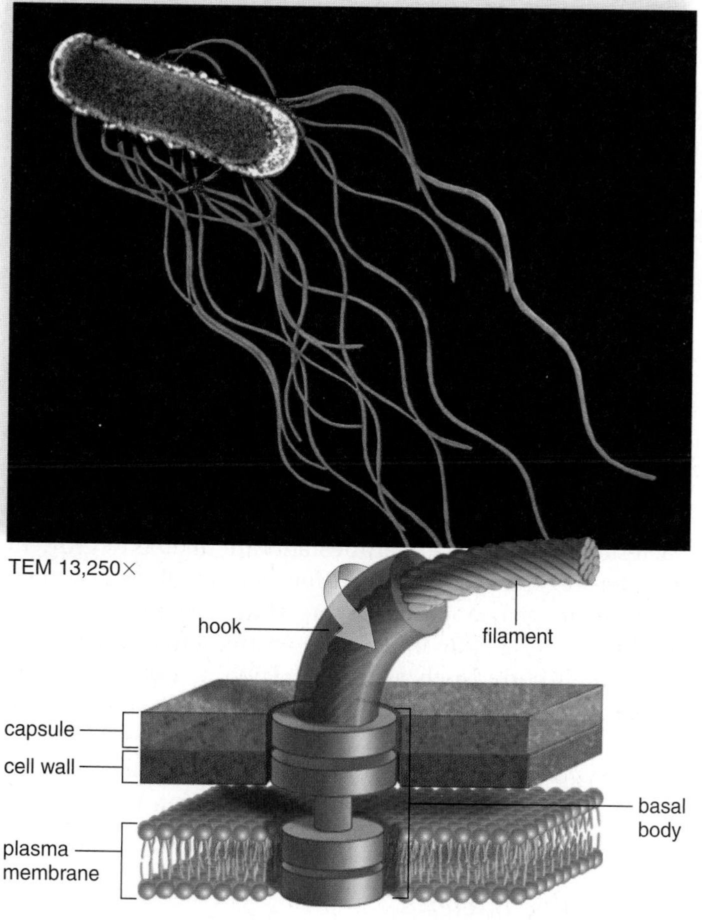


**FIGURE 20.6** **Flagella.**

Each flagellum of a bacterium contains a basal body, a hook, and a filament. The arrow indicates that the basal body, hook, and filament turn 360°. The flagellum of an archaean is more slender and may lack a basal body.

A prokaryotic cell lacks the membranous organelles of a eukaryotic cell, and various metabolic pathways are located on the inside of the plasma membrane. Although prokaryotes do not have a nucleus, they do have a dense area called a **nucleoid** where a single chromosome consisting largely of a circular strand of DNA is found. Many prokaryotes also have accessory rings of DNA called **plasmids.** Plasmids can be extracted and used as vectors to carry foreign DNA into host bacteria during genetic engineering processes. Protein synthesis in a prokaryotic cell is carried out by thousands of ribosomes, which are smaller than eukaryotic ribosomes. The following diagram summarizes prokaryotic cell structure, which is depicted in Figure 4.4:

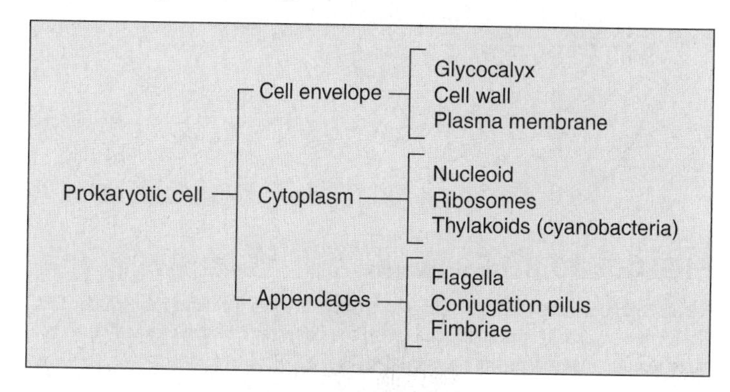

## Reproduction in Prokaryotes

Mitosis, which requires the formation of a spindle apparatus, does not occur in prokaryotes. Instead, prokaryotes reproduce asexually by means of **binary fission** [L. *binarius*, of two, and *fissura*, cleft, break] (Fig. 20.8).

The single circular chromosome replicates, and then two copies separate as the cell enlarges. Newly formed plasma membrane and cell wall separate the cell into two cells. Prokaryotes have a generation time as short as 12 minutes under favorable conditions. Mutations are generated and passed on to offspring more quickly than in eukaryotes. Also, prokaryotes are haploid, and so mutations are immediately subjected to natural selection, which determines any possible adaptive benefit.

In eukaryotes, genetic recombination occurs as a result of sexual reproduction. Sexual reproduction does not occur among prokaryotes, but three means of genetic recombination have been observed in prokaryotes. During **conjugation,** two bacteria are temporarily linked together, often by means of a **conjugation pilus** (see Fig. 20.7*b*). While they are linked, the donor cell passes DNA to a recipient cell. **Transformation** occurs when a cell picks up (from the surroundings) free pieces of DNA secreted by live prokaryotes or released by dead prokaryotes. During **transduction,** bacteriophages carry portions of DNA from one bacterial cell to another. Viruses have also been found to infect archaeal cells, and so transduction may play an important role in gene transfer for both domains of prokaryotes.

**Check Your Progress 20.2**

1. How is a prokaryotic cell structurally different from a eukaryotic cell?
2. Where is the cell wall located relative to the plasma membrane in a typical prokaryotic cell?
3. How is conjugation different from sexual reproduction?

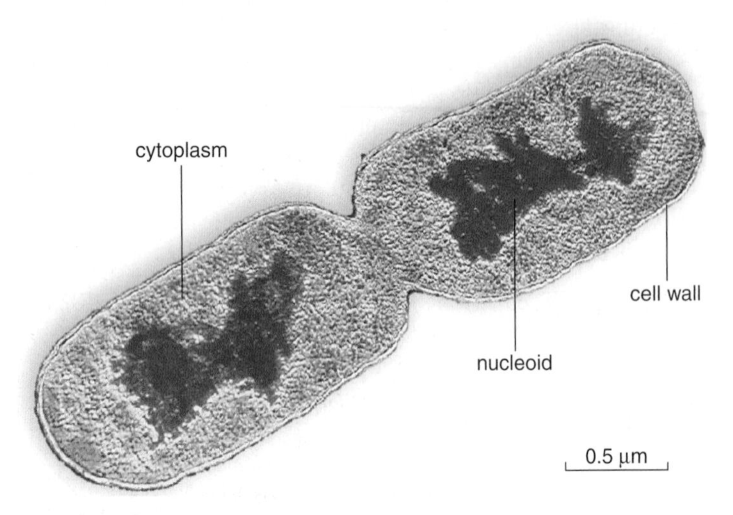


**FIGURE 20.8 Binary fission.**

When conditions are favorable for growth, prokaryotes divide to reproduce. This is a form of asexual reproduction because the daughter cells have exactly the same genetic material as the parent cell.

# 20.3 The Bacteria

**Bacteria** (domain Bacteria) are the more common type of prokaryote. The number of species of bacteria is amazing: To date, over 2,000 different bacteria have been named. They are found in practically every kind of environment on Earth.

## Characteristics of Bacterial Cells

Most bacterial cells are protected by a cell wall that contains the unique molecule **peptidoglycan.** Peptidoglycan is a complex of polysaccharides linked by amino acids. Groups of bacteria are commonly differentiated from one another by using the Gram stain procedure, which was developed in the late 1880s by Hans Christian Gram, a Danish bacteriologist. Even today, the Gram stain is usually the first test used in the identification of unknown bacteria. Gram-positive bacteria retain a dye-iodine complex and appear purple under the light microscope, while Gram-negative bacteria do not retain the complex and appear pink. This difference is dependent on the construction of the cell wall—that is, the Gram-positive bacteria have a thick layer of peptidoglycan in their cell walls, whereas Gram-negative bacteria have only a thin layer. *Clostridium tetani,* which causes tetanus, is an example of a Gram-positive bacterium. *Vibrio cholerae,* which infects the small intestine and causes cholera, is an example of a Gram-negative bacterium.

Bacteria (and archaea) can also be described in terms of their three basic cell shapes (Fig. 20.9): spirilli (sing., spirillum), spiral-shaped or helical-shaped; bacilli (sing., bacillus), rod-shaped; and cocci (sing., coccus), round or spherical. These three basic shapes may be augmented by particular arrangements or shapes of cells. For example, rod-shaped prokaryotes may appear as very short rods (coccobacilli) or as very long filaments (fusiform). Cocci may form pairs (diplococci), chains (streptococci), or clusters (staphylococci).

## Bacterial Metabolism

Bacteria are astoundingly diverse in terms of their metabolic lifestyles. With respect to basic nutrient requirements, bacteria are not much different from other organisms. One difference, however, concerns the need for oxygen. Some bacteria are **obligate anaerobes** and are unable to grow in the presence of free oxygen. A few serious illnesses—such as botulism, gas gangrene, and tetanus—are caused by anaerobic bacteria that infect oxygen-free environments in the human body, such as the intestine or deep puncture wounds. Other bacteria, called **facultative anaerobes,** are able to grow in either the presence or the absence of gaseous oxygen. Most bacteria, however, are aerobic and, like animals, require a constant supply of oxygen to carry out cellular respiration.

### *Autotrophic Bacteria*

Bacteria called **photoautotrophs** [Gk. *photos,* light, *auto,* self, and *trophe,* food] are photosynthetic. They use solar energy to reduce carbon dioxide to organic compounds. There are two types of photoautotrophic bacteria: those that evolved first and do not give off oxygen ($O_2$) and those that evolved later and do give off oxygen. Their characteristics are shown here:

| Do Not Give Off $O_2$ | Do Give Off $O_2$ |
|---|---|
| Photosystem I only | Photosystems I and II |
| Unique type of chlorophyll called bacteriochlorophyll | Type of chlorophyll *a* found in plants |

Green sulfur bacteria and some purple bacteria carry on the first type of photosynthesis. These bacteria usually live in anaerobic conditions, such as the muddy bottom of a marsh, and they cannot photosynthesize in the presence of oxygen. They do not give off oxygen because they do not use water as an electron donor; instead, they can, for example, use hydrogen sulfide ($H_2S$).

$$CO_2 + 2\,H_2S \longrightarrow (CH_2O)_n + 2\,S$$

In contrast, the cyanobacteria (see Fig. 20.12) contain chlorophyll *a* and carry on photosynthesis in the second way, just as algae and plants do.

$$CO_2 + H_2O \longrightarrow (CH_2O)_n + O_2$$

Bacteria called **chemoautotrophs** [Gk. *chemo*, pertaining to chemicals, *auto*, self, and *trophe*, food] carry out chemosynthesis. They oxidize inorganic compounds such as hydrogen gas, hydrogen sulfide, and ammonia to obtain the necessary energy to reduce $CO_2$ to an organic compound. The nitrifying bacteria oxidize ammonia ($NH_3$) to nitrites ($NO_2^-$) and nitrites to nitrates ($NO_3^-$). Their metabolic abilities keep nitrogen cycling through ecosystems. Other bacteria oxidize sulfur compounds. They live in environments such as deep-sea vents 2.5 km below sea level. The organic compounds produced by such bacteria and also archaea (discussed on page 369) support the growth of a community of organisms found at vents. This discovery lends support to the suggestion that the first cells originated at deep-sea vents.

### *Heterotrophic Bacteria*

Bacteria called **chemoheterotrophs** [Gk. *chemo*, pertaining to chemicals, *hetero*, different, and *trophe*, food] take in organic nutrients. They are aerobic **saprotrophs** that decompose almost any large organic molecule to smaller ones that can be absorbed. There is probably no natural organic molecule that cannot be digested by at least one prokaryotic species. In ecosystems, saprotrophic bacteria are called decomposers. They play a critical role in recycling matter and making inorganic molecules available to photosynthesizers.

The metabolic capabilities of chemoheterotrophic bacteria have long been exploited by human beings. Bacteria are used commercially to produce chemicals, such as ethyl alcohol, acetic acid, butyl alcohol, and acetones. Bacterial action is also involved in the production of butter, cheese, sauerkraut, rubber, cotton, silk, coffee, and cocoa. Even antibiotics are produced by some bacteria.

## Symbiotic Relationships

Bacteria (and archaea) form **symbiotic relationships** [Gk. *sym*, together, and *bios*, life] in which two different species live together in an intimate way. When the relationship is *mutualistic*, both species benefit. In *commensalistic* relationships, only one species benefits, and when it is *parasitic*, one species benefits but the other is harmed.

**Commensalism** often occurs when one population modifies the environment in such a way that a second population benefits. Obligate anaerobes can live in our intestines only because the bacterium *Escherichia coli* uses up the available oxygen. The parasitic bacteria cause disease, including human diseases.

Mutualistic bacteria live in human intestines, where they release vitamins K and $B_{12}$, which we can use to help produce blood components. In the stomachs of cows and goats, special mutualistic prokaryotes digest cellulose, enabling these animals to feed on grass. Mutualistic bacteria live in the root nodules of soybean, clover, and alfalfa plants where they reduce atmospheric

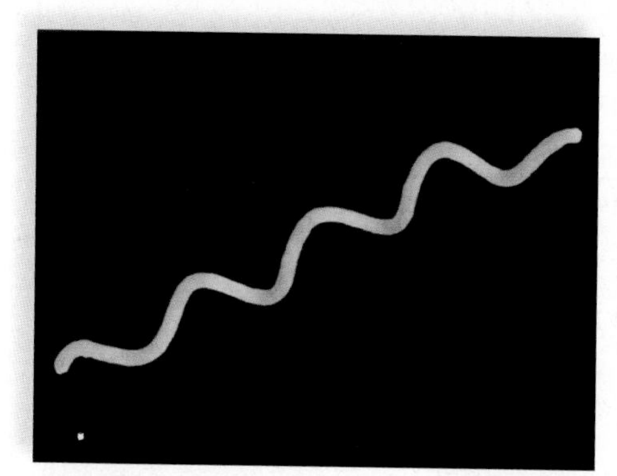

a. Spirillum: *Spirillum volutans* SEM 3,520×

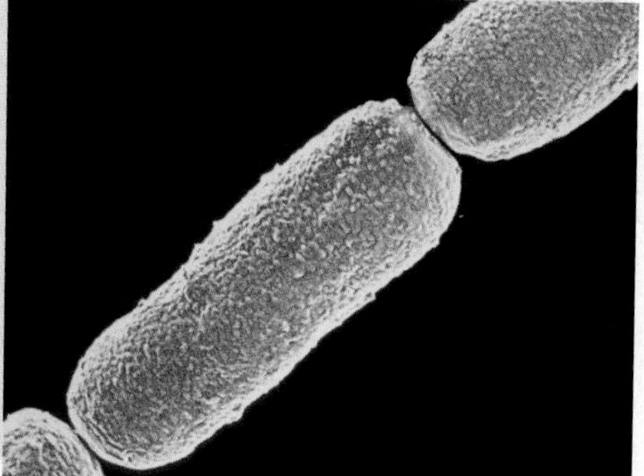
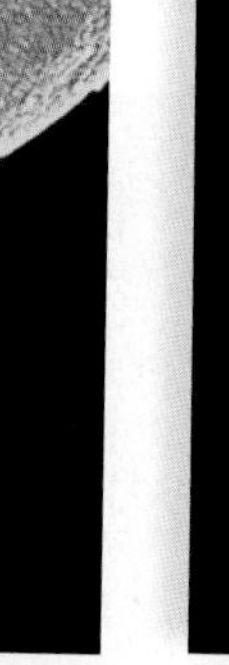

b. Bacilli: *Bacillus anthracis* SEM 35,000×

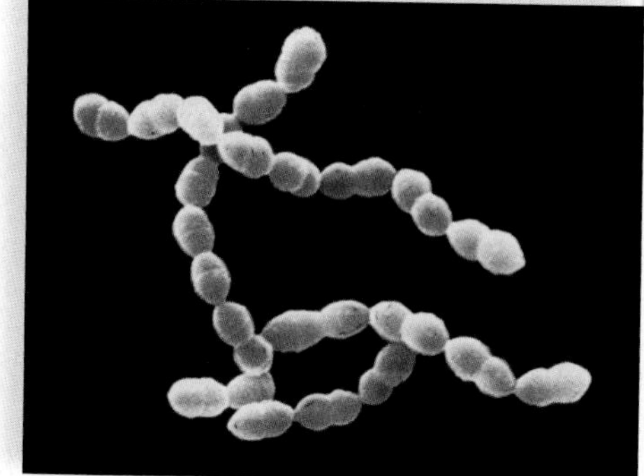

c. Cocci: *Streptococcus thermophilus* SEM 6,250×

**FIGURE 20.9 Diversity of bacteria.**

**a.** Spirillum, a spiral-shaped bacterium. **b.** Bacilli, rod-shaped bacteria. **c.** Cocci, round bacteria.

nitrogen ($N_2$) to ammonia, a process called nitrogen fixation (Fig. 20.10). Plants are unable to fix atmospheric nitrogen, and those without nodules take up nitrate and ammonia from the soil.

Parasitic bacteria cause diseases, and therefore are called **pathogens.** Some of the deadliest pathogens form **endospores** [Gk. *endon*, within, and *spora*, seed] when faced with unfavorable environmental conditions. A portion of the cytoplasm and a copy of the chromosome dehydrate and are then encased by a heavy, protective endospore coat (Fig. 20.11). In some bacteria, the rest of the cell deteriorates, and the endospore is released. Endospores survive in the harshest of environments—desert heat and dehydration, boiling temperatures, polar ice, and extreme ultraviolet radiation. They also survive for very long periods. When anthrax endospores 1,300 years old germinate, they can still cause a severe infection (usually seen in cattle and sheep). Humans also fear a deadly but uncommon type of food poisoning called botulism that is caused by the germination of endospores inside cans of food. To germinate, the endospore absorbs water and grows out of the endospore coat. In a few hours' time, it becomes a typical bacterial cell, capable of reproducing once again by binary fission. Endospore formation is not a means of reproduction, but it does allow survival and dispersal of bacteria to new places.

Many other bacteria cause diseases in humans: A few are listed in Table 20.2. In almost all cases, the growth of microbes themselves does not cause disease; the poisonous substances they release, called **toxins,** cause disease. When Gram-negative bacteria are killed by an antibiotic, the cell wall releases toxins called lipopolysaccharide fragments, and the result may be a high fever and a drop in blood pressure. Other bacteria secrete toxins when they are living. Some of these have a needle-shaped secretion apparatus they can use to inject toxins directly into host cells!

When someone steps on a rusty nail, bacteria can be injected deep into damaged tissue. The damaged area does not have a good blood flow and can become anaerobic. The endospores of *Clostridium tetani* germinate and produce a toxin that causes tetanus. The bacteria never leave the site of the wound, but the tetanus toxin they produce does move throughout the body. This toxin prevents the relaxation of muscles. In time, the body contorts because all the muscles have contracted. Eventually, suffocation occurs.

Fimbriae allow a pathogen to bind to certain cells, and this determines which organs or cells of the body will be its host. Like many bacteria that cause dysentery (severe diarrhea), *Shigella dysenteriae* is able to stick to the intestinal wall. In addition, *S. dysenteriae* produces a toxin called Shiga toxin that increases the potential for fatality. Also, invasive mechanisms that give a pathogen the ability to move through tissues and into the bloodstream result in a more medically significant disease than if it were localized. Usually a person can recover from food poisoning caused by *Salmonella.* But some strains of *Salmonella* have virulence factors—including a needle-shaped

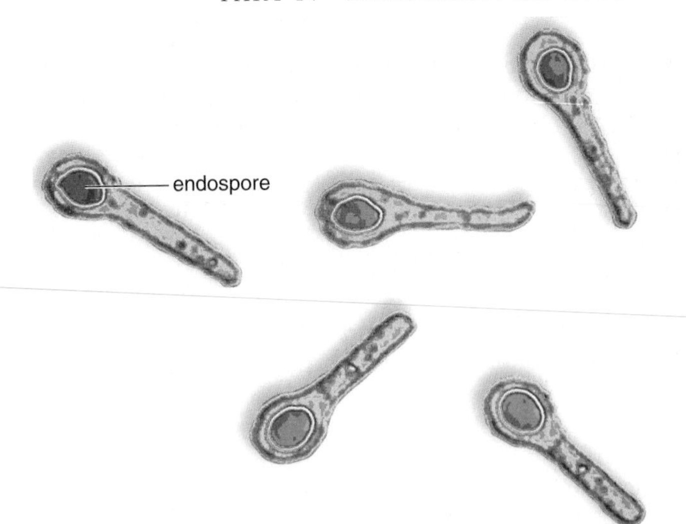


**FIGURE 20.11 The endospore of *Clostridium tetani.***

*C. tetani* produces a terminal endospore that causes it to have a drumstick appearance. If endospores gain access to a wound, they germinate and release bacteria that produce a neurotoxin. The patient develops tetanus, a progressive rigidity that can result in death; immunization can prevent tetanus.

**FIGURE 20.10 Nodules of a legume.**

While some free-living bacteria carry on nitrogen fixation, those of the genus *Rhizobium* invade the roots of legumes, with the resultant formation of nodules. Here the bacteria convert atmospheric nitrogen to an organic nitrogen that the plant can use. These are nodules on the roots of a soybean plant, *Glycine.*

**TABLE 20.2**

**Bacterial Diseases in Humans**

| *Category* | *Disease* |
|---|---|
| Sexually transmitted diseases | Syphilis, gonorrhea, chlamydia |
| Respiratory diseases | Strep throat, scarlet fever, tuberculosis, pneumonia, Legionnaires disease, whooping cough, inhalation anthrax |
| Skin diseases | Erysipelas, boils, carbuncles, impetigo, acne, infections of surgical or accidental wounds and burns, leprosy (Hansen disease) |
| Digestive tract diseases | Gastroenteritis, food poisoning, dysentery, cholera, peptic ulcers, dental caries |
| Nervous system diseases | Botulism, tetanus, leprosy, spinal meningitis |
| Systemic diseases | Plague, typhoid fever, diphtheria |
| Other diseases | Tularemia, Lyme disease |

toxin secretion apparatus—that allow the bacteria to penetrate the lining of the colon and move beyond this organ. Typhoid fever, a life-threatening disease, can then result.

### Antibiotics

Because bacteria are cells in their own right, a number of antibiotic compounds are active against bacteria and are widely prescribed. Most antibacterial compounds fall within two classes, those that inhibit protein biosynthesis and those that inhibit cell wall biosynthesis. Erythromycin and tetracyclines can inhibit bacterial protein synthesis because bacterial ribosomes function somewhat differently than eukaryotic ribosomes. Cell wall biosynthesis inhibitors generally block the formation of peptidoglycan, required to maintain bacterial integrity. Penicillin, ampicillin, and fluoroquinolone (like Cipro) inhibit bacterial cell wall biosynthesis without harming animal cells.

One problem with antibiotic therapy has been increasing bacterial resistance to antibiotics. Genes conferring resistance to antibiotics can be transferred between infectious bacteria by transformation, conjugation, or transduction. When penicillin was first introduced, less than 3% of *Staphylococcus aureus* strains were resistant to it. Now, due to selective advantage, 90% or more are resistant to penicillin and, increasingly, to methicillin, an antibiotic developed in 1957 (see page 283).

## Cyanobacteria

**Cyanobacteria** [Gk. *kyanos,* blue, and *bacterion,* rod] are Gram-negative bacteria with a number of unusual traits. They photosynthesize in the same manner as plants and are believed to be responsible for first introducing oxygen into the primitive atmosphere. Formerly, the cyanobacteria were called blue-green algae and were classified with eukaryotic algae, but now they are classified as prokaryotes. Cyanobacteria can have other pigments that mask the color of chlorophyll so that they appear red, yellow, brown, or black, rather than only blue-green.

Cyanobacterial cells are rather large, ranging from 1 to 50 μm in width. They can be unicellular, colonial, or filamentous. Cyanobacteria lack any visible means of locomotion, although some glide when in contact with a solid surface and others oscillate (sway back and forth). Some cyanobacteria have a special advantage because they possess heterocysts, which are thick-walled cells without nuclei, where nitrogen fixation occurs. The ability to photosynthesize and also to fix atmospheric nitrogen ($N_2$) means that their nutritional requirements are minimal. They can serve as food for heterotrophs in ecosystems.

Cyanobacteria (Fig. 20.12) are common in fresh and marine waters, in soil, and on moist surfaces, but they are also found in harsh habitats, such as hot springs. They are symbiotic with a number of organisms, including liverworts, ferns, and even at times invertebrates such as corals. In association with fungi, they form **lichens** that can grow on rocks. In a lichen, cyanobacterium mutualistically provides organic nutrients to the fungus, while the fungus possibly protects and furnishes inorganic nutrients to the cyanobacterium. It is also possible that the fungus is parasitic on the cyanobacterium. Lichens help transform rocks into soil; other forms of life then may follow. It is hypothesized that cyanobacteria were the first colonizers of land during the course of evolution.

Cyanobacteria are ecologically important in still another way. If care is not taken in disposing of industrial, agricultural, and human wastes, phosphates drain into lakes and ponds, resulting in a "bloom" of these organisms. The surface of the water becomes turbid, and light cannot penetrate to lower levels. When a portion of the cyanobacteria die off, the decomposing prokaryotes use up the available oxygen, causing fish to die from lack of oxygen.

**Check Your Progress 20.3**

1. How is the peptidoglycan layer different in Gram-positive and Gram-negative cells?
2. What is the function of bacterial endospores?
3. Which bacteria produce much of the oxygen we breathe, and what metabolic process gives off this oxygen?

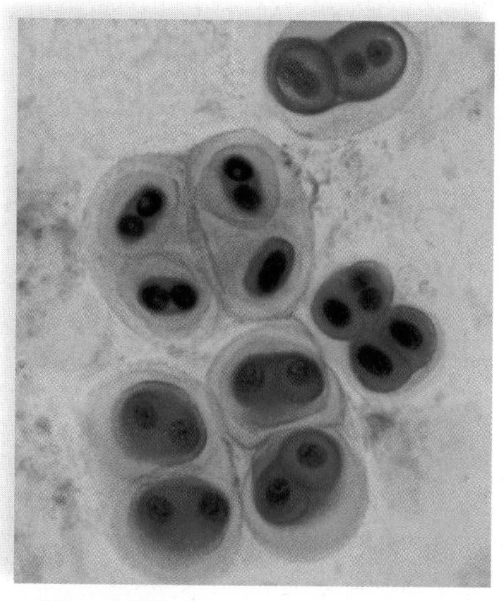
a. *Gloeocapsa* LM 250×

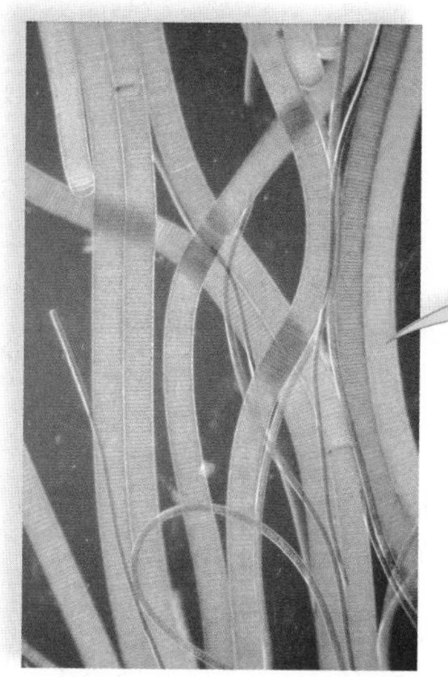

b. *Oscillatoria* LM 100×

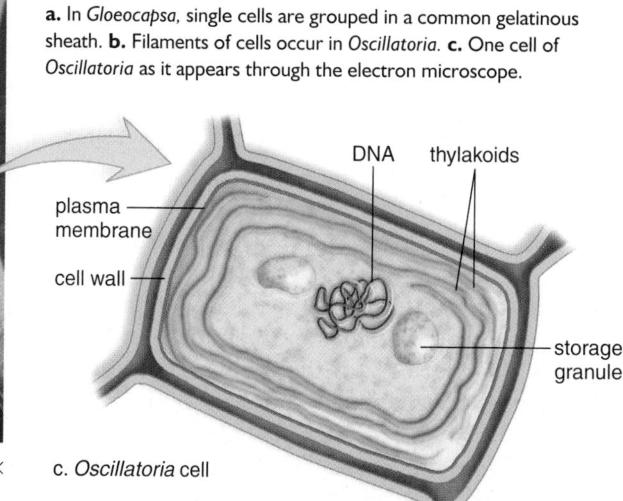

c. *Oscillatoria* cell

**FIGURE 20.12 Diversity among the cyanobacteria.**

**a.** In *Gloeocapsa,* single cells are grouped in a common gelatinous sheath. **b.** Filaments of cells occur in *Oscillatoria.* **c.** One cell of *Oscillatoria* as it appears through the electron microscope.

# 20.4 The Archaea

At one time, **archaea** (domain Archaea) were considered to be a unique group of bacteria. Archaea came to be viewed as a distinct domain of organisms in 1977, when Carl Woese and George Fox discovered that the rRNA of archaea has a different sequence of bases than the rRNA of bacteria. He chose rRNA because of its involvement in protein synthesis—any changes in rRNA sequence probably occur at a slow, steady pace as evolution occurs. As discussed in Chapter 19, it is proposed that the tree of life contains three domains: Archaea, Bacteria, and Eukarya. Because archaea and some bacteria are found in extreme environments (hot springs, thermal vents, salt basins), they may have diverged from a common ancestor relatively soon after life began. Then later, the eukarya diverged from the archaeal line of descent. In other words, the eukarya are more closely related to the archaea than to the bacteria. Archaea and eukarya share some of the same ribosomal proteins (not found in bacteria), initiate transcription in the same manner, and have similar types of tRNA.

## Structure of Archaea

Archaea are prokaryotes with biochemical characteristics that distinguish them from both bacteria and eukaryotes. The plasma membranes of archaea contain unusual lipids that allow many of them to function at high temperatures. The lipids of archaea contain glycerol linked to branched-chain hydrocarbons in contrast to the lipids of bacteria, which contain glycerol linked to fatty acids. The archaea also evolved diverse cell wall types, which facilitate their survival under extreme conditions. The cell walls of archaea do not contain peptidoglycan as do the cell walls of bacteria. In some archaea, the cell wall is largely composed of polysaccharides, and in others, the wall is pure protein. A few have no cell wall.

## Types of Archaea

Archaea were originally discovered living in extreme environmental conditions. Three main types of archaea are still distinguished based on their unique habitats: methanogens, halophiles, and thermoacidophiles (Fig. 20.13).

### *Methanogens*

The **methanogens** (methane makers) are obligate anaerobes found in environments such as swamps, marshes, and the intestinal tracts of animals. Methanogenesis, the ability to form methane ($CH_4$), is a type of metabolism performed only by some archaea. Methanogens are chemoautotrophs, using hydrogen gas ($H_2$) to reduce carbon dioxide ($CO_2$) to methane and couple the energy released to ATP production.

$$CO_2 + 4\,H_2 \longrightarrow CH_4 + 2\,H_2O$$

Methane, also called biogas, is released into the atmosphere, where it contributes to the greenhouse effect and global warming. About 65% of the methane found in our atmosphere is produced by these methanogenic archaea.

Methanogenic archaea may help us anticipate what life may be like on other celestial bodies. Consider, for instance, the unusual microbial community residing in the Lidy Hot Springs of eastern Idaho. The springs, which originate 200 m (660 feet) beneath the Earth's surface, are lacking in organic nutrients, but rich in $H_2$. Scientists have found the springs to be inhabited by vast numbers of microorganisms; over 90%

a.

b.

c.

**FIGURE 20.13 Extreme habitats.**

**a.** Halophilic archaea can live in salt lakes. **b.** Thermoacidophilic archaea can live in the hot springs of Yellowstone National Park. **c.** Methanogens live in swamps and in the guts of animals.

are archaea, and the overwhelming majority are methanogens. The researchers who first investigated the Lidy Hot Springs microbes point out that similar methanogenic communities may someday be found beneath the surfaces of Mars and Europa (one of Jupiter's moons). Since hydrogen is the most abundant element in the universe, it would be readily available for use by methanogens everywhere.

### *Halophiles*

The **halophiles** require high salt concentrations (usually 12–15%—the ocean is about 3.5% salt) for growth. They have been isolated from highly saline environments in which few organisms are able to survive, such as the Great Salt Lake in Utah, the Dead Sea, solar salt ponds, and hypersaline soils. These archaea have evolved a number of mechanisms to survive in environments that are high in salt. This survival ability benefits the halophiles, as they do not have to compete with as many microorganisms as they would encounter in a more moderate environment. The proteins of halophiles have unique chloride pumps that use halorhodopsin (related to the rhodopsin pigment found in our eyes) to pump chloride to the inside of the cell, and this prevents water loss. These organisms are aerobic chemoheterotrophs. However, some species can carry out a unique form of photosynthesis if their oxygen supply becomes scarce, as commonly occurs in highly saline conditions. Instead of chlorophyll, these halophiles use a purple pigment called bacteriorhodopsin to capture solar energy for use in ATP synthesis. Interestingly, most halophiles are so adapted to a high-saline environment that they perish if placed in a solution with a low salt concentration (such as pure water).

### *Thermoacidophiles*

A third major type of archaea are the **thermoacidophiles.** These archaea are isolated from extremely hot, acidic environments such as hot springs, geysers, submarine thermal vents, and around volcanoes. They are chemoautotrophic anaerobes that use hydrogen ($H_2$) as the electron donor, and sulfur (S) or sulfur compounds as terminal electron acceptors, for their electron transport chains. Hydrogen sulfide ($H_2S$) and protons ($H^+$) are common products.

$$S + H_2 \longrightarrow H_2S + H^+$$

Recall that the greater the concentration of protons, the lower (and more acidic) the pH. Thus, it is not surprising that thermoacidophiles grow best at extremely low pH levels, between pH 1 and 2. Due to the unusual lipid composition of their plasma membranes, thermoacidophiles survive best at temperatures above 80°C; some can even grow at 105°C (remember that water boils at 100°C)!

### *Archaea in Moderate Habitats*

Although archaea are capable of living in extremely stressful conditions, they are found in all moderate environments as well. For example, some archaea have been found living in symbiotic relationships with animals, including sponges and sea cucumbers. Such relationships are sometimes mutualistic or even commensalistic, but there are no parasitic archaea—that is, they are not known to cause infectious diseases.

The roles of archaea in activities such as nutrient cycling are still being explored. For example, a group of nitrifying marine archaea has recently been discovered. Some scientists think that these archaea may be major contributors to the supply of nitrite in the oceans. Nitrite can be converted by certain bacteria to nitrate, a form of nitrogen that can be used by plants and other producers to construct amino acids and nucleic acids. Archaea have also been found inhabiting lake sediments, rice paddies, and soil, where they are likely to be involved in nutrient cycling.

**Check Your Progress 20.4**

1. How are archaea different from bacteria?
2. List the three types of archaea distinguished by their unique habitats.
3. Archaea are thought to be closely related to eukaryotes. What evidence supports this possibility?
4. In a recent United Nations report, the practice of maintaining large herds of livestock was blamed for contributing to greenhouse gases. What is the basis for this claim?

## Connecting the Concepts

Microbiology began when Leeuwenhoek first used his microscope to observe microorganisms. Significant advances occurred with the discovery that bacteria and viruses cause disease, and again when microbes were first used in genetic studies. Although there are significant structural differences between prokaryotes and eukaryotes, many biochemical similarities exist between the two. Thus, the details of protein synthesis, first worked out in bacteria, are applicable to all cells, including human cells. Today, transgenic bacteria routinely make products and otherwise serve the needs of human beings.

Many prokaryotes can live in environments that may represent the kinds of habitats available when the Earth first formed. We find prokaryotes in such hostile habitats as swamps, the Dead Sea, and hot sulfur springs. The fossil record suggests that the prokaryotes evolved before the eukaryotes. Not only do all living things trace their ancestry to the prokaryotes, but prokaryotes are believed to have contributed to the evolution of the eukaryotic cell. The mitochondria and chloroplasts of the eukaryotic cell are probably derived from bacteria that took up residence inside a nucleated cell.

Cyanobacteria are credited with introducing oxygen into the Earth's ancestral atmosphere, and they may have been the first colonizers of the terrestrial environment. Some bacteria are decomposers that recycle nutrients in both aquatic and terrestrial environments. Bacteria and archaea play significant roles in the carbon, nitrogen, and phosphorus cycles. Mutualistic bacteria also fix nitrogen in plant nodules, enabling herbivores to digest cellulose, and release certain vitamins in the human intestine. Clearly, humans are dependent on the past and present activities of prokaryotes.

## summary

### 20.1 Viruses, Viroids, and Prions

Viruses are noncellular, while prokaryotes are fully functioning organisms. All viruses have at least two parts: an outer capsid composed of protein subunits and an inner core of nucleic acid, either DNA or RNA, but not both. Some also have an outer membranous envelope.

Viruses are obligate intracellular parasites that can be maintained only inside living cells, such as those of a chicken egg, or those propagated in cell (tissue) culture.

The lytic cycle of a bacteriophage consists of attachment, penetration, biosynthesis, maturation, and release. In the lysogenic cycle of a bacteriophage, viral DNA is integrated into bacterial DNA for an indefinite period of time, but it can undergo the lytic cycle when stimulated.

The reproductive cycle differs for animal viruses. Uncoating is needed to free the genome from the capsid, and either budding or lysis releases the viral particles from the cell. Retroviruses have an enzyme, reverse transcriptase, that carries out reverse transcription. This enzyme produces one strand of DNA (cDNA) using viral RNA as a template, and then another DNA strand that is complementary to the first one. The resulting double-strand DNA becomes integrated into host DNA. The AIDS virus is a retrovirus.

Viruses cause various diseases in plants and animals, including human beings. Viroids are naked strands of RNA (not covered by a capsid) that can cause disease in plants. Prions are protein molecules that have a misshapen tertiary structure. Prions cause diseases such as CJD in humans and mad cow disease in cattle when they cause other proteins of their own kind also to become misshapen.

### 20.2 The Prokaryotes

The bacteria (domain Bacteria) and archaea (domain Archaea) are prokaryotes. Prokaryotic cells lack a nucleus and most of the other cytoplasmic organelles found in eukaryotic cells. Prokaryotes reproduce asexually by binary fission. Their chief method for achieving genetic variation is mutation, but genetic recombination by means of conjugation, transformation, and transduction has been observed.

Prokaryotes differ in their need (and tolerance) for oxygen. There are obligate anaerobes, facultative anaerobes, and aerobic prokaryotes. Some prokaryotes are autotrophic, and some are heterotrophic.

### 20.3 The Bacteria

Bacteria (domain Bacteria) are the more prevalent type of prokaryote. The classification of bacteria is still being developed. Of primary importance at this time are the shape of the cell and the structure of the cell wall, which affects Gram staining. Bacteria occur in three basic shapes: spiral-shaped (spirillum), rod-shaped (bacillus), and round (coccus).

Some prokaryotes are autotrophic—either photoautotrophs (photosynthetic) or chemoautotrophs (chemosynthetic). Some photosynthetic bacteria (cyanobacteria) give off oxygen, and some (purple and green sulfur bacteria) do not. Chemoautotrophs oxidize inorganic compounds such as hydrogen gas, hydrogen sulfide, and ammonia to acquire energy to make their own food. Surprisingly, chemoautotrophs support communities at deep-sea vents.

Many bacteria are chemoheterotrophs (aerobic heterotrophs) and are saprotrophic decomposers that are absolutely essential to the cycling of nutrients in ecosystems. Their metabolic capabilities are so vast that they are used by humans both to dispose of and to produce substances. Many heterotrophic bacteria are symbiotic. The mutualistic nitrogen-fixing bacteria live in nodules on the roots of legumes. Of special interest are the cyanobacteria, which were the first organisms to photosynthesize in the same manner as plants. When cyanobacteria are symbionts with fungi, they form lichens. Some bacterial symbionts, however, are parasitic and cause plant and animal, including human, diseases. Certain bacteria form endospores, which are extremely resistant to destruction. Their genetic material can thereby survive unfavorable conditions.

### 20.4 The Archaea

The archaea (domain Archaea) are a second type of prokaryote. On the basis of rRNA sequencing, it is thought that there are three evolutionary domains: Bacteria, Archaea, and Eukarya. In addition, the archaea appear to be more closely related to the eukarya than to the bacteria. Archaea do not have peptidoglycan in their cell walls, as do the bacteria, and they share more biochemical characteristics with the eukarya than do bacteria.

Three types of archaea live under harsh conditions, such as anaerobic marshes (methanogens), salty lakes (halophiles), and hot sulfur springs (thermoacidophiles). Archaea are also found in moderate environments.

## understanding the terms

archaea 368
bacteria 364
bacteriophage 358
binary fission 364
capsid 357
chemoautotroph 365
chemoheterotroph 365
commensalism 365
conjugation 364
conjugation pilus 364
cyanobacteria 367
emerging virus 361
endospore 366
facultative anaerobe 364
fimbriae 363
flagellum (pl., flagella) 363
halophile 369
host specific 358
lichen 367
lysogenic cell 359
lysogenic cycle 358
lytic cycle 358
methanogen 368
nucleoid 363
obligate anaerobe 364
pathogen 366
peptidoglycan 364
photoautotroph 364
plasmid 363
prion 362
prokaryote 362
retrovirus 361
reverse transcriptase 361
saprotroph 365
symbiotic relationship 365
thermoacidophile 369
toxin 366
transduction 364
transformation 364
viroid 361
virus 356

Match the terms to these definitions:

a. ______________ Bacteriophage life cycle in which the virus incorporates its DNA into that of the bacterium; only later does it begin a lytic cycle, which ends with the destruction of the bacterium.

b. ______________ Organism that contains chlorophyll and uses solar energy to produce its own organic nutrients.

c. ______________ Organism that secretes digestive enzymes and absorbs the resulting nutrients back across the plasma membrane.

d. ______________ Relationship that could be mutualistic, commensalistic, or parasitic.

e. ______________ Type of prokaryote that is most closely related to the Eukarya.

## reviewing this chapter

1. Describe the general structure of viruses, and describe both the lytic cycle and the lysogenic cycle of bacteriophages. 356–59
2. Contrast viruses with cells in terms of the characteristics of life. 356–57
3. How do animal viruses differ in structure and reproductive cycle? 359
4. How do retroviruses differ from other animal viruses? Describe the reproductive cycle of retroviruses in detail. 361
5. Explain Pasteur's experiment, which showed that bacteria do not arise spontaneously. 362
6. Provide a diagram of prokaryotic cell structure and discuss. 362–64
7. How do all prokaryotes introduce variations? How does genetic recombination occur in bacteria? 364
8. How do all prokaryotes differ in their tolerance of and need for oxygen? 364–65
9. Compare photosynthesis between the green sulfur bacteria and the cyanobacteria. 365, 367
10. What are chemoautotrophic prokaryotes, and where have they been found to support whole communities? 365
11. What role do endospores play in disease? 366
12. Discuss the importance of cyanobacteria in ecosystems and in the history of the Earth. 367
13. How do archaea differ from bacteria? 368–69
14. What three different types of archaea may be distinguished based on their different habitats? 368–69

## testing yourself

Choose the best answer for each question.

1. Label this condensed version of bacteriophage reproductive cycles using these terms: penetration, maturation, release, prophage, attachment, biosynthesis, and integration.

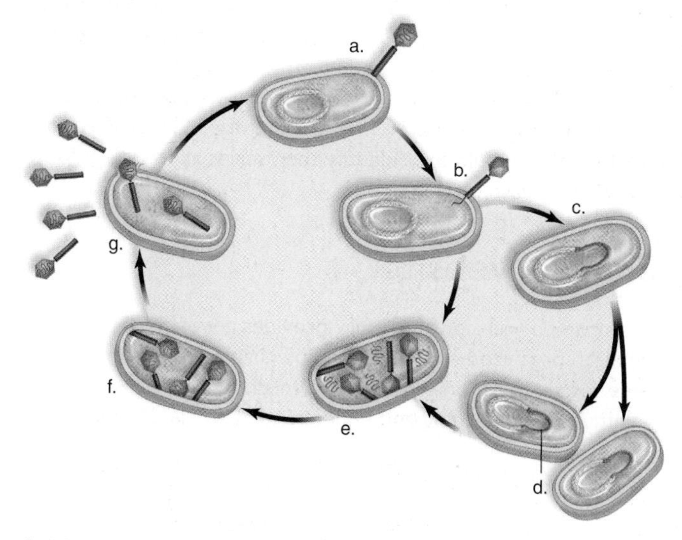


2. Viruses are considered nonliving because
   a. they do not locomote.
   b. they cannot reproduce independently.
   c. their nucleic acid does not code for protein.
   d. they are noncellular.
   e. Both b and d are correct.
3. Which of the following statements about viruses is incorrect?
   a. The nucleic acid may be either DNA or RNA, but not both.
   b. The capsid may be polyhedral or helical.
   c. Viruses do not fit into the current system for naming organisms.
   d. The nucleic acid may be either single stranded or double stranded.
   e. Viruses are rarely, if ever, host specific.
4. The envelope of an animal virus is derived from the ______________ of its host cell.
   a. cell wall
   b. membrane
   c. glycocalyx
   d. receptors
5. A prophage occurs during the reproduction cycle of
   a. a lysogenic virus.
   b. a poxvirus.
   c. a lytic virus.
   d. an enveloped virus.
6. Which of these are found in all viruses?
   a. envelope, nucleic acid, capsid
   b. DNA, RNA, and proteins
   c. proteins and a nucleic acid
   d. proteins, nucleic acids, carbohydrates, and lipids
   e. tail fibers, spikes, and rod shape
7. Which would be the worst choice for cultivation of viruses?
   a. tissue culture
   b. bird embryos
   c. live mammals
   d. sterile broth
8. A pathogen would most accurately be described as
   a. a parasite.
   b. a commensal.
   c. a saprobe.
   d. a symbiont.
9. RNA retroviruses have a special enzyme that
   a. disintegrates host DNA.
   b. polymerizes host DNA.
   c. transcribes viral RNA to cDNA.
   d. translates host DNA.
   e. produces capsid proteins.
10. Which is not true of prokaryotes? They
    a. are living cells.
    b. lack a nucleus.
    c. all are parasitic.
    d. are both archaea and bacteria.
    e. evolved early in the history of life.
11. Facultative anaerobes
    a. require a constant supply of oxygen.
    b. are killed in an oxygenated environment.
    c. do not always need oxygen.
    d. are photosynthetic but do not give off oxygen.
    e. All of these are correct.
12. Which of these is most apt to be a prokaryotic cell wall function?
    a. transport
    b. motility
    c. support
    d. adhesion

13. Cyanobacteria, unlike other types of bacteria that photosynthesize, do
    a. give off oxygen.
    b. not have chlorophyll.
    c. not have a cell wall.
    d. need a fungal partner.
14. Chemoautotrophic prokaryotes
    a. are chemosynthetic.
    b. use the rays of the sun to acquire energy.
    c. oxidize inorganic compounds to acquire energy.
    d. are always bacteria, not archaea.
    e. Both a and c are correct.
15. Archaea differ from bacteria in that
    a. some can form methane.
    b. they have different rRNA sequences.
    c. they do not have peptidoglycan in their cell walls.
    d. they rarely photosynthesize.
    e. All of these are correct.
16. Which of these archaea would live at a deep-sea vent?
    a. thermoacidophile
    b. halophile
    c. methanogen
    d. parasitic forms
    e. All of these are correct.
17. A prokaryote that can synthesize all its required organic components from $CO_2$ using energy from the sun is a
    a. photoautotroph.
    b. photoheterotroph.
    c. chemoautotroph.
    d. chemoheterotroph.
18. While testing some samples of marsh mud, you discover a new microbe that has never been described before. You examine it closely and find that, although it is cellular, there is no nucleus. Biochemical analysis reveals that the plasma membrane is made up of glycerol linked to branched-chain hydrocarbons, and the cell wall contains no peptidoglycan. This microbe is most likely a(n)
    a. enveloped virus.
    b. archaean.
    c. prion.
    d. bacterium.
    e. bacteriophage.
19. The Nobel laureate Peter Medawar called a certain type of microbe "a piece of bad news wrapped up in protein." Which of the following was he describing?
    a. archaea
    b. bacteria
    c. prion
    d. virus
    e. viroid

## thinking scientifically

1. While a few drugs are effective against some viruses, they often impair the function of body cells and thereby have a number of side effects. Most antibiotics (antibacterial drugs) do not cause side effects. Why would antiviral medications be more likely to produce side effects?
2. Model organisms are those widely used by researchers who wish to understand basic processes that are common to many species. Bacteria such as *Escherichia coli* are model organisms for modern geneticists. Compare the characteristics of bacteria to those of peas (see p. 287) and give three reasons why bacteria would be useful in genetic experiments.

## bioethical issue

### Identifying Carriers

Carriers of disease are persons who do not appear to be ill but can nonetheless pass on an infectious disease. The only way society can protect itself is to identify carriers and remove them from areas or activities where transmission of the pathogen is most likely. Sometimes it's difficult to identify all activities that might pass on a pathogen—for example, HIV. A few people believe that they have acquired HIV from their dentists, and while this is generally believed to be unlikely, medical personnel are still required to identify themselves when they are carriers of HIV.

Transmission of HIV is believed to be possible in certain sports. In a statistical study, the Centers for Disease Control and Prevention figured that the odds of acquiring HIV from another football player were 1 in 85 million. But the odds might be higher for boxing, a bloody sport. When two brothers, one of whom had AIDS, got into a vicious fight, the infected brother repeatedly bashed his head against his brother's. Both men bled profusely, and soon after, the previously uninfected brother tested positive for the virus. The possibility of transmission of HIV in the boxing ring has caused several states to require boxers to undergo routine HIV testing. If they are HIV positive, they can't fight.

Should all people who are HIV positive always be required to identify themselves, no matter what the activity? Why or why not? By what method would they identify themselves at school, at work, and in other places?

## *Biology* website

The companion website for *Biology* provides a wealth of information organized and integrated by chapter. You will find practice tests, animations, videos, and much more that will complement your learning and understanding of general biology.

**http://www.mhhe.com/maderbiology10**

# 21

# Protist Evolution and Diversity

*Protists are an astoundingly diverse collection of eukaryotic organisms that vary in form, lifestyle, nutrition, locomotion, and reproduction. The photosynthetic algae are ecological producers responsible for introducing copious amounts of oxygen into the atmosphere. Algae also have economic importance; for instance, the "nori" used to wrap sushi rolls is a red alga. Algin, another seaweed product, is a thickening ingredient in foods such as ice cream; it is also used in the production of pharmaceuticals, cosmetics, paper, and textiles—even the tungsten filaments of lightbulbs. Protozoans, which are heterotrophic and usually have a form of locomotion, serve as food for some animals but also cause diseases. Malaria is just one of several medically and economically important protozoan illnesses of humans.*

*The photo below shows a slime mold familiarly known as dog vomit, which creeps along the forest floor engulfing its food. In the nineteenth century, water molds caused a famine in Ireland by attacking potatoes and devastated the vineyards of France. In this chapter, we will see how diverse protists play essential roles in ecosystems, and how they directly affect our own species.*

The so-called "dog vomit" slime mold (*Fuligo septica*) is a protist.

## *concepts*

# 21.1 General Biology of Protists

**Protists** have eukaryotic cells characterized by membranous organelles. The endosymbiotic theory tells how an eukaryotic cell acquires mitochondria and chloroplasts. Mitochondria are derived from aerobic bacteria, and chloroplasts are derived from cyanobacteria that were engulfed on two separate occasions (Fig. 21.1*a, b*).

Protists vary in size from microscopic algae and protozoans to kelp that can exceed 200 m in length. Kelp, a brown alga, is multicellular; *Volvox,* a green alga, is colonial; while *Spirogyra,* also a green alga, is filamentous. Most protists are unicellular, but despite their small size they have attained a high level of complexity. The amoeboids and ciliates possess unique organelles—their contractile vacuole is an organelle that assists in water regulation.

Protists acquire nutrients in a number of different ways. We will see that the algae are photosynthetic and gather energy from sunlight. Many protozoans are heterotrophic, and some ingest food by endocytosis, thereby forming food vacuoles. A slime mold creeps along the forest floor ingesting decaying plant material in the same manner. Other protozoans are parasitic and absorb nutrient molecules meant for their host. Some protozoans, such as *Euglena,* are **mixotrophic,** meaning they are able to combine autotrophic and heterotrophic nutritional modes.

Asexual reproduction by mitosis is the norm in protists. Sexual reproduction involving meiosis and spore formation generally occurs only in a hostile environment. Spores are resistant to adverse conditions and can survive until favorable conditions return once more. Some protozoans form **cysts,** another type of resting stage. In parasites, a cyst often serves as a means of transfer to a new host.

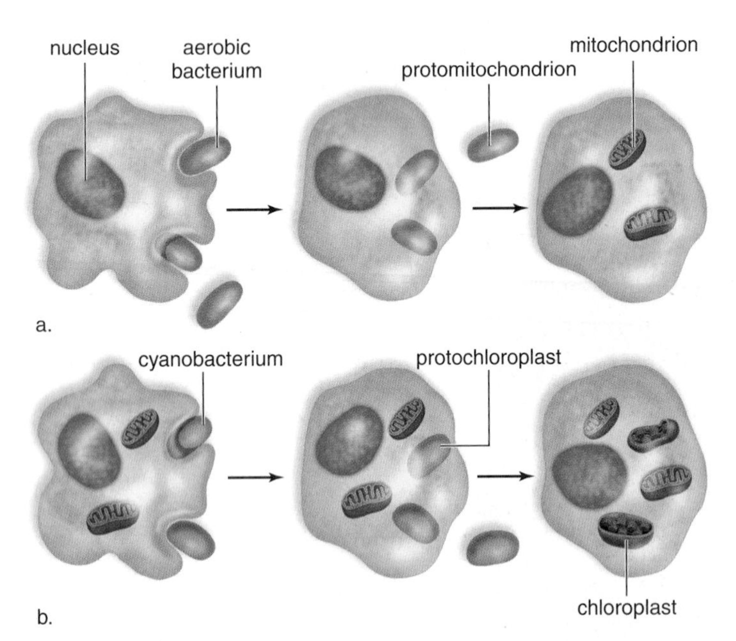


**FIGURE 21.1 Origin of the eukaryotic cell.**

The endosymbiotic theory tells us that mitochondria and chloroplasts were once free-living bacteria. **a.** A nucleated cell takes in aerobic bacteria, which become its mitochondria. **b.** A nucleated cell with mitochondria takes in photosynthetic bacteria (cyanobacteria), which become its chloroplasts.

While the protists have great medical importance because several cause diseases in humans, they also are of enormous ecological importance. Being aquatic, the photosynthesizers give off oxygen and function as producers in both freshwater and saltwater ecosystems. They are a part of **plankton** [Gk. *plankt,* wandering], organisms that are suspended in the water and serve as food for heterotrophic protists and animals.

Protists enter symbiotic relationships ranging from parasitism to mutualism. Coral reef formation is greatly aided by the presence of a symbiotic photosynthetic protist that lives in the tissues of coral animals, for example.

## Evolution and Diversity of Protists

For convenience, the protists can be characterized according to modes of nutrition. The protists known as **algae** are autotrophic, as are land plants. Certain of the algae share a common ancestor with land plants, as is stressed in Chapter 23. The protozoans and slime molds tend to be heterotrophic by ingestion, as are animals. Certain of the protozoans are closely related to animals, as is discussed in Chapter 28. Water molds are heterotrophic by absorption, as are fungi, but they are probably not closely related to fungi.

The term **protozoan** can be somewhat ambiguous; some prefer to restrict the term to unicellular heterotrophic organisms that ingest their food. Or, the term can mean any unicellular (or colonial) eukaryote. In that case, protozoans include photosynthetic, heterotrophic, and mixotrophic organisms. Most protozoans have some form of locomotion, either by flagella, pseudopods, or cilia, and those that don't locomote tend to be parasitic. At one time, zoologists classified protozoans as animals.

New research in the evolution of protists is overturning traditional taxonomical schemes. At this time, the most widely accepted formal approach for categorizing protists is to assign these diverse organisms to supergroups, as noted in the table in Figure 21.2. A **supergroup** is a major eukaryotic group, and the table designates six supergroups that encompass all members of the domain Eukarya, including the protists, plants, fungi, and animals. The relationship of the supergroups based on molecular studies is shown in Figure 21.3.

### Check Your Progress 21.1

1. Which major group of organisms was the first to have eukaryotic cells?
2. How are algae and protozoans nutritionally different from one another?
3. **a.** Which supergroup includes both plants and protists? **b.** Which includes fungi and animals, as well as protists?

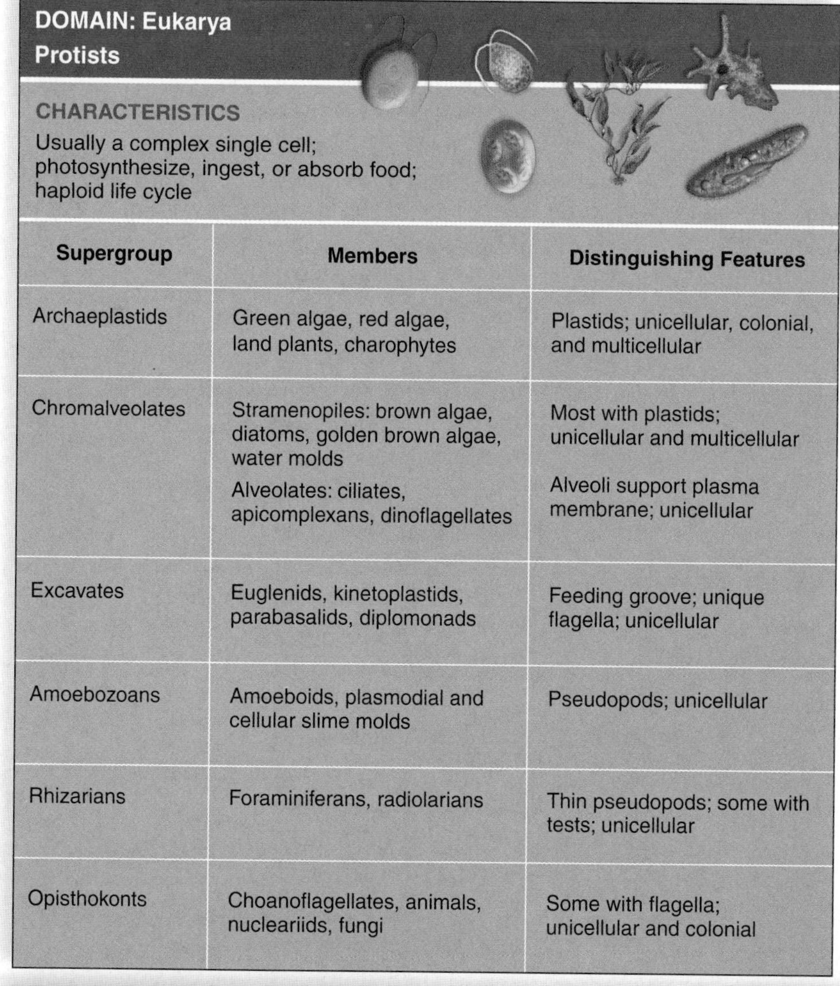

**DOMAIN: Eukarya**
**Protists**

CHARACTERISTICS
Usually a complex single cell; photosynthesize, ingest, or absorb food; haploid life cycle

| Supergroup | Members | Distinguishing Features |
|---|---|---|
| Archaeplastids | Green algae, red algae, land plants, charophytes | Plastids; unicellular, colonial, and multicellular |
| Chromalveolates | Stramenopiles: brown algae, diatoms, golden brown algae, water molds<br>Alveolates: ciliates, apicomplexans, dinoflagellates | Most with plastids; unicellular and multicellular<br>Alveoli support plasma membrane; unicellular |
| Excavates | Euglenids, kinetoplastids, parabasalids, diplomonads | Feeding groove; unique flagella; unicellular |
| Amoebozoans | Amoeboids, plasmodial and cellular slime molds | Pseudopods; unicellular |
| Rhizarians | Foraminiferans, radiolarians | Thin pseudopods; some with tests; unicellular |
| Opisthokonts | Choanoflagellates, animals, nucleariids, fungi | Some with flagella; unicellular and colonial |

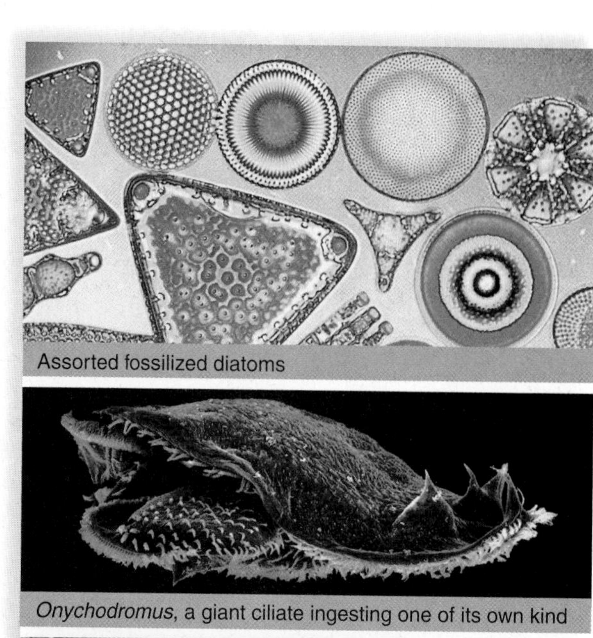

Assorted fossilized diatoms

*Onychodromus*, a giant ciliate ingesting one of its own kind

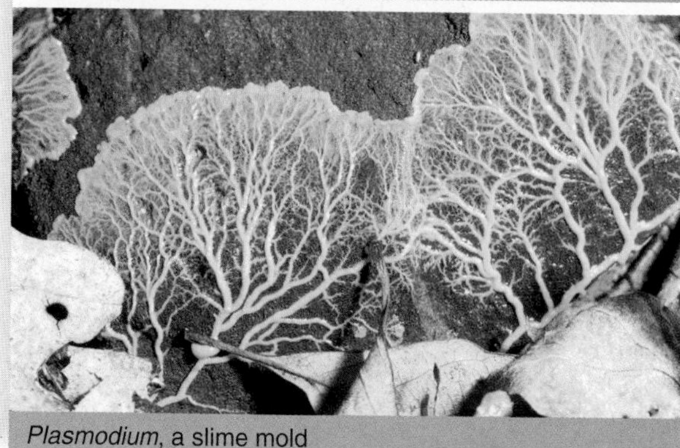

*Plasmodium*, a slime mold

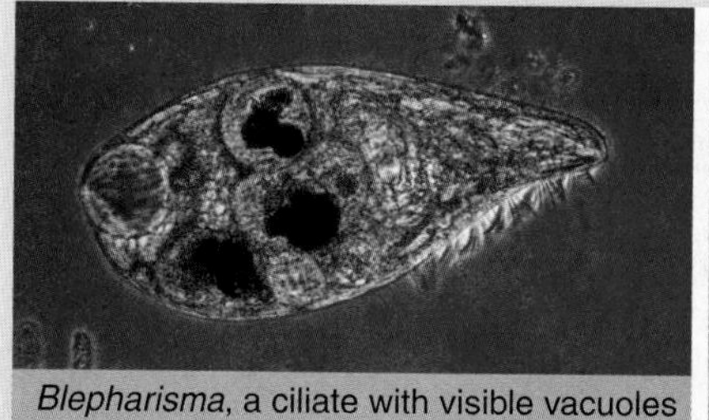

*Blepharisma*, a ciliate with visible vacuoles

*Licmorpha*, a stalked diatom

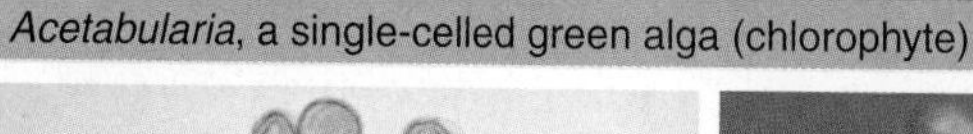

*Acetabularia*, a single-celled green alga (chlorophyte)

*Nonionina*, a foraminiferan

*Ceratium*, an armored dinoflagellate

*Bossiella*, a coralline red alga

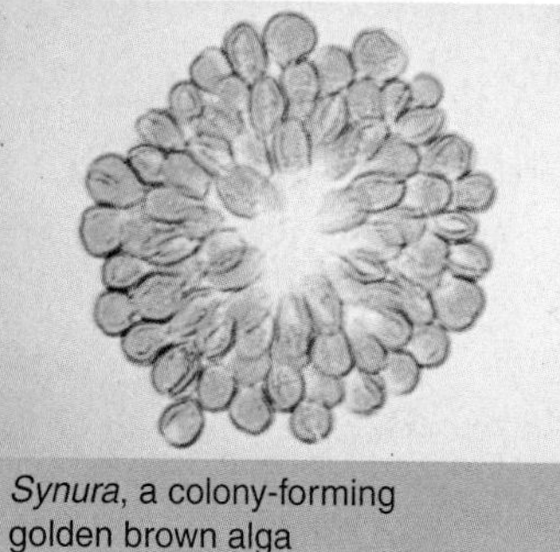

*Synura*, a colony-forming golden brown alga

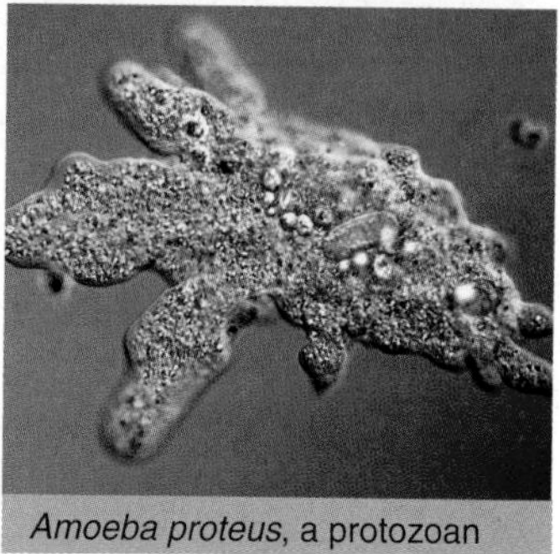

*Amoeba proteus*, a protozoan

**FIGURE 21.2** **Protist diversity.**

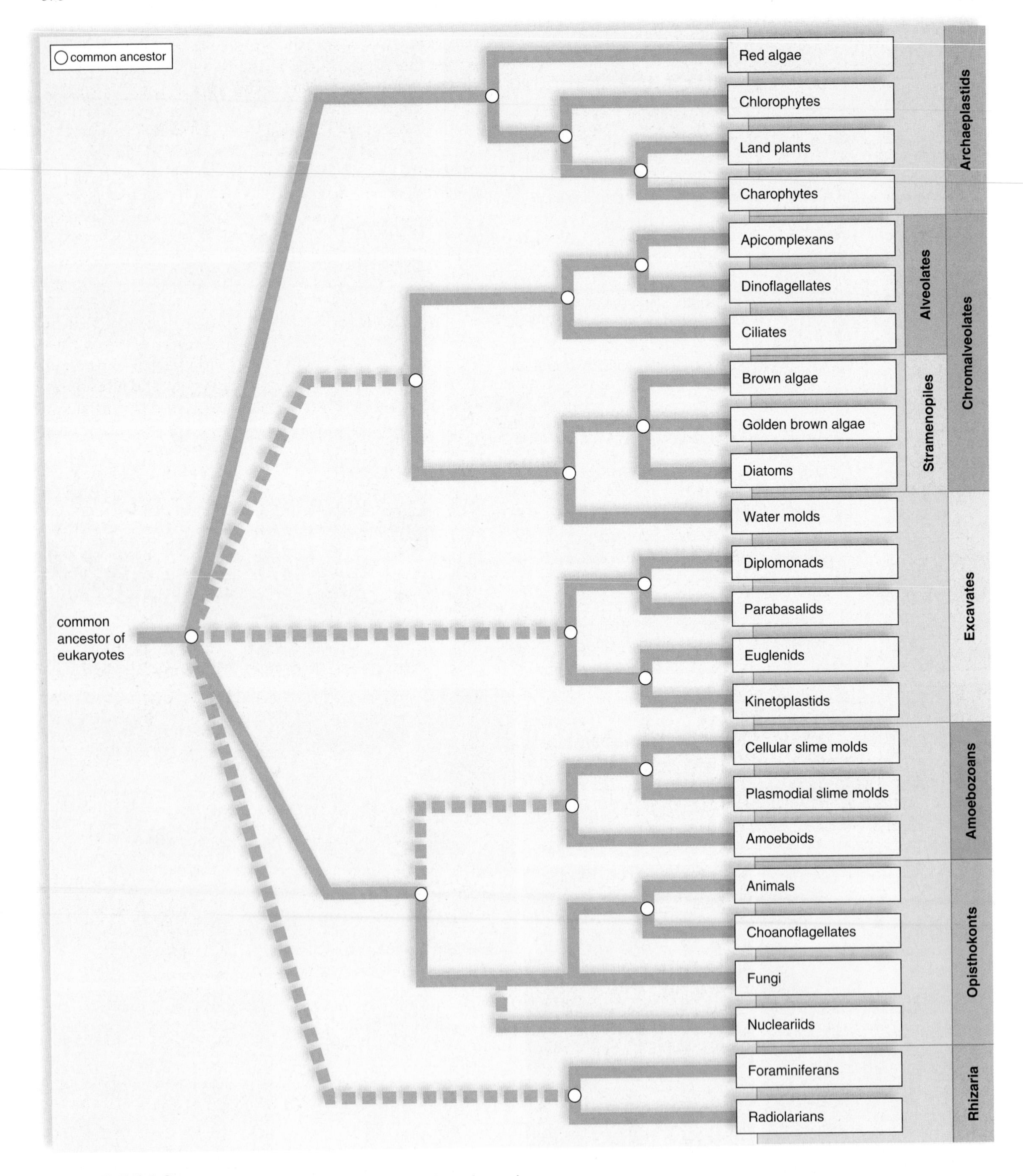


**FIGURE 21.3 Evolutionary relationships between the eukaryotic supergroups.**

Molecular data are used to determine the relatedness of the supergroups and their constituents. The dashed lines indicate relationships that are not certain at this time. This is a simplified tree that does not include all members of each supergroup.

# 21.2 Diversity of Protists

## Supergroup Archaeplastids

The **archaeplastids** (ARE key PLAS tids) [Gk. *archeos*, ancient, *plastikos*, moldable] include land plants and other photosynthetic organisms, such as green and red algae that have plastids derived from endosymbiotic cyanobacteria (see Fig. 21.1*b*).

### Green Algae

The **green algae** are protists that contain both chlorophylls *a* and *b*. They inhabit a variety of environments, including oceans, freshwater environments, snowbanks, the bark of trees, and the backs of turtles. The green algae, which number 17,000 species, also form symbiotic relationships with fungi, plants, and animals. As discussed in Chapter 22, they associate with fungi in lichens. Green algae occur in an abundant variety of forms. The majority of green algae are unicellular; however, filamentous and colonial forms exist. Some multicellular green algae are seaweeds that resemble lettuce leaves. Despite the name, green algae are not always green; some possess additional pigments that give them an orange, red, or rust color.

Biologists have long suggested that land plants are closely related to the green algae because, in addition to chlorophylls *a* and *b*, both land plants and green algae have a cell wall that contains cellulose and store reserve food as starch. Based on molecular data, the green algae may be subdivided into the **chlorophytes** and the **charophytes.** As discussed in Chapter 23, the latter is thought to be the green algae group most closely related to land plants.

**Chlorophytes.** *Chlamydomonas* is a minute, actively moving chlorophyte that inhabits still, freshwater pools. Its fossil ancestors date back over a billion years. Because the alga *Chlamydomonas* is less than 25 μm long, its anatomy is best seen in an electron micrograph (Fig. 21.4). It has a definite cell wall and a single, large, cup-shaped chloroplast that contains a *pyrenoid,* a dense body where starch is synthesized. In many species, a bright red eyespot is sensitive to light and helps bring the organism to locations favorable to photosynthesis. Two long, whiplike flagella projecting from the anterior end operate with a breaststroke motion.

*Chlamydomonas* most often reproduces asexually; mitosis produces as many as 16 daughter cells still within the parent cell wall (Fig. 21.5). Each daughter cell then secretes a cell wall and acquires flagella. The daughter cells escape by secreting an enzyme that digests the parent cell wall.

*Chlamydomonas* occasionally reproduces sexually when growth conditions are unfavorable. Gametes of two different mating types come into contact and join to form a zygote. A heavy wall forms around the zygote, and it becomes a resistant zygospore that undergoes a period of dormancy. When a zygospore germinates, it produces four zoospores by meiosis. **Zoospores,** which are flagellated spores typical of aquatic species, grow to become the adult.

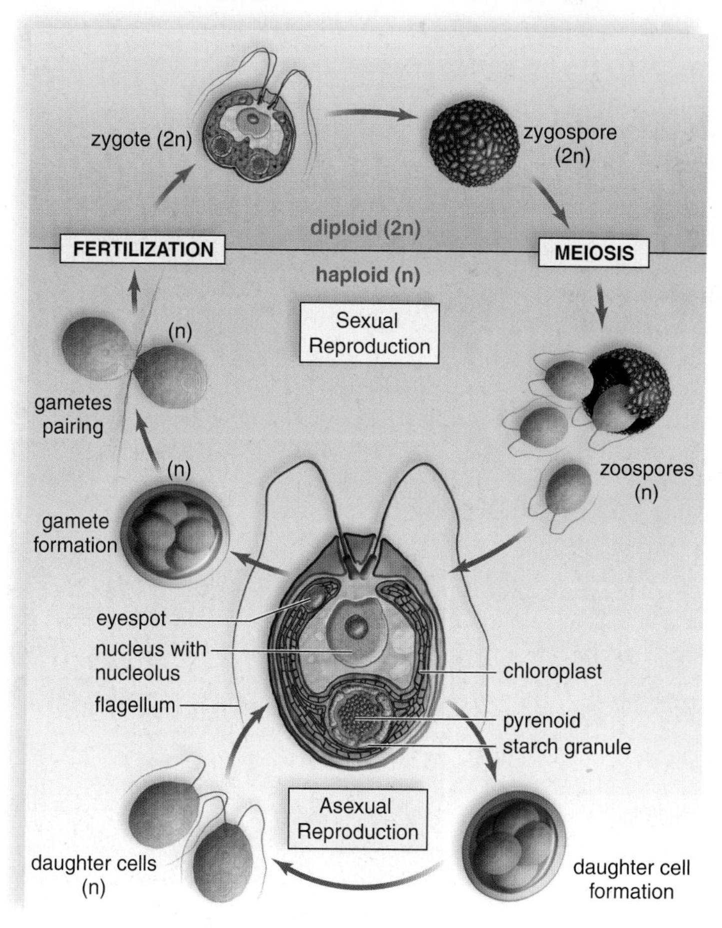


**FIGURE 21.5 Reproduction in *Chlamydomonas*.**

*Chlamydomonas* is motile because it has flagella. During asexual reproduction, all structures are haploid; during sexual reproduction, meiosis follows the zygospore stage, which is the only diploid part of the cycle.

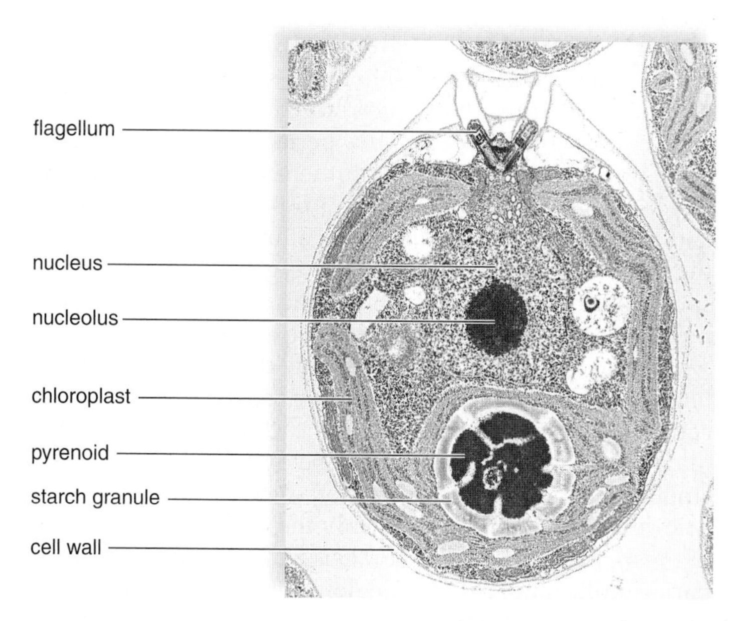


**FIGURE 21.4 Electron micrograph of *Chlamydomonas*.**

*Chlamydomonas* is a microscopic unicellular chlorophyte.

A number of colonial forms occur among the flagellated chlorophytes. *Volvox* is a well-known colonial green alga. A **colony** is a loose association of independent cells. A *Volvox* colony is a hollow sphere with thousands of cells arranged in a single layer surrounding a watery interior. Each cell of a *Volvox* colony resembles a *Chlamydomonas* cell—perhaps it is derived from daughter cells that fail to separate following zoospore formation. In *Volvox,* the cells cooperate in that the flagella beat in a coordinated fashion. Some cells are specialized for reproduction, and each of these can divide asexually to form a new daughter colony (Fig. 21.6). This daughter colony resides for a time within the parent colony, but then it leaves by releasing an enzyme that dissolves away a portion of the parent colony, allowing it to escape.

*Ulva* is a multicellular chlorophyte, commonly called sea lettuce because it lives in the sea and has a leafy appearance (Fig. 21.7). The thallus (body) is two cells thick and can be as much as a meter long. *Ulva* has an alternation of generations life cycle (see Fig. 21A*b*, p. 380) like that of land plants, except that both generations look exactly alike, and the gametes all look the same.

a. *Ulva*, several individuals

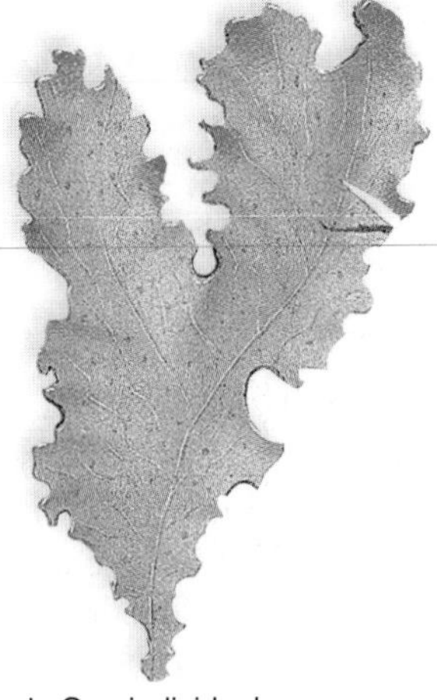

b. One individual

**FIGURE 21.7** ***Ulva.***

*Ulva* is a multicellular chlorophyte known as sea lettuce.

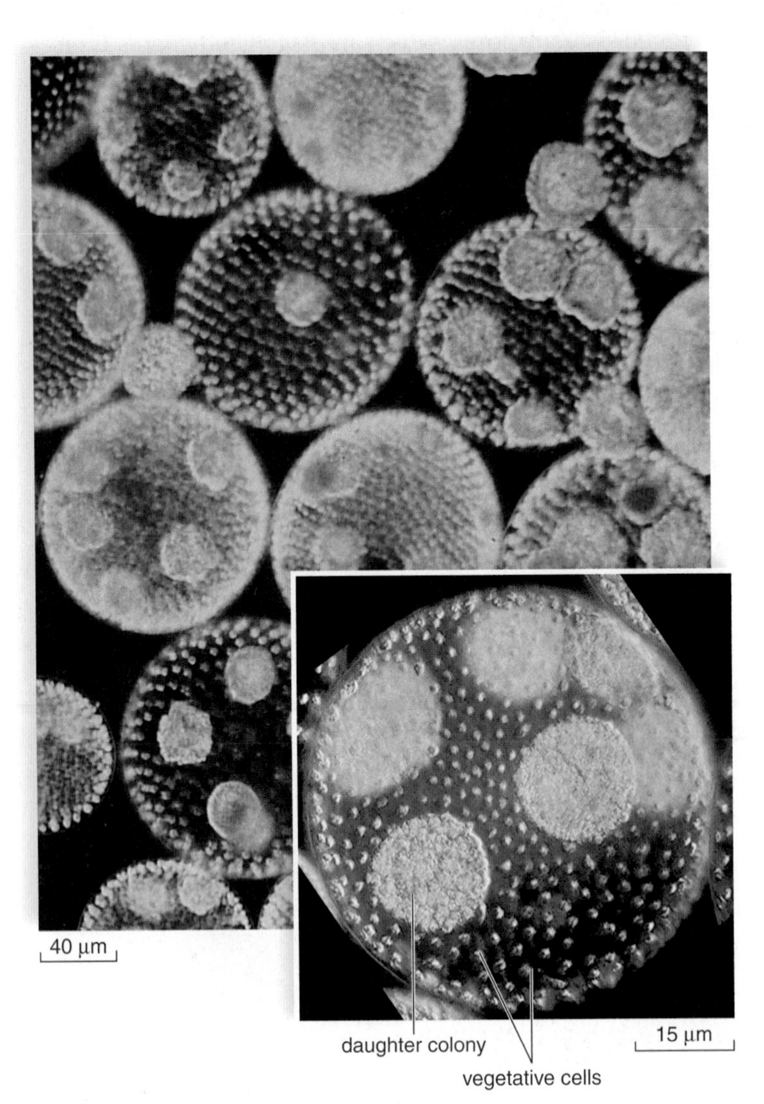


**FIGURE 21.6** ***Volvox.***

*Volvox* is a colonial chlorophyte. The adult *Volvox* colony often contains daughter colonies, which are asexually produced by special cells.

**Charophytes.** **Filaments** [L. *filum,* thread] are end-to-end chains of cells that form after cell division occurs in only one plane. The charophytes are filamentous. In some, the filaments are branched, and in others the filaments are unbranched. *Spirogyra* is an unbranched charophyte. Charophytes often grow epiphytically on (but not taking nutrients from) aquatic flowering plants; they also attach to rocks or other objects underwater. Some are suspended in the water.

*Spirogyra* is found in green masses on the surfaces of ponds and streams. It has ribbonlike, spiralled chloroplasts (Fig. 21.8). During sexual reproduction, *Spirogyra* undergoes **conjugation** [L. *conjugalis,* pertaining to marriage], a temporary union during which the cells exchange genetic material. The two filaments line up parallel to each other, and the cell contents of one filament move into the cells of the other filament, forming diploid zygotes. Resistant zygospores survive the winter, and in the spring they undergo meiosis to produce new haploid filaments.

*Chara* (Fig. 21.9) is a charophyte that lives in freshwater lakes and ponds. It is commonly called a stonewort because it is encrusted with calcium carbonate deposits. The main axis of the alga, which can be over a meter long, is a single file of very long cells anchored by rhizoids, which are colorless, hairlike filaments. Only the cell at the upper end of the main axis produces new cells. Whorls of branches occur at multicellular nodes, regions between the giant cells of the main axis. Each of the branches is also a single file of cells (Fig. 21.9*b*).

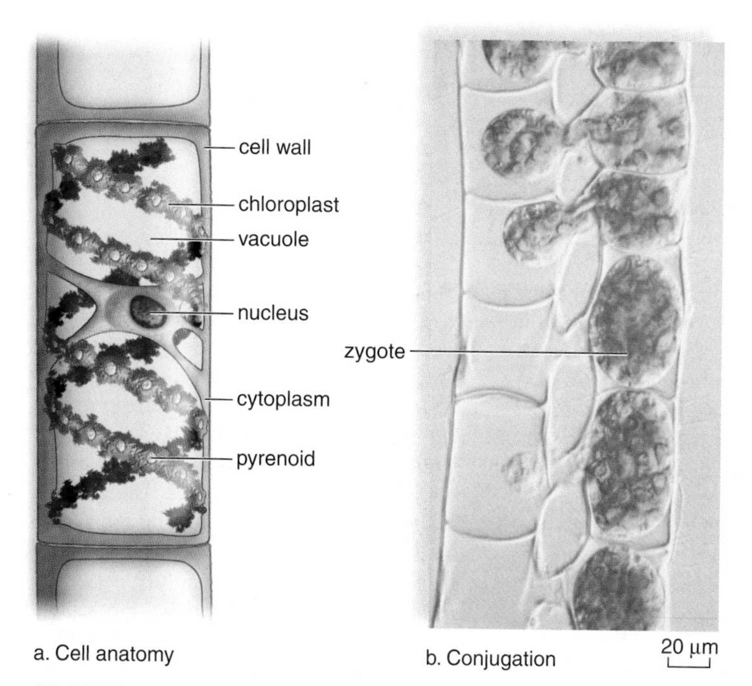


**FIGURE 21.8** ***Spirogyra.***

**a.** *Spirogyra* is an unbranched charophyte in which each cell has a ribbonlike chloroplast. **b.** During conjugation, the cell contents of one filament enter the cells of another filament. Zygote formation follows.

Male and female multicellular reproductive structures grow at the nodes, and in some species they occur on separate individuals. The male structure produces flagellated sperm, and the female structure produces a single egg. The zygote is retained until it is enclosed by tough walls. DNA sequencing data suggest that among green algae, the stoneworts are most closely related to land plants, as is thoroughly discussed on pages 414–15.

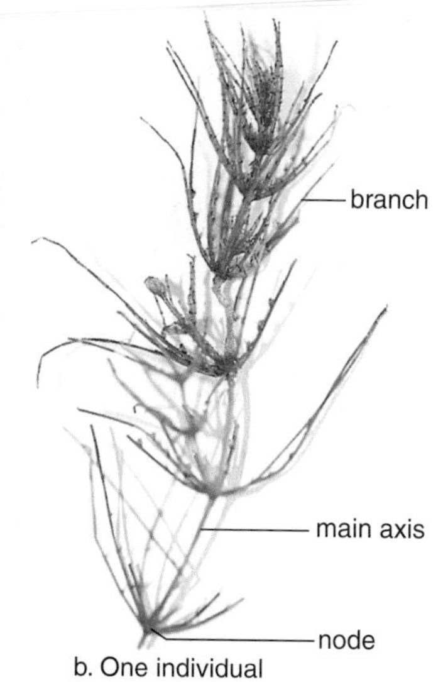


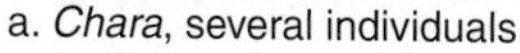

a. *Chara*, several individuals b. One individual

**FIGURE 21.9** ***Chara.***

*Chara* is an example of a stonewort, a charophyte that shares a common ancestor with land plants.

**FIGURE 21.10** **Red alga.**

Red algae are multicellular seaweeds, represented by *Rhodoglossum affine*.

## *Red Algae*

The **red algae** are multicellular seaweeds that possess a red pigment (phycoerythrin) and a blue pigment (phycocyanin) in addition to chlorophyll (Fig. 21.10). These algae live primarily in warm seawater, growing in both shallow and deep waters. Some grow at depths exceeding 70 m, where their unique pigments allow them to absorb the few light rays that penetrate so deep.

The more than 5,000 species are quite variable; most are usually much smaller and more delicate than brown algae, but some species can exceed a meter in length. Red algae can be filamentous but more often they are complexly branched, with the branches having a feathery, flat, or expanded, ribbonlike appearance. Coralline algae are red algae that have cell walls impregnated with calcium carbonate. In some instances, coralline algae contribute as much to the growth of coral reefs as do coral animals.

Red algae are economically important. Agar is a gelatinlike product made primarily from the algae *Gelidium* and *Gracilaria*. Agar is used commercially to make capsules for vitamins and drugs, as a material for making dental impressions, and as a base for cosmetics. In the laboratory, agar is a solidifying agent for a bacterial culture medium. When purified, it becomes the gel for electrophoresis, a procedure that separates proteins or nucleotides. Agar is also used in food preparation—as an antidrying agent for baked goods and to make jellies and desserts set rapidly. Carrageenin, extracted from various red algae, is an emulsifying agent for the production of chocolate and cosmetics. *Porphyra*, a red alga, is the basis of a billion-dollar aquaculture industry in Japan. The reddish-black wrappings around sushi rolls consist of processed *Porphyra* blades.

### Check Your Progress 21.2A

1. What feature of the haploid life cycle causes the adult *Chlamydomonas* to be haploid?
2. Of the archaeplastids studied, which are multicellular?

*science focus*

## Life Cycles Among the Algae

Both asexual and sexual reproduction occur in algae, depending on species and environmental conditions. The types of life cycles seen in algae occur in other protists, and also in plants or animals.

Asexual reproduction is a frequent mode of reproduction among protists when the environment is favorable to growth. Asexual reproduction requires only one parent. The offspring are identical to this parent because the offspring receive a copy of only this parent's genes. The new individuals are likely to survive and flourish if the environment is steady. Various modes of asexual reproduction occur, but growth alone produces a new individual.

Sexual reproduction, with its genetic recombination due in part to fertilization and independent assortment of chromosomes, is more likely to occur among protists when the environment is changing and is unfavorable to growth. Recombination of genes might produce individuals that are more likely to survive extremes in the environment—such as high or low temperatures, acidic or basic pH, or the lack of a particular nutrient.

Sexual reproduction requires two parents, each of which contributes chromosomes (genes) to the offspring by way of gametes. The gametes fuse to produce a diploid zygote. A reproductive cycle is isogamous when the gametes look alike—called isogametes—and a cycle is oogamous when the gametes are dissimilar—called heterogametes. Usually a small flagellated sperm fertilizes a large egg with plentiful cytoplasm.

Meiosis occurs during sexual reproduction—just *when* it occurs makes the sexual life cycles diagrammed in Figure 21A differ from one another. In these diagrams, the diploid phase is shown in blue, and the haploid phase is shown in tan. The haploid life cycle (Fig. 21A*a*) most likely evolved first. In the haploid cycle, the zygote divides by meiosis to form haploid spores that develop into haploid individuals. In algae, the spores are typically zoospores. The zygote is only the diploid stage in this life cycle, and the haploid individual gives rise to gametes. This life cycle is seen in *Chlamydomonas* and a number of other algae.

In alternation of generations, the sporophyte (2n) produces haploid spores by meiosis (Fig. 21A*b*). A spore develops into a haploid gametophyte that produces gametes. The gametes fuse to form a diploid zygote, and the zygote develops into the sporophyte. This life cycle is characteristic of some algae (e.g., *Ulva* and *Laminaria*) and all plants. In *Ulva*, the haploid and diploid generations have the same appearance. In plants, they are noticeably different from each other.

In the diploid life cycle, typical of animals, a diploid individual produces gametes by meiosis (Fig. 21A*c*). Gametes are the only haploid stage in this cycle. They fuse to form a zygote that develops into the diploid individual. This life cycle is rare in algae but does occur in a few species of the brown alga *Fucus*.

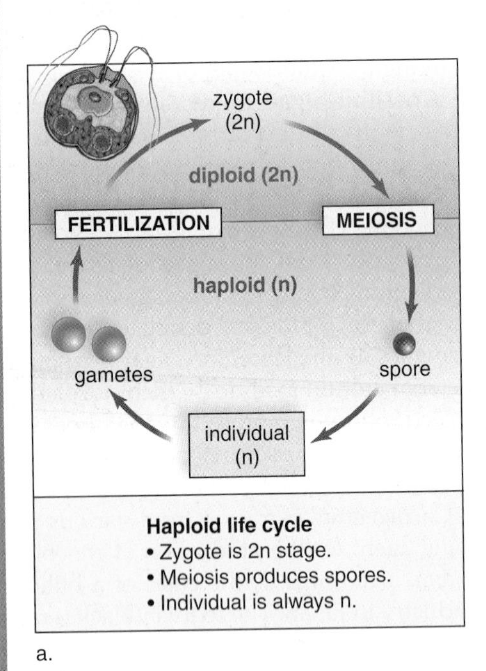


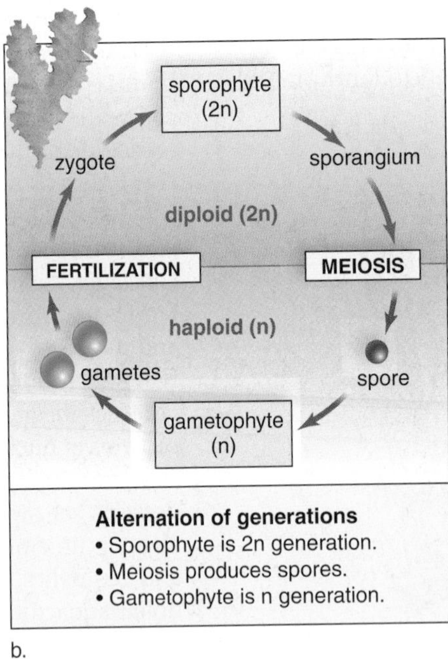


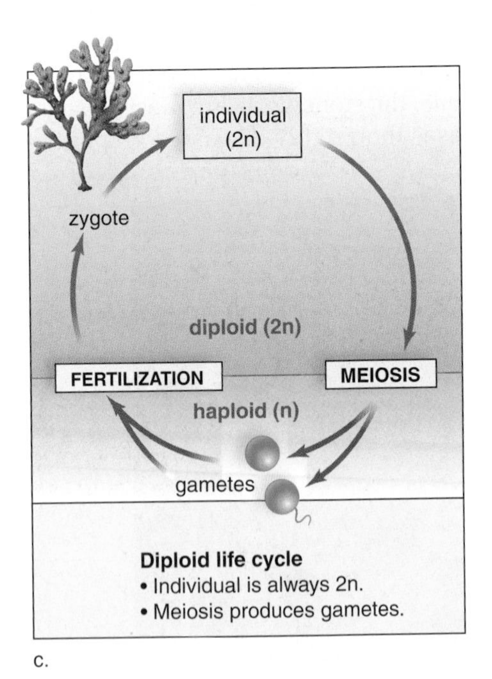


**FIGURE 21A Common life cycles in sexual reproduction.**

## Supergroup Chromalveolates

The **chromalveolates** [Gk. *chroma,* color, L. *alveolus,* hollow] include two large subgroups: the stramenopiles and the alveolates.

### *Stramenopiles*

The **stramenopiles** either have flagella or are descended from an ancestor that had flagella, one of which is longer than the other and covered with hairlike projections.

**Brown Algae.** The **brown algae** are stramenopiles that have chlorophylls *a* and *c* in their chloroplasts and a type of carotenoid pigment (fucoxanthin) that gives them their characteristic color. The reserve food is stored as a carbohydrate called *laminarin.* The brown algae range from small forms with simple filaments to large multicellular forms that may reach 100 m in length (Fig. 21.11). The vast majority of the 1,500 species live in cold ocean waters.

The multicellular brown algae are seaweeds often observed along the rocky coasts in the north temperate zone. They are pounded by waves as the tide comes in and are exposed to dry air as the tide goes out. They dry out slowly, however, because their cell walls contain a mucilaginous, water-retaining material.

Both *Laminaria,* commonly called *kelp,* and *Fucus,* known as rockweed, are examples of brown algae that grow along the shoreline. In deeper waters, the giant kelps (*Macrocystis* and *Nereocystis*) often grow extensively in vast beds. Individuals of the genus *Sargassum* sometimes break off from their holdfasts and form floating masses. Brown algae not only provide food and habitat for marine organisms, they are harvested for human food and for fertilizer in several parts of the world. *Macrocystis* is the source of algin, a pectinlike material that is added to ice cream, sherbet, cream cheese, and other products to give them a stable, smooth consistency.

*Laminaria* is unique among the protists because members of this genus show tissue differentiation—that is, they transport organic nutrients by way of a tissue that resembles phloem in land plants. Most brown algae have the alternation of generations life cycle, but some species of *Fucus* are unique in that meiosis produces gametes, and the adult is always diploid, as in animals.

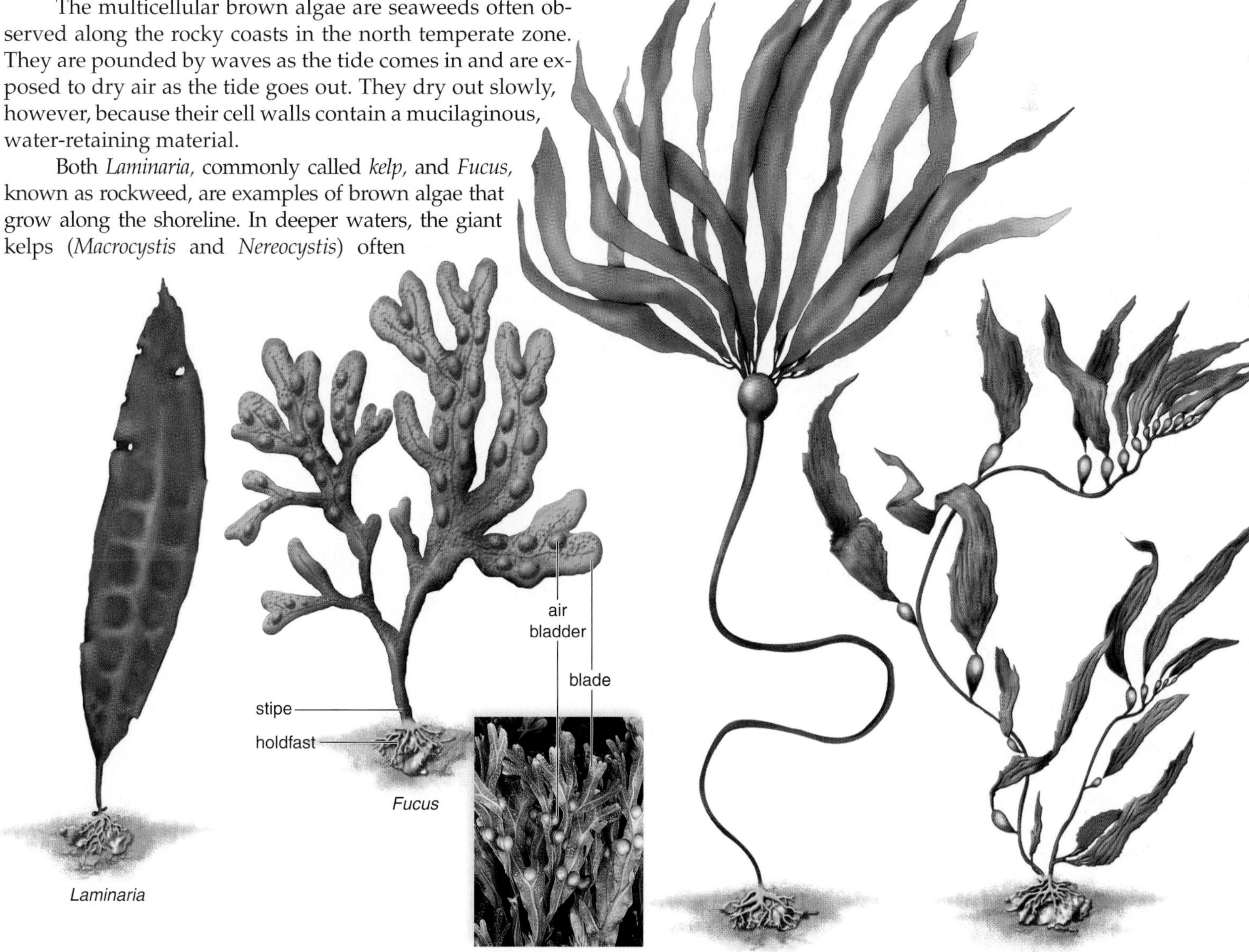


**FIGURE 21.11 Brown algae.**

*Laminaria* and *Fucus* are seaweeds known as kelps. They live along rocky coasts of the north temperate zone. The other brown algae featured, *Nereocystis* and *Macrocystis,* form spectacular underwater "forests" at sea.

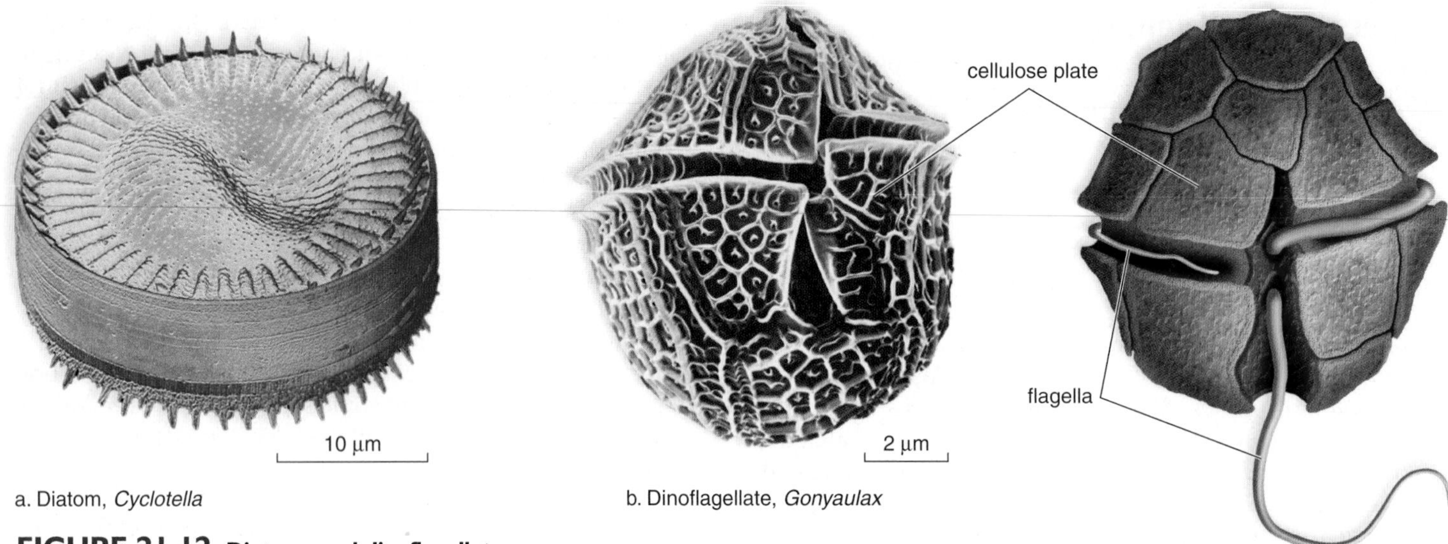


**FIGURE 21.12 Diatoms and dinoflagellates.**

**a.** Diatoms are stramenopiles of various colors, but even so their chloroplasts contain a unique golden brown pigment (*fucoxanthin*), in addition to chlorophylls *a* and *c*. The beautiful pattern results from markings on the silica-embedded wall. **b.** Dinoflagellates are alveolates, such as *Gonyaulax*, with cellulose plates.

**Diatoms.** A **diatom** [Gk. *dia,* through, and *temno,* cut] is a stramenopile that has an ornate silica shell resembling a petri dish—a top shell half, called a valve, fits over a lower valve (Fig. 21.12*a*). Diatoms have an orange-yellow color because they contain a carotenoid pigment in addition to chlorophyll. The approximately 100,000 species are a significant part of plankton serving as a source of oxygen and food for heterotrophs in both freshwater and marine ecosystems.

When diatoms reproduce asexually, each receives one old valve. The new valve fits inside the old one; therefore, new diatoms are smaller than the original ones. This continues until diatoms are about 30% of the original size. Then they reproduce sexually. The zygote becomes a structure that grows and then divides mitotically to produce diatoms of normal size.

The valves are covered with a great variety of striations and markings that form beautiful patterns when observed under the microscope. These are actually depressions or pores through which the organism makes contact with the outside environment. The remains of diatoms, called diatomaceous earth, accumulate on the ocean floor and are mined for use as filtering agents, soundproofing materials, and gentle polishing abrasives, such as those found in silver polish and toothpaste.

**Golden Brown Algae.** The **golden brown algae** derive their distinctive color from yellow-brown carotenoid pigments. The cells of these unicellular or colonial protists typically have two flagella that bear tubular hairs, which is characteristic of stramenopiles. Among the 1,000 species, the cells may be naked, covered with organic or silica scales, or enclosed in a secreted cagelike structure called a lorica. Many golden brown algae, such as *Ochromonas* (Fig. 21.13), are mixotrophs, a term that means they are both autotrophic and heterotropic. *Ochromonas* is capable of photosynthesis as well as phagocytosis. Like diatoms, the golden brown algae contribute to freshwater and marine phytoplankton.

**Water Molds.** The **water molds** usually live in the water, where they form furry growths when they parasitize fishes or insects and decompose remains. In spite of their common name, some water molds live on land and parasitize insects and plants. Nearly 700 species of water molds have been described. A water mold, *Phytophthora infestans,* was responsible for the 1840s potato famine in Ireland, and another, *Plasmopara viticola,* for the downy mildew of grapes that ravaged the vineyards of France in the 1870s. However, most

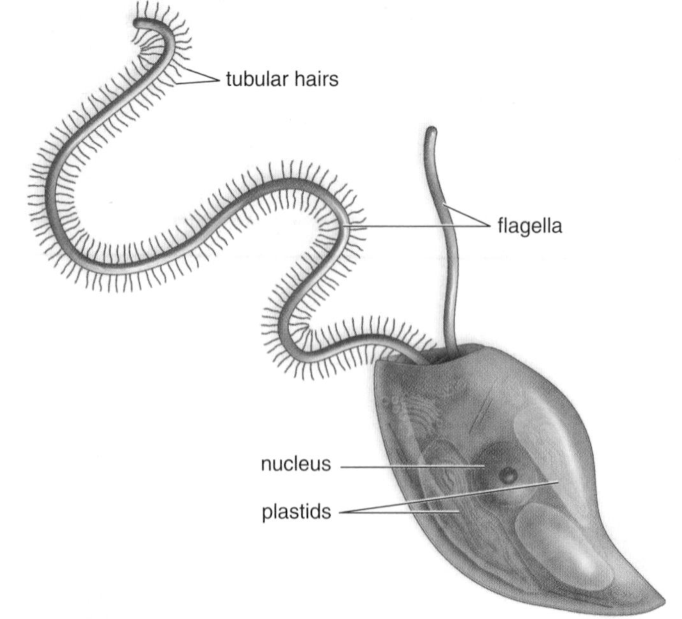


**FIGURE 21.13** ***Ochromonas*, a golden brown alga.**

Golden brown algae have a type of flagella that characterizes the stramenopiles. The longer of the two flagella has rows of tubular hairs.

**FIGURE 21.14 Water mold.**

*Saprolegnia*, a water mold, feeding on a dead insect, is not a fungus.

water molds are saprotrophic and live off dead organic matter. Another well-known water mold is *Saprolegnia*, which is often seen as a cottonlike white mass on dead organisms (Fig. 21.14).

Water molds have a filamentous body as do fungi, but their cell walls are largely composed of cellulose, whereas fungi have cell walls of chitin. The life cycle of water molds differs from that of fungi. During asexual reproduction, water molds produce motile spores (2n zoospores), which are flagellated. The organism is diploid (not haploid as in the fungi), and meiosis produces gametes. Their alternate name "oomycetes" refers to enlarged tips (called oogonia) where eggs are produced.

**Check Your Progress 21.2B**

1. Brown algae, diatoms, and water molds are dissimilar in appearance and way of life. Why are they all stramenopiles?
2. Which of the stramenopiles are you apt to find living on land?

## *Alveolates*

The **alveolates** have alveoli (small sacs) lying just beneath their plasma membranes; the alveoli are thought to lend support to the cell surface. The alveolates are unicellular.

**Dinoflagellates.** The single cell of a **dinoflagellate** is usually bounded by protective cellulose plates impregnated with silicates (see Fig. 21.12*b*). Typically, they have two flagella; one lies in a longitudinal groove with its distal end free, and the other lies in a transverse groove that encircles the organism. The longitudinal flagellum acts as a rudder, and the beating of the transverse flagellum causes the cell to spin as it moves forward. The chloroplasts of a dinoflagellate vary in color from yellow-green to brown because, in addition to chlorophylls *a* and *c*, they contain carotenoids.

The approximately 4,000 species are diverse. Some species, such as *Noctiluca*, are capable of bioluminescence (producing light). Being a part of the plankton, the dinoflagellates are an important source of food for small animals in the ocean. They also live within the bodies of some invertebrates as symbionts. Symbiotic dinoflagellates lack cellulose plates and flagella and are called zooxanthellae. Corals (see Chapter 28), members of the animal kingdom, usually contain large numbers of zooxanthellae. They provide their hosts with organic nutrients, and the corals provide wastes that fertilize the algae. Photosynthetic dinoflagellates are often mixotrophs that can take in available food particles by phagocytosis.

Like the diatoms, dinoflagellates are one of the most important groups of producers in marine environments. Occasionally, however, they undergo a population explosion and become more numerous than usual. At these times, their density can equal 30,000 in a single milliliter. When dinoflagellates such as *Gymnodinium brevis* increase in number, they may cause a phenomenon called **red tide.** Massive fish kills can occur as the result of a powerful neurotoxin produced by these dinoflagellates (Fig. 21.15*a*). Humans who consume shellfish that have fed during a *Gymnodinium* outbreak may suffer from paralytic shellfish poisoning, in which the respiratory organs are paralyzed.

a.

b.

**FIGURE 21.15 Fish kill and dinoflagellate bloom.**

**a.** Fish kills, such as this one in University Lake in Baton Rouge, Louisiana, can be the result of a dinoflagellate bloom. **b.** This bloom, often called a red tide after the color of the water, appeared near California's central coast.

Dinoflagellates usually reproduce asexually using mitosis and longitudinal division of the cell. A daughter cell gets half the cellulose plates unless they were shed before reproduction began. During sexual reproduction the daughter cells act as gametes. The zygote enters a resting stage broken when meiosis produces only one haploid cell because the others disintegrate.

**Ciliates.** The **ciliates** are unicellular protists that move by means of cilia. They are the most structurally complex and specialized of all protozoa. Members of the genus *Paramecium* are classic examples of ciliates (Fig. 21.16*a*). Hundreds of cilia, which beat in a coordinated rhythmic manner, project through tiny holes in a semirigid outer covering, or pellicle. Numerous oval capsules lying in the cytoplasm just beneath the pellicle contain **trichocysts.** Upon mechanical or chemical stimulation, trichocysts discharge long, barbed threads that are useful for defense and for capturing prey. Some trichocysts release poisons.

Most ciliates ingest their food. For example, when a paramecium feeds, food particles are swept down a gullet, below which food vacuoles form. Following digestion, the soluble nutrients are absorbed by the cytoplasm, and the nondigestible residue is eliminated at the anal pore.

During asexual reproduction, ciliates divide by transverse binary fission. Ciliates have two types of nuclei: a large macronucleus and one or more small micronuclei. The macronucleus controls the normal metabolism of the cell, while the micronuclei are concerned with reproduction. Sexual reproduction involves conjugation (Fig. 21.16*b*). The macronucleus disintegrates and the micronuclei undergo meiosis. Two ciliates exchange haploid micronuclei. Then the micronuclei give rise to a new macronucleus, which contains copies of only certain housekeeping genes.

The ciliates are a diverse group of protozoans which range in size from 10 to 3,000 μm and number approximately 8,000 species. The majority of ciliates are free-living; however, several parasitic, sessile, and colonial forms exist. The barrel-shaped didiniums expand to consume paramecia much larger than themselves. *Suctoria* have an even more dramatic way of getting food. They rest quietly on a stalk until a hapless victim comes along. Then they promptly paralyze it and use their tentacles like straws to suck it dry. *Stentor* may be the most elaborate ciliate, resembling a giant blue vase decorated with stripes (Fig. 21.16*c*). *Ichthyophthirius* is responsible for a common disease in fishes called "ick." If left untreated, it can be fatal.

**Apicomplexans.** The **apicomplexans,** also known as *sporozoans,* are nearly 3,900 species of nonmotile, parasitic, sporeforming protozoans. All apicomplexans are parasites; the "apical complex" for which they are named is a unique collection of

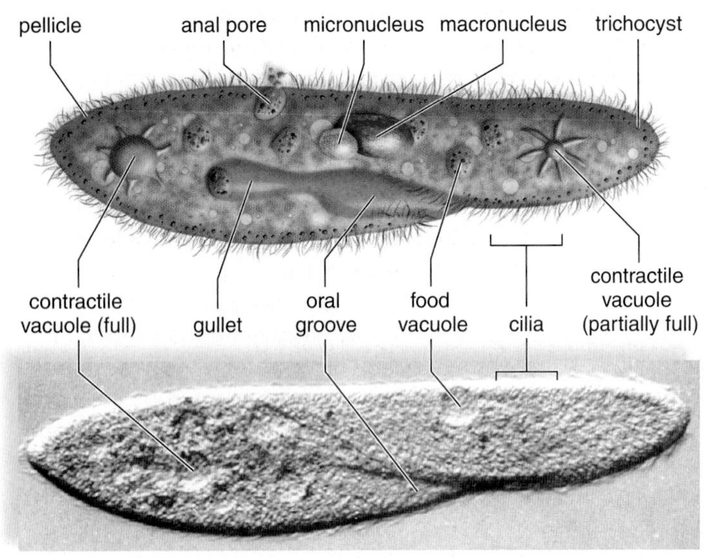


a. *Paramecium*

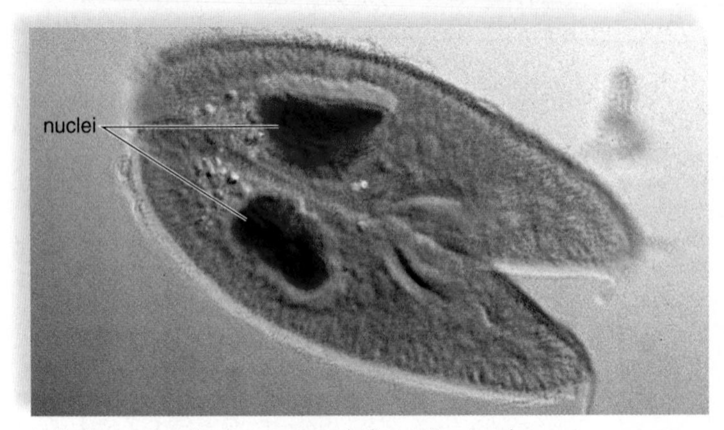


b. During conjugation two paramecia first unite at oral areas

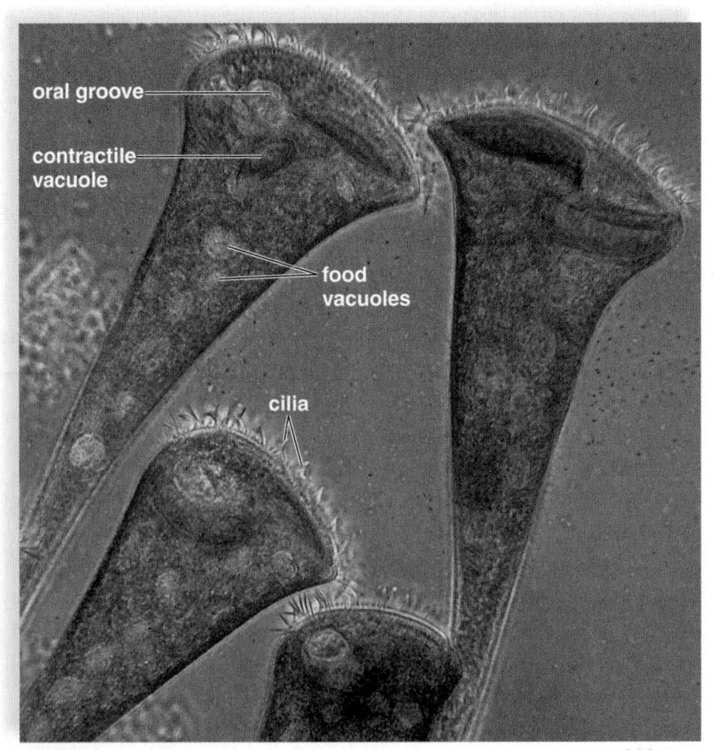


c. *Stentor*

**FIGURE 21.16 Ciliates.**

Ciliates are the most complex of the protists. **a.** Structure of *Paramecium,* adjacent to an electron micrograph. Note the oral groove, the gullet, and anal pore. **b.** A form of sexual reproduction called conjugation occurs periodically. **c.** *Stentor,* a large, vase-shaped, freshwater ciliate.

organelles at one end of the infective stage. Many apicomplexans have multiple hosts.

According to the CDC (Centers for Disease Control and Prevention), there are approximately 350–500 million cases of malaria worldwide; every year, more than a million people die of the infection. In humans, malaria is caused by four distinct members of the genus *Plasmodium*. *Plasmodium vivax*, the cause of one type of malaria, is the most common human parasite. The life cycle alternates between a sexual and an asexual phase in different hosts. Female *Anopheles* mosquitoes acquire protein for production of eggs by biting humans and other animals. When a human is bitten, the parasite eventually invades the red blood cells. The chills and fever of malaria appear when the infected cells burst and release toxic substances into the blood (Fig. 21.17). Despite efforts to control the malaria parasite, as well as the mosquito vector, malaria is making a resurgence. International travel coupled with new resistant forms of both the parasite and the *Anopheles* mosquito is presenting health professionals with formidable problems.

Although the malarial parasite *Plasmodium* is the best known of the sporozoans, the apicomplexans include many examples of human parasites. *Pneumocystis carinii* causes the type of pneumonia seen primarily in AIDS patients. During sexual reproduction, thick-walled cysts form in the lining of pulmonary air sacs. The cysts contain spores that successively divide until the cyst bursts and the spores are released. Each spore becomes a new mature organism that can reproduce asexually but may also enter the sexual stage and form cysts. *Toxoplasma gondii*, another apicomplexan, causes toxoplasmosis, particularly in cats but also in people. In pregnant women, the parasite can infect the fetus and cause birth defects and mental retardation; in AIDS patients, it can infect the brain and cause neurological symptoms.

### Check Your Progress 21.2C

1. Of the alveolates studied, which can locomote (by what means) and which cannot locomote?
2. How do dinoflagellates differ from those typically designated as protozoans?

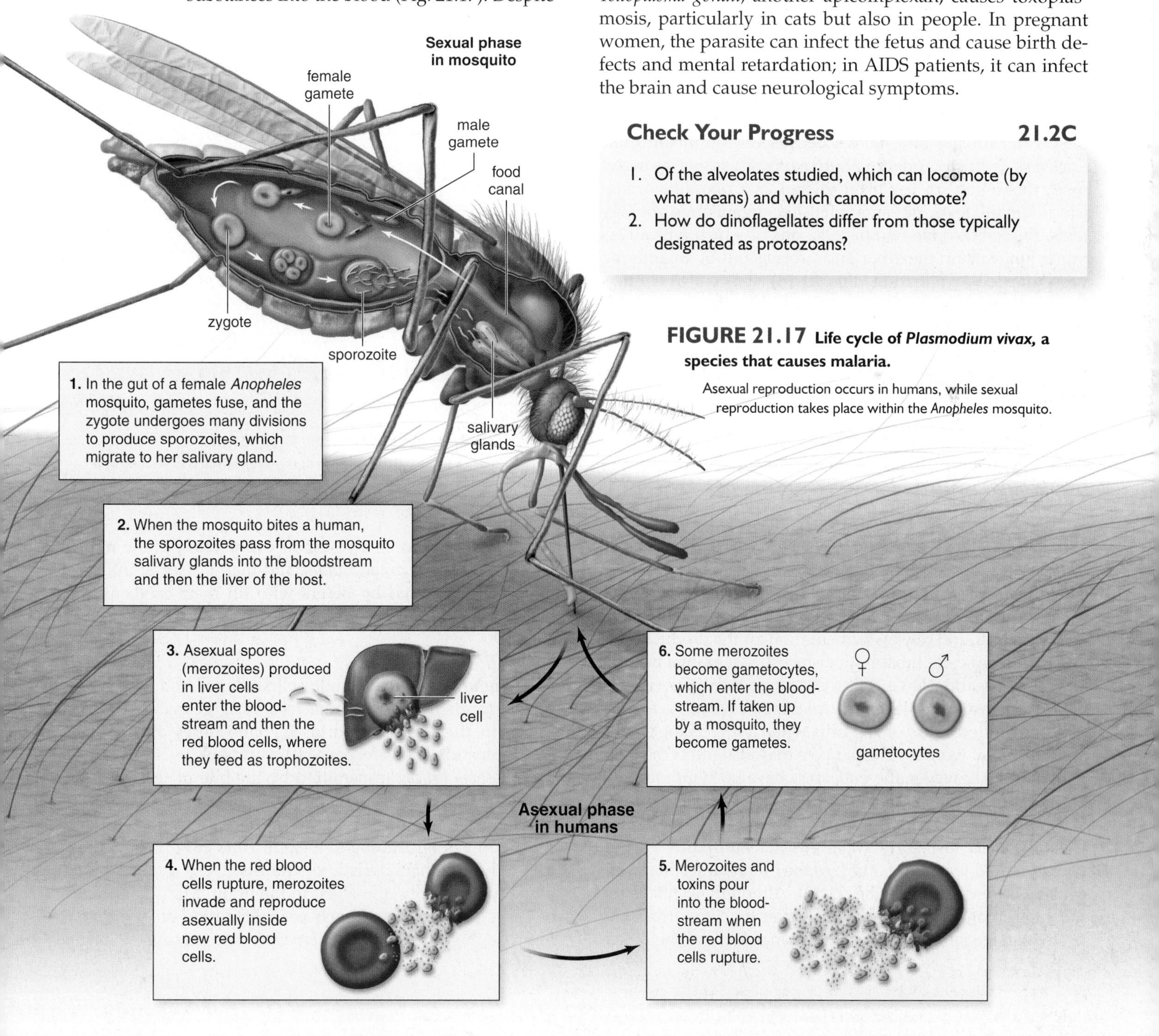

**FIGURE 21.17 Life cycle of *Plasmodium vivax*, a species that causes malaria.**

Asexual reproduction occurs in humans, while sexual reproduction takes place within the *Anopheles* mosquito.

## Supergroup Excavates

The **excavates** [L. *cavus,* hollow] include **zooflagellate** with atypical or absent mitochondria and distinctive flagella and/or deep (excavated) oral grooves.

### *Euglenids*

The **euglenids** are small (10–500 µm), freshwater unicellular organisms. Attempts at classifying the approximately 1,000 species of euglenids typify the problem of classifying excavates, and indeed protists in general. One-third of all genera have chloroplasts; the rest do not. Those that lack chloroplasts ingest or absorb their food. This may not be surprising when one knows that their chloroplasts are like those of green algae and are probably derived from them through endosymbiosis. The chloroplasts are surrounded by three rather than two membranes. A pyrenoid is a special region of the chloroplast where carbohydrate formation occurs. Euglenids produce an unusual type of carbohydrate called paramylon.

A common euglenid is *Euglena deces,* an inhabitant of freshwater ditches and ponds. Because of the mixotrophic *Euglena's* ability to undergo photosynthesis, as well as ingest food, this organism has been treated as a plantlike organism in botany texts and an animal-like organism in zoology texts. Euglenids have two flagella, one of which typically is much longer than the other and projects out of an anterior, vase-shaped invagination (Fig. 21.18). It is called a tinsel flagellum because it has hairs on it. Near the base of this flagellum is an eyespot, which shades a photoreceptor for detecting light. Because euglenids are bounded by a flexible *pellicle* composed of protein bands lying side by side, they can assume different shapes as the underlying cytoplasm undulates and contracts. As in certain other protists, there is a contractile vacuole for ridding the body of excess water. Euglenids reproduce by longitudinal cell division, and sexual reproduction is not known to occur.

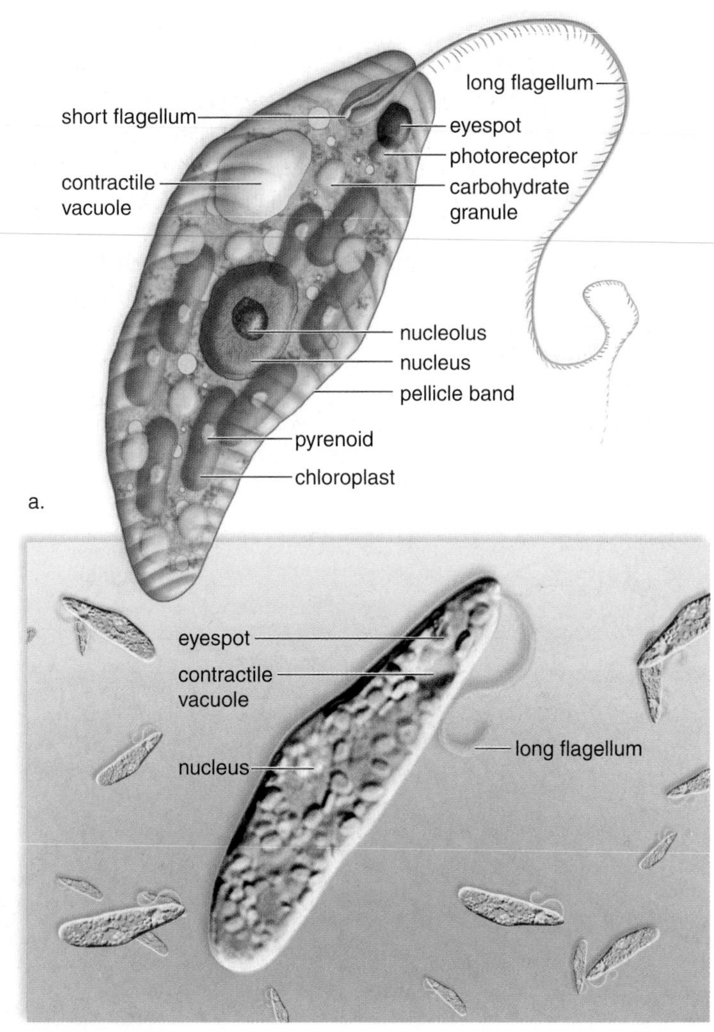


**FIGURE 21.18** ***Euglena.***

**a.** In *Euglena*, a very long flagellum propels the body, which is enveloped by a flexible pellicle. **b.** Micrograph of several specimens.

### *Parabasalids and Diplomonads*

Parabasalids and diplomonads are excavates that are single-celled, flagellated protozoans able to survive in environments with very low oxygen levels. This is because they lack mitochondria, and rely on fermentation for the production of ATP. They typically inhabit inner recesses of animal hosts.

**Parabasalids** have a fibrous connection between the Golgi apparatus and flagella. *Trichomonas vaginalis* is a sexually transmitted cause of vaginitis when it infects the vagina and urethra of women. The parasite may also infect the male genital tract; however, the male may have no symptoms.

A **diplomonad** [Gk. *diplo,* double, and *monas,* unit] cell has two nuclei and two sets of flagella. The diplomonad *Giardia lamblia* forms cysts that are transmitted by way of contaminated water and attach to the human intestinal wall, causing severe diarrhea (Fig. 21.19). Although *Giardia* is the most common flagellate of the human digestive tract, the protozoan lives in a variety of other mammals as well. Beavers are known to be a reservoir of infection in the mountains of western United States, and many cases of infection have been acquired by hikers who fill their canteens at a beaver pond.

### *Kinetoplastids*

The **kinetoplastids** are single-celled, flagellated protozoans named for their distinctive *kinetoplasts,* large masses of DNA found in their mitochondria. Trypanosomes are parasitic kinetoplastids that are passed to humans by insect bites. *Trypanosoma brucei,* transmitted by the bite of the tsetse fly, is the cause of African sleeping sickness (Fig. 21.20). The white blood cells in an infected human accumulate around the blood vessels leading to the brain and cut off circulation. The lethargy characteristic of the disease is caused by an inadequate supply of oxygen to the brain. Many thousands of cases are diagnosed each year. Fatalities or permanent brain damage are common. *Trypanosoma cruzi* causes Chagas

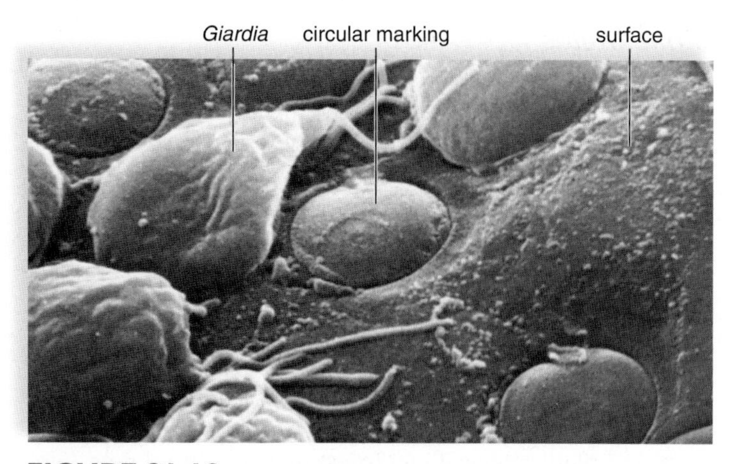


**FIGURE 21.19** ***Giardia lamblia.***

The protist adheres to any surface, including epithelial cells, by means of a sucking disk. Characteristic markings can be seen after the disk detaches.

disease in humans in Central and South America. Approximately 45,000 people die yearly from this parasite.

**Check Your Progress 21.2D**

1. Among the excavates studied, only *Euglena* is photosynthetic. How might you account for *Euglena* having chloroplasts?
2. What anatomical feature of the other excavates studied might make you think they were free-living?

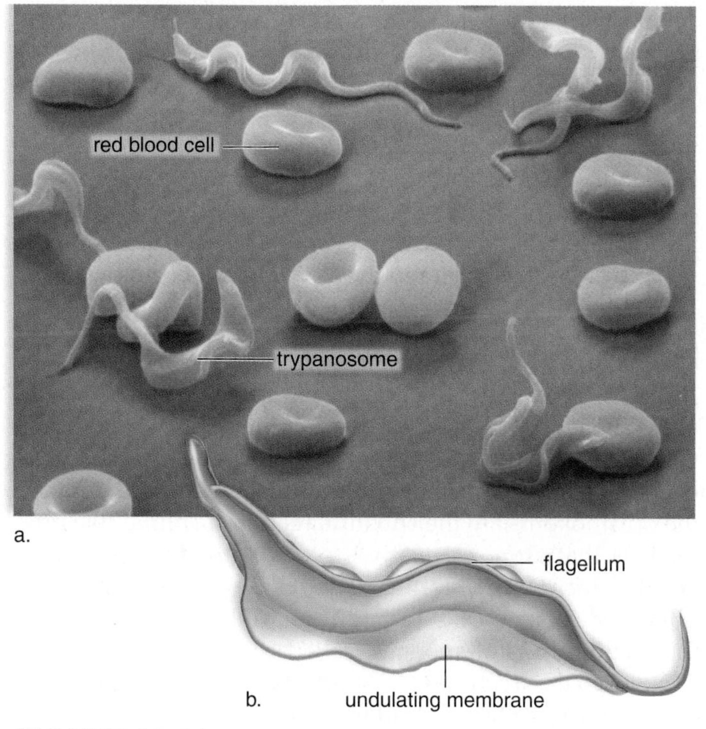


**FIGURE 21.20** ***Trypanosoma brucei.***

**a.** Micrograph of *Trypanosoma brucei,* a causal agent of African sleeping sickness, among red blood cells. **b.** The drawing shows its general structure.

## Supergroup Amoebozoans

The **amoebozoans** [Gk. *ameibein,* to change, *zoa,* animal] is comprised of protozoans that move by **pseudopods** [Gk. *pseudes,* false, and *podos,* foot], processes that form when cytoplasm streams forward in a particular direction. Amoebozoans usually live in aquatic environments, in oceans and freshwater lakes and ponds, and they are often a part of the plankton.

### *Amoeboids*

The **amoeboids** are protists that move and also ingest their food with pseudopods. Hundreds of species of amoeboids have been classified. *Amoeba proteus* is a commonly studied freshwater member of this group (Fig. 21.21). When amoeboids feed, the pseudopods surround and **phagocytize** [Gk. *phagein,* eat, and *kytos,* cell] their prey, which may be algae, bacteria, or other protists. Digestion then occurs within a *food vacuole.* Freshwater amoeboids, including *Amoeba proteus,* have *contractile vacuoles* where excess water from the cytoplasm collects before the vacuole appears to "contract," releasing the water through a temporary opening in the plasma membrane.

*Entamoeba histolytica* is a parasitic amoeboid that lives in the human large intestine and causes amoebic dysentery. The ability of the organism to form cysts makes amoebic dysentery infectious. Complications arise when this parasite invades the intestinal lining and reproduces there. If the parasites enter the body proper, liver and brain involvement can be fatal.

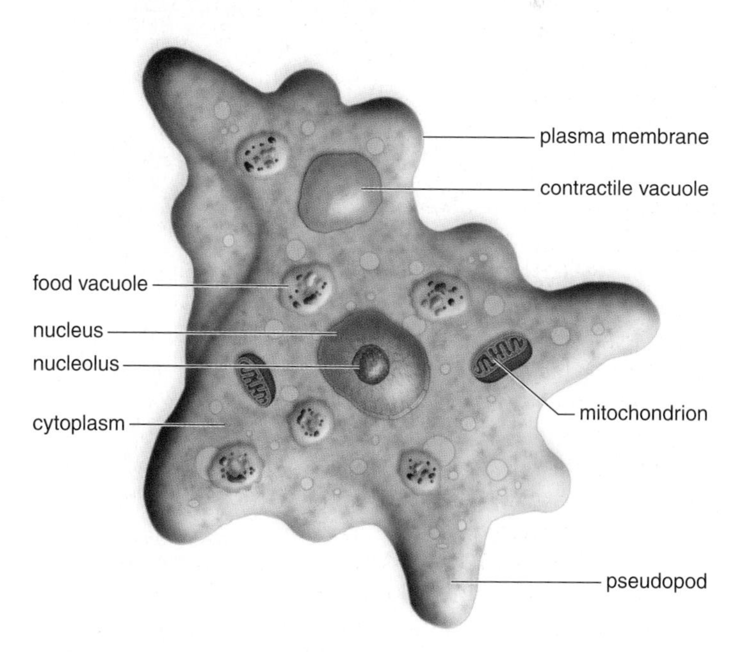


**FIGURE 21.21** ***Amoeba proteus.***

This amoeboid is common in freshwater ponds. Bacteria and other microorganisms are digested in food vacuoles, and contractile vacuoles rid the body of excess water.

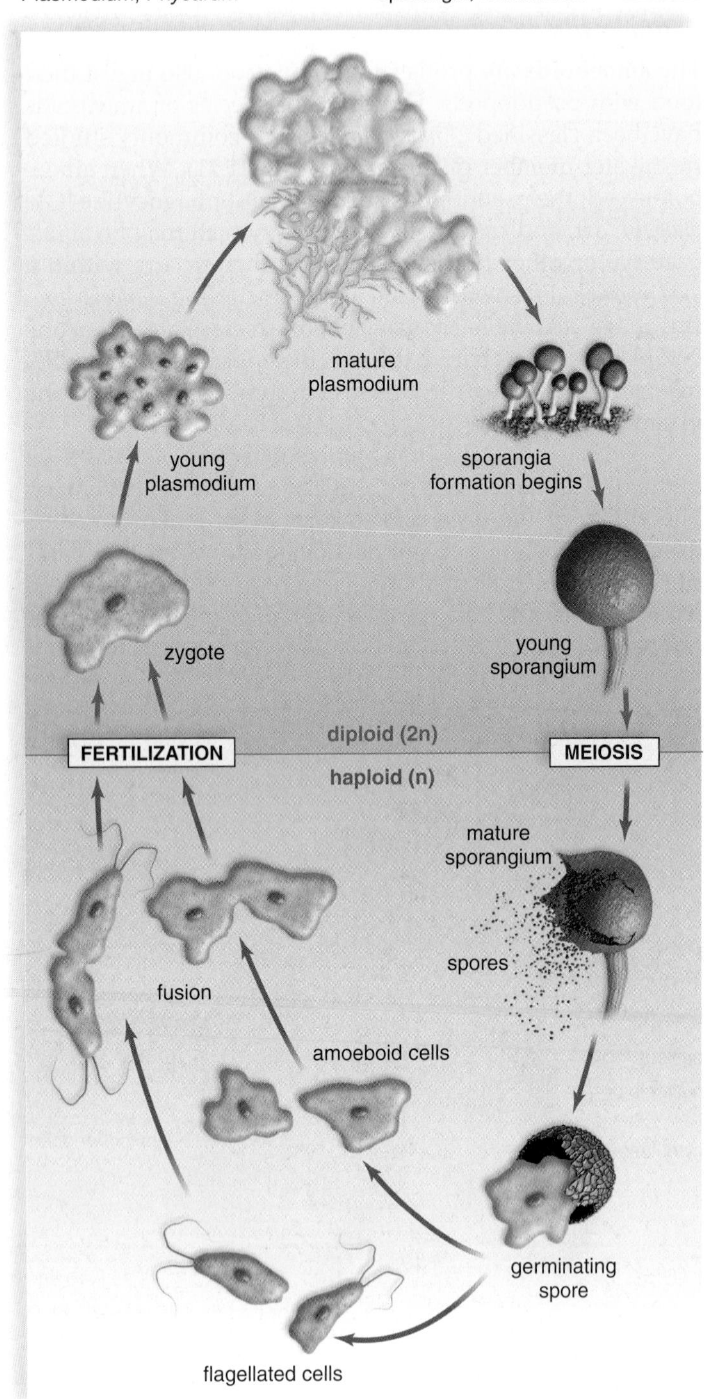


**FIGURE 21.22 Plasmodial slime molds.**

The diploid adult forms sporangia during sexual reproduction, when conditions are unfavorable to growth. Haploid spores germinate, releasing haploid amoeboid or flagellated cells that fuse.

### *Slime Molds*

In forests and woodlands, slime molds contribute to ecological balance when they phagocytize, and therefore help dispose of, dead plant material. They also feed on bacteria, keeping their population under control. Slime molds were once classified as fungi, but unlike fungi, they lack cell walls, and have flagellated cells at some time during their life cycle. The vegetative state of the slime molds is mobile and amoeboid. Slime molds produce spores by meiosis; the spores germinate to form gametes.

Usually, **plasmodial slime molds** exist as a plasmodium, a diploid, multinucleated, cytoplasmic mass enveloped by a slime sheath that creeps along, phagocytizing decaying plant material in a forest or agricultural field (Fig. 21.22). Approximately 700 species of plasmodial slime molds have been described. Many species are brightly colored. At times unfavorable to growth, such as during a drought, the plasmodium develops many sporangia. A **sporangium** [Gk. *spora*, seed, and *angeion* (dim. of *angos*), vessel] is a reproductive structure that produces spores. An aggregate of sporangia is called a fruiting body.

The spores produced by a plasmodial slime mold sporangium can survive until moisture is sufficient for them to germinate. In plasmodial slime molds, spores release a haploid flagellated cell or an amoeboid cell. Eventually, two of them fuse to form a zygote that feeds and grows, producing a multinucleated plasmodium once again.

**Cellular slime molds** are called such because they exist as individual amoeboid cells. They are common in soil, where they feed on bacteria and yeasts. Their small size prevents them from being seen. Nearly 70 species of cellular slime molds have been described.

As the food supply runs out or unfavorable environmental conditions develop, the cells release a chemical that causes them to aggregate into a pseudoplasmodium. The pseudoplasmodium stage is temporary and eventually gives rise to a fruiting body in which sporangia produce spores. When favorable conditions return, the spores germinate, releasing haploid amoeboid cells, and the asexual cycle begins again.

## Supergroup Opisthokonts

Animals and fungi are included in supergroup **opisthokonts** (op is thoe KONTs) [Gk. *opisthos*, behind, *kontos*, pole] along with several closely related protists. This supergroup includes both unicellular and multicellular protozoans. Among the opisthokonts are the **choanoflagellates,** animal-like protozoans that are near relatives of sponges. The choanoflagellates, including unicellular as well as colonial forms, are filter-feeders with cells that bear a striking resemblance to the feeding cells of sponges, called choanocytes (see page 523). The cells each have a single posterior flagellum surrounded by a collar of slender microvilli. Beating of the flagellum creates a water current that flows through the collar, where food particles are taken in by phagocytosis. Colonial choanoflagellates such as *Codonosiga* (Fig. 21.23*a*) commonly attach to surfaces with a stalk, but sometimes float freely like *Proterospongia* (Fig. 21.23*b*).

**Nucleariids** are opisthokonts with a rounded or slightly flattened cell body and threadlike pseudopods called filopodia.

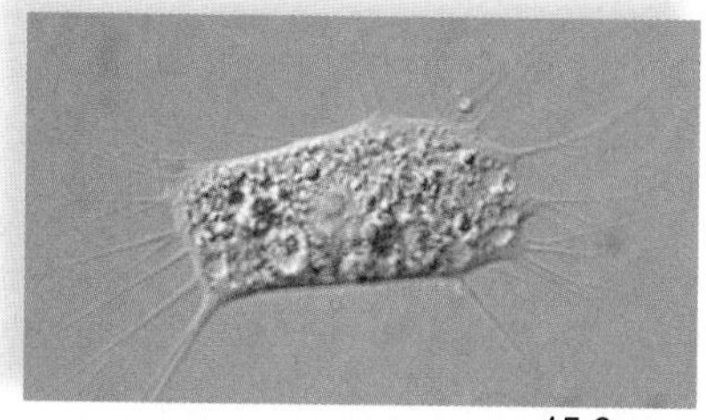


Most feed on algae or cyanobacteria. Although they lack the characteristic cell walls found in fungi, nucleariids appear to be close fungal relatives due to molecular similarities.

## Supergroup Rhizarians

The **rhizarians** [Gk. *rhiza,* root] consist of the **foraminiferans** and the **radiolarians,** organisms with fine, threadlike pseudopods. Although they were previously classified along with amoebozoans, the rhizarians are now assigned to a different supergroup because molecular data indicate the two groups are not very closely related. Foraminiferans and radiolarians both have a skeleton called a **test.** The tests of foraminiferans and radiolarians are intriguing and beautiful. In the foraminiferans, the calcium carbonate test is often multichambered. The pseudopods extend through openings in the test, which covers the plasma membrane (Fig. 21.24*a*). In the radiolarians, the glassy silicon test is internal and usually has a radial arrangement of spines (Fig. 21.24*b*). The pseudopods are external to the test.

The tests of dead foraminiferans and radiolarians form a deep layer (700–4,000 m) of sediment on the ocean floor. The radiolarians lie deeper than the foraminiferans because their glassy test is insoluble at greater pressures. The presence of either or both is used as an indicator of oil deposits on land and sea. Their fossils date even as far back as to Precambrian times and are evidence of the antiquity of the protists. Because each geological period has a distinctive form of foraminiferan, they can be used as index fossils to date sedimentary rock. Deposits of foraminiferans for millions of years, followed by geological upheaval, formed the White Cliffs of Dover along the southern coast of England. Also, the great Egyptian pyramids are built of foraminiferan limestone. One foraminiferan test found in the pyramids is about the size of a silver dollar. This species, known as *Nummulites,* has been found in deposits worldwide, including central eastern Mississippi. The shells of fromainiferans and radiolarians are abundant in the ocean.

### Check Your Progress 21.2E

1. How do the opisthokonts differ from the amoebozoans and the rhizarians?
2. What link is visual evidence between animals and the opisthokonts?

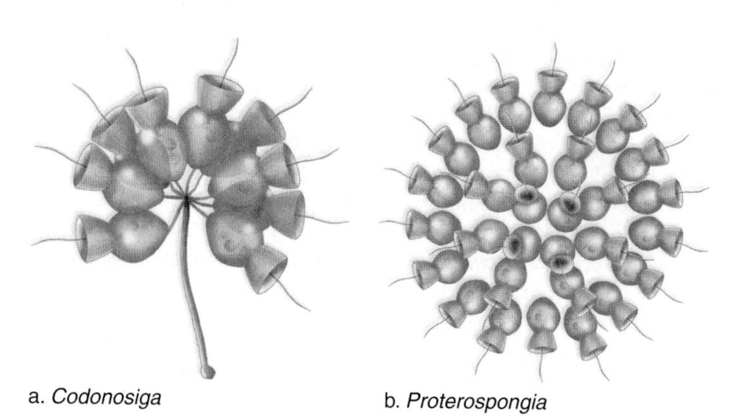

a. *Codonosiga* b. *Proterospongia*

**FIGURE 21.23 Colonial choanoflagellates.**

**a.** A *Codonosiga* colony can anchor itself with a slender stalk. **b.** A *Proterospongia* colony is unattached.

a. Foraminiferan, *Globigerina,* and the White Cliffs of Dover, England

b. Radiolarian tests SEM 200×

**FIGURE 21.24 Foraminiferans and radiolarians.**

**a.** Pseudopods of a live foraminiferan project through holes in the calcium carbonate shell. Fossilized shells were so numerous they became a large part of the White Cliffs of Dover when a geological upheaval occurred. **b.** Skeletal test of a radiolarian. In life, pseudopods extend outward through the openings of the glassy silicon shell.

# Connecting the Concepts

It's possible that the origin of sexual reproduction played a role in fostering the diversity of protists and their ability to inhabit water, soil, food, and even air. Unicellular *Chlamydomonas* usually reproduces asexually, but when it does reproduce sexually, the gametes look alike. Colonial forms of algae, such as *Volvox*, and multicellular forms, such as the brown alga *Fucus*, have a more specialized form of gametogenesis. They undergo oogamy: One gamete is larger and immobile, and the other is flagellated and motile. The green alga *Ulva* and the foraminiferans have an alternation of generations life cycle in which only the haploid stage produces gametes.

Conjugation is another way to introduce genetic variation among offspring. *Spirogyra* undergoes conjugation, but the ciliates have taken conjugation to another level. In ciliates, one haploid nucleus in each mating cell survives meiosis and divides again by mitosis. An exchange of haploid nuclei through a cytoplasmic channel restores diploidy and ensures diversity of genes.

Reproduction in protists is adaptive in still another way. As a part of their life cycle, some protists form spores or cysts that can survive in a hostile environment. The cell emerging from a spore or cyst is often a re-formed haploid or diploid individual, depending on the protist. Spores and cysts can survive inclement conditions for years on end.

The diversity of reproduction among protists exemplifies why it is difficult to discover their evolutionary relationships to the fungi, land plants, and animals without molecular genetics, which can detect evolutionary relationships by using nucleotide sequences.

# summary

### 21.1 General Biology of Protists

Protists are in the domain Eukarya. Independent endosymbiotic events may account for the presence of mitochondria and chloroplasts in eukaryotic cells. Protists are generally unicellular, but still they are quite complex because they (1) have a variety of characteristics, (2) acquire nutrients in a number of ways, and (3) have complicated life cycles that include the ability to withstand hostile environments.

Protists are of great ecological importance because in largely aquatic environments they are the producers that support communities of organisms. Protists also enter into various types of symbiotic relationships. Their great diversity makes it difficult to classify protists, and as yet, there is no general agreement about their categorization. In this chapter, they have been arranged in six supergroups.

### 21.2 Diversity of Protists

The archaeplastids include the plants and all green and red algae. Green algae possess chlorophylls *a* and *b*, store reserve food as starch, and have cell walls of cellulose as do land plants. Green algae are divided into chlorophytes and charophytes. Chlorophytes include forms that are unicellular (*Chlamydomonas*), colonial (*Volvox*), and multicellular (*Ulva*). Charophytes include filamentous forms (*Spirogyra*), as well as the stoneworts (*Chara*) and are thought to be the closest living relatives of land plants. The life cycle varies among the green algae. In most, the zygote undergoes meiosis, and the adult is haploid. *Ulva* has an alternation of generations like land plants, but the sporophyte and gametophyte generations are similar in appearance; the gametes look alike. Red algae are filamentous or multicellular seaweeds that are more delicate than brown algae and are usually found in warmer waters. Red algae have notable economic importance.

The supergroup chromalveolates consists of stramenopiles such as brown algae, diatoms, golden brown algae, and water molds; and alveolates, including dinoflagellates, ciliates, and apicomplexans. Brown algae have chlorophylls *a* and *c* plus a brownish carotenoid pigment. The large, complex brown algae, commonly called seaweeds, are well known and economically important. Diatoms, which have an outer layer of silica, are extremely numerous in both marine and freshwater ecosystems, as are golden brown algae, which may have coverings of silica or organic material. Water molds, which are filamentous and heterotrophic by absorption, are unlike fungi in that they produce flagellated 2n zoospores. Dinoflagellates usually have cellulose plates and two flagella, one at a right angle to the other. They are extremely numerous in the ocean and, on occasion, produce a neurotoxin when they form red tides. The ciliates move by their many cilia. They are remarkably diverse in form, and as exemplified by *Paramecium*, they show how complex a protist can be despite being a single cell. The apicomplexans are nonmotile parasites that form spores; *Plasmodium* causes malaria.

Supergroup excavates is a diverse collection of single-celled, motile protists, including euglenids, parabasalids, diplomonads, and kinetoplastids. Euglenids are flagellated cells with a pellicle instead of a cell wall. Their chloroplasts are most likely derived from a green alga through endosymbiosis. Many of the kinetoplastids are parasites, including trypanosomes, such as those that cause Chagas disease and African sleeping sickness in humans. Parabasalids, such as *Trichomonas vaginalis*, and, diplomonads, such as *Giardia lamblia*, are common inhabitants of animal hosts. They thrive in low-oxygen conditions due to their lack of aerobic respiration.

The amoebozoans supergroup contains amoeboids and slime molds, protists that use pseudopods for motility and feeding. Amoeboids move and feed by forming pseudopods. In *Amoeba proteus*, food vacuoles form following phagocytosis of prey. Contractile vacuoles discharge excess water. Slime molds, which produce nonmotile spores, are unlike fungi in that they have an amoeboid stage and are heterotrophic by ingestion.

The opisthokonts supergroup includes kingdom Animalia, the animal-like protists known as choanoflagellates, kingdom Fungi, and the funguslike protists called nucleariids.

The rhizarians are a supergroup that includes the foraminiferans and radiolarians, protists with threadlike pseudopods and skeletons called tests. The tests of foraminiferans and radiolarians form a deep layer of sediment on the ocean floor. The tests of foraminiferans are responsible for the limestone deposits of the White Cliffs of Dover.

## understanding the terms

algae 374
alveolate 383
amoeboid 387
amoebozoan 387
apicomplexan 384
archaeplastid 377
brown algae 381
cellular slime mold 388
charophyte 377
chlorophyte 377
choanoflagellate 388
chromalveolate 381
ciliate 384
colony 378
conjugation 378
cyst 374
diatom 382
dinoflagellate 383
diplomonad 386
euglenid 386
excavate 386
filament 378
foraminiferan 389
golden brown algae 382
green algae 377
kinetoplastid 386
mixotrophic 374
nucleariid 388
opisthokont 388
parabasalid 386
phagocytize 387
plankton 374
plasmodial slime mold 388
protist 374
protozoan 374
pseudopod 387
radiolarian 389
red algae 379
red tide 383
rhizarian 389
sporangium 388
stramenopile 381
supergroup 374
test 388
trichocyst 384
water mold 382
zooflagellate 386
zoospore 377

Match the terms to these definitions:

a. ______________ Cytoplasmic extension of amoeboid protists; used for locomotion and engulfing food.
b. ______________ Flexible freshwater unicellular organism that usually contains chloroplasts and is flagellated.
c. ______________ Freshwater or marine unicellular protist with a cell wall consisting of two silica-impregnated valves; extremely numerous in phytoplankton.
d. ______________ Causes severe diseases in human beings and domestic animals, including a condition called sleeping sickness.
e. ______________ Freshwater and marine organisms that are suspended on or near the surface of the water.

## reviewing this chapter

1. List and discuss ways that protists are varied. 374
2. Describe the structures of *Chlamydomonas* and *Volvox*, and contrast how they reproduce. 377–78
3. Describe the structures of *Ulva* and *Spirogyra*, and explain how they reproduce. 378
4. Describe the structure of red algae, and discuss their economic importance. 379
5. Describe the structure of brown algae, and discuss their ecological and economic importance. 381
6. Describe the structures of diatoms and dinoflagellates. What is a red tide? 382–84
7. Describe the life cycle of *Plasmodium vivax*, the most common causative agent of malaria. 385
8. Describe the unique structure of euglenids. 386
9. How are ciliates like and different from amoeboids? 384–85, 387–88
10. What features distinguish slime molds and water molds from fungi? Describe the life cycle of a plasmodial slime mold. 383, 388
11. Distinguish between amoeboids, foraminiferans, and radiolarians. 387–89

## testing yourself

Choose the best answer for each question.

For questions 1–6, match each item to those in the key.

**KEY:**

a. Amoebozoans
b. Archaeoplastids
c. Chromalveolates
d. Excavates
e. Opisthokonts
f. Rhizarians

1. foraminiferans
2. ciliates
3. brown algae
4. amoeboids
5. green algae
6. choanoflagellates
7. Which of these is not a green alga?
   a. *Volvox*
   b. *Fucus*
   c. *Spirogyra*
   d. *Chlamydomonas*
   e. *Ulva*
8. Which is not a characteristic of brown algae?
   a. multicellular
   b. chlorophylls *a* and *b*
   c. live along rocky coast
   d. harvested for commercial reasons
   e. contain a brown pigment
9. In *Chlamydomonas*,
   a. the adult is haploid.
   b. the zygospore survives times of stress.
   c. sexual reproduction occurs.
   d. asexual reproduction occurs.
   e. All of these are correct.
10. *Ulva*
   a. undergoes alternation of generations.
   b. is sea lettuce.
   c. is multicellular.
   d. is an archaeplastid.
   e. All of these are correct.
11. Which of these protists are not flagellated?
   a. *Volvox*
   b. *Spirogyra*
   c. dinoflagellates
   d. *Chlamydomonas*
   e. trypanosomes
12. Which pair is mismatched?
   a. diatoms—silica shell, resemble a petri dish, free-living
   b. euglenids—flagella, pellicle, eyespot
   c. *Fucus*—adult is diploid, seaweed, chlorophylls *a* and *c*
   d. *Paramecium*—cilia, calcium carbonate shell, gullet
   e. foraminiferan—test, pseudopod, digestive vacuole
13. Which is a false statement?
   a. Only heterotrophic and not photosynthetic protists are flagellated.
   b. Apicomplexans are parasitic protozoans.
   c. Among protists, the haploid cycle is common.
   d. Ciliates exchange genetic material during conjugation.
   e. Slime molds have an amoeboid stage.

14. Which pair is mismatched?
    a. trypanosome—African sleeping sickness
    b. *Plasmodium vivax*—malaria
    c. amoeboid—severe diarrhea
    d. AIDS—*Giardia lamblia*
    e. dinoflagellates—coral
15. Which is found in slime molds but not in fungi?
    a. nonmotile spores
    b. flagellated cells
    c. zygote formation
    d. photosynthesis
    e. chitin in cell walls
16. Which is a false statement?
    a. Slime molds and water molds are protists.
    b. There are flagellated algae and flagellated protozoans.
    c. Among protists, some flagellates are photosynthetic.
    d. Among protists, only green algae ever have a sexual life cycle.
    e. Conjugation occurs among green algae.
17. Which pair is properly matched?
    a. water mold—flagellate
    b. trypanosome—zooflagellate
    c. *Plasmodium vivax*—mold
    d. amoeboid—algae
    e. golden brown algae—kelp
18. Which is an incorrect statement?
    a. Unicellular protists can be quite complex.
    b. Euglenids are motile but have chloroplasts.
    c. Plasmodial slime molds are amoeboid but have sporangia.
    d. *Volvox* is colonial but has a boxed shape.
    e. Both b and d are incorrect.
19. All are correct about brown algae except that they
    a. range in size from small to large.
    b. are a type of seaweed.
    c. live on land.
    d. are photosynthetic.
    e. are usually multicellular.
20. In the haploid life cycle (e.g., *Chlamydomonas*),
    a. meiosis occurs following zygote formation.
    b. the adult is diploid.
    c. fertilization is delayed beyond the diploid stage.
    d. the zygote produces sperm and eggs.
21. Dinoflagellates
    a. usually reproduce sexually.
    b. have protective cellulose plates.
    c. are insignificant producers of food and oxygen.
    d. have cilia instead of flagella.
    e. tend to be larger than brown algae.
22. Ciliates
    a. move by pseudopods.
    b. are not as varied as other protists.
    c. have a gullet for food gathering.
    d. do not divide by binary fission.
    e. are closely related to the radiolarians.
23. A(n) _______________ is a collared, flagellated, heterotrophic protist that is closely related to animals such as sponges.
    a. radiolarian
    b. kinetoplastid
    c. choanoflagellate
    d. amoeboid
    e. trypanosome
24. Label this diagram of the *Chlamydomonas* life cycle.

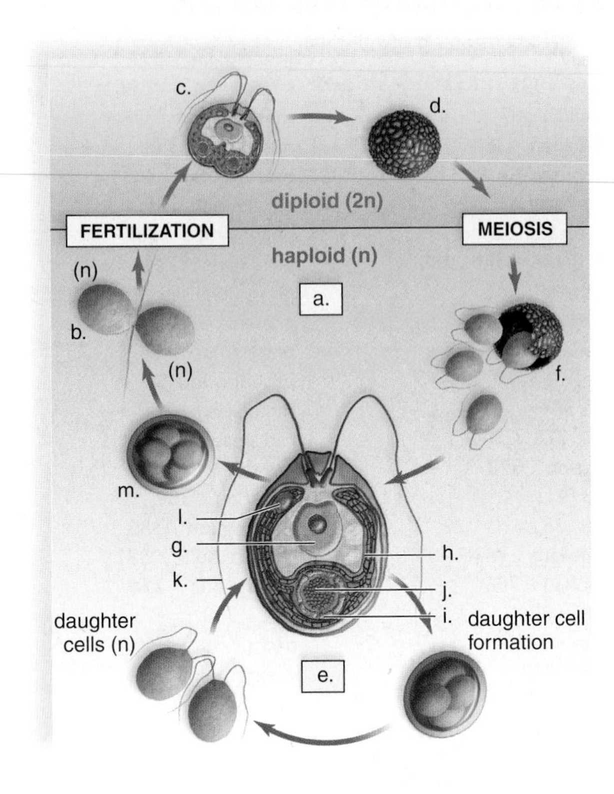


# thinking scientifically

1. While studying a unicellular alga, you discover a mutant in which the daughter cells do not separate after mitosis. This gives you an idea about how filamentous algae may have evolved. You hypothesize that the mutant alga is missing a protein or making a new form of a protein. How might each possibly lead to a filamentous appearance?
2. You are trying to develop a new antitermite chemical that will not harm environmentally beneficial insects. Since termites are adapted to eat only wood, they will starve if they cannot digest this food source. Termites have two symbiotic partners: the protozoan *Trichonympha collaris* and the bacteria it harbors that actually produce the enzyme that digests the wood. Knowing this, how might you prevent termite infestations without targeting the termites directly?

# *Biology* website

The companion website for *Biology* provides a wealth of information organized and integrated by chapter. You will find practice tests, animations, videos, and much more that will complement your learning and understanding of general biology.

**http://www.mhhe.com/maderbiology10**

# 22

# Fungi Evolution and Diversity

## concepts

**22.1 EVOLUTION AND CHARACTERISTICS OF FUNGI**

- Fungi are saprotrophic decomposers that aid the cycling of inorganic nutrients in ecosystems. The body of most fungi is multicellular; it is composed of thin filaments called hyphae. 394–95

**22.2 DIVERSITY OF FUNGI**

- Five widely recognized groups of fungi are chytrids, zygospore fungi, AM fungi, sac fungi, and club fungi. 396–403
- Chytrids are the only group of aquatic fungi; the other groups are adapted to living on land particularly by producing windblown spores during both asexual and sexual life cycles. 396–403
- The terrestrial fungi impact the lives of both plants and animals, including humans. 396–403

**22.3 SYMBIOTIC RELATIONSHIPS OF FUNGI**

- Lichens are an association between a fungus and a cyanobacterium or a green alga. Mycorrhizae are an association between a fungus and the roots of a plant. 404–5

*The largest living organism known is a honey mushroom, whose meshlike body spreads underground for 2,200 acres in the Blue Mountains of eastern Oregon. The reproductive structures of mushrooms and other fungi, such as puffballs, morels, and truffles, appear above ground only occasionally. Most of the time, fungi are busy decomposing leaf litter, fallen tree trunks, and the carcasses of dead animals. It's even been said that without the ability of fungi to coat and help plant roots take up nutrients, plants would not have been able to invade land. Some symbiotic fungi, however, cause diseases in plants and animals, including athlete's foot and ringworm in humans. Still, humans have found many uses for fungi. They are a food source and they help us produce beer, wine, bread, and cheeses.*

*Fungi were once classified as plants, because they have a cell wall and their bodies are nonmotile. However, molecular data now indicate that fungi are much more closely related to animals than they are to plants. Like animals, fungi are heterotrophs. Unlike animals, which ingest and then digest their food, fungi digest their food externally and then absorb the nutrients. In this chapter, you will explore the characteristics and sample the diversity of kingdom Fungi.*

Honey mushroom, *Armillaria ostoyae,* a common cause of tree root rot in the northeastern United States and British Columbia (Canada).

# 22.1 Evolution and Characteristics of Fungi

The **fungi** include over 80,000 species of mostly multicellular eukaryotes that share a common mode of nutrition. Mycologists, scientists that study fungi, expect this number of species to increase to over 1.5 million in the future. Like animals, fungi are heterotrophic and consume preformed organic matter. Animals, however, are heterotrophs that ingest food, while fungi are **saprotrophs** that absorb food. Their cells send out digestive enzymes into the immediate environment and then, when organic matter is broken down, the cells absorb the resulting nutrient molecules.

## Evolution of Fungi

Figure 22.1 lists the groups of fungi we will be discussing and some of the criteria used to distinguish each group. The evolutionary tree shows how these groups are believed to be related. As you can see, the chytrids are different from all other fungi because they are aquatic and have flagellated spores and gametes. Our description of fungal structure applies best to the zygospore fungi, sac fungi, and the club fungi. The AM fungi exist only as mycorrhizae in association with plant roots!

Protists evolved some 1,500 MYA (million years ago). Plants, animals, and fungi can all trace their ancestry to protists, but molecular data tells that animals and fungi shared a common ancestor after plants evolved. Therefore, animals and fungi are more closely related to each other than either is to plants. The common ancestor of animals and fungi was most likely a flagellated unicellular protist, and each became multicellular sometime after they split from one another. The ancestor was also aquatic and flagellated. Animals have retained flagellated cells but most groups of fungi do not have flagella today.

Fungal anatomy doesn't lend itself to becoming fossilized, so fungi probably evolved a lot earlier than the earliest known fungal fossil dated 450 MYA. We do know that while animals were still swimming in the seas during the Silurian, plants were beginning to live on the land, and they brought fungi with them. Mycorrhizae are evident in plant fossils, also some 450 MYA. Perhaps fungi were instrumental in the colonization of land by plants. Much of the fungal diversity we observed most likely had its origin in an adaptive radiation when organisms began to colonize land.

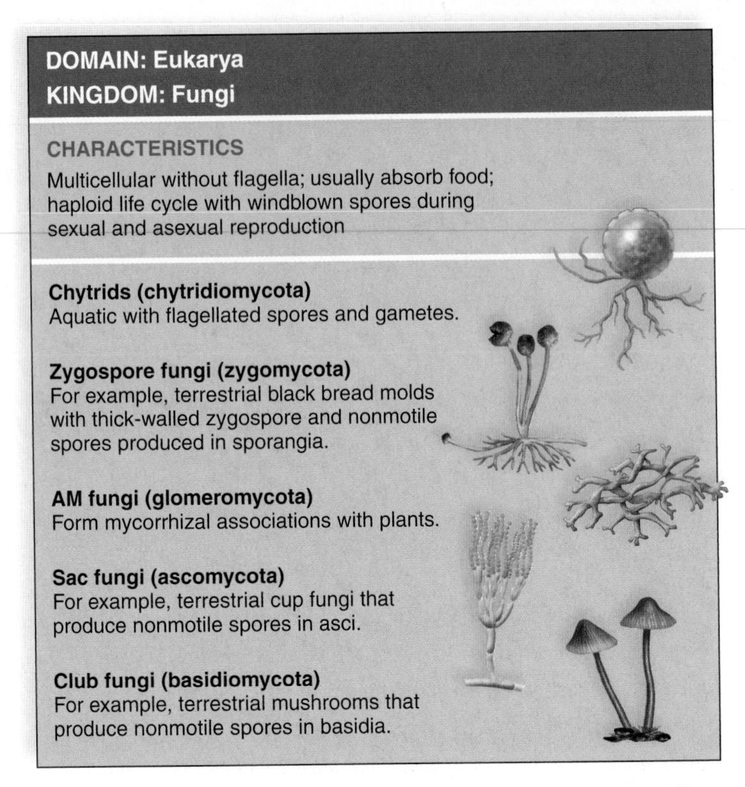


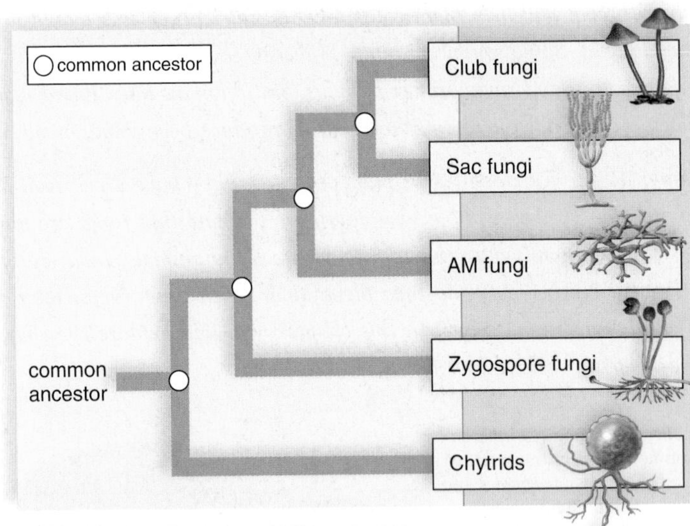


**FIGURE 22.1 Evolutionary relationships among the fungi.**

The common ancestor for the fungi was a flagellated saprotroph with chitin in the cell walls. All but chytrids lost the flagella at some point. They also became multicellular. Only the AM, sac, and club fungi are monophyletic, as is necessary to be a clade.

## Structure of Fungi

Some fungi, including the yeasts, are unicellular; however, the vast majority of species are multicellular. The thallus or body of most fungi is a multicellular structure known as a mycelium (Fig. 22.2*a*). A **mycelium** [Gk. *mycelium*, fungus filaments] is a network of filaments called **hyphae** [Gk. *hyphe*, web]. Hyphae give the mycelium quite a large surface area per volume of cytoplasm, and this facilitates absorption of nutrients into the body of a fungus. Hyphae grow at their tips, and the mycelium absorbs and then passes nutrients on to the growing tips. When a fungus reproduces, a specific portion of the mycelium becomes a reproductive structure that is then nourished by the rest of the mycelium (Fig. 22.2*b*).

Fungal cells are quite different from plant cells, not only by lacking chloroplasts but also by having a cell wall that contains **chitin** and not cellulose. Chitin, like cellulose, is a polymer of glucose organized into microfibrils. In chitin, however, each glucose molecule has a nitrogen-containing amino group attached to it. Chitin is also found in the exoskeleton of arthropods, a major phylum of animals that includes the insects and crustaceans. The energy reserve of fungi is not starch but glycogen, as in animals.

Except for the aquatic chytrids (discussed on page 396), fungi lack motility. The terrestrial fungi lack basal bodies and do not have flagella at any stage in their life cycle. They move toward a food source by growing toward it. Hyphae can cover as much as a kilometer a day!

Some fungi have cross walls, or septa, in their hyphae. These hyphae are called **septate** [L. *septum*, fence, wall]. Actually, the presence of septa makes little difference because pores allow cytoplasm and sometimes even organelles to pass freely from one cell to the other. The septa that separate reproductive cells, however, are complete in all fungal groups. **Nonseptate** fungi are multinucleated; they have many nuclei in the cytoplasm of a hypha (Fig. 22.2*c*).

## Reproduction of Fungi

Both sexual and asexual reproduction occur in fungi. Terrestrial fungal sexual reproduction involves these stages:

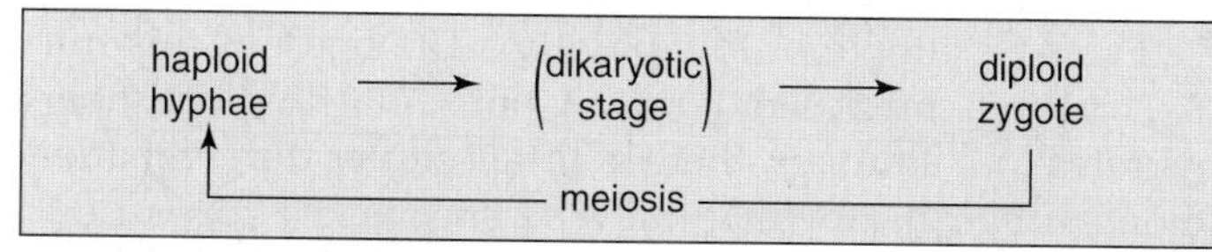


The relative length of time of each phase varies with the species.

During sexual reproduction, hyphae (or a portion thereof) from two different mating types make contact and fuse. It would be expected that the nuclei from the two mating types would also fuse immediately, and they do in some species. In other species, the nuclei pair but do not fuse for days, months, or even years. The nuclei continue to divide in such a way that every cell (in septate hyphae) has at least one of each nucleus. A hypha that contains paired haploid nuclei is said to be n + n or **dikaryotic** [Gk. *dis*, two, and *karyon*, nucleus, kernel]. When the nuclei do eventually fuse, the zygote undergoes meiosis prior to spore formation. Fungal spores germinate directly into haploid hyphae without any noticeable embryological development.

How can the terrestrial and nonmotile fungi ensure that the offspring will be dispersed to new locations? As an adaptation to life on land, fungi usually produce nonmotile, but normally windblown, spores during both sexual and asexual reproduction. A **spore** is a reproductive cell that develops into a new organism without the need to fuse with another reproductive cell. A large mushroom may produce billions of spores within a few days. When a spore lands upon an appropriate food source, it germinates and begins to grow.

Asexual reproduction usually involves the production of spores by a specialized part of a single mycelium. Alternately, asexual reproduction can occur by fragmentation—a portion of a mycelium begins a life of its own. Also, unicellular yeasts reproduce asexually by **budding;** a small cell forms and gets pinched off as it grows to full size (see Fig. 22.5).

### Check Your Progress 22.1

1. How do animals and fungi differ with respect to heterotrophy?
2. How are fungal cell walls different from plant cell walls?
3. Describe the function of a fungal spore.

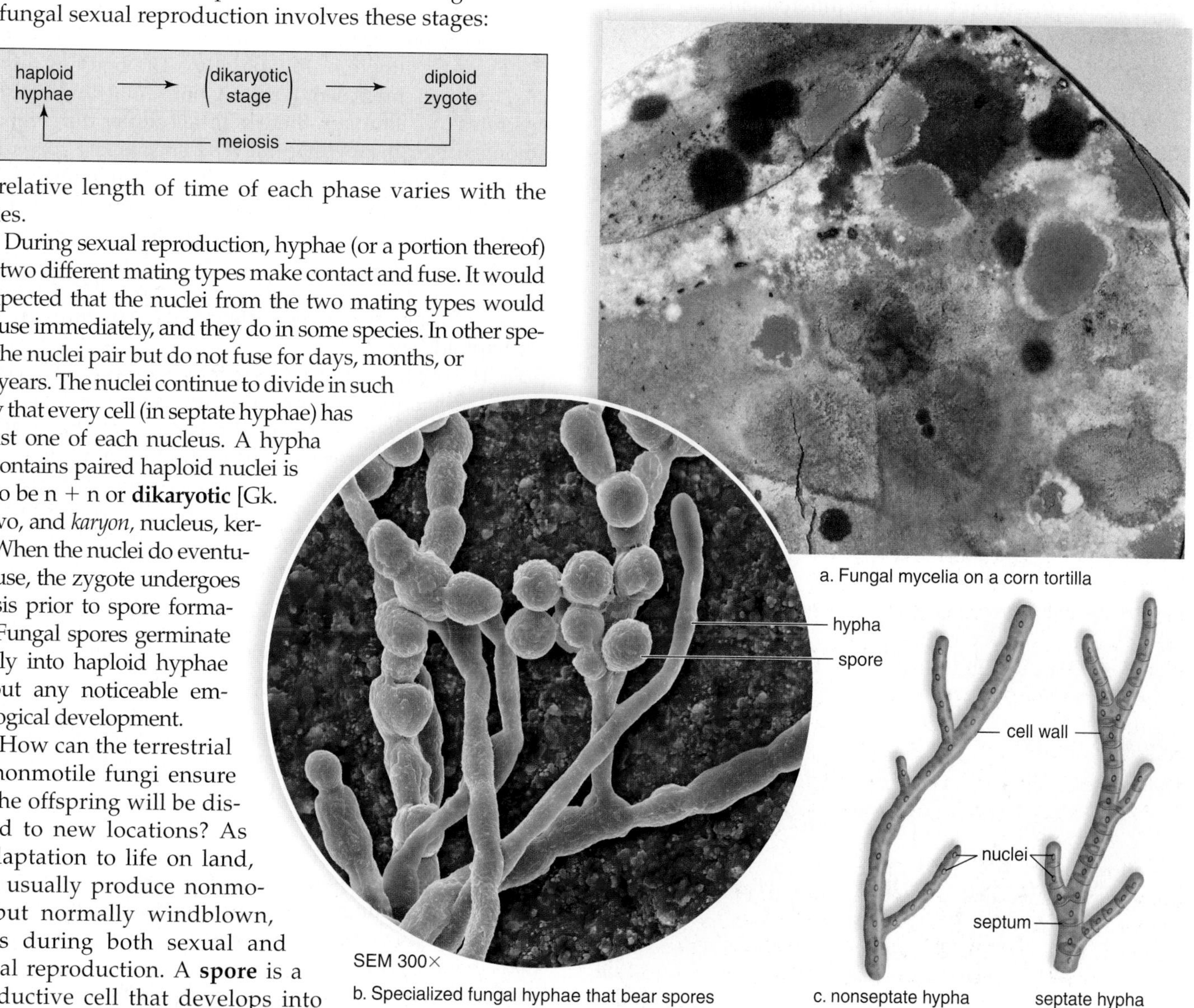


**FIGURE 22.2 Mycelia and hyphae of fungi.**

**a.** Each mycelium grown from a different spore on a corn tortilla is quite symmetrical. **b.** Scanning electron micrograph of specialized aerial fungal hyphae that bear spores. **c.** Hyphae are either nonseptate (do not have cross walls) or septate (have cross walls).

## 22.2 Diversity of Fungi

R. H. Whittaker was the first to say, in 1969, that fungi should be classified as a separate group from protists, plants, and animals, and they remain so today. He based his reasoning on the observation that fungi are the only type of multicellular organism to be saprotrophic. However, as discussed in the previous chapter, some experts now place fungi in the supergroup opisthokonts, which includes animals and certain heterotrophic protists (see page 389).

Without an adequate fossil record, we cannot be certain about the relationship between different types of fungi. However, with the ever more common use of comparative molecular data to decipher evolutionary relationships, we may one day know how the fungi are related. In the meantime, the fungal groups chytrids, zygospore fungi, AM fungi, sac fungi, and club fungi are differentiated according to their life cycle and the type of structure they use to produce spores.

### Chytrids

The **chytrids** (chytridiomycota) include about 790 species of the simplest fungi that may resemble the first fungi to have evolved. Some chytrids are single cells; others form branched nonseptate hyphae. Chytrids are unique among fungi because they are the only fungi to still have flagellated cells. This feature is consistent with their aquatic lifestyle, although some also live in moist soil. They produce flagellated gametes and spores. The placement of the flagella in their spores, called **zoospores,** links fungi to choanoflagellates and animals and helps place fungi in the supergroup opisthokonts (see page 389). Most chytrids reproduce asexually through the production of zoospores within a single cell. The zoospores grow into new chytrids. However, some have an alternation of generations life cycle, much like that of green plants and certain algae, which is quite uncommon among fungi.

Most chytrids play a role in the decay and digestion of dead aquatic organisms, but some are parasitic on plants, animals, and protists (Fig. 22.3). They are also known to cause diseases such as brown spot of corn and black wart of potato. The parasitic chytrid *Batrachochytrium dendrobatidis* has recently decimated populations of harlequin frogs (*Atelopus*) in Central and South America. They grow inside skin cells and disrupt the ability of frogs to acquire oxygen through their skin.

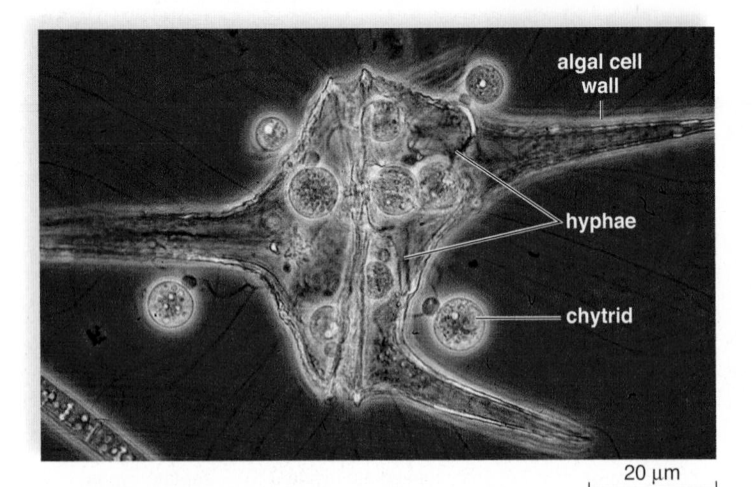


**FIGURE 22.3 Chytrids parasitizing a protist.**

These aquatic chytrids (*Chytriomyces hyalinus*) have penetrated the cell walls of this dinoflagellate and are absorbing nutrients meant for their host. They will produce flagellated zoospores that will go on to parasitize other protists.

### Zygospore Fungi

The **zygospore fungi** (zygomycota) include approximately 1,050 species of fungi. These organisms are saprotrophic, living off plant and animal remains in the soil or in bakery goods in the pantry. Some are parasites of minute soil protists, worms, and insects such as a housefly.

The black bread mold, *Rhizopus stolonifer,* is commonly used as an example of this phylum. The body of this fungus, which is composed of mostly nonseptate hyphae, demonstrates that although there is little cellular differentiation among fungi, the hyphae may be specialized for various purposes. In *Rhizopus,* stolons are horizontal hyphae that exist on the surface of the bread; rhizoids grow into the bread, anchor the mycelium, and carry out digestion; and sporangiophores are aerial hyphae that bear sporangia. A **sporangium** is a capsule that produces spores called sporangiospores. During asexual reproduction, all structures involved are haploid (Fig. 22.4).

The phylum name refers to the zygospore, which is seen during sexual reproduction. The hyphae of opposite mating types, termed plus (+) and minus (−), are chemically attracted, and they grow toward each other until they touch. The ends of the hyphae swell as nuclei enter; then cross walls develop a short distance behind each end, forming **gametangia.** The gametangia merge, and the result is a large multinucleate cell in which the nuclei of the two mating types pair and then fuse. A thick wall develops around the cell, which is now called a **zygospore.** The zygospore undergoes a period of dormancy before meiosis and germination take place. One or more sporangiophores with sporangia at their tips develop, and many spores are released. The spores, dispersed by air currents, give rise to new haploid mycelia. Spores from black bread mold have been found in the air above the North Pole, in the jungle, and far out at sea.

### AM Fungi

The **AM fungi** (glomeromycota) are a relatively small group (160 species) of fungi whose common name stands for arbuscular mycorrhizal fungi. Arbuscules are branching invaginations that the fungus makes when it invades plant roots. Mycorrhizae, which are discussed on page 404, are a mutualistic association that benefit both the fungus and the plant. Long classified as zygospore fungi, the AM fungi are now beginning to receive recognition as a separate group based on molecular data.

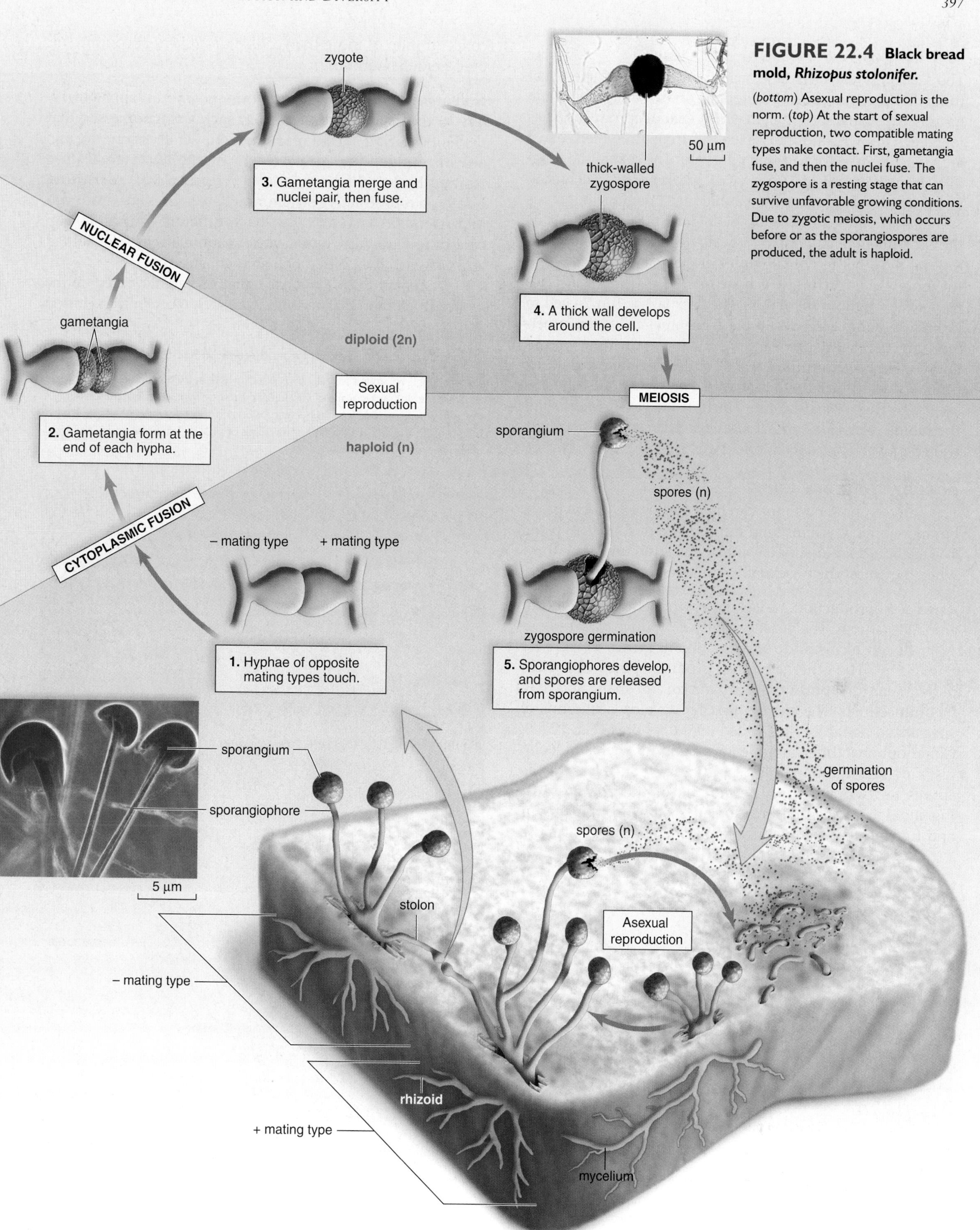


**FIGURE 22.4 Black bread mold, *Rhizopus stolonifer*.**

(*bottom*) Asexual reproduction is the norm. (*top*) At the start of sexual reproduction, two compatible mating types make contact. First, gametangia fuse, and then the nuclei fuse. The zygospore is a resting stage that can survive unfavorable growing conditions. Due to zygotic meiosis, which occurs before or as the sporangiospores are produced, the adult is haploid.

## Sac Fungi

The **sac fungi** (ascomycota) consist of about 50,000 species of fungi. The sac fungi can be thought of as having two main groups: the sexual sac fungi, in which sexual reproduction has long been known, and the asexual sac fungi, in which sexual reproduction has not yet been observed. The sexual sac fungi include such organisms as the **yeast** *Saccharomyces,* the unicellular fungi important in the baking and brewing industries and also in various molecular biological studies. *Neurospora,* the experimental material for the one-gene–one-enzyme studies, and the other **red bread molds** are sexual sac fungi. So are the **morels** and **truffles,** which are famous gourmet delicacies revered throughout the world. The asexual sac fungi used to be in the phylum Deuteromycota, sometimes called the imperfect fungi because their means of sexual reproduction was unknown. However, on the basis of molecular data and structural characteristics, these fungi have now been identified as sac fungi. The asexual sac fungi include the yeast *Candida* and the **molds** *Aspergillus* and *Penicillium.* The asci of *Penicillium* were discovered and, therefore, it was renamed *Talaromyces.*

### *Biology of the Sac Fungi*

The body of the sac fungus can be a single cell, as in yeasts, but more often it is a mycelium composed of septate hyphae. The sac fungi are distinguished by the structures they form when they reproduce asexually and sexually.

**Asexual Reproduction.** Asexual reproduction is the norm among sac fungi. The yeasts usually reproduce by budding. A small cell forms and pinches off as it grows to full size (Fig. 22.5*a*). The other sac fungi produce spores called conidia or **conidiospores** that vary in size and shape and may be multicellular. The conidia usually develop at the tips of specialized aerial hyphae called conidiophores (Fig. 22.5*b*). Conidiophores differ in appearance and this helps mycologists identify the particular sac fungi. When released, the spores are windblown. The conidia of the allergy-causing mold *Cladosporium* are carried easily through the air and transported even over oceans. One researcher found a concentration of more than 35,000 *Cladosporium* conidia/m$^3$ over Leiden (Germany).

**Sexual Reproduction.** Their formal name, ascomycota, refers to the **ascus** [Gk. *askos,* bag, sac], a fingerlike sac that develops during sexual reproduction. On occasion, the asci are surrounded and protected by sterile hyphae within a fruiting body called an *ascocarp* (Fig. 22.6*a*, *b*). A **fruiting body** is a reproductive structure where spores are produced and released. Ascocarps can have different shapes; in cup fungi they are cup shaped and in morels they are stalked and crowned by a pitted bell-shaped ascocarp.

Ascus-producing hyphae remain dikaryotic except in the walled-off portion that becomes the ascus, where nuclear fusion

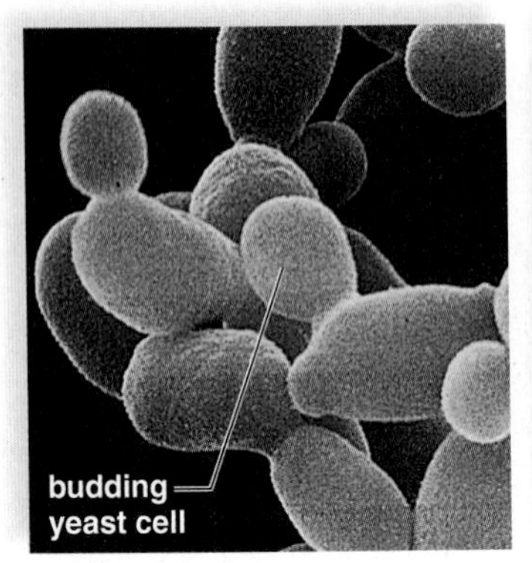


a.

b.

**FIGURE 22.5 Asexual reproduction in sac fungi.**

**a.** Yeasts, unique among fungi, reproduce by budding. **b.** The sac fungi usually reproduce asexually by producing spores called conidia or conidiospores.

**FIGURE 22.6 Sexual reproduction in sac fungi.**

The sac fungi reproduce sexually by producing asci, in fruiting bodies called ascocarps. **a.** In the ascocarp of cup fungi, dikaryotic hyphae terminate forming the asci, where meiosis follows nuclear fusion and spore formation takes place. **b.** In morels, the asci are borne on the ridges of pits. **c.** Peach leaf curl, a parasite of leaves, forms asci as shown.

a. Ascocarp of the cup fungus, *Sarcoscypha*

and meiosis take place. Because mitosis follows meiosis, each ascus contains eight haploid nuclei and produces eight spores. In most ascomycetes, the asci become swollen as they mature, and then they burst, expelling the ascospores. If released into the air, the spores are then windblown.

### The Benefits and Drawbacks of Sac Fungi

The sac fungi play an essential role in recycling by digesting resistant (not easily decomposed) materials containing cellulose, lignin, or collagen. Species are also known that can even consume jet fuel and wall paint. Some are symbiotic with algae, forming lichens, and plant roots, forming mycorrhizae. They also account for most of the known fungal pathogens causing various plant diseases. Powdery mildews grow on leaves, as do leaf curl fungi (Fig. 22.6c); chestnut blight and Dutch elm disease destroy the trees named. Ergot, a parasitic sac fungus that infects rye and (less commonly) other grains, is discussed in the Health Focus on page 401.

A sac fungus produces the drug penicillin, which cures bacterial infections; cyclosporine, which suppresses the immune system leading to the success of transplantation operations; and the steroids, which are present in the birth control pill. They are also used during the production of various foods such as blue cheese and Coke. Many human diseases caused by sac fungi are acquired from the environment. Ringworm comes from soil fungi, rose gardener's disease from thorns, Chicago disease from old buildings, and basketweaver's disease from grass cuttings.

**Yeasts.** Yeasts can be both beneficial and harmful to humans. In the wild, yeasts grow on fruits, and historically the yeasts already present on grapes were used to produce wine. Today, selected yeasts, such as *Saccharomyces cerevisiae*, are added to relatively sterile grape juice to make wine. Also, this yeast is added to prepared grains to make beer. When *Saccharomyces* ferments, it produces ethanol and also carbon dioxide. Both the ethanol and the carbon dioxide are retained for beers and sparkling wines; carbon dioxide is released for still wines. In baking, the carbon dioxide given off is the leavening agent that causes bread to rise. *Saccharomyces* is serviceable to humans in another way. It is sometimes used in genetic engineering experiments requiring a eukaryote.

b. Ascocarp of the morel, *Morchella*

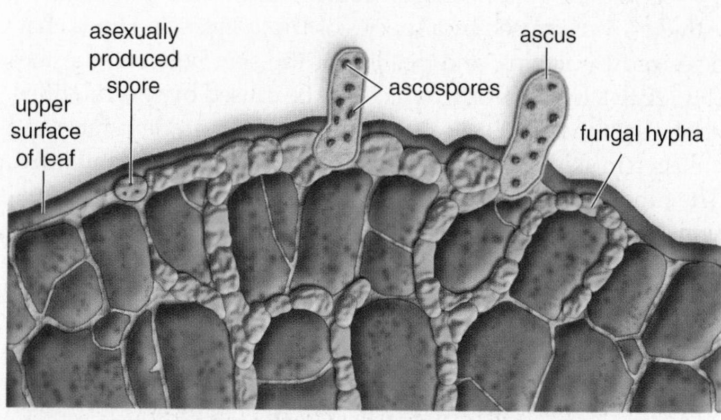


c. Peach leaf curl, *Taphrina*

Yeasts can be harmful to humans. *Candida albicans* is a yeast that causes the widest variety of fungal infections. Disease occurs when the normal balance of microbes in an organ, such as the vagina, is disturbed, particularly by antibiotic therapy. Then *Candida* proliferates and symptoms result. A vaginal infection results in inflammation, itching, and discharge. Oral thrush is a *Candida* infection of the mouth, common in newborns and AIDS patients. In immunocompromised individuals, *Candida* can move through the body, causing a systemic infection that can damage the heart, brain, and other organs.

**Molds.** Molds can be helpful to humans. *Aspergillus* is a group of green molds recognized by the bottle-shaped structure that bears their conidiospores. It is used to produce soy sauce by fermentation of soybeans. A Japanese food called miso is made by fermenting soybeans and rice with *Aspergillus*. In the United States, *Aspergillus* is used to produce citric and gallic acids, which serve as additives during the manufacture of a wide variety of products, including foods, inks, medicines, dyes, plastics, toothpaste, soap, and even chewing gum.

A species of *Penicillium* (blue molds now classified as *Talaromyces*) is the source of the familiar antibiotic called penicillin, which was first manufactured during World War II and still has many applications today. Since the discovery of penicillin by Sir Alexander Fleming, it has saved countless lives. Other *Penicillium* species give the characteristic flavor and aroma to cheeses such as Roquefort and Camembert. The bluish streaks in blue cheese are patches of conidiospores.

Molds can also be harmful to humans. Commonly isolated from soil, plant debris, and house dust, *Aspergillus* is sometimes pathogenic to humans. *Aspergillus flavus*, which grows on moist seeds, secretes a toxin that is the most potent natural carcinogen known. Therefore, in humid climates such as that in the southeastern United States, care must be taken to store grains properly. *Aspergillus* also causes a potentially deadly disease of the respiratory tract that arises after spores have been inhaled.

The mold *Stachybotrys chartarum* (Fig. 22.7) grows well on building materials. It is known as black mold and is responsible for the "sick-building" syndrome. Individuals with chronic exposure to toxins produced by this fungus have reported cold and flulike symptoms, fatigue, and dermatitis. The toxins may suppress and could destroy the immune system, affecting the lymphoid tissue and the bone marrow.

Moldlike fungi cause infections of the skin called tineas. Athlete's foot, caused by a species of *Trichophyton*, is a tinea characterized by itching and peeling of the skin between the toes (Fig. 22.8*a*). In ringworm, which can be caused by several different fungi, the fungus releases enzymes that degrade keratin and collagen in skin. The area of infection becomes red and inflamed. The fungal colony grows outward, forming a ring of inflammation. The center of the lesion begins to heal, thereby giving ringworm its characteristic appearance, a red ring surrounding an area of healed skin (Fig. 22.8*b*). Tinea infections of the scalp are rampant among school-age children, affecting as much as 30% of the population, with the possibility of permanent hair loss.

The vast majority of people living in the eastern and central United States have been infected with *Histoplasma capsulatum*, a thermally dimorphic fungus that grows in mold form at 25°C and in yeast form at 37°C. This common soil fungus, often associated with bird droppings, leads in most cases to a mild "fungal flu." Less than half of those infected notice any symptoms, with 3,000 showing severe disease and about 50 dying each year from histoplasmosis. The fungal pathogen lives and grows within cells of the immune system and causes systemic illness. Lesions are formed in the lungs that leave calcifications, visible in X-ray images, that resemble those of tuberculosis.

**FIGURE 22.7 Black mold.**

*Stachybotrys chartarum*, or black mold, grows well in moist areas, including the walls of homes. It represents a potential health risk.

**Control of Fungal Infections.** The strong similarities between fungal and human cells make it difficult to design fungal medications that do not also harm humans. Researchers exploit any biochemical differences they can discover. The biosynthesis of steroids in fungi differs somewhat from the same pathways in humans. A variety of fungicides are directed against steroid biosynthesis, including some that are applied to fields of grain. Fungicides based on heavy metals are applied to seeds of sorghum and other crops. Topical agents are available for the treatment of yeast infections and tineas, and systemic medications are available for systemic sac fungi infections.

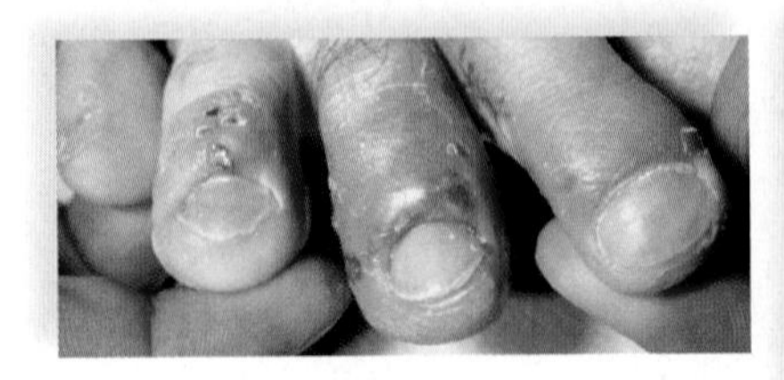


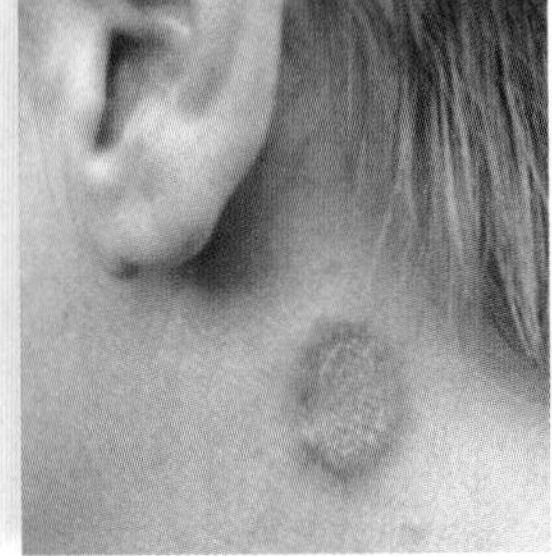


**FIGURE 22.8 Tineas.**

**a.** Athlete's foot and (**b**) ringworm are termed tineas.

*health focus*

## Deadly Fungi

It is unwise for amateurs to collect mushrooms in the wild because certain mushroom species are poisonous. The red and yellow *Amanitas* are especially dangerous. These species are also known as fly agaric because they were once thought to kill flies (the mushrooms were gathered and then sprinkled with sugar to attract flies). Its toxins include muscarine and muscaridine, which produce symptoms similar to those of acute alcoholic intoxication. In one to six hours, the victim staggers, loses consciousness, and becomes delirious, sometimes suffering from hallucinations, manic conditions, and stupor. Luckily, it also causes vomiting, which rids the system of the poison, so death occurs in less than 1% of cases. The death angel mushroom (*Amanita phalloides*, Fig. 22A) causes 90% of the fatalities attributed to mushroom poisoning. When this mushroom is eaten, symptoms don't begin until 10–12 hours later. Abdominal pain, vomiting, delirium, and hallucinations are not the real problem; rather, a poison interferes with RNA (ribonucleic acid) transcription by inhibiting RNA polymerase, and the victim dies from liver and kidney damage.

Some hallucinogenic mushrooms are used in religious ceremonies, particularly among Mexican Indians. *Psilocybe mexicana* contains a chemical called psilocybin that is a structural analogue of LSD and mescaline. It produces a dreamlike state in which visions of colorful patterns and objects seem to fill up space and dance past in endless succession. Other senses are also sharpened to produce a feeling of intense reality.

The only reliable way to tell a nonpoisonous mushroom from a poisonous one is to be able to correctly identify the species. Poisonous mushrooms cannot be identified with simple tests, such as whether they peel easily, have a bad odor, or blacken a silver coin during cooking. Only consume mushrooms identified by an expert!

Like club fungi, some sac fungi also contain chemicals that can be dangerous to people. *Claviceps purpurea*, the ergot fungus, infects rye and replaces the grain with ergot—hard, purple-black bodies consisting of tightly cemented hyphae (Fig. 22B). When ground with the rye and made into bread, the fungus releases toxic alkaloids that cause the disease ergotism. In humans, vomiting, feelings of intense heat or cold, muscle pain, a yellow face, and lesions on the hands and feet are accompanied by hysteria and hallucinations. Ergotism was common in Europe during the Middle Ages. During this period, it was known as St. Anthony's Fire and was responsible for 40,000 deaths in an epidemic in AD 994. We now know that ergot contains lysergic acid, from which LSD is easily synthesized. Based on recorded symptoms, historians believe that those individuals who were accused of practicing witchcraft in Salem, Massachusetts, during the seventeenth century were actually suffering from ergotism. It is also speculated that ergotism is to blame for supposed demonic possessions throughout the centuries. As recently as 1951, an epidemic of ergotism occurred in Pont-Saint-Esprit, France. Over 150 persons became hysterical, and four died.

Because the alkaloids that cause ergotism stimulate smooth muscle and selectively block the sympathetic nervous system, they can be used in medicine to cause uterine contractions and to treat certain circulatory disorders, including migraine headaches. Although the ergot fungus can be cultured in petri dishes, no one has succeeded in inducing it to form ergot in the laboratory. So far, the only way to obtain ergot, even for medical purposes, is to collect it in an infected field of rye.

**FIGURE 22A** **Poisonous mushroom, *Amanita phalloides*.**

**FIGURE 22B** **Ergot infection of rye, caused by *Claviceps purpurea*.**

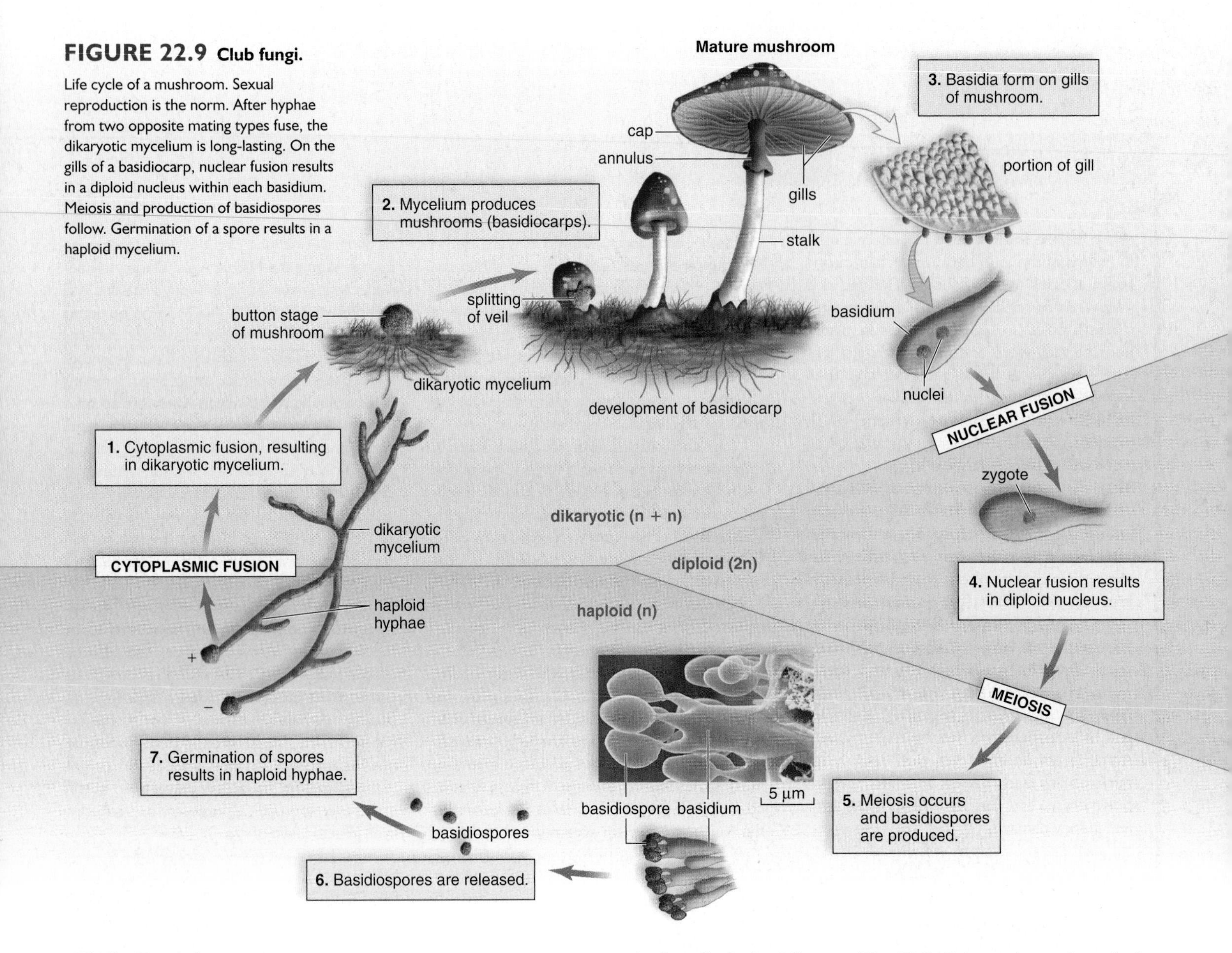


**FIGURE 22.9 Club fungi.**

Life cycle of a mushroom. Sexual reproduction is the norm. After hyphae from two opposite mating types fuse, the dikaryotic mycelium is long-lasting. On the gills of a basidiocarp, nuclear fusion results in a diploid nucleus within each basidium. Meiosis and production of basidiospores follow. Germination of a spore results in a haploid mycelium.

# Club Fungi

**Club fungi** (basidiomycota) consist of over 22,000 species. Mushrooms, toadstools, puffballs, shelf fungi, jelly fungi, bird's-nest fungi, and stinkhorns are basidiomycetes. In addition, fungi that cause plant diseases such as the smuts and rusts are placed in this phylum. Several mushrooms such as the portabella and shiitake mushrooms are savored as foods by humans. Approximately 75 species of basidiomycetes are considered poisonous. The poisonous death angel mushroom is discussed in the Health Focus on page 401.

## *Biology of Club Fungi*

The body of a basidomycota is a mycelium composed of septate hyphae. Most members of this phylum are saprotrophs, although several parasitic species exist.

**Reproduction.** Although club fungi occasionally do produce conidia asexually, they usually reproduce sexually. Their formal name, basidiomycota, refers to the **basidium** [L. *basidi,* small pedestal], a club-shaped structure in which spores called basidiospores develop. Basidia are located within a fruiting body called a basidiocarp (Fig. 22.9). Prior to formation of a basidiocarp, haploid hyphae of opposite mating types meet and fuse, producing a dikaryotic (n + n) mycelium. The dikaryotic mycelium continues its existence year after year, even for hundreds of years on occasion. In many species of mushrooms, the dikaryotic mycelium often radiates out and produces mushrooms in an ever larger, so-called fairy ring (Fig. 22.10*a*).

Mushrooms are composed of nothing but tightly packed hyphae whose walled-off ends become basidia. In gilled mushrooms, the basidia are located on radiating lamellae, the gills. In shelf fungi and pore mushrooms (Fig. 22.10*b, c*), the basidia terminate in tubes. In any case, the extensive surface area of a basidiocarp is lined by basidia, where nuclear fusion, meiosis, and spore production occur. A basidium has four projections in which cytoplasm and a haploid nucleus enter as the basidiospore forms. Basidiospores are windblown; when they germinate, a new haploid mycelium forms. It is estimated that some large mushrooms can produce up to 40 million spores per hour.

In puffballs, spores are produced inside parchmentlike membranes, and the spores are released through a pore or

a. Fairy ring

b. Shelf fungus

c. Pore mushroom, *Boletus*

d. Puffball, *Calvatiga gigantea*

**FIGURE 22.10 Club fungi.**
**a.** Fairy ring. Mushrooms develop in a ring on the outer living fringes of a dikaryotic mycelium. The center has used up its nutrients and is no longer living. **b.** A shelf fungus. **c.** Fruiting bodies of *Boletus*. This mushroom is not gilled; instead, it has basidia-lined tubes that open on the undersurface of the cap. **d.** In puffballs, the spores develop inside an enclosed fruiting body. Giant puffballs are estimated to contain 7 trillion spores.

when the membrane breaks down (Fig. 22.10*d*). In bird's-nest fungi, falling raindrops provide the force that causes the nest's basidiospore-containing "eggs" to fly through the air and land on vegetation. Stinkhorns resemble a mushroom with a spongy stalk and a compact, slimy cap. The long stalk bears the elongated basidiocarp. Stinkhorns emit an incredibly disagreeable odor; flies are attracted by the odor, and when they linger to feed on the sweet jelly, the flies pick up spores that they later distribute.

## *Smuts and Rusts*

Smuts and rusts are club fungi that parasitize cereal crops such as corn, wheat, oats, and rye. They are of great economic importance because of the crop losses they cause every year. Smuts and rusts don't form basidiocarps, and their spores are small and numerous, resembling soot. Some smuts enter seeds and exist inside the plant, becoming visible only near maturity. Other smuts externally infect plants. In corn smut, the mycelium grows between the corn kernels and secretes substances that cause the development of tumors on the ears of corn (Fig. 22.11*a*).

The life cycle of rusts requires alternate hosts, and one way to keep them in check is to eradicate the alternate host. Wheat rust (Fig. 22.11*b*) is also controlled by producing new and resistant strains of wheat. The process is continuous, because rust can mutate to cause infection once again.

a. Corn smut, *Ustilago*

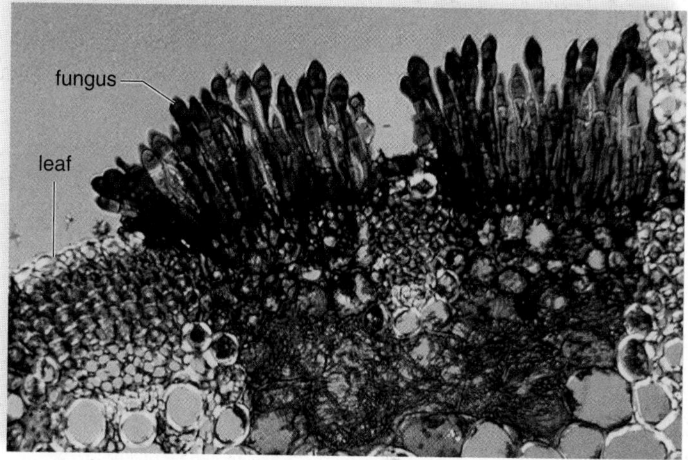

b. Wheat rust, *Puccinia*

**FIGURE 22.11 Smuts and rusts.**
**a.** Corn smut. **b.** Micrograph of wheat rust.

### Check Your Progress 22.2

1. What makes members of the chytrids different from all other fungi?
2. What are fungal infections called? Which type of fungus produces most of the known fungal pathogens?
3. Name the type of fungi for each of these: puffballs, ergots, athlete's foot, and black bread mold.

# 22.3 Symbiotic Relationships of Fungi

Several instances in which fungi are parasites of plants and animals have been mentioned. Two other symbiotic associations are of interest.

## Lichens

**Lichens** are an association between a fungus, usually a sac fungus, and a cyanobacterium or a green alga. The body of a crustose lichen has three layers: The fungus forms a thin, tough upper layer and a loosely packed lower layer that shield the photosynthetic cells in the middle layer (Fig. 22.12*a*). Specialized fungal hyphae, which penetrate or envelop the photosynthetic cells, transfer nutrients directly to the rest of the fungus. Lichens can reproduce asexually by releasing fragments that contain hyphae and an algal cell. In fruticose lichens, the sac fungus reproduces sexually (Fig. 22.12*b*).

In the past, lichens were assumed to be mutualistic relationships in which the fungus received nutrients from the algal cells, and the algal cells were protected from desiccation by the fungus. Actually, lichens may involve a controlled form of parasitism of the algal cells by the fungus, with the algae not benefiting at all from the association. This is supported by experiments in which the fungal and algal components are removed and grown separately. The algae grow faster when they are alone than when they are part of a lichen. On the other hand, it is difficult to cultivate the fungus, which does not naturally grow alone. The different lichen species are identified according to the fungal partner.

Three types of lichens are recognized. Compact crustose lichens are often seen on bare rocks or on tree bark; fruticose lichens are shrublike; and foliose lichens are leaflike (Fig. 22.12*c*). Lichens are efficient at acquiring nutrients and moisture, and therefore they can survive in areas of low moisture and low temperature as well as in areas with poor or no soil. They produce and improve the soil, thus making it suitable for plants to invade the area. Unfortunately, lichens also take up pollutants and cannot survive where the air is polluted. Therefore, lichens can serve as air pollution sensors.

## Mycorrhizae

**Mycorrhizae** [Gk. *mykes*, fungus, and *rhizion*, dim. for root] are mutualistic relationships between soil fungi and the roots of most plants. Plants whose roots are invaded by mycorrhizae grow more successfully in poor soils—particularly soils deficient in phosphates—than do plants without mycorrhizae (Fig. 22.13). The fungal partner, either a glomerulomycete or a sac fungus, may enter the cortex of roots but does not enter the cytoplasm of plant cells. Ectomycorrhizae form a mantle that is exterior to the root, and they grow between cell walls. Endomycorrhizae, such as the AM fungi, penetrate only the cell walls. In any case, the presence of the fungus gives the plant a greater absorptive surface for the intake of minerals. The fungus also benefits from the association by receiving carbohydrates from the plant. As mentioned, even the earli-

reproductive unit

fungal hyphae

algal cell

fungal hyphae

a. Crustose lichen, *Xanthoria*

b. Fruticose lichen, *Lobaria*

c. Foliose lichen, *Xanthoparmelia*

**FIGURE 22.12 Lichen morphology.**

**a.** A section of a compact crustose lichen shows the placement of the algal cells and the fungal hyphae, which encircle and penetrate the algal cells. **b.** Fruticose lichens are shrublike. **c.** Foliose lichens are leaflike.

**FIGURE 22.13 Plant growth experiment.**

A soybean plant (*left*) without mycorrhizae grows poorly compared to two other (*right*) plants infected with different strains of mycorrhizae.

est fossil plants have mycorrhizae associated with them. It would appear, then, that mycorrhizae helped plants adapt to and flourish on land.

It is of interest to know that the truffle, a gourmet delight whose ascocarp is somewhat prunelike in appearance, is a mycorrhizal sac fungus living in association with oak and beech tree roots. In the past, the French used pigs (truffle-hounds) to sniff out and dig up truffles, but now they have succeeded in cultivating truffles by inoculating the roots of seedlings with the proper mycelium.

### Check Your Progress 22.3

1. What type of symbiotic relationship exists in mycorrhizae?
2. How does a lichen reproduce?
3. What is the relationship between air pollution and the presence of lichens?

## Connecting the Concepts

At one time, fungi were considered a part of the plant kingdom, and then later, they were considered a protist. Whittaker argued that, because of their multicellular nature and mode of nutrition, fungi should be in their own kingdom. Like animals, they are heterotrophic, but they do not ingest food—they absorb nutrients. Like plants, they have cell walls, but their cell walls contain chitin instead of cellulose. Fungal cells store energy in the form of glycogen, as do animal cells.

Fungi have been around for at least 570 million years and may be descendants of flagellated protists. The chytrids are fungi that produce spores and gametes with flagella. Flagella have been lost by other fungal phyla over the course of evolution.

Today, the range of species within the kingdom Fungi is broad. Some fungi are unicellular (such as yeasts), some are multicellular parasites (such as those that cause athlete's foot), and some form mutualistic relationships with other species (such as those found in lichens). Most, however, are saprotrophic decomposers that play a vital role in all ecosystems. The decomposers work with bacteria to break down the waste products and dead remains of plants and animals so that organic materials are recycled.

In addition to their various ways of life, fungi have been successful because they have diverse reproductive strategies. During sexual reproduction, the filaments of different mating types typically fuse. Asexual reproduction via spores, fragmentation, and budding is common.

## summary


### 22.1 Evolution and Characteristics of Fungi

Fungi are multicellular eukaryotes that are heterotrophic by absorption. After external digestion, they absorb the resulting nutrient molecules. Most fungi act as saprotrophic decomposers that aid the cycling of chemicals in ecosystems by decomposing dead remains. Some fungi are parasitic, especially on plants, and others are mutualistic with plant roots and algae.

The body of a fungus is composed of thin filaments called hyphae, which collectively are termed a mycelium. The cell wall contains chitin, and the energy reserve is glycogen. With the notable exception of the chytrids, which have flagellated spores and gametes, fungi do not have flagella at any stage in their life cycle. Nonseptate hyphae have no cross walls; septate hyphae have cross walls, but there are pores that allow the cytoplasm and even organelles to pass through.

Fungi produce spores during both asexual and sexual reproduction. During sexual reproduction, hyphae tips fuse so that dikaryotic (n + n) hyphae usually result, depending on the type of fungus. Following nuclear fusion, zygotic meiosis occurs during the production of the sexual spores.

### 22.2 Diversity of Fungi

Five significant groups of fungi are the chytrids (chytridiomycota), zygospore fungi (zygomycota), AM fungi (glomeromycota), sac fungi (ascomycota), and club fungi (basidiomycota).

The chytrids are unique because they produce motile zoospores and gametes. Some chytrids have an alternation of generations life cycle similar to that of plants and certain algae. There are unicellular as well as hyphae-forming chytrids. When hyphae form, they are nonseptate. *Chytriomyces* is an example of a chytrid.

The zygospore fungi are nonseptate, and during sexual reproduction they have a dormant stage consisting of a thick-walled zygospore. When the zygospore germinates, sporangia produce windblown spores. Asexual reproduction occurs when nutrients are plentiful and sporangia again produce spores. An example of a zygomycete is black bread mold.

The AM (arbuscular mycorrhizal) fungi were once classified with the zygospore fungi but are now viewed as a distinct group. AM fungi exist in mutualistic associations with the roots of land plants.

The sac fungi are septate, and during sexual reproduction saclike cells called asci produce spores. Asci are sometimes located in fruiting bodies called ascocarps. Asexual reproduction, which is dependent on the production of conidiospores, is more common. Sexual reproduction is unknown in some sac fungi. Sac fungi include *Talaromyces*, *Aspergillus*, *Candida*, morels and truffles, and various yeasts and molds, some of which cause disease in plants and animals, including humans.

The club fungi are septate, and during sexual reproduction club-shaped structures called basidia produce spores. Basidia are located in fruiting bodies called basidiocarps. Club fungi have a prolonged dikaryotic stage, and asexual reproduction by conidiospores is rare. A dikaryotic mycelium periodically produces fruiting bodies. Mushrooms and puffballs are examples of club fungi.

### 22.3 Symbiotic Relationships of Fungi

Lichens are an association between a fungus, usually a sac fungus, and a cyanobacterium or a green alga. Traditionally, this association was considered mutualistic, but experimentation suggests a controlled parasitism by the fungus on the alga. Lichens may live in extreme environments and on bare rocks; they allow other organisms that will eventually form soil to establish.

The term *mycorrhizae* refers to an association between a fungus and the roots of a plant. The fungus helps the plant absorb minerals, and the plant supplies the fungus with carbohydrates.

## understanding the terms

AM fungi 396
ascus 398
basidium 402
budding 395
chitin 394
chytrid 396
club fungi 402
conidiospore 398
dikaryotic 395
fruiting body 398
fungus (pl., fungi) 394
gametangia 396
hypha 394
lichen 404
mold 398
morel 398
mycelium 394
mycorrhizae 404
nonseptate 395
red bread mold 398
sac fungi 398
saprotroph 394
septate 395
sporangium 396
spore 395
truffle 398
yeast 398
zoospore 396
zygospore 396
zygospore fungi 396

Match the terms to these definitions:

a. _______________ Clublike structure in which nuclear fusion and meiosis occur during sexual reproduction of club fungi.
b. _______________ Tangled mass of hyphal filaments composing the vegetative body of a fungus.
c. _______________ Spore produced by sac and club fungi during asexual reproduction.
d. _______________ Spore-producing and spore-disseminating structure found in sac and club fungi.
e. _______________ Symbiotic relationship between fungal hyphae and roots of vascular plants. The fungus allows the plant to absorb more mineral ions and obtain carbohydrates from the plant.
f. _______________ Fungi with motile spores and gametes.

## reviewing this chapter

1. Which characteristics best define fungi? Describe the body of a fungus and how fungi reproduce. 394–95
2. Discuss the evolution and classification of fungi. 394
3. Explain how chytrids are different from the other phyla of fungi. 396
4. Explain the term *zygospore fungi*. How does black bread mold reproduce asexually? Sexually? 396–97
5. Explain the terms *sexual sac fungi* and *asexual sac fungi*. 398–99
6. Explain the term *sac fungi*. How do sac fungi reproduce asexually? Describe the structure of an ascocarp. 398–99
7. Describe the structure of yeasts, and explain how they reproduce. How are yeasts and molds useful/harmful to humans? 399–400
8. Explain the term *club fungi*. Draw and explain a diagram of the life cycle of a typical mushroom. 402
9. What is the economic importance of smuts and rusts? How can their numbers be controlled? 403
10. Describe the structure of a lichen, and name the three different types. What is the nature of this fungal association? 404
11. Describe the association known as mycorrhizae, and explain how each partner benefits. 404–5

## testing yourself

Choose the best answer for each question.
For questions 1–3, match the fungi to the phyla in the key.

**KEY:**
a. chytridiomycota
b. zygomycota
c. ascomycota
d. basidiomycota

1. club fungi
2. zygospore fungi
3. sac fungi
4. During sexual reproduction, the zygospore fungi produce
   a. an ascus.
   b. a basidium.
   c. a sporangium.
   d. a conidiophore.
5. An organism that decomposes remains is most likely to use which mode of nutrition?
   a. parasitic
   b. saprotrophic
   c. ingestion
   d. chemosynthesis
   e. Both a and b are correct.
6. Hyphae are generally characterized by
   a. strong, impermeable walls.
   b. rapid growth.
   c. large surface area.
   d. pigmented cells.
   e. Both b and c are correct.
7. A fungal spore
   a. contains an embryonic organism.
   b. germinates directly into an organism.
   c. is always windblown, because none are motile.
   d. is most often diploid.
   e. Both b and c are correct.
8. Fungal groups have different
   a. sexual reproductive structures.
   b. sporocarp shapes.
   c. modes of nutrition.
   d. types of cell wall.
   e. levels of organization.
9. In the life cycle of black bread mold, the zygospore
   a. undergoes meiosis and produces zoospores.
   b. produces spores as a part of asexual reproduction.
   c. is a thick-walled dormant stage.
   d. is equivalent to asci and basidia.
   e. All of these are correct.
10. In an ascocarp,
   a. there are fertile and sterile hyphae.
   b. hyphae fuse, forming the dikaryotic stage.
   c. a sperm fertilizes an egg.

d. hyphae do not have chitinous walls.
e. conidiospores form.

11. In which fungus is the dikaryotic stage longer lasting?
    a. zygospore fungus
    b. sac fungus
    c. club fungus
    d. chytrids
12. Conidiospores are formed
    a. asexually at the tips of special hyphae.
    b. during sexual reproduction.
    c. by all types of fungi except water molds.
    d. when it is windy and dry.
    e. as a way to survive a harsh environment.
13. The asexual sac fungi are so called because
    a. they have no zygospore.
    b. they cause diseases.
    c. they form conidiospores.
    d. sexual reproduction has not been observed.
    e. All of these are correct.
14. Lichens
    a. cannot reproduce.
    b. need a nitrogen source to live.
    c. are parasitic on trees.
    d. are able to live in extreme environments.
15. Label this diagram of black bread mold structure and asexual reproduction.

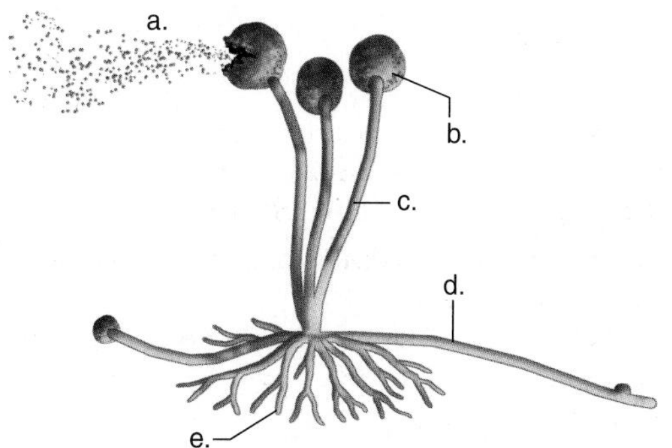


16. Why is it challenging to treat fungal infections of the human body?
    a. Human and fungal cells share common characteristics.
    b. Fungal cells share no features in common with human cells.
    c. All fungal cells are highly resistant to drug treatments.
    d. Both b and c are correct.
17. Mycorrhizae
    a. are a type of lichen.
    b. are mutualistic relationships.
    c. help plants gather solar energy.
    d. help plants gather inorganic nutrients.
    e. Both b and d are correct.
18. Which stage(s) in the chytrid life cycle is/are motile?
    a. diploid zoospores
    b. haploid zoospores
    c. male gametes
    d. female gametes
    e. All of these are correct.
19. Yeasts are what type of fungi?
    a. zygospores
    b. single cell
    c. sac fungi
    d. Both b and c are correct.
    e. All of these are correct.
20. Which statement is incorrect?
    a. Some fungi are parasitic on plants, and others are mutualistic with plant roots and algae.
    b. There are no fungi with motile cells.
    c. The cell walls of fungi contain chitin.
    d. Following nuclear fusion, zygotic meiosis occurs during the production of spores.
    e. Lichens are an association between a fungus and a bacterium or a green alga.
21. Symbiotic relationships of fungi include
    a. athlete's foot.
    b. lichens.
    c. mycorrhizae.
    d. Only b and c are correct.
    e. All three examples are correct.
22. Label this diagram of the life cycle of a mushroom.

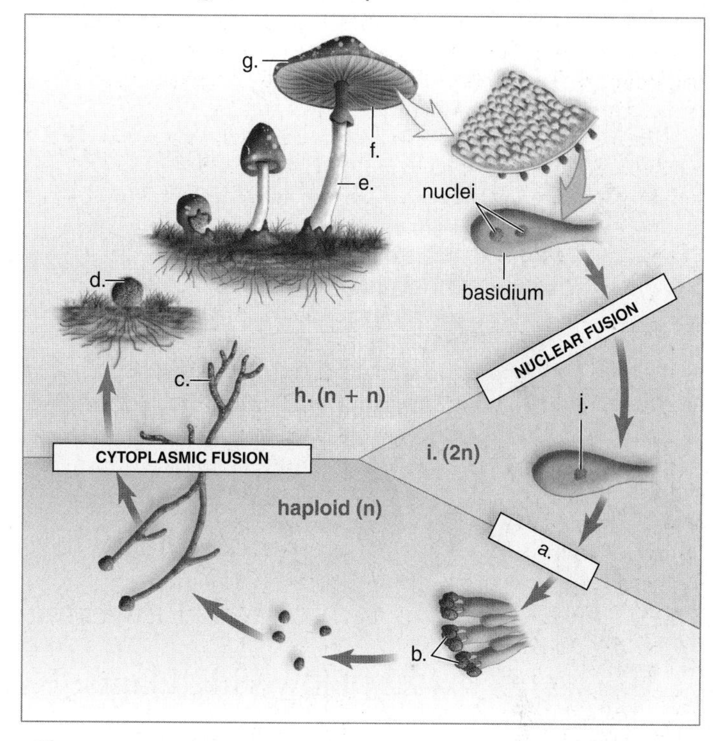


## thinking scientifically

1. The very earliest bakers observed that dough left in the air would rise. Unknown to them, yeast from the air "contaminated" the bread, began to grow, and produced carbon dioxide. Carbon dioxide caused the bread to rise. Later, cooks began to save some soft dough (before much flour was added) from the previous loaf to use in the next loaf. The saved portion was called the mother. What is in the mother, and why was it important to save it in a cool place?
2. There seems to be a fine line between symbiosis and parasitism when you examine the relationships between fungi and plants. What hypotheses could explain how different selective pressures may have caused particular fungal species to adopt one or the other relationship? Under what circumstances might a mutualistic relationship evolve between fungi and plants? Under what circumstances might a parasitic relationship evolve?

## *Biology* website

The companion website for *Biology* provides a wealth of information organized and integrated by chapter. You will find practice tests, animations, videos, and much more that will complement your learning and understanding of general biology.

**http://www.mhhe.com/maderbiology10**

part V

# Plant Evolution and Biology

*Life would be impossible without plants. Plants provide all living things with food and the oxygen needed to break down nutrient molecules and produce ATP. They are a source of fuel to keep us warm, and we use their fibers to make our clothes. The frame of our houses and the furniture therein is often made of wood and other plant products. We even use plants to make the paper for our newspapers and textbooks.*

*Altogether, it would be a good idea to know the principles of plant biology—plant structure and function. Plants have the same characteristics of life as animals do and most reproduce sexually, as we do. Life began in the sea and plants invaded the terrestrial environment before animals, as you would expect since animals are so dependent on plants. When our ancestors evolved, they lived in trees and fed on fruit and only later did they come down from the trees and begin to walk erect. These chapters introduce you to the biology of plants, those incredible organisms that keep the biosphere functioning as it should.*

# 23 Plant Evolution and Diversity

Most likely, when you think of plants, you envision a scene much like that shown below. But plants have a very long evolutionary history, and many other types of plants preceded the evolution of flowering plants. Representatives of each of these groups are alive today and functioning well within their own particular environment. The closest relatives to land plants still live today in an aquatic environment, as do most animals.

*This chapter attempts to trace the evolutionary history of plants by presenting and discussing many other types of plants, both living and extinct, aside from the flowering plants. In other words, you will be studying the diversity of plants within an evolutionary context. We will also give reasons why the most successful and abundant members of the plant kingdom are the flowering plants. Much of their success is due to coevolution between flowering plants and their pollinators, various terrestrial animals with which they share the land environment.*

Grape hyacinth, *Muscaris*, and tulips, *Tulipa*, bloom in spring.

## concepts

# 23.1 The Green Algal Ancestor of Plants

**Plants** are multicellular, photosynthetic eukaryotes, whose evolution is marked by adaptations to a land existence. A land environment does offer certain advantages to plants. For example, there is plentiful light for photosynthesis—water, even if clear, filters light. Also, carbon dioxide is present in higher concentrations and diffuses more readily in air than in water.

The land environment, however, requires adaptations, in particular, to deal with the constant threat of desiccation (drying out). The most successful land plants are those that protect all phases of reproduction (sperm, egg, embryo) from drying out and have an efficient means of dispersing offspring on land. Seed plants disperse their embryos within the seed, which provides the embryo with food within a protective seed coat.

The water environment not only provides plentiful water, it also offers support for the body of a plant. To conserve water, the land plant body, at the very least, is covered by a waxy cuticle that prevents loss of water while still allowing carbon dioxide to enter so that photosynthesis can continue. In many land plants, the roots absorb water from the soil, and a vascular system transports water in the body of the land plant. The vascular system that evolved in land plants allows them to stand tall and support expansive broad leaves that efficiently collect sunlight. The flowering plants, the last type to evolve, employ animals to assist with reproduction and dispersal of seed. The evolutionary history of plants given in Figure 23.1 shows the sequence in which land plants evolved adaptive features for an existence on land.

## The Ancestry of Plants

The plants (listed in Table 23.1) evolved from a freshwater green algal species some 450 million years ago. As evidence for a green algal ancestry, scientists have known for some time that all green algae and plants contain chlorophylls *a* and *b* and various accessory pigments; store excess carbohydrates as starch; and have cellulose in their cell walls.

In recent years, molecular systematists have compared the sequence of ribosomal RNA bases between organisms. The results suggest that among the green algae, land plants are most closely related to freshwater green algae, known as **charophytes.** Fresh water, of course, exists in bodies of water on land, and natural selection would have favored those specimens best able to make the transition to the land itself. The land environment, at the time, was barren and represented a vast opportunity for any photosynthetic plants that were able to leave the water and take advantage of the new environment.

There are several types of charophytes—*Spirogyra*, for example, is a charophyte. But botantists tell us that

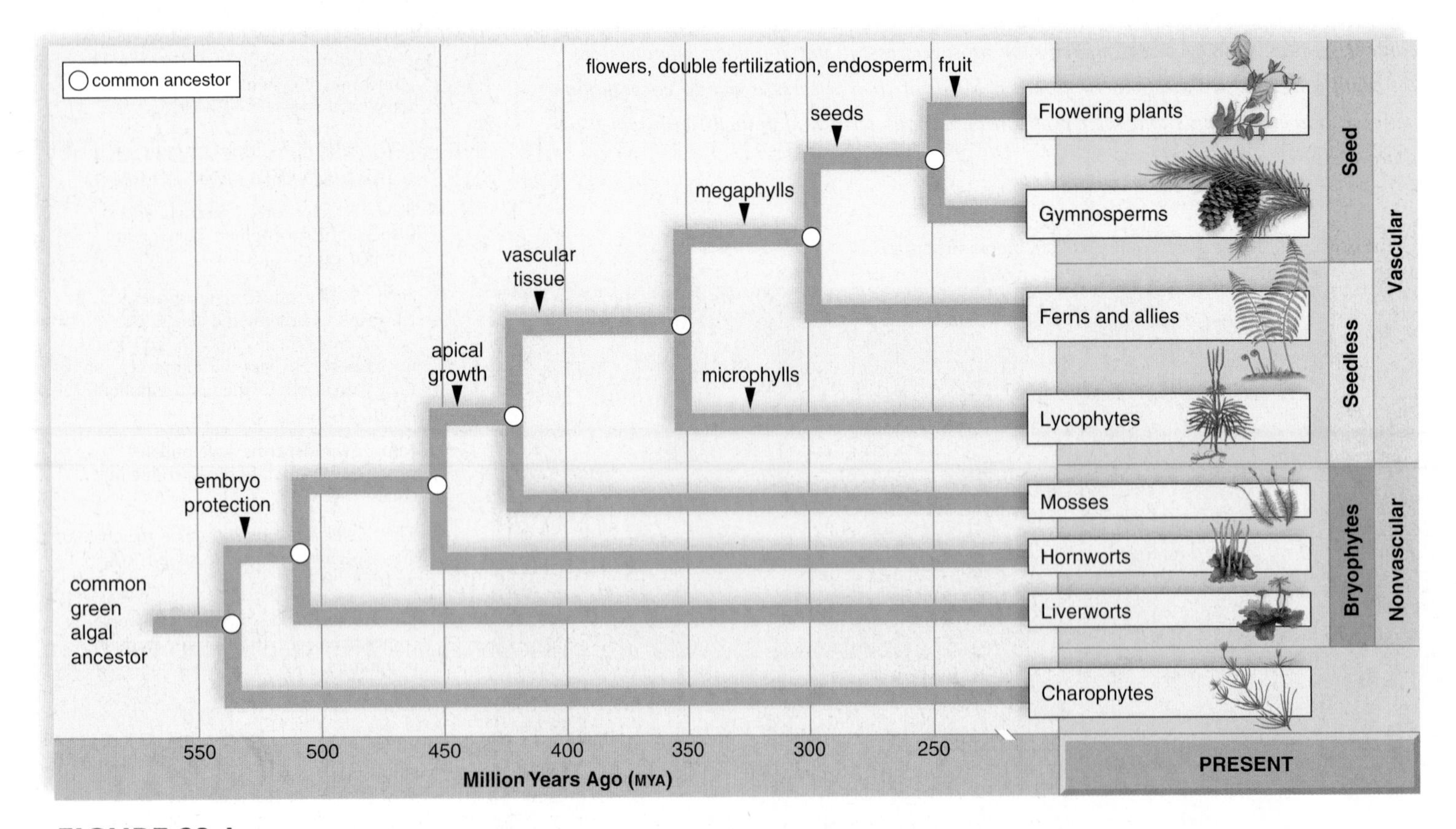


**FIGURE 23.1 Evolutionary history of plants.**

The evolution of plants involves these significant innovations. In particular, protection of a multicellular embryo was seen in the first plants to live on land. Vascular tissue permits the transport of water and nutrients. The evolution of the seed increased the chance of survival for the next generation.

among living charophytes, Charales (an order with 300 macroscopic species) and the *Coleochaete* (a genus with 30 microscopic species), featured in Figure 23.2, are most like land plants. Charophytes and land plants are in the same clade and form a monophyletic group (see Fig. 23.1). Their common ancestor no longer exists, but if it did, it would have features that resemble those of the Charales and *Coleochaete*.

First, let's take a look at these filamentous green algae (Fig. 23.2). The Charales (e.g., *Chara*) are commonly known as stoneworts because some species are encrusted with calcium carbonate deposits. The body consists of a single file of very long cells anchored in mud by thin filaments. Whorls of branches occur at multicellular nodes, regions between the giant cells of the main axis. Male and female reproductive structures grow at the nodes. The zygote is retained until it is enclosed by tough walls. A *Coleochaete* looks like a flat pancake, but the body is actually composed of elongated branched filaments of cells that spread flat across the substrate or form a three-dimensional cushion. The zygote is also retained in *Coleochaete*.

These two groups of charophytes have several features that would have promoted the evolution of multicellular land plants listed in Table 23.1, which have complex tissues and organs. And these features are present in charophytes and/or improved upon in land plants today:

1. *The cellulose cell walls of charophytes* and the land plant lineage are laid down by the same unique type of cellulose synthesizing complexes. Charophytes also have a mechanism of cell-wall formation during cytokinesis that is nearly identical to that of land plants. In land plants, a strong cell wall assists staying erect.
2. *The apical cells of charophytes* produce cells that allow their filaments to increase in length. At the nodes, other cells can divide asymmetrically to produce reproductive structures. Land plants are noted for their apical tissue that produces specialized tissues that add to or develop into new organs, such as new branches and leaves.
3. *The plasmodesmata of charophytes* provide a means of communication between neighboring cells, otherwise separated by cell walls. Perhaps plasmodesmata play a role in the evolution of specialized tissues in land plants, but this is not known for certain.
4. *The placenta* (designated cells) *of charophytes* transfers nutrients from haploid cells of the previous generation to the diploid zygote. Both charophytes and land plants retain and care for the zygote.

**FIGURE 23.2 Charophytes.**

The charophytes (represented here by *Chara* and *Coleochaete*) are the green algae most closely related to the land pants.

## TABLE 23.1

**DOMAIN: Eukarya**
**KINGDOM: Plants**

**CHARACTERISTICS**
Multicellular, usually with specialized tissues; photosynthesizers that became adapted to living on land; most have alternation of generations life cycle.

**Charophytes**
Live in water; haploid life cycle; share certain traits with the land plants

**LAND PLANTS (embryophytes)**
Alternation of generation life cycle; protect a multicellular sporophyte embryo; gametangia produce gametes; apical tissue produces complex tissues; waxy cuticle prevents water loss.

**Bryophytes (liverworts, hornworts, mosses)**
Low-lying, nonvascular plants that prefer moist locations: Dominant gametophyte produces flagellated sperm; unbranched, dependent sporophyte produces windblown spores.

**VASCULAR PLANTS (lycophytes, ferns and their allies, seed plants)**
Dominant, branched sporophyte has vascular tissue: Lignified xylem transports water, and phloem transports organic nutrients; typically has roots, stems, and leaves; and gametophyte is eventually dependent on sporophyte.

**Lycophytes (club mosses)**
Leaves are microphylls with a single, unbranched vein; sporangia borne on sides of leaves produce windblown spores; independent and separate gametophyte produces flagellated sperm.

**Ferns and Allies (pteridophytes)**
Leaves are megaphylls with branched veins; dominant sporophyte produces windblown spores in sporangia borne on leaves; and independent and separate gametophyte produces flagellated sperm.

**SEED PLANTS (gymnosperms and angiosperms)**
Leaves are megaphylls; dominant sporophyte produces heterospores that become dependent male and female gametophytes. Male gametophyte is pollen grain and female gametophyte occurs within ovule, which becomes a seed.

**Gymnosperms (cycads, ginkgoes, conifers, gnetophytes)**
Usually large; cone-bearing; existing as trees in forests. Sporophyte bears pollen cones, which produce windblown pollen (male gametophyte), and seed cones, which produce seeds.

**Angiosperms (flowering plants)**
Diverse; live in all habitats. Sporophyte bears flowers, which produce pollen grains, and bear ovules within ovary. Following double fertilization, ovules become seeds that enclose a sporophyte embryo and endosperm (nutrient tissue). Fruit develops from ovary.

## Alternation of Generations Life Cycle

All land plants have an **alternation of generations life cycle.** Study the alternation of generation life cycle in Figure 23.3 to realize that the sporophyte (2n) is so named for its production of spores by meiosis. A **spore** is a haploid reproductive cell that develops into a new organism without the need to fuse with another reproductive cell. In the plant life cycle, a spore undergoes mitosis and becomes a gametophyte.

The gametophyte (n) is so named for its production of gametes. In plants, eggs and sperm are produced by mitotic cell division. A sperm and egg fuse, forming a diploid zygote that undergoes mitosis—becoming the sporophyte embryo, and then the 2n generation.

Two observations are in order. First, meiosis produces haploid spores. This is consistent with the realization that the sporophyte is the diploid generation and spores are haploid reproductive cells. Second, mitosis occurs as a spore becomes a gametophyte, and mitosis occurs as a zygote becomes a sporophyte. Indeed, it is the occurrence of mitosis at these times that results in two generations.

As a contrast to the alternation of generation life cycle, consider the haploid life cycle of charophytes. In the haploid life cycle (see page 380), the zygote undergoes meiosis, and therefore only four zoospores are produced per zygote. In the plant life cycle, the zygote becomes a multicellular sporophyte with one or more sporangia that produces many windblown spores. The production of so many spores would most likely have assisted land plants in colonizing the land environment.

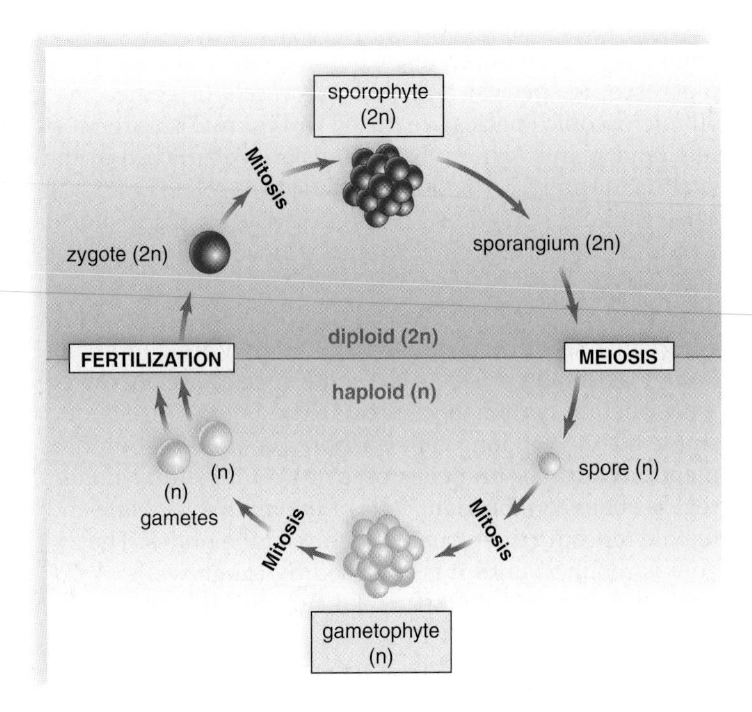


**FIGURE 23.3 Alternation of generations in land plants.**

The zygote develops into a multicellular 2n generation, and meiosis produces spores in multicellular sporangia. The gametophyte generation produces gametes within multicellular gametangia.

### *Dominant Generation*

Land plants differ as to which generation is dominant—that is, more conspicuous. In a moss, the gametophyte is dominant, but in ferns, pine trees, and peach trees, the sporophyte is dominant (Fig. 23.4). In the history of land plants, only the sporophyte evolves vascular tissue; therefore, the shift to sporophyte dominance is an adaptation to life on land. Notice that as the sporophyte gains in dominance, the gametophyte becomes microscopic. It also becomes dependent on the sporophyte.

## Other Derived Traits of Land Plants

Also, in land plants,

1. Not just the zygote but also the multicellular 2n embryo are retained and protected from drying out. Because they protect the embryo, an alternate name for the land plant clade is **embryophyta.**
2. This 2n generation, called a **sporophyte,** produces at least one, and perhaps several, multicellular, sporangia.
3. **Sporangia** (sing., sporangium) produce spores by meiosis. Spores (and pollen grains, if present) have a wall that contains **sporopollenin,** a molecule that prevents drying out.
4. Spores become an n generation, called a **gametophyte,** that bears multicellular gametangia, which have an outer layer of sterile cells and an inner mass of cells that become the gametes:

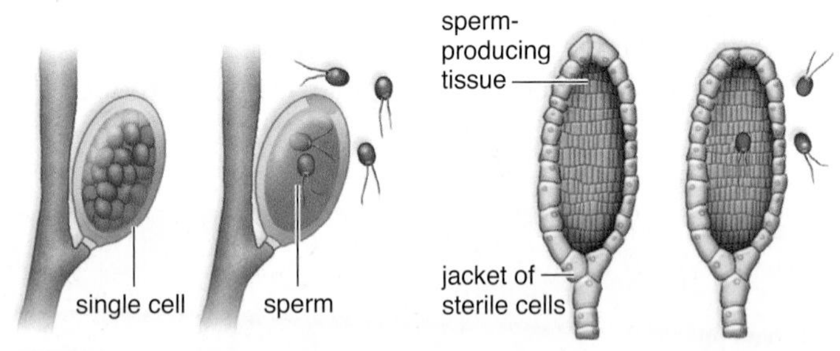


A male gametangium is called an **antheridium,** and a female gametangium is called an **archegonium.**

Aside from innovations associated with the life cycle, the exposed parts of land plants are covered by an impervious waxy **cuticle,** which prevents loss of water. Most land plants also have **stomata** (sing., **stoma**), little openings, that allow gas exchange, despite the plant being covered by a cuticle (Fig. 23.5).

Another trait we see in most land plants is the presence of apical tissue, which has the ability to produce complex tissues and organs.

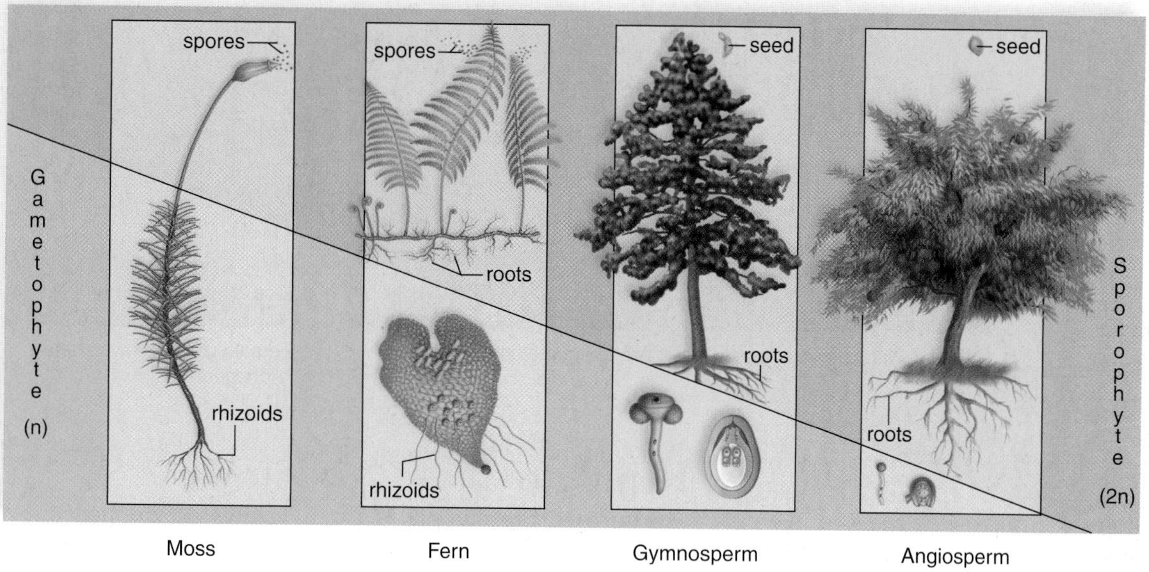

**FIGURE 23.4 Reduction in the size of the gametophyte.**

Notice the reduction in the size of the gametophyte and the increase in the size of the sporophyte among these representatives of today's land plants. This trend occurred as these plants became adapted for life on land. In the moss and fern, spores disperse the gametophyte. In gymnosperms and angiosperms, seeds disperse the sporophyte.

**Check Your Progress 23.1**

1. What are the benefits of a land existence for plants?
2. What traits are shared by both charophytes and land plants?
3. What is the role of each generation in the alternation of generations life cycle?



a. Stained photomicrograph of a leaf cross section

b. Falsely colored scanning electron micrograph of leaf surface

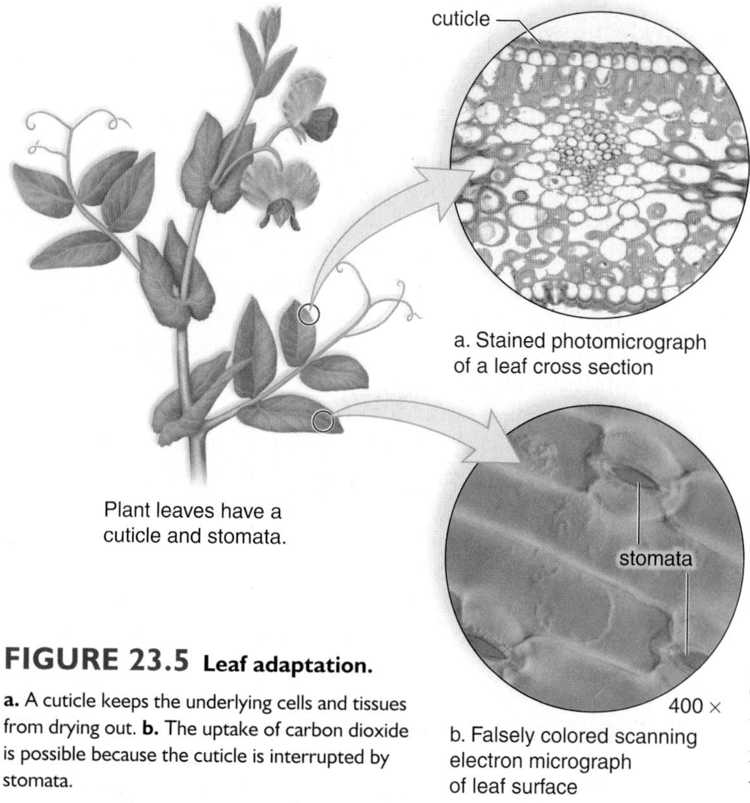

**FIGURE 23.5 Leaf adaptation.**

**a.** A cuticle keeps the underlying cells and tissues from drying out. **b.** The uptake of carbon dioxide is possible because the cuticle is interrupted by stomata.

## 23.2 Evolution of Bryophytes: Colonization of Land

The **bryophytes**—the liverworts, hornworts, and mosses—are the first plants to colonize land. The suffix wort is an Anglo-Saxon term meaning herb. They only superficially appear to have the roots, stems, and leaves because, by definition, true roots, stems, and leaves must contain vascular tissue, which the bryophytes lack. Therefore, bryophytes are often called the **nonvascular plants.** Lacking **vascular tissue,** which is specialized for the transport of water and organic nutrients throughout the body of a plant, bryophytes remain low-lying, and even mosses, which do stand erect, only reach a maximum height of about 20 cm.

The fossil record contains some evidence that the various bryophytes evolved during the Ordovician period. An incomplete fossil record makes it difficult to tell how closely related the various bryophytes are. Molecular data, in particular, suggest that these plants have individual lines of descent, as shown in Figure 23.1, and that they do not form a monophyletic group. The observation that today's mosses have a rudimentary form of vascular tissue suggests that they are more closely related to vascular plants than the hornworts and liverworts.

Bryophytes do share other traits with the vascular plants. For example, they have an alternation of generations life cycle and they have the numbered traits listed on page 412. Their bodies are covered by a cuticle that is interrupted in hornworts and mosses by stomata and they have apical tissue that produces complex tissues. However, bryophytes are the only land plants in which the gametophyte is dominant (see Fig. 23.4). Their gametangia are called antheridia and archegonia. Antheridia produce flagellated sperm, which means they need a film of moisture in order for sperm to swim to eggs located inside archegonia. The lack of vascular tissue and the presence of flagellated sperm means that you are apt to find bryophytes in moist locations. Some bryophytes compete well in harsh environments because they reproduce asexually.

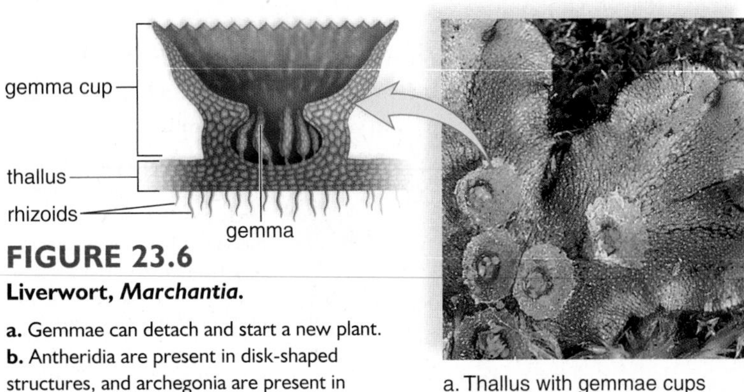


a. Thallus with gemmae cups

b. Male gametophytes bear antheridia

c. Female gametophytes bear archegonia

**FIGURE 23.6**

**Liverwort, *Marchantia*.**

**a.** Gemmae can detach and start a new plant. **b.** Antheridia are present in disk-shaped structures, and archegonia are present in umbrella-shaped structures.

## Liverworts

**Liverworts** are divided into two groups—the thallose liverworts with flattened bodies, known as a thallus; and the leafy liverworts, which superficially resemble mosses. The name liverwort refers to the lobes of a thallus, which to some resemble those of the liver. The majority of liverwort species are the leafy types.

The liverworts in the genus *Marchantia* have a thin thallus, about 30 cells thick in the center. Each branched lobe of the thallus is approximately a centimeter in length; the upper surface is divided into diamond-shaped segments with a small pore, and the lower surface bears numerous hairlike extensions called **rhizoids** [Gk., *rhizion*, dim. of root] that project into the soil (Fig. 23.6). Rhizoids serve in anchorage and limited absorption. *Marchantia* reproduces both asexually and sexually. Gemma cups on the upper surface of the thallus contain gemmae, groups of cells that detach from the thallus and can start a new plant. Sexual reproduction depends on disk-headed stalks that bear antheridia and on umbrella-headed stalks that bear archegonia. Following fertilization, tiny sporophytes composed of a foot, a short stalk, and a capsule begin growing within archegonia. Windblown spores are produced within the capsule.

**FIGURE 23.7 Hornwort, *Anthoceros* sp.**

The "horns" of a hornwort are sporophytes that grow continuously from a base anchored in gametophyte tissue.

## Hornworts

The **hornwort** gametophyte usually grows as a thin rosette or ribbonlike thallus between 1 and 5 cm in diameter. Although some species of hornworts live on trees, the majority of species live in moist, well-shaded areas. They photosynthesize, but they also have a symbiotic relationship with cyanobacteria, which, unlike plants, can fix nitrogen from the air.

The small sporophytes of a hornwort resemble tiny, green broom handles rising from a thin gametophyte, usually less than 2 cm in diameter (Fig. 23.7). Like the gametophyte, a sporophyte can photosynthesize, although it has only one chloroplast per cell. A hornwort can bypass the alternation of generations life cycle by producing asexually through fragmentation.

## Mosses

**Mosses** are the largest phyla of nonvascular plants, with over 15,000 species. The term bryophytes is sometime used to refer only to mosses. There are three distinct groups of mosses: peat mosses, granite mosses, and true mosses. Although most prefer damp, shaded locations in the temperate zone, some survive in deserts, and others inhabit bogs and streams. In forests, they frequently form a mat that covers the ground and rotting logs. In dry environments, they may become shriveled, turn brown, and look completely dead. As soon as it rains, however, the plant becomes green and resumes metabolic activity.

Figure 23.8 describes the life cycle of a typical temperate-zone moss. The gametophyte of mosses begins as an algalike branching filament of cells, the protonema, which precedes and produces upright leafy shoots that sprout rhizoids. The shoots bear either antheridia or archegonia. The dependent sporophyte consists of a foot, which is enclosed in female gametophyte tissue; a stalk; and an upper capsule, the sporangium, where spores are produced. A moss sporophyte can be likened to a child that never leaves home—it is always attached to the gametophyte. At first, the sporophyte is green and photosynthetic; at maturity, it is brown and nonphotosynthetic. In some species, the sporangium can produce as many as 50 million spores. The spores disperse the gametophyte generation.

### Check Your Progress 23.2

1. Name an advantage and disadvantage to the manner in which bryophytes reproduce on land.

**FIGURE 23.8 Moss life cycle, *Polytrichum* sp.**

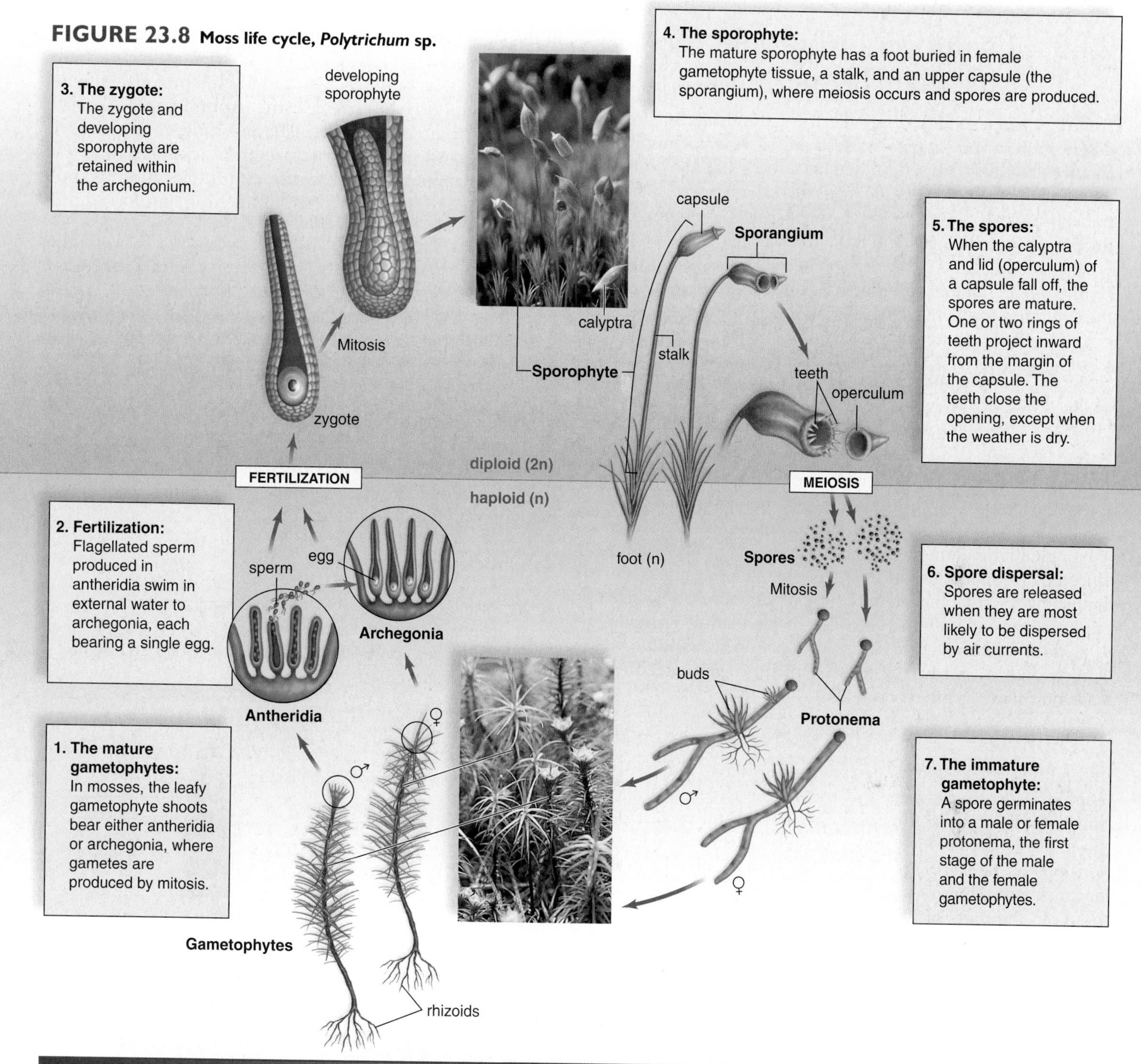


# The Uses of Bryophytes

Mosses, and also liverworts and hornworts, are of great ecological significance. They contribute to the lush beauty of rain forests, the stability of dunes, and conversion of mountain rocks to soil because of their ability to trap and hold moisture, retain metals in the soil, and tolerate desiccation. An effort to conserve bryophytes will most likely center on *Sphagnum* (peat moss) because of its commercial and ecological importance. The cell walls of peat moss have a tremendous ability to absorb water, which is why peat moss is often used in gardening to improve the water-holding capacity of the soil. One percent of the Earth's surface is peatlands, where dead *Sphagnum*, in particular, accumulates and does not decay. This material called peat can be extracted and used for various purposes, such as fuel and building materials. Peat holds much $CO_2$, and there is concern that when peat is extracted, we are losing this depository for $CO_2$, the gas that contributes most to global warming.

Bryophytes are expected to become valuable to genetic research and applications. Once scientists know which genes give bryophytes the ability to resist chemical reagents and decay, as well as animal attacks, these qualities could be transferred to other plants through genetic engineering.

## 23.3 Evolution of Lycophytes: Vascular Tissue

Today, **vascular plants** dominate the natural landscape in nearly all terrestrial habitats. Trees are vascular plants that achieve great height because they have roots that absorb water from the soil and a vascular tissue called **xylem,** which transports water through the stem to the leaves. (Another conducting tissue called **phloem** transports nutrients in a plant.) Further, the cell walls of the conducting cells in xylem contain **lignin,** a material that strengthens plant cell walls; therefore, the evolution of xylem was essential to the evolution of trees.

However, we are getting ahead of our story because the fossil record tells us that the first vascular plants, such as *Cooksonia,* were more likely a bush than a tree. *Cooksonia* is a rhyniophyte, a group of vascular plants that flourished during the Silurian period, but then became extinct by the mid-Devonian period. The rhyniophytes were only about 6.5 cm tall and had no roots or leaves. They consisted simply of a stem that forked evenly to produce branches ending in sporangia (Fig. 23.9). The branching of *Cooksonia* was significant because instead of the single sporangium produced by a bryophyte, the plant produced many sporangia, and therefore many more spores. For branching to occur, meristem has to be positioned at the apex (tip) of stems and its branches, as it is in vascular plants today.

The sporangia of *Cooksonia* produced windblown spores, and since it was not a seed plant, it was a **seedless vascular plant,** as are lycophytes.

### Lycophytes

In addition to the stem of early vascular plants, the first **lycophytes** also had leaves and roots. The leaves are called **microphylls** because they had only one strand of vascular tissue. Microphylls most likely evolved as simple side extensions of the stem (see Fig 23.11*a*). Roots evolved simply as lower extensions of the stem; the organization of vascular tissue in the roots of lycophytes today is much like it was in the stems of fossil vascular plants—the vascular tissue is centrally placed.

Today's lycophytes, also called club mosses, include three groups of 1,150 species: the ground pines (*Lycopodium*), spike mosses (*Selaginella*), and quillworts (*Isoetes*). Figure 23.10 shows the structure of *Lycopodium;* note the structure of the roots and leaves and the location of sporangia. The roots come off a branching, underground stem called a **rhizome.** The microphylls that bear sporangia are called **sporophylls,** and they are grouped into club-shaped **strobili,** accounting for their common name club mosses.

The sporophyte is dominant in lycophytes, as it is in all vascular plants. (This is the generation that has vascular tissue!) Ground pines are homosporous; the spores germinate into inconspicuous and independent gametophytes, as they do in a fern (see Fig. 23.15). The sperm are flagellated in bryophytes, lycophytes, and ferns. (The spike mosses and quillworts are heterosporous, as are seed plants; microspores develop into male gametophytes, and megaspores develop into female gametophytes.)

#### Check Your Progress 23.3

1. Name two features of lycophytes significant to the evolution of land plants.
2. How does the structure of xylem contribute to helping a plant stay erect?

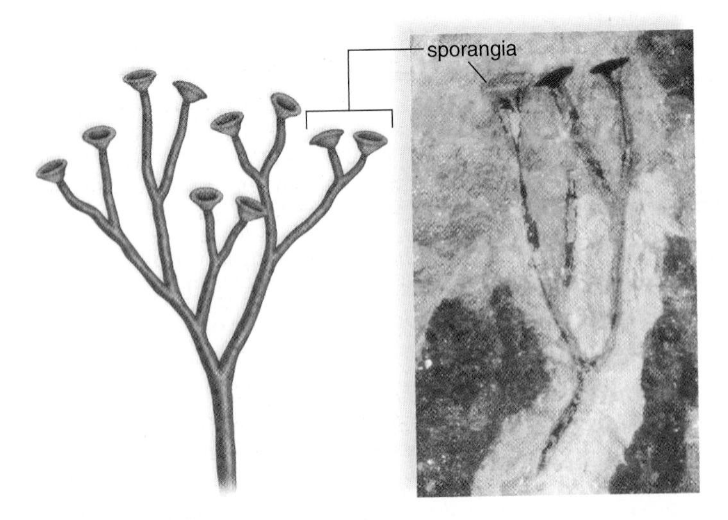


**FIGURE 23.9 A *Cooksonia* fossil.**

The upright branches of a *Cooksonia* fossil, no more than a few centimeters tall, terminated in sporangia as seen here in the drawing and photo.

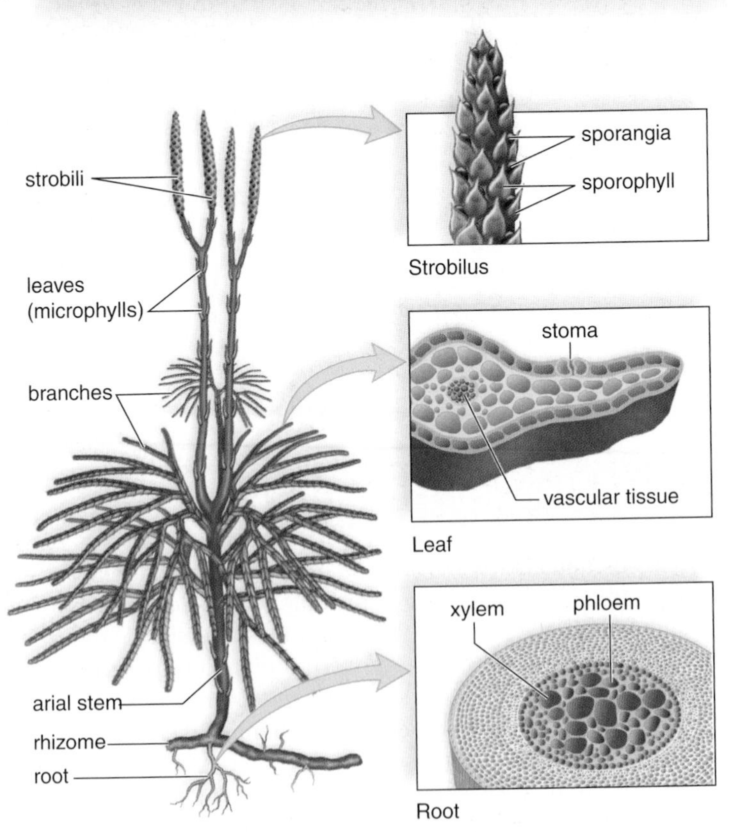


**FIGURE 23.10 Ground pine, *Lycopodium*.**

The *Lycopodium* sporophyte develops an underground rhizome system. A rhizome is an underground stem. This rhizome produces true roots along its length.

# 23.4 Evolution of Pteridophytes: Megaphylls

**Pteridophytes,** consisting of ferns and their allies—horsetails, and whisk ferns—are seedless vascular plants. However, both pteridophytes and seed plants, which are studied in Section 23.5, have megaphylls. **Megaphylls** are broad leaves with several strands of vascular tissue. Figure 23.11*a* shows the difference between microphylls and megaphylls, and Figure 23.11*b* shows how megaphylls could have evolved.

Megaphylls, which evolved about 370 MYA (million years ago), allow plants to efficiently collect solar energy, leading to the production of more food and the possibility of producing more offspring than plants without megaphylls. Therefore, the evolution of megaphylls made plants more fit. Recall that fitness, in an evolutionary sense, is judged by the number of living offspring an organism produces in relationship to others of its own kind.

The pteridophytes, like the lycophytes, were dominant from the late Devonian period through the Carboniferous period. Today, the lycophytes are quite small, but some of the extinct relatives of today's club mosses were 35 m tall and dominated by the Carboniferous swamps. The horsetails, at 18 m, and ancient tree ferns, at 8 m, also contributed significantly to the great swamp forests of the time (see the Ecology Focus on page 423).

## Horsetails

Today, **horsetails** consist of one genus, *Equisetum,* and approximately 25 species of distinct seedless vascular plants. Most horsetails inhabit wet, marshy environments around the globe. About 300 MYA, horsetails were dominant plants and grew as large as modern trees. Today, horsetails have a rhizome that produces hollow, ribbed aerial stems and reaches a height of 1.3 m (Fig. 23.12). The whorls of slender, green side branches at the joints (nodes) of the stem make the plant bear a resemblance to a horse's tail. The leaves may have been megaphylls at one time but now they are reduced and form whorls at the nodes. Many horsetails have strobili at the tips of all stems; others send up special buff-colored stems that bear the strobili. The spores germinate into inconspicuous and independent gametophytes.

The stems are tough and rigid because of silica deposited in cell walls. Early Americans, in particular, used horsetails for scouring pots and called them "scouring rushes." Today, they are still used as ingredients in a few abrasive powders.

**FIGURE 23.12 Horsetail, *Equisetum*.**

Whorls of branches and tiny leaves are at the nodes of the stem. Spore-producing sporangia are borne in strobili.

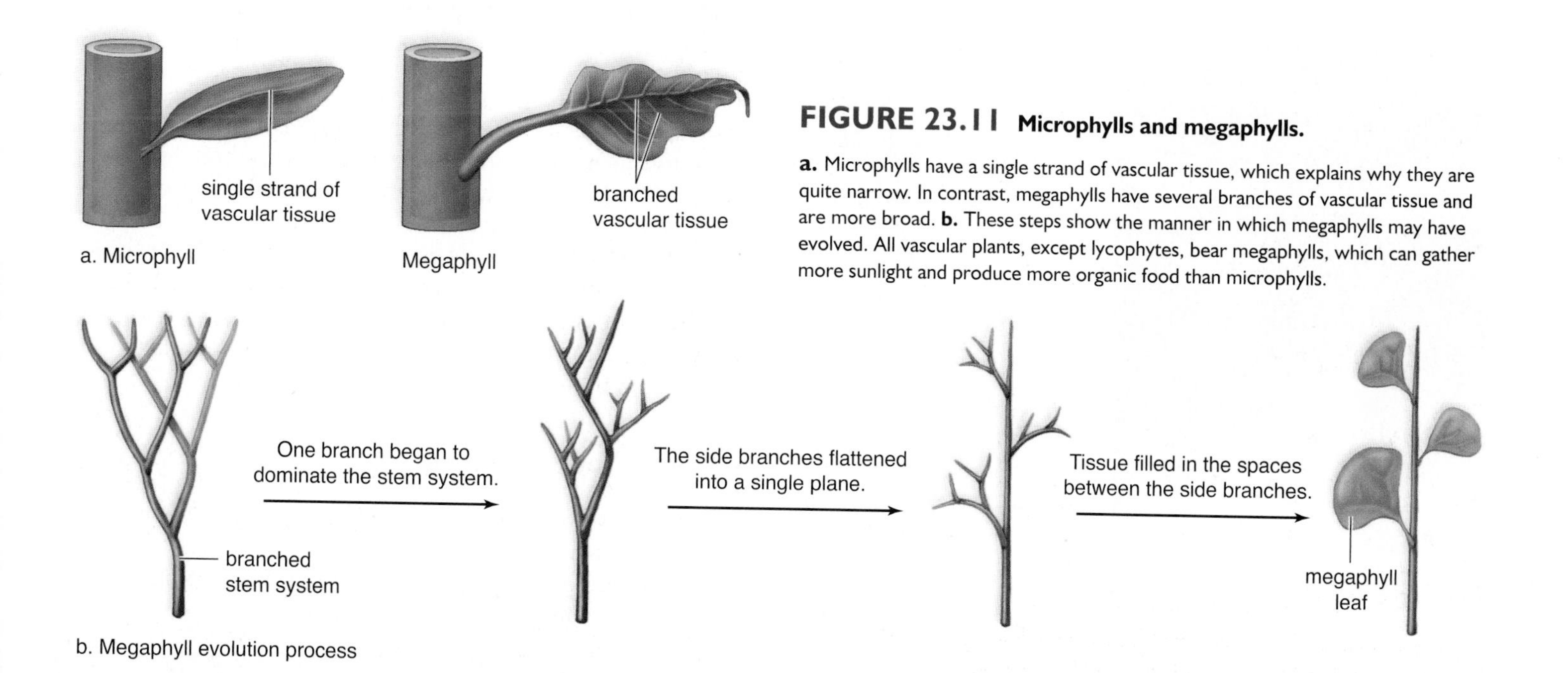


**FIGURE 23.11 Microphylls and megaphylls.**

**a.** Microphylls have a single strand of vascular tissue, which explains why they are quite narrow. In contrast, megaphylls have several branches of vascular tissue and are more broad. **b.** These steps show the manner in which megaphylls may have evolved. All vascular plants, except lycophytes, bear megaphylls, which can gather more sunlight and produce more organic food than microphylls.

## Whisk Ferns

**Whisk ferns** are represented by the genera *Psilotum* and *Tmesipteris*. Both genera live in southern climates as epiphytes (plants that live in/on trees), or they can also be found on the ground. The two *Psilotum* species resemble a whisk broom (Fig. 23.13) because they have no leaves. A horizontal rhizome gives rise to an aerial stem that repeatedly forks. The sporangia are borne on short side branches. The two to three species of *Tmesipteris* have appendages that some maintain are reduced megaphylls.

## Ferns

**Ferns** include approximately 11,000 species. Ferns are most abundant in warm, moist, tropical regions, but they can also be found in temperate regions and as far north as the Arctic Circle. Several species live in dry, rocky places and others have adapted to an aquatic life. Ferns range in size from minute aquatic species less than 1 cm in diameter to modern giant tropical tree ferns that exceed 20 m in height.

The megaphylls of ferns, called **fronds,** are commonly divided into leaflets. The royal fern has fronds that stand about 1.8 m tall; those of the hart's tongue fern are straplike and leathery; and those of the maidenhair fern are broad, with subdivided leaflets (Fig. 23.14). In nearly all ferns, the leaves first appear in a curled-up form called a fiddlehead, which unrolls as it grows.

The life cycle of a typical temperate zone fern, shown in Figure 23.15, applies in general to the other types of vascular seedless plants. The dominant sporophyte produces windblown spores by meiosis within sporangia. In a fern, the sporangia can be located within sori on the underside of the leaflets. The windblown spores disperse the gametophyte, the generation that lacks vascular tissue. The separate heart-shaped gametophyte produces flagellated sperm that swim in a film of water from the antheridium to the egg within the archegonium, where fertilization occurs. Eventually, the gametophyte disappears and the sporophyte is independent. This can be likened to a child that grows up in its parents' house, and then goes out on its own. The two generations of a fern are considered to be separate and independent of one another. Many ferns can reproduce asexually by fragmentation of the fern rhizome.

Cinnamon fern, *Osmunda cinnamomea*

Hart's tongue fern, *Campyloneurum scolopendrium*

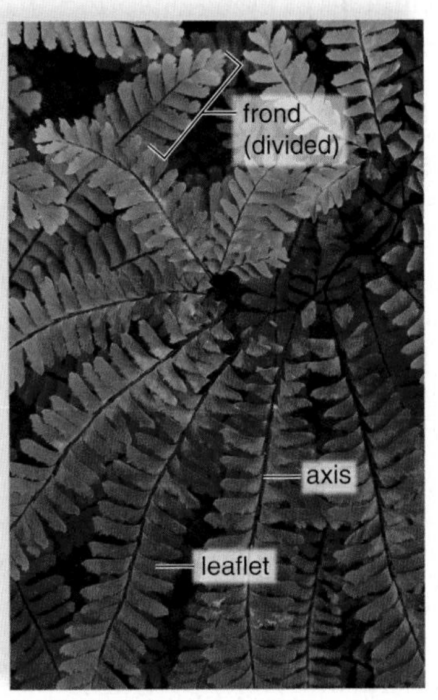

Maidenhair fern, *Adiantum pedatum*

**FIGURE 23.14** **Diversity of ferns.**

**FIGURE 23.13** **Whisk fern, *Psilotum*.**

*Psilotum* has no leaves—the branches carry on photosynthesis. The sporangia are yellow.

### Check Your Progress 23.4

1. In what two ways is the fern life cycle dependent on external water?
2. How is the life cycle of a fern different from the life cycle of a moss?

**FIGURE 23.15**
**Fern life cycle.**

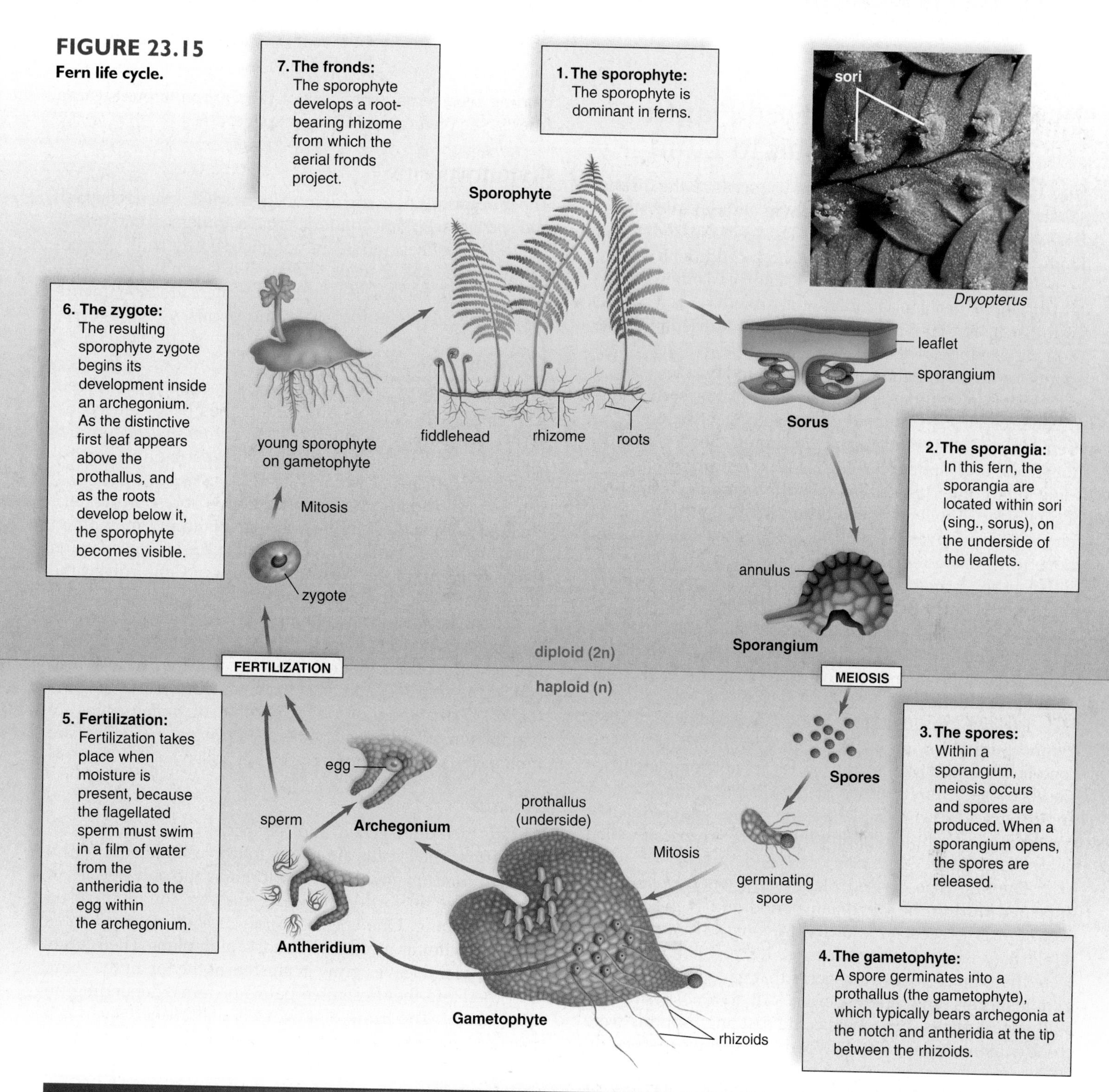


## The Uses of Ferns

Ostrich fern (*Matteuccia truthiopteris*) is the only edible fern to be traded as a food, and it comes to the table in North America as "fiddleheads." In tropical regions of Asia, Africa, and the western Pacific, dozens of types of ferns are taken from the wild and used as food. The fern *Azolla* harbors *Anabaena,* a nitrogen-fixing cyanobacteria, and *Azolla* is grown in rice paddies, where it fertilizes rice plants. It is estimated that each year, *Azolla* converts more atmospheric nitrogen into a form available for plant growth than all the legumes. And like legumes, its use avoids the problems associated with artificial fertilizer applications.

Several authorities have noted that ferns and their allies are used as medicines in China. Among the conditions being treated are boils and ulcers, whooping cough, and dysentery. Extracts from ferns have also been used to kill insects because they inhibit insect molting.

Ferns beautify gardens and horticulturists may use them in floral arrangements. Vases, small boxes and baskets, and also jewelry are made from the trunk of tree ferns. Their black and very hard vascular tissue provides many interesting patterns.

# 23.5 Evolution of Seed Plants: Full Adaptation to Land

Seed plants are vascular plants that use seeds as the dispersal stage. **Seeds** contain a sporophyte embryo and stored food within a protective seed coat. The seed coat and stored food allow an embryo to survive harsh conditions during long periods of dormancy (arrested state) until environmental conditions become favorable for growth. When a seed germinates, the stored food is a source of nutrients for the growing seedling. The survival value of seeds largely accounts for the dominance of seed plants today.

Like a few of the seedless vascular plants, seed plants are **heterosporous** (have microspores and megaspores) but their innovation was to retain the spores. (Seed plants do not release their spores! See Fig. 23.17.) The **microspores** become male gametophytes, called **pollen grains. Pollination** occurs when a pollen grain is brought to the vicinity of the female gametophyte by wind or a pollinator. Then, sperm move toward the female gametophyte through a growing **pollen tube.** A **megaspore** develops into a female gametophyte within an **ovule,** which becomes a seed following fertilization. Note that because the whole male gametophyte, rather than just the sperm as in seedless plants, moves to the female gametophyte, no external water is needed to accomplish fertilization.

The two types of seed plants alive today are called gymnosperms and angiosperms. In **gymnosperms** (mostly cone-bearing seed plants), the ovules are not completely enclosed by sporophyte tissue at the time of pollination. In **angiosperms** (flowering plants), the ovules are completely enclosed within diploid sporophyte tissue (ovaries), which becomes a fruit.

The first type of seed plant was a woody plant that appeared during the Devonian period and is miscalled a seed fern. The seed ferns of the Devonian were not ferns at all, they were progymnosperms. It's possible that these were the type of progymnosperm that gave rise to today's gymnosperms and angiosperms. All gymnosperms are still woody plants, but whereas the first angiosperms were woody, many today are nonwoody. Progymnosperms, including seed ferns, were part of the Carboniferous swamp forests (see the Ecology Focus on page 423).

## Gymnosperms

The four groups of living gymnosperms [Gk. *gymnos,* naked, and *sperma,* seed] are conifers, cycads, ginkgoes, and gnetophytes. Since their seeds are not enclosed by fruit, gymnosperms have "naked seeds." Today, living gymnosperms are classified into 780 species. Still, the conifers are more plentiful today than other types of gymnosperms.

### *Conifers*

**Conifers** consist of about 575 species of trees, many evergreen, including pines, spruces, firs, cedars, hemlocks, redwoods, cypresses, yews, and junipers. The name *conifers* signifies plants that bear **cones,** but other gymnosperm phyla are also cone-bearing. The coastal redwood (*Sequoia sempervirens*), a conifer native to northwestern California and southwestern Oregon, is the tallest living vascular plant and may attain nearly 100 m in height. Another conifer, the bristlecone pine (*Pinus longaeva*) of the White Mountains of California, is the oldest living tree; one is 4,900 years of age.

Vast areas of northern temperate regions are covered in evergreen coniferous forests (Fig. 23.16). The tough, needlelike leaves of pines conserve water because they have a thick cuticle and recessed stomata. Note in the life cycle of the pine (Fig. 23.17) that the sporophyte is dominant, pollen grains are windblown, and the seed is the dispersal stage. Conifers are **monoecious,** since they produce both pollen and seed cones.

### *Cycads*

**Cycads** include 10 genera and 140 species of distinctive gymnosperms. The cycads are native to tropical and subtropical forests. *Zamia pumila* found in Florida is the only species of cycad native to North America. Cycads are commonly used in landscaping. One species, *Cycas revoluta,* referred to as the sago palm, is a common landscaping plant. Their large, finely divided leaves grow in clusters at the top of the stem, and therefore they resemble palms or ferns, depending on their height. The trunk of a cycad is unbranched, even if it reaches a height of 15–18 m, as is possible in some species.

a. A northern coniferous forest of evergreen trees

b. Cones of lodgepole pine, *Pinus contorta*

c. Fleshy seed cones of juniper, *Juniperus*

**FIGURE 23.16** **Conifers.**

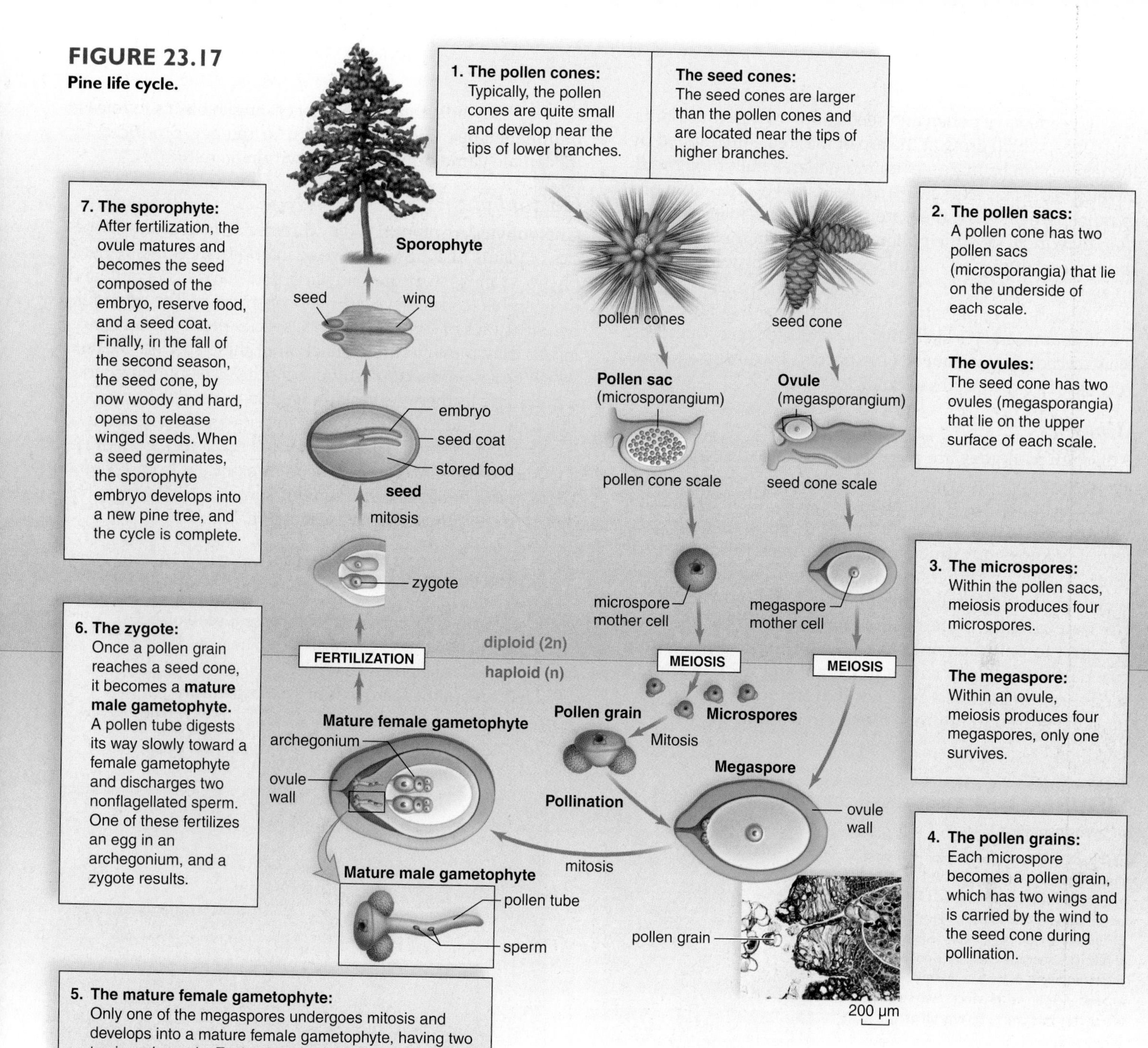


**FIGURE 23.17**
**Pine life cycle.**

## The Uses of Pines

Today, pines are grown in dense plantations for the purpose of providing wood for construction of all sorts. Although technically a softwood, some pinewoods are actually harder than so-called hardwoods. The foundations of the 125-year-old Brooklyn Bridge are made of Southern yellow pine.

Pines are well known for their beauty and pleasant smell; they make attractive additions to parks and gardens, and a number of dwarf varieties are now available for smaller gardens. Some pines are used as Christmas trees or for Christmas decorations.

Pine needles and the inner bark are rich in vitamins A and C. Pine needles can be boiled to make a tea to ease the symptoms of a cold, and the inner bark of white pines can be used in wound dressings or to provide a medicine for colds and coughs. Today, large pine seeds, called pine nuts, are sometimes harvested for use in cooking and baking.

Pine oil is distilled from the twigs and needles of Scotch pines and used to scent a number of household and personal care products, such as room sprays and masculine perfumes. Resin, made by pines as an insect and fungal deterrent, is harvested commercially for a derived product called turpentine.

Cycads have pollen and seed cones on separate plants. The cones, which grow at the top of the stem surrounded by the leaves, can be huge—even more than a meter long with a weight of 40 kg (Fig. 23.18*a*). Cycads have a life cycle of a gymnosperm, except they are pollinated by insects rather than by wind. Also, the pollen tube bursts in the vicinity of the archegonium, and multiflagellated sperm swim to reach an egg.

Cycads were plentiful in the Mesozoic era at the time of the dinosaurs, and it's likely that dinosaurs fed on cycad seeds. Now, cycads are in danger of extinction because they grow very slowly, a distinct disadvantage.

### *Ginkgoes*

Although **ginkgoes** are plentiful in the fossil record, they are represented today by only one surviving species, *Ginkgo biloba*, the ginkgo or maidenhair tree. It is called the maiden hair tree because its leaves resemble those of a maiden hair fern. Ginkgoes are **dioecious**—some trees produce seeds (Fig. 23.18*b*), and others produce pollen. The fleshy seeds, which ripen in the fall, give off such a foul odor that male trees are usually preferred for planting. Ginkgo trees are resistant to pollution and do well along city streets and in city parks. Ginkgo is native to China, and in Asia, ginkgo seeds are considered a delicacy. Extracts from ginkgo trees have been used to improve blood circulation.

Like cycads, the pollen tube of gingko bursts to release multiflagellated sperm that swim to the egg produced by the female gametophyte located within an ovule.

### *Gnetophytes*

**Gnetophytes** are represented by three living genera and 70 species of plants that are very diverse in appearance. In all gnetophytes, xylem is structured similarly, none have archegonia, and their strobili (cones) have a similar construction. The reproductive structures of some gnetophyte species produce nectar, and insects play a role in the pollination of these species. *Gnetum*, which occurs in the tropics, consists of trees or climbing vines with broad, leathery leaves arranged in pairs. *Ephedra*, occurring only in southwestern North America and southeast Asia, is a shrub with small, scalelike leaves (Fig. 23.18*c*). Ephedrine, a medicine with serious side effects, is extracted from *Ephedra*. *Welwitschia*, living in the deserts of southwestern Africa, has only two enormous, straplike leaves (Fig. 23.18*d*).

**Check Your Progress 23.5A**

1. Cite life cycle changes that represent seed plant adaptations, as exemplified by a pine tree, for reproducing on land.
2. What are the four types of gymnosperms?

**FIGURE 23.18 Three groups of gymnosperms.**

**a.** Cycads may resemble ferns or palms but they are cone-producing gymnosperms. **b.** A ginkgo tree has broad leaves and fleshy seeds borne at the end of stalklike megasporophylls. **c.** *Ephedra*, a type of gnetophyte, is a branched shrub. This specimen produces pollen in microsporangia. **d.** *Welwitschia miribilis*, another type of gnetophyte, produces two straplike leaves that split one to several times.

a. *Encephalartos transvenosus*, an African cycad

b. *Ginkgo biloba*, a native of China

c. *Ephedra*, a type of gnetophyte

d. *Welwitschia mirabilis*, a type of gnetophyte

# ecology focus

## Carboniferous Forests

Our industrial society runs on fossil fuels such as coal. The term *fossil fuel* might seem odd at first until one realizes that it refers to the remains of organic material from ancient times. During the Carboniferous period more than 300 million years ago, a great swamp forest (Fig. 23A) encompassed what is now northern Europe, the Ukraine, and the Appalachian Mountains in the United States. The weather was warm and humid, and the trees grew very tall. These are not the trees we know today; instead, they are related to today's seedless vascular plants: the lycophytes, horsetails, and ferns! Lycophytes today may stand as high as 30 cm, but their ancient relatives were 35 m tall and 1 m wide. The stroboli were up to 30 cm long, and some had leaves more than 1 m long. Horsetails too—at 18 m tall—were giants compared to today's specimens. Tree ferns were also taller than tree ferns found in the tropics today. The progymnosperms, including "seed ferns," were significant plants of a Carboniferous swamp. Seed ferns are misnamed because they were actually progymnosperms.

The amount of biomass in a Carboniferous swamp forest was enormous, and occasionally the swampy water rose and the trees fell. Trees under water do not decompose well, and their partially decayed remains became covered by sediment that sometimes changed to sedimentary rock. Exposed to pressure from sedimentary rock, the organic material then became coal, a fossil fuel. This process continued for millions of years, resulting in immense deposits of coal. Geological upheavals raised the deposits to the level where they can be mined today.

With a change of climate, the trees of the Carboniferous period became extinct, and only their herbaceous relatives survived to our time. Without these ancient forests, our life today would be far different because they helped bring about our industrialized society.

**FIGURE 23A Swamp forest of the Carboniferous period.**
*Nonvascular plants and early gymnosperms dominated the swamp forests of the Carboniferous period. Among the early gymnosperms were the seed ferns, so named because their leaves looked like fronds, as shown in a micrograph of fossil remains in the upper right.*

## Angiosperms

**Angiosperms** [Gk. *angion*, dim. of *angos*, vessel, and *sperma*, seed] are the flowering plants. They are an exceptionally large and successful group of plants, with 240,000 known species—six times the number of all other plant groups combined. Angiosperms live in all sorts of habitats, from fresh water to desert, and from the frigid north to the torrid tropics. They range in size from the tiny, almost microscopic duckweed to *Eucalyptus* trees over 100 m tall. It would be impossible to exaggerate the importance of angiosperms in our everyday lives. Angiosperms include all the hardwood trees of temperate deciduous forests and all the broadleaved evergreen trees of tropical forests. Also, all herbaceous (nonwoody plants, such as grasses), and most garden plants are flowering plants. This means that all fruits, vegetables, nuts, herbs, and grains that are the staples of the human diet are angiosperms. As discussed in the Ecology Focus on pages 428–29, they provide us with clothing, food, medicines, and other commercially valuable products.

The flowering plants are called angiosperms because their ovules, unlike those of gymnosperms, are always enclosed within diploid tissues. In the Greek derivation of their name, *angio* ("vessel") refers to the ovary, which develops into a fruit, a unique angiosperm feature.

## Origin and Radiation of Angiosperms

Although the first fossils of angiosperms are no older than about 135 million years (see Fig. 19.8), the angiosperms probably arose much earlier. Indirect evidence suggests the possible ancestors of angiosperms may have originated as long ago as 200 MYA. But their exact ancestral past has remained a mystery since Charles Darwin pondered it.

To find the angiosperm of today that might be most closely related to the first angiosperms, botanists have turned to DNA comparisons. Gene-sequencing data singled out *Amborella trichopoda* (Fig. 23.19) as having ancestral traits. This small woody shrub, with small cream-colored flowers, lives only on the island of New Caledonia in the South Pacific. Its flowers are only about 4–8 mm and the petals and sepals look the same; therefore, they are called tepals. Plants bear either male or female flowers, with a variable number of stamens or carpels.

**FIGURE 23.19** ***Amborella trichopoda.***

Molecular data suggest this plant is most closely related to the first flowering plants.

Although *A. trichopoda* may not be the original angiosperm species, it is sufficiently close that much may be learned from studying its reproductive biology. Botanists hope that this knowledge will help them understand the early adaptive radiation of angiosperms during the Tertiary period. The gymnosperms were abundant during the Mesozoic era but declined during the mass extinction that occurred at the end of the Cretaceous. Angiosperms survived and went on to become the dominant plants during modern times.

### *Monocots and Eudicots*

Most flowering plants belong to one of two classes. These classes are the **Monocotyledones,** often shortened to simply the **monocots** (about 65,000 species), and the **Eudicotyledones,** shortened to **eudicots** (about 175,000 species). The term *eudicot* (meaning true dicot) is more specific than the term *dicot*. It was discovered that some of the plants formerly classified as dicots diverged before the evolutionary split that gave rise to the two major classes of angiosperms. These earlier-evolving plants are not included in the designation eudicots.

Monocots are so called because they have only one cotyledon (seed leaf) in their seeds. Several common monocots include corn, tulips, pineapples, bamboos, and sugarcane. Eudicots are so called because they possess two cotyledons in their seeds. Several common eudicots include cactuses, strawberries, dandelions, poplars, and beans. **Cotyledons** are the seed leaves that contain nutrients that nourish the plant embryo. Table 23.2 lists several fundamental differences between monocots and eudicots.

### *The Flower*

Although **flowers** vary widely in appearance (Fig. 23.20), most have certain structures in common. The **peduncle,** a flower stalk, expands slightly at the tip into a **receptacle,** which bears the other flower parts. These parts, called sepals, petals, stamens, and carpels, are attached to the receptacle in whorls (circles) (Fig. 23.21).

1. The **sepals,** collectively called the calyx, protect the flower bud before it opens. The sepals may

**TABLE 23.2**

**Monocots and Eudicots**

| *Monocots* | *Eudicots* |
|---|---|
| One cotyledon | Two cotyledons |
| Flower parts in threes or multiples of three | Flower parts in fours or fives or multiples of four or five |
| Pollen grain with one pore | Pollen grain with three pores |
| Usually herbaceous | Woody or herbaceous |
| Usually parallel venation | Usually net venation |
| Scattered bundles in stem | Vascular bundles in a ring |
| Fibrous root system | Taproot system |

Beavertail cactus, *Opuntia basilaris*

Water lily, *Nymphaea odorata*

Blue flag iris, *Iris versicolor*

Snow trillium, *Trillium nivale*

Apple blossom, *Malus domestica*

Butterfly weed, *Asclepias tuberosa*

**FIGURE 23.20** **Flower diversity.**

Regardless of size and shape, flowers share certain features, as mentioned in text.

drop off or may be colored like the petals. Usually, however, sepals are green and remain attached to the receptacle.

2. The **petals,** collectively called the corolla, are quite diverse in size, shape, and color. The petals often attract a particular pollinator.
3. Next are the **stamens.** Each stamen consists of two parts: the anther, a saclike container, and the filament, a slender stalk. Pollen grains develop from microspores produced in the anther.
4. At the very center of a flower is the **carpel,** a vaselike structure with three major regions: the **stigma,** an enlarged sticky knob; the **style,** a slender stalk; and the **ovary,** an enlarged base that encloses one or more ovules. The ovule becomes the seed, and the ovary becomes the fruit. Fruit is instrumental in the distribution of seeds.

**TABLE 23.3**

**Other Flower Terminology**

| Term | Type of Flower |
|---|---|
| Complete | All four parts (sepals, petals, stamens, and carpels) present |
| Incomplete | Lacks one or more of the four parts |
| Perfect | Has both stamens and (a) carpel(s) |
| Imperfect | Has stamens or (a) carpel(s), but not both |
| Inflorescence | A cluster of flowers |
| Composite | Appears to be a single flower but consists of a group of tiny flowers |

It can be noted that not all flowers have all these parts (Table 23.3). A flower is said to be complete if it has all four parts; otherwise it is incomplete.

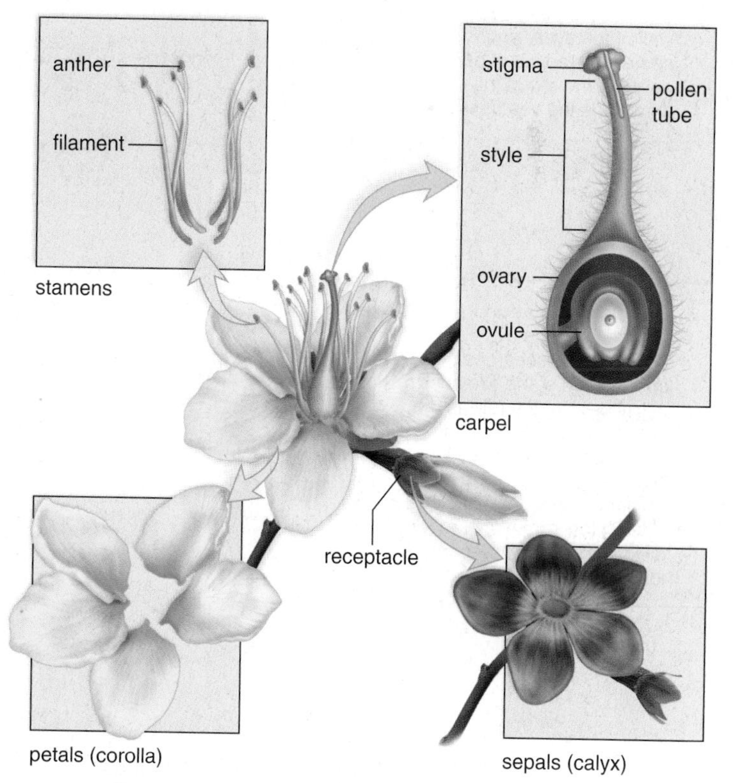


**FIGURE 23.21** **Generalized flower.**

A flower has four main parts: sepals, petals, stamens, and carpels. A stamen has an anther and filament. A carpel has a stigma, style, and ovary. An ovary contains ovules.

## Flowering Plant Life Cycle

Figure 23.22 depicts the life cycle of a typical flowering plant. Like the gymnosperms, flowering plants are heterosporous, producing two types of spores. A **megaspore** located in an ovule within an ovary of a carpel develops into an egg-bearing female gametophyte called the embryo sac. In most angiosperms, the embryo sac has seven cells; one of these is an egg, and another contains two polar nuclei. (These two nuclei are called the polar nuclei because they came from opposite ends of the embryo sac.)

**Microspores,** produced within anthers, become pollen grains that, when mature, are sperm-bearing male gametophytes. The full-fledged mature male gametophyte consists of only three cells: the tube cell and two sperm

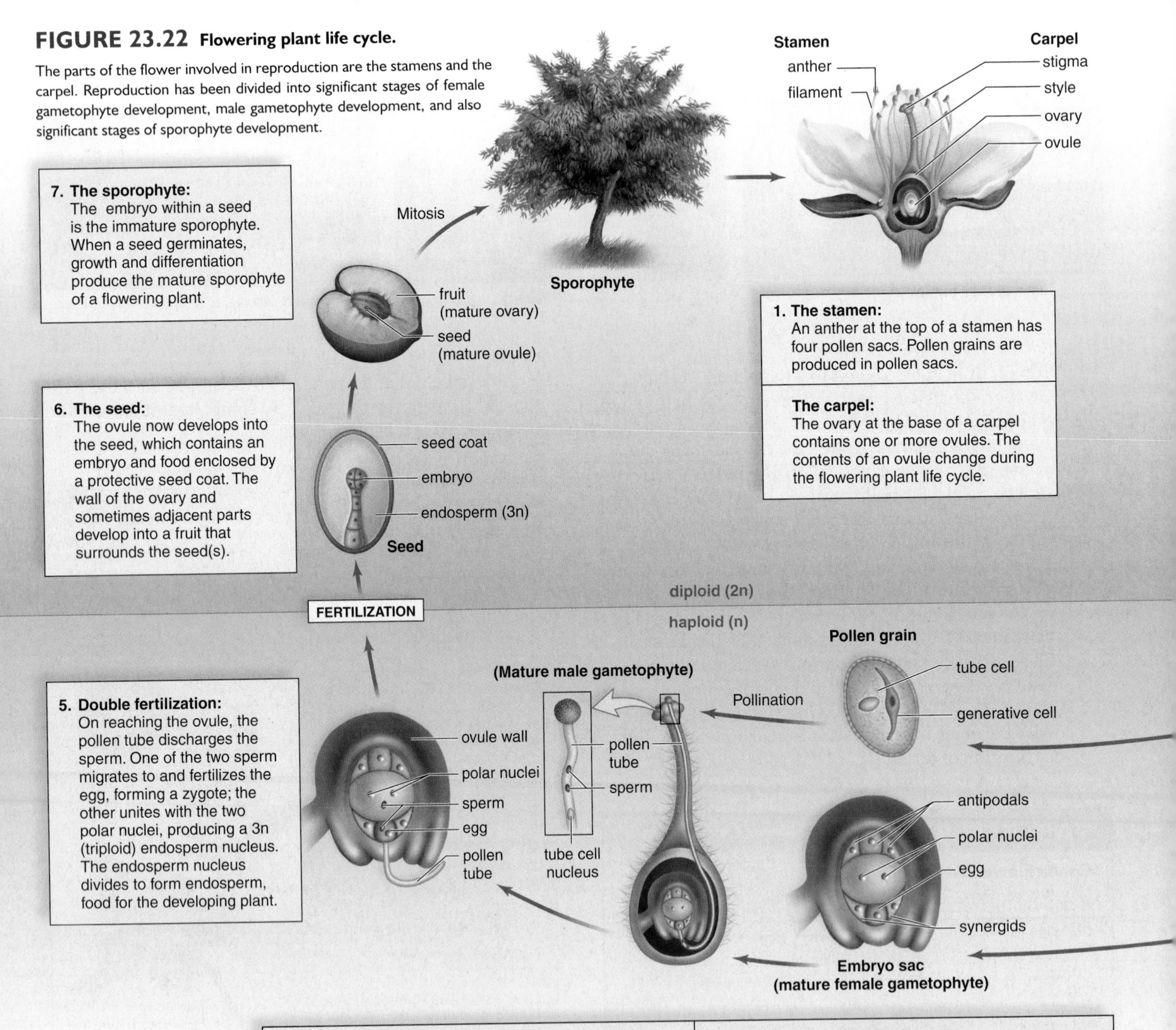


**FIGURE 23.22 Flowering plant life cycle.**
The parts of the flower involved in reproduction are the stamens and the carpel. Reproduction has been divided into significant stages of female gametophyte development, male gametophyte development, and also significant stages of sporophyte development.

cells. During pollination, a pollen grain is transported by various means from the anther to the stigma of a carpel, where it germinates. During germination, the tube cell produces a pollen tube. The **pollen tube** carries the two sperm to the micropyle (small opening) of an ovule. During double fertilization, one sperm unites with an egg, forming a diploid zygote, and the other unites with polar nuclei, forming a triploid endosperm nucleus.

| **2. The pollen sacs:** In pollen sacs (microsporangia) of the anther, meiosis produces microspores. | **The ovules:** In an ovule (megasporangium) within an ovary, meiosis produces four megaspores. |
| --- | --- |

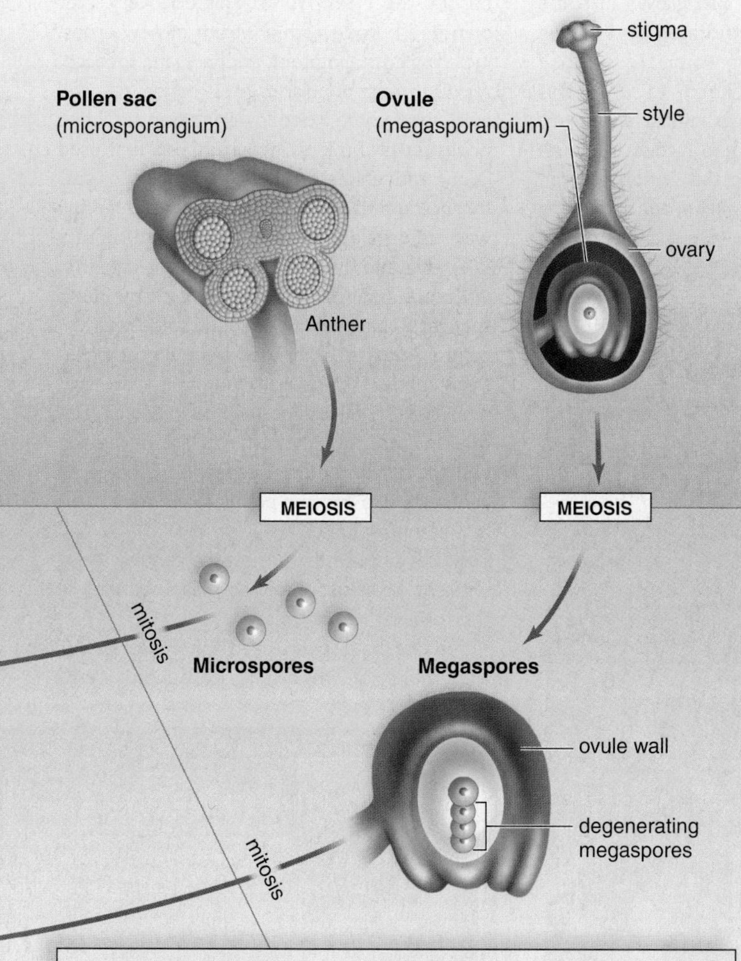


**3. The microspores:**
Each microspore in a pollen sac undergoes mitosis to become an immature pollen grain with two cells: the tube cell and the generative cell. The pollen sacs open, and the pollen grains are windblown or carried by an animal carrier, usually to other flowers. This is pollination.

**The megaspores:**
Inside the ovule of an ovary, three megaspores disintegrate, and only the remaining one undergoes mitosis to become a female gametophyte.

Ultimately, the ovule becomes a seed that contains the embryo (the sporophyte of the next generation) and stored food enclosed within a seed coat. Endosperm in some seeds is absorbed by the cotyledons, whereas in other seeds endosperm is digested as the seed matures.

A **fruit** is derived from an ovary and possibly accessory parts of the flower. Some fruits such as apples and tomatoes provide a fleshy covering, and other fruits such as pea pods and acorns provide a dry covering for seeds.

## Flowers and Diversification

Flowers are involved in the production and development of spores, gametophytes, gametes, and embryos enclosed within seeds. Successful completion of sexual reproduction in angiosperms requires the effective dispersal of pollen and then seeds. The various ways pollen and seeds can be dispersed have resulted in many different types of flowers (see Chapter 27).

Wind-pollinated flowers are usually not showy, whereas insect-pollinated flowers and bird-pollinated flowers are often colorful. Night-blooming flowers attract nocturnal mammals or insects; these flowers are usually aromatic and white or cream-colored. Although some flowers disperse their pollen by wind, many are adapted to attract specific pollinators such as bees, wasps, flies, butterflies, moths, and even bats, which carry only particular pollen from flower to flower. For example, glands located in the region of the ovary produce nectar, a nutrient that is gathered by pollinators as they go from flower to flower. Bee-pollinated flowers are usually blue or yellow and have ultraviolet shadings that lead the pollinator to the location of nectar. The mouthparts of bees are fused into a long tube that is able to obtain nectar from the base of the flower.

The fruits of flowers protect and aid in the dispersal of seeds. Dispersal occurs when seeds are transported by wind, gravity, water, and animals to another location. Fleshy fruits may be eaten by animals, which transport the seeds to a new location and then deposit them when they defecate. Because animals live in particular habitats and/or have particular migration patterns, they are apt to deliver the fruit-enclosed seeds to a suitable location for seed germination (when the embryo begins to grow again) and development of the plant.

### Check Your Progress 23.5B

1. List the components of the stamen. Where is pollen formed?
2. List the components of the carpel. Which part becomes a seed? The fruit?
3. Which groups of plants produce seeds? Give examples of each group.
4. What features of the flowering plant life cycle are not found in any other group?
5. Which are more showy, wind-pollinated flowers or animal-pollinated flowers? Why?

# ecology focus

## Plants: Could We Do Without Them?

Plants define and are the producers in most ecosystems. Humans derive most of their sustenance from three flowering plants: wheat, corn, and rice (Fig. 23B). All three of these plants are in the grass family and are collectively, along with other species, called grains. Most of the Earth's 6.7 billion people live a simple way of life, growing their food on family plots. The continued growth of these plants is essential to human existence. A virus or other disease could hit any one of these three plants and cause massive loss of life from starvation.

Wheat, corn, and rice originated and were first cultivated in different parts of the globe. Wheat is commonly used in the United States to produce flour and bread. It was first cultivated in the Near East (Iran, Iraq, and neighboring countries) about 8000 BC; hence, it is thought to be one of the earliest cultivated plants. Wheat was brought to North America in 1520 by early settlers; now the United States is one of the world's largest producers of wheat. Corn, or what is properly called maize, was first cultivated in Central America about 7,000 years ago. Maize developed from a plant called teosinte, which grows in the highlands of central Mexico. By the time Europeans were exploring Central America, over 300 varieties were already in existence—growing from Canada to Chile. We now commonly grow six major varieties of corn: sweet, pop, flour, dent, pod, and flint. Rice had its origin in southeastern Asia several thousand years ago, where it grew in swamps. Today we are familiar with white and brown rice, which differ in the extent of processing. Brown rice results when the seeds are threshed to remove the hulls—the seed coat and complete embryo remain. If the seed coat and embryo are removed, leaving only the starchy endosperm, white rice results. Unfortunately, the seed coat and embryo are a good source of vitamin B and fat-soluble vitamins. Today, rice is grown throughout the tropics and subtropics where water is abundant. It is also grown in some parts of western United States by flooding diked fields with irrigation water.

Do you have an "addiction" to sugar? This simple carbohydrate comes almost exclusively from two plants—sugarcane (grown in South America, Africa, Asia, and the Caribbean) and sugar beets (grown mostly in Europe and North America). Each provides about 50% of the world's sugar.

Many foods are bland or tasteless without spices. In the Middle Ages, wealthy Europeans spared no cost to obtain spices from the Near and Far East. In the fifteenth and sixteenth centuries, major expeditions were launched in an attempt to find better and cheaper routes for spice importation. The explorer Columbus convinced the Queen of Spain that he could find a shorter route to the Far East by traveling west by ocean rather than east by land. Columbus's idea was sound, but he encountered a little barrier, the New World. This discovery later provided Europe with a wealth of new crops, including corn, potatoes, peppers, and tobacco.

Our most popular drinks—coffee, tea, and cola—also come from flowering plants. Coffee originated in Ethiopia, where it was first used (along with animal fat) during long trips for sustenance and to relieve fatigue. Coffee as a drink was not developed until the thirteenth century in Arabia and Turkey, and it did not catch on in Europe until the seventeenth century. Tea is thought to have been developed somewhere in central Asia. Its earlier uses were almost exclusively medicinal, especially among the Chinese, who still drink tea for medical reasons. The drink as we now know it was not developed until the fourth century. By the mid-seventeenth century it had become popular in Europe. Cola is a common ingredient in tropical drinks and was used around the turn of the century, along with the drug coca (used to make cocaine), in the "original" Coca-Cola.

Plants have been used for centuries for a number of important household items

grain head

Wheat plants, *Triticum*

Corn plants, *Zea*

Rice plants, *Oryza*

**FIGURE 23B Cereal grains.**

*These three cereal grains are the principal source of calories and protein for our civilization.*

a. Dwarf fan palms, *Chamaerops,* for basket weaving

b. Rubber, *Hevea,* for auto tires

d. Tulips, *Tulipa,* for beauty

c. Cotton, *Gossypium,* for cloth

**FIGURE 23C Uses of plants.**
***a.*** *Dwarf fan palms can be used to make baskets.* ***b.*** *A rubber plant provides latex for making tires.* ***c.*** *Cotton can be used for clothing.* ***d.*** *Plants have aesthetic value.*

(Fig. 23C*a*), including the house itself. We are most familiar with lumber being used as the major structural portion in buildings. This wood comes mostly from a variety of conifers: pine, fir, and spruce, among others. In the tropics, trees and even herbs provide important components for houses. In rural parts of Central and South America, palm leaves are preferable to tin for roofs, since they last as long as ten years and are quieter during a rainstorm. In the Near East, numerous houses along rivers are made entirely of reeds.

Rubber is another plant that has many uses today (Fig. 23C*b*). The product had its origin in Brazil from the thick, white sap (latex) of the rubber tree. Once collected, the sap is placed in a large vat, where acid is added to coagulate the latex. When the water is pressed out, the product is formed into sheets or crumbled and placed into bales. Much stronger rubber, such as that in tires, was made by adding sulfur and heating in a process called vulcanization; this produces a flexible material less sensitive to temperature changes. Today, though, much rubber is synthetically produced.

Before the invention of synthetic fabrics, cotton and other natural fibers were our only source of clothing (Fig. 23C*c*). The 5,200-year-old remains of Ice Man found in the Alps had a cape made of grass. China is now the largest producer of cotton. The cotton fiber itself comes from filaments that grow on the seed. In sixteenth-century Europe, cotton was a little-understood fiber known only from stories brought back from Asia. Columbus and other explorers were amazed to see the elaborately woven cotton fabrics in the New World. But by 1800, Liverpool was the world's center of cotton trade. (Interestingly, when Levi Strauss wanted to make a tough pair of jeans, he needed a stronger fiber than cotton, so he used hemp. Hemp is now known primarily as a hallucinogenic drug—marijuana—though there has been a resurgence in its use in clothing.) Over 30 species of native cotton now grow around the world, including the United States.

An actively researched area of plant use today is that of medicinal plants. Currently about 50% of all pharmaceutical drugs have their origins from plants. The treatment of cancers appears to rest in the discovery of new plants. Indeed, the National Cancer Institute (NCI) and most pharmaceutical companies have spent millions (or, more likely, billions) of dollars to send botanists out to collect and test plant samples from around the world. Tribal medicine men, or shamans, of South America and Africa have already been of great importance in developing numerous drugs.

Over the centuries, malaria has caused far more human deaths than any other disease. After European scientists became aware that malaria can be treated by quinine, which comes from the bark of the cinchona tree, a synthetic form of the drug, chloroquine, was developed. But by the late 1960s, it was found that some of the malaria parasites, which live in red blood cells, had become resistant to the synthetically produced drug. Resistant parasites were first seen in Africa but are now showing up in Asia and the Amazon. Today, the only 100% effective drug for malaria treatment must come directly from the cinchona tree, common to northeastern South America.

Numerous plant extracts continue to be misused for their hallucinogenic or other effects on the human body: coca for cocaine and crack, opium poppy for morphine, and yam for steroids. In addition to all these uses of plants, we should not forget or neglect their aesthetic value (Fig. 23C*d*). Flowers brighten any yard, ornamental plants accent landscaping, and trees provide cooling shade during the summer and break the wind of winter days. Plants also produce oxygen, which is so necessary for all plants and animals.

## Connecting the Concepts

Land plants share a common ancestor with the charophytes, who have traits that would have been useful to the first plants to invade land. Through the evolutionary process, land plants adapted to a dry environment and the threat of water loss (desiccation). Today, the bryophytes are generally small, and they are usually found in moist habitats. However, mosses can store large amounts of water and even become dormant during dry spells. The vascular plants, on the other hand, have specialized tissues to transport water and organic nutrients from one part of the body to another. Therefore, they can grow tall. Small pores in their leaves and waxy cuticles open and close to control water loss.

Reproductive strategies in plants are also adapted to a land environment. Mosses and ferns produce flagellated sperm that require external water, but their windblown spores disperse offspring. In seed plants, pollen grains protect sperm until they fertilize an egg. The sporophyte even retains the egg-producing female gametophyte and a seed contains the resulting zygote. The seed protects the sporophyte embryo from drying out until it germinates in a new location. Dispersal of plants by spores or seeds reduces competition for resources.

Humans use land plants for various purposes, including fuel. Massive amounts of biomass were submerged by swamps and covered by sediment during the Carboniferous period. Due to extreme pressure, this organic material became the fossil fuel coal, which, along with the other fuels, makes our way of life today possible. Also, we must not forget that plants produce food and oxygen, two resources that keep our biosphere functioning.

## summary

### 23.1 The Green Algal Ancestor of Plants

Land plants evolved from a common ancestor with multicellular, freshwater green algae about 450 MYA. During the evolution of plants, protecting the embryo, apical growth, vascular tissue for transporting water and organic nutrients, possession of megaphylls, using seeds to disperse offspring, and having flowers were all adaptations to a land existence.

All plants have a life cycle that includes an alternation of generations. In this life cycle, a haploid gametophyte alternates with a diploid sporophyte. During the evolution of plants, the sporophyte gained in dominance, while the gametophyte became microscopic and dependent on the sporophyte.

### 23.2 Evolution of Bryophytes: Colonization of Land

Ancient bryophytes were the first plants to colonize land. Today, the bryophytes consist of liverworts, hornworts, and mosses that lack well-developed vascular tissue. Sporophytes of bryophytes are nutritionally dependent on the gametophyte, which is more conspicuous and photosynthetic. The life cycle of the moss (see Fig. 23.8) demonstrates reproductive strategies such as flagellated sperm and dispersal by means of windblown spores. The bryophytes are not a monophyletic group.

### 23.3 Evolution of Lycophytes: Vascular Tissue

Vascular plants, such as the rhyniophytes, evolved during the Silurian period. The sporophyte has two kinds of well-defined conducting tissues. Xylem is specialized to conduct water and dissolve minerals, and phloem is specialized to conduct organic nutrients. The lycophytes are descended from these first plants and they have vascular tissue. Ancient lycophytes also had the first leaves, which were microphylls; their life cycle is similar to that of the fern.

### 23.4 Evolution of Pteridophytes: Megaphylls

In pteridophytes (ferns and their allies, horsetails and whisk ferns), and also lycophytes, the sporophyte is dominant and is separate from the tiny gametophyte, which produces flagellated sperm. Windblown spores are dispersal agents. Today's ferns have obvious megaphylls—horsetails and whisk ferns have reduced megaphylls.

Seedless vascular plant is a description that applies to lycophytes, ferns, and fern allies that grew to enormous sizes during the Carboniferous period when the climate was warm and wet. Today, seedless vascular plants that live in the temperate zone use asexual propagation to spread into environments that are not favorable to a water-dependent gametophyte generation.

### 23.5 Evolution of Seed Plants: Full Adaptation to Land

Seed plants also have an alternation of generations, but they are heterosporous, producing both microspores and megaspores. Gametophytes are so reduced that the female gametophyte is retained within an ovule. Microspores become the windblown or animal-transported male gametophytes—the pollen grains. Pollen grains carry sperm to the egg-bearing female gametophyte. Following fertilization, the ovule becomes the seed. A seed contains a sporophyte embryo, and therefore seeds disperse the sporophyte generation. Fertilization no longer uses external water, and sexual reproduction is fully adapted to the terrestrial environment.

The gymnosperms (cone-bearing plants) and also possibly angiosperms (flowering plants) evolved from woody seed ferns during the Devonian period. The conifers, represented by the pine tree, exemplify the traits of these plants. Gymnosperms have "naked seeds" because the seeds are not enclosed by fruit, as are those of flowering plants.

A woody shrub, *Amborella trichopoda,* has been identified as most closely related to the common ancestor for the angiosperms. In angiosperms, the reproductive organs are found in flowers; the ovules, which become seeds, are located in the ovary, which becomes the fruit. Therefore, angiosperms have "covered seeds." In many angiosperms, pollen is transported from flower to flower by insects and other animals. Both flowers and fruits are found only in angiosperms and may account for the extensive colonization of terrestrial environments by the flowering plants. (See also Chapter 27.)

## understanding the terms

alternation of generations life cycle 412
angiosperm 420, 424
antheridium 412
archegonium 412
bryophyte 413
carpel 425
charophyte 410
coal 423
cone 420
conifer 420
cotyledon 424
cuticle 412
cycad 420
dioecious 422
embryophyta 412
eudicot 424
Eudicotyledone 424

fern 418
flower 424
frond 418
fruit 427
gametophyte 412
ginkgo 422
gnetophyte 422
gymnosperm 420
heterosporous 420
hornwort 414
horsetail 417
lignin 416
liverwort 414
lycophyte 416
megaphyll 417
megaspore 420, 426
microphyll 416
microspore 420, 426
monocot 424
Monocotyledone 424
monoecious 420
moss 414
nonvascular plant 413
ovary 425
ovule 420
peduncle 424
petal 425
phloem 416
plant 410
pollen grain 420
pollen tube 420, 427
pollination 420
pteridophyte 417
receptacle 424
rhizoid 414
rhizome 416
seed 420
seedless vascular plant 416
sepal 424
sporangia (sing., sporangium) 412
spore 412
sporophyll 416
sporophyte 412
sporopollenin 412
stamen 425
stigma 425
stomata (sing., stoma) 412
strobilus (pl., strobili) 416
style 425
vascular plant 416
vascular tissue 413
whisk fern 418
xylem 416

Match the terms to these definitions:

a. _______________ Diploid generation of the alternation of generations life cycle of a plant; meiosis produces haploid spores that develop into the gametophyte.

b. _______________ Flowering plant group; members have one embryonic leaf, parallel-veined leaves, scattered vascular bundles, and other characteristics.

c. _______________ Male gametophyte in seed plants.

d. _______________ Rootlike hair that anchors a nonvascular plant and absorbs minerals and water from the soil.

## reviewing this chapter

1. Refer to Figure 23.1, and trace the evolutionary history of land plants. What traits do charophytes have that are shared by land plants? 410–11
2. What is meant when it is said that a plant alternates generations? Distinguish between a sporophyte and a gametophyte. 412–13
3. Describe the various types of bryophytes and the life cycle of mosses. Discuss the ecological and commercial importance of mosses. 413–15
4. When do vascular plants appear in the fossil record, and why do lycophytes perhaps resemble the first vascular plants? Mention the importance of branching. 416–18
5. Draw a diagram to describe the life cycle of a fern, pointing out significant features. What are the human uses of ferns? 419
6. What features do all seed plants have in common? When do seed plants appear in the fossil record? 420
7. List and describe the four phyla of gymnosperms. What are the human uses of gymnosperms? 420–22
8. Use a diagram of the pine life cycle to point out significant features, including those that distinguish a seed plant's life cycle from that of a seedless vascular plant. 421
9. What is known about the ancestry of flowering plants? 424
10. How do monocots and eudicots differ? What are the parts of a flower? 424–25
11. Use a diagram to explain and point out significant features of the flowering plant life cycle. 426–27
12. Offer an explanation as to why flowering plants are the dominant plants today. 427

## testing yourself

Choose the best answer for each question.

1. Which of these are characteristics of land plants?
   a. multicellular with specialized tissues and organs
   b. photosynthetic and contain chlorophylls *a* and *b*
   c. protect the developing embryo from desiccation
   d. have an alternation of generations life cycle
   e. All of these are correct.
2. In bryophytes, sperm usually move from the antheridium to the archegonium by
   a. swimming.
   b. flying.
   c. insect pollination.
   d. worm pollination.
   e. bird pollination.
3. Ferns have
   a. a dominant gametophyte generation.
   b. vascular tissue.
   c. seeds.
   d. Both a and b are correct.
   e. Choices a, b, and c are correct.
4. The spore-bearing structure that gives rise to a female gametophyte in seed plants is called a
   a. microphyll.
   b. spore.
   c. megasporangium.
   d. microsporangium.
   e. sporophyll.
5. A small, upright plant that resembles a tiny upright pine tree with club-shaped strobilii and microphylls is a
   a. whisk fern.
   b. lycophyte.
   c. conifer.
   d. horsetail.
   e. fern.
6. Trends in the evolution of plants include all of the following except
   a. from homospory to heterospory.
   b. from less to more reliance on water for life cycle.
   c. from nonvascular to vascular.
   d. from nonwoody to woody.
7. Gymnosperms
   a. have flowers.
   b. are eudicots.
   c. are monocots.
   d. do not have spores in their life cycle.
   e. reproduce by seeds.
8. In the moss life cycle, the sporophyte
   a. consists of leafy green shoots.
   b. is the heart-shaped prothallus.
   c. consists of a foot, a stalk, and a capsule.
   d. is the dominant generation.
   e. All of these are correct.
9. Microphylls
   a. have a single strand of vascular tissue.
   b. evolved before megaphylls.
   c. evolved as extensions of the stem.
   d. are found in lycophytes.
   e. All of these are correct.

10. How are ferns different from mosses?
    a. Only ferns produce spores as dispersal agents.
    b. Ferns have vascular tissue.
    c. In the fern life cycle, the gametophyte and sporophyte are both independent.
    d. Ferns do not have flagellated sperm.
    e. Both b and c are correct.
11. Which of these pairs is mismatched?
    a. pollen grain—male gametophyte
    b. ovule—female gametophyte
    c. seed—immature sporophyte
    d. pollen tube—spores
    e. tree—mature sporophyte
12. In the life cycle of the pine tree, the ovules are found on
    a. needle-like leaves.
    b. seed cones.
    c. pollen cones.
    d. root hairs.
    e. All of these are correct.
13. Monocotyledonous plants often have
    a. parallel leaf venation.
    b. flower parts in units of four or five.
    c. leaves with petioles only.
    d. flowers with stipules.
    e. Choices b, c, and d are correct.
14. Which of these pairs is mismatched?
    a. anther—produces microspores
    b. carpel—produces pollen
    c. ovule—becomes seed
    d. ovary—becomes fruit
    e. flower—reproductive structure
15. Which of these plants contributed the most to our present-day supply of coal?
    a. bryophytes
    b. seedless vascular plants
    c. gymnosperms
    d. angiosperms
    e. Both b and c are correct.
16. Which of these is found in seed plants?
    a. complex vascular tissue
    b. pollen grains that are not flagellated
    c. retention of female gametophyte within the ovule
    d. roots, stems, and leaves
    e. All of these are correct.
17. Which of these is a seedless vascular plant?
    a. gymnosperm
    b. angiosperm
    c. fern
    d. monocot
    e. eudicot
18. Label this diagram of alternation of generations life cycle.

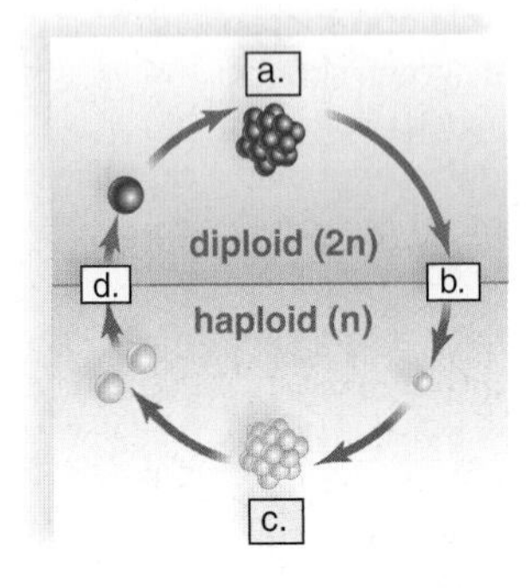


## thinking scientifically

1. Using as many terms as necessary (from both X and Y axes), fill in the proposed phylogenetic tree for vascular plants.

| | ferns | conifers | ginkgos | monocots | eudicots |
|---|---|---|---|---|---|
| vascular tissue | X | X | X | X | X |
| seed plants | | X | X | X | X |
| naked seeds | | X | X | | |
| needle-like leaves | | X | | | |
| fan-shaped leaves | | | X | | |
| enclosed seeds | | | | X | X |
| one embryonic leaf | | | | X | |
| two embryonic leaves | | | | | X |

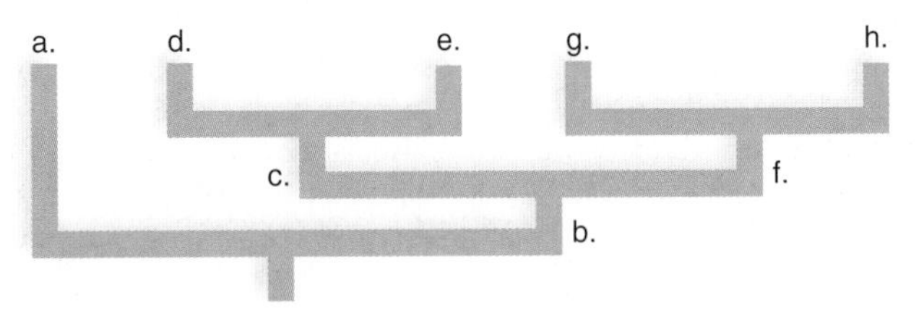


2. Using Figure 23.1, distinguish between the **(a)** microphyll and the **(b)** megaphyll clade.

## bioethical issue

### Saving Plant Species

Pollinator populations have been decimated by pollution, pesticide use, and destruction or fragmentation of natural areas. Belatedly, we have come to realize that various types of bees are responsible for pollinating such cash crops as blueberries, cranberries, and squash and are partly responsible for pollinating apple, almond, and cherry trees.

Why are we so shortsighted when it comes to protecting the environment and living creatures like pollinators? Because pollinators are a resource held in common. The term *commons* originally meant a piece of land where all members of a village were allowed to graze their cattle. The farmer who thought only of himself and grazed more cattle than his neighbor was better off. The difficulty is, of course, that eventually the resource is depleted, and everyone loses.

So, a farmer or property owner who uses pesticides is only thinking of his or her field or lawn and not the good of the whole. The commons can only be protected if citizens have the foresight to enact rules and regulations by which all abide. DDT was outlawed in this country in part because it led to the decline of birds of prey. Similarly, we may need legislation to protect pollinators from factors that kill them off. Legislation to protect pollinators would protect the food supply for all of us.

## *Biology* website

The companion website for *Biology* provides a wealth of information organized and integrated by chapter. You will find practice tests, animations, videos, and much more that will complement your learning and understanding of general biology.

**http://www.mhhe.com/maderbiology10**

# 24

# Flowering Plants: Structure and Organization

*a stunning array of plant life covers the Earth, and over 80% of all living plants are flowering plants, or angiosperms. Therefore, it is fitting that we set aside a chapter of this text to examine the structure and the function of flowering plants. The organization of flowering plants allows them to photosynthesize on land. The elevated leaves have a shape that facilitates absorption of solar energy and carbon dioxide. Strong stems conduct water up to the leaves from the roots, which not only anchor the plant but also absorb water and minerals. All the vegetative organs of flowering plants have a role to play in photosynthesis.*

*When plants photosynthesize, they take in $CO_2$. Therefore, keeping our world green by preserving plants, particularly forests, is a way to remove $CO_2$ from the atmosphere and reduce the dangers of global warming. Think of this when you study this chapter about the structure of roots, stems, and leaves.*

The vegetative organs of a flowering plant consist of root, stems, and leaves.

## concepts

# 24.1 Organs of Flowering Plants

From cacti living in hot deserts to water lilies growing in a nearby pond, the flowering plants, or angiosperms, are extremely diverse. Despite their great diversity in size and shape, flowering plants share many common structural features. Most flowering plants possess a root system and a shoot system (Fig. 24.1). The **root system** simply consists of the roots, while the **shoot system** consists of the stem and leaves. A typical plant features three vegetative **organs** (structures that contain different tissues and perform one or more specific functions) that allow them to live and grow. The roots, stems, and leaves are the vegetative organs. **Vegetative organs** are concerned with growth and nutrition and not reproduction. Flowers, seeds, and fruits are structures involved in reproduction.

## Roots

The root system in the majority of plants is located underground. As a rule of thumb, the root system is at least equivalent in size and extent to the shoot system. An apple tree has a much larger root system than a corn plant, for example. A single corn plant may have roots as deep as 2.5 m and spread out over 1.5 m, while a mesquite tree that lives in the desert may have roots that penetrate to a depth of over 20 m.

The extensive root system of a plant anchors it in the soil and gives it support (Fig. 24.2*a*). The root system absorbs water and minerals from the soil for the entire plant. The cylindrical shape of a root allows it to penetrate the soil as it grows and permits water to be absorbed from all sides. The absorptive capacity of a root is dependent on its many branches, which all bear root hairs in a special zone near the tip. Root hairs, which are projections from epidermal root-hair cells, are the structures that absorb water and minerals. Root hairs are so numerous that they tremendously increase the absorptive surface of a root. It has been estimated that a single rye plant has about 14 billion hair cells, and if placed end to end, the root hairs would stretch 10,626 km. Root-hair cells are constantly being replaced, so this same rye plant forms about 100 million new root-hair cells every day. A plant roughly pulled out of the soil will not fare well when transplanted; this is because small lateral roots and root hairs are torn off. Transplantation is more apt to be successful if you take a part of the surrounding soil along with the plant, leaving as much of the lateral roots and the root hairs intact as possible.

Roots have still other functions. Roots produce hormones that stimulate the growth of stems and coordinate their size with the size of the root. It is most efficient for a plant to have root and stem sizes that are appropriate to each other. **Perennial** plants have vegetative structures that live year after year. Herbaceous perennials, which live in temperate areas and die back, store the products of photosynthesis in their roots. Carrots and sweet potatoes are the roots of such plants.

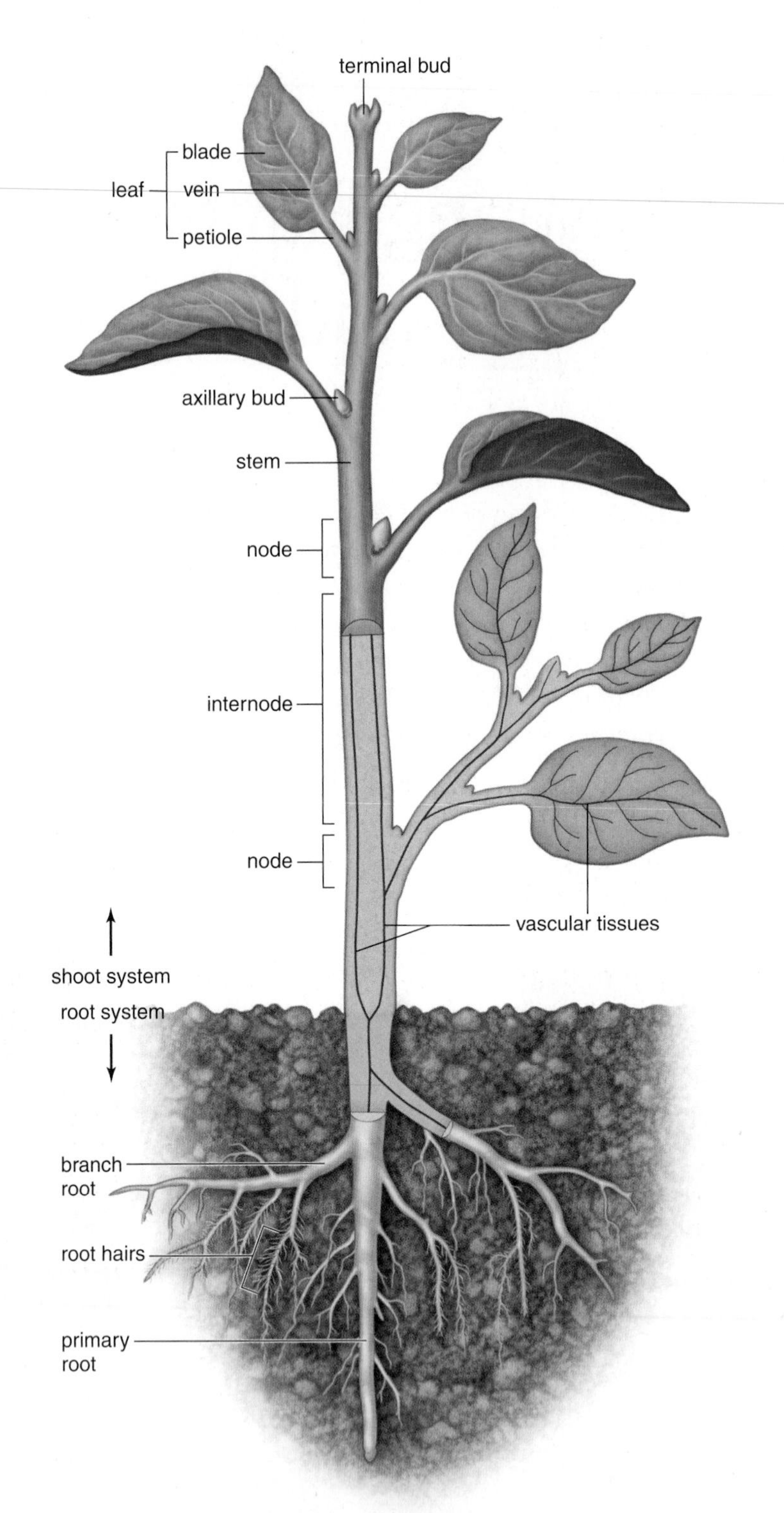


**FIGURE 24.1 Organization of a plant body.**

The body of a plant consists of a root system and a shoot system. The shoot system contains the stem and leaves, two types of plant vegetative organs. Axillary buds can develop into branches of stems or flowers, the reproductive structures of a plant. The root system is connected to the shoot system by vascular tissue (brown) that extends from the roots to the leaves.

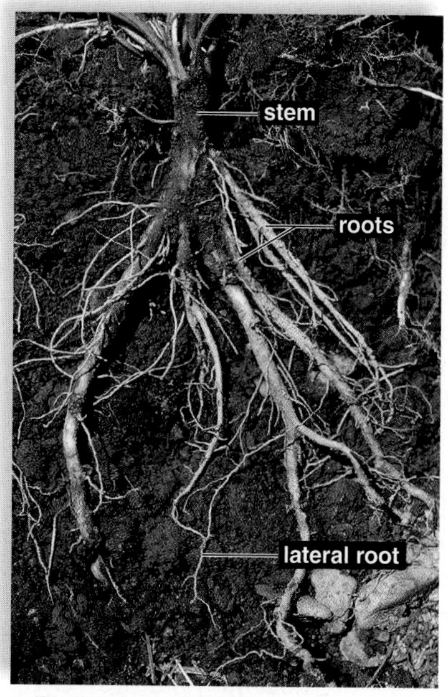


a. Root system, dandelion

b. Shoot system, bean seedling

c. Leaves, pumpkin seedling

**FIGURE 24.2 Vegetative organs of several eudicots.**

**a.** The root system anchors the plant and absorbs water and minerals. **b.** The shoot system consists of a stem and its branches, which support the leaves and transport water and organic nutrients. **c.** The leaves, which may be broad and thin, carry on photosynthesis.

## Stems

The shoot system of a plant is composed of the stem, the branches, and the leaves. A **stem,** the main axis of a plant, has a terminal bud that allows the stem to elongate and produce new leaves (Fig. 24.2*b*). If upright, as most are, a stem supports leaves in a way that exposes each one to as much sunlight as possible. A **node** occurs where leaves are attached to the stem; the region between nodes is called an **internode** (see Fig. 24.1). An **axillary bud,** located at a node in the upper angle between the leaf and the stem can produce new branches of the stem (or flowers). The presence of nodes and internodes is used to identify a stem, even if it happens to be an underground stem. A horizontal underground stem, called a rhizome, sends out roots below and shoots above at the nodes as it grows. Therefore, a rhizome allows a plant, such as ginger and bamboo, to increase its territory.

In addition to supporting the leaves, a stem has vascular tissue that transports water and minerals from the roots through the stem to the leaves and transports the products of photosynthesis, usually in the opposite direction. Nonliving cells form a continuous pipeline for water and mineral transport, while living cells join end to end for organic nutrient transport. A cylindrical stem can sometimes expand in girth as well as length. As trees grow taller each year, they accumulate woody tissue that adds to the strength of their stems.

Stems may have functions other than those mentioned: increasing the length of shoot system and transporting water and nutrients. In some plants (e.g., cactus), the stem is the primary photosynthetic organ. The stem is also a water reservoir in succulent plants. Some underground branches of a stem, or a portion of the root called a tuber, store nutrients.

## Leaves

**Leaves** are the major part of a plant that carries on photosynthesis, a process that requires water, carbon dioxide, and sunlight. Leaves receive water from the root system by way of the stem. Stems and leaves function together to bring about water transport from the roots.

The size, shape, color, and texture of leaves are highly variable. These characteristics are fundamental in plant identification. The leaves of some aquatic duckweeds may be less than 1 mm in diameter, while some palms may have leaves that exceed 6 m in length. The shape of leaves can vary from cactus spines to deeply lobed white oak leaves. Leaves can exhibit a variety of colors from various shades of green to deep purple. The texture of leaves varies from smooth and waxy like a magnolia to coarse like a sycamore. Plants that bear leaves the entire year are called **evergreens** and those that lose their leaves every year are called **deciduous.**

Broad and thin plant leaves have the maximum surface area for the absorption of carbon dioxide and the collection of solar energy needed for photosynthesis. Also unlike stems, leaves are almost never woody. With few exceptions, their cells are living, and the bulk of a leaf contains tissue specialized to carry on photosynthesis.

The wide portion of a foliage leaf is called the **blade.** The **petiole** is a stalk that attaches the blade to the stem (Fig. 24.2*c*). The upper acute angle between the petiole and stem is the leaf axil where the axillary bud is found. Not all leaves are foliage leaves. Some are specialized to protect buds, attach to objects (tendrils), store food (bulbs), or even capture insects. Specialized leaves are discussed on page 451.

## Monocot Versus Eudicot Plants

Flowering plants are divided into two groups, depending on the number of **cotyledons,** or seed leaves, in the embryonic plant (Fig. 24.3). Some have one cotyledon, and these plants are known as monocotyledons, or **monocots.** Other embryos have two cotyledons, and these plants are known as eudicotyledons, or **eudicots.** Cotyledons [Gk. *cotyledon,* cup-shaped cavity] of eudicots supply nutrients for seedlings, but the cotyledon of monocots acts as a transfer tissue, and the nutrients are derived from the endosperm before the true leaves begin photosynthesizing.

The vascular (transport) tissue is organized differently in monocots and eudicots. In the monocot root, vascular tissue occurs in a ring. In the eudicot root, the xylem, which transports water and minerals, is star-shaped; and the phloem, which transports organic nutrients, is located between the points of the star. In the monocot stem, the vascular bundles, which contain vascular tissue surrounded by a bundle sheath, are scattered. In a eudicot stem, the vascular bundles occur in a ring.

Leaf veins are vascular bundles within a leaf. Monocots exhibit parallel venation, and eudicots exhibit netted venation, which may be either pinnate or palmate. Pinnate venation means that major veins originate from points along the centrally placed main vein, and palmate venation means that the major veins all originate at the point of attachment of the blade to the petiole:

Adult monocots and eudicots have other structural differences, such as the number of flower parts and the number of pores in the wall of their pollen grains. Monocots have their flower parts arranged in multiples of three, and eudicots have their flower parts arranged in multiples of four or five. Eudicot pollen grains usually have three pores, and monocot pollen grains usually have one pore.

Although the distinctions between monocots and eudicots may seem of limited importance, they do in fact affect many aspects of their structure. The eudicots are the larger group and include some of our most familiar flowering plants—from dandelions to oak trees. The monocots include grasses, lilies, orchids, and palm trees, among others. Some of our most significant food sources are monocots, including rice, wheat, and corn.

### Check Your Progress 24.1

1. List the three vegetative organs in a plant and state their major functions.
2. List significant differences between monocots and eudicots.

**FIGURE 24.3 Flowering plants are either monocots or eudicots.**

Five features illustrated here are used to distinguish monocots from eudicots: number of cotyledons; the arrangement of vascular tissue in roots, stems, and leaves; and the number of flower parts.

| | Seed | Root | Stem | Leaf | Flower |
|---|---|---|---|---|---|
| Monocots | One cotyledon in seed | Root xylem and phloem in a ring | Vascular bundles scattered in stem | Leaf veins form a parallel pattern | Flower parts in threes and multiples of three |
| Eudicots | Two cotyledons in seed | Root phloem between arms of xylem | Vascular bundles in a distinct ring | Leaf veins form a net pattern | Flower parts in fours or fives and their multiples |

# 24.2 Tissues of Flowering Plants

A flowering plant has the ability to grow its entire life because it possesses meristematic (embryonic) tissue. Apical **meristems** are located at or near the tips of stems and roots, where they increase the length of these structures. This increase in length is called primary growth. In addition to apical meristems, monocots have a type of meristem called intercalary [L. *intercalare,* to insert] meristem, which allows them to regrow lost parts. Intercalary meristems occur between mature tissues, and they account for why grass can so readily regrow after being grazed by a cow or cut by a lawnmower.

Apical meristem continually produces three types of meristem, and these develop into the three types of specialized primary tissues in the body of a plant: Protoderm gives rise to epidermis; ground meristem produces ground tissue; and procambium produces vascular tissue. The functions of these three specialized tissues include:

1. **Epidermal tissue** forms the outer protective covering of a plant.
2. **Ground tissue** fills the interior of a plant.
3. **Vascular tissue** transports water and nutrients in a plant and provides support.

## Epidermal Tissue

The entire body of both nonwoody (herbaceous) and young woody plants is covered by a layer of **epidermis** [Gk. *epi,* over, and *derma,* skin], which contains closely packed epidermal cells. The walls of epidermal cells that are exposed to air are covered with a waxy **cuticle** [L. *cutis,* skin] to minimize water loss. The cuticle also protects against bacteria and other organisms that might cause disease.

In roots, certain epidermal cells have long, slender projections called **root hairs** (Fig. 24.4*a*). As mentioned, the hairs increase the surface area of the root for absorption of water and minerals; they also help anchor the plant firmly in place.

On stems, leaves, and reproductive organs, epidermal cells produce hairs called **trichomes** [Gk. *trichos,* hair] that have two important functions: to protect the plant from too much sun and to conserve moisture. Sometimes trichomes, particularly glandular ones, help protect a plant from herbivores by producing a toxic substance. Under the slightest pressure the stiff trichomes of the stinging nettle lose their tips, forming "hypodermic needles" that inject an intruder with a stinging secretion.

In leaves, the lower epidermis of eudicots and both surfaces of monocots contain specialized cells called guard cells (Fig. 24.4*b*). Guard cells, which are epidermal cells with chloroplasts, surround microscopic pores called **stomata** (sing., stoma). When the stomata are open, gas exchange and water loss occur.

In older woody plants, the epidermis of the stem is replaced by **periderm** [Gk. *peri,* around; *derma,* skin]. The majority component of periderm is boxlike **cork** cells. At maturity, cork cells can be sloughed off (Fig. 24.4*c*). New cork cells are made by a meristem called **cork cambium.** As the new cork cells mature, they increase slightly in volume, and their walls become encrusted with suberin, a lipid material, so that they are waterproof and chemically inert. These nonliving cells protect the plant and make it resistant to attack by fungi, bacteria, and animals. Some cork tissues are commercially used for bottle corks and other products.

The cork cambium overproduces cork in certain areas of the stem surface; this causes ridges and cracks to appear. These features on the surface are called **lenticels.** Lenticels are important in gas exchange between the interior of a stem and the air.

a. Root hairs

guard cell
chloroplasts
epidermal cell
nucleus
stoma

b. Stoma of leaf

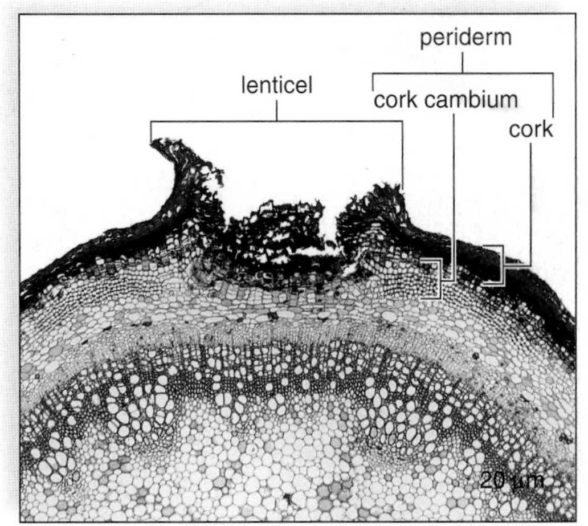


c. Cork of older stem

**FIGURE 24.4 Modifications of epidermal tissue.**

**a.** Root epidermis has root hairs to absorb water. **b.** Leaf epidermis contains stomata (sing., stoma) for gas exchange. **c.** Periderm includes cork and cork cambium. Lenticels in cork are important in gas exchange.

**FIGURE 24.5**
**Ground tissue cells.**
**a.** Parenchyma cells are the least specialized of the plant cells. **b.** Collenchyma cells. Notice how much thicker and irregular the walls are compared to those of parenchyma cells.
**c.** Sclerenchyma cells have very thick walls and are nonliving—their only function is to give strong support.

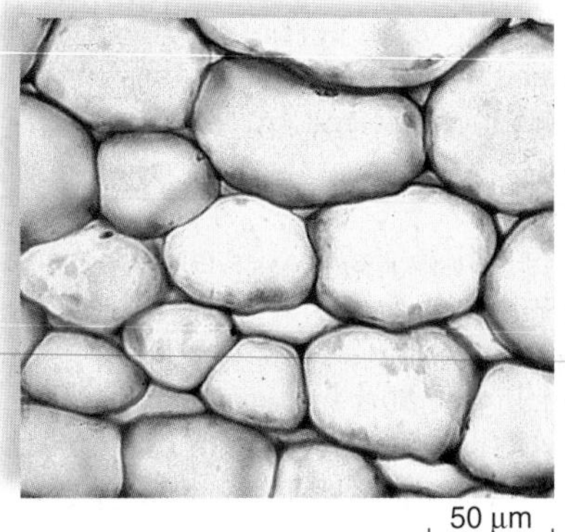
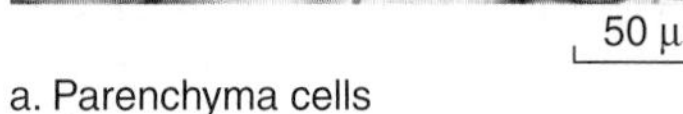


a. Parenchyma cells

b. Collenchyma cells

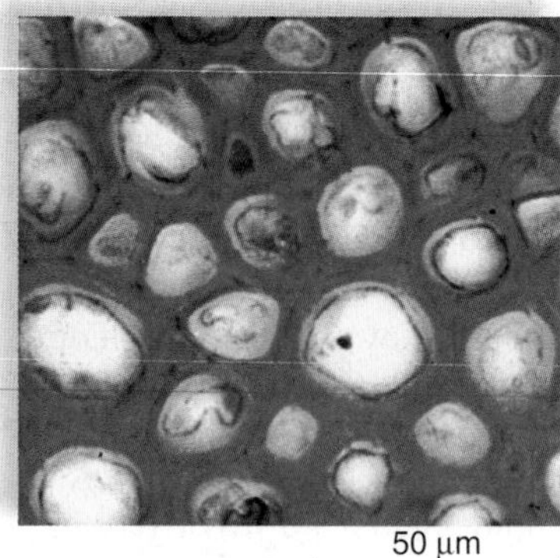


c. Sclerenchyma cells

## Ground Tissue

Ground tissue forms the bulk of a flowering plant and contains parenchyma, collenchyma, and sclerenchyma cells (Fig. 24.5). **Parenchyma** [Gk. *para,* beside, and *enchyma,* infusion] cells are the most abundant and correspond best to the typical plant cell. These are the least specialized of the cell types and are found in all the organs of a plant. They may contain chloroplasts and carry on photosynthesis (chlorenchyma), or they may contain colorless plastids that store the products of photosynthesis. A juicy bite from an apple yields mostly storage parenchyma cells. Parenchyma cells line the connected air spaces of a water lily and other aquatic plants. Parenchyma cells can divide and give rise to more specialized cells, such as when roots develop from stem cuttings placed in water.

**Collenchyma** cells are like parenchyma cells except they have thicker primary walls. The thickness is uneven and usually involves the corners of the cell. Collenchyma cells often form bundles just beneath the epidermis and give flexible support to immature regions of a plant body. The familiar strands in celery stalks (leaf petioles) are composed mostly of collenchyma cells.

**Sclerenchyma** cells have thick secondary cell walls impregnated with **lignin,** which is a highly resistant organic substance that makes the walls tough and hard. If we compare a cell wall to reinforced concrete, cellulose fibrils would play the role of steel rods, and lignin would be analogous to the cement. Most sclerenchyma cells are nonliving; their primary function is to support the mature regions of a plant. Two types of sclerenchyma cells are *fibers* and *sclereids.* Although

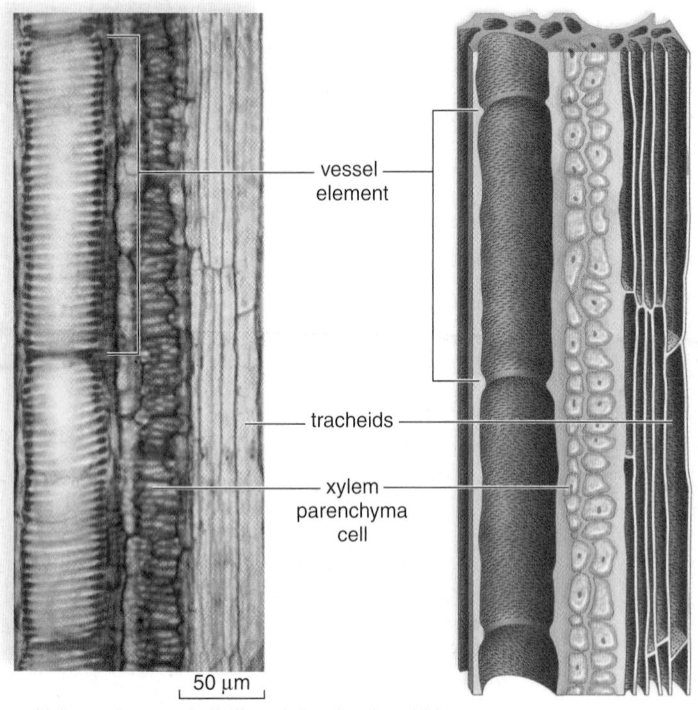


a. Xylem micrograph (*left*) and drawing (*to side*)

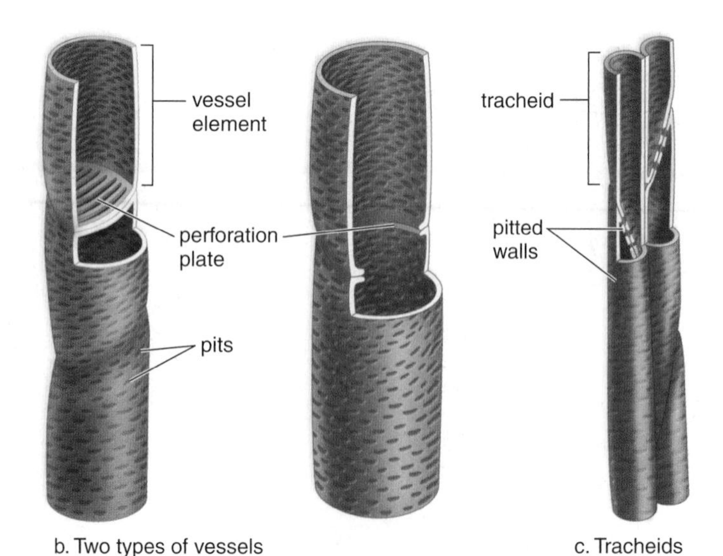


b. Two types of vessels

c. Tracheids

**FIGURE 24.6 Xylem structure.**

**a.** Photomicrograph of xylem vascular tissue and drawing showing general organization of xylem tissue. **b.** Drawing of two types of vessels (composed of vessel elements)—the perforation plates differ. **c.** Drawing of tracheids.

fibers are occasionally found in ground tissue, most are in vascular tissue, which is discussed next. Fibers are long and slender and may be grouped in bundles that are sometimes commercially important. Hemp fibers can be used to make rope, and flax fibers can be woven into linen. Flax fibers, however, are not lignified, which is why linen is soft. Sclereids, which are shorter than fibers and more varied in shape, are found in seed coats and nutshells. Sclereids, or "stone cells," are responsible for the gritty texture of pears. The hardness of nuts and peach pits is due to sclereids.

## Vascular Tissue

There are two types of vascular (transport) tissue. **Xylem** transports water and minerals from the roots to the leaves, and **phloem** transports sucrose and other organic compounds, including hormones, usually from the leaves to the roots. Both xylem and phloem are considered **complex tissues** because they are composed of two or more kinds of cells. Xylem contains two types of conducting cells: tracheids and vessel elements (VE), which are modified sclerenchyma cells (Fig. 24.6). Both types of conducting cells are hollow and nonliving, but the **vessel elements** are larger, may have perforation plates in their end walls, and are arranged to form a continuous vessel for water and mineral transport. The elongated **tracheids,** with tapered ends, form a less obvious means of transport, but water can move across the end walls and side walls because there are **pits,** or depressions, where the secondary wall does not form. In addition to vessel elements and tracheids, xylem contains additional sclerenchyma fibers that lend additional support and parenchyma cells that store various substances. Vascular rays, which are flat ribbons or sheets of parenchyma cells located between rows of tracheids, conduct water and minerals across the width of a plant.

The conducting cells of phloem are specialized parenchyma cells called **sieve-tube members** arranged to form a continuous sieve tube (Fig. 24.7). Sieve-tube members contain cytoplasm but no nuclei. The term *sieve* refers to a cluster of pores in the end walls, which is known as a sieve plate. Each sieve-tube member has a companion cell, which does have a nucleus. The two are connected by numerous plasmodesmata, and the nucleus of the companion cell may control and maintain the life of both cells. The companion cells are also believed to be involved in the transport function of phloem. Sclerenchyma fibers also lend support to phloem.

It is important to realize that vascular tissue (xylem and phloem) extends from the root through stems to the leaves and vice versa (see Fig. 24.1). In the roots, the vascular tissue is located in the **vascular cylinder;** in the stem, it forms **vascular bundles;** and in the leaves, it is found in **leaf veins.**

**Check Your Progress 24.2**

1. List the three specialized tissues in a plant and the cells that make up these tissues.
2. Compare the transport function of xylem and phloem.

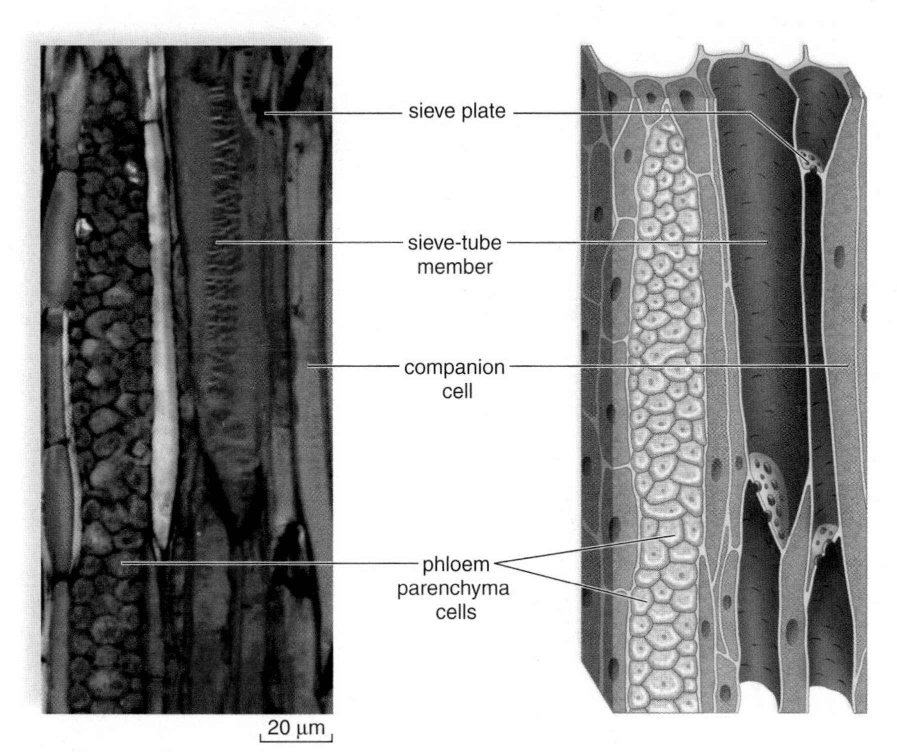


a. Phloem micrograph (*left*) and drawing (*to side*)

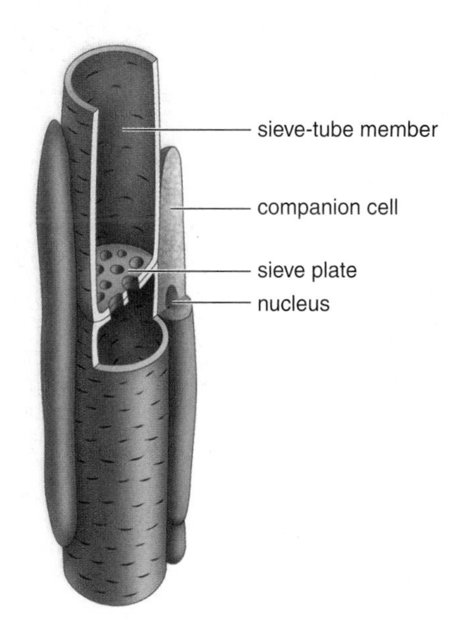


b. Sieve-tube member and companion cells

**FIGURE 24.7 Phloem structure.**

**a.** Photomicrograph of phloem vascular tissue and drawing showing general organization of phloem tissue. **b.** Drawing of sieve tube (composed of sieve-tube members) and companion cells.

# 24.3 Organization and Diversity of Roots

Figure 24.8*a*, a longitudinal section of a eudicot root, reveals zones where cells are in various stages of differentiation as primary growth occurs. The root **apical meristem** is in the region protected by the **root cap.** Root cap cells have to be replaced constantly because they get ground off by rough soil particles as the root grows. The primary meristems are in the zone of cell division, which continuously provides new cells to the zone of elongation. In the zone of elongation, the cells lengthen as they become specialized. The zone of maturation, which contains fully differentiated cells, is

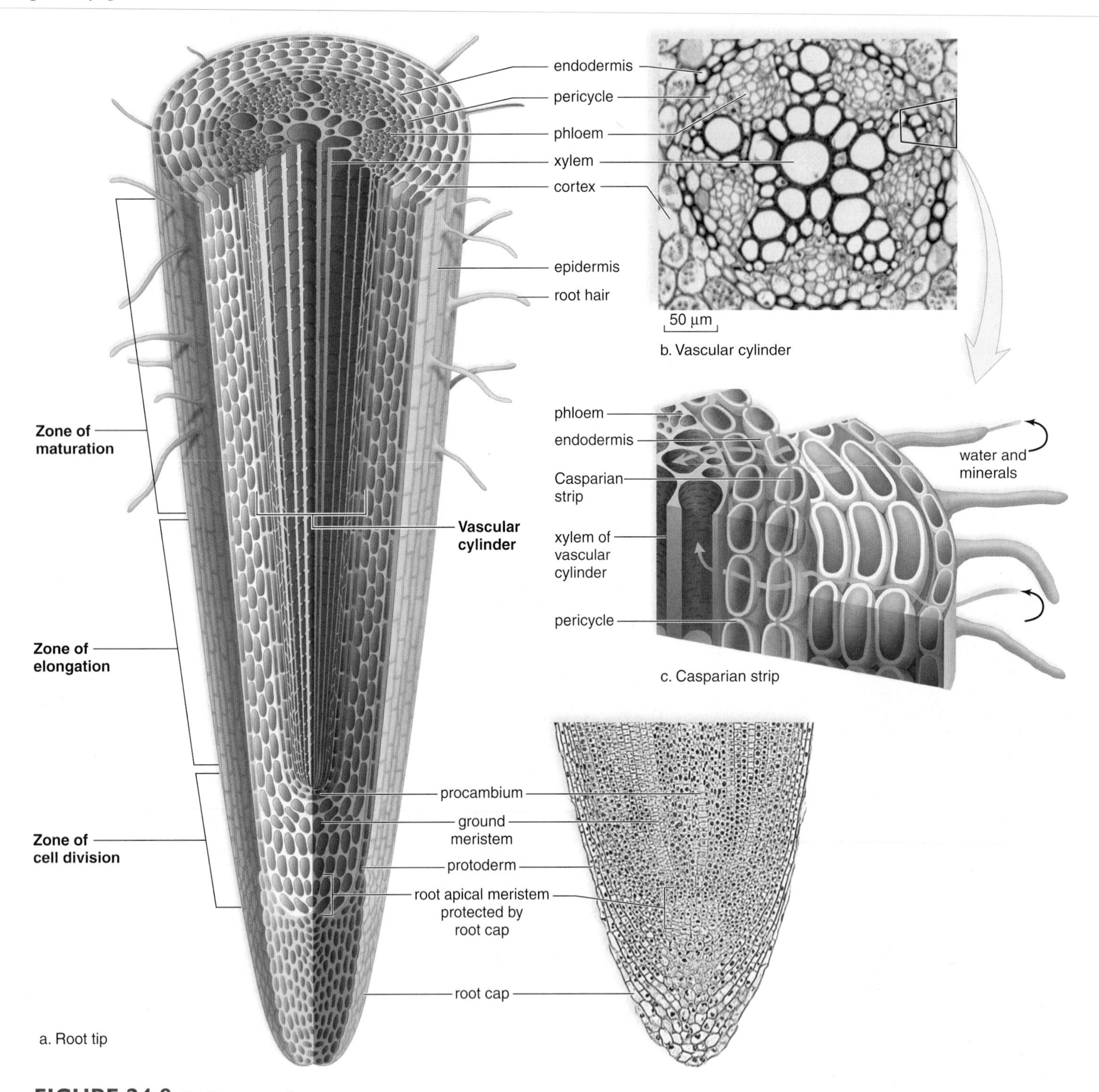


**FIGURE 24.8 Eudicot root tip.**

**a.** The root tip is divided into three zones. **b.** The vascular cylinder of a eudicot root contains the vascular tissue. **c.** Because of the Casparian strip, water and minerals must pass through the cytoplasm of endodermal cells in order to enter the xylem.

recognizable because here root hairs are borne by many of the epidermal cells.

## Tissues of a Eudicot Root

Figure 24.8*a* also shows a cross section of a root at the region of maturation. These specialized tissues are identifiable:

**Epidermis** The epidermis, which forms the outer layer of the root, consists of only a single layer of cells. The majority of epidermal cells are thin-walled and rectangular, but in the zone of maturation, many epidermal cells have root hairs. These project as far as 5–8 mm into the soil particles.

**Cortex** Moving inward, next to the epidermis, large, thin-walled parenchyma cells make up the **cortex** of the root. These irregularly shaped cells are loosely packed, and it is possible for water and minerals to move through the cortex without entering the cells. The cells contain starch granules, and the cortex functions in food storage.

**Endodermis** The **endodermis** [Gk. *endon*, within, and *derma*, skin] is a single layer of rectangular cells that forms a boundary between the cortex and the inner vascular cylinder. The endodermal cells fit snugly together and are bordered on four sides (but not the two sides that contact the cortex and the vascular cylinder) by a layer of impermeable lignin and suberin known as the **Casparian strip** (Fig. 24.8*c*). This strip prevents the passage of water and mineral ions between adjacent cell walls. Therefore, the only access to the vascular cylinder is through the endodermal cells themselves, as shown by the arrow in Figure 24.8*c*. This arrangement regulates the entrance of minerals into the vascular cylinder.

**Vascular tissue** The **pericycle,** the first layer of cells within the vascular cylinder, has retained its capacity to divide and can start the development of branch, or lateral, roots (Fig. 24.9). The main portion of the vascular cylinder contains xylem and phloem. The xylem appears star-shaped in eudicots because several arms of tissue radiate from a common center (see Fig. 24.8*b*). The phloem is found in separate regions between the arms of the xylem.

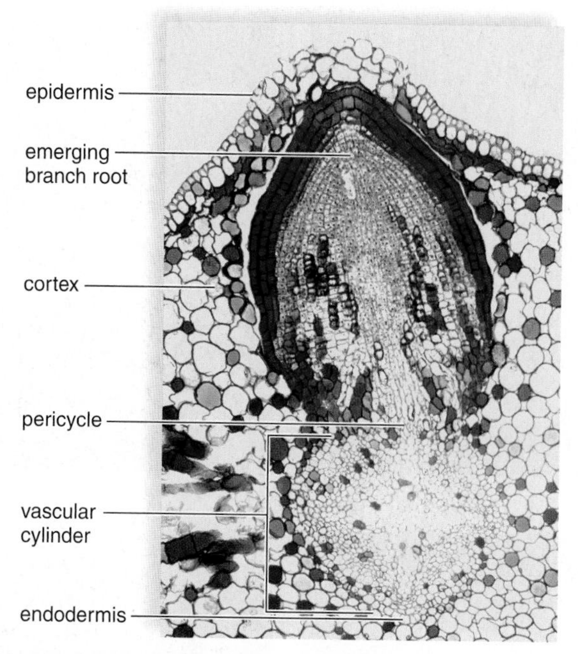


**FIGURE 24.9 Branching of eudicot root.**

This cross section of a willow, *Salix*, shows the origination and growth of a branch root from the pericycle.

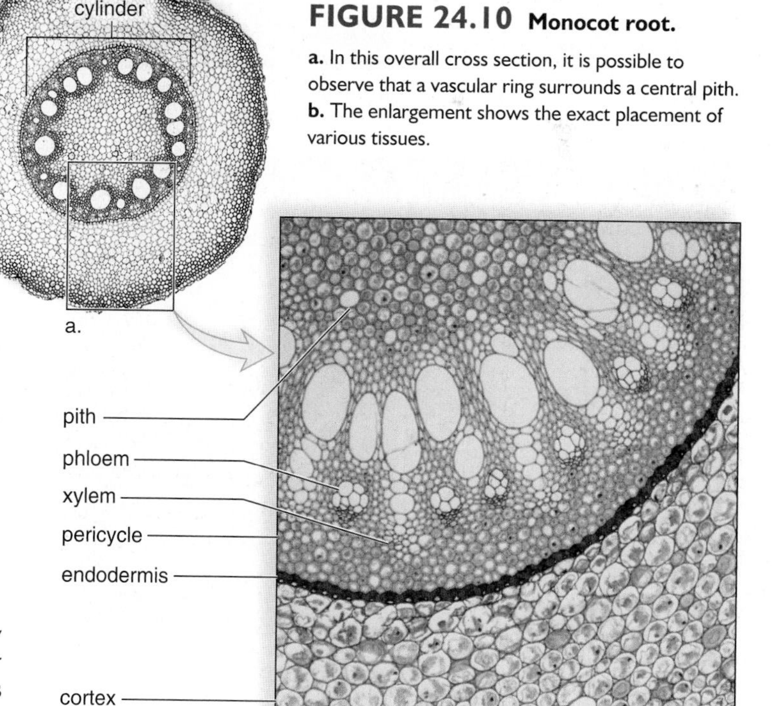


**FIGURE 24.10 Monocot root.**

**a.** In this overall cross section, it is possible to observe that a vascular ring surrounds a central pith. **b.** The enlargement shows the exact placement of various tissues.

## Organization of Monocot Roots

Monocot roots have the same growth zones as eudicot roots, but they do not undergo secondary growth as many eudicot roots go through. Also, the organization of their tissues is slightly different. The ground tissue of a monocot root's **pith,** which is centrally located, is surrounded by a vascular ring composed of alternating xylem and phloem bundles (Fig. 24.10). Monocot roots also have pericycle, endodermis, cortex, and epidermis.

a. Taproot

b. Fibrous root system

c. Prop roots, a type of adventitious root

d. Pneumatophores of black mangrove trees

e. Aerial roots of English ivy clinging to tree trunks

**FIGURE 24.11 Root diversity.**

**a.** A taproot may have branch roots in addition to a main root. **b.** A fibrous root has many slender roots with no main root. **c.** Prop roots are specialized for support. **d.** The pneumatophores of a black mangrove tree allow it to acquire oxygen even though it lives in swampy water. **e.** (*left*) English ivy climbs up the trunk because (*right*) it has aerial roots that cling to tree bark.

## Root Diversity

Roots have various adaptations and associations to better perform their functions: anchorage, absorption of water and minerals, and storage of carbohydrates.

In some plants, notably eudicots, the first or **primary root** grows straight down and remains the dominant root of the plant. This so-called **taproot** is often fleshy and stores food (Fig. 24.11*a*). Carrots, beets, turnips, and radishes have taproots that we consume as vegetables. Sweet potato plants don't have taproots, but they do have roots that expand to store starch. We call these storage areas sweet potatoes.

In other plants, notably monocots, there is no single, main root; instead, there are a large number of slender roots. These grow from the lower nodes of the stem when the first (primary) root dies. These slender roots and their lateral branches make up a **fibrous root system** (Fig. 24.11*b*). Many have observed the fibrous root systems of grasses and have noted how these roots strongly anchor the plant to the soil.

### *Root Specializations*

When roots develop from organs of the shoot system instead of the root system, they are known as **adventitious roots.** Some adventitious roots emerge above the soil line, as they do in corn plants, in which their main function is to help anchor the plant. If so, they are called prop roots (Fig. 24.11*c*). Other examples of adventitious roots are those found on horizontal stems (see Fig. 24.19*a*) or at the nodes of climbing English ivy (Fig. 24.11*e*). As the vines climb, the rootlets attach the plant to any available vertical structure.

Black mangroves live in swampy water and have pneumatophores, root projections that rise above the water and acquire oxygen for cellular respiration (Fig. 24.11*d*).

Some plants, such as dodders and broomrapes, are parasitic on other plants. Their stems have rootlike projections called haustoria (sing., haustorium) that grow into the host plant and make contact with vascular tissue from which they extract water and nutrients (see Fig. 25.8*a*). **Mycorrhizae** are associations between roots and fungi that can extract water and minerals from the soil better than roots that lack a fungus partner. This is a mutualistic relationship because the fungus receives sugars and amino acids from the plant, while the plant receives water and minerals via the fungus.

Peas, beans, and other legumes have **root nodules** where nitrogen-fixing bacteria live. Plants cannot extract nitrogen from the air, but the bacteria within the nodules can take up and reduce atmospheric nitrogen. This means that the plant is no longer dependent on a supply of nitrogen (i.e., nitrate or ammonium) from the soil, and indeed these plants are often planted just to bolster the nitrogen supply of the soil.

**Check Your Progress 24.3**

1. Describe the relationship between the root apical meristem and the root cap.
2. List the function of the cortex, the endodermis, and the pericycle in a root.

# ecology focus

## Paper Comes from Plants

The word *paper* takes its origin from papyrus, the plant Egyptians used to make the first form of paper. The Egyptians manually placed thin sections cut from papyrus at right angles and pressed them together to make a sheet of writing material. From that beginning some 5,500 years ago, the production of paper is now a worldwide industry of major importance (Fig. 24A). The process is fairly simple. Plant material is ground up mechanically to form a pulp that contains "fibers," which biologists know come from vascular tissue. The fibers automatically form a sheet when they are screened from the pulp.

If wood is the source of the fibers, the pulp must be chemically treated to remove lignin. If only a small amount of lignin is removed, the paper is brown, as in paper bags. If more lignin is removed, the paper is white but not very durable, and it crumbles after a few decades. Paper is more durable when it is made from cotton or linen because the fibers from these plants are lignin-free.

Among the other major plants used to make paper are:

**FIGURE 24A Paper production.**
*Today, a revolving wire-screen belt is used to deliver a continuous wet sheet of paper to heavy rollers and heated cylinders, which remove most of the remaining moisture and press the paper flat.*

***Eucalyptus* trees.** In recent years, Brazil has devoted huge areas of the Amazon region to the growing of cloned *Eucalyptus* seedlings, specially selected and engineered to be ready for harvest after about seven years.

**Temperate hardwood trees.** Plantation cultivation in Canada provides birch, beech, chestnut, poplar, and particularly aspen wood for paper making. Tropical hardwoods, usually from Southeast Asia, are also used.

**Softwood trees.** In the United States, several species of pine trees have been genetically improved to have a higher wood density and to be harvestable five years earlier than ordinary pines. Southern Africa, Chile, New Zealand, and Australia also devote thousands of acres to growing pines for paper pulp production.

**Bamboo.** Several Asian countries, especially India, provide vast quantities of bamboo pulp for the making of paper. Because bamboo is harvested without destroying the roots, and the growing cycle is favorable, this plant, which is actually a grass, is expected to be a significant source of paper pulp despite high processing costs to remove impurities.

**Flax and cotton rags.** Linen and cotton cloth from textile and garment mills are used to produce rag paper, whose flexibility and durability are desirable in legal documents, high-grade bond paper, and high-grade stationery.

It has been known for some time that paper largely consists of the cellulose within plant cell walls. It seems reasonable to suppose, then, that paper could be made from synthetic polymers (e.g., rayon). Indeed, synthetic polymers produce a paper that has qualities superior to those of paper made from natural sources, but the cost thus far is prohibitive. Another consideration, however, is the ecological impact of making paper from trees. Plantations containing stands of uniform trees replace natural ecosystems, and when the trees are clear-cut, the land is laid bare. Paper mill wastes, which include caustic chemicals, add significantly to the pollution of rivers and streams.

The use of paper for packaging and making all sorts of products has increased dramatically in the last century. Each person in the United States uses about 318 kg of paper products per year, and this compares to only 2.3 kg of paper per person in India. It is clear, then, that we should take the initiative in recycling paper. When newspaper, office paper, and photocopies are soaked in water, the fibers are released, and they can be used to make a new batch of paper. It's estimated that recycling the Sunday newspapers alone would save approximately 500,000 trees each week!

## 24.4 Organization and Diversity of Stems

The anatomy of a woody twig ready for next year's growth reviews for us the organization of a stem (Fig. 24.12). The **terminal bud** contains the shoot tip protected by bud scales, which are modified leaves. Each spring when growth resumes, bud scales fall off and leave a scar. You can tell the age of a stem by counting these bud scale scars because there is one for each year's growth. Leaf scars and bundle scars mark the location of leaves that have dropped. Dormant axillary buds that will give rise to branches or flowers are also found here.

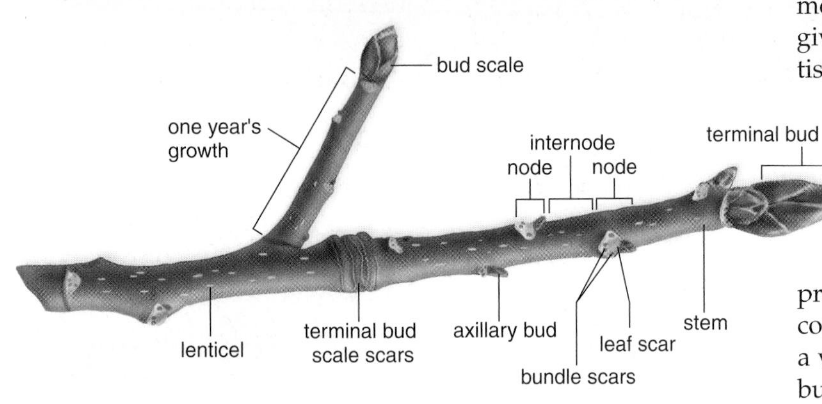


**FIGURE 24.12 Woody twig.**

The major parts of a stem are illustrated by a woody twig collected in winter.

When growth resumes, primary growth continues. The apical meristem at the shoot tip produces new cells that elongate and thereby increase the height of the stem. The **shoot apical meristem** is protected within the terminal bud, where leaf primordia (immature leaves) envelop it (Fig. 24.13). The leaf primordia mark the location of a node; the portion of stem in between nodes is an internode. As a stem grows, the internodes increase in length.

In addition to leaf primordia, the three specialized types of primary meristem, mentioned earlier (see page 437), develop from a shoot apical meristem (Fig. 24.13*b*). These primary meristems contribute to the length of a shoot. As mentioned, the *protoderm*, the outermost primary meristem, gives rise to epidermis. The *ground meristem* produces two tissues composed of parenchyma cells. The parenchyma tissue in the center of the stem is the pith, and the parenchyma tissue between the epidermis and the vascular tissue is the cortex.

The *procambium*, seen as an obvious strand of tissue in Figure 24.13*a*, produces the first xylem cells, called primary xylem, and the first phloem cells, called primary phloem. Differentiation continues as certain cells become the first tracheids or vessel elements of the xylem within a vascular bundle. The first sieve-tube members of a vascular bundle do not have companion cells and are short-lived (some live only a day before being replaced). Mature vascular bundles contain fully differentiated xylem, phloem, and a lateral meristem called **vascular cambium** [L. *vasculum*, dim. of *vas*,

**FIGURE 24.13 Shoot tip and primary meristems.**

**a.** The shoot apical meristem within a terminal bud is surrounded by leaf primordia. **b.** The shoot apical meristem produces the primary meristems: Protoderm gives rise to epidermis; ground meristem gives rise to pith and cortex; and procambium gives rise to vascular tissue, including primary xylem, primary phloem, and vascular cambium.

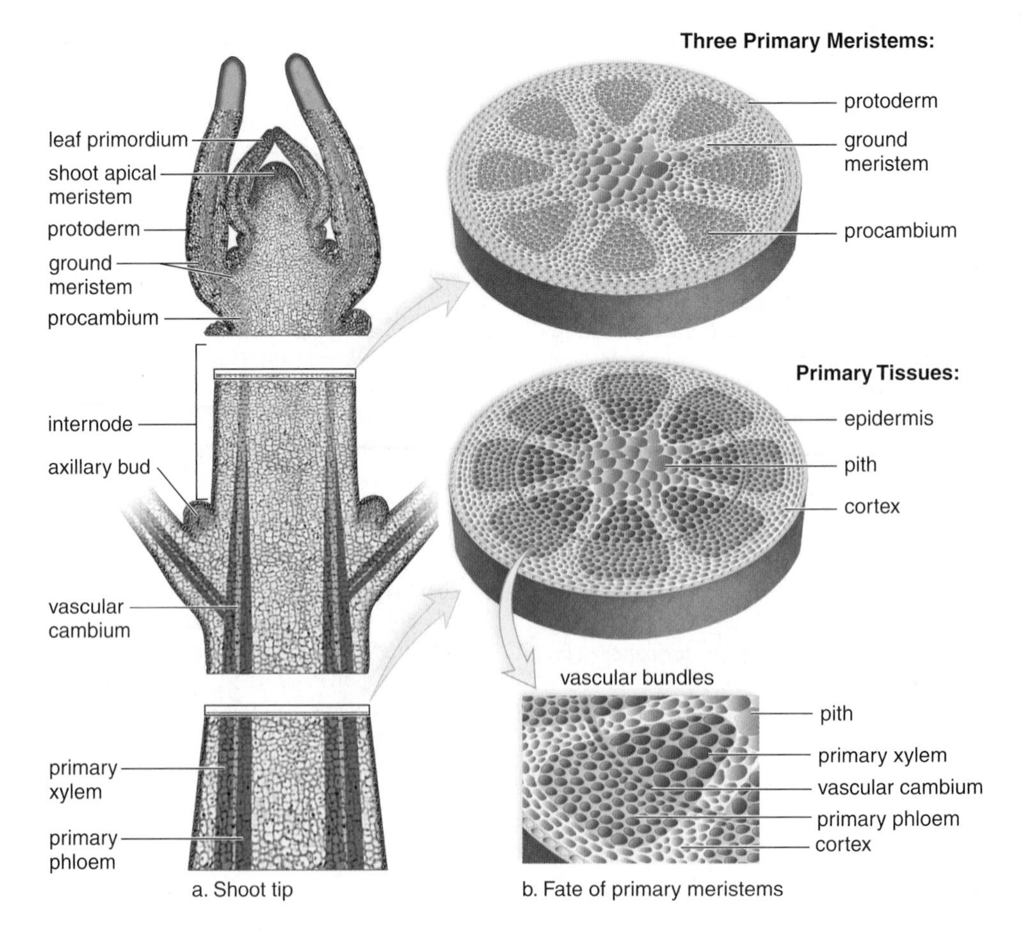

vessel, and *cambio,* exchange]. Vascular cambium is discussed more fully in the next section.

## Herbaceous Stems

Mature nonwoody stems, called **herbaceous stems** [L. *herba,* vegetation, plant], exhibit only primary growth. The outermost tissue of herbaceous stems is the epidermis, which is covered by a waxy cuticle to prevent water loss. These stems have distinctive vascular bundles, where xylem and phloem are found. In each bundle, xylem is typically found toward the inside of the stem, and phloem is found toward the outside.

In the herbaceous eudicot stem such as a sunflower, the vascular bundles are arranged in a distinct ring that separates the cortex from the central pith, which stores water and products of photosynthesis (Fig. 24.14). The cortex is sometimes green and carries on photosynthesis. In a monocot stem such as corn, the vascular bundles are scattered throughout the stem, and often there is no well-defined cortex or well-defined pith (Fig. 24.15).

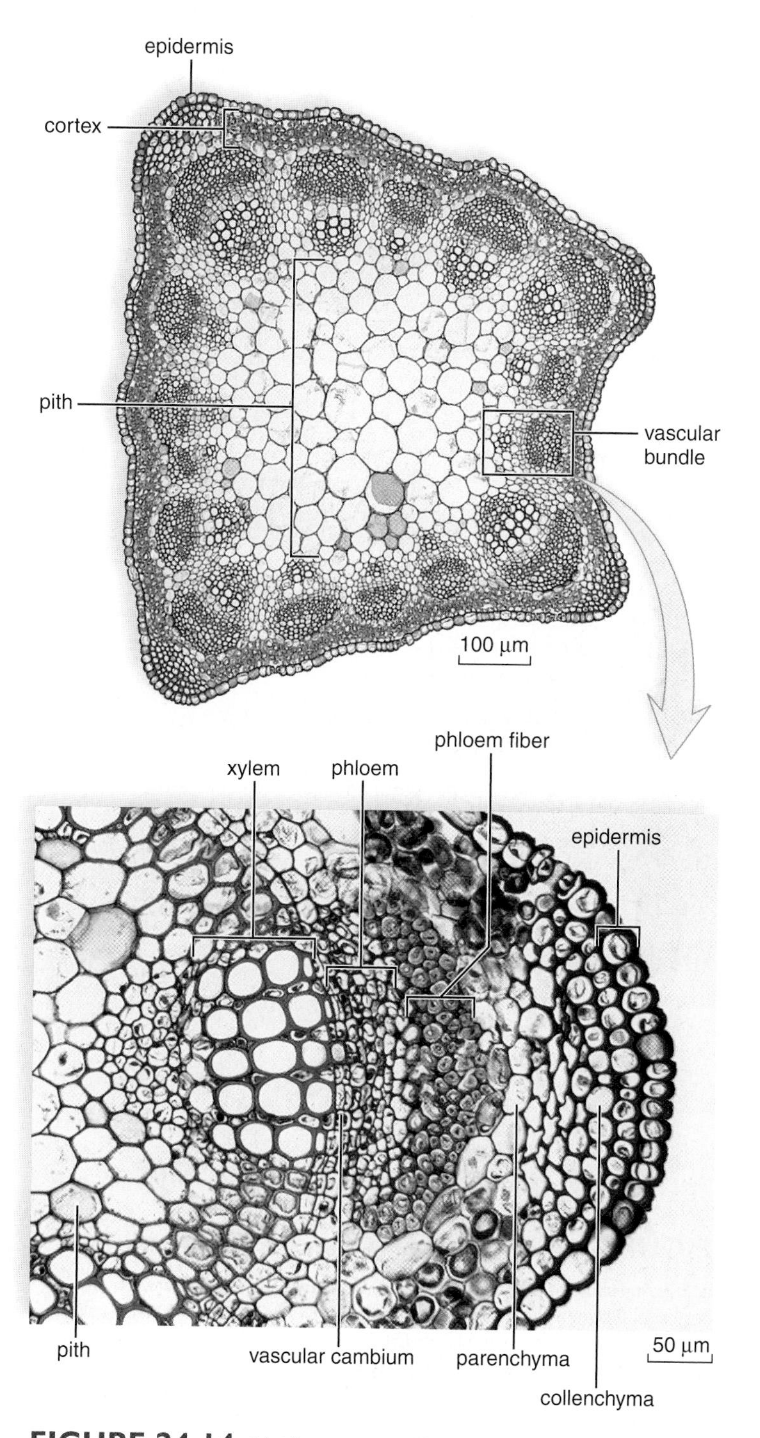


**FIGURE 24.14 Herbaceous eudicot stem.**

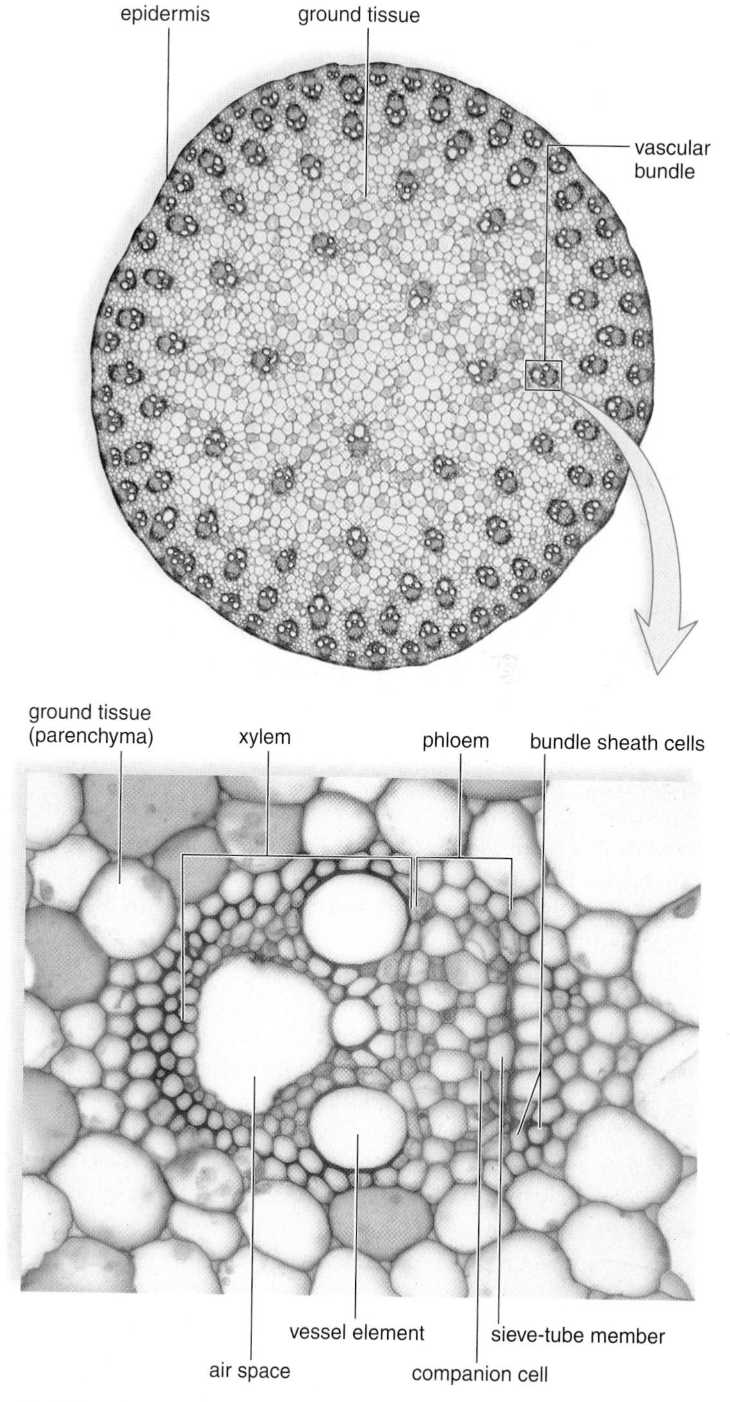


**FIGURE 24.15 Monocot stem.**

## Woody Stems

A woody plant such as an oak tree has both primary and secondary tissues. Primary tissues are those new tissues formed each year from primary meristems right behind the shoot apical meristem. Secondary tissues develop during the first and subsequent years of growth from lateral meristems: vascular cambium and cork cambium. *Primary growth,* which occurs in all plants, increases the length of a plant, and *secondary growth,* which occurs only in conifers and woody eudicots, increases the girth of trunks, stems, branches, and roots.

Trees and shrubs undergo secondary growth because of a change in the location and activity of vascular cambium (Fig. 24.16). In herbaceous plants, vascular cambium is present between the xylem and phloem of each vascular bundle. In woody plants, the vascular cambium develops to form a ring of meristem that divides parallel to the surface of the plant, and produces new xylem toward the inside and phloem toward the outside each year.

Eventually, a woody eudicot stem has an entirely different organization from that of a herbaceous eudicot stem. A woody stem has no distinct vascular bundles and instead has three distinct areas: the bark, the wood, and the pith. Vascular cambium occurs between the bark and the wood, and it causes woody plants to increase in girth. Cork cambium, occurring first beneath the epidermis, is instrumental in the production of cork in woody plants.

You will also notice in Figure 24.16 the *xylem rays* and *phloem rays* that are visible in the cross section of a woody stem. Rays consist of parenchyma cells that permit lateral conduction of nutrients from the pith to the cortex and some storage of food. A phloem ray is actually a continuation of an xylem ray. Some phloem rays are much broader than other phloem rays.

### *Bark*

The **bark** of a tree contains periderm (cork and cork cambium), and phloem. Although secondary phloem is produced each year by vascular cambium, phloem does not build up from season to season. The bark of a tree can be removed; however, this is very harmful because, without phloem, organic nutrients cannot be transported. Girdling, removing a ring of bark from around a tree, can be lethal to the tree. Some herbivores girdle trees, and it can also be deliberately done by humans who want to thin out a forest or fruit tree stand. Girdling kills trees.

At first, cork cambium is located beneath the epidermis, and then later, it is found beneath the periderm. When cork cambium first begins to divide, it produces tissue that disrupts the epidermis and replaces it with cork cells. Cork cells are impregnated with suberin, a waxy layer that makes them waterproof but also causes them to die. This is protective because it makes the stem less edible. But an impervious barrier means that gas exchange is impeded except at lenticels, which are pockets of loosely arranged cork cells not impregnated with suberin.

### *Wood*

**Wood** is secondary xylem that builds up year after year, thereby increasing the girth of trees. In trees that have a growing season, vascular cambium is dormant during the winter. In the spring, when moisture is plentiful and leaves require much water for growth, the secondary xylem contains wide vessel elements with thin walls. In this so-called *spring wood,* wide vessels transport sufficient water to the growing leaves. Later in the season, moisture is scarce, and the wood at this time, called *summer wood,* has a lower proportion of vessels (Fig. 24.17). Strength is required because the tree is growing larger and summer wood contains numerous thick-walled tracheids. At the end of the

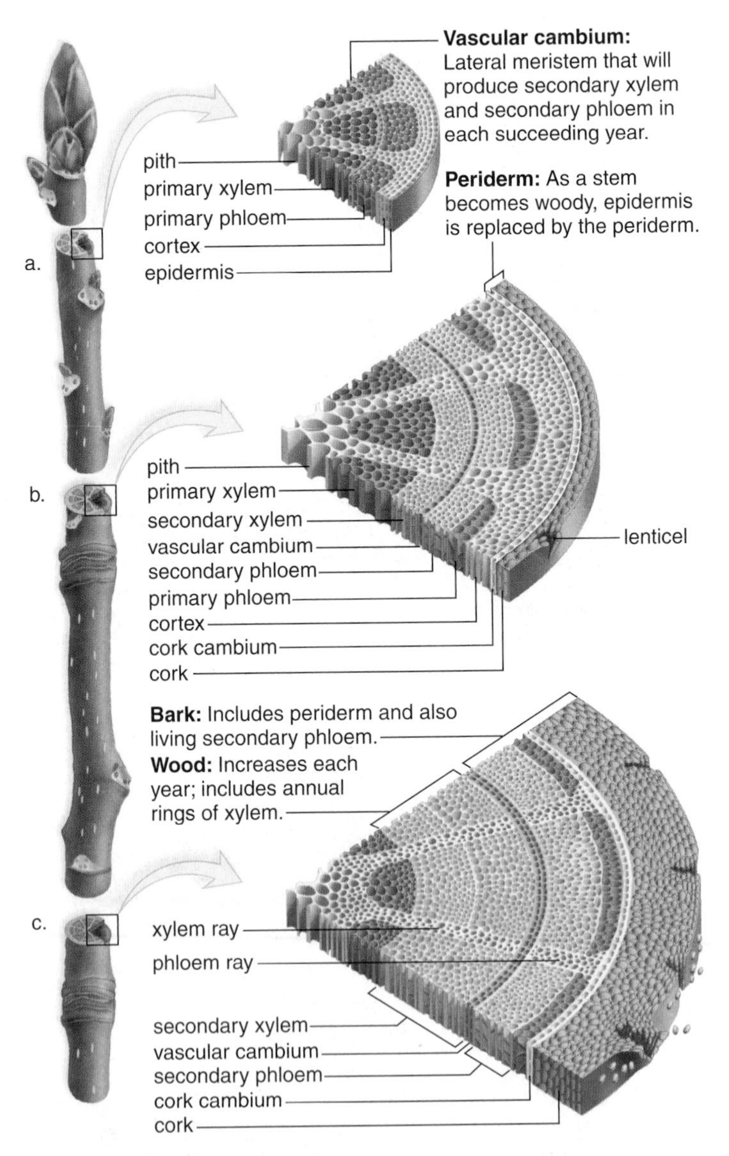


**FIGURE 24.16 Diagrams of secondary growth of stems.**

**a.** Diagram showing eudicot herbaceous stem just before secondary growth begins. **b.** Diagram showing that secondary growth has begun. Periderm has replaced the epidermis. Vascular cambium produces secondary xylem and secondary phloem each year. **c.** Diagram showing a two-year-old stem. The primary phloem and cortex will eventually disappear, and only the secondary phloem (within the bark) produced by vascular cambium will be active that year. Secondary xylem builds up to become the annual rings of a woody stem.

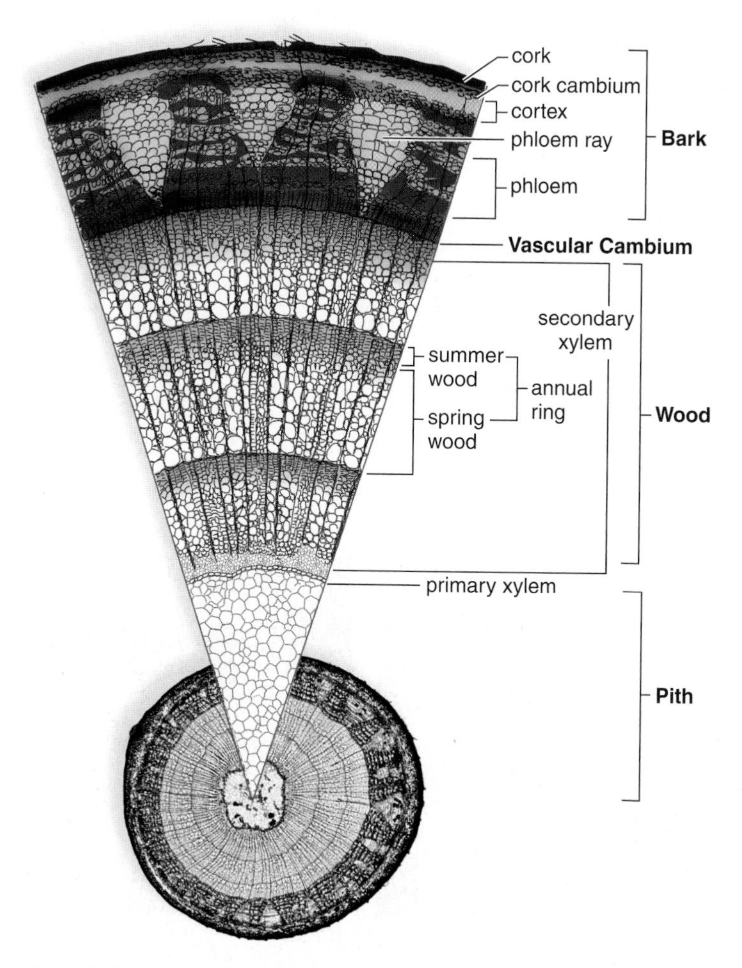


**FIGURE 24.17 Three-year-old woody twig.**

The buildup of secondary xylem in a woody stem results in annual rings, which tell the age of the stem. The rings can be distinguished because each one begins with spring wood (large vessel elements) and ends with summer wood (smaller and fewer vessel elements).

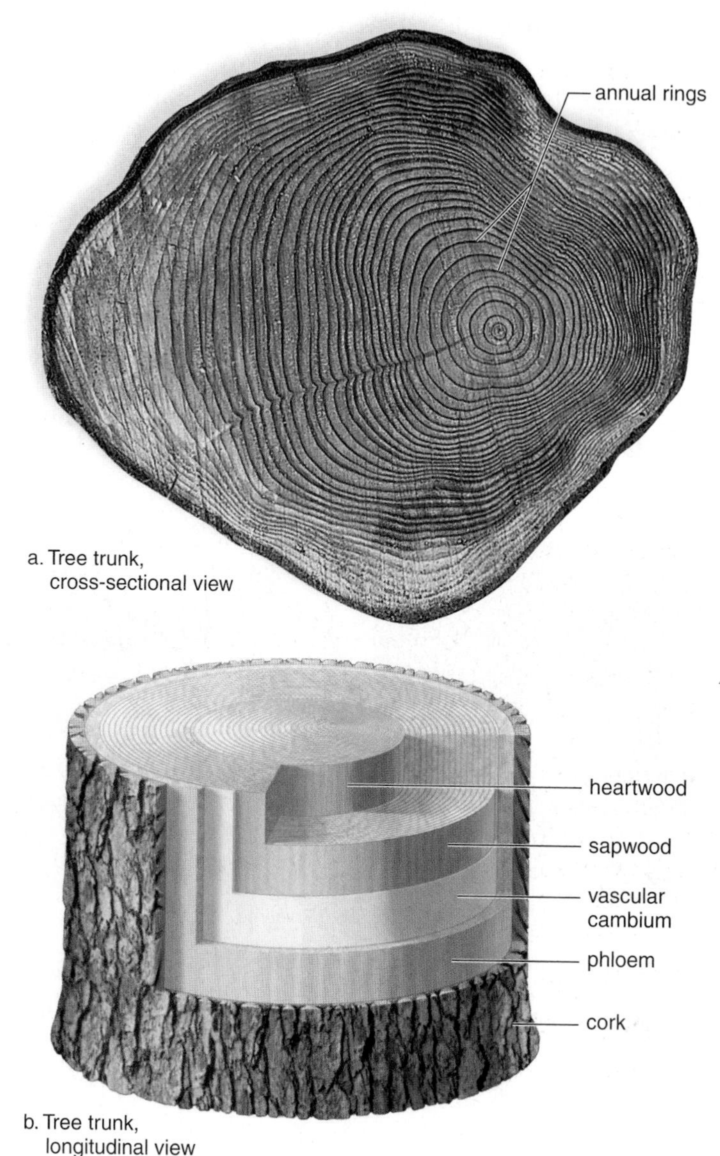


**FIGURE 24.18 Tree trunk.**

**a.** A cross section of a 39-year-old larch, *Larix decidua*. The xylem within the darker heartwood is inactive; the xylem within the lighter sapwood is active. **b.** The relationship of bark, vascular cambium, and wood is retained in a mature stem. The pith has been buried by the growth of layer after layer of new secondary xylem.

growing season, just before the cambium becomes dormant again, only heavy fibers with especially thick secondary walls may develop. When the trunk of a tree has spring wood followed by summer wood, the two together make up one year's growth, or an **annual ring.** You can tell the age of a tree by counting the annual rings (Fig. 24.18*a*). The outer annual rings, where transport occurs, are called sapwood.

In older trees, the inner annual rings, called the heartwood, no longer function in water transport. The cells become plugged with deposits, such as resins, gums, and other substances that inhibit the growth of bacteria and fungi. Heartwood may help support a tree, although some trees stand erect and live for many years after the heartwood has rotted away. Figure 24.18*b* shows the layers of a woody stem in relation to one another.

The annual rings are not only important in telling the age of a tree, they can serve as a historical record of tree growth. For example, if rainfall and other conditions were extremely favorable during a season, the annual ring may be wider than usual. If the tree were shaded on one side by another tree or building, the rings may be wider on the favorable side.

**Woody Plants.** Is it advantageous to be woody? With adequate rainfall, woody plants can grow taller and have more growth because they have adequate vascular tissue to support and service their leaves. However, it takes energy to produce secondary growth and prepare the body for winter if the plant lives in the temperate zone. Also, woody plants need more defense mechanisms because a long-lasting plant that stays in one spot is likely to be attacked by herbivores and parasites. Then, too, trees don't usually reproduce until they have grown several seasons, by which time they may have succumbed to an accident or disease. In certain habitats, it is more advantageous for a plant to put most of its energy into simply reproducing rather than being woody.

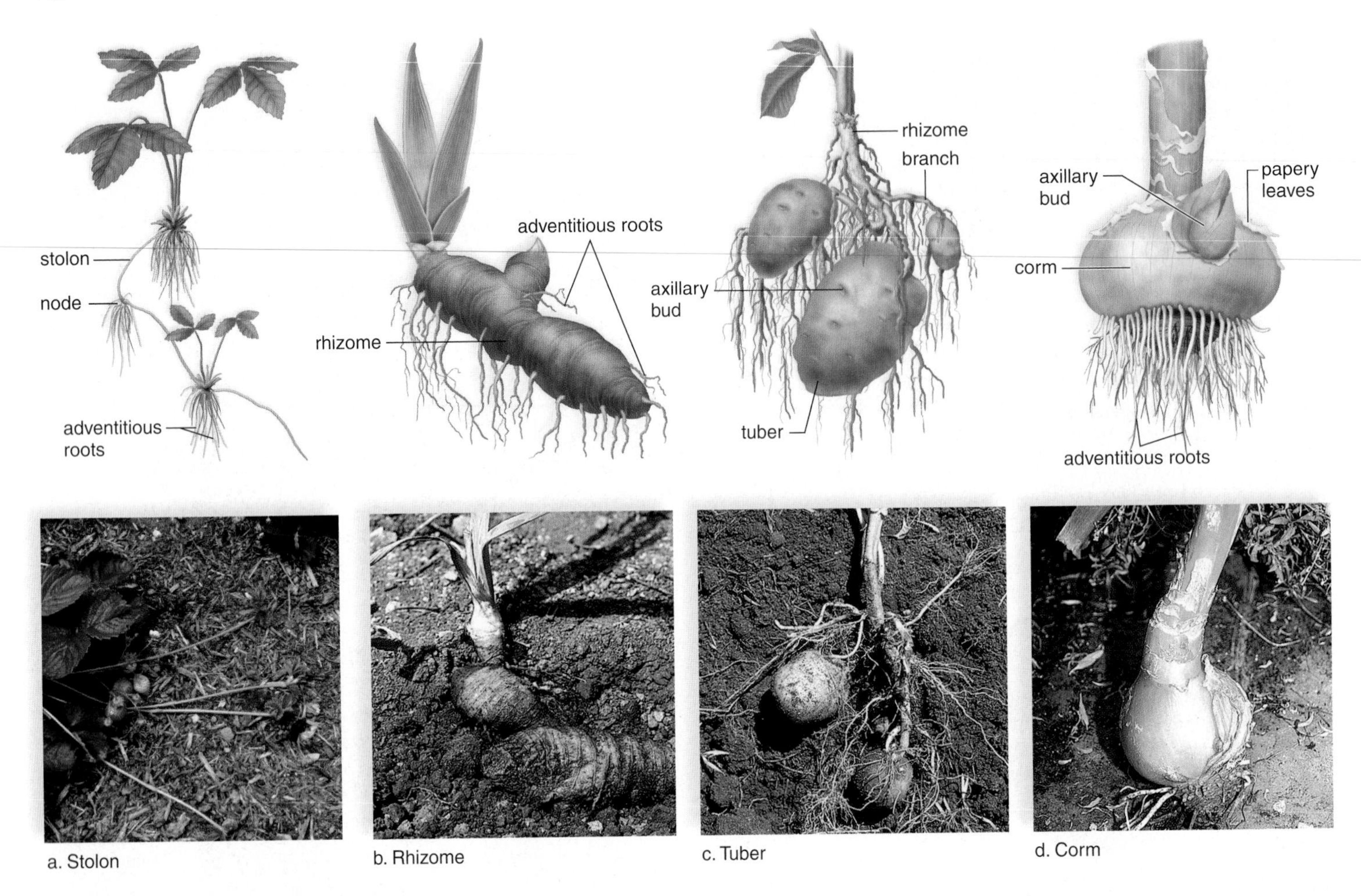


**FIGURE 24.19 Stem diversity.**

**a.** A strawberry plant has aboveground horizontal stems called stolons. Every other node produces a new shoot system. **b.** The underground horizontal stem of an iris is a fleshy rhizome. **c.** The underground stem of a potato plant has enlargements called tubers. We call the tubers potatoes. **d.** The corm of a gladiolus is a stem covered by papery leaves.

## Stem Diversity

Stem diversity is illustrated in Figure 24.19. Aboveground horizontal stems, called **stolons** [L. *stolo,* shoot] or runners, produce new plants where nodes touch the ground. The strawberry plant is a common example of this type of stem, which functions in vegetative reproduction.

Aboveground vertical stems can also be modified. For example, cacti have succulent stems specialized for water storage, and the tendrils of grape plants (which are stem branches) allow them to climb. Morning glory and relatives have stems that twine around support structures. Such tendrils and twining shoots help plants expose their leaves to the sun.

Underground horizontal stems, **rhizomes** [Gk. *rhiza,* root], may be long and thin, as in sod-forming grasses, or thick and fleshy, as in irises. Rhizomes survive the winter and contribute to asexual reproduction because each node bears a bud. Some rhizomes have enlarged portions called tubers, which function in food storage. Potatoes are tubers, in which the eyes are buds that mark the nodes.

Corms are bulbous underground stems that lie dormant during the winter, just as rhizomes do. They also produce new plants the next growing season. Gladiolus corms are referred to as bulbs by laypersons, but the botanist reserves the term *bulb* for a structure composed of modified leaves attached to a short vertical stem. An onion is a bulb.

Humans make use of stems in many ways. The stem of the sugarcane plant is a primary source of table sugar. The spice cinnamon and the drug quinine are derived from the bark of *Cinnamomum verum* and various *Cinchona* species, respectively. And wood is necessary for the production of paper, as discussed in the Ecology Focus on page 443.

### Check Your Progress 24.4

1. What transport tissues are in a vascular bundle?
2. How are vascular bundles arranged in monocot stems? In eudicot stems?
3. Contrast primary growth with secondary growth.
4. List the components of bark.
5. Relate spring wood and summer wood to annual rings.

science focus

## Defense Strategies of Trees

Rainstorms, ice, snow, animals, wind, excess weight, temperature extremes, and chemicals can all injure a tree. So can improper pruning. Pruning, which requires the cutting away of tree parts, can benefit a tree by improving its appearance and helping maintain its balance. But removing the top of a tree, called topping, removing a portion of the roots, and flush-cutting a number of branches at one time is injurious to trees. This is known because a tree reacts to improper pruning in the same manner it reacts to all injuries, no matter what the cause. The wounding of a tree subjects it to disease. Trees, like humans, have defensive strategies against bacterial and fungal invasions that occur when a tree is wounded.

A defense strategy is a mechanism that has arisen through the evolutionary process. In other words, members of the group with the strategy compete better than those that do not. We expect defense strategies to be beneficial—and they are—but the manner in which trees react to disease, called compartmentalization of decay, can still weaken them. Therefore, improper pruning practices should be avoided at all cost if you care about a tree!

Just as with humans, trees have a series of defense strategies against infection. Each one is better than the other at stopping the progress of disease organisms. First, when a tree is injured, the tracheids and vessel elements of xylem immediately plug up with chemicals that block them off above and below the site of the injury. In trees that fail to effectively close off vessel elements, long columns of rot (decay) run up and down the trunk and into branches, which eventually become hollow.

The second defense strategy is a result of tree trunk structure. As you know, a tree trunk has annual rings that tell its age. A dark region at the edge of an annual ring in the cross section of a trunk tells you that this tree was injured, and that disease organisms were unable to advance inward on their way to the pith. It appears, therefore, that disease organisms have a harder time moving across a trunk due to annual ring construction than they do moving through the trunk in vessel elements.

The third defense strategy involves rays. Rays take their name from the fact that they project radially from vascular cambium. Just like the slices of a pie, rays divide the trunk of a tree. Disease organisms can't cross rays either, and this keeps them in a small pie piece of the trunk and prevents them from moving completely around the trunk.

Turkey oak, *Quercus laevis*

Weeping willow, *Salix babylonica*

Cross section of a damaged tree trunk.

**FIGURE 24B Defense strategies.** *An oak tree is better at defense against infection than is a weeping willow tree. The oak tree never has to employ the defense strategy (right) that resulted in this dark ring in the trunk, which can lead to cracks and collapse of the tree.*

The fourth defense strategy is a so-called reaction zone that develops in the region of the injury along the inner portion of the cambium next to the youngest annual ring. The reaction zone can extend from a few inches to a few feet above and below the injury, and partway or all the way around the trunk. The reaction zone doesn't wall off any annual rings that develop after the injury, but it does wall off any annual rings that were present before the injury occurred. Figure 24B (*right*) shows a cross section of a tree that was topped seven years before it was cut down. The reaction zone is seen as a dark circle that, in this case, extends from the top of the tree to the root system.

Although the fourth defense strategy more effectively retards disease, it has a severe disadvantage. Cracks can develop along the reaction zone, and radial cracks also occur from the reaction zone to and through the bark. Cracks can severely weaken a tree and make it more susceptible to breaking. A closure crack is one that occurs at the site of the wound. Sometimes this crack never actually closes.

Some trees are better defenders against disease than others. Trees that effectively carry out strategies 1–3 need never employ strategy 4, which can lead to cracking. Oak trees, *Quercus* (Fig. 24B), are examples of trees that are good at defending themselves, while willows, *Salix*, are not as good.

# 24.5 Organization and Diversity of Leaves

Leaves are the organs of photosynthesis in vascular plants such as flowering plants. As mentioned earlier, a leaf usually consists of a flattened blade and a petiole connecting the blade to the stem. The blade may be single or composed of several leaflets. Externally, it is possible to see the pattern of the leaf veins, which contain vascular tissue. Leaf veins have a net pattern in eudicot leaves and a parallel pattern in monocot leaves (see Fig. 24.3).

Figure 24.20 shows a cross section of a typical eudicot leaf of a temperate zone plant. At the top and bottom are layers of epidermal tissue that often bear trichomes, protective hairs often modified as glands that secrete irritating substances. These features may prevent the leaf from being eaten by insects. The epidermis characteristically has an outer, waxy cuticle that helps keep the leaf from drying out. The cuticle also prevents gas exchange because it is not gas permeable. However, the lower epidermis of eudicot and both surfaces of monocot leaves contain stomata that allow gases to move into and out of the leaf. Water loss also occurs at stomata, but each stoma has two guard cells that regulate its opening and closing, and stomata close when the weather is hot and dry.

The body of a leaf is composed of **mesophyll** [Gk. *mesos*, middle, and *phyllon*, leaf] tissue. Most eudicot leaves have two distinct regions: **palisade mesophyll,** containing elongated cells, and **spongy mesophyll,** containing irregular cells bounded by air spaces. The parenchyma cells of these layers have many chloroplasts and carry on most of the photosynthesis for the plant. The loosely packed arrangement of the cells in the spongy layer increases the amount of surface area for gas exchange.

## Leaf Diversity

The blade of a leaf can be simple or compound (Fig. 24.21). A simple leaf has a single blade in contrast to a compound leaf, which is divided in various ways into leaflets. An example of a plant with simple leaves is a magnolia, and a tree with compound leaves is a pecan tree. Pinnately compound leaves have the leaflets occurring in pairs, such as in a black walnut tree, while palmately compound leaves have all of the leaflets attached to a single point, as in a buckeye tree.

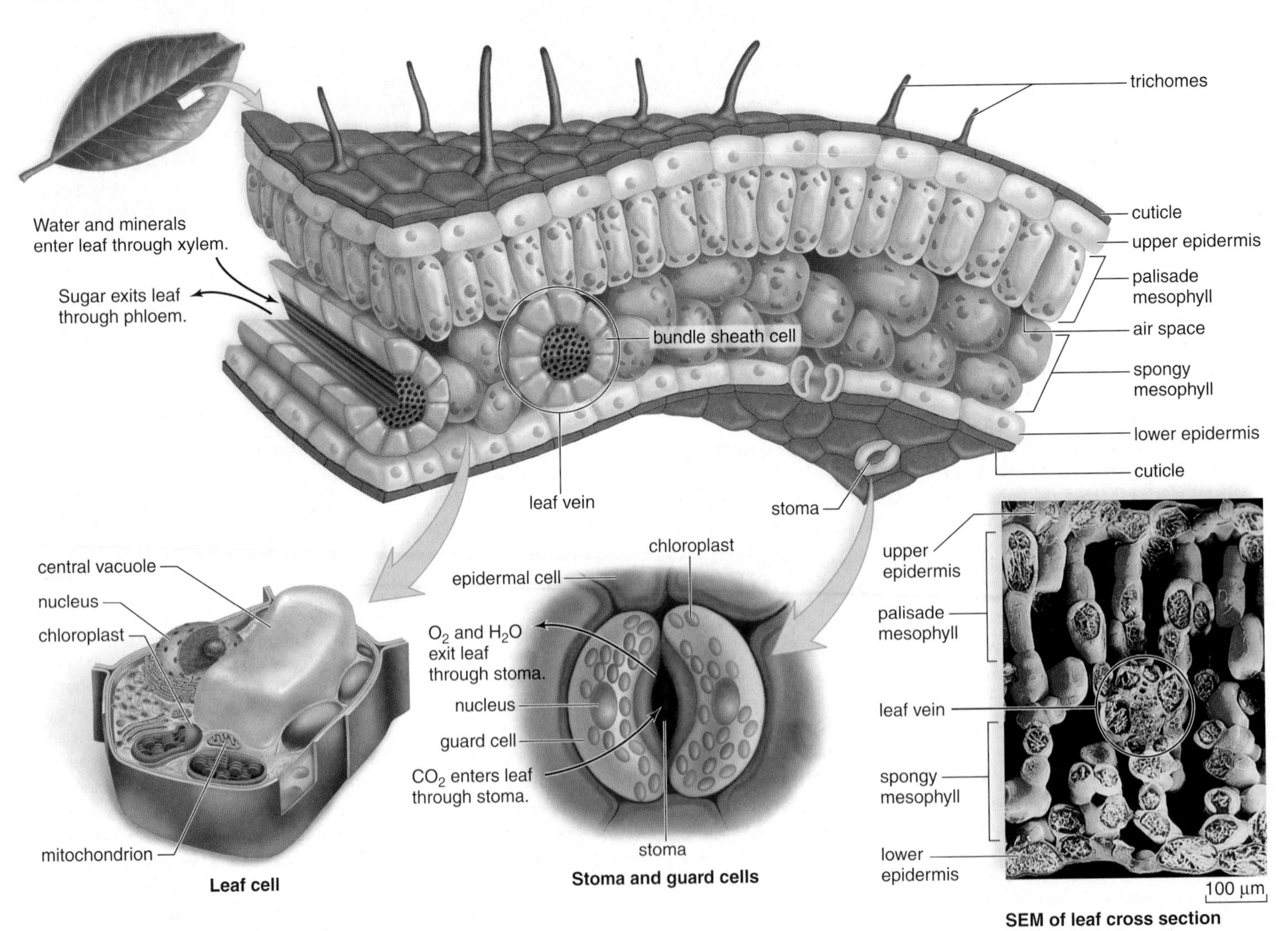


**FIGURE 24.20 Leaf structure.**

Photosynthesis takes place in mesophyll tissue of leaves. The leaf is enclosed by epidermal cells covered with a waxy layer, the cuticle. Leaf hairs are also protective. The veins contain xylem and phloem for the transport of water and solutes. A stoma is an opening in the epidermis that permits the exchange of gases.

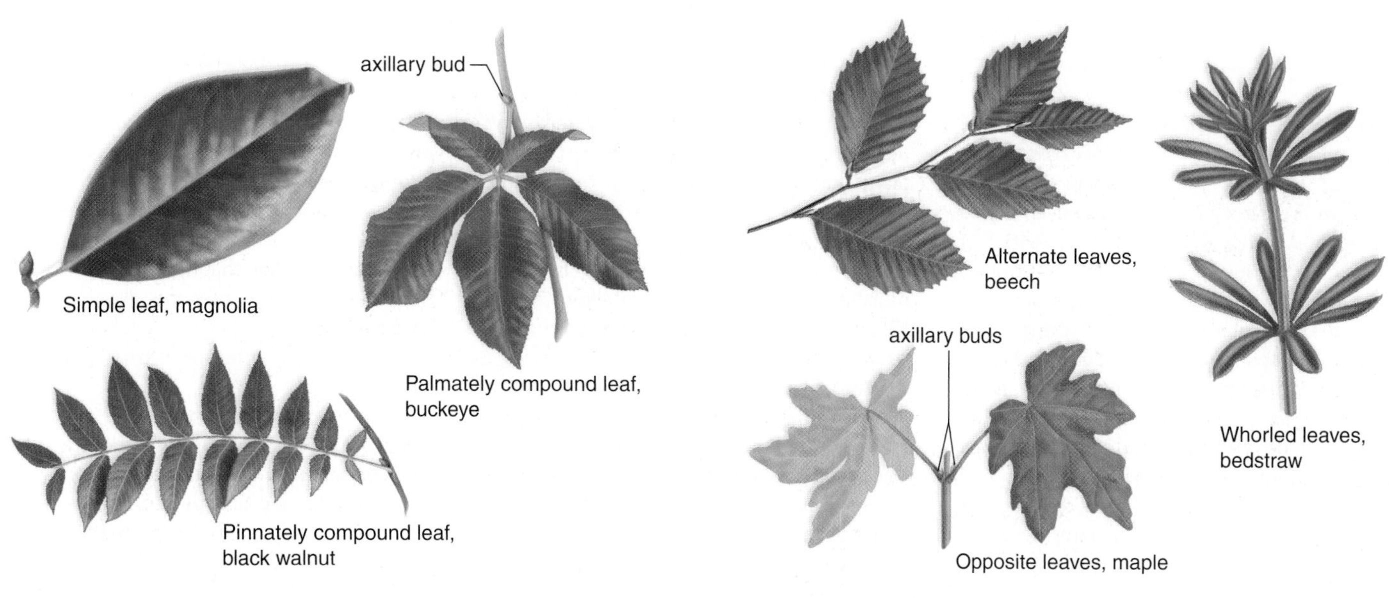


**FIGURE 24.21 Classification of leaves.**

**a.** Leaves are either simple or compound, being either pinnately compound or palmately compound. Note the one axillary bud per compound leaf. **b.** Leaf arrangement on stem can be alternate, opposite, or whorled.

Leaves can be arranged on a stem in three ways: alternate, opposite, or whorled. The leaves are alternate in the American beech; in a maple, the leaves are opposite, being attached to the same node. Bedstraw has a whorled leaf arrangement with several leaves originating from the same node. Leaves are adapted to environmental conditions. Shade plants tend to have broad, wide leaves, and desert plants tend to have reduced leaves with sunken stomata. The leaves of a cactus are the spines attached to the succulent (water-containing) stem (Fig. 24.22*a*).

An onion bulb is made up of leaves surrounding a short stem. In a head of cabbage, large leaves overlap one another. The petiole of a leaf can be thick and fleshy, as in celery and rhubarb. Climbing leaves, such as those of peas and cucumbers, can be modified into tendrils that can attach to nearby objects (Fig. 24.22*b*). The leaves of a few plants are specialized for catching insects. A sundew has sticky trichomes that trap insects and others that secrete digestive enzymes. The Venus's flytrap has hinged leaves that snap shut and interlock when an insect triggers sensitive trichomes that project from inside the leaves (Fig. 24.22*c*). Certain leaves of a pitcher plant resemble a pitcher and have downward-pointing hairs that lead insects into a pool of digestive enzymes secreted by trichomes. Insectivorous plants commonly grow in marshy regions, where the supply of soil nitrogen is severely limited. The digested insects provide the plants with a source of organic nitrogen.

### Check Your Progress 24.5

1. What is the importance of the leaf tissue called mesophyll to a plant?

a. Cactus, *Opuntia*

b. Cucumber, *Cucumis*

c. Venus's flytrap, *Dionaea*

**FIGURE 24.22 Leaf diversity.**

**a.** The spines of a cactus plant are modified leaves that protect the fleshy stem from animal predation. **b.** The tendrils of a cucumber are modified leaves that attach the plant to a physical support. **c.** The modified leaves of the Venus's flytrap serve as a trap for insect prey. When triggered by an insect, the leaf snaps shut. Once shut, the leaf secretes digestive juices that break down the soft parts of the insect's body.

# Connecting the Concepts

In Chapter 23, we saw how land plants became adapted to reproducing in a terrestrial environment. Many other types of adaptations are also required to live on land. Because even humid air is drier than a living cell, the prevention of water loss is critical for land plants. The epidermis, the trichomes, and the cuticle it produces help prevent water loss and overheating in sunlight. (The epidermis and glandular trichomes also protect against invasion by bacteria, fungi, and small insects.) Gas exchange in leaves depends on the presence of stomata, which close when a plant is water-stressed. The cork of woody plants is especially protective against water loss, but when cork is interrupted by lenticels, gas exchange is still possible.

In an aquatic environment, water buoys up plants and keeps them afloat, but land plants had to evolve a way to oppose the force of gravity. The stems of land plants contain strong-walled sclerenchyma fibers, tracheids, and vessel elements. The accumulation of secondary xylem allows a tree to grow in diameter and offers even more support.

Also, in an aquatic environment, water is available to all cells, but on land it is adaptive to have a means of water uptake and transport. In land plants, the roots absorb water and have special extensions called root hairs that facilitate water uptake. Xylem transports water to all plant parts, including the leaves. In Chapter 25, we will see how the drying effect of air allows water to move from the roots to the leaves in the conducting cells of xylem. Roots are buried in soil, where, with the help of mycorrhizae, they can absorb water. In that chapter, we will also see how the properties of water allow phloem to transport sugars from the leaves to the roots and to any other plant part in need of sustenance.

The many adaptations of land plants allow them to carry on photosynthesis and be homeostatic in the terrestrial environment. Homeostatic mechanisms also involve regulation by hormones and defense mechanisms, discussed in Chapter 26.

## summary

### 24.1 Organs of Flowering Plants

A flowering plant has three vegetative organs. A root anchors a plant, absorbs water and minerals, and stores the products of photosynthesis. Stems produce new tissue, support leaves, conduct materials to and from roots and leaves, and help store plant products. Leaves are specialized for gas exchange, and they carry on most of the photosynthesis in the plant.

Flowering plants are divided into the monocots and eudicots according to the number of cotyledons in the seed, the arrangement of vascular tissue in roots, stems, and leaves, and the number of flower parts.

### 24.2 Tissues of Flowering Plants

Flowering plants have apical meristem plus three types of primary meristem. Protoderm produces epidermal tissue. In the roots, epidermal cells bear root hairs; in the leaves, the epidermis contains guard cells. In a woody stem, epidermis is replaced by periderm.

Ground meristem produces ground tissue. Ground tissue is composed of parenchyma cells, which are thin-walled and capable of photosynthesis when they contain chloroplasts. Collenchyma cells have thicker walls for flexible support. Sclerenchyma cells are hollow, nonliving support cells with secondary walls fortified by lignin.

Procambium produces vascular tissue. Vascular tissue consists of xylem and phloem. Xylem contains two types of conducting cells: vessel elements and tracheids. Vessel elements, which are larger and have perforation plates, form a continuous pipeline from the roots to the leaves. In elongated tracheids with tapered ends, water must move through pits in end walls and side walls. Xylem transports water and minerals. In phloem, sieve tubes are composed of sieve-tube members, each of which has a companion cell. Phloem transports sucrose and other organic compounds including hormones.

### 24.3 Organization and Diversity of Roots

A root tip has a zone of cell division (containing the primary meristems), a zone of elongation, and a zone of maturation.

A cross section of a herbaceous eudicot root reveals the epidermis, which protects; the cortex, which stores food; the endodermis, which regulates the movement of minerals; and the vascular cylinder, which is composed of vascular tissue. In the vascular cylinder of a eudicot, the xylem appears star-shaped, and the phloem is found in separate regions, between the points of the star. In contrast, a monocot root has a ring of vascular tissue with alternating bundles of xylem and phloem surrounding the pith.

Roots are diversified. Taproots are specialized to store the products of photosynthesis; a fibrous root system covers a wider area. Prop roots are adventitious roots specialized to provide increased anchorage.

### 24.4 Organization and Diversity of Stems

The activity of the shoot apical meristem within a terminal bud accounts for the primary growth of a stem. A terminal bud contains internodes and leaf primordia at the nodes. When stems grow, the internodes lengthen.

In a cross section of a nonwoody eudicot stem, epidermis is followed by cortex tissue, vascular bundles in a ring, and an inner pith. Monocot stems have scattered vascular bundles, and the cortex and pith are not well defined.

Secondary growth of a woody stem is due to vascular cambium, which produces new xylem and phloem every year, and cork cambium, which produces new cork cells when needed. Cork, a part of the bark, replaces epidermis in woody plants. In a cross section of a woody stem, the bark is all the tissues outside the vascular cambium. It consists of secondary phloem, cork cambium, and cork. Wood consists of secondary xylem, which builds up year after year and forms the annual rings.

Stems are diverse. There are horizontal aboveground and underground stems. Corms and some tendrils are also modified stems.

### 24.5 Organization and Diversity of Leaves

The bulk of a leaf is mesophyll tissue bordered by an upper and lower layer of epidermis; the epidermis is covered by a cuticle and may bear trichomes. Stomata tend to be in the lower layer. Vascular tissue is present within leaf veins. Leaves are diverse: The spines of a cactus are leaves. Other succulents have fleshy leaves. An onion is a bulb with fleshy leaves, and the tendrils of peas are leaves. The Venus's flytrap has leaves that trap and digest insects.

## understanding the terms

adventitious root 442
annual ring 447
apical meristem 440
axillary bud 435
bark 446
blade 435
Casparian strip 441
collenchyma 438
complex tissue 439
cork 437
cork cambium 437
cortex 441
cotyledon 436
cuticle 437
deciduous 435
endodermis 441
epidermal tissue 437
epidermis 437, 441
eudicot 436
evergreen 435
fibrous root system 442
ground tissue 437
herbaceous stem 445
internode 435
leaf 435
leaf vein 439
lenticel 437
lignin 438
meristem 437
mesophyll 450
monocot 436
mycorrhiza 442
node 435
organ 434
palisade mesophyll 450
parenchyma 438
perennial 434
pericycle 441
periderm 437
petiole 435
phloem 439
pit 439
pith 441
primary root 442
rhizome 448
root cap 440
root hair 437
root nodule 442
root system 434
sclerenchyma 438
shoot apical meristem 444
shoot system 434
sieve-tube member 439
spongy mesophyll 450
stem 435
stolon 448
stomata (sing., stoma) 437
taproot 442
terminal bud 444
tracheid 439
trichome 437
vascular bundle 439
vascular cambium 444
vascular cylinder 439
vascular tissue 437
vegetative organ 434
vessel element 439
wood 446
xylem 439

Match the terms to these definitions:

a. ______________ Inner, thickest layer of a leaf; the site of most photosynthesis.
b. ______________ Lateral meristem that produces secondary phloem and secondary xylem.
c. ______________ Seed leaf for embryonic flowering plant; provides nutrient molecules before the leaves begin to photosynthesize.
d. ______________ Stem that grows horizontally along the ground and establishes plantlets periodically when it contacts the soil (e.g., the runners of a strawberry plant).
e. ______________ Vascular tissue that contains vessel elements and tracheids.

## reviewing this chapter

1. Name and discuss the vegetative organs of a flowering plant. 434–35
2. List five differences between monocots and eudicots. 436
3. Epidermal cells are found in what type of plant tissue? Explain how epidermis is modified in various organs of a plant. Contrast an epidermal cell with a cork cell. 437
4. Contrast the structure and function of parenchyma, collenchyma, and sclerenchyma cells. These cells occur in what type of plant tissue? 438
5. Contrast the structure and function of xylem and phloem. Xylem and phloem occur in what type of plant tissue? 439
6. Name and discuss the zones of a root tip. Trace the path of water and minerals across a root from the root hairs to xylem. Be sure to mention the Casparian strip. 440–41
7. Contrast a taproot with a fibrous root system. What are adventitious roots? 442
8. Describe the primary growth of a stem. 444–45
9. Describe cross sections of a herbaceous eudicot, a monocot, and a woody stem. 445–47
10. Discuss the diversity of stems by giving examples of several adaptations. 448
11. Describe the structure and organization of a typical eudicot leaf. 450
12. Note the diversity of leaves by giving examples of several adaptations. 450–51

## testing yourself

Choose the best answer for each question.

1. Which of these is an incorrect contrast between monocots (stated first) and eudicots (stated second)?
   a. one cotyledon—two cotyledons
   b. leaf veins parallel—net veined
   c. vascular bundles in a ring—vascular bundles scattered
   d. flower parts in threes—flower parts in fours or fives
   e. All of these are correct contrasts.
2. Which of these types of cells is most likely to divide?
   a. parenchyma
   b. meristem
   c. epidermis
   d. xylem
   e. sclerenchyma
3. Which of these cells in a flowering plant is apt to be nonliving?
   a. parenchyma
   b. collenchyma
   c. sclerenchyma
   d. epidermal cells
   e. guard cells
4. Root hairs are found in the zone of
   a. cell division.
   b. elongation.
   c. maturation.
   d. apical meristem.
   e. All of these are correct.
5. Cortex is found in
   a. roots, stems, and leaves.
   b. roots and stems.
   c. roots and leaves.
   d. stems and leaves.
   e. roots only.
6. Between the bark and the wood in a woody stem, there is a layer of meristem called
   a. cork cambium.
   b. vascular cambium.
   c. apical meristem.
   d. the zone of cell division.
   e. procambium preceding bark.
7. Which part of a leaf carries on most of the photosynthesis of a plant?
   a. epidermis
   b. mesophyll
   c. epidermal layer
   d. guard cells
   e. Both a and b are correct.
8. Annual rings are the
   a. internodes in a stem.
   b. rings of vascular bundles in a monocot stem.
   c. layers of xylem in a woody stem.
   d. bark layers in a woody stem.
   e. Both b and c are correct.

9. The Casparian strip is found
   a. between all epidermal cells.
   b. between xylem and phloem cells.
   c. on four sides of endodermal cells.
   d. within the secondary wall of parenchyma cells.
   e. in both endodermis and pericycle.
10. Which of these is a stem?
    a. taproot of carrots
    b. stolon of strawberry plants
    c. spine of cactuses
    d. prop roots
    e. Both b and c are correct.
11. Meristem tissue that gives rise to epidermal tissue is called
    a. procambium.
    b. ground meristem.
    c. epiderm.
    d. protoderm.
    e. periderm.
12. New plant cells originate from the
    a. parenchyma.
    b. collenchyma.
    c. sclerenchyma.
    d. base of the shoot.
    e. apical meristem.
13. Ground tissue does not include
    a. collenchyma cells.
    b. sclerenchyma cells.
    c. parenchyma cells.
    d. chlorenchyma cells.
14. Evenly thickened cells that function to support mature regions of a flowering plant are called
    a. guard cells.
    b. aerenchyma cells.
    c. parenchyma cells.
    d. sclerenchyma cells.
    e. xylem cells.
15. Roots
    a. are the primary site of photosynthesis.
    b. give rise to new leaves and flowers.
    c. have a thick cuticle to protect the epidermis.
    d. absorb water and nutrients.
    e. contain spores.
16. Monocot stems have
    a. vascular bundles arranged in a ring.
    b. vascular cambium.
    c. scattered vascular bundles.
    d. a cork cambium.
    e. a distinct pith and cortex.
17. Secondary thickening of stems occurs in
    a. all angiosperms.
    b. most monocots.
    c. many eudicots.
    d. few eudicots.
18. All of these may be found in heartwood except
    a. tracheids.
    b. vessel elements.
    c. parenchyma cells.
    d. sclerenchyma cells.
    e. companion cells.
19. How are compound leaves distinguished from simple leaves?
    a. Compound leaves do not have axillary buds at the base of leaflets.
    b. Compound leaves are smaller than simple leaves.
    c. Simple leaves are usually deciduous.
    d. Compound leaves are found only in pine trees.
    e. Simple leaves are found only in gymnosperms.
20. Label this root using these terms: endodermis, phloem, xylem, cortex, and epidermis.

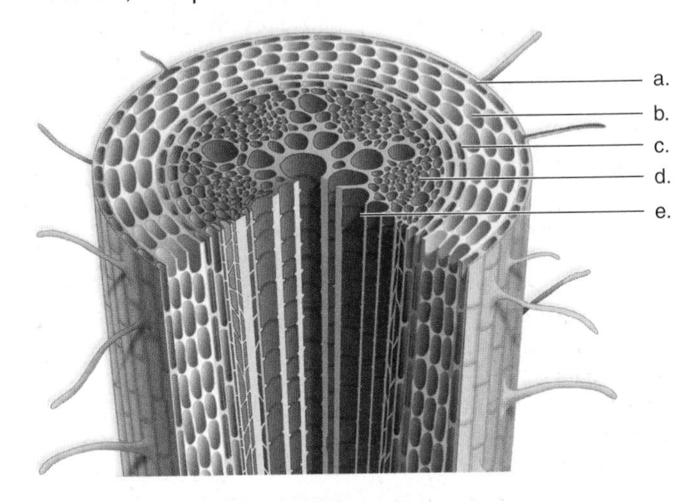


21. Label this leaf using these terms: leaf vein, lower epidermis, palisade mesophyll, spongy mesophyll, and upper epidermis.

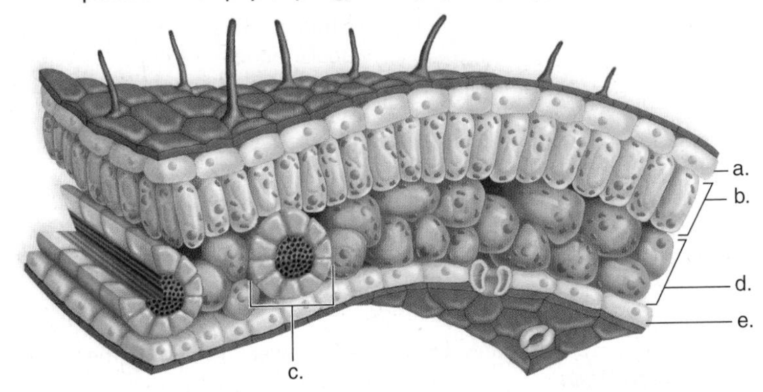


## thinking scientifically

1. Utilizing an electron microscope, how might you confirm that a companion cell communicates with its sieve-tube member?
2. Design an experiment that tests the hypothesis that new plants arise at the nodes of a stolon according to environmental conditions (temperature, water, and sunlight).

## *Biology* website

The companion website for *Biology* provides a wealth of information organized and integrated by chapter. You will find practice tests, animations, videos, and much more that will complement your learning and understanding of general biology.

**http://www.mhhe.com/maderbiology10**

# 25

# Flowering Plants: Nutrition and Transport

*Plants have nutrient requirements just as animals do. They use carbon, hydrogen, oxygen, nitrogen, potassium, calcium, phosphorus, magnesium, and sulfur in relatively large amounts to make all the substances they need to carry out life functions. Plant leaves absorb the gas carbon dioxide and their roots take up oxygen. Water and dissolved minerals move into root hairs often covered by mycorrhizae. Plants have an amazing ability to concentrate minerals in their tissues to a much higher level than they occur in the soil. Some plants, such as those in the legume family (peanuts, clovers, beans), have roots colonized by bacteria that can convert atmospheric nitrogen to a form usable by plants.*

*This chapter discusses the nutrient requirements of plants and how they are absorbed and distributed within the body of a plant. Plants have no central pumping mechanism, yet materials move throughout the body of the plant. The unique properties of water account for the movement of water and minerals in xylem, while osmosis plays an essential role in phloem transport of sugars. The same mechanisms account for transport in very tall redwood trees and in dwarf gardenias.*

Redwoods, *Sequoia sempervirens.*

## concepts

# 25.1 Plant Nutrition and Soil

The ancient Greeks believed that plants were "soil-eaters" and somehow converted soil into plant material. Apparently to test this hypothesis, a seventeenth-century Dutchman named Jean-Baptiste Van Helmont planted a willow tree weighing 5 lb in a large pot containing 200 lb of soil. He watered the tree regularly for five years and then reweighed both the tree and the soil. The tree weighed 170 lb, and the soil weighed only a few ounces less than the original 200 lb. Van Helmont concluded that the increase in weight of the tree was due primarily to the addition of water.

Water is a vitally important nutrient for a plant, but Van Helmont was unaware that water and carbon dioxide (taken in at the leaves) combine in the presence of sunlight to produce carbohydrates, the chief organic matter of plants. Much of the water entering a plant evaporates at the leaves. Roots, like all plant organs, carry on cellular respiration, a process that uses oxygen and gives off carbon dioxide (Fig. 25.1).

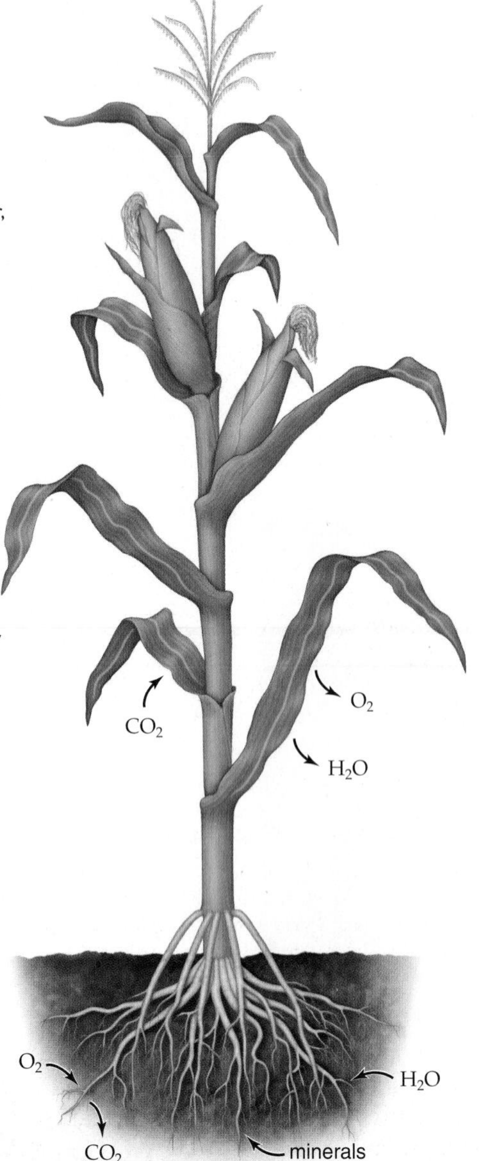


**FIGURE 25.1**
**Overview of plant nutrition.**
Carbon dioxide, which enters leaves, and water, which enters roots, are combined during photosynthesis to form carbohydrates, with the release of oxygen from the leaves. Root cells, and all other plant cells, carry on cellular respiration, which uses oxygen and gives off carbon dioxide. Aside from the elements carbon, hydrogen, and oxygen, plants require nutrients that are absorbed as minerals by the roots.

## Essential Inorganic Nutrients

Approximately 95% of a typical plant's dry weight (weight excluding free water) is carbon, hydrogen, and oxygen. Why? Because these are the elements that are found in most organic compounds, such as carbohydrates. Carbon dioxide ($CO_2$) supplies carbon, and water ($H_2O$) supplies hydrogen and oxygen found in the organic compounds of a plant.

**TABLE 25.1**

**Some Essential Inorganic Nutrients in Plants**

| *Elements* | *Symbol* | *Form* | *Major Functions* |
|---|---|---|---|
| ***Macronutrients*** | | | |
| Carbon<br>Hydrogen<br>Oxygen | C<br>H<br>O | $CO_2$<br>$H_2O$<br>$O_2$ | Major component of organic molecules |
| Phosphorus | P | $H_2PO_4^-$<br>$HPO_4^{2-}$ | Part of nucleic acids, ATP, and phospholipids |
| Potassium | K | $K^+$ | Cofactor for enzymes; water balance and opening of stomata |
| Nitrogen | N | $NO_3^-$<br>$NH_4^+$ | Part of nucleic acids, proteins, chlorophyll, and coenzymes |
| Sulphur | S | $SO_4^{2-}$ | Part of amino acids, some coenzymes |
| Calcium | Ca | $Ca^{2+}$ | Regulates responses to stimuli and movement of substances through plasma membrane; involved in formation and stability of cell walls |
| Magnesium | Mg | $Mg^{2+}$ | Part of chlorophyll; activates a number of enzymes |
| ***Micronutrients*** | | | |
| Iron | Fe | $Fe^{2+}$<br>$Fe^{3+}$ | Part of cytochrome needed for cellular respiration; activates some enzymes |
| Boron | B | $BO_3^{3-}$<br>$B_4O_7^{2-}$ | Role in nucleic acid synthesis, hormone responses, and membrane function |
| Manganese | Mn | $Mn^{2+}$ | Required for photosynthesis; activates some enzymes such as those of the citric acid cycle |
| Copper | Cu | $Cu^{2+}$ | Part of certain enzymes, such as redox enzymes |
| Zinc | Zn | $Zn^{2+}$ | Role in chlorophyll formation; activates some enzymes |
| Chlorine | Cl | $Cl^-$ | Role in water-splitting step of photosynthesis and water balance |
| Molybdenum | Mo | $MoO_4^{2-}$ | Cofactor for enzyme used in nitrogen metabolism |

In addition to carbon, hydrogen, and oxygen, plants require certain other nutrients that are absorbed as minerals by the roots. A **mineral** is an inorganic substance usually containing two or more elements. Why are minerals from the soil needed by a plant? In plants, nitrogen is a major component of nucleic acids and proteins, magnesium is a component of chlorophyll, and iron is a building block of cytochrome molecules. The major functions of various **essential nutrients** for plants are listed in Table 25.1. A nutrient is essential if (1) it has an identifiable role, (2) no other nutrient can substitute and fulfill the same role, and (3) a deficiency of this nutrient causes a plant to die without completing its life cycle. Essential nutrients are divided into **macronutrients** and **micronutrients** according to their relative concentrations in plant tissue. The following diagram and slogan helps us remember which are the macronutrients and which are the micronutrients for plants:

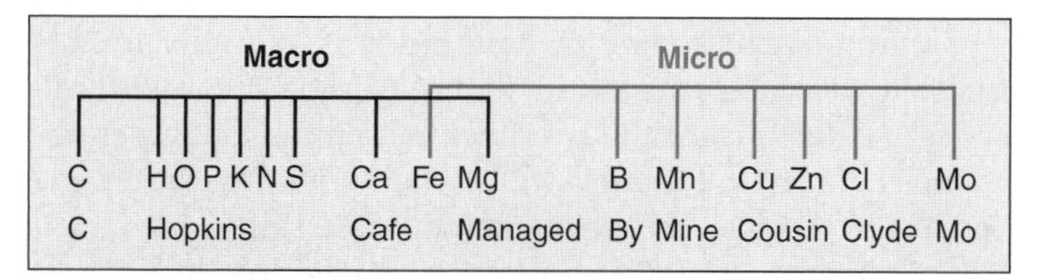


**Beneficial nutrients** are another category of elements taken up by plants. Beneficial nutrients either are required for or enhance the growth of a particular plant. Horsetails require silicon as a mineral nutrient and sugar beets show enhanced growth in the presence of sodium. Nickel is a beneficial mineral nutrient in soybeans when root nodules are present. Aluminum is used by some ferns, and selenium, which is often fatally poisonous to livestock, is used by locoweeds.

## Determination of Essential Nutrients

When a plant is burned, its nitrogen component is given off as ammonia and other gases, but most other essential minerals remain in the ash. The presence of a mineral in the ash, however, does not necessarily mean that the plant normally requires it. The preferred method for determining the mineral requirements of a plant was developed at the end of the nineteenth century by the German plant physiologists Julius von Sachs and Wilhem Knop. This method is called water culture, or **hydroponics** [Gk. *hydrias,* water, and *ponos,* hard work]. Hydroponics allows plants to grow well if they are supplied with all the nutrients they need. The investigator omits a particular mineral and observes the effect on plant growth. If growth suffers, it can be concluded that the omitted mineral is an essential nutrient (Fig. 25.2). This method has been more successful for macronutrients than for micronutrients. For studies involving the latter, the water and the mineral salts used must be absolutely pure, but purity is difficult to attain, because even instruments and glassware can introduce micronutrients. Then, too, the element in question may already be present in the seedling used in the experiment. These factors complicate the determination of essential plant micronutrients by means of hydroponics.

a. Solution lacks nitrogen　　Complete nutrient solution

b. Solution lacks phosphorus　　Complete nutrient solution

c. Solution lacks calcium　　Complete nutrient solution

**FIGURE 25.2 Nutrient deficiencies.**

The nutrient cause of poor plant growth is diagnosed when plants are grown in a series of complete nutrient solutions except for the elimination of just one nutrient at a time. These experiments show that sunflower plants respond negatively to a deficiency of (**a**) nitrogen, (**b**) phosphorus, and (**c**) calcium.

### *Hydroponics*

Hydroponics is of interest as a way to grow crops in the future. Plant pests and diseases are eliminated, and there are no weeds. Water is reused in a pipeline system and little is lost through runoff.

## Soil

Plants acquire carbon when carbon dioxide diffuses into leaves through stomata. Oxygen can enter from the air, but all of the other essential nutrients are absorbed by roots from the soil. It would not be an exaggeration to say that terrestrial life is dependent on the quality of the soil and the ability of soil to provide plants with the nutrients they need.

### *Soil Formation*

Soil formation begins with the weathering of rock in the Earth's crust. Weathering first gradually breaks down rock to rubble and then to soil particles. Some weathering mechanisms, such as the freeze-thaw cycle of ice or the grinding of rock on rock by the action of glaciers or river flow, are purely mechanical. Other forces include a chemical effect, as when acidic rain leaches (washes away) soluble components of rock or when oxygen combines with the iron of rocks.

In addition to these forces, organisms also play a role in the formation of soil. Lichens and mosses grow on pure rock and trap particles that later allow grasses, herbs, and soil animals to follow. When these die, their remains are decomposed, notably by bacteria and fungi. Decaying organic matter, called **humus,** begins to accumulate. Humus supplies nutrients to plants, and its acidity also leaches minerals from rock.

Building soil takes a long time. Under ideal conditions, depending on the type of parent material (the original rock) and the various processes at work, a centimeter of soil may develop within 15 years.

### *The Nutritional Function of Soil*

**Soil** is defined as a mixture of mineral particles, decaying organic material, living organisms, air, and water, which together support the growth of plants. In a good agricultural soil, the first three components come together in such a way that there are spaces for air and water (Fig. 25.3). It's best if the soil contains particles of different sizes because only then will there be spaces for air. Roots take up oxygen from air spaces. Ideally, water clings to particles by capillary action and does not fill the spaces. That's why you shouldn't overwater your houseplants!

**Mineral Particles.** Mineral particles vary in size: Sand particles are the largest (0.05–2.0 mm in diameter); silt particles have an intermediate size (0.002–0.05 mm); and clay particles are the smallest (less than 0.002 mm). Soils are a mixture of these three types of particles. Because sandy soils have many large particles, they have large spaces, and the water drains readily through the particles. In contrast to sandy soils, a soil composed mostly of clay particles has small spaces that fill completely with water. Most likely, you have experienced the feel of sand and clay in your hand: Sand having no moisture flows right through your fingers, while clay clumps together in one large mass because of its water content.

Clay particles have another benefit that sand particles do not have. As Table 25.1 indicates, some minerals are negatively charged and others are positively charged. Clay particles are negative, and they can retain positively charged minerals such as calcium ($Ca^{2+}$) and potassium ($K^{+}$), preventing these minerals from being washed away by leaching. Plants exchange hydrogen ions for these minerals when they take them up (Fig. 25.3). If rain is acidic, its hydrogen ions displace positive mineral ions and cause them to drain away; this is one reason acid rain kills trees. Because clay particles are unable to retain negatively charged $NO_3^-$, the nitrogen content of soil is apt to be low. Legumes (see Fig. 1.13) are sometimes planted to replenish the nitrogen in the soil in preference to relying solely on the addition of fertilizer.

The type of soil called loam is composed of roughly one-third sand, silt, and clay particles. This combination sufficiently retains water and nutrients while still allowing the drainage necessary to provide air spaces. Some of the most productive soils are loam.

**Humus.** Humus, which mixes with the top layer of soil particles, increases the benefits of soil. Plants do well in soils that contain 10–20% humus.

Humus causes soil to have a loose, crumbly texture that allows water to soak in without doing away with air spaces. After a rain, the presence of humus decreases the chances of runoff. Humus swells when it absorbs water and shrinks as it dries. This action helps aerate soil.

Soil that contains humus is nutritious for plants. Humus is acidic; therefore, it retains positively charged minerals until plants take them up. When the organic matter in humus is broken down by bacteria and fungi, inorganic nutrients are returned to plants. Although soil particles are the original source of minerals in soil, recycling of nutrients, as you know, is a major characteristic of ecosystems.

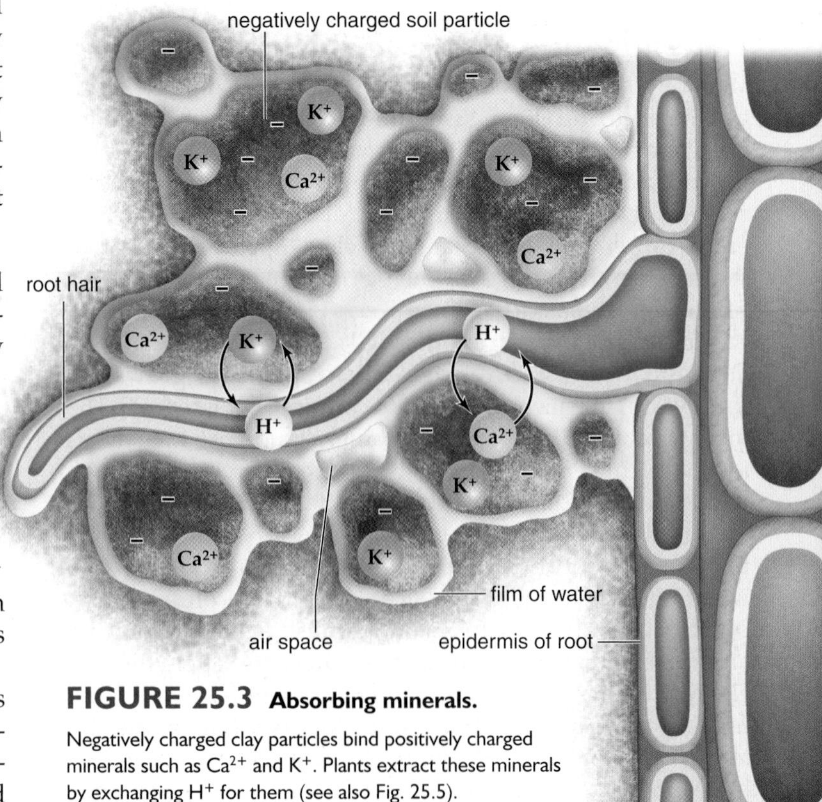


**FIGURE 25.3 Absorbing minerals.**

Negatively charged clay particles bind positively charged minerals such as $Ca^{2+}$ and $K^{+}$. Plants extract these minerals by exchanging $H^{+}$ for them (see also Fig. 25.5).

**Living Organisms.** Small plants play a major role in the formation of soil from bare rock. Due to the process of succession (see Fig. 45.14), larger plants eventually become dominant in certain ecosystems. The roots of larger plants penetrate soil even to the cracks in bedrock. This action slowly opens up soil layers, allowing water, air, and animals to follow.

There are many different types of soil animals. The largest of them, such as toads, snakes, moles, badgers, and rabbits, disturb and mix soil by burrowing. Smaller animals like earthworms ingest fine soil particles and deposit them on the surface as worm casts. Earthworms also loosen and aerate the soil. A range of soil animals, including mites, springtails, and millipedes, help break down leaves and other plant remains by eating them. Soil-dwelling ants construct tremendous colonies with massive chambers and tunnels. These ants also loosen and aerate the soil.

The microorganisms in soil, such as protozoans, fungi, algae, and bacteria, are responsible for the final decomposition of organic remains in humus to inorganic nutrients. Recall that plants are unable to make use of atmospheric nitrogen ($N_2$) and that soil bacteria play an important nutrient role because they make nitrate available to plants.

Insects may improve the properties of soil, but they are also major crop pests when they feed on plant roots. Certain soil organisms, such as some roundworms, can severely impact golf course turf, for example.

## Soil Profiles

A **soil profile** is a vertical section from the ground surface to the unaltered rock below. Usually, a soil profile has parallel layers known as **soil horizons.** Mature soil generally has three horizons (Fig. 25.4). The A horizon is the uppermost (or topsoil) layer that contains litter and humus, although most of the soluble chemicals may have been leached away. The B horizon has little or no organic matter but does contain the inorganic nutrients leached from the A horizon. The C horizon is a layer of weathered and shattered rock.

Because the parent material (rock) and climate (e.g., temperature and rainfall) differ in various parts of the biosphere, the soil profile varies according to the particular ecosystem. Soils formed in grasslands tend to have a deep A horizon built up from decaying grasses over many years, but because of limited rain, there has been little leaching into the B horizon. In forest soils, both the A and B horizons have enough inorganic nutrients to allow for root growth. In tropical rain forests, the A horizon is more shallow than the generalized profile, and the B horizon is deeper, signifying that leaching is more extensive. Since the topsoil of a rain forest lacks nutrients, it can only support crops for a few years.

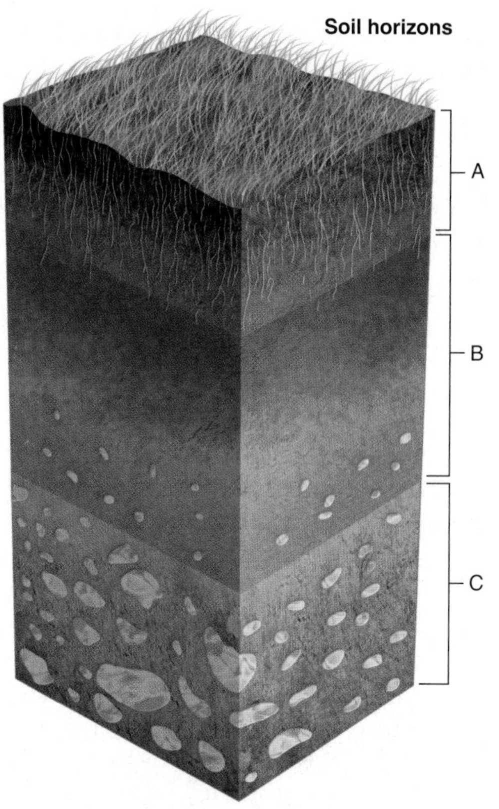


**FIGURE 25.4 Simplified soil profile.**

The top layer (A horizon) contains most of the humus; the next layer (B horizon) accumulates materials leached from the A horizon; and the lowest layer (C horizon) is composed of weathered parent material. Erosion removes the A horizon, a primary source of humus and minerals in soil.

## Soil Erosion

**Soil erosion** occurs when water or wind carry soil away to a new location. Erosion removes about 25 billion tons of topsoil yearly, worldwide. If this rate of loss continues, some scientists predict that the Earth will lose practically all of its topsoil by the middle of the next century. Deforestation (removal of trees) and desertification (increase in deserts due to overgrazing and overfarming marginal lands) contribute to the occurrence of erosion, and so do poor farming practices in general.

In the United States, soil is eroding faster than it is being formed on about one-third of all cropland. Fertilizers and pesticides, carried by eroding soil into groundwater and rivers, are threatening human health. To make up for the loss of soil due to erosion, more energy is used to apply more fertilizers and pesticides to crops. Instead, it would be best to stop erosion before it occurs by following sound agricultural practices.

The coastal wetlands are losing soil at a tremendous rate. These wetlands are important as nurseries for many species of organisms, such as shrimp and redfish, and as protection against storm surge from hurricanes. In Louisiana, 24 $mi^2$ of wetlands are lost each year. This equates to one football field being lost every 38 minutes.

### Check Your Progress 25.1

1. What element(s) in particular, aside from C, H, and O, is/are needed to form (**a**) proteins and (**b**) nucleic acids? How does a plant acquire these elements?
2. Some farmers do not remove the remains of last year's crops from agricultural lands. What are the benefits of this practice?
3. What are the benefits of humus in soil?

## 25.2 Water and Mineral Uptake

The pathways for water and mineral uptake and transport in a plant are the same. As Figure 25.5*a* shows, water along with minerals can enter the root of a flowering plant from the soil simply by passing between the porous cell walls. Eventually, however, the **Casparian strip,** a band of suberin and lignin bordering four sides of root endodermal cells, forces water to enter endodermal cells. Alternatively, water can enter epidermal cells at their **root hairs** and then progress through cells across the cortex and endodermis of a root by means of cytoplasmic strands within plasmodesmata (see Fig. 5.15). Regardless of the pathway, water enters root cells when they have a lower osmotic pressure than does the soil solution.

### Mineral Uptake

In contrast to water, minerals are actively taken up by plant cells. Plants possess an astonishing ability to concentrate minerals—that is, to take up minerals until they are many times more concentrated in the plant than in the surrounding medium. The concentration of certain minerals in roots is as much as 10,000 times greater than in the surrounding soil. Following their uptake by root cells, minerals move into xylem and are transported into leaves by the upward movement of water. Along the way, minerals can exit xylem and enter those cells that require them. Some eventually reach leaf cells. In any case, minerals must again cross a selectively permeable plasma membrane when they exit xylem and enter living cells. By what mechanism do minerals cross plasma membranes?

Recall that plant cells absorb minerals in the ionic form: Nitrogen is absorbed as nitrate ($NO_3^-$), phosphorus as phosphate ($HPO_4^{2-}$), potassium as potassium ions ($K^+$), and so forth. Ions cannot cross the plasma membrane because they are unable to enter the nonpolar phase of the lipid bilayer. It has long been known that plant cells expend energy to actively take up and concentrate mineral ions. If roots are deprived of oxygen or are poisoned so that cellular respiration cannot occur, mineral ion uptake is diminished. The energy of ATP is required for mineral ion transport, but not directly (Fig. 25.5*b*). A plasma membrane pump, called a proton pump, hydrolyzes ATP and uses the energy released to transport hydrogen ions ($H^+$) out of the cell. This sets up an electrochemical gradient that drives positively charged ions such as $K^+$ through a channel protein into the cell. Negatively charged mineral ions are transported, along with $H^+$, by carrier proteins. Since $H^+$ is moving down its concentration gradient, no energy is required. Notice that this model of mineral ion transport in plant cells is based on chemiosmosis, the establishment of an electrochemical gradient to perform work.

#### *Adaptations of Roots for Mineral Uptake*

Two mutualistic relationships assist roots in obtaining mineral nutrients. Root nodules involve a mutualistic relationship with bacteria, and mycorrhizae are a mutualistic relationship with fungi.

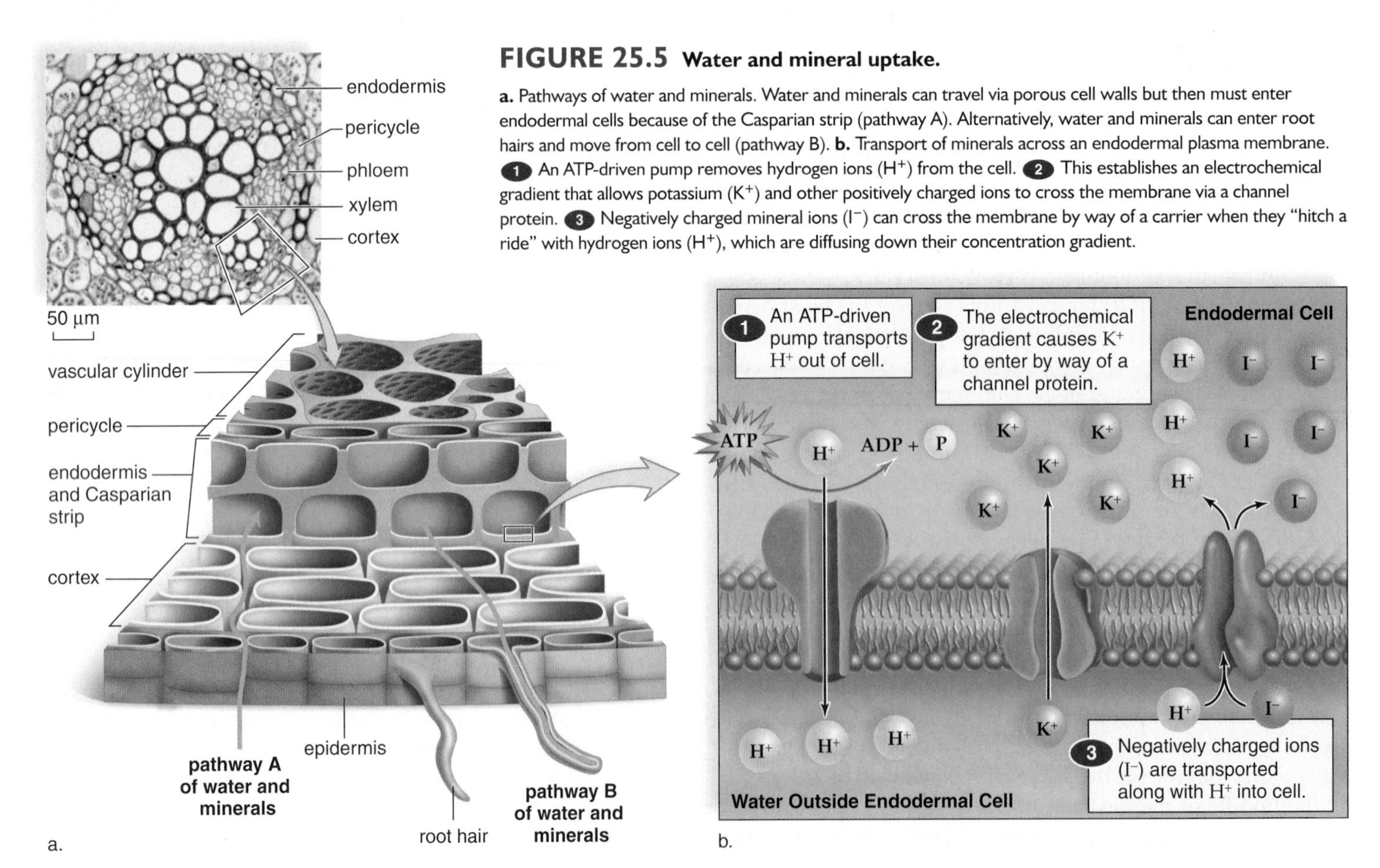


**FIGURE 25.5 Water and mineral uptake.**

**a.** Pathways of water and minerals. Water and minerals can travel via porous cell walls but then must enter endodermal cells because of the Casparian strip (pathway A). Alternatively, water and minerals can enter root hairs and move from cell to cell (pathway B). **b.** Transport of minerals across an endodermal plasma membrane. (1) An ATP-driven pump removes hydrogen ions ($H^+$) from the cell. (2) This establishes an electrochemical gradient that allows potassium ($K^+$) and other positively charged ions to cross the membrane via a channel protein. (3) Negatively charged mineral ions ($I^-$) can cross the membrane by way of a carrier when they "hitch a ride" with hydrogen ions ($H^+$), which are diffusing down their concentration gradient.

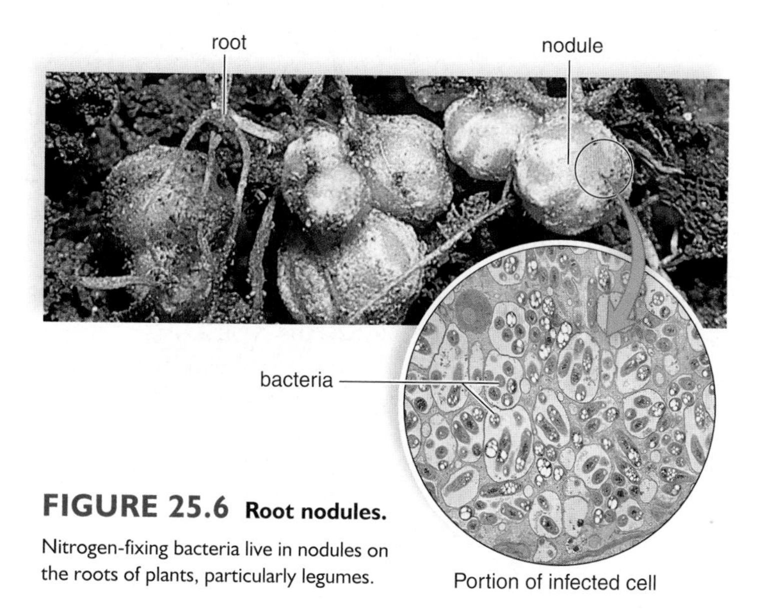


**FIGURE 25.6 Root nodules.**
Nitrogen-fixing bacteria live in nodules on the roots of plants, particularly legumes.

Some plants, such as members of the legume family including bean, clover, and alfalfa, have roots colonized by *Rhizobium* bacteria, which can fix atmospheric nitrogen ($N_2$). They break the $N \equiv N$ bond and reduce nitrogen to $NH_4^+$ for incorporation into organic compounds. The bacteria live in **root nodules** [L. *nodulus,* dim. of *nodus,* knot] and are supplied with carbohydrates by the host plant (Fig. 25.6). The bacteria, in turn, furnish their host with nitrogen compounds.

The second type of mutualistic relationship, called mycorrhizae, involves fungi and almost all plant roots (Fig. 25.7). Only a small minority of plants do not have **mycorrhizae** [Gk. *mykes,* fungus, and *rhiza,* root], and these plants are most often limited as to the environment in which they can grow. The fungus increases the surface area available for mineral and water uptake and breaks down organic matter in the soil, releasing nutrients that the plant can use. In return, the root furnishes the fungus with sugars and amino acids. Plants are extremely dependent on mycorrhizae. Orchid seeds, which are quite small and contain limited nutrients, do not germinate until a mycorrhizal fungus has invaded their cells.

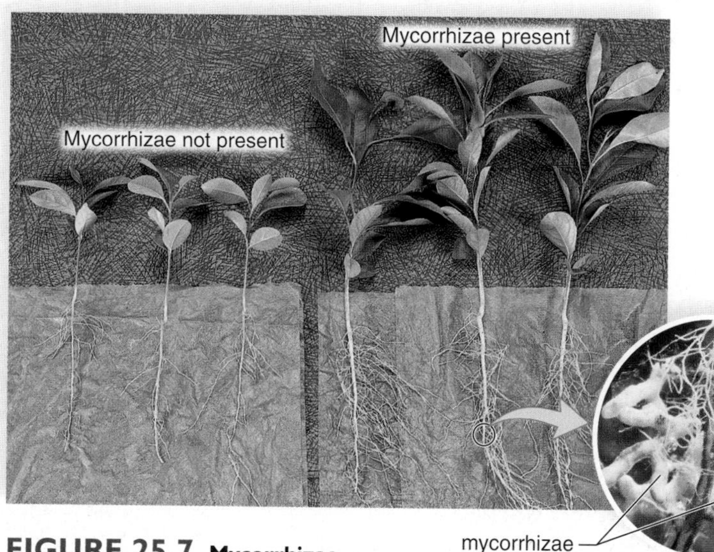


**FIGURE 25.7 Mycorrhizae.**
Plant growth is better when mycorrhizae are present.

a. Dodder, *Cuscuta* sp.

b. Cape sundew, *Drosera capensis*

**FIGURE 25.8 Other ways to acquire nutrients.**
**a.** Some plants, such as the dodder, are parasitic. **b.** Some plants, such as the sundew, are carnivorous.

Nonphotosynthetic plants, such as Indian pipe, use their mycorrhizae to extract nutrients from nearby trees.

Other means of acquiring nutrients also occur. Parasitic plants such as dodders, broomrapes, and pinedrops send out rootlike projections called haustoria that tap into the xylem and phloem of the host stem (Fig. 25.8*a*). Carnivorous plants such as the Venus's flytrap and sundews obtain some nitrogen and minerals when their leaves capture and digest insects (Fig. 25.8*b*).

## Check Your Progress 25.2

1. Review the structure of the plasma membrane on page 460, and explain why the center of the plasma membrane is nonpolar, making it difficult for ions to cross the plasma membrane.
2. Explain the significance to plants of nitrogen-fixing bacteria in the soil.
3. Explain how both partners benefit from a mycorrhizal association.

# 25.3 Transport Mechanisms in Plants

Flowering plants are well adapted to living in a terrestrial environment. Their leaves, which carry on photosynthesis, are positioned to catch the rays of the sun because they are held aloft by the stem (Fig. 25.9). Carbon dioxide enters leaves at the stomata, but water, the other main requirement for photosynthesis, is absorbed by the roots. Water must be transported from the roots through the stem to the leaves.

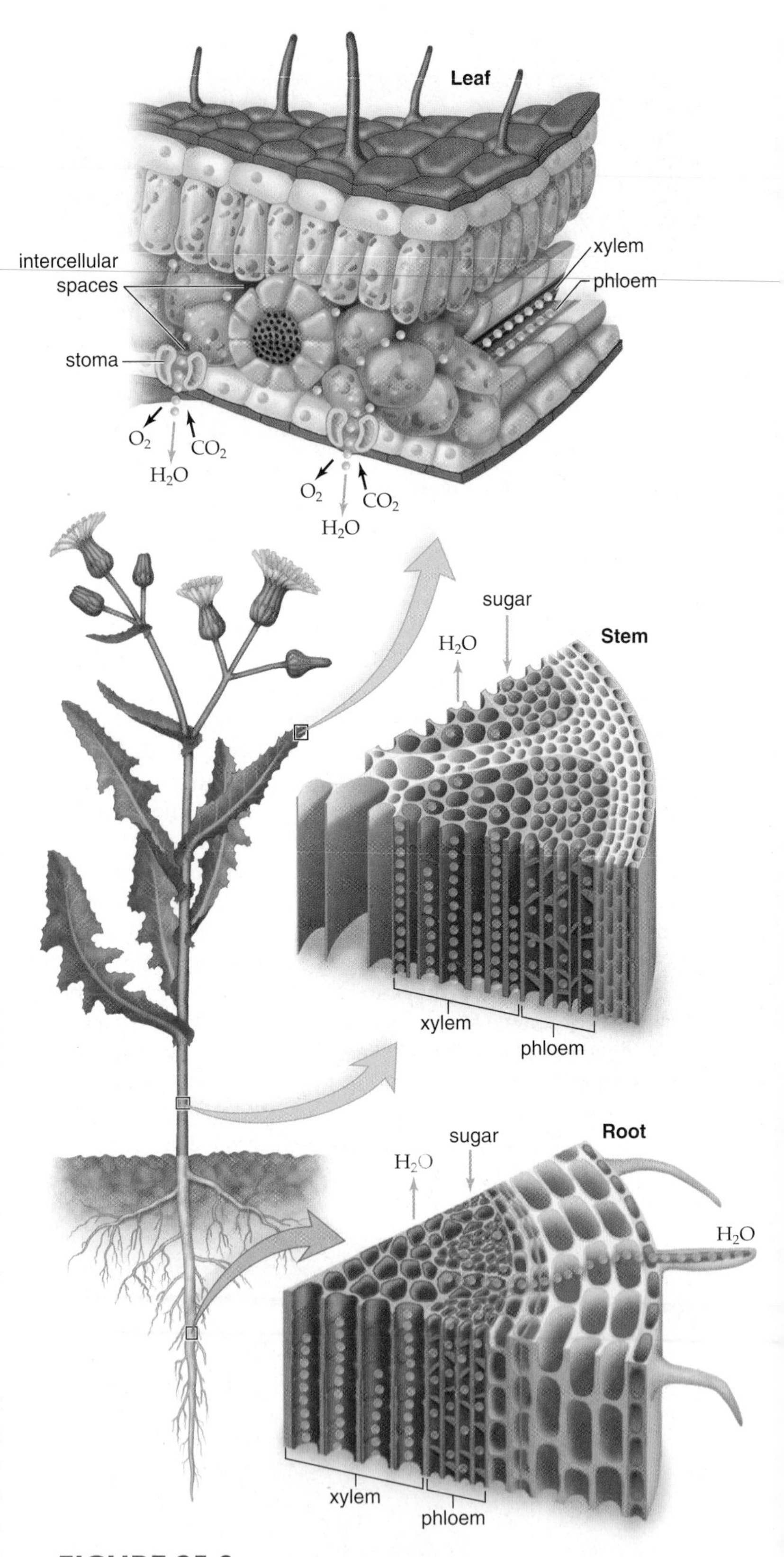


**FIGURE 25.9 Plant transport system.**

Vascular tissue in plants includes xylem, which transports water and minerals from the roots to the leaves, and phloem, which transports organic nutrients oftentimes in the opposite direction. Notice that xylem and phloem are continuous from the roots through the stem to the leaves, which are the vegetative organs of a plant.

## Reviewing Xylem and Phloem Structure

Vascular plants have a transport tissue, called **xylem,** that moves water and minerals from the roots to the leaves. Xylem contains two types of conducting cells: tracheids and vessel elements. **Tracheids** are tapered at both ends. The ends overlap with those of adjacent tracheids (see Fig. 24.6). Pits located in adjacent tracheids allow water to pass from cell to cell. **Vessel elements** are long and tubular with perforation plates at each end (see Fig. 24.6). Vessel elements placed end to end form a completely hollow pipeline from the roots to the leaves. Xylem, with its strong-walled, nonliving cells, gives trees much-needed internal support.

The process of photosynthesis results in sugars, which are used as a source of energy and building blocks for other organic molecules throughout a plant. **Phloem** is the type of vascular tissue that transports organic nutrients to all parts of the plant. Roots buried in the soil cannot possibly carry on photosynthesis, but they still require a source of energy in order to carry on cellular metabolism. Vascular plants are able to transport the products of photosynthesis to regions that require them and/or that will store them for future use. In flowering plants, the conducting cells of phloem are **sieve-tube members,** each of which typically has a companion cell (see Fig. 24.7). **Companion cells** can provide proteins to sieve-tube members, which contain cytoplasm but have no nucleus. The end walls of sieve-tube members are called sieve plates because they contain numerous pores. The sieve-tube members are aligned end to end, and strands of cytoplasm within plasmodesmata extend from one cell to the other through the sieve plates. In this way, sieve-tube members form a continuous **sieve tube** for organic nutrient transport throughout the plant.

## Determining Xylem and Phloem Function

Knowing that vascular plants are structured in a way that allows materials to move from one part to another does not tell us the mechanisms by which they move. Plant physiologists have performed numerous experiments to determine how water and minerals rise to the tops of very tall trees in xylem and how organic nutrients move in the opposite direction in phloem. It would be expected that these processes are mechanical in nature and based on the properties of water because water is a large part of both **xylem sap** and **phloem sap,** as the watery contents of these vessels are called. In living systems, water molecules diffuse

*science focus*

## The Concept of Water Potential

Potential energy is stored energy due to the position of an object. A boulder placed at the top of a hill has potential energy. When pushed, the boulder moves down the hill as potential energy is converted into kinetic (motion) energy. Once it's at the bottom of the hill, the boulder has lost much of its potential energy.

**Water potential** is defined as the energy of water. Just like the boulder, water at the top of a waterfall has a higher water potential than water at the bottom of the waterfall. As illustrated by this example, water moves from a region of higher potential to a region of lower water potential.

In terms of cells, two factors usually determine water potential, which in turn determines the direction in which water will move across a plasma membrane. These factors concern differences in:

1. Water pressure across a membrane
2. Solute concentration across a membrane

*Pressure potential* is the effect that pressure has on water potential. With regard to pressure, it is obvious that water will move across a membrane from the area of higher pressure to the area of lower pressure. The higher the water pressure, the higher the water potential. The lower the water pressure, the lower the water potential, and the more likely it is that water will flow in that direction. Pressure potential is the concept that best explains the movement of sap in xylem and phloem.

To fully explain the movement of water into plant cells, the concept of *osmotic potential* is also required. Osmotic potential takes into account the effects of solutes on the movement of water. The presence of solutes restricts the movement of water because water tends to interact with solutes. Indeed, water tends to move across a membrane from the area of lower solute concentration to the area of higher solute concentration. The lower the concentration of solutes (osmotic potential), the higher the water potential. The higher the concentration of solutes, the lower the water potential and the more likely it is that water will flow in that direction.

Not surprisingly, increasing water pressure will counter the tendency of water to enter a cell because of the presence of solutes. A common situation exists in plant cells. As water enters a plant cell by osmosis, water pressure will increase inside the cell—a plant cell has a strong cell wall that allows water pressure to build up. When will water stop entering the cell? When the pressure potential inside the cell increases and balances the osmotic potential outside the cell.

Pressure potential that increases due to the process of osmosis is often called *turgor pressure*. Turgor pressure is critical, since plants depend on it to maintain the turgidity of their bodies (Fig. 25A). The cells of a wilted plant have insufficient turgor pressure, and the plant droops as a result.

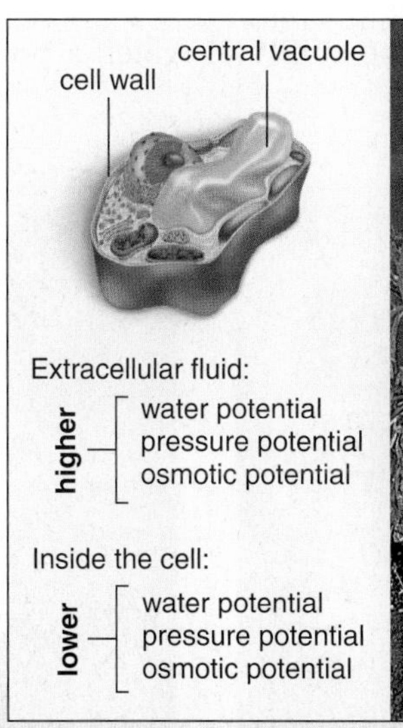

a. Plant cells need water.

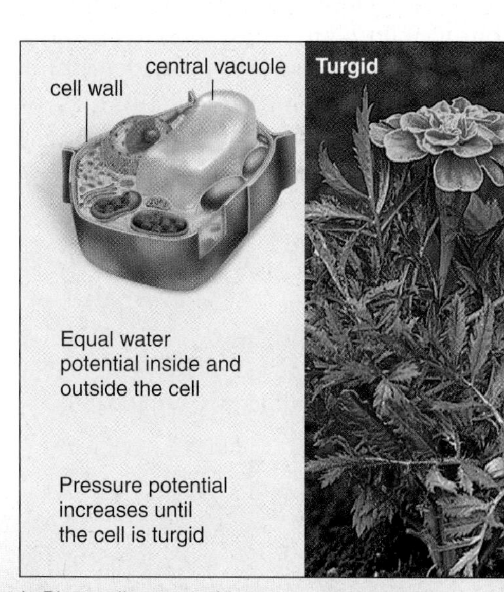


b. Plant cells are turgid.

**FIGURE 25A Water potential and turgor pressure.** Water flows from an area of higher water potential to an area of lower water potential. ***a.*** The cells of a wilted plant have a lower water potential; therefore, water enters the cells. ***b.*** Equilibrium is achieved when the water potential is equal inside and outside the cell. Cells are now turgid, and the plant is no longer wilted.

freely across plasma membranes from the area of higher concentration to the area of lower concentration. Botanists favor describing the movement of water in terms of water potential: Water always flows passively from the area of higher water potential to the area of lower water potential. As can be seen in the Science Focus above, the concept of water potential has the benefit of considering water pressure in addition to osmotic pressure.

Chemical properties of water are also important in movement of xylem sap. The polarity of water molecules and the hydrogen bonding between water molecules allow water to fill xylem cells.

## Water Transport

Figure 25.5 traces the path of water from the root hairs to the xylem. As you know, xylem vessels constitute an open pipeline because the vessel elements have perforation plates separating one from the other (Fig. 25.10*a*, *b*). The tracheids, which are elongated with tapered ends, form a less obvious means of transport, but water can move across the end and side walls of tracheids because of pits, or depressions, where the secondary wall does not form (Fig. 25.10*c*).

Water entering root cells creates a positive pressure called **root pressure.** Root pressure, which primarily occurs at night, tends to push xylem sap upward. Root pressure may be responsible for **guttation** [L. *gutta*, drops, spots] when drops of water are forced out of vein endings along the edges of leaves (Fig. 25.11). Although root pressure may contribute to the upward movement of water in some instances, it is not believed to be the mechanism by which water can rise to the tops of very tall trees. After an injury or pruning, especially in spring, some plants appear to "bleed" as water exudes from the site. This phenomenon is the result of root pressure.

**FIGURE 25.11 Guttation.**

Drops of guttation water on the edges of a strawberry leaf. Guttation, which occurs at night, may be due to root pressure. Root pressure is a positive pressure potential caused by the entrance of water into root cells. Often guttation is mistaken for early morning dew.

### *Cohesion-Tension Model of Xylem Transport*

Once water enters xylem, it must be transported to all parts of the plant. Transporting water can be a daunting task, especially for some plants, such as redwood trees, which can exceed 90 m (almost 300 ft) in height.

The **cohesion-tension model** of xylem transport, outlined in Figure 25.12 describes a mechanism for xylem transport that requires no expenditure of energy by the plant and is dependent on the properties of water. The term *cohesion* refers to the tendency of water molecules to cling together. Because of hydrogen bonding, water molecules interact with one another and form a continuous **water column** in xylem, from the leaves to the roots, that is not easily broken. In addition to cohesion, another property of water called *adhesion* plays a role in xylem transport. Adhesion refers to the ability of water, a polar molecule, to interact

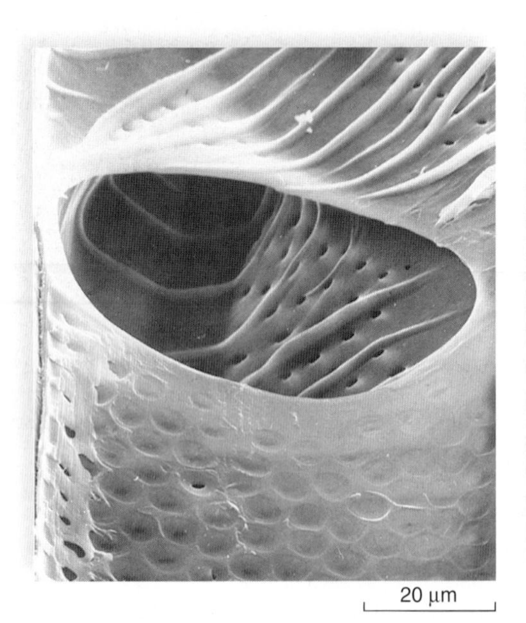


a. Perforation plate with a single, large opening

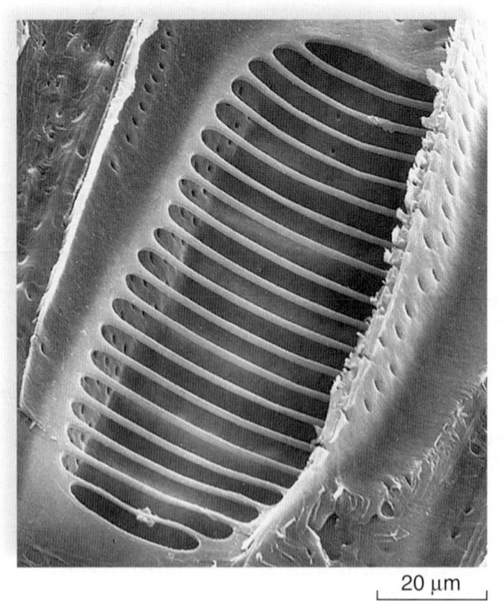


b. Perforation plate with a series of openings

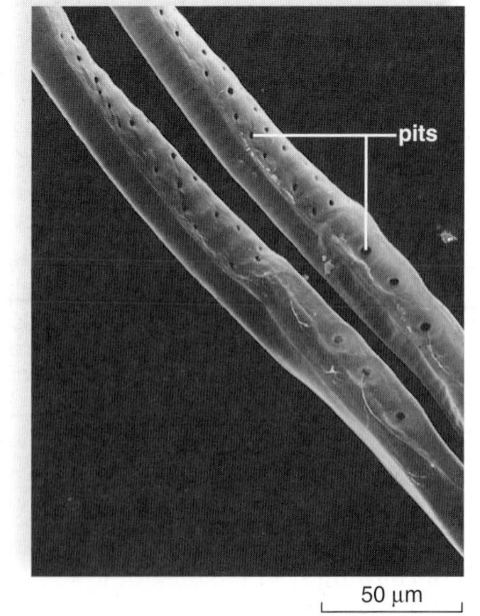


c. Tracheids

**FIGURE 25.10 Conducting cells of xylem.**

Water can move from vessel element to vessel element through perforation plates (**a** and **b**). Vessel elements can also exchange water with tracheids through pits. **c.** Tracheids are long, hollow cells with tapered ends. Water can move into and out of tracheids through pits only.

with the molecules making up the walls of the vessels in xylem. Adhesion gives the water column extra strength and prevents it from slipping back.

**The Leaves.** When the stomata of a leaf are open, the cells of the spongy layer are exposed to the air, which can be quite dry. Water then evaporates as a gas or vapor from the spongy layer into the intercellular spaces. Evaporation of water through leaf stomata is called **transpiration.** At least 90% of the water taken up by the roots is eventually lost by transpiration. This means that the total amount of water lost by a plant over a long period of time is surprisingly large. A single *Zea mays* (corn) plant loses somewhere between 135 and 200 liters of water through transpiration during a growing season. An average-sized birch tree with over 200,000 leaves will transpire up to 3,700 liters of water per day during the growing season.

The water molecules that evaporate from cells into the intercellular spaces are replaced by other water molecules from the leaf veins. Because the water molecules are cohesive, transpiration exerts a *pulling force,* or *tension,* that draws the water column through the xylem to replace the water lost by leaf cells.

Note that the loss of water by transpiration is the mechanism by which minerals are transported throughout the plant body. Also, evaporation of water moderates the temperature of leaf tissues.

There is an important consequence to the way water is transported in plants. When a plant is under water stress, the stomata close. Now the plant loses little water because the leaves are protected against water loss by the waxy **cuticle** of the upper and lower epidermis. When stomata are closed, however, carbon dioxide cannot enter the leaves, and many plants are unable to photosynthesize efficiently. Photosynthesis, therefore, requires an abundant supply of water so that stomata remain open, allowing carbon dioxide to enter.

**The Stem.** The tension in xylem created by evaporation of water at the leaves pulls the water column in the stem upward. Usually, the water column in the stem is continuous because of the cohesive property of water molecules. The water molecules also adhere to the sides of the vessels. What happens if the water column within xylem breaks? The water column "snaps back" down the xylem vessel away from the site of breakage, making it more difficult for conduction to occur. Next time you use a straw to drink a soda, notice that pulling the liquid upward is fairly easy, as long as there is liquid at the end of the straw. When the soda runs low and you begin to get air, it takes considerably more suction to pull up the remaining liquid. When preparing a vase of flowers, you should always cut the stems under water to preserve an unbroken water column and the life of the flowers.

**The Roots.** In the root, water enters xylem passively by osmosis because xylem sap always has a greater concentration of solutes than do the root cells. The water column in xylem extends from the leaves down to the root. Water is pulled upward from the roots due to the tension in xylem created by the evaporation of water at the leaves.

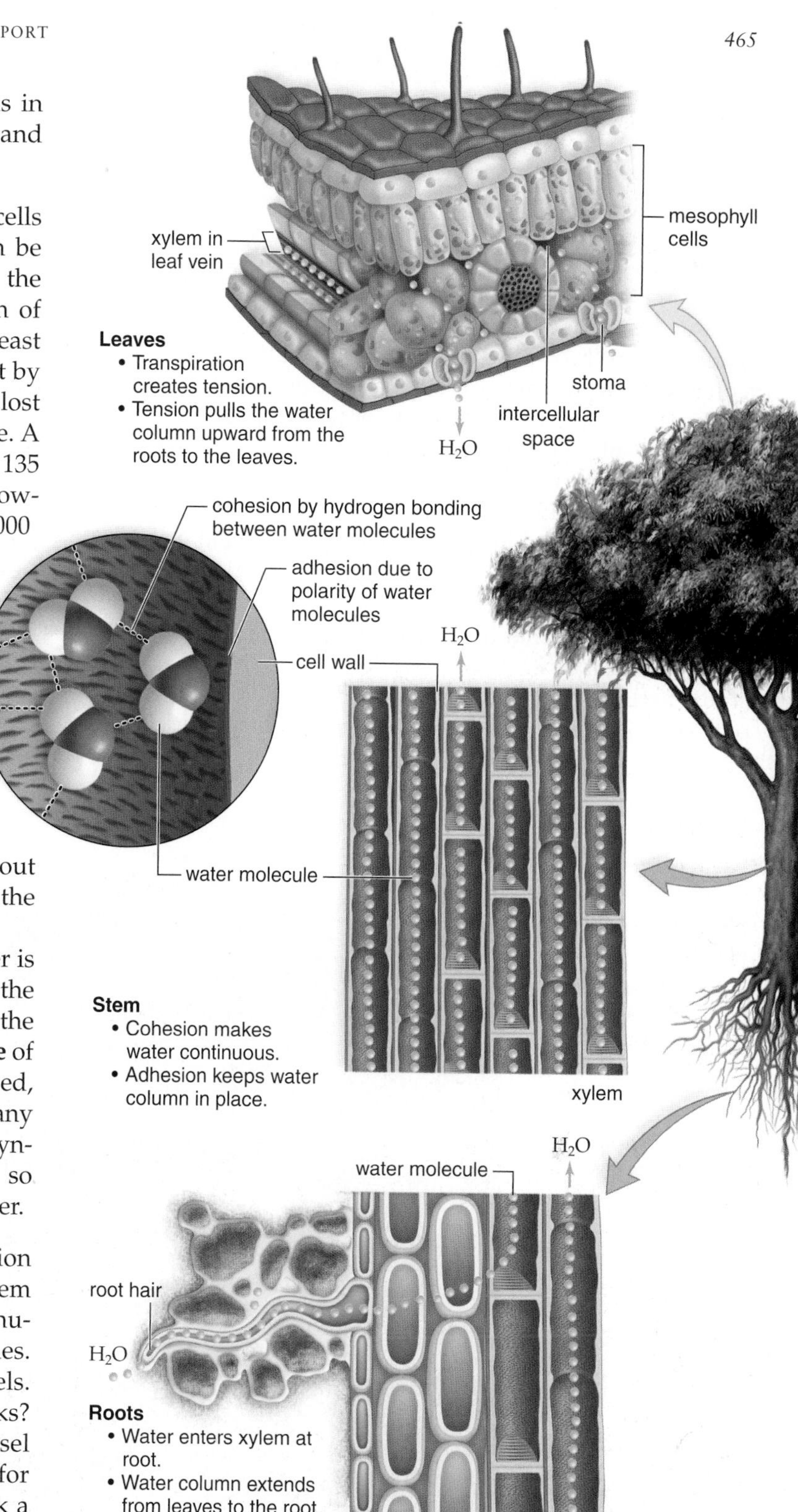


**FIGURE 25.12 Cohesion-tension model of xylem transport.**
Tension created by evaporation (transpiration) at the leaves pulls water along the length of the xylem—from the roots to the leaves.

## Opening and Closing of Stomata

Each **stoma,** a small pore in leaf epidermis, is bordered by **guard cells.** When water enters the guard cells and turgor pressure increases, the stoma opens; when water exits the guard cells and turgor pressure decreases, the stoma closes. Notice in Figure 25.13 that the guard cells are attached to each other at their ends and that the inner walls are thicker than the outer walls. When water enters, a guard cell's radial expansion is restricted because of cellulose microfibrils in the walls, but lengthwise expansion of the outer walls is possible. When the outer walls expand lengthwise, they buckle out from the region of their attachment, and the stoma opens.

Since about 1968, it has been clear that potassium ions ($K^+$) accumulate within guard cells when stomata open. In other words, active transport of $K^+$ into guard cells causes water to follow by osmosis and stomata to open. Also interesting is the observation that hydrogen ions ($H^+$) accumulate outside guard cells as $K^+$ moves into them. A proton pump run by the hydrolysis of ATP transports $H^+$ to the outside of the cell. This establishes an electrochemical gradient that allows $K^+$ to enter by way of a channel protein (see Fig. 25.5*b*).

What regulates the opening and closing of stomata? It appears that the blue-light component of sunlight is a signal that can cause stomata to open. Evidence suggests that a flavin pigment absorbs blue light, and then this pigment sets in motion the cytoplasmic response that leads to activation of the proton pump. Similarly, there could be a receptor in the plasma membrane of guard cells that brings about inactivation of the pump when carbon dioxide ($CO_2$) concentration rises, as might happen when photosynthesis ceases. Abscisic acid (ABA), which is produced by cells in wilting leaves, can also cause stomata to close (see page 480). Although photosynthesis cannot occur, water is conserved.

If plants are kept in the dark, stomata open and close just about every 24 hours, just as if they were responding to the presence of sunlight in the daytime and the absence of sunlight at night. This means that some sort of internal biological clock must be keeping time. Circadian rhythms (a behavior that occurs nearly every 24 hours) and biological clocks are areas of intense investigation at this time. Other factors that influence the opening and closing of stoma include temperature, humidity, and stress.

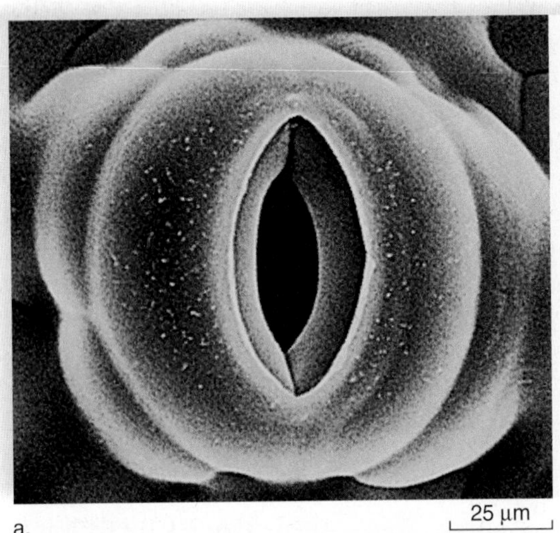


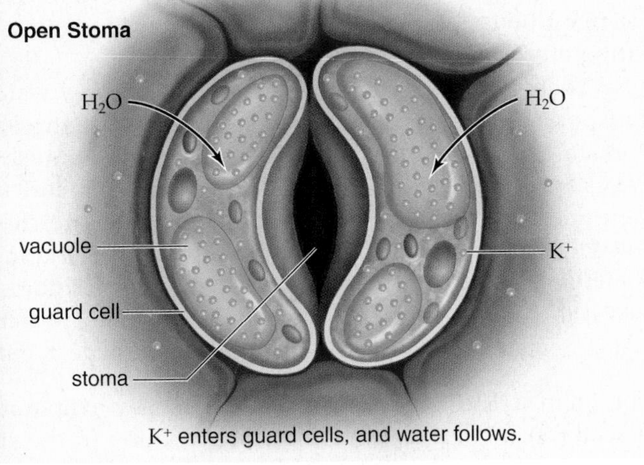


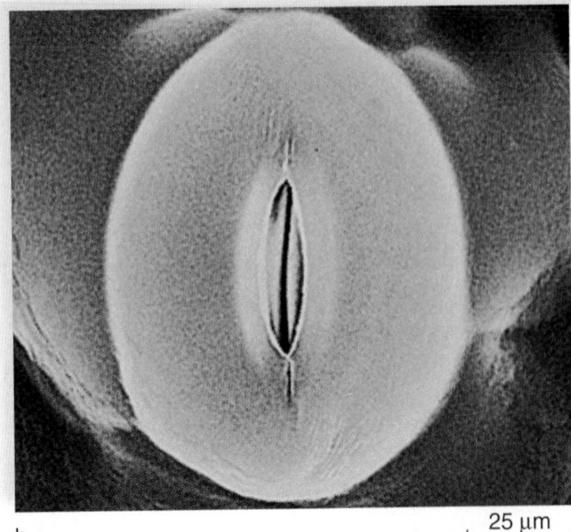


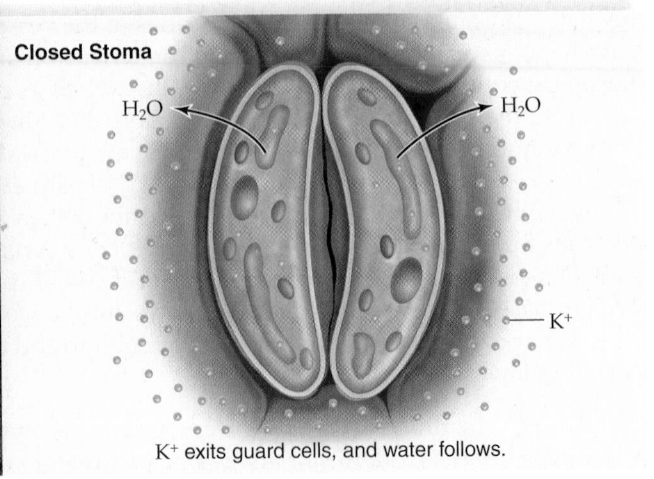


**FIGURE 25.13 Opening and closing of stomata.**

**a.** A stoma opens when turgor pressure increases in guard cells due to the entrance of $K^+$ followed by the entrance of water. **b.** A stoma closes when turgor pressure decreases due to the exit of $K^+$ followed by the exit of water.

*ecology focus*

## Plants Can Clean Up Toxic Messes

**Phytoremediation** uses plants—many of them common species such as poplar, mustard, and mulberry—that have an appetite for lead, uranium, and other pollutants. These plants' genetic makeups allow them to absorb and to store, degrade, or transform substances that kill or harm other plants and animals. "It's an elegantly simple solution to pollution problems" says Louis Licht, who runs Ecolotree, an Iowa City phytoremediation company.

The idea behind phytoremediation is not new; scientists have long recognized certain plants' abilities to absorb and tolerate toxic substances. But the idea of using these plants on contaminated sites has just gained support in the last decade. Different plants work on different contaminants. The mulberry bush, for instance, is effective on industrial sludge; some grasses attack petroleum wastes; and sunflowers (together with soil additives) remove lead. The plants clean up sites in two basic ways, depending on the substance involved. If it is an organic contaminant, such as spilled oil, the plants or microbes around their roots break down the substance. The remainders can either be absorbed by the plant or left in the soil or water. For an inorganic contaminant such as cadmium or zinc, the plants absorb the substance and trap it. The plants must then be harvested and disposed of, or processed to reclaim the trapped contaminant.

### Poplars Take Up Excess Nitrates

Most trees planted along the edges of farms are intended to break the wind. But a mile-long stand of spindly poplars outside Amana, Iowa, is involved in phytoremediation.

The poplars act like vacuum cleaners, sucking up nitrate-laden runoff from a fertilized cornfield before this runoff reaches a nearby brook—and perhaps other waters. Nitrate runoff into the Mississippi River from Midwest farms, after all, is a major cause of the large "dead zone" of oxygen-depleted water that develops each summer in the Gulf of Mexico.

Before the trees were planted, the brook's nitrate levels were as much as ten times the amount considered safe. But then Licht, a University of Iowa graduate student, had the idea that poplars, which absorb lots of water and tolerate pollutants, could help. In 1991, Licht tested his hunch by planting the trees along a field owned by a corporate farm. The brook's nitrate levels subsequently dropped more than 90%, and the trees have thrived.

### Canola Plants Take Up Selenium

Canola plants (*Brassica rapa* and *B. napa*), meanwhile, are grown in California's San Joaquin Valley to soak up excess selenium in the soil to help prevent an environmental catastrophe like the one that occurred there in the 1980s.

Back then, irrigated farming caused naturally occurring selenium to rise to the soil surface. When excess water was pumped onto the fields, some selenium would flow off into drainage ditches, eventually ending up in Kesterson National Wildlife Refuge. The selenium in ponds at the refuge accumulated in plants and fish and subsequently deformed and killed waterfowl, says Gary Bañuelos, a plant scientist with the U.S. Department of Agriculture who helped remedy the problem. He recommended that farmers add selenium-accumulating canola plants to their crop rotations (Fig. 25B). As a result, selenium levels in runoff are being managed. Although the underlying problem of excessive selenium in soils has not been solved, says Bañuelos, "this is a tool to manage mobile selenium and prevent another unlikely selenium-induced disaster."

**FIGURE 25B Canola plants.**
*Scientist Gary Bañuelos recommended planting canola to pull selenium out of the soil.*

### Mustard Plants Take Up Uranium

Phytoremediation has also helped clean up badly polluted sites, in some cases at a fraction of the usual cost. Edenspace Systems Corporation of Reston, Virginia, just concluded a phytoremediation demonstration at a Superfund site on an Army firing range in Aberdeen, Maryland. The company successfully used mustard plants to remove uranium from the firing range, at as little as 10% of the cost of traditional cleanup methods. Depending on the contaminant involved, traditional cleanup costs can run as much as $1 million per acre, experts say.

### Limitations of Phytoremediation

Phytoremediation does have its limitations, however. One of them is its slow pace. Depending on the contaminant, it can take several growing seasons to clean a site—much longer than conventional methods. "We normally look at phytoremediation as a target of one to three years to clean a site," notes Edenspace's Mike Blaylock. "People won't want to wait much longer than that."

Phytoremediation is also only effective at depths that plant roots can reach, making it useless against deep-lying contamination unless the contaminated soils are excavated. Phytoremediation will not work on lead and other metals unless chemicals are added to the soil. In addition, it is possible that animals may ingest pollutants by eating the leaves of plants in some projects.

Despite its shortcomings, experts see a bright future for this technology because, for one reason, the costs are relatively small compared to those of traditional remediation technologies. Traditional methods of cleanup require much energy input and therefore have higher cost. In general, phytoremediation is a low-cost alternative to traditional methods because less energy is required for operation and maintenance. Phytoremediation is a promising solution to pollution problems but, says the EPA's Walter W. Kovalick, "it's not a panacea. It's another arrow in the quiver. It takes more than one arrow to solve most problems."

## Organic Nutrient Transport

Not only do plants transport water and minerals from the roots to the leaves, but they also transport organic nutrients to the parts of plants that need them. This includes young leaves that have not yet reached their full photosynthetic potential; flowers that are in the process of making seeds and fruits; and the roots, whose location in the soil prohibits them from carrying on photosynthesis

### *Role of Phloem*

As long ago as 1679, Marcello Malpighi suggested that bark is involved in translocating sugars from leaves to roots. He observed the results of removing a strip of bark from around a tree, a procedure called **girdling.** If a tree is girdled below the level of the majority of leaves, the bark swells just above the cut, and sugar accumulates in the swollen tissue. We know today that when a tree is girdled, the phloem is removed, but the xylem is left intact. Therefore, the results of girdling suggest that phloem is the tissue that transports sugars.

a. An aphid feeding on a plant stem

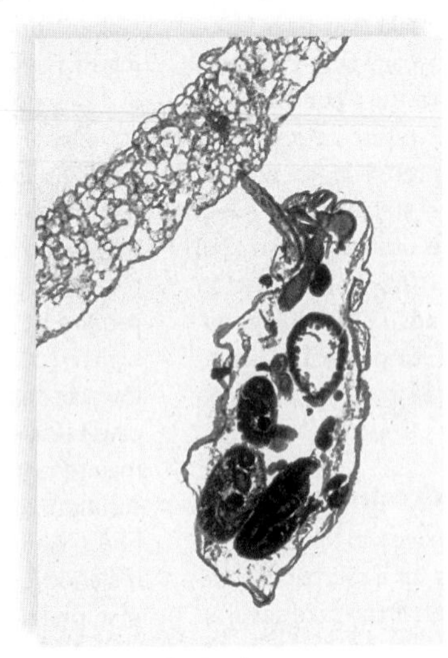

b. Aphid stylet in place

**FIGURE 25.14**
**Acquiring phloem sap.**
Aphids are small insects that remove nutrients from phloem by means of a needle-like mouthpart called a stylet. **a.** Excess phloem sap appears as a droplet after passing through the aphid's body. **b.** Micrograph of stylet in plant tissue. When an aphid is cut away from its stylet, phloem sap becomes available for collection and analysis.

Radioactive tracer studies with carbon 14 ($^{14}C$) have confirmed that phloem transports organic nutrients. When $^{14}C$-labeled carbon dioxide ($CO_2$) is supplied to mature leaves, radioactively labeled sugar is soon found moving down the stem into the roots. It's difficult to get samples of sap from phloem without injuring the phloem, but this problem is solved by using aphids, small insects that are phloem feeders. The aphid drives its stylet, which is a sharp mouthpart that functions like a hypodermic needle, between the epidermal cells, and sap enters its body from a sieve-tube member (Fig. 25.14). If the aphid is anesthetized using ether, its body can be carefully cut away, leaving the stylet. Phloem can then be collected and analyzed by a researcher. By the use of radioactive tracers and aphids, it is known that the movement through phloem can be as fast as 60–100 cm per hour and possibly up to 300 cm per hour.

### *Pressure-Flow Model of Phloem Transport*

The **pressure-flow model** is a current explanation for the movement of organic materials in phloem (Fig. 25.15). Consider the following experiment in which two bulbs are connected by a glass tube. The first bulb contains solute at a higher concentration than the second bulb. Each bulb is bounded by a differentially permeable membrane, and the entire apparatus is submerged in distilled water.

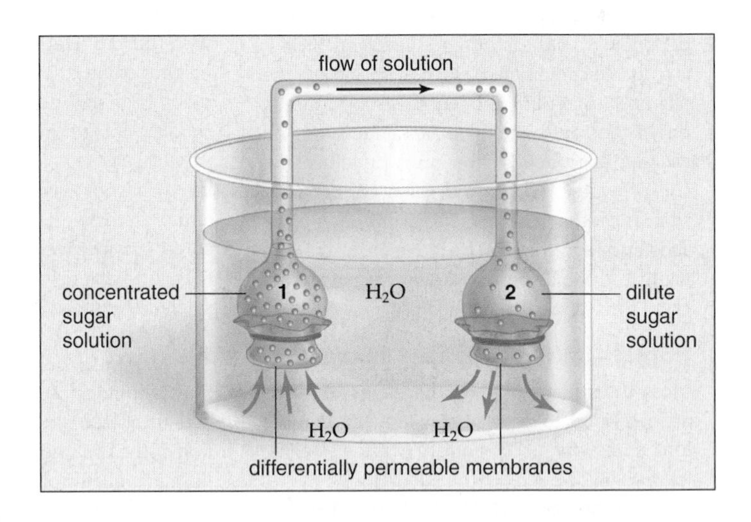


Distilled water flows into the first bulb because it has the higher solute concentration. The entrance of water creates a positive *pressure*, and water *flows* toward the second bulb. This flow not only drives water toward the second bulb, but it also provides enough force for water to move out through the membrane of the second bulb—even though the second bulb contains a higher concentration of solute than the distilled water.

In plants, sieve tubes are analogous to the glass tube that connects the two bulbs. Sieve tubes are composed of sieve-tube members, each of which has a companion cell. It is possible that the companion cells assist the sieve-tube members in some way. The sieve-tube members align end

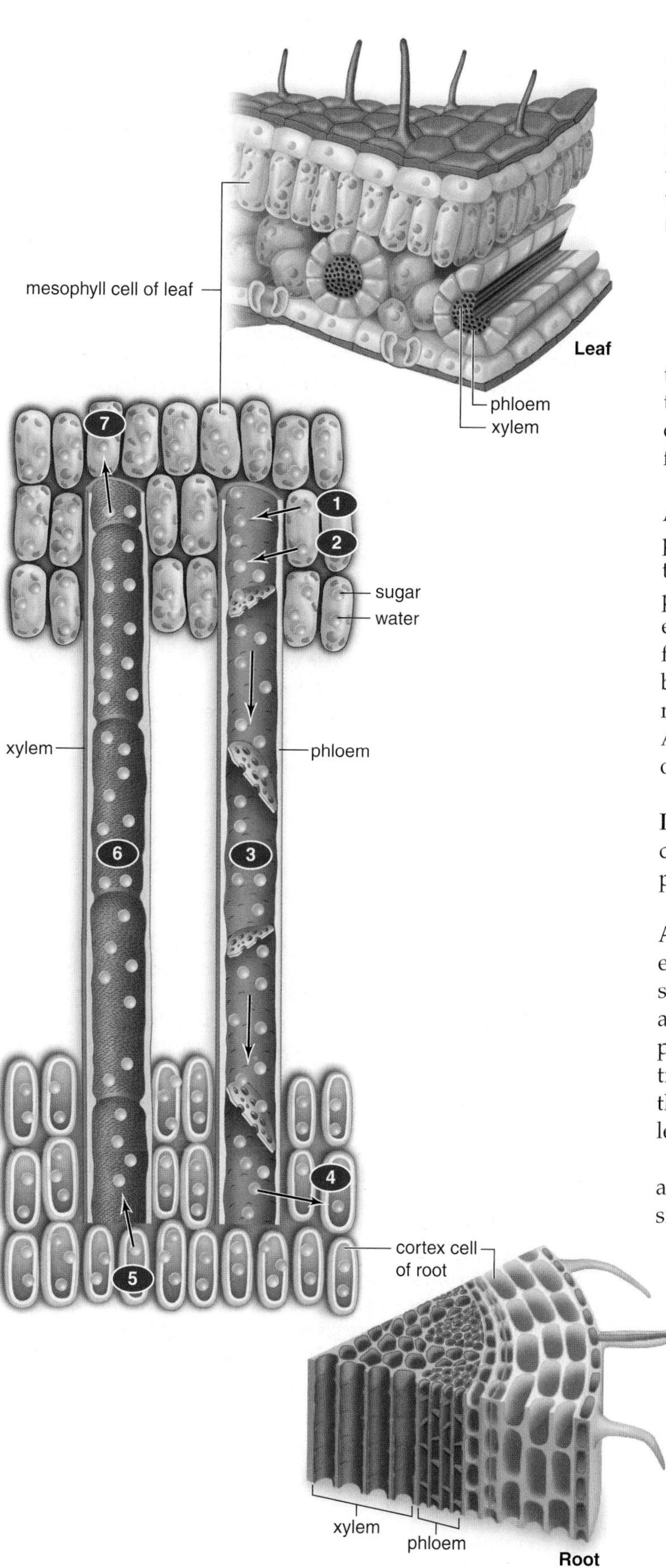


**FIGURE 25.15 Pressure-flow model of phloem transport.**

At a source, 1 sugar (pink) is actively transported into sieve tubes. 2 Water (blue) follows by osmosis. 3 A positive pressure causes phloem contents to flow from source to a sink. At a sink, 4 sugar is actively transported out of sieve tubes and cells use it for cellular respiration. Water exits by osmosis. 5 Some water returns to the xylem, where it mixes with more water absorbed from the soil. 6 Xylem transports water to the mesophyll of the leaf. 7 Most water is transpired, some is used for photosynthesis, and some reenters phloem by osmosis.

to end, and strands of plasmodesmata (cytoplasm) extend through sieve plates from one sieve-tube member to the other. Sieve tubes, therefore, form a continuous pathway for organic nutrient transport throughout a plant.

**At the Source (e.g., leaves).** During the growing season, photosynthesizing leaves are producing sugar. Therefore, they are a source of sugar. This sugar is actively transported into phloem. Again, transport is dependent on an electrochemical gradient established by a proton pump, a form of active transport. Sugar is carried across the membrane in conjunction with hydrogen ions ($H^+$), which are moving down their concentration gradient (see Fig. 25.5). After sugar enters sieve tubes, water follows passively by osmosis.

**In the Stem.** The buildup of water within sieve tubes creates the positive pressure that accounts for the flow of phloem contents.

**At the Sink (e.g., roots).** The roots (and other growth areas) are a sink for sugar, meaning that they are removing sugar and using it for cellular respiration. After sugar is actively transported out of sieve tubes, water exits phloem passively by osmosis and is taken up by xylem, which transports water to leaves, where it is used for photosynthesis. Now, phloem contents continue to flow from the leaves (source) to the roots (sink).

The pressure-flow model of phloem transport can account for any direction of flow in sieve tubes if we consider that the direction of flow is always from **source** to **sink.** For example, recently formed leaves can be a sink, and they will receive sucrose until they begin to maximally photosynthesize.

## Check Your Progress 25.3

1. Explain why water is under tension in stems.
2. Explain the significance of the cohesion and adhesion properties of water to water transport.
3. Explain how sugars move from source to sink in a plant.

# Connecting the Concepts

The land environment offers many advantages for plants, such as greater availability of light and carbon dioxide for photosynthesis. (Water, even if clear, filters out light, and carbon dioxide concentration and rate of diffusion is less in water.) The evolution of a transport system was critical, however, for plants to make full use of these advantages. Only if a transport system is present can plants elevate the leaves so that they are better exposed to solar energy and carbon dioxide in the air. A transport system brings water, a raw material of photosynthesis, from the roots to the leaves and also brings the products of photosynthesis down to the roots. Roots lie beneath the soil, and their cells depend on an input of organic food from the leaves to remain alive. An efficient transport system allows roots to penetrate deeply into the soil to absorb water and minerals.

The presence of a transport system also allows materials to be distributed to those parts of the plant body that are growing most rapidly. New leaves and flower buds would grow rather slowly if they had to depend on their own rate of photosynthesis, for example. Height in vascular plants, due to the presence of a transport system, has other benefits aside from elevation of leaves. It is also adaptive to have reproductive structures located where the wind can better distribute pollen and seeds. Once animal pollination came into existence, it was beneficial for flowers to be located where they are more easily seen by animals.

Clearly, plants with a transport system have a competitive edge in the terrestrial environment.

## summary

### 25.1 Plant Nutrition and Soil

Plants need both essential and beneficial inorganic nutrients. Carbon, hydrogen, and oxygen make up 95% of a plant's dry weight. The other necessary nutrients are taken up by the roots as mineral ions. Even nitrogen (N), which is present in the atmosphere, is most often taken up as $NO_3^-$.

You can determine mineral requirements by hydroponics, in which plants are grown in a solution. The solution is varied by the omission of one mineral. If the plant dies, then the missing mineral must be essential for growth. If it grows poorly, than the mineral is beneficial.

Plant life is dependent on soil, which forms by the weathering of rock. Organisms contribute to the formation of humus and soil. Soil is a mixture of mineral particles, humus, living organisms, air, and water. Soil particles are of three types from the largest to the smallest: sand, silt, and clay. Loam, which contains about equal proportions of all three types, retains water but still has air spaces. Humus contributes to the texture of soil and its ability to provide inorganic nutrients to plants. Topsoil (A horizon of a soil profile) contains humus, and this is the layer that is lost by erosion, a worldwide problem.

### 25.2 Water and Mineral Uptake

Water, along with minerals, can enter a root by passing between the porous cell walls, until it reaches the Casparian strip, after which it passes through an endodermal cell before entering xylem. Water can also enter root hairs and then pass through the cells of the cortex and endodermis to reach xylem.

Mineral ions cross plasma membranes by a chemiosmotic mechanism. A proton pump transports $H^+$ out of the cell. This establishes an electrochemical gradient that causes positive ions to flow into the cells. Negative ions are carried across in conjunction with $H^+$, which is moving along its concentration gradient.

Plants have various adaptations that assist them in acquiring nutrients. Legumes have nodules infected with the bacterium *Rhizobium*, which makes nitrogen compounds available to these plants. Many other plants have mycorrhizae, or fungus roots. The fungus gathers nutrients from the soil, and the root provides the fungus with sugars and amino acids. Some plants have poorly developed roots. Most epiphytes live on, but do not parasitize, trees, whereas dodder and some other plants parasitize their hosts.

### 25.3 Transport Mechanisms in Plants

As an adaptation to life on land, plants have a vascular system that transports water and minerals from the roots to the leaves and must also transport the products of photosynthesis in the opposite direction. Vascular tissue includes xylem and phloem.

In xylem, vessels composed of vessel elements aligned end to end form an open pipeline from the roots to the leaves. Particularly at night, root pressure can build in the root. However, this does not contribute significantly to xylem transport.

The cohesion-tension model of xylem transport states that transpiration creates a tension that pulls water upward in xylem from the roots to the leaves. This means of transport works only because water molecules are cohesive with one another and adhesive with xylem walls.

Most of the water taken in by a plant is lost through stomata by transpiration. Only when there is plenty of water do stomata remain open, allowing carbon dioxide to enter the leaf and photosynthesis to occur.

Stomata open when guard cells take up water. The guard cells are anchored at their ends. They can only stretch lengthwise because microfibrils in their walls prevent lateral expansion. Therefore, guard cells buckle out when water enters. Water enters the guard cells after potassium ions ($K^+$) have entered. Light signals stomata to open, and a high carbon dioxide ($CO_2$) level may signal stomata to close. Abscisic acid produced by wilting leaves also signals for closure.

In phloem, sieve tubes composed of sieve-tube members aligned end to end form a continuous pipeline from the leaves to the roots. Sieve-tube members have sieve plates through which plasmodesmata (strands of cytoplasm) extend from one to the other. The pressure-flow model of phloem transport proposes that a positive pressure drives phloem contents in sieve tubes. Sucrose is actively transported into sieve tubes—by a chemiosmotic mechanism—at a source, and water follows by osmosis. The resulting increase in pressure creates a flow that moves water and sucrose to a sink. A sink can be at the roots or any other part of the plant that requires organic nutrients.

## understanding the terms

beneficial nutrient 457
Casparian strip 460
cohesion-tension model 464
companion cell 462
cuticle 465
essential nutrient 457
girdling 468
guard cell 466

guttation 464
humus 458
hydroponics 457
macronutrient 457
micronutrient 457
mineral 457
mycorrhizae 461
phloem 462
phloem sap 462
phytoremediation 467
pressure-flow model 468
root hair 460
root nodule 461
root pressure 464
sieve tube 462
sieve-tube member 462
sink 469
soil 458
soil erosion 459
soil horizon 459
soil profile 459
source 469
stoma 466
tracheid 462
transpiration 465
vessel element 462
water column 464
water potential 463
xylem 462
xylem sap 462

Match the terms to these definitions:

a. _______________ Explanation for transport in sieve tubes of phloem.
b. _______________ Major layer of soil visible in vertical profile.
c. _______________ Plant's loss of water to the atmosphere, mainly through evaporation at leaf stomata.
d. _______________ Layer of impermeable lignin and suberin bordering four sides of root endodermal cells; causes water and minerals to enter endodermal cells before entering vascular tissue.
e. _______________ Type of plant cell that is found in pairs, with one on each side of a leaf stoma.

## reviewing this chapter

1. Name the elements that make up most of a plant's body. What are essential mineral nutrients and beneficial mineral nutrients? 456–57
2. Briefly describe the use of hydroponics to determine the mineral nutrients of a plant. 457
3. How is soil formed, and how does humus provide nutrients to plants? Describe a generalized soil profile and how a profile is affected by erosion. 458–59
4. Give two pathways by which water and minerals can cross the epidermis and cortex of a root. What feature allows endodermal cells to regulate the entrance of molecules into the vascular cylinder? 460
5. Describe the chemiosmotic mechanism by which mineral ions cross plasma membranes. 460
6. Name two symbiotic relationships that assist plants in taking up minerals and two types of plants that have other means of acquiring nutrients. 460–65
7. A vascular system is adaptive for a land existence. Explain. Describe the composition of a plant's vascular system. 462–63
8. What is root pressure, and why can't it account for the transport of water in xylem? 464
9. Describe and give evidence for the cohesion-tension model of water transport. 464–65
10. Describe the structure of stomata and explain how they can open and close. By what mechanism do guard cells take up potassium ($K^+$) ions? 466
11. What data are available to show that phloem transports organic compounds? Explain the pressure-flow model of phloem transport. 468–69

## testing yourself

Choose the best answer for each question.

1. Which of these molecules is not a nutrient for plants?
   a. water
   b. carbon dioxide gas
   c. mineral ions
   d. nitrogen gas
   e. None of these are nutrients.
2. Which is a component of soil?
   a. mineral particles
   b. humus
   c. organisms
   d. air and water
   e. All of these are correct.
3. The Casparian strip affects
   a. how water and minerals move into the vascular cylinder.
   b. vascular tissue composition.
   c. how soil particles function.
   d. how organic nutrients move into the vascular cylinder.
   e. Both a and d are correct.
4. Which of these is not a mineral ion?
   a. NO3–
   b. Mg+
   c. $CO_2$
   d. Al3+
   e. All of these are correct.
5. What role do cohesion and adhesion play in xylem transport?
   a. Like transpiration, they create a tension.
   b. Like root pressure, they create a positive pressure.
   c. Like sugars, they cause water to enter xylem.
   d. They create a continuous water column in xylem.
   e. All of these are correct.
6. The pressure-flow model of phloem transport states that
   a. phloem content always flows from the leaves to the root.
   b. phloem content always flows from the root to the leaves.
   c. water flow brings sucrose from a source to a sink.
   d. water pressure creates a flow of water toward the source.
   e. Both c and d are correct.
7. Root hairs do not play a role in
   a. oxygen uptake.
   b. mineral uptake.
   c. water uptake.
   d. carbon dioxide uptake.
   e. the uptake of any of these.
8. Xylem includes all of these except
   a. companion cells.
   b. vessels.
   c. tracheids.
   d. dead tissue.
9. After sucrose enters sieve tubes,
   a. it is removed by the source.
   b. water follows passively by osmosis.
   c. it is driven by active transport to the source, which is usually the roots.
   d. stomata open so that water flows to the leaves.
   e. All of these are correct.
10. An opening in the leaf that allows gas and water exchange is called
    a. the lenticel.
    b. the hole.
    c. the stoma.
    d. the guard cell.
    e. the accessory cell.

11. What main force drives absorption of water, creates tension, and draws water through the plant?
    a. adhesion
    b. cohesion
    c. tension
    d. transpiration
    e. absorption
12. A nutrient element is considered essential if
    a. plant growth increases with a reduction in the concentration of the element.
    b. plants die in the absence of the element.
    c. plants can substitute a similar element for the missing element with no ill effects.
    d. the element is a positive ion.
13. Humus
    a. supplies nutrients to plants.
    b. is basic in its pH.
    c. is found in the deepest soil horizons.
    d. is inorganic in origin.
14. Which sequence represents the size of soil particles from largest to smallest?
    a. sand, clay, silt
    b. silt, clay, sand
    c. sand, silt, clay
    d. clay, silt, sand
    e. silt, sand, clay
15. Soils rich in which type of mineral particle will have a high water-holding capacity?
    a. sand
    b. silt
    c. clay
    d. All soil particles hold water equally well.
16. Stomata are usually open
    a. at night, when the plant requires a supply of oxygen.
    b. during the day, when the plant requires a supply of carbon dioxide.
    c. day or night if there is excess water in the soil.
    d. during the day, when transpiration occurs.
    e. Both b and d are correct.
17. Why might the water column in tracheids be less susceptible to breakage than in vessels?
    a. Tracheids are more narrow, giving more opportunity for adhesion to play a role in maintaining the water column.
    b. The end walls of tracheids are more slanted than the end walls of vessel elements.
    c. Tracheids receive support from vessel elements, but not vice versa.
    d. All of these are correct.
18. Explain why this experiment supports the hypothesis that transpiration can cause water to rise to the tops of tall trees.

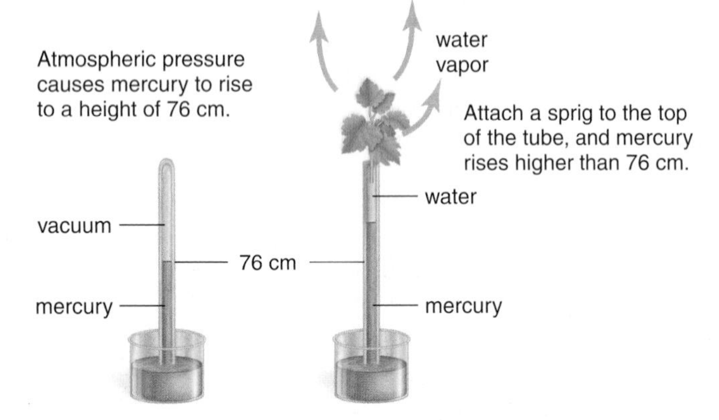


19. Negatively charged clay particles attract
    a. $K^+$.
    b. $NO_3^-$.
    c. $Ca^+$.
    d. Both a and b are correct.
    e. Both a and c are correct.
20. a. Label water ($H_2O$) and potassium ions ($K^+$) appropriately in these diagrams. b. What is the role of $K^+$ in the opening and closing of stomata?

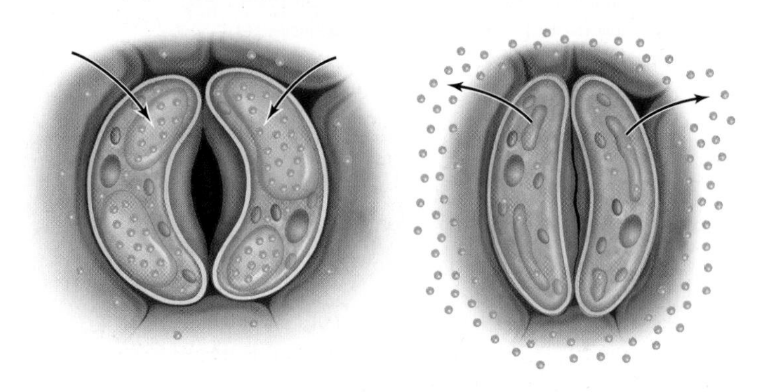

21. Explain why solution flows from the left bulb to the right bulb.

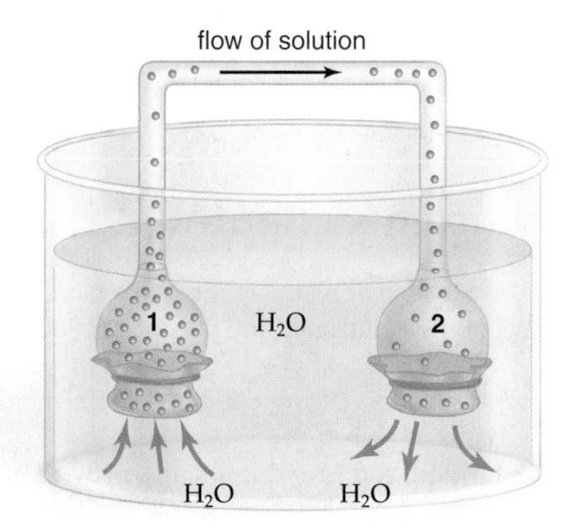


## thinking scientifically

1. Using hydroponics, design an experiment to determine if calcium is an essential plant nutrient. State the possible results.
2. *Welwitschia* is a genus of plant that lives in the Namib and Mossamedes deserts in Africa. Annual rainfall averages only 2.5 cm (1 inch) per year. *Welwitschia* plants contain a large number of stomata (22,000 $cm^2$), which remain closed most of the time. Can you suggest how a large number of stomata would be beneficial to these desert plants?

## *Biology* website

The companion website for *Biology* provides a wealth of information organized and integrated by chapter. You will find practice tests, animations, videos, and much more that will complement your learning and understanding of general biology.

**http://www.mhhe.com/maderbiology10**

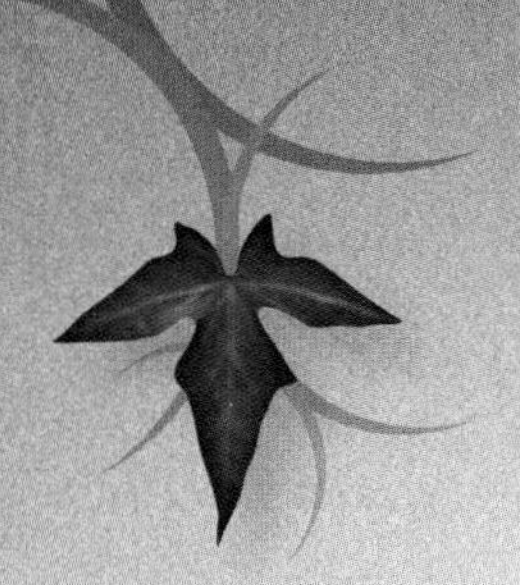

# 26

# Flowering Plants: Control of Growth Responses

*The observation that buttercups track the sun as it moves through the sky is a striking example of a flowering plant's ability to respond to environmental stimuli. Other responses to light can take longer than sun tracking because they involve hormones and an alteration in growth. For example, flowering plants will bend toward the light within a few hours because a hormone produced by the growing tip has moved from the sunny side to the shady side of the stem. Hormones also help flowering plants respond to stimuli in a coordinated manner. In the spring, seeds germinate and growth begins if the soil is warm enough to contain liquid water. In the fall, when temperatures drop, shoot- and root-apical growth ceases. Some plants also flower according to the season. The pigment phytochrome is instrumental in detecting the photoperiod and bringing about genetic changes, which determine whether a plant flowers or does not flower.*

*Plant defenses include physical barriers, chemical toxins, and even mutualistic animals. This chapter discusses the variety of ways flowering plants can respond to their environment, including other organisms.*

Time-lapse photograph of a buttercup, *Ranunculus ficaria,* curving toward and tracking a source of light.

## *concepts*

## 26.1 Plant Hormones

All organisms are capable of responding to environmental stimuli, as when you withdraw your hand from a hot stove. It is adaptive for organisms to respond to stimuli because it leads to their longevity and ultimately to the survival of the species. Flowering plants perceive and react to a variety of environmental stimuli. Some examples include light, gravity, carbon dioxide levels, pathogen infection, drought, and touch. Their responses can be short term, as when stomata open and close in response to light levels, or long term, as when they respond to gravity by the downward growth of the root and the upward growth of the stem.

Although we think of responses in terms of a plant part, the mechanism that brings about a response occurs at the cellular level. In the same manner as animal cells, researchers now know that plant cells utilize signal transduction when they respond to stimuli. Notice in Figure 26.1 that signal transduction involves:

**Receptors**—proteins activated by a specific signal. Receptors can be located in the plasma membrane, the cytoplasm, the nucleus, or even the endoplasmic reticulum. A receptor that responds to light has a pigment component. Fo example, phytochrome has a region that is sensitive to red light, and phototropin has a region that is sensitive to blue light.

**Transduction pathway**—series of relay proteins or enzymes that amplify and transform the signal to one understood by the machinery of the cell. In some instances, a stimulated receptor immediately communicates with the transduction pathway, and in other instances, a second messenger, such as $Ca^{2+}$, initiates the response. As an analogy, consider a mother at work who wants a sitter to fix lunch for her children. The mother (the stimulus) calls home (receptor of cell), and the sitter (the second messenger) fixes lunch (activates transduction pathway).

**Cellular response**—occurs as a result of the transduction pathway. Very often, the response is either the transcription of particular genes or the end product of an activated metabolic pathway. The cellular response brings about the observed macroscopic response, such as stomata closing or a stem that turns toward the light.

How do hormones fit into this model for the ability of flowering plants to respond to both abiotic and biotic stimuli? Coordination between cells is required for a macroscopic response to become evident. Coordination is often dependent on plant **hormones** [Gk. *hormao*, instigate], chemical signals produced in very low concentrations and active in another part of the organism. Hormones, such as auxin, are synthesized or stored in one part of the plant, but they travel within phloem or from cell to cell in response to the appropriate stimulus.

**FIGURE 26.1 Signal transduction in plants.**

1 The hormone auxin enters the cell and is received by a receptor in the nucleus. This complex alters gene expression. 2 A light receptor in the plasma membrane is sensitive to and activated by blue light. Activation leads to stimulation of a transduction pathway that ends with gene expression changes. 3 When attacked by a herbivore, the flowering plant produces defense hormones that bind to a plasma membrane receptor. Again, the transduction pathway results in a change in gene expression.

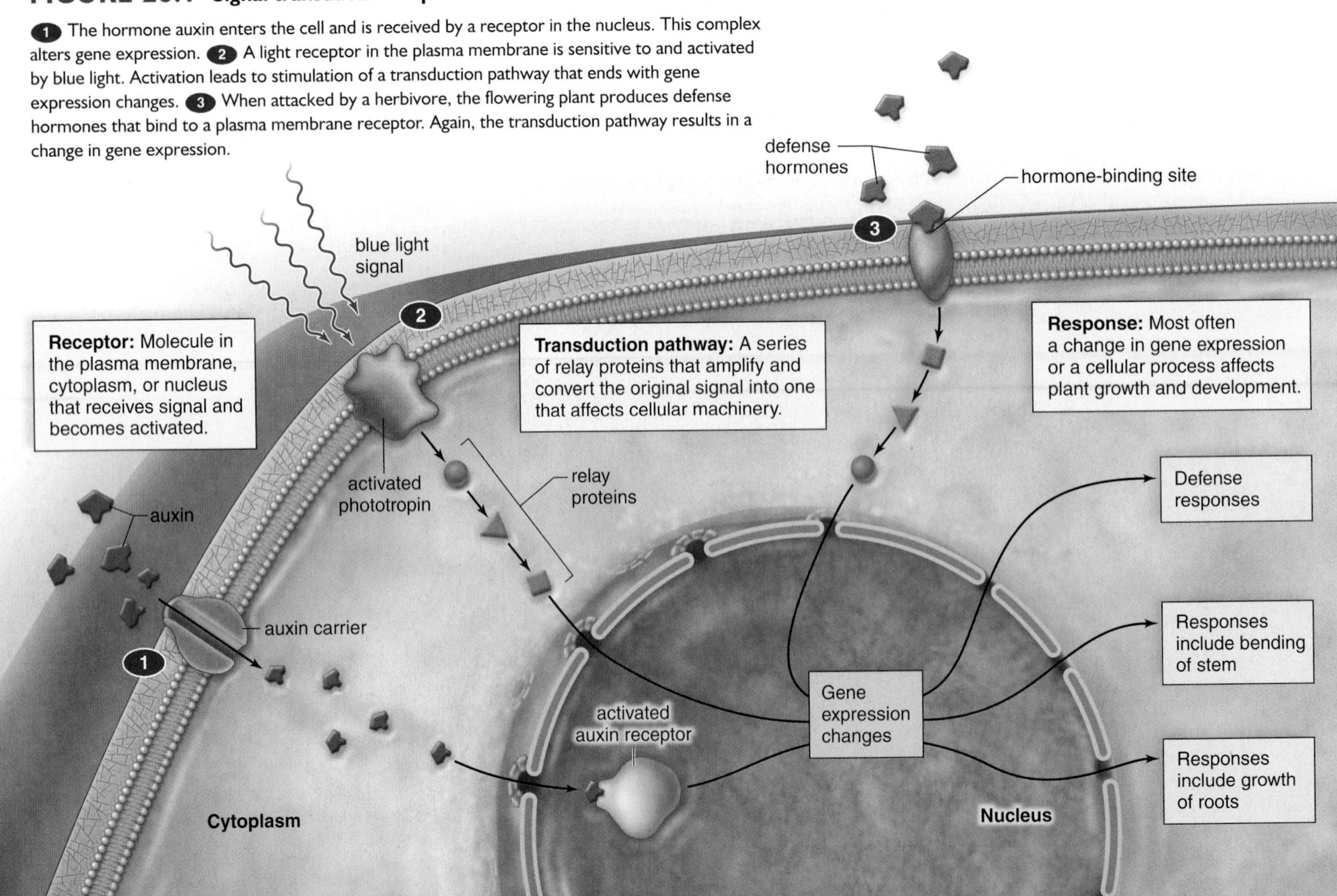

## Auxins

**Auxins** [Gk. *auximos,* promoting growth] are produced in shoot apical meristem and are found in young leaves and in flowers and fruits. The most common naturally occurring auxin is indoleacetic acid (IAA):

$-CH_2-COOH$

N

H

Structure of indoleacetic acid (IAA)

### *Auxins Effect Growth and Development*

Auxins affect many aspects of plant growth and development. Auxins, or more simply, auxin is responsible for **apical dominance,** which occurs when the terminal bud produces new growth instead of the axillary buds. When a terminal bud is removed deliberately or accidentally, the nearest axillary buds begin to grow, and the plant branches. Therefore, pruning the top of a flowering plant generally achieves a fuller look. This removes apical dominance and causes more branching of the main body of the plant.

Auxin causes the growth of roots and fruits and prevents the loss of leaves and fruit. The application of an auxin paste to a stem cutting causes adventitious roots to develop more quickly than they would otherwise. Auxin production by seeds promotes the growth of fruit. As long as auxin is concentrated in leaves or fruits rather than in the stem, leaves and fruits do not drop off. Therefore, trees can be sprayed with auxin to keep mature fruit from falling to the ground.

Synthetic auxins are used today in a number of applications. These auxins are sprayed on plants such as tomatoes to induce the development of fruit without pollination. Thus, seedless tomatoes can be commercially developed. Synthetic auxins such as 2,4-D and 2,4,5-T have been used as herbicides to control broadleaf weeds, such as dandelions and other plants. These substances have little effect on grasses. 2,4-D is still used, but 2,4,5-T was banned in 1979 because of its detrimental effects on human and animal life. A mixture of 2,4-D and 2,4,5-T is best known as the defoliant Agent Orange, used in the Vietnam War.

**Gravitropism and Phototropism.** After gravity has been perceived by a flowering plant, auxin moves to the lower surface of roots and stems. Thereafter, roots curve downward and stems curve upward. Gravitropism is discussed at more length on page 482. The role of auxin in the positive phototropism of stems has been studied for quite some time. The experimental material of choice has been oat seedlings with coleoptiles intact. A **coleoptile** is a protective sheath for the young leaves of the seedling. In 1881, Charles Darwin and his son found that phototropism will not occur if the tip of the seedling is cut off or covered by a black cap. They concluded that some influence that causes curvature is transmitted from the coleoptile tip to the rest of the shoot.

In 1926, Frits W. Went cut off the tips of coleoptiles and placed them on agar (a gelatin-like material). Then he placed an agar block to one side of a tipless coleoptile and found that the shoot would curve away from that side. The bending occurred even though the seedlings were not exposed to light (Fig. 26.2). Went concluded that the agar block contained a chemical that had been produced by the coleoptile tips. This chemical, he decided, had caused the shoots to bend. He named the chemical substance auxin after the Greek word *auximos,* which means promoting growth.

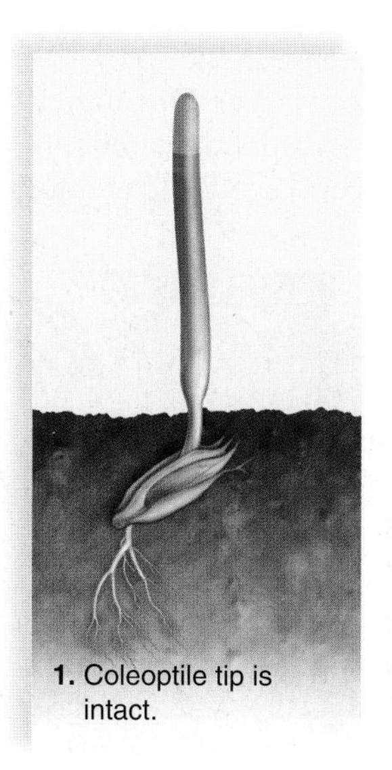


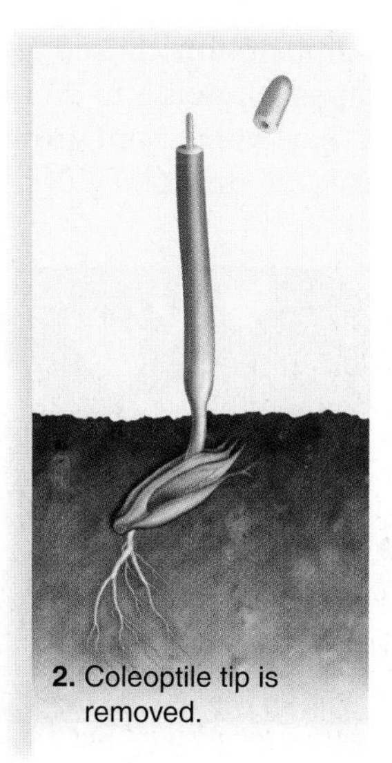


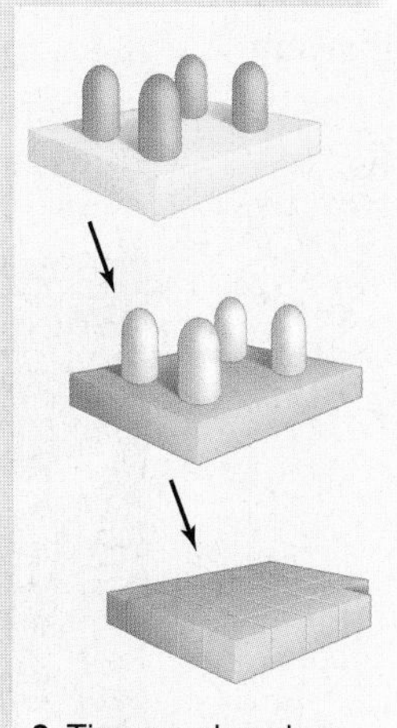

5. Curvature occurs beneath the block.

**FIGURE 26.2 Auxin and phototropism.**

Oat seedlings are protected by a hollow sheath called a coleoptile. After coleoptile tips are removed and placed on agar, a block of the agar to one side of the cut coleoptile can cause it to curve due to the presence of auxin (pink) in the agar. This shows that auxin causes the coleoptile to bend, as it does when exposed to a light source.

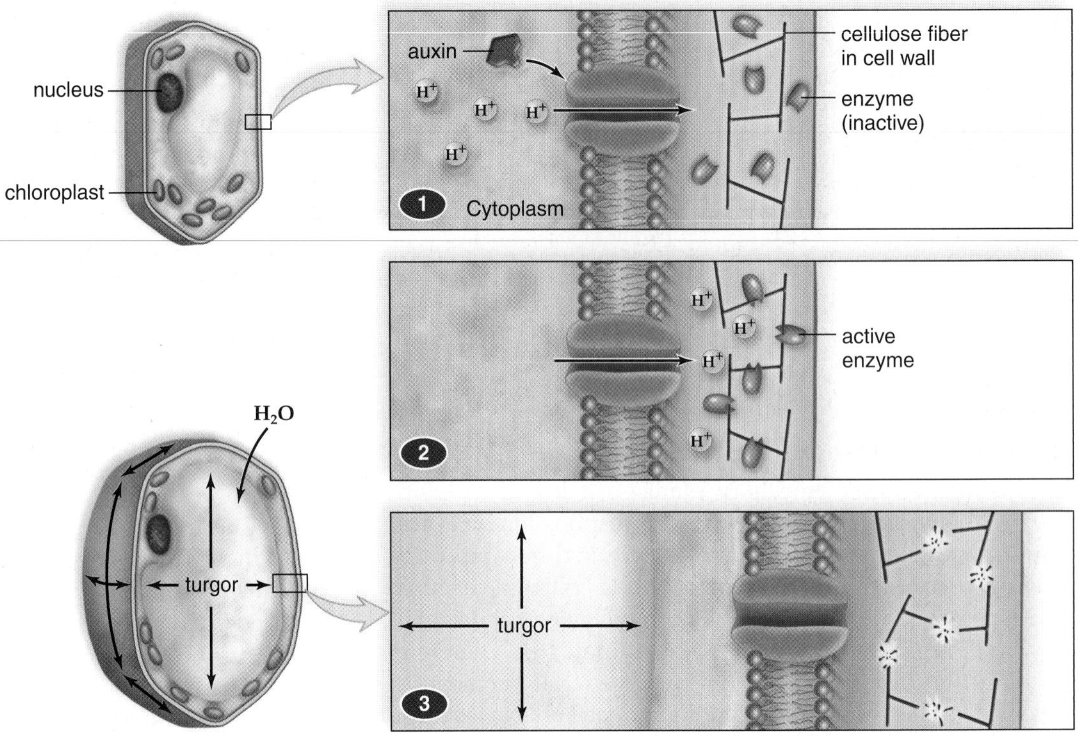


**FIGURE 26.3 Expansion of the cell wall.**

**1** Auxin leads to activation of a proton pump and entrance of hydrogen ions in the cell wall. **2** As the pH decreases, enzymes are activated and break down cellulose fibers in the cell wall. **3** Cellulose fibers burst and the cell expands as turgor pressure inside cell increases.

### *How Auxins Cause Stems to Bend*

When a stem is exposed to unidirectional light, auxin moves to the shady side, where it enters the nucleus and attaches to a receptor. The complex leads to he activation of a proton ($H^+$) pump, and the resulting acidic conditions loosen the cell wall because hydrogen bonds are broken and cellulose fibrils are weakened by enzymatic action. The end result of these activities is elongation of the stem on the shady side so that it bends toward the light (Fig. 26.3).

## Gibberellins

We know of about 70 **Gibberellins** [L. *gibbus,* bent], and they differ chemically only slightly. The most common of these is gibberellic acid, $GA_3$ (the subscript designation distinguishes it from other gibberellins):

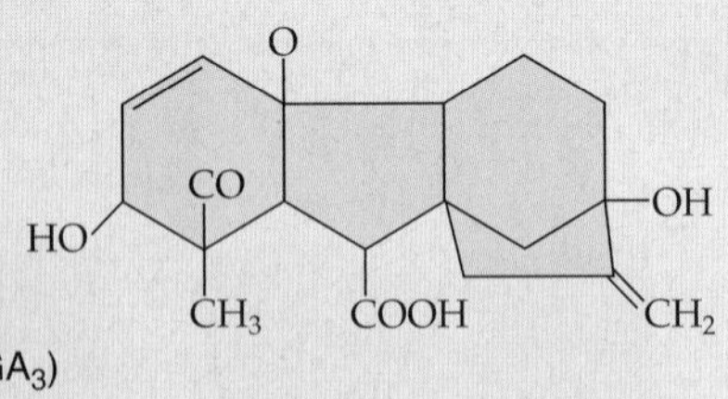


Structure of gibberellic acid ($GA_3$)

### *Gibberellins Promote Stem Elongation*

When gibberellins are applied externally to plants, the most obvious effect is stem elongation (Fig. 26.4*a*). Gibberellins can cause dwarf plants to grow, cabbage plants to become 2 m tall, and bush beans to become pole beans.

Gibberellins were discovered in 1926, the same year that Went performed his classic experiments with auxin. Ewiti Kurosawa, a Japanese scientist, was investigating a fungal disease of rice plants called "foolish seedling disease." The plants elongated too quickly, causing the stem to weaken and the plant to collapse. Kurosawa found that the fungus infecting the plants produced an excess of a chemical he called gibberellin, named after the fungus *Gibberella fujikuroi.* It wasn't until 1956 that gibberellic acid was isolated from a flowering plant rather than from a fungus. Sources of gibberellin in flowering plant parts are young leaves, roots, embryos, seeds, and fruits.

**Commercial Uses.** Commercially, gibberellins are helpful in a number of ways. Gibberellins induce the growth of plants and increase the size of flowers. Gibberellins have also been successfully used to produce larger seedless grapes. In Figure 26.4b, gibberellins caused an increase in the space between the grapes, allowing them to grow larger. **Dormancy** is a period of time when plant growth is suspended. Gibberellins can break the dormancy of buds and seeds. Thereore,

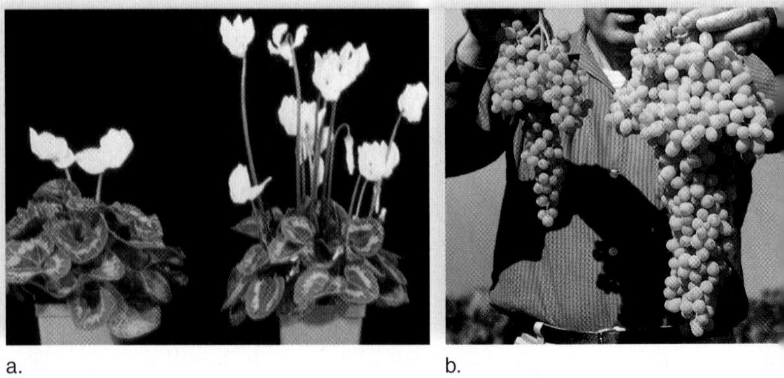

a. b.

**FIGURE 26.4 Gibberellins cause stem elongation.**

**a.** The *Cyclamen* plant on the right was treated with gibberellins; the plant on the left was not treated. **b.** The grapes are larger on the right because gibberellins caused an increase in the space between the grapes, allowing them to grow larger.

application of gibberellins is one way to hasten the development of a flower bud. When Gibberellins break the dormancy of barley seeds, a large, starchy endosperm is broken down into sugars to provide energy for growth. This occurs because amylase, an enzyme that breaks down starch, makes its appearance. It would seem, then, that gibberellins are involved in a transduction pathway that leads to the production of amylase.

## Cytokinins

The **cytokinins** [Gk. *kytos*, cell, and *kineo*, move] are derivatives of adenine, one of the purine bases in DNA and RNA. A naturally occurring cytokinin was not isolated until 1967. Because it came from the kernels of maize (*Zea*), it was called zeatin:

Structure of zeatin

### *Cytokinins Promote Cell Division*

Cytokinins were discovered as a result of attempts to grow plant tissue and organs in culture vessels in the 1940s. It was found that cell division occurred when coconut milk (a liquid endosperm) and yeast extract are added to the culture medium. The effective components were collectively called cytokinins because cytokinesis means cell division. Since then, cytokinins have been isolated from various seed plants, where they occur in the actively dividing tissues of roots and also in seeds and fruits. Cytokinins have been used to prolong the life of flower cuttings as well as vegetables in storage.

Plant tissue culturing is now common practice, and researchers are well aware that the ratio of auxin to cytokinin and the acidity of the culture medium determine whether the plant tissue forms an undifferentiated mass, called a callus, or differentiates to form roots, vegetative shoots, leaves, or floral shoots (Fig. 26.5). These effects illustrate that a plant hormone rarely acts alone, it is the relative concentrations of hormones and their interactions that produce an effect. Researchers have reported that chemicals they call oligosaccharins (chemical fragments released from the cell wall) are also effective in directing differentiation. They hypothesize that auxin and cytokinins are a part of a reception-transduction-response pathway, which leads to the activation of enzymes that release these fragments from the cell wall.

### *Cytokinins Prevent Senescence*

When a plant organ, such as a leaf, loses its natural color, it is most likely undergoing an aging process called **senescence.** During senescence, large molecules within the leaf are broken down and transported to other parts of the plant. Senescence does not always affect the entire plant at once; for example, as some plants grow taller, they naturally lose their lower leaves. It has been found that senescence of leaves can be prevented by the application of cytokinins. Also, axillary buds begin to grow, despite apical dominance, when cytokinin is applied to them.

a.

b.

c.

d.

**FIGURE 26.5 Interaction of hormones.**

Tissue culture experiments have revealed that auxin and cytokinin interact to affect differentiation during development. **a.** In tissue culture that has the usual amounts of these two hormones, tobacco strips develop into a callus of undifferentiated tissue. **b.** If the ratio of auxin to cytokinin is appropriate, the callus produces roots. **c.** Change the ratio, and vegetative shoots and leaves are produced. **d.** Yet another ratio causes floral shoots. It is now clear that each plant hormone rarely acts alone; it is the relative concentrations of hormones that produce an effect. The modern emphasis is to look for an interplay of hormones when a growth response is studied.

science focus

## *Arabidopsis* Is a Model Organism

*Arabidopsis thaliana* is a small flowering plant related to cabbage and mustard plants (Fig. 26A). *Arabidopsis* has no commercial value—in fact, it is a weed! However, it has become a model organism for the study of plant molecular genetics, including signal transduction. Unlike crop plants used formerly, these characteristics make *Arabidopsis* a model organism.

- It is small, so many hundreds of plants can be grown in a small amount of space. *Arabidopsis* consists of a flat rosette of leaves from which grows a short flower stalk.
- Generation time is short. It only takes 5–6 weeks for plants to mature, and each one produces about 10,000 seeds!
- It normally self-pollinates, but it can easily be cross-pollinated. This feature facilitates gene mapping and the production of strains with multiple mutations.
- The number of base pairs in its DNA is relatively small: 125 million base pairs are distributed in 5 chromosomes ($2n = 10$) and 25,500 genes.

In contrast to *Arabidopsis*, crop plants, such as corn, have generation times of at least several months, and they require a great deal of field space for a large number to grow. Crop plants have much larger genomes than *Arabidopsis*. For comparison, the genome sizes for rice (*Oryza sativa*), wheat (*Triticum aestivum*), and corn (*Zea mays*) are 420 million, 16 billion, and 2.5 billion base pairs, respectively. However, crop plants have about the same number of functional genes as *Arabidopsis*, and they occur in the same sequence. Therefore, knowledge of the *Arabidopsis* genome can be used to help locate specific genes in the genomes of other plants. Now that the *Arabidopsis* genome has been sequenced, genes of interest can be cloned from the *Arabidopsis* genome and then used as probes for the isolation of the homologous genes from plants of economic value. Also, cellular processes controlled by a family of genes in other plants require only a single gene or fewer genes in *Arabidopsis*. This, too, facilitates molecular biological studies of the plant.

The creation of *Arabidopsis* mutants plays a significant role in discovering what each of its genes do. For example, if a mutant plant lacks stomata (openings in leaves), then we know that the affected gene influences the formation of stomata. Transformation has emerged as a powerful way to create *Arabidopsis* mutants. The transforming DNA often gets inserted directly within a particular gene sequence. This usually destroys the function of the disrupted gene, resulting in a "knockout mutant." Furthermore, the piece of transformed DNA (T-DNA) that is inserted in the disrupted plant gene can serve as a flag for tracking down the gene by molecular biology methods. Large-scale projects using this T-DNA insertion technique are under way to mutate, identify, and characterize every gene in the *Arabidopsis* genome.

a. *Arabidopsis thaliana*

b. *Arabidopsis thaliana (enlarged drawing)*

**FIGURE 26A Overall appearance of *Arabidopsis thaliana*.**
*Many investigators have turned to this weed as an experimental material to study the actions of genes, including those that control growth and development.* **a.** *Photograph of actual plant.* **b.** *Enlarged drawing.*

Researchers have discovered three classes of genes that are essential to normal floral pattern formation. These are homeotic genes because they cause sepals, petals, stamens, or carpels to appear in place of one another (Fig. 26B, *top*). Triple mutants that lack all three types of genetic activities have flowers that consist entirely of leaves arranged in whorls. And a mutation of a regulatory gene results in flowers that have three whorls of petals. These floral-organ-identity genes appear to be regulated by transcription factors that are expressed and required for extended periods.

The application of *Arabidopsis* genetics to other plants has been shown. For example, one of the mutant genes that alters the development of flowers has been cloned and reintroduced into tobacco plants, where, as expected, it caused sepals and stamens to appear where petals would ordinarily be. The investigators commented that knowledge about the development of flowers in *Arabidopsis* can have far-ranging applications. It will undoubtedly lead someday to more productive crops.

A study of the *Arabidopsis* genome will undoubtedly promote plant molecular genetics in general. And because *Arabidopsis* has been found to be a model organism, its genetics is expected to apply to humans, just as Mendel's laws were discovered by working with pea plants. It's far easier to study signal transduction in *Arabidopsis* cells than in human cells.

*Arabidopsis* flower

Mutated flower

Mutated flower

**FIGURE 26B *Arabidopsis* mutants.**
*Creation of flower mutants (top) and other types of mutants has led to a knowledge of how signal transduction occurs in plant cells. A modern investigator makes use of a computer to analyze data.*

A flat of *Arabidopsis*

Lab

## Abscisic Acid

**Abscisic acid (ABA)** is produced by any "green tissue" (that contains chloroplasts). ABA is also produced in monocot endosperm and roots, where it is derived from carotenoid pigments:

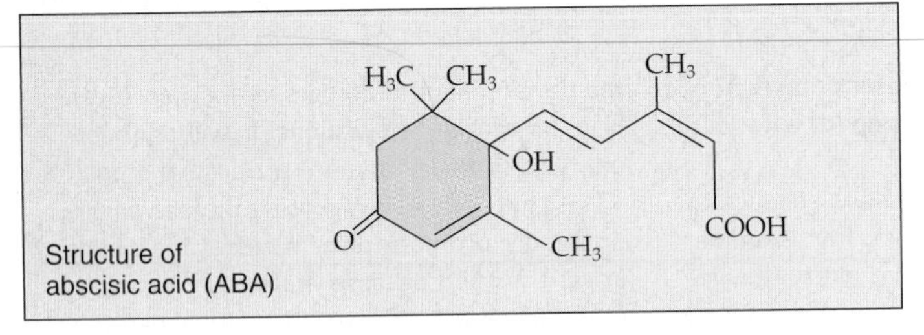


Abscisic acid is sometimes called the stress hormone because it initiates and maintains seed and bud dormancy and brings about the closure of stomata. It was once believed that ABA functioned in **abscission**, the dropping of leaves, fruits, and flowers from a plant. But although the external application of ABA promotes abscission, this hormone is no longer believed to function naturally in this process. Instead, the hormone ethylene seems to bring about abscission.

### ABA Promotes Dormancy

Recall that dormancy is a period of low metabolic activity and arrested growth. Dormancy occurs when a plant organ readies itself for adverse conditions by ceasing to grow (even though conditions at the time may be favorable for growth). For example, it is believed that ABA moves from leaves to vegetative buds in the fall, and thereafter these buds are converted to winter buds. A winter bud is covered by thick, hardened scales (Fig. 26.6). A reduction in the level of ABA and an increase in the level of gibberellins are believed to break seed and bud dormancy. Then seeds germinate, and buds send forth leaves. In Figure 26.7, corn kernels have begun to germinate on the developing cob because this maize mutant is deficient in ABA. Abscisic acid is needed to maintain the dormancy of seeds.

**FIGURE 26.6**
**Dormancy and winter buds.**
Abscisic acid promotes the formation of winter buds.

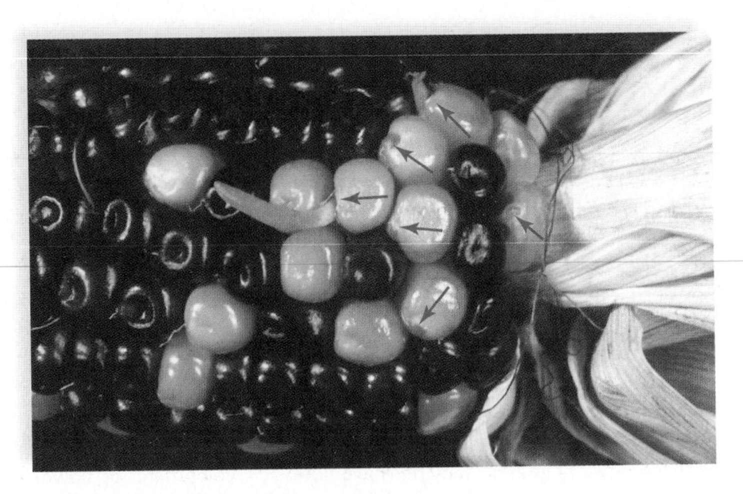

**FIGURE 26.7 Dormancy and germination.**
Corn kernels start to germinate on the cob (see arrows) due to low abscisic acid.

### ABA Closes Stomata

The reception of abscisic acid brings about the closing of stomata when a plant is under water stress, as described in Figure 26.8. Investigators have also found that ABA induces rapid depolymerization of actin filaments and formation of a new type of actin that is randomly oriented throughout the cell. This change in actin organization may also be part of the transduction pathways involved in stomata closure.

## Ethylene

**Ethylene** ($H_2C = CH_2$) is a gas formed from the amino acid methionine. This hormone is involved in abscission and the ripening of fruits.

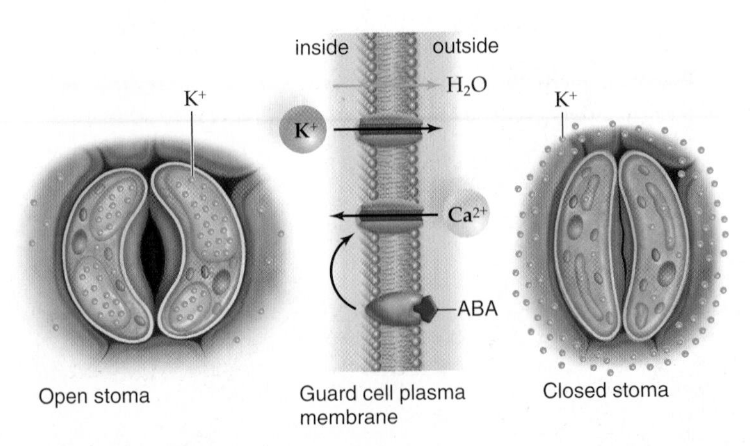


**FIGURE 26.8 Abscisic acid promotes closure of stomata.**
The stoma is open (*left*). When ABA (the first messenger) binds to its receptor in the guard cell plasma membrane, the second messenger ($Ca^{2+}$) enters (*middle*). Now, $K^+$ channels open, and $K^+$ exits the guard cells. After $K^+$ exits, so does water. The stoma closes (*right*).

### *Ethylene Causes Abscission*

The absence of auxin, and perhaps gibberellin, probably initiates abscission. But once abscission has begun, ethylene stimulates certain enzymes, such as cellulase, which helps cause leaf, fruit, or flower drop. In Figure 26.9, a ripe apple, which gives off ethylene, is under the bell jar on the right, but not under the bell jar on the left. As a result, only the holly plant on the right loses it leaves.

### *Ethylene Ripens Fruit*

In the early 1900s, it was common practice to prepare citrus fruits for market by placing them in a room with a kerosene stove. Only later did researchers realize that an incomplete combustion product of kerosene, namely ethylene, ripens fruit. It does so by increasing the activity of enzymes that soften fruits. For example, it stimulates the production of cellulase, which weakens plant cell walls. It also promotes the activity of enzymes that produce the flavor and smell of ripened fruits. And it breaks down chlorophyll, inducing the color changes associated with fruit ripening.

Ethylene moves freely through a plant by diffusion, and because it is a gas, ethylene also moves freely through the air. That is why a barrel of ripening apples can induce ripening of a bunch of bananas some distance away. Ethylene is released at the site of a plant wound due to physical damage or infection (which is why one rotten apple spoils the whole bushel).

The use of ethylene in agriculture is extensive. It is used to hasten the ripening of green fruits, such as melons and honeydews, and is also applied to citrus fruits to attain pleasing colors before marketing. Normally, tomatoes ripen on the vine because the plants produce ethylene. Today, tomato plants can be genetically modified to not produce ethylene. This facilitates shipping because green tomatoes are not subject to as much damage (Fig. 26.10). Once the tomatoes have arrived at their destination, they can be exposed to ethylene so that they ripen.

**FIGURE 26.9 Ethylene and abscission.**

Normally, there is no abscission when a holly twig is placed under a glass jar for a week. When an ethylene-producing ripe apple is also under the jar, abscission of the holly leaves occurs.

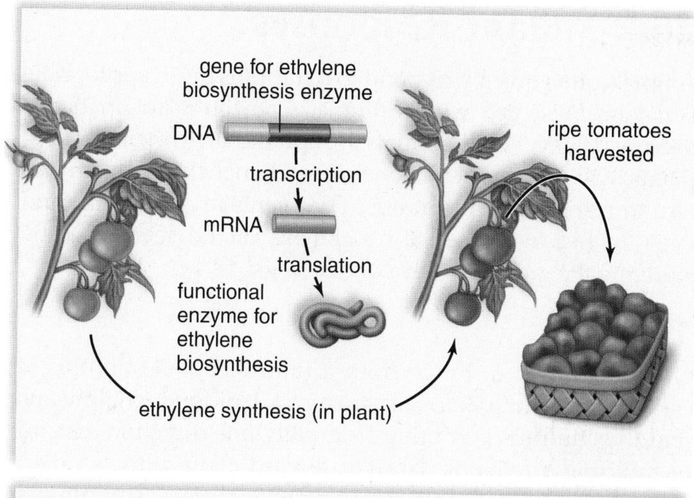


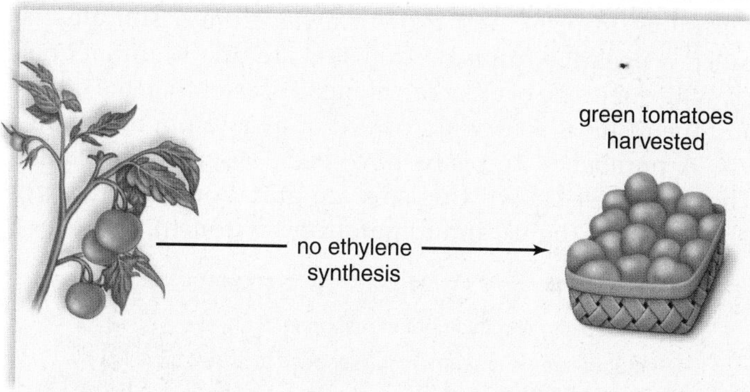


**FIGURE 26.10 Ethylene and fruit ripening.**

Wild-type tomatoes (*above*) ripen on the vine after producing ethylene. Tomatoes (*below*) are genetically modified to produce no ethylene and stay green for shipping.

### *Other Effects of Ethylene*

Ethylene is involved in axillary bud inhibition. Auxin, transported down from the apical meristem of the stem, stimulates the production of ethylene, and this hormone suppresses axillary bud development. Ethylene also suppresses stem and root elongation, even in the presence of other hormones.

This completes our discussion of plant hormones. The next part of the chapter explores plant responses to environmental stimuli.

**Check Your Progress 26.1**

1. In general, how do hormones assist in bringing about responses to stimuli?
2. If you wanted to increase the size of a plant organ, you might apply gibberellins and cytokinins. Explain.
3. **a.** Why is abscisic acid sometimes referred to as an inhibitory hormone? **b.** What hormone has the opposite effect of ABA on seed and bud dormancy?

# 26.2 Plant Responses

Animals often quickly respond to a stimulus by an appropriate behavior. Presented with a nipple, a newborn automatically begins sucking. While animals are apt to change their location, plants, which are rooted in one place, change their growth pattern in response to a stimulus. The events in a tree's life, and even the history of the Earth's climate, can be determined by studying the growth pattern of tree rings!

## Tropisms

Growth toward or away from a unidirectional stimulus is called a **tropism** [Gk. *tropos,* turning]. Unidirectional means that the stimulus is coming from only one direction instead of multiple directions. Growth toward a stimulus is called a positive tropism, and growth away from a stimulus is called a negative tropism. Tropisms are due to differential growth—one side of an organ elongates faster than the other, and the result is a curving toward or away from the stimulus. A number of tropisms have been observed in plants. The three best-known tropisms are gravitropism (gravity), phototropism (light), and thigmotropism (touch).

> Gravitropism: a movement in response to gravity
> Phototropism: a movement in response to a light stimulus
> Thigmotropism: a movement in response to touch

Several other tropisms include chemotropism (chemicals), traumotropism (trauma), skototropism (darkness), and aerotropism (oxygen).

What mechanism permits flowering plants to respond to stimuli? When humans respond to light, the stimulus is first received by a pigment in the retina at the back of the eyes, and then nerve impulses are generated that go to the brain. Only then do humans perform an appropriate behavior. As shown in Figure 26.1, the first step toward a response is *reception* of the stimulus. The next step is *transduction,* meaning that the stimulus has been changed into a form that is meaningful to the organism. (In our example, the light stimulus was changed to nerve impulses.) Finally, there is a *response* by the organism to light. Animals and plants go through this same sequence of events when they respond to a stimulus.

### *Gravitropism*

As is expected from our previous discussion on page 475, when an upright plant is placed on its side, the stem displays negative **gravitropism** [L. *gravis,* heavy; Gk. *tropos,* turning] because it grows upward, opposite the pull of gravity (Fig. 26.11*a*). Again, Charles Darwin and his son were among the first to say that roots, in contrast to stems, show positive gravitropism (Fig. 26.11*b*). Further, they discovered that if the root cap is removed, roots no longer respond to gravity. Later investigators came up with an explanation. Root cap cells contain sensors called **statoliths,** which are starch grains located within amyloplasts, a type of plastid (Fig. 26.11*c*). Due to gravity, the amyloplasts settle to a lower

a.

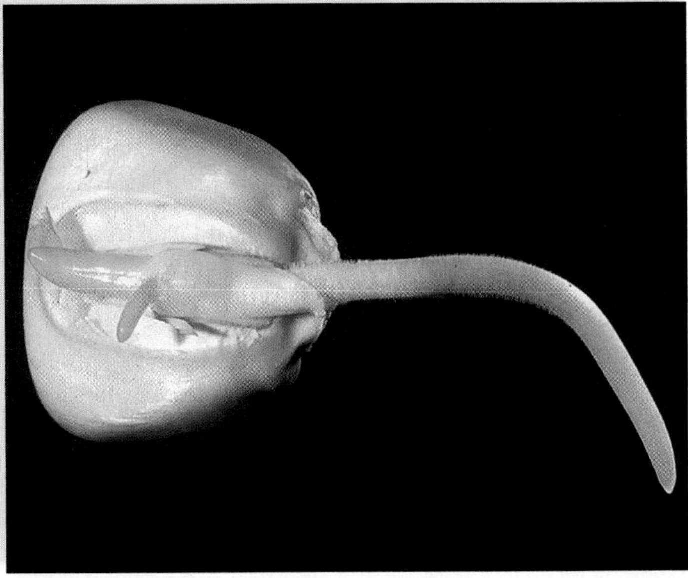

b.

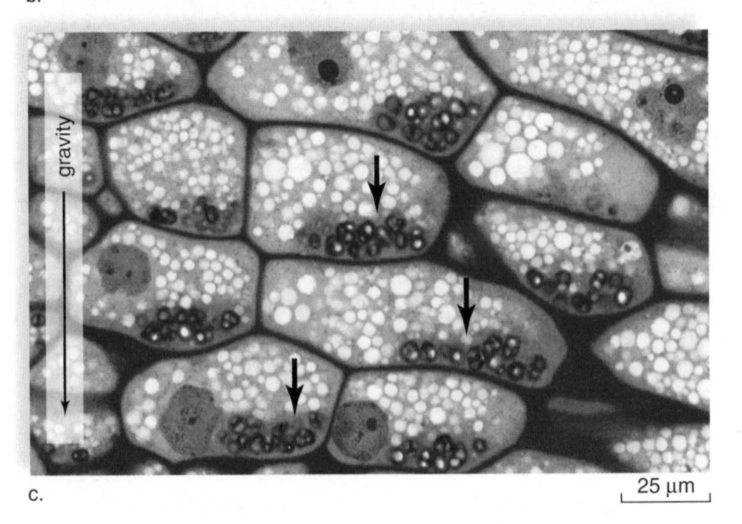


c.

**FIGURE 26.11 Gravitropism.**

**a.** Negative gravitropism of the stem of a *Coleus* plant 24 hours after the plant was placed on its side. **b.** Positive gravitropism of a root emerging from a corn kernel. **c.** Sedimentation of statoliths (see arrows), which are amyloplasts containing starch granules, is thought to explain how roots perceive gravity.

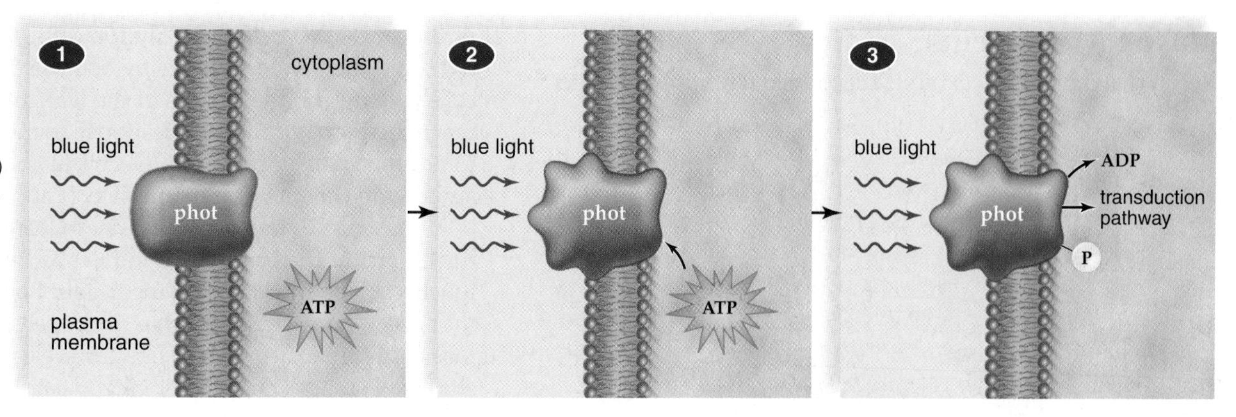


**FIGURE 26.12**
**Phototropin.**
In the presence of blue light, a photoreceptor called phototropin (phot) is activated and becomes phosphorylated. A transduction pathway begins.

part of the cell where they come in contact with cytoskeletal elements. The connection between amyloplasts and auxin is uncertain. Perhaps they cause stems and roots to respond differently to auxin: The upper surface of a root elongates so that the root curves downward and the lower surface of a stem elongates, so that the stem curves upwards.

## *Phototropism*

As discussed previously, positive **phototropism** of stems occurs because the cells on the shady side of the stem elongate due to the presence of auxin. Curving away from light is called negative phototropism. Roots, depending on the species examined, are either insensitive to light or exhibit negative phototropism.

Through the study of mutant *Arabidopsis* plants (see page 479), it is now known that phototropism occurs because plants respond to blue light (Fig. 26.12). ❶ When blue light is absorbed, the pigment portion of a photoreceptor, called phototropin (phot), undergoes a conformation change. ❷ This change results in the transfer of a phosphate group from ATP (adenosine triphosphate) to a protein portion of the photoreceptor. ❸ The phosphorylated photoreceptor triggers a transduction pathway that, in some unknown way, leads to the entry of auxin into the cell.

## *Thigmotropism*

Unequal growth due to contact with solid objects is called **thigmotropism** [Gk. *thigma*, touch, and *tropos*, turning]. An example of this response is the coiling of tendrils or the stems of plants, such as the stems of pea and morning glory plants (Fig. 26.13).

A flowering plant grows straight until it touches something. Then the cells in contact with an object, such as a pole, grow less while those on the opposite side elongate. Thigmotropism can be quite rapid; a tendril has been observed to encircle an object within 10 minutes. The response endures; a couple of minutes of touching can bring about a response that lasts for several days. The response can also be delayed; tendrils touched in the dark will respond once they are illuminated. ATP (adenosine triphosphate) rather than light can cause the response; therefore, the need for light may simply be a need for ATP. Also, the hormones auxin and ethylene may be involved since they can induce curvature of tendrils even in the absence of touch.

Thigmomorphogenesis is a touch response related to thigmotropism. In the former case, however, the entire plant responds to the presence of environmental stimuli, such as wind or rain. The same type of tree growing in a windy location often has a shorter, thicker trunk than one growing in a more protected location. Even simple mechanical stimulation—rubbing a plant with a stick, for example—can inhibit cellular elongation and produce a sturdier plant with increased amounts of support tissue.

**FIGURE 26.13** **Coiling response.**
The stem of a morning glory plant, *Ipomoea*, coiling around a pole illustrates thigmotropism.

## Nastic Movements

Recall that a plant cell exhibits turgor when it fills with water:

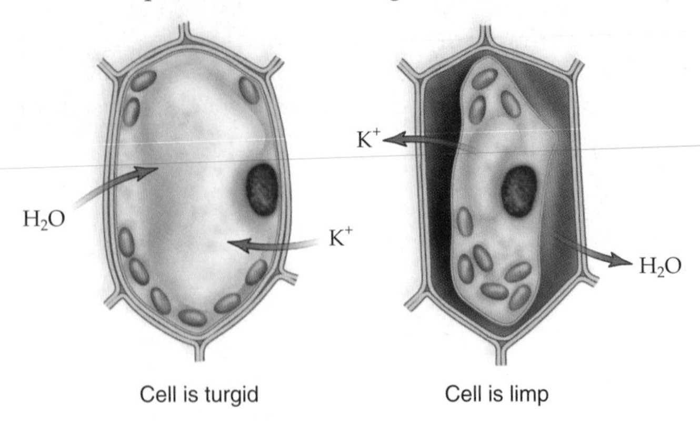


In general, if water exits the many cells of a leaf, the leaf goes limp. Conversely, if water enters a limp leaf, and cells exhibit turgor, the leaf moves as it regains its former position. **Turgor movements** are dependent on turgor pressure changes in plant cells. In contrast to tropisms, turgor movements do not involve growth and are not related to the source of the stimulus.

Turgor movements can result from touch, shaking, or thermal stimulation. The sensitive plant, *Mimosa pudica*, has compound leaves, meaning that each leaf contains many leaflets. Touching one leaflet collapses the whole leaf (Fig. 26.14). *Mimosa* is remarkable because the progressive response to the stimulus takes only a second or two.

The portion of a flowering plant involved in controlling turgor movement is a thickening called a pulvinus at the base of each leaflet. A leaf folds when the cells in the lower half of the pulvinus, called the motor cells, lose potassium ions ($K^+$), and then water follows by osmosis. When the pulvinus cells lose turgor, the leaflets of the leaf collapse. An electrical mechanism may cause the response to move from one leaflet to another. The speed of an electrical charge has been measured, and the rate of transmission is about 1 cm/sec.

A Venus flytrap closes its trap in less than 1 second when three hairs at the base of the trap, called the trigger hairs, are touched by an insect. When the trigger hairs are stimulated by the insect, an electrical charge is propagated throughout the lobes of a leaf. Exactly what causes this electrical charge is being studied. Perhaps (1) the cells located near the outer region of the lobes rapidly secrete hydrogen ions into their cell walls, loosening them, and allowing the walls to swell rapidly by osmosis; or (2) perhaps the cells in the inner portion of the lobes and the midrib rapidly lose ions, leading to a loss of water by osmosis and collapse of these cells. In any case, it appears that turgor movements are involved.

Venus flytrap, *Dionaea*

### *Sleep Movements and Circadian Rhythms*

Leaves that close at night are said to exhibit sleep movements. Activities such as sleep movements that occur regularly in a 24-hour cycle are called **circadian rhythms.** One of the most common examples occurs in a houseplant called the prayer plant (*Maranta leuconeura*) because at night the leaves fold upward

**FIGURE 26.14 Turgor movement.**

A leaf of the sensitive plant, *Mimosa pudica*, before and after it is touched.

into a shape resembling hands at prayer (Fig. 26.15, *top*). This movement is also due to changes in the turgor pressure of motor cells in a pulvinus located at the base of each leaf.

To take a few other examples, morning glory (*Ipomoea leptophylla*) is a plant that opens its flowers in the early part of the day and closes them at night (Fig. 26.15, *bottom*). In most plants, stomata open in the morning and close at night, and some plants secrete nectar at the same time of the day or night.

To qualify as a circadian rhythm, the activity must (1) occur every 24 hours; (2) take place in the absence of external stimuli, such as in dim light; and (3) be able to be reset if external cues are provided. For example, if you take a transcontinental flight, you will likely suffer jet lag because your body will still be attuned to the day/night pattern of its previous environment. But after several days, you will most likely have adjusted and will be able to go to sleep and wake up according to your new time.

### *Biological Clock*

The internal mechanism by which a circadian rhythm is maintained in the absence of appropriate environmental stimuli is termed a **biological clock.** If organisms are sheltered from environmental stimuli, their biological clock keeps the circadian rhythms going, but the cycle extends. In prayer plants, for example, the sleep cycle changes to 26 hours when the plant is kept in constant dim light, as opposed to 24 hours when in traditional day/night conditions. Therefore, it is suggested that biological clocks are synchronized by external stimuli to 24-hour rhythms. The length of daylight compared to the length of darkness, called the photoperiod, sets the clock. Temperature has little or no effect. This is adaptive because the photoperiod indicates seasonal changes better than temperature changes. Spring and fall, in particular, can have both warm and cold days.

Work with *Arabidopsis* (see Science Focus, pages 478–79) and other organisms suggests that the biological clock involves the transcription of a small number of "clock genes." One model proposes that the information-transfer system from DNA to RNA to enzyme to metabolite, with all its feedback controls, is intrinsically cyclical and could be the basis for biological clocks. In *Arabidopsis*, the biological clock involves about 5% of the genome. These genes control sleep movements, the opening and closing of stomata, the discharge of floral fragrances, and the metabolic activities associated with photosynthesis. The biological clock also influences seasonal cycles that depend on day/night lengths, including the regulation of flowering.

While circadian rhythms are outwardly very similar in all species, the clock genes that have been identified are not the same in all species. It would seem, then, that biological clocks have evolved several times to perform similar tasks.

#### Check Your Progress 26.2A

1. Roots grow toward water. Explain why this is adaptive.
2. If a plant is in a horizontal position and rotated horizontally, would the stem or the root exhibit gravitropism?
3. Many bat- and moth-pollinated plants open every 24 hours at night and often produce scent during the evening only. Explain why this is adaptive.

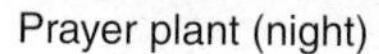

Prayer plant (morning) Prayer plant (night)

a.

Morning glory (morning) Morning glory (night)

b.

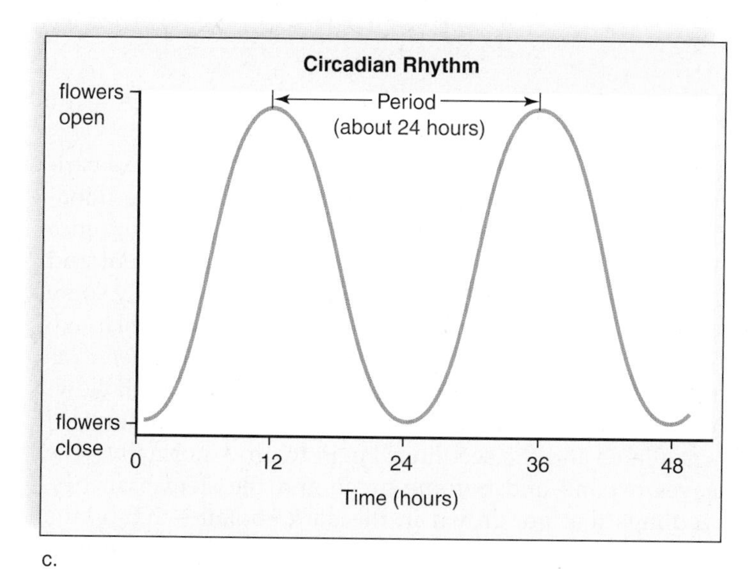


c.

**FIGURE 26.15 Sleep movements and circadian rhythms.**

**a.** The leaves of a prayer plant, *Maranta leuconeura*, fold every 24 hours at night. **b.** The flowers of the morning glory, *Ipomoea leptophylla*, close at night. **c.** Graph of circadian rhythm exhibited by morning glory plant.

## Photoperiodism

Many physiological changes in flowering plants are related to a seasonal change in day length. Such changes include seed germination, the breaking of bud dormancy, and the onset of senescence. A physiological response prompted by changes in the length of day or night in a 24-hour daily cycle is called **photoperiodism** [Gk. *photos,* light, and *periodus,* completed course]. In some plants, photoperiodism influences flowering; for example, violets and tulips flower in the spring, and asters and goldenrod flower in the fall. Photoperiodism requires the participation of a biological clock, which can measure time, and it also requires the activity of a plant photoreceptor called phytochrome.

### *Phytochrome*

**Phytochrome** [Gk. *phyton,* plant, and *chroma,* color] is a blue-green leaf pigment that is present in the cytoplasm of plant cells. A phytochrome molecule is composed of two identical proteins (Fig. 26.16). Each protein has a larger portion where a light-sensitive region is located. The smaller portion is a kinase that can link light absorption with a transduction pathway within the cytoplasm. Phytochrome can be said to act like a light switch because, like a light switch, it can be in the down (inactive) position or in the up (active) position. Red light prevalent in daylight activates phytochrome, and it assumes its active conformation known as $P_{fr}$. When $P_{fr}$ moves into the nucleus, it interacts with specific proteins, such as a transcription factor. The complex activates certain genes and inactivates others. The active form of phytochrome $P_{fr}$ is so called because it absorbs far-red light. Far-red light is prevalent in the evening and it serves to change $P_{fr}$ to $P_r$, which is the inactive form of phytochrome.

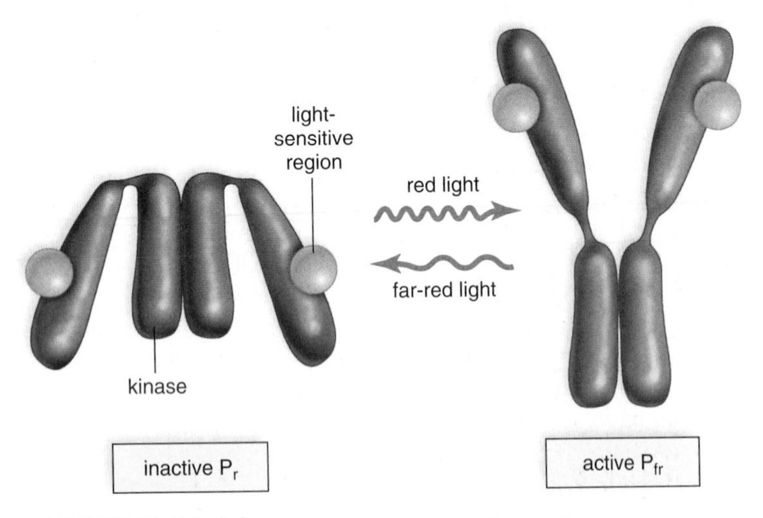


**FIGURE 26.16 Phytochrome conversion cycle.**

The inactive form of phytochrome ($P_r$) is converted to the active form $P_{fr}$ in the presence of red light, which is prevalent in daylight. $P_{fr}$, the active form of phytochrome, is involved in various plant responses such as seed germination, shoot elongation, and flowering. $P_{fr}$ is converted to $P_r$ whenever light is limited, such as in the shade or during the night.

**FIGURE 26.17 Phytochrome control of shoot elongation.**

**a.** If red light is prevalent, as it is in bright sunlight, normal growth occurs. **b.** If far-red light is prevalent, as it is in the shade, etiolation occurs. These effects are due to phytochrome.

### *Functions of Phytochrome*

The $P_r \rightarrow P_{fr}$ conversion cycle is now known to control various growth functions in plants. $P_{fr}$ promotes seed germination and inhibits shoot elongation, for example. The presence of $P_{fr}$ indicates to some seeds that sunlight is present and conditions are favorable for germination. This is why some seeds must be only partly covered with soil when planted. Germination of other seeds, such as those of *Arabidopsis,* is inhibited by light, so they must be planted deeper. Following germination, the presence of $P_{fr}$ indicates that sunlight is available and the seedlings begin to grow normally—the leaves expand and become green and the stem branches. Seedlings that are grown in the dark etiolate—that is, the shoot increases in length, and the leaves remain small (Fig. 26.17). Only when $P_r$ is converted to $P_{fr}$ does the seedling grow normally.

**Flowering.** Flowering plants can be divided into three groups on the basis of their flowering status:

1. **Short-day plants** flower when the day length is shorter than a *critical length.* (Examples are cocklebur, goldenrod, poinsettia, and chrysanthemum.)
2. **Long-day plants** flower when the day length is longer than a critical length. (Examples are wheat, barley, rose, iris, clover, and spinach.)
3. **Day-neutral plants** are not dependent on day length for flowering. (Examples are tomato and cucumber.)

The criterion for designating plants as short-day or long-day is not an absolute number of hours of light, but a critical number that either must be or cannot be exceeded. Spinach is a long-day plant that has a critical length of 14 hours; ragweed is a short-day plant with the same critical length. Spinach, however, flowers in the summer when the day length increases to 14 hours or more, and ragweed flowers in the fall, when the day length shortens to 14 hours or less. In addition, we now know that some plants require a specific sequence of day lengths in order to flower.

Soon after the three groups of flowering plants were discovered, researchers began to experiment with artificial lengths of light and dark that did not necessarily correspond to a normal 24-hour day. These investigators discovered that the cocklebur, a short-day plant, will not flower if a required long dark period is interrupted by a brief flash of white light. (Interrupting the light period with darkness has no effect.) On the other hand, a long-day plant will flower if an overly long dark period is interrupted by a brief flash of white light. They concluded that the length of the dark period, not the length of the light period, controls flowering. Of course, in nature, short days always go with long nights, and vice versa.

To recap, let's consider Figure 26.18:

- Cocklebur is a short-day plant (Fig. 26.18, *left*). (1) When the night is longer than a critical length, cocklebur flowers. (2) The plant does not flower when the night is shorter than the critical length. (3) Cocklebur also does not flower if the longer-than-critical-length night is interrupted by a flash of light.
- Clover is a long-day plant (Fig. 26.18, *right*). (4) When the night is shorter than a critical length, clover flowers. (5) The plant does not flower when the night is longer than a critical length. (6) Clover does flower when a slightly longer-than-critical-length night is interrupted by a flash of light.

### Check Your Progress 26.2B

1. Describe the structure of phytochrome and how it functions in plant cells.
2. A plant is a long-day plant. Explain why the plant will still flower if the long day is interrupted by a period of darkness.

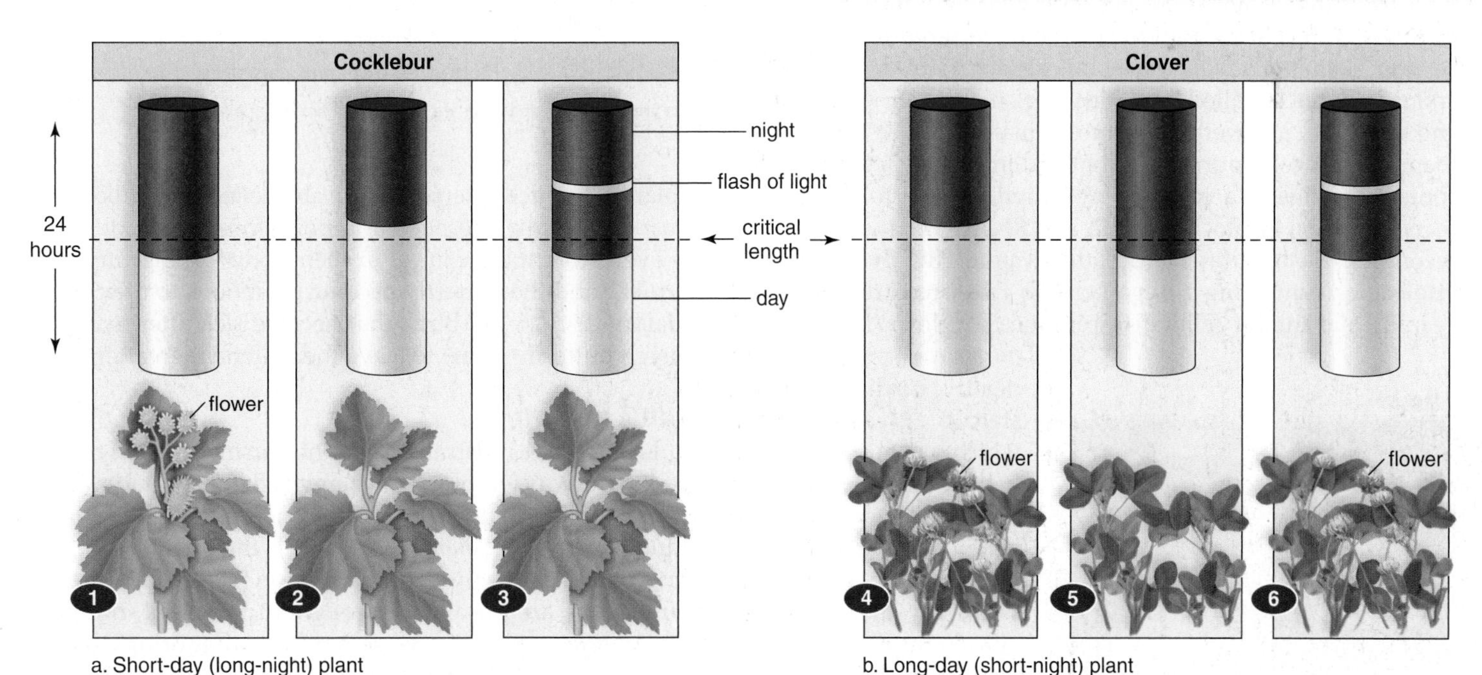


**FIGURE 26.18 Photoperiodism and flowering.**

**a.** Short-day plant. When the day is shorter than a critical length, this type of plant flowers. The plant does not flower when the day is longer than the critical length. It also does not flower if the longer-than-critical-length night is interrupted by a flash of light. **b.** Long-day plant. The plant flowers when the day is longer than a critical length. When the day is shorter than a critical length, this type of plant does not flower. However, it does flower if the slightly longer-than-critical-length night is interrupted by a flash of light.

## Flowering Plants Respond to the Biotic Environment

Plants are always under attack by herbivores (animals that eat plants) and parasites. Fortunately, they have an arsenal of defense mechanisms to deal with insects and fungi, for example (Fig. 26.19).

### *Physical and Chemical Defenses*

A plant's cuticle-covered epidermis and bark, if present, do a good job of discouraging attackers. But, unfortunately, herbivores have ways around a plant's first line of defense. A fungus can invade a leaf by way of the stomata and set up shop inside a leaf, where it feeds on nutrients meant for the plant. Underground nematodes have sharp mouthparts to break through the epidermis of a root and establish a parasitic relationship, sometimes by way of a single cell, which enlarges and transfers carbohydrates to the animal. Similarly, the tiny insects called aphids have styletlike mouthparts that allow them to tap into the phloem of a nonwoody stem. These examples illustrate why plants need several other types of defenses not dependent on the outer surface.

The primary metabolites of plants, such as sugars and amino acids, are necessary to the normal workings of a cell, but plants also produce so-called **secondary metabolites** as a defense mechanism. Secondary metabolites were once thought to be waste products, but now we know that they are part of a plant's arsenal to prevent predation. Tannins, present in or on the epidermis of leaves, are defensive compounds that interfere with the outer proteins of bacteria and fungi. They also deter herbivores because of their astringent effect on the mouth and their interference with digestion. Some secondary metabolites, such as bitter nitrogenous substances called **alkaloids** (e.g., morphine, nicotine, and caffeine), are well-known to humans because we use them for our own purposes. The seedlings of coffee plants contain caffeine at a concentration high enough to kill insects and fungi by blocking DNA and RNA synthesis. Other secondary metabolites include the **cyanogenic glycosides** (molecules containing a sugar group) that break down to cyanide and inhibit cellular respiration. Foxglove (*Digitalis purpurea*) produces deadly cardiac and steroid glycosides, which cause nausea, hallucinations, convulsions, and death in animals that ingest them. Taxol, an unsaturated hydrocarbon, from the Pacific yew (*Taxus brevifolia*), is now a well-known cancer-fighting drug.

Foxglove

Even with regard to secondary metabolites, predators can be one step ahead of the plant. Monarch caterpillars are able to feed on milkweed plants, despite the presence of a poisonous glycoside, and they even store the chemical in their body. In this way, the caterpillar and the butterfly become poisonous to their own predators (Fig. 26.19). Birds that become sick after eating a monarch butterfly know to leave them alone thereafter.

Alfalfa plant bug

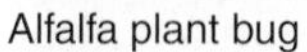

Fungus infection

Monarch caterpillar and butterfly

**FIGURE 26.19 Plant predators and parasites.**

Insects are predators and fungi are parasites of flowering plants.

### *Wound Responses*

Wound responses illustrate that plants can make use of transduction pathways to produce chemical defenses only when they are needed. After a leaf is chewed or injured, a plant produces *proteinase inhibitors*, chemicals that destroy the digestive enzymes of a predator feeding on them. The proteinase inhibitors are produced throughout the plant, not just at the wound site. The defense hormone that brings about this effect is a small peptide called **systemin** (Fig. 26.20). Systemin is produced in the wound area in response to the predator's saliva, but then it travels between cells to reach phloem, which distributes it about the plant. A transduction pathway is activated in cells with systemin receptors, and the cells produce proteinase inhibitors. A chemical called jasmonic acid, and also possibly a chemical called salicylic acid, are part of this transduction pathway. Salicylic acid (a chemical also found in aspirin) has been known since the 1930s to bring about a phenomenon called systemic acquired resistance (SAR), the production of antiherbivore chemicals by defense genes. Recently, companies have begun marketing salicylic acid and other similar compounds as a way to activate SAR in crops, including tomato, spinach, lettuce, and tobacco.

### Hypersensitive Response

On occasion, plants produce a specific gene product that binds (like a key fits a lock) to a viral, bacterial, or fungal gene product made within the cell. This combination offers a way for the plant to "recognize" a particular pathogen. A transduction pathway now ensues, and the final result is a **hypersensitive response (HR)** that seals off the infected area and will also initiate the wound response just discussed.

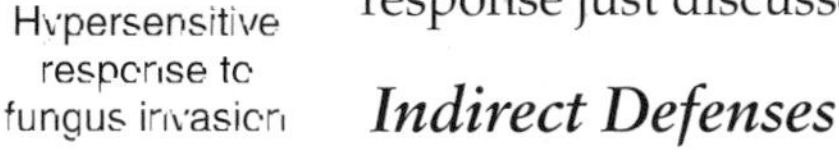

Hypersensitive response to fungus invasion

### Indirect Defenses

Some defenses do not kill or discourage a herbivore outright. For example, female butterflies are less likely to lay their eggs on plants that already have butterfly eggs. So, because the leaves of some passion flowers display physical structures resembling the yellow eggs of *Heliconius* butterflies, these butterflies do not lay eggs on this plant. Other plants produce hormones that prevent caterpillars from metamorphosing into adults and laying more eggs.

Certain plants attract the natural enemies of caterpillars feeding on them. They produce volatile molecules that diffuse into the air and advertise that food is available for a carnivore (an animal that eats other animals). For example, lima beans produce volatiles that attract carnivore mites only when they are being damaged by a spider mite. Corn and cotton plants release volatiles that attract wasps, which then inject their eggs into caterpillars munching on their leaves. The eggs develop into larvae that eat the caterpillars, not the leaves.

### Relationships with Animals

Mutualism is a relationship between two species in which both species benefit. As evidence that a mutualistic relationship can help protect a plant from predators, consider the bullhorn acacia tree, which provides a home for ants of the species *Pseudomyrmex ferruginea.* Unlike other acacias, this species has swollen thorns with a hollow interior where ant larvae can grow and develop.

Mutualism between ants and a plant

In addition to housing the ants, acacias provide them with food. The ants feed from nectaries at the base of leaves and eat fat- and protein-containing nodules called Beltian bodies, which are found at the tips of the leaves. In return, the ants constantly protect the plant by attacking and stinging any would-be herbivores because, unlike other ants, they are active 24 hours a day. Indeed, when the ants on experimental trees were removed, the acacia trees died.

**Check Your Progress 26.2C**

1. What are several ways that flowering plants protect themselves from insect predators?

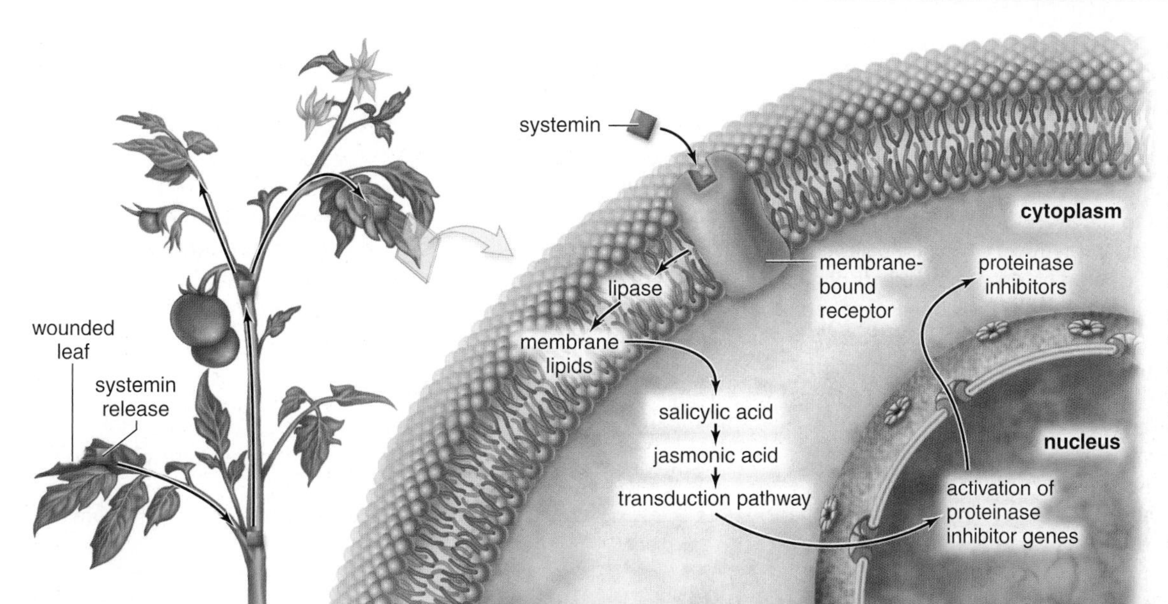


**FIGURE 26.20 Wound response in tomato.**

After wounded leaves produce systemin, it travels in phloem to all parts of a plant where it binds to cells that have a systemin receptor. These cells then produce jasmonic acid, a molecule that initiates a transduction pathway and, in that way, leads to production of proteinase inhibitors, which limit insect feeding.

## Connecting the Concepts

Behavior in plants can be understood in terms of three different levels of organization. On the species level, plant responses that promote survival and reproductive success have evolved through natural selection. At the organismal level, some part or the entire plant responds to a stimulus. And at the cellular level, receptors receive signals, transduction pathways transform them, and genes or metabolic pathways react to them.

We can illustrate these three perspectives by answering the question, why do plants bend toward the light? On the species level, those plants that bend toward the light will be able to produce more organic food and will have more offspring. On the organismal level, elongation of the stem on the shady side causes the stem to bend toward the light. On the cellular level, after auxin is received by a plant cell, cellular activities cause its walls to expand.

The response of both the organism and the cell involve three steps: (1) reception of the stimulus, (2) transduction of the stimulus, and (3) response to the stimulus. Can you designate these steps on the cellular level? A light stimulus causes auxin to enter cells on the shady side, cellular activities is the second step, and stretching of the cell wall is the third step.

If animals were being considered instead of plants, the same type of biological explanations would apply. Plants and animals, and indeed all organisms, share common ancestors, even back to the very first cell(s). Recall that evolution explains both the unity and diversity of living things. Organisms are similar because they share common ancestors; they are different because they are adapted to different ways of life.

## summary

### 26.1 Plant Hormones

Like animals, flowering plants use a reception-transduction-response pathway when they respond to a stimulus. The process involves receptor activation, transduction of the signal by relay proteins, and a cellular response, which can consist of the turning on of a gene or an enzymatic pathway.

Auxin-controlled cell elongation is involved in phototropism and gravitropism. When a plant is exposed to light, auxin moves laterally from the bright to the shady side of a stem.

Gibberellin causes stem elongation between nodes. After this hormone binds to a plasma membrane receptor, a DNA-binding protein activates a gene leading to the production of amylase. Amylase is an enzyme that speeds the breakdown of amylase.

Cytokinins cause cell division, the effects of which are especially obvious when plant tissues are grown in culture. Abscisic acid (ABA) and ethylene are two plant growth inhibitors. ABA is well known for causing stomata to close, and ethylene is known for causing fruits to ripen.

### 26.2 Plant Responses

When flowering plants respond to stimuli, growth and/or movement occurs. Tropisms are growth responses toward or away from unidirectional stimuli. The positive phototropism of stems results in a bending toward light, and the negative gravitropism of stems results in a bending away from the direction of gravity. Roots that bend toward the direction of gravity show positive gravitropism. Thigmotropism occurs when a plant part makes contact with an object, as when tendrils coil about a pole.

Nastic movements are not directional. Due to turgor pressure changes, some plants respond to touch and some perform sleep movements. Plants exhibit circadian rhythms, which are believed to be controlled by a biological clock. The sleep movements of prayer plants, the closing of stomata, and the daily opening of certain flowers have a 24-hour cycle.

Phytochrome is a pigment that is involved in photoperiodism, the ability of plants to sense the length of the day and night during a 24-hour period. This sense leads to seed germination, shoot elongation, and flowering during favorable times of the year. Daylight causes phytochrome to exist as $P_{fr}$, but during the night, it is reconverted to $P_r$ by metabolic processes. Phytochrome in the $P_{fr}$ form leads to a biological response such as flowering. Short-day plants flower only when the days are shorter than a critical length, and long-day plants flower only when the days are longer than a critical length. Actually, research has shown that it is the length of darkness that is critical. Interrupting the dark period with a flash of white light prevents flowering in a short-day plant and induces flowering in a long-day plant.

Flowering plants have defenses against predators and parasites. The first line of defense is their outer covering. They also routinely produce secondary metabolites that protect them from herbivores, particularly insects. Wounding causes plants to produce systemin, which travels about the plant and causes cells to produce proteinase inhibitors that destroy an insects digestive enzymes. During a hypersensitive response, an infected area is sealed off. As an indirect response, plants temporarily attract animals that will destroy predators, and going one step further, plants have permanent relationships with animals, such as ants, that will attack predators.

## understanding the terms

abscisic acid (ABA) 480
abscission 480
alkaloid 488
apical dominance 475
auxin 475
biological clock 485
cellular response 474
circadian rhythm 484
coleoptile 475
cyanogenic glycoside 488
cytokinin 477
day-neutral plant 487
dormancy 477
ethylene 480
gibberellin 476
gravitropism 482
hormone 474
hypersensitive response (HR) 489
long-day plant 487
photoperiodism 486
phototropism 483
phytochrome 486
receptor 474
secondary metabolite 488
senescence 477
short-day plant 487
statolith 482
systemin 488
thigmotropism 483
transduction pathway 474
tropism 482
turgor movement 484

Match the terms to these definitions:

a. ______________ Biological rhythm with a 24-hour cycle.
b. ______________ Directional growth of plants in response to the Earth's gravity.
c. ______________ Dropping of leaves, fruits, or flowers from a plant.

d. ______________ Plant hormone producing increased stem growth between nodes; also involved in flowering and seed germination.

e. ______________ Relative lengths of daylight and darkness that affect the physiology and behavior of an organism.

## reviewing this chapter

1. Name and describe the three stages of signal transduction in plant cells. 474
2. Why does removing a terminal bud cause a plant to get bushier? 475
3. What experiments led to knowledge that a hormone is involved in phototropism? Explain the mechanism by which auxin brings about elongation of cells. 475–76
4. Gibberellin research supports the hypothesis that plant hormones initiate a reception-transduction-response pathway. Explain. 476–77
5. What is the function of cytokinins? Discuss experimental evidence to suggest that hormones interact when they bring about an effect. 477
6. What are some of the primary effects of abscisic acid and how does it bring about these effects? 480
7. What hormones are involved in abscission? How does ethylene bring about ripening of fruits? 480–81
8. Tropisms are responses to stimuli. Why are stems said to exhibit positive phototropism but negative gravitropism? 482–83
9. What are nastic movements, and how do turgor pressure changes bring about movements of plants? 484–85
10. What is a biological clock, how does it function, and what is its primary usefulness in plants? 485
11. Define photoperiodism, and discuss its relationship to flowering in certain plants. 486–87
12. Describe the structure of phytochrome and its response to red and far-red light. 486
13. What mechanisms allow a plant to defend itself? 488–89

## testing yourself

Choose the best answer for each question.

1. During which step of signal transduction is a second messenger released into the cytoplasm?
   a. reception
   b. response
   c. transduction
   d. final step
2. Which of the following plant hormones causes apical dominance?
   a. auxin
   b. gibberellins
   c. cytokinins
   d. abscisic acid
   e. ethylene
3. Internode elongation is stimulated by
   a. abscisic acid.
   b. ethylene.
   c. cytokinin.
   d. gibberellin.
   e. auxin.
4. Which of these is related to gibberellin activity?
   a. initiation of bud dormancy
   b. amylase production
   c. fruit ripening.
   d. Both b and c are correct.
5. ______ always promotes cell division.
   a. Auxin
   b. Phytochrome
   c. Cytokinin
   d. None of these are correct.
6. In the absence of abscisic acid, plants may have difficulty
   a. forming winter buds.
   b. closing the stomata.
   c. Both a and b are correct.
   d. Neither a nor b is correct.
7. Ethylene
   a. is a gas.
   b. causes fruit to ripen.
   c. is produced by the incomplete combustion of fuels such as kerosene.
   d. All of these are correct.
8. Which of the following plant hormones is responsible for a plant losing its leaves?
   a. auxin
   b. gibberellins
   c. cytokinins
   d. abscisic acid
   e. ethylene
9. Which is not a plant hormone?
   a. auxin
   b. cytokinin
   c. gibberellin
   d. All of these are plant hormones.

For questions 10–14, match each statement with a hormone in the key.

**KEY:**

a. auxin
b. gibberellin
c. cytokinin
d. ethylene
e. abscisic acid

10. One rotten apple can spoil the barrel.
11. Cabbage plants bolt (grow tall).
12. Stomata close when a plant is water-stressed.
13. Stems bend toward the sun.
14. Coconut milk causes plant tissues to undergo cell division.
15. You bought green bananas at the grocery store this morning. However, you want a ripe banana for breakfast tomorrow morning. What could you do to accomplish this?
16. Which of the following statements is correct?
    a. Both stems and roots show positive gravitropism.
    b. Both stems and roots show negative gravitropism.
    c. Only stems show positive gravitropism.
    d. Only roots show positive gravitropism.
17. The sensors in the cells of the root cap are called
    a. mitochondria.
    b. central vacuoles.
    c. statoliths.
    d. chloroplasts.
    e. intermediate filaments.
18. A student places 25 morning glory (see Fig. 26.13) seeds in a large pot and allows the seeds to germinate in total darkness. Which of the following growth or movement activities would the seedlings exhibit?
    a. gravitropism, as the roots grow down and the shoots grow up
    b. phototropism, as the shoots search for light
    c. thigmotropism, as the tendrils coil around other seedlings
    d. Both a and c are correct.
19. Circadian rhythms
    a. require a biological clock.
    b. do not exist in plants.
    c. are involved in the tropisms.
    d. are involved in sleep movements.
    e. Both a and d are correct.

20. Plants that flower in response to long nights are
    a. day-neutral plants.
    b. long-day plants.
    c. short-day plants.
    d. impossible.
21. Short-day plants
    a. are the same as long-day plants.
    b. are apt to flower in the fall.
    c. do not have a critical photoperiod.
    d. will not flower if a short day is interrupted by bright light.
    e. All of these are correct.
22. A plant requiring a dark period of at least 14 hours will
    a. flower if a 14-hour night is interrupted by a flash of light.
    b. not flower if a 14-hour night is interrupted by a flash of light.
    c. not flower if the days are 14 hours long.
    d. not flower if the nights are longer than 14 hours.
    e. Both b and c are correct.
23. Phytochrome plays a role in
    a. flowering.
    b. stem growth.
    c. leaf growth.
    d. All of these are correct.
24. Phytochrome
    a. is a plant pigment.
    b. is present as $P_{fr}$ during the day.
    c. activates DNA-binding proteins.
    d. is a photoreceptor.
    e. All of these are correct.
25. Primary metabolites are needed for ______ while secondary metabolites are produced for ______.
    a. growth, signal transduction
    b. normal cell functioning, defense
    c. defense, growth
    d. signal transduction, normal cell functioning
26. Which of the following is a plant secondary metabolite used by humans to treat disease?
    a. morphine
    b. codeine
    c. quinine
    d. penicillin
    e. All but d are correct.
27. Which of these is mismatched?
    a. cuticle—first line of defense
    b. secondary metabolite—growth response
    c. systemin—wound response
    d. acadia and ants—mutualism
28. Which of these would account for sleep movements in prayer plants?
    a. turgor movements
    b. biological clock
    c. daylight
    d. phytochrome
29. Plants etoliate when they need
    a. fertilizer.
    b. water.
    c. daylight.
    d. phytochrome.
30. Which is part of phytochrome structure?
    a. protein kinase
    b. photoreceptor
    c. $P_r$ and $P_{fr}$
    d. All of these are correct.
31. Which of these is an indirect defense?
    a. secondary chemical
    b. making a predator think that butterfly eggs are already on the leaves
    c. inviting ants to live on a plant
    d. All of these are correct.

## thinking scientifically

1. You hypothesize that abscisic acid (ABA) is responsible for the turgor pressure changes that permit a plant to track the sun (see photograph on page 473). What observations could you make to support your hypothesis?
2. You formulate the hypothesis that the negative gravitropic response of stems is greater than the positive phototropism of stems. How would you test your hypothesis?

## bioethical issue

### Environmental Activism

The government paid the owner of Pacific Lumber some $500 million so that 3,500 acres of redwood trees along the coast of the Pacific Northwest would be preserved. Activists thought the deal was unfair—they wanted 60,000 acres preserved. Any less and there would not be enough habitat to keep the marbled murrelets, an endangered seabird, from sinking into oblivion. Besides, it would take hundreds of years for the trees to regrow to their original size.

Activists feel betrayed by their own representatives in the government. After all, the owner of Pacific Lumber is a big contributor to the reelection campaigns of elected officials who approved the deal. Under the circumstances, what would you do to save the trees? Some activists climbed the trees and refused to come down even while the trees were being cut. In September of 1998, David Chain, 24, lost his life when a tree fell on him and crushed his skull. What is the proper action for activist groups when they feel their government is letting them down? Should they defy the law, and if so, what is the proper response of government officials?

## *Biology* website

The companion website for *Biology* provides a wealth of information organized and integrated by chapter. You will find practice tests, animations, videos, and much more that will complement your learning and understanding of general biology.

**http://www.mhhe.com/maderbiology10**

# 27

# Flowering Plants: Reproduction

*The flowering plants, or angiosperms, are the most diverse and widespread of all the plants, and their means of sexual reproduction, centered in the flower, is well adapted to life on land. The flower shields the female gametophyte and produces pollen grains that protect a male gametophyte until fertilization takes place. The presence of an ovary allows angiosperms to produce seeds within fruits. Seeds guard the embryo until conditions are favorable for germination.*

*The evolution of the flower permits pollination not only by wind but also by animals. Flowering plants that rely on animals for pollination have a mutualistic relationship with them. The flower provides nutrients for a pollinator such as a bee, a fly, a beetle, a bird, or even a bat. The animals, in turn, inadvertently carry pollen from one flower to another, allowing pollination to occur. Similarly, animals help flowering plants disperse their fruits and, therefore, seeds. As we learn in this chapter, the diversity of flowering plants can, in part, be attributed to their relationships with a great variety of animals.*

Honeybee on landing platform of a yellow aster.

## concepts

# 27.1 Sexual Reproductive Strategies

As in animals, sexual reproduction in plants is advantageous because it can generate variety among the offspring due to the process of meiosis and fertilization. In a changing environment, a new variety may be better adapted for survival and reproduction than either parent.

## Life Cycle Overview

When plants reproduce sexually, they alternate between two multicellular stages. As also described on page 412, a diploid stage alternates with a haploid one. (1) In flowering plants, the diploid sporophyte is dominant, and it is the generation that bears flowers (Fig. 27.1) (2) A **flower,** which is the reproductive structure of angiosperms, produces two types of spores by meiosis, microspores and megaspores. (3) A **microspore** [Gk. *mikros*, small, little] undergoes mitosis and becomes a pollen grain, which is either windblown or carried by an animal to the vicinity of the female gametophyte. (4) In the meantime, the **megaspore** [Gk. *megas*, great, large] has undergone mitosis to become the female gametophyte, an embryo sac located within an ovule found within an ovary. (5) At maturity, a pollen grain contains nonflagellated sperm, which travel by way of a pollen tube to the embryo sac. (6) Once a sperm fertilizes an egg, the zygote becomes an embryo, still within an ovule. (7) The ovule develops into a **seed,** which contains the embryo and stored food surrounded by a seed coat. The ovary becomes a fruit, which aids in dispersing the seeds. (8) When a seed germinates, a new sporophyte emerges and through mitosis and growth becomes a mature organism.

Notice that the sexual life cycle of flowering plants is adapted to a land existence. The microscopic female gametophytes develop completely within the sporophyte and are thereby protected from desiccation. Pollen grains (male gametophytes) are not released until they develop a thick wall. No external water is needed to bring about fertilization in flowering plants. Instead, the pollen tube provides passage for a sperm to reach an egg. Following fertilization, the embryo and its stored food are enclosed within a protective seed coat until external conditions are favorable for germination.

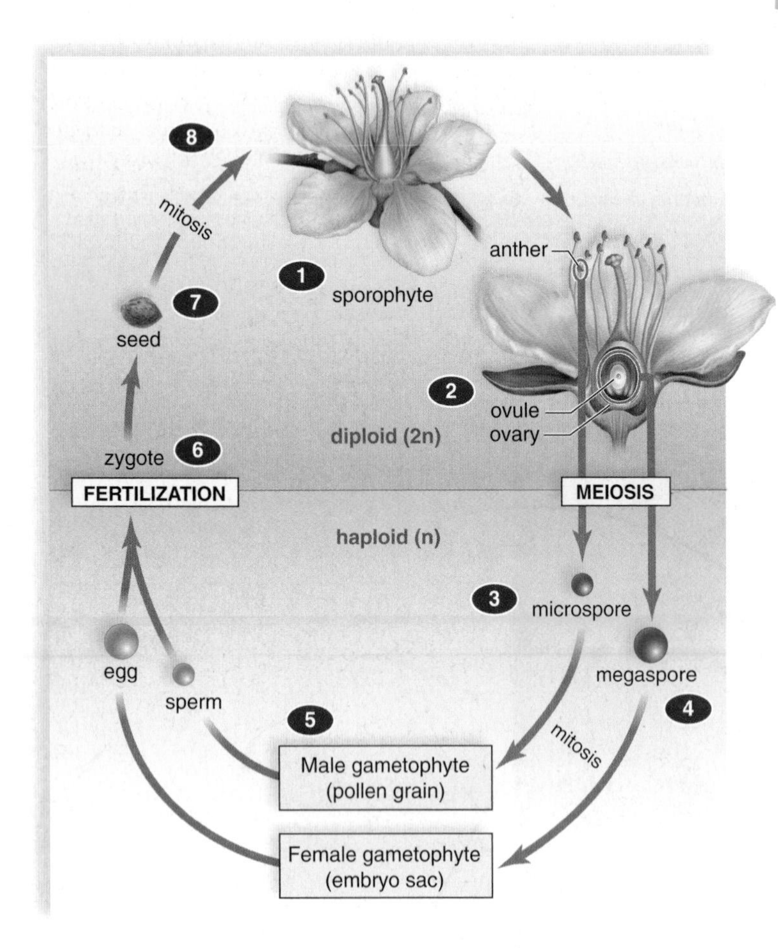


**FIGURE 27.1 Sexual reproduction in flowering plants.**

The sporophyte bears flowers. The flower produces microspores within anthers and megaspores within ovules by meiosis. A megaspore becomes a female gametophyte, which produces an egg within an embryo sac, and a microspore becomes a male gametophyte (pollen grain), which produces sperm. Fertilization results in a zygote and sustenance for the embryo. A seed contains an embryo and stored food within a seed coat. After dispersal, a seed becomes a new sporophyte plant.

## Flowers

The flower is unique to angiosperms (Fig. 27.2). Aside from producing the spores and protecting the gametophytes, flowers often attract pollinators, which aid in transporting pollen from plant to plant. Flowers also produce the fruits that enclose the seeds. The evolution of the flower was a major factor leading to the success of angiosperms, with over 240,000 species. Flowering is often a response to environmental signals such as the length of the day (see page 486). In many plants, a flower develops when shoot apical meristem that previously formed leaves suddenly stops producing leaves and starts producing a

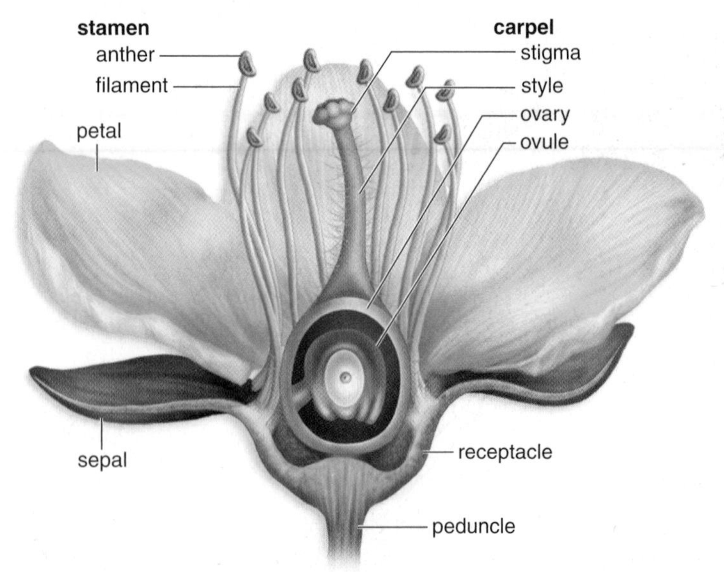


**FIGURE 27.2 Anatomy of a flower.**

A complete flower has all flower parts: sepals, petals, stamens, and at least one carpel.

flower enclosed within a bud. In other plants, axillary buds develop directly into flowers. In monocots, flower parts occur in threes and multiples of three; in eudicots, flower parts are in fours or fives and multiples of four or five (Fig. 27.3).

## *Flower Structure*

A typical flower has four whorls of modified leaves attached to a receptacle at the end of a flower stalk called a peduncle.

1. The **sepals,** which are the most leaflike of all the flower parts, are usually green, and they protect the bud as the flower develops within. Collectively, the sepals are called the **calyx.**
2. An open flower next has a whorl of **petals,** whose color accounts for the attractiveness of many flowers. The size, the shape, and the color of petals are attractive to a specific pollinator. Wind-pollinated flowers may have no petals at all. Collectively, the petals are called the **corolla.**
3. **Stamens** are the "male" portion of the flower. Each stamen has two parts: the **anther,** a saclike container, and the **filament** [L. *filum,* thread], a slender stalk. Pollen grains develop from the microspores produced in the anther.
4. At the very center of a flower is the **carpel,** a vaselike structure that represents the "female" portion of the flower. A carpel usually has three parts: the **stigma,** an enlarged sticky knob; the **style,** a slender stalk; and the **ovary,** an enlarged base that encloses one or more **ovules** (see Fig. 27.2).

Ovules [L. *ovulum,* little egg] play a significant role in the production of megaspores and, therefore, female gametophytes. A flower can have a single carpel or multiple carpels. Sometimes several carpels are fused into a single structure, in which case the ovary is compound and has several chambers, each of which contains ovules. For example, an orange develops from a compound ovary, and every section of the orange is a chamber.

## *Variations in Flower Structure*

We have space to mention only a few variations in flower structure. Not all flowers have sepals, petals, stamens, or carpels. Those that do are said to be *complete* and those that do not are said to be *incomplete.* Flowers that have both stamens and carpels are called *perfect* (bisexual) flowers; those with only stamens and those that have only carpels are *imperfect* (unisexual) flowers. If staminate flowers and carpellate flowers are on one plant, the plant is *monoecious* [Gk. *monos,* one, and *oikos,* home, house] (Fig. 27.4). Corn is an example of a plant that is monoecious. If staminate and carpellate flowers are on separate plants, the plant is *dioecious.* Holly trees are dioecious, and if red berries are a priority, it is necessary to acquire a plant with staminate flowers and another plant with carpellate flowers.

a. Daylily, *Hemerocallis* sp.

b. Festive azalea, *Rhododendron* sp.

**FIGURE 27.3 Monocot versus eudicot flowers.**

**a.** Monocots, such as daylilies, have flower parts usually in threes. In particular, note the three petals and three sepals, both of which are colored in this flower. **b.** Azaleas are eudicots. They have flower parts in fours or fives; note the five petals of this flower. p = petal, s = sepal

a. Staminate flowers

b. Carpellate flowers

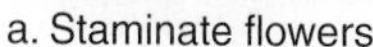

**FIGURE 27.4 Corn plants are monoecious.**

A corn plant has clusters of staminate flowers (**a**) and carpellate flowers (**b**). Staminate flowers produce the pollen that is carried by wind to the carpellate flowers, where ears of corn develop.

## Life Cycle in Detail

In all land plants, the sporophyte produces haploid spores by meiosis. The haploid spores grow and develop into haploid gametophytes, which produce gametes by mitotic division. Flowering plants, however, are *heterosporous*—they produce microspores and megaspores. Microspores become mature male gametophytes (sperm-bearing pollen grains), and megaspores become mature female gametophytes (egg-bearing embryo sacs).

### Development of Male Gametophyte

Microspores are produced in the anthers of flowers (Fig. 27.5). An anther has four pollen sacs, each containing many microspore mother cells. A microspore mother cell undergoes meiosis to produce four haploid microspores. In each, the haploid nucleus divides mitotically, followed by unequal cytokinesis, and the result is two cells enclosed by a finely sculptured wall. This structure, called the **pollen grain,** is at first an immature **male gametophyte** that consists of a tube cell and a generative cell. The larger tube cell will eventually produce a *pollen tube.* The smaller generative cell divides mitotically either now or later to produce two sperm. Once these events take place, the pollen grain has become the mature male gametophyte.

### Development of Female Gametophyte

The ovary contains one or more ovules. An ovule has a central mass of parenchyma cells almost completely covered by layers of tissue called integuments except where there is an opening, the micropyle. One parenchyma cell enlarges to become a megaspore mother cell, which undergoes meiosis, producing four haploid megaspores (Fig. 27.5). Three of these megaspores are nonfunctional, and one is functional. In a typical pattern, the nucleus of the functional megaspore

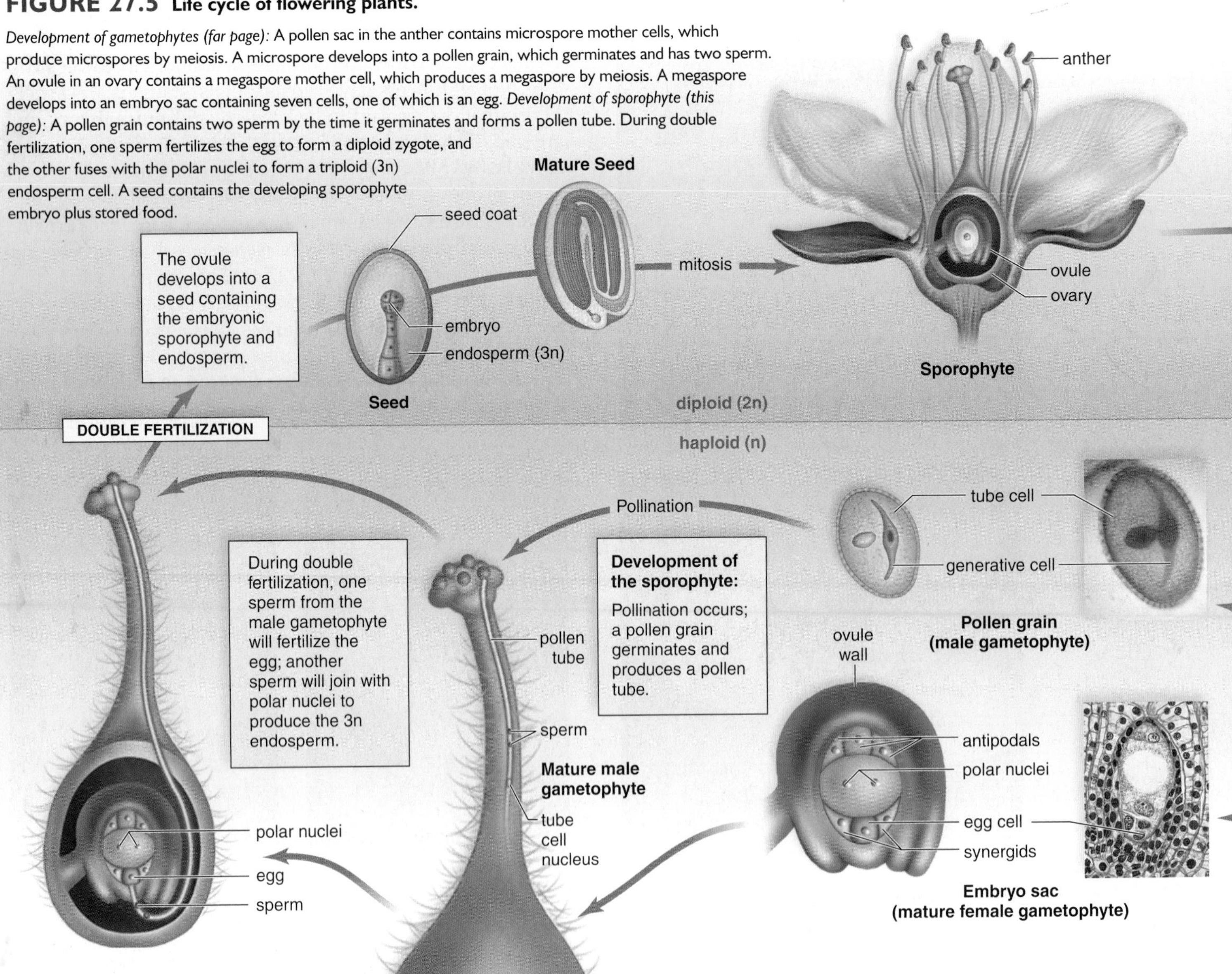


**FIGURE 27.5 Life cycle of flowering plants.**

*Development of gametophytes (far page):* A pollen sac in the anther contains microspore mother cells, which produce microspores by meiosis. A microspore develops into a pollen grain, which germinates and has two sperm. An ovule in an ovary contains a megaspore mother cell, which produces a megaspore by meiosis. A megaspore develops into an embryo sac containing seven cells, one of which is an egg. *Development of sporophyte (this page):* A pollen grain contains two sperm by the time it germinates and forms a pollen tube. During double fertilization, one sperm fertilizes the egg to form a diploid zygote, and the other fuses with the polar nuclei to form a triploid (3n) endosperm cell. A seed contains the developing sporophyte embryo plus stored food.

divides mitotically until there are eight nuclei in the **female gametophyte** [Gk. *megas*, great, large]. When cell walls form later, there are seven cells, one of which is binucleate. The female gametophyte, also called the **embryo sac,** consists of these seven cells:

- one egg cell, associated with two synergid cells;
- one central cell, with two polar nuclei;
- three antipodal cells

## *Development of New Sporophyte*

The walls separating the pollen sacs in the anther break down when the pollen grains are ready to be released. **Pollination** is simply the transfer of pollen from an anther to the stigma of a carpel. Self-pollination occurs if the pollen is from the same plant, and cross-pollination occurs if the pollen is from a different plant of the same species.

**FIGURE 27.6 Pollination.**

**a.** Cocksfoot grass, *Dactylus glomerata*, releasing pollen. **b.** Pollen grains of Canadian goldenrod, *Solidago canadensis*. **c.** Pollen grains of pussy willow, *Salix discolor*. The shape and pattern of pollen grain walls are quite distinctive, and experts can use them to identify the genus, and even sometimes the species, that produced a particular pollen grain. Pollen grains have strong walls resistant to chemical and mechanical damage; therefore, they frequently become fossils.

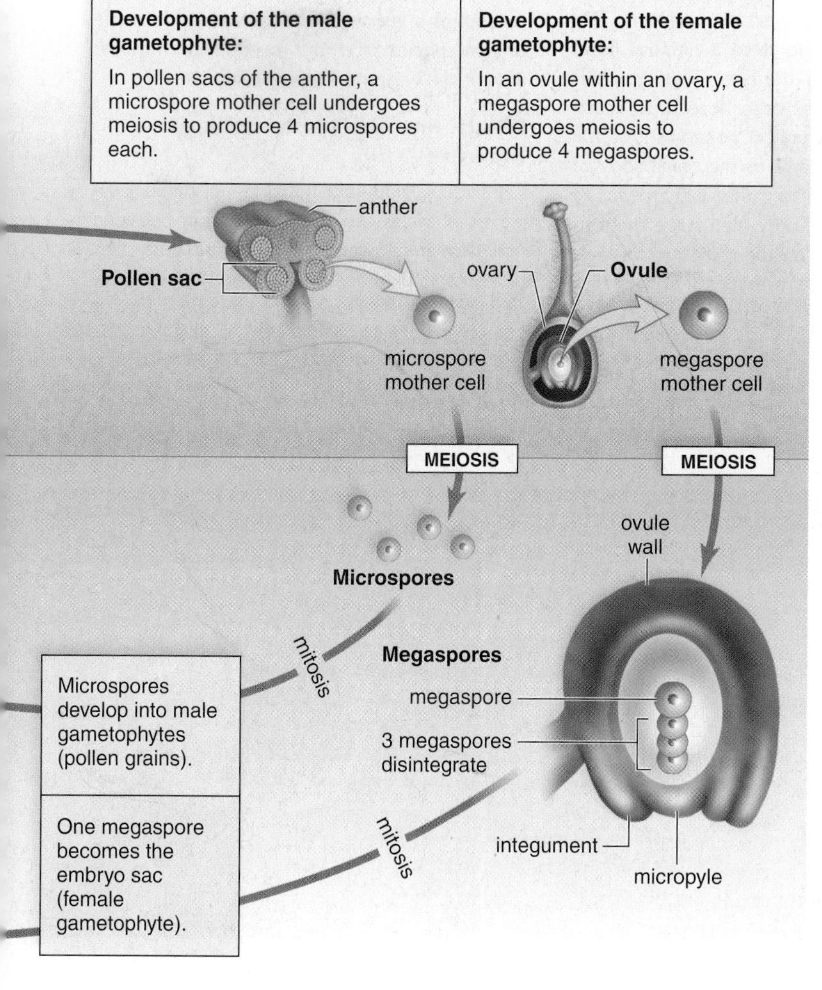


When a pollen grain lands on the stigma of the same species, it germinates, forming a pollen tube (see Fig. 27.5). The germinated pollen grain, containing a tube cell and two sperm, is the mature male gametophyte. As it grows, the pollen tube passes between the cells of the stigma and the style to reach the micropyle, a pore of the ovule. When the pollen tube reaches the micropyle, **double fertilization** occurs. As expected, one of the sperm unites with the egg, forming a 2n zygote. Unique to angiosperms, however, the other sperm unites with two polar nuclei centrally placed in the embryo sac, forming a 3n endosperm nucleus. This endosperm nucleus eventually develops into the **endosperm** [Gk. *endon*, within, and *sperma*, seed], a nutritive tissue that the developing embryonic sporophyte will use as an energy source. Now the ovule begins to develop into a seed. One important aspect of seed development is formation of the seed coat from the ovule wall. A mature seed contains (1) the embryo, (2) stored food, and (3) the seed coat (see Fig. 27.8).

## *Cross Pollination*

Some species of flowering plants—for example, the grasses and grains—rely on wind pollination (Fig. 27.6), as do the gymnosperms, the other type of seed plant. Much of the plant's energy goes into making pollen to ensure that some pollen grains actually reach a stigma. Even the amount successfully transferred is staggering: A single corn plant may produce from 20 to 50 million grains a season. In corn, the flowers tend to be monoecious, and clusters of tiny male flowers move in the wind, freely releasing pollen into the air. Most angiosperms rely on animals—be they insects (e.g., bumblebees, flies, butterflies, and moths), birds (e.g., hummingbirds), or mammals (e.g., bats)—to carry out pollination. The use of animal pollinators is unique to flowering plants, and it helps account for why these plants are so successful on land. By the time flowering plants appear in the fossil record some 135 MYA, insects had long been present. For millions of years, then, plants and their animal pollinators have coevolved. **Coevolution** means that as one species changes, the other changes too, so that in the end, the two species are suited to one another. Plants with flowers that attracted a pollinator enjoyed an advantage because, in the end, they produced more seeds. Similarly, pollinators that were able to find and remove food from the flower were more successful. Today, we see that the reproductive parts of the flower are positioned so that the pollinator can't help but pick up pollen from one flower and deliver it to another. On the other hand, the mouthparts of the pollinator are suited to gathering the nectar from these particular plants.

*science focus*

## Plants and Their Pollinators

A plant and its pollinator(s) are adapted to one another. They have a mutualistic relationship in which each benefits—the plant uses its pollinator to ensure that cross-pollination takes place, and the pollinator uses the plant as a source of food. This mutualistic relationship came about through the process of coevolution—that is, the codependency of the plant and the pollinator is the result of suitable changes in the structure and function of each. The evidence for coevolution is observational. For example, floral coloring and odor are suited to the sense perceptions of the pollinator; the mouthparts of the pollinator are suited to the structure of the flower; the type of food provided is suited to the nutritional needs of the pollinator; and the pollinator forages at the time of day that specific flowers are open. The following are examples of such coevolution.

### Bee-Pollinated Flowers

There are 20,000 known species of bees that pollinate flowers. The best-known pollinators are the honeybees (Fig. 27A*a*). Bee eyes see a spectrum of light that is different from the spectrum seen by humans. The bees' visible spectrum is shifted so that they do not see red wavelengths but do see ultraviolet wavelengths. Bee-pollinated flowers are usually brightly colored and are predominantly blue or yellow; they are not entirely red. They may also have ultraviolet shadings called nectar guides, which highlight the portion of the flower that contains the reproductive structures. The mouthparts of bees are fused into a long tube that contains a tongue. This tube is an adaptation for sucking up nectar provided by the plant, usually at the base of the flower.

Bee flowers are delicately sweet and fragrant to advertise that nectar is present. The nectar guides often point to a narrow floral tube large enough for the bee's feeding apparatus but too small for other insects to reach the nectar. Bees also collect pollen as food for their larvae. Pollen clings to the hairy body of a bee, and the bees also gather it by means of bristles on their legs. They then store the pollen in pollen baskets on the third pair of legs. Bee-pollinated flowers are sturdy and irregular in shape because they often have a landing platform where the bee can alight. The landing platform requires the bee to brush up against the anther and stigma as it moves toward the floral tube to feed. One type of orchid, *Ophrys*, has evolved a unique adaptation. The flower resembles a female wasp, and when the male of that species attempts to copulate with the flower, the wasp receives pollen.

### Moth- and Butterfly-Pollinated Flowers

Contrasting moth- and butterfly-pollinated flowers emphasizes the close adaptation between pollinator and flower. Both moths and butterflies have a long, thin, hollow proboscis, but they differ in other characteristics. Moths usually feed at night and have a well-developed sense of smell. The flowers they visit are visible at night because they are lightly shaded (white, pale yellow, or pink), and they have strong, sweet perfume,

a.

b.

**FIGURE 27A Pollinators.**
***a.*** *A bee-pollinated flower is a color other than red (bees cannot detect this color) and has a landing platform where the reproductive structures of the flower brush up against the bee's body.* ***b.*** *A butterfly-pollinated flower is often a composite, containing many individual flowers. The broad expanse provides room for the butterfly to land, after which it lowers its proboscis into each flower in turn.*

Many examples of the coevolution between plants and their pollinators are given in the Science Focus on these pages. Here, we can note that bee-pollinated flowers are usually yellow, blue, or white because these are the colors bees can see. Bees respond to ultraviolet markings called nectar guides that help them locate nectar. Humans do not use ultraviolet light in order to see, but bees are sensitive to ultraviolet light. A bee has a feeding proboscis of the right length to collect nectar from certain flowers and a pollen basket on its hind legs that allows it to carry pollen back to the hive. Because many fruits and vegetables are dependent on bee pollination, there is much concern today that the number of bees is declining due to disease and the use of pesticides.

### Check Your Progress 27.1

1. What specific part of a flower produces male gametophytes? Female gametophytes?
2. Contrast the development and structure of the male gametophyte and the female gametophyte.
3. What are the products of double fertilization in angiosperms?

which helps attract moths. Moths hover when they feed, and their flowers have deep tubes with open margins that allow the hovering moths to reach the nectar with their long proboscis. Butterflies are active in the daytime and have good vision but a weak sense of smell. Their flowers have bright colors—even red because butterflies can see the color red—but the flowers tend to be odorless. Unable to hover, butterflies need a place to land. Flowers that are visited by butterflies often have flat landing platforms (Fig. 27A*b*). Composite flowers (composed of a compact head of numerous individual flowers) are especially favored by butterflies. Each flower has a long, slender floral tube, accessible to the long, thin butterfly proboscis.

#### Bird- and Bat-Pollinated Flowers

In North America, the most well-known bird pollinators are the hummingbirds. These tiny animals have good eyesight but do not have a well-developed sense of smell. Like moths, they hover when they feed. Typical flowers pollinated by hummingbirds are red, with a slender floral tube and margins that are curved back and out of the way. And although they produce copious amounts of nectar, the flowers have little odor. As a hummingbird feeds on nectar with its long, thin beak, its head comes into contact with the stamens and pistil (Fig. 27B*a*).

Bats are adapted to gathering food in various ways, including feeding on the nectar and pollen of plants. Bats are nocturnal and have an acute sense of smell. Those that are pollinators also have keen vision and a long, extensible, bristly tongue. Typically, bat-pollinated flowers open only at night and are light-colored or white. They have a strong, musty smell similar to the odor that bats produce to attract one another. The flowers are generally large and sturdy and are able to hold up when a bat inserts part of its head to reach the nectar. While the bat is at the flower, its head is dusted with pollen (Fig. 27B*b*).

#### Coevolution

There are many examples of plant and animal coevolution, but how did this coevolution come about? Some 200 million years ago, when seed plants were just beginning to evolve and insects were not as diverse as they are today, wind alone was used to carry pollen. Wind pollination, however, is a hit-or-miss affair. Perhaps beetles feeding on vegetative leaves were the first insects to carry pollen directly from plant to plant by chance. This use of animal motility to achieve cross-fertilization no doubt resulted in the evolution of flowers, which have features, such as the production of nectar, to attract pollinators. Then, if beetles developed the habit of feeding on flowers, other features, such as the protection of ovules within ovaries, may have evolved.

As cross-fertilization continued, more and more flower variations likely developed, and pollinators became increasingly adapted to specific angiosperm species. Today, there are some 240,000 species of flowering plants and over 900,000 species of insects. This diversity suggests that the success of angiosperms has contributed to the success of insects, and vice versa.

a.

b.

**FIGURE 27B More pollinators.**
**a.** *Hummingbird-pollinated flowers are curved back, allowing the bird to insert its beak to reach the rich supply of nectar. While doing this, the bird's forehead and other body parts touch the reproductive structures.* **b.** *Bat-pollinated flowers are large, sturdy flowers that can take rough treatment. Here the head of the bat is positioned so that its bristly tongue can lap up nectar.*

# 27.2 Seed Development

Development of the embryo within the seed is the next event in the life cycle of the angiosperm. Plant growth and development involves cell division, cell elongation, and differentiation of cells into tissues and then organs. Development is a programmed series of stages from a simple to a more complex form. Cellular differentiation, or specialization of structure and function, occurs as development proceeds.

## Stages of Eudicot Development

Figure 27.7 shows the stages of development for a eudicot embryo.

### *Zygote and Proembryo Stages*

1 Immediately after double fertilization, we can make out the zygote and the endosperm. The zygote (true green) is small with dense cytoplasm. 2 The zygote divides repeatedly in different planes, forming several cells called a proembryo. Also present is an elongated structure called a suspensor. The suspensor transfers and produces nutrients from the endosperm and these allow the embryo to grow.

### *Globular Stage*

3 During the globular stage, the proembryo is largely a ball of cells. The root-shoot axis of the embryo is already established at this stage because the embryonic cells near the suspensor will become a root, while those at the other end will ultimately become a shoot.

The outermost cells of the plant embryo will become dermal tissue. These cells divide with their cell plate perpendicular to the surface; therefore, they produce a single outer layer of cells. Recall that dermal tissue protects the plant from desiccation and includes the stomata, which open and close to facilitate gas exchange and minimize water loss.

### *The Heart Stage and Torpedo Stage Embryos*

4 The embryo has a heart shape when the **cotyledons,** or seed leaves, appear because of local, rapid cell division. 5 As the embryo continues to enlarge and elongate, it takes on a torpedo shape. Now the root and shoot apical meristems are distinguishable. The shoot apical meristem is responsible for aboveground growth, and the root apical meristem is responsible for underground growth. Ground meristem, which gives rise to the bulk of the embryonic interior, is also now present.

### *The Mature Embryo*

6 In the mature embryo, the epicotyl is the portion between the cotyledon(s) that contributes to shoot development. The plumule is found at the tip of the epicotyl and consists of the shoot tip and a pair of small leaves. The hypocotyl is the portion below the cotyledon(s). It contributes to stem development and terminates in the radicle or embryonic root.

The cotyledons are quite noticeable in a eudicot embryo and may fold over. Procambium is visible at the core of the embryo and is destined to form the future vascular tissue responsible for water and nutrient transport.

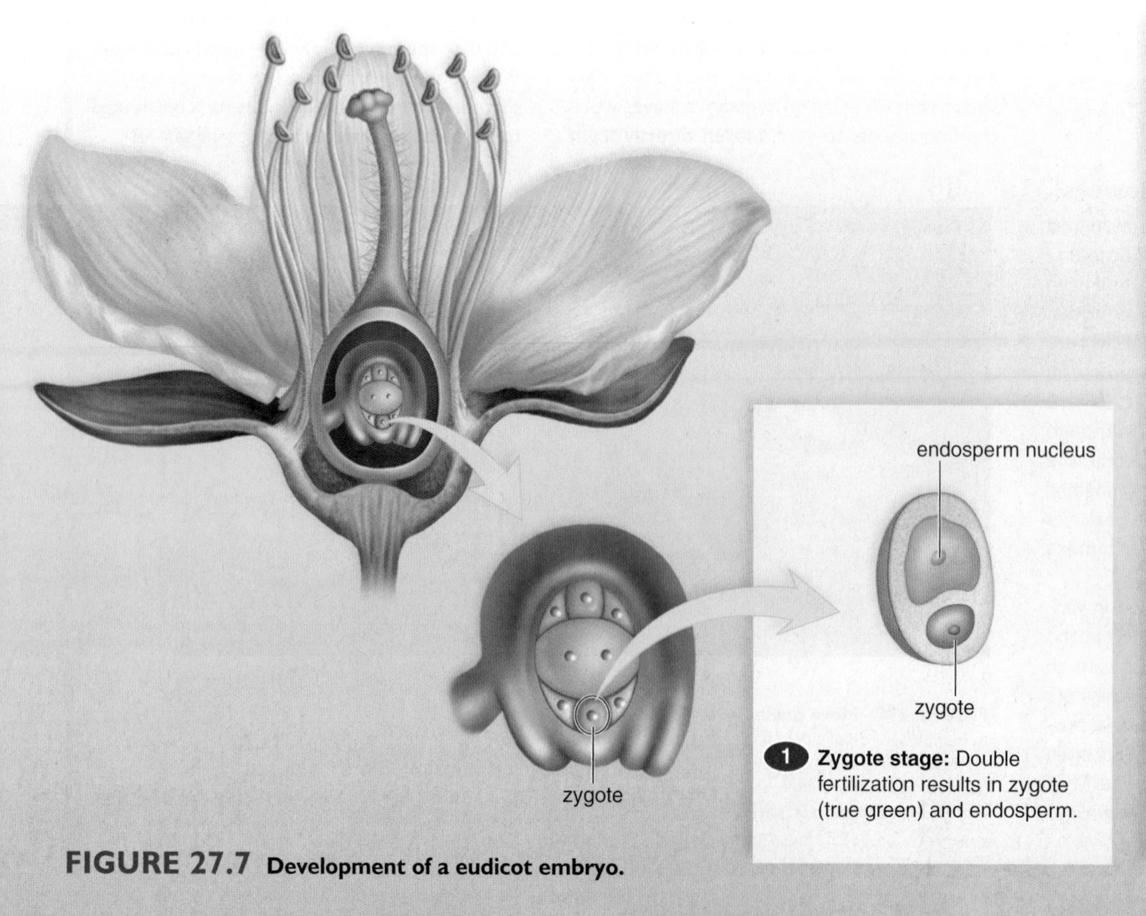


1 **Zygote stage:** Double fertilization results in zygote (true green) and endosperm.

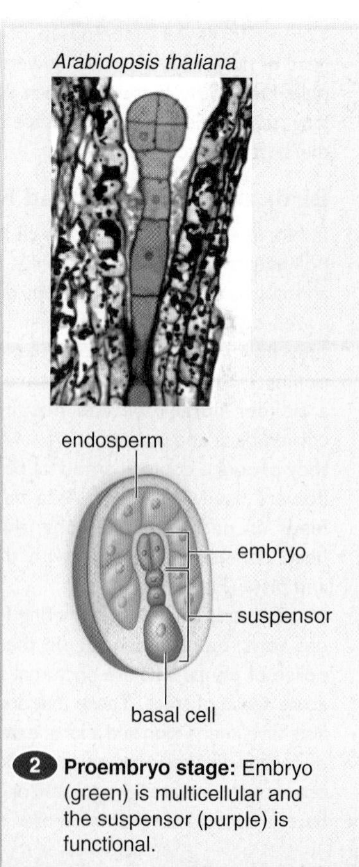


2 **Proembryo stage:** Embryo (green) is multicellular and the suspensor (purple) is functional.

**FIGURE 27.7 Development of a eudicot embryo.**

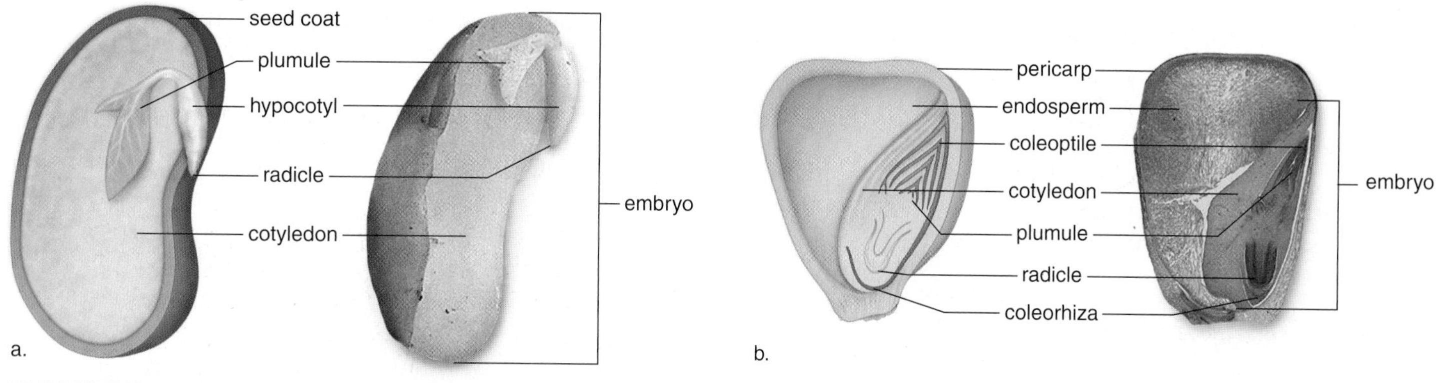


**FIGURE 27.8 Monocot versus eudicot.**

**a.** In a bean seed (eudicot), the endosperm has disappeared; the bean embryo's cotyledons take over food storage functions. The hypocotyl becomes the shoot system, which will include the plumule (first leaves). The radicle becomes the root system. **b.** The corn kernel (monocot) has endosperm that is still present at maturity. The coleoptile is a protective sheath for the shoot system; the coleorhiza similarly protects the future root system. The term pericarp is explained in the next section.

As the embryo develops, the integuments of the ovule become the seed coat. The seed coat encloses and protects the embryo and its food supply.

## Monocots Versus Eudicot Seeds

Monocots, unlike eudicots, have only one cotyledon. Another important difference between monocots and eudicots is the manner in which nutrient molecules are stored in the seed. In monocots, the cotyledon, in addition to storing certain nutrients, absorbs other nutrient molecules from the endosperm and passes them to the embryo. In eudicots, the cotyledons usually store all the nutrient molecules that the embryo uses. Therefore, in Figure 27.7 we can see that the endosperm seemingly disappears. Actually, it has been taken up by the two cotyledons.

Figure 27.8 contrasts the structure of a bean seed (eudicot) and a corn kernel (monocot). The size of seeds may vary from the dust-sized seeds of orchids to the 27-kg seed of the double coconut.

### Check Your Progress 27.2

1. Identify the origin of each of the three parts of a seed.
2. Why are both the seed coat and the embryo 2n?
3. What are cotyledons and what role do they play?

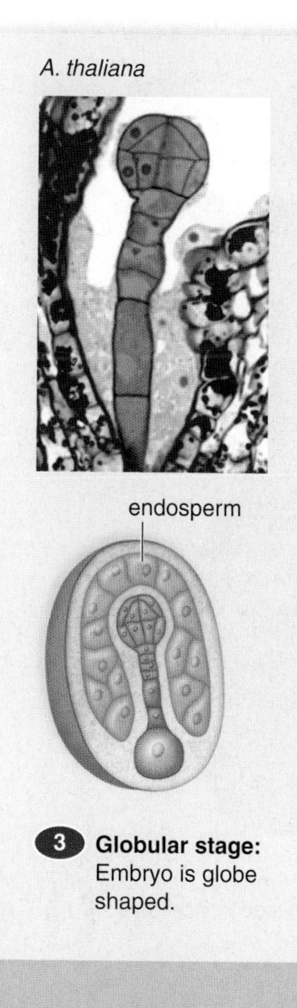


**3 Globular stage:** Embryo is globe shaped.

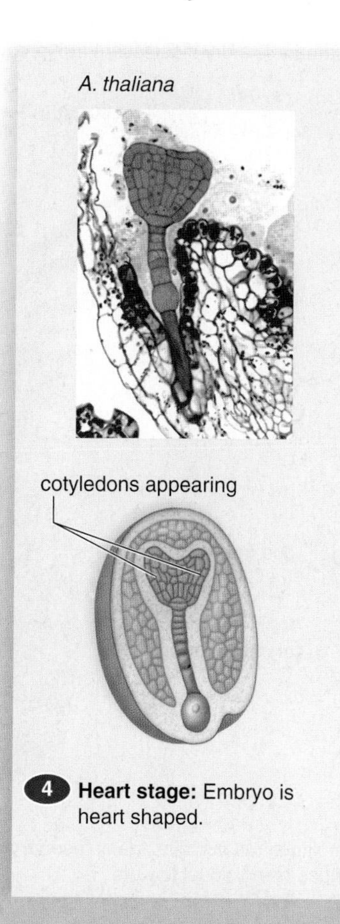


**4 Heart stage:** Embryo is heart shaped.

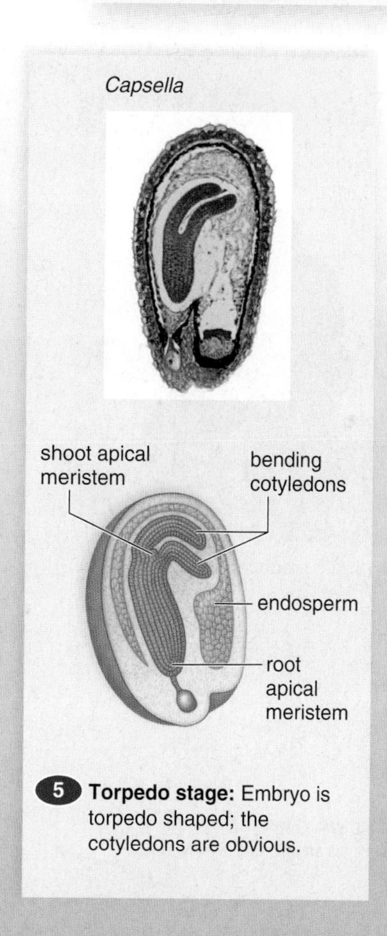


**5 Torpedo stage:** Embryo is torpedo shaped; the cotyledons are obvious.

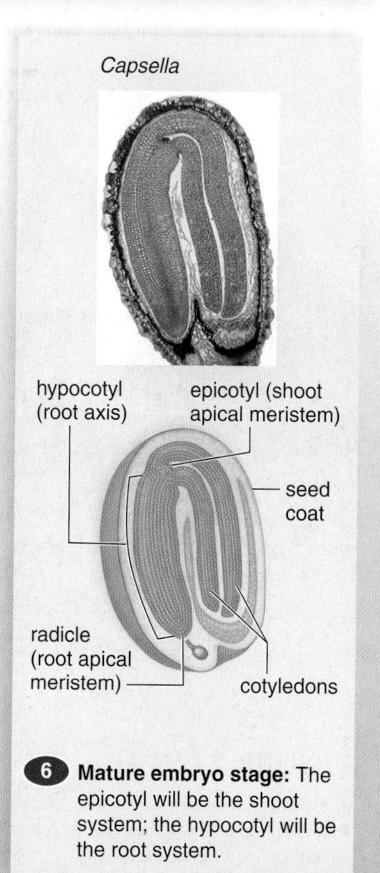


**6 Mature embryo stage:** The epicotyl will be the shoot system; the hypocotyl will be the root system.

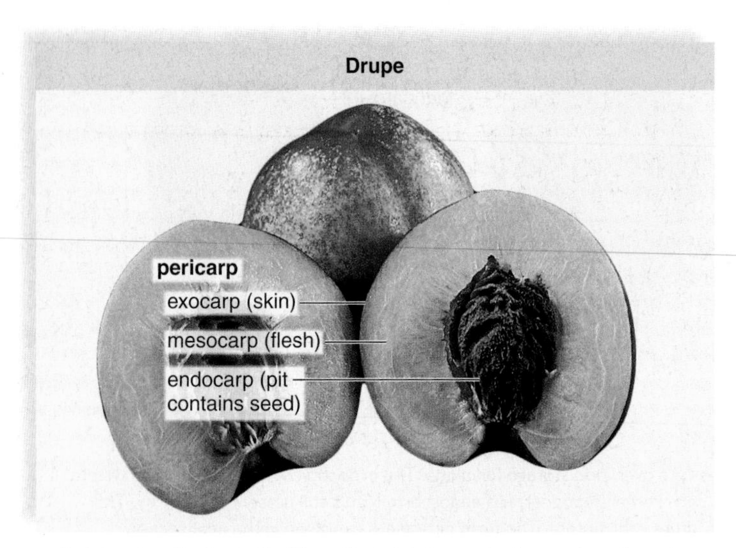


a. A drupe is a fleshy fruit with a pit containing a single seed produced from a simple ovary.

b. A berry is a fleshy fruit having seeds and pulp produced from a compound ovary.

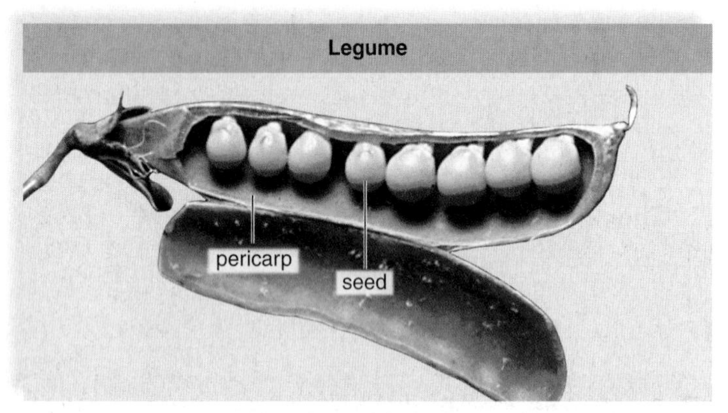


c. A legume is a dry dehiscent fruit produced from a simple ovary.

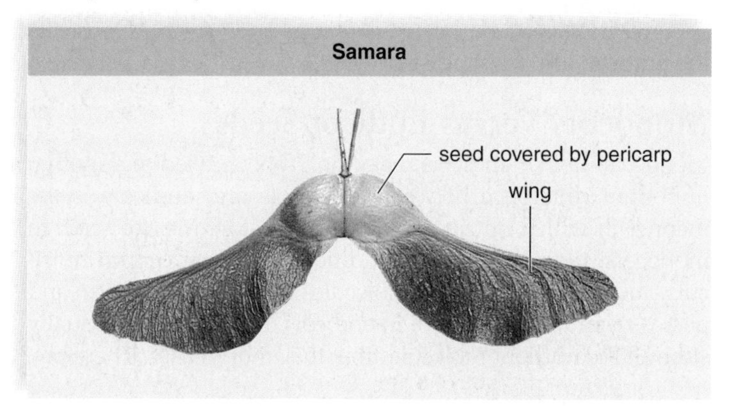


d. A samara is a dry indehiscent fruit produced from a simple ovary.

e. An aggregate fruit contains many fleshy fruits produced from simple ovaries of the same flower.

f. A multiple fruit contains many fused fruits produced from simple ovaries of individual flowers.

**FIGURE 27.9 Fruits.**

**a.** A peach is a drupe. **b.** A tomato is a true berry. **c.** A pea is a legume. **d.** The fruit of a maple tree is a samara. **e.** A blackberry is an aggregate fruit. **f.** A pineapple is a multiple fruit.

# 27.3 Fruit Types and Seed Dispersal

Most of us have to become accustomed to the botanical definition of a fruit (Table 27.1). A **fruit** is a mature ovary and sometimes, in addition, other flower parts, such as the receptacle. This means that pea pods, tomatoes, and what is usually called winged maple seeds are actually fruits (Fig. 27.9). Fruits protect and help disperse seeds. Some fruits are better at one of these functions than the other. The fruit of a peach protects the seed well, but the pit may make it difficult for germination to occur. Peas easily escape from pea pods, but once they are free, they are protected only by the seed coat.

## Kinds of Fruits

Fruits can be simple or compound (Fig. 27.9). A *simple fruit* is derived from a single ovary that can have one or several chambers (Fig. 27.9*a–d*). A *compound fruit* is derived from several groups of ovaries (Fig. 27.9*e–f*). If a single flower had many ovaries, as in a blackberry, then the fruit is an *aggregate* one (Fig. 27.9*e*). In contrast, a pineapple comes from many individual ovaries. Because the flowers had only one receptacle, the ovaries fused to form a large, *multiple fruit* (Fig. 27.9*f*).

As a fruit develops, the ovary wall thickens to become the pericarp, which can have as many as three layers: exocarp, mesocarp, and endocarp.

- The exocarp forms the outermost skin of a fruit.
- The mesocarp is often the fleshy tissue between the exocarp and endocarp of the fruit.
- The endocarp serves as the boundary around the seed(s). The endocarp may be hard, as in peach pits, or papery, as in apples.

Some fruits such as legumes and cereal grains of wheat, rice, and corn are *dry fruits.* The fruits of grains can be mistaken for seeds because a dry pericarp adheres to the seed within. Legumes such as the pea pod (Fig. 27.9*c*) are dehiscent because they split open when ripe. Grains are indehiscent—they don't split open. Humans gather grains before they are released from the plant and then process them to acquire their nutrients. You might be more familiar with fleshy fruits, such as the peach and tomato. In these fruits, the mesocarp is well developed.

**TABLE 27.1**

**Fruit Classification Based on Composition and Texture**

| |
|---|
| *Composition (based on type and arrangement of ovaries and flowers)* |
| Simple: develops from a simple ovary or compound ovary |
| Compound: develops from a group of ovaries |
| Aggregate: ovaries are from a single flower on one receptacle |
| Multiple: ovaries are from separate flowers on a common receptacle |
| *Texture (based on mature pericarp)* |
| Fleshy: the entire pericarp or portions of it are soft and fleshy at maturity |
| Dry: the pericarp is papery, leathery, or woody when the fruit is mature |
| Dehiscent: the fruit splits open when ripe |
| Indehiscent: the fruit does not split open when ripe |

## Dispersal of Fruits

Many dry fruits are dispersed by wind. Woolly hairs, plumes, and wings are all adaptations for this type of dispersal. The somewhat heavier dandelion fruit uses a tiny "parachute" for dispersal. Milkweed pods split open to release seeds that float away on puffy white threads. The winged fruit of a maple tree has been known to travel up to 10 km from its parent. Other fruits depend on animals for dispersal.

### *Dispersal by Animals*

When ripe, *fleshy colorful fruits,* such as peaches and cherries, often attract animals and provide them with food (Fig. 27.10*a*). Their hard endocarp protects the seed so it can pass through the digestive system of an animal and remain unharmed. As the flesh of a tomato is eaten, the small size of the seeds and the slippery seed coat means that tomato seeds rarely get crushed by the teeth of animals. The seeds swallowed by birds and mammals are defecated (passed out of the digestive tract with the feces) some distance from the parent plant. Squirrels and other animals that gather seeds and fruits bury them some distance away and may even forget where they have been stored. The hooks and spines of clover, bur, and cocklebur attach a dry fruit to the fur of animals and the clothing of humans (Fig. 27.10*b*).

a.

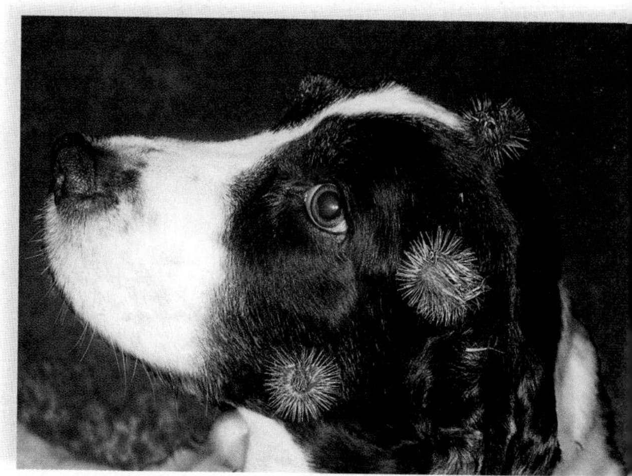

b.

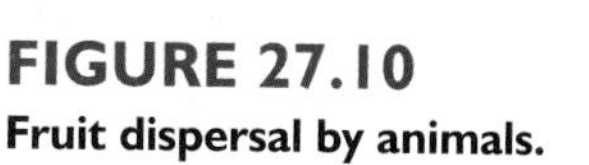

**FIGURE 27.10**

**Fruit dispersal by animals.**

**a.** When birds eat fleshy fruits, seeds pass through their digestive system. **b.** Burdock, a dry fruit, clings to the fur of animals.

## Seed Germination

Following dispersal, if conditions are right, seeds may **germinate** to form a seedling. Germination doesn't usually take place until there is sufficient water, warmth, and oxygen to sustain growth. These requirements help ensure that seeds do not germinate until the most favorable growing season has arrived. Some seeds do not germinate until they have been dormant for a period of time. For seeds, *dormancy* is the time during which no growth occurs, even though conditions may be favorable for growth. In the temperate zone, seeds often have to be exposed to a period of cold weather before dormancy is broken. Fleshy fruits (e.g., apples, pears, oranges, and tomatoes) contain inhibitors so that germination does not occur while the fruit is still on the plant. For seeds to take up water, bacterial action and even fire may be needed. Once water enters, the seed coat bursts and the seed germinates.

If the two cotyledons of a bean seed are parted, the rudimentary plant with immature leaves is exposed (Fig. 27.11*a*). As the eudicot seedling starts to form, the root emerges first. The shoot is hook-shaped to protect the immature leaves as they emerge from the soil. The cotyledons provide the new seedlings with enough energy for the stem to straighten and the leaves to grow. As the mature leaves of the plant begin photosynthesizing, the cotyledons shrivel up.

A corn kernel is actually a fruit, and therefore its outer covering is the pericarp and seed coat combined (Fig. 27.11*b*). Inside is the single cotyledon. Also, the immature leaves and the radicle are covered, respectively, by a coleoptile and a coleorhiza. These sheaths are discarded as the root grows directly downward into the soil and the shoot of the seedling begins to grow directly upward. This completes our discussion of sexual reproduction in flowering plants.

### Check Your Progress 27.3

1. How does a dry fruit differ in structure and dispersal from a fleshy fruit?
2. Contrast how a eudicot seedling and a monocot seedling protect the first true leaves.

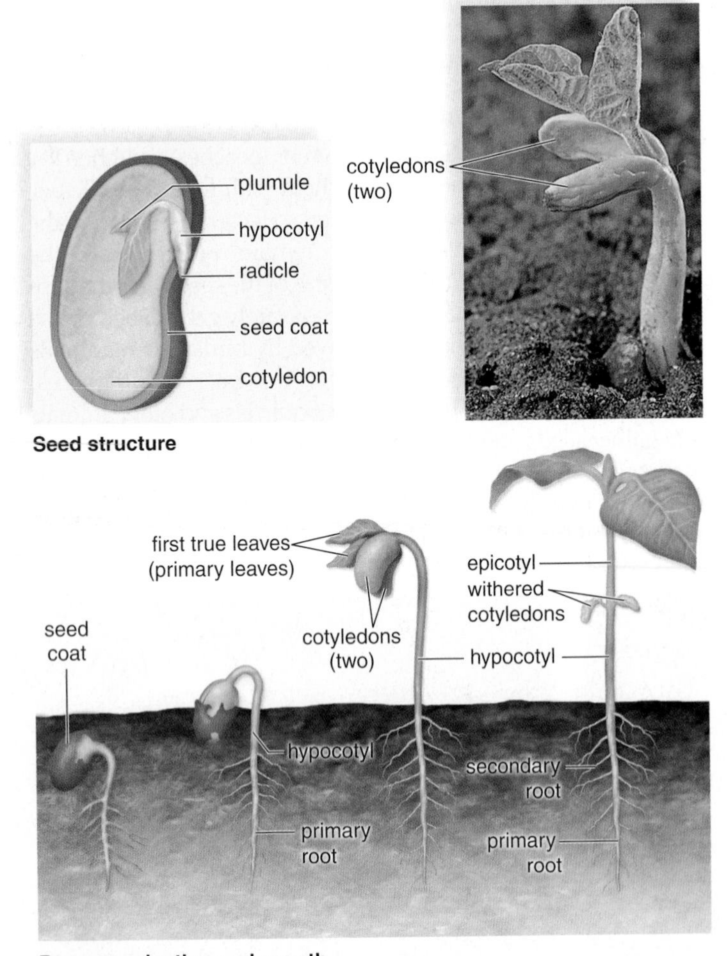


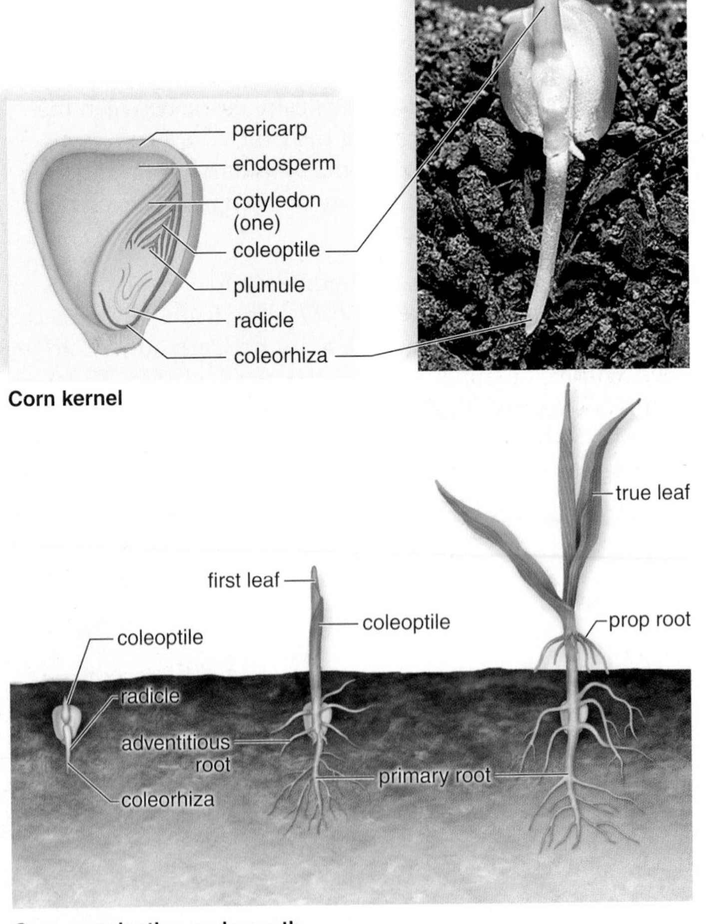


**FIGURE 27.11 Eudicot and monocot seed structure and germination.**

**a.** Bean (eudicot) seed structure and germination. **b.** Corn (monocot) kernel structure and germination.

# 27.4 Asexual Reproductive Strategies

Asexual reproduction is the production of an offspring identical to a single parent. Asexual reproduction is less complicated in plants because pollination and seed production is not required. Therefore, it can be advantageous when the parent is already well adapted to a particular environment and the production of genetic variations is not an apparent necessity.

Figure 27.12 features a strawberry plant that has produced a stolon. **Stolons** are horizontal stems that can be seen because they run aboveground. As you know, the nodes of stems are regions where new growth can occur. In the case of stolons, a new shoot system appears above the node, and a new root system appears below the node. The larger plant on the left is the parent plant, and the smaller plant on the right is the asexual offspring that has arisen at a node. The characteristics of new offspring produced by stolons are identical to those of the parent plant.

**Rhizomes** are underground stems that produce new plants asexually. Irises are examples of plants that have no aboveground stem because their main stem is a rhizome that grows horizontally underground. As with stolons, new plants arise at the nodes of a rhizome. White potatoes are expanded portions of a rhizome branch, and each eye is a bud that will produce a new potato plant if it is planted with a portion of the swollen tuber. Sweet potatoes are modified roots; they can be propagated by planting sections of the root.

You may have noticed that the roots of some fruit trees, such as cherry and apple trees, produce "suckers," small plants that can be used to grow new trees. In addition, pineapple, sugarcane, azalea, gardenia, and many other food and ornamental plants have been propagated from stem cuttings. In these plants, a cut stem will automatically produce roots. The discovery that the plant hormone auxin can cause roots to develop has expanded the list of plants that can be propagated from stem cuttings.

## Tissue Culture of Plants

**Tissue culture** is the growth of a tissue in an artificial liquid or solid culture medium. Somatic embryogenesis, meristem tissue culture, and anther tissue culture are three methods of cloning plants due to the ability of plants to grow from single cells. Many plant cells are **totipotent,** which means that each one has the genetic capability of becoming an entire plant.

During *somatic embryogenesis,* hormones are added to the medium and they cause leaf or other tissue cells to generate small masses of cells, which can be genetically engineered before being allowed to become many new identical plants.

**FIGURE 27.12**
**Asexual reproduction in plants.**
Meristem tissues at nodes can generate new plants, as when the stolons of strawberry plants, *Fragaria*, give rise to new plants.

Thousands of little "plantlets" can be produced by using this method of plant tissue culture (Fig. 27.13). Many important crop plants, such as tomato, rice, celery, and asparagus, as well as ornamental plants such as lilies, begonias, and African violets, have been produced using somatic embryogenesis. Plants generated from somatic embryos are not always genetically identical clones. They can vary because of mutations that arise spontaneously during the production process. These mutations, called *somaclonal variations*, are another way to produce new plants with desirable traits. Somatic embryos can be encapsulated in hydrated gel, creating artificial "seeds" that can be shipped anywhere.

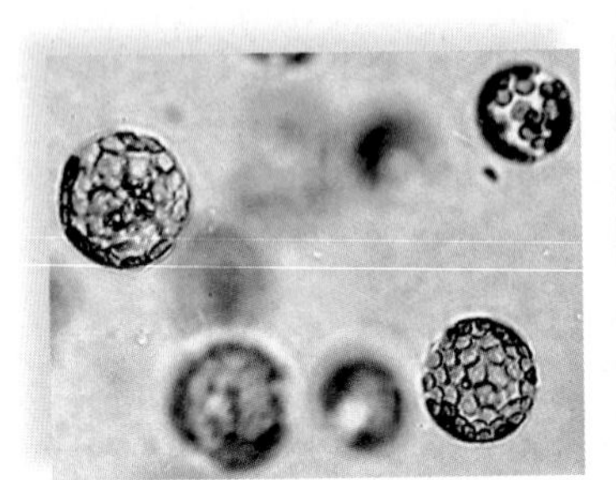

a. Protoplasts, naked cells

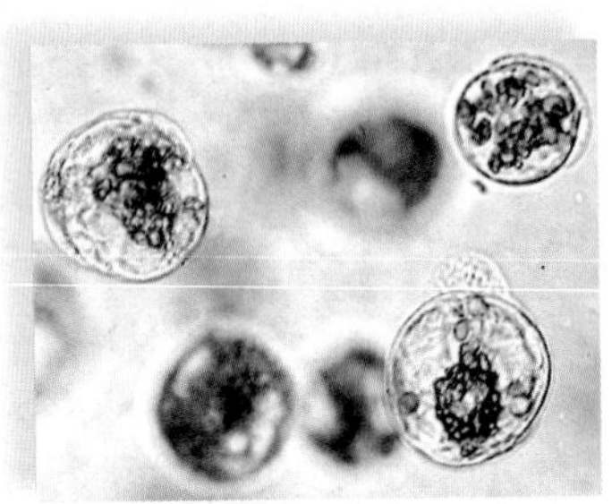

b. Cell wall regeneration

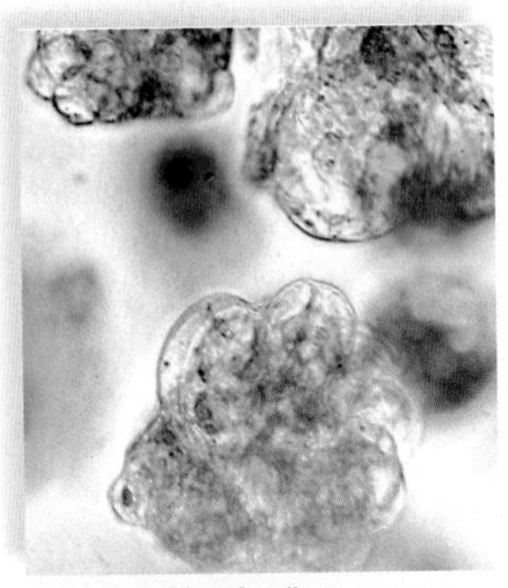

c. Aggregates of cells

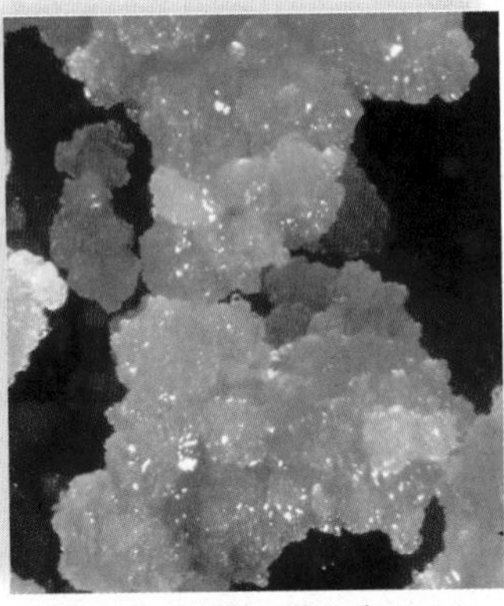

d. Callus, undifferentiated mass

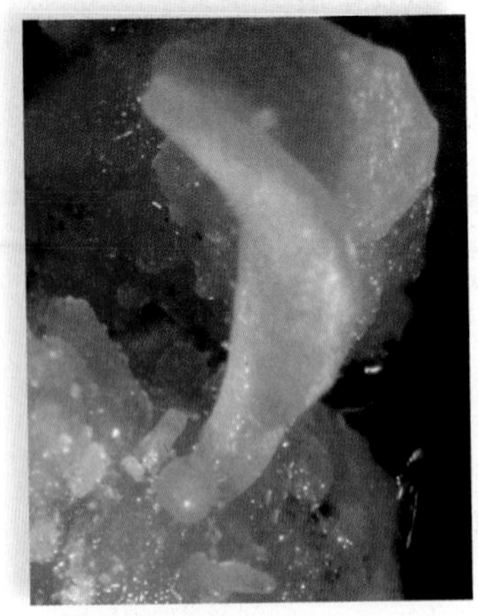

e. Somatic embryo

f. Plantlet

**FIGURE 27.13 Asexual reproduction through tissue culture.**

**a.** When plant cell walls are removed by digestive enzyme action, the result is naked cells, or protoplasts. **b.** Regeneration of cell walls and the beginning of cell division. **c.** Cell division produces aggregates of cells. **d.** An undifferentiated mass, called a callus. **e.** Somatic cell embryos such as this one appear. **f.** The embryos develop into plantlets that can be transferred to soil for growth into adult plants.

**FIGURE 27.14 Producing whole plants from meristem tissue.**

*Meristem tissue* can also be used as a source of plant cells. In this case, the resulting products are clonal plants that always have the same traits. In Figure 27.14, culture flasks containing meristematic orchid tissue are rotated under lights. If the correct proportions of hormones are added to the liquid medium, many new shoots develop from a single shoot tip. When these are removed, more shoots form. Another advantage to producing identical plants from meristem tissue is that the plants are virus-free. (The presence of plant viruses weakens plants and makes them less productive.)

*Anther tissue culture* is a technique in which the haploid cells within pollen grains are cultured in order to produce haploid plantlets. Conversely, a diploid (2n) plantlet can be produced if chemical agents, to encourage chromosomal doubling, are added to the anther culture. Anther tissue culture is a direct way to produce plants that are certain to have the same characteristics.

## Cell Suspension Culture

A technique called **cell suspension culture** allows scientists to extract chemicals (i.e., secondary metabolites) from plant cells, which may have been genetically modified (see page 255), in high concentration and without having to over-collect wild-type plants growing in their natural environments. These cells produce the same chemicals the entire plant produces. For example, cell suspension cultures of *Cinchona ledgeriana* produce quinine, which is used to treat leg cramping, a major symptom of malaria. And those of several *Digitalis* species produce digitalis, digitoxin, and digoxin, which are useful in the treatment of heart disease.

### Check Your Progress 27.4

1. What are the possible benefits of asexual reproduction?
2. How are new plants produced asexually in the wild?
3. How are new plants produced asexually in the laboratory?

# Connecting the Concepts

Life as we know it would not be possible without vascular plants, and specifically flowering plants, which now dominate the biosphere. *Homo sapiens* evolved in a world dominated by flowering plants and therefore does not know a world without them. The earliest humans were mostly herbivores; they relied on foods they could gather for survival—fruits, nuts, seeds, tubers, roots, and so forth. Plants also provided protection from the environment, offering shelter from heavy rains and noonday sun. Later on, human civilizations could not have begun without the development of agriculture. The majority of the world's population still relies primarily on three flowering plants—corn, wheat, and rice—for the majority of its sustenance. Sugar, coffee, spices of all kinds, cotton, rubber, and tea are plants that have even promoted wars because of their importance to a country's economy. Although we now live in an industrialized society, we are still dependent on plants and have put them to even more uses. To take a couple of examples, plants produce substances needed to lubricate the engines of supersonic jets and to make cellulose acetate for films. We grow them not only for the products they produce, but also for their simple beauty. For millions of urban dwellers, plants are their major contact with the natural world.

Most people fail to appreciate the importance of plants, but plants may be even more critical to our lives today than they were to our early ancestors on the African plains. Currently, half of all pharmaceutical drugs have their origin in plants. The world's major drug companies are engaged in a frantic rush to collect and test plants from the rain forests for their drug-producing potential. Why the rush? Because the rain forests may be gone before all the possible cures for cancer, AIDS, and other killers have been found. Wild plants cannot only help cure human ills, but they can also serve as a source of genes for improving the quality of the plants that support our way of life.

## summary

### 27.1 Sexual Reproductive Strategies

Flowering plants exhibit an alternation of generations life cycle. Flowers borne by the sporophyte produce microspores and megaspores by meiosis. Microspores develop into a male gametophyte, and megaspores develop into the female gametophyte. The gametophytes produce gametes by mitotic cell division. Following fertilization, the sporophyte is enclosed within a seed covered by fruit.

The flowering plant life cycle is adapted to a land existence. The microscopic gametophytes are protected from desiccation by the sporophyte; the pollen grain has a protective wall and fertilization does not require external water. The seed has a protective seed coat, and seed germination does not occur until conditions are favorable.

A typical flower has several parts: Sepals, which are usually green in color, form an outer whorl; petals, often colored, are the next whorl; and stamens, each having a filament and anther, form a whorl around the base of at least one carpel. The carpel, in the center of a flower, consists of a stigma, style, and ovary. The ovary contains ovules.

Each ovule contains a megaspore mother cell, which divides meiotically to produce four haploid megaspores, only one of which survives. This megaspore divides mitotically to produce the female gametophyte (embryo sac), which usually has seven cells. One is an egg cell and another is a central cell with two polar nuclei.

The anthers contain microspore mother cells, each of which divides meiotically to produce four haploid microspores. Each of these divides mitotically to produce a two-celled pollen grain. One cell is the tube cell, and the other is the generative cell. The generative cell later divides mitotically to produce two sperm cells. The pollen grain is the male gametophyte. After pollination, the pollen grain germinates, and as the pollen tube grows, the sperm cells travel to the embryo sac. Pollination is simply the transfer of pollen from anther to stigma.

Flowering plants experience double fertilization. One sperm nucleus unites with the egg nucleus, forming a 2n zygote, and the other unites with the polar nuclei of the central cell, forming a 3n endosperm cell.

After fertilization, the endosperm cell divides to form multicellular endosperm. The zygote becomes the sporophyte embryo. The ovule matures into the seed (its integuments become the seed coat). The ovary becomes the fruit.

### 27.2 Seed Development

As the ovule is becoming a seed, the zygote is becoming an embryo. After the first several divisions, it is possible to discern the embryo and the suspensor. The suspensor attaches the embryo to the ovule and supplies it with nutrients. The eudicot embryo becomes first heart-shaped and then torpedo-shaped. Once you can see the two cotyledons, it is possible to distinguish the shoot tip and the root tip, which contain the apical meristems. In eudicot seeds, the cotyledons frequently take up the endosperm.

### 27.3 Fruit Types and Seed Dispersal

The seeds of flowering plants are enclosed by fruits. There are different types of fruits. Simple fruits are derived from a single ovary (which can be simple or compound). Some simple fruits are fleshy, such as a peach or an apple. Others are dry, such as peas, nuts, and grains. Compound fruits consist of aggregate fruits, which develop from a number of ovaries of a single flower, and multiple fruits develop from a number of ovaries of separate flowers.

Flowering plants have several ways to disperse seeds. Seeds may be blown by the wind, attached to animals that carry them away, eaten by animals that defecate them some distance away, or adapted to water transport.

Prior to germination, you can distinguish a bean (eudicot) seed's two cotyledons and plumule, which is the shoot that bears leaves. Also present are the epicotyl, the hypocotyl, and the radicle. In a corn kernel (monocot), the endosperm, the cotyledon, the plumule, and the radicle are visible.

### 27.4 Asexual Reproductive Strategies

Many flowering plants reproduce asexually, as when the nodes of stems (either aboveground or underground) give rise to entire plants, or when roots produce new shoots.

Somatic embryogenesis is the development of adult plants from protoplasts in tissue culture. Micropropagation, the production of clonal plants as a result of meristem culture in particular, is now a commercial venture. Flower meristem culture results in somatic embryos that can be packaged in gel for worldwide distribution. Anther culture results in homozygous plants that express recessive genes. Leaf, stem, and root culture can result in cell suspensions that allow plant chemicals to be produced in large tanks.

## understanding the terms

anther 495
calyx 495
carpel 495
cell suspension culture 506
coevolution 498
corolla 495
cotyledon 500
double fertilization 497
embryo sac 497
endosperm 497
female gametophyte 497
filament 495
flower 494
fruit 503
germinate 504
male gametophyte 496
megaspore 494
microspore 494
ovary 495
ovule 495
petal 495
pollen grain 496
pollination 497
rhizome 505
seed 494
sepal 495
stamen 495
stigma 495
stolon 505
style 495
tissue culture 505
totipotent 505

Match the terms to these definitions:

a. ______________ Flower structure consisting of an ovary, a style, and a stigma.
b. ______________ Flowering plant structure consisting of one or more ripened ovaries that usually contain seeds.
c. ______________ The gametophyte that produces an egg; an embryo sac in flowering plants.
d. ______________ Mature ovule that contains an embryo, with stored food enclosed in a protective coat.
e. ______________ Mature male gametophyte in seed plants.

## reviewing this chapter

1. Draw a diagram of alternation of generations in flowering plants, and indicate which structures are protected by the sporophyte. Explain. 494
2. Draw a diagram of a flower, and name the parts. 494–95
3. Describe the development of a male gametophyte, from the microsporocyte to the production of sperm. 496
4. Describe the development of a female gametophyte, from the megasporocyte to the production of an egg. 496–97
5. What is the difference between pollination and fertilization? Why doesn't fertilization require any external water? What is double fertilization? 497–99
6. Describe methods of cross-pollination, including the use of animal pollinators. 498–99
7. Describe the sequence of events as a eudicot zygote becomes an embryo enclosed within a seed. 500–501
8. Distinguish between simple dry fruits and simple fleshy fruits. Give an example of each type. What is an aggregate fruit? A multiple fruit? 502–3
9. Name several mechanisms of seed and/or fruit dispersal. 503
10. What are the requirements for seed germination? Contrast the germination of a bean seed with that of a corn kernel. 504
11. In what ways do plants ordinarily reproduce asexually? What is the importance of totipotency with regard to tissue culture? 505–6

## testing yourself

Choose the best answer for each question.

1. In plants,
   a. a gamete becomes a gametophyte.
   b. a spore becomes a sporophyte.
   c. both sporophyte and gametophyte produce spores.
   d. only a sporophyte produces spores.
   e. Both a and b are correct.
2. The flower part that contains ovules is the
   a. carpel.
   b. stamen.
   c. sepal.
   d. petal.
   e. seed.
3. The megaspore and the microspore mother cells
   a. both produce pollen grains.
   b. both divide meiotically.
   c. both divide mitotically.
   d. produce pollen grains and embryo sacs, respectively.
   e. All of these are correct.
4. A pollen grain is
   a. a haploid structure.
   b. a diploid structure.
   c. first a diploid and then a haploid structure.
   d. first a haploid and then a diploid structure.
   e. the mature gametophyte.
5. Which of these pairs is incorrectly matched?
   a. polar nuclei—plumule
   b. egg and sperm—zygote
   c. ovule—seed
   d. ovary—fruit
   e. stigma—carpel
6. Which of these is not a fruit?
   a. walnut
   b. pea
   c. green bean
   d. peach
   e. All of these are fruits.
7. Animals assist with
   a. pollination and seed dispersal.
   b. control of plant growth and response.
   c. translocation of organic nutrients.
   d. asexual propagation of plants.
   e. germination of seeds.
8. A seed contains
   a. a seed coat.
   b. an embryo.
   c. stored food.
   d. cotyledon(s).
   e. All of these are correct.
9. Which of these is mismatched?
   a. plumule—leaves
   b. cotyledon—seed leaf
   c. epicotyl—root
   d. pericarp—corn kernel
   e. carpel—ovule
10. Which of these is not a common procedure in the tissue culture of plants?
   a. shoot tip culture for the purpose of micropropagation
   b. meristem culture for the purpose of somatic embryos
   c. leaf, stem, and root culture for the purpose of cell suspension cultures
   d. culture of hybridized mature plant cells

11. In the life cycle of flowering plants, a microspore develops into
    a. a megaspore.
    b. a male gametophyte.
    c. a female gametophyte.
    d. an ovule.
    e. an embryo.
12. Carpels
    a. are the female part of a flower.
    b. contain ovules.
    c. are the innermost part of a flower.
    d. may be absent in a flower.
    e. All of these are correct.
13. Which of these is part of a male gametophyte?
    a. synergid cells
    b. the central cell
    c. polar nuclei
    d. a tube nucleus
    e. antipodal cells
14. Bat-pollinated flowers
    a. are colorful.
    b. are open throughout the day.
    c. are strongly scented.
    d. have little scent.
    e. Both b and c are correct.
15. Heart, torpedo, and globular refer to
    a. embryo development.
    b. sperm development.
    c. female gametophyte development.
    d. seed development.
    e. Both b and d are correct.
16. Fruits
    a. nourish embryo development.
    b. help with seed dispersal.
    c. signal gametophyte maturity.
    d. attract pollinators.
    e. signal when they are ripe.
17. Asexual reproduction in flowering plants
    a. is unknown.
    b. is a rare event.
    c. is common.
    d. occurs in all plants.
    e. is no fun.
18. Plant tissue culture takes advantage of
    a. a difference in flower structure.
    b. sexual reproduction.
    c. gravitropism.
    d. phototropism.
    e. totipotency.
19. In plants, meiosis directly produces
    a. new xylem.
    b. phloem.
    c. spores.
    d. egg.
    e. sperm.
20. Label this diagram of alternation of generations in flowering plants.

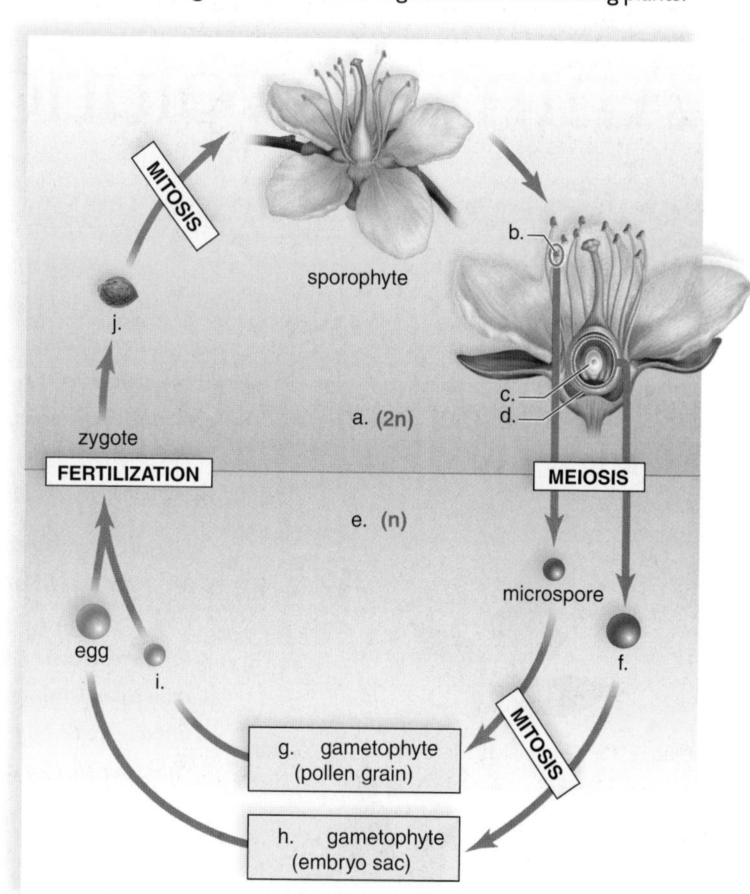


## thinking scientifically

1. You notice that a type of wasp has been visiting a flower type in your garden. What web/library research would allow you to hypothesize that this wasp is a pollinator for this flower type?
2. The pollinator for a very rare plant has become extinct. **a.** What laboratory technique would you use to prevent the plant from also becoming extinct? **b.** How might you improve the hardiness of the plant?

## *Biology* website

The companion website for *Biology* provides a wealth of information organized and integrated by chapter. You will find practice tests, animations, videos, and much more that will complement your learning and understanding of general biology.

**http://www.mhhe.com/maderbiology10**

part VI

# Animal Evolution and Diversity

*In his book* On the Origin of Species, *Charles Darwin says that, while the planet Earth cycles year after year around the sun, "endless forms most beautiful and most wonderful" have appeared and will keep on appearing. This phraseology can certainly be applied to the evolution of animals, whose variety seems without bounds. The fossil record even reveals a myriad of animals that are extinct today, as are the dinosaurs. The search for food, shelter, and mates under a variety of environmental conditions can explain why the diversity of animals is so great.*

*Despite their diversity, evolution from a common ancestor has provided animal life with an unbroken thread of unity. At the biosphere level, animals are heterotrophic consumers that require a constant supply of food by way of autotrophs. At the organismal level, their eukaryotic cells usually form tissues and organs with specialized functions. At the molecular level, animals share a common chemistry, including a genetic code that we now know reveals how the many groups of animals are related. This part concentrates on the characteristics of animals, their origin, and evolution as revealed by molecular genetics.*

# 28

# Invertebrate Evolution

When we think of animals, we tend to imagine birds, dogs, fishes, squirrels, and other vertebrates. However, the animals that lack a backbone—invertebrates—are far more diverse and numerous than the vertebrates. They range in size from tiny gnats to the giant and colorful squid; some are as familiar as a housefly—while others, say, the comb jellies—may be unknown to you. Many invertebrates enhance the quality of our lives. For example, coral reefs, built by stony corals, act as storm barriers for the shoreline and provide safe harbors for ships. Bees are insects that pollinate fruit trees—otherwise, they would not be able produce their delectable products.*

*Still, other invertebrates are harmful to us. Insects called pests feed on and damage crops, making them less valuable or unusable. Working in pairs, certain invertebrates bring about human diseases. For example, mosquitoes serve as vectors for roundworms, whose presence accounts for elephantiasis in humans and heartworm in dogs. Parasites, as well as free-living invertebrates, are adapted to live as they do. In this chapter, you will read about the major groups of invertebrates and the evolutionary ties that bind them together, while allowing them to adapt to particular habitats.*

Dirofilaria immitis *is an invertebrate that causes heartworm in dogs.*

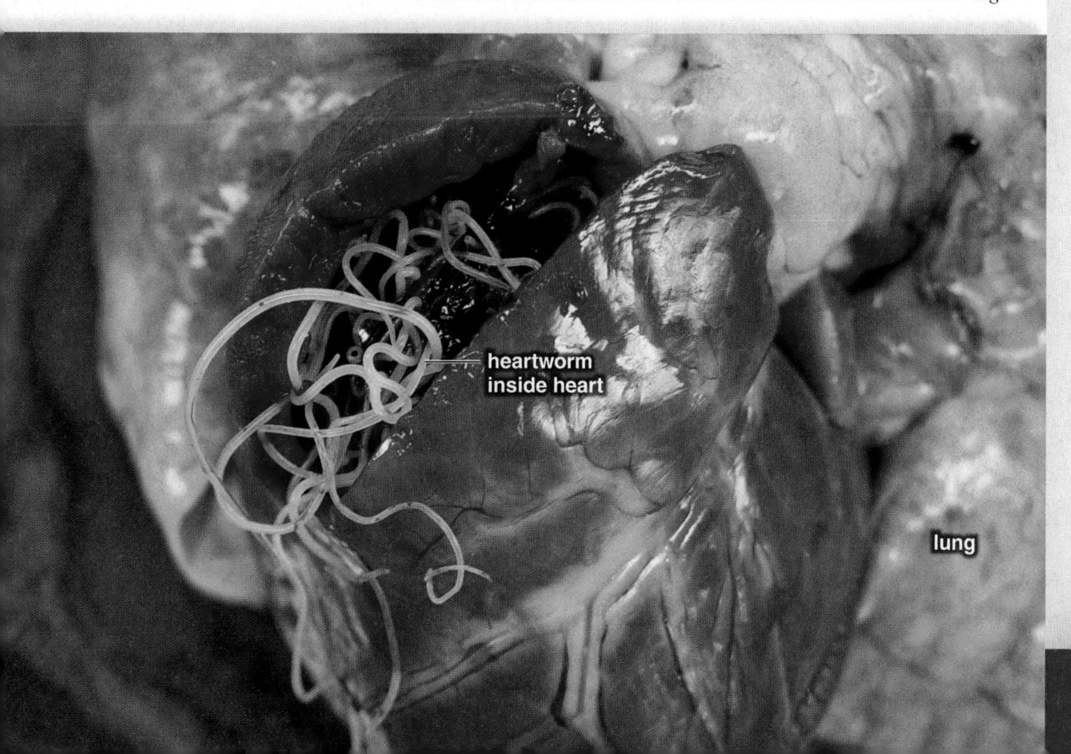


## concepts

# 28.1 Evolution of Animals

Animals can be contrasted with plants and fungi. Like both of these, animals are multicellular eukaryotes, but unlike plants, which make their food through photosynthesis, they are heterotrophs, and must acquire nutrients from an external source. Fungi digest their food externally and absorb the breakdown products. Animals ingest (eat) whole food and digest it internally.

Animals have the diploid life cycle (see Fig. 21A*c*). They usually carry on sexual reproduction and begin life only as a fertilized diploid egg. From this starting point, they undergo a series of developmental stages to produce an organism that has specialized tissues, usually within organs that carry on specific functions. Two types of tissues in particular—muscle and nerve—characterize animals. The evolution of these tissues allows an animal to exhibit motility and a variety of flexible movements. They enable many types of animals to search actively for their food and to prey on other organisms. Coordinated movements also allow animals to seek mates, shelter, and a suitable climate—behaviors that have resulted in the vast diversity of animals. Animals are monophyletic; and both invertebrates and vertebrates can trace their ancestry to the same ancestor. Adult **vertebrates** have a spinal cord (or backbone) running down the center of the back, while **invertebrates,** the topic of this chapter, do not have a backbone. The frog in Figure 28.1 is a vertebrate, while the damselfly it is devouring is an invertebrate.

Figure 28.1 illustrates the characteristics of a complex animal, using the frog as an example. A frog goes through a number of embryonic stages to become a larval form (the tadpole) with many specialized tissues. A larva is an immature stage that typically lives in a different habitat and feeds on different foods than the adult. By means of a change in body form called metamorphosis, the larva, which typically swims, turns into a sexually mature adult frog that swims and hops. The tadpole lives on tiny aquatic organisms, and the terrestrial adult typically feeds on insects and worms. A large African bullfrog will try to eat just about anything, including other frogs, as well as small fish, reptiles, and mammals.

**FIGURE 28.1 Animals—multicellular, heterotrophic eukaryotes.**

Animals begin life as a 2n zygote that undergoes development to produce a multicellular organism that has specialized tissues. Animals depend on a source of external food to carry on life's processes. This series of images shows the development and metamorphosis of the frog, a complex animal.

Adult frog

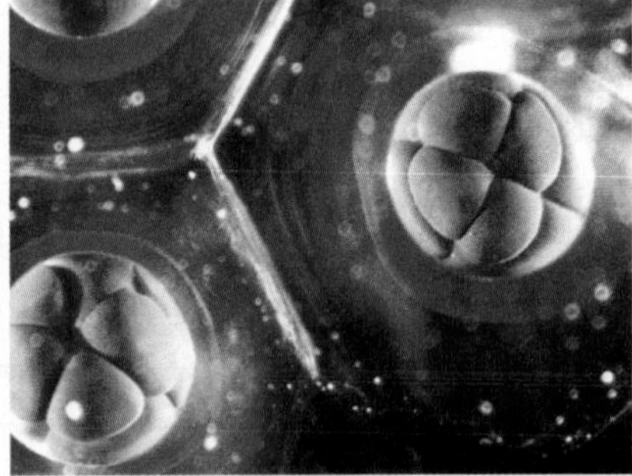

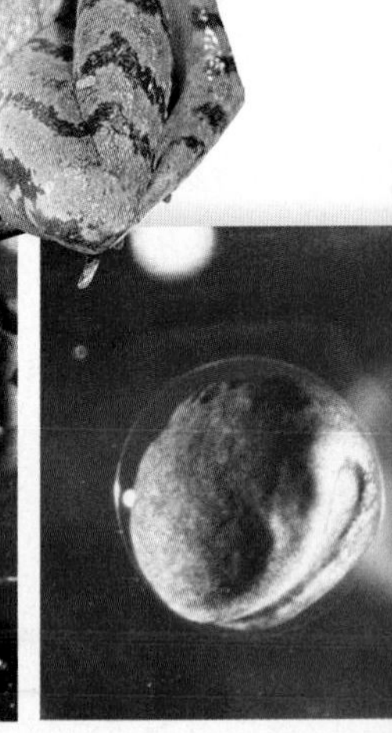

Stages in development, from zygote to embryo.

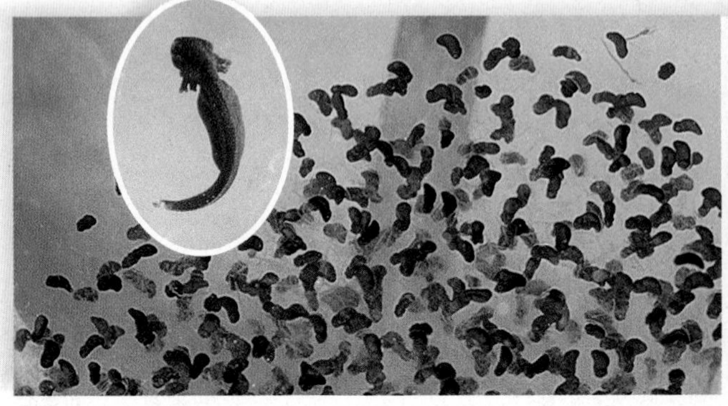

Stages in metamorphosis, from hatching to tadpole.

## Ancestry of Animals

In Chapter 23, we discussed evidence that plants most likely share a green algal ancestor with the charophytes. What about animals? Did they also evolve from a protist, most likely a particular motile protozoan? The *colonial flagellate hypothesis* states that animals are descended from an ancestor that resembled a hollow spherical colony of flagellated cells. Figure 28.2 shows how the process would have begun with an aggregate of a few flagellated cells. From there a larger number of cells could have formed a hollow sphere. Individual cells within the colony would have become specialized for particular functions, such as reproduction. Two tissue layers could have arisen by an infolding of certain cells into a hollow sphere. Tissue layers do arise in this manner during the development of animals today. The colonial flagellate hypothesis is also attractive because it implies that **radial symmetry** preceded **bilateral symmetry** in the history of animals, as is probably the case. In a radially symmetrical animal, any longitudinal cut produces two identical halves; in a bilaterally symmetrical animal, only one longitudinal cut yields two identical halves:

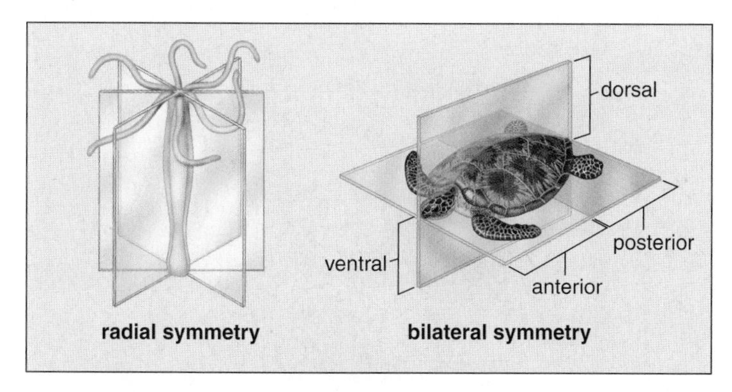


Among the protists, the choanoflagellates (collared flagellates) most likely resemble the last unicellular ancestor of animals, and molecular data tells us that they are the closest living relatives of animals! A choanoflagellate is a single cell, 3–10 $\mu$m in diameter, with a flagellum surrounded by a collar of 30–40 microvilli. Movement of the flagellum creates water currents that pull the protist along. As the water moves through the microvilli, they engulf bacteria and debris from the water. Interesting to our story, choanoflagellates also exist as a colony of cells (Fig. 28.3). Several can be found together at the end of a stalk or simply clumped together like a bunch of grapes.

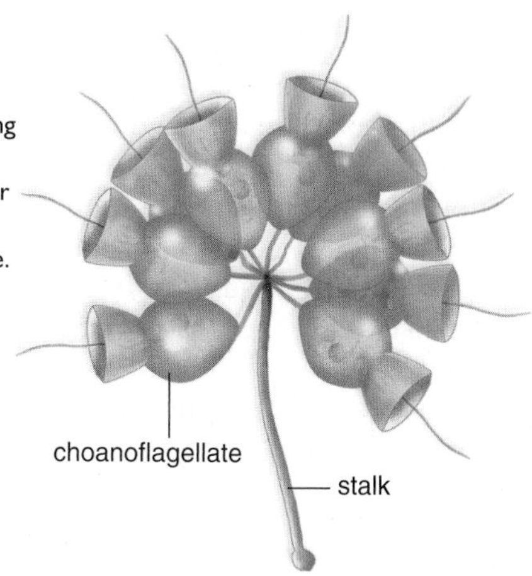


**FIGURE 28.3**
**Choanoflagellate.**
The choanoflagellates are the living protozoans most closely related to animals and may resemble their immediate unicellular ancestor. Some live as a colony like this one.

### *Evolution of Body Plans*

As we discussed in the Science Focus on page 308, all of the various animal body plans were present by the Cambrian period. How could such diversity have arisen within a relatively short period of geological time? As an animal develops, there are many possibilities regarding the number, position, size, and patterns of its body parts. Different combinations could have led to the great variety of animal forms in the past and present. We now know that slight shifts in genes called *Hox* (homeotic) genes are responsible for the major differences between animals that arise during development. Perhaps changes in the expression of *Hox* developmental genes explain why all the animal groups of today had representatives in the Cambrian seas.

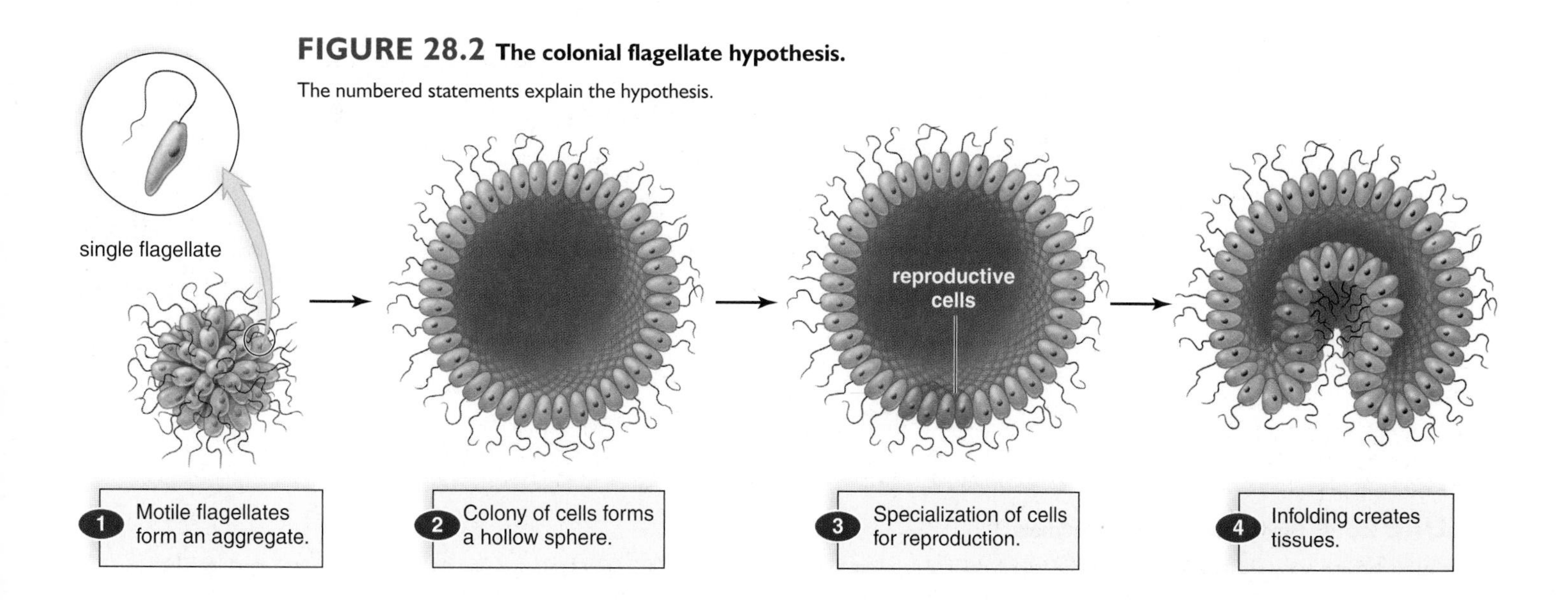


**FIGURE 28.2** **The colonial flagellate hypothesis.**
The numbered statements explain the hypothesis.

## The Phylogenetic Tree of Animals

There is no adequate fossil record by which to trace the early evolution of animals. Therefore, the phylogenetic tree of animals shown in Figure 28.4 is based on molecular and morphological data, including homologies that become apparent during the development of animals. When utilizing molecular data, it is assumed that the more closely related two organisms, the more rRNA nucleotide sequences they will have in common. Molecular data have resulted in a phylogenetic tree that is quite different from one based only on morphological characteristics.

### *Morphological Data*

Refer to the tree in Figure 28.4 as we discuss the anatomical characteristics substantiating the molecular data used to construct the tree.

**Type of Symmetry.** Three types of symmetry exist in the animal world. *Asymmetrical symmetry* is seen in sponges that have no particular body shape. The cnidarians and comb jellies are *radially symmetrical*—they are organized circularly, similar to a wheel, and two identical halves are obtained, no matter how the animal is sliced longitudinally. The rest of the

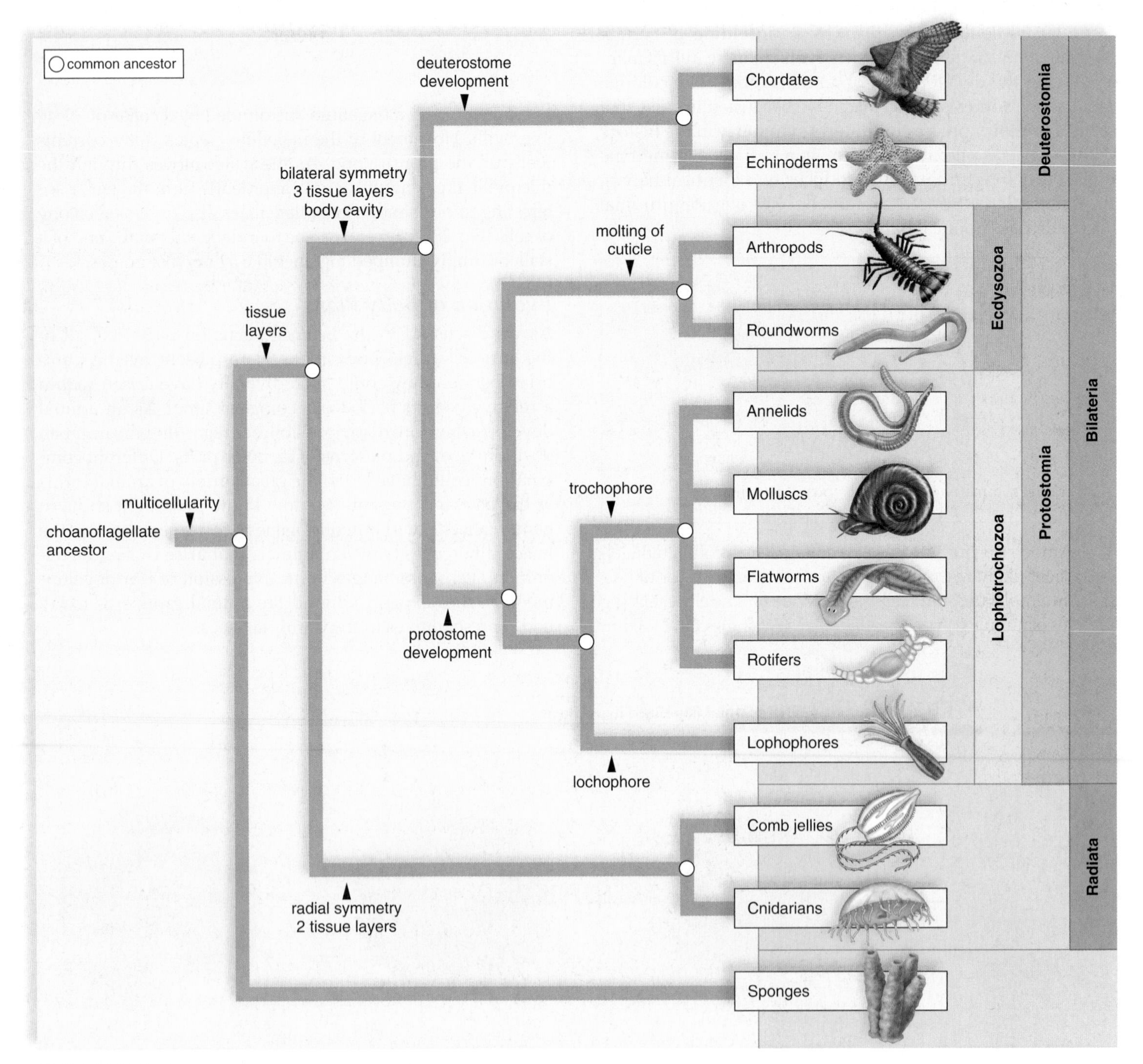


**FIGURE 28.4 Phylogenetic tree of animals.**

All animal phyla living today are most likely descended from a colonial flagellated protist living about 600 million years ago, and this accounts for their motility. This phylogenetic (evolutionary) tree uses morphological and molecular data (rRNA sequencing) to determine which phyla are most closely related to one another.

animals are *bilaterally symmetrical* as adults, and they have a definite left and right half, and only a longitudinal cut down the center of the animal will produce two equal halves.

Radially symmetrical animals are sometimes attached to a substrate—that is, they are **sessile.** This type of symmetry is useful because it allows these animals to reach out in all directions from one center. This advantage also applies to floating animals with radial symmetry, such as jellyfish. Bilaterally symmetrical animals tend to be active and to move forward with an anterior end. During the evolution of animals, bilateral symmetry is accompanied by **cephalization,** localization of a brain and specialized sensory organs at the anterior end of an animal.

**Embryonic Development.** Like all animals, sponges are multicellular but they do not have true tissues. Therefore, sponges have the cellular level of organization. True tissues appear in the other animals as they undergo embryological development. The first three tissue layers are often called **germ layers** because they give rise to the organs and organ systems of complex animals. Animals such as the cnidarians, which have only two tissue layers (ectoderm and endoderm) as embryos, are *diploblastic* with the tissue level of organization. Those animals that develop further and have all three tissue layers (ectoderm, mesoderm, and endoderm) as embryos are *triploblastic* and have the organ level of organization. Notice in the phylogenetic tree that the animals with three tissue layers are either **protostomes** [Gk. *proto,* first; *stoma,* mouth] or **deuterostomes** [Gk. *deuter,* second; *stoma,* mouth].

**DOMAIN: Eukarya**
**KINGDOM: Animals**

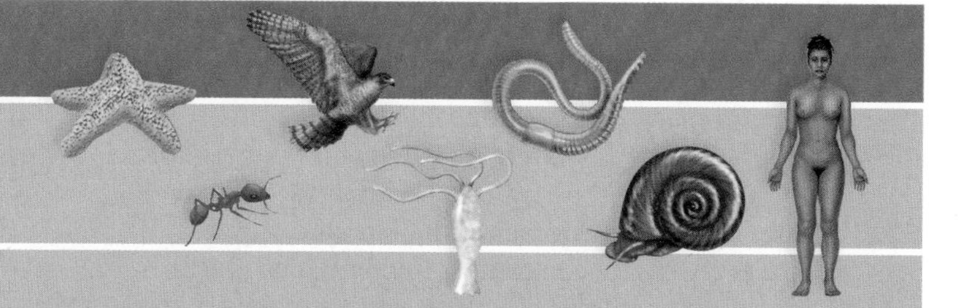

**CHARACTERISTICS**
Multicellular, usually with specialized tissues; ingest or absorb food; diploid life cycle

**INVERTEBRATES**

Sponges (bony, glass, spongin):* Asymmetrical, saclike body perforated by pores; internal cavity lined by choanocytes; spicules serve as internal skeleton. 5,150⁺

**Radiata**

Cnidarians (hydra, jellyfish, corals, sea anemones): Radially symmetrical with two tissue layers; sac body plan; tentacles with nematocysts. 10,000⁺

Comb jellies: Have the appearance of jellyfish; the "combs" are eight visible longitudinal rows of cilia that can assist locomotion; lack the nematocysts of cnidarians but some have two tentacles. 150⁺

**Protostomia (Lophotrochozoa)**

Lophophorates (lampshells, bryozoa): Filter feeders with a circular or horseshoe-shaped ridge around the mouth that bears feeding tentacles. 5,935⁺

Flatworms (planarians, tapeworms, flukes): Bilateral symmetry with cephalization; three tissue layers and organ systems; acoelomate with incomplete digestive tract that can be lost in parasites; hermaphroditic. 20,000⁺

Rotifers (wheel animals): Microscopic animals with a corona (crown of cilia) that looks like a spinning wheel when in motion. 2,000⁺

Molluscs (chitons, clams, snails, squids): Coelom, all have a foot, mantle, and visceral mass; foot is variously modified; in many, the mantle secretes a calcium carbonate shell as an exoskeleton; true coelom and all organ systems. 110,000⁺

Annelids (polychaetes, earthworms, leeches): Segmented with body rings and setae; cephalization in some polychaetes; hydroskeleton; closed circulatory system. 16,000⁺

**Protostomia (Ecdysozoa)**

Roundworms (*Ascaris*, pinworms, hookworms, filarial worms): Pseudocoelom and hydroskeleton; complete digestive tract; free-living forms in soil and water; parasites common. 25,000⁺

Arthropods (crustaceans, spiders, scorpions, centipedes, millipedes, insects): Chitinous exoskeleton with jointed appendages undergoes molting; insects—most have wings—are most numerous of all animals. 1,000,000⁺

**Deuterostomia**

Echinoderms (sea stars, sea urchins, sand dollars, sea cucumbers): Radial symmetry as adults; unique water-vascular system and tube feet; endoskeleton of calcium plates. 7,000⁺

Chordates (tunicates, lancelets, vertebrates): All have notochord, dorsal tubular nerve cord, pharyngeal pouches, and postanal tail at some time; contains mostly vertebrates in which notochord is replaced by vertebral column. 56,000⁺

**VERTEBRATES**

Fishes (jawless, cartilaginous, bony): Endoskeleton, jaws, and paired appendages in most; internal gills; single-loop circulation; usually scales. 28,000⁺

Amphibians (frogs, toads, salamanders): Jointed limbs; lungs; three-chambered heart with double-loop circulation; moist, thin skin. 5,383⁺

Reptiles (snakes, turtles, crocodiles): Amniotic egg; rib cage in addition to lungs; three- or four-chambered heart typical; scaly, dry skin; copulatory organ in males and internal fertilization. 8,000⁺ Birds (songbirds, waterfowl, parrots, ostriches): Endothermy, feathers, and skeletal modifications for flying; lungs with air sacs; four-chambered heart. 10,000⁺

Mammals (monotremes, marsupials, placental): Hair and mammary glands. 4,800⁺

*After a character is listed, it is present in the rest, unless stated otherwise.
⁺Number of species.

Figure 28.5 shows that protostome and deuterostome development are differentiated by three major events:

1. Cleavage, the first event of development, is cell division without cell growth. In protostomes, spiral cleavage occurs, and daughter cells sit in grooves formed by the previous cleavages. The fate of these cells is fixed and determinate in protostomes; each can contribute to development in only one particular way. In deuterostomes, radial cleavage occurs, and the daughter cells sit right on top of the previous cells. The fate of these cells is indeterminate—that is, if they are separated from one another, each cell can go on to become a complete organism.
2. As development proceeds, a hollow sphere of cells, or blastula, forms and the indentation that follows produces an opening called the blastopore. In protostomes, the mouth appears at or near the blastopore, hence the origin of their name; in deuterostomes, the anus appears at or near the blastopore, and only later does a second opening form the mouth, hence the origin of their name.
3. Certain of the protostomes and all deuterostomes have a body cavity completely lined by mesoderm, called a **true coelom** [Gk. *koiloma,* cavity]. However, a true coelom develops differently in the two groups. In protostomes, the mesoderm arises from cells located near the embryonic blastopore, and a splitting occurs that produces the coelom. In deuterostomes, the coelom arises as a pair of mesodermal pouches from the wall of the primitive gut. The pouches enlarge until they meet and fuse.

The deuterostomes include the echinoderms and the chordates, two groups of animals that will be examined in detail later. The protostomes are divided into the two groups: the **ecdysozoa** [Gk. *ecdysis,* stripping off] and the **lophotrochozoa.** The ecdysozoans include the roundworms and arthropods. Both of these types of animals **molt;** they shed their outer covering as they grow. Ecdysozoa means molting animals. The lophotrochozoa contain two groups: the **lophophores** [Gk. *lophos,* crest; *phoros,* bearing] and the **trochophores** [Gk. *trochos,* wheel]. All the lophophores have the same type of feeding apparatus. The trochophores either have presently or their ancestors had a trochophore larva (see page 520).

## Check Your Progress 28.1

1. State three characteristics that all animals have in common.
2. Explain the colonial flagellate hypothesis about the origin of animals.
3. Refer to Figure 28.4 and state the characteristics that pertain to arthropods.

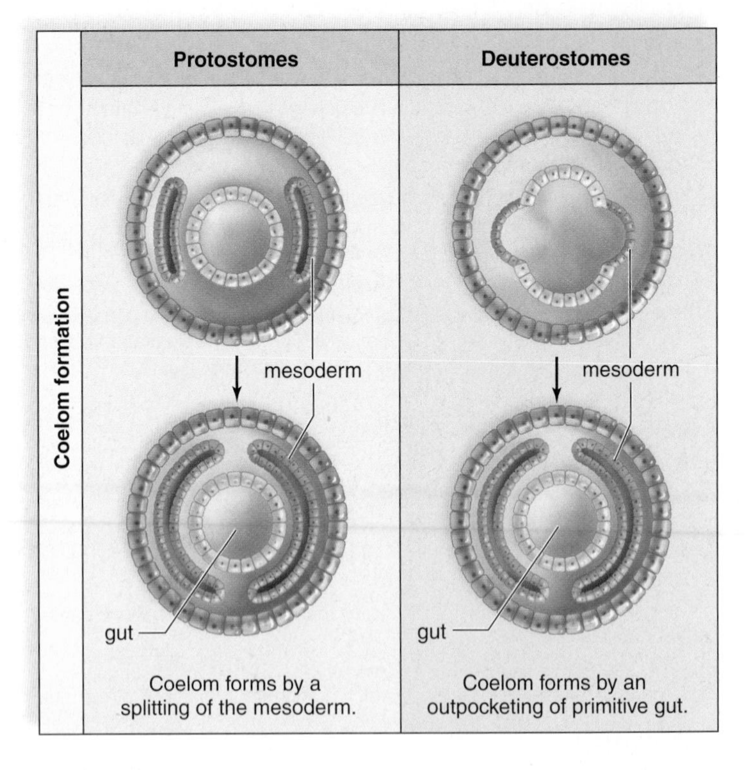


**FIGURE 28.5 Protostomes compared to deuterostomes.**

*Left:* In the embryo of protostomes, cleavage is spiral—new cells are at an angle to old cells—and each cell has limited potential and cannot develop into a complete embryo; the blastopore is associated with the mouth; and the coelom, if present, develops by a splitting of the mesoderm. *Right:* In deuterostomes, cleavage is radial—new cells sit on top of old cells—and each one can develop into a complete embryo; the blastopore is associated with the anus; and the coelom, if present, develops by an outpocketing of the primitive gut.

# 28.2 Introducing the Invertebrates

Sponges, cnidarians, and comb jellies are the least complex of the animals we will study.

## Sponges

While all animals are multicellular, **sponges** (phylum Porifera [L. *porus,* pore, and *ferre,* to bear]) are the only animals to lack true tissues and to have a cellular level of organization. Actually, they have few cell types and no nerve or muscle cells to speak of. Still, molecular data show that they are at the base of the evolutionary tree of animals.

The saclike body of a sponge is perforated by many pores (Fig. 28.6). Sponges are aquatic, largely marine animals that vary greatly in size, shape, and color. But, they all have a canal system of varying complexity that allows water to move through their bodies.

The interior of the canals is lined with flagellated cells that resemble a choanoflagellate. In a sponge, these cells are called collar cells, or choanocytes. The beating of the flagella produces water currents that flow through the pores into the central cavity and out through the osculum, the upper opening of the body. Even a simple sponge only 10 cm tall is estimated to filter as much as 100 L of water each day. It takes this much water to supply the needs of the sponge. A sponge is a stationary filter feeder, also called a suspension feeder, because it filters suspended particles from the water by means of a straining device—in this case, the pores of the walls and the microvilli making up the collar of collar cells. Microscopic food particles that pass between the microvilli are engulfed by the collar cells and digested by them in food vacuoles.

The skeleton of a sponge prevents the body from collapsing. All sponges have fibers of spongin, a modified form of collagen; a bath sponge is the dried spongin skeleton from which all living tissue has been removed. Today, however, commercial "sponges" are usually synthetic. Typically, the endoskeleton of sponges also contains **spicules**—small, needle-shaped structures with one to six rays. Traditionally, the type of spicule has been used to classify sponges, in which case there are bony, glass, and spongin sponges. The success of sponges—they have existed longer than many other animal groups—can be attributed to their spicules. They have few predators because a mouth full of spicules is an unpleasant experience. Also, they produce a number of foul-smelling and toxic substances that discourage predators.

Sponges can reproduce both asexually and sexually. They reproduce asexually by fragmentation or by budding. During budding, a small protuberance appears and gradually increases in size until a complete organism forms. Budding produces colonies of sponges that can become quite large. During sexual reproduction, eggs and sperm are released into the central cavity, and the zygote develops into a flagellated larva that may swim to a new location. If the cells of a sponge are mechanically separated, they will reassemble into a complete and functioning organism! Like many less specialized organisms, sponges are also capable of regeneration, or growth of a whole from a small part.

**FIGURE 28.6 Simple sponge anatomy.**

a. Yellow tube sponge, *Aplysina fistularis*

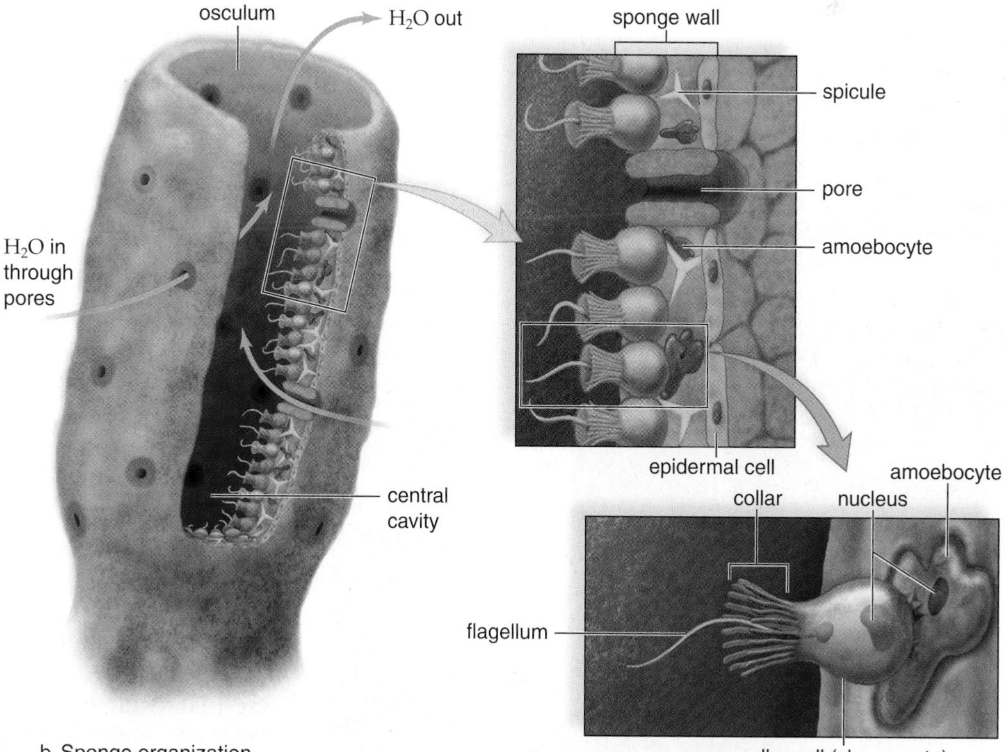

b. Sponge organization

## Comb Jellies and Cnidarians

These two groups of animals (Fig. 28.7) have true tissues and, as embryos, they have the two germ layers ectoderm and endoderm. They are radially symmetrical as adults, a type of symmetry with the advantages discussed on page 515.

### *Comb Jellies*

**Comb jellies** (phylum Ctenophora) are solitary, mostly free-swimming marine invertebrates that are usually found in warm waters. Ctenophores represent the largest animals propelled by beating cilia and range in size from a few centimeters to 1.5 m in length. Their body is made up of a transparent jellylike substance called **mesoglea.** Most ctenophores do not have stinging cells and capture their prey by using sticky adhesive cells called colloblasts. Some ctenophores are bioluminescent, capable of producing their own light.

### *Cnidarians*

**Cnidarians** (phylum Cnidaria) are tubular or bell-shaped animals that reside mainly in shallow coastal waters; however, there are some freshwater, brackish, and oceanic forms. The term *cnidaria* is derived from the presence of specialized stinging cells called cnidocytes. Each cnidocyte has a fluid-filled capsule called a **nematocyst** [Gk. *nema,* thread, and *kystis,* bladder] that contains a long, spirally coiled hollow thread. When the trigger of the cnidocyte is touched, the nematocyst is discharged. Some threads merely trap a prey, and others have spines that penetrate and inject paralyzing toxins.

The body of a cnidarian is a two-layered sac. The outer tissue layer is a protective epidermis derived from ectoderm. The inner tissue layer, which is derived from endoderm, secretes digestive juices into the internal cavity, called the **gastrovascular cavity** [Gk. *gastros,* stomach; L. *vasculum,* dim. of *vas,* vessel] because it serves for digestion of food and circulation of nutrients. The fluid-filled gastrovascular cavity also serves as a supportive **hydrostatic skeleton,** so called because it offers some resistance to the contraction of muscle but permits flexibility. The two tissue layers are separated by mesoglea.

Two basic body forms are seen among cnidarians. The mouth of a **polyp** is directed upward, while the mouth of a jellyfish, or **medusa,** is directed downward. The bell-shaped medusa has more mesoglea than a polyp, and the tentacles are concentrated on the margin of the bell. At one time, both body forms may have been a part of the life cycle of all cnidarians. When both are present, the animal is dimorphic: The sessile polyp stage produces medusae by asexual budding, and the motile medusan stage produces egg and sperm. In some cnidarians, one stage is dominant and the other is reduced; in other species, one form is absent altogether.

a. Sea anemone, *Corynactis*

b. Cup coral, *Tubastrea*

c. Portuguese man-of-war, *Physalia*

d. Jellyfish, *Crambionella*

**FIGURE 28.8 Cnidarian diversity.**

**a.** The anemone, which is sometimes called the flower of the sea, is a solitary polyp. **b.** Corals are colonial polyps residing in a calcium carbonate or proteinaceous skeleton. **c.** The Portuguese man-of-war is a colony of modified polyps and medusae. **d.** True jellyfishes undergo the complete life cycle; this is the medusa stage. The polyp is small.

a.

b.

**FIGURE 28.7 Comb jelly compared to cnidarian.**

**a.** *Pleurobrachia pileus,* a comb jelly. **b.** *Polyorchis penicillatus,* medusan form of a cnidarian. Both animals have similar symmetry, diploblastic organization, and gastrovascular cavities.

## *Cnidarian Diversity*

Sea anemones (Fig. 28.8*a*) are sessile polyps that live attached to submerged rocks, timbers, or other substrate. Most sea anemones range in size from 0.5–20 cm in length and 0.5–10 cm in diameter and are often colorful. Their upward-turned oral disk that contains the mouth is surrounded by a large number of hollow tentacles containing nematocysts.

Corals (Fig. 28.8*b*) resemble sea anemones encased in a calcium carbonate (limestone) house. The coral polyp can extend into the water to feed on microorganisms and retreat into the house for safety. Some corals are solitary, but the vast majority live in colonies that vary in shape from rounded to branching. Many corals exhibit elaborate geometric designs and stunning colors and are responsible for the building of coral reefs. The slow accumulation of limestone can result in massive structures, such as the Great Barrier Reef along the eastern coast of Australia. Coral reef ecosystems are very productive, and a diverse group of marine life call the reef home.

The hydrozoans have a dominant polyp. *Hydra* (see Fig. 28.9) is a hydrozoan, and so is a Portuguese man-of war. You might think the Portuguese man-of-war is an odd-shaped medusa, but actually it is a colony of polyps (Fig. 28.8*c*). The original polyp becomes a gas-filled float that provides buoyancy, keeping the colony afloat. Other polyps, which bud from this one, are specialized for feeding or for reproduction. A long, single tentacle armed with numerous nematocysts arises from the base of each feeding polyp. Swimmers who accidentally come upon a Portuguese man-of-war can receive painful, even serious, injuries from these stinging tentacles.

In true jellyfishes (Fig. 28.8*d*), the medusa is the primary stage, and the polyp remains small. Jellyfishes are zooplankton and depend on tides and currents for their primary means of movement. They feed on a variety of invertebrates and fishes and are themselves food for marine animals.

**Hydra.** The body of a hydra [Gk. *hydra*, a many-headed serpent] is a small tubular polyp about one-quarter inch in length. Hydras are often studied as an example of a cnidarian. Hydras are likely to be found attached to underwater plants or rocks in most lakes and ponds. The only opening (the mouth) is in a raised area surrounded by four to six tentacles that contain a large number of nematocysts.

Figure 28.9 shows the microscopic anatomy of *Hydra*. The cells of the epidermis are termed epitheliomuscular cells because they contain muscle fibers. Also present in the epidermis are nematocyst-containing cnidocytes and sensory cells that make contact with the nerve cells within a **nerve net.** These interconnected nerve cells allow transmission of impulses in several directions at once. The body of a hydra can contract or extend, and the tentacles that ring the mouth can reach out and grasp prey and discharge nematocysts.

Hydras reproduce asexually by forming buds, small outgrowths that develop into a complete animal and then detach. Interstitial cells of the epidermis are capable of becoming other types of cells such as an ovary and/or a testis. When hydras reproduce sexually, sperm from a testis swim to an egg within an ovary. The embryo is encased within a hard, protective shell that allows it to survive until conditions are optimum for it to emerge and develop into a new polyp. Like the sponges, cnidarians have great regenerative powers, and hydras can grow an entire organism from a small piece.

**Check Your Progress 28.2**

1. In what ways are cnidarians more complex than the sponges?

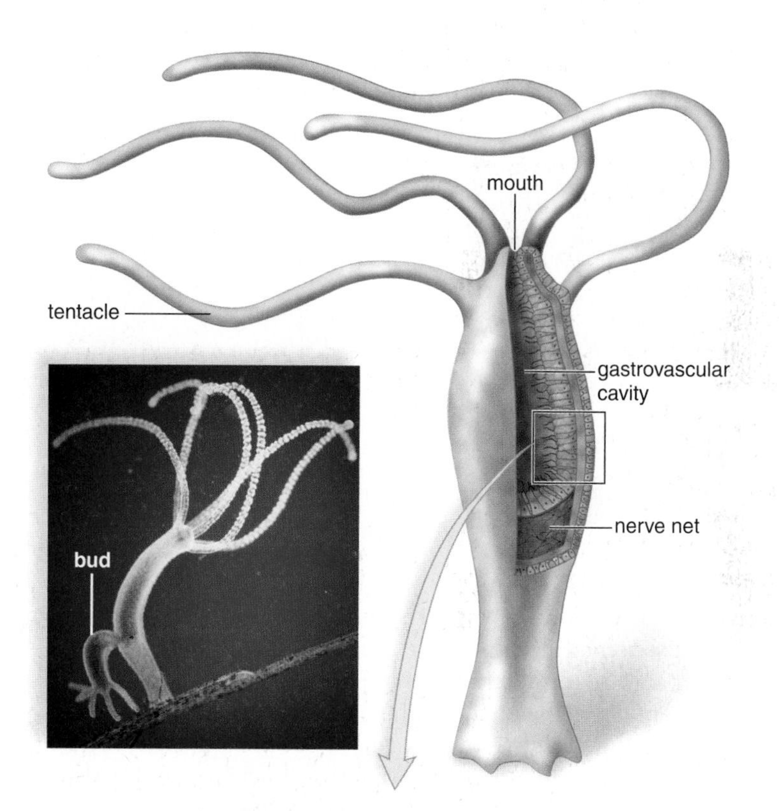


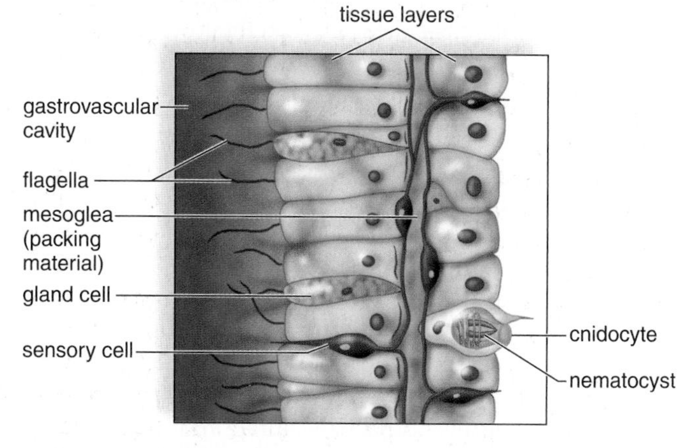


**FIGURE 28.9 Anatomy of *Hydra*.**

*Above:* The body of *Hydra* is a small tubular polyp that reproduces asexually by forming outgrowths called buds. The buds develop into a complete animal. *Below:* The body wall contains two tissue layers separated by mesoglea. Cnidocytes are cells that contain nematocysts.

# 28.3 Variety Among the Lophotrochozoans

The lophotrochozoa encompass several groups of animals. These animals are bilaterally symmetrical at least in some stage of their development. As embryos, they have three germ layers, and as adults, they have the organ level of organization. As discussed on page 516, lophotrochozoans have the protostome pattern of development. Some have a true coelom, as exemplified best in the annelids.

The lophotrochozoans can be divided into two groups: the lophophores (e.g., bryozoans and brachiopods) and the trochophores (flatworms, rotifers, molluscs, and annelids). The lophophores may not be closely related, but they all have a feeding apparatus called the lophophore. The trochophores either have a trochophore larva today (molluscs and annelids), or an ancestor had one some time in the past (flatworms and rotifers).

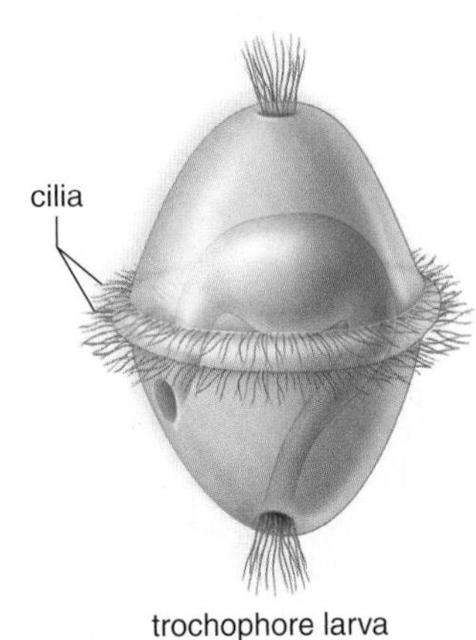


## Flatworms

**Flatworms** (phylum Platyhelminthes) are trochozoans rightly named because they are worms with an extremely flat body. Like the cnidarians, flatworms have a sac body plan and only one opening, the mouth. When one opening is present, the digestive tract is said to be *incomplete,* and when two openings are present, the digestive tract is *complete.* Also, flatworms have no body cavity, and instead the third germ layer, mesoderm, fills the space between their organs.

Among flatworms, planarians are free-living; flukes and tapeworms are parasitic.

### *Free-living Flatworms*

*Dugesia* is a planarian that lives in freshwater lakes, streams, and ponds, where it feeds on small living or dead organisms. A planarian captures food by wrapping itself around the prey, entangling it in slime, and pinning it down. Then a muscular pharynx is extended through the mouth and a sucking motion takes pieces of the prey into the pharynx. The pharynx leads into a three-branched gastrovascular cavity in which digestion is both extracellular and intracellular (Fig. 28.10*a*). The digestive system delivers nutrients and oxygen to the cells, and there is no circulatory system nor respiratory system. Waste molecules exit through the mouth.

Planarians have a well-developed excretory system (Fig. 28.10*b*). The excretory organ functions in osmotic regulation,

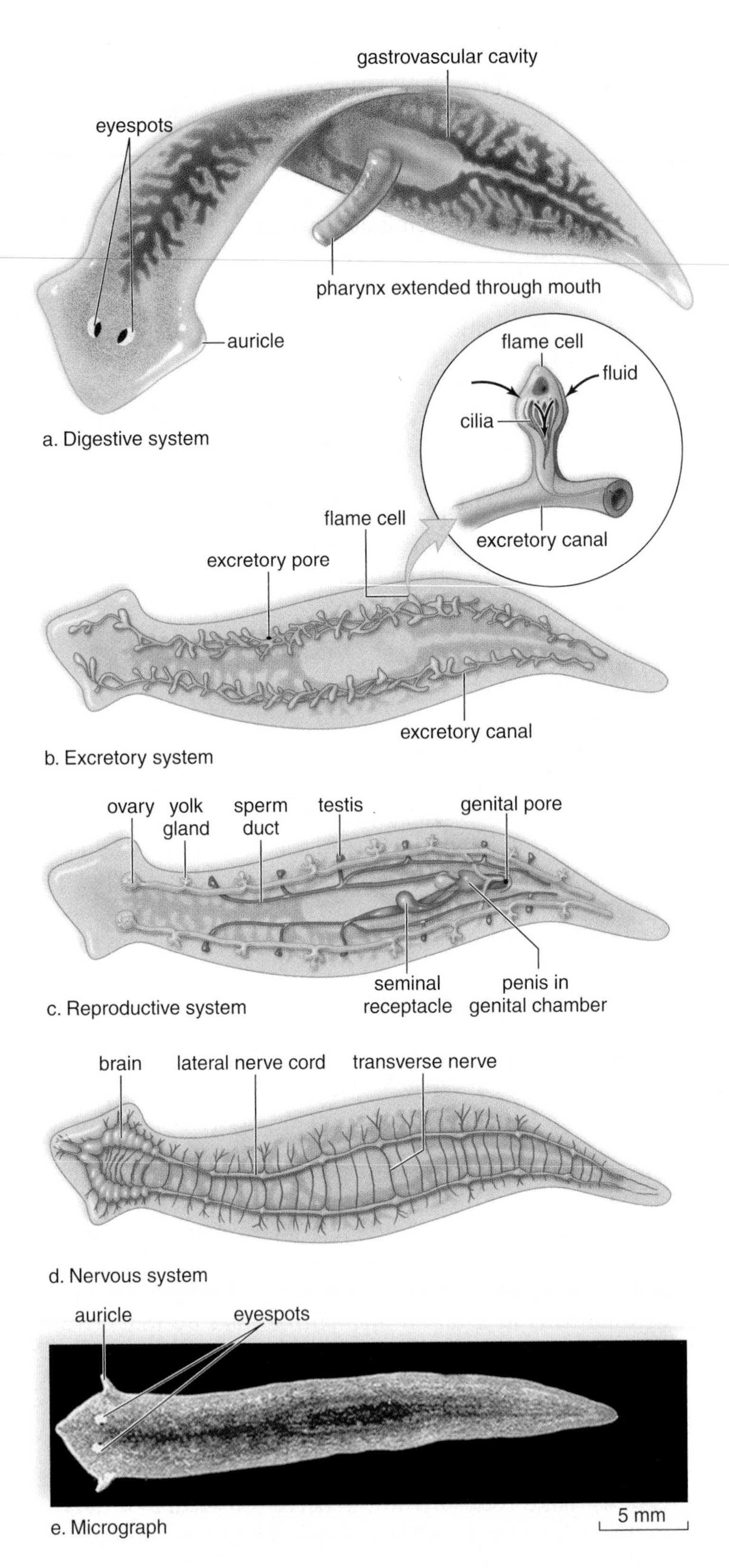


**FIGURE 28.10 Planarian anatomy.**

**a.** When a planarian extends the pharynx, food is sucked up into a gastrovascular cavity that branches throughout the body. **b.** The excretory system with flame cells is shown in detail. **c.** The reproductive system (shown in pink and blue) has both male and female organs. **d.** The nervous system has a ladderlike appearance. **e.** The photograph shows that a planarian, *Dugesia*, is bilaterally symmetrical and has a head region with eyespots.

as well as in water excretion. The organ consists of a series of interconnecting canals that run the length of the body on each side. Bulblike structures containing cilia are at the ends of the side branches of the canals. The cilia move back and forth, bringing water into the canals that empty at pores. The excretory system often functions as an osmotic-regulating system. The beating of the cilia reminded an early investigator of the flickering of a flame, and so the excretory organ of the flatworm is called a flame cell.

Planarians can reproduce asexually. They constrict beneath the pharynx, and each part grows into a whole animal again. Because planarians have the ability to regenerate, they have been the subject of much research. Planarians also reproduce sexually. They are **hermaphroditic** or monoecious, which means that they possess both male and female sex organs (Fig. 28.10*c*). The worms practice cross-fertilization when the penis of one is inserted into the genital pore of the other. The fertilized eggs are enclosed in a cocoon and hatch in two or three weeks as tiny worms.

The nervous system is called a *ladder-type* because the two lateral nerve cords plus transverse nerves look like a ladder (Fig. 28.10*d*). Paired ganglia (collections of nerve cells) function as a brain in the head region. Planarians are bilaterally symmetrical and they undergo cephalization. The head of a planarian is bluntly arrow shaped, with lateral extensions called auricles that contain chemosensory cells and tactile cells used to detect potential food sources and enemies. The pigmentation of the two light-sensitive eyespots causes the worm to look cross-eyed (Fig. 28.10*e*).

Three kinds of muscle layers—an outer circular layer, an inner longitudinal layer, and a diagonal layer—allow for quite varied movement. In larger forms, locomotion is accomplished by the movement of cilia on the ventral and lateral surfaces. Numerous gland cells secrete a mucus upon which the animal moves.

## *Parasitic Flatworms*

Flukes (trematodes) and tapeworms (cestodes) are parasitic flatworms. Both are highly modified for the parasitic mode of life, losing some of the attributes of free-living flatworms and gaining others. Flukes and tapeworms are covered by a protective tegument, which is a specialized body covering resistant to host digestive juices. Associated with the loss of a predatory lifestyle is an absence of cephalization. The head with sensory structures is replaced by an anterior end with hooks and/or suckers for attachment to the host. They no longer hunt for prey and the nervous system is not well developed. On the other hand, a well-developed reproductive system helps ensure transmission to a new host.

Both flukes and tapeworms utilize a secondary, or intermediate, host to transmit offspring from primary host to primary host. The primary host is infected with the sexually mature adult; the secondary host contains the larval stage or stages.

**Flukes.** Flukes are named for the organ they inhabit; for example, there are liver, lung, and blood flukes (Fig. 28.11). The almost 11,000 species have an oval to more elongated

**FIGURE 28.11 Life cycle of a blood fluke, *Schistosoma*.**

**a.** Micrograph of *Schistosoma*.
**b.** *Schistosomiasis*, an infection of humans caused by the blood fluke *Schistosoma*, is an extremely prevalent disease in Egypt—especially since the building of the Aswan High Dam. Standing water in irrigation ditches, combined with unsanitary practices, has created the conditions for widespread infection.

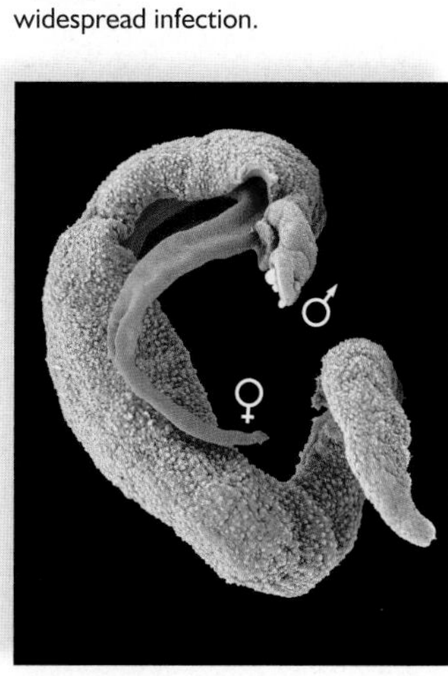


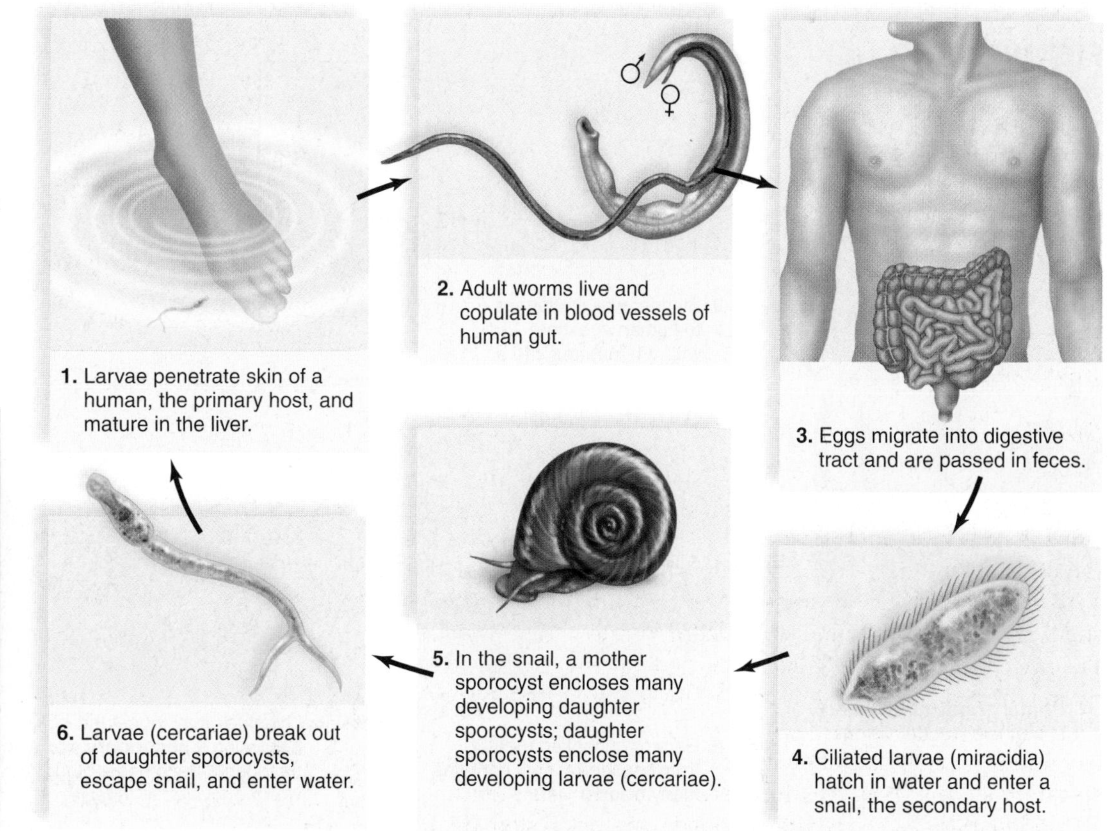

flattened body about 2.5 cm long. At the anterior end is an oral sucker surrounded by sensory papilla and at least one other sucker used for attachment to a host. The blood fluke (*Schistosoma*) occurs predominantly in the Middle East, Asia, and Africa where 200,000 infected persons die each year from *schistosomiasis.* In this disease, female flukes deposit their eggs in small blood vessels close to the lumen of the intestine, and the eggs make their way into the digestive tract by a slow migratory process (Fig. 28.11). After the eggs pass out with the feces, they hatch into tiny larvae that swim about in rice paddies and elsewhere until they enter a particular species of snail. Within the snail, asexual reproduction occurs; sporocysts, which are spore-containing sacs, eventually produce new larval forms that leave the snail. If the larvae penetrate the skin of a human, they begin to mature in the liver and implant themselves in the blood vessels of the small intestine. The flukes and their eggs can cause dysentery, anemia, bladder inflammation, brain damage, and severe liver complications. Infected persons usually die of secondary diseases brought on by their weakened condition.

The Chinese liver fluke, *Clonorchis sinensis,* is a parasite of cats, dogs, pigs, and humans and requires two secondary hosts: a snail and a fish. The adults reside in the liver and deposit their eggs in bile ducts, which carry them to the intestines for elimination in feces. Nonhuman species generally become infected through the fecal route, but humans usually become infected by eating raw fish. A heavy *Clonorchis* infection can cause severe cirrhosis of the liver and death.

**Tapeworms.** Tapeworms vary in length from a few millimeters to nearly 20 m. They have a highly modified head region called the scolex that contains hooks for attachment to the intestinal wall of the host and suckers for feeding. Behind the **scolex,** a series of reproductive units called **proglottids** are found that contain a full set of female and male sex organs. The number of proglottids may vary depending on the species. After fertilization, the organs within a proglottid disintegrate and the proglottids become filled with mature eggs. These egg-filled proglottids are called gravid, and in some species may contain 100,000 eggs. Once mature, depending on the species, the eggs may be released through a pore into the host's intestine, where they exit with the feces. In other species, the gravid proglottids break off and are eliminated with the feces.

Most tapeworms have complicated life cycles that usually involve several hosts. Figure 28.12 illustrates the life cycle of the pork tapeworm, *Taenia solium,* which involves the human as the primary host and the pig as the secondary host. After a pig feeds on feces-contaminated food, the larvae are released. They burrow through the intestinal wall and travel in the bloodstream to finally lodge and encyst in muscle. This **cyst** is a small, hard-walled structure that contains a larva called a bladder worm. When humans eat infected meat that has not been thoroughly cooked, the bladder worms break out of the cysts, attach themselves to the intestinal wall, and grow to adulthood. Then the cycle begins again. Generally, tapeworm infections cause diarrhea, weight loss, and fatigue in the primary host.

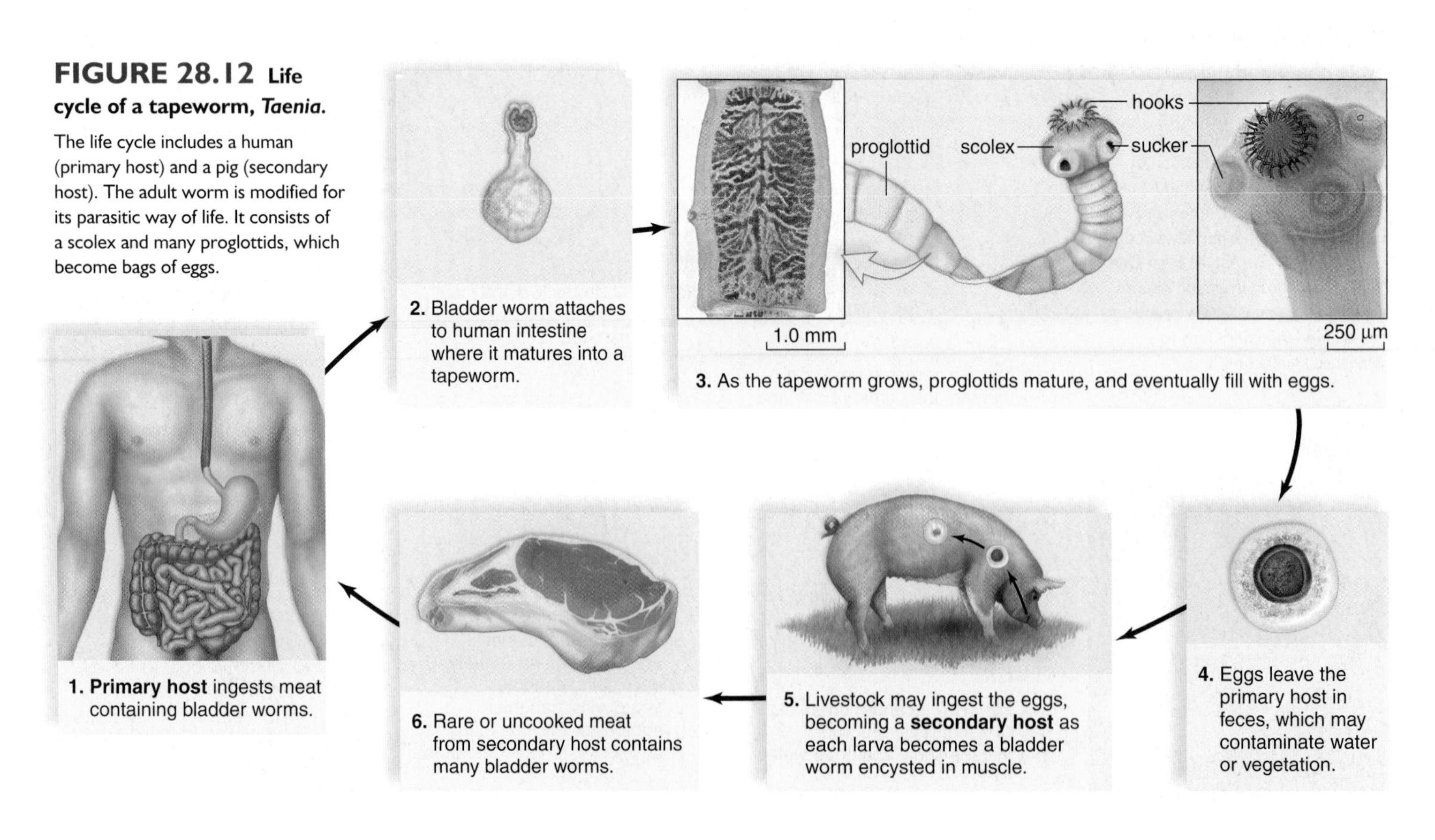


**FIGURE 28.12 Life cycle of a tapeworm, *Taenia*.** The life cycle includes a human (primary host) and a pig (secondary host). The adult worm is modified for its parasitic way of life. It consists of a scolex and many proglottids, which become bags of eggs.

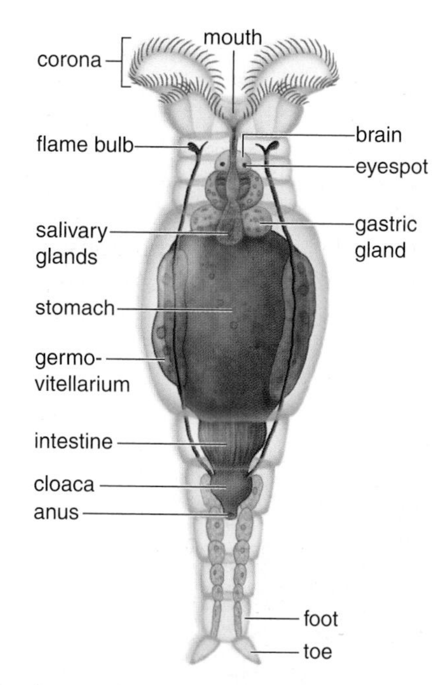


**FIGURE 28.13**
**Rotifer.**

Rotifers are microscopic animals only 0.1–3 mm in length. The beating of cilia on two lobes at the anterior end of the animal gives the impression of a pair of spinning wheels

## Rotifers

**Rotifers** (phylum Rotifera) are related to the flatworms and both are trochozoans. Anton von Leeuwenhoek viewed rotifers through his microscope and called them the "wheel animacules." Rotifers have a crown of cilia, known as the corona, on their heads (Fig. 28.13). When in motion, the corona, which looks like a spinning wheel, serves as an organ of locomotion and also directs food into the mouth. The approximately 2,000 species primarily live in fresh water; however, some marine and terrestrial forms exist. The majority of rotifers are transparent, but some are very colorful. Many species of rotifers can desiccate during harsh conditions and remain dormant for lengthy periods of time. This characteristic has earned them the title "resurrection animacules."

## Molluscs

The **molluscs** (phylum Mollusca), the second most numerous group of animals, inhabit a variety of environments, including marine, freshwater, and terrestrial habitats. This diverse phylum includes chitons, limpets, slugs, snails, abalones, conchs, nudibranchs, clams, scallops, squid, and octopuses. Molluscs vary in size from microscopic to the giant squid, which can attain lengths of over 20 m and weigh over 450 kg. The group includes herbivores, carnivores, filter feeders, and parasites.

Although diverse, molluscs share a three-part body plan consisting of the visceral mass, mantle, and foot (Fig. 28.14*a*). The visceral mass contains the internal organs, including a highly specialized digestive tract, paired kidneys, and reproductive organs. The **mantle** is a covering that lies to either side of, but does not completely enclose, the visceral mass. It may secrete a shell and/or contribute to the development of gills or lungs. The space between the folds of the mantle is called the mantle cavity. The foot is a muscular organ that may be adapted for locomotion, attachment, food capture, or a combination of functions. Another feature often present in molluscs is a rasping, tonguelike **radula,** an organ that bears many rows of teeth and is used to obtain food (Fig. 28.14*b*).

The true coelom is reduced and largely limited to the region around the heart in molluscs. Most molluscs have an open circulatory system. The heart pumps blood, more properly called hemolymph, through vessels into sinuses (cavities) collectively called a **hemocoel** [Gk. *haima,* blood, and *koiloma,* cavity]. Blue hemocyanin, rather than red hemoglobin, is the respiratory pigment. Nutrients and oxygen diffuse into the tissues from these sinuses instead of being carried into the tissues by capillaries, microscopic blood vessels present in animals with closed circulatory systems.

The nervous system of molluscs consists of several ganglia connected by nerve cords. The amount of cephalization and sensory organs varies from nonexistent in clams to complex in squid and octopi. The molluscs also exhibit variations in mobility. Oysters are sessile, snails are extremely slow moving, and squid are fast-moving, active predators.

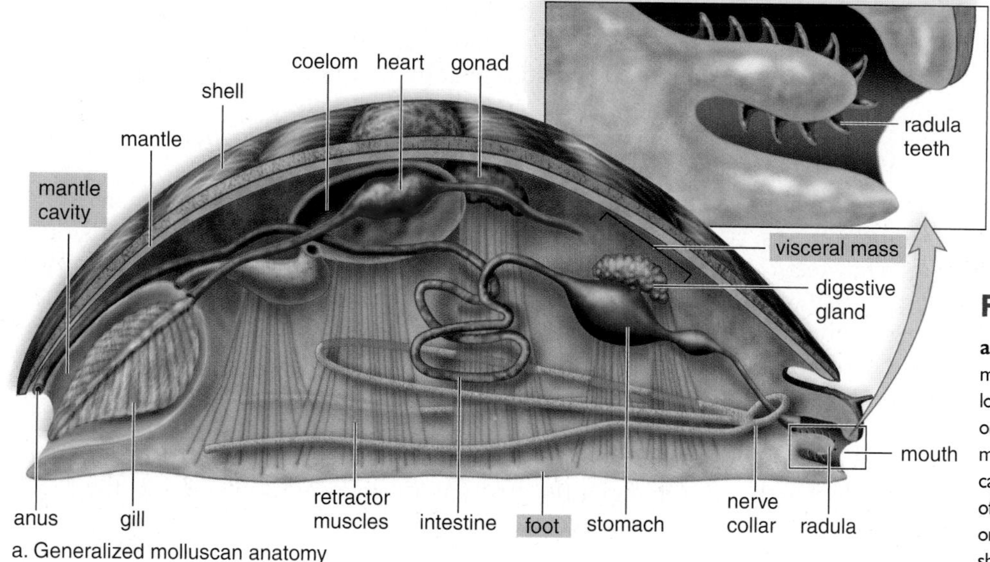


a. Generalized molluscan anatomy

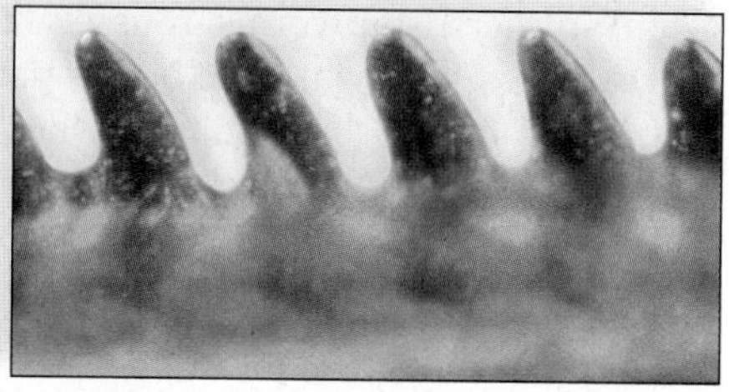

b. Radula

**FIGURE 28.14 Body plan of molluscs.**

**a.** Molluscs have a three-part body consisting of a ventral, muscular foot that is specialized for various means of locomotion; a visceral mass that includes the internal organs; and a mantle that covers the visceral mass and may secrete a shell. Ciliated gills may lie in the mantle cavity and direct food toward the mouth. **b.** In the mouth of many molluscs, such as snails, the radula is a tonguelike organ that bears rows of tiny teeth that point backward, shown here in a drawing and a micrograph.

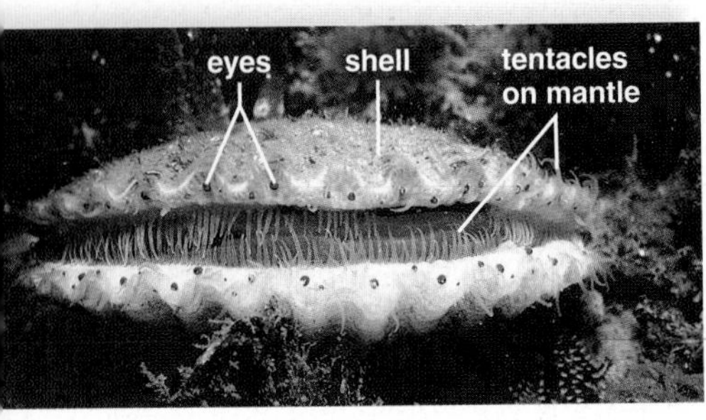


a. Scallop, *Pecten* sp.

b. Mussels, *Mytilus edulis*

**FIGURE 28.15 Bivalve diversity.**

Bivalves have a two-part shell. **a.** Scallops clap their valves and swim by jet propulsion. This scallop has sensory organs consisting of blue eyes and tentacles along the mantle edges. **b.** Mussels form dense beds in the intertidal zone of northern shores. **c.** In this drawing of a clam, the mantle has been removed from one side. Follow the path of food from the incurrent siphon to the gills, the mouth, the stomach, the intestine, the anus, and the excurrent siphon. Locate the three ganglia: anterior, foot, and posterior. The heart lies in the reduced coelom.

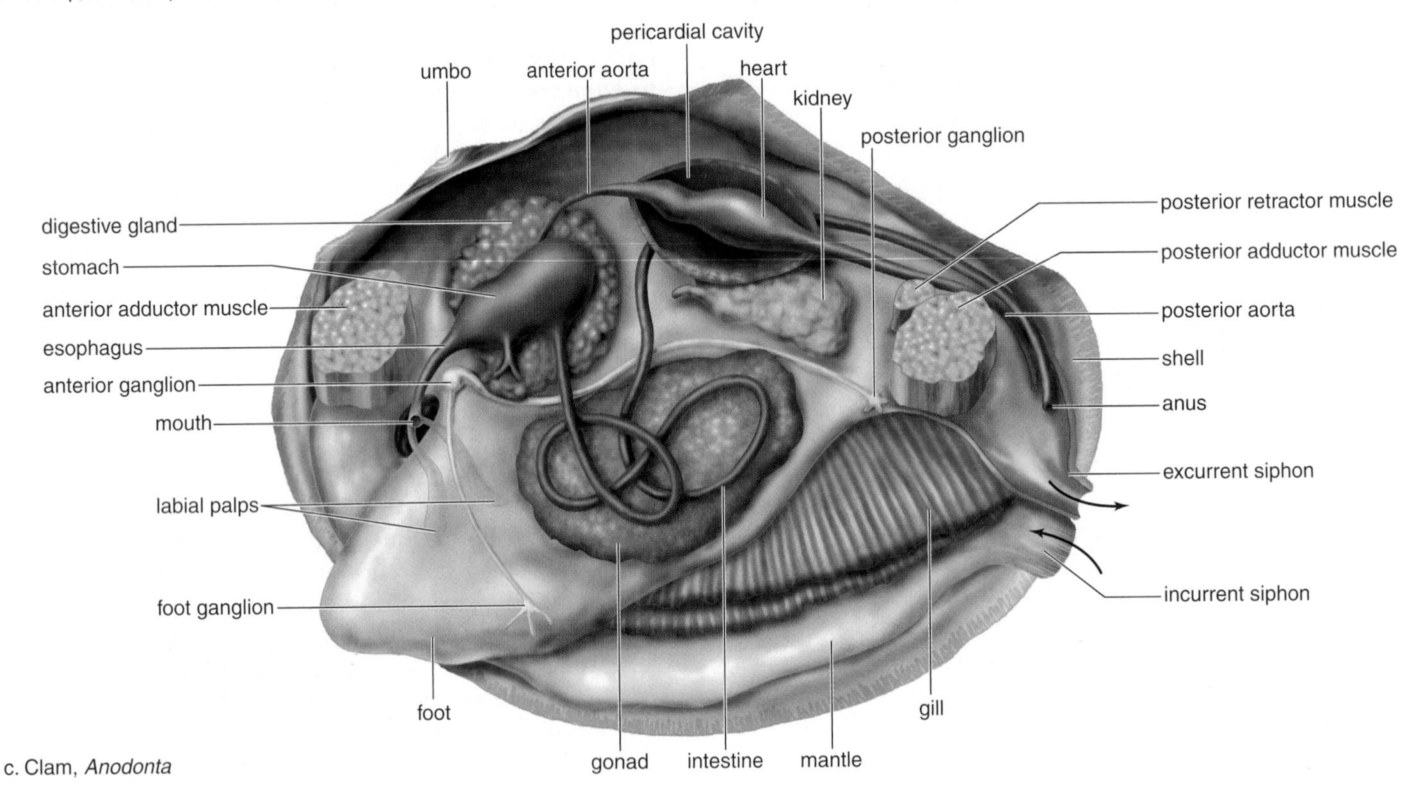


c. Clam, *Anodonta*

## *Bivalves*

Clams, oysters, shipworms, mussels, and scallops are all **bivalves** (class Bivalvia) with a two-part shell that is hinged and closed by powerful muscles (Fig. 28.15). They have no head, no radula, and very little cephalization. Clams use their hatchet-shaped foot for burrowing in sandy or muddy soil, and mussels use their foot to produce threads that attach them to nearby objects. Scallops both burrow and swim; rapid clapping of the valves releases water in spurts and causes the animal to move forward in a jerky fashion for a few feet.

In freshwater clams such as *Anodonta* (Fig. 28.15c), the shell, secreted by the mantle, is composed of protein and calcium carbonate with an inner layer, called mother of pearl. If a foreign body is placed between the mantle and the shell, pearls form as concentric layers of shell are deposited about the particle. The compressed muscular foot of a clam projects ventrally from the shell; by expanding the tip of the foot and pulling the body after it, the clam moves forward.

Within the mantle cavity, the ciliated gills hang down on either side of the visceral mass. The beating of the cilia causes water to enter the mantle cavity by way of the incurrent siphon and to exit by way of the excurrent siphon. The clam is a filter feeder; small particles in this constant stream of water adhere to the gills, and ciliary action sweeps them toward the mouth.

The mouth leads to a stomach and then to an intestine, which coils about in the visceral mass before going right through the heart and ending in an anus. The anus empties at the excurrent siphon. There is also an accessory organ of digestion called a digestive gland. The heart lies just below the hump of the shell within the pericardial cavity, the only remains of the coelom. The circulatory system is open; the heart pumps hemolymph into vessels that open into the *hemocoel*. The nervous system is composed of three pairs of ganglia (located anteriorly, posteriorly, and in the foot), which are connected by nerves.

There are two excretory kidneys, which lie just below the heart and remove waste from the pericardial cavity for excretion into the mantle cavity. The clam excretes ammonia ($NH_3$), a toxic substance that requires the excretion of water at the same time.

In freshwater clams, the sexes are separate, and fertilization is internal. Fertilized eggs develop into specialized larvae and are released from the clam. Some larvae attach to the gills of a fish and become a parasite before they sink to the bottom and develop into a clam. Certain clams and annelids have the same type of larva (see page 520), and this reinforces their evolutionary relationship between molluscs and annelids.

## *Other Molluscs*

The **gastropods** [Gk. *gastros,* stomach, and *podos,* foot], the largest class of molluscs, include slugs, snails, whelks, conchs, limpets, and nudibranchs. While most are marine, slugs and garden snails are adapted to terrestrial environments (Fig. 28.16*c*).

Gastropods have an elongated, flattened foot and most, except for slugs and nudibranchs, have a one-piece coiled shell that protects the visceral mass. The anterior end bears a well-developed head region with a cerebral ganglion and eyes on the ends of tentacles. Land snails are hermaphroditic; when two snails meet, they shoot calcareous darts into each other's body wall as a part of premating behavior. Then each inserts a penis into the vagina of the other to provide sperm for the future fertilization of eggs, which are deposited in the soil. Development proceeds directly without the formation of larvae.

**Cephalopods** [Gk. *kaphale,* head, and *podos,* foot] range in length from 2 cm to 20 m as in the giant squid, *Architeuthis.* Cephalopod means head footed; both squids and octopi can squeeze their mantle cavity so that water is forced out through a funnel, propelling them by jet propulsion (Fig. 28.16). Also, the tentacles and arms that circle the head capture prey by adhesive secretions or by suckers. A powerful, parrotlike beak is used to tear prey apart. They have well-developed sense organs, including eyes that are similar to those of vertebrates and focus like a camera. Cephalopods, particularly octopi, have well-developed brains and show a remarkable capacity for learning.

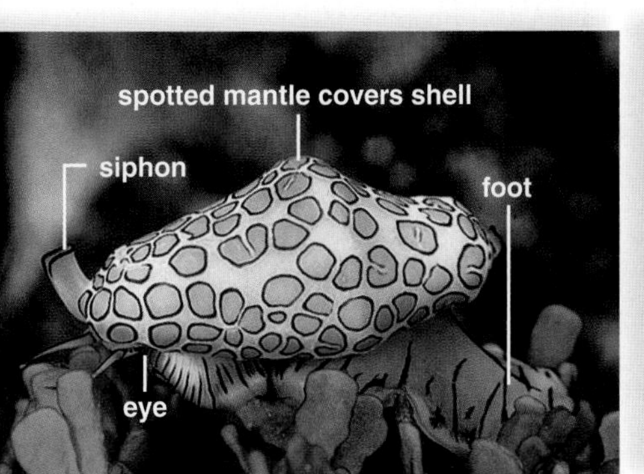


a. Flamingo tongue shell, *Cyphoma gibbosum*

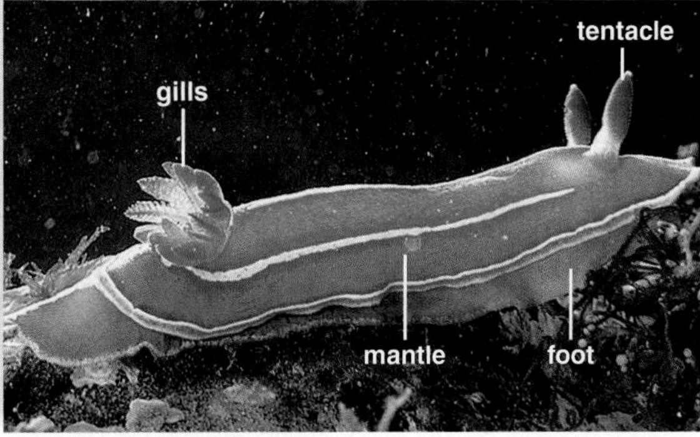


b. Nudibranch, *Glossodoris macfarlandi*

c. Land snail, *Helix aspersa*

d. Two-spotted octopus, *Octopus bimaculatus*

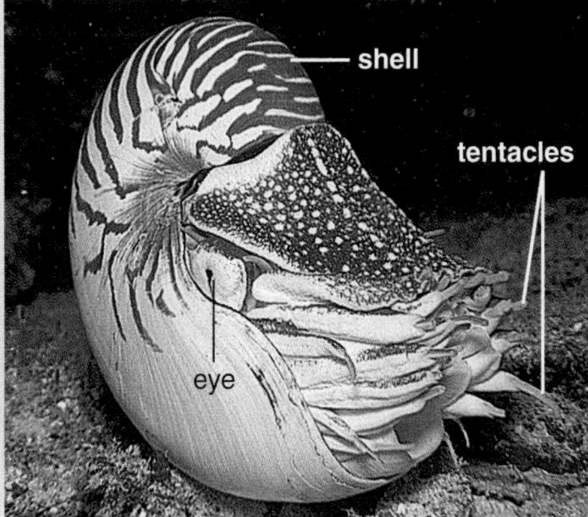


e. Chambered nautilus, *Nautilus belauensis*

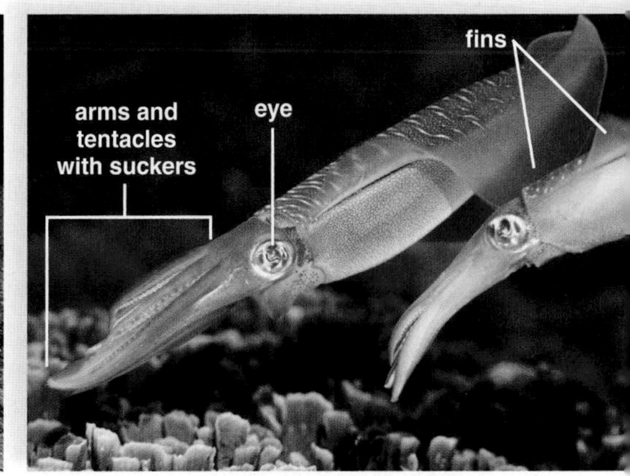


f. Bigfin reef squid, *Sepioteuthis lessoniana*

**FIGURE 28.16 Gastropod and cephalopod diversity.**

**a–c.** Gastropods have the three parts of a mollusc, and the foot is muscular, elongated, and flattened. Nudibranchs have no shell, as do the other two shown.
**d–f.** Cephalopods have tentacles and/or arms in place of a head. Speed suits their predatory lifestyle, and only the chambered nautilus has a shell among those shown.

# Annelids

**Annelids** (phylum Annelida [L. *anellus,* little ring]), which are sometimes called the segmented worms, vary in size from microscopic to tropical earthworms that can be over 4 m long. The most familiar members of this group are earthworms, marine worms, and leeches.

Annelids are the only trochozoan with segmentation and a well-developed coelom. **Segmentation** is the repetition of body parts along the length of the body. The well-developed coelom is fluid-filled and serves as a supportive *hydrostatic skeleton.* A hydrostatic skeleton, along with partitioning of the coelom, permits independent movement of each body segment. Instead of just burrowing in the mud, an annelid can crawl on a surface.

**Setae** [L. *seta,* bristle] are bristles that protrude from the body wall, can anchor the worm, and help it move. The *oligochaetes* are annelids with few setae, and the *polychaetes* are annelids with many setae.

## *Earthworms*

The common earthworm, *Lumbricus terrestris,* is an oligochaete (Fig. 28.17). Earthworm setae protrude in pairs directly from the surface of the body. Locomotion, which is accomplished section by section, uses muscle contraction and the setae. When longitudinal muscles contract, segments bulge and their setae protrude into the soil; then, when circular muscles contract, the setae are withdrawn, and these segments move forward.

Earthworms reside in soil where there is adequate moisture to keep the body wall moist for gas exchange. They are scavengers that feed on leaves or any other organic matter conveniently taken into the mouth along with dirt. Segmentation and a complete digestive tract have led to increased specialization of parts. Food drawn into the mouth by the action of the muscular pharynx is stored in a crop and ground up in a thick, muscular gizzard. Digestion and absorption occur in a long intestine whose dorsal surface has an expanded region called a **typhlosole** that increases the surface for absorption. Waste is eliminated through the anus.

Earthworm segmentation, which is obvious externally, is also internally evidenced by septa that occur between segments. The long, ventral nerve cord leading from the brain has ganglionic swellings and lateral nerves in each segment. The excretory system consists of paired **nephridia** [Gk. *nephros,* kidney], which are coiled tubules in each segment. A nephridium has two openings: One is a ciliated funnel that collects coelomic fluid, and the other is an exit in the body wall. Between the two openings is a convoluted region where waste material is removed from the blood vessels about the tubule.

Annelids have a *closed circulatory system,* which means that the blood is always contained in blood vessels that run the length of the body. Red blood moves anteriorly in the dorsal blood vessel, which connects to the ventral blood vessel by five pairs of muscular vessels called "hearts." Pulsations of the dorsal blood vessel and the five pairs of hearts are responsible for blood flow. As the ventral vessel takes the blood toward the posterior regions of the worm's body, it gives off branches in every segment.

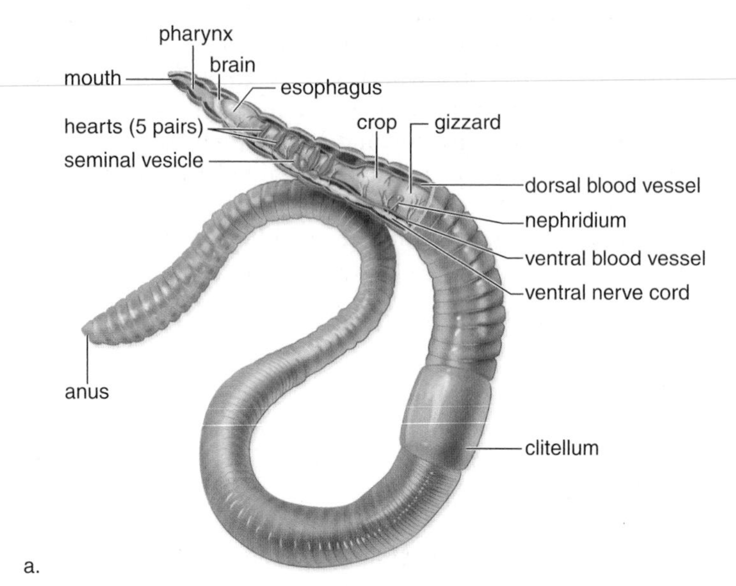


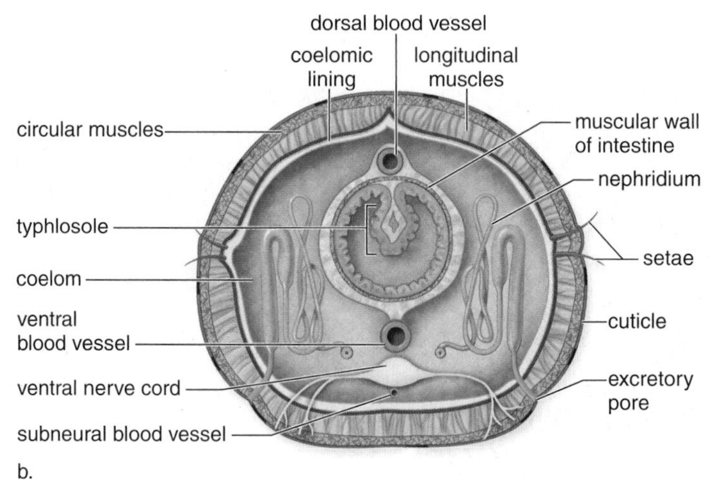

**FIGURE 28.17 Earthworm, *Lumbricus terrestris.***

**a.** In the longitudinal section, note the specialized parts of the digestive tract. **b.** In cross section, note the spacious coelom, the paired setae and nephridia, and a ventral nerve cord that has branches in each segment. **c.** When earthworms mate, they are held in place by a mucus secreted by the clitellum. The worms are hermaphroditic, and when mating, sperm pass from the seminal vesicles of each to the seminal receptacles of the other.

Earthworms are *hermaphroditic;* the male organs are the testes, the seminal vesicles, and the sperm ducts, and the female organs are the ovaries, the oviducts, and the seminal receptacles. During mating, two worms lie parallel to each other facing in opposite directions (Fig. 28.17*c*). The fused midbody segment, called a clitellum, secretes mucus, protecting the sperm from drying out as they pass between the worms. After the worms separate, the clitellum of each produces a slime tube, which is moved along over the anterior end by muscular contractions. As it passes, eggs and the sperm received earlier are deposited, and fertilization occurs. The slime tube then forms a cocoon to protect the worms as they develop. There is no larval stage in earthworms.

## *Other Annelids*

Approximately two-thirds of annelids are marine polychaetes. In polychaetes, the setae are in bundles on parapodia [Gk. *para,* beside, and *podos,* foot], which are paddlelike appendages found on most segments. These are used not only in swimming but also as respiratory organs, where the expanded surface area allows for exchange of gases. Some polychaetes are free-swimming, but the majority live in crevices or burrow into the ocean bottom. Clam worms, such as *Nereis* (Fig. 28.18*a*), are predators. They prey on crustaceans and other small animals, which are captured by a pair of strong chitinous jaws that extend with a part of the pharynx when the animal is feeding. Associated with its way of life, *Nereis* is cephalized, having a head region with eyes and other sense organs.

Other marine polychaetes are sedentary (sessile) tube worms, with radioles (ciliated mouth appendages) used to gather food (Fig 28.18*b*). Christmas tree worms, fan worms, and featherduster worms all have radioles. In featherduster worms, the beautiful radioles cause the animal to look like an old fashioned feather duster.

Polychaetes have breeding seasons, and only during these times do the worms have sex organs. In *Nereis,* many worms simultaneously shed a portion of their bodies containing either eggs or sperm, and these float to the surface where fertilization takes place. The zygote rapidly develops into the trochophore larva, similar to that of the marine clam.

Leeches are annelids that normally live in freshwater habitats. They range in size from less than 2 cm to the medicinal leech, which can be 20 cm in length. They exhibit a variety of patterns and colors but most are brown or olive green. The body of a leech is flattened dorsoventrally. They have the same body plan as other annelids, but they have no setae and each body ring has several transverse grooves.

While some are free-living, most leeches are fluid feeders that attach themselves to open wounds. Among their modifications are two suckers, a small oral one around the mouth and a posterior one. Some bloodsuckers, such as the medicinal leech, can cut through tissue. Leeches are able to keep blood flowing by means of hirudin, a powerful anticoagulant in their saliva. Medicinal leeches have been used for centuries in blood-letting and other procedures. Today, they are used in reconstructive surgery for severed digits and in plastic surgery (Fig. 28.18*c*).

**Check Your Progress 28.3**

1. Even though flatworms, molluscs, and annelids are very different, they are all considered lophotrochozoans. What characteristic(s) unite(s) these three phyla?
2. Compare these three groups with regard to body cavity, digestive tract, and circulatory system.
3. Briefly describe any parasites that occur among these three groups.

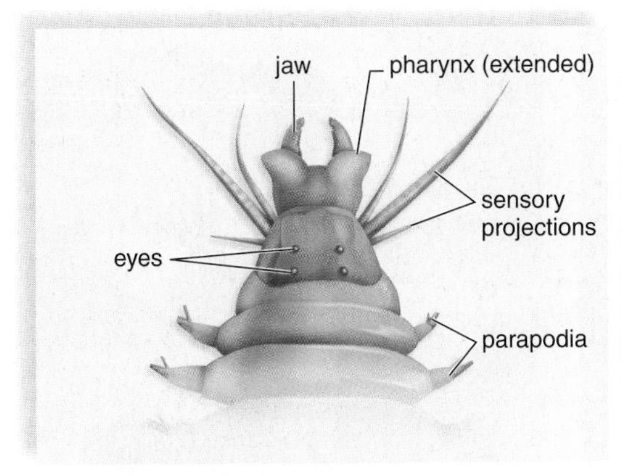


a. Clam worm, *Nereis succinea*

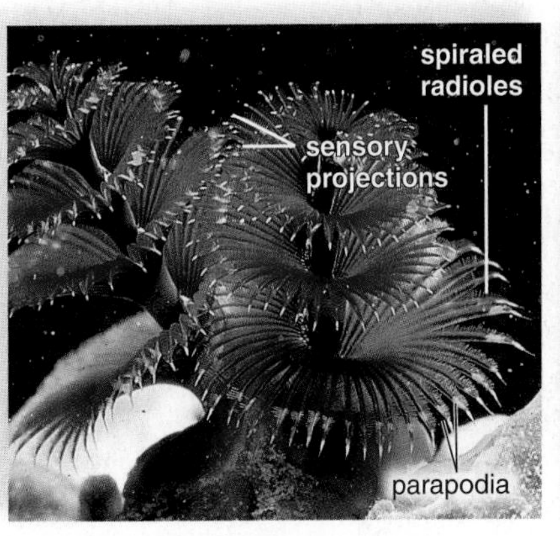


b. Christmas tree worm, *Spirobranchus giganteus*

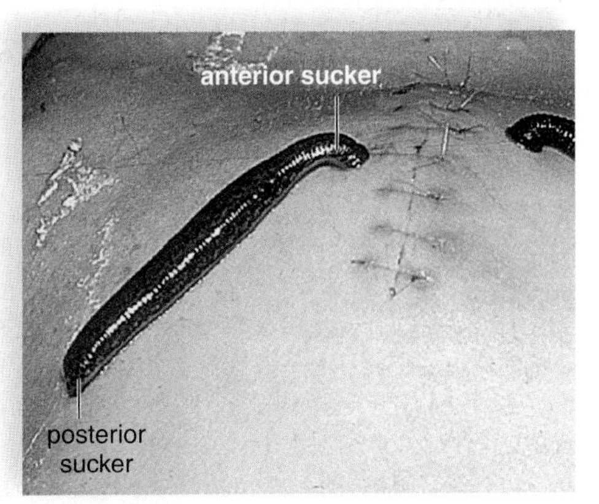


c. Medicinal leech, *Hirudo medicinalis*

**FIGURE 28.18 Polychaete diversity.**

**a.** Clam worms are predaceous polychaetes that undergo cephalization. Note also the parapodia, which are used for swimming and as respiratory organs. **b.** Christmas tree worms (a type of tube worm) are sessile feeders whose radioles (ciliated mouth appendages) spiral as shown here. **c.** Medical leech is a blood sucker.

# 28.4 Quantity Among the Ecdysozoans

The ecdysozoans are protostomes, as are the lophotrochozoans. The term *ecdysis* means molting, and both roundworms and arthropods, which belong to this group, periodically shed their outer covering.

## Roundworms

**Roundworms** (phylum Nematoda) are nonsegmented worms that are prevalent in almost any environment. Generally, nematodes are colorless and range in size from microscopic to exceeding 1 m in length. The internal organs, including the tubular reproductive organs, lie within the pseudocoelom. A **pseudocoelom** [Gk. *Pseudes,* false, and *koiloma,* cavity] is a body cavity that is incompletely lined by mesoderm. In other words, mesoderm occurs inside the body wall but not around the digestive cavity (gut).

Nematodes have developed a variety of lifestyles from free-living to parasitic. One species, *Caenorhabditis elegans,* a free-living nematode, is a model animal used in genetics and developmental biology.

### *Parasitic Roundworms*

An *Ascaris* infection is common in humans, cats, dogs, pigs, and a number of other vertebrates. As in other nematodes, *Ascaris* males tend to be smaller (15–31 cm long) than females (20–49 cm long) (Fig. 28.19*a*). In males, the posterior end is curved and comes to a point. Both sexes move by means of a characteristic whiplike motion because only longitudinal muscles lie next to the body wall. A typical female *Ascaris* is very prolific, producing over 200,000 eggs daily. The eggs are eliminated with the host's feces and can remain viable in the soil for many months. Eggs enter the body via uncooked vegetables, soiled fingers, or ingested fecal material and hatch in the intestines. The juveniles make their way into the veins and lymphatic vessels and are carried to the heart and lungs. From the lungs, the larvae travel up the trachea, where they are swallowed and eventually reach the intestines. There, the larvae mature and begin feeding on intestinal contents.

The symptoms of an *Ascaris* infection depend on the stage of the infection. Larval *Ascaris* in the lungs can cause pneumonia-like symptoms. In the intestines, *Ascaris* can cause malnutrition; blockage of the bile duct, pancreatic duct, and appendix; and poor health.

*Trichinosis,* caused by *Trichinella spiralis,* is a serious infection that humans can contract when they eat rare pork containing encysted *Trichinella* larvae. After maturation, the female adult burrows into the wall of the small intestine and produces living offspring that are carried by the bloodstream to the skeletal muscles, where they encyst (Fig. 28.19*b*). Heavy infections can be painful and lethal.

Filarial worms, a type of roundworm, cause various diseases. In the United States, mosquitoes transmit the larvae of a parasitic filarial worm to dogs. Because the worms live in the heart and the arteries that serve the lungs, the infection is called *heartworm disease.* The condition can be fatal; therefore, heartworm medicine is recommended as a preventive measure for all dogs.

Elephantiasis is a disease of humans caused by the filarial worm, *Wuchereria bancrofti.* Restricted to tropical areas of Africa, this parasite also uses a mosquito as a secondary host. Because the adult worms reside in lymphatic vessels, collection of fluid is impeded, and the limbs of an infected person may swell to a monstrous size (Fig. 28.19*c*). Elephantiasis is treatable in its early stages but usually not after scar tissue has blocked lymphatic vessels.

Pinworms are the most common nematode parasite in the United States. The adult parasites live in the cecum and large intestine. Females migrate to the anal region at night and lay their eggs. Scratching the resultant itch can contaminate hands, clothes, and bedding. The eggs are swallowed, and the life cycle begins again.

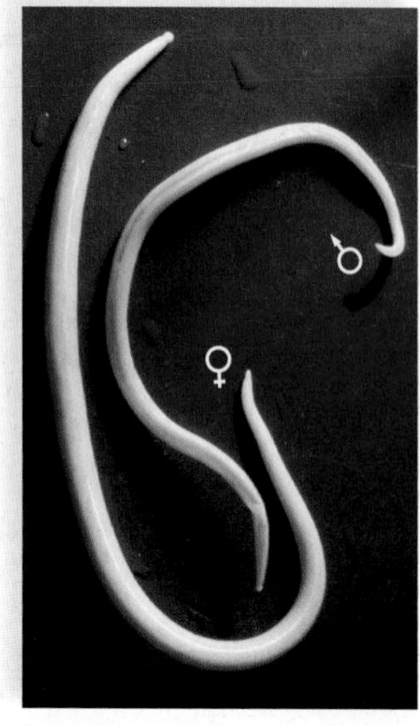


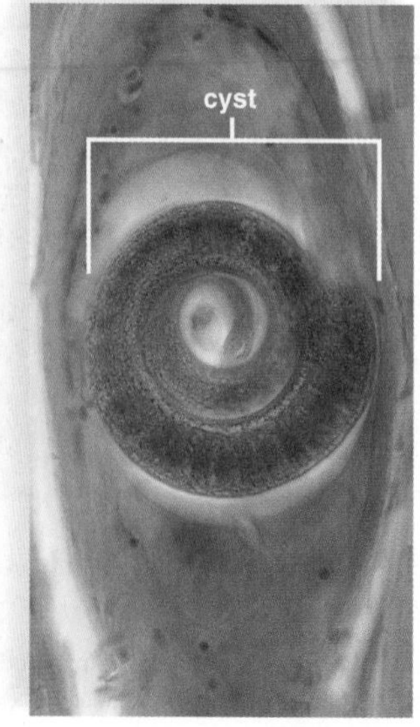

**FIGURE 28.19**
**Roundworm diversity.**
**a.** *Ascaris,* a common cause of a roundworm infection in humans. **b.** Encysted *Trichinella* larva in muscle. **c.** A filarial worm infection causes elephantiasis, which is characterized by a swollen body part when the worms block lymphatic vessels.

## Arthropods

The **arthropods** (phylum Arthropoda [Gk. *arthron,* joint, and *podos,* foot]) vary greatly in size. The parasitic mite measures less than 0.1 mm in length, while the Japanese crab measures up to 4 m in length. Arthropods, which also occupy every type of habitat, are considered the most successful group of all the animals. The remarkable success of arthropods is dependent on five characteristics:

1. A rigid but jointed **exoskeleton** (Fig. 28.20*a, b*). The exoskeleton is composed primarily of **chitin** [Gk. *chiton,* tunic], a strong, flexible, nitrogenous polysaccharide. The exoskeleton serves many functions, including protection, attachment for muscles, locomotion, and prevention of desiccation. However, because it is hard and nonexpandable, arthropods must molt, or shed, the exoskeleton as they grow larger. Arthropods have this in common with other ecdysozoans. Before molting, the body secretes a new, larger exoskeleton, which is soft and wrinkled, underneath the old one. After enzymes partially dissolve and weaken the old exoskeleton, the animal breaks it open and wriggles out. The new exoskeleton then quickly expands and hardens (Fig. 28.20*c*).
2. Segmentation. Segmentation is readily apparent because each segment has a pair of jointed appendages, even though certain segments are fused into a head, thorax, and abdomen. The jointed appendages of arthropods are basically hollow tubes moved by muscles. Typically, the appendages are highly adapted for a particular function, such as food gathering, reproduction, and locomotion. In addition, many appendages are associated with sensory structures and used for tactile purposes.
3. Well-developed nervous system. Arthropods have a brain and a ventral nerve cord. The head bears various types of sense organs, including eyes of two types—simple and compound. The compound eye is composed of many complete visual units, each of which operates independently (Fig. 28.20*d*). The lens of each visual unit focuses an image on the light-sensitive membranes of a small number of photoreceptors within that unit. The simple eye, like that of vertebrates, has a single lens that brings the image to focus onto many receptors, each of which receives only a portion of the image. In addition to sight, many arthropods have well-developed touch, smell, taste, balance, and hearing. Arthropods display many complex behaviors and methods of communication.
4. Variety of respiratory organs. Marine forms use gills, which are vascularized, highly convoluted, thin-walled tissue specialized for gas exchange. Terrestrial forms have book lungs (e.g., spiders) or air tubes called **tracheae** [L. *trachia,* windpipe]. Tracheae serve as a rapid way to transport oxygen directly to the cells.
5. Reduced competition through **metamorphosis** [Gk. *meta,* implying change, and *morphe,* shape, form]. Many arthropods undergo a drastic change in form and physiology that occurs as an immature stage, called a larva, becomes an adult. Among arthropods, the larva eats different food and lives in a different environment than the adult. For example, larval crabs live among and feed on plankton, while adult crabs are bottom dwellers that catch live prey or scavenge dead organic matter. Among insects, such as butterflies, the caterpillar feeds on leafy vegetation, while the adult feeds on nectar.

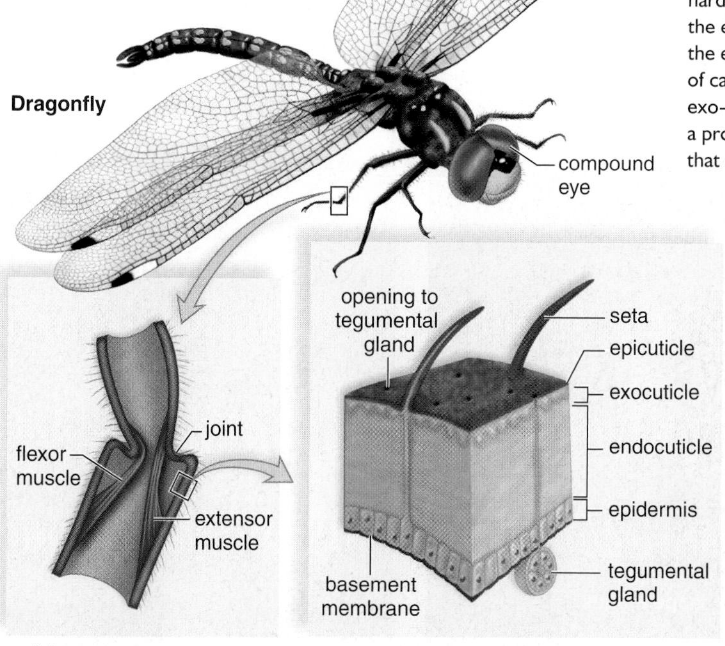

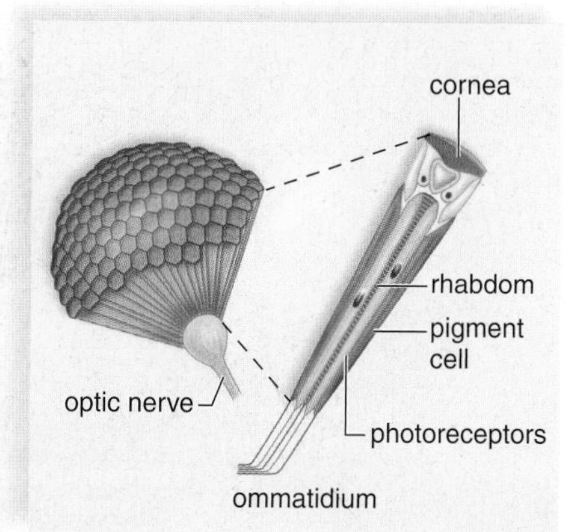


**FIGURE 28.20 Arthropod skeleton and eye.**

**a.** The joint in an arthropod skeleton is a region where the cuticle is thinner and not as hard as the rest of the cuticle. The direction of movement is toward the flexor muscle or the extensor muscle, whichever one has contracted. **b.** The exoskeleton is secreted by the epidermis and consists of the endocuticle; the exocuticle, hardened by the deposition of calcium carbonate; and the epicuticle, a waxy layer. Chitin makes up the bulk of the exo- and endocuticles. **c.** Because the exoskeleton is nonliving, it must be shed through a process called molting for the arthropod to grow. **d.** Arthropods have a compound eye that contains many individual units, each with its own lens and photoreceptors.

## *Crustaceans*

The name **crustacean** is derived from their hard, crusty exoskeleton, which contains calcium carbonate in addition to the typical chitin. Although crustaceans are extremely diverse, the head usually bears a pair of compound eyes and five pairs of appendages. The first two pairs, called antennae and antennules, lie in front of the mouth and have sensory functions. The other three pairs (mandibles, first and second maxillae) lie behind the mouth and are mouthparts used in feeding. Biramous [Gk. *bis,* two, and *ramus,* a branch] appendages on the thorax and abdomen are segmentally arranged; one branch is the gill branch, and the other is the leg branch.

The majority of crustaceans live in marine and aquatic environments (Fig. 28.21). **Decapods,** which are the most familiar and numerous crustaceans, include lobsters, crabs, crayfish, hermit crabs, and shrimp. These animals have a thorax that bears five pairs of walking appendages. Typically, the gills are positioned above the walking legs. The first pair of walking legs may be modified as claws.

Copepods and krill are small crustaceans that live in the water, where they feed on algae. In the marine environment, they serve as food for fishes, sharks, and whales. They are so numerous that, despite their small size, some believe they are harvestable as food. Barnacles are also crustaceans, but they have a thick, heavy shell as befits their inactive lifestyle. Barnacles can live on wharf pilings, ship hulls, seaside rocks, and even the bodies of whales. They begin life as free-swimming larvae, but they undergo a metamorphosis that transforms their swimming appendages to cirri, feathery structures that are extended and allow them to filter feed when they are submerged.

**Anatomy of a Crayfish.** Figure 28.22*a* gives a view of the external anatomy of the crayfish. The head and thorax are fused into a cephalothorax, which is covered on the top and sides by a nonsegmented carapace. The abdominal segments are equipped with swimmerets, small paddlelike structures. The first two pairs of swimmerets in the male, known as claspers, are quite strong and are used to pass sperm to the female.

The last two segments bear the uropods and the telson, which make up a fan-shaped tail. Ordinarily, a crayfish lies in wait for prey. It faces out from an enclosed spot with the claws extended, and the antennae moving about. The claws seize any small animal, either dead or living that happens by, and carry it to the mouth. When a crayfish moves about, it generally crawls slowly, but may swim rapidly by using its heavy abdominal muscles and tail.

The respiratory system consists of gills that lie above the walking legs protected by the carapace. As shown in Figure 28.22*b*, the digestive system includes a stomach, which is divided into two main regions: an anterior portion called the gastric mill, equipped with chitinous teeth to grind coarse food, and a posterior region, which acts as a filter to prevent coarse particles from entering the digestive glands, where absorption takes place. Green glands lying in the head region, anterior to the esophagus, excrete metabolic wastes through a duct that opens externally at the base of the antennae. The coelom, which

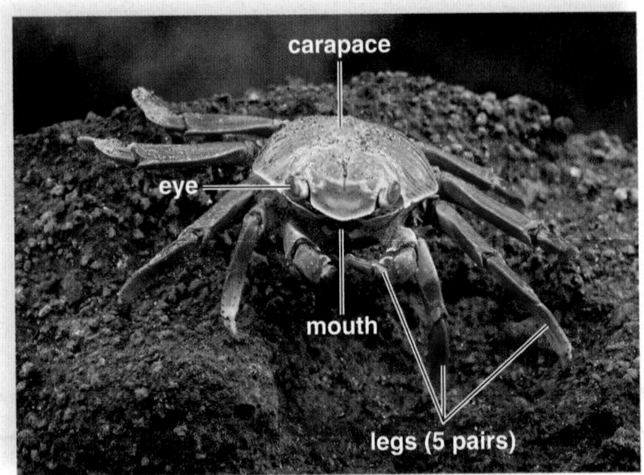

a. Sally lightfoot crab, *Grapsus grapsus*

**FIGURE 28.21 Crustacean diversity.**

The crayfish on the next page, crabs (**a**), and shrimp (**b**) are decapods—they have five pairs of walking legs. Shrimp resemble crayfish more closely than crabs, which have a reduced abdomen. Pelagic shrimp feed on copepods, such as the one seen from below in (**c**). A copepod has long antennae used for floating, and feathery maxillae used for filter feeding. Barnacles have no abdomen and a reduced head; the thoracic legs project through a shell to filter feed. Barnacles often live on human-made objects such as ships, buoys, and cables. The gooseneck barnacle (**d**) is attached to an object by a long stalk.

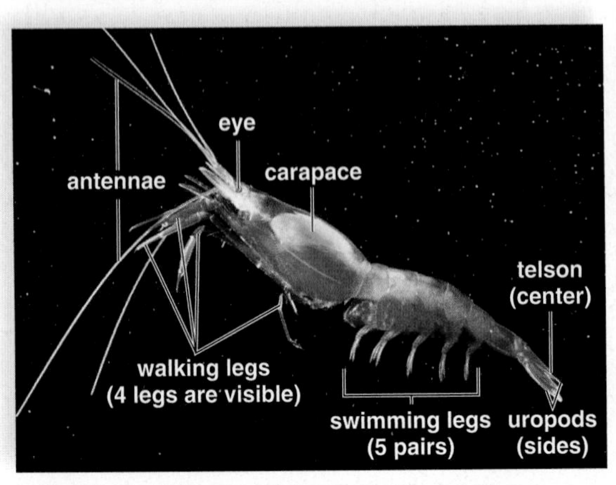

b. Red-backed cleaning shrimp, *Lysmata grasbhami*

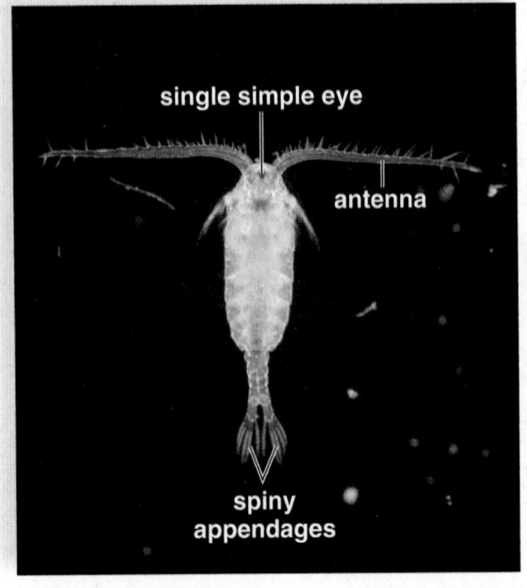

c. Copepod, *Diaptomus*

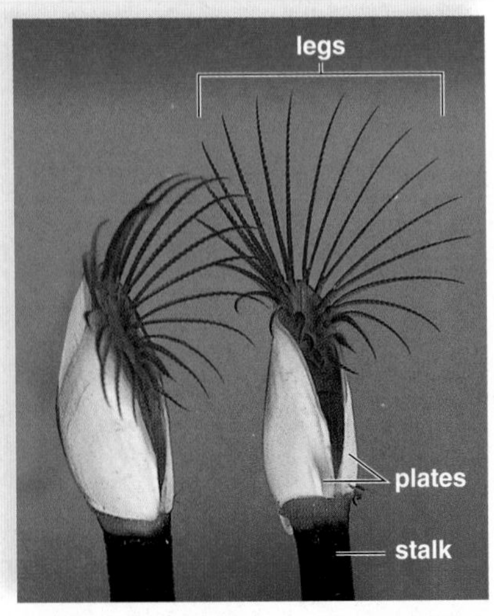

d. Gooseneck barnacles, *Lepas anatifera*

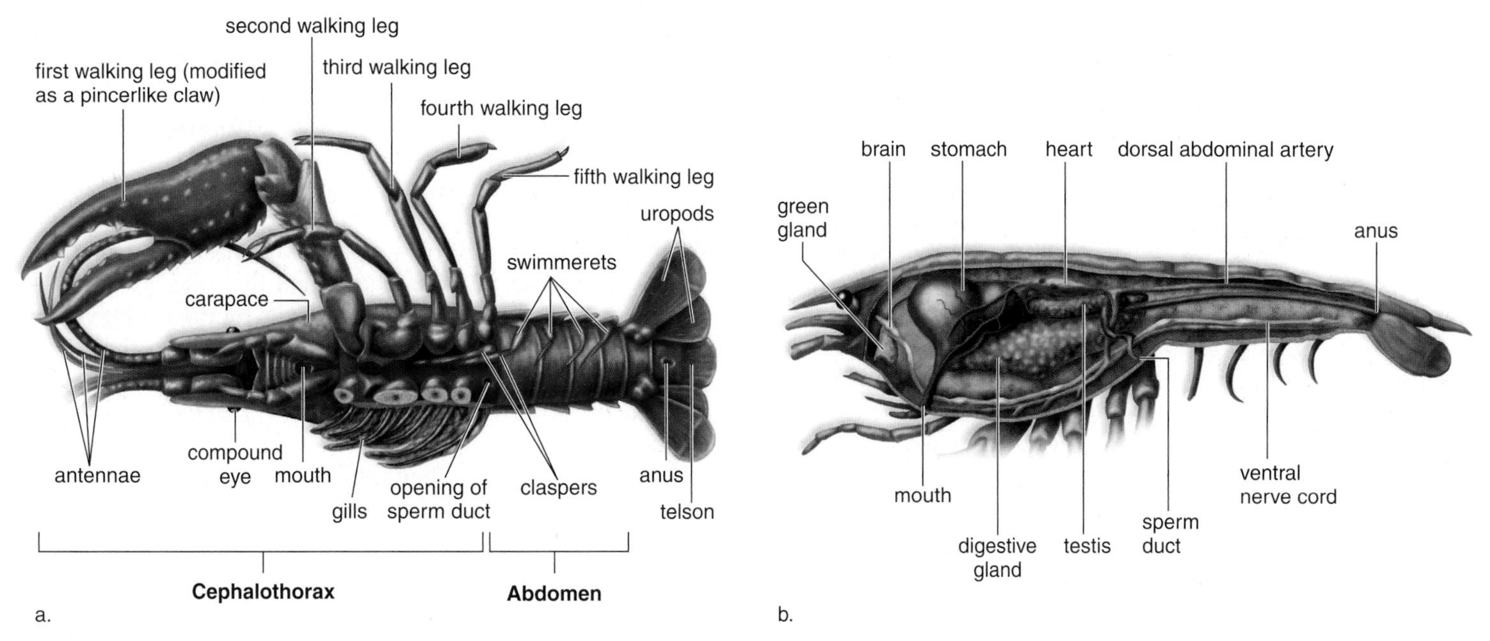


**FIGURE 28.22 Male crayfish, *Cambarus*.**

**a.** Externally, it is possible to observe the jointed appendages, including the swimmerets, and the walking legs, which include the claws. These appendages, plus a portion of the carapace, have been removed from the right side so that the gills are visible. **b.** Internally, the parts of the digestive system are particularly visible. The circulatory system can also be clearly seen. Note the ventral nerve cord.

is so well developed in the annelids, is reduced in the arthropods and is composed chiefly of the space about the reproductive system. A heart pumps hemolymph containing the blue respiratory pigment hemocyanin into a *hemocoel* consisting of sinuses (open spaces), where the hemolymph flows about the organs. As in the molluscs, this is an open circulatory system because blood is not contained within blood vessels.

The crayfish nervous system is well developed. Crayfish have a brain and a ventral nerve cord that passes posteriorly. Along the length of the nerve cord, periodic ganglia give off lateral nerves. Sensory organs are well developed. The compound eyes are found on the ends of movable eyestalks. These eyes are accurate and can detect motion and respond to polarized light. Other sensory organs include tactile antennae and chemosensitive setae. Crayfish also have statocysts that serve as organs of equilibrium.

The sexes are separate in the crayfish, and the gonads are located just ventral to the pericardial cavity. In the male, a coiled sperm duct opens to the outside at the base of the fifth walking leg. Sperm transfer is accomplished by the modified first two swimmerets of the abdomen. In the female, the ovaries open at the bases of the third walking legs. A stiff fold between the bases of the fourth and fifth pairs serves as a seminal receptacle. Following fertilization, the eggs are attached to the swimmerets of the female. Young hatchlings are miniature adults, and no metamorphosis occurs.

## *Centipedes and Millipedes*

The centipedes and millipedes are known for their many legs (Fig. 28.23*a*). In **centipedes** ("hundred-leggers"), each of their many body segments has a pair of walking legs. The approximately 3,000 species prefer to live in moist environments such as under logs, in crevices, and in leaf litter, where they are active predators on worms, small crustaceans, and insects. The head of a centipede includes paired antennae and jawlike mandibles. Appendages on the first trunk segment are clawlike venomous jaws that kill or immobilize prey, while mandibles chew.

In **millipedes** ("thousand leggers"), each of four thoracic segments bears one pair of legs (Fig. 28.23*b*), while abdominal segments have two pairs of legs. Millipedes live under stones or burrow in the soil as they feed on leaf litter. Their cylindrical bodies have a tough chitinous exoskeleton. Some secrete hydrogen cyanide, a poisonous substance.

a.

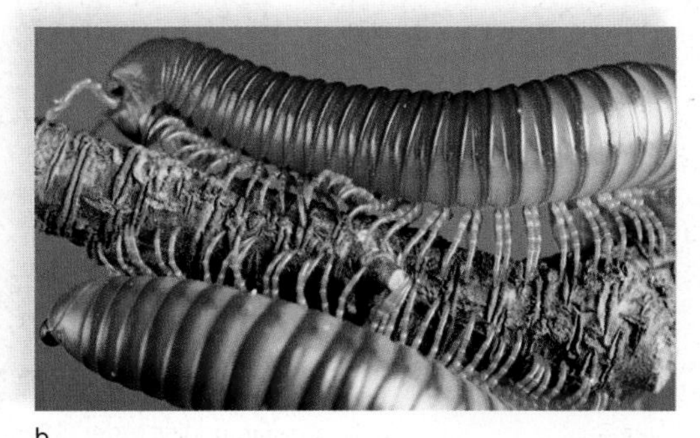

b.

**FIGURE 28.23**
**Centipede and millipede.**

**a.** A centipede has a pair of appendages on almost every segment. **b.** A millipede has two pairs of legs on most segments.

### Insects

**Insects** are adapted for an active life on land, although some have secondarily invaded aquatic habitats. The body of an insect is divided into a head, a thorax, and an abdomen. The head bears the sense organs and mouthparts (Fig. 28.24). The thorax bears three pairs of legs and possibly one or two pairs of wings; and the abdomen contains most of the internal organs. Wings enhance an insect's ability to survive by providing a way of escaping enemies, finding food, facilitating mating, and dispersing offspring.

In the grasshopper (Fig. 28.25), the third pair of legs is suited to jumping. There are two pairs of wings. The forewings are tough and leathery, and when folded back at rest, they protect the broad, thin hindwings. On each lateral surface, the first abdominal segment bears a large tympanum for the reception of sound waves. The posterior region of the exoskeleton in the female has an ovipositor, used to dig a hole in which eggs are laid.

The digestive system is suitable for a herbivorous diet. In the mouth, food is broken down mechanically by mouthparts and enzymatically by salivary secretions. Food is temporarily stored in the crop before passing into the gizzard, where it is finely ground. Digestion is completed in the stomach, and nutrients are absorbed into the hemocoel from outpockets called gastric ceca (*cecum,* a cavity open at one end only). The excretory system consists of **Malpighian tubules,** which extend into a hemocoel and collect nitrogenous wastes that are concentrated and excreted into the digestive tract. The formation of a solid nitrogenous waste, namely uric acid, conserves water.

The respiratory system begins with openings in the exoskeleton called spiracles. From here, air enters small tubules called tracheae (Fig. 28.25*a*). The tracheae branch and rebranch, finally ending in moist areas where the actual exchange of gases takes place. No individual cell is very far from a site of gas exchange. The movement of air through this complex of tubules is not a passive process; air is pumped through by a series of bladderlike structures (air sacs) attached to the tracheae near the spiracles. Air enters the anterior four spiracles and exits by the posterior six spiracles. Breathing by tracheae may account for the small size of insects (most are less than 6 cm in length), since the tracheae are so tiny and fragile that they would be crushed by any amount of weight.

The circulatory system contains a slender, tubular heart that lies against the dorsal wall of the abdominal exoskeleton and pumps hemolymph into the hemocoel, where it circulates before returning to the heart again. The hemolymph is colorless and lacks a respiratory pigment, and so transports nutrients and wastes. The highly efficient tracheal system transports respiratory gases.

Grasshoppers undergo *incomplete metamorphosis,* a gradual change in form as the animal matures. The immature grasshopper, called a nymph, is recognizable as a grasshopper, even though it differs in body proportions from the adult. Other insects, such as butterflies, undergo *complete metamorphosis,* involving drastic changes in form. At first, the animal is a wormlike larva (caterpillar) with chewing mouthparts. It then forms a case, or cocoon, about itself and becomes a pupa. During this stage, the body parts are completely reorganized;

**FIGURE 28.24** **Insect diversity.**

Mealybug, order Homoptera

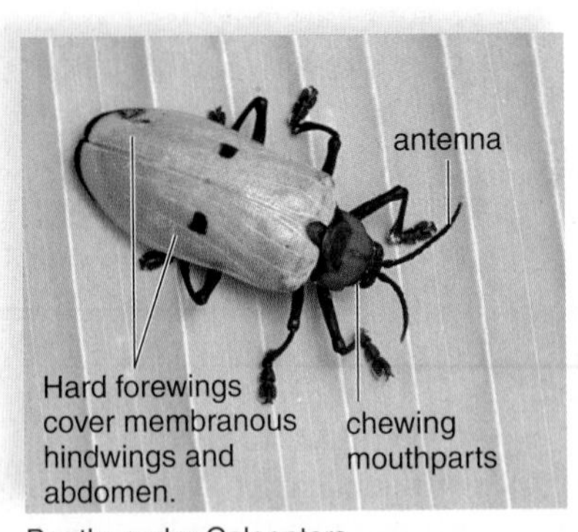

Beetle, order Coleoptera

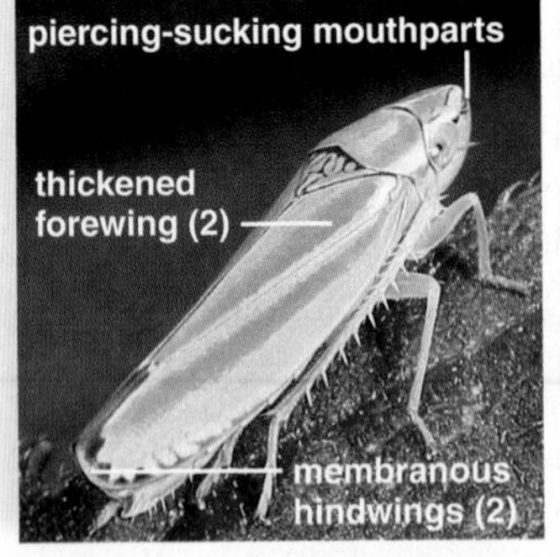

Leafhopper, order Homoptera

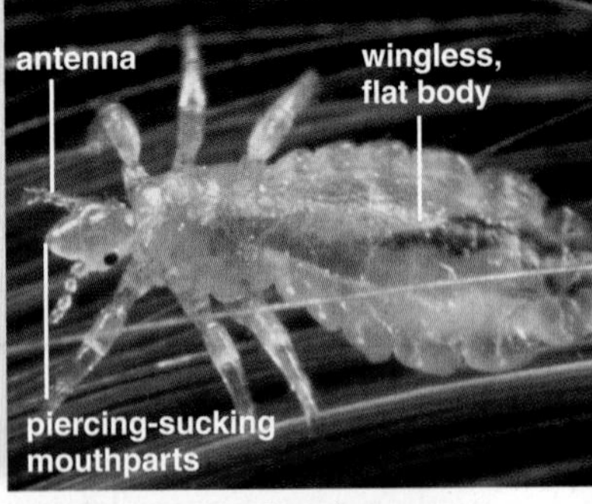

Head louse, order Anoplura

Wasp, order Hymenoptera

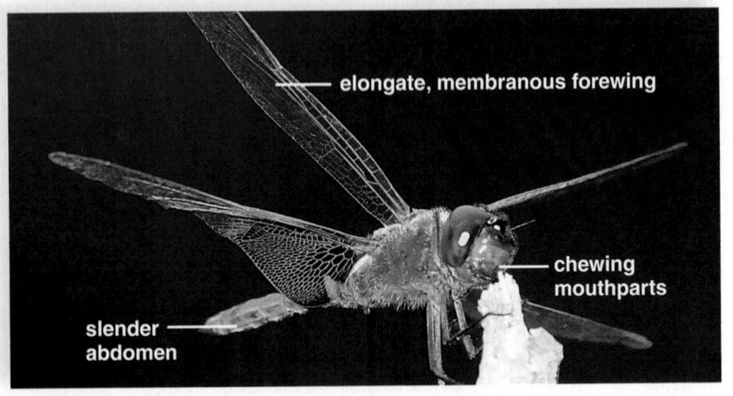

Dragonfly, order Odonata

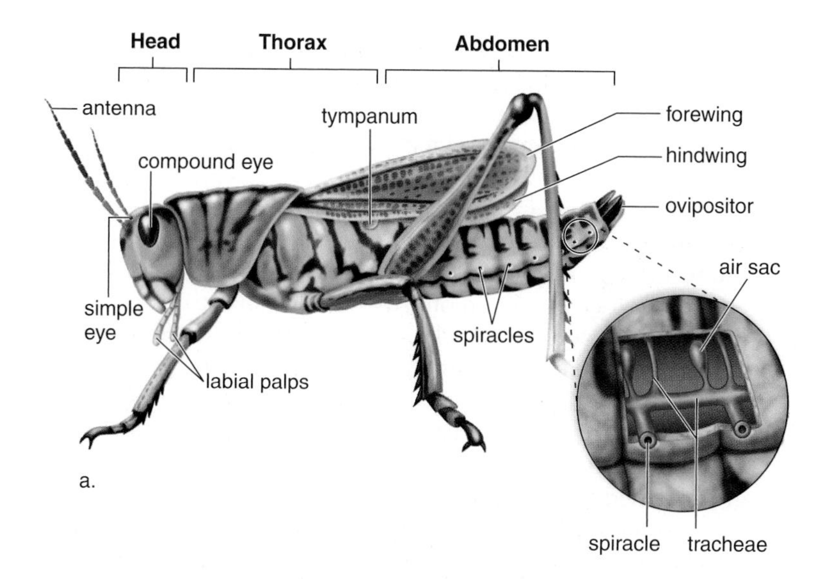


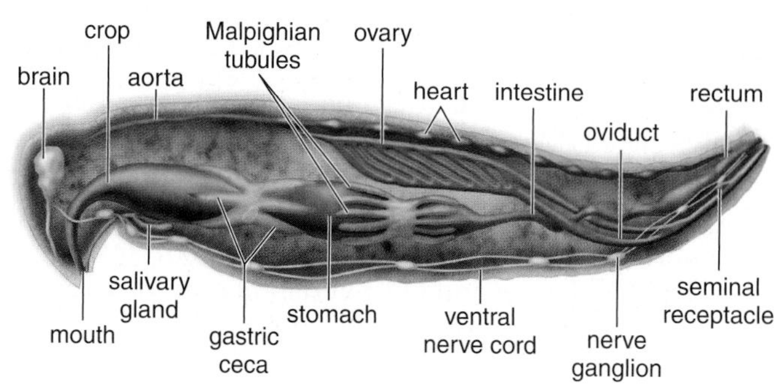


**FIGURE 28.25 Female grasshopper, *Romalea*.**

**a.** Externally, the body of a grasshopper is divided into three sections and has three pairs of legs. The tympanum receives sound waves, and the jumping legs and the wings are for locomotion. **b.** Internally, the digestive system is specialized. The Malpighian tubules excrete a solid nitrogenous waste (uric acid). A seminal receptacle receives sperm from the male, which has a penis.

the adult then emerges from the cocoon. This life cycle allows the larvae and adults to use different food sources.

Insects show remarkable behavior adaptations. Bees, wasps, ants, termites, and other colonial insects have complex societies.

## *Chelicerates*

The **chelicerates** live in terrestrial, aquatic, and marine environments. The first pair of appendages is the pincerlike chelicerae, used in feeding and defense. The second pair is the pedipalps, which can have various functions. A *cephalothorax* [Gk. *kephale*, head, and *thorax*, breastplate] (fused head and thorax) is followed by an abdomen that contains internal organs.

Horseshoe crabs of the genus *Limulus* are familiar along the east coast of North America. The body is covered by exoskeletal shields. The anterior shield is a horseshoe-shaped carapace, which bears two prominent compound eyes (Fig. 28.26*a*). Ticks, mites, scorpions, spiders, and harvestmen are all arachnids. Over 25,000 species of mites and ticks have been classified, some of which are parasitic on a variety of other animals. Ticks are ectoparasites of various vertebrates, and they are carriers for such diseases as Rocky Mountain spotted fever and Lyme disease. When not attached to a host, ticks hide on plants and in the soil.

Scorpions can be found on all continents except Antarctica (Fig. 28.26*b*). North America is home to approximately 1,500 species. Scorpions are nocturnal and spend most of the day hidden under a log or a rock. Their pedipalps are large pincers, and their long abdomen ends with a stinger that contains venom.

Presently, over 35,000 species of spiders have been classified (Fig. 28.26*c*). Spiders, the most familiar chelicerates, have a narrow waist that separates the cephalothorax from the abdomen. Spiders do not have compound eyes; instead, they have numerous simple eyes that perform a similar function. The chelicerae are modified as fangs, with ducts from poison glands, and the pedipalps are used to hold, taste, and chew food. The abdomen often contains silk glands, and spiders spin a web in which to trap their prey.

### Check Your Progress 28.4

1. Name two ways the roundworms are anatomically similar to the arthropods.
2. Name two ways crustaceans are adapted to an aquatic life and insects are adapted to living on land.
3. What feature do the chelicerates have in common?

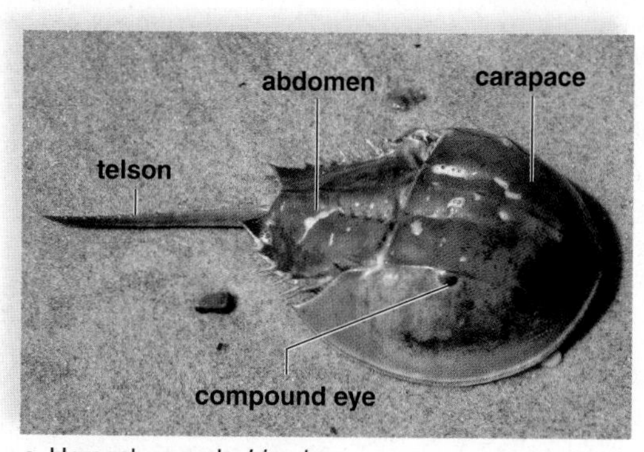


a. Horseshoe crab, *Limulus*

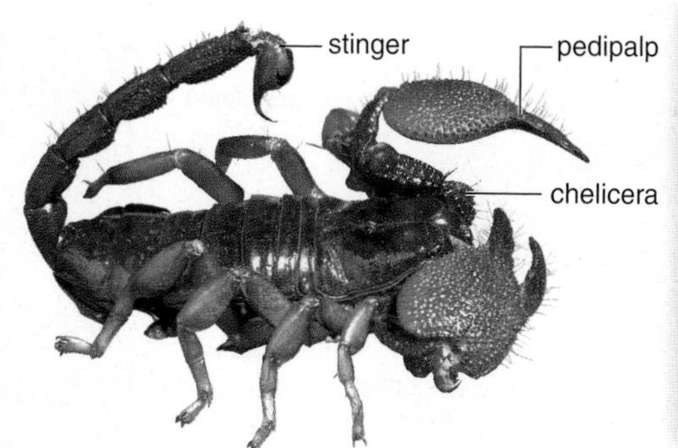


b. Kenyan giant scorpion, *Pandinus*

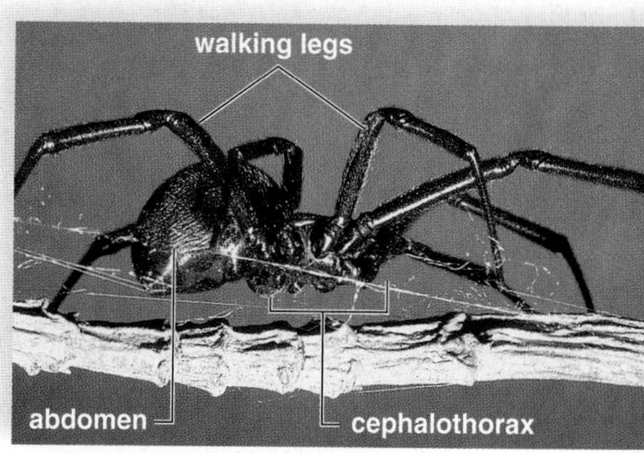


c. Black widow spider, *Latrodectus*

**FIGURE 28.26 Chelicerate diversity.**

**a.** Horseshoe crabs are common along the east coast. **b.** Scorpions are more common in tropical areas. **c.** The black widow spider is a poisonous spider that spins a web.

# 28.5 Invertebrate Deuterostomes

Molecular data tell us that echinoderms and chordates are closely related. Morphological data indicate that these two groups share the deuterostome pattern of development, which was described on page 516. The echinoderms and a few chordates are invertebrates; the echinoderms are discussed in this chapter and the invertebrate chordates are discussed in Chapter 29. Most of the chordates are vertebrates, as will become apparent.

## Echinoderms

**Echinoderms** (phylum Echinodermata [Gk. *echinos*, spiny, and *derma*, skin]) are primarily bottom-dwelling marine animals. They range in size from brittle stars less than 1 cm in length to giant sea cucumbers over 2 m long. The most striking feature of echinoderms is their 5-pointed radial symmetry, as illustrated by a sea star. Although echinoderms are radially symmetrical as adults, their larvae are free-swimming filter feeders with bilateral symmetry. Echinoderms have an endoskeleton of spiny calcium-rich plates called ossicles. The spines protruding from their skin account for the phylum name Echinodermata. Another innovation is their unique **water vascular system** consisting of canals and appendages that function in locomotion, feeding, gas exchange, and sensory reception.

The more familiar of the echinoderms are the Asteroidea containing the sea stars (Fig. 28.27*a*, *b*), which are studied here; the Holothurians, including the sea cucumbers, which have long leathery bodies and resemble a cucumber (Fig. 28.27*c*); and the Echinoidea, including the sea urchin and sand dollar, both of which use their spines for locomotion, defense, and burrowing (Fig. 28.27*d*). Less familiar are the Ophiuroidea, which includes the brittle stars, with a central disk surrounded by radially flexible arms, and the Crinoidea, the oldest group, which includes the stalked feather stars and the motile feather stars.

### Sea Stars

**Sea stars** number about 1,600 species that are commonly found along rocky coasts, where they feed on clams, oysters, and other bivalve molluscs. Various structures project through the body wall: (1) spines from the endoskeletal plates offer some protection; (2) pincerlike structures around the bases of spines keep the surface free of small particles; and (3) skin gills, tiny fingerlike extensions of the skin, are used for respiration. On the oral surface, each arm has a groove lined by little **tube feet** (Fig. 28.27).

To feed, a sea star positions itself over a bivalve and attaches some of its tube feet to each side of the shell. By working its tube feet in alternation, it pulls the shell open. A very small crack is enough for the sea star to evert its cardiac stomach and push it through the crack, so that it contacts the soft parts of the bivalve. The stomach secretes enzymes, and digestion begins, even while the bivalve is attempting to close its shell. Later, partly digested food is taken into the sea star's body, where digestion continues in the pyloric stomach using enzymes from the digestive glands found in each arm. A short intestine opens at the anus on the aboral side (side opposite the mouth).

In each arm, the well-developed coelom contains a pair of digestive glands and gonads (either male or female) that open on the aboral surface by very small pores. The nervous system consists of a central nerve ring that gives off radial nerves in each arm. A light-sensitive eyespot is at the tip of each arm.

**FIGURE 28.27 Echinoderms.**

**a.** Sea star (starfish) anatomy. Like other echinoderms, sea stars have a water vascular system that begins with the sieve plate and ends with expandable tube feet. **b.** The red sea star, *Mediastar*, uses the suction of its tube feet to open a clam, a primary source of food. **c.** Sea cucumber, *Pseudocolochirus*. **d.** Sea urchin, *Strongylocentrotus*.

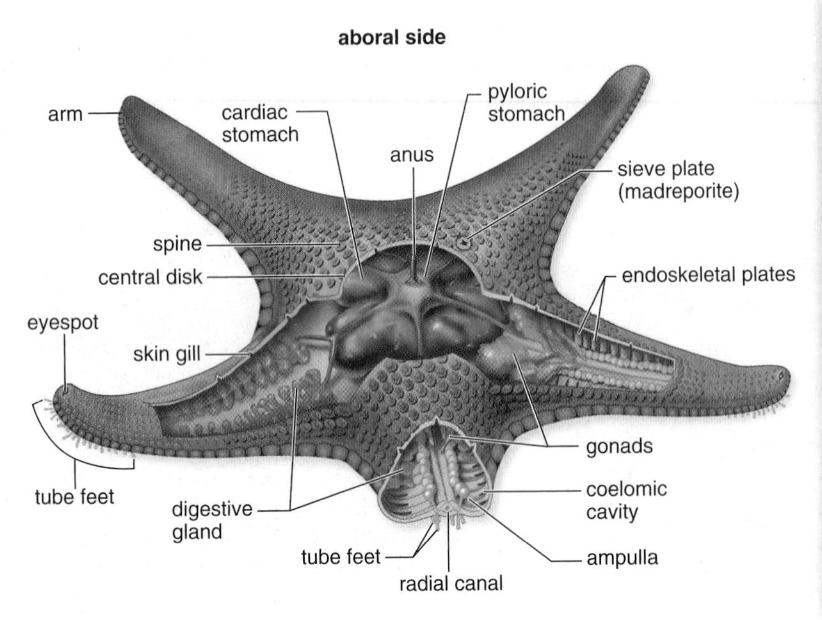


a.

b. Red sea star, *Mediastar*

Locomotion depends on the water vascular system. Water enters this system through a structure on the aboral side called the *sieve plate,* or *madreporite.* From there it passes through a stone canal to a ring canal, which circles around the central disc, and then to a radial canal in each arm. From the radial canals, many lateral canals extend into the tube feet, each of which has an ampulla. Contraction of an ampulla forces water into the tube foot, expanding it. When the foot touches a surface, the center is withdrawn, giving it suction so that it can adhere to the surface. By alternating the expansion and contraction of the tube feet, a sea star moves slowly along.

Echinoderms do not have a respiratory, excretory, or circulatory system. Fluids within the coelomic cavity and the water vascular system carry out many of these functions. For example, gas exchange occurs across the skin gills and the tube feet. Nitrogenous wastes diffuse through the coelomic fluid and the body wall. Cilia on the peritoneum lining the coelom keep the coelomic fluid moving.

Sea stars reproduce asexually and sexually. If the body is fragmented, each fragment can regenerate a whole animal as long as a portion of the central disc is present. Sea stars spawn and release either eggs or sperm at the same time. The bilateral larva will undergo a metamorphosis to become the radially symmetrical adult.

**Check Your Progress 28.5**

1. What evidence do we have that echinoderms evolved from bilaterally symmetrical animals?
2. Describe the functions of the water vascular system in sea stars.

## Connecting the Concepts

About 90% of all animal species have a coelom and are in the phyla Mollusca, Annelida, Arthropoda, Echinodermata, and Chordata. (The chordates, which include the vertebrates, will be studied in Chapter 29.) This indicates that there are some adaptive advantages to having a coelom, or perhaps a combination of bilateral symmetry and the complexity made possible by the presence of a coelom. Most animals with a coelom have both circulatory and respiratory systems, which permit them to be active. The echinoderms are an anomaly because they lack these systems and have radial symmetry.

Some of the coelomate phyla are protostomes (molluscs, annelids, and arthropods) and some are deuterostomes (echinoderms and chordates) on the basis of embryological development. However, it is interesting that both lines of descent have representatives that are segmented. Based on the current phylogenetic tree, segmentation must have evolved three separate times in the annelids, arthropods, and chordates. Segmentation of the annelids and many arthropods is obvious because the body has demarcations that can be seen externally. The repeating vertebrae of the backbone in vertebrates signal that they, too, are segmented. Was a coelom present before the evolution of segmentation? Perhaps it was. Later, a partitioned coelom may have provided a hydrostatic skeleton that enabled a worm to burrow more efficiently in the soil. In other words, there was a selective advantage to having a segmented coelom in organisms that lacked limbs. The later evolution of limbs in jointed arthropods (e.g., insects) and vertebrates (e.g., frogs, lizards, horses) was even more adaptive for locomotion on land.

Animals that live on land must also have a means of reproduction that allows fertilization and development to take place without requiring external water.

c. Sea cucumber, *Pseudocolochirus*

d. Purple sea urchin, *Strongylocentrotus*

## summary

### 28.1 Evolution of Animals

Animals are multicellular organisms that are heterotrophic and ingest their food. They have the diploid life cycle. Typically, they have the power to move by means of contracting fibers. It is hypothesized that animals evolved from a protist that resembles the choanoflagellates of today. Molecular data used to construct an evolutionary tree of the animals can be substantiated by morphological data.

### 28.2 Introducting the Invertebrates

Sponges resemble colonial protozoans. They have the cellular level of organization, lack tissues, and have various symmetries. Sponges are sessile filter feeders and depend on a flow of water through the body to acquire food, which is digested in vacuoles within collar cells that line a central cavity.

Comb jellies and cnidarians have two tissue layers derived from the germ layers ectoderm and endoderm. They are radially symmetrical.

Cnidarians have a sac body plan. They exist as either polyps or medusae, or they can alternate between the two. Hydras and their relatives—sea anemones and corals—are polyps; in jellyfishes, the medusan stage is dominant. In *Hydra* and other cnidarians, an outer epidermis is separated from an inner gastrodermis by mesoglea. They possess tentacles to capture prey and cnidocytes armed with nematocysts to stun it. A nerve net coordinates movements.

### 28.3 Variety Among the Lophotrochozoans

Free-living flatworms (planarians) exemplify that flatworms have three tissue layers and no coelom. Planarians have muscles and a ladder-type nervous system, and they show cephalization. They take in food through an extended pharynx leading to a gastrovascular cavity, which extends throughout the body. There is an osmotic-regulating organ that contains flame cells. Flukes and tapeworms are parasitic. Flukes have two suckers by which they attach to and feed from their hosts. Tapeworms have a scolex with hooks and suckers for attaching to the host intestinal wall. The body of a tapeworm is made up of proglottids, which, when mature, contain thousands of eggs. If these eggs are taken up by pigs or cattle, larvae become encysted in their muscles. If humans eat this meat, they too may become infected with a tapeworm.

Rotifers are microscopic and have a corona that resembles a spinning wheel when in motion.

The body of a mollusc typically contains a visceral mass, a mantle, and a foot. Many also have a head and a radula. The nervous system consists of several ganglia connected by nerve cords. There is a reduced coelom and an open circulatory system. Clams (bivalves) are adapted to a sedentary coastal life, squids (cephalopods) to an active life in the sea, and snails (gastropods) to life on land.

Annelids are segmented worms. They have a well-developed coelom divided by septa, a closed circulatory system, a ventral solid nerve cord, and paired nephridia. Earthworms are oligochaetes ("few bristles") that use the body wall for gas exchange. Polychaetes ("many bristles") are marine worms that have parapodia. They may be predators, with a definite head region, or they may be filter feeders, with ciliated tentacles to filter food from the water. Leeches also belong to this phylum.

### 28.4 Quantity Among the Ecdysozoans

Roundworms have a pseudocoelom. Roundworms are usually small and very diverse; they are present almost everywhere in great numbers. Many are significant parasites of humans. The parasite *Ascaris* is representative of the group. Infections can also be caused by *Trichinella,* whose larval stage encysts in the muscles of humans. Elephantiasis is caused by a filarial worm that blocks lymphatic vessels.

Arthropods are the most varied and numerous of animals. Their success is largely attributable to a flexible exoskeleton, specialized body regions, and jointed appendages. Also important are a high degree of cephalization, a variety of respiratory organs, and reduced competition through metamorphosis. Crustaceans (crayfish, lobsters, shrimps, copepods, krill, and barnacles) have a head that bears compound eyes, antennae, antennules, and mouthparts. Crayfish illustrate other features, such as an open circulatory system, respiration by gills, and a ventral solid nerve cord. Insects include butterflies, grasshoppers, bees, and beetles. The anatomy of the grasshopper illustrates insect anatomy and the ways they are adapted to life on land. Like other insects, grasshoppers have wings and three pairs of legs attached to the thorax. Grasshoppers have a tympanum for sound reception, a digestive system specialized for a grass diet, Malpighian tubules for excretion of solid nitrogenous waste, tracheae for respiration, internal fertilization, and incomplete metamorphosis.

Chelicerates (horseshoe crabs, spiders, scorpions, ticks, and mites) have chelicerae, pedipalps, and four pairs of walking legs attached to a cephalothorax.

### 28.5 Invertebrate Deuterostomes

Echinoderms (e.g., sea stars, sea urchins, sea cucumbers, and sea lilies) have radial symmetry as adults (not as larvae) and internal calcium-rich plates with spines. Typical of echinoderms, sea stars have tiny skin gills, a central nerve ring with branches, and a water vascular system for locomotion. Each arm of a sea star contains branches from the nervous, digestive, and reproductive systems.

## understanding the terms

Match the terms to these definitions:

a. _____________ Blind digestive cavity that also serves a circulatory (transport) function in animals lacking a circulatory system.
b. _____________ Body cavity lying between the digestive tract and body wall that is completely lined by mesoderm.
c. _____________ Change in shape and form that some animals, such as insects, undergo during development.
d. _____________ System of canals and appendages used for movement in echinoderms.

## reviewing this chapter

1. What does the phylogenetic tree (see Fig. 28.4) tell you about the evolution of the animals studied in this chapter? 514–15
2. What features make sponges different from the other organisms placed in the animal kingdom? 517
3. What are the two body forms found in cnidarians? Explain how they function in the life cycle of various types of cnidarians. 518–19
4. Describe the anatomy of *Hydra,* pointing out those features that typify cnidarians. 519
5. Describe the anatomy of a free-living planarian, and how it differs from the parasitic flatworms. 520–22
6. What are the general characteristics of molluscs and the specific features of bivalves, cephalopods, and gastropods? 523–25
7. What are the general characteristics of annelids and the specific features of earthworms? 526–27
8. Describe the anatomy of *Ascaris,* pointing out those features that typify roundworms. 528
9. What are the general characteristics of arthropods, specifically crustaceans and insects? 529–33
10. What other types of arthropods were discussed in the chapter? 531–33
11. What are the general characteristics of echinoderms? Explain how the water vascular system works in sea stars. 534–35

## testing yourself

Choose the best answer for each question.

1. Which of these is not a characteristic of animals?
   a. heterotrophic
   b. diploid life cycle
   c. have contracting fibers
   d. single cells or colonial
   e. lack of chlorophyll
2. The phylogenetic tree of animals shows that
   a. three germ layers evolved before a coelom.
   b. both molluscs and annelids are protostomes.
   c. some animals have radial symmetry.
   d. sponges were the first to evolve from an ancestral protist.
   e. All of these are correct.
3. Which of these descriptions does not pertain to both protostomes and deuterostomes?
   a. three germ layers, bilateral symmetry, first opening is mouth
   b. bilateral symmetry, first opening is mouth, all have a true coelom
   c. spiral cleavage, first opening is anus, true coelom develops by a splitting of mesoderm
   d. bilateral symmetry, three germ layers, second opening is mouth
   e. None pertain to both protostomes and deuterostomes.
4. Which of these sponge characteristics is not typical of animals?
   a. They practice sexual reproduction.
   b. They have the cellular level of organization.
   c. They have various symmetries.
   d. They have flagellated cells.
   e. Both b and c are not typical.
5. Which of these pairs is mismatched?
   a. sponges—spicules
   b. tapeworms—proglottids
   c. cnidarians—nematocysts
   d. roundworms—cilia
   e. cnidarians—polyp and medusa
6. Flukes and tapeworms
   a. show cephalization.
   b. have well-developed reproductive systems.
   c. have well-developed nervous systems.
   d. are free-living.
7. *Ascaris* is a parasitic
   a. roundworm.
   b. flatworm.
   c. hydra.
   d. sponge.
   e. comb jelly.
8. The phylogenetic tree of animals shows that
   a. cnidarians evolved directly from sponges.
   b. flatworms evolved directly from roundworms.
   c. rotifers are closely related to flatworms.
   d. coelomates gave rise to the acoelomates.
   e. All of these are correct.
9. Comb jellies are most closely related to
   a. cnidarians.
   b. sponges.
   c. flatworms.
   d. roundworms.
   e. Both a and b are correct.
10. Write the correct type of animal beside each of the following terms.
    a. proglottids:
    b. mantle cavity:
    c. collar cells:
    d. cnidocytes:
    e. crown of cilia:
    f. branched gastrovascular cavity:
    g. exoskeleton contains chitin:
11. Which of these does not pertain to a protostome?
    a. spiral cleavage
    b. blastopore is associated with the anus
    c. coelom, splitting of mesoderm
    d. annelids, arthropods, and molluscs
    e. mouth is associated with first opening
12. Which of these best shows that snails are not closely related to crayfish?
    a. Snails are terrestrial, and crayfish are aquatic.
    b. Snails have a broad foot, and crayfish have jointed appendages.
    c. Snails are hermaphroditic, and crayfish have separate sexes.
    d. Snails are insects, but crayfish are fishes.
    e. Snails are bivalves, and crayfish are chelicerates.
13. Which of these pairs is mismatched?
    a. clam—gills
    b. lobster—gills
    c. grasshopper—book lungs
    d. polychaete—parapodia
14. A radula is a unique organ for feeding found in
    a. molluscs.
    b. annelids.
    c. arthropods.
    d. only insects.
    e. All of these are correct.

15. Which of these pairs is mismatched?
    a. crayfish—walking legs
    b. clam—hatchet foot
    c. grasshopper—wings
    d. earthworm—many cilia
    e. squid—jet propulsion
16. Associate each of these terms to annelids, to arthropods, and/or to molluscs.

**Terms:**
    a. organ system level of organization ______________
    b. segmentation ______________
    c. true coelom ______________
    d. cephalization in some representatives ______________
    e. bilateral symmetry ______________
    f. complete gut ______________
    g. jointed appendages ______________
    h. three-part body plan ______________
17. Associate each of these terms to clams and/or to earthworms.

**Terms:**
    a. annelid ______________
    b. mollusc ______________
    c. three ganglia ______________
    d. gills ______________
    e. closed circulatory system ______________
    f. setae ______________
    g. open circulatory system ______________
    h. hatchet foot ______________
    i. hydrostatic skeleton ______________
    j. ventral nerve cord ______________
18. Label this diagram.

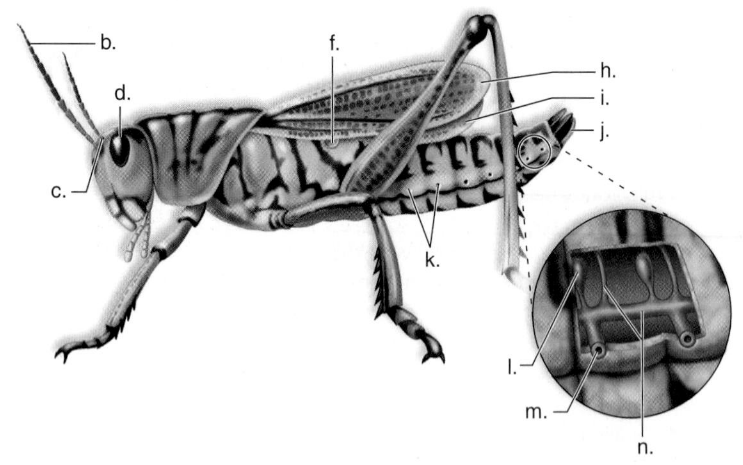


For questions 19–23, match the description to a body system in the key.

**KEY:**
    a. respiratory system
    b. digestive system
    c. cardiovascular system
    d. nervous system

19. particles adhere to labial palps
20. passes through the heart
21. has three pairs of ganglia
22. hemocoel
23. gills

For questions 24–27, match the classification to an animal in the key.

**KEY:**
    a. shrimp
    b. squid
    c. earthworm
    d. spider

24. molluscan cephalopod
25. annelid oligochaete
26. arthropod crustacean
27. arthropod chelicerate
28. Segmentation in the earthworm is not exemplified by
    a. body rings.
    b. coelom divided by septa.
    c. setae on most segments.
    d. nephridia interior in most segments.
    e. tympanum exterior to segments.
29. Which of these is an incorrect difference between clam and squid?
    Clam — Squid
    a. filter feeder—active predator
    b. hatchet foot—jet propulsion
    c. brain and nerves—three separate ganglia
    d open circulation—closed circulation
    e. no cephalization—marked cephalization
30. Which characteristic accounts for the success of arthropods?
    a. jointed exoskeleton
    b. well-developed nervous system
    c. segmentation
    d. respiration adapted to environment
    e. All of these are correct.
31. Sea stars
    a. have no respiratory, excretory, or circulatory systems.
    b. usually reproduce asexually.
    c. are protostomes.
    d. can evert their stomach to digest food.
    e. Both a and d are correct.
32. The tube feet of echinoderms
    a. are their head.
    b. are a part of the water vascular system.
    c. are found in the coelom.
    d. help pass sperm to females during reproduction.
    e. All of these are correct.

## thinking scientifically

1. How is the lifestyle of radially symmetrical animals different from that of bilaterally symmetrical animals? Explain.
2. Animals, like other life-forms, evolved in the ocean. What made water a better nursery than land for the evolution of animals?

## *Biology* website

The companion website for *Biology* provides a wealth of information organized and integrated by chapter. You will find practice tests, animations, videos, and much more that will complement your learning and understanding of general biology.

**http://www.mhhe.com/maderbiology10**

# 29

# Vertebrate Evolution

*The New Guinea (NG) singing dog lives on the island of New Guinea, but it doesn't actually sing. It yelps, whines, and howls at different pitches unlike other dogs. The howl is eerie to the point of causing goose bumps when one tone blends with the next. The evolutionary history of the NG singing dog is a matter of debate. Is it a unique species, a type of gray wolf, or is it related to the dingo, the nonbarking canine that lives in Australia? Some suggest that the NG singing dog might be the most primitive known "breed" of domestic dog because Stone Age people brought it to New Guinea 6,000 years ago! Living on an island, it has remained geographically and reproductively isolated ever since. Therefore, it could be a living fossil, an organism with the same genes and characteristics as its original ancestor. If so, the NG singing dog offers an opportunity to study not only why it "sings" but also what the first domesticated dog was like.*

*However, like many other vertebrates, the topic of this chapter, the NG singing dog is threatened with extinction. Expeditions into the highlands of New Guinea have only yielded a few droppings, tracks, and haunting howls in the distance.*

New Guinea singing dog, *Canis familiaris hallstromi.*

## concepts

# 29.1 The Chordates

Chordates (phylum Chordata), like echinoderms, have the deuterostome pattern of development. However, the **chordates** have a different type of skeleton compared to the invertebrates. In invertebrates, the skeleton is external, and the muscles are attached to its inner surface. In chordates, the skeleton is internal, and the muscles are attached to the outer surface. This allows the chordates to enjoy freedom of movement and attainment of a larger size than invertebrates. These four basic characteristics occur during the life history of all chordates:

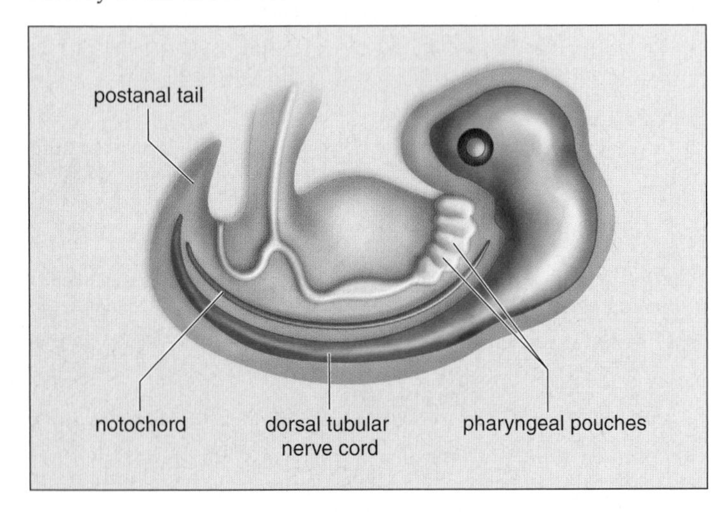


**Notochord** [Gk. *notos,* back, and *chorde,* string]. A dorsal supporting rod; the notochord is located just below the nerve cord. The majority of vertebrates have an embryonic notochord that is replaced by the vertebral column during development.

**Dorsal tubular nerve cord.** The anterior portion becomes the brain in most chordates. In vertebrates, the nerve cord, often called the spinal cord, is protected by vertebrae.

**Pharyngeal pouches.** These are seen only during embryonic development in most vertebrates. In the nonvertebrate chordates, the fishes, and amphibian larvae, the pharyngeal pouches become functioning **gills** (respiratory organs of aquatic vertebrates). Water passing into the mouth and the pharynx goes through the gill slits, which are supported by gill arches. In terrestrial vertebrates, the pouches are modified for various purposes. For example, in humans the first pair of pouches become the auditory tubes.

**Postanal tail.** A tail—in the embryo, if not in the adult—extends beyond the anus. In other groups of animals, the anus is terminal.

## Nonvertebrate Chordates

The nonvertebrate chordates are divided into two groups: the cephalochordates and the urochordates. Lancelets (genus *Branchiostoma*) are the **cephalochordates.** (The former genus for these organisms, Amphioxus, is now sometimes used as a common name also.) These marine chordates, which are only a few centimeters long, are named for their resemblance to a lancet—a small, two-edged surgical knife (Fig. 29.1). Lancelets are found in the shallow water along most coasts, where they usually lie partly buried in sandy or muddy substrates with only their anterior mouth and gill apparatus exposed. They feed on microscopic particles filtered out of the constant stream of water that enters the mouth and passes through the gill slits into an atrium that opens at the atriopore.

Lancelets retain the four chordate characteristics as an adult and for that reason lancelets are important in comparative anatomy and evolutionary studies. In lancelets, the notochord extends from the tail to the head, and this accounts for their group name, the cephalochordates. In addition to the four features of all chordates, segmentation is present, as witnessed by the fact that the muscles are segmentally arranged, and the dorsal tubular nerve cord has periodic branches. Segmentation may not be an important feature in lancelets, but it is in the vertebrates where it leads

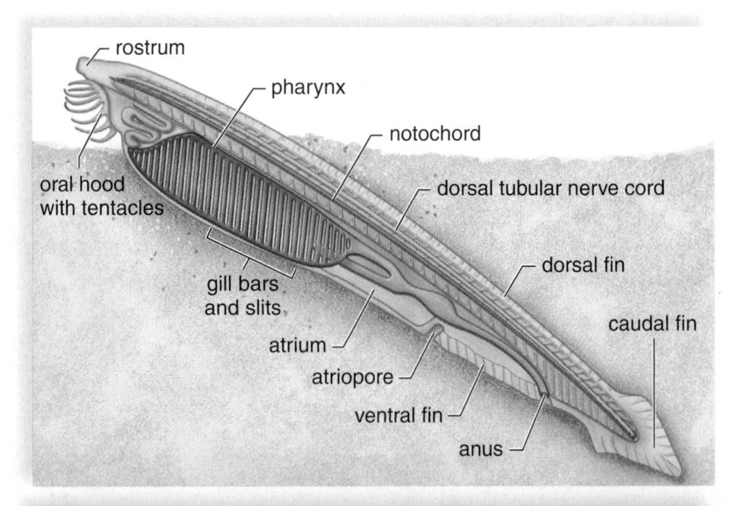

**FIGURE 29.1** **Lancelet, *Branchiostoma.***

Lancelets are filter feeders. Water enters the mouth and exits at the atriopore after passing through the gill slits.

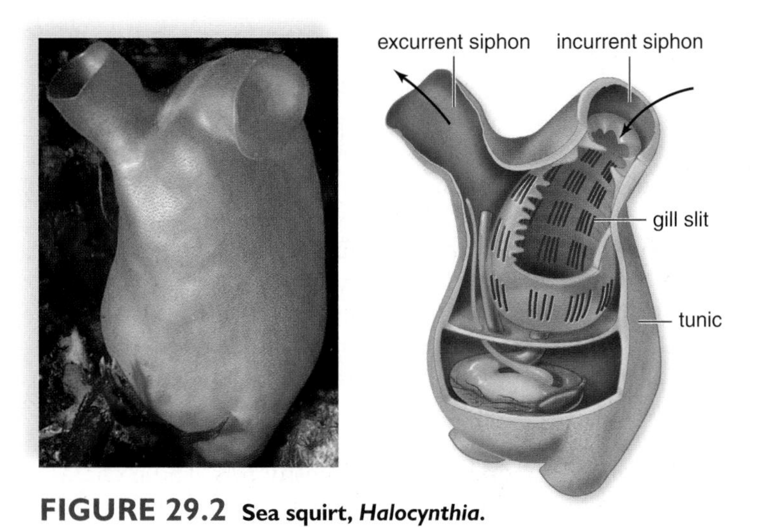


**FIGURE 29.2 Sea squirt, *Halocynthia*.**

Note that the only chordate characteristic remaining in the adult is gill slits.

to specialization of parts, as we also witnessed in the annelids and arthropods.

Sea squirts, the **urochordates,** live on the ocean floor, where they squirt out water from their excurrent siphon when disturbed. They are also called tunicates because they have a tunic that makes them look like thick-walled, squat sacs. The sea squirt larva is bilaterally symmetrical and has the four chordate characteristics. Metamorphosis produces the sessile adult with an incurrent and excurrent siphon (Fig. 29.2). The pharynx is lined by numerous cilia whose beating creates a current of water that moves into the pharynx and out the numerous gill slits, the only chordate characteristic that remains in the adult.

Many evolutionary biologists hypothesize that the sea squirts are directly related to the vertebrates. It has been suggested that a larva with the four chordate characteristics may have become sexually mature without developing the other adult sea squirt characteristics. Then it may have evolved into a fishlike vertebrate. Figure 29.3 shows how the main groups of chordates may have evolved.

### Check Your Progress 29.1

1. Would you predict that humans have the four chordate characteristics as embryos? Why or why not?
2. Adult sea squirts look like thick-walled, squat sacs. Why are they classified as chordates?

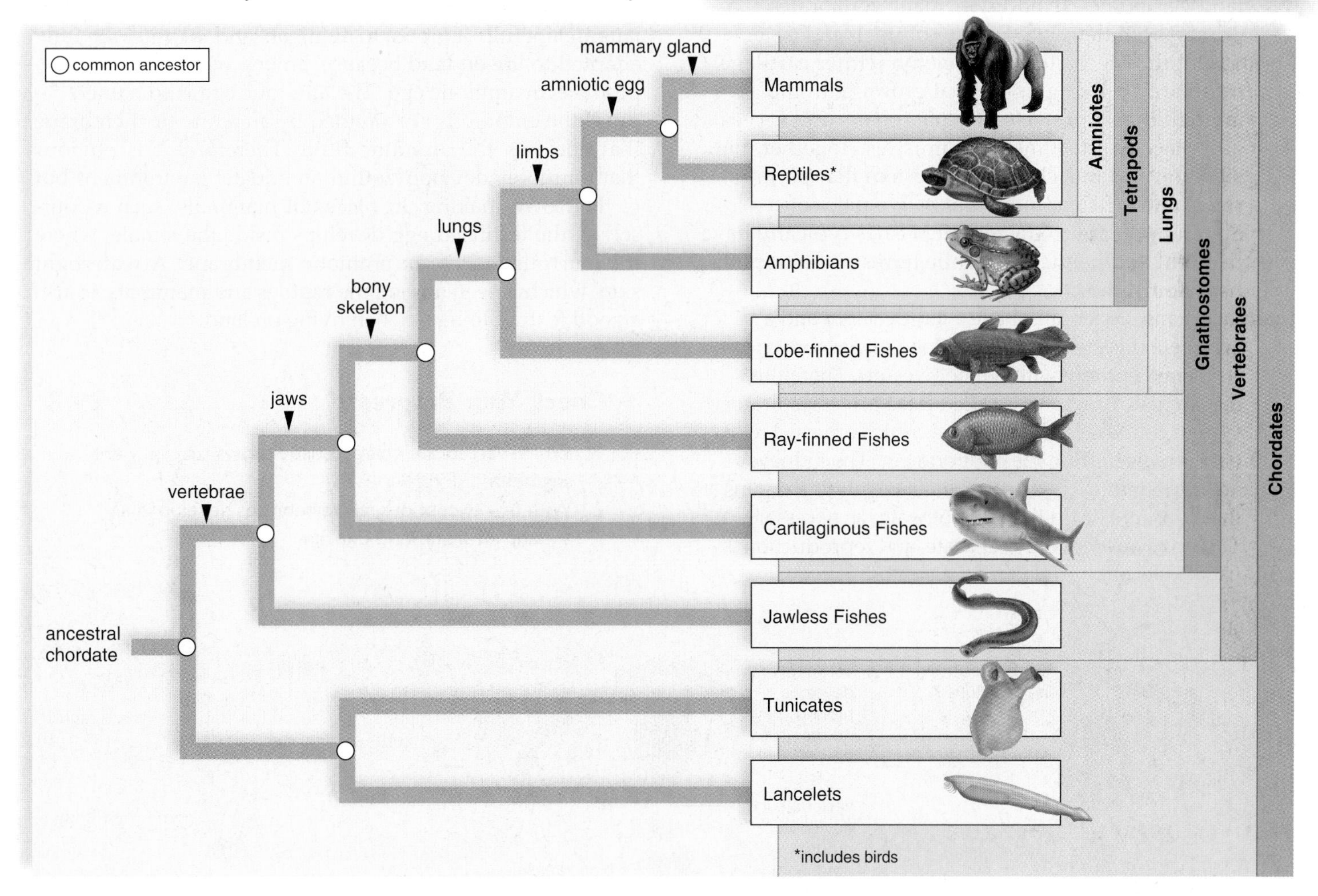


**FIGURE 29.3 Phylogenetic tree of the chordates.**

The evolution of vertebrates is marked by at least seven derived characteristics.

# 29.2 The Vertebrates

As embryos, vertebrates have the four chordate characteristics. In addition, vertebrates have these features:

**Vertebral column.** The embryonic notochord is generally replaced by a vertebral column composed of individual vertebrae (Fig. 29.4). Remnants of the notochord are seen in the intervertebral discs, which are compressible cartilaginous pads found between the vertebrae. The vertebral column, which is a part of the flexible but strong endoskeleton, gives evidence that vertebrates are segmented.

**Skull.** The main axis of the internal skeleton consists of not only the vertebral column, but also a skull that encloses and protects the brain. During vertebrate evolution, the brain increases in complexity, and specialized regions developed to carry out specific functions. The high degree of cephalization is accompanied by complex sense organs. The eyes develop as outgrowths of the brain. The ears are primarily equilibrium devices in aquatic vertebrates, but they also function as sound-wave receivers in land vertebrates. In addition, many vertebrates possess well-developed senses of smell and taste.

**Endoskeleton.** The vertebrate skeleton (either cartilage or bone) is a living tissue that grows with the animal. It also protects internal organs and serves as a place of attachment for muscles. Together, the skeleton and muscles form a system that permits rapid and efficient movement. Two pairs of appendages are characteristic. Fishes typically have pectoral and pelvic fins, while terrestrial tetrapods have four limbs.

**Internal organs.** Vertebrates have a large coelom and a complete digestive tract. The blood in vertebrates is contained entirely within blood vessels. Therefore, the circulatory system is called closed. The respiratory system consists of gills or lungs, which are used to obtain oxygen from the environment. The kidneys are important excretory and water-regulating organs that conserve or rid the body of water as necessary. The sexes are generally separate, and reproduction is usually sexual.

## Vertebrate Evolution

Chordates, including vertebrates, appear on the scene suddenly at the start of the Cambrian period (see page 327). Even though we do not know the origin of vertebrates, we can trace their later evolution, as shown in Figure 29.3. The earliest vertebrates were fishes, organisms that are abundant both in marine and freshwater habitats. A few of today's fishes lack jaws and have to suck and, otherwise, engulf their prey. Most fishes have jaws, which are a more efficient means of gasping and eating prey. The jawed fishes and all the other vertebrates are **gnathostomes**—animals with jaws.

Certain of the early fishes not only had jaws, they also had a bony skeleton, lungs, and fleshy fins. These characteristics were preadaptive for a land existence, and the amphibians, the first vertebrates to live on land, evolved from these fishes. The amphibians were the first vertebrates to have limbs. The terrestrial vertebrates are **tetrapods** because they have four limbs. Some, such as the snakes, no longer have four limbs, but their evolutionary ancestors had four limbs.

Many amphibians, such as the frog, reproduce in an aquatic environment. This means that, in general, amphibians are not fully adapted to living on land. Reptiles are fully adapted to life on land because, among other features, they produce an amniotic egg. The amniotic egg is so named because the embryo is surrounded by an amniotic membrane that encloses the amniotic fluid. Therefore, it is obvious that **amniotes** develop within an aquatic environment but of their own making. In placental mammals, such as ourselves, the fertilized egg develops inside the female, where it is surrounded by an amniotic membrane. A watertight skin, which is seen among the reptiles and mammals, is also a good feature to have when living on land.

### Check Your Progress 29.2

1. Which vertebrate characteristic shows that they are segmented? Explain.
2. With few exceptions, all vertebrates develop in an aquatic environment. Explain.

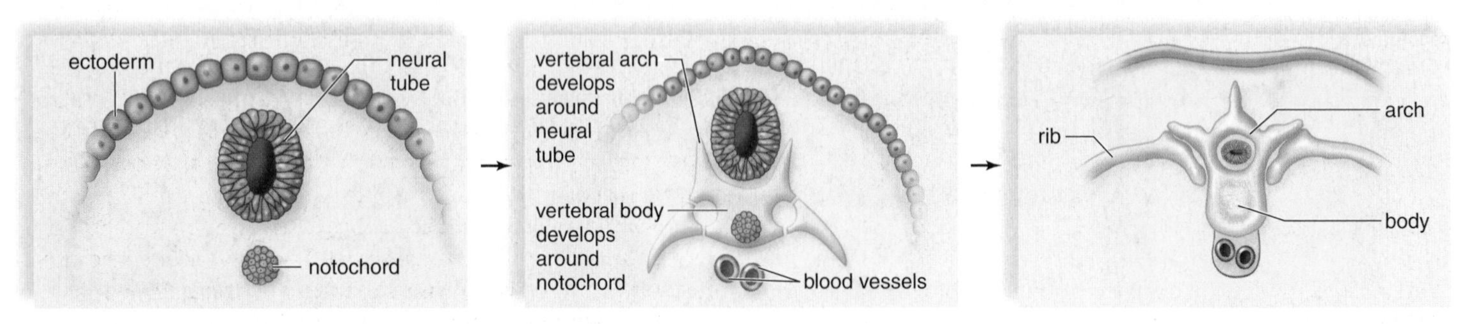


**FIGURE 29.4 Replacement of notochord by the vertebrae.**

During vertebrate development, the vertebrae replace the notochord and surround the neural tube. The result is the flexible vertebral column, which protects the nerve cord. The term spine refers to the vertebral column plus the nerve cord.

# 29.3 The Fishes

**Fishes** are the largest group of vertebrates with nearly 28,000 recognized species. They range in size from a few millimeters in length to the whale shark, which may reach lengths of 12 m. The fossil record of the fishes is extensive.

## Jawless Fishes

The earliest fossils of Cambrian origin were the small, filter-feeding, jawless and finless **ostracoderms.** Several groups of ostracoderms developed heavy dermal armor for protection.

Today's **jawless fishes,** or **agnathans,** have a cartilaginous skeleton and persistent notochord. They are cylindrical, up to a meter long, and have smooth, nonscaly skin (Fig. 29.5). The hagfishes are exclusively marine scavengers that feed on soft-bodied invertebrates and dead fishes. Many species of lampreys are filter feeders like their ancestors. Parasitic lampreys have a round, muscular mouth used to attach themselves to another fish and suck nutrients from the host's cardiovascular system. The parasitic sea lamprey gained access to the Great Lakes in 1829 and by 1950 had almost demolished the resident trout and whitefish populations.

## Fishes with Jaws

Fishes with jaws have these characteristics:

**Ectothermy.** Like all fishes, jawed fishes are **ectotherms** [Gk. *ekto,* outer, and *therme,* heat], which means that they depend on the environment to regulate their temperature.

**Gills.** Like all fishes, jawed fishes breathe with gills and have a single-looped, closed circulatory system with a heart that pumps the blood first to the gills. Then, oxygenated blood passes to the rest of the body.

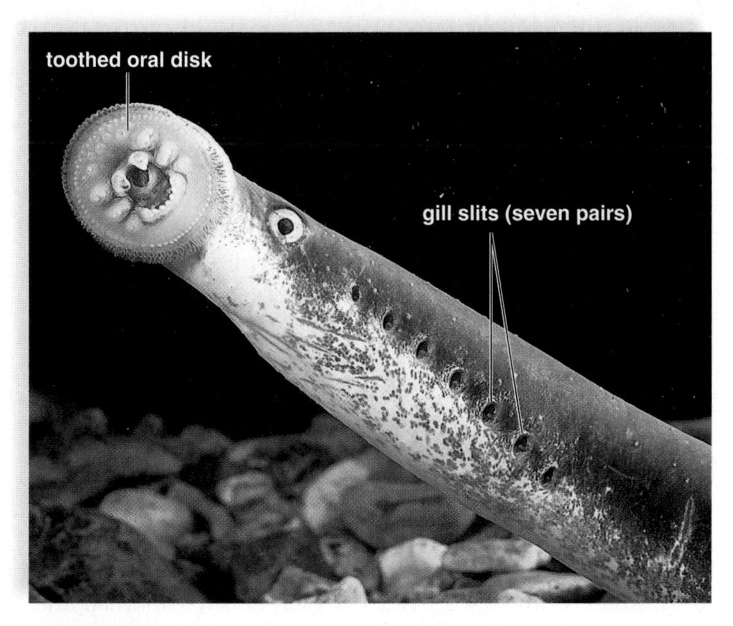


**FIGURE 29.5** **Lamprey, *Petromyzon.***

Lampreys, which are agnathans, have an elongated, rounded body and nonscaly skin. Note the lamprey's toothed oral disk, which is attached to the aquarium glass.

**Cartilaginous or bony endoskeleton.** The endoskeleton of jawed fishes includes the vertebral column, a skull with jaws, and paired pectoral and pelvic fins. The large muscles of the body actually do most of the work of locomotion, but the fins help with balance and turning.

Jaws evolved from the first pair of gill arches present in ancestral agnathans. The second pair of gill arches became support structures for the jaws:

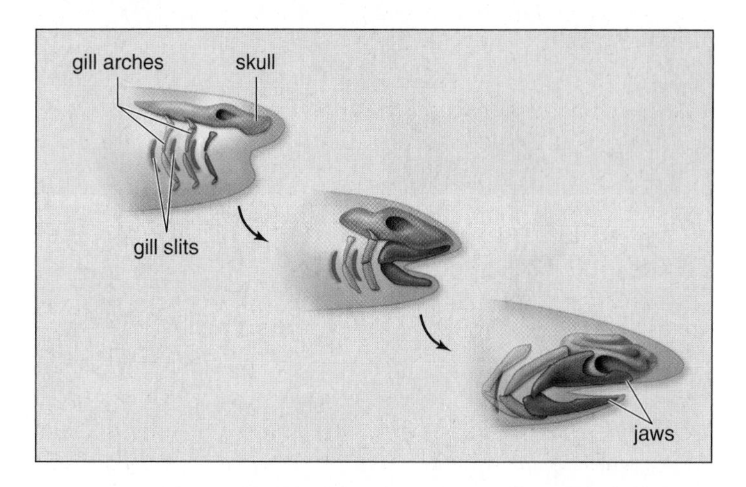


**Scales.** The skin of the jawed fishes is covered by scales, therefore, it is not exposed to the environment. A scientist can tell the age of a fish from examining the growth of the scales.

The **placoderms,** extinct jawed fishes of the Devonian period, are probably the ancestors of early sharks and bony fishes. Placoderms were armored with heavy, bony plates and had strong jaws. Like modern-day fishes, they also had paired pectoral and pelvic fins (see Fig. 29.6).

### *Cartilaginous Fishes*

Sharks, rays, skates, and chimaeras are marine **cartilaginous fishes (Chondrichthyes).** The cartilaginous fishes have a skeleton composed of cartilage instead of bone; have five to seven gill slits on both sides of the pharynx; and lack the gill cover of bony fishes. In addition, many have openings to the gill chamber located behind the eyes called spiracles. Their body is covered with dermal denticles, tiny teethlike scales that project posteriorly, which is why a shark's skin feels like sandpaper. The menacing teeth of sharks and their relatives are simply larger, specialized versions of these scales. At any one time, a shark such as the great white shark may have up to 3,000 teeth in its mouth, arranged in 6 to 20 rows. Only the first row or two are actively used for feeding; the other rows are replacement teeth.

Three well-developed senses enable sharks and rays to detect their prey. They have the ability to sense electric currents in water—even those generated by the muscle movements of animals. They have a lateral line system, a series of pressure-sensitive cells that lie within canals along both sides of the body, which can sense pressure caused by a fish or other animal swimming nearby. They also have a very

a. Sand tiger shark, *Carcharias taurus*

b. Blue-spotted stingray, *Taeniura lymma*

**FIGURE 29.6 Cartilaginous fishes.**

**a.** Sharks are predators or scavengers that move gracefully through open ocean waters. **b.** Most stingrays grovel in the sand, feeding on bottom-dwelling invertebrates. This blue-spotted stingray is protected by the spine on its whiplike tail.

keen sense of smell; the part of the brain associated with this sense is very well developed. Sharks can detect about one drop of blood in 115 liters of water.

The largest sharks are filter feeders, not predators. The basking sharks and whale sharks ingest tons of small crustaceans, collectively called krill. Many sharks are fast-swimming predators in the open sea (Fig. 29.6*a*). The great white shark, about 7 m in length, feeds regularly on dolphins, sea lions, and seals. Humans are normally not attacked except when mistaken for sharks' usual prey. Tiger sharks, so named because the young have dark bands, reach 6 m in length and are unquestionably one of the most predaceous sharks. As it swims through the water, a tiger shark will swallow anything, including rolls of tar paper, shoes, gasoline cans, paint cans, and even human parts. The number of bull shark attacks has increased in the Gulf of Mexico in recent years. This is due to more swimmers in the shallow waters and loss of the sharks' natural food supply.

In rays and skates (Fig. 29.6*b*), the pectoral fins are greatly enlarged into a pair of large, winglike **fins,** and the bodies are dorsoventrally flattened. The spiracles are enlarged and allow them to move water over the gills while resting on the bottom, where they feed on organisms such as crustaceans, small fishes, and molluscs.

Stingrays have a whiplike tail that has serrated spines with venom glands at the base. These animals can deliver a painful "sting." Manta rays are harmless oceanic filter-feeding giants with a fin span of up to 6 m and a weight of 2,000 kg. Sawfish rays are named for their large, protruding anterior "saw." Some species of stingrays can deliver an electric shock. Their large electric organs, located at the base of their pectoral fins, can discharge over 300 volts. Skates resemble stingrays but possess two dorsal fins and a caudal fin.

The chimaeras, or ratfishes, are a group of cartilaginous fishes that live in cold marine waters. They are known for their unusual shape and iridescent colors.

## *Bony Fishes*

The majority of living vertebrates, approximately 25,000 species, are **bony fishes (Osteichthyes).** The bony fishes range in size from gobies, which are less than 7.5 mm long, to the giant sturgeons, which can obtain a length of 4 m.

The majority of fish species are **ray-finned bony fishes** with fan-shaped fins supported by a thin, bony ray. These fishes are the most successful and diverse of all the vertebrates (Fig. 29.7). Some, such as herrings, are filter feeders; others, such as trout, are opportunists; and still others are predaceous carnivores, such as piranhas and barracudas.

Despite their diversity, bony fishes have many features in common. They lack external gill slits and, instead, their gills are covered by an **operculum.** Many bony fishes have a **swim bladder,** a gas-filled sac into which they can secrete gases or from which they can absorb gases, altering its pressure. This results in a change in the fishes' buoyancy and, therefore, their depth in the water. Bony fishes have a single-loop, closed cardiovascular system (see Fig. 29.9*a*). The nervous system and brain in bony fishes are well developed, and complex behaviors are common. Bony fishes have separate sexes, and the majority of species undergo external fertilization after females deposit eggs and males deposit sperm into the water.

**Lobe-Finned Fishes** **Lobe-finned fishes** possess fleshy fins supported by bones. Ancestral lobe-finned fishes gave rise to modern-day lobe-finned fishes, the lungfishes, and the amphibians. **Lungfishes** have lungs and also gills for gas exchange. The lobe-finned fishes and lungfishes are grouped together as the **Sarcopterygii.** Today, there are only two species of lobe-finned fishes (the coelacanths) and six species of lungfish. Lungfishes live in Africa, South America, and Australia, either in stagnant fresh water or in ponds that dry up annually.

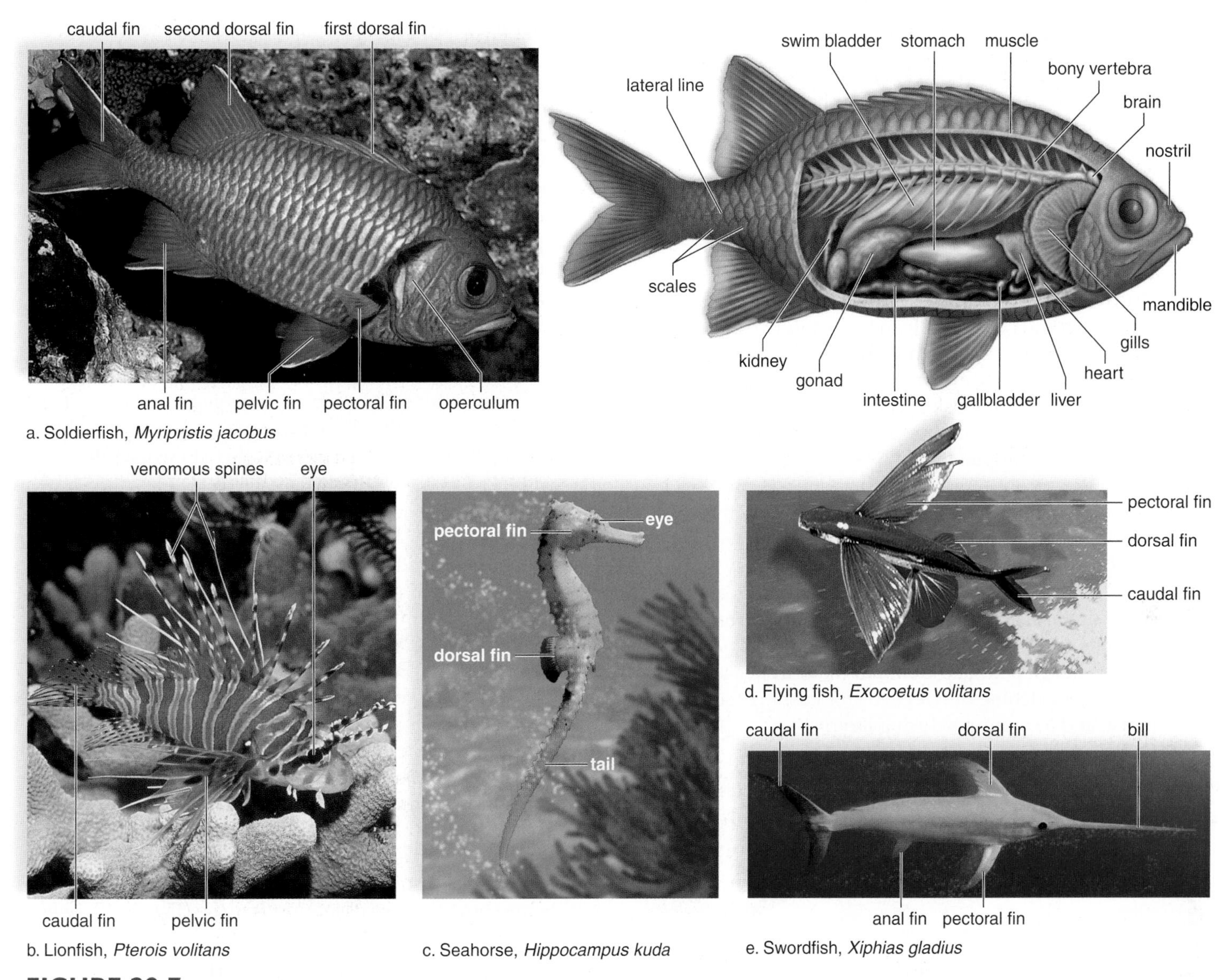


a. Soldierfish, *Myripristis jacobus*

b. Lionfish, *Pterois volitans*

c. Seahorse, *Hippocampus kuda*

d. Flying fish, *Exocoetus volitans*

e. Swordfish, *Xiphias gladius*

**FIGURE 29.7 Ray-finned fishes.**

**a.** A soldierfish has the typical appearance and anatomy of a ray-finned fish. A lionfish (**b**), a seahorse (**c**), a flying fish (**d**), and a swordfish (**e**) show how diverse ray-finned fishes can be.

In 1938, a coelacanth was caught from the deep waters of the Indian Ocean off the eastern coast of South Africa. It took the scientific world by surprise because these animals were thought to be extinct for 70 million years. Approximately 200 coelacanths have been captured in recent years (Fig. 29.8).

### Check Your Progress 29.3

1. List and describe the characteristics that fishes have in common.
2. What characteristics distinguish cartilaginous fishes?

**FIGURE 29.8 Coelacanth, *Latimeria chalumnae*.**

A coelacanth is a lobe-finned fish once thought to be extinct.

## 29.4 The Amphibians

**Amphibians** (class Amphibia [Gk. *amphibios,* living both on land and in water]), which have these characteristics, were abundant during the Carboniferous period:

**Limbs.** Typically, amphibians are tetrapods [Gk. *tetra,* four, and *podos,* foot], meaning that they have four limbs. The skeleton, particularly the pelvic and pectoral girdles, is well developed to promote locomotion.

**Smooth and nonscaly skin.** The skin, which is kept moist by mucous glands, plays an active role in water balance and respiration and can also help in temperature regulation when on land through evaporative cooling. A thin, moist skin does mean, however, that most amphibians stay close to water, or else risk drying out.

**Lungs.** If lungs are present, they are relatively small, and respiration is supplemented by exchange of gases across the porous skin called cutaneous respiration.

**Double-loop circulatory pathway.** (Fig. 29.9*b, c*). A three-chambered heart with a single ventricle and two atria pumps blood to both the lungs and to the body.

**Sense organs.** Special senses, such as sight, hearing, and smell, are fine-tuned for life on land. Amphibian brains are larger than those of fish, and the cerebral cortex is more developed. These animals have a specialized tongue for catching prey, eyelids for keeping their eyes moist, and a sound-producing larynx.

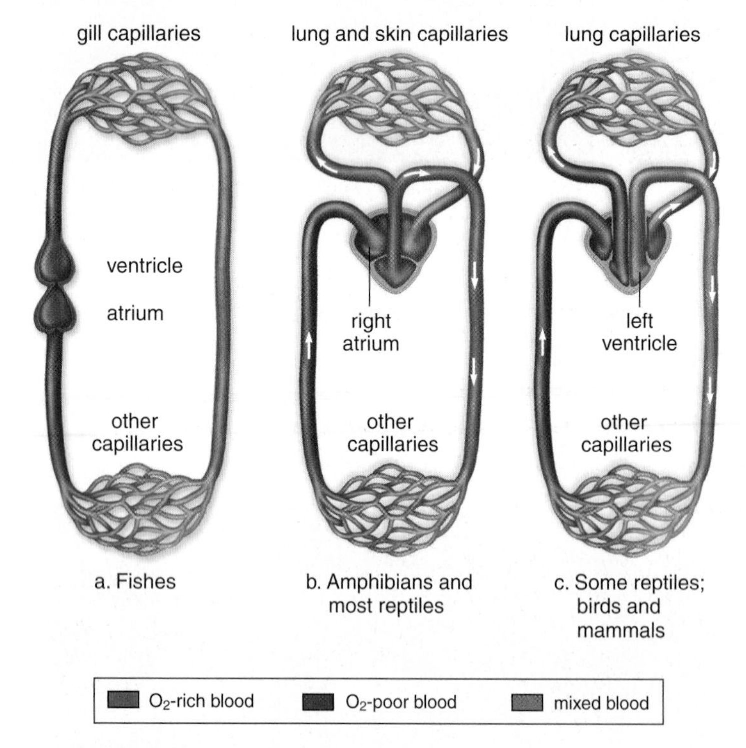


**FIGURE 29.9 Vertebrate circulatory pathways.**

**a.** The single-loop pathway of fishes has a two-chambered heart. **b.** The double-loop pathway of other vertebrates sends blood to the lungs and to the body. In amphibians and most reptiles, there is limited mixing of $O_2$-rich and $O_2$-poor blood in the single ventricle of their three-chambered heart. **c.** The four-chambered heart of some reptiles (crocodilians and birds) and mammals sends only $O_2$-poor blood to the lungs and $O_2$-rich blood to the body.

**Ectothermy.** Like fishes, amphibians are ectotherms but are able to live in environments where the temperature fluctuates greatly. During winters in the temperate zone, they become inactive and enter torpor. The European common frog can survive in temperatures dropping to as low as −6°C.

**Aquatic reproduction.** Their name, amphibians, is appropriate because many return to water for the purpose of reproduction. They deposit their eggs and sperm into the water, where external fertilization takes place. Generally, the eggs are protected only by a jelly coat and not by a shell. When the young hatch, they are tadpoles (aquatic larvae with gills) that feed and grow in the water. After they undergo a **metamorphosis** (change in form), amphibians emerge from the water as adults that breathe air. Some amphibians, however, have evolved mechanisms that allow them to bypass this aquatic larval stage and reproduce on land.

### Evolution of Amphibians

Amphibians evolved from the lobe-finned fishes with lungs by way of transitional forms. Two hypotheses have been suggested to account for the evolution of amphibians from lobe-finned fishes. Perhaps lobe-finned fishes had an advantage over others because they could use their lobed fins to move from pond to pond. Or, perhaps the supply of food on land in the form of plants and insects—and the absence of predators—promoted further adaptations to the land environment. Paleontologists have recently found a well-preserved transitional fossil from the late Devonian period in Arctic Canada that represents an intermediate between lobe-finned fishes and tetrapods with limbs. This fossil, named *Tiktaalik roseae* (Fig. 29.10, *left*), provides unique insights into how the legs of tetrapods arose (Fig. 29.10, *right*).

### Diversity of Living Amphibians

The amphibians of today occur in three groups: salamanders and newts; frogs and toads; and caecilians. Salamanders and newts have elongated bodies, long tails, and usually two pairs of limbs (Fig. 29.11*a*). Salamanders and newts range in size from less than 15 cm to the giant Japanese salamander, which exceeds 1.5 m in length. Most have limbs that are set at right angles to the body and resemble the earliest fossil amphibians. They move like a fish, with side-to-side, sinusoidal (S-shaped) movements:

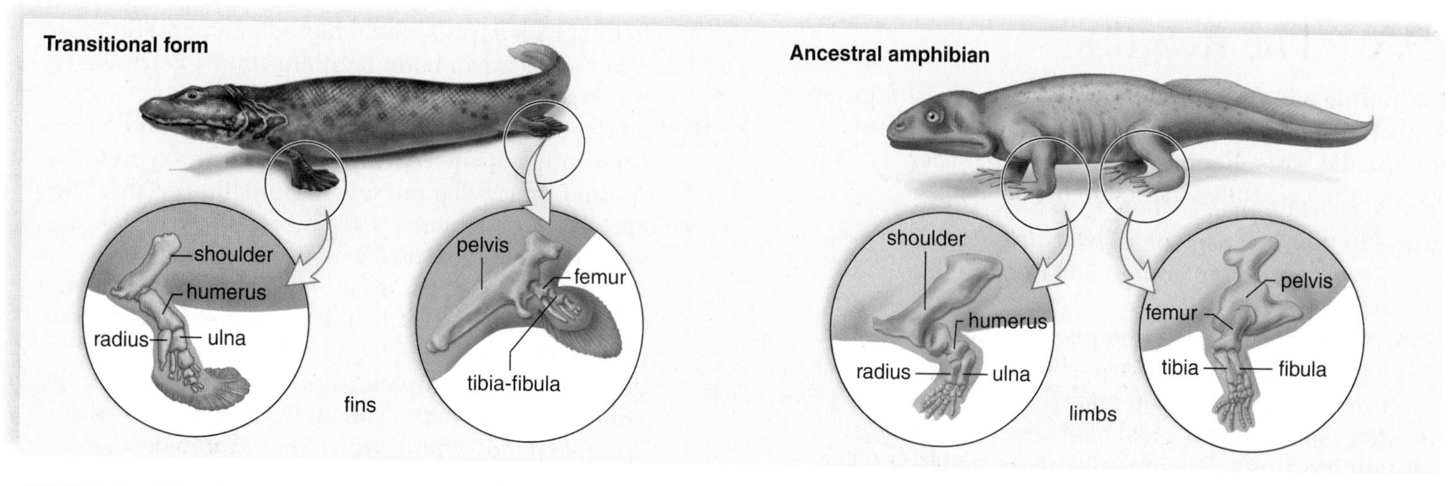


**FIGURE 29.10 Lobe-finned fishes to amphibians.**

This transitional form links the lobes of lobe-finned fishes to the limbs of ancestral amphibians. Compare the fins of the transitional form (*left*) to the limbs of the ancestral amphibian (*right*).

Both salamanders and newts are carnivorous, feeding on small invertebrates such as insects, slugs, snails, and worms. Salamanders practice internal fertilization; in most, males produce a sperm-containing spermatophore that females pick up with their **cloaca** (the terminal chamber common to the urinary, digestive, and genital tracts). Then the fertilized eggs are laid in water or on land, depending on the species. Some amphibians, such as the mudpuppy of eastern North America, remain in the water and retain the gills of the larva.

Frogs and toads, which range in length from less than 1 cm to 30 cm, are common in subtropical to temperate to desert climates around the world. In these animals, which lack tails as adults, the head and trunk are fused, and the long hindlimbs are specialized for jumping (Fig. 29.11*b*). All species are carnivorous and have a tremendous array of specializations depending on their habitats. Glands in the skin secrete poisons that make the animal distasteful to eat and protect them from microbial infections. Some tropical species with brilliant fluorescent green and red coloration are particularly poisonous (see Fig. 29A*a*). Colombian Indians dip their darts in the deadly secretions of these frogs, aptly called poison-dart frogs. The tree frogs have adhesive toepads that allow them to climb trees, while others, the spadefoots, have hardened spades that act as shovels enabling them to dig into the soil.

Caecilians are legless, often sightless, worm-shaped amphibians that range in length from about 10 cm to more than 1 m (Fig. 29.11*c*). Most burrow in moist soil, feeding on worms and other soil invertebrates. Some species have folds of skin that make them look like a segmented earthworm.

### Check Your Progress 29.4

1. **a.** List and describe the characteristics that amphibians have in common. **b.** What evidence links lobe-finned fishes to the amphibians?
2. Describe the usual life cycle of amphibians.

a. Barred tiger salamander, *Ambystoma tigrinum*

b. Tree frog, *Hyla andersoni*

sightless head
smooth skin

c. Caecilian, *Caecilia nigricans*

**FIGURE 29.11 Amphibians.**

Living amphibians are divided into three orders: **a.** Salamanders and newts. Members of this order have a tail throughout their lives and, if present, unspecialized limbs. **b.** Frogs and toads. Like this frog, members of this order are tailless and have limbs specialized for jumping. **c.** The caecilians are wormlike burrowers.

## 29.5 The Reptiles

The **reptiles** (class Reptilia) are a very successful group of terrestrial animals consisting of more than 17,000 species, including the birds. Reptiles have these characteristics showing that they are fully adapted to life on land:

**Paired limbs.** Two pairs of limbs usually with five toes. Reptiles are adapted for climbing, running, paddling, or flying.

**Skin.** A thick and dry skin is impermeable to water. Therefore, the skin prevents water loss. In reptiles, the skin is wholly or in part scaly (Fig. 29.12). Many reptiles (e.g., snakes and lizards) molt several times a year.

**Efficient breathing.** The lungs are more developed than in amphibians. Also in many reptiles, an expandable rib cage assists breathing.

**Efficient circulation.** The heart prevents mixing of blood. A septum divides the ventricle either partially or completely. If it partially divides the ventricle, the mixing of $O_2$-poor blood and $O_2$-rich blood is reduced. If the septum is complete, $O_2$-poor blood is completely separate from $O_2$-rich blood (see Fig. 29.9*c*).

**Efficient excretion.** The kidneys are well developed. The kidneys excrete uric acid, and therefore less water is required to rid the body of nitrogenous wastes.

**Ectothermy.** Most reptiles are ectotherms, and this allows them to survive on a fraction of the food per body weight required by birds and mammals. Ectothermic reptiles are adapted behaviorally to maintain a warm body temperature by warming themselves in the sun.

**Well-adapted reproduction.** Sexes are separate and fertilization is internal. Internal fertilization prevents sperm from drying out when copulation occurs. The **amniotic egg** contains extraembryonic membranes, which protect the embryo, remove nitrogenous wastes, and provide the embryo with oxygen, food, and water (see Fig. 29.14*e*). These membranes are not part of the embryo itself and are disposed of after development is complete. One of the membranes, the amnion, is a sac that fills with fluid and provides a "private pond" within which the embryo develops.

### Evolution of Amniotes

An ancestral amphibian gave rise to the amniotes, which includes animals now classified as the reptiles (including birds) and the mammals. The embryo of an amniote has extracellular membranes, including an amnion. Figure 29.13 shows that the amniotes consist of three lineages: (1) the turtles, in which the skull has no openings behind the orbit—eye socket; (2) all the other reptiles including the birds, in which the skull has two openings behind the orbit; and (3) the mammals, in which the skull has one opening behind the orbit.

This chapter concerns the reptiles, an artificial grouping because it has no common ancestor. In other words, reptiles are a paraphyletic group and not a monophyletic group.

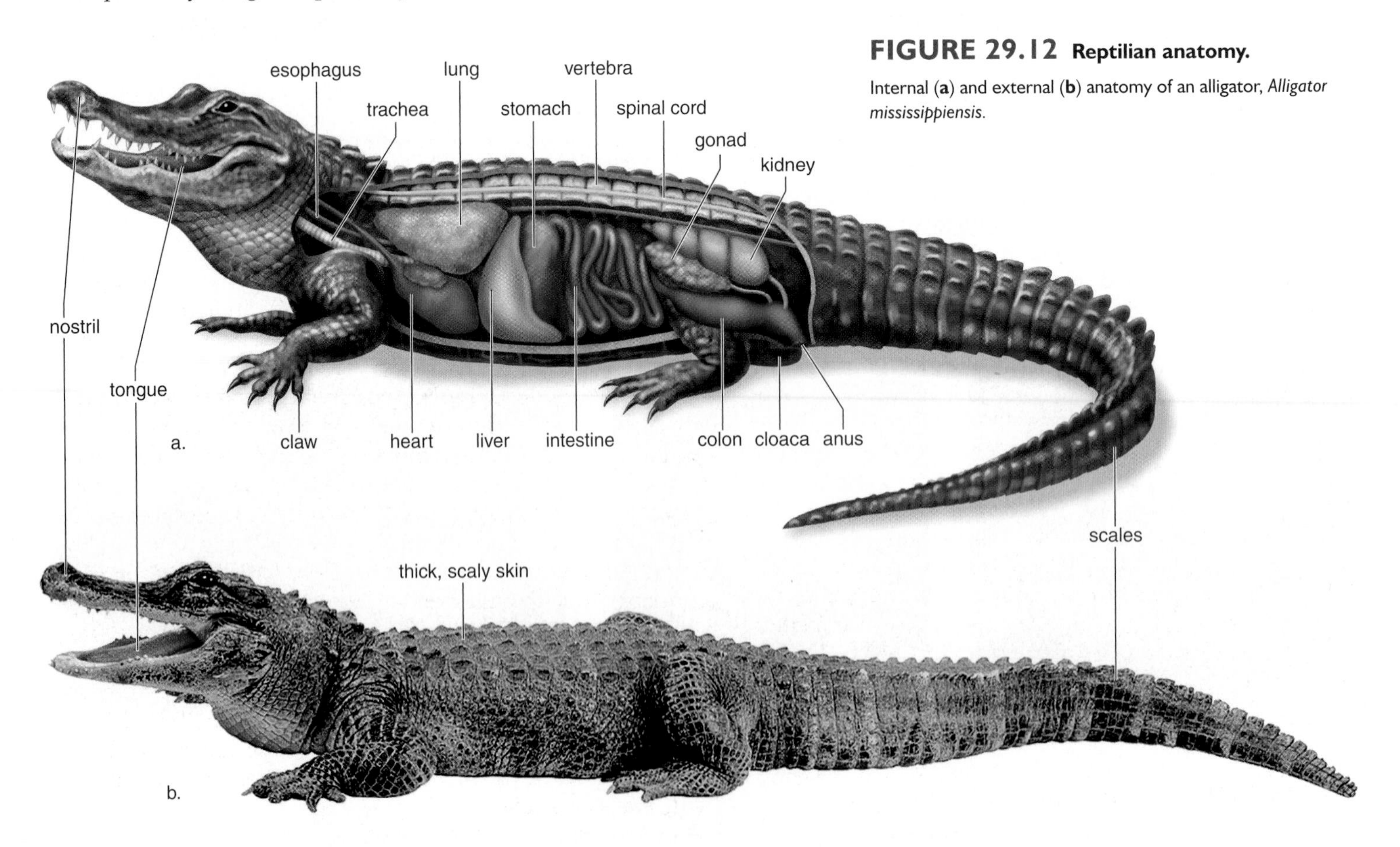


**FIGURE 29.12 Reptilian anatomy.**
Internal (**a**) and external (**b**) anatomy of an alligator, *Alligator mississippiensis*.

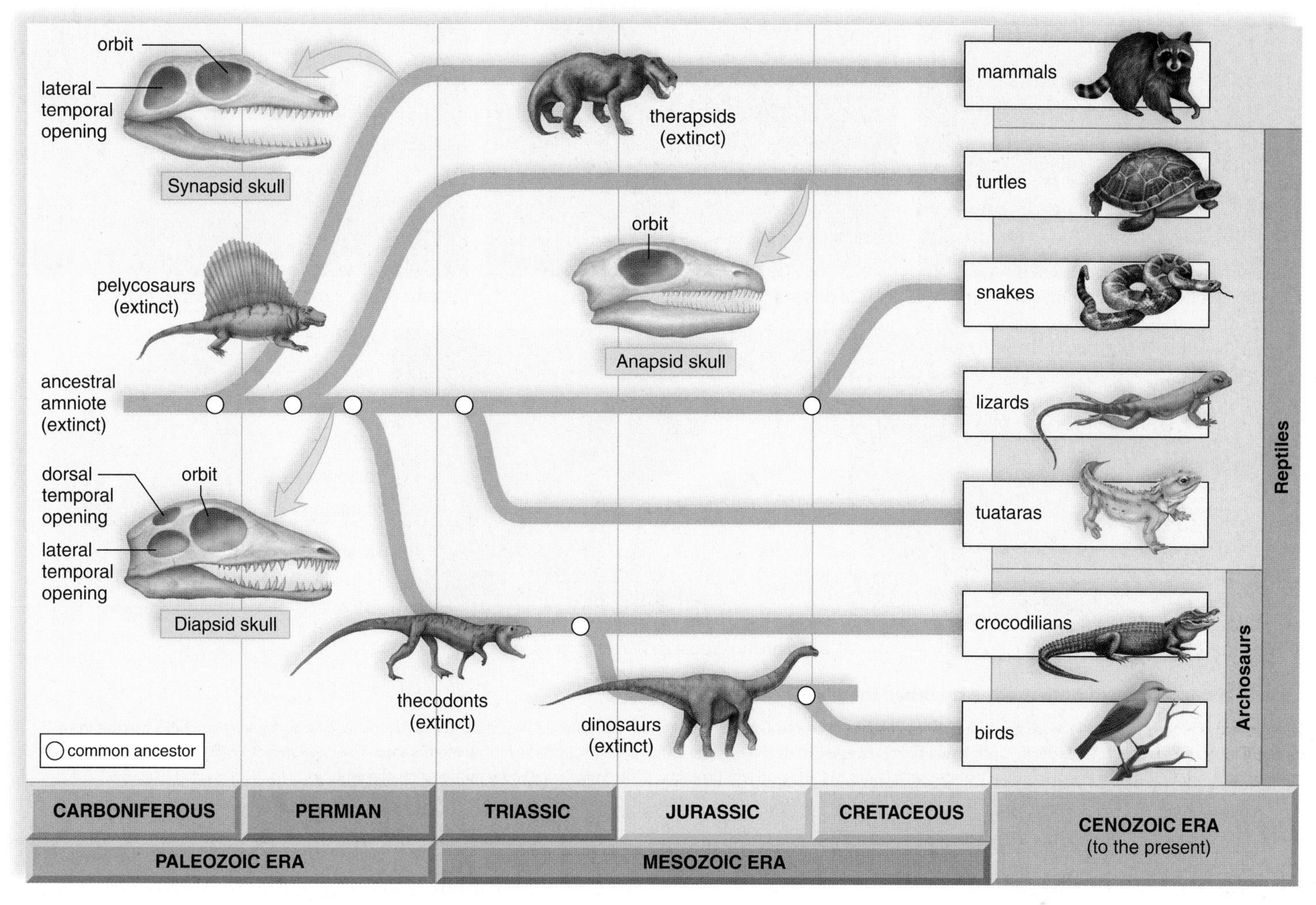


**FIGURE 29.13 Phylogenetic tree of reptiles.**

This phylogenetic tree shows the presumed evolutionary relationships among major groups of reptiles, starting with an amniote ancestor in the Paleozoic era. Openings in the skull are evidence that there are two major groups of reptiles. The turtles have a different ancestry from the other reptiles.

Therefore, authorities are in the process of dividing the reptiles into a number of monophyletic groups.

The other reptiles are *diapsids* because they have a skull with two openings behind the eyes. The thecodonts are diapsids that gave rise to the ichthyosaurs, which returned to the aquatic environment, and the pterosaurs of the Jurassic period, which had a keel for the attachment of large flight muscles and air spaces in their bones to reduce weight. Their wings were membranous and supported by elongated bones of the fourth finger. *Quetzalcoatlus,* the largest flying animal ever to live, had an estimated wingspan of nearly 13.5 m.

Of interest to us, the thecodonts gave rise to the crocodiles and dinosaurs. A sequence of now known transitional forms occurs between the dinosaurs and the birds. The crocodilians and birds share derived features, such as skull openings in front of the eyes and clawed feet. It is customary now to use the designation archosaurs for the crocodilians, dinosaurs, and birds. This means that these animals are more closely related to each other than they are to snakes and lizards.

The **dinosaurs** varied greatly in size and behavior. The average size of a dinosaur was about the size of a chicken. Some of the dinosaurs, however, were the largest land animals ever to live. *Brachiosaurus,* a herbivore, was about 23 m long and about 17 m tall. *Tyrannosaurus rex,* a carnivore, was 5 m tall when standing on its hind legs. A bipedal stance freed the forelimbs and allowed them to be used for purposes other than walking, such as manipulating prey. It was also preadaptive for the evolution of wings in the birds.

Dinosaurs dominated the Earth for about 170 million years before they died out at the end of the Cretaceous period. One hypothesis for this mass extinction is that a massive meteorite struck the Earth near the Yucatán Peninsula. The resultant cataclysmic events disrupted existing ecosystems, destroying many living things. This hypothesis is supported by the presence of a layer of the mineral iridium, which is rare on Earth but common in meteorites in the late Cretaceous strata.

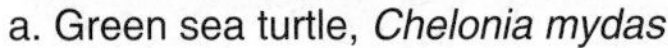
a. Green sea turtle, *Chelonia mydas*

b. Gila monster, *Heloderma suspectum*

c. Diamondback rattlesnake, *Crotalus atrox*

d. Tuatara, *Sphenodon punctatus*

e. American crocodile, *Crocodylus acutus*

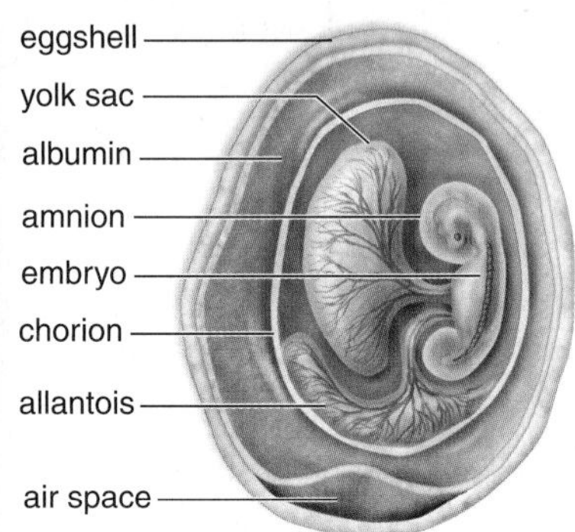


**FIGURE 29.14 Reptilian diversity other than birds.**

Representative living reptiles include (**a**) green sea turtles, (**b**) the venomous Gila monster, (**c**) the diamondback rattlesnake, and (**d**) the tuatara. **e.** A young crocodile hatches from an egg. The eggshell is leathery and flexible, not brittle like birds' eggs. Inside the egg, the embryo is surrounded by three membranes. The chorion aids in gas exchange, the allantois stores waste, and the amnion encloses a fluid that prevents drying out and provides protection. The yolk sac provides nutrients for the embryo.

### *Diversity of Living Reptiles*

Living reptiles are represented by turtles, lizards, snakes, tuataras, crocodilians, and birds. Figure 29.14 shows representatives of all but the birds.

Along with tortoises, turtles can be found in marine, freshwater, and terrestrial environments. Most turtles have ribs and thoracic vertebrae that are fused into a heavy shell. They lack teeth but have a sharp beak. The legs of sea turtles are flattened and paddlelike (Fig. 29.14*a*), while terrestrial tortoises have strong limbs for walking.

Lizards have four clawed feet and resemble their prehistoric ancestors in appearance (Fig. 29.14*b*), although some species have lost their limbs and superficially resemble snakes. Typically, they are carnivorous and feed on insects and small animals, including other lizards. Marine iguanas of the Galápagos Islands are adapted to spending time each day at sea, where they feed on sea lettuce and other algae. Chameleons are adapted to live in trees and have long, sticky tongues for catching insects some distance away. They can change color to blend in with their background. Geckos are primarily nocturnal lizards with adhesive pads on their toes. Skinks are common elongated lizards with reduced limbs and shiny scales. Monitor lizards and Gila monsters, despite their names, are not a dangerous threat to humans.

Although most snakes (Fig. 29.14*c*) are harmless, several venomous species, including rattlesnakes, cobras, mambas, and copperheads, have given the whole group a reputation of being dangerous. Snakes evolved from lizards and have lost their limbs as an adaptation to burrowing. A few species such as pythons and boas still possess the vestiges of pelvic girdles. Snakes are carnivorous and have a jaw that is loosely attached to the skull; therefore, they can eat prey that is much larger than their head size. When snakes and lizards flick out their tongue, it is collecting airborne molecules and transferring them to a Jacobson's organ at the roof of the mouth and sensory cells on the floor of the mouth. A Jacobson's organ is an olfactory organ for the analysis of airborne chemicals. Snakes possess internal ears that are capable of detecting low-frequency sounds and vibrations. Their ears lack external ear openings

Two species of tuataras are found in New Zealand (Fig. 29.14*d*). They are lizardlike animals that can attain a length of 66 cm and can live for nearly 80 years. These animals possess a well-developed "third" eye, known as a pineal eye, which is light sensitive and buried beneath the skin in the upper part of the head. The tuataras are the only member of an ancient group of reptiles that included the common ancestor of modern lizards and snakes.

The majority of crocodilians (including alligators and crocodiles) live in fresh water feeding on fishes, turtles, and terrestrial animals that venture too close to the water. They have long, powerful jaws (Fig. 29.14*e*) with numerous teeth and a muscular tail that serves as both a weapon and a paddle. Male crocodiles and alligators bellow to attract mates. In some species, the male protects the eggs and cares for the young.

science focus

## Vertebrates and Human Medicine

Hundreds of pharmaceutical products come from other vertebrates, and even those that produce poisons and toxins give us medicines that benefit us. The Thailand cobra paralyzes its victim's nerves and muscles with a potent venom that eventually leads to respiratory arrest. However, that venom is also the source of the drug Immunokine, which has been used for ten years in multiple sclerosis patients. Immunokine, which is almost without side effects, actually protects the patient's nerve cells from destruction by their immune system. A compound known as ABT-594, derived from the skin of the poison-dart frog, is approximately 50 times more powerful than morphine in relieving chronic and acute pain without the addictive properties. The southern copperhead snake and the fer-de-lance pit viper are two of the unlikely vertebrates that either serve as the source of pharmaceuticals or provide a chemical model for the synthesis of effective drugs in the laboratory. These drugs include anticoagulants ("clot busters"), painkillers, antibiotics, and anticancer drugs.

A variety of friendlier vertebrates produce proteins that are similar enough to human proteins to be used for medical treatment. Until 1978, when recombinant DNA human insulin was produced, diabetics injected insulin purified from pigs. Currently, the flu vaccine is produced in fertilized chicken eggs. The production of these drugs, however, is often time-consuming, labor intensive, and expensive. In 2003, pharmaceutical companies used 90 million chicken eggs and took nine months to produce the flu vaccine.

Some of the most powerful applications of genetic engineering can be found in the development of drugs and therapies for human diseases. In fact, this new biotechnology has actually led to a new industry: animal pharming. Animal pharming uses genetically altered vertebrates, such as mice, sheep, goats, cows, pigs, and chickens, to produce medically useful pharmaceutical products. The human gene for some useful product is inserted into the embryo of the vertebrate. That embryo is implanted into a foster mother, which gives birth to the transgenic animal, so called because it contains genes from two sources. An adult transgenic vertebrate produces large quantities of the pharmed product in its blood, eggs, or milk, from which the product can be easily harvested and purified. A pharmed product advanced in development and the FDA approval process is ATIII, a bioengineered form of human antithrombin. This medication is important in the treatment of individuals who have a hereditary deficiency of this protein and so are at high risk for life-threatening blood clots, especially during such events as surgery or childbirth procedures.

Xenotransplantation, the transplantation of nonhuman vertebrate tissues and organs into humans, is another benefit of genetically altered animals. There is an alarming shortage of human donor organs to fill the need for hearts, kidneys, and livers. The first animal–human transplant occurred in 1984 when a team of surgeons implanted a baboon heart into an infant, who unfortunately lived only a short while before dying of circulatory complications. In the late 1990s, two patients were kept alive using a pig liver outside of their body to filter their blood until a human organ was available for transplantation. Although baboons are phylogenetically closer to humans than pigs, pigs are generally healthier, produce more offspring in a shorter time, and are already farmed for food. Despite the fears of some, scientists think that viruses unique to pigs are unlikely to cross the species barrier and infect the human recipient. Currently, pig heart valves and skin are routinely used for treatment of humans. Miniature pigs, whose heart size is similar to humans, are being genetically engineered to make their tissues less foreign to the human immune system, in order to avoid rejection.

The use of transgenic vertebrates for medical purposes does raise health and ethical concerns. What viral AIDS-like epidemic might be unleashed by cross-species transplantation? What other unseen health consequences might there be? Is it ethical to change the genetic makeup of vertebrates, in order to use them as drug or organ factories? Are we redefining the relationship between humans and other vertebrates to the detriment of both? These questions will continue to be debated as the research goes forward. Meanwhile, several U.S. regulatory bodies, including the FDA, have adopted voluntary guidelines for this new technology.

a. Poison-dart frogs, source of a medicine

b. Pigs, source of organs

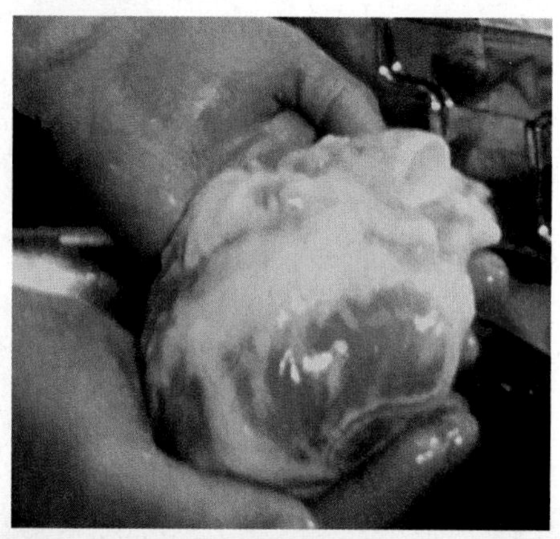

c. Heart for transplantation

**FIGURE 29A Use of other vertebrates for medical purposes.**
***a.*** *The poison-dart frog is the source of a pain medication.* ***b.*** *Pigs are now being genetically altered to provide a supply of* ***(c)*** *hearts for heart transplant operations.*

## Birds

Birds share a common ancestor with crocodilians and have traits, such as the presence of scales (feathers are modified scales), a tail with vertebrae, and clawed feet, that show they are indeed reptiles.

To many people, **birds** are the most conspicuous, melodic, beautiful, and fascinating group of vertebrates. Birds range in size from the tiny "bee" hummingbird at 1.8 g (less than a penny) and 5 cm long to the ostrich at a maximum weight of 160 kg and a height of 2.7 m.

Nearly every anatomical feature of a bird can be related to its ability to fly (Fig. 29.15). These features are involved in the action of flight, providing energy for flight or the reduction of the bird's body weight, making flight less energetically costly:

**Feathers.** Soft down keeps birds warm, wing feathers allow flight, and tail feathers are used for steering. A feather is a modified reptilian scale with the complex structure shown in Figure 29.15*a*. Nearly all birds molt (lose their feathers) and replace their feathers about once a year.

**Modified skeleton.** Unique to birds, the collarbone is fused (the wishbone), and the sternum has a keel (Fig. 29.15*b*). Many other bones are fused, making the skeleton more rigid than the reptilian skeleton. The breast muscles are attached to the keel, and their action accounts for a bird's ability to fly.

A horny beak has replaced jaws equipped with teeth, and a slender neck connects the head to a rounded, compact torso.

**Modified respiration.** In birds, unlike other reptiles, the lobular lungs connect to anterior and posterior air sacs. The presence of these sacs means the air circulates one way through the lungs, and gases are continuously exchanged across respiratory tissues. Another benefit of air sacs is that they lighten the body and bones for flying. Some of the air sacs are present in cavities within the bones.

**Endothermy.** Birds, unlike other reptiles, generate internal heat. Many **endotherms** can use metabolic heat to maintain a constant internal temperature. Endothermy may be associated with their efficient nervous, respiratory, and circulatory systems.

**Well-developed sense organs and nervous system.** Birds have particularly acute vision and well-developed brains. Their muscle reflexes are excellent. An enlarged portion of the brain seems to be the area

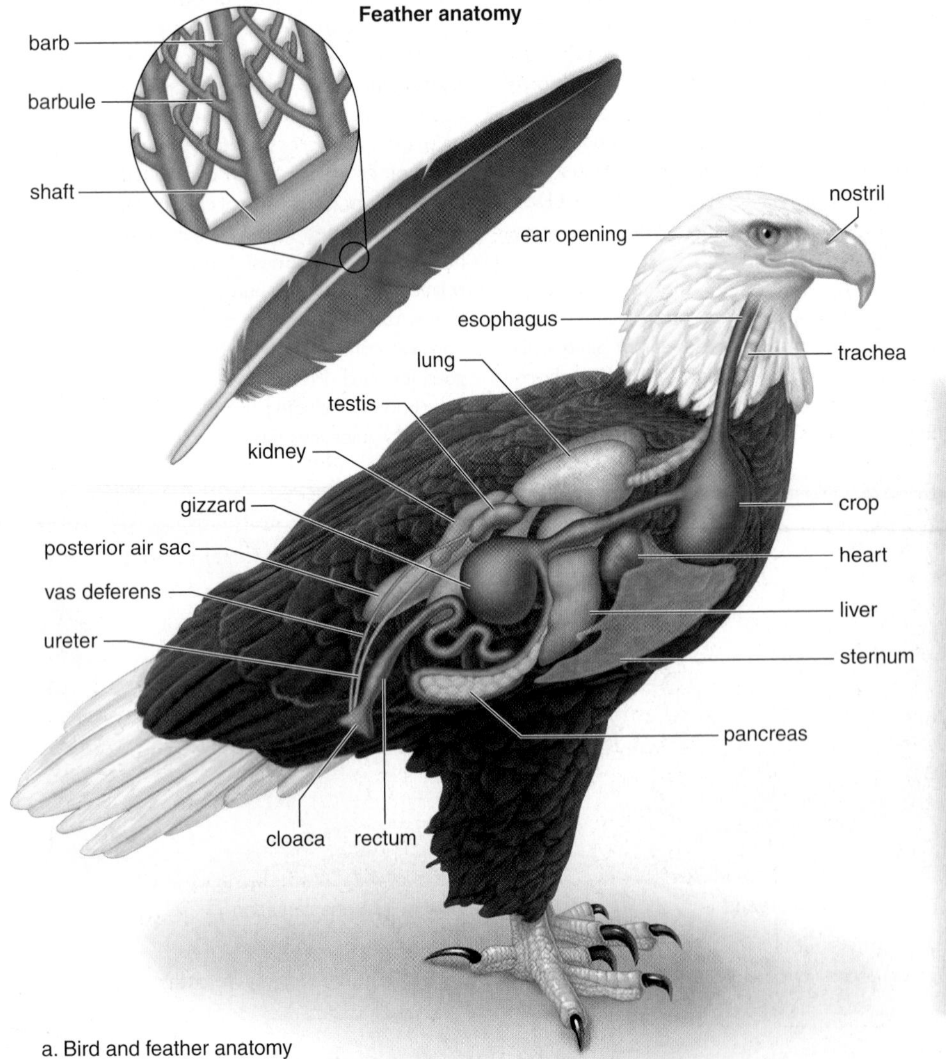


a. Bird and feather anatomy

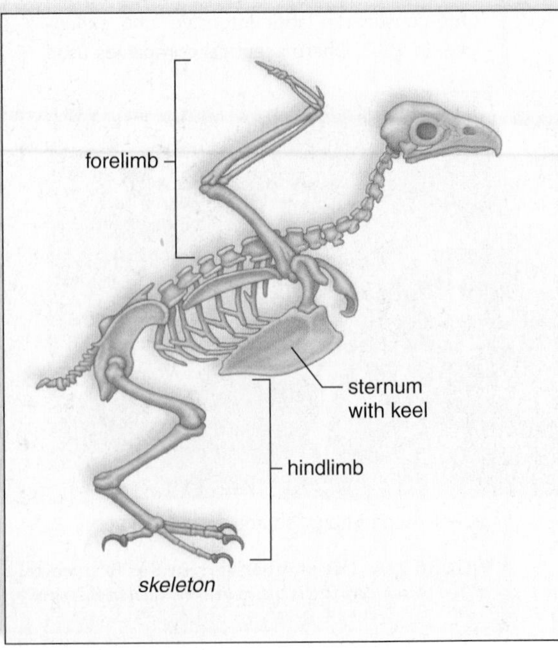


b. Bald eagle, *Haliaetus*

**FIGURE 29.15 Bird anatomy and flight.**

**a.** Bird anatomy. *Top:* In feathers, a hollow central shaft gives off barbs and barbules, which interlock in a latticelike array. *Bottom:* The anatomy of an eagle is representative of bird anatomy. **b.** Bird flight. *Left:* The skeleton of an eagle shows that birds have a large, keeled sternum to which flight muscles attach. The bones of the forelimb help support the wings. *Right:* Birds fly by flapping their wings. Bird flight requires an airstream and a powerful wing downstroke for lift, a force at right angles to the airstream.

a. Bald eagle, *Haliaetus leucocephalus*

b. Pileated woodpecker, *Dryocopus pileatus*

**FIGURE 29.16**
**Bird beaks.**

**a.** A bald eagle's beak allows it to tear apart prey. **b.** A woodpecker's beak is used to chisel in wood. **c.** A flamingo's beak strains food from the water with bristles that fringe the mandibles. **d.** A parrot's beak is modified to pry open nuts. **e.** A cardinal's beak allows it to crack tough seeds.

c. Flamingo, *Phoenicopterus ruber*

d. Blue-and-yellow macaw, *Ara ararauna*

e. Cardinal, *Cardinalis cardinalis*

responsible for instinctive behavior. A ritualized courtship often precedes mating. Many newly hatched birds require parental care before they are able to fly away and seek food for themselves. A remarkable aspect of bird behavior is the seasonal migration of many species over long distances. Birds navigate by day and night, whether it is sunny or cloudy, by using the sun and stars and even the Earth's magnetic field to guide them. Birds are very vocal animals. Their vocalizations are distinctive and so convey an abundance of information.

### *Diversity of Living Birds*

The majority of birds, including eagles, geese, and mockingbirds, have the ability to fly. However, some birds, such as emus, penguins, kiwis, and ostriches, are flightless. Traditionally, birds have been classified according to beak and foot type (Fig. 29.16) and, to some extent, on their habitat and behavior. The birds of prey have notched beaks and sharp talons; shorebirds have long, slender, probing beaks and long, stiltlike legs; woodpeckers have sharp, chisel-like beaks and grasping feet; waterfowl have broad beaks and webbed toes; penguins have wings modified as paddles; songbirds have perching feet; and parrots have short, strong "plierlike" beaks and grasping feet.

**Check Your Progress 29.5**

1. How are reptiles adapted to a land environment?
2. Contrast the characteristics of alligators to those of snakes.
3. The common ancestor for birds was a dinosaur. Does this make birds a reptile? Explain.

## 29.6 The Mammals

The **mammals** include the largest animal ever to live, the blue whale (130 metric tons); the smallest mammal, the Kitti's bat (1.5 g); and the fastest land animal, the cheetah (110 km/hr). These characteristics distinguish mammals:

**Hair.** The most distinguishing characteristics of mammals are the presence of hair and milk-producing mammary glands. Hair provides insulation against heat loss, and being endothermic allows mammals to be active even in cold weather. The color of hair can camouflage a mammal and help the animal blend into its surroundings. In addition, hair can be ornamental and can serve sensory functions.

**Mammary glands.** These glands enable females to feed (nurse) their young without leaving them to find food. Nursing also creates a bond between mother and offspring that helps ensure parental care while the young are helpless and provides antibodies to the young from the mother through the milk.

**Skeleton.** The mammalian skull accommodates a larger brain relative to body size than does the reptilian skull. Also, mammalian cheek teeth are differentiated as premolars and molars. The vertebrae of mammals are highly differentiated, typically the middle region of the vertebral column is arched, and the limbs are under the body.

**Internal organs.** Efficient respiratory and circulatory systems ensure a ready oxygen supply to muscles whose contraction produces body heat. Like birds, mammals have a double-loop circulatory pathway and a four-chambered heart. The kidneys are adapted to conserving water in terrestrial mammals. The nervous system of mammals is highly developed. Special senses in mammals are well developed, and mammals exhibit complex behavior.

**Internal development.** In most mammals, the young are born alive after a period of development in the uterus, a part of the female reproductive tract. Internal development shelters the young and allows the female to move actively about while the young are maturing.

### Evolution of Mammals

Mammals share an amniote ancestor with reptiles (see Fig. 29.13). Their more immediate ancestors in the Mesozoic era had a synapsid skull (two openings behind the eyes). The first true mammals appeared during the Triassic period, about the same time as the first dinosaurs, and were similar in size to mice. During the reign of the dinosaurs (170 million years), mammals were a minor group that changed little. The two earliest mammalian groups, represented today by the monotremes and marsupials, are not abundant today. The marsupials probably originated in the Americas and then spread through South America and Antarctica to Australia before these continents separated. Placental mammals, the third branch of the mammalian lineage, originated in Eurasia and spread to the Americas also by land connections that existed between the continents during the Mesozoic. The placental mammals underwent an adaptive radiation into the habitats previously occupied by the dinosaurs.

### *Monotremes*

**Monotremes** [Gk. *monos*, one, and *trema*, hole] are egg-laying mammals that include only the duckbill platypus (Fig. 29.17*a*) and two species of spiny anteaters. The term monotreme refers to the presence of a single opening, the cloaca. Monotremes unlike other mammals lay hard-shelled amniotic eggs. No embryonic development occurs inside the female's body. The female duckbill platypus lays her eggs in a burrow in the ground. She incubates the eggs, and after hatching, the young lick up milk that seeps from modified sweat glands on the mother's abdomen. Spiny anteaters, which actually feed mainly on termites and not ants, have pores that seep milk in a shallow belly pouch formed by skin folds on each side. The egg moves from the cloaca to this pouch, where hatching takes place and the young remain for about 53 days. Then they stay in a burrow, where the mother periodically visits and nurses them.

### *Marsupials*

The **marsupials** [Gk. *marsupium*, pouch] are also known as the pouched mammals. Marsupials include kangaroos, koalas, Tasmanian devils, wombats, sugar gliders, and opossums. The young of marsupials begin their development inside the female's body, but they are born in a very immature condition. Newborns are typically hairless, have yet to

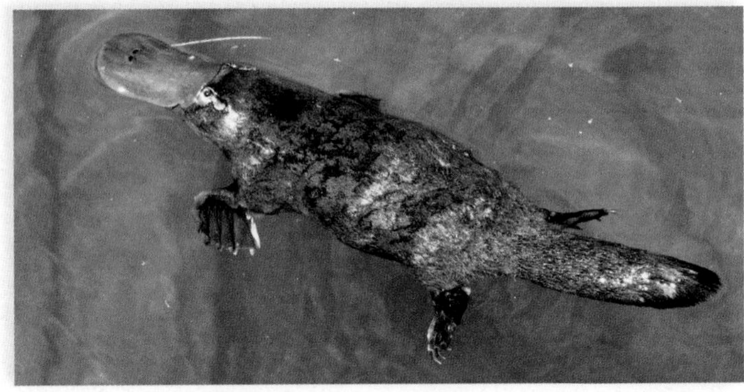

a. Duckbill platypus, *Ornithorhynchus anatinus*

b. Koala, *Phascolarctos cinereus*

c. Virginia opossum, *Didelphis virginianus*

**FIGURE 29.17 Monotremes and marsupials.**

**a.** The duckbill platypus is a monotreme that inhabits Australian streams. **b.** The koala is an Australian marsupial that lives in trees. **c.** The opossum is the only marsupial in North America. The Virginia opossum is found in a variety of habitats.

a. White-tailed deer, *Odocoileus virginianus*

b. African lioness, *Panthera leo*

c. Squirrel monkey, *Saimiri sciureus*

d. Killer whale, *Orcinus orca*

**FIGURE 29.18**
**Placental mammals.**

Placental mammals have adapted to various ways of life. **a.** Deer are herbivores that live in forests. **b.** Lions are carnivores on the African plain. **c.** Monkeys inhabit tropical forests. **d.** Whales are sea-dwelling placental mammals.

open their eyes, yet crawl up into a pouch on their mother's abdomen. Inside the pouch, they attach to nipples of mammary glands and continue to develop. Frequently, more are born than can be accommodated by the number of nipples, and it's "first come, first served."

Today, marsupial mammals are most abundant in Australia and New Guinea, filling all the typical roles of placental mammals on other continents. For example, among herbivorous marsupials in Australia today, koalas are tree-climbing browers (Fig. 29.17*b*), and kangaroos are grazers. A significant number of marsupial species are also found in South and Central America. The opossum is the only North American marsupial (Fig. 29.17*c*).

## *Placental Mammals*

The **placental mammals** are the dominant group of mammals on Earth. Developing placental mammals are dependent on the **placenta,** an organ of exchange between maternal blood and fetal blood. Nutrients are supplied to the growing offspring, and wastes are passed to the mother for excretion. While the fetus is clearly parasitic on the female, in exchange, she is able to freely move about while the fetus develops.

Placental mammals lead an active life. The senses are acute, and the brain is enlarged due to the convolution and expansion of the foremost part—the cerebral hemispheres. The brain is not fully developed for some time after birth, and there is a long period of dependency on the parents, during which the young learn to take care of themselves.

Most mammals live on land, but some (e.g., whales, dolphins, seals, sea lions, and manatees) are secondarily adapted to live in water, and bats are able to fly. While bats are the only mammal that can fly, three types of placentals can glide: the flying squirrels, scaly-tailed squirrels, and the flying lemurs. These are the main types of placental mammals:

The *ungulates* are hoofed mammals, which comprise about a third of all living and extinct mammalian groups. The hoofed mammals have a reduced number of toes and are divided according to whether an odd number remain (e.g., horses, zebras, tapirs, rhinoceroses) or whether a even number of toes remain (e.g., pigs, cattle, deer, hippopotamuses, buffaloes, giraffes) (Fig. 29.18*a*). Many of the hoofed animals have elongated limbs and are adapted for running, often across open grasslands. Both groups of animals are herbivorous and have large, grinding teeth.

The *carnivores* (e.g., dogs, cats, bears, raccoons, and skunks) are predaceous meat eaters with large and conical-shaped canine teeth (Fig. 29.18*b*). Some carnivores are aquatic (e.g., seals, sea lions, and walruses) and must return to land to reproduce.

The *primates* are tree-dwelling fruit eaters (e.g., lemurs, monkeys, gibbons, chimpanzees, gorillas, and humans) (Fig. 29.18*c*). Humans are ground dwellers, well known for their opposable thumb and well-developed brain.

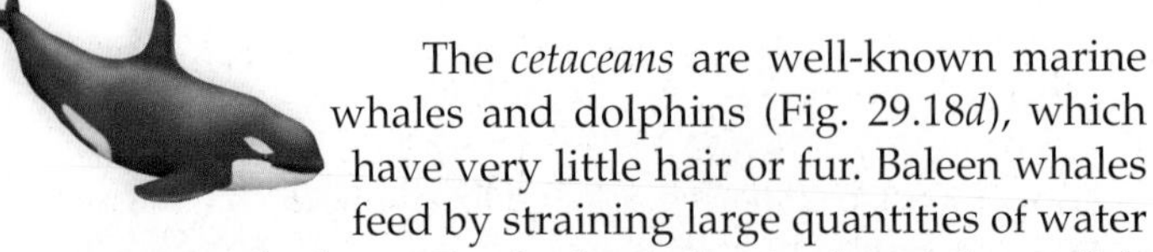

The *cetaceans* are well-known marine whales and dolphins (Fig. 29.18*d*), which have very little hair or fur. Baleen whales feed by straining large quantities of water containing plankton. Toothed whales feed mainly on fish and squid.

The *chiroptera* are the flying mammals (e.g., bats), whose wings consist of two layers of skin and connective tissue stretched between the elongated bones of all fingers but the first. Many species use echolocation to navigate at night and to locate their usual insect prey. But there are also bird-, fish-, frog-, plant-, and blood-eating bats.

The *rodents* are most often small plant eaters (e.g., mice, rats, squirrels, beavers, and porcupines). The incisors of these gnawing animals suffer heavy wear and tear, and they grow continuously.

The *proboscideans*, the herbivorous elephants, are the largest living land mammals, whose upper lip and nose have become elongated and muscularized to form a trunk.

The *lagomorphans* are the rodentlike jumpers (e.g., rabbits, hares, and pikas). These herbivores have two pairs of continually growing incisors, and their hind legs are longer than their front legs.

The *insectivores* are the small burrowing mammals (e.g., shrews and moles), which have short snouts and live primarily underground.

At one time, it was thought that insectivores were most like the original placentals. More recent analysis suggests that the *edentates* (anteaters) and pangolins (scaly anteaters) are the more primitive groups of living placentals.

**Check Your Progress 29.6**

1. What are the major characteristics of mammals? What are the evolutionary advantages of these characteristics?
2. Describe the three groups of mammals and several groups of the placental mammals.

## Connecting the Concepts

How do you measure success? As human beings, we may assume that vertebrate chordates, such as ourselves, are the most successful organisms. But depending on the criteria used, organisms that are in some ways less complex may come out on top!

For example, vertebrates are eukaryotes, which have been assigned to one domain, while the prokaryotes are now divided into two domains. In fact, the total number of prokaryotes is greater than the number of eukaryotes, and there are possibly more types of prokaryotes than any other living form. Therefore, the unseen world is much larger than the seen world. Furthermore, prokaryotes are adapted to use most energy sources and to live in almost any type of environment.

As terrestrial mammals, humans might assume that terrestrial species are more successful than aquatic ones. However, if not for the myriad types of terrestrial insects, there would be more aquatic species than terrestrial ones on Earth. The adaptative radiation of mammals has taken place on land, and this might seem impressive to some. But actually, the number of mammalian species (4,800) is small compared to, say, the molluscs (110,000 species), which radiated in the sea.

The size and complexity of the brain is also sometimes cited as a criterion by which vertebrates are more successful than other living things. However, this very characteristic has been linked to others that make an animal prone to extinction. Studies have indicated that large animals have a long life span, are slow to mature, have few offspring, expend much energy caring for their offspring, and tend to become extinct if their normal way of life is destroyed. And finally, vertebrates, in general, are more threatened than other types of organisms by our present biodiversity crisis—a crisis brought on by the activities of the vertebrate with the most complex brain of all, *Homo sapiens*.

Chapter 30 traces the increase in complexity of the human brain by exploring the evolution of the primates, a group that includes humans.

## summary

### 29.1 The Chordates

Chordates (sea squirts, lancelets, and vertebrates) have a notochord, a dorsal tubular nerve cord, pharyngeal pouches, and a postanal tail at some time in their life history.

Lancelets and sea squirts are the nonvertebrate chordates. Lancelets are the only chordate to have the four characteristics in the adult stage. Sea squirts lack chordate characteristics (except gill slits) as adults, but they have a larva that could be ancestral to the vertebrates.

### 29.2 The Vertebrates

Vertebrates have the four chordate characteristics as embryos. As adults, the notochord is replaced by the vertebral column. Vertebrates undergo cephalization, and have an endoskeleton, paired appendages, and well-developed internal organs.

Vertebrate evolution is marked by the evolution of vertebrae, jaws, a bony skeleton, lungs, limbs, and the amniotic egg.

### 29.3 The Fishes

The first vertebrates lacked jaws and paired appendages. They are represented today by the hagfishes and lampreys. Ancestral bony fishes, which had jaws and paired appendages, gave rise during the Devonian period to two groups: today's cartilaginous fishes (skates, rays, and sharks) and the bony fishes, including the ray-finned fishes and the lobe-finned fishes. The ray-finned fishes (actinopterygii) became the most diverse group among the vertebrates. Ancient lobe-finned fishes (sarcopterygii) gave rise to the coelacanths and amphibians.

### 29.4 The Amphibians

Amphibians are tetrapods represented primarily today by frogs and salamanders. Most frogs and some salamanders return to the water to reproduce and then metamorphose into terrestrial adults.

### 29.5 The Reptiles

Reptiles (today's alligators and crocodiles, birds, turtles, tuataras, lizards, and snakes) lay a shelled amniotic egg, which allows them to reproduce on land. Turtles with an anapsid skull have a separate ancestry from the other reptiles mentioned. The other reptiles with a diapsid skull include the crocodilians, dinosaurs, and the birds. Birds have reptilian features, including scales (feathers are modified scales), tail with vertebrae, and clawed feet.

The feathers of birds help them maintain a constant body temperature. Birds are adapted for flight: Their bones are hollow, their shape is compact, their breastbone is keeled, and they have well-developed sense organs.

### 29.6 The Mammals

Mammals share an amniote ancestor with reptiles but they have a synapsid skull (two openings behind the eyes). Mammals remained small and insignificant while the dinosaurs existed, but when dinosaurs became extinct at the end of the Cretaceous period, mammals became the dominant land organisms.

Mammals are vertebrates with hair and mammary glands. Hair helps them maintain a constant body temperature, and the mammary glands allow them to feed and establish an immune system in their young. Monotremes lay eggs, while marsupials have a pouch in which the newborn crawls and continues to develop. The placental mammals, which are the most varied and numerous, retain offspring inside the female until birth.

## understanding the terms

agnathan 543
amniote 542
amniotic egg 548
amphibian 546
bird 552
bony fish (Osteichthyes) 544
cartilaginous fish (Chondrichthyes) 543
cephalochordate 540
chordate 540
cloaca 547
dinosaur 549
ectotherm 543
endotherm 552
fin 544
fishes 543
gills 540
gnathostome 542
jawless fishes 543
lobe-finned fishes (Sarcopterygii) 544
lungfishes 544
mammal 554
marsupial 554
metamorphosis 546
monotreme 554
notochord 540
operculum 544
ostracoderm 543
placenta 555
placental mammal 555
placoderm 543
ray-finned bony fishes 544
reptile 548
sarcopterygii 544
swim bladder 544
tetrapod 542
urochordate 541

Match the terms to these definitions:

a. ______________ Animal (bird or mammal) that maintains a uniform body temperature independent of the environmental temperature.

b. ______________ Egg-laying mammal—for example, duckbill platypus and spiny anteater.

c. ______________ Terrestrial vertebrates with internal fertilization, scaly skin, and a shelled egg; includes turtles, lizards, snakes, crocodilians, and birds.

d. ______________ Dorsal supporting rod that exists in all chordates sometime in their life history; replaced by the vertebral column in vertebrates.

## reviewing this chapter

1. What four characteristics do all chordates have at some time in their development? 540
2. Describe the two groups of nonvertebrate chordates, and explain how the sea squirts might be ancestral to vertebrates. 540–41
3. Discuss the distinguishing characteristics and the evolution of vertebrates. 542
4. Describe the jawless fishes, including ancient ostracoderms. 543
5. Describe the characteristics of fishes with jaws. What is the significance of having jaws? Describe today's cartilaginous and bony fishes. The amphibians evolved from what type of ancestral fish? 543–45
6. Discuss the characteristics of amphibians, stating which ones are especially adaptive to a land existence. Explain how their name (amphibians) characterizes these animals. 546–47
7. What is the significance of the amniotic egg? What other characteristics make reptiles less dependent on a source of external water? 548
8. Draw a simplified phylogenetic tree that includes the anapsid, diapsid, and synapsid skulls. 548–50
9. What is the significance of wings? In what other ways are birds adapted to flying? 552–53
10. What are the three major groups of mammals, and what are their primary characteristics? 554–56

## testing yourself

Choose the best answer for each question.

1. Which of these is not a chordate characteristic?
   a. dorsal supporting rod, the notochord
   b. dorsal tubular nerve cord
   c. pharyngeal pouches
   d. postanal tail
   e. vertebral column
2. Adult sea squirts
   a. do not have all five chordate characteristics.
   b. are also called tunicates.
   c. are fishlike in appearance.
   d. are the first chordates to be terrestrial.
   e. All of these are correct.
3. Cartilaginous fishes and bony fishes are different in that only
   a. bony fishes have paired fins.
   b. bony fishes have a keen sense of smell.
   c. bony fishes have an operculum.
   d. cartilaginous fishes have a complete skeleton.
   e. cartilaginous fishes are predaceous.
4. Amphibians evolved from what type of ancestral fish?
   a. sea squirts and lancelets
   b. cartilaginous fishes
   c. jawless fishes
   d. ray-finned fishes
   e. lobe-finned fishes

5. Which of these is not a feature of amphibians?
   a. dry skin that resists desiccation
   b. metamorphosis from a swimming form to a land form
   c. small lungs and a supplemental means of gas exchange
   d. reproduction in the water
   e. a single ventricle
6. Reptiles
   a. were dominant during the Mesozoic era.
   b. include the birds.
   c. lay shelled eggs.
   d. are ectotherms, except for birds.
   e. All of these are correct.
7. Which of these is a true statement?
   a. In all mammals, offspring develop completely within the female.
   b. All mammals have hair and mammary glands.
   c. All mammals have one birth at a time.
   d. All mammals are land-dwelling forms.
   e. All of these are true.
8. Which of these is not an invertebrate? Choose more than one answer if correct.
   a. tunicate
   b. frog
   c. lancelet
   d. squid
   e. roundworm
9. Which of these is not a characteristic of vertebrates? Choose more than one answer if correct.
   a. All vertebrates have a complete digestive system.
   b. Vertebrates have a closed circulatory system.
   c. Sexes are usually separate in vertebrates.
   d. Vertebrates have a jointed endoskeleton.
   e. Most vertebrates never have a notochord.
10. Bony fishes are divided into which two groups?
   a. hagfishes and lampreys
   b. sharks and ray-finned fishes
   c. ray-finned fishes and lobe-finned fishes
   d. jawless fishes and cartilaginous fishes
11. Which of these is an incorrect difference between reptiles and birds?

| Reptiles | Birds |
|---|---|
| a. shelled egg | partial internal development |
| b. scales | feathers |
| c. tetrapods | wings |
| d. ectothermy | endothermy |
| e. no air sacs | air sacs |

12. Which of these does not produce an amniotic egg? Choose more than one answer if correct.
   a. bony fishes
   b. duckbill platypus
   c. snake
   d. robin
   e. frog
13. Label the following diagram of a chordate embryo.

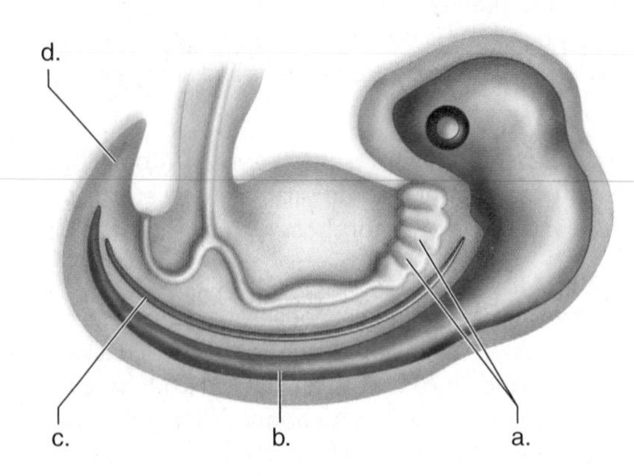


14. The amniotes include all but the
   a. birds.
   b. mammals.
   c. reptiles.
   d. amphibians.
15. Which of the following groups has a three-chambered heart?
   a. all birds
   b. all reptiles
   c. all mammals
   d. all amphibians
16. Ancestors to the mammals known only in the fossil record are
   a. synapsids.
   b. marsupials.
   c. monotremes.
   d. placentals.

## thinking scientifically

1. *Archaeopteryx* was a birdlike reptile that had a toothed beak. Give an evolutionary explanation for the elimination of teeth in a bird's beak.
2. While amphibians have rudimentary lungs, skin is also a respiratory organ. Why would a thin skin be more sensitive to pollution than lungs?

## *Biology* website

The companion website for *Biology* provides a wealth of information organized and integrated by chapter. You will find practice tests, animations, videos, and much more that will complement your learning and understanding of general biology.

**http://www.mhhe.com/maderbiology10**

# 30 Human Evolution

*Sometimes you hear people say that evolutionists believe humans evolved from apes. This is a mistake; instead, evolutionists tell us that modern humans and certain of the apes followed their own evolutionary pathways after evolving from a common ancestor. Among the apes, gorillas and chimpanzees are our cousins, and we couldn't have evolved from our cousins because we are all contemporaries—living on Earth at the same time. The pattern of descent is just like that between you and your cousin in that cousins are descended from the same grandparents.*

*Scientists have discovered that the same patterns of evolution characterize human evolution as any other group of organisms. Various prehuman groups died out, migrated, or interbred with other groups all within a very short period of time, making the evolutionary descent of humans very complex. Additional fossils are always being found; the photo below shows how scientists have reconstructed the appearance of* Sahelanthropus tchadensis *from his fossil remains. This fossil has been dated somewhat later than the time when humans and apes parted, but the skull opening for the spine indicates that* Sahelanthropus tchadensis *walked erect, just as we do. This chapter traces the ancestry of primates, including humans, from their origins.*

Reconstruction of *Sahelanthropus tchadensis,* a possible human ancestor that lived 7 million years ago (MYA).

## concepts

### 30.1 EVOLUTION OF PRIMATES

- Primate characteristics include an enlarged brain, an opposable thumb, stereoscopic vision, and an emphasis on learned behavior. 560–63

### 30.2 EVOLUTION OF HUMANLIKE HOMININS

- The hominins include modern humans and extinct species most closely related to humans. 564–65
- The evolutionary split between the ape lineage and the human lineage occurred about 7 MYA (million years ago). Several fossils, dated about this time, may be the earliest hominins. 564–65

### 30.3 EVOLUTION OF LATER HUMANLIKE HOMININS

- About 4 MYA, australopiths were prevalent in Africa. The australopiths had a relatively small brain, but they could walk upright. 566–67

### 30.4 EVOLUTION OF EARLY *HOMO*

- About 2 MYA, early *Homo* types evolved that had a larger brain than the australopiths and were able to make primitive stone tools. 568
- *Homo ergaster* in Africa and *Homo erectus* in Asia had knowledge of fire and made more advanced tools. 568–69

### 30.5 EVOLUTION OF LATER *HOMO*

- The replacement model says that modern humans evolved in Africa, and after migrating to Asia and Europe about 100,000 years BP (before the present), they replaced the archaic *Homo* species. 570
- Cro-Magnon is the name given to modern humans. They made sophisticated tools and definitely had culture. 570–71
- Today, humans have various ethnic backgrounds. Even so, genetic evidence suggests that they share a fairly recent common ancestor and that noticeable differences are due to adaptations to local environmental conditions or genetic drift. 572

## 30.1 Evolution of Primates

**Primates** [L. *primus,* first] include prosimians, monkeys, apes, and humans (Fig. 30.1). In contrast to other types of mammals, primates are adapted for an **arboreal** life—that is, for living in trees. The evolution of primates is characterized by trends toward mobile limbs; grasping hands; a flattened face; stereoscopic vision; a large, complex brain; and a reduced reproductive rate. These traits are particularly useful for living in trees.

### Mobile Forelimbs and Hindlimbs

Primates have evolved grasping hands and feet, which have five digits each. In most primates, flat nails have replaced the claws of ancestral primates, and sensitive pads on the undersides of fingers and toes assist the grasping of objects. All primates have a thumb, but it is only truly opposable in Old World monkeys, great apes, and humans. Because an **opposable thumb** can touch each of the other fingers, the grip is both powerful and precise (Fig. 30.2). In all but humans, primates with an opposable thumb also have an opposable toe.

The evolution of the primate limb was a very important adaptation for their life in trees. Mobile limbs with clawless opposable digits allow primates to freely grasp and release tree limbs. They also allow primates to easily reach out and bring food, such as fruit, to the mouth.

### Stereoscopic Vision

A foreshortened snout and a relatively flat face are also evolutionary trends in primates. These may be associated with a general decline in the importance of smell and an increased reliance on vision. In most primates, the eyes are located in the front, where they can focus on the same object from slightly different angles (Fig. 30.3). The **stereoscopic** (three-dimensional) **vision** and good depth perception that result permit primates to make accurate judgments about the distance and position of adjoining tree limbs.

Some primates, humans in particular, have color vision and greater visual acuity because the retina contains cone cells in addition to rod cells. Rod cells are activated in dim light, but the blurry image is in shades of gray. Cone cells require bright light, but the image is sharp and in color. The lens of the eye focuses light directly on the fovea, a region of the retina where cone cells are concentrated.

PROSIMIANS

Ring-tailed lemur, *Lemus catta*

Tarsier, *Tarsius bancanus*

a.

NEW WORLD MONKEY

White-faced monkey, *Cebus capucinus*

OLD WORLD MONKEY

Anubis baboon, *Papio anubis*

b.

ASIAN APES

Orangutan, *Pongo pygmaeus*

White-handed gibbon, *Hylobates lar*

c.

## Large, Complex Brain

Sense organs are only as beneficial as the brain that processes their input. The evolutionary trend among primates is toward a larger and more complex brain. This is evident when comparing the brains of prosimians, such as lemurs and tarsiers, with that of apes and humans. In apes and humans, the portion of the brain devoted to smell is smaller, and the portions devoted to sight have increased in size and complexity. Also, more of the brain is devoted to controlling and processing information received from the hands and the thumb. The result is good hand-eye coordination. A larger portion of the brain is devoted to communication skills, which supports primates' tendency to live in social groups.

## Reduced Reproductive Rate

One other trend in primate evolution is a general reduction in the rate of reproduction, associated with increased age of sexual maturity and extended life spans. Gestation is lengthy, allowing time for forebrain development. One birth at a time is the norm in primates; it is difficult to care for several offspring while moving from limb to limb. The juvenile period of dependency is extended, and there is an emphasis on learned behavior and complex social interactions.

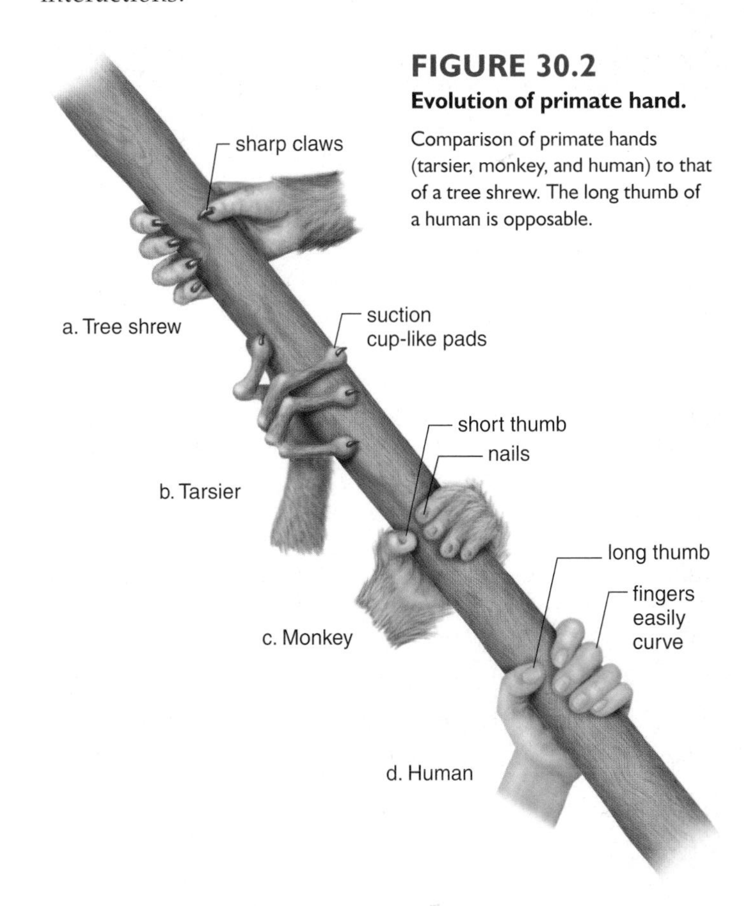


**FIGURE 30.2**
**Evolution of primate hand.**
Comparison of primate hands (tarsier, monkey, and human) to that of a tree shrew. The long thumb of a human is opposable.

AFRICAN APES

Chimpanzee, *Pan troglodytes*

estern lowland gorilla, *Gorilla gorilla*

Humans, *Homo sapiens*

d.

**FIGURE 30.1**
**Primate diversity.**
**a.** Today's prosimians may resemble the first group of primates to evolve. **b.** Today's monkeys are divided into the New World monkeys and the Old World monkeys. **c.** Today's apes can be divided into the Asian apes (orangutans and gibbons) and the African apes (chimpanzees and gorillas). **d.** Humans are also primates.

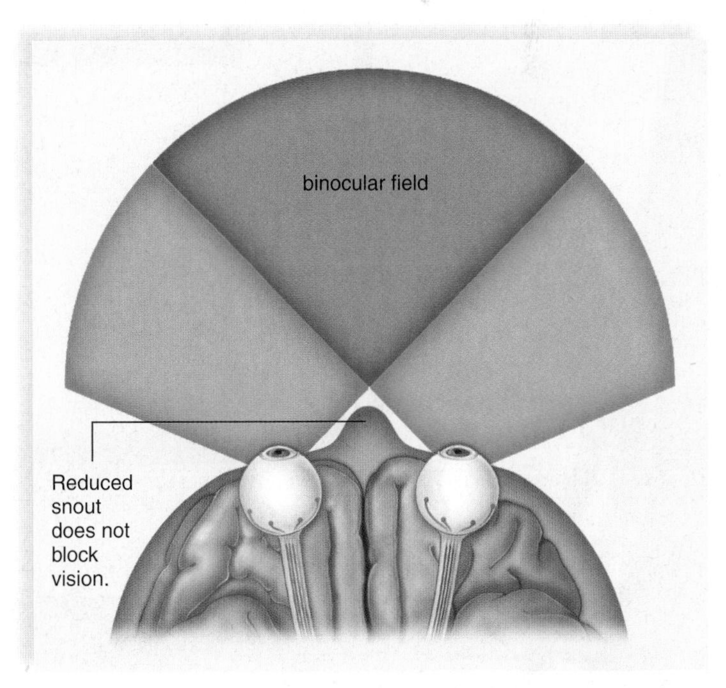


**FIGURE 30.3 Stereoscopic vision.**
In primates, the snout is reduced, and the eyes are at the front of the head. The result is a binocular field that aids depth perception and provides stereoscopic vision.

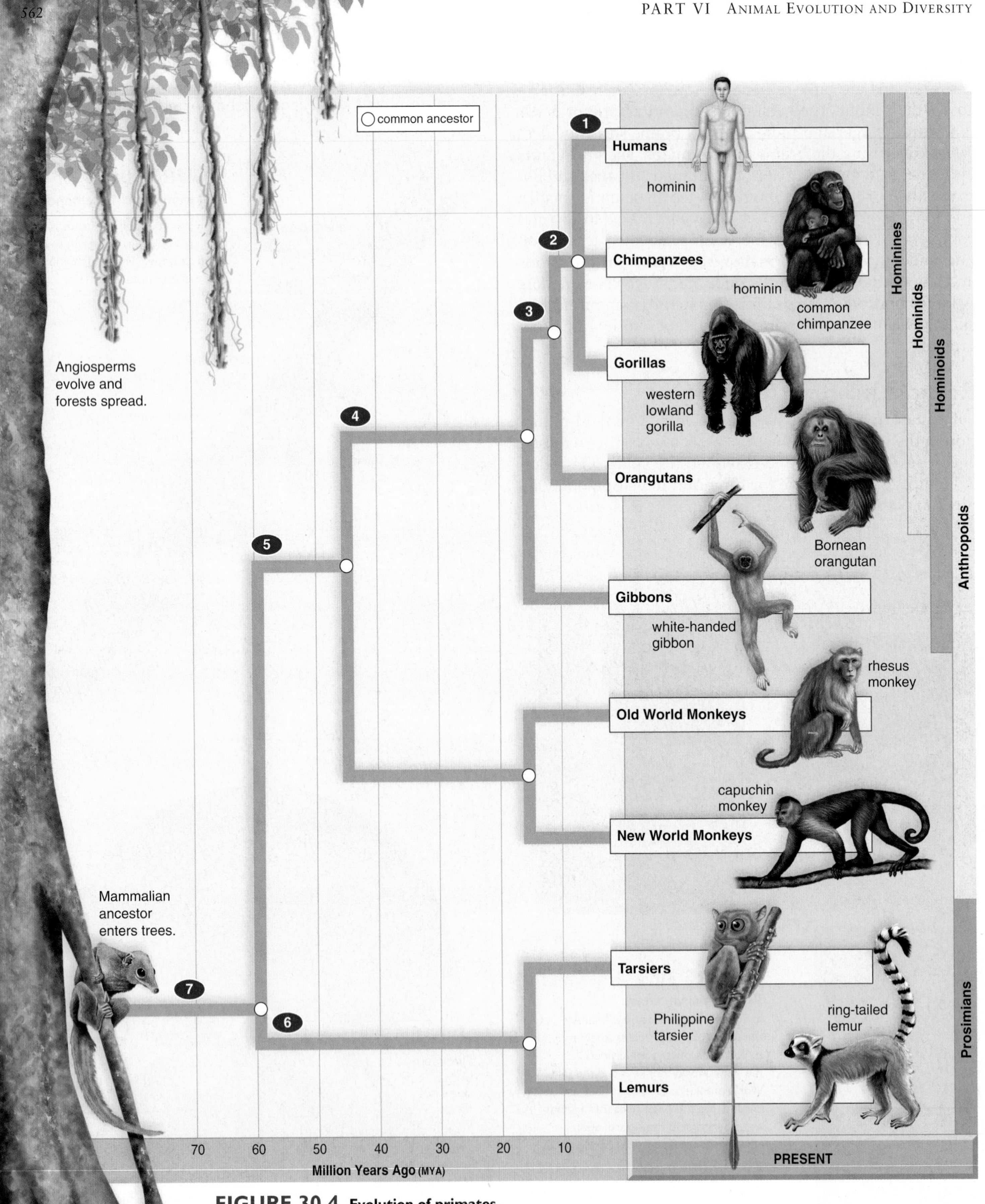


**FIGURE 30.4 Evolution of primates.**

Primates are descended from an ancestor that may have resembled a tree shrew. The descendants of this ancestor adapted to the new way of life and developed traits such as a shortened snout and nails instead of claws. The time when each type of primate diverged from the main line of descent is known from the fossil record. A common ancestor was living at each point of divergence; for example, there was a common ancestor for hominines about 7 MYA, for the hominoids about 15 MYA, and one for anthropoids about 45 MYA.

## Sequence of Primate Evolution

Figure 30.4 traces the evolution of primates during the Cenozoic era. 1 **Hominins** (the designation that includes chimpanzees, humans, and species very closely related to humans) first evolved about 5 MYA. 2 Molecular data shows that hominins and gorillas are closely related and these two groups must have shared a common ancestor sometime during the Miocene. Hominins and gorillas are now grouped together as **hominines.** 3 The **hominids** [L. *homo,* man; Gk. *eides,* like] include the hominines and the orangutan. 4 The **hominoids** include the gibbon and the hominids. The hominoid common ancestor first evolved at the beginning of the Miocene about 23 MYA.

5 The **anthropoids** [Gk. *anthropos,* man, and *eides,* like] include the hominoids and the Old World monkeys and New World monkeys. Old World monkeys lack tails and have protruding noses. Some of the better-known Old World monkeys are the baboon, a ground dweller, and the rhesus monkey, which has been used in medical research. The New World monkeys often have long prehensile (grasping) tails and flat noses. Two of the well-known New World monkeys are the spider monkey and the capuchin, the "organ grinder's monkey."

Primate fossils similar to monkeys are first found in Africa, dated about 45 MYA. At that time, the Atlantic Ocean would have been too expansive for some of them to have easily made their way to South America, where the New World monkeys live today. It is hypothesized that a common ancestor to both the New World and Old World monkeys arose much earlier when a narrower Atlantic would have made crossing much more reasonable. The New World monkeys evolved in South America, and the Old World monkeys evolved in Africa.

6 Notice that **prosimians** [L. *pro,* before, and *simia,* ape, monkey], represented by lemurs and tarsiers, were the first type of primate to diverge from the common ancestor for all the primates. 7 All primates share one common mammalian ancestor, which lived about 55 MYA. This ancestor may have resembled today's tree shrews.

### *Hominoid Evolution*

About 15 MYA, there were dozens of hominoid species, but the anatomy of a fossil classified as *Proconsul* makes it a probable transitional link between the monkeys and the hominoids. *Proconsul* was about the size of a baboon, and the size of its brain (165 cc) was also comparable. This fossil species didn't have the tail of a monkey (Fig. 30.5), and its elbow is similar to that of modern apes, but its limb proportions suggest that it walked as a quadruped on top of tree limbs as monkeys do. Although primarily a tree dweller, *Proconsul* may have also spent time exploring nearby grasslands for food.

*Proconsul* was probably ancestral to the **dryopithecines,** from which the hominoids arose. About 10 MYA, Africarabia (Africa plus the Arabian Peninsula) joined with Asia, and the apes migrated into Europe and Asia. In 1966, Spanish paleontologists announced the discovery of a specimen of *Dryopithecus* dated at 9.5 MYA near Barcelona. The anatomy of these bones clearly indicates that *Dryopithecus* was a tree dweller and locomoted by swinging from branch to branch as gibbons do today. They did not walk along the top of tree limbs as *Proconsul* did.

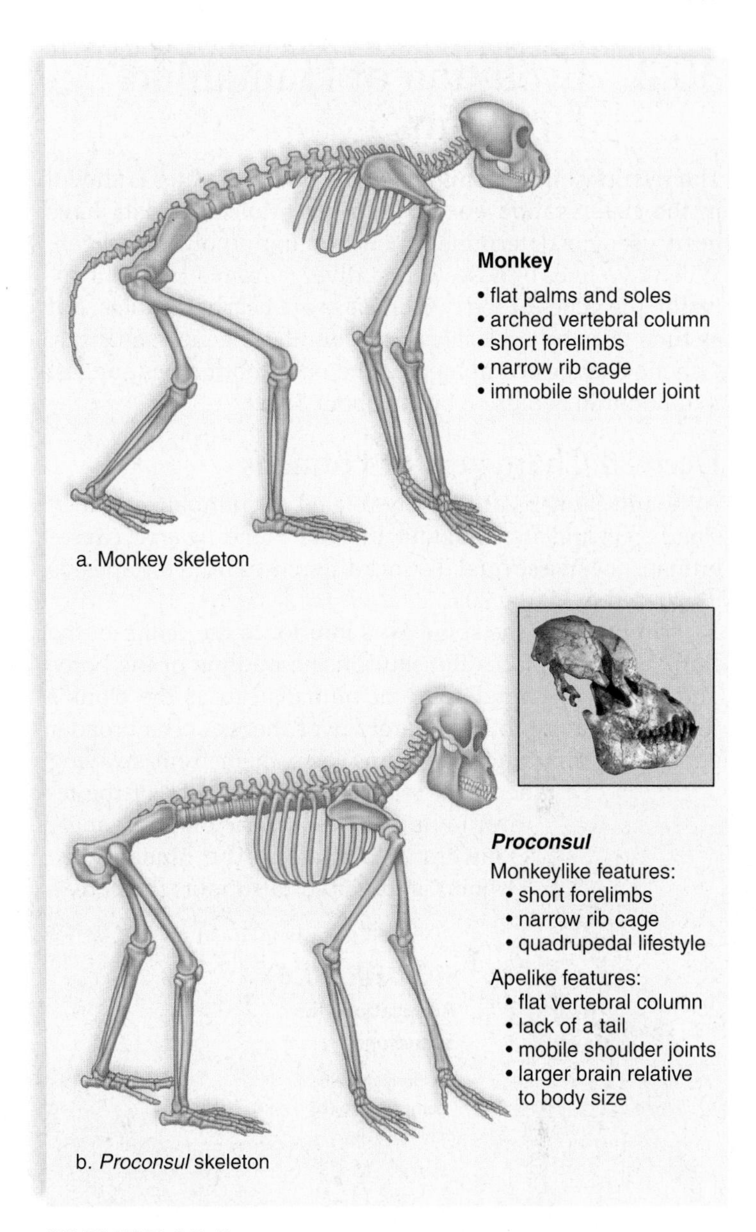


**FIGURE 30.5 Monkey skeleton compared to *Proconsul* skeleton.**

Comparison of a monkey skeleton (**a**) with that of *Proconsul* (**b**) shows various dissimilarities, indicating that *Proconsul* is more related to today's apes than to today's monkeys.

**Check Your Progress 30.1**

1. Match these terms to a number in Figure 30.4: prosimians, anthropoids, hominoids, hominins, hominines.
2. What type of evidence tells us that we humans are very closely related to the chimpanzees and gorillas?

# 30.2 Evolution of Humanlike Hominins

The relationship of hominins to the other primates is shown in the classification box at the right. Molecular data have been used to determine when hominin evolution began. When two lines of descent first diverge from a common ancestor, the genes of the two lineages are nearly identical. But as time goes by, each lineage accumulates genetic changes. Genetic changes compared to the other hominines suggest that hominin evolution began about 5 MYA.

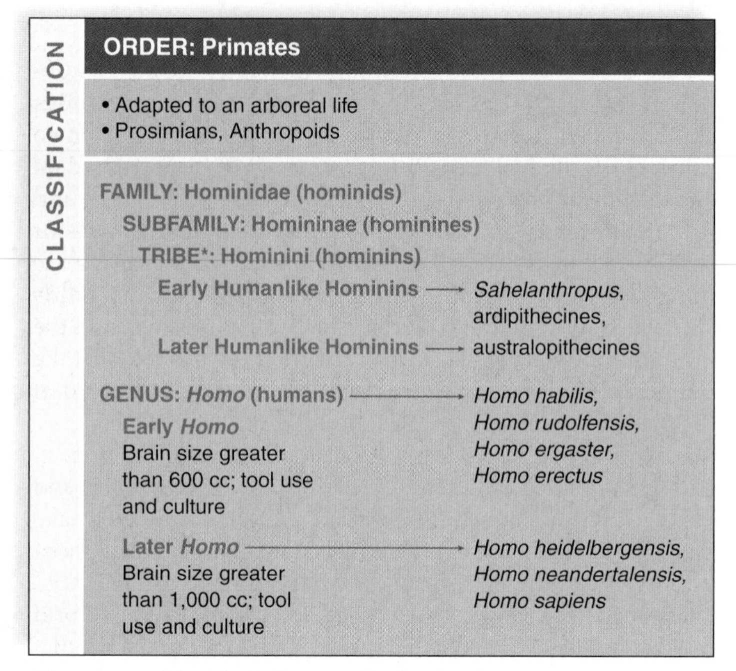

CLASSIFICATION

**ORDER: Primates**

- Adapted to an arboreal life
- Prosimians, Anthropoids

**FAMILY: Hominidae (hominids)**
**SUBFAMILY: Homininae (hominines)**
**TRIBE*: Hominini (hominins)**
**Early Humanlike Hominins** → *Sahelanthropus*, ardipithecines,
**Later Humanlike Hominins** → australopithecines

**GENUS: *Homo* (humans)** → *Homo habilis*, *Homo rudolfensis*, *Homo ergaster*, *Homo erectus*
**Early *Homo***
Brain size greater than 600 cc; tool use and culture

**Later *Homo*** → *Homo heidelbergensis*, *Homo neandertalensis*, *Homo sapiens*
Brain size greater than 1,000 cc; tool use and culture

* A new taxonomic level that lies between subfamily and genus.

## Derived Characters of Humans

Although humans are closely related to chimpanzees, they stand erect and are, therefore, bipedal. Standing erect causes humans to have several distinct differences from the apes, as illustrated in Figure 30.6.

In humans, the spine exits inferior to the center of the skull, and this places the skull in the midline of the body. The longer, S-shaped spine of humans causes the trunk's center of gravity to be squarely over the feet. The broader pelvis and hip joint of humans keep them from swaying when they walk. The longer neck of the femur in humans causes the femur to angle inward at the knees. The human knee joint is modified to support the body's weight—that is, the femur is larger at the bottom, and the tibia is larger at the top. The human toe is not opposable; instead, the foot has an arch, which enables humans to walk long distances and run with less chance of injury.

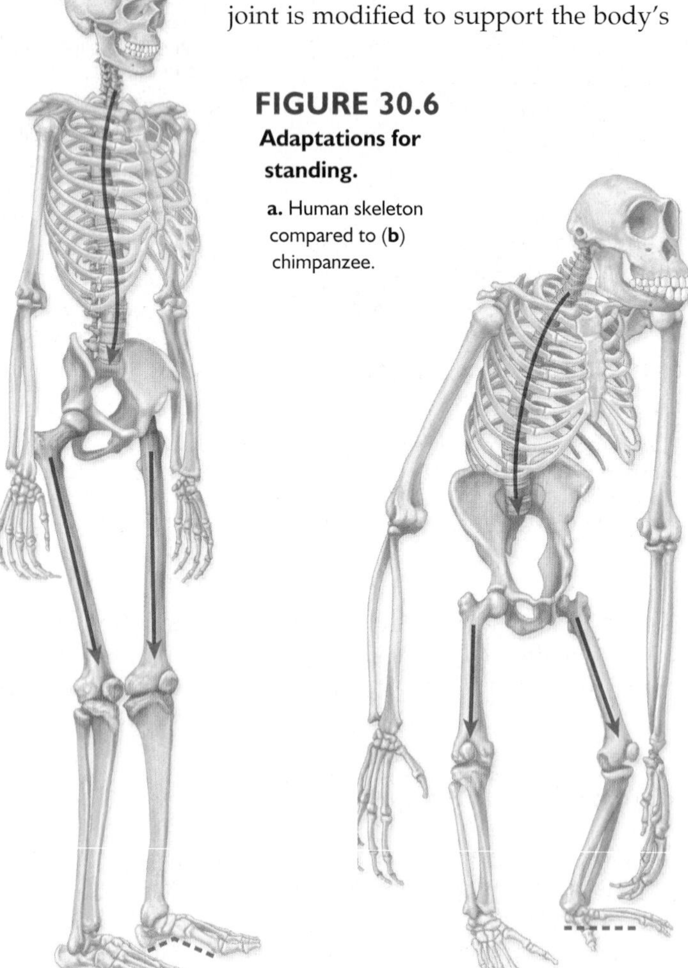


**FIGURE 30.6 Adaptations for standing.**
**a.** Human skeleton compared to (**b**) chimpanzee.

## The Early Humanlike Hominins

Paleontologists use evidence of bipedalism to identify the early humanlike hominins. Until recently, many scientists thought that hominins began to stand upright in response to a dramatic change in climate that caused forests to be replaced by grassland. Now, some biologists suggest that the first humanlike hominins evolved even while they lived in trees, because they see no evidence of a dramatic shift in vegetation about 7 MYA. Their environment is now thought to have included some forest, some woodland, and some grassland. While still living in trees, the first humanlike hominins may have walked upright on large branches as they collected fruit from overhead. Then, when they began to forage on the ground, an upright stance would have made it easier for them to travel from woodland to woodland and/or to forage among bushes. Bipedalism may have had the added advantage of making it easier for males to carry food back to females. Or, bipedalism may be associated with the need to carry a helpless infant from place to place.

In Figure 30.7, early humanlike hominins are represented by orange-colored bars. The bars extend from the date of a species' appearance in the fossil record to the date it became extinct. Paleontologists have now found several fossils dated around the time the ape lineage and the human lineage are believed to have split, and one of these is *Sahelanthropus tchadensis*. Only the braincase has been found and dated at 7 MYA. Although the braincase is very apelike, the location of the opening for the spine at the back of the skull suggests bipedalism. Also, the canines are smaller and the tooth enamel is thicker than those of an ape.

Another early humanlike hominin, *Ardipithecus ramidus*, is representative of the ardipithecines of 4.5 MYA. So far, only skull fragments of *A. ramidus* have been described. Indirect evidence suggests that the species was possibly bipedal, and that some individuals may have been 122 cm tall. The teeth seem intermediate between those of earlier apes and later humanlike hominins, which are discussed next. Recently, fossils dated 4 MYA show a direct link between *A. ramidus* and the australopiths, discussed next.

**Check Your Progress 30.2**

1. What environmental influence may have caused bipedalism to evolve?
2. What is the strongest evidence that humanlike hominin evolution began around 7 MYA?

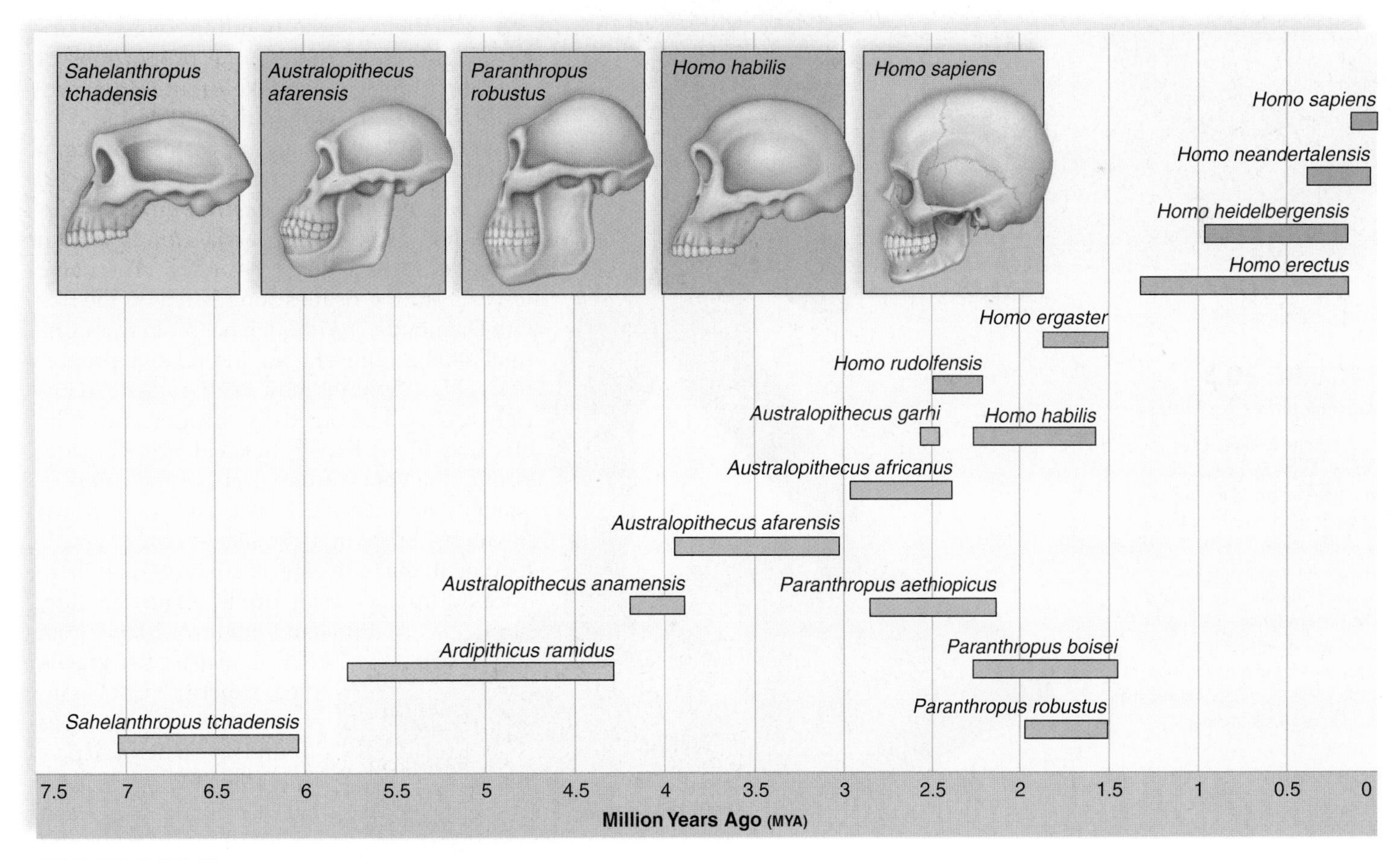


**FIGURE 30.7 Human Evolution.**

Several groups of extinct hominins preceded the evolution of modern humans. These groups have been divided into the early humanlike hominins (orange), later humanlike hominins (green), early *Homo* species (lavender), and finally the later *Homo* species (blue). Only modern humans are classified as *Homo sapiens*.

## 30.3 Evolution of Later Humanlike Hominins

The **australopithecines** (called **australopiths** for short) are a group of hominins that evolved and diversified in Africa from 4 MYA until about 1 MYA. In Figure 30.7, the australopiths are represented by green-colored bars. The australopiths had a small brain (an apelike characteristic) and walked erect (a humanlike characteristic). Therefore, it seems that human characteristics did not evolve all together at the same time. Australopiths give evidence of **mosaic evolution,** meaning that different body parts change at different rates and, therefore, at different times.

Australopiths stood about 100–115 cm in height and had relatively small brains averaging from about 370–515 cc—slightly larger than that of a chimpanzee. The forehead was low and the face projected forward (Fig. 30.8). Tool use is not in evidence except for later appearing *A. garhi.*

Some australopiths were slight of frame and termed *gracile* (slender). Others were *robust* (powerful) and tended to have massive jaws because of their large grinding teeth. The larger species, now placed in the genus *Paranthropus,* had well-developed chewing muscles that were anchored to a prominent bony crest along the top of the skull. Their diet included seeds and roots. The gracile types of genus *Australopithecus* most likely fed on soft fruits and leaves. Therefore, the australopiths show an adaptation to different ways of life. Fossil remains of australopiths have been found in both southern Africa and in eastern Africa.

### The Fossils

The first australopith to be discovered was unearthed in southern Africa by Raymond Dart in the 1920s. This hominin, named ***Australopithecus africanus,*** is a gracile type. A second specimen from southern Africa, *Paranthropus robustus,* is a robust type. Both *A. africanus* and *P. robustus* had a brain size of about 500 cc; variations in their skull anatomy are essentially due to their different diets. These hominins walked upright. Nevertheless, the proportions of their limbs are apelike—that is, the forelimbs are longer than the hindlimbs.

More than 20 years ago in eastern Africa, a team led by Donald Johanson unearthed nearly 250 fossils of a hominin called ***Australopithecus afarensis.*** A now-famous female skeleton is known worldwide by its field name, Lucy. (The name derives from the Beatles song "Lucy in the Sky with Diamonds.") Although her brain was quite small (400 cc), the shapes and relative proportions of Lucy's limbs indicate that she stood upright and walked bipedally at least some of the time (Fig. 30.8*a*). Even better evidence of bipedal locomotion comes from a trail of footprints in Laetoli dated about 3.7 MYA. The larger prints are double, as though a smaller-sized being was stepping in the footprints of another—and there are additional small prints off to the side, within hand-holding distance (Fig. 30.8*b*). *A. afarensis,* a gracile type, is believed to be ancestral to the robust types found in eastern Africa, including *P. aethiopicus* and *P. boisei. P. boisei* had a powerful upper body and the largest molars of any humanlike hominin.

b.

a.

**FIGURE 30.8**
***Australopithecus afarensis.***

**a.** A reconstruction of Lucy on display at the St. Louis Zoo. **b.** These fossilized footprints occur in ash from a volcanic eruption some 3.7 MYA. The larger footprints are double, and a third, smaller individual was walking to the side. (A female holding the hand of a youngster may have been walking in the footprints of a male.) The footprints suggest that *A. afarensis* walked bipedally.

In 2000, a team of scientists from the Max Planck Institute unearthed the fossilized remains of a 3.3-million-year-old juvenile *A. afarensis* just 4 km from where Lucy had been discovered. Dubbed Salem by her discoverer, she is often called "Lucy's baby," even though she is tens of thousands of years older than Lucy. Not only is this fossil exceptional because the remains of infants and juveniles rarely fossilize, but it represents the most complete *A. afarensis* fossil to date.

An earlier find called *A. garhi* may be the transitional link between the australopiths and the next group of fossils we will be discussing, namely, the early *Homo* species, represented in Figure 30.7 by lavender-colored bars. *A garhi* is an australopith, but it made tools.

**Check Your Progress 30.3**

1. In general terms, compare the gracile australopith species with the robust species. What accounts for the differences in their anatomies?
2. Is a southern or eastern australopith fossil likely to be an ancestor of early *Homo*? Explain.

*science focus*

## Origins of the Genus *Homo*

Remains of australopiths indicate that they spent part of their time climbing trees and that they retained many apelike traits. Most likely, the australopiths climbed trees for the same reason that chimpanzees do today: to gather fruits and nuts in trees and to sleep aboveground at night so as to avoid predatory animals, such as lions and hyenas.

Whereas our brain is about the size of a grapefruit, that of the australopiths was about the size of an orange. Their brain was only slightly larger than that of a chimpanzee.

We know that the genus *Homo* evolved from the genus *Australopithecus*, but it seemed to me [Stephen Stanley] they could not have done so as long as the australopiths climbed trees every day. The obstacle relates to the way we, members of *Homo*, develop our large brain. Unlike other primates, we retain the high rate of fetal brain growth through the first year after birth. (That is why a one-year-old child has a very large head in proportion to the rest of its body.) The brain of other primates, including monkeys and apes, grows rapidly before birth, but immediately after birth, their brain grows more slowly. As a result, an adult human brain is more than three times as large as that of an adult chimpanzee.

A continuation of the high rate of fetal brain growth eventually allowed the genus *Homo* to evolve from the genus *Australopithecus*. But there was a problem in that continued brain growth is linked to underdevelopment of the entire body. Although the human brain becomes much larger, human babies are remarkably weak and uncoordinated. Such helpless infants must be carried about and tended. Human babies are unable to cling to their mothers the way chimpanzee babies can (Fig. 30A).

The origin of the *Homo* genus entailed a great evolutionary compromise. Humans gained a large brain, but they were saddled with the largest interval of infantile helplessness of all the mammals. The positive value of a large brain must have outweighed the negative aspects of infantile helplessness, such as the inability of adults to climb trees while holding a helpless infant, or else genus *Homo* would not have evolved. Having a larger brain meant that humans were able to outsmart or ward off predators with weapons they were clever enough to manufacture.

Probably very few genetic changes were required to delay the maturation of *Australopithecus* and produce the large brain of *Homo*. The mutation of a *Hox* developmental gene could have delayed early maturation, allowing the brain to enlarge under the selection forces resulting from the increased social nature of *Homo*. As we learn more about the human genome, we will eventually uncover the particular gene or gene combinations that cause us to have a large brain, and this will be a very exciting discovery.

*Steven Stanley*
*Johns Hopkins University*

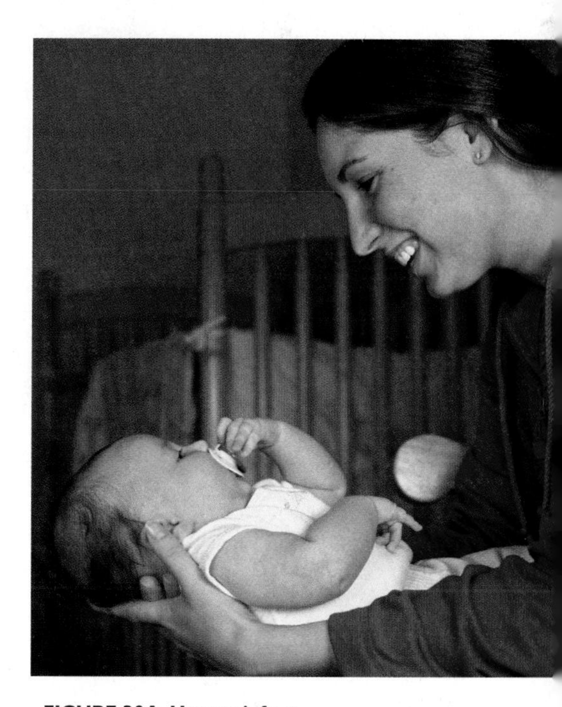

**FIGURE 30A Human infant.**
*A human infant is often cradled and has no means to cling to its mother when she goes about her daily routine.*

## 30.4 Evolution of Early *Homo*

Early *Homo* species appear in the fossil record somewhat earlier or later than 2 MYA. They all have a brain size that is 600 cc or greater, their jaw and teeth resemble those of humans, and tool use is in evidence.

### *Homo habilis* and *Homo rudolfensis*

*Homo habilis* and *Homo rudolfensis* are closely related and will be considered together. *Homo habilis* means handyman, and these two species are credited by some as being the first peoples to use stone tools, as discussed in the Ecology Focus on page 569. Most believe that while they were socially organized, they were probably scavengers rather than hunters. The cheek teeth of these hominins tend to be smaller than even those of the gracile australopiths. This is also evidence that they were omnivorous and ate meat, in addition to plant material.

Compared to australopiths, the protrusion of the face was less, and the brain was larger. Although the height of *H. rudolfensis* did not exceed that of the australopiths, some of this species' fossils have a brain size as large as 800 cc, which is considerably larger than that of *A. afarensis*.

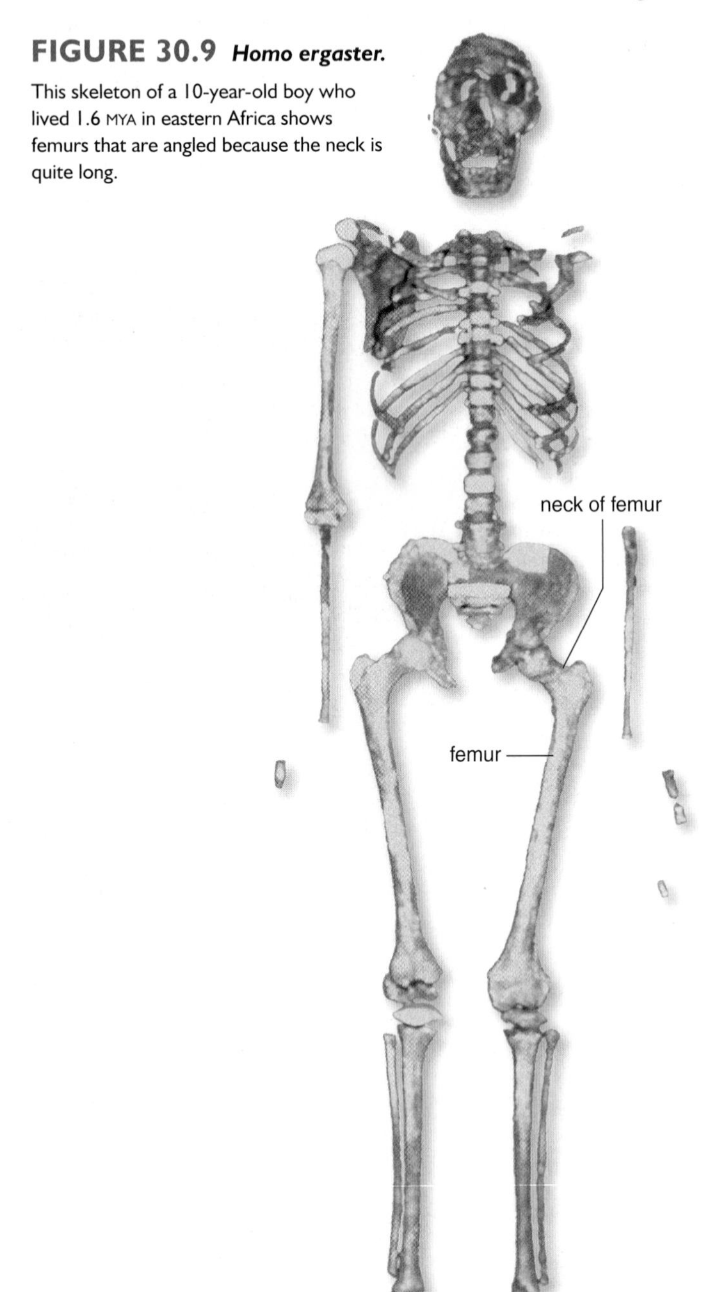


**FIGURE 30.9** ***Homo ergaster.*** This skeleton of a 10-year-old boy who lived 1.6 MYA in eastern Africa shows femurs that are angled because the neck is quite long.

### *Homo ergaster* and *Homo erectus*

***Homo ergaster*** evolved in Africa perhaps from *H. rudolfensis.* Similar fossils found in Asia are different enough to be classified as ***Homo erectus*** [L. *homo,* man, and *erectus,* upright]. These fossils span the dates between 1.9 and 0.3 MYA, and many other fossils belonging to both species have been found in Africa and Asia.

Compared to *H. habilis, H. ergaster* had a larger brain (about 1,000 cc) and a flatter face with a projecting nose. This type of nose is adaptive for a hot, dry climate because it permits water to be removed before air leaves the body. The recovery of an almost complete skeleton of a 10-year-old boy indicates that *H. ergaster* was much taller than the hominins discussed thus far (Fig. 30.9). Males were 1.8 m tall, and females were 1.55 m tall. Indeed, these hominins stood erect and, most likely, had a *striding gait* like that of modern humans. The robust and most likely heavily muscled skeleton still retained some australopithecine features. Even so, the size of the birth canal indicates that infants were born in an immature state that required an extended period of care.

*H. ergaster* first appeared in Africa but then migrated into Europe and Asia sometime between 2 MYA and 1 MYA. Most likely, *H. erectus* evolved from *H. ergaster* after *H. ergaster* arrived in Asia. In any case, such an extensive population movement is a first in the history of humankind and a tribute to the intellectual and physical skills of these peoples. They also had a knowledge of fire and may have been the first to cook meat.

### *Homo floresiensis*

In 2004, scientists announced the discovery of the fossil remains of *Homo floresiensis.* The 18,000-year-old fossil of a 1 m tall, 25 kg adult female was discovered on the island of Flores in the South Pacific. The specimen was the size of a three-year-old *Homo sapien* but possessed a braincase only one-third the size of modern humans. A 2007 study supports the hypothesis that this diminutive hominin and her peers evolved from normal-sized, island hopping *Homo erectus* populations that reached Flores about 840,000 years ago. Apparently, *H. floresiensis* used tools and fire.

**Check Your Progress 30.4**

1. What is significant about the migration of *H. ergaster* out of Africa?
2. What are the cultural advancements that early *Homo* made over the australopiths?

*ecology focus*

## Biocultural Evolution Began with *Homo*

Culture encompasses human activities and products that are passed on from one generation to another outside of direct biological inheritance. *Homo habilis* (and *Homo rudolfensis*) could make the simplest of stone tools, called Oldowan tools after a location in Africa where the tools were first found. The main tool could have been used for hammering, chopping, and digging. A flake tool was a type of knife sharp enough to scrape away hide and remove meat from bones. The diet of *H. habilis* most likely consisted of collected plants. But they probably had the opportunity to eat meat scavenged from kills abandoned by lions, leopards, and other large predators in Africa.

*Homo erectus,* who lived in Eurasia, also made stone tools, but the flakes were sharper and had straighter edges. They are called Acheulian tools for a location in France where they were first found. Their so-called multipurpose hand axes were large flakes with an elongated oval shape, a pointed end, and sharp edges on the sides. Supposedly they were handheld, but no one knows for sure. *H. erectus* also made the same core and flake tools as *H. habilis*. In addition, *H. erectus* could have also made many other implements out of wood or bone, and even grass, which can be twisted together to make string and rope. Excavation of *H. erectus* campsites dated 400,000 years ago have uncovered literally tens of thousands of tools.

*H. erectus,* like *H. habilis,* also gathered plants as food. However, *H. erectus* may have also harvested large fields of wild plants. The members of this species were not master hunters, but they gained some meat through scavenging and hunting. The bones of all sorts of animals litter the areas where they lived. Apparently, they ate pigs, sheep, rhinoceroses, buffalo, deer, and many other smaller animals. *H. erectus* lived during the last Ice Age, but even so, moved northward. No wonder *H. erectus* is believed to have used fire. A campfire would have protected them from wild beasts and kept them warm at night. And the ability to cook would have made meat easier to eat. In order for the humans to survive during the winter in northern climates, meat must have become a substantial part of the diet since plant sources are not available in the dead of winter. It is even possible that the campsites of *H. erectus* were "home bases" where the women stayed behind with the children while the men went out to hunt. If so, these people may have been the first **hunter-gatherers** (Fig. 30B)—that is, they hunted animals and gathered plants. This was a successful way of life that allowed the hominin populations to increase from a few thousand australopiths in Africa 2 MYA to hundreds of thousands of *H. erectus* by 300,000 years ago.

Hunting most likely encourages the development and spread of culture between individuals and generations. Those who could speak a language would have been able to cooperate better as they hunted and even as they sought places to gather food. Among animals, only humans have a complex language that allows them to communicate their experiences symbolically. Words stand for objects and events that can be pictured in the mind. The cultural achievements of *H. erectus* essentially began a new phase of human evolution, called **biocultural evolution,** in which natural selection is influenced by cultural achievements rather than by anatomic phenotype. *H. erectus* succeeded in new, colder environments because these individuals occupied caves, used fire, and became more capable of obtaining and eating meat as a substantial part of their diet.

**FIGURE 30B** ***Homo erectus.*** *The Homo erectus people may have been hunter-gatherers.*

## 30.5 Evolution of Later *Homo*

Later *Homo* species are represented by blue-colored bars in Figure 30.7. The evolution of these species from older *Homo* species has been the subject of much debate. Most researchers believe that modern humans (*Homo sapiens*) evolved from *H. ergaster,* but they differ as to the details. Many disparate early *Homo* species in Europe are now classified as *Homo heidelbergensis.* Just as *H. erectus* is believed to have evolved from *H. ergaster* in Asia, so *H. heidelbergensis* is believed to have evolved from *H. ergaster* in Europe. Further, for the sake of discussion, *H. ergaster* in Africa, *H. erectus* in Asia, and *H. heidelbergensis* (and *H. neandertalensis*) in Europe can be grouped together as archaic humans who lived as long as a million years ago. The most widely accepted hypothesis for the evolution of modern humans from archaic humans is referred to as the **replacement model** or **out-of-Africa hypothesis,** which proposes that modern humans evolved from archaic humans only in Africa, and then modern humans migrated to Asia and Europe, where it replaced the archaic species about 100,000 years BP (before the present) (Fig. 30.10).

The replacement model is supported by the fossil record. The earliest remains of modern humans (Cro-Magnon), dating at least 130,000 years BP, have been found only in Africa. Modern humans are found in Asia until 100,000 years BP and not in Europe until 60,000 years BP. Until earlier modern human fossils are found in Asia and Europe, the replacement model is supported.

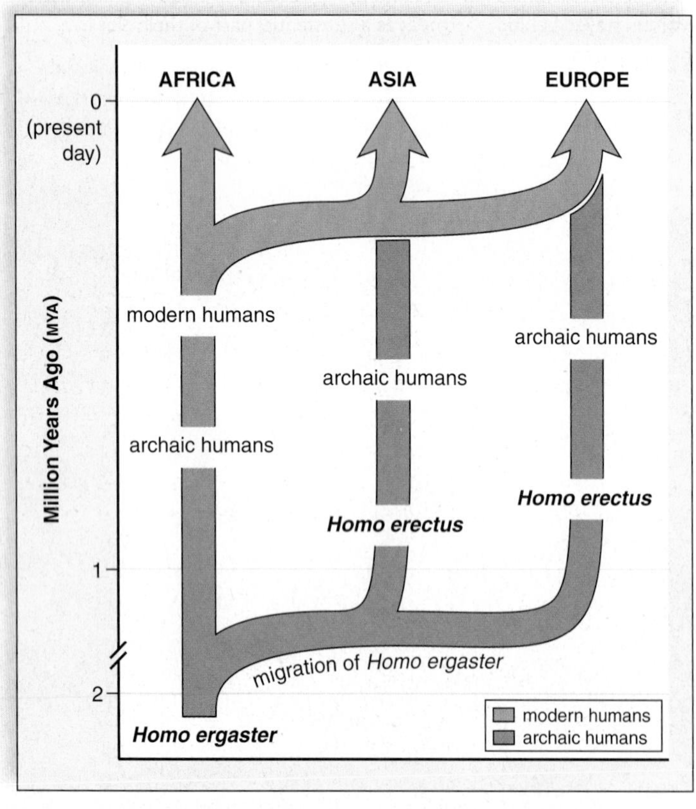


**FIGURE 30.10 Replacement model.**

Modern humans evolved in Africa and then replaced archaic humans in Asia and Europe.

The replacement model is also supported by DNA data. Several years ago, a study showed that the mitochondrial DNA of Africans is more diverse that the DNA of the people in Europe (and the world). This is significant because if mitochondrial DNA has a constant rate of mutation, Africans should show the greatest diversity, since modern humans have existed the longest in Africa. Called the "mitochondrial Eve" hypothesis by the press (note that this is a misnomer because no single ancestor is proposed), the statistics that calculated the date of the African migration were found to be flawed. Still, the raw data—which indicate a close genetic relationship among all Europeans—support the replacement model.

An opposing hypothesis to the out-of-Africa hypothesis does exist. This hypothesis, called the multiregional continuity hypothesis, proposes that modern humans arose from archaic humans in essentially the same manner in Africa, Asia, and Europe. The hypothesis is multiregional because it applies equally to Africa, Asia, and Europe, and it supposes that in these regions, genetic continuity will be found between modern populations and archaic populations. This hypothesis has sparked many innovative studies to test which hypothesis is correct.

### Neandertals

The **Neandertals,** *Homo neandertalensis,* are an intriguing species of archaic humans that lived between 200,000 and 28,000 years ago. Neandertal fossils are known from the Middle East and throughout Europe. Neandertals take their name from Germany's Neander Valley, where one of the first Neandertal skeletons, dated some 200,000 years ago, was discovered.

According to the replacement model, the Neandertals were also supplanted by modern humans. Surprisingly, however, the Neandertal brain was, on the average, slightly larger than that of *Homo sapiens* (1,400 cc, compared with 1,360 cc in most modern humans). The Neandertals had massive brow ridges and wide, flat noses. They also had a forward-sloping forehead and a receding lower jaw. Their nose, jaws, and teeth protruded far forward. Physically, the Neandertals were powerful and heavily muscled, especially in the shoulders and neck (Fig. 30.11). The bones of Neandertals were shorter and thicker than those of modern humans. New fossils show that the pubic bone was long compared to that of modern humans. The Neandertals lived in Europe and Asia during the last Ice Age, and their sturdy build could have helped conserve heat.

Archaeological evidence suggests that Neandertals were culturally advanced. Some Neandertals lived in caves; however, others probably constructed shelters. They manufactured a variety of stone tools, including spear points, which could have been used for hunting, and scrapers and knives, which would have helped in food preparation. They most likely successfully hunted bears, woolly mammoths, rhinoceroses, reindeer, and other contemporary animals. They used and could control fire, which probably helped in cooking frozen meat and in keeping warm. They even buried their dead with flowers and tools and may have had a religion.

**FIGURE 30.11 Neandertals.**

This drawing shows that the nose and the mouth of the Neandertals protruded from their faces, and their muscles were massive. They made stone tools and were most likely excellent hunters.

**FIGURE 30.12 Cro-Magnons.**

Cro-Magnon people are the first to be designated *Homo sapiens*. Their tool-making ability and other cultural attributes, such as their artistic talents, are legendary.

## Cro-Magnons

**Cro-Magnons** are the oldest fossils to be designated *Homo sapiens*. In keeping with the replacement model, the Cro-Magnons, who are named after a fossil location in France, were the modern humans who entered Asia from Africa about 100,000 years BP and then spread to Europe. They probably reached western Europe about 40,000 years ago. Cro-Magnons had a thoroughly modern appearance (Fig. 30.12). They had lighter bones, flat high foreheads, domed skulls housing brains of 1,590 cc, small teeth, and a distinct chin. They were hunter-gatherers, as was *H. erectus,* but they hunted more efficiently.

### *Tool Use in Cro-Magnons*

During the last Ice Age, *Homo sapiens* had colonized all of the continents except Antarctica. Glaciation had caused a significant drop in sea level and, as a result, land bridges to the New World and Australia were available. No doubt, colonization was fostered by the combination of a larger brain and free hands with opposable thumbs that made it possible for Cro-Magnons to draft and manipulate tools and weapons of increasing sophistication. They made advanced stone tools, including compound tools, as when stone flakes were fitted to a wooden handle. They may have been the first to make knifelike blades and to throw spears, enabling them to kill animals from a distance. They were such accomplished hunters that some researchers believe they may have been responsible for the extinction of many larger mammals, such as the giant sloth, the mammoth, the saber-toothed tiger, and the giant ox, during the late Pleistocene epoch. This event is known as the Pleistocene overkill.

### *Language and Cro-Magnons*

A more highly developed brain may have also allowed Cro-Magnons to perfect a language composed of patterned sounds. Language greatly enhanced the possibilities for cooperation and a sense of cohesion within the small bands that were the predominant form of human social organization, even for the Cro-Magnons. They combined hunting and fishing with the gathering of fruits, berries, grains, and root crops that grew in the wild.

The Cro-Magnons were extremely creative. They sculpted small figurines and jewelry out of reindeer bones and antlers. These sculptures could have had religious significance or been seen as a way to increase fertility. The most impressive artistic achievements of the Cro-Magnons were cave paintings, realistic and colorful depictions of a variety of animals, from woolly mammoths to horses, that have been discovered deep in caverns in southern France and Spain. These paintings suggest that Cro-Magnons had the ability to think symbolically, as would be needed in order to speak.

**Check Your Progress 30.5A**

1. What evidence is there to support the replacement model (out-of-Africa hypothesis) for the evolution of modern humans from archaic humans?
2. Describe the tools of Cro-Magnons. How are they advanced over those seen in archaic humans?
3. What is the significance of the development of art by Cro-Magnons?

## Human Variation

Human beings have been widely distributed about the globe ever since they evolved. As with any other species that has a wide geographic distribution, phenotypic and genotypic variations are noticeable between populations. Today, we say that people have different ethnicities (Fig. 30.13*a*).

It has been hypothesized that human variations evolved as adaptations to local environmental conditions. One obvious difference among people is skin color. A darker skin is protective against the high UV intensity of bright sunlight. On the other hand, a white skin ensures vitamin D production in the skin when the UV intensity is low. Harvard University geneticist Richard Lewontin points out, however, that this hypothesis concerning the survival value of dark and light skin has never been tested.

Two correlations between body shape and environmental conditions have been noted since the nineteenth century. The first, known as Bergmann's rule, states that animals in colder regions of their range have a bulkier body build. The second, known as Allen's rule, states that animals in colder regions of their range have shorter limbs, digits, and ears. Both of these effects help regulate body temperature by increasing the surface-area-to-volume ratio in hot climates and decreasing the ratio in cold climates. For example, Figure 30.13*b, c* shows that the Massai of East Africa tend to be slightly built with elongated limbs, while the Eskimos, who live in northern regions, are bulky and have short limbs.

Other anatomic differences among ethnic groups, such as hair texture, a fold on the upper eyelid (common in Asian peoples), or the shape of lips, cannot be explained as adaptations to the environment. Perhaps these features became fixed in different populations due simply to genetic drift. As far as intelligence is concerned, no significant disparities have been found among different ethnic groups.

a.

b.

c.

**FIGURE 30.13 Ethnic groups.**

**a.** Some of the differences between the various prevalent ethnic groups in the United States may be due to adaptations to the original environment. **b.** The Massai live in East Africa. **c.** Eskimos live near the Arctic Circle.

### *Genetic Evidence for a Common Ancestry*

The replacement model for the evolution of humans, discussed on page 570, pertains to the origin of ethnic groups. This hypothesis proposes that all modern humans have a relatively recent common ancestor, that is, Cro-Magnon, who evolved in Africa and then spread into other regions. Paleontologists tell us that the variation among modern populations is considerably less than among archaic human populations some 250,000 years ago. If so, all ethnic groups evolved from the same single, ancestral population.

A comparative study of mitochondrial DNA shows that the differences among human populations are consistent with their having a common ancestor no more than a million years ago. Lewontin has also found that the genotypes of different modern populations are extremely similar. He examined variations in 17 genes, including blood groups and various enzymes, among seven major geographic groups: Caucasians, black Africans, mongoloids, south Asian Aborigines, Amerinids, Oceanians, and Australian Aborigines. He found that the great majority of genetic variation—85%—occurs within ethnic groups, not among them. In other words, the amount of genetic variation between individuals of the same ethnic group is greater than the variation between ethnic groups.

**Check Your Progress 30.5B**

1. What data support the hypothesis that humans are one species?
2. Where is the greater amount of variability in modern human populations, within ethnic groups or between them?

## Connecting the Concepts

Aside from various anatomical differences related to human bipedalism and intelligence, a cultural evolution separates us from the apes. A hunter-gatherer society evolved when humans became able to make and use tools. That society then gave way to an agricultural economy about 12,000 to 15,000 years ago, perhaps because we were too efficient at killing big game so that a food shortage arose. The agricultural period extended from that time to about 200 years ago, when the Industrial Revolution began. Now most people live in urban areas. Perhaps as a result, modern humans are for the most part divorced from nature and often endowed with the philosophy of exploiting and controlling nature.

Our cultural evolution has had far-reaching effects on the biosphere, especially since the human population has expanded to the point that it is crowding out many other species. Our degradation and disruption of the environment threaten the continued existence of many species, including our own. As discussed in Chapter 47, however, we have recently begun to realize that we must work with, rather than against, nature if biodiversity is to be maintained and our own species is to continue to exist.

Before we examine the environment and the role of humans in ecosystems, we will study the various organ systems of the human body. Humans need to keep themselves and the environment fit so that they and their species can endure.

## summary

### 30.1 Evolution of Primates

Primates, in contrast to other types of mammals, are adapted for an arboreal life. The evolution of primates is characterized by trends toward mobile limbs; grasping hands; a flattened face; stereoscopic vision; a large, complex brain; and one birth at a time. These traits are particularly useful for living in trees.

The term hominin is now used for chimpanzees, humans, and their closely related, but extinct, relatives. A hominin is a member of the group hominines that also includes the gorilla, which are the apes most closely related to hominins on the basis of molecular data. There follows ever increasing sized groups.

*Proconsul* is a transitional link between monkeys and the hominoids, which include the gibbons, organgutans, and the hominines.

### 30.2 Evolution of Humanlike Hominins

Fossil and molecular data tell us humanlike hominins shared a common ancestor with chimpanzees until about 5 MYA, and the split between their lineage and the human lineage occurred around this time.

Humans walk erect, and this causes our anatomy to differ from the apes. In humans, the spinal cord curves and exits from the center of the skull, rather than from the rear of the skull. The human pelvis is broader and more bowl-shaped to place the weight of the body over the legs. Humans use only the longer, heavier lower limbs for walking; in apes, all four limbs are used for walking, and the upper limbs are longer than the lower limbs.

To be a humanlike hominin, a fossil must have an anatomy suitable to standing erect. Perhaps bipedalism developed when humanlike hominins stood on branches to reach fruit overhead, and then they continued to use this stance when foraging among bushes. An upright posture reduces exposure of the body to the sun's rays, and leaves the hands free to carry food, perhaps as a gift to receptive females.

Several early humanlike hominin fossils, such as *Sahelanthropus tchadensis*, have been dated around the time of a shared ancestor for apes and humans (7 MYA). The ardipithecines appeared about 4.5 MYA. All the early humanlike hominins have a chimp-sized braincase but are believed to have walked erect.

### 30.3 Evolution of Later Humanlike Hominins

It is possible that an australopith (4 MYA–1 MYA) is a direct ancestor for humans. These hominins walked upright and had a brain size of 370–515 cc. In southern Africa, hominins classified as australopiths include *Australopithecus africanus*, a gracile form, and *Paranthropus robustus*, a robust form. In eastern Africa, hominins classified as australopiths include, *A. afarensis* (Lucy), a gracile form, and also robust forms. Many of the australopiths coexisted, and the species *A. garhi* is the probable ancestor to the genus *Homo*.

### 30.4 Evolution of Early *Homo*

Early *Homo*, such as *Homo habilis* and *Homo rudolfensis*, dated around 2 MYA, is characterized by a brain size of at least 600 cc, a jaw with teeth that resembled those of modern humans, and the use of tools.

*Homo ergaster* and *Homo erectus* (1.9–0.3 MYA) had a striding gait, made well-fashioned tools, and could control fire. *Homo ergaster* migrated into Asia and Europe from Africa between 2 and 1 MYA. *Homo erectus* evolved in Asia and gave rise to *H. floresiensis*.

### 30.5 Evolution of Later *Homo*

The replacement model of human evolution says that modern humans originated only in Africa and, after migrating into Europe and Asia, replaced the archaic *Homo* species found there.

The Neandertals, a group of archaic humans, lived in Europe and Asia. Their chinless faces, squat frames, and heavy muscles are apparently adaptations to the cold. Cro-Magnon is a name often given to modern humans. Their tools were sophisticated, and they definitely had a culture, as witnessed by the paintings on the walls of caves. The human ethnic groups of today differ in ways that can be explained in part by adaptation to the environment. Genetic studies tell us that there are more genetic differences between people of the same ethnic group than between ethnic groups. We are one species.

## understanding the terms

anthropoid 563
arboreal 560
australopithecine (australopith) 566
*Australopithecus afarensis* 566
*Australopithecus africanus* 566
biocultural evolution 569
Cro-Magnon 571
dryopithecine 563
hominid 563
hominin 563
hominine 563
hominoid 563
*Homo erectus* 568
*Homo ergaster* 568
hunter-gatherer 569
mosaic evolution 566
Neandertal 570
opposable thumb 560
out-of-Africa hypothesis 570
primate 560
prosimian 563
replacement model 570
stereoscopic vision 560

Match the terms to these definitions:

a. ______________ Group of primates that includes monkeys, apes, and humans.
b. ______________ The common name for the first fossils generally accepted as being modern humans.
c. ______________ Hominin with a sturdy build who lived during the last Ice Age in Eurasia; hunted large game and lived together in a kind of society.
d. ______________ Type of early *Homo* to first have a striding gait similar to that of modern humans.
e. ______________ Member of a group that does not include prosiminans nor monkeys, nor gibbons.

## reviewing this chapter

1. List and discuss various evolutionary trends among primates, and state how they would be beneficial to animals with an arboreal life. 560–61
2. What is the significance of the fossils known as *Proconsul?* 563
3. How does an upright stance cause human anatomy to differ from that of chimpanzees? 564–65
4. Discuss the possible benefits of bipedalism in early hominins. 565
5. Why does the term *mosaic evolution* apply to the australopiths? 566
6. Why are the early *Homo* species classified as humans? If these hominins did make tools, what does this say about their probable way of life? 568–69
7. What role might *H. ergaster* have played in the evolution of modern humans according to the replacement model? 570
8. Who were the Neandertals and the Cro-Magnons, and what is their place in the evolution of humans according to the replacement model mentioned in question 7? 570–71

## testing yourself

Choose the best answer for each question.

1. Which of these gives the correct order of divergence from the main primate line of descent?
   a. prosimians, monkeys, gibbons, orangutans, African apes, humanlike hominins
   b. gibbons, orangutans, prosimians, monkeys, African apes, humanlike hominins
   c. monkeys, gibbons, prosimians, African apes, orangutans, humanlike hominins
   d. African apes, gibbons, monkeys, orangutans, prosimians, humanlike hominins
   e. *H. habilis, H. ergaster, H. neandertalensis,* Cro-Magnon
2. Lucy is a(n)
   a. early *Homo*.
   b. australopith.
   c. ardipithecine.
   d. modern human.
3. What possibly influenced the evolution of bipedalism?
   a. Humans wanted to stand erect in order to use tools.
   b. With bipedalism, it's possible to reach food overhead.
   c. With bipedalism, sexual intercourse is facilitated.
   d. An upright stance exposes more of the body to the sun, and vitamin D production requires sunlight.
   e. All of these are correct.
4. Which of these is an incorrect association with robust types?
   a. massive chewing muscles attached to bony skull crest
   b. some australopiths
   c. diet included fibrous foods
   d. lived during an Ice Age
   e. Both a and c are incorrect associations.
5. *H. ergaster* could have been the first to
   a. use and control fire.
   b. migrate out of Africa.
   c. make axes and cleavers.
   d. have a brain of about 1,000 cc.
   e. All of these are correct.
6. Which of these characteristics is not consistent with the others?
   a. brow ridges
   b. small cheek teeth (molars)
   c. high forehead
   d. projecting face
   e. stereoscopic vision
7. Which of these statements is correct? The last common ancestor for chimpanzees and hominins
   a. has been found, and it resembles a gibbon.
   b. was probably alive around 5 MYA.
   c. has been found, and it has been dated at 30 MYA.
   d. is not expected to be found because there was no such common ancestor.
   e. is now believed to have lived in Asia, not Africa.
8. Which of these pairs is incorrectly matched?
   a. gibbon—hominoid
   b. *A. africanus*—hominin
   c. tarsier—anthropoid
   d. *H. erectus*—*H. ergaster*
   e. early *Homo*—*H. habilis*
9. If the out-of Africa hypothesis is correct, then
   a. human fossils in China after 100,000 years BP would not be expected to resemble earlier fossils.
   b. human fossils in China after 100,000 years BP would be expected to resemble earlier fossils.
   c. humans did not migrate out of Africa.
   d. Both b and c are correct.
   e. Both a and c are correct.
10. Which of these pairs is incorrectly matched?
    a. *H. erectus*—made tools
    b. Neandertal—good hunter
    c. *H. habilis*—controlled fire
    d. Cro-Magnon—good artist
    e. *A. robustus*—fibrous diet
11. Which hominins could have inhabited the Earth at the same time?
    a. australopiths and Cro-Magnons
    b. *Paranthropus robustus* and *Homo habilis*
    c. *Homo habilis* and *Homo sapiens*
    d. gibbons and humans
12. Which of these is an incorrect statement?
    a. *H. habilis* and *H. rudolfensis* were omnivores with a brain size of about 800 cc.
    b. *H. ergaster* had a brain size larger than that of *H. erectus*.
    c. *H. floresiensis,* discovered in 2004, used tools and fire.
    d. All of these are correct.

For questions 13–17, indicate whether the statement is true (T) or false (F).

13. Australopiths were adapted to different diets. ______________

14. *Homo habilis* made stone tools. ______________
15. The human pelvis is bowl-shaped, and the ape pelvis is long and narrow. ______________
16. The gibbon is an Asian ape, while the chimpanzee is an African ape. ______________
17. Mitochondrial DNA differences are inconsistent with the existence of a recent human common ancestor for all ethnic groups. ______________

For questions 18–22, fill in the blanks.

18. Along with monkeys and all apes, humans are ______________.
19. The out-of-Africa hypothesis proposes that modern humans evolved in ______________ only.
20. The australopiths could probably walk ______________, but they had a ______________ brain.
21. The only fossil rightly called *Homo sapien* is that of ______________.
22. Modern humans evolved ______________ (choose billions, millions, thousands) of years ago.
23. Which human characteristic is not thought to be an adaptation to the environment?
    a. bulky bodies of Eskimos
    b. long limbs of Africans
    c. light skin of northern Europeans
    d. hair texture of Asians
    e. Both a and b are correct.
24. Complete this diagram of the replacement model by filling in the blanks.

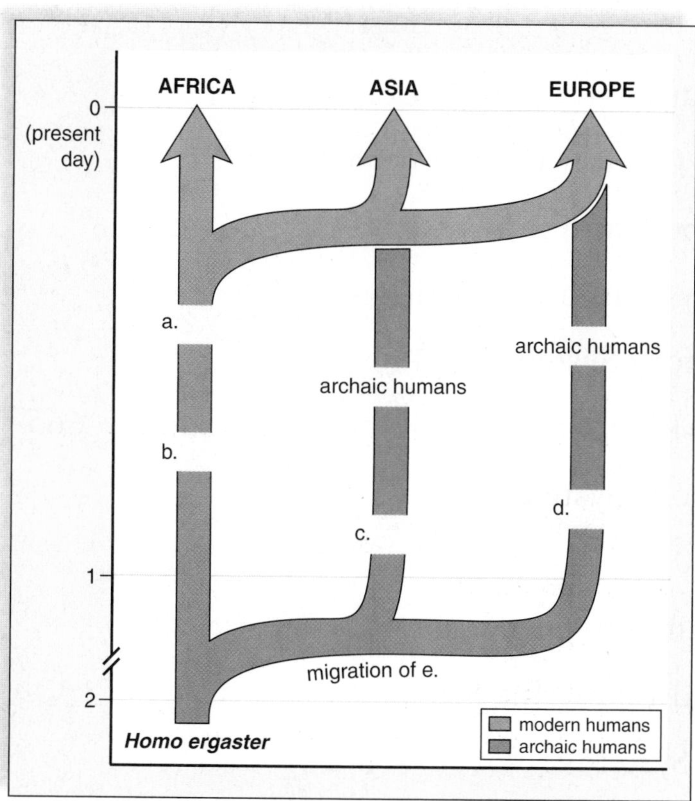


**Replacement Model**

## thinking scientifically

1. Bipedalism has many selective advantages. However, there is one particular disadvantage to walking on two feet: Giving birth to an offspring with a large head through the smaller pelvic opening that is necessitated by upright posture is very difficult. This situation results in a high percentage of deaths (of both mother and child) during birth compared to other primates. How do you explain the selection of a trait that is both positive and negative?
2. How might you use biotechnology to show that humans today have Neandertal genes, and therefore, Cro-Magnons and Neandertals interbred with one another?

## bioethical issue

### Manipulation of Evolution

Since the dawn of civilization, humans have carried out cross-breeding programs to develop plants and animals of use to them. With the advent of DNA technology, we have entered a new era in which even greater control can be exerted over the evolutionary process. We can manipulate genes and give organisms traits that they would not ordinarily possess. Some plants today produce human proteins that can be extracted from their seeds, and some animals grow larger because we have supplied them with an extra gene for growth hormone. Does this type of manipulation seem justifiable?

What about the possibility that we are manipulating our own evolution? Should doctors increase the fitness of certain couples by providing them with a means to reproduce that they cannot achieve on their own? Is the use of alternate means of reproduction bioethically justifiable? In the near future, it may be possible for parents to choose the phenotypic traits of their offspring; in effect, this might enable humans to ensure that their offspring are stronger and brighter than their parents. Does this choosing of "designer babies" seem ethical to you?

## *Biology* website

The companion website for *Biology* provides a wealth of information organized and integrated by chapter. You will find practice tests, animations, videos, and much more that will complement your learning and understanding of general biology.

**http://www.mhhe.com/maderbiology10**

part VII

# Comparative Animal Biology

*In contrast to plants, which are autotrophic and make their own organic food, animals are heterotrophic and feed on organic molecules made by other organisms. Their mobility, which is dependent upon nerve and muscle fibers, is essential to escaping predators, finding a mate, and acquiring food. Food is then digested, and the nutrients are distributed to the body's cells. Finally, waste products are expelled.*

*In complex animals, a distinct division of labor exists, and each of the organ systems is specialized to carry out specific functions. A cardiovascular system transports materials from one body part to another; a respiratory system carries out gas exchange; and a urinary system filters the blood and removes its wastes. The lymphatic system, along with the immune system, protects the body from infectious diseases. The nervous system and endocrine system coordinate the activities of the other systems.*

*Our comparative study will show how the systems evolved and how they function to maintain homeostasis, the relative constancy of the internal environment.*

# 31

# Animal Organization and Homeostasis

*The need for a space suit to take a space walk reminds us that organ systems function best if the internal environment stays within normal limits. For example, a warm temperature speeds enzymatic reactions, a moderate blood pressure helps circulate blood, and a sufficient oxygen concentration facilitates ATP production. Working in harmony and under the coordination of the nervous and endocrine systems, healthy organ systems are capable of maintaining homeostasis, a dynamic equilibrium of the internal environment. Swim the English Channel, cross the Sahara Desert by camel, hike to the South Pole, or take a space walk—your body temperature will stay at just about 37°C as long as you take proper precautions. An astronaut depends on artificial systems in addition to natural systems to maintain homeostasis.*

*This chapter discusses homeostasis after a look at the body's organization. Just as in other complex animals, each organ system of the human body contains a particular set of organs. The circulatory system contains the heart and blood vessels and the nervous system contains the brain and nerves. Organs are composed of tissues, and each type of tissue has like cells that perform specific functions. We will begin the chapter by examining several of the major types of tissues.*

An astronaut needs a special suit to take a walk outside his spaceship.

## concepts

# 31.1 Types of Tissues

Like all living things, animals are highly organized. Animals begin life as a single cell, the fertilized egg or zygote. The zygote undergoes cell division, and the cells differentiate into a variety of tissues that go on to become parts of organs. Several organs are found in an organ system. In this chapter, we consider the tissue, organ, and organ system levels of organization.

A **tissue** is composed of specialized cells of the same type that perform a common function in the body. The tissues of the human body can be categorized into four major types:

1. *Epithelial tissue* covers body surfaces, lines body cavities, and forms glands.
2. *Connective tissue* binds and supports body parts.
3. *Muscular tissue* moves the body and its parts.
4. *Nervous tissue* receives stimuli and transmits nerve impulses.

## Epithelial Tissue

**Epithelial tissue,** also called epithelium (pl., epithelia), consists of tightly packed cells that form a continuous layer. Epithelial tissue covers surfaces and lines body cavities. Usually, it has a protective function, but it can also be modified to carry out secretion, absorption, excretion, and filtration.

Epithelial cells may be connected to one another by three types of junctions composed of proteins (see Fig. 5.14). Regions where proteins join them together are called tight junctions. In the intestine, the gastric juices stay out of the body, and in the kidneys, the urine stays within kidney tubules because epithelial cells are joined by tight junctions. For example, adhesion junctions in the skin allow epithelial cells to stretch and bend, while gap junctions are protein channels that permit the passage of molecules between two adjacent cells.

Epithelial cells are exposed to the environment on one side, but on the other side they have a **basement membrane.** The basement membrane is simply a thin layer of various types of proteins that anchors the epithelium to the extracellular matrix, which is often a type of connective tissue. The basement membrane should not be confused with the plasma membrane or the body membranes we will be discussing.

### *Simple Epithelia*

Epithelial tissue is either simple or complex. Simple epithelia have only a single layer of cells (Fig. 31.1) and are classified according to cell type. **Squamous epithelium,** which is composed of flattened cells, is found lining blood vessels and the air sacs of lungs. **Cuboidal epithelium** contains cube-shaped cells and is found lining the kidney tubules and various glands. **Columnar epithelium** has cells resembling rectangular pillars or columns, with nuclei usually located near the bottom of each cell. This epithelium is found lining the digestive tract, where it efficiently absorbs nutrients from the small intestine because of minute cellular extensions called microvilli. Ciliated columnar epithelium is found lining the oviducts, where it propels the egg toward the uterus.

When an epithelium is pseudostratified, it appears to be layered, but true layers do not exist because each cell touches the basement membrane. The lining of the windpipe, or trachea, is pseudostratified ciliated columnar epithelium. A secreted covering of mucus traps foreign particles, and the upward motion of the cilia carries the mucus to the back of the throat, where it may be either swallowed or expectorated. Smoking can cause a change in mucus secretion and inhibit

**FIGURE 31.1 Types of epithelial tissues in vertebrates.**

Basic epithelial tissues found in vertebrates are shown, along with locations of the tissue and the primary function of the tissue at these locations.

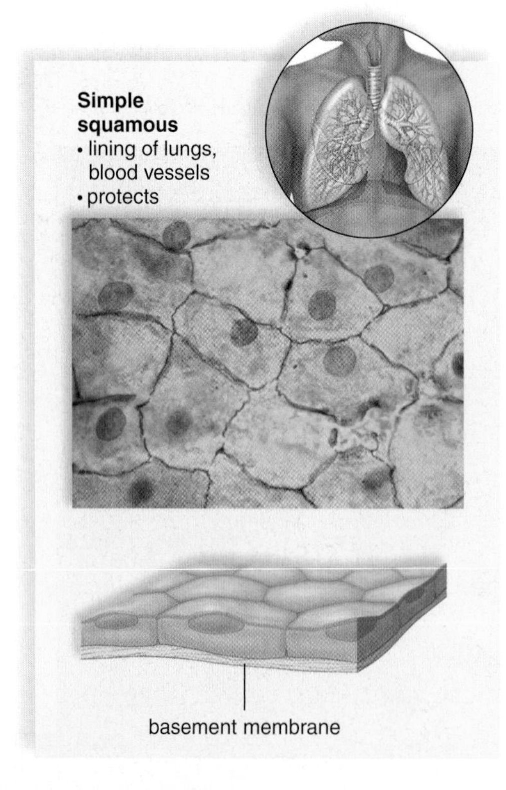


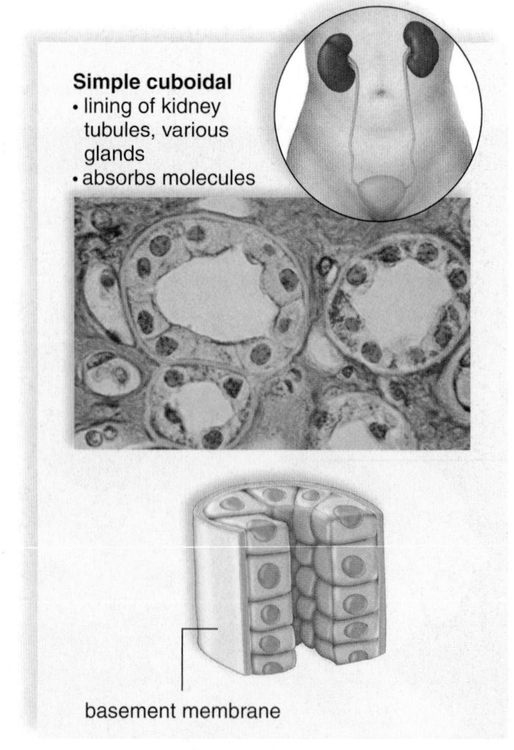

ciliary action, resulting in a chronic inflammatory condition called bronchitis.

### *Stratified Epithelia*

Stratified epithelia have layers of cells piled one on top of the other. Only the bottom layer touches the basement membrane. The nose, mouth, esophagus, anal canal, and vagina are all lined with stratified squamous epithelium. As we shall see, the outer layer of skin is also stratified squamous epithelium, but the cells have been reinforced by keratin, a protein that provides strength. Stratified cuboidal and stratified columnar epithelia also occur in the body.

### *Glandular Epithelia*

When an epithelium secretes a product, it is said to be glandular. A **gland** can be a single epithelial cell, as in the case of mucus-secreting goblet cells within the columnar epithelium lining the digestive tract, or a gland can contain many cells. Glands that secrete their product into ducts are called **exocrine glands.**

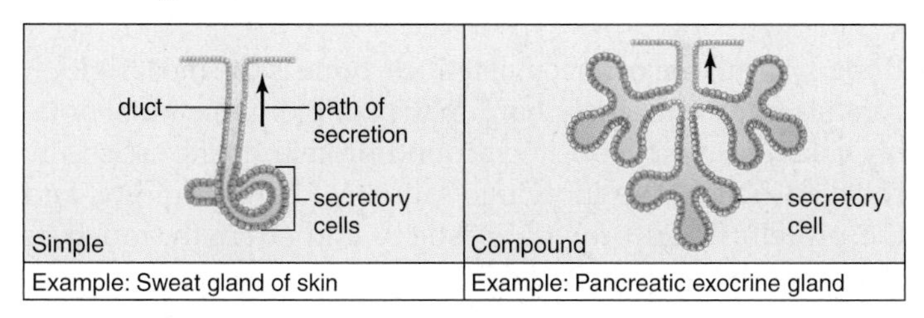


Glands that have no duct are appropriately known as **endocrine glands.** Endocrine glands (e.g., pituitary gland and thyroid) secrete hormones internally, so they are transported by the bloodstream.

## Connective Tissue

**Connective tissue** is the most abundant and widely distributed tissue in complex animals. It is quite diverse in structure and function, but, even so, all types have three components: specialized cells, ground substance, and protein fibers (Fig. 31.2). The ground substance is a noncellular material that separates the cells and varies in consistency from solid to semifluid to fluid. The fibers are of three possible types. White **collagen fibers** contain collagen, a protein that gives them flexibility and strength. **Reticular fibers** are very thin collagen fibers that are highly branched and form delicate supporting networks. Yellow **elastic fibers** contain elastin, a protein that is not as strong as collagen but is more elastic.

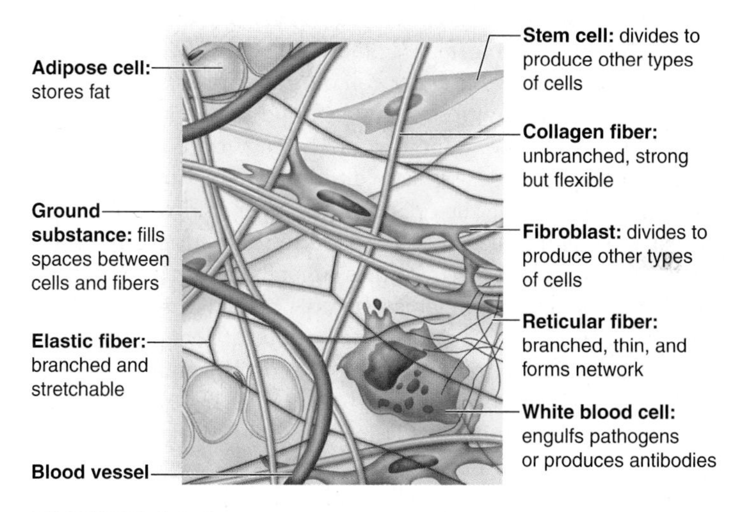


**FIGURE 31.2** **Diagram of fibrous connective tissue.**

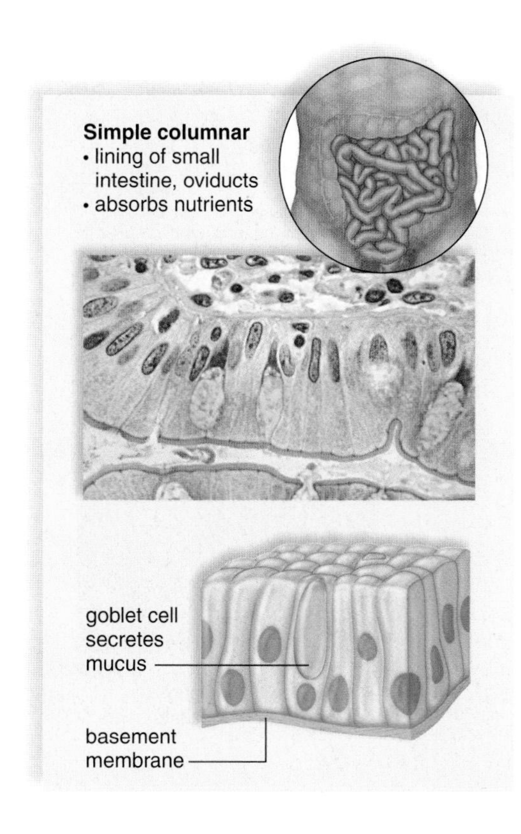


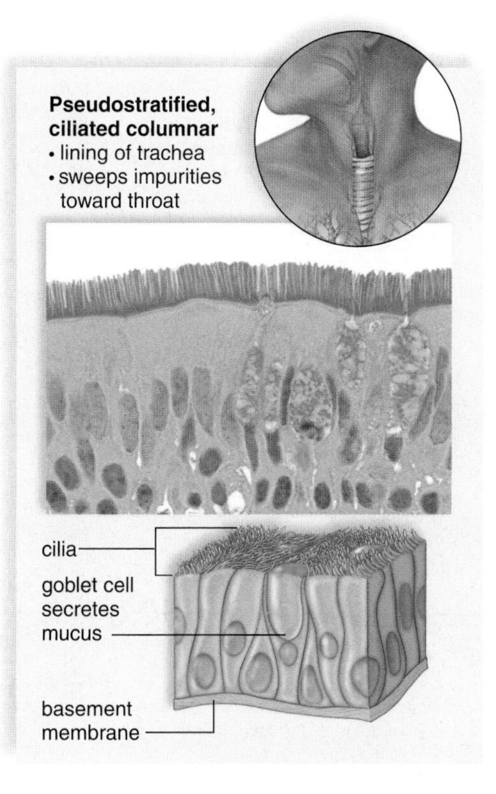


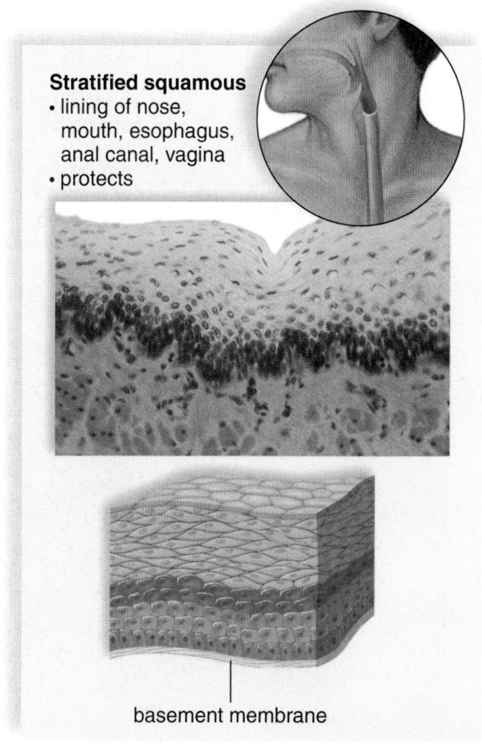

Connective tissue is classified into three categories: fibrous, supportive, and fluid. Each category includes a number of different types of tissues.

## *Fibrous Connective Tissue*

Both loose fibrous and dense fibrous connective tissues have cells called **fibroblasts** [L. *fibra,* thread, and Gk. *blastos,* bud] that are located some distance from one another and are separated by a jellylike matrix containing white collagen fibers and yellow elastic fibers.

**Loose fibrous connective tissue** supports epithelium and also many internal organs (Fig. 31.3*a*). Its presence in lungs, arteries, and the urinary bladder allows these organs to expand. It forms a protective covering enclosing many internal organs, such as muscles, blood vessels, and nerves.

**Adipose tissue** [L. *adipalis,* fatty] serves as the body's primary energy reservoir (Fig. 31.3*b*). Adipose tissue also insulates the body, contributes to body contours, and provides cushioning. In mammals, adipose tissue is found particularly beneath the skin, around the kidneys, and on the surface of the heart. The number of adipose cells in an individual is fixed. When a person gains weight, the cells become larger, and when weight is lost, the cells shrink. In obese people, the individual cells called adipocytes may be up to five times larger than normal. Most adipocytes are white, but in fetuses, infants, and children, they may be brown due to mitochondria and be good at heat production.

**Dense fibrous connective tissue** contains many collagen fibers that are packed together (Fig. 31.3*c*). This type of tissue has more specific functions than does loose connective tissue. For example, dense fibrous connective tissue is found in **tendons** [L. *tendo,* stretch], which connect muscles to bones, and in **ligaments** [L. *ligamentum,* band], which connect bones to other bones at joints.

## *Supportive Connective Tissue*

In **cartilage,** the cells lie in small chambers called lacunae (sing., **lacuna**), separated by a matrix that is solid yet flexible. Unfortunately, because this tissue lacks a direct blood supply, it heals very slowly. There are three types of cartilage, distinguished by the type of fiber in the matrix.

**Hyaline cartilage** (Fig. 31.3*d*), the most common type of cartilage, contains only very fine collagen fibers. The matrix has a white, translucent appearance. Hyaline cartilage is found in the nose and at the ends of the long bones and the ribs, and it forms rings in the walls of respiratory passages. The fetal skeleton also is made of this type of cartilage. Later, the cartilaginous fetal skeleton is replaced by bone.

**Elastic cartilage** has more elastic fibers than hyaline cartilage. For this reason, it is more flexible and is found, for example, in the framework of the outer ear.

**Fibrocartilage** has a matrix containing strong collagen fibers. Fibrocartilage is found in structures that withstand tension and pressure, such as the pads between the vertebrae in the backbone and the wedges in the knee joint.

**Bone.** Of all the connective tissues, **bone** is the most rigid. It consists of an extremely hard matrix of inorganic salts, notably calcium salts, deposited around protein fibers, especially collagen fibers. The inorganic salts give bones rigidity, and the protein fibers provide elasticity and strength, much as steel rods do in reinforced concrete.

**Compact bone** makes up the shaft of a long bone (Fig. 31.3*e*). It consists of cylindrical structural units called osteons (Haversian systems). The central canal of each osteon is surrounded by rings of hard matrix. Bone cells are located in spaces called lacunae between the rings of matrix. Blood vessels in the central canal carry nutrients that allow bone to

**FIGURE 31.3**
**Types of connective tissue in vertebrates.**
Pertinent information about each type of connective tissue is given.

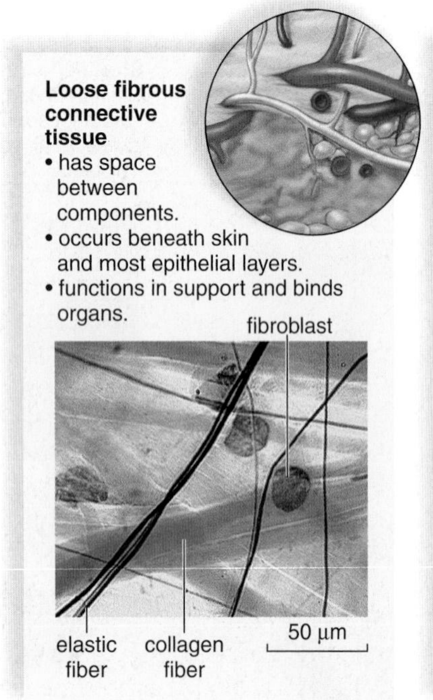


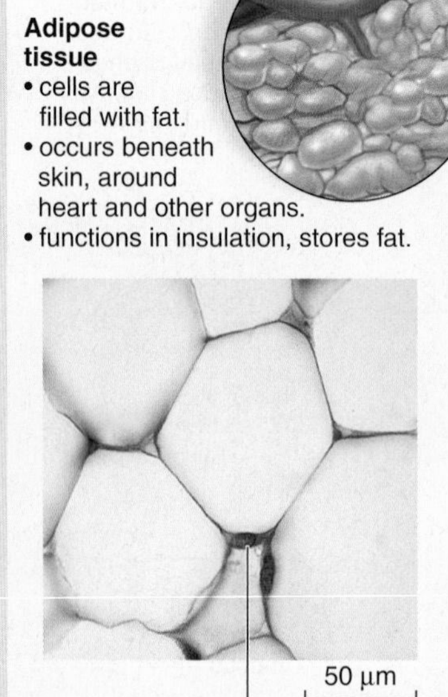


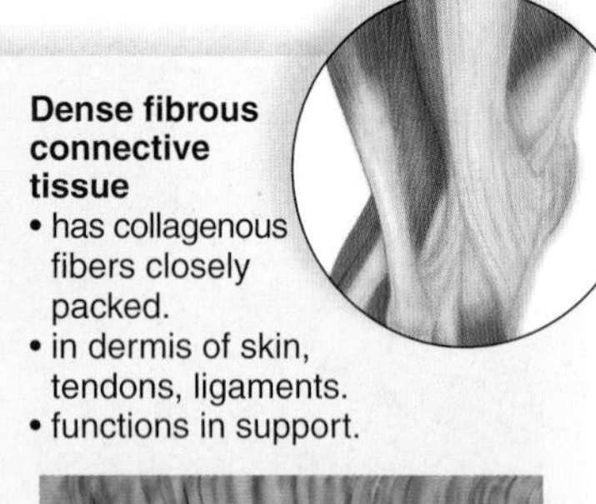


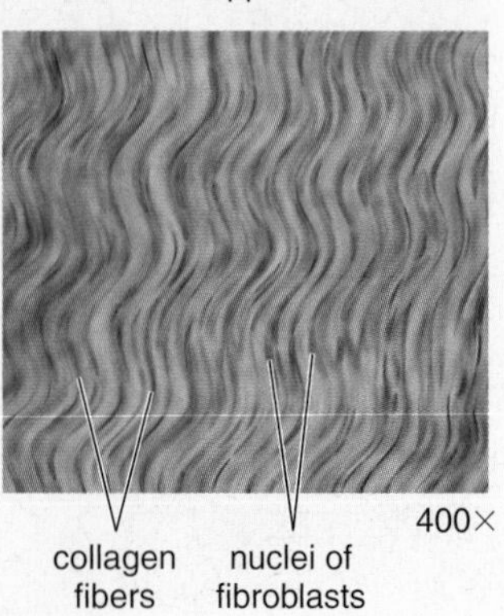

renew itself. Thin extensions of bone cells within canaliculi (minute canals) connect the cells to each other and to the central canal.

The ends of a long bone contain spongy bone, which has an entirely different structure. **Spongy bone** contains numerous bony bars and plates, separated by irregular spaces. Although lighter than compact bone, spongy bone is still designed for strength. Just as braces are used for support in buildings, the solid portions of spongy bone follow lines of stress.

### *Fluid Connective Tissues*

**Blood,** which consists of formed elements and plasma, is a fluid connective tissue located in blood vessels (Fig. 31.4). In adults, the production of blood cells, known as hematopoiesis, occurs in the red bone marrow.

The internal environment of the body consists of blood and **tissue fluid.** The systems of the body help keep blood composition and chemistry within normal limits, and blood in turn creates tissue fluid. Blood transports nutrients and oxygen to tissue fluid and removes carbon dioxide and other wastes. It helps distribute heat and also plays a role in fluid, ion, and pH balance. The formed elements, discussed following, each have specific functions.

The **red blood cells** are small, biconcave, disk-shaped cells without nuclei. The absence of a nucleus makes the cells biconcave. The presence of the red pigment hemoglobin makes them red and, in turn, makes the blood red. Hemoglobin is composed of four units; each unit is composed of the protein globin and a complex iron-containing structure called heme. The iron forms a loose association with oxygen, and in this way red blood cells transport oxygen and readily give it up in the tissues.

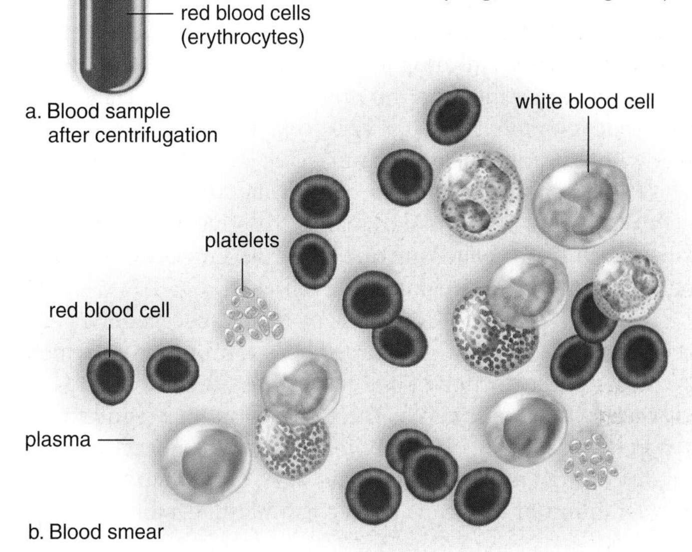


**FIGURE 31.4**

**Blood, a liquid tissue.**

**a.** Blood is classified as connective tissue because the cells are separated by a matrix—plasma. Plasma, the liquid portion of blood, usually contains several types of cells. **b.** Drawing of the components of blood: red blood cells, white blood cells, and platelets (which are actually fragments of a larger cell).

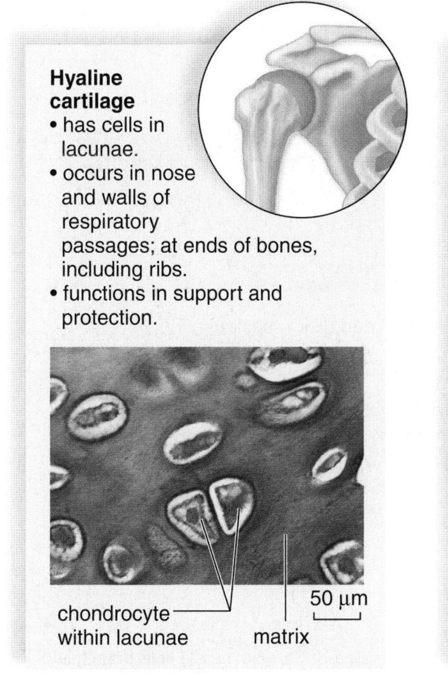


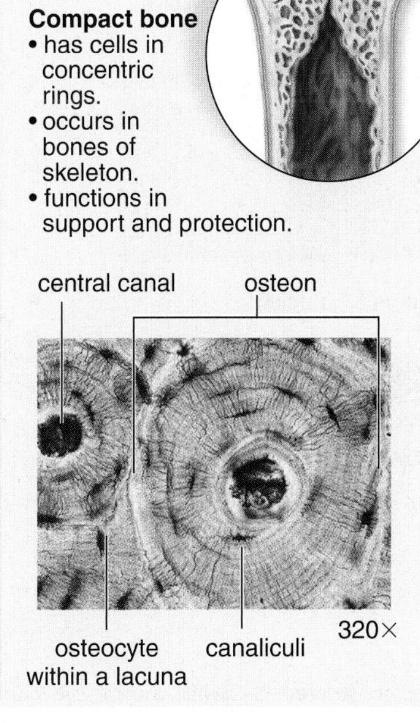


**White blood cells** may be distinguished from red blood cells by the fact that they are usually larger, have a nucleus, and without staining would appear translucent. White blood cells characteristically look bluish because they have been stained that color. White blood cells fight infection, primarily in two ways. Some white blood cells are phagocytic and engulf infectious **pathogens,** while other white blood cells either produce antibodies, molecules that combine with foreign substances to inactivate them, or they kill cells outright.

**Platelets** are not complete cells; rather, they are fragments of giant cells present only in bone marrow. When a blood vessel is damaged, platelets form a plug that seals the vessel, and injured tissues release molecules that help the clotting process.

**Lymph** is a fluid connective tissue located in lymphatic vessels. Lymphatic vessels absorb excess tissue fluid and various dissolved solutes in the tissues and transport them to particular vessels of the cardiovascular system. Special lymphatic capillaries, called lacteals, absorb fat molecules from the small intestine. Lymph nodes, composed of fibrous connective tissue, occur along the length of lymphatic vessels. In particular, lymph is cleansed as it passes through lymph nodes because white blood cells congregate there. Lymph nodes enlarge when you have an infection.

## Muscular Tissue

**Muscular (contractile) tissue** is composed of cells called muscle fibers. Muscle fibers contain actin filaments and myosin filaments, whose interaction accounts for movement. The muscles are also important in the generation of body heat. There are three distinct types of muscle tissue: skeletal, smooth, and cardiac. Each type differs in appearance, physiology, and function.

**Skeletal muscle,** also called voluntary muscle (Fig. 31.5*a*), is attached by tendons to the bones of the skeleton, and when it contracts, body parts move. Contraction of skeletal muscle is under voluntary control and occurs faster than in the other muscle types. Skeletal muscle fibers are cylindrical and quite long—sometimes they run the length of the muscle. They arise during development when several cells fuse, resulting in one fiber with multiple nuclei. The nuclei are located at the periphery of the cell, just inside the plasma membrane. The fibers have alternating light and dark bands that give them a **striated** appearance. These bands are due to the placement of actin filaments and myosin filaments in the cell.

**Smooth (visceral) muscle** is so named because the cells lack striations. The spindle-shaped cells form layers in which the thick middle portion of one cell is opposite the thin ends of adjacent cells. Consequently, the nuclei form an irregular pattern in the tissue (Fig. 31.5*b*). Smooth muscle is not under voluntary control and therefore is said to be involuntary. Smooth muscle, found in the walls of viscera (intestine, stomach, and other internal organs) and blood vessels, contracts more slowly than skeletal muscle but can remain contracted for a longer time. When the smooth muscle of the intestine contracts, food moves along its lumen (central cavity). When the smooth muscle of the blood vessels contracts, blood vessels constrict, helping to raise blood pressure. Small amounts of smooth muscle are also found in the iris of the eye and in the skin.

**Cardiac muscle** (Fig. 31.5*c*) makes up the walls of the heart. Its contraction pumps blood and accounts for the heartbeat. Cardiac muscle combines features of both smooth muscle and skeletal muscle. Like skeletal muscle, it has striations, but the contraction of the heart is involuntary for the most part. Cardiac muscle cells also differ from skeletal muscle cells in that they have a single, centrally placed nucleus. The cells are branched and seemingly fused one with the other, and the heart appears to be composed of one large interconnecting mass of muscle cells. Actually, cardiac muscle cells are separate and individual, but they are bound end to end at **intercalated disks,** areas where folded plasma membranes between two cells contain adhesion junctions and gap junctions.

## Nervous Tissue

**Nervous tissue** contains nerve cells called neurons. An average person has about 1 trillion neurons. A **neuron** is a specialized cell that has three parts: dendrites, a cell body, and an axon (Fig. 31.6, *top*). A dendrite is a process that conducts signals toward

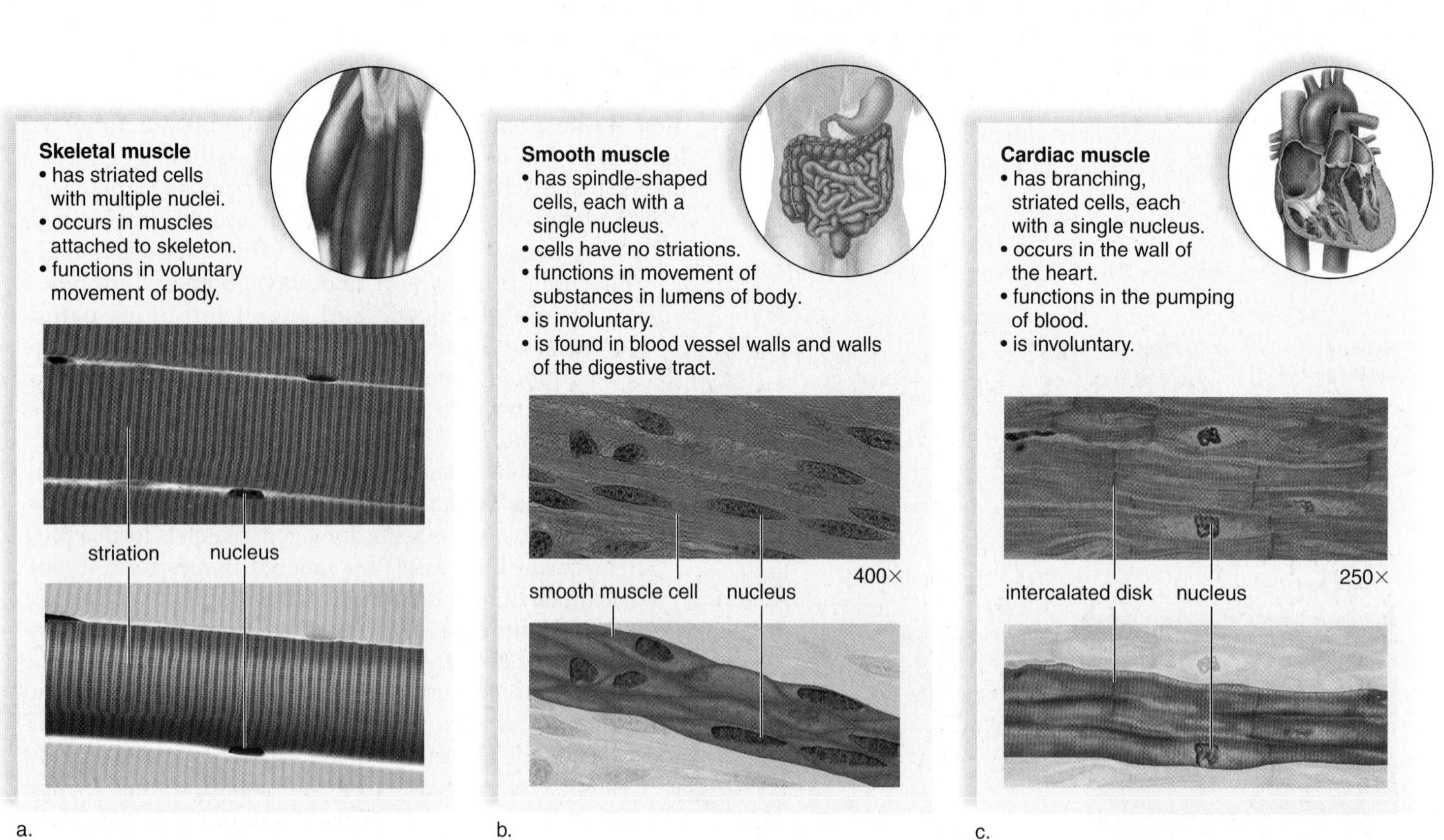


**FIGURE 31.5 Muscular tissue.**

**a.** Skeletal muscle is voluntary and striated. **b.** Smooth muscle is involuntary and nonstriated. **c.** Cardiac muscle is involuntary and striated. Cardiac muscle cells branch and fit together at intercalated disks.

the cell body. The cell body contains the major concentration of the cytoplasm and the nucleus of the neuron. An axon is a process that typically conducts nerve impulses away from the cell body. Long axons are covered by myelin, a white, fatty substance. The term *fiber*[1] is used here to refer to an axon along with its myelin sheath if it has one. Outside the brain and spinal cord, fibers bound by connective tissue form **nerves.**

The nervous system has just three functions: sensory input, integration of data, and motor output. Nerves conduct impulses from sensory receptors to the spinal cord and the brain, where integration occurs. The phenomenon called sensation occurs only in the brain, however. Nerves also conduct nerve impulses away from the spinal cord and brain to the muscles and glands, causing them to contract and secrete, respectively. In this way, a coordinated response to the stimulus is achieved.

In addition to neurons, nervous tissue contains neuroglia.

## *Neuroglia*

**Neuroglia** are cells that outnumber neurons as much as 50 to 1, and take up more than half the volume of the brain. Although the primary function of neuroglia is to support and nourish neurons, research is currently being conducted to determine how much they directly contribute to brain function. Various types of neuroglia are found in the brain. Microglia, astrocytes, and oligodendrocytes are shown in Figure 31.6*a*. Microglia, in addition to supporting neurons, engulf bacterial and cellular debris. Astrocytes provide nutrients to neurons and produce a hormone known as glia-derived growth factor, which someday might be used as a cure for Parkinson disease and other diseases caused by neuron degeneration. Oligodendrocytes form myelin. Neuroglia do not have a long process, but even so, researchers are now beginning to gather evidence that they do communicate among themselves and with neurons!

Mature neurons have little capacity for cell division and seldom form tumors. The majority of brain tumors in adults involve actively dividing neuroglia cells. Most brain tumors have to be treated with surgery or radiation therapy because of a blood-brain barrier.

### Check Your Progress 31.1

1. Distinguish between the three types of simple epithelium and give a location for each.
2. Compare and contrast the three types of connective tissue.
3. How does the structure and function of skeletal, smooth, and cardiac muscle differ?
4. What are the three parts of a neuron, and what does each part do?

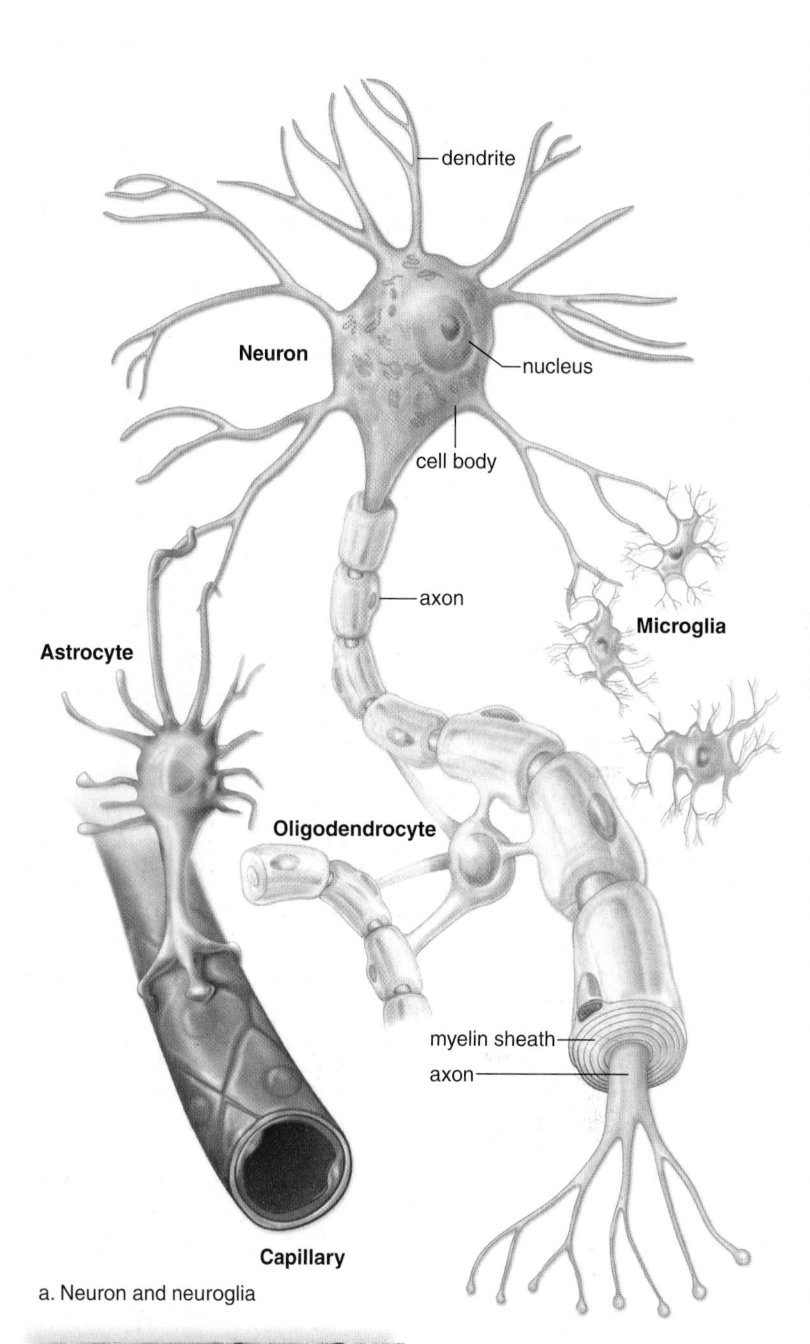

a. Neuron and neuroglia

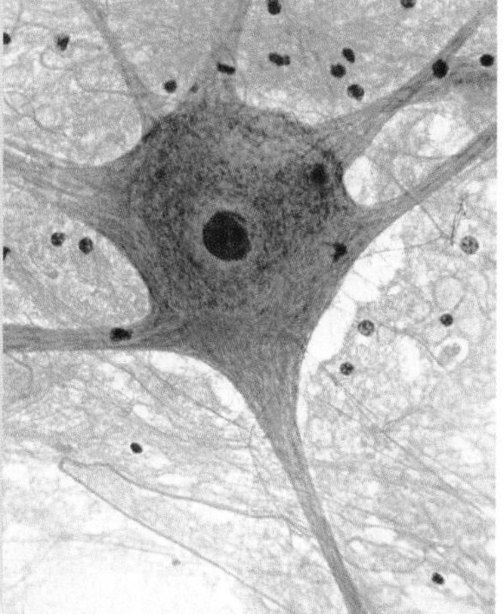
b. Micrograph of a neuron 200×

**FIGURE 31.6**
**Neurons and neuroglia.**
Neurons conduct nerve impulses. Neuroglia consist of cells that support and service neurons and have various functions: Microglia are a type of neuroglia that become mobile in response to inflammation and phagocytize debris. Astrocytes lie between neurons and a capillary; therefore, nutrients entering neurons from the blood must first pass through astrocytes. Oligodendrocytes form the myelin sheaths around fibers in the brain and spinal cord.

[1] In connective tissue, a fiber is a component of the matrix; in muscular tissue, a fiber is a muscle cell; in nervous tissue, a nerve fiber is an axon and its myelin sheath.

*health focus*

## Nerve Regeneration

In humans, axons outside the brain and spinal cord can regenerate, but not those inside these organs. After injury, axons in the human central nervous system (CNS) degenerate, resulting in permanent loss of nervous function. Not so in cold-water fishes and amphibians, where axon regeneration in the CNS does occur. So far, investigators have identified several proteins that seem to be necessary to axon regeneration in the CNS of these animals (Fig. 31A), but it will be a long time before biochemistry can offer a way to bring about axon regeneration in the human CNS. It's possible, though, that one day these proteins will become drugs or that gene therapy might be used to cause humans to produce the same proteins when CNS injuries occur.

In the meantime, some accident victims are trying other ways to bring about a cure. In 1995, Christopher Reeve, best known for his acting role as "Superman," was thrown headfirst from his horse, crushing the spinal cord just below the neck's top two vertebrae. Immediately, his brain lost almost all communication with the portion of his body below the site of damage and he could not move his arms and legs. Many years later, Reeve could move his left index finger slightly and could take tiny steps while being held upright in a pool. He had sensation throughout his body and could feel his wife's touch.

Reeve's improvement was not the result of cutting-edge drugs or gene therapy—it was due to exercise (Fig. 31B)! Reeve exercised as much as five hours a day, especially using a recumbent bike outfitted with electrodes that made his leg muscles contract and relax. The bike cost him $16,000. It could cost less if commonly used by spinal cord injury patients in their own homes. Reeve, who was an activist for the disabled, was pleased that insurance would pay for the bike about 50% of the time.

It's possible that Reeve's advances were the result of improved strength and bone density, which lead to stronger nerve signals. Normally, nerve cells are constantly signaling one another, but after a spinal cord injury, the signals cease. Perhaps Reeve's intensive exercise brought back some of the normal communication between nerve cells. Reeve's physician, John McDonald, a neurologist at Washington University in St. Louis, is convinced that his axons were regenerating. The neuroscientist Fred Gage at the Salk Institute in La Jolla, California, has shown that exercise does enhance the growth of new cells in adult brains.

For himself, Reeve was convinced that stem cell therapy would one day allow him to be off his ventilator and functioning normally; however, Reeve died in 2004. So far, researchers have shown that both embryonic stem cells and bone marrow stem cells can differentiate into neurons in the laboratory. Bone marrow stem cells apparently can also become neurons when injected into the body.

a.

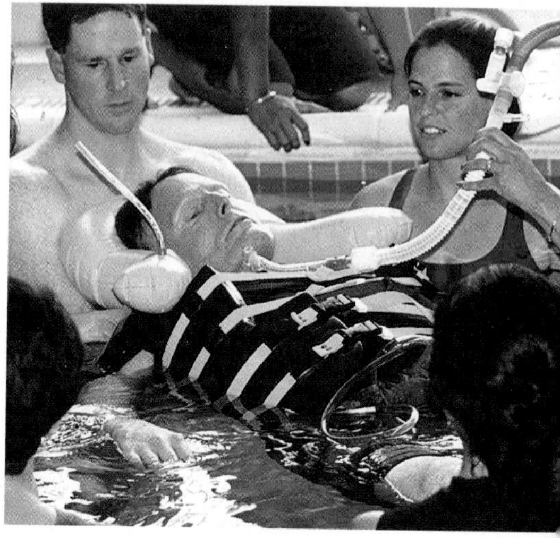

b.

**FIGURE 31B Treatment today for spinal cord injuries.** ***a.*** *Reeve suffered a spinal cord injury when horseback riding in 1995.* ***b.*** *He exercised many hours a day. Here he receives aqua therapy. Reeve died in 2004.*

**FIGURE 31A Researchers at work.** ***a.*** *Some researchers are studying the activity of proteins that allow cold-bodied animals to regenerate axons in the CNS.* ***b.*** *Others are doing stem cell research. Stem cells might one day be used to cure people with spinal cord injuries.*

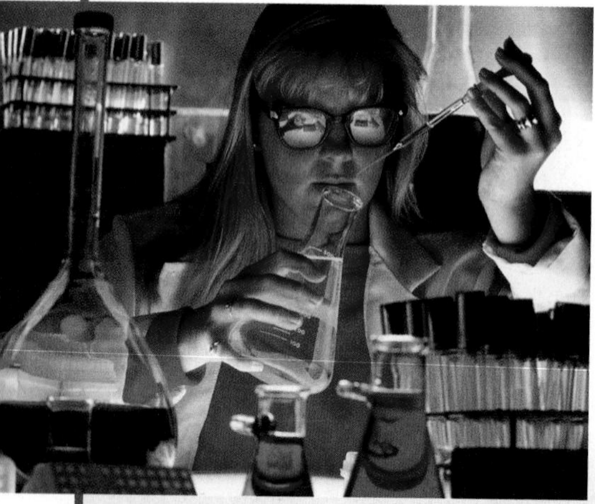

a.

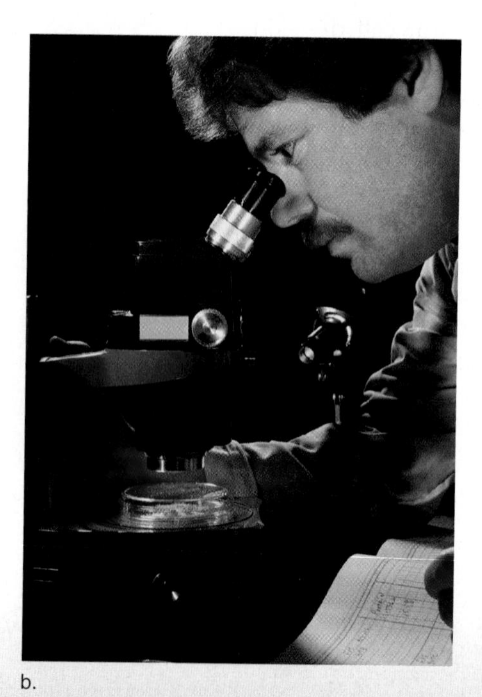

b.

## 31.2 Organs and Organ Systems

Specific tissues are associated with particular organs. For example, nervous tissue is associated with the brain. In actuality, an **organ** is composed of two or more types of tissues working together to perform particular functions. An **organ system** contains many different organs that cooperate to carry out a process, such as the digestion of food.

The integumentary system, consisting of the skin and its derivatives (hair, nails, and cutaneous glands), is the most conspicuous system in the body. Being the largest organ, the skin covers an area of 1.5–2 $m^2$ and accounts for nearly 15% of the weight of an average human.

Derivatives of the skin differ throughout the vertebrate world. Fishes possess a number of bony scales. Amphibians have smooth skin covered with mucous glands. Reptiles possess epidermal scales that vary in color and shape. Birds have scales on their legs, but most of the body is covered with feathers. Mammals are characterized by hair, nails, glands, and a number of sensory detectors.

### Skin as an Organ

Human skin covers the body, protecting underlying parts from physical trauma, pathogen invasion, and water loss. The skin is also important in thermoregulation or regulating body temperature. Skin is equipped with a variety of sensory structures that monitor touch, pressure, temperature, and pain. In addition, skin cells manufacture precursor molecules that are converted to vitamin D after exposure to UV (ultraviolet) light.

### Regions of Skin

The **skin** has two regions: the epidermis and the dermis (Fig. 31.7). A **subcutaneous layer** known as the hypodermis is found between the skin and any underlying structures, such as muscle or bone.

#### *The Epidermis*

The **epidermis** [Gk. *epi,* over, and *derma,* skin] is made up of stratified squamous epithelium. Skin can be described as thin skin or thick skin based on the thickness of the epidermis. Thin skin covers most of the body and is associated with hair follicles, sebaceous (oil) glands, and sweat glands. Thick skin appears in regions of wear and tear, such as the palms of the hands and soles of the feet. Thick skin has sweat glands but no sebaceous glands or hair follicles. In both types of skin, new cells derived from stem (basal) cells become flattened and hardened as they push to the surface (Fig. 31.8*a*). Hardening takes place because the cells produce keratin, a waterproof protein. Dandruff occurs when the rate of keratinization in the skin of the scalp is two or three times the normal rate. A thick layer of dead keratinized cells, arranged

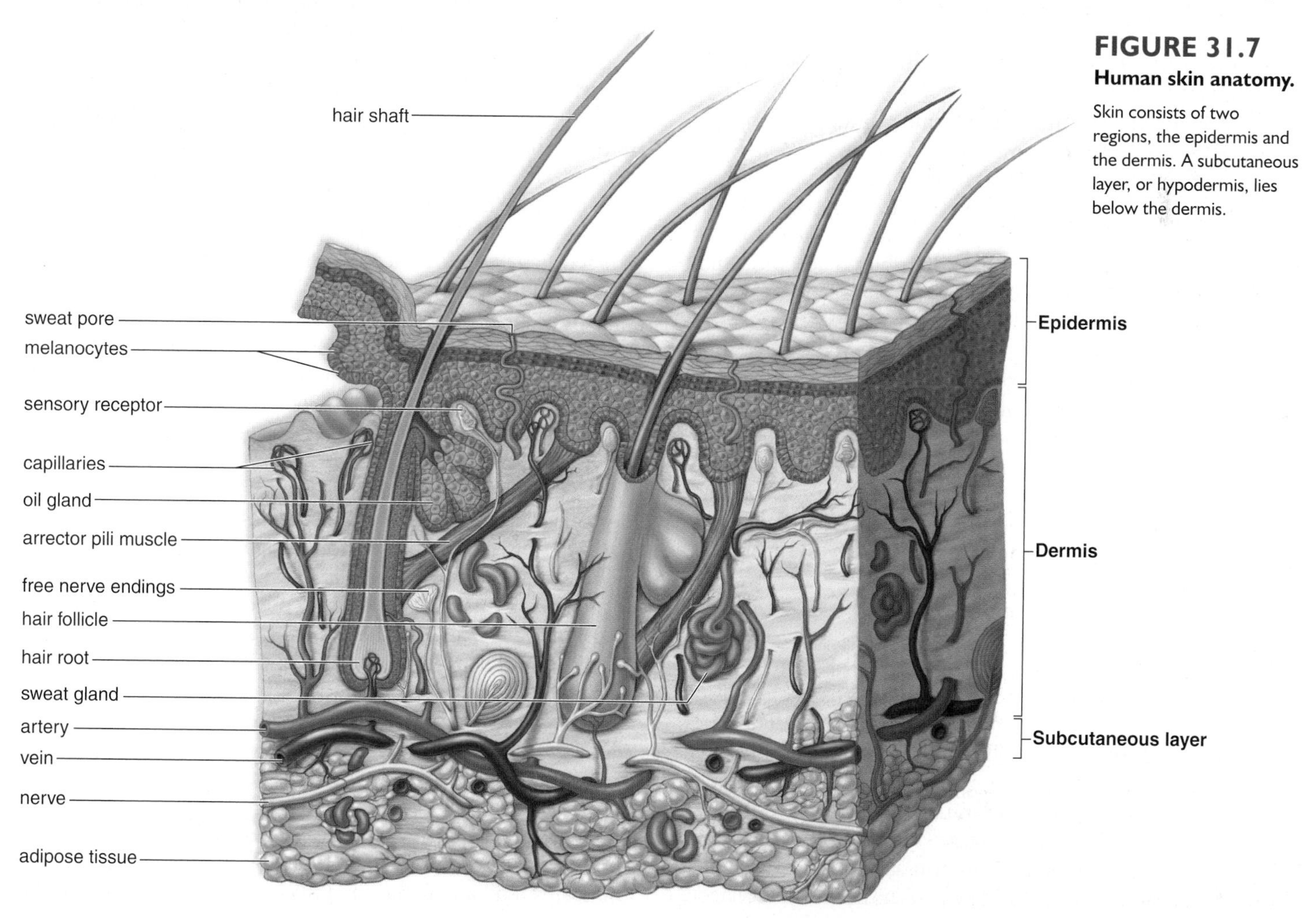


**FIGURE 31.7**
**Human skin anatomy.**
Skin consists of two regions, the epidermis and the dermis. A subcutaneous layer, or hypodermis, lies below the dermis.

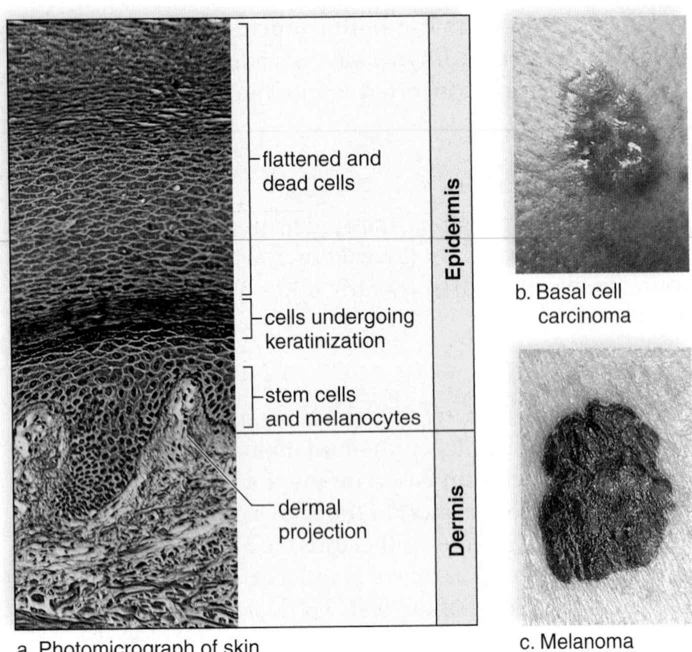


**FIGURE 31.8** **The epidermis.**

**a.** Epidermal ridges following dermal projections are clearly visible. Stem cells and melanocytes are in this region. **b.** Basal cell carcinoma derived from stem cells and melanoma (**c**) derived from melanocytes are types of skin cancer. Remember the A, B, C, D rule when examining a questionable freckle or mole: A–asymmetrical shape; B–border irregularity; C–color change; D–diameter or sudden change in size.

in spiral and concentric patterns, forms fingerprints and footprints.

Specialized cells in the epidermis called **melanocytes** produce melanin, the pigment responsible for skin color. The amount of melanin varies throughout the body. It is concentrated in freckles and moles. Tanning occurs after a light-skinned person is exposed to sunlight because melanocytes produce more melanin, which is distributed to epidermal cells before they rise to the surface. While we tend to associate a tan with health, actually it signifies that the body is trying to protect itself from the dangerous rays of the sun. Some ultraviolet radiation does serve a purpose, however. As mentioned, certain cells in the epidermis convert a steroid related to cholesterol into **vitamin D** only with the aid of ultraviolet radiation. Vitamin D is required for proper bone growth.

Too much ultraviolet radiation is dangerous and can lead to skin cancer. Basal cell carcinoma (Fig. 31.8*b*) derived from stem cells gone awry is the more common type of skin cancer and the most curable. Melanoma (Fig. 31.8*c*), the type of skin cancer derived from melanocytes, is extremely serious.

### *The Dermis*

The **dermis** [Gk. *derma,* skin] is a region of dense fibrous connective tissue beneath the epidermis. The dermis contains collagen and elastic fibers. The collagen fibers are flexible but offer great resistance to overstretching; they prevent the skin from being torn. Stretching of the dermis, as occurs in obesity and pregnancy, can produce stretch marks, or striae. The elastic fibers maintain normal skin tension but also stretch to allow movement of underlying muscles and joints. (The number of collagen and elastic fibers decreases with age and with exposure to the sun, causing the skin to become less supple and more prone to wrinkling.) The dermis also contains blood vessels that nourish the skin. When blood rushes into these vessels, a person blushes, and when blood is minimal in them, a person turns "blue."

Sensory receptors are specialized nerve endings in the dermis that respond to external stimuli. There are sensory receptors for touch, pressure, pain, and temperature. The fingertips contain the most touch receptors, and these add to our ability to use our fingers for delicate tasks.

### *The Subcutaneous Layer*

Technically speaking, the subcutaneous layer (the hypodermis) beneath the dermis is not a part of skin. It is composed of loose connective tissue and adipose tissue, which stores fat. Fat is a stored source of energy in the body. Adipose tissue helps to thermally insulate the body from either gaining heat from the outside or losing heat from the inside. A well-developed subcutaneous layer gives the body a rounded appearance and provides protective padding against external assaults. Excessive development of the subcutaneous layer accompanies obesity.

## Accessory Organs of the Skin

Nails, hair, and glands are structures of epidermal origin, even though some parts of hair and glands are largely found in the dermis.

**Nails** are a protective covering of the distal part of fingers and toes, collectively called digits. Nails grow from special epithelial cells at the base of the nail in the portion called the nail root. The cuticle is a fold of skin that hides the nail root. The whitish color of the half-moon-shaped base, or lunula, results from the thick layer of cells in this area. The cells of a nail become keratinized as they grow out over the nail bed. The appearance of nails can be medically important. For example, in clubbing of the nails, the nails turn down instead of lying flat. This condition is associated with a deficiency of oxygen in the blood.

**Hair follicles** begin in the dermis and continue through the epidermis, where the hair shaft extends beyond the skin. Contraction of the arrector pili muscles attached to hair follicles causes the hairs to "stand on end" and goose bumps to develop. Epidermal cells form the root of hair, and their division causes a hair to grow. The cells become keratinized and die as they are pushed farther from the root.

Hair, except for the root, is formed of dead, hardened epidermal cells; the root is alive and resides at the base of a follicle in the dermis. An average person has about 100,000 hair follicles on the scalp. The number of follicles varies from one body region to another. The texture of hair is dependent on the shape of the shaft. In wavy hair, the shaft is oval, and in straight hair, the shaft is round. Hair color is determined by pigmentation. Dark hair is due to melanin concentration, and blond hair has scanty amounts of melanin. Red hair is caused by an iron-containing pigment called trichosiderin. Gray or white hair results from a lack of pigment. A hair on the scalp grows about 1 mm every three days.

Each hair follicle has one or more **oil glands,** also called sebaceous glands, which secrete sebum, an oily substance that lubricates the hair within the follicle and the skin itself. If the sebaceous glands fail to discharge the secretions collect and form "whiteheads" or "blackheads." The color of blackheads is due to oxidized sebum. Acne is an inflammation of the sebaceous glands that most often occurs during adolescence due to hormonal changes.

**Sweat glands,** also called sudoriferous glands, are quite numerous and are present in all regions of skin. A sweat gland is a tubule that begins in the dermis and either opens into a hair follicle, or more often opens onto the surface of the skin. Sweat glands located all over the body play a role in modifying body temperature. When the body temperature starts to rise, sweat glands become active. Sweat absorbs body heat as it evaporates. Once the body temperature lowers, sweat glands are no longer active. Other sweat glands occur in the groin and axillary regions and are associated with distinct scents.

## Organ Systems

In most animals, individual organs function as part of an organ system, the next higher level of animal organization. These same systems are found in all vertebrate animals. The organ systems of vertebrates carry out the life processes that are common to all animals, and indeed to all organisms.

| Life Processes | Human Systems |
|---|---|
| Coordinate body activities | Nervous system<br>Endocrine system |
| Acquire materials and energy (food) | Skeletal system<br>Muscular system<br>Digestive system |
| Maintain body shape | Skeletal system<br>Muscular system |
| Exchange gases | Respiratory system |
| Transport materials | Cardiovascular system |
| Excrete wastes | Urinary system |
| Protect the body from disease | Lymphatic system<br>Immune system |
| Produce offspring | Reproductive system |

### *Body Cavities*

Each organ system has a particular distribution within the human body. There are two main body cavities: the smaller dorsal cavity and the larger ventral cavity (Fig. 31.9*a*). The brain and the spinal cord are in the dorsal cavity.

During development, the ventral cavity develops from the coelom. In humans and other mammals, the coelom is divided by a muscular diaphragm that assists breathing. The heart (a pump for the cardiovascular system) and the lungs are located in the upper (thoracic or chest) cavity (Fig. 31.9*b*). The major portions of the digestive system, including the accessory organs (e.g., the liver and pancreas) are located in the abdominal cavity, as are the kidneys of the urinary system. The urinary bladder, the female reproductive organs, or certain of the male reproductive organs, are located in the pelvic cavity.

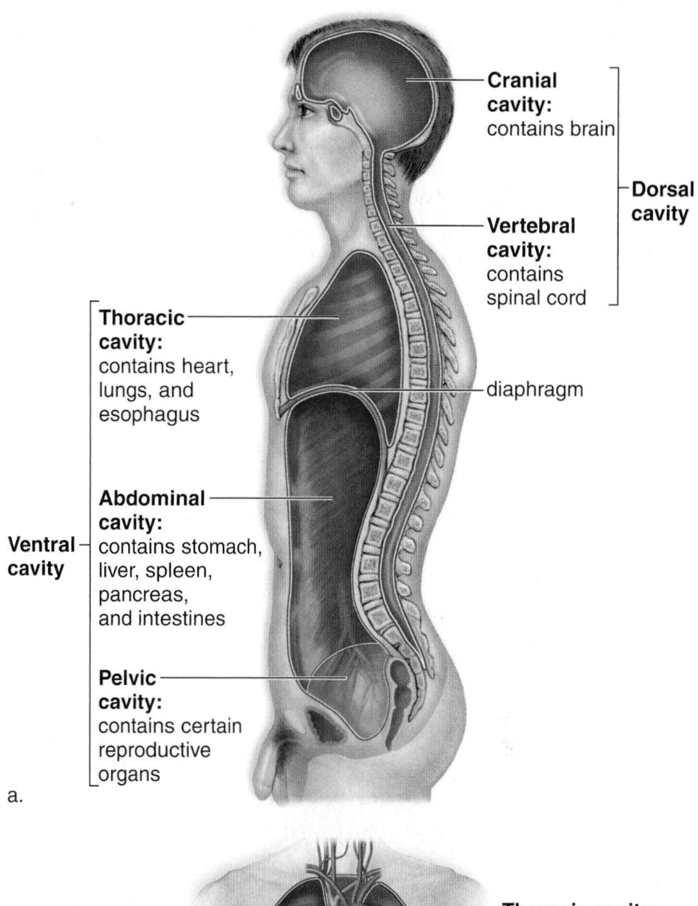


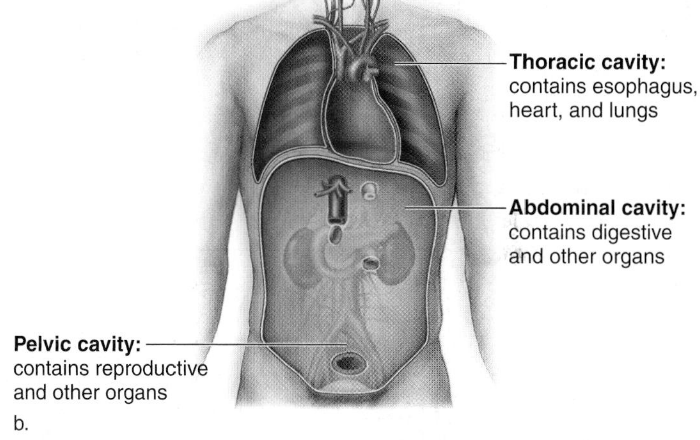


**FIGURE 31.9 Mammalian body cavities.**

**a.** Side view. The dorsal (toward the back) cavity contains the cranial cavity and the vertebral canal. The brain is in the cranial cavity, and the spinal cord is in the vertebral canal. The well-developed ventral (toward the front) cavity is divided by the diaphragm into the thoracic cavity and the abdominopelvic cavity (abdominal cavity and pelvic cavity). The heart and lungs are in the thoracic cavity, and most other internal organs are in the abdominal cavity. **b.** Frontal view of the thoracic cavity.

**Check Your Progress 31.2**

1. Compare the structure and function of the epidermis and dermis of the skin.
2. Contrast the location and function of sweat glands and oil glands.
3. What are the two major body cavities? What two cavities are in each of these?

# 31.3 Homeostasis

All type systems of the body contribute to **homeostasis.** The digestive system takes in and digests food, providing nutrient molecules that enter the blood and replace the nutrients that are constantly being used by the body cells. The respiratory system adds oxygen to the blood and removes carbon dioxide. The amount of oxygen taken in and carbon dioxide given off can be increased to meet body needs. The liver and the kidneys contribute greatly to homeostasis. For example, immediately after glucose enters the blood, it can be removed by the liver and stored as glycogen. Later, glycogen is broken down to replace the glucose used by the body cells; in this way, the glucose composition of the blood remains constant. The hormone insulin, secreted by the pancreas, regulates glycogen storage. The kidneys are also under hormonal control as they excrete wastes and salts, substances that can affect the pH level of the blood.

Although homeostasis is, to a degree, controlled by hormones, it is ultimately controlled by the nervous system. In humans, the brain contains regulatory centers that control the function of other organs, maintaining homeostasis. These regulatory centers are often a part of a negative feedback system.

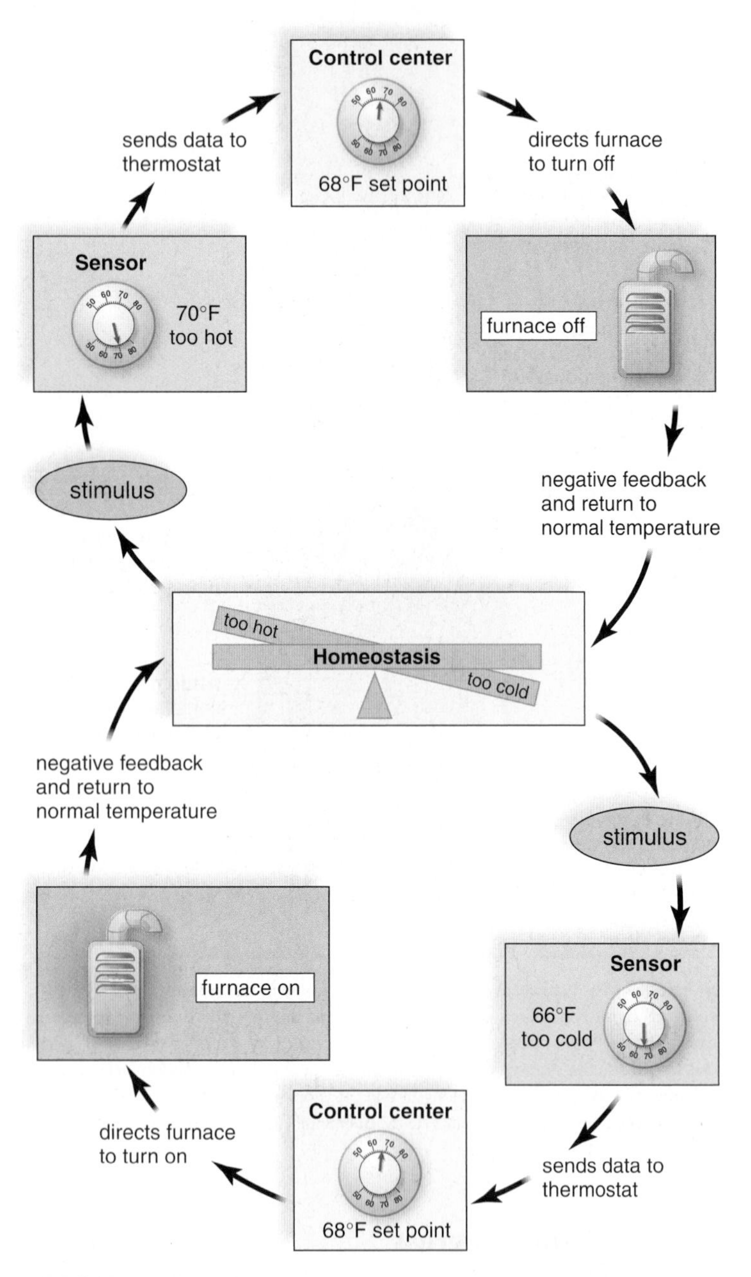


**FIGURE 31.10 Regulation of room temperature.**

This diagram shows how room temperature is returned to normal when the room becomes too hot (*above*) or too cold (*below*). The thermostat contains both the sensor and the control center. *Above:* The sensor detects that the room is too hot, and the control center turns the furnace off. The stimulus is no longer present when the temperature returns to normal. *Below:* The sensor detects that the room is too cold, and the control center turns the furnace on. The stimulus is no longer present when the temperature returns to normal.

## Negative Feedback

**Negative feedback** is the primary homeostatic mechanism that keeps a variable, such as the blood glucose level, close to a particular value, or set point. A homeostatic mechanism has at least two components: a sensor and a control center. The sensor detects a change in the internal environment; the control center then brings about an effect to bring conditions back to normal again. Now, the sensor is no longer activated. In other words, a negative feedback mechanism is present when the output of the system dampens the original stimulus. For example, when blood pressure rises, sensory receptors signal a control center in the brain. The center stops sending nerve impulses to the arterial walls, and they relax. Once the blood pressure drops, signals no longer go to the control center. A home heating system is often used to illustrate how a more complicated negative feedback mechanism works (Fig. 31.10). You set the thermostat at, say, 68°F. This is the *set point*. The thermostat contains a thermometer, a sensor that detects when the room temperature is above or below the set point. The thermostat also contains a control center; it turns the furnace off when the room is warm and turns it on when the room is cool. When the furnace is off, the room cools a bit, and when the furnace is on, the room warms a bit. In other words, typical of negative feedback mechanisms, there is a fluctuation above and below normal.

### *Human Example: Regulation of Body Temperature*

The sensor and control center for body temperature is located in a part of the brain called the hypothalamus. Notice that a negative feedback mechanism prevents change in the same direction; body temperature does not get warmer and warmer because warmth brings about a change toward a lower body temperature. Also, body temperature does not get colder and colder because a body temperature below normal brings about a change toward a warmer body temperature.

**Above Normal Temperature.** When the body temperature is above normal, the control center directs the blood vessels of the skin to dilate. This allows more blood to flow near the surface of the body, where heat can be lost to the environment. In addition, the nervous system activates the sweat glands, and the evaporation of sweat helps lower body temperature. Gradually, body temperature decreases to 98.6°F.

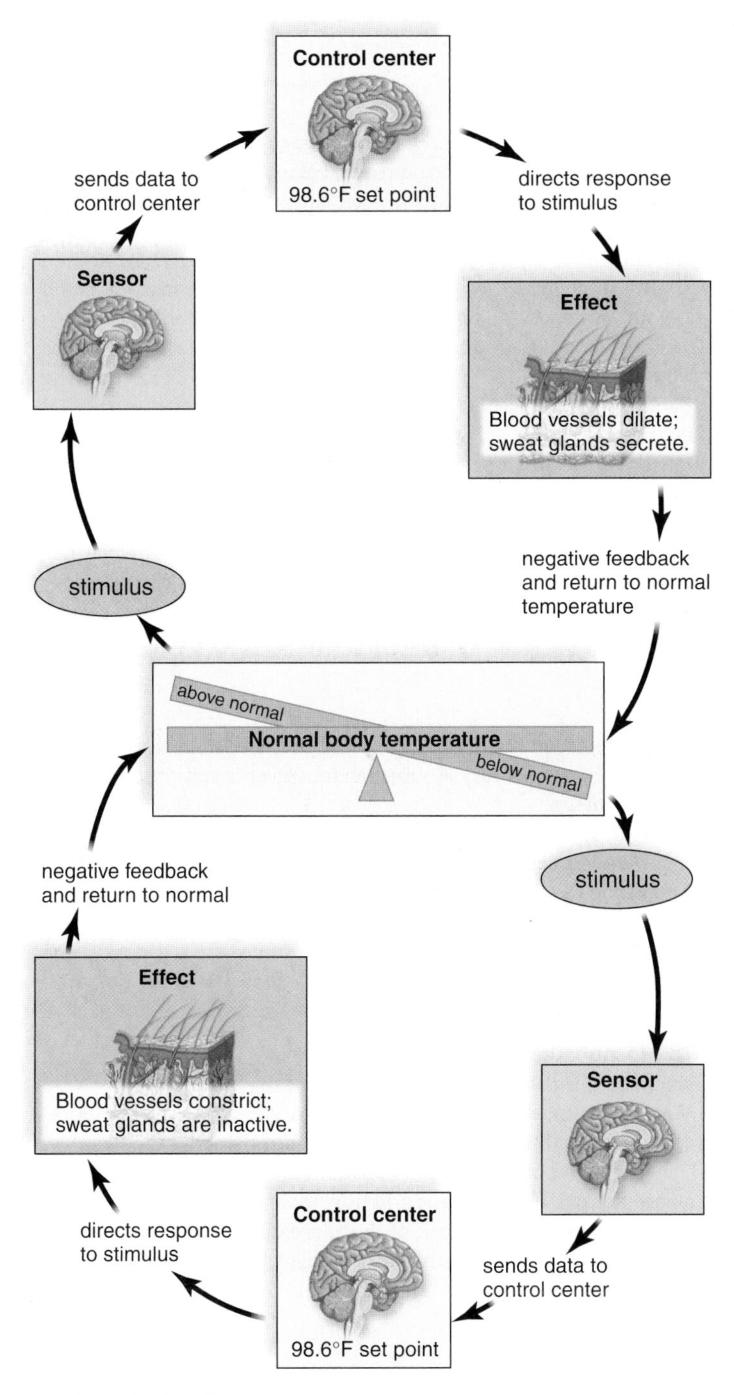


**FIGURE 31.11 Regulation of body temperature.**

*Above:* When body temperature rises above normal, the hypothalamus senses the change and causes blood vessels to dilate and sweat glands to secrete so that body temperature returns to normal. *Below:* When body temperature falls below normal, the hypothalamus senses the change and causes blood vessels to constrict. In addition, shivering may occur to bring body temperature back to normal. In this way, the original stimulus was removed.

**Below Normal Temperature.** When the body temperature falls below normal, the control center directs (via nerve impulses) the blood vessels of the skin to constrict (Fig. 31.11). This conserves heat. If body temperature falls even lower, the control center sends nerve impulses to the skeletal muscles, and shivering occurs. Shivering generates heat, and gradually body temperature rises to 98.6°C.

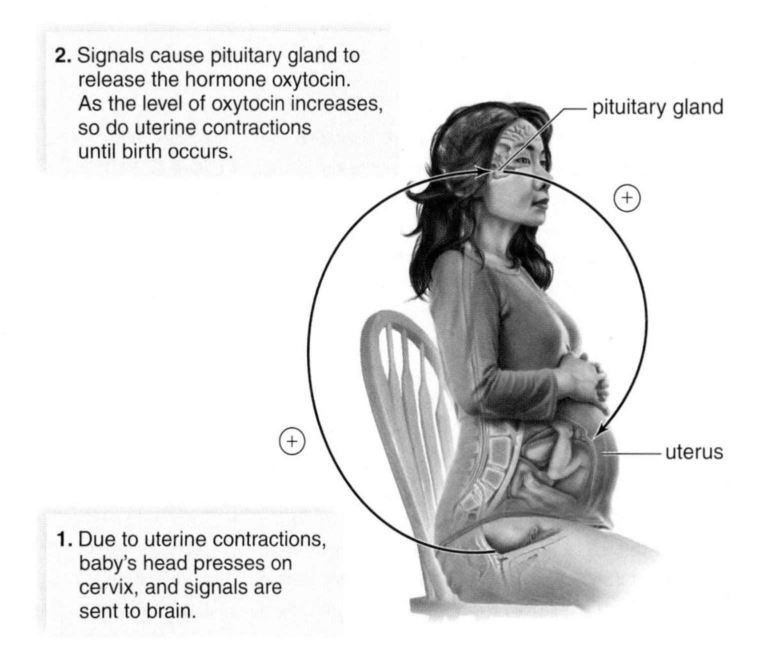


**FIGURE 31.12 Positive feedback.**

This diagram shows how positive feedback works. The signal causes a change in the same direction until there is a definite cutoff point, such as birth of a child.

When the temperature rises to normal, the control center is inactivated.

## Positive Feedback

**Positive feedback** is a mechanism that brings about an ever greater change in the same direction (Fig. 31.12). When a woman is giving birth, the head of the baby begins to press against the cervix (opening to the birth canal) stimulating sensory receptors there. When nerve impulses reach the brain, the brain causes the pituitary gland to secrete the hormone oxytocin. Oxytocin travels in the blood and causes the uterus to contract. As labor continues, the cervix is ever more stimulated and uterine contractions become ever stronger until birth occurs.

A positive feedback mechanism can be harmful, as when a fever causes metabolic changes that push the fever still higher. Death occurs at a body temperature of 113°F because cellular proteins denature at this temperature and metabolism stops. Still, positive feedback loops such as those involved in childbirth, blood clotting, and the stomach's digestion of protein assist the body in completing a process that has a definite cutoff point.

### Check Your Progress 31.3

1. What is homeostasis, and why is it important to body function?
2. How do the circulatory, respiratory, and urinary systems contribute to homeostasis?
3. How does negative feedback work?

# Connecting the Concepts

In this chapter, we have concentrated on humans, but homeostasis occurs even in the simplest of animals. Homeostasis in unicellular organisms and thin invertebrate animals occurs only at the cellular level, and each cell must carry out its own exchanges with the external environment to maintain a relative constancy of cytoplasm.

In complex animals, such as humans, there are localized boundaries where materials are exchanged with the external environment. Gas exchange usually occurs within lungs, food is digested within a digestive tract, and kidneys collect and excrete metabolic wastes. Exchange boundaries are an effective way to regulate the internal environment if there is a transport system to carry materials from one body part to another. Circulation in complex animals carries out this function (see p. 610).

Regulating mechanisms occur at all levels of organization. At the cellular level, the actions of enzymes are often controlled by feedback mechanisms. However, in animals with organ systems, the nervous and endocrine systems regulate the actions of organs. In addition, an animal's nervous system gathers and processes information about the external environment. Sensory receptors act as specialized boundaries through which external stimuli are received and converted into a form that can be processed by the nervous system. The information received and processed by the nervous system may then influence the organism's behavior in a way that contributes to homeostasis.

Homeostasis is so critical that without it the organism dies. Let us take a familiar example in humans. After eating, when the hormone insulin is present, glucose is removed from blood and stored in the liver as glycogen. In between eating, glycogen breakdown keeps the blood glucose level at just about 0.1%. When a person has type I diabetes, the pancreas fails to secrete insulin, and glucose is not stored in the liver. Worse yet, cells are unable to take up glucose even after eating, when there is a plentiful supply in the blood. Lacking glucose for cellular respiration, cells begin to break down fats with the result that acids are released in cells and enter the bloodstream. Now the person has acidosis—a low pH, which may hinder enzymatic activity to the point that cellular metabolism falters and the person dies.

## summary

### 31.1 Types of Tissues

During development, the zygote divides to produce cells that go on to become tissues composed of similar cells specialized for a particular function. Tissues make up organs, and organ systems make up the organism. This sequence describes the levels of organization within an organism.

Tissues are categorized into four groups. Epithelial tissue, which covers the body and lines cavities, is of three types: squamous, cuboidal, and columnar epithelium. Each type can be simple or stratified; it can also be glandular or have modifications, such as cilia. Epithelial tissue protects, absorbs, secretes, and excretes.

Connective tissue has a matrix between cells. Loose fibrous connective tissue and dense fibrous connective tissue contain fibroblasts and fibers. Loose fibrous connective tissue has both collagen and elastic fibers. Dense fibrous connective tissue, like that of tendons and ligaments, contains closely packed collagen fibers. In adipose tissue, the cells enlarge and store fat.

Both cartilage and bone have cells within lacunae, but the matrix for cartilage is more flexible than that for bone, which contains calcium salts. In bone, the lacunae lie in concentric circles within an osteon (or Haversian system) about a central canal. Blood is a connective tissue in which the matrix is a liquid called plasma.

Muscular (contractile) tissue can be smooth or striated (skeletal and cardiac), and involuntary (smooth and cardiac) or voluntary (skeletal). In humans, skeletal muscle is attached to bone, smooth muscle is in the wall of internal organs, and cardiac muscle makes up the heart.

Nervous tissue has one main type of conducting cell, the neuron, and several types of neuroglia. The majority of neurons have dendrites, a cell body, and an axon. The brain and spinal cord contain complete neurons, while nerves contain only axons. Axons are specialized to conduct nerve impulses.

### 31.2 Organs and Organ Systems

Organs contain various tissues. Skin is an organ that has two regions. Epidermis (stratified squamous epithelium) overlies the dermis (fibrous connective tissue containing sensory receptors, hair follicles, blood vessels, and nerves). A subcutaneous layer is composed of loose connective tissue.

Organ systems contain several organs. The organ systems of humans have specific functions and carry out the life processes that are common to all organisms.

The human body has two main cavities. The dorsal cavity contains the brain and spinal cord. The ventral cavity is divided into the thoracic cavity (heart and lungs) and the abdominal cavity (most other internal organs).

### 31.3 Homeostasis

Homeostasis is the dynamic equilibrium of the internal environment (blood and tissue fluid). All organ systems contribute to homeostasis, but special contributions are made by the liver, which keeps the blood glucose constant, and the kidneys, which regulate the pH. The nervous and hormonal systems regulate the other body systems. Both of these are controlled by negative feedback mechanisms, which result in slight fluctuations above and below desired levels. Body temperature is regulated by a center in the hypothalamus.

## understanding the terms

adipose tissue 580
basement membrane 578
blood 581
bone 580
cardiac muscle 582
cartilage 580
collagen fiber 579
columnar epithelium 578
compact bone 580
connective tissue 579
cuboidal epithelium 578
dense fibrous connective tissue 580
dermis 586
elastic cartilage 580
elastic fiber 579
endocrine gland 579
epidermis 585
epithelial tissue 578
exocrine gland 579
fibroblast 580
fibrocartilage 580
gland 579
hair follicle 586
homeostasis 588
hyaline cartilage 580
intercalated disk 582
lacuna 580

ligament 580
loose fibrous connective tissue 580
lymph 581
melanocyte 586
muscular (contractile) tissue 582
nail 586
negative feedback 588
nerve 583
nervous tissue 582
neuroglia 583
neuron 582
oil gland 586
organ 585
organ system 585
pathogen 581
platelet 581
positive feedback 589
red blood cell 581
reticular fiber 579
skeletal muscle 582
skin 585
smooth (visceral) muscle 582
spongy bone 581
squamous epithelium 578
striated 582
subcutaneous layer 585
sweat gland 587
tendon 580
tissue 578
tissue fluid 581
vitamin D 586
white blood cell 581

Match the terms to these definitions:

a. ______________ Fibrous connective tissue that joins bone to bone at a joint.
b. ______________ Outer region of the skin composed of stratified squamous epithelium.
c. ______________ Having striations, such as in cardiac and skeletal muscle.
d. ______________ Self-regulatory state in which imbalances result in a fluctuation above and below a mean.
e. ______________ Porous bone found at the ends of long bones where blood cells are formed.

## reviewing this chapter

1. Name the four major types of tissues. 578
2. Describe the structure and the functions of three types of simple epithelial tissue. 578–79
3. Describe the structure and the functions of six major types of connective tissue. 579–81
4. Describe the structure and the functions of three types of muscular tissue. 582
5. Nervous tissue contains what types of cells? 582–83
6. Describe the structure of skin, and state at least two functions of this organ. 585–87
7. In general terms, describe the locations of the human organ systems. 587
8. Tell how the various systems of the body contribute to homeostasis. 588–89
9. What is the function of sensors, the regulatory center, and effectors in a negative feedback mechanism? Why is it called negative feedback? 588–89

## testing yourself

Choose the best answer for each question.

1. Which of these pairs is incorrectly matched?
   a. tissues—like cells
   b. epithelial tissue—protection and absorption
   c. muscular tissue—contraction and conduction
   d. connective tissue—binding and support
   e. nervous tissue—conduction and message sending
2. Which of these is not a type of epithelial tissue?
   a. simple cuboidal and stratified columnar
   b. bone and cartilage
   c. stratified squamous and simple squamous
   d. pseudostratified and ciliated
   e. All of these are epithelial tissue.
3. Which tissue is more apt to line a lumen?
   a. epithelial tissue
   b. connective tissue
   c. nervous tissue
   d. muscular tissue
   e. only smooth muscle
4. Tendons and ligaments
   a. are connective tissue.
   b. are associated with the bones.
   c. are found in vertebrates.
   d. contain collagen.
   e. All of these are correct.
5. Which tissue has cells in lacunae?
   a. epithelial tissue
   b. cartilage
   c. bone
   d. smooth muscle
   e. Both b and c are correct.
6. Cardiac muscle is
   a. striated.
   b. involuntary.
   c. smooth.
   d. many fibers fused together.
   e. Both a and b are correct.
7. Which of these components of blood fights infection?
   a. red blood cells
   b. white blood cells
   c. platelets
   d. hydrogen ions
   e. All of these are correct.
8. Which of these body systems contribute to homeostasis?
   a. digestive and urinary systems
   b. respiratory and nervous systems
   c. nervous and endocrine systems
   d. immune and cardiovascular systems
   e. All of these are correct.
9. In a negative feedback mechanism,
   a. the output cancels the input.
   b. there is a fluctuation above and below the average.
   c. there is self-regulation.
   d. a regulatory center communicates with other body parts.
   e. All of these are correct.
10. When a human being is cold, the superficial blood vessels
    a. dilate, and the sweat glands are inactive.
    b. dilate, and the sweat glands are active.
    c. constrict, and the sweat glands are inactive.
    d. constrict, and the sweat glands are active.
    e. contract so that shivering occurs.

11. Give the name, the location, and the function for each of these tissues in the human body.
    a. type of epithelial tissue ______________
    b. type of muscular tissue ______________
    c. type of connective tissue ______________

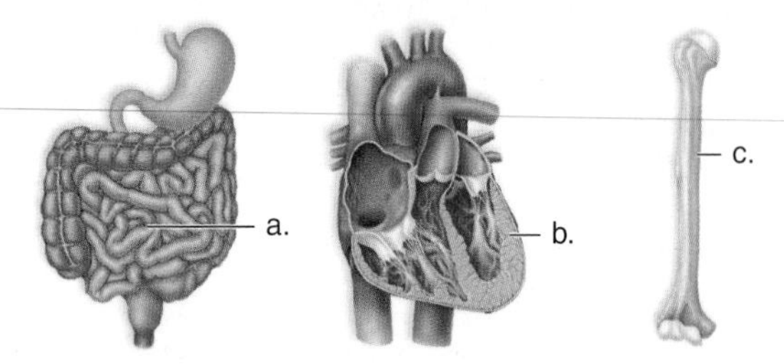


12. Which of these is a function of skin?
    a. temperature regulation
    b. manufacture of vitamin D
    c. collection of sensory input
    d. protection from invading pathogens
    e. All of these are correct.
13. Which of these is an example of negative feedback?
    a. Air conditioning goes off when room temperature lowers.
    b. Insulin decreases blood sugar levels after eating a meal.
    c. Heart rate increases when blood pressure drops.
    d. All of these are examples of negative feedback.
14. Which of these correctly describes a layer of the skin?
    a. The epidermis is simple squamous epithelium in which hair follicles develop and blood vessels expand when we are hot.
    b. The subcutaneous layer lies between the epidermis and the dermis. It contains adipose tissue, which keeps us warm.
    c. The dermis is a region of connective tissue that contains sensory receptors, nerve endings, and blood vessels.
    d. The skin has a special layer, still unnamed, in which there are all the accessory structures such as nails, hair, and various glands.
15. The ______________ separates the thoracic cavity from the abdominal cavity.
    a. liver
    b. pancreas
    c. diaphragm
    d. pleural membrane
    e. intestines

In questions 16–18, match each type of muscle tissue to as many terms in the key as possible.

**KEY:**

a. voluntary
b. involuntary
c. striated
d. nonstriated
e. spindle-shaped cells
f. branched cells
g. long, cylindrical cells

16. skeletal muscle
17. smooth muscle
18. cardiac muscle

In questions 19–22, match each description to the tissues in the key.

**KEY:**

a. loose fibrous connective tissue
b. hyaline cartilage
c. adipose tissue
d. compact bone

19. occurs in nose and walls of respiratory passages
20. occurs only within bones of skeleton
21. occurs beneath most epithelial layers
22. occurs beneath skin, and around organs, including the heart

## thinking scientifically

1. Many cancers develop from epithelial tissue. These include lung, colon, and skin cancers. What are two attributes of this tissue type that make cancer more likely to develop?
2. Bacterial or viral infections can cause a fever. Fevers occur when the hypothalamus changes its temperature set point. Signaling of the hypothalamus could be direct (from the infectious agent itself) or indirect (from the immune system). Which of these would enable the hypothalamus to respond to the greatest variety of infectious agents? Is there any disadvantage to such a signaling system?

## bioethical issue

### Organ Transplants

Despite widespread efforts to convince people of the need for organ donors, supply continues to lag far behind demand. One proposed strategy to bring supply and demand into better balance is to develop an "insurance" program for organs. In this program, participants would pay the "premium" by promising to donate their organs at death. They, in turn, would receive priority for transplants as their "benefit." To avoid the problem of too many high-risk people applying for this type of insurance, a medical exam would be required so that only people with a normal risk of requiring a transplant would be accepted. Do you suppose this system would be an improvement over the current system in which organ donation is voluntary, with no tangible benefit? Can you think of any other strategy that would be more effective for increasing organ donations?

## *Biology* website

The companion website for *Biology* provides a wealth of information organized and integrated by chapter. You will find practice tests, animations, videos, and much more that will complement your learning and understanding of general biology.

**http://www.mhhe.com/maderbiology10**

# 32

# Circulation and Cardiovascular Systems

*blood is a lifeline for human beings because it is the means by which needed supplies are transported to the cells of the body. Ordinarily, red blood cells, which transport oxygen, are far more numerous than the white blood cells, which defend the body from infection. But when leukemia is present, white blood cells proliferate wildly (see inset) and are so numerous that blood can't carry out its many functions, even though the heart is still functioning as it should.*

*We shall see in this chapter that many animals have a cardiovascular system in which the heart pumps blood about the body to all organs. The pumping of the heart is merely an auxiliary function because it is blood that transports gases and nutrients to and carries waste away from the cells of all organs, including those such as the lungs, digestive tract, and kidneys, which carry out exchanges between the external environment and blood. These exchanges keep the composition of blood relatively constant.*

Normally, red blood cells far outnumber white blood cells. When a cancer called leukemia is present, the blood cell composition is reversed (see inset).

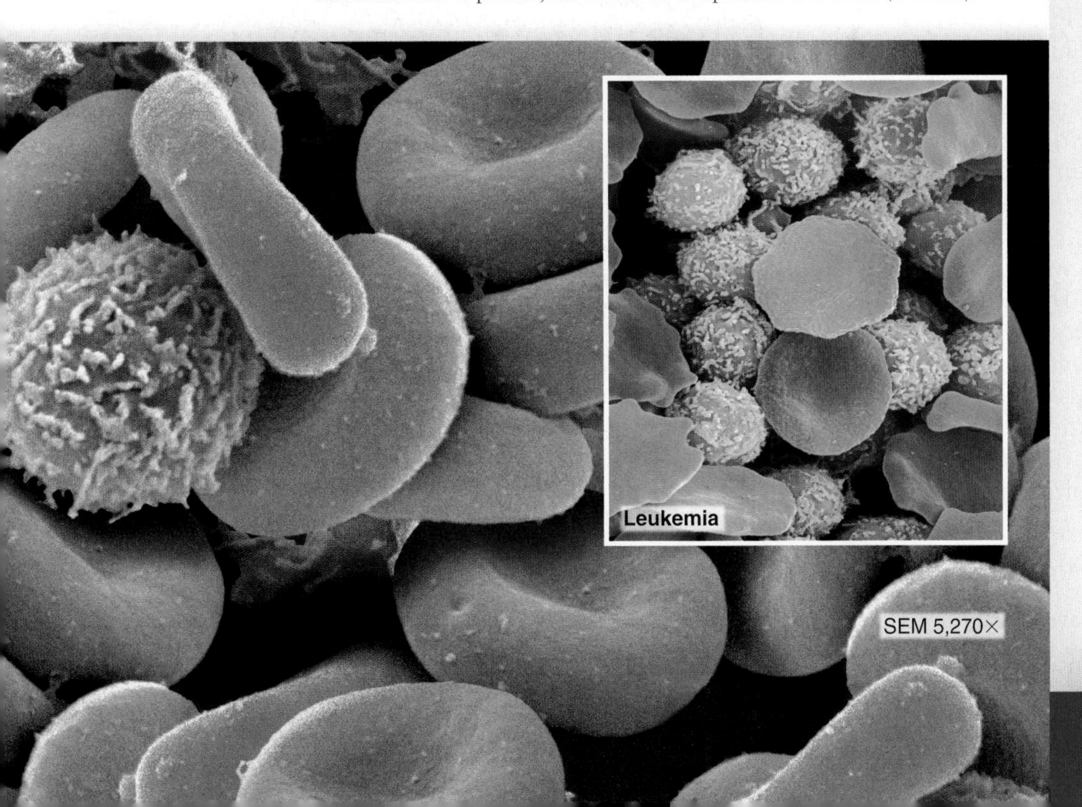


## concepts

# 32.1 Transport in Invertebrates

Some invertebrates such as sponges, cnidarians (e.g., hydras, sea aneomones), and flatworms (e.g., planarians) do not have a circulatory system (Fig. 32.1*a, b*). Their thin body wall makes a circulatory system unnecessary. In hydras, cells are either part of an external layer, or they line the gastrovascular cavity. Each cell is exposed to water and can independently exchange gases and rid itself of wastes. The cells that line the gastrovascular cavity are specialized to complete the digestive process. They pass nutrient molecules to other cells by diffusion. In a planarian, a trilobed gastrovascular cavity branches throughout the small, flattened body. No cell is very far from one of the three digestive branches, so nutrient molecules can diffuse from cell to cell. Similarly, diffusion meets the respiratory and excretory needs of the cells.

Pseudocoelomate invertebrates, such as nematodes, use the coelomic fluid of their body cavity for transport purposes. The coelomate echinoderms also rely on movement of coelomic fluid within a body cavity as a circulatory system (Fig. 32.1*c*).

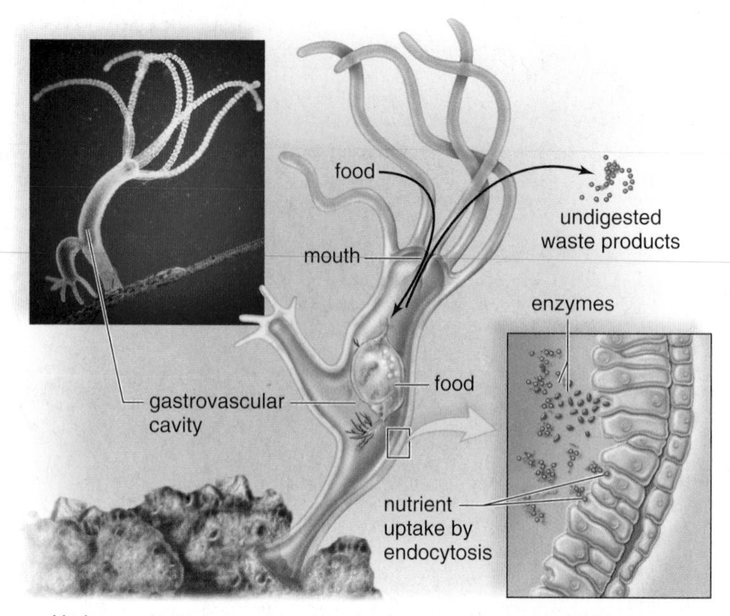


a. Hydra

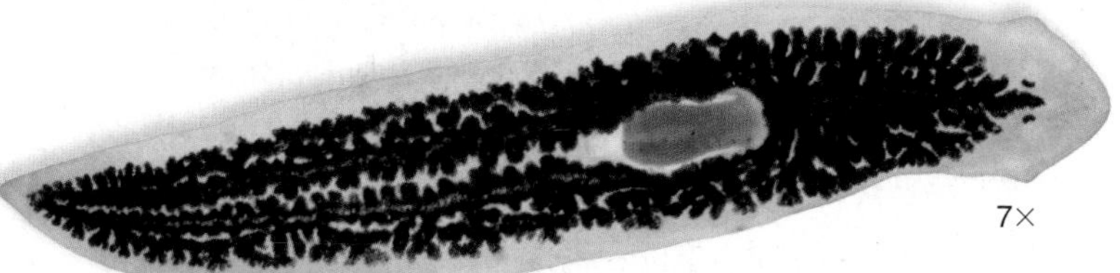


b. Flatworm

c. Red sea star, *Mediastar*

**FIGURE 32.1 Aquatic organisms without a circulatory system.**

**a.** In a hydra, a cnidarian, the gastrovascular cavity makes digested material available to the cells that line the cavity. These cells can also acquire oxygen from the watery contents of the cavity and discharge their wastes there. **b.** In a planarian, a flatworm, the gastrovascular cavity branches throughout the body, bringing nutrients to body cells. **c.** In a sea star, the coelomic fluid distributes oxygen and picks up wastes.

## Invertebrates with a Circulatory System

Most animals have a **circulatory system** that serves the needs of their cells. The circulatory system transports oxygen and nutrients, such as glucose and amino acids, to the cells. There it picks up wastes, which are later excreted from the body by the lungs or kidneys. There are two types of circulatory fluids: **blood,** which is always contained within blood vessels, and **hemolymph,** which flows into a body cavity called a hemocoel. Hemolymph is a mixture of blood and tissue fluid.

### *Open Circulatory System*

Hemolymph is seen in animals that have an **open circulatory system** that consists of blood vessels plus open spaces. For example, in most molluscs and arthropods, the heart pumps hemolymph via vessels into tissue spaces that are sometimes enlarged into saclike sinuses (Fig. 32.2*a*). Eventually, hemolymph drains back to the heart. In the grasshopper, an arthropod, the dorsal tubular heart pumps hemolymph into a dorsal aorta, which empties into the hemocoel. When the heart contracts, openings called ostia (sing., ostium) are closed; when the heart relaxes, the hemolymph is sucked back into the heart by way of the ostia. The hemolymph of a grasshopper is colorless because it does not contain hemoglobin or any other respiratory pigment. It carries nutrients but no oxygen. Oxygen is taken to cells, and carbon dioxide is removed from them by way of air tubes, called tracheae, which are found throughout the body. The tracheae provide efficient transport and delivery of respiratory gases, while at the same time restricting water loss.

### *Closed Circulatory System*

Blood is seen in animals that have a **closed circulatory system,** which consists of blood vessels only. For example, in annelids such as earthworms and in some molluscs such as squid and octopuses, blood, consisting of cells and plasma, (a liquid) is pumped by the heart into a system of blood vessels (Fig. 32.2*b*). Valves prevent the backward flow of blood. In the segmented earthworm, five pairs of anterior hearts (aortic arches) pump blood into the ventral blood vessel (an artery), which has a branch called a lateral vessel in every segment of the worm's body. Blood moves through these branches into capillaries, the thinnest of the blood vessels, where exchanges with tissue fluid take place. Both gas exchange and nutrient-for-waste exchange occur across the capillary walls. (No cell in the body of an animal with a closed circulatory system is far from a capillary.) In an earthworm, after leaving a capillary, blood moves from small veins into the dorsal blood vessel (a vein). This dorsal blood vessel returns blood to the heart for repumping.

The earthworm has red blood that contains the respiratory pigment hemoglobin. Hemoglobin is dissolved in the blood and is not contained within cells. The earthworm has no specialized organ, such as lungs, for gas exchange with the external environment. Gas exchange takes place across the body wall, which must always remain moist for this purpose.

### Check Your Progress 32.1

1. What is the primary function of a circulatory system?
2. Compare and contrast an open circulatory system of a grasshopper with a closed circulatory system of an earthworm. Why might it be advantageous for a grasshopper to use tracheae to transport oxygen to cells?

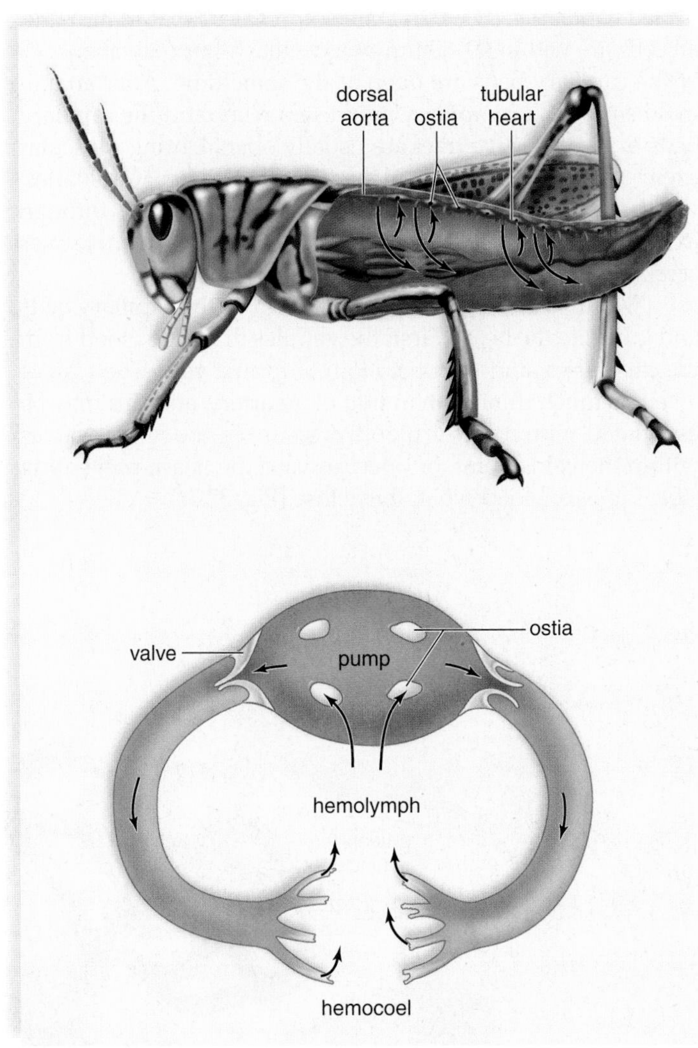


a. Open circulatory system

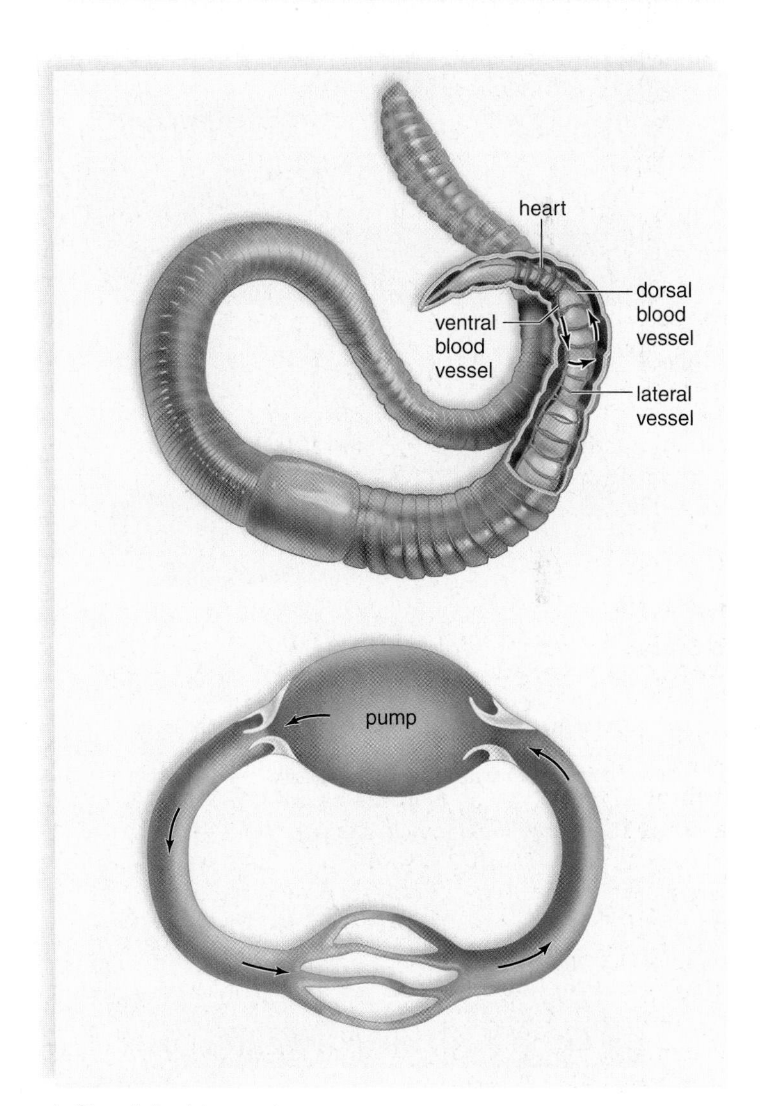


b. Closed circulatory system

**FIGURE 32.2 Open versus closed circulatory systems.**

**a.** (*above*) The grasshopper, an arthropod, has an open circulatory system. (*below*) A hemocoel is a body cavity filled with hemolymph, which freely bathes the internal organs. The heart, a pump, sends hemolymph out through vessels and collects it through ostia (openings). This open system probably could not supply oxygen to wing muscles rapidly enough. These muscles receive oxygen directly from tracheae (air tubes). **b.** (*above*) The earthworm, an annelid, has a closed circulatory system. The dorsal and ventral blood vessels are joined by five pairs of anterior hearts, which pump blood. (*below*) The lateral vessels distribute blood to the rest of the worm.

## 32.2 Transport in Vertebrates

All vertebrate animals have a closed circulatory system, which is called a **cardiovascular system** [Gk. *kardia,* heart; L. *vascular,* vessel]. It consists of a strong, muscular heart in which the atria (sing., atrium) receive blood and the muscular ventricles pump blood through the blood vessels. There are three kinds of blood vessels: **arteries,** which carry blood away from the heart; **capillaries** [L. *capillus,* hair], which exchange materials with tissue fluid; and **veins** [L. *vena,* blood vessel], which return blood to the heart (Fig. 32.3).

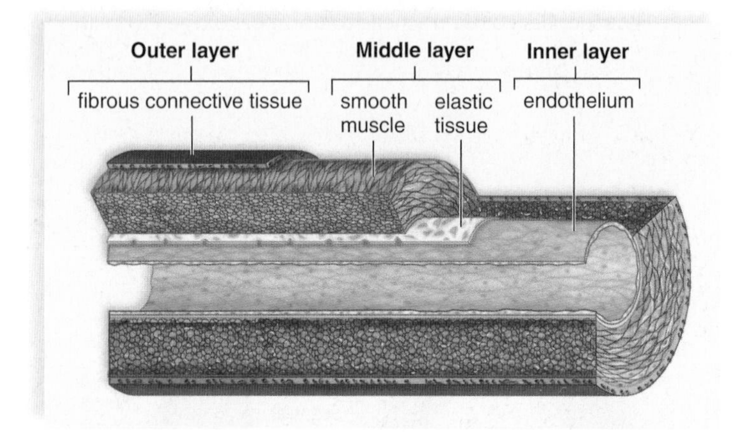


a. Artery

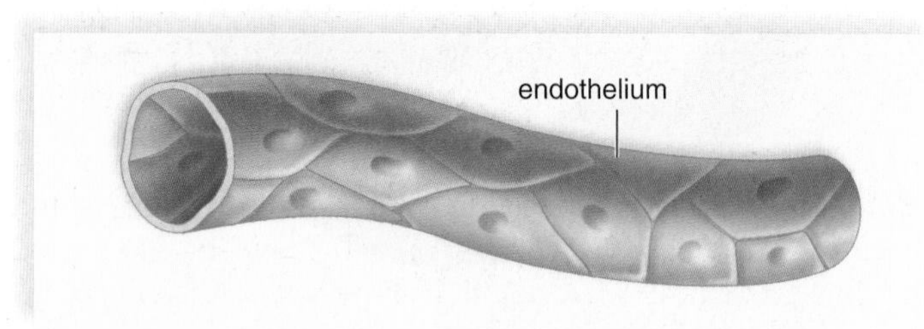


b. Capillary

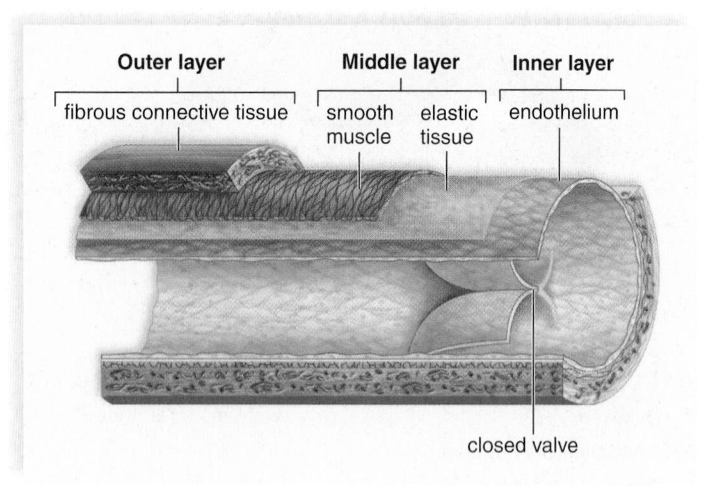


c. Vein

**FIGURE 32.3 Transport in vertebrates.**

**a.** Arteries have well-developed walls with a thick middle layer of elastic tissue and smooth muscle. **b.** Capillary walls are only one cell thick. **c.** Veins have flabby walls, particularly because the middle layer is not as thick as in arteries. Veins have valves, which ensure one-way flow of blood back to the heart.

An artery or a vein has three distinct layers (Fig. 32.3*a, c*). The outer layer consists of fibrous connective tissue, which is rich in elastic and collagen fibers. The middle layer is composed of smooth muscle and elastic tissue. The innermost layer, called the endothelium, is similar to squamous epithelium.

Arteries have thick walls, and those attached to the heart are resilient, meaning that they are able to expand and accommodate the sudden increase in blood volume that results after each heartbeat. **Arterioles** are small arteries whose diameter can be regulated by the nervous system. Arteriole constriction and dilation affect blood pressure in general. The greater the number of vessels dilated, the lower the blood pressure.

Arterioles branch into capillaries, which are extremely narrow, microscopic tubes with a wall composed of only one layer of cells. Capillary beds, which consist of many interconnected capillaries (Fig. 32.4), are so prevalent that in humans, all cells are within 60–80 μm of a capillary. But only about 5% of the capillary beds are open at the same time. After an animal has eaten, precapillary sphincters relax, and the capillary beds in the digestive tract are usually open. During muscular exercise, the capillary beds of the muscles are open. Capillaries, which are usually so narrow that red blood cells pass through in single file, allow exchange of nutrient and waste molecules across their thin walls.

**Venules** and veins collect blood from the capillary beds and take it to the heart. First, the venules drain the blood from the capillaries, and then they join to form a vein. The wall of a vein is much thinner than that of an artery, and this may be associated with a lower blood pressure in the veins. Valves within the veins point, or open, toward the heart, preventing a backflow of blood when they close (Fig. 32.3*c*).

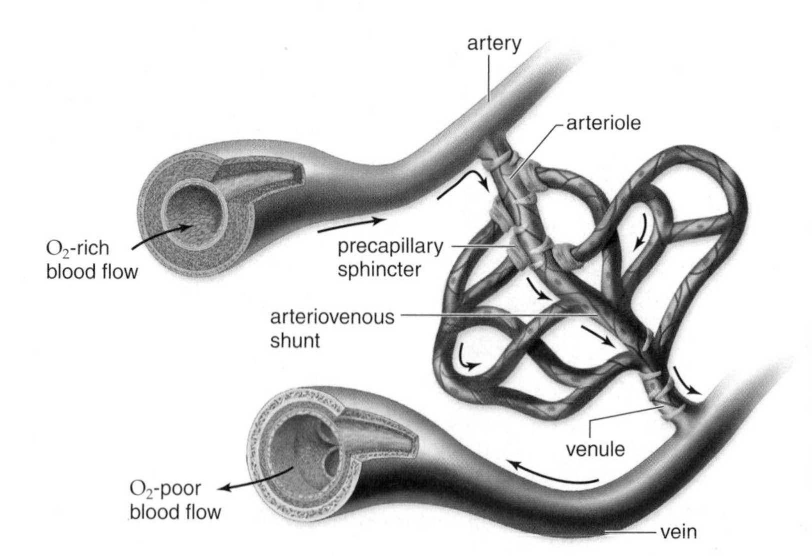


**FIGURE 32.4 Anatomy of a capillary bed.**

When a capillary bed is open, sphincter muscles are relaxed and blood flows through the capillaries. When precapillary sphincter muscles are contracted, the bed is closed and blood flows through an arteriovenous shunt that carries blood directly from an arteriole to a venule.

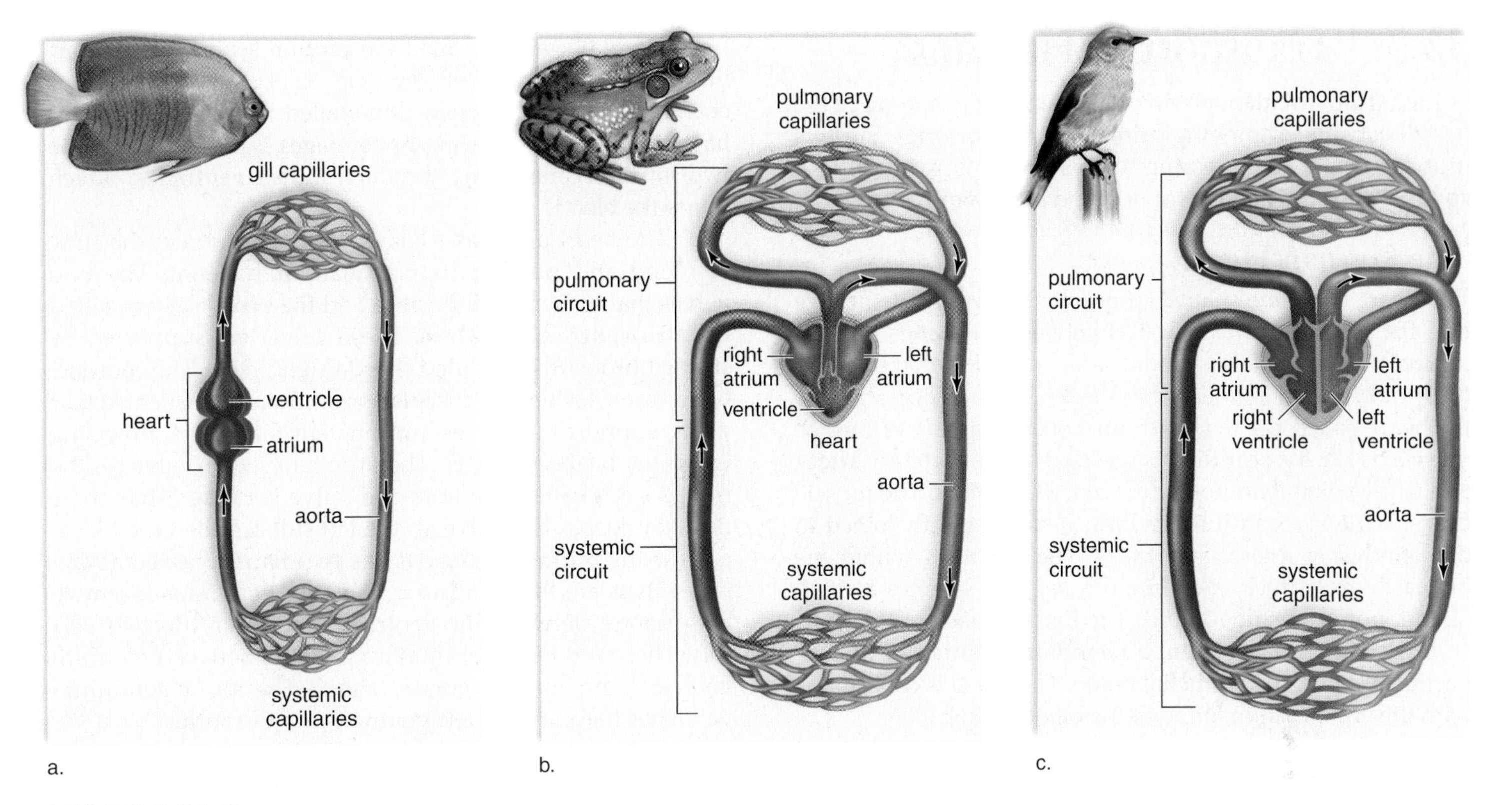


**FIGURE 32.5 Comparison of circulatory circuits in vertebrates.**

**a.** In fishes, the blood moves in a single circuit. Blood pressure created by the pumping of the heart is dissipated after the blood passes through the gill capillaries. This is a disadvantage of this one-circuit system. **b.** Amphibians and most reptiles have a two-circuit system in which the heart pumps blood to both the pulmonary capillaries in the lungs and the systemic capillaries in the body itself. Although there is a single ventricle, there is little mixing of $O_2$-rich and $O_2$-poor blood. **c.** The pulmonary and systemic circuits are completely separate in crocodiles (a reptile) and in birds and mammals, because the heart is divided by a septum into right and left halves. The right side pumps blood to the lungs, and the left side pumps blood to the body proper.

## Comparison of Circulatory Pathways

Two different types of circulatory pathways are seen among vertebrate animals. In fishes, blood follows a one-circuit (single-loop) circulatory pathway through the body. The heart has a single atrium and a single ventricle (Fig. 32.5*a*). The pumping action of the ventricle sends blood under pressure to the gills, where gas exchange occurs. After passing through the gills, blood returns to the dorsal aorta, which distributes blood throughout the body. Veins return $O_2$-poor blood to an enlarged chamber called the sinus venosus that leads to the atrium. The atrium pumps blood back to the ventricle. This single circulatory loop has an advantage in that the gill capillaries receive $O_2$-poor blood and the capillaries of the body, called systemic capillaries, receive fully $O_2$-rich blood. It is disadvantageous in that after leaving the gills the blood is under reduced pressure.

As a result of evolutionary changes, the other vertebrates have a two-circuit (double-loop) circulatory pathway. The heart pumps blood to the tissues, called a **systemic circuit,** and also pumps blood to the lungs, called a **pulmonary circuit** [L. *pulmonarius*, of the lungs]. This double pumping action is an adaptation to breathing air on land.

In amphibians, the heart has two atria and a single ventricle (Fig. 32.5*b*). The sinus venosus collects $O_2$-poor blood returning via the veins and pumps it to the right atrium. $O_2$-rich blood returning from the lungs passes to the left atrium. Both of the atria empty into the single ventricle. $O_2$-rich and $O_2$-poor blood are kept somewhat separate because $O_2$-poor blood is pumped out of the ventricle before $O_2$-rich blood enters. The $O_2$-rich blood is pumped out of the ventricle for distribution to the body. The $O_2$-poor blood is delivered to the lungs and, perhaps, to the skin for recharging with oxygen.

In most reptiles, a septum partially divides the ventricle. In these animals, mixing of $O_2$-rich and $O_2$-poor blood is kept to a minimum. In crocodilians (alligators and crocodiles), the septum completely separates the ventricle. These reptiles have a four-chambered heart.

In all birds and mammals, as well as crocodilians, the heart is divided into left and right halves (Fig. 32.5*c*). The right ventricle pumps blood to the lungs, and the larger left ventricle pumps blood to the rest of the body. This arrangement provides adequate blood pressure for both the pulmonary and systemic circuits.

### Check Your Progress 32.2

1. List and describe the functions of the three types of vessels in a cardiovascular system.
2. Contrast a one-circuit circulatory pathway with a two-circuit pathway.

# 32.3 Transport in Humans

In the cardiovascular system of humans, the pumping of the heart keeps blood moving primarily in the arteries. Skeletal muscle contraction pressing against veins is primarily responsible for the movement of blood in the veins.

## The Human Heart

The **heart** is a cone-shaped, muscular organ about the size of a fist (Fig. 32.6). It is located between the lungs directly behind the sternum (breastbone) and is tilted so that the apex (the pointed end) is oriented to the left. The major portion of the heart, called the myocardium, consists largely of cardiac muscle tissue. Myocardium is serviced by the coronary artery and cardiac vein and not by the blood it pumps. The muscle fibers of the myocardium are branched and tightly joined to one another at intercalated disks. The heart lies within the pericardium, a thick, membranous sac that secretes a small quantity of lubricating liquid. The inner surface of the heart is lined with endocardium, a membrane composed of connective tissue and endothelial tissue. The lining is continuous with the endothelium lining of the blood vessels.

Internally, a wall called the **septum** separates the heart into a right side and a left side (Fig. 32.7). The heart has four chambers. The two upper, thin-walled atria (sing., **atrium**) have wrinkled, protruding appendages called auricles. The two lower chambers are the thick-walled **ventricles,** which pump the blood.

The heart also has four valves, which direct the flow of blood and prevent its backward movement. The two valves that lie between the atria and the ventricles are called the **atrioventricular valves.** These valves are supported by strong fibrous strings called chordae tendineae. The chordae, which are attached to muscular projections of the ventricular walls, support the valves and prevent them from inverting when the heart contracts. The atrioventricular valve on the right side is called the tricuspid valve because it has three flaps, or cusps. The valve on the left side is called the bicuspid (or the mitral) because it has two flaps. The remaining two valves are the **semilunar valves,** whose flaps resemble half-moons, between the ventricles and their attached vessels. The pulmonary semilunar valve lies between the right ventricle and the pulmonary trunk. The aortic semilunar valve lies between the left ventricle and the aorta.

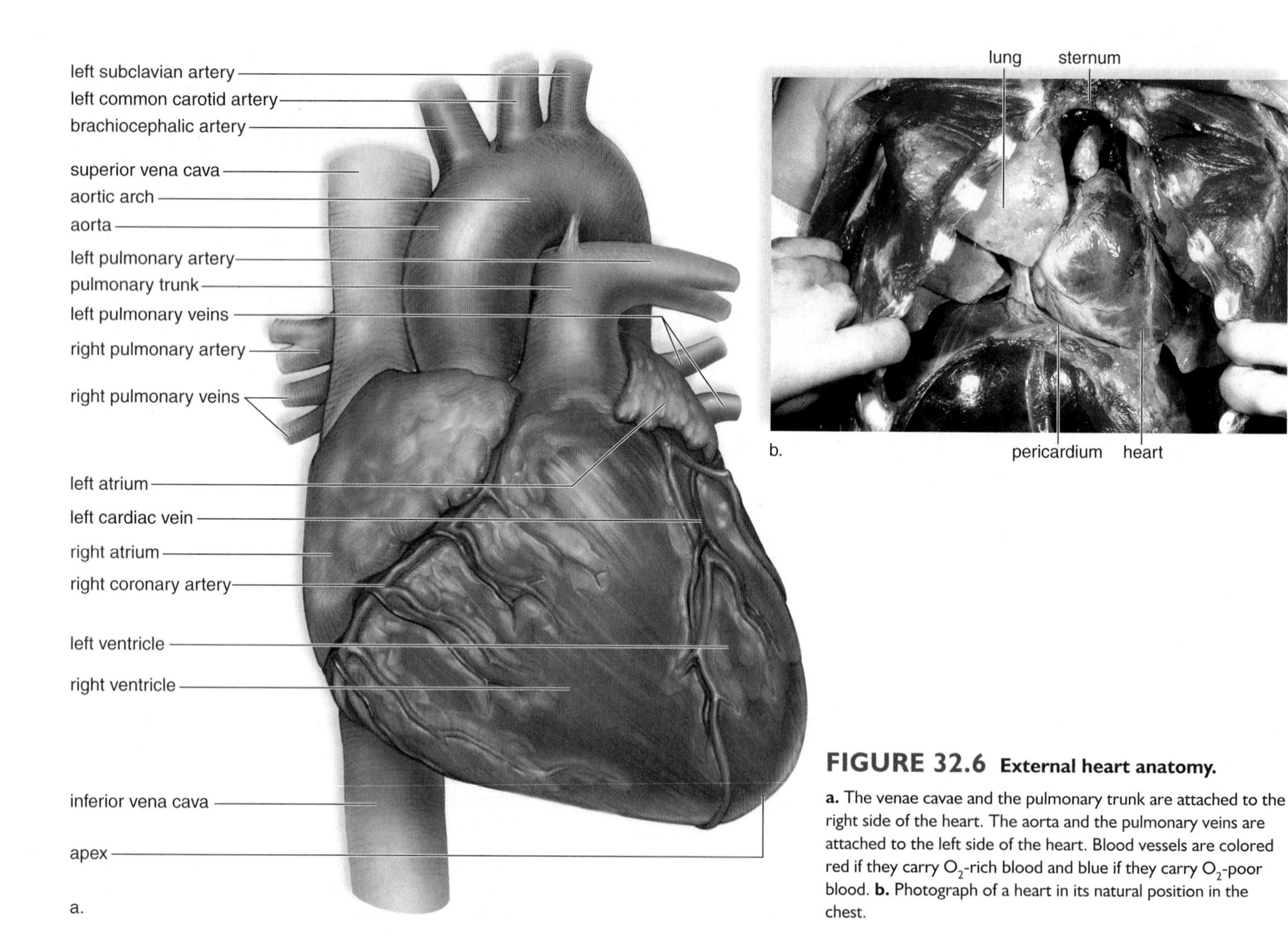


**FIGURE 32.6 External heart anatomy.**

**a.** The venae cavae and the pulmonary trunk are attached to the right side of the heart. The aorta and the pulmonary veins are attached to the left side of the heart. Blood vessels are colored red if they carry $O_2$-rich blood and blue if they carry $O_2$-poor blood. **b.** Photograph of a heart in its natural position in the chest.

## *Path of Blood Through the Heart*

Even though the presence of intercalated disks (Fig. 32.7*b*) between cardiac muscle cells allows both atria and then both ventricles to contract simultaneously, we can travel the path of blood through the heart in the following manner:

- The superior vena cava and the inferior vena cava, which carry $O_2$-poor blood that is relatively high in carbon dioxide, enter the right atrium.
- The right atrium sends blood through an atrioventricular valve (the tricuspid valve) to the right ventricle.
- The right ventricle sends blood through the pulmonary semilunar valve into the pulmonary trunk and the two pulmonary arteries to the lungs.
- Four pulmonary veins, which carry $O_2$-rich blood, enter the left atrium.
- The left atrium sends blood through an atrioventricular valve (the bicuspid or mitral valve) to the left ventricle.
- The left ventricle sends blood through the aortic semilunar valve into the aorta to the body proper.

From this description, it is obvious that $O_2$-poor blood never mixes with $O_2$-rich blood and that blood must go through the lungs in order to pass from the right side to the left side of the heart. In fact, the heart is a double pump because the right ventricle of the heart sends blood into the pulmonary circuit, and the left ventricle sends blood into the systemic circuit. Since the left ventricle has the harder job of pumping blood to the entire body, its walls are thicker than those of the right ventricle, which pumps blood a relatively short distance to the lungs. Some people associate $O_2$-rich blood with all arteries and $O_2$-poor poor blood with all veins, but this idea is incorrect: Pulmonary arteries and pulmonary veins are just the reverse. That is why pulmonary arteries are colored blue and pulmonary veins are colored red in Figures 32.6 and 32.7.

The pumping of the heart sends blood out under pressure into the arteries. Because the left side of the heart is the stronger pump, blood pressure is greatest in the aorta. Blood pressure then decreases as the cross-sectional area of arteries and then arterioles increases. Therefore, a different mechanism is needed to move blood in the veins, as we shall discuss.

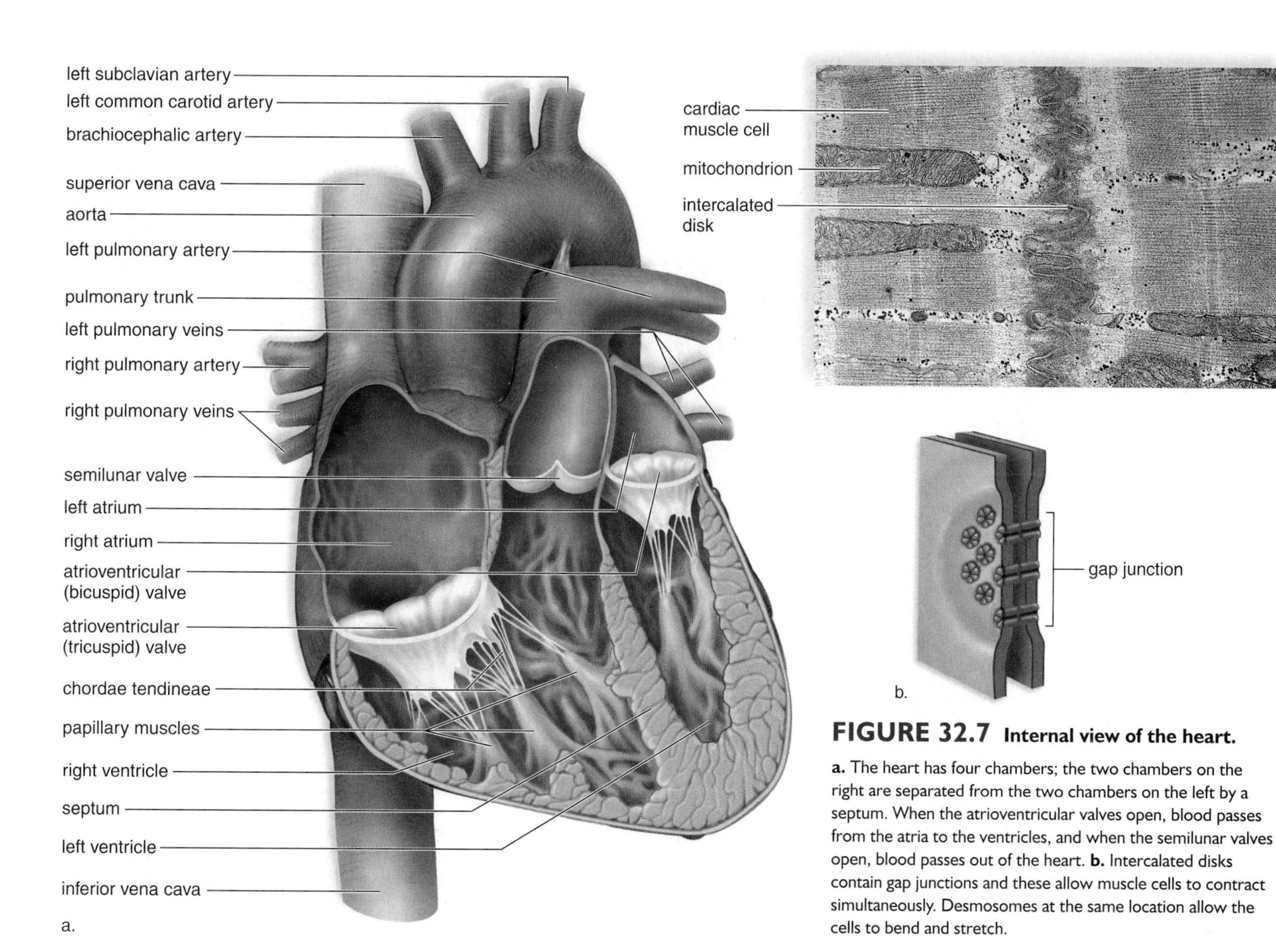


**FIGURE 32.7 Internal view of the heart.**

**a.** The heart has four chambers; the two chambers on the right are separated from the two chambers on the left by a septum. When the atrioventricular valves open, blood passes from the atria to the ventricles, and when the semilunar valves open, blood passes out of the heart. **b.** Intercalated disks contain gap junctions and these allow muscle cells to contract simultaneously. Desmosomes at the same location allow the cells to bend and stretch.

### *The Heartbeat*

The average human heart contracts, or beats, about 70 times a minute, or 2.5 billion times in a lifetime, and each heartbeat lasts about 0.85 second. The term **systole** [Gk. *systole*, contraction] refers to contraction of the heart chambers, and the word **diastole** [Gk. *diastole*, dilation, spreading] refers to relaxation of these chambers. Each heartbeat, or **cardiac cycle,** consists of the following phases, which are also depicted in Figure 32.8:

| Cardiac Cycle | | |
|---|---|---|
| **Time** | **Atria** | **Ventricles** |
| 0.15 sec | Systole | Diastole |
| 0.30 sec | Diastole | Systole |
| 0.40 sec | Diastole | Diastole |

First the atria contract (while the ventricles relax), then the ventricles contract (while the atria relax), and then all chambers rest. Note that the heart is in diastole about 50% of the time. The short systole of the atria is appropriate since the atria send blood only into the ventricles. It is the muscular ventricles that actually pump blood out into the cardiovascular system proper. When the word *systole* is used alone, it usually refers to the left ventricular systole. The volume of blood that the left ventricle pumps per minute into the systemic circuit is called the **cardiac output.** A person with a heartbeat of 70 beats per minute has a cardiac output of 5.25 liters a minute. This is almost equivalent to the amount of blood in the body. During heavy exercise, the cardiac output can increase manyfold.

When the heart beats, the familiar lub-dub sound is heard as the valves of the heart close. The longer and lower-pitched *lub* is caused by vibrations of the heart when the atrioventricular valves close due to ventricular contraction. The shorter and sharper *dub* is heard when the semilunar valves close due to back pressure of blood in the arteries. A heart murmur, a slight slush sound after the lub, is often due to ineffective valves, which allow blood to pass back into the atria after the atrioventricular valves have closed.

The **pulse** is a wave effect that passes down the walls of the arterial blood vessels when the aorta expands and then recoils following ventricular systole. Because there is one arterial pulse per ventricular systole, the arterial pulse rate can be used to determine the heart rate.

The rhythmic contraction of the heart is due to the **cardiac conduction system.** Nodal tissue, which has both

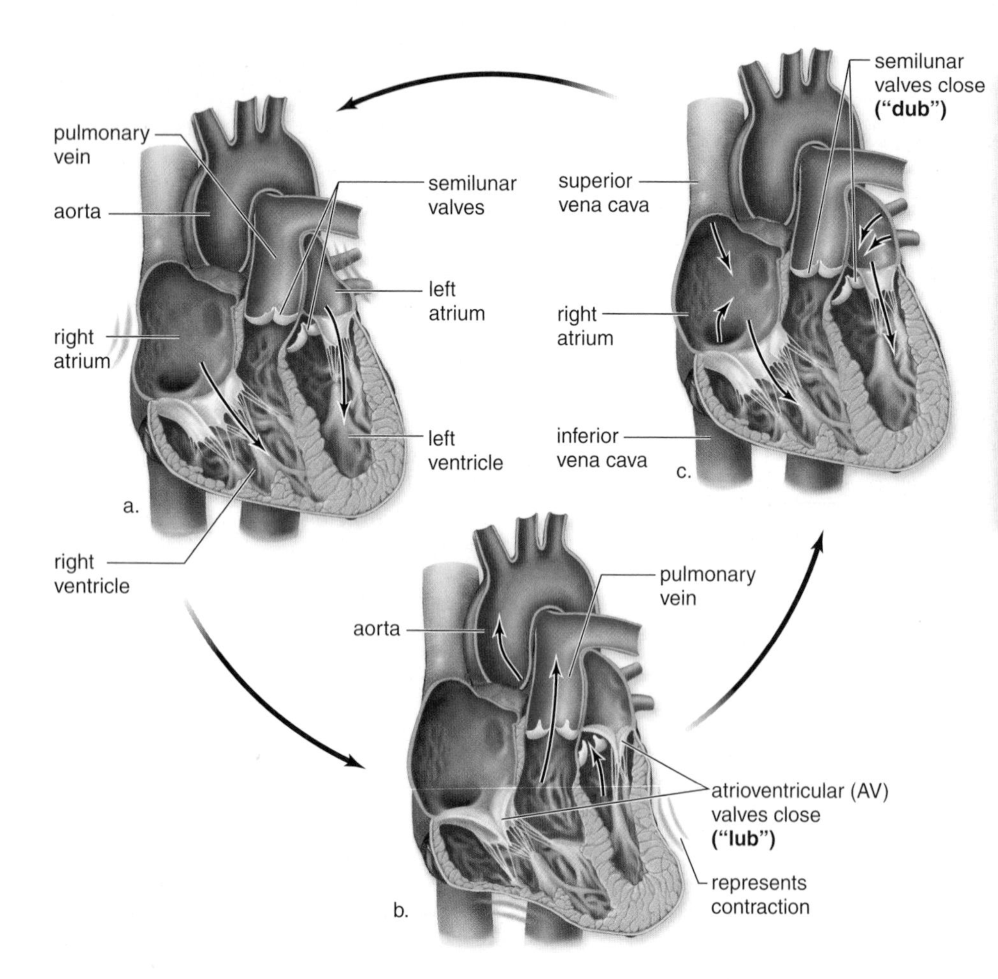


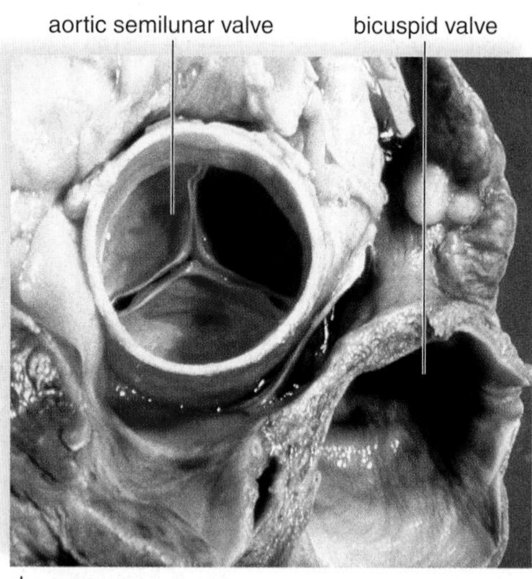


**FIGURE 32.8 Stages in the cardiac cycle.**

**a.** When the atria contract, the ventricles are relaxed and filling with blood. The atrioventricular valves are open, and the semilunar valves are closed. **b.** When the ventricles contract, the atrioventricular valves are closed, the semilunar valves are open, and the blood is pumped into the pulmonary trunk and aorta. **c.** When the heart is relaxed, both the atria and the ventricles are filling with blood. The atrioventricular valves are open, and the semilunar valves are closed. **d.** Aortic semilunar valve and bicuspid valve, an atrioventricular valve on left.

muscular and nervous characteristics, is a unique type of cardiac muscle located in two regions of the heart. The SA (sinoatrial) node is found in the upper dorsal wall of the right atrium; the AV (atrioventricular) node is found in the base of the right atrium very near the septum (Fig. 32.9*a*). The SA node initiates the heartbeat and every 0.85 second automatically sends out an excitation impulse, which causes the atria to contract. Therefore, the SA node is called the **cardiac pacemaker** because it usually keeps the heartbeat regular. When the impulse reaches the AV node, the AV node signals the ventricles to contract by way of large fibers terminating in the more numerous and smaller Purkinje fibers. Although the beat of the heart is intrinsic, it is regulated by the nervous system, which can increase or decrease the heartbeat rate. The hormones epinephrine and norepinephrine, which are released by the adrenal glands, which are endocrine glands, also stimulate the heart. During exercise, for example, the heart pumps faster and stronger due to nervous stimulation and due to the release of epinephrine and norepinephrine.

**The Electrocardiogram.** An **electrocardiogram (ECG)** is a recording of the electrical changes that occur in the myocardium during a cardiac cycle. Body fluids contain ions that conduct electrical currents, and therefore the electrical changes in the myocardium can be detected on the skin's surface. When an electrocardiogram is being taken, electrodes placed on the skin are connected by wires to an instrument that detects the myocardium's electrical changes. Thereafter, a pattern appears that reflects the contractions of the heart. Figure 32.9*b* depicts the pattern that results from a normal cardiac cycle.

When the SA node triggers an impulse, the atrial fibers produce an electrical change called the P wave. The P wave indicates that the atria are about to contract. After that, the QRS complex signals that the ventricles are about to contract and the atria are relaxing. The electrical changes that occur as the ventricular muscle fibers recover produce the T wave.

Various types of abnormalities can be detected by an electrocardiogram. One of these, called ventricular fibrillation, is caused by uncoordinated contraction of the ventricles (Fig. 32.9*c*). Ventricular fibrillation is of special interest because it can be caused by an injury or drug overdose. It is the most common cause of sudden cardiac death in a seemingly healthy person. Once the ventricles are fibrillating, they have to be defibrillated by applying a strong electric current for a short period of time. Then the SA node may be able to reestablish a coordinated beat.

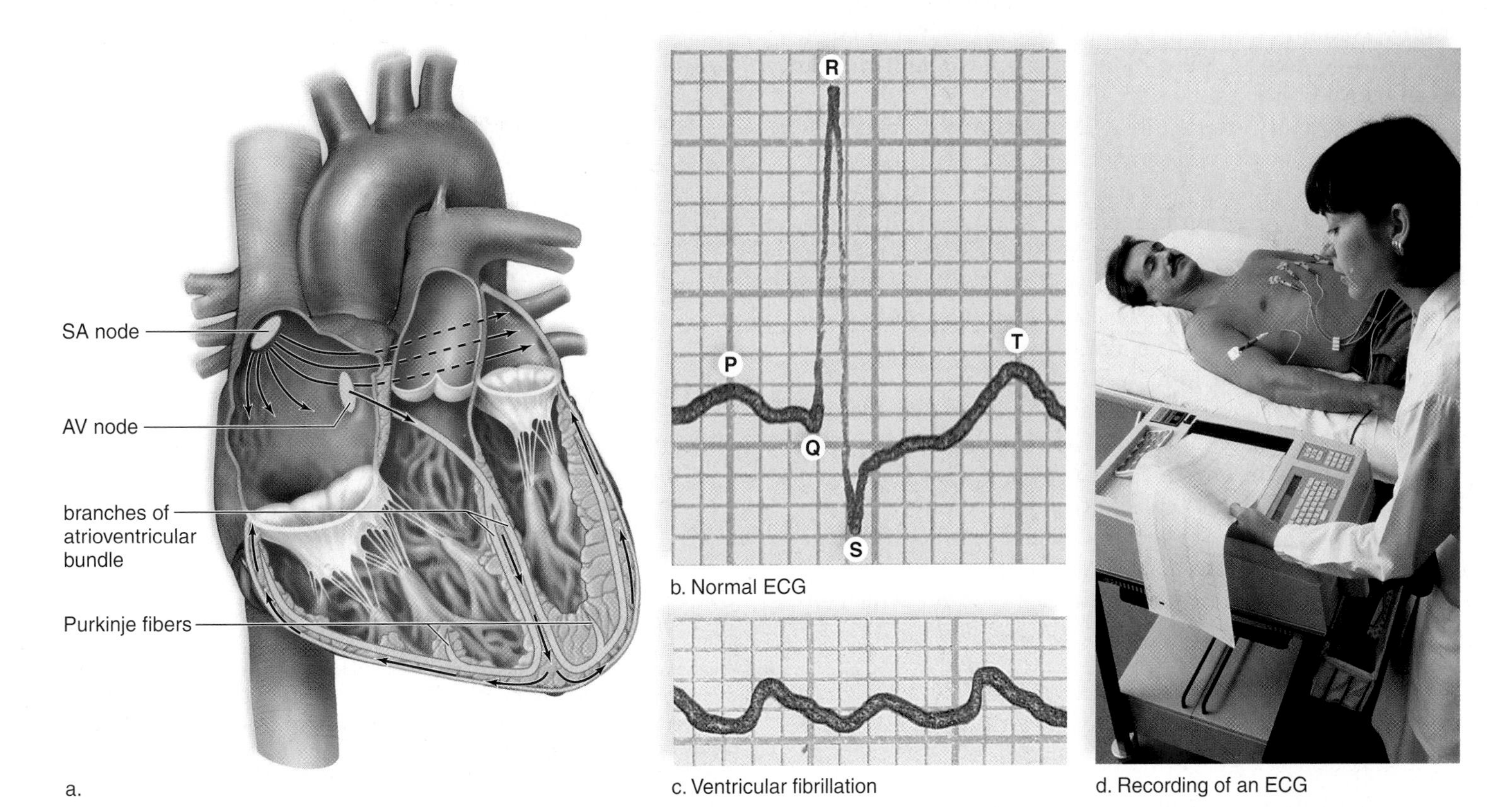


**FIGURE 32.9 Conduction system of the heart.**

**a.** The SA node sends out a stimulus (black arrows), which causes the atria to contract. When this stimulus reaches the AV node, it signals the ventricles to contract. Impulses pass down the two branches of the atrioventricular bundle to the Purkinje fibers, and thereafter the ventricles contract. **b.** A normal ECG usually indicates that the heart is functioning properly. The P wave occurs just prior to atrial contraction; the QRS complex occurs just prior to ventricular contraction; and the T wave occurs when the ventricles are recovering from contraction. **c.** Ventricular fibrillation produces an irregular electrocardiogram due to irregular stimulation of the ventricles. **d.** The recording of an ECG.

## Vascular Pathways

The human cardiovascular system includes two major circular pathways, the pulmonary circuit and the systemic circuit (Fig. 32.10).

### *The Pulmonary Circuit*

In the pulmonary circuit, the path of blood can be traced as follows: $O_2$-poor blood from all regions of the body collects in the right atrium and then passes into the right ventricle, which pumps it into the pulmonary trunk. The pulmonary trunk divides into the right and left pulmonary arteries, which carry blood to the lungs. As blood passes through pulmonary capillaries, carbon dioxide is given off and oxygen is picked up. $O_2$-rich blood returns to the left atrium of the heart, through pulmonary venules that join to form pulmonary veins.

### *The Systemic Circuit*

The **aorta** [L. *aorte*, great artery] and the **venae cavae** (sing., vena cava) [L. *vena*, blood vessel, and *cavus*, hollow] are the major blood vessels in the systemic circuit. To trace the path of blood to any organ in the body, you need only start with the left ventricle, mention the aorta, the proper branch of the aorta, the organ, and the vein returning blood to the vena cava, which enters the right atrium. In the systemic circuit, arteries contain $O_2$-rich blood and have a bright red color, but veins contain $O_2$-poor blood and appear dull red or, when viewed through the skin, blue.

The coronary arteries are extremely important because they serve the heart muscle itself (see Fig. 32.6). The coronary arteries arise from the aorta just above the aortic semilunar valve. They lie on the exterior surface of the heart, where they branch into arterioles and then capillaries. In the capillary beds, nutrients, wastes, and gases are exchanged between the blood and the tissues. The capillary beds enter venules, which join to form the cardiac veins, and these empty into the right atrium.

A **portal system** [L. *porto*, carry, transport] is one that begins and ends in capillaries. The hepatic portal system takes blood from the intestines to the liver. The liver, an organ of homeostasis, modifies substances absorbed by the intestines, removes toxins and bacteria picked up from the intestines, and monitors the normal composition of the blood. Blood leaves the liver by way of the hepatic vein, which enters the inferior vena cava.

**Tracing the Path of Blood.** Branches from the aorta go to the organs and major body regions. For example, this is the path of blood to and from the lower legs:

> left ventricle—aorta—common iliac artery—femoral artery—lower leg capillaries—femoral vein—common iliac vein—inferior vena cava—right atrium

In most instances, the artery and the vein that serve the same region are given the same name. What happens in between the artery and the vein? Arterioles from the artery branch into capillaries, where exchange takes place, and then venules join to form the vein that enters a vena cava. An exception occurs between the digestive tract and the liver, because blood must pass through two sets of capillaries because of the hepatic portal system.

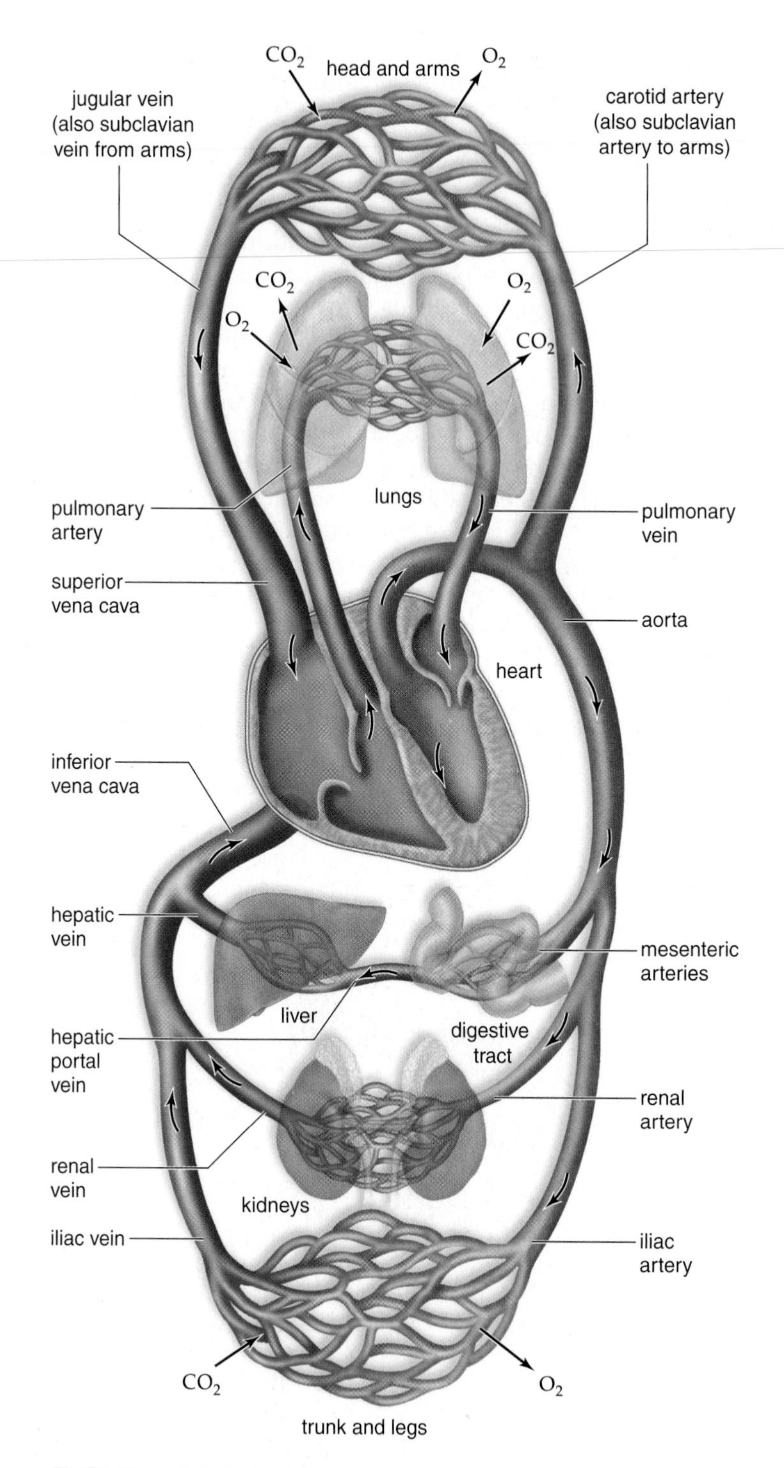


**FIGURE 32.10 Path of blood.**

When tracing blood from the right to the left side of the heart in the pulmonary circuit, you must mention the pulmonary vessels. When tracing blood from the digestive tract to the right atrium in the systemic circuit, you must mention the hepatic portal vein, the hepatic vein, and the inferior vena cava. The blue-colored vessels carry $O_2$-poor blood, and the red-colored vessels carry $O_2$-rich blood; the arrows indicate the flow of blood.

## Blood Pressure

When the left ventricle contracts, blood is forced into the aorta and then other systemic arteries under pressure. Systolic pressure results from blood being forced into the arteries during ventricular systole, and diastolic pressure is the pressure in the arteries during ventricular diastole. Human **blood pressure** can be measured with a **sphygmomanometer,** which has a pressure cuff that determines the amount of pressure required to stop the flow of blood through an artery. Blood pressure is normally measured on the brachial artery, an artery in the upper arm. Today, digital manometers are often used to take one's blood pressure instead. Blood pressure is given in millimeters of mercury (mm Hg). A blood pressure reading consists of two numbers—for example, 120/80—that represent systolic and diastolic pressures, respectively.

As blood flows from the aorta into the various arteries and arterioles, blood pressure falls. Also, the difference between systolic and diastolic pressure gradually diminishes. In the capillaries, there is a slow, fairly even flow of blood. This may be related to the very high total cross-sectional area of the capillaries (Fig. 32.11). It has been calculated that if all the blood vessels in a human were connected end to end, the total distance would reach around the Earth at the equator two times. A large portion of this distance would be due to the quantity of capillaries.

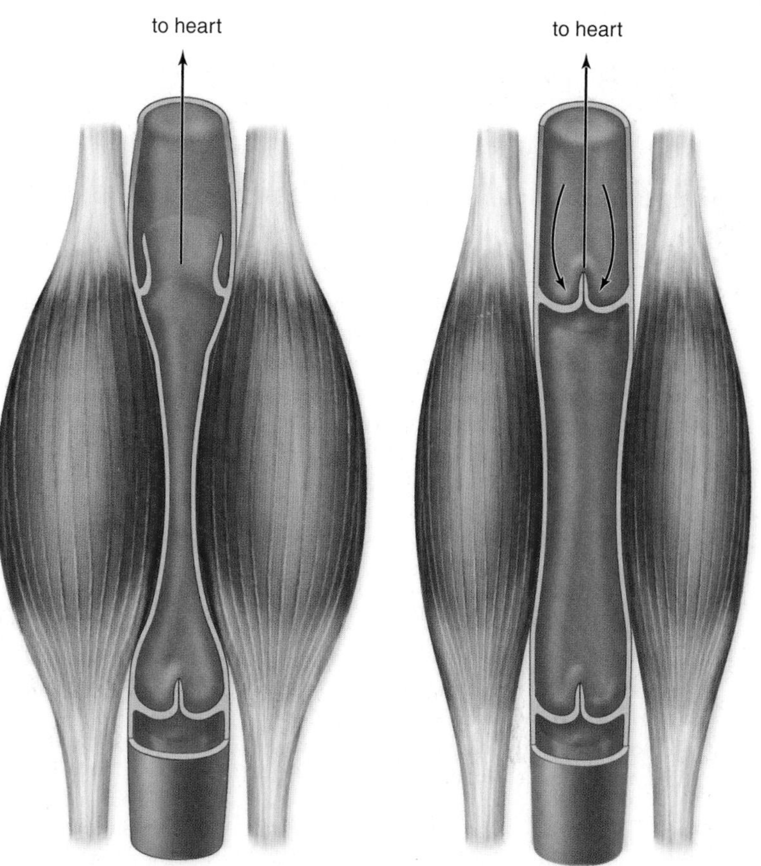


**FIGURE 32.12 Cross section of a valve in a vein.**

**a.** Pressure on the walls of a vein, exerted by skeletal muscles, increases blood pressure within the vein and forces a valve open. **b.** When external pressure is no longer applied to the vein, blood pressure decreases, and back pressure forces the valve closed. Closure of the valves prevents the blood from flowing in the opposite direction.

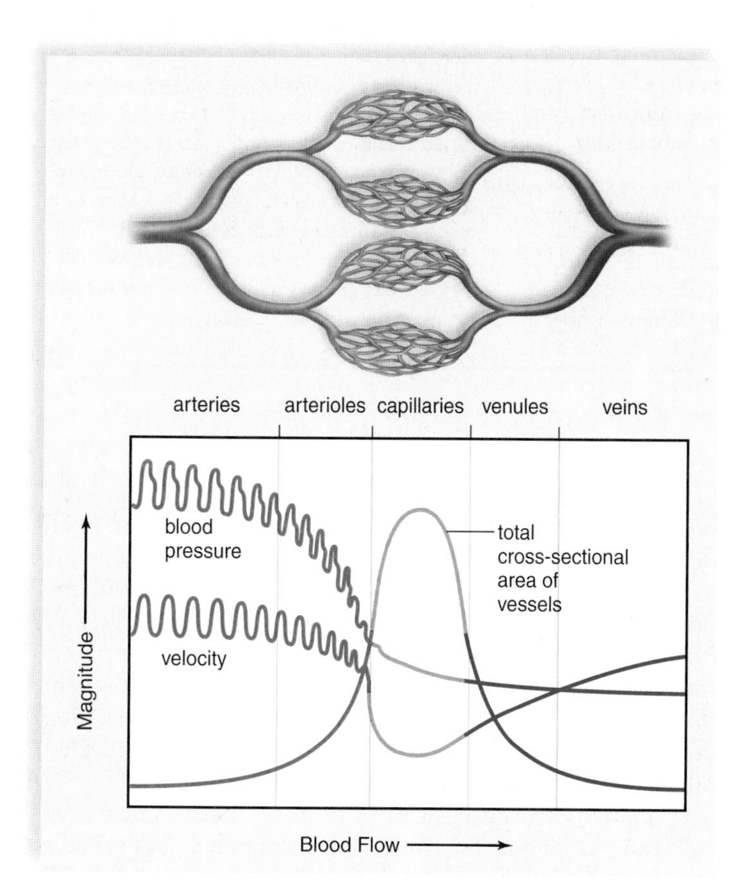


**FIGURE 32.11 Velocity and blood pressure related to vascular cross-sectional area.**

In capillaries, blood is under minimal pressure and has the least velocity. Blood pressure and velocity drop off because capillaries have a greater total cross-sectional area than arterioles.

Blood pressure in the veins is low and cannot move blood back to the heart, especially from the limbs of the body. Venous return is dependent on these factors: (1) When skeletal muscles near veins contract, they put pressure on the collapsible walls of the veins and on blood contained in these vessels. (2) Veins have valves that prevent the backward flow of blood, and therefore pressure from muscle contraction is sufficient to move blood through veins toward the heart (Fig. 32.12). Varicose veins, abnormal dilations in superficial veins, develop when the valves of the veins become weak and ineffective due to a backward pressure of the blood. Varicose veins of the anal canal are known as hemorrhoids. (3) A respiratory pump works like this: When we inhale, the chest expands and this reduces pressure in the thoracic cavity. Blood will flow from the higher pressure (in the abdominal cavity) to lower pressure (in the thoracic cavity).

*health focus*

## Prevention of Cardiovascular Disease

All of us can take steps to prevent cardiovascular disease, the most frequent cause of death in the United States. Certain genetic factors predispose an individual to cardiovascular disease, such as family history of heart attack under age 55, male gender, and ethnicity (African Americans are at greater risk). People with one or more of these risk factors need not despair, however. It means only that they should pay particular attention to the following guidelines for a heart-healthy lifestyle.

### The Don'ts

*Smoking*

Hypertension is well recognized as a major contributor to cardiovascular disease. When a person smokes, the drug nicotine, present in cigarette smoke, enters the bloodstream. Nicotine causes the arterioles to constrict and the blood pressure to rise. Restricted blood flow and cold hands are associated with smoking in most people. More serious is the need for the heart to pump harder to propel the blood through the lungs at a time when the oxygen-carrying capacity of the blood is reduced.

*Drug Abuse*

Stimulants, such as cocaine and amphetamines, can cause an irregular heartbeat and lead to heart attacks and strokes in people who are using drugs even for the first time. Intravenous drug use may result in a cerebral embolism.

Too much alcohol can destroy just about every organ in the body, the heart included. But investigators have discovered that people who take an occasional drink have a 20% lower risk of heart disease than do teetotalers. Two to four drinks a week is the recommended limit for men; one to three drinks for women.

*Weight Gain*

Hypertension is prevalent in persons who are more than 20% above the recommended weight for their height. In those who are overweight, more tissues require servicing, and the heart sends the extra blood out under greater pressure. It may be harder to lose weight once it is gained, and therefore it is recommended that weight control be a lifelong endeavor. Even a slight decrease in weight can bring with it a reduction in hypertension. A 4.5-kg weight (about 10 lbs) loss doubles the chance that blood pressure can be normalized without drugs.

### The Dos

*Healthy Diet*

Diet influences the amount of cholesterol in the blood. Cholesterol is ferried by two types of plasma proteins, called LDL (low-density lipoprotein) and HDL (high-density lipoprotein). LDL (called "bad" lipoprotein) takes cholesterol from the liver to the tissues, and HDL (called "good" lipoprotein) transports cholesterol out of the tissues to the liver. When the LDL level in blood is high or the HDL level is abnormally low, plaque, which interferes with circulation, accumulates on arterial walls (Fig. 32A).

Eating foods high in saturated fat (red meat, cream, and butter) and foods containing so-called trans-fats (most margarines, commercially baked goods, and deep-fried foods) raises the LDL-cholesterol level. Replacement of these harmful fats with healthier ones, such as monounsaturated fats (olive and canola oils) and polyunsaturated fats (corn, safflower, and soybean oils), is recommended. Cold water fish (e.g., halibut, sardines, tuna, and salmon) contain polyunsaturated fatty acids and especially omega-3 polyunsaturated fatty acids, which can reduce plaque.

Evidence is mounting to suggest a role for antioxidant vitamins (A, E, and C) in preventing cardiovascular disease. Antioxidants protect the body from free radicals that oxidize cholesterol and damage the lining of an artery, leading to a blood clot that can block blood vessels. Nutritionists believe that consuming at least five servings of fruits and vegetables a day may protect against cardiovascular disease.

## Cardiovascular Disease

Cardiovascular disease (CVD) is the leading cause of untimely death in the Western countries. The Health Focus on this page emphasizes how to prevent CVD from developing.

### *Hypertension*

It is estimated that about 20% of all Americans suffer from hypertension, which is high blood pressure. Under age 45, a reading above 130/90 is hypertensive, and beyond age 45, a reading above 140/95 is hypertensive. While both systolic and diastolic pressures are important, it is the diastolic pressure that is emphasized when medical treatment is being considered.

### *Atherosclerosis*

Hypertension is also seen in individuals who have atherosclerosis (formerly called arteriosclerosis), an accumulation of soft masses of fatty materials, particularly cholesterol, beneath the inner linings of arteries (see Fig. 32A). Such deposits are called plaque. As deposits occur, plaque tends to protrude into the lumen of the vessel, interfering with the flow of blood. Atherosclerosis begins in early adulthood and develops progressively through middle age, but symptoms may not appear until an individual is 50 or older. To prevent its onset and development, the American Heart Association and other organizations recommend a diet low in saturated fat and cholesterol and rich in fruits and vegetables.

Plaque can cause a clot to form on the irregular arterial wall. As long as the clot remains stationary, it is called a thrombus, but when and if it dislodges and moves along with the blood, it is called an embolus. If thromboembolism is not treated, complications can arise (see the following section).

In certain families, atherosclerosis is due to an inherited condition such as familial hypercholesterolemia. The presence

*Cholesterol Profile*

Starting at age 20, all adults are advised to have their cholesterol levels tested at least every five years. Even in healthy individuals, an LDL level above 160 mg/100 ml and an HDL level below 40 mg/100 ml are matters of concern. If a person has heart disease or is at risk for heart disease, an LDL level below 100 mg/100 ml is now recommended. Medications will most likely be prescribed for individuals who do not meet these minimum guidelines.

*Exercise*

People who exercise are less apt to have cardiovascular disease. One study found that moderately active men who spent an average of 48 minutes a day on a leisure-time activity such as gardening, bowling, or dancing had one-third fewer heart attacks than peers who spent an average of only 16 minutes each day being active. Exercise helps keep weight under control, may help minimize stress, and reduces hypertension. The heart beats faster when exercising, but exercise slowly increases its capacity. This means that the heart can beat slower when we are at rest and still do the same amount of work. One physician recommends that cardiovascular patients walk for one hour, three times a week, and, in addition, practice meditation and yogalike stretching and breathing exercises to reduce stress.

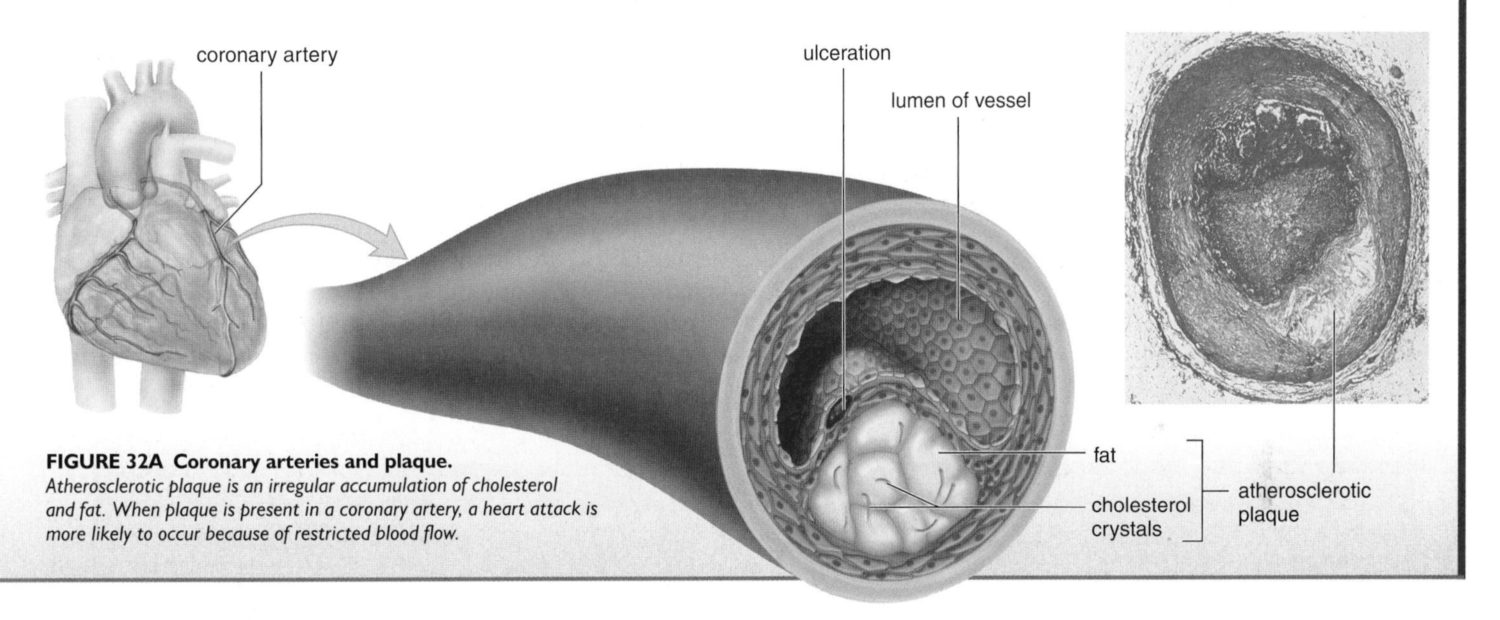


**FIGURE 32A Coronary arteries and plaque.** *Atherosclerotic plaque is an irregular accumulation of cholesterol and fat. When plaque is present in a coronary artery, a heart attack is more likely to occur because of restricted blood flow.*

of the disease-associated mutation can be detected, and this information is helpful if measures are taken to prevent the occurrence of the disease.

## *Stroke and Heart Attack*

Stroke, heart attack, and aneurysm are associated with hypertension and atherosclerosis. A cardiovascular accident, also called a **stroke,** often results when a small cranial arteriole bursts or is blocked by an embolus. A lack of oxygen causes a portion of the brain to die, and paralysis or death can result. A person is sometimes forewarned of a stroke by a feeling of numbness in the hands or the face, difficulty in speaking, or temporary blindness in one eye.

When a coronary artery is completely blocked, perhaps because of a thromboembolism, a portion of the heart muscle dies due to a lack of oxygen. This is a myocardial infarction, also called a **heart attack.** If a coronary artery becomes partially blocked, the individual may suffer from **angina pectoris,** characterized as a squeezing sensation or a flash of burning. Nitroglycerin or related drugs dilate blood vessels and help relieve the pain.

### Check Your Progress 32.3

1. Contrast the structure and function of the right and left ventricles.
2. Trace the path of blood through the heart from the venae cavae to the aorta.
3. Explain what happens during a heartbeat and what makes the familiar *lub-dub* sounds.
4. What conditions might occur as a result of hypertension and plaque?

# 32.4 Blood, a Transport Medium

The blood of mammals has numerous functions that help maintain homeostasis. Blood (1) transports gases, nutrients, waste products, and hormones throughout the body; (2) helps destroy pathogenic microorganisms; (3) distributes antibodies that are important in immunity; (4) aids in maintaining water balance and pH; (5) helps regulate body temperature; and (6) carries platelets and factors that ensure clotting and prevent blood loss.

In humans, blood has two main portions: the liquid portion, called plasma, and the formed elements, consisting of various cells and platelets (Fig. 32.13). **Plasma** [Gk. *plasma*, something molded] contains many types of molecules, including nutrients, wastes, salts, and proteins. The salts and proteins are involved in buffering the blood, effectively keeping the pH near 7.4. They also maintain the blood's osmotic pressure so that water has an automatic tendency to enter blood capillaries. Several plasma proteins are involved in blood clotting, and others transport large organic molecules in the blood. Albumin, the most plentiful of the plasma proteins, transports bilirubin, a breakdown product of hemoglobin. Globulins have various functions; among the globulins are the lipoproteins, which transport cholesterol.

## Formed Elements

The formed elements are of three types: red blood cells, or erythrocytes [Gk. *erythros*, red, and *kytos*, cell]; white blood cells, or leukocytes [Gk. *leukos*, white, and *kytos*, cell]; and platelets, or thrombocytes [Gk. *thrombos*, blood clot, and *kytos*, cell].

### *Red Blood Cells*

**Red blood cells** are small, biconcave disks that at maturity lack a nucleus and contain the respiratory pigment hemoglobin. Approximately 25 trillion red blood cells exist in an average adult. There are 6 million red blood cells per cubic millimeter ($mm^3$) of whole blood, and each one of these cells contains about 250 million hemoglobin molecules. **Hemoglobin** [Gk. *haima*, blood; L. *globus*, ball] contains four globin protein chains, each associated with heme, an iron-containing group. Iron combines loosely with oxygen, and in this way oxygen is carried in the blood. If there is an insufficient number of red blood cells, or if the cells do not have enough hemoglobin, the individual suffers from anemia and has a tired, run-down feeling.

Red blood cells are manufactured continuously in the red bone marrow of the skull, the ribs, the vertebrae, and the ends of the long bones. The hormone erythropoietin, which is produced by the kidneys, stimulates the production of red blood cells. Now available as a drug, erythropoietin is helpful to persons with anemia and is also sometimes abused by athletes who want to enhance their performance.

Before they are released from the bone marrow into blood, red blood cells lose their nuclei and synthesize hemoglobin. After living about 120 days, they are destroyed chiefly in the liver and the spleen, where they are engulfed by large phagocytic cells. When red blood cells are destroyed, hemoglobin is released. The iron is recovered and returned

| Plasma | |
|---|---|
| **Type** | **Function** |
| **Water** (90–92% of plasma) | Maintains blood volume; transports molecules |
| **Plasma proteins** (7–8% of plasma) | Maintain blood osmotic pressure and pH |
| Globulins | Transport; fight infection |
| Fibrinogen | Blood clotting |
| **Salts** (less than 1% of plasma) | Maintain blood osmotic pressure and pH; aid metabolism |
| **Gases** ($O_2$ and $CO_2$) | Cellular respiration |
| **Nutrients** (lipids, glucose, and amino acids) | Food for cells |
| **Wastes** (urea and uric acid) | End product of metabolism; excretion by kidneys |
| **Hormones** | Aid metabolism |

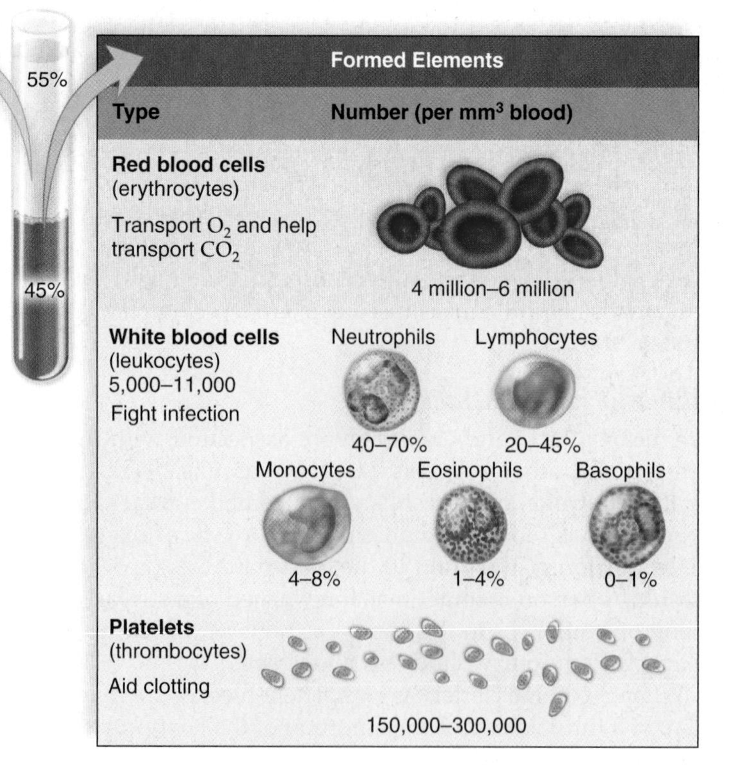


**FIGURE 32.13** Composition of blood.

to the red bone marrow for reuse. The heme portions of the molecules undergo chemical degradation and are excreted by the liver as bile pigments in the bile. The bile pigments are primarily responsible for the color of feces.

## *White Blood Cells*

**White blood cells** differ from red blood cells in that they are usually larger and have a nucleus, they lack hemoglobin, and, without staining, they appear translucent. With staining, white blood cells appear light blue unless they have granules that bind with certain stains. The following are granular leukocytes with a lobed nucleus: neutrophils, which have granules that stain slightly pink; eosinophils, which have granules that take up the red dye eosin; and basophils, which have granules that take up a basic dye, staining them a deep blue. The agranular leukocytes with no granules and a circular or indented nucleus are the larger monocytes and the smaller lymphocytes. There are approximately 5,000–11,000 white blood cells per $mm^3$. Stem cell growth factor can be used to increase the production of all white blood cells, and various other growth factors are also available to stimulate the production of specific stem cells. These growth factors are helpful to people with low immunity, such as AIDS patients.

When microorganisms enter the body due to an injury, the response is called an inflammatory reaction because swelling and reddening occur at the injured site. Cells in the vicinity release substances that cause vasodilation (increase in the diameter of a vessel), and increased capillary permeability. **Neutrophils** [Gk. *neuter*, neither, and *phileo*, love], which are amoeboid, squeeze through the capillary wall and enter the tissue fluid, where they phagocytize foreign material. **Monocytes** appear and are transformed into **macrophages** [Gk. *makros*, long, and *phagein*, to eat], which are large phagocytizing cells that release white blood cell growth factors. Soon there is an explosive increase in the number of leukocytes. The thick, yellowish fluid called pus contains a large proportion of dead white blood cells that have fought the infection.

**Lymphocytes** [L. *lympha*, clear water; Gk. *kytos*, cell] also play an important role in fighting infection. Certain lymphocytes called T cells attack infected cells that contain viruses. Other lymphocytes called B cells produce antibodies. Each B cell produces just one type of antibody, which is specific for one type of antigen. An **antigen** [Gk. *anti*, against; L. *genitus*, forming, causing], which is most often a protein but sometimes a polysaccharide, causes the body to produce an antibody because the antigen doesn't belong to the body. Antigens are present in the outer covering of parasites or in their toxins. When **antibodies** [Gk. *anti*, against] combine with antigens, the complex is often phagocytized by a macrophage. An individual is actively immune when a large number of B cells are all producing the specific antibody needed for a particular infection.

The **eosinophils** are granular leukocytes that are important in releasing enzymes used in fighting parasites and destroying allergens. **Basophils** are the least common leukocyte. They contain the anticoagulant heparin, which prevents blood from clotting too quickly. Basophils also contain the vasodilator histamine, which promotes blood flow to tissues.

## *Platelets*

**Platelets** (thrombocytes) result from fragmentation of certain large cells, called megakaryocytes, in the red bone marrow. Platelets are produced at a rate of 200 billion a day, and the blood contains 150,000–300,000 per $mm^3$. These formed elements are involved in blood clotting, or coagulation.

**Blood Clotting.** When a blood vessel in the body is damaged, platelets clump at the site of the puncture and partially seal the leak. Platelets and the injured tissues release a clotting factor called prothrombin activator that converts prothrombin to thrombin. This reaction requires calcium ions ($Ca^{2+}$). *Thrombin*, in turn, acts as an enzyme that severs two short amino acid chains from each fibrinogen molecule. These activated fragments then join end to end, forming long threads of fibrin. Fibrin threads wind around the platelet plug in the damaged area of the blood vessel and provide the framework for the clot. Red blood cells also are trapped within the fibrin threads; these cells make a clot appear red (Fig. 32.14). A fibrin clot is present only temporarily. As soon as blood vessel repair is initiated, an enzyme called plasmin destroys the fibrin network and restores the fluidity of plasma.

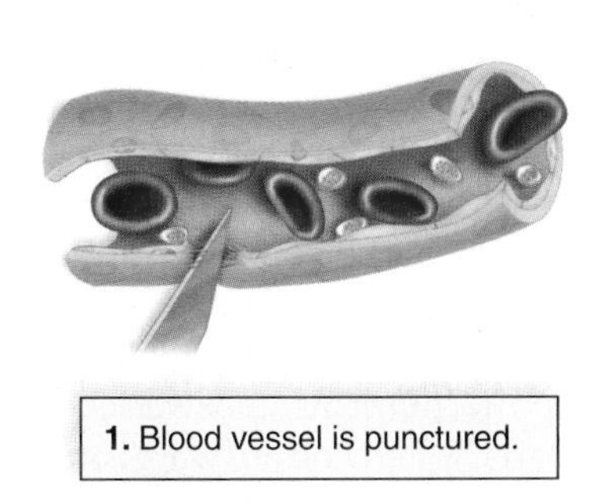


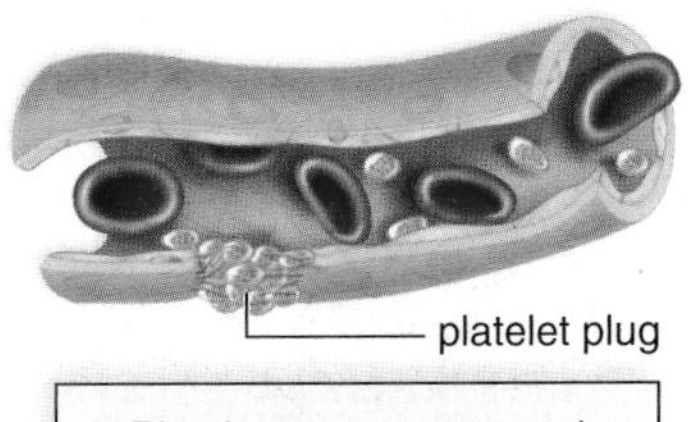


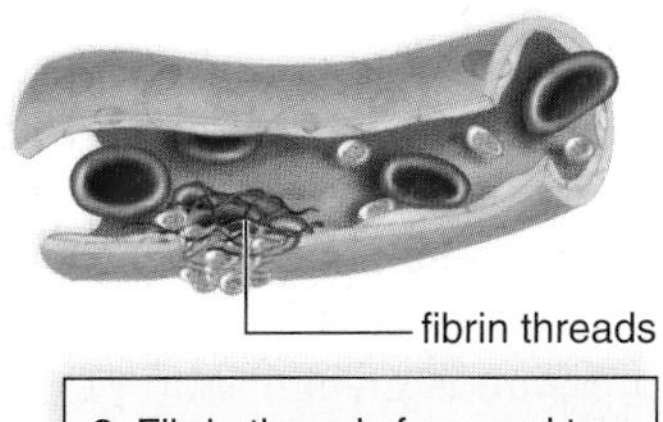


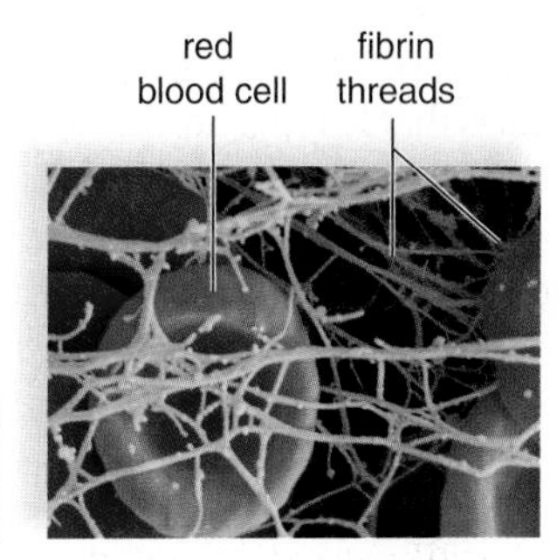


**FIGURE 32.14 Blood clotting.**

A number of plasma proteins participate in a series of enzymatic reactions that lead to the formation of fibrin threads.

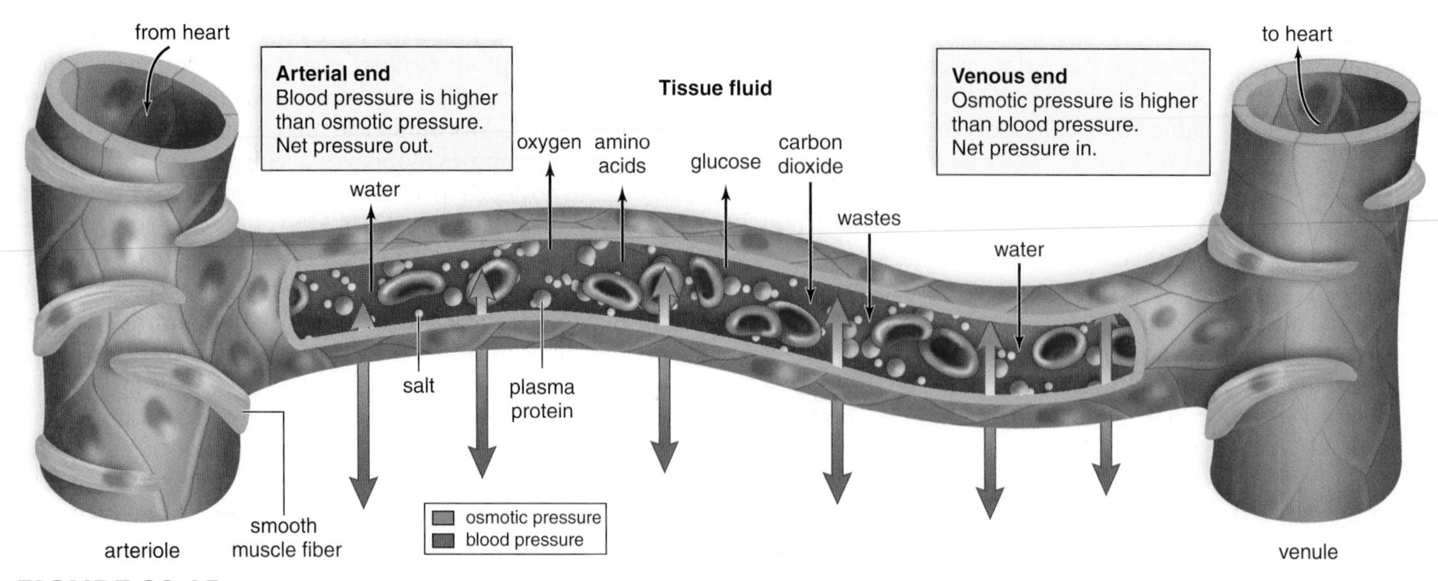


**FIGURE 32.15 Capillary exchange.**

A capillary, illustrating the exchanges that take place and the forces that aid the process. At the arterial end of a capillary, the blood pressure is higher than the osmotic pressure; therefore, water ($H_2O$) tends to leave the bloodstream. In the midsection, molecules, including oxygen ($O_2$) and carbon dioxide ($CO_2$), follow their concentration gradients. At the venous end of a capillary, the osmotic pressure is higher than the blood pressure; therefore, water tends to enter the bloodstream. Notice that the red blood cells and the plasma proteins are too large to exit a capillary.

## Capillary Exchange

Figure 32.15 illustrates capillary exchange between a systemic capillary and tissue fluid, the fluid between the body's cells. Blood that enters a capillary at the arterial end is rich in oxygen and nutrients, and it is under pressure created by the pumping of the heart. Two forces primarily control movement of fluid through the capillary wall: osmotic pressure, which tends to cause water to move from tissue fluid to blood, and blood pressure, which tends to cause water to move in the opposite direction. At the arterial end of a capillary, blood pressure (30 mm Hg) is higher than the osmotic pressure of blood (21 mm Hg). Osmotic pressure is created by the presence of salts and the plasma proteins. Because blood pressure is higher than osmotic pressure at the arterial end of a capillary, water exits a capillary at this end.

Midway along the capillary, where blood pressure is lower, the two forces essentially cancel each other, and there is no net movement of water. Solutes now diffuse according to their concentration gradient: Oxygen and nutrients (glucose and amino acids) diffuse out of the capillary; carbon dioxide and wastes diffuse into the capillary. Red blood cells and almost all plasma proteins remain in the capillaries. The substances that leave a capillary contribute to **tissue fluid,** the fluid between the body's cells. Since plasma proteins are too large to readily pass out of the capillary, tissue fluid tends to contain all components of plasma except much lesser amounts of protein.

At the venule end of a capillary, where blood pressure has fallen even more, osmotic pressure is greater than blood pressure, and water tends to move into the capillary. Almost the same amount of fluid that left the capillary returns to it, although some excess tissue fluid is always collected by the lymphatic capillaries (Fig. 32.16). Tissue fluid contained within lymphatic vessels is called **lymph.** Lymph is returned to the systemic venous blood when the major lymphatic vessels enter the subclavian veins in the shoulder region.

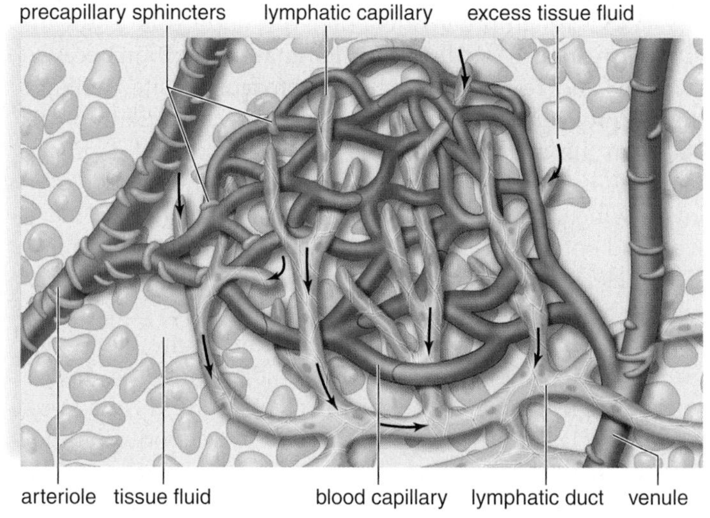


**FIGURE 32.16 Capillary bed.**

A lymphatic capillary bed lies near a blood capillary bed. When lymphatic capillaries take up excess tissue fluid, it becomes lymph. Precapillary sphincters can shut down a blood capillary, and blood then flows through the shunt.

Not all capillary beds are open at the same time. When the precapillary sphincters (circular muscles) shown in Figure 32.4 are relaxed, the capillary bed is open and blood flows through the capillaries. When precapillary sphincters are contracted, blood flows through a shunt that carries blood directly from an arteriole to a venule. In addition to nutrients and wastes, the blood distributes heat to body parts. When you are warm, many capillaries that serve the skin are open, and your face is flushed. This helps rid the body of excess heat. When you are cold, skin capillaries close, conserving heat, and your skin turns blue.

## Blood Types

Many early blood transfusions resulted in illness and even death of some recipients. Eventually, it was discovered that only certain types of blood are compatible because red blood cell membranes carry specific proteins or carbohydrates that are antigens to blood recipients. An antigen is a foreign molecule, usually a protein, that the body reacts to. Several groups of red blood cell antigens exist, the most significant being the ABO system. Clinically, it is very important that the blood groups be properly cross-matched to avoid a potentially deadly transfusion reaction. In such a reaction, the recipient may die of kidney failure within a week.

### *ABO System*

In the ABO system, the presence or absence of type A and type B antigens on red blood cells determines a person's blood type. For example, if a person has type A blood, the A antigen is on his or her red blood cells. This molecule is not an antigen to this individual, although it can be an antigen to a recipient who does not have type A blood.

In the ABO system, there are four types of blood: A, B, AB, and O. Within the plasma are antibodies to the antigens that are not present on the person's red blood cells. These antibodies are called anti-A and anti-B. This chart tells you what antibodies are present in the plasma of each blood type:

| Blood Type | Antigen on Red Blood Cells | Antibody in Plasma |
|---|---|---|
| A | A | Anti-B |
| B | B | Anti-A |
| AB | A, B | None |
| O | None | Anti-A and anti-B |

Because type A blood has anti-B and not anti-A antibodies in the plasma, a donor with type A blood can give blood to a recipient with type A blood (Fig. 32.17). However, if type A blood is given to a type B recipient, **agglutination** (Fig. 32.18), the clumping of red blood cells, occurs or it can cause blood to stop circulating in small blood vessels, and this leads to organ damage.

Theoretically, which type blood would be accepted by all recipients? Type O blood has no antigens on the red blood cells and is sometimes called the universal donor. Which type blood could receive blood from any other blood type? Type AB blood has no anti-A or anti-B antibodies in the plasma and is sometimes called the universal recipient. In practice, however, it is not safe to rely solely on the ABO system when matching blood. Instead, samples of the two types of blood are physically mixed, and the result is microscopically examined before blood transfusions are done.

### *Rh System*

Another important antigen in matching blood types is the Rh factor. Eighty-five percent of the U.S. population has this particular antigen on the red blood cells and is called Rh positive. Fifteen percent does not have the antigen and is Rh negative. Rh-negative individuals normally do not have antibodies to the Rh factor, but they may make them when exposed to the Rh factor. The designation of blood type usually also includes whether the person has or does not have the Rh factor on the red blood cells.

During pregnancy, if the mother is Rh negative and the father is Rh positive, the child may be Rh positive. If the Rh-positive red blood cells leak across the placenta, the mother will produce anti-Rh antibodies. In this or a subsequent pregnancy with another Rh-positive baby, these antibodies may cross the placenta and destroy the child's red blood cells. This condition, called hemolytic disease of the newborn (HDN), can be fatal without an immediate blood transfusion.

**Check Your Progress 32.4**

1. List the functions of blood.
2. Contrast red blood cells with white blood cells.
3. Describe how a blood clot forms.
4. Why can't you give type A blood to a type B recipient?

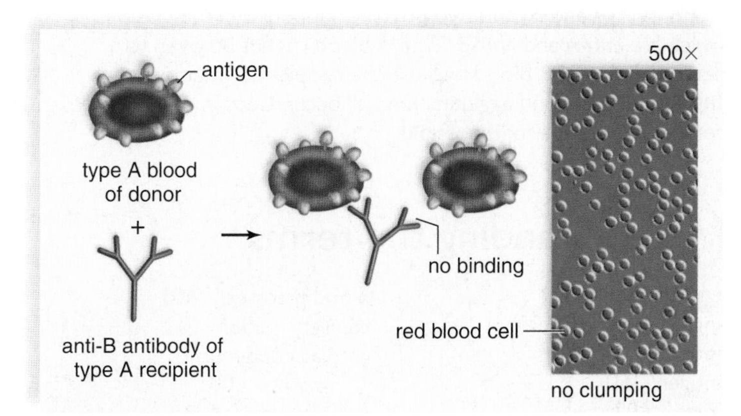

No agglutination

**FIGURE 32.17 No agglutination.**

No agglutination occurs when the donor and recipient have the same type blood.

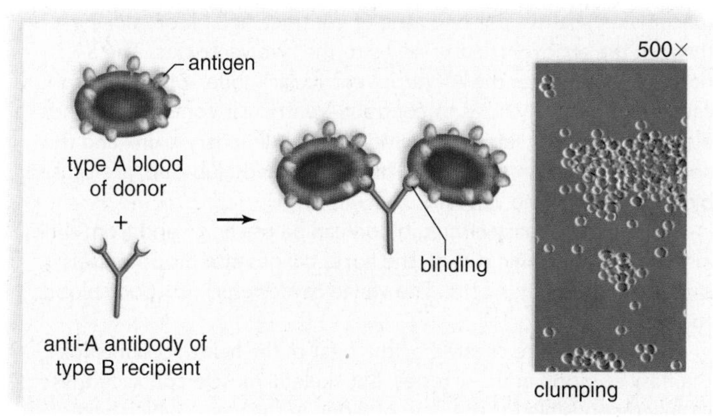

Agglutination

**FIGURE 32.18 Agglutination.**

Agglutination occurs because blood type B has anti-A antibodies in the plasma.

# Connecting the Concepts

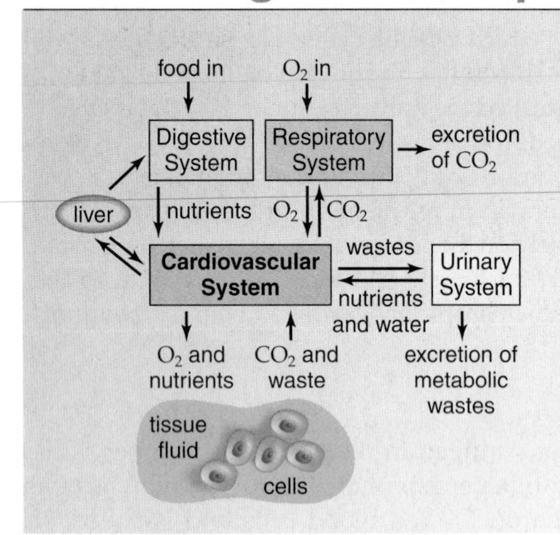


The cardiovascular system consists of the heart and the blood vessels, which can be likened to the streets of a town. Imagine that you have received birthday gifts and have decided to return some and pick up items you prefer. In a similar way, the cardiovascular system picks up nutrients at the intestinal tract, exchanges carbon dioxide for oxygen at the lungs, and deposits metabolic wastes at the kidneys. Therefore, the composition of the blood and the tissue fluid stays relatively constant. Tissue fluid, which surrounds the cells, makes exchanges with the blood, and if the composition of the blood stays constant, tissue fluid will be able to supply the cells with the nutrients and oxygen they require and receive their metabolic wastes. In this way homeostasis, the relative constancy of the internal environment, will be maintained.

The pumping of the heart merely helps keep the blood moving so that it can continue to service the cells. While we tend to think of the body in terms of organs, it is the cells of the organs that do the work of keeping us alive and only if the cells are well cared for do we remain healthy. In Chapter 34. we consider the contribution of the digestive system to homeostasis.

## summary

### 32.1 Transport in Invertebrates

Some invertebrates do not have a transport system. The presence of a gastrovascular cavity allows diffusion alone to supply the needs of cells in cnidarians and flatworms. Roundworms make use of their pseudocoelom in the same way that echinoderms use their coelom to circulate materials.

Other invertebrates do have a transport system. Insects have an open circulatory system, and earthworms have a closed one.

### 32.2 Transport in Vertebrates

Vertebrates have a closed system in which arteries carry blood away from the heart to capillaries, where exchange takes place, and veins carry blood to the heart.

Fishes have a one-circuit circulatory pathway because the heart, with the single atrium and ventricle, pumps blood only to the gills. The other vertebrates have both pulmonary and systemic circuits. Amphibians have two atria but a single ventricle. Crocodilians, birds, and mammals, including humans, have a heart with two atria and two ventricles, in which $O_2$-rich blood is always separate from $O_2$-poor blood.

### 32.3 Transport in Humans

The heartbeat in humans begins when the SA (sinoatrial) node (pacemaker) causes the two atria to contract, and blood moves through the atrioventricular valves to the two ventricles. The SA node also stimulates the AV (atrioventricular) node, which in turn causes the two ventricles to contract. Ventricular contraction sends blood through the semilunar valves to the pulmonary trunk and the aorta. Now all chambers rest. The heart sounds, lub-dub, are caused by the closing of the valves.

In the pulmonary circuit, blood can be traced to and from the lungs. In the systemic circuit, the aorta divides into blood vessels that serve the body's cells. The venae cavae return $O_2$-poor blood to the heart.

Blood pressure created by the beat of the heart accounts for the flow of blood in the arteries, but skeletal muscle contraction is largely responsible for the flow of blood in the veins, which have valves preventing a backward flow.

Hypertension and atherosclerosis are two circulatory disorders that lead to heart attack and to stroke. Following a heart-healthy diet, getting regular exercise, maintaining a proper weight, and not smoking cigarettes are protective against the development of these conditions.

### 32.4 Blood, a Transport Medium

Blood has two main parts: plasma and formed elements. Plasma contains mostly water (90–92%) and proteins (7–8%) but also contains nutrients and wastes.

The red blood cells contain hemoglobin and function in oxygen transport. Defense against disease depends on the various types of white blood cells. Neutrophils and monocytes are phagocytic and are very active during the inflammatory reaction. Lymphocytes are involved in the development of immunity to disease. Eosinophils are important in allergic reactions and parasitic infections, and basophils contain the anticoagulant heparin.

The platelets and two plasma proteins, prothrombin and fibrinogen, function in blood clotting, an enzymatic process that results in fibrin threads. Blood clotting is a complex process that includes three major events: (1) Platelets and injured tissue release prothrombin activator, which (2) enzymatically changes prothrombin to thrombin; (3) thrombin is an enzyme that causes fibrinogen to be converted to fibrin threads.

When blood reaches a capillary, water moves out at the arterial end, due to blood pressure. At the venous end, water moves in, due to osmotic pressure. In between, nutrients diffuse out of, and wastes diffuse into, the capillary.

The ABO system recognizes two possible antigens on the red blood cells (A and B) and two possible antibodies in the plasma, which are anti-A and anti-B. Type A blood cannot be given to a person with type B blood because the recipient's blood contains anti-A antibodies and agglutination will occur. Certain other combinations are also impossible.

## understanding the terms

agglutination 609
angina pectoris 605
antibody 607
antigen 607
aorta 602
arteriole 596
artery 596
atrioventricular valve 598
atrium 598
basophil 607
blood 594
blood pressure 603
capillary 596
cardiac conduction system 600
cardiac cycle 600
cardiac output 600
cardiac pacemaker 601
circulatory (or cardiovascular) system 594, 596
closed circulatory system 594
diastole 600

electrocardiogram (ECG) 601
eosinophil 607
heart 598
heart attack 605
hemoglobin 606
hemolymph 594
lymph 608
lymphocyte 607
macrophage 607
monocyte 607
neutrophil 607
open circulatory system 594
plasma 606
platelet 607
portal system 602
pulmonary circuit 597
pulse 600
red blood cell 606
semilunar valve 598
septum 598
sphygmomanometer 603
stroke 605
systemic circuit 597
systole 600
tissue fluid 608
vein 596
vena cava 602
ventricle 598
venule 596
white blood cell 607

Match the terms to these definitions:

a. ______________ Blood vessel that transports blood away from the heart.
b. ______________ Cell fragment that is necessary to blood clotting.
c. ______________ The liquid portion of blood; contains nutrients, wastes, salts, and proteins.
d. ______________ The major systemic veins that take blood to the heart from the tissues.
e. ______________ Iron-containing respiratory pigment occurring in vertebrate red blood cells and in the blood plasma of many invertebrates.

## reviewing this chapter

1. Describe transport in invertebrates that have no circulatory system; in those that have an open circulatory system; and in those that have a closed circulatory system. 594–95
2. Compare the circulatory systems of a fish, an amphibian, and a mammal. 597
3. Trace the path of blood in humans from the right ventricle to the left atrium; from the left ventricle to the kidneys and to the right atrium; from the left ventricle to the small intestine and to the right atrium. 599
4. Describe the mechanism of a heartbeat, mentioning all the factors that account for this repetitive process. Describe how the heartbeat affects blood flow. What other factors are involved in blood flow? 600–601
5. Define these terms: pulmonary circuit, systemic circuit, and portal system. 601–2
6. Discuss the life cycle and function of red blood cells. 606–7
7. How are white blood cells classified? What are the functions of neutrophils, monocytes, and lymphocytes? 607
8. Name the steps that take place when blood clots. Which substances are present in blood at all times, and which appear during the clotting process? 607
9. What forces facilitate exchange of molecules across the capillary wall? 608
10. Explain the ABO system of typing blood. 609

## testing yourself

Choose the best answer for each question.

1. Which one of these would you expect to be part of a closed, but not an open, circulatory system?
   a. ostia
   b. capillary beds
   c. hemocoel
   d. heart
   e. All of these are correct.
2. In a one-circuit circulatory pathway, blood pressure
   a. is constant throughout the system.
   b. drops significantly after gas exchange has taken place.
   c. is higher at the intestinal capillaries than at the gill capillaries.
   d. brings $O_2$-rich blood directly to the heart.
   e. does not occur in the animal kingdom.
3. In which animal does aortic blood have less oxygen than blood in the pulmonary vein?
   a. frog
   b. chicken
   c. monkey
   d. fish
   e. All of these are correct.
4. Which of these factors has little effect on blood flow in arteries?
   a. heartbeat
   b. blood pressure
   c. total cross-sectional area of vessels
   d. skeletal muscle contraction
   e. the amount of blood leaving the heart
5. In humans, blood returning to the heart from the lungs returns to
   a. the right ventricle.
   b. the right atrium.
   c. the left ventricle.
   d. the left atrium.
   e. both the right and left sides of the heart.
6. Systole refers to the contraction of the
   a. major arteries.
   b. SA node.
   c. atria and ventricles.
   d. major veins.
   e. All of these are correct.
7. Which of these is an incorrect association?
   a. white blood cells—infection fighting
   b. red blood cells—blood clotting
   c. plasma—water, nutrients, and wastes
   d. red blood cells—hemoglobin
   e. platelets—blood clotting
8. Water enters capillaries on the venous end as a result of
   a. active transport from tissue fluid.
   b. an osmotic pressure gradient.
   c. higher blood pressure on the venous end.
   d. higher blood pressure on the arterial side.
   e. higher red blood cell concentration on the venous end.
9. The last step in blood clotting
   a. is the only step that requires calcium ions.
   b. occurs outside the bloodstream.
   c. is the same as the first step.
   d. converts prothrombin to thrombin.
   e. converts fibrinogen to fibrin.
10. Macrophages are derived from
    a. basophils.
    b. eosinophils.
    c. neutrophils.
    d. lymphocytes.
    e. monocytes.
11. Which of the following is not a formed element of blood?
    a. leukocyte
    b. eosinophil
    c. fibrinogen
    d. platelet
12. Which of these is an incorrect statement concerning the heartbeat?
    a. The atria contract at the same time.
    b. The ventricles relax at the same time.
    c. The atrioventricular valves open at the same time.
    d. The semilunar valves open at the same time.
    e. First, the right side contracts, and then the left side contracts.

13. All arteries in the body contain $O_2$-rich blood, with the exception of the
    a. aorta.
    b. pulmonary arteries.
    c. renal arteries.
    d. coronary arteries.
14. The cardiac veins directly enter
    a. the inferior vena cava.
    b. the superior vena cava.
    c. the right atrium.
    d. the left atrium.
15. The "lub," the first heart sound, is produced by the closing of
    a. the aortic semilunar valve.
    b. the pulmonary semilunar valve.
    c. the right (atrioventricular) tricuspid valve.
    d. the left (atrioventricular) bicuspid valve, or mitral valve.
    e. both atrioventricular valves.

For questions 16–19, indicate whether the statement is true (T) or false (F).

16. Carbon dioxide exits the arterial end of the capillary, and oxygen enters the venous end of the capillary. ______
17. Platelets form a plug by sticking to each other. ______
18. Another term for blood clotting is agglutination. ______
19. SA node impulses cause the atria to contract. ______
20. Label portions of these arrows a–d as either blood pressure or osmotic pressure.

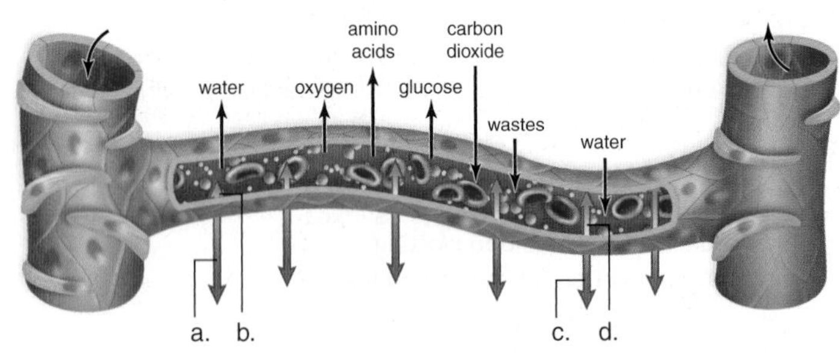


21. Label this diagram of the heart.

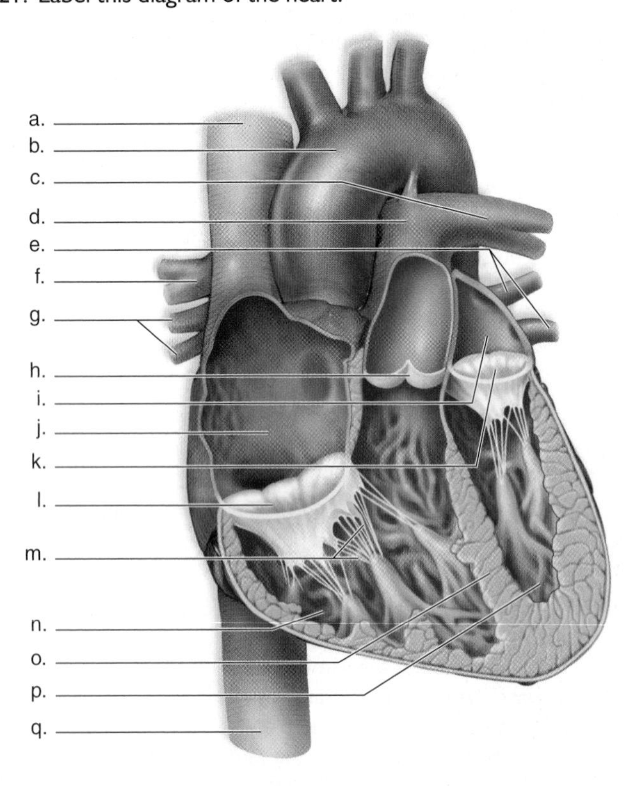


## thinking scientifically

1. For several years, researchers have attempted to produce an artificial blood for transfusions. Artificial blood would most likely be safer and more readily available than human blood. While artificial blood might not have all the characteristics of human blood, it would be useful on the battlefield and in emergency situations. Which characteristics of normal blood must artificial blood have to be useful, and which would probably be too difficult to reproduce?
2. You have to stand in front of the class to give a report. You are nervous, and your heart is pounding. How would your ECG appear?

## bioethical issue

### A Healthy Lifestyle

Many deaths a year could be prevented if people adopted the healthy lifestyle described in the Health Focus on page 604. Tobacco, lack of exercise, and a high-fat diet probably cost the nation about $200 billion per year in health-care costs. To what lengths should we go to prevent these deaths and reduce health-care costs?

E. A. Miller, a meatpacking entity of ConAgra in Hyrum, Utah, charges extra for medical coverage of employees who smoke. Eric Falk, Miller's director of human resources, says, "We want to teach employees to be responsible for their behavior." Anthem Blue Cross–Blue Shield of Cincinnati, Ohio, takes a more positive approach. They give insurance plan participants $240 a year in extra benefits, such as additional vacation days, if they get good scores in five out of seven health-related categories. The University of Alabama, Birmingham, School of Nursing has a health-and-wellness program that counsels employees about how to get into shape in order to keep their insurance coverage. Audrey Brantley, who is in the program, has mixed feelings. She says, "It seems like they are trying to control us, but then, on the other hand, I know of folks who found out they had high blood pressure or were borderline diabetics and didn't know it."

Does it really work? Turner Broadcasting System in Atlanta has a policy that affects all employees hired after 1986. They will be fired if caught smoking—whether at work or at home—but some admit they still manage to sneak a smoke. What steps do you think are ethical to encourage people to adopt a healthy lifestyle?

## *Biology* website

The companion website for *Biology* provides a wealth of information organized and integrated by chapter. You will find practice tests, animations, videos, and much more that will complement your learning and understanding of general biology.

**http://www.mhhe.com/maderbiology10**

# 33

# Lymph Transport and Immunity

*the immunodeficiency virus (HIV) attacks the immune system and prevents it from mounting an organized defense. An HIV infection terminates in AIDS (acquired immunodeficiency syndrome), an illness characterized by weight loss, innumerable infections including herpes, and life-threatening illnesses from pneumonia to cancer. HIV is sexually transmitted, and everyone should follow certain guidelines to prevent an HIV infection because presently there is no cure: Either abstain or have a monogamous relationship with a person free of STDs (sexually transmitted diseases); don't inject drugs and be wary of a relationship with an intravenous drug user; always use a condom and avoid both anal intercourse and oral sex because HIV can be transmitted in these ways also.*

*This chapter begins with the lymphatic system, which has an intimate connection with immunity, and then it reviews the mechanisms the immune system ordinarily uses to keep us healthy. Finally, we will consider various illnesses that beset us when immunity falters.*

A patient with AIDS. These photos show how the health of a patient with AIDS deteriorates.

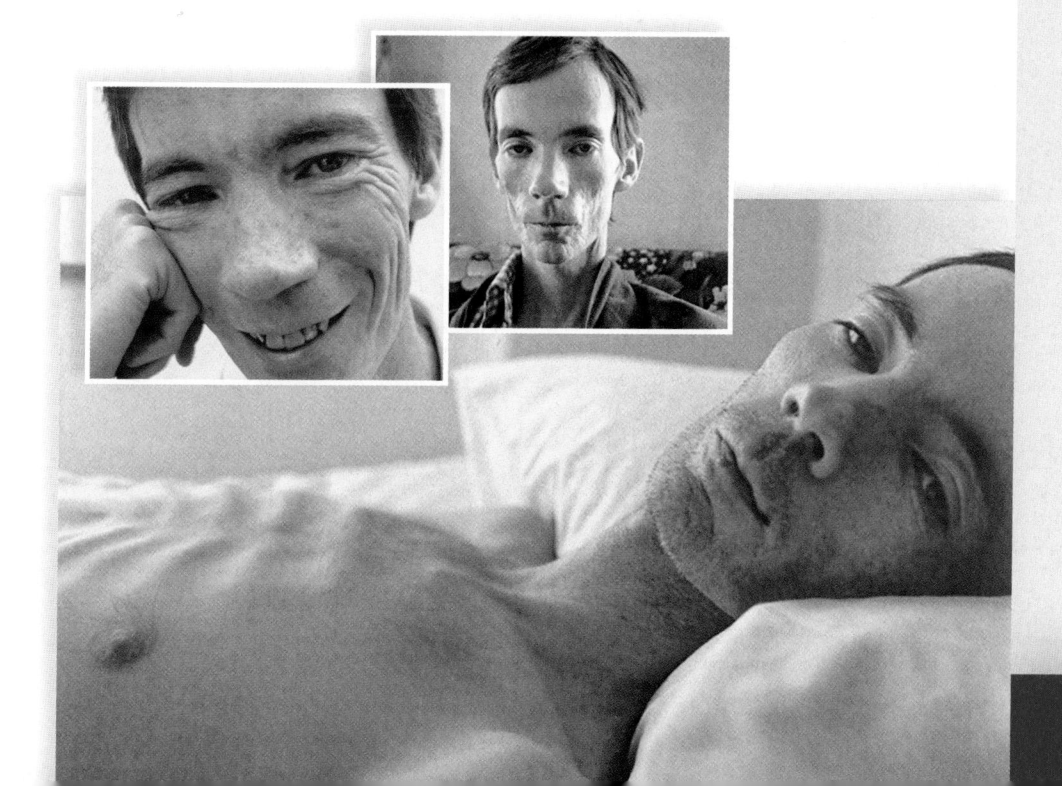

## concepts

# 33.1 The Lymphatic System

The **lymphatic system,** which is closely associated with the cardiovascular system, has four main functions that contribute to homeostasis:

- Lymphatic capillaries absorb excess tissue fluid and return it to the bloodstream.
- In the small intestines, lymphatic capillaries called lacteals absorb fats in the form of lipoproteins and transport them to the bloodstream.
- The lymphatic system is responsible for the production, maintenance, and distribution of lymphocytes.
- The lymphatic system helps defend the body against pathogens.

## Lymphatic Vessels

**Lymphatic vessels** form a one-way system that begins with lymphatic capillaries (Fig. 33.1). Most regions of the body are richly supplied with lymphatic capillaries—tiny, closed-ended vessels. Lymphatic capillaries take up excess tissue fluid. The fluid inside lymphatic capillaries is called **lymph.** In addition to water and fat molecules, lymph contains the same ions, nutrients, gases, and proteins that are present in tissue fluid. It also contains defense molecules called antibodies, which are produced by lymphocytes.

The lymphatic capillaries join to form lymphatic vessels that merge before entering one of two ducts: the thoracic duct or the right lymphatic duct. The larger thoracic duct returns lymph to the left subclavian vein. The right lymphatic duct returns lymph to the right subclavian vein.

The construction of the larger lymphatic vessels is similar to that of cardiovascular veins, including the presence of valves. Skeletal muscle contraction forces lymph through lymphatic vessels, and then it is prevented from flowing backward by the one-way valves.

## Lymphatic Organs

The **lymphatic (lymphoid) organs** are included in Figure 33.1. **Red bone marrow** is a spongy, semisolid red tissue where stem cells divide and produce all the various types of blood cells, including lymphocytes (Fig. 33.2*a*). Some of

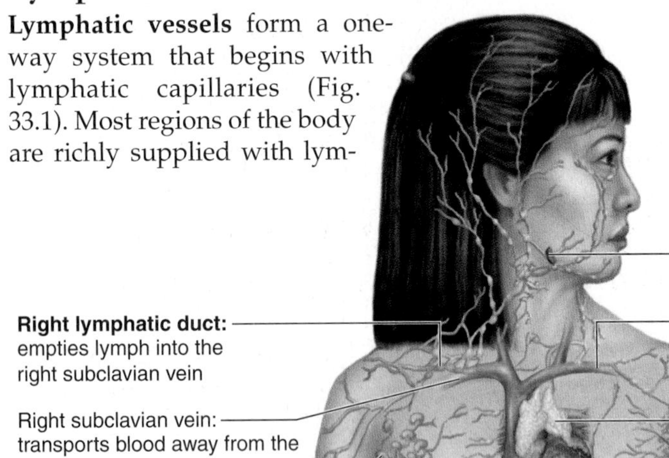


**FIGURE 33.1 Lymphatic system.**

Lymphatic vessels drain excess fluid from the tissues and return it to the cardiovascular system. The enlargement shows that lymphatic vessels, like cardiovascular veins, have valves to prevent backward flow. The lymph nodes, spleen, thymus gland, and red bone marrow are the main lymphatic organs that assist immunity.

these become mature **B cells,** a major type of lymphocyte, in the bone marrow. In a child, most of the bones have red bone marrow, but in an adult, it is present only in the bones of the skull, the sternum (breastbone), the ribs, the clavicle (collarbone), the pelvic bones, the vertebral column, and the proximal heads of the femur and humerus.

The red bone marrow consists of a network of connective tissue fibers that supports the stem cells and their progeny. They are packed around thin-walled sinuses filled with venous blood. Differentiated blood cells enter the bloodstream at these sinuses.

The soft, bilobed **thymus gland** is located in the thoracic cavity between the trachea and the sternum ventral to the heart (see Fig. 33.1). Immature **T cells,** the other major type of lymphocyte, migrate from the bone marrow through the bloodstream to the thymus, where they mature. The thymus also produces thymic hormones, such as thymosin, that are thought to aid in the maturation of T cells. The thymus varies in size, but it is largest in children and shrinks as we get older. In the elderly, it is barely detectable. When well developed, it contains many lobules (Fig. 33.2*b*).

**Lymph nodes** are small (about 1–25 mm in diameter), ovoid structures occurring along lymphatic vessels. Lymph nodes are named for their location. For example, inguinal lymph nodes are in the groin, and axillary lymph nodes are in the armpits. Physicians often feel for the presence of swollen, tender lymph nodes in the neck as evidence that the body is fighting an infection.

As lymph courses through the sinuses (open spaces) of the medulla (Fig. 33.2*c*), it is cleansed by macrophages, which engulf debris and pathogens. The many B and T cells also present help the immune system destroy pathogens.

Unfortunately, cancer cells sometimes enter lymphatic vessels and congregate in lymph nodes. Therefore, when a person undergoes surgery for cancer, it is a routine procedure to remove some lymph nodes and examine them to determine whether the cancer has spread to other regions of the body.

The **spleen,** an oval organ with a dull purplish color, is located in the upper left side of the abdominal cavity posterior to the stomach. Most of the spleen is red pulp that filters and cleanses the blood. Red pulp consists of blood vessels and sinuses, where macrophages remove old and defective blood cells. The spleen also has white pulp that is inside the red pulp and consists of little lumps of lymphatic tissue where B and T cells congregate (Fig. 33.2*d*).

The spleen's outer capsule is relatively thin, and an infection or a blow can cause the spleen to burst. Although the spleen's functions can be largely replaced by other organs, a person without a spleen is often slightly more susceptible to infections and may require antibiotic therapy indefinitely.

Patches of lymphatic tissue in the body include the **tonsils,** located in the pharynx; **Peyer patches,** located in the intestinal wall; and the vermiform appendix, attached to the cecum. These structures encounter pathogens and antigens that enter the body by way of the mouth.

### Check Your Progress 33.1

1. Give a brief description of the lymphatic system.
2. Summarize the functions of the lymphatic system.
3. Briefly describe the general appearance, location, and function of the lymphatic organs.

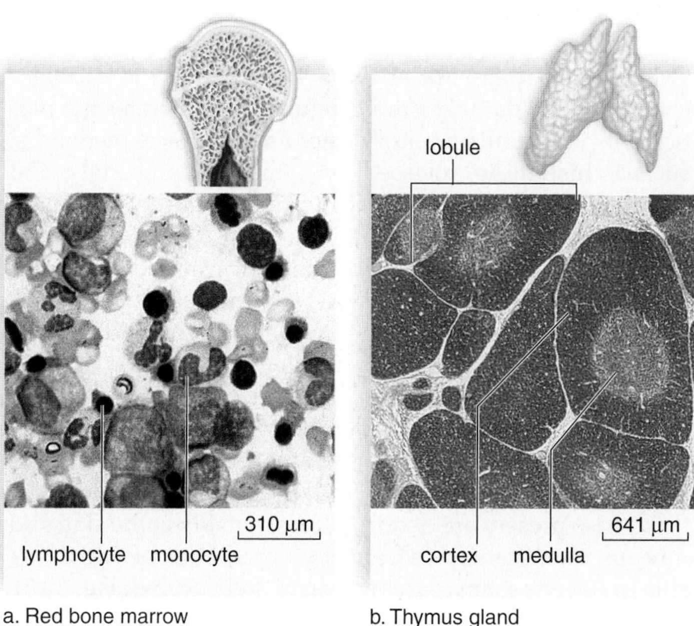


a. Red bone marrow

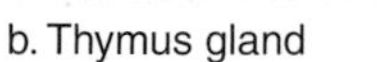

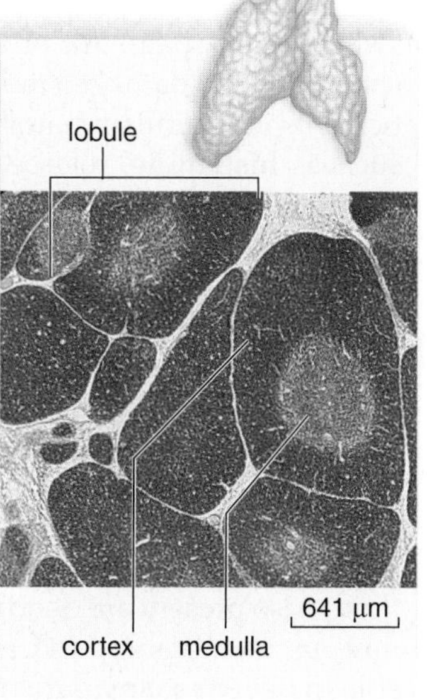


b. Thymus gland

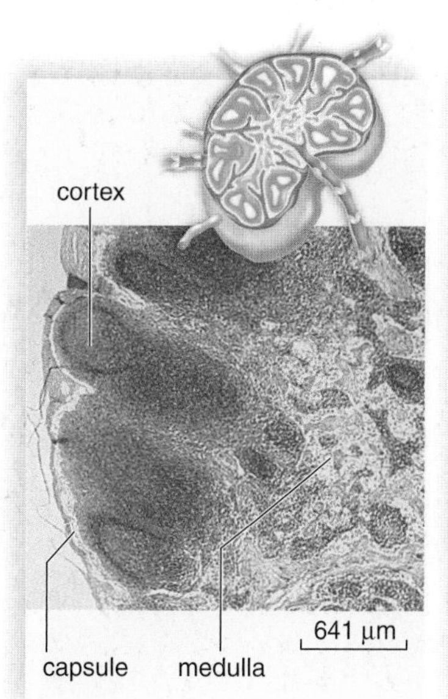

c. Lymph node

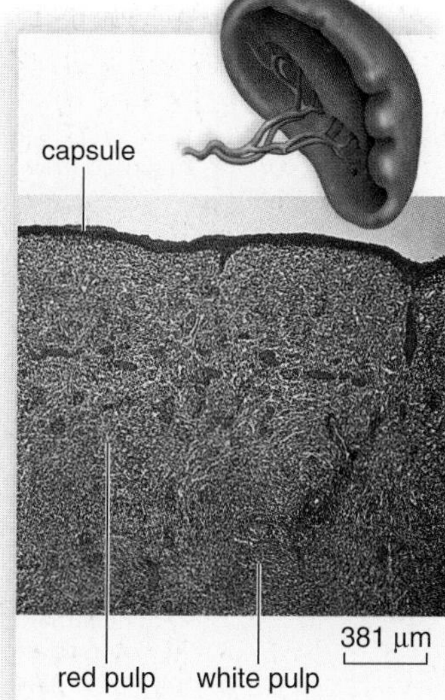


d. Spleen

**FIGURE 33.2 The lymphatic organs.**

**a.** Blood cells, including lymphocytes, are produced in red bone marrow. B cells mature in the bone marrow, but (**b**) T cells mature in the thymus. **c.** Lymph is cleansed in lymph nodes, while (**d**) blood is cleansed in the spleen.

# 33.2 Nonspecific Defense Against Disease

We are constantly exposed to pathogens in our food and drink and as we breathe air or touch objects, both living and inanimate, in our environment. The warm temperature and constant supply of nutrients in our bodies make them an ideal place for pathogens to flourish. Without a means of defense, we would be unable to prevent invasions by all sorts of pathogens and would soon die. Fortunately, our body provides us with nonspecific defenses and specific defenses. Other organisms also have this ability to defend themselves, but none have been studied as much as humans because our defense mechanisms are very important to the field of medicine.

**Immunity** begins with the nonspecific defenses, which are summarized in Figure 33.3. They include (1) barriers to entry such as the skin, (2) protective proteins such as complement and interferons, (3) phagocytes and natural killer cells, and (4) the inflammatory response. Because nonspecific defenses occur automatically, the term *innate immunity* is preferred by some. With this type of immunity, there is no recognition that an intruder has attacked before. Later, we will discuss specific defenses, which are directed against particular pathogens and do exhibit memory.

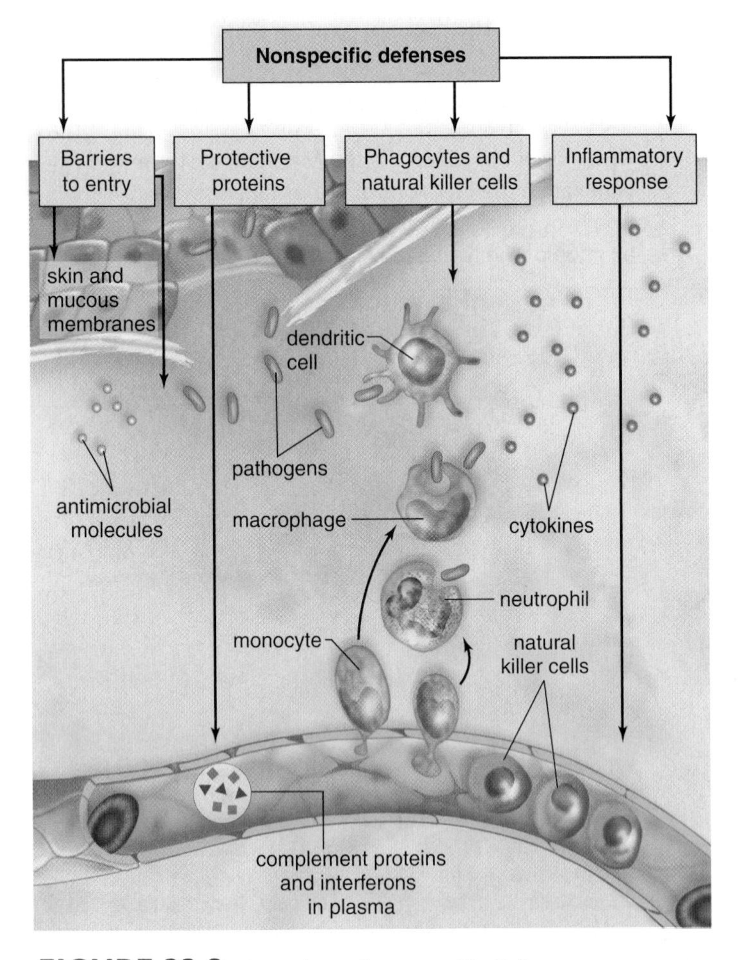


**FIGURE 33.3** **Overview of nonspecific defenses.**

Nonspecific defenses act rapidly to detect and respond to an infection by any and all pathogens and cancer cells. Nonspecific defenses include barriers to entry, protective proteins, phagocytes and natural killer cells, and the inflammatory response.

## Barriers to Entry

Barriers to entry by pathogens include non-chemical, mechanical barriers, such as the skin and the mucous membranes lining the respiratory, digestive, and urinary tracts. For example, the upper layers of our skin are composed of dead, keratinized cells that form an impermeable barrier. But when the skin has been injured, one of the first concerns is the possibility of an infection. We are also familiar with the importance of sterility before an injection is given. The injection needle and the skin must be free of pathogens. This testifies to the importance of skin as a defense against invasion by a pathogen.

The mucus of mucous membranes physically ensnares microbes. The upper respiratory tract is lined by ciliated cells that sweep mucus and trapped particles up into the throat, where they can be swallowed or expectorated (coughed out). In addition, the various bacteria that normally reside in the intestine and other areas, such as the vagina, prevent pathogens from taking up residence.

Barriers to entry also include antimicrobial molecules. Oil gland secretions contain chemicals that weaken or kill certain bacteria on the skin; mucous membranes secrete lysozyme, an enzyme that can lyse bacteria; and the stomach has an acidic pH, which inhibits the growth of many types of bacteria, or may even kill them.

## Inflammatory Response

Whenever tissue is damaged by physical or chemical agents or by pathogens, a series of events occurs that is known as the **inflammatory response.**

An inflamed area has four outward signs: redness, heat, swelling, and pain. All of these signs are due to capillary changes in the damaged area. Figure 33.4 illustrates the participants in the inflammatory response. Chemical mediators, such as **histamine,** released by damaged tissue cells, and mast cells cause the capillaries to dilate and become more permeable. Excess blood flow, due to enlarged capillaries, causes the skin to redden and become warm. Increased permeability of the capillaries allows proteins and fluids to escape into the tissues, resulting in swelling. The swollen area stimulates free nerve endings, causing the sensation of pain.

Migration of phagocytes, namely neutrophils and monocytes, also occurs during the inflammatory response. Neutrophils and monocytes are amoeboid and can change shape to squeeze through capillary walls and enter tissue fluid. Also present are dendritic cells, notably in the skin and mucous membranes, and macrophages, both of which are able to devour many pathogens and still survive (Fig. 33.5). Macrophages also release colony-stimulating factors, cytokines that pass by way of the blood to the red bone marrow, where they stimulate the production and release of white blood cells, primarily neutrophils. As the infection is being

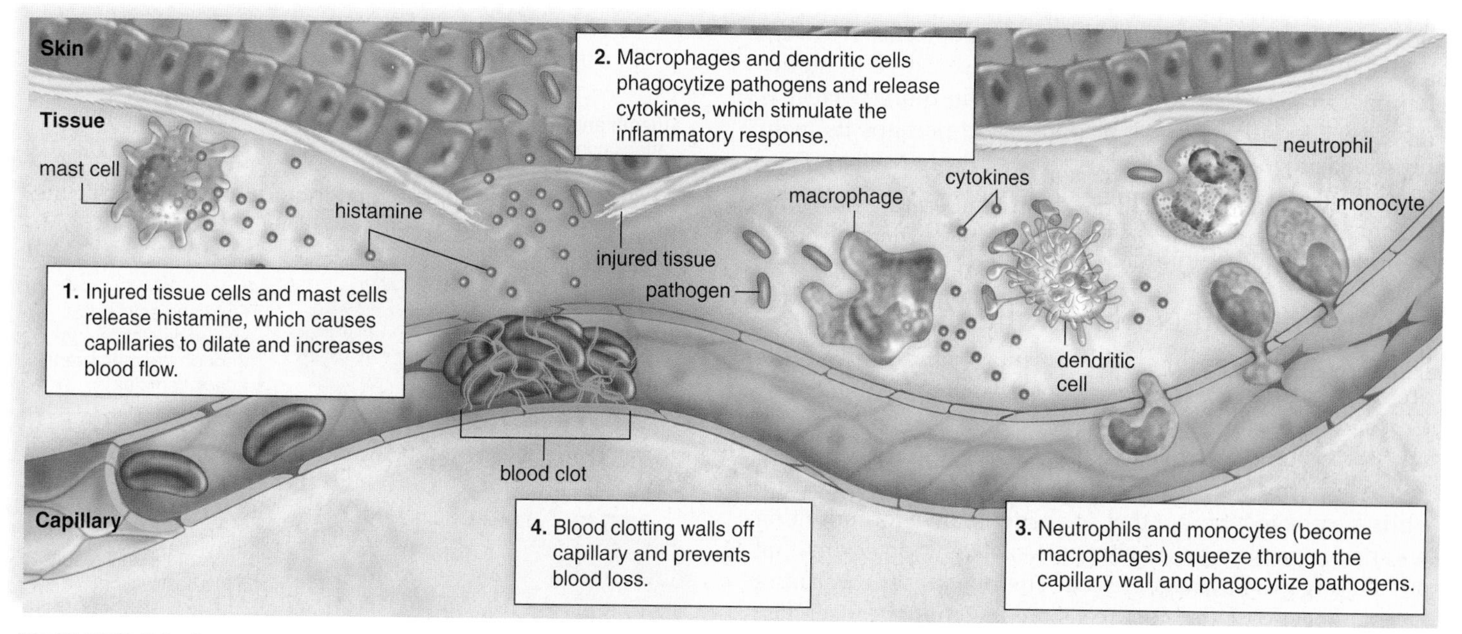


**FIGURE 33.4 Inflammatory response.**

Due to capillary changes in a damaged area and the release of chemical mediators, such as histamine by mast cells, an inflamed area exhibits redness, heat, swelling, and pain. The inflammatory response can be accompanied by other reactions to the injury. Macrophages and dendritic cells, present in the tissues, phagocytize pathogens, as do neutrophils, which squeeze through capillary walls from the blood. Macrophages and dendritic cells release cytokines, which stimulate the inflammatory and other immune responses. A blood clot can form to seal a break in a blood vessel.

overcome, some phagocytes die. These—along with dead tissue cells, dead bacteria, and living white blood cells—form pus, a whitish material. The presence of pus indicates that the body is trying to overcome an infection.

The inflammatory response can be accompanied by other responses to the injury. A blood clot can form to seal a break in a blood vessel. Antigens, chemical mediators, dendritic cells, and macrophages move through the tissue fluid and lymph to the lymph nodes. There, B cells and T cells are activated to mount a specific defense to the infection.

Sometimes an inflammation persists, and the result is chronic inflammation that is often treated by administering anti-inflammatory agents such as aspirin, ibuprofen, or cortisone. These medications act against the chemical mediators released by the white blood cells in the damaged area.

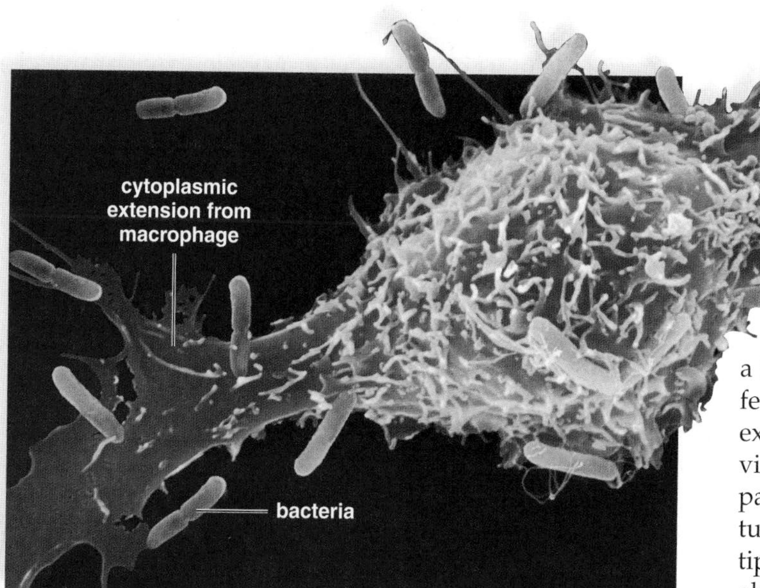


**FIGURE 33.5 Macrophage engulfing bacteria.**

Monocyte-derived macrophages are the body's scavengers. They engulf microbes and debris in the body's fluids and tissues, as illustrated in this colorized scanning electron micrograph.

### *Fever*

An illness may result in a fever, an elevated body temperature. To treat or not to treat a fever is controversial within the medical field for, in some instances, a fever may be beneficial. Perhaps a fever is the body's way of informing us that something is wrong. Alternatively, a fever could be part of our first line of defense. At times, a fever may directly participate in overcoming an illness. For example, a fever can contribute to the host's defense by providing an unfavorable environment for the invader. Some pathogens have very strict temperature requirements, and turning up the heat on them can slow their ability to multiply and thrive. In fact, supporting this hypothesis is the observation that increasing the body temperature in mice has been shown to decrease death rates and shorten the recovery time linked to many infectious agents.

Other medical experts believe that the main function of a fever is to stimulate immunity. In support of this hypothesis, a fever has been shown to limit the growth of tumor cells more severely than that of normal body cells. This suggests either that tumor cells are directly sensitive to higher temperatures or that a fever stimulates immunity. Heat is a part of the inflammatory response, and perhaps its main function is to jump-start the response of the body.

While data concerning the benefits of fever are inconclusive, the general consensus is that an extreme fever should be treated but milder cases may be best left alone.

## Phagocytes and Natural Killer Cells

Several types of white blood cells are phagocytic. **Neutrophils** are cells that are able to leave the bloodstream and phagocytize (engulf) bacteria in connective tissues. They have various other ways of killing bacteria also. For example, their granules release antimicrobial peptides called defensins. **Eosinophils** are phagocytic, but they are better known for mounting an attack against animal parasites such as tapeworms that are too large to be phagocytized. The two most powerful of the phagocytic white blood cells are **macrophages** (see Fig. 33.5) and macrophage-derived **dendritic cells.** They engulf pathogens, which are then destroyed by enzymes when their endocytic vesicles combine with lysosomes. Dendritic cells are found in the skin; once they devour pathogens, they travel to lymph nodes, where they stimulate natural killer cells or lymphocytes. Macrophages are found in all sorts of tissues, where they voraciously devour pathogens and then stimulate lymphocytes to carry on specific immunity.

**Natural killer (NK) cells** are large, granular lymphocytes that kill virus-infected cells and cancer cells by cell-to-cell contact. NK cells do their work while specific defenses are still mobilizing, and they produce cytokines that promote specific defenses.

What makes NK cells attack and kill a cell? First, they normally congregate in the tonsils, lymph nodes, and spleen, where they are stimulated by dendritic cells before they travel forth. Then, NK cells look for a self-protein on the body's cells. As may happen, if a virus-infected cell or a cancer cell has lost its self-proteins, the NK cell kills it in the same manner used by cytotoxic T cells. Unlike cytotoxic T cells, NK cells are not specific; they have no memory; and their numbers do not increase after stimulation.

## Protective Proteins

**Complement** is composed of a number of blood plasma proteins that "complement" certain immune responses, which accounts for their name. These proteins are continually present in the blood plasma but must be activated by pathogens to exert their effects. Complement helps destroy pathogens in three ways:

1. Enhanced inflammation. Complement proteins are involved in and amplify the inflammatory response because certain ones can bind to **mast cells** (type of white blood cell in tissues) and trigger histamine release, and others can attract phagocytes to the scene.
2. Some complement proteins bind to the surface of pathogens already coated with antibodies, which ensures that the pathogens will be phagocytized by a neutrophil or macrophage.
3. Certain other complement proteins join to form a membrane attack complex that produces holes in the surface of some bacteria and viruses. Fluids and salts then enter the bacterial cell or virus to the point that it bursts (Fig. 33.6).

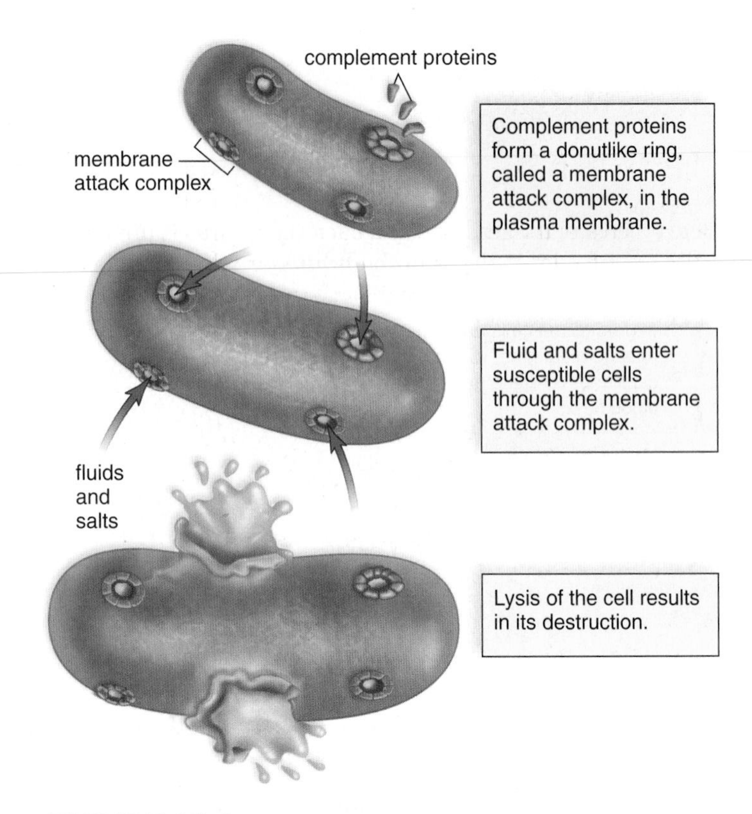


**FIGURE 33.6 Action of the complement system against a bacterium.**

When complement proteins in the blood plasma are activated by an immune response, they form a membrane attack complex that makes holes in bacterial cell walls and plasma membranes, allowing fluids and salts to enter until the cell eventually bursts.

**Interferons** are **cytokines,** soluble proteins that affect the behavior of other cells. Interferons are made by virus-infected cells. They bind to the receptors of noninfected cells, causing them to produce substances that interfere with viral replication. Interferons, now available as a biotechnology product, are used to treat certain viral infections, such as hepatitis C.

### Check Your Progress 33.2

1. What are some examples of the body's nonspecific defenses?
2. Which blood cells, in particular, should you associate with nonspecific defenses, and how do they function?
3. Why are the complement proteins so named?

# 33.3 Specific Defense Against Disease

When nonspecific defenses have been inadequate to stop an infection, specific defense comes into play. Immunity is complete when a pathogen such as a bacteria or virus is unable to cause an infection now or in the future, or when people are able to avoid cancer. These steps accomplish a specific defense and from them you can see why the term *acquired immunity* is preferred by some. During the first step, *recognition* of a particular molecule, called an **antigen,** occurs. Some antigens are termed **foreign antigens** because the body does not produce them. Pathogens, cancer cells, and transplanted tissues and organs bear antigens the immune system usually recognizes as foreign. Other antigens are termed **self-antigens** because the body itself produces them. (The self-proteins mentioned on the previous page are self-antigens.) It is unfortunate when the immune system reacts to the body's own pancreatic cells (causing diabetes mellitus) or to nerve fiber sheaths (causing multiple sclerosis), but fortunate when the immune system can destroy the cancerous cells of a tumor. Second, a *response* occurs. Unlike the nonspecific defenses, which occur immediately, it usually takes five to seven days to mount a specific defense. Third, the immune system can *remember* antigens it has met before. This is the reason, for example, that once we recover from the measles, we usually do not get the disease a second time. The reaction time for a nonspecific defense is always about the same, but if the body is immune, the reaction time for a specific defense is quite short the next time it encounters the same antigen.

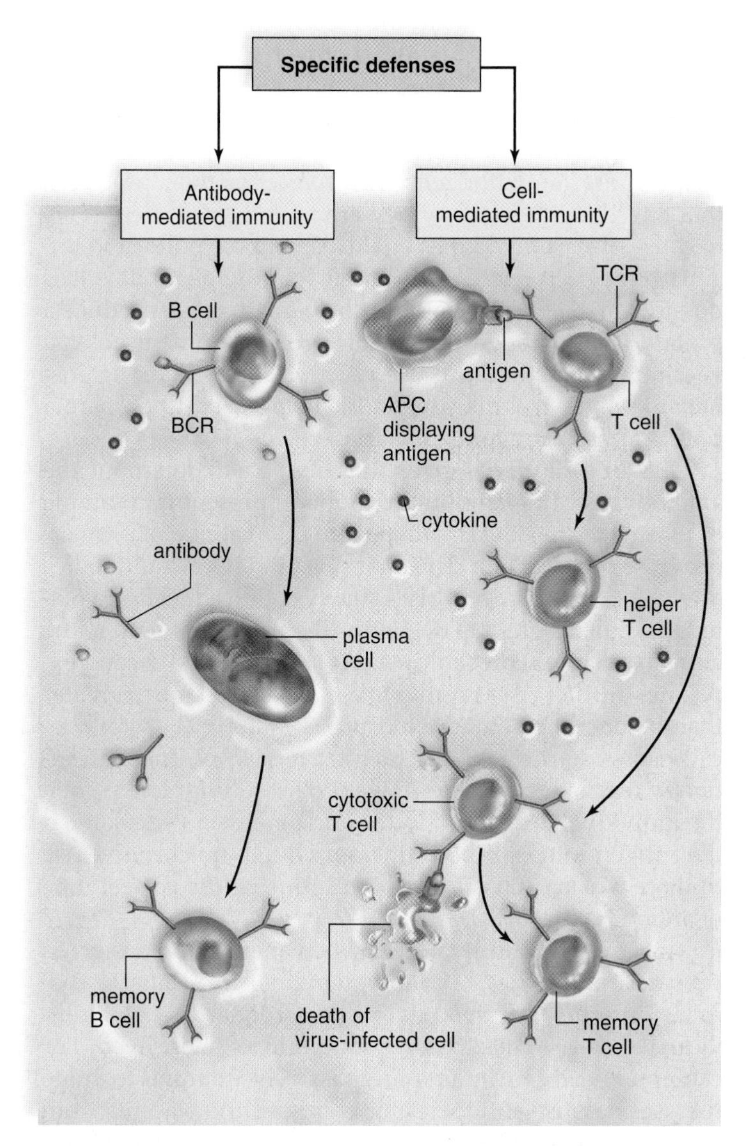


**FIGURE 33.7 Overview of specific defenses.**

If nonspecific defenses are inadequate, specific defenses come into play, and attack is then directed against a particular antigen (colored green). The antigen is a part of a pathogen or cancer cell. On the *left*, B cells are responsible for antibody-mediated immunity, which involves the production of antibodies by plasma cells and the production of membory B cells. On the *right*, T cells are responsible for cell-mediated immunity. They recognize an antigen only after it is presented to them by an APC. Then, T cells kill virus-infected or cancer cells on contact.

Specific defenses primarily depend on the two types of lymphocytes, called B cells and T cells (Fig. 33.7). Both B cells and T cells are manufactured in the red bone marrow. B cells mature there, but T cells mature in the thymus. These cells are capable of recognizing antigens because they have specific **antigen receptors** that combine with antigens. B cells have **B-cell receptors (BCRs),** and T cells have **T-cell receptors (TCRs).** Each lymphocyte has receptors that will combine with only one type of antigen. If a particular B cell could respond to an antigen—a molecule projecting from the bacterium *Streptococcus pyogenes* for example—it would not react to any other antigen. It is often said that the receptor and the antigen fit together like a lock and a key. Remarkably, diversification occurs to such an extent during maturation that there are specific B cells and/or T cells for any possible antigen we are likely to encounter during a lifetime.

B cells are responsible for **antibody-mediated immunity** (Fig. 33.7, *left*). When B cells encounter an antigen, a BCR recognizes it directly and combines with it right away. Then, the B cell gives rise to **plasma cells,** which produce specific antibodies. These antibodies can react to the same antigen as the original B cell. Therefore, an antibody has the same specificity as the BCR. Some progeny of the activated B cells become memory B cells, so called because these cells always "remember" a particular antigen and make us immune to a particular illness, but not to any other illness.

In contrast to B cells, T cells are responsible for **cell-mediated immunity** (Fig. 33.7, *right*). T cells do not recognize an antigen directly and instead the antigen must be presented to them by an **antigen-presenting cell (APC).** Macrophages and dendritic cells are APCs. T cells exist primarily as **helper T cells,** which regulate specific immunity, and **cytotoxic T cells,** which attack and kill virus-infected cells and cancer cells. Some cytotoxic T cells become memory T cells, ever ready to defend against the same virus or kill the same type of cancer cell again.

Because B and T cells defend us from disease by specifically reacting to antigens, they can be likened to special forces that can attack selected targets without harming nearby residents (cells).

## B Cells and Antibody-Mediated Immunity

The **clonal selection model** describes what happens when a *B-cell receptor (BCR)* combines with an antigen (Fig. 33.8). ❶ Many B cells are present, but only the one that has BCRs that can combine with the specific antigen clone goes on to divide and produce many new cells. Therefore, the antigen is said to "select" the B cell that will clone. Also, at this time, cytokines secreted by helper T cells stimulate B cells to clone. ❷ Defense by B cells is called *antibody-mediated immunity* because most members of the clone become plasma cells that produce specific antibodies. It is also called humoral immunity because these antibodies are present in blood and lymph. (A *humor* is any fluid normally occurring in the body.) ❸ Some progeny of activated B cells become **memory B cells,** which are the means by which long-term immunity is possible. Once the threat of an infection has passed, the development of new plasma cells ceases, and those present undergo apoptosis.

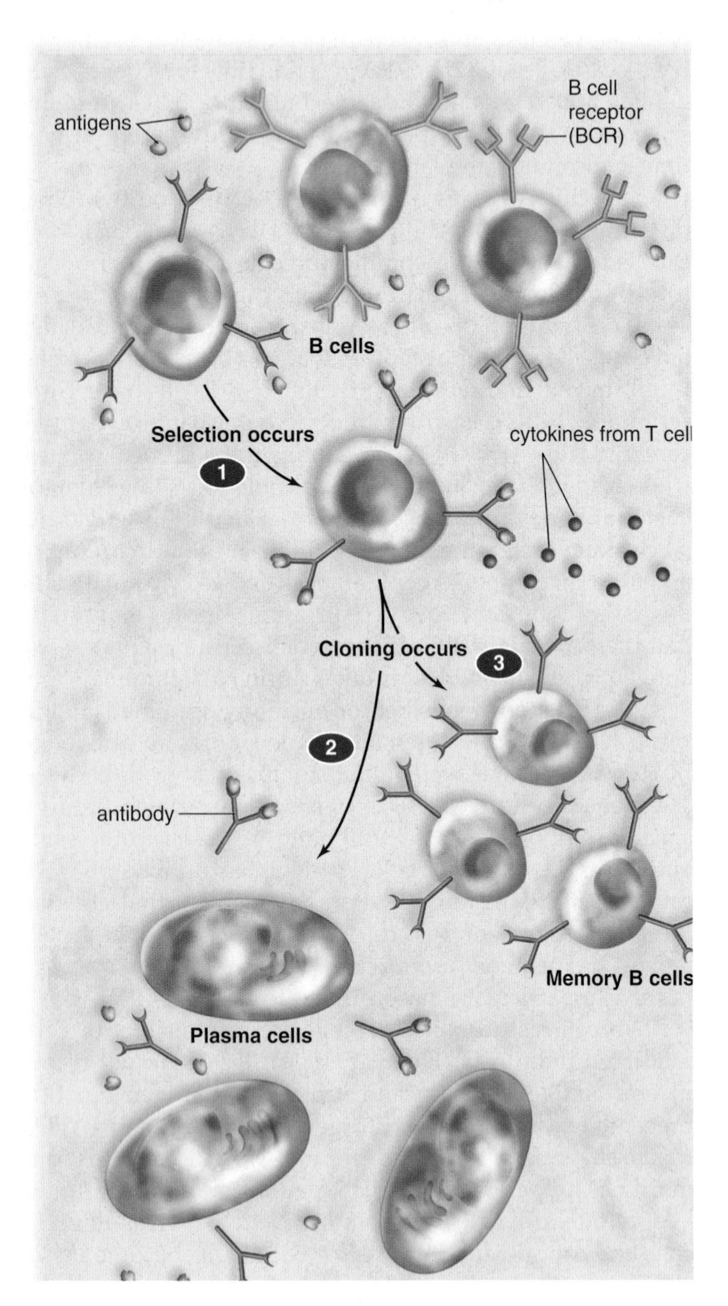


**FIGURE 33.8 Clonal selection model as it applies to B cells.**

Activation of a B cell occurs when its B-cell receptor (BCR) can combine with an antigen (colored green). In the presence of cytokines, this B cell undergoes clonal expansion, producing many plasma cells that secrete antibodies specific to the antigen.

### *Active Versus Passive Immunity*

**Active immunity** occurs when an individual produces a supply of antibodies. For example, when you catch the measles, you recover because your body produces the antibodies that combine with the measle-causing viruses and bring about their destruction. Another way to become actively immune is to undergo immunization. **Immunization** involves the use of vaccines to bring about clonal expansion, not only of B cells, but also of T cells. Traditionally, vaccines are the pathogens themselves, or their products, that have been treated so they are no longer virulent (able to cause disease). Vaccines against smallpox, polio, and tetanus have been successfully used worldwide. Today, it is possible to genetically engineer bacteria to mass-produce a protein from pathogens, and this protein can be used as a vaccine. This method was used to produce a vaccine against hepatitis B, a viral disease, and is being used to prepare a potential vaccine against malaria.

After a vaccine is given, it is possible to determine the antibody titer (the amount of antibody present in a sample of plasma). After the first exposure to a vaccine, a primary response occurs. For a period of several days, no antibodies are present; then the titer rises slowly, followed by first a plateau and then a gradual decline as the antibodies bind to the antigen or simply break down (Fig. 33.9). After a second exposure, the titer rises rapidly to a plateau level much greater than before. The second exposure is called a "booster" because it boosts the antibody titer to a high level. The high antibody titer is expected to prevent disease symptoms when the individual is exposed to the antigen. Even years later, if the antigen enters the body, memory B cells quickly give rise to more plasma cells capable of producing the correct type of antibody.

**Passive immunity** occurs when an individual is given prepared antibodies (immunoglobulins) to combat a disease. Since these antibodies are not produced by the individual's plasma cells, passive immunity is short-lived. For example, newborn infants are passively immune to some diseases because antibodies have crossed the placenta from the mother's blood (Fig. 33.10). These antibodies soon disappear, however, so that within a few months, infants become more susceptible to infections. Breast-feeding prolongs the natural passive immunity an infant receives from its mother because antibodies are present in the mother's milk.

Even though passive immunity does not last, it is sometimes used to prevent illness in a patient who has been

**FIGURE 33.9**
**Antibody titers.**

During immunization, the primary response after the first injection of a vaccine is minimal, but the secondary response, which may occur after the second injection, shows a dramatic rise in the amount of antibody present in plasma.

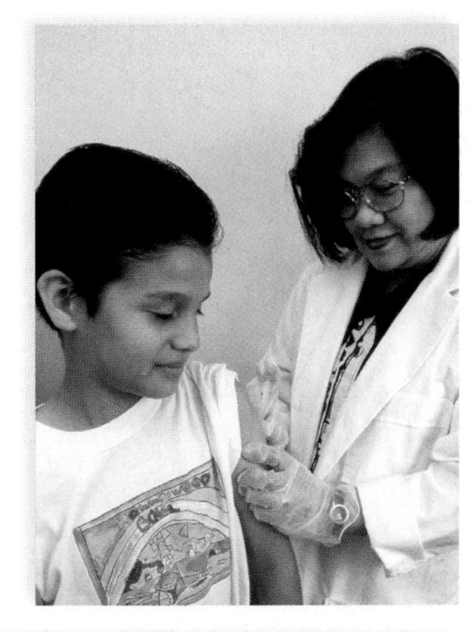

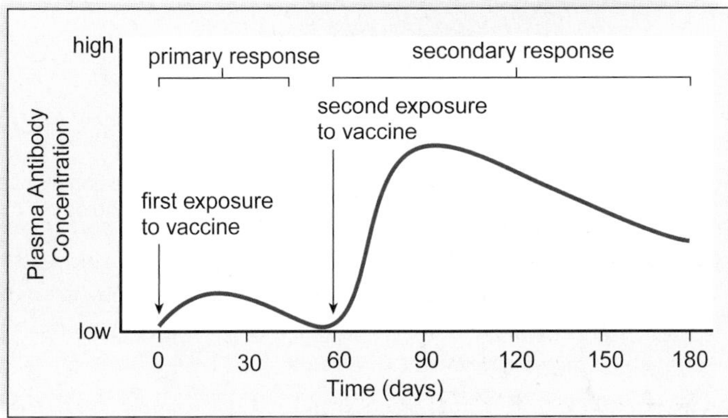


unexpectedly exposed to an infectious disease. Artificial passive immunity is used in the emergency treatment of rabies, measles, tetanus, diphtheria, botulism, hepatitis A, and snakebites. Usually, the patient receives a gamma globulin injection (serum that contains antibodies), extracted from a large, diverse adult population.

## *Structure of Antibodies*

The basic unit that composes antibody structure is a Y-shaped protein molecule with two arms. Each arm has a "heavy" (long) polypeptide chain and a "light" (short) polypeptide chain (Fig. 33.11). These chains have constant regions, located at the trunk of the Y, where the sequence of amino acids is set. The class of antibody is determined by the structure of the antibody's constant region. The variable regions form an antigen-binding site, and their shape is specific to a particular antigen. The antigen combines with the antibody at the antigen-binding site in a lock-and-key manner.

Another name for an antibody is **immunoglobulin (Ig).** The most typical antibody, called **IgG,** is one Y-shaped molecule (Fig. 33.11). IgG antibodies are the major type in blood, and lesser amounts are found in lymph and tissue fluid. IgG antibodies bind to pathogens and their toxins (poisons they secrete). IgG antibodies can cross the placenta from a mother to her fetus, so the newborn has a temporary, particular immune response. Other types of antibodies contain two or more Y-shaped molecules. Antibodies belonging to the M class are pentamers—that is, clusters of five Y-shaped molecules linked together. IgM antibodies are the first antibodies produced by a newborn's body.

The antigen-antibody reaction can take several forms. When antigens are a part of a pathogen, antibodies may coat them completely, a process called neutralization. Often, the reaction produces a clump of antigens combined with antibodies, termed an immune complex. The antibodies in an immune complex are like a beacon that attracts white blood cells that move in for the kill.

**FIGURE 33.10**
**Passive immunity.**

Breast-feeding is believed to prolong the passive immunity an infant receives from the mother during pregnancy because antibodies are present in the mother's milk.

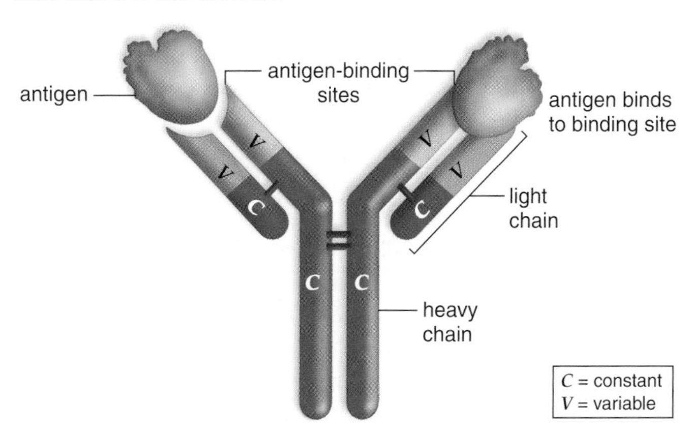


**FIGURE 33.11 Structure of antibodies.**

An antibody contains two heavy (long) polypeptide chains and two light (short) chains arranged so there are two variable regions, where a particular antigen is capable of binding with the antibody. The shape of the antigen fits the shape of the binding site.

## Monoclonal Antibodies

Every plasma cell derived from a single B cell secretes antibodies against a specific antigen. These are called **monoclonal antibodies** because all of them are the same type. One method of producing monoclonal antibodies in vitro (outside the body in the laboratory) is depicted in Figure 33.12. B cells are removed from an animal (usually mice are used) and exposed to a particular antigen. (1) The resulting plasma cells are fused with myeloma cells (malignant plasma cells that live and divide indefinitely). The fused cells are called hybridomas—*hybrid-* because they result from the fusion of two different cells, and *-oma* because one of the cells is a cancer cell. (2) The hybridomas then secrete large amounts of the specified monoclonal antibody, which recognizes only a single antigen of interest.

**Research Uses for Monoclonal Antibodies.** The ability to quickly produce monoclonal antibodies in the laboratory has made them an important tool for academic research. Monoclonal antibodies are very useful because of their extreme specificity for only a particular molecule. A monoclonal antibody can be used to select out a specific molecule among many others, much like finding needles in a haystack. Now the molecule can be purified from all the others that are also present in a sample. In this way, monoclonal antibodies have simplified formerly tedious laboratory tasks, allowing investigators more time to focus on other priorities.

**Medical Uses for Monoclonal Antibodies.** Monoclonal antibodies also have many applications in medicine. For instance, they can now be used to make quick and certain diagnoses of various conditions. Today, a monoclonal antibody is used to signify pregnancy by detecting a particular hormone in the urine of a woman after she becomes pregnant. Thanks to this technology, pregnancy tests that once required a visit to a doctor's office and the use of expensive laboratory equipment can now be performed in the privacy and comfort of a woman's own home, at minimal expense.

Monoclonal antibodies can be used not only to diagnose infections and illnesses but also to fight them. Many bacteria and viruses possess unique proteins on their cell surfaces that make them easily recognized by an appropriate monoclonal antibody. When binding occurs with a certain monoclonal antibody but not another, a physician knows what type of infection is present. And because monoclonal antibodies can distinguish some cancer cells from normal tissue cells, they may also be used to identify cancers at very early stages when treatment can be most effective.

Finally, monoclonal antibodies have also shown promise as potential drugs to help fight disease. RSV, a common virus that causes serious respiratory tract infections in very young children, is now being successfully treated with a monoclonal antibody drug. The antibody recognizes a protein on the viral surface, and when it binds very tightly to the surface of the virus, the patient's own immune system can easily recognize the virus and destroy it before it has a chance to cause serious illness.

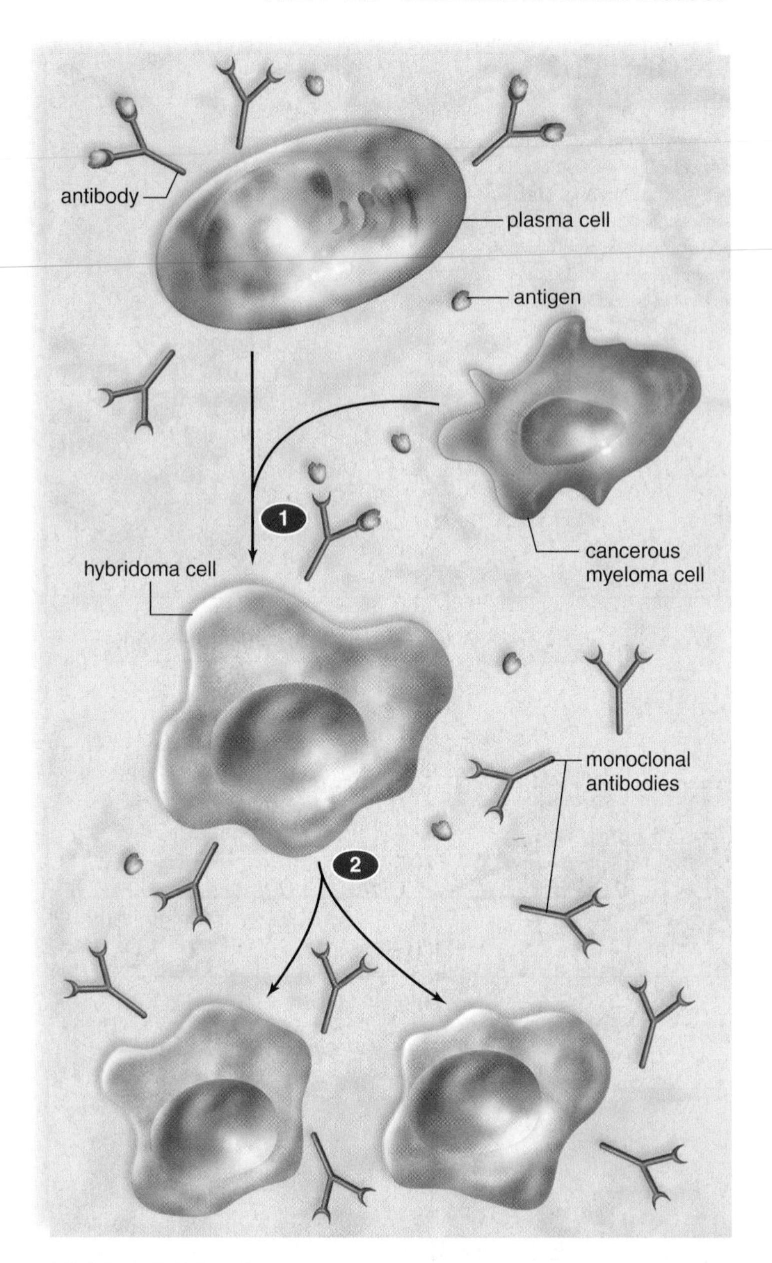


**FIGURE 33.12 Production of monoclonal antibodies.**

Plasma cells of the same type (derived from immunized mice) are fused with myeloma (cancerous) cells, producing hybridoma cells that are "immortal." Hybridoma cells divide and continue to produce the same type of antibody, called monoclonal antibodies.

Other illnesses, such as cancer, are also being successfully treated with monoclonal antibodies. Since these antibodies are able to distinguish between cancerous and normal cells, they have been engineered to carry radioisotopes or toxic drugs to tumors so that cancer cells can be selectively destroyed without damaging other body cells. In short, monoclonal antibodies are helping scientists in their research and physicians in their attempts to diagnose and cure patients.

*science focus*

# Antibody Diversity

In 1987, Susumu Tonegawa (Fig. 33A*a*) became the first Japanese scientist to win the Nobel Prize in Physiology or Medicine, after dedicating himself to finding the solution to an engrossing puzzle. Immunologists and geneticists knew that each B cell makes an antibody especially equipped to recognize the specific shape of a particular antigen. But they did not know how the human genome contained enough genetic information to permit the production of up to 2 million different antibody types needed to combat all of the pathogens we are likely to encounter during our lives.

## The Puzzle

An antibody is composed of two light and two heavy polypeptide chains, which are divided into constant and variable regions. The constant region determines the antibody class, and the variable region determines the specificity of the antibody, because this is where an antigen binds to a specific antibody (see Fig. 33.11). Each B cell must have a genetic way to code for the variable regions of both the light and heavy chains.

## The Experiment

Tonegawa's colleagues say that he is a creative genius who intuitively knows how to design experiments to answer specific questions. In this instance, he examined the DNA sequences of lymphoblasts (immature lymphocytes) and compared them to mature B cells. He found that the DNA segments coding for the variable and constant regions were scattered throughout the genome in lymphoblasts, and that only certain of these segments were present in each mature antibody-secreting B cell, where they randomly came together and coded for a specific variable region. Later, the variable and constant regions are joined to give a specific antibody (Fig. 33A*b*). As an analogy, consider that each person entering a supermarket chooses various items for purchase, and that the possible combination of items in any particular grocery bag is astronomical. Tonegawa also found that mutations occur as the variable segments are undergoing rearrangements. Such mutations are another source of antibody diversity.

## A Prize Winner

Tonegawa received his BS in chemistry in 1963 at Kyoto University and earned his PhD in biology from the University of California at San Diego (UCSD) in 1969. After that, he worked as a research fellow at UCSD and the Salk Institute. In 1971, he moved to the Basel Institute for Immunology and began the experiments that eventually led to his Nobel Prize–winning discovery. Tonegawa also contributed to the effort to decipher the receptors of T cells. This was an even more challenging area of research than the diversity of antibodies produced by B cells. Since 1981, he has been a full professor at Massachusetts Institute of Technology (MIT), where he has a reputation for being an "aggressive, determined researcher" who often works late into the night.

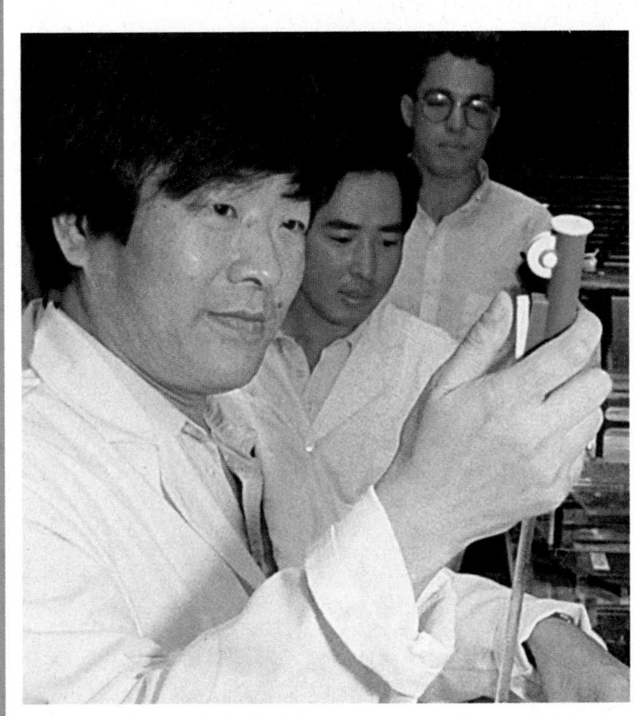

a. Susumu Tonegawa in the laboratory

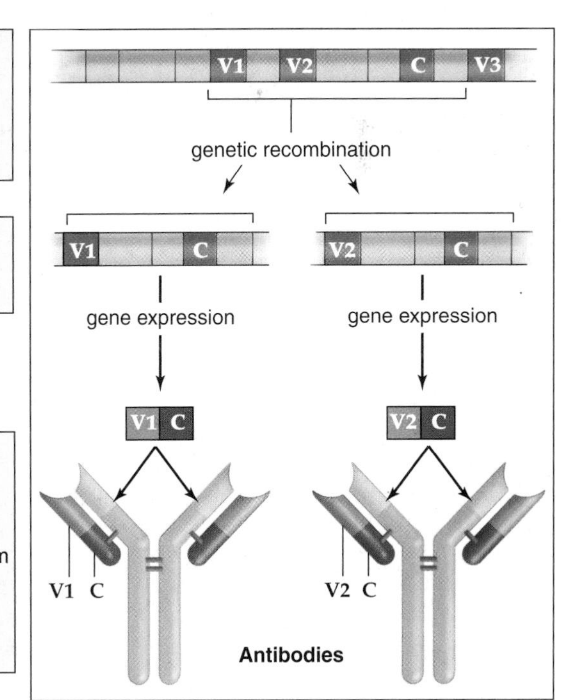


b. Antibody variable regions

**FIGURE 33A Antibody diversity.**
***a.*** *Susumu Tonegawa received a Nobel Prize for his findings regarding antibody diversity.* ***b.*** *Different genes for the variable regions of heavy and light chains are brought together during the production of B lymphocytes so that the antigen receptors of each one can combine with only a particular antigen.*

## T Cells and Cell-Mediated Immunity

T cells do not produce antibodies. Cell-mediated immunity is named for the action of cytotoxic T cells that directly attack diseased cells and cancer cells. Helper T cells, however, release cytokines that regulate both innate and acquired immunity and still other T cells are memory T cells that are ever ready to fight an infection.

### *Life History of T cells*

T cells are formed in red bone marrow before they migrate to the thymus, a gland that secretes thymic hormones. These hormones stimulate T cells to develop *T-cell receptors (TCRs).* When a T cell leaves the thymus, it has a unique TCR, just as B cells have a BCR. Unlike B cells, however, T cells are unable to recognize an antigen without help. The antigen must be displayed to them by an antigen-presenting cell (APC), such as a dendritic cell or a macrophage. After phagocytizing a pathogen, APCs travel to a lymph node or the spleen, where T cells also congregate. In the meantime, the APC has broken the pathogen apart in a lysosome. A piece of the pathogen is then displayed in an **MHC (major histocompatibility complex) protein** on the cell's surface. MHC proteins are self-antigens because they mark cells as belonging to a particular individual and, therefore, make transplantation of organs difficult. MHC proteins differ by the sequence of their amino acids, and the immune system will attack as foreign any tissue that bears MHC antigens different from those of the individual.

In Figure 33.13, the different types of T cells have specific TCRs represented by their different shapes and colors. (1) A macrophage presents an antigen only to a T cell that has a TCR capable of combining with a particular antigen (colored green). (2) A major difference in recognition of an antigen by helper T cells and cytotoxic T cells is that helper T cells only recognize an antigen in combination with MHC class II molecules while cytotoxic T cells only recognize an antigen in combination with MHC class I molecules. In Figure 33.13 a cytotoxic T cell binds to an antigen displayed in combination with an MHC I and undergoes cloning during which many copies of the cytotoxic T cell are produced. (3) Some of the cloned cells become **memory T cells.**

(4) As the illness disappears, the immune response wanes, and active T cells become susceptible to apoptosis, which is programmed cell death (see Fig. 9.2). As mentioned previously, apoptosis contributes to homeostasis by regulating the number of cells present in an organ, or in this case, in the immune system. When apoptosis does not occur as it should, T cell cancers (e.g., lymphomas and leukemias) can result. Also, in the thymus, any T cell that has the potential to destroy the body's own cells undergoes apoptosis.

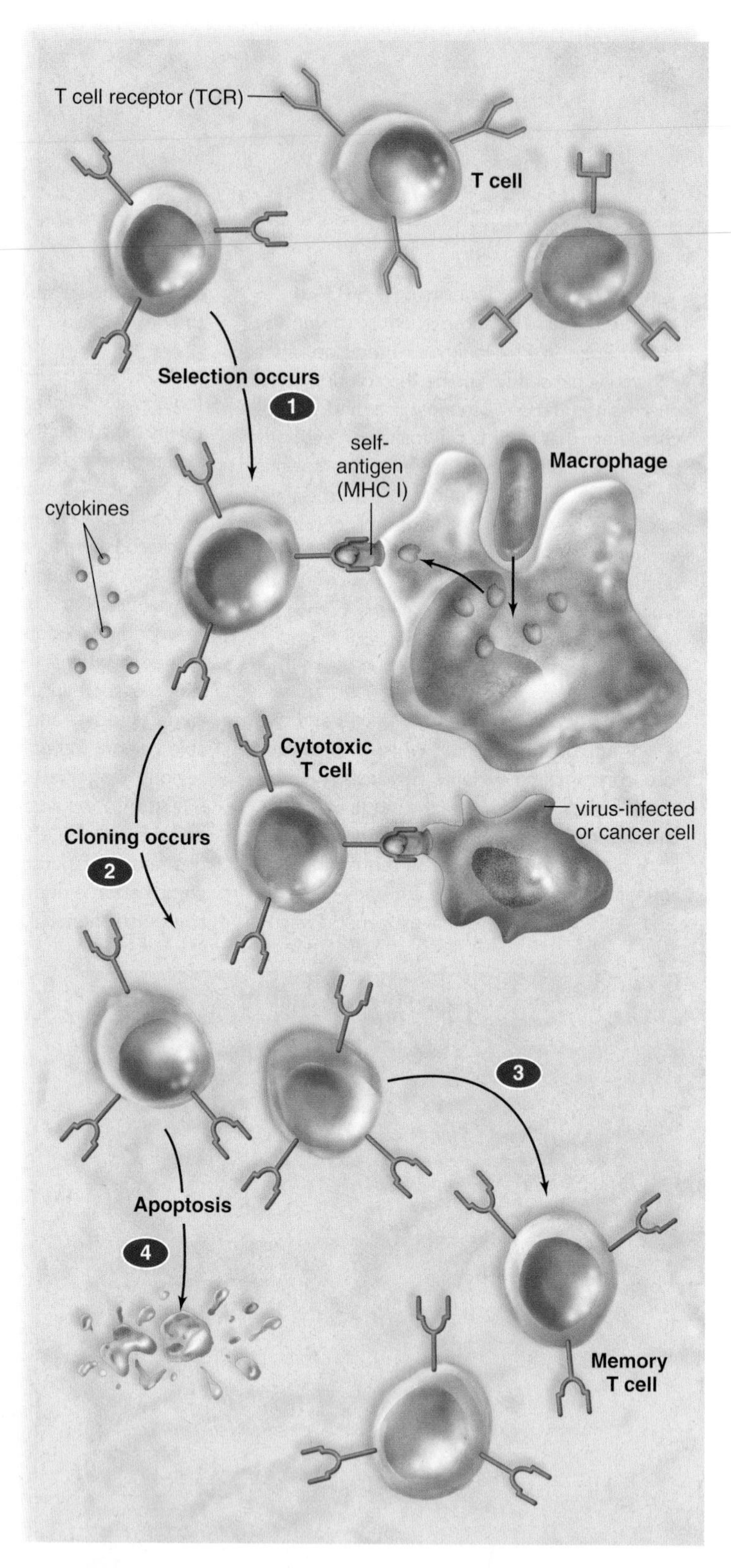


**FIGURE 33.13 Clonal selection model as it applies to T cells.**

Each T cell has a T-cell receptor (TCR) designated by a shape and color that will combine only with a specific antigen. Activation of a T cell occurs when its TCR can combine with an antigen. A macrophage presents the antigen (colored green) in the groove of an MHC I. Thereafter, the T cell undergoes clonal expansion, and many copies of the same type T cell are produced. After the immune response has been successful, the majority of T cells undergo apoptosis, but a small number are memory T cells. Memory T cells provide protection should the same antigen enter the body again at a future time.

### *Functions of Cytotoxic T Cells and Helper T Cells*

Cytotoxic T cells specialize in cell to cell combat. They have storage vacuoles containing perforins and storage vacuoles

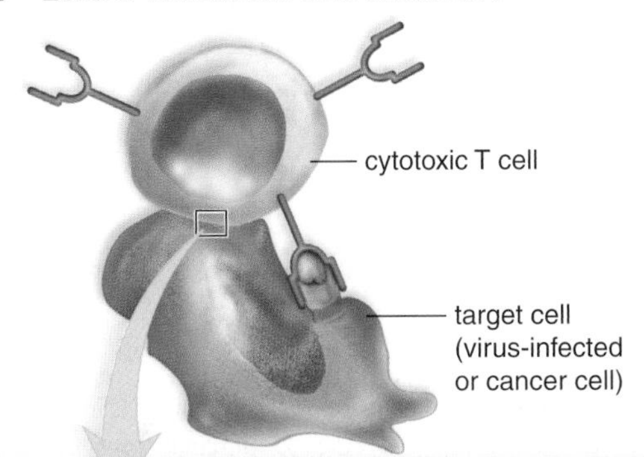


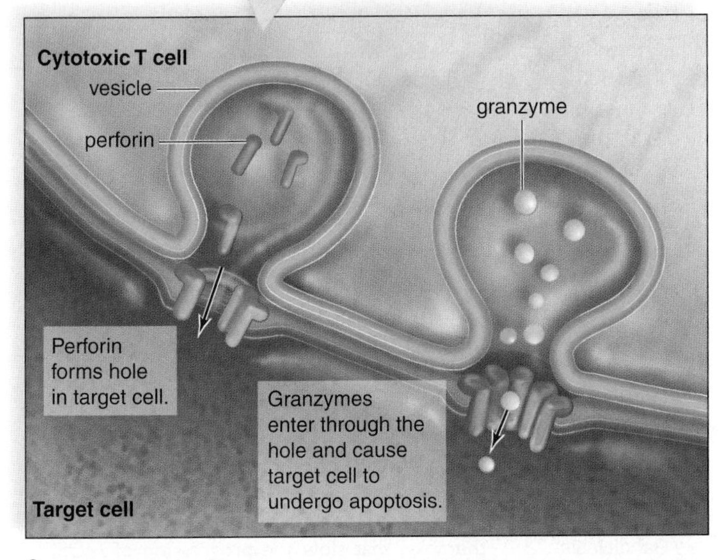


a.

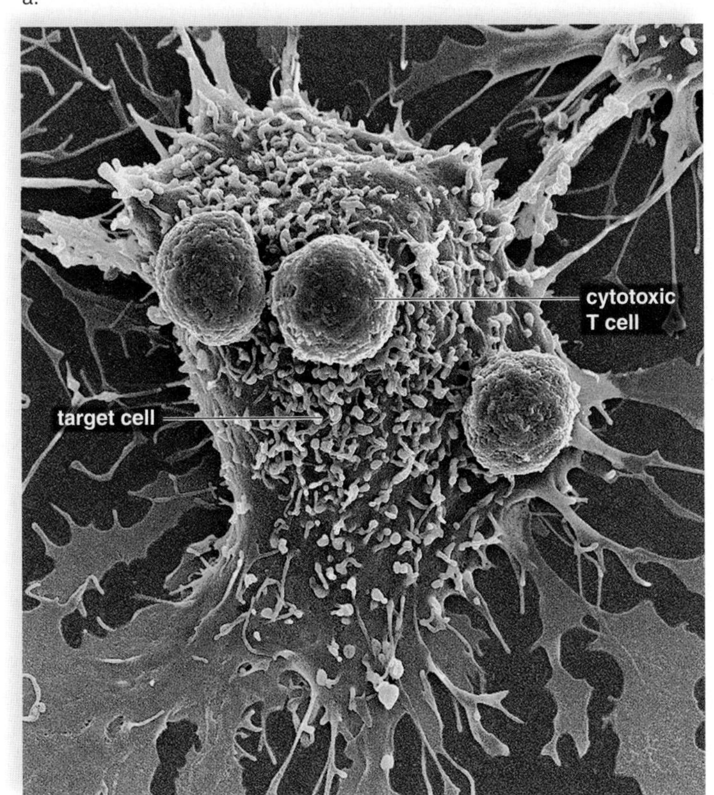


b. SEM 1,250×

**FIGURE 33.14 Cell-mediated immunity.**

**a.** How a T cell destroys a virus-infected cell or cancer cell. **b.** The scanning electron micrograph shows cytotoxic T cells attacking and destroying a cancer cell (target cell).

containing enzymes called granzymes (Fig. 33.14). After a cytotoxic T cell is activated, it binds to a virus-infected cell or cancer cell, it releases perforin molecules, which perforate the plasma membrane, forming a pore. Cytotoxic T cells then deliver granzymes into the pore, and these cause the cell to undergo apoptosis and die. Once cytotoxic T cells have released the perforins and granzymes, they move on to the next target cell. Cytotoxic T cells are responsible for so-called cell-mediated immunity.

Helper T cells play a critical role in coordinating nonspecific defenses (innate immunity) and specific defenses (acquired immunity), including both cell-mediated immunity and also antibody-mediated immunity. How do helper T cells perform this function? After a helper T cell is activated, it secretes cytokines. Cytokines are chemical mediators that attract neutrophils, natural killer cells, and macrophages to where they are needed. Cytokines stimulate phagocytosis of pathogens, and they stimulate the clonal expansion of T and B cells. Cytokines are used for immunotherapy purposes also, as we shall discuss in the next section.

Because more and more immune cells are recruited by helper T cells, the number of pathogens eventually begins to wane. But by now the body is fully prepared to deal with this pathogen again. Therefore, it is said that the immune system possesses memory—it can often remember former antigens. Memory T cells, like memory B cells, are long-lived, and their number is far greater than the original number of T cells that could recognize a specific antigen. Therefore, when the same antigen enters the body later on, the immune response may occur so rapidly that no detectable illness occurs.

## *HIV Infections*

The primary host for an HIV (human immunodeficiency virus) is a helper T cell, but macrophages are also under attack. After HIV enters a host cell, it reproduces as described in Figure 20.4, and many progeny bud from the cell. In other words, the host helper T cell produces viruses that go on to destroy more helper T cells. Figure 33B in the Health Focus on page 626 describes the progression of an HIV infection over time. At first an individual is able to stay ahead of the virus by producing enough helper T cells to keep their number within the normal range. But gradually, as the HIV count rises, the helper T cell count drops to way below normal. Then the person comes down with what are called opportunistic infections—infections that would be unable to take hold in a person with a healthy immune system. Now the individual has AIDS (acquired immunodeficiency syndrome). An HIV infection is presently a treatable disease, but the regimen of medicines is very difficult to maintain, and viral resistance to these medications is becoming apparent. Therefore, it is much wiser for individuals to prevent becoming infected by following the recommendations given on page 613.

*health focus*

## Opportunistic Infections and HIV

AIDS (acquired immunodeficiency syndrome) is caused by the destruction of the immune system, following an HIV (human immunodeficiency virus) infection. An HIV infection leads to the eventual destruction of immune system cells, known as helper T lymphocytes or, simply, helper T cells. Then, the individual succumbs to many unusual types of infections that would not cause disease in a person with a healthy immune system. Such infections are known as "opportunistic infections" (OIs).

AIDS victim: Kaposi sarcoma is evident

HIV not only kills helper T cells by directly infecting them; it also causes many uninfected T cells to die by a variety of mechanisms, including apoptosis (programmed cell death or "cell suicide"). Many helper T cells are also killed by the person's own immune system as it tries to overcome the HIV infection. After initial infection with HIV, it may take up to ten years for an individual's helper T cells to become so depleted that the immune system can no longer organize a specific response to OIs (Fig. 33B). In a healthy individual, the number of helper T cells typically ranges from 800–1,000 cells per $mm^3$ of blood. The appearance of specific OIs can be associated with the helper T-cell count.

- Shingles. Painful infection with varicella-zoster (chickenpox) virus. Helper T-cell count of less than 500/$mm^3$.
- Candidiasis. Fungal infection of the mouth, throat, or vagina. Helper T-cell count of about 350/$mm^3$.
- *Pneumocystis* pneumonia. Fungal infection causing the lungs to become useless as they fill with fluid and debris. Helper T-cell count of less than 200/$mm^3$.

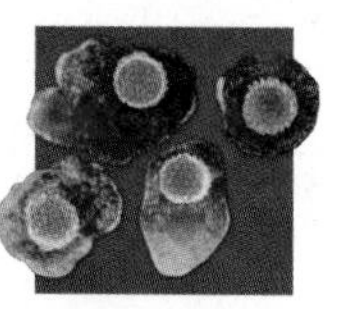

Chickenpox virus

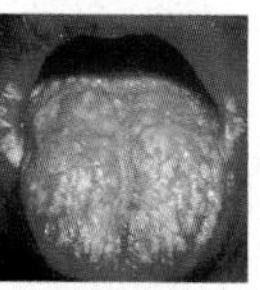

Candidiasis

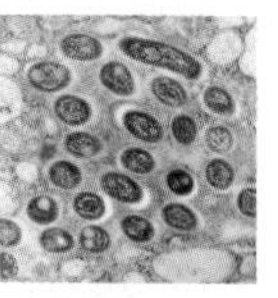

Pneumocystic pneumonia

- Kaposi's sarcoma. Cancer of blood vessels due to human herpesvirus 8 gives rise to reddish purple, coin-sized spots and lesions on the skin. Helper T-cell count of less than 200/$mm^3$.
- Toxoplasmic encephalitis. Protozoan infection characterized by severe headaches, fever, seizures, and coma. Helper T-cell count of less than 100/$mm^3$.
- *Mycobacterium avium* complex (MAC). Bacterial infection resulting in persistent fever, night sweats, fatigue, weight loss, and anemia. Helper T-cell count of less than 75/$mm^3$.
- Cytomegalovirus. Viral infection that leads to blindness; inflammation of the brain, throat ulcerations. Helper T-cell count of less than 50/$mm^3$.

Due to development of powerful drug therapies that slow the progression of AIDS, people infected with HIV in the United States are suffering lower incidence of OIs than in the 1980s and 1990s. It is hoped that a vaccine will be developed for AIDS one day.

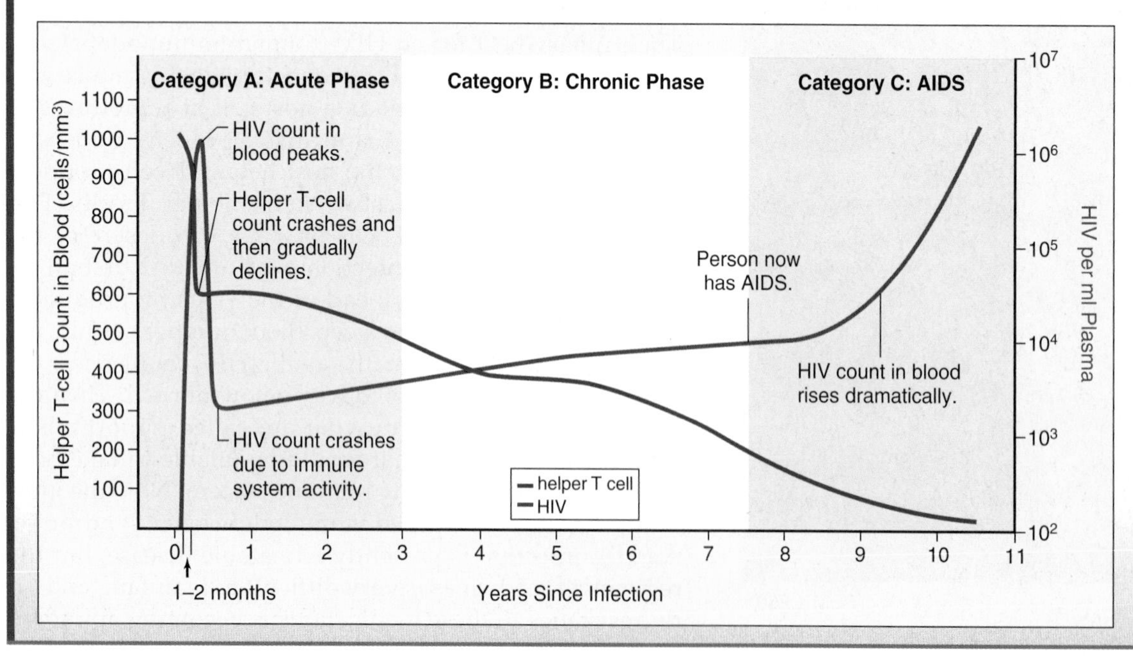


**FIGURE 33B Progression of HIV infection during its three stages, called categories A, B, and C.**
*In category A, the individual may have no symptoms or very mild symptoms associated with the infection. By category B, opportunistic infections have begun to occur, such as candidiasis, shingles, and diarrhea. Category C is characterized by more severe opportunistic infections and is clinically described as AIDS.*

## Cytokines and Cancer Therapy

The term cytokine simply means a soluble protein that acts as a signaling molecule. Because cytokines stimulate white blood cells, they have been studied as a possible adjunct therapy for cancer. We have already mentioned that interferons are cytokines. Interferons are produced by virus-infected cells, and they signal other cells of the need to prevent infection. Interferon has been investigated as a possible cancer drug, but so far it has proven to be effective only in certain patients, and the exact reasons for this, as yet, cannot be discerned. Also, interferon has a number of side effects that limit its use.

Cytokines called interleukins are produced by white blood cells, and they act to stimulate other white blood cells. Scientists actively engaged in interleukin research believe that interleukins will soon be used in addition to vaccines, for the treatment of chronic infectious diseases, and perhaps for the treatment of cancer. Interleukin antagonists may also prove helpful in preventing skin and organ rejection, autoimmune diseases, and allergies.

Because cancer cells carry an altered protein on their cell surface, they should be attacked and destroyed by cytotoxic T cells. Whenever cancer does develop, it is hypothesized that the cytotoxic T cells have not been activated. In that case, cytokines such as interleukins might awaken the immune system and lead to the destruction of the cancer. In one technique being investigated, researchers bioengineer APC cells withdrawn from the patient and activate the cells by culturing them in the presence of an interleukin. The engineered cells are then reinjected into the patient, who is given doses of interleukin to maintain the activity of the APC cells (Fig. 33.15).

Tumor necrosis factor (TNF) is a cytokine produced by macrophages that has the ability to promote the inflammatory response and to cause the death of cancer cells. Like the interferons and interleukins, TNF stimulates the body's immune cells to fight cancer. TNF also directly affects tumor cells, damaging them and the blood vessels within the tumor. Without an adequate blood supply, a cancerous tumor cannot thrive. However, researchers are still uncertain about exactly how TNF destroys tumors. Researchers have found that TNF therapy is most effective and least toxic when directed at a specific tumor site, rather than used as a general medicine. Clinical trials are under way.

### Check Your Progress 33.3

1. What steps accomplish a specific defense?
2. What are some examples of the immune system's use of a specific defense?
3. Which blood cells are mainly responsible for specific defense, and how do they function?

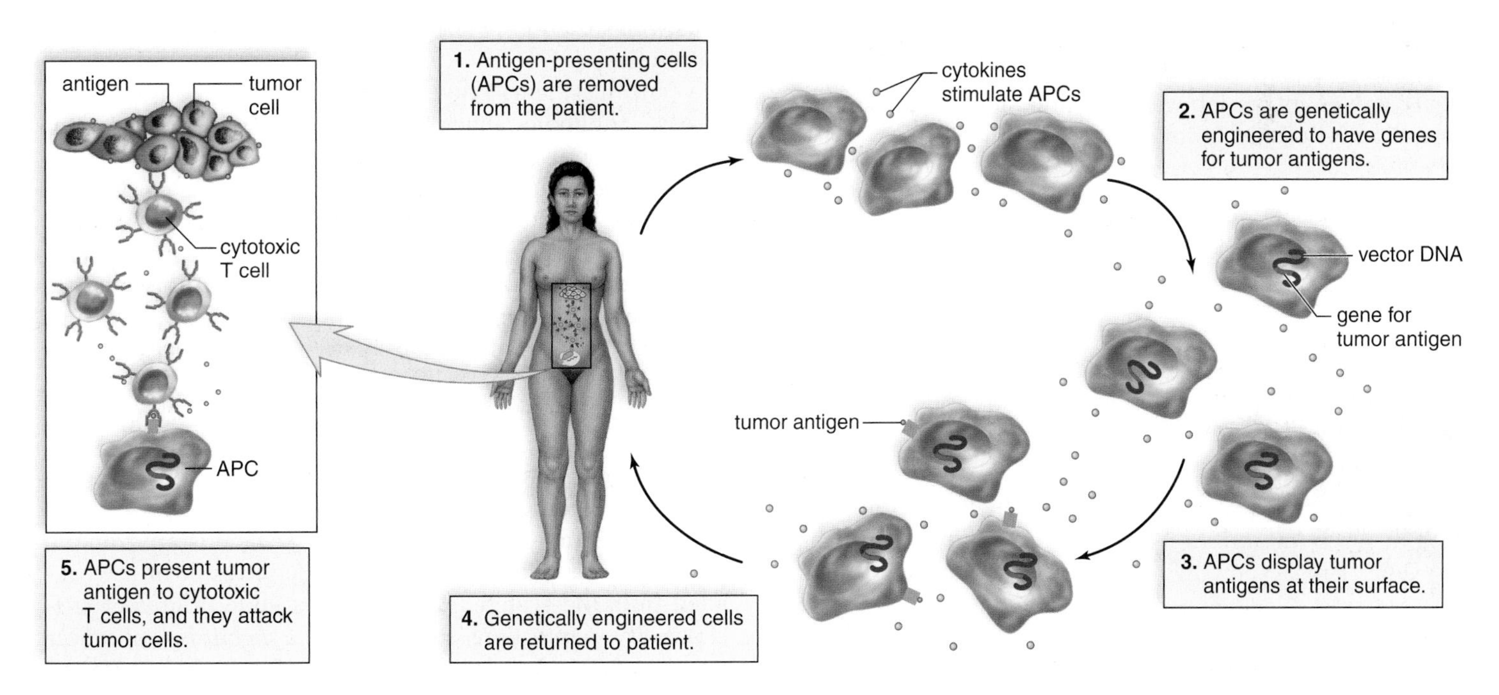


**FIGURE 33.15 Immunotherapy.**

1. Antigen-presenting cells (APCs) are removed and (2) are genetically engineered to (3) display tumor antigens. 4. After APCs are returned to the patient, (5) they present the antigen to cytotoxic T cells, which then kill tumor cells.

# 33.4 Immunity Side Effects

In certain instances, immunity works against the best interests of the body, as when it makes it more difficult to transplant organs from a donor to a recipient, causes autoimmune disorders, or leads to allergic reactions.

## Tissue Rejection

Certain organs, such as skin, the heart, and the kidneys, could be transplanted easily from one person to another if the body did not attempt to reject them. Rejection occurs because antibodies and cytotoxic T cells bring about the destruction of foreign tissues in the body. When rejection occurs, the immune system is correctly distinguishing between self and nonself.

Organ rejection can be controlled by carefully selecting the organ to be transplanted and administering immunosuppressive drugs. It is best if the transplanted organ has the same type of MHC antigens as those of the recipient, because cytotoxic T cells recognize foreign MHC antigens. Two well-known immunosuppressive drugs, cyclosporine and tacrolimus, both act by inhibiting the production of certain cytokines that stimulate cytotoxic T cells.

*Xenotransplantation*, the transplantation of animal tissues and organs into human beings, is another way to solve the problem of rejection. Genetic engineering can make pig organs less antigenic by removing the MHC antigens. The ultimate goal is to make pig organs as widely accepted as blood type O. Other researchers hope that tissue engineering, including the production of human organs by using stem cells, will one day do away with the problem of rejection. Scientists have recently grown new heart valves in the laboratory using stem cells gathered from amniotic fluid following amniocentesis.

## Disorders of the Immune System

We will discuss disorders in which the immune system mistakenly attacks the body's own cells as if they bear foreign antigens and other disorders in which immunity is lacking and infections are common.

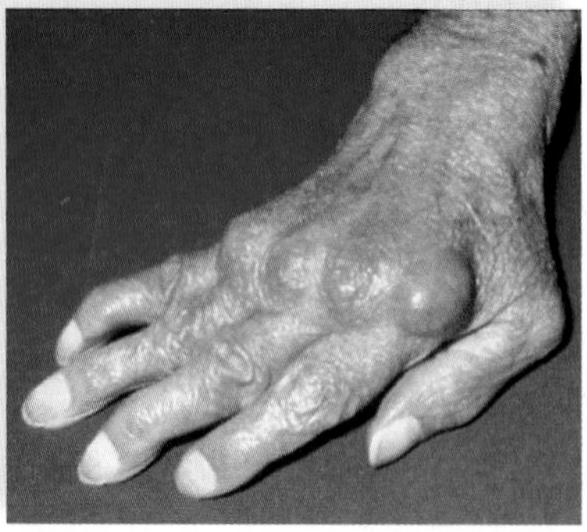

**FIGURE 33.16 Rheumatoid arthritis.**

Rheumatoid arthritis is due to recurring inflammation in skeletal joints, due to immune system attack.

**Autoimmune diseases** can be characterized by the failure of the immune system to distinguish between foreign antigens and the self-antigens that mark the body's own tissues. In an autoimmune disease, chronic inflammation occurs, and cytotoxic T cells or antibodies mistakenly attack the body's own cells.

The exact events that trigger an autoimmune disorder are not known. Some autoimmune disorders set in following a noticeable infection; for example, the heart damage following rheumatic fever is thought to be due to an autoimmune disorder triggered by the illness. Otherwise, most autoimmune disorders probably start after an undetected inflammatory response. However, the tendency to develop autoimmune disorders is known to be inherited in some cases, so there may be genetic causes as well.

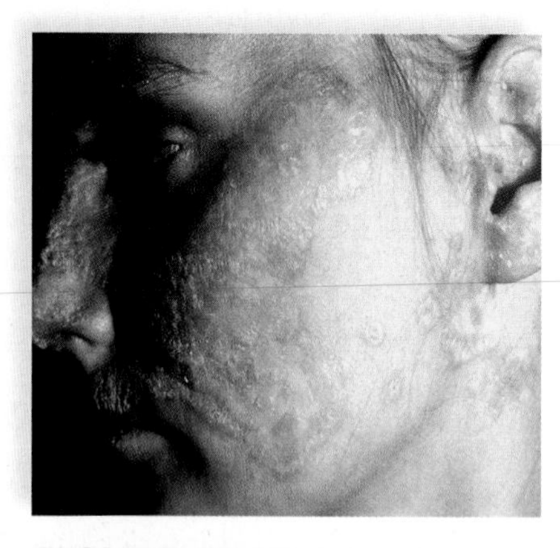

**FIGURE 33.17 Systemic lupus.**

A butterfly-shaped rash appears on the face.

Since little is known about the origin of autoimmune disorders, no cures are currently available. Regardless, most of these conditions can be managed over the long term with immunosuppressive drugs that control the various symptoms.

Rheumatoid arthritis is a common autoimmune disorder that causes recurring inflammation in synovial joints (Fig. 33.16). Complement proteins, T cells, and B cells all participate in deterioration of the joints, which eventually become immobile. This chronic inflammation gradually causes destruction of the delicate membrane and cartilage within the joint.

In myasthenia gravis, a well-understood autoimmune disease, antibodies attach to and interfere with the functioning of neuromuscular junctions, causing muscular weakness.

Systemic lupus erythematosus (lupus) is a chronic autoimmune disorder characterized by the presence of antibodies to the nuclei of the body's cells. Unlike rheumatoid arthritis or myasthenia gravis, lupus affects multiple tissues and organs, and is still very poorly understood. The symptoms vary somewhat, but most patients experience a characteristic skin rash (Fig. 33.17), joint pain, and kidney damage. Lupus typically progresses to include many life-threatening complications.

When a person has an immune deficiency, the immune system is unable to protect the body against disease. Acquired immunodeficiency syndrome (AIDS) is an example of an acquired immune deficiency. As a result of a weakened immune system, AIDS patients show a greater susceptibility to infections and also have a higher risk of cancer (see page 626). Infrequently, a child may be born with an impaired immune system, caused by a defect in lymphocyte development. In severe combined immunodeficiency disease (SCID), both antibody- and cell-mediated immunity are lacking or inadequate. Without treatment, even common infections can be fatal. Gene therapy has been successful in SCID patients.

## Allergies

**Allergies** are hypersensitivities to substances, such as pollen, food, or animal hair, that ordinarily would do no harm to the body. The response to these antigens, called allergens, usually includes some degree of tissue damage.

An **immediate allergic response** can occur within seconds of contact with the antigen. The response is caused by antibodies known as IgE (Table 33.1). IgE antibody receptors have mast cells in the tissues. When an allergen attaches to the IgE antibodies and the antibodies attach to their receptors on mast cells, the cells release histamine and other substances that bring about the allergic symptoms (Fig. 33.18). If the allergen is pollen, histamine stimulates the mucous membranes of the nose and eyes to release fluid, causing the runny nose and watery eyes typical of **hay fever.** If a person has **asthma,** the airways leading to the lungs constrict; labored breathing is accompanied by wheezing. Nausea, vomiting, and diarrhea typically occur when food contains an allergen.

Drugs called antihistamines are used to treat allergies. These drugs compete for the same receptors on the nose, eyes, airways, and lining of the digestive tract cells that ordinarily combine with histamine. In this way, histamine is prevented from binding and causing its unpleasant symptoms. However, antihistamines are only partially effective because mast cells release other molecules, in addition to histamine, that cause allergic symptoms.

**Anaphylactic shock** is an immediate allergic response that occurs after an allergen has entered the bloodstream. Bee stings and penicillin shots are known to cause this reaction in some individuals because both inject the allergen into the blood. Anaphylactic shock is characterized by a sudden and life-threatening drop in blood pressure, due to increased permeability of the capillaries by histamine.

People with allergies produce ten times more IgE than people without allergies. A new treatment using injections of monoclonal IgG antibodies for IgEs is currently being tested in individuals with severe food allergies. More routinely, injections of the allergen are given so that the body will build up high quantities of IgG antibodies. The hope is that these will combine with allergens received from the environment before they have a chance to combine with the IgE antibodies.

A **delayed allergic response** is initiated by memory T cells at the site of allergen contact in the body. The allergic response is regulated by the cytokines secreted by both T cells and macrophages. A classic example of a delayed allergic response is the skin test for tuberculosis (TB). When the test result is positive, the tissue where the antigen was injected becomes red and hardened. This indicates prior exposure to tubercle bacilli, the cause of TB. Contact dermatitis, which occurs when a person is allergic to poison ivy, jewelry, cosmetics, and many other substances that touch the skin, is another example of a delayed allergic response.

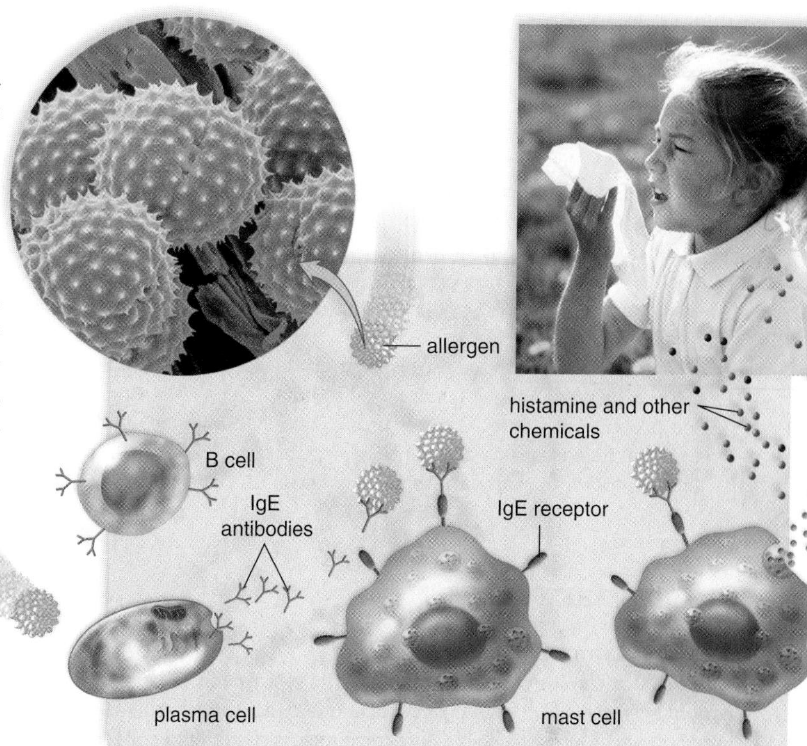


**FIGURE 33.18 An allergic reaction.**

An allergen attaches to IgE antibodies, which then cause mast cells to release histamine and other chemicals that are responsible for the allergic reaction.

### Check Your Progress 33.4

1. What types of complications and disorders are associated with the functioning of the immune system? Explain each of these.

**TABLE 33.1**
**Comparison of Immediate and Delayed Allergic Responses**

| | *Immediate Response* | *Delayed Response* |
|---|---|---|
| **Onset of Symptoms** | Takes several minutes | Takes 1 to 3 days |
| **Lymphocytes Involved** | B cells | T cells |
| **Immune Reaction** | IgE antibodies | Cell-mediated immunity |
| **Type of Symptoms** | Hay fever, asthma, and many other allergic responses | Contact dermatitis (e.g., poison ivy) |
| **Therapy** | Antihistamine and adrenaline | Cortisone |

# Connecting the Concepts

The role of the lymphatic system in homeostasis cannot be overemphasized. Excess tissue fluid, which is collected within lymphatic vessels, is called lymph. Because this fluid is returned to the cardiovascular system, blood volume and pressure are maintained. The lymphatic system is also intimately involved in immunity.

Levels of defense against invasion of the body by pathogens can be compared to how we protect our homes. Homes usually have external defenses such as a fence, a dog, or locked doors. Similarly, the body has barriers, such as the skin and mucous membranes, that prevent pathogens from entering the blood and lymph. Like a home alarm system, if invasion does occur, a signal goes off. First, nonspecific defense mechanisms such as the complement system and phagocytosis by white blood cells come into play. Finally, specific defense, which is dependent on the activities of B and T cells, occurs.

A strong connection exists between the immune, endocrine, and nervous systems. The thymus gland produces hormones that influence the immune response. Cortisone has the ability to mollify the inflammatory response. Cytokines help the body recover from disease in part by affecting the brain's temperature control center. A fever is thought to create an unfavorable environment for foreign invaders. Also, cytokines bring about a feeling of sluggishness, sleepiness, and loss of appetite. These behaviors tend to make us take care of ourselves until we feel better.

We continue our discussion of homeostasis in the next chapter, which considers the structure and function of the digestive system in both invertebrates and vertebrates.

## summary


### 33.1 The Lymphatic System

The lymphatic system consists of lymphatic vessels and organs. The lymphatic vessels (1) receive glycerol and fatty acids packaged as lipoproteins at intestinal villi; (2) receive excess tissue fluid collected by lymphatic capillaries; and (3) carry these to the bloodstream.

Lymphocytes are produced and accumulated in the lymphatic organs (lymph nodes, tonsils, spleen, thymus gland, and red bone marrow). Lymph is cleansed of pathogens in lymph nodes, and blood is cleansed of pathogens in the spleen. T lymphocytes mature in the thymus, while B lymphocytes mature in the red bone marrow where all blood cells are produced. White blood cells such as these lymphocytes are necessary for nonspecific and specific defenses.

### 33.2 Nonspecific Defense Against Disease

Immunity involves nonspecific and specific defenses. Nonspecific defenses (also called innate immunity) include barriers to entry, the inflammatory response, phagocytes and natural killer cells, and protective proteins.

### 33.3 Specific Defense Against Disease

Specific defenses (also called acquired immunity) require B lymphocytes and T lymphocytes, also called B cells and T cells. B cells undergo clonal selection with production of plasma cells and memory B cells after their B-cell receptor combines with a specific antigen. Plasma cells secrete antibodies and are responsible for antibody-mediated immunity. Some progeny of activated B cells become memory B cells, which remain in the body and produce antibodies if the same antigen enters the body at a later date. Active (long-lived) immunity occurs as a response to an illness or the administration of vaccines when a person is well and in no immediate danger of contracting an infectious disease. Passive immunity is needed when an individual is in immediate danger of succumbing to an infectious disease. Passive immunity is short-lived because the antibodies are administered to and not made by the individual.

An antibody is usually a Y-shaped molecule that has two binding sites for a specific antigen. Monoclonal antibodies, which are produced by the same plasma cell, have various functions, from detecting infections to treating cancer.

T cells have T-cell receptors and are responsible for cell-mediated immunity. The two main types of T cells are cytotoxic T cells and helper T cells. Cytotoxic T cells kill virus-infected or cancer cells on contact. Helper T cells produce cytokines and regulate both nonspecific defenses (innate immunity) and specific defenses (acquired immunity). Cytokines, including interferon, are used in attempts to treat AIDS and to promote the body's ability to recover from cancer.

For a T cell to recognize an antigen, the antigen must be presented by an antigen-presenting cell (APC), a dendritic cell, or a macrophage, along with an MHC (major histocompatibility complex) protein. Thereafter, the activated T cell undergoes clonal expansion until the illness has been stemmed. Then, most of the activated T cells undergo apoptosis. A few cells remain, however, as memory T cells.

### 33.4 Immunity Side Effects

Immune side effects also include tissue rejection and autoimmune diseases. Allergic responses occur when the immune system reacts vigorously to substances not normally recognized as foreign. Immediate allergic responses, usually consisting of coldlike symptoms, are due to the activity of antibodies. Delayed allergic responses, such as contact dermatitis, are due to the activity of T cells.


## understanding the terms

Match the terms to these definitions:

a. _____________ Antigens prepared in such a way that they can promote active immunity without causing disease.
b. _____________ Fluid, derived from tissue fluid, that is carried in lymphatic vessels.
c. _____________ Foreign substance, usually a protein or a polysaccharide, that stimulates the immune system to react, such as by producing antibodies.
d. _____________ Process of programmed cell death involving a cascade of specific cellular events leading to the death and destruction of the cell.
e. _____________ Lymphocyte that matures in the thymus and exists in three varieties, one of which kills antigen-bearing cells outright.

## reviewing this chapter

1. Which functions of the lymphatic system are not assisted by another system? Explain. 614
2. Describe the microscopic structure and the function of lymph nodes, the spleen, the thymus gland, and red bone marrow. 614–15
3. Discuss the body's nonspecific defense mechanisms. 616–18
4. Describe the inflammatory response, and give a role for each type of cell and molecule that participates in the response. 616–18
5. Describe the clonal selection model as it applies to B cells. B cells are responsible for which type of immunity? 620
6. How is active immunity artificially achieved? How is passive immunity achieved? 620–21
7. Describe the structure of an antibody, and define the terms variable regions and constant regions. 621
8. How are monoclonal antibodies produced, and what are their applications? 622
9. Discuss the clonal selection model as it applies to T cells. 624
10. Name the two main types of T cells, and state their functions. 619, 624–25
11. What are cytokines, and how are they used in immunotherapy? 627
12. Discuss autoimmune diseases and allergies as they relate to the immune system. 628–29

## testing yourself

Choose the best answer for each question.

1. Both veins and lymphatic vessels
   a. have thick walls of smooth muscle.
   b. contain valves for one-way flow of fluids.
   c. empty directly into the heart.
   d. are fed fluids from arterioles.
2. Complement
   a. is a nonspecific defense mechanism.
   b. is involved in the inflammatory response.
   c. is a series of proteins present in the plasma.
   d. plays a role in destroying bacteria.
   e. All of these are correct.
3. Which of these pertain(s) to T cells?
   a. have specific receptors
   b. are more than one type
   c. are responsible for cell-mediated immunity
   d. stimulate antibody production by B cells
   e. All of these are correct.
4. Which one of these does not pertain to B cells?
   a. have passed through the thymus
   b. have specific receptors
   c. are responsible for antibody-mediated immunity
   d. become plasma cells that synthesize and liberate antibodies
5. The clonal selection model says that
   a. an antigen selects certain B cells and suppresses them.
   b. an antigen stimulates the multiplication of B cells that produce antibodies against it.
   c. T cells select those B cells that should produce antibodies, regardless of antigens present.
   d. T cells suppress all B cells except the ones that should multiply and divide.
   e. Both b and c are correct.
6. Plasma cells are
   a. the same as memory cells.
   b. formed from blood plasma.
   c. B cells that are actively secreting antibody.
   d. inactive T cells carried in the plasma.
   e. a type of red blood cell.
7. Which of these pairs is incorrectly matched?
   a. cytotoxic T cells—help complement react
   b. cytotoxic T cells—active in tissue rejection
   c. macrophages—activate T cells
   d. memory T cells—long-living T cells
   e. T cells—mature in thymus
8. Vaccines are
   a. the same as monoclonal antibodies.
   b. treated bacteria or viruses, or one of their proteins.
   c. short-lived.
   d. MHC proteins.
   e. All of these are correct.
9. The inflammatory process involves
   a. complement, lymphocytes, and antigens.
   b. increased blood flow, phagocytes, and blood clotting.
   c. barriers to entry, tonsils, and fever.
   d. passive immunity, MHC proteins, and interferon.
   e. All of these are correct.
10. Label a–c on this IgG molecule using these terms: antigen-binding sites, light chain, heavy chain.

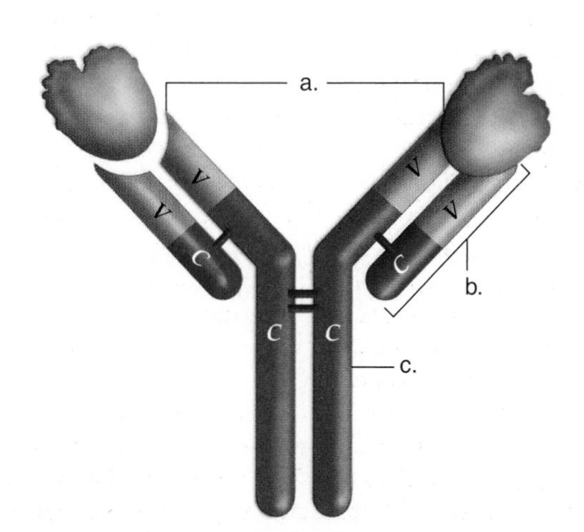


    d. What do *V* and *C* stand for in the diagram?

11. The lymphatic system does not
    a. transport tissue fluid back to the blood.
    b. transport absorbed lipoproteins from the small intestine to the blood.
    c. play a role in immunological defense.
    d. filter metabolic wastes, such as urea.
12. Which is a nonspecific defense against a pathogen?
    a. skin
    b. gastric juice
    c. complement
    d. interferons
    e. All of these are correct.
13. Which cell does not phagocytize?
    a. neutrophil
    b. lymphocyte
    c. monocyte
    d. macrophage
14. B cells mature within
    a. the lymph nodes.
    b. the spleen.
    c. the thymus.
    d. the bone marrow.
15. Plasma cells secrete
    a. antibodies.
    b. perforins.
    c. lysosomal enzymes.
    d. histamine.
    e. lymphokines.
16. Mast cell secretion occurs after an allergen combines with
    a. IgG antibodies.
    b. IgE antibodies.
    c. IgM antibodies.
    d. IgA antibodies.
17. After a second exposure to a vaccine,
    a. antibodies are made quickly and in greater amounts.
    b. immunity lasts longer than after the first exposure.
    c. antibodies of the IgG class are produced.
    d. plasma cells are active.
    e. All of these are correct.
18. Active immunity may be produced by
    a. having a disease.
    b. receiving a vaccine.
    c. receiving gamma globulin injections.
    d. Both a and b are correct.
    e. Both b and c are correct.
19. T cells do not
    a. promote the activity of B cells.
    b. undergo apoptosis.
    c. secrete cytokines.
    d. produce antibodies.
20. MHC proteins
    a. are present only on the surface of certain cells.
    b. help present the antigen to T cells.
    c. are unnecessary to the immune response.
    d. Both b and c are correct.
21. Lymph nodes
    a. block the flow of lymph.
    b. contain B cells and T cells.
    c. decrease in size during an illness.
    d. filter blood.
22. Which of the following is a function of the spleen?
    a. produces T cells
    b. removes worn-out red blood cells
    c. produces immunoglobulins
    d. produces macrophages
    e. regulates the immune system

## thinking scientifically

1. The transplantation of organs from one person to another was impossible until the discovery of immunosuppressant drugs. Now, with the use of drugs such as cyclosporine, organs can be transplanted without rejection. Transplant patients must take immunosuppressant drugs for the remainder of their lives. How can a person do this and not eventually succumb to disease?
2. Laboratory mice are immunized with a measles vaccine. When the mice are challenged with measles virus to test the strength of their immunity, the memory cells do not completely prevent replication of the measles virus. The virus undergoes a few rounds of replication before the immune response is observed. You have developed a strain of mice with a much faster response to a viral challenge, but these mice often develop an autoimmune disease. Speculate on the connection between speed of response and an autoimmune disease.

## bioethical issue

### Cost of Drugs to Treat AIDS

Over 36 million people worldwide are living with AIDS. This disease is deadly without proper medical care but a chronic disease if treated. Drug companies typically charge a high price for AIDS medications because Americans and their insurance companies can afford them. However, these drugs are out of reach in many countries, such as those in Africa, where AIDS is a widespread problem. Some people argue that drug companies should use the profits from other drugs (such as those for heart disease, depression, and impotence) to make AIDS drugs affordable to those who need them. This has not happened yet. In some countries, governments have allowed companies to infringe on foreign patents held by major drug companies so that they can produce affordable AIDS drugs. Do drug companies have a moral obligation to provide low-cost AIDS drugs, even if they have to do so at a loss of revenue? Is it right for governments to ignore patent laws in order to provide their citizens with affordable drugs?

## *Biology* website

The companion website for *Biology* provides a wealth of information organized and integrated by chapter. You will find practice tests, animations, videos, and much more that will complement your learning and understanding of general biology.

**http://www.mhhe.com/maderbiology10**

# 34

# Digestive Systems and Nutrition

## concepts

### 34.1 DIGESTIVE TRACTS

- An incomplete digestive tract with only one opening has little specialization of parts; a complete digestive tract with two openings does have specialization of parts. 634
- Discontinuous feeders, rather than continuous feeders, have a storage area for food. 634–35
- The dentition of herbivores, carnivores, and omnivores is adapted to the type of food they eat. 635–36

### 34.2 HUMAN DIGESTIVE TRACT

- The human digestive tract has many specialized parts and several accessory organs of digestion, which contribute in their own way to the digestion of food. 636–42

### 34.3 DIGESTIVE ENZYMES

- The digestive enzymes are specific to the particular type of food digested. Each functions in a particular part of the digestive tract. 642–43
- The products of digestion are small molecules, such as amino acids and glucose, that can cross plasma membranes. 642–43

### 34.4 NUTRITION

- Proper nutrition supplies the body with nutrients, including all vitamins and minerals. 643–46

*a great blue heron stands at the water's edge with its long neck cocked in the familiar S shape. To feed, its swordlike bill is launched forward to grasp a fish. Most birds, like the heron, are discontinuous feeders and feed periodically. Some birds, like a hummingbird, are continuous feeders and feed almost every waking moment. In keeping with feeding discontinuously, the esophagus of a great blue heron has a crop, a storage area that allows the bird to delay digestion of the meal. The bird's stomach has a gizzard that crushes hard materials with its muscular walls and abrades them with sand swallowed sometime in the past. Similar to humans, the bulk of enzymatic digestion and also absorption occur in the small intestine.*

*In this chapter, we will compare the digestive system of various animals before examining how the organs of the human digestive system function. This is the first step toward an appreciation of dietary concerns, which are also discussed.*

This great blue heron has caught a fish. Digestion will produce nutrient molecules small enough to enter the bloodstream.

# 34.1 Digestive Tracts

Not all animals have a digestive tract. Consider, for example, that in sponges, digestion occurs in food vacuoles, as it does in protozoa. Digestion in hydras, which are cnidarians, begins in their gastrovascular cavity, but is finished in food vacuoles.

The majority of animals have some sort of gut, or digestive tract, where food is digested into small nutrient molecules that can cross plasma membranes. Digestion contributes to homeostasis by providing the body with the nutrients needed to sustain the life of cells. A digestive tract (1) ingests food, (2) breaks food down into small molecules that can cross plasma membranes, (3) absorbs these nutrient molecules, and (4) eliminates undigestible remains.

## Incomplete Versus Complete Tracts

An **incomplete digestive tract** has a single opening, usually called a mouth. However, the single opening is used both as an entrance for food and an exit for wastes. Planarians, which are flatworms, have an incomplete tract (Fig. 34.1). It begins with a mouth and muscular pharynx and then the tract, a gastrovascular cavity, branches throughout the body. Planarians are primarily carnivorous and feed largely on smaller aquatic animals, as well as bits of organic debris. When a planarian is feeding, the pharynx actually extends beyond the mouth. The body is wrapped about the prey and the pharynx sucks up minute quantities at a time. Digestive enzymes present in the tract allow some extracellular digestion to occur. Digestion is finished intracellularly by the cells that line the tract. No cell in the body is far from the digestive tract; therefore, diffusion alone is sufficient to distribute nutrient molecules.

The digestive tract of a planarian is notable for its lack of specialized parts. It is saclike because the pharynx serves not only as an entrance for food but also as an exit for undigestible material. Specialization of parts does not occur under these circumstances.

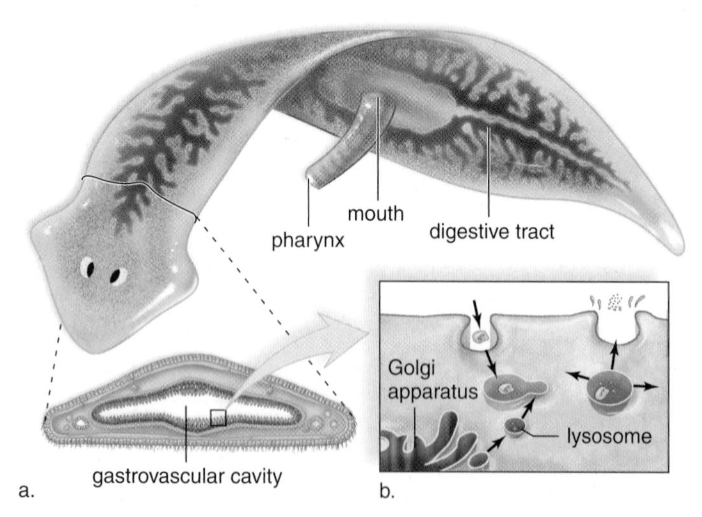


**FIGURE 34.1 Incomplete digestive tract of a planarian.**

**a.** Planarians, which are flatworms, have a gastrovascular cavity with a single opening that acts as both an entrance and an exit. Like hydra, planarians rely on intracellular digestion to complete the digestive process. **b.** Phagocytosis produces a vacuole, which joins with an enzyme-containing lysosome. The digested products pass from the vacuole into the cytoplasm before any undigestible material is eliminated at the plasma membrane.

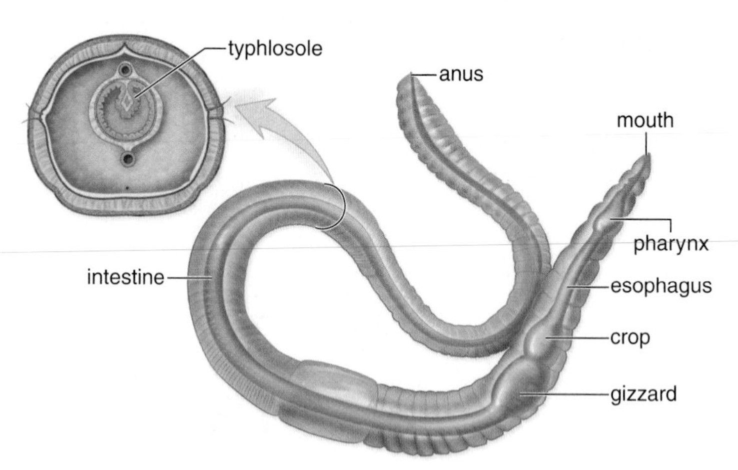


**FIGURE 34.2 Complete digestive tract of an earthworm.**

Complete digestive tracts have both a mouth and an anus and can have many specialized parts, such as those labeled in this drawing. Also in earthworms, which are annelids, the absorptive surface of the intestine is increased by an internal fold called the typhlosole.

Planarians have some modified parasitic relatives. Tapeworms, which are parasitic flatworms, lack a digestive system. Nutrient molecules are absorbed by the tapeworm from the intestinal juices of the host, which surround the tapeworm's body. The integument and body wall of the tapeworm are highly modified for this purpose. They have millions of microscopic, fingerlike projections that increase the surface area for absorption.

In contrast to planarians, earthworms, which are annelids, have a **complete digestive tract,** meaning that the tract has a mouth and an anus (Fig. 34.2). Earthworms feed mainly on decayed organic matter in soil. The muscular pharynx draws in food with a sucking action. Food then enters the crop, which is a storage area with thin, expansive walls. From there, food goes to the gizzard, where thick, muscular walls crush and sand grinds the food. Digestion is extracellular within an intestine. The surface area of digestive tracts is often increased for absorption of nutrient molecules, and in earthworms, this is accomplished by an intestinal fold called the **typhlosole.** Undigested remains pass out of the body at the anus. Specialization of parts is obvious in the earthworm because the pharynx, the crop, the gizzard, and the intestine each has a particular function in the digestive process.

## Continuous Versus Discontinuous Feeders

Clams, which are molluscs, are continuous feeders, called filter feeders (Fig. 34.3*a*). Water is always moving into the mantle cavity by way of the incurrent siphon (slitlike opening) and depositing particles, including algae, protozoans, and minute invertebrates, on the gills. The size of the incurrent siphon permits the entrance of only small particles, which adhere to the gills. Ciliary action moves suitably sized particles to the labial palps, which force them through the mouth into the stomach. Digestive enzymes are secreted by a large digestive gland, but amoeboid cells present throughout the tract are believed to complete the digestive process by intracellular digestion.

**FIGURE 34.3 Nutritional mode of a clam compared to a squid.**

Clams and squids are molluscs. A clam burrows in the sand or mud, where it filter feeds, whereas a squid swims freely in open waters and captures prey. In keeping with their lifestyles, a clam (**a**) is a continuous feeder and a squid (**b**) is a discontinuous feeder. Digestive system labels are shaded green.

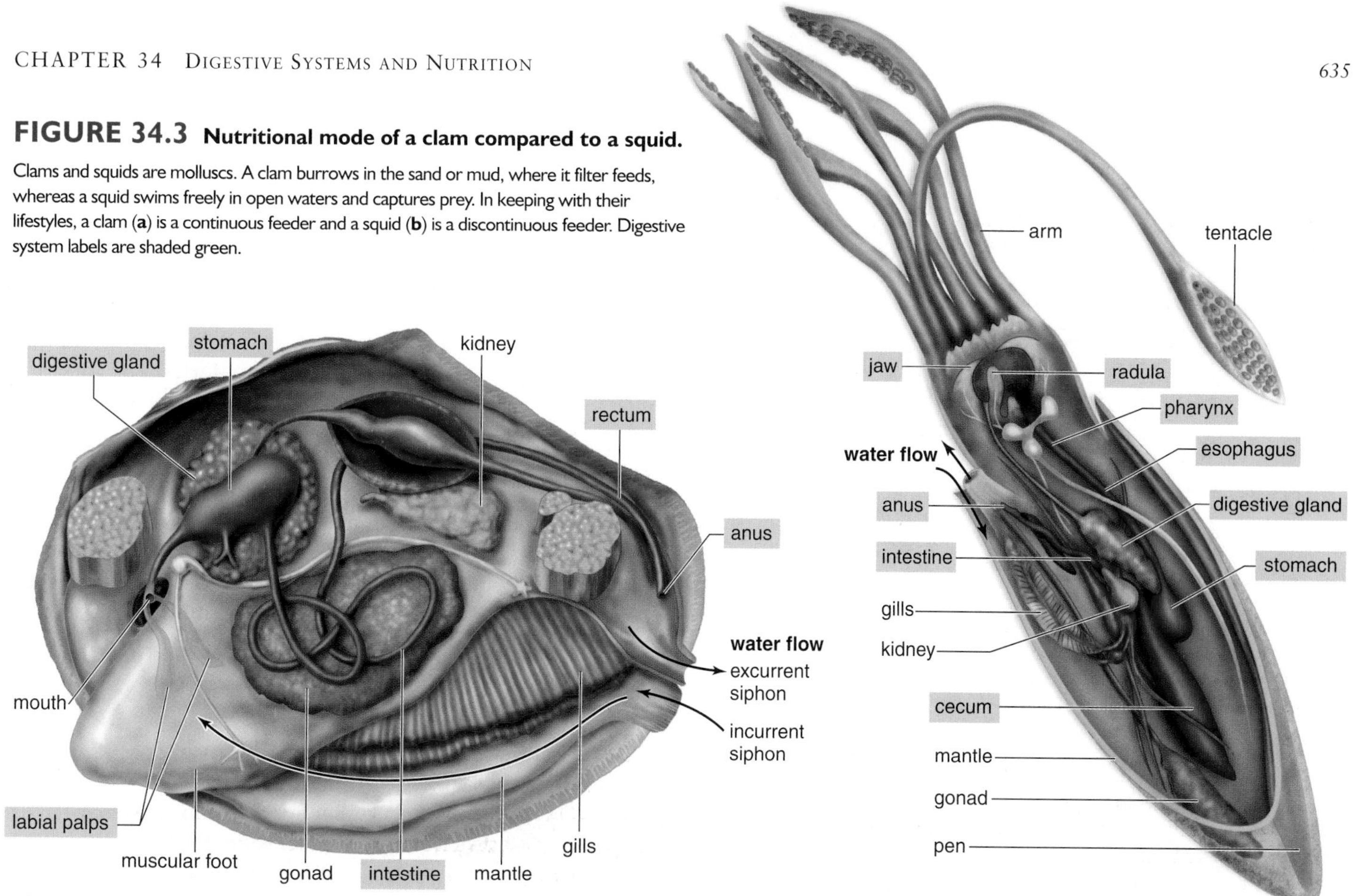

a. Digestive system (green) of clam



b. Digestive system (green) of squid

Marine fanworms, which are annelids, are continuous filter feeders. The ciliated appendages of these worms are specialized for gathering fine particles and microscopic plankton from the water. They allow only small particles to enter the mouth. Larger particles are rejected. A baleen whale, such as a blue whale, is an active filter feeder. Baleen, a keratinized curtainlike fringe, hangs from the roof of the mouth and filters small shrimp called krill from the water. A baleen whale filters up to a ton of krill every few minutes.

Squids, which are molluscs, are discontinuous feeders (Fig. 34.3*b*). The body of a squid is streamlined, and the animal moves rapidly through the water using jet propulsion (forceful expulsion of water from a tubular funnel). The head of a squid is surrounded by ten arms, two of which have developed into long, slender tentacles whose suckers have toothed, horny rings. These tentacles seize prey (fishes, shrimps, and worms) and bring it to the squid's beaklike jaws, which bite off pieces pulled into the mouth by the action of a **radula,** a tonguelike structure. An esophagus leads to a stomach and a cecum (blind sac), where digestion occurs. The stomach, supplemented by the cecum, retains food until digestion is complete. Discontinuous feeders, whether they are carnivores, like blue herons, or herbivores, like elephants, require a storage area for food, which can be a crop, where no digestion occurs, or a stomach, where digestion begins.

## Adaptation to Diet

Some animals are omnivores; they eat both plants and animals. Others are herbivores; they feed only on plants. Still others are carnivores; they eat only other animals. Among invertebrates, filter feeders such as clams and tube worms are omnivores. Land snails, which are terrestrial molluscs, and some insects, such as grasshoppers and locusts, are herbivores. Spiders (arthropods) are carnivores, as are sea stars (echinoderms), which feed on clams. A sea star positions itself above a clam and uses its tube feet to pull the valves of the shell apart (see Fig. 28.27). Then, it everts a part of its stomach to start the digestive process, even while the clam is trying to close its shell. Some invertebrates are cannibalistic. A female praying mantis (an insect), if starved, will feed on her mate as the reproductive act is taking place!

Among mammals, the dentition differs according to mode of nutrition. Among herbivores, the koala of Australia is famous for its diet of only eucalyptus leaves, and likewise many other mammals are browsers, feeding off bushes and trees. Grazers, like the horse, feed off grasses. The horse has sharp, even incisors for neatly clipping off blades of grass and large, flat premolars and molars for grinding and crushing the grass (Fig. 34.4*a*). Extensive grinding and crushing disrupts plant cell walls, allowing bacteria located in a part of the digestive tract called the cecum to digest cellulose. Other mammalian grazers, such as cattle, sheep, and deer, are ruminants with a large, four-chambered stomach. In contrast to horses, they graze quickly and swallow partially chewed grasses into a special part of the stomach called a **rumen.** Here, microorganisms start the digestive process, and the result, called cud, is regurgitated at a later time when the animal is no longer feeding. The cud is chewed again before being swallowed for complete digestion.

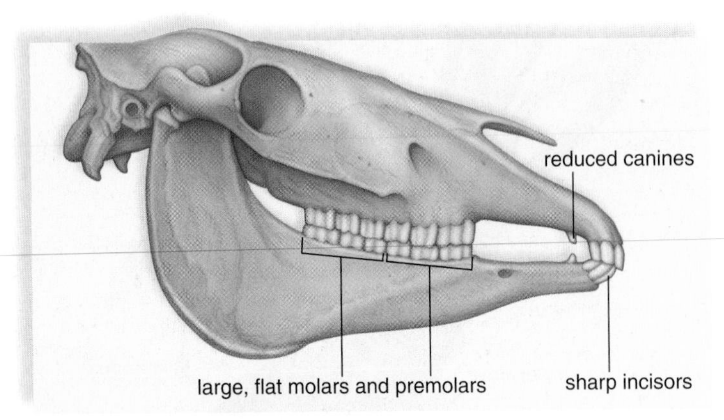


a. Horses are herbivores.

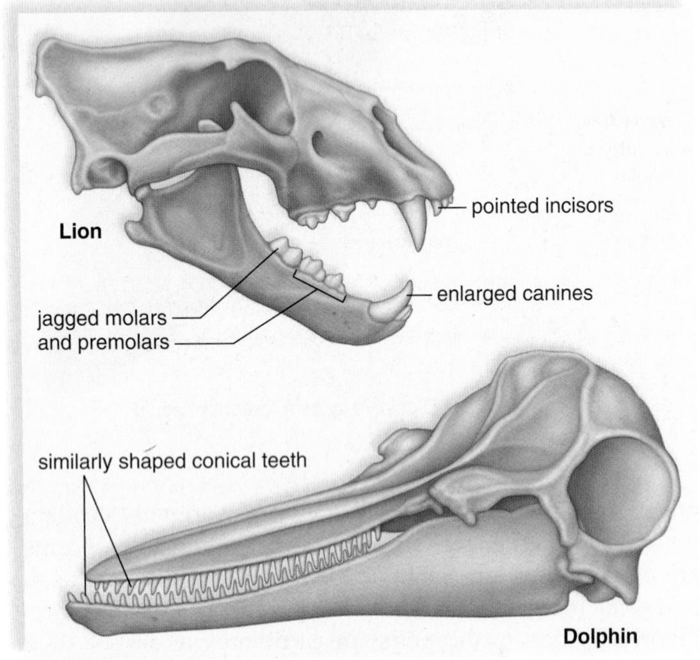


b. Lions and dolphins are carnivores.

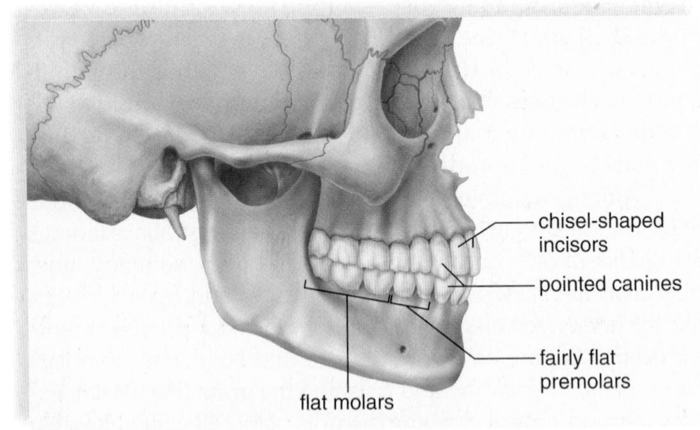


c. Humans are omnivores.

**FIGURE 34.4 Dentition among mammals.**

**a.** Horses are herbivores and have teeth suitable to clipping and chewing grass. **b.** Lions and dolphins are carnivores. Dentition in a lion is suitable for killing large animals such as zebras and wildebeests and tearing apart their flesh. Dentition in a dolphin is suitable to grasping small animals like fishes, which are swallowed whole. **c.** Humans are omnivores and have teeth suitable to a mixed diet of vegetables and meat.

Many mammals, including dogs, lions, toothed whales, and dolphins, are carnivores. Lions use pointed canine teeth for killing, short incisors for scraping bones, and pointed molars for slicing flesh (Fig. 34.4*b*, *top*). Dolphins and toothed whales swallow food whole without chewing it first; they are equipped with many identical, conical teeth that are used to catch and grasp their slippery prey before swallowing (Fig. 34.4*b*, *bottom*). Meat is rich in protein and fat and is easier to digest than plant material. The intestine of a rabbit, a herbivore, is much longer than that of a similarly sized cat, a carnivore.

Humans, as well as raccoons, rats, and brown bears, are omnivores. Therefore, the dentition has a variety of specializations to accommodate both a vegetable diet and a meat diet. An adult human has 32 teeth. One-half of each jaw has teeth of four different types: two chisel-shaped incisors for shearing; one pointed canine for tearing; two fairly flat premolars for grinding; and three molars, well flattened for crushing (Fig. 34.4*c*).

**Check Your Progress 34.1**

1. A complete digestive tract can have many specialized parts. Explain.
2. Discontinuous feeders, but not continuous feeders, tend to have a storage area for food. Explain.
3. Compare the teeth of carnivores to those of herbivores.

# 34.2 Human Digestive Tract

Humans have a tube-within-a-tube body plan and a complete digestive tract, which begins with a mouth and ends in an anus. The major structures of the human digestive tract are illustrated in Figure 34.5. The pancreas, liver, and gallbladder are accessory organs of digestion, and they produce secretions that aid digestion.

The digestion of food in humans is an extracellular event. Digestion requires a cooperative effort between different parts of the body. Digestion consists of two major stages: mechanical digestion and chemical digestion. Mechanical digestion involves the physical breakdown of food into smaller particles. This is accomplished through the chewing of food in the mouth and the physical churning and mixing of food in the stomach and small intestine. Chemical digestion requires enzymes that are secreted by the digestive tract or by accessory glands that lie nearby. Specific enzymes break down particular macromolecules into smaller molecules that can be absorbed.

## Mouth

The **mouth,** or oral cavity, serves as the beginning of the digestive tract. The palate, or roof of the mouth, separates the oral cavity from the nasal cavity. It consists of the anterior hard palate and the posterior soft palate. The fleshy uvula is the posterior extension of the soft palate (see Fig. 34.6). The cheeks and lips retain food while it is chewed by the teeth and mixed with saliva.

There are three major pairs of **salivary glands** that send their juices by way of ducts to the mouth. Saliva contains

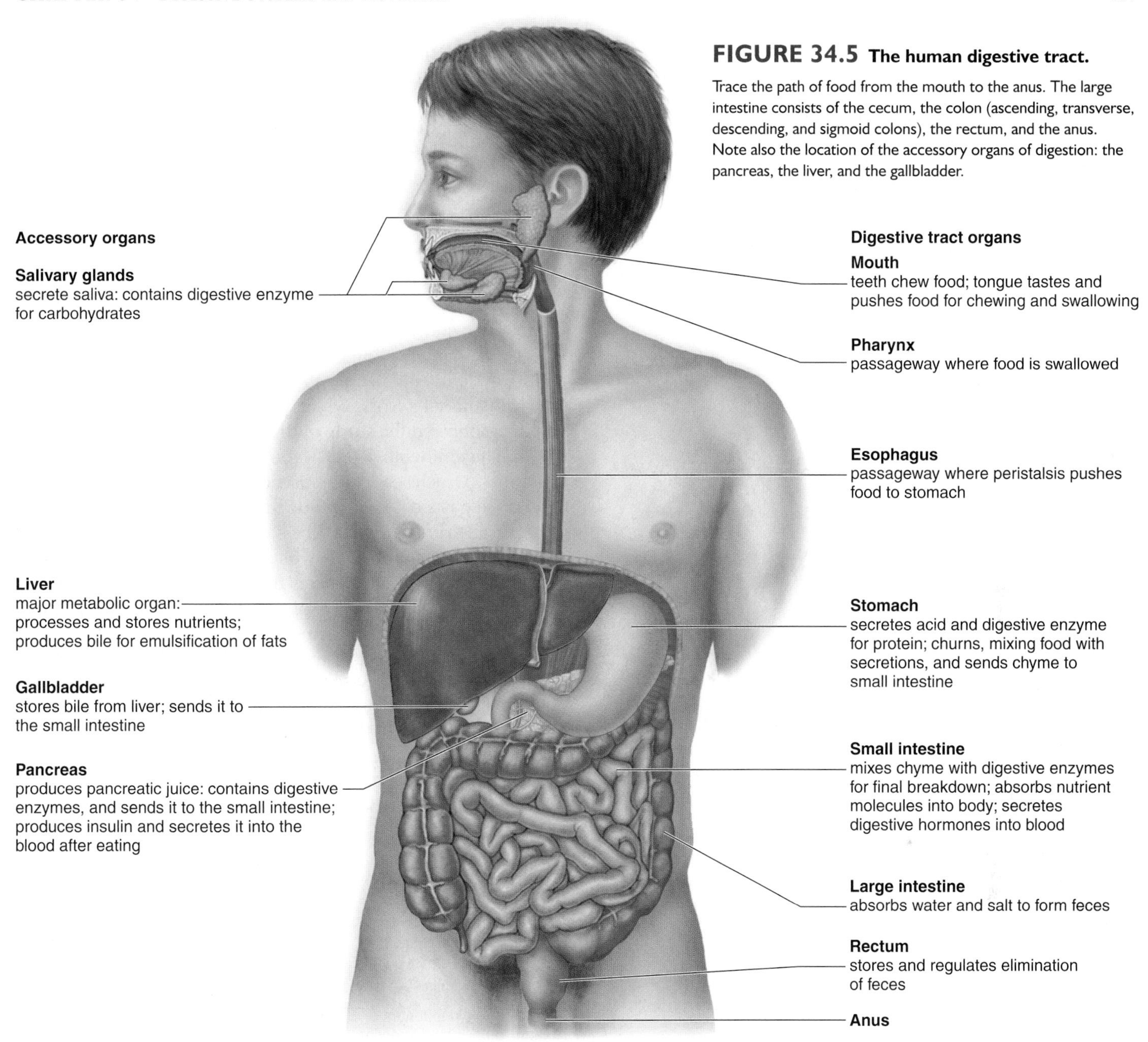


**FIGURE 34.5 The human digestive tract.**

Trace the path of food from the mouth to the anus. The large intestine consists of the cecum, the colon (ascending, transverse, descending, and sigmoid colons), the rectum, and the anus. Note also the location of the accessory organs of digestion: the pancreas, the liver, and the gallbladder.

the enzyme **salivary amylase,** which begins the process of starch digestion. The disaccharide maltose is a typical end product of salivary amylase digestion.

$$\text{starch} + H_2O \xrightarrow{\text{salivary amylase}} \text{maltose}$$

While in the mouth, food is manipulated by a muscular tongue, which has touch and pressure receptors similar to those in the skin. Taste buds, sensory receptors that are stimulated by the chemical composition of food, are also found primarily on the tongue as well as on the surface of the mouth. The tongue, which is composed of striated muscle and an outer layer of mucous membrane, mixes the chewed food with saliva. It then forms this mixture into a mass called a bolus in preparation for swallowing.

## The Pharynx and the Esophagus

The digestive and respiratory passages come together in the **pharynx** and then separate. The **esophagus** [Gk. *eso,* within, and *phagein,* eat] is a tubular structure, of about 25 cm in length, that takes food to the stomach. Sphincters are muscles that encircle tubes and act as valves; tubes close when sphincters contract, and they open when sphincters relax. The lower gastroesophageal sphincter is located where the esophagus enters the stomach. When food enters the stomach, the sphincter relaxes for a few seconds and then it closes again. Heartburn occurs due to acid reflux, when some of the stomach's contents escape into the esophagus. When vomiting occurs, the abdominal muscles and the diaphragm, a muscle that separates the thoracic and abdominal cavities, contract.

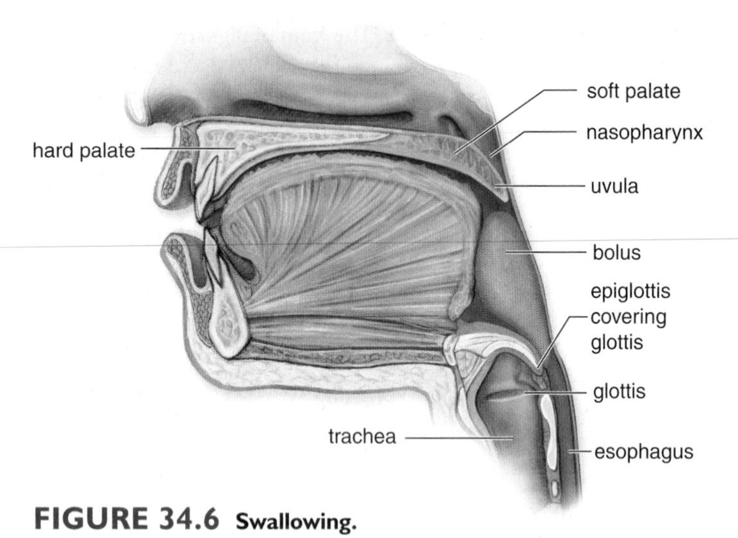


**FIGURE 34.6 Swallowing.**

Respiratory and digestive passages converge and diverge in the pharynx. When food is swallowed, the soft palate closes off the nasopharynx, and the epiglottis covers the glottis, forcing the bolus to pass down the esophagus. Therefore, a person does not breathe when swallowing.

When food is swallowed, the soft palate, the rear portion of the mouth's roof, moves back to close off the nasopharynx. A flap of tissue called the epiglottis covers the glottis, an opening into the trachea. Now the bolus must move through the pharynx into the esophagus because the air passages are blocked (Fig. 34.6). When food enters the esophagus, peristalsis begins. **Peristalsis** [Gk. *peri*, around, and *stalsis*, compression] is a rhythmical contraction that serves to move the contents along in tubular organs, such as the digestive tract (Fig. 34.7).

## Stomach

The **stomach** is a thick-walled, J-shaped organ that lies on the left side of the body beneath the diaphragm. The wall of the stomach has deep folds (rugae) that disappear as the stomach fills to an approximate capacity of 1 liter. Therefore, humans can periodically eat relatively large meals and spend the rest of their time at other activities.

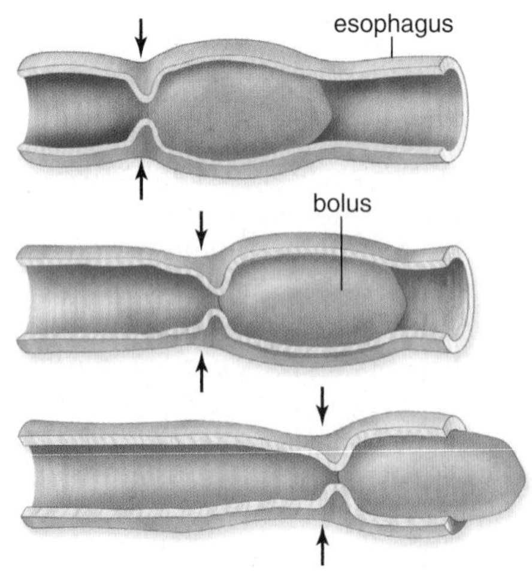


**FIGURE 34.7 Peristalsis in the digestive tract.**

These three drawings show how a peristaltic wave moves through a single section of the esophagus over time. The arrows point to areas of contraction.

The stomach (Fig. 34.8) is much more than a storage organ, as was discovered by William Beaumont in the mid-nineteenth century. Beaumont, an American doctor, had a French Canadian patient, Alexis St. Martin. St. Martin had been shot in the stomach, and when the wound healed, he was left with a fistula, or opening, that allowed Beaumont to look inside the stomach and to collect gastric (stomach) juices produced by gastric glands. Beaumont was able to determine that the muscular walls of the stomach contract vigorously and mix food with juices that are secreted whenever food enters the stomach. He found that gastric juice contains hydrochloric acid (HCl) and a substance, now called pepsin, that is active in digestion. He also found that the gastric juices are produced independently of the protective mucous secretions of the stomach. Beaumont's work, which was carefully and painstakingly done, pioneered the study of digestive physiology.

The epithelial lining of the stomach has millions of gastric pits, which lead into gastric glands. The gastric glands produce gastric juice. So much hydrochloric acid is secreted by the gastric glands that the stomach routinely has a pH of about 2. Such a high acidity usually is sufficient to kill bacteria and other microorganisms that might be in food. This low pH also stops the activity of salivary amylase, which functions optimally at the near-neutral pH of saliva.

As with the rest of the digestive tract, a thick layer of mucus protects the wall of the stomach from enzymatic action. Still, an ulcer, which is an open sore in the wall caused by the gradual destruction of tissues, does occur in some

**FIGURE 34.8**
**Anatomy of the stomach.**

**a.** The stomach, which has thick walls, expands as it fills with food. **b.** The mucous membrane layer of its walls secretes mucus and contains gastric glands, which secrete a gastric juice active in the digestion of protein.

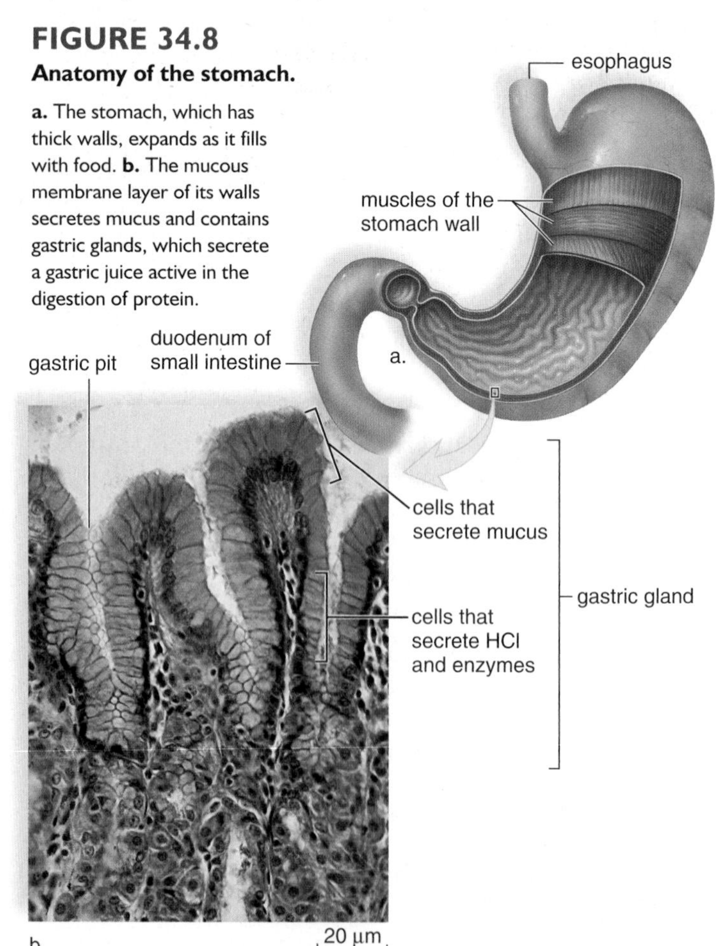

individuals. Ulcers can be caused by an infection by an acid-resistant bacterium, *Helicobacter pylori,* which is able to attach to the epithelial lining. Wherever the bacterium attaches, the lining stops producing mucus, and the area becomes exposed to digestive action. As a result, an ulcer develops.

Eventually, food mixing with gastric juice in the stomach contents become **chyme,** which has a thick, creamy consistency. At the base of the stomach is a narrow opening controlled by a sphincter. A sphincter is a muscle that surrounds a tube and closes or opens the tube by contracting and relaxing. Whenever the sphincter relaxes, a small quantity of chyme passes through the opening into the small intestine. When chyme enters the small intestine, it sets off a neural reflex that causes the muscles of the sphincter to contract vigorously and to close the opening temporarily. Then the sphincter relaxes again and allows more chyme to enter. The slow manner in which chyme enters the small intestine allows for thorough digestion.

## The Small Intestine

The **small intestine** is named for its small diameter (compared to that of the large intestine), but perhaps it should be called the long intestine. The small intestine averages about 6 m in length, compared to the large intestine, which is about 1.5 m in length.

The first 25 cm of the small intestine is called the **duodenum.** A duct brings bile from the liver and gallbladder, and pancreatic juice from the pancreas, into the small intestine (see Fig. 34.11*a*). **Bile** emulsifies fat—emulsification causes fat droplets to disperse in water. The intestine has a slightly basic pH because pancreatic juice contains sodium bicarbonate ($NaHCO_3$), which neutralizes chyme. The enzymes in pancreatic juice and enzymes produced by the intestinal wall complete the process of food digestion.

It has been suggested that the surface area of the small intestine is approximately that of a tennis court. What factors contribute to increasing its surface area? The wall of the small intestine contains fingerlike projections called villi (sing., **villus**), which give the intestinal wall a soft, velvety appearance (Fig. 34.9). A villus has an outer layer of columnar epithelial cells, and each of these cells has thousands of microscopic extensions called microvilli. Collectively, in electron micrographs, microvilli give the villi a fuzzy border, known as a "brush border." Since the microvilli bear the intestinal enzymes, these enzymes are called brush-border enzymes. The microvilli greatly increase the surface area of the villus for the absorption of nutrients.

Nutrients are absorbed into the vessels of a villus, which contains blood capillaries and a lymphatic capillary, called a **lacteal.** Sugars (digested from carbohydrates) and amino acids (digested from proteins) enter the blood capillaries of a villus. Glycerol and fatty acids (digested from fats) enter the epithelial cells of the villi, and within these cells they are joined and packaged as lipoprotein droplets, which enter a lacteal. After nutrients are absorbed, they are eventually carried to all the cells of the body by the bloodstream.

## Large Intestine

The **large intestine,** which includes the cecum, the colon, the rectum, and the anus, is larger in diameter (6.5 cm) but shorter in length (1.5 m) than the small intestine. The large intestine absorbs water, salts, and some vitamins. It also stores undigestible material until it is eliminated at the anus. No digestion takes place in the large intestine.

The cecum, which lies below the junction with the small intestine, is the blind end of the large intestine. The cecum has a small projection called the vermiform **appendix** [L. *verm,* worm, and *form,* shape, and *append,* an addition] (Fig. 34.10). In humans, the appendix may play a role in fighting infections. In the case of appendicitis, the appendix becomes infected and so filled with fluid that it may burst. If an infected appendix bursts before it can be removed, it can lead to a serious, generalized infection of the abdominal lining, called peritonitis.

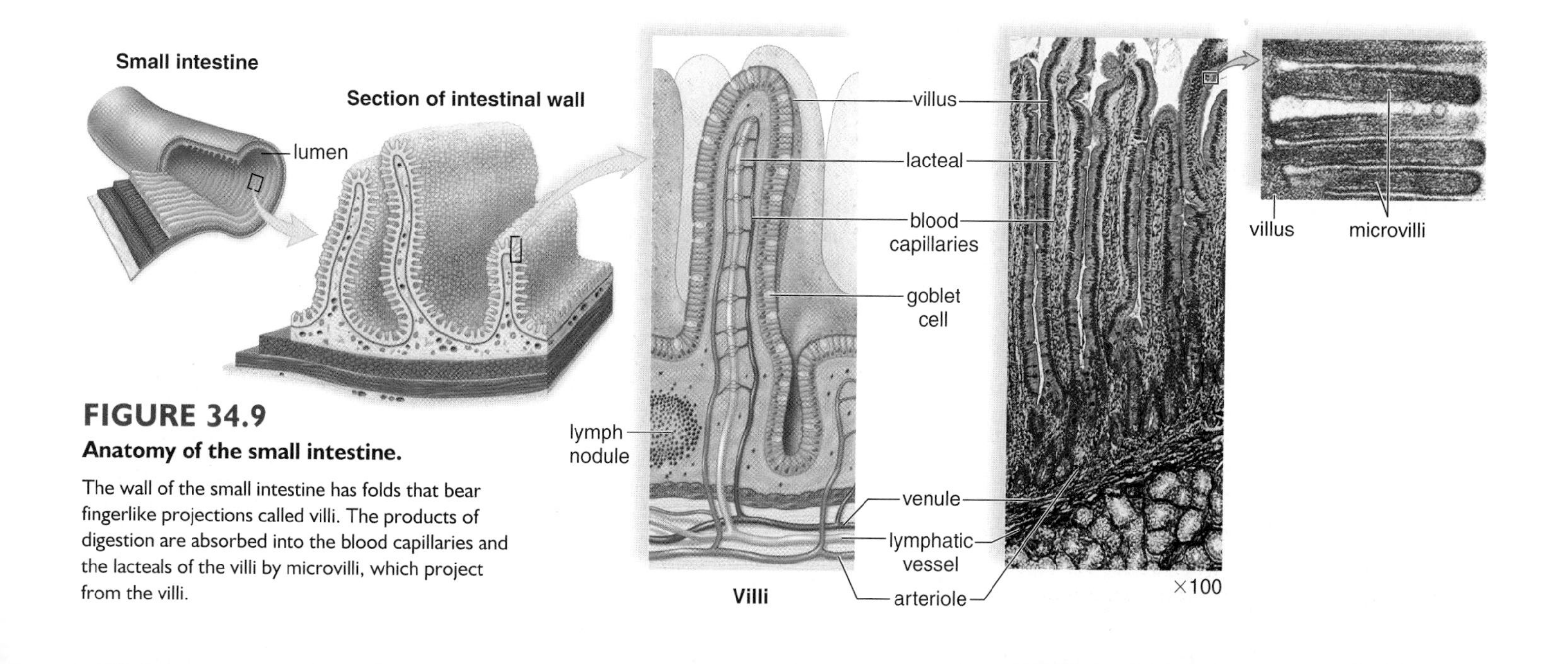


**FIGURE 34.9**
**Anatomy of the small intestine.**

The wall of the small intestine has folds that bear fingerlike projections called villi. The products of digestion are absorbed into the blood capillaries and the lacteals of the villi by microvilli, which project from the villi.

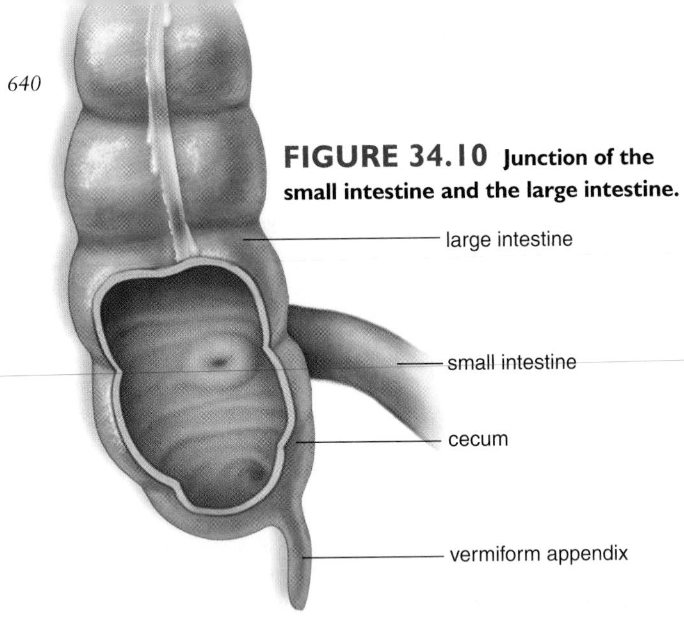


**FIGURE 34.10 Junction of the small intestine and the large intestine.**

The colon joins the rectum, the last 20 cm of the large intestine. About 1.5 liters of water enters the digestive tract daily as a result of eating and drinking. An additional 8.5 liters enter the digestive tract each day carrying the various substances secreted by the digestive glands. About 95% of this water is absorbed by the small intestine, and much of the remaining portion is absorbed by the colon. If this water is not reabsorbed, **diarrhea** can lead to serious dehydration and ion loss, especially in children.

The large intestine has a large population of bacteria, notably *Escherichia coli.* The bacteria break down undigestible material, and they also produce some vitamins, such as vitamin K. Vitamin K is necessary to blood clotting. Digestive wastes (feces) eventually leave the body through the **anus,** the opening of the anal canal. Feces are about 75% water and 25% solid matter. Almost one-third of this solid matter is made up of intestinal bacteria. The remainder is undigested plant material, fats, waste products (such as bile pigments), inorganic material, mucus, and dead cells from the intestinal lining. The color of feces is the result of bilirubin breakdown and the presence of oxidized iron. The foul odor is the result of bacterial action.

The colon is subject to the development of **polyps,** which are small growths arising from the mucosa. Polyps, whether they are benign or cancerous, can be removed surgically. Some investigators believe that dietary fat increases the likelihood of colon cancer. Dietary fat causes an increase in bile secretion, and it could be that intestinal bacteria convert bile salts to substances that promote the development of colon cancer. Dietary fibers absorb water and add bulk, thereby diluting the concentration of bile salts and facilitating the movement of substances through the intestine. Regular elimination reduces the time that the colon wall is exposed to any cancer-promoting agents in feces.

### Check Your Progress 34.2A

1. Trace the path of food from the mouth to the large intestine.
2. What are the functions of the small intestine, and how is the wall of the small intestine modified to perform these functions?

## Three Accessory Organs

The pancreas, liver, and gallbladder are accessory digestive organs. Figure 34.11*a* shows how the pancreatic duct from the pancreas and the common bile duct from the liver and gallbladder enter the duodenum.

### *The Pancreas*

The **pancreas** lies deep in the abdominal cavity, resting on the posterior abdominal wall. It is an elongated and somewhat flattened organ that has both an endocrine and an exocrine function. As an endocrine gland, it secretes insulin and glucagon, hormones that help keep the blood glucose level within normal limits. In this chapter, however, we are interested in its exocrine function. Most pancreatic cells produce pancreatic juice, which contains sodium bicarbonate ($NaHCO_3$) and digestive enzymes for all types of food. Sodium bicarbonate neutralizes acid chyme from the stomach. Pancreatic amylase digests starch, trypsin digests protein, and lipase digests fat.

### *The Liver*

The **liver,** which is the largest gland in the body, lies mainly in the upper right section of the abdominal cavity, under the diaphragm (see Fig. 34.5). The liver contains approximately 100,000 lobules that serve as its structural and functional units (Fig. 34.11*b*). Triads, located between the lobules, consist of a bile duct, which takes bile away from the liver; a branch of the hepatic artery, which brings $O_2$-rich blood to the liver; and a

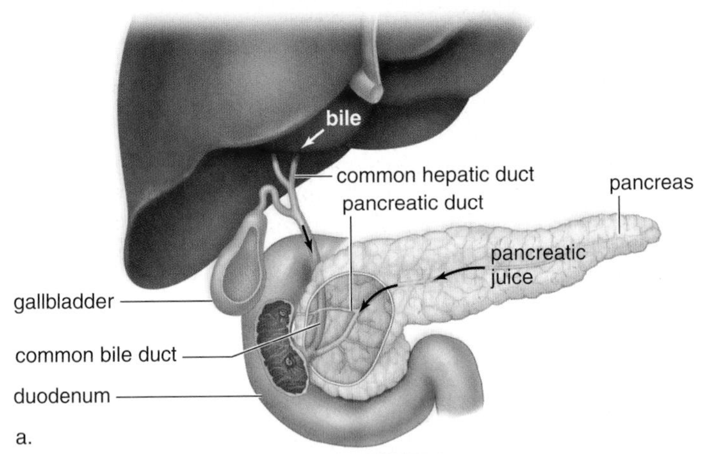


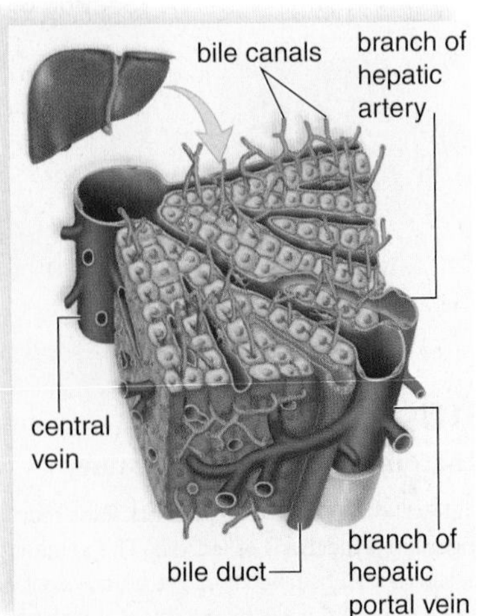


**FIGURE 34.11 Liver, gallbladder, and pancreas.**

**a.** The liver makes bile, which is stored in the gallbladder and sent (black arrow) to the small intestine by way of the common bile duct. The pancreas produces digestive enzymes that are sent (black arrows) to the small intestine by way of the pancreatic duct. **b.** The liver contains over 100,000 lobules. Each lobule contains many cells that perform the various functions of the liver. They remove and add materials to the blood and deposit bile in a duct.

*health focus*

## Wall of the Digestive Tract

We can compare the human digestive tract to a garden hose that has a beginning (mouth) and an end (anus). The so-called **lumen** is the central space where food is digested. The wall of the digestive tract has four layers (Fig. 34A), and we will associate each layer with a particular disorder.

The first layer of the wall next to the lumen is called the **mucosa.** The mucosa is more familiarly called a mucous membrane and, of course, it produces mucus, which protects the wall from the digestive enzymes inside the lumen. In the mouth, stomach, and small intestine, the mucosa either contains glands that secrete and/or receive digestive enzymes from glands that secrete digestive enzymes.

Diverticulosis is a condition in which portions of the mucosa literally have pushed through the other layers and formed pouches, where food can collect. The pouches can be likened to an inner tube that pokes through weak places in a tire. When the pouches become infected or inflamed, the condition is called *diverticulitis.* This happens in 10–25% of people with diverticulosis.

The second layer in the digestive wall is called the **submucosa.** The submucosal layer is a broad band of loose connective tissue that contains blood vessels, lymphatic vessels, and nerves. These are the vessels that will carry the nutrients absorbed by the mucosa. Lymph nodules, called Peyer's patches, are also in the submucosa. Like the tonsils, they help protect us from disease. Because the submucosa contains blood vessels, it can be the site of an inflammatory response that leads to *inflammatory bowel disease (IBD)*, characterized by chronic diarrhea, abdominal pain, fever, and weight loss.

The third layer is termed the **muscularis,** and it contains two layers of smooth muscle. The inner, circular layer encircles the tract; the outer, longitudinal layer lies in the same direction as the tract. The contraction of these muscles, which are under nervous control, accounts for movement of digested food from the esophagus to the anus. The muscularis can be associated with *irritable bowel syndrome (IBS)*, in which contractions of the wall cause abdominal pain, constipation, and/or diarrhea. The underlying cause of IBS is not known, although some suggest stress as an underlying cause.

The fourth layer of the tract is the **serosa** (serous membrane layer), which secretes a serous fluid. The serosa is a part of the peritoneum, the internal lining of the abdominal cavity. The appendix is a worm-shaped blind tube projecting from the first part of the large intestine on the right side of the abdomen. An inflamed appendix (*appendicitis*) has to be removed because, should the appendix burst, the result can be *peritonitis,* a life-threatening infection of the peritoneum.

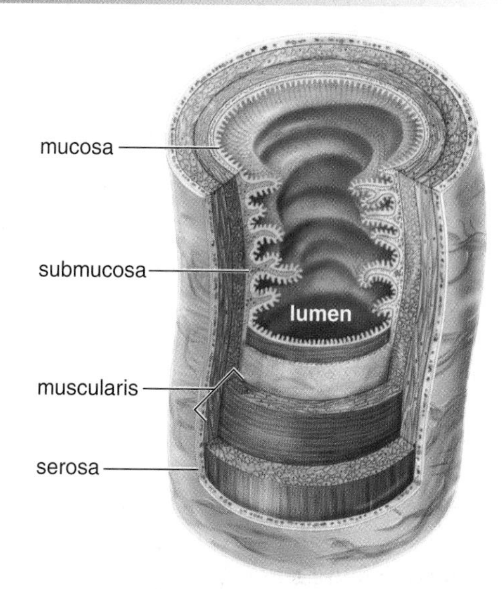


**FIGURE 34A Wall of the digestive tract.** *The wall of the digestive tract contains the four layers noted.*

branch of the hepatic portal vein, which transports nutrients from the intestines (see Fig. 32.10). The central veins of lobules enter a hepatic vein. Blood moves from the intestines to the liver via the hepatic portal vein and from the liver to the inferior vena cava via the hepatic veins.

In some ways, the liver acts as the gatekeeper to the blood (Table 34.1). As blood in the hepatic portal vein passes through the liver, it removes poisonous substances and detoxifies them. The liver also removes and stores iron and the vitamins A, $B_{12}$, D, E, and K. The liver makes the plasma proteins and helps regulate the quantity of cholesterol in the blood.

The liver maintains the blood glucose level at about 100 mg/100 ml (0.1%), even though a person eats intermittently. When insulin is present, any excess glucose present in blood is removed and stored by the liver as glycogen. Between meals, glycogen is broken down to glucose, which enters the hepatic veins, and in this way, the blood glucose level remains constant.

If the supply of glycogen is depleted, the liver converts glycerol (from fats) and amino acids to glucose molecules. The conversion of amino acids to glucose necessitates deamination, the removal of amino groups. By a complex metabolic pathway, the liver then combines ammonia with carbon dioxide to form urea. Urea is the usual nitrogenous waste product from amino acid breakdown in humans.

The liver produces bile, which is stored in the gallbladder. Bile has a yellowish green color because it contains the bile pigment *bilirubin,* derived from the breakdown of hemoglobin, the red pigment of red blood cells. Bile also contains bile salts.

**TABLE 34.1**

**Functions of the Liver**

1. Detoxifies blood by removing and metabolizing poisonous substances
2. Stores iron and the vitamins A, $B_{12}$, D, E, and K
3. Makes plasma proteins, such as albumins and fibrinogen, from amino acids
4. Stores glucose as glycogen after a meal, and breaks down glycogen to glucose to maintain the glucose level of blood between eating periods
5. Produces urea after breaking down amino acids
6. Removes bilirubin, a breakdown product of hemoglobin from the blood, and excretes it in bile, a liver product
7. Helps regulate blood cholesterol level, converting some to bile salts

Bile salts are derived from cholesterol, and they emulsify fat in the small intestine. When fat is emulsified, it breaks up into droplets, providing a much larger surface area, which can be acted upon by a digestive enzyme from the pancreas.

**Liver Disorders.** Hepatitis and cirrhosis are two serious diseases that affect the entire liver and hinder its ability to repair itself. Therefore, they are life-threatening diseases. When a person has a liver ailment, jaundice may occur. **Jaundice** is present when the skin and the whites of the eyes have a yellowish tinge. Jaundice occurs because bilirubin is deposited in the skin, due to an abnormally large amount in the blood. Jaundice can also result from **hepatitis,** inflammation of the liver. Viral hepatitis occurs in several forms. Hepatitis A is usually acquired from sewage-contaminated drinking water. Hepatitis B, which is usually spread by sexual contact, can also be spread by blood transfusions or contaminated needles. The hepatitis B virus is more contagious than the AIDS virus, which is spread in the same way. Thankfully, however, a vaccine is now available for hepatitis B. Hepatitis C, which is usually acquired by contact with infected blood and for which there is no vaccine, can lead to chronic hepatitis, liver cancer, and death.

**Cirrhosis** is another chronic disease of the liver. First, the organ becomes fatty, and then liver tissue is replaced by inactive fibrous scar tissue. Cirrhosis of the liver is often seen in alcoholics, due to malnutrition and to the excessive amounts of alcohol (a toxin) the liver is forced to break down.

The liver has amazing regenerative powers and can recover if the rate of regeneration exceeds the rate of damage. During liver failure, however, there may not be enough time to let the liver heal itself. Liver transplantation is usually the preferred treatment for liver failure, but artificial livers have been developed and tried in a few cases. One type is a cartridge that contains liver cells. The patient's blood passes through the cellulose acetate tubing of the cartridge and is serviced in the same manner as with a normal liver.

### *The Gallbladder*

The **gallbladder** is a pear-shaped, muscular sac attached to the surface of the liver (see Fig. 34.5). About 1,000 ml of bile are produced by the liver each day, and any excess is stored in the gallbladder. Water is reabsorbed by the gallbladder so that bile becomes a thick, mucuslike material. When needed, bile leaves the gallbladder and proceeds to the duodenum via the common bile duct.

The cholesterol content of bile can come out of solution and form crystals. If the crystals grow in size, they form gallstones. The passage of the stones from the gallbladder may block the common bile duct and cause obstructive jaundice. Then, the gallbladder must be removed.

**Check Your Progress** **34.2B**

1. What are the three main accessory organs that assist with the digestive process?
2. How does each accessory organ contribute to the digestion of food?

## 34.3 Digestive Enzymes

The various digestive enzymes present in the digestive juices, mentioned previously, help break down carbohydrates, proteins, nucleic acids, and fats, the major components of food. Starch is a polysaccharide, and its digestion begins in the mouth. Saliva from the salivary glands has a neutral pH and contains **salivary amylase,** the first enzyme to act on starch.

$$\text{starch} + H_2O \xrightarrow{\text{salivary amylase}} \text{maltose}$$

Maltose molecules cannot be absorbed by the intestine; additional digestive action in the small intestine converts maltose to glucose, which can be absorbed.

Protein digestion begins in the stomach. Gastric juice secreted by gastric glands has a very low pH—about 2—because it contains hydrochloric acid (HCl). Pepsinogen, a precursor that is converted to **pepsin** when exposed to HCl, is also present in gastric juice. Pepsin acts on protein to produce peptides.

$$\text{protein} + H_2O \xrightarrow{\text{pepsin}} \text{peptides}$$

Peptides are usually too large to be absorbed by the intestinal lining, but later they are broken down to amino acids in the small intestine.

Starch, proteins, nucleic acids, and fats are all enzymatically broken down in the small intestine. Pancreatic juice, which enters the duodenum, has a basic pH because it contains sodium bicarbonate ($NaHCO_3$). One pancreatic enzyme, **pancreatic amylase,** digests starch (Fig. 34.12*a*).

$$\text{starch} + H_2O \xrightarrow{\text{pancreatic amylase}} \text{maltose}$$

Another pancreatic enzyme, **trypsin,** digests protein (Fig. 34.12*b*).

$$\text{protein} + H_2O \xrightarrow{\text{trypsin}} \text{peptides}$$

Trypsin is secreted as trypsinogen, which is converted to trypsin in the duodenum.

Maltase and peptidases, enzymes produced by the small intestine, complete the digestion of starch to glucose and protein to amino acids, respectively. Glucose and amino acids are small molecules that cross into the cells of the villi and enter the blood (Fig. 34.12*a*, *b*).

Maltose, a disaccharide that results from the first step in starch digestion, is digested to glucose by **maltase.**

$$\text{maltose} + H_2O \xrightarrow{\text{maltase}} \text{glucose} + \text{glucose}$$

Other disaccharides have their own enzyme and are digested in the small intestine. The absence of any one of these enzymes can cause illness.

Peptides, which result from the first step in protein digestion, are digested to amino acids by **peptidases.**

$$\text{peptides} + H_2O \xrightarrow{\text{peptidases}} \text{amino acids}$$

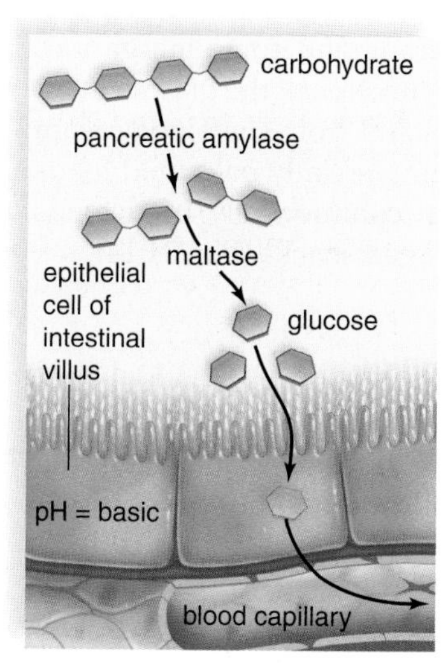


a. Carbohydrate digestion

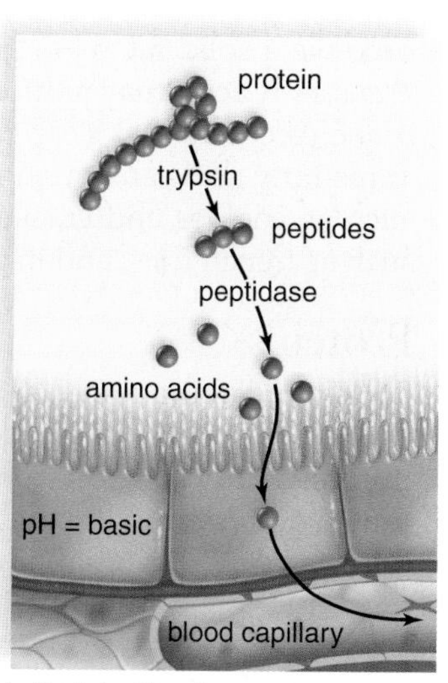


b. Protein digestion

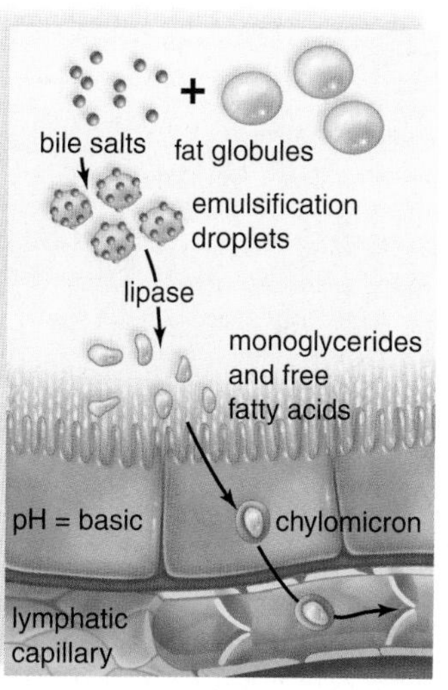


c. Fat digestion

**FIGURE 34.12 Digestion and absorption of nutrients.**
**a.** Starch is digested to glucose, which is actively transported into the epithelial cells of intestinal villi. From there, glucose moves into the bloodstream. **b.** Proteins are digested to amino acids, which are actively transported into the epithelial cells of intestinal villi. From there, amino acids move into the bloodstream. **c.** Fats are emulsified by bile and digested to monoglycerides and fatty acids. These diffuse into epithelial cells, where they recombine and join with proteins to form lipoproteins, called chylomicrons. Chylomicrons enter a lacteal.

**Lipase,** a third pancreatic enzyme, digests fat molecules in fat droplets after they have been emulsified by bile salts.

$$\text{fat} \xrightarrow{\text{bile salts}} \text{fat droplets}$$

$$\text{fat droplets} + H_2O \xrightarrow{\text{lipase}} \text{glycerol} + 3 \text{ fatty acids}$$

Specifically, the end products of lipase digestion are monoglycerides (glycerol + one fatty acid) and fatty acids. These enter the cells of the villi, and within these cells, they are rejoined and packaged as lipoprotein droplets, called chylomicrons. Chylomicrons enter the lacteals (Fig. 34.12*c*).

**Check Your Progress 34.3**

1. Describe the breakdown and absorption of starch and protein.

# 34.4 Nutrition

This part of the chapter discusses the benefits of a balanced diet, which should include all necessary minerals and vitamins.

## Carbohydrates

**Carbohydrates** are present in food in the form of sugars, starch, and fiber. Fruits, vegetables, milk, and honey are natural sources of sugars. Glucose and fructose are monosaccharide sugars, and lactose (milk sugar) and sucrose (table sugar) are disaccharides. After being absorbed from the digestive tract into the bloodstream, all sugars are converted to glucose for transport in the blood and use by cells. Glucose is the preferred direct energy source in cells.

Plants store glucose as starch, and animals store glucose as glycogen. Good sources of starch are beans, peas, cereal grains, and potatoes. Starch is digested to glucose in the digestive tract, and any excess glucose is stored as glycogen. Although other animals likewise store glucose as glycogen in liver or muscle tissue (meat), little is left by the time an animal is eaten for food. Except for honey and milk, which contain sugars, animal foods do not contain carbohydrates.

**Fiber** includes various undigestible carbohydrates derived from plants. Food sources rich in fiber include beans, peas, nuts, fruits, and vegetables. Whole-grain products are also a good source of fiber, and are therefore more nutritious than food products made from refined grains. During *refinement*, fiber and also vitamins and minerals are removed from grains, so that primarily starch remains. For example, a slice of bread made from whole-wheat flour contains 3 g of fiber; a slice of bread made from refined wheat flour contains less than 1 g of fiber.

Technically, fiber is not a nutrient for humans because it cannot be digested to small molecules that enter the bloodstream. Insoluble fiber, however, adds bulk to fecal material, which stimulates movement in the large intestine, preventing constipation. Soluble fiber combines with bile acids and cholesterol in the small intestine and prevents them from being absorbed. In this way, high-fiber diets may protect against heart disease. The typical American consumes only about 15 g of fiber each day; the recommended daily intake of fiber is 25 g for women and 38 g for men. To increase your fiber intake, eat whole-grain foods, snack on fresh fruits and raw vegetables, and include nuts and beans in your diet (Fig. 34.13).

If you, or someone you know, has lost weight by following the Atkins or South Beach diet, you may think "carbs" are unhealthy and should be avoided. According to nutritionists, however, carbohydrates should supply a large portion of your energy needs. Evidence suggests that Americans are not eating the right kind of carbohydrates. In some countries, the traditional diet is 60–70% high-fiber carbohydrates, and these people have a low incidence of the diseases that plague Americans.

Some nutritionists hypothesize that the high intake of foods that are rich in refined carbohydrates and fructose sweeteners processed from cornstarch may be responsible

**FIGURE 34.13** **Fiber-rich foods.**

Plants provide a good source of carbohydrates. They also provide a good source of vitamins, minerals, and fiber when they are not processed (refined).

for the prevalence of obesity in the United States. Because certain foods, such as donuts, cakes, pies, and cookies, are high in both refined carbohydrates and fat, it is difficult to determine which dietary component is responsible for the current epidemic of obesity among Americans. In any case, they are empty-calorie foods that provide sugars but no vitamins or minerals. Nutritionists also point out that consuming too much energy from any source contributes to body fat, which increases a person's risk of obesity and associated illnesses.

## Lipids

Like carbohydrates, **triglycerides** (fats and oils) supply energy for cells, but **fat** is stored for the long term in the body. Nutritionists generally recommend that people include unsaturated, rather than saturated, fats in their diets. Two unsaturated fatty acids (alpha-linolenic and linoleic acids) are *essential* in the diet. They can be supplied by eating fatty fish and by including plant oils, such as canola and soybean oils, in the diet. Delayed growth and skin problems can develop in people whose diets lack these essential unsaturated *fatty* acids.

Animal-derived foods, such as butter, meat, whole milk, and cheeses, contain saturated fatty acids. Plant oils contain unsaturated fatty acids; each type of oil has a particular percentage of monounsaturated and polyunsaturated fatty acids.

**Cholesterol,** a lipid, can be synthesized by the body. Cells use cholesterol to make various compounds, including bile, steroid hormones, and vitamin D. Plant foods do not contain cholesterol; only animal foods such as cheese, egg yolks, liver, and certain shellfish (shrimp and lobster) are rich in cholesterol. Elevated blood cholesterol levels are associated with an increased risk of cardiovascular disease, the number-one killer of Americans. A diet rich in cholesterol and saturated fats increases the risk of cardiovascular disease.

Statistical studies suggest that trans-fatty acids (trans-fats) are even more harmful than saturated fatty acids. Trans-fatty acids arise when unsaturated oils are hydrogenated to produce a solid fat, as in shortening and some margarines. Trans-fatty acids may reduce the function of the plasma membrane receptors that clear cholesterol from the bloodstream. Trans-fatty acids are found in commercially packaged foods, such as cookies and crackers; in commercially fried foods, such as french fries; and in packaged snacks (Fig. 34.14).

## Proteins

Dietary **proteins** are digested to amino acids, which cells use to synthesize hundreds of cellular proteins. Of the 20 different amino acids, nine are *essential amino acids* that must be present in the diet. Children will not grow if their diets lack the essential amino acids. Eggs, milk products, meat, poultry, and most other foods derived from animals contain all nine essential amino acids and are considered "complete" or "high-quality" protein sources.

Foods derived from plants generally do not have as much protein per serving as those derived from animals, and each type of plant food generally lacks one or more of the essential amino acids. Therefore, most plant foods are "incomplete" or "low-quality" protein sources. Vegetarians, however, do not have to rely on animal sources of protein. To meet their protein needs, total vegetarians (vegans) can eat grains, beans, and nuts in various combinations. Also, tofu, soymilk, and other foods made from processed soybeans are complete protein sources. A balanced vegetarian diet is quite possible with a little planning.

According to nutritionists, protein should not supply the bulk of dietary calories. The average American eats about twice as much protein as he or she needs, and some people may be on a diet that encourages the intake of proteins, instead of carbohydrates, as an energy source. Also, bodybuilders should realize that excess amino acids are not always converted into muscle tissue. When amino acids are broken down, the liver removes the nitrogen portion (*deamination*) and uses it to form urea, which is excreted in urine.

**FIGURE 34.14** **Food high in trans-fats.**

Most people enjoy sweets, but cookies and cakes are apt to contain saturated fats and trans-fats.

The water needed for excretion of urea can cause dehydration when a person is exercising and losing water by sweating. High-protein diets can also increase calcium loss in the urine and encourage the formation of kidney stones. Furthermore, high-protein foods often contain a high amount of fat.

## Diet and Obesity

Nutritionists point out that consuming too many Calories from any source contributes to body fat, which increases a person's risk of obesity and associated illnesses. (Obesity is defined as weighing 30% more than the ideal body weight for your height and body build.) Still, foods such as donuts, cakes, pies, cookies, and white bread, which are high in refined carbohydrates (starches and sugars), and fried foods, which are high in fat, may very well be responsible for the current epidemic of obesity among Americans. Also implicated is the lack of exercise because of a sedentary lifestyle. Type 2 diabetes and cardiovascular disease are often seen in people who are obese.

### *Type 2 Diabetes*

Diabetes mellitus is indicated by the presence of glucose in the urine. Glucose has spilled over into the urine because there is too high a level of glucose in the blood. Diabetes occurs in two forms. Type 1 diabetes is not associated with obesity. When a person has type 1 diabetes, the pancreas does not produce insulin, and the patient must have daily insulin injections. In contrast, children, and more often adults, with type 2 diabetes are usually obese and display impaired insulin production and insulin resistance. In a person with insulin resistance, the body's cells fail to take up glucose, even when insulin is present. Therefore, the blood glucose level exceeds the normal level, and glucose appears in the urine.

Type 2 diabetes is increasing rapidly in most industrialized countries of the world. A healthy diet, increased physical activity, and weight loss have been seen to improve the ability of insulin to function properly in type 2 diabetics (Fig. 34.15). It is well worth the effort to control type 2 diabetes because all diabetics, whether type 1 or type 2, are at risk for blindness, kidney disease, as well as cardiovascular disease.

**FIGURE 34.15 Exercising for good health.**

Regular exercise helps prevent and control type 2 diabetes.

### *Cardiovascular Disease*

In the United States, cardiovascular disease, which includes hypertension, heart attack, and stroke, is among the leading causes of death. Cardiovascular disease is often due to arteries blocked by plaque, which contains saturated fats and cholesterol. Cholesterol is carried in the blood by two types of lipoproteins: low-density lipoprotein (LDL) and high-density lipoprotein (HDL). LDL molecules are considered "bad" because they are like delivery trucks that carry cholesterol from the liver to the cells and to the arterial walls. HDL molecules are considered "good" because they are like garbage trucks that dispose of cholesterol. HDL transports cholesterol from the cells to the liver, which converts it to bile salts that enter the small intestine.

Consuming saturated fats, including trans-fats, tends to raise LDL cholesterol levels, while eating unsaturated fats lowers LDL cholesterol levels. Beef, dairy foods, and coconut oil are rich sources of saturated fat. Foods containing partially hydrogenated oils (e.g., vegetable shortening and stick margarine) are sources of trans-fats. Unsaturated fatty acids in olive and canola oils, most nuts, and coldwater fish tend to lower LDL cholesterol levels. Furthermore, coldwater fish (e.g., herring, sardines, tuna, and salmon) contain polyunsaturated fatty acids and especially *omega-3 unsaturated fatty acids,* which are believed to reduce the risk of cardiovascular disease. However, taking fish oil supplements to obtain omega-3s is not recommended without a physician's approval, because too much of these fatty acids can interfere with normal blood clotting.

The American Heart Association recommends limiting total cholesterol intake to 300 mg per day. This requires careful selection of the foods we include in our daily diets. For example, an egg yolk contains about 210 mg of cholesterol, which would be two-thirds of the recommended daily intake. Still, this doesn't mean eggs should be eliminated from a healthy diet, since the proteins in them are very nutritious; in fact, most healthy people can eat a couple of whole eggs each week without experiencing an increase in their blood cholesterol levels.

A physician can determine whether patients' blood lipid levels are normal. If a person's cholesterol and triglyceride levels are elevated, modifying the fat content of the diet, losing excess body fat, and exercising regularly can reduce them. If lifestyle changes do not lower blood lipid levels enough to reduce the risk of cardiovascular disease, a physician may prescribe special medications.

## Vitamins and Minerals

**Vitamins** are organic compounds (other than carbohydrates, fats, and proteins) that regulate various metabolic activities and must be present in the diet. Many vitamins are part of coenzymes; for example, niacin is the name for a portion of the coenzyme $NAD^+$, and riboflavin is a part of FAD. Coenzymes are needed in small amounts because they are used over and over again in cells. Not all vitamins are coenzymes; vitamin A, for example, is a precursor for the pigment that prevents night blindness. It has been known for some time that the absence of a vitamin can be associated with a particular disorder. Vitamins are especially abundant in fruits and vegetables, and so it is suggested that we eat about 4 1/2 cups of fruits and vegetables per day. Although many foods are now enriched or fortified with vitamins, some individuals are still at risk for vitamin deficiencies, generally as a result of poor food choices.

The body also needs about 20 elements called **minerals** for various physiological functions, including regulation of biochemical reactions, maintenance of fluid balance, and incorporation into certain structures and compounds. Occasionally individuals (especially women) do not receive enough iron, calcium, magnesium, or zinc in their diets. Adult females need more iron in the diet than males (18 mg compared to 10 mg) if they are menstruating each month. Many people take calcium supplements, as directed by a physician, to counteract osteoporosis, a degenerative bone disease that especially affects older men and women. Many people consume too much sodium, even double the amount needed. Excess sodium can cause water retention and contribute to hypertension.

### Check Your Progress 34.4

1. Why is a diet that includes plentiful vegetables better for you than a diet that includes excess protein?
2. How can you decrease your chances of acquiring type 2 diabetes and cardiovascular disease?

## Connecting the Concepts

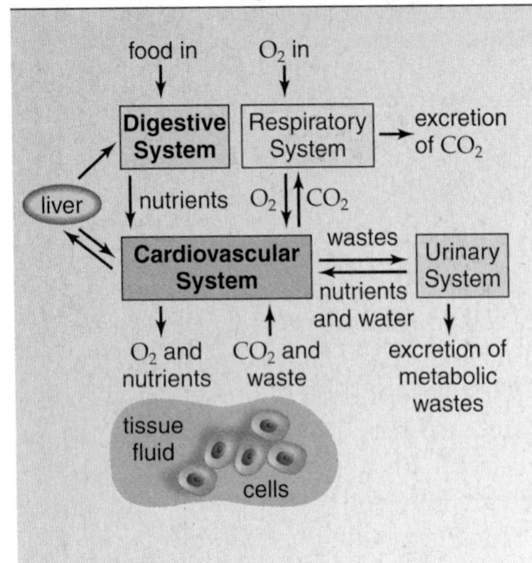


The digestive system consists of the digestive tract with the mouth at one end and the anus at the other end. Food enters the mouth and is digested to small nutrient molecules, which enter the bloodstream at the small intestine. Undigestible residue (feces) exits at the anus.

Digestive juices enter the tract from the salivary glands and the pancreas. The gallbladder also sends bile to the tract for emulsification of fats. Bile is made in the liver, which makes a significant contribution to homeostasis by helping to keep the composition of the blood constant. For example, after eating and under the influence of insulin, the liver stores glucose from the digestive tract as glycogen and releases glucose to the blood in between eating. The liver also removes nitrogenous and other types of injurious molecules from the blood and metabolizes them to excretory products that the bloodstream takes to the kidneys.

Nutrient molecules, such as amino acids, glucose, and vitamins, are absorbed from the small intestine into the blood. Fatty acids and glycerol enter the lacteals, a part of the lymphatic system. Body cells receive these molecules from blood and then build the body's own proteins and carbohydrates. In Chapter 35, we consider the contribution of the respiratory system to homeostasis.

## summary

### 34.1 Digestive Tracts

Some animals (e.g., planarians) have an incomplete digestive tract, so called because there is only one opening. An incomplete tract has little specialization of parts. Other animals (e.g., earthworms) have a complete digestive tract, so called because the tract has both a mouth and an anus. A complete tract tends to have specialization of parts.

Some animals are continuous feeders (e.g., clams, which are filter feeders); others are discontinuous feeders (e.g., squid). Discontinuous feeders need a storage area for food.

Most mammals have teeth. Herbivores need teeth that can clip off plant material and grind it up. Also, the herbivore's stomach contains bacteria that can digest cellulose.

Carnivores need teeth that can tear and rip meat into pieces. Meat is easier to digest than plant material, so the digestive system of carnivores has fewer specialized regions and the intestine is shorter than that of herbivores.

### 34.2 Human Digestive Tract

In the human digestive tract, food is chewed and manipulated in the mouth, where salivary glands secrete saliva. Saliva contains salivary amylase, which begins carbohydrate digestion.

Food then passes to the pharynx and down the esophagus by peristalsis to the stomach. The stomach stores and mixes food with mucus and gastric juices to produce chyme. Pepsin begins protein digestion in the stomach.

The duodenum of the small intestine receives bile from the liver and pancreatic juice from the pancreas. Bile emulsifies fat and readies it for digestion by an enzyme produced by the pancreas. The pancreas also produces enzymes that digest starch and protein. The intestinal enzymes finish the process of chemical digestion.

The walls of the small intestine have fingerlike projections called villi where small nutrient molecules are absorbed. Amino acids and glucose enter the blood vessels of a villus. Glycerol and fatty acids are joined and packaged as lipoproteins before entering lymphatic vessels called lacteals in a villus.

The large intestine consists of the cecum, the colon, and the rectum, which ends at the anus. The large intestine does not produce digestive enzymes; it does absorb water, salts, and some vitamins. Reduced water absorption results in diarrhea. The intake of water and fiber help prevent constipation.

Three accessory organs of digestion—the pancreas, liver, and gallbladder—send secretions to the duodenum via ducts. The pancreas produces pancreatic juice, which contains digestive enzymes for carbohydrates, protein, and fat.

The liver produces bile, which is stored in the gallbladder. The liver receives blood from the small intestine by way of the hepatic portal vein. It has numerous important functions, and any malfunction of the liver is a matter of considerable concern.

### 34.3 Digestive Enzymes

Digestive enzymes are present in digestive juices and break down food into the nutrient molecules glucose, amino acids, fatty acids, and glycerol. Salivary amylase and pancreatic amylase begin the digestion of starch. Pepsin and trypsin digest protein to peptides. Following emulsification by bile, lipase digests fat to glycerol and fatty acids. Intestinal enzymes finish the digestion of starch and protein.

Each digestive enzyme is present in a particular part of the digestive tract. Salivary amylase functions in the mouth; pepsin functions in the stomach; trypsin, lipase, and pancreatic amylase occur in the intestine along with the various enzymes that digest disaccharides and peptides.

### 34.4 Nutrition

The nutrients released by the digestive process should provide us with an adequate amount of energy, essential amino acids and fatty acids, and all necessary vitamins and minerals.

Carbohydrates are necessary in the diet, simple sugars and refined starches are not helpful because they provide calories but no fiber, vitamins, or minerals. Proteins supply us with essential amino acids, but it is wise to avoid meats that are fatty because fats from animal sources are saturated. While unsaturated fatty acids, particularly the omega-3 fatty acids, are protective against cardiovascular disease, the saturated fatty acids lead to plaque, which blocks blood vessels. Obesity is to be avoided particularly because obesity is now known to be associated with the development of type 2 diabetes and cardiovascular disease.

## understanding the terms

anus 640
appendix 639
bile 639
carbohydrate 643
cholesterol 644
chyme 639
cirrhosis 642
complete digestive tract 634
diarrhea 640
duodenum 639
esophagus 637
fat 644
fiber 643
gallbladder 642
hepatitis 642
incomplete digestive tract 634
jaundice 642
lacteal 639
large intestine 639
lipase 643
liver 640
lumen 641
maltase 642
mineral 646
mouth 636
mucosa 641
muscularis 641
pancreas 640
pancreatic amylase 642
pepsin 642
peptidase 642
peristalsis 638
pharynx 637
polyp 640
protein 644
radula 635
rumen 635
salivary amylase 637, 642
salivary gland 636
serosa 641
small intestine 639
stomach 638
submucosa 641
triglyceride 644
trypsin 642
typhlosole 634
villus 639
vitamin 646

Match the terms to these definitions:

a. ______________ Essential requirement in the diet, needed in small amounts. They are often part of coenzymes.
b. ______________ Fat-digesting enzyme secreted by the pancreas.
c. ______________ Lymphatic vessel in an intestinal villus; it aids in the absorption of fats.
d. ______________ Muscular tube for moving swallowed food from the pharynx to the stomach.
e. ______________ Organ attached to the liver that serves to store and concentrate bile.

## reviewing this chapter

1. Contrast the incomplete digestive tract with the complete digestive tract, using the planarian and earthworm as examples. 634
2. Contrast a continuous feeder with a discontinuous feeder, using the clam and squid as examples. 634–35
3. Contrast the dentition of the mammalian herbivore with that of the mammalian carnivore, using the horse and lion as examples. 635–36
4. List the parts of the human digestive tract, anatomically describe them, and state the contribution of each to the digestive process. 636–40
5. Discuss the absorption of the products of digestion into the lymphatic and cardiovascular systems. 639
6. Describe the structure and function of the large intestine. Name several medical conditions associated with the large intestine. 639–40
7. State the location and describe the functions of the pancreas, the liver, and the gallbladder. 640–42
8. Name and discuss three serious illnesses of the liver. 642
9. Assume that you have just eaten a ham sandwich. Discuss the digestion of the contents of the sandwich. Mention all the necessary enzymes. 642–43
10. Explain why good nutrition is important to human health. Explain why obesity can lead to type 2 diabetes and cardiovascular disease. 643–46

## testing yourself

Choose the best answer for each question.

1. Animals that feed discontinuously
   a. have digestive tracts that permit storage.
   b. are always filter feeders.
   c. exhibit extremely rapid digestion.
   d. have a nonspecialized digestive tract.
   e. usually eat only meat.
2. In which of these types of animals would you expect the digestive tract to be more complex?
   a. those with a single opening for the entrance of food and exit of wastes
   b. those with two openings, one serving as an entrance and the other as an exit
   c. only those complex animals that also have a respiratory system
   d. those with two openings that use the digestive tract to help fight infections
   e. All but a are correct.

3. The typhlosole within the gut of an earthworm compares best to which of these organs in humans?
   a. teeth in the mouth
   b. esophagus in the thoracic cavity
   c. folds in the stomach
   d. villi in the small intestine
   e. the large intestine because it absorbs water
4. Which of these animals is a continuous feeder with a complete digestive tract?
   a. planarian
   b. clam
   c. squid
   d. lion
   e. human
5. Tracing the path of food in humans, which step is out of order first?
   a. mouth
   b. pharynx
   c. esophagus
   d. small intestine
   e. stomach
   f. large intestine
6. The products of digestion are
   a. large macromolecules needed by the body.
   b. enzymes needed to digest food.
   c. small nutrient molecules that can be absorbed.
   d. regulatory hormones of various kinds.
   e. the foods we eat.
7. Why can a person not swallow food and talk at the same time?
   a. To swallow, the epiglottis must close off the trachea.
   b. The brain cannot control two activities at once.
   c. To speak, air must come through the larynx to form sounds.
   d. A swallowing reflex is only initiated when the mouth is closed.
   e. Both a and c are correct.
8. Which of these could be absorbed directly without need of digestion?
   a. glucose
   b. fat
   c. protein
   d. nucleic acid
   e. All of these are correct.
9. Which association is incorrect?
   a. protein—trypsin
   b. fat—lipase
   c. maltose—pepsin
   d. starch—amylase
   e. protein—pepsin
10. Most of the absorption of the products of digestion takes place in humans across
    a. the squamous epithelium of the esophagus.
    b. the convoluted walls of the stomach.
    c. the fingerlike villi of the small intestine.
    d. the smooth wall of the large intestine.
    e. the lacteals of the lymphatic system.
11. The hepatic portal vein is located between
    a. the hepatic vein and the vena cava.
    b. two capillary beds.
    c. the pancreas and the small intestine.
    d. the small intestine and the liver.
    e. Both b and d are correct.
12. Bile in humans
    a. is an important enzyme for the digestion of fats.
    b. is made by the gallbladder.
    c. emulsifies fat.
    d. must be activated during the first few weeks of life.
    e. All of these are correct.
13. Which of these is not a function of the liver in adults?
    a. produces bile
    b. stores glucose
    c. produces urea
    d. makes red blood cells
    e. produces proteins needed for blood clotting
14. The large intestine in humans
    a. digests all types of food.
    b. is the longest part of the intestinal tract.
    c. absorbs water.
    d. is connected to the stomach.
    e. All of these are correct.
15. The appendix connects to the
    a. cecum.
    b. small intestine.
    c. esophagus.
    d. large intestine.
    e. liver.
    f. All of these are correct.
16. Which association is incorrect?
    a. mouth—starch digestion
    b. esophagus—protein digestion
    c. small intestine—starch, lipid, protein digestion
    d. stomach—food storage
    e. liver—production of bile
17. Predict and explain the expected digestive results per test tube for this experiment.

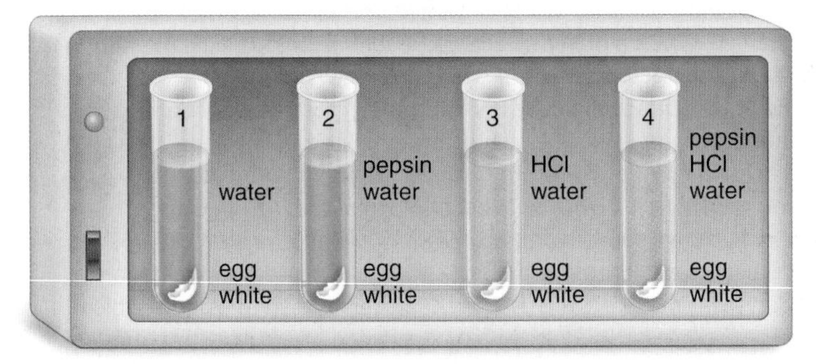


## thinking scientifically

1. A drug for leukemia is not broken down in the stomach and is well absorbed by the intestine. However, the molecular form of the drug collected from the blood is not the same as the form that was swallowed by the patient. What explanation is most likely?
2. Snakes often swallow whole animals, a process that takes a long time. Then snakes spend some time digesting their food. What structural modifications would allow slow swallowing and storage of a whole animal to occur? What chemical modifications would be necessary to digest a whole animal?

## *Biology* website

The companion website for *Biology* provides a wealth of information organized and integrated by chapter. You will find practice tests, animations, videos, and much more that will complement your learning and understanding of general biology.

**http://www.mhhe.com/maderbiology10**

# 35 Respiratory Systems

Elephant seals can dive to a depth of 1,500 m and stay submerged for two hours, even though they breathe air like we do. Their secret includes these adaptations: aquatic mammals (seals, whales, dolphins) have more blood cells and more blood per body weight than we do; their muscles are chock full of myoglobin, a respiratory pigment that stores oxygen in the muscles; and they have a special diving response. The heart rate slows down, the peripheral blood vessels constrict, and the blood circulates to only the heart and lungs. Now the spleen releases its supply of fully-oxygenated red blood cells, keeping the heart and brain going for a while longer.

*Humans sometimes free-dive—we can't store oxygen, but we can bring on the diving response. You can practice by submerging your face in cold water. If you hold your breath long enough, the spleen will discharge its red blood cells. Even so, free-diving is not recommended for humans; it can have disasterous results because humans are highly adapted to live on land. In this chapter, we will learn how animals in the water and on land ordinarily breathe and transport gases to and from their cells.*

An elephant seal can free-dive and suspend breathing for two hours.

## concepts

# 35.1 Gas Exchange Surfaces

**Respiration** is the sequence of events that results in gas exchange between the body's cells and the environment. In terrestrial vertebrates, respiration includes these steps:

- **Ventilation** (i.e., breathing) includes inspiration (entrance of air into the lungs) and expiration (exit of air from the lungs).
- **External respiration** is gas exchange between the air and the blood within the lungs. Blood then transports oxygen from the lungs to the tissues.
- **Internal respiration** is gas exchange between the blood and the tissue fluid. (The body's cells exchange gases with the tissue fluid.) The blood then transports carbon dioxide to the lungs.

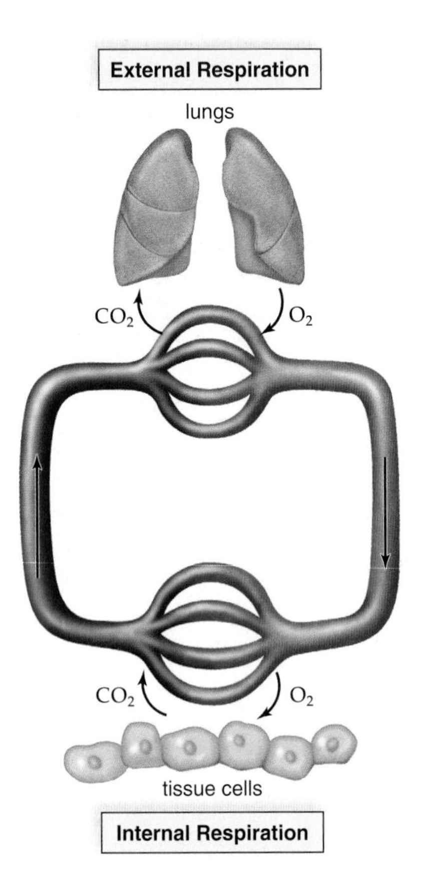


Gas exchange takes place by the physical process of diffusion. For external respiration to be effective, the gas-exchange region must be (1) moist, (2) thin, and (3) large in relation to the size of the body. Some animals, such as planarians, are small and shaped in a way that allows the surface of the animal to be the gas-exchange surface. Most complex animals have specialized external respiration surfaces, such as gills in aquatic animals and lungs in terrestrial animals. The effectiveness of diffusion is enhanced by vascularization (the presence of many capillaries), and delivery of oxygen to the cells is promoted when the blood contains a respiratory pigment, such as hemoglobin.

Regardless of the particular external respiration surface and the manner in which gases are delivered to the cells, in the end, oxygen enters mitochondria, where cellular respiration takes place. Without internal respiration, ATP production does not take place, and life ceases.

## External Gas-Exchange Surfaces

It is more difficult for animals to obtain oxygen from water than from air. Water fully saturated with air contains only a fraction of the amount of oxygen that would be present in the same volume of air. Also, water is more dense than air. Therefore, aquatic animals expend more energy carrying out gas exchange than do terrestrial animals. Fishes use as much as 25% of their energy output to respire, while terrestrial mammals use only 1–2% of their energy output for that purpose.

Hydras, which are cnidarians, and planarians, which are flatworms, have a large surface area in comparison to their size. This makes it possible for most of their cells to exchange gases directly with the environment. In hydras, the outer layer of cells is in contact with the external environment, and the inner layer can exchange gases with the water in the gastrovascular cavity (Fig. 35.1). In planarians, the flattened body permits cells to exchange gases with the external environment.

The tubular shape of annelids (segmented worms) also provides a surface area adequate for external respiration. The earthworm, an annelid, is an example of a terrestrial invertebrate that is able to use its body surface for respiration because the capillaries come close to the surface (Fig. 35.2).

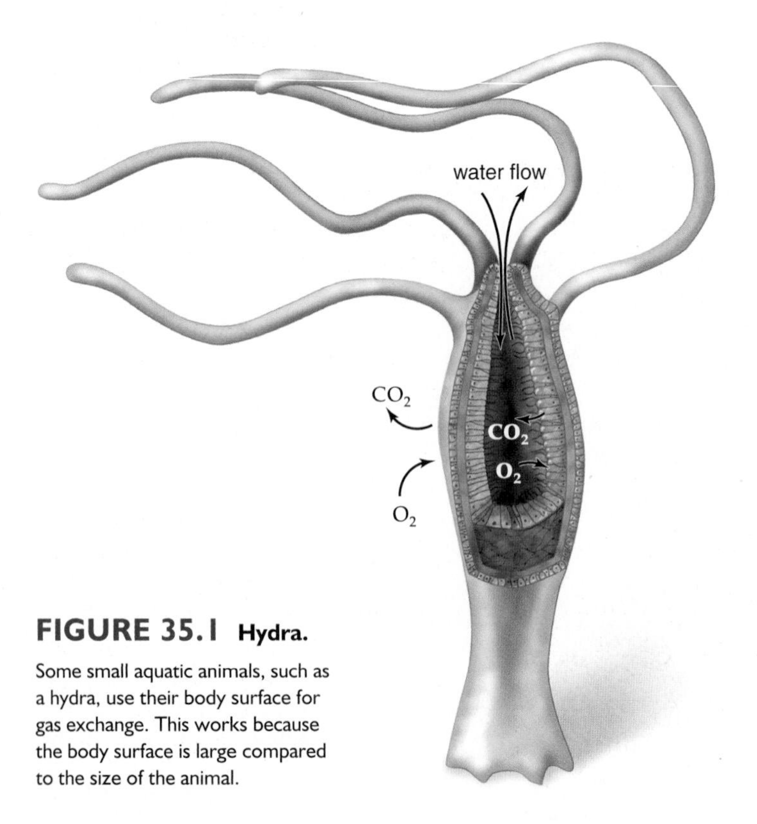


**FIGURE 35.1 Hydra.**

Some small aquatic animals, such as a hydra, use their body surface for gas exchange. This works because the body surface is large compared to the size of the animal.

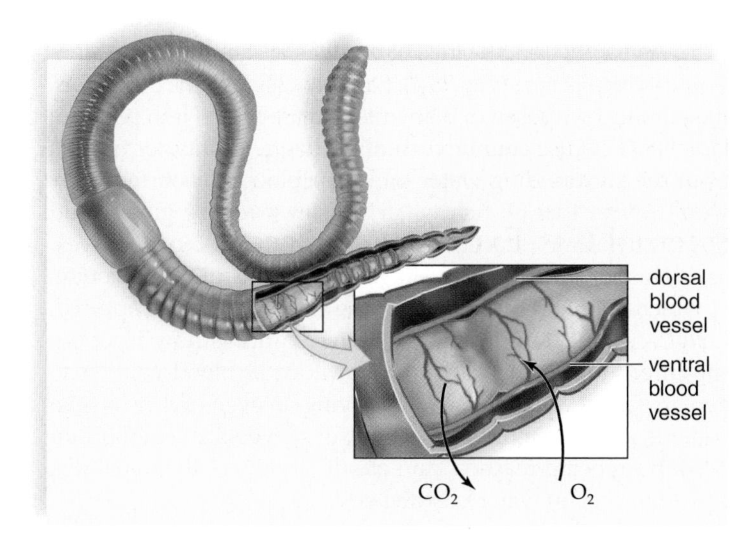


**FIGURE 35.2 Earthworm.**

An earthworm's entire external surface functions in external respiration.

An earthworm keeps its body surface moist by secreting mucus and by releasing fluids from excretory pores. Further, the worm is behaviorally adapted to remain in damp soil during the day, when the air is driest. In addition to a tubular shape, aquatic polychaete worms have extensions of the body wall called parapodia, which are vascularized and used for gas exchange.

Aquatic invertebrates (e.g., clams and crayfish) and aquatic vertebrates (e.g., fish and tadpoles) have gills that extract oxygen from a watery environment. **Gills** are finely divided, vascularized outgrowths of the body surface or the pharynx (Fig. 35.3*a*). Various mechanisms are used to pump water across the gills, depending on the organism.

Insects have a system of air tubes called tracheae through which oxygen is delivered directly to the cells without entering the blood (Fig. 35.3*b*). Tracheole fluid occurs at the end of the tracheae. Air sacs located near the wings, legs, and abdomen act as bellows to help move the air into the tubes through external openings.

Terrestrial vertebrates usually have **lungs,** which are vascularized outgrowths from the lower pharyngeal region. The tadpoles of frogs live in the water and have gills as external respiratory organs, but adult amphibians possess simple, saclike lungs. Most amphibians respire to some extent through the skin, and some salamanders depend entirely on the skin, which is kept moist by mucus produced by numerous glands on the surface of the body.

The lungs of birds and mammals are elaborately subdivided into small passageways and spaces, respectively (Fig. 35.3*c*). It has been estimated that human lungs have a total surface area that is at least 50 times the skin's surface area. Air is a rich source of oxygen compared to water; however, it does have a drying effect on external respiratory surfaces. A human loses about 350 ml of water per day when the air has a relative humidity of only 50%. To keep the lungs from drying out, air is moistened as it moves through the passageways leading to the lungs.

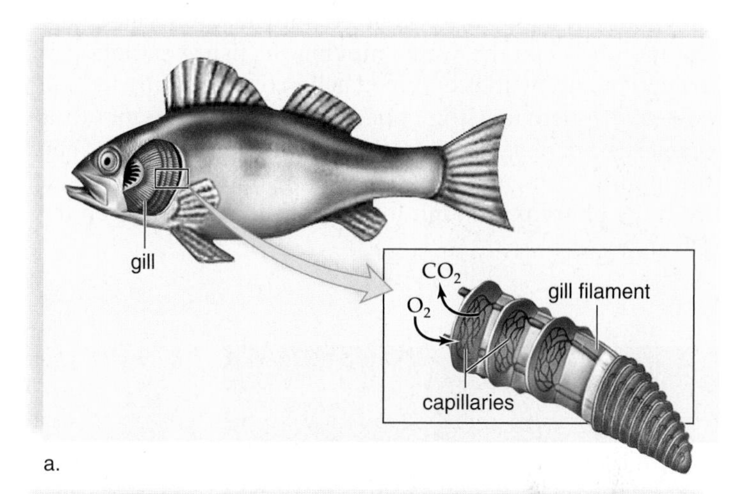


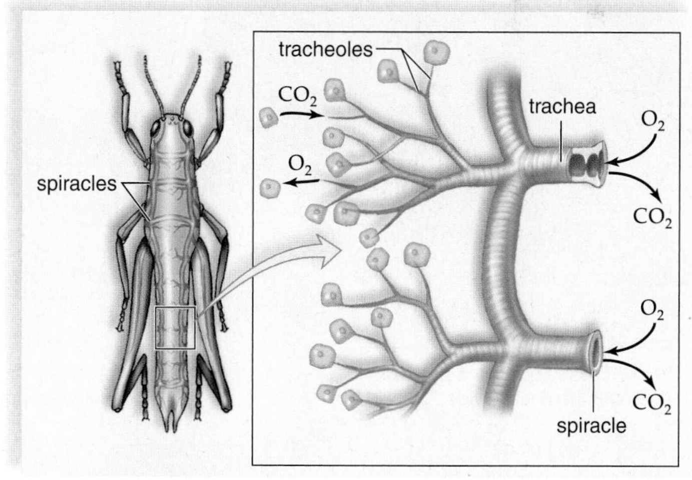


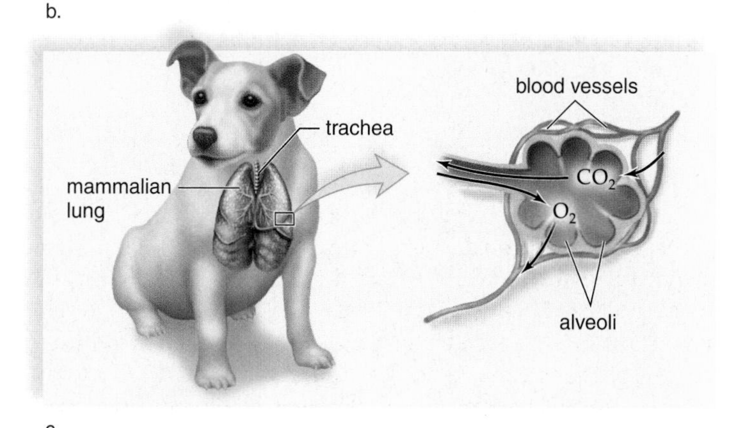


**FIGURE 35.3 Respiratory organs.**

**a.** Fish have gills to assist external respiration. **b.** Insects have a tracheal system that delivers oxygen directly to their cells. **c.** Vertebrates have lungs with a large total external respiration surface.

## The Gills of a Fish

Animals with gills use various means of ventilation. Among molluscs, such as a clam or a squid, water is drawn into the mantle cavity, where it passes through the gills. In crustaceans, such as crabs and shrimps, which are arthropods, the gills are located in thoracic chambers covered by the exoskeleton. The action of specialized appendages located near the mouth keeps the water moving. In fish, ventilation is brought about by the combined action of the mouth and gill covers, or opercula (sing., operculum). When the mouth is open, the opercula are closed and water is drawn in. Then the mouth closes, and the opercula open, drawing the water from the pharynx through the gill slits located between the gill arches.

As mentioned, the gills of bony fishes are outward extensions of the pharynx (Fig. 35.4). On the outside of the gill arches, the gills are composed of filaments that are folded into platelike lamellae. Fish use **countercurrent exchange** to transfer oxygen from the surrounding water into their blood. *Con*current flow would mean that $O_2$-rich water passing over the gills would flow in the same direction as $O_2$-poor blood in the blood vessels. This arrangement would result in an equilibrium point, at which only half the oxygen in the water would be captured. *Counter*current flow means that the two fluids flow in opposite directions. With countercurrent flow, as blood gains oxygen, it always encounters water having an even higher oxygen content. A countercurrent mechanism prevents an equilibrium point from being reached, and about 80–90% of the initial dissolved oxygen in water is extracted.

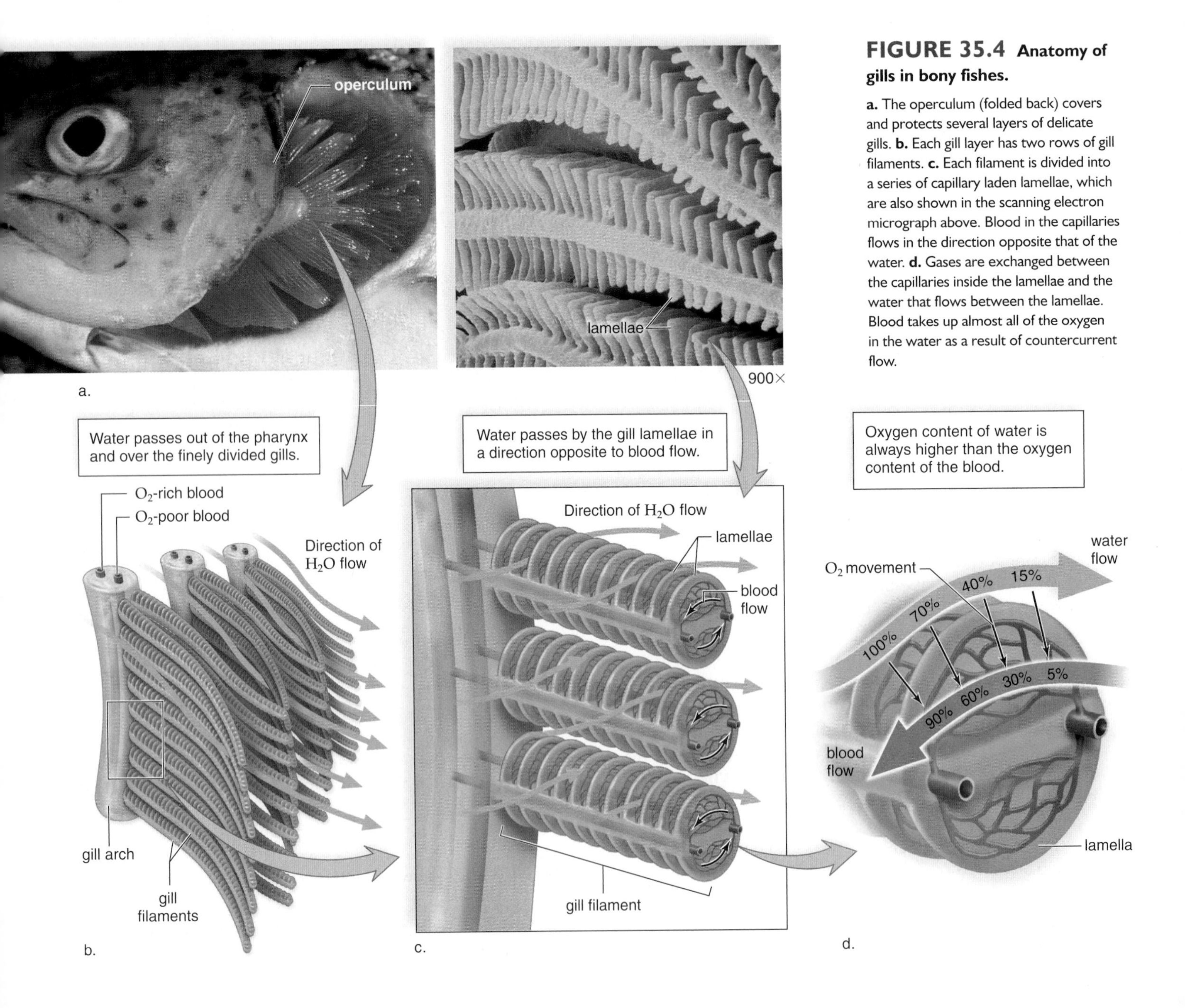


**FIGURE 35.4 Anatomy of gills in bony fishes.**

**a.** The operculum (folded back) covers and protects several layers of delicate gills. **b.** Each gill layer has two rows of gill filaments. **c.** Each filament is divided into a series of capillary laden lamellae, which are also shown in the scanning electron micrograph above. Blood in the capillaries flows in the direction opposite that of the water. **d.** Gases are exchanged between the capillaries inside the lamellae and the water that flows between the lamellae. Blood takes up almost all of the oxygen in the water as a result of countercurrent flow.

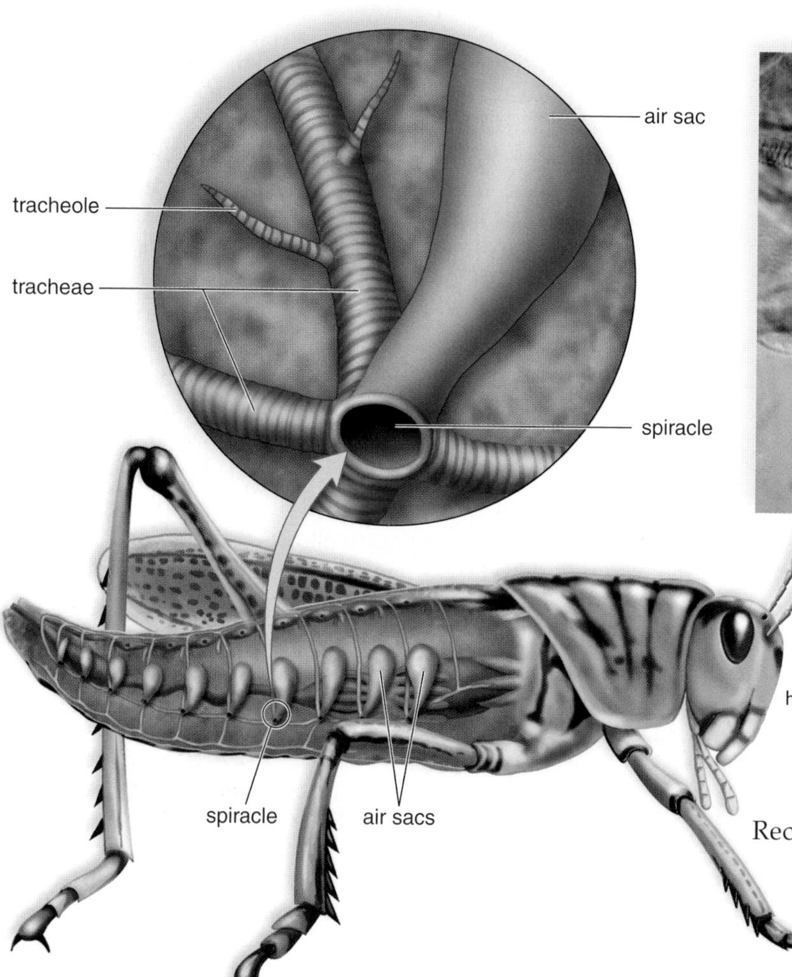


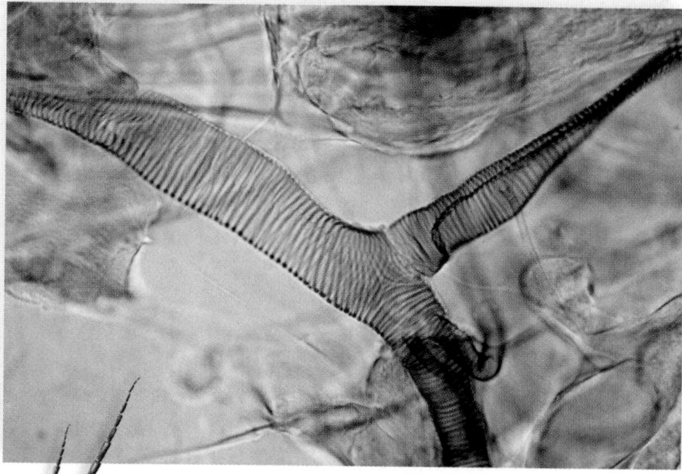

**FIGURE 35.5 Tracheae of insects.**
A system of air tubes extends throughout the body of an insect, and the tubes carry oxygen to the cells. Air enters the tracheae at openings called spiracles. From here, the air moves to the smaller tracheoles, which take it to the cells, where gas exchange takes place. The photomicrograph shows how the walls of the trachea are stiffened with bands of chitin.

## The Tracheal System of Insects

Arthropods are coelomate animals, but the coelom is reduced and the internal organs lie within a cavity called the hemocoel because it contains hemolymph, a mixture of blood and lymph. Hemolymph flows freely through the hemocoel, making circulation in arthropods inefficient. Many insects are adapted for flight, and their flight muscles require a steady supply of oxygen. Insects overcome the inefficiency of their blood flow by having a respiratory system that consists of **tracheae,** tiny air tubes that take oxygen directly to the cells (Fig. 35.5). Tracheae have a single layer of cells supported by spiral thickenings of chitin. The tracheae branch into even smaller tubules called tracheoles, which also branch and rebranch until finally the air tubes are only about 0.1 μm in diameter. There are so many fine tracheoles that almost every cell is near one. Also, the tracheoles indent the plasma membrane so that they terminate close to mitochondria. Therefore, $O_2$ can flow more directly from a tracheole to mitochondria, where cellular respiration occurs. The tracheae also dispose of $CO_2$.

The tracheoles are fluid-filled, but the larger tracheae contain air and open to the outside by way of spiracles (Fig. 35.5). Usually, the spiracle has some sort of closing device that reduces water loss, and this may be why insects have no trouble inhabiting drier climates. Recently, investigators presented evidence that tracheae actually expand and contract, thereby drawing air into and out of the system. This method is comparable to the way the human lungs expand to draw air into them. Otherwise, a tracheal system consisting of an expansive network of thin-walled tubes seems to be an entirely different mechanism of respiration from those used by other animals.

It's been suggested that a tracheal system that has no efficient method to improve flow is sufficient only in small insects. Larger insects have still another mechanism to ventilate—keep the air moving in and out—tracheae. Many larger insects have air sacs, which are thin-walled and flexible, located near major muscles. Contraction of these muscles causes the air sacs to empty, and relaxation causes the air sacs to expand and draw in air. Even so, insects still lack the efficient circulatory system of birds and mammals that is able to pump $O_2$-rich blood through arteries to all the cells of the body. This may be why insects remain small, despite the attempts of movies to make us think otherwise.

A tracheal system is an adaptation to breathing air, and yet, some insect larval stages and even some adult insects live in the water. In these instances, the tracheae do not receive air by way of spiracles. Instead, diffusion of oxygen across the body wall supplies the tracheae with oxygen. Mayfly and stonefly nymphs have thin extensions of the body wall called tracheal gills—the tracheae are particularly numerous in this area. This is an interesting adaptation because it dramatizes that tracheae function to deliver oxygen in the same manner as vertebrate blood vessels.

## The Lungs of Humans

The human respiratory system includes all of the structures that conduct air in a continuous pathway to and from the lungs (Fig. 35.6*a*). The lungs lie deep within the thoracic cavity, where they are protected from drying out. As air moves through the nose, the pharynx, the trachea, and the bronchi to the lungs, it is filtered so that it is free of debris, warmed, and humidified. By the time the air reaches the lungs, it is at body temperature and saturated with water. In the nose, hairs and cilia act as a screening device. In the trachea and the bronchi, cilia beat upward, carrying mucus, dust, and occasional small bits of food that "went down the wrong way" into the throat, where the accumulation may be swallowed or expectorated (Fig. 35.6*b*).

The hard and soft palates separate the nasal cavities from the mouth, but the air and food passages cross in the **pharynx.** This may seem inefficient, and there is danger of choking if food accidentally enters the trachea; however, this arrangement does have the advantage of letting you breathe through your mouth in case your nose is plugged up. In addition, it permits greater intake of air during heavy exercise, when greater gas exchange is required.

Air passes from the pharynx through the **glottis,** an opening into the **larynx,** or voice box. At the edges of the glottis, embedded in mucous membrane, are the **vocal cords.** These flexible and pliable bands of connective tissue vibrate and produce sound when air is expelled past them through the glottis from the larynx.

The larynx and the trachea remain open to receive air at all times. The larynx is held open by a complex of nine cartilages, among them the Adam's apple. Easily seen in many men, the Adam's apple resembles a small, rounded apple just under the skin in the front of the neck. The **trachea** is held open by a series of C-shaped, cartilaginous rings that do not completely meet in the rear. When food is being swallowed, the larynx rises, and the glottis is closed by a flap of tissue called the **epiglottis.** A backward movement of the soft palate covers the entrance of the nasal passages into the pharynx. The food then enters the esophagus, which lies behind the larynx.

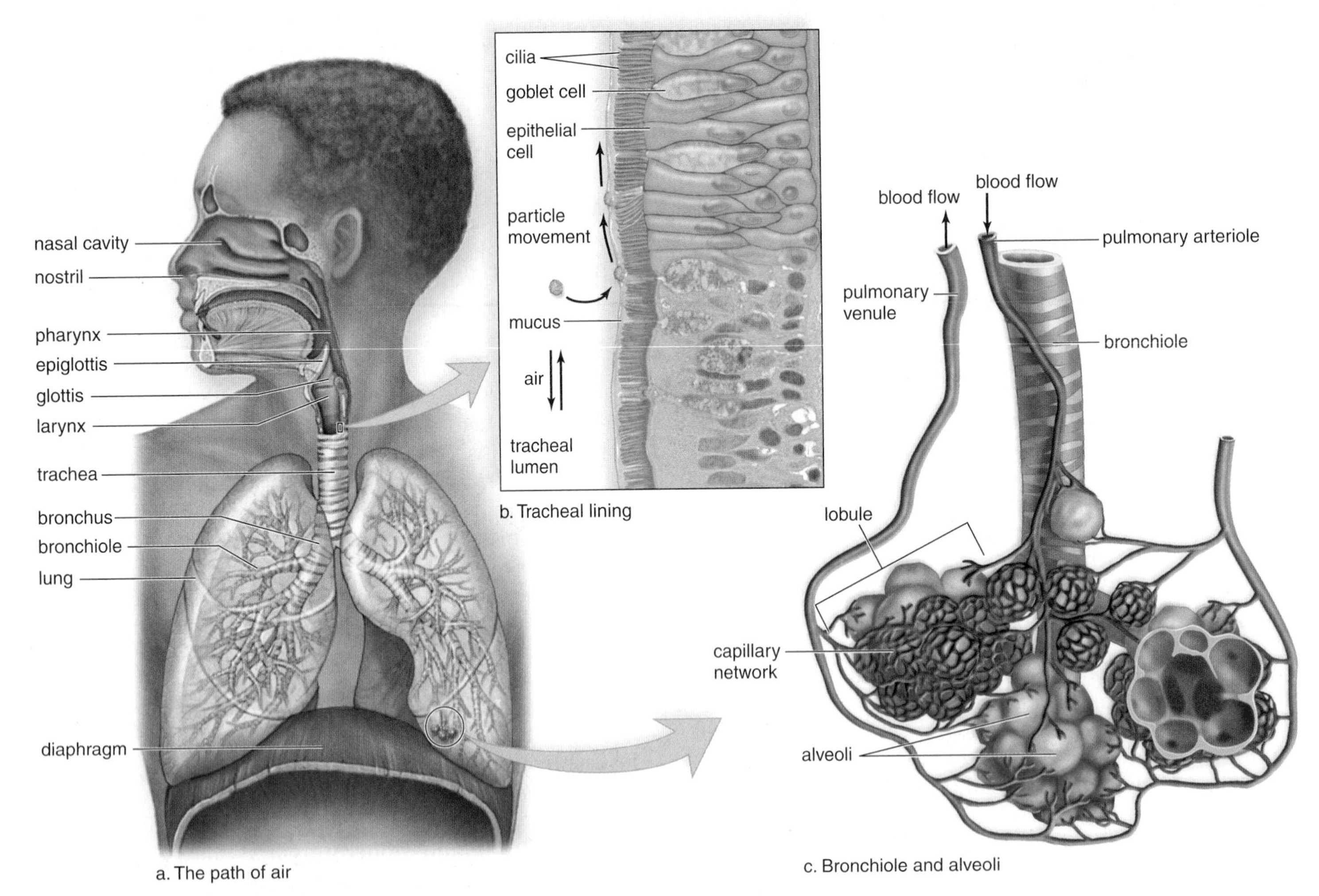


**FIGURE 35.6 The human respiratory tract.**

**a.** The respiratory tract extends from the nose to the lungs, which are composed of air sacs called alveoli. **b.** The lining of the trachea is a ciliated epithelium (art above, micrograph below) with mucus-producing goblet cells. The lining prevents inhaled particles from reaching the lungs: the mucus traps the particles, and the cilia help move the mucus toward the throat, where it can be swallowed. **c.** Gas exchange occurs between air in the alveoli and blood within a capillary network that surrounds the alveoli.

*science focus*

## Questions About Tobacco, Smoking, and Health

**Is there a safe way to smoke?** No. All cigarettes can damage the human body. Any amount of smoke is dangerous.

**Is cigarette smoking really addictive?** Yes. The nicotine in cigarette smoke causes addiction to smoking. Nicotine is an addictive drug (just like heroin and cocaine) for three main reasons: Small amounts make the smoker want to smoke more; smokers usually suffer withdrawal symptoms when they stop; and nicotine can affect the mood and nature of the smoker.

**Does smoking cause cancer?** Yes. Tobacco use accounts for about one-third of all cancer deaths in the United States. Smoking causes almost 90% of lung cancers. Smoking also causes cancers of the larynx (voice box), oral cavity, pharynx (throat), and esophagus, and contributes to the development of cancers of the bladder, pancreas, cervix, kidney, and stomach.

**How does cigarette smoke affect the lungs?** Cigarette smoking causes several lung diseases that can be just as dangerous as lung cancer: chronic bronchitis, in which the airways produce excess mucus, forcing the smoker to cough more; emphysema, a disease that slowly destroys a person's ability to breathe; and chronic obstructive pulmonary disease, a name that encompasses both chronic bronchitis and emphysema.

**Why do smokers have "smoker's cough"?** Cigarette smoke contains chemicals that irritate the air passages and lungs. When a smoker inhales these substances, the body tries to protect itself by producing mucus and stimulating coughing. Normally, cilia (tiny, hairlike formations that line the airways) beat outward and "sweep" harmful material out of the lungs. Smoke, however, decreases this sweeping action, so some of the poisons in the smoke remain in the lungs.

**If you smoke but do not inhale, is there any danger?** Yes. Even if smokers don't inhale, they are breathing the smoke secondhand and are still at risk for lung cancer. Pipe and cigar smokers, who often do not inhale, are at increased risk for lip, mouth, tongue, and several other cancers.

**Does cigarette smoking affect the heart?** Yes. Smoking increases the risk of heart disease, which is the number-one cause of death in the United States. Smoking is a risk factor for heart disease, but cigarette smoking is the biggest risk factor for sudden heart death. Smokers are more likely to die from a heart attack within an hour of the attack than nonsmokers. Cigarette smoke can cause harm to the heart at very low levels, much lower than the amount that causes lung disease.

**How does smoking affect pregnant women and their babies?** Smoking during pregnancy is linked to a greater chance of miscarriage, premature delivery, stillbirth, infant death, low birth weight, and sudden infant death syndrome (SIDS). When a pregnant woman smokes, she is really smoking for two because the nicotine, carbon monoxide, and other dangerous chemicals in smoke enter her bloodstream and then pass into the baby's body, preventing the baby from getting essential nutrients and oxygen for growth.

**What are some of the short-term and long-term effects of smoking cigarettes?** Short-term effects include shortness of breath and a nagging cough, diminished ability to smell and taste, premature aging of the skin, and increased risk of sexual impotence in men. Long-term effects include many types of cancer, heart disease, aneurysms, bronchitis, emphysema, and stroke. Smoking contributes to the severity of pneumonia and asthma.

**What are the dangers of environmental tobacco smoke (ETS)?** In nonsmokers, ETS causes about 3,000 lung cancer deaths and about 35,000 to 40,000 deaths from heart disease each year. Children whose parents smoke are more likely to suffer from asthma, pneumonia or bronchitis, ear infections, coughing, wheezing, and increased mucus production in the first two years of life than children who come from smoke-free households.

**Are chewing tobacco and snuff safe alternatives to cigarette smoking?** No. The juice from smokeless tobacco is absorbed directly through the lining of the mouth. This creates sores and white patches that often lead to cancer of the mouth. Smokeless tobacco users also greatly increase their risk of other cancers, including those of the pharynx (throat). Other effects of smokeless tobacco include harm to teeth and gums.

The trachea divides into two primary **bronchi,** which enter the right and left lungs. Branching continues, eventually forming a great number of smaller passages called **bronchioles.** The two bronchi resemble the trachea in structure, but as the bronchial tubes divide and subdivide, their walls become thinner, and rings of cartilage are no longer present. Each bronchiole terminates in an elongated space enclosed by a multitude of air pockets, or sacs, called **alveoli,** which make up the lungs (Fig. 35.6c). Internal gas exchange occurs between the air in the alveoli and the blood in the capillaries.

### Check Your Progress 35.1

1. Per unit volume, air contains more oxygen than water, but breathing air causes one problem easily preventable in the water. What is it?
2. One page 650, we said that respiration has three events, but this is not so in insects. Explain.
3. Name six illnesses associated with smoking cigarettes.

# 35.2 Breathing and Transport of Gases

During breathing, the lungs are ventilated. $O_2$ moves into the blood, and $CO_2$ moves out of the blood into the lungs. Blood transports $O_2$ to the body's cells and $CO_2$ from the cells to the lungs.

## Breathing

Terrestrial vertebrates ventilate their lungs by moving air into and out of the respiratory tract. Amphibians use positive pressure to force air into the respiratory tract. With the mouth and nostrils firmly shut, the floor of the mouth rises and pushes the air into the lungs. Reptiles, birds, and mammals use negative pressure to move air into the lungs and positive pressure to move it out. **Inspiration** (or inhalation) is the act of moving air into the lungs, and **expiration** (or exhalation) is the act of moving air out of the lungs.

Reptiles have jointed ribs that can be raised to expand the lungs, but mammals have both a rib cage and a diaphragm. The **diaphragm** is a horizontal muscle that divides the thoracic cavity (above) from the abdominal cavity (below). During inspiration in mammals, the rib cage moves up and out, and the diaphragm contracts and moves down (Fig. 35.7). As the thoracic (chest) cavity expands and lung volume increases, air flows into the lungs due to decreased air pressure in the thoracic cavity and lungs. Also, inspiration is the active phase of breathing in reptiles and mammals.

During expiration in mammals, the rib cage moves down, and the diaphragm relaxes and moves up to its former position (Fig. 35.8). No muscle contraction is required and expiration is the inactive phase of breathing in reptiles and mammals. During expiration, air flows out as a result of increased pressure in the thoracic cavity and lungs.

We can liken ventilation in reptiles and mammals to the way a bellows, used to fan a fire, functions (Fig. 35.9). First, the handles of the bellows are pulled apart, decreasing the air pressure inside the bellows. This causes air to automatically flow into the bellows, just as air automatically enters the lungs because the rib cage moves up and out during inspiration. Then, when the handles of the bellows are pushed together, air automatically flows out because the air pressure increases inside the bellows. Similarly, air

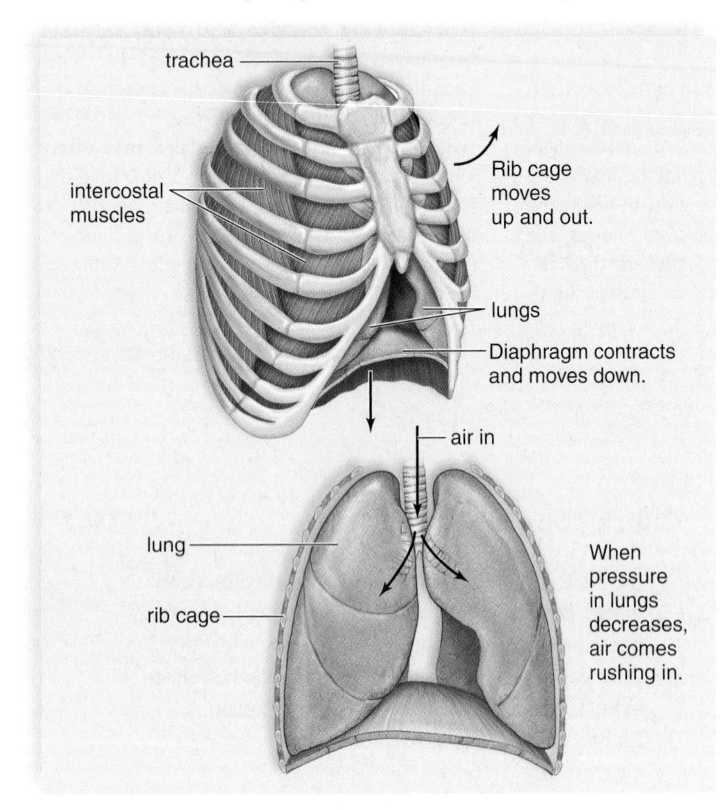


**FIGURE 35.7 Inspiration.**

During inspiration, the thoracic cavity and lungs expand so that air is drawn in.

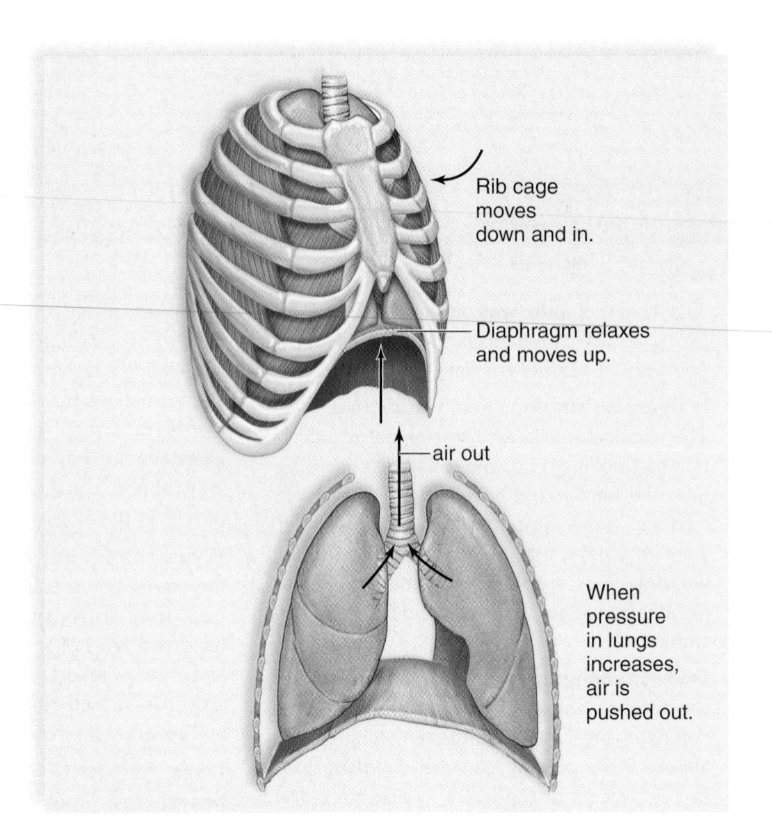


**FIGURE 35.8 Expiration.**

During expiration, the thoracic cavity and lungs resume their original positions and pressures. Now, air is forced out.

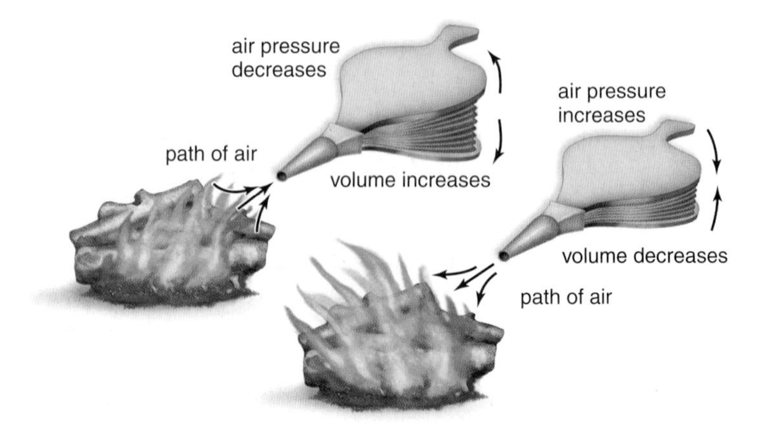


**FIGURE 35.9 Bellows.**

Lungs function much as bellows do.

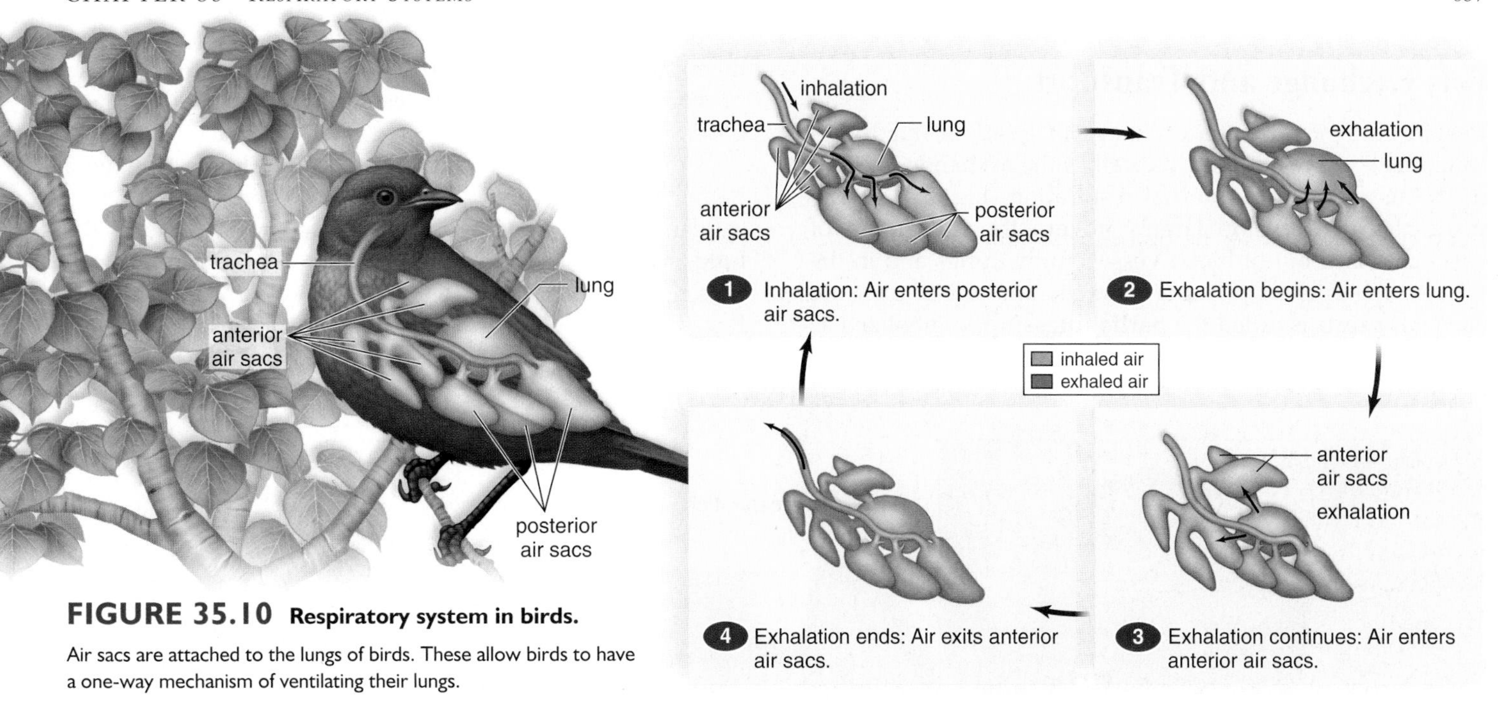


**FIGURE 35.10 Respiratory system in birds.**
Air sacs are attached to the lungs of birds. These allow birds to have a one-way mechanism of ventilating their lungs.

automatically exits the lungs when the rib cage moves down and in during expiration. The analogy is not exact, however, because no force is required for the rib cage to move down, and inspiration is the only active phase of breathing. Forced expiration can occur if we so desire, however.

All terrestrial vertebrates, except birds, use a *tidal ventilation mechanism,* so called because the air moves in and out by the same route. This means that the lungs of amphibians, reptiles, and mammals are not completely emptied and refilled during each breathing cycle. Because of this, the air entering mixes with used air remaining in the lungs. While this does help conserve water, it also decreases gas-exchange efficiency. In contrast, birds use a *one-way ventilation mechanism* (Fig. 35.10). Incoming air is carried past the lungs by a trachea, which takes it to a set of posterior air sacs. The air then passes forward through the lungs into a set of anterior air sacs. From here, it is finally expelled. Notice that fresh air never mixes with used air in the lungs of birds, thereby greatly improving gas-exchange efficiency.

## *Modifications of Breathing in Humans*

Normally, adults have a breathing rate of 12 to 20 ventilations per minute. The rhythm of ventilation is controlled by a **respiratory center** in the medulla oblongata of the brain. The respiratory center automatically sends out impulses by way of a spinal nerve to the diaphragm (phrenic nerve) and intercostal nerves to the intercostal muscles of the rib cage (Fig. 35.11). Now inspiration occurs. When the respiratory center stops sending neuronal signals to the diaphragm and the rib cage, expiration occurs.

Although the respiratory center automatically controls the rate and depth of breathing, its activity can also be influenced by nervous input and chemical input. Following forced inhalation, stretch receptors in the alveolar walls initiate inhibitory nerve impulses that travel from the inflated lungs to the respiratory center. This stops the respiratory center from sending out nerve impulses.

The respiratory center is directly sensitive to the levels of hydrogen ions ($H^+$). However, when carbon dioxide ($CO_2$) enters the blood, it reacts with water and releases hydrogen ions. In this way, $CO_2$ participates in regulating the breathing rate. When hydrogen ions rise in the blood and the pH decreases, the respiratory center increases the rate and depth of breathing. The chemoreceptors in the **carotid bodies,** located in the carotid arteries, and in the **aortic bodies,** located in the aorta, will stimulate the respiratory center during intense exercise due to a reduction in pH and also if and when arterial oxygen decreases to 50% of normal.

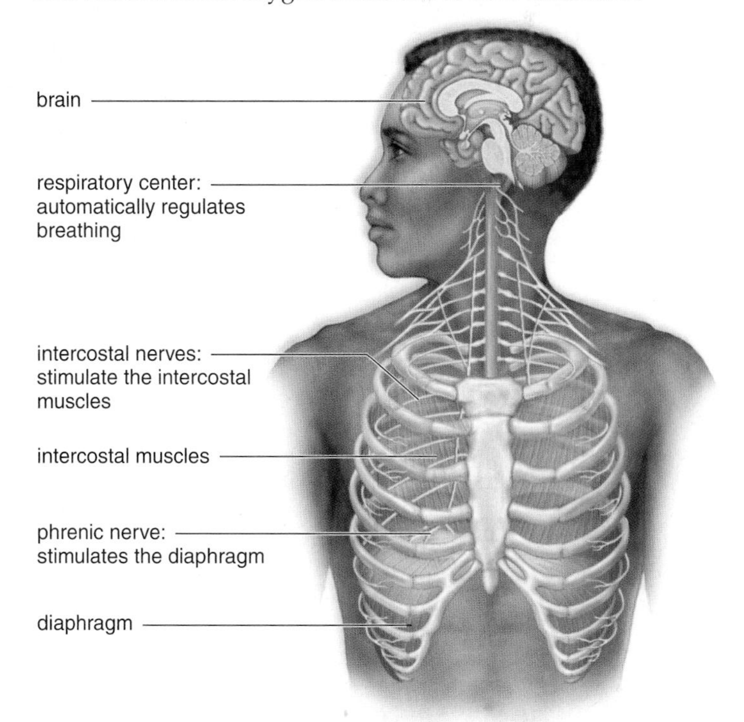


**FIGURE 35.11 Nervous control of breathing.**
The breathing rate can be modified by nervous stimulation of the intercostal muscles and diaphragm.

## Gas Exchange and Transport

Respiration includes the exchange of gases in our lungs, called external respiration, as well as the exchange of gases in the tissues, called internal respiration (Fig. 35.12). The principles of diffusion largely govern the movement of gases into and out of blood vessels in the lungs and in the tissues. Gases exert pressure, and the amount of pressure each gas exerts is called the **partial pressure,** symbolized as $P_{O_2}$ and $P_{CO_2}$. If the partial pressure of oxygen differs across a membrane, oxygen will diffuse from the higher to the lower pressure. Similarly, $CO_2$ diffuses from the higher to the lower partial pressure.

Ventilation causes the alveoli of the lungs to have a higher $P_{O_2}$ and a lower $P_{CO_2}$ than the blood in pulmonary capillaries, and this accounts for the exchange of gases in the lungs. When blood reaches the tissues, cellular respiration in cells causes the tissue fluid to have a lower $P_{O_2}$ and a higher $P_{CO_2}$ than the blood in the systemic capillaries and this accounts for the exchange of gases in the tissues.

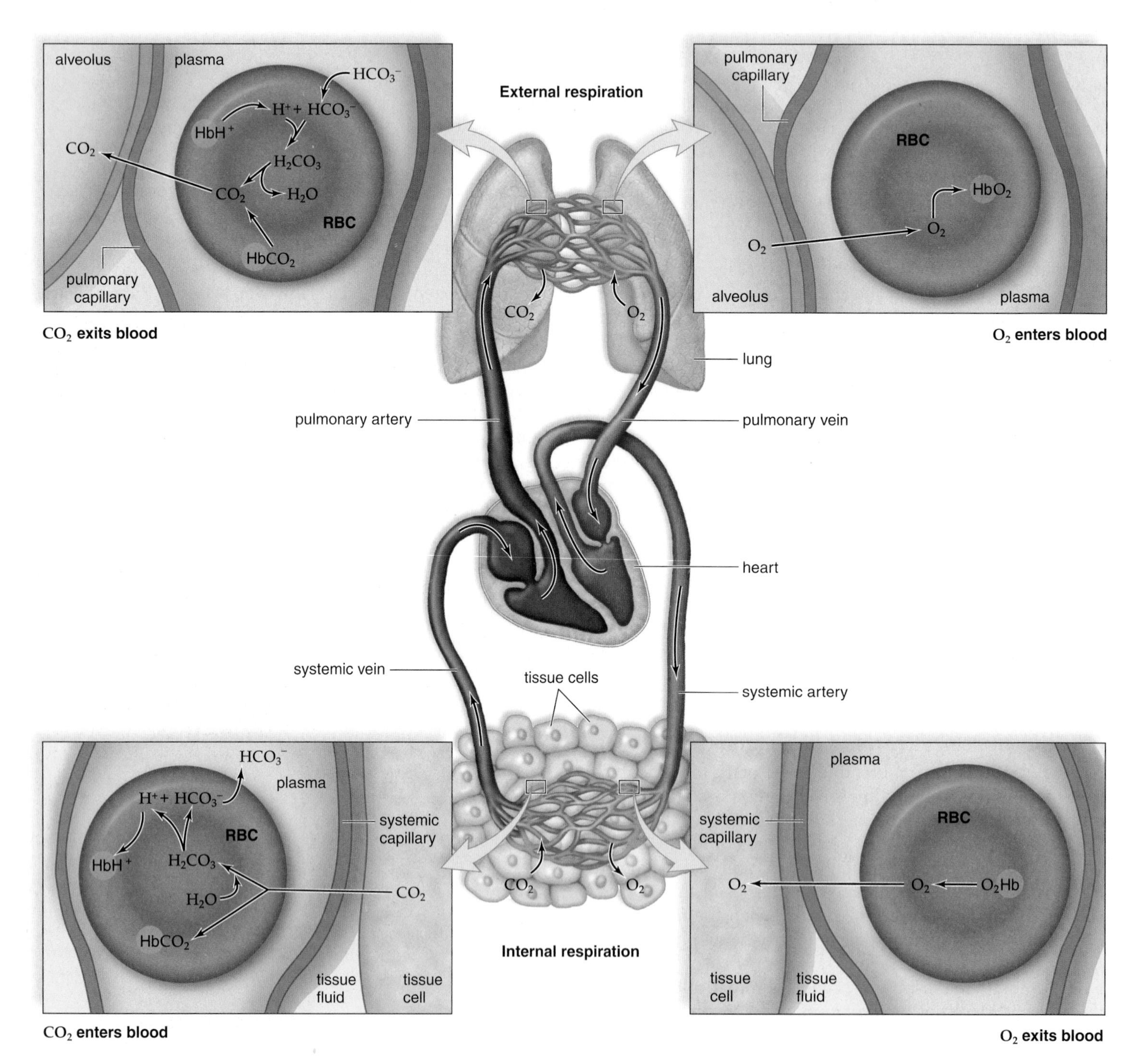


**FIGURE 35.12 External and internal respiration.**

During external respiration (*above*) in the lungs, carbon dioxide ($CO_2$) leaves blood, and oxygen ($O_2$) enters blood. During internal respiration (*below*) in the tissues, oxygen leaves blood, and carbon dioxide enters blood.

## *Transport of Oxygen and Carbon Dioxide*

**External Respiration.** As blood enters the lungs, a small amount of $CO_2$ is being carried by hemoglobin with the formula $HbCO_2$. Also, some hemoglobin is carrying hydrogen ions with the formula $HbH^+$. Most of the $CO_2$ in the pulmonary capillaries is carried as bicarbonate ions ($HCO_3^-$) in the plasma. As the free $CO_2$ from the following equation begins to diffuse out, this reaction is driven to the right:

$$\underset{\text{hydrogen ion}}{H^+} + \underset{\text{bicarbonate ion}}{HCO_3^-} \longrightarrow \underset{\text{carbonic acid}}{H_2CO_3} \longrightarrow \underset{\text{water}}{H_2O} + \underset{\text{carbon dioxide}}{CO_2}$$

The reaction occurs in red blood cells, where the enzyme **carbonic anhydrase** speeds the breakdown of carbonic acid (see Fig. 35.12, *upper left*). Pushing this equation to the far right by breathing fast can cause you to stop breathing for a time; pushing this equation to the left by not breathing is even more temporary because breathing will soon resume due to the rise in $H^+$.

Most oxygen entering the pulmonary capillaries from the alveoli of the lungs combines with **hemoglobin (Hb)** in red blood cells (RBCs) to form **oxyhemoglobin** (Fig. 35.12, *upper right*):

$$\underset{\text{deoxyhemoglobin}}{Hb} + \underset{\text{oxygen}}{O_2} \longrightarrow \underset{\text{oxyhemoglobin}}{HbO_2}$$

At the normal $P_{O_2}$ in the lungs, hemoglobin is practically saturated with oxygen. Each hemoglobin molecule contains four polypeptide chains, and each chain is folded around an iron-containing group called **heme** (Fig. 35.13). It is actually the iron that forms a loose association with oxygen. Since there are about 250 million hemoglobin molecules in each red blood cell, each red blood cell is capable of carrying at least one billion molecules of oxygen.

Unfortunately, carbon monoxide (CO) is an air pollutant that comes from the incomplete combustion of natural gas and gasoline. Because CO is a colorless, odorless gas, people can be unaware that they are breathing it. But once CO is in the bloodstream, it combines with the iron of hemoglobin 200 times more tightly than oxygen, and the result can be death. This is the reason that homes are equipped with CO detectors.

**Internal Respiration.** Blood entering the systemic capillaries is a bright red color because RBCs contain oxyhemoglobin. Because the temperature in the tissues is higher and the pH is lower than in the lungs, oxyhemoglobin has a tendency to give up oxygen:

$$HbO_2 \longrightarrow Hb + O_2$$

Oxygen diffuses out of the blood into the tissues because the $P_{O_2}$ of tissue fluid is lower than that of blood (see Fig. 35.12, *lower right*). The lower $P_{O_2}$ is due to cells continuously using up oxygen in cellular respiration. After oxyhemoglobin gives up $O_2$, $O_2$ leaves the blood and enters tissue fluid, where it is taken up by cells.

Carbon dioxide, on the other hand, enters blood from the tissues because the $P_{CO_2}$ of tissue fluid is higher than that of blood. Carbon dioxide, produced continuously by cells, collects in tissue fluid. After $CO_2$ diffuses into the blood, it enters the red blood cells, where a small amount combines with the protein portion of hemoglobin to form **carbaminohemoglobin** ($HbCO_2$). Most of the $CO_2$, however, is transported in the form of the **bicarbonate ion** ($HCO_3^-$). First, $CO_2$ combines with water, forming carbonic acid, and then this dissociates to a hydrogen ion ($H^+$) and $HCO_3^-$:

$$\underset{\text{carbon dioxide}}{CO_2} + \underset{\text{water}}{H_2O} \longrightarrow \underset{\text{carbonic acid}}{H_2CO_3} \longrightarrow \underset{\text{hydrogen ion}}{H^+} + \underset{\text{bicarbonate ion}}{HCO_3^-}$$

Carbonic anhydrase also speeds this reaction. The $HCO_3^-$ diffuses out of the red blood cells to be carried in the plasma (see Fig. 35.12, *lower left*).

The release of $H^+$ from this reaction could drastically change the pH of the blood, which is highly undesirable because cells require a normal pH in order to remain healthy. However, the $H^+$ is absorbed by the globin portions of hemoglobin. Hemoglobin that has combined with $H^+$ is called reduced hemoglobin and has the formula $HbH^+$. $HbH^+$ plays a vital role in maintaining the normal pH of the blood. Blood that leaves the systemic capillaries is a dark maroon color because red blood cells contain reduced hemoglobin.

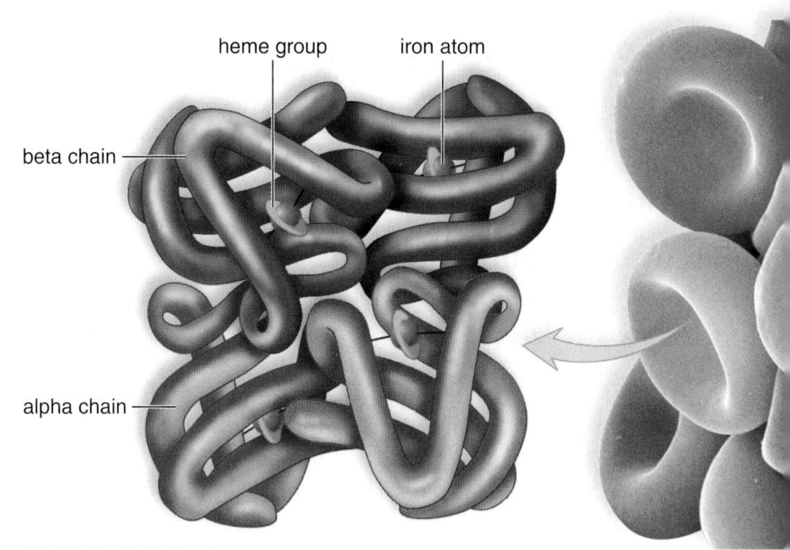


**FIGURE 35.13 Hemoglobin.**

Hemoglobin consists of four polypeptide chains (two alpha—red; two beta—purple), each associated with a heme group. Each heme group contains an iron atom, which can bind to $O_2$.

### Check Your Progress 35.2

1. Explain the process of inspiration and expiration in humans.
2. Describe the functions of hemoglobin in the transport of $O_2$ and $CO_2$.

## 35.3 Respiration and Health

We have seen that the entire respiratory tract has a warm, wet, mucous membrane lining that is constantly exposed to environmental air. The quality of this air, determined by the pollutants and the pathogens therein, can affect our health.

### Lower Respiratory Tract Disorders

Lower respiratory tract infections and disorders are shown in Figure 35.14.

#### *Infections*

*Acute bronchitis* is an infection of the primary and secondary bronchi. Usually it is preceded by a viral upper respiratory infection that has led to a secondary bacterial infection. Most likely, a nonproductive cough has become a deep cough that expectorates mucus and perhaps pus.

*Pneumonia* is a viral, bacterial, or fungal infection of the lungs in which bronchi and alveoli fill with a discharge, such as pus and fluid (Fig. 35.14). Most often it is preceded by influenza. Rather than being a generalized lung infection, pneumonia may be localized in specific lobules of the lungs. Obviously the more lobules involved, the more serious the infection. Pneumonia can be caused by a microbe that is usually held in check but has gained the upper hand due to stress and/or reduced immunity. AIDS patients are subject to a particularly rare form of pneumonia caused by a fungal-like organism called *Pneumocystis carinii*. Pneumonia of this type is almost never seen in individuals with a healthy immune system. High fever and chills with headache and chest pain are symptoms of pneumonia.

*Pulmonary tuberculosis* is caused by the tubercle bacillus, a type of bacterium. It is possible to tell if a person has ever been exposed to tuberculosis with a skin test in which a highly diluted extract of the bacillus is injected into the skin of the patient. A person who has never been in contact with the tubercle bacillus shows no reaction, but one who has developed immunity to the organism shows an area of inflammation that peaks in about 48 hours. When tubercle bacilli invade the lung tissue, the cells build a protective capsule about the organisms, isolating them from the rest of the body. This tiny capsule is called a tubercle. If the resistance of the body is high, the imprisoned organisms die, but if the resistance is low, the organisms eventually can be liberated. If a chest X ray detects active tubercles, the individual is put on appropriate drug therapy to ensure the localization of the disease and the eventual destruction of any live bacterial organisms.

Tuberculosis was a major killer in the United States before the middle of the twentieth century, after which antibiotic therapy brought it largely under control. In recent years, the incidence of tuberculosis has been on the rise, particularly among AIDS patients, the homeless, and the rural poor. Worse, the new strains are resistant to the usual antibiotic therapy.

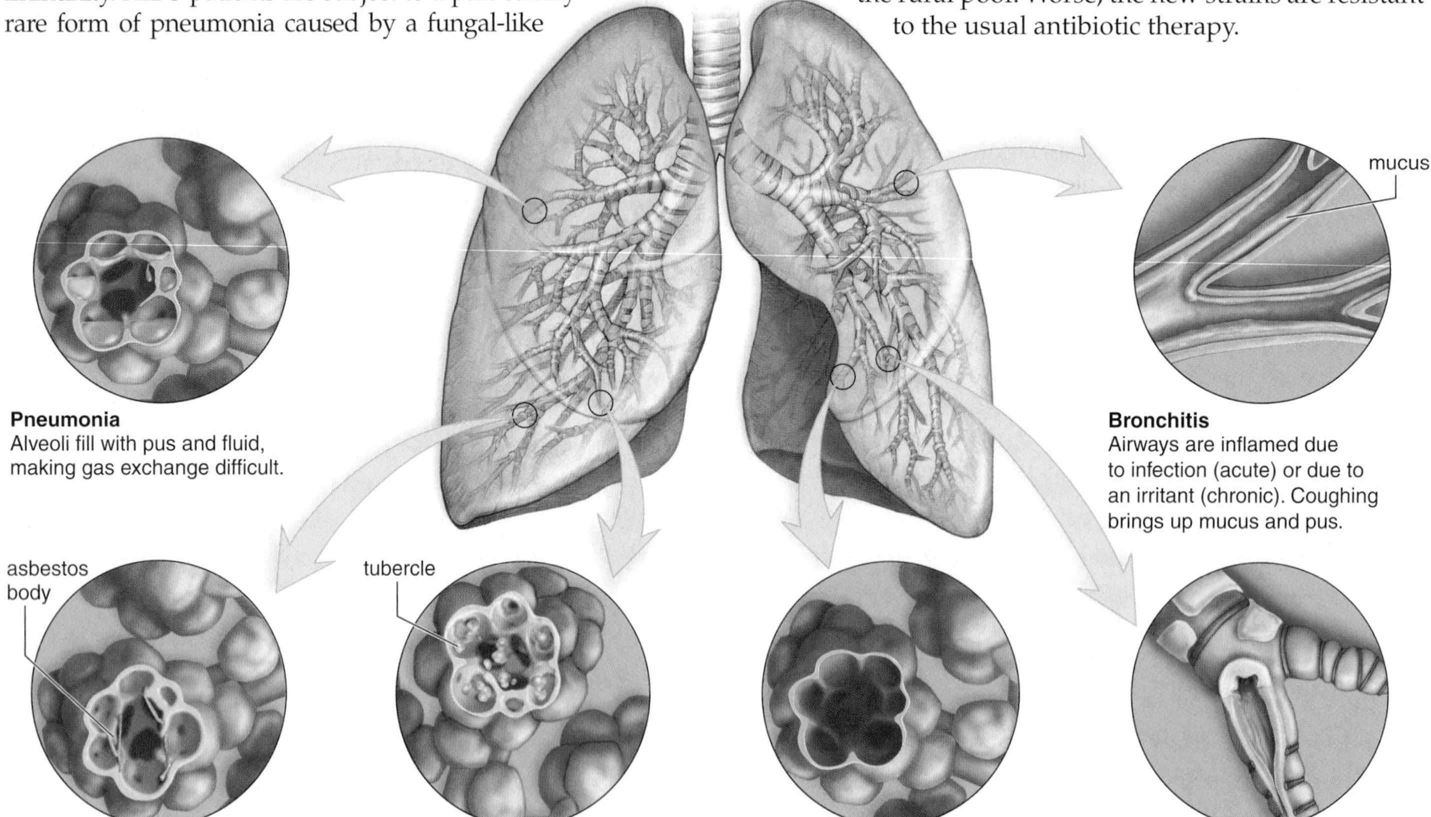


**FIGURE 35.14 Common bronchial and pulmonary diseases.**

Exposure to infectious pathogens and/or polluted air, including tobacco smoke, causes the diseases and disorders shown here.

**FIGURE 35.15 Smoking and lung disorders.**

Smoking causes almost 90% of all lung cancers and is also a major cause of emphysema. **a.** Normal lungs. **b.** The lungs of a person who died from emphysema are shrunken and blackened from trapped smoke. **c.** The lungs of a person who died from lung cancer are blackened from smoke except for the presence of the tumor, which is a mass of malformed soft tissue.

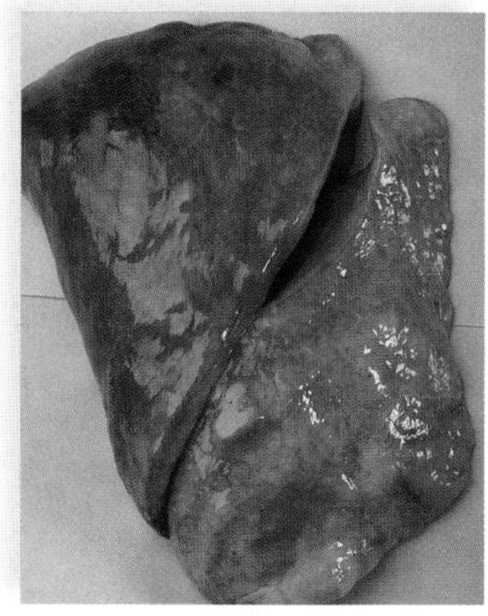

a. Normal lungs

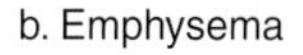

b. Emphysema

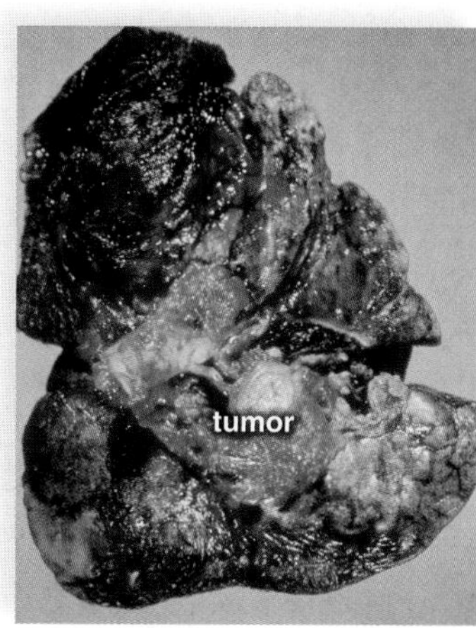


c. Lung cancer

## *Disorders*

Inhaling particles such as silica (sand), coal dust, asbestos, and, now it seems, fiberglass can lead to *pulmonary fibrosis,* a condition in which fibrous connective tissue builds up in the lungs. The lungs cannot inflate properly and are always tending toward deflation. Breathing asbestos is also associated with the development of cancer. Since asbestos has been used so widely as a fireproofing and insulating agent, unwarranted exposure has occurred.

In *chronic bronchitis,* the airways are inflamed and filled with mucus. A cough that brings up mucus is common. The bronchi have undergone degenerative changes, including the loss of cilia and their normal cleansing action. Under these conditions, an infection is more likely to occur. Smoking cigarettes and cigars and exposure to other pollutants, such as was experienced by workers and bystanders at Ground Zero after the September 11, 2001, attack, is the most frequent cause of chronic bronchitis.

*Emphysema* is a chronic and incurable lung disorder in which the alveoli are distended and their walls damaged so that the surface area available for gas exchange is reduced. Emphysema is often preceded by chronic bronchitis. Air trapped in the lungs leads to alveolar damage and a noticeable ballooning of the chest. The elastic recoil of the lungs is reduced, so not only are the airways narrowed, but the driving force behind expiration is also reduced. The patient is breathless and may have a cough. Because the surface area for gas exchange is reduced, oxygen reaching the heart and the brain is reduced. Even so, the heart works furiously to force more blood through the lungs, and an increased workload on the heart can result. Lack of oxygen to the brain can make the person feel depressed, sluggish, and irritable. Exercise, drug therapy, and supplemental oxygen, along with giving up smoking, may relieve the symptoms and possibly slow the progression of emphysema. Upon death, the lungs are decidedly abnormal (Fig. 35.15*b*).

*Asthma* is a disease of the bronchi and bronchioles that is marked by wheezing, breathlessness, and sometimes cough and expectoration of mucus. The airways are unusually sensitive to specific irritants, which can include a wide range of allergens such as pollen, animal dander, dust, cigarette smoke, and industrial fumes. Even cold air can be an irritant. When exposed to the irritant, the smooth muscle in the bronchioles undergoes spasms. It now appears that chemical mediators given off by immune cells in the bronchioles result in the spasms. Most asthma patients have some degree of bronchial inflammation that reduces the diameter of the airways and contributes to the seriousness of an attack. Asthma is not curable, but it is treatable. Special inhalers can control the inflammation and possibly prevent an attack, while other types of inhalers can stop the muscle spasms should an attack occur.

## *Lung Cancer*

*Lung cancer* used to be more prevalent in men than in women, but recently it has surpassed breast cancer as a cause of death in women. This can be linked to an increase in the number of women who smoke today. Nearly 150,000 people in the United States die of lung cancer each year (Fig. 35.15*c*). The American Cancer Society links over 85% of these deaths to smoking. Autopsies on smokers have revealed the progressive steps by which the most common form of lung cancer develops. The first event appears to be thickening of the lining of the airways. Then there is a loss of cilia so that it is impossible to prevent dust and dirt from settling in the lungs. Following this, cells with atypical nuclei appear in the thickened lining. A tumor consisting of disordered cells with atypical nuclei is considered to be cancer in situ (at one location). A final step occurs when some of these cells break loose and penetrate other tissues, a process called metastasis. Now the cancer has spread. The original tumor may grow until a bronchus is blocked, cutting off the supply of air to that lung. The entire lung then collapses, the secretions trapped in the lung spaces become infected, and pneumonia or a lung abscess (localized area of pus) results. The only treatment that offers a possibility of cure is to remove a lobe or the whole lung before metastasis has had time to occur. This operation is called a *pneumonectomy.*

### Check Your Progress 35.3

1. Which of the lower respiratory tract disorders are largely due to infections and which are largely environmental in origin?

## Connecting the Concepts

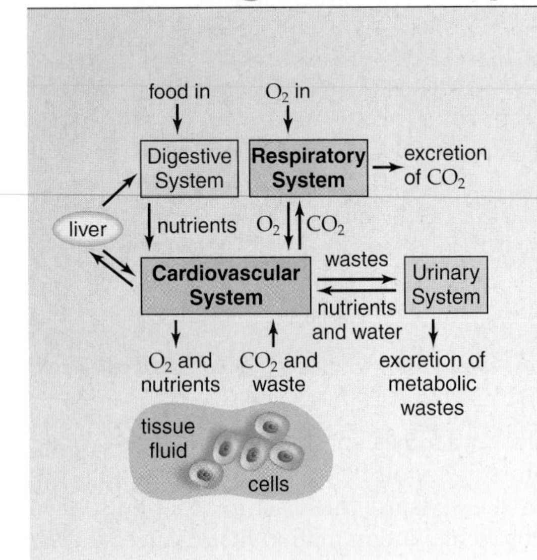


In mammals, the respiratory system consists of the respiratory tract with the nasal passages (or mouth) at one end and the lungs at the other end. Inspired air is 20% $O_2$ and 0.04% $CO_2$, while expired air is about 14% $O_2$ and 6% $CO_2$. Gas exchange in the lungs accounts for the difference in composition of inspired and expired air.

In the lungs, oxygen is absorbed into the bloodstream and from there it is transported by red blood cells to the capillaries, where it exits and enters tissue fluid. On the other hand, carbon dioxide enters capillaries at the tissues and is transported largely as the bicarbonate ion to the lungs, where it is converted to carbon dioxide and exits the body. Diffusion alone accounts for gas exchange in the lungs, called external respiration, and gas exchange in the tissues, called internal respiration. Energy is not needed, as gases follow their concentration gradients according to their partial pressures.

Internal gas exchange is extremely critical because cells use oxygen and release carbon dioxide as a result of cellular respiration, the process that generates ATP in cells. External gas exchange has the benefit of helping to keep the pH of the blood constant as required for homeostasis. When carbon dioxide exits, the blood pH returns to normal. In Chapter 36, we consider the contribution of the kidneys to homeostasis.

## summary

### 35.1 Gas Exchange Surfaces

Some aquatic animals, such as hydras and planarians, use their entire body surface for gas exchange. Most animals have a specialized gas-exchange area. Large aquatic animals usually pass water through gills. In bony fishes, blood in the capillaries flows in the direction opposite that of the water. Blood takes up almost all of the oxygen in the water as a result of this countercurrent flow. On land, insects use tracheal systems, and vertebrates have lungs. In insects, air enters the tracheae at openings called spiracles. From there, the air moves to ever smaller tracheoles until gas exchange takes place at the cells themselves. Lungs are found inside the body, where water loss is reduced. To ventilate the lungs, some vertebrates use positive pressure, but most inhale, using muscular contraction to produce a negative pressure that causes air to rush into the lungs. When the breathing muscles relax, air is exhaled.

Birds have a series of air sacs attached to the lungs. When a bird inhales, air enters the posterior air sacs, and when a bird exhales, air moves through the lungs to the anterior air sacs before exiting the respiratory tract. The one-way flow of air through the lungs allows more fresh air to be present in the lungs with each breath, and this leads to greater uptake of oxygen from one breath of air.

### 35.2 Breathing and Transport of Gases

During inspiration, air enters the body at nasal cavities and then passes from the pharynx through the glottis, larynx, trachea, bronchi, and bronchioles to the alveoli of the lungs, where exchange occurs, and during expiration air passes in the opposite direction. Humans breathe by negative pressure, as do other mammals. During inspiration, the rib cage goes up and out, and the diaphragm lowers. The lungs expand and air comes rushing in. During expiration, the rib cage goes down and in, and the diaphragm rises. Therefore, air rushes out.

The rate of breathing increases when the amount of $H^+$ and carbon dioxide in the blood rises, as detected by chemoreceptors such as the aortic and carotid bodies.

Gas exchange in the lungs and tissues is brought about by diffusion. Hemoglobin transports oxygen in the blood; carbon dioxide is mainly transported in plasma as the bicarbonate ion. Excess hydrogen ions are transported by hemoglobin. The enzyme carbonic anhydrase found in red blood cells speeds the formation of the bicarbonate ion.

### 35.3 Respiration and Health

The respiratory tract is subject to infections such as pneumonia and pulmonary tuberculosis. New strains of tuberculosis are resistant to the usual antibiotic therapy.

Major lung disorders are usually due to cigarette smoking. In chronic bronchitis the air passages are inflamed, mucus is common, and the cilia that line the respiratory tract are gone. Emphysema and lung cancer are two of the most serious consequences of smoking cigarettes. When the lungs of these patients are removed upon death, they are blackened by smoke.

## understanding the terms

alveolus (pl., alveoli) 654
aortic body 657
bicarbonate ion 659
bronchiole 655
bronchus (pl., bronchi) 655
carbaminohemoglobin 659
carbonic anhydrase 659
carotid body 657
countercurrent exchange 652
diaphragm 656
epiglottis 654
expiration 656
external respiration 650
gills 651
glottis 654
heme 659
hemoglobin (Hb) 659
inspiration 656
internal respiration 650
larynx 654
lungs 651
oxyhemoglobin 659
partial pressure 658
pharynx 654
respiration 650
respiratory center 657
trachea (pl., tracheae) 653, 654
ventilation 650
vocal cord 654

Match the terms to these definitions:

a. _______________ In terrestrial vertebrates, the mechanical act of moving air in and out of the lungs; breathing.

b. _______________ Dome-shaped muscularized sheet separating the thoracic cavity from the abdominal cavity in mammals.

c. _______________ Fold of tissue within the larynx; creates vocal sounds when it vibrates.

d. ______________ Respiratory organ in most aquatic animals; in fish, an outward extension of the pharynx.
e. ______________ Stage during breathing when air is pushed out of the lungs.

## reviewing this chapter

1. Compare the respiratory organs of aquatic animals to those of terrestrial animals. 650–54
2. How does the countercurrent flow of blood within gill capillaries and water passing across the gills assist respiration in fishes? 652
3. Why is it beneficial for the body wall of earthworms to be moist? Why don't insects require circulatory system involvement in air transport? 653
4. Name the parts of the human respiratory system, and list a function for each part. How is air reaching the lungs cleansed? 654
5. Explain the phrase "breathing by using negative pressure." 656
6. Contrast the tidal ventilation mechanism in humans with the one-way ventilation mechanism in birds, and explain the benefits of the ventilation mechanism in birds. 656–57
7. The concentration of what substances in blood controls the breathing rate in humans? Explain. 658
8. How are oxygen and carbon dioxide transported in blood? What does carbonic anhydrase do? 659
9. Which conditions depicted in Figure 35.14 are due to infection? Which are due to behavioral or environmental factors? Explain. 660–61

## testing yourself

Choose the best answer for each question.

1. Label the following diagram depicting respiration.

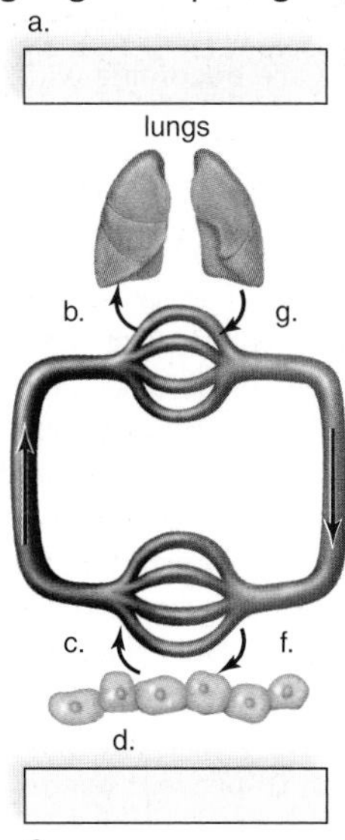


2. One problem faced by terrestrial animals with lungs, but not by aquatic animals with gills, is that
   a. gas exchange involves water loss.
   b. breathing requires considerable energy.
   c. oxygen diffuses very slowly in air.
   d. the concentration of oxygen in water is greater than that in air.
   e. All of these are correct.
3. In which animal is the circulatory system not involved in gas transport?
   a. mouse
   b. dragonfly
   c. trout
   d. sparrow
   e. human
4. Birds have more efficient lungs than humans because the flow of air
   a. is the same during both inspiration and expiration.
   b. travels in only one direction through the lungs.
   c. never backs up as it does in human lungs.
   d. is not hindered by a larynx.
   e. enters their bones.
5. Which animal breathes by positive pressure?
   a. fish
   b. human
   c. bird
   d. frog
   e. planarian
6. Which of these is a true statement?
   a. In lung capillaries, carbon dioxide combines with water to produce carbonic acid.
   b. In tissue capillaries, carbonic acid breaks down to carbon dioxide and water.
   c. In lung capillaries, carbonic acid breaks down to carbon dioxide and water.
   d. In tissue capillaries, carbonic acid combines with hydrogen ions to form the carbonate ion.
   e. All of these statements are true.
7. Air enters the human lungs because
   a. atmospheric pressure is less than the pressure inside the lungs.
   b. atmospheric pressure is greater than the pressure inside the lungs.
   c. although the pressures are the same inside and outside, the partial pressure of oxygen is lower within the lungs.
   d. the residual air in the lungs causes the partial pressure of oxygen to be less than it is outside.
   e. the process of breathing pushes air into the lungs.
8. If the digestive and respiratory tracts were completely separate in humans, there would be no need for
   a. swallowing.
   b. a nose.
   c. an epiglottis.
   d. a diaphragm.
   e. All of these are correct.
9. In tracing the path of air in humans, you would list the trachea
   a. directly after the nose.
   b. directly before the bronchi.
   c. before the pharynx.
   d. directly before the lungs.
   e. Both a and c are correct.
10. In humans, the respiratory control center
   a. is stimulated by carbon dioxide.
   b. is located in the medulla oblongata.
   c. controls the rate of breathing.
   d. is stimulated by hydrogen ion concentration.
   e. All of these are correct.
11. Carbon dioxide is carried in the plasma
   a. in combination with hemoglobin.
   b. as the bicarbonate ion.
   c. combined with carbonic anhydrase.
   d. only as a part of tissue fluid.
   e. All of these are correct.
12. Which of these is anatomically incorrect?
   a. The nose has two nasal cavities.
   b. The pharynx connects the nasal and oral cavities to the larynx.
   c. The larynx contains the vocal cords.
   d. The trachea enters the lungs.
   e. The lungs contain many alveoli.

13. How is inhaled air modified before it reaches the lungs?
    a. It must be humidified.
    b. It must be warmed.
    c. It must be filtered and cleansed.
    d. All of these are correct.
14. Internal respiration refers to
    a. the exchange of gases between the air and the blood in the lungs.
    b. the movement of air into the lungs.
    c. the exchange of gases between the blood and tissue fluid.
    d. cellular respiration, resulting in the production of ATP.
15. The chemical reaction that converts carbon dioxide to a bicarbonate ion takes place in
    a. the blood plasma.
    b. red blood cells.
    c. the alveolus.
    d. the hemoglobin molecule.
16. Which of these would affect hemoglobin's $O_2$-binding capacity?
    a. pH
    b. partial pressure of oxygen
    c. blood pressure
    d. temperature
    e. All of these except c are correct.
17. The enzyme carbonic anhydrase
    a. causes the blood to be more basic in the tissues.
    b. speeds the conversion of carbonic acid to carbon dioxide and water.
    c. actively transports carbon dioxide out of capillaries.
    d. is active only at high altitudes.
    e. All of these are correct.
18. Which of these is incorrect concerning inspiration?
    a. Rib cage moves up and out.
    b. Diaphragm contracts and moves down.
    c. Pressure in lungs decreases, and air comes rushing in.
    d. The lungs expand because air comes rushing in.
19. Label this diagram of the human respiratory system.

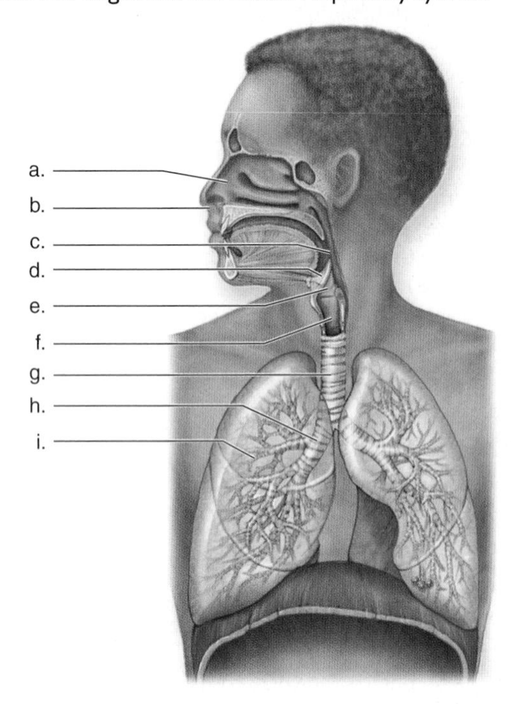


## thinking scientifically

1. You are a physician who witnessed Christopher Reeve's riding accident. Why might you immediately use mouth to mouth resuscitation until mechanical ventilation becomes available?
2. Fetal hemoglobin picks up oxygen from the maternal blood. If the oxygen-binding characteristics of hemoglobin in the fetus were identical to the hemoglobin of the mother, oxygen could never be transferred at the placenta to fetal circulation. What hypothesis about the oxygen-binding characteristics of fetal hemoglobin would explain how fetuses get the oxygen they need?

## bioethical issue

### Antibiotic Therapy

Antibiotics cure respiratory infections, but there are problems associated with antibiotic therapy. Aside from a possible allergic reaction, antibiotics not only kill off disease-causing bacteria, but they also reduce the number of beneficial bacteria in the intestinal tract and other locations. These beneficial bacteria hold in check the growth of other pathogens that now begin to flourish. Diarrhea can result, as can a vaginal yeast infection. The use of antibiotics can also prevent natural immunity from occurring, leading to the need for recurring antibiotic therapy. Especially alarming at this time is the occurrence of resistance. Resistance takes place when vulnerable bacteria are killed off by an antibiotic, and this allows resistant bacteria to become prevalent. The bacteria that cause ear, nose, and throat infections as well as scarlet fever and pneumonia are becoming widely resistant because we have not been using antibiotics properly. Tuberculosis is on the rise, and the new strains are resistant to the usual combined antibiotic therapy.

Every citizen needs to be aware of our present crisis situation. Stuart Levy, a Tufts University School of Medicine microbiologist, says that we should do what is ethical for society and ourselves. What is needed? Antibiotics kill bacteria, not viruses—therefore, we shouldn't take antibiotics unless we know for sure we have a bacterial infection. And we shouldn't take them prophylactically—that is, just in case we might need one. If antibiotics are taken in low dosages and intermittently, resistant strains are bound to take over. Animal and agricultural use should be pared down, and household disinfectants should no longer be spiked with antibacterial agents. Perhaps then, Levy says, vulnerable bacteria will begin to supplant the resistant ones in the population. Are you doing all you can to prevent bacteria from becoming resistant?

## *Biology* website

The companion website for *Biology* provides a wealth of information organized and integrated by chapter. You will find practice tests, animations, videos, and much more that will complement your learning and understanding of general biology.

**http://www.mhhe.com/maderbiology10**

# 36

# Body Fluid Regulation and Excretory Systems

## concepts

**36.1 EXCRETION AND THE ENVIRONMENT**

- The amounts of water and energy required to excrete various nitrogenous waste products differ. 666
- Most animals have organs of excretion that maintain the water balance of the body and rid the body of metabolic waste molecules. 667
- The mechanism for maintaining water balance differs according to the environment of the organism. 668–69

**36.2 URINARY SYSTEM IN HUMANS**

- The urinary system of humans consists of organs that produce, store, and rid the body of urine. 670
- The work of an organ is dependent on its microscopic anatomy; nephrons within the human kidney produce urine. 670–71
- Like many physiological processes, urine formation in humans is a multistep process. 671–73
- In addition to ridding the body of waste molecules and maintaining water balance, the human kidneys adjust the pH of the blood. 674–75

*If the salt concentration in body fluids is too high, cells shrink and die. If it is too low, cells explode and die. Yet animals are found in all sorts of environments, including marine environments that are too salty, freshwater environments that don't have enough salt, and even terrestrial environments that are simply too dry. Animals clearly spend a lot of energy regulating the composition of their body fluids, and chief among the organs that help are the kidneys of the urinary system. Sometimes animals, such as marine birds and reptiles, get a little help from accessory glands. Sea turtles have salt glands located above their eye that, true to their name, rid the body of salt. When the glands excrete a salty solution collected from body fluids, sea turtles appear to cry. We don't have salt glands and cannot survive after drinking salty water because the kidneys alone can't handle all the salt.*

*In this chapter, you will learn how animals excrete various metabolic wastes while maintaining their normal saltwater balance and their pH balance. The latter functions are of primary importance to homeostasis and continued good health.*

Marine organisms rid the body of excess salt; fishes extrude salt at their gills, and turtles do so near their eyes.

# 36.1 Excretion and the Environment

The particular nitrogenous end product differs among animals according to the environment, but an excretory organ tends to be tubular, as we will see first among the invertebrates and then among the vertebrates.

## Nitrogenous Waste Products

The breakdown of various molecules, including amino acids and nucleic acids, results in nitrogenous wastes. For simplicity's sake, we will limit our discussion to amino acid metabolism. When amino acids are broken down by the body to generate energy, or are converted to fats or carbohydrates, the amino groups ($—NH_2$) must be removed because they are not needed. Once the amino groups have been removed, they may be excreted from the body in the form of ammonia, urea, or uric acid, depending on the species. Removal of amino groups from amino acids requires a fairly set amount of energy. However, the amount of energy required to convert amino groups to ammonia, urea, or uric acid differs, as indicated in Figure 36.1.

### *Ammonia*

Amino groups removed from amino acids immediately form **ammonia** ($NH_3$) by the addition of a third hydrogen ion. Little or no energy is required to convert an amino group to ammonia by adding a hydrogen ion. Ammonia is quite toxic and can be a nitrogenous excretory product if a good deal of water is available to wash it from the body. Ammonia is excreted by most fishes and other aquatic animals whose gills and skin surfaces are in direct contact with the water of the environment.

### *Urea*

Production of urea requires the expenditure of energy because it is produced in the liver by a set of energy-requiring enzymatic reactions, known as the urea cycle. In this cycle, carrier molecules take up carbon dioxide and two molecules of ammonia, finally releasing urea. **Urea** is much less toxic than ammonia and can be excreted in a moderately concentrated solution. This allows body water to be conserved, an important advantage for terrestrial animals with limited access to water. Sharks, adult amphibians, and mammals usually excrete urea as their main nitrogenous waste.

### *Uric Acid*

Uric acid is synthesized by a long, complex series of enzymatic reactions that requires expenditure of even more ATP than does urea synthesis. **Uric acid** is not very toxic, and it is poorly soluble in water. Poor solubility is an advantage if water conservation is needed, because uric acid can be concentrated even more readily than can urea. Uric acid is routinely excreted by insects, reptiles, and birds. In reptiles and birds, a dilute solution of uric acid passes from the kidneys to the *cloaca,* a common reservoir for the products of the digestive, urinary, and reproductive systems. The cloacal contents are refluxed into the large intestine, where water is reabsorbed. The white substance in bird feces is uric acid. Embryos of reptiles and birds develop inside completely enclosed shelled eggs. The production of insoluble, relatively nontoxic uric acid is advantageous for shelled embryos because all nitrogenous wastes are stored inside the shell until hatching takes place. Here again, the advantage of water conservation seems to counterbalance the disadvantage of energy expenditure for synthesis of an excretory molecule.

In general, it is possible to predict which metabolic waste product an animal excretes based on its anatomy and environment—but not always. For example, unlike most other birds, hummingbirds excrete more ammonia than uric acid. **Excretion** of ammonia, as stated, requires a lot of fluid to keep its toxicity under control. Recall that hummingbirds are fluid feeders, and as such, they have plenty of fluid available for excreting ammonia.

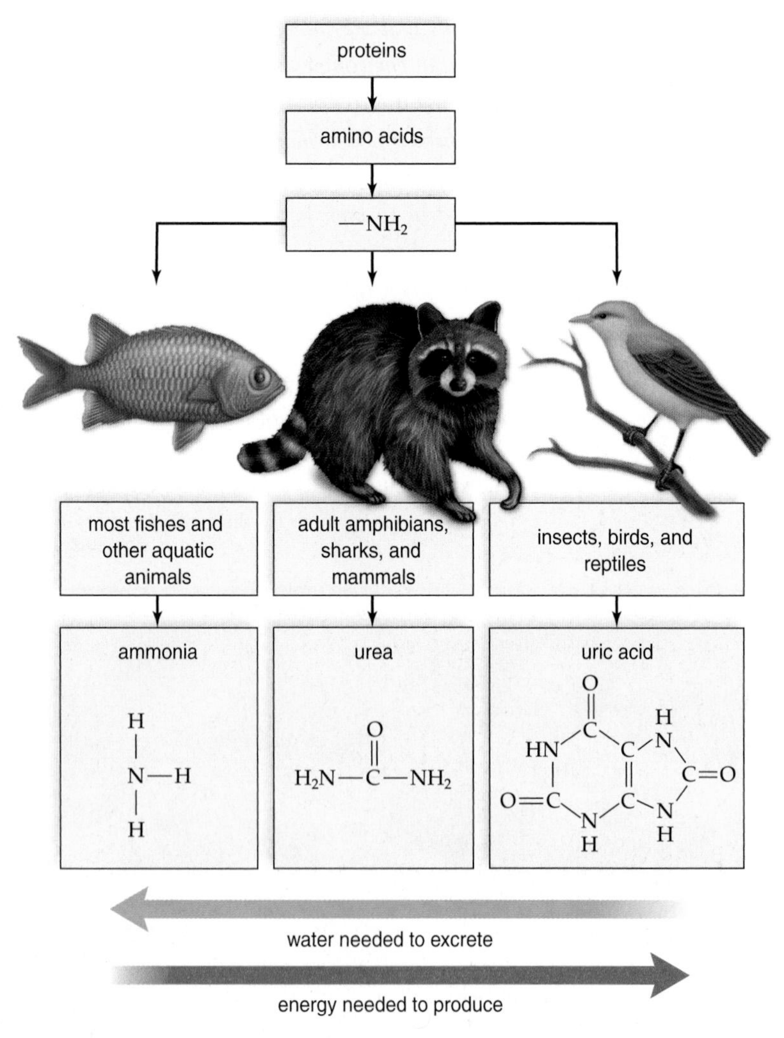


**FIGURE 36.1 Nitrogenous wastes.**

Proteins are hydrolyzed to amino acids, whose breakdown results in carbon chains and amino groups ($—NH_2$). The carbon chains can be used as an energy source, but the amino groups must be excreted as ammonia, urea, or uric acid.

## Excretory Organs Among Invertebrates

Most animals have tubular excretory organs that regulate the water-salt balance of the body and excrete metabolic wastes into the environment. Here we give three examples among the invertebrates.

The planarians, which are flatworms living in fresh water, have two strands of branching excretory tubules that open to the outside of the body through excretory pores (Fig. 36.2*a*). Located along the tubules are bulblike **flame cells,** each of which contains a cluster of beating cilia that looks like a flickering flame under the microscope. The beating of flame-cell cilia propels fluid through the excretory tubules and out of the body. The system is believed to function in ridding the body of excess water and in excreting wastes.

The body of an earthworm is divided into segments, and nearly every body segment has a pair of excretory structures called **nephridia.** Each nephridium is a tubule with a ciliated opening and an excretory pore (Fig. 36.2*b*). As fluid from the coelom is propelled through the tubule by beating cilia, its composition is modified. For example, nutrient substances are reabsorbed and carried away by a network of capillaries surrounding the tubule. The urine of an earthworm contains metabolic wastes, salts, and water. Although the earthworm is considered a terrestrial animal, it excretes a very dilute urine. Each day, an earthworm may produce a volume of urine equal to 60% of its body weight. The excretion of ammonia is consistent with these data.

Insects have a unique excretory system consisting of long, thin tubules called **Malpighian tubules** attached to the gut. Uric acid is actively transported from the surrounding hemolymph into these tubules, and water follows a salt gradient established by active transport of $K^+$. Water and other useful substances are reabsorbed at the rectum, but the uric acid leaves the body at the anus. Insects that live in water, or eat large quantities of moist food, reabsorb little water. But insects in dry environments reabsorb most of the water and excrete a dry, semisolid mass of uric acid.

The excretory organs of other arthropods are given different names, although they function similarly. In crustaceans (e.g., crabs, crayfish), nitrogenous wastes are generally removed by diffusion across the gills—for those species that have gills. Even so, crustaceans possess excretory organs called *green glands* located in the ventral portion of the head region. Fluid collects within the tubules from the surrounding blood of the hemocoel, but this fluid is modified by the time it leaves the tubules. The secretion of salts into the tubule regulates the amount of urine excreted. In shrimp and pillbugs, the excretory organs are located in the maxillary segments and are called *maxillary glands.* Spiders, scorpions, and other arachnids possess *coxal glands,* which are located near one or more appendages and used for excretion. Coxal glands are spherical sacs resembling annelid nephridia. Wastes are collected from the surrounding blood of the hemocoel and discharged through pores at one to several pairs of appendages.

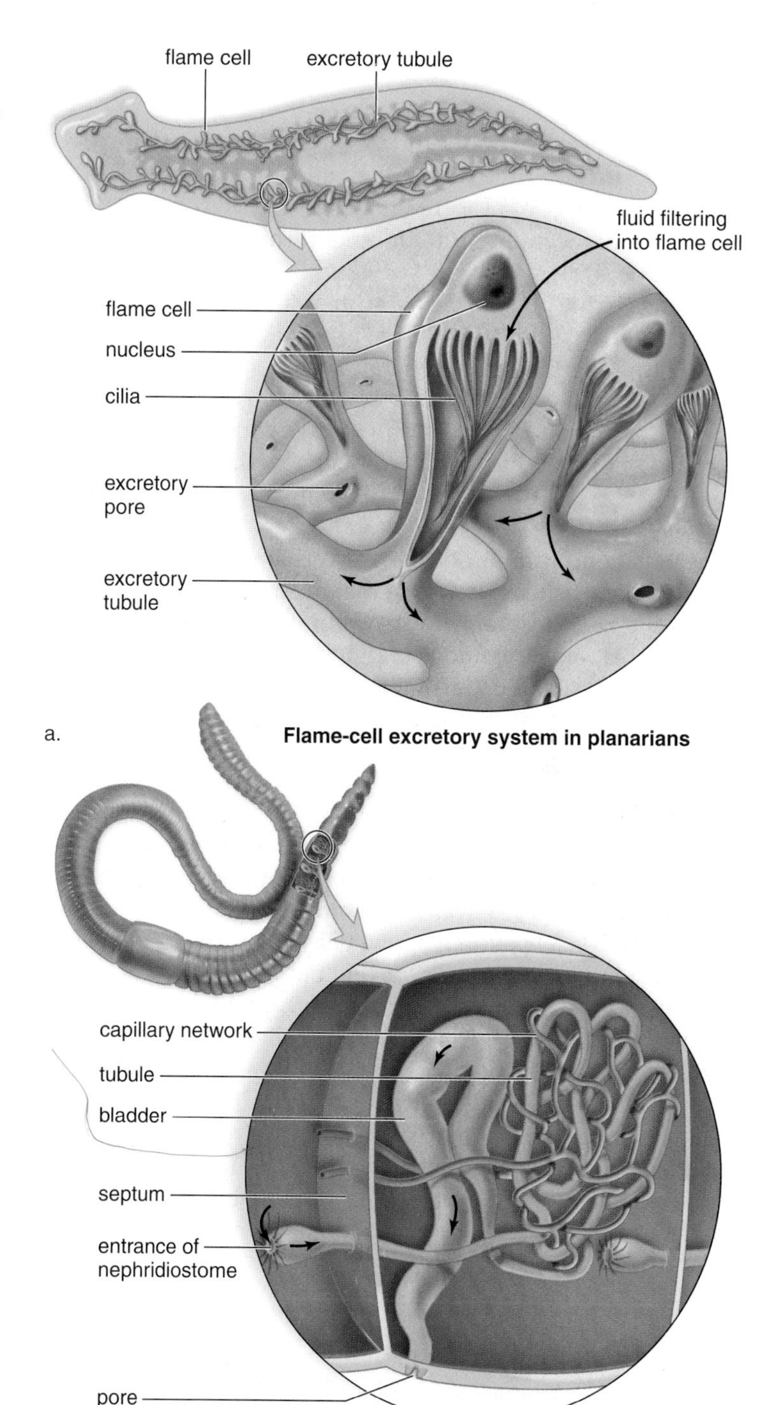


**FIGURE 36.2 Excretory organs in animals.**

**a.** Two or more tracts of branching tubules run the length of the body and open to the outside by pores. At the ends of side branches are small bulblike cells called flame cells. **b.** The nephridium has a ciliated opening, the nephridiostome, that leads to a coiled tubule surrounded by a capillary network. Urine can be temporarily stored in the bladder before being released to the outside via a pore termed a nephridiopore.

In conclusion, invertebrates utilize tubules to rid the body of wastes and maintain a water-salt balance. On occasion, excretion also involves other organs, such as the rectum in the earthworm and the gills in crayfish.

## Osmoregulation Among Aquatic Vertebrates

Most vertebrates **osmoregulate**—that is, maintain particular ion concentrations in their blood. Osmoregulation is absolutely essential to maintain *homeostasis,* the relative constancy of the internal environment. This is a necessity because ions such as $Na^+$, $Ca^{2+}$, $K^+$, and $PO_4^{3-}$ greatly affect the workings of the body systems, such as the skeletal, nervous, and muscular systems.

Osmoregulation in general is necessary because few vertebrates have blood that is isotonic to seawater. Not so for the cartilaginous fishes (Fig. 36.3), whose blood is isotonic to seawater, for reasons we will now discuss.

### Cartilaginous Fishes

The total concentration of the various ions in the blood of cartilaginous fishes is less than that in seawater. Their blood plasma is nearly isotonic to seawater because they pump it full of urea, and this molecule gives their blood the same tonicity as seawater. Cartilaginous fishes do regulate the concentration of other solutes in their blood and have rectal glands that rid the body of excess salt.

### Marine Bony Fishes

A marine environment, which is high in salts, is hypertonic to the blood plasma of bony fishes. Apparently, their common ancestor evolved in fresh water, and only later did some groups invade the sea. Therefore, marine bony fishes must avoid the tendency to become dehydrated (Fig. 36.4*a*). As the sea washes over their gills, marine bony fishes lose water by osmosis. To counteract this, they drink seawater almost constantly. On the average, marine bony fishes swallow an amount of water equal to 1% of their body weight every hour. This is equivalent to a human drinking about 700 ml of water every hour around the clock. But while they get water by drinking, this habit also causes these fishes to acquire salt. To rid the body of excess salt, they actively transport it into the surrounding seawater at the gills. The kidneys conserve water, and marine bony fishes produce a scant amount of isotonic urine.

### Freshwater Bony Fishes

The osmotic problems of freshwater bony fishes and the response to their environment are exactly opposite those of marine bony fishes (Fig. 36.4*b*). Freshwater fishes tend to gain water by osmosis across the gills and the body surface. As a consequence, these fishes never drink water. They actively transport salts into the blood across the membranes of their gills. They eliminate excess water by producing large quantities of dilute (hypotonic) urine. They discharge a quantity of urine equal to one-third their body weight each day.

## Osmoregulation Among Terrestrial Vertebrates

Desert mammals, such as the kangaroo rat, and seabirds, such as a seagull, illustrate different strategies for dealing with extreme terrestrial environments.

**FIGURE 36.3 Osmoregulation in a shark.**

The blood of a shark is isotonic to seawater because the blood contains a high concentration of urea.

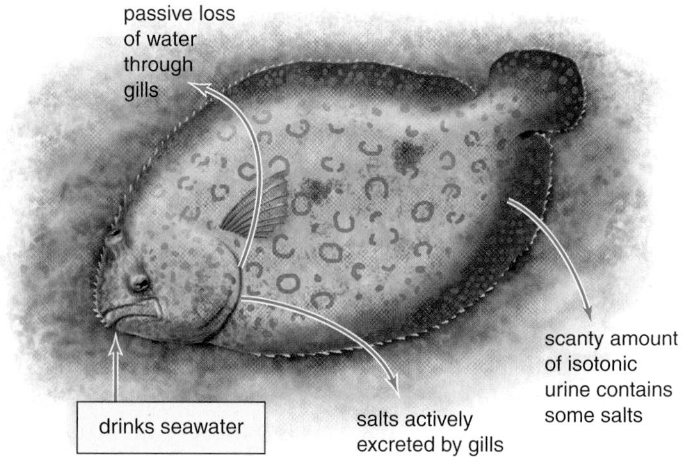


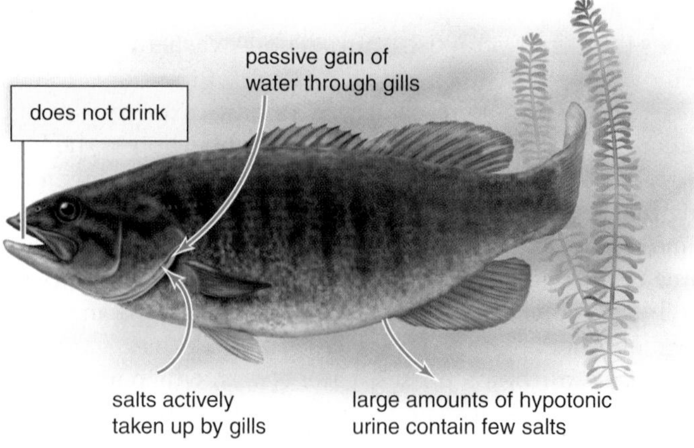


**FIGURE 36.4 Body fluid regulation in bony fishes.**

**a.** Marine bony fishes employ different mechanisms compared to (**b**) freshwater fishes in order to osmoregulate their body fluids.

**FIGURE 36.5 Adaptations of a kangaroo rat to a dry environment.**

A kangaroo rat minimizes water loss in the many ways noted.

## *Kangaroo Rat*

Dehydration threatens all terrestrial animals, especially those that live in a desert, as does the kangaroo rat. Its fur prevents loss of water to the air, and during the day, it remains in a cool burrow. In addition, the kangaroo rat's nasal passage has a highly convoluted mucous membrane surface that captures condensed water from exhaled air. Exhaled air is usually full of moisture, which is why you can see it on cold winter mornings—the moisture in exhaled air is condensing.

As we shall see, humans mainly conserve water by producing urine that is hypertonic to blood plasma. The kangaroo rat forms a very concentrated urine—20 times more concentrated than its blood plasma. Also, its fecal material is almost completely dry.

Most terrestrial animals need to drink water occasionally to make up for the water lost from the skin and respiratory passages and through urination. However, the kangaroo rat is so adapted to conserving water that it can survive by using metabolic water derived from cellular respiration, and it never drinks water (Fig. 36.5). The adaptations of the kangaroo rat allow it to remain in water-salt balance, even under desert conditions.

## *Seagulls, Reptiles, and Mammals*

Birds, reptiles, and mammals evolved on land, and their kidneys are especially good at conserving water. However, some animals have become secondarily adapted to living near or in the sea. They drink seawater and still manage to survive. If humans drink seawater, we lose more water than we take in just ridding the body of all that salt. Little is known about how whales manage to get rid of extra salt, but we know that their kidneys are enormous. Other animals have been studied, and we have learned that seabirds and reptiles have salt glands that pump out salt (Fig. 36.6). In the two types of animals we will mention, each has commandeered a gland meant for another purpose and used it to pump out the salt from blood plasma and leave behind the water, just as in a desalination plant.

In birds, salt-excreting glands are located near the eyes. The glands produce a salty solution that is excreted through the nostrils and moves down grooves on their beaks until it drips off. In marine turtles, the salt gland is a modified tear (lacrimal) gland, and in sea snakes, a salivary sublingual gland beneath the tongue gets rid of excess salt. The work of the gland is regulated by the nervous system. Osmoreceptors, perhaps located near the heart, are thought to stimulate the brain, which then orders the gland to excrete salt until the salt concentration in the blood decreases to a tolerable level.

### Check Your Progress 36.1

1. What is the advantage of excreting urea instead of ammonia or uric acid?
2. Earthworms have a thin skin for respiration. If they had thicker skin, would it affect the function of the nephridia?
3. What evidence do we have that excess salt does not enter the body of a shark?
4. Would the tonicity of the urine produced by a seagull be greater than that produced by a human? Explain.

**FIGURE 36.6 Adaptations of marine birds to a high salt environment.**

Many marine birds and reptiles have glands that pump salt out of the body.

# 36.2 Urinary System in Humans

The urinary system of humans contains excretory organs called the kidneys (Fig. 36.7). The kidneys are the chief organs of homeostasis in the human body because they are the ultimate regulators of blood composition, as we shall see.

The human **kidneys** are bean-shaped, reddish-brown organs, each about the size of a fist. They are located on either side of the vertebral column just below the diaphragm, in the lower back, where they are partially protected by the lower rib cage. The right kidney is slightly lower than the left kidney. **Urine** [Gk. *urina*, urine] made by the kidneys is conducted from the body by the other organs in the urinary system. Each kidney is connected to a **ureter,** a duct that takes urine from the kidney to the **urinary bladder,** where it is stored until it is voided from the body through the single **urethra.** In males, the urethra passes through the penis, and in females, the opening of the urethra is ventral to that of the vagina. There is no connection between the genital (reproductive) and urinary systems in females, but there is a connection in males. In males, the urethra also carries sperm during ejaculation.

## Kidneys

If a kidney is sectioned longitudinally, three major parts can be distinguished (Fig. 36.8). The **renal cortex,** which is the outer region of a kidney, has a somewhat granular appearance. The **renal medulla** consists of six to ten cone-shaped renal pyramids that lie on the inner side of the renal cortex. The innermost part of the kidney is a hollow chamber called the **renal pelvis.** Urine collects in the renal pelvis and then is carried to the bladder by a ureter.

A kidney stone or renal calculus is a hard granule of phosphate, calcium, protein, or uric acid that forms in the renal pelvis. Many are passed unnoticed. However, larger and jagged stones can block the renal pelvis or ureter causing intense pain and damage.

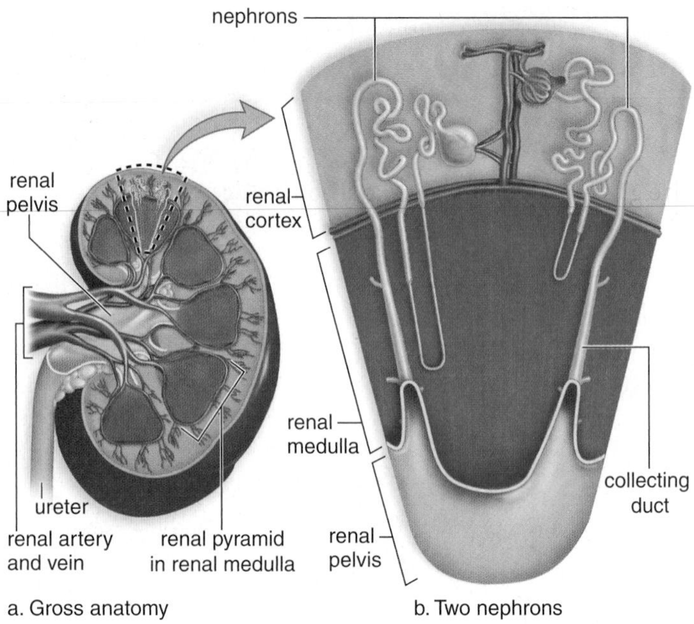

nephrons
renal pelvis
renal cortex
renal medulla
collecting duct
ureter
renal artery and vein
renal pyramid in renal medulla
renal pelvis

a. Gross anatomy

b. Two nephrons

**FIGURE 36.8 Macroscopic and microscopic anatomy of the kidney.**

**a.** Longitudinal section of a kidney, showing the location of the renal cortex, the renal medulla, and the renal pelvis. **b.** An enlargement of one renal lobe, showing the placement of nephrons.

### *Nephrons*

Microscopically, each kidney is composed of over 1 million tiny tubules called **nephrons** [Gk. *nephros*, kidney]. The nephrons of a kidney produce urine. Some nephrons are located primarily in the renal cortex, but others dip down into the renal medulla, as shown in Figure 36.8*b*. Each nephron is made of several parts (Fig. 36.9). The blind end of a nephron is

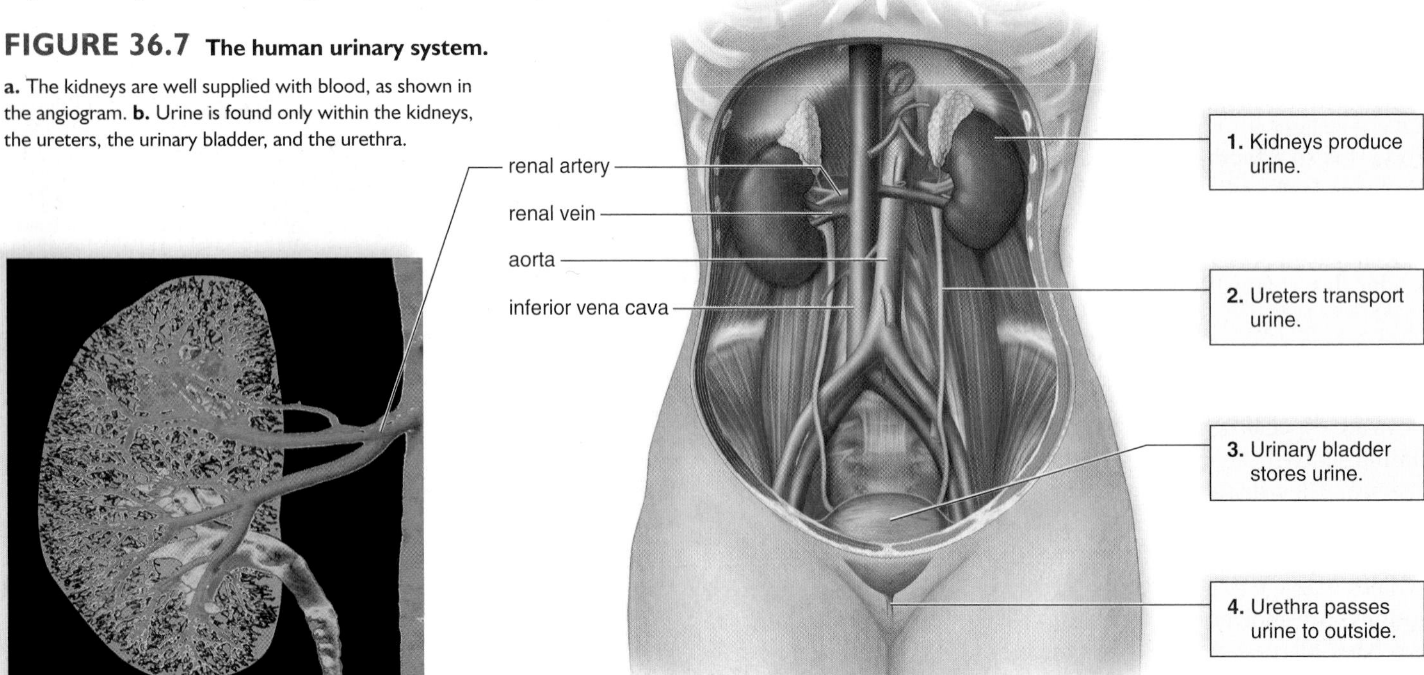


**FIGURE 36.7 The human urinary system.**

**a.** The kidneys are well supplied with blood, as shown in the angiogram. **b.** Urine is found only within the kidneys, the ureters, the urinary bladder, and the urethra.

pushed in on itself to form a cuplike structure called the **glomerular capsule** [L. *glomeris,* ball] (Bowman's capsule). The outer layer of the glomerular capsule is composed of squamous epithelial cells; the inner layer is composed of specialized cells that allow easy passage of molecules. Leading from the glomerular capsule is a portion of the nephron known as the **proximal convoluted tubule** [L. *proximus,* nearest], which is lined by cells with many mitochondria and tightly packed microvilli. Then, simple squamous epithelium appears in the **loop of the nephron** (loop of Henle), which has a descending limb and an ascending limb. This is followed by the **distal convoluted tubule** [L. *distantia,* far]. Several distal convoluted tubules enter one **collecting duct.** The collecting duct transports urine down through the renal medulla and delivers it to the renal pelvis. The loop of the nephron and the collecting duct give the pyramids of the renal medulla their striped appearance (see Fig. 36.8).

Each nephron has its own blood supply (Fig. 36.9). The renal artery branches into numerous small arteries, which branch into arterioles, one for each nephron. Each arteriole, called an afferent arteriole, divides to form a capillary bed, the **glomerulus** [L. *glomeris,* ball], which is surrounded by the glomerular capsule. The glomerulus drains into an efferent arteriole, which subsequently branches into a second capillary bed around the tubular parts of the nephron. These capillaries, called peritubular capillaries, lead to venules that join to form veins leading to the renal vein, a vessel that enters the inferior vena cava.

## Urine Formation

An average human produces between 1 and 2 liters of urine daily. Urine production requires three distinct processes (see Fig. 36.11), and, as you can see, the entire tubule portion of a nephron participates in the last two steps in urine formation:

1. glomerular filtration at the glomerular capsule;
2. tubular reabsorption at the convoluted tubules; and
3. tubular secretion at the convoluted tubules.

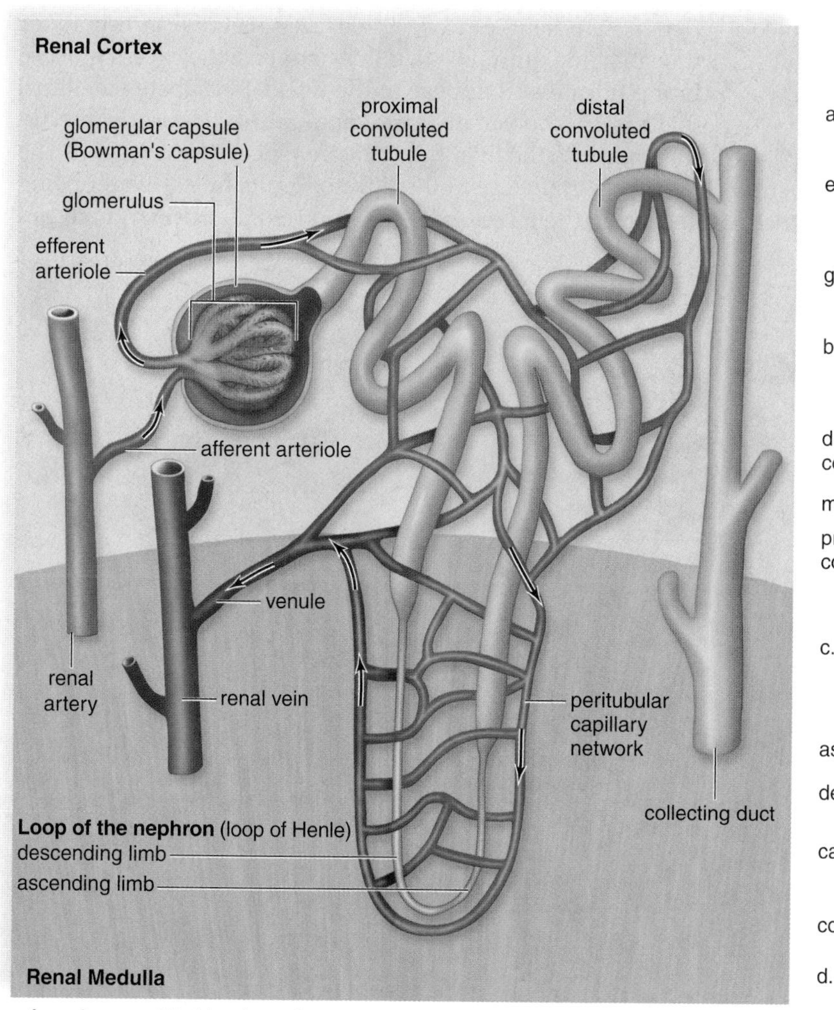


a. A nephron and its blood supply

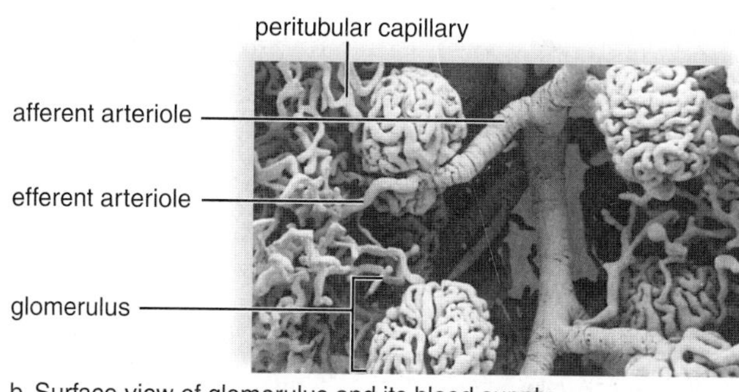


b. Surface view of glomerulus and its blood supply

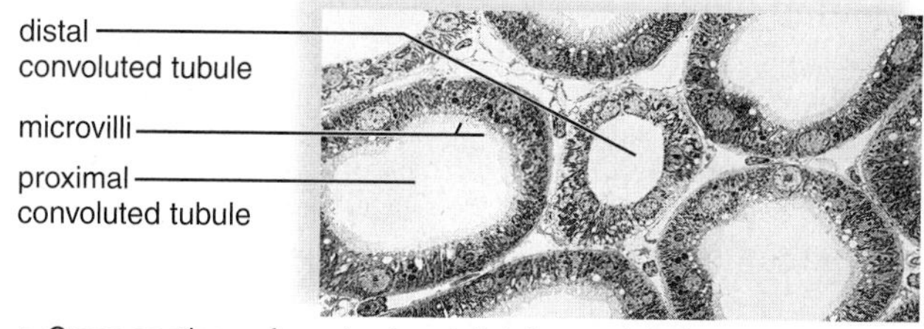


c. Cross sections of proximal and distal convoluted tubules

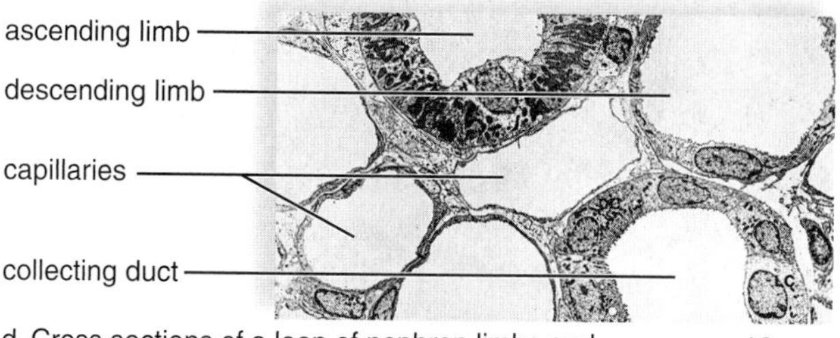


d. Cross sections of a loop of nephron limbs and collecting duct. (The other cross sections are those of capillaries.)

**FIGURE 36.9 Nephron anatomy.**

**a.** You can trace the path of blood about a nephron by following the arrows. A nephron is made up of a glomerular capsule, the proximal convoluted tubule, the loop of the nephron, the distal convoluted tubule, and the collecting duct. The micrographs in (**b**), (**c**), and (**d**) show these structures.

## Glomerular Filtration

**Glomerular filtration** (see Fig. 36.11) is the movement of small molecules across the glomerular wall into the glomerular capsule as a result of blood pressure. When blood enters the glomerulus, blood pressure is sufficient to cause small molecules, such as water, nutrients, salts, and wastes, to move from the glomerulus to the inside of the glomerular capsule, especially since the glomerular walls are 100 times more permeable than the walls of most capillaries elsewhere in the body. The molecules that leave the blood and enter the glomerular capsule are called the glomerular filtrate. Plasma proteins and blood cells are too large to be part of this filtrate, so they remain in the blood as it flows into the efferent arteriole.

Glomerular filtrate is essentially protein free, but otherwise it has the same composition as blood plasma. If this composition were not altered in other parts of the nephron, death from loss of nutrients (starvation) and loss of water (dehydration) would quickly follow. The total blood volume averages about 5 liters, and this amount of fluid is filtered every 40 minutes. Thus 180 liters of filtrate is produced daily, some 60 times the amount of blood plasma in the body. Most of the filtered water is obviously quickly returned to the blood, or a person would actually die from urination. Tubular reabsorption prevents this from happening.

## Tubular Reabsorption

**Tubular reabsorption** (see Fig. 36.11) takes place when substances move across the walls of the tubules into the associated peritubular capillary network (Fig. 36.10). The osmolarity of the blood is essentially the same as that of the filtrate within the glomerular capsule, and therefore osmosis of water from the filtrate into the blood cannot yet occur. However, sodium ions ($Na^+$) are actively pumped into the peritubular capillary, and then chloride ions ($Cl^-$) follow passively. Now the osmolarity of the blood is such that water moves passively from the tubule into the blood. About 60–70% of salt and water are reabsorbed at the proximal convoluted tubule.

Nutrients, such as glucose and amino acids, also return to the blood at the proximal convoluted tubule. This is a selective process, because only molecules recognized by carrier proteins in plasma membranes are actively reabsorbed. The cells of the proximal convoluted tubule have numerous microvilli, which increase the surface area, and numerous mitochondria, which supply the energy needed for active transport (Fig. 36.10). Glucose is an example of a molecule that ordinarily is reabsorbed completely because there is a plentiful supply of carrier molecules for it. However, if there is more glucose in the filtrate than there are carriers to handle it, glucose will exceed its renal threshold, or transport maximum. When this happens, the excess glucose in the filtrate will appear in the urine. In diabetes mellitus, there is an abnormally large amount of glucose in the filtrate because the liver fails to store glucose as glycogen. The presence of glucose in the filtrate results in less water being absorbed; the increased thirst and frequent urination in un-

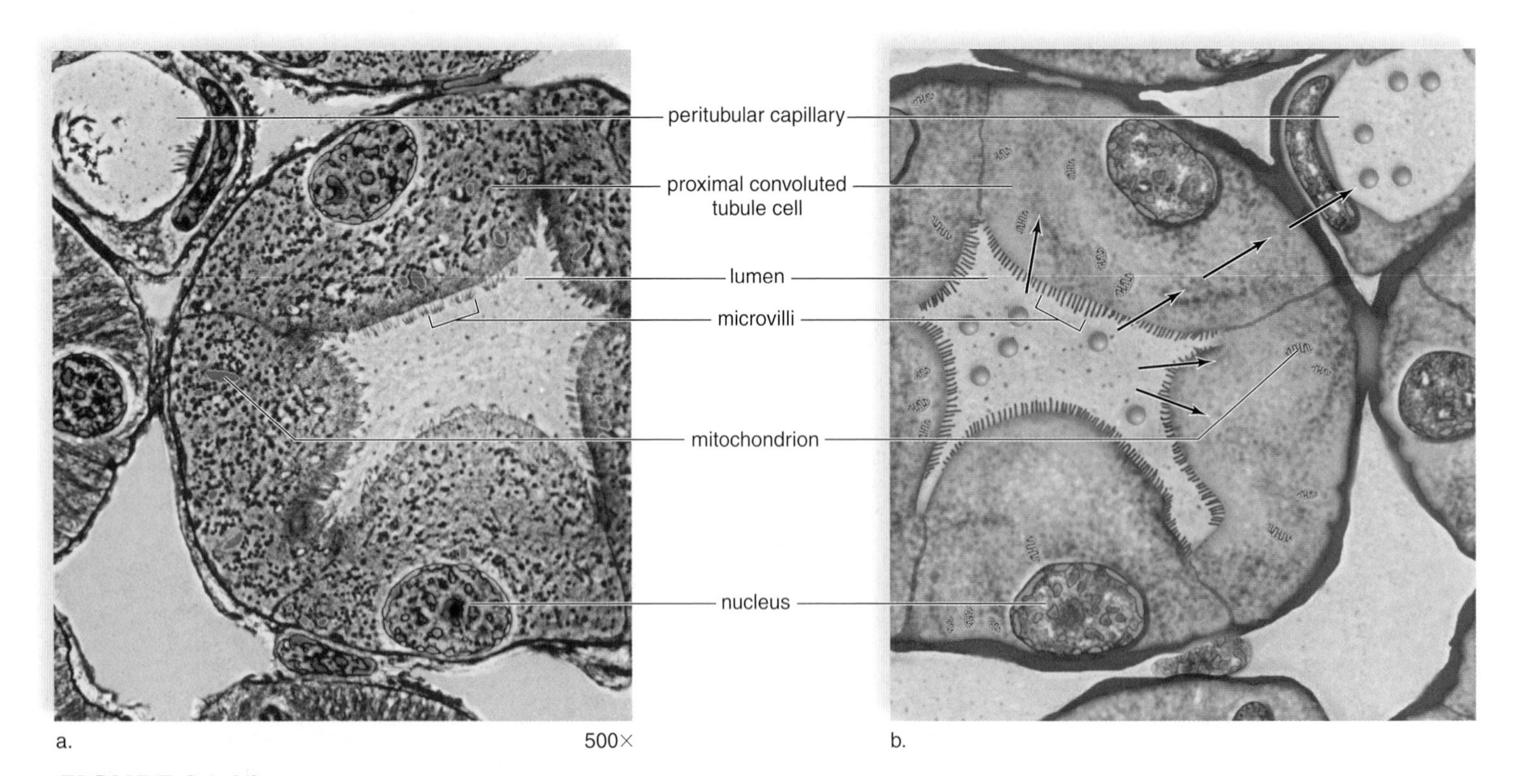


**FIGURE 36.10 Proximal convoluted tubule.**

**a.** This photomicrograph shows that the cells lining the proximal convoluted tubule have a brush border composed of microvilli, which greatly increases the surface area exposed to the lumen. The peritubular capillary adjoins the cells. **b.** Diagrammatic representation of (**a**) shows that each cell has many mitochondria, which supply the energy needed for active transport, the process that moves molecules (green) from the lumen of the tubule to the capillary, as indicated by the arrows.

treated diabetics are due to less water being reabsorbed into the peritubular capillary network.

Urea is an example of a substance that is passively reabsorbed from the filtrate. At first, the concentration of urea within the filtrate is the same as that in blood plasma. But after water is reabsorbed, the urea concentration is greater than that of peritubular plasma. In the end, about 50% of the filtered urea is reabsorbed.

### *Tubular Secretion*

**Tubular secretion** is the second way substances are removed from blood and added to tubular fluid (Fig. 36.11). Substances such as uric acid, hydrogen ions, ammonia, creatinine, histamine, and penicillin are eliminated by tubular secretion. The process of tubular secretion may be viewed as helping to rid the body of potentially harmful compounds that were not filtered into the glomerulus.

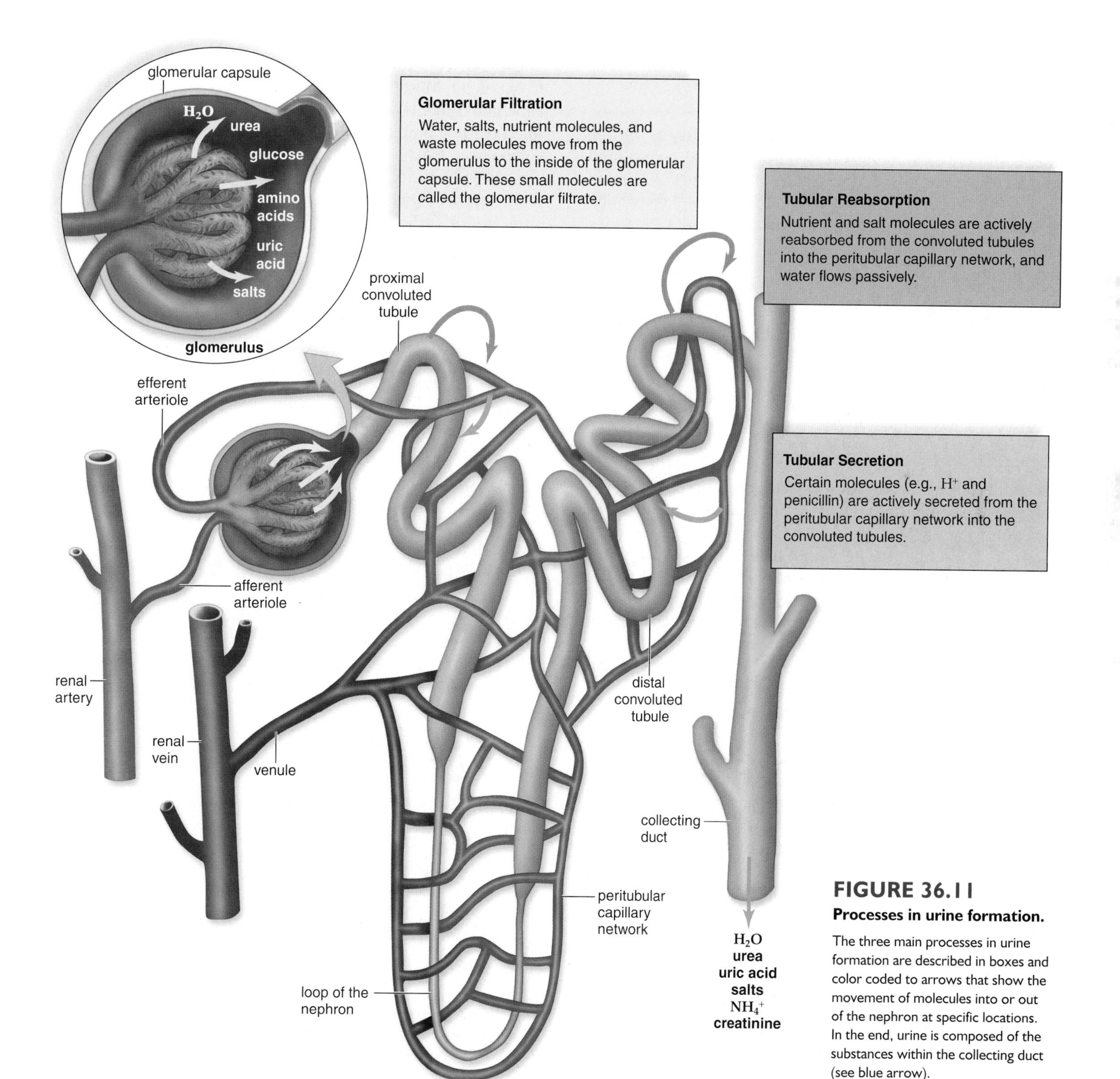


**FIGURE 36.11**

**Processes in urine formation.**

The three main processes in urine formation are described in boxes and color coded to arrows that show the movement of molecules into or out of the nephron at specific locations. In the end, urine is composed of the substances within the collecting duct (see blue arrow).

## The Kidneys and Homeostasis

The kidneys are organs of homeostasis for four main reasons: The kidneys (1) excrete metabolic wastes such as urea, which is the primary nitrogenous waste of humans; (2) maintain the water-salt balance in a way to be described; (3) maintain the acid-base balance and, therefore, the pH balance; and (4) secrete hormones. One of the hormones secreted by the kidneys boosts the number of red blood cells when insufficient oxygen is being delivered to its cells. This hormone, called **erythropoietin,** stimulates the stem cells in the bone marrow to produce more red blood cells. Another hormone produced by the kidneys, called renin, will be discussed in this part of the chapter.

### *Maintaining the Salt-Water Balance*

Most of the water and salt (NaCl) present in the filtrate is reabsorbed across the wall of the proximal convoluted tubule. The excretion of a hypertonic urine (one that is more concentrated than blood) is dependent on the reabsorption of water from the loop of the nephron and the collecting duct. During the process of reabsorption, water passes through recently discovered water channels called **aquaporins.**

**Loop of the Nephron.** A long loop of the nephron, which typically penetrates deep into the renal medulla, is made up of a descending (going down) limb and an ascending (going up) limb. Salt (NaCl) passively diffuses out of the lower portion of the ascending limb, but the upper, thick portion of the limb actively extrudes salt out into the tissue of the outer renal medulla (Fig. 36.12). Less and less salt is available for transport as fluid moves up the thick portion of the ascending limb. Because of these circumstances, there is an osmotic gradient within the tissues of the renal medulla: The concentration of salt is greater in the direction of the inner medulla. (Note that water cannot leave the ascending limb because the limb is impermeable to water.)

The innermost portion of the inner medulla has the highest concentration of solutes. This cannot be due to salt because active transport of salt does not start until fluid reaches the thick portion of the ascending limb. Urea is believed to leak from the lower portion of the collecting duct, and it is this molecule that contributes to the high solute concentration of the inner medulla.

Because of the osmotic gradient within the renal medulla, water leaves the descending limb along its entire length. This is a countercurrent mechanism: As water diffuses out of the descending limb, the remaining fluid within the limb encounters an even greater osmotic concentration of solute; therefore, water will continue to leave the descending limb from the top to the bottom. Filtrate within the collecting duct also encounters the same osmotic gradient mentioned earlier (Fig. 36.12). Therefore, water diffuses out of the collecting duct into the renal medulla, and the urine within the collecting duct becomes hypertonic to blood plasma.

**Antidiuretic hormone (ADH)** [Gk. *anti,* against; L. *ouresis,* urination] released by the posterior lobe of the pituitary plays a role in water reabsorption at the collecting duct. To understand the action of this hormone, consider its name. *Diuresis* means increased amount of urine, and *antidiuresis* means decreased amount of urine. When ADH is present, more water is reabsorbed (blood volume and pressure rise), and a decreased amount of urine results. In practical terms, if an individual does not drink much water on a certain day, the posterior lobe of the pituitary releases ADH, causing more water to be reabsorbed and less urine to form. On the other hand, if an individual drinks a large amount of water and does not perspire much, ADH is not released. Now more water is excreted, and more urine forms. Diuretics, such as caffeine and alcohol, increase the flow of urine by interfering with the action of ADH.

**Hormones Control the Reabsorption of Salt.** Usually, more than 99% of sodium ($Na^+$) filtered at the glomerulus is returned to the blood. Most sodium (67%) is reabsorbed at the proximal convoluted tubule, and a sizable amount (25%) is extruded by the ascending limb of the loop of the nephron. The rest is reabsorbed from the distal convoluted tubule and collecting duct.

Blood volume and pressure is, in part, regulated by salt reabsorption. When blood volume, and therefore blood pressure, is not sufficient to promote glomerular filtration, the kidneys secrete renin. **Renin** is an enzyme that changes angiotensinogen (a large plasma protein produced by the liver) into

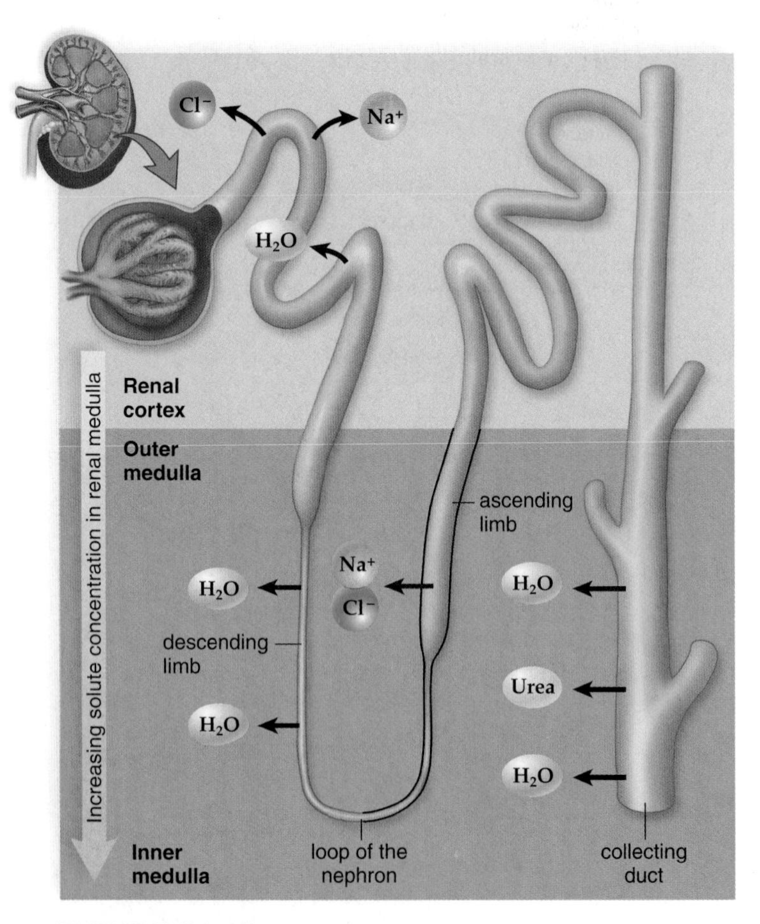


**FIGURE 36.12 Reabsorption of salt and water.**

Salt (NaCl) diffuses and is actively transported out of the ascending limb of the loop of the nephron into the renal medulla; also, urea leaks from the collecting duct and enters the tissues of the renal medulla. This creates a hypertonic environment, which draws water out of the descending limb and the collecting duct. This water is returned to the cardiovascular system.

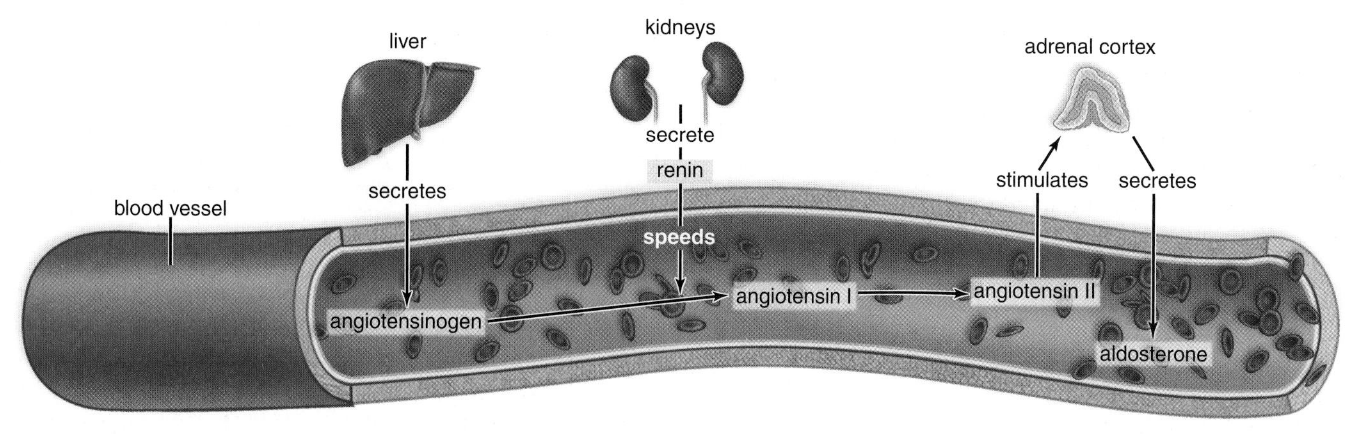


**FIGURE 36.13 The renin-angiotensin-aldosterone system.**

The liver secretes angiotensinogen into the bloodstream. Renin from the kidneys initiates the chain of events that results in angiotensin II. Angiotensin II acts on the adrenal cortex to secrete aldosterone, which causes reabsorption of sodium ions by the kidneys and a subsequent rise in blood pressure.

angiotensin I. Later, angiotensin I is converted to **angiotensin II,** a powerful vasoconstrictor that also stimulates the adrenal glands, which lie on top of the kidneys, to release aldosterone (Fig. 36.13). **Aldosterone** is a hormone that promotes the excretion of potassium ions ($K^+$) and the reabsorption of sodium ions ($Na^+$) at the distal convoluted tubule. The reabsorption of sodium ions is followed by the reabsorption of water. Therefore, blood volume and blood pressure increase.

**Atrial natriuretic hormone (ANH)** is a hormone secreted by the atria of the heart when cardiac cells are stretched due to increased blood volume. ANH inhibits the secretion of renin by the juxtaglomerular apparatus and the secretion of aldosterone by the adrenal cortex. Its effect, therefore, is to promote the excretion of $Na^+$—that is, natriuresis. When $Na^+$ is excreted, so is water, and therefore blood volume and blood pressure decrease.

These examples show that the kidneys regulate the water balance in blood by controlling the excretion and the reabsorption of ions. Sodium ($Na^+$) is an important ion in plasma that must be regulated, but the kidneys also excrete or reabsorb other ions, such as potassium ions ($K^+$), bicarbonate ions ($HCO_3^-$), and magnesium ions ($Mg^{2+}$), as needed.

## *Maintaining the Acid-Base Balance*

The functions of cells are influenced by pH. Therefore regulation of pH is extremely important to good health. The *bicarbonate ($HCO_3^-$) buffer system* and breathing work together to help maintain the pH of the blood. Central to the mechanism is this reaction, which you have seen before:

$$H^+ + HCO_3^- \rightleftharpoons H_2CO_3 \rightleftharpoons H_2O + CO_2$$

The *excretion of carbon dioxide ($CO_2$) by the lungs* helps keep the pH within normal limits, because when carbon dioxide is exhaled, this reaction is pushed to the right, and hydrogen ions are tied up in water. Indeed, when blood pH decreases, chemoreceptors in the carotid bodies (located in the carotid arteries) and in aortic bodies (located in the aorta) stimulate the respiratory control center, and the rate and depth of breathing increase. On the other hand, when blood pH begins to rise, the respiratory control center is depressed, and the amount of bicarbonate ion increases in the blood.

As powerful as this system is, only the kidneys can rid the body of a wide range of acidic and basic substances. The kidneys are slower acting than the buffer/breathing mechanism, but they have a more powerful effect on pH. For the sake of simplicity, we can think of the kidneys as reabsorbing bicarbonate ions and excreting hydrogen ions as needed to maintain the normal pH of the blood:

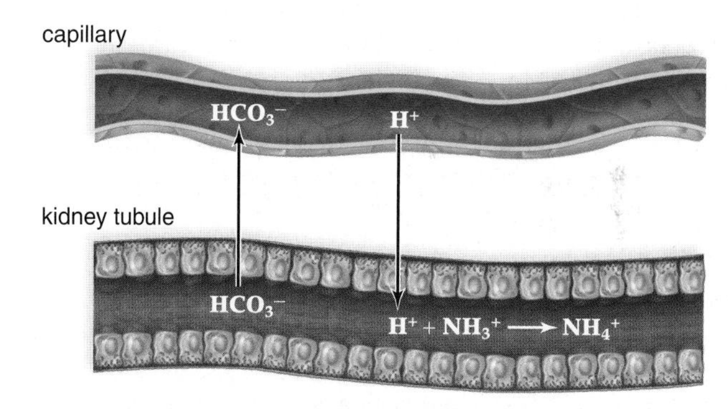


If the blood is acidic, hydrogen ions are excreted and bicarbonate ions are reabsorbed. If the blood is basic, hydrogen ions are not excreted and bicarbonate ions are not reabsorbed. The fact that urine is usually acidic (pH about 6) shows that usually an excess of hydrogen ions are excreted. Ammonia ($NH_3$) provides a means for buffering these hydrogen ions in urine: ($NH_3 + H^+ \longrightarrow NH_4^+$). Ammonia (whose presence is quite obvious in the diaper pail or kitty litter box) is produced in tubule cells by the deamination of amino acids. Phosphate provides another means of buffering hydrogen ions in urine.

### Check Your Progress 36.2

1. Which of the organs shown in Figure 36.7 are organs of homeostasis involved in osmoregulation and excretion?
2. The kidneys function on a take-back system. Explain.
3. What does the renin-angiotensin-aldosterone system accomplish?

## Connecting the Concepts

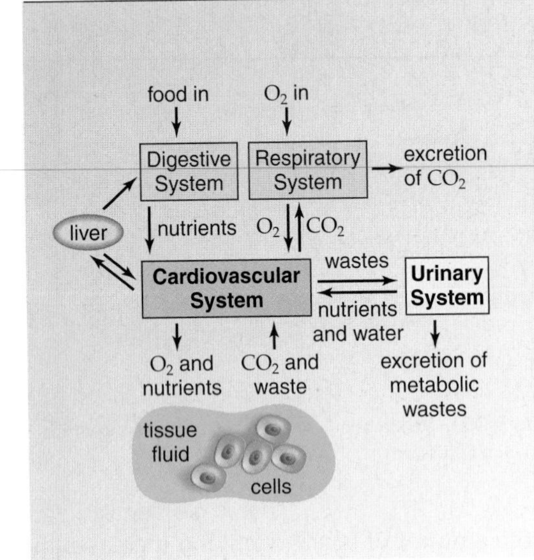


We have seen that the cardiovascular system works with the digestive and respiratory systems to maintain homeostasis, and now we wish to consider the contribution of the urinary system. The kidneys are the chief regulators of the internal environment because they have ultimate control over what is removed and what is retained in blood. They remove nitrogenous wastes such as urea (produced by the liver) and also uric acid. Even more important, the kidneys maintain the water-salt balance and the pH balance of blood. Hormones affect the workings of the kidneys. Too low a concentration of $Na^+$ in the blood causes blood pressure to lower and activates the renin-aldosterone sequence, and then the kidneys increase $Na^+$ reabsorption. Take in too much salt and ADH from the pituitary gland causes the kidneys to reabsorb more water. The kidneys work with the respiratory system to maintain pH. The respiratory system excretes $CO_2$ and this helps lower pH, but the kidneys can retain $HCO_3^-$, which helps raise pH. The kidneys can also retain or excrete $H^+$ ions. Ion composition of the blood affects osmolarity and the workings of other body systems.

The kidneys affect homeostasis another way. They produce erythropoietin, a hormone that stimulates red blood cell formation, and in this way, they help the cardiovascular and respiratory systems.

## summary

### 36.1 Excretion and the Environment

Animals excrete nitrogenous wastes. The amount of water and energy required to excrete nitrogenous wastes differs. Aquatic animals usually excrete ammonia (needs much water to excrete), and land animals excrete either urea or uric acid (needs much energy to produce).

Animals often have an excretory organ. The flame cells of planarians rid the body of excess water. Earthworm nephridia exchange molecules with the blood in a manner similar to that of vertebrate kidneys. Malpighian tubules in insects take up metabolic wastes and water from the hemolymph. Later, the water is absorbed by the gut.

Osmotic regulation is important to animals. Most have to balance their water and salt intake and excretion to maintain normal solute and water concentration in body fluids. Marine fishes constantly drink water, excrete salts at the gills, and pass an isotonic urine. Freshwater fishes never drink water; they take in salts at the gills and excrete a hypotonic urine. Terrestrial animals that live in extreme environments also have adaptations. For example, the kangaroo rat can survive on metabolic water because of its many ways of conserving water; marine birds and reptiles have glands that extrude salt.

### 36.2 Urinary System in Humans

The kidneys, excretory organs, are part of the human urinary system. Microscopically, each kidney is made up of nephrons, each of which has several parts and its own blood supply.

Urine formation by a nephron requires three steps: glomerular filtration, when nutrients, water, and wastes enter the nephron's glomerular capsule; tubular reabsorption, when nutrients and most water are reabsorbed into the peritubular capillary network; and tubular secretion, when additional wastes are added to the convoluted tubules.

Humans excrete a hypertonic urine. The ascending limb of the loop of the nephron actively extrudes salt so that the renal medulla is increasingly hypertonic relative to the contents of the descending limb and the collecting duct. Since urea leaks from the lower end of the collecting duct, the inner renal medulla has the highest concentration of solute. Therefore, a countercurrent mechanism ensures that water will diffuse out of the descending limb and the collecting duct.

Three hormones are involved in maintaining the water content of the blood. The hormone ADH (antidiuretic hormone), which makes the collecting duct more permeable, is secreted by the posterior pituitary in response to an increase in the osmotic pressure of the blood. The hormone aldosterone is secreted by the adrenal cortex after low blood pressure has caused the kidneys to release renin. The presence of renin leads to the formation of angiotensin II, which causes the adrenal cortex to release aldosterone. Aldosterone acts on the kidneys to retain $Na^+$; therefore, water is reabsorbed and blood pressure rises. The atrial natriuretic hormone prevents the secretion of renin and aldosterone.

The kidneys keep blood pH within normal limits. They reabsorb $HCO_3^-$ and excrete $H^+$ as needed to maintain the pH at about 7.4.

## understanding the terms



Match the terms to these definitions:

a. ______________ Blind, threadlike excretory tubule near the anterior end of an insect hindgut.

b. ______________ Cuplike structure that is the initial portion of a nephron; where glomerular filtration occurs.

c. _______________ Main nitrogenous waste of terrestrial amphibians and most mammals.
d. _______________ Hormone secreted by the adrenal cortex that regulates the sodium and potassium ion balance of the blood.
e. _______________ Main nitrogenous waste of insects, reptiles, and birds.

## reviewing this chapter

1. Relate the three primary nitrogenous wastes to the habitat of animals. 666
2. Describe how the excretory organs of the earthworm and the insect function. 667
3. Contrast the osmotic regulation of a marine bony fish with that of a freshwater bony fish. 668
4. Give examples of how other types of animals regulate their water and salt balance. 668–69
5. Describe the path of urine in humans, and give a function for each structure mentioned. 670
6. Describe the macroscopic anatomy of a human kidney, and relate it to the placement of nephrons. 670–71
7. List the parts of a nephron, and give a function for each structure mentioned. 670–71
8. Describe how urine is made by outlining what happens at each part of the nephron. 671–73
9. Describe the reabsorption of water and salt along the length of the nephron. Include the contribution of the loop of the nephron. 674–75
10. Name and describe the action of antidiuretic hormone (ADH), the renin-aldosterone connection, and the atrial natriuretic hormone (ANH). 674–75
11. How does the nephron regulate the pH of the blood? 675

## testing yourself

Choose the best answer for each question.

1. Which of these pairs is mismatched?
   a. insects—excrete uric acid
   b. humans—excrete urea
   c. fishes—excrete ammonia
   d. birds—excrete ammonia
   e. All of these are correct.
2. One advantage of urea excretion over uric acid excretion is that urea
   a. requires less energy than uric acid to form.
   b. can be concentrated to a greater extent.
   c. is not a toxic substance.
   d. requires no water to excrete.
   e. is a larger molecule.
3. Freshwater bony fishes maintain water balance by
   a. excreting salt across their gills.
   b. periodically drinking small amounts of water.
   c. excreting a hypotonic urine.
   d. excreting wastes in the form of uric acid.
   e. Both a and c are correct.
4. Animals with which of these are most likely to excrete a semisolid nitrogenous waste?
   a. nephridia
   b. Malpighian tubules
   c. human kidneys
   d. flame cells
   e. All of these are correct.
5. In which of these human structures are you least apt to find urine?
   a. large intestine
   b. urethra
   c. collecting duct
   d. bladder
   e. Both a and c are correct.
6. Excretion of a hypertonic urine in humans is associated best with
   a. the glomerular capsule.
   b. the proximal convoluted tubule.
   c. the loop of the nephron.
   d. the collecting duct.
   e. Both c and d are correct.
7. The presence of ADH (antidiuretic hormone) causes an individual to excrete
   a. less salt.
   b. less water.
   c. more water.
   d. more salt.
   e. Both a and c are correct.
8. In humans, water is
   a. found in the glomerular filtrate.
   b. reabsorbed from the nephron.
   c. in the urine.
   d. reabsorbed from the collecting duct.
   e. All of these are correct.
9. Which of these is out of order first?
   a. glomerular capsule
   b. proximal convoluted tubule
   c. distal convoluted tubule
   d. loop of the nephron
   e. collecting duct
10. Normally in humans, glucose
    a. is always in the filtrate and urine.
    b. is always in the filtrate with little or none in urine.
    c. undergoes tubular secretion and is in urine.
    d. undergoes tubular secretion and is not in urine.
    e. is not in the filtrate and is not in the urine.
11. Which of these causes blood pressure to decrease?
    a. aldosterone
    b. antidiuretic hormone (ADH)
    c. renin
    d. atrial natriuretic hormone (ANH)
12. If a drug inhibits the kidneys' ability to reabsorb bicarbonate so that bicarbonate is excreted in the urine, the blood will become
    a. acidic.
    b. alkaline.
    c. first acidic and then alkaline.
    d. first alkaline and then acidic.
13. Which of these materials is not filtered from the blood at the glomerulus?
    a. water
    b. urea
    c. protei
    d. glucose
    e. sodium ions
14. The renal medulla has a striped appearance due to the presence of which structures?
    a. loop of the nephron
    b. collecting ducts
    c. peritubular capillaries
    d. Both a and b are correct.

15. By what process are most molecules secreted from the blood into the convoluted tubules?
    a. osmosis
    b. diffusion
    c. active transport
    d. facilitated diffusion
16. Which of these is not correct?
    a. Uric acid is produced from the breakdown of amino acids.
    b. Urea is produced from the breakdown of proteins.
    c. Ammonia results from the deamination of amino acids.
    d. All of these are correct.
17. When tracing the path of blood, the blood vessel that follows the renal artery is the
    a. peritubular capillary.
    b. efferent arteriole.
    c. afferent arteriole.
    d. renal vein.
    e. glomerulus.
18. Absorption of the glomerular filtrate occurs at
    a. the convoluted tubules.
    b. only the distal convoluted tubule.
    c. the loop of the nephron.
    d. the collecting duct.
19. Label this diagram of a nephron.

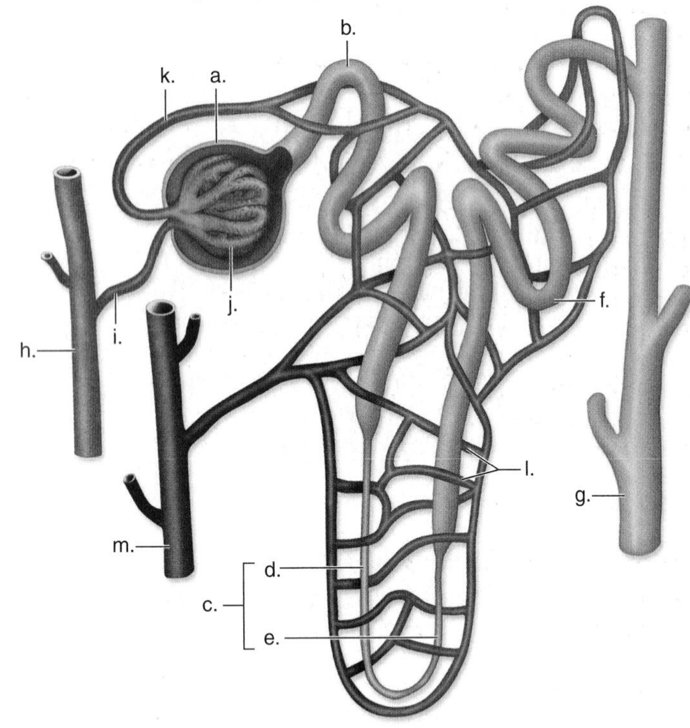


## thinking scientifically

1. High blood pressure often is accompanied by kidney damage. In some people, the kidney damage is subsequent to the high blood pressure, but in others the kidney damage is what caused the high blood pressure. Explain how a low-salt diet would enable you to determine whether the high blood pressure or the kidney damage came first?
2. The renin-angiotensin-aldosterone system can be inhibited in order to reduce high blood pressure. Usually, the angiotensin-converting enzyme, which converts angiotensin I to angiotensin II, is inhibited by drug therapy. Why would this enzyme be an effective point to disrupt the system?

## bioethical issue

### Increasing Life Span

As a society, we are accustomed to thinking that as we grow older, diseases such as urinary disorders will begin to occur. Almost everyone is aware that most males are subject to enlargement of the prostate as they age, and that cancer of the prostate is not uncommon among older men. However, as with many illnesses associated with aging, medical science now knows how to treat or even cure prostate problems. Because of these successes, our life span has lengthened. A child born in the United States in 1900 lived to, say, the age of 47. If that same child were born today, he or she would probably live to at least 76. Even more exciting is the probability that scientists will improve the life span still further. People could live beyond 100 years and have the same vigor and vitality they had when they were young.

Most people are appreciative of living longer, especially if they can expect to be free of the illnesses and inconveniences associated with aging. But have we examined how we feel about longevity as a society? We are accustomed to considering that if the birthrate increases, so does the size of a population. But what about the death rate? If the birthrate stays constant and the death rate decreases, obviously population size also increases. Most experts agree that population growth depletes resources and increases environmental degradation. Having more people in the older population can also put a strain on the economy if they are unable to meet their financial needs, including medical expenses, without government assistance.

What is the ethical solution to this problem? Should we just allow the population to increase as older people live longer? Should we decrease the birthrate? Should we reduce government assistance to older people so they realize that they must be able to take care of themselves? Or should we call a halt to increasing the life span through advancements in medical science?

## *Biology* website

The companion website for *Biology* provides a wealth of information organized and integrated by chapter. You will find practice tests, animations, videos, and much more that will complement your learning and understanding of general biology.

**http://www.mhhe.com/maderbiology10**

# 37

# Neurons and Nervous Systems

*Through input from* **sensory receptors,** *the nervous system receives a continuous barrage of information, which it integrates before it stimulates* **effectors,** *such as muscles and glands. An impairment of these operations can have serious consequences for the individual. Spinal cord injuries can result in paralysis when commands from the brain and spinal cord fail to reach the nerves that bring about muscle contraction. Similarly, disease can cause paralysis. Amyotrophic lateral sclerosis (ALS), also known as Lou Gehrig disease (for a famous baseball player with ALS), is a fatal degenerative disease characterized by the death of neurons, which signal muscle contraction. People with ALS gradually lose the ability to move, and eventually cannot even breathe on their own; however, their intellectual ability is not impaired. Professor Stephen Hawking, pictured below, is a renowned physicist and author who is afflicted with ALS.*

*In this chapter, you will explore the structure, evolution, and function of nervous systems in invertebrate and vertebrate animals.*

Stephen Hawking suffers from amyotrophic lateral sclerosis, or Lou Gehrig disease.

## concepts

# 37.1 Evolution of the Nervous System

The nervous system is vitally important in complex animals, enabling them to seek food and avoid danger. It ceaselessly monitors internal and external conditions and makes appropriate changes to maintain homeostasis. A comparative study of animal nervous system organization indicates the evolutionary trends that may have led to the nervous system of vertebrates.

## Invertebrate Nervous Organization

Simple animals, such as sponges, which have the cellular level of organization, can respond to stimuli; the most common observable response is closure of the osculum (central opening). Hydras, which are cnidarians with the tissue level of organization, can contract and extend their bodies, move their tentacles to capture prey, and even turn somersaults. They have a **nerve net** that is composed of neurons (nerve cells) in contact with one another and with contractile cells in the body wall (Fig. 37.1*a*). Sea anemones and jellyfishes, which are also cnidarians, seem to have two nerve nets. A fast-acting one allows major responses, particularly in times of danger, and the other coordinates slower and more delicate movements.

Planarians, which are flatworms, have a nervous organization that reflects their bilateral symmetry. They have a **ladderlike nervous system,** with two ventrally located lateral or longitudinal nerve cords (bundles of nerves) that extend from the cerebral ganglia to the posterior end of their body. Transverse nerves connect the nerve cords, as well as the cerebral ganglia, to the eyespots. **Cephalization** has occurred, as evidenced by a concentration of neurons and sensory receptors in a head region. A cluster of neurons is called a **ganglion** (pl., ganglia), and the anterior cerebral ganglia receive sensory information from photoreceptors in the eyespots and sensory cells in the auricles (Fig. 37.1*b*). The two lateral nerve cords allow a rapid transfer of information from the cerebral ganglia to the posterior end, and the transverse nerves between the nerve cords keep the movement of the two sides coordinated. Bilateral symmetry plus cephalization are two significant trends in the development of a nervous organization that is adaptive for an active way of life. Also, the nervous organization in planarians is a foreshadowing of the central nervous system and peripheral nervous system seen in vertebrates.

Annelids (e.g., earthworm, Fig. 37.1*c*), arthropods (e.g., crab, Fig. 37.1*d*), and molluscs (e.g., squid, Fig. 37.1*e*) are complex animals with true nervous systems. The annelids and arthropods have the typical invertebrate nervous system. There is a brain and a ventral nerve cord having a ganglion in each segment. The brain, which normally receives sensory information, controls the activity of the ganglia and assorted nerves so that the muscle activity of the entire animal is coordinated. The crab and squid show marked cephalization—the anterior end has a well-defined brain, and there are well-developed sense organs, such as eyes. The presence of a brain and other ganglia in the body of all these animals indicates an increase in the number of neurons among more complex invertebrates.

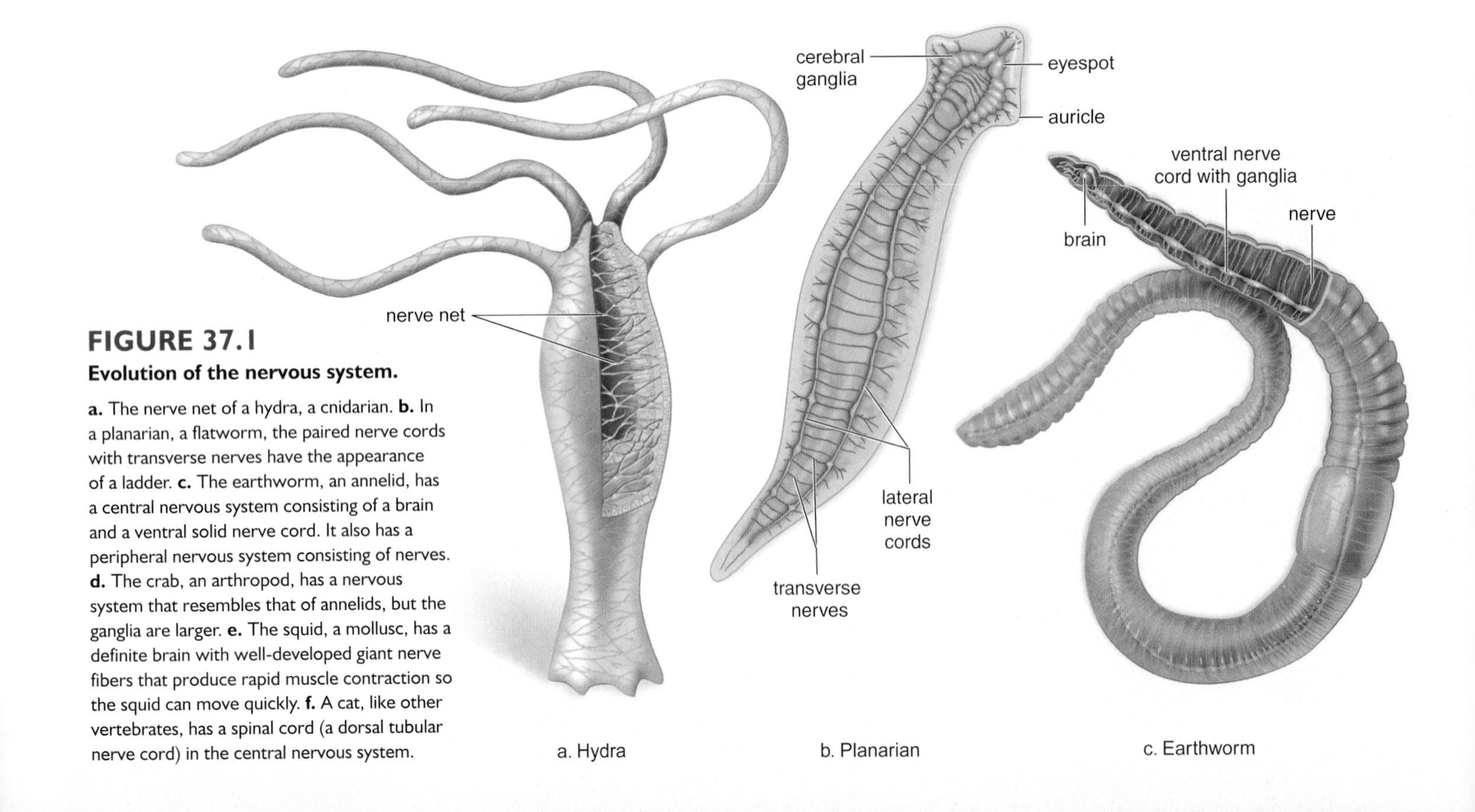


**FIGURE 37.1**

**Evolution of the nervous system.**

**a.** The nerve net of a hydra, a cnidarian. **b.** In a planarian, a flatworm, the paired nerve cords with transverse nerves have the appearance of a ladder. **c.** The earthworm, an annelid, has a central nervous system consisting of a brain and a ventral solid nerve cord. It also has a peripheral nervous system consisting of nerves. **d.** The crab, an arthropod, has a nervous system that resembles that of annelids, but the ganglia are larger. **e.** The squid, a mollusc, has a definite brain with well-developed giant nerve fibers that produce rapid muscle contraction so the squid can move quickly. **f.** A cat, like other vertebrates, has a spinal cord (a dorsal tubular nerve cord) in the central nervous system.

## Vertebrate Nervous Organization

In vertebrates (e.g., cat), cephalization, coupled with bilateral symmetry, results in several types of paired sensory receptors, including the eyes, ears, and olfactory structures that allow the animal to gather information from the environment. Paired cranial and spinal nerves contain numerous nerve fibers. Vertebrates have many more neurons than do invertebrates. For example, an insect's entire nervous system may contain a total of about 1 million neurons, while a vertebrate's nervous system may contain many thousand to many billion times that number. A vertebrate's **central nervous system (CNS),** consisting of a spinal cord and brain, develops from an embryonic dorsal neural tube. The spinal cord is continuous with the brain because the embryonic neural tube becomes the spinal cord posteriorly, while the vertebrate brain is derived from the enlarged anterior end of the neural tube. Ascending tracts carry sensory information to the brain, and descending tracts carry motor commands to the neurons in the spinal cord that control the muscles.

It is customary to divide the vertebrate brain into the hindbrain, midbrain, and forebrain (Fig. 37.2). The hindbrain is the most ancient part of the brain. Nearly all vertebrates have a well-developed hindbrain that regulates motor activity below the level of consciousness. In humans, for example, the lungs and heart function even when we are sleeping. The medulla oblongata contains control centers for breathing and heart rate. Coordination of motor activity associated with limb movement, posture, and balance eventually became centered in the cerebellum.

The optic lobes are part of the midbrain, which was originally a center for coordinating reflexes involving the eyes and ears. Starting with the amphibians and continuing in the other vertebrates, the forebrain processes sensory information. Originally, the forebrain was concerned mainly with the sense of smell. Later, the thalamus evolved to receive sensory input from the midbrain and the hindbrain and to pass it on to the cerebrum, the anterior part of the forebrain in vertebrates. In the forebrain, the hypothalamus is particularly concerned with homeostasis, and in this capacity, the hypothalamus communicates with the medulla oblongata and the pituitary gland.

The cerebrum, which is highly developed in mammals, integrates sensory and motor input and is particularly associated with higher mental capabilities. In humans, the outer layer of the cerebrum, called the cerebral cortex, is especially large and complex.

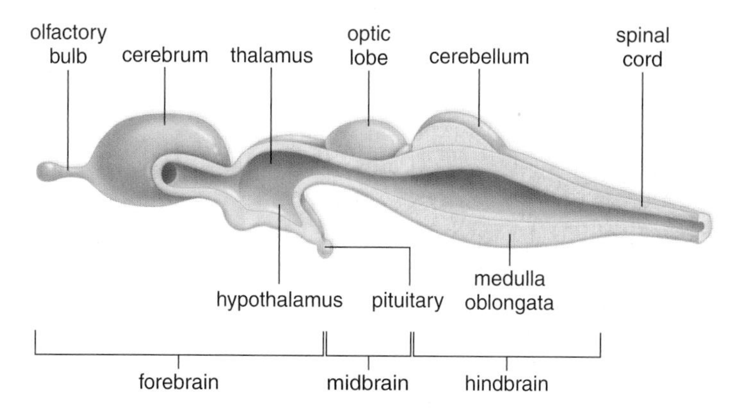


**FIGURE 37.2 Organization of the vertebrate brain.**

The vertebrate brain is divided into the forebrain, the midbrain, and the hindbrain.

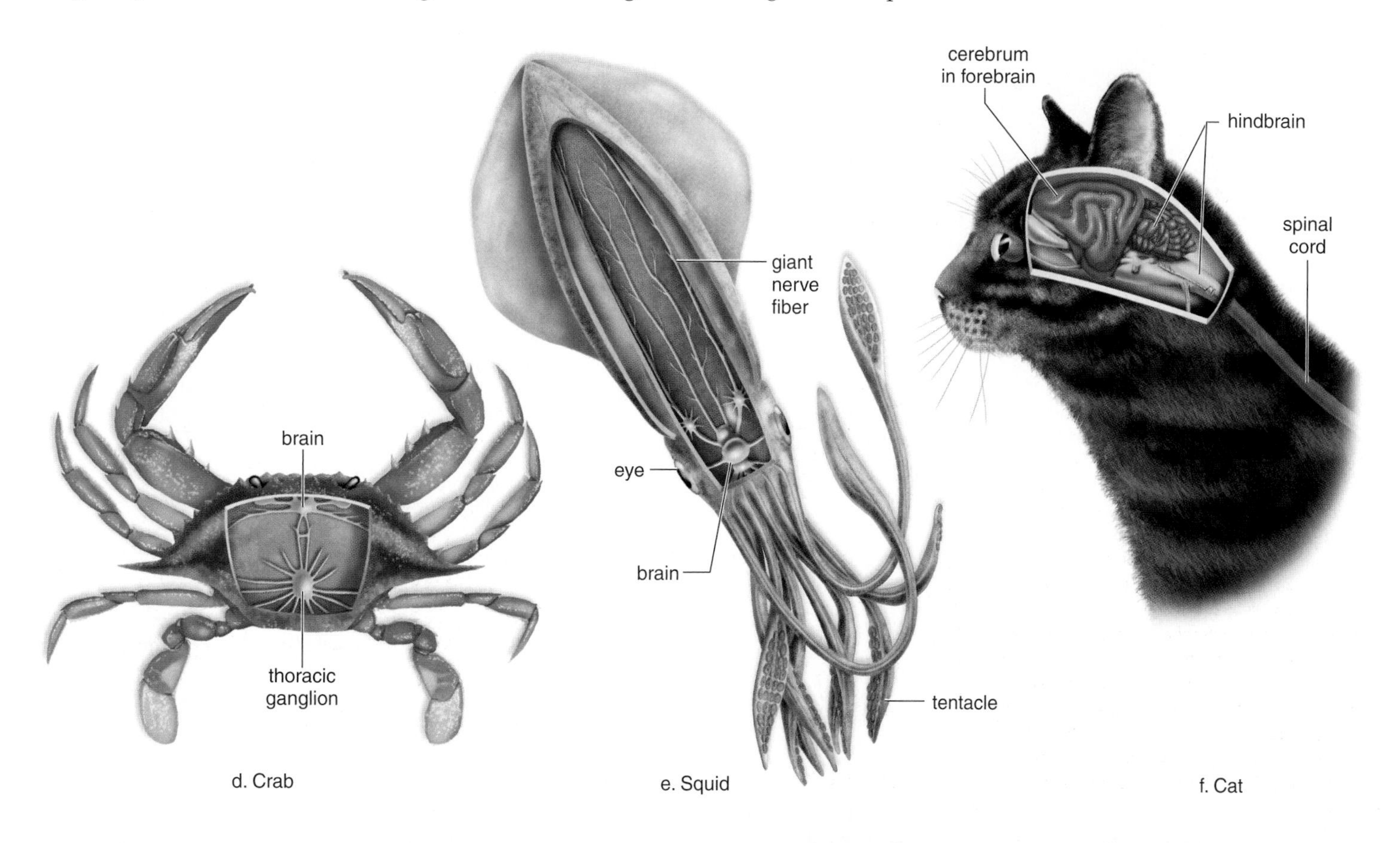

## The Human Nervous System

The human nervous system has three specific functions: (1) it receives sensory input—sensory receptors in skin and other organs respond to external and internal stimuli by generating nerve impulses that travel to the central nervous system (CNS); (2) it performs integration—the CNS sums up the input it receives from all over the body; and (3) it generates motor output—nerve impulses from the CNS go to the muscles and glands. Muscle contractions and gland secretions are responses to stimuli received by sensory receptors. As an example, consider the events that occur as a person raises a glass to the lips. Continual sensory input to the CNS from the eyes and hand informs the CNS of the position of the glass, and the CNS continually sums up the incoming data before commanding the hand to proceed. At any time, integration with other sensory data might cause the CNS to command a different motion instead.

In humans, the central nervous system (CNS) consists of the brain and spinal cord (Fig. 37.3). The brain is housed in the skull and the spinal cord is housed in the vertebral column. The **peripheral nervous system (PNS)** [Gk. *periphereia,* circumference] consists of all the nerves and ganglia that lie outside the central nervous system. The paired cranial and spinal nerves are part of the PNS. In the PNS, the somatic nervous system has sensory and motor functions that control the skeletal muscles. The autonomic nervous system controls smooth muscle, cardiac muscle, and the glands. It is further divided into the sympathetic and parasympathetic divisions.

The components and functions of the central and peripheral nervous systems are complex. For an organism to maintain homeostasis, both systems have to work in harmony.

### Check Your Progress 37.1

1. What is a ganglion?
2. What is the advantage for an animal having cephalization in addition to bilateral symmetry?
3. Distinguish between the CNS and the PNS.

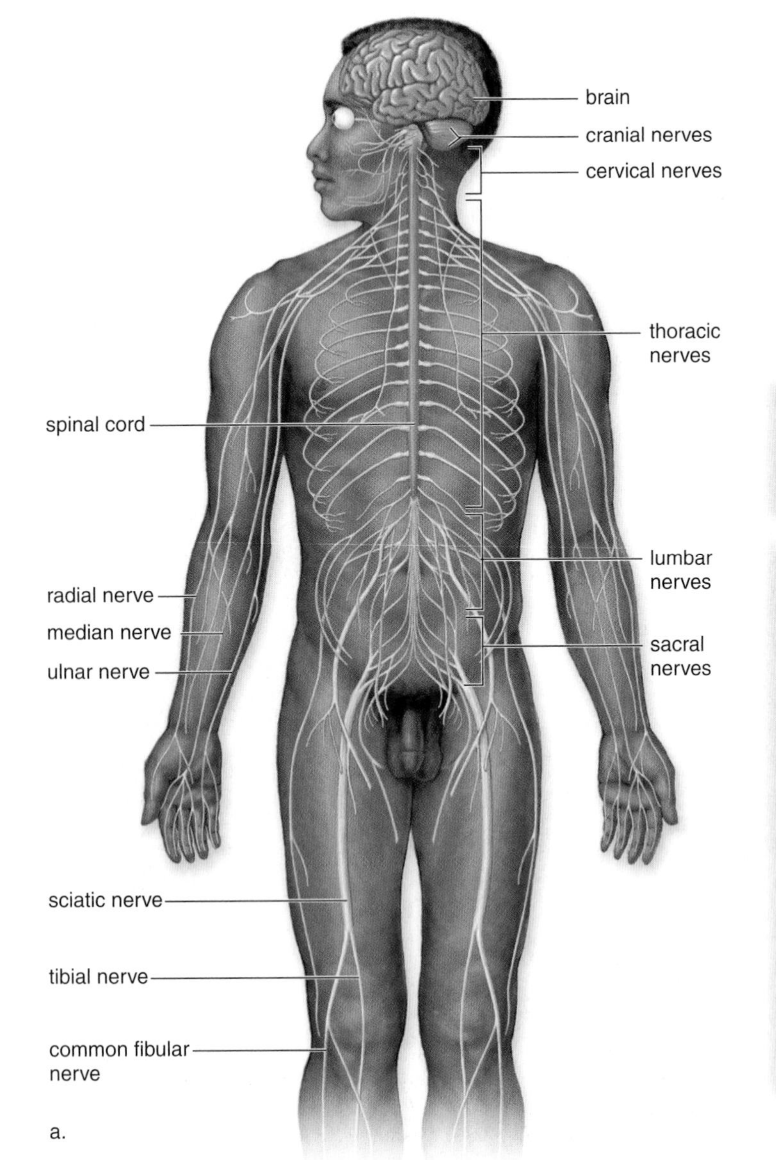


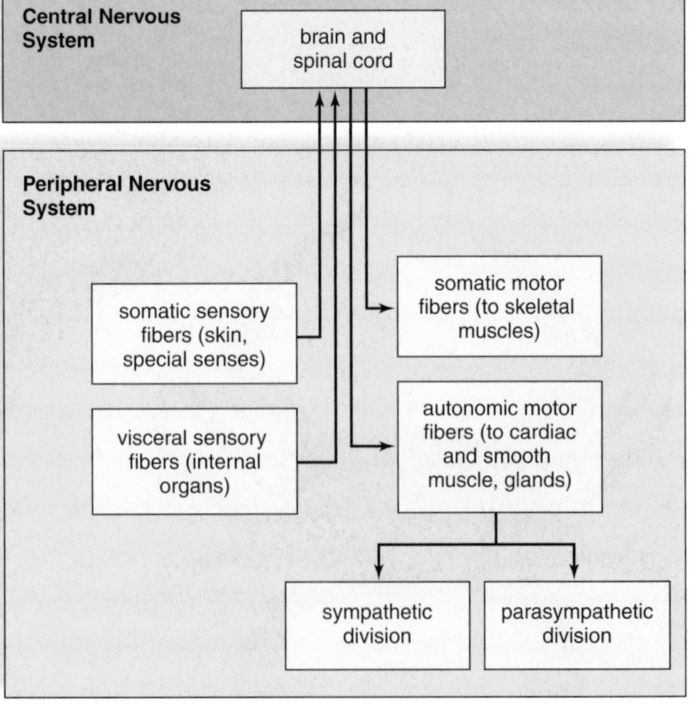


**FIGURE 37.3 Organization of the nervous system in humans.**

**a.** The central nervous system (CNS) is composed of brain and spinal cord; the peripheral nervous system (PNS) consists of nerves. **b.** In the somatic system of the PNS, nerves conduct impulses from sensory receptors located in the skin and internal organs to the CNS and motor impulses from the CNS to the skeletal muscles. In the autonomic system, consisting of the sympathetic and parasympathetic divisions, motor impulses travel to smooth muscle, cardiac muscle, and glands.

# 37.2 Nervous Tissue

Although complex, nervous tissue is composed of two principal types of cells. **Neurons,** also known as nerve cells, are the functional units of the nervous system. They receive sensory information, convey the information to an integration center such as the brain, and conduct signals from the integration center to effector structures such as the glands and muscles. **Neuroglia** serve as supporting cells, providing support and nourishment to the neurons.

## Neurons and Neuroglia

Neurons [Gk. *neuron,* nerve] vary in appearance depending on their function and location. They consist of three major parts: a cell body, dendrites, and an axon (Fig. 37.4). The **cell body** contains a nucleus and a variety of organelles. The **dendrites** [Gk. *dendron,* tree] are short, highly branched processes that receive signals from the sensory receptors or other neurons and transmit them to the cell body. The **axon** [Gk. *axon,* axis] is the portion of the neuron that conveys information to another neuron or to other cells. Axons can be bundled together to form nerves. For this reason, axons are often called **nerve fibers.** Many axons are covered by a white insulating layer called the **myelin sheath** [Gk. *myelos,* spinal cord].

Neuroglia, or glial cells, which were discussed on page 583, greatly outnumber neurons in the brain. There are several different types in the CNS, each with specific functions. Some (microglia) help remove bacteria and debris, some (astrocytes) provide metabolic and structural support directly to the neurons. The myelin sheath is formed from the membranes of tightly spiraled neuroglia. In the PNS, **Schwann cells** perform this function, leaving gaps called **nodes of Ranvier.** In the CNS, neuroglial cells called **oligodendrocytes** perform this function.

### *Types of Neurons*

Neurons can be described in terms of their function and shape. **Motor (efferent) neurons** take nerve impulses from the CNS to muscles or glands. Motor neurons are said to have a multipolar shape because they have many dendrites and a single axon (Fig. 37.4*a*). Motor neurons cause muscle fibers to contract or glands to secrete, and therefore they are said to innervate these structures.

**Sensory (afferent) neurons** take nerve impulses from sensory receptors to the CNS. The sensory receptor, which is the distal end of the long axon of a sensory neuron, may be as simple as a naked nerve ending (a pain receptor), or may be built into a highly complex organ, such as the eye or ear. Almost all sensory neurons have a structure that is termed unipolar (Fig. 37.4*b*). In unipolar neurons, the process that extends from the cell body divides into a branch that extends to the periphery and another that extends to the CNS. Since both of these extensions are long and myelinated and transmit nerve impulses, it is now generally accepted to refer to them as an axon.

**Interneurons** [L. *inter,* between; Gk. *neuron,* nerve] occur entirely within the CNS. Interneurons, which are typically multipolar in shape (Fig. 37.4*c*), convey nerve impulses between various parts of the CNS. Some lie between sensory neurons and motor neurons; some take messages from one side of the spinal cord to the other or from the brain to the cord, and vice versa. They also form complex pathways in the brain where processes accounting for thinking, memory, and language occur.

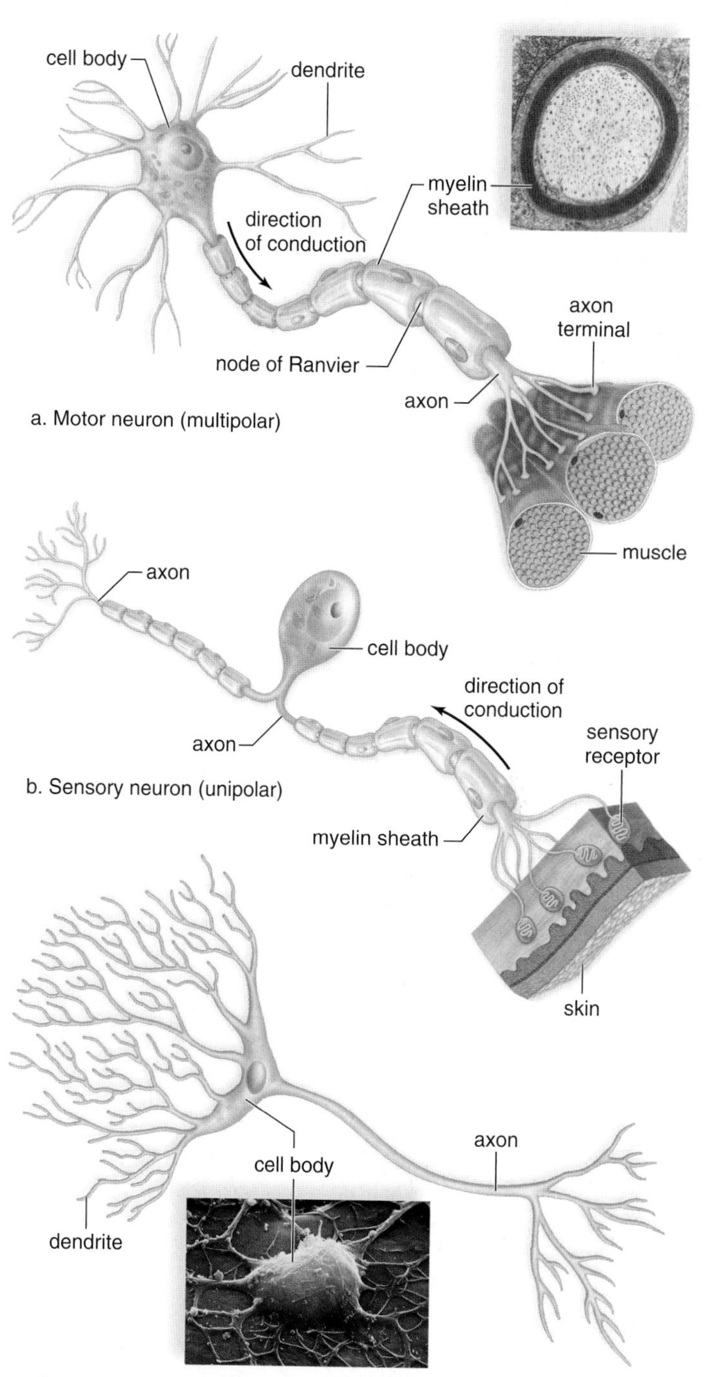


**FIGURE 37.4 Neuron anatomy.**

**a.** Motor neuron. Note the branched dendrites and the single, long axon, which branches only near its tip. **b.** Sensory neuron with dendritelike structures projecting from the peripheral end of the axon. **c.** Interneuron (from the cortex of the cerebellum) with very highly branched dendrites.

## Transmission of the Nerve Impulses

In the early 1900s, Julius Bernstein at the University of Halle, Germany, suggested that the nerve impulse is an electrochemical phenomenon involving the movement of unequally distributed ions on either side of an axonal membrane, the plasma membrane of an axon. It was not until later, however, that investigators developed a technique that enabled them to support this hypothesis. A. L. Hodgkin and A. F. Huxley, English neurophysiologists, received the Nobel Prize in 1963 for their work in this field. They and a group of researchers, headed by K. S. Cole and J. J. Curtis at Woods Hole, Massachusetts, managed to insert a tiny electrode into the giant axon of the squid *Loligo*. This internal electrode was then connected to a voltmeter, an instrument with a screen that shows voltage differences over time (Fig. 37.5). Voltage is a measure of the electrical potential difference between two points, which in this case is the difference between two electrodes—one placed inside and another placed outside the axon. (An electrical potential difference across a membrane is called the membrane potential.) When a membrane potential exists, we can say that a plus pole and a minus pole exist; therefore, the voltmeter indicates the existence of polarity and records polarity changes.

### *Resting Potential*

When the axon is not conducting an impulse, the voltmeter records a membrane potential equal to about −65 mV (millivolts), indicating that the inside of the neuron is more negative than the outside (Fig. 37.5*a*). This is called the **resting potential** because the axon is not conducting an impulse.

The existence of this polarity can be correlated with a difference in ion distribution on either side of the axonal membrane. As Figure 37.5*a* shows, there is a higher concentration of sodium ions ($Na^+$) outside the axon and a higher concentration of potassium ions ($K^+$) inside the axon. The unequal distribution of these ions is in part due to the activity of the sodium-potassium pump (see page 94). This pump is an active transport system in the plasma membrane that pumps three sodium ions out of and two potassium ions into the axon. The pump is always working because the membrane is somewhat permeable to these ions and they tend to diffuse toward their lesser concentration. Since the membrane is more permeable to potassium ions than to sodium ions, there are always more positive ions outside the membrane than inside; this accounts for some of the polarity recorded by the voltmeter. There are also large, negatively charged proteins in the cytoplasm of the axon; altogether, then, the

**FIGURE 37.5 Resting and action potential of the axonal membrane.**

**a.** Resting potential. A voltmeter that records voltage changes indicates the axonal membrane has a resting potential of −65 mV. There is a preponderance of $Na^+$ outside the axon and a preponderance of $K^+$ inside the axon. The permeability of the membrane to $K^+$ compared to $Na^+$, and the presence of large, negatively charged proteins (not shown) within the axon, causes the inside to be negative compared to the outside. **b.** Action potential. Depolarization occurs when $Na^+$ gates open and $Na^+$ moves inside the axon, and (**c**) repolarization occurs when $K^+$ gates open and $K^+$ moves outside the axon. **d.** Graph of the action potential.

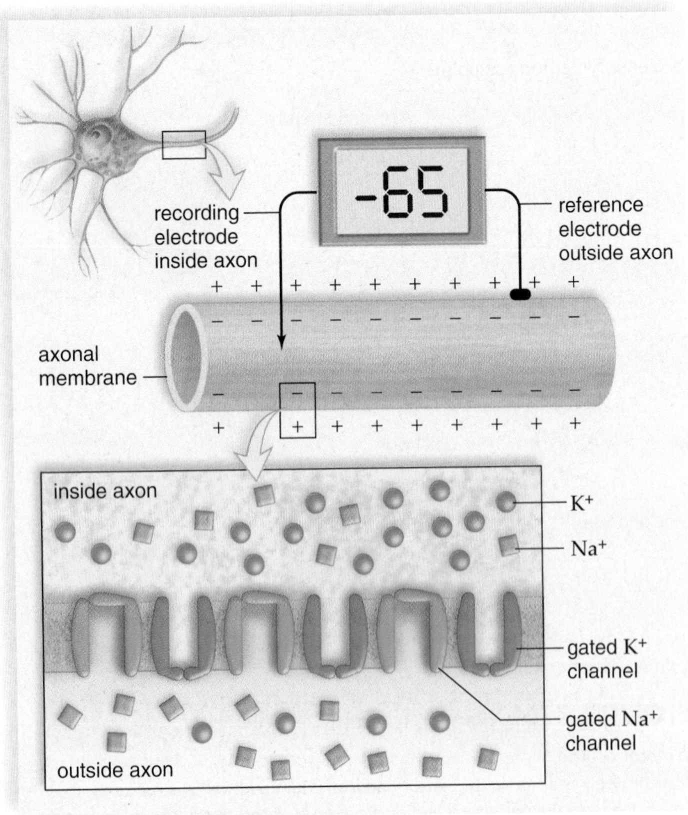


a. Resting potential: more $Na^+$ outside the axon and more $K^+$ inside the axon causes polarization.

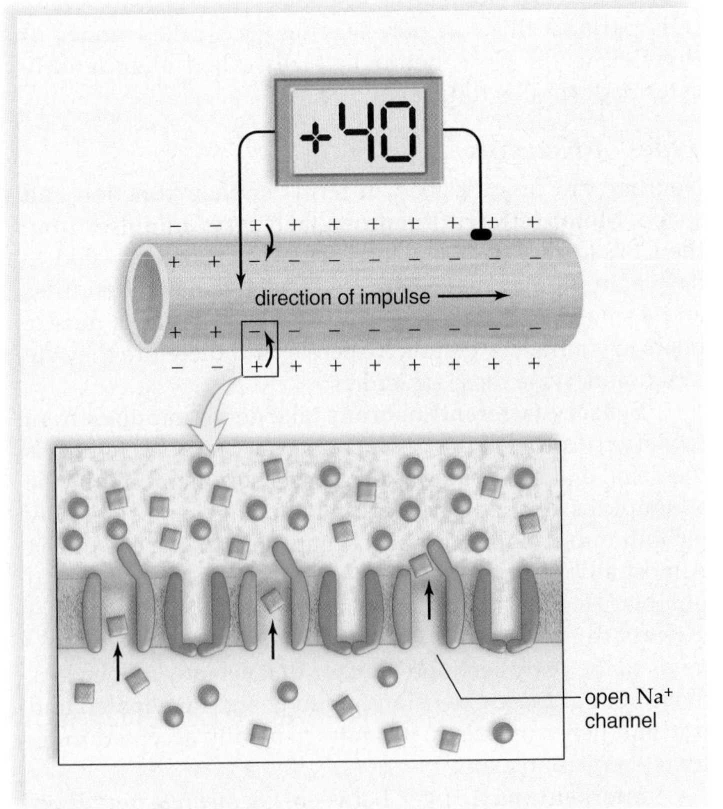


b. Action potential begins: depolarization occurs when $Na^+$ gates open and $Na^+$ moves to inside the axon.

voltmeter records that the inside is −65 mV compared to the outside. This is the resting potential.

## *Action Potential*

An **action potential** is a rapid change in polarity across a portion of an axonal membrane as the nerve impulse occurs. An action potential uses two types of gated ion channels in the axonal membrane. In the axonal membrane, a gated ion channel allows sodium ($Na^+$) to pass through the membrane, and another allows potassium ($K^+$) to pass through the membrane. In contrast to ungated ion channels, which constantly allow ions across the membrane, gated ion channels open and close in response to a stimulus such as a signal from another neuron.

Threshold is the minimum change in polarity across the axonal membrane that is required to generate an action potential. Therefore, the action potential is an all-or-none event. During *depolarization,* the inside of a neuron becomes positive because of the sudden entrance of sodium ions. If threshold is reached, many more sodium channels open, and the action potential begins. As sodium ions rapidly move across the membrane to the inside of the axon, the action potential swings up from −65 mV to +40 mV (Fig. 37.5*b*). This reversal in polarity causes the sodium channels to close and the potassium channels to open. As potassium ions leave the axon, the action potential swings down from +40 mV to −65 mV. In other words, a *repolarization* occurs (Fig. 37.5*c*). An action potential only takes 2 milliseconds. In order to visualize such rapid fluctuations in voltage across the axonal membrane, researchers generally find it useful to plot the voltage changes over time (Fig. 37.5*d*).

## *Propagation of Action Potentials*

In nonmyelinated axons, the action potential travels down an axon one small section at a time, at a speed of about 1 m/second. In myelinated axons, the gated ion channels that produce an action potential are concentrated at the nodes of Ranvier. Just as taking giant steps during a game of "Simon Says" is more efficient, so ion exchange only at the nodes makes the action potential travel faster. *Saltar* in Spanish means "to jump," and so this mode of conduction, called **saltatory conduction,** means that the action potential "jumps" from node to node:

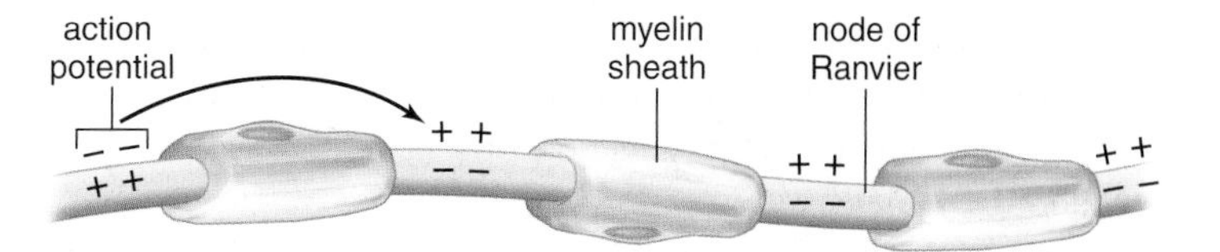


Speeds of 200 m/second (450 miles per hour) have been recorded.

The intensity of a message traveling down a nerve fiber is determined by how many nerve impulses are generated within a given time span. A fiber can conduct a volley of nerve impulses because only a small number of ions are exchanged with each impulse. As soon as an action potential has moved on, the previous section undergoes a **refractory period,** during which the $Na^+$ gates are unable to open. Notice, therefore, that the action potential cannot move backward and instead always moves down an axon toward its terminals.

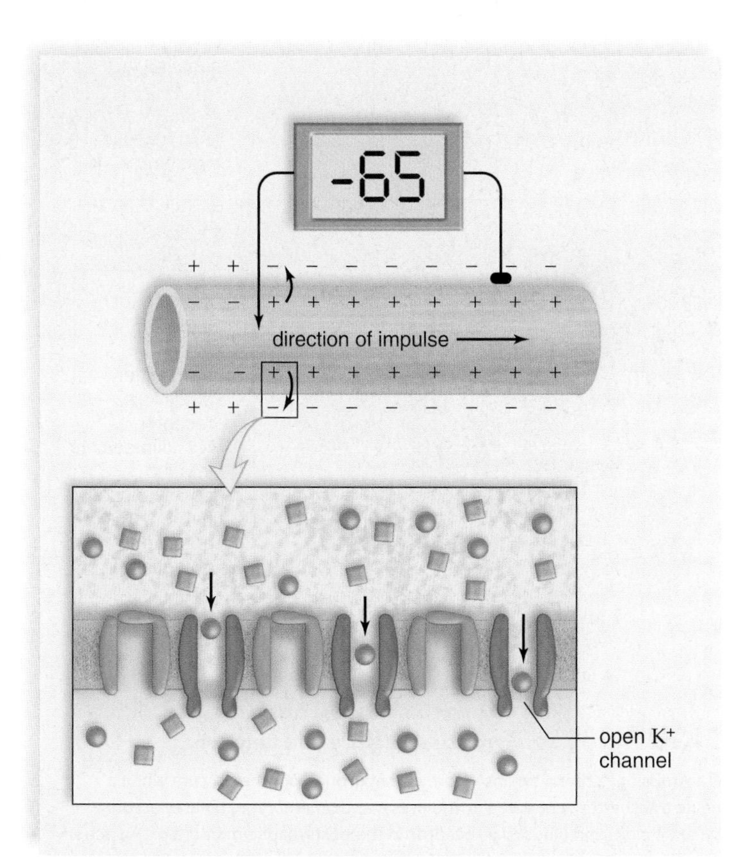


c. Action potential ends: repolarization occurs when $K^+$ gates open and $K^+$ moves to outside the axon.

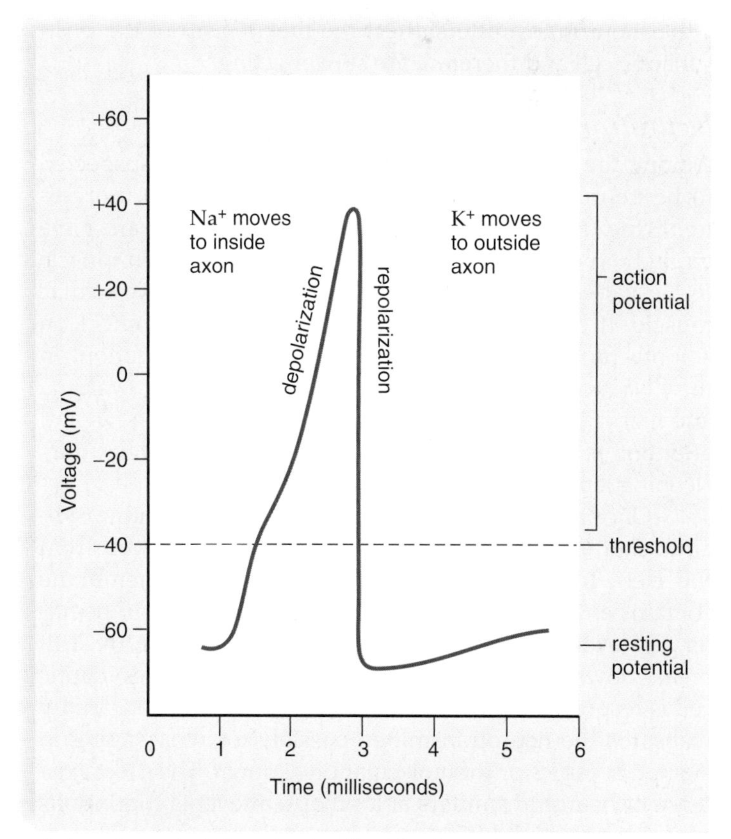


d. An action potential can be visualized if voltage changes are graphed over time.

## Transmission Across a Synapse

Every axon branches into many fine endings, each tipped by a small swelling, called an axon terminal (Fig. 37.6). Each terminal lies very close to the dendrite (or the cell body) of another neuron. This region of close proximity is called a **synapse.** At a synapse, the membrane of the first neuron is called the *pre*synaptic membrane, and the membrane of the next neuron is called the *post*synaptic membrane. The small gap between the neurons is called the **synaptic cleft.**

A nerve impulse cannot cross a synaptic cleft. Transmission across a synapse is carried out by molecules called **neurotransmitters,** which are stored in synaptic vesicles. When nerve impulses traveling along an axon reach an axon terminal, gated channels for calcium ions ($Ca^{2+}$) open, and calcium enters the terminal. This sudden rise in $Ca^{2+}$ stimulates synaptic vesicles to merge with the presynaptic membrane, and neurotransmitter molecules are released into the synaptic cleft. They diffuse across the cleft to the postsynaptic membrane, where they bind with specific receptor proteins.

Depending on the type of neurotransmitter and/or the type of receptor, the response of the postsynaptic neuron can be toward excitation or toward inhibition. Excitatory neurotransmitters that use gated ion channels are fast acting. Other neurotransmitters affect the metabolism of the postsynaptic cell and therefore are slower acting.

### *Neurotransmitters and Neuromodulators*

Among the more than 100 substances known or suspected to be neurotransmitters are **acetylcholine (ACh), norepinephrine (NE), dopamine,** and **serotonin,** which are present in both the CNS and PNS. The effect of ACh on muscle tissue varies. It excites skeletal muscle but inhibits cardiac muscle. It has either an excitatory or inhibitory effect on smooth muscle or glands, depending on their location. In the CNS, norepinephrine is important to dreaming, waking, and mood. Dopamine is involved in emotions, learning, and attention while serotonin is involved in thermoregulation, sleeping, emotions, and perception.

Once a neurotransmitter has been released into a synaptic cleft and has initiated a response, it is removed from the cleft. In some synapses, the postsynaptic membrane contains enzymes that rapidly inactivate the neurotransmitter. For example, the enzyme **acetylcholinesterase (AChE)** breaks down acetylcholine. In other synapses, the presynaptic cell is responsible for reuptake, a process in which it rapidly reabsorbs the neurotransmitter, possibly for repackaging in synaptic vesicles or for molecular breakdown. The short existence of neurotransmitters at a synapse prevents continuous stimulation (or inhibition) of postsynaptic membranes.

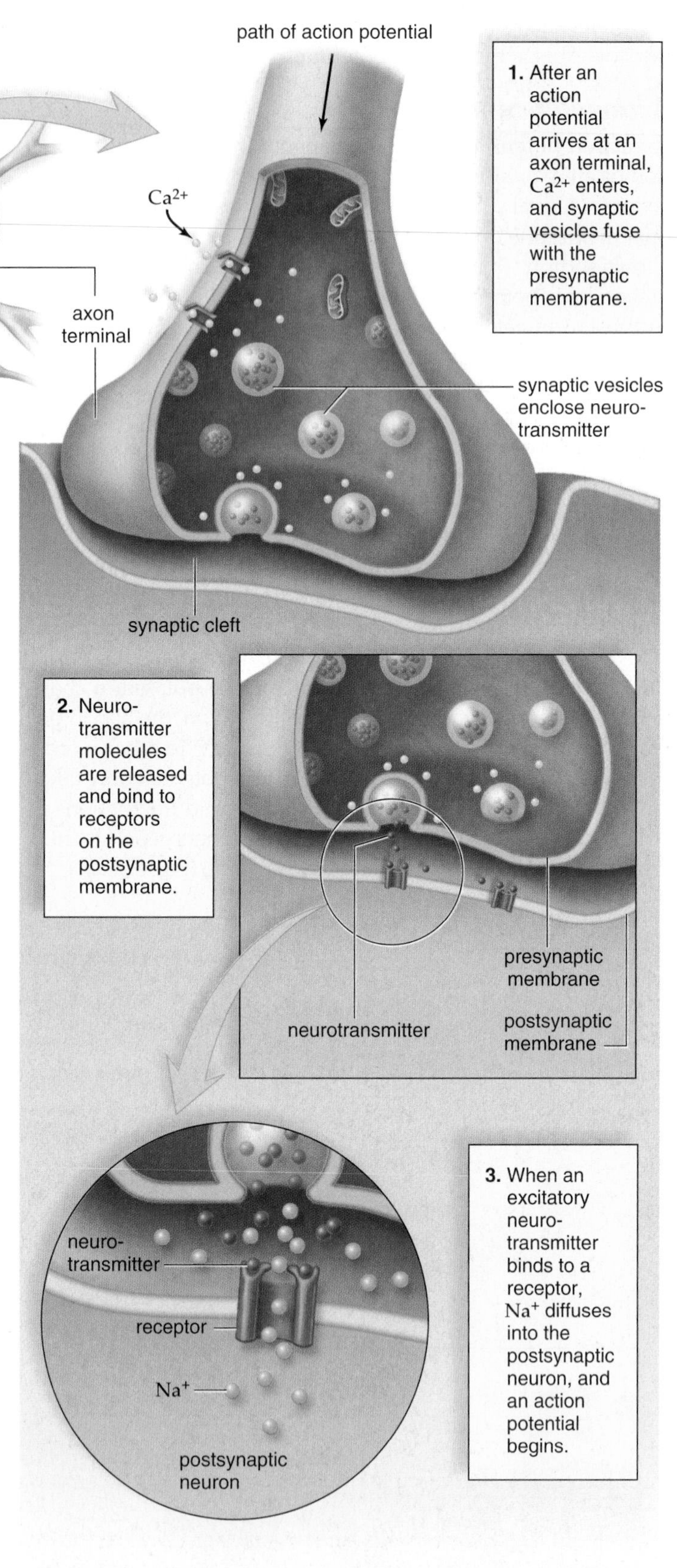


**FIGURE 37.6 Synapse structure and function.**

Transmission across a synapse from one neuron to another occurs when a neurotransmitter is released at the presynaptic membrane, diffuses across a synaptic cleft, and binds to a receptor in the postsynaptic membrane. An action potential may begin.

Neurotransmitter imbalances are associated with a number of disorders. Parkinson disease is associated with a lack of dopamine in the brain. Levodopa, one of the drugs used to treat Parkinson disease, serves as a precursor in the synthesis of dopamine, thereby boosting dopamine levels in the brain.

It is of interest to note here that many drugs affecting the nervous system act by interfering with or potentiating the action of neurotransmitters. Such drugs can enhance or block the release of a neurotransmitter, mimic the action of a neurotransmitter or block the receptor, or interfere with the removal of a neurotransmitter from a synaptic cleft. For instance, Alzheimer disease is associated with a deficiency of ACh; some of the drugs that slow the progression of the disease are cholinesterase inhibitors that block AChE and slow the degradation of ACh. Likewise, depression, a common mood disorder, appears to involve imbalances in norepinephrine and serotonin. Some antidepressant drugs, such as fluoxetine (Prozac), prevent the reuptake of serotonin, and others, including bupropion hydrochloride (Wellbutrin), prevent the reuptake of both serotonin and norepinephrine. Blocking reuptake prolongs the effects of these two neurotransmitters in networks of neurons within the brain that are involved in the emotional state. Several so-called "recreational drugs" (used for enjoyment rather than medical reasons) also affect neurotransmitter activity, as described in the Science Focus on pages 696–97.

**Neuromodulators** are molecules that block the release of a neurotransmitter or modify a neuron's response to a neurotransmitter. Substance P and the endorphins are well-known neuromodulators. Substance P is released by sensory neurons when pain is present. Endorphins block the release of substance P and, therefore, serve as natural painkillers. They are associated with the "runner's high" of joggers because they also produce a feeling of tranquility. Endorphins are produced by the brain not only when there is physical stress but also when emotional stress is present.

## Synaptic Integration

A single neuron has many dendrites plus the cell body, and both can have synapses with many other neurons. One thousand to 10,000 synapses per a single neuron is not uncommon. Therefore, a neuron is on the receiving end of many excitatory and inhibitory signals. An excitatory neurotransmitter produces a potential change called a signal that drives the neuron closer to an action potential, and an inhibitory neurotransmitter produces a signal that drives the neuron farther from an action potential. Excitatory signals have a depolarizing effect, and inhibitory signals have a hyperpolarizing effect.

Neurons integrate these incoming signals. **Integration** is the summing up of excitatory and inhibitory signals (Fig. 37.7). If a neuron receives many excitatory signals (either from different synapses or at a rapid rate from one synapse), chances are the axon will transmit a nerve impulse. On the other hand, if a neuron receives both inhibitory and excitatory signals, the summing up of these signals may prohibit the axon from firing.

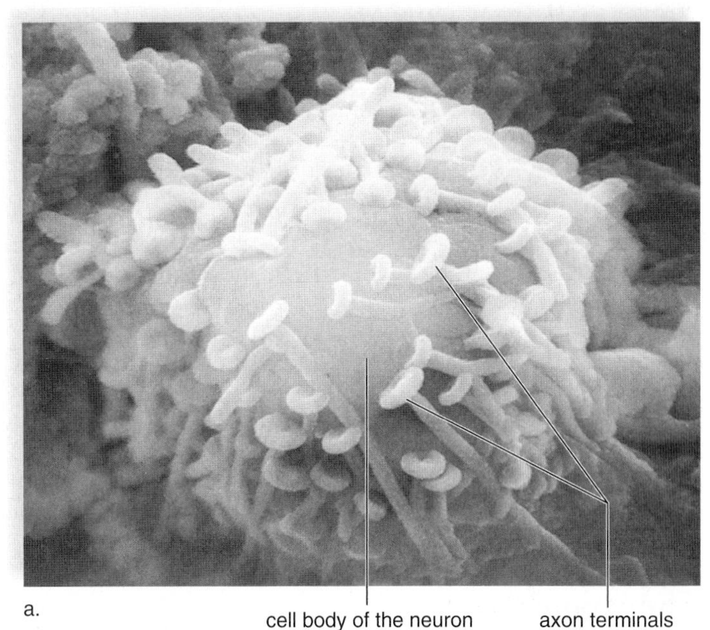


**FIGURE 37.7 Synaptic integration.**

**a.** Many neurons synapse with a cell body. **b.** Both inhibitory signals (blue) and excitatory signals (red) are summed up in the dendrite and cell body of the postsynaptic neuron. Only if the combined signals cause the membrane potential to rise above threshold does an action potential occur. In this example, threshold was not reached.

### Check Your Progress 37.2

1. Would a nerve impulse travel more quickly down an unmyelinated axon or a myelinated axon? Why?
2. A nerve impulse has two parts. **a.** During the first part, which ion moves where? **b.** During the second part, which ion moves where?
3. How are neurotransmitter molecules removed from synaptic clefts?

# 37.3 Central Nervous System: Brain and Spinal Cord

The central nervous system (CNS) consists of the spinal cord and the brain, where sensory information is received and motor control is initiated. The spinal cord and the brain are both protected by bone; the spinal cord is surrounded by vertebrae, and the brain is enclosed by the skull. Both the spinal cord and the brain are wrapped in three protective membranes known as **meninges. Meningitis** (inflammation of the meninges) is a serious disorder caused by a number of bacteria or viruses that invade the meninges. The spaces between the meninges are filled with **cerebrospinal fluid,** which cushions and protects the CNS. Cerebrospinal fluid is contained in the central canal of the spinal cord and within the **ventricles** of the brain, which are interconnecting spaces that produce and serve as reservoirs for cerebrospinal fluid.

## The Spinal Cord

The **spinal cord** is a bundle of nervous tissue enclosed in the vertebral column (see Fig. 37.12); it extends from the base of the brain to the vertebrae just below the rib cage. The spinal cord has two main functions: (1) it is the center for many **reflex actions,** which are automatic responses to external stimuli, and (2) it provides a means of communication between the brain and the spinal nerves, which leave the spinal cord.

A cross section of the spinal cord reveals that it is composed of a central portion of **gray matter** and a peripheral region of white matter. The gray matter consists of cell bodies and unmyelinated fibers. It is shaped like a butterfly, or the letter H, with two dorsal (posterior) horns and two ventral (anterior) horns surrounding a central canal. The gray matter contains portions of sensory neurons and motor neurons, as well as short interneurons that connect sensory and motor neurons.

Myelinated long fibers of interneurons that run together in bundles called **tracts** give **white matter** its color. These tracts connect the spinal cord to the brain. These tracts are like a busy superhighway, by which information continuously passes between the brain and the rest of the body. Dorsally, the tracts are primarily ascending, taking information *to* the brain; ventrally, the tracts are primarily descending, carrying information *from* the brain. Because the tracts at one point cross over, the left side of the brain controls the right side of the body, and the right side of the brain controls the left side of the body.

If the spinal cord is severed as the result of an injury, paralysis results. If the injury occurs in the cervical (neck) region, all four limbs are usually paralyzed, a condition known as quadriplegia. If the injury occurs in the thoracic region, the lower body may be paralyzed, a condition called paraplegia.

## The Brain

The **brain** contains four interconnected chambers called ventricles (Fig. 37.8*a*). The two lateral ventricles are inside the cerebrum. The third ventricle is surrounded by the diencephalon, and the fourth ventricle lies between the cerebellum and the pons. Cerebrospinal fluid is continuously produced in the ventricles and circulates through them; it then flows out of the brain between the meninges.

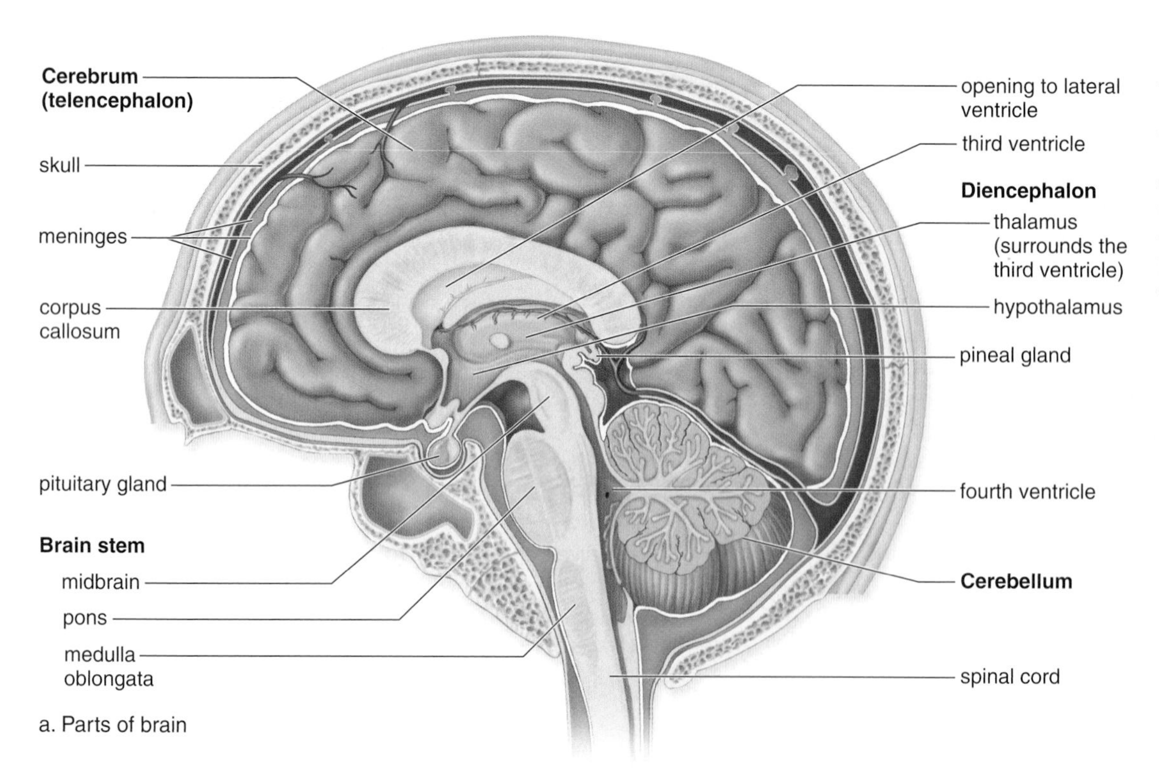


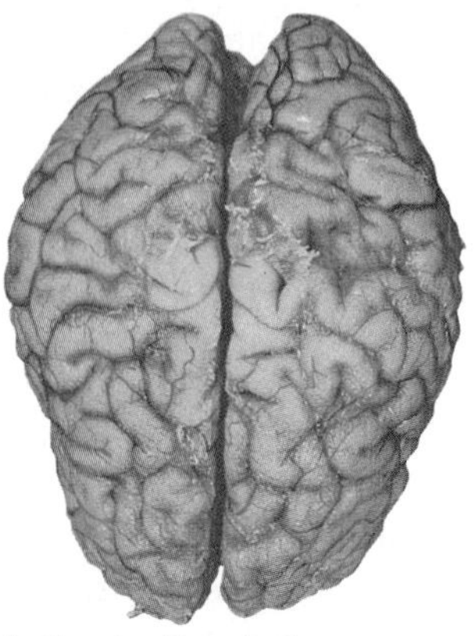


**FIGURE 37.8**
**The human brain.**
**a.** The right cerebral hemisphere is shown here, along with other, closely associated structures. The hemispheres are connected by the corpus callosum. **b.** The cerebrum is divided into the right and left cerebral hemispheres.

## *The Cerebrum*

The **cerebrum,** also called the telencephalon, is the largest portion of the brain in humans. The cerebrum is the last center to receive sensory input and carry out integration before commanding voluntary motor responses. It communicates with and coordinates the activities of the other parts of the brain.

**Cerebral Hemispheres.** The cerebrum is divided into two halves, called **cerebral hemispheres** (see Fig. 37.8*b*). A deep groove called the longitudinal fissure divides the cerebrum into the right and left hemispheres. Each hemisphere receives information from and controls the opposite side of the body. Although the hemispheres appear the same, the right hemisphere is associated with artistic and musical ability, emotion, spatial relationships, and pattern recognition. The left hemisphere is more adept at mathematics, language, and analytical reasoning. The two cerebral hemispheres are connected by a bridge of tracts within the corpus callosum.

Shallow grooves called sulci (sing., sulcus) divide each hemisphere into lobes (Fig. 37.9). The *frontal lobe* lies toward the front of the hemispheres and is associated with motor control, memory, reasoning, and judgment. For example, if a fire occurs, the frontal lobe enables you to decide whether to exit via the stairs or the window, or how to dress if the temperature plummets to subzero. The frontal lobe on the left side contains the *Broca area,* which organizes motor commands to produce speech.

The *parietal lobes* lie posterior to the frontal lobe and are concerned with sensory reception and integration, as well as taste. A *primary taste area* in the parietal lobe accounts for taste sensations.

The *temporal lobe* is located laterally. A primary auditory area in the temporal lobe receives information from our ears. The *occipital lobe* is the most posterior lobe. A *primary visual area* in the occipital lobe receives information from our eyes.

**The Cerebral Cortex.** The **cerebral cortex** is a thin (less than 5 mm thick), but highly convoluted, outer layer of gray matter that covers the cerebral hemispheres. The convolutions increase the surface area of the cerebral cortex. The cerebral cortex contains tens of billions of neurons and is the region of the brain that accounts for sensation, voluntary movement, and all the thought processes required for learning, memory, language, and speech.

Two regions of the cerebral cortex are of particular interest. The **primary motor area** is in the frontal lobe just ventral to (before) the central sulcus. Voluntary commands to skeletal muscles begin in the primary motor area, and each part of the body is controlled by a certain section. The size of the section indicates the precision of motor control. For example, the face and hand take up a much larger portion of the primary motor area than does the entire trunk. The **primary somatosensory area** is just dorsal to the central sulcus in the parietal lobe. Sensory information from the skin and skeletal muscles arrives here, where each part of the body is sequentially represented in a manner similar to the primary motor area.

**Basal Nuclei.** While the bulk of the cerebrum is composed of white matter (i.e., tracts), masses of gray matter are located deep within the white matter. These so-called **basal nuclei** (formerly termed basal ganglia) integrate motor commands, ensuring that proper muscle groups are activated or inhibited. **Huntington disease** and **Parkinson disease,** which are both characterized by uncontrollable movements, are due to malfunctioning basal nuclei.

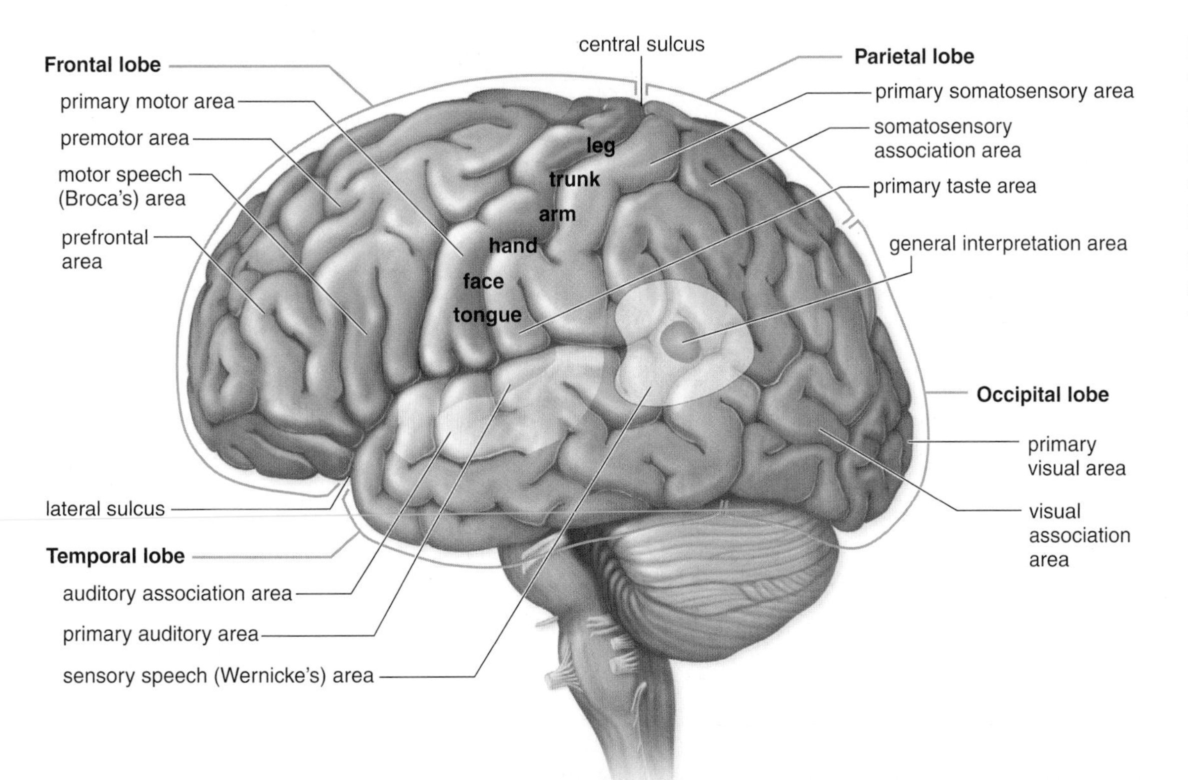


**FIGURE 37.9**

**The lobes of a cerebral hemisphere.**

Each cerebral hemisphere is divided into four lobes: frontal, parietal, temporal, and occipital. These lobes contain centers for reasoning and movement, somatic sensing, hearing, and vision, respectively.

### *Other Parts of the Brain*

The hypothalamus and the thalamus are in the **diencephalon,** a region that encircles the third ventricle. The **hypothalamus** forms the floor of the third ventricle. It is an integrating center that helps maintain homeostasis by regulating hunger, sleep, thirst, body temperature, and water balance. The hypothalamus controls the pituitary gland and, thereby, serves as a link between the nervous and endocrine systems.

The **thalamus** consists of two masses of gray matter located in the sides and roof of the third ventricle. It is on the receiving end for all sensory input except smell. Visual, auditory, and somatosensory information arrives at the thalamus via the cranial nerves and tracts from the spinal cord. The thalamus integrates this information and sends it on to the appropriate portions of the cerebrum. For this reason, the thalamus is often referred to as the "gatekeeper" for sensory information en route to the cerebral cortex. The thalamus is involved in arousal of the cerebrum, and it also participates in higher mental functions such as memory and emotions.

The **pineal gland,** which secretes the hormone melatonin, is located in the diencephalon. Presently, there is much interest in the role of melatonin in our daily rhythms; some researchers believe it may be involved in jet lag and insomnia. Scientists are also interested in the possibility that the hormone regulates the onset of puberty.

The **cerebellum** lies under the occipital lobe of the cerebrum and is separated from the brain stem by the fourth ventricle. It is the largest part of the hindbrain. The cerebellum has two portions that are joined by a narrow central portion. Each portion is primarily composed of white matter, which in longitudinal section has a treelike pattern. Overlying the white matter is a thin layer of gray matter that forms a series of complex folds.

The cerebellum receives sensory input from the eyes, ears, joints, and muscles about the present position of body parts, and it also receives motor output from the cerebral cortex about where these parts should be located. After integrating this information, the cerebellum sends motor impulses by way of the brain stem to the skeletal muscles. In this way, the cerebellum maintains posture and balance. It also ensures that all of the muscles work together to produce smooth, coordinated voluntary movements. The cerebellum assists the learning of new motor skills such as playing the piano or hitting a baseball. New evidence indicates that the cerebellum is important in judging the passage of time.

The **brain stem** contains the midbrain, the pons, and the medulla oblongata (see Fig. 37.8). The **midbrain** acts as a relay station for tracts passing between the cerebrum and the spinal cord or cerebellum. The tracts cross in the brain stem so that the right side of the body is controlled by the left portion of the brain, and the left portion of the body is controlled by the right portion of the brain.

The brain stem also has reflex centers for visual, auditory, and tactile responses. The word **pons** means "bridge" in Latin, and true to its name, the pons contains bundles of axons traveling between the cerebellum and the rest of the CNS. In addition, the pons functions with the medulla oblongata to regulate breathing rate, and has reflex centers concerned with head movements in response to visual and auditory stimuli.

The **medulla oblongata** contains a number of reflex centers for regulating heartbeat, breathing, and blood pressure. It also contains the reflex centers for vomiting, coughing, sneezing, hiccuping, and swallowing. The medulla oblongata lies just superior to the spinal cord, and it contains tracts that ascend or descend between the spinal cord and higher brain centers.

**The Reticular Activating System.** The reticular formation is a complex network of nuclei (masses of gray matter) and nerve fibers that extend the length of the brain stem (Fig. 37.10). The reticular formation is a major component of the reticular activating system (RAS), which receives sensory signals that it sends up to higher centers, and motor signals that it sends to the spinal cord.

The RAS arouses the cerebrum via the thalamus and causes a person to be alert. Apparently, the RAS can filter out unnecessary sensory stimuli, explaining why you can study with the TV on. If you want to awaken the RAS, surprise it with a sudden stimulus, like splashing your face with cold water; if you want to deactivate it, remove visual and auditory stimuli. General anesthetics function by artificially suppressing the RAS. A severe injury to the RAS can cause a person to be comatose, from which recovery may be impossible.

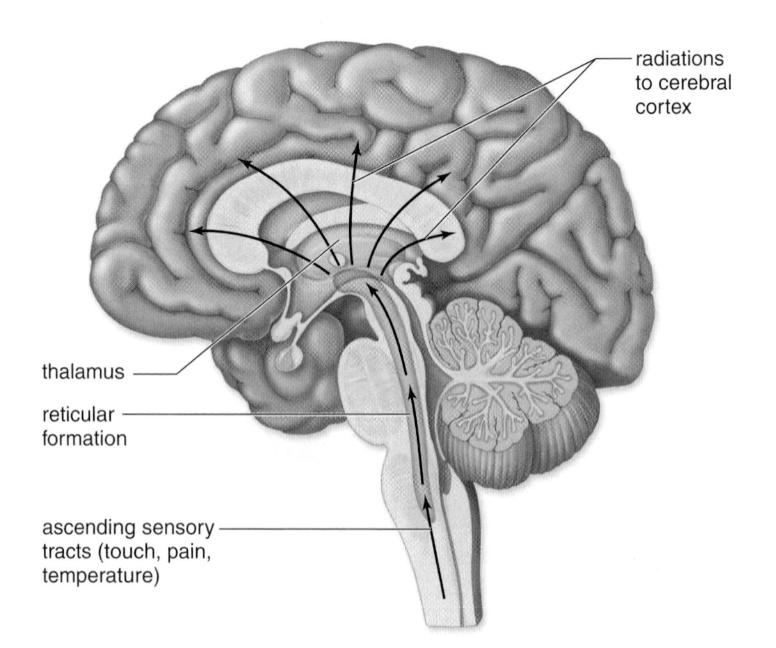


**FIGURE 37.10 The reticular activating system.**

The reticular formation receives and sends on motor and sensory information to various parts of the CNS. One portion, the reticular activating system (RAS; see arrows), arouses the cerebrum and, in this way, controls alertness versus sleep.

### *The Limbic System*

The **limbic system** is a complex network of tracts and nuclei that incorporates medial portions of the cerebral lobes, the basal nuclei, and the diencephalon (Fig. 37.11). The limbic system blends higher mental functions and primitive emotions into a united whole. It accounts for why activities like sexual behavior and eating seem pleasurable and also why, say, mental stress can cause high blood pressure.

Two significant structures within the limbic system are the hippocampus and the amygdala, which are essential for learning and memory. The hippocampus, a seahorse-shaped structure that lies deep in the temporal lobe, is well situated in the brain to make the prefrontal area aware of past experiences stored in sensory association areas. The amygdala, in particular, can cause these experiences to have emotional overtones. For example, the smell of smoke may serve as an alarm to search for fire in the house. The inclusion of the frontal lobe in the limbic system means that reason can keep us from acting out strong feelings.

**Learning and Memory.** **Memory** is the ability to hold a thought in mind or recall events from the past, ranging from a word we learned only yesterday to an early emotional experience that has shaped our lives. Learning takes place when we retain and use past memories.

The prefrontal area in the frontal lobe is active during short-term memory as when we temporarily recall a telephone number. Some telephone numbers go into long-term memory. Think of a telephone number you know by heart, and see if you can bring it to mind without also thinking about the place or person associated with that number. Most likely you cannot, because typically long-term memory is a mixture of what is called semantic memory (numbers, words, etc.) and episodic memory (persons, events, etc.). Skill memory is a type of memory that can exist independent of episodic memory. Skill memory is being able to perform motor activities like riding a bike or playing ice hockey.

What parts of the brain are functioning when you remember something from long ago? The hippocampus gathers our long-term memories, which are stored in bits and pieces throughout the sensory association areas, and makes them available to the frontal lobe. Why are some memories so emotionally charged? The amygdala is responsible for fear conditioning and associating danger with sensory information received from the thalamus and the cortical sensory areas.

Long-term potentiation (LTP) is an enhanced response at synapses seen particularly within the hippocampus. During LTP, glutamate binds to the postsynaptic membrane, and calcium may rush in too fast due to a malformed receptor. The apoptosis that follows may contribute to Alzheimer disease (AD). AD neurons have neurofibrillary tangles (bundles of fibrous protein) surrounding the nucleus and protein-rich accumulations called amyloid plaques enveloping the axon branches. Although it is not yet known how excitotoxicity is related to structural abnormalities of AD neurons, some researchers are trying to develop neuroprotective drugs that can possibly guard brain cells against damage due to glutamate.

**Check Your Progress 37.3**

1. The brain is very dependent on the spinal cord. Explain.
2. The hypothalamus, which has sleep centers, communicates with the RAS. What might cause narcolepsy, the disorder characterized by brief periods of unexpected sleep?
3. Brain injury can cause a disconnect between the amygdala and the portion of the cortex devoted to recognizing faces. People with this ailment can identify the faces of family members, but have no feelings for them. This is so disturbing that sufferers come to believe their "real" families have been replaced with "imposters." Explain.

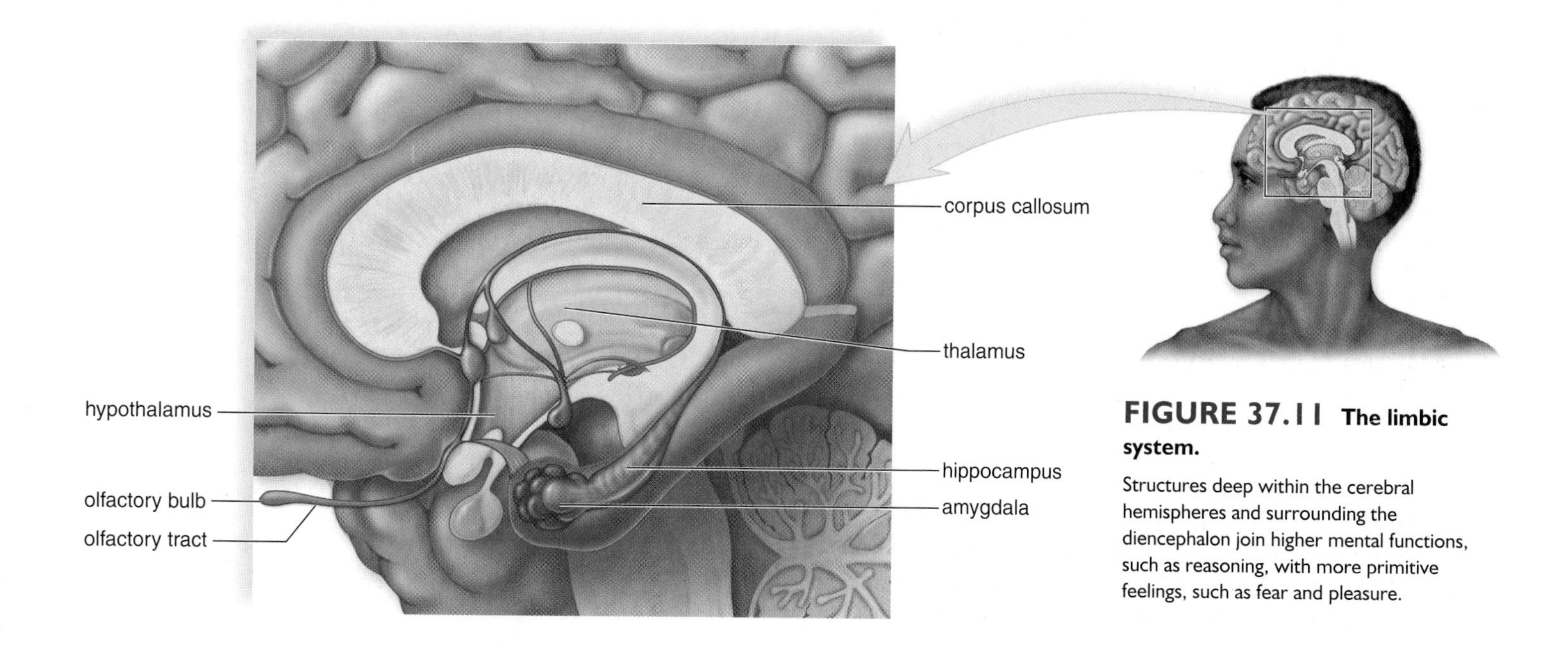


**FIGURE 37.11 The limbic system.**

Structures deep within the cerebral hemispheres and surrounding the diencephalon join higher mental functions, such as reasoning, with more primitive feelings, such as fear and pleasure.

# 37.4 Peripheral Nervous System

The peripheral nervous system (PNS) lies outside the central nervous system and contains **nerves,** which are bundles of axons. Axons that occur in nerves are also called *nerve fibers.* The cell bodies of neurons are found in the CNS and in ganglia, collections of cell bodies outside the CNS.

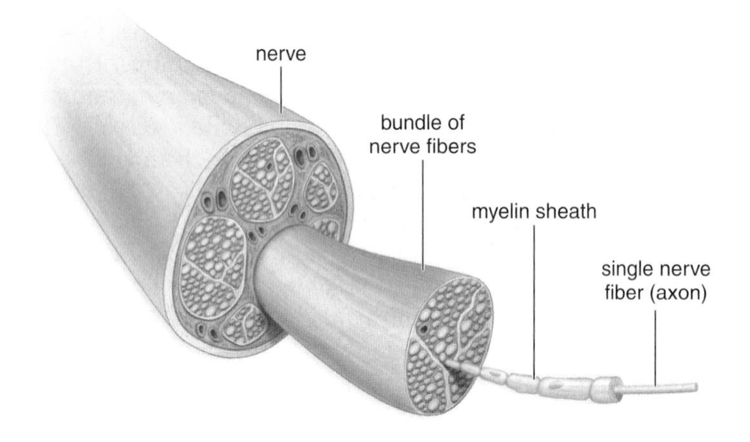


Humans have 12 pairs of **cranial nerves** attached to the brain (Fig. 37.12*a*). Some of these are sensory nerves; that is, they contain only sensory nerve fibers. Some are motor nerves that contain only motor fibers, and others are mixed nerves that contain both sensory and motor fibers. Cranial nerves are largely concerned with the head, neck, and facial regions of the body. However, the vagus nerve has branches not only to the pharynx and larynx but also to most of the internal organs.

Humans have 31 pairs of **spinal nerves** (Figs. 37.12*b* and 37.13). The paired spinal nerves emerge from the spinal cord by two short branches, or roots. The dorsal root contains the axons of sensory neurons, which conduct impulses to the spinal cord from sensory receptors. The cell body of a sensory neuron is in the **dorsal root ganglion.** The ventral root contains the axons of motor neurons, which conduct impulses away from the spinal cord to effectors. These two roots join to form a spinal nerve. All spinal nerves are mixed nerves that contain many sensory and motor fibers. Each spinal nerve serves the particular region of the body in which it is located.

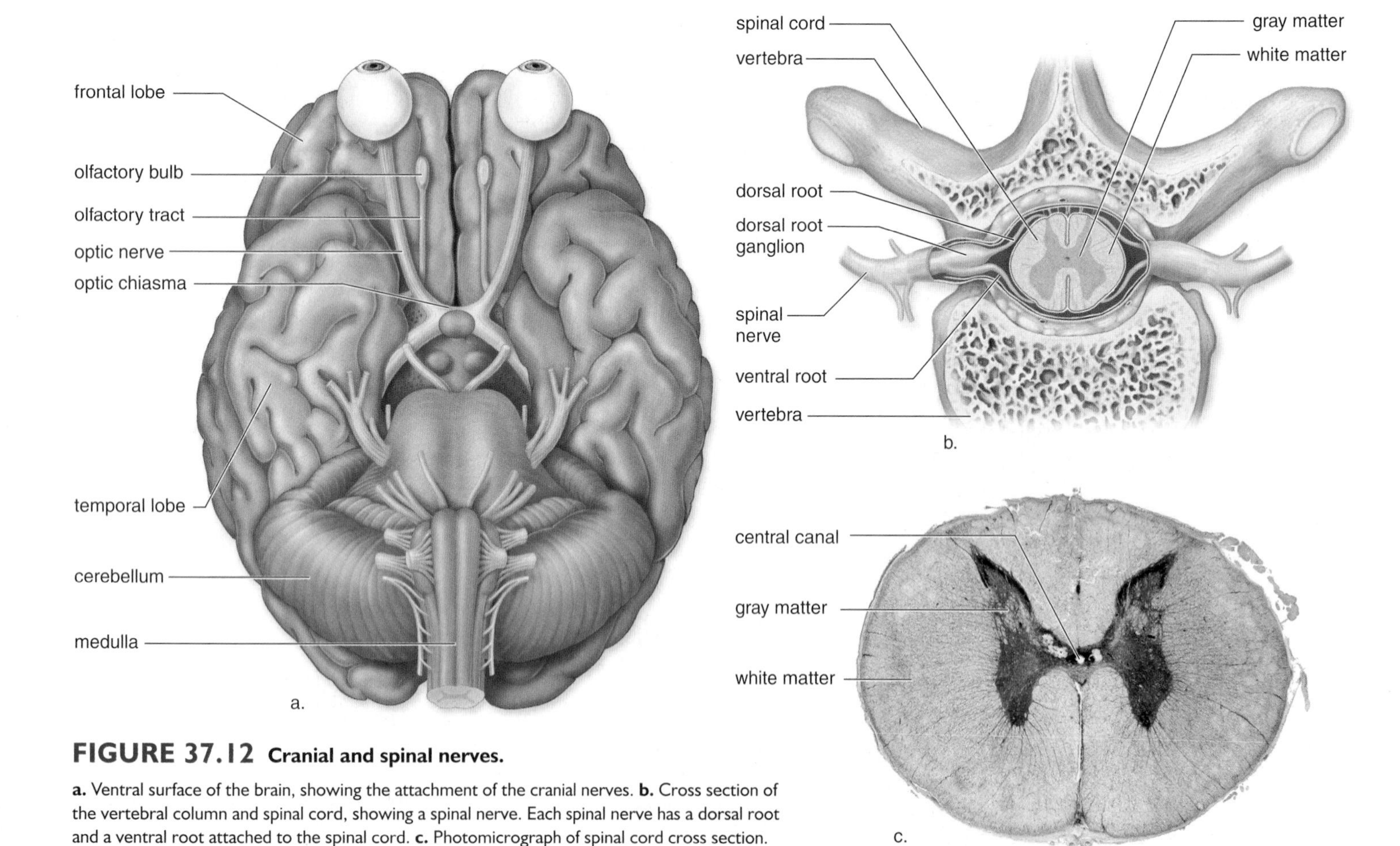


**FIGURE 37.12 Cranial and spinal nerves.**

**a.** Ventral surface of the brain, showing the attachment of the cranial nerves. **b.** Cross section of the vertebral column and spinal cord, showing a spinal nerve. Each spinal nerve has a dorsal root and a ventral root attached to the spinal cord. **c.** Photomicrograph of spinal cord cross section.

## Somatic System

The PNS has two divisions—somatic and autonomic—and we are going to consider the somatic system first. The nerves in the **somatic system** serve the skin, joints, and skeletal muscles. Therefore, the somatic system includes nerves that take (1) sensory information from external sensory receptors in the skin and joints to the CNS, and (2) motor commands away from the CNS to the skeletal muscles. The neurotransmitter acetylcholine (ACh) is active in the somatic system.

Voluntary control of skeletal muscles always originates in the brain. Involuntary responses to stimuli, called reflex actions, can involve either the brain or just the spinal cord. Reflexes enable the body to react swiftly to stimuli that could disrupt homeostasis. Flying objects cause our eyes to blink, and sharp pins cause our hands to jerk away, even without us having to think about it.

**The Reflex Arc.** Figure 37.13 illustrates the path of a reflex that involves only the spinal cord. If your hand touches a sharp pin, **sensory receptors** generate nerve impulses that move along sensory axons through a dorsal root ganglion toward the spinal cord. Sensory neurons that enter the cord dorsally pass signals on to many interneurons in the gray matter of the spinal cord. Some of these interneurons synapse with motor neurons. The short dendrites and the cell bodies of motor neurons are also in the spinal cord, but their axons leave the cord ventrally. Nerve impulses travel along motor axons to an **effector,** which brings about a response to the stimulus. In this case, a muscle contracts so that you withdraw your hand from the pin. (Sometimes an effector is a gland.) Various other reactions are possible—you will most likely look at the pin, wince, and cry out in pain. This whole series of responses is explained by the fact that some of the interneurons in the white matter of the cord carry nerve impulses in tracts to the brain. The brain makes you aware of the stimulus and directs subsequent reactions to the situation. You don't feel pain until the brain receives the information and interprets it! Visual information received directly by way of a cranial nerve may make you aware that your finger is bleeding. Then you might decide to look for a Band-Aid.

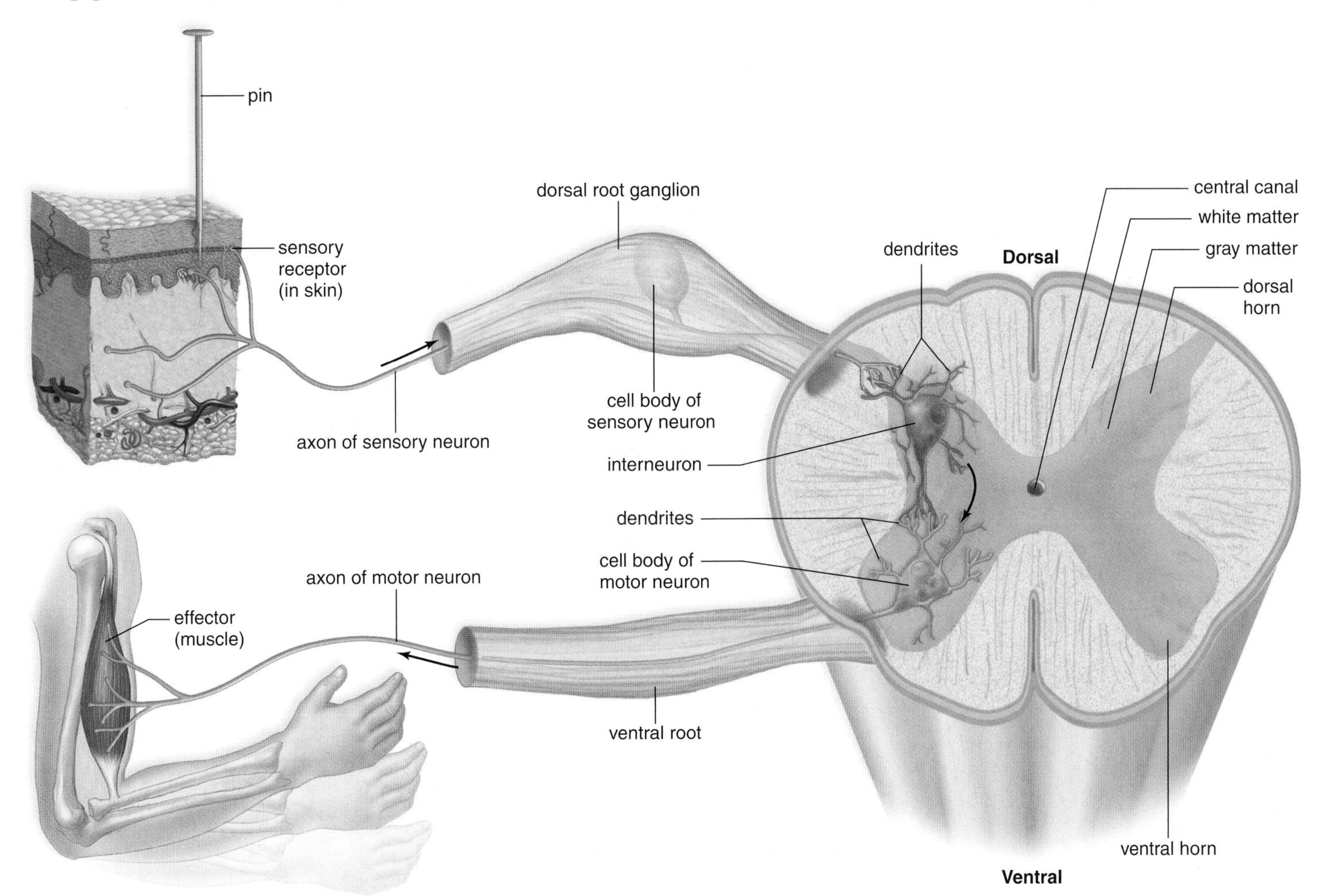


**FIGURE 37.13 A reflex arc showing the path of a spinal reflex.**

A stimulus (e.g., sharp pin) causes sensory receptors in the skin to generate nerve impulses that travel in sensory axons to the spinal cord. Interneurons integrate data from sensory neurons and then relay signals to motor axons. Motor axons convey nerve impulses from the spinal cord to a skeletal muscle, which contracts. Movement of the hand away from the pin is the response to the stimulus.

**FIGURE 37.14 Autonomic system structure and function.**

Sympathetic preganglionic fibers (*left*) arise from the cervical, thoracic, and lumbar portions of the spinal cord; parasympathetic preganglionic fibers (*right*) arise from the cranial and sacral portions of the spinal cord. Each system innervates the same organs but has contrary effects.

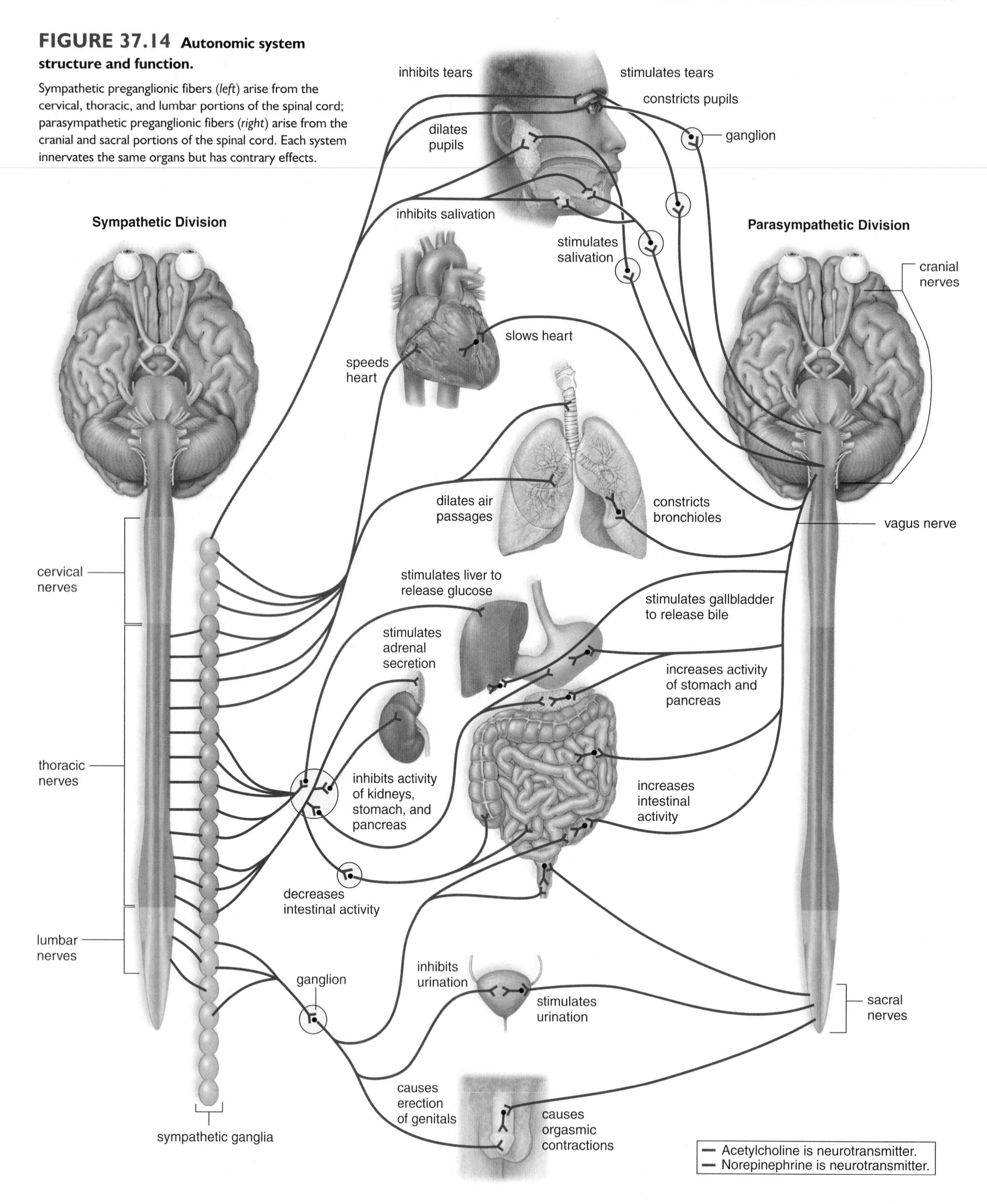

## Autonomic System

The **autonomic system** of the PNS regulates the activity of cardiac and smooth muscle and glands. It carries out its duties without our awareness or intent. The system is divided into the sympathetic and parasympathetic divisions (Fig. 37.14 and Table 37.1). Both of these divisions (1) function automatically and usually in an involuntary manner; (2) innervate all internal organs; and (3) use two neurons and one ganglion for each impulse. The first neuron has a cell body within the CNS and a preganglionic fiber. The second neuron has a cell body within the ganglion and a postganglionic fiber.

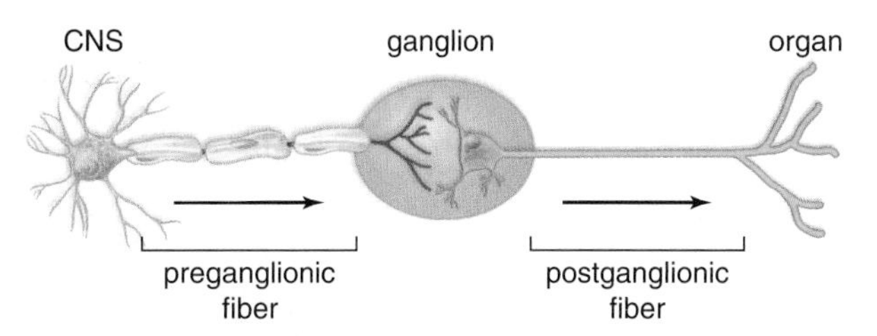


Reflex actions, such as those that regulate blood pressure and breathing rate, are especially important to the maintenance of homeostasis. These reflexes begin when the sensory neurons in contact with internal organs send information to the CNS. They are completed by motor neurons within the autonomic system.

### *Sympathetic Division*

Most preganglionic fibers of the **sympathetic division** arise from the middle, or thoracolumbar, portion of the spinal cord and almost immediately terminate in ganglia that lie near the cord. Therefore, in this division, the preganglionic fiber is short, but the postganglionic fiber that makes contact with an organ is long.

The sympathetic division is especially important during emergency situations and is associated with "fight or flight." If you need to fend off a foe or flee from danger, active muscles require a ready supply of glucose and oxygen. The sympathetic division accelerates the heartbeat and dilates the bronchi. On the other hand, the sympathetic division inhibits the digestive tract, since digestion is not an immediate necessity if you are under attack. The neurotransmitter released by the postganglionic axon is primarily norepinephrine (NE). The structure of NE is like that of epinephrine (adrenaline), an adrenal medulla hormone that usually increases heart rate and contractility.

### *Parasympathetic Division*

The **parasympathetic division** includes a few cranial nerves (e.g., the vagus nerve) and also fibers that arise from the sacral (bottom) portion of the spinal cord. Therefore, this division often is referred to as the craniosacral portion of the autonomic system. In the parasympathetic division, the preganglionic fiber is long, and the postganglionic fiber is short because the ganglia lie near or within the organ.

The parasympathetic division, sometimes called the "housekeeper" or "rest and digest division," promotes all the internal responses we associate with a relaxed state; for example, it causes the pupil of the eye to contract, promotes digestion of food, and slows the heartbeat. The neurotransmitter used by the parasympathetic division is acetylcholine (ACh).

**Check Your Progress 37.4**

1. What are two ways in which cranial nerves and spinal nerves differ from one another?
2. **a.** What part of the CNS is active when a reflex action involving the limbs occurs? **b.** What part of the CNS is active when we override a reflex action and do not react automatically?
3. You are sitting quietly, enjoying a slice of pizza and a soft drink. Your mischievous best friend sneaks up behind you, then dumps ice down the back of your shirt. Describe the shift in autonomic system activity that ensues.

**TABLE 37.1**

**Comparison of Somatic Motor and Autonomic Motor Pathways**

| | *Somatic Motor Pathway* | *Autonomic Motor Pathways* | |
|---|---|---|---|
| | | ***Sympathetic*** | ***Parasympathetic*** |
| **Type of control** | Voluntary/involuntary | Involuntary | Involuntary |
| **Number of neurons per message** | One | Two (preganglionic shorter than postganglionic) | Two (preganglionic longer than postganglionic) |
| **Location of motor fiber** | Most cranial nerves and all spinal nerves | Thoracolumbar spinal nerves | Cranial (e.g., vagus) and sacral spinal nerves |
| **Neurotransmitter** | Acetylcholine | Norepinephrine | Acetylcholine |
| **Effectors** | Skeletal muscles | Smooth and cardiac muscle, glands | Smooth and cardiac muscle, glands |

science focus

# Drugs of Abuse

Drug abuse is apparent when a person takes a drug at a dose level and under circumstances that increase the potential for a harmful effect. Addiction is present when more of the drug is needed to get the same effect, and withdrawal symptoms occur when the user stops taking the drug. This is true not only for teenagers and adults, but also for newborn babies of mothers who abuse and are addicted to drugs. Alcohol, drugs, and tobacco can all adversely affect the developing embryo, fetus, or newborn.

## Alcohol

Alcohol consumption is the most socially accepted form of drug use worldwide. The approximate number of adults who consume alcohol in the United States on a regular basis is 65%. Of those, 5% say they are "heavy drinkers." Notably, 80% of college-age young adults drink. Unfortunately, so-called binge drinking has resulted in the deaths of many college students.

Alcohol (ethanol) acts as a *depressant* on many parts of the brain where it affects neurotransmitter release or uptake. For example, alcohol increases the action of GABA, which inhibits motor neurons, and it also increases the release of endorphins, which are natural painkillers. Depending on the amount consumed, the effects of alcohol on the brain can lead to a feeling of relaxation, lowered inhibitions, impaired concentration and coordination, slurred speech, and vomiting. If the blood level of alcohol becomes too high, coma or death can occur.

Chronic alcohol consumption can damage the frontal lobes, decrease overall brain size, and increase the size of the ventricles. Brain damage is manifested by permanent memory loss, amnesia, confusion, apathy, disorientation, or lack of motor coordination. Prolonged alcohol use can also permanently damage the liver, the major detoxification organ of the body, to the point that a liver transplant may be required.

## Nicotine

When tobacco is smoked, nicotine is rapidly delivered to the CNS, especially the midbrain. There it binds to neurons, causing the release of dopamine, the neurotransmitter that promotes a sense of pleasure and is involved in motor control. In the PNS, nicotine also acts as a *stimulant* by mimicking acetylcholine and increasing heart rate, blood pressure, and muscle activity. Fingers and toes become cold because blood vessels have constricted. Increased digestive tract motility may account for the weight loss sometimes seen in smokers.

The physiologically and psychologically addictive nature of nicotine is well known. The addiction rate of smokers is about 70%. The failure rate in those who try to quit smoking is about 80–90% of smokers. Withdrawal symptoms include irritability, headache, insomnia, poor cognitive performance, the urge to smoke, and weight gain. Ways to quit smoking include applying nicotine skin patches, chewing nicotine gum, or taking oral drugs that block the actions of acetylcholine. The effectiveness of these therapies is variable. An experimental therapy involves "immunizing" the brain of smokers against nicotine. Injections cause the production of antibodies that bind to nicotine and prevent it from entering the brain. The effectiveness of this new therapy is not yet known.

## Club and Date Rape Drugs

Methamphetamine and Ecstasy are considered club or party drugs. Methamphetamine (commonly called meth or speed) is a synthetic drug made by the addition of a methyl group to amphetamine. Because the addition of the methyl group is fairly simple, methamphetamine is often produced from amphetamine in clandestine, makeshift laboratories in homes, motel rooms, or campers. The number of toxic chemicals used to prepare the drug makes a former meth lab site hazardous to humans and to the environment. Over 9 million people in the United States have used methamphetamine at least once in their lifetime. It is available as a powder or as crystals (crystal meth or ice).

The structure of methamphetamine is similar to that of dopamine, and its *stimulatory* effect mimics that of cocaine. It reverses the effects of fatigue, maintains wakefulness, and temporarily elevates the user's mood. The initial rush is typically followed by a state of high agitation that, in some individuals, leads to violent behavior. Chronic use can result in what is called an amphetamine psychosis, characterized by paranoia, auditory and visual hallucinations, self-absorption, irritability, and aggressive, erratic behavior. Excessive intake can lead to hyperthermia, convulsions, and death.

Ecstasy is the street name for MDMA (methylenedioxymethamphetamine), a drug with effects similar to those of methamphetamine. Also referred to as E, X, or the love drug, it is taken as a pill that looks like an aspirin or candy. Many people using Ecstasy believe that it is totally safe if used with lots of water to counter its effect on body temperature. A British teen, Lorna Spinks, died after taking two high-strength Ecstasy pills, which caused her body temperature to rise to a fatal level. Also, Ecstasy has an overstimulatory effect on neurons that produce serotonin, which, like dopamine, elevates our mood. Most of the damage to these neurons can be repaired when the use of Ecstasy is discontinued, but some damage appears to be permanent.

Drugs with sedative effects, known as date rape or predatory drugs, include Rohypnol (roofies), gamma-hydroxybutyric acid (GHB), and ketamine (special K). These drugs can be given to an unsuspecting person, who then becomes vulnerable to sexual assault after the drug takes effect. Relaxation, amnesia, and disorientation occur after taking these drugs, which are popular at clubs or raves because they enhance the effect of heroin and Ecstasy.

## Cocaine

Cocaine is an alkaloid derived from the shrub *Erythroxylon coca.* Approximately 35 million Americans have used cocaine by sniffing/snorting, injecting, or smoking. Cocaine is a powerful *stimulant* in the CNS that interferes with the re-uptake of dopamine at synapses. The result is a rush of well-being that lasts from 5 to 30 minutes. People on cocaine sprees (or binges) take the drug repeatedly and at ever-higher doses. The result is sleeplessness, lack of appetite, increased sex drive, tremors, and "cocaine psychosis," a condition that resembles paranoid schizophrenia. During the crash period, fatigue, depression, and irritability are common, along with memory loss and confused thinking.

"Crack" is the street name given to cocaine that is processed to a free base for smoking. The term *crack* refers to the crackling sound heard when smoking. Smoking allows extremely high doses of the drug to reach the brain rapidly, providing an intense and immediate high, or "rush." Approximately 8 million Americans use crack. Long-term use is expected to cause brain damage (Fig. 37A).

Cocaine is highly addictive; related deaths are usually due to cardiac and/or respiratory arrest. The combination of cocaine and alcohol dramatically increases the risk of sudden death.

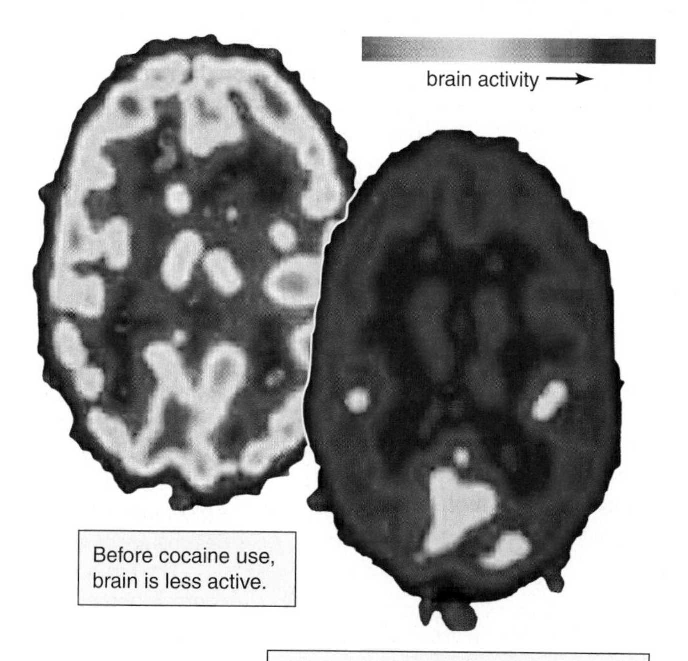


**FIGURE 37A**
**Drug use.**
*Brain activity before and after the use of cocaine.*

## Heroin

Heroin is derived from the resin or sap of the opium poppy plant, which is widely grown—from Turkey to Southeast Asia and in parts of Latin America. Heroin is a highly addictive drug that acts as a *depressant* in the nervous system. Drugs derived from opium are called opiates, a class that also includes morphine and codeine, both of which have painkilling effects.

Heroin is the most abused opiate—it travels rapidly to the brain, where it is converted to morphine, and the result is a rush sensation and a feeling of euphoria. Opiates depress breathing, block pain pathways, cloud mental function, and sometimes cause nausea and vomiting. Long-term effects of heroin use are addiction, hepatitis, HIV/AIDS, and various bacterial infections due to the use of shared needles. As with other drugs of abuse, addiction is common, and heavy users may experience convulsions and death by respiratory arrest.

Heroin can be injected, snorted, or smoked. Abusers typically inject heroin up to four times a day. It is estimated that 4 million Americans have used heroin some time in their lives, and over 300,000 people use heroin annually.

## Marijuana

The dried flowering tops, leaves, and stems of the Indian hemp plant, *Cannabis sativa,* contain and are covered by a resin that is rich in THC (tetrahydrocannabinol). The names *cannabis* and *marijuana* apply to either the plant or THC. Marijuana can be consumed, but usually it is smoked in a cigarette called a "joint." An estimated 22 million Americans use marijuana. Although the drug was banned in the United States in 1937, several states have legalized its use for medical purposes, such as lessening the effects of chemotherapy.

It seems that THC may mimic the actions of anandamide, a neurotransmitter that was recently discovered. Both THC and anandamide belong to a class of chemicals called cannabinoids. Receptors that bind cannabinoids are located in the hippocampus, cerebellum, basal nuclei, and cerebral cortex, brain areas that are important for memory, orientation, balance, motor coordination, and perception.

When THC reaches the CNS, the person experiences mild euphoria, along with alterations in vision and judgment. Distortions of space and time can also occur in occasional users. In heavy users, hallucinations, anxiety, depression, rapid flow of ideas, body image distortions, paranoia, and psychotic symptoms can result. The terms cannabis psychosis and cannabis delirium describe such reactions to marijuana's influence on the brain. Regular usage of marijuana can cause cravings that make it difficult to stop.

## Treatment for Addictive Drugs

Presently, treatment for addiction to drugs consists mainly of behavior modification. Heroin addiction can be treated with synthetic opiate compounds, such as methadone or suboxone, that decrease withdrawal symptoms and block heroin's effects. Unfortunately, inappropriate methadone use can be dangerous, as demonstrated by celebrity deaths associated with methadone overdose or taking methadone along with other drugs.

New treatment techniques include the administration of antibodies to block the effects of cocaine and methamphetamine. These antibodies would make relapses by former drug abusers impossible and could be used to treat overdoses. A vaccine for cocaine that would stimulate antibody production is being tested.

# Connecting the Concepts

Like the wiring of a modern office building, the peripheral nervous system of humans contains nerves that reach to all parts of the body. There is a division of labor among the nerves. The cranial nerves serve the head region, with the exception of the vagus nerve. All body movements are controlled by spinal nerves, and this is why paralysis may follow a spinal injury. Except for the vagus nerve, only spinal nerves make up the autonomic system, which controls the internal organs. As in most other animals, much of the work of the nervous system in humans is below the level of consciousness.

The nervous system has just three functions: sensory input, integration, and motor output. Sensory input would be impossible without sensory receptors, which are sensitive to external and internal stimuli. You might even argue that sense organs like the eyes and ears should be considered a part of the nervous system, since there would be no sensory nerve impulses without their ability to generate them.

Nerve impulses are the same in all neurons, so how is it that stimulation of eyes causes us to see, and stimulation of ears causes us to hear? Essentially, the central nervous system carries out the function of integrating incoming data. The brain allows us to perceive our environment, reason, and remember. After sensory data have been processed by the CNS, motor output occurs. Muscles and glands are the effectors that allow us to respond to the original stimuli. Without the musculoskeletal system, we would never be able to respond to a danger detected by our eyes and ears.

## summary

### 37.1 Evolution of the Nervous System

A comparative study of the invertebrates shows a gradual increase in the complexity of the nervous system. The vertebrate nervous system, like that of the earthworm, is divided into the central and peripheral nervous systems.

### 37.2 Nervous Tissue

The anatomical unit of the nervous system is the neuron, of which there are three types: sensory, motor, and interneuron. Each of these is made up of a cell body, an axon, and dendrites.

When an axon is not conducting an action potential (nerve impulse), the resting potential indicates that the inside of the fiber is negative compared to the outside. The sodium-potassium pump helps maintain a concentration of $Na^+$ outside the fiber and $K^+$ inside the fiber. When the axon is conducting a nerve impulse, an action potential (i.e., a change in membrane potential) travels along the fiber. Depolarization occurs (inside becomes positive) due to the movement of $Na^+$ to the inside, and then repolarization occurs (inside becomes negative again) due to the movement of $K^+$ to the outside of the fiber.

Transmission of the nerve impulse from one neuron to another takes place across a synapse. In humans, synaptic vesicles release a chemical, known as a neurotransmitter, into the synaptic cleft. The binding of neurotransmitters to receptors in the postsynaptic membrane can either increase the chance of an action potential (stimulation) or decrease the chance of an action potential (inhibition) in the next neuron. A neuron usually transmits several nerve impulses, one after the other.

### 37.3 Central Nervous System: Brain and Spinal Cord

The CNS consists of the spinal cord and brain, which are both protected by bone. The CNS receives and integrates sensory input and formulates motor output. The gray matter of the spinal cord contains neuron cell bodies; the white matter consists of myelinated axons that occur in bundles called tracts. The spinal cord sends sensory information to the brain, receives motor output from the brain, and carries out reflex actions.

In the brain, the cerebrum has two cerebral hemispheres connected by the corpus callosum. Sensation, reasoning, learning and memory, and language and speech take place in the cerebrum. The cerebral cortex is a thin layer of gray matter covering the cerebrum. The cerebral cortex of each cerebral hemisphere has four lobes: a frontal, parietal, occipital, and temporal lobe. The primary motor area in the frontal lobe sends out motor commands to lower brain centers, which pass them on to motor neurons. The primary somatosensory area in the parietal lobe receives sensory information from lower brain centers in communication with sensory neurons. Association areas for vision are in the occipital lobe, and those for hearing are in the temporal lobe.

The brain has a number of other regions. The hypothalamus controls homeostasis, and the thalamus specializes in sending sensory input on to the cerebrum. The cerebellum primarily coordinates skeletal muscle contractions. The medulla oblongata and the pons have centers for vital functions such as breathing and the heartbeat.

### 37.4 Peripheral Nervous System

The peripheral nervous system contains the somatic system and the autonomic system. Reflexes are automatic, and some do not require involvement of the brain. A simple reflex requires the use of neurons that make up a reflex arc. In the somatic system, a sensory neuron conducts nerve impulses from a sensory receptor to an interneuron, which in turn transmits impulses to a motor neuron, which stimulates an effector to react.

While the motor portion of the somatic system of the PNS controls skeletal muscle, the motor portion of the autonomic system controls smooth muscle of the internal organs and glands. The sympathetic division, which is often associated with reactions that occur during times of stress, and the parasympathetic division, which is often associated with activities that occur during times of relaxation, are both parts of the autonomic system.

## understanding the terms

Match the terms to these definitions:

a. ______________ Automatic, involuntary responses of an organism to a stimulus.
b. ______________ Chemical stored at the ends of axons that is responsible for transmission across a synapse.
c. ______________ System within the peripheral nervous system that regulates internal organs.
d. ______________ Collection of neuron cell bodies usually outside the central nervous system.
e. ______________ Neurotransmitter active in the somatic system of the peripheral nervous system.

## reviewing this chapter

1. Trace the evolution of the nervous system by contrasting its organization in hydras, planarians, earthworms, and humans. 680–82
2. Describe the structure of a neuron, and give a function for each part mentioned. Name three types of neurons, and give a function for each. 683
3. What are the major events of an action potential, and what ion changes are associated with each event? 685
4. Describe the mode of action of a neurotransmitter at a synapse, including how it is stored and how it is destroyed. 686–87
5. Name the major parts of the human brain, and give a principal function for each part. 688–90
6. Describe the limbic system, and discuss its possible involvement in learning and memory. 691
7. Discuss the structure and function of the peripheral nervous system. 692
8. Trace the path of a spinal reflex. 693
9. Contrast the sympathetic and parasympathetic divisions of the autonomic system. 694–95

## testing yourself

Choose the best answer for each question.

1. Which is the most complete list of animals that have a central nervous system (CNS) and a peripheral nervous system (PNS)?
   a. hydra, planarian, earthworm, rabbit, human
   b. planarian, earthworm, rabbit, human
   c. earthworm, rabbit, human
   d. rabbit, human
2. Which of these are the first and last elements in a spinal reflex?
   a. axon and dendrite
   b. sense organ and muscle effector
   c. ventral horn and dorsal horn
   d. motor neuron and sensory neuron
   e. sensory receptor and the brain
3. A spinal nerve takes nerve impulses
   a. to the CNS.
   b. away from the CNS.
   c. both to and away from the CNS.
   d. only inside the CNS.
   e. only from the cerebrum.
4. Which of these correctly describes the distribution of ions on either side of an axon when it is not conducting a nerve impulse?
   a. more sodium ions ($Na^+$) outside and fewer potassium ions ($K^+$) inside
   b. $K^+$ outside and $Na^+$ inside
   c. charged proteins outside; $Na^+$ and $K^+$ inside
   d. charged proteins inside
   e. Both a and d are correct.
5. When the action potential begins, sodium gates open, allowing $Na^+$ to cross the membrane. Now the polarity changes to
   a. negative outside and positive inside.
   b. positive outside and negative inside.
   c. There is no difference in charge between outside and inside.
   d. Any one of these could be correct.
6. Transmission of the nerve impulse across a synapse is accomplished by
   a. the release of $Na^+$ at the presynaptic membrane.
   b. the release of neurotransmitters at the postsynaptic membrane.
   c. the reception of neurotransmitters at the postsynaptic membrane.
   d. Only a and c are correct.
7. The autonomic system has two divisions, called the
   a. CNS and PNS.
   b. somatic and skeletal systems.
   c. efferent and afferent systems.
   d. sympathetic and parasympathetic divisions.
8. Synaptic vesicles are
   a. at the ends of dendrites and axons.
   b. at the ends of axons only.
   c. along the length of all long fibers.
   d. at the ends of interneurons only.
   e. Both b and d are correct.
9. Which of these pairs is mismatched?
   a. cerebrum—thinking and memory
   b. thalamus—motor and sensory centers
   c. hypothalamus—internal environment regulator
   d. cerebellum—motor coordination
   e. medulla oblongata—fourth ventricle

10. Repolarization of an axon during an action potential is produced by
    a. inward diffusion of $Na^+$.
    b. active extrusion of $K^+$.
    c. outward diffusion of $K^+$.
    d. inward active transport of $Na^+$.
11. Which two parts of the brain are least likely to work directly together?
    a. thalamus and cerebrum
    b. cerebrum and cerebellum
    c. hypothalamus and medulla oblongata
    d. cerebellum and medulla oblongata
12. Which of the following is not part of the spinal cord?
    a. central canal
    b. dorsal horn
    c. association areas
    d. tracts
    e. ventral horn
13. A drug that inactivates acetylcholinesterase
    a. stops the release of ACh from presynaptic endings.
    b. prevents the attachment of ACh to its receptor.
    c. increases the ability of ACh to stimulate postsynaptic cells.
    d. All of these are correct.
14. Which of these statements about autonomic neurons is correct?
    a. They are motor neurons.
    b. Preganglionic neurons have cell bodies in the CNS.
    c. Postganglionic neurons innervate smooth muscles, cardiac muscle, and glands.
    d. All of these are correct.
15. Which of these fibers release norepinephrine?
    a. preganglionic sympathetic axons
    b. postganglionic sympathetic axons
    c. preganglionic parasympathetic axons
    d. postganglionic parasympathetic axons
16. Sympathetic nerve stimulation does not cause
    a. the liver to release glycogen.
    b. dilation of bronchioles.
    c. the gastrointestinal tract to digest food.
    d. an increase in the heart rate.
17. The limbic system
    a. involves portions of the cerebral lobes, basal nuclei, and the diencephalon.
    b. is responsible for our deepest emotions, including pleasure, rage, and fear.
    c. is not responsible for reason and self-control.
    d. All of these are correct.
18. Which of these would be covered by a myelin sheath?
    a. short dendrites
    b. globular cell bodies
    c. long axons
    d. interneurons
    e. All of these are correct.
19. Label this diagram of a reflex arc.

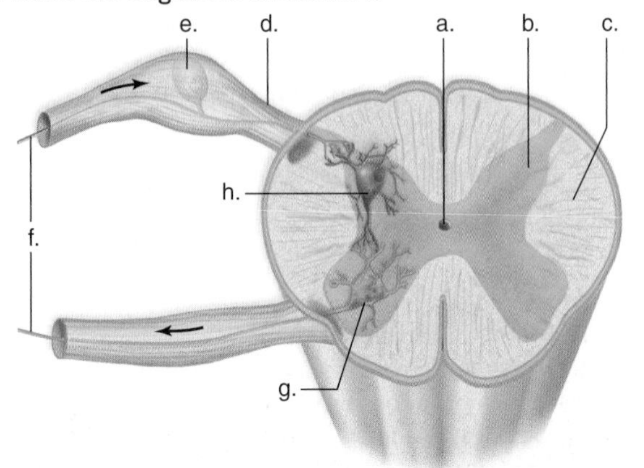


## thinking scientifically

1. In individuals with panic disorder, the fight-or-flight response is activated by inappropriate stimuli. How might it be possible to directly control this response in order to treat panic disorder? Why is such control often impractical?
2. A man who lost his leg several years ago continues to experience pain as though it were coming from the missing limb. What hypothesis could explain the neurological basis of this pain?

## bioethical issue

### The Terri Schiavo Case

On March 31, 2005, news outlets across the country were dominated by the death of a 41-year-old Florida woman named Terri Schiavo. Mrs. Schiavo had required a feeding tube to provide her with fluid and nutrients for 15 years, ever since cardiac arrest deprived her brain of oxygen, causing what doctors diagnosed as a permanent vegetative state (PVS). She died about two weeks after her feeding tube was withdrawn under court order.

PVS is different from brain death, in which all brain function is destroyed. A brain-dead patient lacks the reflexes needed to breathe and maintain blood pressure, and therefore needs more extensive life-support than a PVS patient. A person in a PVS is not aware of his or her surroundings and is unable to communicate or respond to commands. However, the individual can usually breathe without a ventilator, make spontaneous movements, respond to stimuli with involuntary reflexes, open and close his or her eyes, and sometimes even laugh or cry. PVS is caused by prolonged oxygen deprivation of the brain, and is diagnosed if a completely unconscious individual remains so for several months. The longer a person is in a vegetative state, the less likely it is that he or she will ever recover.

The decision to remove Terri Schiavo's feeding tube was sought by Mrs. Schiavo's husband, who stated that his wife would not choose to exist in such a condition. However, his decision met with strong opposition from her parents, who insisted their daughter could be in an ongoing, minimally conscious state (MCS). There is evidence that individuals in a MCS may be aware, but are severely limited in their ability to respond to stimuli and communicate. Even if Mrs. Schiavo could never regain full consciousness, her parents maintained their desire to continue care and support for her.

Although an autopsy following Terri Schiavo's death showed that her brain was less than half the mass that it should have been, it was impossible to determine if she was in a PVS or MCS state. The Terri Schiavo case raises a multitude of difficult questions. What evidence should a family require before making the decision to cease all life-sustaining measures for a loved one? Does it make a difference if the patient is in a MCS versus a PVS? Who should bear the cost of diagnosing and caring for patients with long-term loss of consciousness? Should everyone sign a document or carry a card stating his or her wishes in the event of a traumatic brain injury?

## *Biology* website

The companion website for *Biology* provides a wealth of information organized and integrated by chapter. You will find practice tests, animations, videos, and much more that will complement your learning and understanding of general biology.

**http://www.mhhe.com/maderbiology10**

# 38 Sense Organs

## concepts

**38.1 CHEMICAL SENSES**

- Chemoreceptors are almost universally found in animals for sensing chemical substances in food and air. 702
- Human taste buds and olfactory cells are chemoreceptors that respond to chemicals in food and air, respectively. 702–3

**38.2 SENSE OF VISION**

- The eye of arthropods is a compound eye made up of many individual units; the human eye is a camera-type eye with a single lens. Photoreceptors contain visual pigments that respond to light rays. In humans, synaptic integration and processing begin in the retina before nerve impulses are sent to the brain. 704–8

**38.3 SENSES OF HEARING AND BALANCE**

- The inner ear of humans contains mechanoreceptors for hearing and for a sense of balance. The mechanoreceptors for hearing are hair cells in the cochlea, and the mechanoreceptors for balance are hair cells in the vestibule and semicircular canals. 709–13

*Dogs, especially bloodhounds, have an astonishing sense of smell and can detect drugs, explosives, human remains, termites, gas leaks, and much more. Dogs have a better sense of smell than humans because they have approximately 40 times more olfactory cells in their nasal cavities. Upon stimulation, olfactory cells generate nerve impulses that travel to the brain where they are interpreted. Impulses arriving at a particular sensory area of the brain can be interpreted in only one way; for example, those arriving at the olfactory area only result in smell sensations.*

*Dogs and other animals, including humans, have a variety of sensory receptors. Chemoreceptors are sensitive to the presence of particular molecules, photoreceptors are sensitive to light, and mechanoreceptors detect touch and pressure. Sensory receptors play a significant role in homeostasis because they gather the information that allows the brain to make decisions about finding prey, escaping a predator, and any number of other adaptative behaviors. In this chapter, you will learn how sense organs work, and how they enable animals to survive in changing environments.*

Once a bloodhound is given an item to "scent" (inset), it can follow a trail that is hours—or even days—old.

## 38.1 Chemical Senses

The sensory receptors responsible for taste and smell are termed **chemoreceptors** [Gk. *chemo,* pertaining to chemicals; L. *receptor,* receiver] because they are sensitive to certain chemical substances in food, including liquids, and air. Chemoreception is found almost universally in animals and is, therefore, believed to be the most primitive sense.

The location and sensitivity of chemoreceptors vary throughout the animal kingdom. They are important in finding food, locating a mate, and detecting potentially dangerous chemicals in the environment. Although chemoreceptors are present throughout the body of planarians, they are concentrated in the auricles located on the sides of the head. Insects, crustaceans, and other arthropods possess a number of chemoreceptors. In the housefly, chemoreceptors are located primarily on the feet. A fly literally tastes with its feet instead of its mouth. Insects also detect airborne pheromones, which are chemical messages passed between individuals. In crustaceans such as lobsters and crabs, chemoreceptors are widely distributed in their appendages and antennae. In vertebrates such as amphibians, chemoreceptors are located in the nose, mouth, and skin. Snakes possess Jacobsen's organs, a pair of sensory pitlike organs located in the roof of the mouth. When a snake flicks its forked tongue, scent molecules are carried to the Jacobsen's organs and sensory information is transmitted to the brain for interpretation. In mammals, the receptors for taste are located in the mouth, and the receptors for smell are located in the nose.

### Sense of Taste

In adult humans, approximately 3,000 **taste buds** are located primarily on the tongue (Fig. 38.1). Many taste buds lie along the walls of the papillae, the small elevations on the tongue that are visible to the unaided eye. Isolated taste buds are also present on the hard palate, the pharynx, and the epiglottis.

Taste buds open at a taste pore. Taste buds have supporting cells and a number of elongated taste cells that end in microvilli. The microvilli, which project into the taste pore, bear receptor proteins for certain molecules. When molecules bind to receptor proteins, nerve impulses are generated in associated sensory nerve fibers. These nerve impulses go to the brain, including cortical areas that interpret them as tastes.

There are five primary types of taste: sweet, sour, salty, bitter, and umami (Japanese savory, delicious). Foods rich in certain amino acids, such as the common seasoning monosodium glutamate (MSG), as well as certain flavors of cheese, beef broth, and some seafood, produce the taste of umami. Taste buds for each of these tastes are located throughout the tongue, although certain regions may be slightly more sensitive to particular tastes. A particular food can stimulate more than one of these types of taste buds. In this way, the response of taste buds can result in a range of sweet, sour, salty, bitter, and umami tastes. The brain appears to survey the overall pattern of incoming sensory impulses and to take a "weighted average" of their taste messages as the perceived taste.

### Sense of Smell

In humans, the sense of smell, or olfaction, is dependent on between 10 and 20 million **olfactory cells.** These structures are located within olfactory epithelium high in the roof of the nasal cavity (Fig. 38.2). Olfactory cells are modified neurons. Each cell ends in a tuft of about five olfactory cilia that bear receptor proteins for odor molecules. Each olfactory cell has only 1 out of 1,000 different types of receptor proteins. Nerve fibers from like olfactory cells lead to the same neuron in the olfactory

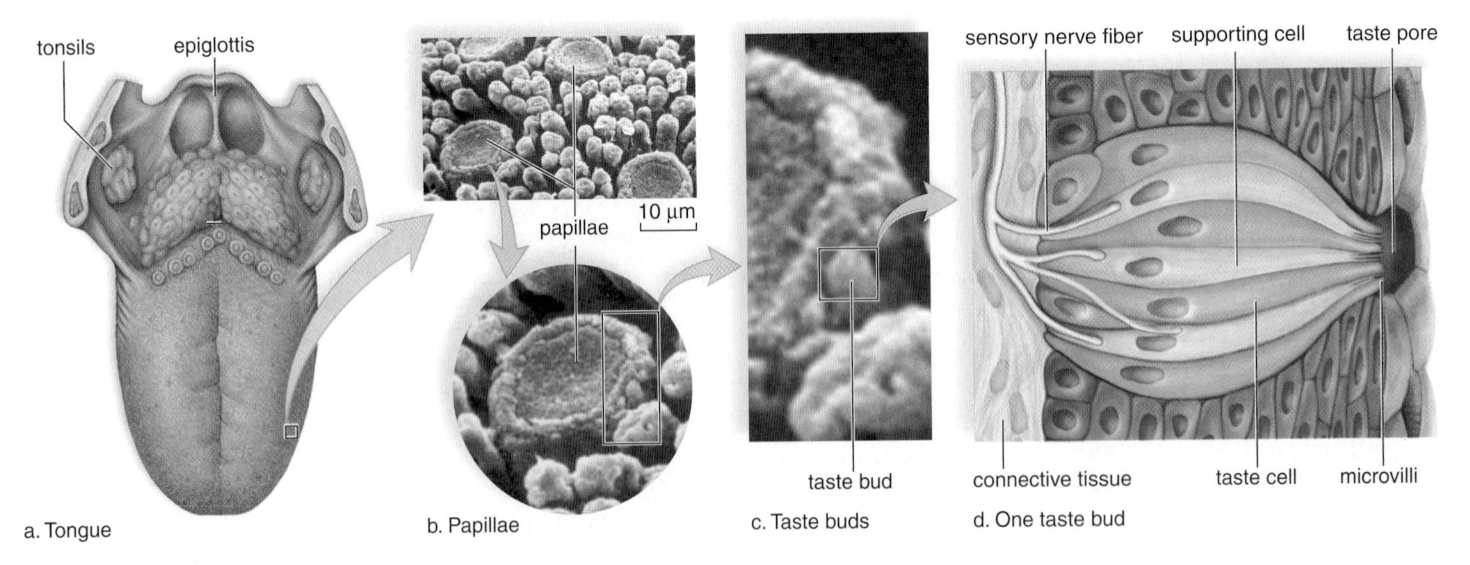


**FIGURE 38.1 Taste buds in humans.**

**a.** Papillae on the tongue contain taste buds that are sensitive to sweet, sour, salty, bitter, and umami. **b.** Photomicrograph and enlargement of papillae. **c.** Taste buds occur along the walls of the papillae. **d.** Taste cells end in microvilli that bear receptor proteins for certain molecules. When molecules bind to the receptor proteins, nerve impulses are generated and go to the brain, where the sensation of taste occurs.

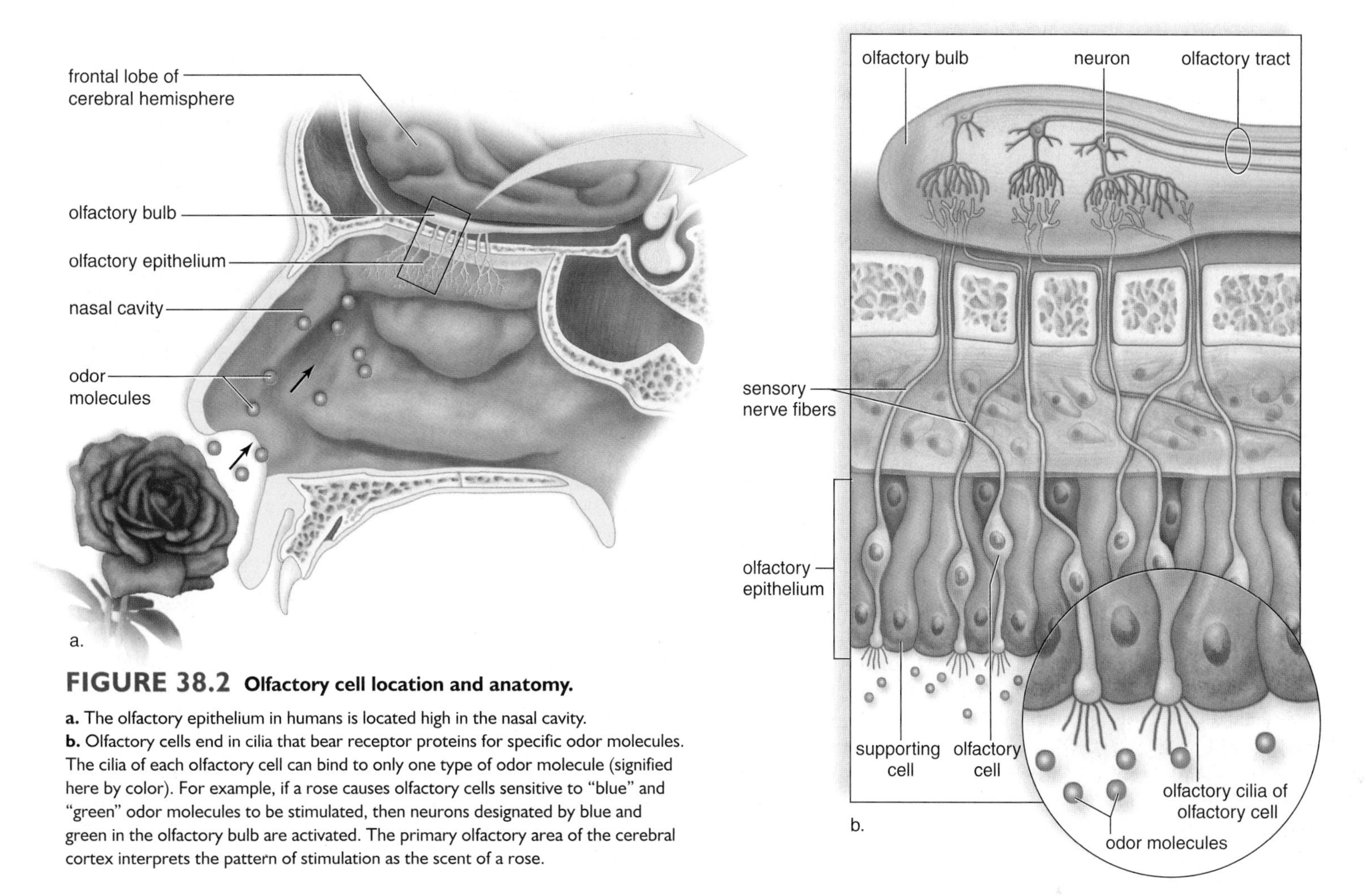


**FIGURE 38.2 Olfactory cell location and anatomy.**

**a.** The olfactory epithelium in humans is located high in the nasal cavity.
**b.** Olfactory cells end in cilia that bear receptor proteins for specific odor molecules. The cilia of each olfactory cell can bind to only one type of odor molecule (signified here by color). For example, if a rose causes olfactory cells sensitive to "blue" and "green" odor molecules to be stimulated, then neurons designated by blue and green in the olfactory bulb are activated. The primary olfactory area of the cerebral cortex interprets the pattern of stimulation as the scent of a rose.

bulb, an extension of the brain. An odor contains many odor molecules that activate a characteristic combination of receptor proteins. A rose might stimulate olfactory cells, designated by blue and green in Figure 38.2, while a gardenia might stimulate a different combination. An odor's signature in the olfactory bulb is determined by which neurons are stimulated. When the neurons communicate this information via the olfactory tract to the olfactory areas of the cerebral cortex, we know we have smelled a rose or a gardenia.

Have you ever noticed that a certain aroma vividly brings to mind a certain person or place? A whiff of perfume may remind you of a specific person, or the smell of boxwood may remind you of your grandfather's farm. The olfactory bulbs have direct connections with the limbic system and its centers for emotions and memory. One investigator showed that when subjects smelled an orange while viewing a painting, they not only remembered the painting when asked about it later, they had many deep feelings about the painting.

The number of olfactory cells declines with age and the remaining population of receptors becomes less sensitive. Thus, older people tend to apply excessive amounts of perfume or aftershave to detect its smell. The ability to smell can also be lost as the result of head trauma, respiratory infection, or brain disease. This can become dangerous if these individuals cannot smell spoiled food, smoke, or a gas leak.

Usually, the sense of taste and the sense of smell work together to create a combined effect when interpreted by the cerebral cortex. For example, when you have a cold, you think food has lost its taste, but most likely you have lost the ability to sense its smell. This method works in reverse also. When you smell something, some of the molecules move from the nose down into the mouth region and stimulate the taste buds there. Therefore, part of what we refer to as smell may in fact be taste.

### Check Your Progress 38.1

1. How are the senses of smell and taste similar to one another?
2. List the five types of taste.
3. The sense of smell has been called the most direct of the senses. Examine Figure 38.2 again, and explain the reason why.

# 38.2 Sense of Vision

**Photoreceptors** [Gk. *photos,* light; L. *receptor,* receiver] are sensory receptors that are sensitive to light. Some animals lack photoreceptors and depend on senses such as smelling and hearing instead; other animals have photoreceptors but live in environments that do not require them. For example, moles live underground and use their senses of smell and touch rather than eyesight.

Not all photoreceptors form images. The "eyespots" of planarians allow these animals to determine the direction of light. Image-forming eyes are found among four invertebrate groups: cnidarians, annelids, molluscs, and arthropods. Arthropods have **compound eyes** composed of many independent visual units called ommatidia [Gk. *ommation,* dim. of *omma,* eye], each possessing all the elements needed for light reception (Fig. 38.3). Both the cornea and crystalline cone function as lenses to direct light rays toward the photoreceptors. The photoreceptors generate nerve impulses, which pass to the brain by way of optic nerve fibers. The outer pigment cells absorb stray light rays so that the rays do not pass from one visual unit to the other. The image that results from all the stimulated visual units is crude because the small size of compound eyes limits the number of visual units, which still might number as many as 28,000. How arthropod brains integrate images from the compound eye to perceive objects is not known.

Insects have color vision, but they make use of a slightly shorter range of the electromagnetic spectrum compared to humans. However, they can see the longest of the ultraviolet rays, and this enables them to be especially sensitive to the particular parts of flowers such as nectar guides that have particular ultraviolet patterns (Fig. 38.4). Some fishes, all reptiles, and most birds are believed to have color vision, but among mammals, only humans and other primates have color vision. It would seem, then, that this trait was adaptive for a diurnal habit (active during the day), which accounts for its retention in a few mammals.

Vertebrates (including humans) and certain molluscs, such as the squid and the octopus, have a **camera-type eye.** Since molluscs and vertebrates are not closely related, this similarity is an example of convergent evolution. A single lens focuses an image of the visual field on photoreceptors, which are closely packed together. In vertebrates, the lens changes shape to aid focusing, but in molluscs the lens moves back and forth. All of the photoreceptors taken together can be compared to a piece of film in a camera. The human eye is more complex than a camera, however, as we shall see.

**FIGURE 38.4 Nectar guides.**

Evening primrose, *Oenothera,* as seen by humans (*left*) and insects (*right*). Humans see no markings, but insects see distinct blotches because their eyes respond to ultraviolet rays. These types of markings, known as nectar guides, often highlight the reproductive parts of flowers, where insects feed on nectar and pick up pollen at the same time.

Animals with two eyes facing forward have three-dimensional, or **stereoscopic, vision.** The visual fields overlap and each eye is able to view an object from a different angle. Predators tend to have stereoscopic vision and so do humans. Animals with eyes facing sideways, such as rabbits, don't have stereoscopic vision, but they do have **panoramic vision,** meaning that the visual field is very wide. Panoramic vision is useful to prey animals because it makes it more difficult for a predator to sneak up on them.

## The Human Eye

The most important parts of the human eye and their functions are listed in Table 38.1. The human eye, which is an elongated sphere about 2.5 cm in diameter, has three layers, or coats: the sclera, the choroid, and the retina (Fig. 38.5). The outer layer, the **sclera** [Gk. *skleros,* hard], is an opaque, white, fibrous layer that covers most of the eye; in front of the

**FIGURE 38.3 Compound eye.**

Each visual unit of a compound eye has a cornea and a lens that focus light onto photoreceptors. The photoreceptors generate nerve impulses that are transmitted to the brain, where interpretation produces a mosaic image.

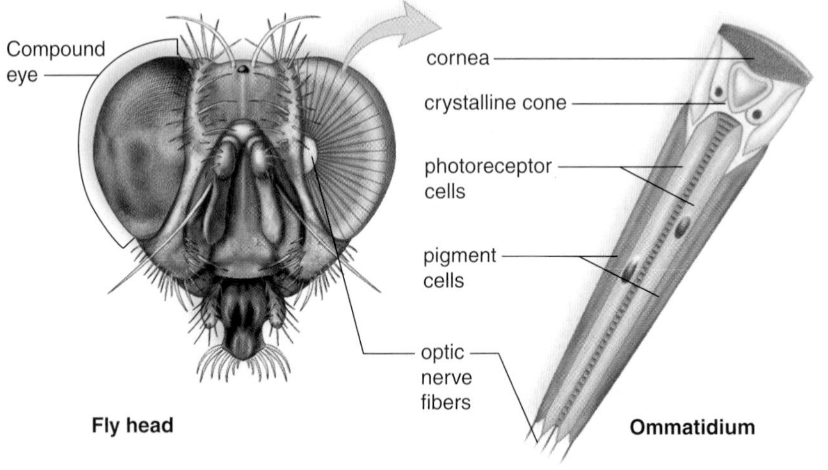

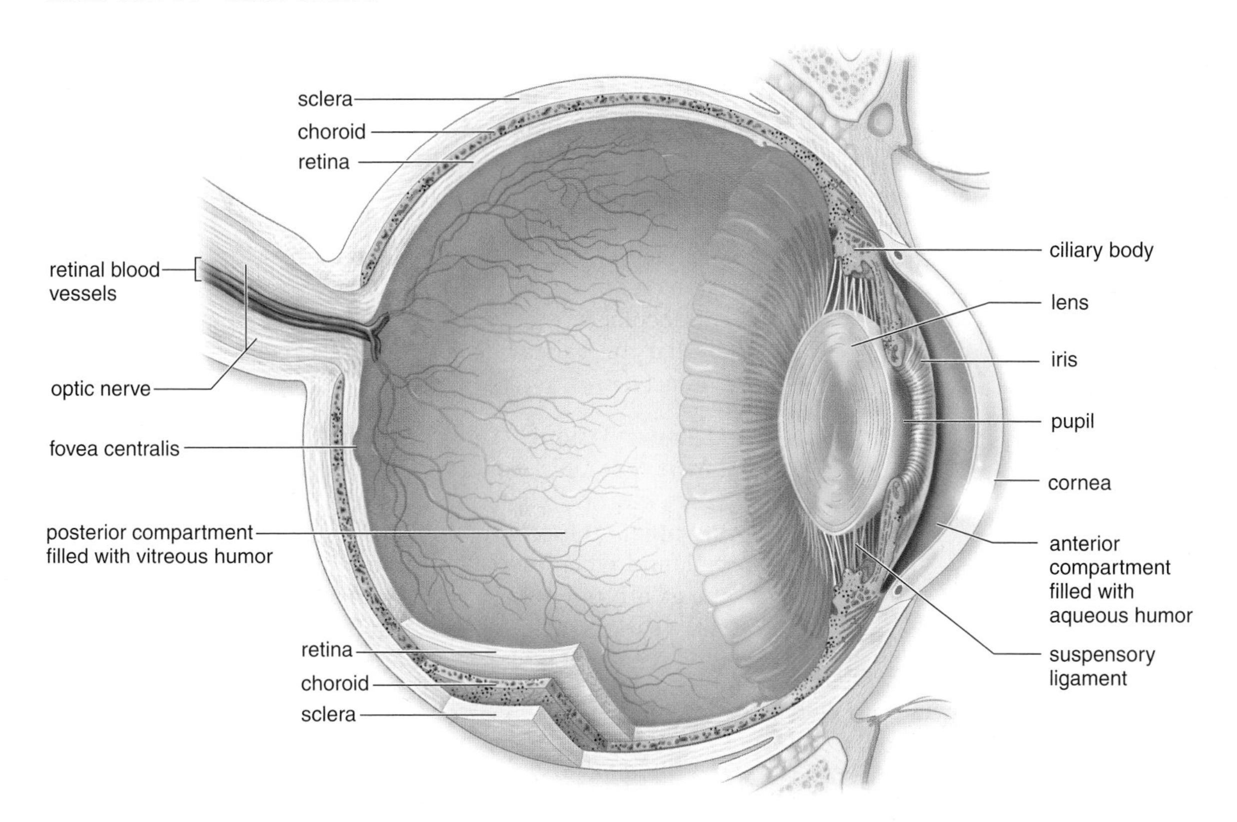


**FIGURE 38.5 Anatomy of the human eye.**

Notice that the sclera, the outer layer of the eye, becomes the cornea and that the choroid, the middle layer, is continuous with the ciliary body and the iris. The retina, the inner layer, contains the photoreceptors for vision. The fovea centralis is the region where vision is most acute.

eye, the sclera becomes the transparent **cornea,** the window of the eye. A thin layer of epithelial cells forms a mucous membrane called the **conjunctiva** that covers the surface of the sclera and keeps the eyes moist. The middle, thin, dark-brown layer, the **choroid** [Gk. *chorion,* membrane], contains many blood vessels and a brown pigment that absorbs stray light rays. Toward the front of the eye, the choroid thickens and forms the ring-shaped ciliary body and a thin, circular, muscular diaphragm, the iris. The **iris** is the colored portion of the eye and regulates the size of an opening called the pupil. The **pupil,** like the aperture of a camera lens, regulates light entering the eye. The **lens,** which is attached to the ciliary body by ligaments, divides the cavity of the eye into two portions and helps form images. A basic, watery solution called aqueous humor fills the anterior compartment between the cornea and the lens. The aqueous humor provides a fluid cushion and nutrient and waste transport for the eye. Glaucoma results when aqueous humor builds up and increases intraocular pressure. A viscous, gelatinous material, the vitreous humor, fills the large posterior compartment behind the lens. The vitreous humor helps stabilize the shape of the eye and supports the retina.

The inner layer of the eye, the **retina** [L. *rete,* net], is located in the posterior compartment. The retina contains photoreceptors called rod cells and cone cells. The rods are very sensitive to light, but they do not see color; therefore, at night or in a darkened room, we see only shades of gray. The cones, which require bright light, are sensitive to different wavelengths of light, and therefore, we have the ability to distinguish colors. The retina has a very special region called the **fovea centralis,** where cone cells are densely packed. Light is normally focused on the fovea when we look directly at an object. This is helpful because vision is most acute in the fovea centralis. Rods are distributed in the peripheral regions of the retina. Sensory fibers form the optic nerve, which takes nerve impulses to the brain.

## TABLE 38.1

**Functions of the Parts of the Eye**

| *Part* | *Function* |
|---|---|
| Sclera | Protects and supports eyeball |
| Conjunctiva | Moistens eye surface |
| Cornea | Refracts light rays |
| Pupil | Admits light |
| Choroid | Absorbs stray light |
| Ciliary body | Holds lens in place, accommodation |
| Iris | Regulates light entrance |
| Retina | Contains sensory receptors for sight |
| Rods | Make black-and-white vision possible |
| Cones | Make color vision possible |
| Fovea centralis | Makes acute vision possible |
| Other | |
| Lens | Refracts and focuses light rays |
| Humors | Transmit light rays and support eyeball |
| Optic nerve | Transmits impulse to brain |

## *Focusing of the Eye*

When we look at an object, light rays pass through the pupil and are focused on the retina. The image produced is much smaller than the object because light rays are bent (refracted) when they are brought into focus. Focusing starts at the cornea and continues as the rays pass through the lens and the humors. The image on the retina is inverted (it is upside down) and reversed from left to right. When information from the retina reaches the brain, it is processed so that we perceive our surroundings in the correct orientation.

The lens provides additional focusing power as **visual accommodation** occurs for close vision. The shape of the lens is controlled by the **ciliary muscle** within the ciliary body. When we view a distant object, the ciliary muscle is relaxed, causing the suspensory ligaments attached to the ciliary body to be taut; therefore, the lens remains relatively flat (Fig. 38.6*a*). When we view a near object, the ciliary muscle contracts, releasing the tension on the suspensory ligaments, and the lens becomes more round due to its natural elasticity (Fig. 38.6*b*). Because close work requires contraction of the ciliary muscle, it very often causes muscle fatigue known as eyestrain. With normal aging, the lens loses its ability to accommodate for near objects (Fig. 38.6*b*); therefore, people frequently need reading glasses once they reach middle age.

Aging, or possibly exposure to the sun, also makes the lens subject to cataracts; the lens can become opaque and therefore incapable of transmitting light rays. Currently, surgery is the only viable treatment for cataracts. First, a surgeon opens the eye near the rim of the cornea. The protein-digesting enzyme zonulysin may be used to dissolve the ligaments holding the lens in place. Most surgeons then use a cryoprobe, which freezes the lens for easy removal. An intraocular lens attached to the iris can then be implanted so that the patient does not need to wear thick glasses or contact lenses.

**Distance Vision.** Those who can easily see a near object but have trouble seeing what is designated as a size 20 letter 20 ft away on an optometrist's chart are said to be *nearsighted* (myopic). These individuals often have an elongated eyeball, and when they attempt to look at a distant object, the image is brought to focus in front of the retina. These people can wear concave lenses, which diverge the light rays so that the image can be focused on the retina (Fig. 38.7*a*). Rather

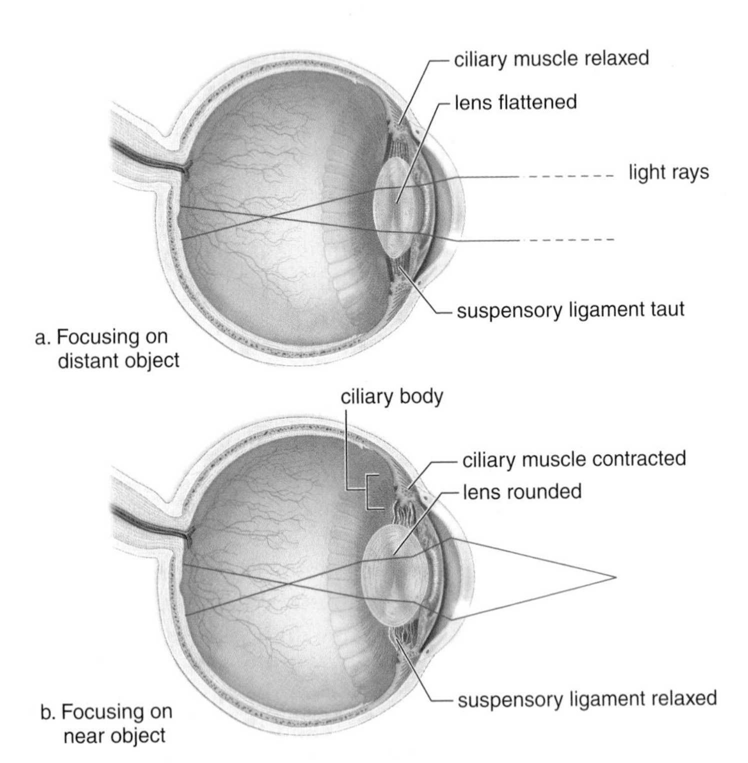


**FIGURE 38.6 Focusing of the human eye.**

Light rays from each point on an object are bent by the cornea and the lens in such a way that an inverted and reversed image of the object forms on the retina. **a.** When focusing on a distant object, the lens is flat because the ciliary muscle is relaxed and the suspensory ligament is taut. **b.** When focusing on a near object, the lens accommodates; that is, it becomes rounded because the ciliary muscle contracts, causing the suspensory ligament to relax.

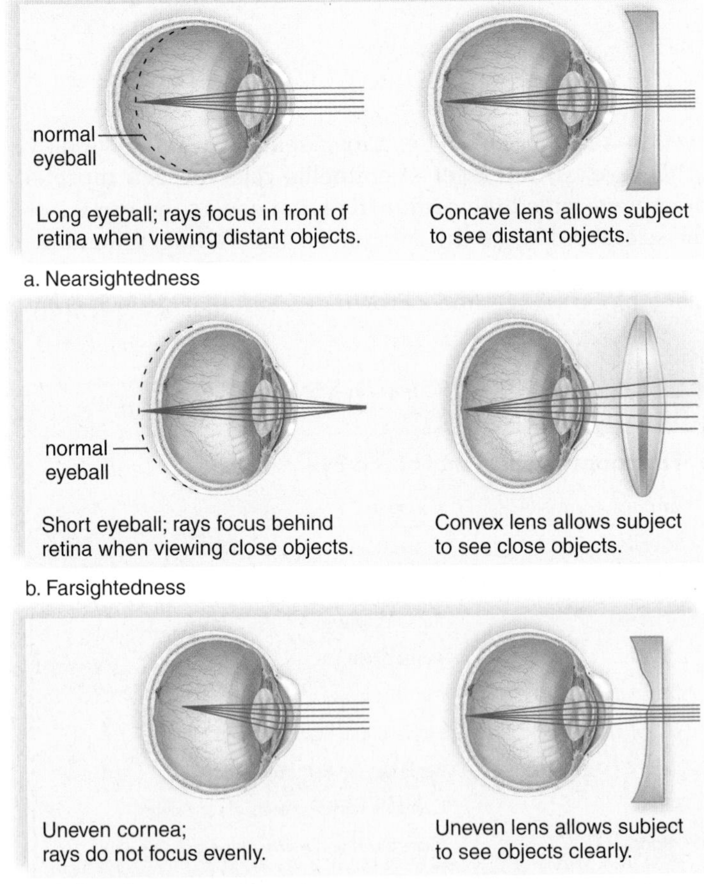


**FIGURE 38.7 Common abnormalities of the eye with possible corrective lenses.**

**a.** A concave lens in nearsighted persons focuses light rays on the retina. **b.** A convex lens in farsighted persons focuses light rays on the retina. **c.** An uneven lens in persons with astigmatism focuses light rays on the retina.

than wear glasses or contact lenses, many nearsighted people are now choosing to undergo laser surgery. First, specialists determine how much the cornea needs to be flattened to achieve visual acuity. Controlled by a computer, the laser then removes this amount of the cornea. Most patients achieve at least 20/40 vision, but a few complain of glare and varying visual acuity.

Those who can easily see the optometrist's chart but cannot easily see near objects are *farsighted* (hyperopic). They often have a shortened eyeball, and when they try to see near objects, the image is focused behind the retina. These individuals must wear a convex lens to increase the bending of light rays so that the image can be focused on the retina (Fig. 38.7*b*). When the cornea or lens is uneven, the image is fuzzy. This condition, called astigmatism, can be corrected by an unevenly ground lens to compensate for the uneven cornea (Fig. 38.7*c*).

## Photoreceptors of the Eye

Vision begins once light has been focused on the photoreceptors in the retina. Figure 38.8 illustrates the structure of the photoreceptors called **rod cells** and **cone cells.** Both rods and cones have an outer segment joined to an inner segment by a stalk. Pigment molecules are embedded in the membrane of the many disks present in the outer segment. Synaptic vesicles are located at the synaptic endings of the inner segment.

The visual pigment in rods is a deep-purple pigment called rhodopsin. **Rhodopsin** is a complex molecule made up of the protein opsin and a light-absorbing molecule called *retinal*, which is a derivative of vitamin A. When a rod absorbs light, rhodopsin splits into opsin and retinal, leading to a cascade of reactions and the closure of ion channels in the rod cell's plasma membrane. The release of inhibitory transmitter molecules from the rod's synaptic vesicles ceases. Thereafter, nerve impulses go to the visual areas of the cerebral cortex. Rods are very sensitive to light and, therefore, are suited to night vision. (Since carrots are rich in vitamin A, it is true that eating carrots can improve your night vision.) Rod cells are plentiful in the peripheral region of the retina; therefore, they also provide us with peripheral vision and perception of motion.

The cones, on the other hand, are located primarily in the fovea centralis and are activated by bright light. They allow us to detect the fine detail and the color of an object. Color vision depends on three different kinds of cones, which contain pigments called the B (blue), G (green), and R (red) pigments. Each pigment is made up of retinal and opsin, but there is a slight difference in the opsin structure of each, which accounts for their individual absorption patterns. Various combinations of cones are believed to be stimulated by in-between shades of color. For example, the color yellow is perceived when green cones are highly stimulated, red cones are partially stimulated, and blue cones are not stimulated. In color blindness, an individual lacks certain visual pigments. As indicated in Chapter 11, color blindness is a hereditary disorder.

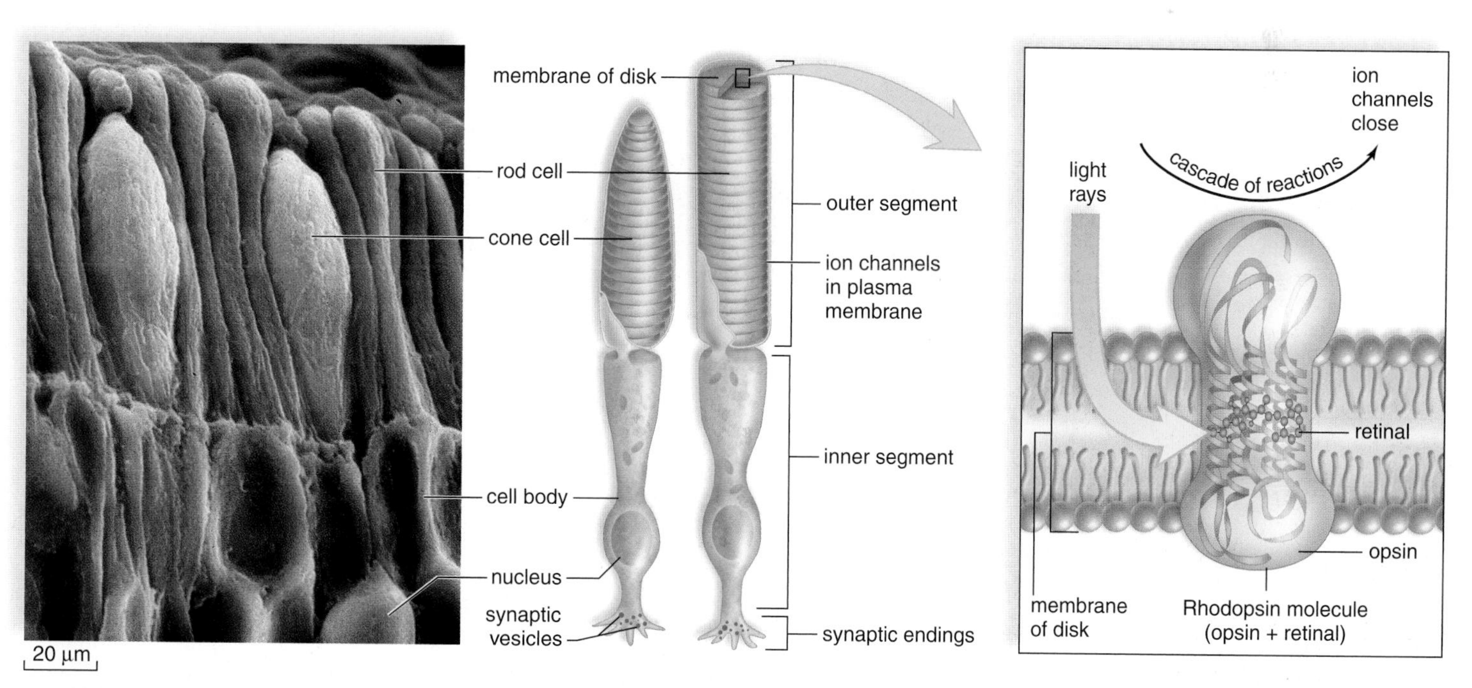


**FIGURE 38.8 Photoreceptors in the eye.**

The outer segment of rods and cones contains stacks of membranous disks, which contain visual pigments. In rods, the membrane of each disk contains rhodopsin, a complex molecule containing the protein opsin and the pigment retinal. When rhodopsin absorbs light energy, it splits, releasing opsin, which sets in motion a cascade of reactions that cause ion channels in the plasma membrane to close. Thereafter, nerve impulses go to the brain.

## Integration of Visual Signals in the Retina

The retina has three layers of neurons (Fig. 38.9). The layer closest to the choroid contains the rod cells and cone cells; the middle layer contains bipolar cells; and the innermost layer contains ganglion cells, whose sensory fibers become the optic nerve. Only the rod cells and the cone cells are sensitive to light, and therefore light must penetrate to the back of the retina before they are stimulated.

The rod cells and the cone cells synapse with the bipolar cells, which in turn synapse with ganglion cells that initiate nerve impulses. Notice in Figure 38.9 that there are many more rod cells and cone cells than ganglion cells. In fact, the retina has as many as 150 million rod cells and 6 million cone cells but only 1 million ganglion cells. The sensitivity of cones versus rods is mirrored by how directly they connect to ganglion cells. As many as 150 rods may excite the same ganglion cell. Stimulation of rods results in vision that is blurred and indistinct. In contrast, some cone cells in the fovea centralis excite only one ganglion cell. This explains why cones, especially in the fovea, provide us with a sharper, more detailed image of an object.

As signals pass to bipolar cells and ganglion cells, integration occurs. Each ganglion cell receives signals from rod cells covering about 1 $mm^2$ of retina (about the size of a thumbtack hole). This region is the ganglion cell's receptive field. Some time ago, scientists discovered that a ganglion cell is stimulated only by messages received from the center of its receptive field; otherwise, it is inhibited. If all the rod cells in the receptive field receive light, the ganglion cell responds in a neutral way—that is, it reacts only weakly or perhaps not at all. This supports the hypothesis that considerable processing occurs in the retina before nerve impulses are sent to the brain. Additional integration occurs in the visual areas of the cerebral cortex.

### *Blind Spot*

Figure 38.9 provides an opportunity to point out that there are no rods and cones where the optic nerve exits the retina. Therefore, no vision is possible in this area. You can prove this to yourself by putting a dot to the right of center on a piece of paper. Use your right hand to move the paper slowly toward your right eye while you look straight ahead. The dot will disappear at one point—this is your **blind spot.**

#### Check Your Progress 38.2

1. Contrast rod and cone photoreceptors in terms of main functions, light conditions, and excitation of ganglion cells.
2. List the three layers (or coats) of the human eye, from the outside in.
3. List the three layers of neurons in the human retina, from the layer facing the inside of the eyeball to the layer that contacts the choroid. Why do you think the arrangement of neuron layers is sometimes described as "backwards?"

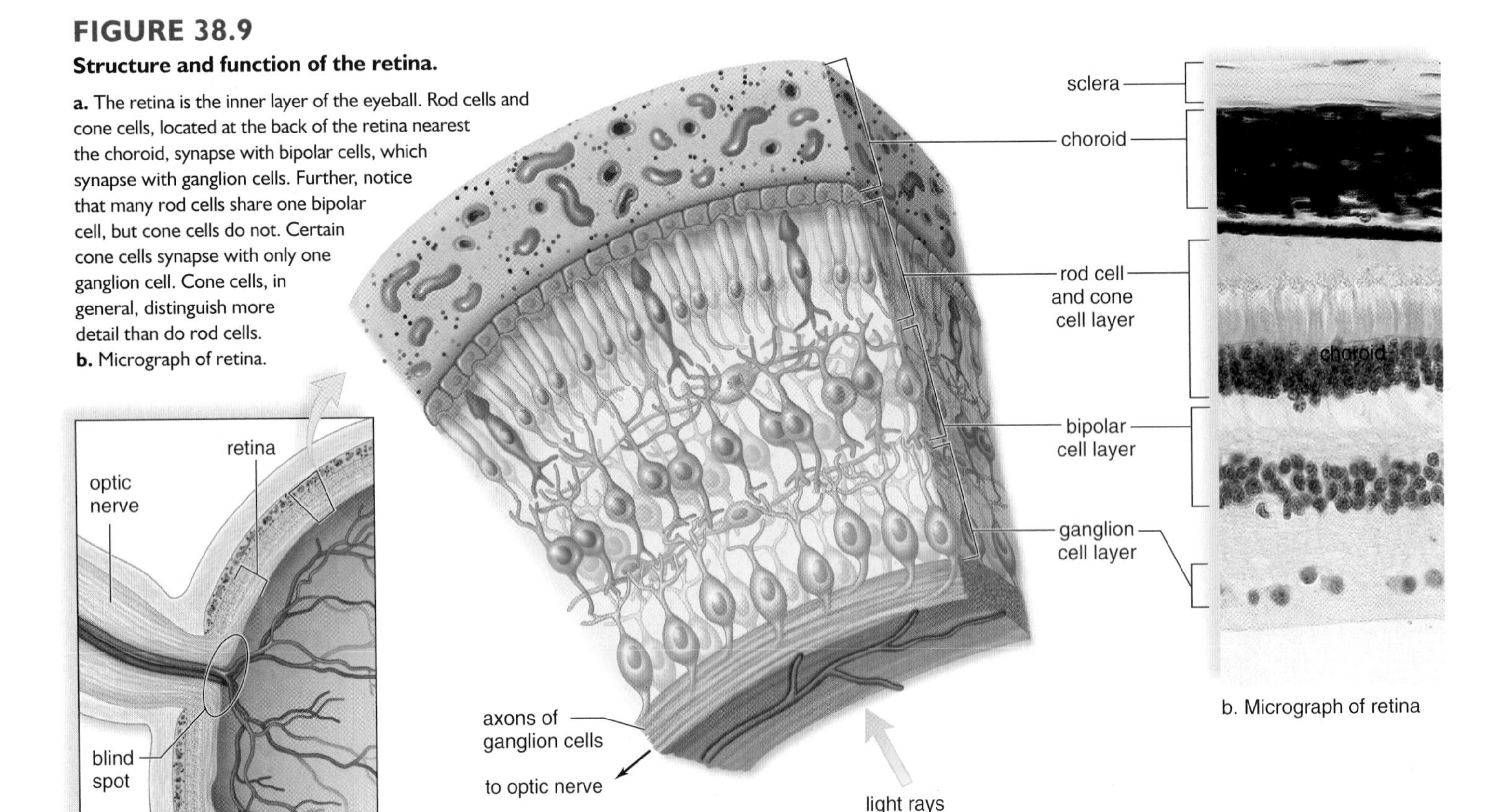


**FIGURE 38.9**

**Structure and function of the retina.**

**a.** The retina is the inner layer of the eyeball. Rod cells and cone cells, located at the back of the retina nearest the choroid, synapse with bipolar cells, which synapse with ganglion cells. Further, notice that many rod cells share one bipolar cell, but cone cells do not. Certain cone cells synapse with only one ganglion cell. Cone cells, in general, distinguish more detail than do rod cells. **b.** Micrograph of retina.

a. Location of retina

b. Micrograph of retina

# 38.3 Senses of Hearing and Balance

The ear has two sensory functions: hearing and balance (equilibrium). The sensory receptors for both of these are located in the inner ear, and each consists of *hair cells* with stereocilia (long microvilli) that are sensitive to mechanical stimulation. They are **mechanoreceptors.**

## Anatomy of the Ear

The ear has three distinct divisions: the outer, inner, and middle ear (Fig. 38.10). The **outer ear** consists of the pinna (external flap) and the auditory canal. The opening of the auditory canal is lined with fine hairs and glands. Glands that secrete earwax are located in the upper wall of the auditory canal. Earwax helps guard the ear against the entrance of foreign materials, such as air pollutants and microorganisms.

The **middle ear** begins at the **tympanic membrane** (eardrum) and ends at a bony wall containing two small openings covered by membranes. These openings are called the *oval window* and the *round window*. Three small bones are found between the tympanic membrane and the oval window. Collectively called the **ossicles,** individually they are the *malleus* (hammer), the *incus* (anvil), and the *stapes* (stirrup) because their shapes resemble these objects. The malleus adheres to the tympanic membrane, and the stapes touches the oval window. An *auditory tube* (eustachian tube), which extends from each middle ear to the nasopharynx, permits equalization of air pressure. Chewing gum, yawning, and swallowing in elevators and airplanes help move air through the auditory tubes upon ascent and descent. As this occurs, we often hear the ears "pop."

Whereas the outer ear and the middle ear contain air, the inner ear is filled with fluids. Anatomically speaking, the **inner ear** has three areas: the **semicircular canals** and the **vestibule** are both concerned with equilibrium; the **cochlea** is concerned with hearing. The cochlea resembles the shell of a snail because it spirals.

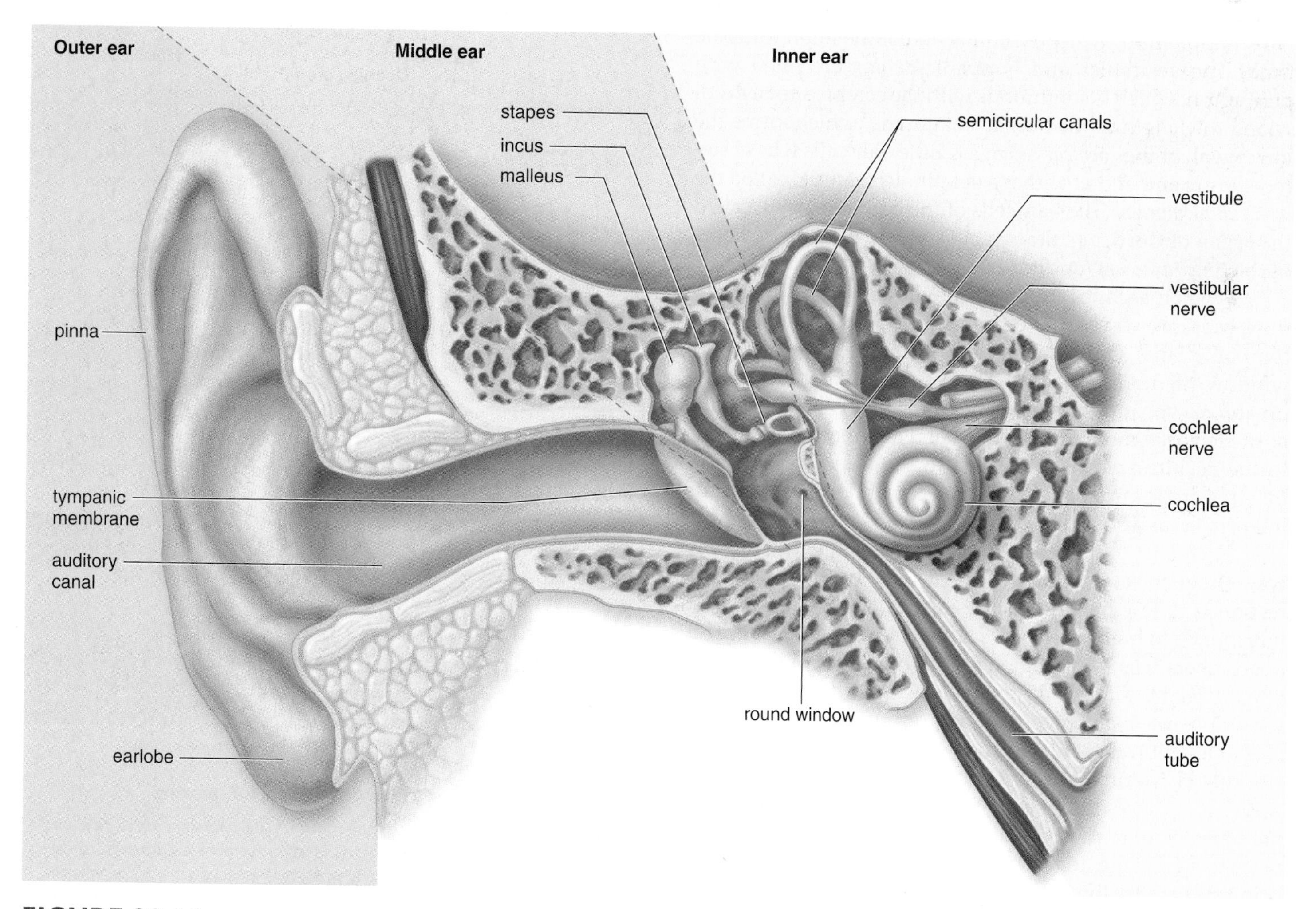


**FIGURE 38.10 Anatomy of the human ear.**

In the middle ear, the malleus (hammer), the incus (anvil), and the stapes (stirrup) amplify sound waves. In the inner ear, the mechanoreceptors for equilibrium are in the semicircular canals and the vestibule, and the mechanoreceptors for hearing are in the cochlea.

## Process of Hearing

The sound pathway travels from the auditory canal to the inner ear.

### *The Auditory Canal and Middle Ear*

The process of hearing begins when sound waves enter the auditory canal. Just as ripples travel across the surface of a pond, sound waves travel by the successive vibrations of molecules. Ordinarily, sound waves do not carry much energy, but when a large number of waves strike the tympanic membrane, it moves back and forth (vibrates) ever so slightly. The malleus then takes the pressure from the inner surface of the tympanic membrane and passes it by means of the incus to the stapes in such a way that the pressure is multiplied about 20 times as it moves. The stapes strikes the membrane of the oval window, causing it to vibrate, and in this way, the pressure is passed to the fluid within the cochlea.

### *Inner Ear*

When the snail-shaped cochlea is examined in cross section (Fig. 38.11), the vestibular canal, the cochlear canal, and the tympanic canal become apparent. The cochlear canal contains endolymph, which is similar in composition to tissue fluid. The vestibular and tympanic canals are filled with perilymph, which is continuous with the cerebrospinal fluid. Along the length of the basilar membrane, which forms the lower wall of the *cochlear canal*, are little hair cells whose stereocilia are embedded within a gelatinous material called the *tectorial membrane*. The hair cells of the cochlear canal, called the **organ of Corti,** or spiral organ, synapse with nerve fibers of the *cochlear nerve* (auditory nerve).

When the stapes strikes the membrane of the oval window, pressure waves move from the vestibular canal to the tympanic canal across the basilar membrane, and the round window membrane bulges. The basilar membrane moves up and down, and the stereocilia of the hair cells embedded in the tectorial membrane bend. Then nerve impulses begin in the cochlear nerve and travel to the brain stem. When they reach the auditory areas of the cerebral cortex, they are interpreted as a sound.

Each part of the organ of Corti is sensitive to different wave frequencies, or pitch. Near the tip, the organ of Corti responds to low pitches, such as a tuba, and near the base, it responds to higher pitches, such as a bell or a whistle. The nerve fibers from each region along the length of the organ of Corti lead to slightly different areas in the brain. The pitch sensation we experience depends on which region of the basilar membrane vibrates and which area of the brain is stimulated. Volume is a function of the amplitude of sound waves. Loud noises cause the fluid within the vestibular canal to exert more pressure and the basilar membrane to vibrate to a greater extent. The resulting increased stimulation is interpreted by the brain as volume. It is believed that the brain interprets the tone of a sound based on the distribution of the hair cells stimulated.

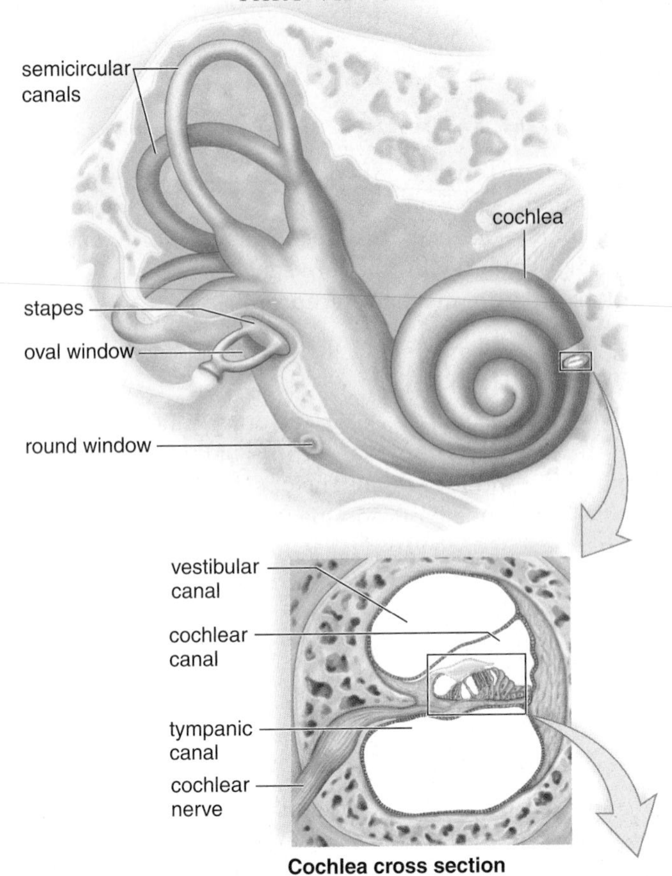


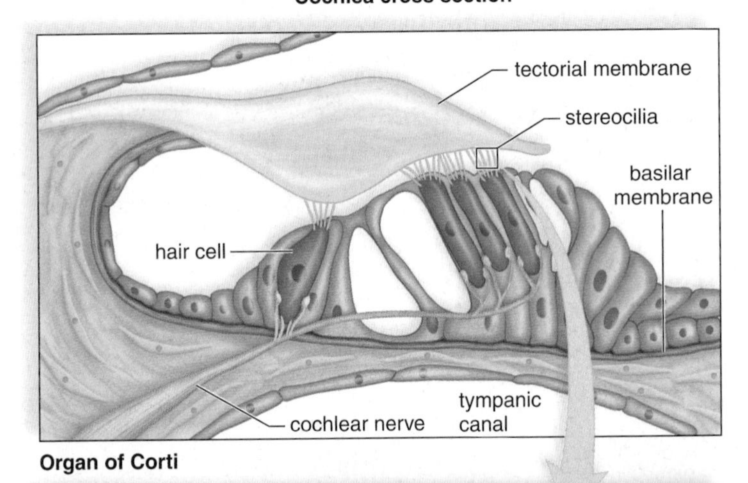


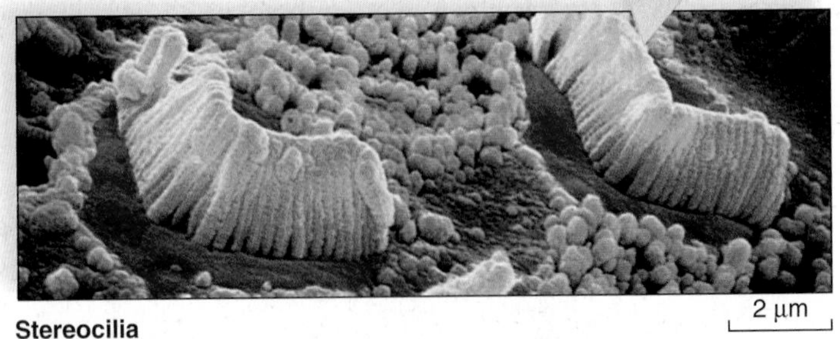


**FIGURE 38.11 Mechanoreceptors for hearing.**

The organ of Corti is located within the cochlea. In the uncoiled cochlea, note that the organ consists of hair cells resting on the basilar membrane, with the tectorial membrane above. Pressure waves move from the vestibular canal to the tympanic canal, causing the basilar membrane to vibrate. This causes the stereocilia (or at least a portion of the more than 20,000 hair cells) embedded in the tectorial membrane to bend. Nerve impulses traveling in the cochlear nerve result in hearing.

*health focus*

## Protecting Vision and Hearing

A serious loss of vision and hearing can occur as we age. Therefore we should start protecting our eyes and ears when we are young.

### Preventing a Loss of Vision

Although flying objects sometimes penetrate the cornea and damage the iris, lens, or retina, careless use of contact lenses is the most common cause of injuries to the eye. Injuries cause only 4% of all cases of blindness; the most frequent causes are retinal disorders, glaucoma, and cataracts, in that order. Retinal disorders are varied. In diabetic retinopathy, which blinds many people between the ages of 20 and 74, capillaries to the retina burst, and blood spills into the vitreous humor. Careful regulation of blood glucose levels in these patients may be protective. In macular degeneration, the cones are destroyed because thickened choroid vessels no longer function as they should. Glaucoma occurs when the drainage system of the eyes fails, so that fluid builds up and destroys nerve fibers responsible for peripheral vision. Eye doctors always check for glaucoma, but it is advisable to be aware of the disorder in case it comes on quickly. Those who have experienced acute glaucoma report that the eyeball feels as heavy as a stone. Cataracts occur in 50% of people between the ages of 65 and 74, and in 70% of those ages 75 or older. In cataracts, cloudy spots on the lens of the eye eventually pervade the whole lens. The milky yellow-white lens scatters incoming light and blocks vision. The extent of visual impairment depends on the size and density of the cataract, and where it is located in the lens. A dense, centrally placed cataract causes severe blurring of vision.

There are preventive measures that we can take to reduce the chance of defective vision as we age. It is recommended, therefore, that everyone, especially those who live in sunny climates or work outdoors, wear glass, not plastic, sunglasses to absorb ultraviolet light. Large lenses worn close to the eyes offer further protection. Special-purpose lenses that block at least 99% of UV-B, 60% of UV-A, and 20–97% of visible light are good for bright sun combined with sand, snow, or water. Health-care providers have found an increased incidence of cataracts in heavy cigarette smokers. In men, smoking 20 cigarettes or more a day, and in women, smoking more than 35 cigarettes a day, doubles the risk of cataracts. It is possible that smoking reduces the delivery of blood and therefore nutrients to the lens.

### Preventing a Loss of Hearing

Especially when we are young, the middle ear is subject to infections that can lead to hearing impairment if they are not treated promptly by a physician. The mobility of ossicles decreases with age, and in the condition called otosclerosis, new filamentous bone grows over the stirrup, impeding its movement. Surgical treatment is the only remedy for this type of conduction deafness. However, age-associated nerve deafness due to stereocilia damage from exposure to loud noises is preventable. Hospitals are now aware that even the ears of the newborn need to be protected from noise and are taking steps to make sure neonatal intensive care units and nurseries are as quiet as possible.

In today's society, exposure to excessive noise is commonplace. Noise is measured in decibels, and any noise above a level of 80 decibels could result in damage to the hair cells of the organ of Corti. Eventually, the stereocilia and then the hair cells disappear completely (Fig. 38A). If listening to city traffic for extended periods can damage hearing, it stands to reason that frequent attendance at rock concerts, constantly playing a stereo loudly, or using earphones at high volume are also damaging to hearing. However, by the time symptoms are noticed, some degree of hearing has already been irreversibly destroyed.

Aside from loud music, noisy indoor or outdoor equipment, such as a rug-cleaning machine or a chain saw, is also troublesome. Even motorcycles and recreational vehicles such as snowmobiles and motocross bikes can contribute to a gradual loss of hearing. Exposure to intense sounds of short duration, such as a burst of gunfire, can result in an immediate hearing loss. The butt of the rifle offers some protection to the ear nearest the gun when it is shot, so it is the opposite ear that often suffers a loss of hearing.

The first hint of noise-induced hearing loss could be temporary hearing loss, a "full" feeling in the ears, muffled hearing, or tinnitus (e.g., ringing in the ears). If you have any of these symptoms, take steps immediately to prevent further damage. If exposure to noise is unavoidable, specially designed noise-reduction earmuffs are available, and it is also possible to purchase earplugs made from a compressible, spongelike material at the drugstore or sporting-goods store. These earplugs are not the same as those worn for swimming, and they should not be used interchangeably.

Finally, people need to be aware that some medicines are ototoxic. Anticancer drugs, most notably cisplatin, and certain antibiotics (e.g., streptomycin, kanamycin, gentamicin) make the ears especially susceptible to hearing loss.

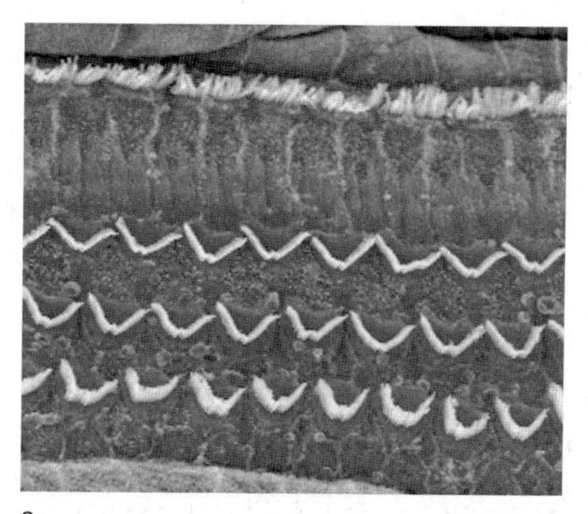

a.

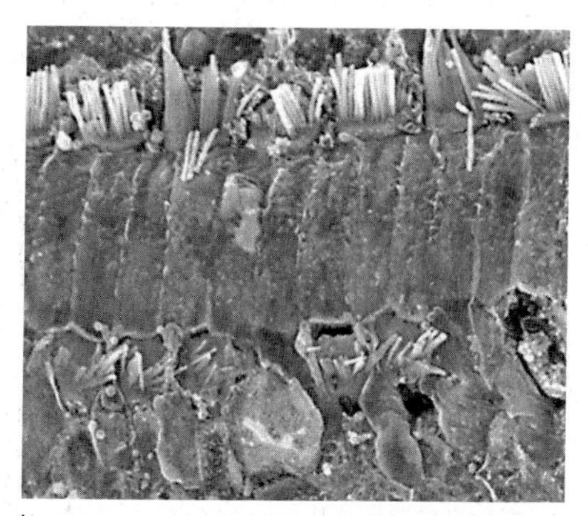

b.

**FIGURE 38A Hearing loss.**
***a.*** *Normal hair cells in the organ of Corti of a guinea pig.* ***b.*** *Damaged hair cells in the organ of Corti of a guinea pig. This damage occurred after 24-hour exposure to a noise level typical of a rock concert.*

## ..se of Balance

Mechanoreceptors in the semicircular canals detect rotational and/or angular movement of the head **(rotational equilibrium),** while mechanoreceptors in the utricle and saccule detect straight-line movement of the head in any direction **(gravitational equilibrium).**

### *Rotational Equilibrium*

Rotational equilibrium (Fig. 38.12*a*) involves the semicircular canals, which are arranged so that there is one in each dimension of space. The base of each of the three canals, called the ampulla, is slightly enlarged. Little hair cells, whose stereocilia are embedded within a gelatinous material called a cupula, are

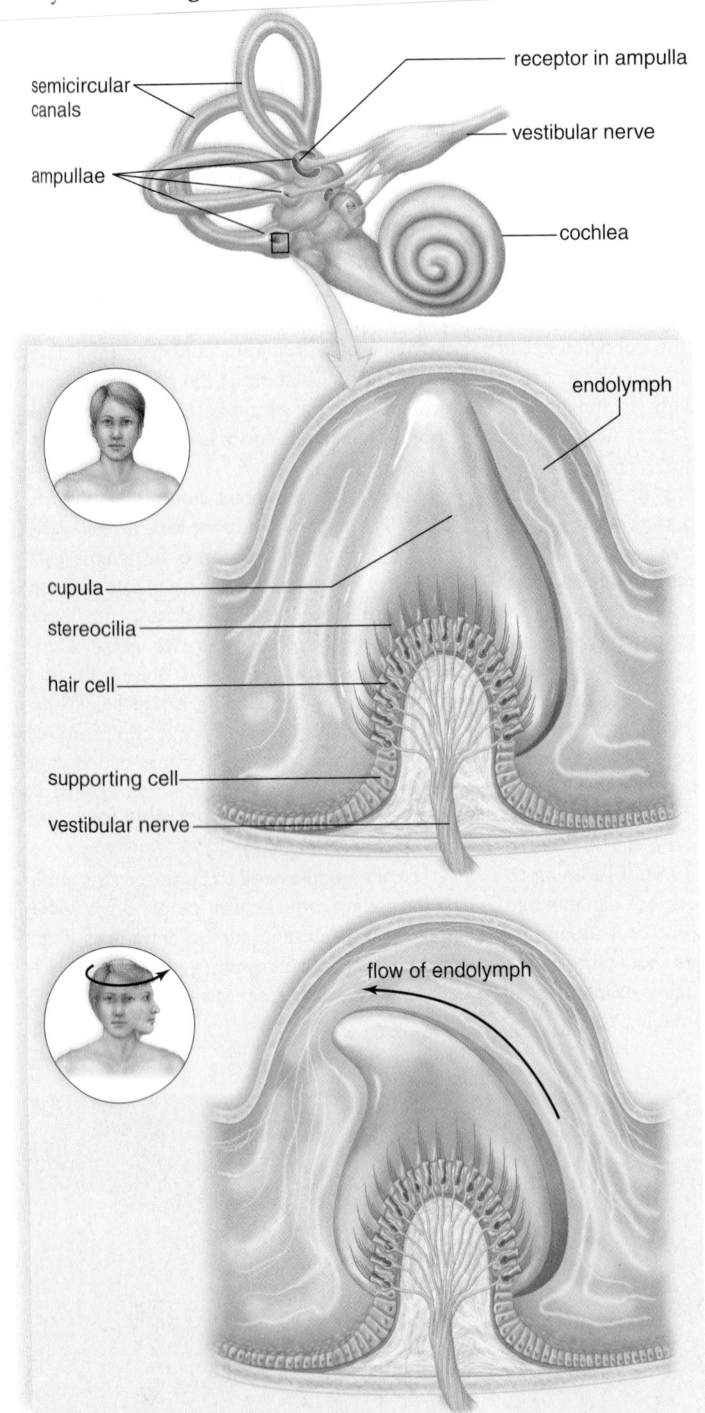


a. Rotational equilibrium: receptors in ampullae of semicircular canal

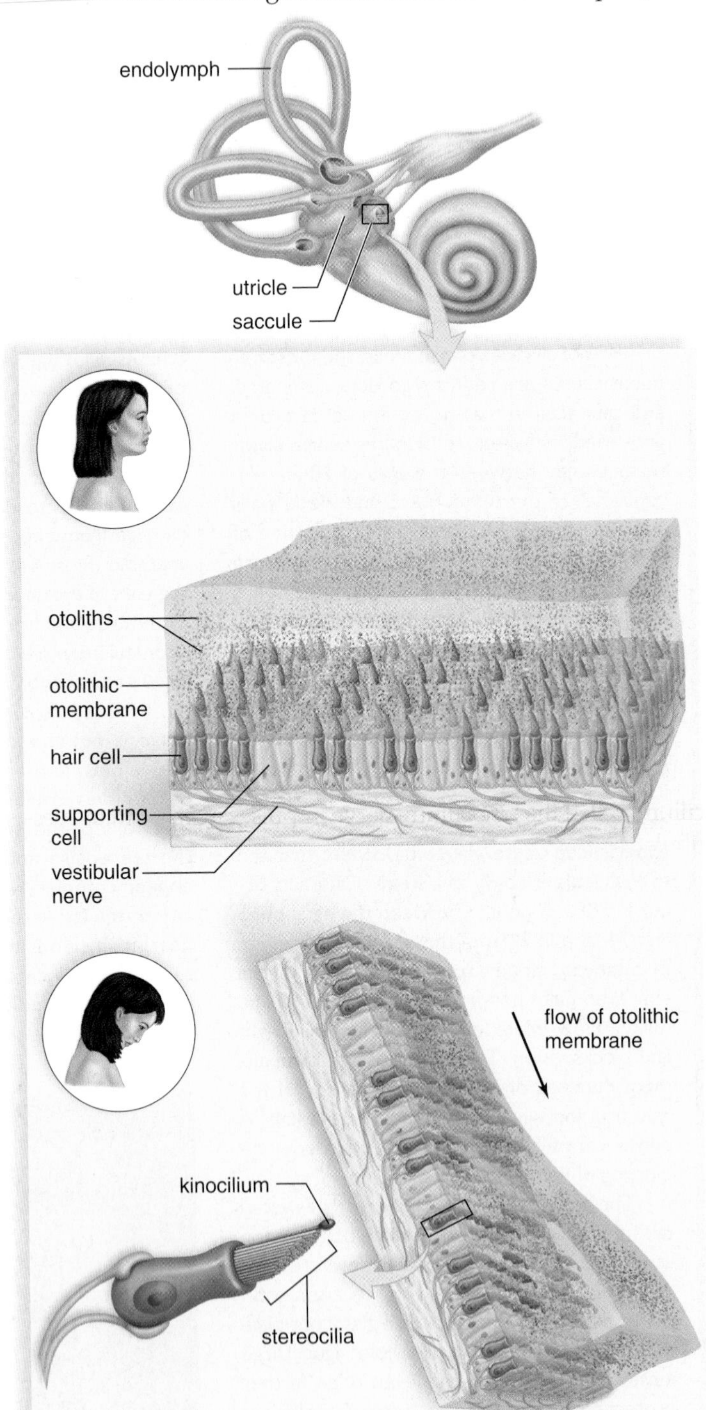


b. Gravitational equilibrium: receptors in utricle and saccule of vestibule

**FIGURE 38.12 Mechanoreceptors for equilibrium.**

**a.** Rotational equilibrium. The ampullae of the semicircular canals contain hair cells with stereocilia embedded in a cupula. When the head rotates, the cupula is displaced, bending the stereocilia. Thereafter, nerve impulses travel in the vestibular nerve to the brain. **b.** Gravitational equilibrium. The utricle and the saccule contain hair cells with stereocilia embedded in an otolithic membrane. When the head bends, otoliths are displaced, causing the membrane to sag and the stereocilia to bend. If the stereocilia bend toward the kinocilium, the longest of the stereocilia, nerve impulses increase in the vestibular nerve. If the stereocilia bend away from the kinocilium, nerve impulses decrease in the vestibular nerve. This difference tells the brain in which direction the head moved.

found within the ampullae. Because there are three semicircular canals, each ampulla responds to head movement in a different plane of space. As fluid (endolymph) within a semicircular canal flows over and displaces a cupula, the stereocilia of the hair cells bend, and the pattern of impulses carried by the vestibular nerve to the brain changes. The brain uses information from the hair cells within ampulla of the semicircular canals to maintain equilibrium through appropriate motor output to various skeletal muscles that can right our present position in space as need be.

Vertigo is dizziness and a sensation of rotation. It is possible to simulate a feeling of vertigo by spinning rapidly and stopping suddenly. When the eyes are rapidly jerked back to a midline position, the person feels like the room is spinning. This shows that the eyes are also involved in our sense of equilibrium.

### *Gravitational Equilibrium*

Gravitational equilibrium (Fig. 38.12*b*) depends on the **utricle** and **saccule,** two membranous sacs located in the vestibule. Both of these sacs contain little hair cells, whose stereocilia are embedded within a gelatinous material called an otolithic membrane. Calcium carbonate ($CaCO_3$) granules, or **otoliths,** rest on this membrane. The utricle is especially sensitive to horizontal (back and forth) movements of the head, while the saccule responds best to vertical (up and down) movements.

When the head is still, the otoliths in the utricle and the saccule rest on the otolithic membrane above the hair cells. When the head moves in a straight line, the otoliths are displaced and the otolithic membrane sags, bending the stereocilia of the hair cells beneath. If the stereocilia move toward the largest stereocilium, called the kinocilium, nerve impulses increase in the vestibular nerve. If the stereocilia move away from the kinocilium, nerve impulses decrease in the vestibular nerve. If you are upside down, nerve impulses in the vestibular nerve cease. These data tell the brain the direction of the movement of the head.

### *Sensory Receptors in Other Animals*

The **lateral line** system of fishes (Fig. 38.13*a*) guides them in their movements and in locating other fishes, including predators and prey and mates. The system detects water currents and pressure waves from nearby objects in the same manner as the sensory receptors in the human ear. In bony fishes, the sensory receptors are located within a canal that has openings to the outside. A lateral line receptor is a collection of hair cells with cilia embedded in a gelatinous cupula. When the cupula bends due to pressure waves, the hair cells initiate nerve impulses.

Gravitational equilibrium organs, called statocysts (Fig. 38.13*b*), are found in cnidarians, molluscs, and crustaceans, which are arthropods. These organs give information only about the position of the head; they are not involved in the sensation of movement. When the head stops moving, a small particle called a statolith stimulates the cilia of the closest hair cells, and these cilia generate impulses, indicating the position of the head.

#### Check Your Progress 38.3

1. Determine whether each of the following belongs to the outer, middle, or inner ear: **a.** ossicles; **b.** pinna; **c.** semicircular canals; **d.** cochlea; **e.** vestibule; **f.** auditory canal.
2. List, in order, the structures that must conduct a sound wave from the time it enters the auditory canal until it reaches the cochlea.
3. Which structures of the inner ear are responsible for gravitational equilibrium? Rotational equilibrium?

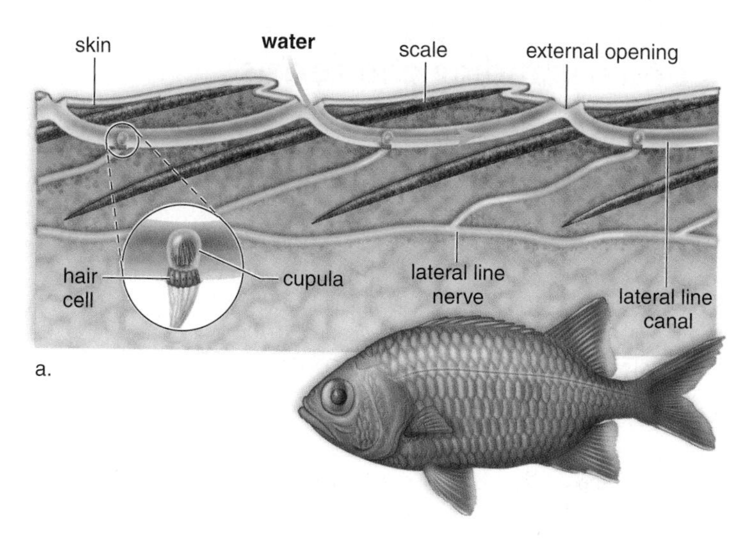


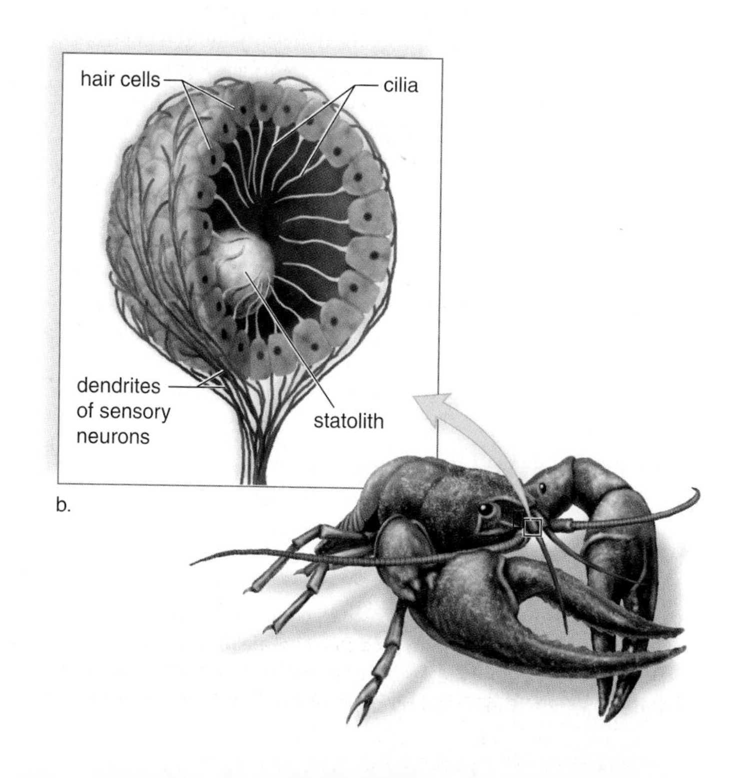


**FIGURE 38.13 Sensory receptors in other animals.**

**a.** Hairs located within cupulas of a lateral line detect wave vibrations and currents, helping to guide fish movements in order to locate predators and prey and mates. **b.** Within a statocyst, a small particle (the statolith) comes to rest on hair cells and tells a crustacean the position of its head.

## Connecting the Concepts

An animal's information exchange with the internal and external environment is dependent on just a few types of sensory receptors. We have examined chemoreceptors, such as taste cells and olfactory cells; photoreceptors, such as eyes; and mechanoreceptors, such as the hair cells for hearing and balance. The senses are not equally developed in all animals. Male moths have chemoreceptors on the filaments of their antennae to detect minute amounts of an airborne sex attractant released by a female. This is certainly a more efficient method than searching for a mate by sight.

Birds that live in forested areas signal that a territory is occupied by singing because it is difficult to see a bird in a tree, as most birders know. On the other hand, hawks have such a keen sense of sight that they are able to locate a small mouse far below them. Insectivorous bats have an unusual adaptation for finding prey in the dark. They send out a series of sound pulses and listen for the echoes that come back. The time it takes for an echo to return indicates the location of an insect. A unique adaptation is found among the so-called electric fishes of Africa and Australia. They have electroreceptors that can detect disturbances in an electrical current they emit into the water. These disturbances indicate the location of obstacles and prey.

Animals that migrate use various senses to find their way. Salmon hatch in a freshwater stream but drift to the ocean as larvae. By the end of the third or fourth year, they migrate back to where they were hatched. Like other migrating animals, they apparently can use the sun as a compass to find their way back to the vicinity of their home river, but then salmon switch to a sense of smell to find the exact location of their hatching. Perhaps the odor of the plants and soil in this stream was imprinted on the nervous system of the larval fish.

Through the evolutionary process, animals tend to rely on those stimuli and senses that are adaptive to their particular environment and way of life. In all cases, sensory receptors generate nerve impulses that travel to the brain, where sensation occurs. In mammals, and particularly human beings, integration of the data received from various sensory receptors results in perception of events occurring in the external environment.

## summary

### 38.1 Chemical Senses

Chemoreception is found universally in animals and is, therefore, believed to be the most primitive sense.

Human olfactory cells and taste buds are chemoreceptors. They are sensitive to chemicals in water and air.

### 38.2 Sense of Vision

Vision is dependent on the eye, the optic nerves, and the visual areas of the cerebral cortex. The eye has three layers. The outer layer, the sclera, can be seen as the white of the eye; it also becomes the transparent bulge in the front of the eye called the cornea. The middle pigmented layer, called the choroid, absorbs stray light rays. The rod cells (sensory receptors for dim light) and the cone cells (sensory receptors for bright light and color) are located in the retina, the inner layer of the eyeball. The cornea, the humors, and especially the lens bring the light rays to focus on the retina. To see a close object, accommodation occurs as the lens rounds up.

When light strikes rhodopsin within the membranous disks of rod cells, rhodopsin splits into opsin and retinal. A cascade of reactions leads to the closing of ion channels in a rod cell's plasma membrane. Inhibitory transmitter molecules are no longer released, and nerve impulses are carried in the optic nerve to the brain.

Integration occurs in the retina, which is composed of three layers of cells: the rod and cone layer, the bipolar cell layer, and the ganglion cell layer. Integration also occurs in the visual areas of the cerebral cortex.

### 38.3 Senses of Hearing and Balance

Hearing in humans is dependent on the ear, the cochlear nerve, and the auditory areas of the cerebral cortex. The ear is divided into three parts. The outer ear consists of the pinna and the auditory canal, which direct sound waves to the middle ear. The middle ear begins with the tympanic membrane and contains the ossicles (malleus, incus, and stapes). The malleus is attached to the tympanic membrane, and the stapes is attached to the oval window, which is covered by membrane. The inner ear contains the cochlea and the semicircular canals, plus the utricle and saccule.

Hearing begins when the outer and middle portions of the ear convey and amplify the sound waves that strike the oval window. Its vibrations set up pressure waves within the cochlea, which contains the organ of Corti, consisting of hair cells whose stereocilia are embedded within the tectorial membrane. When the stereocilia of the hair cells bend, nerve impulses begin in the cochlear nerve and are carried to the brain.

The ear also contains receptors for our sense of equilibrium. Rotational equilibrium is dependent on the stimulation of hair cells within the ampullae of the semicircular canals. Gravitational equilibrium relies on the stimulation of hair cells within the utricle and the saccule.

## understanding the terms

Match the terms to these definitions:

a. ______________ Photoreceptor, composed of many independent units, which is typical of arthropods.

b. ______________ Inner layer of the eyeball containing the photoreceptors—rod cells and cone cells.

c. ______________ Outer, white, fibrous layer of the eye that surrounds the eye except for the transparent cornea.

d. ______________ Receptor that is sensitive to chemical stimulation—for example, receptors for taste and smell.

e. ______________ Specialized region of the cochlea containing the hair cells for sound detection and discrimination.

## reviewing this chapter

1. Discuss the structure and the function of human chemoreceptors. 702–3
2. In general, how does the eye in arthropods differ from that in humans? 704–5
3. What types of animals have eyes that are constructed like the human eye? 704
4. Name the parts of the human eye, and give a function for each part. 704–5
5. Explain focusing and accommodation in terms of the anatomy of the human eye. 706–7
6. Contrast the location and the function of rod cells to those of cone cells. 707
7. Explain the process of integration in the retina and the brain. 708
8. Describe the structure of the human ear. 709
9. Describe how we hear. 710
10. Describe the role of the semicircular canals, utricle, and saccule in balance. 712–13

## testing yourself

Choose the best answer for each question.

1. A sensory receptor
   a. is the first portion of a reflex arc.
   b. initiates nerve impulses.
   c. is specific for one type of stimulus.
   d. is associated with a sensory neuron.
   e. All of these are correct.
2. Which of these gives the correct path for light rays entering the human eye?
   a. sclera, retina, choroid, lens, cornea
   b. fovea centralis, pupil, aqueous humor, lens
   c. cornea, pupil, lens, vitreous humor, retina
   d. optic nerve, sclera, choroid, retina, humors
   e. All of these are correct.
3. Which gives an incorrect function for the structure?
   a. lens—focusing
   b. iris—regulation of amount of light
   c. choroid—location of cones
   d. sclera—protection
   e. fovea centralis—acute vision
4. Retinal is
   a. a derivative of vitamin A.
   b. sensitive to light energy.
   c. a part of rhodopsin.
   d. found in rod cells.
   e. All of these are correct.
5. Which association is incorrect?
   a. taste buds—humans
   b. compound eye—arthropods
   c. camera-type eye—squid
   d. statocysts—sea stars
   e. chemoreceptors—planarians
6. Which one of these wouldn't you mention if you were tracing the path of sound vibrations?
   a. auditory canal
   b. tympanic membrane
   c. semicircular canals
   d. cochlea
   e. ossicles
7. Which one of these correctly describes the location of the organ of Corti?
   a. between the tympanic membrane and the oval window in the inner ear
   b. in the utricle and saccule within the vestibule
   c. between the tectorial membrane and the basilar membrane in the cochlear canal
   d. between the outer and inner ear within the semicircular canals
8. Which of these pairs is mismatched?
   a. semicircular canals—inner ear
   b. utricle and saccule—outer ear
   c. auditory canal—outer ear
   d. ossicles—middle ear
   e. cochlear nerve—inner ear
9. Both olfactory receptors and sound receptors have cilia, and they both
   a. are chemoreceptors.
   b. are a part of the brain.
   c. are mechanoreceptors.
   d. initiate nerve impulses.
   e. All of these are correct.
10. Which of these is an incorrect difference between olfactory receptors and equilibrium receptors?

| | **Olfactory receptors** | **Equilibrium receptors** |
|---|---|---|
| a. | located in nasal cavities | located in the inner ear |
| b. | chemoreceptors | mechanoreceptors |
| c. | respond to molecules in air | respond to movements of the body |
| d. | communicate with brain via a tract | communicate with brain via vestibular nerve |

   e. All of these contrasts are correct.

11. Stimulation of hair cells in the semicircular canals results from the movement of
    a. endolymph.
    b. aqueous humor.
    c. basilar membrane.
    d. otoliths.
12. To focus on objects that are close to the viewer,
    a. the suspensory ligaments must be pulled tight.
    b. the lens needs to become more rounded.
    c. the ciliary muscle will be relaxed.
    d. the image must focus on the area of the optic nerve.
13. Which abnormality of the eye is incorrectly matched?
    a. astigmatism—either the lens or cornea is not even
    b. farsightedness—eyeball is shorter than usual
    c. nearsightedness—image focuses behind the retina
    d. color blindness—genetic disorder in which certain types of cones may be missing

14. Which of these would allow you to know that you were upside down, even if you were in total darkness?
    a. utricle and saccule
    b. cochlea
    c. semicircular canals
    d. tectorial membrane
15. The thin, darkly pigmented layer that underlies most of the sclera is the
    a. conjunctiva.
    b. cornea.
    c. retina.
    d. choroid.
16. Adjustment of the lens to focus on objects close to the viewer is called
    a. convergence.
    b. accommodation.
    c. focusing.
    d. constriction.
17. The middle ear is separated from the inner ear by
    a. the oval window.
    b. the tympanic membrane.
    c. the round window.
    d. Both a and c are correct.
18. Label this diagram of the human eye. State a function for each structure labeled.

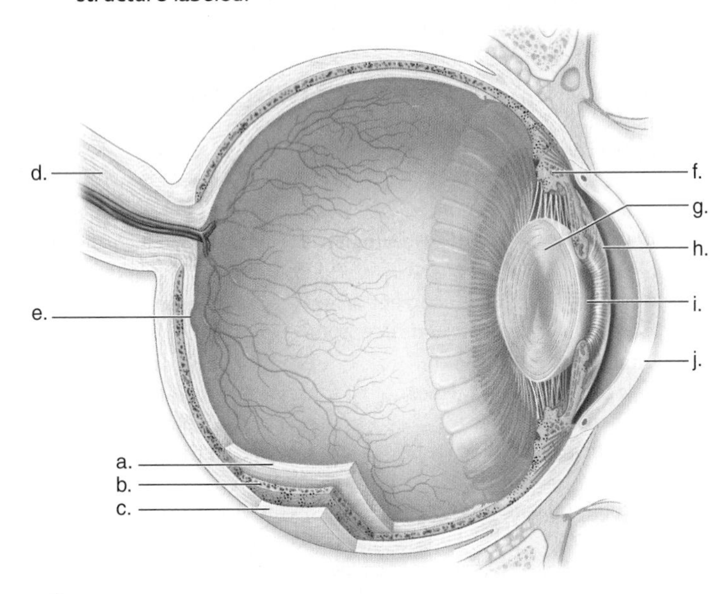


## thinking scientifically

1. The density of taste buds on the tongue can vary. Some obese individuals have a lower density of taste buds than usual. Assume that taste perception is related to taste bud density. If so, what hypothesis would you test to see if there is a relationship between taste bud density and obesity?
2. A man who has spent many years serving on submarines complains of hearing loss, particularly the inability to hear high tones. When a submarine submerges, the inside air pressure intensifies. What hypothesis or hypotheses might explain hearing loss in this individual?

## bioethical issue

### Preventable Hearing Loss

The National Institute on Deafness and Other Communication Disorders (NIDCD) tells us that, of the approximately 28 million people in the United States with hearing loss, over one-third have been affected to some degree by noise exposure. As you learned in the Health Focus on page 711, there are simple steps one can take to prevent noise-induced hearing loss, such as wearing ear protection when engaged in loud activities such as mowing the lawn. Portable music devices are common culprits in hearing loss: If you can hear sound coming from someone's earphones from a distance of 3 feet, that person is inflicting a damaging level of noise on him- or herself.

It's clear that for many of us, our own behavior contributes to the occurrence of hearing loss. Should we be responsible in our actions and take all possible steps to safeguard our hearing? Or, should we simply rely on medical science to help us once our hearing has been lost? Those with hearing loss may cope by using hearing aids, which can be costly and do not restore one's hearing to "normal." A new device, called a cochlear implant, can compensate for damaged hair cells by directly stimulating the auditory nerve. However, the device must be surgically implanted, requires extensive postsurgical rehabilitation, does not replicate unassisted hearing, and is extremely expensive. Today, most health insurance provides at least partial assistance for hearing aids and cochlear implants. However, it is estimated that over the past 30 years, the number of people in the United States with hearing loss has doubled. Therefore, health insurance costs must rise in order to assist those with hearing loss.

What is your responsibility as an individual when it comes to protecting your own hearing? Does this responsibility change if you are in a group, say, taking a road trip in a car with the stereo turned all the way up? If you are aware that a friend is damaging his or her own hearing in preventable ways, should you say anything? What is society's responsibility for educating its members about the importance of hearing protection?

## *Biology* website

The companion website for *Biology*, provides a wealth of information organized and integrated by chapter. You will find practice tests, animations, videos, and much more that will complement your learning and understanding of general biology.

**http://www.mhhe.com/maderbiology10**

# 39

# Locomotion and Support Systems

*When Paul Hamm, a champion gymnast, does handstands on rings, his muscular and skeletal systems are working together under the control of his nervous system. The same is true when eagles fly and fish swim or when animals feed, escape prey, reproduce, or simply play. Although not all animals have muscles and bones, they all use contractile fibers to move about. In planarians, hydras, and earthworms, muscles push against body fluids located inside either a gastrovascular cavity or a coelom.*

*Only in vertebrates are muscles attached to a bony endoskeleton. Both the skeletal system and the muscular system contribute to homeostasis. Aside from giving the body shape and protecting internal organs, the skeleton serves as a storage area for inorganic calcium and produces blood cells. The skeleton also protects internal organs while supporting the body against the pull of gravity. While contributing to body movement, the skeletal muscles give off heat, which warms the body. This chapter compares locomotion in animals and reviews the musculoskeletal system of vertebrates.*

Gymnastics requires coordination between the nervous and support systems.

## *concepts*

# 39.1 Diversity of Skeletons

Skeletons serve as support systems for animals, providing rigidity, protection, and surfaces for muscle attachment. Several different kinds of skeletons occur in the animal kingdom. Cnidarians, flatworms, roundworms, and annelids have a hydrostatic skeleton. Typically, molluscs and arthropods have an **exoskeleton** (external skeleton) composed of calcium carbonate or chitin, respectively. Sponges, echinoderms, and vertebrates possess an **endoskeleton** (internal skeleton). The endoskeleton of sponges consists of mineralized spicules and spongin. In echinoderms, the endoskeleton is composed of calcareous plates and in vertebrates, the endoskeleton is comprised of cartilage, bone, or both.

## Hydrostatic Skeleton

In animals that lack a hard skeleton, a fluid-filled gastrovascular cavity or a fluid-filled coelom can act as a hydrostatic skeleton. A **hydrostatic skeleton** [Gk. *hydrias*, water, and *stasis*, standing] offers support and resistance to the contraction of muscles so that mobility results. As analogies, consider that a garden hose stiffens when filled with water, and that a water-filled balloon changes shape when squeezed at one end. Similarly, an animal with a hydrostatic skeleton can change shape and perform a variety of movements.

Hydras and planarians use their fluid-filled gastrovascular cavity as a hydrostatic skeleton. When muscle fibers at the base of epidermal cells in a hydra contract, the body or tentacles shorten rapidly. Planarians usually glide over a substrate with the help of muscular contractions that control the body wall and cilia. Roundworms have a fluid-filled pseudocoelom and move in a whiplike manner when their longitudinal muscles contract. Annelids, such as earthworms, are segmented and have septa that divide the coelom into compartments (Fig. 39.1). Each segment has its own set of longitudinal and circular muscles and its own nerve supply, so each segment or group of segments may function independently. When circular muscles contract, the segments become thinner and elongate. When longitudinal muscles contract, the segments become thicker and shorten. By alternating circular muscle contraction and longitudinal muscle contraction and by using its setae to hold its position during contractions, the animal moves forward.

### *Muscular Hydrostat*

Even animals that have an exoskeleton or an endoskeleton move selected body parts by means of muscular hydrostats, meaning that fluid contained within certain muscle fibers assists movement of that part. Muscular hydrostats are used by clams to extend their muscular foot and by sea stars to extend their tube feet. Spiders depend on them to move their legs, and moths depend on them to extend their proboscis. In vertebrates, movement of an elephant's trunk involves a muscular hydrostat that allows the animal to reach as high as 23 feet, or pick up a morsel of food, or pull down a tree.

## Exoskeletons and Endoskeletons

Molluscs, arthropods, and vertebrates have rigid skeletons. The exoskeleton of molluscs and arthropods protects and supports these animals and provides a location for muscle attachment. Strength of an exoskeleton can be improved by increasing its thickness and weight, but this leaves less room for internal organs.

In molluscs, such as snails and clams, a thick and nonmobile calcium carbonate shell is primarily used for protection against the environment and predators. A mollusc's shell can grow as the animal grows.

The exoskeleton of arthropods, such as insects and crustaceans, is composed of chitin, a strong, flexible nitrogenous polysaccharide. Their exoskeleton protects them against wear and tear, predators, and drying out—an important feature for arthropods that live on land. The jointed and movable exoskeleton of arthropods is particularly suitable for terrestrial life in another way. The jointed and movable appendages allow

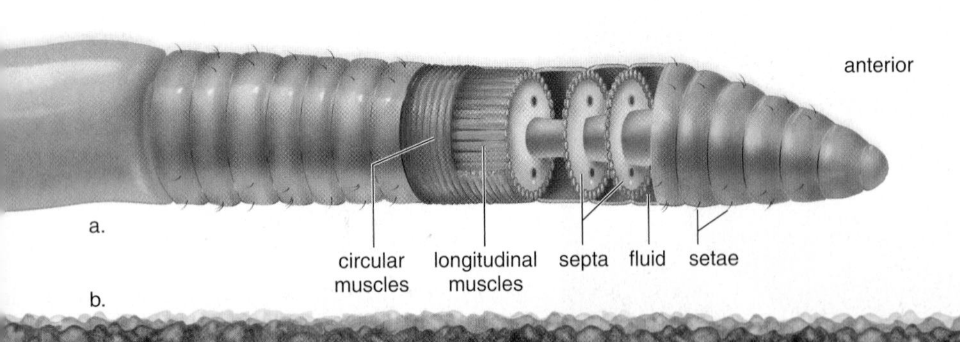


**FIGURE 39.1 Locomotion in an earthworm.**

**a.** The coelom is divided by septa, and each body segment is a separate locomotor unit. There are both circular and longitudinal muscles. **b.** As circular muscles contract, a few segments extend. The worm is held in place by setae, needlelike chitinous structures on each segment of the body. Then, as longitudinal muscles contract, a portion of the body is brought forward. This series of events occurs down the length of the worm.

**FIGURE 39.2** **Exoskeleton.**

Exoskeletons support muscle contraction and prevent drying out. The chitinous exoskeleton of an arthropod is shed as the animal molts; until the new skeleton dries and hardens, the animal is more vulnerable to predators, and muscle contractions may not translate into body movements. In this photo, a dog-day cicada, *Tibicen*, has just finished molting.

flexible movements. To grow, however, arthropods must molt to rid themselves of an exoskeleton that has become too small (Fig. 39.2). Until the new exoskeleton dries and hardens, the animals are more vulnerable to predators.

Both echinoderms and vertebrates have an endoskeleton. The skeleton of echinoderms consists of spicules and plates of calcium carbonate embedded in the living tissue of the body wall. In contrast, the vertebrate endoskeleton is living tissue. Sharks and rays have skeletons composed only of cartilage. Other vertebrates, such as bony fishes, amphibians, reptiles, birds, and mammals, have endoskeletons composed of bone and cartilage. The advantages of the jointed vertebrate endoskeleton are listed in Figure 39.3. An endoskeleton grows with the animal, and molting is not required. It supports the weight of a large animal without limiting the space for internal organs. An endoskeleton also offers protection to vital internal organs, but it is protected by the soft tissues around it. Injuries to soft tissue are usually easier to repair than injuries to a hard skeleton. The vertebrate endoskeleton is also jointed, allowing for complex movements such as swimming, jumping, flying, and running.

### Check Your Progress 39.1

1. Identify the type of skeleton of each of the following animals: butterfly, crow, shrimp, oyster, and sea urchin.
2. Stick out your tongue. What type of muscular support system makes this possible?
3. A dead earthworm loses its cylindrical shape, becoming limp and flattened. Why?

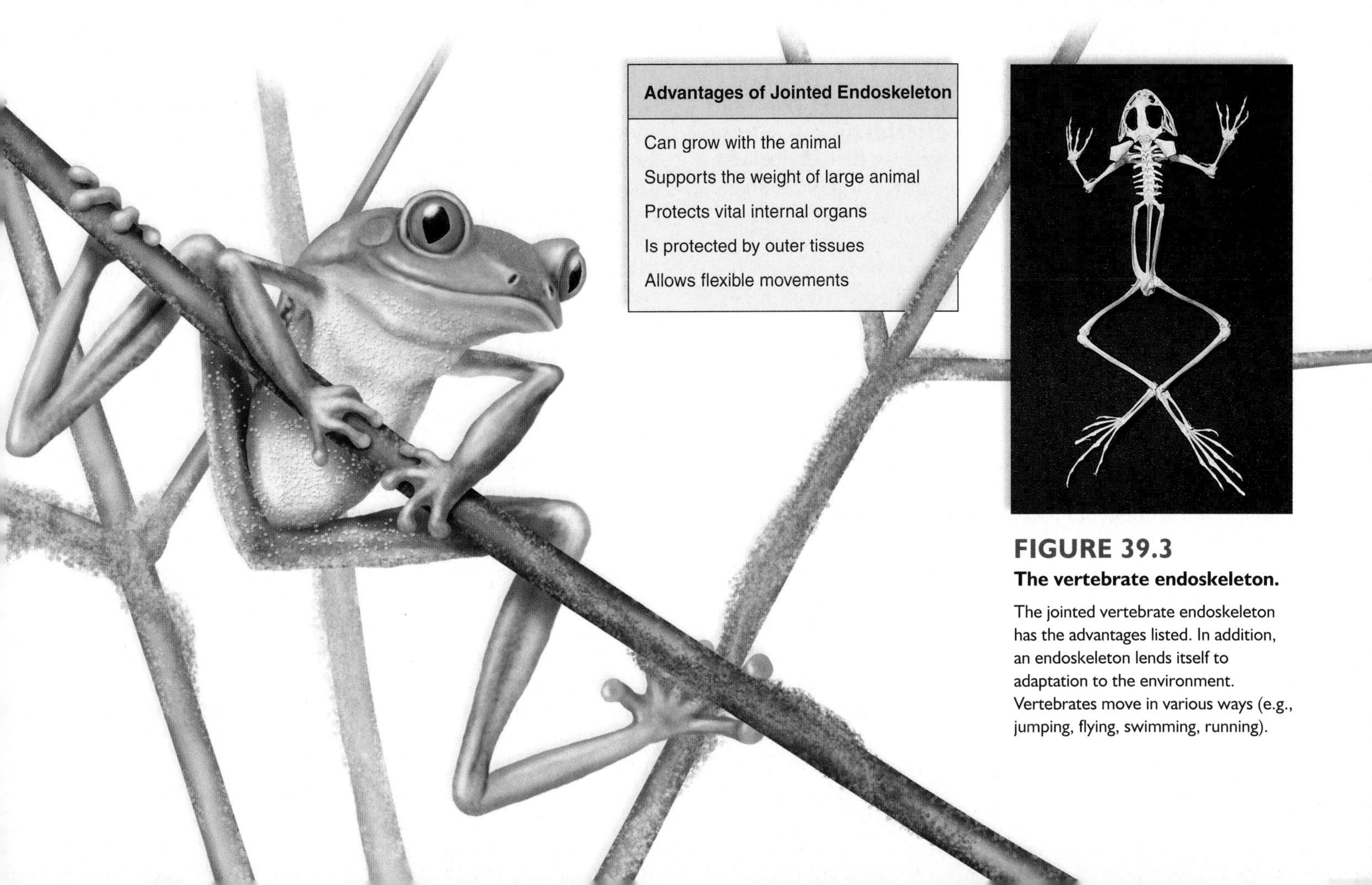


**FIGURE 39.3**
**The vertebrate endoskeleton.**

The jointed vertebrate endoskeleton has the advantages listed. In addition, an endoskeleton lends itself to adaptation to the environment. Vertebrates move in various ways (e.g., jumping, flying, swimming, running).

# 39.2 The Human Skeletal System

The human skeletal system has many functions that contribute to homeostasis.

*The rigid skeleton supports the body.* An endoskeleton grows with the body, and molting, as seen in arthropods, is not needed.

*The skeleton protects vital internal organs,* such as the brain, heart, and lungs. The bones of the skull protect the brain, the rib cage protects the heart and lungs. The vertebrae protect the spinal cord, which makes nervous connections to all the muscles of the limbs.

*The skeleton provides sites for muscle attachment,* making movement possible. While articulations (joints) occur between all the bones, we associate body movement particularly with jointed appendages.

*The skeleton serves as an important storage reservoir for ions,* such as calcium and phosphorus. All bones have a matrix that contains calcium phosphate, a source of calcium ions and phosphate ions in the blood.

*The skeleton produces blood cells.* Blood cells and other blood elements are produced within the red bone marrow of the skull, ribs, sternum, pelvis, and long bones.

## Bone Growth and Renewal

Most of the bones of the human skeleton are composed of cartilage during prenatal development. Because the cartilaginous structures are shaped like the future bones, they provide "models" of these bones. The cartilaginous models are converted to bones when calcium salts are deposited in the matrix (nonliving material), first by the cartilage cells and later by bone-forming cells called **osteoblasts** [Gk. *osteon,* bone, and *blastos,* bud]. This type of bone is called endochondral bone, or replacement bone. The conversion of cartilaginous models to bones is called endochondral ossification.

There are also examples of ossification that have no previous cartilaginous model. This type of ossification occurs in the dermis and forms bones called dermal bones. Several dermal bones include the mandible (lower jaw), certain bones of the skull, and the clavicle (collarbone). During intramembranous ossification, fibrous connective tissue membranes give support as ossification begins.

Endochondral ossification of a long bone begins in a region called a primary ossification center, located in the middle of the cartilaginous model. In the primary ossification center, the cartilage is broken down and invaded by blood vessels, and cells in the area mature into bone-forming osteoblasts. Later, secondary ossification centers form at the ends of the model. A cartilaginous *growth plate* remains between the primary ossification center and each secondary center. As long as these plates remain, growth is possible. The rate of growth is controlled by hormones, particularly growth hormone (GH) and the sex hormones. Eventually, the plates become ossified, causing the primary and secondary centers of ossification to fuse, and the bone stops growing.

In the adult, bone is continually being broken down and built up again. Bone-absorbing cells, called **osteoclasts** [Gk. *osteon,* bone, and *klastos,* broken in pieces], break down bone, remove worn cells, and deposit calcium in the blood. In this way, osteoclasts help maintain the blood calcium level and contribute to homeostasis. Among other functions, calcium ions play a major role in muscle contraction and nerve conduction. The blood calcium level is closely regulated by the antagonistic hormones parathyroid hormone (PTH) and calcitonin. PTH promotes the activity of osteoclasts, and calcitonin inhibits their activity to keep the blood calcium level within normal limits.

Assuming that the blood calcium level is normal, bone destruction caused by the work of osteoclasts is repaired by osteoblasts. As they form bone, osteoblasts take calcium from the blood. Eventually, some of these cells get caught in the matrix (nonliving material) they secrete and are converted to **osteocytes** [Gk. *osteon,* bone, and *kytos,* cell], the cells found within the lacunae of osteons. Adults are thought to require more calcium in the diet than do children to promote the work of osteoblasts.

Through this process of remodeling, old bone tissue is replaced by new bone tissue. Osteoclasts and osteoblasts work together to heal broken bones by breaking down and building bone at the site of the damage. Therefore, the thickness of bones can change, depending on exercise and hormone balances. As discussed in the Health Focus on page 726, a thinning of the bones called osteoporosis can occur as we age if proper precautions are not taken.

## Anatomy of a Long Bone

A long bone, such as the humerus, illustrates principles of bone anatomy. When the bone is split open, as in Figure 39.4, the longitudinal section shows that it is not solid but has a cavity called the medullary cavity bounded at the sides by compact bone and at the ends by spongy bone. Beyond the spongy bone, there is a thin shell of compact bone and finally a layer of hyaline cartilage. The cavity of an adult long bone usually contains yellow bone marrow, which is a fat-storage tissue.

**Compact bone** contains many osteons (Haversian systems), where osteocytes lie in tiny chambers called lacunae. The lacunae are arranged in concentric circles around central canals that contain blood vessels and nerves. The lacunae are separated by a matrix of collagen fibers and mineral deposits, primarily calcium and phosphorus salts.

**Spongy bone** has numerous bony bars and plates separated by irregular spaces. Although lighter than compact bone, spongy bone is still designed for strength. Just as braces are used for support in buildings, the solid portions of spongy bone follow lines of stress. The spaces in spongy bone are often filled with **red bone marrow,** a specialized tissue that produces blood cells. This is an additional way the skeletal system assists homeostasis. As you know, red blood cells transport oxygen, and white blood cells are a part of the immune system, which fights infection.

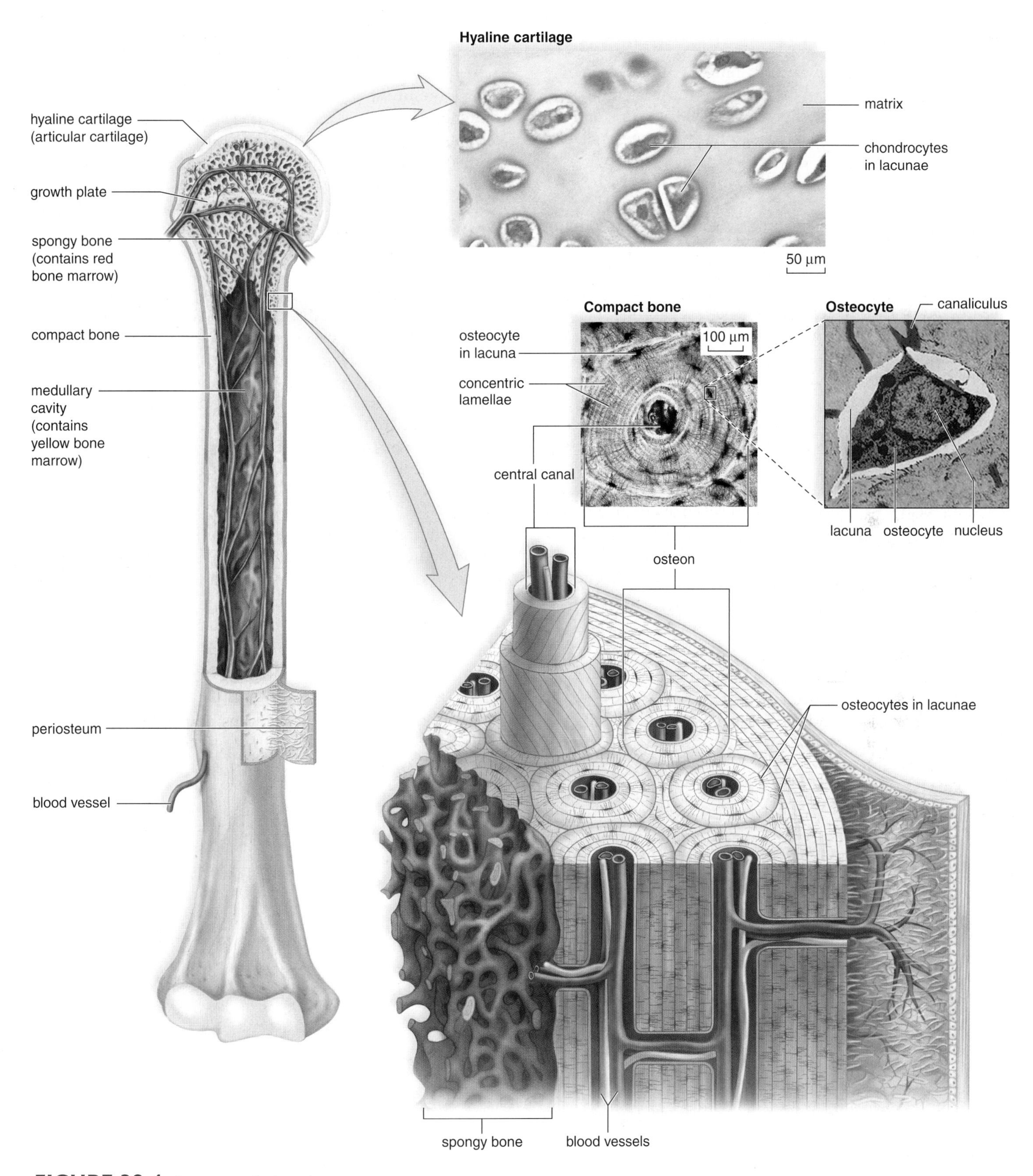


**FIGURE 39.4 Anatomy of a long bone.**

*Left:* A long bone is encased by fibrous membrane (periosteum) except where it is covered at the ends by hyaline cartilage (see micrograph). Spongy bone located beneath the cartilage may contain red bone marrow. The central shaft contains yellow bone marrow and is bordered by compact bone, which is shown in the enlargement and micrograph (*right*).

## The Axial Skeleton

The **axial skeleton** [L. *axis,* axis, hinge; Gk. *skeleton,* dried body] lies in the midline of the body and consists of the skull, the vertebral column, the thoracic cage, the sacrum, and the coccyx (blue labels in Fig. 39.5). A total of 80 bones make up the axial skeleton.

### *The Skull*

The skull, which protects the brain, is formed by the cranium and the facial bones (Fig. 39.6). In newborns, certain bones of the cranium are joined by membranous regions called **fontanels** (or "soft spots"), all of which usually close and become **sutures** by the age of two years. The bones of the cranium contain the sinuses [L. *sinus,* hollow], air spaces lined by mucous membrane that reduce the weight of the skull and give a resonant sound to the voice. Two sinuses, called the mastoid sinuses, drain into the middle ear. Mastoiditis, a condition that can lead to deafness, is an inflammation of these sinuses.

The major bones of the cranium have the same names as the lobes of the brain. On the top of the cranium, the frontal bone forms the forehead, and the parietal bones extend to the sides. Below the much larger parietal bones, each temporal bone has an opening that leads to the middle ear. In the rear of the skull, the occipital bone curves to form the base of the skull. At the base of the skull, the spinal cord passes upwards through a large opening called the **foramen magnum** [L. *foramen,* hole, and *magnus,* great, large] and becomes the brain stem.

The temporal and frontal bones are cranial bones that contribute to the face. The sphenoid bones account for the flattened areas on each side of the forehead, which we call the temples. The frontal bone not only forms the forehead, but it also has supraorbital ridges where the eyebrows are located. Glasses sit where the frontal bone joins the nasal bones.

The most prominent of the facial bones are the mandible [L. *mandibula,* jaw], the maxillae, the zygomatic bones, and the nasal bones. The mandible, or lower jaw, is the only freely movable portion of the skull (Fig. 39.6), and its action permits us to chew our food. It also forms the "chin." Tooth sockets are located in the mandible and on the maxillae, which form the upper jaw and a portion of the hard palate. The zygomatic bones are the cheekbone prominences, and the nasal bones form the bridge of the nose. Other bones make up the nasal septum, which divides the nose cavity into two regions.

Whereas the ears are formed only by cartilage and not by bone, the nose is a mixture of bones, cartilage, and connective tissues. The lips and cheeks have a core of skeletal muscle.

### *The Vertebral Column and Rib Cage*

The **vertebral column** [L. *vertebra,* bones of backbone] supports the head and trunk and protects the spinal cord and the roots of the spinal nerves. It is a longitudinal axis that

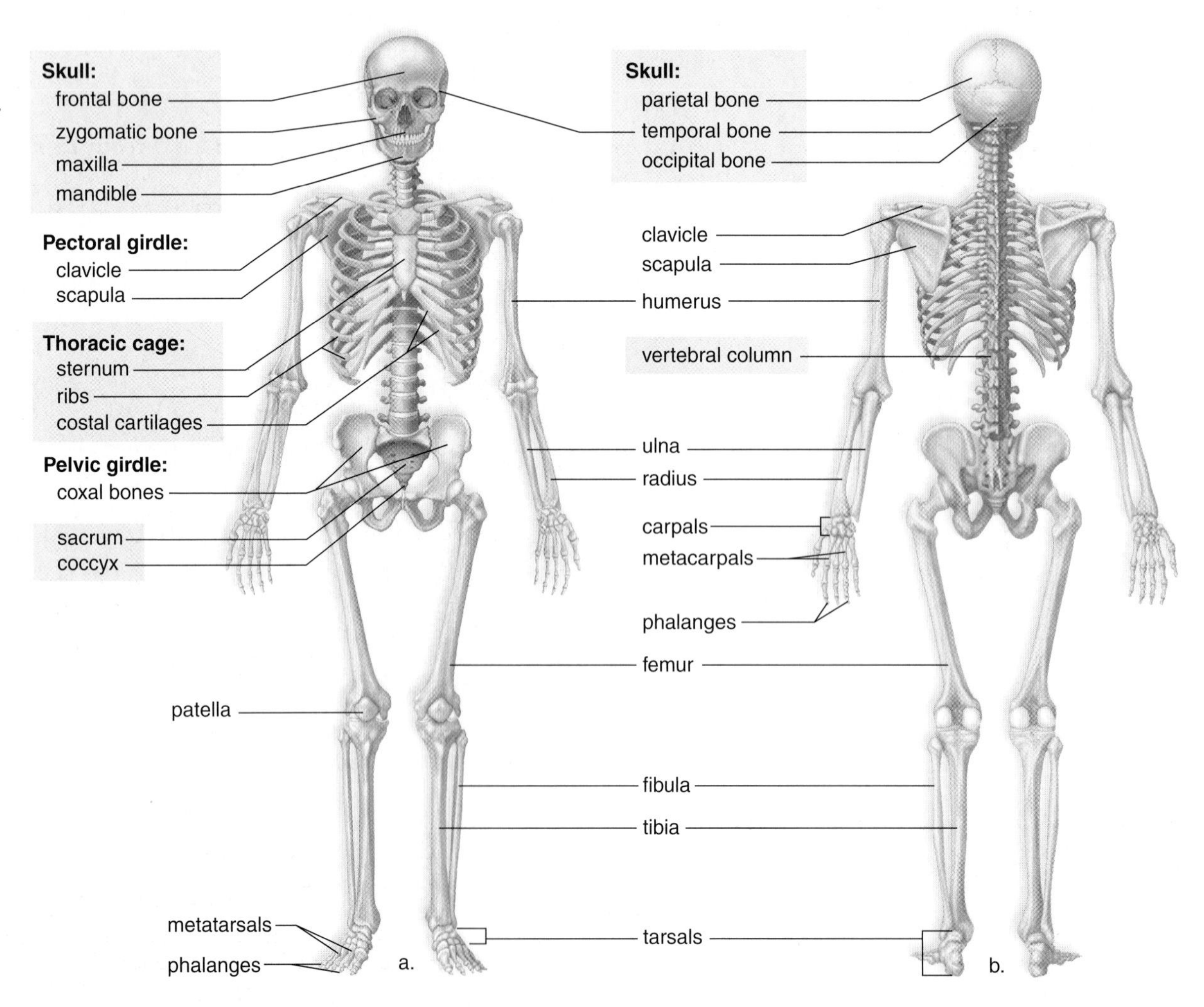


**FIGURE 39.5** **The human skeleton.**
**a.** Anterior view. **b.** Posterior view. The bones of the axial skeleton are in blue and the rest is the appendicular skeleton.

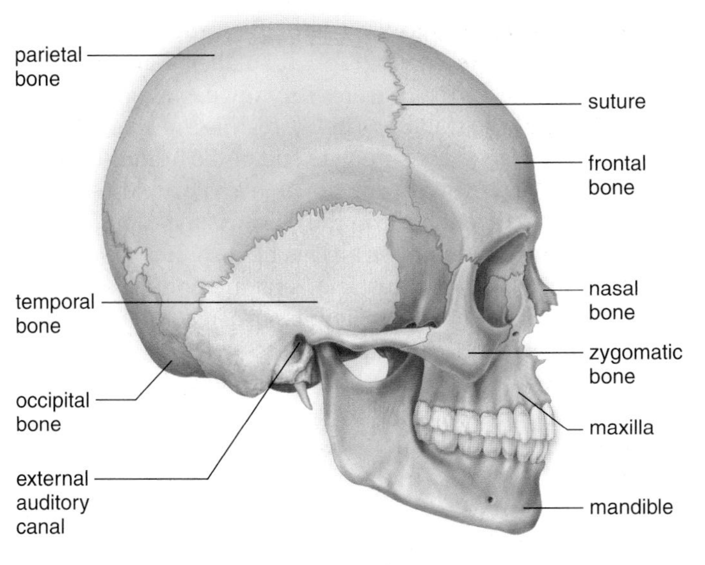


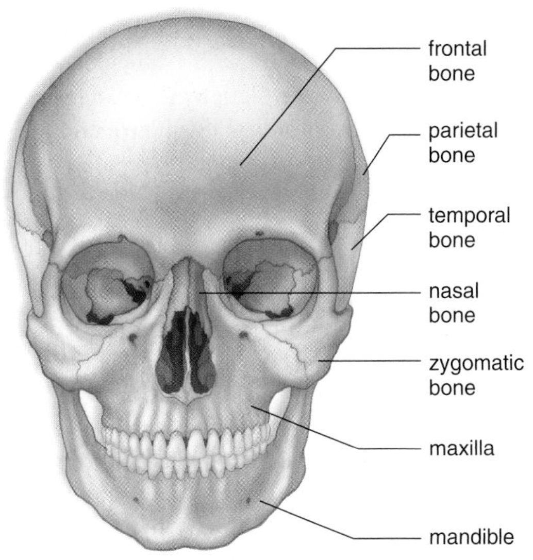


**FIGURE 39.6 The skull.**

The skull consists of the cranium and the facial bones. The frontal bone is the forehead; the zygomatic arches form the cheekbones, and the maxillae form the upper jaw. The mandible has a projection we call the chin.

serves either directly or indirectly as an anchor for all the other bones of the skeleton.

Twenty-four vertebrae make up the vertebral column. Seven cervical vertebrae are located in the neck; 12 thoracic vertebrae are in the thorax; 5 lumbar vertebrae are in the lower back; 5 fused sacral vertebrae form a single sacrum; and several fused vertebrae are in the coccyx, or tailbone. Normally, the vertebral column has four curvatures that provide more resilience and strength for an upright posture than could a straight column. Abnormal curvatures also occur (Fig. 39.7).

Intervertebral disks, composed of fibrocartilage between the vertebrae, provide padding. They prevent the vertebrae from grinding against one another and absorb shock caused by movements such as running, jumping, and even walking. The presence of the disks allows the vertebrae to move as we bend forward, backward, and from side to side. Unfortunately, these disks become weakened with age and can herniate and rupture. Pain results if a disk presses against the spinal cord and/or spinal nerves. The body may heal itself, or the disk can be removed surgically. If so, the vertebrae can be fused together, but this limits the flexibility of the body.

**Rib Cage.** The thoracic vertebrae are a part of the rib cage. The rib cage also contains the ribs, the costal cartilages, and the sternum, or breastbone (Fig. 39.8).

There are 12 pairs of ribs. The upper 7 pairs are "true ribs" because they attach directly to the sternum. The lower 5 pairs do not connect directly to the sternum and are called the "false ribs." Three pairs of false ribs attach by means of

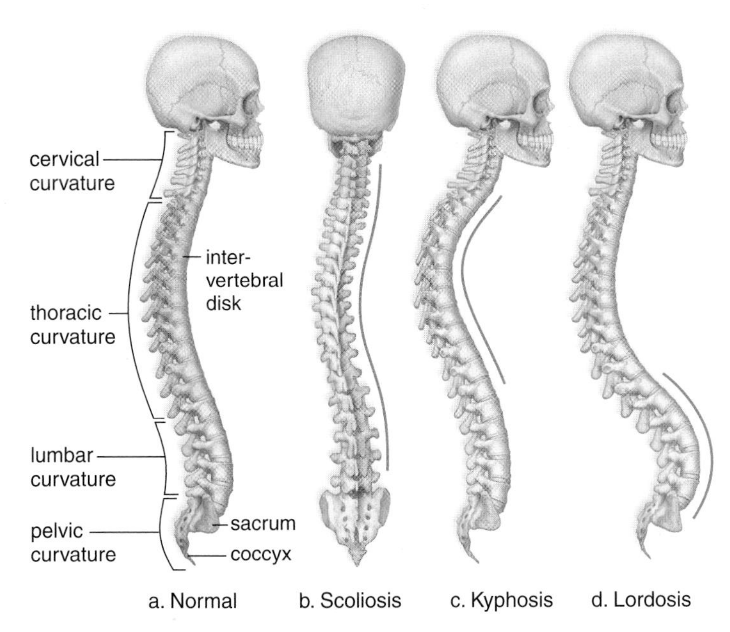


**FIGURE 39.7 Spinal curvatures.**

**a.** The vertebral column normally has four curvatures. **b.** Scoliosis is abnormal lateral (sideways) curvature. **c.** Kyphosis is abnormal thoracic curvature. **d.** Lordosis is abnormal lumbar curvature.

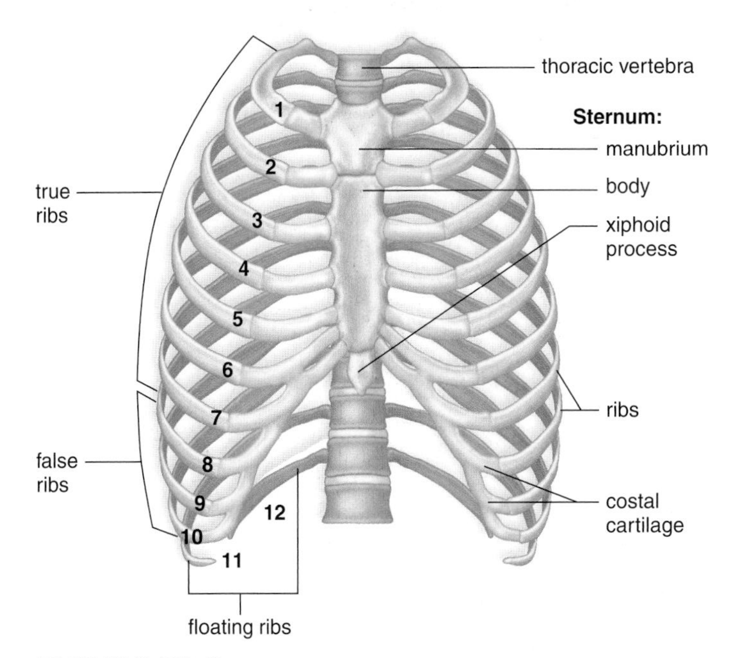


**FIGURE 39.8 The rib cage.**

The rib cage consists of the thoracic vertebrae, the 12 pairs of ribs, the costal cartilages, and the sternum, or breastbone.

a common cartilage, and 2 pairs are "floating ribs" because they do not attach to the sternum at all.

The rib cage demonstrates how the skeleton is protective but also flexible. The rib cage protects the heart and lungs; yet it swings outward and upward upon inspiration and then downward and inward upon expiration.

## The Appendicular Skeleton

The **appendicular skeleton** [L. *appendicula,* dim. of *appendix,* appendage] consists of the bones within the pectoral and pelvic girdles and the attached limbs (see Fig. 39.5). The pectoral (shoulder) girdle and upper limbs are specialized for flexibility, but the pelvic girdle (hipbones) and lower limbs are specialized for strength. A total of 126 bones make up the appendicular skeleton.

### *The Pectoral Girdle and Upper Limb*

The components of the **pectoral girdle** [Gk. *pechys,* forearm] are only loosely linked together by ligaments (Fig. 39.9). Each clavicle (collarbone) connects with the sternum and the scapula (shoulder blade), but the scapula is held in place only by muscles. This allows it to glide and rotate on the clavicle. The single long bone in the arm, the humerus, has a smoothly rounded head that fits into a socket of the scapula. The socket, however, is very shallow and much smaller than the head. Although this means that the arm can move in almost any direction, there is little stability. Therefore, this is the joint that is most apt to dislocate. The opposite end of the humerus meets the two bones of the lower arm, the ulna and the radius, at the elbow. (The prominent bone in the elbow is the topmost part of the ulna.) When the upper limb is held so that the palm is turned frontward, the radius and ulna are about parallel to one another. When the upper limb is turned so that the palm is next to the body, the radius crosses in front of the ulna, a feature that contributes to the easy twisting motion of the forearm.

The many bones of the hand increase its flexibility. The wrist has eight carpal bones, which look like small pebbles. From these, five metacarpal bones fan out to form a framework for the palm. The metacarpal bone that leads to the thumb is placed in such a way that the thumb can reach out and touch the other digits. (Digits is a term that refers to either fingers or toes.) Beyond the metacarpals are the phalanges, the bones of the fingers and the thumb. The phalanges of the hand are long, slender, and lightweight.

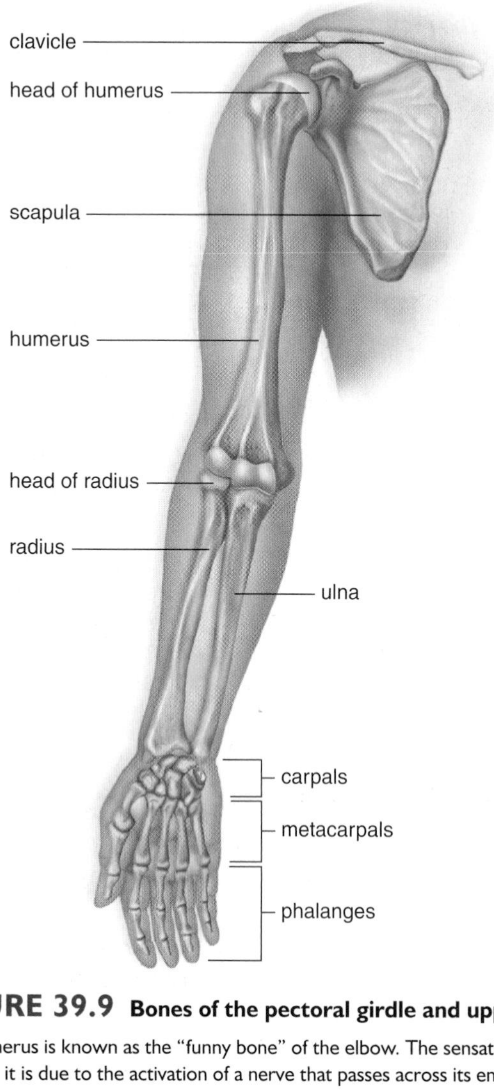

clavicle
head of humerus
scapula
humerus
head of radius
radius
ulna
carpals
metacarpals
phalanges

**FIGURE 39.9 Bones of the pectoral girdle and upper limb.**

The humerus is known as the "funny bone" of the elbow. The sensation upon bumping it is due to the activation of a nerve that passes across its end.

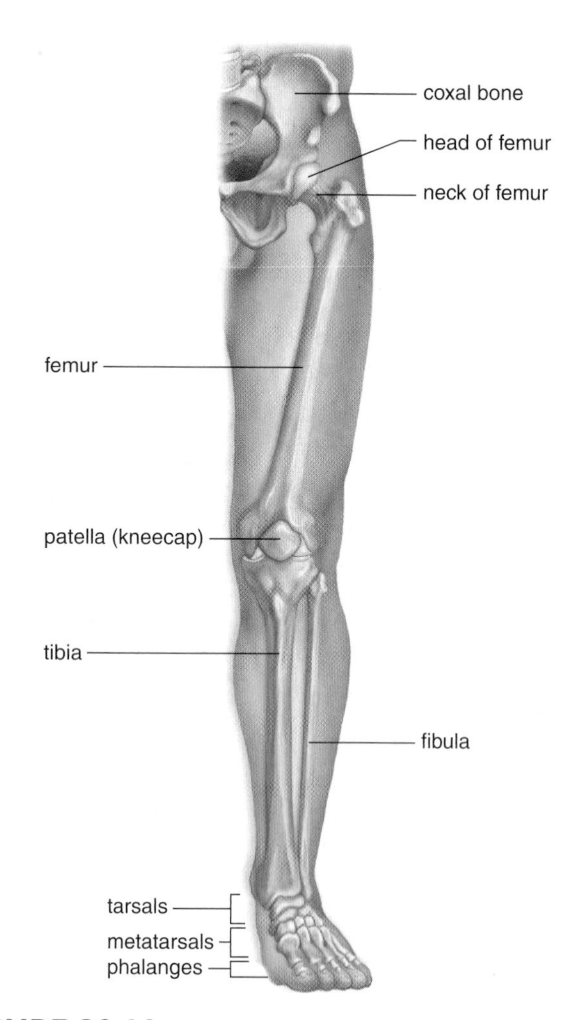


**FIGURE 39.10 Bones of the pelvic girdle and lower limb.**

The femur is our strongest bone—it withstands a pressure of 540 kg per 2.5 $cm^3$ when we walk.

### *The Pelvic Girdle and Lower Limb*

The **pelvic girdle** [L. *pelvis,* basin] (Fig. 39.10) consists of two heavy, large coxal bones (hipbones). The coxal bones are anchored to the sacrum, and together these bones form a hollow cavity called the pelvic cavity. The wider pelvic cavity in females than males accommodates pregnancy and childbirth. The weight of the body is transmitted through the pelvis to the lower limbs and then onto the ground. The largest bone in the body is the femur, or thighbone.

In the leg, the larger of the two bones, the tibia, has a ridge we call the shin. Both of the bones of the leg have a prominence that contributes to the ankle—the tibia on the inside of the ankle and the fibula on the outside of the ankle. Although there are seven tarsal bones in the ankle, only one receives the weight and passes it on to the heel and the ball of the foot. If you wear high-heeled shoes, the weight is thrown toward the front of your foot. The metatarsal bones participate in forming the arches of the foot. There is a longitudinal arch from the heel to the toes and a transverse arch across the foot. These provide a stable, springy base for the body. If the tissues that bind the metatarsals together become weakened, "flat feet" are apt to result. The bones of the toes are called phalanges, just as are those of the fingers, but in the foot the phalanges are stout and extremely sturdy.

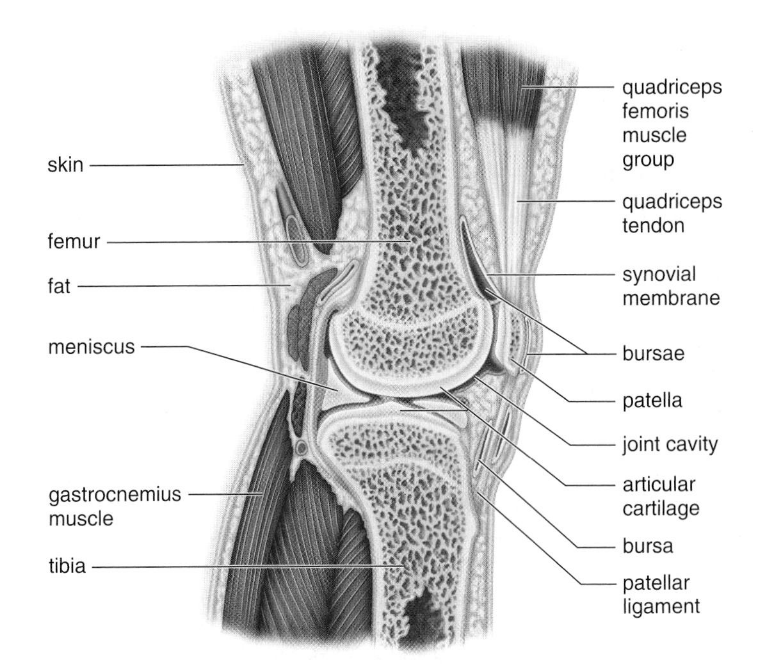


**FIGURE 39.11 Knee joint.**

The knee joint is an example of a synovial joint. The cavity between the bones is encased by ligaments and lined by synovial membrane. The patella (kneecap) serves to guide the quadriceps tendon over the joint when flexion or extension occurs.

## Classification of Joints

Bones are connected at the **joints,** which are classified as fibrous, cartilaginous, and synovial. Fibrous joints, such as the sutures that exist between the cranial bones, are immovable. Cartilaginous joints, such as those between the vertebrae, are slightly movable. The vertebrae are also separated by disks, which increase their flexibility. The two hipbones are slightly movable because they are ventrally joined by cartilage. Owing to hormonal changes, this joint becomes more flexible during late pregnancy, allowing the pelvis to expand during childbirth.

In freely movable **synovial joints,** the two bones are separated by a cavity. **Ligaments,** composed of fibrous connective tissue, bind the two bones to each other, holding them in place as they form a capsule. In a "double-jointed" individual, the ligaments are unusually loose. The joint capsule is lined by synovial membrane, which produces synovial fluid, a lubricant for the joint.

The knee is an example of a synovial joint (Fig. 39.11). In the knee, as in other freely movable joints, the bones are capped by a layer of articular cartilage. In addition, there are crescent-shaped pieces of cartilage between the bones called menisci [Gk. *meniscus,* crescent]. These give added stability, helping to support the weight placed on the knee joint. Unfortunately, athletes often suffer injury of the meniscus, known as torn cartilage. Thirteen fluid-filled sacs called bursae (sing., **bursa**) occur around the knee joint. Bursae ease friction between tendons and ligaments and between tendons and bones. Inflammation of the bursae is called bursitis. Tennis elbow is a form of bursitis.

There are different types of movable joints. The knee and elbow joints are hinge joints because, like a hinged door, they largely permit movement in one direction only. In a pivot joint, a small, cylindrical projection of one bone pivots within the ring formed of bone and ligament of another bone, making rotation possible. The joint between the first two cervical vertebrae, which permits side-to-side movement of the head, is an example of a pivot joint. More movable are the ball-and-socket joints; for example, the ball of the femur fits into a socket on the hipbone. Ball-and-socket joints allow movement in all planes and even rotational movement.

Synovial joints are subject to arthritis. In rheumatoid arthritis, the synovial membrane becomes inflamed and thickened. Degenerative changes take place that make the joint almost immovable and painful to use. There is evidence that these effects are brought on by an autoimmune reaction. In old-age arthritis, or osteoarthritis, the cartilage at the ends of the bones disintegrates so that the two bones become rough and irregular. This type of arthritis is apt to affect the joints that have received the greatest use over the years.

### Check Your Progress 39.2

1. What are bone-forming and bone-absorbing cells called? Name the cells that reside in the lacunae of osteons.
2. Determine whether each of the following bones belongs to the axial or appendicular skeleton: sacrum, frontal bone, humerus, tibia, vertebra, coxal bone, temporal bone, scapula, and sternum.

*health focus*

## You Can Avoid Osteoporosis

Osteoporosis is a condition in which the bones are weakened due to a decrease in the bone mass that makes up the skeleton. Throughout life, bones are continuously remodeled. While a child is growing, the rate of bone formation is greater than the rate of bone breakdown. The skeletal mass continues to increase until ages 20 to 30. After that, there is an equal rate of formation and breakdown of bone mass until ages 40 to 50. Then, reabsorption begins to exceed formation, and the total bone mass slowly decreases.

Over time, men are apt to lose 25% and women to lose 35% of their bone mass. But we have to consider that men tend to have denser bones than women anyway, and their testosterone (male sex hormone) level generally does not begin to decline significantly until after age 65. In contrast, the estrogen (female sex hormone) level in women begins to decline at about age 45. Since sex hormones play an important role in maintaining bone strength, this difference means that women are more likely than men to suffer a higher incidence of fractures, involving especially the hip vertebrae, long bones, and pelvis. Although osteoporosis may at times be the result of various disease processes, it is essentially a disease that occurs as we age. An estimated 10 million people in the United States have osteoporosis, which results in 1.5 million fractures each year.

### How to Avoid Osteoporosis

Everyone can take measures to avoid having osteoporosis when they get older. Adequate dietary calcium throughout life is an important protection against osteoporosis. The U.S. National Institutes of Health recommend a calcium intake of 1,200–1,500 mg per day during puberty. Males and females require 1,000 mg per day until the age of 65 and 1,500 mg per day after age 65. In older women 1,500 mg per day is especially desirable.

A small daily amount of vitamin D is also necessary for the body to use calcium correctly. Exposure to sunlight is required to allow skin to synthesize a precursor to vitamin D. If you reside on or north of a "line" drawn from Boston to Milwaukee, to Minneapolis, to Boise, chances are you're not getting enough vitamin D during the winter months. Therefore, you should avail yourself of vitamin D present in fortified foods such as low-fat milk and cereal.

Very inactive people, such as those confined to bed, lose bone mass 25 times faster than people who are moderately active. On the other hand, a regular, moderate, weight-bearing exercise like walking or jogging is another good way to maintain bone strength (Fig. 39A).

Postmenopausal women with any of the following risk factors should have an evaluation of their bone density:

- white or Asian race
- thin body type
- family history of osteoporosis
- early menopause (before age 45)
- smoking
- a diet low in calcium, or excessive alcohol consumption and caffeine intake
- sedentary lifestyle

Presently bone density is measured by a method called dual energy X-ray absorptiometry (DEXA). This test measures bone density based on the absorption of photons generated by an X-ray tube. Soon there may be blood and urine tests to detect the biochemical markers of bone loss. Then it will be made easier for physicians to screen older women and at-risk men for osteoporosis.

If the bones are thin, it is worthwhile to take all possible measures to gain bone density because even a slight increase can significantly reduce fracture risk. Hormone therapy includes black cohosh, which is a phytoestrogen (estrogen made by a plant as opposed to an animal). Calcitonin is a naturally occurring hormone whose main site of action is the skeleton, where it inhibits the action of osteoclasts, the cells that break down bone. Also, alendronate is a drug that acts similarly to calcitonin. After three years of alendronate therapy, an increase in spinal density by about 8% and hip density by about 7% is obtained. Promising new drugs include slow-release fluoride therapy and certain growth hormones. These medications stimulate the formation of new bone.

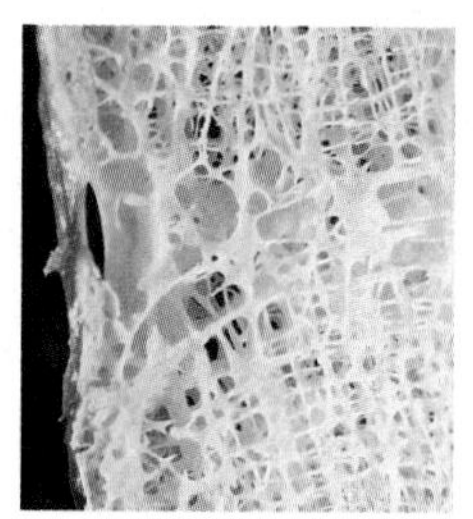

a. normal bone

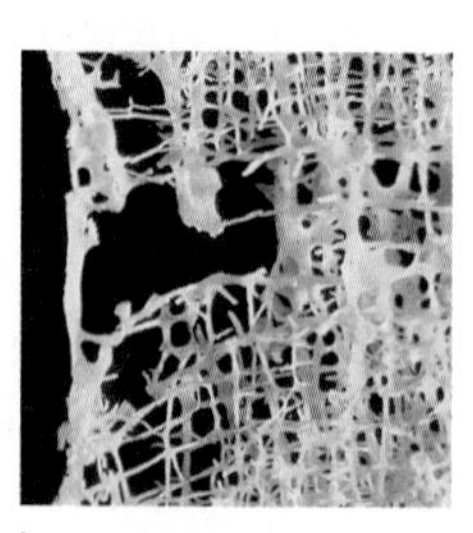

b. osteoporosis

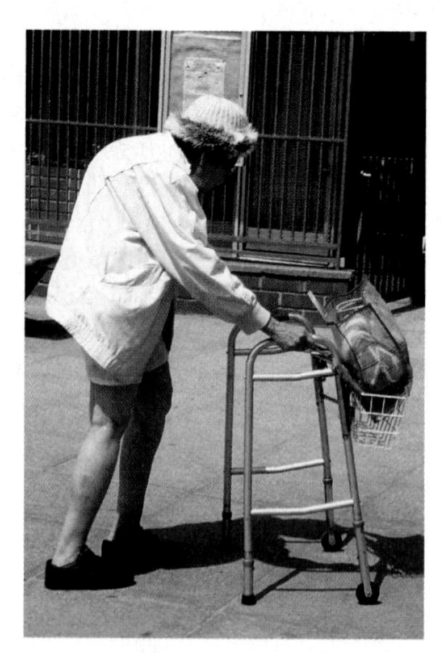

c.

**FIGURE 39A Preventing osteoporosis.**
*Weight-lifting exercise, when we are young, can help prevent osteoporosis when we are older.* ***a.*** *Normal bone growth compared with bone from a person with* ***(b)*** *osteoporosis.* ***c.*** *An elderly person with osteoporosis.*

## 39.3 The Human Muscular System

Three types of muscle tissue can be found in humans: smooth, cardiac, and skeletal muscle. Skeletal muscle, or striated voluntary muscle, is important in maintaining posture, providing support, and allowing for movement. In addition, skeletal muscle is important in homeostasis by helping maintain a constant body temperature. The contraction of skeletal muscle causes ATP breakdown, releasing heat that is distributed about the body.

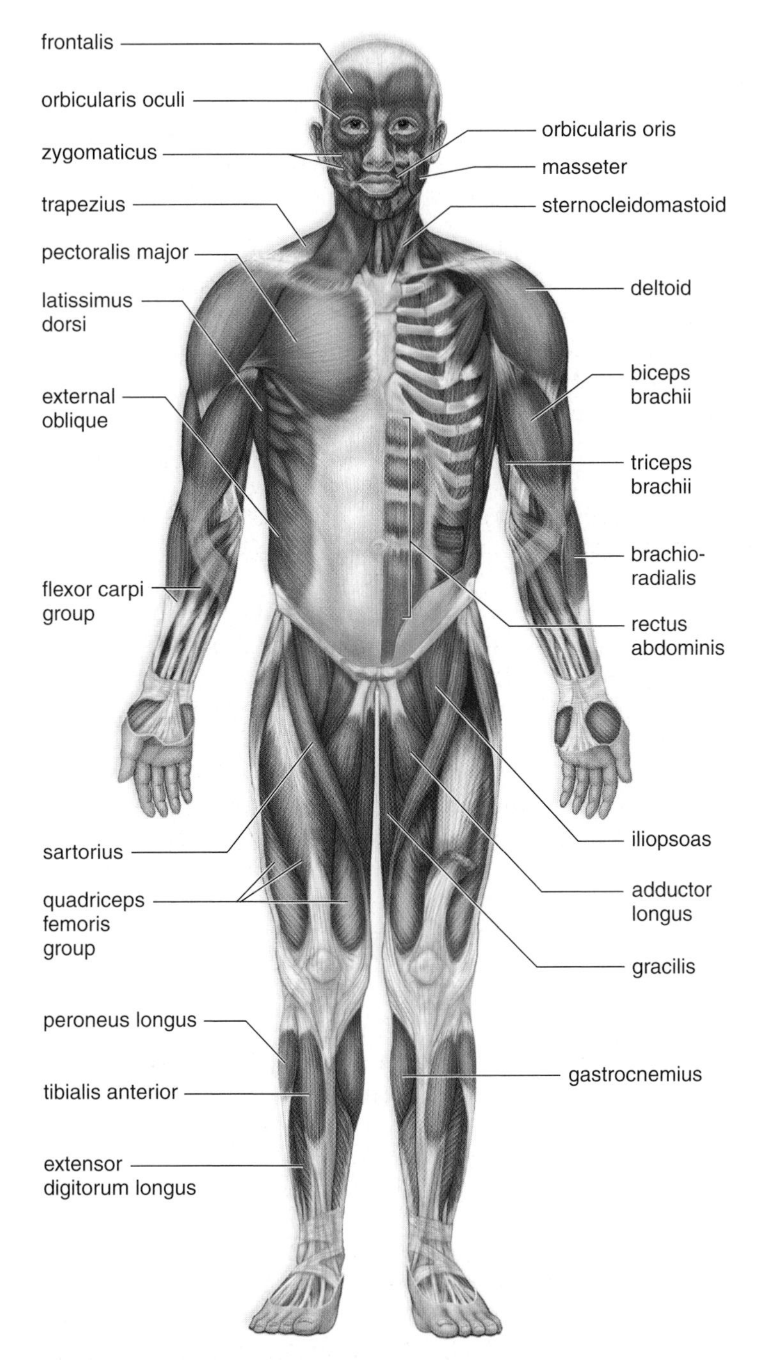


**FIGURE 39.12 Human musculature.**

Anterior view of some of the major superficial skeletal muscles.

### Macroscopic Anatomy and Physiology

The nearly 700 skeletal muscles and their associated tissues make up approximately 40% of the weight of an average human. Several of the major superficial muscles are illustrated in Figure 39.12. Skeletal muscles are attached to the skeleton by bands of fibrous connective tissue called **tendons** [L. *tendo,* stretch]. When muscles contract, they shorten. Therefore, muscles can only pull; they cannot push. Because of this, skeletal muscles must work in antagonistic pairs. If one muscle of an antagonistic pair flexes the joint and bends the limb, the other one extends the joint and straightens the limb. Figure 39.13 illustrates this principle.

In the laboratory, if a muscle is given a rapid series of threshold stimuli (strong enough to bring about action potentials, as described on page 685), it can respond to the next stimulus without relaxing completely. In this way, muscle contraction summates until maximal sustained contraction, called **tetanus,** is achieved. Tetanic contractions ordinarily occur in the body's muscles whenever skeletal muscles are actively used. Even when muscles appear to be at rest, they exhibit **tone,** in which some of their fibers are contracting. Muscle tone is particularly important in maintaining posture. If all the fibers within the muscles of the neck, trunk, and legs were to suddenly relax, the body would collapse.

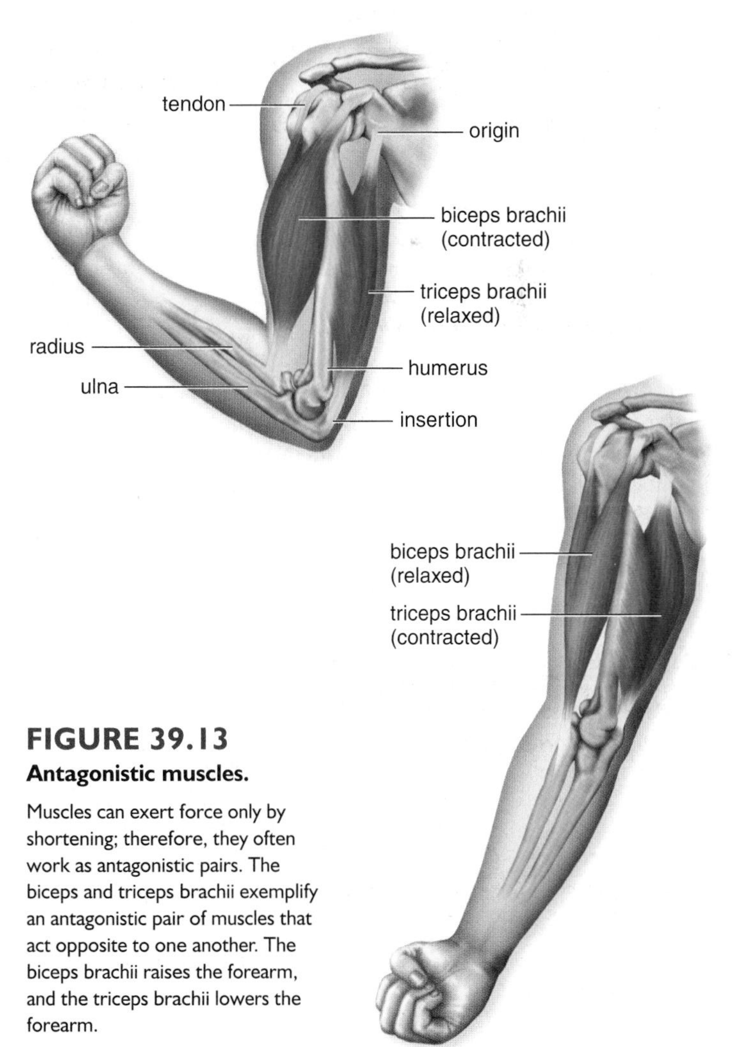


**FIGURE 39.13**
**Antagonistic muscles.**

Muscles can exert force only by shortening; therefore, they often work as antagonistic pairs. The biceps and triceps brachii exemplify an antagonistic pair of muscles that act opposite to one another. The biceps brachii raises the forearm, and the triceps brachii lowers the forearm.

## Microscopic Anatomy and Physiology

A vertebrate skeletal muscle is composed of a number of muscle fibers in bundles. Each muscle fiber is a cell containing the usual cellular components, but some components have special features (Fig. 39.14). The **sarcolemma,** or plasma membrane [Gk. *sarkos,* flesh, and *lemma,* sheath], forms a T (for transverse) system. The T tubules penetrate, or dip down, into the cell so that they come into contact—but do not fuse—with expanded portions of modified endoplasmic reticulum, called the **sarcoplasmic reticulum.** These expanded portions serve as storage sites for calcium ions ($Ca^{2+}$), which are essential for muscle contraction. The sarcoplasmic reticulum encases hundreds and sometimes even thousands of **myofibrils** [Gk. *myos,* muscle; L. *fibra,* thread], which are the contractile portions of muscle fiber.

A myofibril has many sarcomeres. Myofibrils are cylindrical in shape and run the length of the muscle fiber. The light microscope shows that a myofibril has light and dark bands called striations. It is these bands that cause skeletal muscle to appear striated. The electron microscope reveals that the striations of myofibrils are formed by the placement of protein filaments within contractile units called **sarcomeres** [Gk. *sarkos,* flesh, and *meros,* part].

Examining sarcomeres when they are relaxed shows that a sarcomere extends between two dark lines called the Z lines. There are two types of protein filaments: thick filaments made up of **myosin,** and thin filaments made up of **actin.** The I band is light colored because it contains only actin filaments attached to a Z line. The dark regions of the A band contain overlapping actin and myosin filaments, and its H zone has only myosin filaments.

### *Sliding Filament Model*

Examining muscle fibers when contracted shows that the sarcomeres within the myofibrils have shortened. When a sarcomere shortens, the actin (thin) filaments slide past the myosin (thick) filaments and approach one another. This causes the I band to shorten and the H zone to nearly or completely disappear. The movement of actin filaments in relation to myosin filaments is called the **sliding filament model** of muscle contraction. During the sliding process, the sarcomere shortens, even though the filaments themselves remain the same length. When you play "tug of war," your hands grasp the rope, pull, let go, attach farther down the rope, and pull again. The myosin heads are like your hands—grasping, pulling, letting go, and then repeating the process.

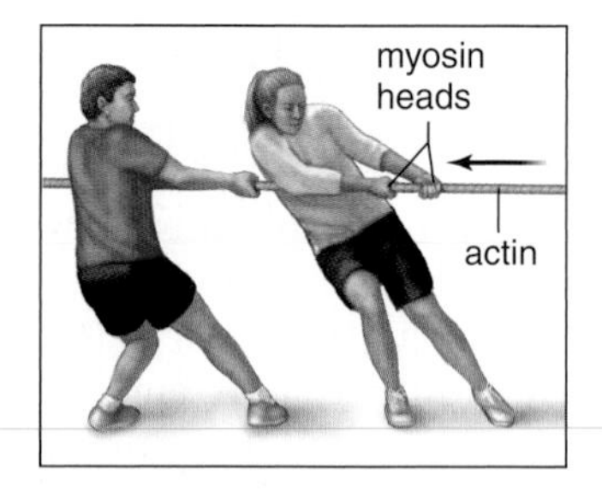


The participants in muscle contraction have the functions listed in Table 39.1. ATP supplies the energy for muscle contraction. Although the actin filaments slide past the myosin filaments, it is the myosin filaments that do the work. Myosin filaments break down ATP and form cross-bridges that attach to and pull the actin filaments toward the center of the sarcomere.

**ATP.** ATP provides the energy for muscle contraction. Although muscle cells contain *myoglobin,* a molecule that stores oxygen, cellular respiration does not immediately supply all the ATP that is needed. In the meantime, muscle fibers rely on *creatine phosphate* (phosphocreatine), a storage form of high-energy phosphate. Creatine phosphate cannot directly participate in muscle contraction. Instead, it anaerobically regenerates ATP by the following reaction:

$$\text{creatine—P} + \text{ADP} \longrightarrow \text{ATP} + \text{creatine}$$

This reaction occurs in the midst of sliding filaments, and therefore, this method of supplying ATP is the speediest energy source available to muscles.

When all of the creatine phosphate is depleted, mitochondria may by then be producing enough ATP for muscle contraction to continue. If not, fermentation is a second way for muscles to supply ATP without consuming oxygen. Fermentation, which is apt to occur when strenuous exercise first begins, supplies ATP for only a short time, and lactate builds up. Whether lactate causes muscle aches and fatigue upon exercising is now being questioned.

We all have had the experience of needing to continue deep breathing following strenuous exercise. This continued intake of oxygen, which is required to complete the metabolism of lactate and restore cells to their original energy state, represents an **oxygen debt.** The lactate is transported to the liver, where 20% of it is completely broken down to carbon dioxide ($CO_2$) and water ($H_2O$). The ATP gained by this respiration is then used to reconvert 80% of the lactate to glucose. In persons who train, the number of mitochondria increases, and there is a greater reliance on them rather than on fermentation to produce ATP. Less lactate is produced, and there is less oxygen debt.

**TABLE 39.1**

**Muscle Contraction**

| *Name* | *Function* |
|---|---|
| Actin filaments | Slide past myosin, causing contraction |
| $Ca^{2+}$ | Needed for myosin to bind to actin |
| Myosin filaments | Pull actin filaments by means of cross-bridges; are enzymatic and split ATP |
| ATP | Supplies energy for muscle contraction |

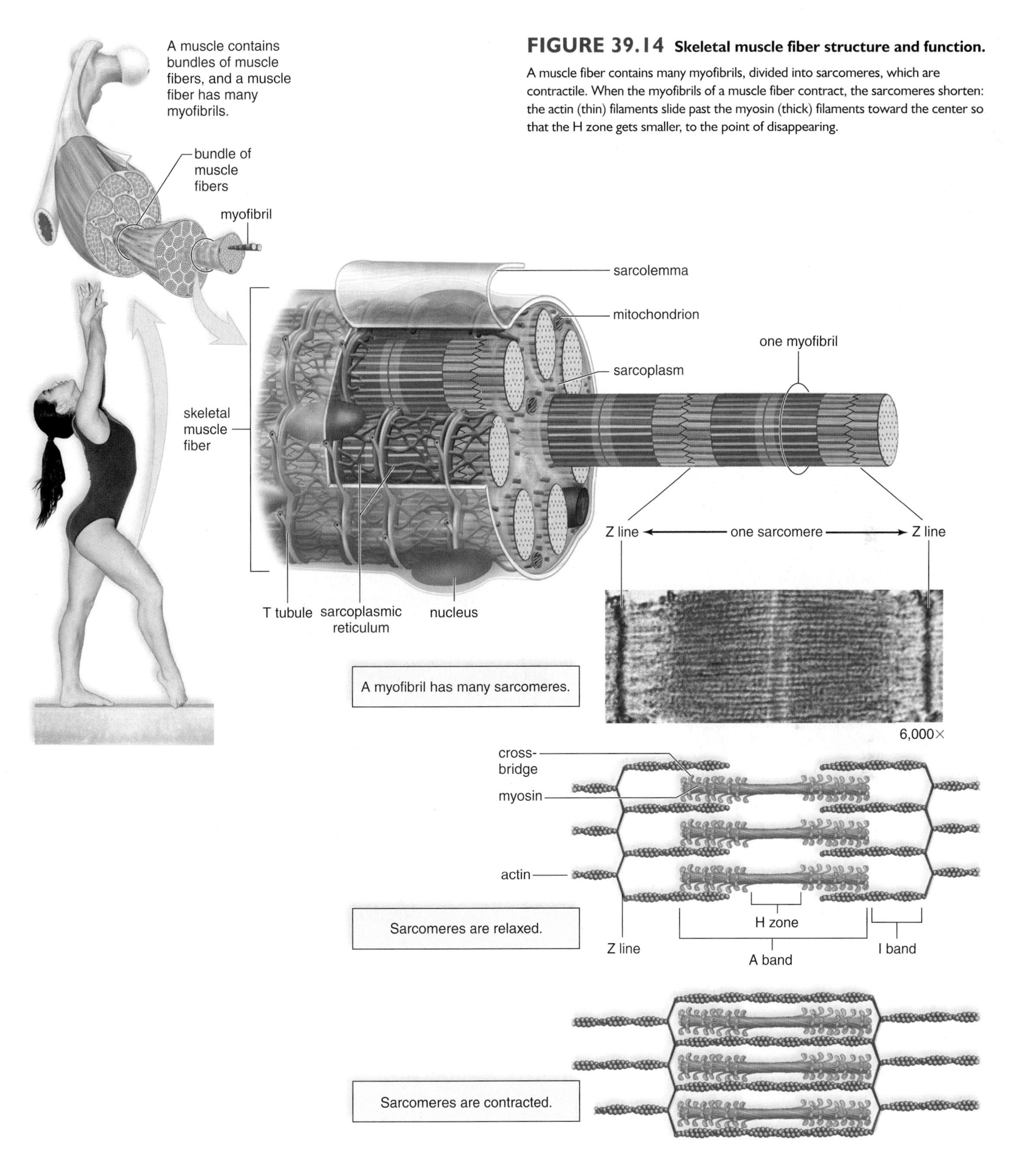


**FIGURE 39.14 Skeletal muscle fiber structure and function.**

A muscle fiber contains many myofibrils, divided into sarcomeres, which are contractile. When the myofibrils of a muscle fiber contract, the sarcomeres shorten: the actin (thin) filaments slide past the myosin (thick) filaments toward the center so that the H zone gets smaller, to the point of disappearing.

## Muscle Innervation

Muscles are stimulated to contract by motor nerve fibers. Nerve fibers have several branches, each of which ends at an axon terminal that lies in close proximity to the sarcolemma of a muscle fiber. A small gap, called a synaptic cleft, separates the axon terminal from the sarcolemma. This entire region is called a **neuromuscular junction** (Fig. 39.15).

Axon terminals contain synaptic vesicles that are filled with the neurotransmitter acetylcholine (ACh). When nerve impulses traveling down a motor neuron arrive at an axon terminal, the synaptic vesicles release ACh into the synaptic cleft. ACh quickly diffuses across the cleft and binds to receptors in the sarcolemma. Now the sarcolemma generates impulses that spread over the sarcolemma and down T tubules to the sarcoplasmic reticulum. The release of calcium from the sarcoplasmic reticulum causes the filaments within sarcomeres to slide past one another. Sarcomere contraction results in myofibril contraction, which in turn results in muscle fiber, and finally muscle, contraction.

Once a neurotransmitter has been released into a neuromuscular junction and has initiated a response, it is removed from the junction. When the enzyme acetylcholinesterase (AChE) breaks down acetylcholine, muscle contraction ceases due to reasons we will discuss next.

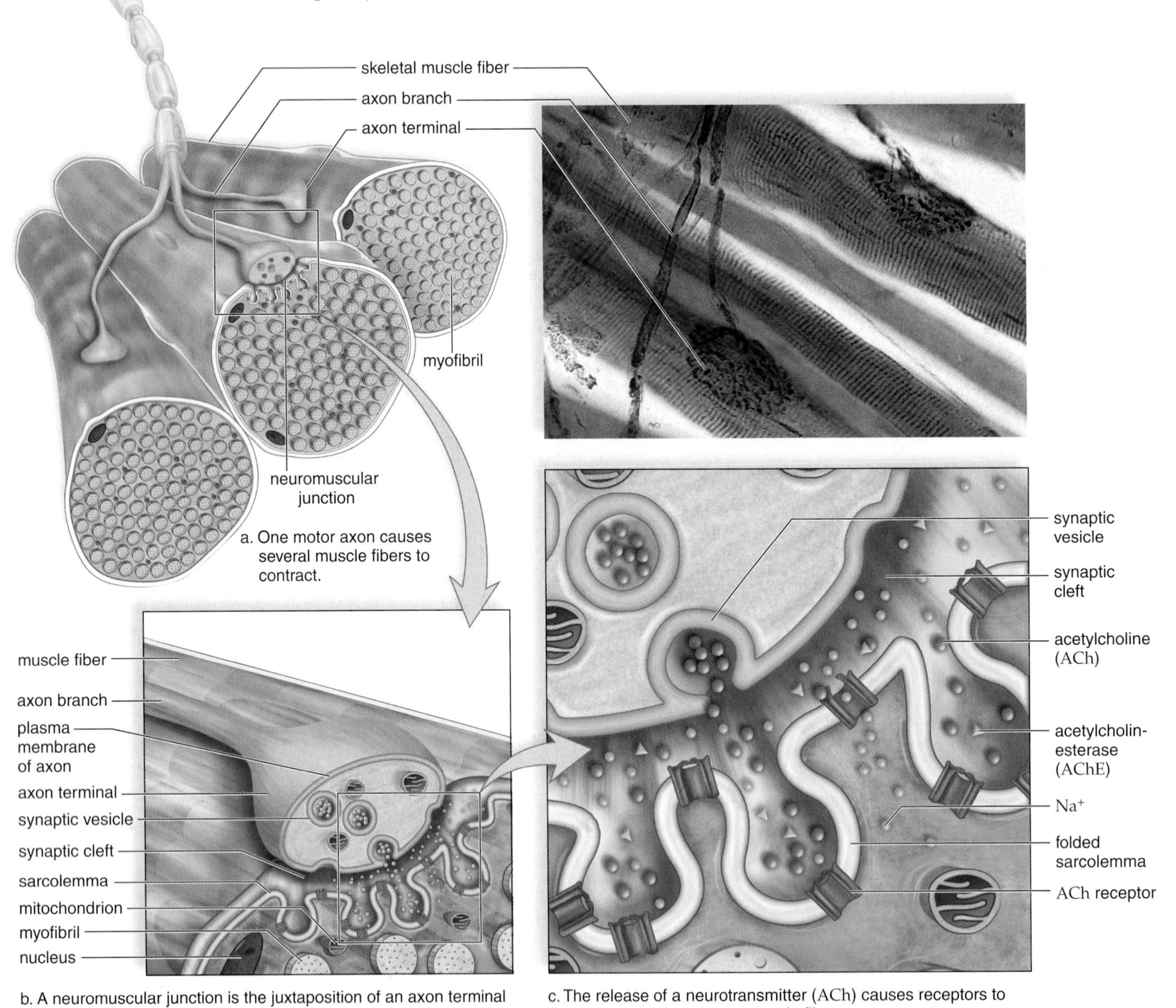


a. One motor axon causes several muscle fibers to contract.

b. A neuromuscular junction is the juxtaposition of an axon terminal and the sarcolemma of a muscle fiber.

c. The release of a neurotransmitter (ACh) causes receptors to open and $Na^+$ to enter a muscle fiber.

**FIGURE 39.15 Neuromuscular junction.**

The branch of a motor nerve fiber ends in an axon terminal that meets but does not touch a muscle fiber. A synaptic cleft separates the axon terminal from the sarcolemma of the muscle fiber. Nerve impulses traveling down a motor fiber cause synaptic vesicles to discharge a neurotransmitter that diffuses across the synaptic cleft. When the neurotransmitter is received by the sarcolemma of a muscle fiber, impulses begin that lead to muscle fiber contraction.

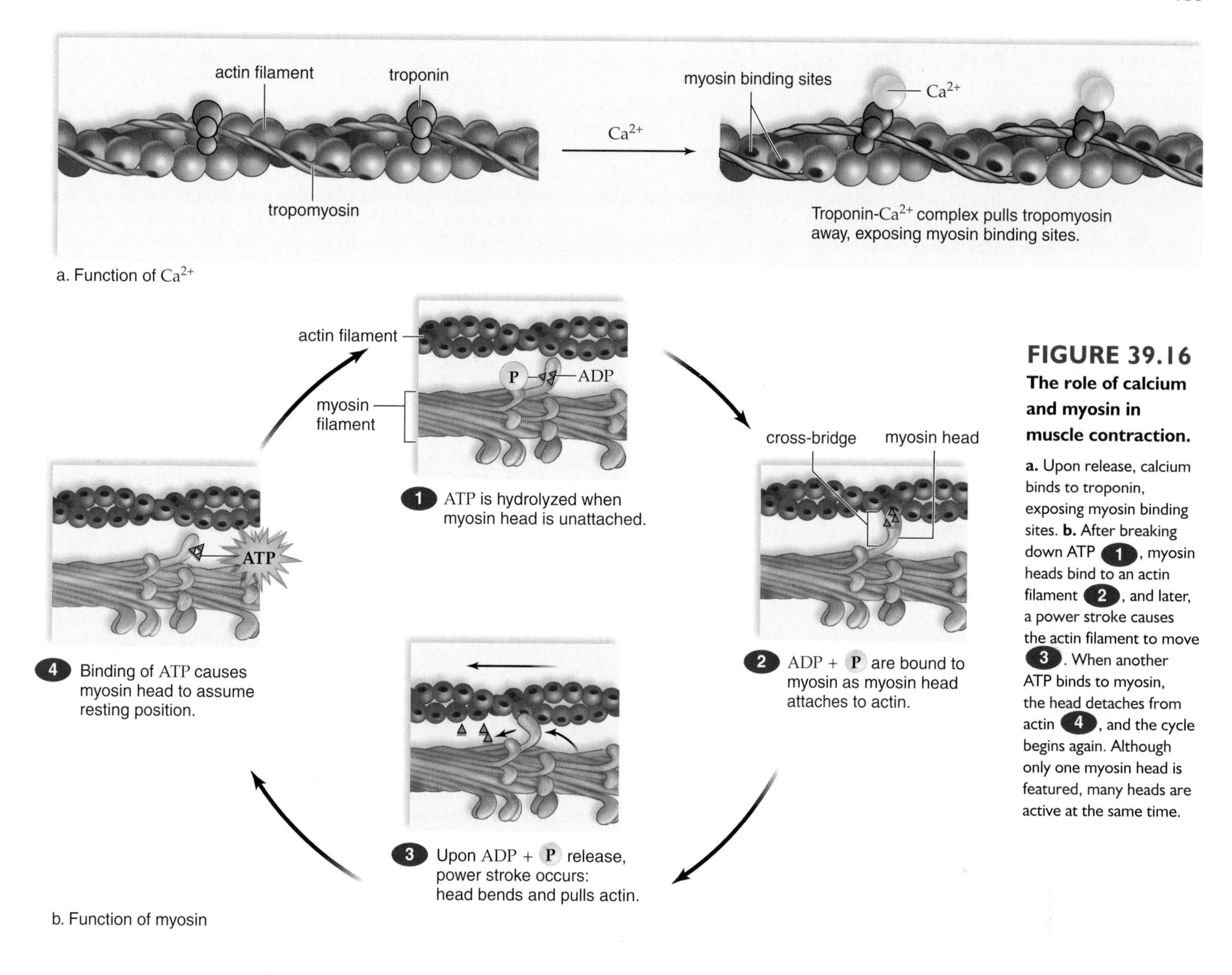


**FIGURE 39.16 The role of calcium and myosin in muscle contraction.**

**a.** Upon release, calcium binds to troponin, exposing myosin binding sites. **b.** After breaking down ATP (1), myosin heads bind to an actin filament (2), and later, a power stroke causes the actin filament to move (3). When another ATP binds to myosin, the head detaches from actin (4), and the cycle begins again. Although only one myosin head is featured, many heads are active at the same time.

## *Role of Calcium in Muscle Contraction*

Figure 39.16 illustrates the placement of two other proteins associated with a thin filament, which is composed of a double row of twisted actin molecules. Threads of tropomyosin wind about an actin filament, and troponin occurs at intervals along the threads. Calcium ions ($Ca^{2+}$) that have been released from the sarcoplasmic reticulum combine with troponin. After binding occurs, the tropomyosin threads shift their position, and myosin binding sites are exposed.

Thick filaments are bundles of myosin molecules with double globular heads. Myosin heads function as ATPase enzymes, splitting ATP into ADP and Ⓟ. This reaction activates the heads so that they bind to actin. The ADP and Ⓟ remain on the myosin heads until the heads attach to actin, forming cross-bridges. Now, ADP and Ⓟ are released, and this causes the cross-bridges to change their positions. This is the power stroke that pulls the thin filaments toward the middle of the sarcomere. When more ATP molecules bind to myosin heads, the cross-bridges are broken as the heads detach from actin. The cycle begins again; the actin filaments move nearer the center of the sarcomere each time the cycle is repeated.

Contraction continues until nerve impulses cease and calcium ions are returned to their storage sites. The membranes of the sarcoplasmic reticulum contain active transport proteins that pump calcium ions back into the calcium storage sites, and muscle relaxation occurs. When a person or animal dies, ATP production ceases. Without ATP, the myosin heads cannot detach from actin, nor can calcium be pumped back into the sarcoplasmic reticulum. As a result, the muscles remain contracted, a phenomenon called *rigor mortis*.

### Check Your Progress 39.3

1. What is an "antagonistic pair" of muscles?
2. What are the microscopic levels of structure in a skeletal muscle?
3. What is the role of ATP in muscle contraction?

## Connecting the Concepts

The adage that structure fits function is evident when one observes how different animals locomote. Among a group of widely diversified animals, such as mammals, various modes of locomotion have evolved. Humans are bipedal and walk on the soles of their feet formed by the tarsal and metatarsal bones. This form of locomotion allows the hands to be free and may have evolved from the habit of monkeys and apes to use only forelimbs as they swing through the branches of trees. Dexterity of hands and feet is actually the ancestral mammalian condition. In humans and apes, the bones of the hands and feet are not fused, and the wrist and ankle can rotate in three dimensions.

Carnivores, such as members of the cat family, walk on their toes. This is an obvious adaptation to running—we also raise the heel and engage the toes in order to run faster. Hoofed mammals, such as horses and deer, have greatly elongated legs and run on the tips of their digits. The hoof of a horse is its third digit only. A cheetah can sprint faster than a horse, but a horse will eventually outdistance the cheetah because it has more endurance.

Mammals that jump, such as kangaroos and rabbits, have a squat shape and elongated hindlimbs that propel them forward. Several groups of arboreal mammals, including flying squirrels and sugar gliders, have membranes attached to their bodies that permit them to glide through the air from tree to tree. Among mammals, only bats truly fly. Their membranous wings are stretched between greatly elongated forelimbs and fingers. In both birds and bats, the wing is moved downward and forward in one motion, and then backward and upward in another.

In terrestrial animals, the skeleton gives the body its shape but also supports it against the pull of gravity. Because water is buoyant, gravity is not much of a problem for aquatic animals, but a shape that reduces friction is quite helpful. Seals, sea lions, whales, and dolphins have a streamlined torpedo shape that facilitates movement through water. A whale has few protruding parts; a male's penis is completely hidden within muscular folds, and the teats of the female lie behind slits on either side of the genital area. Thus, we see that while locomotion chiefly involves musculoskeletal adaptations, other body systems are involved as well.

## summary

### 39.1 Diversity of Skeletons

Three types of skeletons are found in the animal kingdom: hydrostatic skeleton (cnidarians, flatworms, and segmented worms); exoskeleton (certain molluscs and arthropods); and endoskeleton (vertebrates). The rigid but jointed skeleton of arthropods and vertebrates helped them colonize the terrestrial environment.

### 39.2 The Human Skeletal System

The human skeleton gives support to the body, helps protect internal organs, provides sites for muscle attachment, and is a storage area for calcium and phosphorus salts, as well as a site for blood cell formation.

Most bones are cartilaginous in the fetus but are converted to bone during development. A long bone undergoes endochondral ossification in which a cartilaginous growth plate remains between the primary ossification center in the middle and the secondary centers at the ends of the bones. Growth of the bone is possible as long as the growth plates are present, but eventually they too are converted to bone. Bone is constantly being renewed; osteoclasts break down bone, and osteoblasts build new bone. Osteocytes are in the lacunae of osteons; a long bone has a shaft of compact bone and two ends that contain spongy bone. The shaft contains a medullary cavity with yellow marrow, and the ends contain red marrow.

The human skeleton is divided into two parts: (1) the axial skeleton, which is made up of the skull, the vertebral column, the sternum, and the ribs; and (2) the appendicular skeleton, which is composed of the girdles and their appendages.

Joints are classified as immovable, like those of the cranium; slightly movable, like those between the vertebrae; and freely movable (synovial joints), like those in the knee and hip. In synovial joints, ligaments bind the two bones together, forming a capsule containing synovial fluid.

### 39.3 The Human Muscular System

Whole skeletal muscles can only shorten when they contract; therefore, they work in antagonistic pairs. For example, if one muscle flexes the joint and brings the limb toward the body, the other one extends the joint and straightens the limb. A muscle at rest exhibits tone, which is dependent on tetanic contractions.

A whole skeletal muscle is composed of muscle fibers. Each muscle fiber is a cell that contains myofibrils in addition to the usual cellular components. Longitudinally, myofibrils are divided into sarcomeres, which display the arrangement of actin and myosin filaments.

The sliding filament model of muscle contraction says that myosin filaments have cross-bridges, which attach to and detach from actin filaments, causing actin filaments to slide and the sarcomere to shorten. (The H zone disappears as actin filaments approach one another.) Myosin breaks down ATP, and this supplies the energy for muscle contraction. Anaerobic creatine phosphate breakdown and fermentation quickly generate ATP. Sustained exercise requires cellular respiration for the generation of ATP.

Nerves innervate muscles. Nerve impulses traveling down motor neurons to neuromuscular junctions cause the release of ACh, which binds to receptors on the sarcolemma (plasma membrane of a muscle fiber). Impulses begin and move down T tubules that approach the sarcoplasmic reticulum (endoplasmic reticulum of muscle fibers), where calcium is stored. Thereafter, calcium ions are released and bind to troponin. The troponin-$Ca^{2+}$ complex causes tropomyosin threads winding around actin filaments to shift their position, revealing myosin binding sites. Myosin filaments are composed of many myosin molecules with double globular heads. When myosin heads break down ATP, they are ready to attach to actin. The release of ADP + Ⓟ causes myosin heads to change their position. This is the power stroke that causes the actin filament to slide toward the center of a sarcomere. When more ATP molecules bind to myosin, the heads detach from actin, and the cycle begins again.

## understanding the terms

actin 728
appendicular skeleton 724
axial skeleton 722
bursa 725
compact bone 720
endoskeleton 718
exoskeleton 718
fontanel 722
foramen magnum 722
hydrostatic skeleton 718
joint 725
ligament 725
myofibril 728
myosin 728
neuromuscular junction 730
osteoblast 720
osteoclast 720
osteocyte 720
oxygen debt 728
pectoral girdle 724
pelvic girdle 725
red bone marrow 720
sarcolemma 728
sarcomere 728
sarcoplasmic reticulum 728
sliding filament model 728
spongy bone 720
suture 722
synovial joint 725
tendon 727
tetanus 727
tone 727
vertebral column 722

Match the terms to these definitions:

a. ______________ Bone-forming cell.
b. ______________ Movement of actin filaments in relation to myosin filaments, which accounts for muscle contraction.
c. ______________ Muscle protein making up the thin filaments in a sarcomere; its movement shortens the sarcomere, yielding muscle contraction.
d. ______________ Part of the skeleton that consists of the pectoral and pelvic girdles and the bones of the arms and legs.
e. ______________ Portion of the skeleton that provides support and attachment for the arms.
f. ______________ Fluid-filled body cavity surrounded by layers of muscle fibers, which functions for body support and movement.

## reviewing this chapter

1. What are the three types of skeletons found in the animal kingdom and how do they differ? Cite animals that have these types of skeletons. 718–19
2. Give several functions of the skeletal system in humans. How does the skeletal system contribute to homeostasis? 720
3. Contrast compact bone with spongy bone. Explain how bone grows and is renewed. 720–21
4. Distinguish between the axial and appendicular skeletons. 722–25
5. List the bones that form the pectoral girdle and upper limb; the pelvic girdle and lower limb. 724–25
6. How are joints classified? Describe the anatomy of a freely movable joint. 725
7. Give several functions of the muscular system in humans. How does the muscular system contribute to homeostasis? 727
8. Describe how muscles are attached to bones. What is accomplished by muscles acting in antagonistic pairs? 727
9. Discuss the microscopic structural features of a muscle fiber and a sarcomere. What is the sliding filament model? 728–29
10. Discuss the availability and the specific role of ATP during muscle contraction. What is oxygen debt, and how is it repaid? 728
11. Describe the structure and function of a neuromuscular junction. 730
12. Describe the cyclical events as myosin pulls actin toward the center of a sarcomere. 731

## testing yourself

Choose the best answer for each question.
For questions 1–4, match each bone to the location in the key.

**KEY:**

a. arm
b. forearm
c. pectoral girdle
d. pelvic girdle
e. thigh
f. leg

1. ulna
2. tibia
3. clavicle
4. femur
5. Spongy bone
   a. is a storage area for fat.
   b. contains red bone marrow, where blood cells are formed.
   c. lends no strength to bones.
   d. contributes to homeostasis.
   e. Both b and d are correct.
6. Which of these pairs is mismatched?
   a. slightly movable joint—vertebrae
   b. hinge joint—hip
   c. synovial joint—elbow
   d. immovable joint—sutures in cranium
   e. ball-and-socket joint—hip
7. The skeletal system does not
   a. produce blood cells.
   b. store minerals.
   c. help produce movement.
   d. store fat.
   e. produce body heat.
8. All blood cells—red, white, and platelets—are produced by which of the following?
   a. yellow bone marrow
   b. red bone marrow
   c. periosteum
   d. medullary cavity
9. Which of the following is not a bone of the appendicular skeleton?
   a. the scapula
   b. a rib
   c. a metatarsal bone
   d. the patella
   e. the ulna
10. The vertebrae that articulate with the ribs are the
    a. lumbar vertebrae.
    b. sacral vertebrae.
    c. thoracic vertebrae.
    d. cervical vertebrae.
    e. coccyx.
11. In a muscle fiber,
    a. the sarcolemma is connective tissue holding the myofibrils together.
    b. the sarcoplasmic reticulum stores calcium.
    c. both myosin and actin filaments have cross-bridges.
    d. there is a T system but no endoplasmic reticulum.
    e. All of these are correct.

12. When muscles contract,
    a. sarcomeres shorten.
    b. myosin heads break down ATP.
    c. actin slides past myosin.
    d. the H zone disappears.
    e. All of these are correct.
13. Which of these is the direct source of energy for muscle contraction?
    a. ATP
    b. creatine phosphate
    c. lactic acid
    d. glycogen
    e. Both a and b are correct.
14. Nervous stimulation of muscles
    a. occurs at a neuromuscular junction.
    b. involves the release of ACh.
    c. results in impulses that travel down the T system.
    d. causes calcium to be released from the sarcoplasmic reticulum.
    e. All of these are correct.
15. A neuromuscular junction occurs between an axon terminal and
    a. a muscle fiber.
    b. a myofibril.
    c. a myosin filament.
    d. a sarcomere only.
    e. Both a and d are correct.
16. When calcium is released from the sarcoplasmic reticulum, it binds to
    a. myosin.
    b. actin.
    c. troponin.
    d. sarcomeres.
    e. Both b and d are correct.
17. At what point is ATP hydrolyzed?
    a. just as myosin attaches to troponin
    b. just before myosin attaches to actin
    c. just when myosin pulls on actin
    d. just when impulses move down a T tubule
18. ACh
    a. is active at somatic synapses but not at neuromuscular junctions.
    b. binds to receptors in the sarcolemma.
    c. precedes the buildup of ATP in mitochondria.
    d. is stored in the sarcoplasmic reticulum.
    e. Both b and d are correct.
19. Label this diagram of a muscle fiber, using these terms: myofibril, Z line, T tubule, sarcomere, sarcolemma, sarcoplasmic reticulum.

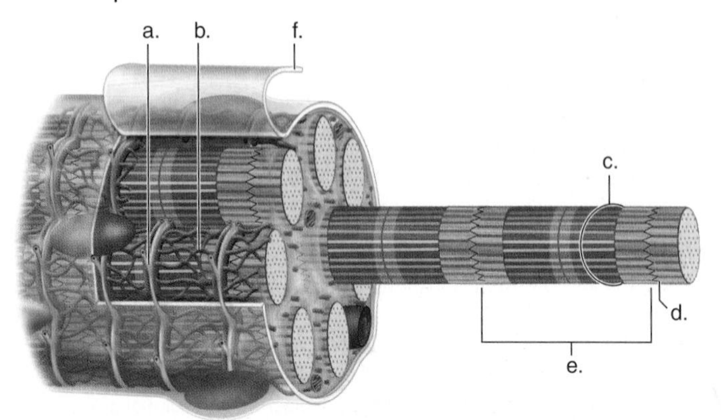


## thinking scientifically

1. It is observed that some motor neurons innervate only a few muscle fibers in the biceps brachii. Other motor neurons each innervate many muscle fibers. How might this observation correlate with our ability to pick up a pencil or a 2-liter soda bottle? On what basis would the brain bring about the correct level of contraction?
2. Some athletes believe that taking oral creatine will increase their endurance because it will increase the amount of phosphate available to their muscles for ATP synthesis. This statement can be regarded as two hypotheses: (a) oral creatine increases endurance, and (b) oral creatine increases the amount of creatine available in muscles for ATP synthesis. How could these two hypotheses be tested?

## bioethical issue

### Support Systems and Locomotion

A natural advantage does not bar an athlete from participating in and winning a medal in a particular sport at the Olympic Games. Nor are athletes restricted to a certain amount of practice or required to eliminate certain foods from their diets.

Athletes are, however, prevented from participating in the Olympic Games if they have taken certain performance-enhancing drugs. There is no doubt that regular use of drugs such as anabolic steroids leads to kidney disease, liver dysfunction, hypertension, increased aggression, and a myriad of other undesirable side effects (see Fig. 40.16, page 750). Even so, shouldn't the individual be allowed to take these drugs if he or she wants to? Anabolic steroids are synthetic forms of the male sex hormone testosterone. Taking large doses, along with strength training, leads to much larger muscles than otherwise. Extra strength and endurance can give an athlete an advantage in certain sports, such as racing, swimming, and weight lifting.

Should the Olympic committee outlaw the taking of anabolic steroids, and if so, on what basis? The basis can't be an unfair advantage, because some athletes naturally have an unfair advantage over other athletes. Should these drugs be outlawed on the basis of health reasons? Excessive practice or a purposeful decrease or increase in weight to better perform in a sport can also be injurious to one's health. In other words, how can you justify allowing some behaviors that enhance performance and not others?

## *Biology* website

The companion website for *Biology* provides a wealth of information organized and integrated by chapter. You will find practice tests, animations, videos, and much more that will complement your learning and understanding of general biology.

**http://www.mhhe.com/maderbiology10**

# 40

# Hormones and Endocrine Systems

*Hormones, chemical messengers of the endocrine system, cause a caterpillar to become a moth. One hormone, ecdysone, initiates molting of the exoskeleton as the caterpillar passes through a series of growth stages. A decline in the production of juvenile hormone triggers the final metamorphosis into an adult moth.*

*The endocrine system, in addition to the nervous system, coordinates the activities of the body's other organ systems and helps maintain homeostasis. In contrast to the nervous system, the endocrine system is not centralized, but consists of several organs scattered throughout the body. The hormones secreted by endocrine glands travel through the bloodstream and tissue fluid to reach their target tissues. The metabolism of a cell changes when it has a plasma membrane or nuclear receptor for that hormone. In this chapter, you will learn how hormones exert their slow but powerful influences on the body. You will see how the endocrine system maintains homeostasis when working properly, as well as some consequences of endocrine malfunction.*

The tobacco hornworm (*Manduca sexta*) begins life as a caterpillar. The caterpillar molts and undergoes metamorphosis, as orchestrated by hormones.

## concepts

## 40.1 Endocrine Glands

The nervous system and the endocrine system work together to regulate the activities of the other systems. Both systems use chemical signals when they respond to changes that might threaten homeostasis, but they have different means of delivering these signals (Fig. 40.1). As discussed, the nervous system is composed of neurons. In this system, sensory receptors (specialized dendrites) detect changes in the internal and external environment. The CNS integrates the information and responds by stimulating muscles and glands. Communication depends on nerve impulses, conducted in axons, and neurotransmitters, which cross synapses. Axon conduction occurs rapidly and so does diffusion of a neurotransmitter across the short distance of a synapse. In other words, the nervous system is organized to respond rapidly to stimuli. This is particularly useful if the stimulus is an external event that endangers our safety—we can move quickly to avoid being hurt.

The **endocrine system** functions differently. The endocrine system is largely composed of glands (Fig. 40.2). These glands secrete **hormones,** which are carried by the bloodstream to target cells throughout the body. If a child is short for his or her age, it is because his or her blood didn't contain enough growth hormone produced by the anterior pituitary. Growth hormone stimulates the growth of long bones, and in this way affects the height of an individual. It takes time to deliver hormones, and it takes time for cells to respond, but the effect is longer lasting. In other words, the endocrine system is organized for a slow but prolonged response.

Endocrine glands can be contrasted with exocrine glands. Exocrine glands have ducts and secrete their products into these ducts, which take them to the lumens of other organs or outside the body. For example, the salivary glands send saliva into the mouth by way of the salivary ducts. **Endocrine glands,** as stated, secrete their products into the bloodstream, which delivers them throughout the body. It must be stressed that only certain cells, called target cells, can respond to certain hormones. If a cell can respond to a hormone, the hormone and receptor proteins bind together as a key fits a lock.

It is of interest to note that both the nervous system and the endocrine system make use of negative feedback mechanisms. If the blood pressure falls, sensory receptors signal a control center in the brain. This center sends out nerve impulses to the arterial walls so that they constrict, and blood pressure rises. Now the sensory receptors are no longer stimulated, and the feedback mechanism is inactivated. Similarly, a rise in blood glucose level causes the pancreas to release insulin, which, in turn, promotes glucose uptake by the liver, muscles, and other cells of the body (Fig. 40.2). When the blood glucose level falls, the pancreas no longer secretes insulin.

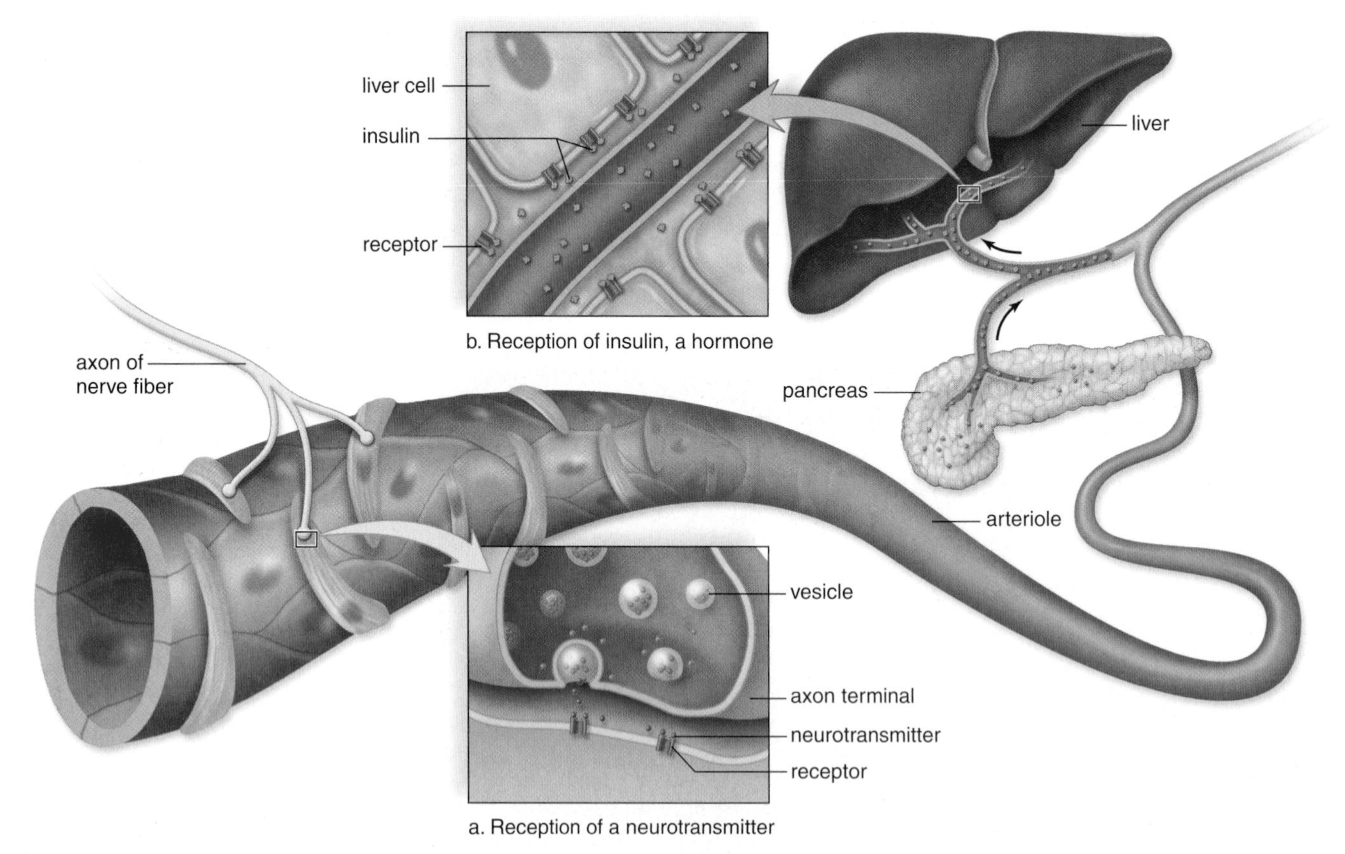


**FIGURE 40.1 Modes of action of the nervous and endocrine systems.**

**a.** Nerve impulses passing along an axon cause the release of a neurotransmitter. The neurotransmitter, a chemical signal, binds to a receptor and causes the wall of an arteriole to constrict. **b.** The hormone insulin, a chemical signal, travels in the cardiovascular system from the pancreas to the liver, where it binds to a receptor and causes liver cells to store glucose as glycogen.

**HYPOTHALAMUS**

Releasing and inhibiting hormones: regulate the anterior pituitary

**PITUITARY GLAND**

**Posterior Pituitary**

Antidiuretic (ADH): water reabsorption by kidneys

Oxytocin: stimulates uterine contraction and milk letdown

**Anterior Pituitary**

Thyroid stimulating (TSH): stimulates thyroid

Adrenocorticotropic (ACTH): stimulates adrenal cortex

Gonadotropic (FSH, LH): egg and sperm production; sex hormone production

Prolactin (PL): milk production

Growth (GH): bone growth, protein synthesis, and cell division

**PINEAL GLAND**

Melatonin: controls circadian and circannual rhythms

**PARATHYROIDS**

Parathyroid hormone (PTH): raises blood calcium level

parathyroid glands (posterior surface of thyroid)

**THYROID**

Thyroxine ($T_4$) and triiodothyronine ($T_3$): increase metabolic rate; regulates growth and development

Calcitonin: lowers blood calcium level

**THYMUS**

Thymosins: production and maturation of T lymphocytes

**ADRENAL GLAND**

**Adrenal cortex**

Glucocorticoids (cortisol): raises blood glucose level; stimulates breakdown of protein

Mineralocorticoids (aldosterone): reabsorption of sodium and excretion of potassium

Sex hormones: reproductive organs and bring about sex characteristics

**Adrenal medulla**

Epinephrine and norepinephrine: active in emergency situations; raise blood glucose level

**PANCREAS**

Insulin: lowers blood glucose level and promotes glycogen buildup

Glucagon: raises blood glucose level and promotes glycogen breakdown

testis (male)

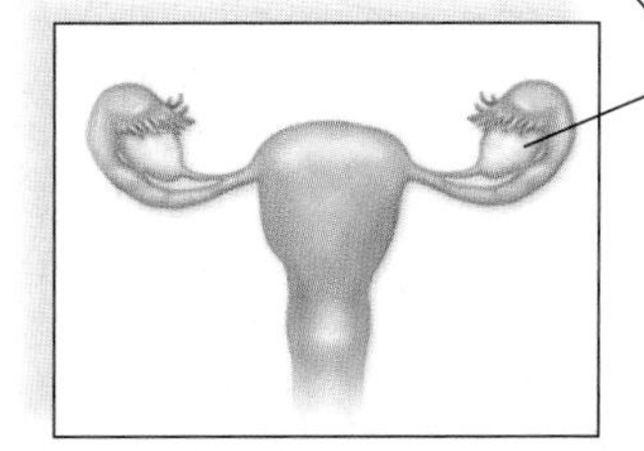

ovary (female)

**GONADS**

**Testes**

Androgens (testosterone): male sex characteristics

**Ovaries**

Estrogens and progesterone: female sex characteristics

**FIGURE 40.2 Major glands of the human endocrine system.**

Major glands and the hormones they produce are depicted. Also, the endocrine system includes other organs such as the kidneys, gastrointestinal tract, and the heart, which also produce hormones but not as a primary function of these organs.

## Hormones Are Chemical Signals

Like other **chemical signals,** hormones are a means of communication between cells, between body parts, and even between individuals. They typically affect the metabolism of cells that have receptors to receive them (Fig. 40.3). In a condition called androgen insensitivity, an individual has X and Y sex chromosomes, and the testes, which remain in the abdominal cavity, produce the sex hormone testosterone. However, the body cells lack receptors that are able to combine with testosterone, and the individual appears to be a normal female.

Like testosterone, most hormones act at a distance between body parts. They travel in the bloodstream from the gland that produced them to their target cells. Also counted as hormones are the secretions produced by neurosecretory cells in the hypothalamus, a part of the brain. They travel in the capillary network that runs between the hypothalamus and the pituitary gland. Some of these secretions stimulate the pituitary to secrete its hormones, and others prevent it from doing so.

Not all hormones act between body parts. As we shall see, prostaglandins are a good example of a *local hormone.* After prostaglandins are produced, they are not carried in the bloodstream; instead, they affect neighboring cells, sometimes promoting pain and inflammation. Also, growth factors are local hormones that promote cell division and mitosis.

Chemical signals that influence the behavior of other individuals are called **pheromones.** Pheromones have been released by other animals and a researcher has isolated one released by men that reduces premenstrual nervousness and tension in women. Women who live in the same household often have menstrual cycles in synchrony, perhaps because the armpit secretions of a woman who is menstruating affects the menstrual cycle of other women in the household.

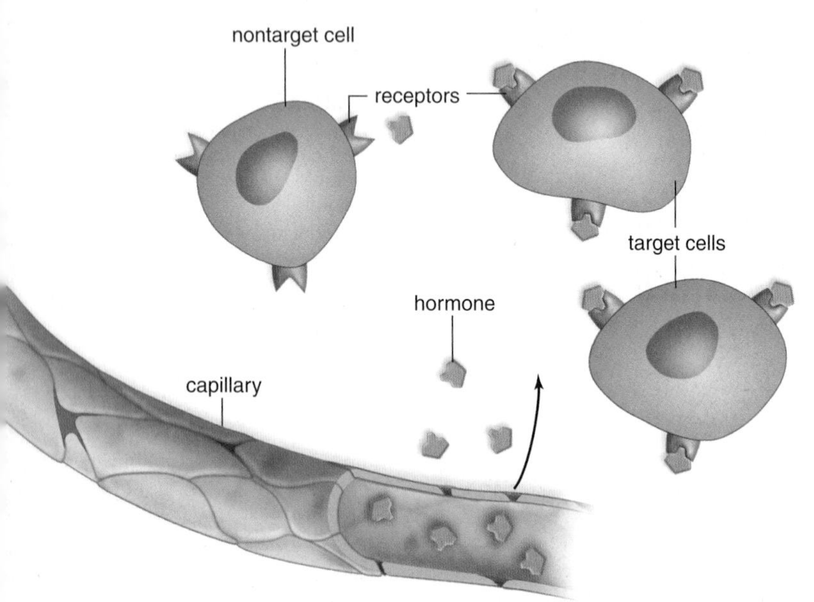


**FIGURE 40.3 Target cell concept.**

Most hormones are distributed by the bloodstream to target cells. Target cells have receptors for the hormone, and the hormone combines with the receptor as a key fits a lock.

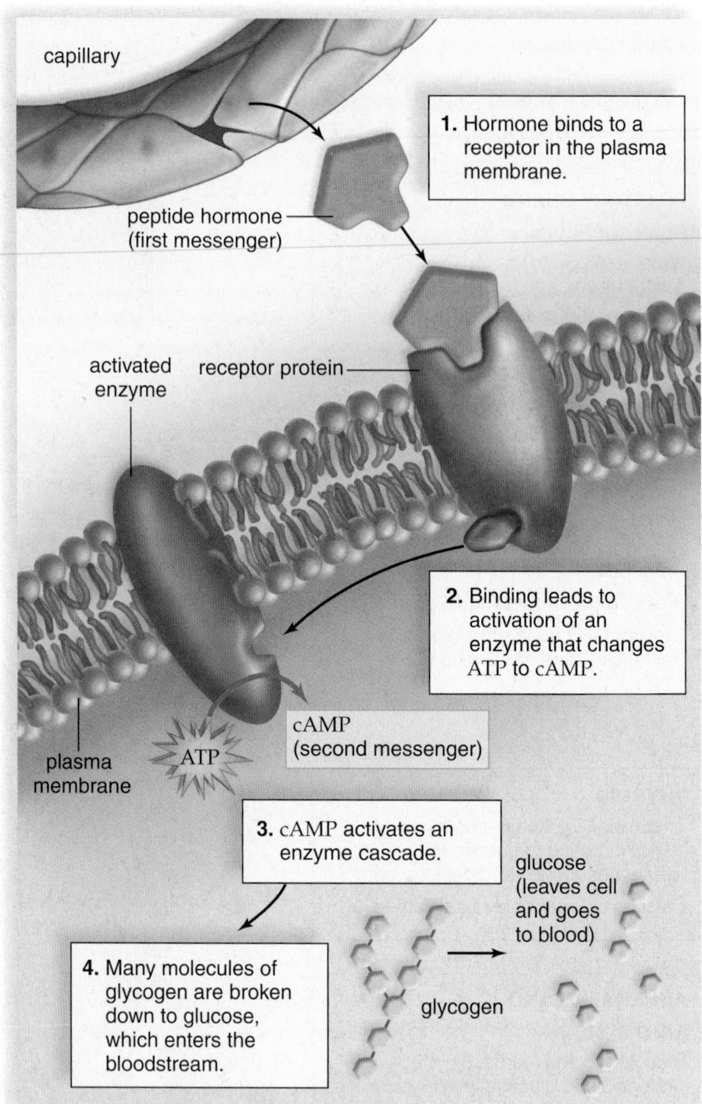


**FIGURE 40.4 Peptide hormone.**

The peptide hormone (first messenger) binds to a receptor in the plasma membrane. Thereafter, cyclic AMP (second messenger) forms and activates an enzyme cascade.

### *The Action of Hormones*

Hormones have a wide range of effects on cells. Some of these effects induce a target cell to increase its uptake of particular molecules, such as glucose, or ions, such as calcium. Some bring about an alteration of the target cell's structure in some way. Some simply influence cell metabolism. Growth hormone is a peptide hormone that influences cell metabolism leading to a change in the structure of bone. This hormone is a **peptide hormone.** The term peptide hormone is used to include hormones that are peptides, proteins, glycoproteins, and modified amino acids. Growth hormone is a protein produced and secreted by the anterior pituitary. **Steroid hormones** have the same complex of four carbon rings because they are all derived from cholesterol.

**The Action of Peptide Hormones.** Most hormonal glands secrete peptide hormones. The actions of peptide hormones can vary, and we will concentrate on what happens in muscle cells after the hormone epinephrine binds to a receptor in the plasma membrane (Fig. 40.4). In muscle cells, the reception of epinephrine leads to the breakdown of glycogen to glucose, which provides energy for ATP production. The immediate result of binding is the formation of **cyclic adenosine monophosphate (cAMP).** Cyclic AMP contains one phosphate group attached to adenosine at two locations. Therefore, the molecule is cyclic. Cyclic AMP activates a protein kinase enzyme in the cell, and this enzyme, in turn, activates another enzyme, and so forth. The series of enzymatic reactions that follows cAMP formation is called an enzyme cascade. Because each enzyme can be used over and over again at every step of the cascade, more enzymes are involved. Finally, many molecules of glycogen are broken down to glucose, which enters the bloodstream.

Typical of a peptide hormone, epinephrine never enters the cell. Therefore, the hormone is called the **first messenger,** while cAMP, which sets the metabolic machinery in motion, is called the **second messenger.** To explain this terminology, let's imagine that the adrenal medulla, which produces epinephrine, is like the home office that sends out a courier (i.e., the hormone epinephrine is the first messenger) to a factory (the cell). The courier doesn't have a pass to enter the factory, so when he arrives at the factory, he tells a supervisor through the screen door that the home office wants the factory to produce a particular product. The supervisor (i.e., cAMP, the second messenger) walks over and flips a switch that starts the machinery (the enzymatic pathway), and a product is made.

**The Action of Steroid Hormones.** Only the adrenal cortex, the ovaries, and the testes produce steroid hormones. Thyroid hormones act similarly to steroid hormones, even though they have a different structure. Steroid hormones do not bind to plasma membrane receptors, and instead they are able to enter the cell because they are lipids (Fig. 40.5). Once inside, a steroid hormone binds to a receptor, usually in the nucleus but sometimes in the cytoplasm. Inside the nucleus, the hormone-receptor complex binds with DNA and activates certain genes. Messenger RNA (mRNA) moves to the ribosomes in the cytoplasm and protein (e.g., enzyme) synthesis follows. To continue our analogy, a steroid hormone is like a courier that has a pass to enter the factory (the cell). Once inside, he makes contact with the plant manager (DNA), who sees to it that the factory (cell) is ready to produce a product.

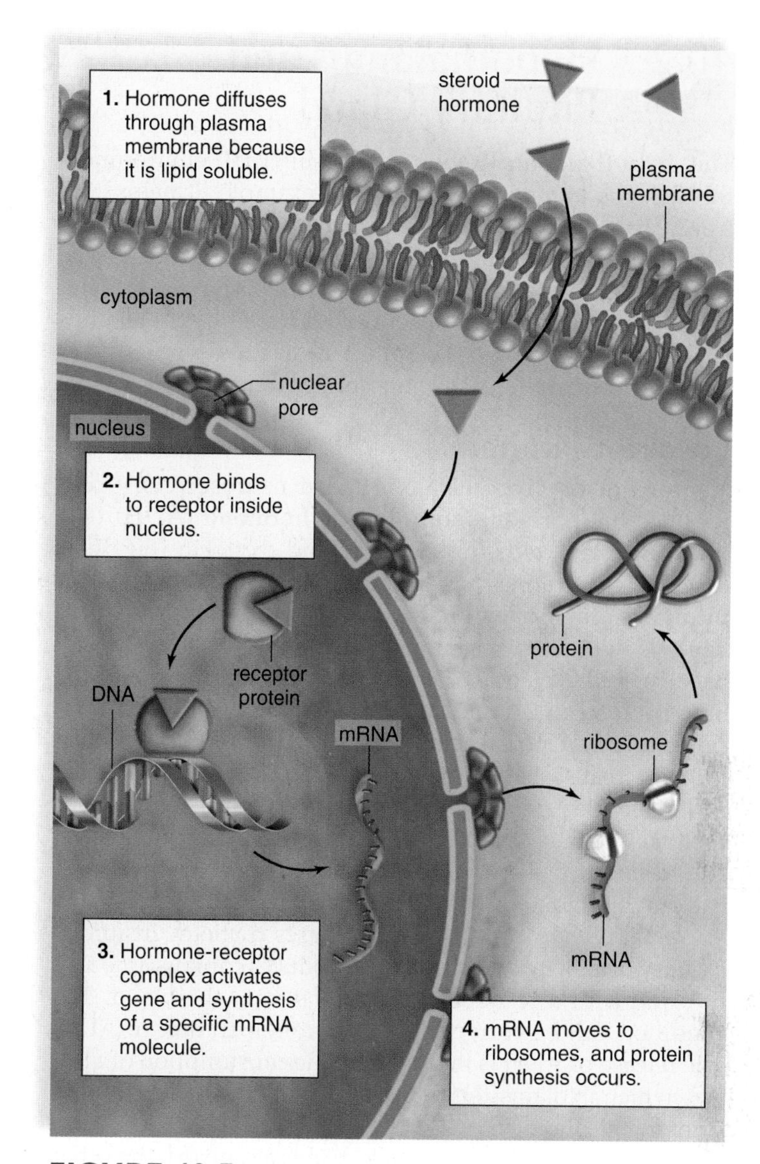


**FIGURE 40.5 Steroid hormone.**

A steroid hormone passes directly through the target cell's plasma membrane before binding to a receptor in the nucleus or cytoplasm. The hormone-receptor complex binds to DNA and gene expression follows.

Steroids act more slowly than peptides because it takes more time to synthesize new proteins than to activate enzymes already present in cells. Their action lasts longer, however.

### Check Your Progress 40.1

1. What is the main difference between an endocrine gland and an exocrine gland?
2. Where are the receptors for peptide and steroid hormones found?
3. Which type of hormone, peptide or steroid, causes the formation of cAMP upon binding its receptor?

# 40.2 Hypothalamus and Pituitary Gland

The **hypothalamus** regulates the internal environment through the autonomic system. For example, it helps control heartbeat, body temperature, and water balance. The hypothalamus also controls the glandular secretions of the **pituitary gland** (hypophysis). The pituitary, a small gland about 1 cm in diameter, is connected to the hypothalamus by a stalklike structure. The pituitary has two portions: the posterior pituitary and the anterior pituitary.

## Posterior Pituitary

Neurons in the hypothalamus called neurosecretory cells produce the hormones **antidiuretic hormone (ADH)** [Gk. *anti*, against; L. *ouresis*, urination] and oxytocin (Fig. 40.6, *left*). These hormones pass through axons into the **posterior pituitary,** where they are stored in axon endings. Certain neurons in the hypothalamus are sensitive to the water-salt balance of the blood. When these cells determine that the blood is too concentrated, ADH is released from the posterior pituitary. Upon reaching the kidneys, ADH causes water to be reabsorbed. As the blood becomes dilute, ADH is no longer released. This is an example of control by **negative feedback** because the *effect* of the hormone (to dilute blood) acts to shut down the *release* of the hormone. Negative feedback maintains stable conditions and homeostasis.

Inability to produce ADH causes diabetes insipidus (watery urine), in which a person produces copious amounts of urine with a resultant loss of ions from the blood. The condition can be corrected by the administration of ADH. The release of ADH is inhibited by the consumption of alcohol, which explains the frequent urination associated with drinking alcohol.

**Oxytocin** [Gk. *oxys*, quick, and *tokos*, birth], the other hormone made in the hypothalamus, causes uterine contractions during childbirth and milk letdown when a baby is nursing. The more the uterus contracts during labor, the more nerve impulses reach the hypothalamus, causing oxytocin to be released. Similarly, the more a baby suckles, the more oxytocin is released. In both instances, the release of oxytocin from the posterior pituitary is controlled by **positive feedback**—that is, the stimulus continues to bring about an effect that ever increases in intensity. Positive feedback is not a way to maintain stable conditions and homeostasis. Oxytocin may also play a role in the propulsion of semen through the male reproductive tract and may affect feelings of sexual satisfaction and emotional bonding.

## Anterior Pituitary

A portal system, consisting of two capillary systems connected by a vein, lies between the hypothalamus and the anterior pituitary (Fig. 40.6, *right*). The hypothalamus controls the anterior pituitary by producing **hypothalamic-releasing hormones** and in some instances **hypothalamic-inhibiting hormones.** For example, there is a hypothalamic-releasing hormone that stimulates the anterior pituitary to secrete a thyroid-stimulating hormone and a hypothalamic-inhibiting hormone that prevents the anterior pituitary from secreting prolactin.

### *Anterior Pituitary Hormones Affecting Other Glands*

Certain hormones produced by the **anterior pituitary** affect other glands. **Gonadotropic hormones** stimulate the gonads—the testes in males and the ovaries in females—to produce gametes and sex hormones. **Adrenocorticotropic hormone (ACTH)** stimulates the adrenal cortex to produce glucocorticoid. **Thyroid-stimulating hormone (TSH)** stimulates the thyroid to produce thyroxine ($T_4$) and triiodothyronine ($T_3$). In each instance, these hormones are involved in a three-tier system and the blood level of the last hormone in the sequence exerts negative feedback control over the secretion of the first two hormones. This is how it works for TSH:

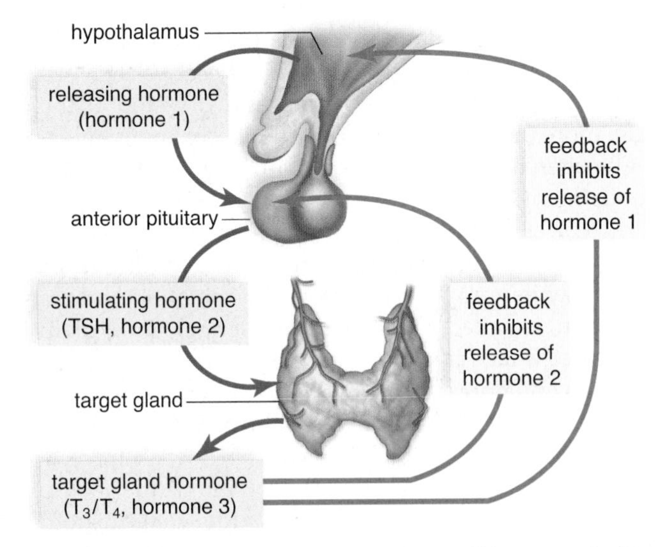


### *Anterior Pituitary Hormones Not Affecting Other Glands*

Other types of hormones produced by the anterior pituitary are under the control of the hypothalamus, but they do not affect other endocrine glands. **Prolactin (PRL)** [L. *pro*, before, and *lactis*, milk] is produced in quantity only after childbirth. It causes the mammary glands in the breasts to develop and produce milk. It also plays a role in carbohydrate and fat metabolism.

**Growth hormone (GH),** or somatotropic hormone, promotes skeletal and muscular growth. It stimulates the rate at which amino acids enter cells and protein synthesis occurs. It also promotes fat metabolism as opposed to glucose metabolism.

**Melanocyte-stimulating hormone (MSH)** [Gk. *melanos*, black, and *kytos*, cell] causes skin-color changes in many fishes, amphibians, and reptiles having melanophores, special skin cells that produce color variations. The concentration of this hormone in humans is very low.

### *Effects of Growth Hormone*

GH is one of the hormones produced by the anterior pituitary after being stimulated to do so by a releasing hormone from the hypothalamus (Fig. 40.6, *right*). The quantity is greatest during childhood and adolescence, when most body growth is occurring. If too little GH is produced during childhood, the individual has **pituitary dwarfism,** characterized by normal proportions but small stature.

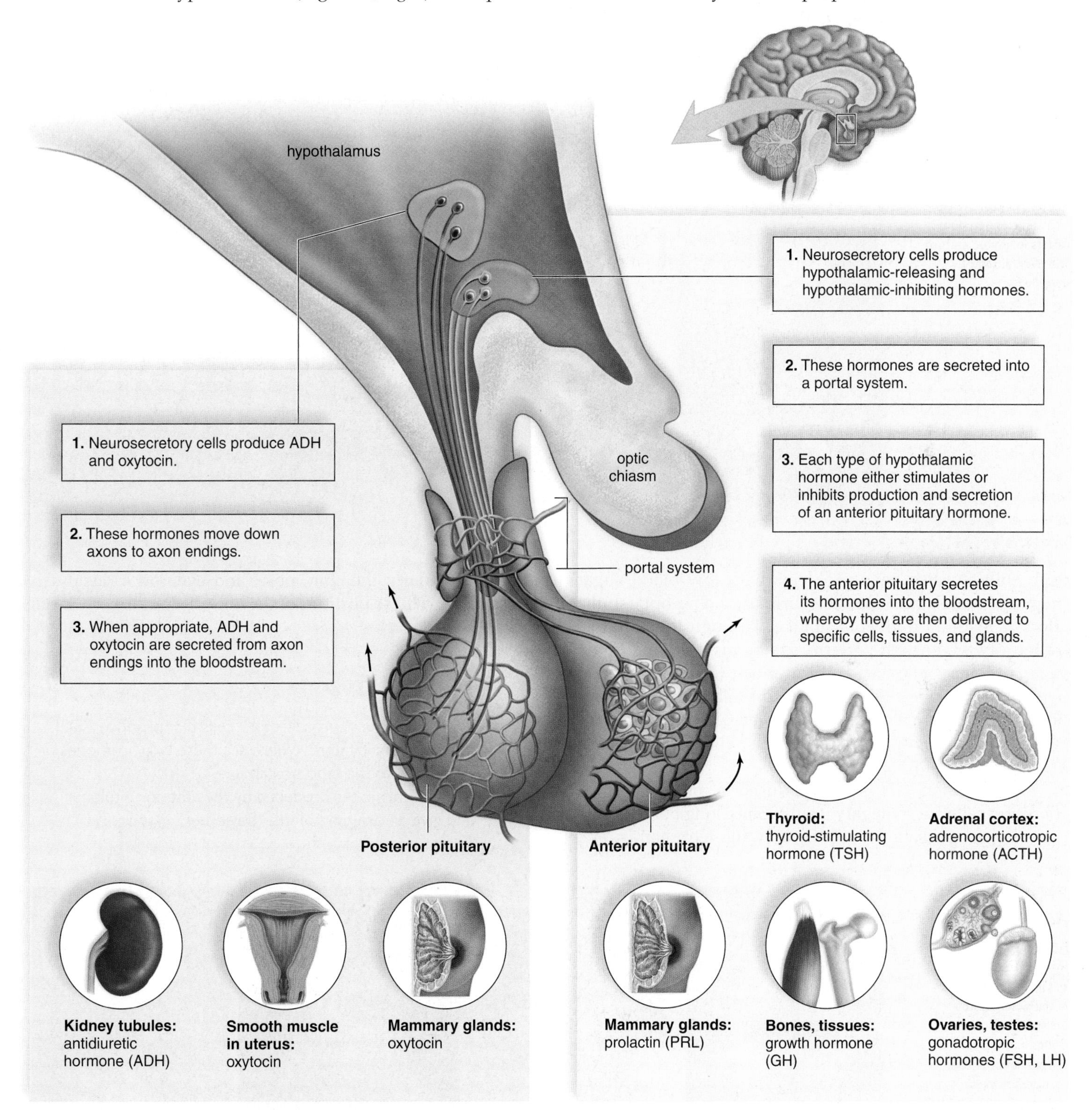


**FIGURE 40.6 Hypothalamus and the pituitary.**

In this diagram, the name of the hormone is given below its target organ, which is depicted in the circle. *Left:* The hypothalamus produces two hormones, ADH and oxytocin, which are stored and secreted by the posterior pituitary. *Right:* The hypothalamus controls the secretions of the anterior pituitary, and the anterior pituitary controls the secretions of the thyroid, adrenal cortex, and gonads, which are also endocrine glands.

a.

b.

**FIGURE 40.7**
**Effect of growth hormone.**

**a.** The amount of growth hormone produced by the anterior pituitary during childhood affects the height of an individual. Plentiful growth hormone produces very tall basketball players. **b.** Too much growth hormone can lead to giantism, while an insufficient amount results in limited stature and even pituitary dwarfism.

Such children also have problems with low blood sugar (hypoglycemia), because GH normally helps oppose the effect of insulin on glucose uptake. Through the administration of GH, growth patterns can be restored and blood sugar problems alleviated. If too much GH is secreted, a person can become a giant (Fig. 40.7*b*). Giants usually have poor health, primarily because elevated GH cancels out the effects of insulin, promoting an illness called diabetes mellitus (see page 748).

On occasion, GH is overproduced in the adult, and a condition called **acromegaly** results. Since long bone growth is no longer possible in adults, only the feet, hands, and face (particularly the chin, nose, and eyebrow ridges) can respond, and these portions of the body become overly large (Fig. 40.8).

**Check Your Progress 40.2**

1. Where are the neurosecretory cells that produce ADH and oxytocin located? Where are these two hormones released into the bloodstream?
2. List the hormones produced by the anterior pituitary and give an example of the "three-tier" system.

Age 9

Age 16

Age 33

Age 52

**FIGURE 40.8**
**Acromegaly.**

Acromegaly is caused by overproduction of GH in the adult. It is characterized by enlargement of the bones in the face, the fingers, and the toes as a person ages.

# 40.3 Other Endocrine Glands and Hormones

## Thyroid and Parathyroid Glands

The **thyroid gland** [Gk. *thyreos*, large, door-shaped shield] is a large gland located in the neck, where it is attached to the trachea just below the larynx (see Fig. 40.2). The parathyroid glands are embedded in the posterior surface of the thyroid gland.

### *Thyroid Gland*

The thyroid gland is the largest endocrine gland, weighing approximately 20 g. It is excessively red in appearance because of its high blood volume. It consists of two distinct lobes connected by a slender isthmus. The thyroid gland is composed of a large number of follicles, each a small spherical structure made of thyroid cells filled with the thyroid hormones triiodothyronine ($T_3$), which contains three iodine atoms, and **thyroxine ($T_4$)**, which contains four iodine atoms.

**Effects of Thyroid Hormones.** To produce triiodothyronine and thyroxine, the thyroid gland actively acquires iodine. The concentration of iodine in the thyroid gland can increase to as much as 25 times that of the blood. If iodine is lacking in the diet, the thyroid gland is unable to produce the thyroid hormones. In response to constant stimulation by the anterior pituitary, the thyroid enlarges, resulting in a **simple goiter** (Fig. 40.9). Some years ago, it was discovered that the use of iodized salt allows the thyroid to produce the thyroid hormones and therefore helps prevent simple goiter.

Thyroid hormones increase the metabolic rate. They do not have a target organ; instead, they stimulate all the cells of the body to metabolize at a faster rate. More glucose is broken down, and more energy is used.

If the thyroid fails to develop properly, a condition called **congenital hypothyroidism** (cretinism) results (Fig. 40.9). Individuals with this condition are short and stocky and have had extreme hypothyroidism (undersecretion of thyroid hormone) since infancy or childhood. Thyroid hormone therapy can initiate growth, but unless treatment is begun within the first two months of life, mental retardation results. The occurrence of hypothyroidism in adults produces the condition known as **myxedema,** which is characterized by lethargy, weight gain, loss of hair, slower pulse rate, lowered body temperature, and thickness and puffiness of the skin. The administration of adequate doses of thyroid hormones restores normal function and appearance.

In the case of hyperthyroidism (oversecretion of thyroid hormone), or Graves disease, the thyroid gland is overactive, and a goiter forms. This type of goiter is called **exophthalmic goiter.** The eyes protrude (exophthalmos) because of edema in eye socket tissues and swelling of the muscles that move the eyes. The patient usually becomes hyperactive, nervous, and irritable and suffers from insomnia. In addition, the individual may have unusual sweating and heat sensitivity. Removal or destruction of a portion of the thyroid by means of radioactive iodine is sometimes effective in curing the condition. Hyperthyroidism can also be caused by a thyroid tumor, which is usually detected as a lump during physical examination. Again, the treatment is surgery in combination with administration of radioactive iodine. The prognosis for most patients is excellent.

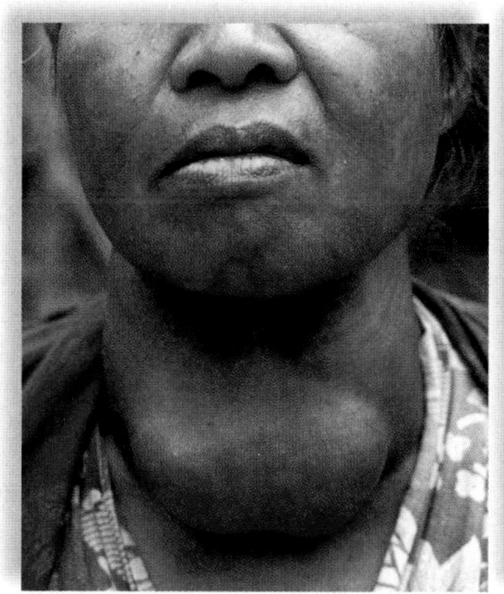

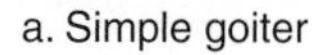

a. Simple goiter

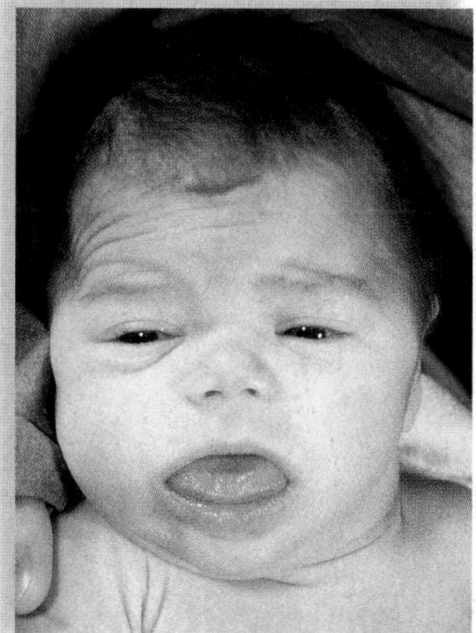

b. Congenital hypothyroidism

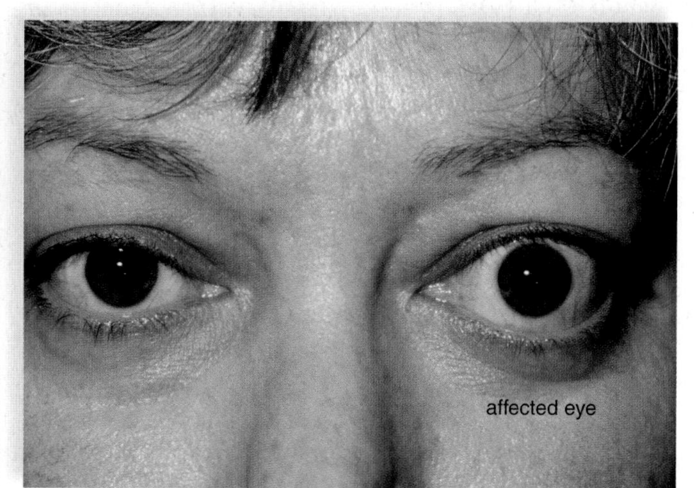


c. Exophthalmic goiter

**FIGURE 40.9 Abnormalities of the thyroid.**

**a.** An enlarged thyroid gland is often caused by a lack of iodine in the diet. Without iodine, the thyroid is unable to produce its hormones and continued anterior pituitary stimulation causes the gland to enlarge. **b.** Individuals who develop hypothyroidism during infancy or childhood do not grow and develop as others do. Unless medical treatment is begun, the body is short and stocky; mental retardation is also likely. **c.** In exophthalmic goiter, a goiter is due to an overactive thyroid, and the eye protrudes because of edema in eye socket tissue.

**Calcitonin.** Calcium ($Ca^{2+}$) plays a significant role in both nervous conduction and muscle contraction. It is also necessary for blood clotting. The blood calcium level is regulated in part by **calcitonin,** a hormone secreted by the thyroid gland when the blood calcium level rises. The primary effect of calcitonin is to bring about the deposit of calcium in the bones (Fig. 40.10, *above*). (It does this by temporarily reducing the activity and number of osteoclasts.) When the blood calcium lowers to normal, the release of calcitonin by the thyroid is inhibited, but a low level stimulates the release of **parathyroid hormone (PTH)** by the parathyroid glands.

Calcitonin is important in growing children but seems to have little effect in adults. It may be helpful in reducing bone loss in osteoporosis and in pregnant women. The deficiency of calcitonin is not linked with any specific disorder.

## *Parathyroid Glands*

Many years ago, the four **parathyroid glands** were sometimes mistakenly removed during thyroid surgery because of their size and location. Parathyroid hormone causes the blood phosphate ($HPO_4^{2-}$) level to decrease and the blood calcium level to increase when the level is low.

PTH corrects a low blood calcium level in a number of ways. PTH promotes the activity of osteoclasts and the release of calcium from the bones. PTH also promotes the reabsorption of calcium by the kidneys so it is not excreted. In the kidneys, PTH also brings about activation of vitamin D. Vitamin D, in turn, stimulates the absorption of calcium from the intestine (Fig. 40.10, *below*). These effects bring the blood calcium level back to the normal range so that the parathyroid glands no longer secrete PTH.

When insufficient parathyroid hormone production (hypoparathyroidism) leads to a dramatic drop in the blood calcium level, tetany results. In **tetany,** the body shakes from continuous muscle contraction. This effect is brought about by increased excitability of the nerves, which initiate nerve impulses spontaneously and without rest.

In hyperparathyroidism, the blood calcium level becomes abnormally high. This disorder can cause the bones to become abnormally soft and fragile. In addition, it can cause the individual to become unusually irritable and prone to kidney stones.

## *Anatagonistic Hormones*

Calcitonin and PTH are antagonistic hormones because their action is opposite to one another, and both hormones work together to regulate the blood calcium level. When the blood calcium level is low, the thyroid secretes calcitonin and when the blood calcium level is low, the parathyroid glands release PTH (Fig. 40.10).

You will have an opportunity to learn of other antagonistic hormones in the pages that follow. For example, the liver secretes both insulin and glucagon, which work together to maintain the blood glucose level.

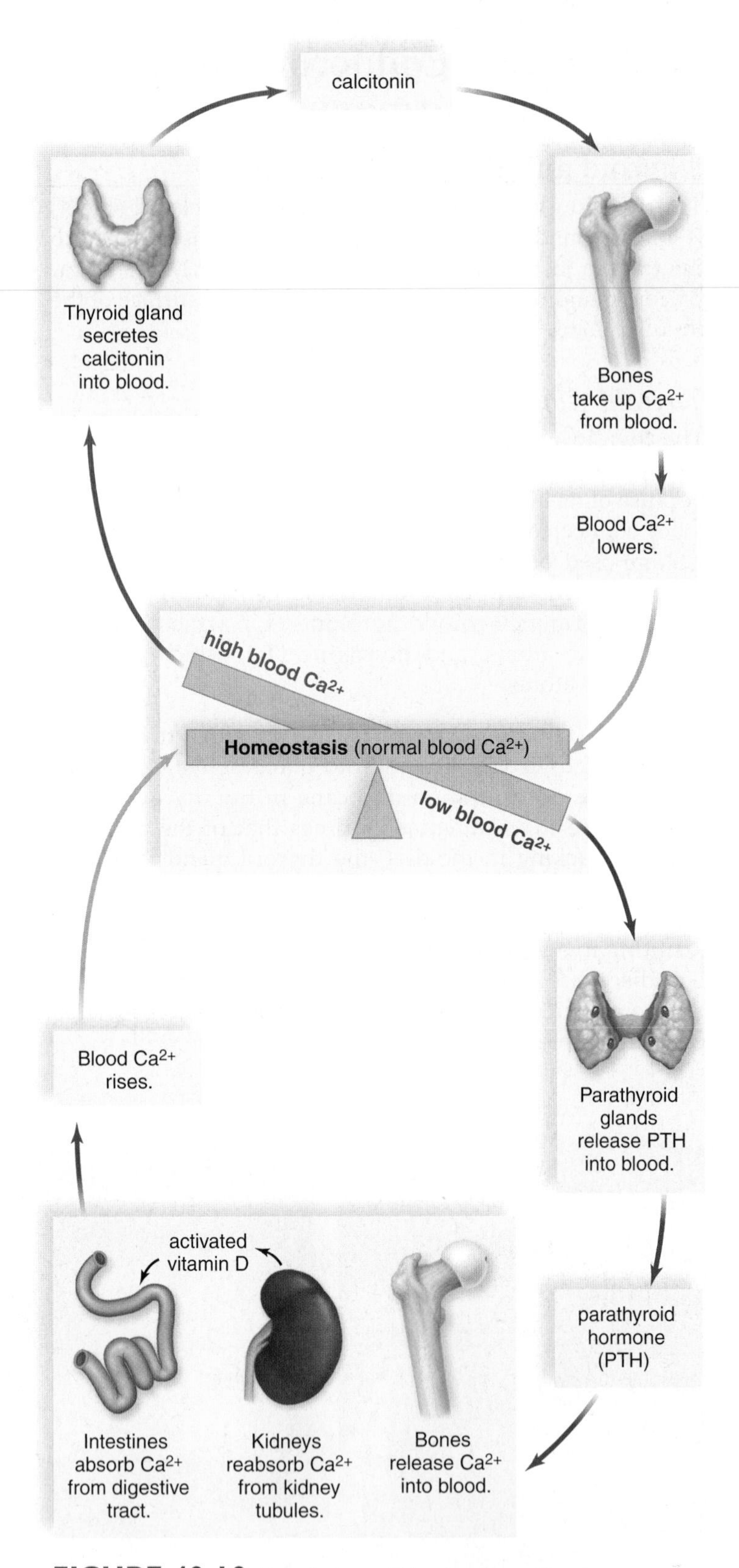


**FIGURE 40.10 Regulation of blood calcium level.**

*Above:* When the blood calcium ($Ca^{2+}$) level is high, the thyroid gland secretes calcitonin. Calcitonin promotes the uptake of $Ca^{2+}$ by the bones, and therefore the blood $Ca^{2+}$ level returns to normal. *Below:* When the blood $Ca^{2+}$ level is low, the parathyroid glands release parathyroid hormone (PTH). PTH causes the bones to release $Ca^{2+}$ and the kidneys to reabsorb $Ca^{2+}$ and activate vitamin D; thereafter, the intestines absorb $Ca^{2+}$. Therefore, the blood $Ca^{2+}$ level returns to normal.

## Adrenal Glands

The **adrenal glands** [L. *ad*, toward, and *renis*, kidney] sit atop the kidneys (see Fig. 40.2). Each gland is about 5 cm long and 3 cm wide and weigh about 5 g. An adrenal gland consists of an inner portion called the **adrenal medulla** and an outer portion called the **adrenal cortex.** These portions, like the anterior pituitary and the posterior pituitary, have no physiological connection with one another.

The hypothalamus exerts control over the activity of both portions of the adrenal glands. It initiates nerve impulses that travel by way of the brain stem, spinal cord, and sympathetic nerve fibers to the adrenal medulla, which then secretes its hormones. The hypothalamus, by means of ACTH-releasing hormone, controls the anterior pituitary's secretion of ACTH, which in turn stimulates the adrenal cortex to secrete glucocorticoids. Stress of all types, including both emotional and physical trauma, prompts the hypothalamus to stimulate the adrenal glands.

**Epinephrine** (adrenaline) and **norepinephrine** (noradrenaline) produced by the adrenal medulla rapidly bring about all the bodily changes that occur when an individual reacts to an emergency situation. The effects of these hormones are short-term (Fig. 40.11). Epinephrine and norepinephrine accelerate the breakdown of glucose to form ATP, trigger the mobilization of glycogen reserves in skeletal muscle, and increase the cardiac rate and force of contraction.

In contrast, the hormones produced by the adrenal cortex provide a long-term response to stress. The two major types of hormones produced by the adrenal cortex are the mineralocorticoids and the glucocorticoids. The **mineralocorticoids** regulate salt and water balance, leading to increases in blood volume and blood pressure. The **glucocorticoids** regulate carbohydrate, protein, and fat metabolism, leading to an increase in blood glucose level.

The adrenal cortex also secretes a small amount of male sex hormones and a small amount of female sex hormones in both sexes—that is, in the male, both male and female sex hormones are produced by the adrenal cortex, and in the female, both male and female sex hormones are also produced by the adrenal cortex.

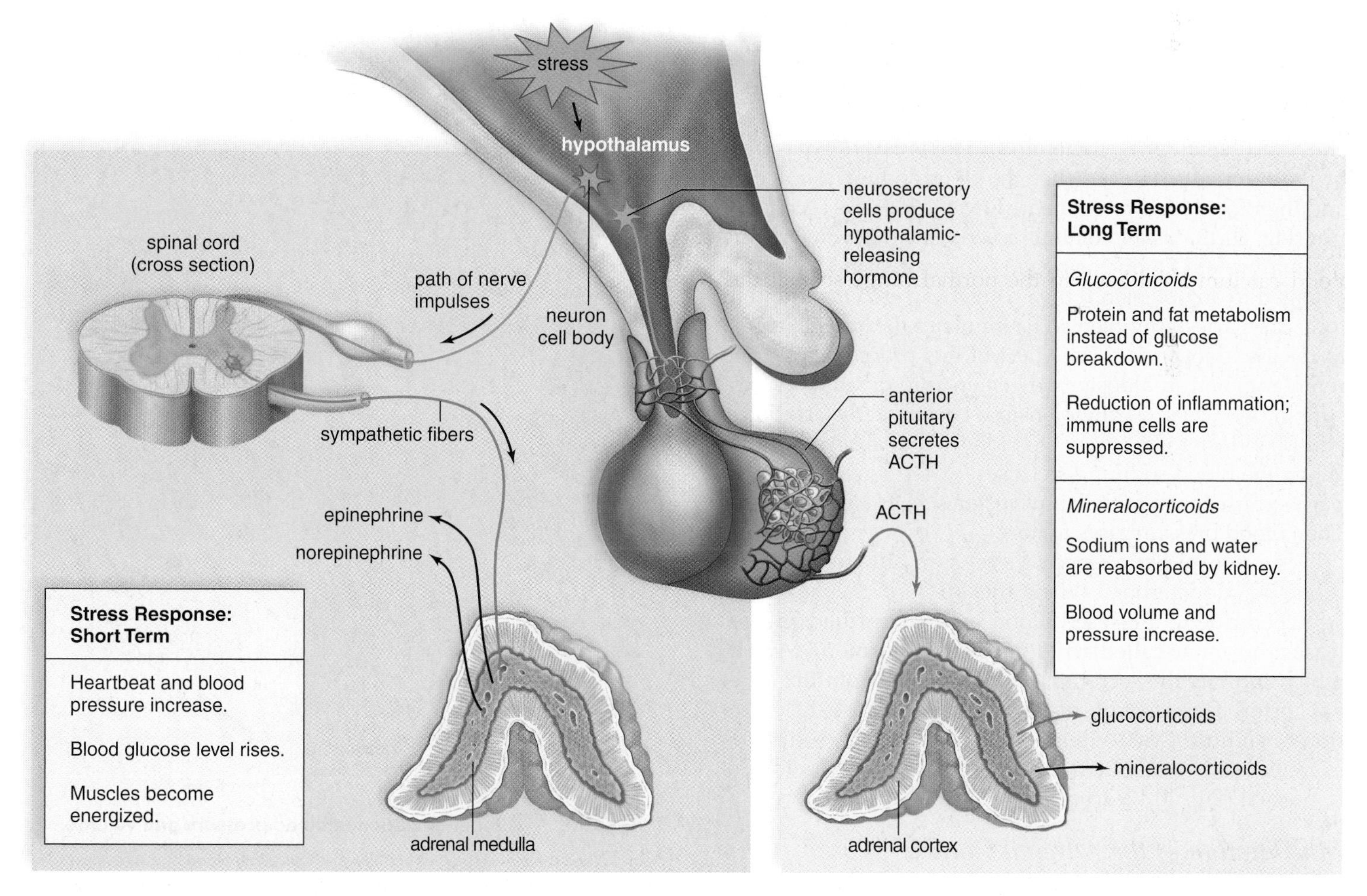


**FIGURE 40.11 Adrenal glands.**

Both the adrenal cortex and the adrenal medulla are under the control of the hypothalamus when they help us respond to stress. *Left:* Nervous stimulation causes the adrenal medulla to provide a rapid, but short-term, stress response. *Right:* ACTH from the anterior pituitary causes the adrenal cortex to release glucocorticoids. Independently, the adrenal cortex releases mineralocorticoids. The adrenal cortex provides a slower, but long-term, stress response.

### Glucocorticoids

**Cortisol** is a biologically significant glucocorticoid produced by the adrenal cortex. Cortisol raises the blood glucose level in at least two ways: (1) It promotes the breakdown of muscle proteins to amino acids, which are taken up by the liver from the bloodstream. The liver then breaks down these excess amino acids to glucose, which enters the blood. (2) Cortisol promotes the metabolism of fatty acids rather than carbohydrates, and this spares glucose.

Cortisol also counteracts the inflammatory response, which leads to the pain and swelling of joints in arthritis and bursitis. The administration of cortisol in the form of cortisone aids these conditions because it reduces inflammation. Very high levels of glucocorticoids in the blood can suppress the body's defense system, including the inflammatory response that occurs at infection sites. Cortisone and other glucocorticoids can relieve swelling and pain from inflammation, but by suppressing pain and immunity, they can also make a person highly susceptible to injury and infection.

### Mineralocorticoids

**Aldosterone** is the most important of the mineralocorticoids. Aldosterone primarily targets the kidney, where it promotes renal absorption of sodium ($Na^+$) and renal excretion of potassium ($K^+$).

The secretion of mineralocorticoids is not controlled by the anterior pituitary. When the blood sodium ($Na^+$) level and therefore blood pressure are low, the kidneys secrete **renin** (Fig. 40.12, *below*). Renin is an enzyme that converts the plasma protein angiotensinogen to angiotensin I, which is changed to angiotensin II by a converting enzyme found in lung capillaries. Angiotensin II stimulates the adrenal cortex to release aldosterone. The effect of this process, called the renin-angiotensin-aldosterone system, is to raise blood pressure in two ways: (1) angiotensin II constricts the arterioles, and (2) aldosterone causes the kidneys to reabsorb sodium. When the blood sodium level rises, water is reabsorbed, in part because the hypothalamus secretes ADH (see page 740). Then blood pressure rises to normal.

As you might suspect, there is an antagonistic hormone to aldosterone. When the atria of the heart are stretched due to increased blood volume, cardiac cells release a hormone called **atrial natriuretic hormone (ANH),** which inhibits the secretion of aldosterone from the adrenal cortex. The effect of this hormone is to cause the excretion of sodium ($Na^+$)—that is, *natriuresis.* When sodium is excreted, so is water, and therefore blood pressure lowers to normal (Fig. 40.12, *above*).

### Malfunction of the Adrenal Cortex

When the level of adrenal cortex hormones is low due to hyposecretion, a person develops **Addison disease.** The presence of excessive but ineffective ACTH causes bronzing of the skin because ACTH, like MSH, can lead to a buildup of melanin (Fig. 40.13). Other symptoms include weight loss,

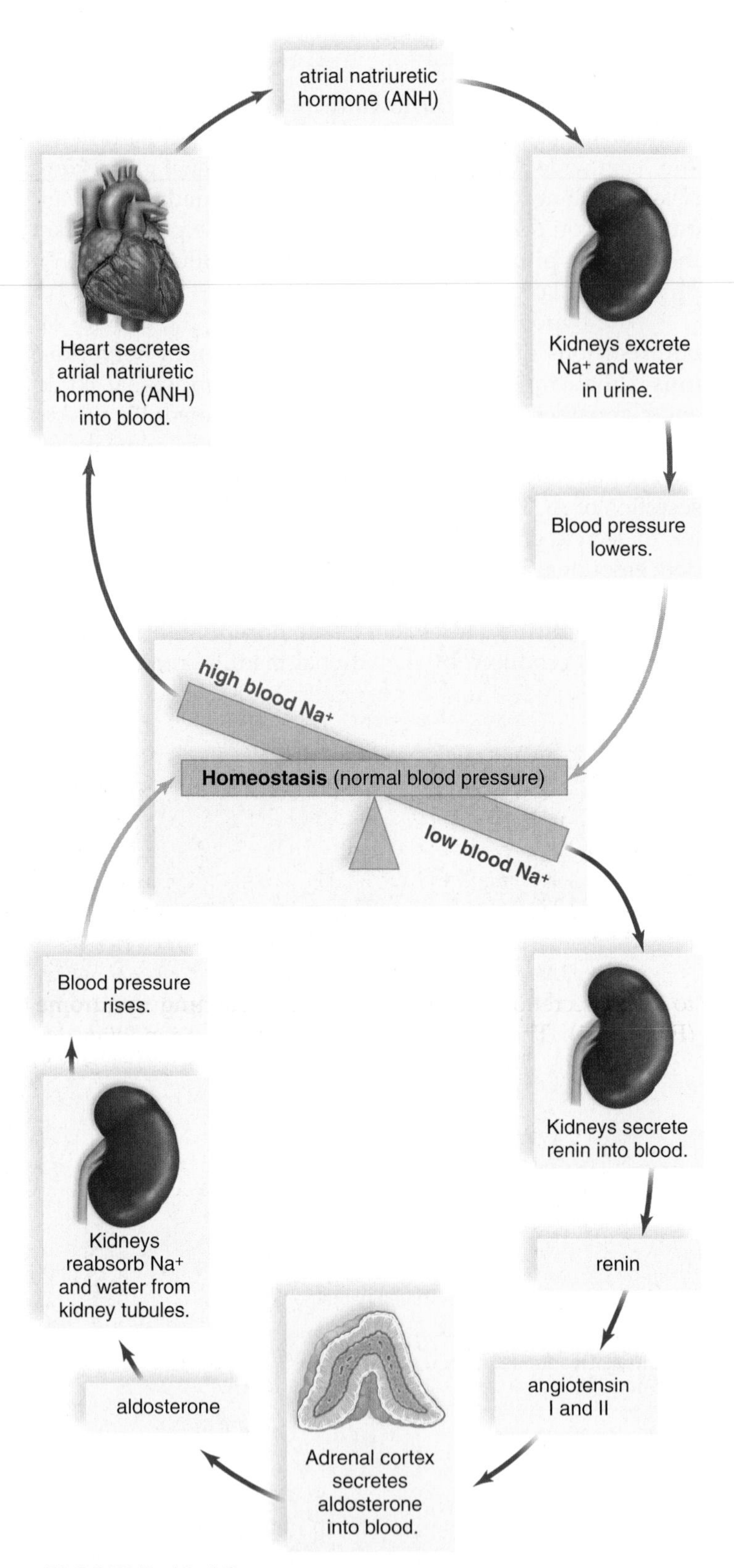


**FIGURE 40.12 Regulation of blood pressure and volume.**

*Below:* When the blood sodium ($Na^+$) level is low, a low blood pressure causes the kidneys to secrete renin. Renin leads to the secretion of aldosterone from the adrenal cortex. Aldosterone causes the kidneys to reabsorb $Na^+$, and water follows, so that blood volume and pressure return to normal. *Above:* When the blood $Na^+$ is high, a high blood volume causes the heart to secrete atrial natriuretic hormone (ANH). ANH causes the kidneys to excrete $Na^+$, and water follows. The blood volume and pressure return to normal.

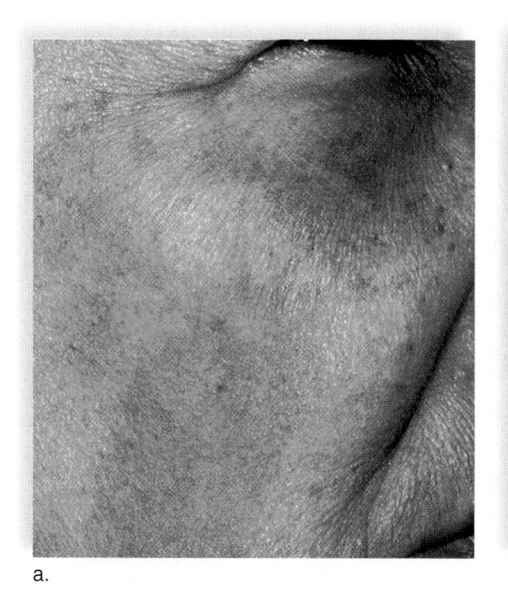
a.

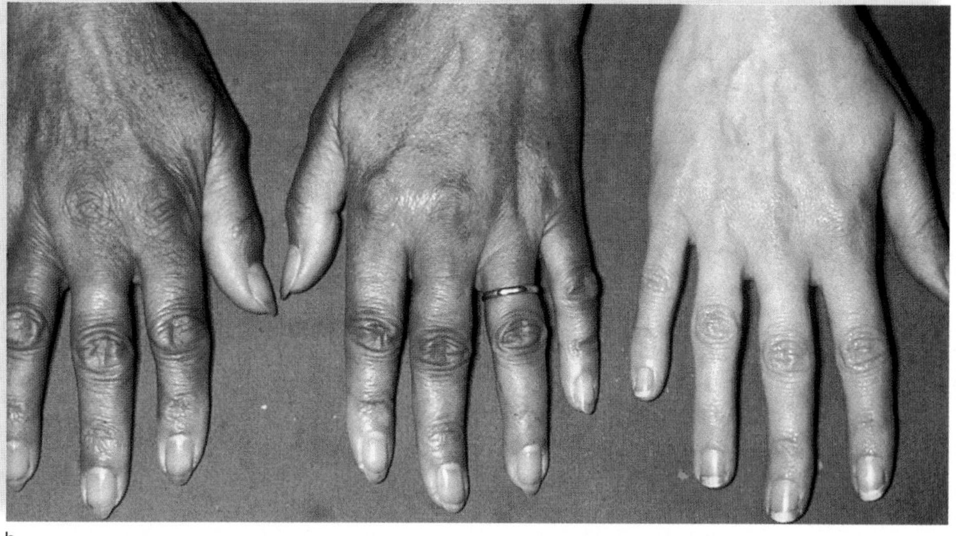
b.

**FIGURE 40.13 Addison disease.**

Addison disease is characterized by a peculiar bronzing of the skin, particularly noticeable in light-skinned individuals. Note the color of (**a**) the face and (**b**) the hands compared to the hand of an individual without the disease.

weakness, and hypotension (low blood pressure). Without cortisol, glucose cannot be replenished when a stressful situation arises. Even a mild infection can lead to death. The lack of aldosterone results in the loss of sodium and water, the development of low blood pressure, and possibly severe dehydration. Left untreated, Addison disease can be fatal.

When the level of adrenal cortex hormones is high due to hypersecretion, a person develops **Cushing syndrome** (Fig. 40.14). The excess cortisol results in a tendency toward diabetes mellitus as muscle protein is metabolized and subcutaneous fat is deposited in the midsection. The trunk is obese, while the arms and legs remain a normal size. An excess of aldosterone and reabsorption of sodium and water by the kidneys lead to a basic blood pH and hypertension. The face is moon-shaped due to edema. Hypertension (high blood pressure) is common in patients with Cushing syndrome. Masculinization may occur in women because of excess adrenal male sex hormones. This is referred to as adrenogenital syndrome (AGS). In women with AGS, an increase in body hair, deepening of the voice, and beard growth may occur.

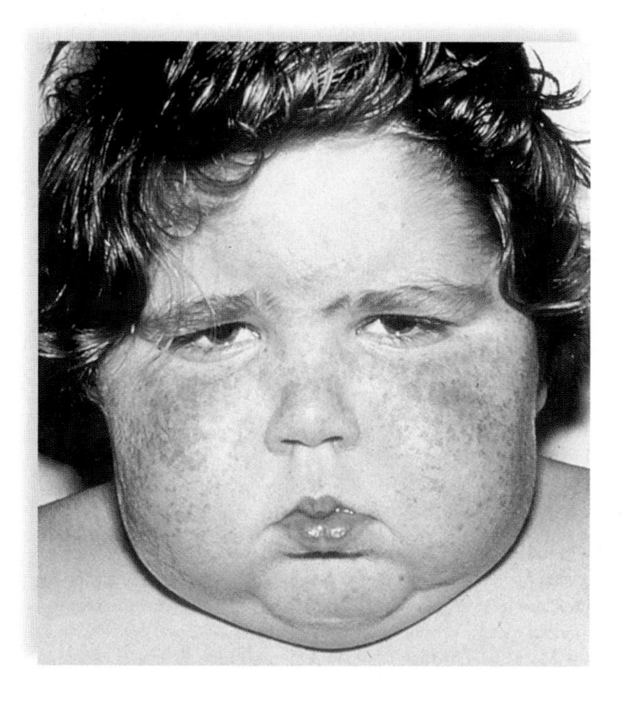

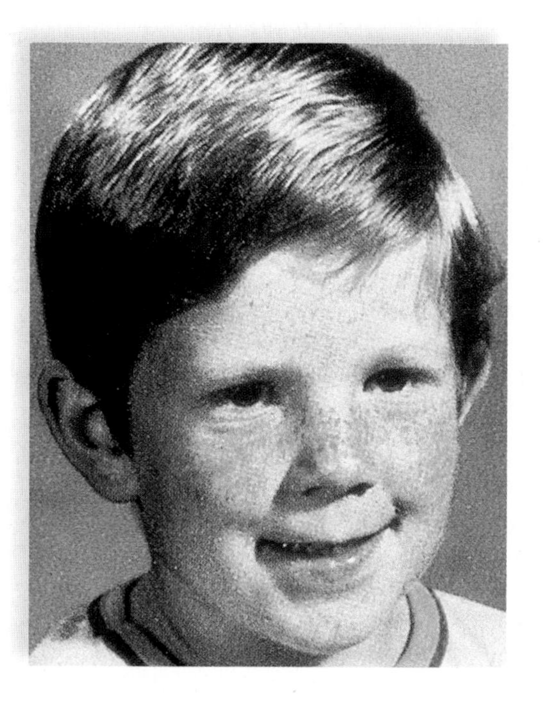

**FIGURE 40.14**
**Cushing syndrome.**

Cushing syndrome results from hypersecretion of hormones due to an adrenal cortex tumor. *Left:* Patient first diagnosed with Cushing syndrome. *Right:* Four months later, after therapy.

## Pancreas

The **pancreas** (see Fig. 40.2) is a slender, pale-colored organ that lies transversely in the abdomen between the kidneys and near the duodenum of the small intestine. It is about 20 cm long and 3 cm thick and weighs about 80 g. The pancreas is rather lumpy in consistency and is composed of two types of tissue. Exocrine tissue produces and secretes digestive juices that go by way of ducts to the small intestine. Endocrine tissue, called the **pancreatic islets** (islets of Langerhans), produces and secretes the hormones insulin and glucagon directly into the blood. The majority of pancreas tissues are exocrine in nature.

Insulin is secreted when there is a high blood glucose level, which usually occurs just after eating (Fig. 40.15, *above*). Insulin stimulates the uptake of glucose by cells, especially liver cells, muscle cells, and adipose tissue cells. In liver and muscle cells, glucose is then stored as glycogen. In muscle cells, the breakdown of glucose supplies energy for protein metabolism, and in fat cells the breakdown of glucose supplies glycerol for the formation of fat. In these ways, insulin lowers the blood glucose level.

Glucagon is secreted from the pancreas, usually in between eating, when there is a low blood glucose level (Fig. 40.15, *below*). The major target tissues of glucagon are the liver and adipose tissue. Glucagon stimulates the liver to break down glycogen to glucose and to use fat and protein in preference to glucose as energy sources. Adipose tissue cells break down fat to glycerol and fatty acids. The liver takes these up and uses them as substrates for glucose formation. In these ways, glucagon raises the blood glucose level.

### *Diabetes Mellitus*

**Diabetes mellitus** is a fairly common hormonal disease in which liver cells, and indeed most body cells, are unable to take up glucose as they should. Therefore, cellular famine exists in the midst of plenty, and the person becomes extremely hungry. As the blood glucose level rises, glucose, along with water, is excreted in the urine. Urination is frequent and the loss of water in this way causes the diabetic to be extremely thirsty.

The glucose tolerance test assists in the diagnosis of diabetes mellitus. After the patient is given 100 g of glucose, the blood glucose concentration is measured at intervals. In a diabetic, the blood glucose level rises greatly and remains elevated for several hours. In the meantime, glucose appears in the urine. In a nondiabetic, the blood glucose level rises somewhat and then returns to normal after about two hours.

**Types of Diabetes.** There are two types of diabetes mellitus. In *type 1 diabetes,* the pancreas is not producing insulin. This condition is believed to be brought on by exposure to an environmental agent, most likely a virus, whose presence causes cytotoxic T cells to destroy the pancreatic islets. The body turns to the metabolism of fat, which leads to the buildup of ketones in the blood, called ketonuria, and, in turn, to acidosis

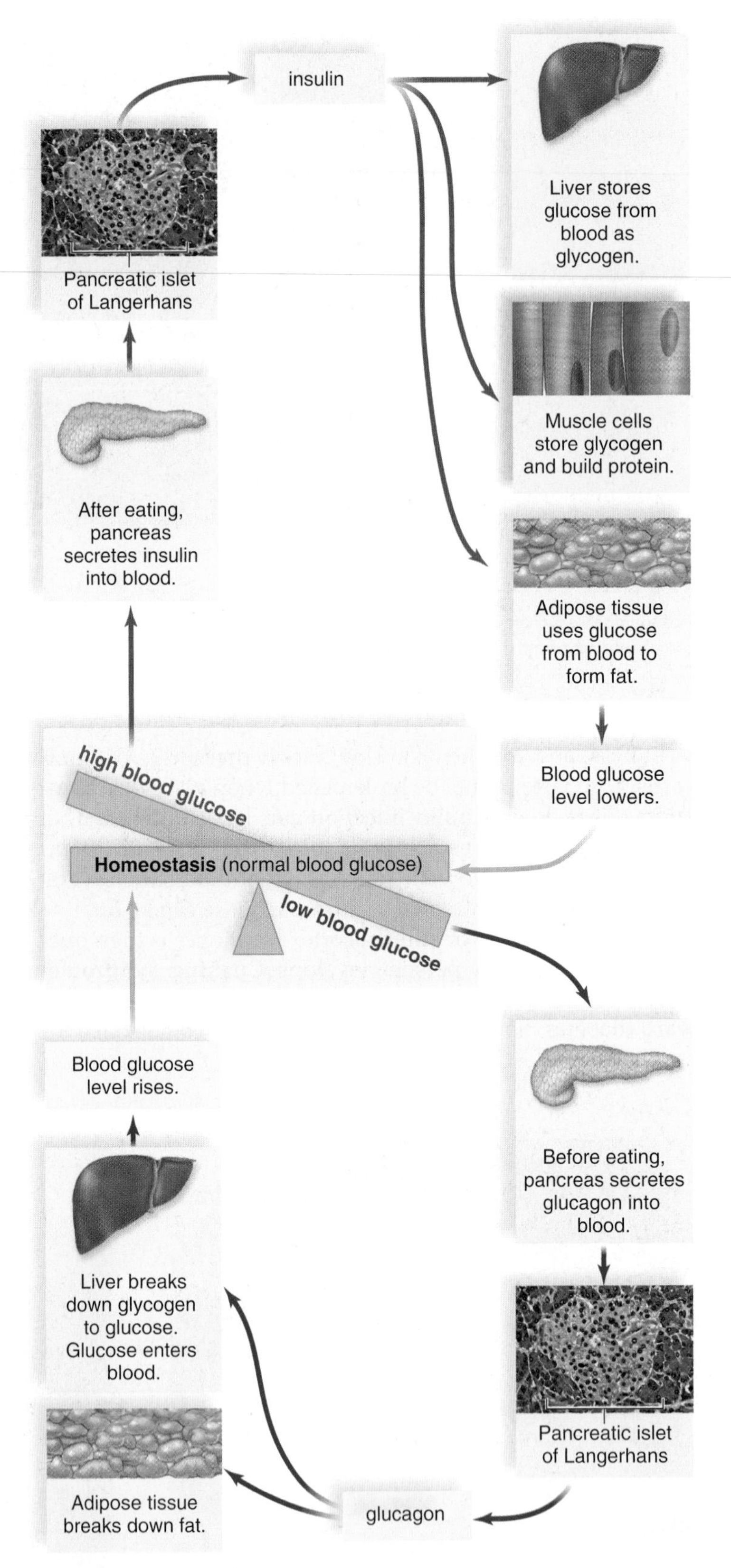


**FIGURE 40.15 Regulation of blood glucose level.**

*Above:* When the blood glucose level is high, the pancreas secretes insulin. Insulin promotes the storage of glucose as glycogen and the synthesis of proteins and fats (as opposed to their use as energy sources). Therefore, insulin lowers the blood glucose level to normal. *Below:* When the blood glucose level is low, the pancreas secretes glucagon. Glucagon acts opposite to insulin; therefore, glucagon raises the blood glucose level to normal.

*science focus*

## Identifying Insulin as a Chemical Messenger

The pancreas is both an exocrine gland and an endocrine gland. It sends digestive juices to the duodenum by way of the pancreatic duct, and it secretes the hormones insulin and glucagon into the bloodstream. In 1920, physician Frederick Banting decided to try to isolate insulin in order to identify it as a chemical messenger. Previous investigators had been unable to do this because the enzymes in the digestive juices destroyed insulin (a protein) during the isolation procedure. Banting hit upon the idea of tying off the pancreatic duct, which he knew from previous research would lead to the degeneration only of the cells that produce digestive juices and not of the pancreatic islets (of Langerhans), where insulin is made. His professor, J. J. Macleod, made a laboratory available to him at the University of Toronto and also assigned a graduate student, Charles Best, to assist him. Banting and Best had limited funds and spent that summer working, sleeping, and eating in the lab. By the end of the summer, they had obtained pancreatic extracts that did lower the blood glucose level in diabetic dogs. Macleod then brought in biochemists, who purified the extract. Insulin therapy for the first human patient began in 1922, and large-scale production of purified insulin from pigs and cattle followed. Banting and Macleod received a Nobel Prize for their work in 1923. The amino acid sequence of insulin was determined in 1953. Insulin is presently synthesized using recombinant DNA technology. Banting and Best performed the required steps given in the chart (lower left) to identify a chemical messenger.

| *Steps to Identify a Chemical Messenger* | *Banting and Best* |
|---|---|
| 1. Identify the source of the chemical | Pancreatic islets are source |
| 2. Identify the effect to be studied | Presence of pancreas in body lowers blood glucose |
| 3. Isolate the chemical | Insulin isolated from pancreatic secretions |
| 4. Show that the chemical has the effect | Insulin lowers blood glucose |

**FIGURE 40A Early insulin experiments.** *Charles H. Best and Sir Frederick Banting in 1921 with the first dog to be kept alive by insulin.*

(acid blood), which can lead to coma and death. As a result, the individual must have daily insulin injections. These injections control the diabetic symptoms but still can cause inconveniences, since either an overdose of insulin or missing a meal can bring on the symptoms of hypoglycemia (low blood sugar). These symptoms include perspiration, pale skin, shallow breathing, and anxiety. Because the brain requires a constant supply of glucose, unconsciousness can result. The cure is quite simple: Immediate ingestion of a sugar cube or fruit juice can very quickly counteract hypoglycemia.

It is possible to transplant a working pancreas into patients with type 1 diabetes. To do away with the necessity of taking immunosuppressive drugs after the transplant, fetal pancreatic islet cells have been injected into patients. Another experimental procedure is to place pancreatic islet cells in a capsule that allows insulin to get out but prevents antibodies and T lymphocytes from getting in. This artificial organ is implanted in the abdominal cavity. Recently, there has been progress in developing a vaccine to block the immune system's attack on islet cells. Such a vaccine, possibly used in conjunction with immunosuppressive drugs, may someday stop or even reverse type 1 diabetes.

Of the over 20.8 million people who now have diabetes in the United States, most have *type 2 diabetes*. Often, the patient is obese—adipose tissue produces a substance that impairs insulin receptor function. Normally, but not in the type 2 diabetic, the binding of insulin to a receptor causes the number of glucose transporters to increase in the plasma membrane. Also, the blood insulin level is low and cells do not have enough insulin receptors.

It is possible to prevent or at least control type 2 diabetes by adhering to a low-fat, low-sugar diet and exercising regularly. If this fails, oral drugs that stimulate the pancreas to secrete more insulin and enhance the metabolism of glucose in the liver and muscle cells are available. It is projected that as many as 5 million Americans may have type 2 diabetes without being aware of it. Yet, the effects of untreated type 2 diabetes are as serious as those of type 1 diabetes.

Long-term complications of both types of diabetes are blindness, kidney disease, and cardiovascular disorders, including atherosclerosis, heart disease, stroke, and reduced circulation. The latter can lead to gangrene in the arms and legs. Pregnancy carries an increased risk of diabetic coma, and the child of a diabetic is somewhat more likely to be stillborn or to die shortly after birth. These complications of diabetes are not expected to appear if the mother's blood glucose level is carefully regulated and kept within normal limits.

## Testes and Ovaries

The **testes** are located in the scrotum, and the **ovaries** are located in the pelvic cavity. The testes produce **androgens** (e.g., **testosterone**), which are the male sex hormones, and the ovaries produce **estrogens** and **progesterone,** the female sex hormones. The anterior pituitary releases the gonadotropic hormones, follicle-stimulating hormone (FSH), and luteinizing hormone (LH), and they control the secretions of the testes and ovaries in a three-tier system, as described on page 740.

Greatly increased testosterone secretion at the time of puberty stimulates the growth of the penis and the testes. Testosterone also brings about and maintains the male secondary sex characteristics that develop during puberty. Testosterone causes growth of a beard, axillary (underarm) hair, and pubic hair. It prompts the larynx and the vocal cords to enlarge, causing the voice to change. It is partially responsible for the muscular strength of males, and this is the reason some athletes take supplemental amounts of **anabolic steroids,** which are either testosterone or related chemicals. The dangerous effects of taking anabolic steroids are listed in Figure 40.16. Testosterone also stimulates oil and sweat glands in the skin; therefore, it is largely responsible for acne and body odor. Another side effect of testosterone is baldness. Genes for baldness are probably inherited by both sexes, but baldness is seen more often in males because of the presence of testosterone.

The female sex hormones, estrogen and progesterone, have many effects on the body. In particular, estrogens secreted at the time of puberty stimulate the growth of the uterus and the vagina. Estrogen [Gk. *oistros,* sexual heat; L. *genitus,* producing] is necessary for egg maturation and is largely responsible for the secondary sex characteristics in females, including female body hair and fat distribution. In general, females have a more rounded appearance than males because of a greater accumulation of fat beneath the skin. Also, the pelvic girdle is wider in females than in males, resulting in a larger pelvic cavity. Both estrogen and progesterone are required for breast development and regulation of the uterine cycle, which includes monthly menstruation (discharge of blood and mucosal tissues from the uterus).

## Pineal Gland

The **pineal gland,** which is located in the brain, produces the hormone **melatonin,** primarily at night. Melatonin is involved in our daily sleep-wake cycle; normally we grow sleepy at night when melatonin levels increase and awaken once daylight returns and melatonin levels are low (Fig. 40.17). Daily 24-hour cycles such as this are called **circadian rhythms** [L. *circum,* around, and *dies,* day], and circadian rhythms are controlled by an internal timing mechanism called a biological clock.

Based on animal research, it appears that melatonin also regulates sexual development. It has been noted that children whose pineal gland has been destroyed due to a brain tumor experience early puberty.

## Thymus Gland

The **thymus gland** is a lobular gland that lies just beneath the sternum (see Fig. 40.2). This organ reaches its largest size and is most active during childhood. With aging, the organ gets smaller and becomes fatty. Lymphocytes that originate in the bone marrow and then pass through the thymus are transformed into T lymphocytes. The lobules of the thymus are lined by epithelial cells that secrete hormones called thymosins. These hormones aid in the differentiation of T lymphocytes packed inside the lobules. Although the hormones secreted by the thymus ordinarily work in the thymus, there is hope that these hormones could be injected into AIDS or cancer patients, where they would enhance T-lymphocyte function.

**FIGURE 40.16** **The effects of anabolic steroid use.**

## Other Hormones

Some hormones in the body, as exemplified by the ones discussed here, are produced by tissues/cells rather than by one of the glands.

### *Leptin*

**Leptin** is a peptide hormone secreted by adipose (fat) tissue throughout the body. It was first described in the 1990s. One of its most interesting functions is its role in the feedback control of appetite. Leptin binds to neurons in the CNS that are concerned with the control of appetite. It can bring about feelings of satiation and can suppress appetite. Early researchers hoped that leptin could be used to control obesity in humans. Unfortunately, the trials have not yielded satisfactory results.

### *Erythropoietin*

**Erythropoietin (EPO)** is a peptide hormone produced by the kidneys. It is released in response to low oxygen levels in kidney tissues. EPO serves to stimulate the production of red blood cells (erythropoiesis) and speed up the maturation of red blood cells. Under the influence of EPO, bone marrow can increase the rate of red blood cell production upward to 30 million per second. People with anemia, common in kidney disease, cancer, and AIDS, may be effectively treated with injections of recombinant EPO. In recent years, some athletes have practiced blood doping, in which EPO is used to improve performance by increasing the oxygen-carrying capacity of the blood. The potential dangers of blood doping far outweigh the temporary advantages. Because EPO increases the number of red blood cells, the blood becomes thicker, blood pressure can become elevated, and the athlete is at increased risk of a heart attack or stroke.

### *Local Hormones*

Local hormones are produced by cells, and they act on neighboring cells. Examples include growth factors, cytokines, and prostaglandins. **Prostaglandins** are potent chemical signals produced within cells from arachidonate, a fatty acid. In the uterus, prostaglandins cause muscles to contract; therefore, they are implicated in the pain and discomfort of menstruation in some women. Also, prostaglandins mediate the effects of pyrogens, chemicals that are believed to reset the temperature regulatory center in the brain. Aspirin reduces body temperature and controls pain because of its effect on prostaglandins.

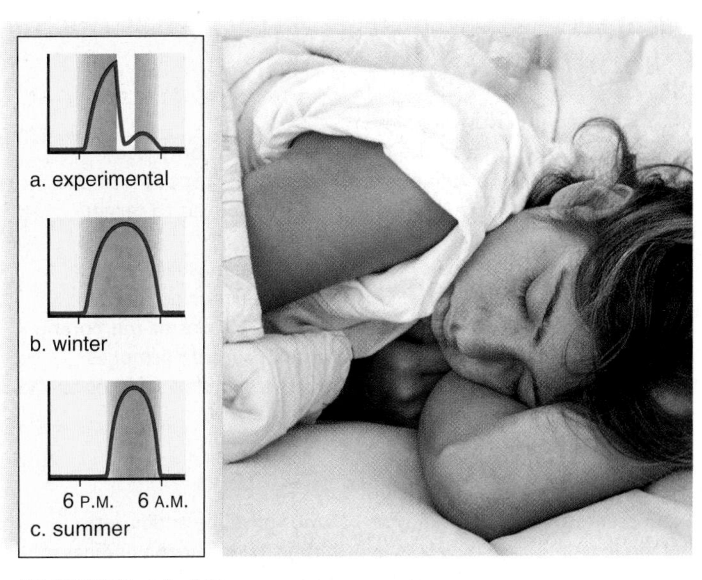


**FIGURE 40.17 Melatonin production.**

Melatonin production is greatest at night when we are sleeping. Light suppresses melatonin production (**a**), so its duration is longer in the winter (**b**) than in the summer (**c**).

Certain prostaglandins reduce gastric secretion and have been used to treat ulcers; others lower blood pressure and have been used to treat hypertension; and yet others inhibit platelet aggregation and have been used to prevent thrombosis. However, different prostaglandins have contrary effects, and it has been very difficult to successfully standardize their use. Therefore, prostaglandin therapy is still considered experimental.

**Check Your Progress 40.3**

1. How does the renin-angiotensin-aldosterone system raise blood pressure?
2. Name the source of each of the following hormones: aldosterone, melatonin, epinephrine, EPO, leptin, glucagon, ANH, cortisol, and calcitonin.
3. Which hormone stimulates osteoclasts? Which inhibits them?

## Connecting the Concepts

The nervous system and the endocrine system are structurally and functionally related. The hypothalamus, a portion of the brain, controls the pituitary, an endocrine gland. The hypothalamus even produces the hormones that are released by the posterior pituitary. It also stimulates the release of hormones by the adrenal medulla. Neurosecretory cells in the hypothalamus also produce releasing hormones that control the activity of the anterior piuitary.

The nervous system is well known for bringing about an immediate response to environmental stimuli, as in the fight-or-flight reaction. Most often, the chemical signals released by the nervous system, called neurotransmitters, help maintain homeostasis. Heart rate, breathing rate, and blood pressure are all regulated to stay relatively constant. Hormones released by endocrine glands also help maintain homeostasis, especially by keeping the levels of calcium, sodium, glucose, and other blood constitutents within normal limits.

The endocrine system is slower acting than the nervous system and often regulates processes that occur over days or even months. Hormones secreted into the bloodstream control whole body processes, such as growth and reproduction. In this chapter, we mentioned the effects of growth hormone, and in Chapter 41, we explore the activity of the hormones that influence reproduction.

## summary

### 40.1 Endocrine Glands

The nervous system and the endocrine system both use chemical signals. Endocrine glands secrete hormones into the bloodstream, and from there they are distributed to target organs or tissues.

Hormones are a type of chemical signal that usually act at a distance between body parts. Hormones are either peptides or steroids. Reception of a peptide hormone at the plasma membrane activates an enzyme cascade inside the cell. Steroid hormones combine with a receptor in the cell, and the complex attaches to and activates DNA. Protein synthesis follows.

### 40.2 Hypothalamus and Pituitary Gland

Neurosecretory cells in the hypothalamus produce antidiuretic hormone (ADH) and oxytocin, which are stored in axon endings in the posterior pituitary until they are released.

The hypothalamus produces hypothalamic-releasing and hypothalamic-inhibiting hormones, which pass to the anterior pituitary by way of a portal system. The anterior pituitary produces several types of hormones, and some of these stimulate other hormonal glands to secrete hormones in a three-tier system.

### 40.3 Other Endocrine Glands and Hormones

The thyroid gland, controlled by TSH, requires iodine to produce thyroxine ($T_4$) and triiodothyronine ($T_3$), which increase the metabolic rate. If iodine is available in insufficient quantities, a simple goiter develops; if the thyroid is overactive, an exophthalmic goiter develops. The thyroid gland also produces calcitonin, which helps lower the blood calcium level. The parathyroid glands secrete parathyroid hormone, which raises the blood calcium and decreases the blood phosphate levels.

The adrenal glands respond to stress: Immediately, the adrenal medulla secretes epinephrine and norepinephrine, which bring about responses we associate with emergency situations. On a long-term basis, the adrenal cortex, controlled by ACTH, produces the glucocorticoids (e.g., cortisol) and the mineralocorticoids (e.g., aldosterone). Cortisol stimulates hydrolysis of proteins to amino acids that are converted to glucose; in this way, it raises the blood glucose level. Aldosterone causes the kidneys to reabsorb sodium ions ($Na^+$) and to excrete potassium ions ($K^+$). Addison disease develops when the adrenal cortex is underactive, and Cushing syndrome develops when the adrenal cortex is overactive.

The pancreatic islets secrete insulin, which lowers the blood glucose level, and glucagon, which has the opposite effect. The most common illness caused by hormonal imbalance is diabetes mellitus, which is due to the failure of the pancreas to produce insulin or the failure of the cells to take it up.

The gonads, controlled by gonadotropic hormones, produce the sex hormones; the pineal gland produces melatonin, which may be involved in circadian rhythms and the development of the reproductive organs; and the thymus secretes thymosins, which stimulate T-lymphocyte production and maturation.

Tissue and organs having other functions also produce hormones. Leptin is a newly described hormone that regulates appetite, and erythropoietin stimulates the production of red blood cells. Cells produce local hormones, for example, prostaglandins are produced and act locally.

## understanding the terms

acromegaly 742
Addison disease 746
adrenal cortex 745
adrenal gland 745
adrenal medulla 745
adrenocorticotropic hormone (ACTH) 740
aldosterone 746
anabolic steroid 750
androgen 750
anterior pituitary 740
antidiuretic hormone (ADH) 740
atrial natriuretic hormone (ANH) 746
calcitonin 744
chemical signal 738
circadian rhythm 750
congenital hypothyroidism 743
cortisol 746
Cushing syndrome 747
cyclic adenosine monophosphate (cAMP) 739
diabetes mellitus 748
endocrine gland 736
endocrine system 736
epinephrine 745
erythropoietin (EPO) 751
estrogen 750
exophthalmic goiter 743
first messenger 739
glucocorticoid 745
gonadotropic hormone 740
growth hormone (GH) 740
hormone 736
hypothalamic-inhibiting hormone 740
hypothalamic-releasing hormone 740
hypothalamus 740
leptin 751
melanocyte-stimulating hormone (MSH) 740
melatonin 750
mineralocorticoid 745
myxedema 743
negative feedback 740
norepinephrine 745
ovary 750
oxytocin 740
pancreas 748
pancreatic islet 748
parathyroid gland 744
parathyroid hormone (PTH) 744
peptide hormone 738
pheromone 738
pineal gland 750
pituitary dwarfism 741
pituitary gland 740
positive feedback 740
posterior pituitary 740
progesterone 750
prolactin (PRL) 740
prostaglandin 751
renin 746
second messenger 739
simple goiter 743
steroid hormone 739
testes 750
testosterone 750
tetany 744
thymus gland 750
thyroid gland 743
thyroid-stimulating hormone (TSH) 740
thyroxine ($T_4$) 743

Match the terms to these definitions:

a. ______________ Organ in the neck; secretes several important hormones, including thyroxine and calcitonin.
b. ______________ Common homeostatic control mechanism in which the output of a system shuts off or reduces the intensity of the original stimulus.
c. ______________ Organ that produces melatonin.
d. ______________ Type of hormone that binds to a plasma membrane receptor; results in activation of enzyme cascade.
e. ______________ Chemical substance secreted by one organism that influences the behavior of another.
f. ______________ Endocrine organ in which immature lymphocytes become T lymphocytes.
g. ______________ Testosterone and related chemicals used by some athletes to build muscle mass.
h. ______________ Appetite-suppressing peptide hormone produced by adipose tissue.

## reviewing this chapter

1. In what ways are the nervous and endocrine systems alike, and how are they different? 736
2. Explain how steroid hormones and peptide hormones affect the metabolism of the cell. 738–39
3. Give an example of the three-tier system and the negative feedback relationship among the hypothalamus, the anterior pituitary, and other endocrine glands. 740–41
4. Explain the relationship of the hypothalamus to the posterior pituitary gland and to the anterior pituitary gland. List the hormones secreted by the posterior and anterior pituitary. 740–41
5. Discuss the effect of there being too much or too little growth hormone when a young person is growing. What is the result if the anterior pituitary produces growth hormone in an adult? 741–42
6. What two types of goiters are associated with a malfunctioning thyroid? Explain each type. 743
7. How do the thyroid and the parathyroid work together to control the blood calcium level? 744
8. How do the adrenal glands respond to stress? What hormones are secreted by the adrenal medulla, and what effects do these hormones have? 745
9. Name the most significant glucocorticoid and mineralocorticoid, and discuss the function of each. Explain the symptoms of Addison disease and Cushing syndrome. 746–47
10. Draw a diagram to explain how insulin and glucagon maintain the blood glucose level. Use your diagram to explain three major symptoms of type I diabetes mellitus. 748–49
11. Name the other endocrine glands cited in this chapter, and discuss the functions of the hormones they secrete. Also, discuss the hormones not produced by endocrine glands. 750–51

## testing yourself

Choose the best answer for each question.
For questions 1–5, match each hormone to a gland in the key.

**KEY:**

a. pancreas
b. anterior pituitary
c. posterior pituitary
d. thyroid
e. adrenal medulla
f. adrenal cortex

1. cortisol
2. growth hormone (GH)
3. oxytocin storage
4. insulin
5. epinephrine
6. Which of the following relies on the activation of cAMP, a second messenger, to stimulate its target cells?
   a. estrogen
   b. aldosterone
   c. glucagon
   d. testosterone
7. The anterior pituitary controls the secretion(s) of both
   a. the adrenal medulla and the adrenal cortex.
   b. the thyroid and the adrenal cortex.
   c. the ovaries and the testes.
   d. Both b and c are correct.
8. Diabetes mellitus is associated with
   a. too much insulin in the blood.
   b. too much glucose in the blood.
   c. blood that is too dilute.
   d. All of these are correct.
9. Which of these is not a pair of antagonistic hormones?
   a. insulin—glucagon
   b. calcitonin—parathyroid hormone
   c. aldosterone—atrial natriuretic hormone (ANH)
   d. thyroxine—growth hormone
10. Which hormone and condition are mismatched?
    a. cortisol—myxedema
    b. growth hormone—acromegaly
    c. thyroxine—goiter
    d. parathyroid hormone—tetany
    e. insulin—diabetes
11. Which of the following hormones could affect fat metabolism?
    a. growth hormone
    b. thyroxine
    c. insulin
    d. glucagon
    e. All of these are correct.
12. The difference between type I and type 2 diabetes is that
    a. for type 2 diabetes, insulin is produced but not used; type I results from lack of insulin production.
    b. treatment for type 2 involves insulin injections, while type I can be controlled, usually by diet.
    c. only type I can result in complications such as kidney disease, reduced circulation, or stroke.
    d. type I can be a result of lifestyle, and type 2 is thought to be caused by a virus or other agent.
13. Which of the following hormones is/are found in females?
    a. estrogen
    b. testosterone
    c. follicle-stimulating hormone
    d. Both a and c are correct.
    e. All of these are correct.
14. Parathyroid hormone causes
    a. the kidneys to excrete more calcium ions.
    b. bone tissue to break down and release calcium into the bloodstream.
    c. fewer calcium ions to be absorbed by the intestines.
    d. more calcium ions to be deposited in bone tissue.
15. Hormones from all but which of the following glands can affect glucose levels in the body?
    a. pancreas
    b. adrenal glands
    c. pituitary
    d. hypothalamus
    e. thymus
16. Tropic hormones are hormones that affect other endocrine tissues. Which of the following would be considered a tropic hormone?
    a. calcitonin
    b. oxytocin
    c. glucagon
    d. melatonin
    e. follicle-stimulating hormone
17. One of the chief differences between endocrine hormones and local hormones is
    a. the distance over which they act.
    b. that one is a chemical signal and the other is not.
    c. only endocrine hormones are made by humans.
    d. All of these are correct.

18. Peptide hormones
    a. are received by a receptor located in the plasma membrane.
    b. are received by a receptor located in the cytoplasm.
    c. bring about the transcription of DNA.
    d. Both b and c are correct.
19. Complete this diagram by filling in blanks a–e.

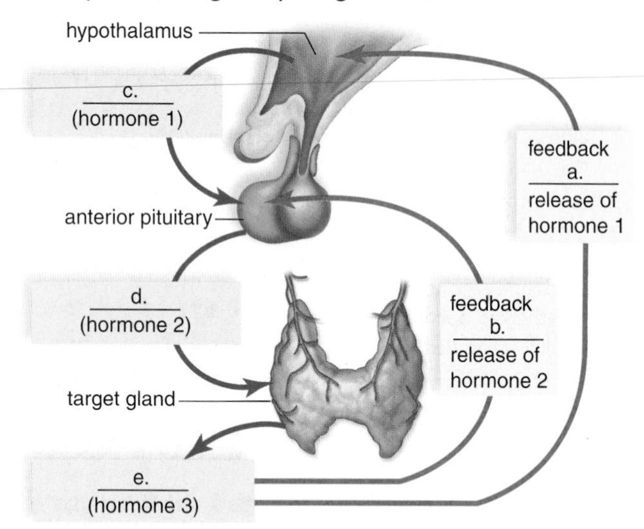


For questions 20–25, match the function to a hormone in the key. Choose more than one answer if correct. Answers may be used more than once.

**KEY:**

a. antidiuretic hormone
b. oxytocin
c. glucocorticoids
d. glucagon
e. parathyroid hormone

20. raises blood glucose level
21. stimulates uterine muscle contraction
22. stimulates water reabsorption by kidneys
23. stimulates release of milk by mammary glands
24. raise blood glucose and stimulate breakdown of protein
25. raises blood calcium level
26. Steroid hormones are secreted by
    a. the adrenal cortex.
    b. the gonads.
    c. the thyroid.
    d. Both a and b are correct.
    e. Both b and c are correct.
27. Steroid hormones
    a. bind to a receptor located in the plasma membrane.
    b. cause the production of cAMP.
    c. activate protein kinase.
    d. stimulate the production of mRNA.
28. Which of the following statements about the pituitary gland is incorrect?
    a. The pituitary lies inferior to the hypothalamus.
    b. Growth hormone and prolactin are secreted by the anterior pituitary.
    c. The anterior pituitary and posterior pituitary communicate with each other.
    d. Axons run between the hypothalamus and the posterior pituitary.
29. Prostaglandins
    a. have a consistent effect.
    b. are useful in the treatment of cancer.
    c. are carried in the blood.
    d. stimulate other glands.
    e. act locally.
30. Erythropoietin
    a. stimulates platelet production.
    b. stimulates leukocyte production.
    c. inhibits red blood cell production.
    d. inhibits leukocyte production.
    e. stimulates red blood cell production.

## thinking scientifically

1. Caffeine inhibits the breakdown of cAMP in the cell. Referring to Figure 40.4, how would this influence a stress response brought about by epinephrine?
2. Both males and females can develop secondary sex characteristics of the opposite sex if they take enough of the appropriate sex hormone. Hypothesize the mechanism that would make this possible after reviewing Figure 40.4.

## bioethical issue

### Growth Hormone

Untreated GH deficiency in childhood results in pituitary dwarfism. If young children with GH deficiency receive daily GH injections through adolescence, most will attain normal stature. Few would argue that treatment of pituitary dwarfism is not justified. But is it justified to use GH therapy to make a child of normal height taller?

GH levels naturally decline with age, and there is evidence that treating older adults with GH boosts muscle mass and reduces body fat. Pills do not work, and GH injections are by prescription only and very expensive, being around $1,000 a month. Furthermore, GH therapy can result in undesirable side effects, such as elevated blood sugar, fluid retention, and joint pain. Nevertheless, there is keen interest in GH therapy not only from those who wish to delay or avoid the effects of aging, but also from athletes in search of performance-enhancing substances. Thus far, like anabolic steroids and erythropoietin (EPO), GH is banned by most competitive sports authorities, including the U.S. and International Olympic Committees.

The existence of GH therapy raises questions about medical treatment for disease versus enhancement of an already healthy body. Under what circumstances is it suitable to increase a child's growth? Should the physical decline of aging be accepted, or should we try to maintain a youthful build using GH? If an athlete wants to make him- or herself more competitive using GH, and is willing to accept the risks, should therapy be permitted? Who decides?

## *Biology* website

The companion website for *Biology* provides a wealth of information organized and integrated by chapter. You will find practice tests, animations, videos, and much more that will complement your learning and understanding of general biology.

**http://www.mhhe.com/maderbiology10**

# 41

# Reproductive Systems

*Sea horses are fishes with an unusual style of sexual reproduction: The males become pregnant and give birth to the young. When sea horses mate, the female deposits eggs in a special brood pouch on the male's abdomen, where they are fertilized by his sperm. The pouch seals and stays closed as the embryos develop, a process that can take 10 to 25 days depending on the species. Then, the pouch opens and muscular contractions expel the young sea horses. Although the evolutionary advantage conferred by male pregnancy in sea horses is not certain, some experts think it increases the efficiency of reproduction. The female does not have to wait until the completion of pregnancy for her eggs to mature and become ready for fertilization, so when the male has given birth, she is ready to mate and impregnate him again immediately.*

*In this chapter, you will learn about the mode of asexual reproduction that requires but a single parent and various modes of sexual reproduction that require production of gametes and fertilization. Most of the chapter focuses on the human female and male reproductive systems, which produce gametes and keep them from drying out—no external water is required for human reproduction.*

Male potbelly sea horses with brood pouches holding 150–300 oocytes each.

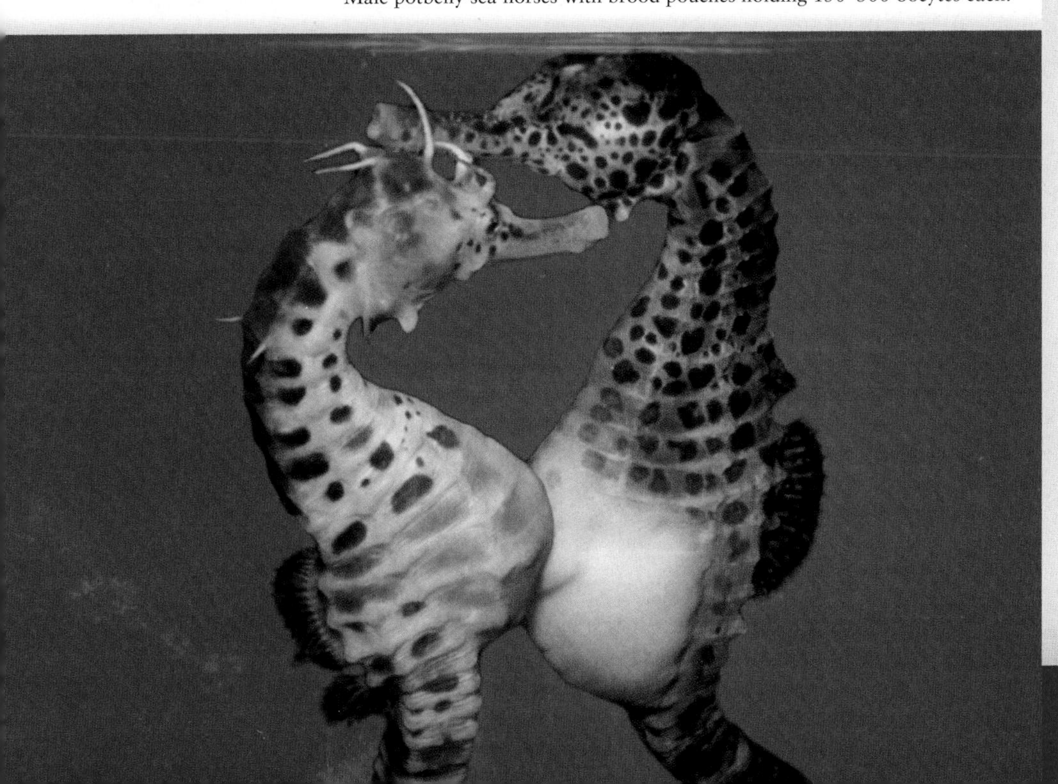

## *concepts*

# 41.1 How Animals Reproduce

Although the majority of animals reproduce sexually, a few groups of animals are also capable of asexual reproduction. In sexual reproduction, sex cells, or gametes, produced by the parents unite to form a genetically unique individual. In asexual reproduction, a single parent gives rise to offspring that are identical to the parent, unless mutations have occurred. The adaptive advantage of asexual reproduction is that organisms can reproduce rapidly and colonize favorable environments quickly.

## Asexual Reproduction

Several types of invertebrates, such as sponges, cnidarians, flatworms, annelids, and echinoderms, can reproduce asexually. Sponges produce asexual gemmules that develop into new individuals. Cnidarians, such as hydras, can reproduce asexually by budding (Fig. 41.1). A new individual arises as an outgrowth (bud) of the parent. *Obelia* is dimorphic and has an asexual colonial stage and a sexually reproducing medusa stage.

You can horizontally cut a planarian into as many as ten pieces in the laboratory and get ten new planarians. The parasitic flatworms reproduce asexually during certain stages of their complicated life cycles. Several annelids, such as earthworms and sandworms, also have the ability to regenerate from fragments. Fragmentation, followed by regeneration, is seen among sponges and echinoderms as well. If a sea star is chopped up, it has the potential to regenerate into several new individuals.

Several types of flatworms, roundworms, crustaceans, annelids, insects, fishes, lizards, and even some turkeys have the ability to reproduce parthenogenetically. **Parthenogenesis** is a modification of sexual reproduction in which an unfertilized egg develops into a complete individual. In honeybees, the queen bee makes and stores sperm she uses to selectively fertilize eggs. Any unfertilized eggs become haploid males.

## Sexual Reproduction

Usually during sexual reproduction, the egg of one parent is fertilized by the sperm of another. The majority of animals are *di*oecious, which means having separate sexes. *Mono*ecious, or hermaphroditic, organisms have both male and female sex organs in the same body. Some hermaphroditic organisms, such as tapeworms, are capable of self-fertilization, but the majority, such as earthworms, practice cross-fertilization. Sequential hermaphroditism, or sex reversal, also occurs. In coral reef fishes called wrasses, a male has a harem of several females. If the male dies, the largest female becomes a male.

Animals usually produce gametes in specialized organs called **gonads.** Sponges are an exception to this rule because the collar cells lining the central cavity of a sponge give rise to sperm and eggs. Hydras and other cnidarians produce only temporary gonads in the fall, when sexual reproduction occurs. Animals in other phyla have permanent reproductive organs. The gonads are **testes,** which produce sperm, and **ovaries,** which produce eggs. Eggs or sperm are derived from **germ cells,** which become specialized for this purpose during early development. Other cells in a gonad support and nourish the developing gametes or produce hormones necessary to the reproductive process. The reproductive system also usually has a number of accessory structures, such as ducts and storage areas, that aid in bringing the gametes together.

Many aquatic animals, such as the fishes in Figure 41.2, practice external fertilization. Palolo worms release their eggs in the water only when the moon moves closer to the Earth, and the tides become somewhat higher than

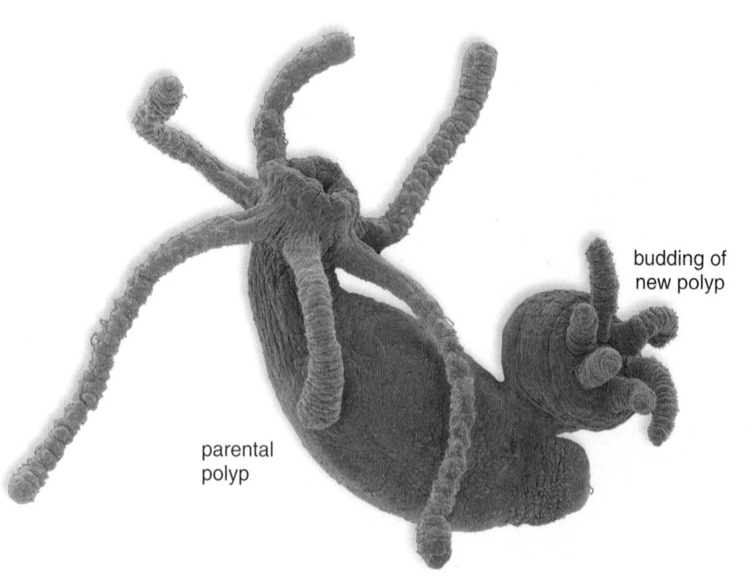


**FIGURE 41.1 Reproduction in *Hydra*.**

Hydras reproduce asexually and sexually. During asexual reproduction, a new polyp buds from the parental polyp. During sexual reproduction, temporary gonads develop in the body wall.

**FIGURE 41.2 Reproduction in anemonefish.**

Fishes such as orange-fin anemonefish, *Amphiprion chrysopterus,* usually reproduce sexually. Sperm from the male fertilize eggs from the female in the water.

usual. Most likely, they have a biological clock that can sense the passage of time so that their reproductive behavior is synchronized. Hundreds of thousands of palolo worms rise to the surface of the sea and release their eggs during a two- to four-hour period on two or three successive days of the year.

**Copulation** is sexual union to facilitate the reception of sperm by a female. In terrestrial vertebrates, males typically have a penis for depositing sperm into the vagina of females. But not so in birds, which lack a penis and vagina. They have a cloaca, a chamber that receives products from the digestive, urinary, and reproductive tracts. A male transfers sperm to a female after placing his cloacal opening against hers. In damselflies, the female curls her abdomen forward to receive sperm previously deposited in a pouch by the male, and copulation doesn't occur (Fig. 41.3).

## Life History Strategies

Any animal that deposits an egg in the external environment is **oviparous.** Most aquatic animals deposit their eggs in the water, where they undergo development. Many have a larval stage, an immature form capable of feeding. Since the **larva** has a different lifestyle, it is able to use a different food source than the adult. In seastars, the bilaterally symmetrical larva undergoes metamorphosis to become a radially symmetrical juvenile. Crayfish, on the other hand, do not have a larval stage; the egg hatches into a tiny juvenile with the same form as the adult. Some aquatic animals retain their eggs in some way and release young able to fend for themselves. These animals are *ovoviviparous.* For example, oysters, which are molluscs, retain their eggs in the mantle cavity, and male sea horses, which are vertebrates, have a special brood pouch in which the eggs develop (see page 755). Even though they are terrestrial, garter snakes retain their eggs inside the abdomen until they hatch.

Usually, reptiles, such as turtles and crocodiles, lay a leathery-shelled egg that contains **extraembryonic membranes** to serve the needs of the embryo and prevent drying out. One membrane surrounds an abundant supply of **yolk,** which is a nutrient-rich material. The shelled egg is a significant adaptation to the terrestrial environment. Birds lay and care for hard-shelled eggs with extraembryonic membranes. The newly hatched birds usually have to be fed before they are able to fly away and seek food for themselves (Fig. 41.4). Complex hormonal and neural regulation are involved in the reproductive behavior of parental birds.

Among mammals, the duckbill platypus and spiny anteater lay shelled eggs. Marsupials and placental mammals do not lay eggs. In marsupials, the yolk sac membrane functions briefly to supply the unborn with nutrients acquired internally from the mother. Immature young finish their development within a pouch where they are nourished on milk. The placental mammals are termed **viviparous,** because they do not lay eggs and development occurs inside the female's body until offspring can live independently. Their **placenta** is a complex structure derived, in part, from the chorion, another of the reptilian extraembryonic membranes. The evolution of this type of placenta allows the developing young to internally exchange materials with the mother until the offspring can function on their own. Viviparity represents the ultimate in caring for the unborn, and in placental mammals, the mother continues to supply the nutrient needs of her offspring after birth.

### Check Your Progress 41.1

1. What is the advantage of asexual reproduction? Sexual reproduction?
2. Distinguish between oviparous, ovoviviparous, and viviparous.
3. How does a shelled egg allow reproduction on land?

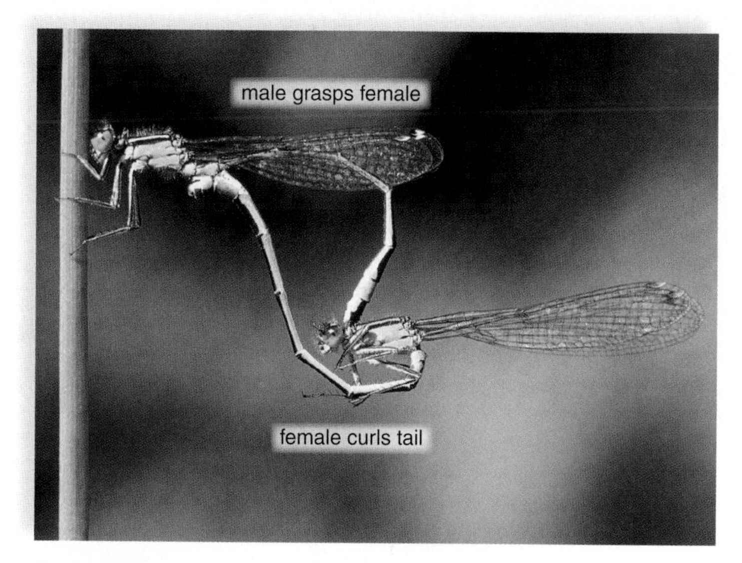


**FIGURE 41.3 Damselflies mating on land.**

On land, animals have various means to ensure that the gametes do not dry out. Here, a female damselfly receives sperm from the male.

**FIGURE 41.4 Parenting in birds.**

Birds, such as the American goldfinch, *Carduelis tristis,* are oviparous and lay hard-shelled eggs. They are well known for incubating their eggs and caring for their offspring after they hatch.

# 41.2 Male Reproductive System

The human male reproductive system includes the organs pictured in Figure 41.5 and listed in Table 41.1. The male gonads are paired testes, which are suspended within the sacs of the scrotum. The testes begin their development inside the abdominal cavity, but they descend into the scrotal sacs as development proceeds. If the testes do not descend—and the male does not receive hormone therapy or undergo surgery to place the testes in the scrotum—sterility (the inability to produce offspring) results. Sterility occurs because undescended testes developing in the body cavity are subject to higher body temperatures than those in the scrotum. A cooler temperature is critical for the normal development of sperm.

Testicular cancer is the most common type of cancer in young men between the ages of 15 and 34. The cure rate is high if early treatment is received. All men should perform regular self-exams to detect unusual lumps or swellings of the testes, and report any changes to their doctors.

Sperm produced in the seminiferous tubules of the testes mature within the epididymides (sing., epididymis), which are tightly coiled tubules lying just outside the testes. Maturation seems to be required for the sperm to swim to the egg. Once the sperm have matured, they are propelled into the vasa deferentia (sing., vas deferens) by muscular contractions. Sperm are stored in both the epididymides and the vasa deferentia. When a male becomes sexually aroused, sperm enter the ejaculatory ducts and then the urethra, part of which is located within the penis. The vasa deferentia are severed or blocked in a surgical form of birth control called a vasectomy. Sperm are unable to complete their journey down the male reproductive tract, and are phagocytized by macrophages.

**TABLE 41.1**

**Male Reproductive System**

| *Organ* | *Function* |
|---|---|
| Testes | Produce sperm and sex hormones |
| Epididymides | Sites of maturation and some storage of sperm |
| Vasa deferentia | Conduct and store sperm |
| Seminal vesicles | Contribute fluid to semen |
| Prostate gland | Contributes fluid to semen |
| Urethra | Conducts sperm (and urine) |
| Bulbourethral glands | Contribute fluid to semen |
| Penis | Organ of copulation |

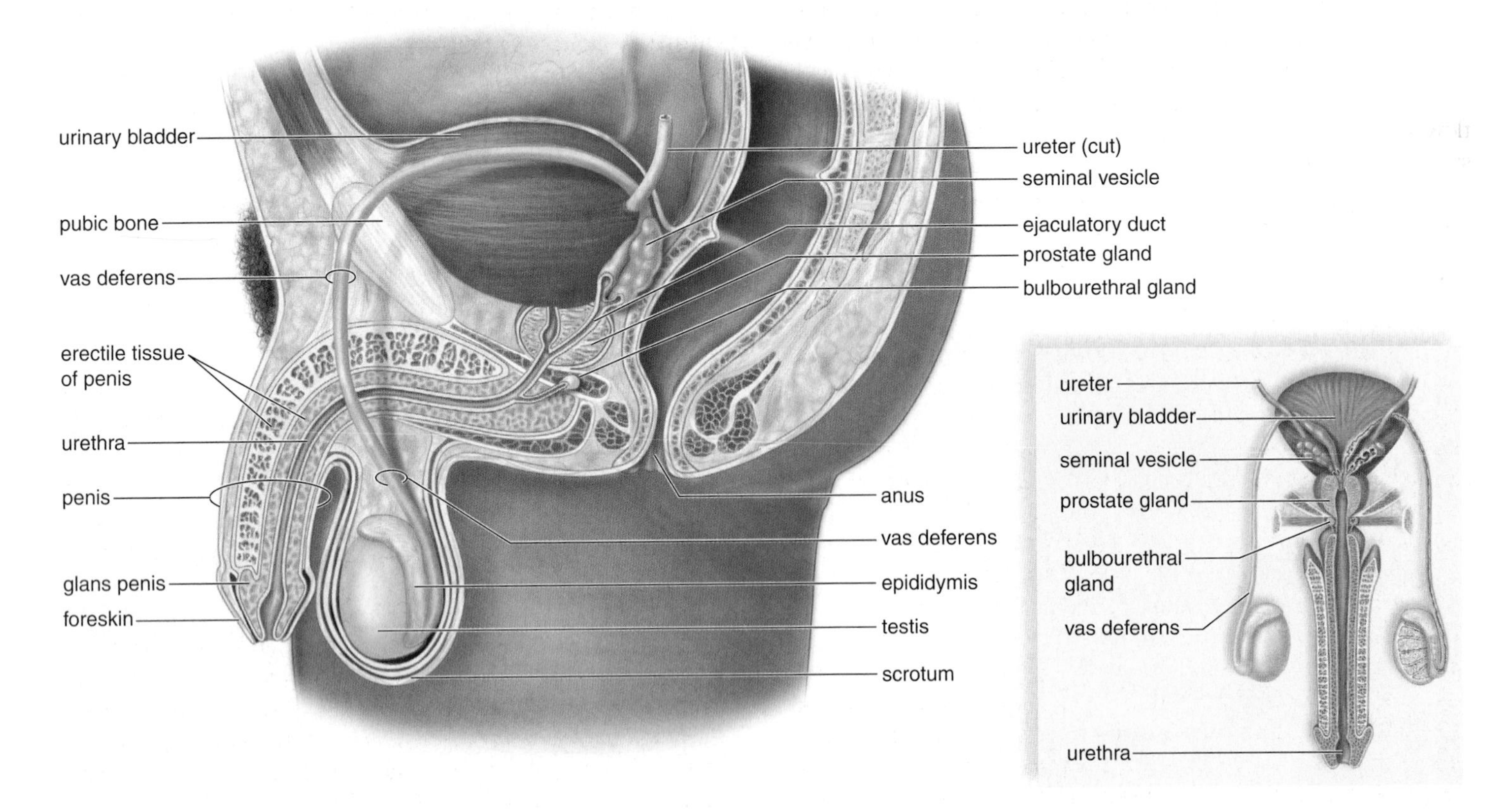


**FIGURE 41.5 The male reproductive system.**

The testes produce sperm. The seminal vesicles, the prostate gland, and the bulbourethral glands provide a fluid medium for the sperm, which move from the vas deferens through the ejaculatory duct to the urethra in the penis. The foreskin (prepuce) is removed when a penis is circumcised.

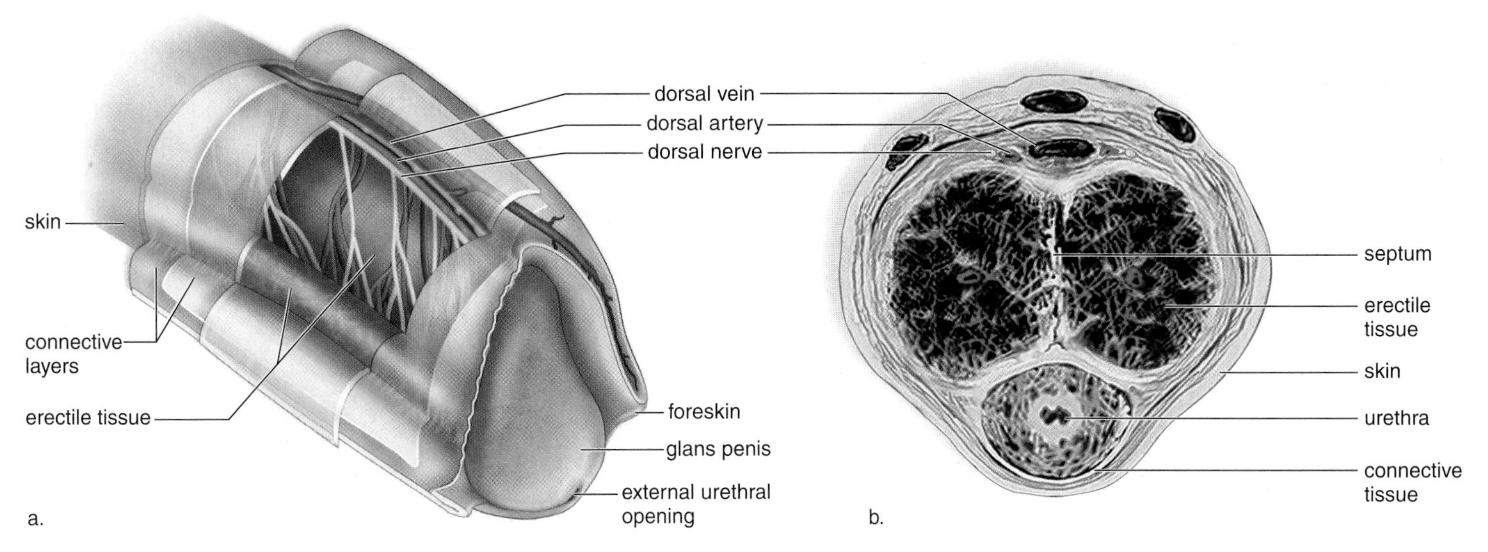


**FIGURE 41.6 Penis anatomy.**

**a.** Beneath the skin and the connective tissue lies the urethra, surrounded by erectile tissue. This tissue expands to form the glans penis, which in uncircumcised males is partially covered by the foreskin (prepuce). **b.** Micrograph of shaft in cross section showing location of erectile tissue. One column surrounds the urethra.

## Orgasm in Males

The **penis** is the male organ of sexual intercourse. The penis has a long shaft and an enlarged tip called the glans penis. The glans penis is normally covered by a layer of skin called the foreskin. Circumcision, the surgical removal of the foreskin, is usually done soon after birth. Three cylindrical columns of spongy, erectile tissue containing distensible blood spaces extend through the shaft of the penis (Fig. 41.6). During sexual arousal, nervous reflexes cause an increase in arterial blood flow to the penis. This increased blood flow fills the blood space in the erectile tissue, and the penis, which is normally limp (flaccid), stiffens and increases in size. These changes are called erection. If the penis fails to become erect, the condition is called erectile dysfunction (formerly called impotency). Medications for the treatment of erectile dysfunction ensure that a full erection will take place.

**Semen (seminal fluid)** [L. *semen*, seed] is a thick, whitish fluid that contains sperm and secretions from three glands (Table 41.1). The seminal vesicles lie at the base of the bladder. As sperm pass from the vasa deferentia into the ejaculatory ducts, these vesicles secrete a thick, viscous fluid containing nutrients (fructose) for possible use by the sperm. The fluid also contains prostaglandins that stimulate smooth muscle contraction along the male and female reproductive tracts. Just below the bladder lies the prostate gland, which secretes a milky alkaline fluid believed to activate or increase the motility of the sperm. Sperm are more viable in a basic solution, and semen, which is milky in appearance, has a slightly basic pH. Prostatic secretions also include an antibiotic called seminalplasmin that helps prevent urinary tract infections in males. Slightly below the prostate gland, on either side of the urethra, is a pair of small glands called bulbourethral glands that produce mucous secretions with a lubricating effect. An average male releases 2–5 ml of semen during orgasm. Typically semen is composed of 60% seminal vesicle secretions, 30% prostatic fluids, less than 10% sperm and spermatic duct secretions, and a trace amount of bulbourethral fluid.

If sexual arousal reaches its peak, ejaculation follows an erection. The first phase of ejaculation is called emission. During emission, the spinal cord sends nerve impulses via appropriate nerve fibers to the epididymides and vasa deferentia. Their muscular walls contract, causing sperm to enter the ejaculatory duct, whereupon the seminal vesicles, prostate gland, and bulbourethral glands release their secretions. Secretions from the bulbourethral glands are the first to enter the urethra, and they function to cleanse the urethra of acidic residue from urine. This fluid does not normally contain sperm, but considering that the young human adult male produces up to 1 billion sperm a day, it may. It is therefore possible, though not probable, for fertilization of an egg and pregnancy of a female to take place even though ejaculation has not occurred.

During the second phase of ejaculation, called expulsion, rhythmical contractions of muscles at the base of the penis and within the urethral wall expel semen in spurts from the opening of the urethra. These rhythmical contractions are an example of release from myotonia, or muscle tenseness. Myotonia is another important sexual response. An erection lasts for only a limited amount of time. The penis now returns to its normal flaccid state. Following ejaculation, a male may typically experience a period of time, called the refractory period, during which stimulation does not bring about an erection. The contractions that expel semen from the penis are a part of male **orgasm** [Gk. *orgasmos*, sexual excitement], the physiological and psychological sensations that occur at the climax of sexual stimulation.

## The Testes

A longitudinal section of a testis shows that it is composed of compartments called lobules, each of which contains one to three tightly coiled **seminiferous tubules** (Fig. 41.7*a*). Altogether, these tubules have a combined length of approximately 250 m (almost two and three-quarters times the length of a football field). A microscopic cross section of a seminiferous tubule shows that it is packed with cells undergoing spermatogenesis (Fig. 41.7*b*), the production of sperm. As seen in Figure 41.7*c*, spermatogonia increase in

**FIGURE 41.7 Testis and sperm.**

**a.** The lobules of a testis contain seminiferous tubules. **b.** Light micrograph of a cross section of the seminiferous tubules, where spermatogenesis occurs. Note the location of interstitial cells in clumps among the seminiferous tubules. **c.** Diagrammatic representation of spermatogenesis, which occurs in wall of tubules. **d.** A sperm has a head, a middle piece, and a tail. The nucleus is in the head, which is capped by the enzyme-containing acrosome.

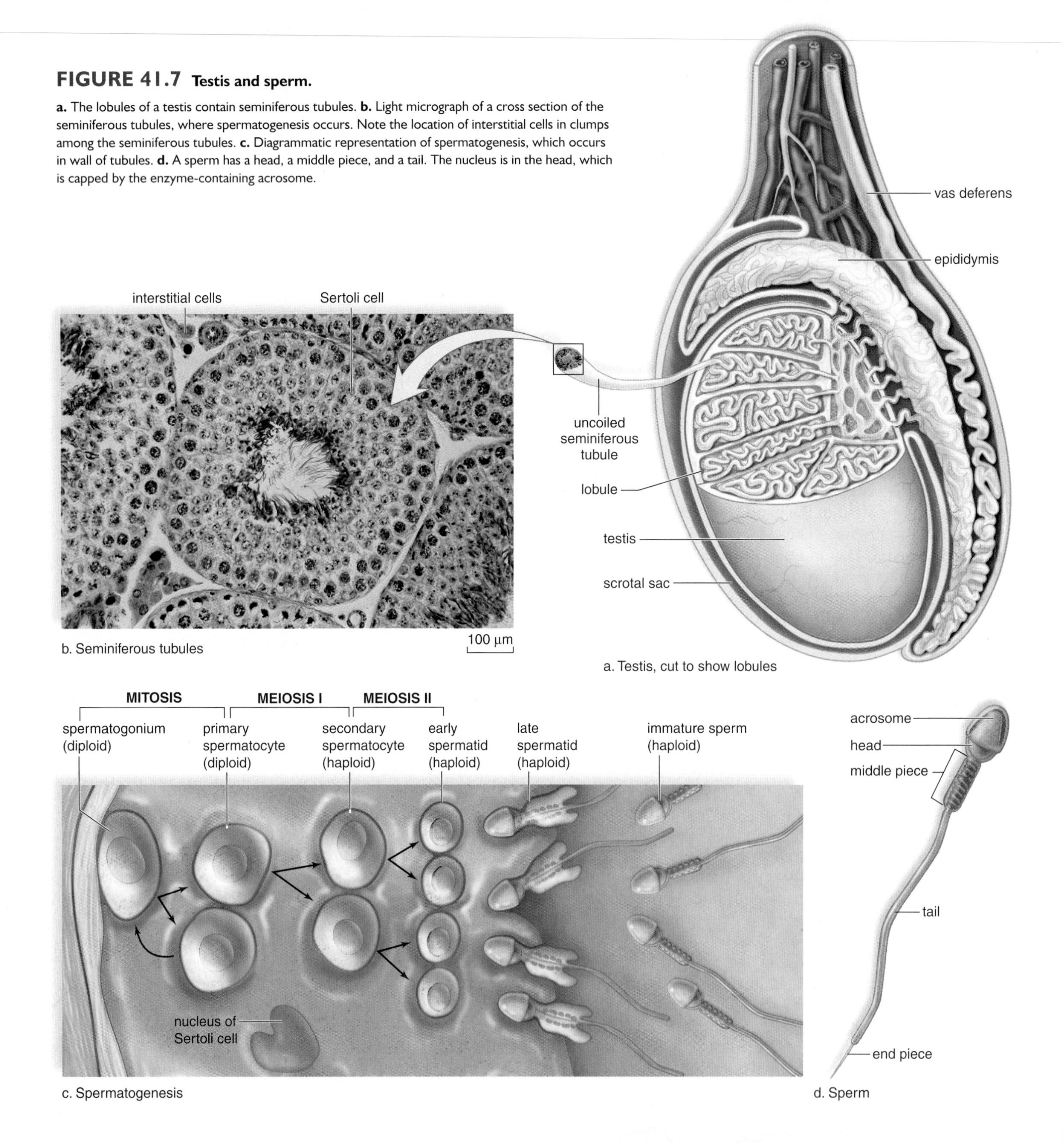

size and divide. One of these diploid cells undergoes meiosis I and II to become four spermatids, which contain 23 chromosomes each. Spermatids then differentiate into sperm. Also present are Sertoli cells, which support, nourish, and regulate spermatogenesis. It takes approximately 74 days for spermatogonia to develop into sperm.

Mature **sperm** have three distinct parts: a head, a middle piece, and a tail (Fig. 41.7*d*). The middle piece and the tail contain microtubules, in the characteristic 9 + 2 pattern of cilia and flagella. In the middle piece, mitochondria are wrapped around the microtubules and provide the energy for movement. The head contains a nucleus covered by a cap called the acrosome [Gk. *akros*, at the tip, and *soma*, body], that stores enzymes needed to penetrate the egg. (Because the human egg is surrounded by several layers of cells and a thick membrane, the acrosomal enzymes play a role in allowing a sperm to penetrate the layers surrounding the egg.) The ejaculated semen of a normal male contains between 50 and 150 million sperm per milliliter. If less than 25 million sperm per milliliter are released, infertility may result. Fewer than 100 ever reach the vicinity of the egg, however, and only one sperm normally enters an egg. With contractions of the female uterus, it takes sperm about two hours to reach the egg traveling approximately 12.5 cm per hour.

## Hormonal Regulation in Males

The hypothalamus has ultimate control of the testes' sexual function because it secretes a hormone called gonadotropic-releasing hormone, or GnRH, that stimulates the anterior pituitary to produce the gonadotropic hormones. There are two gonadotropic hormones—follicle-stimulating hormone (FSH) and luteinizing hormone (LH)—in both males and females. In males, FSH promotes spermatogenesis in the seminiferous tubules.

LH in males controls the production of the androgen testosterone by the interstitial cells (Leydig cells), which are scattered in the spaces between the seminiferous tubules (Fig. 41.7*b*). All these hormones, including inhibin, a hormone released by the seminiferous tubules, are involved in a negative feedback relationship that maintains the fairly constant production of sperm and testosterone (Fig. 41.8).

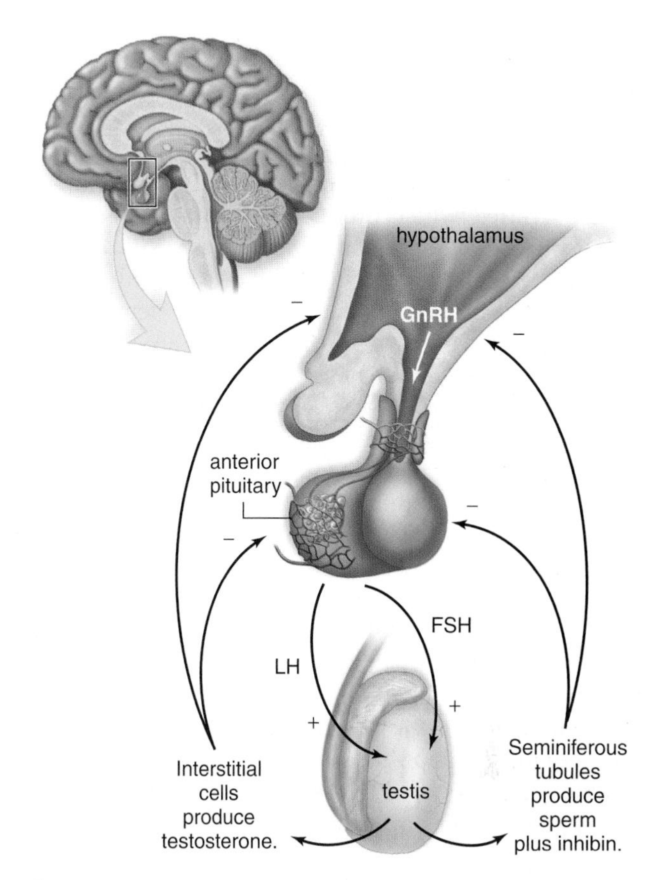


**FIGURE 41.8 Hormonal control of testes.**

GnRH (gonadotropic-releasing hormone) stimulates the anterior pituitary to produce FSH and LH. FSH stimulates the testes to produce sperm, and LH stimulates the testes to produce testosterone. Testosterone from interstitial cells and inhibin from the seminiferous tubules exert negative feedback control over the hypothalamus and the anterior pituitary, and this ultimately regulates the level of testosterone in the blood.

### *Functions of Testosterone*

**Testosterone** is the main sex hormone in males. It is essential for the normal development and functioning of the organs listed in Table 41.1. Testosterone is also necessary for the maturation of sperm.

In addition, testosterone brings about and maintains the male **secondary sex characteristics** that develop at the time of **puberty.** Males are generally taller than females and have broader shoulders and longer legs relative to trunk length. The deeper voice of males compared to females is due to the fact they have a larger larynx with longer vocal cords. Since the so-called Adam's apple is a part of the larynx, it is usually more prominent in males than in females.

Testosterone is responsible for the greater muscular development in males. Because of this effect, males and females sometimes take anabolic steroids (either the natural or synthetic form of testosterone) to build up their muscles (see page 750). Testosterone causes males to develop noticeable hair on the face, chest, and occasionally other regions of the body, such as the back. Testosterone also leads to the receding hairline and pattern baldness that occur in males. Testosterone levels peak at about 20 years of age and then decline as a male ages.

**Check Your Progress 41.2**

1. Trace the pathway a sperm must follow from its origin to its exit from the male reproductive tract.
2. List the glands that contribute fluids to the semen.
3. Which gonadotropic hormone stimulates the interstitial cells to produce testosterone? Which stimulates sperm production in the seminiferous tubules?

# 41.3 Female Reproductive System

The human female reproductive system includes the ovaries, the oviducts, the uterus, and the vagina (Fig. 41.9 and Table 41.2). The ovaries, which produce a secondary **oocyte** each month, lie in shallow depressions, one on each side of the upper pelvic cavity. The oviducts [L. *ovum,* egg, and *duco,* lead out], also called uterine or fallopian tubes, extend from the ovaries to the uterus; however, the oviducts are not attached to the ovaries. Instead, they have fingerlike projections called fimbriae (sing., fimbria) that sweep over the ovaries. When an oocyte bursts from an ovary during ovulation, it usually is swept into an oviduct by the combined action of the fimbriae and the beating of cilia that line the oviducts. Fertilization, if it occurs, normally takes place in an oviduct. The developing embryo is propelled slowly by ciliary movement and tubular muscle contraction to the uterus. Some women opt to permanently prevent pregnancy by having a surgery called tubal ligation, in which the oviducts are either burned or clipped shut so that sperm cannot reach the oocyte.

The **uterus** [L. *uterus,* womb] is a thick-walled muscular organ about the size and shape of an inverted pear. The narrow end of the uterus is called the **cervix.** The embryo completes its development after embedding itself in the uterine lining, called the **endometrium** [Gk. *endon,* within, and *metra,* womb]. If, by chance, the embryo should embed itself in another location, such as an oviduct, a so-called ectopic pregnancy results.

**TABLE 41.2**

**Female Reproductive Organs**

| *Organ* | *Function* |
|---|---|
| Ovaries | Produce egg and sex hormones |
| Oviducts (fallopian tubes) | Conduct egg; location of fertilization |
| Uterus (womb) | Houses developing embryo and fetus |
| Vagina | Receives penis during copulation and serves as birth canal |

A small opening in the cervix leads to the vaginal canal. The vagina [L. *vagina,* sheath] is a tube at a 45-degree angle to the small of the back. The mucosal lining of the vagina lies in folds and the vagina can distend. This is especially important when the vagina serves as the birth canal, and it also can facilitate intercourse, when the vagina receives the penis during copulation. Several different types of bacteria normally reside in the vagina and create an acidic environment. This environment is protective against the possible growth of pathogenic bacteria, but sperm prefer the basic environment provided by semen.

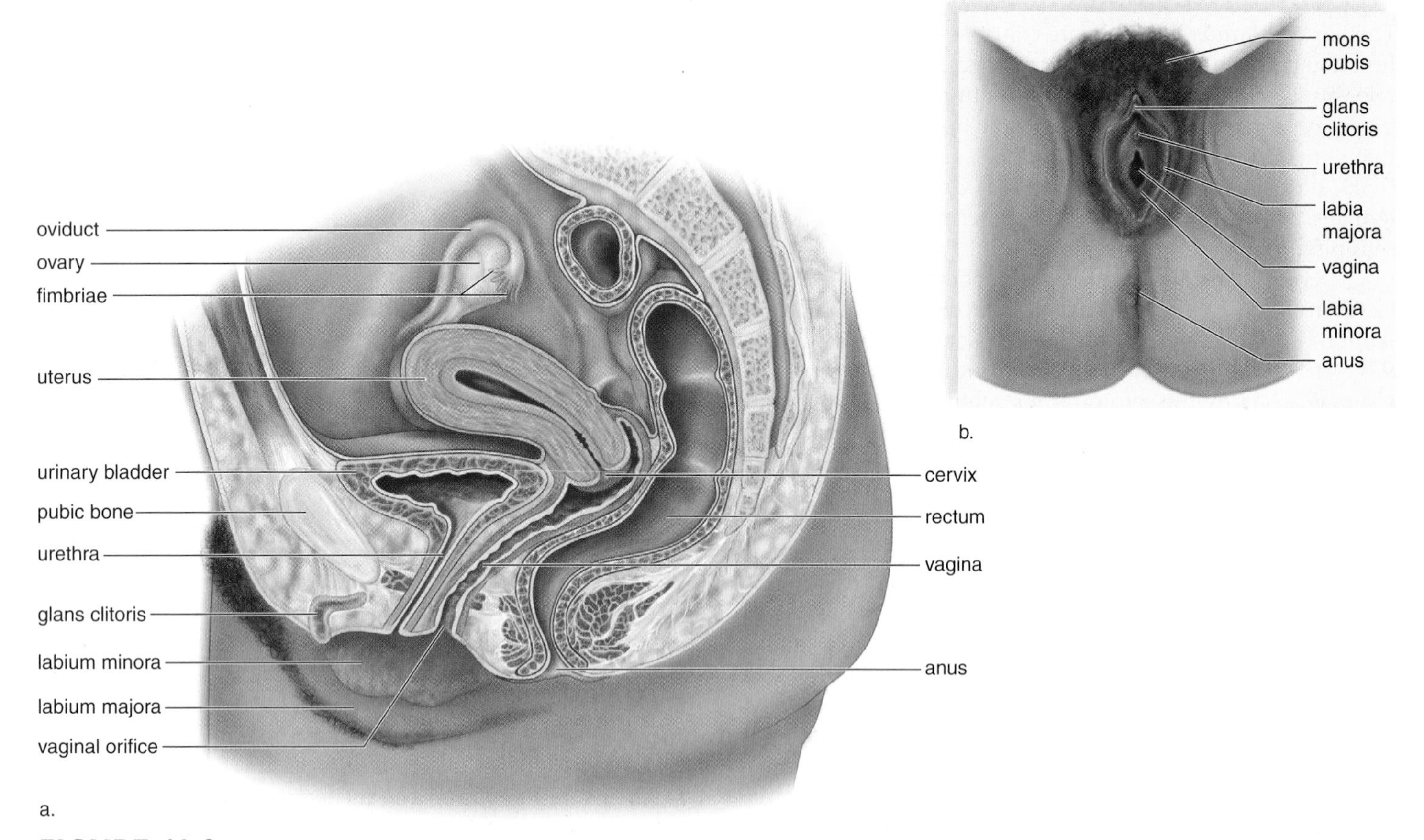


**FIGURE 41.9 Female reproductive system.**

**a.** The ovaries produce one oocyte per month. Fertilization occurs in the oviduct, and development occurs in the uterus. The vagina is the birth canal and organ of sexual intercourse. **b.** Vulva. At birth, the opening of the vagina is partially occluded by a membrane called the hymen. Physical activities and sexual intercourse disrupt the hymen.

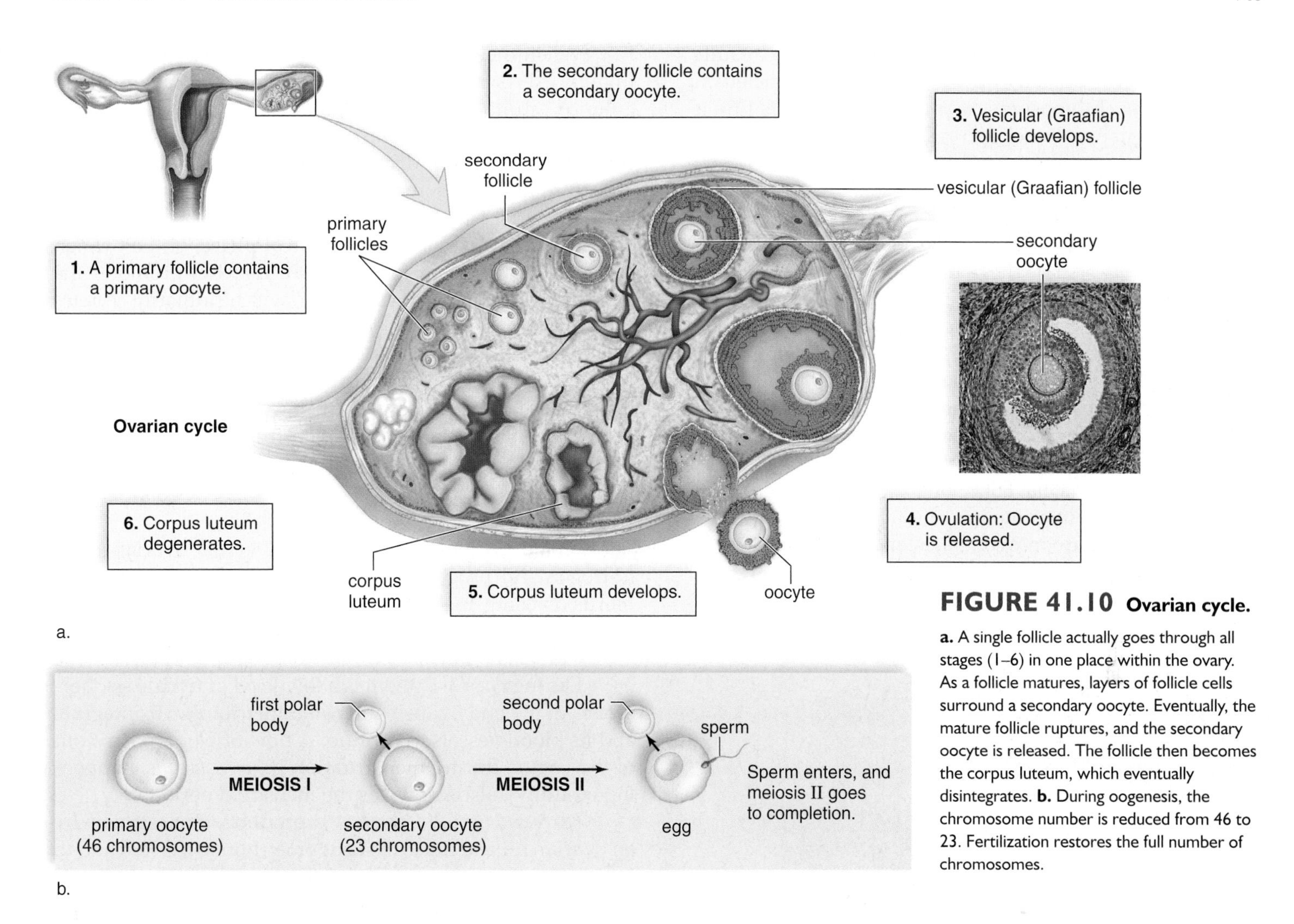


**FIGURE 41.10 Ovarian cycle.**

**a.** A single follicle actually goes through all stages (1–6) in one place within the ovary. As a follicle matures, layers of follicle cells surround a secondary oocyte. Eventually, the mature follicle ruptures, and the secondary oocyte is released. The follicle then becomes the corpus luteum, which eventually disintegrates. **b.** During oogenesis, the chromosome number is reduced from 46 to 23. Fertilization restores the full number of chromosomes.

## Orgasm in Females

The external genital organs of a female are known collectively as the vulva (Fig. 41.9*b*). The mons pubis and two folds of skin called labia minora and labia majora are on either side of the urethral and vaginal openings. Beneath the labia majora, pea-sized greater vestibular glands (Bartholin glands) open on either side of the vagina. They keep the vulva moist and lubricated during intercourse.

At the juncture of the labia minora is the clitoris, which is homologous to the penis in males. The clitoris has a shaft of erectile tissue and is capped by a pea-shaped glans. The many sensory receptors of the clitoris allow it to function as a sexually sensitive organ. The clitoris has twice as many nerve endings as the penis. Orgasm in the female is a release of neuromuscular tension in the muscles of the genital area, vagina, and uterus.

## The Ovaries

Normally, the ovaries alternate in producing one oocyte each month. For the sake of convenience, the released oocyte is often called an **egg.** The ovaries also produce the female sex hormones, **estrogens,** collectively called estrogen, and **progesterone,** during the ovarian cycle.

### *The Ovarian Cycle*

The **ovarian cycle** occurs as a **follicle** [L. dim. of *folliculus,* bag] changes from a primary to a secondary to a vesicular (Graafian) follicle (Fig. 41.10*a*) under the influence of follicle-stimulating hormone (FSH) and luteninizing hormone (LH) from the anterior pituitary (see Fig. 41.11). Epithelial cells of a primary follicle surround a primary oocyte. Pools of follicular fluid surround the oocyte in a secondary follicle. In a vesicular follicle, a fluid-filled cavity increases to the point that the follicle wall balloons out on the surface of the ovary.

As a follicle matures, oogenesis, a form of meiosis depicted in Figure 41.10*b*, is initiated and continues. The primary oocyte divides, producing two haploid cells. One cell is a secondary oocyte, and the other is a polar body. The vesicular follicle bursts, releasing the secondary oocyte (often called an egg for convenience) surrounded by a clear membrane. This process is referred to as **ovulation.** Once a vesicular follicle has lost the secondary oocyte, it develops into a **corpus luteum** [L. *corpus,* body, and *luteus,* yellow], a glandlike structure.

The secondary oocyte enters an oviduct. If fertilization occurs, a sperm enters the secondary oocyte and then the oocyte completes meiosis and becomes an egg. An egg with 23 chromosomes and a second polar body results. When the sperm nucleus unites with the egg nucleus, a zygote with 46 chromosomes is produced. If zygote formation and pregnancy do not occur, the corpus luteum begins to degenerate after about ten days.

### *Phases of the Ovarian Cycle*

The ovarian cycle is controlled by the gonadotropic hormones, FSH and LH (Fig. 41.11). The gonadotropic hormones are not present in constant amounts and instead are secreted at different rates during the cycle. For simplicity's sake, it is convenient to emphasize that during the first half, or **follicular phase,** of the cycle, FSH promotes the development of a follicle that primarily secretes estrogen. As the estrogen level in the blood rises, it exerts negative feedback control over the anterior pituitary secretion of FSH so that the follicular phase comes to an end. At the same time that FSH along with LH release is being dampened by moderate amounts of estrogen, the synthesis of the gonadotropic hormones continues, and they build up in the anterior pituitary.

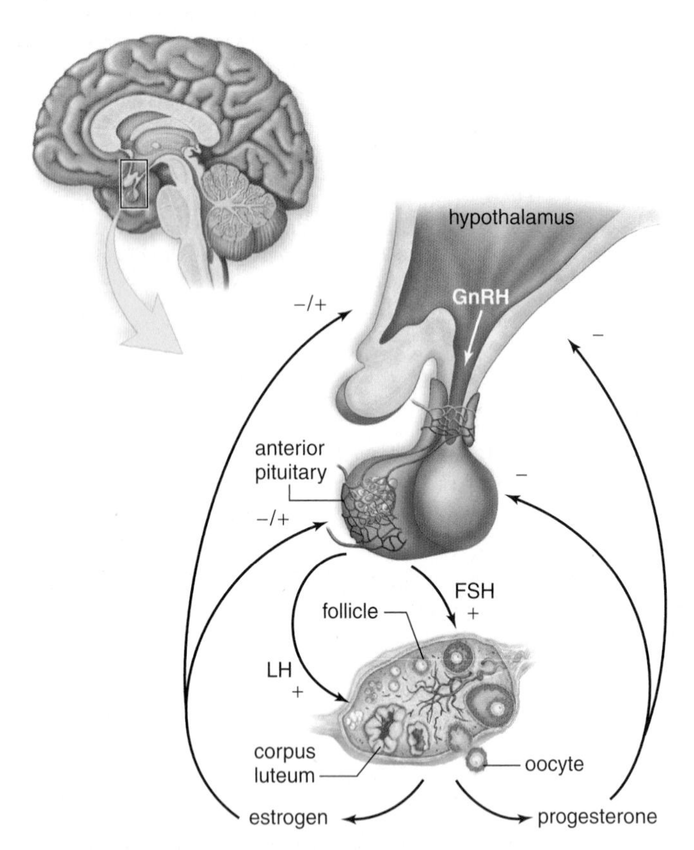


**FIGURE 41.11 Hormonal control of ovaries.**

The hypothalamus produces GnRH (gonadotropic-releasing hormone). GnRH stimulates the anterior pituitary to produce FSH (follicle-stimulating hormone) and LH (luteinizing hormone). FSH stimulates the follicle to produce primarily estrogen, and LH stimulates the corpus luteum to produce primarily progesterone. Estrogen and progesterone maintain the sex organs (e.g., uterus) and the secondary sex characteristics, and they exert feedback control over the hypothalamus and the anterior pituitary. Feedback control regulates the relative amounts of estrogen and progesterone in the blood.

When the level of estrogen in the blood becomes very high, it exerts positive feedback on the hypothalamus and anterior pituitary. The hypothalamus is stimulated to suddenly secrete a large amount of GnRH. This leads to a surge of LH (and to a lesser degree, FSH) by the anterior pituitary and to ovulation at about the fourteenth day of a 28-day cycle (Fig. 41.12, *top*).

During the second half, or **luteal phase,** of the ovarian cycle, it is convenient to emphasize that LH promotes the development of the corpus luteum, which primarily secretes progesterone. As the blood level of progesterone rises, it exerts negative feedback control over anterior pituitary secretion of LH so that the corpus luteum begins to degenerate. As the luteal phase comes to an end, the low levels of progesterone and estrogen in the body cause menstruation to begin, as discussed next.

## The Uterine Cycle

The female sex hormones produced in the ovarian cycle (estrogen and progesterone) affect the endometrium of the uterus, causing the cyclical series of events known as the **uterine cycle** (Fig. 41.12, *bottom*). Twenty-eight-day cycles are divided as follows.

During *days 1–5,* there is a low level of female sex hormones in the body, causing the endometrium to disintegrate and its blood vessels to rupture. A flow of blood passes out of the vagina during **menstruation** [L. *menstrualis,* happening monthly], also known as the menstrual period.

During *days 6–13,* increased production of estrogen by an ovarian follicle causes the endometrium to thicken and to become vascular and glandular. This is called the proliferative phase of the uterine cycle.

Ovulation usually occurs on *day 14* of the 28-day cycle.

During *days 15–28,* increased production of progesterone by the corpus luteum causes the endometrium to double in thickness and the uterine glands to mature, producing a thick mucoid secretion. This is called the secretory phase of the uterine cycle. The endometrium now is prepared to receive the developing embryo. If pregnancy does not occur, the corpus luteum degenerates, and the low level of sex hormones in the female body causes the endometrium to break down as menstruation occurs. Table 41.3 compares the stages of the uterine cycle with those of ovarian cycle.

### *Menstruation*

Seven to ten days before the start of menstruation, some women suffer from premenstrual syndrome (PMS). During this time a woman may exhibit breast enlargement, achiness, headache, and irritability. The exact cause for PMS has yet to be discovered.

During menstruation, arteries that supply the lining constrict and the capillaries weaken. Blood spilling from the damaged vessels detaches layers of the lining, not all at once but in random patches. Mucus, blood, and degenerating endometrium descend from the uterus, through the vagina, creating menstrual flow. Fibrinolysin, an enzyme released by dying cells, prevents the blood from clotting. Menstruation

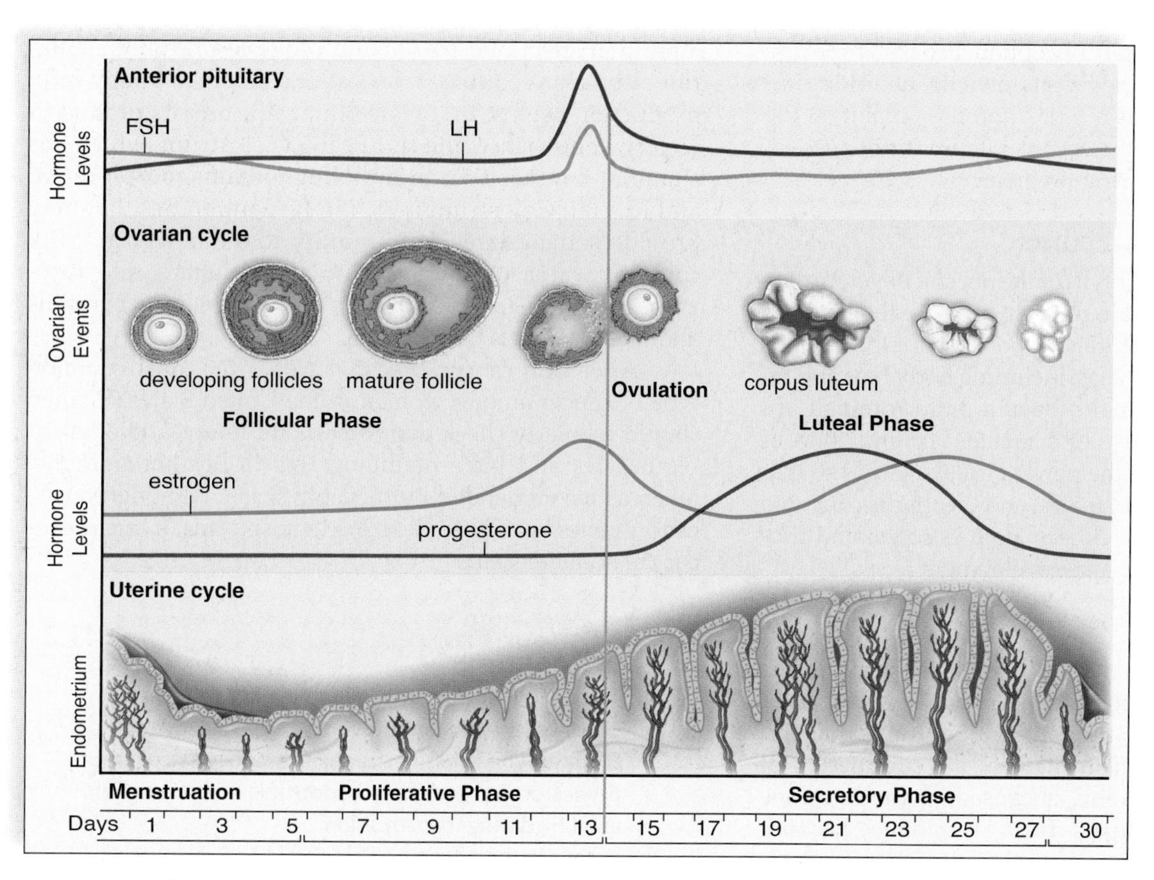


**FIGURE 41.12**
**Female hormone levels during the ovarian and uterine cycles.**
During the follicular phase of the ovarian cycle (*top*), FSH released by the anterior pituitary promotes the maturation of a follicle in the ovary. The ovarian follicle produces increasing levels of estrogen, which causes the endometrium to thicken during the proliferative phase of the uterine cycle (*bottom*). After ovulation and during the luteal phase of the ovarian cycle, LH promotes the development of the corpus luteum. This structure produces increasing levels of progesterone, which causes the endometrium to become secretory. Menstruation and the proliferative phase begin when progesterone production declines to a low level.

lasts from three to five days, as the uterus sloughs off the thick lining that was three weeks in the making.

The first menstrual period, called **menarche,** typically occurs between the ages of 11 and 13. Menarche signifies that the ovarian and uterine cycles have begun. If menarche does not occur by age 16, or normal uterine cycles are interrupted for six months or more, amenorrhea exists. Primary amenorrhea is usually caused by nonfunctional ovaries or developmental abnormalities. Secondary amenorrhea may be caused by weight loss and/or excessive exercise.

**Menopause,** which usually occurs between ages 45 and 55, is the time in a woman's life when menstruation ceases because the ovaries are no longer functioning. Menopause is not complete until menstruation is absent for a year.

## Fertilization and Pregnancy

If fertilization does occur, an embryo begins development even as it travels down the oviduct to the uterus. The endometrium is now prepared to receive the developing embryo, which becomes embedded in the lining several days following fertilization. The placenta originates from both maternal and embryonic tissues. It is shaped like a large, thick pancake and is the site of the exchange of gases and nutrients between fetal and maternal blood, although the two rarely mix. At first, the placenta produces **human chorionic gonadotropin (HCG),** which maintains the corpus luteum until the placenta begins its own production of progesterone and estrogen. HCG is the hormone detected in pregnancy tests, since it is present in the mother's blood and

**TABLE 41.3**

**Ovarian and Uterine Cycles (Simplified)**

| *Ovarian Cycle* | *Events* | *Uterine Cycle* | *Events* |
|---|---|---|---|
| Follicular phase—Days 1–13 | FSH<br>Follicle maturation<br>Estrogen | Menstruation—Days 1–5<br>Proliferative phase—Days 6–13 | Endometrium breaks down<br>Endometrium rebuilds |
| **Ovulation—Day 14*** | LH spike | | |
| Luteal phase—Days 15–28 | LH<br>Corpus luteum<br>Progesterone | Secretory phase—Days 15–28 | Endometrium thickens and glands are secretory |

** Assuming a 28-day cycle*

urine as soon as ten days after conception. Progesterone and estrogen have two effects. They shut down the anterior pituitary so that no new follicles mature, and they maintain the lining of the uterus so that the corpus luteum is not needed. No menstruation occurs during pregnancy.

## Estrogen and Progesterone

Estrogen in particular is essential for the normal development and functioning of the female reproductive organs listed in Table 41.2. Estrogen is also largely responsible for the secondary sex characteristics in females, including body hair and fat distribution. In general, females have a more rounded appearance than males because of a greater accumulation of fat beneath their skin. Also, the pelvic girdle enlarges so that females have wider hips than males, and the thighs converge at a greater angle toward the knees. Both estrogen and progesterone are required for breast development as well.

## The Female Breast

A female breast contains between 15 and 24 lobules, each with its own mammary duct (Fig. 41.13). A duct begins at the nipple and divides into numerous other ducts that end in blind sacs called alveoli. **Lactation,** the production of milk by the cells of the alveoli, is caused by the hormone prolactin. Milk is not produced during pregnancy because production of prolactin is suppressed by the feedback inhibition effect of estrogen and progesterone on the anterior pituitary. A couple of days after delivery of a baby, milk production begins. In the meantime, the breasts produce a watery, yellowish-white fluid called colostrum, which has a similar composition to milk but contains more protein and less fat. Colostrum is rich in IgA antibodies that may provide some degree of immunity to the newborn. Milk contains water, proteins, amino acids, sugars, and lysozymes (enzymes with antibiotic properties). Milk contains about 750 calories per liter.

After skin cancer, breast cancer is the most common type of cancer among women in the United States. Women should regularly check their breasts for lumps and other irregularities and have mammograms (X-ray photographs) taken as recommended by their physician. Although breast cancer genes have been described, most forms of breast cancer are nonhereditary.

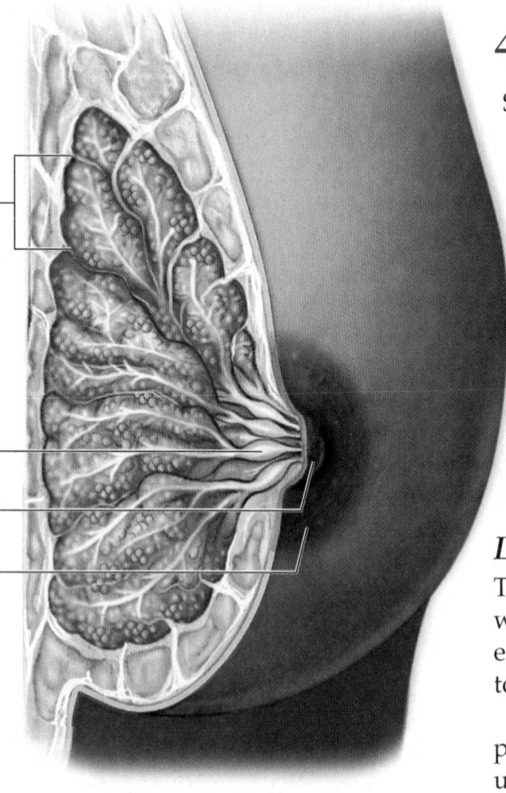

lobule containing alveoli
mammary duct
nipple
areola

**FIGURE 41.13 Anatomy of the breast.**

The female breast contains lobules consisting of ducts and alveoli. In the lactating breast, cells lining the alveoli have been stimulated to produce milk by the hormone prolactin.

**Check Your Progress 41.3**

1. Trace the pathway an oocyte must follow from its origin to its exit from the female reproductive tract.
2. Describe the levels of the sex hormones estrogen and progesterone, and the gonadotropic hormones, FSH and LH, during menstruation.
3. What are the effects of FSH and LH on the ovarian cycle?

# 41.4 Control of Reproduction

Several means are available to dampen or enhance the human reproductive potential. Contraceptives are medications and devices that reduce the chance of pregnancy.

## Birth Control Methods

The most reliable method of birth control is abstinence—that is, not engaging in sexual intercourse. This form of birth control has the added advantage of preventing transmission of a sexually transmitted disease. The male and female condoms also offer some protection against sexually transmitted diseases. These and other common means of birth control used in the United States are shown in Figure 41.14.

### *Less Common Birth Control Methods*

The expression "morning-after pill" refers to a medication that will prevent pregnancy after unprotected intercourse. The expression is a misnomer, in that medication can begin one to several days after unprotected intercourse.

A kit called Preven is made up of four synthetic progesterone pills; two are taken up to 72 hours after unprotected intercourse, and two more are taken 12 hours later. The medication upsets the normal uterine cycle, making it difficult for the embryo to implant itself in the endometrium. A recent study estimated that the medication was 85% effective in preventing unintended pregnancies.

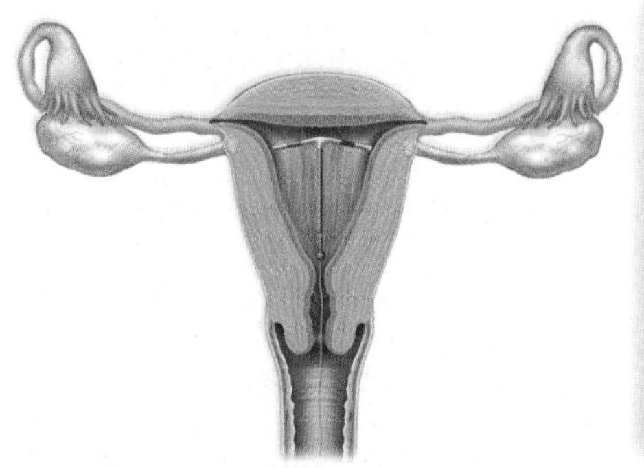
a. Intrauterine device placement

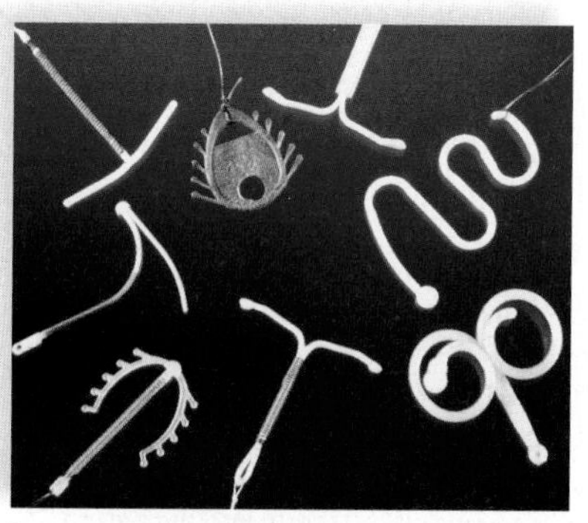
Intrauterine devices

b. Hormone skin patch

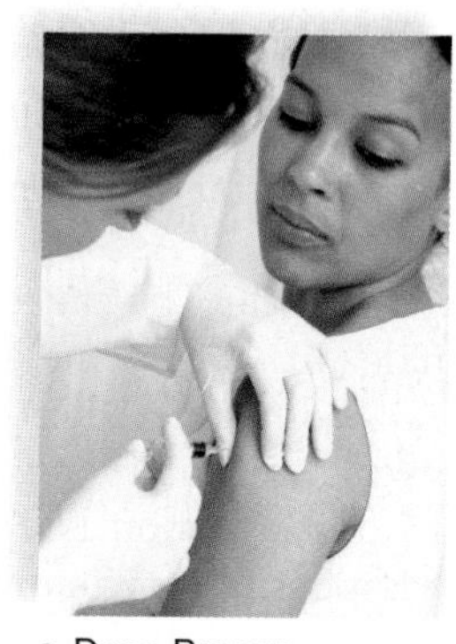
c. Depo-Provera (a progesterone injection)

d. Diaphragm and spermicidal jelly

e. Female condom

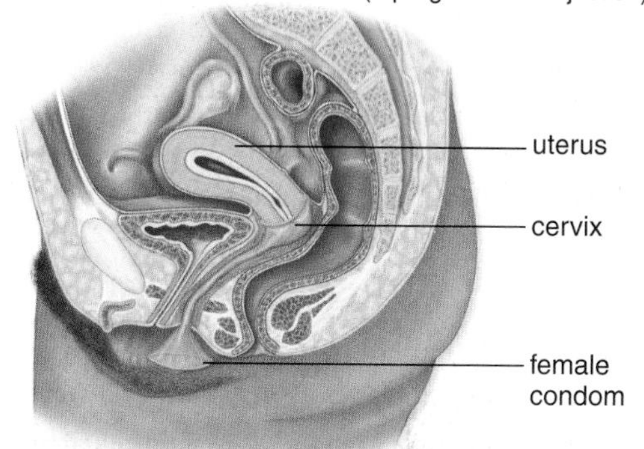

Female condom placement

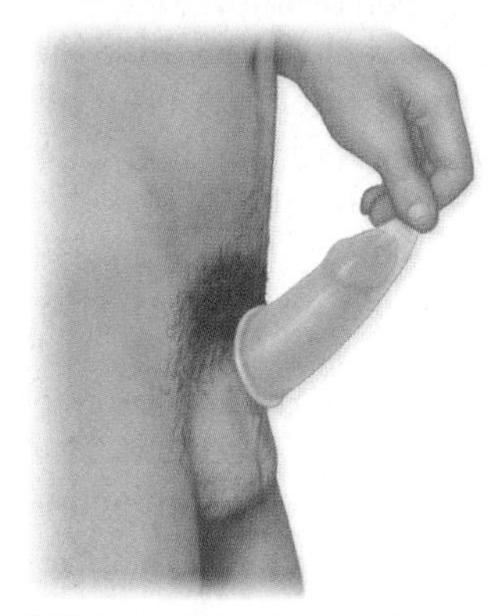
f. Male condom placement

Male condom

g. Implant

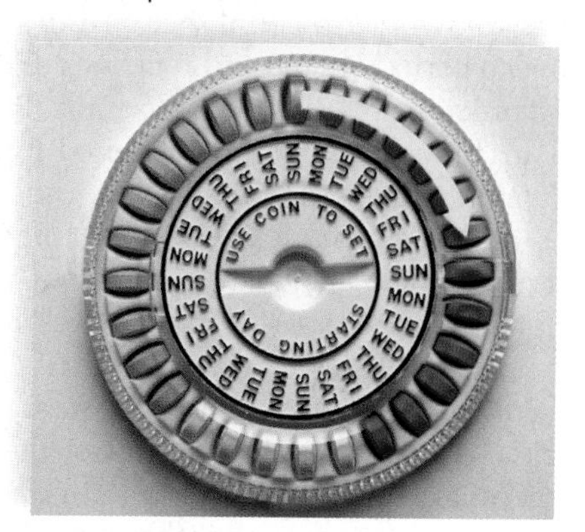

h. Oral contraception (birth control pills)

**FIGURE 41.14 Various birth control methods.**

**a.** Intrauterine devices mechanically prevent implantation and can contain progesterone to prevent ovulation, prevent implantation, and thicken cervical mucus. **b.** Skin patches eliminate need to take oral contraceptives. **c.** Depo-Provera injection of progesterone every three months. **d.** Diaphragm, a latex cup that covers cervix, (**e**) female condom fitted inside vagina, and (**f**) male condom that fits over penis prevent access to uterus by sperm. **g.** Implant inserted under skin contains progesterone. **h.** Birth control pill prevents ovulation by combined action of estrogen and progesterone.

Mifepristone, better known as RU-486, is a pill that is presently used to cause the loss of an implanted embryo by blocking the progesterone receptors of endometrial cells. Without functioning receptors for progesterone, the endometrium sloughs off, carrying the embryo with it. When taken in conjunction with a prostaglandin to induce uterine contractions, RU-486 is 95% effective. It is possible that some day this medication will also be a "morning-after pill," taken when menstruation is late without evidence that pregnancy has occurred.

**Contraceptive vaccines** are now being developed. For example, a vaccine intended to immunize women against HCG, the hormone so necessary to maintaining the implantation of the embryo, was successful in a limited clinical trial. Since HCG is not normally present in the body, no autoimmune reaction is expected, but the immunization does wear off with time. Others believe that it would also be possible to develop a safe antisperm vaccine that could be used in women.

A great deal of research is being devoted to developing safe and effective hormonal birth control for males. Implants, pills, patches, and injections are being explored as ways to deliver testosterone and/or progesterone at adequate levels to suppress sperm production. Even the most successful formulations are still in the experimental stage and are unlikely to be available outside of clinical trials for at least a few more years.

## Reproductive Technologies

**Infertility** is the inability of a couple to achieve pregnancy after one year of regular, unprotected intercourse. The American Medical Association estimates that 15% of all couples are infertile. The cause of infertility can be attributed to the male (40%), the female (40%), or both (20%).

Sometimes the causes of infertility may be corrected by medical intervention so that couples can have children. If no obstruction is apparent and body weight is normal, it is possible for females to take fertility drugs, which are gonadotropic hormones that stimulate the ovaries and bring about ovulation. Such hormone treatments may cause multiple ovulations and multiple births.

When reproduction does not occur in the usual manner, many couples adopt a child. Others sometimes try one of the assisted reproductive technologies (ARTs) developed to increase the chances of pregnancy. In these cases, sperm and/or oocytes are often retrieved from the testes and ovaries, and fertilization takes place in a clinical or laboratory setting.

### *Artificial Insemination by Donor (AID)*

During artificial insemination, sperm are placed in the vagina by a physician. Sometimes a woman is artificially inseminated by her partner's sperm. This is especially helpful if the partner has a low sperm count, because the sperm can be collected over a period of time and concentrated so that the sperm count is sufficient to result in fertilization. Often, however, a woman is inseminated by sperm acquired from a donor. At times, a combination of partner and donor sperm is used.

A variation of AID is *intrauterine insemination (IUI)*. In IUI, fertility drugs are given to stimulate the ovaries, and then the donor's sperm is placed in the uterus, rather than in the vagina.

If the prospective parents wish, sperm can be sorted into those believed to be X-bearing or Y-bearing to increase the chances of having a child of the desired sex. First, the sperm are dosed with a DNA-staining chemical. Because the X chromosome has slightly more DNA than the Y chromosome, it takes up more dye. When a laser beam shines on the sperm, the X-bearing sperm shine a little more brightly than the Y-bearing sperm. A machine sorts the sperm into two groups on this basis. Parents can expect about a 65% success rate for males and about 85% for females.

### *In Vitro Fertilization (IVF)*

During IVF, fertilization, the union of a sperm and an egg to form a zygote, occurs in laboratory glassware. Ultrasound machines can now spot follicles in the ovaries that hold immature oocytes; therefore, the latest method is to forgo the administration of fertility drugs and retrieve immature oocytes by using a needle. The immature oocytes are then brought to maturity in glassware before concentrated sperm are added. After about two to four days, the embryos are ready to be transferred to the uterus of the woman, who is now in the secretory phase of her uterine cycle. If desired, the embryos can be tested for a genetic disease, as discussed in the Health Focus on page 769. If implantation is successful, development continues to term.

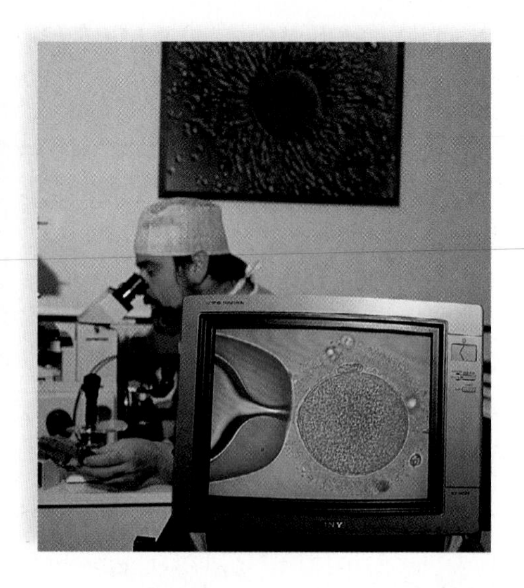

**FIGURE 41.15 Intracytoplasmic sperm injection (ICSI).**
A microscope connected to a television screen is used to carry out in vitro fertilization. A pipette holds the oocyte steady while a needle (not visible) introduces the sperm into the oocyte.

### *Gamete Intrafallopian Transfer (GIFT)*

Recall that the term **gamete** refers to a sex cell, either a sperm or an oocyte. Gamete intrafallopian transfer was devised to overcome the low success rate (15–20%) of in vitro fertilization. The method is exactly the same as for in vitro fertilization, except the oocytes and the sperm are placed in the oviducts immediately after they have been brought together. GIFT has the advantage of being a one-step procedure for the woman—the oocytes are removed and reintroduced all in the same time period. A variation on this procedure is to fertilize the eggs in the laboratory and then place the zygotes in the oviducts.

### *Surrogate Mothers*

In some instances, women are contracted and paid to have babies. These women are called surrogate mothers. The sperm and even the oocyte can be contributed by the contracting parents.

### *Intracytoplasmic Sperm Injection (ICSI)*

In this highly sophisticated procedure, a single sperm is injected into an oocyte (Fig. 41.15). It is used effectively when a man has severe infertility problems.

If all the assisted reproductive technologies discussed were employed simultaneously, it would be possible for a baby to have five parents: (1) sperm donor, (2) oocyte donor, (3) surrogate mother, (4) contracting mother, and (5) contracting father.

**Check Your Progress 41.4**

1. In Figure 41.14, which three birth control methods mechanically block sperm from entering the uterus? Which methods can utilize progesterone?
2. Briefly describe each of the following assisted reproductive technologies: artificial insemination by donor (AID), in vitro fertilization (IVF), gamete intrafallopian transfer (GIFT), and intracytoplasmic sperm injection (ICSI).

*health focus*

## Preimplantation Genetic Diagnosis

If prospective parents are heterozygous for a genetic disorder, they may want the assurance that their offspring will be free of the disorder. Determining the genotype of the embryo will provide this assurance. For example, if both parents are *Aa* for a recessive disorder, the embryo will develop normally if it has the genotype *AA* or *Aa*. On the other hand, if one of the parents is *Aa* for a dominant disorder, the embryo will develop normally only if it has the genotype *aa*.

Following in vitro fertilization (IVF), the zygote (fertilized egg) divides. When the embryo has six to eight cells (Fig. 41A*a*), removal of one of these cells for testing purposes has no effect on normal development. Only embryos with a cell that tests negative for the genetic disorder of interest are placed in the uterus to continue developing.

So far, about 1,000 children with normal genotypes for genetic disorders that run in their families have been born worldwide following embryo testing. In the future, it's possible that embryos who test positive for a disorder could be treated by gene therapy, so that they, too, would be allowed to continue to term.

Testing the oocyte is possible if the condition of concern is recessive. Recall that meiosis in females results in a single egg and at least two polar bodies[1]. Polar bodies later disintegrate and receive very little cytoplasm, but they do receive a haploid number of chromosomes. When a woman is heterozygous for a recessive genetic disorder, about half the first polar bodies will have received the genetic defect, and in these instances, the egg will receive the normal allele. Therefore, if a polar body tests positive for a recessive defect, the egg will receive the normal dominant allele. Only normal eggs are then used for IVF. Even if the sperm should happen to carry the mutation, the zygote will, at worst, be heterozygous. But the phenotype will appear normal (Fig. 41A*b*).

If, in the future, gene therapy becomes routine, it's possible that an egg could be given genes that control traits desired by the parents, such as musical or athletic ability, prior to IVF.

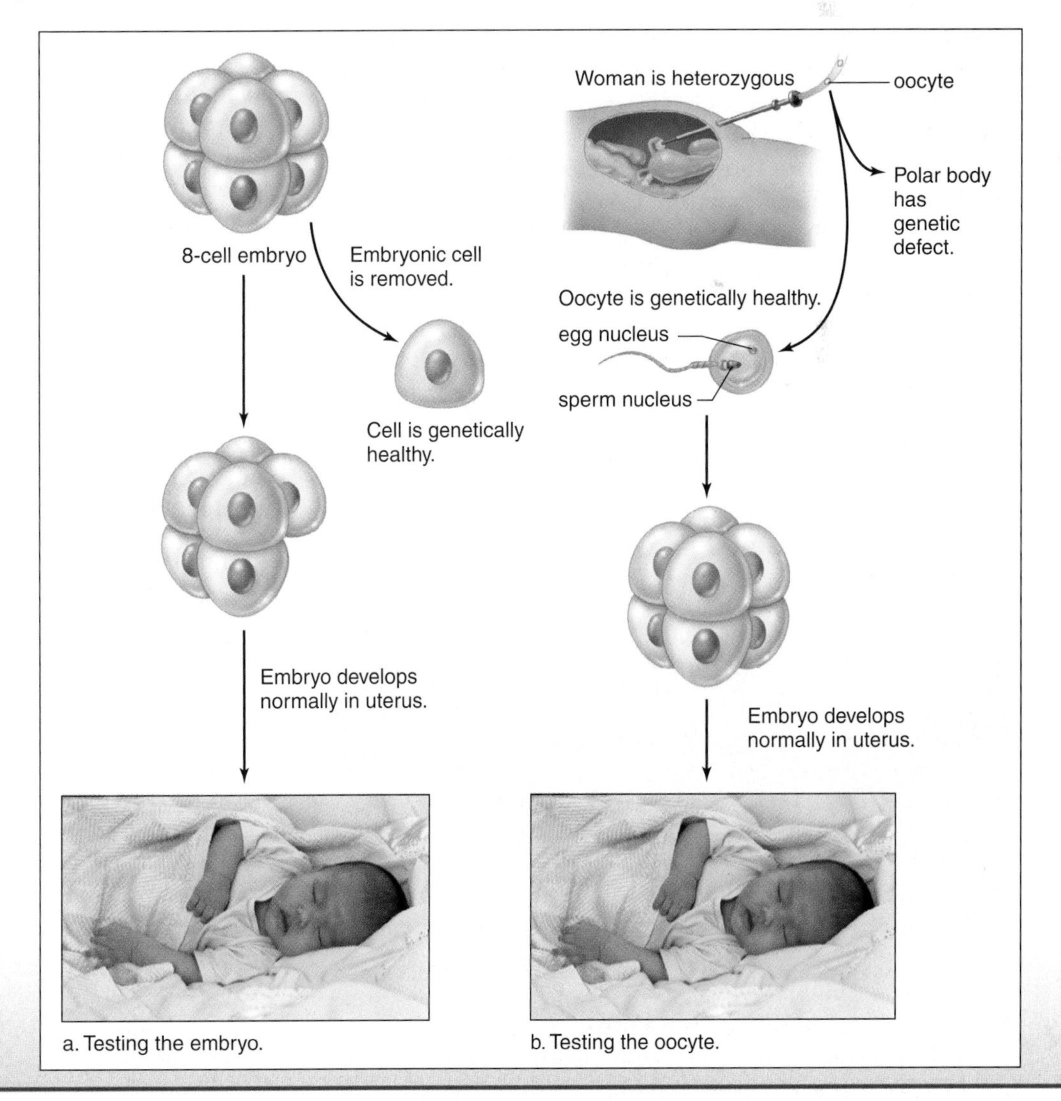


**FIGURE 41A Preimplantation genetic diagnosis.**
***a.*** *Following IVF and cleavage, genetic analysis is performed on one cell removed from an eight-celled embryo. If it is found to be free of the genetic defect of concern, the seven-celled embryo is implanted in the uterus and develops into a newborn phenotype.* ***b.*** *Chromosome and genetic analysis is performed on a polar body attached to an oocyte. If the oocyte is free of a genetic defect, the egg is used for IVF, and the embryo is implanted in the uterus for further development.*

[1] Once meiosis is complete, the oocyte is an egg.

# 41.5 Sexually Transmitted Diseases

Sexually transmitted diseases (STDs) are caused by organisms ranging from viruses to arthropods; however, we will discuss only certain STDs caused by viruses and bacteria. Unfortunately, for unknown reasons, humans cannot develop good immunity to any STD. Therefore, prompt medical treatment should be sought when exposed to an STD. To prevent the spread of STDs, a latex condom can be used.

## AIDS

The organism that causes acquired immunodeficiency syndrome (AIDS) is a virus called the **human immunodeficiency virus (HIV).** HIV attacks the type of lymphocyte known as helper T cells (Fig. 41.16). Helper T cells, you will recall, stimulate the activities of B lymphocytes, which produce antibodies. After an HIV infection sets in, helper T cells begin to decline in number, and the person becomes more susceptible to other types of infections.

### *Symptoms*

AIDS has three stages of infection, called categories A, B, and C. During the category A stage, which may last about a year, the individual is an asymptomatic carrier. He or she may exhibit no symptoms but can pass on the infection. Immediately after infection and before the blood test becomes positive, a large number of infectious viruses are present in the blood, and these could be passed on to another person. Even after the blood test becomes positive, the person remains well as long as the body produces sufficient helper T lymphocytes to keep the count higher than 500 per $mm^3$. During the category B stage, or AIDS-related complex (ARC), which may last six to eight years, the lymph nodes swell, and the person may experience weight loss, night sweats, fatigue, fever, and diarrhea. Infections such as thrush (white sores on the tongue and in the mouth) and herpes recur. Finally, the person may progress to category C, which is full-blown AIDS, characterized by nervous disorders and the development of an opportunistic disease, such as an unusual type of pneumonia or skin cancer. Opportunistic diseases are those that occur only in individuals who have little or no capability of fighting an infection. Without intensive medical treatment, the AIDS patient dies about seven to nine years after infection. Now, with a combination therapy of several drugs, AIDS patients are beginning to live longer in the United States.

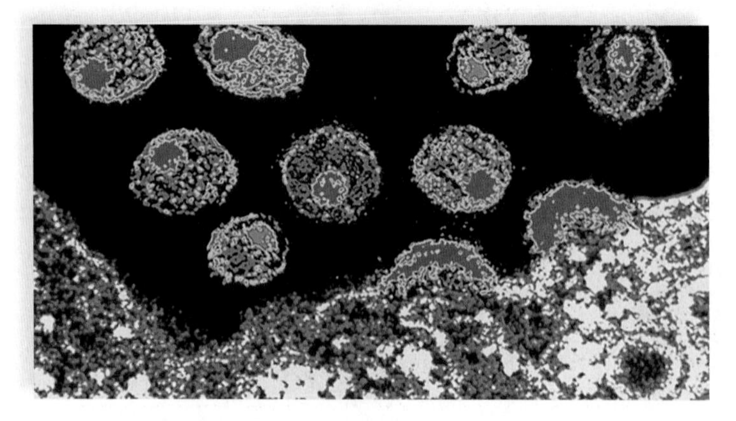

**FIGURE 41.16 HIV, the AIDS virus.**

False-colored micrograph showing HIV particles budding from an infected helper T cell. These viruses can infect other helper T cells and also macrophages, which work with helper T cells to stem infection.

### *Transmission*

AIDS is transmitted by sexual contact with an infected person, including vaginal or rectal intercourse and oral/genital contact. Also, needle-sharing among intravenous drug users is high-risk behavior. A less common mode of transmission (now occurring only rarely in countries where donated blood is screened for HIV) is through transfusions of infected blood or blood-clotting factors. Transmission through tears, saliva, and sweat is thought to be highly unlikely and has never been documented.

HIV first spread through the homosexual community, and male-to-male sexual contact still accounts for the largest percentage of new AIDS cases in the United States. But the largest increases in HIV infections are occurring through heterosexual contact or by intravenous drug use. Now, women account for 25% of all newly diagnosed cases of AIDS.

### *Treatment*

There is no cure for AIDS, but a treatment called highly active antiretroviral therapy (HAART) is usually able to stop HIV replication to such an extent that the viral load becomes undetectable. HAART uses a combination of drugs that interfere with the life cycle of HIV. Entry inhibitors stop HIV from entering a cell by, for example, preventing the virus from binding to a receptor in the plasma membrane. Reverse transcriptase inhibitors, such as AZT, interfere with the operation of the reverse transcriptase enzyme. Integrase inhibitors prevent HIV from inserting its own genetic material into that of the host cells. Protease inhibitors prevent protease from processing newly created polypeptides. Assembly and budding inhibitors are in the experimental stage, and none are available as yet. The hope is that by giving a patient a combination of these drugs, the virus is less likely to replicate, successfully mutate to make drug therapy ineffective, and/or become resistant to drug therapy.

An HIV-positive pregnant woman who takes reverse transcriptase inhibitors during her pregnancy reduces the chances of HIV transmission to her newborn. If possible, drug therapy should be delayed until the tenth to twelfth week of pregnancy to minimize any adverse effects of AZT on fetal development. If treatment begins at this time and delivery is by cesarean section, the chance of transmission from mother to infant is very slim (about 1%).

There is a general consensus that control of the AIDS epidemic will not occur until a vaccine is available. Many different approaches have thus far been tried with limited success. Perhaps a combination of these vaccines will work best, as in drug therapy.

## Genital Warts

Genital warts are caused by the human papillomaviruses (HPVs) (Fig. 41.17). Many times, carriers either do not have

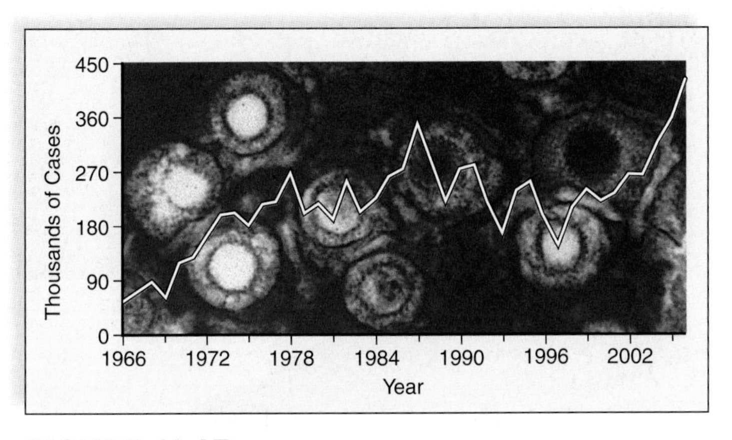


**FIGURE 41.17 Genital warts.**

A graph depicting the incidence of new cases of genital warts reported in the United States from 1966–2006 is superimposed on a photomicrograph of human papillomaviruses.

any sign of warts or merely have flat lesions. When present, flat or raised warts can be found on the penis and the foreskin of males and at the vaginal orifice and cervix in females. Warts on the cervix are always flat and can be hard to detect.

Presently, there is no cure for an HPV infection, but it can be treated effectively by surgery, freezing, application of an acid, or laser burning, depending on severity. If visible warts are removed, they may recur. HPV infection is associated with cancer of the cervix, as well as tumors of the vulva, the vagina, the anus, the penis, and the upper throat. A vaccine against the most common HPV has been developed, and it is recommended that girls be vaccinated before they become sexually active. (Research is under way to see if the HPV vaccine is equally effective for boys.)

## Genital Herpes

Genital herpes is caused by herpes simplex virus (Fig. 41.18). Type 1 usually causes cold sores and fever blisters, while type 2 more often causes genital herpes. Crossover infections do occur, however. That is, type 1 has been known to cause a genital infection, while type 2 has been known to cause cold sores and fever blisters.

Genital herpes is one of the more prevalent sexually transmitted diseases today. At any one time, millions of persons may be having recurring symptoms. After infection, some people exhibit no symptoms; others may experience a tingling or itching sensation before blisters appear on the genitals. Once the blisters rupture, they leave painful ulcers that may take as long as three weeks or as little as five days to heal. The blisters may be accompanied by fever, pain on urination, swollen lymph nodes in the groin, and in women, a copious discharge. At this time, the individual has an increased risk of acquiring an AIDS infection.

After the ulcers heal, the disease is only latent, and blisters can recur, although usually at less frequent intervals and with milder symptoms. Fever, stress, sunlight, and menstruation are associated with recurrence of symptoms. Exposure to herpes in the birth canal can cause an infection in the newborn, which leads to neurological disorders and even death. Birth by cesarean section prevents this possibility.

## Hepatitis

There are several types of hepatitis and each type infects the liver and can cause cancer and death. The type of hepatitis and the virus that causes it are designated by the same letter. Hepatitis A is usually acquired from sewage-contaminated drinking water, but this infection can also be sexually transmitted through oral/anal contact. Hepatitis B, which can be spread by blood transfusions and bodily fluids, can also be spread in the same manner as AIDS and is more infectious. Hepatitis C is called posttransfusion hepatitis but can be transmitted through sexual contact. Fortunately, a vaccine is now available for hepatitis B. No vaccines are available for hepatitis C.

## Chlamydia

Chlamydia is named for the tiny bacterium that causes it (*Chlamydia trachomatis*). The rate of new chlamydial infections has steadily increased since 1987 (Fig. 41.19).

Chlamydial infections of the lower reproductive tract are usually mild or asymptomatic, especially in women. About 8–21 days after infection, men may experience a mild burning sensation on urination and a mucoid discharge.

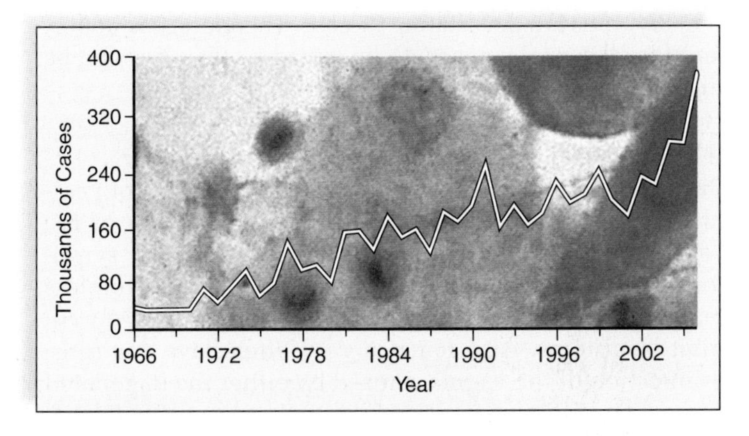


**FIGURE 41.18 Genital herpes.**

A graph depicting the incidence of new reported cases of genital herpes in the United States from 1966–2006 is superimposed on a photomicrograph of cells infected with the herpes simplex virus.

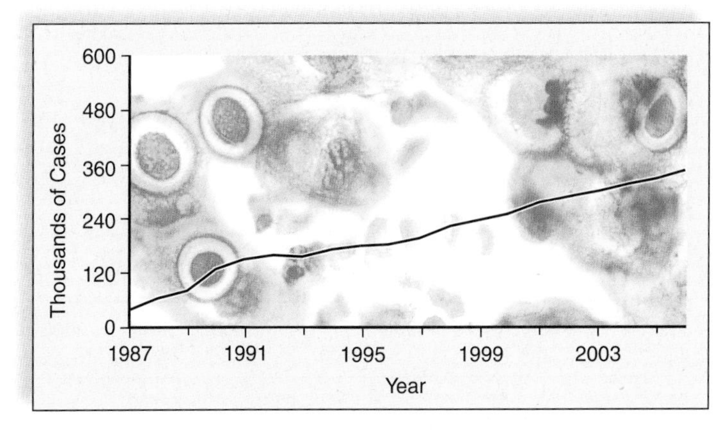


**FIGURE 41.19 Chlamydial infection.**

A graph depicting the incidence of new cases of chlamydia in the United States from 1987–2006 is superimposed on a photomicrograph of a cell containing different stages of the organism.

Women may have a vaginal discharge along with the symptoms of a urinary tract infection. Chlamydia also causes cervical ulcerations, which increase the risk of acquiring AIDS.

If the infection is misdiagnosed or if a woman does not seek medical help, there is a particular risk of the infection spreading from the cervix to the oviducts so that pelvic inflammatory disease (PID) results. This painful condition can result in blockage of the oviducts with the possibility of sterility and infertility. If a baby comes in contact with chlamydia during birth, inflammation of the eyes or pneumonia can result.

## Gonorrhea

Gonorrhea is caused by the bacterium *Neisseria gonorrhoeae* (Fig. 41.20). Diagnosis in the male is not difficult, since typical symptoms are pain upon urination and a thick, greenish yellow urethral discharge. In males and females, a latent infection leads to pelvic inflammatory disease (PID), which affects the vasa deferentia or oviducts. As the inflamed tubes heal, they may become partially or completely blocked by scar tissue, resulting in sterility or infertility. If a baby is exposed during birth, an eye infection leading to blindness can result. All newborns are given eyedrops to prevent this possibility.

Gonorrhea proctitis, an infection of the anus characterized by anal pain and blood or pus in the feces, also occurs in patients. Oral/genital contact can cause infection of the mouth, throat, and tonsils. Gonorrhea can spread to internal parts of the body, causing heart damage or arthritis. If, by chance, the person touches infected genitals and then touches his or her eyes, a severe eye infection can result. Up to now, gonorrhea was curable by antibiotic therapy, but resistance to antibiotics is becoming more and more common, and 40% of all strains are now known to be resistant to therapy.

## Syphilis

Syphilis, which is caused by the bacterium *Treponema pallidum,* has three stages that are typically separated by latent periods. In the primary stage, a hard chancre (ulcerated sore with hard edges) appears (Fig. 41.21*a*). In the secondary stage, a rash appears all over the body—even on the palms of the hands and the soles of the feet (Fig. 41.21*b*). During the tertiary stage, syphilis may affect the cardiovascular and/or nervous system. An infected person may become mentally retarded, become blind, walk with a shuffle, or show signs of insanity. Gummas, which are large, destructive ulcers, may develop on the skin or within the internal organs (Fig. 41.21*c*). Syphilitic bacteria can cross the placenta, causing birth defects or a stillbirth. Unlike the other STDs discussed, there is a blood test to diagnose syphilis.

Syphilis is a devastating disease. Therefore, it is important for all sexual contacts to be traced so that they can be treated with antibiotic therapy.

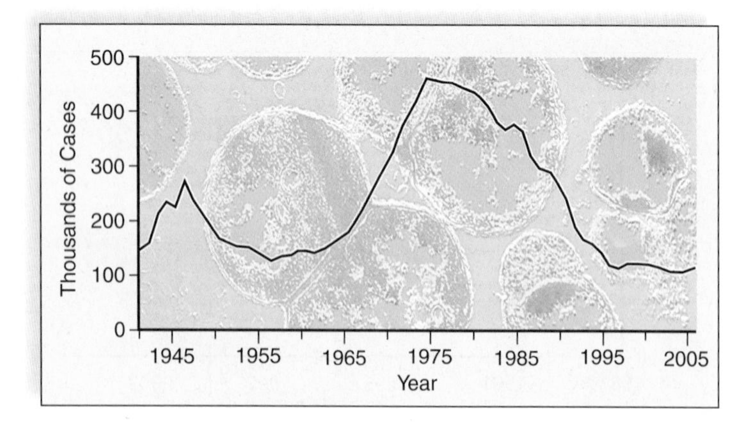


**FIGURE 41.20 Gonorrhea.**

A graph depicting the incidence of new cases of gonorrhea in the United States from 1941–2006 is superimposed on a photomicrograph of a urethral discharge from an infected male. Gonorrheal bacteria (*Neisseria gonorrhoeae*) occur in pairs; for this reason, they are called diplococci.

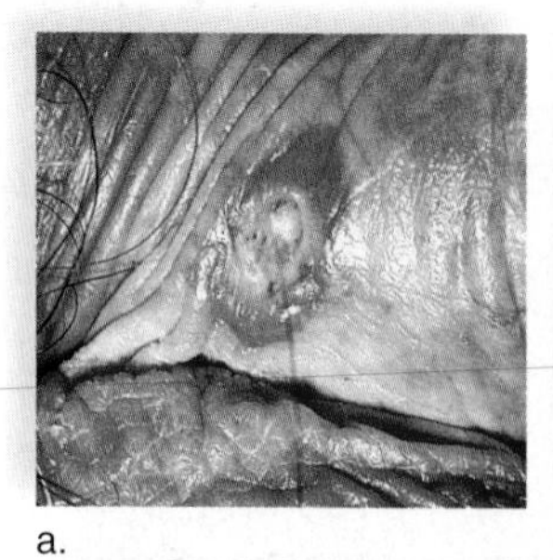


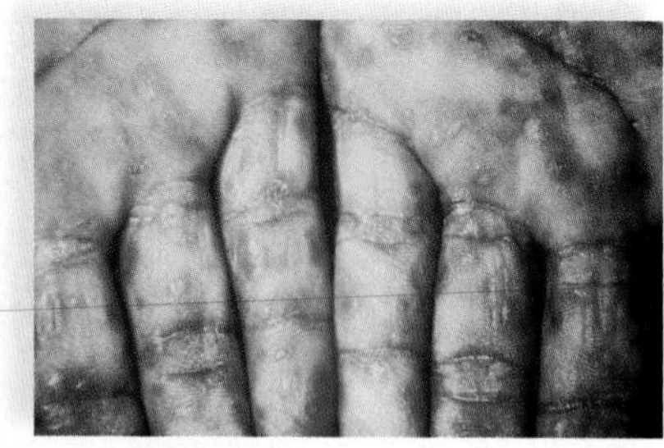


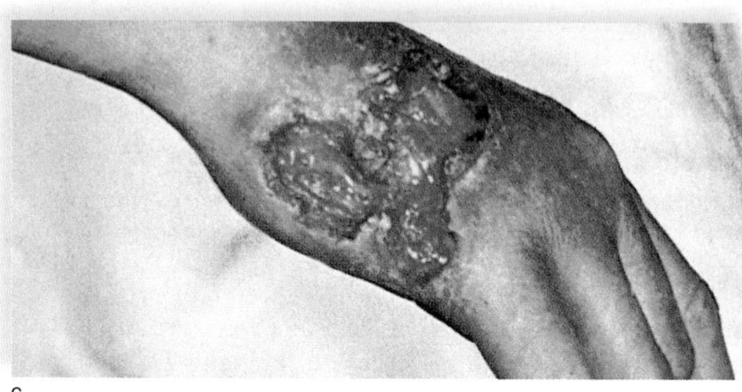


**FIGURE 41.21 Syphilis.**

**a.** The primary stage of syphilis is a chancre at the site where the bacterium enters the body. **b.** The secondary stage is a body rash that occurs even on the palms of the hands and soles of the feet. **c.** In the tertiary stage, gummas may appear on the skin or internal organs.

## Two Other Infections

**Bacterial vaginosis** is believed to account for 50% of vaginitis cases in American women. The overgrowth of the bacterium *Gardnerella vaginalis* and consequent symptoms can occur for nonsexual reasons, but symptomless males may pass on the bacteria to women, who do exhibit symptoms. Also females very often have vaginitis, or infection of the vagina, caused by either the flagellated protozoan *Trichomonas vaginalis* or the yeast *Candida albicans.* The protozoan infection causes a frothy white or yellow, foul-smelling discharge accompanied by itching, and the yeast infection causes a thick, white, curdy discharge, also accompanied by itching. **Trichomoniasis** is

*health focus*

## Preventing Transmission of STDs

It is wise to protect yourself from getting a sexually transmitted disease (STD). Some of the STDs, such as gonorrhea, syphilis, and chlamydia, can be cured by taking an antibiotic, but medication for the ones transmitted by viruses is much more problematic. In any case, it is best to prevent the passage of STDs from person to person so that treatment becomes unnecessary.

### Sexual Activities Transmit STDs

***Abstain from sexual intercourse or develop a long-term monogamous*** (always the same partner) sexual relationship with a partner who is free of STDs (Fig. 41B).

***Refrain from multiple sex partners or having relations with someone who has multiple sex partners.*** If you have sex with two other people and each of these has sex with two people and so forth, the number of people who are relating is quite large.

***Remember that, although the prevalence of AIDS is presently higher among homosexuals and bisexuals,*** the highest rate of increase is now occurring among heterosexuals. The lining of the uterus is generally only one cell thick, and it does allow infected cells from a sexual partner to enter.

***Be aware that having relations with an intravenous drug user is risky*** because the behavior of this group risks hepatitis and an HIV infection. Be aware that anyone who already has another sexually transmitted disease is more susceptible to an HIV infection.

***Uncircumcised males are more likely to become infected*** with an STD than circumcised males because vaginal secretions can remain under the foreskin for a long period of time.

***Avoid anal-rectal intercourse*** (in which the penis is inserted into the rectum) because the lining of the rectum is thin and cells infected with HIV can easily enter the body there.

### Unsafe Sexual Practices Transmit STDs

***Always use a latex condom during sexual intercourse*** if you do not know for certain that your partner has been free of STDs for some time. Be sure to follow the directions supplied by the manufacturer.

***Avoid fellatio*** (kissing and insertion of the penis into a partner's mouth) ***and cunnilingus*** (kissing and insertion of the tongue into the vagina) because they may be a means of transmission. The mouth and gums often have cuts and sores that facilitate the entrance of infected cells.

***Be cautious about the use of alcohol or any drug*** that may prevent you from being able to control your behavior. Females have to be particularly aware of the "date-rape" drug GHB (gamma-hydroxy-butyramine), a drug that puts a person in an uninhibited state with no memory of what transpired.

### Drug Use Transmits Hepatitis and HIV

***Stop, if necessary, or do not start the habit of injecting drugs into your veins.*** Be aware that hepatitis and HIV can be spread by blood-to-blood contact.

***Always use a new sterile needle for injection or one that has been cleaned in bleach*** if you are a drug user and cannot stop your behavior (Fig. 41C).

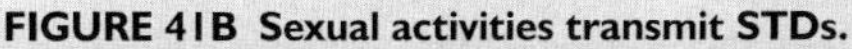

**FIGURE 41B Sexual activities transmit STDs.**

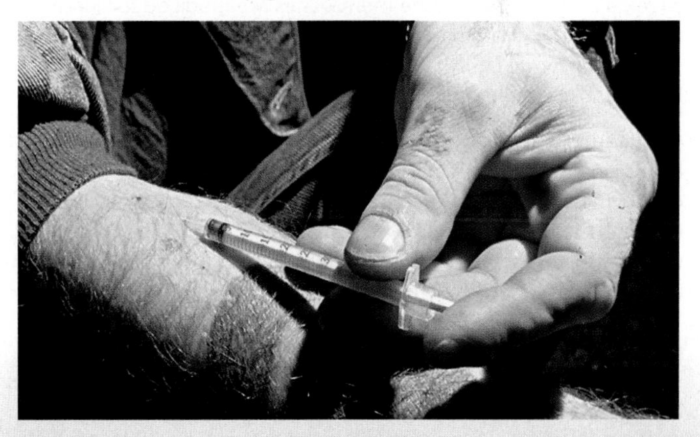

**FIGURE 41C Sharing needles transmits STDs.**

most often acquired through sexual intercourse, and the asymptomatic male is usually the reservoir of infection. *Candida albicans,* however, is a normal organism found in the vagina; its growth simply increases beyond normal under certain circumstances. For example, women taking birth control pills are sometimes prone to yeast infections. Also, the indiscriminate use of antibiotics can alter the normal balance of organisms in the vagina so that a yeast infection flares up.

### Check Your Progress 41.5

1. Which of the STDs described in this chapter are caused by bacteria and, therefore, can be effectively treated with antibiotics?
2. How does HIV infection harm the immune system?
3. Which of the STDs described in this chapter are likely to cause PID-related infertility?

## Connecting the Concepts

The dizzying array of reproductive technologies has resulted in many legal complications. Questions range from which mother has first claim to the child—the surrogate mother, the woman who donated the egg, or the primary caregiver—to which partner has first claim to frozen embryos following a divorce. Legal decisions about who has the right to use what techniques have rarely been discussed, much less decided upon. Some clinics will help anyone, male or female, no questions asked, as long as he or she has the ability to pay. And most clinics are heading toward doing any type of procedure, including guaranteeing the sex of the child or making sure the child will be free from a particular genetic disorder. It would not be surprising if, in the future, zygotes could be engineered to have any particular trait desired by the parents.

Even today, eugenic (good gene) goals are evidenced by the fact that reproductive clinics advertise for egg and sperm donors, primarily in elite college newspapers. Is it too late for us as a society to make ethical decisions about reproductive issues? Should we come to a consensus about what techniques should be allowed and who should be able to use them? We all want to avoid, if possible, what happened to Jonathan Alan Austin. Jonathan, who was born to a surrogate mother, later died from injuries inflicted by his father. Should background checks be legally required? Should surrogate mothers only make themselves available to individuals or couples who possess certain psychological characteristics?

## summary

### 41.1 How Animals Reproduce

Ordinarily, asexual reproduction may quickly produce a large number of offspring genetically identical to the parent. Sexual reproduction involves gametes and produces offspring that are genetically slightly different from the parents. The gonads are the primary sex organs, but there are also accessory organs. The accessory organs consist of storage areas for sperm and ducts that conduct the gametes. They also contribute to formation of the semen.

Animals typically protect their eggs and embryos. The egg of oviparous animals contains yolk, and in terrestrial animals a shelled egg prevents drying out. The amount of yolk is dependent on whether there is a larval stage.

Reptiles have extraembryonic membranes that allow them to develop on land; these same membranes are modified for internal development in mammals. Ovoviviparous animals retain their eggs until the offspring have hatched, and viviparous animals retain the embryo. Placental mammals exemplify viviparous animals.

### 41.2 Male Reproductive System

In human males, sperm are produced in the testes, mature in the epididymides, and may be stored in the vasa deferentia before entering the urethra, along with seminal fluid (produced by seminal vesicles, the prostate gland, and bulbourethral glands). Sperm are ejaculated during male orgasm, when the penis is erect.

Spermatogenesis occurs in the seminiferous tubules of the testes, which also produce testosterone in interstitial cells. Testosterone brings about the maturation of the primary sex organs during puberty and promotes the secondary sex characteristics of males, such as low voice, facial hair, and increased muscle strength.

Follicle-stimulating hormone (FSH) from the anterior pituitary stimulates spermatogenesis, and luteinizing hormone (LH) stimulates testosterone production. A hypothalamic-releasing hormone, gonadotropic-releasing hormone (GnRH), controls anterior pituitary production and FSH and LH release. The level of testosterone in the blood controls the secretion of GnRH and the anterior pituitary hormones by a negative feedback system.

### 41.3 Female Reproductive System

In females, an oocyte produced by an ovary enters an oviduct, which leads to the uterus. The uterus opens into the vagina. The external genital area of women includes the vaginal opening, the clitoris, the labia minora, and the labia majora.

In either ovary, one follicle a month matures, produces a secondary oocyte, and becomes a corpus luteum. This is called the ovarian cycle. The follicle and the corpus luteum produce estrogens, collectively called estrogen, and progesterone, the female sex hormones.

The uterine cycle occurs concurrently with the ovarian cycle. In the first half of these cycles (days 1–13, before ovulation), the anterior pituitary produces FSH and the follicle produces estrogen. Estrogen causes the endometrium to increase in thickness. In the second half of these cycles (days 15–28, after ovulation), the anterior pituitary produces LH and the follicle produces progesterone. Progesterone causes the endometrium to become secretory. Feedback control of the hypothalamus and anterior pituitary causes the levels of estrogen and progesterone to fluctuate. When they are at a low level, menstruation begins.

If fertilization occurs, a zygote is formed, and development begins. The resulting embryo travels down the oviduct and implants itself in the prepared endometrium. A placenta, which is the region of exchange between the fetal blood and the mother's blood, forms. At first, the placenta produces HCG, which maintains the corpus luteum; later, it produces progesterone and estrogen.

The female sex hormones, estrogen and progesterone, also affect other traits of the body. Primarily, estrogen brings about the maturation of the primary sex organs during puberty and promotes the secondary sex characteristics of females, including less body hair than males, a wider pelvic girdle, a more rounded appearance, and development of breasts.

### 41.4 Control of Reproduction

Numerous birth control methods and devices, such as the birth control pill, diaphragm, and condom, are available for those who wish to prevent pregnancy. A morning-after pill, RU-486, is now available. Some couples are infertile, and if so, they may use assisted reproductive technologies to have a child. Artificial insemination and in vitro fertilization have been followed by more sophisticated techniques, such as intracytoplasmic sperm injection.

### 41.5 Sexually Transmitted Diseases

Sexually transmitted diseases include AIDS, an epidemic disease; genital warts, which lead to cancer of the cervix; genital herpes, which repeatedly flares up; hepatitis, especially types A and B; chlamydia and gonorrhea, which cause pelvic inflammatory disease (PID); and syphilis, which has cardiovascular and neurological complications if untreated.

## understanding the terms



Match the terms to these definitions:

a. _______________ Release of an oocyte from the ovary.
b. _______________ Development of an egg into a whole organism without fertilization.
c. _______________ Female sex hormone that causes the endometrium of the uterus to become secretory during the uterine cycle; along with estrogen, it maintains secondary sex characteristics in females.
d. _______________ Thick, whitish fluid consisting of sperm and secretions from several glands of the male reproductive tract.
e. _______________ Organ that produces gametes; the ovary, which produces eggs, and the testis, which produces sperm.

## reviewing this chapter

1. Contrast asexual reproduction with sexual reproduction, reproduction in water with reproduction on land, and the life history of an insect with that of a bird. 756–57
2. Trace the path of sperm in a human male. What glands contribute fluids to semen? 758–59
3. Discuss the anatomy and physiology of the testes. Describe the structure of sperm. 760–61
4. Name the endocrine glands involved in maintaining the sex characteristics of males and the hormones produced by each. 761
5. Trace the path of an oocyte in a human female. Where do fertilization and implantation occur? When does the oocyte become an egg? Name two functions of the vagina. 762
6. Describe the external genital organs in females. 762–63
7. Discuss the anatomy and physiology of the ovaries. Describe the ovarian cycle and ovulation. 763–64
8. Describe the uterine cycle, and relate it to the ovarian cycle. In what way is menstruation prevented if pregnancy occurs? 764–65
9. What events occur at fertilization? Name three functions of the female sex hormones, aside from their involvement in the uterine cycle. 765–66
10. Describe the anatomy and physiology of the breast. 766
11. Which means of birth control require surgery, use hormones, use barrier methods, or are dependent on none of these? 766–67
12. If couples are infertile, what assisted reproductive technologies are available? 768
13. List the cause, symptoms, and treatment for the most common types of sexually transmitted diseases. 770–73

## testing yourself

1. Label this diagram of the male reproductive system and trace the path of sperm.

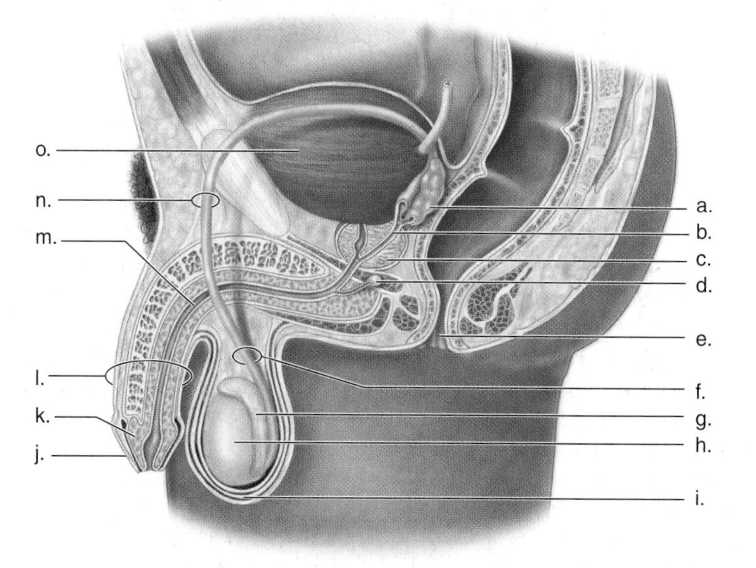


Choose the best answer for each question.

2. Which of these is a requirement for sexual reproduction?
   a. male and female parents
   b. production of gametes
   c. optimal environmental conditions
   d. aquatic habitat
   e. All of these are correct.
3. Internal fertilization
   a. can prevent the drying out of gametes and zygotes.
   b. must take place on land.
   c. is practiced by humans.
   d. requires that males have a penis.
   e. Both a and c are correct.
4. Which of these pairs is mismatched?
   a. interstitial cells—testosterone
   b. seminiferous tubules—sperm production
   c. vasa deferentia—seminal fluid production
   d. urethra—conducts sperm
   e. Both c and d are mismatched.
5. Follicle-stimulating hormone (FSH)
   a. is secreted by females but not males.
   b. stimulates the seminiferous tubules to produce sperm.
   c. secretion is controlled by gonadotropin-releasing hormone (GnRH).
   d. is the same as luteinizing hormone.
   e. Both b and c are correct.

6. Which of these combinations is most likely to be present before ovulation occurs?
   a. FSH, corpus luteum, estrogen, secretory uterine lining
   b. luteinizing hormone (LH), follicle, progesterone, thick endometrium
   c. FSH, follicle, estrogen, endometrium becoming thick
   d. LH, corpus luteum, progesterone, secretory endometrium
   e. Both c and d are correct.
7. In tracing the path of sperm, you would mention vasa deferentia before
   a. testes.
   b. epididymides.
   c. urethra.
   d. seminiferous tubules.
   e. All of these are correct.
8. An oocyte is fertilized in
   a. the vagina.
   b. the uterus.
   c. the oviduct.
   d. the ovary.
   e. All of these are correct.
9. During pregnancy,
   a. the ovarian and uterine cycles occur more quickly than before.
   b. GnRH is produced at a higher level than before.
   c. the ovarian and uterine cycles do not occur.
   d. the female secondary sex characteristics are not maintained.
   e. Both b and c are correct.
10. Which means of birth control is most effective in preventing sexually transmitted diseases?
   a. condom
   b. pill
   c. diaphragm
   d. spermicidal jelly
   e. vasectomy
11. Which of these is a sexually transmitted disease caused by a bacterium?
   a. gonorrhea
   b. hepatitis B
   c. genital warts
   d. genital herpes
   e. HIV
12. The HIV virus has a preference for binding to
   a. B lymphocytes.
   b. cytotoxic T lymphocytes.
   c. helper T lymphocytes.
   d. All of these are correct.

For questions 13–15, match the descriptions with the sexually transmitted diseases in the key.

**KEY:**

a. AIDS
b. hepatitis B
c. genital herpes
d. genital warts
e. gonorrhea
f. chlamydia
g. syphilis

13. blisters, ulcers, pain on urination, swollen lymph nodes
14. flulike symptoms, jaundice; eventual liver failure possible
15. males have a thick, greenish-yellow discharge; no symptoms in female; can lead to PID
16. Which of these is the primary sex organ of the male?
   a. penis
   b. scrotum
   c. testis
   d. prostate
   e. vasectomy
17. The luteal phase of the uterine cycle is characterized by
   a. high levels of LH and progesterone.
   b. low levels of estrogen and progesterone.
   c. increasing estrogen and little or no progesterone.
   d. high levels of LH only.
18. The secretory phase of the uterine cycle occurs during which ovarian phase?
   a. follicular phase
   b. ovulation
   c. luteal phase
   d. menstrual phase
19. Which of these is the correct path of an oocyte?
   a. oviduct, fimbriae, uterus, vagina
   b. ovary, fimbriae, oviduct, uterine cavity
   c. oviduct, fimbriae, abdominal cavity
   d. fimbriae, uterine tube, uterine cavity, ovary
20. Following ovulation, the corpus luteum develops under the influence of
   a. progesterone.
   b. FSH.
   c. LH.
   d. estradiol.

## thinking scientifically

1. Female athletes who train intensively often stop menstruating. The important factor appears to be the reduction of body fat below a certain level. Give a possible evolutionary explanation for a relationship between body fat in females and reproductive cycles.
2. The average sperm count in males is now lower than it was several decades ago. The reasons for the lower sperm count usually seen today are not known. What data might be helpful in order to formulate a testable hypothesis?

## bioethical issue

### Fertility Drugs

Higher-order multiple births (triplets or more) in the United States increased 19% between 1980 and 1994. During these years, it became customary to use fertility drugs (gonadotropic hormones) to stimulate the ovaries. Although a variety of assisted reproductive technologies have become commonplace in recent years, fertility drugs are still routinely prescribed for women who are infertile due to ovulation disorders. However, fertility drugs frequently result in the release and subsequent fertilization of multiple oocytes. The risks for premature delivery, low birth weight, and developmental abnormalities rise sharply for higher-order multiple births. And the physical and emotional burden placed on the parents is extraordinary. They face endless everyday chores and find it difficult to maintain normal social relationships, if only because they get insufficient sleep. Finances are strained to provide for the children's needs, including housing and child-care assistance. About one-third report that they received no help from relatives, friends, or neighbors in the first year after the birth. Trips to the hospital for accidental injury are more frequent because parents with only two arms and two legs cannot keep so many children safe at one time.

A higher-order multiple pregnancy can be terminated, or selective reduction can be done. During selective reduction, one or more of the fetuses is killed by an injection of potassium chloride. Selective reduction could very well result in psychological and social complications for the mother and surviving children.

Many clinicians are now urging that all possible steps be taken to ensure that the chance of higher-order multiple births be reduced. For example, pre-ovulation oocytes may be obtained and subjected to in vitro fertilization (in which the eggs are fertilized in the lab), as shown in Figure 41A on page 769. Even this option poses an ethical dilemma: In vitro fertilization may be carried out with the intent that only one or two zygotes will be placed in the woman's womb. But then any leftover zygotes may never have an opportunity to continue development.

## *Biology* website

The companion website for *Biology* provides a wealth of information organized and integrated by chapter. You will find practice tests, animations, videos, and much more that will complement your learning and understanding of general biology.

**http://www.mhhe.com/maderbiology10**

# 42

# Animal Development

Many of the fundamental processes that guide development, from prefertilization onward, are shared throughout most of the animal kingdom. Whether the embryo is that of a worm, insect, amphibian, bird, or mammal, it must first establish an anterior end and a posterior end. As the embryo continues to grow, the various body segments form and give rise to structures such as appendages. The coordination of these developmental events is quite intricate, and easily disrupted by harmful environmental factors such as chemicals, radiation, and pathogens.

*Although the main focus of this chapter is on events that occur early in an animal's life span, it is worth noting that, once an animal has reached maturity, development does not cease. If an animal's body is injured, the wound must be repaired. Not surprisingly, many of the same genes involved in wound repair are those which were active during earlier stages in development. In this chapter, you will examine the processes of early animal development, which begin before the sperm encounters the egg and continue until birth.*

As development occurs, complexity increases.

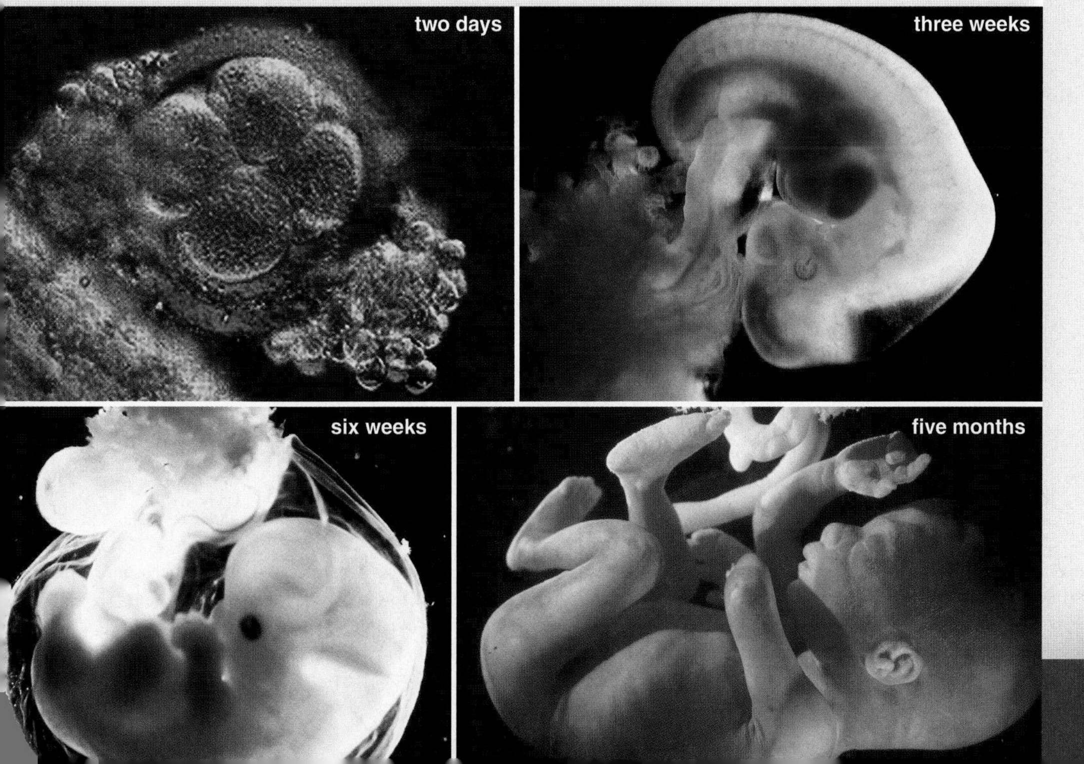


## concepts

# 42.1 Early Developmental Stages

**Fertilization** is the union of a sperm and an egg to form a zygote. Figure 42.1 illustrates how fertilization occurs in mammals, including humans. Human sperm have three distinct parts: a tail, a middle piece, and a head. The tail is a flagellum, which allows the sperm to swim toward the egg, and the middle piece contains energy-producing mitochondria. The head contains the sperm nucleus and is capped by a membrane-bounded acrosome. The egg of a mammal (actually the secondary oocyte) is surrounded by a few layers of adhering follicular cells, collectively called the *corona radiata.* These cells nourished the oocyte when it was in a follicle of the ovary. Next the oocyte has an extracellular matrix termed the *zona pellucida* just outside the plasma membrane.

Fertilization requires a series of events that will result in the diploid zygote. Although only one sperm will actually fertilize the oocyte, many sperm begin this journey and a number succeed in reaching the oocyte. The sperm cover the surface of the oocyte and secrete enzymes that help weaken the corona radiata. They squeeze through the corona radiata and bind to the zona pellucida. After a sperm head binds tightly to the zona pellucida, the acrosome releases digestive enzymes that forge a pathway for the sperm through the zona pellucida to the oocyte plasma membrane.

When the first sperm binds to the oocyte plasma membrane the next few events prevent *polyspermy* (entrance of more than one sperm). As soon as a sperm touches the plasma membrane of an oocyte, the oocyte's plasma membrane depolarizes and this change in charge, known as the "fast block," serves to repel sperm only for a few seconds. Then vesicles in the oocyte called *cortical granules* secrete enzymes that turn the zona pellucida into an impenetrable *fertilization membrane.* The longer-lasting cortical reaction is known as the "slow block."

The last of the events includes formation of the diploid zygote. Microvilli extending from the plasma membrane of the oocyte (Fig. 42.1) bring the entire sperm into the oocyte. The sperm nucleus releases its chromatin, which re-forms into chromosomes enclosed within the so-called sperm pronucleus. In the meantime, the secondary oocyte completes meiosis, becoming an egg whose chromosomes are also enclosed in a pronucleus. A single nuclear envelope soon surrounds both sperm and egg pronuclei. Cell division is imminent, and the

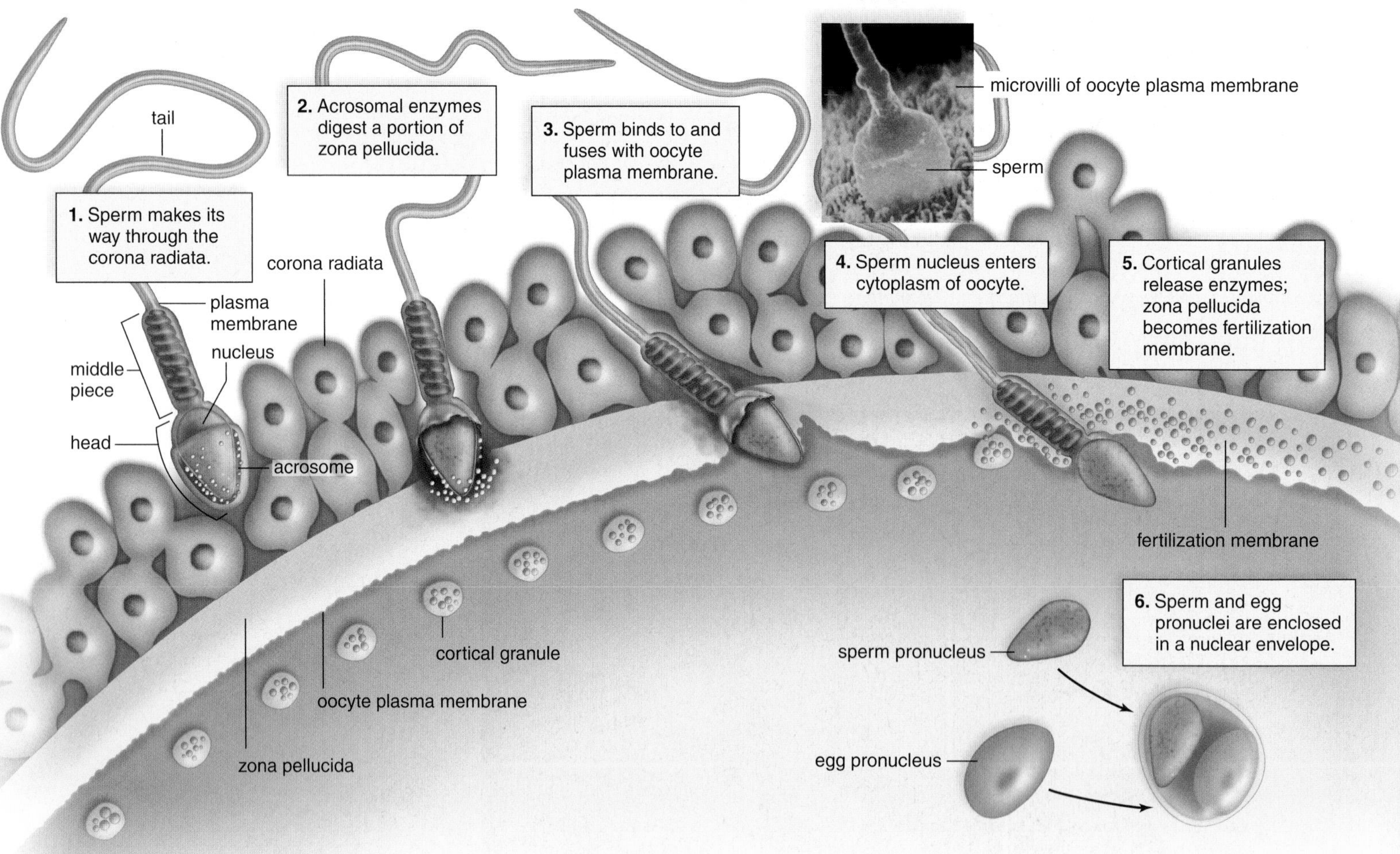


**FIGURE 42.1 Fertilization.**

During fertilization, a single sperm is drawn into the oocyte by microvilli of its plasma membrane (micrograph). The head of a sperm has a membrane-bounded acrosome filled with enzymes. When released, these enzymes digest a pathway for the sperm through the zona pellucida. After a sperm binds to the plasma membrane of the oocyte, changes occur that prevent other sperm from entering the oocyte. The oocyte finishes the second meiotic division and is now an egg. Fertilization is complete when the sperm pronucleus and the egg pronucleus contribute chromosomes to the zygote.

centrosomes that give rise to a spindle apparatus are derived from the basal body of the sperm's flagellum! The two haploid sets of chromosomes share the first spindle apparatus of the fertilized egg, also called a zygote.

## Embryonic Development

Development is all the changes that occur during the life cycle of an organism. During the first stages of development, an organism is called an **embryo.**

### *Cellular Stages of Development*

The cellular stages of development are (1) cleavage resulting in a multicellular embryo and (2) formation of the blastula. **Cleavage** is cell division without growth. DNA replication and mitotic cell division occur repeatedly, and the cells get smaller with each division. In other words, cleavage increases only the number of cells; it does not change the original volume of the egg's cytoplasm.

As shown in Figure 42.2, cleavage in a lancelet is equal and results in uniform cells that form a **morula,** which is a ball of cells. The 16-cell morula resembles a mulberry and continues to divide forming a blastula. A **blastula** is a hollow ball of cells having a fluid-filled cavity called a **blastocoel.** The blastocoel forms when the cells of the morula extrude $Na^+$ into extracellular spaces and water follows by osmosis. The water collects in the center, and the result is a hollow ball of cells.

The zygotes of other animals, such as a frog, chick, or human, which are vertebrates, also undergo cleavage and form a morula. In frogs, cleavage is not equal because of the presence of **yolk,** a dense nutrient material. When yolk is present, the zygote and embryo exhibit polarity, and the embryo has an animal pole and a vegetal pole. The animal pole of a frog embryo has a deep gray color because the cells contain melanin granules, and the vegetal pole has a yellow color because the cells contain yolk.

Similarly, all vertebrates have a blastula stage, but the appearance of the blastula can be different from that of a lancelet. A chick is a vertebrate animal that develops on land and lays a hard-shelled egg containing plentiful yolk. Because yolk-filled cells do not participate in cleavage, the blastula is a layer of cells that spreads out over the yolk. The blastocoel is a space that separates these cells from the yolk:

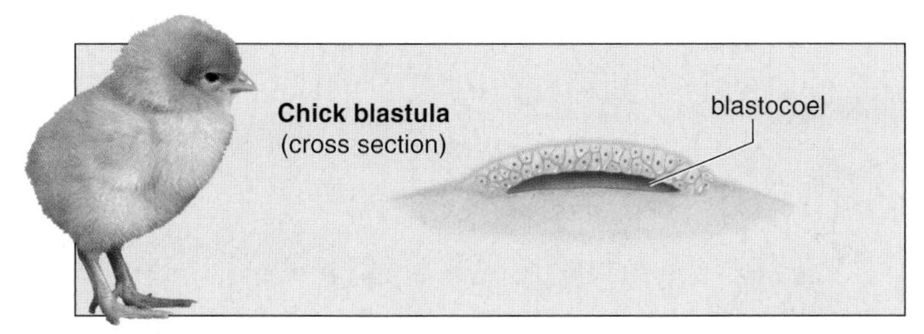


The blastula of humans resembles that of the chick embryo, yet this resemblance cannot be related to the amount of yolk because the human egg contains little yolk. But the evolutionary history of these two animals can provide an explanation for this similarity. Both birds

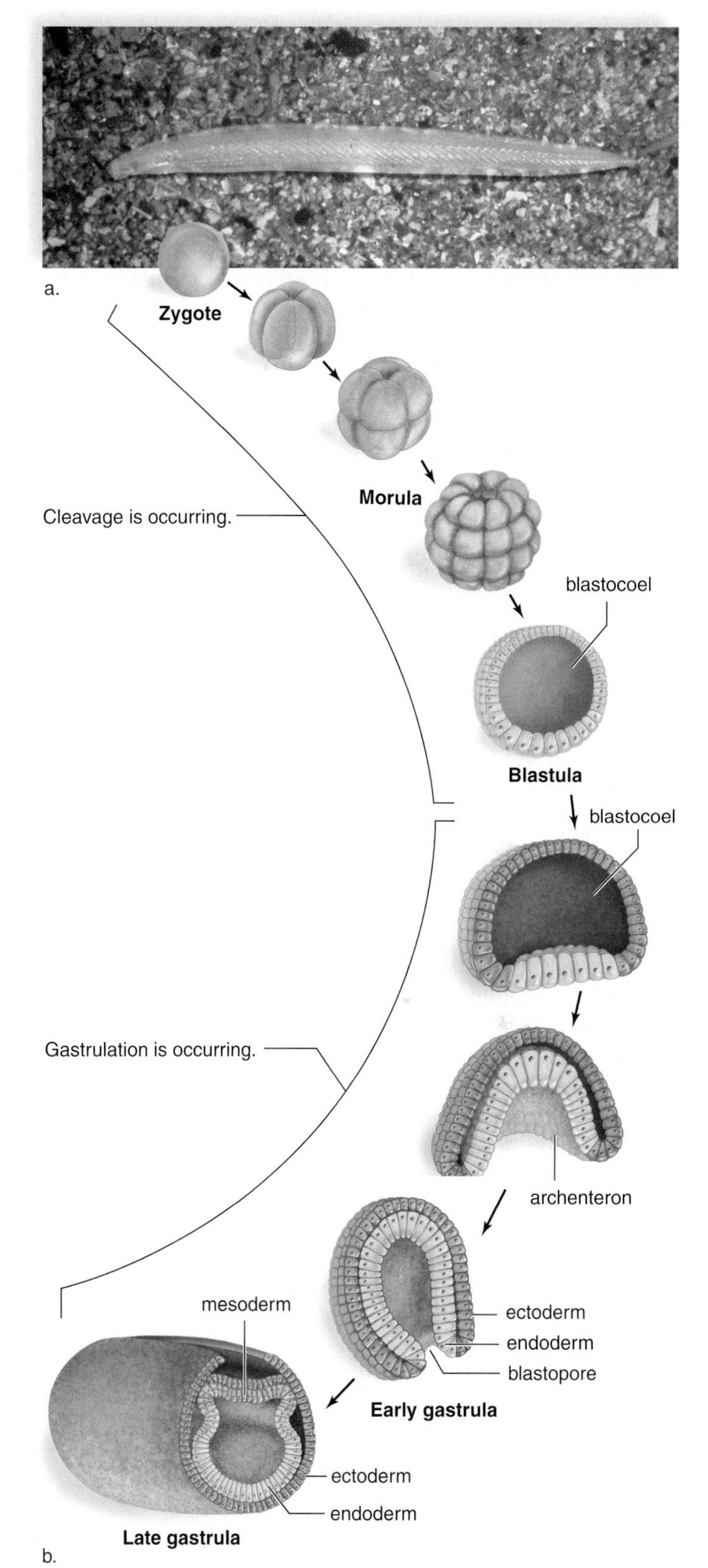


**FIGURE 42.2 Lancelet early development.**

**a.** A lancelet. **b.** The early stages of development are exemplified in the lancelet. Cleavage produces a number of cells that form a cavity, the blastocoel. Invagination during gastrulation produces the germ layers ectoderm and endoderm. Then the mesoderm arises.

and mammals are related to reptiles. This explains why all three groups develop similarly, despite a difference in the amount of yolk in their eggs.

### *Tissue Stages of Development*

The tissue stages of development are (1) the early gastrula and (2) the late gastrula. The early **gastrula** stage begins when certain cells begin to push, or invaginate, into the blastocoel, creating a double layer of cells (see Fig. 42.2). Cells migrate during this and other stages of development, sometimes traveling quite a distance before reaching a destination, where they continue developing. Extracellular proteins and cytoskeletal elements participate in allowing migration to occur. As cells migrate, they "feel their way" by changing their pattern of adhering to extracellular proteins.

An early gastrula has two layers of cells. The outer layer of cells is called the **ectoderm,** and the inner layer is called the **endoderm.** The endoderm borders the gut, but at this point, it is termed either the archenteron or the primitive gut. The pore, or hole, created by invagination (inward folding) is the **blastopore,** and in a lancelet, the blastopore eventually becomes the anus. **Gastrulation** is not complete until three layers of cells that will develop into adult organs are produced. In addition to ectoderm and endoderm, the late gastrula has a middle layer of cells called the **mesoderm.**

Figure 42.2 illustrates gastrulation in a lancelet and Figure 42.3 compares the lancelet, frog, and chick late gastrula stages. In the lancelet, mesoderm formation begins as outpocketings from the primitive gut (Fig. 42.3). These outpocketings will grow in size until they meet and fuse forming two layers of mesoderm. The space between them is the coelom. The coelom is a body cavity lined by mesoderm that contains internal organs. (In humans, the coelom becomes the thoracic and abdominal cavities of the body.)

In the frog, the cells containing yolk do not participate in gastrulation and, therefore, they do not invaginate. Instead, a slitlike blastopore is formed when the animal pole cells begin to invaginate from above, forming endoderm. Animal pole cells also move down over the yolk, to invaginate from below. Some yolk cells, which remain temporarily in the region of the blastopore, are called the yolk plug. Mesoderm forms when cells migrate between the ectoderm and endoderm. Later, a splitting of the mesoderm creates the coelom.

A chick egg contains so much yolk that endoderm formation does not occur by invagination. Instead, an upper layer of cells becomes ectoderm, and a lower layer becomes endoderm. Mesoderm arises by an invagination of cells along the edges of a longitudinal furrow in the midline of the embryo. Because of its appearance, this furrow is called the *primitive streak.* Later, the newly formed mesoderm splits to produce a coelomic cavity.

Ectoderm, mesoderm, and endoderm are called the embryonic **germ layers.** No matter how gastrulation takes place, the result is the same: three germ layers are formed. It is possible to relate the development of future organs to these germ layers (Table 42.1).

**TABLE 42.1**

**Embryonic Germ Layers**

| *Embryonic Germ Layer* | *Vertebrate Adult Structures* |
|---|---|
| Ectoderm (outer layer) | Nervous system; epidermis of skin and derivatives of the epidermis (hair, nail s, glands); tooth enamel, dentin, and pulp; epithelial lining of oral cavity and rectum |
| Mesoderm (middle layer) | Musculoskeletal system; dermis of skin; cardiovascular system; urinary system; lymphatic system; reproductive system—including most epithelial linings; outer layers of respiratory and digestive systems |
| Endoderm (inner layer) | Epithelial lining of digestive tract and respiratory tract, associated glands of these systems; epithelial lining of urinary bladder; thyroid and parathyroid glands |

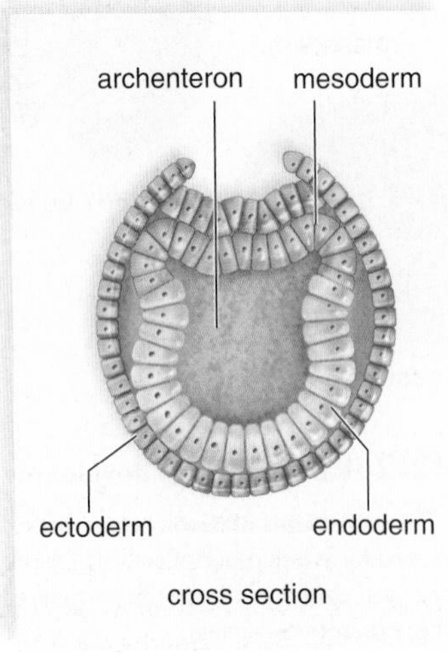
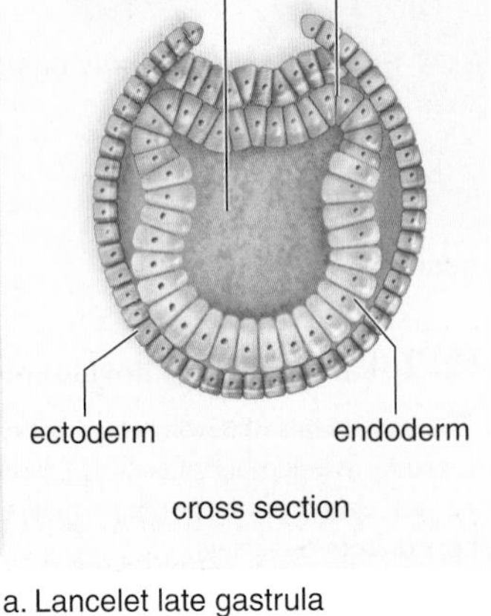


a. Lancelet late gastrula

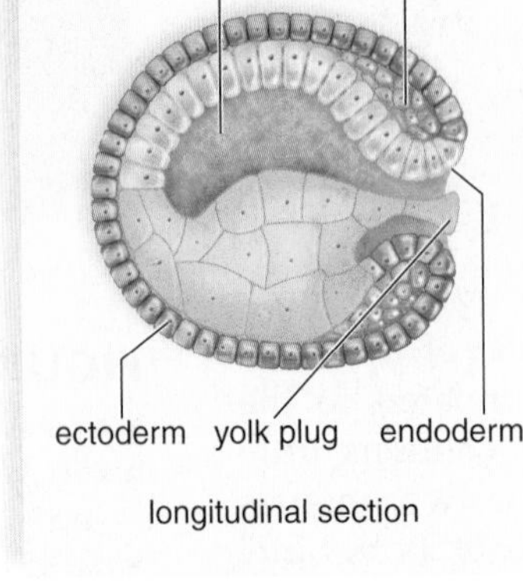


b. Frog late gastrula

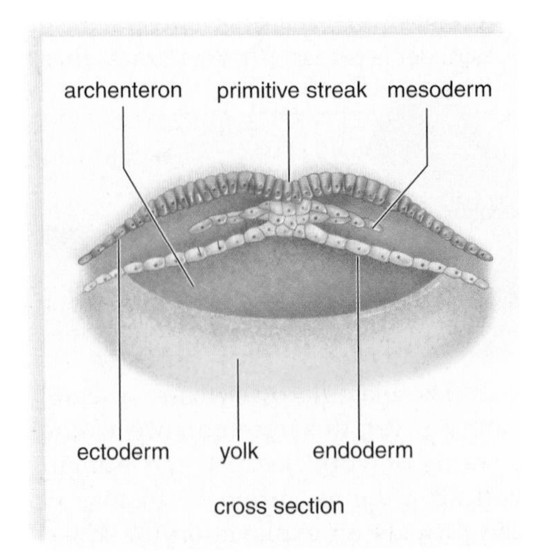


c. Chick late gastrula

**FIGURE 42.3** **Comparative development of mesoderm.** **a.** In the lancelet, mesoderm forms by an outpocketing of the archenteron. **b.** In the frog, mesoderm forms by migration of cells between the ectoderm and endoderm. **c.** In the chick, mesoderm also forms by invagination of cells.

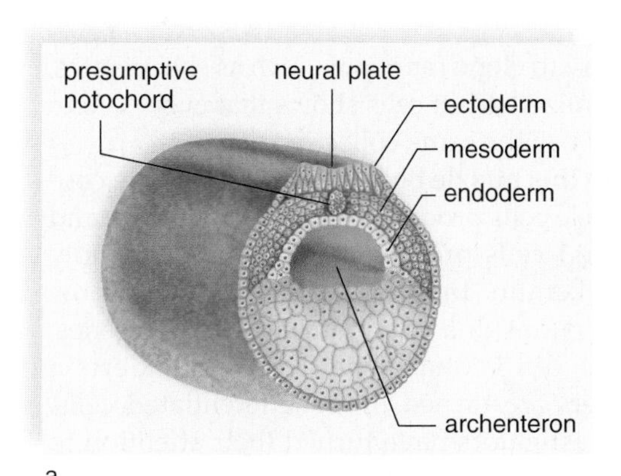


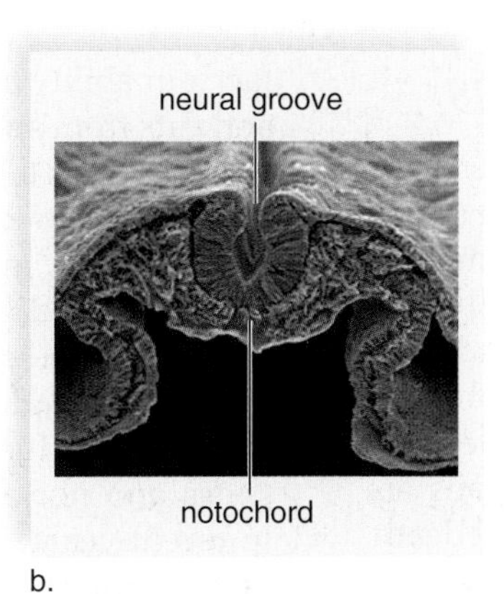


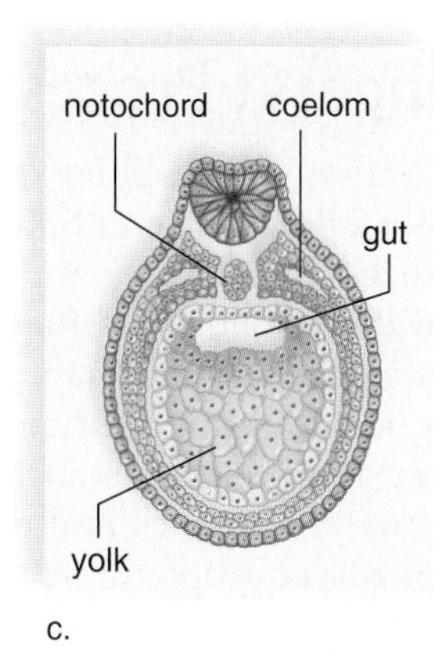


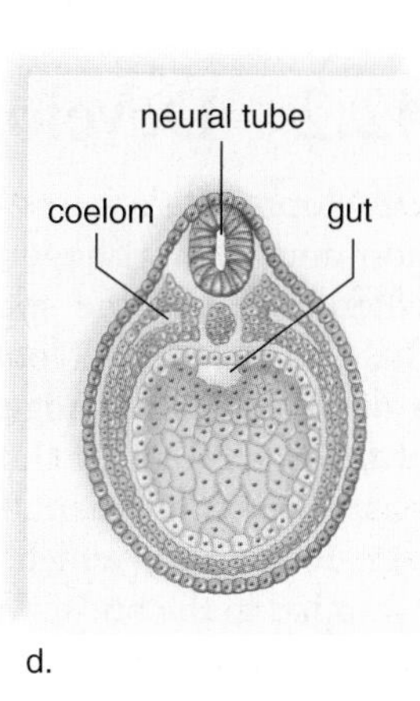


**FIGURE 42.4 Development of neural tube and coelom in a frog embryo.**

**a.** Ectodermal cells that lie above the future notochord (called presumptive notochord) thicken to form a neural plate. **b.** The neural groove and folds are noticeable as the neural tube begins to form. **c.** A splitting of the mesoderm produces a coelom, which is completely lined by mesoderm. **d.** A neural tube and a coelom have now developed.

## *Organ Stages of Development*

The organs of an animal's body develop from the three embryonic germ layers. Much study has been devoted to how the nervous system develops.

The newly formed mesoderm cells lie along the main longitudinal axis of the animal and coalesce to form a dorsal supporting rod called the **notochord.** The notochord persists in lancelets, but in frogs, chicks, and humans, it is later replaced by the vertebral column. Therefore, these animals are called vertebrates.

The nervous system develops from midline ectoderm located just above the notochord. At first, a thickening of cells, called the **neural plate,** is seen along the dorsal surface of the embryo. Then, neural folds develop on either side of a neural groove, which becomes the **neural tube** when these folds fuse. Figure 42.4 shows cross sections of frog development to illustrate the formation of the neural tube. At this point, the embryo is called a **neurula.** Later, the anterior end of the neural tube develops into the *brain,* and the rest becomes the *spinal cord.* In addition, the neural crest is a band of cells that develops where the neural tube pinches off from the ectoderm. Neural crest cells migrate to various locations, where they contribute to formation of skin and muscles, in addition to the adrenal medulla and the ganglia of the peripheral nervous system.

Midline mesoderm cells that did not contribute to the formation of the notochord now become two longitudinal masses of tissue. These two masses become blocked off into somites, which are serially arranged along both sides along the length of the notochord. Somites give rise to muscles associated with the axial skeleton and to the vertebrae. The serial origin of axial muscles and the vertebrae testify that vertebrates are segmented animals. Lateral to the somites, the mesoderm splits, forming the mesodermal lining of the coelom.

A primitive gut tube is formed by endoderm as the body itself folds into a tube. The heart, too, begins as a simple tubular pump. Organ formation continues until the germ layers have given rise to the specific organs listed in Table 42.1. Figure 42.5 will help you relate the formation of vertebrate structures and organs to the three embryonic layers of cells: the ectoderm, the mesoderm, and the endoderm.

### Check Your Progress 42.1

1. Describe the fast and slow blocks to polyspermy.
2. A mature gastrula has three germ layers. Name the germ layer that gives rise to each of the following: notochord, thyroid and parathyroid glands, nervous system, epidermis, skeletal muscles, kidneys, bones, and pancreas.
3. At which stage of embryonic development do cross-sections of all chordate embryos closely resemble one another?

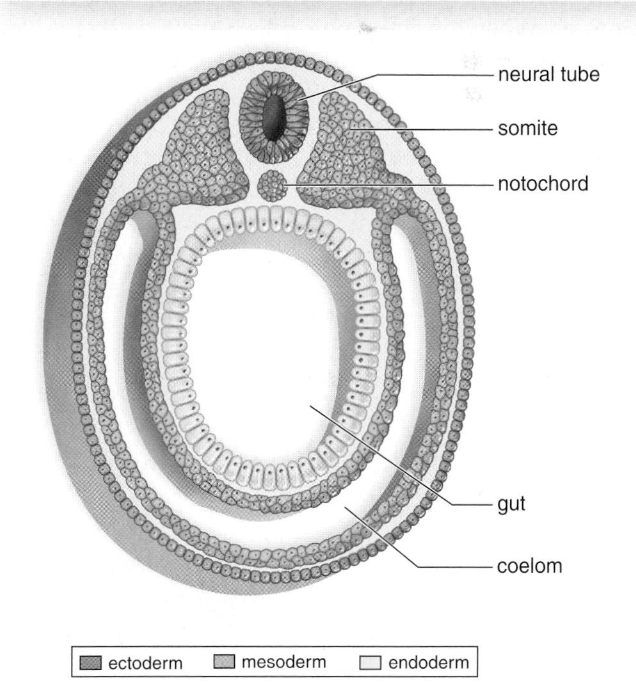


**FIGURE 42.5 Vertebrate embryo, cross section.**

At the neurula stage, each of the germ layers, indicated by color (see key), can be associated with the later development of particular parts. The somites give rise to the muscles of each segment and to the vertebrae, which replace the notochord in vertebrates.

# 42.2 Developmental Processes

Development requires (1) growth, (2) cellular differentiation, and (3) morphogenesis. **Cellular differentiation** occurs when cells become specialized in structure and function; that is, a muscle cell looks different and acts differently than a nerve cell. **Morphogenesis** produces the shape and form of the body. One of the earliest indications of morphogenesis is cell movement. Later, morphogenesis includes **pattern formation,** which means how tissues and organs are arranged in the body. **Apoptosis,** or programmed cell death, which was first discussed on page 153, is an important part of pattern formation.

Developmental genetics has benefited from research using the roundworm, *Caenorhabditis elegans,* and the fruit fly, *Drosophila melanogaster.* These organisms are referred to as model organisms because the study of their development produced concepts that help us understand development in general.

## Cellular Differentiation

At one time, investigators mistakenly believed that irreversible genetic changes must account for differentiation and morphogenesis. Perhaps, they speculated, the genes are parceled out as development occurs, and that is why cells of the body have a different structure and function. We now know that is not the case; rather, every cell in the body has a full complement of genes.

The zygote is **totipotent;** it has the ability to generate the entire organism and, therefore, must contain all the instructions needed by any other specialized cell in the body. For the first few days of cell division, all the embryonic cells are totipotent. When the embryonic cells begin to specialize and lose their totipotency, they do not lose genetic information. In fact, our ability today to clone mammals such as sheep, mice, and cats from specialized adult cells shows that every cell in an organism's body has the same collection of genes.

The answer to this puzzle becomes clear when we consider that only muscle cells produce the proteins myosin and actin; only red blood cells produce hemoglobin; and only skin cells produce keratin. In other words, we now know that specialization is not due to a parceling out of genes; rather, it is due to differential gene expression. Certain genes and not others are turned on in differentiated cells. In recent years, investigators have turned their attention to discovering the mechanisms that lead to differential gene expression. Two mechanisms—cytoplasmic segregation and induction—seem to be especially important.

### *Cytoplasmic Segregation*

Differentiation must begin long before we can recognize specialized types of cells. Ectodermal, endodermal, and mesodermal cells in the gastrula look quite similar, but they must be different because they develop into different organs. The egg is now known to contain substances called **maternal determinants,** which influence the course of development. Cytoplasmic segregation is the parceling out of maternal determinants as mitosis occurs:

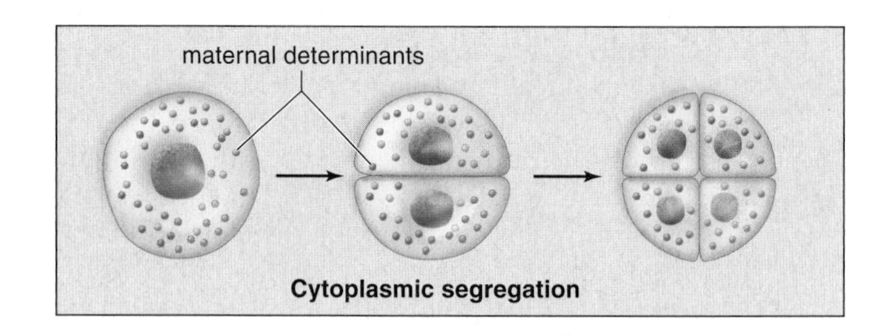


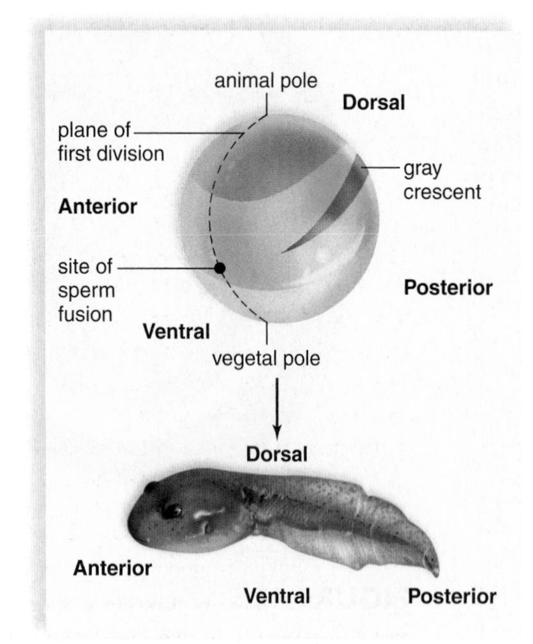


a. Zygote of a frog is polar and has axes.

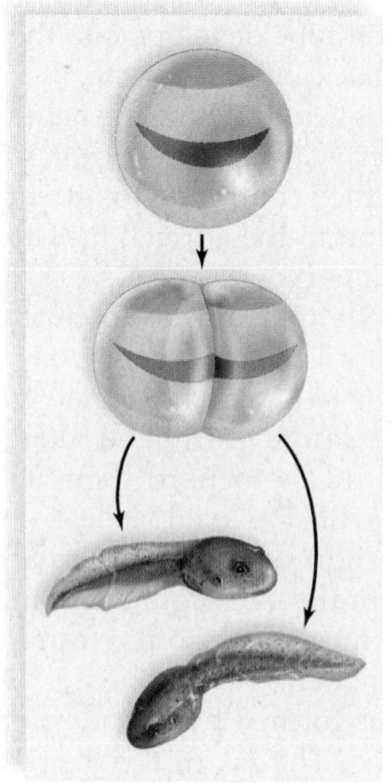

b. Each cell receives a part of the gray crescent.

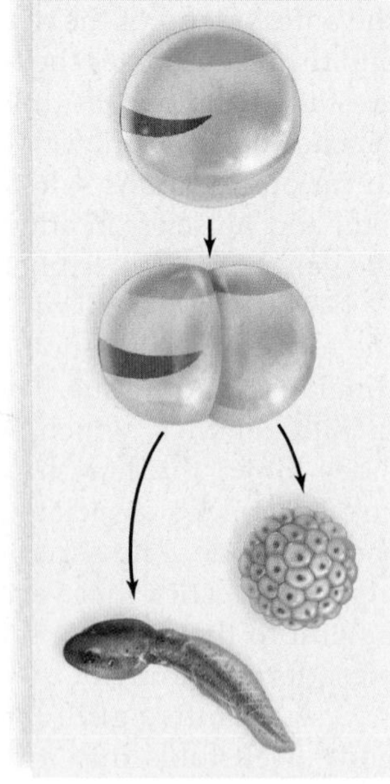

c. Only the cell on the left receives the gray crescent.

**FIGURE 42.6**
**Cytoplasmic influence on development.**
**a.** The zygote of a frog has anterior/posterior and dorsal/ventral axes that correlate with the position of the gray crescent. **b.** The first cleavage normally divides the gray crescent in half, and each daughter cell is capable of developing into a complete tadpole. **c.** But if only one daughter cell receives the gray crescent, then only that cell can become a complete embryo. This shows that maternal determinants are present in the cytoplasm of a frog's egg.

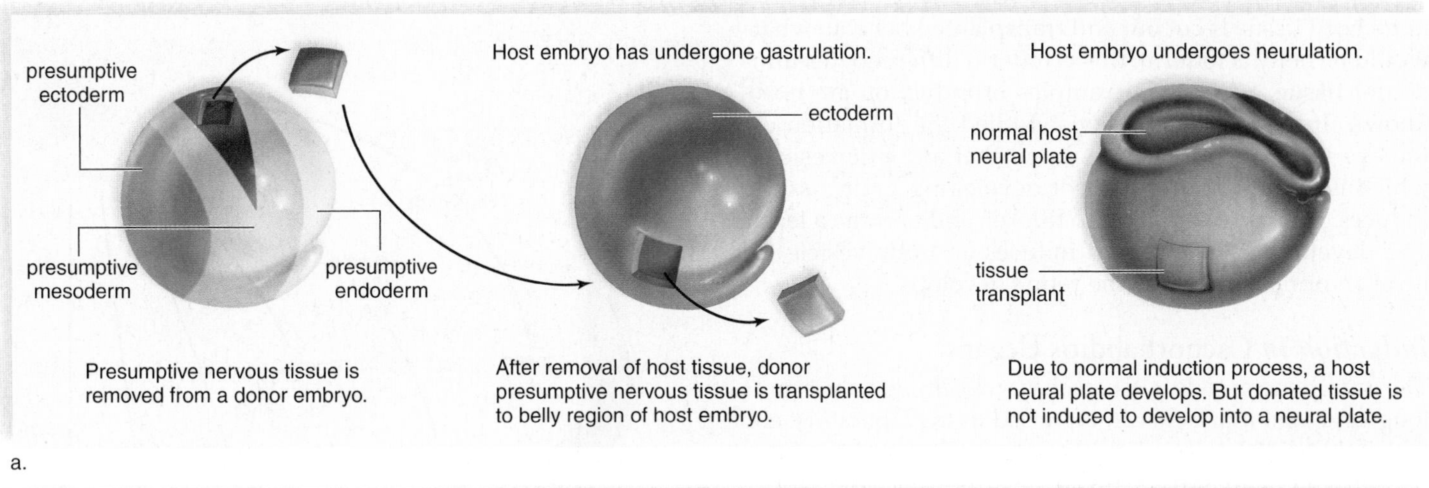


a.

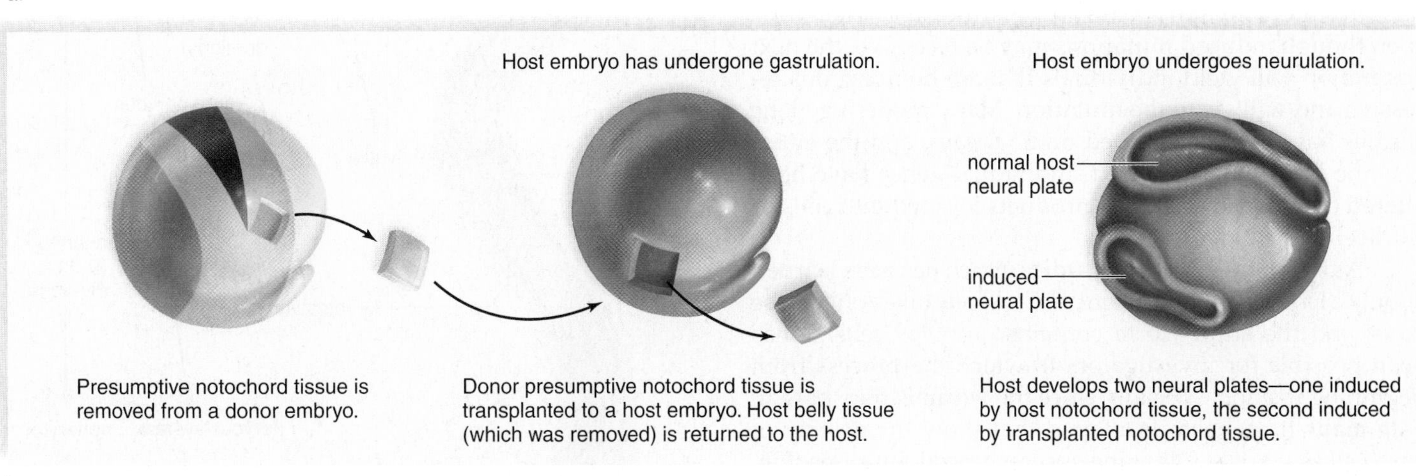


b.

**FIGURE 42.7 Control of nervous system development.**

**a.** In this experiment, the presumptive nervous system (blue) does not develop into the neural plate if moved from its normal location. **b.** In this experiment, the presumptive notochord (pink) can cause even belly ectoderm to develop into the neural plate (blue). This shows that the notochord induces ectoderm to become a neural plate, most likely by sending out chemical signals.

Cytoplasmic segregation helps determine how the various cells of the morula will develop.

An early experiment showed that the cytoplasm of a frog's egg is not uniform. It is polar and has both an anterior/posterior axis and a dorsal/ventral axis, which can be correlated with the **gray crescent,** a gray area that appears after the sperm fertilizes the egg (Fig. 42.6*a*). Hans Spemann, who received a Nobel Prize in 1935 for his extensive work in embryology, showed that if the gray crescent is divided equally by the first cleavage, each experimentally separated daughter cell develops into a complete embryo (Fig. 42.6*b*). However, if the zygote divides so that only one daughter cell receives the gray crescent, only that cell becomes a complete embryo (Fig. 42.6*c*). This experiment allows us to speculate that the gray crescent must contain particular chemical signals that are needed for development to proceed normally.

## *Induction and Frog Experiments*

As development proceeds, specialization of cells and formation of organs are influenced not only by maternal determinants but also by signals given off by neighboring cells. **Induction** [L. *in,* into, and *duco,* lead] is the ability of one embryonic tissue to influence the development of another tissue.

Spemann showed that a frog embryo's gray crescent becomes the dorsal lip of the blastopore, where gastrulation begins. Since this region is necessary for complete development, he called the dorsal lip of the blastopore the *primary organizer.* The cells closest to Spemann's primary organizer become endoderm, those farther away become mesoderm, and those farthest away become ectoderm. This suggests that there may be a molecular concentration gradient that acts as a chemical signal to induce germ layer differentiation.

The gray crescent in the zygote of a frog marks the dorsal side of the embryo where the mesoderm becomes notochord and ectoderm becomes nervous system. In a classic experiment Spemann and his colleague Hilde Mangold showed that presumptive (potential) notochord tissue induces the formation of the nervous system (Fig. 42.7). If presumptive nervous system tissue, located just above the presumptive notochord, is cut out and transplanted to the belly region of the embryo, it does not form a neural tube. On the other hand, if presumptive notochord tissue is cut out and transplanted beneath what would be belly ectoderm, this ectoderm differentiates into neural tissue. Still other examples of induction are now known. In 1905, Warren Lewis studied the formation of the eye in frog embryos. He found that an optic vesicle, which is a lateral outgrowth of developing brain tissue, induces overlying ectoderm to thicken and become a lens. The developing lens in turn induces an optic vesicle to form an optic cup, where the retina develops.

### *Induction in* Caenorhabditis elegans

The minute nematode, *Caenorhabditis elegans,* is only 1 mm long, and vast numbers can be raised in the laboratory either in petri dishes or a liquid medium. The worm is hermaphroditic, and self-fertilization is the rule. Therefore, even though induced mutations may be recessive, the next generation will yield individuals that are homozygous recessive and will show the mutation. Many modern genetic studies have been performed on *C. elegans,* and the entire genome has been sequenced. Individual genes have been altered and cloned and their products injected into cells or extracellular fluid.

As the result of genetic studies, much has been learned about *C. elegans.* Development of *C. elegans* takes only three days, and the adult worm contains only 959 cells. It has been possible for investigators to watch the process from beginning to end, especially since the worm is transparent. **Fate maps** have been developed that show the destiny of each cell as it arises following successive cell divisions (Fig. 42.8). Some investigators have studied in detail the development of the vulva, a pore through which eggs are laid. A cell called the anchor cell induces the vulva to form. The cell closest to the anchor cell receives the most inducer and becomes the inner vulva. This cell in turn produces another inducer, which acts on its two neighboring cells, and they become the outer vulva. The inducers are growthlike factors that alter the metabolism of the receiving cell and activate particular genes. Work with *C. elegans* has shown that induction requires the transcriptional regulation of genes in a particular sequence. This diagram shows how induction can occur sequentially:

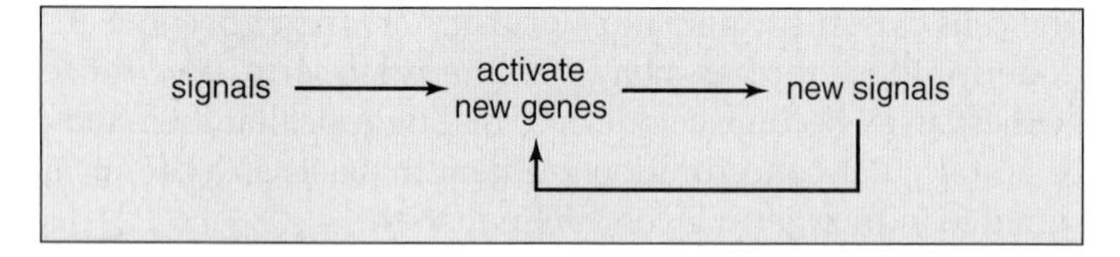


## Morphogenesis

An animal achieves its ordered and complex body form through morphogenesis, which requires that cells associate to form tissues, and tissues give rise to organs. Pattern formation is the process that enables morphogenesis. In pattern formation, cells of the embryo divide and differentiate, taking up orderly positions in tissues and organs. Although animals display an amazingly diverse array of morphologies, or body forms, most share common sets of genes that direct

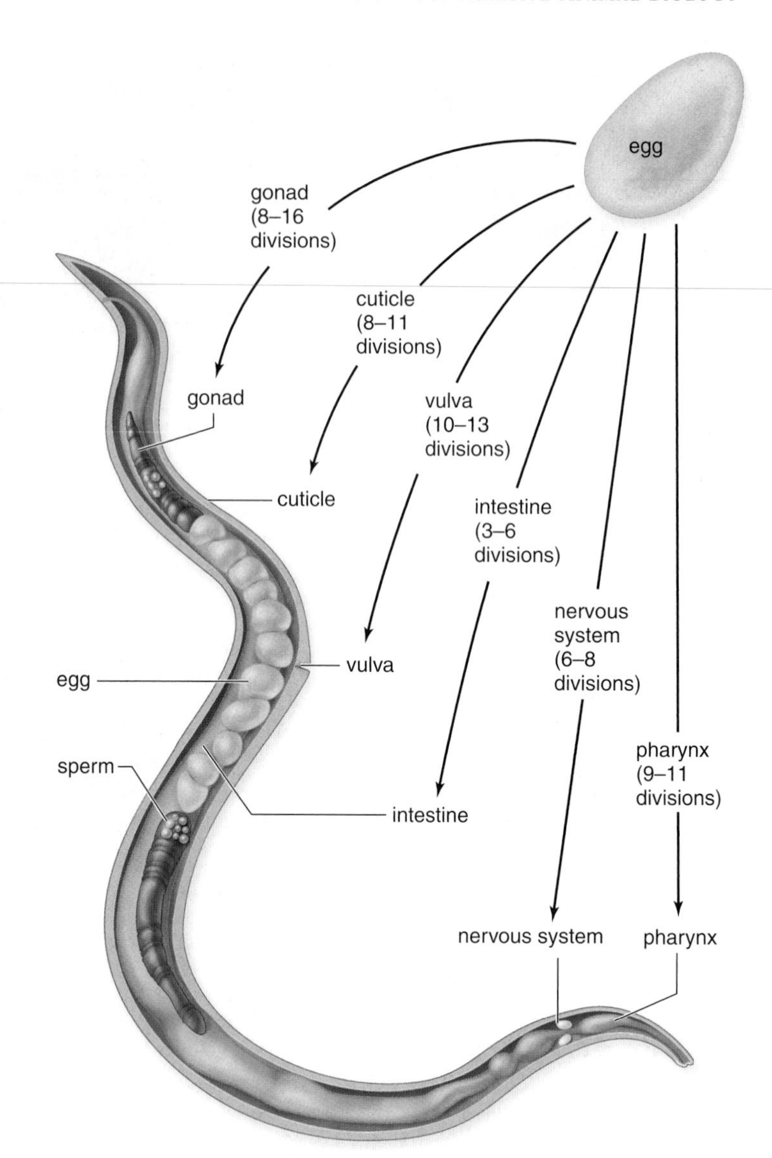


**FIGURE 42.8 Development of *C. elegans*, a nematode.**

A fate map of the worm showing that as cells arise by cell division, they are destined to become particular structures.

pattern formation. When pattern formation has ensured that key cells are properly arranged, then morphogenesis, the construction of the ultimate body form, can take place.

### *Morphogenesis in* Drosophila melanogaster

Much of the study of pattern formation and morphogenesis has been done using relatively simple models such as the minuscule fruit fly *Drosophila melanogaster.* A *Drosophila* egg is only about 0.5 mm long, and develops into an adult in about two weeks. The fly's genome is considerably smaller than that of humans or even mice. However, the genes that direct pattern formation appear to be highly conserved among animals with segmented body morphologies. Thus, *Drosophila* research increases our understanding of human morphogenetic processes.

Perhaps the most striking aspect of pattern formation in any animal is that all the cells in the embryo, beginning with the zygote, contain the same genetic information. The differences that arise are due to differences in gene expression, not DNA content. Various genes become activated in different regions of the embryo. These differences are instigated as embryonic cells communicate with one another through the products of gene expression. As pattern formation occurs in *Drosophila,* the embryonic cells begin to express genes differently in graded, periodic, and eventually striped arrangements. Boundaries between large body regions are established first, before the refinement of smaller, subdivided parts can take place.

**The Anterior/Posterior Axes.** The first step in *Drosophila* pattern formation and morphogenesis begins with the establishment of anteroposterior polarity, meaning that the anterior (head) and posterior (abdomen) ends are different from one another. Such polarity is present in the egg before it is fertilized by a sperm. Egg polarity results from maternal determinants, mRNAs that are deposited in specific positions within the egg while it is still in the ovary. The protein products of these genes diffuse away from the areas of their highest concentration in the embryo, forming gradients.

Two of the most important maternal determinants are *bicoid* and *nanos.* Proteins such as those that result when *bicoid* and *nanos* mRNAs are translated are known as **morphogens** due to their crucial influence in morphogenesis. Bicoid protein is most concentrated anteriorly, where it prevents the formation of the posterior region. A mutation in the *bicoid* gene results in an embryo with two posterior ends. (*Bicoid* means "two-tailed.") Nanos protein is most concentrated posteriorly, and is required for abdomen formation. A mutation in the *nanos* gene results in an embryo without an abdomen.

**The Segmentation Pattern.** Once the anterior and posterior ends of the embryo have been established by the expression of maternal determinants, a group of zygotic genes called *gap* genes come into play. The task of gap genes is to divide the anteroposterior axis into broad regions (Fig. 42.9*a*). They are so-named because mutations in gap genes result in gaps in the embryo, where large blocks of segments are missing. Gap genes are temporarily activated by the gradients of anterior and posterior morphogens, and in turn activate the *pair-rule* genes.

The pair-rule genes are expressed periodically, in alternating stripes (Fig. 42.9*b*). They serve to "rough out" a preliminary segmentation pattern along the anteroposterior axis. The products of pair-rule genes may stimulate or suppress the expression of other genes, particularly the *segment-polarity* genes. These genes ensure that each segment has boundaries, with distinct anterior and posterior halves. The segment-polarity genes are also expressed in a striped fashion, but with twice as many stripes as the pair-rule genes (Fig. 42.9*c*). Mutations in segment-polarity genes result in the loss of one part of each segment, and the duplication of another portion of the same segment.

**Homeotic Genes.** The **homeotic genes** are often referred to as *selector genes* because they select for segmental identity—in other words, they dictate which body parts arise from the segments. Mutations in homeotic genes may result in the development of body parts in inappropriate areas, such as legs instead of antennae, or wings instead of tiny balancing organs called halteres (Fig. 42.10*a*). Such alterations in morphology are known as *homeotic transformations.*

Interestingly, homeotic genes in *Drosophila* and other organisms have all been found to share a structural feature called a **homeobox.** (*Hox,* the term used for mammalian homeotic genes, is a shortened form of homeobox.) A

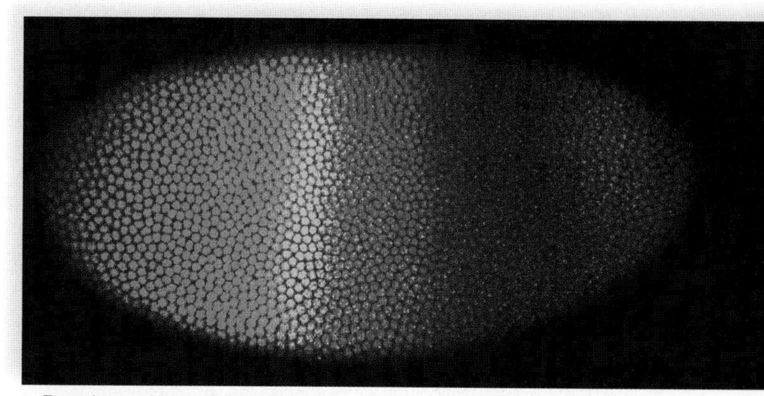

a. Protein products of gap genes

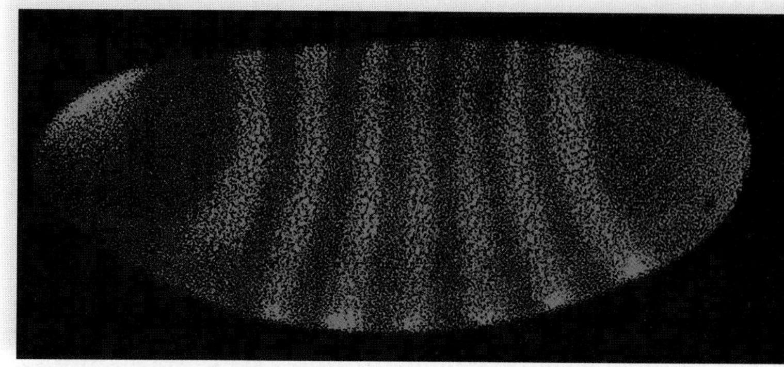

b. Protein products of pair-rule genes

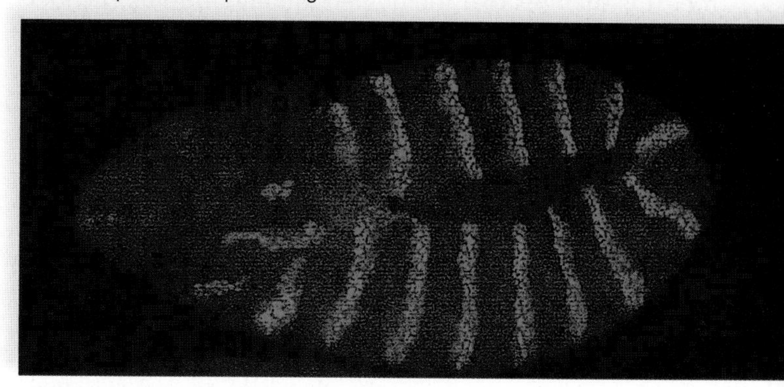

c. Protein products of segment-polarity genes

**FIGURE 42.9 Development in *Drosophila,* a fruit fly.**
**a.** The different colors show that two different gap gene proteins are present from the anterior to the posterior end of an embryo. **b.** The green stripes show that a pair-rule gene is being expressed as segmentation of the fly occurs. **c.** Now segment-polarity genes help bring about division of each segment into an anterior and posterior end.

homeobox is a sequence of nucleotides that encodes a 60 amino acid sequence called a **homeodomain:**

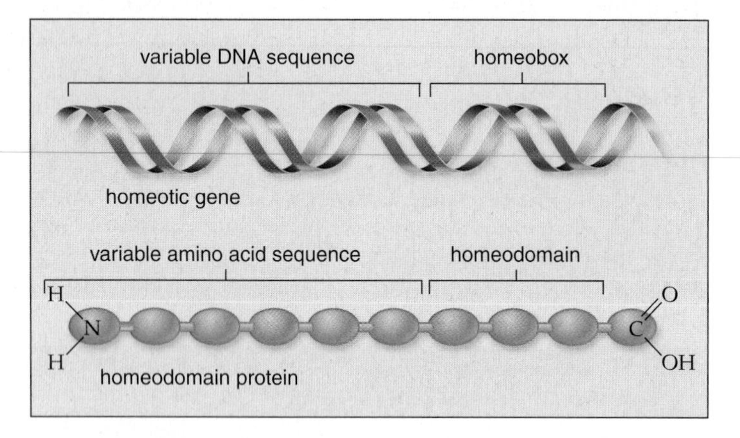


The homeodomain is a functionally important part of the protein encoded by a homeotic gene. Homeotic genes code for transcription factors, proteins that bind to regulatory regions of DNA and determine whether or not specific target genes are turned on. The homeodomain is the DNA-binding portion of the transcription factor, although the other, more variable sequences of a transcription factor determine which target genes are turned on. It is thought that the homeodomain proteins encoded by homeotic genes direct the activities of various target genes involved in morphogenesis, such as those involved in cell-to-cell adhesion. This orderly process in the end determines the morphology of particular segments.

The importance of homeotic genes is underscored by the finding that the homeotic genes are highly conserved, being present in the genomes of many organisms, including mammals such as mice and even humans. Most homeotic genes have their loci on the same chromosome in *Drosophila,* while in mammals there are four clusters that reside on different chromosomes. Notice that, in both flies and mammals, the position of the homeotic genes on the chromosome matches their anterior-to-posterior expression pattern in the body (Fig. 42.10*b*). The first gene clusters determine the final development of anterior segments, while those later in the sequence determine the final development of posterior segments of the animal's body.

Mutations in homeotic genes have similar effects in the mammalian body to the homeotic transformations observed in *Drosophila.* For instance, mutations in two adjacent *Hox* genes in the mouse result in shortened forelimbs that are missing the radius and ulna bones. In humans, mutations in a different *Hox* gene cause synpolydactyly, a rare condition in which there are extra digits (fingers and toes), some of which are fused to their neighbors.

**Apoptosis.** We have already discussed the importance of apoptosis (programmed cell death) in the normal day-to-day operation of the immune system and in preventing the occurrence of cancer. Apoptosis is also an important part of morphogenesis. During development of humans, we know that apoptosis is necessary to the shaping of the hands and feet; if it does not occur, the child is born with webbing between its fingers and toes.

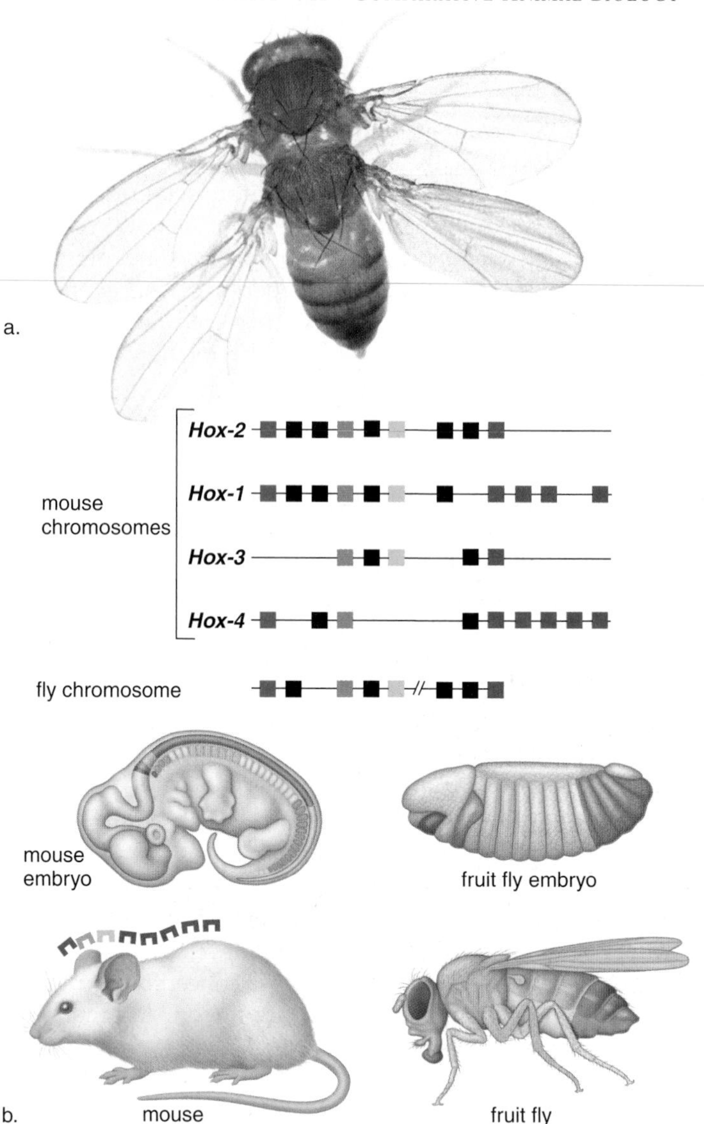


**FIGURE 42.10 Pattern formation in *Drosophila*.**

Homeotic genes control pattern formation, an aspect of morphogenesis. **a.** If homeotic genes are activated at inappropriate times, abnormalities such as a fly with four wings occur. **b.** The green, blue, yellow, and red colors show that homologous homeotic genes occur on four mouse chromosomes and on a fly chromosome in the same order. These genes are color coded to the region of the embryo, and therefore the adult, where they regulate pattern formation. The black boxes are homeotic genes that are not identical between the two animals. In mammals, homeotic genes are called *Hox* genes.

The fate maps of *C. elegans* (see Fig. 42.8) indicate that apoptosis occurs in 131 cells as development takes place. When a cell-death signal is received, an inhibiting protein becomes inactive, allowing a cell-death cascade to proceed that ends in enzymes destroying the cell.

## Check Your Progress 42.2

1. Cellular differentiation occurs as the result of two mechanisms. Name and define both of them.
2. Define the term "morphogen."
3. What is the function of the homeobox sequence in a homeotic gene?

# 42.3 Human Embryonic and Fetal Development

In humans, the length of time from conception (fertilization followed by **implantation**) to birth (parturition) is approximately nine months (266 days). It is customary to calculate the time of birth by adding 280 days to the start of the last menstruation, because this date is usually known, whereas the day of fertilization is usually unknown. Because the time of birth is influenced by so many variables, only about 5% of babies actually arrive on the forecasted date.

In humans, pregnancy, or gestation, is the time in which the developing embryo is carried by the mother. Human development is often divided into embryonic development (months 1 and 2) and fetal development (months 3–9). During the **embryonic period,** the major organs are formed, and during fetal development, these structures are refined.

Development can also be divided into trimesters. Each trimester can be characterized by specific developmental accomplishments. During the first trimester, embryonic and early fetal development occur. The second trimester is characterized by the development of organs and organ systems. By the end of the second trimester, the fetus appears distinctly human. In the third trimester, the fetus grows rapidly and the major organ systems become functional. An infant born one or perhaps two months premature has a reasonable chance of survival.

## *Extraembryonic Membranes*

Before we consider human development chronologically, we must understand the placement of **extraembryonic membranes** [L. *extra,* on the outside]. Extraembryonic membranes are best understood by considering their function in reptiles and birds. In reptiles, these membranes made development on land first possible. If an embryo develops in the water, the water supplies oxygen for the embryo and takes away waste products. The surrounding water prevents desiccation, or drying out, and provides a protective cushion. For an embryo that develops on land, all these functions are performed by the extraembryonic membranes.

In the chick, the extraembryonic membranes develop from extensions of the germ layers, which spread out over the yolk. Figure 42.11 shows the chick surrounded by the membranes. The **chorion** [Gk. *chorion,* membrane] lies next to the shell and carries on gas exchange. The **amnion** [Gk. *amnion,* membrane around fetus] contains the protective amniotic fluid, which bathes the developing embryo. The **allantois** [Gk. *allantos,* sausage] collects nitrogenous wastes, and the **yolk sac** surrounds the remaining yolk, which provides nourishment.

The function of the extraembryonic membranes in humans has been modified to suit internal development. Their presence, however, shows that we are related to the reptiles.

The chorion develops into the fetal half of the placenta, the organ that provides the embryo/fetus with nourishment and oxygen and takes away its waste. Blood vessels within the chorionic villi are continuous with the umbilical blood vessels. The blood vessels of the allantois become the umbilical blood vessels and the allantois accumulates the small amount of urine produced by the fetal kidneys and later gives rise to the urinary bladder. The yolk sac, which lacks yolk, is the first site of blood cell formation. The amnion contains fluid to cushion and protect the embryo, which develops into a fetus. It is interesting to note that all chordate animals develop in water—either in bodies of water or surrounded by amniotic fluid within a shell or uterus.

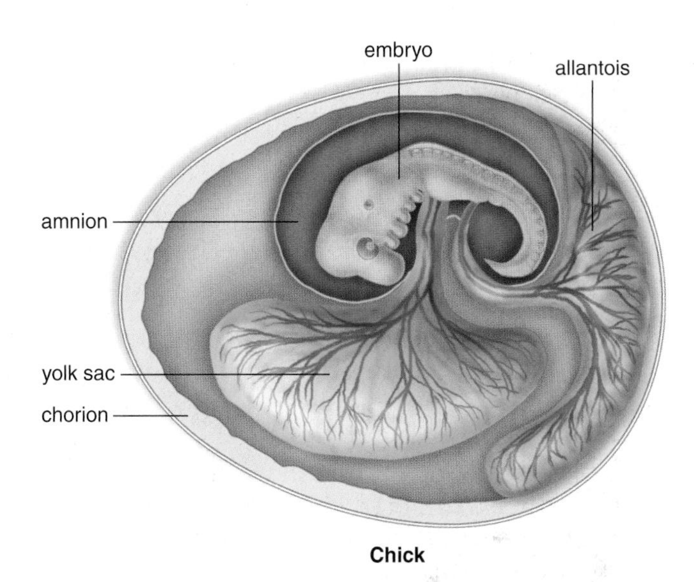


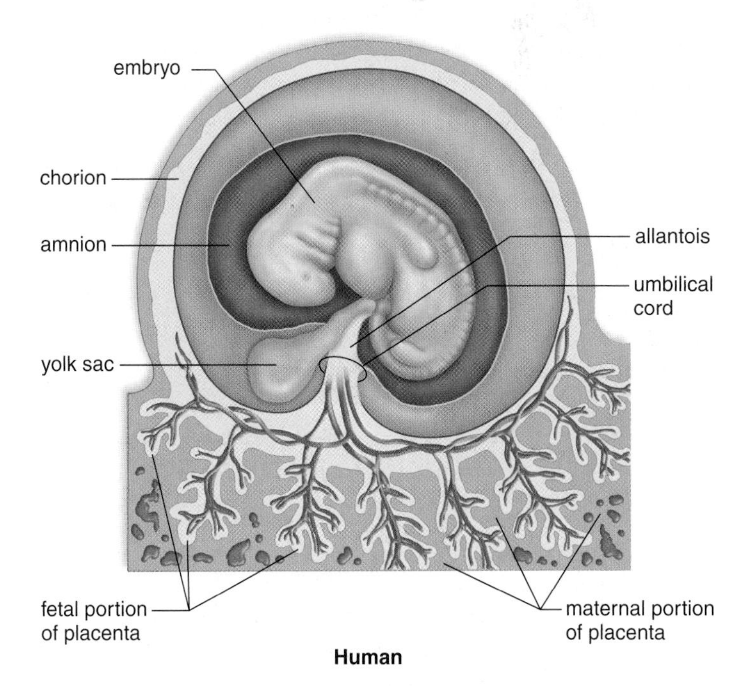


**FIGURE 42.11 Extraembryonic membranes.**

Extraembryonic membranes, which are not part of the embryo, are found during the development of chicks and humans. Each has a specific function.

## Embryonic Development

Embryonic development includes the first two months of development.

### *The First Week*

Fertilization occurs in the upper third of the oviduct (Fig. 42.12). Cleavage begins 30 hours after fertilization and continues as the embryo passes through the oviduct to the uterus. By the time the embryo reaches the uterus on the third day, it is a morula. The morula is not much larger than the zygote because, even though multiple cell divisions have occurred, there has been no growth of these newly formed cells. By about the fifth day, the morula is transformed into the blastocyst. The **blastocyst** has a fluid-filled cavity, a single layer of outer cells called the **trophoblast** [Gk. *trophe,* food, and *blastos,* bud], and an inner cell mass. The early function of the trophoblast is to provide nourishment for the embryo. Later, the trophoblast, reinforced by a layer of mesoderm, gives rise to the chorion, one of the extraembryonic membranes (see Fig. 42.11). The inner cell mass eventually becomes the embryo, which develops into a fetus.

### *The Second Week*

At the end of the first week, the embryo begins the process of implanting in the wall of the uterus. The trophoblast secretes enzymes to digest away some of the tissue

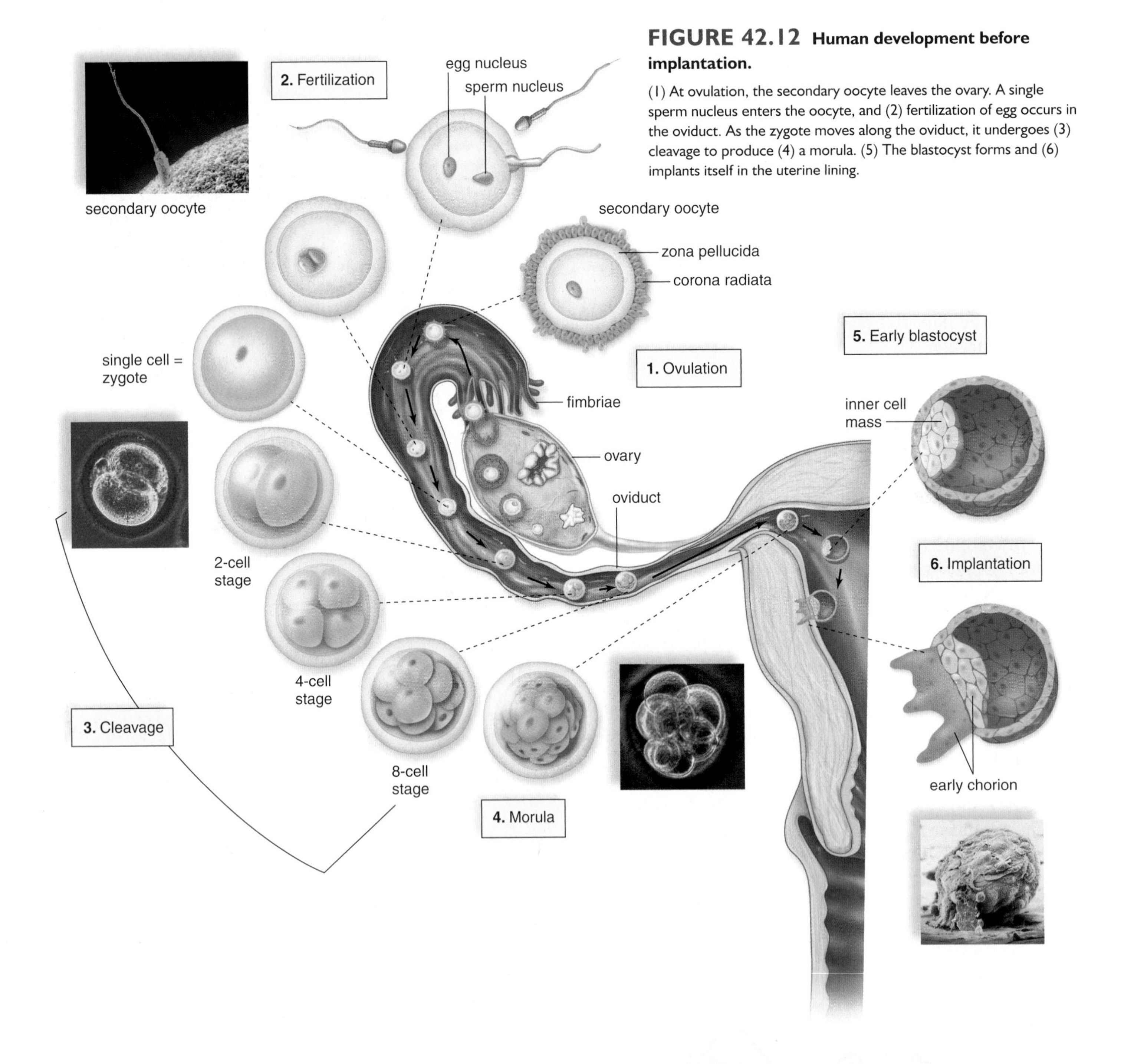


**FIGURE 42.12 Human development before implantation.**

(1) At ovulation, the secondary oocyte leaves the ovary. A single sperm nucleus enters the oocyte, and (2) fertilization of egg occurs in the oviduct. As the zygote moves along the oviduct, it undergoes (3) cleavage to produce (4) a morula. (5) The blastocyst forms and (6) implants itself in the uterine lining.

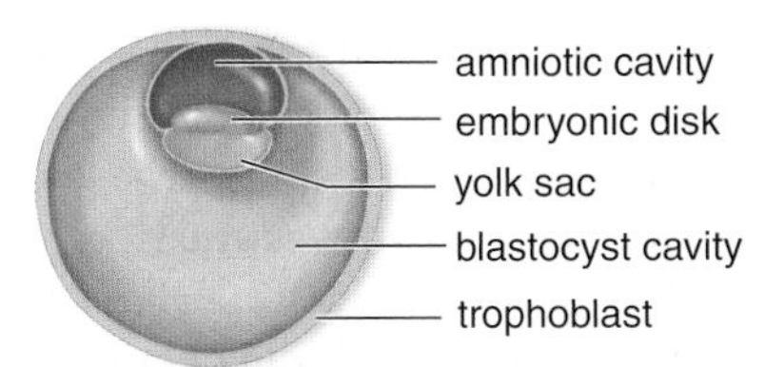


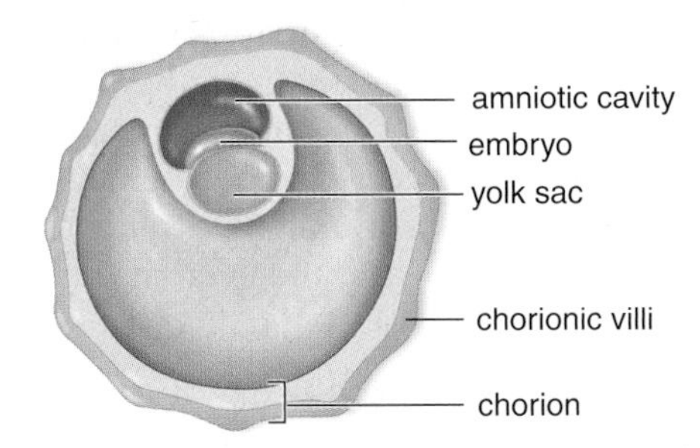


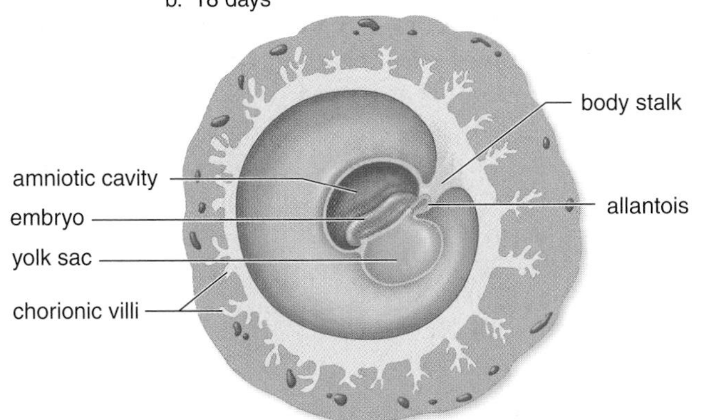


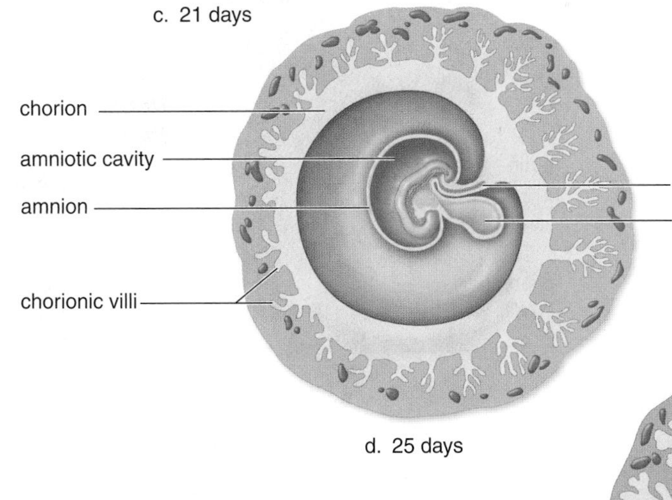


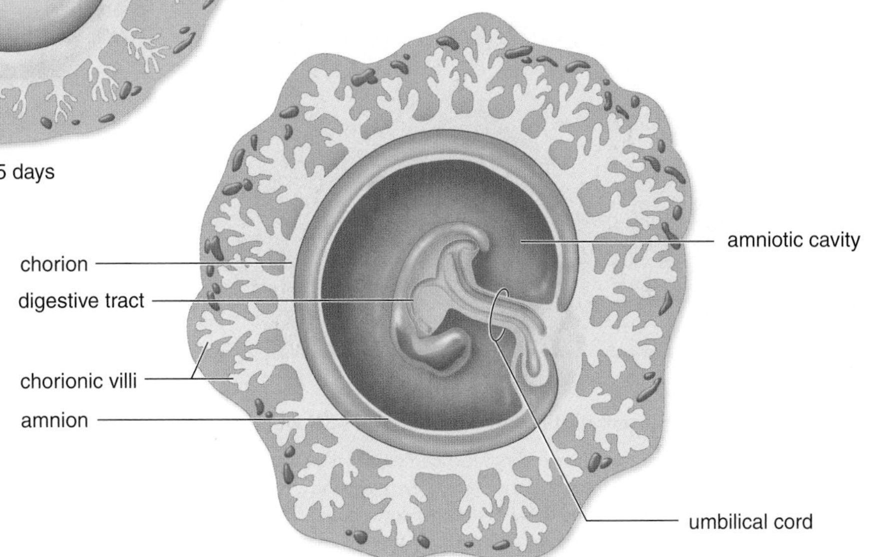


**FIGURE 42.13 Human embryonic development.**

**a.** At first, the embryo contains no organs, only tissues. The amniotic cavity is above the embryo, and the yolk sac is below. **b.** The chorion develops villi, the structures so important to the exchange between mother and child. **c., d.** The allantois and yolk sac, two more extraembryonic membranes, are positioned inside the body stalk as it becomes the umbilical cord. **e.** At 35+ days, the embryo has a head region and a tail region. The umbilical cord takes blood vessels between the embryo and the chorion (placenta).

and blood vessels of the endometrium of the uterus (Fig. 42.12). The embryo is now about the size of the period at the end of this sentence. The trophoblast begins to secrete **human chorionic gonadotropin (HCG),** the hormone that is the basis for the pregnancy test and that serves to maintain the corpus luteum past the time it normally disintegrates. (Recall that the corpus luteum is a yellow body formed in the ovary from a follicle that has discharged its secondary oocyte.) Because of this, the endometrium is maintained, and menstruation does not occur.

As the week progresses, the inner cell mass detaches itself from the trophoblast, and two more extraembryonic membranes form (Fig. 42.13*a*). The yolk sac, which forms below the embryonic disk, has no nutritive function as in chicks, but it is the first site of blood cell formation. However, the amnion and its cavity are where the embryo (and then the fetus) develops. In humans, amniotic fluid acts as an insulator against cold and heat and also absorbs shock, such as that caused by the mother exercising.

Gastrulation occurs during the second week. The inner cell mass now has flattened into the **embryonic disk,** composed of two layers of cells: ectoderm above and endoderm below. Once the embryonic disk elongates to form the primitive streak, the third germ layer, mesoderm, forms by invagination of cells along the streak. The trophoblast is reinforced by mesoderm and becomes the chorion (Fig. 42.13*b*). It is possible to relate the development of future organs to these germ layers (see Table 42.1).

### The Third Week

Two important organ systems make their appearance during the third week. The nervous system is the first organ system to be visually evident. At first, a thickening appears along the entire dorsal length of the embryo; then the neural folds appear. When the neural folds meet at the midline, the neural tube, which later develops into the brain and the nerve cord, is formed (see Fig. 42.4). After the notochord is replaced by the vertebral column, the nerve cord is called the spinal cord.

Development of the heart begins in the third week and continues into the fourth week. At first, there are right and left heart tubes; when these fuse, the heart begins pumping blood, even though the chambers of the heart are not fully formed. The veins enter posteriorly, and the arteries exit anteriorly from this largely tubular heart, but later the heart twists so that all major blood vessels are located anteriorly.

### The Fourth and Fifth Weeks

At four weeks, the embryo is barely larger than the height of this print. A bridge of mesoderm called the body stalk connects the caudal (tail) end of the embryo with the chorion, which has treelike projections called **chorionic villi** [Gk. *chorion,* membrane; L. *villus,* shaggy hair] (Fig. 42.13*c, d*). The chorionic villi eventually form the placental sinus. The fourth extraembryonic membrane, the allantois, is contained within this stalk, and its blood vessels become the umbilical blood vessels. The head and the tail then lift up, and the body stalk moves anteriorly by constriction. Once this process is complete, the **umbilical cord** [L. *umbilicus,* navel], which connects the developing embryo to the placenta, is fully formed (Fig. 42.13*e*).

Little flippers called limb buds appear (Fig. 42.14); later, the arms and the legs develop from the limb buds, and even the hands and the feet become apparent. At the same time—during the fifth week—the head enlarges, and the sense organs become more prominent. It is possible to make out the developing eyes, ears, and even the nose.

### The Sixth Through Eighth Weeks

During the sixth to eighth weeks of development, the embryo becomes easily recognizable as human. Concurrent with brain development, the head achieves its normal relationship with the body as a neck region develops. The nervous system is developed well enough to permit reflex actions, such as a startle response to touch. At the end of this period, the embryo is about 38 mm long and weighs no more than an aspirin tablet, even though all organ systems are established.

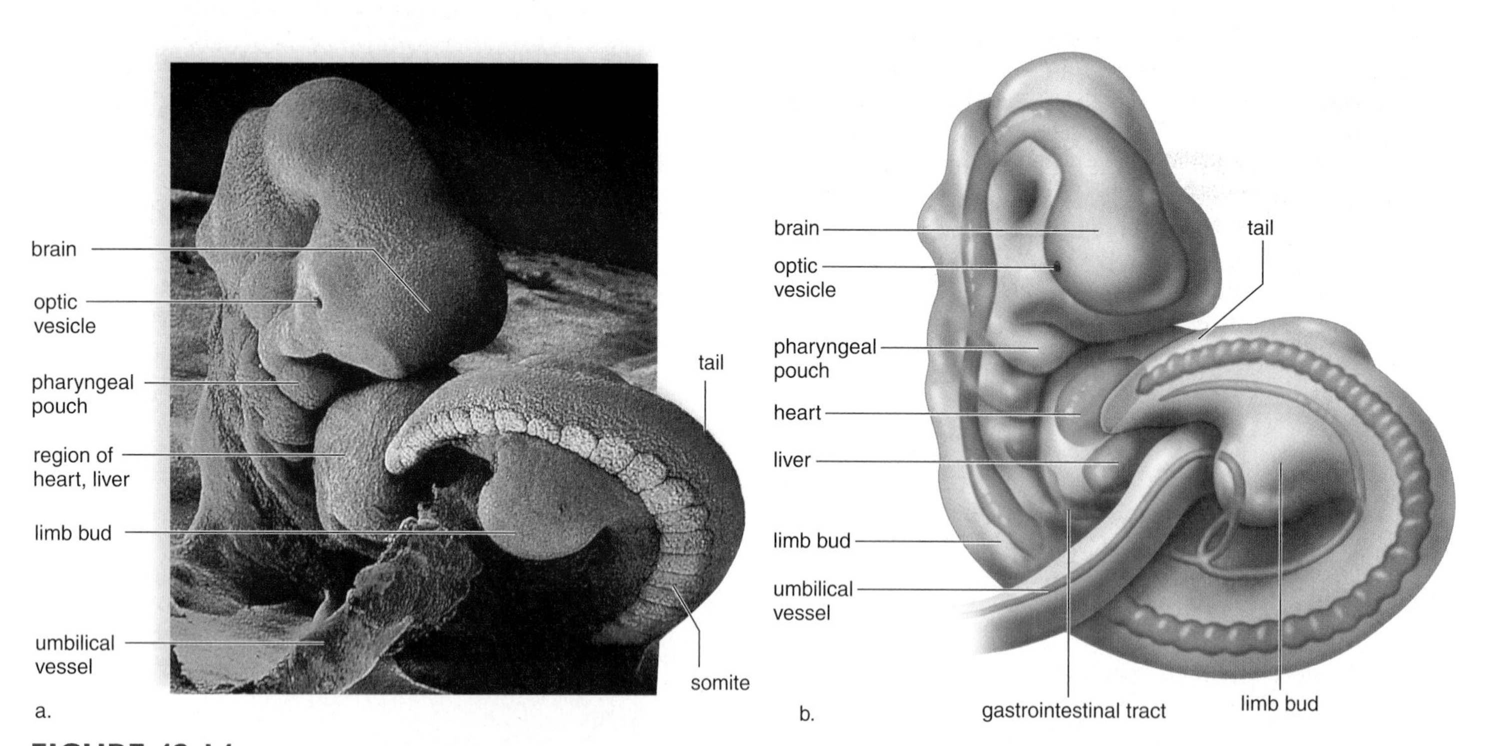


**FIGURE 42.14 Human embryo at beginning of fifth week.**

**a.** Scanning electron micrograph. **b.** The embryo is curled so that the head touches the heart, the two organs whose development is farther along than the rest of the body. The organs of the gastrointestinal tract are forming, and the arms and the legs develop from the bulges that are called limb buds. The tail is an evolutionary remnant; its bones regress and become those of the coccyx (tailbone). The pharyngeal arches become functioning gills only in fishes and amphibian larvae; in humans, the first pair of pharyngeal pouches becomes the auditory tubes. The second pair becomes the tonsils, while the third and fourth become the thymus gland and the parathyroid glands.

## The Structure and Function of the Placenta

The **placenta** is a mammalian structure that functions in gas, nutrient, and waste exchange between embryonic (later fetal) and maternal cardiovascular systems. The placenta begins formation once the embryo is fully implanted. At first, the entire chorion has chorionic villi that project into endometrium. Later, these disappear in all areas except where the placenta develops. By the tenth week, the placenta (Fig. 42.15) is fully formed and is producing progesterone and estrogen. These hormones have two effects: (1) due to their negative feedback control of the hypothalamus and the anterior pituitary, they prevent any new follicles from maturing, and (2) they maintain the lining of the uterus, so now the corpus luteum is not needed. No menstruation occurs during pregnancy.

The placenta has a fetal side contributed by the chorion and a maternal side consisting of uterine tissues. Notice in Figure 42.15 how the chorionic villi are surrounded by maternal blood; yet maternal and fetal blood do not mix under normal conditions because exchange always takes place across plasma membranes. Carbon dioxide and other wastes move from the fetal side to the maternal side of the placenta and nutrients and oxygen move from the maternal side to the fetal side. The umbilical cord stretches between the placenta and the fetus. Although it may seem that the umbilical cord travels from the placenta to the intestine, actually the umbilical cord is simply taking fetal blood to and from the placenta. The umbilical cord is the lifeline of the fetus because it contains the umbilical arteries and vein, which transport waste molecules (carbon dioxide and urea) to the placenta for disposal into the maternal blood and take oxygen and nutrient molecules from the placenta to the rest of the fetal circulatory system. If the placenta prematurely tears from the uterine wall, the lives of the fetus and mother are endangered.

Harmful chemicals can also cross the placenta as discussed in the Health Focus on pages 792–93. This is of particular concern during the embryonic period, when various structures are first forming. Each organ or part seems to have a sensitive period during which a substance can alter its normal development. For example, if a woman takes the drug thalidomide, a tranquilizer, between days 27 and 40 of her pregnancy, the infant is likely to be born with deformed limbs. After day 40, however, the infant is born with normal limbs.

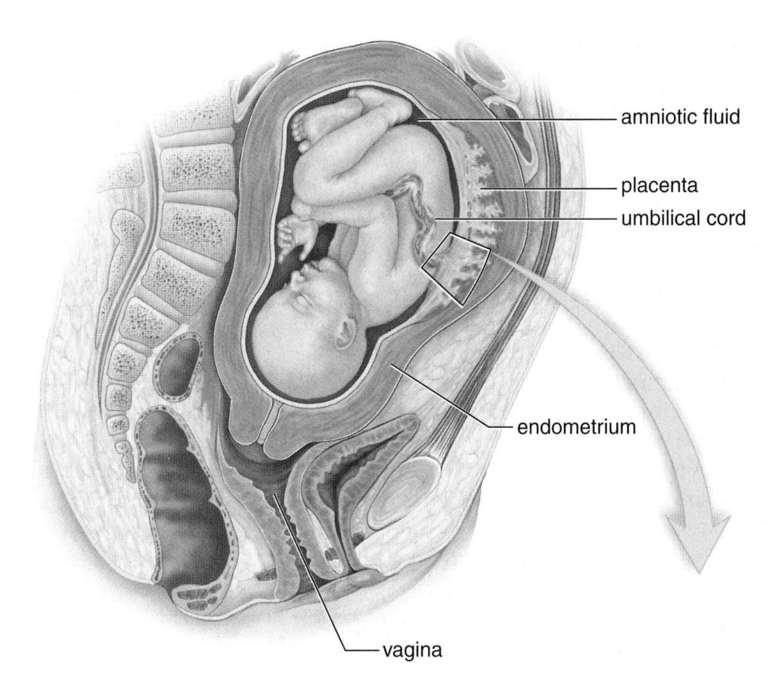


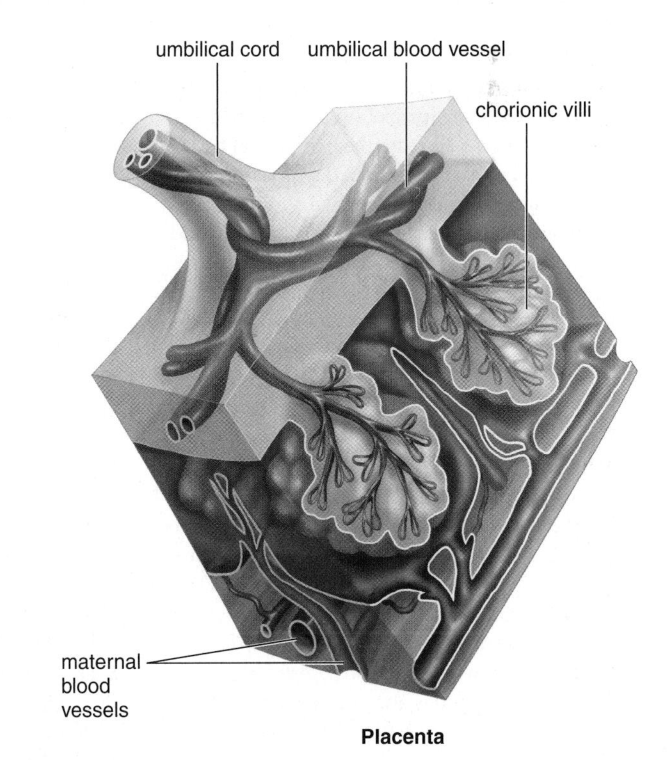


**FIGURE 42.15 Anatomy of the placenta in a fetus at six to seven months.**

The placenta is composed of both fetal and maternal tissues. Chorionic villi penetrate the uterine lining and are surrounded by maternal blood. Exchange of molecules between fetal and maternal blood takes place across the walls of the chorionic villi.

## Fetal Development and Birth

Fetal development (months 3–9) is marked by an extreme increase in size. Weight multiplies 600 times, going from less than 28 g to 3 kg. During this time, too, the fetus grows to about 50 cm in length. The genitalia appear in the third month, so it is possible to tell if the fetus is male or female.

Soon, hair, eyebrows, and eyelashes add finishing touches to the face and head. In the same way, fingernails and toenails complete the hands and feet. A fine, downy hair (lanugo) covers the limbs and trunk, only to later disappear. The fetus looks very old because the skin is growing so fast that it wrinkles. A waxy, almost cheese-like substance (vernix caseosa) [L. *vernix*, varnish, and

*health focus*

# Preventing and Testing for Birth Defects

It is believed that at least 1 in 16 newborns has a birth defect, either minor or serious, and the actual percentage may be even higher. Most likely, only 20% of all birth defects are due to heredity. Those that are hereditary can sometimes be detected before birth. Amniocentesis allows the fetus to be tested for abnormalities of development; chorionic villi sampling allows the embryo to be tested; and during preimplantation genetic diagnosis (see also Fig. 41A*b*), eggs (actually oocytes) are screened prior to in vitro fertilization (Fig. 42A). These guidelines can help prevent most other birth defects.

## Have Good Health Habits

Women who consume a nutritious diet and avoid potentially harmful substances, radiation, and pathogens, increase their chances of having healthy babies. For example, women of childbearing age are urged to make sure they consume adequate amounts of the vitamin folic acid in order to prevent neural tube defects such as spina bifida and anencephaly in their children. In spina bifida, part of the vertebral column fails to develop properly and cannot adequately protect the delicate spinal cord. In the most severe cases, fluid-filled meninges, or even part of the spinal cord itself, protrude from an opening in the back. Anencephaly may be even more devastating than spina bifida, as most of the fetal brain fails to develop. Anencephalic infants are stillborn or survive for only a few days after birth. Fortunately, it is easy to get enough folic acid to greatly reduce the chances of neural tube defects. Folic acid is present in multivitamins, plus many breads and cereals are fortified with it. Folic acid is also plentiful in leafy green vegetables, nuts, and citrus fruits. The CDC recommends that all women of childbearing age get at least 400 micrograms of folic acid every day through vitamin supplements and fortified foods, in addition to following a healthy, folic acid–rich diet. You should be aware that neural tube birth defects occur just a few weeks after conception, when many women are still unaware that they are pregnant—especially if the pregnancy is unplanned.

## Avoid Alcohol, Smoking, and Drugs of Abuse

Alcohol consumption during pregnancy is a leading cause of birth defects. In severe instances, the baby is born with fetal alcohol syndrome (FAS), estimated to occur in 0.2 to 1.5 of every 1,000 live births in the United States. Children with FAS are frequently underweight, have an abnormally small head, abnormal facial development, and mental retardation. Children with FAS often exhibit short attention span, impulsiveness, and poor judgment, as well as serious difficulties with learning and memory. Vision, hearing, and physical coordination may also be adversely affected. In addition to its directly damaging effects, alcohol reduces a woman's folic acid level, increasing the risk of the neural tube defects just described. Because no level of alcohol consumption has been deemed safe for an unborn child, women who are pregnant should avoid alcohol altogether.

Many preventable birth defects are caused by cigarette smoking. Babies born to smoking mothers typically have low birth weight and may continue to lag in growth. Smoking mothers are also at increased risk of premature membrane rupture, placental displacement or detachment from the fetus, and premature birth. Infants born to smoking mothers are more likely to have defects of the face, heart, and brain than those born to nonsmokers. Illegal drugs should be avoided. For example, cocaine causes blood pressure fluctuations that deprive the fetus of oxygen. Cocaine babies have visual problems, lack coordination, and are mentally retarded.

## Avoid Certain Medications and Supplements

Even prescription drugs may cause birth defects. In the late 1950s and early 1960s, thalidomide was commonly prescribed as a sleeping pill, and many pregnant women were given the drug to treat morning sickness. At the time, no one knew of thalidomide's effects on developing fetuses. As a result of thalidomide exposure during the period of limb bud development, thousands of children were born with flipperlike structures instead of limbs. Thalidomide was promptly taken off the market. Similarly, a drug called DES (diethylstilbestrol) was prescribed from 1938 to 1971 to pregnant women to prevent premature labor and miscarriage. Daughters of women who took DES while pregnant have since been found to have an increased risk of vaginal and cervical cancer, and also a higher incidence of structural abnormalities of the reproductive tract, complications of their own pregnancies, and infertility. The sons of women who took DES have an increased frequency of noncancerous cysts of the epididymis. Research is currently under way to assess the impact of this drug on the grandchildren of women who took DES while pregnant.

## Avoid Infections That Cause Birth Defects

Pathogens such as the virus that causes rubella (German measles) can cause birth defects. In the past, this virus caused many birth defects, specifically mental retardation, deafness, blindness, and heart defects. Viral-induced birth defects occur when the tiny viral particles cross the placenta and infect the fetus. Rubella is much less of a danger in developed countries today because most women were vaccinated against rubella as children. HIV, the virus that causes AIDS, is another virus that affects many newborns every year. Infants born with HIV often show signs of infection early in life and do not survive for more than a few years. If a pregnant woman is known to be positive for HIV, her physician may avoid amniocentesis and other procedures that could expose the fetus to maternal blood. The mother may also be given antiviral drugs to reduce the chances of fetal infection.

## Avoid Having X-rays

Penetrating forms of radiation such as X-rays can hinder cell division and damage DNA, a particular concern for the rapidly dividing and differentiating cells of a fetus. Damage from X-ray exposure may contribute to birth defects, and increase the risk of a child developing leukemia later on. Pregnant women should avoid unnecessary X-rays, especially those targeting the lower back or abdomen. If a medical X-ray is unavoidable, the woman should notify the X-ray technician so that her fetus can be protected as much as possible, for example, by draping a lead apron over her abdomen if this area is not targeted for X-ray examination.

Now that physicians and laypeople are aware of the various ways birth defects can be prevented, it is hoped that the incidence of birth defects will decrease in the future.

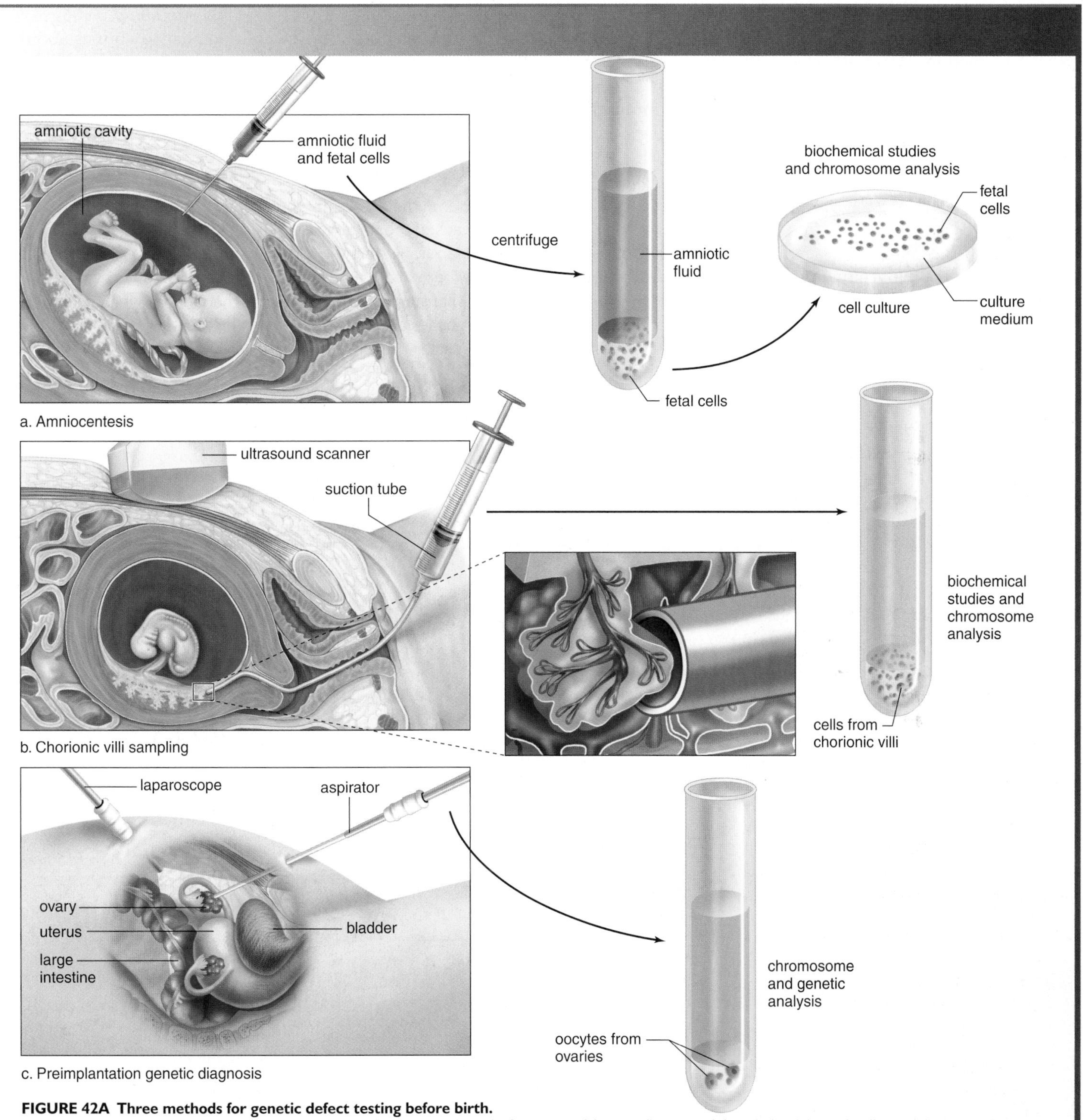


**FIGURE 42A Three methods for genetic defect testing before birth.**
***a.*** *Amniocentesis is usually performed from the fifteenth to the seventeenth week of pregnancy. A long needle is passed through the abdominal wall to withdraw a small amount of amniotic fluid, along with fetal cells. Since there are only a few cells in the amniotic fluid, testing may be delayed as long as four weeks until cell culture produces enough cells for testing purposes. About 40 tests are available for different defects.* ***b.*** *Chorionic villi sampling is usually performed from the eighth to the twelfth week of pregnancy. The doctor inserts a long, thin tube through the vagina into the uterus. With the help of ultrasound, which gives a picture of the uterine contents, the tube is placed between the lining of the uterus and the chorion. Then a sampling of the chorionic villi cells is obtained by suction. Chromosome analysis and biochemical tests for genetic defects can be done immediately on these cells.* ***c.*** *Preimplantation genetic diagnosis is performed prior to in vitro fertilization. Preovulatory oocytes are removed by aspiration after a laparoscope (optical telescope) is inserted into the abdominal cavity through a small incision in the region of the navel. The first polar body is tested. If the woman is heterozygous (Aa) and the defective gene (a) is found in the polar body, then the oocyte must have received the normal gene (A). Normal oocytes then undergo in vitro fertilization and are placed in the prepared uterus.*

*caseus*, cheese] protects the wrinkly skin from the watery amniotic fluid.

The fetus at first only flexes its limbs and nods its head, but later it can move its limbs vigorously to avoid discomfort. The mother feels these movements from about the fourth month on. The other systems of the body also begin to function. After 16 weeks, the fetal heartbeat is heard through a stethoscope. A fetus born at 24 weeks has a chance of surviving, although the lungs are still immature and often cannot capture oxygen adequately. Weight gain during the last couple of months increases the likelihood of survival.

### *The Stages of Birth*

The latest findings suggest that when the fetal brain is sufficiently mature, the hypothalamus causes the pituitary to stimulate the adrenal cortex so that androgens are released into the bloodstream. The placenta uses androgens as a precursor for estrogens, hormones that stimulate the production of prostaglandin (a molecule produced by many cells that acts as a local hormone) and oxytocin. All three of these molecules cause the uterus to contract and expel the fetus.

The process of birth (parturition) includes three stages. During the first stage, the cervix dilates to allow passage of the baby's head and body. The amnion usually bursts about this time. During the second stage, the baby is born and the umbilical cord is cut. During the third stage, the placenta is delivered (Fig. 42.16).

**Check Your Progress 42.3**

1. Name the extraembryonic membrane that gives rise to each of the following: umbilical blood vessels, the first blood cells, and the fetal half of the placenta.
2. Where in the human body does fertilization occur? At what stage of development is the embryo when it first reaches the uterus?
3. What is the function of the placenta?

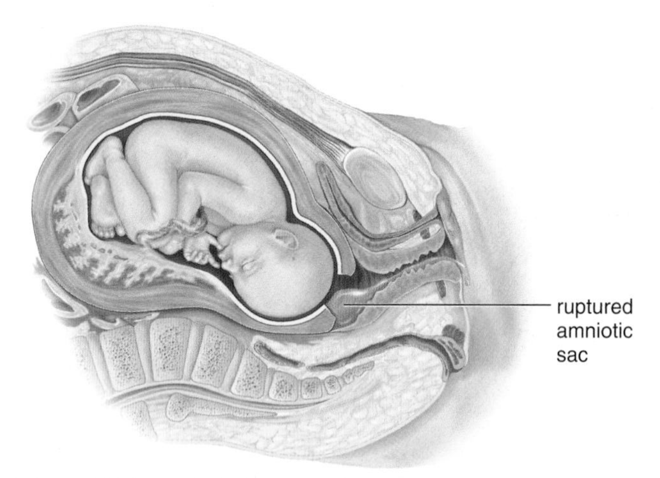


a. First stage of birth: cervix dilates

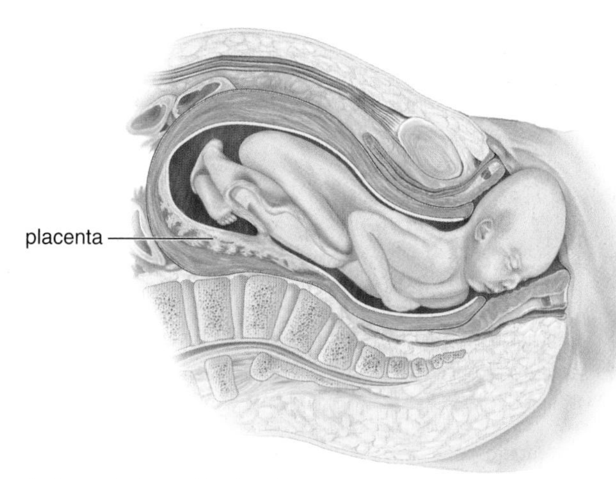


b. Second stage of birth: baby emerges

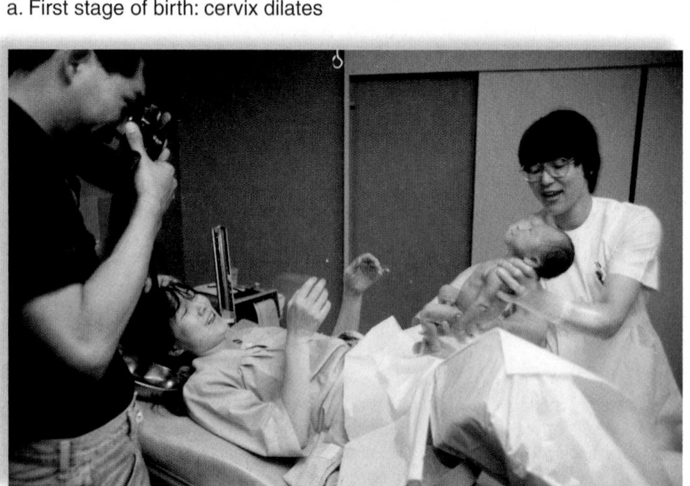

c. Baby has arrived

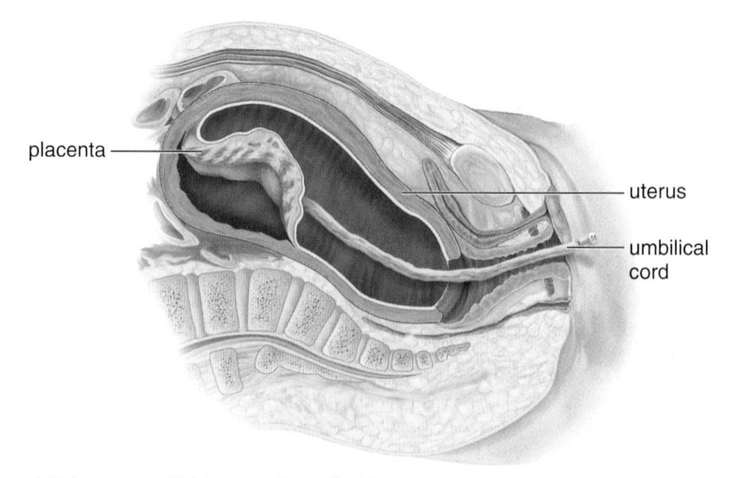


d. Third stage of birth: expelling afterbirth

**FIGURE 42.16 Three stages of parturition.**

**a.** Position of fetus just before birth begins. **b.** Dilation of cervix. **c.** Birth of baby. **d.** Expulsion of afterbirth.

# Connecting the Concepts

We have come full circle. We began our study of biology by considering the structure of the cell and its genetic machinery, including how the expression of genes is regulated. In this chapter, we have observed that animals go through the same early embryonic stages of morula, blastula, gastrula, and so forth. The set sequence of these stages is due to the expression of genes that bring about cellular changes. Therefore, once again, we are called upon to study organisms at the cellular level of organization.

We have seen that hormones are signals that affect cellular metabolism. Like gibberellins in plants, steroid hormones in animals turn on the expression of genes. When a steroid hormone binds to a specific hormone receptor, a gene is transcribed, and then translation produces the corresponding protein. This same type of signal transduction pathway occurs during development. Transduction means that the signal has been transformed into an event that has an effect on the organism.

A set sequence of signaling molecules is produced as development occurs. Each new signal in the sequence turns on a specific gene, or more likely, a sequence of genes. Gene expression cascades are common during development. Homeotic genes are arranged in sets on the chromosomes, and a protein product from one set of genes acts as a transcription factor to turn on another set, and so forth. During the development of flies, it is possible to observe that first one region of a chromosome and then another puffs out, indicating that genes are transcribed in sequence as development occurs.

Developmental biology is now making a significant contribution to the field of evolution. Homeotic genes with the same homeoboxes (sequence of about 180 base pairs) have been discovered in many different types of organisms. This suggests that homeotic genes arose early in the history of life, and mutations in these genes could possibly account for macroevolution—the appearance of new species or even higher taxons.

## summary

### 42.1 Early Developmental Stages

Development begins at fertilization. Only one sperm actually enters the oocyte. Both sperm and egg contribute chromosomes to the diploid zygote.

The early developmental stages in animals proceed from cellular stages to tissue stages to organ stages. During the cellular stage, cleavage (cell division) occurs, but there is no overall growth. The result is a morula, which becomes the blastula when an internal cavity (the blastocoel) appears.

During the tissue stage, gastrulation (invagination of cells into the blastocoel) results in formation of the germ layers: ectoderm, mesoderm, and endoderm. Both the cellular and tissue stages can be affected by the amount of yolk. The human blastula stage resembles that of the chick whose egg has a large amount of yolk.

Organ formation can be related to germ layers. For example, during neurulation, the nervous system develops from midline ectoderm, just above the notochord. At this point, it is possible to draw a typical cross section of a vertebrate embryo in which the notochord has not been replaced by the vertebral column (see Fig. 42.5).

### 42.2 Developmental Processes

Cellular differentiation begins with cytoplasmic segregation in the egg. After the first cleavage of a frog embryo, only a daughter cell that receives a portion of the gray crescent is able to develop into a complete embryo. Therefore, cytoplasmic segregation of maternal determinants occurs during early development of a frog. Induction is also part of cellular differentiation. For example, the notochord induces the formation of the neural tube in frog embryos. The reciprocal induction that occurs between the lens and the optic vesicle is another good example of induction. In *C. elegans*, investigators have shown that induction is an ongoing process in which one tissue after the other regulates the development of another, through chemical signals coded for by particular genes.

Some morphogen genes determine the axes of the body, and others regulate the development of segments. An important concept has emerged: During development, sequential sets of master genes code for morphogen gradients that activate the next set of master genes, in turn. Morphogens are transcription factors that bind to DNA.

Homeotic genes control pattern formation such as the presence of antennae, wings, and limbs on the segments of *Drosophila*. Homeotic genes code for proteins that contain a homeodomain, a particular sequence of 60 amino acids. These proteins are also transcription factors, and the homeodomain is the portion of the protein that binds to DNA. Homologous homeotic genes have been found in a wide variety of organisms, and therefore they must have arisen early in the history of life and been conserved.

### 42.3 Human Embryonic and Fetal Development

Human development can be divided into embryonic development (months 1 and 2) and fetal development (months 3–9). The early stages in human development resemble those of the chick. The similarities are probably due to their evolutionary relationship, not to the amount of yolk the eggs contain, because the human egg has little yolk.

The extraembryonic membranes appear early in human development. The trophoblast of the blastocyst is the first sign of the chorion, which goes on to become the fetal part of the placenta. Exchange occurs between fetal and maternal blood at the placenta. The amnion contains amniotic fluid, which cushions and protects the embryo. The yolk sac and allantois are also present.

Fertilization occurs in the oviduct, and cleavage occurs as the embryo moves toward the uterus. The morula becomes the blastocyst before implanting in the endometrium of the uterus. Organ development begins with neural tube and heart formation. There follows a steady progression of organ formation during embryonic development. During fetal development, refinement of organ systems occurs, and the fetus adds weight.

## understanding the terms

allantois 787
amnion 787
apoptosis 782
blastocoel 779
blastocyst 788
blastopore 780
blastula 779
cellular differentiation 782
chorion 787
chorionic villus 790
cleavage 779
ectoderm 780
embryo 779
embryonic disk 789
embryonic period 787
endoderm 780
extraembryonic membrane 787
fate map 784
fertilization 778

gastrula 780
gastrulation 780
germ layer 780
gray crescent 783
homeobox 785
homeodomain 786
homeotic genes 785
human chorionic gonadotropin (HCG) 789
implantation 787
induction 783
maternal determinant 782
mesoderm 780
morphogen 785
morphogenesis 782
morula 779
neural plate 781
neural tube 781
neurula 781
notochord 781
pattern formation 782
placenta 791
totipotent 782
trophoblast 788
umbilical cord 790
yolk 779
yolk sac 787

Match the terms to these definitions:

a. _______________ Ability of a chemical or a tissue to influence the development of another tissue.
b. _______________ Primary tissue layer of a vertebrate embryo—namely, ectoderm, mesoderm, or endoderm.
c. _______________ Extraembryonic membrane of birds, reptiles, and mammals that forms an enclosing, fluid-filled sac.
d. _______________ A 180-nucleotide sequence located in nearly all homeotic genes.
e. _______________ Stage of early animal development during which the germ layers form, at least in part, by invagination.

## reviewing this chapter

1. Describe the events of fertilization that (a) allow the sperm to reach the plasma membrane of the secondary oocyte, (b) prevent polyspermy, and (c) result in a diploid zygote. 778–79
2. What happens during the cellular stages, the tissue stages, and the organ stages of early development in animals? 779–81
3. What are the germ layers, and which organs are derived from each of the germ layers? 780
4. Draw a cross section of a typical vertebrate embryo at the neurula stage, and label your drawing. 781
5. Describe two mechanisms that are known to be involved in the processes of cellular differentiation. 782–84
6. Describe an experiment performed by Spemann suggesting that the notochord induces formation of the neural tube. Give another well-known example of induction between tissues. 783–84
7. With regard to *C. elegans,* what is a fate map? How does induction occur? 784
8. With regard to *Drosophila,* what is a morphogen gradient, and what does such a gradient do to bring about morphogenesis? 784–85
9. What is the function of homeotic genes, and what is the significance of the homeobox within these genes? 785–86
10. List the human extraembryonic membranes, give a function for each, and compare their functions to those in the chick. 787
11. Tell where fertilization, cleavage, the morula stage, and the blastocyst stage occur in humans. What happens to the embryo in the uterus? 788–90
12. Describe the structure and the function of the placenta in humans. 791
13. List and describe the stages of birth. 794

## testing yourself

Choose the best answer for each question.

1. Which of these stages is the first one out of sequence?
   a. cleavage
   b. blastula
   c. morula
   d. gastrula
   e. neurula
2. Which of these stages is mismatched?
   a. cleavage—cell division
   b. blastula—gut formation
   c. gastrula—three germ layers
   d. neurula—nervous system
   e. Both b and c are mismatched.
3. A cell that is capable of giving rise to a complete organism is said to be
   a. totipotent.
   b. homeotic.
   c. ectodermal.
   d. differentiated.
   e. induced.
4. Morphogenesis is associated with
   a. protein gradients.
   b. induction.
   c. transcription factors.
   d. homeotic genes.
   e. All of these are correct.
5. In humans, the placenta develops from the chorion. This indicates that human development
   a. resembles that of the chick.
   b. is dependent on extraembryonic membranes.
   c. cannot be compared to that of lower animals.
   d. begins only upon implantation.
   e. Both a and b are correct.
6. In humans, the fetus
   a. is surrounded by four extraembryonic membranes.
   b. has developed organs and is recognizably human.
   c. is dependent on the placenta for excretion of wastes and acquisition of nutrients.
   d. is embedded in the endometrium of the uterus.
   e. Both b and c are correct.
7. Developmental changes
   a. require growth, differentiation, and morphogenesis.
   b. stop occurring when one is grown.
   c. are dependent on a parceling out of genes into daughter cells.
   d. are dependent on activation of master genes in an orderly sequence.
   e. Both a and d are correct.
8. Which of these pairs is mismatched?
   a. brain—ectoderm
   b. gut—endoderm
   c. bone—mesoderm
   d. lens—endoderm
   e. heart—mesoderm
9. Label this diagram illustrating the placement of the extraembryonic membranes, and give a function for each membrane in humans.

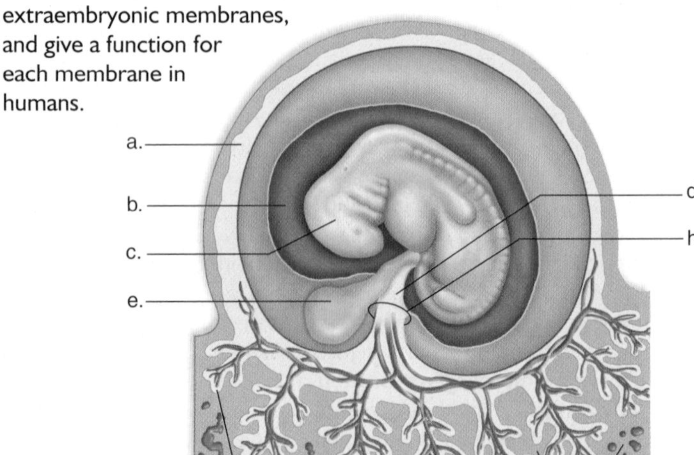

For questions 10–13, match the statement with the terms in the key.

**KEY:**

a. apoptosis
b. homeotic genes
c. fate maps
d. morphogen gradients
e. segment-polarity genes

10. have been developed that show the destiny of each cell that arises through cell division during the development of *C. elegans*
11. occurs when a cell-death cascade is activated
12. can have a range of effects, depending on its concentration in a particular portion of the embryo
13. genes that control pattern formation, the organization of differentiated cells into specific three-dimensional structures
14. The ability of one embryonic tissue to influence the growth and development of another tissue is termed
    a. morphogenesis.
    b. pattern formation.
    c. apoptosis.
    d. cellular differentiation.
    e. induction.
15. Which hormone is administered to begin the process of birth?
    a. estrogen
    b. oxytocin
    c. prolactin
    d. testosterone
    e. Both b and d are correct.
16. Only one sperm enters and fertilizes a human ovum because
    a. sperm have an acrosome.
    b. the corona radiata gets larger.
    c. the zona pellucida lifts up.
    d. the plasma membrane hardens.
    e. All of these are correct.
17. Which is a correct sequence in humans that ends with the stage that implants?
    a. morula, blastocyst, embryonic disk, gastrula
    b. ovulation, fertilization, cleavage, morula, early blastocyst
    c. embryonic disk, gastrula, primitive streak, neurula
    d. primitive streak, neurula, extraembryonic membranes, chorion
    e. cleavage, neurula, early blastocyst, morula

## thinking scientifically

1. A mutant gene is known to disrupt the earliest stages of development in sea urchins. Individuals with two copies of the mutant gene seem to develop normally. However, when normal males are crossed with mutant (normal-appearing) females, none of the offspring develop at all. How could this pattern of expression be explained?
2. *Drosophila* belongs to the insect order Diptera, in which one pair of wings is the norm. In *Drosophila, Ubx* mutation converts the halteres into an extra pair of wings located posterior to the normal pair (see Figure 42.10*a*). Butterflies belong to the order Lepidoptera in which two pairs of wings is normal. The ancestor of both flies and butterflies is thought to have possessed two pairs of wings. However, in butterflies, mutations in *Ubx* have been found to alter pigmentation, spot pattern, and scale morphology in the hindwings, but not the number of wing pairs. What does this tell us about the role of *Ubx* in insect wing development?

## bioethical issue

### Human Embryonic Stem Cell Research

Human embryonic stem cell lines (also called ES cell lines) are obtained by removing the inner cell mass from a blastocyst and growing the cells in a petri dish. The cells are valued by researchers because they are pluripotent, meaning they have the potential to differentiate into a wide range of different types of cells if properly stimulated. At this time, the federal government funds research on ES cell lines that were started before 2001, obtained from extra embryos left over from in vitro fertilization (IVF), and voluntarily donated for scientific study. However, funding may be obtained from certain state or private sources, and researchers overseas are not subject to U.S. regulations.

Advocates of ES cell research say that such cells could be used to develop cures for conditions such as Parkinson disease, diabetes, heart disease, Alzheimer disease, cystic fibrosis, and spinal cord injuries. In addition, ES cells could be studied to help scientists understand the basic processes of human development and used to test new drugs.

ES cell research opponents say that it should be restricted because it requires the destruction of human life. Many believe that embryos, even those so early in development, should be considered human beings. If so, then producing an excess of them for IVF and then discarding them would be wrong. It might also be wrong to benefit from their destruction.

Stem cells may be obtained from other sources besides embryos. Some, known as embryonic germ cells, are harvested from aborted or miscarried fetuses, but this source is subject to the same sort of controversy as ES cells. Research using stem cells taken from the umbilical cord and placenta has yielded some very promising results; adult tissues such as bone marrow and parts of the brain are other stem cell sources. And as discussed on page 160, genetically modified fibroblasts can serve as stem cells. In fact, some of these nonembryonic cells have already been used to treat medical conditions, including blood disorders, spinal cord injury, and heart attack damage. Such stem cells are obtained without the use of embryos or fetuses. However, they appear to be more limited in their ability to differentiate than ES cells.

Should ES cell research continue to be permitted? If so, should it be supported by government funding? Do the origins of the ES cell lines make a difference? These are all decisions that have been faced by voters in recent years and will certainly continue to generate vigorous debate.

## *Biology* website

The companion website for *Biology* provides a wealth of information organized and integrated by chapter. You will find practice tests, animations, videos, and much more that will complement your learning and understanding of general biology.

**http://www.mhhe.com/maderbiology10**

part VIII

# Behavior and Ecology

*Most everyone is instinctively drawn to the study of ecology. It is the science that describes how the environment works, our place in the great scheme of things, and what we can do to improve environmental conditions for ourselves and all organisms on planet Earth.*

*Ecology is defined as the study of interactions among organisms and the physical environment, but this definition doesn't give a hint as to the importance of understanding these relationships. Ecologists have two statements that can help: "Everything is affected by everything else" and "There is no free lunch." For example, when tall smokestacks were constructed in the Midwest, the pollutants didn't go away, they caused acid rain in the Northeast. As is so often the case, the only way to solve an ecological problem is to stop the behavior that caused the problem. There is no getting around the problem, and it must be dealt with head-on.*

*Ecology describes the biosphere (the portions of the sea, land, and air that contain living things), but today ecology is also an experimental science. Ecologists use the scientific process to determine how ecosystems function and how they will respond to our manipulations. It is a goal of conservation biology, an active field of current research, to help preserve species and manage ecosystems for sustainable human welfare.*

# 43

# Behavioral Ecology

*Naked mole rats spend almost their entire lives underground and are extremely social in the same manner as ants, which also have underground chambers. Like the ants, mole rats have a rigid social hierarchy. At the top is the queen, the only female who reproduces. At the bottom are workers of both sexes, who do not reproduce, but instead work tirelessly to excavate long intricate tunnels and to help the queen reproduce. Behaviorists ask: Why are mole rats social? And they can give a two-part answer. Mole rats have to dig underground tunnels in order to obtain the roots and tubers that they rely on for food and moisture. Without efficient tunnel digging, a mole rat can't survive, and it takes several mole rats working together to efficiently dig a tunnel. The other answer has to do with their common genes. The queen keeps a harem of very closely related males with whom she mates. The result of this inbreeding is that all members of the colony share about 80% of the same genes. Members of the group largely see to the propagation of their own genes when they help the queen reproduce. Behavioral ecology, as discussed in this chapter, is dedicated to the principle that natural selection shapes behavior just as it does the anatomy and physiology of an animal.*

Among naked mole rats, *Heterocephalus glaber,* workers help the queen reproduce.

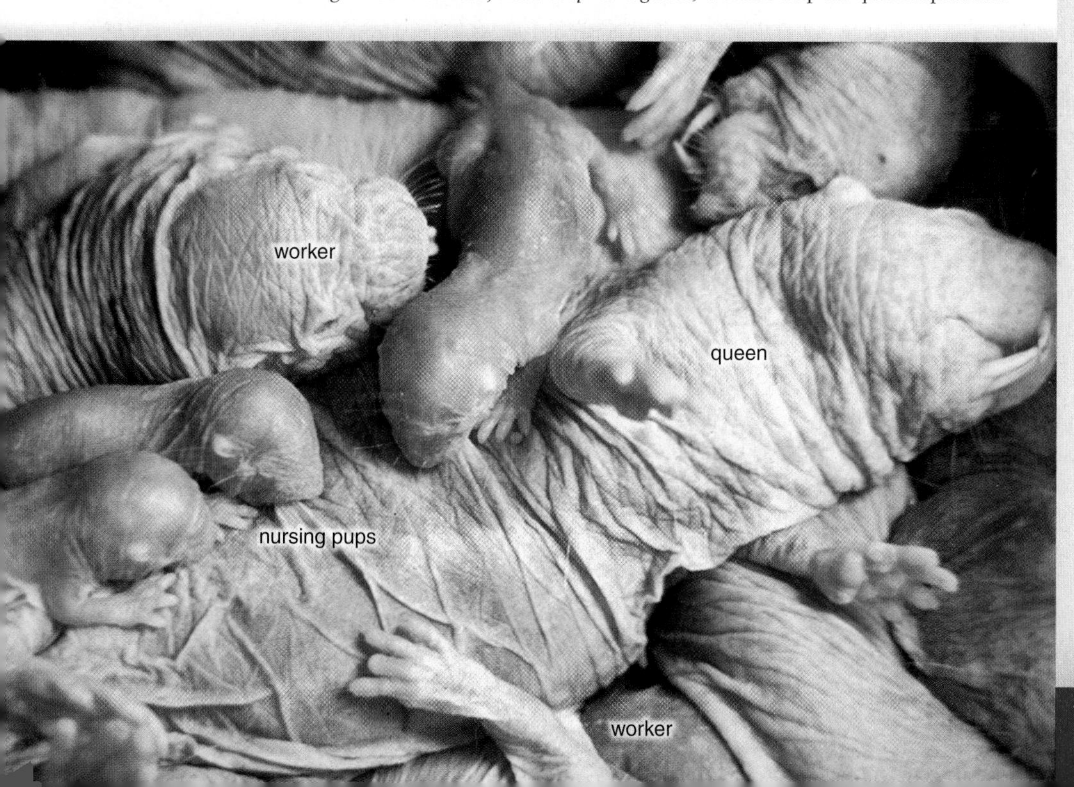


## concepts

# 43.1 Inheritance Influences Behavior

**Behavior** encompasses any action that can be observed and described. Figure 1.15 described the aggressive behavior of mountain bluebirds toward other males during mating. An investigator was able to center in on this behavior, observe it, and record it objectively. In the same manner, we wish to pose the question if anatomy determines the behavior a bluebird or any type animal is capable of performing.

The nature versus nurture question asks to what extent both our genes (nature) and environmental influences (nurture) affect behavior. We would expect that genes, which control the development of neural and hormonal mechanisms, also influence the behavior of an animal. The results of experiments done to discover the degree to which genetics controls behavior support the hypothesis that most behaviors have, at least in part, a genetic basis.

## Experiments That Suggest Behavior Has a Genetic Basis

Among the many behavior studies, we will study two that suggest that behavior has a genetic basis. The first one concerns lovebirds. Lovebirds are small, green and pink African parrots that nest in tree hollows. There are several closely related species of lovebirds in the genus *Agapornis* that build their nests differently. Fischer lovebirds cut large leaves (or in the laboratory, pieces of paper) into long strips with their bills. They use their bills to carry the strips (Fig. 43.1*a*) to the nest, where they weave them in with others to make a deep cup. Peach-faced lovebirds cut somewhat shorter strips and they carry them to the nest in a very unusual manner. They pick up the strips in their bills and then insert them into their feathers (Fig. 43.1*b*). In this way, they can carry several of these short strips with each trip to the nest, while Fischer lovebirds can carry only one of the longer strips at a time.

Researchers hypothesized that if the behavior for obtaining and carrying nesting material is inherited, then hybrids might show intermediate behavior. When the two species of birds were mated, it was observed that the hybrid birds have difficulty carrying nesting materials. They cut strips of intermediate length and then attempt to tuck the strips into their rump feathers. They do not push the strips far enough into the feathers, however, and when they walk or fly, the strips always come out. Hybrid birds eventually (about three years in this study) learn to carry the cut strips in their beak, but they still briefly turn their head toward their rump before flying off. These studies support the hypothesis that behavior has a genetic basis.

a. Fischer lovebird with nesting material in its beak.

b. Peach-faced lovebird with nesting material in its rump feathers.

**FIGURE 43.1 Nest building behavior in lovebirds.**

**a.** Fischer lovebirds, *Agapornis fischeri*, carry strips of nesting material in the bill, as do most other birds. **b.** Peach-faced lovebirds, *Agapornis roseicollis*, tuck strips of nesting material into their rump feathers before flying back to the nest.

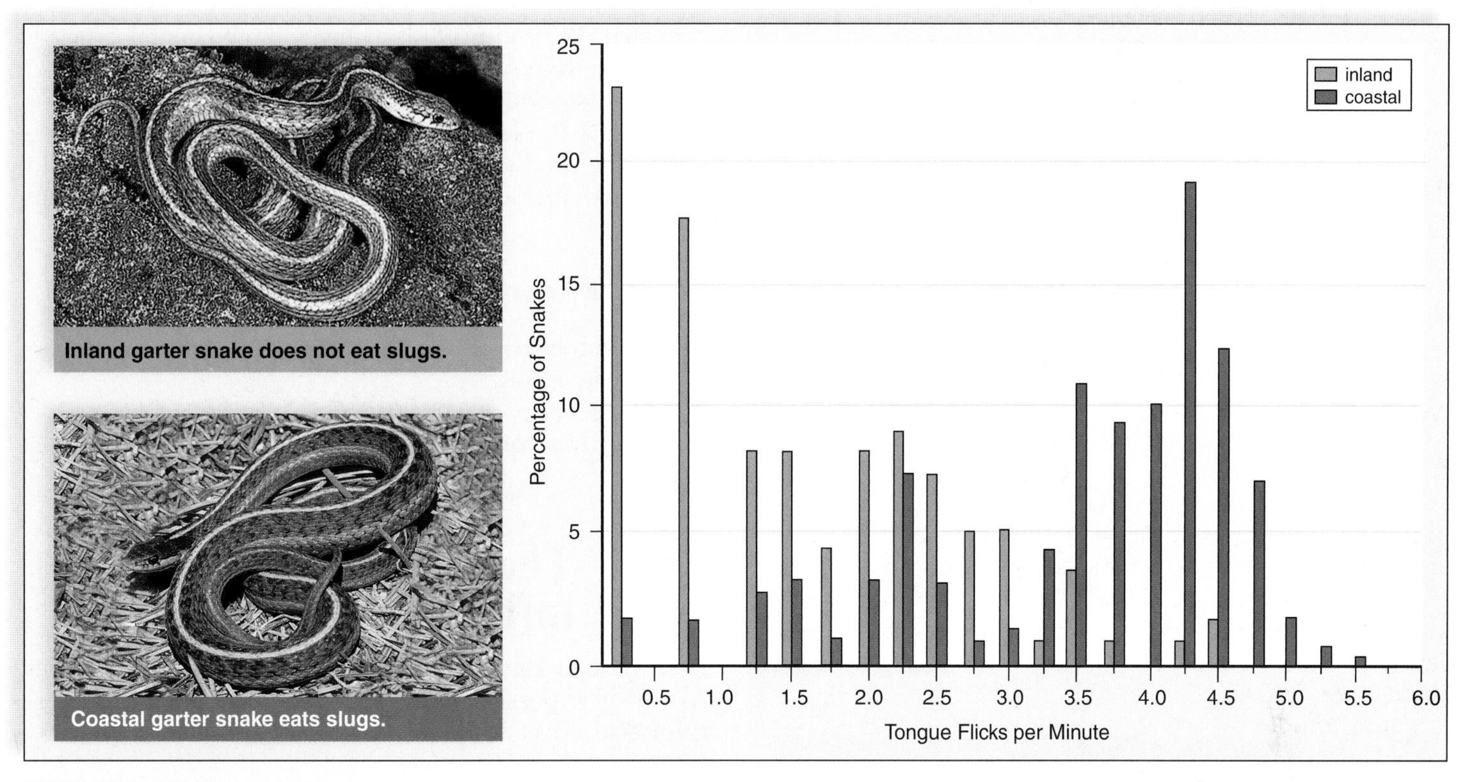


**FIGURE 43.2 Feeding behavior in garter snakes.**

The number of tongue flicks by inland and coastal garter snakes, *Thamnophis elegans*, is measured in terms of their response to slug extract on cotton swabs. Coastal snakes tongue-flicked more than inland snakes.

Several experiments using the garter snake to determine if food preference has a genetic basis have been conducted. There are two different types of garter snake populations in California. Inland populations are aquatic and commonly feed underwater on frogs and fish. Coastal populations are terrestrial and feed mainly on slugs. In the laboratory, inland adult snakes refused to eat slugs, while coastal snakes readily did so. The experimental results of matings between snakes from the two populations (inland and coastal) show their newborns have an overall intermediate incidence of slug acceptance.

Differences between slug acceptors and slug rejecters appear to be inherited, but what physiological difference is there between the two populations? A clever experiment answered this question. When snakes eat, their tongues carry chemicals to an odor receptor in the roof of the mouth. They use tongue flicks to recognize their prey. Even newborns will flick their tongues at cotton swabs dipped in fluids of their prey. Swabs were dipped in slug extract, and the number of tongue flicks were counted for newborn inland snakes and coastal snakes. Coastal snakes had a higher number of tongue flicks than inland snakes (Fig. 43.2). Apparently, inland snakes do not eat slugs because they are not sensitive to their smell. A genetic difference between the two populations of snakes results in a physiological difference in their nervous systems. Although hybrids showed a great deal of variation in the number of tongue flicks, they were generally intermediate, as predicted by the genetic hypothesis.

### *Experiments with Humans*

Human twins, on occasion, have been separated at birth and raised under different environmental conditions. Studies of such twins show that they have similar food preferences, activity patterns, and even select mates with similar characteristics. These twin studies lend support to the hypothesis that at least certain types of behavior are primarily influenced by nature (i.e., the genes).

## Experiments That Show Behavior Has a Genetic Basis

The nervous and endocrine systems are both responsible for the coordination of body systems. Is the endocrine system also involved in behavior? Research studies answer this question in the affirmative. For example, the egg-laying behavior in the marine snail *Aplysia* involves a set sequence of movements. Following copulation, the animal extrudes long strings of more than a million egg cases. It takes the egg case string in its mouth, covers it with mucus, waves its head back and forth to wind the string into an irregular mass, and attaches the mass to a solid object, such as a rock. Several years ago, scientists isolated and analyzed an egg-laying hormone (ELH) that causes the snail to lay eggs, even if it has not mated. ELH was found to be a small protein of 36 amino acids that diffuses into the circulatory system and excites the smooth muscle cells of the reproductive duct, causing them to contract and expel the egg string. Using

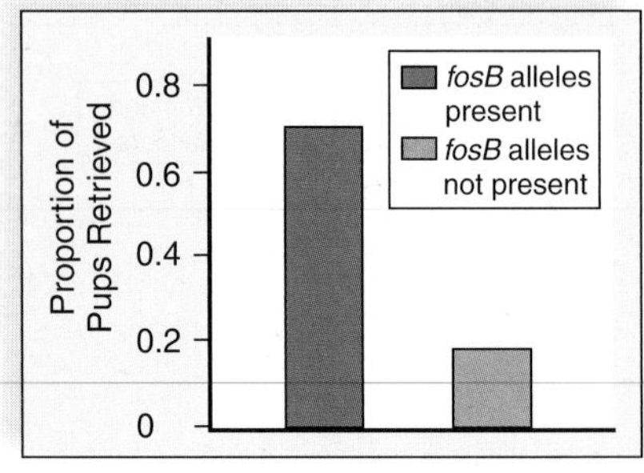


a.

b. *fosB* alleles present.

c. *fosB* alleles not present.

**FIGURE 43.3 Maternal care in mice.**

**a.** A mouse with *fosB* alleles spends time retrieving and crouching over its young, whereas mice that lack these alleles do not display these maternal behaviors. **b.** This mouse shows good maternal care by retrieving her young and crouching over them. **c.** This mouse does not retrieve her young and does not crouch over them.

recombinant DNA techniques, the investigators isolated the ELH gene. The gene's product turned out to be a protein with 271 amino acids. The protein can be cleaved into as many as 11 possible products, and ELH is one of these. ELH alone, or in conjunction with these other products, is thought to control all the components of egg-laying behavior in *Aplysia*.

In another study, investigators found that maternal behavior in mice was dependent on the presence of a gene called *fosB*. Normally, when mothers first inspect their newborn, various sensory information from their eyes, ears, nose, and touch receptors travel to the hypothalamus. This incoming information causes *fosB* alleles to be activated, and a particular protein is produced. The protein begins a process during which cellular enzymes and other genes are activated. The end result is a change in the neural circuitry within the hypothalamus, which manifests itself in good maternal behavior. Mice that lack good maternal behavior also lack *fosB* alleles, and the hypothalamus fails to make any of the products, nor activate any of the enzymes, and other genes that lead to good maternal behavior (Fig. 43.3).

**Check Your Progress 43.1**

1. Describe two studies that suggest behavior has a genetic basis.
2. Describe two studies that show in more detail how behavior has a genetic bases.

# 43.2 The Environment Influences Behavior

Even though genetic inheritance serves as a basis for behavior, it is possible that environmental influences (nurture) also affect behavior. For example, behaviorists originally believed that some behaviors were **fixed action patterns (FAP)** elicited by a sign stimulus. For example, male stickleback fish aggressively defend a territory against other males. In laboratory studies, the male reacts more aggressively to any model that has a red belly like he has, rather than to a model that otherwise looks like a stickleback fish. In this instance, the color red is called a sign stimulus for the aggressive behavior.

male stickleback

Investigators discovered, however, that many behaviors, formerly thought to be FAPs, improve with practice, for example, by learning. In this context, **learning** is defined as a durable change in behavior brought about by experience. For example, habituation is a form of learning because, with experience, an animal no longer responds to a particular stimuli. Deer grazing on the side of a busy highway, ignoring traffic, is an example of habituation.

## Instinct and Learning

Laughing gull chicks' begging behavior appears to be a FAP, because it is always performed the same way in response to the parent's red bill (the sign stimulus). A chick directs a pecking motion toward the parent's bill, grasps it, and strokes it downward (Fig. 43.4). Parents can bring about the begging behavior by swinging their bill gently from side to side. After the chick responds, the parent regurgitates food onto the floor of the nest. If need be, the parent then encourages the chick to eat. This interaction between the chicks and their parents suggests that the begging behavior involves learning. To test this hypothesis, diagrammatic pictures of gull heads were painted on small cards, and then eggs were collected in the field. The eggs were hatched in a dark incubator to eliminate visual stimuli before the test. On the day of hatching, each chick was allowed to make about a dozen pecks at the model. The chicks were returned to the nest, and then each was retested. The tests showed that on the average, only one-third of the pecks by a newly hatched chick strike the model. But one day after hatching, more than half of the pecks are accurate, and two days after hatching, the accuracy reaches a level of more than 75%. Investigators concluded that improvement in motor skills, as well as visual experience, strongly affect development of chick begging behavior.

## Imprinting

**Imprinting** is considered a form of learning. Imprinting was first observed in birds when chicks, ducklings, and goslings followed the first moving object they saw after hatching. This object is ordinarily their mother. Imprinting in the wild, rather than in the laboratory, has survival value and leads to reproductive success, because it leads to being able to recognize one's species and, therefore, an appropriate mate.

But in the laboratory, investigators found that birds can seemingly be imprinted on any object—a human or a red ball—if it is the first moving object they see during a sensitive period of two to three days after hatching. The term *sensitive period* means that the behavior develops only during this time.

A chick imprinted on a red ball follows it around and chirps whenever the ball is moved out of sight. Social interactions between parent and offspring during the sensitive period seem key to normal imprinting. For example, female mallards cluck during the entire time imprinting is occurring, and it could be that vocalization before and after hatching is necessary to normal imprinting.

Goslings imprinted on an investigator

## Social Interactions and Learning

White-crowned sparrows sing a species-specific song, but males of a particular region have their own dialect. Birds were caged in order to test the hypothesis that young white-crowned sparrows learn how to sing from older members of their species.

Three groups of birds were tested. Birds in the first group *heard no songs at all.* When grown, these birds sang a song, but it was not fully developed. Birds in the second group *heard tapes of white-crowns singing.* When grown, they sang in that dialect, as long as the tapes had been played during a sensitive period from about age 10–50 days. White-crowned sparrows' dialects (or other species' songs) played before or after this sensitive period had no effect on the birds. Birds in a third group did not hear tapes and instead were *given an adult tutor.* These birds sang a song of a different species—no matter when the tutoring began—showing that social interactions apparently assist learning in birds.

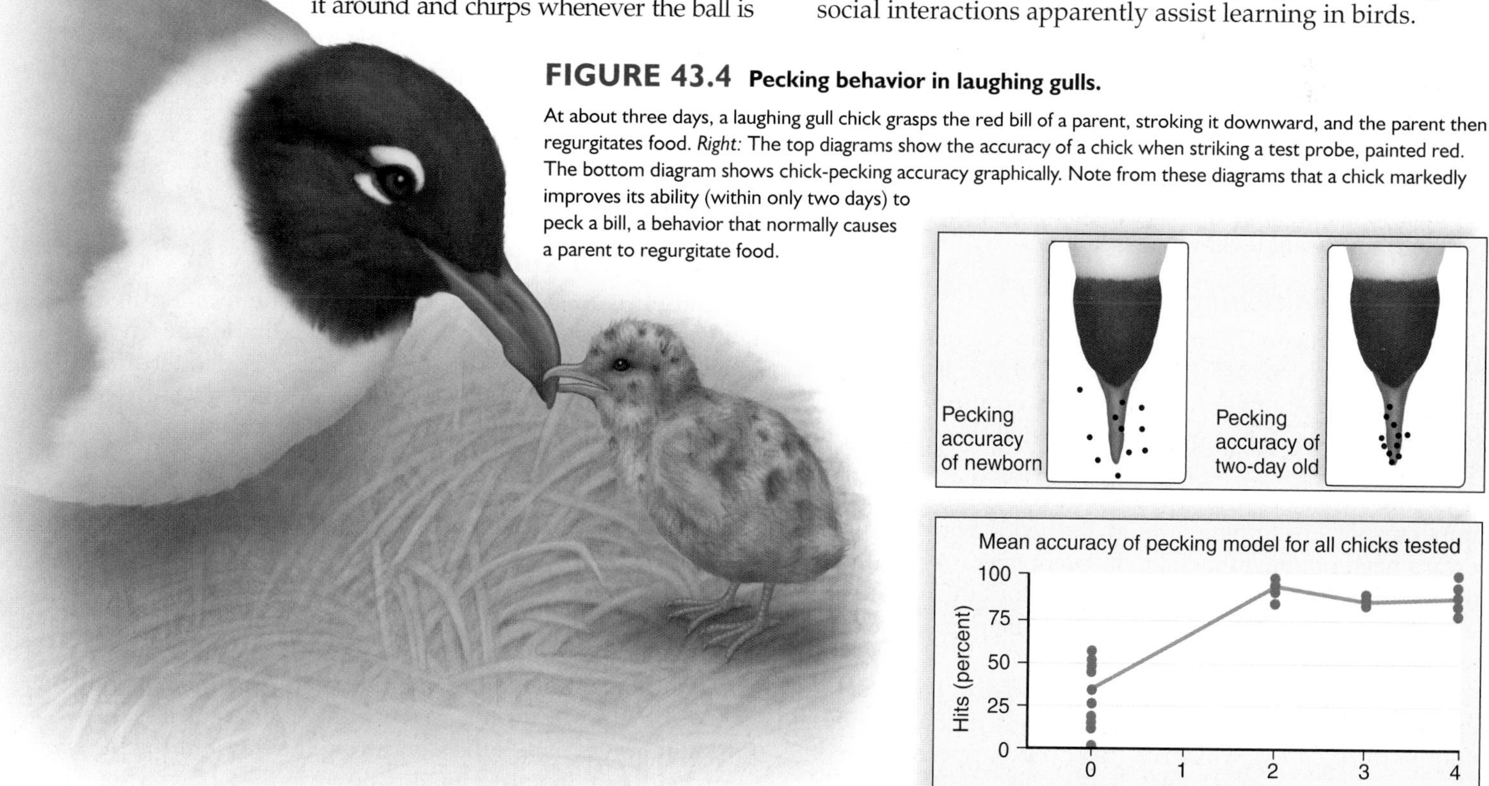

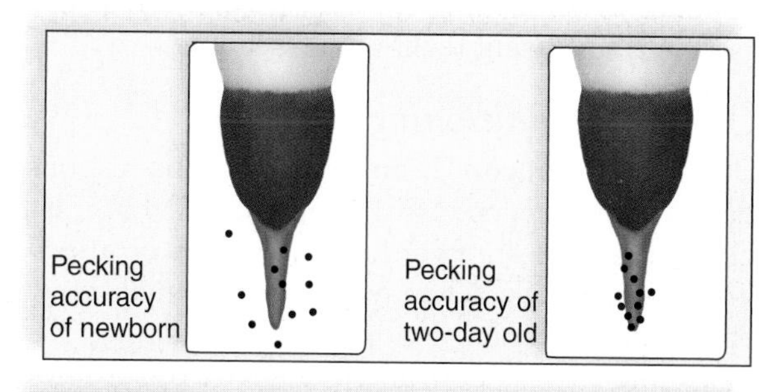


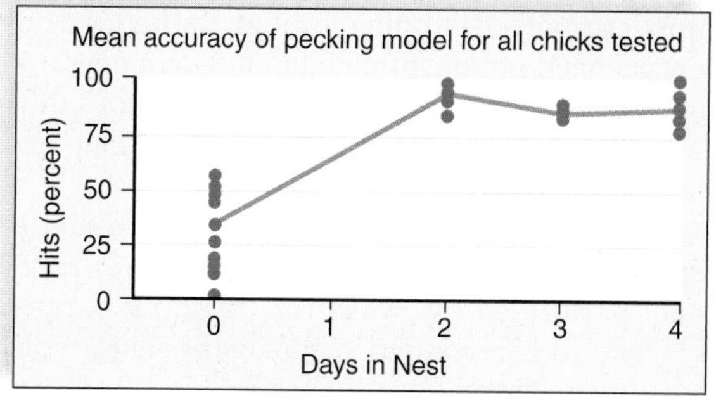


**FIGURE 43.4 Pecking behavior in laughing gulls.**

At about three days, a laughing gull chick grasps the red bill of a parent, stroking it downward, and the parent then regurgitates food. *Right:* The top diagrams show the accuracy of a chick when striking a test probe, painted red. The bottom diagram shows chick-pecking accuracy graphically. Note from these diagrams that a chick markedly improves its ability (within only two days) to peck a bill, a behavior that normally causes a parent to regurgitate food.

## Associative Learning

A change in behavior that involves an association between two events is termed **associative learning.** For example, birds that get sick after eating a monarch butterfly no longer prey on monarch butterflies, even though they may be readily available. Or the smell of fresh baked bread may entice you, even though you may have just eaten. If so, perhaps you associate the taste of bread with a pleasant memory, such as being at home. Both classical conditioning and operant conditioning are examples of associative learning.

### *Classical Conditioning*

In **classical conditioning,** the paired presentation of two different types of stimuli (at the same time) causes an animal to form an association between them. The best-known laboratory example of classical conditioning is that of an experiment done by the Russian psychologist Ivan Pavlov. First, Pavlov observed that dogs salivate when presented with food. Then, he rang a bell whenever the dogs were fed. Eventually, the dogs would salivate whenever the bell was rung, regardless of whether food was present (Fig. 43.5).

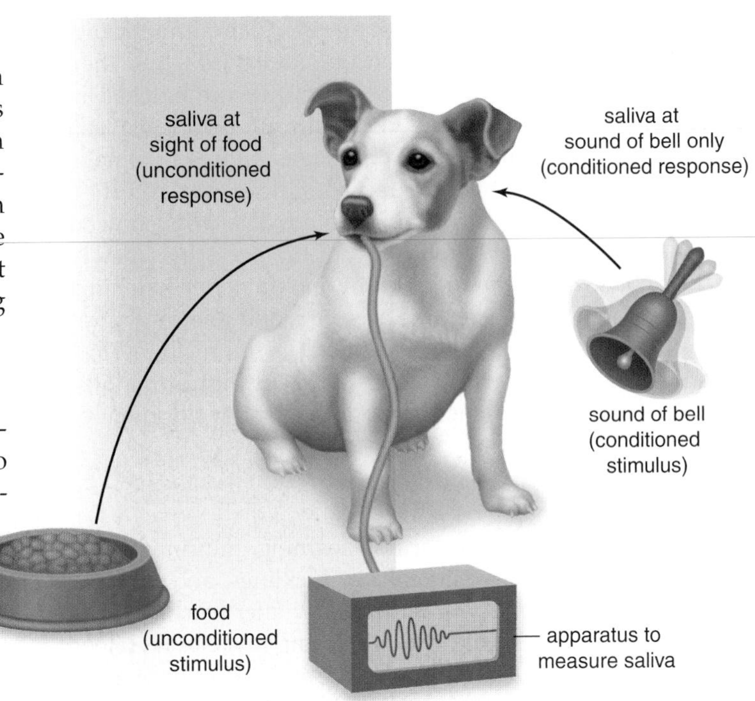


**FIGURE 43.5 Classical conditioning.**

Ivan Pavlov discovered classical conditioning by performing this experiment with dogs. A bell is rung when a dog is fed food. Salivation is noted. Eventually, the dog salivates when the bell is rung, even though no food is presented. Food is an unconditioned stimulus, and the sound of the bell is a conditioned stimulus that brings about the response—that is, salivation.

Classical conditioning suggests that an organism can be trained—that is, conditioned—to associate any response to any stimulus. Unconditioned responses are those that occur naturally, as when salivation follows the presentation of food. Conditioned responses are those that are learned, as when a dog learns to salivate when it hears a bell. Advertisements attempt to use classical conditioning to sell products. Why do commercials pair attractive people with a product being advertised? The hope is that viewers will associate attractiveness with the product. This pleasant association may cause them to buy the product.

Some types of classical conditioning can be helpful. For example, it's been suggested that you hold a child on your lap when reading to them. Why? Because the child will associate a pleasant feeling with reading.

### *Operant Conditioning*

During **operant conditioning,** a stimulus-response connection is strengthened. Most people know that it is helpful to give an animal an award, such as food or affection, when teaching it a trick. When we go to an animal show, it is quite obvious that trainers use operant conditioning. They present a stimulus, say, a hoop, and then give a reward (food) for the proper response (jumping through the hoop). Sometimes the reward need not be immediate. In latent operant conditioning, an animal makes an association without the immediate reward, as when squirrels make a mental map of where they have hidden nuts.

B. F. Skinner is well known for studying this type of learning in the laboratory. In the simplest type of experiment performed by Skinner, a caged rat happens to press a lever and is rewarded with sugar pellets, which it avidly consumes. Thereafter, the rat regularly presses the lever whenever it wants a sugar pellet. In more sophisticated experiments, Skinner even taught pigeons to play Ping-Pong by reinforcing desired responses to stimuli.

As an example in child rearing again, it's been suggested that parents who give a positive reinforcement for good behavior will be more successful than parents who punish behaviors they believe are undesirable.

## Orientation and Migratory Behavior

**Migration** is long-distance travel from one location to another. Loggerhead sea turtles hatch on a Florida beach and then travel across the Atlantic Ocean to the Mediterranean Sea, which offers an abundance of food. After several years, pregnant females return to the same beaches to lay their eggs. Every year, monarch butterflies fly from North America to Mexico where they can continue breeding because milkweed plants are still available there to serve as a source of food for their larvae.

At the very least, migration requires **orientation,** the ability to travel in a particular direction, such as south in the winter and north in the spring. Most of the work regarding orientation has been done in birds. Many birds can use the sun during the day or the stars at night to orient themselves. The sun moves across the sky during the day, but the birds are able to compensate for this because they have a sense of time. They are presumed to have a biological clock that

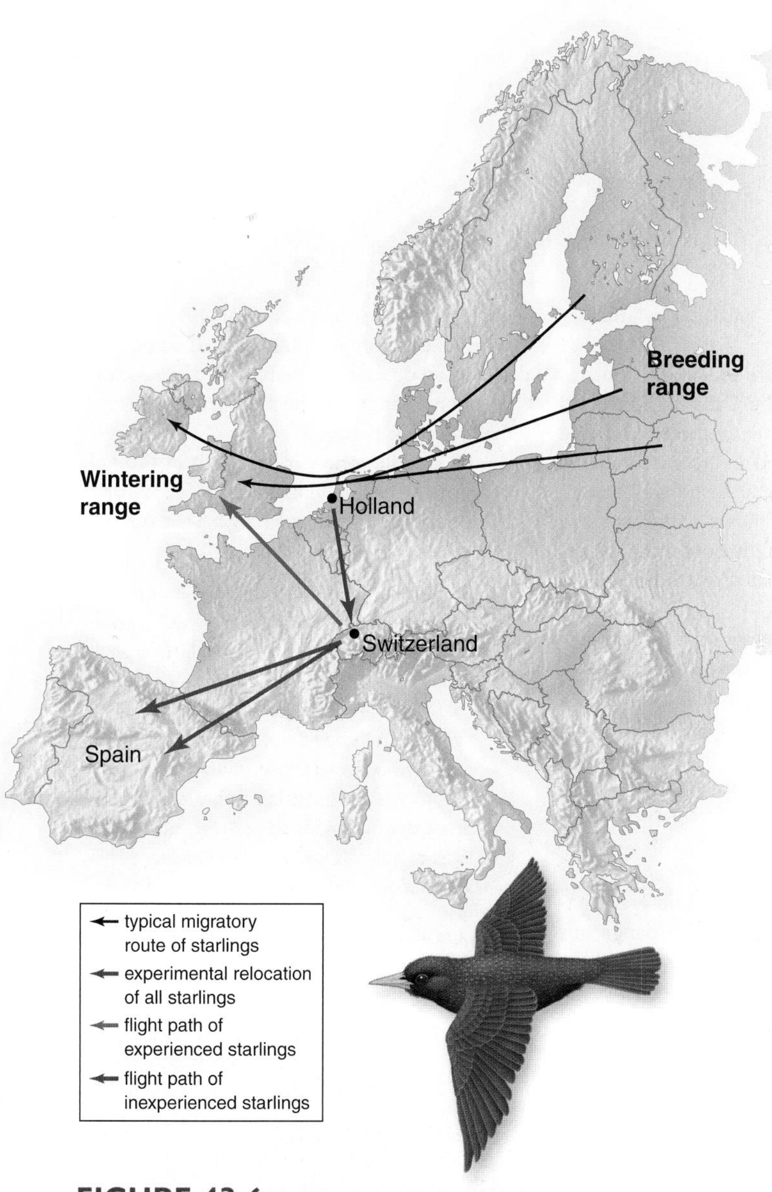


**FIGURE 43.6 Starling migratory experiment.**

Starlings, *Sturnus vulgaris*, on their way from the Baltics to Great Britain were captured and released in Switzerland. Inexperienced birds kept flying in the same direction and ended up in Spain. Experienced birds had learned to navigate, as witnessed by the fact that they still arrived in Great Britain.

allows them to know where the sun will be in relation to the way they should be going any time of the day.

Experienced birds can also **navigate.** They are able to change their direction in response to other environmental clues that tell them they are currently headed in the wrong direction. These clues are apt to come from the Earth's magnetic field. Figure 43.6 shows a study that was done with starlings, which typically migrate from the Baltics to Great Britain and return. Test starlings were captured in Holland and transported to Switzerland. Experienced birds corrected their flight pattern and still got to Great Britain. Young, inexperienced birds ended up in Spain instead of Great Britain!

Migratory behavior has a proximate cause and an ultimate cause. The proximate cause is due to environmental stimuli that tell the birds it is time to travel. The ultimate cause is due to the possibility of reaching a more favorable environment for survival and reproduction. Is the benefit worth the cost—the dangers of the journey? It must be, or else the behavior would not have arisen.

## Cognitive Learning

In addition to the modes of learning already discussed, animals may learn through imitation and insight. Many organisms learn through observation and imitation. For example, Japanese macaques learn to wash sweet potatoes before eating them by imitating others.

**Insight learning** occurs when an animal suddenly solves a problem without any prior experience with the situation. The animal appears to call upon prior experience with other circumstances to solve the problem. For example, chimpanzees have been observed stacking boxes to reach bananas in laboratory settings. Other animals too, aside from primates, seem to be able to reason things out. In one experiment, ravens were offered meat that was attached to string hanging from a branch in a confined aviary. The ravens were accustomed to eating meat, but had no knowledge of how strings work. It took several hours but eventually one raven flew to the branch, reached down, grabbed the string with its beak, and pulled the string up over and over again, each time securing the string with its foot. Eventually, the meat was within reach, and the raven was able to grab the meat with its beak. Other ravens were then able to also perform this behavior.

Ravens learn to retrieve food.

Can animals also plan ahead? It seems so. A sea otter saves a particular rock to act as a hard surface against which to bash open clams. A chimpanzee strips leaves off a twig, which it then uses to secure termites from a termite nest.

If animals can think, do they have emotions? This too is an unexplored area that is now an area of interest. The Science Focus reading on page 806 explores the possibility that animals have emotions and they can feel pain.

### Check Your Progress 43.2

1. What type of learning occurs when an animal no longer tries to eat bumblebees after being stung by one?
2. Give an example that shows how instinct and learning may interact as behavior develops.
3. Cite evidence that animals have cognitive abilities.

*science focus*

## Do Animals Have Emotions?

In recent years, investigators have become interested in determining whether animals have emotions. The body language of animals can be interpreted to suggest that animals do have feelings. When wolves reunite, they wag their tails to and fro, whine, and jump up and down; elephants vocalize—emit their "greeting rumble"—flay their ears, and spin about. Many young animals play with one another or even with themselves, as when dogs chase their own tails. On the other hand, upon the death of a friend or parent, chimps are apt to sulk, stop eating, and even die. It seems reasonable to hypothesize that animals are "happy" when they reunite, "enjoy" themselves when they play, and are "depressed" over the loss of a close friend or relative. Even people who rarely observe animals usually agree about what an animal must be feeling when it exhibits certain behaviors (Fig. 43A).

In the past, scientists found it expedient to collect data only about observable behavior and to ignore the possible mental state of the animal. Why? Because emotions are personal, and no one can ever know exactly how another animal is feeling. B. F. Skinner, whose research method is described earlier in this chapter, regarded animals as robots that become conditioned to respond automatically to a particular stimulus. He and others never considered that animals might have feelings. But now, some scientists believe they have sufficient data to suggest that at least other vertebrates and/or mammals do have feelings, including fear, joy, embarrassment, jealousy, anger, love, sadness, and grief. And they believe that those who hypothesize otherwise should have to present the opposing data.

Perhaps it would be reasonable to consider the suggestion of Charles Darwin, the father of evolution, who said that animals are different in degree rather than in kind. This means that all animals can, say, feel love, but perhaps not to the degree that humans can. B. Würsig watched the courtship of two baleen whales. They touched, caressed, rolled side-by-side, and eventually swam off together. He wondered if their behavior indicated they felt love for one another. When you think about it, it is unlikely that emotions first appeared in humans with no evolutionary homologies in animals.

Iguanas, but not fish and frogs, tend to stay where it is warm. M. Cabanac has found that warmth makes iguanas experience a rise in body temperature and an increase in heart rate. These are biological responses associated with emotions in humans. Perhaps the ability of animals to feel pleasure and displeasure is a mental state that rises to the level of consciousness.

Neurobiological data support the hypothesis that other animals, aside from humans, are capable of enjoying themselves when they perform an activity such as play. Researchers have found a high level of dopamine in the brain when rats play, and the dopamine level increases even when rats anticipate the opportunity to play. Certainly even the staunchest critic is aware that many different species of animals have limbic systems and are capable of fight-or-flight responses to dangerous situations. Can we go further and suggest that animals feel fear even when no physiological response has yet occurred?

Laboratory animals may be too stressed to provide convincing data on emotions; we have to consider that emotions evolved under an animal's normal environmental conditions. This makes field research more useful. It is possible to fit animals with devices that transmit information on heart rate, body temperature, and eye movements as they go about their daily routine. Such information will help researchers learn how animal emotions might correlate with their behavior, just as emotions influence human behavior. One possible definition describes emotion as a psychological phenomenon that helps animals direct and manage their behavior.

M. Bekoff, who is prominent in the field of animal behavior, encourages us to be open to the possibility that animals have emotions. He states:

> By remaining open to the idea that many animals have rich emotional lives, even if we are wrong in some cases, little truly is lost. By closing the door on the possibility that many animals have rich emotional lives, even if they are very different from our own or from those of animals with whom we are most familiar, we will lose great opportunities to learn about the lives of animals with whom we share this wonderous planet.[1]

[1]Bekoff, M. Animal emotions: Exploring passionate natures. October 2000. *Bioscience 50:10*, page 869.

Is the snowshoe hare afraid?

Is the young chimp comforted?

**FIGURE 43A Emotions in animals.**

# 43.3 Animal Communication

Animals exhibit a wide diversity of social behaviors. Some animals are largely solitary and join with a member of the opposite sex only for the purpose of reproduction. Others pair, bond, and cooperate in raising offspring. Still others form a **society** in which members of a species are organized in a cooperative manner, extending beyond sexual and parental behavior. We have already mentioned the social groups of naked mole rats, baboons, and red deer (see page 292). Social behavior in these and other animals requires that they communicate with one another.

## Communicative Behavior

**Communication** is an action by a sender that may influence the behavior of a receiver. The communication can be purposeful, but it does not have to be. Bats send out a series of sound pulses and listen for the corresponding echoes to find their way through dark caves and locate food at night. Some moths have an ability to hear these sound pulses, and they begin evasive tactics when they sense that a bat is near. Are the bats purposefully communicating with the moths? No, bat sounds are simply a cue to the moths that danger is near.

### *Chemical Communication*

Chemical signals have the advantage of being effective both night and day. The term **pheromone** [Gk. *phero*, bear, carry, and *monos*, alone] designates chemical signals in low concentration that are passed between members of the same species. Some animals are capable of secreting different pheromones, each with a different meaning. Female moths secrete chemicals from special abdominal glands, which are detected downwind by receptors on male antennae. The antennae are especially sensitive, and this ensures that only male moths of the correct species (not predators) will be able to detect them.

**FIGURE 43.7 Use of a pheromone.**

This male cheetah is spraying urine onto a tree to mark its territory.

Ants and termites mark their trails with pheromones. Cheetahs and other cats mark their territories by depositing urine, feces, and anal gland secretions at the boundaries (Fig. 43.7). Klipspringers (small antelope) use secretions from a gland below the eye to mark twigs and grasses of their territory. Pheromones are known to control the behavior of social insects, as when workers slavishly care for the offspring produced by a queen.

To what degree do pheromones, in addition to hormones, affect the behavior of mammals, even humans, determining whether they carry out parental care, become agressive, or engage in courtship behavior? Some researchers maintain that human behavior is influenced by undetectable pheromones wafting through the air. They have discovered that like the mouse, humans have an organ in the nose, called the vomeronasal organ (VNO), that can detect not only odors, but also pheromones. The neurons from this organ lead to the hypothalamus, the part of the brain that controls the release of many hormones in the body. One investigator has isolated and plans to market a perfume containing a pheromone released by men that appears to reduce premenstrual nervousness and tension in women.

### *Auditory Communication*

**Auditory** (sound) **communication** has some advantages over other kinds of communication. It is faster than chemical communication, and it too is effective both night and day. Further, auditory communication can be modified not only by loudness but also by pattern, duration, and repetition. In an experiment with rats, a researcher discovered that an intruder can avoid attack by increasing the frequency with which it makes an appeasement sound.

Male crickets have calls, and male birds have songs for a number of different occasions. For example, birds may have one song for distress, another for courting, and still another for marking territories. Sailors have long heard the songs of humpback whales transmitted through the hull of a ship. But only recently has it been shown that the song has six basic themes, each with its own phrases, that can vary in length and be interspersed with sundry cries and chirps. The purpose of the song is probably sexual, serving to advertise the availability of the singer. Bottlenose dolphins have one of the most complex languages in the animal kingdom.

Language is the ultimate auditory communication. Only humans have the biological ability to produce a large number of different sounds and to put them together in many different ways. Nonhuman primates have different vocalizations, each having a definite meaning, such as when vervet monkeys give alarm calls (Fig. 43.8). Although chimpanzees can be taught to use an artificial language, they never progress beyond the capability level of a two-year-old child. It has also been difficult to prove that chimps understand the concept of grammar or can use their language to reason. It still seems as if humans possess a communication ability unparalleled by other animals.

a.

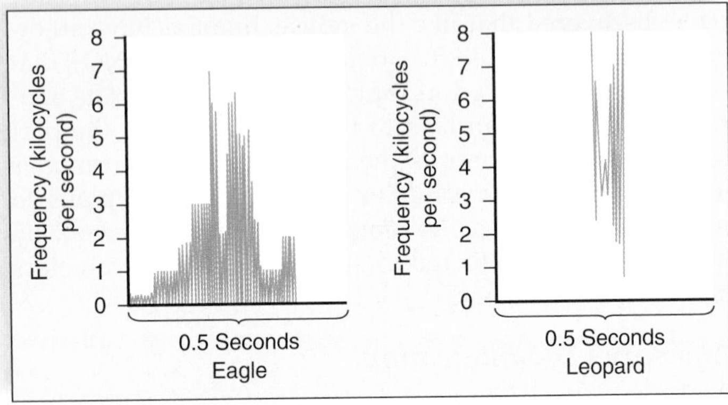


b.

**FIGURE 43.8 Auditory communication.**

**a.** Vervet monkeys, *Cercopithecus aethiops*, are responding to an alarm call. Vervet monkeys can give different alarm calls according to whether a troop member sights an eagle or a leopard, for example. **b.** The frequency per second of the sound differs for each type call.

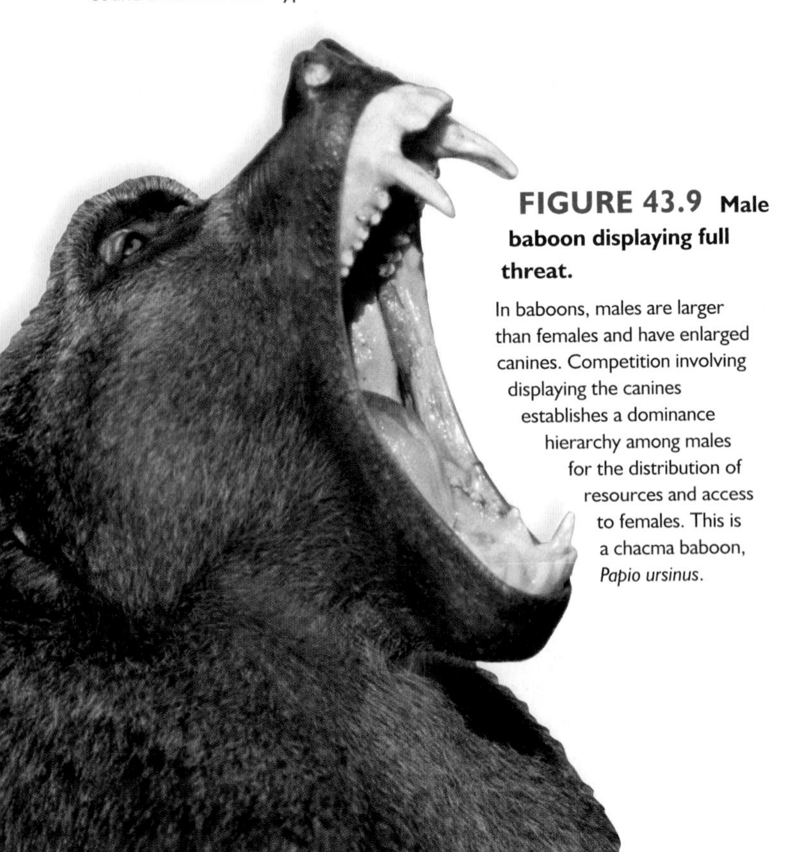

**FIGURE 43.9 Male baboon displaying full threat.**

In baboons, males are larger than females and have enlarged canines. Competition involving displaying the canines establishes a dominance hierarchy among males for the distribution of resources and access to females. This is a chacma baboon, *Papio ursinus*.

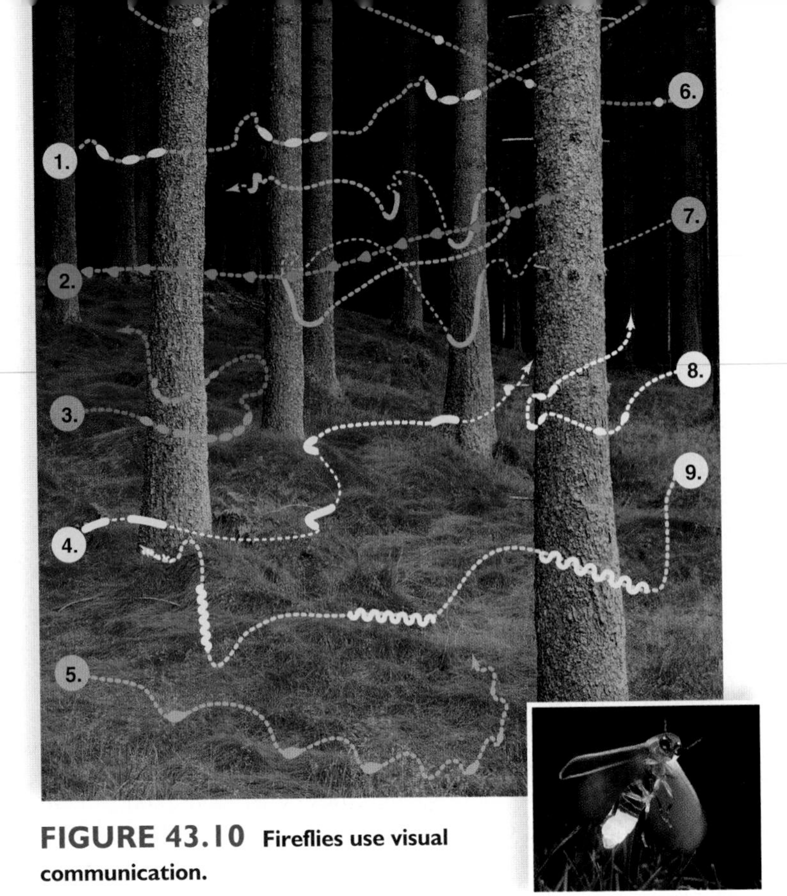


**FIGURE 43.10 Fireflies use visual communication.**

Each number represents the male flash pattern of a different species. The patterns are a behavorial reproductive isolation mechanism.

### *Visual Communication*

Visual signals are most often used by species that are active during the day. Contests between males make use of threat postures and possibly prevent outright fighting, a behavior that might result in reduced fitness. A male baboon displaying full threat is an awesome sight that establishes his dominance and keeps peace within the baboon troop (Fig. 43.9). Hippopotamuses perform territorial displays that include mouth opening.

Many animals use complex courtship behaviors and displays. The plumage of a male Raggiana Bird of Paradise allows him to put on a spectacular courtship dance to attract a female, giving her a basis on which to select a mate. Defense and courtship displays are exaggerated and always performed in the same way so that their meaning is clear. Fireflies use a flash pattern to signal females of the same species (Fig. 43.10).

**Visual communication** allows animals to signal others of their intentions without the need to provide any auditory or chemical messages. The body language of students during a lecture provides an example. Some students lean forward in their seats and make eye contact with the instructor. They want the instructor to know they are interested and find the material of value. Others lean back in their chairs and look at the floor or doodle. These students indicate they are not interested in the material. Teachers can use students' body language to determine if they are effectively presenting the material and make changes accordingly.

a.

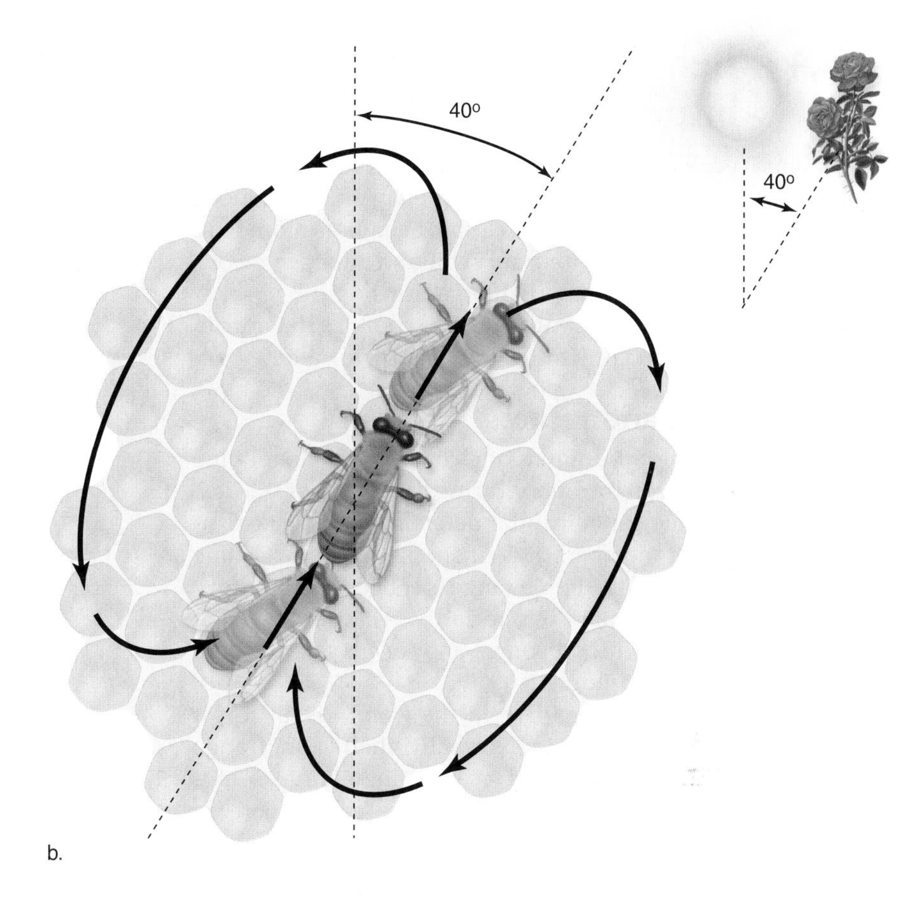


b.

**FIGURE 43.11 Communication among bees.**

**a.** Honeybees do a waggle dance to indicate the direction of food. **b.** If the dance is done outside the hive on a horizontal surface, the straight run of the dance will point to the food source. If the dance is done inside the hive on a vertical surface, the angle of the straightaway to that of the direction of gravity is the same as the angle of the food source to the sun.

Other human behaviors also send visual clues to others. The hairstyle and dress of a person or the way he or she walks and talks are ways to send messages to others. Some studies have suggested that women are apt to dress in an appealing manner and be sexually inviting when they are ovulating. People who dress in black, move slowly, fail to make eye contact, and sit alone may be telling others that they are unhappy. Psychologists have long tried to understand how visual clues can be used to better understand human emotions and behavior. Similarly, body language in animals is being used by researchers to suggest that they, as well as humans, have emotions.

### *Tactile Communication*

**Tactile communication** occurs when one animal touches another. For example, laughing gull chicks peck at the parent's bill to induce the parent to feed them (see Fig. 43.4). A male leopard nuzzles the female's neck to calm her and to stimulate her willingness to mate. In primates, grooming—one animal cleaning the coat and skin of another—helps cement social bonds within a group.

Honeybees use a combination of communication methods, but especially tactile ones, to impart information about the environment. When a foraging bee returns to the hive, it performs a waggle dance that indicates the distance and the direction of a food source (Fig. 43.11). As the bee moves between the two loops of a figure 8, it buzzes noisily and shakes its entire body in so-called waggles. Outside the hive, the dance is done on a horizontal surface, and the straight run indicates the direction of the food. Inside the hive, the angle of the straight run to that of the direction of gravity is the same as the angle of the food source with the sun. In other words, a 40° angle to the left of vertical means that food is 40° to the left of the sun.

Bees can use the sun as a compass to locate food because they have a biological clock, which allows them to compensate for the movement of the sun in the sky. A biological clock is an internal means of telling time. Today, we know that the ticking of the clock, in both insects and mammals (including humans), requires alterations in the expression of a gene called *period.*

**Check Your Progress 43.3**

1. Give evidence that communication is meant to affect the behavior of the receiver.
2. Give an advantage and disadvantage to each type of communication discussed.
3. State the human receptor for each type of communication discussed.

# 43.4 Behaviors That Increase Fitness

**Behavioral ecology** assumes that behavior is subject to natural selection. We have established that behavior has a genetic basis, and we would expect that certain behaviors more than others will lead to increased survival and number of offspring. Therefore, the behavior of organisms we can observe today must have adaptive value. These types of behaviors in particular have been studied for their adaptive value.

## Territoriality and Fitness

In order to gather food, animals often have a particular home range where they can be found during the course of the day. One portion of the range can be defended for their exclusive use and other members of their species are not welcome there. This portion of the home range is called their **territory** and the behavior is called **territoriality.** An animal's territory may have a good food source and/or may be the area in which they will reproduce.

For example, gibbons live in the tropical rain forest of South and Southeast Asia. Normally, their home range can be covered in about 3–4 days, and they are also monogamous and territorial. Territories are maintained by loud singing (Fig. 43.12). Males sing just before sunrise, and mated pairs sing duets during the morning. Males, but not females, show evidence of fighting to defend their territory in the form of broken teeth and scars. Obviously, defense of a territory has a certain cost; it takes energy to sing and fight off others. Also, you might get hurt. So, what is the adaptive value of being territorial? Chief among the benefits of territoriality are to ensure a source of food, exclusive rights to one or more females, and to have a place to rear young and possibly to protect yourself from predators. The territory has to be the right size for the animal. Too large a territory cannot be defended, and too small a territory may not contain enough food. Cheetahs require a large territory in order to hunt for their prey and, therefore, they use means of marking their territory that will last for a while. As shown in Figure 43.7, cheetahs, like many dogs, use urine to mark their territory. Hummingbirds are known to defend a very small territory because they depend on only a small patch of flowers as their food source.

**FIGURE 43.12 Male and female gibbons.**

Siamang gibbons, *Hylobates syndactylus*, are monogamous, and they both share the task of raising offspring. They also share the task of marking their territory by singing. As is often the case in monogamous relationships, the sexes are similar in appearance. Male is above and female is below.

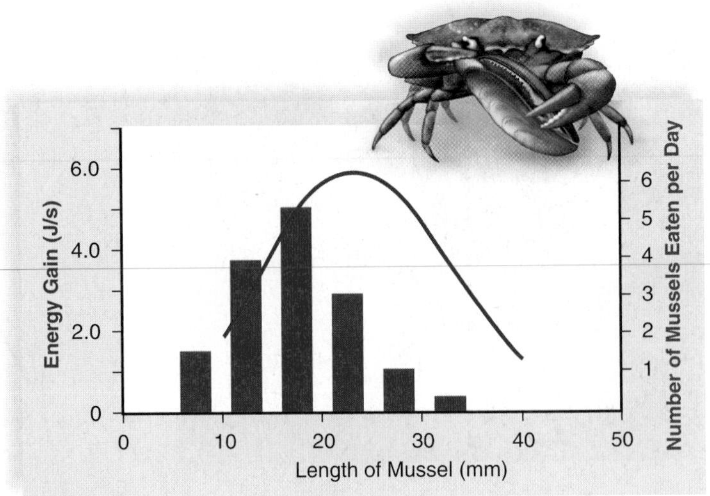

Energy Gain (J/s): 0, 2.0, 4.0, 6.0
Length of Mussel (mm): 0, 10, 20, 30, 40, 50
Number of Mussels Eaten per Day: 1, 2, 3, 4, 5, 6

**FIGURE 43.13 Foraging for food.**

When offered a choice of an equal number of each size of mussel, the shore crab, *Carcinus maenas*, prefers the intermediate size. Their size provides the highest rate of net energy return. Net energy is determined by the yield per time used in breaking open the shell.

Territoriality is more likely to occur during times of reproduction. Seabirds have very large home ranges consisting of hundreds of kilometers of open ocean, but when they reproduce they become fiercely territorial. Each bird has a very small territory consisting of only a small patch of beach where they place their nest.

### *Foraging for Food*

Food gathering is technically called foraging for food. A concern for an animal is to acquire a food source that will provide more energy than the effort of acquiring the food. In one study, it was shown that shore crabs eat intermediate-sized mussels because the net energy gain was more than if they ate larger-sized mussels (Fig. 43.13). The large mussels take too much energy to open per the amount of energy they provide. The **optimal foraging model** states that it is adaptive for foraging behavior to be as energetically efficient as possible.

Even though it can be shown that animals that take in more energy are the ones that are likely to have more offspring, animals often have to consider other factors such as escaping from predation. If an animal is killed and eaten, it has no chance at all of having offspring. Animals often face trade-offs that cause them to modify their behavior or even stop foraging for a while.

**FIGURE 43.14 Hamadryas baboons.**

Among Hamadryas baboons, *Papio hamadryas*, a male, which is silver-white and twice the size of a female, keeps and guards a harem of females with whom he mates exclusively.

**FIGURE 43.15 Competition.**

During the mating season, bull elk, *Cervus elaphus*, males may find it necessary to engage in antler wrestling in order to have sole access to females in a territory.

## Reproductive Strategies and Fitness

Usually, primates are **polygamous,** and males monopolize multiple females. Because of gestation and lactation, females invest more in their offspring than do males and may not always be available for mating. Under these circumstances, it is adaptive for females to be concerned with a good food source. When food sources are clumped, females congregate in small groups. Because only a few females are expected to be receptive at a time, males will likely be able to defend these few from other males. Males are expected to compete with other males for the limited number of receptive females available (Fig. 43.14).

A limited number of primates are **polyanthrus.** Tamarins are squirrel-sized New World monkeys that live in Central or South America. Tamarins live together in groups of one or more families in which one female mates with more than one male. The female normally gives birth to twins of such a large size that the fathers, and not the mother, carry it about. This may be the reason these animals are polyanthrus. Polyandry also occurs when the environment does not have sufficient resources to support several young at a time.

We have already mentioned the reproductive strategy of gibbons. They are **monogamous,** which means that they pair bond, and both male and female help with the rearing of the young. Males are active fathers, frequently grooming and handling infants. Monogamy is relatively rare in primates, which includes prosimians, monkeys, and apes (only about 18% are monogamous). In primates, monogamy occurs when males have limited mating opportunities, territoriality exists, and the male is fairly certain the offspring are his. In gibbons, females are evenly distributed in the environment, most likely because they are aggressive to one another. Investigators note that females do attack a speaker when it plays female sounds in their territory.

### *Sexual Selection*

**Sexual selection** is a form of natural selection that favors features that increase an animal's chances of mating. In other words, these features are adaptive in the sense that they lead to increased fitness.

Sexual selection often results in female choice and male competition. Because females produce only one egg a month, it is adaptive for them to be choosy about their mate. If they choose a mate that passes on features to a male offspring that will cause him to be chosen by females, their fitness has increased. Whether females choose features that are adaptive to the environment is in question. For example, peahens are likely to choose peacocks that have the most elaborate tails. Such a fancy tail could otherwise be detrimental to the male and make him more likely to be captured by a predator. In one study, an extra ornament was attached to a father zebra finch and the daughters of this bird underwent the process of imprinting (see page 803). Now, these females were more likely to choose a mate that also had the same artificial ornament.

While females can always be sure an offspring is theirs, males do not have this certainty. However, males produce a plentiful supply of sperm. The best strategy for males to increase their fitness, therefore, is to have as many offspring as possible. Competition may be required for them to gain access to females, and ornaments, such as antlers, can enhance a male's ability to fight (Fig. 43.15). When bull elk compete, they issue a loud number of screams that gives way to a series of grunts. If still necessary, the two bulls walk in parallel to show each other their physique. If this doesn't convince one or the other to back off, the pair resorts to ramming each other with their antlers. Rarely is either bull actually hurt. Whereas a peacock cannot shed his tail, elk shed their antlers as soon as mating season is over.

science focus

## Sexual Selection in Male Bowerbirds

At the start of the breeding season, male bowerbirds use small sticks and twigs to build elaborate display areas called bowers (Fig. 43B). They clear the space around the bower and decorate the area with fresh flowers, fruits, pebbles, shells, bits of glass, tinfoil, and any bright baubles they can find. The Satin Bowerbird of eastern Australia prefers blue objects, a color that harmonizes with the male's glossy blue-back plumage. Males collect blue parrot feathers, flowers, berries, ballpoint pens, clothespins, and even toothbrushes from researchers' cabins.

After the bower is complete, a male bowerbird spends most of his time near this bower, calling to females, renewing his decorations, and guarding his work against possible raids by other males. After inspecting many bowers and their owners, a female approaches one, and the male begins a display. He faces her, fluffs up his feathers, and flaps his wings to the beat of a call. The female enters the bower, and if she crouches, the two mate.

Female bowerbirds build their own nests and raise the young without help from their mates, so attractive males can mate with multiple females. The reproductive advantage gained by attractive male bowerbirds is quite large; the most attractive males may mate with up to 25 females per year, but most males mate rarely or not at all. As already discussed, Dr. [Gerald] Borgia found that the males most often chosen by females have well-built bowers with well-decorated platforms. In addition, it is possible that the ability of a male bowerbird to respond appropriately to the female during courtship might influence his success. I [Gail Patricelli] and my colleagues studied this interactive component of Satin Bowerbird mating behavior as part of my doctoral dissertation research at the University of Maryland, in collaboration with Dr. Borgia, my graduate advisor.

**FIGURE 43B Male and female bowerbird.**
*Among Satin Bowerbirds,* Ptilonorhynchus violaceus, *the male bowerbird (right) has prepared this bower and decorated its platform with particularly blue objects. A female bowerbird (left) has entered the bower and a male courtship display will now begin.*

Male bowerbirds are not gaudy in appearance, but their displays are highly intense and aggressive. Their courting displays are similar to those used by males to intimidate each other in aggressive encounters—with males puffing their feathers, rapidly extending their wings, and running, while making a loud, buzzing vocalization. Analysis of natural courtships has shown that males must display intensely to be attractive, *but males that are too intense too soon can startle females*. Females may benefit from preferentially mating with the most intensely displaying males (e.g., if these displays indicate male health or vigor), but when females are startled repeatedly by male displays, they may not be able to efficiently assess male traits. Thus both sexes can maximize the potential benefits of intense male courtship displays—and minimize the potential costs—by communicating. Indeed, a female behavior (degree of crouching) reflects the level of display intensity that the female will tolerate without being startled. By giving higher-intensity displays only when females increase their crouching, males could increase their courtship success by displaying intensely enough to be attractive without threatening females with displays more intense than they are ready to tolerate.

With this information in mind, we specifically tested the hypothesis that males respond to female crouching signals by adjusting their intensity, and that a particular male's ability to respond to female signals is related to his success in courtship. A male's ability to modify his courtship display according to the rapidity with which the female crouches was difficult

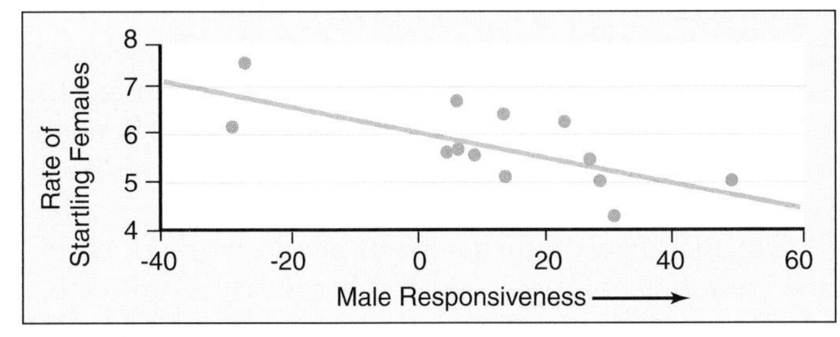


a.

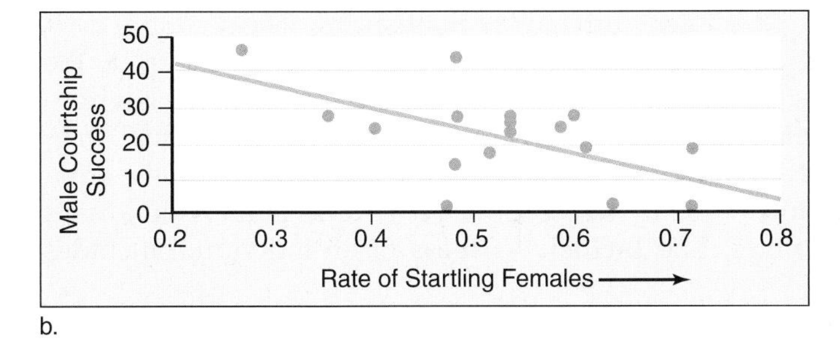


b.

**FIGURE 43C Courtship success of males.**
*Some males have more courtship success than others. This experiment may explain why.* ***a.*** *An experimenter using a remote control (photograph) can regulate the crouch rate of a robotic female. Some males are better able than others to vary the intensity of their courtship depending on the crouch rate of the robotic female (top graph). In other words, only if the robotic female crouches more do they respond more. Therefore, they are predicted to startle females less.* ***b.*** *Experimenters found that males who respond best under experimental conditions do startle live females less and do have better courtship success.*

to measure in natural courtships, since it was not clear whether males were responding to females, or vice versa. To solve this problem, we collaborated with an engineer to develop robotic female bowerbirds, which used tiny servo motors to mimic female movements (Fig. 43C *right*).

Using these "fembots," we were able to control female signals and measure male response in experimental courtships. During the bowerbird mating season at our field site in Wallaby Creek, Australia, we worked with student volunteers to test each male in our population with robots that crouched at four different rates: no crouch, slow, moderate, and fast. These experiments showed that male Satin Bowerbirds in general modulate their displays in response to robotic female crouching (Fig. 43C*a*). This supports the prediction that males are able to respond to female signals by giving their highest-intensity courtship displays for females who crouched the fastest and are least likely to be startled.

Utilizing automatically triggered video cameras that monitor behaviors at bowers, it was possible to measure each male's courtship/mating success with no difficulty. We found that males who modulate their displays more effectively in response to robotic female signals startle real females less often in natural courtships, and are thus more successful in courting females (Fig. 43C*b*).

Our results suggest that females prefer intensely displaying males as mates, but that successful males do not always display at maximum intensity; they modulate their intensity in response to female signals, thus producing displays attractive to females without threatening them.

Male responsiveness to female signals may be an important part of successful courtship in many species—even if males do not dance aggressively during courtship like male bowerbirds. For instance, when females choose their mates based on bright coloration, successful males may respond to female signals by altering their position relative to the sun, their distance from the female, or the way they shake their tail when displaying their colors. So, along with extreme male traits—such as gaudy colors and aggressive dances—sexual selection may favor the ability of males to read female signals and adjust courtship displays accordingly.

*Courtesy of Gail Patricelli*
*University of Maryland*

## Societies and Fitness

The principles of evolutionary biology can be applied to the study of social behavior in animals. Sociobiologists hypothesize that living in a society has a greater reproductive benefit than reproductive cost. A cost-benefit analysis can help determine if this hypothesis is supported.

Group living does have its benefits. It can help an animal avoid predators, rear offspring, and find food. A group of impalas is more likely to hear an approaching predator than a solitary one. Many fish moving rapidly in many directions might distract a would-be predator. Weaver birds form giant colonies that help protect them from predators, but the birds may also share information about food sources. Primate members of the same baboon troop signal to one another when they have found an especially bountiful fruit tree. Lions working together are able to capture large prey, such as zebra and buffalo.

Group living also has its disadvantages. When animals are crowded together into a small area, disputes can arise over access to the best feeding places and sleeping sites. Dominance hierarchies are one way to apportion resources, but this puts subordinates at a disadvantage. Among red deer, sons are preferable because, as a harem master, sons will result in a greater number of grandchildren. However, sons, being larger than daughters, need to be nursed more frequently and for a longer period of time. Subordinate females do not have access to enough food resources to adequately nurse sons and, therefore, they tend to rear daughters, not sons. Still, like the subordinate males in a baboon troop, subordinate females in a red deer harem may be better off in terms of fitness if they stay with a group, despite the cost involved.

Living in close quarters exposes individuals to illness and parasites that can easily pass from one animal to another. Social behavior helps to offset some of the proximity disadvantages. For example, baboons and other types of social primates invest much time in grooming one another, and this most likely helps them remain healthy. Humans use extensive medical care to help offset the health problems that arise from living in the densely populated cities of the world.

## Altruism Versus Self-Interest

**Altruism** [L. *alter,* the other] is a behavior that has the potential to decrease the lifetime reproductive success of the altruist, while benefiting the reproductive success of another member of the society. In insect societies, especially, reproduction is limited to only one pair, the queen and her mate. For example, among army ants, the queen is inseminated only during her nuptial flight, and thereafter she spends her time reproducing (Fig. 43.16). The society has three different sizes of sterile female workers. The smallest workers (3 mm), called the nurses, take care of the queen and larvae, feeding them and keeping them clean. The intermediate-sized workers, constituting most of the population, go out on raids to collect food. The soldiers (14 mm), with huge heads and powerful jaws, run along the sides and rear of raiding parties and protect the column of ants from attack by intruders. While rare in mammals, the introduction to this chapter describes how mole rats have a similar type of societal structure.

Can the altruistic behavior of sterile workers be explained in terms of fitness, which is judged by reproductive success? Genes are passed from one generation to the next in two quite different ways. The first way is direct: A parent can pass a gene directly to an offspring. The second way is indirect: A relative that reproduces can pass the gene to the next generation. *Direct selection* is adaptation to the environment due to the reproductive success of an individual. *Indirect selection,* called **kin selection,** is adaptation to the environment due to the reproductive success of the individual's relatives. The **inclusive fitness** of an individual includes

**FIGURE 43.16 Queen ant.**

A queen ant, *Solenopsis geminata,* has a large abdomen for egg production and is cared for by small ants, called nurses. The idea of inclusive fitness suggests that relatives, in addition to offspring, increase an individual's reproductive success. Therefore, sterile nurses are being altruistic when they help the queen produce offspring to whom they are closely related.

personal reproductive success and the reproductive success of relatives.

Among social bees, social wasps, and ants, the queen is diploid (2n), but her mate is haploid (n). If the queen has had only one mate, sister workers are more closely related to each other. They share, on average, 75% of their genes because they inherit 100% of their father's alleles. Their potential offspring would share, on average, only 50% of their genes with the queen. Therefore, a worker can achieve a greater inclusive fitness by helping her mother (the queen) produce additional sisters than by directly reproducing. Under these circumstances, behavior that appears altruistic is more likely to evolve.

Indirect selection can also occur among animals whose offspring receive only a half set of genes from both parents. Consider that your brother or sister shares 50% of your genes, your niece or nephew shares 25%, and so on. Therefore, the survival of two nieces (or nephews) is worth the survival of one sibling, assuming they both go on to reproduce.

Among chimpanzees in Africa, a female in estrus frequently copulates with several members of the same group, and the males make no attempt to interfere with each other's matings. How can they be acting in their own self-interest? Genetic relatedness appears to underlie their apparent altruism. Members of a group share more than 50% of their genes in common because members never leave the territory in which they are born.

**FIGURE 43.17 Inclusive fitness.**

A meerkat is acting as a babysitter for its young sisters and brothers while their mother is away. Researchers point out that the helpful behavior of the older meerkat can lead to increased inclusive fitness.

### *Reciprocal Altruism*

In some bird species, offspring from a previous clutch of eggs may stay at the nest to help parents rear the next batch of offspring. In a study of Florida scrub jays, the number of fledglings produced by an adult pair doubled when they had helpers. Mammalian offspring are also observed to help their parents (Fig. 43.17). Among jackals in Africa, solitary pairs managed to rear an average of 1.4 pups, whereas pairs with helpers reared 3.6 pups. What are the benefits of staying behind to help? First, a helper is contributing to the survival of its own kin. Therefore, the helper actually gains a fitness benefit. Second, a helper is more likely than a nonhelper to inherit a parental territory—including other helpers. Helping, then, involves making a minimal, short-term reproductive sacrifice in order to maximize future reproductive potential. Therefore, helpers at the nest are also practicing a form of **reciprocal altruism.** Reciprocal altruism also occurs in animals that are not necessarily closely related. In this event, an animal helps or cooperates with another animal with no immediate benefit. However, the animal that was helped will repay the debt at some later time. Reciprocal altruism usually occurs in groups of animals that are mutually dependent. Cheaters in reciprocal altruism are recognized and not reciprocated in future events. Reciprocal altruism occurs in vampire bats that live in the tropics. Bats returning to the roost after a feeding activity share their blood meal with other bats in the roost. If a bat fails to share blood with one that had previously shared blood with it, the cheater bat will be excluded from future blood sharing.

Vampire bat, *Desmodus rotundus*

**Check Your Progress 43.4**

1. In what way is territoriality related to foraging for food?
2. In what way is an animal's reproductive strategy related to sexual selection.
3. Why can it be said that altruistic behavior is probably in the self-interest of the animal.

# Connecting the Concepts

Birds build nests, dogs bury bones, cats chase quick-moving objects, and snakes bask in the sun. All animals, including humans, behave—they respond to stimuli, both from the physical environment and from other individuals of the same or different species. It can be shown that behaviors have a genetic basis and yet behaviors can be modified by experience. Regardless, behaviorists use an evolutionary approach to generate hypotheses that can be tested to better understand how the behavior increases individual fitness (i.e., the capacity to produce surviving offspring).

Inheritance produces the behavioral variations that are subject to natural selection, resulting in the most adaptive responses to stimuli. Still, behaviorists concede that both nature and nurture determine behavior. Genetics determines, for example, that hawks hunt by using vision rather than smell, and that cats, not dogs, climb trees. Songbirds are born with the ability to sing, but which song or dialect they sing is strongly dependent on the songs they hear from their parents and siblings. A new chimpanzee mother naturally cares for her young, but she is a better mother if she has observed other females in her troop raising young.

A behavior such as territoriality or foraging has its trade-offs, but this behavior would not have evolved unless the benefit outweighed the cost. Sexual selection explains much about mating behavior. Males are apt to compete and females to choose because this type of behavior is most apt to lead to reproductive success. Also, there is survival value, after all, in the ability of male baboons to react aggressively when the troop is under attack. There is a cost to certain behaviors. Why should older offspring help younger offspring? The evolutionary answer is that, in the end, there are benefits that outweigh the costs. Otherwise, the behavior would not continue. Similarly, group living in which animals communicate must have some benefits. The evolutionary approach to studying behavior has proved fruitful in helping us understand why birds sing their melodious songs, dolphins frolic in groups, and wondrous male Raggiana Birds of Paradise display to females.

## summary

### 43.1 Inheritance Influences Behavior

Investigators have long been interested in the degree to which nature (genetics) or nurture (environment) influences behavior. Studies with birds, snakes, humans, snails, and mice have been done, among many others. Hybrid studies with lovebirds produce results consistent with the hypothesis that behavior has a genetic basis. Garter snake experiments indicate that the nervous system controls behavior. Twin studies in humans show that certain types of behavior are apparently inherited. *Aplysia* DNA studies indicate that the endocrine system also controls behavior.

### 43.2 The Environment Influences Behavior

Even behaviors formerly thought to be fixed action patterns (FAPs), or otherwise inflexible, sometimes can be modified by learning. The red bill of laughing gulls initiates chick begging behavior. However, with experience, chick begging behavior improves and the chicks demonstrate an increased ability to recognize parents.

Other studies suggest that learning is involved in behaviors. Imprinting in birds, during a sensitive period, causes them to follow the first moving object they see. Song learning in birds involves various elements—including the existence of a sensitive period, during which an animal is primed to learn—and the positive benefit of social interactions.

Associative learning includes classical conditioning and operant conditioning. In classical conditioning, the pairing of two different types of stimuli causes an animal to form an association between them. In this way, dogs will salivate at the sound of a bell. In operant conditioning, animals learn behaviors because they are rewarded when they perform them.

Orientation and migratory behavior occur in several groups of animals. Orientation is the ability to move in a certain direction, but migration can require navigation, and is a learned ability to change direction if need be. Animals use the sun, stars, and the Earth's magnetic field in order to migrate.

Imitation and insight learning does occur in animals. Insight learning has occurred when an animal can solve a new and different problem without prior experience.

### 43.3 Animal Communication

Communication is an action by a sender that affects the behavior of a receiver. Chemical, auditory, visual, and tactile signals are forms of communication that foster cooperation that benefits both the sender and the receiver. Pheromones are chemical signals that are passed between members of the same species. Auditory communication includes language, which may occur between other types of animals and not just humans. Visual communication allows animals to signal others without the need of auditory or chemical messages. Tactile communication is especially associated with sexual behavior.

### 43.4 Behaviors That Increase Fitness

Traits that promote reproductive success are expected to be advantageous overall, despite any possible disadvantage. Some animals are territorial and defend a territory where they have food resources and can reproduce. When animals choose those foods that return the most net energy, they have more energy left over for reproduction.

Reproductive strategies include monogamy, polygamy, and polyandry. Which strategy is employed depends on the animal and its environment. Sexual selection is a form of natural selection that selects for traits that increase an animal's fitness. Males produce many sperm and are expected to compete to inseminate females. Females produce few eggs and are expected to be selective about their mates.

Living in a social group can have its advantages (e.g., ability to avoid predators, raise young, and find food). It also has disadvantages (e.g., tension between members, spread of illness and parasites, and reduced reproductive potential). When animals live in groups, the benefits must outweigh the costs or the behavior would not exist.

In most instances, the individuals of a society act to increase their own fitness (ability to produce surviving offspring). In this context, it is necessary to consider inclusive fitness, which includes personal reproductive success and the reproductive success of relatives, also. Sometimes, animals perform altruistic acts, as when individuals help their parents rear siblings. Social insects help their mother reproduce, but this behavior seems reasonable when we consider that siblings share 75% of their genes. Among mammals, a parental helper may be likely to inherit the parent's territory. In reciprocal altruism, animals aid one another for future benefits.

## understanding the terms

| | |
|---|---|
| altruism 814 | monogamous 811 |
| associative learning 804 | navigate 805 |
| auditory communication 807 | operant conditioning 804 |
| behavior 800 | optimal foraging model 810 |
| behavioral ecology 810 | orientation 804 |
| classical conditioning 804 | pheromone 807 |
| communication 807 | polyanthrus 811 |
| fixed action pattern (FAP) 802 | polygamous 811 |
| imprinting 803 | reciprocal altruism 815 |
| inclusive fitness 814 | sexual selection 811 |
| insight learning 805 | society 807 |
| kin selection 814 | tactile communication 809 |
| learning 802 | territoriality 810 |
| migration 804 | territory 810 |
| | visual communication 808 |

Match the terms to these definitions:

a. ______________ Behavior related to defending a particular area, which is often used for the purpose of feeding, mating, and caring for young.

b. ______________ Social interaction that benefits others and has the potential to decrease the lifetime reproductive success of the member exhibiting the behavior.

c. ______________ Signal by a sender that may influence the behavior of a receiver.

d. ______________ Chemical substance secreted into the environment by one organism that may influence the behavior of another.

## reviewing this chapter

1. Describe two studies that suggest behavior has a genetic basis. 800–801
2. Describe two studies that show that behavior has a genetic basis. 801–2
3. How does an experiment with laughing gull chicks support the hypothesis that environment (nurture) influences behavior? 802–3
4. Describe two types of associative learning. 804
5. Give examples of the different types of communication among members of a social group. 807–9
6. What is territoriality, and how might it increase fitness? 810
7. Name three types of reproductive strategies, and describe how they can be related to the environment. 811
8. What is sexual selection, and why does it foster female choice and male competition during mating? 811
9. Give examples of behaviors that appear to be altruistic but actually increase the inclusive fitness of an individual. 814–15

## testing yourself

Choose the best answer for each question.
For questions 1–4, match the type of learning in the key with its description. Answers may be used more than once.

**KEY:**

a. classical conditioning
b. insight learning
c. imprinting
d. migration

1. Ducks follow the first moving object they see after hatching.
2. A dog salivates when a bell is rung.
3. Starlings fly between the Balkins and Great Britain in the fall and spring.
4. Chimpanzees pile up boxes to reach bananas.
5. Behavior is
   a. any action that is learned.
   b. all responses to the environment.
   c. any action that can be observed and described.
   d. all activity that is controlled by hormones.
   e. unique to birds and humans.
6. A behaviorist would most likely study which one of the following items?
   a. the flow of energy through an ecosystem
   b. the way a bird digests seeds
   c. the number of times a fiddler crab flashes its claw to attract a mate
   d. the structure of a horse's leg
   e. All of the above are correct.
7. Which one of the following is considered, by behaviorists, to control (in part or in whole) animal behavior?
   a. circulatory and respiratory systems
   b. respiratory and digestive systems
   c. digestive and nervous systems
   d. nervous and endocrine systems
   e. All systems of the body control behavior.
8. Which of the following is not an example of a genetically based behavior?
   a. Inland garter snakes do not eat slugs, while coastal populations do.
   b. One species of lovebird carries nesting strips one at a time, while another carries several.
   c. One species of warbler migrates, while another one does not.
   d. Snails lay eggs in response to egg-laying hormone.
   e. Wild foxes raised in captivity are not capable of hunting for food.
9. How would the following graph differ if pecking behavior in laughing gulls was a fixed action pattern?
   a. It would be a diagonal line with an upward incline.
   b. It would be a diagonal line with a downward incline.
   c. It would be a horizontal line.
   d. It would be a vertical line.
   e. None of these is correct.

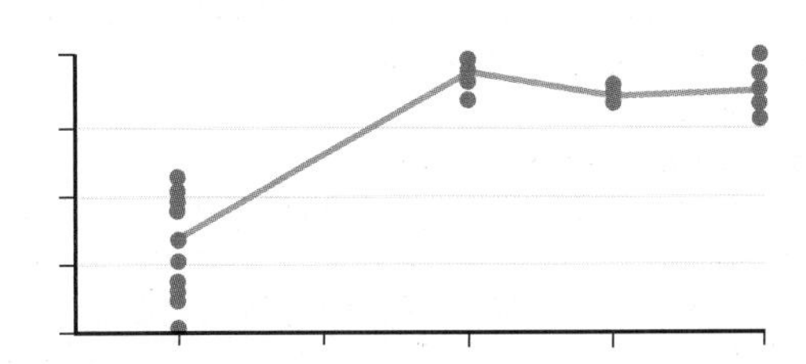

10. Egg-laying hormone causes snails to lay eggs, implicating which system in this behavior?
    a. digestive
    b. endocrine
    c. nervous
    d. lymphatic
    e. respiratory
11. Using treats to train a dog to do a trick is an example of
    a. imprinting.
    b. tutoring.
    c. vocalization.
    d. operant conditioning.

12. The benefits of imprinting generally outweigh the costs because
    a. an animal that has been imprinted on the wrong object can be reimprinted on the mother.
    b. imprinting behavior never lasts more than a few months.
    c. animals in the wild rarely imprint on anything other than their mother.
    d. animals that imprint on the wrong object generally die before they pass their genes on.
13. In white-crowned sparrows, social experience exhibits a very strong influence over the development of singing patterns. What observation led to this conclusion?
    a. Birds only learned to sing when they were trained by other birds.
    b. The window in which birds learn from other birds is wider than that when birds learn from tape recordings.
    c. Birds can learn different dialects only from other birds.
    d. Birds that learned to sing from a tape recorder could change their song when they listened to another bird.
14. Which of the following best describes classical conditioning?
    a. The gradual strengthening of stimulus-response connections that seemingly are unrelated.
    b. A type of associative learning in which there is no contingency between response and reinforcer.
    c. The learning behavior in which an organism follows the first moving object it encounters.
    d. The learning behavior in which an organism exhibits a fixed action pattern from the time of birth.
15. The observation that male bowerbirds decorate their nests with blue objects favored by females can best be associated with
    a. insight learning.
    b. imprinting.
    c. sexual selection.
    d. altruism
16. Foraging
    a. is best done at night.
    b. should result in net energy to the animal.
    c. is an unnecessary activity.
    d. is a part of sexual selection.
17. Migratory behavior
    a. has a higher cost than a benefit.
    b. affects the reproductive strategy.
    c. pertains only to birds.
    d. involves the ability to navigate.

For problems 18–22, match the type of communication in the key with its description. Answers may be used more than once.

**KEY:**

a. chemical communication
b. auditory communication
c. visual communication
d. tactile communication

18. Aphids (insects) release an alarm pheromone when they sense they are in danger.
19. Male peacocks exhibit an elaborate display of feathers to attract females.
20. Ground squirrels give an alarm call to warn others of the approach of a predator.
21. Male silk moths are attracted to females by a sex attractant released by the female moth.
22. Sage grouses perform an elaborate courtship dance.
23. After migrating south, birds usually
    a. reproduce.
    b. fight to win the right to mate.
    c. learn a new song.
    d. All of these are correct.
24. Bees that do a waggle dance are teaching other bees
    a. how to dance.
    b. where to find food.
    c. how to find and use the sun for navigation.
    d. how to use auditory communication.

## thinking scientifically

1. Meerkats are said to exhibit altruistic behavior because certain members of a population act as sentries. How would you test the hypothesis that sentries are engaged in altruistic behavior?
2. You are testing the hypothesis that human infants instinctively respond to higher-pitched voices. Your design is to record head turns toward speakers when they play voices in different pitches. When you do the experiment using several different infants, your data support your hypothesis. However, prior learning by infants is still a serious criticism. What is the basis of this criticism?

## bioethical issue

### Putting Animals in Zoos

Is it ethical to keep animals in zoos, where they are not free to behave as they would in the wild? If we keep animals in zoos, are we depriving them of their freedom? Some point out that freedom is never absolute. Even an animal in the wild is restricted in various ways by its abiotic and biotic environment. Many modern zoos keep animals in habitats that nearly match their natural ones so that they have some freedom to roam and behave naturally. Perhaps, too, we should consider the education and enjoyment of the many thousands of human visitors to a zoo compared to the freedom lost by a much smaller number of animals kept in a zoo.

Today, reputable zoos rarely go out and capture animals in the wild—they usually get their animals from other zoos. Most people feel it is not a good idea to take animals from the wild except for very serious reasons. Certainly, zoos should not be involved in the commercial and often illegal trade of wild animals that still goes on today. Many zoos today are involved in the conservation of animals. They provide the best home possible while animals are recovering from injury or increasing their numbers until they can be released to the wild. Can we perhaps look at zoos favorably if they show that they are keeping animals under good conditions and are also involved in preserving animals? What is your opinion?

## *Biology* website

The companion website for *Biology* provides a wealth of information organized and integrated by chapter. You will find practice tests, animations, videos, and much more that will complement your learning and understanding of general biology.

**http://www.mhhe.com/maderbiology10**

# 44 Population Ecology

*Elephants attain a large size, are social, live a long time, even up to 70 years, and produce few offspring. Females live in social family units, and the much larger males visit them only during breeding season. Females give birth about every five years to a single calf that is well cared for and has a good chance of meeting the challenges of its lifestyle. Normally, an elephant population exists at the carrying capacity of the environment.*

*Elephants are currently threatened because of human population growth, and because males, in preference to females, are killed for their ivory tusks. A moratorium on killing male elephants has increased the estimate of their numbers. So, the hope is that we do have enough time to learn what factors are most critical for elephants to sustain a healthy population before it is too late. This chapter previews the principles of population ecology, a field that is extremely important to the preservation of species and the maintenance of the diversity of life on Earth.*

Social unit of female elephants, *Loxodonta africana*.

## concepts

# 44.1 Scope of Ecology

In 1866, the German zoologist Ernst Haeckel coined the word *ecology* from two Greek roots [Gk. *oikos*, home, house, and *-logy*, "study of" from *logikos*, rational, sensible]. He said that **ecology** is the study of the interactions among all organisms and with their physical environment. Haeckel also pointed out that ecology and evolution are intertwined because ecological interactions are selection pressures that result in evolutionary change, which in turn affects ecological interactions.

Ecology, like so many biological disciplines, is wide-ranging. At one of its lowest levels, ecologists study how the individual organism is adapted to its environment. For example, they study how a fish is adapted to and survives in its **habitat** (the place where the organism lives) (Fig. 44.1). Most organisms do not exist singly; rather, they are part of a population, the functional unit that interacts with the environment and on which natural selection operates. A **population** is defined as all the organisms belonging to the same species within an area at the same time. At this level of study, ecologists are interested in factors that affect the growth and regulation of population size.

A **community** consists of all the various populations of multiple species interacting at a locale. In a coral reef, there are numerous populations of algae, corals, crustaceans, fishes, and so forth. At this level, ecologists want to know how interactions such as predation and competition affect the organization of a community. An **ecosystem** contains a community of populations and also the abiotic environment (e.g., the availability of sunlight for plants). Energy flow and chemical cycling are significant aspects of understanding how an ecosystem functions. Ecosystems rarely have distinct boundaries and are not totally self-sustaining. Usually, a transition zone called an ecotone, which has a mixture of organisms from adjacent ecosystems, exists between ecosystems. The **biosphere** encompasses the zones of the Earth's soil, water, and air where living organisms are found. Table 44.1 summarizes the levels of biological study.

Modern ecology is not just descriptive, it is predictive. It analyzes levels of organization and develops models and hypotheses that can be tested. A central goal of modern ecology is to develop models that explain and predict the distribution and abundance of organisms. Ultimately, ecology considers not one particular area, but the distribution and abundance of

**TABLE 44.1**

**Ecological Terms**

| *Term* | *Definition* |
|---|---|
| Ecology | Study of the interactions of organisms with each other and with the physical environment |
| Population | All the members of the same species that inhabit a particular area |
| Community | All the populations found in a particular area |
| Ecosystem | A community and its physical environment, including both nonliving (abiotic) and living (biotic) components |
| Biosphere | All the communities on Earth whose members exist in air and water and on land |

**FIGURE 44.1** **Ecological levels.**

The study of ecology encompasses levels of organization that proceed from the individual organism to the population, to the community, and finally to an ecosystem.

populations in the biosphere. For example, what factors have selected for the mix of plants and animals in a tropical rain forest at one latitude and in a desert at another? While modern ecology is useful in and of itself, it also has unlimited application possibilities, including the proper management of plants and wildlife, the identification of and efficient use of renewable and nonrenewable resources, the preservation of habitats and natural cycles, the maintenance of food resources, and the ability to predict the impact and course of a disease such as malaria or AIDS.

**Check Your Progress 44.1**

1. What is the difference between a population and a community?
2. What is a central goal of modern ecological studies today?
3. What is meant by the "abiotic environment"?

# 44.2 Demographics of Populations

**Demography** is the statistical study of a population, such as its density, its distribution, and its rate of growth, which is dependent on such factors as its mortality pattern and age distribution.

## Density and Distribution

**Population density** is the number of individuals per unit area, such as there are 73 persons per square mile in the United States. Population density figures make it seem as if individuals are uniformly distributed, but this often is not the case. For example, we know full well that most people in the United States live in cities, where the number of people per unit area is dramatically higher than in the country. And even within a city, more people live in particular neighborhoods than others, and such distributions can change over time. Therefore, basing ecological models solely on population density, as has often been done in the past, can lead to misleading results.

**Population distribution** is the pattern of dispersal of individuals across an area of interest. The availability of resources can affect where populations live. **Resources** are nonliving (abiotic) and living (biotic) components of an environment that support living organisms. Light, water, space, mates, and food are some important resources for populations. **Limiting factors** are those environmental aspects that particularly determine where an organism lives. For example, trout live only in cool mountain streams, where the oxygen content is high, but carp and catfish are found in rivers near the coast because they can tolerate warm waters, which have a low concentration of oxygen. The timberline is the limit of tree growth in mountainous regions or in high latitudes. Trees cannot grow above the high timberline because of low temperatures and the fact that water remains frozen most of the year. The distribution of organisms can also be due to biotic factors. In Australia, the red kangaroo does not live outside arid inland areas because it is adapted to feeding on the grasses that grow there.

Three descriptions—*clumped, random,* and *uniform*—are often used to characterize observed patterns of distribution. Suppose you considered the distribution of a species across its full range. A range is that portion of the globe where the species can be found; red kangaroos live in Australia. On that scale, you would expect to find a clumped distribution. However, organisms are located in areas suitable to their adaptations; as mentioned, red kangaroos live in grasslands, and catfish live in warm river water near the coast.

Within a smaller area such as a single body of water or a single forest, the availability of resources again influences which of the patterns of distribution is common for a particular population. For example, a study of the distribution of hard clams in a bay on the south shore of Long Island, New York, showed that clam abundance is associated with sediment shell content. Investigators hope to use this information to transform areas that have few clams into high-abundance areas. Distribution patterns need not be constant. In a study of desert shrubs, it was found that the distribution changed from clumped to random to a uniform distribution pattern as the plants matured. As time passed, it was found that competition for belowground resources caused the distribution pattern to become uniform (Fig. 44.2).

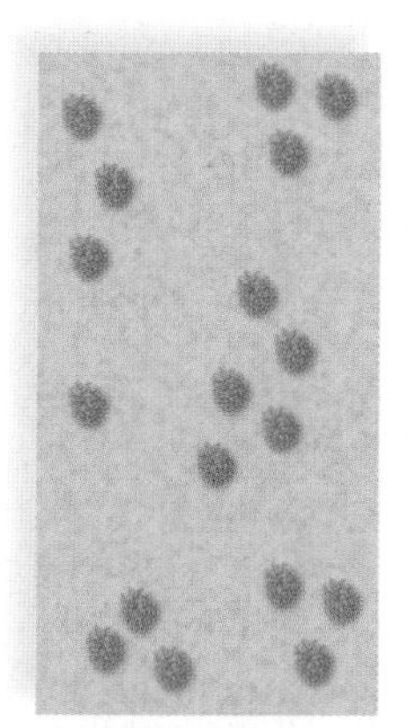
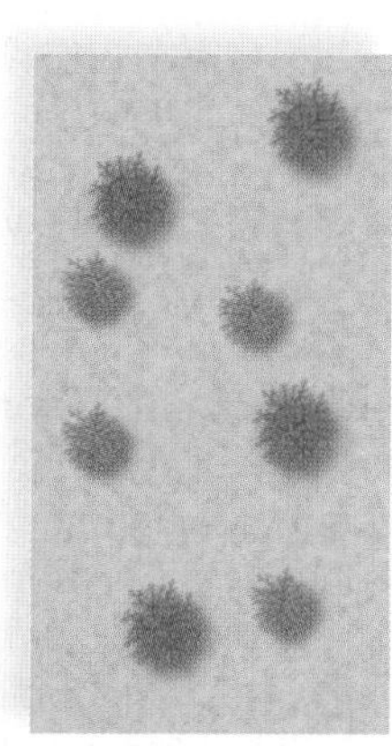
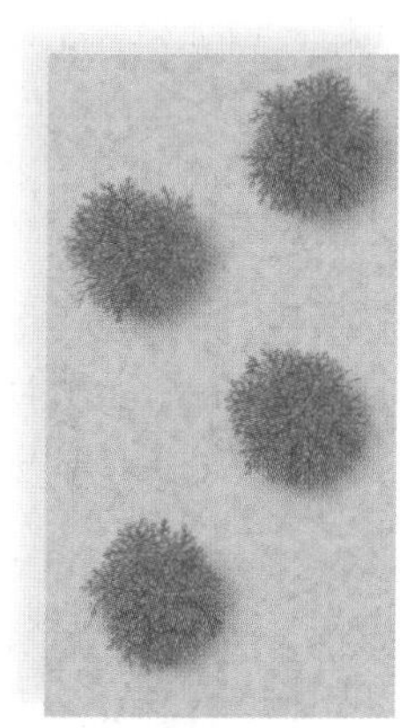
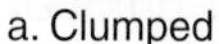

a. Clumped b. Random c. Uniform

d. Mature desert shrubs

**FIGURE 44.2 Distribution patterns of the creosote bush.**

**a.** Young, small desert shrubs are clumped. **b.** Medium shrubs are randomly distributed. **c.** Mature shrubs are uniformly distributed. **d.** Photograph of mature shrub distribution.

a.

b.

**FIGURE 44.3 Biotic potential.**

A population's maximum growth rate under ideal conditions—that is, its biotic potential—is greatly influenced by the number of offspring produced in each reproductive event. **a.** Pigs, which produce many offspring that quickly mature to produce more offspring, have a much higher biotic potential than (**b**) the rhinoceros, which produces only one or two offspring per infrequent reproductive event.

Other factors besides resource availability can influence distribution patterns. Breeding golden eagles, like many other birds, exhibit territoriality, and this behavioral characteristic discourages a clumped distribution at this time. On the other hand, cedar trees tend to be clumped near the parent plant because seeds are not widely dispersed.

## Population Growth

The **rate of natural increase (*r*),** which for our purposes is the same as the growth rate, is dependent on the number of individuals born each year and the number of individuals that die each year. Usually it is possible to assume that immigration and emigration are equal and need not be considered in the calculation of the growth rate. Populations grow when the number of births exceeds the number of deaths. If the number of births is 30 per year and the number of deaths is 10 per year per 1,000 individuals, the growth rate would be:

$$(30 - 10)/1{,}000 = 0.02 = 2.0\%$$

The highest possible rate of natural increase for a population when resources are unlimited is called its **biotic potential** (Fig. 44.3). Whether the biotic potential is high or low depends on characteristics of the population that reduce or slow its potential reproduction, such as the following:

- Usual number of offspring per reproductive event
- Chances of survival until age of reproduction
- How often each individual reproduces
- Age at which reproduction begins

### *Mortality Patterns*

Population growth patterns assume that populations are made up of identical individuals. Actually the individuals of a population are in different stages of their life span. A **cohort** is all the members of a population born at the same time. Some investigators study population dynamics and construct life tables that show how many members of a cohort are still alive after certain intervals of time. For example, Table 44.2 is a life table for a bluegrass cohort. The cohort contains 843 individuals. The table tells us that after three months, 121 individuals have died, and therefore the mortality rate is 0.143 per capita. Another way to express this same statistic, however, is to consider that 722 individuals are still alive—have survived—after three months. **Survivorship** is the probability of newborn individuals of a cohort surviving to particular ages. If we plot the number surviving at each age, a survivorship curve is produced (see Fig. 44.4*b*). The results of such investigations show that each species tends to have one of the typical survivorship curves.

**TABLE 44.2**

**A Life Table for a Bluegrass Cohort**

| *Age (months)* | *Number Observed Alive* | *Number Dying* | *Mortality Rate per Capita* | *Avg. Number of Seeds per Individual* |
|---|---|---|---|---|
| 0–3 | 843 | 121 | 0.143 | 0 |
| 3–6 | 722 | 195 | 0.271 | 300 |
| 6–9 | 527 | 211 | 0.400 | 620 |
| 9–12 | 316 | 172 | 0.544 | 430 |
| 12–15 | 144 | 95 | 0.626 | 210 |
| 15–18 | 54 | 39 | 0.722 | 60 |
| 18–21 | 15 | 12 | 0.800 | 30 |
| 21–24 | 3 | 3 | 1.000 | 10 |
| 24 | 0 | — | — | — |

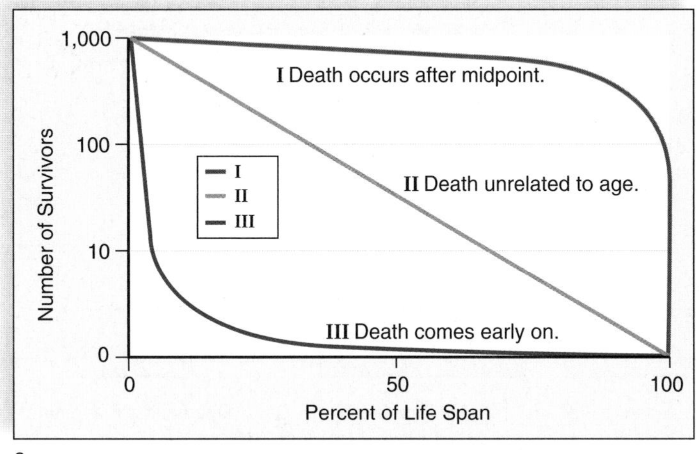


a.

b. Bluegrasses

c. Lizards

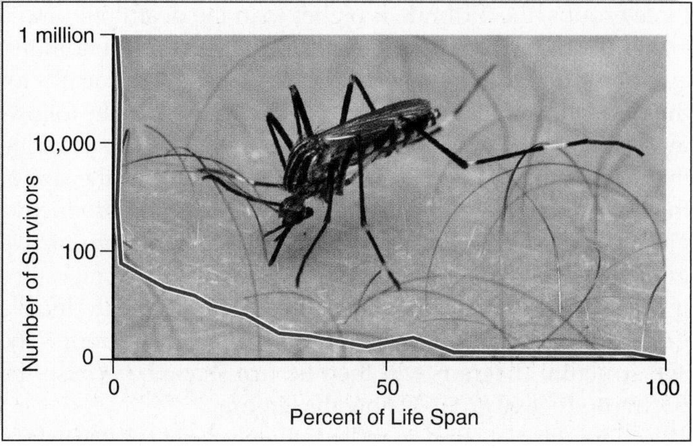


d. Mosquitoes

**FIGURE 44.4 Survivorship curves.**

Survivorship curves show the number of individuals of a cohort that are still living over time. **a.** Three generalized survivorship curves. **b.** The survivorship curve for bluegrasses seems to be a combination of the type I and type II curves. **c.** The survivorship curve for lizards fits the type II curve somewhat. **d.** The survivorship curve for mosquitoes is a type III curve.

Three types of idealized survivorship curves, numbered I, II, and III, are usually recognized (Fig. 44.4*a*). The type I curve is characteristic of a population in which most individuals survive well past the midpoint of the life span, and death does not come until near the end of the life span. Animals that have this type of survivorship curve include large mammals and humans in more-developed countries. On the other hand, the type III curve is typical of a population in which most individuals die very young. This type of survivorship curve occurs in many invertebrates, fishes, and humans in less-developed countries. In the type II curve, survivorship decreases at a constant rate throughout the life span. In many songbirds and small mammals, death is usually unrelated to age; thus they represent a type II survivorship curve.

The survivorship curves of natural populations do not fit these three idealized curves exactly. In a bluegrass cohort, as shown in Table 44.2 for example, most individuals survive till six to nine months, and then the chances of survivorship diminish at an increasing rate (Fig. 44.4*b*). Statistics for a lizard cohort are close enough to classify the survivorship curve in the type II category (Fig. 44.4*c*), while a mosquito cohort has a type III curve (Fig. 44.4*d*).

Much can be learned about the life history of a species by studying its life table and the survivorship curve that can be constructed based on this table. Would you predict that natural selection would favor the most or the fewest members of a population with a type III survivorship curve to contribute offspring to the next generation? Obviously, since death comes early for most members, only a few are living long enough to reproduce. What about the other two types of survivorship curves?

Other types of information are also available from studying life tables. Look again at Table 44.2, the bluegrass life table. It tells us that per capita seed production increases as plants mature, and then seed production drops off. How do you predict this would compare to a cohort of human beings?

## *Age Distribution*

When the individuals in a population reproduce repeatedly, several generations may be alive at any given time. From the perspective of population growth, a population contains three major age groups: prereproductive, reproductive, and postreproductive. Populations differ according to what proportion of the population falls in each age group. At least three **age structure diagrams** are possible (Fig. 44.5).

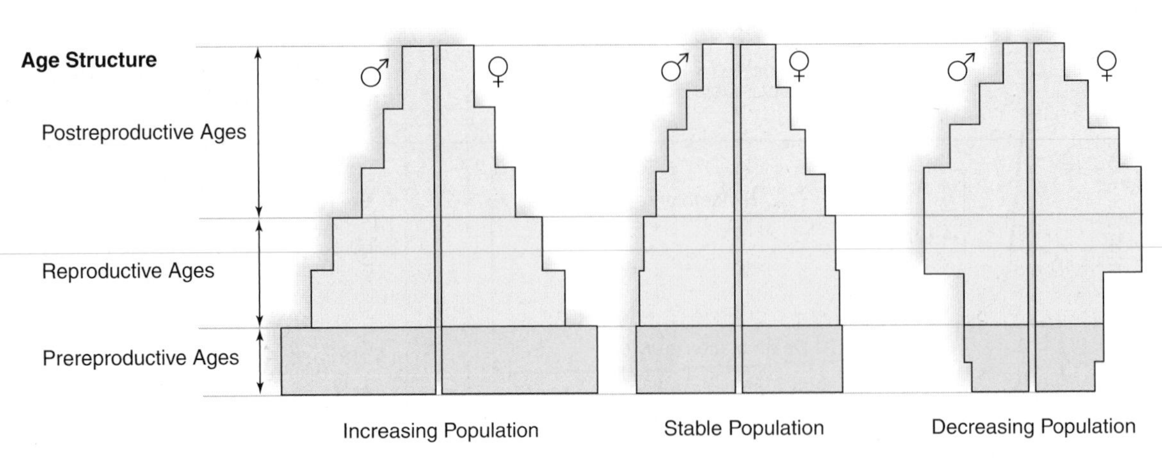


**FIGURE 44.5 Age structure diagrams.**

Typical age structure diagrams for hypothetical populations that are increasing, stable, or decreasing. Different numbers of individuals in each age class create these distinctive shapes. In each diagram, the left half represents males while the right half represents females.

When the prereproductive group is the largest of the three groups, the birthrate is higher than the death rate, and a pyramid-shaped diagram is expected. Under such conditions, even if the growth for that year were matched by the deaths for that year, the population would continue to grow in the following years. Why? Because there are more individuals entering than leaving the reproductive years. Eventually, as the size of the reproductive group equals the size of the prereproductive group, a bell-shaped diagram will result. The postreproductive group will still be the smallest, however, because of mortality. If the birthrate falls below the death rate, the prereproductive group will become smaller than the reproductive group. The age structure diagram will then be urn-shaped, because the postreproductive group is now the largest.

The age distribution reflects the past and future history of a population. Because a postwar baby boom occurred in the United States between 1946 and 1964, the postreproductive group will soon be the largest group.

### Check Your Progress 44.2

1. Describe the difference between population density and population distribution.
2. Describe the differences among type I, II, and III survivorship curves.
3. Why does a bell-shaped age pyramid indicate a growing population?

## 44.3 Population Growth Models

Based on observation and natural selection principles, ecologists have developed two working models for population growth. In the pattern called **semelparity** [Gk. *seme,* once, and L. *parous,* to bear or bring forth], the members of the population have only a single reproductive event in their lifetime. When the time for reproduction draws near, the mature adults cease to grow and expend all their energy in reproduction, and then die. Many insects, such as winter moths, and annual plants, such as zinnias, follow this pattern of reproduction growth. They produce a resting stage of development such as eggs or seeds that survive unfavorable conditions and resume growth the next favorable season. In other words, semelparity is an adaptation to an unstable environment. In the pattern called **iteroparity** [Gk. *itero,* repeat], members of the population experience many reproductive events throughout their lifetime. They continue to invest energy in their future survival and this increases their chances of reproducing again. Iteroparity is an adaptation to a stable environment when the offsprings' chances of survival are relatively high. Most vertebrates, shrubs, and trees have this pattern of reproduction.

Figure 44.6 exemplifies that reproduction does not always fit these two patterns. However, ecologists have found it useful to develop mathematical models of

a.

b.

**FIGURE 44.6 Patterns of reproduction.**

Although we can assume that members of populations either have a single sacrificial reproductive event, or they reproduce repeatedly, actually both are simplifications. **a.** Aphids reproduce repeatedly by asexual reproduction during the summer and then reproduce sexually only once, right before the onset of winter. Therefore, aphids use both patterns of reproduction. **b.** The offspring of annual plants can germinate several seasons later. Under these circumstances, population size of these organisms could fluctuate according to environmental conditions.

population growth based on these two very different patterns of reproduction. Although the mathematical models we will be describing are simplifications, they still may predict how best to control the distribution and abundance of organisms, or how to predict the responses of populations when their environment is altered in some way. Testing predictions permits the development of new hypotheses that can then be tested.

## Exponential Growth

As an example of semelparous reproduction, we will consider a population of insects in which females reproduce only once a year and then the adult population dies. Each female produces on the average 2.4 eggs per generation that will survive the winter and become offspring the next year. In the next generation, each female will again produce 2.4 eggs. In the case of discrete breeding, $R$ = net reproductive rate.[1] Why net reproductive rate? Because it is the observed rate of natural increase after deaths have occurred.

Figure 44.7*a* shows how the population would grow year after year for ten years, assuming that $R$ stays constant from generation to generation. This growth is equal to the size of the population because all members of the previous generation have died. Mayflies, featured in Figure 44.7, have one reproductive event usually in the spring—hence their name—and then development of the next generation requires as many as 50 molts during the winter. Figure 44.7*b* shows the growth curve for such a population. This growth curve, which has a J shape, depicts exponential growth. With **exponential growth,** the number of individuals added each generation increases as the total number of females increases.

Notice that the curve has these phases:

*Lag phase* During this phase, growth is slow because the population is small.

*Exponential growth phase* During this phase, growth is accelerating.

Figure 44.7*c* gives the mathematical equation that allows you to calculate growth and size for any population that has discrete (nonoverlapping) generations. In other words, as discussed, all members of the previous generation die off before the new generation appears. To use this equation to determine future population size, it is necessary to know $R$, which is the net reproductive rate determined after gathering mathematical data regarding past population increases. Notice that even though $R$ remains constant, growth is exponential because the number of individuals added each year is increasing. Therefore, the growth of the population is accelerating.

For exponential growth to continue unchecked, each member of the population has to have unlimited resources. Plenty of room, food, shelter, and any other requirements necessary to sustain growth must be available. But in reality, environmental conditions prevent exponential growth. Eventually, any further growth is impossible because the food supply runs out and waste products begin to accumulate. Also, as the population increases in size so do the effects of competition between members, predation, parasites, and disease.

| Generation | Population Size |
|---|---|
| 0 | 10.0 |
| 1 | 24.0 |
| 2 | 57.6 |
| 3 | 138.2 |
| 4 | 331.7 |
| 5 | 796.1 |
| 6 | 1,910.6 |
| 7 | 4,585.4 |
| 8 | 11,005.0 |
| 9 | 26,412.0 |
| 10 | 63,388.8 |

a.

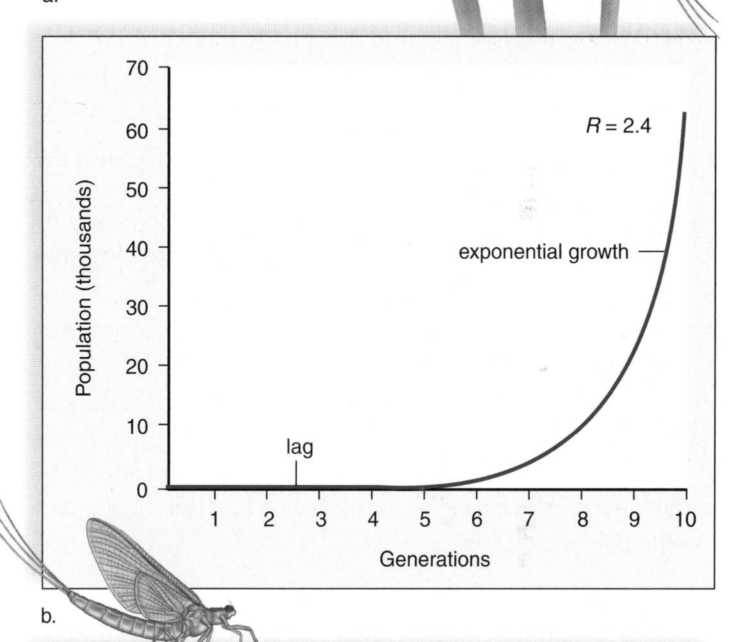


b.

To calculate population size from year to year, use this formula:

$$N_{t+1} = RN_t$$

$N_t$ = number of females already present

$R$ = net reproductive rate

$N_{t+1}$ = population size the following year

c.

**FIGURE 44.7 Model for exponential growth.**
When the data for discrete reproduction in (**a**) are plotted, the exponential growth curve in (**b**) results. **c.** This formula produces the same results as (**a**) and generates the same graph.

[1] The change of *r* to *R* is simply customary in discrete breeding calculations; both coefficients deal with the same thing (birth minus death).

## Logistic Growth

What type of growth curve results when environmental factors opposing growth come into play? In 1930, Raymond Pearl developed a method for estimating the number of yeast cells accruing every two hours in a laboratory culture vessel. His data are shown in Figure 44.8*a*. When the data are plotted, the growth curve has the appearance shown in Figure 44.8*b*. This type of growth curve is a sigmoidal (S) or S-shaped curve.

Notice that this so-called **logistic growth** has these phases:

*Lag phase* During this phase, growth is slow because the population is small.
*Exponential growth phase* During this phase, growth is accelerating.
*Deceleration phase* During this phase, growth slows down.
*Stable equilibrium phase* During this phase, there is little if any growth because births and deaths are about equal.

Figure 44.8*c* gives the mathematical equation that allows us to calculate logistic growth (so-called because the exponential portion of the curve would produce a straight line if the log of $N$ were plotted). The entire equation for logistic growth is:

$$\frac{N}{t} = rN\frac{(K-N)}{K}$$

but let's consider each portion of the equation separately.

Because the population has repeated reproductive events, we need to consider growth as a function of change in time ($\Delta$):

$$\frac{\Delta N}{\Delta t} = rN$$

If the change in time is very small, then we can turn to differential calculus, and the instantaneous population growth ($d$) is given by:

$$\frac{dN}{dt} = rN$$

This portion of the equation applies to the first two phases of growth—the lag phase and the exponential growth phase. Here, also, we do not expect exponential growth to continue. Charles Darwin calculated that a single pair of elephants could have over 19 million live descendants after 750 years. Others have calculated that a single female housefly could produce over 5 trillion flies in one year! Such explosive growth does not occur because environmental conditions, both abiotic and biotic, cause population growth to slow. The yeast population was grown in a vessel in which food could run short and waste products could accumulate. These environmental conditions prevent exponential growth from continuing. Look again at Figure 44.8. Following exponential growth, a population is expected to enter a deceleration phase and then a stable equilibrium phase of the logistic growth curve. Now the population is at the carrying capacity of the environment.

**Growth of Yeast Cells in Laboratory Culture**

| Time ($t$) (hours) | Number of individuals ($N$) | Number of individuals added per 2-hour period $\left(\frac{\Delta N}{\Delta t}\right)$ |
|---|---|---|
| 0 | 9.6 | 0 |
| 2 | 29.0 | 19.4 |
| 4 | 71.1 | 42.1 |
| 6 | 174.6 | 103.5 |
| 8 | 350.7 | 176.1 |
| 10 | 513.3 | 162.6 |
| 12 | 594.4 | 81.1 |
| 14 | 640.8 | 46.4 |
| 16 | 655.9 | 15.1 |
| 18 | 661.8 | 5.9 |

a.

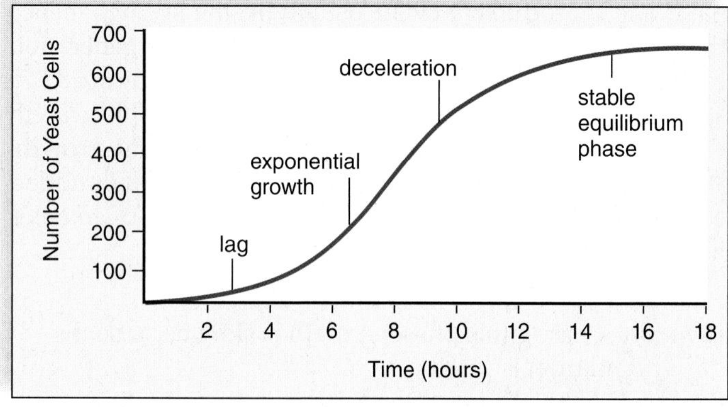


b.

To calculate population growth as time passes, use this formula:

$$\frac{N}{t} = rN\left(\frac{K-N}{K}\right)$$

$N$ = population size

$N/t$ = change in population size

$r$ = rate of natural increase

$K$ = carrying capacity

$\frac{K-N}{K}$ = effect of carrying capacity on population growth

c.

**FIGURE 44.8 Model for logistic growth.**

When the data for repeated reproduction (**a**) are plotted, the logistic growth curve in (**b**) results. **c.** This formula produces the same results as (**a**) and generates the same graph.

### *Carrying Capacity*

The environmental **carrying capacity** ($K$) is the maximum number of individuals of a given species the environment can support. The closer population size nears the carrying capacity of the environment, the more likely resources will become scarce and biotic effects such as competition and predation will become evident. The birthrate is expected to decline and the death rate is expected to increase. This will result in a decrease in population growth; eventually, the population stops growing and its size remains stable. Carrying capacity in any given environment can vary throughout time depending on fluctuating conditions, for example, amount of rainfall from one year to the next.

How does our mathematical model for logistic growth take this process into account? To our equation for growth under conditions of exponential growth we add the term:

$$\frac{(K - N)}{K}$$

In this expression, $K$ is the carrying capacity of the environment. The easiest way to understand the effects of this term is to consider two extreme possibilities. First, consider a time at which the population size is well below carrying capacity. Resources are relatively unlimited, and we expect rapid, nearly exponential growth to take place. Does the model predict this? Yes, it does. When $N$ is very small relative to $K$, the term $(K-N)/K$ is very nearly $(K-0)/K$, or approximately 1. Therefore, $(dN)/(dt)$ = approximately $rN$.

Similarly, consider what happens when the population reaches carrying capacity. Here, we predict that growth will stop and the population will stabilize. What happens in the model? When $N = K$, the term $(K - N)/K$ declines from nearly 1 to 0, and the population growth slows to zero.

As mentioned, the model we have developed predicts that exponential growth will occur only when population size is much lower than the carrying capacity. So, as a practical matter, if we are using a fish population as a continuous food source, it would be best to maintain the population size in the exponential phase of the logistic growth curve. Biotic potential can have its full effect, and the birthrate is the highest it can be during this phase. If we overfish, the population will sink into the lag phase, and it will be years before exponential growth recurs. On the other hand, if we are trying to limit the growth of a pest, it is best, if possible, to reduce the carrying capacity rather than reduce the population size. Reducing the population size only encourages exponential growth to begin once again. Farmers can reduce the carrying capacity for a pest by alternating rows of different crops rather than growing one type of crop throughout the entire field.

**Check Your Progress 44.3**

1. What ecological factors might result in iteroparity rather than semelparity?
2. How does carrying capacity (K) limit exponential growth?

# 44.4 Regulation of Population Size

In a study of winter moth population dynamics, it was discovered that a large proportion of eggs did not survive the winter and exponential growth never occurred. Perhaps the low number of individuals at the start of each season helps prevent the occurrence of exponential growth. This observation raises the question, "How well do the models for exponential and logistic growth predict population growth in natural populations?"

Is it possible, for example, that exponential growth may cause population size to rise above the carrying capacity of the environment, and as a consequence a population crash may occur? For example, in 1911, 4 male and 21 female reindeer were released on St. Paul Island in the Bering Sea off Alaska. St. Paul Island had a completely undisturbed environment—there was little hunting pressure, and there were no predators. The herd grew exponentially to about 2,000 reindeer in 1938, overgrazed the habitat, and then abruptly declined to only 8 animals in 1950 (Fig. 44.9). This pattern of a population explosion eventually followed by a population crash is called irruptive, or Malthusian, growth. It is named in honor of the eighteenth-century economist who had a great influence on Charles Darwin. Populations do not ordinarily undergo Malthusian growth because of factors that regulate population growth.

**FIGURE 44.9 Density-dependent effect.**

On St. Paul Island, Alaska, reindeer, *Rangifer*, grew exponentially for several seasons and then underwent a sharp decline as a result of overgrazing the available range.

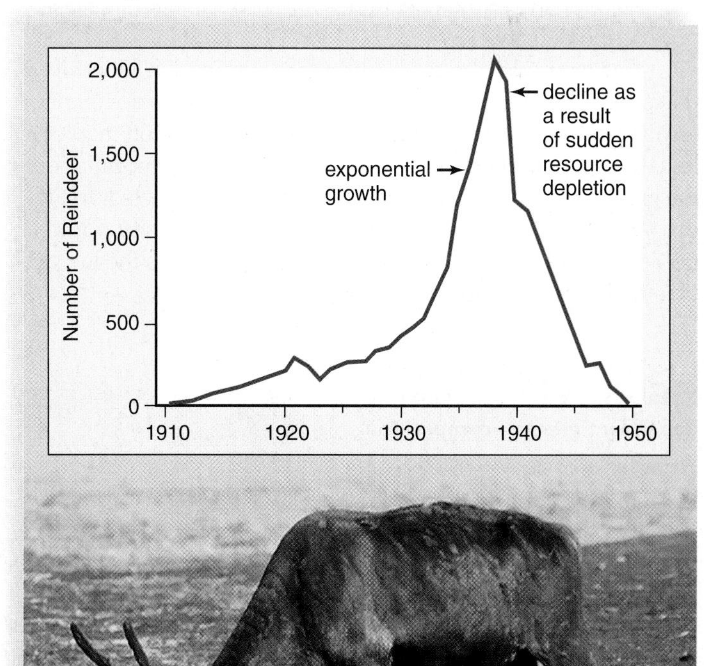

a. Low density of mice

b. High density of mice

**FIGURE 44.10 Density-independent effects.**

The impact of a density-independent factor, such as weather or natural disasters, is not influenced by population density. The impact of a flash flood on (**a**) a low-density population (mortality rate of 3/5, or 60%) is similar to the impact on (**b**) a high-density population (mortality rate of 12/20, or 64%).

## Factors That Regulate Population Growth

Ecologists have long recognized that both abiotic conditions and biotic conditions play an important role in regulating population size in nature.

### *Density-Independent Factors*

Abiotic factors include droughts (lack of rain), freezes, hurricanes, floods, and forest fires. Any one of these natural disasters can cause individuals to die and lead to a sudden and catastrophic reduction in population size. However, such an event does not necessarily kill a larger percentage of a dense population compared to a less dense population. Therefore, an abiotic factor is usually a **density-independent factor,** meaning that the intensity of the effect does not increase with increased population density.

For example, the proportion of a population killed in a flash flood event is independent of density—floods do not necessarily kill a larger percentage of a dense population than of a less dense population. Nevertheless, the larger the population, the greater the number of individuals probably affected. In Figure 44.10, the impact of a flash flood on a low-density population of mice living in a field was 3 out of 5, whereas the impact on a high-density population was 12 out of 20.

### *Density-Dependent Factors*

Biotic factors are considered **density-dependent factors** because the percentage of the population affected does increase as the density of the population increases. Competition, predation, and parasitism are all biotic factors that increase in intensity as the density increases.

*Competition* can occur when members of a species attempt to use resources (such as light, food, space) that are in limited supply. As a result, not all members of the population can have access to the resource to the degree necessary to ensure survival or reproduction or some other aspect of their life history. As an example, let's consider a woodpecker population in which members have to compete for nesting sites. Each pair of birds requires a tree hole to raise offspring. If there are more holes than breeding pairs, each pair can have a hole in which to lay eggs and rear young birds (Fig. 44.11*a*). But if there are fewer holes than there are breeding pairs, then each pair must compete to acquire a nesting site (Fig. 44.11*b*). Pairs that fail to gain access to holes will be unable to contribute new members to the population.

Competition for food also controls population growth. However, resource partitioning among different age groups is a way to reduce competition for food. The life cycle of butterflies includes caterpillars, which require a different food from the adults. The caterpillars graze on leaves, while

**FIGURE 44.11 Density-dependent effects—competition.**

The impact of competition on a population is directly proportional to the density of the population. When density is low (**a**), every member of the population has access to the resource. But when the density is high (**b**), there is competition between members of the population to gain access to available resources.

a. Low density of birds

b. High density of birds

the adults feed on nectar produced by flowers. Therefore, parents do not compete with their offspring for food.

*Predation* occurs when one living organism, the predator, eats another, the prey. In the broadest sense, predation can include not only animals such as lions, which kill zebras, but also filter-feeding blue whales, which strain krill from the ocean waters; and even herbivorous deer, which browse on trees and bushes. The effect of predation on a prey population generally increases as the prey population grows denser, because prey is easier to find when hiding places are limited. Consider a field inhabited by a population of mice (Fig. 44.12). Each mouse must have a hole in which to hide to avoid being eaten by a hawk. If there are 100 holes, and a low density of 102 mice, then only 2 mice will be left out in the open. It might be hard for the hawk to find only 2 mice in the field. If neither mouse is caught, then the predation rate is 0/2 = 0%. However, if there are 100 holes, and a high density of 120 mice, then there is a greater chance that the hawk will be able to find some of these 100 mice without holes. If half of the exposed mice are caught, the predation rate is 50/100 = 50%. Therefore, increasing the density of the available prey has increased the proportion of the population preyed upon.

Parasites, such as blood-sucking ticks, are generally much smaller than their host. Although parasites do not always kill their hosts, they do usually weaken them over time. A highly parasitized individual is less apt to produce as many offspring than if it were healthy. In this way, parasitism also plays a role in regulating population size.

### *Other Considerations*

Density-independent and density-dependent factors are extrinsic to the organism. Is it possible that intrinsic factors—those based on the anatomy, physiology, or behavior of the organism—might affect population size and growth rates? Territoriality and dominance hierarchies are behaviors that affect population size and growth rates. Recruitment and migration are other intrinsic social means by which the population sizes of more complex organisms are regulated.

Outside of any regulating factors, it could be that some populations have an innate instability. Ecologists have developed models that predict complex, erratic changes in even simple systems. For example, a computer model of Dungeness crab populations assumed that adults produce many larvae and then die. Most of the larvae do not survive, and those that do stay close to home. Under these circumstances, the model predicted wild fluctuations in population size with no recurring pattern, which is now termed *chaos*.

Population growth regulating factors can serve as selective agents. Some members of a population may possess traits that make it more likely that they, rather than other members of the population, will survive and reproduce when these particular density-independent or density-dependent factors are present in the environment. Therefore, these traits will be more prevalent in the next generation whenever these factors are a part of the environment (see Fig. 15.11).

**Check Your Progress 44.4**

1. What effect does population density have on competition and predation?
2. Give an example to show that a density-independent factor can be a selective agent.

a. Low density of mice

b. High density of mice

**FIGURE 44.12**
**Density-dependent effects—predation.**

The impact of predation on a population is directly proportional to the density of the population. In a low-density population (**a**), the chances of a predator finding the prey are low, resulting in little predation. But in the higher density population (**b**), there is a greater likelihood of the predator locating potential prey, resulting in a greater predation rate.

# 44.5 Life History Patterns

Populations vary on such particulars as the number of births per reproduction, the age of reproduction, the life span, and the probability of living the entire life span. Such particulars are part of a species' life history. Life histories contain characteristics that can be thought of as trade-offs. Each population is able to capture only so much of the available energy, and how this energy is distributed between its life span (short versus long), reproduction events (few versus many), care of offspring (little versus much), and so forth has evolved over time. Natural selection shapes the final life history of individual species, and therefore it is not surprising that related species, such as frogs and toads, may have different life history patterns, if they occupy different types of environments (Fig. 44.13).

The logistic population growth model has been used to suggest that members of some populations are subject to *r*-selection and members of other populations are subject to *K*-selection. In fluctuating and/or unpredictable environments, density-independent factors will keep populations in the lag or exponential phase of population growth. Population size is low relative to *K*, and ***r*-selection** favors *r*-strategists. As a consequence of this pattern of energy allocation, small individuals that mature early and have a short life span are favored. Most energy goes into producing many relatively small offspring, and no energy goes into parental care. The more offspring, the more likely it is that some of them will survive any future population crash. Because of low population densities, density-dependent mechanisms such as predation and intraspecific competition are unlikely to play a major role in regulating population size

a. Mouth-brooding frog, *Rhinoderma darwinii*

b. Strawberry poison arrow frog, *Dendrobates pumilio*

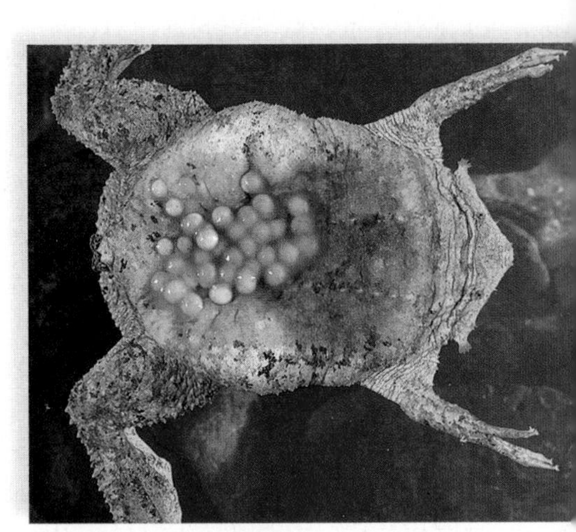

c. Surinam toad, *Pipa pipa*

d. Wood frog, *Rana sylvatica*

e. Midwife toad, *Alyces obstetricans*

**FIGURE 44.13 Parental care among frogs and toads.**

**a.** In mouth-brooding frogs of South America, the male carries the larvae in a vocal pouch (brown area), which elongates the full length of his body before the froglets are released. **b.** In poison arrow frogs of Costa Rica, after the eggs hatch, the tadpoles wiggle onto the parent's back (at white arrow) and are then carried to water. **c.** In the Surinam toads of South America, males fertilize the eggs during a somersaulting bout because the eggs are on the female's back. Each egg develops in a separate pocket, where the tail of the tadpole acts as a placenta to take nourishment from the female's circulatory system. **d.** Wood frogs live mainly in wooded areas, but they breed in temporary ponds arising from spring snowmelt. Toads and any frogs that lay their eggs on land exhibit various forms of parental care. **e.** The midwife toad of Europe carries strings of eggs entwined around his hind legs and takes them to water when they are ready to hatch.

**FIGURE 44.14 Life history strategies.**

Are dandelions *r*-strategists with the characteristics noted, and are bears *K*-strategists with the characteristics noted? Most often the distinctions between these two possible life strategies are not as clear-cut as they may seem.

and growth rates most of the time. Such organisms are often very good dispersers and colonizers of new habitats. Classic examples of such *opportunistic species* are bacteria, some fungi, many insects, rodents, and annual plants (Fig. 44.14, *left*).

In contrast, some environments are relatively stable and/or predictable, where populations tend to be near *K*, with minimal fluctuations in size. Resources such as food and shelter will be relatively scarce for these individuals, and those that are best able to compete will have the largest number of offspring. ***K*-selection** favors *K*-strategists, which allocate energy to their own growth and survival and to the growth and survival of their offspring. Therefore, they are usually fairly large, late to mature, have low fecundity and parity, and have a fairly long life span. Because these organisms, termed *equilibrium species*, are strong competitors, they can become established and exclude opportunistic species. They are specialists rather than colonizers and tend to become extinct when their normal way of life is destroyed. The best possible examples of *K*-strategists include long-lived plants (saguaro cacti, oaks, cypresses, and pines), birds of prey (hawks and eagles), and large mammals (whales, elephants, bears, and humans) (Fig. 44.14, *right*). Another example of a *K*-strategist is the Florida panther, the largest mammal in the Florida Everglades. It requires a very large range and produces few offspring that require parental care. Currently, the Florida panther is unable to compensate for a reduction in its range and is therefore on the verge of extinction.

Nature is actually more complex than these two possible life history patterns suggest. It now appears that our descriptions of *r*-strategist and *K*-strategist populations are at the ends of a continuum, and most populations lie somewhere in between these two extremes. For example, recall that plants have a two-generation life cycle, the sporophyte and the gametophyte. Ferns, which could be classified as *r*-strategists, distribute many spores and leave the gametophyte to fend for itself, but gymnosperms (e.g., pine trees) and angiosperms (e.g., oak trees), which could be classified as *K*-strategists, retain and protect the gametophyte. They produce seeds that contain the next sporophyte generation plus stored food. The added investment is significant, but these plants still release copious numbers of seeds.

Also, adult size is not always a determining factor for the life history pattern. For example, a cod is a rather large fish weighing up to 12 kg and measuring nearly 2 m in length—but cod release gametes in vast numbers, the zygotes form in the sea, and the parents make no further investment in developing offspring. Of the 6–7 million eggs released by a single female cod, only a few will become adult fish. Cod are considered *r*-strategists.

### Check Your Progress 44.5

1. Describe the general characteristics of a *K*- and *r*-strategist.

*ecology focus*

## When a Population Grows Too Large

White-tailed deer, which live from southern Canada to below the equator in South America, are prolific breeders. In one study, investigators found that two male and four female deer produced 160 offspring in six years. Theoretically, the number could have been 300 because a large proportion of does (female deer) breed their first year, and once they start breeding, produce about two young each year of life.

A century ago, the white-tailed deer population across eastern United States was less than half a million. Today, it is well over 200 million deer—even more than existed when Europeans first arrived to colonize America. This dramatic increase in population size can probably be attributed to a lack of predators. For one thing, hunting is tightly controlled by government agencies, and in some areas, it is banned altogether because of the danger it poses to the general public. Similarly, the natural predators of deer, such as wolves and mountain lions, are now absent from most regions. This too can be traced to a large human population that is fearful of large predators because they could possibly attack humans and domestic animals.

We like to see a mother with her fawns by the side of the road or scampering off into the woods with tails raised to show off the white underside. Or, we find it thrilling to see a large buck (male deer) with majestic antlers partially hidden in the woods (Fig. 44A). But the sad reality is that, in those areas where deer populations have become too large, the deer suffer from starvation as they deplete their own food supply. For example, after deer hunting was banned on Long Island, New York, the deer population quickly outgrew available food resources. The animals became sickly and weak and weighed so little that their ribs, vertebrae, and pelvic bones were visible through their skin.

Then, too, a very large deer population causes humans many problems. A homeowner is dismayed to see new plants decimated and evergreen trees damaged due to the munching of deer. The economic damage that large deer populations cause to agriculture, landscaping, and forestry exceeds a billion dollars per year. More alarming, a million deer-vehicle collisions take place in the U.S. each year, resulting in over a billion dollars in insurance claims, thousands of human injuries, and hundreds of human deaths. Lyme disease, transmitted by deer ticks to humans, infects over 3,000 people annually. Untreated Lyme disease can lead to debilitating arthritic symptoms.

Deer overpopulation hurts not only deer and humans, but other species as well. The forested areas that are overpopulated by deer have fewer understory plants. Furthermore, the deer selectively eat certain species of plants, while leaving others alone. This can cause long-lasting changes in the number and diversity of trees in forests, leading to a negative economic impact on logging and forestry. The number of songbirds, insects, squirrels, mice, and other animals declines with an increasing deer population. It behooves us, therefore, to learn to manage deer populations. And the good news is that in some states, such as Texas, large landowners now set aside a portion of their property for a deer herd. They improve the nutrition of the herd and restrict the harvesting of young bucks, but allow the harvesting of does. The result is a self-sustaining herd that brings them economic benefits because they charge others for the privilege of hunting on their land.

a.

b. c.

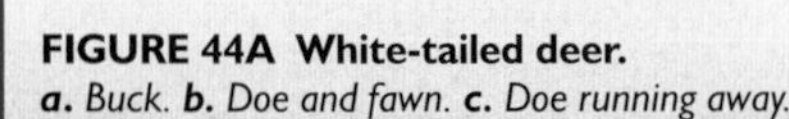

**FIGURE 44A White-tailed deer.**
***a.*** *Buck.* ***b.*** *Doe and fawn.* ***c.*** *Doe running away.*

## 44.6 Human Population Growth

The world's population has risen steadily to a present size of about 6.7 billion people (Fig. 44.15). Prior to 1750, the growth of the human population was relatively slow, but as more reproducing individuals were added, growth increased, until the curve began to slope steeply upward, indicating that the population was undergoing exponential growth. The number of people added annually to the world population peaked at about 87 million around 1990, and currently it is a little over 79 million per year. This is roughly equal to the current populations of Argentina, Ecuador, and Peru combined.

The potential for future population growth can be appreciated by considering the **doubling time,** the length of time it takes for the population size to double. Currently, the doubling time is estimated to be 52 years. Such an increase in population size would put extreme demands on our ability to produce and distribute resources. In 52 years, the world would need double the amount of food, jobs, water, energy, and so on just to maintain the present standard of living.

Many people are gravely concerned that the amount of time needed to add each additional billion persons to the world population has become shorter and shorter. The first billion didn't occur until 1800; the second billion was attained in 1930; the third billion in 1960; and today there are over 6 billion. The world's population may level off at 8, 10.5, or 14.2 billion, depending on the speed at which the growth rate declines. Zero population growth cannot be achieved until on the average each couple has only two children and the number of women entering their reproductive years is the same as those leaving them behind.

### More-Developed Versus Less-Developed Countries

The countries of the world can be divided into two groups. In the **more-developed countries (MDCs),** typified by countries in North America, Europe, Japan, and Australia, population growth is low, and the people enjoy a good standard of living. In the **less-developed countries (LDCs),** such as certain countries in Latin America, Africa, and Asia, population growth is expanding rapidly, and the majority of people live in poverty.

The MDCs doubled their populations between 1850 and 1950. This was largely due to a decline in the death rate, the development of modern medicine, and improved socioeconomic conditions. The decline in the death rate was followed shortly thereafter by a decline in the birthrate, so that populations in the MDCs experienced only modest growth between 1950 and 1975. This sequence of events (i.e., decreased death rate followed by decreased birthrate) is termed a **demographic transition.** Yearly growth of the MDCs has now stabilized at about 0.1%.

In contrast to the other MDCs, there is no end in sight to the U.S. population growth. Although yearly growth of the U.S. population is only 0.6%, many people immigrate to the United States each year. In addition, an unusually large number of births occurred between 1946 and 1964 (called a baby boom), resulting in a large number of women still of reproductive age.

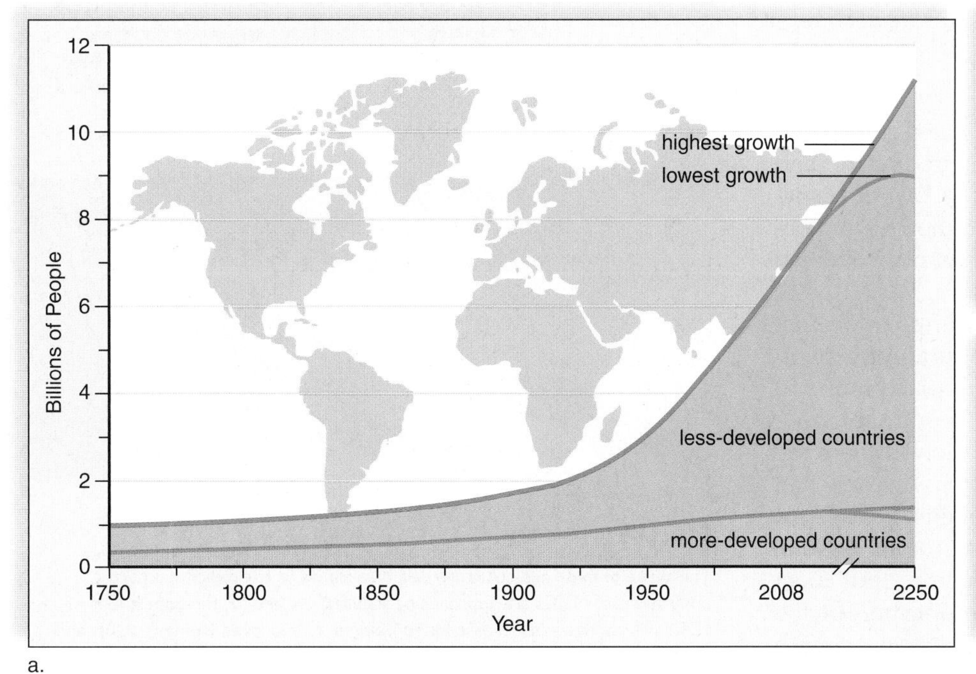


a.

b.

**FIGURE 44.15 World population growth.**

**a.** The world's' population size is now about 7 billion. It is predicted that the world's population size may level off at 9 billion or increase to more than 11 billion by 2250, depending on the speed with which the growth rate declines. **b.** Lifestyle of most individuals in the MDCs (*above*) is contrasted with most individuals in the LDCs (*below*).

Most of the world's population (80%) now lives in LDCs. Although the death rate began to decline steeply in the LDCs following World War II because of the importation of modern medicine from the MDCs, the birthrate remained high. The yearly growth of the LDCs peaked at 2.5% between 1960 and 1965. Since that time, a demographic transition has occurred and the death rate and the birthrate have fallen. The collective growth rate for the LDCs is now 1.5%, but many countries in sub-Saharan Africa have not participated in this decline. In some of these, women average more than five children each.

The population of the LDCs may explode from 5.5 billion today to 8 billion in 2050. Some of this increase will occur in Africa, but most will occur in Asia because the Asian population is now about 4 times the size of the African population. Asia already has 56% of the world's population living on 31% of its arable (farmable) land. Therefore, Asia is expected to experience acute water scarcity, a significant loss of biodiversity, and more urban pollution. Twelve of the world's 15 most polluted cities are in Asia.

These are suggestions about how to reduce the expected population increase in the LDCs:

1. Establish and/or strengthen family planning programs. A decline in growth is seen in countries with good family planning programs supported by community leaders.
2. Use social progress to reduce the desire for large families. Providing education, raising the status of women, and reducing child mortality are social improvements that could reduce this desire.
3. Delay the onset of childbearing. This, along with wider spacing of births, could help birthrate decline.

## Age Distributions

The populations of the MDCs and LDCs can be divided into three age groups: prereproductive, reproductive, and postreproductive (Fig. 44.16). Currently, the LDCs are experiencing a population momentum because they have more women entering the reproductive years than older women leaving them.

Laypeople are sometimes under the impression that if each couple has two children, **zero population growth** (no increase in population size) will take place immediately. However, most countries today will continue growing due to the age structure of the population. If there are more young women entering the reproductive years than there are older women leaving them, so-called **replacement reproduction** will still result in population growth.

Many MDCs have a stable age structure, but most LDCs have a youthful profile—a large proportion of the population is younger than 15. For example, in Nigeria, 49% of the population is under age 15. On average, Nigerian women have 5.9 children. As a result of these rates, the population of Nigeria is expected to increase from 148 million presently to 282 million in 2050. The population of the continent of Africa is projected to increase from 957 million to 2 billion between 2008 and 2050. This means that the LDC populations will still expand, even after replacement reproduction is attained. The more quickly replacement reproduction is achieved, however, the sooner zero population growth will result.

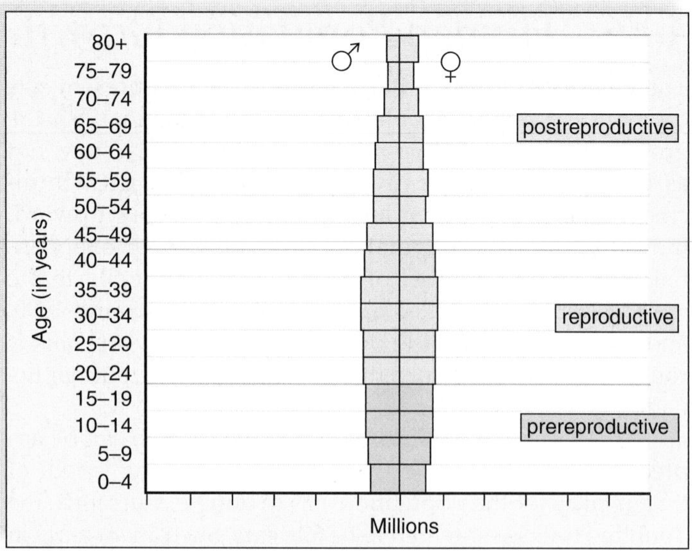


a. More-developed countries (MDCs)

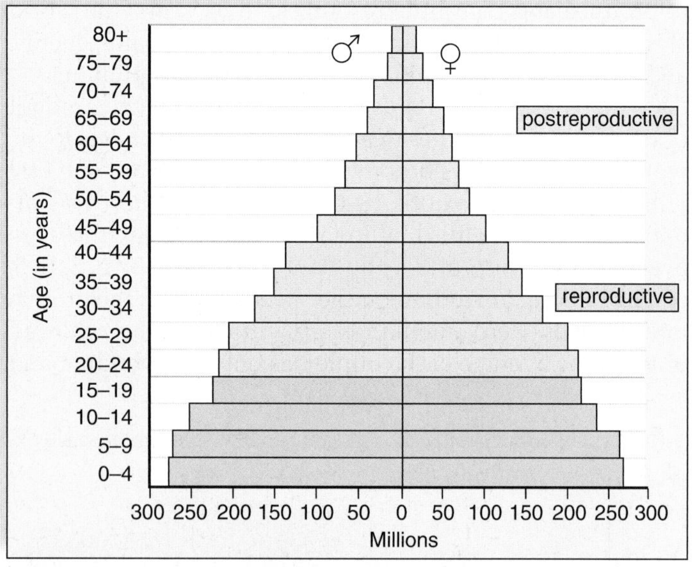


b. Less-developed countries (LDCs)

c.

**FIGURE 44.16 Age-structure diagrams.**

The shape of these age-structure diagrams allows us to predict that (a) the populations of MDCs are approaching stabilization, and (**b**) the populations of LDCs will continue to increase for some time. **c.** Improved women's rights and increasing contraceptive use could change this scenario. Here a community health worker is instructing women in Bangladesh about the use of contraceptives.

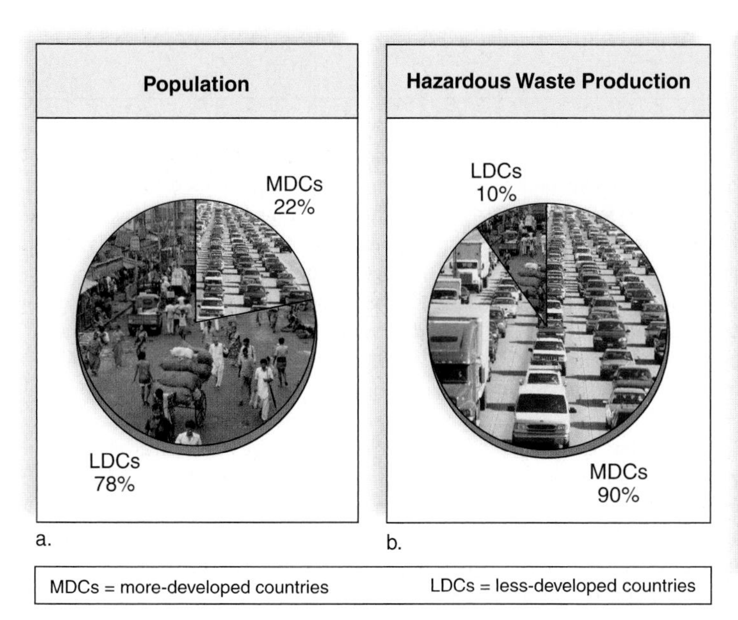


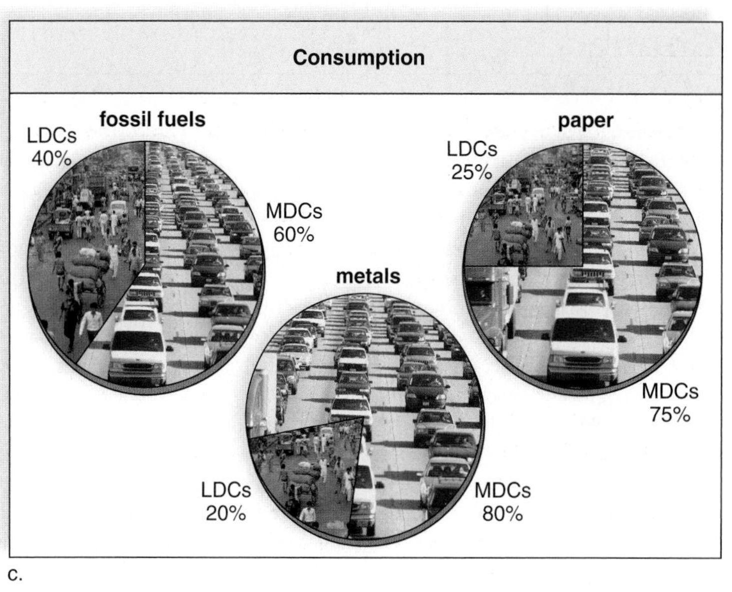


**FIGURE 44.17 Environmental impact caused by MDCs and LDCs.**

**a.** The combined population of MDCs is smaller than that of LDCs: The MDCs account for 22% of the world's population, and the LDCs account for 78%. **b.** MDCs produce most of the world's hazardous wastes—90% for MDCs compared with 10% for LDCs. The production of hazardous waste is tied to (**c**), the consumption of fossil fuels, metals, and paper, among other resources. The MDCs consume 60% of fossil fuels compared with 40% for the LDCs, 80% of metals compared with 20% for the LDCs, and 75% of paper compared with 25% for the LDCs.

# Population Growth and Environmental Impact

Population growth is putting extreme pressure on each country's social organization, the Earth's resources, and the biosphere. Since the population of the LDCs is still growing at a significant rate, it might seem that their population increase will be the greater cause of future environmental degradation. But this is not necessarily the case because the MDCs consume a much larger proportion of the Earth's resources than do the LDCs. This consumption leads to environmental degradation, which is a major concern in itself, but the usage of these resources is disproportionate to human population growth.

## *Environmental Impact*

The environmental impact (E.I.) of a population is measured not only in terms of population size, but also in terms of resource consumption per capita and the pollution that results because of this consumption. In other words:

$$\begin{aligned} \text{E.I.} &= \text{population size} \times \text{resource consumption per capita} \\ &= \text{pollution per unit of resource used} \end{aligned}$$

Therefore, there are two possible types of overpopulation: The first is simply due to population growth, and the second is due to increased resource consumption caused by population growth. The first type of overpopulation is more obvious in LDCs, and the second type is more obvious in MDCs because the per capita consumption is so much higher in the MDCs. For example, an average family in the United States, in terms of per capita resource consumption and waste production, is the equivalent of 30 people in India. We need to realize, therefore, that only a limited number of people can be sustained anywhere near the standard of living in the MDCs.

The current comparative environmental impact of MDCs and LDCs is shown in Figure 44.17. The MDCs account for only about one-fourth of the world population (Fig. 44.17*a*). But the MDCs account for 90% of the hazardous waste production (pollution) (Fig. 44.17*b*). Why should that be? MDCs account for much more pollution than LDCs because they consume much greater amounts of fossil fuel, metal, and paper than do LDCs, as shown in Figure 44.17*c*.

As the LDCs become more industrialized, their per capita consumption will also rise and, in some LDCs, it may nearly equal that of a more developed country. For example, China's economy is growing rapidly and, because China has such a large population (1.3 billion), it is already competing with the United States for oil and metals on the world markets. It could be that as developing countries consume more, people in the United States will adjust and learn to consume less.

## *Results of Environmental Impact*

Consumption of resources has a negative effect on the environment. In Chapter 45, you will learn how resource consumption leads to various pollution problems and, in Chapter 47, you will learn how our increasing environmental impact may cause a mass extinction of wildlife greater than any other since the evolution of life began.

**Check Your Progress 44.6**

1. Compare the population growth of the LDCs with that of the MDCs.
2. Under what circumstances can replacement reproduction still cause continued growth?
3. How will increased LDC consumption affect MDC consumption? Explain.

# Connecting the Concepts

The science of ecology began when nineteenth-century naturalists studied and described the population parameters of various species. However, modern ecology has grown into much more than a simple descriptive field. Ecology, which is integral to the evolution of species, is now very much an experimental and predictive science.

Much of the success in the development of ecology as a predictive science has come from studies of populations and the creation of models that examine how populations change over time. The simplest models are based on population growth when there are unlimited resources. This results in exponential growth, a type of population growth that is rarely seen in nature for extended lengths of time. Pest species may exhibit exponential growth until they run out of resources. Because few populations exhibit exponential growth, population ecologists incorporated resource limitation into their models. The simplest models that account for limited resources result in logistic growth. Populations that exhibit logistic growth will cease growth when they reach the environmental carrying capacity.

Many modern ecological studies are concerned with identifying the factors that place limits on population growth and that set the environmental carrying capacity. A combination of careful, descriptive studies, experiments done in nature, and sophisticated models has allowed ecologists to make good predictions about which factors have the greatest influence on population growth.

The next step in the development of modern ecology has been determining how populations of different species affect one another. This is known as community ecology. Because each population in a community responds to environmental changes in slightly different ways, developing predictive models that explain how communities change has been challenging. However, ecologists are better able to predict how communities will change through time and to understand what factors influence community properties such as species number, abundance of individuals, and species interactions.

## summary

### 44.1 Scope of Ecology

Ecology is the study of the interactions of organisms with other organisms and with the physical environment. Ecology encompasses several levels of study: organism, population, community, ecosystem, and finally the biosphere. Ecologists are particularly interested in how interactions affect the distribution, abundance, and life history strategies of organisms.

### 44.2 Demographics of Populations

Demographics include statistical data about a population. For example, population density is the number of individuals per unit area or volume. Distribution of individuals can be uniform, random, or clumped. A population's distribution is often determined by resources, that is, abiotic factors such as water, temperature, and availability of nutrients.

Population growth is dependent on number of births and number of deaths, immigration, and emigration. The number of births minus the number of deaths results in the rate of natural increase (growth rate). Mortality (deaths per capita) within a population can be recorded in a life table and illustrated by a survivorship curve. The pattern of population growth is reflected in the age distribution of a population, which consists of prereproductive, reproductive, and postreproductive segments. Populations that are growing exponentially have a pyramid-shaped age distribution pattern.

### 44.3 Population Growth Models

One model for population growth assumes that the environment offers unlimited resources. In the example given, the members of the population have discrete reproductive events, and therefore the size of next year's population is given by the equation $N_{t+1} = RN_t$. Under these conditions, exponential growth results in a J-shaped curve.

Most environments restrict growth, and exponential growth cannot continue indefinitely. Under these circumstances, an S-shaped, or logistic, growth curve results. The growth of the population is given by the equation $N/t = rN\,(K - N)/K$ for populations in which individuals have repeated reproductive events. The term $(K - N)/K$ represents the unused portion of the carrying capacity ($K$). When the population reaches carrying capacity, the population stops growing because environmental conditions oppose biotic potential, the maximum rate of natural increase (growth rate) for a population.

### 44.4 Regulation of Population Size

Population growth is limited by density-independent factors (e.g., abiotic factors such as weather) and density-dependent factors (biotic factors such as predation and competition). Other means of regulating population growth exist in some populations. For example, territoriality is possibly a means of population regulation. Other populations seem to not be regulated, and their population size fluctuates widely.

### 44.5 Life History Patterns

The logistic growth model has been used to suggest that the environment promotes either *r*-selection or *K*-selection. So-called *r*-selection occurs in unpredictable environments where density-independent factors affect population size. Energy is allocated to producing as many small offspring as possible. Adults remain small and do not invest in parental care of offspring. *K*-selection occurs in environments that remain relatively stable, where density-dependent factors affect population size. Energy is allocated to survival and repeated reproductive events. The adults are large and invest in parental care of offspring. Actual life histories contain trade-offs between these two patterns, and such trade-offs are subjected to natural selection.

### 44.6 Human Population Growth

The human population is still expanding, but deceleration has begun. It is unknown when the population size will level off, but it may occur by 2050. Substantial increases are expected in certain LDCs (less-developed countries) of Asia and also Africa. Support for family planning, human development, and delayed childbearing could help prevent the increase.

## understanding the terms

age structure diagram 823
biosphere 820
biotic potential 822
carrying capacity 827
cohort 822
community 820
demographic transition 833
demography 821
density-dependent factor 828
density-independent factor 828
doubling time 833
ecology 820
ecosystem 820
exponential growth 825
habitat 820
iteroparity 824
*K*-selection 831
less-developed country (LDC) 833
limiting factor 821
logistic growth 826
more-developed country (MDC) 833
population 820
population density 821
population distribution 821
rate of natural increase (*r*) 822
replacement reproduction 834
resource 821
*r*-selection 830
semelparity 824
survivorship 822
zero population growth 834

Match the terms to these definitions:

a. _______________ Due to industrialization, a decline in the birthrate following a reduction in the death rate so that the population growth rate is lowered.
b. _______________ Group of organisms of the same species occupying a certain area and sharing a common gene pool.
c. _______________ Growth, particularly of a population, in which the increase occurs in the same manner as compound interest.
d. _______________ Largest number of organisms of a particular species that can be maintained indefinitely in a given environment.
e. _______________ Maximum population growth rate under ideal conditions.

## reviewing this chapter

1. What are the various levels of ecological study? 820–21
2. What largely determines a population's density, distribution, and growth rate? 821–22
3. How does the mortality pattern and age distribution affect the growth rate? 822–24
4. What type of growth curve indicates that exponential growth is occurring? What are the environmental conditions for exponential growth? 825
5. What type of growth curve levels off because of environmental conditions? What environmental conditions are involved? 826
6. What is the carrying capacity of an area? 827
7. Are density-independent or density-dependent factors more likely to regulate population size? Why? 827–29
8. What other types of factors are involved in regulating population size? 829
9. Why would you expect the life histories of natural populations to vary and contain some characteristics that are so-called *r*-selected and some that are so-called *K*-selected? 830–31
10. What type of growth curve presently describes the growth of the human population? In what types of countries is most of this growth occurring, and how might it be curtailed? 833–34

## testing yourself

Choose the best answer for each question.

1. Which of these levels of ecological study involves an interaction between abiotic and biotic components?
   a. organisms
   b. populations
   c. communities
   d. ecosystem
   e. All of these are correct.
2. When phosphorus is made available to an aquatic community, the algal populations suddenly bloom. This indicates that phosphorus is
   a. a density-dependent regulating factor.
   b. a reproductive factor.
   c. a limiting factor.
   d. an *r*-selection factor.
   e. All of these are correct.
3. A J-shaped growth curve can be associated with
   a. exponential growth.
   b. biotic potential.
   c. unlimited resources.
   d. rapid population growth.
   e. All of these are correct.
4. An S-shaped growth curve
   a. occurs when resources become limited.
   b. includes an exponential growth phase.
   c. occurs in natural populations but not in laboratory ones.
   d. is subject to a sharp decline.
   e. All of these are correct.
5. If a population has a type I survivorship curve (most have a long life span), which of these would you also expect?
   a. a single reproductive event per adult
   b. overlapping generations
   c. reproduction occurring near the end of the life span
   d. a very low birthrate
   e. None of these are correct.
6. A pyramid-shaped age distribution means that
   a. the prereproductive group is the largest group.
   b. the population will grow for some time in the future.
   c. the country is more likely an LDC than an MDC.
   d. fewer women are leaving the reproductive years than entering them.
   e. All of these are correct.
7. Which of these is a density-independent regulating factor?
   a. competition
   b. predation
   c. weather
   d. resource availability
   e. the average age when childbearing begins
8. Fluctuations in population growth can correlate to changes in
   a. predation.
   b. weather.
   c. resource availability.
   d. parasitism.
   e. All of these are correct.
9. A species that has repeated reproductive events, lives a long time, but suffers a crash due to the weather is exemplifying
   a. *r*-selection.
   b. *K*-selection.
   c. a mixture of both *r*-selection and *K*-selection.
   d. density-dependent and density-independent regulation.
   e. a pyramid-shaped age structure diagram.

10. In which pair is the first included in the second?
    a. habitat—population
    b. population—community
    c. community—ecosystem
    d. ecosystem—biosphere
    e. All of these except a are correct.
11. How does population density differ from population distribution?
    a. Actually, population density is the same as population distribution.
    b. The greater the population density, the more likely that the population distribution will be clumped.
    c. Population density has nothing to do with population distribution.
    d. The less dense the population, the more likely that the population distribution will be equally spaced.
    e. Both b and c are correct.
12. Which one of these has nothing to do with a population's growth curve?
    a. exponential growth
    b. biotic potential
    c. environmental conditions
    d. rate of natural increase
    e. All of these pertain to a population's growth curve.
13. When a population is undergoing logistic growth, environmental conditions determine
    a. whether to expect exponential growth.
    b. the carrying capacity.
    c. the length of the lag phase.
    d. whether the population is subject to density-independent regulation.
    e. All of these are correct.
14. Which of these statements is correct?
    a. A life table is based on whether a population has cohorts or not.
    b. The life table supplies the information for constructing the survivorship curve.
    c. A life table, like an ideal weight table, supplies information that can keep members of a population healthy.
    d. A life table tells which ideal survivorship curve is most appropriate to a particular population.
    e. Both b and d are correct.
15. Because of the postwar baby boom, the U.S. population
    a. can never level off.
    b. presently has a pyramid-type age distribution.
    c. is subject to regulation by density-dependent factors.
    d. is more conservation-minded than before.
    e. None of these is correct.

## thinking scientifically

1. In the winter moth life cycle, parasites were found to be a less important cause of mortality than cold winter weather and predators. Give an evolutionary explanation for the inefficiency of parasites in controlling population size.
2. You are a river manager charged with maintaining the flow through the use of dams so that trees, which have equilibrium life histories, can continue to grow along the river. What would you do?

## bioethical issue

### Population Control

The answer to how to curb the expected increase in the world's population lies in discovering how to curb the rapid population growth of the less-developed countries. In these countries, population experts have discovered what they call the "virtuous cycle." Family planning leads to healthier women, and healthier women have healthier children, and the cycle continues. Women no longer need to have many babies for only a few to survive. More education is also helpful because better-educated people are more interested in postponing childbearing and promoting women's rights. Women who have equal rights with men tend to have fewer children.

"There isn't any place where women have had the choice that they haven't chosen to have fewer children," says Beverly Winikoff at the Population Council in New York City. "Governments don't need to resort to force." Bangladesh is a case in point. Bangladesh is one of the densest and poorest countries in the world. In 1990, the birthrate was 4.9 children per woman, and now it is 3.3. This achievement was due in part to the Dhaka-based Grameen Bank, which loans small amounts of money mostly to destitute women to start a business. The bank discovered that when women start making decisions about their lives, they also start making decisions about the size of their families. Family planning within Grameen families is twice as common as the national average; in fact, those women who get a loan promise to keep their families small! Also helpful has been the network of village clinics that counsel women who want to use contraceptives. The expression "contraceptives are the best contraceptives" refers to the fact that you don't have to wait for social changes to get people to use contraceptives—the two feed back on each other.

Recently, some of the less-developed countries, faced with economic crises, have cut back on their family planning programs, and the more-developed countries have not taken up the slack. Indeed, some foreign donors have also cut back on aid—the United States by one-third. Are you in favor of foreign aid to help countries develop family planning programs? Why or why not?

## *Biology* website

The companion website for *Biology* provides a wealth of information organized and integrated by chapter. You will find practice tests, animations, videos, and much more that will complement your learning and understanding of general biology.

**http://www.mhhe.com/maderbiology10**

# 45

# Community and Ecosystem Ecology

*The narrow green leaves of a Venus flytrap (*Dionaea muscipula*) end with two reddish spiked lobes on either side of a midrib. An insect, most likely a fly, is lured to the leaves because they are lined by a band of sweet-smelling nectar glands. When the fly touches a trigger hair, the trap is sprung, and the spikes of the lobes become interlocked, enclosing the insect like the bars of a jail cell. Digestive enzymes pour forth from glands on the leaf surface, breaking down the helpless victim. Carnivorous plants carry on photosynthesis, but they live in mineral-poor wetlands. Nitrogen and phosphate are absorbed from the bodies of their animal prey and not from the soil.*

*The organic nutrients a Venus flytrap produces are transferred to any animal that feeds on them. Ordinarily, upon death and decomposition of organisms, inorganic nutrients are made available to plants once again. This chapter examines various types of interactions that occur among the populations of a community. It also looks at interactions with the physical environment.*

The Venus flytrap, *Dionaea muscipula*, is a plant that feeds on insects.

## concepts

## 45.1 Ecology of Communities

Populations do not occur as single entities; they are part of a community. A **community** is an assemblage of populations of different species interacting with one another in the same environment. Communities come in different sizes, and it is sometimes difficult to decide where one community ends and another begins.

A fallen log can be considered a community because the various populations living on and within a fallen log, such as plants, fungi, worms, and insects, interact with one another and form a community. The fungi break down the log and provide food for the earthworms and insects living in and on the log. Those insects may feed on one another, too. If birds flying throughout the entire forest feed on the insects and worms living in and on the log, then they are also part of the larger forest community.

### Community Structure

Two characteristics—species composition and diversity—allow us to compare communities. The species composition of a community, also known as **species richness,** is simply a listing of the various species found in that community. **Species diversity** includes both species richness and species evenness, or the relative abundance of the different species.

#### *Species Composition*

It is apparent, by comparing the photographs in Figure 45.1, that a coniferous forest has a composition different from a tropical rain forest. The narrow-leaved evergreen trees are prominent in a coniferous forest, and broad-leaved evergreen trees are numerous in a tropical rain forest. As the list of mammals demonstrates, a coniferous community and a tropical rain forest community contain different types of mammals. Ecologists comparing these two communities

a.

b.

**FIGURE 45.1 Community structure.**

Communities differ in their composition, as witnessed by their predominant plants and animals. The diversity of communities is described by the richness of species and their relative abundance. **a.** A coniferous forest. Some mammals found here are listed to the *right*. **b.** A tropical rain forest. Some mammals found here are listed to the *left*.

would go on to find differences in other plants and animals too. In the end, ecologists would conclude that not only are the species compositions of these two communities different, but the tropical rain forest has more species and, therefore, higher species richness.

### *Species Diversity*

The *diversity* of a community goes beyond composition because it includes not only a listing of all the species in a community but also the relative abundance of each species. To take an extreme example: A forest sampled in West Virginia has 76 yellow poplar trees but only one American elm (among other species). If we were simply walking through this forest, we might miss seeing the American elm. If, instead, the forest had 36 poplar trees and 41 American elms, the forest would seem more diverse to us, and indeed it would be more diverse. The greater the species richness and the more even the distribution of the species in a community, the greater the diversity.

## Community Interactions

This chapter examines the various types of community interactions and their importance to the structure of a community. Such interactions illustrate some of the most important evolutionary selection pressures acting on individuals. They also help us develop an understanding of how biodiversity can be preserved.

### *Habitat and Ecological Niche*

Each species occupies a particular position in the community, both in a spatial sense (where it lives) and in a functional sense (what role it plays). A particular place where a species lives and reproduces is its **habitat.** The habitat might be the forest floor, a swift stream, or the ocean's edge. The **ecological niche** is the role a species plays in its community, including its habitat and its interactions with other organisms. The niche includes the resources used to meet energy, nutrient, and survival demands. For a dragonfly larva, its habitat is a pond or lake where it eats other insects. The pond must contain vegetation where the dragonfly larva can hide from its predators, such as fish and birds. In addition, the water must be clear enough for it to see its prey and warm enough for it to be in active pursuit. Since it is difficult to study all aspects of niche, some observations focus only on one aspect, as with the birds featured in Figure 45.2. Each of these birds are specialist species because they have a limited diet, tolerate only small changes in environmental conditions, and live in a specific habitat. Other specialist species are pandas, spotted owls, and freshwater dolphins. Some species, such as raccoons, roaches, and house sparrows, are generalist with a broad range of niches. These organisms have a diversified diet, tolerate a wide range of environmental conditions, and can live in a variety of places.

Because a species' niche is affected by both abiotic factors (such as climate and habitat) and biotic factors (such as competitors, parasites, and predators), ecologists like to distinguish between the fundamental and the realized niche. The *fundamental niche* comprises all the abiotic conditions under which a species could survive when adverse biotic conditions are absent. The *realized niche* comprises those conditions under which a species does survive when adverse biotic interactions, such as competition and predation, are present. Therefore, a species' fundamental niche tends to be larger than its realized niche.

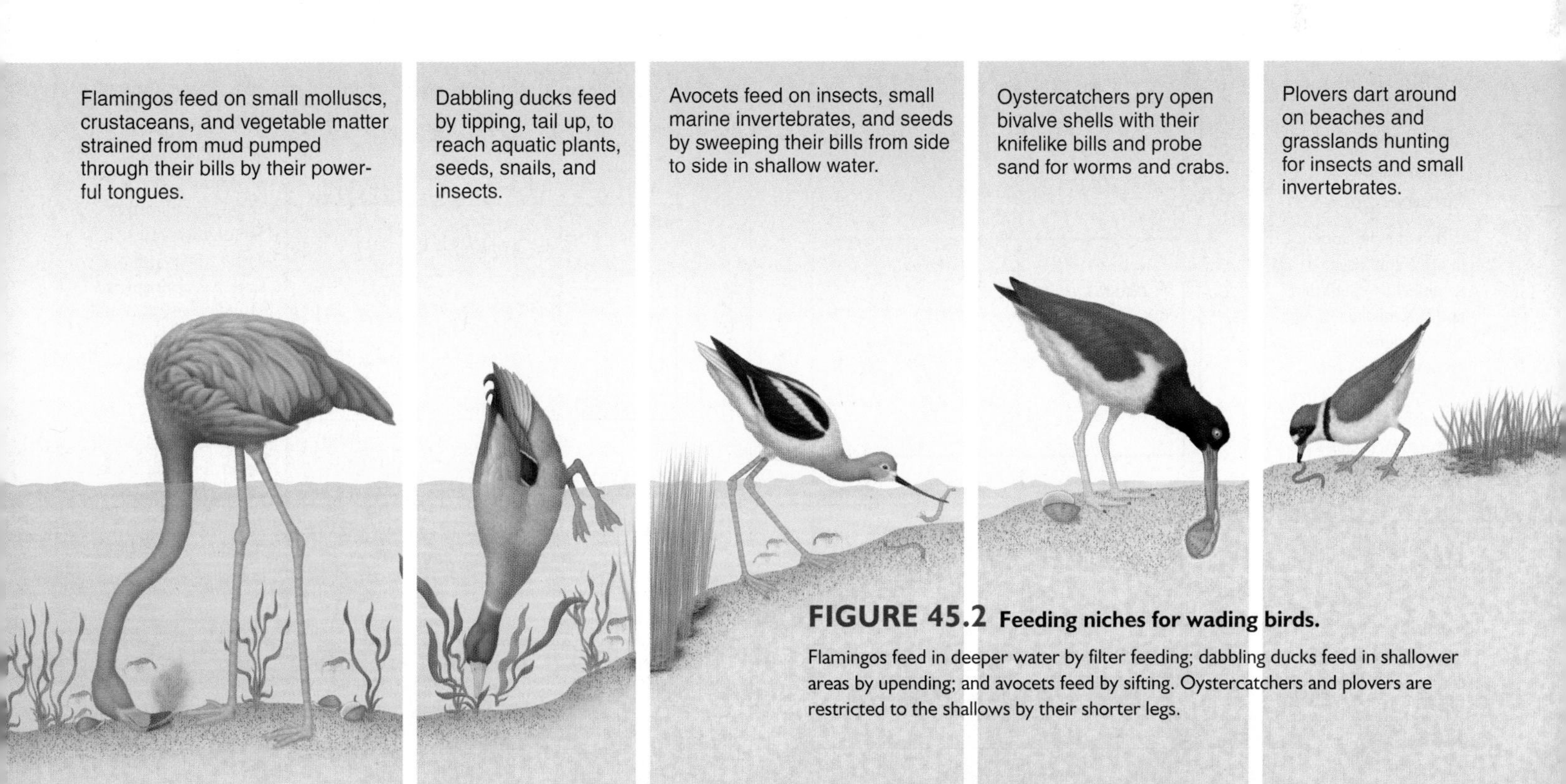


**FIGURE 45.2 Feeding niches for wading birds.**

Flamingos feed in deeper water by filter feeding; dabbling ducks feed in shallower areas by upending; and avocets feed by sifting. Oystercatchers and plovers are restricted to the shallows by their shorter legs.

### *Competition Between Populations*

Competition occurs when members of different species try to use a resource (such as light, space, or nutrients) that is in limited supply. If the resource is not in limited supply, there is no competition. In the 1930s, Russian ecologist G. F. Gause grew two species of *Paramecium* in one test tube containing a fixed amount of bacterial food. Although each population survived when grown separately, only one survived when they were grown together (Fig. 45.3). The successful *Paramecium* population had a higher biotic potential than the unsuccessful population. After observing the outcome of other similar experiments in the laboratory, ecologists formulated the **competitive exclusion principle,** which states that no two species can indefinitely occupy the same niche at the same time.

What does it take to have different ecological niches so that extinction of one species is avoided? In another laboratory experiment, Gause found that two species of *Paramecium* did continue to occupy the same tube when one species fed on bacteria at the bottom of the tube and the other fed on bacteria suspended in solution. Individuals of each species that can avoid competition have a reproductive advantage. **Resource partitioning** decreases competition between two species, leading to increased niche specialization and less niche overlap. An example of resource partitioning involves owl and hawk populations. Owls and hawks feed on similar prey (small rodents), but owls are nocturnal hunters and hawks are diurnal hunters. What could have been one niche became two more specialized niches because of a divergence of behavior.

It is possible to observe the process of niche specialization in nature. When three species of ground finches of the Galápagos Islands occur on separate islands, their beaks tend to be the same intermediate size, enabling each to feed on a wider range of seeds (Fig. 45.4). Where they co-occur, selection has favored divergence in beak size because the size of the beak affects the kinds of seeds that can be eaten. In other words, competition has led to resource partitioning and, therefore, niche specialization. The tendency for characteristics to be more divergent when populations belong to the same community than when they are isolated is termed **character displacement.** Character displacement is often used as evidence that competition and resource partitioning have taken place.

Niche specialization can be subtle. Five different species of warblers that occur in North American forests are all nearly the same size, and all feed on a type of spruce tree caterpillar. For all these species to exist, they must be avoiding direct competition. In a famous study, ecologist Robert MacArthur recorded the amount of time each of five warbler species spent in different regions of spruce canopies to determine where each species did most of its feeding (Fig. 45.5). He discovered that each species primarily used different parts of the tree canopy and, in that way, had a more specialized niche.

As another example, consider that swallows, swifts, and martins all eat flying insects and parachuting spiders. These birds even frequently fly together in mixed flocks. But each type of bird has a different nesting site and migrates at a slightly different time of year. In doing so, they are not competing for the same food source when they are feeding their young.

In all these cases of niche partitioning, we have merely assumed that what we observe today is due to competition

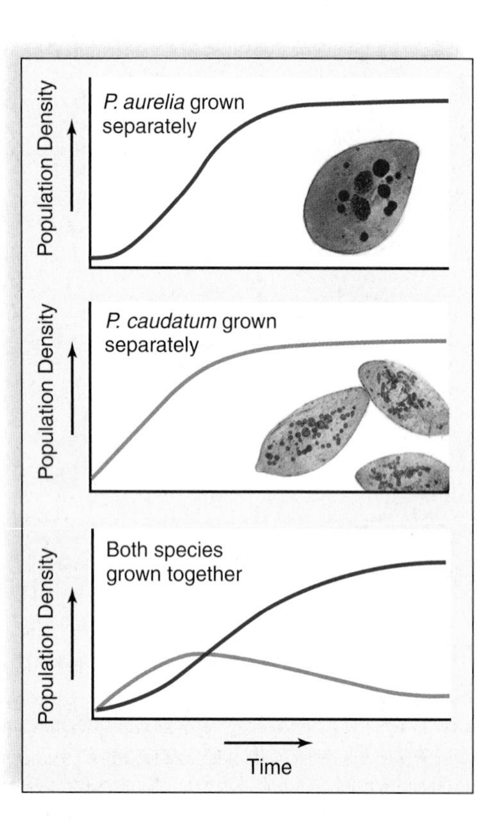


**FIGURE 45.3 Competition between two laboratory populations of *Paramecium*.**

When grown alone in pure culture, *Paramecium caudatum* and *Paramecium aurelia* exhibit sigmoidal growth. When the two species are grown together in mixed culture, *P. aurelia* is the better competitor, and *P. caudatum* dies out. Both attempted to exploit the same resources, which led to competitive exclusion.

Source: Data from G. F. Gause, *The Struggle for Existence*, 1934, Williams & Wilkins Company, Baltimore, MD.

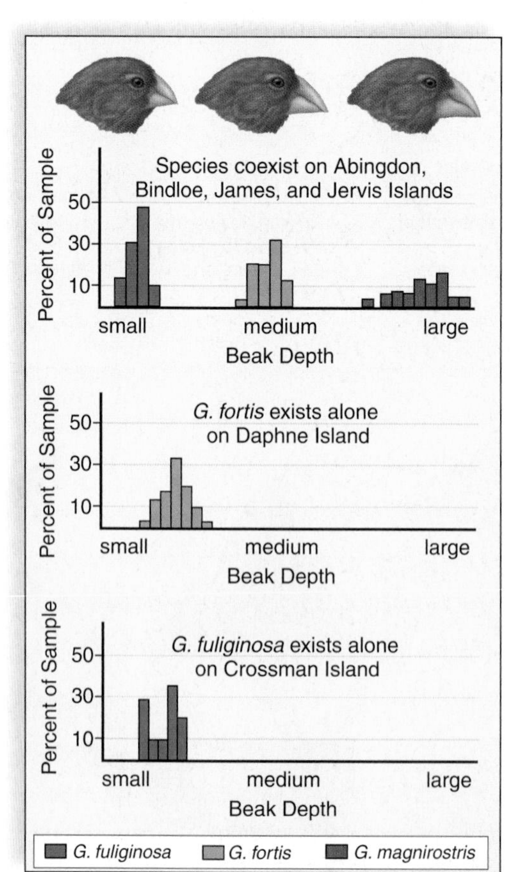


**FIGURE 45.4 Character displacement in finches on the Galápagos Islands.**

When *Geospiza fuliginosa*, *G. fortis*, and *G. magnirostris* are on the same island, their beak sizes are appropriate to eating small-, medium-, and large-sized seeds. When *G. fortis* and *G. fuliginosa* are on separate islands, their beaks have the same intermediate size, which allows them to eat seeds that vary in size.

**FIGURE 45.5 Niche specialization among five species of coexisting warblers.**

The diagrams represent spruce trees. The time each species spent in various portions of the trees was determined; each species spent more than half its time in the blue regions.

in the past. Some ecologists are fond of saying that in doing so we have invoked the "ghosts of competition past." Are there any instances in which competition has actually been observed? Joseph Connell has studied the distribution of barnacles on the Scottish coast, where a small barnacle lives on the high part of the intertidal zone, and a large barnacle lives on the lower part (Fig. 45.6). Free-swimming larvae of both species attach themselves to rocks randomly throughout in the intertidal zone; however, in the lower zone, the large *Balanus* barnacles seem to either force the smaller *Chthamalus* individuals off the rocks or grow over them. To test his observation, Connell removed the larger barnacles and found that the smaller barnacles grew equally well on all parts of the rock. The entire intertidal zone is the fundamental niche for *Chthamalus,* but competition is restricting the range of *Chthamalus* on the rocks. *Chthamalus* is more resistant to drying out than is *Balanus;* therefore, it has an advantage that permits it to grow in the upper intertidal zone where it is exposed more often to the air. In other words, the upper intertidal zone becomes the realized niche for *Chthamalus*.

## Predator-Prey Interactions

**Predation** occurs when one living organism, called the **predator,** feeds on another, called the **prey.** In its broadest sense, predaceous consumers include not only animals such as lions, which kill zebras, but also filter-feeding blue

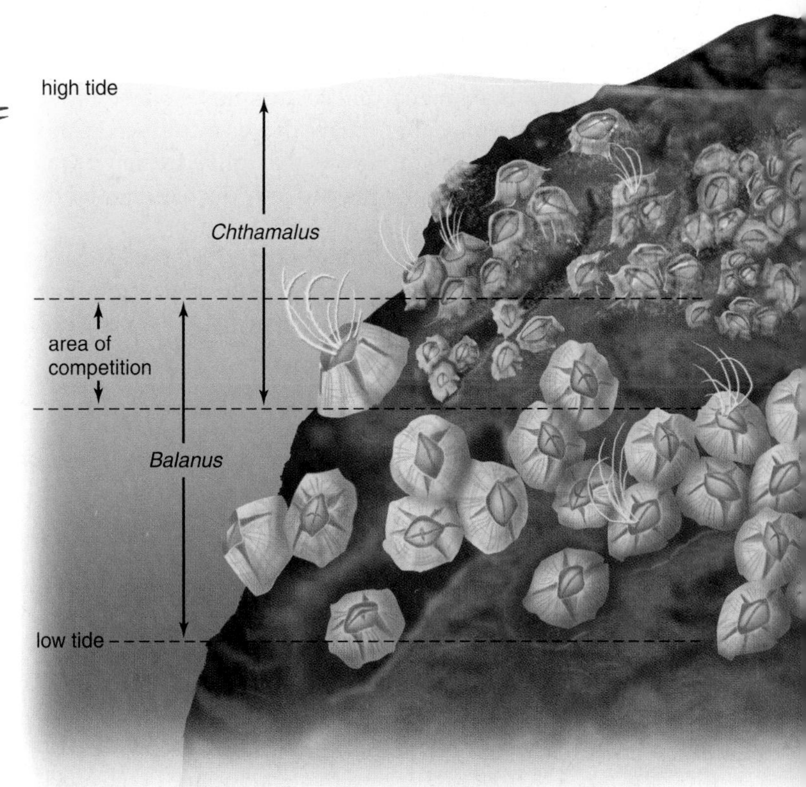


**FIGURE 45.6 Competition between two species of barnacles.**

Competition prevents two species of barnacles from occupying as much of the intertidal zone as possible. Both exist in the area of competition between *Chthamalus* and *Balanus*. Above this area only *Chthamalus*, a small barnacle, survives, and below it only *Balanus*, a large barnacle, survives.

whales, which strain krill from ocean waters; herbivorous deer, which browse on trees and shrubs; parasitic ticks, which suck blood from their victims; and *parasitoids,* which are wasps that lay their eggs inside the body of a host. The resulting larvae feed on the host, sometimes causing death. Parasitism can be considered a type of predation because one individual obtains nutrients from another (see page 846). Predation and parasitism are expected to increase the abundance of the predator and parasite at the expense of the abundance of the prey or host.

### Predator-Prey Population Dynamics

Do predators reduce the population density of prey? In another classic experiment, G. F. Gause reared the ciliated protozoans *Paramecium caudatum* (prey) and *Didinium nasutum* (predator) together in a culture medium. He observed that *Didinium* ate all the *Paramecium* and then died of starvation. In nature, we can find a similar example. When a gardener brought prickly-pear cactus to Australia from South America, the cactus spread out of control until millions of acres were covered with nothing but cacti. The cacti were brought under control when a moth from South America, whose caterpillar feeds only on the cactus, was introduced. The caterpillar was a voracious predator on the cactus, efficiently reducing the cactus population. Now both cactus and moth are found at greatly reduced densities in Australia.

This raises an interesting point: The population density of the predator can be affected by the prevalence of the prey. In other words, the predator-prey relationship is actually a two-way street. In that context, consider that at first the biotic potential (maximum reproductive rate) of the prickly-pear cactus was maximized, but factors that oppose biotic potential came into play after the moth was introduced. And the biotic potential of the moth was maximized when it was first introduced, but the carrying capacity decreased after its food supply was diminished.

Sometimes, instead of remaining in a steady state, predator and prey populations first increase in size and then decrease. We can appreciate that an increase in predator population size is dependent on an increase in prey population size. But what causes a decrease in population size instead of the establishment of a steady population size? At least two possibilities account for the reduction: (1) Perhaps the biotic potential (reproductive rate) of the predator is so great that its increased numbers overconsume the prey, and then as the prey population declines, so does the predator population; or (2) perhaps the biotic potential of the predator is unable to keep pace with the prey, and the prey population overshoots the carrying capacity and suffers a crash. Now the predator population follows suit because of a lack of food. In either case, the result will be a series of peaks and valleys, with the predator population size lagging slightly behind that of the prey population.

A famous example of predator-prey cycles occurs between the snowshoe hare and the Canadian lynx, a type of small predatory cat (Fig. 45.7). The snowshoe hare is a common herbivore in the coniferous forests of North America, where it feeds on terminal twigs of various shrubs and small trees. The Canadian lynx feeds on snowshoe hares but also on ruffed grouse and spruce grouse, two types of birds. Studies have revealed that the hare and lynx populations cycle regularly, as graphed in Figure 45.7. Investigators at first assumed that the lynx brings about a decline in the hare population and that this accounts for the cycling. But others have noted that the decline in snowshoe hare abundance was accompanied by low growth and reproductive rates, which could be signs of a food shortage. Experiments were done to test whether (1) predation or (2) lack of food caused the decline in the hare population. The results suggest that both factors combined to produce a low hare population and the cycling effect.

### Prey Defenses

Prey defenses are mechanisms that thwart the possibility of being eaten by a predator. Prey species have evolved a variety of mechanisms that enable them to avoid predators, including heightened senses, speed, protective armor, protective spines or thorns, tails and appendages that break off, and chemical defenses.

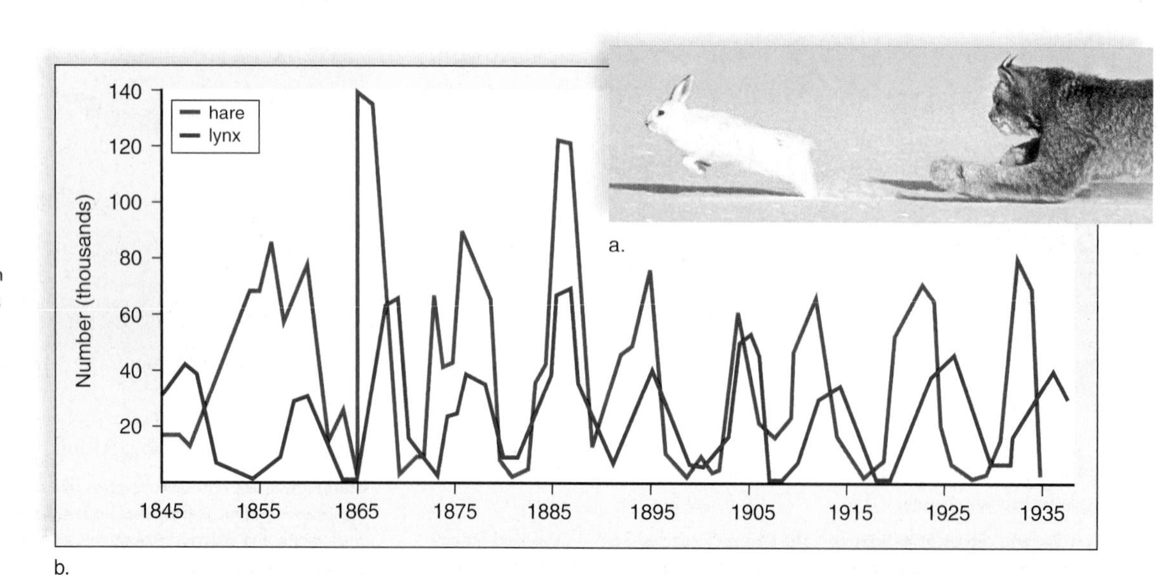


**FIGURE 45.7 Predator-prey interaction between a lynx and a snowshoe hare.**

**a.** A Canadian lynx, *Felis canadensis*, is a solitary predator. A long, strong forelimb with sharp claws grabs its main prey, the snowshoe hare, *Lepus americanus*. The lynx lives in northern forests. Its brownish gray coat blends in well against tree trunks, and its long legs enable it to walk through deep snow. **b.** The number of pelts received yearly by the Hudson Bay Company for almost 100 years shows a pattern of ten-year cycles in population densities. The snowshoe hare population reaches a peak abundance before that of the lynx by a year or more.

a. Camouflage

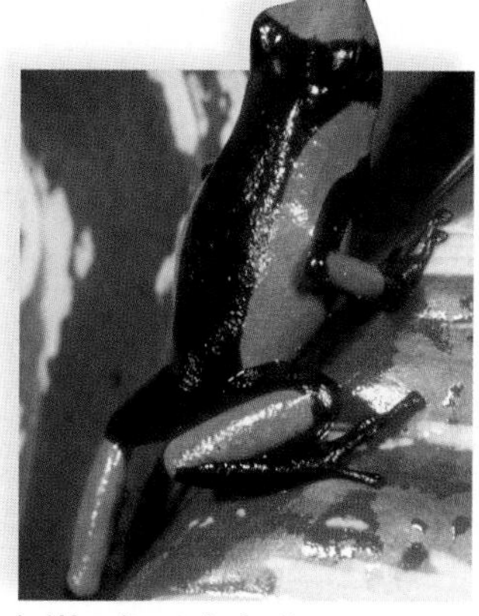

b. Warning colorization

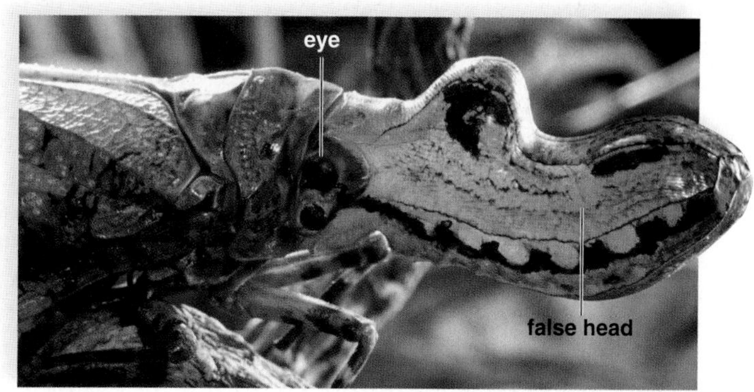


c. Fright

**FIGURE 45.8 Antipredator defenses.**

**a.** Flounder is a fish that blends in with its background. **b.** The skin secretions of poison arrow frogs make them dangerous to touch. **c.** The large false head of the South American lantern fly may startle a potential predator.

One common strategy to avoid capture by a predator is **camouflage,** or the ability to blend into the background. Some animals have cryptic coloration that blends them into their surroundings. For example, flounders can take on the same coloration as their background (Fig. 45.8*a*). Many examples of protective camouflage are known: Walking sticks look like twigs; katydids look like sprouting green leaves; some caterpillars resemble bird droppings; and some insects and moths blend into the bark of trees.

Another common antipredator defense among animals is *warning coloration,* which tells the predator that the prey is potentially dangerous. As a warning to possible predators, poison arrow frogs are brightly colored (Fig. 45.8*b*). Also, many animals, including caterpillars, moths, and fishes, possess false eyespots that confuse or startle another animal. Other animals have elaborate anatomic structures that cause the *startle response.* The South American lantern fly has a large false head with false eyes, making it resemble the head of an alligator (Fig. 45.8*c*). However, antipredator defenses are not always false. A porcupine certainly looks formidable, and for good reason. Its arrowlike quills have barbs that dig into the predator's flesh and penetrate even deeper as the enemy struggles after being impaled. In the meantime, the porcupine runs away.

Association with other prey is another common strategy that may help avoid capture. Flocks of birds, schools of fish, and herds of mammals stick together as protection against predators. Baboons that detect predators visually, and antelopes that detect predators by smell, sometimes forage together, gaining double protection against stealthy predators. The gazellelike springboks of southern Africa jump stiff-legged 2–4 m into the air when alarmed. Such a jumble of shapes and motions might confuse an attacking lion, allowing the herd to escape.

## *Mimicry*

**Mimicry** occurs when one species resembles another that possesses an overt antipredator defense. A mimic that lacks the defense of the organism it resembles is called a Batesian mimic (named for Henry Bates, who described the phenomenon). Once an animal experiences the defense of the model, it remembers the coloration and avoids all animals that look similar. Figure 45.9*a, b* shows two insects (flower fly and longhorn beetle) that resemble a yellow jacket wasp but lack the wasp's ability to sting. Classic examples of Batesian mimicry include the scarlet kingsnake mimicking the venomous coral snake and the viceroy butterfly mimicking the foul-tasting monarch butterfly.

There are also examples of species that have the same defense and resemble each other. Many stinging insects—bees, wasps, hornets, and bumblebees—have the familiar black and yellow bands. Once a predator has been stung by a black and yellow insect, it is wary of that color pattern in the future. Mimics that share the same protective defense are called Müllerian mimics, after Fritz Müller, who suggested that this, too, is a form of mimicry. The bumblebee in Figure 45.9*c* is a Müllerian mimic of the yellow jacket wasp because both of them can sting.

a. Flower fly

b. Longhorn beetle

c. Bumblebee

d. Yellow jacket

**FIGURE 45.9 Mimicry among insects.**

**a.** A flower fly, *Chrysotoxum*, and (**b**) a longhorn beetle, *Strophiona*, are Batesian mimics because they are incapable of stinging another animal, yet they have the same appearance as the yellow jacket wasp (**d**). **c.** The bumblebee, *Bombus*, and (**d**) the yellow jacket, *Vespula*, are Müllerian mimics because they both use stinging as a defense.

## Symbiotic Relationships

Symbiotic relations are of three types, as shown in Table 45.1. Is this categorization artificial? Some biologists argue that the amount of harm or good two species do one another is dependent on what the investigator chooses to measure.

**Parasitism** is similar to predation in that an organism, called the **parasite,** derives nourishment from another, called the **host.** Parasitism is an example of a **symbiosis,** an association in which at least one of the species is dependent on the other (Table 45.1). Viruses, such as HIV, that reproduce inside human lymphocytes are always parasitic, and parasites occur in all of the kingdoms of life as well. Bacteria (e.g., strep infection), protists (e.g., malaria), fungi (e.g., rusts and smuts), plants (e.g., indian pipe), and animals (e.g., tapeworms and fleas) all have parasitic members. While small parasites can be endoparasites (heartworms) (Fig. 45.10), larger ones are more likely to be ectoparasites (leeches), which remain attached to the exterior of the body by means of specialized organs and appendages. The effects of parasites on the health of the host can range from slightly weakening them to actually killing them over time. When host populations are at a high density, parasites readily spread from one host to the next, causing intense infestations and a subsequent decline in host density. Parasites that do not kill their host can still play a role in reducing the host's population density because an infected host is less fertile and becomes more susceptible to another cause of death.

In addition to nourishment, host organisms also provide their parasites with a place to live and reproduce, as well as a mechanism for dispersing offspring to new hosts. Many parasites have both a primary and a secondary host. The secondary host may be a vector that transmits the parasite to the next primary host. Usually both hosts are required in order to complete the life cycle. The association between parasite and host is so intimate that parasites are often specific and even require certain species as hosts.

### *Commensalism*

**Commensalism** is a symbiotic relationship between two species in which one species is benefited, and the other is neither benefited nor harmed.

**TABLE 45.1**

**Symbiotic Relationships**

| *Interaction* | *Expected Outcome* |
|---|---|
| Parasitism | Abundance of parasite increases, and abundance of host decreases. |
| Commensalism | Abundance of one species increases, and the other is not affected. |
| Mutualism | Abundance of both species increases. |

Instances are known in which one species provides a home and/or transportation for the other species. Barnacles that attach themselves to the backs of whales and the shells of horseshoe crabs are provided with both a home and transportation. It is possible though that the movement of the host is impeded by the presence of the attached animals, and therefore some are reluctant to use these as instances of commensalism.

Epiphytes, such as Spanish moss and some species of orchids and ferns, grow in the branches of trees, where they receive light, but they take no nourishment from the trees. Instead, their roots obtain nutrients and water from the air. Clownfishes live within the waving mass of tentacles of sea anemones (Fig. 45.11). Because most fishes avoid the stinging tentacles of the anemones, clownfishes are protected from predators. Perhaps this relationship borders on mutualism, because the clownfishes actually may attract other fishes on which the anemone can feed or provide some cleaning services for the anemone.

Commensalism often turns out, on closer examination, to be an instance of either mutualism or parasitism. Cattle egrets are so named because these birds stay near cattle, which flush out their prey—insects and other animals—from vegetation. The relationship becomes mutualistic when egrets remove ectoparasites from the cattle. Remoras are fishes that attach themselves to the bellies of sharks by means of a modified dorsal fin acting as a suction cup. Remoras benefit by getting a free ride and feeding on a shark's leftovers. However, the shark benefits when remoras remove its ectoparasites.

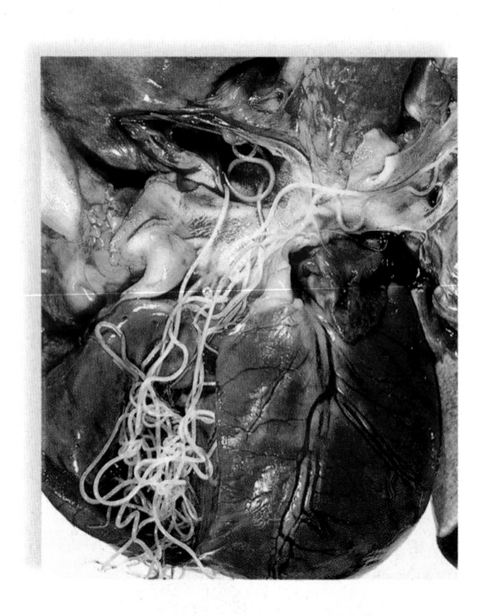

**FIGURE 45.10**
**Heartworm.**
*Dirofilaria immitis* is a parasitic nematode spread by mosquitoes. The worms, which live in the heart and pulmonary blood artery, can cause death of the host.

**FIGURE 45.11** **Clownfish among sea anemone's tentacles.**
If the clownfish, *Premnasa biaculeatus,* performs no service for the sea anemone, this association is a case of commensalism. If the clownfish lures other fish to be eaten by the sea anemone, this is a case of mutualism.

*ecology focus*

## Interactions and Coevolution

**Coevolution** is present when two species adapt in response to selective pressure imposed by the other. Symbiosis (close association between two species), which includes parasitism, commensalism, and mutualism, is especially prone to the process of coevolution. Flowers pollinated by animals have features that attract them (see the Science Focus on pages 498–99). As an example of this type of coevolution, a butterfly-pollinated flower is often a composite, containing many individual flowers; the broad expanse provides a platform for the butterfly to land, and the butterfly has a proboscis that it in turn inserts into each tiny flower. In this case, natural selection has selected both for flat flowers and butterfly mouthparts to feed on such flowers. However, neither species would have evolved this way without the influence of the other.

Coevolution also occurs between predators and prey. For example, a cheetah sprints forward to catch its prey, and this behavior selects for those gazelles that are fast enough to avoid capture. Over generations, the adaptation of the prey (a great running speed, in this case) may very well put selective pressure on the predator for an adaptation to the prey's defense mechanism. Hence, an evolutionary "arms race" can develop. The process of coevolution has been studied in the cuckoo, a social parasite that reproduces at the expense of other bird species by laying its eggs in their nests. It is a strange sight to see a small bird feeding a cuckoo nestling several times its size. How did this strange happening develop? Investigators discovered that in order to "trick" a host bird, the adult cuckoo has to (1) lay an egg that mimics the host's egg, (2) lay its egg very rapidly (only 10 seconds are required) in the afternoon while the host is away from the nest, and (3) leave most of the hosts' eggs in the nest because hosts will desert a nest that has only one egg in it. (The cuckoo chick hatches first and is adapted to removing any other eggs in the nest [Fig. 45A*a*].) At this stage in the arms race, the cuckoo appears to have the upper hand, but the host birds may very well next evolve a way to distinguish the cuckoo from their own young.

Coevolution can take many forms. In the case of *Plasmodium,* the cause of malaria, the sexual portion of the life cycle occurs within mosquitoes (the vector), and the asexual portion occurs in humans. The human immune system uses surface proteins to detect pathogens, and *Plasmodium* has numerous genes for surface proteins, and it is capable of changing these surface proteins repeatedly. In this way, it stays one step ahead of the host's immune system. A similar capability of HIV has added to the difficulty of producing an AIDS vaccine.

The relationship between parasite and host can even include the ability of parasites to seemingly manipulate the behavior of their hosts in self-serving ways. Ants infected with the lance fluke (but not those uninfected) mysteriously cling to blades of grass with their mouthparts. There, the infected ants are eaten by grazing sheep, and the flukes are transmitted to the next host in their life cycle. Similarly, when snails of the genus *Succinea* are parasitized by worms of the genus *Leucochloridium,* they are eaten by birds. As the worms mature, they invade the snail's eyestalks, making them resemble edible caterpillars. Now the birds eat the snails, and the parasites release their eggs, which complete development inside the urinary tracts of birds.

The traditional view was that as host and parasite coevolved, each would become more tolerant of the other since, if the opposite occurred, the parasite would soon run out of hosts. Parasites could first become commensal, or harmless to the host. Then, given enough time, the parasite and host might even become mutualists. In fact, the evolution of the eukaryotic cell by endosymbiosis is predicated on the supposition that bacteria took up residence inside a larger cell, and then the parasite and cell became mutualists.

However, this argument is too teleological for some; after all, no organism is capable of "looking ahead" at its evolutionary fate. Rather, if an aggressive parasite could transmit more of itself in less time than a benign one, aggressiveness would be favored by natural selection. On the other hand, other factors, such as the life cycle of the host, can determine whether aggressiveness is beneficial or not. For example, a benign parasite of newts will do better than an aggressive one. Why? Because newts take up solitary residence outside ponds in the forest for six years, and parasites have to wait that long before they are likely to meet up with another potential host. If a parasite kills its host before it can reach another, it not only has lost its food source, but also its home.

b.

a.

**FIGURE 45A Social parasitism.**
***a.*** *The cuckoo,* Cuculus, *is a social parasite of the reed warbler,* Acrocephalus. *A cuckoo chick is heaving the eggs of its host out of the nest.* ***b.*** *Its own egg (see inset) mimics and is accepted by the host as its own.*

## *Mutualism*

**Mutualism** is a symbiotic relationship in which both members benefit. As with other symbiotic relationships, it is possible to find numerous examples among all organisms. Bacteria that reside in the human intestinal tract acquire food, but they also provide us with vitamins, molecules we are unable to synthesize for ourselves. Termites would not be able to digest wood if not for the protozoans that inhabit their intestinal tracts and digest cellulose. Mycorrhizae are mutualistic associations between the roots of plants and fungal hyphae. The hyphae improve the uptake of nutrients for the plant, protect the plant's roots against pathogens, and produce plant growth hormones. In return, the plant provides the fungus with carbohydrates. Some sea anemones make their home on the backs of crabs. The crab uses the stinging tentacles of the sea anemone to gather food and to protect itself; the sea anemone gets a free ride that allows it greater access to food than other anemones. Lichens can grow on rocks because their fungal member conserves water and leaches minerals that are provided to the algal partner, which photosynthesizes and provides organic food for both populations. However, it's been suggested that the fungus is parasitic, at least to a degree, on the algae.

In tropical America, the bullhorn acacia tree is adapted to provide a home for ants of the species *Pseudomyrmex ferruginea*. Unlike other acacias, this species has swollen thorns with a hollow interior, where ant larvae can grow and develop. In addition to housing the ants, acacias provide them with food. The ants feed from nectaries at the base of the leaves and eat fat- and protein-containing nodules called Beltian bodies, found at the tips of the leaves. The ants constantly protect the plant from herbivores and other plants that might shade it because, unlike other ants, they are active 24 hours a day.

The relationship between plants and their pollinators, mentioned previously, is a good example of mutualism. Perhaps the relationship began when herbivores feasted on pollen. The provision of nectar by the plant may have spared the pollen and, at the same time, allowed the animal to become an agent of pollination. By now, pollinator mouthparts are adapted to gathering the nectar of a particular plant species, and this species is dependent on the pollinator for dispersing pollen (see pages 498–99). The mutualistic relationships between flowers and their pollinators are examples of coevolution.

The outcome of mutualism is an intricate web of species interdependencies critical to the community. For example, in areas of the western United States, the branches and cones of whitebark pine are turned upward, meaning that the seeds do not fall to the ground when the cones open. Birds called Clark's nutcrackers eat the seeds of whitebark pine trees and store them in the ground (Fig. 45.12). Therefore, Clark's nutcrackers are critical seed dispersers for the trees. Also, grizzly bears find the stored seeds and consume them. Whitebark pine seeds do not germinate unless their seed coats are exposed to fire. When natural forest fires in the area are suppressed, whitebark pine trees decline in number, and so do Clark's nutcrackers and grizzly bears. When lightning-ignited fires are allowed to burn, or prescribed burning is used in the area, the whitebark pine populations increase, as do the populations of Clark's nutcrackers and grizzly bears.

*Cleaning symbiosis* is a symbiotic relationship in which crustaceans, fish, and birds act as cleaners for a variety of vertebrate clients. Large fish in coral reefs line up at cleaning stations and wait their turn to be cleaned by small fish that even enter the mouths of the large fish (Fig. 45.13). Whether cleaning symbiosis is an example of mutualism has been questioned because of the lack of experimental data. If clients respond to tactile stimuli by remaining immobile while cleaners pick at them, then cleaners may be exploiting this response by feeding on host tissues, as well as on ectoparasites.

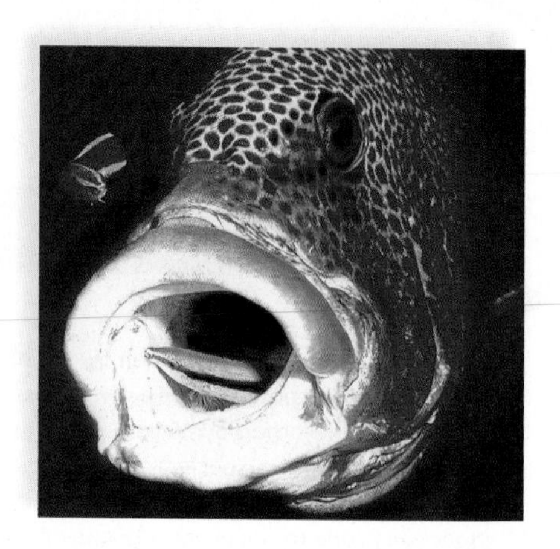

**FIGURE 45.13**
**Cleaning symbiosis.**
A cleaner wrasse, *Labroides dimidiatus*, in the mouth of a spotted sweetlip, *Plectorhincus chaetodontoides*, is feeding off parasites. Does this association improve the health of the sweetlip, or is the sweetlip being exploited? Investigation is under way.

**FIGURE 45.12**
**Clark's nutcrackers.**
Mutualism can take many forms, such as when birds like Clark's nutcrackers feed on but also disperse the seeds of whitebark pine trees.

### Check Your Progress 45.1

1. What is the difference between an organism's habitat and niche?
2. What two factors can cause predator and prey populations to cycle in a predictable manner?
3. Give examples to show mutualism can be involved when one population feeds off another.

# science focus

## Island Biogeography Pertains to Biodiversity

Would you expect larger coral reefs to have a greater number of species, called species richness, than smaller coral reefs? The area (space) occupied by a community can have a profound effect on its biodiversity. American ecologists Robert MacArthur and E. O. Wilson developed a general **model of island biogeography** to explain and predict the effects of (1) distance from the mainland and (2) size of an island on community diversity.

Imagine two new islands that, as yet, contain no species at all. One of these islands is near the mainland, and one is far from the mainland (Fig. 45B*a*). Which island will receive more immigrants from the mainland? Most likely, the near one because it's easier for immigrants to get there. Similarly, imagine two islands that differ in size (Fig. 45B*b*). Which island will be able to support a greater number of species? The large one, because its greater amount of resources can support more populations, while species on the smaller island may eventually face extinction due to scarce resources. MacArthur and Wilson studied the biodiversity on many island chains, including the West Indies, and discovered that species richness does correlate positively with island distance from mainland and island size. They developed a model of island biogeography that takes into account both factors. An equilibrium is reached when the rate of species immigration matches the rate of species extinction due to limited space (Fig. 45B*c*). Notice that the equilibrium point is highest for a large island that is near the mainland. The equilibrium could be dynamic (new species keep on arriving, and new extinctions keep on occurring), or the composition of the community could remain steady unless disturbed.

### Biodiversity

Conservationists note that the trends graphed in Figure 45B*c* in particular apply to their work because humans often create preserved areas surrounded by farms, towns, and cities, or even water. For example, in Panama, Barro Colorado Island (BCI) was created in the 1910s when a river was dammed to form a lake. As predicted by the model of island biogeography, BCI lost species because it was a small island that had been cut off from the mainland. Among the species that became extinct were the top predators on the island, namely the jaguar, puma, and ocelot. Thereafter, medium-sized terrestrial mammals, such as the coatimundi, increased in number. Because the coatimundi is an avid predator of bird eggs and nestlings, soon there were fewer bird species on BCI, even though the island is large enough to support them.

The model of island biogeography suggests that the larger the conserved area, the better the chance of preserving more species. Is it possible to increase the amount of space without using more area? Two possibilities come to mind. If the environment has patches, it has a greater number of habitats—and thus greater diversity. As gardeners, we are urged to create patches in our yards if we wish to attract more butterflies and birds! One way to introduce patchiness is through stratification, the use of layers. Just as a high-rise apartment building allows more human families to live in an area, so can stratification within a community provide more and different types of living space for different species.

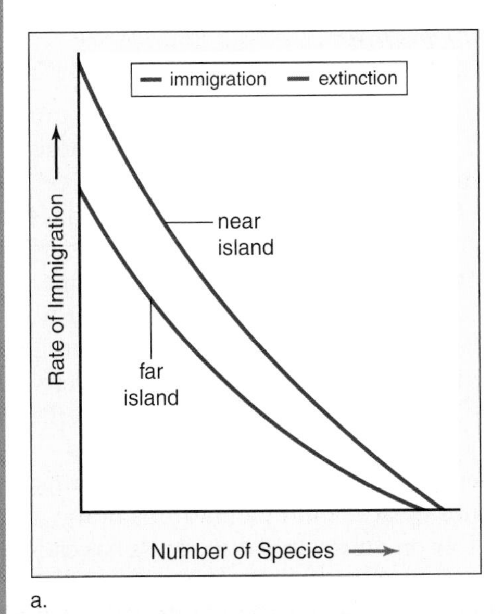


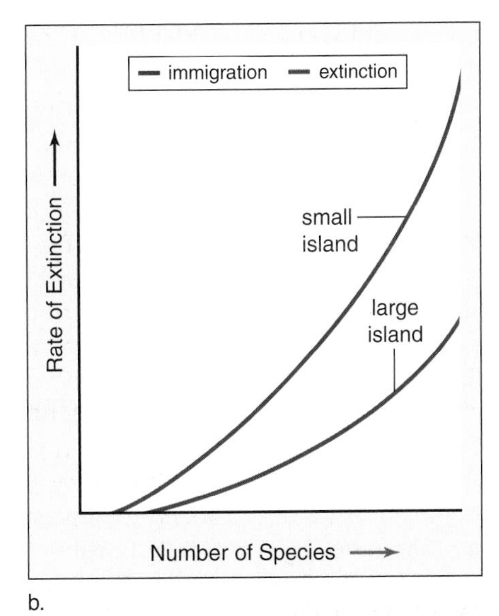


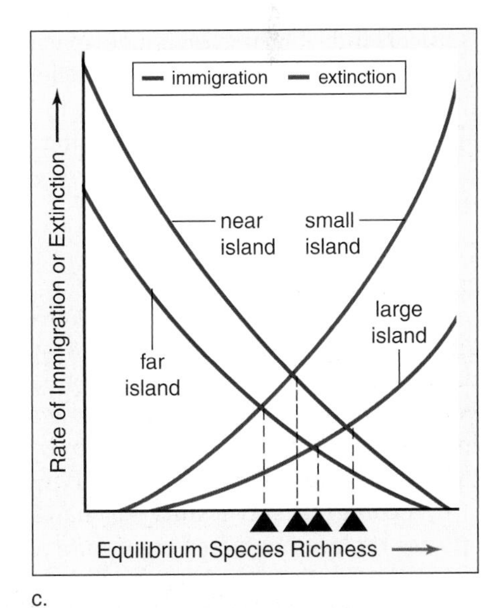


**FIGURE 45B Species richness.**
***a.*** *An island near the mainland will have a higher immigration rate than an island far from the mainland.* ***b.*** *A small island will have a higher extinction rate than a large island.* ***c.*** *The balance between immigration and extinction for four possible types of islands. A large island that is near the mainland will have a higher equilibrium species richness than the other types of islands.*

**FIGURE 45.14**
**Secondary succession.**
This example of secondary succession occurred in a former cornfield in New Jersey on the east coast of the United States. **a.** During the first year, only the remains of corn plants are seen. **b.** During the second year, wild grasses have invaded the area. **c.** By the fifth year, the grasses look more mature, and sedges have joined them. **d.** During the tenth year, there are goldenrod plants, shrubs (blackberry), and juniper trees. **e.** After 20 years, the juniper trees are mature, and there are also birch and maple trees in addition to the blackberry shrubs.

a. First year

b. Second year

# 45.2 Community Development

Each community has a history that can be surveyed over a short time period or even over geological time. We know that the distribution of life has been influenced by dynamic changes occurring during the history of Earth. We have previously discussed how continental drift contributed to various mass extinctions that have occurred in the past. For example, when the continents joined to form the supercontinent Pangaea, many forms of marine life became extinct. Or, when the continents drifted toward the poles, immense glaciers drew water from the oceans and even chilled once-tropical lands. During an ice age, glaciers moved southward, and then in between ice ages, glaciers retreated, changing the environment and allowing life to colonize the land once again. Over time, complex communities evolved. Many ecologists, however, try to observe changes as they occur during much shorter timescales.

## Ecological Succession

Communities are subject to disturbances that can range in severity from a storm blowing down a patch of trees, to a beaver damming a pond, to a volcanic eruption. We know from observation that following these disturbances, we'll see changes in the plant and animal communities over time; often, we'll wind up with the same kind of community with which we started.

**Ecological succession** is a change involving a series of species replacements in a community. *Primary succession* occurs in areas where there is no soil formation, such as following a volcanic eruption or a glacial retreat. Wind, water, and other abiotic factors start the formation of soil from exposed rock. *Secondary succession* occurs in areas where soil is present, as when a cultivated field, such as a cornfield in New Jersey, returns to a natural state (Fig. 45.14). This is disturbance-based succession. Notice that the progression changes from grasses to shrubs to a mixture of shrubs and trees.

The first species to begin secondary succession are called **pioneer species**—that is, plants that are invaders of disturbed areas—and then the area progresses through the series of stages described in Figure 45.15. Again, we observe a series that begins with small, short-lived species and proceeds through stages of species of mixed sizes and life spans, until finally there are only large, long-lived species of trees. Ecologists have tried to determine the processes and mechanisms by which the changes described in Figures 45.14 and 45.15 take place—and whether these processes always have the same "end point" of community composition and diversity.

### Models About Succession

In 1916, F. E. Clements proposed that succession in a particular area will always lead to the same type of community, which he called a **climax community.** He hypothesized that climate, in particular, determined whether succession resulted in a desert, a type of grassland, or a particular type of forest. This is the reason, he said, that coniferous forests occur in northern latitudes, deciduous forests in temperate zones, and tropical rain forests in the tropics. Secondarily, he hypothesized that soil conditions might also affect the results. Shallow, dry soil might produce a grassland where otherwise a forest might be expected, or the rich soil of a riverbank might produce a woodland where a prairie is expected.

Further, Clements hypothesized that each stage facilitated the invasion and replacement by organisms of the next stage. Shrubs can't grow on dunes until dune grass has caused soil to develop. Similarly, in the example given in Figure 45.15, shrubs can't arrive until grasses have made the soil suitable for them. Each successive community prepares the way for the next, so that grass-shrub-forest development occurs in a sequential way. Therefore, in what is sometimes called "climax theory," this is known as the *facilitation model* of succession.

Aside from this facilitation model, there is also an *inhibition model*. That model predicts that colonists hold onto

c. Fifth year

d. Tenth year

e. Twentieth year

their space and inhibit the growth of other plants until the colonists die or are damaged. Still another model, called the *tolerance model*, predicts that different types of plants can colonize an area at the same time. Sheer chance determines which seeds arrive first, and successional stages may simply reflect the length of time it takes species to mature. This alone could account for the herb-shrub-forest development that is often seen (Fig. 45.15). The length of time it takes for trees to develop might give the impression that there is a recognizable series of plant communities, from the simple to the complex. In reality, succession does occur, and the models mentioned here are probably not mutually exclusive, but a mixture of multiple, complex processes.

Although it may not have been apparent to early ecologists, we now recognize that the most outstanding characteristic of natural communities is their dynamic nature. Also, it seems obvious to us now that the most complex communities most likely consist of habitat patches that are at various stages of succession. Each successional stage has its own mix of plants and animals, and if a sample of all stages is present, community diversity is greatest. Further, we do not know if succession continues to certain end points, because the process may not be complete anywhere on the face of the Earth.

**Check Your Progress 45.2**

1. Several different hypothesis (models) are available to explain succession. Does this mean that ecological succession doesn't occur?

**FIGURE 45.15 Secondary succession in a forest.**

In secondary succession in a large conifer plantation in central New York State, certain species are common to particular stages. However, the process of regrowth shows approximately the same stages as secondary succession from a cornfield (see Fig. 45.14).

# 45.3 Dynamics of an Ecosystem

In an **ecosystem,** populations interact among themselves and with the physical environment. The abiotic components of an ecosystem are the nonliving components such as the atmosphere, water, and soil. The biotic components are living things that can be categorized according to their food source (Fig. 45.16).

## Autotrophs

**Autotrophs** require only inorganic nutrients and an outside energy source to produce organic nutrients for their own use and for all the other members of a community. They are called **producers** because they produce food. Autotrophs include photosynthetic organisms such as land plants and algae. They possess chlorophyll and carry on photosynthesis in freshwater and marine habitats. Algae make up the phytoplankton, which are photosynthesizing organisms suspended in water. Green plants are the dominant photosynthesizers on land.

Some autotrophic bacteria are chemosynthetic. They obtain energy by oxidizing inorganic compounds such as ammonia, nitrites, and sulfides, and they use this energy to synthesize organic compounds. Chemoautotrophs have been found to support communities in some caves and also at hydrothermal vents along deep-sea oceanic ridges where sunlight is unavailable.

## Heterotrophs

**Heterotrophs** need a preformed source of organic nutrients as they acquire food from a different (*hetero*) source. They are called **consumers** because they consume food that was generated by a producer. **Herbivores** are animals that graze directly on plants or algae. In terrestrial habitats, insects are small herbivores; antelopes and bison are large herbivores. In aquatic habitats, zooplankton are small herbivores; fishes and manatees are large herbivores. **Carnivores** feed on other animals; birds that feed on insects are carnivores, and so are hawks that feed on birds and small mammals. **Omnivores** are animals that feed on both plants and animals. Chickens, raccoons, and humans are omnivores. Some animals are scavengers, such as vultures and jackals, which eat the carcasses of dead animals.

**Detritivores** are organisms that feed on detritus, which is decomposing particles of organic matter. Marine fan worms filter detritus from the water, while clams take it from the substratum. Earthworms, some beetles, termites, and ants are all terrestrial detritivores. Bacteria and fungi, including mushrooms, are decomposers; they acquire nutrients by breaking down dead organic matter, including animal wastes. **Decomposers** perform a valuable service because they release inorganic substances that are taken up by plants once more. Otherwise, plants would be completely dependent only on physical processes, such as the release of minerals from rocks, to supply them with inorganic nutrients.

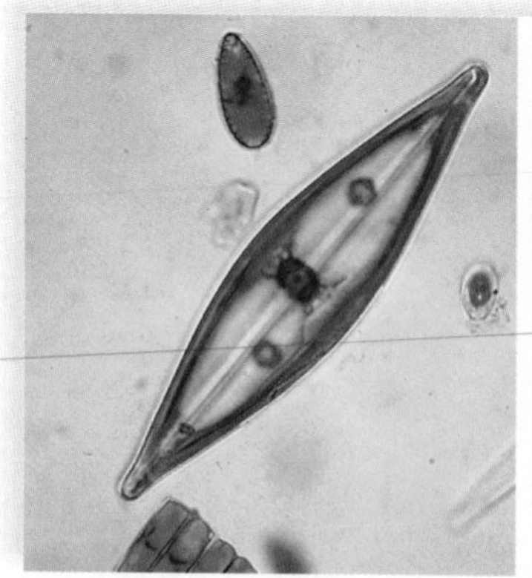

a. Producers

b. Herbivores

c. Carnivores

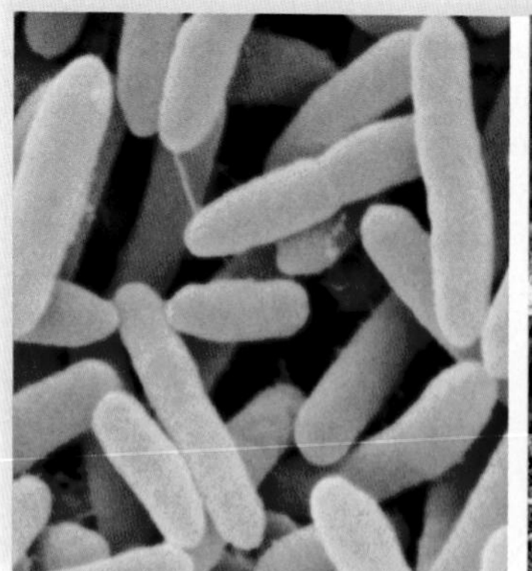

d. Decomposers

**FIGURE 45.16 Biotic components.**

**a.** Diatoms and green plants are photoautotrophs. **b.** Caterpillars and rabbits are herbivores. **c.** Spiders and osprey are carnivores. **d.** Bacteria and mushrooms are decomposers.

## Energy Flow and Chemical Cycling

A diagram of all the biotic components of an ecosystem illustrates that every ecosystem is characterized by two fundamental phenomena: energy flow and chemical cycling (Fig. 45.17). Energy flow begins when producers absorb solar energy, and chemical cycling begins when producers take in inorganic nutrients from the physical environment. Thereafter, via photosynthesis, producers make organic nutrients (food) directly for themselves and indirectly for the other populations of the ecosystem. Energy flows through an ecosystem via photosynthesis because as organic nutrients pass from one component of the ecosystem to another, such as when an herbivore eats a plant or a carnivore eats an herbivore, only a portion of the original amount of energy is transferred. Eventually, the energy dissipates into the environment as heat. Therefore, the vast majority of ecosystems cannot exist without a continual supply of solar energy.

Only a portion of the organic nutrients made by producers is passed on to consumers because plants use organic molecules to fuel their own cellular respiration. Similarly, only a small percentage of nutrients consumed by lower-level consumers, such as herbivores, is available to higher-level consumers, or carnivores. As Figure 45.18 demonstrates, a certain amount of the food eaten by an herbivore is never digested and is eliminated as feces.

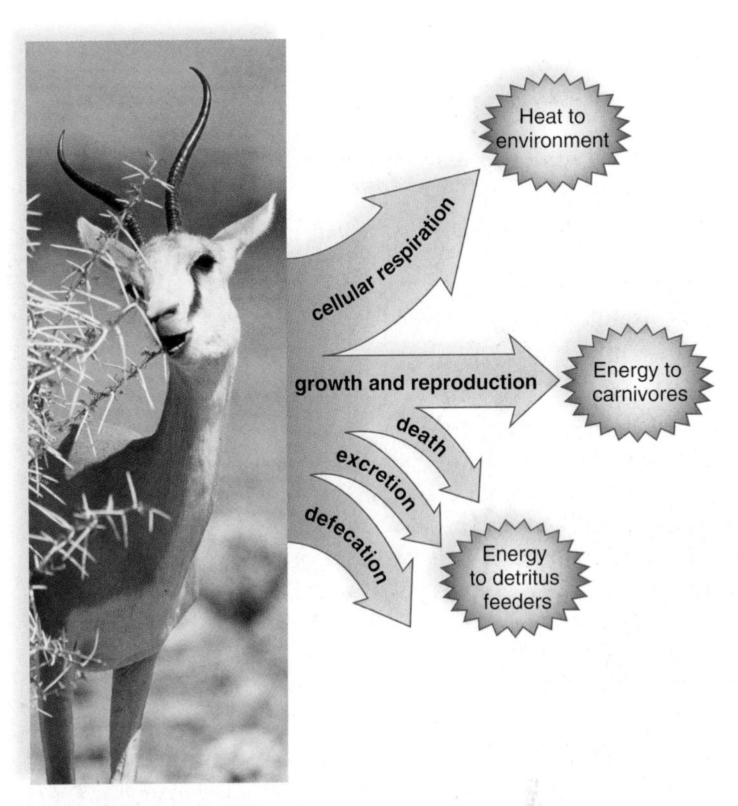


**FIGURE 45.18 Energy balances.**

Only about 10% of the food energy taken in by an herbivore is passed on to carnivores. A large portion goes to detritus feeders via defecation, excretion, and death, and another large portion is used for cellular respiration.

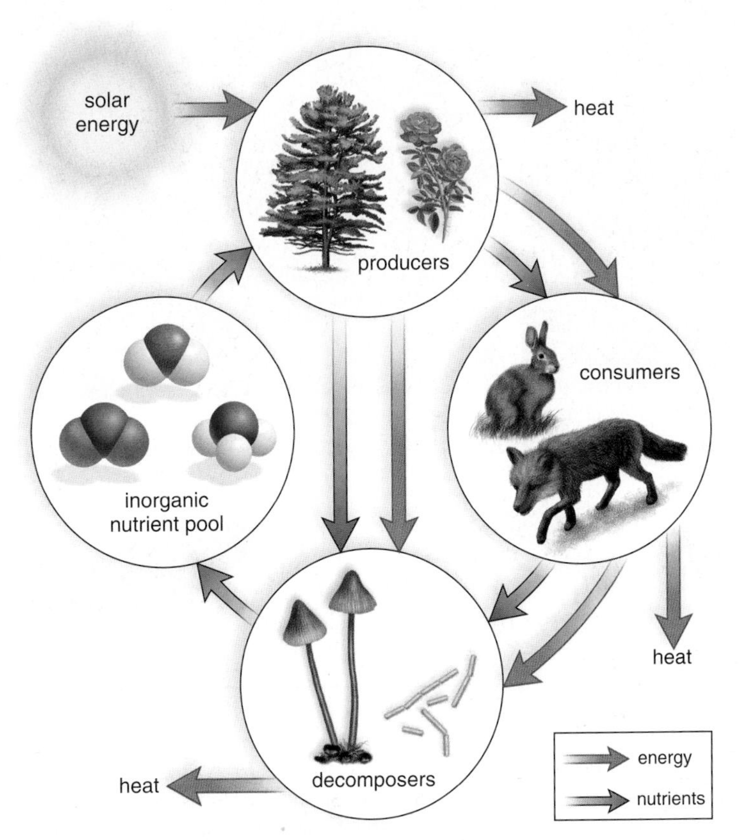


**FIGURE 45.17 Nature of an ecosystem.**

Chemicals cycle, but energy flows through an ecosystem. As energy transformations repeatedly occur, all the energy derived from the sun eventually dissipates as heat.

Metabolic nitrogenous wastes are excreted as urine. Of the assimilated energy, a large portion is used during cellular respiration for the production of ATP and thereafter becomes heat. Only the remaining energy, which is converted into increased body weight or additional offspring, becomes available to carnivores.

The elimination of feces and urine by a heterotroph, and indeed the death of all organisms, does not mean that organic nutrients are lost to an ecosystem. Instead, they represent the organic nutrients made available to decomposers. Decomposers convert the organic nutrients, such as glucose, back into inorganic chemicals, such as carbon dioxide and water, and release them to the soil or atmosphere. When these inorganic chemicals are absorbed by producers from the atmosphere or soil, these chemicals have completed their cycle within an ecosystem.

The laws of thermodynamics support the concept that energy flows through an ecosystem. The first law states that energy can neither be created nor destroyed. This explains why ecosystems are dependent on a continual outside source of energy, usually solar energy, which is used by photosynthesizers to produce organic nutrients. The second law states that, with every transformation, some energy is degraded into a less available form such as heat. Because plants carry on cellular respiration, for example, only about 55% of the original energy absorbed by plants is available to an ecosystem.

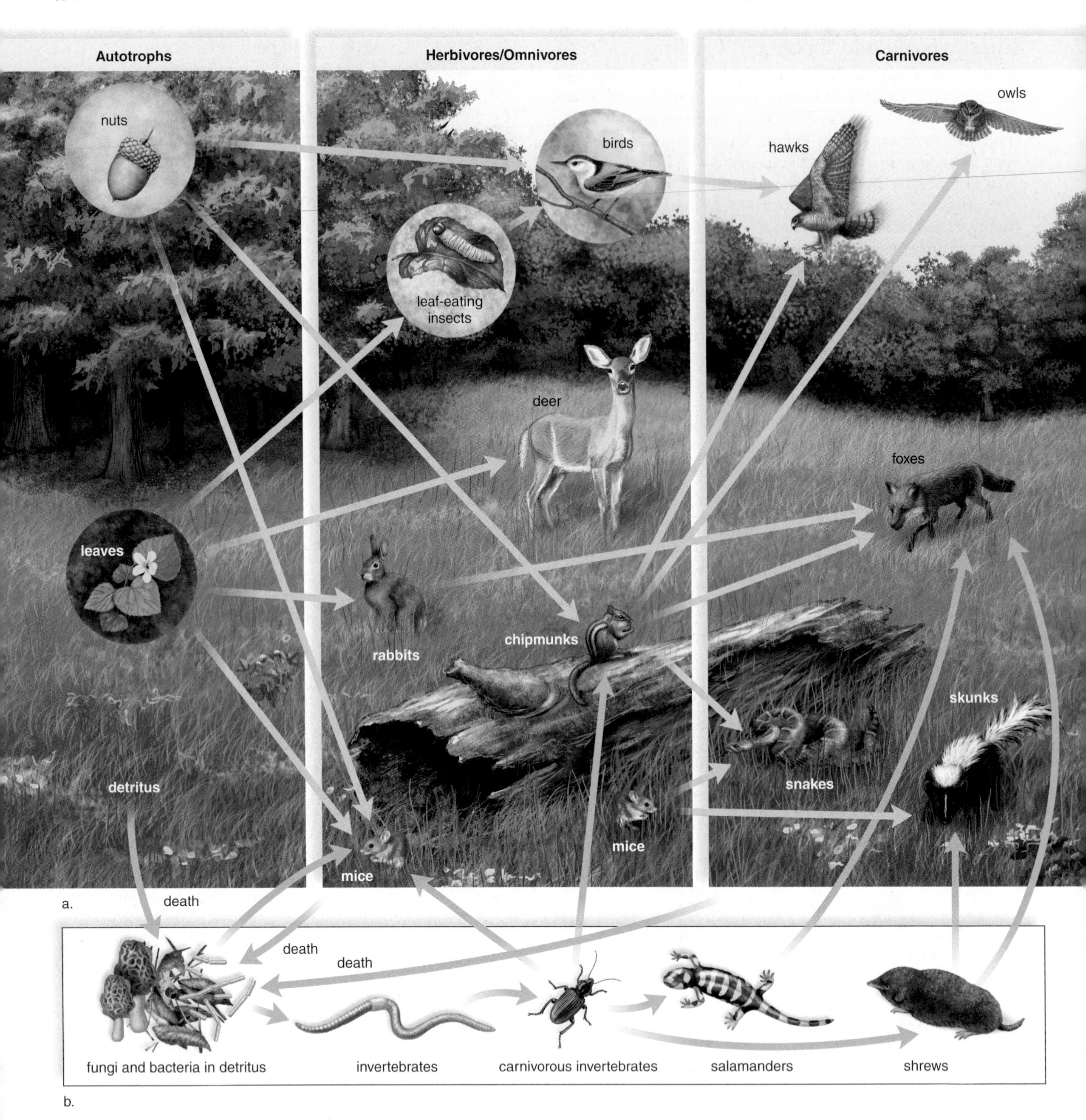


**FIGURE 45.19 Grazing and detrital food web.**

Food webs are descriptions of who eats whom. **a.** Tan arrows illustrate possible grazing food webs. For example, birds, which feed on nuts, may be eaten by a hawk. Autotrophs such as the tree are producers (first trophic, or feeding, level), the first series of animals are primary consumers (second trophic level), and the next group of animals are secondary consumers (third trophic level). **b.** Green arrows illustrate possible detrital food webs, which begin with detritus—the bacteria and fungi of decay and the remains of dead organisms. A large portion of these remains are from the grazing food web illustrated in (**a**). The organisms in the detrital food web are sometimes fed on by animals in the grazing food web, as when robins feed on earthworms. Thus, the grazing food web and the detrital food web are connected to one another.

## Energy Flow

The principles discussed previously in this chapter can now be applied to an actual ecosystem—a forest of 132,000 $m^2$ in New Hampshire. The various interconnecting paths of energy flow are represented by a **food web,** a diagram that describes trophic (feeding) relationships. Figure 45.19*a* is a grazing food web because it begins with a producer, specifically the oak tree depicted. Insects in the form of caterpillars feed on leaves, while mice, rabbits, and deer feed on leaf tissue at or near the ground. Birds, chipmunks, and mice feed on fruits and nuts, but they are in fact omnivores because they also feed on caterpillars. These herbivores and omnivores all provide food for a number of different carnivores.

Figure 45.19*b* is a detrital food web, which begins with detritus. Detritus is food for soil organisms such as earthworms. Earthworms are in turn fed on by carnivorous invertebrates, and they may be eaten by shrews or salamanders. Because the members of a detrital food web may become food for aboveground carnivores, the detrital and grazing food webs are joined.

We naturally tend to think that aboveground plants such as trees are the largest storage form of organic matter and energy, but this is not necessarily the case. In this particular forest, the organic matter lying on the forest floor and mixed into the soil contains over twice as much energy as the leaf matter of living trees. Therefore, more energy in a forest may be funneling through the detrital food web than through the grazing food web.

### *Trophic Levels*

The arrangement of the species in Figure 45.19 suggests that organisms are linked to one another in a straight line, according to feeding relationships, or who eats whom. Diagrams that show a single path of energy flow in an ecosystem are called **food chains.** For example, in the grazing food web, we could find this grazing food chain:

leaves ⟶ caterpillars ⟶ birds ⟶ hawks

And in the detrital food web, we could find this detrital food chain:

detritus ⟶ earthworms ⟶ salamanders

A **trophic level** is a level of nourishment within a food web or chain. In the grazing food web in Figure 45.19*a*, going from left to right, the green plants are producers (first trophic level), the first series of animals are primary consumers (second trophic level), and the next group of animals are secondary consumers (third trophic level).

### *Ecological Pyramids*

The shortness of food chains can be attributed to the loss of energy between trophic levels. In general, only about 10% of the energy of one trophic level is available to the next trophic level. Therefore, if an herbivore population consumes 1,000 kg of plant material, only about 100 kg is converted to body tissue of an herbivore, 10 kg to first-level carnivores, and 1 kg to second-level carnivores. The so-called 10% rule explains why few carnivores can be supported in a food web. The flow of energy with large losses between successive trophic levels is sometimes depicted as an **ecological pyramid** (Fig. 45.20).

Energy losses between trophic levels also result in pyramids based on the number of organisms or the amount of biomass at each trophic level. When constructing such pyramids, problems arise, however. For example, in Figure 45.19, each tree would contain numerous caterpillars; therefore, there would be more herbivores than autotrophs! The explanation, of course, has to do with size. An autotroph can be as tiny as a microscopic alga or as big as a beech tree; similarly, an herbivore can be as small as a caterpillar or as large as an elephant.

Pyramids of biomass eliminate size as a factor because **biomass** is the number of organisms multiplied by the dry weight of the organic matter within one organism. The biomass of the producers is expected to be greater than the biomass of the herbivores, and that of the herbivores is expected to be greater than that of the carnivores. In aquatic ecosystems, such as some lakes and open seas where algae are the only producers, the herbivores may have a greater biomass than the producers when their measurements are taken because the algae are consumed at a high rate. Such

**FIGURE 45.20 Ecological pyramid.**

The biomass, or dry weight ($g/m^2$), for trophic levels in a grazing food web in a bog at Silver Springs, Florida. There is a sharp drop in biomass between the producer level and herbivore level, which is consistent with the common knowledge that the detrital food web plays a significant role in bogs.

pyramids, which have more herbivores than producers, are called inverted pyramids:

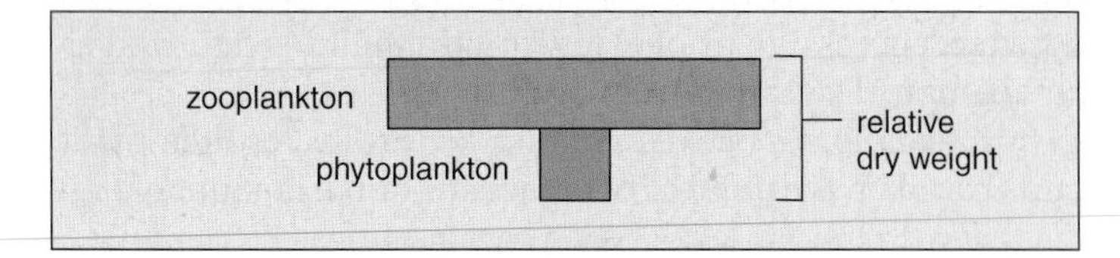


These kinds of problems are making some ecologists hesitant about using pyramids to describe ecological relationships. One more problem is what to do with the decomposers, which are rarely included in pyramids, even though a large portion of energy becomes detritus in many ecosystems.

## Chemical Cycling

The pathways by which chemicals circulate through ecosystems involve both living (biotic) and nonliving (geologic) components; therefore, they are known as **biogeochemical cycles.** In this chapter, we describe four of the biogeochemical cycles: the water, carbon, phosphorus, and nitrogen cycles. A biogeochemical cycle may be sedimentary or gaseous. The phosphorus cycle is a sedimentary cycle; the chemical is absorbed from the soil by plant roots, passed to heterotrophs, and eventually returned to the soil by decomposers. The carbon and nitrogen cycles are gaseous, meaning that the chemical returns to and is withdrawn from the atmosphere as a gas.

Chemical cycling involves the components of ecosystems shown in Figure 45.21. A *reservoir* is a source normally unavailable to producers, such as the carbon present in calcium carbonate shells on ocean bottoms. An *exchange pool* is a source from which organisms do generally take chemicals, such as the atmosphere or soil. Chemicals move along food chains in a *biotic community*, perhaps never entering an exchange pool.

Human activities (purple arrows) remove chemicals from reservoirs and exchange pools and make them available to the biotic community. In this way, human activities result in pollution because they upset the normal balance of nutrients for producers in the environment.

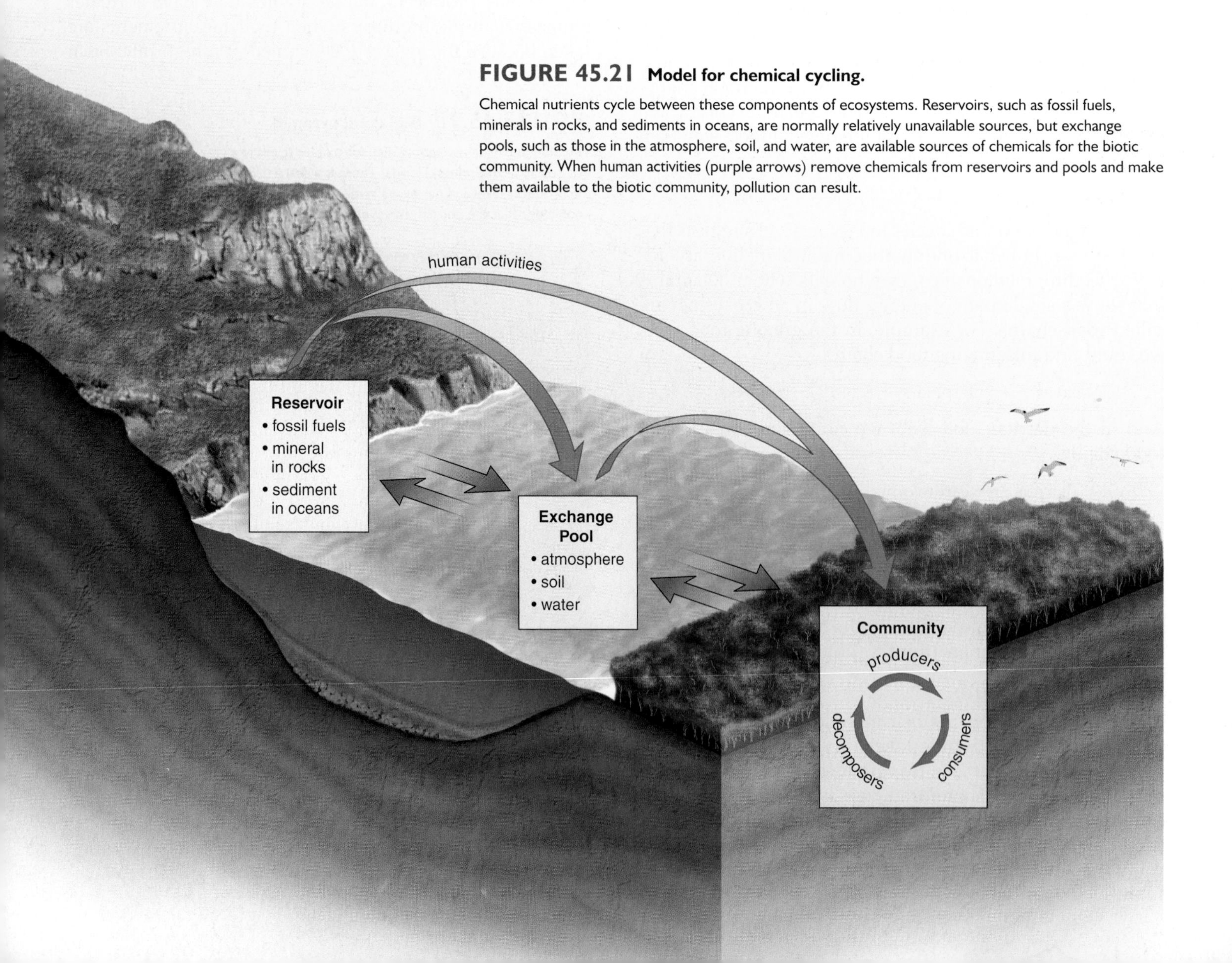


**FIGURE 45.21 Model for chemical cycling.**

Chemical nutrients cycle between these components of ecosystems. Reservoirs, such as fossil fuels, minerals in rocks, and sediments in oceans, are normally relatively unavailable sources, but exchange pools, such as those in the atmosphere, soil, and water, are available sources of chemicals for the biotic community. When human activities (purple arrows) remove chemicals from reservoirs and pools and make them available to the biotic community, pollution can result.

## The Water Cycle

The **water (hydrologic) cycle** is described in Figure 45.22. A **transfer rate** is defined as the amount of a substance that moves from one component of the environment to another within a specified period of time. The width of the arrows in Figure 45.22 indicates the transfer rate of water.

During the water cycle, fresh water is first distilled from salt water through evaporation. During evaporation, a liquid, in this case water, changes to a gaseous state. The sun's rays cause fresh water to evaporate from the seawater, and the salts are left behind. Next, condensation occurs. During condensation, a gas is converted into a liquid. For example, vaporized fresh water rises into the atmosphere, is stored in clouds, cools, and falls as rain over the oceans and the land.

Water evaporates from land and from plants (evaporation from plants is called transpiration) and also from bodies of fresh water. Because land lies above sea level, gravity eventually returns all fresh water to the sea. In the meantime, water is contained within standing waters (lakes and ponds), flowing water (streams and rivers), and groundwater.

Some of the water from precipitation (e.g., rain, snow, sleet, hail, and fog) sinks, or percolates, into the ground and saturates the Earth to a certain level. The top of the saturation zone is called the groundwater table, or simply, the water table. Because water infiltrates through the soil and rock layers, sometimes groundwater is also located in aquifers, rock layers that contain water and release it in appreciable quantities to wells or springs. Aquifers are recharged when rainfall and melted snow percolate into the soil.

### *Human Activities*

In some parts of the United States, especially the arid West and southern Florida, withdrawals from aquifers exceed any possibility of recharge. This is called "groundwater mining." In these locations, the groundwater is dropping, and residents may run out of groundwater, at least for irrigation purposes, within a few years.

Fresh water, which makes up only about 3% of the world's supply of water, is called a renewable resource because a new supply is always being produced because of the water cycle. But it is possible to run out of fresh water when the available supply is not adequate or is polluted so that it is not usable.

**FIGURE 45.22 The hydrologic (water) cycle.**

Evaporation from the ocean exceeds precipitation, so there is a net movement of water vapor onto land, where precipitation results in surface water and groundwater that flow back to the sea. On land, transpiration by plants contributes to evaporation. The numbers in this diagram indicate water flow in cubic kilometers per year.

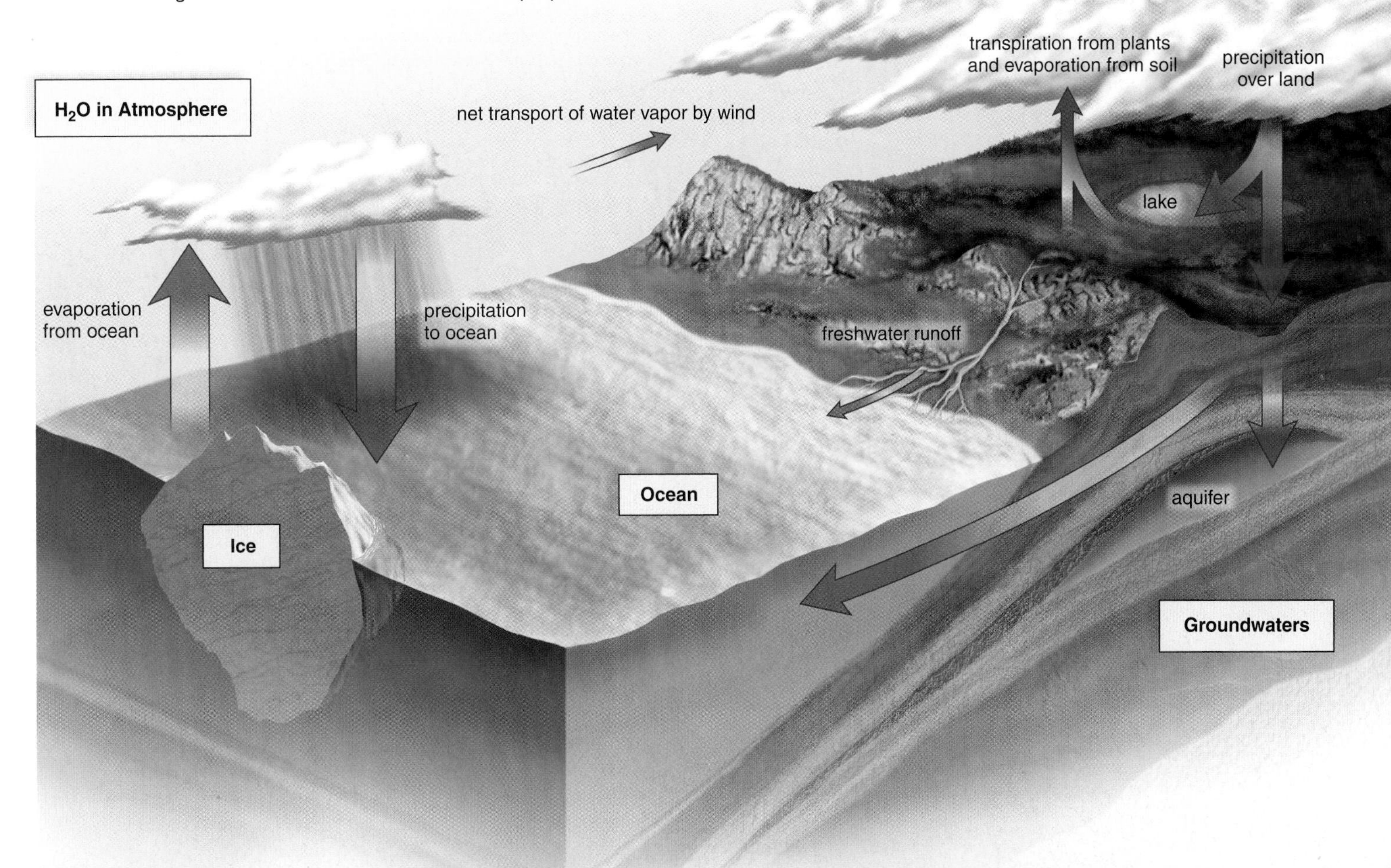

## The Carbon Cycle

In the carbon cycle, organisms in both terrestrial and aquatic ecosystems exchange carbon dioxide ($CO_2$) with the atmosphere (Fig. 45.23). Therefore, the $CO_2$ in the atmosphere is the exchange pool for the carbon cycle. On land, plants take up $CO_2$ from the air and, through photosynthesis, they incorporate carbon into nutrients that are used by autotrophs and heterotrophs alike. (1) When organisms, including plants, respire, carbon is returned to the atmosphere as $CO_2$. (2) $CO_2$ then recycles to plants by way of the atmosphere.

In aquatic ecosystems, the exchange of $CO_2$ with the atmosphere is indirect. (3) Carbon dioxide from the air combines with water to produce bicarbonate ion ($HCO_3^-$), a source of carbon for algae that produce food for themselves and for heterotrophs. Similarly, when aquatic organisms respire, the $CO_2$ they give off becomes $HCO_3^-$. (4) The amount of bicarbonate in the water is in equilibrium with the amount of $CO_2$ in the air.

### Reservoirs Hold Carbon

Living and dead organisms contain organic carbon and serve as one of the reservoirs for the carbon cycle. The world's biotic components, particularly trees, contain 800 billion tons of organic carbon, and an additional 1,000–3,000 billion metric tons are estimated to be held in the remains of plants and animals in the soil. (5) Ordinarily, decomposition of organisms returns $CO_2$ to the atmosphere.

Some 300 MYA, plant and animal remains were transformed into coal, oil, and natural gas, the materials we call fossil fuels. Another reservoir for carbon is the inorganic carbonate that accumulates in limestone and in calcium carbonate shells. Many marine organisms have calcium carbonate shells that remain in bottom sediments long after the organisms have died. Geologic forces change these sediments into limestone.

### Human Activities and the Carbon Cycle

(6) More $CO_2$ is being deposited in the atmosphere than is being removed, largely due to the burning of fossil fuels and the destruction of forests to make way for farmland and pasture. When we humans do away with forests, we reduce a reservoir and also the very organisms that take up excess carbon dioxide. Today, the amount of $CO_2$ released into the atmosphere is about twice the amount that remains in the atmosphere. Much of the $CO_2$ dissolves into the ocean.

Other gases, as well as $CO_2$, are excess, emitted into the atmosphere due to human activities. The other gases include nitrous oxide ($N_2O$) from fertilizers and animal wastes and methane ($CH_4$) from bacterial decomposition that takes place particularly in the guts of animals, in sediments, and in flooded rice paddies. These gases are known

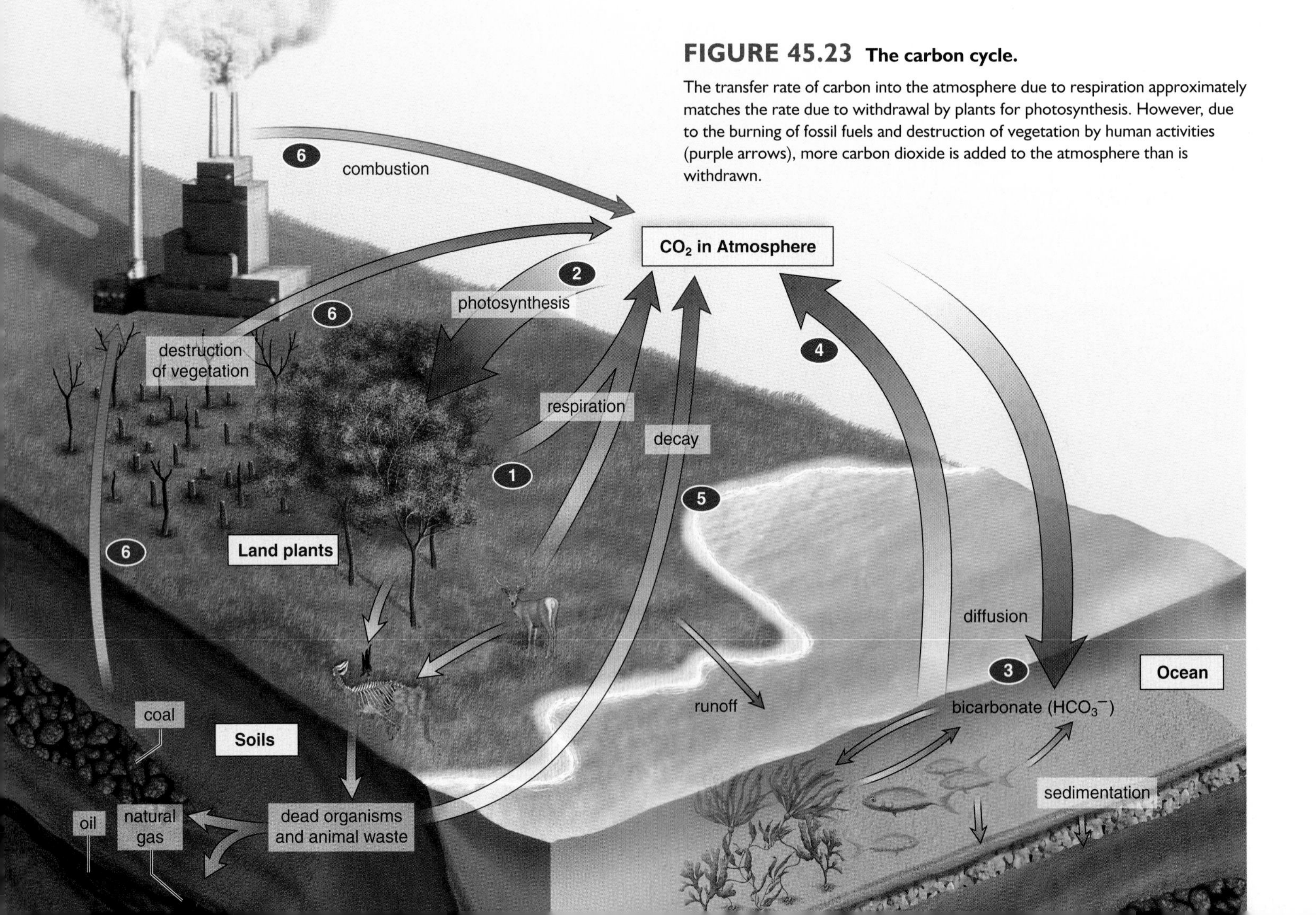


**FIGURE 45.23 The carbon cycle.**

The transfer rate of carbon into the atmosphere due to respiration approximately matches the rate due to withdrawal by plants for photosynthesis. However, due to the burning of fossil fuels and destruction of vegetation by human activities (purple arrows), more carbon dioxide is added to the atmosphere than is withdrawn.

as **greenhouse gases** because, just like the panes of a greenhouse, they allow solar radiation to pass through but hinder the escape of infrared rays (heat) back into space. This phenomenon has come to be known as the **greenhouse effect.** The greenhouse gases are contributing significantly to an overall rise in the Earth's ambient temperature, a trend called **global warming.** The global climate has already warmed about 0.6°C since the Industrial Revolution. Computer models are unable to consider all possible variables, but the Earth's temperature may rise 1.5–4.5°C by 2100 if greenhouse emissions continue at the current rates.

It is predicted that, as the oceans warm, temperatures in the polar regions will rise to a greater degree than in other regions. If so, glaciers will melt, and sea level will rise, not only due to this melting but also because water expands as it warms. Increased rainfall is likely along the coasts, while dryer conditions are expected inland. Coastal agricultural lands, such as the deltas of Bangladesh and China, will be inundated with seawater, and billions of dollars will have to be spent to keep coastal cities such as New Orleans, New York, Boston, Miami, and Galveston from disappearing into the sea.

## The Phosphorus Cycle

Figure 45.24 depicts the phosphorus cycle. (1) Phosphorus, trapped in oceanic sediments, moves onto land due to a geologic uplift. (2) On land, the very slow weathering of rocks places (3) phosphate ions ($PO_3^-$ and $HPO_4^+$) in the soil. (4) Some of these become available to plants, which use phosphate in a variety of molecules, including phospholipids, ATP, and the nucleotides that become a part of DNA and RNA. (5) Animals eat producers and incorporate some of the phosphate into their teeth, bones, and shells, which take many years to decompose. (6) However, eventually the death and decay of all organisms and also the decomposition of animal wastes make phosphate ions available to producers once again. Because the available amount of phosphate is already being used within food chains, phosphate is usually a limiting inorganic nutrient for plants—that is, the lack of it limits the size of populations in ecosystems.

(7) Some phosphate naturally runs off into aquatic ecosystems, where algae acquire phosphate from the water before it becomes trapped in sediments. Phosphate in marine sediments does not become available to producers on land again until a geologic upheaval exposes sedimentary rocks on land. Now, the cycle begins again.

### *Human Activities and the Phosphorus Cycle*

(8) Human beings boost the supply of phosphate by mining phosphate ores for producing fertilizer and detergents. Runoff of phosphate and nitrogen due to fertilizer use, animal wastes from livestock feedlots, and discharge from sewage treatment plants results in **eutrophication** (overenrichment) of waterways.

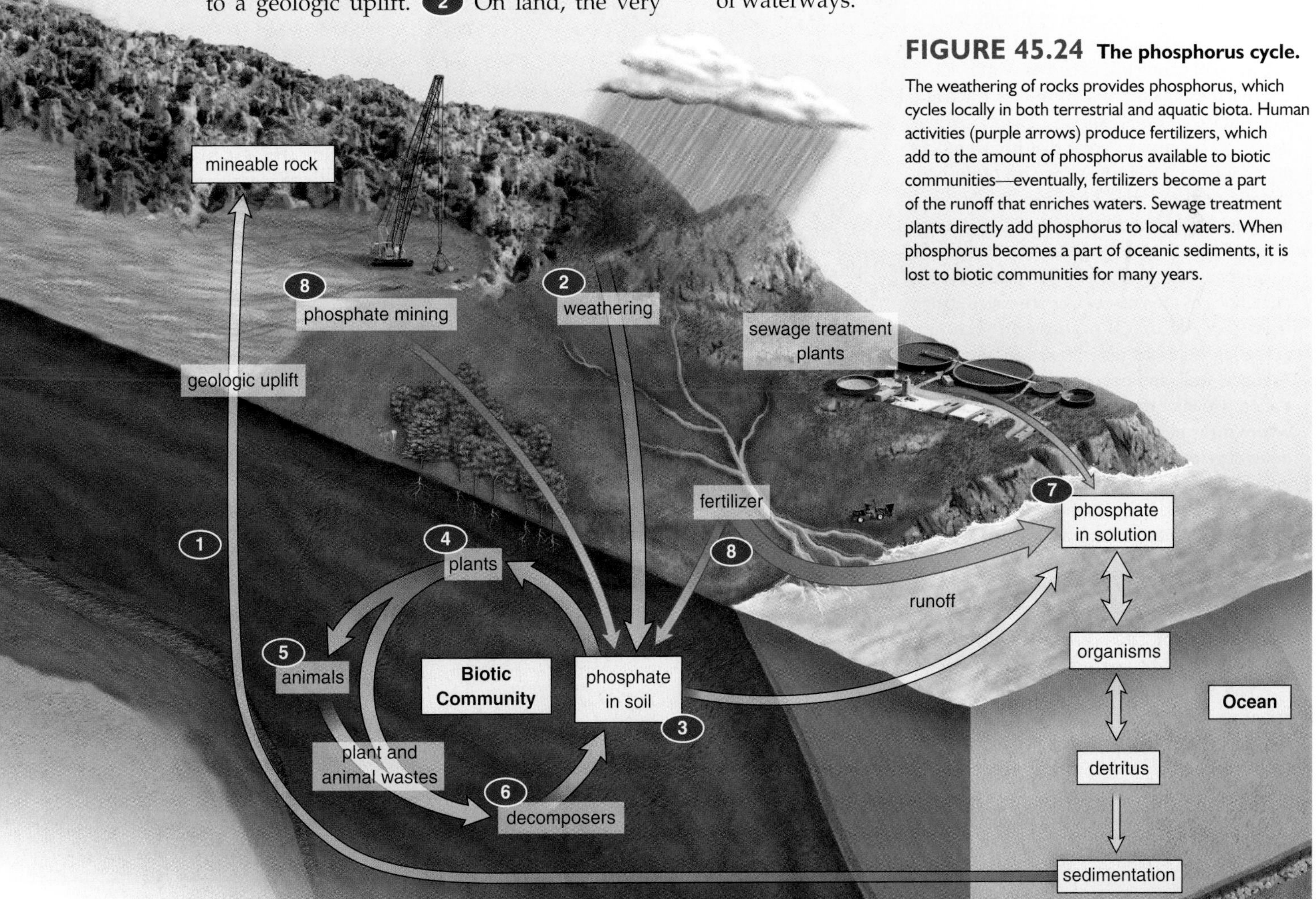


**FIGURE 45.24 The phosphorus cycle.**

The weathering of rocks provides phosphorus, which cycles locally in both terrestrial and aquatic biota. Human activities (purple arrows) produce fertilizers, which add to the amount of phosphorus available to biotic communities—eventually, fertilizers become a part of the runoff that enriches waters. Sewage treatment plants directly add phosphorus to local waters. When phosphorus becomes a part of oceanic sediments, it is lost to biotic communities for many years.

science focus

# Ozone Shield Depletion

In the stratosphere, some 50 km above the Earth, ozone forms the **ozone shield,** a layer of ozone that absorbs most of the ultraviolet (UV) rays of the sun so that fewer rays strike the Earth. Ozone forms when ultraviolet radiation from the sun splits oxygen molecules ($O_2$), and then the oxygen atoms (O) combine with other oxygen molecules to produce ozone ($O_3$).

## Cause of Depletion

The absorption of UV radiation by the ozone shield is critical for living things. In humans, UV radiation causes mutations that can lead to skin cancer and can make the lens of the eye develop cataracts. In addition, it adversely affects the immune system and our ability to resist infectious diseases. UV radiation also impairs crop and tree growth and kills off algae and tiny shrimplike animals (krill) that sustain oceanic life. Without an adequate ozone shield, therefore, our health and food sources are threatened.

It became apparent in the 1980s that depletion of ozone had occurred worldwide and that the depletion was most severe above the Antarctic every spring. There, ozone depletion became so great that it covered an area two and a half times the size of Europe, and exposed not only Antarctica but also the southern tip of South America and vast areas of the Pacific and Atlantic oceans to harmful ultraviolet rays. In the popular press, severe depletions of the ozone layer are called ozone holes (Fig. 45C*a*). Of even greater concern, an ozone hole has now appeared above the Arctic as well, and ozone holes were also detected within northern and southern latitudes, where many people live. Whether or not these holes develop in the spring depends on prevailing winds, weather conditions, and the type of particles in the atmosphere. A United Nations Environmental Program report predicts a 26% rise in cataracts and nonmelanoma skin cancers for every 10% drop in the ozone level. A 26% increase translates into 1.75 million additional cases of cataracts and 300,000 more skin cancers every year, worldwide.

The seriousness of the situation caused scientists around the globe to begin studying the cause of ozone depletion. The cause was found to be chlorine atoms (Cl), which can destroy up to 100,000 molecules of ozone before settling to the Earth's surface as chloride years later.

## Control of CFCs

The chlorine atoms that enter the troposphere and eventually reach the stratosphere come primarily from the breakdown of **chlorofluorocarbons (CFCs),** chemicals much in use by humans. The best-known CFC is Freon, a coolant found in refrigerators and air conditioners. CFCs are also used as cleaning agents and as foaming agents during the production of Styrofoam coffee cups, egg cartons, insulation, and paddings. Formerly, CFCs were used as propellants in spray cans, but this application is now banned in the United States and several European countries. Other molecules, such as the cleaning solvent methyl chloroform, are also sources of harmful chlorine atoms.

Most of the countries of the world have stopped using CFCs, and the United States halted production in 1995. Since that time, satellite measurements indicate that the amount of harmful chlorine pollution in the stratosphere has started to decline. It is clear, however, that recovery of the ozone shield may take several more years and involve other pollution-fighting approaches, aside from lowering chlorine pollution. Researchers report that currently, there were more and longer-lasting polar clouds than previously. Why might that be? As the Earth's surface warms due to global warming, less heat reradiates into the stratosphere. Mathematical modeling suggests that stratospheric clouds could last twice as long over the Arctic before the year 2010, when the coldest winter ever is expected.

Cloud cover contributes to the breakdown of the ozone shield by chlorine pollution (Fig. 45C*b*). It is speculated that once polar stratospheric clouds become twice as persistent, there could still be an ozone loss of 30%.

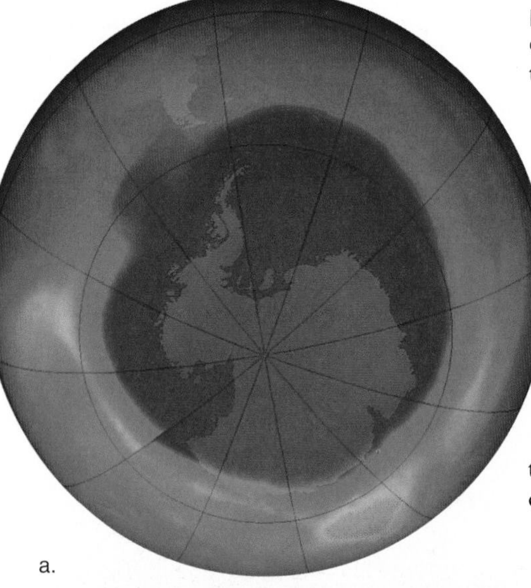

a.

b.

**FIGURE 45C Ozone shield depletion.**
***a.*** *Map of ozone levels in the atmosphere of the Southern Hemisphere, September 2007. The ozone depletion is larger than the size of Europe.* ***b.*** *Global warming is contributing to a cloud cover in the stratosphere, which contributes to the breakdown of the ozone shield by chlorine pollution.*

## The Nitrogen Cycle

Nitrogen gas ($N_2$) makes up about 78% of the atmosphere, but plants cannot make use of nitrogen in its gaseous form. Therefore, nitrogen can be a nutrient that limits the amount of growth in an ecosystem. First, let's consider that (1) **$N_2$ (nitrogen) fixation** occurs when nitrogen gas ($N_2$) is converted to ammonium ($NH_4^+$), a form plants can use (Fig. 45.25). Some cyanobacteria in aquatic ecosystems and some free-living bacteria in soil are able to fix atmospheric nitrogen in this way. Other nitrogen-fixing bacteria live in nodules on the roots of legumes, such as beans, peas, and clover. They make organic compounds containing nitrogen available to the host plants so that the plant can form proteins and nucleic acids. (2) Plants can also use nitrates ($NO_3^-$) as a source of nitrogen. The production of nitrates during the nitrogen cycle is called **nitrification.** Nitrification can occur in two ways: (1) Nitrogen gas ($N_2$) is converted to $NO_3^-$ in the atmosphere when cosmic radiation, meteor trails, and lightning provide the high energy needed for nitrogen to react with oxygen. (2) Ammonium ($NH_4^+$) in the soil from various sources, including decomposition of organisms and animal wastes, is converted to $NO_3^-$ by nitrifying bacteria in soil. Specifically, $NH_4^+$ (ammonium) is converted to $NO_2^-$ (nitrite), and then $NO_2^-$ is converted to $NO_3^-$ (nitrate). (3) During the process of assimilation, plants take up $NH_4^+$ and $NO_3^-$ from the soil and use these ions to produce proteins and nucleic acids. Notice in Figure 45.25 that the subcycle involving the biotic community, which occurs on land and in the ocean, need not depend on the presence of nitrogen gas at all. Finally, (4) **denitrification** is the conversion of nitrate back to nitrogen gas, which then enters the atmosphere. Denitrifying bacteria living in the anaerobic mud of lakes, bogs, and estuaries carry out this process as a part of their own metabolism. In the nitrogen cycle, denitrification would counterbalance nitrogen fixation if not for human activities.

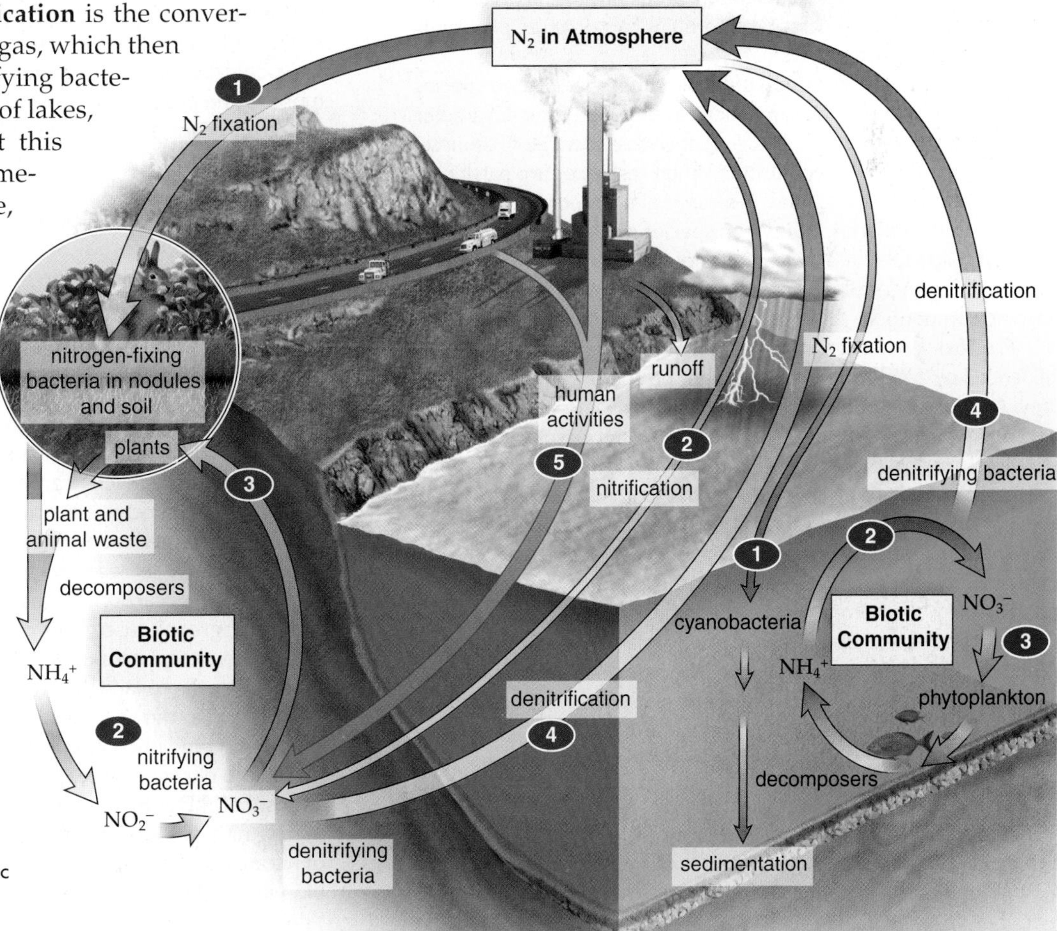


**FIGURE 45.25 The nitrogen cycle.**

Nitrogen is primarily made available to biotic communities by internal cycling of the element. Without human activities, the amount of nitrogen returned to the atmosphere (denitrification in terrestrial and aquatic communities) exceeds withdrawal from the atmosphere ($N_2$ fixation and nitrification). Human activities (purple arrows) result in an increased amount of $NO_3^-$ in terrestrial communities with resultant runoff to aquatic biotic communities.

### *Human Activities and the Nitrogen Cycle*

(5) Humans significantly alter the transfer rates in the nitrogen cycle by producing fertilizers from $N_2$—in fact, they nearly double the fixation rate. Fertilizer, which also contains phosphate, runs off into lakes and rivers and results in an overgrowth of algae and rooted aquatic plants. When the algae die off, enlarged populations of decomposers use up all the oxygen in the water, and the result is a massive fish kill.

**Acid deposition** occurs because nitrogen oxides ($NO_x$) and sulfur dioxide ($SO_2$) enter the atmosphere from the burning of fossil fuels. Both these gases combine with water vapor to form acids that eventually return to the Earth. Acid deposition has drastically affected forests and lakes in northern Europe, Canada, and the northeastern United States because their soils are naturally acidic and their surface waters are only mildly alkaline (basic). Acid deposition reduces agricultural yields and corrodes marble, metal, and stonework.

#### Check Your Progress 45.3

1. What type of population is at the base of an ecological pyramid and the start of a food chain?
2. How does the passage of energy differ from that of chemicals in an ecosystem?
3. Human activities intensify what natural process associated with the carbon cycle?

# Connecting the Concepts

Community ecology is concerned with how populations interact with each other in a particular locale. Historically, some ecologists felt that communities functioned as "superorganisms," with each population a vital and integral part of the community. At the other extreme, some ecologists felt that each population simply occurred where conditions were best for it and that communities were chance aggregations of species. Most likely, the truth lies somewhere in between these two extreme views. We know that a population of any species is distributed where conditions are best for it. However, communities do exhibit emergent properties that result from certain groupings of species co-occurring.

The number of individuals in many populations within ecological communities is determined by interactions such as interspecific competition, predation, and parasitism. The sizes of all populations are negatively affected by at least one of those interactions. Positive interactions such as mutualisms are fairly common in nature (especially for plants), but their effect on population size is not well understood yet.

Perhaps one of the most important recent discoveries about communities is that they are highly dynamic. The number of species, kinds of species, and sizes of populations within most communities are constantly changing due to disturbances and climatic variability.

Because the species composition of communities can be so variable, many ecologists study the movement of energy and nutrients through communities. The physical environment has a large influence on energy and nutrient flow, and thus the nonliving world must be incorporated into our studies of communities and ecosystems.

## summary

### 45.1 Ecology of Communities

A community is an assemblage of populations interacting with one another within the same environment. Communities differ in their composition (species found there) and their diversity (species richness and relative abundance).

An organism's habitat is where it lives in the community. An ecological niche is defined by the role an organism plays in its community, including its habitat and how it interacts with other species in the community. Competition, predator-prey, parasite-host, commensalistic, and mutualistic relationships help organize populations into an intricate dynamic system.

The competitive exclusion principle states that no two species can indefinitely occupy the same niche at the same time. Character displacement is a structural change that gives evidence of resource partitioning and niche specialization When resources are partitioned between two or more species, increased niche specialization occurs. But the difference between species can be more subtle, as when warblers feed at different parts of the tree canopy. Barnacles competing on the Scottish coast may be an example of present ongoing competition.

Predator-prey interactions between two species are especially influenced by amount of predation and the amount of food for the prey. A cycling of population densities may occur. Prey defenses take many forms: Camouflage, use of fright, and warning coloration are three possible mechanisms. Batesian mimicry occurs when one species has the warning coloration but lacks the defense. Müllerian mimicry occurs when two species with the same warning coloration have the same defense. We would expect coevolution to occur within a community. For example, the better the predator becomes at catching prey, the better the prey becomes at escaping the predator.

Like predators, parasites take nourishment from their host. Whether parasites are aggressive (kill their host) or benign probably depends on which results in the highest fitness. Symbiotic relationships are classified as commensalistic, parasitic, or mutualistic. Mutualistic relationships as when Clark's nuthatches feed on but disperse whitebark pine seeds are critical to the cohesiveness of a community.

### 45.2 Community Development

Ecological succession involves a series of species replacements in a community. Primary succession occurs where there is no soil present. Secondary succession occurs where soil is present and certain plant species can begin to grow. A climax community forms when stages of succession lead to a particular type of community.

### 45.3 Dynamics of an Ecosystem

Ecosystems have biotic and abiotic components. The biotic components are autotrophs, heterotrophs, detritus feeders, and decomposers. Abiotic components are resources such as nutrients and conditions such as type of soil and temperature.

Ecosystems are characterized by energy flow and chemical cycling. Energy flows because as food passes from one population to the next, each population makes energy conversions that result in a loss of usable energy. Chemicals cycle because they pass from one population to the next until decomposers return them once more to the producers. Ecosystems contain food webs in which the various organisms are connected by trophic relationships. In grazing food webs, food chains begin with a producer. In a detrital food web, food chains begin with detritus. Ecological pyramids are graphic representations of the number of organisms, biomass, or energy content of trophic levels.

Biogeochemical cycles may be sedimentary (phosphorus cycle) or gaseous (carbon and nitrogen cycles). Chemical cycling involves a reservoir, an exchange pool, and a biotic community.

In the water cycle, evaporation over the ocean is not compensated for by precipitation. Precipitation over land results in bodies of fresh water plus groundwater, including aquifers. Eventually, all water returns to the oceans.

In the carbon cycle, carbon dioxide in the atmosphere is an exchange pool; both terrestrial and aquatic plants and animals exchange carbon dioxide with the atmosphere. Living and dead organisms serve as reservoirs for the carbon cycle because they contain organic carbon. Human activities increase the level of $CO_2$ and other greenhouse gases contributing to global warming.

In the phosphorus cycle, geological upheavals move phosphorus from the ocean to land. Slow weathering of rocks returns phosphorus to the soil. Most phosphorus is recycled within a community, and phosphorus is a limiting nutrient.

In the nitrogen cycle, plants cannot use nitrogen gas from the atmosphere. During nitrogen fixation, $N_2$ converts to ammonium, making nitrogen available to plants. Nitrification is the production of nitrates while denitrification is the conversion of nitrate back to $N_2$, which enters the atmosphere. Human activities increase transfer rates in the nitrogen cycle. Acid deposition occurs when nitrogen oxides enter the atmosphere, combine with water vapor, and return to Earth in precipitation.

## understanding the terms

acid deposition 861
autotroph 852
biogeochemical cycle 856
biomass 855
camouflage 845
carnivore 852
character displacement 842
chlorofluorocarbon (CFC) 860
climax community 850
coevolution 847
commensalism 846
community 840
competitive exclusion principle 842
consumer 852
decomposer 852
denitrification 861
detritivore 852
ecological niche 841
ecological pyramid 855
ecological succession 850
ecosystem 852
eutrophication 859
food chain 855
food web 855
global warming 859
greenhouse effect 859
greenhouse gas 858
habitat 841
herbivore 852
heterotroph 852
host 846
mimicry 845
model of island biogeography 849
mutualism 848
$N_2$ (nitrogen) fixation 861
nitrification 861
omnivore 852
ozone shield 860
parasite 846
parasitism 846
pioneer species 850
predation 843
predator 843
prey 843
producer 852
resource partitioning 842
species diversity 840
species richness 840
symbiosis 846
transfer rate 857
trophic level 855
water (hydrologic) cycle 857

Match the terms to these definitions:

a. ______________ Complex pattern of interlocking and crisscrossing food chains.
b. ______________ Place where an organism lives and is able to survive and reproduce.
c. ______________ Directional pattern of change in which one community replaces another until a community typical of the area results.
d. ______________ Process by which atmospheric nitrogen gas is changed to forms that plants can use.

## reviewing this chapter

1. What data do you need to describe a community's composition and diversity? 840–41
2. Describe the habitat and ecological niche of a particular species. 841
3. What is the competitive exclusion principle? How does the principle relate to character displacement and niche specialization? 842–43
4. Explain the observation that some predator-prey population densities cycle. Give examples of prey defenses. What is mimicry, and why does it work as a prey defense? 843–45
5 Give examples of parasitism, commensalism, and mutualism, and examples of coevolution that occur within a community. 846–48
6. What are the two types of ecological succession? What is the present controversy surrounding the concept? 850–51
7. Give examples of autotrophs and heterotrophs in an ecosystem. 852
8. Distinguish between energy flow and chemical cycling in an ecosystem. Describe two types of food webs and give examples of food chains. 853–54
9. Explain the appearance of an ecological pyramid and the expression trophic level. 855–56
10. Draw diagrams to illustrate the water, carbon, phosphorous, and nitrogen cycles. 857–61
11. List and explain human activities that affect each of these cycles and contribute to a degraded environment for all organisms. 857–61

## testing yourself

Choose the best answer for each question.

1. According to the competitive exclusion principle,
   a. one species is always more competitive than another for a particular food source.
   b. competition excludes multiple species from using the same food source.
   c. no two species can occupy the same niche at the same time.
   d. competition limits the reproductive capacity of species.
2. Resource partitioning pertains to
   a. niche specialization.
   b. character displacement.
   c. increased species diversity.
   d. the development of mutualism.
   e. All but d are correct.

For statements 3–7, indicate the type of interaction in the key that is described in each scenario.

**KEY:**
a. competition
b. predation
c. parasitism
d. commensalism
e. mutualism

3. An alfalfa plant gains fixed nitrogen from the bacterial species *Rhizobium* in its root system, while *Rhizobium* gains carbohydrates from the plant.
4. Both foxes and coyotes in an area feed primarily on a limited supply of rabbits.
5. Roundworms establish a colony inside a cat's digestive tract.
6. A fungus captures nematodes as a food source.
7. An orchid plant lives in the treetops, gaining access to sun and pollinators, but not harming the trees.
8. A bullhorn acacia provides a home and nutrients for ants. Which statement is likely?
   a. The plant is under the control of pheromones produced by the ants.
   b. The ants protect the plant.
   c. The plant and the ants compete with each other.
   d. The plant and the ants have coevolved to occupy different ecological niches.
   e. All of these are correct.
9. The frilled lizard of Australia suddenly opened its mouth wide and unfurled folds of skin around its neck. Most likely, this was a way to
   a. conceal itself.
   b. warn that it was noxious to eat.
   c. scare a predator.
   d. scare its prey.
   e. All of these are correct.

10. When one species mimics another species, the mimic sometimes
    a. lacks the defense of the model.
    b. possesses the defense of the model.
    c. is brightly colored.
    d. All of these are correct.
11. The species within a community are
    a. used to compare communities.
    b. present due to their abiotic requirements.
    c. more diverse as the size of the area increases.
    d. present due to their biotic interactions.
    e. All of these are correct.
12. Label this diagram of an ecosystem:

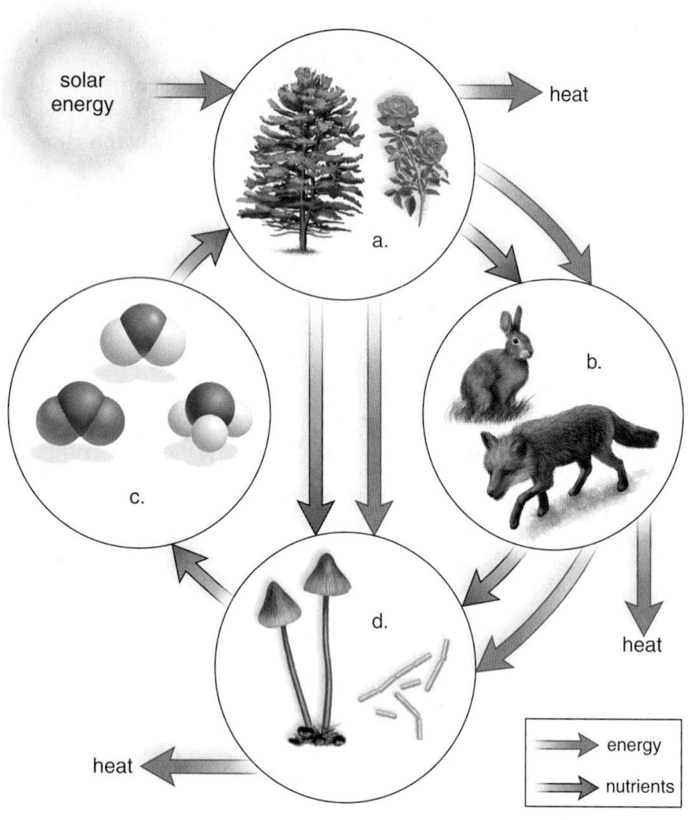


13. Mosses growing on bare rock will eventually help to create soil. These mosses are involved in ______________ succession.
    a. primary
    b. secondary
    c. tertiary
14. The ecological niche of an organism
    a. is the same as its habitat.
    b. includes how it competes and acquires food.
    c. is specific to the organism.
    d. is usually occupied by another species.
    e. Both b and c are correct.
15. In what way are decomposers like producers?
    a. Either may be the first member of a grazing or a detrital food chain.
    b. Both produce oxygen for other forms of life.
    c. Both require nutrient molecules and energy.
    d. Both are present only on land.
    e. Both produce organic nutrients for other members of ecosystems.
16. When a heterotroph takes in food, only a small percentage of the energy in that food is used for growth because
    a. some food is not digested and is eliminated as feces.
    b. some metabolites are excreted as urine.
    c. some energy is given off as heat.
    d. All of these are correct.
    e. None of these are correct.
17. During chemical cycling, inorganic nutrients are typically returned to the soil by
    a. autotrophs.
    b. detritivores.
    c. decomposers.
    d. tertiary consumers.
18. In a grazing food web, carnivores that eat herbivores are
    a. producers.
    b. primary consumers.
    c. secondary consumers.
    d. tertiary consumers.
19. Choose the statement that is true concerning this food chain: grass → rabbits → snakes → hawks
    a. Each predator population has a greater biomass than its prey population.
    b. Each prey population has a greater biomass than its predator population.
    c. Each population is omnivorous.
    d. Each population returns inorganic nutrients and energy to the producer.
    e. Both a and c are correct.
    f. Both a and b are correct.
20. Which of the following is a sedimentary biogeochemical cycle?
    a. carbon
    b. nitrogen
    c. phosphorus
21. Which of the following could not be a component of the nitrogen cycle?
    a. proteins
    b. ammonium
    c. decomposers
    d. photosynthesis
    e. bacteria in root nodules
22. How do plants contribute to the carbon cycle?
    a. When plants respire, they release $CO_2$ into the atmosphere.
    b. When plants photosynthesize, they consume $CO_2$ from the atmosphere.
    c. When plants photosynthesize, they provide oxygen to heterotrophs.
    d. Both a and b are correct.

## thinking scientifically

1. As per Figure 17.2, you observe three species of *Empidonax* flycatchers in the same general area, and you hypothesize that they occupy different niches. How could you substantiate your hypothesis?
2. In order to improve species richness, you decide to add phosphate to a pond. How might you determine how much phosphate to add in order to avoid eutrophication?

## *Biology* website

The companion website for *Biology* provides a wealth of information organized and integrated by chapter. You will find practice tests, animations, videos, and much more that will complement your learning and understanding of general biology.

**http://www.mhhe.com/maderbiology10**

# 46

# Major Ecosystems of the Biosphere

## concepts

**46.1 CLIMATE AND THE BIOSPHERE**

- The unequal distribution of solar radiation around the Earth produces variations in climate. 866
- Global air circulation patterns and physical geographic features help produce various patterns of temperature and rainfall about the globe. 867

**46.2 TERRESTRIAL ECOSYSTEMS**

- The Earth's major terrestrial biomes are tundra, forests (coniferous, temperate deciduous, and tropical), shrublands, grasslands (temperate grasslands and tropical savannas), and deserts. 868–78

**46.3 AQUATIC ECOSYSTEMS**

- The Earth's major aquatic ecosystems are of two types: freshwater and saltwater. 879–84
- Ocean currents also affect the climate and the weather over the continents. 884

*From space, the Earth is a pristine aqua globe, hovering against a vast backdrop of darkness. Get somewhat closer and the Earth's churning atmosphere, immense water systems, and seven continents come into view. Not until we see the surface can we make out the vast differences in land formations and physical features. Rainfall and temperature largely account for the great terrestrial ecosystems of the world, whether the freezing, snow-covered Arctic tundra, the hot scorched deserts, the lush tropical rain forests, or the sea of grass savanna.*

*In this chapter, we will study the mix of species in the major ecosystems already mentioned and also the oceans. Each species has a particular way of life and is adapted to living under particular environmental conditions. Through biogeochemical cycles driven by solar energy, natural ecosystems transformed the Earth's crust, its waters, and the atmosphere into a life-supporting environment.*

Plants and animals from ecosystems of planet Earth.

# 46.1 Climate and the Biosphere

**Climate** refers to the prevailing weather conditions in a particular region. Climate is dictated by temperature and rainfall, which are influenced by the following factors: (1) variations in solar radiation distribution due to the tilt of the Earth as it orbits about the sun; and (2) other effects, such as topography and whether a body of water is nearby.

## Effect of Solar Radiation

Because the Earth is a sphere, it receives a direct hit of the sun's rays at the equator but a glancing blow at the poles (Fig. 46.1*a*). The region between latitudes approximately 26.5° north and south of the equator is considered the tropics. The tropics are warmer than the areas north of 23.5°N and south of 23.5°S, known as the temperate regions. The tilt of the Earth as it orbits around the sun causes one pole or the other to be closer to the sun (except at the spring and fall equinoxes, when the sun aims directly at the equator), and this accounts for the seasons that occur in all parts of the Earth except at the equator (Fig. 46.1*b*). When the Northern Hemisphere is having winter, the Southern Hemisphere is having summer, and vice versa.

If the Earth were standing still and were a solid, uniform ball, all air movements—which we call winds—would be in two directions. Air at the equator warmed by the sun would rise and move toward colder air at the poles. Rising air creates zones of lower air pressure. However, because the Earth rotates on its axis daily and its surface consists of continents and oceans, the flows of warm and cold air are modified into three large circulation cells in each hemisphere (Fig. 46.2). At the equator, the sun heats the air and evaporates water. The warm, moist air rises, cools, and loses most of its moisture as rain. The greatest amounts of rainfall on Earth are near the equator. The rising air flows toward the poles, but at about 30° north and south latitude, it cools before it sinks toward the Earth's surface and reheats. As the dry air descends and warms, areas of high pressure are generated. High-pressure regions are zones of low rainfall. The great deserts of Africa, Australia, and the Americas occur at these latitudes. At the Earth's surface, the air flows both toward the poles and the equator. As dry air moves across the Earth, moisture from both land and water gets absorbed. At about 60° north and south latitude, the warmed air rises and cools, producing another low-pressure area with high rainfall. This moisture supports the great forests of the temperate zone. Part of this rising air flows toward the equator, and part continues toward the poles, where it descends. The poles are high pressure areas and have low amounts of precipitation.

Besides affecting precipitation, the spinning of the Earth also affects the winds (Fig. 46.2). In the Northern Hemisphere, large-scale winds generally bend clockwise, and in the Southern Hemisphere, they bend counterclockwise. The curving pattern of the winds, ocean currents, and cyclones is the result of the fact that the Earth rotates in an eastward direction. At about 30° north latitude and 30° south latitude, the winds blow from the east-southeast in the Southern Hemisphere and from the east-northeast in the Northern Hemisphere (the east coasts of continents at these latitudes are wet). The doldrums, regions of calm, occur at the equator. The winds blowing from the doldrums toward the poles are called trade winds because sailors depended on them to fill the sails of their trading ships. Between 30° and 60° north and south latitude, strong winds, called the prevailing westerlies, blow from west to east. The west coasts of the continents at these latitudes are wet, as is the Pacific Northwest, where a massive evergreen forest is located. Weaker winds, called the polar easterlies, blow from east to west at still higher latitudes of their respective hemispheres.

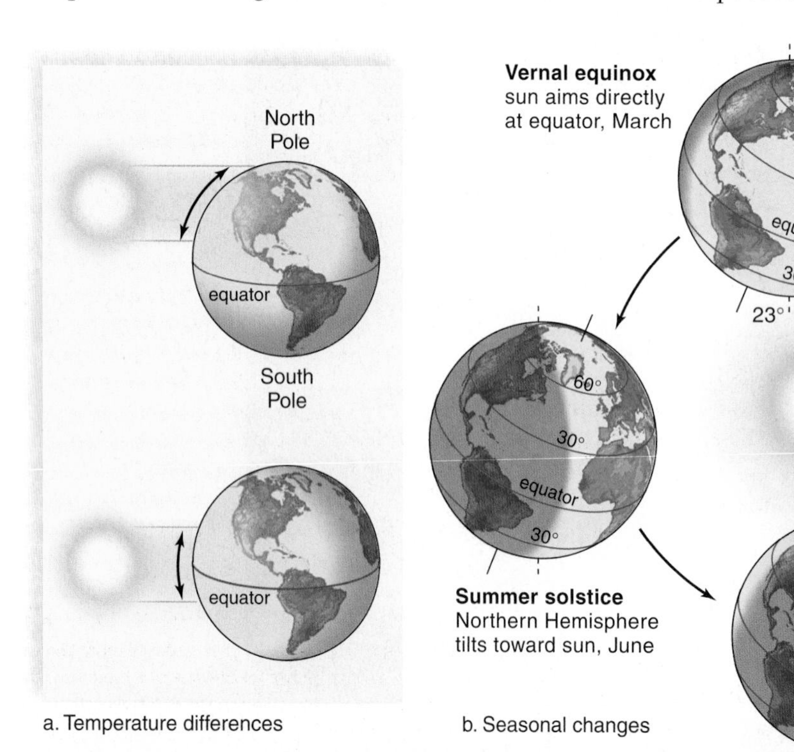


**FIGURE 46.1 Distribution of solar energy.**

**a.** Since the Earth is a sphere, beams of solar energy striking the Earth near one of the poles are spread over a wider area than similar beams striking the Earth at the equator. **b.** The seasons of the Northern and Southern Hemispheres are due to the tilt of the Earth on its axis as it rotates about the sun.

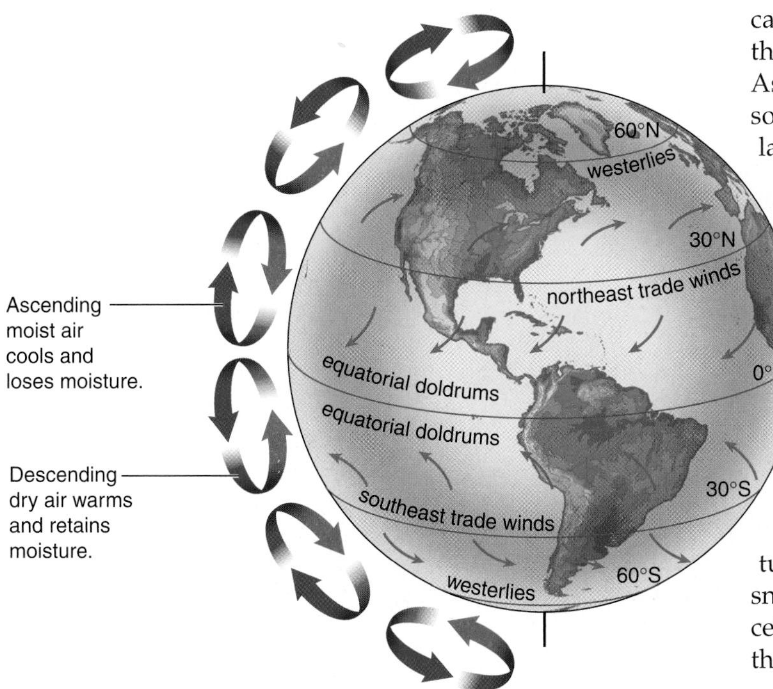


**FIGURE 46.2 Global wind circulation.**

Air ascends and descends as shown because the Earth rotates on its axis. Also, the trade winds move from the northeast to the west in the Northern Hemisphere, and from the southeast to the west in the Southern Hemisphere. The westerlies move toward the east.

## Other Effects

**Topography** means the physical features, or "the lay," of the land. One physical feature that affects climate is the presence of mountains. As air blows up and over a coastal mountain range, it rises and cools. One side of the mountain, called the windward side, receives more rainfall than the other side, called the leeward side. On the leeward side, the air descends, absorbs moisture from the ground, and produces clear weather (Fig. 46.3). The difference between the windward side and the leeward side can be quite dramatic. In the Hawaiian Islands, for example, the windward side of the mountains receives more than 750 cm of rain a year, while the leeward side, which is in a **rain shadow,** gets on the average only 50 cm of rain and is generally sunny. In the United States, the western side of the Sierra Nevada Mountains is lush, while the eastern side is a semidesert.

The temperature of the oceans is more stable than that of landmasses. Oceanic water gains or loses heat more slowly than terrestrial environments. This causes coasts to have a unique weather pattern that is not observed inland. During the day, the land warms more quickly than the ocean, and the air above the land rises, pulling a cool sea breeze in from the ocean. At night, the reverse happens; the breeze blows from the land toward the sea.

India and some other countries in southern Asia have a **monsoon** climate, in which wet ocean winds blow onshore for almost half the year. The land heats more rapidly than the waters of the Indian Ocean during spring. The difference in temperature between the land and the ocean causes an enormous circulation of air: Warm air rises over the land, and cooler air comes in off the ocean to replace it. As the warm air rises, it loses its moisture, and the monsoon season begins. As just discussed, rainfall is particularly heavy on the windward side of hills. Cherrapunji, a city in northern India, receives an annual average of 1,090 cm of rain a year because of its high altitude. This weather pattern has reversed by November. The land is now cooler than the ocean; therefore, dry winds blow from the Asian continent across the Indian Ocean. In the winter, the air over the land is dry, the skies cloudless, and temperatures pleasant. The chief crop of India is rice, which starts to grow when the monsoon rains begin.

In the United States, people often speak of the "lake effect," meaning that in the winter, arctic winds blowing over the Great Lakes become warm and moisture-laden. When these winds rise and lose their moisture, snow begins to fall. Places such as Buffalo, New York, receive heavy snowfalls due to the lake effect, and snow is on the ground there for an average of 90–140 days every year.

### Check Your Progress 46.1

1. What accounts for a warm climate at the equator?
2. Name two physical features that can affect rainfall.

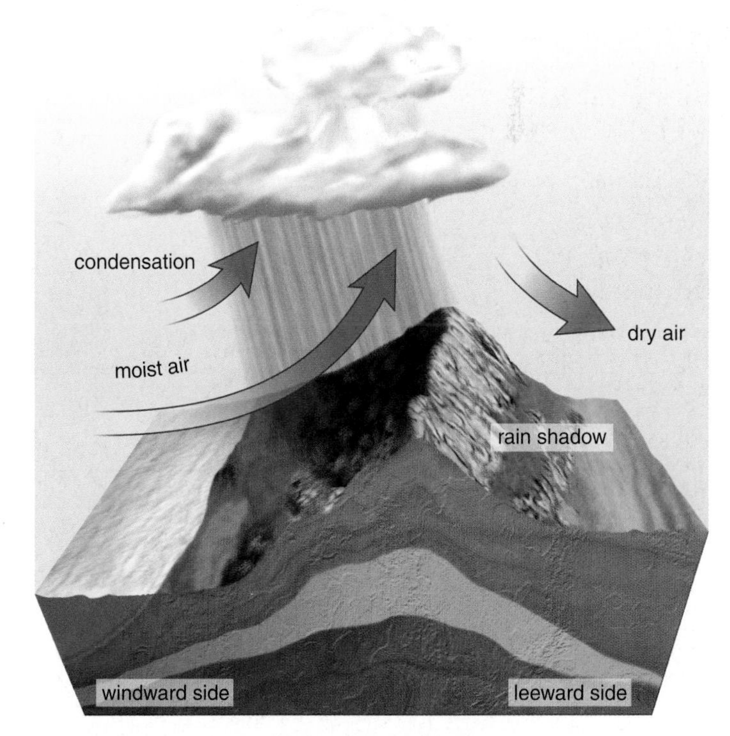


**FIGURE 46.3 Formation of a rain shadow.**

When winds from the sea cross a coastal mountain range, they rise and release their moisture as they cool this side of a mountain, called the windward side. The leeward side of a mountain receives relatively little rain and is therefore said to lie in a "rain shadow."

**FIGURE 46.4 Pattern of biome distribution.**

**a.** Pattern of world biomes in relation to temperature and moisture. The dashed line encloses a wide range of environments in which either grasses or woody plants can dominate the area, depending on the soil type.
**b.** The same type of biome can occur in different regions of the world, as shown on this global map.

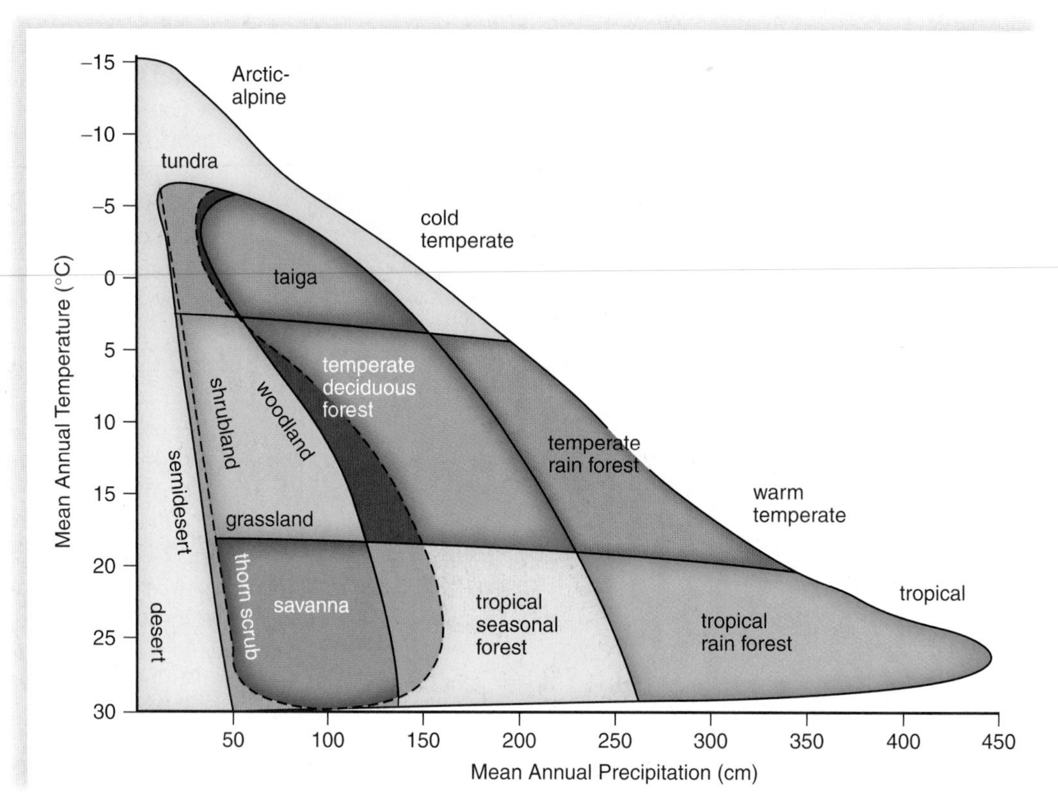


a. Biome pattern of temperature and precipitation

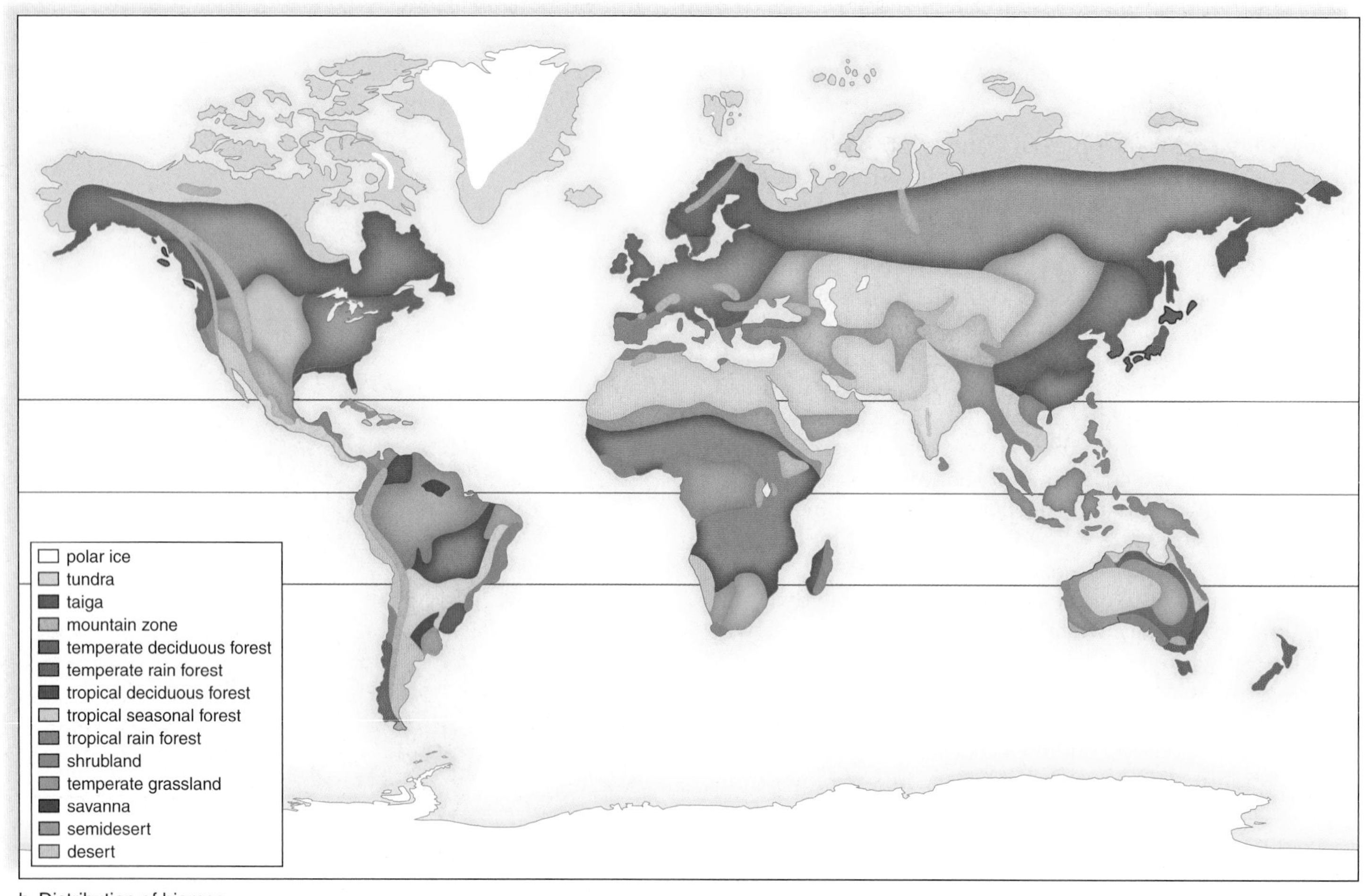


b. Distribution of biomes

## 46.2 Terrestrial Ecosystems

Major terrestrial ecosystems called a **biome** are characterized by their climate and geography (Table 46.1). A biome has a particular mix of plants and animals that are adapted to living under certain environmental conditions, of which climate is an overriding influence. When terrestrial biomes are plotted according to their mean annual temperature and mean annual precipitation, a particular pattern results (Fig. 46.4*a*). The distribution of biomes is shown in Figure 46.4*b*. Even though Figure 46.4 shows definite demarcations, keep in mind that the biomes gradually change from one type to the other. Also, although each type of biome will be described separately, remember that each biome has inputs from and outputs to all the other terrestrial and aquatic ecosystems of the biosphere.

The distribution of the biomes and their corresponding organismal populations are determined principally by differences in climate due to the distribution of solar radiation and defining topographical features. Both latitude and altitude are responsible for temperature gradients. If one travels from the equator to the North Pole, it is possible to observe first a tropical rain forest, followed by a temperate deciduous forest, a coniferous forest, and tundra, in that order, and this sequence is also seen when ascending a mountain (Fig. 46.5). The coniferous forest of a mountain is called a **montane coniferous forest,** and the tundra near the peak of a mountain is called an **alpine tundra.** When going from the equator to the South Pole, one would not reach a region corresponding to a coniferous forest and tundra of the Northern Hemisphere because the majority of the landmasses is shifted toward the north.

**TABLE 46.1**

**Selected Biomes**

| *Name* | *Characteristics* |
|---|---|
| Tundra | Around North Pole; average annual temperature is −12°C to −6°C; low annual precipitation (less than 25 cm); permafrost (permanent ice) year-round within a meter of surface. |
| Taiga (coniferous forest) | Large northern biome that circles just below the Arctic Circle; temperature is below freezing for half the year; moderate annual precipitation (30–85 cm); long nights in winter and long days in summer. |
| Temperate deciduous forest | Eastern half of United States, Canada, Europe, and parts of Russia; four seasons of the year with hot summers and cold winters; goodly annual precipitation (75–150 cm) |
| Grasslands | Called prairies in North America, savannas in Africa, pampas in South America, steppes in Europe; hot in summer and cold in winter (United States); moderate annual precipitation (25–50 cm); good soil for agriculture. |
| Tropical rain forests | Located near the equator in Latin America, Southeast Asia, and West Africa; warm (20–25°C) and wet (190 cm/year); has wet/dry season. |
| Deserts | Northern and Southern Hemispheres at 30° latitude; hot (38°C) days and cold (7°C) nights; low annual precipitation (less than 25 cm). |

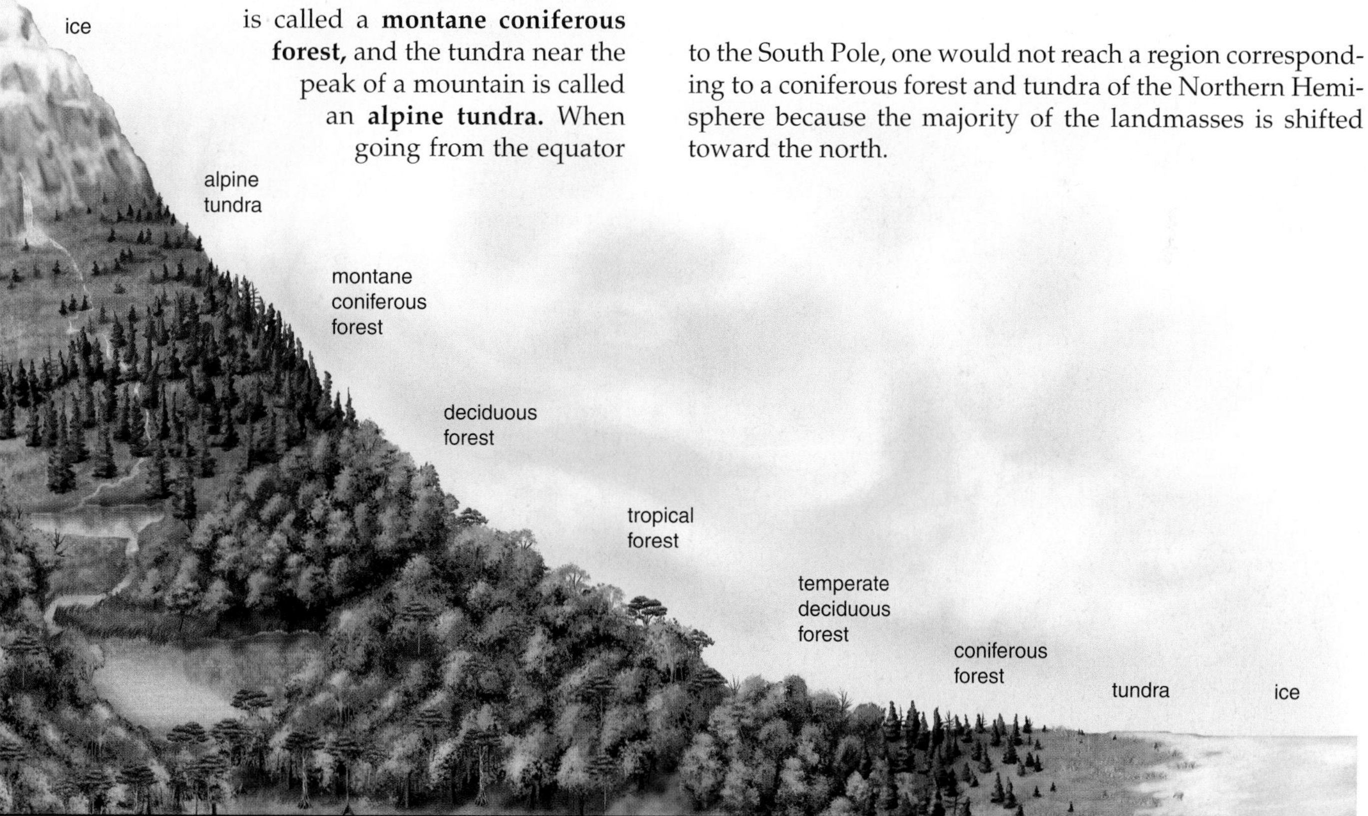


**FIGURE 46.5 Climate and biomes.**

Biomes change with altitude just as they do with latitude because vegetation is partly determined by temperature. Precipitation also plays a significant role, which is one reason grasslands, instead of tropical or deciduous forests, are sometimes found at the base of mountains.

## Tundra

The **Arctic tundra** biome, which encircles the Earth just south of ice-covered polar seas in the Northern Hemisphere, covers about 20% of the Earth's land surface (Fig. 46.6). (A similar ecosystem, called the alpine tundra, occurs above the timberline on mountain ranges.) The Arctic tundra is cold and dark much of the year. Arctic tundra has extremely long, cold, harsh winters and short summers (6–8 weeks). Because rainfall amounts to only about 20 cm a year, the tundra could possibly be considered a desert, but melting snow creates a landscape of pools and bogs in the summer, especially because so little evaporates. Only the topmost layer of soil thaws; the **permafrost** beneath this layer is always frozen, and therefore, drainage is minimal. The available soil in the tundra is nutrient-poor.

Trees are not found in the tundra because the growing season is too short, their roots cannot penetrate the permafrost, and they cannot become anchored in the shallow boggy soil of summer. In the summer, the ground is covered with short grasses and sedges, as well as numerous patches of lichens and mosses. Dwarf woody shrubs, such as dwarf birch, flower and seed quickly while there is plentiful sun for photosynthesis.

A few animals live in the tundra year-round. For example, the mouselike lemming stays beneath the snow; the ptarmigan, a grouse, burrows in the snow during storms; and the musk ox conserves heat because of its thick coat and short, squat body. Other animals that live in the tundra include snowy owls, lynx, voles, Arctic foxes, and snowshoe hares. In the summer, the tundra is alive with numerous insects and birds, particularly shorebirds and waterfowl that migrate inland. Caribou in North America and reindeer in Asia and Europe also migrate to and from the tundra, as do the wolves that prey upon them. Polar bears are common near the coastal regions. All species have adaptations for living in extreme cold and short growing or feeding seasons.

tundra

a. Tundra vegetation and location (light blue)

b. Plentiful bird life

c. Caribou, *Rangifer taradanus*, a large mammal

**FIGURE 46.6 The tundra.**

**a.** In this biome, which is nearest the polar regions, the vegetation consists principally of lichens, mosses, grasses, and low-growing shrubs. **b.** Pools of water that do not evaporate or drain into the permanently frozen ground attract many birds. **c.** Caribou, more plentiful in the summer than in the winter, feed on lichens, grasses, and shrubs.

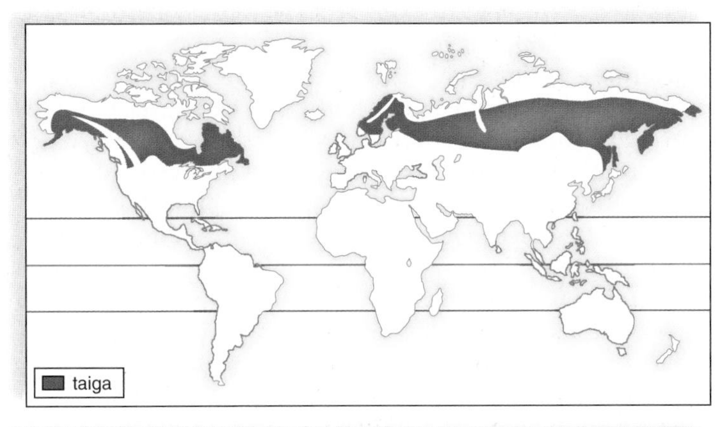

a. Taiga vegetation and location (dark blue)

b. Bull moose, *Alces americanus*, a large mammal

**FIGURE 46.7 The taiga.**

The taiga, which means swampland, spans northern Europe, Asia, and North America. The appellation "spruce-moose" refers to the (**a**) dominant presence of spruce trees and (**b**) moose, which frequent the ponds.

## Coniferous Forests

Coniferous forests are found in three locations: in the **taiga,** which extends around the world in the northern part of North America and Eurasia; near mountaintops (where it is called a montane coniferous forest); and along the Pacific coast of North America, as far south as northern California.

The taiga, or boreal forest, exists south of the tundra and covers approximately 11% of the Earth's landmasses (Fig. 46.7). There are no comparable biomes in the Southern Hemisphere because no large landmasses exist at that latitude. The taiga typifies the coniferous forest with its cone-bearing trees, such as spruce, fir, and pine. These trees are well adapted to the cold because both the leaves and bark have thick coverings. Also, the needlelike leaves can withstand the weight of heavy snow. There is a limited understory of plants, but the floor is covered by low-lying mosses and lichens beneath a layer of needles. Birds harvest the seeds of the conifers, and bears, deer, moose, beavers, and muskrats live around the cool lakes and along the streams. Wolves prey on these larger mammals. A montane coniferous forest also harbors the wolverine and the mountain lion.

The coniferous forest that runs along the west coast of Canada and the United States is sometimes called a **temperate rain forest.** The prevailing winds moving in off the Pacific Ocean lose their moisture when they meet the coastal mountain range. The plentiful rainfall and rich soil have produced some of the tallest conifer trees ever in existence, including the coastal redwoods. This forest is also called an old-growth forest because some trees are as old as 800 years. It truly is an evergreen forest because mosses, ferns, and other plants grow on all the tree trunks. Squirrels, lynx, and numerous species of amphibians, reptiles, and birds inhabit the temperate rain forest. The northern spotted owl is an endangered species of this particular ecosystem that has received recent conservation efforts.

## Temperate Deciduous Forests

**Temperate deciduous forests** are found south of the taiga in eastern North America, eastern Asia, and much of Europe (Fig. 46.8). The climate in these areas is moderate, with relatively high rainfall (75–150 cm per year). The seasons are well defined, and the growing season ranges between 140 and 300 days. The trees, such as oak, beech, sycamore, and maple, have broad leaves and are termed deciduous trees; they lose their leaves in the fall and grow them in the spring. In the southern temperate deciduous forests, evergreen magnolia trees can be found.

The tallest trees form a canopy, an upper layer of leaves that are the first to receive sunlight. Even so, enough sunlight penetrates to provide energy for another layer of trees, called understory trees. Beneath these trees are shrubs that may flower in the spring before the trees have put forth their leaves. Still another layer of plant growth—mosses, lichens, and ferns—resides beneath the shrub layer. This stratification provides a variety of habitats for insects and birds. Ground life is also plentiful. Squirrels, rabbits, woodchucks, and chipmunks are small herbivores. These and ground birds such as turkeys, pheasants, and grouse are preyed on by red foxes. White-tailed deer and black bears have increased in number in recent years. In contrast to the taiga, amphibians and reptiles occur in this biome because the winters are not as cold. Frogs and turtles prefer an aquatic existence, as do the beaver and muskrat, which are mammals.

Autumn fruits, nuts, and berries provide a supply of food for the winter, and the leaves, after turning brilliant colors and falling to the ground, contribute to the rich layer of humus. The minerals within the rich soil are washed far into the ground by spring rains, but the deep tree roots capture these and bring them back up into the forest system again.

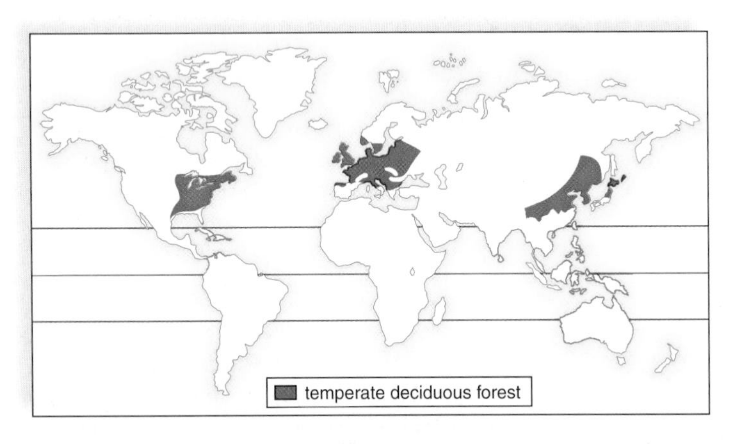

a. Temperate deciduous vegetation and location (red)

Millipede, *Marceus* sp.

Eastern chipmunk, *Tamias striatus*

Marsh marigolds, *Caltha howellii*

Bobcat, *Felis rufus*

b. Animal life of forest biome

**FIGURE 46.8 Temperate deciduous forest.**

**a.** A temperate deciduous forest is home to many varied plants and animals. **b.** Millipedes can be found among leaf litter, chipmunks feed on acorns, and bobcats prey on these and other small mammals.

*ecology focus*

## Wildlife Conservation and DNA

After DNA analysis, scientists were amazed to find that some 60% of loggerhead turtles drowning in the nets and hooks of fisheries in the Mediterranean Sea were from beaches in the southeastern United States. Since the unlucky creatures were a good representative sample of the turtles in the area, that meant more than half of the young turtles living in the Mediterranean Sea had hatched from nests on beaches in Florida, Georgia, and South Carolina (Fig. 46A*a*). Some 20,000–50,000 loggerheads die each year due to the Mediterranean fisheries, which may partly explain the decline in loggerheads nesting on southeastern U.S. beaches for the last 25 years.

The sequencing of DNA from Alaskan brown bears allowed Sandra Talbot (a graduate student at the University of Alaska's Institute of Arctic Biology) and wildlife geneticist Gerald Shields to conclude that there are two types of brown bears in Alaska. One type resides only on southeastern Alaska's Admiralty, Baranof, and Chichagof Islands, known as the ABC Islands. The other brown bear in Alaska is found throughout the rest of the state, as well as in Siberia and western Asia (Fig. 46A*b*).

A third distinct type of brown bear, known as the Montana grizzly, resides in other parts of North America. These three types comprise all of the known brown bears in the New World.

The ABC bears' uniqueness may be bad news for the timber industry, which has expressed interest in logging parts of the ABC Islands. Says Shields, "Studies show that when roads are built and the habitat is fragmented, the population of brown bears declines. Our genetic observations suggest they are truly unique, and we should consider their heritage. They could never be replaced by transplants."

In what will become a classic example of how DNA analysis might be used to protect endangered species from future ruin, scientists from the United States and New Zealand carried out discreet experiments in a Japanese hotel room on whale sushi bought in local markets. Sushi, a staple of the Japanese diet, is a rice and meat concoction wrapped in seaweed. Armed with a miniature DNA sampling machine, the scientists found that, of the 16 pieces of whale sushi they examined, many were from whales that are endangered or protected under an international moratorium on whaling. "Their findings demonstrated the true power of DNA studies," says David Woodruff, a conservation biologist at the University of California, San Diego.

One sample was from an endangered humpback, four were from fin whales, one was from a northern minke, and another from a beaked whale. Stephen Palumbi of the University of Hawaii says the technique could be used for monitoring and verifying catches. Until then, he says, "no species of whale can be considered safe."

Meanwhile, Ken Goddard, director of the unique U.S. Fish and Wildlife Service Forensics Laboratory in Ashland, Oregon, is already on the watch for wildlife crimes in the United States and 122 other countries that send samples to him for analysis. "DNA is one of the most powerful tools we've got," says Goddard, a former California police crime-lab director.

The lab has blood samples, for example, for all of the wolves being released into Yellowstone National Park—"for the obvious reason that we can match those samples to a crime scene," says Goddard. The lab has many cases currently pending in court that he cannot discuss. But he likes to tell the story of the lab's first DNA-matching case. Shortly after the lab opened in 1989, California wildlife authorities contacted Goddard. They had seized the carcass of a trophy-sized deer from a hunter. They believed the deer had been shot illegally on a 3,000-acre preserve owned by actor Clint Eastwood. The agents found a gut pile on the property but had no way to match it to the carcass. The hunter had two witnesses to deny the deer had been shot on the preserve.

Goddard's lab analysis made a perfect match between tissue from the gut pile and tissue from the carcass. Says Goddard: "We now have a cardboard cutout of Clint Eastwood at the lab saying 'Go ahead: Make my DNA.' "

**FIGURE 46A DNA studies.** ***a.*** *Many loggerhead turtles found in the Mediterranean Sea are from the southeastern United States.* ***b.*** *These two brown bears appear similar, but one type, known as an ABC bear, resides only on southeastern Alaska's Admiralty, Baranof, and Chichagof islands.*

a. Loggerhead turtle, *Caretta caretta*

b. Brown bears, *Ursus arctos*

## Tropical Forests

In the **tropical rain forests** of South America, Africa, and the Indo-Malayan region near the equator, the weather is always warm (between 20° and 25°C), and rainfall is plentiful (with a minimum of 190 cm per year). This may be the richest biome, in terms of both number of different kinds of species and their abundance.

A tropical rain forest has a complex structure, with many levels of life, including the forest floor, understory, and canopy. The vegetation of the forest floor is very sparse because much of the sunlight is filtered out by the canopy. The understory consists of smaller plants that are specialized for life in the shade. The canopy, topped by the crowns of tall trees, is the most productive level of the tropical rain forest (Fig. 46.9). Some of the broadleaf evergreen trees grow from 15–50 m or more. These tall trees often have trunks buttressed at ground level to prevent their toppling over. Lianas, or woody vines, which encircle the tree as it grows, also help strengthen the trunk. The diversity of species is enormous—a 10-km$^2$ area of tropical rain forest may contain 750 species of trees and 1,500 species of flowering plants.

Although some animals live on the forest floor (e.g., pacas, agoutis, peccaries, and armadillos), most live in the trees (Fig. 46.10). Insect life is so abundant that the majority of species have not been identified yet. Termites play a vital role in the decomposition of woody plant material, and ants are found everywhere, particularly in the trees. The various birds, such as hummingbirds, parakeets, parrots, and toucans, are often beautifully colored. Amphibians and reptiles are well represented by many types of frogs, snakes, and lizards. Lemurs, sloths, and monkeys are well-known primates that feed on the fruits of the trees. The largest carnivores are the big cats—the jaguars in South America and the leopards in Africa and Asia.

Many animals spend their entire life in the canopy, as do some plants. **Epiphytes** are plants that grow on other plants but usually have roots of their own that absorb moisture and minerals leached from the canopy; others catch rain and debris by forming vases of overlapping leaves. The most common epiphytes are related to pineapples, orchids, and ferns.

While we usually think of tropical forests as being nonseasonal rain forests, tropical forests that have wet and

**FIGURE 46.9 Levels of life in a tropical rain forest.**

The primary levels within a tropical rain forest are the canopy, the understory, and the forest floor. But the canopy (solid layer of leaves) contains levels as well, and some organisms spend their entire life in one particular level. Long lianas (hanging vines) climb into the canopy, where they produce leaves. Epiphytes are air plants that grow on the trees but do not parasitize them.

dry seasons are found in India, Southeast Asia, West Africa, South and Central America, the West Indies, and northern Australia. Here, there are deciduous trees, with many layers of growth beneath them.

Whereas the soil of a temperate deciduous forest biome is rich enough for agricultural purposes, the soil of a tropical rain forest biome is not. Nutrients are cycled directly from the litter to the plants again. Productivity is high because of high temperatures, a yearlong growing season, and the rapid recycling of nutrients from the litter. (In humid tropical forests, iron and aluminum oxides occur at the surface, causing a reddish residue known as laterite. When the trees are cleared, laterite bakes in the hot sun to a bricklike consistency that will not support crops.) Swidden agriculture, often called slash-and-burn agriculture, has been successful, but also destructive, in the tropics. Trees are felled and burned, and the ashes provide enough nutrients for several harvests. Thereafter, the forest must be allowed to regrow, and a new section must be cut and burned.

It is estimated that 2.4 acres of rain forest are destroyed per second. This rate equates to nearly 78 million acres annually. Unless conservation strategies are employed soon, rain forests will be destroyed beyond recovery, taking with them unique and interesting life-forms. Ecologists estimate that an average of 137 species are driven to extinction every day in rain forests.

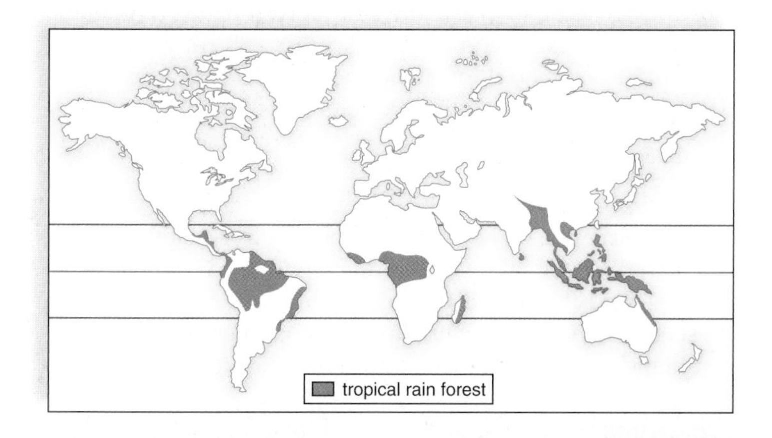

Poison arrow frog,
*Dendrobates azureus*

Cone-headed katydid,
*Panacanthus cuspidatus*

Ocelot,
*Felis pardalis*

Blue and gold macaw,
*Ara ararauna*

Brush-footed butterfly,
*Anartia amalthea linnaeus*

Lemur,
*Propithecus verreauxi*

Arboreal lizard,
*Calotes calotes*

**FIGURE 46.10 Location (green) and representative animals of the tropical rain forests of the world.**

## Shrublands

It is difficult to define a shrub, but in general, shrubs are shorter than trees (4.5–6 m) with a woody, persistent stem and no central trunk. Shrubs have small but thick evergreen leaves, which are often coated with a waxy material that prevents loss of moisture from the leaves. Their thick underground roots can survive dry summers and frequent fires and take deep moisture from the soil. Shrubs are adapted to withstand arid conditions and can also quickly sprout new growth after a fire. As a point of interest, you will recall from Chapter 45 that a shrub stage is part of the process of both primary and secondary succession.

**Shrublands** tend to occur along coasts that have dry summers and receive most of their rainfall in the winter. Shrublands are found along the cape of South Africa, the western coast of North America, and the southwestern and southern shores of Australia, as well as around the Mediterranean Sea and in central Chile. The dense shrubland that occurs in California is known as **chaparral** (Fig. 46.11). This type of shrubland, called Mediterranean, lacks an understory and ground litter and is highly flammable. The seeds of many species require the heat and scarring action of fire to induce germination. Other shrubs sprout from the roots after a fire. Typical animals of the chaparral include mule deer, rodents, lizards, and scrub jays.

There is also a northern shrub area that lies west of the Rocky Mountains. This area is sometimes classified as a cold desert, but the region is dominated by sagebrush. Some of the birds found there are dependent on sagebrush for their existence.

## Grasslands

**Grasslands** occur where annual rainfall is greater than 25 cm but generally insufficient to support trees. For example, in temperate areas, where rainfall is between 25 and 75 cm, it is too dry for forests and too wet for deserts to form.

Grasses are well adapted to a changing environment and can tolerate a high degree of grazing, flooding, drought, and sometimes fire. Where rainfall is high, tall grasses that reach more than 2 m in height (e.g., pampas grass) can flourish. In drier areas, shorter grasses (between 5 and 10 cm) are dominant. Low-growing bunch grasses (e.g., grama grass) grow in the United States near deserts. The growth of grasses is seasonal. As a result, grassland animals such as bison migrate, and others such as ground squirrels hibernate, when there is little grass for them to eat.

### *Temperate Grasslands*

The temperate grasslands include the Russian steppes, the South American pampas, and the North American prairies (Fig. 46.12). In these grasslands, winters are bitterly cold and summers are hot and dry. When traveling across the United States from east to west, the line between the temperate deciduous forest and a tall-grass prairie is roughly along the border between Illinois and Indiana. The tall-grass prairie receives more rainfall than does the short-grass prairie, which occurs near deserts. Large herds of bison—estimated at hundreds of thousands—once roamed the prairies, as did herds of pronghorn antelope. Now, small mammals, such as mice, prairie dogs, and rabbits, typically live belowground, but usually feed aboveground. Hawks, snakes, badgers, coyotes, and foxes feed on these mammals. Virtually all of these grasslands, however, have been converted to agricultural lands because of their fertile soils.

### *Savannas*

**Savannas** occur in regions where a relatively cool dry season is followed by a hot rainy season (Fig. 46.13). The largest savannas are in central and southern Africa. Other savannas exist in Australia, southeast Asia, and South America. The savanna is characterized by large expanses of grasses with sparse populations of trees. The plants of the savanna have extensive and deep root systems that enable them to survive drought and fire. One tree that can survive the severe dry season is the thorny flat-topped *Acacia,* which sheds its leaves during a drought. The African savanna supports

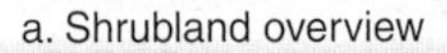
a. Shrubland overview

b. Scrub jay, *Aphelocoma californica*

c. Chemise, *Adenostema faciculatum*

**FIGURE 46.11 Shrubland.**

**a.** Shrublands, such as chaparral in California, are subject to raging fires, but the shrubs are adapted to quickly regrow. **b.** Scrub jays find a home here as does (**c**) a plant commonly called chemise.

the greatest variety and number of large herbivores of all the biomes. Elephants and giraffes are browsers that feed on tree vegetation. Antelopes, zebras, wildebeests, water buffalo, and rhinoceroses are grazers that feed on grasses. Any plant litter that is not consumed by grazers is attacked by a variety of small organisms, among them termites. Termites build towering nests in which they tend fungal gardens, their source of food. The herbivores support a large population of carnivores. Lions and hyenas sometimes hunt in packs, cheetahs hunt singly by day, and leopards hunt singly by night.

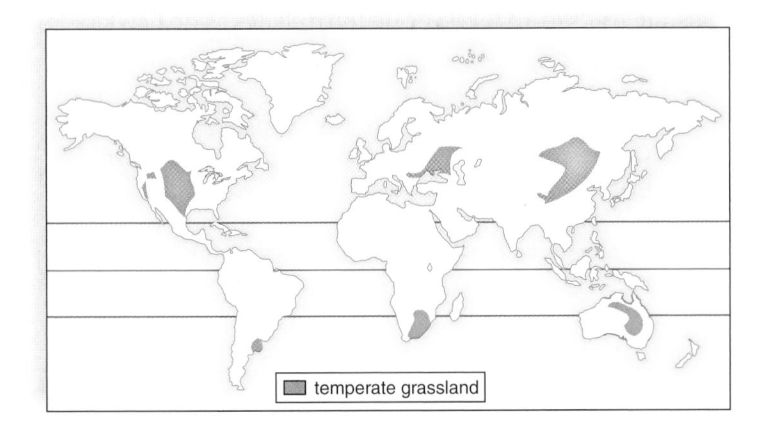

a. Temperate grassland vegetation and location (purple)

b. American bison, *Bison bison*, a large mammal

**FIGURE 46.12 Temperate grassland.**

**a.** Tall-grass prairies are seas of grasses dotted by pines and junipers. **b.** Bison, once abundant, are now being reintroduced into certain areas.

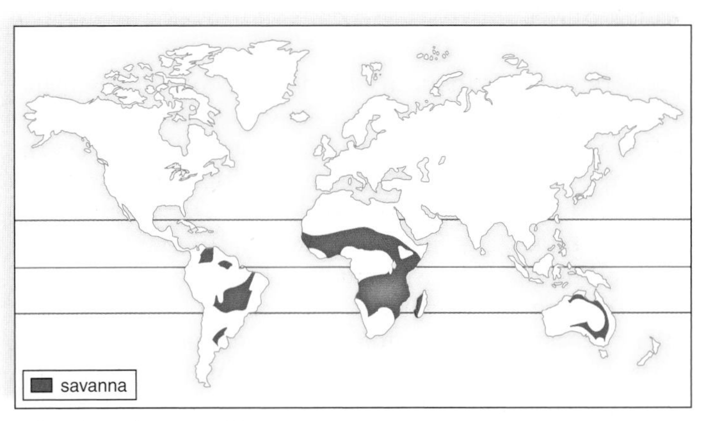


a. Location of savanna biome

Cheetah, *Acinonyx jubatus*

Giraffe, *Giraffa camelopardalis*

Wildebeest, *Connochaetes* sp.

Zebra (in foreground), *Equus quagga*

b. Animal life of savanna biome

**FIGURE 46.13 The savanna.**

**a.** The African savanna, located as shown, varies from grassland to widely spaced shrubs and trees. **b.** This biome supports a large assemblage of herbivores (e.g., zebras, wildebeests, and giraffes). Carnivores (e.g., cheetahs) prey on these.

## Deserts

**Deserts** are usually found at latitudes of about 30°, in both the Northern and Southern Hemispheres. Deserts cover nearly 30% of the Earth's land surface. The winds that descend in these regions lack moisture. Therefore, the annual rainfall is less than 25 cm. Days are hot because a lack of cloud cover allows the sun's rays to penetrate easily, but nights are cold because heat escapes easily into the atmosphere.

The Sahara, which stretches all the way from the Atlantic coast of Africa to the Arabian peninsula, and a few other deserts have little or no vegetation. However, most deserts have a variety of plants (Fig. 46.14*a*) that are highly adapted to survive long droughts, extreme heat, and extreme cold. Adaptations to these conditions include thick epidermal layers, water-storing succulent stems and leaves, and the ability to set seeds quickly in the spring. The best-known desert perennials in North America are the spiny cacti, which have stems that store water and carry on photosynthesis. Also common are nonsucculent shrubs, such as the many-branched sagebrush with silvery gray leaves and the spiny-branched ocotillo, which produces leaves during wet periods and sheds them during dry periods.

Some animals are adapted to the desert environment. To conserve water, many desert animals such as reptiles and insects are nocturnal or burrowing and have a protective outer body covering. A desert has numerous insects, which pass through the stages of development in synchrony with the periods of rain. Reptiles, especially lizards and snakes, are perhaps the most characteristic group of vertebrates found in deserts, but running birds (e.g., the roadrunner) and rodents (e.g., the kangaroo rat) are also well known (Fig. 46.14*b*). Larger mammals, such as the kit fox, prey on the rodents, as do hawks.

**Check Your Progress 46.2**

1. Contrast the vegetation of the tropical rain forest with that of a temperate deciduous forest.
2. Account for why there are more predaceous carnivores on the African savanna than in the tundra.

**FIGURE 46.14 The desert.**

Plants and animals that live in a desert are adapted to arid conditions. **a.** The plants are either succulents, which retain moisture, or shrubs with woody stems and small leaves, which lose little moisture. **b.** Among the animal life, the kangaroo rat feeds on seeds and other vegetation; the roadrunner preys on insects, lizards, and snakes. The kit fox is a desert carnivore.

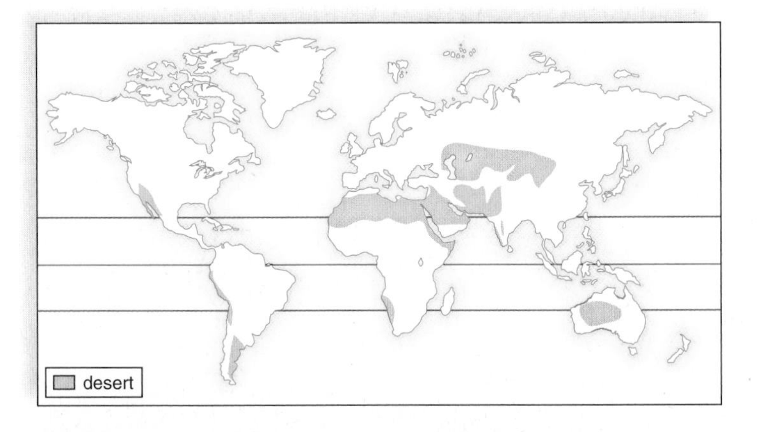


a. Desert vegetation and location (light brown)

Bannertail kangaroo rat, *Dipodomys spectabilis*

Greater roadrunner, *Geococcyx californianus*

Kit fox, *Vulpes velox*

b. Animal life of desert biome

# 46.3 Aquatic Ecosystems

Aquatic ecosystems are classified as two types: freshwater (inland) or saltwater. Brackish water, however, is a mixture of fresh and salt water. Figure 46.15 shows how these ecosystems are joined physically and discusses some of the organisms that are adapted to live in them.

In the water cycle, the sun's rays cause seawater to evaporate, and the salts are left behind. As discussed in Chapter 45, the evaporated fresh water rises into the atmosphere, cools, and falls as rain. When rain falls, some of the water sinks, or percolates, into the ground and saturates the Earth to a certain level. The top of the saturation zone is called the groundwater table, or simply the water table.

Since land lies above sea level, gravity eventually returns all fresh water to the sea, but in the meantime, it is contained as standing water within basins, called lakes and ponds, or as flowing water within channels, called streams or rivers. Sometimes groundwater is also located in underground rivers called aquifers. Whenever the Earth contains basins or channels, water will appear to the level of the water table.

**Wetlands** are areas that are wet for at least part of the year. Generally, wetlands are classified by their vegetation. **Marshes** are wetlands that are frequently or continually inundated by water and characterized by the presence of rushes, reeds, and other grasses. They provide excellent habitat for waterfowl and small mammals. Marshes are one of the most productive ecosystems on Earth. **Swamps** are wetlands that are dominated by either woody plants or shrubs. Common swamp trees include cypress, red maple, and tupelo. The American alligator is a top predator in many swamp ecosystems. **Bogs** are wetlands that are characterized by acidic waters, peat deposits, and sphagnum moss. Bogs receive most of their water from precipitation and are nutrient-poor. Several species of plants thrive in bogs, including cranberries, orchids, and insectivorous plants such as Venus flytraps and pitcher plants. Moose and a number of other animals are inhabitants of bogs in the northern United States and Canada.

Humans have historically channeled aboveground rivers and filled in wetlands with the attitude that useless land was being improved. However, these activities degrade ecosystems, can cause seasonal flooding, and wetlands provide food and habitats for many unique fishes, waterfowl, and other wildlife. They also purify waters by filtering them and by diluting and breaking down toxic wastes and excess nutrients. Wetlands directly absorb storm waters and also absorb overflows from lakes and rivers. In this way, they protect farms, cities, and towns from the devastating effects of floods. Federal and local laws have been enacted for the protection of wetlands, and the current attitude of many has changed.

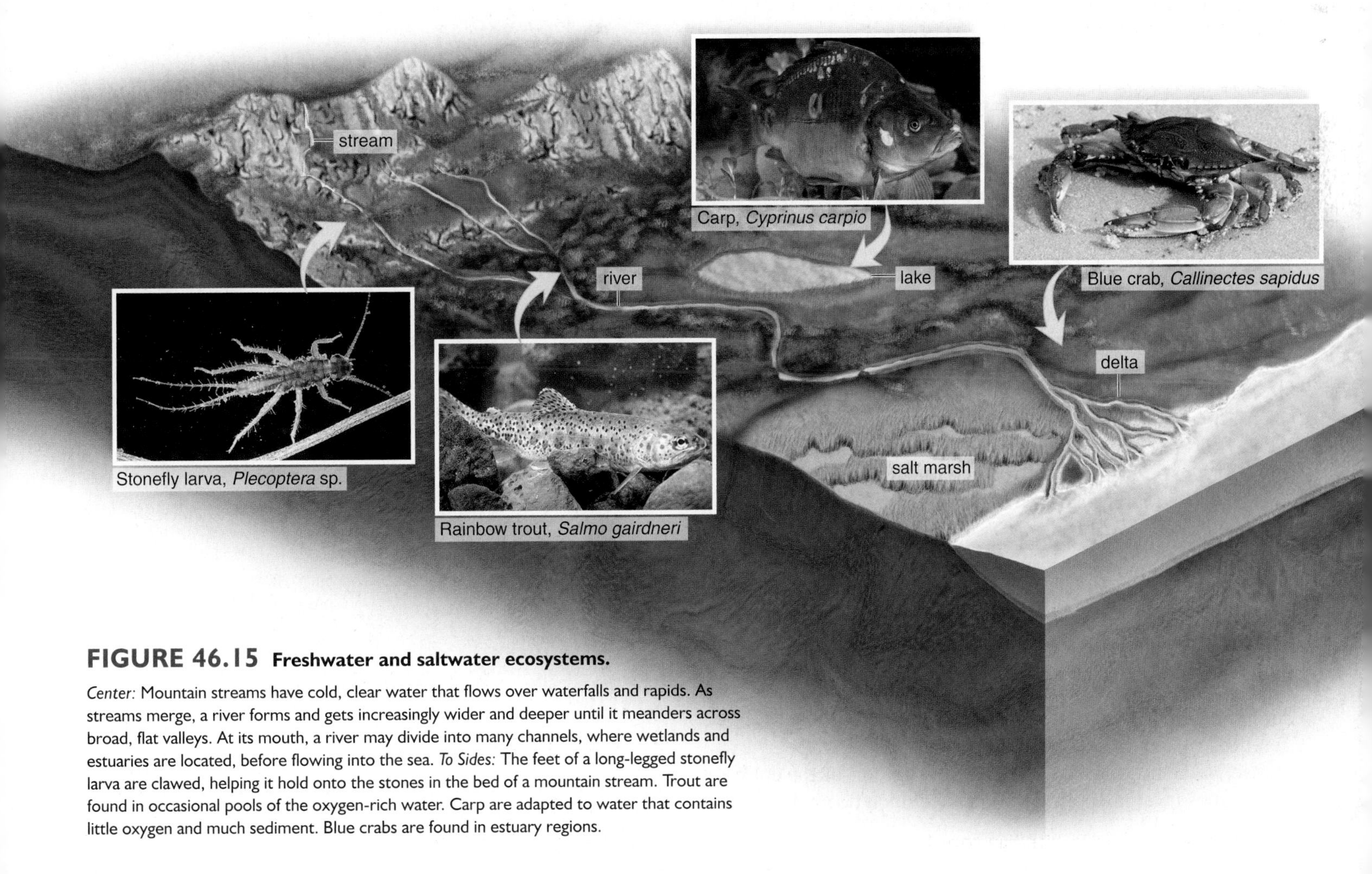


**FIGURE 46.15 Freshwater and saltwater ecosystems.**

*Center:* Mountain streams have cold, clear water that flows over waterfalls and rapids. As streams merge, a river forms and gets increasingly wider and deeper until it meanders across broad, flat valleys. At its mouth, a river may divide into many channels, where wetlands and estuaries are located, before flowing into the sea. *To Sides:* The feet of a long-legged stonefly larva are clawed, helping it hold onto the stones in the bed of a mountain stream. Trout are found in occasional pools of the oxygen-rich water. Carp are adapted to water that contains little oxygen and much sediment. Blue crabs are found in estuary regions.

a. Oligotrophic lake

b. Eutrophic lake

**FIGURE 46.16** **Types of lakes.**

Lakes can be classified according to whether they are (**a**) oligotrophic (nutrient-poor) or (**b**) eutrophic (nutrient-rich). Eutrophic lakes tend to have large populations of algae and rooted plants, resulting in a large population of decomposers that use up much of the oxygen and leave little oxygen for fishes.

## Lakes

**Lakes** are bodies of fresh water often classified by their nutrient status. Oligotrophic (nutrient-poor) lakes are characterized by a small amount of organic matter and low productivity (Fig. 46.16*a*). Eutrophic (nutrient-rich) lakes are characterized by plentiful organic matter and high productivity (Fig. 46.16*b*). Such lakes are usually situated in naturally nutrient-rich regions or are enriched by agricultural or urban and suburban runoff. Oligotrophic lakes can become eutrophic through large inputs of nutrients. This process is called **eutrophication.**

In the temperate zone, deep lakes are stratified in the summer and winter and have distinct vertical zones. In summer, lakes in the temperate zone have three layers of water that differ in temperature (Fig. 46.17). The surface layer, the epilimnion, is warm from solar radiation; the middle layer, the thermocline, experiences an abrupt drop in temperature; and the lowest layer, the hypolimnion, is cold. These differences in temperature prevent mixing. The warmer, less dense water of the epilimnion "floats" on top of the colder, more dense water of the thermocline, which floats on top of the hypolimnion.

Phytoplankton found in the sunlit epilimnion use up nutrients as they photosynthesize. Photosynthesis releases oxygen, giving this layer a ready supply. Detritus naturally falls by gravity to the bottom of the lake, and there oxygen is used up as decomposition occurs. Decomposition releases nutrients, however. As the season progresses, the epilimnion becomes nutrient-poor, while the hypolimnion begins to be depleted of oxygen.

In the fall, as the epilimnion cools, and in the spring, as it warms, an overturn occurs. In the fall, the upper epilimnion waters become cooler than the hypolimnion waters. This causes the surface water to sink and the deep water to rise. This **fall overturn** continues until the temperature is uniform throughout the lake. At this point, wind aids in the circulation of water so that mixing occurs. Eventually, oxygen and nutrients become evenly distributed.

As winter approaches, the water cools. Ice formation begins at the top, and the ice remains there because ice is less dense than cool water. Ice has an insulating effect, preventing further cooling of the water below. This permits aquatic organisms to live through the winter in the water beneath the surface of the ice.

In the spring, as the ice melts, the cooler water on top sinks below the warmer water on the bottom. This

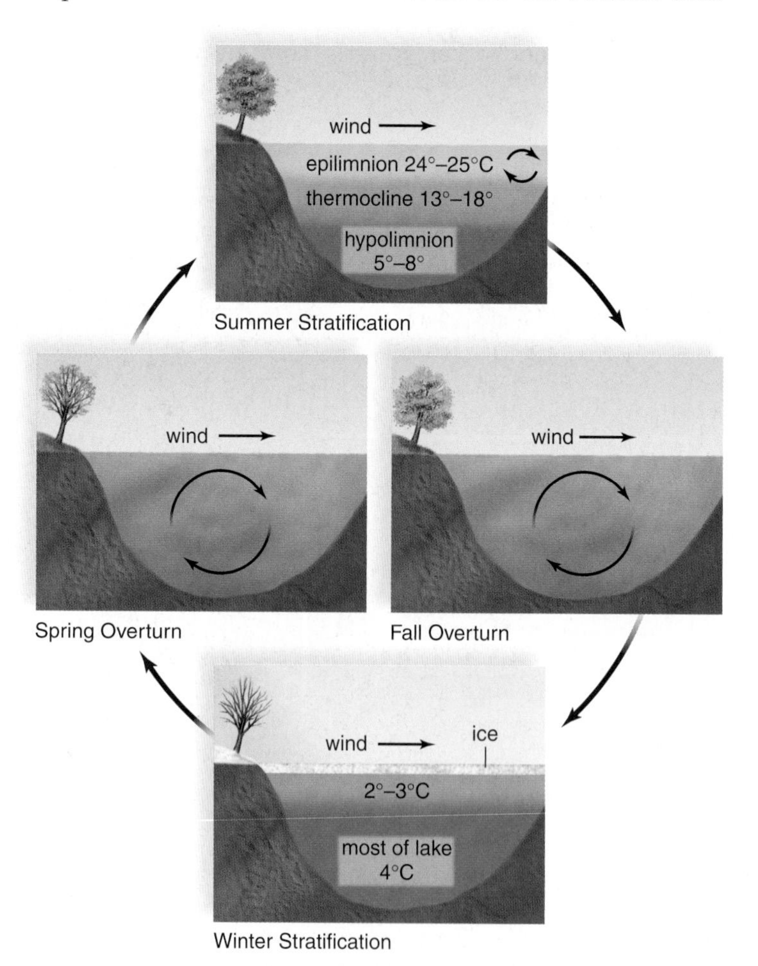


**FIGURE 46.17** **Lake stratification in a temperate region.**

Temperature profiles of a large oligotrophic lake in a temperate region vary with the season. During the spring and fall overturns, the deep waters receive oxygen from surface waters, and surface waters receive inorganic nutrients from deep waters.

**spring overturn** continues until the temperature is uniform throughout the lake. At this point, wind aids in the circulation of water as before. When the surface waters absorb solar radiation, thermal stratification occurs once more.

The vertical stratification and seasonal change of temperatures in a lake influence the seasonal distribution of fish and other aquatic life in the lake basin. For example, coldwater fish move to the deeper water in summer and inhabit the upper water in winter. In the fall and spring just after mixing occurs, phytoplankton growth at the surface is most abundant.

## *Life Zones*

In both fresh and salt water, free-drifting microscopic organisms, called *plankton* [Gk. *planktos,* wandering], are important components of the ecosystem. **Phytoplankton** [Gk. *phyton,* plant, and *planktos,* wandering] are photosynthesizing algae that become noticeable when a green scum or red tide appears on the water. **Zooplankton** [Gk. *zoon,* animal, and *planktos,* wandering] are minute animals that feed on the phytoplankton. Lakes and ponds can be divided into several life zones (Fig. 46.18). The *littoral zone* is closest to the shore, the *limnetic zone* forms the sunlit body of the lake, and the *profundal zone* is below the level of light penetration. The *benthic zone* includes the sediment at the soil-water interface. Aquatic plants are rooted in the shallow littoral zone of a lake, providing habitat for numerous protozoans, invertebrates, fishes, and some reptiles. Pike are largemouth bass that are "lurking perdators." They wait among vegetation around the margins of lakes and surge out to capture passing prey. Wading birds are commonly seen feeding in the littoral zone. Some organisms, such as the water strider, live at the water-air interface and can literally walk on water. In the limnetic zone, small fishes, such as minnows and killifish, feed on plankton and also serve as food for larger fish, such as bass. In the profundal zone, zooplankton, invertebrates, and fishes such as catfish and whitefish feed on debris that falls from higher zones.

The bottom of the lake is known as the benthic zone. Primarily, the benthic zone is composed of silt, sand, inorganic sediment, and dead organic material (detritus). Bottom-dwelling organisms are known as benthic species and include worms, snails, clams, crayfishes, and some insect larvae. Decomposers, such as bacteria, are also found in the benthic zone and serve to break down wastes and dead organisms into nutrients that are eventually used by producers.

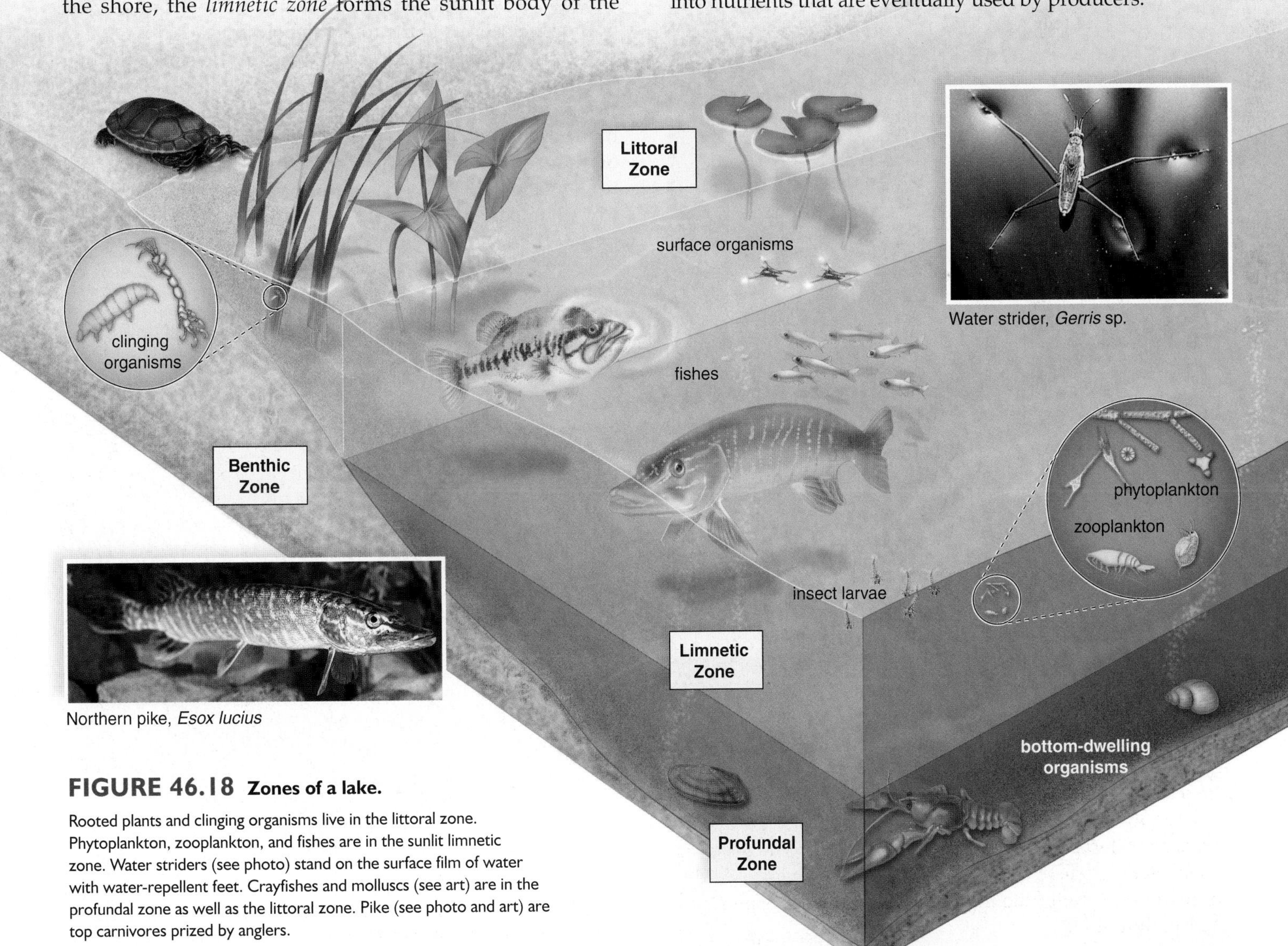


**FIGURE 46.18 Zones of a lake.**

Rooted plants and clinging organisms live in the littoral zone. Phytoplankton, zooplankton, and fishes are in the sunlit limnetic zone. Water striders (see photo) stand on the surface film of water with water-repellent feet. Crayfishes and molluscs (see art) are in the profundal zone as well as the littoral zone. Pike (see photo and art) are top carnivores prized by anglers.

## Coastal Ecosystems Border the Oceans

Salt marshes, discussed previously, and also mudflats and mangrove swamps, featured in Figure 46.19, are ecosystems that occur at a delta. Mangrove swamps develop in subtropical and tropical zones, while marshes and mudflats occur in temperate zones. These ecosystems are often designated as an estuary. So are coastal bays, fjords (an inlet of water between high cliffs), and some lagoons (a body of water separated from the sea by a narrow strip of land). Therefore, the term estuary has a very broad definition. An **estuary** is a partially enclosed body of water where fresh water and seawater meet and mix as a river enters the ocean.

Organisms living in an estuary must be able to withstand constant mixing of waters and rapid changes in salinity. But those organisms adapted to the estuarine environment find an abundance of nutrients. An estuary acts as a nutrient trap because the sea prevents the rapid escape of nutrients brought by a river. As the result of usually warm, calm waters and plentiful nutrients, estuaries are biologically diverse and highly productive.

Phytoplankton and shore plants thrive in the nutrient-rich estuaries, providing an abundance of food and habitat for animals. It is estimated that nearly two-thirds of marine fishes and shellfish spawn and develop in the protective and rich environment of estuaries, making the estuarine environment the nursery of the sea. An abundance of larval, juvenile, and mature fish and shellfish attract a number of predators, such as reptiles, birds, and fishes of various types.

*Rocky shores* (Fig. 46.19*c*) and sandy shores are constantly bombarded by the sea as the tides roll in and out. The **intertidal zone** lies between the high- and low-tide marks (Fig. 46.20). In the upper portion of the intertidal zone, barnacles are glued so tightly to the stone by their own secretions that their calcareous outer plates remain in place, even after the enclosed shrimplike animal dies. In the midportion of the intertidal zone, brown algae, known as rockweed, may overlie the barnacles. Below the intertidal zone, macroscopic seaweeds, which are the main photosynthesizers, anchor themselves to the rocks by holdfasts.

Organisms cannot attach themselves to shifting, unstable sands on a sandy beach; therefore, nearly all the permanent residents dwell underground. They either burrow during the day and surface to feed at night, or they remain permanently within their burrows and tubes. Ghost crabs and sandhoppers (amphipods) burrow themselves above the high-tide mark and feed at night when the tide is out. Sandworms and sand (ghost)

**FIGURE 46.19** **Coastal ecosystems.**

**a.** Mudflats are frequented by migrant birds. **b.** Mangrove swamps skirt the coastlines of many tropical and subtropical lands. **c.** Some organisms of a rocky coast live in tidal pools.

a. Mudflat

b. Mangrove swamp

c. Rocky shore

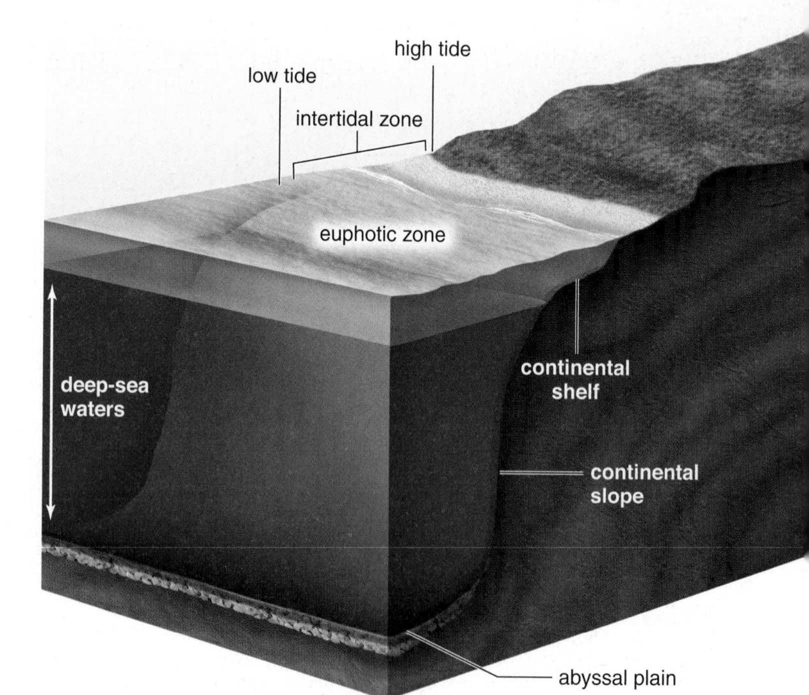


**FIGURE 46.20** **Ocean ecosystems.**

Organisms live in the well-lit waters of the euphotic zone and in the increasing darkness of the deep-sea waters of the pelagic zones (see Figure 46.21).

shrimp remain within their burrows in the intertidal zone and feed on detritus whenever possible. Still lower in the sand, clams, cockles, and sand dollars are found. A variety of shorebirds visit the beaches and feed on various invertebrates and fishes.

## Oceans

Shallow ocean waters (called the *euphotic zone*) contain a greater concentration of organisms than the rest of the sea (see Fig. 46.20). Here, phytoplankton, (i.e., algae) is food not only for zooplankton (i.e., protozoans and microscopic animals) but also for small fishes. These attract a number of predatory and commercially valuable fishes. On the continental shelf, seaweed can be found growing, even on outcroppings as the water gets deeper. Clams, worms, and sea urchins are preyed upon by sea stars, lobsters, crabs, and brittle stars.

**Coral reefs** are areas of biological abundance just below the surface in shallow, warm, tropical waters. Their chief constituents are stony corals, animals that have a calcium carbonate (limestone) exoskeleton, and calcareous red and green algae. Corals provide a home for microscopic algae called *zooxanthellae*. The corals, which feed at night, and the algae, which photosynthesize during the day, are mutualistic and share materials and nutrients. The algae need sunlight, and this may be the reason coral reefs form only in shallow, sunlit waters.

A reef is densly populated with life. The large number of crevices and caves provide shelter for filter feeders (sponges, sea squirts, and fanworms) and for scavangers (crabs and sea urchins). The barracuda, moray eel, and shark are top predators in coral reefs. There are many types of small, beautifully colored fishes. These become food for larger fishes, including snappers that are caught for human consumption.

Most of the ocean lies within the **pelagic zones,** as noted in Figure 46.21. The *epipelagic zone* lacks the inorganic nutrients of shallow waters, and therefore it does not have as high a concentration of phytoplankton, even though the surface is sunlit. Still, the photosynthesizers are food for a large assembly of zooplankton, which then become food for schools of various fishes. A number of porpoise and dolphin species visit and feed in the epipelagic zone. Whales, too, are mammals found in this zone. Baleen whales strain krill (small crustaceans) from the water, and toothed sperm whales feed primarily on the common squid.

Animals in the deeper waters of the *mesopelagic zone* are carnivores, which are adapted to the absence of light, and tend to be translucent, red colored, or even luminescent. There are luminescent shrimps, squids, and fishes, including lantern and hatchet fishes. Various species of zooplankton, invertebrates, and fishes migrate from the mesopelagic zone to the surface to feed at night.

The deepest waters of the *bathypelagic zone* are in complete darkness except for an occasional flash of bioluminescent light. Carnivores and scavengers are found in this zone. Strange-looking fishes with distensible mouths and abdomens and small, tubular eyes feed on infrequent prey.

It once was thought that few vertebrates exist on the *abyssal plain* beneath the bathypelagic zone because of the intense pressure and the extreme cold. Yet, many invertebrates survive there by feeding on debris floating down from the mesopelagic zone. Sea lilies (crinoids) rise above the seafloor; sea cucumbers and sea urchins crawl around on the sea bottom; and tube worms burrow in the mud.

The flat abyssal plain is interrupted by enormous underwater mountain chains called oceanic ridges. Along the axes of the ridges, crustal plates spread apart, and molten magma rises to fill the gap. At **hydrothermal vents,** seawater

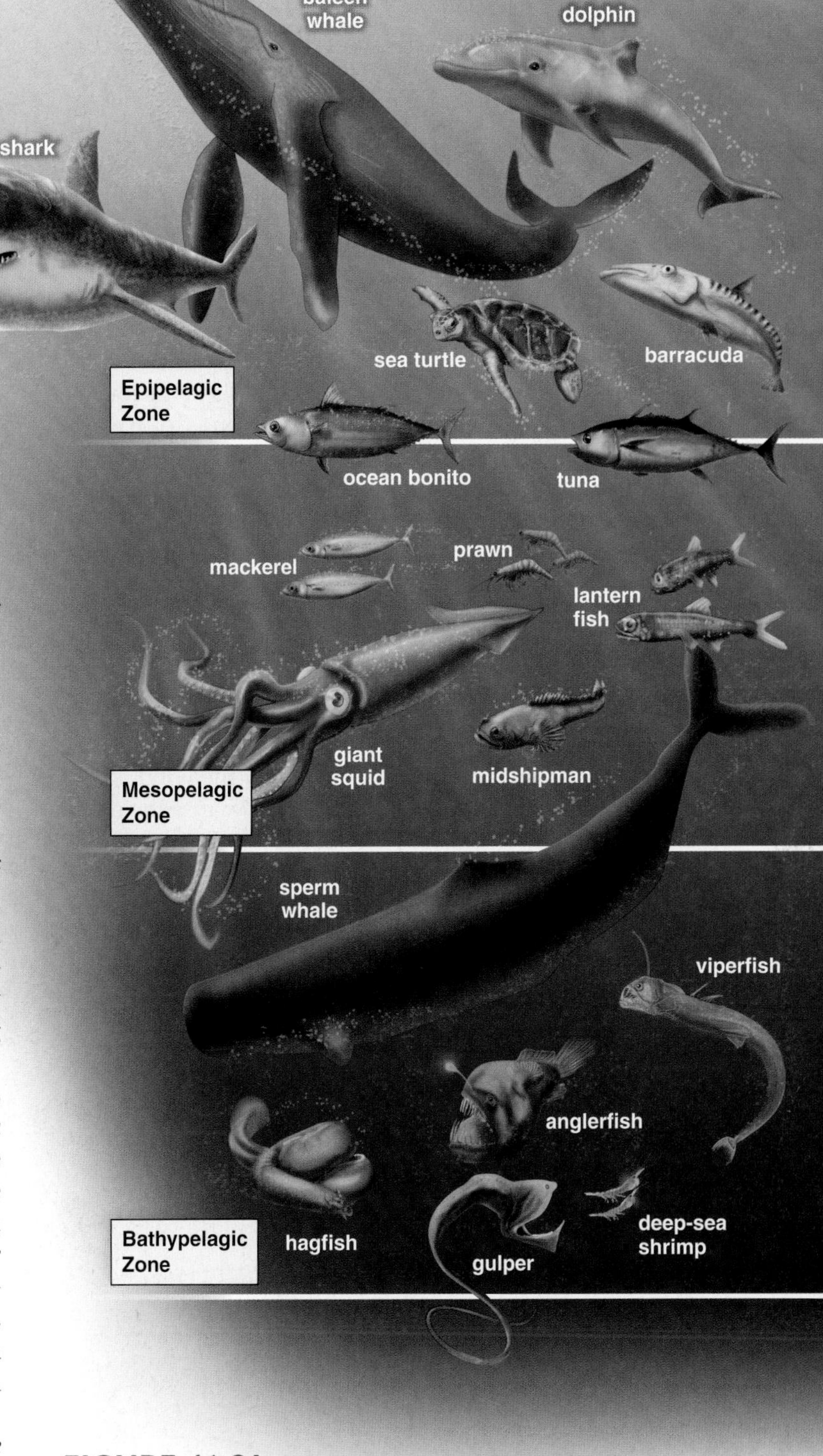


**FIGURE 46.21 Ocean inhabitants of pelagic zones.**
Different organisms are characteristic of the epipelagic, mesopelagic, and bathypelagic zones.

percolates through cracks and is heated to about 350°C, causing sulfate to react with water and form hydrogen sulfide ($H_2S$). Chemoautotrophic bacteria that obtain energy from oxidizing hydrogen sulfide exist freely or mutualistically within the tissues of organisms. They are the start of food chains for an ecosystem that includes huge tube worms, clams, crustaceans, echinoderms, and fishes. This ecosystem can exist where light never penetrates because, unlike photosynthesis, chemosynthesis does not require light energy.

## Ocean Currents

Climate is driven by the sun, but the oceans play a major role in redistributing heat in the biosphere. Water tends to be warm at the equator and much cooler at the poles because of the distribution of the sun's rays, as discussed earlier (see Fig. 46.1*a*). Air takes on the temperature of the water below, and warm air moves from the equator to the poles. In other words, the oceans make the winds blow. (Landmasses also play a role, but the oceans hold heat longer and remain cool longer during periods of changing temperature than do continents.)

When wind blows strongly and steadily across a great expanse of ocean for a long time, friction from the moving air begins to drag the water along with it. Once the water has been set in motion, its momentum, aided by the wind, keeps it moving in a steady flow called a current. Because the ocean currents eventually strike land, they move in a circular path—clockwise in the Northern Hemisphere and counterclockwise in the Southern Hemisphere (Fig. 46.22). As the currents flow, they take warm water from the equator to the poles. One such current, called the Gulf Stream, brings tropical Caribbean water to the east coast of North America and the higher latitudes of western Europe. Without the Gulf Stream, Great Britain, which has a relatively warm temperature, would be as cold as Greenland. In the Southern Hemisphere, another major ocean current warms the eastern coast of South America.

Also in the Southern Hemisphere, a current called the Humboldt Current flows toward the equator. The Humboldt Current carries phosphorus-rich cold water northward along the west coast of South America. During a process called **upwelling,** cold offshore winds cause cold nutrient-rich waters to rise and take the place of warm nutrient-poor waters. In South America, the enriched waters cause an abundance of marine life that supports the fisheries of Peru and northern Chile. Birds feeding on these organisms deposit their droppings on land, where they are mined as guano, a commercial source of phosphorus. When the Humboldt Current is not as cool as usual, upwelling does not occur, stagnation results, the fisheries decline, and climate patterns change globally. This phenomenon, which is discussed in the Ecology Focus on page 885, is called an **El Niño–Southern Oscillation.**

### Check Your Progress 46.3

1. Describe the zones of the open ocean.

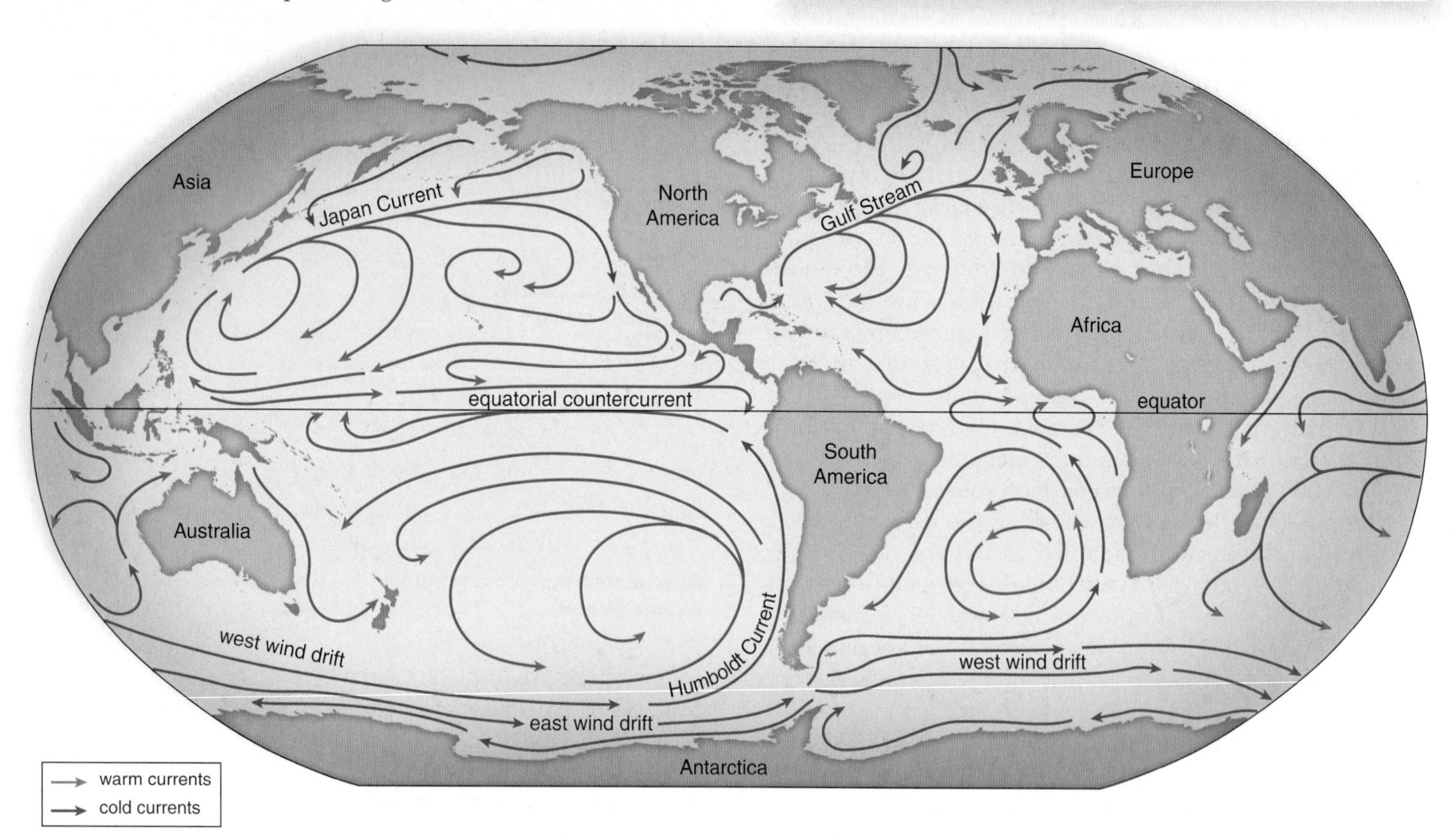


**FIGURE 46.22 Ocean currents.**

The arrows on this map indicate the locations and directions of the major ocean currents set in motion by the global wind circulation. By carrying warm water to cool latitudes (e.g., the Gulf Stream) and cool water to warm latitudes (e.g., the Humboldt Current), these currents have a major effect on the world's climates.

*ecology focus*

## El Niño–Southern Oscillation

Climate largely determines the distribution of life on Earth. Short-term variations in climate, which we call weather, also have a pronounced effect on living things. There is no better example than an El Niño. Originally, El Niño referred to a warming of the seas off the coast of Peru at Christmastime—hence, the name El Niño, "the boy child," for the Christ child Jesus.

Now scientists prefer the term El Niño–Southern Oscillation (ENSO) for a severe weather change brought on by an interaction between the atmosphere and ocean currents. Ordinarily, the southeast trade winds move along the coast of South America and turn west because of the Earth's daily rotation on its axis. As the winds drag warm ocean waters from east to west, there is an upwelling of nutrient-rich cold water from the ocean's depths, resulting in a bountiful Peruvian harvest of anchovies. When the warm ocean waters reach their western destination, the monsoons bring rain to India and Indonesia. Scientists have noted that these events correlate with a difference in the barometric pressure over the Indian Ocean and the southeastern Pacific—that is, the barometric pressure is low over the Indian Ocean and high over the southeastern Pacific. But when a "southern oscillation" occurs and the barometric pressures switch, an El Niño begins.

During an El Niño, both the northeast and the southeast trade winds slacken. Upwelling no longer occurs, and the anchovy catch off the coast of Peru plummets. During a severe El Niño, waters from the east never reach the west, and the winds lose their moisture in the middle of the Pacific instead of over the Indian Ocean. The monsoons fail, and drought occurs in India, Indonesia, Africa, and Australia. Harvests decline, cattle must be slaughtered, and famine is likely in highly populated India and Africa, where funds to import replacement supplies of food are limited.

A backward movement of winds and ocean currents may even occur so that the waters warm to more than 14° above normal along the west coast of the Americas. This is a sign that a severe El Niño has occurred, and the weather changes are dramatic in the Americas also. Southern California is hit by storms and even hurricanes, and the deserts of Peru and Chile receive so much rain that flooding occurs. A jet stream (strong wind currents) can carry moisture into Texas, Louisiana, and Florida, with flooding a near certainty. Or the winds can turn northward and deposit snow in the mountains along the west coast so that flooding occurs in the spring. Some parts of the United States, however, benefit from an El Niño. The Northeast is warmer than usual, few if any hurricanes hit the east coast, and there is a lull in tornadoes throughout the Midwest. Altogether, a severe El Niño affects the weather over three-quarters of the globe.

Eventually, an El Niño dies out, and normal conditions return. The normal cold-water state off the coast of Peru is known as La Niña (the girl). Figure 46B contrasts the weather conditions of a La Niña with those of an El Niño. Since 1991, El Niños have varied in magnitude, and two record-breaking El Niños have occurred. As our overall climate changes, the severity of El Niños remains somewhat unpredictable.

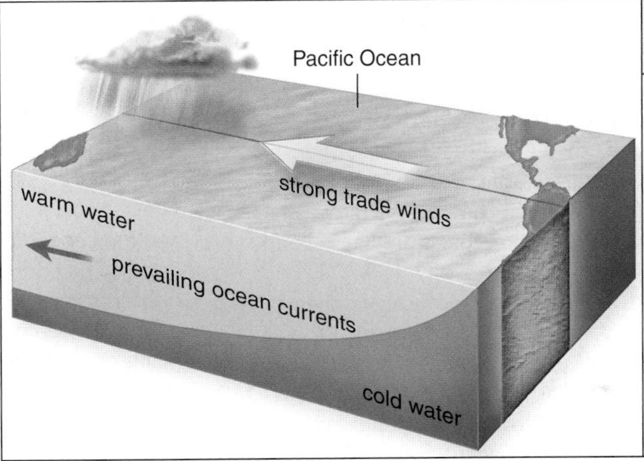


**La Niña**

- Upwelling off the west coast of South America brings cold waters to the surface.
- Barometric pressure is high over the southeastern Pacific.
- Monsoons associated with the Indian Ocean occur.
- Hurricanes occur off the east coast of the United States.

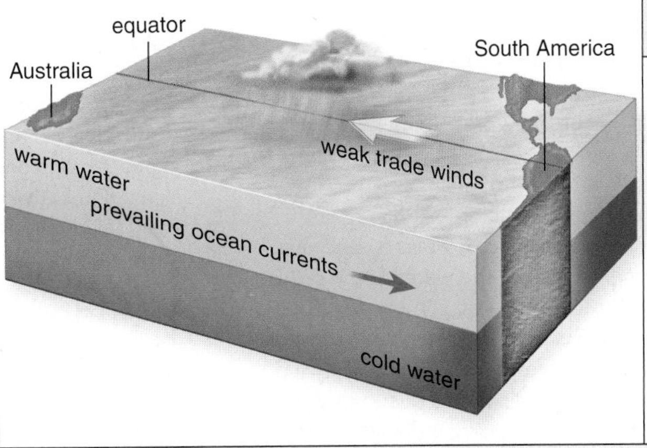


**El Niño**

- Great ocean warming occurs off the west coast of the Americas.
- Barometric pressure is low over the southeastern Pacific.
- Monsoons associated with the Indian Ocean fail.
- Hurricanes occur off the west coast of the United States.

**FIGURE 46B La Niña (*above*) and El Niño (*below*).**

# Connecting the Concepts

The biosphere is the product of interactions of living organisms with the Earth's physical and chemical environment over billions of years. Life has affected the atmosphere, modifying its composition and influencing global climate. Organisms have helped establish the chemical and physical conditions of streams, lakes, and oceans. The soils of terrestrial ecosystems and the sediments of aquatic ecosystems are structured largely by the activities of organisms. Reef-building plants and animals have helped build the islands on which millions of humans live. The Earth's diverse biomes also result from interactions of the biotic communities and the abiotic environment.

Over geological time, the biosphere has been changing constantly. The biomes of the age of dinosaurs were strikingly different from those of today. Astrophysical events have triggered some of this change. Changes in the sun's radiation output and in the tilt of the Earth's axis have altered the pattern of solar energy reaching the Earth's surface. Geological processes have also modified conditions for life. The drifting of continents has changed the arrangement of continents and oceans. Mountain ranges have been thrust up and eroded down. Through these changing conditions, life has evolved, and the structure of the Earth's biomes has evolved as well. In the last few million years, humans appeared and learned to exploit the Earth's biomes.

Humans have transformed vast areas of many of the terrestrial biomes into farmland, cities, highways, and other developments. Still, people depend on the biodiversity that exists in the Earth's biomes, and on the interactions of other organisms within the biosphere. These interactions influence nutrient cycling, waste processing, and basic biological productivity. However, the Earth's biotic diversity also provides enjoyment and inspiration to millions of people, who spend billions of dollars to visit coral reefs, deserts, rain forests, and even the Arctic tundra.

## summary

### 46.1 Climate and the Biosphere

Because the Earth is a sphere, the sun's rays at the poles are distributed out over a larger area than the direct rays at the equator. The temperature at the surface of the Earth therefore decreases from the equator to each pole. The Earth is tilted on its axis, and the seasons change as the Earth revolves annually around the sun.

Warm air rises near the equator, loses its moisture, and then descends at about 30° north and south latitude to the poles. When the air descends, it absorbs moisture from the land, and therefore the great deserts of the world are formed at 30° latitudes. Because the Earth rotates on its axis daily, the winds blow in opposite directions above and below the equator. Topography also plays a role in the distribution of moisture. Air rising over coastal ranges loses its moisture on the windward side, making the leeward side arid.

### 46.2 Terrestrial Ecosystems

A biome is a major type of terrestrial community. Biomes are distributed according to climate—that is, temperature and rainfall influence the pattern of biomes about the world. The effect of temperature causes the same sequence of biomes when traveling to northern latitudes as when traveling up a mountain.

The Arctic tundra is the northernmost biome and consists largely of short grasses, sedges, and dwarf woody plants. Because of cold winters and short summers, most of the water in the soil is frozen year-round. This is called the permafrost.

The taiga, a coniferous forest, has less rainfall than other types of forests. The temperate deciduous forest has trees that gain and lose their leaves because of the alternating seasons of summer and winter. Tropical rain forests are continually warm and wet. These are the most complex and productive of all biomes.

Shrublands usually occur along coasts that have dry summers and receive most of their rainfall in the winter. Among grasslands, the savanna, a tropical grassland, supports the greatest number of different types of large herbivores. Temperate grasslands, such as that found in the central United States, have a limited variety of vegetation and animal life.

Deserts are characterized by a lack of water—they are usually found in places with less than 25 cm of precipitation per year. Some desert plants, such as cacti, are succulents with thick stems and leaves, and others are shrubs that are deciduous during dry periods.

### 46.3 Aquatic Ecosystems

Streams, rivers, lakes, and wetlands are different freshwater ecosystems. In deep lakes of the temperate zone, the temperature and the concentration of nutrients and gases in the water vary with depth. The entire body of water is cycled twice a year, distributing nutrients from the bottom layers. Lakes and ponds have three life zones. Rooted plants and clinging organisms live in the littoral zone, plankton and fishes live in the sunlit limnetic zone, and bottom-dwelling organisms such as crayfishes and molluscs live in the profundal zone.

Marine ecosystems are divided into coastal ecosystems and the oceans. The coastal ecosystems, especially estuaries, are more productive than the oceans. Estuaries (and associated salt marshes, mudflats, and mangrove forests) are near the mouth of a river. Estuaries are considered the nurseries of the sea.

An ocean is divided into the pelagic zone and the ocean floor. The pelagic zone (open waters) has three zones. The epipelagic zone receives adequate sunlight and supports the most life. The mesopelagic zone contains organisms adapted to minimum or no light. The bathypelagic zone is in complete darkness. The ocean floor includes the continental shelf, the continental slope, and the abyssal plain.

## understanding the terms

alpine tundra 869
Arctic tundra 870
biome 869
bog 879
chaparral 876
climate 866
coral reef 883
desert 878
El Niño–Southern Oscillation 884
epiphyte 874
estuary 882
eutrophication 880
fall overturn 880
grassland 876
hydrothermal vent 883
intertidal zone 882
lake 880
marsh 879
monsoon 867

Match the terms to these definitions:

a. ______________ End of a river where fresh water and salt water mix as they meet.
b. ______________ Open portion of the sea.
c. ______________ Terrestrial biome that is a coniferous forest extending in a broad belt across northern Eurasia and North America.
d. ______________ Terrestrial biome that is a grassland in Africa, characterized by few trees and a severe dry season.
e. ______________ Oxygen-rich top waters mix with nutrient-rich bottom waters in stratified lakes.

## reviewing this chapter

1. Describe how a spherical Earth and the path of the Earth about the sun affect climate. 866
2. Describe the air circulation about the Earth, and tell why deserts are apt to occur at 30° north and south of the equator. 866
3. How does a coastal mountain range affect climate? What causes a monsoon climate? 867
4. Name the terrestrial biomes you would expect to find when going from the base to the top of a mountain. 869
5. Describe the location, the climate, and the populations of the Arctic tundra, coniferous forests (both taiga and temperate rain forest), temperate deciduous forests, tropical rain forests, shrublands, grasslands (both temperate grasslands and savanna), and deserts. 870–78
6. Describe the importance of wetlands and the major types of wetlands. 879
7. Describe the overturn of a temperate lake, the life zones of a lake, and the organisms you would expect to find in each life zone. 880–81
8. Describe the coastal communities, and discuss the importance of estuaries to the productivity of the ocean. 882–83
9. Describe the zones of the open ocean and the organisms you would expect to find in each zone. 883–84
10. Describe the ocean currents and how the Gulf Stream accounts for Great Britain having a mild temperature. 884

## testing yourself

Choose the best answer for each question.

1. The seasons are best explained by
   a. the distribution of temperature and rainfall in biomes.
   b. the tilt of the Earth as it orbits about the sun.
   c. the daily rotation of the Earth on its axis.
   d. the fact that the equator is warm and the poles are cold.
2. The mild climate of Great Britain is best explained by
   a. the winds called the westerlies.
   b. the spinning of the Earth on its axis.
   c. Great Britain being a mountainous country.
   d. the flow of ocean currents.
3. Which of these pairs is mismatched?
   a. tundra—permafrost
   b. savanna—*Acacia* trees
   c. prairie—epiphytes
   d. coniferous forest—evergreen trees
4. All of these phrases describe the tundra except
   a. low-lying vegetation.
   b. northernmost biome.
   c. short growing season.
   d. many different types of species.
5. The forest with a multilevel canopy is the
   a. tropical rain forest.
   b. coniferous forest.
   c. tundra.
   d. temperate deciduous forest.
6. Why are lush evergreen forests present in the Pacific Northwest of the United States?
   a. The rotation of the Earth causes this.
   b. Winds blow from the ocean, bringing moisture.
   c. They are located on the leeward side of a mountain range.
   d. Both b and c are correct.
   e. All of these are correct.
7. Which of these influences the location of a particular biome?
   a. latitude
   b. average annual rainfall
   c. average annual temperature
   d. altitude
   e. All of these are correct.
8. Which of these is a function of a wetland?
   a. purifies water
   b. is an area where toxic wastes can be broken down
   c. helps absorb overflow and prevents flooding
   d. is a home for organisms that are links in food chains
   e. All of these are correct.
9. The area of a lake closest to shore and where rooted plants are found is the
   a. littoral zone.
   b. limnetic zone.
   c. profundal zone.
   d. benthic zone.
10. An estuary acts as a nutrient trap because of the
    a. action of rivers and tides.
    b. depth at which photosynthesis can occur.
    c. amount of rainfall received.
    d. height of the water table.
11. Which area of an ocean has the most light and nutrients?
    a. epipelagic zone and abyssal plain
    b. epipelagic zone only
    c. benthic zone only
    d. neritic province
12. Which area of the pelagic zone is completely dark?
    a. epipelagic zone
    b. mesopelagic zone
    c. bathypelagic zone
13. Runoff of fertilizer and animal wastes from a large farm that drains into a lake would be an example of which process?
    a. fall overturn
    b. eutrophication
    c. spring overturn
    d. upwelling

14. Phytoplankton are more likely to be found in which life zone of a lake?
    a. limnetic zone
    b. profundal zone
    c. benthic zone
    d. All of these are correct.
15. Which area of the seashore is only exposed during low tide?
    a. upper littoral zone
    b. mid-littoral zone
    c. lower littoral zone
    d. None of these are correct.
16. Which of these would normally be found in the benthic zone of a lake?
    a. minnows
    b. clams
    c. zooplankton
    d. plants
    e. large fish
17. All of the following phrases describe a tropical rain forest, except
    a. nutrient-rich soil.
    b. many arboreal plant and animals.
    c. canopy composed of many layers.
    d. broad-leaved evergreen trees.
18. An oligotrophic lake
    a. is nutrient-rich.
    b. is cold.
    c. is likely to be found in an agricultural or urban area.
    d. has poor productivity.
19. Energy for the food chain near hydrothermal vents comes from
    a. dead organisms that fall down from above.
    b. highly efficient photosynthetic phytoplankton.
    c. chemoautotrophic bacteria.
    d. heat given off by the vents.

## thinking scientifically

1. Pharmaceutical companies are interested in "bioprospecting" in tropical rain forests. These companies are looking for naturally occurring compounds in plants or animals that can be used as drugs for a variety of diseases. The most promising compounds act as antibacterial or antifungal agents. Even discounting the fact that the higher density of species in tropical rain forests would produce a wider array of compounds than another biome, why would antibacterial and antifungal compounds be more likely to evolve in this particular biome?
2. When hurricanes come ashore, the ocean tide may surge dozens of feet above normal into coastal communities. Recently, hurricane Katrina had devastating effects on New Orleans, Louisiana, a city actually built below sea level following the draining of the natural wetlands. How would the presence of wetlands possibly decrease the damage of seawater brought in by an incoming hurricane?

## bioethical issue

### Water Pollution

Agricultural fertilizers are the chief cause of nitrate contamination of drinking-water wells. Excessive nitrates in a baby's bloodstream can lead to slow suffocation, known as blue-baby syndrome. Agricultural herbicides are suspected carcinogens in the tap water of scattered ecosystems coast to coast. What can be done?

Some farmers are already using irrigation methods that deliver water directly to plant roots, no-till agriculture that reduces the loss of topsoil and cuts back on herbicide use, and integrated pest management, which relies heavily on good bugs to kill bad bugs. Perhaps more should do so. Encouraged—in some cases, compelled—by state and federal agents, dairy farmers have built sheds, concrete containments, and underground liquid storage tanks to hold wastes when it rains. Later, the manure is trucked to fields and spread as fertilizer.

Homeowners, like business golf clubs and ski resorts, also contribute to the problem. The manicuring of lawns, the use of motor vehicles, and the construction and use of roads and buildings all add contaminants to streams, lakes, and aquifers. Citizens around Grand Traverse Bay on the eastern shore of Lake Michigan have also gotten the message, especially because they want to keep on enjoying water-dependent activities such as boating, swimming, and fishing. James Haverman, a concerned member of the Traverse Bay Watershed Initiative says, "If we can't change the way people live their everyday lives, we are not going to be able to make a difference." Builders in Traverse County are already required to control soil erosion with filter fences, steer rainwater away from exposed soil, build sediment basins, and plant protective buffers. Presently, homeowners must have a 25-foot setback from wetlands and a 50-foot setback from lakes and creeks. They are also encouraged to pump out their septic systems every two years. Do you approve of legislation that requires farmers and homeowners to protect freshwater supplies? Why or why not?

## *Biology* website

The companion website for *Biology* provides a wealth of information organized and integrated by chapter. You will find practice tests, animations, videos, and much more that will complement your learning and understanding of general biology.

**http://www.mhhe.com/maderbiology10**

# 47 Conservation of Biodiversity

*e*ven schoolchildren on a field trip in the woods may come across many deformed frogs. These frogs often have extra limbs arising from their midsection. Scientists are working to discover why this problem is so widespread and, so far, two hypotheses have been suggested. One hypothesis being studied is that a small trematode burrows into a tadpole and prevents the normal development of legs. The other hypothesis is that a chemical called methoprene, which is used by many farmers as a pesticide, is affecting development and causing these abnormalities. Once the cause is known, something can be done about it.*

*In this chapter, we discuss the emergence of conservation biology as a branch of environmental science directed at actively improving environmental conditions for wildlife and ourselves. Conservation biology examines the nature and evolutionary origin of biodiversity. As an applied science, it studies the human activities that are presently reducing biodiversity and what steps to take to reverse the situation. Conservation biology has an ultimate goal: the preservation and management of ecosystems for human welfare.*

Deformed frogs are undoubtedly due to environmental effects.

## concepts

## 47.1 Conservation Biology and Biodiversity

**Conservation biology** [L. *conservatio,* keep, save] is a relatively new discipline of biology that studies all aspects of biodiversity with the goal of conserving natural resources for this generation and all future generations. Conservation biology is unique in that it is concerned with both the development of scientific concepts and the application of these concepts to the everyday world. A primary goal is the management of biodiversity for sustainable use by humans. To achieve this goal, conservation biologists are interested in, and come from, many subfields of biology that only now have been brought together into a cohesive whole.

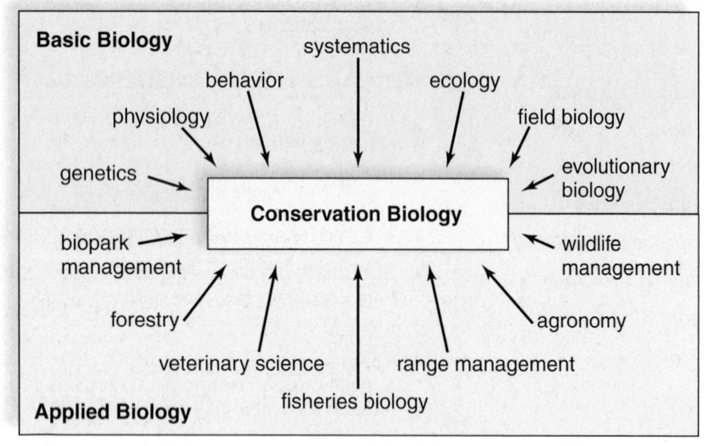


Like a physician, a conservation biologist must be aware of the latest findings, both theoretical and practical, and be able to use this knowledge to diagnose the source of trouble and suggest a suitable treatment. Often, it is necessary for conservation biologists to work with government officials at both the local and federal levels. Public education is another important duty of conservation biologists.

Conservation biology is a unique science in another way. It blatantly supports the following ethical principles: (1) Biodiversity is desirable for the biosphere and therefore for humans; (2) extinctions, due to human actions, are therefore undesirable; (3) the complex interactions in ecosystems support biodiversity and are desirable; and (4) biodiversity generated by evolutionary change has value in and of itself, regardless of any practical benefit. The consequences of disrupting ecosystem interactions through biodiversity loss are potentially very large and unpredictable. Therefore, ecosystem disruption should be avoided.

Conservation biology has emerged in response to a crisis—never before in the history of the Earth are so many extinctions expected in such a short period of time. Estimates vary, but at least 10–20% of all species now living most likely will become extinct in the next 20–50 years unless planned coordinated actions are taken. It is urgently important, then, that all citizens understand the concept of biodiversity, the value of biodiversity, the likely causes of present-day extinctions, what could be done to prevent extinctions from occurring, and the potential consequences of decreased biodiversity.

To protect biodiversity, **bioinformatics** is applied. Bioinformatics is the science of collecting, analyzing, and making readily available biological information. Throughout the world, molecular, descriptive, and biogeographical information on organisms is being collected. This information is used in understanding and protecting biodiversity and will become more useful as more data are accumulated.

### Biodiversity

At its simplest level, **biodiversity** [Gk. *bio,* life; L. *diversus,* various] is the variety of life on Earth. It is common practice to describe biodiversity in terms of the number of species among various groups of organisms. Figure 47.1 only accounts for the species that have so far been described. It has been estimated that there may be between 10 and 50 million species in all; if so, many species are still to be found and described.

Of these, nearly 1,200 species in the United States and 30,000 species worldwide are in danger of extinction. An **endangered species** is one that is in peril of immediate extinction throughout all or most of its range. Examples of endangered species include the black lace cactus, armored snail, hawksbill sea turtle, California condor, West Indian manatee, and snow leopard. **Threatened species** are organisms that are likely to become endangered species in the foreseeable future. Examples of threatened species include the Navaho sedge, puritan tiger beetle, gopher tortoise, bald eagle, gray wolf, and Louisiana black bear.

To develop a meaningful understanding of life on Earth, we need to know more about species than their total number. Ecologists describe biodiversity as an attribute of three other levels of biological organization: genetic diversity, community diversity, and landscape diversity.

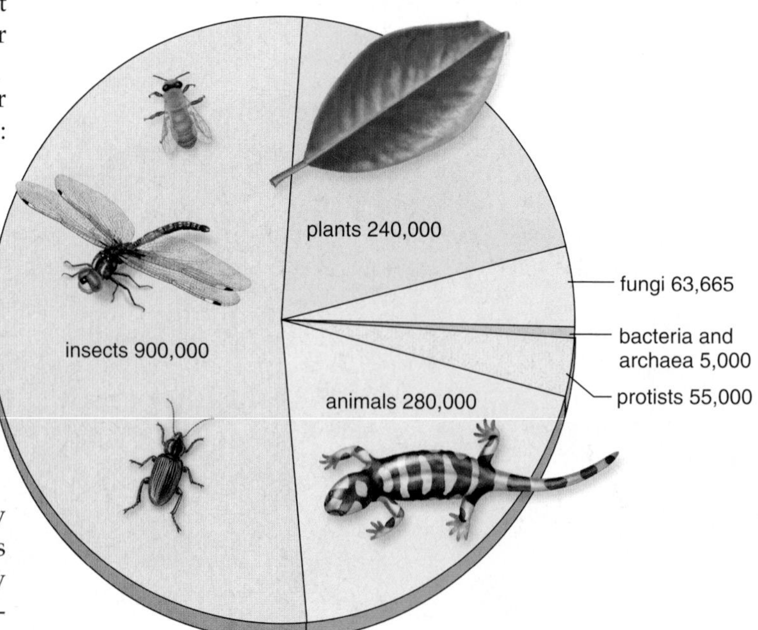


**FIGURE 47.1 Number of described species.**

There are only about 1.75 million described species; insects are far more prevalent than organisms in other groups. Undescribed species probably number far more than those species that have been described.

**Genetic diversity** refers to variations among the members of a population. Populations with high genetic diversity are more likely to have some individuals that can survive a change in the structure of their ecosystem. For example, the 1846 potato blight in Ireland, the 1922 wheat failure in the Soviet Union, and the 1984 outbreak of citrus canker in Florida were all made worse by limited genetic variation among these crops. If a species' population is quite small and isolated, it is more likely to eventually become extinct because of a loss of genetic diversity. As organisms become endangered and threatened, they lose their genetic diversity.

**Ecosystem diversity** is dependent on the interactions of species at a particular locale. One community's species composition can be completely different from that of other communities. Community composition, therefore, increases the levels of biodiversity in the biosphere. Although past conservation efforts frequently concentrated on saving particular charismatic species, such as the California condor, the black-footed ferret, or the spotted owl, this is a weak, shortsighted approach. A more effective approach is to conserve species that have a critical role to play in an ecosystem. Saving an entire community can save many species, and the contrary is also true—disrupting a community threatens the existence of more than one species. Opossum shrimp, *Mysis relicta,* were introduced into Flathead Lake in Montana and its tributaries as food for salmon. The shrimp ate so much zooplankton that there was in the end far less food for the fish and ultimately for the grizzly bears and bald eagles as well (Fig. 47.2).

**Landscape diversity** involves a group of interacting ecosystems; within one **landscape,** for example, there may be plains, mountains, and rivers. Any of these ecosystems can be so fragmented that they are connected by only patches (remnants) or strips of land that allow organisms to move from one ecosystem to the other. Fragmentation of the landscape reduces reproductive capacity and food availability and can disrupt seasonal behaviors.

## *Distribution of Biodiversity*

Biodiversity is not evenly distributed throughout the biosphere; therefore, protecting some areas will save more species than protecting other areas. Biodiversity is highest at the tropics, and it declines toward each pole on land, in fresh water, and in the ocean. Also, more species are found in the coral reefs of the Indonesian archipelago than in other coral reefs as one moves westward across the Pacific.

Some regions of the world are called **biodiversity hotspots** because they contain unusually large concentrations of species. Biodiversity in these hotspots account for about 44% of all known higher plant species and 35% of all terrestrial vertebrate species but cover only about half of 1.4% of the Earth's land area. The island of Madagascar, the Cape region of South Africa, Indonesia, the coast of California, and the Great Barrier Reef of Australia are all biodiversity hotspots.

One surprise of late has been the discovery that rain forest canopies and the deep-sea benthos have many more species than formerly thought. Some conservationists refer to these two areas as biodiversity frontiers.

### Check Your Progress 47.1

1. What is conservation biology?
2. What is biodiversity?

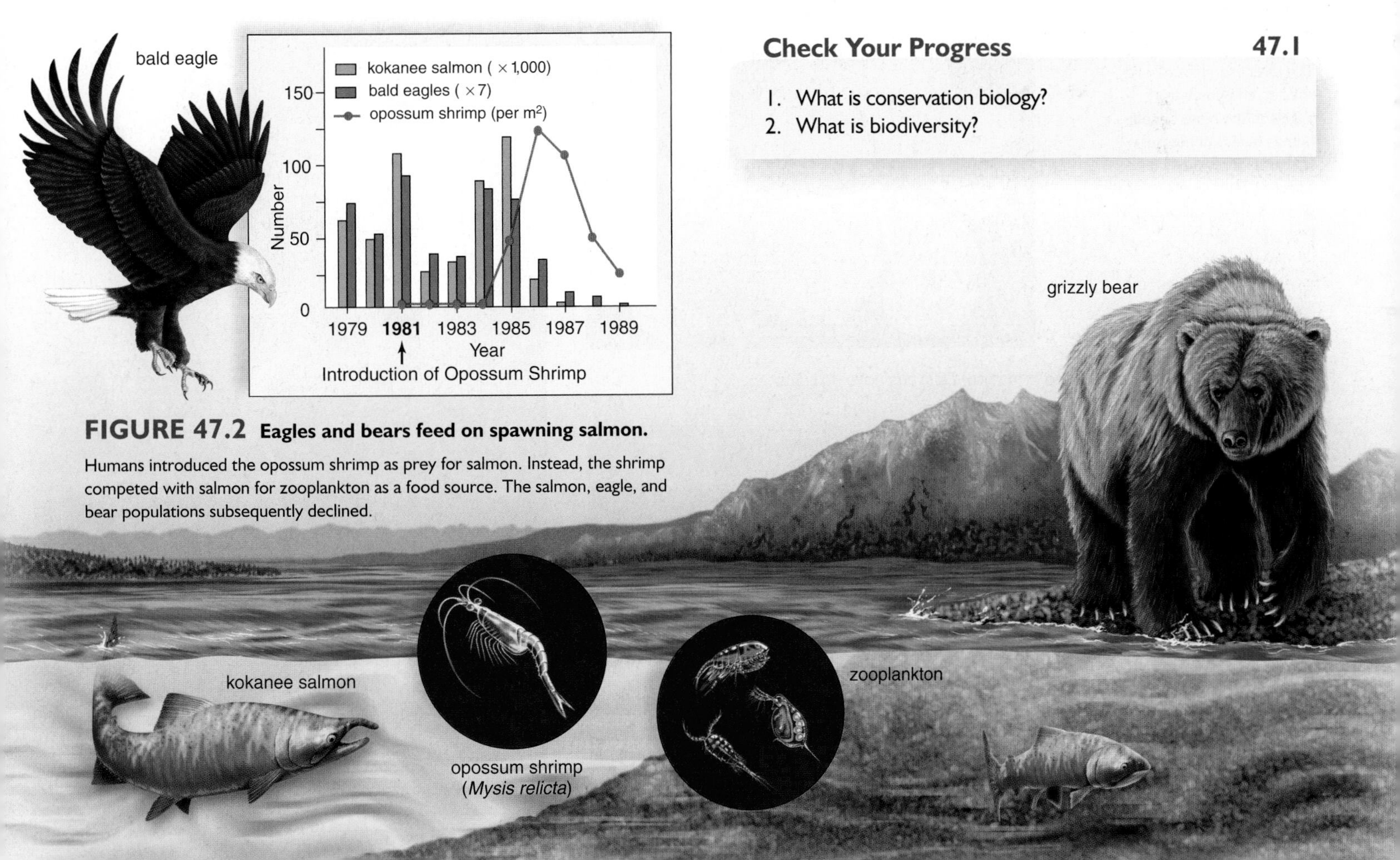


**FIGURE 47.2 Eagles and bears feed on spawning salmon.**

Humans introduced the opossum shrimp as prey for salmon. Instead, the shrimp competed with salmon for zooplankton as a food source. The salmon, eagle, and bear populations subsequently declined.

# 47.2 Value of Biodiversity

Conservation biology strives to reverse the trend toward the possible extinction of living species. To bring this about, it is necessary to make all people aware that biodiversity is a resource of immense value.

## Direct Value

Various individual species perform services for human beings and contribute greatly to the value we should place on biodiversity. Only some of the most obvious values are discussed here and illustrated in Figure 47.3.

### *Medicinal Value*

Most of the prescription drugs used in the United States, valued at over $200 billion, were originally derived from living organisms. The rosy periwinkle from Madagascar is an excellent example of a tropical plant that has provided us with useful medicines. Potent chemicals from this plant are now used to treat two forms of cancer: leukemia and Hodgkin disease. Because of these drugs, the survival rate for childhood leukemia has gone from 10% to 90%, and Hodgkin disease is usually curable. Although the value of saving a life cannot be calculated, it is still sometimes easier for us to appreciate the worth of a resource if it is explained in monetary terms. Thus, researchers tell us that, judging from the success rate in the past, an additional 328 types of drugs are yet to be found in tropical rain forests, and the value of this resource to society is probably $147 billion.

You may already know that the antibiotic penicillin is derived from a fungus and that certain species of bacteria produce the antibiotics tetracycline and streptomycin. These drugs have proven to be indispensable in the treatment of diseases, including certain sexually transmitted diseases.

Leprosy is among those diseases for which there is as yet no cure. The bacterium that causes leprosy will not grow in the laboratory, but scientists discovered that it grows naturally in the nine-banded armadillo. Having a source for the bacterium may make it possible to find a cure for leprosy. The blood of horseshoe crabs, *Limulus*, contains a substance called limulus amoebocyte lysate, which is used to ensure that medical devices such as pacemakers, surgical implants, and prosthetic devices are free of bacteria. Hemolymph (the bloodlike fluid in arthropods) is taken from 250,000 crabs a year, and then they are returned to the sea unharmed.

### *Agricultural Value*

Crops such as wheat, corn, and rice are derived from wild plants that have been modified to be high producers. The same high-yield, genetically similar strains tend to be grown worldwide. When rice crops in Africa were being devastated by a virus, researchers grew wild rice plants from thousands

**FIGURE 47.3**
**Direct value of wildlife.**
The direct services of wild species, shown on this page and the next, benefit human beings immensely. It is sometimes possible to calculate the monetary value, which is always surprisingly large.

Wild species, like the rosy periwinkle, *Catharanthus roseus*, are sources of many medicines.

Wild species, like many marine species, provide us with food.

Wild species, like the nine-banded armadillo, *Dasypus novemcinctus*, play a role in medical research.

of seed samples until they found one that contained a gene for resistance to the virus. These wild plants were then used in a breeding program to transfer the gene into high-yield rice plants. If this variety of wild rice had become extinct before it could be discovered, rice cultivation in Africa might have collapsed.

Biological pest controls—natural predators and parasites—are often preferable to using chemical pesticides. When a rice pest called the brown planthopper became resistant through natural selection to pesticides, farmers began to use natural brown planthopper enemies instead. The economic savings were calculated at well over $1 billion. Similarly, cotton growers in Cañete Valley, Peru, found that pesticides were no longer working against the cotton aphid because of the resistance the aphids evolved. Research identified natural predators that are now being used to an ever greater degree by cotton farmers. Again, savings have been enormous.

Most flowering plants are pollinated by animals, such as bees, wasps, butterflies, beetles, birds, and bats. The domesticated honeybee, *Apis mellifera,* pollinates almost $10 billion worth of food crops annually in the United States. The danger of this dependency on a single species is exemplified by mites, which have now wiped out more than 20% of the commercial honeybee population in the United States. Where can we get resistant bees? From the wild, of course. The value of wild pollinators to the U.S. agricultural economy has been calculated at $4.1–$6.7 billion a year.

### *Consumptive Use Value*

Humans have had much success cultivating crops, keeping domesticated animals, growing trees in plantations, and so forth. But so far, aquaculture, the growing of fish and shellfish for human consumption, has contributed only minimally to human welfare—instead, most freshwater and marine harvests depend on the catching of wild animals, such as fishes (e.g., trout, cod, tuna, and flounder), crustaceans (e.g., lobsters, shrimps, and crabs), and mammals (e.g., whales). Obviously, these aquatic organisms are an invaluable biodiversity resource.

The environment provides a variety of other products that are sold in the marketplace worldwide, including wild fruits and vegetables, skins, fibers, beeswax, and seaweed. Also, by hunting and fishing, some people obtain their meat directly from the environment. In one study, researchers calculated that the economic value of wild pig in the diet of native hunters in Sarawak, East Malaysia, was approximately $40 million per year.

Similarly, many trees are still felled in the natural environment for their wood. Researchers have calculated that a species-rich forest in the Peruvian Amazon is worth far more if the forest is used for fruit and rubber production than for timber production. Fruit and the latex needed to produce rubber can be brought to market for an unlimited number of years, whereas once the trees are gone, no more timber can be harvested.

Wild species, like the lesser long-nosed bat, *Leptonycteris curasoae*, are pollinators of agricultural and other plants.

Wild species, like ladybugs, *Coccinella*, play a role in biological control of agricultural pests.

Wild species, like rubber trees, *Hevea*, can provide a product indefinitely if the forest is not destroyed.

## Indirect Value

The wild species we have been discussing play a role in their respective ecosystems. If we want to preserve them, it is more economical to save the ecosystems, and subsequently all the species within, than individual species. Ecosystems perform many services for modern humans, who increasingly live in cities. Humans evolved outdoors and today, we still have the innate need to "get away" from our indoor lifestyles. It is thought that the function of houseplants is "bringing the outdoors and its biodiversity inside" since that is what we require, but often do not make the time to fill this need. These services are said to be indirect because they are pervasive and not easily discernible (Fig. 47.4). Even so, our very survival depends on the functions that ecosystems perform for us. Think of the alternative if there was no wild ecosystem to which to escape the rat race.

### *Biogeochemical Cycles*

Ecosystems are characterized by energy flow and chemical cycling. The biodiversity within ecosystems contributes to the workings of the water, carbon, nitrogen, phosphorus, and other biogeochemical cycles. We are dependent on these cycles for fresh water, removal of carbon dioxide from the atmosphere, uptake of excess soil nitrogen, and provision of phosphate. When human activities upset the usual workings of biogeochemical cycles, the environmental consequences include the release of excess pollutants that are harmful to us. Technology is currently unable to artificially contribute to or create any of the biogeochemical cycles.

### *Waste Disposal*

Decomposers break down dead organic matter and other types of wastes to inorganic nutrients that are used by the producers within ecosystems. This function aids humans immensely because we dump millions of tons of waste material into natural ecosystems each year. If it were not for decomposition, waste would soon cover the entire surface of our planet. We can build sewage treatment plants, but they are expensive, and few of them break down solid wastes completely to inorganic nutrients. It

a.

b.

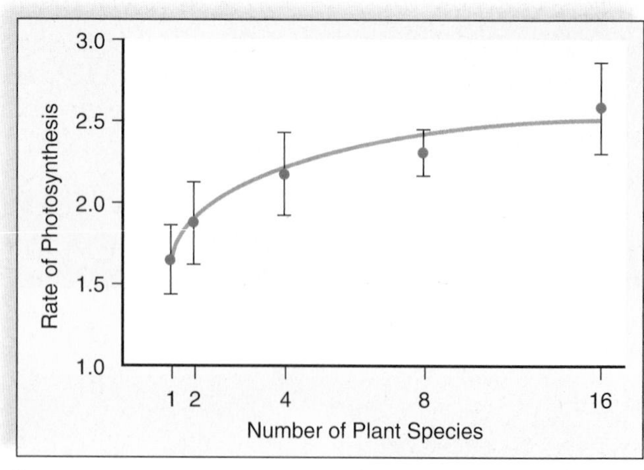


c.

**FIGURE 47.4 Indirect value of ecosystems.**

**a.** Natural ecosystems provide (**b**) human-impacted ecosystems with many ecological services. **c.** Research results show that the higher the biodiversity (measured by number of plant species), the greater the rate of photosynthesis of an experimental community.

is less expensive and more efficient to provide plants and trees with partially treated wastewater and let soil bacteria cleanse it completely.

Biological communities are also capable of breaking down and immobilizing pollutants, such as heavy metals and pesticides, that humans release into the environment. A review of wetland functions in Canada assigned a value of \$50,000 per hectare (100 acres, or 10,000 $m^2$) per year to the ability of natural areas to purify water and take up pollutants.

### *Provision of Fresh Water*

Few terrestrial organisms are adapted to living in a salty environment—they need fresh water. The water cycle continually supplies fresh water to terrestrial ecosystems. Humans use fresh water in innumerable ways, including drinking it and irrigating their crops. Freshwater ecosystems, such as rivers and lakes, also provide us with fishes and other organisms for food.

Unlike other commodities, there is no substitute for fresh water. We can remove salt from seawater to obtain fresh water, but the cost of desalination is about four to eight times the average cost of fresh water acquired via the water cycle.

Forests and other natural ecosystems exert a "sponge effect." They soak up water and then release it at a regular rate. When rain falls in a natural area, plant foliage and dead leaves lessen its impact, and the soil slowly absorbs it, especially if the soil has been aerated by organisms. The water-holding capacity of forests reduces the possibility of flooding. The value of a marshland outside Boston, Massachusetts, has been estimated at \$72,000 per hectare per year solely on its ability to reduce floods. Forests release water slowly for days or weeks after the rains have ceased. Rivers flowing through forests in West Africa release twice as much water halfway through the dry season, and between three and five times as much at the end of the dry season, as do rivers from coffee plantations.

### *Prevention of Soil Erosion*

Intact ecosystems naturally retain soil and prevent soil erosion. The importance of this ecosystem attribute is especially observed following deforestation. In Pakistan, the world's largest dam, the Tarbela Dam, is losing its storage capacity of 12 billion $m^3$ many years sooner than expected because silt is building up behind the dam due to deforestation. At one time, the Philippines were exporting \$100 million worth of oysters, mussels, clams, and cockles each year. Now, silt carried down rivers following deforestation is smothering the mangrove ecosystem that serves as a nursery for the sea. Most coastal ecosystems are not as bountiful as they once were because of deforestation and a myriad of other assaults.

### *Regulation of Climate*

At the local level, trees provide shade and reduce the need for fans and air conditioners during the summer. Proper placement of shade trees near a home can reduce energy bills.

Globally, forests restore the climate because they take up carbon dioxide. The leaves of trees use carbon dioxide when they photosynthesize, the bodies of the trees store carbon, and oxygen is released as a by-product. When trees are cut and burned, carbon dioxide is released into the atmosphere. The reduction in forests reduces the carbon dioxide uptake and the oxygen output. This change in the atmospheric gases, especially greenhouse gases such as $CO_2$, affects the amount of solar radiation retained on Earth's surface. Large scale deforestation may affect the global atmosphere and, in turn, the climate.

### *Ecotourism*

Almost everyone prefers to vacation in the natural beauty of an ecosystem. In the United States, nearly 100 million people enjoy vacationing in a natural setting. To do so, they spend \$4 billion each year on fees, travel, lodging, and food. Many tourists want to go sport fishing, whale watching, boat riding, hiking, birdwatching, and the like. Others want to merely immerse themselves in the beauty and serenity of a natural environment. Many underdeveloped countries in tropical regions are taking advantage of this by offering "ecotours" of the local biodiversity. Providing guided tours of forests is often more profitable than destroying them.

## Biodiversity and Natural Ecosystems

Massive changes in biodiversity, such as deforestation, have a significant impact on ecosystems. Researchers are interested in determining whether a high degree of biodiversity also helps ecosystems function more efficiently. To test the benefits of biodiversity in a Minnesota grassland habitat, researchers sowed plots with seven levels of plant diversity. Their study found that ecosystem performance improves with increasing diversity. A similar study in California also showed greater overall resource use in more diverse plots because of resource partitioning among the plants.

Another group of experimenters tested the effects of an increase in diversity at four levels: producers, herbivores, parasites, and decomposers. They found that the rate of photosynthesis increased as diversity increased (Fig. 47.4*c*). A computer simulation has shown that the response of a deciduous forest to elevated carbon dioxide is a function of species diversity. The more complex community, composed of nine tree species, exhibited a 30% greater amount of photosynthesis than a community composed of a single species.

More studies are needed to test whether biodiversity maximizes resource acquisition and retention within an ecosystem. Also, are more diverse ecosystems better able to withstand environmental changes and invasions by other species, including pathogens? Then, too, how does fragmentation affect the distribution of organisms within an ecosystem and the functioning of an ecosystem?

**Check Your Progress 47.2**

1. Explain the difference between a direct value of biodiversity and an indirect value of biodiversity.

## 47.3 Causes of Extinction

To stem the tide of extinction due to human activities, it is first necessary to identify its causes. Researchers examined the records of 1,880 threatened and endangered wild species in the United States and found that habitat loss was involved in 85% of the cases (Fig. 47.5*a*). Exotic species had a hand in nearly 50%, pollution was a factor in 24%, overexploitation in 17%, and disease in 3%. The percentages add up to more than 100% because most of these species are imperiled for more than one reason. Macaws are a good example that a combination of factors can lead to a species decline (Fig. 47.5*b*). Not only has their habitat been reduced by encroaching timber and mining companies, but macaws are also hunted for food and collected for the pet trade.

### Habitat Loss

Habitat loss has occurred in all ecosystems, but concern has now centered on tropical rain forests and coral reefs because they are particularly rich in species. A sequence of events in Brazil offers a fairly typical example of the manner in which rain forest is converted to land uninhabitable for wildlife. The construction of a major highway into the forest first provided a way to reach the interior of the forest (see Fig. 47.5*b*). Small towns and industries sprang up along the highway, and roads branching off the main highway gave rise to even more roads. The result was fragmentation of the once immense forest. The government offered subsidies to anyone willing to take up residence in the forest, and the people who came cut and burned trees in patches (see Fig. 47.5*b*). Tropical soils contain limited nutrients, but when the trees are burned, nutrients are released that support a lush growth for the grazing of cattle for about three years. However, once the land was degraded (see Fig. 47.5*b*), the farmers moved on to another portion of the forest to start over again.

Loss of habitat also affects freshwater and marine biodiversity. Coastal degradation is mainly due to the large concentration of people living on or near the coast. Already, 60% of coral reefs have been destroyed or are on the verge of destruction; it is possible that all coral reefs may disappear during the next 40 years unless our behaviors drastically change. Mangrove forest destruction is also a problem; Indonesia, with the most mangrove acreage, has lost 45% of its mangroves, and the percentage is even higher for other tropical countries. Wetland areas, estuaries, and seagrass beds are also being rapidly destroyed by the actions of humans.



b. Macaws on salt lick

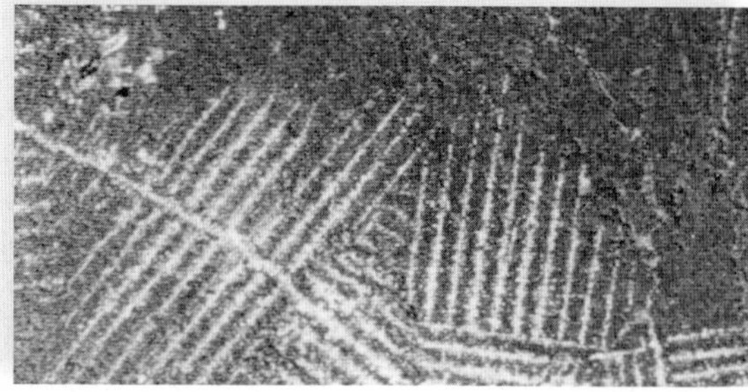

Roads cut through forest

Forest occurs in patches

Destroyed areas

c. Wildlife habitat is reduced.

**FIGURE 47.5 Habitat loss.**

**a.** In a study that examined records of imperiled U.S. plants and animals, habitat loss emerged as the greatest threat to wildlife. **b.** Macaws that reside in South American tropical rain forests are endangered for some of the reasons listed in the graph in (**a**). **c.** Habitat loss due to road construction in Brazil. (*above*) Road construction opened up the rain forest and subjected it to fragmentation. (*middle*) The result was patches of forest and degraded land. (*below*) Wildlife could not live in destroyed portions of the forest.

## Exotic Species

**Exotic species,** sometimes called alien species, are nonnative members of an ecosystem. Ecosystems around the globe are characterized by unique assemblages of organisms that have evolved together in one location. Migrating to a new location is not usually possible because of barriers such as oceans, deserts, mountains, and rivers. Humans, however, have introduced exotic species into new ecosystems in the following ways:

*Colonization* Europeans, in particular, brought various familiar species with them when they colonized new places. For example, the pilgrims brought the dandelion to the United States as a familiar salad green. In addition, they introduced pigs to North America that have since become feral, reverting to their wild state. In some parts of the United States, feral pigs are very destructive.

*Horticulture and agriculture* Some exotics now taking over vast tracts of land have escaped from cultivated areas. Kudzu is a vine from Japan that the U.S. Department of Agriculture thought would help prevent soil erosion. The plant now covers much landscape in the South, including even walnut, magnolia, and sweet gum trees (Fig. 47.6*a*). The water hyacinth was introduced to the United States from South America because of its beautiful flowers. Today, it clogs up waterways and diminishes natural diversity.

*Accidental transport* Global trade and travel accidentally bring many new species from one country to another. Researchers found that the ballast water released from ships into Coos Bay, Oregon, contained 367 marine species from Japan. The zebra mussel from the Caspian Sea was accidentally introduced into the Great Lakes in 1988. It now forms dense beds that squeeze out native mussels. Other organisms accidentally introduced into the United States include the Formosan termite, the Argentinian fire ant, and the nutria, a type of large rodent.

Exotic species can disrupt food webs. As mentioned earlier, opossum shrimp introduced into a lake in Montana added a trophic level that in the end meant less food for bald eagles and grizzly bears (see Fig. 47.2).

### *Exotics on Islands*

Islands are particularly susceptible to environmental discord caused by the introduction of exotic species. Islands have unique assemblages of native species that are closely adapted to one another and cannot compete well against exotics. Myrtle trees, *Myrica faya,* introduced into the Hawaiian Islands from the Canary Islands, are symbiotic with a type of bacterium that is capable of nitrogen fixation. This feature allows the species to establish itself on nutrient-poor volcanic soil, a distinct advantage in Hawaii. Once established, myrtle trees call a halt to the normal succession of native plants on volcanic soil.

The brown tree snake has been introduced onto a number of islands in the Pacific Ocean. The snake eats adult birds, their eggs, and nestlings. On Guam, it has reduced ten native bird species to the point of extinction. On the Galápagos Islands, black rats have reduced populations of giant tortoise, while goats and feral pigs have changed the vegetation from highland forest to pampaslike grasslands and destroyed stands of cacti. In Australia, mice and rabbits have stressed native marsupial populations. Mongooses introduced into the Hawaiian Islands to control rats also prey on native birds (Fig. 47.6*b*).

## Pollution

In the present context, **pollution** can be defined as any environmental change that adversely affects the lives and health of living things. Pollution has been identified as the third main cause of extinction. Pollution can also weaken organisms and lead to disease, the fifth main cause of extinction. Biodiversity is particularly threatened by the following types of environmental pollution:

*Acid deposition* Both sulfur dioxide from power plants and nitrogen oxides in automobile exhaust are converted to acids when they combine with water vapor in the atmosphere. These acids return to Earth as either wet deposition (acid rain or snow) or dry deposition (sulfate and nitrate salts). Sulfur dioxide and nitrogen oxides are emitted in one locale, but deposition occurs across state and national boundaries. Acid deposition causes trees to weaken and increases their susceptibility to disease and insects. It also kills small invertebrates and decomposers so that the entire ecosystem is threatened. Many lakes in the northern United States are now lifeless because of the effects of acid deposition.

*Eutrophication* Lakes are also under stress due to over-enrichment. When lakes receive excess nutrients due to runoff from agricultural fields and wastewater from sewage treatment, algae begin to grow in

a.

b.

**FIGURE 47.6 Exotic species.**

**a.** Kudzu, a vine from Japan, was introduced in several southern states to control erosion. Today, kudzu has taken over and displaced many native plants.

**b.** Mongooses were introduced into Hawaii to control rats, but they also prey on native birds.

abundance. An algal bloom is apparent as a green scum or excessive mats of filamentous algae. Upon death, the decomposers break down the algae, but in so doing, they use up oxygen. A decreased amount of oxygen is available to fish, leading sometimes to a massive fish kill.

*Ozone depletion* The ozone shield is a layer of ozone ($O_3$) in the stratosphere, some 50 km above the Earth. The ozone shield absorbs most of the wavelengths of harmful ultraviolet (UV) radiation so that they do not strike the Earth. The cause of ozone depletion can be traced to chlorine atoms ($Cl^-$) that come from the breakdown of chlorofluorocarbons (CFCs). The best-known CFC is Freon, a heat transfer agent still found in refrigerators and air conditioners today. Severe ozone shield depletion can impair crop and tree growth and also kill plankton (microscopic plant and animal life) that sustain oceanic life. The immune system and the ability of all organisms to resist infectious diseases will most likely be weakened.

*Organic chemicals* Our modern society uses organic chemicals in all sorts of ways. Organic chemicals called nonylphenols are used in products ranging from pesticides to dishwashing detergents, cosmetics, plastics, and spermicides. These chemicals mimic the effects of hormones, and in that way most likely harm wildlife. Salmon are born in fresh water but mature in salt water. After investigators exposed young fish to nonylphenol, they found that 20–30% were unable to make the transition between fresh and salt water. Nonylphenols cause the pituitary to produce prolactin, a hormone that may prevent saltwater adaptation.

*Global warming* The expression **global warming** refers to an expected increase in average temperature during the twenty-first century. You may recall from Chapter 46 that carbon dioxide is a gas that comes from the burning of fossil fuels, and methane is a gas that comes from oil and gas wells, rice paddies, and animals. These gases are known as greenhouse gases because, just like the panes of a greenhouse, they allow solar radiation to pass through but hinder the escape of its heat back into space. Data collected around the world show a steady rise in $CO_2$ concentration. These data are used to generate computer models that predict the Earth may warm to temperatures higher than currently experienced (Fig. 47.7*a*). An upward shift in temperatures could influence everything from growing seasons in plants to migratory patterns in animals.

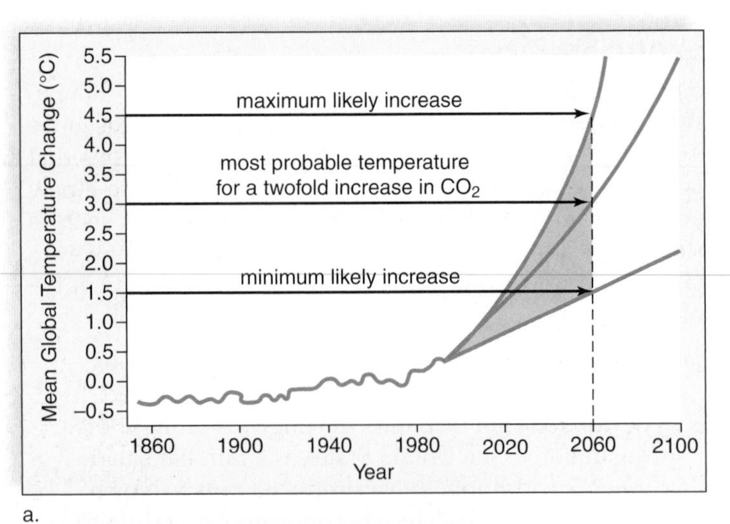


a.

b.

**FIGURE 47.7 Global warming.**

**a.** Mean global temperature is expected to rise due to the introduction of greenhouse gases into the atmosphere. **b.** Global warming has the potential to significantly affect the world's biodiversity distribution. A temperature rise of only a few degrees causes coral reefs to "bleach" and become lifeless. If, in the meantime, migration occurs, coral reefs could move northward.

As temperatures rise, regions of suitable climate for various terrestrial species may shift toward the poles and higher elevations. Extinctions are expected because the present assemblages of species in ecosystems will be disrupted as some species migrate northward, leaving others behind. Plants migrate when seeds disperse, and growth occurs in a new locale. For example, to remain in a favorable habitat, it's been calculated that the rate of beech tree migration would have to be 40 times faster than has ever been observed. It seems unlikely that beech or any other type of tree would be able to meet the pace required. Then, too, many species of organisms are confined to relatively small habitat patches that are surrounded by agricultural or urban areas they would not be able to cross. And even if they have the capacity to disperse to new sites, suitable habitats may not be available. If the global climate changes faster than organisms can migrate, extinction of such species is likely. Other species may experience population increase. For example, parasites and pests that are usually killed by cold winters will now be able to survive in greater numbers. The tropics may very well expand, and whether present-day temperate-zone agriculture will survive is questionable.

*ecology focus*

# Overexploitation of Asian Turtles

## Conservation Alert

Collection and trade of terrestrial tortoises and freshwater turtles for human consumption and other uses has surged in Asia over the past two decades and is now spreading to areas around the globe (Fig. 47A). With 40 to 60% of all types of turtles already endangered, these practices have virtually wiped out many tortoise and freshwater turtle populations from wide areas of Asia and have brought others to the brink of extinction in a matter of years. The wild-collection trade started in Bangladesh in order to supply consumption demands in South China and then it quickly spread across tropical Asia as one area after another became depleted. Currently, the practice has spread to the United States.

Tortoises and turtles mature late (at the order of 10–20+ years), experience great longevity (measured in decades), and have low annual reproductive rates. Traders prefer to collect larger size, and therefore breeding age, individuals. This means that wild populations are not likely to recover after they have been plundered. Presently, the stocking of thousands of new turtle farms (ranging from backyard to industrial scale operations) is also causing a further run on the last remaining wild stocks of many endangered species. Then, too, there is a very active illegal pet trade in rare tortoises and freshwater turtles for wealthy patrons. This illegal pet trade in rare tortoise and freshwater turtle species is now vexing enforcement authorities in Asia, Europe, and the United States.

## Major Challenges Today and in the Future

Basic scientific knowledge about the range, natural history, and conservation needs of individual species of tortoises and turtles is lacking. In fact, some traded species are so poorly known, and so endangered, that specimens have sometimes only been found by researchers in wildlife markets. Wildlife inspectors and enforcement officials are often unaware of conservation concerns and unable to identify turtle species and to associate them with pertinent regulations. Because the wild-collection trade is now worldwide, it is essential that all nations and states with (remaining) wild turtle populations ensure that their domestic legislation is adequate to secure the future of their turtle populations. Too often, trade starts up and occurs faster than regulatory measures can be put in place to prevent local populations from being decimated.

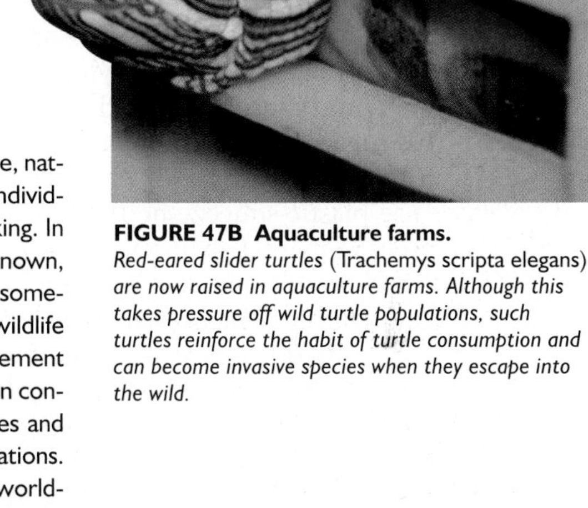

**FIGURE 47B Aquaculture farms.**
*Red-eared slider turtles* (Trachemys scripta elegans) *are now raised in aquaculture farms. Although this takes pressure off wild turtle populations, such turtles reinforce the habit of turtle consumption and can become invasive species when they escape into the wild.*

**FIGURE 47A Turtles for sale.**
*Since the 1990s, great numbers of wild-caught tortoises and freshwater turtles can be seen offered for sale at markets in East Asia. Shown here are steppe tortoises* (Testudo horsefieldii) *(back) and elongated tortoises* (Indotestudo elongata) *(front).*

Another issue that needs attention is the threat of invasive species and the spread of diseases from aquaculture facilities to wild populations. It is known that Chinese soft-shell turtles and red-eared slider turtles have an extreme ability to adapt to new environments (Fig. 47B). Reports abound from around the globe that turtles have either escaped or have been deliberately released from aquaculture facilities. Ecologists can only speculate about how they will impact wild populations when they become competitors, predators, hybridizers, and disease vectors. Impact studies are desirable, as is a regulatory framework governing the transport and handling of non-local turtles.

The public needs to become aware and concerned about the plight of tortoises and freshwater turtle populations because serious conservation research and legislative action is immediately required in order to save local populations from extinction.

Written by Peter Paul van Dijk with the assistance of Bruce J. Weissgold.

## Overexploitation

**Overexploitation** occurs when the number of individuals taken from a wild population is so great that the population becomes severely reduced in numbers. A positive feedback cycle explains overexploitation: The smaller the population, the more valuable its members, and the greater the incentive to capture the few remaining organisms. Poachers are very active in the collecting and sale of endangered and threatened species because it has become so lucrative. The overall international value of trading wildlife species is $20 billion, of which $8 billion is attributed to the illegal sale of rare species.

Markets for rare plants and exotic pets support both legal and illegal trade in wild species. Rustlers dig up rare cacti, such as the crested saguaros, and sell them to gardeners for as much as $15,000 each. Parrots are among birds taken from the wild for sale to pet owners. For every bird delivered alive, many more have died in the process. The same holds true for tropical fish, which often come from the coral reefs of Indonesia and the Philippines. Divers dynamite reefs or use plastic squeeze-bottles of cyanide to stun them; in the process, many fish and valuable corals die.

The Convention of International Trade of Endangered Species (CITES) was an agreement established in 1973 to ensure that international trade of species does not threaten their survival. Today, over 30,000 species of plants and animals receive some level of protection from over 172 countries worldwide.

Poachers still hunt for hides, claws, tusks, horns, or bones of many endangered mammals. Because of its rarity, a single Siberian tiger is now worth more than $500,000—its bones are pulverized and used as a medicinal powder. The horns of rhinoceroses become ornate carved daggers, and their bones are ground up to sell as a medicine. The ivory of an elephant's tusk is used to make art objects, jewelry, or piano keys. The fur of a Bengal tiger sells for as much as $100,000 in Tokyo.

The U.N. Food and Agricultural organization tells us that we have now overexploited 11 of 15 major oceanic fishing areas. Fish are a renewable resource if harvesting does not exceed the ability of the fish to reproduce. Our society uses larger and more efficient fishing fleets to decimate fishing stocks. Pelagic species such as tuna are captured by purse-seine fishing, in which a very large net surrounds a school of fish, and then the net is closed in the same manner as a drawstring purse. Up to thousands of dolphins that swim above schools of tuna are often captured and then killed in this type of net. However, many tuna suppliers advertise their product as "dolphin safe." Other fishing boats drag huge trawling nets, large enough to accommodate 12 jumbo jets, along the seafloor to capture bottom-dwelling fish (Fig. 47.8*a*). Only large fish are kept; undesirable small fish and sea turtles are discarded, dying, back into the ocean. Trawling has been called the marine equivalent of clear-cutting trees because after the net goes by, the sea bottom is devastated (Fig. 47.8*b*). Today's fishing practices don't allow fisheries to recover. Cod and haddock, once the most abundant bottom-dwelling fish along the northeast coast of the United States, are now often outnumbered by dogfish and skate.

a. Fishing by use of a drag net

b. Result of drag net fishing

**FIGURE 47.8 Trawling.**

**a.** These Alaskan pollock were caught by dragging a net along the seafloor.
**b.** Appearance of the seafloor after the net passed.

A marine ecosystem can be disrupted by overfishing, as exemplified on the U.S. west coast. When sea otters began to decline in numbers, investigators found that they were being eaten by orcas (killer whales). Usually orcas prefer seals and sea lions to sea otters, but they began eating sea otters when few seals and sea lions could be found. What caused a decline in seals and sea lions? Their preferred food sources—perch and herring—were no longer plentiful due to overfishing. Ordinarily, sea otters keep the population of sea urchins, which feed on kelp, under control. But with fewer sea otters around, the sea urchin population exploded and decimated the kelp beds. Thus, overfishing set in motion a chain of events that detrimentally altered the food web of an ecosystem.

### Check Your Progress 47.3

1. What are the five main causes of extinction?
2. Explain why the introduction of exotic species can be detrimental to biodiversity with reference to Figures 47.6 and 47B.

# 47.4 Conservation Techniques

Despite the value of biodiversity to our very survival, human activities are causing the extinction of thousands of species a year. Clearly, we need to reverse this trend and preserve as many species as possible. Habitat preservation and restoration are important in preserving biodiversity.

## Habitat Preservation

Preservation of a species' habitat is of primary concern, but first we must prioritize which species to preserve. As mentioned previously, the biosphere contains biodiversity hotspots, relatively small areas having a concentration of endemic (native) species not found anyplace else. In the tropical rain forests of Madagascar, 93% of the primate species, 99% of the frog species, and over 80% of the plant species are endemic to Madagascar. Preserving these forests and other hotspots will save a wide variety of organisms.

**Keystone species** are species that influence the viability of a community, although their numbers may not be excessively high. The extinction of a keystone species can lead to other extinctions and a loss of biodiversity. For example, bats are designated a keystone species in tropical forests of the Old World. They are pollinators that also disperse the seeds of trees. When bats are killed off and their roosts destroyed, the trees fail to reproduce. The grizzly bear is a keystone species in the northwestern United States and Canada (Fig. 47.9*a*). Bears disperse the seeds of berries; as many as 7,000 seeds may be in one dung pile. Grizzly bears kill the young of many hoofed animals and thereby keep their populations under control. Grizzly bears are also a principal mover of soil when they dig up roots and prey upon hibernating ground squirrels and marmots. Other keystone species are beavers in wetlands, bison in grasslands, alligators in swamps, and elephants in grasslands and forests.

Keystone species should not be confused with **flagship species,** which evoke a strong emotional response in humans. Flagship species are considered charismatic and are treasured for their beauty, cuteness, and regal nature. These species can motivate the public to preserve biodiversity. Flagship species include lions, tigers, dolphins, and the giant panda.

a. Grizzly bear, *Ursus arctos horribilis*

b. Old-growth forest; northern spotted owl, *Strix occidentalis caurina* (inset)

**FIGURE 47.9**
**Habitat preservation.**

When particular species are protected, other wildlife benefits. **a.** The Greater Yellowstone Ecosystem has been delineated in an effort to save grizzly bears, which need a very large habitat. **b.** Currently, the remaining portions of old-growth forests in the Pacific Northwest are not being logged in order to save the northern spotted owl (inset).

### *Metapopulations*

The grizzly bear population is actually a **metapopulation** [Gk. *meta,* between; L. *populus,* people], a population subdivided into several small, isolated populations due to habitat fragmentation. Originally there were probably 50,000–100,000 grizzly bears south of Canada, but this number has been reduced because human housing communities have encroached on their home range and bears are killed by frightened homeowners. Now there are six virtually isolated subpopulations totaling about 1,000 individuals. The Yellowstone National Park population numbers 200, but the others are even smaller.

Saving metapopulations sometimes requires determining which of the populations is a source and which are sinks. A **source population** is one that most likely lives in a favorable area, and its birthrate is most likely higher than its death rate. Individuals from source populations move into **sink populations,** where the environment is not as favorable and where the birthrate equals the death rate at best. When trying to save the northern spotted owl, conservationists determined that it was best to avoid having owls move into sink habitats. The northern spotted owl reproduces successfully in old-growth rain forests of the Pacific Northwest (Fig. 47.9*b*) but not in nearby immature forests that are in the process of recovering from logging. Distinct boundaries that hindered the movement of owls into these sink habitats proved to be beneficial in maintaining source populations.

### *Landscape Preservation*

Grizzly bears inhabit a number of different types of ecosystems, including plains, mountains, and rivers. Saving any one of these types of ecosystems alone would not be sufficient to preserve the grizzly bears. Instead, it is necessary to save diverse ecosystems that are at least connected. You will recall that a landscape encompasses different types of ecosystems. An area called the Greater Yellowstone Ecosystem, where bears are free to roam, has now been defined. It contains millions of acres in Yellowstone National Park; state lands in Montana, Idaho, and Wyoming; five different national forests; various wildlife refuges; and even private lands.

Landscape protection for one species is often beneficial for other wildlife that share the same space. The last of the contiguous 48 states' harlequin ducks, bull trout, westslope cutthroat trout, lynx, pine martens, wolverines, mountain caribou, and great gray owls are found in areas occupied by grizzly bears. Gray wolves have also recently returned to this territory. Then, too, grizzly bear range overlaps with 40% of Montana's vascular plants of special conservation concern.

**The Edge Effect.** When preserving landscapes, it is necessary to consider the **edge effect.** An edge reduces the amount of habitat typical of an ecosystem because the edges around a patch have a habitat slightly different from the interior of the patch. For example, forest edges are brighter, warmer, drier, and windier, with more vines, shrubs, and weeds than the forest interior. Also, Figure 47.10*a* shows that a small and a large patch of habitat have the same amount of edge; therefore, the effective habitat shrinks as a patch gets smaller.

Many popular game animals, such as turkeys and white-tailed deer, are more plentiful in the edge region of a particular area. However, today it is known that creating edges can be detrimental to wildlife because of habitat fragmentation.

The edge effect can also have a serious impact on population size. Songbird populations west of the Mississippi have been declining of late, and ornithologists have noticed that the nesting success of songbirds is quite low at the edge of a forest. The cause turns out to be the brown-headed cowbird, a social parasite of songbirds. Adult cowbirds prefer to feed in open agricultural areas, and they only briefly enter the forest when searching for a host nest in which to lay their eggs (Fig. 47.10*b*). Cowbirds are therefore benefited, while songbirds are disadvantaged, by the edge effect.

## Habitat Restoration

**Restoration ecology** is a new subdiscipline of conservation biology that seeks scientific ways to return ecosystems to their state prior to habitat degradation. Three principles have so far emerged. First, it is best to begin as soon as possible before remaining fragments of the original habitat are lost. These fragments are sources of wildlife and seeds from which to restock the restored habitat. Second, once the natural histories of the species in the habitat are understood, it is best to use biological techniques that mimic natural processes to bring about restoration. This might take the form of using controlled burns to bring back grassland habitats, biological pest controls to rid the area of exotic species, or bioremediation techniques to clean up pollutants. Third, the goal is **sustainable development,** the ability of an ecosystem to maintain itself while providing services to human beings. The Everglades ecosystem is used here to illustrate these principles. Although habitat restoration is good, there is some concern that the restored areas may not be functionally equivalent to the natural regions.

### *The Everglades*

Originally, the Everglades encompassed the whole of southern Florida from Lake Okeechobee down to Florida Bay (Fig. 47.11*a*). This ecosystem is a vast sawgrass prairie, interrupted occasionally by a cypress dome or hardwood tree island. Within these islands, both temperate and tropical evergreen trees grow amongst dense and tangled vegetation. Mangroves are found along sloughs (creeks) and at the shoreline. The

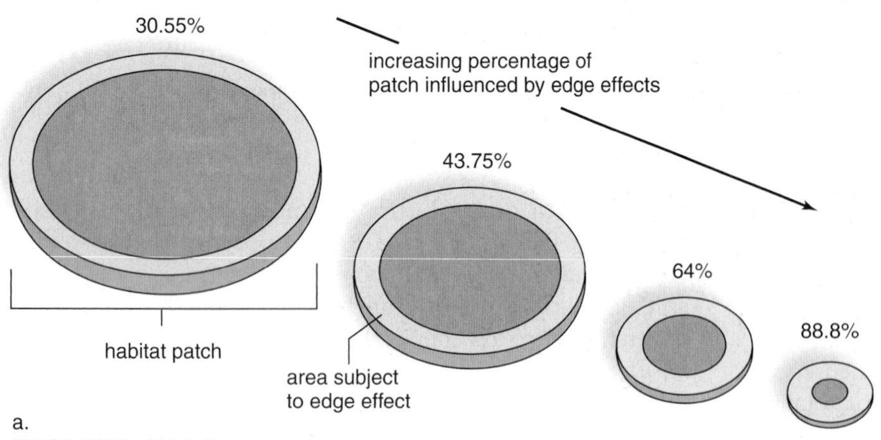

**FIGURE 47.10 Edge effect.**

**a.** The smaller the patch, the greater the proportion that is subject to the edge effect. **b.** Cowbirds lay their eggs in the nests of songbirds (yellow warblers). A cowbird is bigger than a warbler nestling and will be able to acquire most of the food brought by the warbler parent.

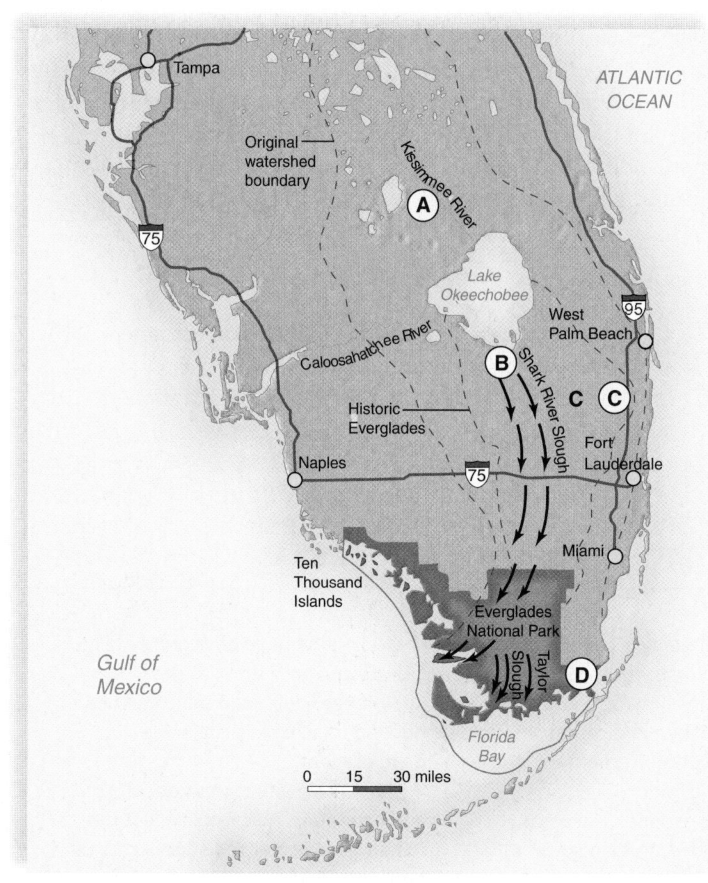


a. Location of Everglades National Park (purple)

Florida panther, *Puma concolor coryi*

American alligator, *Alligator mississippiensis*

White ibis, *Eudocimus albus*

Roseate spoonbill, *Ajaia ajaja*

Wood stork, *Mycteria americana*

b. Wildlife in Everglades

**FIGURE 47.11**
**Restoration of the Everglades.**

**a.** Restoration plans call for adding curves and habitat back to the Kissimmee River (A); creating large marshes and making the Shark River Slough free-flowing again (B); creating a buffer zone of wetlands between urban development along Florida's eastern coast (C); and reducing salinity by letting fresh water flow into and through Taylor Slough (D). **b.** Wildlife of the Florida Everglades.

prop roots of red mangroves protect over 40 different types of juvenile fishes as they grow to maturity. During a wet season, from May to November, animals disperse throughout the region, but in the dry season, from December to April, they congregate wherever pools of water are found. Alligators are famous for making "gator holes," where water collects and fishes, shrimp, crabs, birds, and a host of living things survive until the rains come again. The Everglades once supported millions of large and colorful birds, including herons, egrets, the white ibis, and the roseate spoonbill (Fig. 47.11*b*).

At the turn of the century, settlers began to drain the land just south of Lake Okeechobee to grow crops in the newly established Everglades Agricultural Area (EAA). A large dike now rings Lake Okeechobee and prevents water from overflowing its banks and moving slowly southward. To provide flood protection for urban development, water is shunted through the St. Lucie Canal to the Atlantic Ocean or through the canalized Caloosahatchee River to the Gulf of Mexico. In times of drought, water is contained not only in the lake but also in three conservation areas established to the south of the lake. Water must be conserved to irrigate the farmland and to recharge the Biscayne aquifer (underground river), which supplies drinking water for the cities on the east coast of Florida. The Central and Southern Florida Flood Control Project (C&SF) included the construction of over 2,250 km of canals, 125 water control stations, and 18 large pumping stations. Now the Everglades National Park receives water only when it is discharged artificially from a conservation area, and the discharge is according to the convenience of the C&SF rather than according to the natural wet/dry season of southern Florida. Largely because of this, the Everglades are now dying, as witnessed by declining bird populations. The birds, which used to number in the millions, now number in the thousands.

**Restoration Plan.** A restoration plan has been developed that will sustain the Everglades ecosystem while maintaining the services society requires. The U.S. Army Corps of Engineers is to redesign the C&SF so that the Everglades receive a more natural flow of water from Lake Okeechobee. This will require flooding the EAA and growing only crops such as sugarcane and rice that can tolerate these wetter conditions. This has the benefit of stopping the loss of topsoil and preventing possible residential development in the area. There will also be an extended buffer zone between an expanded Everglades and the urban areas on Florida's east coast. The buffer zone will contain a contiguous system of interconnected marsh areas, detention reservoirs, seepage barriers, and water treatment areas. This plan is expected to stop the decline of the Everglades and its biodiversity, while still allowing agriculture to continue and providing water and flood control to the eastern coast. Sustainable development will maintain the ecosystem indefinitely and still meet human needs.

### Check Your Progress 47.4

1. Why is landscape preservation, rather than ecosystem preservation, of primary importance?
2. What are the three principles of habitat restoration?

## Connecting the Concepts

Our industrial societies are overusing the environment to the point of exhaustion. Forests throughout tropical, temperate, and subarctic regions are being harvested and cut for timber at unsustainable rates. Urban sprawl is completely replacing natural ecosystems in highly populated regions. Fresh waters are being diverted for agricultural and urban uses to the extent that riverbeds and lake beds are becoming dry in places. Dams are being constructed for hydropower and irrigation with little consideration for their impact on aquatic life. Exotic species of plants, animals, and microbes are being released into new environments with little or no restraint. Marine fisheries are being exploited by major fishing nations at unsustainable levels.

All these actions, and others, are reducing biodiversity, which we now realize is a resource of enormous economic value. If properly managed, sustainable yields of food and fiber can be obtained from many natural lands and waters. Modern genetic engineering technologies make the genes of millions of wild species available for use in breeding improved crops, domestic animals, and biological control agents. Enjoyment of nature appears to be an innate requirement of a healthy human life.

As natural forests, grasslands, streams, lakes, and seas are degraded, human society must expend greater amounts of nonrenewable energy and materials to substitute for benefits that biodiversity provides at no cost. Lost species, and ultimately, lost ecosystems, cannot be replaced. Biodiversity is therefore a nonrenewable resource. The goal of conservation biology is to protect, restore, and use this resource wisely. To that end, the vision of conservation biology is:

> A world where leaders are committed to long-term environmental protection and to international leadership and cooperation in addressing the world's environmental problems.
>
> A world with an environmentally literate citizenry that has the knowledge, skills, and ethical values needed to achieve sustainable development.
>
> A world in which market prices and economic indicators reflect the full environmental and social costs of human activities.
>
> A world in which a new generation of technologies contributes to the conservation of resources and the protection of the environment.
>
> A world landscape that sustains natural systems, maximizes biological diversity, and uplifts the human spirit.
>
> A world in which human numbers are stabilized, all people enjoy a decent standard of living through sustainable development, and the global environment is protected for future generations.

*Modified from The Report of the National Commission on the Environment, 1993.*


## summary

### 47.1 Conservation Biology and Biodiversity

Conservation biology is the scientific study of biodiversity and its management for sustainable human welfare. The unequaled present rate of extinctions has drawn together scientists and environmentalists in basic and applied fields to address the problem.

Biodiversity is the variety of life on Earth; the exact number of species is not known, but there are assuredly many species yet to be discovered and recognized. Biodiversity must also be preserved at the genetic, community (ecosystem), and landscape levels of organization.

Conservationists have discovered that biodiversity is not evenly distributed in the biosphere, and therefore saving particular areas may protect more species than saving other areas.

### 47.2 Value of Biodiversity

The direct value of biodiversity is seen in the observable services of individual wild species. Wild species are our best source of new medicines to treat human ills, and they help meet other medical needs. For example, the bacterium that causes leprosy grows naturally in armadillos, and horseshoe crab blood contains a bacteria-fighting substance.

Wild species have agricultural value. Domesticated plants and animals are derived from wild species, and they use wild species as a source of genes for the improvement of their phenotypes. Instead of pesticides, wild species can be used as biological controls, and most flowering plants benefit from animal pollinators. Much of our food, particularly fish and shellfish, is still caught in the wild. Hardwood trees from natural forests supply us with lumber for various purposes, such as making furniture.

The indirect services provided by ecosystems are largely unseen and difficult to quantify but absolutely necessary to our well-being. These services include the workings of biogeochemical cycles, waste disposal, provision of fresh water, prevention of soil erosion, and regulation of climate. Many people enjoy vacationing in natural settings. Various studies show that more diverse ecosystems function better than less diverse systems.

### 47.3 Causes of Extinction

Researchers have identified the major causes of extinction. Habitat loss is the most frequent cause, followed by introduction of exotic species, pollution, overexploitation, and disease. (Pollution often leads to disease, so these were discussed at the same time.) Habitat loss has occurred in all parts of the biosphere, but concern has now centered on tropical rain forests and coral reefs, where biodiversity is especially high. Exotic species have been introduced into foreign ecosystems through colonization, horticulture and agriculture, and accidental transport. Various causes of pollution include fertilizer runoff, industrial emissions, and improper disposal of wastes, among others. Overexploitation is exemplified by commercial fishing, which is so efficient that fisheries of the world are collapsing.

### 47.4 Conservation Techniques

To preserve species, it is necessary to preserve their habitat. Some emphasize the need to preserve biodiversity hotspots because of their richness. Often today it is necessary to save metapopulations because of past habitat fragmentation. If so, it is best to determine the source populations and save those instead of the sink populations. A keystone species such as the grizzly bear requires the preservation of a landscape consisting of several types of ecosystems over millions of acres of territory. Obviously, in the process, many other species will also be preserved.

Conservation today is assisted by two types of computer analysis. A gap analysis tries for a fit between biodiversity concentrations and land still available to be preserved. A population viability analysis indicates the minimum size of a population needed to prevent extinction from happening.

Since many ecosystems have been degraded, habitat restoration may be necessary before sustainable development is possible. Three principles of restoration are (1) start before sources of wildlife and seeds are lost; (2) use simple biological techniques that mimic natural processes; and (3) aim for sustainable development so that the ecosystem fulfills the needs of humans.

## understanding the terms

biodiversity 890
biodiversity hotspot 891
bioinformatics 890
conservation biology 890
ecosystem diversity 891
edge effect 902
endangered species 890
exotic species 897
flagship species 901
genetic diversity 891
global warming 898
keystone species 901
landscape 891
landscape diversity 891
metapopulation 901
overexploitation 900
pollution 897
restoration ecology 902
sink population 901
source population 901
sustainable development 902
threatened species 890

Match the terms to these definitions:

a. ______________ A rather small area with an unusually large concentration of species.
b. ______________ A subdivided population in isolated patches of habitat.
c. ______________ A population that has a positive growth rate and net emigration rate to other locations.
d. ______________ Species that are likely to become endangered in the foreseeable future.

## reviewing this chapter

1. Explain these attributes of conservation biology: (a) both academic and applied; (b) supports ethical principles; and (c) is responding to a biodiversity crisis. 890
2. Discuss the conservation of biodiversity at the species, genetic, community, and landscape levels. 890–91
3. Describe the uneven distribution of diversity in the biosphere. What is the implication of uneven distribution for conservation biologists? 891
4. List various ways in which individual wild species provide us with valuable services. 892–93
5. List various ways in which ecosystems provide us with indispensable services. 894–95
6. List and discuss the five major causes of extinction, starting with the most frequent cause. 896–900
7. Introduction of exotic species is usually due to what events? 897
8. List and discuss five major types of pollution that particularly affect biodiversity. 897–98
9. Use the positive feedback cycle to explain why overexploitation occurs. 900
10. Using the grizzly bear population as an example, explain keystone species, metapopulations, and landscape preservation. 901–2
11. Explain the three principles of habitat restoration with reference to the Everglades. 902–4

## testing yourself

Choose the best answer for each question.

1. Which of these would not be within the realm of conservation biology?
   a. helping to manage a national park
   b. a government board charged with restoring an ecosystem
   c. writing textbooks and/or popular books on the value of biodiversity
   d. introducing endangered species back into the wild
   e. All of these are concerns of conservation biology.
2. Which of these pairs does not show a contrast in the number of species?
   a. temperate zone—tropical zone
   b. hotspots—cold spots
   c. rain forest canopy—rain forest floor
   d. pelagic zone—abyssal plain
3. The value of wild pollinators to the U.S. agricultural economy has been calculated to be $4.1–$6.7 billion a year. What is the implication?
   a. Society could easily replace wild pollinators by domesticating various types of pollinators.

b. Pollinators may be valuable, but that doesn't mean any other species also provide us with valuable services.
c. If we did away with all natural ecosystems, we wouldn't be dependent on wild pollinators.
d. Society doesn't always appreciate the services that wild species provide naturally and without any fanfare.
e. All of these statements are correct.

4. The services provided to us by ecosystems are unseen. This means
a. they are not valuable.
b. they are noticed particularly when the service is disrupted.
c. biodiversity is not needed for ecosystems to keep functioning as before.
d. we should be knowledgeable about them and protect them.
e. Both b and d are correct.

5. Which of these is a true statement?
a. Habitat loss is the most frequent cause of extinctions today.
b. Exotic species are often introduced into ecosystems by accidental transport.
c. Climate change may cause many extinctions but also expand the ranges of other species.
d. Overexploitation of fisheries could very well lead to a complete collapse of the fishing industry.
e. All of these statements are true.

6. Which of these is expected if the average global temperature increases?
a. the inability of some species to migrate to cooler climates as environmental temperatures rise
b. the possible drowning of coral reefs
c. an increase in the number of parasites in the temperate zone
d. some species will experience a population decline and others will experience an increase.
e. All of these are expected.

7. Why is a grizzly bear a keystone species existing as a metapopulation?
a. Grizzly bears require many thousands of miles of preserved land because they are large animals.
b. Grizzly bears have functions that increase biodiversity, but presently the population is subdivided into isolated subpopulations.
c. When grizzly bears are present, so are many other types of species within a diverse landscape.
d. Grizzly bears are a source population for many other types of organisms across several population types.
e. All of these statements are correct.

8. Sustainable development of the Everglades will mean that
a. the various populations that make up the Everglades will continue to exist indefinitely.
b. human needs will also be met while successfully managing the ecosystem.
c. the means used to maintain the Everglades will mimic the processes that naturally maintain the Everglades.
d. the restoration plan is a workable plan.
e. All of these statements are correct.

9. Which statement accepted by conservation biologists best shows that they support ethical principles?
a. Biodiversity is the variety of life observed at various levels of biological organization.
b. Wild species directly provide us with all sorts of goods and services.
c. New technologies can help determine conservation plans.
d. There are three principles of restoration biology that need to be adhered to in order to restore ecosystems.
e. Biodiversity is desirable and has value in and of itself, regardless of any practical benefit.

10. A population in an unfavorable area with a high infant mortality rate would be
a. a metapopulation.
b. a source population.
c. a sink population.
d. a new population.

11. What is the edge effect?
a. More species live near the edge of an ecosystem, where more resources are available to them.
b. New species originate at the edge of ecosystems due to interactions with other species.
c. The edge of an ecosystem is not a typical habitat and may be an area where survival is more difficult.
d. More species are found at the edge of a rain forest due to deforestation of the forest interior.

12. Eagles and bears feed on spawning salmon. If shrimp are introduced that compete with salmon for food,
a. the salmon population will decline.
b. the eagle and bear populations will decline.
c. only the shrimp population will decline.
d. all populations will increase in size.
e. Both a and b are correct.

13. Biodiversity hotspots
a. have few populations because the temperature is too hot.
b. contain about 20% of the Earth's species even though their area is small.
c. are always found in tropical rain forests and coral reefs.
d. are sources of species for the ecosystems of the world.
e. All except a are correct.

14. Consumptive use value
a. means we should think of conservation in terms of the long run.
b. means we are placing too much emphasis on living things that are useful to us.
c. means some organisms, other than crops and farm animals, are valuable as products.
d. is a type of direct value.
e. Both c and d are correct.

15. Which of these is not an indirect value of species?
a. participates in biogeochemical cycles
b. participates in waste disposal
c. helps provide fresh water
d. prevents soil erosion
e. All of these are indirect values.

16. Most likely, ecosystem performance improves
a. the more diverse the ecosystem.
b. as long as selected species are maintained.
c. as long as species have both direct and indirect value.
d. if extinctions are diverse.
e. Both a and b are correct.

17. Efforts made to preserve species include all except
a. catching tuna in dolphin-safe nets.
b. international regulation of the trade of endangered species.
c. captive breeding of popular plants and animals.
d. fragmenting habitats, producing metapopulations.
e. All of these are examples of conservation efforts.

18. Sea urchins feed on kelp beds, and sea otters feed on sea urchins. If sea otters are killed off, which of these statement(s) is true? Choose more than one answer if correct.
    a. Kelp beds will increase.
    b. Kelp beds will decrease.
    c. Sea urchins will increase.
    d. Sea urchins will decrease.
19. Complete the following graph by labeling each bar (a–e) with a cause of extinction, from the most influential to the least influential.

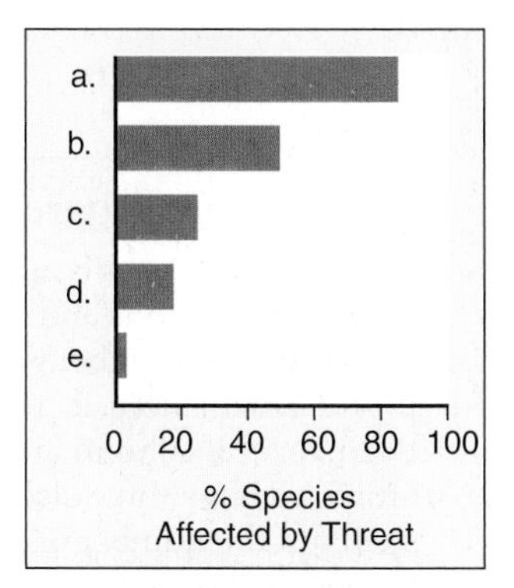


## thinking scientifically

1. The scale at which conservation biologists work often makes direct experimentation difficult. But computer models can assist in predicting the future of populations or ecosystems. Some scientists feel that these models are inadequate because they cannot reproduce all the variables found in the real world. If you were trying to predict the impact on songbirds of clear-cutting a portion of a forest, what information would you need to develop a good model?
2. Bioprospecting is the search for medically useful molecules derived from living things. The desire for monetary and medicinal gains from such discoveries deters the preservation of endangered habitats. However, bioprospecting does protect ecosystems and, in this way, saves many species rather than individual species. What types of living things would bioprospectors be most interested in?

## bioethical issue

### Protecting Bighorn Sheep

To protect the state's declining bighorn sheep populations, the New Mexico Game Commission approved a plan that calls for killing scores of mountain lions over several years. The commission pointed out that the lions have killed 36 of 43 radio-collared bighorn sheep released into the wild since 1996 and have increasingly turned to killing sheep as the state's deer herd has declined.

Mountain lions and bighorns have long coexisted as part of a common ecosystem. Rather than killing mountain lions to increase the bighorn sheep population, it would be better to consider a restoration management plan that would improve the ecosystem. The state recently engaged the Hornocker Wildlife Institute of Idaho to study the situation. The institute's report concluded that the number of lions does not necessarily affect the size of the sheep population. Rather, diseases from domestic livestock grazing in the area are responsible for more than 50% of the deaths in some bighorn sheep populations.

Opposition to the restoration plan claims that the plan to kill mountain lions was solidly based in science. The report said that killing lions in selected areas will give the sheep population a chance to rebound because there are 2,000 lions and only 760 bighorn sheep in two separate populations. Do you approve of taking steps to protect bighorn sheep from mountain lions? Why or why not? If you do approve, how would you proceed?

## *Biology* website

The companion website for *Biology* provides a wealth of information organized and integrated by chapter. You will find practice tests, animations, videos, and much more that will complement your learning and understanding of general biology.

**http://www.mhhe.com/maderbiology10**

# APPENDIX A

## Answer Key

### CHAPTER 1

Figure 2.6: Because carbon has only 6 electrons, while phosphorus has 15 and sulfur has 16.

#### Check Your Progress

**1.1:** 1. Acquisition and use of energy, responding to the environment, reproduction, and adaptation. 2. Viruses are not "alive" because they do not meet all the criteria for a living thing. 3. Cacti possess adaptations, such as a thick cuticle, that help prevent it from drying out from transpiration, and have deep root systems designed to tap water deep underground. **1.2:** 1. Domain, Kingdom, Phylum, Class, Order, Family, Genus, Species. 2. Protists are generally unicellular organisms; fungi are multicellular filamentous organisms that absorb food; plants are multicellular organisms that are usually photosynthetic; animals are multicellular organisms that must ingest and process their food. 3. Members of a population that have new adaptive traits are selected to reproduce more than other members and pass on these traits to the next generation. **1.3:** 1. Individuals of a community interact with members of the same species, with other populations, and with the physical environment. 2. Human activities may destroy ecosystems by altering the physical environment, much like agricultural runoff may fill in streams, ponds, and lakes; 3. They are more vulnerable, because high biodiversity ecosystems often depend on species with very specific roles that are more vulnerable to changes in the environment. **1.4:** 1. A control experiment provides a comparison for the investigator to use in determining whether or not manipulation of the independent variable in the experiment affects the outcome. 2. A model may simplify a complex living organism, so there may be other factors in the living organism that would affect the experiment that are not accounted for in the model. Sometimes lax peer review occurs, but in other cases new and novel ideas may be hindered.

#### Understanding the Terms

**a.** metabolism; **b.** evolution; **c.** experimental variable; **d.** photosynthesis; **e.** control

#### Testing Yourself

**1.** c; **2.** Brain thinks but nerve cells do not think; **3.** b; **4.** b; **5.** b; **6.** d; **7.** b; **8.** b; **9.** b; **10.** a; **11.** c; **12.** e; **13.** b; **14.** b; **15.** Energy is brought into an ecosystem for the first time through photosynthesis. The sun provides energy for photosynthesis. **16.** Excess carbon dioxide emitted may alter the amount of it available from its reservoirs, such as from the atmosphere. This excess carbon dioxide is more readily available to photosynthetic plants and protists, possibly causing them to overpopulate. **17.** Not necessarily, if the new organism does not serve as prey for one species, allowing them to overpopulate, nor prey on or compete with existing species in the ecosystem.

#### Thinking Scientifically

**1. a.** After dye is spilled on culture plate, bacteria live despite exposure to sunlight; **b.** Dye protects bacteria against death by UV light; **c.** Expose two sets of plates to UV light: one set of plates contains bacteria and dye, and the other set contains only bacteria. The bacteria in both sets die; **d.** Rejects hypothesis because dye does not protect bacteria against death by UV light. **2.** Treat both groups the same, except one group receives name-brand fertilizer and another group receives the generic brand. At the end of the growing season, weigh the fruit that you harvest from each plot and compare the total weight of the name-brand treatment with that of the generic treatment. **3.** Experimental variable is the drug: one group of patients receives the drug and one doesn't. Responding variable is the results of the experiment.

### CHAPTER 2

#### Check Your Progress

**2.1:** 1. Atomic mass is approximately the sum of the protons and neutrons in an atom. Atomic number is the number of protons in an atom. 2. a. In Group 3 (the third vertical column in the table), the atoms all have three electrons in the outer shell but the number of shells increases by one; b. In a period, such as Period III, the number of shells remains the same but the number of electrons in the outer shell sequentially increase by one. 3. Uses include imaging body parts (e.g., PET scans, thyroid imaging), sterilization of medical equipment, cancer therapy, and increased storage life of produce.
**2.2:** 1. An ionic bond is created when one atom gives up electron(s) and another gains electron(s) so that both atoms have outer shells filled. A covalent bond is formed when two atoms share electrons to fill their outer shells. 2. Calcium gives away its two outer electrons in order to have a complete outer shell. This will give calcium two more protons than electrons. 3. By sharing with four hydrogen atoms, carbon acquires the four electrons it needs to fill its outer shell. Hydrogen has only one shell, which is complete with two electrons; therefore, each of four hydrogens can acquire a needed electron by sharing with carbon.
**2.3:** 1. When water changes to a gas, it takes much energy to break hydrogen bonds. 2. Body heat is being used to evaporate water. 3. Water freezes from the top down because ice is less dense than water. It's less dense because hydrogen bonds become more open when water freezes. **2.4:** 1. An acid dissociates in water to release hydrogen ions. A base either takes up hydrogen ions or releases hydroxide ions. 2. Both trees and fish die when their environment becomes too acid. 3. The more $H^+$, the lower the pH; the fewer $H^+$, the higher pH.

#### Understanding the Terms

**a.** polar covalent bond; **b.** ion; **c.** acid; **d.** molecule; **e.** buffer

## Testing Yourself

**1.** d; **2.** e; **3.** c; **4.** d; **5.** a; **6.** d; **7.** e; **8.** d; **9.** b; **10.** c; **11.**

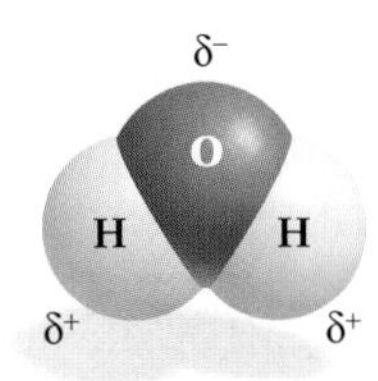


**12.** c; **13.** b; **14.** a; **15.** a; **16.** 7p and 7n in nucleus; two electrons in first shell, and five electrons in outer shell. This means that nitrogen needs three more electrons in the outer shell to be stable; because each hydrogen contributes one electron, the formula for ammonia in $NH_3$; **17.** d; **18.** e; **19.** e; **20.** b; **21.** c; **22.** d; **23.** a; **24.** c; **25.** b

## Thinking Scientifically

**1.** $Na^+Cl^-$ interrupts hydrogen bonding enough to prevent the formation of the ice lattice that forms during freezing. **2.** Chemical behavior is dependent on the number of electrons in the outer shell, not the number of neutrons in the nucleus.

# CHAPTER 3

## Check Your Progress

**3.1:** 1. The carbon atom bonds with up to four different elements. Carbon-to-carbon bonds are stable, so long chains can be built, chains can have branching patterns, and chains can form rings. Isomers are possible. 2. a. Same because both are made of pearls; different because the pearls can vary in color and source (e.g., fresh water, marine). b. Same because both are composed of amino acids; different because particular types of amino acids can be used. **3.2:** 1. Humans have no digestive juices capable of breaking the bonds in cellulose. 2. The monomer in cellulose is glucose; in chitin, an amino group is attached to each glucose. Cellulose is found in plant cell walls, while chitin is in the exoskeleton of some animals. **3.3:** 1. a. A saturated fatty acid contains no double bonds between carbon atoms, while an unsaturated fatty acid contains one or more double bonds; b. Unsaturated because it is not associated with high blood pressure, as is saturated. 2. In a bilayer, phospholipids arrange themselves so that their hydrophilic heads are adjacent to water, while the hydrophobic tails point inward toward each other. **3.4:** 1. Metabolism, transport (protein carrier and channels in plasma membrane), regulation (hormones). 2. The protein's sequence of amino acids is its primary structure. 3. a. The peptide bond includes a partially negative oxygen and a partially negative hydrogen. Therefore, hydrogen bonding between peptide bonds accounts for a protein's secondary structure; b. Covalent bonding between *R* groups, such as disulfide linkages. **3.5:** 1. Phosphate, 5-carbon sugar, and nitrogen-containing base. 2. Between the two strands of DNA, the base A is hydrogen bonded to the base T, and the base C is bonded to the base G. Between a strand of DNA and RNA, the base A is hydrogen bonded to the base U, and the base C is bonded to the base G. 3. When ATP breaks down to ADP + P, energy is released.

## Understanding the Terms

**a.** carbohydrate; **b.** lipid; **c.** polymer; **d.** isomer; **e.** peptide

## Testing Yourself

**1.** c; **2.** a; **3.** e; **4.** b; **5.** c; **6.** b; **7.** c; **8.** c; **9.** d; **10.** c; **11.** e; **12.** c; **13.** **a.** monomer; **b.** monomer; **c.** dehydration reaction; **d.** $H_2O$; **e.** monomer; **f.** monomer; **g.** polymer; **14.** c; **15.** d; **16.** c; **17.** d; **18.** a; **19.** a; **20.** d; **21.** a; **22.** c; **23.** b; **24.** c; **25.** b; **26.** d; **27.** a; **28.** c; **29.** b; **30.** a; **31.** d; **32.** b; **33.** d; **34.** b; **35.** c; **36.** a; **37.** e; **38.** d

## Thinking Scientifically

**1. a.** Subject the seeds of temperate and tropical plants, for which you know the amount and kind of fatty acid, to a range of temperatures from above freezing to below freezing for an extended length of time. Plant the seeds and compare the percentage of survivals per type of plant. **b.** The presence of unsaturated fatty acids in temperate plant seeds may be an adaptation to the environment. **2.** Possible hypothesis: (1) The abnormal enzyme will not produce as much product per unit time as the normal enzyme. (2) The abnormal enzyme will have a different shape from the normal enzyme due to altered levels of organization.

# CHAPTER 4

## Check Your Progress

**4.1:** 1. A cell needs a relatively large surface area for absorption of nutrients and secretion of wastes. **4.2:** 1. A prokaryotic cell lacks a membrane-bounded nucleus, while a eukaryotic cell has one. 2. Cell envelope: mesosome, plasma membrane, cell wall, glycocalyx; cytoplasm: inclusion bodies, nucleoid, ribosomes; appendages: conjugation pilus, fimbriae, flagellum. **4.3:** 1. Separates various metabolic processes; localizes enzymes, substrates, and products; and allows cells to become specialized. 2. Mitochondria are derived from aerobic bacteria, and chloroplasts are derived from photosynthetic bacteria that were taken into a eukaryotic cell. **4.4:** 1. nuclear envelope—defines the nucleus; nuclear pore—allows substances to move into and out of nucleus; nucleolus—formation of ribosomal RNA (rRNA); chromatin—becomes chromosomes and contains DNA. 2. Ribosomes are found attached to the ER and in the cytoplasm. In the cytoplasm, they occur either singly or as polyribosomes. Ribosomes carry out protein synthesis. **4.5:** 1. Rough ER contains ribosomes, while smooth ER does not. Rough ER synthesizes proteins and modifies them, while smooth ER synthesizes lipids, among other activities. 2. Transport vesicles from the ER proceed to the Golgi apparatus. The Golgi apparatus modified their contents and repackages them in new vesicles, some of which carry out secretion, and some of which are lysosomes. **4.6:** 1. Both peroxisomes and lysosomes enclose enzymes; a peroxisome is a metabolic assistant to other organelles; a lysosome digests molecules and can break down cell parts and their component molecules. 2. Both the plant central vauole and the lysosome break down cell parts but vacuoles also store molecules. **4.7:** 1. Structure: The two main parts of a chloroplast are the thylakoids and the stroma; the two main parts of a mitochondrion are the cristae and the matrix. Function: Chloroplasts are larger than mitochondria and capture energy from the sun to build carbohydrates. Mitochondria break down carbohydrates to release energy for ATP production. **4.8:** 1. Microtubules, intermediate filaments, and actin filaments. 2. Cilia and flagella are both composed of microtubules arranged in a particular pattern and enclosed by the plasma membrane. Cilia are shorter than flagella. 3. Cells lining the respiratory tract have cilia that sweep mucus and debris back up into the throat where it can be swallowed or ejected; sperm have flagella that allow them to swim to the egg.

### Understanding the Terms

**a.** Golgi apparatus; **b.** peroxisome; **c.** nucleolus; **d.** cytoskeleton; **e.** fimbria

### Testing Yourself

**1.** c; **2.** c; **3.** d; **4.** a; **5.** c; **6.** c; **7.** a; **8.** d; **9. a.** rough ER—produces proteins; **b.** chromatin—DNA specifies the order of amino acids in proteins; **c.** nucleolus—forms ribosomal RNA, which participates in protein synthesis; **d.** is not involved in protein synthesis; **e.** Golgi apparatus—processes and packages proteins for distribution; **10.** b; **11.** c; **12.** e; **13.** c; **14.** a; **15. a.** example; **b.** Mitochondria and chloroplasts are a pair because they are both membranous structures involved in energy metabolism; **c.** Centrioles and flagella are a pair because they both contain microtubules; centrioles give rise to the basal bodies of flagella; **d.** ER and ribosomes are a pair because together they are rough ER, which produces proteins.

### Thinking Scientifically

**1.** The plastid must be derived from an independent prokaryote that was taken up by a protist sometime in the past. **2.** Uracil is a base found in RNA; therefore, you expect to find it in the nucleus and at the ribosomes.

## CHAPTER 5

### Check Your Progress

**5.1:** 1. The plasma membrane is a phospholipid bilayer with many embedded proteins. The polar heads are exposed and the fatty acid tail of phospholipids point inward. 2. Channel, transport, cell recognition, receptor, enzymatic, and junction proteins. **5.2:** 1. A hypotonic solution has less solute and more solvent than the cell. A hypertonic solution has more solute and less solvent than the cell. 2. Both move molecules from high to low concentration. Diffusion does not require a membrane or transport proteins, while facilitated transport does. **5.3:** 1. Both require a carrier protein, but active transport also requires energy. 2. Both use vesicles to transport materials across the plasma membrane. Molecules are transported out by exocytosis and in by endocytosis. **5.4:** 1. The extracellular matrix is composed of polysaccharides and proteins. In addition, the matrix in bone contains mineral salts. 2. Adhesion junctions: Intercellular filaments run between the two cells. Tight junctions: Plasma membranes are joined, and the result is an impermeable barrier. Gap junctions: Membrane channels allow molecules and ions to pass from one cell to the other. 3. A primary wall is composed of cellulose and other molecules, such as pectin. Secondary cell walls contain cellulose fibrils at right angles and also lignin that adds rigidity and strength.

### Understanding the Terms

**a.** differentially permeable; **b.** osmosis; **c.** hypertonic solution; **d.** glycoprotein; **e.** phagocytosis

### Testing Yourself

**1. a.** hypertonic—cell shrinks due to loss of water; **b.** hypotonic—central vacuole expands due to gain of water; **2.** b; **3.** b; **4.** b; **5.** c; **6.** c; **7.** e; **8.** e; **9.** b; **10.** b; **11.** See Figure 5.1, page 86; **12.** b; **13.** e; **14.** d; **15.** a; **16.** e

### Thinking Scientifically

**1.** $Cl^-$ moves across the plasma membrane from the cytoplasm of cells into the bronchial tube. This creates a hypertonic solution that causes water to follow. In cystic fibrosis, the channels that allow $Cl^-$ movement do not function, and the result is debilitating lung infections. **2.** Fluidity of the plasma membrane is inversely correlated with the percent of saturated fatty acids in the phospholipids. Decreasing saturation of the fatty acid portion of the phospholipids would keep the membrane fluid enough to allow substances to enter and exit cells, even in colder weather.

## CHAPTER 6

### Check Your Progress

**6.1:** 1. Potential energy is stored energy, while kinetic energy is energy of motion. 2. The second energy law tells us that every energy transformation increases disorder. Entropy is the tendency toward disorder. **6.2:** 1. ATP holds energy but easily gives it up because the last phosphate group can be removed, releasing energy. 2. ATP can donate a phosphate to energize a compound for a reaction. Alternatively, it causes a molecule to change its shape and in that way bring about a change. **6.3:** 1. They bring reactants together at their active site or position a substrate so it is ready to react. 2. Optimum amount of substrate and enzyme. Also, optimum pH and temperature. **6.4:** 1. a. glucose; b. It gets broken down to carbon dioxide and water. 2. a. reduced; b. oxidized.

### Understanding the Terms

**a.** metabolism; **b.** potential energy; **c.** vitamin; **d.** entropy; **e.** coenzyme; **f.** oxidation

### Testing Yourself

**1.** e; **2.** e; **3.** e; **4.** e; **5.** d; **6.** d; **7.** b; **8. a.** active site; **b.** substrates; **c.** product; **d.** enzyme; **e.** enzyme-substrate complex; **f.** enzyme. The shape of an enzyme is important to its activity because it allows an enzyme-substrate complex to form. **9.** a; **10.** a; **11.** c; **12.** b; **13.** c; **14.** c; **15.** c; **16.** d; **17.** a; **18.** c; **19.** a; **20.** b; **21.** d; **22.** d; **23.** e; **24.** c; **25.** e; **26.** c; **27.** e; **28.** See Figure 6.13, page 113.

### Thinking Scientifically

**1.** Sugar breakdown releases a lot more energy per molecule than does breaking down ATP. Further, in converting sugar to ATP, energy has already been lost, and therefore, is metabolically wasteful to break down ATP. **2.** Flower relies on enzymes, which lower the energy of activation, to break down glucose.

## CHAPTER 7

### Check Your Progress

**7.1:** 1. Plants, algae, and cyanobacteria. 2. Thylakoid membrane absorbs solar energy, and carbohydrate forms in stroma. **7.2:** 1. $O_2$ is reduced to $CH_2O$ and when the oxygen in water gives its $H_2$ ($2\ H^+ + 2\ e^-$), it is oxidized. 2. Light reactions use solar energy to split water and produce ATP and NADPH; Calvin cycle reactions use ATP and NADPH to reduce $CO_2$ to a carbohydrate. **7.3:** 1. Visible light, specifically blue and red light. 2. ATP and NADPH. **7.4:** 1. Carbon dioxide fixation, carbon dioxide reduction, and regeneration of RuBP. 2. Glucose, sucrose, starch, cellulose, fatty acids, glycerol, and amino acids. **7.5:** 1. $C_4$ plants include many grasses, sugarcane, and corn; CAM plants include cacti, stonecrops, orchids, and bromeliads. 2. $C_4$ plants prevent oxygen from competing with carbon dioxide for an active site on the enzyme rubisco.

### Understanding the Terms

**a.** light reaction; **b.** photosystem; **c.** photosynthesis; **d.** Calvin cycle

### Testing Yourself

**1.** e; **2.** d; **3.** a; **4.** a, b; **5.** c; **6.** a, b, c; **7.** b; **8.** d; **9.** T; **10.** F; **11.** F; **12.** e; **13.** e; **14.** c; **15.** See Figure 7. 4, page 121; **f.** thylakoid membranes; **g.** stroma; **16.** a; **17. a.** water; **b.** oxygen; **c.** carbon dioxide; **d.** carbohydrate; **e.** ADP + P; **f.** ATP; **g.** $NADP^+$; **h.** NADPH; **18.** e; **19.** e; **20.** c; **21.** c; **22.** e

### Thinking Scientifically

**1.** The bacteria were clustered where solar energy was being absorbed, and photosynthesis was occurring because it produces oxygen needed by the bacteria. **2.** It is possible to extract the pigments from a leaf and test for which ones are present because each type pigment has its own absorption spectrum. Test for the pigments in both summer leaves and fall leaves to tell if the pigments were there all season or appear only in the fall.

## CHAPTER 8

### Check Your Progress

**8.1:** 1. Step by step breakdown allows much of the released energy to be captured and utilized by the cell. 2. Glycolysis, the preparatory reaction, the citric acid cycle, and the electron transport chain. The prep reaction and the citric acid cycle release $CO_2$. ETC produces $H_2O$. **8.2:** 1. During the energy-investment steps, ATP breakdown provides the phosphate groups to activate substrates. During the energy-harvesting steps, NADH and ATP are produced. 2. Fermentation occurs when oxygen is not available. Pyruvate enters the mitochondria for further breakdown when oxygen is available. **8.3:** 1. Drawbacks: Most of the energy in a glucose molecule is unused and it results in a toxic end product. Benefits: The 2 ATP gained can be used as a burst of energy. **8.4:** 1. The $C_2$ acetyl group comes from the prep reaction. 2. Per glucose molecule, the citric acid cycle produces 4 $CO_2$, 6 NADH, 2 $FADH_2$, 2 ATP. 3. A dam holds back water, just as the inner membrane holds back hydrogen ions. As water flows over a dam, electricity is produced. As hydrogen ions flow down their concentration gradient through an ATP synthase complex, ATP is produced. **8.5:** 1. a. Hydrolytic reactions are catabolic. Dehydration reactions are anabolic. b. ATP breakdown is catabolic. 2. Both have an inner membrane (in chloroplasts, results in thylakoids; in mitochondria, results in cristae) where complexes form an ETC and ATP is produced by chemiosmosis. Both have a fluid-filled interior (in stroma of chloroplasts, NADPH help reduce $CO_2$ to a carbohydrate, and in matrix of mitochondria, NAD helps oxidize glucose products with the release of $CO_2$).

### Understanding the Terms

**a.** glycolysis; **b.** oxygen debt; **c.** catabolism; **d.** metabolic pool; **e.** chemiosmosis

### Testing Yourself

**1.** b; **2.** c; **3.** a; **4.** c; **5.** c; **6.** c; **7.** a; **8.** b; **9.** e; **10.** c; **11.** a; **12.** c; **13.** b; **14.** b; **15.** c; **16.** d; **17.** d; **18.** c; **19.** a; **20.** b; **21.** b; **22.** b; **23.** d; **24.** a, c, d; **25.** b, d; **26.** a, b, d; **27.** d; **28. a.** cristae, contains electron transport chain and ATP synthase complex; **b.** matrix, location of prep reaction and citric acid cycle; **c.** outer membrane, defines the boundary of the mitochondrion; **d.** intermembrane space, accumulation of $H^+$; **e.** inner membrane, partitions the mitochondrion into the intermembrane space and the matrix.

### Thinking Scientifically

**1.** Acid, because for ATP to be produced, $H^+$ must flow through the ATP synthase complex. **2.** The two radioactive carbons were taken up by a component of the cycle and later released as $CO_2$.

## CHAPTER 9

### Check Your Progress

**9.1:** 1. G1, S, G2, and M stage. DNA is replicated during S stage, and cell division occurs during M stage. 2. DNA damage, failure to properly replicate the DNA, and failure of chromosomes to attach properly to spindle. **9.2:** 1. During prophase, the nuclear membrane fragments, the nucleus disappears, the chromosomes condense, and the spindle begins to form. Without these events, mitosis could not occur. 2. Animal cells furrow and plant cells have a cell plate. Because plant cells have rigid cell walls, they cannot furrow. **9.3:** 1. Cancer cells lack differentiation, have abnormal nuclei, fail to undergo apoptosis, may form tumors, undergo angiogenesis, and may also metastasize throughout the body. 2. Accumulating mutations allow the tumor to escape the capsule containing it, to attract new blood vessels, and ultimately allow tumor cells to enter circulation and lodge in other tissues of the body. 3. a. Cell commits to cell division even in the absence of proper stimuli. b. Cell fails to stop dividing because the proper stimuli to stop are absent. **9.4:** 1. Binary fission involves inward growth of plasma membrane and cell wall concomitant with the separation of the duplicated chromosome attached to the plasma membrane. Mitosis always involves a mitotic spindle to distribute the daughter chromosomes. 2. Prokaryotes usually have a single, small circular chromosome with a few genes and only a few associated proteins. Eukaryotes have many long, linear chromosomes, which contain many thousands of genes and many more proteins.

### Understanding the Terms

**a.** centrosome; **b.** centromere; **c.** spindle; **d.** sister chromatid; **e.** apoptosis

### Testing Yourself

**1.** b; **2.** e; **3.** c; **4.** a; **5.** b; **6.** b; **7.** b; **8.** e; **9.** b; **10.** c; **11.** e; **12.** b; **13.** c; **14.** c; **15.** b; **16.** d; **17.** a; **18.** c; **19.** d; **20.** e; **21. a.** chromatid of chromosome; **b.** centriole; **c.** spindle fiber or aster; **d.** nuclear envelope (fragment); Early prophase.

### Thinking Scientifically

**1.** Since histones are only needed during S stage, one would expect to see high amounts made then, and no synthesis at all during $G_1$ and $G_2$. **2.** The radiation caused mutations to occur that can lead to cancer. The number of mutations required varies according to the type of cancer. Cancers that occurred earlier required fewer mutations than those that occurred later.

## CHAPTER 10

### Check Your Progress

**10.1:** 1. Homologous chromosomes are two copies of the same kind of chromosome, judged by length and location of the centromere. Also, they contain genes for the same traits in the same order. 2. Homologous chromosomes pair during synapsis and then separate so each daughter cell receives one from each pair. During mitosis sister chromatids separate becoming daughter chromosomes so each daughter cell receives the same number and kinds of chromosomes as the parent cell. **10.2:** 1. Independent assortment of chromosomes creates an increased number of possible combinations of chromosomes in each gamete. Crossing-over shuffles the alleles between homologous chromosomes to create even more variation. 2. $2^4$ or 16. 3. Genetic variability ensures that at least some individuals have traits that will allow a species to survive adverse conditions. **10.3:** 1. Two daughter cells that share the same parent cell from meiosis I are identical unless crossing-over occurred. 2. Interkinesis is the intervening cell cycle between meiotic divisions. It differs from interphase in that the stages of the cell cycle do not occur. The chromosomes are already duplicated. **10.4:** 1. In metaphase I of meiosis, homologous chromosomes are paired at the metaphase plate with each homologue facing opposite spindle poles. In metaphase II and mitotic metaphase, homologous chromosomes are not paired, and sister chromatids are attached to spindle fibers from opposite spindle poles. 2. Meiosis II resembles mitosis because sister chromatids are separated during both processes. Meiosis II differs from mitosis because the cells are haploid and not diploid. **10.5:** 1. In males, the primary spermatocytes located within the testes; in females, the primary oocytes located within the ovaries. 2. The bulk of the cytoplasm and other cellular contents are in the one cell that will undergo embryonic development. **10.6:** 1. Nondisjunction in meiosis may cause aneuploidy, an extra or missing chromosome. 2. Sex chromosome aneuploidy is more common because only one of the X chromosomes is active. Any extra X chromosomes become Barr bodies. 3. An inversion involves the reversal of a piece of a chromosome from within, and normally does not cause symptoms. A translocation is swapping of two chromosome fragments from one to the other, and while not usually troublesome, may cause severe problems in offspring if the two chromosomes go into separate cells.

## Understanding the Terms

**a.** spermatogenesis; **b.** bivalent; **c.** polar body; **d.** secondary oocyte; **e.** homologue

## Testing Yourself

**1.** b; **2.** d; **3.** e; **4.** b; **5.** a; **6.** d; **7.** c; **8.** a; **9.** c; **10.** d; **11.** b; **12.** d; **13.** 24, 12; **14.** spermatogenesis, oogenesis; **15.** fertilization; **16.** gametes, spore; **17.** diploid, haploid; **18.** e; **19.** b; **20.** b; **21.** c; **22.** d; **23.** a; **24.** a

## Thinking Scientifically

**1.** The homologous chromosomes separate during anaphase I. **2.** It is possible because this checkpoint functions to ensure that DNA is not damaged before the second meiotic division occurs. One way that this could possibly be tested experimentally would be to induce DNA damage with radiation in cells that have completed meiosis I, and then examine the cells to see if meiosis II still occurs. **3.** There is a 50% chance that the man will pass his abnormal copy of chromosome 2 to his children; likewise, the chance he will pass the abnormal copy of chromosome 6 to his children is 50%. To calculate the chance of passing both abnormal chromosomes to his child (maintaining the balanced translocation), multiply the odds of both events together (as described in Chapter 11). Thus, the chance he will pass the balanced translocation to his child is 0.25, or 25%.

# CHAPTER 11

## Check Your Progress

**Figure 11.8:** This person is heterozygous because she has a child that is affected. **Figure 11.9:** This person is heterozygous because he has a child that is unaffected. **11.1:** 1. Mendel was successful because he chose a good subject, always followed the same well-planned procedure, kept careful records, and used mathematical analysis to analyze his data. 2. The garden pea was a good choice because it has many easily observed traits, a relatively short generation time, each plant produces many offspring (peas), and cross pollination is only possible by hand. **11.2A:** 1.a. all *W*; b. ½ *W*, ½ *w*; c. ½ *T*, ½ *t*; d. all *T*. 2. *bb*. 3. 3:1, 40. **11.2B:** 1. *LG*, *Lg*, *lG*, *lg*. 2. 9:3:3:1. **11.2C:** 1. 75% yellow; 25% green. 2. 75%. 3. a. 9:3:3:1; b. freckles, short fingers; freckles, long fingers; no freckles, short fingers; no freckles, long fingers; c. 1/16. **11.2D:** 1. 25%. 2. *LlGg* x *llgg*; *LlGg* x *LlGg*. 3. *Tt* x *tt*, *t*. **11.2E:** 1. *cc*, *Cc*. 2. 25%. 3. woman: *Hh*; husband: *hh*. 4. 50%. **11.3A:** 1. $A_1A_2$. 2. child: *ii*; mother: $I^Ai$; father: $I^Ai$, $I^Bi$, or *i*. 3. See Figure 11.15. **11.3B:** 3. Mother $X^bX^b$; father, $X^BY$; female offspring are $X^BX^b$, and males are $X^bY$. 1. $X^RX^R$ and $X^RY$. 2. a. 100%; b. none; c. 100%.

## Understanding the Terms

**a.** recessive allele; **b.** allele; **c.** dominant allele; **d.** testcross; **e.** genotype

## Testing Yourself

**1.** b; **2.** a; **3.** c; **4.** d; **5.** c; **6.** d; **7.** a; **8.** b; **9.** b; **10.** b; **11.** a; **12.** d; **13.** d; **14.** c; **15.** c; **16.** b; **17.** a; **18.** c; **19.** autosomal dominant

## Additional Genetic Problems

**1.** 100% chance for widow's peak and 0% chance for straight hairline; **2.** 25%; **3.** *Bb x bb;* **4.** 210 gray bodies and 70 black bodies; 140 = heterozygous; **5.** 50%; **6.** 9/16; **7.** $I^AI^B$; *yes, mother could be AB, A, B, O;* **8.** No, because the son does not inherit alleles on the X from his father.; **9.** mother, $X^HX^h$, $X^HY$, $X^hY$.

## Thinking Scientifically

**1.** Cross it now with a fly that lacks the characteristic. Most likely, the fly is heterozygous and only a single autosomal mutation has occurred. Therefore, the cross will be *Aa* x *aa* with 1:1 results. If the characteristic disappears in males, cross two $F_1$ flies to see if it reappears; it could be X-linked. **2.** Give plants with a particular leaf pattern different amounts of fertilizer from none (your control) to over-enriched, and observe the results. Keep other conditions, such as amount of water, the same for all.

# CHAPTER 12

## Check Your Progress

**12.1:** 1. DNA must be able store information about the development of an organism, must be stable enough so that it can be replicated accurately, and must be able to undergo changes that provide genetic variability within a population. 2. DNA is a right-handed double helix with two strands that run in opposite directions. The backbone is composed of alternating sugar-phosphate groups, and the molecule is held together in the center by hydrogen bonds between

interacting bases. A always hydrogen bonds to T, and G to C. **12.2:** 1. (1) The DNA strands are separated by DNA helicase, (2) new nucleotides are positioned by complementary base pairing, and (3) the new nucleotides are joined together by DNA polymerase to form a new DNA strand. 2. When a DNA molecule is replicated, each copy contains one pre-existing strand and one newly made strand. 3. Prokaryotic DNA replication begins at a single origin of replication and usually proceeds in both directions towards a termination region on the opposite side of the chromosome. Eukaryotic DNA replication begins at multiple origins of replication and continues until the replication forks meet. **12.3:** 1. mRNA carries information from DNA to direct the synthesis of a protein. rRNA makes up part of the ribosomes that are used to translate messenger RNAs. tRNA transfers amino acids to the ribosome during protein synthesis. 2. Several different codons may specify the same amino acid.
**12.4:** 1. Transcription proceeds along the template strand in the 3′ to 5′ direction. The RNA molecule is therefore built in the 5′ to 3′ direction. 3. The introns are spliced out and the exons joined together, and a 5′ guanosine cap and a 3′ poly-A tail are added as the mRNA is processed. **12.5:** 1. Transfer RNA delivers amino acids to the ribosome by binding to the appropriate codon on the mRNA being translated. 2. A ribosome consists of a small and large subunit. Each subunit is composed of a mixture of protein and ribosomal RNA. 3. Initiation of all components of the translational complex, including the first tRNA carrying methionine, are assembled. Elongation: Amino acids are delivered one by one as tRNA molecules pair with the codons on the mRNA.
Termination: A stop codon is reached, a release factor binds to it, and the completed protein is cleaved from the last tRNA as the ribosomal subunits dissociate. **12.6:** 1. Euchromatin consists of the 30 nm zigzag structure that is folded into radial loops, whereas heterchromatin is further compacted by further association with scaffold proteins.

### Understanding the Terms

**a.** intron; **b.** DNA polymerase; **c.** interspersed repeats; **d.** translation

### Testing Yourself

**1.** c; **2.** a; **3.** e; **4.** c; **5.** b; **6.** d; **7.** e; **8.** e; **9.** b; **10. a.** GGA GGA CUU ACG UUU; **b.** CCU CCU GAA UGC AAA; **c.** glycine-glycine-leucine-threonine-phenylalanine; **11.** d; **12.** a; **13.** a; **14.** a; **15.** b, d; **16.** a; **17.** a; **18.** c; **19.** d; **20.** d; **21.** e; **22.** e

### Thinking Scientifically

**1.** Sequence the gene and determine if a transposon sequence is present in the sequence. **2.** Isolate a plant cell and insert into the cell the gene that codes for green fluorescent protein (GFP). Allow the cell to develop into a mature plant.

## CHAPTER 13

### Check Your Progress

**13.1:** 1. An operon is a group of genes that are regulated in a coordinated manner. 2. A gene under positive control is transcribed when it is regulated by a protein that is an activator and not a repressor, whereas one under negative control is not transcribed when it is regulated by a protein that is a repressor. **13.2:** 1. Chromatin, transcriptional, posttranscriptional, translational, and after posttranslational. 2. Packing genes into heterochromatin inactivates a gene, while it is genetically active when held in loosely packed euchromatin.
3. Alternative processing of mRNA allows organisms to produce multiple types of mRNAs, and thus proteins, from a single gene, allowing cells to fine tune gene activity. **13.3:** 1. Errors in DNA replication and natural chemical changes in the bases in DNA may lead to spontaneous mutations; organic chemicals and physical mutagens like x-rays and UV radiation may cause induced mutations. 2. A frameshift mutation may shift the reading frame of a gene so that all following codons encode different amino acids, rendering the protein nonfunctional.

### Understanding the Terms

**a.** posttranscriptional control; **b.** operon; **c.** Barr body; **d.** induced mutation; **e.** carcinogen

### Testing Yourself

**1.** a; **2.** e; **3.** a; **4.** b; **5.** d; **6.** e; **7.** b; **8.** b; **9.** b; **10.** b; **11.** b; **12.** e; **13.** e; **14.** e; **15. a.** DNA; **b.** regulator gene; **c.** promoter; **d.** operator; **e.** active repressor; **16.** c; **17.** d; **18.** d; **19.** d

### Thinking Scientifically

**1.** Translocation may cause portions of two genes to become one gene. If so, the resulting protein could have a new activity with respect to cell cycle regulation. Alternatively, the regulatory sequences of one gene could be controlling the other gene leading to higher or lower levels of expression. Either way, the normal regulation of the cell cycle could be lost.
**2.** A mutation outside a gene may alter its expression if it, for example, (1) disrupts an enhancer or silencer that regulates a nearby gene or (2) affects the chromatin structure so that it changes from euchromatin to heterochromatin or the reverse in the vicinity of the gene.

## CHAPTER 14

### Check Your Progress

**14.1:** 1. To create an rDNA molecule, a piece of foreign DNA is cut with with restriction enzymes and mixed with a plasmid vector cut with the same restriction enzyme. The DNAs are mixed together and DNA ligase is added to seal the molecule. Then, it can be given to bacteria. 2. DNA molecules amplified by PCR can be used to create fingerprints, enabling paternity testing and forensic DNA analysis, among other applications. **14.2:** 1. Animals are multicellular, requiring that the genes be introduced into a fertilized egg, whereas bacteria are unicellular. 2. A transgenic animal contains recombinant DNA molecules in addition to its genome, while a cloned animal is genetically identical to the one from which it was created, but does not contain rDNA. **14.3:** 1. Liposomes, nasal sprays, and adenoviruses are currently being used to deliver genes to cells for gene therapy. 2. Ex vivo gene therapy is being used to treat SCID and familial hypercholesterolemia by adding genes to isolated bone marrow stem cells and liver cells, respectively, before returning them to the patient. In vivo gene therapy is being used to treat cystic fibrosis by introducing genes to cells in the respiratory tract, and genes are being delivered to tumors to make them more susceptible to chemotherapy. **14.4a:** 1. A tandem repeat consists of repeated sequences that are one next to the other, whereas interspersed repeats may be spread across different portions of the same or different chromosomes. 2. We now know that much of eukaryotic DNA has functions other than coding for amino acids in proteins. **14.4b:** 1. Scientists are discovering individual differences in base sequence from the human genome sequence and how some of these differences may be linked to health and disease. 2. Proteomics gives information regarding which genes are expressed as proteins in a particular cell type or tissue, and can help us to understand what makes some cells or tissues unique. 3. Comparative genomics reveals the overall similarity in base sequence from one species to another, giving a rough idea of exactly how similar the organisms are at the molecular level.

## Understanding the Terms

**a.** restriction enzyme; **b.** transgenic organism; **c.** xenotransplantation; **d.** cloning; **e.** polymerase chain reaction (PCR)

## Testing Yourself

**1.** c; **2.** c; **3.** e; **4.** d; **5.** e; **6.** c; **7.** e; **8.** a; **9.** c; **10.** a; **11.** b; **12.** d; **13.** e; **14.** d; **15.** e; **16.** a; **17.** d; **18.** left: AATT; right: TTAA; **19.** a; **20.** e

## Thinking Scientifically

**1.** If a researcher has both a genomic clone and a cDNA clone of a gene, they may compare the sequence of the two clones to determine where the introns and exons occur within the genomic clone. **2.** For example, if an individual has an oncogene, intergenic DNA might contain a silencer that decreases expression of the mutant allele. But if this intergenic DNA contained an enhancer, the additional copies of the intergenic DNA could be harmful because it could increase expression of the oncogene.

# CHAPTER 15

## Check Your Progress

**15.1:** 1. Black. 2. They would be created anew because it would be impossible to get replacement species from the surrounding area. **15.2:** 1. Variations. 2. Long-fur rabbits at top of mountain and short-fur rabbits at bottom of mountain. Longer fur would keep the rabbits warmer at the top of the mountain. **15.3:** 1. Fossils are direct evidence. Dating fossils allows you to trace the history of evolution. 2. In South America, marsupials had to compete with placental mammals, and they became extinct. 3. No, because biogeography gives evidence of evolution when unrelated species are similarly adapted to same type of environment.

## Understanding the Terms

**a.** biogeography; **b.** paleontology; **c.** vestigial structure; **d.** adaptation; **e.** inheritance of acquired characteristics

## Testing Yourself

**1.** d; **2.** e; **3.** b; **4.** e; **5.** e; **6.** e; **7.** e; **8.** e; **9.** e; **10.** b; **11.** a; **12.** b, d; **13.** c; **14.** a; **15.** d; **16.** c; **17.** b, d; **18.** d; **19.** d; **20.** Life has a history, and it's possible to trace the history of individual organisms. **21.** Two different continents can have similar environments, and therefore unrelated organisms that are similarly adapted. **22.** All vertebrate forelimbs contain the same sets of bones organized in similar ways. **23.** All vertebrates share a common ancestor, who had pharyngeal pouches during development. **24.** Similarities are expected because all species share recent and distant common ancestors. Base difference through the occurrence of mutations account for the diversity of life.

## Thinking Scientifically

**1.** Because of the frequent rate of mutations, the virus' recognition proteins would change making it less likely the virus would be detected by the immune system. **2.** Yes; due to natural selection of boll weevils resistant to the insecticide.

# CHAPTER 16

## Check Your Progress

**16.1:** 1. Over time, the allele frequency differences between the two populations will tend to disappear. 2. The offspring represent only a fraction of the genetic diversity of the original gene pool. **16.2:** 1. Directional selection because the result is a shifting of traits in one direction. 2. Sexual selection increases the ability of an organism to reproduce. **16.3:** 1. Aside from noting that mutations, gene flow, and genetic drift still occur, natural selection only acts on certain types of traits; not all traits are exposed to natural selection; the environment and selective agents can be changeable; and the heterozygote can be favored so that all three genotypes are maintained. 2. The dominant allele (p) and the recessive allele (q) must be present in the previous generation in order for the heterozygote (pq) to appear in the next generation.

## Understanding the Terms

**a.** stabilizing selection; **b.** territoriality; **c.** genetic drift; **d.** gene pool; **e.** gene flow

## Testing Yourself

**1.** c; **2.** c; **3.** c; **4.** c; **5.** e; **6.** b; **7.** c; **8.** e; **9.** b; **10.** See Figure 16.8, page 289; **11.** a; **12.** d; **13.** c; **14.** c; **15.** a; **16.** e; **17.** b; **18.** e; **19.** e; **20.** c; **21.** c

## Additional Genetics Problems

**1.** 36%; **2.** 99%; **3.** 0.10

## Thinking Scientifically

**1.** Some insects are naturally resistant to the pesticide. On the first farm, the dose was low enough that even some nonresistant insects could survive to reproduce. Their offspring are also sensitive and so the pesticide retains its effectiveness for several years. The second farmer attempted to kill all pests, and the naturally resistant insects were able to survive. Therefore, in just a few seasons, the pesticide lost its effectiveness. **2.** Hypothesis: females prefer bright-feathered males because their plumage indicates they are in good health and are probably more fit.

# CHAPTER 17

## Check Your Progress

**17.1:** 1. No, because ligers share the ancestry of both lions and tigers. 2. a. Habitat isolation; b. $F_2$ fitness. **17.2:** 1. Show that each of the cats is adapted to a different environment. 2. The fossil record would have to show fossils of the different types of cats existing in the same location at the same time—before they begin to appear in various locations. **17.3:** 1. A gradualistic model because the liger could be a transitional link. 2. *Hox* genes have a powerful affect on development and offer a mechanism by which evolution could occur rapidly. 3. Ligers are bigger than lions and tigers. 4. No, evolution is not goal oriented.

## Understanding the Terms

**a.** postzygotic isolating mechanism; **b.** adaptive radiation; **c.** speciation; **d.** allopatric speciation

### Testing Yourself

**1.** c; **2.** c; **3.** b; **4.** f; **5.** a; **6.** e; **7.** h; **8.** e; **9.** c; **10.** e; **11. a.** species 1; **b.** geographic barrier; **c.** genetic changes; **d.** species 2; **e.** genetic changes; **f.** species 3; **12.** b; **13.** b; **14.** b; **15.** d; **16.** d; **17.** b; **18.** b; **19.** c; **20.** b; **21.** b; **22.** d; **23.** b; **24.** c; **25.** c

### Thinking Scientifically

**1.** Both the biological species concept and DNA sequences provide a way to identify species without the need to examine them anatomically, but the DNA sequences method is faster and unequivocal. The evolutionary species concept allows you to trace the history of an organism in the fossil record, and the biological species concept allows you to determine how species are kept separate. **2.** Their chromosomes are compatible, and the two species are very closely related. It's doubtful they should be considered different species.

## CHAPTER 18

### Check Your Progress

**18.1:** 1. Serve as a template for RNA/DNA synthesis. 2. Fermentation.
**18.2A:** 1. Chemical evolution, evolution of first cells, evolution of eukaryotic cells by endosymbiosis, and first heterotrophic protists before photosynthetic protists. 2. Evolution of multicellularity.
**18.2B:** 1. During the Carboniferous period, plants and animals invaded land and the plants became the fossil fuel we burn today. **18.2C:** 1. Cenozoic era. 2. Cycads and dinosaurs. **18.3:** 1. Perhaps not because the continents separated during the Mesozoic period. 2. Humans did not evolve until after the last mass extinction discussed.

### Understanding the Terms

**a.** molecular clock; **b.** protocell; **c.** liposome; **d.** ocean ridge; **e.** ozone shield

### Testing Yourself

**1.** c; **2.** d; **3.** a; **4.** b; **5.** e; **6.** b; **7.** c; **8.** d; **9.** b; **10.** b; **11.** b; **12.** e; **13.** e; **14.** c; **15.** b; **16.** c; **17.** d; **18.** a; **19.** c; **20.** c; **21.** e; **22.** c; **23. a.** oldest eukaryotic fossils; **b.** $O_2$ accumulates; **c.** oldest known fossils; **d.** protists diversify; **e.** Ediacaran animals; **f.** Cambrian animals; **24.** liposomes, microspheres; **25.** true; **26.** photosynthesizing; **27.** Carboniferous; **28.** Cenozoic; **29.** meteorite, drift; **30.** b; **31.** e; **32.** d

### Thinking Scientifically

**1.** The tree shows that all life forms have a common source and how they are related, despite the occurrence of divergence, which gives rise to different groups of organisms. **2.** The specialized environmental niche of these organisms is the same as it was when they first evolved.

## CHAPTER 19

### Check Your Progress

**19.1:** 1. More inclusive: class, phylum, kingdom, domain; less inclusive: family, genus, species. 2. Yes, a genus usually contains more than one species.
**19.2A:** 1. Determining the common ancestors indicates how many clades. Both the common ancestor and its descendants have the same derived traits. 2. The immediate common ancestor for birds is also the immediate common ancestor for certain reptiles.
**19.2B:** 1. See if their wings are constructed similarly. Use DNA analysis to see how closely related are insects and bats. 2. The snake and the bird have fewer differences than either has with a monkey. **19.3:** 1. RNA sequence data, along with morphological data, suggested that prokaryotes were not all the same, and further, the eukarya are more closely related to archaea than they are to bacteria. 2. Molecular data (RNA/DNA sequence data) indicates that fungi are related to animals. We have no structural data for a close association.

### Understanding the Terms

**a.** taxonomy; **b.** phylogenetic tree; **c.** taxon; **d.** cladistics; **e.** homology

### Testing Yourself

**1.** e; **2.** e; **3.** d; **4.** a, b, c; **5.** c, d, e; **6.** b, c, d, e; **7.** a; **8.** c; **9.** a; **10.** d; **11.** b; **12.** e; **13.** b; **14.** e; **15.** a; **16.** b; **17.** e; **18.** a; **19.** b; **20.** b; **21.** c

### Thinking Scientifically

**1.** New systems are invariably slow to spread. In the meantime, those who use the new system find it difficult to communicate with those who use the old system. On the other hand, if classification were never revised, we would still be learning Aristotle's system. **2.** If you found significant differences in rRNA sequences, you might conclude that the eukaryotes belong in different domains. Most likely, the fungi, plants, and animals would be in separate domains. Perhaps each type protist could be assigned to one of these domains since protists are ancestral to these multicellular groups.

## CHAPTER 20

### Check Your Progress

**20.1:** 1. All viruses have a nucleic acid and a capsid. 2. Viroids and prions are nonliving because, like viruses, they are noncellular and unable to reproduce without a host. 3. If a virus' host survives, many more copies of the virus will be produced and spread to other hosts than if the host dies.
**20.2:** 1. Prokaryotic cells lack a nucleus and membranous organelles. 2. The cell wall lies outside the plasma membrane. 3. In conjugation, the recipient prokaryotic cell acquires new DNA from the donor cell. Sexual reproduction occurs in eukaryotes and results in a new individual with a haploid set of chromosomes from each parent.
**20.3:** 1. The peptidoglycan layer is much thicker in Gram-positive cells than in Gram-negative cells. 2. Endospores permit survival when environmental conditions are harsh. 3. Cyanobacteria produce by photosynthesis much of the oxygen we breathe. **20.4:** 1. Archaea and bacteria differ in rRNA base sequences, and their plasma membranes and cell walls are biochemically distinct. 2. Methogens, halophiles, and thermoacidophiles. 3. Archaea and eukaryotes share some of the same ribosomal proteins, initiate transcription in the same way, and have similar tRNA. 4. Archaea that inhabit livestock intestines generate methane, a greenhouse gas.

### Understanding the Terms

**a.** lysogenic cycle; **b.** photoautotroph; **c.** saprotroph; **d.** symbiotic; **e.** archaea

### Testing Yourself

**1. a.** attachment; **b.** penetration; **c.** integration; **d.** prophage; **e.** biosynthesis; **f.** maturation; **g.** release; **2.** e; **3.** e; **4.** b; **5.** a; **6.** c; **7.** d; **8.** a; **9.** c; **10.** c; **11.** c; **12.** c; **13.** a; **14.** e; **15.** e; **16.** a; **17.** a; **18.** b; **19.** d

### Thinking Scientifically

**1.** For the most part, viruses use host enzymes, which can cause side effects. **2.** Bacteria are very small and reproduce very rapidly, so it is possible to produce and keep many generations in a small test tube or petri dish. Having only one set of genes means that any new mutations show immediately and can be more easily analyzed. Lastly, plasmids can be used as a highly effective vector in genetic engineering experiments.

## CHAPTER 21

### Check Your Progress

**21.1:** 1. Protists. 2. Algae photosynthesize and protozoans ingest their food. 3. a. Archaeplastids; b. Opisthokonts. **21.2A:** 1. The zygote undergoes meiosis. 2. *Ulva, Chara,* and red algae. **21.2B:** 1. Their DNA base sequences are similar, and their ancestor had a flagellum. 2. Water molds. **21.2C:** 1. Dinoflagellates locomote by flagella, and ciliates locomote by cilia. Apicomplexans cannot locomote. 2. Dinoflagellates can photosynthesize. **21.2D:** 1. By endosymbiosis of an algae. 2. They can locomote by flagella. **21.2E:** 1. The amoebozoans and rhizaria have pseudopods. 2. The feeding cells of sponges resemble choanoflagellates, which are opisthokonts.

### Understanding the Terms

**a.** pseudopod; **b.** euglenid; **c.** diatom; **d.** apicomplexan; **e.** plankton

### Testing Yourself

**1.** f; **2.** c; **3.** c; **4.** a; **5.** b; **6.** e; **7.** b; **8.** b; **9.** e; **10.** e; **11.** b; **12.** d; **13.** a; **14.** d; **15.** b; **16.** d; **17.** b; **18.** d; **19.** c; **20.** a; **21.** b; **22.** c; **23.** c; **24. a.** sexual reproduction; **b.** gametes pairing; **c.** zygote (2n); **d.** zygospore (2n); **e.** asexual reproduction; **f.** zoospores (n); **g.** nucleus with nucleolus; **h.** chloroplast; **i.** starch granule; **j.** pyrenoid; **k.** flagellum; **l.** eyespot; **m.** gamete formation. See also Figure 21.5, page 376.

### Thinking Scientifically

**1.** The mutant might be missing a protein that is responsible for completing cell wall synthesis during cytokinesis. Without complete cell wall synthesis, perhaps daughter cells cannot separate. Alternatively, if a new, unusually sticky cell wall protein is being made in the mutant, then daughter cells may not be able to separate. Either would result in a filamentous phenotype. **2.** If either the protozoan or the bacteria are killed, the termite should also die since it will be unable to digest its food. The bacteria could be killed by treating the termite's food source wth an antibiotic, or the protozoan could be killed with an antiprotozoan drug (similar to those used to treat protozoan infections of humans).

## CHAPTER 22

### Check Your Progress

**22.1:** 1. Animals are heterotrophs by ingestion, and fungi are heterotrophs by absorption. 2. Fungal cell walls contain chitin, and those of plants contain cellulose. 3. A fungal spore can grow into a new organism without fusing with another cell. **22.2:** 1. Chytrids have flagellated gametes and spores; other fungi are nonmotile at all stages of their life cycle. 2. Mycoses. Sac fungi. 3. Club fungi, sac fungi, sac fungi, and zygospore fungi. **22.3:** 1. Mutualism. 2. Asexually through fragmentation. 3. Lichens become scarce when air is highly polluted.

### Understanding the Terms

**a.** basidium; **b.** mycelium; **c.** conidiospore; **d.** fruiting body; **e.** mycorrhizae; **f.** chytrids

### Testing Yourself

**1.** d; **2.** b; **3.** c; **4.** c; **5.** b; **6.** e; **7.** e; **8.** a; **9.** c; **10.** a; **11.** c; **12.** a; **13.** d; **14.** d; **15. a.** spores; **b.** sporangium; **c.** sporangiophore; **d.** stolon; **e.** rhizoid; **16.** a; **17.** e; **18.** e; **19.** d; **20.** b; **21.** e; **22. a.** meiosis; **b.** basidiospores; **c.** dikaryotic mycelium; **d.** button stage of the mushroom (basidiocarp); **e.** stalk; **f.** gill; **g.** cap; **h.** dikaryotic; **i.** diploid; **j.** zygote. See also Figure 22.9, page 402.

### Thinking Scientifically

**1.** While yeast is usually available in the air, the kinds and abundance of yeast would be expected to vary with different weather and climate conditions. By reserving some of the dough of bread that rose in a preferred way, a cook could be sure of having enough yeast of the correct variety for the next loaf. By keeping the mother in a cool place between baking days, the yeast would divide slowly, and not accumulate levels of waste products that would start to kill the yeast. **2.** A mutualistic relationship between a fungus and a plant might evolve when environmental conditions are harsh enough that neither can survive well alone. During the transition period between free-living and symbiosis, it would be essential that fungus not harm the plant. Varieties that did kill plants would be less successful (leave fewer offspring) than varieties that did not. A parasitism lifestyle might have evolved if there were competition for free space on the ground. An additional requirement would be an abundance of host plants. Since the parasite often kills the host, there must be other hosts available for the fungus to survive.

## CHAPTER 23

### Check Your Progress

**23.1:** 1. Plentiful light and $CO_2$. 2. A cellulose cell wall produced in same way; apical cells that produce new tissue; plasmodesmata between cells; transfer of nutrients from haploid cells of previous generation to zygote of new generation. 3. The diploid sporophyte produces haploid spores by meiosis. The haploid gametophyte produces gametes. **23.2:** 1. Advantages: The sporophyte embryo is protected from drying out, and the sporophyte produces windblown spores that are resistant to drying out. Disadvantage: The sperm are flagellated and need an outside source of moisture in order to swim to the egg. **23.3:** 1. In lycophytes, the dominant sporophyte has vascular tissue, and therefore roots, stems, and leaves. 2. The walls of xylem contains lignin, a strengthening agent. **23.4:** 1. The independent gametophyte generation lacks vascular tissue, and it produces flagellated sperm. 2. In ferns, but not mosses, the sporophyte is dominant and separate from the gametophyte. **23.5A:** 1. (1) Water is not required for fertilization because pollen grains (male gametophytes) are windblown, and (2) ovules protect female gametophytes and become seeds that disperse the sporophyte, the generation that has vascular tissue. 2. Conifers, cycads, ginkgoes, gnetophytes. **23.5B:** 1. The stamen contains the anther and the filament. Pollen forms in the pollen sac of the anther. 2. The carpel contains the stigma, style, and ovary. An ovule in the ovary becomes a seed, and the ovary becomes the fruit. 3. Gymnosperms (cone-bearing, such as cycads and pine trees) and angiosperms (flowering plants, such as fruit trees and garden plants) produce seeds. 4. Presence of an ovary leads to production of seeds enclosed by a fruit. Animals are often used as pollinators. 5. Animal-pollinated flowers are showy, and in different ways, such as color and fragrance, attract their particular pollinators.

### Understanding the Terms

**a.** sporophyte; **b.** monocotyledon; **c.** pollen grain; **d.** rhizoid

### Testing Yourself

**1.** e; **2.** a; **3.** b; **4.** c; **5.** b; **6.** b; **7.** e; **8.** c; **9.** c; **10.** e; **11.** d; **12.** b; **13.** a; **14.** b; **15.** b; **16.** e; **17.** c; **18. a.** sporophyte (2n); **b.** meiosis; **c.** gametophyte (n); **d.** fertilization. See also Figure 23.3, page 412.

### Thinking Scientifically

**1. a.** ferns; **b.** seed plants; **c.** naked seeds; **d.** needlelike leaves, Conifers; **e.** fan-shaped leaves, Gingkos; **f.** enclosed seeds; **g.** one embryonic leaf, Monocots; **h.** two embryonic leaves, Eudicots.
**2. a.** Lycophytes evolved from a common ancestor that had microphylls. **b.** Ferns, gymnosperms, and angiosperms evolved from a common ancestor that has megaphylls.

## CHAPTER 24

### Check Your Progress

**24.1:** 1. Vegetative organs are the leaves (photosynthesis), the stem (support, new growth, transport), and the root (absorb water and minerals). 2. Monocots: embryo with single cotyledon; xylem and phloem in a ring in the root; scattered vascular bundles in the stem; parallel leaf veins; flower parts in multiples of three. Eudicots embryo with two cotyledons; phloem located between arms of xylem in the root; vascular bundles in a ring in the stem; netted leaf veins; flower parts in multiples of fours or fives.
**24.2:** 1. Epidermal tissue: epidermal cells; ground tissue: parenchyma, collenchyma, and sclerenchyma cells; vascular tissue: xylem (vessel elements and tracheids) and phloem (sieve-tube members). 2. Xylem transports water and minerals usually from roots to leaves. Phloem transports organic compounds throughout the plant. **24.3:** 1. The root apical meristem is located at the tip of the root and is covered by the root cap. 2. Cortex: food storage; endodermis: control of mineral uptake; pericycle: formation of branch roots. **24.4:** 1. A vascular bundle contains xylem and phloem. 2. Vascular bundles are scattered in monocot stems and form a ring in eudicot stems. 3. Primary growth is growth in length and is nonwoody; secondary growth is growth in girth and is woody. 4. Bark is composed of cork, cork cambium, cortex, and phloem. 5. An annual ring is composed of one year's growth of wood—one layer of spring wood followed by one layer of summer wood. **24.5:** 1. Photosynthesis, which produces organic food for a plant, occurs in the mesophyll.

### Understanding the Terms

**a.** mesophyll; **b.** vascular cambium; **c.** cotyledon; **d.** stolon; **e.** xylem

### Testing Yourself

**1.** c; **2.** b; **3.** c; **4.** c; **5.** b; **6.** b; **7.** b; **8.** c; **9.** c; **10.** b; **11.** d; **12.** e; **13.** d; **14.** d; **15.** d; **16.** c; **17.** c; **18.** e; **19.** a; **20. a.** epidermis; **b.** cortex; **c.** endodermis; **d.** phloem; **e.** xylem. See also Figure 24. 8, page 440; **21. a.** upper epidermis; **b.** palisade mesophyll; **c.** leaf vein; **d.** spongy mesophyll; **e.** lower epidermis. See also Figure 24.20, page 450.

### Thinking Scientifically

**1.** Use tissue autoradiography: allow plants to take up radioactive amino acids for a short time; at increasing intervals of time, prepare thin sections of stem; radiation will expose a photographic film, and microscopic examination of films allows the experimenter to follow the path of protein. **2.** Grow stolons under various environmental conditions. Control group: These stolons are provided with a warm temperature, plentiful water, and sunlight. Make sure the nodes are touching the ground: it should be observed that new plants are arising from the nodes. Test groups: Deprive each test group of only one variable, either the mechanical stimuli of having the nodes touch the ground, or warm temperature, or water, or sunlight. Most likely, all these conditions are requirements for the growth of new plants from nodes.

## CHAPTER 25

### Check Your Progress

**25.1:** 1. a. Nitrogen and sulfur are needed to form protein. All plant roots take up nitrate ($NO_3^-$) and sulfate ($SO_4^{2-}$) from the soil. b. Nitrogen and phosphate ($HPO_4^{2-}$) are needed to make nucleic acids. Plant roots also take up phosphate from the soil. 2. (1) Helps prevent soil erosion; (2) helps retain moisture; and (3) as the remains decompose, nutrients are returned to the soil. 3. Humus improves soil aeration, soil texture, increases water-holding capacity, decomposes to release nutrients for plant growth, and helps retain positively charged minerals and make them available for plant uptake. **25.2:** 1. The nonpolar tails of phospholipid molecules make the center of the plasma membrane nonpolar. 2. The bacteria convert atmospheric nitrogen to nitrate or ammonium, which can be taken up by plant roots. 3. The fungus obtains sugars and amino acids from the plant. The plant obtains inorganic nutrients and water from the fungus.
**25.3:** 1. Evaporation of water from leaf surfaces causes water to be under tension in stems. 2. When water molecules are pulled upward during transpiration, their cohesiveness creates a continuous water column. Adhesion allows water molecules to cling to the sides of xylem vessels, so the column of water does not slip down. 3. Sugars enter sieve tubes at sources, creating pressure as water flows in as well. The pressure is relieved at the other end when sugars and water are removed at the sink.

### Understanding the Terms

**a.** pressure-flow model; **b.** soil horizon; **c.** transpiration; **d.** Casparian strip; **e.** guard cell

### Testing Yourself

**1.** d; **2.** e; **3.** a; **4.** c; **5.** d; **6.** c; **7.** d; **8.** a; **9.** b; **10.** c; **11.** d; **12.** b; **13.** a; **14.** c; **15.** c; **16.** e; **17.** a; **18.** The diagram shows that air pressure pushing down on mercury in the pan can raise a column of mercury only to 76 cm. When water above the column is transpired, it pulls on the mercury and raises it higher than 76 cm. This suggests that transpiration would be able to raise water to the tops of trees. **19.** e; **20. a.** See Figure 25.13, page 466. **b.** After $K^+$ enters guard cells, water follows by osmosis and the stoma opens. **21.** There is more solute in bulb 1 than in bulb 2, therefore water enters bulb 1. This creates a positive pressure that causes water, along with solute, to flow toward bulb 2. See also illustration on page 468.

### Thinking Scientifically

**1.** Divide a large number of identical plants into control and experimental groups. Both groups are to receive the same treatment, including all necessary nutrients, but the experimental group will not be given any calcium. It is expected that only the experimental group will suffer any ill effects. If only the control or if both groups do poorly, some unknown variable is affecting the results. **2.** The plants get most of their water from fog that rolls off the nearby ocean at night. Therefore at night, the plants open all their stomata and take in both moisture and carbon dioxide for photosynthesis. The stomata are closed during the day. The large number of stomata, in this unusual case, actually helps the plant to survive in a very dry environment.

## CHAPTER 26

### Check Your Progress

**26.1:** 1. Hormones coordinate the responses of plants to stimuli. 2. You could apply gibberellins to induce growth and cytokinins to increase the number of cells. 3. a. ABA maintains dormancy and closes stomata. b. Gibberellins have the opposite effect. **26.2A:** 1. It is adaptive for roots to grow toward water because it enhances their ability to extract water and dissolve minerals from the soil for plant tissues. 2. Rotating horizontally will prevent the statoliths from settling and triggering differential growth. Therefore, neither the root nor the shoot is expected to curve up or down. 3. These animals are nocturnal, so it would be a waste of energy to open their flowers and produce scent during the day. **26.2B:** 1. Red light converts Pr to $P_{fr}$; $P_{fr}$ binds to a transcription factor; and the complex moves to the nucleus, where it binds to DNA so that genes are turned on or off. 2. The plant is responding to a short night, not to the length of the day. **26.2C:** 1. Plants have (1) physical and chemical defenses (e.g., secondary metabolites); (2) wound responses (e.g., proteinase inhibitors); (3) hypersensitive responses (e.g., sealing off of infected areas); and (4) relationships with animals (e.g., acacia and ants).

### Understanding the Terms

**a.** circadian rhythm; **b.** gravitropism; **c.** abscission; **d.** gibberellin; **e.** photoperiodism

### Testing Yourself

**1.** c; **2.** a; **3.** d; **4.** b; **5.** c; **6.** c; **7.** d; **8.** e; **9.** d; **10.** d; **11.** b; **12.** e; **13.** a; **14.** c; **15.** Place the banana in a closed container with a ripened fruit. **16.** d; **17.** c; **18.** d; **19.** e; **20.** c; **21.** b; **22.** e; **23.** a; **24.** e; **25.** b; **26.** e; **27.** b; **28.** a.; **29.** c; **30.** d; **31.** d

### Thinking Scientifically

**1.** Use a plant that tracks the sun as your experimental material. Make tissue slides to confirm the presence of a pulvinus, as in Figure 26.14. Apply ABA to live pulvinus tissue under the microscope to test for the results described in Figure 26.14. **2.** Shine a light underneath a plant growing on its side. If the stem now curves down, the phototropic response is greater than the gravitropic response and your hypothesis is not supported.

## CHAPTER 27

### Check Your Progress

**27.1:** 1. Male gametophytes are produced in the anther of the stamen. The female gametophyte is produced in an ovule within the ovary of the carpel. 2. Each microspore produces a two-celled pollen grain. The generative cell produces two sperm, and the tube cell produces a pollen tube. One of the four megaspores produces a seven-celled female gametophyte, called the embryo sac, within the ovule. 3. When one sperm fertilizes the egg, a zygote results. When the second sperm joins with two other nuclei of the embryo sac, the endosperm results. **27.2:** 1. The embryo is derived from the zygote; the stored food is derived from the endosperm; and the seed coat is derived from the ovule wall. 2. The ovule is a sporophyte structure produced by the female parent. Therefore, the wall (becomes seed coat) is 2n. The embryo inside the ovule is the product of fertilization and is, therefore, 2n. 3. Cotyledons are embryonic leaves that are present in seeds. Cotyledons store nutrients derived from endosperm (in eudicots). **27.3:** 1. Dry fruits, with a dull, thin, and dry covering derived from the ovary, are more apt to be windblown. Fleshy fruits, with a juicy covering derived from the ovary and possibly other parts of the flower, are more apt to be eaten by animals. 2. Eudicot seedlings have a hook shape, and monocot seedlings have a sheath to protect the first true leaves. **27.4:** 1. Advantages to asexual reproduction include: (1) the newly formed plant is often supported nutritionally by the parent plant until it is established; (2) if the parent is ideally suited for the environment, the offspring will be as well; and (3) if distance between individuals make cross-pollination unlikely, asexual reproduction is a good alternative. 2. For example, stolons and rhizomes produce new shoots and roots; fruit trees produce suckers; and stem cuttings grow new roots and become a shoot system.
3. Tissue from leaves, meristem, and anthers can become whole plants in tissue culture.

### Understanding the Terms

**a.** carpel; **b.** fruit; **c.** female gametophyte; **d.** seed; **e.** pollen grain

### Testing Yourself

**1.** d; **2.** a; **3.** b; **4.** a; **5.** a; **6.** e; **7.** a; **8.** e; **9.** c; **10.** d; **11.** b; **12.** e; **13.** d; **14.** c; **15.** a; **16.** b; **17.** c; **18.** e; **19.** c; **20. a.** diploid ; **b.** anther; **c.** ovule; **d.** ovary; **e.** haploid; **f.** megaspore; **g.** male; **h.** female; **i.** sperm; **j.** seed. See also Figure 27.1, page 496.

### Thinking Scientifically

**1.** You could study (a) the anatomy of the wasp and flower, trying to determine if the mouth parts of the wasp are suitable for collecting nectar from this flower; (b) the appearance of the flower in sunlight/ultraviolet light to determine if the result is suitable to the vision of the wasp; and (c) the behavior of the wasp to see if it is suitable as a pollinator of this flower.
**2. a.** Use asexual reproduction through tissue culture (see Fig. 27.13). **b.** Continue to propagate in this manner only the most hardy plants.

## CHAPTER 28

### Check Your Progress

**28.1:** 1. Multicellular, usually with specialized tissues, ingest food, diploid life cycle. 2. Animals are descended from an ancestor that resembles a hollow spherical colony of flagellated cells. Individual cells became specialized for reproduction. Two tissue layers arose by invagination. 3. Multicellular; bilateral symmetry, three tissue layers, body cavity, deuterostome development.
**28.2:** 1. Sponges are multicellular; no symmetry; no digestive cavity.

Cnidarians have true tissues; radial symmetry; have a gastrovascular cavity. **28.3:** 1. They all have bilateral symmetry, three tissue layers, and protostome development. They have no body plan, coelom, or any sort of nervous tissue. 2. Annelids and molluscs have a complete digestive tract, a true coelom, and a circulatory system (closed in annelids and open in molluscs). Flatworms have a gastrovascular cavity with only one opening, no coelom, and no circulatory system. 3. Flukes and tapeworms are parasitic flatworms. Their head region now contains hooks and/or suckers for attaching to the digestive tract (tapeworms) or blood vessel (fluke). Leeches (annelids) are external parasites. **28.4:** 1. Roundworms and arthropods are the molting protostomes. They both have a true coelom. 2. Crustaceans breathe by gills and have swimmerets. Insects breathe by tracheae and they have wings. 3. The first pair of appendages is the chelicerae (modified fangs), and the second pair is the pedipalps (hold, taste, chew food). **28.5:** 1. The larval stage is bilaterally symmetrical. 2. The water vascular system functions in locomotion, feeding, gas exchange, and sensory reception.

### Understanding the Terms

**a.** gastrovascular cavity; **b.** true coelom; **c.** metamorphosis; **d.** water vascular

### Testing Yourself

**1.** d; **2** e; **3.** e; **4.** e; **5.** d; **6.** b; **7.** a; **8.** c; **9.** a; **10. a.** tapeworm; **b.** mollusc; **c.** sponge; **d.** cnidaria; **e.** rotifer; **f.** flatworm; **g.** arthropod; **11.** b; **12.** b; **13.** c; **14.** a; **15.** d; **16. a.** all; **b.** annelids and arthropods; **c.** all; **d.** all; **e.** all; **f.** all; **g.** arthropods; **h.** mollusc; **17. a.** earthworms; **b.** clams; **c.** clams; **d.** clams; **e.** earthworms; **f.** earthworms; **g.** clams; **h.** clams; **i.** earthworms; **j.** earthworms; **18. a.** head; **b.** antenna; **c.** simple eye; **d.** compound eye; **e.** thorax; **f.** tympanum; **g.** abdomen; **h.** forewing; **i.** hindwing; **j.** ovipositor; **k.** spiracles; **l.** air sac; **m.** spiracle; **n.** tracheae; **19.** b; **20.** b; **21.** d; **22.** c; **23.** a; **24.** b; **25.** c; **26.** a; **27.** d; **28.** e; **29.** c; **30.** e; **31.** e; **32.** b

### Thinking Scientifically

**1.** Animals that are sessile tend to be radially symmetrical because their food comes to them from all directions. There is no need to have anterior and posterior body regions. Animals that move through their environment are bilaterally symmetrical, with the anterior portion containing sensory organs. This allows the animal to sense and respond to the environment as it travels through it. **2.** (1) Drying out is not a danger. (2) Water facilitates metabolic reactions and moderates temperatures. (3) Diffusion in water helps distribute nutrients in cells. (4) Water supports animals.

## CHAPTER 29

### Check Your Progress

**29.1:** 1. Humans are chordates, and therefore, they have the four characteristics at some point in their life cycle. Humans have all four chordate characteristics as embryos. 2. A sea squirt larva has the four characteristics as a larva, then undergoes metamorphosis to become an adult, which has gill slits but none of the other characteristics. **29.2:** 1. The vertebral column shows vertebrates are segmented because it is composed of repeating units. 2. Vertebrates evolved in the water where the environment prevents drying out. The terrestrial vertebrates practice internal fertilization and development with extraembryonic membranes. Amniotic fluid surrounds the embryo. **29.3:** 1. All fishes are aquatic vertebrates and ectothermic. They all live in water, breathe by gills, and have a single circulatory loop (Fig. 29.9*a*). 2. Cartilagenous fish have jaws, two pairs of paired fins, gill slits, dermal denticles, and a skeleton made up of cartilage. **29.4:** 1. a. Paired limbs, smooth, nonscaly skin that stays moist, lungs, a three-chambered heart with a double-loop circulatory pathway, sense organs adapted for a land environment, ectothermic, and have aquatic reproduction. b. Lobe-finned fish and amphibians both have lungs and internal nares that allow them to breathe air. The same bones are present in the front fins of the lobe-finned fish as in the forelimbs of early amphibians. 2. Usually, amphibians carry out external fertilization in the water. The embryos develop in the eggs until the tadpoles emerge. They then undergo metamorphosis, growing legs and reabsorbing the tail, and become adults. **29.5:** 1. Paired limbs allow reptiles to locomote on land; a thick, dry skin prevents water loss; they breathe air and have a double circulatory path; they lay a shelled egg that contains extraembryonic membranes. 2. Alligators live in fresh water and have a thick skin, two pairs of legs, powerful jaws, and a long muscular tail that allow them to capture and eat other animals that are in or come to the water's edge. Snakes have no limbs and have relatively thin skin. They live close to or in the ground and can escape detection. They use smell (Jacobson's organ) and vibrations to detect prey. Some use venom to subdue prey, which they eat whole because their jaws are distendable. 3. Yes, birds are reptiles: feathers are modified scales; they have clawed feet and a tail that contains vertebrae. If their common ancestor was a dinosaur, they are dinosaurs. **29.6:** 1. Mammals have hair or fur and mammary glands, endothermy, limbs under body, differentiated teeth, and an enlarged brain. 2. Three groups of mammals are monotremes (have a cloaca and lay eggs), marsupials (young are born immature and finish development in a pouch), and placental mammals (development occurs internally and the fetus is nourished by placenta). Placental mammals include bats (chiroptera) that can fly, primates that live in trees, whales and dolphins (cetaceans) that live in the sea, and elephants (proboscidea) that have a long trunk.

### Understanding the Terms

**a.** endothermic; **b.** monotreme; **c.** reptile; **d.** notochord

### Testing Yourself

**1.** e; **2.** a; **3.** c; **4.** e; **5.** a; **6.** e; **7.** b; **8.** b; **9.** e; **10.** c; **11.** a; **12.** a; **13. a.** pharyngeal pouches; **b.** dorsal tubular nerve cord; **c.** notochord; **d.** postanal tail; **14.** d; **15.** d; **16.** a

### Thinking Scientifically

**1.** Teeth in the jaw bones are much heavier than a beak. As with many of the characteristics of birds, beaks probably evolved due to selection pressures to decrease the body weight of birds, making flight possible. Being lower in weight also makes flight less energetically costly. **2.** The skin in amphibians is highly vascularized (many blood vessels) because of its role in respiration. Since the skin is on the outside of the body, it would come in contact with pollutants in the soil and water, not just the air, like the lungs.

## CHAPTER 30

### Check Your Progress

**30.1:** 1. prosimians: 6, anthropoid: 5, 4, 3, 2, 1, hominoids: 4, 3, 2, 1, hominines: 3, 2, 1, hominins: 2, 1; 2. Molecular data. **30.2:** 1. Standing on tree limbs to reach fruit overhead; traveling and foraging on ground. 2. Molecular data. **30.3:** 1. Gracile: slight of frame and smaller teeth. Robust: larger frame, massive jaws with large teeth. Diet because robust fed on tough plant material. 2. Eastern because their arm length is proportioned as in *Homo.* **30.4:** 1. Shows they are physically and intellectually competent. 2. *Homo* has the use of tools and fire. Most likely, *Homo erectus* men could hunt, which means they had to work together and perhaps even speak to each other. The woman may have gathered edible plants and plant products. **30.5A:** 1. Cro-Magnon made knife blades and combined them with wooden handles. They made spears that could be thrown from a distance. 2. Fossil evidence shows that humans evolved in Africa. Molecular data shows that there are few genetic differences between people today. 3. It represents symbolic thinking. **30.5B:** 1. People from any two ethnic groups produce fertile offspring. 2. Within ethnic groups.

### Understanding the Terms

**a.** anthropoid; **b.** Cro-Magnon; **c.** Neandertal; **d.** *Homo ergaster;* **e.** hominids

### Testing Yourself

**1.** a; **2.** b; **3.** b; **4.** d; **5.** e; **6.** d; **7.** b; **8.** c; **9.** a; **10.** c; **11.** b; **12.** d; **13.** T; **14.** T; **15.** T; **16.** T; **17.** F; **18.** anthropoids; **19.** Africa; **20.** erect, small; **21.** Cro-Magnon; **22.** thousands; **23.** d; **24. a.** modern humans; **b.** archaic humans; **c.** *Homo erectus;* **d.** *Homo erectus.* See also Figure 30.10, page 570.

### Thinking Scientifically

**1.** The fact that the trait has survived is usually taken as definition that the trait is more advantageous than not, even though there are "trade-offs."
**2.** Sequence the Neandertal genome using DNA from Neandertal bones, and compare the sequence to the human genome of today. Look for sequences present in both genomes.

## CHAPTER 31

### Check Your Progress

**31.1:** 1. Squamous epithelium: flat cells that line the blood vessels and air sacs of lungs; cuboidal epithelium: cube-shaped cells that line the kidney tubules and various glands; columnar epithelium: rectangular cells that line the digestive tract. 2. Fibrous connective tissue has collagen and elastic fibers in a jellylike matrix between fibroblasts; supportive connective tissue has protein fibers in a solid matrix between collagen or bone cells; fluid connective tissue lacks fibers and has a fluid matrix between blood cells or lymphatic cells. 3. Skeletal muscle, which is striated with multiple nuclei, causes bones to move when contracted. Smooth muscle, which is spindle-shaped with a single nucleus, causes the walls of internal organs to constrict. Cardiac muscle, which has branching, striated cells each with a single nucleus, causes the heart to beat. 4. Dendrites conduct signals toward the cell body; cell body contains most of the cytoplasm and the nucleus, it carries on the usual functions of the cell; and the axon conducts nerve impulses. **31.2:** 1. The epidermis is stratified squamous epithelium, and it protects and prevents water loss. The dermis is dense fibrous connective tissue, and it helps regulate body temperature and provides sensory reception. 2. Sweat glands are located in all regions of the skin. They help modify body temperature. Oil glands are associated with hair follicles and lubricate the hair within the follicle and the skin. 3. The dorsal cavity contains the cranial cavity and the vertebral cavity; the ventral cavity contains the thoracic cavity and the abdominopelvic cavity. **31.3:** 1. Homeostasis, the dynamic equilibrium of the internal environment, maintains body conditions within a range appropriate for cells to continue living. 2. Circulatory system brings nutrients and removes waste from tissue fluid. Respiratory system carries out gas exchange. Urinary system excretes metabolic wastes and maintains salt-water balance and pH of blood. 3. When conditions go beyond or below a set point, a correction is made to bring conditions back to normality again.

### Understanding the Terms

**a.** ligament; **b.** epidermis; **c.** striated; **d.** homeostasis; **e.** spongy bone

### Testing Yourself

**1.** c; **2.** b; **3.** a; **4.** e; **5.** e; **6.** e; **7.** b; **8.** e; **9.** e; **10.** c; **11. a.** columnar epithelium, lining of intestine (digestive tract), protection and absorption; **b.** cardiac muscle, wall of heart, pumps blood; **c.** compact bone, skeleton, support and protection. **12.** e; **13.** d; **14.** c; **15.** c; **16.** c, a, g; **17.** e, d, b; **18.** b, c, f; **19.** b; **20.** d; **21.** a; **22.** c

### Thinking Scientifically

**1.** Epithelial cells are the outer layer of protection of the body. Therefore, any mutagens in the environment contact epithelial cells first. A second factor is the high rate of cell division in these cells. Epithelial cells are constantly being sloughed off and replaced. This high rate of cell division means that spontaneous mutations arising from errors in DNA replication are more likely to occur here.
**2.** The immune system responds to a wide variety of infectious agents. If the hypothalamus were to also respond it would be duplicating a function of the immune system. It is economical for the immune system to recognize infectious agents and signal the hypothalamus when the body's temperature set point needs to be changed. A potential disadvantage arises if the immune system signal is sent in error or cannot be turned off. Inappropriate or uncontrollable fevers could result.

## CHAPTER 32

### Check Your Progress

**32.1:** 1. To carry nutrients and oxygen to cells and to carry away their wastes. 2. Both use a heart to pump fluid. An open system pumps hemolymph through channels and cavities. The hemolymph eventually drains back to the heart. A closed system pumps blood through vessels that carry blood both away from and back to the heart. The trachea, but not an open circulatory system, is more efficient because it takes oxygen directly to the muscles. **32.2:** 1. Arteries carry blood away from the heart, capillaries exchange their contents with tissue fluid, and veins return blood back to the heart. 2. The one-circuit pathway utilizes a heart with one atrium and one ventricle to send blood to the gill capillaries and then the systemic capillaries in a single loop. The two-circuit pathway pumps blood to both the pulmonary and systemic capillaries simultaneously. **32.3:** 1. The wall of the left ventricle is thicker than the wall of the right ventricle, and it generates a greater pressure than the right ventricle. The right ventricle pumps blood into the pulmonary circuit, which takes blood only to the lungs for gas exchange, while the left ventricle pumps blood into the systemic circuit, which take blood to all the cells of the body. 2. From the body: venae cavae, right atrium, tricuspid valve, right ventricle, pulmonary semilunar valve, pulmonary trunk and arteries. From the lungs: pulmonary veins, left atrium, bicuspid valve, left ventricle, aortic semilunar valve, aorta. 3. First the atria contract, then the ventricles contract, and then they both rest. The *lub* sound occurs when the atrioventricular valves close, and the *dub* sound occurs when the semiluner valves close. 4. Thromboembolism, stroke, heart attack. **32.4:** 1. Blood transports substances to and from the capillaries, defends against pathogen invasion, helps regulate body temperature, and forms clots to prevent excessive blood loss. 2. Red blood cells are smaller, lack a nucleus, contain hemoglobin, and are red in color. White blood cells are larger, have a nucleus, do not contain hemoglobin, and are translucent in appearance. 3. Platelets accumulate at the site of injury and release a clotting factor that results in the synthesis of thrombin. Thrombin synthesizes fibrin threads that provide a framework for the clot. 4. A type B recipient has anti-A antibodies in the plasma, and they will react with the donor's red blood cells, causing agglutination.

### Understanding the Terms

**a.** artery; **b.** platelet; **c.** plasma; **d.** venae cavae; **e.** hemoglobin

### Testing Yourself

**1.** b; **2.** b; **3.** a; **4.** d; **5.** d; **6.** c; **7.** b; **8.** b; **9.** e; **10.** e; **11.** c; **12.** e; **13.** b; **14.** c; **15.** e; **16.** F; **17.** T; **18.** F; **19.** T; **20. a.** blood pressure; **b.** osmotic pressure; **c.** blood pressure; **d.** osmotic pressure; **21.** See Figure 32.7, page 599.

### Thinking Scientifically

**1.** Artificial blood must be able to carry oxygen from the lungs to the tissues. Therefore it must contain a molecule like hemoglobin which binds oxygen and still releases it. The ion composition ($K^+$, $Ca^{2+}$, and $Na^+$) of artificial blood should be similar to human blood. Most likely artificial blood would not contain formed elements because they would be too difficult to replicate. Most likely it would not be possible to reproduce any part of the clotting mechanism also. **2.** SA rapid heartbeat is not an electrically abnormal heartbeat. The space between the T wave and the P wave would be very short, and the amplitude of the QRS complex might be higher but the overall pattern should be similar to a normal ECG.

## CHAPTER 33

### Check Your Progress

**33.1:** 1. The lymphatic system consists of the lymphatic vessels, which have the same structure as cardiovascular veins, and the lymphatic organs: red bone marrow, lymph nodes, and spleen. 2. The lymphatic system absorbs fats, returns excess tissue fluid to the bloodstream, produces lymphocytes, and helps defend the body against pathogens. 3. Red bone marrow is a spongy, semisolid red tissue located in certain bones (e.g., ribs, clavicle, vertebral column, heads of femur, and humerus), which produces all the blood cells of the body. The thymus is a soft, bilobed gland located in the thoracic cavity between the trachea and the sternum where T lymphocytes mature. A lymph node is a small ovoid structure located along lymphatic vessels where lymph is cleansed. The spleen is an oval organ with a dull purplish color that cleanses the blood. **33.2:** 1. Barriers to entry (e.g., skin); inflammatory response; phagocytes and natural killer cells; protective proteins (e.g., complement). 2. Phagocytes (dendritic cells, macrophages, neutrophils) devour pathogens. Natural killer cells kill virus-infected cells and cancer cells by cell-to-cell contact. 3. Complement proteins complement (assist) the other nonspecific defenses by enhancing inflammation, binding to the surface of pathogens, and forming a membrane attack complex. **33.3:** 1. Specific defense requires that the immune system be able to (1) recognize, (2) respond to, and (3) remember foreign antigens. 2. Once we recover from an infection, the immune system remembers the antigen and we are immune to it. The immune system also reacts to foreign tissues and cancer cells. 3. B cells are responsible for antibody-mediated immunity, and T cells are responsible for cell mediated immunity. B cells produce antibodies; cytotoxic T cells attack viral-infected or cancer cells, and helper T cells produce cytokines that stimulate the immune response. **33.4:** 1. Tissue rejection occurs when foreign tissues and organs are rejected by the body. Autoimmune diseases (e.g., rheumatoid arthritis) occur when cytotoxic T cells or antibodies mistakenly attack the body's own cells. Immune deficiency occurs when the immune system is deficient, as in AIDS. Allergies occur when the body responds to environmental substances such as pollen that normally do not provoke a response.

### Understanding the Terms

**a.** vaccine; **b.** lymph; **c.** antigen; **d.** apoptosis; **e.** T lymphocyte

### Testing Yourself

**1.** b; **2.** e; **3.** e; **4.** a; **5.** b; **6.** c; **7.** a; **8.** b; **9.** b; **10. a.** antigen-binding sites; **b.** light chain; **c.** heavy chain; **d.** *V stands for variable region; C stands for constant region;* **11.** d; **12.** e **13.** b; **14.** d; **15.** a; **16.** b; **17.** e; **18.** d; **19.** d; **20.** d; **21.** b; **22.** b

### Thinking Scientifically

**1.** Drugs such as cyclosporine inhibit IL-2, therefore suppressing the production of natural killer cells and cytotoxic T cells. However, it does not affect other components of the immune of healing systems, including the production of other types of white blood cells. **2.** Your results indicate than an immune system that can mount an immediate response to a viral challenge tends to be indiscriminate in its response. It reacts against the virus and the body's own cells. Therefore, an autoimmune disease frequently occurs. Your results also suggest that a slow specific response is better than a fast specific response because the chance of a crippling autoimmune response is less.

## CHAPTER 34

### Check Your Progress

**34.1:** 1. When a digestive tract has both a mouth and anus, each part of the tract can become specialized as in the earthworm where the pharynx, crop, gizzard, and intestine have specialized functions. 2. Discontinuous feeders tend to eat large meals and often have a storage area, such as a crop or stomach, to hold their food before the start of the digestive process. Continuous feeders take in small amounts all the time and do not require a storage area. 3. Carnivores tend to have pointed

incisors and enlarged canine teeth to tear off pieces small enough to quickly swallow. The molars are jagged for efficient chewing of meat. Herbivores have reduced canines but sharp even incisors to clip grasses. The large flat molars grind and crush tough grasses. **34.2A:** 1. Mouth, pharynx, esophagus, stomach, small intestine, large intestine, rectum, anus. 2. The small intestine finishes the digestion of proteins, fats, carbohydrates, and nucleic acids. Bile from the liver (emulsifies fat) and pancreatic juice assist digestion but so do the brush-border enzymes, so called because they are on the microvilli that extend from intestinal villi. The villi and microvilli greatly enhance the surface area of the intestinal wall, thereby assisting the small intestine's second
function: absorption of the final products of digestion. **34.2B:** 1. The pancreas, the liver, and the gallbladder. 2. Pancreatic juice, which enters the duodenum, contains pancreatic amylase for the digestion of starch, trypsin for the digestion of protein, and lipase for the digestion of fat. The liver makes bile, which is stored in the gallbladder. Bile enters the duodenum where it emulsifies fat preparatory to its digestion by lipase. The liver has many other functions, such as the storage of glucose as glycogen. **34.3:** 1. Starch digestion begins in the mouth where salivary amylase digests starch to maltose and pancreatic amylase continues this same process in the small intestine. Maltase and brush-border enzyme digests maltose to glucose, which enters a blood capillary. Protein digestion starts in the stomach where pepsin digests protein to peptides and continues in the small intestine where trypsin carries out this same process. The intestinal enzyme called peptidase digest peptides to amino acids, which enter a blood capillary.
**34.4:** 1. Vegetables, if properly chosen, can supply limited calories but all necessary amino acids and vitamins. Much urea results when excess amino acids from proteins are metabolized. The loss of water needed to excrete urea can result in dehydration and loss of calcium ions.
2. Eat well-balanced meals (limit saturated fats and instead consume unsaturated fats). Keep body weight within the normal range and exercise regularly.

## Understanding the Terms

**a.** vitamins; **b.** lipase; **c.** lacteal; **d.** esophagus; **e.** gallbladder

## Testing Yourself

**1.** a; **2.** b; **3.** d; **4.** b; **5.** d; **6.** c; **7.** e; **8.** a; **9.** c; **10.** c; **11.** e; **12.** c; **13.** d; **14.** c; **15.** a; **16.** b; **17.** Test tube 1: no digestion—no enzyme and no HCl; Test tube 2: some digestion—no HCl; Test tube 3: no digestion—no enzyme; Test tube 4: digestion—both enzyme and HCl are present.

## Thinking Scientifically

**1.** The drug was probably modified by the liver. All the blood from the intestine goes to the liver where any actual or potentially poisonous compound is metabolized, or changed, in order to make it less toxic or easier to excrete via the kidneys. **2.** The jaws of snakes that swallow whole animals unhinge to make this possible. The teeth curve backwards to retain the prey. Extrusion of the trachea allows breathing while slow swallowing occurs. The esophagus and stomach expand to allow passage and storage of the animal. The intestines are short because the food source is mostly protein and fat. Chemical digestion of a whole animal would require powerful digestive enzymes.

# CHAPTER 35

## Check Your Progress

**35.1:** 1. Air has a drying effect, and respiratory surfaces have to be moist. The body of a terrestrial animal provides this moisture. 2. Insects have many tracheae that branch into ever smaller tubes, which deliver oxygen to the cells. The two steps of respiration (breathing, external exchange) are not necessary.
3. Chronic bronchitis, emphysema, cancer, aneurysms, stroke, miscarriage, and many more. **35.2:** 1. During inspiration, the rib cage moves up and out, and the diaphragm contracts and moves down. As the thoracic cavity expands, air flows into the lungs due to decreased air pressure in the lungs. During expiration, the rib cage moves down and the diaphragm relaxes and moves up to its former position. Air flows out as a result of increased pressure in the lungs. 2. In the lungs, oxygen entering pulmonary capillaries combines with hemoglobin (Hb) in red blood cells to form oxyhemoglobin ($HbO_2$. In the tissues, Hb gives up $O_2$. $CO_2$ enters the blood and the red blood cells. Some combines with Hb to form carbaminohemoglobin $HbCO_2$. Most $CO_2$ combines with water to form carbonic acid, which dissociates into $H^+$ and $HCO_3^-$. The $H^+$ is absorbed by the globin portions of hemoglobin to form reduced hemoglobin $HbH^+$. This helps stabilize the pH of the blood. The $HCO_3^-$ is carried in the plasma. **35.3:** 1. Infections: pneumonia, pulmonary tuberculosis. Environmental: pulmonary fibrosis, emphysema, bronchitis, lung cancer

## Understanding the Terms

**a.** ventilation; **b.** diaphragm; **c.** vocal cord; **d.** gill; **e.** expiration

## Testing Yourself

**1. a.** external respiration**; b.** $CO_2$; **c.** $CO_2$; **d.** tissue cells; **e.** internal respiration; **f.** $O_2$; **g.** $O_2$; **2.** a; **3.** b; **4.** b; **5.** d; **6.** c; **7.** b; **8.** c; **9.** b; **10.** e; **11.** b; **12.** d; **13.** d; **14.** c; **15.** b; **16.** e; **17.** b; **18.** d; **19. a.** nasal cavity; **b.** nostril; **c.** pharynx; **d.** epiglottis; **e.** glottis; **f.** larynx; **g.** trachea; **h.** bronchus; **i.** bronchiole. See also Figure 35. 6*a*, page 654.

## Thinking Scientifically

**1.** A severed spinal cord prevents the medulla oblongata from communicating with the rib cage and diaphragm via the phrenic nerve and intercostal nerves.
**2.** Fetal hemoglobin must have a higher affinity of oxygen than maternal hemoglobin. Therefore it will bind to oxygen at a lower partial pressure and a lower pH than does maternal hemoglobin.

# CHAPTER 36

## Check Your Progress

**36.1:** 1. Urea is not as toxic as ammonia, and it does not require as much water to excrete; uric acid takes more energy to prepare than urea. 2. No, the workings of the nephridia stay the same, regardless of the thickness of the skin. 3. The blood of a shark is isotonic to seawater. 4. Most likely, the tonicity of a seagull's urine is about the same as that of a human because they rid the body of salt using a salt gland, not kidneys. **36.2:** 1. The kidneys. 2. All small molecules enter the filtrate, and the blood takes back what it needs. 3. It fine-tunes the reabsorption of sodium ions.

## Understanding the Terms

**a.** Malpighian tubule; **b.** glomerular capsule; **c.** urea; **d.** aldosterone; **e.** uric acid

### Testing Yourself

**1.** d; **2.** a; **3.** c; **4.** b; **5.** a; **6.** e; **7.** b; **8.** e; **9.** c; **10.** b; **11.** d; **12.** a; **13.** c; **14.** d; **15.** c; **16.** d; **17.** c; **18.** a; **19. a.** glomerular capsule; **b.** proximal convoluted tubule; **c.** Loop of the nephron; **d.** descending limb; **e.** ascending limb; **f.** distal convoluted tubule; **g.** collecting duct; **h.** renal artery; **i.** afferent arteriole; **j.** glomerulus; **k.** efferent arteriole; **l.** peritubular capillary network; **m.** renal vein. See also Figure 36.9, page 671.

### Thinking Scientifically

**1.** A low-salt diet should by itself reduce blood pressure. Kidney damage, if it is not too serious, should then repair itself. If kidney damage came first, the low-salt diet may not by itself lower the blood pressure. **2.** Angiotensin II is a vasoconstrictor, and in addition, it stimulates aldosterone secretion. Therefore, this treatment causes a large decrease in blood pressure.

## CHAPTER 37

### Check Your Progress

**37.1:** 1. A ganglion is a cluster of neuron (nerve cell) bodies. In animals with a CNS and a PNS, it is a cluster of neurons located outside the CNS. 2. Bilateral symmetry plus cephalization leads to paired sensory organs for sight, hearing, and smell, that are useful for obtaining information about the animal's environment. 3. The CNS (central nervous system) consists of the brain and spinal cord. The PNS (peripheral nervous system) is composed of nerves and ganglia. **37.2:** 1. The nerve impulse would travel more quickly down the myelinated axon due to saltatory conduction. 2. a. $Na^+$ moves from the outside of the axon membrane to the inside. b. $K^+$ moves from the inside of the axon membrane to the outside. 3. Neurotransmitter molecules may be degraded by enzymes, or be taken up by the presynaptic cell. **37.3:** 1. The spinal cord contains important pathways for communication between the brain and the spinal nerves which serve the rest of the body. 2. Output from the RAS functions keeps us awake. A malfunctioning RAS may stop signaling to the sleep centers in the hypothalamus, enabling them to temporarily take over and cause uncontrollable sleepiness. 3. We normally experience positive feelings when we recognize the familiar faces of our loved ones. A disconnect between the amygdala and the cortex disables this emotional response, and the injured person, desperate for an explanation, adopts the "imposter" belief. **37.4:** 1. Cranial nerves emerge from the brain; some are sensory, some are motor, and others are mixed. Spinal nerves emerge from the spinal cord; all are mixed. 2. a. The spinal cord; b. The brain. 3. The parasympathetic ("rest and digest") division dominates as you enjoy your meal, but your friend's "surprise" causes a sudden increase in sympathetic ("fight or flight") activity.

### Understanding the Terms

**a.** reflex; **b.** neurotransmitter; **c.** autonomic system; **d.** ganglion; **e.** acetylcholine

### Testing Yourself

**1.** b; **2.** b; **3.** c; **4.** a; **5.** a; **6.** c; **7.** d; **8.** b; **9.** b; **10.** c; **11.** d; **12.** c; **13.** c; **14.** d; **15.** b; **16.** c; **17.** b; **18.** c; **19. a.** central canal; **b.** gray matter; **c.** white matter; **d.** dorsal root; **e.** cell body of sensory neuron in dorsal root ganglion; **f.** spinal nerve; **g.** cell body of motor neuron; **h.** interneuron

### Thinking Scientifically

**1.** The most direct cause of the fight-or-flight response is norepinephrine released by the sympathetic nervous system. If norepinephrine is being released inappropriately it could be that the postganglionic fibers are being triggered unnecessarily. If so, control could be attempted by blocking production of norepinephrine or interaction of norepinephrine with its receptor. However, the harm caused by disabling the sympathetic nervous system would probably outweigh the benefit of reducing the panic response. **2.** The portion of the brain's somatosensory cortex originally devoted to sensation from an amputated limb gradually reorganizes itself; as a result, sensory input from different areas of the body is often perceived as pain in the missing limb. In addition, some portion of the sensory neurons serving the amputated leg will still be present. The axon portion of these neurons in the spinal cord would still be able to release neurotransmitter substances. The release of neurotransmitter by sensory neurons coming from an amputated limb is being perceived as pain by the brain.

## CHAPTER 38

### Check Your Progress

**38.1:** 1. Both are chemical senses that use chemoreceptors to detect molecules in the environment. 2. Sweet, sour, salty, bitter, and umami. 3. The olfactory cells that bind odor molecules are neurons, and they convey impulses directly to the olfactory bulbs of the brain. **38.2:** 1. Rods are for peripheral vision and motion detection; they are well-suited for dim light. Cones are for color perception and fine detail, and are best-suited for bright light. Many rods may excite a single ganglion cell, but much smaller numbers of cones excite individual ganglion cells. 2. Sclera, choroid, retina. 3. Ganglion cell layer, bipolar cell layer, photoreceptor (rod and cone) layer. Light must pass through the ganglion and bipolar cell layers before it reaches the photoreceptor cells. **38.3:** 1. a. middle; b. outer; c. inner; d. inner; e. inner; f. outer. 2. Auditory canal, tympanic membrane (eardrum), ossicles (malleus, incus, and stapes), oval window, cochlea. 3. Utricle and saccule. Semicircular canals.

### Understanding the Terms

**a.** compound eye; **b.** retina; **c.** sclera; **d.** chemoreceptor; **e.** organ of Corti

### Testing Yourself

**1.** e; **2.** c; **3.** c; **4.** e; **5.** d; **6.** c; **7.** c; **8.** b; **9.** d; **10.** e; **11.** a; **12.** b; **13.** c; **14.** a; **15.** d; **16.** b; **17.** d; **18. a.** retina—contains sensory receptors; **b.** choroid—absorbs stray light; **c.** sclera—protects and supports eyeball; **d.** optic nerve—transmits impulses to brain; **e.** fovea centralis—makes acute vision possible; **f.** muscle in ciliary body—holds lens in place, accommodation; **g.** lens—refracts and focuses light rays; **h.** iris—regulates light entrance; **i.** pupil—admits light; **j.** cornea—refracts light rays. See also Figure 38.5, page 705.

### Thinking Scientifically

**1.** Taste perception in the brain may be less in obese individuals with low density of taste buds. Measuring eating-associated brain activity may indicate taste perception. Hypothesis: If low density of taste buds causes obesity then brain activity associated with taste perception would be less in obese individuals compared to those who are not obese. While quantity of taste perception may be related to obesity, many other factors may also be involved because eating must have various levels of control. **2.** Perhaps the increased air pressure upon submersion intensifies volume (loudness) leading to hearing loss. Hypothesis 1: If increased air pressure causes hearing loss, then hair cells of the organ of Corti will be damaged in individuals subjected to increased air pressure. Hypothesis 2: If increased air pressure causes the inability to hear high tones, then the organ of Corti at the base of the cochlea will show the greatest damage.

## CHAPTER 39

### Check Your Progress

**39.1:** Exoskeleton, endoskeleton, exoskeleton, exoskeleton, endoskeleton. 1. The tongue is a muscular hydrostat. 2. Because the muscle layers surrounding the coelom no longer contract, the hydrostatic skeleton cannot provide support for the body. **39.2:** 1. Osteoblasts build bone and osteoclasts break it down. Osteocytes occupy lacunae. 2. Axial, axial, appendicular, appendicular, axial, appendicular, axial, appendicular, axial. **39.3:** 1. Pair of muscles that work opposite to one another; for example, if one muscle flexes (bends) the joint the other extends (straightens) it. 2. Myofibrils are tubular contractile units that are divided into sarcomeres. Each sarcomere contains actin (thin filaments) and myosin (thick filaments). 3. The movement of myosin heads, triggered by the release of ADP and P, that pulls thin filaments toward the center of the sarcomere.

### Understanding the Terms

**a.** osteoblast; **b.** sliding filament model; **c.** actin; **d.** appendicular skeleton; **e.** pectoral girdle; **f.** hydrostatic skeleton

### Testing Yourself

**1.** b; **2.** f; **3.** c; **4.** e; **5.** e; **6.** b; **7.** e; **8.** b; **9.** b; **10.** c; **11.** b; **12.** e; **13.** a; **14.** e; **15.** a; **16.** c; **17.** b; **18.** b; **19.** **a.** T tubule; **b.** sarcoplasmic reticulum; **c.** myofibril; **d.** Z line; **e.** sarcomere; **f.** sarcolemma of muscle fiber

### Thinking Scientifically

**1.** Neurons that cause the contraction of many muscle fibers would produce more lifting power than neurons that cause the contraction of only a few fibers. Previous experience may be the basis on which the brain "decides" how much lifting power is needed and appropriately innervates the correct number of muscle fibers. If the brain does not perceive the situation correctly, as when a large box is empty, the box or you could go flying. **2.** Hypothesis 2 can be tested by determining if oral creatine is absorbed into the blood; if so, does absorbed creatine reach the inside of muscle fibers and result in a measurably greater amount of creatine phosphate? Hypothesis 1 is probably more difficult to test since it is subject to a serious placebo effect: "Endurance" can be affected by a great many emotional factors. It would be difficult, if not impossible, to control these factors in an experimental investigation.

## CHAPTER 40

### Check Your Progress

**40.1:** 1. Exocrine glands secrete their products through ducts, while endocrine glands generally secrete hormones into the bloodstream. 2. Peptide hormones have receptors in the plasma membrane. Steroid hormones have receptors that are generally in the nucleus, sometimes in the cytoplasm. 3. Peptide hormones. **40.2:** 1. PRL stimulates the mammary glands to produce milk, and oxytocin triggers milk letdown so that milk is released from the breasts. 2. ADH and oxytocin are produced in the hypothalamus, and released from the posterior pituitary. 3. TSH, ACTH, PRL, GH, FSH, LH, and MSH. **40.3:** 1. Angiotensin II causes arterioles to constrict; aldosterone causes reabsorption of $Na^+$, accompanied by water, in the kidneys. 2. Adrenal cortex, pineal gland, adrenal medulla, kidneys, adipose tissue, pancreas, heart, adrenal cortex, thyroid gland. 3. PTH stimulates osteoclasts and calcitonin inhibits them

### Understanding the Terms

**a.** thyroid gland; **b.** negative feedback; **c.** pineal gland; **d.** peptide hormone; **e.** pheromone; **f.** thymus; **g.** anabolic steroids; **h.** leptin

### Testing Yourself

**1.** f; **2.** b; **3.** c; **4.** a; **5.** e; **6.** c; **7.** d; **8.** b; **9.** d; **10.** a; **11.** e; **12.** a; **13.** e; **14.** b; **15.** e; **16.** e; **17.** a; **18.** a; **19.** **a.** inhibits; **b.** inhibits; **c.** releasing hormone; **d.** stimulating hormone; **e.** target gland hormone; **20.** d; **21.** b; **22.** a; **23.** b; **24.** c; **25.** e; **26.** d; **27.** d; **28.** c; **29.** e; **30.** e

### Thinking Scientifically

**1.** Caffeine would have the effect of increasing the effect of epinephrine. cAMP is normally broken down quickly, but in the presence of caffeine, its slower breakdown would be equivalent to increasing the amount of epinephrine in the blood, and a longer or stronger response would result. **2.** The genes for all possible sexual characteristics must be present in both males and females. Like any other genes, one allele of each pair of genes is inherited from each parent. Which genes are expressed is dependent on the sex hormones present, which ultimately is dependent on the inheritance of sex chromosomes.

## CHAPTER 41

### Check Your Progress

**41.1:** 1. Asexual reproduction allows organisms to reproduce rapidly and colonize favorable environments quickly. Sexual reproduction produces offspring with a new combination of genes that may be more adaptive to a changed environment. 2. An oviparous animal lays eggs which hatch outside the body. A viviparous animal gives birth after the offspring have developed within the mother's body. Ovoviviparous animals retain fertilized eggs within a parent's body until they hatch; the parent then gives birth to the young. 3. A shelled egg contains extraembryonic membranes which keep the embryo moist, carries out gas exchange, collect wastes, and provide yolk as food. **41.2:** 1. Seminiferous tubule, epididymis, vas deferens, ejaculatory duct, urethra. 2. Seminal vesicles, prostate gland, and bulbourethral glands. 3. LH, FSH. **41.3:** 1. Ovary, oviduct, uterus, cervix, vagina. 2. All four hormones are at their lowest or nearly lowest levels. 3. FSH stimulates ovarian follicles to produce primarily estrogen. LH stimulates the corpus luteum to produce primarily progesterone. **41.4:** 1. Male and female condom and the diaphragm prevent sperm from coming in contact with the egg. 2. In AID, sperm are placed in the vagina or sometimes the uterus. In IVF, conception takes place in laboratory glassware and embryos are transferred to the woman's uterus. In GIFT, eggs and sperm are brought together in laboratory glassware, and placed in the oviducts immediately afterward. In ICSI, one sperm is injected directly into an egg. **41.5:** 1. Chlamydia, gonorrhea, syphilis, and bacterial vaginosis. 2. HIV infects helper T lymphocytes, which are important to the immune system because they stimulate B lymphocytes and cause them to produce antibodies. 3. Chlamydia and gonorrhea.

### Understanding the Terms

**a.** ovulation; **b.** parthenogenesis; **c.** progesterone; **d.** semen; **e.** gonad

### Testing Yourself

**1. a.** seminal vesicle; **b.** ejaculatory duct; **c.** prostate gland; **d.** bulbourethral gland; **e.** anus; **f.** vas deferens; **g.** epididymis; **h.** testis; **i.** scrotum; **j.** foreskin; **k.** glans penis; **l.** penis; **m.** urethra; **n.** vas deferens; **o.** urinary bladder; **2.** b; **3.** e; **4.** c; **5.** e; **6.** c; **7.** c; **8.** c; **9.** c; **10.** a; **11.** a; **12.** c; **13.** c; **14.** b; **15.** e; **16.** c; **17.** a; **18.** c; **19.** b; **20.** c

### Thinking Scientifically

**1.** Fetuses take a high caloric toll on their mothers. In modern times food supply is usually sufficient for our needs, but earlier in human history the food supply may not have been as dependable. Mothers with low percentages of body fat might starve during pregnancy or the prolonged period of breast-feeding that early human infants presumably required, and thus would not be able to reproduce at all. By stopping the reproductive cycle when the body senses insufficient food reserves, the woman is perhaps better able to survive to a time when food is more plentiful. **2.** The possibility exists that spermatogenesis is not occurring as it should. Therefore, it would be helpful to see if the number of cells undergoing spermatogenesis is normal. Data concerning the levels of various sex hormones in the blood would also be useful. A low testosterone level would be a significant find. It's also been suggested that many organic pollutants such as pesticides have hormonal effects that could be interfering with the normal stimulation of spermatogenesis by testosterone. Therefore, comparative sperm count data between men exposed to organic pollutants and men not exposed would be helpful.

## CHAPTER 42

### Check Your Progress

**42.1:** 1. The fast block is the depolarization of the egg's plasma membrane that occurs upon initial contact with a sperm. The slow block occurs when the secretion of cortical granules converts the zona pellucida into the fertilization membrane. 2. Mesoderm, endoderm, ectoderm, ectoderm, mesoderm, mesoderm, mesoderm, endoderm. 3. Neurulation.
**42.2:** 1. Cytoplasmic segregation is the parceling out of maternal determinants as mitosis occurs. Induction is the influence of one embryonic tissue on the development of another. 2. A morphogen is a transcription factor that is distributed along a concentration gradient in the embryo and helps direct morphogenesis. 3. The homeobox encodes the homeodomain region of the protein product of the gene. The homeodomain is the DNA-binding region of the protein, which is a transcription factor.
**42.3:** 1. Allantois, yolk sac, and chorion. 2. The upper third of the oviduct. Morula. 3. The placenta provides gas exchange, nutrient delivery, and waste removal for the embryo and later the fetus.

### Understanding the Terms

**a.** induction; **b.** germ layer; **c.** amnion; **d.** homeobox; **e.** gastrula

### Testing Yourself

**1.** b; **2.** b; **3.** a; **4.** e; **5.** b; **6.** e; **7.** e; **8.** d; **9. a.** chorion (contributes to forming placenta where wastes are exchanged for nutrients and oxygen); **b.** amnion (protects and prevents desiccation); **c.** embryo; **d.** allantois (blood vessels become umbilical blood vessels); **e.** yolk sac (first site of blood cell formation); **f.** chorionic villi (embryonic portion of placenta); **g.** maternal portion of placenta; **h.** umbilical cord (connects developing embryo to the placenta). See also Figure 42. 11, page 807. **10.** c; **11.** a; **12.** d; **13.** b; **14.** e; **15.** b; **16.** c; **17.** b

### Thinking Scientifically

**1.** The gene is for a maternal determinant and such genes are only expressed as the egg is maturing. Individuals can have mutant genes and still develop normally because genes for maternal determinates are not expressed in the present generation; They are expressed in the next generation. The mutant female is sterile because none of her eggs contain the maternal determinant in question. **2.** These findings tell us that *Ubx* is important in wing morphogenesis for both orders of insects, but as a homeotic gene its role is that of a selector for target gene activity (not a simple controller of wing number). Since the effects of *Ubx* mutation are so different in flies versus butterflies, there must have been a shift in the target genes of *Ubx* since these two orders of insects diverged from their common ancestor.

## CHAPTER 43

### Check Your Progress

**43.1:** 1. Fisher lovebird (carry nesting material in beak) mated to Peach-faced lovebird (carry nesting material in rump feathers) result in offspring with intermediate behavior. Offspring of inland garter snakes (do not eat slugs) and coastal garter snakes (eat slugs) show an intermediate liking for slugs. 2. Gene for egg-laying hormone in *Aplysia* was isolated and its protein product controls egg-laying behavior. The gene fosB has been found to control maternal behavior in mice. **43.2:** 1. Associative learning. 2. Just hatched, laughing gull chicks instinctively peck at parents bill to be fed but their accuracy improves after a few days. 3. Chimpanzees pile up boxes to reach food and ravens use their beak and feet to bring up food attached to a string.
**43.3:** 1. Pheromones are used to mark a territory so other animals of that species will stay away; honeybees do a waggle dance to guide other bees to a food source; vervet monkeys have calls that make other vervets run away. 2. Chemical (effective all the time, not as fast as auditory); auditory (can be modified but the recipient has to be present when message is sent); visual (need not be accompanied by chemical or auditory, needs light in order to receive); tactile (permits bonding; recipient must be close). 3. chemical: taste buds and olfactory receptors; auditory: ears; visual: eyes; tactile: touch receptors in skin.
**43.4:** 1. One benefit of territoriality is to ensure a source of food. 2. Both an animal's reproductive strategy and sexual selection favors features that increase an animal's chance of leaving offspring. 3. Altruistic behavior is supposed to be selfless but when, for example, a child helps its parents raise siblings, the child is helping to increase some of its own genes in the next generation.

### Understanding the Terms

**a.** territoriality; **b.** altruism; **c.** communication; **d.** pheromone

## Testing Yourself

**1.** c; **2.** a; **3.** d; **4.** b; **5.** c; **6.** c; **7.** d; **8.** e; **9.** c; **10.** b; **11.** d; **12.** c; **13.** c; **14.** a; **15.** c; **16.** b; **17.** d; **18.** a; **19.** c; **20.** b; **21.** a; **22.** c; **23.** a; **24.** b

## Thinking Scientifically

**1.** Evidence supporting the hypothesis would be that the sentries reproduce less than nonsentries, and that reproduction of others is enhanced by the activity of the sentry. **2.** Infants could have been conditioned to turn their head toward their mother's voice.

# CHAPTER 44

## Check Your Progress

**44.1:** 1. A population is all the members of a one species that inhibit a particular area and a community is all the populations that interact within that area. 2. To develop models that explain and predict the distribution and abundance of organisms. 3. Abiotic means the nonliving aspects of an environment such as rainfall and temperature. **44.2:** 1. Population density is the number of individuals per unit area and population distribution is the pattern of dispersal of individuals across an area of interest. 2. In type I survivorship curve, most individuals survive well past the midpoint of the life span and death does not come until near the end of the life span. In type II, survivorship decreases at a constant rate throughout the life span. In type III, most individuals die young. 3. In a bell-shaped age pyramid, the pre-reproductive members represent the largest portion of the population. **44.3:** 1. An environment in which the weather, food supply etc. remains stable favors iteroparity. 2. Exponential growth ceases when the environment cannot support a larger population size, that is when the size of the population has reached the environment's carrying capacity.
**44.4:** 1. As population density increases, competition and predation become more intense. 2. If a flash flood occurs, mice that can stay afloat will survive and reproduce whereas those that quickly sink will not survive and will not reproduce. In this way the ability to stay afloat will be more prevalent in the next generation. **44.5:** 1. K-strategist: allocate energy to their own growth and survival and to the growth and survival of their limited number of offspring. r-strategist: allocate energy to producing a large number of offspring and little or no energy goes into parental care.
**44.6:** 1. More-developed countries have a low rate of population growth while the less-developed countries have a high population growth. 2. When there are more women entering the reproductive years than those that are leaving them behind. 3. Since resources are in limited supply, consumption in the MDC will have to decrease.

## Understanding the Terms

**a.** demographic transition; **b.** population; **c.** exponential growth; **d.** carrying capacity; **e.** biotic potential

## Testing Yourself

**1.** d; **2.** c; **3.** e; **4.** b; **5.** b; **6.** e; **7.** c; **8.** e; **9.** c; **10.** e; **11.** c; **12.** e; **13.** b; **14.** e; **15.** e

## Thinking Scientifically

**1.** If the parasites killed a significant number of moths, there would be fewer moth hosts the next year. Evolution would seem to favor a parasite that was not too efficient in killing its host, since that would ensure an adequate supply of hosts for future generations of parasites. **2.** Determine the original normal flow of the river and maintain the flow as close to normal as possible.

# CHAPTER 45

## Check Your Progress

**45.1:** 1. An organism's habitat is the place where it lives and reproduces. The niche is the role it plays in its community such as whether it is a producer or consumer. 2. The two factors are (1) the predator causes the prey population to decline leading to a decline in the predator population; later when the prey population recovers so does the predator population; (2) lack of food causes the prey population to decline followed by the prey population; later when food is available to the prey population they both recover. 3. Acacias feed the ants that protect them from herbivores; Clark's nutcrackers feed on the seeds of whitebark pine trees but also disperse the seeds; pollinators take nectar from flowers and carry their pollen to other flowers of the same species. **45.2:** 1. No, ecological succession is observable (see Figure 45.14) but ecologists want to provide an explanation for succession and decide if it results in a "climax community". **45.3:** 1. A producer of food (photosynthesizer) is at the base of an ecological pyramid. 2. Energy passes from one population to the next and at each step more is converted to heat until all of the original input is heat. Therefore energy flows though an ecosystem. Chemicals pass from one population to the next and then recycle back to the producer populations again. 3. Return of $CO_2$ to the atmosphere because humans burn fossil fuels and destroy forests that take up $CO_2$.

## Understanding the Terms

**a.** food web; **b.** habitat; **c.** ecological succession; **d.** nitrogen fixation

## Testing Yourself

**1.** c; **2.** e; **3.** e; **4.** a; **5.** c; **6.** b; **7.** d; **8.** b; **9.** c; **10.** d; **11.** e; **12.** **a.** producers; **b.** consumers; **c.** inorganic nutrient pool; **d.** decomposers; **13.** a; **14.** e; **15.** c; **16.** d; **17.** c; **18.** c; **19.** b; **20.** c; **21.** d; **22.** d

## Thinking Scientifically

**1.** Observe the birds carefully to see if they differ in habitat and food requirements, relationships with other organisms, time of day for feeding and season of year for reproduction, and effect on abiotic environment. **2.** Fill a large container with water from the pond. Add phosphate slowly over several days or months, and when you see growth, calculate the amount of phosphate you need for the pond.

# CHAPTER 46

## Check Your Progress

**46.1:** 1. Because the Earth is a sphere, the sun's rays hit the equator straight on but are angled to reach the poles. 2. The windward side of the mountain receives more rainfall than the other side. Winds blowing over bodies of water collect moisture that they lose when they reach land. **46.2:** 1. A tropical rain forest has a canopy (tops of great variety of tall evergreen hardwood trees) with buttressed trunks at ground level. Long lianas (hanging vines) limb into the canopy. Epiphytes grow on the trees. The understory consists of smaller plants and the forest floor is very sparse. A temperate deciduous forest contains trees (oak, beech, sycamore, and maple) that lose their leaves in the fall. Enough light penetrates the canopy to allow a layer of understory trees. Shrubs, mosses, and ferns grow at ground level. 2. The savanna is an expansive grassland that has a moderate climate. Therefore, the grasses keep producing throughout the year and provide plentiful food for a great variety of and number of herbivores that provide food for carnivores. The tundra is cold much of the year and has a limited growing season; therefore, its productivity is low and it supports only small populations of a few types of herbivores. **46.3:** 1. Most of the open ocean is the pelagic zone (open waters) divided into the epipelagic zone (contains phytoplankton, zooplankton, many types of fishes and also dolphins and whales). The mesopelagic zone contains only carnivores adapted to the absence of light. The bathypelagic zone is incomplete darkness that contains strange-looking fishes and invertebrates. Few vertebrates but many invertebrates (echinoderms, tube worms) exist on the abyssal plain and feed on debris that floats down from above.

### Understanding the Terms

**a.** estuary; **b.** pelagic zone; **c.** taiga; **d.** savanna; **e.** spring turnover

### Testing Yourself

**1.** b; **2.** d; **3.** c; **4.** d; **5.** a; **6.** b; **7.** e; **8.** e; **9.** a; **10.** a; **11.** b; **12.** d; **13.** b; **14.** a; **15.** c; **16.** b; **17.** a; **18.** d; **19.** c

### Thinking Scientifically

**1.** Bacteria and fungi grow in warm, moist environments, such as tropical rain forests. **2.** Wetlands act as buffers between the land and sea. They can absorb much of the water coming in from the ocean.

## CHAPTER 47

### Check Your Progress

**47.1:** 1. Conservation biology studies all aspects of biodiversity with the goal of conserving natural resources for all generations. 2. Biodiversity includes the number of species on Earth; genetic diversity (variations in a species); ecosystem diversity (interactions of species); landscape diversity (interactions of ecosystems). **47.2:** 1. Direct value is a service that is immediately recognizable such as producing a medicine, food, or commercial product. Indirect value may not be as noticeable as in assisting biogeochemical cycles waste disposal, providing fresh water, preventing soil erosion, regulating climate or providing a place to vacation. **47.3:** 1. Habitat loss, exotic species, pollution, overexploitation, disease. 2. Exotic plants displace native plants, predators introduced to kill pests also kill native animals; escaped animals may compete with, prey on, hybridize with or introduce diseases into native populations. **47.4:** 1. A landscape involves more than one ecosystem and sometimes keystone species move between ecosystems. 2. Begin as soon as possible, mimic natural processes, strive for sustainable development while providing services to humans.

### Understanding the Terms

**a.** biodiversity hotspot; **b.** metapopulation; **c.** source population; **d.** threatened species

### Testing Yourself

**1.** e; **2.** b; **3.** d; **4.** e; **5.** e; **6.** e; **7.** b; **8.** e; **9.** e; **10.** c; **11.** c; **12.** e; **13.** b; **14.** e; **15.** e; **16.** a; **17.** e; **18.** b, c; **19. a.** habitat loss; **b.** introduction of exotic species; **c.** pollution; **d.** overexploitation; **e.** disease

### Thinking Scientifically

**1.** Natural history data on all species in the forest. **2.** Species that show natural resistance to disease.

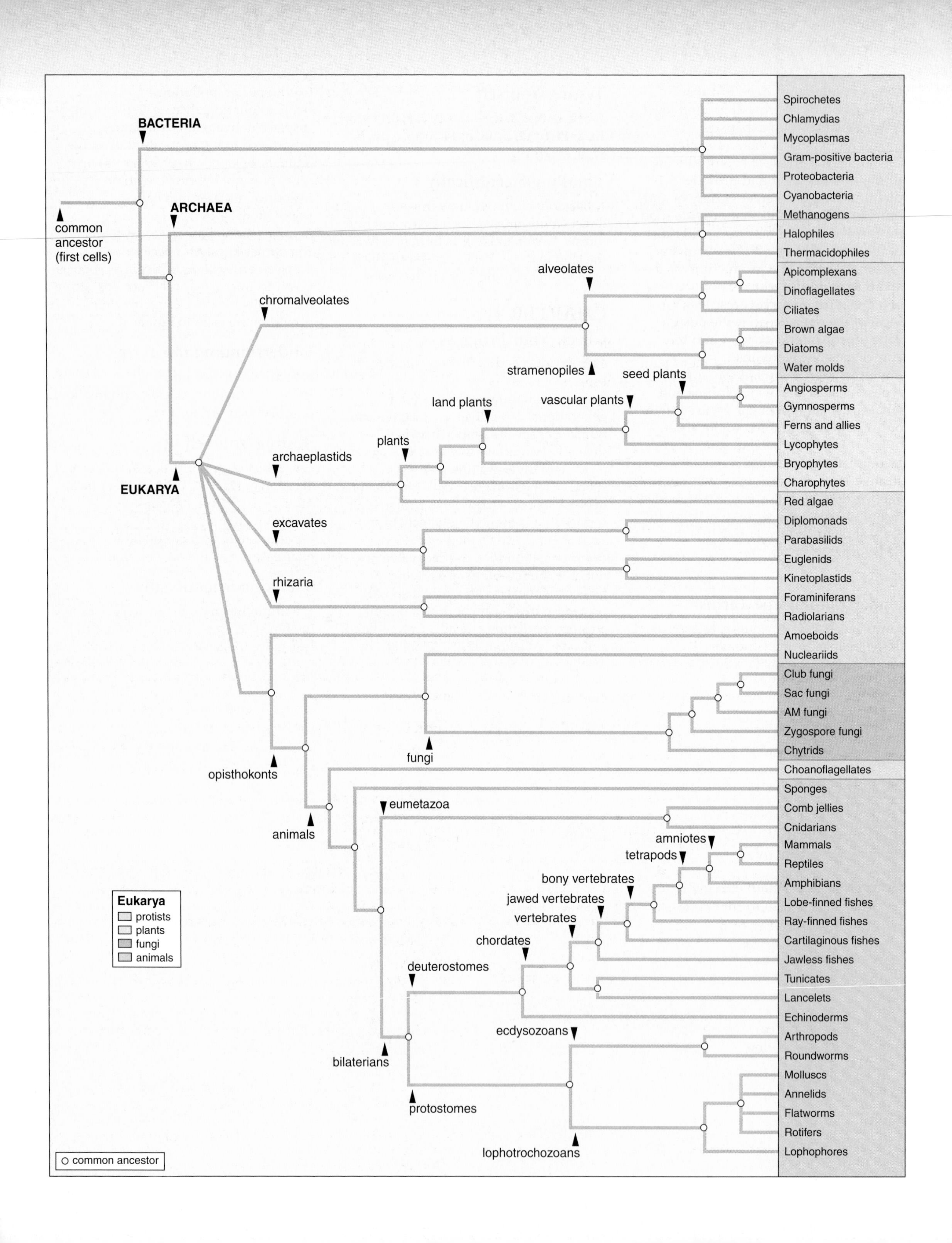

BACTERIA
ARCHAEA
EUKARYA
common ancestor (first cells)
Spirochetes
Chlamydias
Mycoplasmas
Gram-positive bacteria
Proteobacteria
Cyanobacteria
Methanogens
Halophiles
Thermacidophiles
alveolates
chromalveolates
stramenopiles
Apicomplexans
Dinoflagellates
Ciliates
Brown algae
Diatoms
Water molds
seed plants
vascular plants
land plants
plants
archaeplastids
Angiosperms
Gymnosperms
Ferns and allies
Lycophytes
Bryophytes
Charophytes
Red algae
excavates
Diplomonads
Parabasilids
Euglenids
Kinetoplastids
rhizaria
Foraminiferans
Radiolarians
Amoeboids
Nucleariids
Club fungi
Sac fungi
AM fungi
Zygospore fungi
Chytrids
fungi
opisthokonts
Choanoflagellates
animals
eumetazoa
Sponges
Comb jellies
Cnidarians
amniotes
tetrapods
bony vertebrates
jawed vertebrates
vertebrates
chordates
deuterostomes
Mammals
Reptiles
Amphibians
Lobe-finned fishes
Ray-finned fishes
Cartilaginous fishes
Jawless fishes
Tunicates
Lancelets
Echinoderms
ecdysozoans
bilaterians
protostomes
lophotrochozoans
Arthropods
Roundworms
Molluscs
Annelids
Flatworms
Rotifers
Lophophores
Eukarya
protists
plants
fungi
animals
common ancestor

# APPENDIX B

## Tree of Life

The Tree of Life depicted in this appendix is based on the phylogenetic (evolutionary) trees presented in the text. Figure 1.5 showed how the three domains of life—Bacteria, Archaea, and Eukarya—are related. This relationship is also apparent in the Tree of Life, which combines the individual trees given in the text for eukaryotic protists, plants, fungi, and animals. In combining these trees, we show how all organisms may be related to one another through the evolutionary process.

The text also described the organisms that are included in the tree. These descriptions are repeated here.

### PROKARYOTES

Domains Bacteria and Archaea (Chapter 20) constitute the prokaryotic organisms that are characterized by their simple structure but a complex metabolism. The chromosome of a prokaryote is not bounded by a nuclear envelope, and therefore, these organisms do not have a nucleus. Prokaryotes carry out all the metabolic processes performed by eukaryotes and many others besides. However, they do not have organelles, except for plentiful ribosomes.

### DOMAIN BACTERIA

Bacteria are the most plentiful of all organisms, capable of living in most habitats, and carry out many different metabolic processes. While most bacteria are aerobic heterotrophs, some are photosynthetic, and some are chemosynthetic. Motile forms move by flagella consisting of a single filament. Their cell wall contains peptidoglycan and they have distinctive RNA sequences.

### DOMAIN ARCHAEA

Their cell walls lack peptidoglycan, their lipids have a unique branched structure, and their ribosomal RNA sequences are distinctive.

**Methanogens.** Obtain energy by using hydrogen gas to reduce carbon dioxide to methane gas. They live in swamps, marshes and intestines of mammals.

**Extremophiles.** Able to grow under conditions that are too hot, too cold, and too acidic for most forms of life to survive.

**Nonextreme archaea.** Grow in wide variety of environments that are considered within the normal range for living organisms.

### DOMAIN EUKARYA

Eukarya have a complex cell structure with a nucleus and several types of organelles that compartmentalize the cell. Mitochondria that produce ATP and chloroplasts that produce carbohydrate are derived from prokaryotes that took up residence in a larger nucleated cell. Protists tend to be unicellular, while plants, fungi, and animals are multicellular with specialized cells. Each multicellular group is characterized by a particular mode of nutrition. Flagella, if present, have a 9 + 2 organization.

### PROTISTS

The protists (Chapter 21) are a catchall group for any eukaryote that is not a plant, fungus, or animal. Division into six supergroups is a working hypothesis that is subject to change as more is known about the evolutionary relationships of the protists. A supergroup is a major eukaryotic group and six supergroups encompass all members of the domain Eukarya including protists, plants, fungi, and animals.

**Archaeplastids.** A supergroup of photosynthesizers with plastids derived from endosymbiotic cyanobacteria. Includes land plants and other photosynthetic organisms, such as green and red algae and charophytes, exemplified by the stoneworts, which share a common ancestor with land plants.

**Chromalveolates.** A supergroup that includes the Stramenopiles, which have a unique flagella, and the Alveolates, which have small sacs under plasma membrane.

*Stramenopiles.* Includes brown algae, such as *Laminaria* and *Fucus*, diatoms, golden brown algae, and water molds.

*Alveolates.* Includes dinoflagellates, ciliates such as *Paramecium*, and apicomplexans, such as *Plasmodium vivax.*

**Excavates.** Have an excavated oral grove and form a supergroup that includes zooflagellates, such as euglenids (e.g., *Euglena*); diplomonads, such as *Giardia lambia;* and kinetoplastids (have a DNA granule called a kinetoplast), such as trypanosomes.

**Amoebozoans.** Supergroup of amoeboid cells that move by pseudopodia. Includes amoeboids, such as *Amoeba proteus,* and slime molds.

**Rhizarians.** Supergroup of amoeboid cells with tests. They form a supergroup that includes the foraminiferans and the radiolarians.

**Opisthokonts.** Supergroup named for members that have a single posterior flagellum (Gk., *opistho,* rear and *kontos,* pole). Includes animals and choanoflagellates that may be related to the common ancestor of animals and the fungi.

# PLANTS

Plants (Chapter 23) are photosynthetic eukaryotes that became adapted to living on land. Includes aquatic green algae called charophytes, which have a haploid life cycle and share certain traits with the land plants.

## Land Plants (embryophytes)

Have an alternation of generation life cycle; protect a multicellular sporophyte embryo; produce gametes in gametangia; possess apical tissue that produces complex tissues; and a waxy cuticle that prevents water loss.

**Bryophytes.** Low-lying, nonvascular plants that prefer moist locations: the dominant gametophyte produces flagellated sperm. The sporophyte is unbranched, and dependent sporophyte produces windblown spores. Includes mosses, liverworts, and hornworts.

## Vascular Plants

Have a dominant, branched sporophyte with vascular tissue: a lignified xylem that transports water, and phloem that transports organic nutrients. Typically produces roots, stems, and leaves; the gametophyte is eventually dependent on the sporophyte.

**Lycophytes (club mosses).** Have leaves called microphylls, which have a single, unbranched vein. The sporangia, which are borne on sides of leaves, produce windblown spores. The independent and separate gametophyte produces flagellated sperm.

**Ferns and their allies.** Have leaves called megaphylls that have branched veins. The dominant sporophyte produces windblown spores in sporangia borne on leaves, and the independent and separate gametophyte produces flagellated sperm. Includes ferns, whisk ferns, and horsetails, see pages 417-418.

### *Seed Plants*

Have leaves that are megaphylls; a dominant sporophyte produces heterospores that become dependent male and female gametophytes. Male gametophyte is pollen grain and female gametophyte develops within ovule, which becomes a seed.

**Gymnosperms.** Large, cone-bearing trees. The sporophyte bears pollen cones, which produce windblown pollen (male gametophyte), and seed cones, which bear ovules. Ovules develop into naked seeds. Includes conifers, gnetophytes, the ginko, and cycads, see pages 420-422.

**Angiosperms (flowering plants).** Nonwoody or woody plants that live in all habitats. The sporophyte bears flowers, which produce pollen grains, and bear ovules within ovary. Following double fertilization, ovules become seeds that enclose a sporophyte embryo and endosperm (nutrient tissue). Fruit develops from ovary.

# FUNGI

Fungi (Chapter 22) have multicellular bodies composed of hyphae; usually absorb food and lack flagella; and produce nonmotile spores during both asexual and sexual reproduction. Chytrids (Chytridiomycota) are aquatic fungi with flagellated spores and gametes.

**Zygospore fungi (Zygomycota).** Exemplified by black bread mold; produce a thick-walled zygospore during sexual reproduction.

**AM fungi (Glomeromycota).** Form a mutualistic relationship with plants called mycorrhizae.

**Sac fungi (Ascomycota).** Exemplified by cup fungi; produce fruiting bodies during sexual reproduction where spores develop fingerlike sacs called asci.

**Club fungi (Basidiomycota).** Exemplified by mushrooms; produce fruiting bodies during sexual reproduction where spores develop in club-shaped structures called basidia.

# ANIMALS

Animals (Chapters 28 and 29) are multicellular, usually with specialized tissues and digestive cavity; ingest or absorb food; and have a diploid life cycle.

**Sponges.** Have an asymmetrical, saclike body perforated by pores internal cavity lined by food-filtering cells called choanocytes; spicules serve as internal skeleton.

**Ctenophores.** Have two tentacles; eight rows of cilia that resemble combs; biradial symmetry. Includes comb jellies.

**Cnidarians.** Radially symmetrical with two tissue layers; sac body plan; and tentacles with nematocysts. Includes hydras, jellyfish, sea anemones, and corals.

## Protostomes

Bilaterally symmetrical with protostome development in which the first opening is the mouth.

## *Lophotrochozoans*

Includes lophophores, which have a specific type of ciliated feeding device (see page 516); and trochophores, which have a trochophore larva or their ancestors had one.

**Flatworms.** Bilaterally symmetrical with cephalization; have three tissue layers and organ systems, including both male and female sex organs. They are acoelomate with an incomplete digestive tract that can be lost in parasites. Planarians are free-living; flukes and tapeworms are parasitic.

**Rotifers (aquatic wheel animals).** Microscopic aquatic animals with a corona (crown of cilia) that looks like a spinning wheel when in motion.

**Annelids. Segmented with body rings and setae.** Cephalization occurs in some polychaetes. They utilize the coelom as a hydroskeleton and have a closed circulatory system. Includes earthworms, polychaetes, and leeches.

**Molluscs. Have a foot, mantel, and visceral mass.** The foot is variously modified; in many, the mantle secretes a calcium carbonate shell. They have a coelom and all organ systems. Includes clams, snails, and squids.

## *Ecdysozoa*

Animals that undergo ecdysis (molting).

**Roundworms.** Have a pseudocoelom, which they use as a hydroskeleton, and a complete digestive tract. Although many are free-living, parasites such as *Ascaris,* pinworms, hookworms, and filarial worms are well known.

**Arthropods.** Have a chitinous exoskeleton with jointed appendages, specialized for particular functions. Insects, many of which are winged, are the most numerous of all arthropods and animals. Includes crustaceans, spiders, scorpions, centipedes, and millipedes, in addition to insects.

## Deuterostomes

Bilaterally symmetrical with deuterstome development in which the second opening is the mouth.

**Echinoderms.** Radial symmetry as adults; unique water-vascular system; and associated tube feet. Their endoskeleton is composed of calcium plates. Includes sea stars, sea urchins, sand dollars, and sea cucumbers.

**Chordates.** Have a notochord, dorsal tubular nerve cord, pharyngeal pouches, and postanal tail at some time; segmentation has led to specialization of parts. Includes tunicates, lancelets, and vertebrates.

*Lancelets (Cephalochordates).* Marine, nonvertebrate chordates shaped like a lance that retain the four chordate characteristics as an adult. Segmentation of muscles is obvious.

*Tunicates (Urochordates).* Marine, nonvertebrate chordates that produce a tunic, a tough sac containing mainly cellulose. Only the larva has the characteristics of chordates; the adult has gill slits. Segmentation is not present.

**Vertebrates (Vertebrata).** Chordates in which the notochord has been replaced with vertebrae. Vertebrae, which make up the spine, are an obvious sign of segmentation.

*Fishes.* Diverse group of marine or freshwater vertebrates that breathe by means of gills and have a single-looped and closed blood circuit. Vertebral column of bone or cartilage; most have jaws and paired appendages. Includes jawless, cartilaginous, ray-finned, and lobe-finned fishes.

*Amphibians.* Vertebrates with lungs, cutaneous respiration, and a three-chambered heart. Frogs and salamanders have legs but caecilians do not.

*Reptiles.* Vertebrates fully adapted to living on land because they have an amniotic-shelled egg, dry, scaly skin, and a rib cage. Turtles, lizards, and snakes have a three-chambered heart but crocodiles and alligators have a four-chambered heart. Birds are crocodilians unique among reptiles because they have feathers and are endothermic.

*Mammals.* Vertebrates characterized by fur and mammary glands. They are endothermic amniotes that, for the most part, practice internal fertilization and development. Monotremes lay shelled eggs; marsupials have a pouch where offspring finish development; and placental mammals produce young capable of independency.

# APPENDIX C

## Metric System

| Unit and Abbreviation | Metric Equivalent | Approximate English-to-Metric Equivalents |
|---|---|---|
| ***Length*** | | |
| nanometer (nm) | $= 10^{-9}$ m ($10^{-3}$ µm) | |
| micrometer (µm) | $= 10^{-6}$ m ($10^{-3}$ mm) | |
| millimeter (mm) | = 0.001 ($10^{-3}$) m | |
| centimeter (cm) | = 0.01 ($10^{-2}$) m | 1 inch = 2.54 cm<br>1 foot = 30.5 cm |
| meter (m) | = 100 ($10^{2}$) cm<br>= 1,000 mm | 1 foot = 0.30 m<br>1 yard = 0.91 m |
| kilometer (km) | = 1,000 ($10^{3}$) m | 1 mi = 1.6 km |
| ***Weight (mass)*** | | |
| nanogram (ng) | $= 10^{-9}$ g | |
| microgram (µg) | $= 10^{-6}$ g | |
| milligram (mg) | $= 10^{-3}$ g | |
| gram (g) | = 1,000 mg | 1 ounce = 28.3 g<br>1 pound = 454 g |
| kilogram (kg) | = 1,000 ($10^{3}$) g | = 0.45 kg |
| metric ton (t) | = 1,000 kg | 1 ton = 0.91 t |
| ***Volume*** | | |
| microliter (µl) | $= 10^{-6}$ l ($10^{-3}$ ml) | |
| milliliter (ml) | $= 10^{-3}$ liter<br>= 1 $cm^3$ (cc)<br>= 1,000 $mm^3$ | 1 tsp = 5 ml<br>1 fl oz = 30 ml |
| liter (l) | = 1,000 ml | 1 pint = 0.47 liter<br>1 quart = 0.95 liter<br>1 gallon = 3.79 liter |
| kiloliter (kl) | = 1,000 liter | |

| °C | °F | |
|---|---|---|
| 100 | 212 | Water boils at standard temperature and pressure. |
| 71 | 160 | Flash pasteurization of milk |
| 57 | 134 | Highest recorded temperature in the United States, Death Valley, July 10, 1913 |
| 41 | 105.8 | Average body temperature of a marathon runner in hot weather |
| 37 | 98.6 | Human body temperature |
| 13.7 | 56.66 | Human survival is still possible at this temperature. |
| 0 | 32.0 | Water freezes at standard temperature and pressure. |

**Units of Temperature**

°F °C

°F scale: 230, 220, 210, 200, 190, 180, 170, 160, 150, 140, 130, 120, 110, 100, 90, 80, 70, 60, 50, 40, 30, 20, 10, 0, -10, -20, -30, -40

°C scale: 110, 100, 90, 80, 70, 60, 50, 40, 30, 20, 10, 0, -10, -20, -30, -40

212° — 100°
160° — 71°
134° — 57°
131°
105.8° — 41°
98.6° — 37°
56.66° — 13.7°
32° — 0°

To convert temperature scales:

$$°C = \frac{(°F - 32)}{1.8}$$

$$°F = 1.8\,(°C) + 32$$

# APPENDIX D

## Periodic Table of Elements

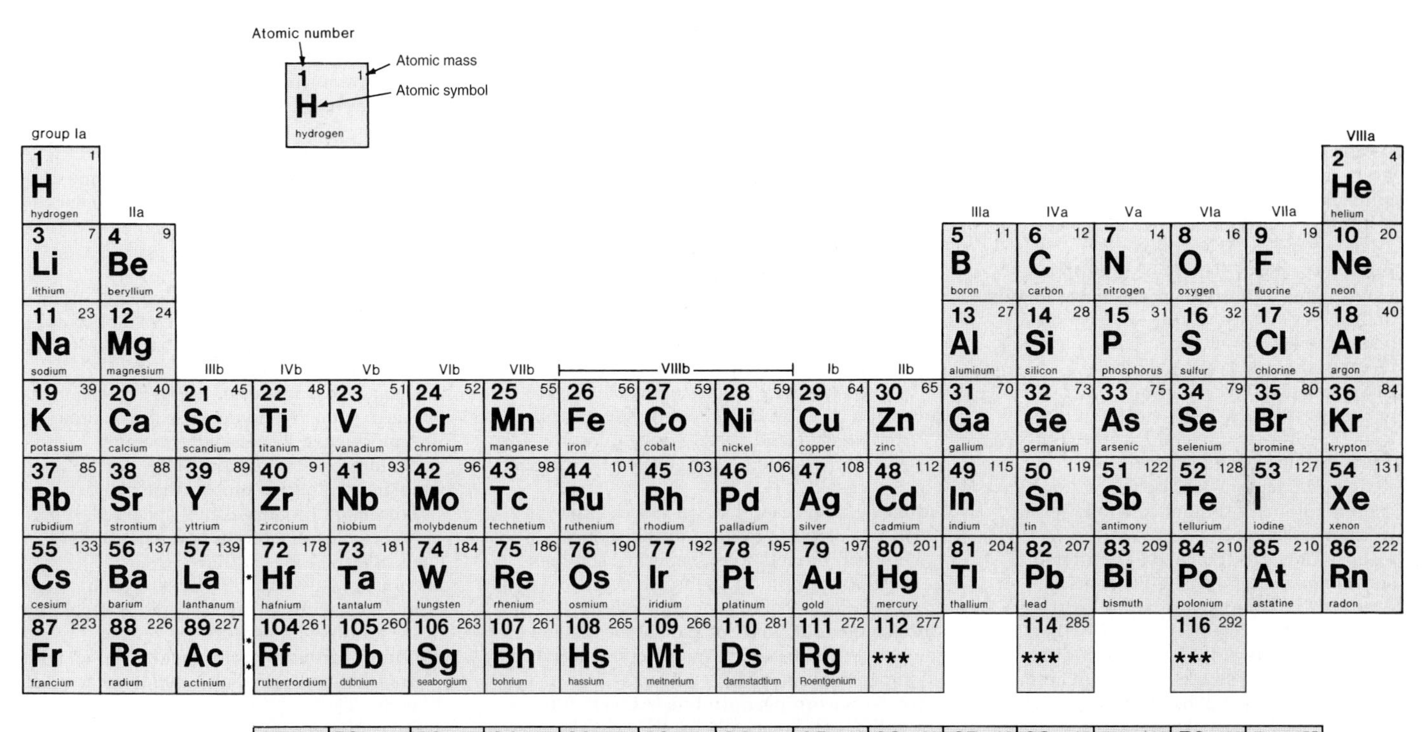

# GLOSSARY

## A

**abscisic acid (ABA)** (ab SIH sick) Plant hormone that causes stomata to close and initiates and maintains dormancy. 480

**abscission** (ab SIH shun) Dropping of leaves, fruits, or flowers from a land plant. 480

**absolute dating (of fossils)** Determining the age of a fossil by direct measurement, usually involving radioisotope decay. 324

**absorption spectrum** Pattern of absorption when pigments take up various wave lengths of light.

**acetylcholine (ACh)** (uh see tuhl KOH lean) Neurotransmitter active in both the peripheral and central nervous systems. 686

**acetylcholinesterase (AChE)** (uh see tuhl KOH lean ESS turr raze) Enzyme that breaks down acetylcholine within a synapse. 686

**acid** Molecules tending to raise the hydrogen ion concentration in a solution and to lower its pH numerically. 32

**acid deposition** The return to Earth in rain or snow of sulfate or nitrate salts of acids produced by commercial and industrial activities. 861

**acromegaly** (ack row MEG uh lee) Condition resulting from an increase in growth hormone production after adult height has been achieved. 742

**actin** (ACK tin) One of two major proteins of muscle; makes up thin filaments in myofibrils of muscle fibers. *See also* myosin. 728

**actin filament** Muscle protein filament in a sarcomere; its movement shortens the sarcomere, yielding muscle contraction. Actin filaments play a role in the movement of the cell and its organelles. 78

**action potential** Electrochemical changes that take place across the axomembrane; the nerve impulse. 685

**active immunity** Ability to produce antibodies due to the immune system's response to a microorganism or a vaccine. 620

**active site** Region on the surface of an enzyme where the substrate binds and where the reaction occurs. 108

**active transport** Use of a plasma membrane carrier protein to move a molecule or ion from a region of lower concentration to one of higher concentration; it opposes equilibrium and requires energy. 94

**adaptation** Organism's modification in structure, function, or behavior suitable that better suits the environment. 5, 273

**adaptive radiation** Rapid evolution of several species from a common ancestor into new ecological or geographical zones. 306

**Addison disease** (ADD dih sun) Condition resulting from a deficiency of adrenal cortex hormones; characterized by low blood glucose, weight loss, and weakness. 746

**adenine (A)** (AD duh neen) One of four nitrogen-containing bases in nucleotides composing the structure of DNA and RNA. Pairs with uracil (U) and thymine (T). 214

**adenosine** (ad DEN oh seen) Portion of ATP and ADP that is composed of the base adenine and the sugar ribose. 53

**adhesion junction** (ad HE shun) Junction between cells in which the adjacent plasma membranes do not touch but are held together by intercellular filaments attached to buttonlike thickenings. 98

**adipose tissue** (AD dip ose) Connective tissue in which fat is stored. 580

**ADP (adenosine diphosphate)** (ad DEN oh seen dye FOSS fate) Nucleotide with two phosphate groups that can accept another phosphate group and become ATP. 53, 106

**adrenal cortex** (uh DREEN uhl) Outer portion of the adrenal gland; secretes mineralocorticoids, such as aldosterone, and glucocorticoids, such as cortisol. 745

**adrenal gland** Gland that lies atop a kidney; the *adrenal medulla* produces the hormones epinephrine and norepinephrine, and the *adrenal cortex* produces the glucocorticoid and mineralocorticoid hormones. 745

**adrenal medulla** Inner portion of the adrenal gland; secretes the hormones epinephrine and norepinephrine. 745

**adrenocorticotropic hormone (ACTH)** (uh DREEN oh core tic oh TROH pick) Hormone secreted by the anterior lobe of the pituitary gland that stimulates activity in the adrenal cortex. 740

**adventitious root** (ad vin TIH shus) Fibrous roots that develop from stems or leaves, such as the prop roots of corn or the holdfast roots of ivy. 442

**aerobic** Phase of cellular respiration that requires oxygen. 135

**age structure diagram** In demographics, a display of the age groups of a population; a growing population has a pyramid-shaped diagram. 823

**agglutination** (ag glue tin NAY shun) Clumping of red blood cells due to a reaction between antigens on red blood cell plasma membranes and antibodies in the plasma. 609

**agnathan** (ag NATH uhn) Fishes that lack jaws; Namely, the lampreys and hagfishes. 543

**aldosterone** (al DOSS turr own) Hormone secreted by the adrenal cortex that regulates the sodium and potassium ion balance of the blood. 675, 746

**alkaloid** Bitter-tasting nitrogenous compounds that have a basic pH (e.g., caffeine). 488

**allantois** (uh LANN toys) Extraembryonic membrane that accumulates nitrogenous wastes the eggs of reptiles, including birds; contributes to the formation of umbilical blood vessels in mammals. 787

**allele** (uh LEEL) Alternative form of a gene—alleles occur at the same locus on homologous chromosomes. 170, 193

**allergy** Immune response to substances that usually are not recognized as foreign. 629

**allopatric speciation** (al low PAT trick spee see AY shun) Origin of new species between populations that are separated geographically. 304

**alloploidy** Polyploid organism that contains the genomes of two or more different species. 307

**allosteric site** Site on an allosteric enzyme that binds an effector molecule; binding alters the activity of the enzyme. 111

**alpine tundra** Tundra near the peak of a mountain. 869

**alternation of generations life cycle** Life cycle, typical of land plants, in which a diploid sporophyte alternates with a haploid gametophyte. 176, 412

**altruism** (AL true is uhm) Social interaction that has the potential to decrease the lifetime reproductive success of the member exhibiting the behavior. 814

**alveolats** A group of protists that includes unicellular dinoflagellates, apicomplexans, and ciliates; alveoli support plasma membrane. 383

**alveolus (pl., alveoli)** (al VEE oh luss) In humans, terminal, microscopic, grapelike air sac found in lungs. 654

**AM fungi (arbuscular mycorrhiza)** Fungi with branching invaginations used to invade plant roots. 396

**amino acid** (uh MEAN no) Organic molecule composed of an amino group and an acid group; covalently bonds to produce peptide molecules. 49

**ammonia** Nitrogenous end product that takes a limited amount of energy to produce but requires much water to excrete because it is toxic. 666

**amnion** (AM nee ahn) Extraembryonic membrane of birds, reptiles, and mammals that forms an enclosing, fluid-filled sac. 787

**amniote** Vertebrate that produces an egg surrounded by four membranes, one of which is the amnion; amniote groups are the reptiles, (including birds), and mammals. 542

**amniotic egg** (am nee AH tick) Egg that has an amnion, as seen during the development of reptiles, (including birds), and mammals. 548

**amoeboid** (uh ME boid) Cell that moves and engulfs debris with pseudopods. 387

**amboebozoan** Supergroup of eukaryotes that includes amoebas and slime molds and is characterized by lobe-shaped pseudopodia. 387

**amphibian** Member of vertebrate class Amphibia that includes frogs, toads, and salamanders; they are still tied to a watery environment for reproduction. 546

**anabolic steroid** (ann uh BAH lick STARE oid) Synthetic steroid that mimics the effect of testosterone. 750

**anaerobic** (ann air OH bick) Growing or metabolizing in the absence of oxygen. 135

**analogous structure** (ann AL oh gus) Structure that has a similar function in separate lineages but differs in anatomy and ancestry. 278, 345

**analogy** (ann AL oh gee) Similarity of function but not of origin. 345
**anaphase** (ANN uh faze) Mitotic phase during which daughter chromosomes move toward the poles of the spindle. 158
**anaphylactic shock** (ann uh fuh LACK tick) Severe systemic form of anaphylaxis involving bronchiolar constriction, impaired breathing, vasodilation, and a rapid drop in blood pressure with a threat of circulatory failure. 629
**androgen** (ANN droh jen) Male sex hormone (e.g., testosterone). 750
**aneuploidy** (ANN you ploid ee) Individual whose chromosome number is not an exact multiple of the haploid number for the species. 180
**angina pectoris** (ann JYE nuh peck TORE iss) Condition characterized by thoracic pain resulting from occluded coronary arteries; precedes a heart attack. 605
**angiogenesis** (ann jee oh JEN uh sis) Formation of new blood vessels; one mechanism by which cancer spreads. 162
**angiosperm** (ANN jee oh sperm) Flowering land plant; the seeds are borne within a fruit. 420, 424
**angiotensin II** Hormone produced from angiotensinogen (a plasma protein) by the kidneys and lungs; raises blood pressure. 675
**animal** Multicellular, heterotrophic eukaryote characterized by the presence of muscular and nervous tissue and undergoing development to achieve its final form. 8
**annelid** (ANN uh lid) The segmented worms, such as the earthworm and the clam worm. 526
**annual ring** Layer of wood (secondary xylem) usually produced during one growing season. 447
**anterior pituitary** (an TEER rih urr pit TWO uh tare ree) Portion of the pituitary gland that is controlled by the hypothalamus and produces six types of hormones, some of which control other endocrine glands. 740
**anther** (ANN thurr) In flowering land plants, pollen-bearing portion of stamen. 495
**antheridia** Sperm-producing structures, as in the moss life cycle. 412
**anthropoid** (ANN throw poid) Group of primates that includes monkeys, apes, and humans. 563
**antibody** Protein produced in response to the presence of an antigen; each antibody combines with a specific antigen. 607
**antibody-mediated immunity** Specific mechanism of defense in which plasma cells derived from B cells produce antibodies that combine with antigens. 619
**anticodon** (ann tie COH don) Three-base sequence in a transfer RNA molecule base that pairs with a complementary codon in mRNA. 224
**antidiuretic hormone (ADH)** (an tee die you REH tick) Hormone secreted by the posterior pituitary that increases the permeability of the collecting ducts in a kidney. 674, 740
**antigen** (ANN tih jen) Foreign substance, usually a protein or a polysaccharide, that stimulates the immune system to react, such as to produce antibodies. 607, 619
**antigen-presenting cell (APC)** Cell that displays the antigen to certain cells of the immune system so they can defend the body against that particular antigen. 619
**antigen receptor** Receptor proteins in the plasma membrane of immune system cells whose shape allows them to combine with a specific antigen. 619
**anus** (AY nuss) Outlet of the digestive tube. 640
**aorta** (ay OR tuh) In humans, the major systemic artery that takes blood from the heart to the tissues. 602
**aortic body** Sensory receptor in the aortic arch sensitive to the $O_2$, $CO_2$, and $H^+$ content of the blood. 657
**apical dominance** (AY pick uhl) Influence of a terminal bud in suppressing the growth of axillary buds. 475
**apical meristem** (AY pick uhl MARE uh stem) In vascular land plants, masses of cells in the root and shoot that reproduce and elongate as primary growth occurs. 440
**apicomplexan** Parasitic protozoans, formerly called sporozoans that lack mobility and form spores; now named for a unique collection of organelles. 384
**apoptosis** (ay pop TOE sis) Programmed cell death involving a cascade of specific cellular events leading to death and destruction of the cell. 153, 782
**appendicular skeleton** (app pen DICK you lurr) Part of the vertebrate skeleton forming the appendages, shoulder girdle, and hip girdle. 724
**appendix** (app PEN dicks) In humans, small, tubular appendage that extends outward from the cecum of the large intestine. 639
**aquaporin** Channel protein through which water can diffuse across a membrane. 90, 674
**arboreal** (are BORE ree uhl) Living in trees. 560
**archaea** Member of the domain Archaea. 368
**archaeplastid** Supergroup of eukaryotes that includes land plants and red and green algae. Developed from endosymbiotic cyanobacteria. 377
**archegonia** Egg-producing structures, as in the moss life cycle. 412
**Arctic tundra** Biome that encircles the Earth just south of ice-covered polar seas in the Northern Hemisphere. 870
**arteriole** (are TEER ree ohl) Vessel that takes blood from an artery to capillaries. 596
**artery** Blood vessel that transports blood away from the heart. 596
**arthropod** (ARTH throw pod) Invertebrates, with an exoskeleton and jointed appendages, such as crustaceans and insects. 529
**artificial selection** Intentional breeding of certain traits, or combinations of traits, over others to produce a desirable outcome. 273
**ascus** (ASK us) Fingerlike sac in which nuclear fusion, meiosis, and ascospore production occur during sexual reproduction of sac fungi. 398
**asexual reproduction** Reproduction that requires only one parent and does not involve gametes. 164
**associative learning** Acquired ability to associate two stimuli or between a stimulus and a response. 804
**assortative mating** (ah Sor tah tive) Mating of individuals with similar phenotypes. 287
**aster** (ASS turr) Short, radiating fibers produced by the centrosomes in animal cells. 156
**asthma** Condition in which bronchioles constrict and cause difficulty in breathing. 629
**atom** Smallest particle of an element that displays the properties of the element. 22
**atomic mass** Average of atom mass units for all the isotopes of an atom. 23
**atomic number** Number of protons within the nucleus of an atom. 23
**atomic symbol** One or two letters that represent the name of an element—e.g., H stands for a hydrogen atom, and Na stands for a sodium atom. 22
**ATP (adenosine triphosphate)** (ad DEN no seen try FOSS fate) Nucleotide with three phosphate groups. The breakdown of ATP into ADP + P makes energy available for energy-requiring processes in cells. 52, 106
**ATP synthase complex** (SIN thaze) Complex of proteins in the cristae of mitochondria and thylakoid membrane of chloroplast that produces ATP as hydrogen flows down a concentration gradient. 113, 124
**atrial natriuretic hormone (ANH)** (AY tree uhl nat tree you RETT tick) Hormone secreted by the heart that increases sodium excretion. 675, 746
**atrioventricular valve** (ay tree oh vinn TRICK you lurr) Heart valve located between an atrium and a ventricle. 598
**atrium** (AY tree uhm) Chamber; particularly an upper chamber of the heart lying above a ventricle. 598
**auditory communication** Sound that an animal makes for the purpose of sending a message to another individual. 807
**australopithecine (australopith)** (oss stray loh PITH ih seen) One of several species of *Australopithecus*, a genus that contains the first generally recognized humanlike hominins. 566
***Australopithecus afarensis*** Bipedal humanlike hominin that lived 3.9 and 3 MYA, e.g., Lucy, discovered at Hadar, Ethiopia, in 1974. 566
***Australopithecus africanus*** Gracile humanlike hominin that lived between 3 and 2.2 MYA, unearthed in southern Africa in the 1920s; one of the first australopiths to be discovered. 566
**autoimmune disease** (ah toe ih MUNE) Disease that results when the immune system mistakenly attacks the body's own tissues. 628
**autonomic system** (ah toe NAHM mick) Portion of the peripheral nervous system that regulates internal organs. 695
**autoploidy** Polyploid organism with chromosome sets from the same species. 307
**autosome** (AH toe sohm) Any chromosome other than the sex-determining pair. 198
**autotroph** (AH toe trofe) Organism that can capture energy and synthesize organic molecules from inorganic nutrients. 118, 852
**auxin** (OX sin) Plant hormone regulating growth, particularly cell elongation; also called indoleacetic acid (IAA). 475
**axial skeleton** (AXE ee uhl) Part of the vertebrate skeleton forming the vertical support or axis, including the skull, the rib cage, and the vertebral column. 722
**axillary bud** (AXE ill air ree) Bud located in the axil of a leaf. 435
**axon** (AXE ahn) Elongated portion of a neuron that conducts nerve impulses, typically from the cell body to the synapse. 683

# B

**bacillus** (buh SILL us) A rod-shaped bacterium; also a genus of bacteria, *Bacillus*. 64
**bacterial vaginosis** Sexually transmitted disease caused by *Gardnerella vaginalis, Mobiluncus* spp., *Mycoplasma hominis*, and various anaerobic bacteria. Although a mild disease, it is a risk factor for obstetric infections and pelvic inflammatory disease. 772
**bacteriophage** (back TEER ree oh fahj) Virus that infects bacteria. 358

**bacterium (pl., bacteria)** Member of the domain Bacteria. 364

**bark** External part of a tree, containing cork, cork cambium, and phloem. 446

**Barr body** Dark-staining body (discovered by M. Barr) in the nuclei of female mammals that contains a condensed, inactive X chromosome. 182, 238

**basal body** (BAY zull) A cytoplasmic structure that is located at the base of—and may organize—cilia or flagella. 80

**basal nuclei** (BAY zull NEW clee eye) Subcortical nuclei deep within the white matter that serve as relay stations for motor impulses and produce dopamine to help control skeletal muscle activities. 689

**base** Molecules tending to lower the hydrogen ion concentration in a solution and raise the pH numerically. 32

**basement membrane** Layer of nonliving material that anchors epithelial tissue to underlying connective tissue. 578

**basidium** (buh SIH dee uhm) Clublike structure in which nuclear fusion, meiosis, and basidiospore production occur during sexual reproduction of club fungi. 402

**basophil** (BASE oh fill) White blood cell with a granular cytoplasm; able to be stained with a basic dye. 607

**B cell** Lymphocyte that matures in the bone marrow and, when stimulated by the presence of a specific antigen, gives rise to antibody-producing plasma cells. 615

**B cell receptor (BCR)** Molecule on the surface of a B cell that binds to a specific antigen. 619

**behavior** Observable, coordinated responses to environmental stimuli. 800

**behavioral ecology** Study of how natural selection shapes behavior. 810

**beneficial nutrient** In plants, element that is either required or enhances the growth and production of a plant. 457

**benign** (buh NINE) Mass of cells derived from a single mutated cell that has repeatedly undergone cell division but has remained at the site of origin. 161

**bicarbonate ion** (by CAR boh nate EYE ahn) Ion that participates in buffering the blood, and the form in which carbon dioxide is transported in the bloodstream. 659

**bilateral symmetry** (by LATT turr uhl SIMM met tree) Body plan having two corresponding or complementary halves. 513

**bile** Secretion of the liver that is temporarily stored and concentrated in the gallbladder before being released into the small intestine, where it emulsifies fat. 639

**binary fission** (BY nuh ree FISH uhn) Splitting of a parent cell into two daughter cells; serves as an asexual form of reproduction in bacteria. 164, 364

**binomial nomenclature** (by NO mee uhl) Scientific name of an organism, the first part of which designates the genus and the second part of which designates the specific epithet. 8, 339

**biocultural evolution** Phase of human evolution in which cultural events affect natural selection. 569

**biodiversity** (by oh die VERSE sit tee) Total number of species, the variability of their genes, and the communities in which they live. 10, 890

**biodiversity hotspot** Region of the world that contains unusually large concentrations of species. 891

**biogeochemical cycle** (by oh jee oh KEM ick cull) Circulating pathway of elements such as carbon and nitrogen involving exchange pools, storage areas, and biotic communities. 856

**biogeography** (by oh jee AH gruh fee) Study of the geographical distribution of organisms. 269

**bioinformatics** (by oh in for MAT ticks) Computer technologies used to study the genome. 259, 890

**biological clock** Internal mechanism that maintains a biological rhythm in the absence of environmental stimuli. 485

**biological species concept** The concept that defines species as groups of populations that have the potential to interbreed and that are reproductively isolated from other groups. 300

**biology** Scientific study of life. 11

**biomass** (BY oh mass) The number of organisms multiplied by their weight. 855

**biome** (BY ohm) One of the biosphere's major communities, characterized in particular by certain climatic conditions and particular types of plants. 869

**biomolecule** Organic molecule (macromolecule as a protein or nucleic acid) in living organisms. 38

**biosphere** (BY ohs fear) Zone of air, land, and water at the surface of the Earth in which living organisms are found. 9, 820

**biotechnology products** Commercial or agricultural products that are made with or derived from transgenic organisms. 252

**biotic potential** (by AH tick) Maximum population growth rate under ideal conditions. 822

**bird** Endothermic reptile that has feathers and wings, is often adapted for flight, and lays hard-shelled eggs. 552

**bivalent** (by VAY lent) Homologous chromosomes, each having sister chromatids that are joined by a nucleoprotein lattice during meiosis; also called a tetrad. 171

**bivalve** (BY valve) Type of mollusc with a shell composed of two valves; includes clams, oysters, and scallops. 524

**blade** Broad, expanded portion of a land plant leaf that may be single or compound leaflets. 435

**blastocoel** (BLAST toe seal) Fluid-filled cavity of a blastula. 779

**blastocyst** (BLAST toe sist) Early stage of human embryonic development that consists of a hollow, fluid-filled ball of cells. 788

**blastopore** Opening into the primitive gut formed at gastrulation. 780

**blastula** (BLAST you luh) Hollow, fluid-filled ball of cells occurring during animal development prior to gastrula formation. 779

**blind spot** Region of the retina, lacking rods or cones, where the optic nerve leaves the eye. 708

**blood** Fluid circulated by the heart through a closed system of vessels. 581, 594

**blood pressure** Force of blood pushing against the inside wall of blood vessels. 603

**bog** Wet, spongy ground in a low-lying area, usually acidic and low in organic nutrients. 879

**bone** Connective tissue having protein fibers and a hard matrix of inorganic salts, notably calcium salts. 580

**bony fishes (Osteichthyes)** A fish that has a bony rather than cartilaginous skeleton. 544

**bottleneck effect** Type of genetic drift; occurs when a majority of genotypes are prevented from participating in the production of the next generation as a result of a natural disaster or human interference. 288

**brain** Ganglionic mass at the anterior end of the nerve cord; in vertebrates, the brain is located in the cranial cavity of the skull. 688

**brain stem** In mammals; portion of the brain consisting of the medulla oblongata, pons, and midbrain. 690

**bronchiole** (BRAHN key ohl) In terrestrial vertebrates, small tube that conducts air from a bronchus to the alveoli. 655

**bronchus (pl., bronchi)** (BRAHN cuss, BRAHN kie) In terrestrial vertebrates, branch of the trachea that leads to the lungs. 655

**brown algae** Marine photosynthetic protist with a notable abundance of xanthophyll pigments; this group includes well-known seaweeds of northern rocky shores. 381

**bryophyte** (BRY oh fite) A nonvascular land plant—the mosses, liverworts, and hornworts— in which the gametophyte is dominant.. 413

**budding** Asexual form of reproduction whereby a new organism develops as an outgrowth of the body of the parent. 395

**buffer** Substance or group of substances that tend to resist pH changes of a solution, thus stabilizing its relative acidity and basicity. 32

**bulk transport** Movement of elements in an organism in large amount. 90

**bursa** (BURR suh) Saclike, fluid-filled structure, lined with synovial membrane, that occurs near a synovial joint. 725

# C

**$C_3$ plant** Plant that fixes carbon dioxide via the Calvin cycle; the first stable product of $C_3$ photosynthesis is a 3-carbon compound. 128

**$C_4$ plant** Plant that fixes carbon dioxide to produce a $C_4$ molecule that releases carbon dioxide to the Calvin cycle. 128

**calcitonin** (cal sit ON in) Hormone secreted by the thyroid gland that increases the blood calcium level. 744

**calorie** Amount of heat energy required to raise the temperature of one gram of water 1°C. 29

**Calvin cycle reaction** Portion of photosynthesis that takes place in the stroma of chloroplasts and can occur in the dark; it uses the products of the light reactions to reduce $CO_2$ to a carbohydrate. 121

**calyx** The sepals collectively; the outermost flower whorl. 495

**camera-type eye** Type of eye found in vertebrates and certain molluscs; a single lens focuses an image on closely packed photoreceptors. 704

**camouflage** (CAM oh flaj) Process of hiding from predators in which the organism's behavior, form, and pattern of coloration allow it to blend into the background and prevent detection. 845

**cancer** Malignant tumor whose nondifferentiated cells exhibit loss of contact inhibition, uncontrolled growth, and the ability to invade tissue and metastasize. 161

**capillary** (CAP pill air ree) Microscopic blood vessel; gases and other substances are exchanged across the walls of a capillary between blood and tissue fluid. 596

**capsid** (CAP sid) Protective protein containing the genetic material of a virus. 357

**capsule** Gelatinous layer surrounding the cells of

blue-green algae and certain bacteria. 64

**carbaminohemoglobin** (car buh meen oh HEE muh glow bin) Hemoglobin carrying carbon dioxide. 659

**carbohydrate** (car boh HI drate) Class of organic compounds that includes monosaccharides, disaccharides, and polysaccharides. 41, 643

**carbon dioxide ($CO_2$) fixation** Photosynthetic reaction in which carbon dioxide is attached to an organic compound. 126

**carbonic anhydrase** (car BAH nick ann HI draze) Enzyme in red blood cells that speeds the formation of carbonic acid from water and carbon dioxide. 659

**carcinogen** (car SIN uh jen) Environmental agent that causes mutations leading to the development of cancer. 243

**cardiac conduction system** System of specialized cardiac muscle fibers that conducts impulses from the SA node to the chambers of the heart, causing them to contract. 600

**cardiac cycle** One complete cycle of systole and diastole for all heart chambers. 600

**cardiac muscle** Striated, involuntary muscle tissue found only in the heart. 582

**cardiac output** Blood volume pumped by each ventricle per minute (not total output pumped by both ventricles). 600

**cardiac pacemaker** Mass of specialized cardiac muscle tissue that controls the rhythm of the heartbeat; the SA node. 601

**cardiovascular system** (car dee oh VASS cue lurr) Organ system in which blood vessels distribute blood under the pumping action of the heart. 594, 596

**carnivore** (CAR nih vore) Consumer in a food chain that eats other animals. 852

**carotenoid** (car RAH ten oid) Yellow or orange pigment that serves as an accessory to chlorophyll in photosynthesis. 122

**carotid body** Structure located at the branching of the carotid arteries; contains chemoreceptors sensitive to the $O_2$, $CO_2$, and $H^+$ content in blood. 657

**carpel** (CAR pull) Ovule-bearing unit that is a part of a pistil. 425, 495

**carrier** Heterozygous individual who has no apparent abnormality but can pass on an allele for a recessively inherited genetic disorder. 198

**carrier protein** Protein that combines with and transports a molecule or ion across the plasma membrane. 88

**carrying capacity** Largest number of organisms of a particular species that can be maintained indefinitely by a given environment. 827

**cartilage** (CAR tih ledge) Connective tissue in which the cells lie within lacunae embedded in a flexible, proteinaceous matrix. 580

**cartilaginous fishes (Chondrichthyes)** (car tih LAJJ jen us) A fish that has a cartilaginous rather than bony skeleton; includes sharks, rays, and skates. 543

**Casparian strip** (cass PAIR ree uhn) Layer of impermeable lignin and suberin bordering four sides of root endodermal cells; prevents water and solute transport between adjacent cells. 441, 460

**caspase** (CASS pase) Cell cycle regulatory enzyme that initiates apoptosis. 153

**catabolism** (cuh TAB uh liz uhm) Metabolic process that breaks down large molecules into smaller ones; catabolic metabolism. 145

**catastrophism** (cuh TASS troh fizz uhm) Belief espoused by Georges Cuvier that periods of catastrophic extinctions occurred, after which repopulation of surviving species took place, giving the appearance of change through time. 268

**cell** Smallest unit that displays the properties of life; composed of organelle-containing cytoplasm surrounded by a plasma membrane. 2, 60

**cell body** Portion of a neuron that contains a nucleus and from which dendrites and an axon extend. 683

**cell cycle** Repeating sequence of events in eukaryotes that involves cell growth and nuclear division; consists of the stages $G_1$, S, $G_2$, and M. 152

**cell envelope** In a prokaryotic cell, the portion composed of the plasma membrane, the cell wall, and the glycocalyx. 64

**cell-mediated immunity** Specific mechanism of defense in which T cells destroy antigen-bearing cells. 619

**cell plate** Structure across a dividing plant cell that signals the location of new plasma membranes and cell walls. 159

**cell recognition protein** Glycoprotein that helps the body defend itself against pathogens. 88

**cell suspension culture** Small clumps of naked plant cells grown in tissue culture that produce drugs, cosmetics, agricultural chemicals, among others. 506

**cell theory** One of the major theories of biology, which states that all organisms are made up of cells; cells are capable of self-reproduction and come only from preexisting cells. 60

**cellular differentiation** Process and developmental stages by which a cell becomes specialized for a particular function. 782

**cellular respiration** Metabolic reactions that use the energy from carbohydrate, fatty acid, or amino acid breakdown to produce ATP molecules. 134

**cellular response** Response to the transduction pathway in which proteins or enzymes change a signal to a format that the cell can understand, resulting in the appropriate response. 474

**cellular slime mold** Free-living amoeboid cells that feed on bacteria and yeasts by phagocytosis and aggregate to form a plasmodium that produces spores. 388

**cellulose** (SELL you lohs) Polysaccharide that is the major complex carbohydrate in plant cell walls. 43

**cell wall** Structure that surrounds a plant, protistan, fungal, or bacterial cell and maintains the cell's shape and rigidity. 64, 99

**centipede** (SEN tih peed) Elongated arthropod characterized by having one pair of legs to each body segment; they may have 15 to 173 pairs of legs. 531

**central nervous system (CNS)** Portion of the nervous system consisting of the brain and spinal cord. 681

**central vacuole** (VACK you ohl) In a plant cell, a large, fluid-filled sac that stores metabolites. During growth, it enlarges, forcing the primary cell wall to expand and the cell surface-area-to-volume ratio to increase. 75

**centriole** (SENT tree ohl) Cell organelle, existing in pairs, that occurs in the centrosome and may help organize a mitotic spindle for chromosome movement during animal cell division. 80, 155

**centromere** (SENT troh meer) Constriction where sister chromatids of a chromosome are held together. 155

**centrosome** (SENT troh sohm) Central microtubule organizing center of cells. In animal cells, it contains two centrioles. 78, 155

**cephalization** (seff full lih ZAY shun) Having a well-recognized anterior head with a brain and sensory receptors. 515, 680

**cephalochordate** Small, fishlike invertebrate that is a member of the phylum Chordata. Probably the closest living relative to vertebrates. 540

**cephalopod** (SEF ful lo pod) Type of mollusc in which the head is prominent and the foot is modified to form two arms and several tentacles; includes squids, cuttlefish, octopuses, and nautiluses. 525

**cerebellum** (sair uh BELL uhm) In terrestrial vertebrates, portion of the brain that coordinates skeletal muscles to produce smooth, graceful motions. 690

**cerebral cortex** (sir REE brull CORE tex) Outer layer of cerebral hemispheres; receives sensory information and controls motor activities. 689

**cerebral hemisphere** Either of the two lobes of the cerebrum in vertebrates. 689

**cerebrospinal fluid** (sir ree broh SPY null) Fluid found in the ventricles of the brain, in the central canal of the spinal cord, and in association with the meninges. 688

**cerebrum** (sir REE brumm) Largest part of the brain in mammals. 689

**cervix** Narrow end of the uterus, which leads into the vagina. 762

**channel protein** Protein that forms a channel to allow a particular molecule or ion to cross the plasma membrane. 88

**chaparral** (shapp purr AL) Biome characterized by broad-leafed evergreen shrubs forming dense thickets. 876

**chaperone protein** (shapp purr OHN) Molecule that interacts with a protein so that it folds into its proper shape. 51

**character displacement** Tendency for characteristics to be more divergent when similar species belong to the same community than when they are isolated from one another. 842

**charophyte** Type of living green algae that on the basis of nucleotide sequencing and cellular features is most closely related to land plants. 377, 410

**chelicerate** (shell lih sir AH tuh) Arthropods (e.g., horseshoe crabs, sea spiders, arachnids), that have a pair of pointed appendages used to manipulate food. 533

**chemical energy** Energy associated with the interaction of atoms in a molecule. 104

**chemical evolution** Increase in the complexity of chemicals over time that could have led to the first cells. 318

**chemical signal** Molecule that brings about a change in a cell, tissue, organ, or individual when it binds to a specific receptor. 738

**chemiosmosis** (kim mee oz MOW sis) Process by which mitochondria and chloroplasts use the energy of an electron transport chain to create a hydrogen ion gradient that drives ATP formation. 113, 124, 143

**chemoautotroph** (key mow AH toe trofe) Organism able to synthesize organic molecules by using carbon dioxide as the carbon source and the oxidation of an inorganic substance (such as hydrogen sulfide) as the energy source. 365

**chemoheterotroph** (key mow HETT turr row trofe) Organism that is unable to produce its own organic molecules, and therefore requires organic nutrients in its diet. 365

**chemoreceptor** (key mow ree SEPP turr) Sensory receptor that is sensitive to chemical stimulation—for example, receptors for taste and smell. 702

**chitin** (KITE in) Strong but flexible nitrogenous polysaccharide found in the exoskeleton of arthropods and in the cell walls of fungi. 43, 394, 529

**chlorofluorocarbons (CFCs)** (klore oh flur oh CAR buns) Organic compounds containing carbon, chlorine, and fluorine atoms. CFCs such as Freon can deplete the ozone shield by releasing chlorine atoms in the upper atmosphere. 860

**chlorophyll** (KLORE uh fill) Green pigment that absorbs solar energy and is important in algal and land plant photosynthesis; occurs as chlorophyll *a* and chlorophyll *b*. 119

**chlorophyte** Most abundant and diverse group of green algae, including freshwater, marine, and terrestrial forms that synthesize. Chlorophytes share chemical and anatomical characteristics with land plants. 377

**chloroplast** (KLORE oh plast) Membrane-bounded organelle in algae and land plants with chlorophyll-containing membranous thylakoids; where photosynthesis takes place. 76, 119

**choanoflagellate** Unicellular choanoflagellates have one and colonial forms have many collar cells like those of sponges; choanoflagellates are the protists most closely related to animals. 389

**cholesterol** (koh LESS turr all) A steroid found in animal plasma membrane and from which other types of steroids are derived. 86, 644

**chordate** (CORE date) Animals that have a dorsal tubular nerve cord, a notochord, pharyngeal gill pouches, and a postanal tail at some point in their life cycle; includes a few types of invertebrates (e.g. sea squirts and lancelets) and the vertebrates. 540

**chorion** (CORE ree ahn) Extraembryonic membrane functioning for respiratory exchange in birds and reptiles; contributes to placenta formation in mammals. 787

**chorionic villus** (core ree AH nick VILL us) In placental mammals treelike extension of the chorion, projecting into the maternal tissues at the placenta. 790

**choroid** (CORE oid) Vascular, pigmented middle layer of the eyeball. 705

**chromalveolate** Supergroup of eukaryotes that includes alveolates and stramenopiles. 381

**chromatid** (CROW muh tid) Following replication a chromrsome consists of a pair of sister chromatids, each a single DNA helix, held together at the centromere. Following separation, each chromosome is a single chromatid. 152

**chromatin** (CROW muh tin) Network of DNA strands and associated proteins observed within a nucleus that is not dividing. 70, 155

**chromoplast** Plastid in land plants responsible for orange, yellow, and red color of plants, including the autumn colors in leaves. 76

**chromosome** (CROW muh sohm) An observable structure that results when chromatin condenses and coils, each species has a particular number of chromosomes that is passed on to the next generation. 70

**chyme** (KIME) Thick, semiliquid food material that passes from the stomach to the small intestine. 639

**chytrid** Mostly aquatic fungi with flagellated spores that may represent the most ancestral fungal lineage. 396

**ciliary muscle** (SILL lee air ree) Within the ciliary body of the vertebrate eye, the ciliary muscle controls the shape of the lens. 706

**ciliate** (SILL lee ate) Complex unicellular protist that moves by means of cilia and digests food in food vacuoles. 384

**cilium** (SILL lee uhm) Short, hairlike projections from the plasma membrane, occurring usually in larger numbers (cilia). 80

**circadian rhythm** (sir KAY dee uhn) Biological rhythm with a 24-hour cycle. 484, 750

**circulatory system** In animals, an organ system that moves substances to and from cells, usually via a heart, blood, and blood vessels. 594, 596

**cirrhosis** Chronic, irreversible injury to liver tissue; commonly caused by frequent alcohol consumption. 642

**citric acid cycle** Cycle of reactions in mitochondria that begins with citric acid. This cycle breaks down an acetyl group and produces $CO_2$, ATP, NADH, and $FADH_2$; also called the Krebs cycle. 135, 141

**clade** Evolutionary lineage consisting of an ancestral species and all of its descendants, forming a distinct branch on a cladogram. 342

**cladistics** (kluh DISS tick) Method of systematics that uses derived characters to determine monophyletic groups and construct cladograms. 342

**cladogram** (CLAYD doe gram) In cladistics, a branching diagram that shows the relationship among species in regard to their shared derived characters. 342

**class** One of the categories are subgroups used by taxonomists to group species; class within a phylim or division. 6, 340

**classical conditioning** Type of learning whereby an unconditioned stimulus that elicits a specific response is paired with a neutral stimulus so that the response becomes conditioned. 804

**cleavage** (CLEAVE edge) Cell division without cytoplasmic addition or enlargement; occurs during the first stage of animal development. 779

**climate** Generalized weather patterns of an area, primarily determined by temperature and average rainfall. 866

**climax community** In ecology, community that results when succession has come to an end. 850

**cloaca** (kloh AY cuh) Common chamber and opening to the digestive, urinary, and genital tracts in certain vertebrates. 547

**clonal selection model** (KLOH null) States that the antigen selects which lymphocyte will undergo clonal expansion and produce more lymphocytes bearing the same type of receptor. 620

**cloning** Production of identical copies. In organisms, the production of organisms with the same genes; in genetic engineering, the production of many identical copies of a gene. 250

**closed circulatory system** A type of circulatory system where blood is confined to vessels and is kept separate from the interstitial fluid. 594

**club fungi** Fungi that produce spores in club-shaped basidia within a fruiting body; includes mushrooms, shelf fungi and puffballs. 402

**cnidarian** (neye DARE ree uhn) Invertebrates existing as either a polyp or medusa with two tissue layers and radial symmetry. 518

**coacervate droplet** (coh AY sir vate) An aggregate of colloidal droplets held together by electrostatic forces. 320

**coal** Fossil fuel formed millions of years ago from plant material that did not decay. 423

**coccus** (COCK us) A spherical-shaped bacterium. 64

**cochlea** (COKE lee uh) Spiral-shaped structure of the vertebrate inner ear containing the sensory receptors for hearing. 710

**codominance** (koh DAH men unce) Inheritance pattern in which both alleles of a gene are equally expressed in a heterozygote. 202

**codon** (KOH dahn ) Three-base sequence in messenger RNA that during translation directs the addition of a particular amino acid into a protein or directs termination of the process. 221

**coenzyme** (koh IN zime) Nonprotein organic molecule that aids the action of the enzyme to which it is loosely bound. 52, 110

**coevolution** Mutual evolution in which two species exert selective pressures on the other species. 498, 847

**cofactor** Nonprotein adjunct required by an enzyme in order to function; many cofactors are metal ions, others are coenzymes. 110

**cohesion-tension model** Explanation for upward transport of water in xylem based upon transpiration-created tension and the cohesive properties of water molecules. 464

**cohort** (KOH hort) Group of individuals having a statistical factor in common, such as year of birth, in a population study. 822

**coleoptile** (koh lee OPP tile) Protective sheath that covers the young leaves of a seedling. 475

**collagen fiber** White fiber in the matrix of connective tissue giving flexibility and strength. 579

**collecting duct** Duct within the kidney that receives fluid from several nephrons; the reabsorption of water occurs here. 671

**collenchyma** (kuh LENN kih muh) Plant tissue composed of cells with unevenly thickened walls; supports growth of stems and petioles. 438

**colony** Loose association of cells each remaining independent for most functions. 378

**columnar epithelium** (kuh LUM nurr epp uih THEE lee uhm) Type of epithelial tissue with cylindrical cells. 578

**comb jelly** Invertebrates that resemble jelly fishes and are the largest animals to be propelled by beating cilia. 518

**commensalism** (kuh MENS suh liz uhm) Symbiotic relationship in which one species is benefited, and the other is neither harmed nor benefited. 365, 846

**common ancestor** Ancestor common to at least two lines of descent. 341

**communication** Signal by a sender that influences the behavior of a receiver. 807

**community** Assemblage of species interacting with one another within the same environment. 9, 820, 840

**compact bone** Type of bone that contains osteons consisting of concentric layers of matrix and osteocytes in lacunae. 580, 720

**companion cell** Cell associated with sieve-tube members in phloem of vascular plants. 462

**comparative genomics** Study of genomes through the direct comparison of their genes and DNA sequences. 258

**competitive exclusion principle** Theory that no two species can occupy the same niche in the same place and at the same time. 842

**competitive inhibition** Form of enzyme inhibition where the substrate and inhibitor are both able to bind to the enzyme's active site. Only when the substrate is at the active site will product form. 111

**complement** Collective name for a series of enzymes and activators in the blood, some of which may bind to antibody and may lead to rupture of a foreign cell. 618

**complementary base pairing** Hydrogen bonding between particular purines and pyrimidines in DNA. 53, 216

**complementary DNA (cDNA)** DNA that has been synthesized from mRNA by the action of reverse transcriptase. 253

**complete digestive tract** Digestive tract that has both a mouth and an anus. 634

**complex tissue** In plants, tissue composed of two or more kinds of cells (e.g., xylem, containing tracheids and vessel elements; phloem, containing sieve-tube members and companion cells). 439

**compound** Substance having two or more different elements united chemically in fixed ratio. 26

**compound eye** Type of eye found in arthropods; it is composed of many independent visual units. 704

**concentration gradient** Gradual change in chemical concentration between two areas of differing concentrations. 90

**conclusion** Statement made following an experiment as to whether or not the results support the hypothesis. 12

**cone** Reproductive structure in conifers comprised of scales bearing sporangia; pollen cones bear microsporangia, and seed cones bear megasporangia. 420

**cone cell** Photoreceptor in vertebrate eyes that responds to bright light and makes color vision possible. 707

**congenital hypothyroidism** Condition that results from the thyroid gland not developing properly; characteristics include stunted growth and possible mental retardation. 743

**conidiospore** (koh NIDD dee uh spore) Spore produced by sac and club fungi during asexual reproduction. 398

**conifer** (KAH nih fur) Member of a group of cone-bearing gymnosperm land plants that includes pine, cedar, and spruce trees. 420

**conjugation** (kahn jew GAY shun) Transfer of genetic material from one cell to another. 364, 378

**conjugation pilus (pl., conjugation pili)** (PIE luss, PIE lie) In a bacterium, elongated, hollow appendage used to transfer DNA to other cells. 65, 364

**conjunctiva** (kahn junk TY vuh) Delicate membrane that lines the eyelid protecting the sclera. 705

**connective tissue** Type of animal tissue that binds structures together, provides support and protection, fills spaces, stores fat, and forms blood cells; adipose tissue, cartilage, bone, and blood are types of connective tissue; living cells in a nonliving matrix. 579

**conservation biology** Discipline that seeks to understand the effects of human activities on species, communities, and ecosystems and to develop practical approaches to preventing the extinction of species and the destruction of ecosystems. 890

**consumer** Organism that feeds on another organism in a food chain generally; primary consumers eat plants, and secondary consumers eat animals. 852

**continental drift** The movement of the Earth's crust by plate tectonics resulting in the movement of continents with respect to one another. 332

**contraceptive vaccine** Under development, this birth control method immunizes against the hormone HCG, crucial to maintaining implantation of the embryo. 767

**contrast** In magnification with a microscope, brightness differences between objects. 63

**control** Sample that goes through all the steps of an experiment but does not contain the variable being tested; a standard against which the results of an experiment are checked. 12

**convergent evolution** (kuhn VERGE ent) Similarity in structure in distantly related groups generally due to similiar selective pressures in like environments. 345

**copulation** (cop you LAY shun) Sexual union between a male and a female. 757

**coral reef** Coral formations in shallow tropical waters that support an abundance of diversity. 883

**corepressor** (koh ree PRESS her) Molecule that binds to a repressor, allowing the repressor to bind to an operator in a repressible operon. 235

**cork** Outer covering of the bark of trees; made of dead cells that may be sloughed off. 437

**cork cambium** Lateral meristem that produces cork. 437

**cornea** (CORE nee uh) Transparent, anterior portion of the outer layer of the eyeball. 705

**corolla** The petals, collectively; usually the conspicuously colored flower whorl. 495

**corpus luteum** (CORE pus LU tee uhm) Follicle that has released an egg and increases its secretion of progesterone. 763

**cortex** (CORE tex) In plants, ground tissue bounded by the epidermis and vascular tissue in stems and roots; in animals, outer layer of an organ, such as the cortex of the kidney or adrenal gland. 441

**cortisol** (CORE tih zahl) Glucocorticoid secreted by the adrenal cortex that responds to stress on a long-term basis; reduces inflammation and promotes protein and fat metabolism. 746

**cost-benefit analysis** A weighing-out of the costs and benefits (in terms of contributions to reproductive success) of a particular strategy or behavior. 291

**cotyledon** (cot tih LEE dunn) Seed leaf for embryo of a flowering plant; provides nutrient molecules for the developing plant before photosynthesis begins. 424, 436, 500

**countercurrent exchange** Fluids flow side-by-side in opposite directions, as in the exchange of fluids in the kidneys. 652

**coupled reactions** Reactions that occur simultaneously; one is an exergonic reaction that releases energy, and the other is an endergonic reaction that requires an input of energy in order to occur. 107

**covalent bond** (koh VALE lunt) Chemical bond in which atoms share one pair of electrons. 27

**cranial nerve** (CRANE nee uhl) Nerve that arises from the brain. 692

**crenation** (krin AY shun) In animal cells, shriveling of the cell due to water leaving the cell when the environment is hypertonic. 93

**cristae (sing., crista)** (KRISS tee, KRISS tuh) Short, fingerlike projections formed by the folding of the inner membrane of mitochondria. 77

**Cro-Magnon** (crow MAG nahn) Common name for the first fossils to be designated *Homo sapiens.* 571

**crossing-over** Exchange of segments between nonsister chromatids of a bivalent during meiosis. 172

**crustacean** (crust TAY shun) Member of a group of marine arthropods that contains, among others, shrimps, crabs, crayfish, and lobsters. 530

**cuboidal epithelium** (cube OID uhl epp pih THEE lee uhm) Type of epithelial tissue with cube-shaped cells. 578

**Cushing syndrome** Condition resulting from hypersecretion of glucocorticoids; characterized by thin arms and legs and a "moon face," and accompanied by high blood glucose and sodium levels. 747

**cuticle** Waxy layer covering the epidermis of plants that protects the plant against water loss and disease-causing organisms. 412, 437, 465

**cyanobacterium (pl., cyanobacteria)** (SIGH uhn no back TEER ree uhm) Photosynthetic bacterium that contains chlorophyll and releases oxygen; formerly called a blue-green alga. 65, 367

**cyanogenic glycoside** Plant compound that contains sugar; produces cyanide. 488

**cycad** (SIGH cad) Type of gymnosperm with palmate leaves and massive cones; cycads are most often found in the tropics and subtropics. 420

**cyclic adenosine monophosphate (cAMP)** ATP-related compound that acts as the second messenger in peptide hormone transduction; it initiates activity of the metabolic machinery. 739

**cyclin** (SIGH klin) Protein that cycles in quantity as the cell cycle progresses; combines with and activates the kinases that function to promote the events of the cycle. 153

**cyst** (SIST) In protists and invertebrates, resting structure that contains reproductive bodies or embryos. 374, 522

**cytochrome** (SIGH toe krome) Any of several iron-containing protein molecules that are members of the electron transport chain in photosynthesis and cellular respiration. 142

**cytokine** (SIGH toe kine) Type of protein secreted by a T lymphocyte that attacks viruses, virally infected cells, and cancer cells. 618

**cytokinesis** (sigh toe kin NEE sis) Division of the cytoplasm following mitosis or meiosis. 152

**cytokinin** (sigh toe KINE ninn) Plant hormone that promotes cell division; often works in combination with auxin during organ development in plant embryos. 477

**cytoplasm** (SIGH toe plaz uhm) Contents of a cell between the nucleus (nucleoid) region of bacteria and the plasma membrane. 64

**cytosine (C)** (SIGH toe zeen) One of four nitrogen-containing bases in the nucleotides composing the structure of DNA and RNA; pairs with guanine. 214

**cytoskeleton** (sigh toe SKELL luh ton) Internal framework of the cell, consisting of microtubules, actin filaments, and intermediate filaments. 67

**cytotoxic T cell** (sigh toe TOX ick) T lymphocyte that attacks and kills antigen-bearing cells. 619

# D

**data (sing., datum)** (DAY tuh, DAY tum) Facts or information collected through observation and/or experimentation. 12

**day-neutral plant** Plant whose flowering is not dependent on day length—e.g., tomato and cucumber. 487

**deamination** (dee am in AY shun) Removal of an amino group (—$NH_2$) from an amino acid or other organic compound. 145

**decapod** (DECK uh pod) Type of crustacean in which the thorax bears five pairs of walking legs; includes shrimps, lobsters, crayfish, and crabs. 530

**deciduous** (dih SIDD you us) Land plant which sheds its leaves annually. 435

**decomposer** Organism, usually a bacterium or fungus, that breaks down organic matter into inorganic nutrients that can be recycled in the environment. 852

**deductive reasoning** Process of logic and reasoning, using "if . . . then" statements. 11

**dehydration reaction** Chemical reaction resulting in a covalent bond with the accompanying loss of a water molecule. 40

**delayed allergic response** Allergic response initiated at the site of the allergen by sensitized T cells, involving macrophages and regulated by cytokines. 629

**deletion** (duh LEE shun) Change in chromosome structure in which the end of a chromosome breaks off or two simultaneous breaks lead to the loss of an internal segment; often causes abnormalities—e.g., cri du chat syndrome. 184

**demographic transition** (dem oh GRAFF ick) Due to industrialization, a decline in the birthrate following a reduction in the death rate so that the population growth rate is lowered. 833

**demography** Properties of the rate of growth and the age structure of populations. 821

**denatured** (dee NATE churd) Loss of an enzyme's normal shape so that it no longer functions; caused by a less than optimal pH and temperature. 51, 110

**dendrite** (DEN drite) Part of a neuron that sends signals toward the cell body. 683

**dendritic cell** Antigen-presenting cell of the epidermis and mucous membranes. 618

**denitrification** (dee nite trih fih KAY shun) Conversion of nitrate or nitrite to nitrogen gas by bacteria in soil. 861

**dense fibrous connective tissue** Type of connective tissue containing many collagen fibers packed together; found in tendons and ligaments, for example. 580

**density-dependent factor** Biotic factor, such as disease or competition, that affects population size in a direct relationship to the population's density. 828

**density-independent factor** Abiotic factor, such as fire or flood, that affects population size independent of the population's density. 828

**deoxyribose** (dee ox ee RYE bohs) Pentose sugar found in DNA. 41

**derived trait** Structural, physiological, or behavioral trait that is present in a specific lineage and is not present in the common ancestor for several lineages. 341

**dermis** (DER miss) In mammals, thick layer of the skin underlying the epidermis. 586

**desert** Ecological biome characterized by a limited amount of rainfall; deserts have hot days and cool nights. 878

**desmosome** (DEZ moh sohm) Intercellular junction that connects cytoskeletons of adjacent cells. 98

**detritivore** Any organism that obtains most of its nutrients from the detritus in an ecosystem. 852

**deuterostome** (DEW turr row stome) Group of coelomate animals in which the second embryonic opening is associated with the mouth; the first embryonic opening, the blastopore, is associated with the anus. 515

**diabetes mellitus** Condition characterized by a high blood glucose level and the appearance of glucose in the urine due to a deficiency of insulin production and failure of cells to take up glucose. 748

**diaphragm** (DIE uh framm) In mammals, dome-shaped muscularized sheet separating the thoracic cavity from the abdominal cavity. 656

**diarrhea** Excessively frequent and watery bowel movements. 640

**diastole** (die ASS tuh lee) Relaxation period of a heart chamber during the cardiac cycle. 600

**diatom** (DIE uh tom) Golden-brown alga with a cell wall in two parts, or valves; significant part of phytoplankton. 382

**diencephalon** (die in SEF uh lahn) In vertebrates, portion of the brain in the region of the third ventricle that includes the thalamus and hypothalamus. 690

**differentially permeable** Ability of plasma membranes to regulate the passage of substances into and out of the cell, allowing some to pass through and preventing the passage of others. 90

**diffusion** Movement of molecules or ions from a region of higher to lower concentration; it requires no energy and tends to lead to an equal distribution. 91

**dihybrid cross** (die HIGH brid) Cross between parents that differ in two traits. 194

**dikaryotic** (die care ree AH tick) Having two haploid nuclei that stem from different parent cells; during sexual reproduction, sac and club fungi have dikaryotic cells. 395

**dinoflagellate** (dine no FLAJ ell ate) Photosynthetic unicellular protist with two flagella, one whiplash and the other located within a groove between protective cellulose plates; significant part of phytoplankton. 383

**dioecious** (dye EESH us) Having unisexual flowers or cones, with the male flowers or cones confined to certain land plants and the female flowers or cones of the same species confined to other different plants. 422

**diploid (2n) number** (DIP loid) Cell condition in which two of each type of chromosome are present. 155, 170

**diplomonad** Protist that has modified mitochondria, two equal-sized nuclei, and multiple flagella. 386

**directional selection** Outcome of natural selection in which an extreme phenotype is favored, usually in a changing environment. 290

**disaccharide** (die SACK uh ride) Sugar that contains two units of a monosaccharide; e.g., maltose. 41

**disruptive selection** Outcome of natural selection in which the two extreme phenotypes are favored over the average phenotype, leading to more than one distinct form. 290

**distal convoluted tubule** (DISS tull KAHN vole loot ted TUBE yule) Final portion of a nephron that joins with a collecting duct; associated with tubular secretion. 671

**DNA (deoxyribonucleic acid)** (dee OX ee RYE bow new CLAY ick) Nucleic acid polymer produced from covalent bonding of nucleotide monomers that contain the sugar deoxyribose; the genetic material of nearly all organisms. 52

**DNA ligase** (LIE gaze) Enzyme that links DNA fragments; used during production of recombinant DNA to join foreign DNA to vector DNA. 250

**DNA microarray** Thousands of different single-stranded DNA fragments arranged in an array (grid) on a glass slide; used to detect and measure gene expression. 258

**DNA polymerase** (pah LIMM urr race) During replication, an enzyme that joins the nucleotides complementary to a DNA template. 217

**DNA repair enzyme** One of several enzymes that restore the original base sequence in an altered DNA strand. 244

**DNA replication** Synthesis of a new DNA double helix prior to mitosis and meiosis in eukaryotic cells and during prokaryotic fission in prokaryotic cells. 217

**domain** Largest of the categories, or taxa, used by taxonomists to group species; the three domains are Archaea, Bacteria, and Eukarya. 6, 340

**domain Archaea** One of the three domains of life; contains prokaryotic cells that often live in extreme habitats and have unique genetic, biochemical, and physiological characteristics; its members are sometimes referred to as *archaea*. 7, 348

**domain Bacteria** One of the three domains of life; contains prokaryotic cells that differ from archaea because they have their own unique genetic, biochemical, and physiological characteristics. 7, 348

**domain Eukarya** One of the three domains of life, consisting of organisms with eukaryotic cells; includes protists, fungi, plants, or animals. 7, 348

**dominance hierarchy** Organization of animals in a group that determines the order in which the animals have access to resources. 291

**dominant allele** (uh LEEL) Allele that exerts its phenotypic effect in the heterozygote; it masks the expression of the recessive allele. 193

**dopamine** (DOPE uh meen) Neurotransmitter in the central nervous system. 686

**dormancy** In plants, a cessation of growth under conditions that seem appropriate for growth. 477

**dorsal root ganglion** (GANG lee uhn) Mass of sensory neuron cell bodies located in the dorsal root of a spinal nerve. 692

**double fertilization** In flowering plants, one sperm nucleus unites with the egg nucleus, and a second sperm nucleus unites with the polar nuclei of an embryo sac. 497

**double helix** Double spiral; describes the three-dimensional shape of DNA. 215

**doubling time** Number of years it takes for a population to double in size. 833

**dryopithecine** Tree dwelling primate existing 12–9 MYA; ancestral to apes. 563

**duodenum** (dew ODD duh num) First part of the small intestine, where chyme enters from the stomach. 639

**duplication** Change in chromosome structure in which a particular segment is present more than once in the same chromosome. 184

# E

**ecdysozoa** A protostome characterized by periodic molting of their exoskeleton. Includes the roundworms and arthropods. 516

**echinoderm** (ee KINE oh derm) Invertebrates such as sea stars, sea urchins, and sand dollars; characterized by radial symmetry and a water vascular system. 534

**ecological niche** Role an organism plays in its community, including its habitat and its interactions with other organisms. 841

**ecological pyramid** Visual depiction of the biomass, number of organisms, or energy content of various trophic levels in a food web—from the producer to the final consumer populations. 855

**ecological succession** The gradual replacement of communities in an area following a disturbance (secondary succession) or the creation of new soil (primary succession). 850

**ecology** Study of the interactions of organisms with other organisms and with the physical and chemical environment. 820

**ecosystem** Biological community together with the associated abiotic environment; characterized by a flow of energy and a cycling of inorganic nutrients. 9, 820, 852

**ecosystem diversity** Variety of species in a particular locale, dependent on the species interactions. 891

**ectoderm** (EK toe derm) Outermost primary tissue layer of an animal embryo; gives rise to the nervous system and the outer layer of the integument. 780

**ectotherm** (ek toe THERM) Organism having a body temperature that varies according to the environmental temperature. 543

**edge effect** Phenomenon in which the edges around a landscape patch provide a slightly different habitat than the favorable habitat in the interior of the patch. 902

**effector** Muscle or gland that receives signals from motor fibers and thereby allows an organism to respond to environmental stimuli. 679

**elastic cartilage** Type of cartilage composed of elastic fibers, allowing greater flexibility. 580

**elastic fiber** Yellow fiber in the matrix of connective tissue, providing flexibility. 579

**electrocardiogram (ECG)** (ee leck troh CARD dee oh gram) Recording of the electrical activity associated with the heartbeat. 601

**electron** Negative subatomic particle, moving about in an energy level around the nucleus of an atom. 22

**electronegativity** The ability of an atom to attract electrons toward itself in a chemical bond. 27

**electron shell** Concentric energy levels in which electrons orbit. 22

**electron transport chain (ETC)** Passage of electrons along a series of membrane-bound electron carrier molecules from a higher to lower energy level; the energy released is used for the synthesis of ATP. 112, 119, 135, 142

**element** Substance that cannot be broken down into substances with different properties; composed of only one type atom. 22

**El Nino–Southern Oscillation** Warming of water in the Eastern Pacific equatorial region such that the Humboldt Current is displaced, with possible negative results such as reduction in marine life. 884

**elongation** Middle stage of translation in which additional amino acids specified by the mRNA are added to the growing polypeptide. 226

**embryo** Stage of a multicellular organism that develops from a zygote before it becomes free-living; in seed plants, the embryo is part of the seed. 779

**embryonic disk** (em bree AHN ick) During human development, flattened area during gastrulation from which the embryo arises. 789

**embryonic period** First eight weeks of human development, during which the major organ systems are organized. 787

**embryophyta** Bryophytes and vascular plants; both of which produce embryos. 412

**embryo sac** Female gametophyte (megagametophyte) of flowering plants. 497

**emergent property** Quality that appears as biological complexity increases. 2

**emerging virus** Newly identified viruses that are becoming more prominent usually because they cause serious disease. 361

**endangered species** A species that is in peril of immediate extinction throughout all or most of its range (e.g., California condor, snow leopard). 890

**endergonic reaction** (en der GONN ick) Chemical reaction that requires an input of energy; opposite of exergonic reaction. 106

**endocrine gland** (EN doe crinn) Ductless organ that secretes hormone(s) into the bloodstream. 579, 736

**endocrine system** Organ system involved in the coordination of body activities; uses hormones as chemical signals secreted into the bloodstream. 736

**endocytosis** (en doe site TOE sis) Process by which substances are moved into the cell from the environment by phagocytosis (cellular eating) or pinocytosis (cellular drinking); includes receptor-mediated endocytosis. 96

**endoderm** (EN doe derm) Innermost primary tissue layer of an animal embryo that gives rise to the linings of the digestive tract and associated structures. 780

**endodermis** (en doe DERM miss) Internal plant root tissue forming a boundary between the cortex and the vascular cylinder. 441

**endomembrane system** (en doe MEM brain) Cellular system that consists of the nuclear envelope, endoplasmic reticulum, Golgi apparatus, and vesicles. 72

**endometrium** (en doe MEET tree uhm) Mucous membrane lining the interior surface of the uterus. 762

**endoplasmic reticulum (ER)** (en doe PLAZ mick ruh TICK you lum) System of membranous saccules and channels in the cytoplasm, often with attached ribosomes. 72

**endoskeleton** (en doe SKELL uh ton) Protective internal skeleton, as in vertebrates. 718

**endosperm** (EN doe sperm) In flowering plants, nutritive storage tissue that is derived from the union of a sperm nucleus and polar nuclei in the embryo sac. 497

**endospore** (EN doe spore) Spore formed within a cell; certain bacteria form endospores. 366

**endosymbiotic theory** (en doe simm bee AH tick) Explanation of the evolution of eukaryotic organelles by phagocytosis of prokaryotes. 66, 325

**endotherm** (en doe THERM) Organism in which maintenance of a constant body temperature is independent of the environmental temperature. 552

**energy** Capacity to do work and bring about change; occurs in a variety of forms. 4, 104

**energy of activation** Energy that must be added in order for molecules to react with one another. 108

**enhancer** DNA sequence that acts as a regulatory element to increase the level of transcription when a transcription factor binds to it. 242

**entropy** (EN truh pee) Measure of disorder or randomness. 105

**enzymatic protein** (en zih MATT tick) Protein that catalyzes a specific reaction. 88

**enzyme** (EN zime) Organic catalyst, usually a protein, that speeds a reaction in cells due to its particular shape. 40, 108

**enzyme inhibition** Means by which cells regulate enzyme activity; may be competitive or noncompetitive inhibition. 111

**eosinophil** (ee oh SIN uh fill) White blood cell containing cytoplasmic granules that stain with acidic dye. 607, 618

**epidermal tissue** Exterior tissue, usually one cell thick, of leaves, young stems, roots, and other parts of plants. 437

**epidermis** (eh pih DERM miss) In mammals, the outer, protective layer of the skin; in plants, tissue that covers roots, leaves, and stems of nonwoody organisms. 437, 585

**epigenetic inheritance** An inheritance pattern in which a nuclear gene has been modified but the changed expression of the gene is not permanent over many generations. 237

**epiglottis** (eh pih GLOTT tiss) Structure that covers the glottis, the air-tract opening, during the process of swallowing. 654

**epinephrine** (eh pih NEFF rinn) Hormone secreted by the adrenal medulla in times of stress; adrenaline. 745

**epiphyte** (EPP pih fite) Plant that takes its nourishment from the air because its placement in other plants gives it an aerial position. 874

**epithelial tissue** (eh pih THEE lee uhl) Tissue that lines hollow organs and covers surfaces. 578

**erythropoietin (EPO)** (eh rih throw poe EE tin) Hormone produced by the kidneys that speeds red blood cell formation. 674, 751

**esophagus** (eh SOFF uh gus) Muscular tube for moving swallowed food from the pharynx to the stomach. 637

**essential nutrient** In plants, substance required for normal growth, development, or reproduction. 457

**estrogen** (ESS truh jen) Female sex hormone that helps maintain sexual organs and secondary sex characteristics. 750, 763

**estuary** (EST you air ree) Portion of the ocean located where a river enters and fresh water mixes with salt water. 882

**ethylene** (ETH uh leen) Plant hormone that causes ripening of fruit and is also involved in abscission. 480

**euchromatin** (you CROW muh tin) Chromatin that is extended and accessible for transcription. 229, 239

**eudicot** (you DIE cot) Abbreviation of eudicotyledon. Flowering plant group; members have two embryonic leaves (cotyledons), net-veined leaves, vascular bundles in a ring, flower parts in fours or fives and their multiples, and other characteristics. 424, 436

**Eudicotyledone** One of two major classes of angiosperms; abbreviated as eudicot. 424

**euglenid** (YOU glen id) Flagellated and flexible freshwater unicellular protist that usually contains chloroplasts and has a semirigid cell wall. 386

**eukaryotic cell (eukaryote)** (you care ree AH tick) Type of cell that has a membrane-bounded nucleus and membranous organelles; found in organisms within the domain Eukarya. 7, 64

**euploidy** (you PLOY dee) Cells containing only complete sets of chromosomes. 180

**eutrophication** (you troh fih KAY shun) Enrichment of water by inorganic nutrients used by phytoplankton. Often, overenrichment caused by human activities leads to excessive bacterial growth and oxygen depletion. 859, 880

**evaporate (evaporation)** Conversion of a liquid or a solid into a gas. 29

**evergreen** Land plant that sheds leaves over a long period, so some leaves are always present. 435

**evolution** Descent of organisms from common ancestors with the development of genetic and phenotypic changes over time that make them more suited to the environment. 5, 267

**evolutionary species concept** Every species has its own evolutionary history, which is partly documented in the fossil record. 300

**excavate** Supergroup of eukaryotes that includes euglenids, kinetoplastids, parabasalids, and diplomonads. 386

**excretion** Elimination of metabolic wastes by an organism at exchange boundaries such as the plasma membrane of unicellular organisms and excretory tubules of multicellular animals. 666

**exergonic reaction** (ex urr GONN ick) Chemical reaction that releases energy; opposite of endergonic reaction. 106

**exocrine gland** (EX oh krinn) Gland that secretes its product to an epithelial surface directly or through ducts. 579

**exocytosis** (ex oh sigh TOE sis) Process in which an intracellular vesicle fuses with the plasma membrane so that the vesicle's contents are released outside the cell. 96

**exon** Segment of mRNA containing the protein-coding portion of a gene that remains within the mRNA after splicing has occurred. 223

**exophthalmic goiter** (ex opp THOWL mick GOI turr) Enlargement of the thyroid gland accompanied by an abnormal protrusion of the eyes. 743

**exoskeleton** (ex oh SKELL uh ton) Protective external skeleton, as in arthropods. 529, 718

**exotic species** Nonnative species that migrate or are introduced by humans into a new ecosystem; also called alien species. 897

**experiment** Artificial situation devised to test a hypothesis. 11

**experimental design** Methodology by which an experiment will seek to support the hypothesis. 11

**experimental variable** Factor of the experiment being tested. 14

**expiration** Act of expelling air from the lungs; exhalation. 656

**exponential growth** Growth, particularly of a population, in which the increase occurs in the same manner as compound interest. 825

**extant** Species, or other levels of taxa, that are still living. 267

**external respiration** Exchange of oxygen and carbon dioxide between alveoli and blood. 650

**extinct; extinction** Total disappearance of a species or higher group. 10, 327

**extracellular matrix (ECM)** Nonliving substance in which animal cells are imbedded; is composed of protein and polysaccharides. 87

**extraembryonic membrane** (ex truh em bree AH nick) Membrane that is not a part of the embryo but is necessary to the continued existence and health of the embryo. 757, 787

**ex vivo gene therapy** Gene therapy in which cells are removed from an organism, and DNA injected to correct a genetic defect; the cells are returned to the organism to treat a disease or disorder. 256

# F

**facilitated transport** Passive transfer of a substance into or out of a cell along a concentration gradient by a process that requires a carrier. 94

**facultative anaerobe** (fac ull TAY tihv ANN air robe) Prokaryote that is able to grow in either the presence or the absence of gaseous oxygen. 364

**FAD** Flavin adenine dinucleotide; a coenzyme of oxidation-reduction that becomes $FADH_2$ as oxidation of substrates occurs, and then delivers electrons to the electron transport chain in mitochondria during cellular respiration. 134

**fall overturn** Mixing process that occurs in fall in stratified lakes, whereby oxygen-rich top waters mix with nutrient-rich bottom waters. 880

**family** One of the categories, or taxa, used by taxonomists to group species; the taxon above the genus level. 6, 340

**family pedigree** Chart of genetic relationship of family individuals the through the generations. 201

**fat** Organic molecule that contains glycerol and fatty acids and is found in adipose tissue of vertebrates. 44, 644

**fate map** Diagram that traces the differentiation of cells during development from their origin to their final structure and function. 784

**fatty acid** Molecule that contains a hydrocarbon chain and ends with an acid group. 44

**fecundity** Potential capacity of an individual to produce offspring. 830

**female gametophyte** (guh MEET oh fite) In seed plants, the gametophyte that produces an egg; in flowering plants, an embryo sac. Sometimes called a megagametophyte. 496

**fermentation** Anaerobic breakdown of glucose that results in a gain of two ATP and end products such as alcohol and lactate. 135, 138

**fern** Member of a group of land plants that have large fronds; in the sexual life cycle, the independent gametophyte produces flagellated sperm, and the vascular sporophyte produces windblown spores. 418

**fertilization** Fusion of sperm and egg nuclei, producing a zygote that develops into a new individual. 172, 778

**fiber** Plant material that is nondigestible and promotes regularity of elimination. 643

**fibroblast** (FIE broh blast) Cell found in loose connective tissue that synthesizes collagen and elastic fibers in the matrix. 580

**fibrocartilage** Cartilage with a matrix of strong collagenous fibers. 580

**fibrous protein** A protein that has only a secondary structure; generally insoluble; includes collagens, elastins, and keratins. 51

**fibrous root system** In most monocots, a mass of similarly sized roots that cling to the soil. 442

**filament** (FILL uh mint) End-to-end chains of cells that form as cell division occurs in only one plane; in plants, the elongated stalk of a stamen. 378, 495

**fimbria (pl., fimbriae)** (FIMM bree uh, FIMM bree ee) Small, bristlelike fiber on the surface of a bacterial cell, which attaches bacteria to a surface; also fingerlike extension from the oviduct near the ovary. 65, 363

**fin** In fish and other aquatic animals, membranous, winglike, or paddlelike process used to propel, balance, or guide the body. 544

**first messenger** Chemical signal such as a peptide hormone that binds to a plasma membrane receptor protein and alters the metabolism of a cell because a second messenger is activated. 739

**fishes** Aquatic, gill-breathing vertebrate that usually has fins and skin covered with scales; fishes were among the earliest vertebrates that evolved. 543

**fitness** Ability of an organism to reproduce and pass its genes to the next fertile generation; measured against the ability of other organisms to reproduce in the same environment. 271, 291

**five-kingdom system** System of classification that contains the kingdoms Monera, Protista, Plantae, Animalia, and Fungi. 348

**fixed action pattern (FAP)** Innate behavior pattern that is stereotyped, spontaneous, independent of immediate control, genetically encoded, and independent of individual learning. 802

**flagellum (pl., flagella)** (fluh JELL uhm) Long, slender extension used for locomotion by some bacteria, protozoans, and sperm. 65, 80, 363

**flagship species** Species that evoke a strong emotional response in humans; charismatic, cute, regal (e.g., lions, tigers, dolphin, panda). 901

**flame cell** Found along excretory tubules of planarians; functions in propulsion of fluid through the excretory canals and out of the body. 667

**flatworm** Invertebrates such as planarians and tapeworms with extremely thin bodies; a three-branched gastrovascular cavity and a ladder type nervous system 520

**flower** Reproductive organ of a flowering plant, consisting of several kinds of modified leaves arranged in concentric rings and attached to a modified stem called the receptacle. 424, 494

**fluid-mosaic model** Model for the plasma membrane based on the changing location and pattern of protein molecules in a fluid phospholipid bilayer. 87

**follicle** (FOLL lick cull) Structure in the ovary of animals that contains an oocyte; site of oocyte production. 763

**follicular phase** (foe LICK you lurr) First half of the ovarian cycle, during which the follicle matures and much estrogen (and some progesterone) is produced. 764

**fontanel** (fahn tuh NELL) Membranous region located between certain cranial bones in the skull of a vertebrate fetus or infant. 722

**food chain** The order in which one population feeds on another in an ecosystem, thereby showing the flow of energy from a detrivore (detrital food chain) or a producer (grazing food chain) to the final consumer. 855

**food web** In ecosystems, a complex pattern of interlocking and crisscrossing food chains. 855

**foramen magnum** (for AY men MAG num) Opening in the occipital bone of the vertebrate skull through which the spinal cord passes. 722

**foraminiferan** (for am men IF furr uhn) A protist bearing a calcium carbonate test with many openings through which pseudopods extend. 388

**foreign antigen** An antigen not produced by the organism and to which it will react 619

**formula** A group of symbols and numbers used to express the composition of a compound. 26

**fossil** Any past evidence of an organism that has been preserved in the Earth's crust. 276, 322

**founder effect** Cause of genetic drift due to colonization by a limited number of individuals who, by chance, have different geneotype and allele frequencies than the parent population. 288

**fovea centralis** (FOE vee uh sen TRAHL liss) Region of the retina consisting of densely packed cones; responsible for the greatest visual acuity. 705

**frameshift mutation** Insertion or deletion of at least one base so that the reading frame of the corresponding mRNA changes. 244

**free energy** Useful energy in a system that is capable of performing work. 106

**frond** Leaf of a fern palm, or cycad. 418

**fruit** Flowering plant structure consisting of one or more ripened ovaries that usually contain seeds. 427, 503

**fruiting body** Spore-producing and spore-disseminating structure found in sac and club fungi. 398

**functional genomics** Study of gene function at the genome level. It involves the study of many genes simultaneously and the use of DNA microarrays. 258

**functional group** Specific cluster of atoms attached to the carbon skeleton of organic molecules that enters into reactions and behaves in a predictable way. 39

**fungus (pl., fungi)** Saprotrophic decomposer; the body is made up of filaments called hyphae that form a mass called a mycelium. 7, 394

# G

**gallbladder** Organ attached to the liver that serves to store and concentrate bile. 642

**gametangia** Cell or multicellular structure in which gametes are formed. 396

**gamete** (GAMM eet) Haploid sex cell; e.g., egg and sperm. 170, 768

**gametogenesis** (gamm eet oh JEN uh sis) Development of the male and female sex gametes. 179

**gametophyte** (guh MEET uh fite) Haploid generation of the alternation of generations life cycle of a plant; produces gametes that unite to form a diploid zygote. 178, 412

**ganglion** (GANG lee ahn) Collection or bundle of neuron cell bodies usually outside the central nervous system. 680

**gap junction** Junction between cells formed by the joining of two adjacent plasma membranes; it lends strength and allows ions, sugars, and small molecules to pass between cells. 99

**gastropod** (gas trah POD) Mollusc with a broad, flat foot for crawling (e.g., snails and slugs). 525

**gastrovascular cavity** (gas troh VASS cue lurr) Blind digestive cavity in animals that have a sac body plan. 518

**gastrula** (GAS true luh) Stage of animal development during which the germ layers form, at least in part, by invagination. 780

**gastrulation** (gas true LAY shun) Formation of a gastrula from a blastula; characterized by an invagination to form cell layers of a caplike structure. 780

**gene** (JEEN) Unit of heredity existing as alleles on the chromosomes; in diploid organisms, typically two alleles are inherited—one from each parent. 5, 70

**gene cloning** DNA cloning to produce many identical copies of the same gene. 250

**gene flow** Sharing of genes between two populations through interbreeding. 287

**gene locus** Specific location of a particular gene on a chromosomes. 193

**gene mutation** Altered gene whose sequence of bases differs from the previous sequence. 243

**gene pharming** Production of pharmaceuticals using transgenic farm animals. 253

**gene pool** Total of all the genes of all the individuals in a population. 285

**gene therapy** Correction of a detrimental mutation by the addition of new DNA and its insertion in a genome. 250

**genetically modified organis (GMO)** Organism that carries the genes of another organism as a result of DNA technology. 252

**genetic code** Universal code that has existed for eons and allows for conversion DNA and RNA's chemical code to a sequence of amino acids in a protein. Each codon consists of three bases that stand for one of the 20 amino acids found in proteins or directs the termination of translation. 221

**genetic diversity** Variety among members of a population. 891

**genetic drift** Mechanism of evolution due to random changes in the allelic frequencies of a population; more likely to occur in small populations or when only a few individuals of a large population reproduce. 287

**genetic profile** An individuals genome, including any possible mutations. 258

**genetic recombination** Process in which chromosomes are broken and rejoined to form novel combinations; in this way offspring receive alleles in combinations different from their parents. 172

**genomics** Study of whole genomes. 255

**genotype** (JEEN oh type) Genes of an organism for a particular trait or traits; often designated by letters—for example, *BB* or *Aa*. 193

**genus** (JEEN us) One of the categories, or taxa, used by taxonomists to group species; contains those species that are most closely related through evolution. 6, 340

**geologic timescale** History of the earth based on the fossil record and divided into eras, periods, and epochs 324

**germ cell** During zygote development, cells that are set aside from the somatic cells and that will eventually undergo meiosis to produce gametes. 756

**germinate** Beginning of growth of a seed, spore, or zygote, especially after a period of dormancy. 504

**germ layer** Primary tissue layer of a vertebrate embryo—namely, ectoderm, mesoderm, or endoderm. 515, 780

**gibberellin** (jib urr ELL uhn) Plant hormone promoting increased stem growth; also involved in flowering and seed germination. 476

**gills** Respiratory organ in most aquatic animals; in fish, an outward extension of the pharynx. 540, 651

**ginkgo** Member of phylum Ginkgophyte; maidenhair tree. 422

**girdling** Removing a strip of bark from around a tree. 468

**gland** Epithelial cell or group of epithelial cells that are specialized to secrete a substance. 579

**global warming** Predicted increase in the Earth's temperature due to human activities that promote the greenhouse effect. 125, 859, 898

**globular protein** Most of the proteins in the body; soluble in water or salt solution; includes albumins, globulins, histones. 51

**glomerular capsule** (glow MARE you lurr) Cuplike structure that is the initial portion of a nephron. 671

**glomerular filtration** Movement of small molecules from the glomerulus into the glomerular capsule due to the action of blood pressure. 672

**glomerulus** (glow MARE you luss) Capillary network within the glomerular capsule of a nephron. 671

**glottis** (GLAH tiss) Opening for airflow in the larynx. 654

**glucocorticoid** (glue koh CORE tih coid) Type of hormone secreted by the adrenal cortex that influences carbohydrate, fat, and protein metabolism; *See* also cortisol. 745

**glucose** (GLUE kohs) Six-carbon sugar that organisms degrade as a source of energy during cellular respiration. 41

**glycerol** (GLISS ur all) Three-carbon carbohydrate with three hydroxyl groups attached; a component of fats and oils. 44

**glycocalyx** (glie koh KAY licks) Gel-like coating outside the cell wall of a bacterium. If compact, it is called a capsule; if diffuse, it is called a slime layer. 64

**glycogen** (GLIE kuh jen) Storage polysaccharide found in animals; composed of glucose molecules joined in a linear fashion but having numerous branches. 42

**glycolipid** (glie koh LIP pidd) Lipid in plasma membranes that bears a carbohydrate chain attached to a hydrophobic tail. 87

**glycolysis** (glie KAH lih sis) Anaerobic breakdown of glucose that results in a gain of two ATP and the end product pyruvate. 135, 136

**glycoprotein** (glie koh PRO teen) Protein in plasma membranes that bears a carbohydrate chain. 87

**gnathostome** Vertebrates with jaws. 542

**gnetophyte** Member of one of the four phyla of gymnosperms; Gnetophyta has only three living genera, which differ greatly from one another—e.g., *Welwitschia* and *Ephedra*. 422

**golden brown algae** Unicellular organism that contains pigments, including chlorophyll *a* and *c* and carotenoids, that produce its color. 382

**Golgi apparatus** (GOAL ghee app uh RAT us) Organelle consisting of sacs and vesicles that processes, packages, and distributes molecules about or from the cell. 72

**gonad** (GO nadd) Organ that produces gametes; the ovary produces eggs, and the testis produces sperm. 756

**gonadotropic hormone** (go nadd oh TROH pick) Substance secreted by the anterior pituitary that regulates the activity of the ovaries and testes; principally, follicle-stimulating hormone (FSH) and luteinizing hormone (LH). 740

**granum (pl., grana)** (GRA numm) Stack of chlorophyll-containing thylakoids in a chloroplast. 76, 119

**grassland** Biome characterized by rainfall greater than 25 cm/yr, grazing animals, and warm summers; includes the prairie of the U.S. midwest and the African savanna. 876

**gravitational equilibrium** Maintenance of balance when the head and body are motionless. 712

**gravitropism** (grav ih TROPE is uhm) Growth response of roots and stems of plants to the Earth's gravity; roots demonstrate positive gravitropism, and stems demonstrate negative gravitropism. 482

**gray crescent** Gray area that appears in an amphibian egg after being fertilized by the sperm; thought to contain chemical signals that turn on the genes that control development. 783

**gray matter** Nonmyelinated axons and cell bodies in the central nervous system. 688

**green algae** Members of a diverse group of photosynthetic protists; contain chlorophylls *a* and *b* and have other biochemical characteristics like those of plants. 377

**greenhouse effect** Reradiation of solar heat toward the Earth, caused by an atmosphere that allows the sun's rays to pass through but traps the heat in the same manner as the glass of a greenhouse. 859

**greenhouse gases** Gases in the atmosphere such as carbon dioxide, methane, water vapor, ozone, and nitrous oxide that are involved in the greenhouse effect. 858

**ground tissue** Tissue that constitutes most of the body of a plant; consists of parenchyma, collenchyma, and sclerenchyma cells that function in storage, basic metabolism, and support. 437

**growth factor** A hormone or chemical, secreted by one cell, that may stimulate or inhibit growth of another cell or cells. 153

**growth hormone (GH)** Substance secreted by the anterior pituitary; controls size of an individual by promoting cell division, protein synthesis, and bone growth. 740

**guanine (G)** (GWAH neen) One of four nitrogen-containing bases in nucleotides composing the structure of DNA and RNA; pairs with cytosine. 214

**guard cell** One of two cells that surround a leaf stoma; changes in the turgor pressure of these cells cause the stoma to open or close. 466

**guttation** (gutt TAY shun) Liberation of water droplets from the edges and tips of leaves. 464

**gymnosperm** (JIM no sperm) Type of woody seed plant in which the seeds are not enclosed by fruit and are usually borne in cones, such as those of the conifers. 420

# H

**habitat** Place where an organism lives and is able to survive and reproduce. 820, 841

**hair follicle** Tubelike depression in the skin in which a hair develops. 586

**halophile** (HAL uh file) Type of archaea that lives in extremely salty habitats. 369

**haploid (n) number** (HAP loid) Cell condition in which only one of each type of chromosome is present. 155, 170

**Hardy-Weinberg principle** Law stating that the gene frequencies in a population remain stable if evolution does not occur due to nonrandom mating, selection, migration, and genetic drift. 286

**hay fever** Seasonal variety of allergic reaction to a specific allergen. Characterized by sudden attacks of sneezing, swelling of nasal mucosa, and often asthmatic symptoms. 629

**heart** Muscular organ whose contraction causes blood to circulate in the body of an animal. 598

**heart attack** Damage to the myocardium due to blocked circulation in the coronary arteries; myocardial infarction. 605

**heat** Type of kinetic energy; captured solar energy eventually dissipates as heat in the environment. 104

**helper T cell** Secretes lymphokines, which stimulate all kinds of immune cells. 619

**heme** (HEEM) Iron-containing group found in hemoglobin. 659

**hemizygous** Possessing only one allele for a gene in a diploid organism; males are hemizygous for genes on the X chromosome. 206

**hemocoel** (HEEM uh seel) Residual coelom found in arthropods, which is filled with hemolymph. 523

**hemoglobin (Hb)** (HEEM uh globe in) Iron-containing respiratory pigment occurring in vertebrate red blood cells and in the blood plasma of some invertebrates. 48, 659, 606

**hemolymph** (HEEM uh limf) Circulatory fluid that is a mixture of blood and interstitial fluid; seen in animals that have an open circulatory system, such as molluscs and arthropods. 594

**hepatitis** Inflammation of the liver. Viral hepatitis occurs in several forms. 642

**herbaceous stem** (her BAY shus) Nonwoody stem. 445

**herbivore** (HER bih vore) Primary consumer in a grazing food chain; a plant eater. 852

**hermaphroditic** Type of animal that has both male and female sex organs. 521

**heterochromatin** (hett turr oh CROW muh tin) Highly compacted chromatin that is not accessible for transcription. 229, 240

**heterosporous** Seed plant that produces two types of spores—microspores and megaspores. A plant that produces only one type of spore is *homosporous*. 420

**heterotroph** (HETT turr uh trofe) Organism that cannot synthesize organic compounds from inorganic substances and therefore must take in organic food. 118, 852

**heterozygote advantage** Situation in which individuals heterozygous for a trait have a selective advantage over those who are homozygous dominant or recessive; an example is sickle-cell anemia disease. 295

**heterozygous** (hett turr oh ZYE guss) Possessing unlike alleles for a particular trait. 193

**hexose** (HEX ohs) Six-carbon sugar. 41

**histamine** (HISS tuh mean) Substance, produced by basophils in blood and mast cells in connective tissue, that causes capillaries to dilate. 616

**histone** A group of proteins involved in forming the nucleosome structure of eukaryote chromatin. 155, 228

**holozoic** (hoe low ZOE ick) Obtaining nourishment by ingesting solid food particles. 384

**homeobox** (HOME me oh box) 180-nucleotide sequence located in all homeotic genes. 785

**homeodomain** Conserved DNA-binding region of transcription factors encoded by the homeobox of homeotic genes. 786

**homeostasis** (home me oh STAY sis) Maintenance of normal internal conditions in a cell or an organism by means of self-regulating mechanisms. 4, 588

**homeotic genes** (home me AH tick) Genes that control the overall body plan by controlling the fate of groups of cells during development. 785

**hominid** (HAH men idd) Member of the family Hominidae, including humans, chimpanzees, gorillas, and orangutans. 563

**hominin** Taxon that includes human and species very closely related to humans and chimpanzees. 563

**hominine** Taxon that includes the hominins and gorillas. 563

**hominoid** (HAH men oid) Member of the superfamily Hominoidea, which includes apes, humans, and their recent ancestors. 563

***Homo erectus*** (HOE mow eh RECK tuss) Hominin related to *H. erectus* that remained in Africa. 568

***Homo ergaster*** Extinct hominin; some paleontologists separate it from *H. erectus*, some do not and consider it a part of the African line of *H. erectus*. 568

**homologous chromosome** (hoe MOLL uh gus) Member of a pair of chromosomes that are alike and come together in synapsis during prophase of the first meiotic division; a *homologue*. 170

**homologous gene** Gene that codes for the same protein, even if the base sequence may be different. 260

**homologous structure** In evolution, a structure that is similar in different types of organisms because these organisms are derived from a common ancestor. 278, 345

**homologue** (HOE mow log) Member of a homologous pair of chromosomes. 170

**homology** (hoe MAH low jee) Similarity of parts or organs of different organisms caused by evolutionary derivation from a corresponding part or organ in a remote ancestor, and usually having a similar embryonic origin. 344

**homozygous** (hoe mow ZYE guss) Possessing two identical alleles for a particular trait. 193

**hormone** Chemical messenger produced in one part of the body that controls the activity of other parts. 474, 736

**hornwort** A broyphyte ( phylum *Anthocerophyta*) with a thin gametophyte and tiny sporophyte that resembles a broom handle. 414

**horsetail** A seedless vascular plant having only one genus (*Equisetum*) in existence today; characterized by rhizomes, scalelike leaves, strobili, and tough, rigid stems. 417

**host** Organism that provides nourishment and/or shelter for a parasite. 846

**host specific** Parasite that can infect only one type of host. 358

**human chorionic gonadotropin (HCG)** (core ree AH nick go nadd uh TROPE in) Gonadotropic hormone produced by the chorion that functions to maintain the uterine lining. 765, 789

**Human Genome Project (HGP)** Initiative to determine the complete sequence of the human genome and to analyze this information. 255

**human immunodeficiency virus (HIV)** (im you no duh FISH ens see) Virus responsible for AIDS. 770

**humus** (HUE muss) Decomposing organic matter in the soil. 458

**hunter-gatherer** Human that hunted animals and gathered plants for food. 569

**Huntington diease** Genetic disease marked by progressive deterioration of the nervous system and resulting in neuromuscular abnormalities. 689

**hyaline cartilage** Cartilage whose cells lie in lacunae separated by a white translucent matrix containing very fine collagen fibers. 580

**hydra** A freshwater cnidaria that only exists as a polyp with tentacles. 519

**hydrogen bond** Weak bond that arises between a slightly positive hydrogen atom of one molecule and a slightly negative atom of another molecule or between parts of the same molecule. 28

**hydrogen ion ($H^+$)** Hydrogen atom that has lost its electron and therefore bears a positive charge. 32

**hydrolysis reaction** (high DRAH lih sis) Splitting of a bond by the addition of water, with the $H^+$ going to one molecule and the $OH^-$ going to the other. 40

**hydrophilic** (high droh FILL ick) Type of molecule that interacts with water by dissolving in water and/or by forming hydrogen bonds with water molecules. 30, 39

**hydrophobic** (high droh FOE bick) Type of molecule that does not interact with water because it is nonpolar. 30, 39

**hydroponics** (high droh PAH nicks) Technique for growing plants by suspending them with their roots in a nutrient solution. 457

**hydrostatic skeleton** (high droh STAT ick) Fluid-filled body compartment that provides support for muscle contraction resulting in movement; seen in cnidarians, flatworms, roundworms, and segmented worms. 518, 718

**hydrothermal vent** (high droh THERM mull) Hot springs in the seafloor along ocean ridges where heated seawater and sulfate react to produce hydrogen sulfide; here, chemosynthetic bacteria support a community of varied organisms. 883

**hydroxide ion ($OH^-$)** (high DROX side EYE ahn) One of two ions that results when a water molecule dissociates; it has gained an electron and therefore bears a negative charge. 32

**hypersensitive response (HR)** Plants respond to pathogens by selectively killing plant cells to block the spread of the pathogen. 489

**hypertonic solution** (high purr TAH nick) Higher solute concentration (less water) than the cytoplasm of a cell; causes cell to lose water by osmosis. 93

**hypha** (HIGH fuh) Filament of the vegetative body of a fungus. 394

**hypothalamic-inhibiting hormone** (high poh THOWL mick) One of many hormones produced by the hypothalamus that inhibits the secretion of an anterior pituitary hormone. 740

**hypothalamic-releasing hormone** One of many hormones produced by the hypothalamus that stimulates the secretion of an anterior pituitary hormone. 740

**hypothalamus** (high poh THOWL uh muss) In vertebrates, part of the brain that helps regulate the internal environment of the body—for example, heart rate, body temperature, and water balance. 690, 740

**hypothesis** (high PAH thuh sis) Supposition established by reasoning after consideration of available evidence; it can be tested by obtaining more data, often by experimentation. 11

**hypotonic solution** (high poh TAH nick) Lower solute (more water) concentration than the cytoplasm of a cell; causes cell to gain water by osmosis. 92

# I

**IgG** Most abundant immunoglobulin; mostly found in the blood, but also in the lymph and tissue fluid. Y shaped with two binding sites. 621

**immediate allergic response** Allergic response that occurs within seconds of contact with an allergen; caused by the attachment of the allergen to IgE antibodies. 629

**immunity** Ability of the body to protect itself from foreign substances and cells, including disease-causing agents. 616

**immunization** Strategy for achieving immunity to the effects of specific disease-causing agents. 620

**immunoglobulin (Ig)** (imm you no GLOB you linn) Globular plasma protein that functions as an antibody. 621

**implantation** In placental mammals, the embedding of an embryo at the blastocyst stage into the endometrium of the uterus. 787

**imprinting** Learning to make a particular response to only one type of animal or object. 803

**inclusion body** In a bacterium, stored nutrients for later use. 65

**inclusive fitness** Fitness that results from personal reproduction and from helping nondescendant relatives reproduce. 814

**incomplete digestive tract** Digestive tract that has a single opening, usually called a mouth. 634

**incomplete dominance** Inheritance pattern in which the offspring has an intermediate phenotype, as when a red-flowered plant and a white-flowered plant produce pink-flowered offspring. 202

**incomplete penetrance** Dominant alleles that are not always expressed, often for unknown reasons. 202

**independent assortment** Alleles of unlinked genes segregate independently of each other during meiosis so that the gametes contain all possible combinations of alleles. 172

**index fossil** Deposits found in certain layers of strata; similar fossils can be found in the same strata around the world. 322

**induced fit model** Change in the shape of an enzyme's active site that enhances the fit between the active site and its substrate(s). 108

**induced mutation** Mutation that is caused by an outside influence, such as organic chemicals or ionizing radiation. 243

**inducer** Molecule that brings about activity of an operon by joining with a repressor and preventing it from binding to the operator. 236

**inducible operon** (in DOO sih bull AH purr ahn) In a catabolic pathway, an operon causes transcription of the genes controlling a group of enzymes. 236

**induction** Ability of a chemical or a tissue to influence the development of another tissue. 783

**inductive reasoning** Using specific observations and the process of logic and reasoning to arrive at a hypothesis. 11

**industrial melanism** (MELL uh nizz uhm) Increased frequency of darkly pigmented (melanic) forms in a population when soot and pollution make lightly pigmented forms easier for predators to see against a pigmented background. 286

**infertility** Inability to have as many children as desired. 768

**inflammatory response** Tissue response to injury that is characterized by redness, swelling, pain, and heat. 616

**ingroup** In a cladistic study of evolutionary relationships among organisms, the group that is being analyzed. 342

**inheritance of acquired characteristics** Lamarckian belief that characteristics acquired during the lifetime of an organism can be passed on to offspring. 268

**initiation** First stage of translation in which the translational machinery binds an mRNA and assembles. 226

**inner ear** Portion of the ear consisting of a vestibule, semicircular canals, and the cochlea where equilibrium is maintained and sound is transmitted. 710

**inorganic chemistry** Branch of science which deals with compounds that are not unique to the plant or animal worlds. 38

**insect** Type of arthropod. The head has antennae, compound eyes, and simple eyes; the thorax has three pairs of legs and often wings; and the abdomen has internal organs. 532

**insight learning** Ability to apply prior learning to a new situation without trial-and-error activity. 805

**inspiration** Act of taking air into the lungs; inhalation. 656

**integration** Summing up of excitatory and inhibitory signals by a neuron or by some part of the brain. 687

**intercalated disk** (in TURK uh lay tidd) Region that holds adjacent cardiac muscle cells together; disks appear as dense bands at right angles to the muscle striations. 585

**interferon** (in turr FEAR ron) Antiviral agent produced by an infected cell that blocks the infection of another cell. 618

**intergenic sequence** (in tur GEN ic) Region of DNA that lies between genes on a chromosome. 256

**interkinesis** (in turr kuh NEE sis) Period of time between meiosis I and meiosis II during which no DNA replication takes place. 176

**intermediate filament** Ropelike assemblies of fibrous polypeptides in the cytoskeleton that provide support and strength to cells; so called because they are intermediate in size between actin filaments and microtubules. 78

**internal respiration** Exchange of oxygen and carbon dioxide between blood and tissue fluid. 650

**interneuron** (in turr NURE ron) Neuron located within the central nervous system that conveys messages between parts of the central nervous system. 683

**internode** (IN turr node) In vascular plants, the region of a stem between two successive nodes. 435

**interphase** (IN turr faze) Stages of the cell cycle ($G_1$, S, $G_2$) during which growth and DNA synthesis occur when the nucleus is not actively dividing. 152

**interspersed repeat** (in tur SPURSED) Repeated DNA sequence that is spread across several regions of a chromosome or across multiple chromosomes. 257

**intertidal zone** Region along a coastline where the tide recedes and returns. 882

**intron** (IN trahn) Intervening sequence found between exons in mRNA that is removed before translation. 223

**inversion** Change in chromosome structure in which a segment of a chromosome is turned around 180°; this reversed sequence of genes can lead to altered gene activity and abnormalities. 184

**invertebrate** (in VURR tuh brate) Animal without a vertebral column or back bone. 512

**in vivo gene therapy** Gene therapy in which normal genes are injected directly into an organism to treat a condition often due to a faulty gene. 256

**ion** (EYE ahn) Charged particle that carries a negative or positive charge. 26

**ionic bond** (eye AH nick) Chemical bond in which ions are attracted to one another by opposite charges. 26

**iris** Muscular ring that surrounds the pupil and regulates the passage of light through this opening. 705

**isomer** (EYE so murr) Molecules with the same molecular formula but a different structure, and therefore a different shape. 39

**isotonic solution** (eye so TAH nick) Solution that is equal in solute concentration to that of the cytoplasm of a cell; causes cell to neither lose nor gain water by osmosis. 92

**isotope** (EYE so tope) Atom of the same element having the same atomic number but a different mass number due to the number of neutrons. 24

**iteroparity** Repeated production of offspring at intervals throughout the life cycle of an organism. 824

# J

**jaundice** Yellowish tint to the skin caused by an abnormal amount of bilirubin (bile pigment) in the blood, indicating liver malfunction. 642

**jawless fishes** Type of fish that has no jaws; includes today's hagfishes and lampreys. 543

**joint** Articulation between two bones of a skeleton. 725

**junction protein** Protein that assists cell-to-cell communication at the plasma membrane. 88

# K

**karyokinesis** (CARE ree oh kin ee sis) Division of the nucleus. 152

**karyotype** (CARE ree oh type) Chromosomes arranged by pairs according to their size, shape, and general appearance in mitotic metaphase. 181

**keystone species** Species whose activities significantly affect community structure. 901

**kidneys** Paired organs of the vertebrate urinary system that regulate the chemical composition of the blood and produce a waste product called urine. 670

**kinetic energy** (kin NET tick) Energy associated with motion. 104

**kinetochore** (kin NET uh core) An assembly of proteins that attaches to the centromere of a chromosome during mitosis. 155, 171

**kinetoplastid** Unicellular, flagellate protist characterized by the presence in their single mitochondrion of a kinetoplast (a structure containing a large mass of DNA). 386

**kingdom** One of the categories, or taxa, used by taxonomists to group species; the taxon above phylum. 6, 340

**kin selection** Indirect selection; adaptation to the environment due to the reproductive success of an individual's relatives. 814

***K*-selection** Favorable life-history strategy under stable environmental conditions characterized by the production of a few offspring with much attention given to offspring survival. 831

# L

**lactation** (lack TAY shun) Secretion of milk by mammary glands, usually for the nourishment of an infant. 766

**lacteal** (LACK tee uhl) Lymphatic vessel in an intestinal villus; aids in the absorption of fats. 639

**lacuna** (luh COON uh) Small pit or hollow cavity, as in bone or cartilage, where a cell or cells are located. 580

**ladderlike nervous system** In planarians, two lateral nerve cords joined by transverse nerves. 680

**lake** Body of fresh water, often classified by nutrient status, such as oligotrophic (nutrient-poor) or eutrophic (nutrient-rich). 880

**landscape** A number of interacting ecosystems. 891

**landscape diversity** Variety of habitat elements within an ecosystem (e.g., plains, mountains, and rivers). 891

**large intestine** In vertebrates, portion of the digestive tract that follows the small intestine; in humans, consists of the cecum, colon, rectum, and anal canal. 639

**larva** (LARR vuh) Immature form in the life cycle of some animals; it sometimes undergoes metamorphosis to become the adult form. 757

**larynx** (LAIR inks) Cartilaginous organ located between the pharynx and the trachea; in humans, contains the vocal cords; sometimes called the voice box. 654

**lateral line** Canal system containing sensory receptors that allow fishes and amphibians to detect water currents and pressure waves from nearby objects. 713

**law** *See* principle 12

**laws of thermodynamics** Two laws explaining energy and its relationships and exchanges. The first, also called the "law of conservation," says that energy cannot be created or destroyed but can only be changed from one form to another; the second says that energy cannot be changed from one form to another without a loss of usable energy. 104

**leaf** Lateral appendage of a stem, highly variable in structure, often containing cells that carry out photosynthesis. 435

**leaf vein** Vascular tissue within a leaf. 439

**learning** Relatively permanent change in an animal's behavior that results from practice and experience. 802

**lens** Clear, membranelike structure found in the vertebrate eye behind the iris; brings objects into focus. 705

**lenticel** (LENN tiss uhl) Frond of usually numerous, lightly raised, somewhat spongy, groups of cells in the bark of woody plants. Permits gas exchange between the interior of a plant and the external atmosphere. 437

**leptin** Hormone produced by adipose tissue that acts on the hypothalamus to signal satiety (fullness). 751

**less-developed country (LDC)** Country that is becoming industrialized; typically, population growth is expanding rapidly, and the majority of people live in poverty. 833

**leucoplast** (LOO coh plast) Plastid, generally colorless, that synthesizes and stores starch and oils. 76

**lichen** (LIKE in) Symbiotic relationship between certain fungi and algae, in which the fungi possibly provide inorganic food or water and the algae provide organic food. 367, 404

**life cycle** Recurring pattern of genetically programmed events by which individuals grow, develop, maintain themselves, and reproduce. 178

**ligament** Tough cord or band of dense fibrous tissue that binds bone to bone at a joint. 580, 725

**light reaction** Portion of photosynthesis that captures solar energy and takes place in thylakoid membranes of chloroplasts; it produces ATP and NADPH. 120

**lignin** (LIGG nihn) Chemical that hardens the cell walls of land plants. 416, 438

**limbic system** (LIMM bick) In humans, functional association of various brain centers, including the amygdala and hippocampus; governs learning and memory and various emotions such as pleasure, fear, and happiness. 691

**limiting factor** Resource or environmental condition that restricts the abundance and distribution of an organism. 821

**lipase** (LIE pace) Fat-digesting enzyme secreted by the pancreas. 643

**lipid** (LIP pid) Class of organic compounds that tends to be soluble in nonpolar solvents; includes fats and oils. 44

**liposome** (LIP uh sohm) Droplet of phospholipid molecules formed in a liquid environment. 320

**liver** Large, dark red internal organ that produces urea and bile, detoxifies the blood, stores glycogen, and produces the plasma proteins, among other functions. 640

**liverwort** Bryophyte with a lobed or leafy gametophyte and a sporophyte composed of a stalk and capsule. 413

**lobe-finned fishes** Type of fishes with limblike fins. 544

**logistic growth** (luh JISS tick) Population increase that results in an S-shaped curve; growth is slow at first, steepens, and then levels off due to environmental resistance. 826

**long-day plant** Plant that flowers when day length is longer than a critical length; e.g., wheat, barley, clover, and spinach. 487

**loop of the nephron** (NEFF ron) Portion of a nephron between the proximal and distal convoluted tubules; functions in water reabsorption. 671

**loose fibrous connective tissue** Tissue composed mainly of fibroblasts widely separated by a matrix containing collagen and elastic fibers. 580

**lophophore** A general term to describe several groups of lophotrochoans that have a feeding structure called a lophophore. 516

**lophotrochozoa** Main group of protostomes; widely diverse. Includes the flatworms, rotifers, annelids, and molluscs. 516

**lumen** Cavity inside any tubular structure, such as the lumen of the digestive tract. 641

**lung fishes** Type of lobe-finned fish that utilizes lungs in addition to gills for gas exchange. 544

**lungs** Internal respiratory organ containing moist surfaces for gas exchange. 651

**luteal phase** (LOO tee uhl) Second half of the ovarian cycle, during which the corpus luteum develops and much progesterone (and some estrogen) is produced. 764

**lycophyte** Club mosses, among the first vascular plants to evolve and to have leaves. The leaves of the lycophytes are mirrophylls. 416

**lymph** (LIMF) Fluid, derived from tissue fluid, that is carried in lymphatic vessels. 581, 608, 614

**lymphatic organ** (limm FAT ick) Organ other than a lymphatic vessel that is part of the lymphatic system; the lymphatic organs are the lymph nodes, tonsils, spleen, thymus gland, and bone marrow. 614

**lymphatic system** (limm FAT ick) Organ system consisting of lymphatic vessels and lymphatic organs; transports lymph and lipids, and aids the immune system. 614

**lymphatic vessel** Vessel that carries lymph. 614

**lymph node** Mass of lymphatic tissue located along the course of a lymphatic vessel. 615

**lymphocyte** (LIMM foe site) Specialized white blood cell that functions in specific defense; occurs in two forms—T lymphocytes and B lymphocytes. 607

**lysogenic cell** Cell that contains a prophage (virus incorporated into DNA), which is replicated when the cell divides. 359

**lysogenic cycle** (lie so JEN ick) Bacteriophage life cycle in which the virus incorporates its DNA into that of a bacterium; occurs preliminary to the lytic cycle. 358

**lysosome** (LIE so sohm) Membrane-bounded vesicle that contains hydrolytic enzymes for digesting macromolecules. 73

**lytic cycle** (LIH tick) Bacteriophage life cycle in which the virus takes over the operation of the bacterium immediately upon entering it and subsequently destroys the bacterium. 358

# M

**macroevolution** (mac crow evv oh LOO shun) Large-scale evolutionary change, such as the formation of new species. 310

**macronutrient** Essential element needed in large amounts for plant growth, such as nitrogen, calcium, or sulfur. 457

**macrophage** (MAC crow fahj) In vertebrates, large phagocytic cell derived from a monocyte that ingests microbes and debris. 607, 618

**magnification** Using a microscope, enlarging an object for viewing. 62

**male gametophyte** (guh MEET toe fite) In seed plants, the gametophyte that produces sperm; a pollen grain. Sometimes called a microgametophyte. 496

**malignant** (muh LIGG nunt) The power to threaten life; cancerous. 161

**Malpighian tubule** (mal PIG ee uhn TUBE yule) Blind, threadlike excretory tubule near the anterior end of an insect's hindgut. 532, 667

**maltase** Enzyme produced in small intestine that breaks down maltose to two glucose molecules. 642

**mammal** Endothermic vertebrate characterized especially by the presence of hair and mammary glands. 554

**mantle** In molluscs, an extension of the body wall that covers the visceral mass and may secrete a shell. 523

**marsh** Soft, wetland, which is treeless. 879

**marsupial** Member of a group of mammals bearing immature young nursed in a marsupium, or pouch—for example, kangaroo and opossum. 554

**mass extinction** Episode of large-scale extinction in which large numbers of species disappear in a few million years or less. 327

**mass number** Mass of an atom equal to the number of protons plus the number of neutrons within the nucleus. 23

**mast cell** Connective tissue cell that releases histamine in allergic reactions. 618

**maternal determinant** One of many substances present in the egg that influences the course of development. 782

**matrix** (MAY tricks) Unstructured semifluid substance that fills the space between cells in connective tissues or inside organelles. 77

**matter** Anything that takes up space and has mass. 22

**maturity** In biology, the age of reproduction. 830

**mechanical energy** A type of kinetic energy, such as walking or running. 104

**mechanoreceptor** (muh can oh ree SEPP turr) Sensory receptor that responds to mechanical stimuli, such as pressure, sound waves, or gravity. 710

**medulla oblongata** (muh DULE uh ahb long AH tuh) In vertebrates, part of the brain stem that is continuous with the spinal cord; controls heartbeat, blood pressure, breathing, and other vital functions. 690

**medusa** Among cnidarians, bell-shaped body form that is directed downward and contains much mesoglea. 518

**megaphyll** Large leaf with several to many veins. 417

**megaspore** (MEG uh spore) One of the two types of spores produced by seed plants; develops into a female gametophyte (embryo sac). 420, 426, 494

**meiosis** (my OH sis) Type of nuclear division that occurs as part of sexual reproduction, in which the daughter cells receive the haploid number of chromosomes in varied combinations. 170

**melanocyte** (mell ANN oh site) Specialized cell in the epidermis that produces melanin, the pigment responsible for skin color. 586

**melanocyte-stimulating hormone (MSH)** Substance that causes melanocytes to secrete melanin in most vertebrates. 740

**melatonin** (mell uh TONE in) Hormone, secreted by the pineal gland, that is involved in biorhythms. 750

**memory** Capacity of the brain to store and retrieve information about past sensations and perceptions; essential to learning. 691

**memory B cell** Forms during a primary immune response but enters a resting phase until a secondary immune response occurs. 620

**memory T cell** T cell that differentiated during an initial infection and responds rapidly during subsequent exposure to the same antigen. 624

**menarche** Onset of menstruation. 765

**meninges** (men IN jeez) Protective membranous coverings around the central nervous system. 688

**meningitis** (men in JIE tuss) A condition that refers to inflammation of the brain or spinal cord meninges (membranes). 688

**menopause** Termination of the ovarian and uterine cycles in older women. 765

**menstruation** (men strew AY shun) Periodic shedding of tissue and blood from the inner lining of the uterus in primates. 764

**meristem** (MARE uh stem) Undifferentiated embryonic tissue in the active growth regions of plants. 437

**mesoderm** (MESS oh derm) Middle primary tissue layer of an animal embryo that gives rise to muscle, several internal organs, and connective tissue layers. 780

**mesoglea** Transparent jellylike substance. 518

**mesophyll** (MESS oh fill) Inner, thickest layer of a leaf consisting of palisade and spongy mesophyll; the site of most of photosynthesis. 450

**mesosome** (MESS oh sohm) In a bacterium, plasma membrane that folds into the cytoplasm and increases surface area. 64

**messenger RNA (mRNA)** Type of RNA formed from a DNA template and bearing coded information for the amino acid sequence of a polypeptide. 220

**metabolic pathway** (met uh BAH lick) Series of linked reactions, beginning with a particular reactant and terminating with an end product. 108

**metabolic pool** Metabolites that are the products of and/or the substrates for key reactions in cells, allowing one type of molecule to be changed into another type, such as carbohydrates converted to fats. 145

**metabolism** (met TAB uh liz uhm) All of the chemical reactions that occur in a cell during growth and repair. 4, 106

**metamorphosis** (met uh MORE foh sis) Change in shape and form that some animals, such as insects, undergo during development. 529, 546

**metaphase** (MET uh faze) Mitotic phase during which chromosomes are aligned at the metaphase plate. 157

**metaphase plate** A disk formed during metaphase in which all of a cell's chromosomes lie in a single plane at right angles to the spindle fibers. 157

**metapopulation** Population subdivided into several small and isolated populations due to habitat fragmentation. 901

**metastasis** (muh TASS tuh sis) Spread of cancer from the place of origin throughout the body; caused by the ability of cancer cells to migrate and invade tissues. 162

**methanogen** (meth THANN uh jen) Type of archaea that lives in oxygen-free habitats, such as swamps, and releases methane gas. 368

**MHC** (major histocompatibility complex) **protein** Protein marker that is a part of cell-surface markers anchored in the plasma membrane, which the immune system uses to identify "self." 624

**microevolution** Change in gene frequencies between populations of a species over time. 285

**micronutrient** Essential element needed in small amounts for plant growth, such as boron, copper, and zinc. 457

**microphyll** Small leaf with one vein. 416

**microRNA** Introns that are processed into smaller signals; after being degraded, they combine with a protein, and the complex binds to mRNAs. These are then destroyed instead of being translated. 243

**microsphere** Formed from proteinoids exposed to water; has properties similar to those of today's cells. 319

**microspore** (MY crow spore) One of the two types of spores produced by seed plants; develops into a male gametophyte (pollen grain). 420, 426, 494

**microtubule** (my crow TUBE yule) Small, cylindrical organelle composed of tubulin protein around an empty central core; present in the cytoplasm, centrioles, cilia, and flagella. 78

**midbrain** In mammals, the part of the brain located below the thalamus and above the pons. 690

**middle ear** Portion of the ear consisting of the tympanic membrane, the oval and round windows, and the ossicles, where sound is amplified. 710

**migration** Regular back-and-forth movement of animals between two geographic areas at particular times of the year. 804

**millipede** (MILL ih peed) More or less cylindrical arthropod characterized by having two pairs of short legs on most of its body segments; may have 13 to almost 200 pairs of legs. 531

**mimicry** (MIMM ick kree) Superficial resemblance of two or more species; a mechanism that avoids predation by appearing to be noxious. 845

**mineral** Naturally occurring inorganic substance containing two or more elements; certain minerals are needed in the diet. 457, 646

**mineralocorticoid** (men urr ull oh CORE tih coid) Hormones secreted by the adrenal cortex that regulate salt and water balance, leading to increases in blood volume and blood pressure. 745

**mitochondrion** (mite oh KAHN dree uhn) Membrane-bounded organelle in which ATP molecules are produced during the process of cellular respiration. 76, 140

**mitosis** (my TOE sis) Process in which a parent nucleus produces two daughter nuclei, each having the same number and kinds of chromosomes as the parent nucleus. 152

**mixotrophic** Organism that can use autotrophic and heterotrophic means of gaining nutrients. 374

**model** Simulation of a process that aids conceptual understanding until the process can be studied firsthand; a hypothesis that describes how a particular process could possibly be carried out. 12

**mitotic spindle** Microtubule structure that brings about chromosomal movement during nuclear division. 153

**model of island biogeography** Model to explain the biodiversity of an island based on distance from the mainland and the island's size. 849

**mold** Various fungi whose body consists of a mass of hyphae (filaments) that grow on and receive nourishment from organic matter such as food and clothing. 398

**molecular clock** Idea that the rate at which mutational changes accumulate in certain genes is constant over time and is not involved in adaptation to the environment. 327, 346

**molecule** Union of two or more atoms of the same element; also, the smallest part of a compound that retains the properties of the compound. 26

**mollusc** Invertebrates such as squids, clams, snails, and chitons; characterized by a visceral mass, a mantle, and a foot. 523

**molt** Periodic shedding of the exoskeleton in arthropods and cuticle in roundworms. 516

**monoclonal antibody** (mah no CLONE uhl) One of many antibodies produced by a clone of hybridoma cells that all bind to the same antigen. 622

**monocot** (MAH no cot) Abbreviation of monocotyledon. Flowering plant group; members have one embryonic leaf (cotyledon), parallel-veined leaves, scattered vascular bundles, flower parts in threes or multiples of three, and other characteristics. 424, 436

**monocotyledone** Plant whose embryo has one cotyledon; one of the two classes of angiosperms; abbreviated as monocot. 424

**monocyte** (MAH no site) Type of a granular leukocyte that functions as a phagocyte, particularly after it becomes a macrophage, which is also an antigen-presenting cell. 607

**monoecious** Having unisexual male flowers or cones and unisexual female flowers or cones both on the same plant. 420

**monogamous** Breeding pair of organisms that only reproduce with each other through their lifetime. 811

**monohybrid cross** Cross between parents that differ in only one trait. 192

**monomer** (MAH nuh murr) Small molecule that is a subunit of a polymer—e.g., glucose is a monomer of starch. 40

**monophyletic group** A group of species including the most recent common ancestor and all its descendants. 344

**monosaccharide** (mah no SACK uh ride) Simple sugar; a carbohydrate that cannot be decomposed by hydrolysis—e.g., glucose. 41

**monosomy** (MAH no sohm mee) One less chromosome than usual. 180

**monotreme** (MAH no treem) Egg-laying mammal—e.g., duckbill platypus and spiny anteater. 554

**monsoon** (mahn SOON) Climate in India and southern Asia caused by wet ocean winds that blow onshore for almost half the year. 867

**montane coniferous forest** (MAHN tane cuh NIFF urr us) Coniferous forest of a mountain. 869

**more-developed country (MDC)** Country that is industrialized; typically, population growth is low, and the people enjoy a good standard of living. 833

**morel** Edible fungi having a conical cap with a highly pitted surface. 398

**morphogen** (MORF uh jen) Protein that is part of a gradient that influences morphogenesis. 785

**morphogenesis** (morf oh JEN uh sis) Emergence of shape in tissues, organs, or entire embryo during development. 782

**morula** (MORE you luh) Spherical mass of cells resulting from cleavage during animal development prior to the blastula stage. 779

**mosaic evolution** Concept that human characteristics did not evolve at the same rate; for example, some body parts are more humanlike than others in early hominins. 566

**moss** Type of bryophyte. 414

**motor molecule** Protein that moves along either actin filaments or microtubules and translocates organelles. 78

**motor (efferent) neuron** Nerve cell that conducts nerve impulses away from the central nervous system and innervates effectors (muscle and glands). 683

**mouth** In humans, organ of the digestive tract where food is chewed and mixed with saliva. 636

**mRNA transcript** mRNA molecule formed during transcription that has a sequence of bases complementary to a gene. 222

**mucosa** Epithelial membrane containing cells that secrete mucus; found in the inner cell layers of the digestive (first layer) and respiratory tracts. 641

**multicellular** Organism composed of many cells; usually has organized tissues, organs, and organ systems. 2

**multifactorial trait** Trait controlled by polygenes subject to environmental influences; each dominant allele contributes to the phenotype in an additive and like manner. 204

**multiple alleles** (uh LEEL) Inheritance pattern in which there are more than two alleles for a particular trait; each individual has only two of all possible alleles. 202

**muscularis** Smooth muscle layer found in the digestive tract. 641

**muscular** (contractile) **tissue** (cunn TRACK tile) Type of animal tissue composed of fibers that shorten and lengthen to produce movements. 582

**mutagen** (MEWT uh jen) Chemical or physical agent that increases the chance of mutation. 245

**mutation** Alternation in chromosome structure or number and also an alteration in a gene due to a change in DNA composition. 287

**mutualism** (mute you uh LIZ uhm) Symbiotic relationship in which both species benefit in terms of growth and reproduction. 848

**mycelium** (my SEE lee uhm) Tangled mass of hyphal filaments composing the vegetative body of a fungus. 394

**mycorrhizae (sing., mycorrhiza)** (my coh RIZE ee) Mutualistic relationship between fungal hyphae and roots of vascular plants. 404, 442, 461

**myelin sheath** (MY uh linn) White, fatty material—derived from the membrane of neurolemmocytes—that forms a covering for nerve fibers. 683

**myofibril** (my oh FIBE rull) Specific muscle cell organelle containing a linear arrangement of sarcomeres, which shorten to produce muscle contraction. 728

**myosin** (MY oh sin) Muscle protein making up the thick filaments in a sarcomere; it pulls actin to shorten the sarcomere, yielding muscle contraction. 728

**myxedema** (mikes uh DEEM uh) Condition resulting from a deficiency of thyroid hormone in an adult. 743

# N

**$N_2$ (nitrogen) fixation** Process whereby free atmospheric nitrogen is converted into compounds, such as ammonium and nitrates, usually by bacteria. 861

**$NAD^+$ (nicotinamide adenine dinucleotide)** (nick coh TIN uh mide ADD uh neen die NUKE klee oh tide) Coenzyme of oxidation-reduction that accepts electrons and hydrogen ions to become $NADH + H^+$ as oxidation of substrates occurs. During cellular respiration, NADH carries electrons to the electron transport chain in mitochondria. 112, 134

**$NADP^+$ (nicotinamide adenine dinucleotide phosphate)** (nick coh TIN uh mide ADD uh neen die NUKE klee oh tide FOSS fate) Coenzyme of oxidation-reduction that accepts electrons and hydrogen ions to become $NADPH + H^+$. During photosynthesis, NADPH participates in the reduction of carbon dioxide to a carbohydrate. 112

**nail** Flattened epithelial tissue from the stratum lucidum of the skin; located on the tips of fingers and toes. 586

**natural killer (NK) cell** Lymphocyte that causes an infected or cancerous cell to burst. 618

**natural selection** Mechanism of evolution caused by environmental selection of organisms most fit to reproduce; results in adaptation to the environment. 8, 271

**navigate** Ability to steer or manage a course by adjusting your bearings and follow the result of the adjustment. 805

**Neandertal** (nee AND urr tall) Hominin with a sturdy build that lived during the last Ice Age in Europe and the Middle East; hunted large game and left evidence of being culturally advanced. 570

**negative feedback** Mechanism of homeostatic response by which the output of a system suppresses or inhibits activity of the system. 588, 740

**nematocyst** (nuh MAT uh sist) In cnidarians, a capsule that contains a threadlike fiber, the release of which aids in the capture of prey. 518

**nephridium (pl., nephridia)** (nuh FRIDD ee uhm, nuh FRIDD ee uh) Segmentally arranged, paired excretory tubules of many invertebrates, as in the earthworm. 526, 667

**nephron** (NEFF rahn) Microscopic kidney unit that regulates blood composition by glomerular filtration, tubular reabsorption, and tubular secretion. 670

**nerve** Bundle of long axons outside the central nervous system. 583, 692

**nerve fiber** Axon; conducts nerve impulses away from the cell. They are classified as either myelinated or unmyelinated based on the presence or absence of a myelin sheath. 683

**nerve net** Diffuse, noncentralized arrangement of nerve cells in cnidarians. 519, 680

**nervous tissue** Tissue that contains nerve cells (neurons), which conduct impulses, and neuroglia, which support, protect, and provide nutrients to neurons. 582

**neural plate** (NURE uhl) Region of the dorsal surface of the chordate embryo that marks the future location of the neural tube. 781

**neural tube** Tube formed by closure of the neural groove during development. In vertebrates, the neural tube develops into the spinal cord and brain. 781

**neuroglia** (nure RAH glee uh) Nonconducting nerve cells that are intimately associated with neurons and function in a supportive capacity. 583, 683

**neuromodulator** (nure oh MAH dew lay turr) Electrical stimulant of a peripheral nerve, the spinal cord, or the brain; used to ease pain. 687

**neuromuscular junction** (nure oh MUSS cue lurr) Region where an axon bulb approaches a muscle fiber; contains a presynaptic membrane, a synaptic cleft, and a postsynaptic membrane. 730

**neuron** (NURE ahn) Nerve cell that characteristically has three parts: dendrites, cell body, and an axon. 582, 683

**neurotransmitter** (nure oh trans MITT urr) Chemical stored at the ends of axons that is responsible for transmission across a synapse. 686

**neurula** The early embryo during the development of the neural tube from the neural plate, marking the first appearance of the nervous system; the next stage after the gastrula. 781

**neutron** (NEW trahn) Neutral subatomic particle, located in the nucleus and assigned one atomic mass unit. 22

**neutrophil** (NEW troh fill) Granular leukocyte that is the most abundant of the white blood cells; first to respond to infection. 607, 618

**nitrification** (nite trih fih KAY shun) Process by which nitrogen in ammonia and organic compounds is oxidized to nitrites and nitrates by soil bacteria. 861

**node** In plants, the place where one or more leaves attach to a stem. 435

**nodes of Ranvier** (RAN veer) Gap in the myelin sheath around a nerve fiber. 683

**noncompetitive inhibition** Form of enzyme inhibition where the inhibitor binds to an enzyme at a location other than the active site; while at this site, the enzyme shape changes, the inhibitor is unable to bind to its substrate, and no product forms. 111

**nonpolar covalent bond** (nahn POH lurr coh VALE lent) Bond in which the sharing of electrons between atoms is fairly equal. 27

**nonrandom mating** Mating among individuals on the basis of their phenotypic similarities or differences, rather than mating on a random basis. 287

**nonseptate** (nahn SEPP tate) Lacking cell walls; some fungal species have hyphae that are nonseptate. 395

**nonvascular plants** Bryophytes, such as mosses and liverworts, that have no vascular tissue and either occur in moist locations or have special adaptations for living in dry locations. 413

**norepinephrine (NE)** (nor epp pin EFF renn) Neurotransmitter of the postganglionic fibers in the sympathetic division of the autonomic system; also, a hormone produced by the adrenal medulla. 686, 745

**notochord** (NO toh cord) Cartilaginous-like supportive dorsal rod in all chordates sometime in their life cycle; replaced by vertebrae in vertebrates. 540, 781

**nuclear envelope** Double membrane that surrounds the nucleus in eukaryotic cells and is connected to the endoplasmic reticulum; has pores that allow substances to pass between the nucleus and the cytoplasm. 70

**nucleariid** Protist that may be related to fungi although nucleariids lack the same type of cell wall and have threadlike pseudopods. 389

**nuclear pore** Opening in the nuclear envelope that permits the passage of proteins into the nucleus and ribosomal subunits out of the nucleus. 70

**nucleic acid** (new CLAY ick) Polymer of nucleotides; both DNA and RNA are nucleic acids. 52

**nucleoid** (NEW klee oid) Region of prokaryotic cells where DNA is located; it is not bounded by a nuclear envelope. 64, 164, 363

**nucleolus** (new KLEE uh luss) Dark-staining, spherical body in the nucleus that produces ribosomal subunits. 70

**nucleoplasm** (NEW klee oh plazz uhm) Semifluid medium of the nucleus containing chromatin. 70

**nucleosome** (NEW klee oh sohm) In the nucleus of a eukaryotic cell, a unit composed of DNA wound around a core of eight histone proteins, giving the appearance of a string of beads. 229

**nucleotide** (NEW klee oh tide) Monomer of DNA and RNA consisting of a 5-carbon sugar bonded to a nitrogenous base and a phosphate group. 52, 212

**nucleus** Membrane-bounded organelle within a eukaryotic cell that contains chromosomes and controls the structure and function of the cell. 64

# O

**obligate anaerobe** (AHB lih gate ANN urr robe) Prokaryote unable to grow in the presence of free oxygen. 364

**observation** Step in the scientific method by which data are collected before a conclusion is drawn. 11

**ocean ridge** Ridge on the ocean floor where oceanic crust forms and from which it moves laterally in each direction. 319

**octet rule** The observation that an atom is most stable when its outer shell is complete and contains eight electrons; an exception is hydrogen which requires only two electrons in its outer shell to have a completed shell. 25

**oil** Triglyceride, usually of plant origin, that is composed of glycerol and three fatty acids and is liquid in consistency due to many unsaturated bonds in the hydrocarbon chains of the fatty acids. 44

**oil gland** Gland of the skin, associated with hair follicle, that secretes sebum; sebaceous gland. 586

**olfactory cell** (ohl FACT toh ree) Modified neuron that is a sensory receptor for the sense of smell. 702

**oligodendrocyte** Type of glial cell that forms myelin sheaths around neurons in the CNS. 683

**omnivore** (AHM nih vore) Organism in a food chain that feeds on both plants and animals. 852

**oncogene** (AHN coh jeen) Cancer-causing gene. Oncogenes code for proteins that stimulate the cell cycle and inhibit apoptosis. 162

**oocyte** (OH oh site) Immature egg that is undergoing meiosis; upon completion of meiosis, the oocyte becomes an egg. 762

**oogenesis** (oh JENN us sis) Production of eggs in females by the process of meiosis and maturation. 179

**open circulatory system** Arrangement of internal transport in which blood bathes the organs directly, and there is no distinction between blood and interstitial fluid. 594

**operant conditioning** (AH purr unt) Learning that results from rewarding or reinforcing a particular behavior. 804

**operator** In an operon, the sequence of DNA that serves as a binding site for a repressor, and thereby regulates the expression of structural genes. 234

**operon** (AH purr rahn) Group of structural and regulating genes that function as a single unit. 234

**opisthokont** Supergroup of eukaryotes that choanoflagellates animals, nucleariids and fungi. 389

**opposable thumb** Fingers arranged in such a way that the thumb can touch the ventral surface of the fingertips of all four fingers. 560

**optimal foraging model** Analysis of behavior as a compromise of feeding costs versus feeding benefits. 810

**order** One of the categories, or taxa, used by taxonomists to group species; the taxon above the family level. 6, 340

**organ** Combination of two or more different tissues performing a common function. 434, 585

**organelle** Small, often membranous structure in the cytoplasm having a specific structure and function. 66

**organic chemistry** Branch of science which deals with organic molecules including those that are unique to living things. 38

**organic molecule** Molecule that always contains carbon and hydrogen, and often contains oxygen as well; organic molecules are associated with living things. 38

**organism** Individual living thing. 2

**organ of Corti** (CORE tie) Structure in the vertebrate inner ear that contains auditory receptors (also called spiral organ). 711

**organ system** Group of related organs working together. 585

**orgasm** (OR gazz uhm) Physiological and psychological sensations that occur at the climax of sexual stimulation. 759

**orientation** In birds, the ability to know present location by tracking stimuli in the environment. 804

**osmoregulate** Regulation of the salt water balance to maintain a normal balance within internal fluids. 668

**osmosis** (oz MOH sis) Diffusion of water through a differentially permeable membrane. 92

**osmotic pressure** (oz MAH tick) Measure of the tendency of water to move across a differentially permeable membrane; visible as an increase in liquid on the side of the membrane with higher solute concentration. 92

**ossicle** (AH sick cull) One of the small bones of the vertebrate middle ear—malleus, incus, and stapes. 710

**osteoblast** (AH stee oh blast) Bone-forming cell. 720

**osteoclast** (AH stee oh clast) Cell that causes erosion of bone. 720

**osteocyte** (AH stee oh site) Mature bone cell located within the lacunae of bone. 720

**ostracoderm** (ah STRAH cuh derm) Earliest vertebrate fossils of the Cambrian and Devonian periods; these fishes were small, jawless, and finless. 543

**otolith** (OH toe lith) Calcium carbonate granule associated with sensory receptors for detecting movement of the head; in vertebrates, located in the utricle and saccule. 713

**outer ear** Portion of the ear consisting of the pinna and the auditory canal. 710

**outgroup** In a cladistic study of evolutionary relationships among organisms, a group that has a known relationship to, but not a member of, the taxa being analyzed. 342

**out-of-Africa hypothesis** Proposal that modern humans originated only in Africa; then they migrated and supplanted populations of *Homo* in Asia and Europe about 100,000 years ago. 570

**ovarian cycle** (oh VAIR ree uhn) Monthly changes occurring in the ovary that determine the level of sex hormones in the blood. 763

**ovary** In flowering plants, the enlarged, ovule-bearing portion of the carpel that develops into a fruit; female gonad in animals that produces an egg and female sex hormones. 425, 495, 750, 756

**overexploitation** When the number of individuals taken from a wild population is so great that the population becomes severely reduced in numbers. 900

**oviparous** Type of reproduction in which development occurs in an egg, laid by mother, in reptiles. 757

**ovulation** (ah view LAY shun) Bursting of a follicle when a secondary oocyte is released from the ovary; if fertilization occurs, the secondary oocyte becomes an egg. 763

**ovule** (OH vule) In seed plants, a structure that contains the female gametophyte and has the potential to develop into a seed. 420, 495

**ovum** (OH vuhm) Haploid egg cell that is usually fertilized by a sperm to form a diploid zygote. 763

**oxidation** Loss of one or more electrons from an atom or molecule; in biological systems, generally the loss of hydrogen atoms. 112

**oxygen debt** Amount of oxygen required to oxidize lactic acid produced anaerobically during strenuous muscle activity. 138, 728

**oxyhemoglobin** (ox zee HEEM uh glow bin) Compound formed when oxygen combines with hemoglobin. 659

**oxytocin** (ox zee TOE sin) Hormone released by the posterior pituitary that causes contraction of the uterus and milk letdown. 740

**ozone shield** Accumulation of $O_3$, formed from oxygen in the upper atmosphere; a filtering layer that protects the Earth from ultraviolet radiation. 325, 860

# P

***p53*** A tumor suppressor gene that (1) attempts to repair DNA damage or (2) stops the cell cycle, or (3) initiates apoptosis. 153

**paleontology** (pale lee uhn TAH loh jee) Study of fossils that results in knowledge about the history of life. 268, 322

**palisade mesophyll** (PAL uh sade MESS oh fill) Layer of tissue in a plant leaf containing elongated cells with many chloroplasts. 450

**pancreas** (PAN kree us) Internal organ that produces digestive enzymes and the hormones insulin and glucagon. 640, 748

**pancreatic amylase** (pan kree AT tick AM uhl laze) Enzyme that digests starch to maltose. 642

**pancreatic islet** (pan kree AT tick EYE lit) Masses of cells that constitute the endocrine portion of the pancreas. 748

**panoramic vision** Vision characterized by having a wide field of vision; found in animals with eyes to the side. 704

**parabasalid** Unicellular protist that lacks mitochondria; possess flagella in clusters near the anterior of the cell. 386

**parasite** Species that is dependent on a host species for survival, usually to the detriment of the host species. 846

**parasitism** (PAIR uh sit tiz uhm) Symbiotic relationship in which one species (the *parasite*) benefits in terms of growth and reproduction to the detriment of the other species (the *host*). 846

**parasympathetic division** (pair uh simm puh THETT ick) Division of the autonomic system that is active under normal conditions; uses acetylcholine as a neurotransmitter. 695

**parathyroid gland** (pair uh THIGH roid) Gland embedded in the posterior surface of the thyroid gland; it produces parathyroid hormone. 744

**parathyroid hormone (PTH)** Hormone secreted by the four parathyroid glands that increases the blood calcium level and decreases the phosphate level. 744

**parenchyma** (puh RENN kih muh) Plant tissue composed of the least-specialized of all plant cells; found in all organs of a plant. 438

**Parkinson disease** Progressive deterioration of the central nervous system due to a deficiency in the neurotransmitter dopamine. 689

**parthenogenesis** (par thin oh JENN uh sis) Development of an egg cell into a whole organism without fertilization. 756

**partial pressure** Pressure exerted by each gas in a mixture of gases. 658

**passive immunity** Protection against infection acquired by transfer of antibodies to a susceptible individual. 621

**pathogen** Disease-causing agent such as viruses, parasitic bacteria, fungi, and animals. 366, 581

**pattern formation** Positioning of cells during development that determines the final shape of an organism. 782

**pectoral girdle** (PECK tore uhl) Portion of the vertebrate skeleton that provides support and attachment for the upper (fore) limbs; consists of the scapula and clavicle on each side of the body. 724

**peduncle** Flower stalk; expands into the receptacle. 424

**pelagic zone** (puh LAJJ ick) Open portion of the sea. 883

**pelvic girdle** Portion of the vertebrate skeleton to which the lower (hind) limbs are attached; consists of the coxal bones. 725
**penis** Male copulatory organ; in humans, the male organ of sexual intercourse. 759
**pentose** (PEN toes) Five-carbon sugar. Deoxyribose is the pentose sugar found in DNA; ribose is the pentose sugar found in RNA. 41
**pepsin** (PEP sin) Enzyme secreted by gastric glands that digests proteins to peptides. 642
**peptidase** Intestinal enzyme that breaks down short chains of amino acids to individual amino acids that are absorbed across the intestinal wall. 642
**peptide** (PEP tide) Two or more amino acids joined together by covalent bonding. 48
**peptide bond** Type of covalent bond that joins two amino acids. 48
**peptide hormone** Type of hormone that is a protein, a peptide, or derived from an amino acid. 738
**peptidoglycan** (pep tih doe GLIKE can) Unique molecule found in bacterial cell walls. 43, 364
**perennial** (purr IN nee uhl) Flowering plant that lives more than one growing season because the underground parts regrow each season. 434
**pericycle** (pair ih SIGH cull) Layer of cells surrounding the vascular tissue of roots; produces branch roots. 441
**periderm** (PAIR ih derm) Protective tissue that replaces epidermis; includes cork, cork cambium. 437
**peripheral nervous system (PNS)** (purr IF fur uhl) Nerves and ganglia that lie outside the central nervous system. 682
**peristalsis** (pair iss STALL sis) Wavelike contractions that propel substances along a tubular structure such as the esophagus. 638
**permafrost** Permanently frozen ground, usually occurring in the tundra, a biome of Arctic regions. 870
**peroxisome** (purr OX ih sohm) Enzyme-filled vesicle in which fatty acids and amino acids are metabolized to hydrogen peroxide that is broken down to harmless products. 75
**petal** A flower part that occurs just inside the sepals; often conspicuously colored to attract pollinators. 425, 494
**petiole** (PET tee ohl) The part of a plant leaf that connects the blade to the stem. 435
**Peyer patches** Lymphatic organs located in small intestine. 615
**phagocytize** (fag OSS sit tize) To ingest extracellular particles by engulfing them, as do amoeboid cells. 387
**phagocytosis** (fag oh site OH sis) Process by which amoeboid-type cells engulf large substances, forming an intracellular vacuole. 96
**pharynx** (FAIR inks) In vertebrates, common passageway for both food intake and air movement; located between the mouth and the esophagus. 637, 654
**phenomenon** (fin NAH men ahn) Observable event. 11
**phenotype** (FEE no type) Visible expression of a genotype—e.g., brown eyes or attached earlobes. 193
**pheromone** (FAIR oh moan) Chemical messenger that works at a distance and alters the behavior of another member of the same species. 738, 807
**phloem** (FLOW emm) Vascular tissue that conducts organic solutes in plants; contains sieve-tube members and companion cells. 416, 439, 462
**phloem sap** Solution of sugars, nutrients, and hormones found in the phloem tissue of a land plant. 462
**phospholipid** (foss foe LIP id) Molecule that forms the bilayer of the cell's membranes; has a polar, hydrophilic head bonded to two nonpolar, hydrophobic tails. 46
**photoautotroph** (foe toe AH toe trofe) Organism able to synthesize organic molecules by using carbon dioxide as the carbon source and sunlight as the energy source. 364
**photoperiodism** Relative lengths of daylight and darkness that affect the physiology and behavior of an organism. 486
**photoreceptor** Sensory receptor that responds to light stimuli. 704
**photorespiration** Series of reactions that occurs in plants when carbon dioxide levels are depleted but oxygen continues to accumulate, and the enzyme RuBP carboxylase fixes oxygen instead of carbon dioxide. 128
**photosynthesis** (foe toe SIN thuh sis) Process occurring usually within chloroplasts whereby chlorophyll-containing organelles trap solar energy to reduce carbon dioxide to carbohydrate. 4, 118
**photosystem** Photosynthetic unit where solar energy is absorbed and high-energy electrons are generated; contains a pigment complex and an electron acceptor; occurs as PS (photosystem) I and PS II. 122
**phototropism** (foe toe TROH piz uhm) Growth response of plant stems to light; stems demonstrate positive phototropism. 483
**pH scale** Measurement scale for hydrogen ion concentration. 32
**phylogenetic tree** (file oh jenn ETT ick) Diagram that indicates common ancestors and lines of descent among a group of organisms. 341
**phylogeny** (file AH jenn ee) Evolutionary history of a group of organisms. 341
**phylum** (FILE uhm) One of the categories, or taxa, used by taxonomists to group species; the taxon above the class level. 6, 340
**phytochrome** (FITE toe chrome) Photoreversible plant pigment that is involved in photoperiodism and other responses of plants, such as etiolation. 486
**phytoplankton** (fite oh PLANK ton) Part of plankton containing organisms that photosynthesize, releasing oxygen to the atmosphere and serving as food producers in aquatic ecosystems. 382, 881
**phytoremediation** (FITE toe ruh mee dee AY shun) The use of plants to restore a natural area to its original condition. 467
**pineal gland** (PIN nee uhl) Gland—either at the skin surface (fish, amphibians) or in the third ventricle of the brain (mammals)—that produces melatonin. 690, 750
**pinocytosis** (pie no site OH sis) Process by which vesicle formation brings macromolecules into the cell. 96
**pioneer species** Early colonizer of barren or disturbed habitats that usually has rapid growth and a high dispersal rate. 850
**pit** Any depression or opening; usually in reference to the small openings in the cell walls of xylem cells that function in providing a continuum between adjacent xylem cells. 439
**pith** Parenchyma tissue in the center of some stems and roots. 441
**pituitary dwarfism** (pit TWO it air ree) Condition caused by inadequate growth hormone in which affected individual has normal proportions but small stature. 741
**pituitary gland** Small gland that lies just inferior to the hypothalamus; consists of the anterior and posterior pituitary, both of which produce hormones. 740
**placenta** (pluh SENT uh) Organ formed during the development of placental mammals from the chorion and the uterine wall; allows the embryo, and then the fetus, to acquire nutrients and rid itself of wastes; produces hormones that regulate pregnancy. 555, 757, 791
**placental mammal** A group of species that rely on internal development whereby the fetus exchanges nutrients and wastes with its mother via a placenta 555
**placoderm** (PLACK uh derm) First jawed vertebrates; heavily armored fishes of the Devonian period. 543
**plankton** (PLANK ton) Freshwater and marine organisms that are suspended on or near the surface of the water; includes phytoplankton and zooplankton. 374
**plant** Multicellular, photosynthetic, eukaryotes that increasingly become adapted to live on land. 7, 410
**plasma** (PLAZZ muh) In vertebrates, the liquid portion of blood; contains nutrients, wastes, salts, and proteins. 606
**plasma cell** Mature B cell that mass-produces antibodies. 619
**plasma membrane** Membrane surrounding the cytoplasm that consists of a phospholipid bilayer with embedded proteins; functions to regulate the entrance and exit of molecules from cell. 64
**plasmid** (PLAZZ mid) Extrachromosomal ring of accessory DNA in the cytoplasm of bacteria. 64, 250, 363
**plasmodesmata** (plazz moh dezz MAH tuh) In plants, cytoplasmic strands that extend through pores in the cell wall and connect the cytoplasm of two adjacent cells. 99
**plasmodial slime mold** (plazz MOH dee uhl) Free-living mass of cytoplasm that moves by pseudopods on a forest floor or in a field, feeding on decaying plant material by phagocytosis; reproduces by spore formation. 388
**plasmolysis** (plazz MOLL ih sis) Contraction of the cell contents due to the loss of water. 93
**plastid** (PLASS tidd) Organelles of plants and algae that are bounded by a double membrane and contain internal membranes and/or vesicles (i.e., chloroplasts, chromoplasts, leucoplasts). 76
**platelet** (PLATE let) Component of blood that is necessary to blood clotting. 581, 607
**plate tectonics** (tec TAH nicks) Concept that the Earth's crust is divided into a number of fairly rigid plates whose movements account for continental drift. 332
**pleiotropy** (ply AH troh pee) Inheritance pattern in which one gene affects many phenotypic characteristics of the individual. 203
**point mutation** Change of one base only in the sequence of bases in a gene. 244
**polar body** In oogenesis, a nonfunctional product; two to three meiotic products are of this type. 179

**polar covalent bond** Bond in which the sharing of electrons between atoms is unequal. 27

**pollen grain** In seed plants, structure that is derived from a microspore and develops into a male gametophyte. 420, 496

**pollen tube** In seed plants, a tube that forms when a pollen grain lands on the stigma and germinates. The tube grows, passing between the cells of the stigma and the style to reach the egg inside an ovule, where fertilization occurs. 420, 427

**pollination** In gymnosperms, the transfer of pollen from pollen cone to seed cone; in angiosperms, the transfer of pollen from anther to stigma. 420, 497

**pollution** Any environmental change that adversely affects the lives and health of living things. 897

**polyandrous** Practice of female animals having several male mates; found in the New World monkeys where the males help in rearing the offspring. 811

**polygamous** Practice of males having several female mates. 811

**polygenic inheritance** (pah lee JENN ick) Pattern of inheritance in which a trait is controlled by several allelic pairs; each dominant allele contributes to the phenotype in an additive and like manner. 204

**polymer** (PAH lee murr) Macromolecule consisting of covalently bonded monomers; for example, a polypeptide is a polymer of monomers called amino acids. 40

**polymerase chain reaction (PCR)** (pah LIMM mare raze) Technique that uses the enzyme DNA polymerase to produce millions of copies of a particular piece of DNA. 252

**polyp** (PAH lip) Among cnidarians, body form that is directed upward and contains much mesoglea; in anatomy; small, abnormal growth that arises from the epithelial lining. 518, 640

**polypeptide** (pah lee PEP tide) Polymer of many amino acids linked by peptide bonds. 48

**polyploidy** (PAH lee ploid) Having a chromosome number that is a multiple greater than twice that of the monoploid number. 307

**polyribosome** (pah lee RIBE uh sohm) String of ribosomes simultaneously translating regions of the same mRNA strand during protein synthesis. 71, 225

**polysaccharide** (pah lee SACK uh ride) Polymer made from sugar monomers; the polysaccharides starch and glycogen are polymers of glucose monomers. 42

**pons** (PAHNS) Portion of the brain stem above the medulla oblongata and below the midbrain; assists the medulla oblongata in regulating the breathing rate. 690

**population** Group of organisms of the same species occupying a certain area and sharing a common gene pool. 9, 284, 820

**population density** The number of individuals per unit area or volume living in a particular habitat. 821

**population distribution** The pattern of dispersal of individuals living within a certain area. 821

**population genetics** The study of gene frequencies and their changes within a population. 284

**portal system** Pathway of blood flow that begins and ends in capillaries, such as the portal system located between the small intestine and liver. 602

**positive feedback** Mechanism of homeostatic response in which the output of the system intensifies and increases the activity of the system. 740

**posterior pituitary** (pit YOU ih tare rree) Portion of the pituitary gland that stores and secretes oxytocin and antidiuretic hormone produced by the hypothalamus. 740

**posttranscriptional control** Gene expression following translation regulated by the way mRNA transcripts are processed. 240

**posttranslational control** Alternation of gene expression by changing a protein's activity after it is translated. 242

**postzygotic isolating mechanism** (post zie GAH tick) Anatomical or physiological difference between two species that prevents successful reproduction after mating has taken place. 303

**potential energy** Stored energy as a result of location or spatial arrangement. 104

**predation** (preh DAY shun) Interaction in which one organism (the *predator*) uses another (the *prey*) as a food source. 843

**predator** Organism that practices predation. 843

**prediction** Step of the scientific process that follows the formulation of a hypothesis and assists in creating the experimental design. 11

**preparatory (prep) reaction** Reaction that oxidizes pyruvate with the release of carbon dioxide; results in acetyl CoA and connects glycolysis to the citric acid cycle. 135, 140

**pressure-flow model** Explanation for phloem transport; osmotic pressure following active transport of sugar into phloem brings a flow of sap from a source to a sink. 468

**prey** Organism that provides nourishment for a predator. 843

**prezygotic isolating mechanism** (pree zie GAH tick) Anatomical or behavioral difference between two species that prevents the possibility of mating. 302

**primary motor area** Area in the frontal lobe where voluntary commands begin; each section controls a part of the body. 689

**primary root** Original root that grows straight down and remains the dominant root of the plant; contrasts with fibrous root system. 442

**primary somatosensory area** (so mat oh SENSE uh ree) Area dorsal to the central sulcus where sensory information arrives from the skin and skeletal muscles. 689

**primate** Member of the order Primate; includes prosimians, monkeys, apes, and hominins, all of whom have adaptations for living in trees. 560

**principle** Theory that is generally accepted by an overwhelming number of scientists; also called a law. 12

**prion** (PRY ahn) Infectious particle consisting of protein only and no nucleic acid. 51, 362

**producer** Photosynthetic organism at the start of a grazing food chain that makes its own food—e.g., green plants on land and algae in water. 852

**product** Substance that forms as a result of a reaction. 106

**progesterone** (pro JEST turr ohn) Female sex hormone that helps maintain sexual organs and secondary sex characteristics. 750, 763

**proglottid** (pro GLAH tid) Segment of a tapeworm that contains both male and female sex organs and becomes a bag of eggs. 522

**prokaryote** (pro CARE ree oat) Organism that lacks the membrane-bounded nucleus and membranous organelles typical of eukaryotes. 7, 362

**prokaryotic cell** (pro care ree AH tick) Lacking a membrane-bounded nucleus and organelles; the cell type within the domains Bacteria and Archaea. 64

**prolactin (PRL)** (pro LACK tin) Hormone secreted by the anterior pituitary that stimulates the production of milk from the mammary glands. 740

**prometaphase** (pro MET uh faze) Phase of mitosis during which the chromosomes are condensed but not fully aligned at the metaphase plate. 157

**promoter** In an operon, a sequence of DNA where RNA polymerase binds prior to transcription. 222, 234

**prophase** (PRO faze) Mitotic phase during which chromatin condenses so that chromosomes appear; chromosomes are scattered. 156

**prosimian** (pro SIMM me uhn) Group of primates that includes lemurs and tarsiers, and may resemble the first primates to have evolved. 563

**prostaglandin** (pro stah GLAN din) Hormone that has various and powerful local effects. 751

**protein** (PRO teen) Molecule consisting of one or more polypeptides. 48, 644

**protein-first hypothesis** In chemical evolution, the proposal that protein originated before other macromolecules and made possible the formation of protocells. 319

**proteinoid** (PRO tin oid) Abiotically polymerized amino acids that, when exposed to water, become microspheres having cellular characteristics. 319

**proteome** Collection of proteins resulting from the translation of genes into proteins. 258

**proteomics** Study of the complete collection of proteins that an organism produces. 227, 258

**protist** (PRO teest) A eukaryotic organism that is not a plant, fungus, or animal. Protists are generally a microscopic complex single cell; they evolved before other types of eukaryotes in the history of Earth 7, 374

**protobiont** Also called protocell, possible first cell. 320

**protocell** (PRO toe cell) In biological evolution, a possible cell forerunner that became a cell once it acquired genes. 320

**proton** (PRO tahn) Positive subatomic particle located in the nucleus and assigned one atomic mass unit. 22

**proto-oncogene** (pro toe AHN coh jeen) Normal gene that can become an oncogene through mutation. 162

**protostome** (PRO toe stome) Group of coelomate animals in which the first embryonic opening (the blastopore) is associated with the mouth. 515

**protozoan** (pro toe ZOH uhn) Heterotrophic, unicellular protist that moves by flagella, cilia, or pseudopodia. 374

**proximal convoluted tubule** Portion of a nephron following the glomerular capsule where tubular reabsorption of filtrate occurs. 671

**pseudocoelom** Body cavity lying between the digestive tract and body wall that is incompletely lined by mesoderm. 528

**pseudopod** (SUE doe pod) Cytoplasmic extension of amoeboid protists; used for locomotion and engulfing food. 78, 387

**pteridophyte** Ferns and their allies (horsetail and whisk ferns). 417

**puberty** Period of life when secondary sex changes occur in humans; marked by the onset of menses in females and sperm production in males. 761

**pulmonary circuit** (PULL moh nair ree) Circulatory pathway between the lungs and the heart. 597

**pulse** Vibration felt in arterial walls due to expansion of the aorta following ventricle contraction. 600

**Punnett square** (PUN net) Grid used to calculate the expected results of simple genetic crosses. 196

**pupil** Opening in the center of the iris of the vertebrate eye. 705

**pyruvate** (pie ROO vate) End product of glycolysis; its further fate, involving fermentation or entry into a mitochondrion, depends on oxygen availability. 135

# R

**radial symmetry** (RAY dee uhl SIM meh tree) Body plan in which similar parts are arranged around a central axis, like spokes of a wheel. 513

**radiolarian** (ray dee oh LAIR ree uhn) Protist that has a glassy silicon test, usually with a radial arrangement of spines; pseudopods are external to the test. 388

**radula** (RADD you luh) Tonguelike organ found in molluscs that bears rows of tiny teeth, which point backward; used to obtain food. 523, 635

**rain shadow** Leeward side (side sheltered from the wind) of a mountainous barrier, which receives much less precipitation than the windward side. 867

**rate of natural increase (*r*)** Growth rate dependent on the number of individuals that are born each year and the number of individuals that die each year. 822

**ray-finned bony fishes** Group of bony fishes with fins supported by parallel bony rays connected by webs of thin tissue. 544

**RB** Tumor suppressor genes whose protein interprets growth signals and nutrient availability before allowing the cell cycle to proceed. 153

**reactant** (ree ACT unt) Substance that participates in a reaction. 104

**receptacle** Area where a flower attaches to a floral stalk. 424

**receptor** Type of membrane protein that binds to specific molecules in the environment, providing a mechanism for the cell to sense and adjust to its surroundings. 474

**receptor-mediated endocytosis** (en doe site TOE sis) Selective uptake of molecules into a cell by vacuole formation after they bind to specific receptor proteins in the plasma membrane. 96

**receptor protein** Protein located in the plasma membrane or within the cell; binds to a substance that alters some metabolic aspect of the cell. 88

**recessive allele** (re SESS ihv uh LEEL) Allele that exerts its phenotypic effect only in the homozygote; its expression is masked by a dominant allele. 193

**reciprocal altruism** The trading of helpful or cooperative acts, such as helping at the nest, by individuals—the animal that was helped will repay the debt at some later time. 815

**recombinant DNA (rDNA)** (ree CAHM bih nunt) DNA that contains genes from more than one source. 250

**red algae** Marine photosynthetic protists with a notable abundance of phycobilin pigments; include coralline algae of coral reefs. 379

**red blood cell** Erythrocyte; contains hemoglobin and carries oxygen from the lungs or gills to the tissues in vertebrates. 581, 606

**red bone marrow** Vascularized, modified connective tissue that is sometimes found in the cavities of spongy bone; site of blood cell formation. 614, 720

**red bread mold** Sac fungus that grows on bread and was the experimental material in the formulation of the one gene, one enzyme hypothesis 398

**red tide** A population bloom of dinoflagellates that causes costal waters to turn red. Releases a toxin that can lead to paralytic shellfish poisoning. 383

**reduction** Gain of electrons by an atom or molecule with a concurrent storage of energy; in biological systems, the electrons are accompanied by hydrogen ions. 112

**reflex action** Automatic, involuntary response of an organism to a stimulus. 688

**refractory period** Time following an action potential when a neuron is unable to conduct another nerve impulse. 685

**regulator gene** In an operon, a gene that codes for a protein that regulates the expression of other genes. 235

**relative dating (of fossils)** Determining the age of fossils by noting their sequential relationships in strata; *absolute dating* relies on radioactive dating techniques to assign an actual date. 322

**renal cortex** (REE null CORE tex) Outer portion of the kidney that more appears granular. 670

**renal medulla** (REE null muh DOO luh) Inner portion of the kidney that consists of renal pyramids. 670

**renal pelvis** Hollow chamber in the kidney that lies inside the renal medulla and receives freshly prepared urine from the collecting ducts. 670

**renin** (REN ninn) Enzyme released by the kidneys that leads to the secretion of aldosterone and a rise in blood pressure. 746

**repetitive DNA element** Sequence of DNA on a chromosome that is repeated several times. 256

**replacement model** Proposal that modern humans originated only in Africa; then they migrated and supplanted populations of *Homo* in Asia and Europe about 100,000 years ago. 570

**replacement reproduction** Population in which each person is replaced by only one child. 834

**replication fork** In eukaryotes, the point where the two parental DNA strands separate to allow replication. 219

**repressible operon** (AH purr ahn) Operon that is normally active because the repressor is normally inactive. 235

**repressor** In an operon, protein molecule that binds to an operator, preventing transcription of structural genes. 234

**reproduce** To produce a new individual of the same kind. 5

**reproductive cloning** Used to create an organism that is genetically identical to the original individual. 159

**reptile** Terrestrial vertebrate with internal fertilization, scaly skin, and an egg with a leathery shell; includes snakes, lizards, turtles, crocodiles, and birds. 548

**resolution** Capability of a microscope to distinguish the separate parts of an object 63

**resource partitioning** Mechanism that increases the number of niches by apportioning the supply of a resource such as food or living space between species. 842

**respiration** Sequence of events that results in gas exchange between the cells of the body and the environment. 650

**respiratory center** Group of nerve cells in the medulla oblongata that send out nerve impulses on a rhythmic basis, resulting in involuntary inspiration on an ongoing basis. 657

**responding variable** Result or change that occurs when an experimental variable is utilized in an experiment. 14

**resting potential** Membrane potential of an inactive neuron. 684

**restoration ecology** Subdiscipline of conservation biology that seeks ways to return ecosystems to their former state. 902

**restriction enzyme** Bacterial enzyme that stops viral reproduction by cleaving viral DNA; used to cut DNA at specific points during production of recombinant DNA. 250

**reticular fiber** (reh TICK cue lurr) Very thin collagen fibers in the matrix of connective tissue, highly branched and forming delicate supporting networks. 579

**retina** (RETT tih nuh) Innermost layer of the vertebrate eyeball containing the photoreceptors—rod cells and cone cells. 705

**retrovirus** (rett troh VIE russ) RNA virus containing the enzyme reverse transcriptase that carries out RNA/DNA transcription. 361

**reverse transcriptase** Viral enzyme found in retroviruses that is capable of converting their RNA genome into a DNA copy. 361

**rhizarian** Supergroup of eukaryotes that includes foraminiferans and radiolarians. 388

**rhizoid** (RYE zoid) Rootlike hair that anchors a plant and absorbs minerals and water from the soil. 413

**rhizome** (RYE zohm) Rootlike underground stem. 416, 448, 505

**rhodopsin** (rode AHP sin) Light-absorbing molecule in rod cells and cone cells that contains a pigment and the protein opsin. 707

**ribose** (RYE bohs) Pentose sugar found in RNA. 41

**ribosomal RNA (rRNA)** (rye boh SOHM uhl) Type of RNA found in ribosomes that translate messenger RNAs to produce proteins. 220

**ribosome** (RYE boh sohm) RNA and protein in two subunits; site of protein synthesis in the cytoplasm. 64, 71

**ribozyme** (RYE boh zime) RNA molecule that can catalyze chemical reactions. 108, 223

**RNA (ribonucleic acid)** (rye boh new CLAY ick) Nucleic acid produced from covalent bonding of nucleotide monomers that contain the sugar ribose; occurs in three forms: messenger RNA, ribosomal RNA, and transfer RNA. 52

**RNA-first hypothesis** In chemical evolution, the proposal that RNA originated before other macromolecules and allowed the formation of the first cell(s). 319

**RNA polymerase** (pah LIMM mare raze) During transcription, an enzyme that creates on mRNA transcript by joining nucleotides complementary to a DNA template. 222

**rod cell** Photoreceptor in vertebrate eyes that responds to dim light. 707

**root cap** Protective cover of the root tip, whose cells are constantly replaced as they are ground off when the root pushes through rough soil particles. 440

**root hair** Extension of a root epidermal cell that collectively increases the surface area for the absorption of water and minerals. 437, 460

**root nodule** (NOD yule) Structure on plant root that contains nitrogen-fixing bacteria. 442, 461

**root pressure** Osmotic pressure caused by active movement of mineral into root cells; serves to elevate water in xylem for a short distance. 464

**root system** Includes the main root and any and all of its lateral (side) branches. 434

**rotational equilibrium** Maintenance of balance when the head and body are suddenly moved or rotated. 711

**rotifer** Microscopic invertebrates characterized by ciliated corona that when beating looks like a rotating wheel. 523

**rough ER (endoplasmic reticulum)** (in doe PLAZZ mick ruh TICK you lumm) Membranous system of tubules, vesicles, and sacs in cells; has attached ribosomes. 72

**roundworm** Invertebrates with nonsegmented cylindrical body covered by a cuticle that molts; some forms are free-living in water and soil, and many are parasitic. 528

***r*-selection** Favorable life history strategy under certain environmental conditions; characterized by a high reproductive rate with little or no attention given to offspring survival. 830

**RuBP carboxylase** (car BOX ill laze) An enzyme that starts the Calvin cycle reactions by catalyzing attachment of the carbon atom from $CO_2$ to RuBP. 126

# S

**saccule** (SACK yule) Saclike cavity in the vestibule of the vertebrate inner ear; contains sensory receptors for gravitational equilibrium. 713

**sac fungi** Fungi that produce spores in fingerlike sacs called asci within a fuiting body; includes morels, truffles, yeasts and molds. 398

**salivary amylase** (SAL lih vair ree AM uh laze) In humans, enzyme in saliva that digests starch to maltose. 637, 642

**salivary gland** In humans, gland associated with the mouth that secretes saliva. 636

**salt** Ionic compound that results from a classical acid-base reaction. 26

**saltatory conduction** (SALT tuh tore ree) Movement of nerve impulses from one neurolemmal node to another along a myelinated axon. 685

**saprotroph** (SAP pro trofe) Organism that secretes digestive enzymes and absorbs the resulting nutrients back across the plasma membrane. 365, 394

**sarcolemma** (sark oh LIMM uh) Plasma membrane of a muscle fiber; also forms the tubules of the T system involved in muscular contraction. 728

**sarcomere** (SARK oh meer) One of many units, arranged linearly within a myofibril, whose contraction produces muscle contraction. 728

**sarcoplasmic reticulum** (sark oh PLAZZ mick ruh TICK you lumm) Smooth endoplasmic reticulum of skeletal muscle cells; surrounds the myofibrils and stores calcium ions. 728

**sarcopterygii** Mesozoic marine reptiles. 544

**saturated fatty acid** Fatty acid molecule that lacks double bonds between the carbons of its hydrocarbon chain. The chain bears the maximum number of hydrogens possible. 44

**savanna** (suh VANN uh) Terrestrial biome that is a grassland in Africa, characterized by few trees and a severe dry season. 876

**Schwann cell** Cell that surrounds a fiber of a peripheral nerve and forms the myelin sheath. 683

**scientific method** Process by which scientists formulate a hypothesis, gather data by observation and experimentation, and come to a conclusion. 11

**scientific theory** Concept supported by a broad range of observations, experiments, and data. 12

**sclera** (SKLARE uh) White, fibrous, outer layer of the eyeball. 704

**sclerenchyma** (skluh RINK ih muh) Plant tissue composed of cells with heavily lignified cell walls; functions in support. 438

**scolex** (SCOLE lex) Tapeworm head region; contains hooks and suckers for attachment to host. 522

**sea star** An echinoderm with noticeable 5-pointed radial symmetry; found along rocky coasts where they feed on bivalves. 534

**seaweed** Multicellular forms of red, green, and brown algae found in marine habitats. 380

**secondary metabolite** Molecule not directly involved in growth, development, or reproduction of an organism; in plants, these molecules, which include nicotine, caffeine, tannins, and menthols, can discourage herbivores. 488

**secondary oocyte** (OH oh site) In oogenesis, the functional product of meiosis I; becomes the egg. 179

**secondary sex characteristic** Trait that is sometimes helpful but not absolutely necessary for reproduction and is maintained by the sex hormones in males and females. 761

**second messenger** Chemical signal such as cyclic AMP that causes the cell to respond to the first messenger—a hormone bound to plasma membrane receptor protein. 739

**secretion** (suh KREE shun) Release of a substance by exocytosis from a cell that may be a gland or part of a gland. 72

**sedimentation** (sed ih men TAY shun) Process by which particulate material accumulates and forms a stratum. 322

**seed** Mature ovule that contains an embryo, with stored food enclosed in a protective coat. 420, 494

**seedless vascular plant** Collective name for club mosses (lycophyte) and ferns (pteridophyte) Characterized by windblown spores. 416

**segmentation** (seg men TAY shun) Repetition of body units as seen in the earthworm. 526

**self-antigen** Antigen that is produced by an organism. 619

**semelparity** Condition of having a single reproductive effort in a lifetime. 824

**semen (seminal fluid)** (SEE men, SIMM in uhl) Thick, whitish fluid consisting of sperm and secretions from several glands of the male reproductive tract. 759

**semicircular canal** One of three half-circle-shaped canals of the vertebrate inner ear; contains sensory receptors for rotational equilibrium. 710

**semiconservative replication** Duplication of DNA resulting in two double helix molecules, each having one parental and one new strand. 217

**semilunar valve** Valve resembling a half moon located between the ventricles and their attached vessels. 598

**seminiferous tubule** (seh men IF furr us TUBE yule) Long, coiled structure contained within chambers of the testis where sperm are produced. 760

**senescence** (seh NESS sense) Sum of the processes involving aging, decline, and eventual death of a plant or plant part. 477

**sensory (afferent) neuron** Nerve cell that transmits nerve impulses to the central nervous system after a sensory receptor has been stimulated. 683

**sensory receptor** Structure that receives either external or internal environmental stimuli and is a part of a sensory neuron or transmits signals to a sensory neuron. 679

**sepal** (SEE pull) Outermost, sterile, leaflike covering of the flower; usually green in color. 425, 494

**septate** (SEPP tate) Having cell walls; some fungal species have hyphae that are septate. 395

**septum** (SEPP tum) Partition or wall that divides two areas; the septum in the heart separates the right half from the left half. 598

**serosa** Outer embryonic membrane of birds and reptiles; chorion. 641

**serotonin** A neurotransmitter. 686

**sessile** (SESS isle) Tending to stay in one place. 515

**seta (pl., setae)** (SEE tuh, SEE tee) A needlelike, chitinous bristle in annelids, arthropods, and others. 526

**sexual reproduction** Reproduction involving meiosis, gamete formation, and fertilization; produces offspring with chromosomes inherited from each parent with a unique combination of genes. 170

**sexual selection** Changes in males and females, often due to male competition and female selectivity, leading to increased fitness. 291, 811

**shoot apical meristem** (AY pick uhl MARE ih stem) Group of actively dividing embryonic cells at the tips of plant shoots. 444

**shoot system** Aboveground portion of a plant consisting of the stem, leaves, and flowers. 434

**short-day plant** Plant that flowers when day length is shorter than a critical length—e.g., cocklebur, poinsettia, and chrysanthemum. 487

**short tandem repeat (STR) profiling** Procedure of analyzing DNA in which PCR and gel electrophoresis are used to create an individuals band pattern with each one being unique because each person has their own number of repeats at different locations. 253

**shrubland** Arid terrestrial biome characterized by shrubs and tending to occur along coasts that have dry summers and receive most of their rainfall in the winter. 876

**sieve-tube member** Member that joins with others in the phloem tissue of plants as a means of transport for nutrient sap. 439, 462

**signal** Molecule that stimulates or inhibits an event in the cell cycle. 153

**signal peptide** Sequence of amino acids that binds with a SRP, causing a ribosome to bind to ER. 71

**simple goiter** Condition in which an enlarged thyroid produces low levels of thyroxine. 743

**single nucleotide polymorphism (SNP)** Site present in at least 1% of the population at which individuals differ by a single nucleotide. These can be used as genetic markers to map unknown genes or traits. 284

**sink** In the pressure-flow model of phloem transport, the location (roots) from which sugar is constantly being removed. Sugar will flow to the roots from the source. 469

**sink population** Population that is found in an unfavorable area where at best the birthrate equals the death rate; sink populations receive new members from source populations. 901

**sister chromatid** (CROW muh tid) One of two genetically identical chromosomal units that are the result of DNA replication and are attached to each other at the centromere. 155

**skeletal muscle** Striated, voluntary muscle tissue that comprises skeletal muscles; also called striated muscle. 582

**skin** Outer covering of the body; can be called the integumentary system because it contains organs such as sense organs. 585

**sliding filament model** An explanation for muscle contraction based on the movement of actin filaments in relation to myosin filaments. 728

**small intestine** In vertebrates, the portion of the digestive tract that precedes the large intestine; in humans, consists of the duodenum, jejunum, and ileum. 639

**smooth (visceral) muscle** Nonstriated, involuntary muscles found in the walls of internal organs. 582

**smooth ER (endoplasmic reticulum)** (in doe PLAZZ mick ruh TICK cue lumm) Membranous system of tubules, vesicles, and sacs in eukaryotic cells; lacks attached ribosomes. 72

**society** Group in which members of species are organized in a cooperative manner, extending beyond sexual and parental behavior. 807

**sodium-potassium pump** Carrier protein in the plasma membrane that moves sodium ions out of and potassium ions into animal cells; important in nerve and muscle cells. 94

**soil** Accumulation of inorganic rock material and organic matter that is capable of supporting the growth of vegetation. 458

**soil erosion** Movement of topsoil to a new location due to the action of wind or running water. 459

**soil horizon** Major layer of soil visible in vertical profile; for example, topsoil is the A horizon. 459

**soil profile** Vertical section of soil from the ground surface to the unaltered rock below. 459

**solute** (SAHL yute) Substance that is dissolved in a solvent, forming a solution. 29, 91

**solution** Fluid (the solvent) that contains a dissolved solid (the solute). 29, 91

**solvent** (SAHL vent) Liquid portion of a solution that serves to dissolve a solute. 91

**somatic cell** (so MAT tick) Body cell; excludes cells that undergo meiosis and become sperm or egg. 153

**somatic system** Portion of the peripheral nervous system containing motor neurons that control skeletal muscles. 693

**source** In the pressure-flow model of phloem transport, the location (leaves) of sugar production. Sugar will flow from the leaves to the sink. 469

**source population** Population that can provide members to other populations of the species because it lives in a favorable area, and the birthrate is most likely higher than the death rate. 901

**speciation** (spee see AY shun) Origin of new species due to the evolutionary process of descent with modification. 304

**species** Group of similarly constructed organisms capable of interbreeding and producing fertile offspring; organisms that share a common gene pool; the taxon at the lowest level of classification. 6, 340

**species diversity** Variety of species that make up a community. 840

**species richness** Number of species in a community. 840

**specific epithet** (spuh SIFF ick EPP pih thett) In the binomial system of taxonomy, the second part of an organism's name; it may be descriptive. 339

**sperm** Male gamete having a haploid number of chromosomes and the ability to fertilize an egg, the female gamete. 761

**spermatogenesis** (sperm mat oh JENN uh sis) Production of sperm in males by the process of meiosis and maturation. 179

**sphygmomanometer** (sfig moh mah NAHM met turr) Device consisting of inflatable cuff and pressure gauge for measuring arterial blood pressure. 603

**spicule** (SPICK yule) Skeletal structure of sponges composed of calcium carbonate or silicate. 517

**spinal cord** In vertebrates, the nerve cord that is continuous with the base of the brain and housed within the vertebral column. 688

**spinal nerve** Nerve that arises from the spinal cord. 692

**spirillum (pl., spirilla)** (spy RILL lumm) Long, rod-shaped bacterium that is twisted into a rigid spiral; if the spiral is flexible rather than rigid, it is called a spirochete. 64

**spirochete** (SPY roe keet) Long, rod-shaped bacterium that is twisted into a flexible spiral; if the spiral is rigid rather than flexible, it is called a spirillum. 64

**spleen** Large, glandular organ located in the upper left region of the abdomen; stores and purifies blood. 615

**sponge** Invertebrates that are pore-bearing filter feeders whose inner body wall is lined by collar cells that resemble a unicellular choanoflagellate. 517

**spongy bone** Type of bone that has an irregular, meshlike arrangement of thin plates of bone. 581, 720

**spongy mesophyll** (MESS oh fill) Layer of tissue in a plant leaf containing loosely packed cells, increasing the amount of surface area for gas exchange. 450

**spontaneous mutation** Mutation that arises as a result of anomalies in normal biological processes, such as mistakes made during DNA replication. 243

**sporangium (pl., sporangia)** (spore RAN jee uhm) Structure that produces spores. 388, 396, 412

**spore** Asexual reproductive or resting cell capable of developing into a new organism without fusion with another cell, in contrast to a gamete. 176, 395, 412

**sporophyll** Modified leaf that bears a sporangium or sporangia. 416

**sporophyte** (SPORE oh fite) Diploid generation of the alternation of generations life cycle of a plant; produces haploid spores that develop into the haploid generation. 178, 412

**sporopollenin** Tough substance that the outer wall of spores and pollen grains is composed. 412

**spring overturn** Mixing process that occurs in spring in stratified lakes whereby oxygen-rich top waters mix with nutrient-rich bottom waters. 881

**squamous epithelium** (SQUAY muss epp pih THEE lee uhm) Type of epithelial tissue that contains flat cells. 578

**stabilizing selection** Outcome of natural selection in which extreme phenotypes are eliminated and the average phenotype is conserved. 289

**stamen** (STAY men) In flowering plants, the portion of the flower that consists of a filament and an anther containing pollen sacs where pollen is produced. 425, 495

**starch** Storage polysaccharide found in plants that is composed of glucose molecules joined in a linear fashion with few side chains. 42

**statistical phylogenetics** System of creating phylogenetic trees using statistical tools rather than parisomy. 343

**statolith** (STAT oh lith) Sensors found in root cap cells that cause a plant to demonstrate gravitropism. 482

**stem** Usually the upright, vertical portion of a plant that transports substances to and from the leaves. 435

**stereoscopic vision** Vision characterized by depth perception and three-dimensionality. 560, 704

**steroid** (STARE oid) Type of lipid molecule having a complex of four carbon rings—e.g., cholesterol, estrogen, progesterone, and testosterone. 46

**steroid hormone** Type of hormone that has the same complex of four carbon rings, but each one has different side chains. 739

**stigma** (STIG muh) In flowering plants, portion of the carpel where pollen grains adhere and germinate before fertilization can occur. 425, 495

**stolon** (STOLE uhn) Stem that grows horizontally along the ground and may give rise to new plants where it contacts the soil—e.g., the runners of a strawberry plant. 448, 505

**stoma (pl., stomata)** (STOME muh, stoh MAH tuh) Small opening between two guard cells on the underside of leaf epidermis through which gases pass. 119, 412, 437, 466

**stomach** In vertebrates, muscular sac that mixes food with gastric juices to form chyme, which enters the small intestine. 638

**stramenopile** Group of protists that includes water molds, diatoms, and golden brown algae and is characterized by a "hairy" flagellum. 381

**stratum** (STRAY tum) Ancient layer of sedimentary rock; results from slow deposition of silt, volcanic ash, and other materials. 322

**striated** (STRY ate ted) Having bands; in cardiac and skeletal muscle, alternating light and dark bands produced by the distribution of contractile proteins. 582

**strobilus** (stroh BILL us) In club mosses, terminal clusters of leaves that bear sporangia. 416

**stroke** Condition resulting when an arteriole in the brain bursts or becomes blocked by an embolism; cerebrovascular accident. 605

**stroma** (STROH muh) Fluid within a chloroplast that contains enzymes involved in the synthesis of carbohydrates during photosynthesis. 76, 119

**stromatolite** (stroh MAT oh lite) Domed structure found in shallow seas consisting of cyanobacteria bound to calcium carbonate. 324

**structural gene** Gene that codes for an enzyme in a metabolic pathway. 234

**structural genomics** Study of the sequence of DNA bases and the amount of genes in organisms. 255

**style** Elongated, central portion of the carpel between the ovary and stigma. 425, 495

**subcutaneous layer** A sheet that lies just beneath the skin and consists of loose connective and adipose tissue. 585

**submucosa** Tissue layer just under the epithelial lining of the lumen of the digestive tract (second layer). 641

**substrate** Reactant in a reaction controlled by an enzyme. 108

**substrate-level ATP synthesis** (foss for ill LAY shun) Process in which ATP is formed by transferring a phosphate from a metabolic substrate to ADP. 136

**supergroup** In this text, refers to the major groups of eukaryotes. 374

**surface-area-to-volume ratio** Ratio of a cell's outside area to its internal volume. 61

**surface tension** Force that holds moist membranes together due to the attraction of water molecules. 30

**survivorship** Probability of newborn individuals of a cohort surviving to particular ages. 822

**sustainable development** Management of an ecosystem so that it maintains itself while providing services to human beings. 902

**suture** Line of union between two nonarticulating bones, as in the skull. 722

**swamp** Wet, spongy land that is saturated and sometimes partially or intermittently covered with water. 879

**sweat gland** Skin gland that secretes a fluid substance for evaporative cooling; sudoriferous gland. 587

**swim bladder** In fishes, a gas-filled sac whose pressure can be altered to change buoyancy. 544

**symbiosis** Relationship that occurs when two different species live together in a unique way; it may be beneficial, neutral, or detrimental to one and/or the other species. 846

**symbiotic relationship** *See* symbiosis. 365

**sympathetic division** Division of the autonomic system that is active when an organism is under stress; uses norepinephrine as a neurotransmitter. 695

**sympatric speciation** (simm PAT trick spee see AY shun) Origin of new species in populations that overlap geographically. 307

**synapomorphy** In systematics, a derived character that is shared by clade members. 343

**synapse** (SIN naps) Junction between neurons consisting of the presynaptic (axon) membrane, the synaptic cleft, and the postsynaptic (usually dendrite) membrane. 686

**synapsis** (sin NAP sis) Pairing of homologous chromosomes during meiosis I. 171

**synaptic cleft** (sin NAP tick) Small gap between presynaptic and postsynaptic membranes of a synapse. 686

**synovial joint** (sin OH vee uhl) Freely moving joint in which two bones are separated by a cavity. 725

**systematics** (sis tim MAT ticks) Study of the diversity of organisms to classify them and determine their evolutionary relationships. 338

**systemic circuit** (sis TIM mick SIR kit) Circulatory pathway of blood flow between the tissues and the heart. 597

**systemin** In plants, an 18-amino-acid peptide that is produced by damaged or injured leaves that leads to the wound response. 488

**systole** (SIS toe lee) Contraction period of the heart during the cardiac cycle. 600

# T

**tactile communication** Communication through touch; for example, when a chick pecks its mother for food, chimpanzees grooming each other, and honeybees "dance." 809

**taiga** (TIE guh) Terrestrial biome that is a coniferous forest extending in a broad belt across northern Eurasia and North America. 871

**tandem repeat** Repetitive DNA sequence in which the repeats occur one after another in the same region of a chromosome. 256

**taproot** Main axis of a root that penetrates deeply and is used by certain plants (such as carrots) for food storage. 442

**taste bud** Structure in the vertebrate mouth containing sensory receptors for taste; in humans, most taste buds are on the tongue. 702

**taxon (pl., taxa)** (TAX ahn, TAX uh) Group of organisms that fills a particular classification category. 340

**taxonomy** (tax AH no mee) Branch of biology concerned with identifying, describing, and naming organisms. 6, 338

**T cell** Lymphocyte that matures in the thymus and exists in four varieties, one of which kills antigen-bearing cells outright. 615

**T cell receptor (TCR)** Molecule on the surface of a T cell that can bind to a specific antigen fragment in combination with an MHC molecule. 619

**telomere** (TELL oh meer) Tip of the end of a chromosome that shortens with each cell division and may thereby regulate the number of times a cell can divide. 162, 218

**telophase** (TELL oh faze) Mitotic phase during which daughter cells are located at each pole. 158

**temperate deciduous forest** (TIM purr utt duh SIDD you us) Forest found south of the taiga; characterized by deciduous trees such as oak, beech, and maple, moderate climate, relatively high rainfall, stratified plant growth, and plentiful ground life. 872

**temperate rain forest** Coniferous forest—e.g., that running along the west coast of Canada and the United States—characterized by plentiful rainfall and rich soil. 871

**template** (TEM plate) Parental strand of DNA that serves as a guide for the complementary daughter strand produced during DNA replication. 217

**tendon** Strap of fibrous connective tissue that connects skeletal muscle to bone. 580, 727

**terminal bud** Bud that develops at the apex of a shoot. 444

**termination** End of translation that occurs when a ribosome reaches a stop codon on the mRNA that it is translating, causing release of the completed protein. 227

**territoriality** Marking and/or defending a particular area against invasion by another species member; area often used for the purpose of feeding, mating, and caring for young. 292, 810

**territory** Area occupied and defended exclusively by an animal or group of animals. 291, 810

**test** Loose-fitting shell of a foraminiferan or a radiolarian; made of calcium carbonate or silicon, respectively. 388

**testcross** Cross between an individual with the dominant phenotype and an individual with the recessive phenotype. The resulting phenotypic ratio indicates whether the dominant phenotype is homozygous or heterozygous. 197

**testes (sing., testis)** (TEST tiss, TEST teez) Male gonad that produces sperm and the male sex hormones. 750, 756

**testosterone** (test TOSS turr ohn) Male sex hormone that helps maintain sexual organs and secondary sex characteristics. 750, 761

**tetanus** (TETT uh nuss) Sustained muscle contraction without relaxation. 727

**tetany** (TETT uh nee) Severe twitching caused by involuntary contraction of the skeletal muscles due to a calcium imbalance. 744

**tetrapod** (TETT truh pod) Four-footed vertebrate; includes amphibians, reptiles, birds, and mammals. 542

**thalamus** (THAL uh muss) In vertebrates, the portion of the diencephalon that passes on selected sensory information to the cerebrum. 690

**therapeutic cloning** Used to create mature cells of various cell types. Also, used to learn about specialization of cells and provide cells and tissue to treat human illnesses. 159

**therapsid** (thurr RAP sid) Mammal-like reptiles appearing in the middle Permian period; ancestral to mammals. 549

**thermoacidophile** (therm moh uh SIDD oh file) Type of archaea that lives in hot, acidic, aquatic habitats, such as hot springs or near hydrothermal vents. 369

**thigmotropism** (thig MAH troh piz uhm) In plants, unequal growth due to contact with solid objects, as the coiling of tendrils around a pole. 483

**threatened species** Species that is likely to become an endangered species in the foreseeable future (e.g., bald eagle, gray wolf, Louisiana black bear). 890

**thylakoid** (THIGH luh koid) Flattened sac within a granum whose membrane contains chlorophyll and where the light reactions of photosynthesis occur. 65, 76, 119

**thymine (T)** (THIGH men) One of four nitrogen-containing bases in nucleotides composing the structure of DNA; pairs with adenine. 214

**thymus gland** (THIGH muss) Lymphoid organ involved in the development and functioning of the immune system; T lymphocytes mature in the thymus gland. 615, 750

**thyroid gland** (THIGH roid) Large gland in the neck that produces several important hormones, including thyroxine, triiodothyronine, and calcitonin. 743

**thyroid-stimulating hormone (TSH)** Substance produced by the anterior pituitary that causes the thyroid to secrete thyroxine and triiodothyronine. 740

**thyroxine ($T_4$)** (thigh ROCKS sin) Hormone secreted from the thyroid gland that promotes growth and development; in general, it increases the metabolic rate in cells. 743

**tight junction** Junction between cells when adjacent plasma membrane proteins join to form an impermeable barrier. 99

**tissue** Group of similar cells combined to perform a common function. 578

**tissue culture** Process of growing tissue artificially, usually in a liquid medium in laboratory glassware. 505

**tissue fluid** Fluid that surrounds the body's cells; consists of dissolved substances that leave the blood capillaries by filtration and diffusion. 581, 608

**tone** Continuous, partial contraction of muscle. 727

**tonicity** (tone ISS ih tee) Osmolarity of a solution compared to that of a cell. If the solution is isotonic to the cell, there is no net movement of water; if the solution is hypotonic, the cell gains water; and if the solution is hypertonic, the cell loses water. 92

**tonsils** Partially encapsulated lymph nodules located in the pharynx. 615

**topography** Surface features of the Earth. 867

**totipotent** (toe TIP uh tent) Cell that has the full genetic potential of the organism, including the potential to develop into a complete organism. 505, 782

**toxin** Poisonous substance produced by living cells or organisms. Toxins are nearly always proteins that are capable of causing disease on contact or absorption with body tissues. 366

**tracer** Substance having an attached radioactive isotope that allows a researcher to track its whereabouts in a biological system. 24

**trachea (pl., tracheae)** (TRAY kee uh, TRAY kee ee) In insects, air tubes located between the spiracles and the tracheoles. In tetrapod vertebrates, air tube (windpipe) that runs between the larynx and the bronchi. 529, 653, 654

**tracheid** (TRAY kee id) In vascular plants, type of cell in xylem that has tapered ends and pits through which water and minerals flow. 439, 462

**tract** Bundle of myelinated axons in the central nervous system. 688

**transcription** Process whereby a DNA strand serves as a template for the formation of mRNA. 220

**transcription activator** Protein that speeds transcription. 240

**transcriptional control** Control of gene expression during the transcriptional phase determined by mechanisms that control whether transcription occurs or the rate at which it occurs. 242

**transcription factor** In eukaryotes, protein required for the initiation of transcription by RNA polymerase. 242

**transduction** (trans DUCK shun) Exchange of DNA between bacteria by means of a bacteriophage. 364

**transduction pathway** Series of proteins or enzymes that change a signal to one understood by the cell. 474

**transfer rate** Amount of a substance that moves from one component of the environment to another within a specified period of time. 857

**transfer RNA (tRNA)** Type of RNA that transfers a particular amino acid to a ribosome during protein synthesis; at one end, it binds to the amino acid, and at the other end it has an anticodon that binds to an mRNA codon. 220

**transformation** Taking up of extraneous genetic material from the environment by bacteria. 364

**transgenic organism** (trans JENN ick) Free-living organism in the environment that has had a foreign gene inserted into it. 252

**transitional fossil** Fossil that bears a resemblance to two groups that in present day are classified separately. 276

**translation** Process whereby ribosomes use the sequence of codons in mRNA to produce a polypeptide with a particular sequence of amino acids. 220

**translational control** Gene expression regulated by the activity of mRNA transcripts. 241

**translocation** (trans low KAY shun) Movement of a chromosomal segment from one chromosome to another nonhomologous chromosome, leading to abnormalities—e.g., Down syndrome. 184, 227

**transpiration** Plant's loss of water to the atmosphere, mainly through evaporation at leaf stomata. 465

**transposon** (trans POSE ahn) DNA sequence capable of randomly moving from one site to another in the genome. 257

**trichocyst** (TRICK oh sist) Found in ciliates; contains long, barbed threads useful for defense and capturing prey. 384

**trichomes** (TRY cohmz) In plants, specialized outgrowth of the epidermis (e.g., root hairs). 437

**trichomoniasis** (trih coh moh NIE uh sis) Sexually transmitted disease caused by the parasitic protozoan *Trichomonas vaginalis*. 772

**triglyceride** (try GLISS suh ride) Neutral fat composed of glycerol and three fatty acids. 44, 644

**triplet code** During gene expression, each sequence of three nucleotide bases stands for a particular amino acid. 221

**trisomy** (try SO mee) Having three of a particular type of chromosome (2n + 1). 180

**trochophore** Type of protostome that produces a trochophore larva; also has two bands of cilia around its middle. 516

**trophic level** (TROFE ick) Feeding level of one or more populations in a food web. 855

**trophoblast** (TROFE oh blast) Outer membrane surrounding the embryo in mammals; when thickened by a layer of mesoderm, it becomes the chorion, an extraembryonic membrane. 788

**tropical rain forest** Biome near the equator in South America, Africa, and the Indo-Malay regions; characterized by warm weather, plentiful rainfall, a diversity of species, and mainly tree-living animal life. 874

**tropism** (TROPE iz uhm) In plants, a growth response toward or away from a directional stimulus. 482

**true coelom** Body cavity completely lined with mesoderm; found in certain protostomes and all deuterostomes. 516

**truffle** Subterranean edible fungi. 398

**trypsin** (TRIP sin) Protein-digesting enzyme secreted by the pancreas. 642

**tube foot** Part of the water vascular system in sea stars, located on the oral surface of each arm; functions in locomotion. 534

**tubular reabsorption** (TUBE yule lurr ree ab SORP shun) Movement of primarily nutrient molecules and water from the contents of the nephron into blood at the proximal convoluted tubule. 672

**tubular secretion** Movement of certain molecules from blood into the distal convoluted tubule of a nephron so that they are added to urine. 673

**tumor** Cells derived from a single mutated cell that has repeatedly undergone cell division; benign tumors remain at the site of origin, while malignant tumors metastasize. 161

**tumor suppressor gene** Gene that codes for a protein that ordinarily suppresses the cell cycle; inactivity due to a mutation can lead to a tumor. 162

**turgor movement** In plant cells, pressure of the cell contents against the cell wall when the central vacuole is full. 484

**turgor pressure** (TURR gurr) Pressure of the cell contents against the cell wall; in plant cells, determined by the water content of the vacuole and provides internal support. 93

**tympanic membrane** (tim PAN ick) Membranous region that receives air vibrations in an auditory organ; in humans, the eardrum. 710

**typhlosole** (TIFE low sole) Expanded dorsal surface of long intestine of earthworms, allowing additional surface for absorption. 526, 634

# U

**umbilical cord** (uhm BILL lick cull) Cord connecting the fetus to the placenta through which blood vessels pass. 790

**unicellular** (you nih SELL you lurr) Made up of but a single cell, as in the bacteria. 2

**uniformitarianism** (you nih form ih TARE ree uhn iz uhm) Belief espoused by James Hutton that geological forces act at a continuous, uniform rate. 269

**unsaturated fatty acid** Fatty acid molecule that carbons of its hydrocarbon chain. The chain bears fewer hydrogens than the maximum number possible. 44

**upwelling** Upward movement of deep, nutrient-rich water along coasts; it replaces surface waters that move away from shore when the direction of prevailing wind shifts. 884

**uracil (U)** (YUR a sill) Pyrimidine base that occurs in RNA, replacing thymine. 220

**urea** (you REE uh) Main nitrogenous waste of terrestrial amphibians and most mammals. 666

**ureter** (you REE turr) Tubular structure conducting urine from the kidney to the urinary bladder. 670

**urethra** (you REE thruh) Tubular structure that receives urine from the bladder and carries it to the outside of the body. 670

**uric acid** (YOUR rick) Main nitrogenous waste of insects, reptiles, and birds. 666

**urinary bladder** (YOUR rinn air ree) Organ where urine is stored. 670

**urine** Liquid waste product made by the nephrons of the vertebrate kidney through the processes of glomerular filtration, tubular reabsorption, and tubular secretion. 670

**uterine cycle** (YOU turr rinn) Cycle that runs concurrently with the ovarian cycle; it prepares the uterus to receive a developing zygote. 764

**uterus** (YOU turr us) In mammals, expanded portion of the female reproductive tract through which eggs pass to the environment or in which an embryo develops and is nourished before birth. 762

**utricle** (YOU trick cull) Cavity in the vestibule of the vertebrate inner ear; contains sensory receptors for gravitational equilibrium. 713

# V

**vacuole** (VAC you ohl) Membrane-bounded sac, larger than a vesicle; usually functions in storage and can contain a variety of substances. In plants, the central vacuole fills much of the interior of the cell. 75

**valence shell** Outer shell of an atom. 25

**vascular bundle** (VASS cue lurr) In plants, primary phloem and primary xylem enclosed by a bundle sheath. 439

**vascular cambium** (VASS cue lurr CAMM bee uhm) In plants, lateral meristem that produces secondary phloem and secondary xylem. 444

**vascular cylinder** In eudicots, the tissues in the middle of a root, consisting of the pericycle and vascular tissues. 439

**vascular plant** Plant that has xylem and phloem. 416

**vascular tissue** Transport tissue in plants, consisting of xylem and phloem. 413, 437

**vector** (VECK turr) In genetic engineering, a means to transfer foreign genetic material into a cell—e.g., a plasmid. 250

**vegetative organ** Nonreproductive plant part. 434

**vein** Blood vessel that arises from venules and transports blood toward the heart. 596

**vena cava** (VEE nuh CAVE uh) Large systemic vein that returns blood to the right atrium of the heart in tetrapods; either the superior or inferior vena cava. 602

**ventilation** (venn tih LAY shun) Process of moving air into and out of the lungs; breathing. 650

**ventricle** (VENT trih cull) Cavity in an organ, such as a lower chamber of the heart or the ventricles of the brain. 598, 688

**venule** (VENN yule) Vessel that takes blood from capillaries to a vein. 596

**vertebral column** (VERT tih brull) Portion of the vertebrate endoskeleton that houses the spinal cord; consists of many vertebrae separated by intervertebral disks. 722

**vertebrate** (VERT tih brate) Chordate in which the notochord is replaced by a vertebral column. 512

**vesicle** (VESS sick cull) Small, membrane-bounded sac that stores substances within a cell. 66

**vessel element** Cell that joins with others to form a major conducting tube found in xylem. 439, 462

**vestibule** (VESS tibb yule) Space or cavity at the entrance to a canal, such as the cavity that lies between the semicircular canals and the cochlea. 710

**vestigial structure** (vest TIH jee uhl) Remains of a structure that was functional in some ancestor but is no longer functional in the organism in question. 267, 278

**villus** (VILL us) Small, fingerlike projection of the inner small intestinal wall. 639

**viroid** (VYE roid) Infectious strand of RNA devoid of a capsid and much smaller than a virus. 361

**virus** Noncellular parasitic agent consisting of an outer capsid and an inner core of nucleic acid. 356

**visible light** Portion of the electromagnetic spectrum that is visible to the human eye. 122

**visual accommodation** Ability of the eye to focus at different distances by changing the curvature of the lens. 706

**visual communication** Form of communication between animals using their bodies, includes fighting. 808

**vitamin** Essential requirement in the diet, needed in small amounts. Vitamins are often part of coenzymes. 110, 646, 666

**vitamin D** Fat-soluble compound; deficiency tends to cause rickets in children. 586

**viviparous** (vie VIP purr us) Animal that gives birth after partial development of offspring within mother. 757

**vocal cord** In humans, fold of tissue within the larynx; creates vocal sounds when it vibrates. 654

# W

**water column** In plants, water molecules joined together in xylem from the leaves to the roots. 464

**water (hydrologic) cycle** (high droh LAH jick) Interdependent and continuous circulation of water from the ocean, to the atmosphere, to the land, and back to the ocean. 857

**water mold** Filamentous organisms having cell walls made of cellulose; typically decomposers of dead freshwater organisms, but some are parasites of aquatic or terrestrial organisms. 382

**water potential** Potential energy of water; a measure of the capability to release or take up water relative to another substance. 463

**water vascular system** Series of canals that takes water to the tube feet of an echinoderm, allowing them to expand. 534

**wax** Sticky, solid, waterproof lipid consisting of many long-chain fatty acids usually linked to long-chain alcohols. 47

**wetland** Wet area. (*See also* bog or swamp.) 879

**whisk fern** Common name for seedless vascular plant that consists only of stems and has no leaves or roots. 418

**white blood cell** Leukocyte, of which there are several types, each having a specific function in protecting the body from invasion by foreign substances and organisms. 581, 607

**white matter** Myelinated axons in the central nervous system. 688

**wobble hypothesis** Ability of the tRNAs to recongnize more than one codon; the condons differ in their third nucleotide. 224

**wood** Secondary xylem that builds up year after year in woody plants and becomes the annual rings. 446

# X

**xenotransplantation** Use of animal organs, instead of human organs, in human transplant patients. 254

**X-linked** Allele that is located on an X chromosome but may control a trait that has nothing to do with the sexual characteristics of an animal. 205

**xylem** (ZIE lumm) Vascular tissue that transports water and mineral solutes upward through the plant body; it contains vessel elements and tracheids. 416, 439, 462

**xylem sap** Solution of inorganic nutrients moves from a plant's roots to its shoots through xylem tissue. 462

# Y

**yeast** Unicellular fungus that has a single nucleus and reproduces asexually by budding or fission, or sexually through spore formation. 398

**yolk** Dense nutrient material in the egg of a bird or reptile. 757, 779

**yolk sac** One of the extraembryonic membranes that, in shelled vertebrates, contains yolk for the nourishment of the embryo, and in placental mammals is the first site for blood cell formation. 787

# Z

**zero population growth** No growth in population size. 834

**zooflagellate** (zoh oh FLAJ jell ate) Nonphotosynthetic protist that moves by flagella; typically zooflagellates enter into symbiotic relationships, and some are parasitic. 386

**zooplankton** (zoe oh PLANK ton) Part of plankton containing protozoans and other types of microscopic animals. 387, 881

**zoospore** Spore that is motile by one or more flagella. 377, 396

**zygospore** (ZIE go spore) Thick-walled resting cell formed during sexual reproduction of zygospore fungi. 396

**zygospore fungi** Fungi such as black bread mold that reproduces by forming windblown spores in sporangia; sexual reproduction involves a thick-walled zygospore. 396

**zygote** (ZIE goat) Diploid cell formed by the union of two gametes; the product of fertilization. 170

# CREDITS

## PHOTOGRAPHS

**History of Biology:** Leeuwenhoek, Darwin, Pasteur, Koch, Lorenz, Pauling: © Bettman/Corbis; Pavlov: © Hulton-Deutsch Collection/Corbis; McClintock: © AP Photo/Middlemiss; Franklin: © Photo Researchers, Inc.

### Chapter 1

**Opener:** Courtesy Ernesto Sandoval, UC Davis Botanical Conservatory; **1.1 (Bacteria):** © Dr. Dennis Kunkel/Phototake; **1.1 (Paramecium):** © M. Abbey/Visuals Unlimited; **1.1 (Morel):** © Royalty-Free Corbis; **1.1 (Sunflower):** © Photodisc Green/Getty Images; **1.1 (Snow goose):** © Charles Bush Photography; **1.3a:** © Niebrugge Images; **1.3b:** © Photodisc Blue/Getty Images; **1.3c:** © Charles Bush Photography; **1.3d:** © Michael Abby/Visuals Unlimited; **1.3e:** © Pat Pendarvis; **1.3f:** National Park Service Photo; **1.4:** © Francisco Erize/Bruce Coleman, Inc.; **1.6:** © Ralph Robinson/Visuals Unlimited; **1.7:** © A.B. Dowsett/SPL/Photo Researchers, Inc.; **1.8 (Protist):** © Michael Abby/Visuals Unlimited; **(Plant):** © Pat Pendarvis; **(Fungi):** © Rob Planck/Tom Stack; **(Animal):** © Royalty-Free/Corbis; **1.11a:** © Frank & Joyce Burek/Getty Images; **1.11b (All):** © Dr. Phillip Dustan; **1.12:** Courtesy Leica Microsystems Inc.; **1A (Left):** © Royalty-Free/Corbis; **1A (Right):** © Jim Craigmyle/Corbis; **1.13:** © Dr. Jeremy Burgess/Photo Researchers, Inc.; **1.14 (All):** Courtesy Jim Bidlack; **1.15:** © Erica S. Leeds.

### Chapter 2

**Opener:** © Peter Weimann/Animals Animals/Earth Scenes; **2.1:** © Gunter Ziesler/Peter Arnold, Inc.; **2.4a:** © Biomed Commun./Custom Medical Stock Photo; **2.4b (Right):** © Hank Morgan/Rainbow; **2.4b (Left):** © Mazzlota et al./Photo Researchers, Inc; **2.5a (Peaches):** © Tony Freeman/PhotoEdit; **2.5b:** © Geoff Tompkinson/SPL/Photo Researchers, Inc.; **2.7 (Crystals):** © Charles M. Falco/Photo Researchers, Inc.; **2.7 (Salt shaker):** © Erica S. Leeds; **2.10:** © Grant Taylor/Getty Images; **2Aa:** © Lionel Delevingue/Phototake; **2Ab:** © Mauritius, GMBH/Phototake.

### Chapter 3

**Opener:** © Sylvia S. Mader; **3.1a:** © Brand X Pictures/PunchStock; **3.1b:** © Ingram Publishing/Alamy; **3.1c:** © H. Pol/CNRI/SPL/Photo Researchers, Inc.; **3.4:** © The McGraw Hill Companies, Inc./John Thoeming, photographer; **3.6:** © Steve Bloom/Taxi/Getty; **3.8a:** © Jeremy Burgess/SPL/Photo Researchers, Inc.; **3.8b:** © Don W. Fawcett/Photo Researchers, Inc.; **3.9:** © Science Source/J.D. Litvay/Visuals Unlimited; **3.10:** © Paul Nicklen/National Geographic/Getty Images; **3.13:** © Ernest A. Janes/Bruce Coleman, Inc.; **3.14a:** © Das Fotoarchiv/Peter Arnold, Inc.; **3.14b:** © Martha Cooper/Peter Arnold, Inc.; **(Runner, p. 48):** © Duomo/Corbis; **3.18a:** © Gregory Pace/Corbis; **3.18b:** © Ronald Siemoneit/Corbis Sygma; **3.18c:** © Kjell Sandved/Visuals Unlimited; **3.21:** © Photodisk Red/Getty Images; **3.22c:** © Jennifer Loomis/Animals Animals/Earth Scenes.

### Chapter 4

**Opener:** © Oliver Meckes/Nicole Ottawa/Photo Researchers, Inc.; **4.1a:** © Geoff Bryant/Photo Researchers, Inc.; **4.1b:** Courtesy Ray F. Evert/University of Wisconsin Madison; **4.1c:** © Barbara J. Miller/Biological Photo Service; **4.1d:** Courtesy O. Sabatakou and E. Xylouri-Frangiadaki; **4Aa:** © Robert Brons/Biological Photo Service; **4Ab:** © M. Schliwa/Visuals Unlimited; **4Ac:** © Kessel/Shih/Peter Arnold, Inc.; **4B (Bright field):** © Ed Reschke; **4B (Bright field stained):** © Biophoto Associates/Photo Researchers, Inc.; **4B (Differential, Phase contrast, Dark field):** © David M. Phillips/Visuals Unlimited; **4.4:** © Howard Sochurek/The Medical File/Peter Arnold, Inc.; **4.6:** © Dr. Dennis Kunkel/Visuals Unlimited; **4.7:** © Newcomb/Wergin/Biological Photo Service; **4.8 (Bottom):** Courtesy Ron Milligan/Scripps Research Institute; **4.8 (Top right):** Courtesy E.G. Pollock; **4.10:** © R. Bolender & D. Fawcett/Visuals Unlimited; **4.11:** Courtesy Charles Flickinger, from *Journal of Cell Biology* 49: 221-226, 1971, Fig. 1 page 224; **4.12a:** Courtesy Daniel S. Friend; **4.12b:** Courtesy Robert D. Terry/Univ. of San Diego School of Medicine; **4.14:** © S.E. Frederick & E.H. Newcomb/Biological Photo Service; **4.15:** © Newcomb/Wergin/Biological Photo Service; **4.16a:** Courtesy Herbert W. Israel, Cornell University; 4.17a: Courtesy Dr. Keith Porter; **4.18a(Actin):** © M. Schliwa/Visuals Unlimited; **4.18b, c (Intermediate, Microtubules):** © K.G. Murti/Visuals Unlimited; **4.18a (Chara):** The McGraw-Hill Companies, Inc./photo by Dennis Strete and Darrell Vodopich; **4.18b(Peacock):** © Vol. 86/Corbis; **4.18c (Chameleon):** © Photodisc/Vol. 6/Getty Images; **4.19 (Top):** Courtesy Kent McDonald, University of Colorado Boulder; **4.19 (Bottom):** *Journal of Structural Biology, Online* by Manley McGill et al. Copyright 1976 by Elsevier Science & Technology Journals. Reproduced with permission of Elsevier Science & Technology Journals in the format Textbook via Copyright Clearance Center; **4.20 (Sperm):** © David M. Phillips/Photo Researchers, Inc.; **4.20 (Flagellum, Basal body):** © William L. Dentler/Biological Photo Service.

### Chapter 5

**Opener:** © Professors P. Motta & T. Naguro/Science Photo Library/Photo Researchers, Inc.; **5Aa (Human egg):** © Anatomical Travelogue/Photo Researchers, Inc.; **5Aa (Embryo):** © Neil Harding/Getty Images; **5Aa (Baby):** © Photodisc Collection/Getty Images; **5.12 (Top):** © Eric Grave/Phototake; **5.12 (Center):** © Don W. Fawcett/Photo Researchers, Inc.; **5.12 (Bottom, both):** Courtesy Mark Bretscher; **5.14a:** From Douglas E. Kelly, *Journal of Cell Biology* 28 (1966): 51. Reproduced by copyright permission of The Rockefeller University Press; **5.14b:** © David M. Phillips/Visuals Unlimited; **5.14c:** Courtesy Camillo Peracchia, M.D.; **5.15:** © E.H. Newcomb/Biological Photo Service.

### Chapter 6

**Opener:** © PhotoAlto/PunchStock; **6.3b:** © Darwin Dale/Photo Researchers, Inc.; **6.8b:** © James Watt/Visuals Unlimited; **6.8c:** © Creatas/PunchStock; **6.9:** © G.K. & Vikki Hart/The Image Bank/Getty Images; **6A:** © Sygma/Corbis.

### Chapter 7

**Opener:** © David Muench/Corbis; **7.1 (Moss):** © Steven P. Lynch; **7.1 (Trees):** © Digital Vision/PunchStock; **7.1 (Kelp):** © Chuck Davis/Stone/Getty Images; **7.1 (Cyanobacteria):** © Sherman Thomas/Visuals Unlimited; **7.1 (Diatoms):** © Ed Reschke/Peter Arnold; **7.1 (Euglena):** © T.E. Adams/Visuals Unlimited; **7.1 (Sunflower):** © Royalty-Free/Corbis; **7.2:** © Dr. George Chapman/Visuals Unlimited; **7.3:** © B. Runk/S. Schoenberger/Grant Heilman Photography; **7.5:** Courtesy Lawrence Berkeley National Lab; **7A:** © George Holton/Photo Researchers, Inc.; **7.1:** © Herman Eisenbeiss/Photo Researchers, Inc.; **7.11a:** © Jim Steinberg/Photo Researchers, Inc.; **7.11b:** © Nigel Cattlin/Photo Researchers, Inc.; **7.12:** © S. Alden/PhotoLink/Getty Images.

## Chapter 8

**Opener:** © Wolfgang Kaehler/Corbis; **8.1:** © E. & P. Bauer/zefa/Corbis; **(Bread, wine, cheese, p. 139):** © The McGraw Hill Companies, Inc./John Thoeming, photographer; **(Yogurt, p. 139):** © The McGraw Hill Companies, Inc./Bruce M. Johnson, photographer; **8A:** © The McGraw Hill Companies, Inc./Bruce M. Johnson, photographer; **8.6:** © Dr. Donald Fawcett and Dr. Porter/Visuals Unlimited; **8.11:** © C Squared Studios/Getty Images.

## Chapter 9

**Opener:** © SPL/Photo Researchers, Inc.; **9.2:** Courtesy Douglas R. Green/LaJolla Institute for Allergy and Immunology; **9.3:** © Andrew Syred/Photo Researchers, Inc.; **9.4 Animal cell (Early prophase, Prophase, Metaphase, Anaphase, Telophase):** © Ed Reschke; **9.4 Animal cell (Prometaphase):** © Michael Abbey/Photo Researchers, Inc.; **9.4 Plant cell (Early prophase, Prometaphse):** © Ed Reschke; **9.4 Plant cell (Prophase, Metaphase, Anaphase):** © R. Calentine/Visuals Unlimited; **9.4 Plant cell (Telophase):** © Jack M. Bostrack/Visuals Unlimited; **9.5 (Both):** © R.G. Kessel and C.Y. Shih, *Scanning Electron Microscopy in Biology: A Students' Atlas on Biological Organization*, 1974 Springer-Verlag, New York; **9.6:** © Katherine Esau; **9.8d:** © Biophoto Associates/Photo Researchers, Inc.; **9.9 (All):** © Stanley C. Holt/Biological Photo Service.

## Chapter 10

**Opener:** Barcroft Media; **10.1:** © L. Willatt/Photo Researchers, Inc.; **10.3a:** Courtesy Dr. D. Von Wettstein; **10.5:** © American Images, Inc/Getty Images; **10.6 (All):** © Ed Reschke; **10.11a:** © Jose Carrilo/PhotoEdit; **10.11b:** © CNRI/SPL/Photo Researchers, Inc.; **10.12a:** Courtesy UNC Medical Illustration and Photography; **10.12b:** Courtesy Stefan D. Schwarz, http://klinefeltersyndrome.org; **10.14b:** Courtesy The Williams Syndrome Association; **10.15b (Both):** *American Journal of Human Genetics* by N.B. Spinner. Copyright 1994 by Elsevier Science & Technology Journals. Reproduced with permission of Elsevier Science & Technology Journals in the format Textbook via Copyright Clearance Center.

## Chapter 11

**Opener:** © Felicia Martinez Photography; **11.1:** © Ned M. Seidler National Geographic Image Collection; **11.10:** Courtesy Division of Medical Toxicology, University of Virginia; **11.11:** © Pat Pendarvis; **11.12:** From Lichtman M. L., Beutler E, Kipps T.J., et al. *Williams Hematology*, 7th edition. Color Plate IX-4. The McGraw-Hill Companies, Inc., 2006; **11Ba:** © Steve Uzzell; **11.14 (Left):** © AP/Wide World Photos; **11.14 Right):** © Ed Reschke; **(Sickled cells, p. 203):** © Phototake, Inc./Alamy; **11.18(Abnormal):** Courtesy Dr. Rabi Tawil, Director, Neuromuscular Pathology Laboratory, University of Rochester Medical Center; **11.18 (Boy):** Courtesy Muscular Dystrophy Association; **11.18 (Normal):** Courtesy Dr. Rabi Tawil, Director, Neuromuscular Pathology Laboratory, University of Rochester Medical Center.

## Chapter 12

**Opener:** (Leopard): © James Martin/Stone/Getty Images; (Flower): © Rosemary Calvert/Stone/Getty Images; (Crab): © Tui DeRoy/Bruce Coleman; (Protozoa): © A. M. Siegelman/Visuals Unlimited; **12.2 (Bacteria):** © Martin Shields/Photo Researchers, Inc.; 12.2 (Jellyfish): © R. Jackman/OSF/Animals Animals/Earth Scenes; **12.2 (Pigs):** Courtesy Norrie Russell, The Roslin Institute; **12.2 (Mouse):** © Eye of Science/Photo Researchers, Inc.; **12.2 (Plant):** © Dr. Neal Stewart; **12.4a:** © Photo Researchers, Inc.; **12.4c:** © Science Source/Photo Researchers, Inc.; **12.5a:** © Kenneth Eward/Photo Researchers, Inc.; **12.5d:** © A. Barrington Brown/Photo Researchers, Inc.; **12.12:** © Oscar L. Miller/Photo Researchers, Inc.; **12.15d:** Courtesy Alexander Rich; **12.20a:** © Ada L. Olins and Donald E. Olins/Biological Photo Service; **12.20b:** Courtesy Dr. Jerome Rattner, Cell Biology and Anatomy, University of Calgary; **12.20c:** *Cell* by Paulson, J.R. & Laemmli, UK. Copyright 1977 by Elsevier Science & Technology Journals. Reproduced with permission of Elsevier Science & Technology Journals in the format Textbook via Copyright Clearance Center; **12.20d, e:** © Peter Engelhardt/Department of Pathology and Virology, Haartman Institute/Centre of Excellence in Computational Complex Systems Research, Biomedical Engineering and Computational Science, Faculty of Information and Natural Sciences, Helsinki University of Technology, Helsinki, Finland.

## Chapter 13

**Opener:** © Eye of Science/Photo Researchers, Inc.; **13.5a:** Courtesy Stephen Wolfe; **13.6:** © Chanan Photo 2004; **13A:** © *Fitzpatrick Color Atlas & Synopsis of Clinical Dermatology* 5/e. Used with permission of The McGraw-Hill Companies, Inc. (Access Medicine); **13B:** Courtesy Heather and Patrick Kelly; **13.11:** © Ken Greer/Visuals Unlimited; **13.12b, c:** © Stan Flegler/Visuals Unlimited.

## Chapter 14

**Opener:** © Ken Lucas/Visuals Unlimited; **14.4 (Both):** Courtesy General Electric Research & Development; **14.8:** © Cindy Charles/PhotoEdit.

## Chapter 15

**Opener:** © Brand X Pictures/PunchStock; **15.1b:** © Wolfgang Kaehler/Corbis; **15.1c:** © Luiz C. Marigo/Peter Arnold; **15.1d:** © Gary J. James/Biological Photo Service; **15.1e:** © Charles Benes/Index Stock Imagery; **15.1f:** © Galen Rowell/Corbis; **15.1g:** © D. Parer & E. Parer-Cook/Ardea; **15.2:** © Carolina Biological/Visuals Unlimited; **15.3a:** © Joseph H. Bailey/National Geographic Image Sales; **15.3b:** © Daryl Balfour/Photo Researchers, Inc.; **15.5 (European hare):** © WILDLIFE/Peter Arnold, Inc.; **15.5 (Patagonian hare):** © Juan & Carmecita Munoz/Photo Researchers, Inc.; **15.6a:** © Kevin Schafer/Corbis; **15.6b:** © Michael Dick/Animals Animals/Earth Scenes; 15.7: © Lisette Le Bon/SuperStock; **15.8 (Wolf):** © Gary Milburn/Tom Stack & Associates; **15.8 (Irish wolfhound):** © Ralph Reinhold/Index Stock Imagery; **15.8 (Boston terrier):** © Robert Dowling/Corbis; **15.8 (Dalmation):** © Alexander Lowry/Photo Researchers, Inc.; **15.8 (Shih tzu):** © Bob Shirtz/SuperStock; **15.8 (Bloodhound):** © Mary Bloom/Peter Arnold, Inc.; **15.8 (Scottish terrier):** © Carolyn A. McKeone/Photo Researchers, Inc.; **15.8 (Beagle):** © Tim Davis/Photo Researchers, Inc.; **15.8 (Red chow):** © Jeanne White/Photo Researchers, Inc.; **15.8 (Shetland sheepdog):** © Ralph Reinhold/Index Stock Imagery; **15.8 (English sheepdog):** © Yann Arthus-Bertrand/Corbis; **15.8 (Chihuahua):** © Kent & Donna Dannen/Photo Researchers, Inc.; **15.8 (Fox):** © Steven J. Kazlowski/Alamy; **15.9 (Cabbage, Brussel sprouts, kohlrabi):** Courtesy W. Atlee Burpee Company; **15.9 (Mustard):** © Jack Wilburn/Animals Animals/Earth Scenes; **15A:** © Stock Montage; **(Moths, p. 275):** © Michael Wilmer Forbes Tweedie/Photo Researchers, Inc.; **15.10a:** © Adrienne T. Gibson/Animals Animals/Earth Scenes; **15.10b:** © Joe McDonald/Animals Animals/Earth Scenes; **15.10c:** © Leonard Lee Rue/Animals Animals/Earth Scenes; **15.12a:** © Jean-Claude Carton/Bruce Coleman Inc.; **15.12b:** © Joe Tucciarone; **15.13 (Fossil):** © J.G.M. Thewissen, http://darla.neoucom.edu/Depts/Anat Thewissen/; **15.14 (Sugar glider):** © ANT Photo Library/Photo Researchers, Inc.; **15.14 (Tasmanian wolf):** © Tom McHugh/Photo Researchers, Inc.; **15.14 (Wombat):** © Photodisc Blue/Getty; **15.14 (Dasyurus):** © Tom McHugh/Photo Researchers, Inc.; **15.14 (Kangaroo):** © George Holton/Photo Researchers, Inc.

## Chapter 16

**Opener:** © 2008 The Associated Press. All rights reserved; **16.1 (Top left, center, right; Bottom 1, 3, 4):** © Vol. 105/PhotoDisc/Getty; **16.1 (Bottom 2):** © Vol. 42/PhotoDisc/Getty Images; **16.1 (Bottom 5):** © Vol. 116/PhotoDisc/Getty Images; **16.3 (Both):** © Michael Wilmer Forbes Tweedie/Photo Researchers, Inc.; **16.7:** Courtesy Victor McKusick; **16.10:** © Helen Rodd; **16.11:** © Bob Evans/Peter Arnold, Inc.; **16.13:** © Barbara Gerlach/Visuals Unlimited; **16.14a:** © Y. Arthus-Bertrand/Peter Arnold, Inc.; **16.14b:** © Neil McIntre/Getty Images; **16A:** © Jody Cobb/National Geographic; **16.15 (E.o. lindheimeri, E.o. quadrivittata**): © Zig Leszczynski/Animals Animals/Earth Scenes; **16.15 (E.o. spiloides):** © Joseph Collins/Photo Researchers, Inc.; **16.15 (E.o. rossalleni):** © Dale Jackson/Visuals Unlimited; **16.15 (E.o. obsoleta):** © William Weber/Visuals Unlimited.

## Chapter 17

**Opener:** © Andy Carvin; **17.2 (Acadian):** © Karl Maslowski/Visuals Unlimited; **17.2 (Willow):** © Ralph Reinhold/Animals Animals/Earth Scenes; **17.2 (Least):** © Stanley Maslowski/Visuals Unlimited; **17.3 (Left):** © Sylvia S. Mader; **17.3 (Right):** © B & C Alexander/Photo Researchers, Inc.; **17.6:** © Barbara Gerlach/Visuals Unlimited; **17.7 (Stallion):** © Superstock, Inc.; **17.7 (Donkey):** © Robert J. Erwin/Photo Researchers, Inc.; **17.7 (Mule):** © Jorg & Petra Wegner/Animals Animals/Earth Scenes; **(Lizard, p. 305):** © Jonathan Losos; **17.12 (C. pulchella):** © J. L. Reveal; **17.12 (C. concinna):** © Gerald & Buff Corsi/Visuals Unlimited; **17.12 (C. virgata):** ©: Dr. Dean Wm. Taylor/Jepson Herbarium, UC Berkeley; **17A:** © Boehm Photography; **17B (Opabina):** © A. J. Copley/Visuals Unlimited; **17B (Thaumaptilon):** © Simon Conway Morris, University of Cambridge; **17B (Wiwaxia):** © Albert Copley/Visuals Unlimited; **17B (Vauxia):** © Alan Siruinikoff/Photo Researchers, Inc.; **17.14 (Left):** © Carolina Biological Supply/Photo Researchers, Inc.; **17.14 (Center):** © Vol. OS02/PhotoDisc/Getty Images; **17.14 (Right):** © Aldo Brando/Peter Arnold, Inc.; **17.15:** Courtesy Walter Gehring, reprinted with permission from *Induction of Ectopic Eyes by Target Expression of the Eyeless Gene in Drosophila,* G. Halder, P. Callaerts, Walter J. Gehring, *Science* Vol. 267, © 24 March 1995 American Association for the Advancement of Science; **17.16 (Both):** © A. C. Burke, 2000.

## Chapter 18

**Opener:** © Joe Tucciarone/Photo Researchers, Inc.; **18.2:** © Ralph White/Corbis; **18.3a:** © Science VU/Visuals Unlimited; **18.3b:** Courtesy Dr. David Deamer; **18.5:** © Henry W. Robinson/Visuals Unlimited; **18.6 (Trilobite):** © Francois Gohier/Photo Researchers, Inc.; **18.6 (Tusks):** © AP Photo/Francis Latreille/Nova Productions; **18.6 (Placoderm):** © The Cleveland Museum of Natural History; **18.6 (Fern):** © George Bernard/Natural History Photo Agency; **18.6 (Petrified wood):** Courtesy National Park Service; **18.6 (Ammonites):** © Sinclair Stammers/SPL/Photo Researchers, Inc.; **18.6 (Scorpion):** © George O. Poinar; **18.6 (Dinosaur footprint):** © Scott Berner/Visuals Unlimited; **18.6 (Ichthyosaur):** © Natural History Museum, London; **18.8a:** Courtesy J. William Schopf; **18.8b:** © Francois Gohier/Photo Researchers, Inc.; **18.9a:** Courtesy James G. Gehling, South Australian Museum; **18.9b:** Courtesy Dr. Bruce N. Runnegar; **18.10 (Opabina):** © A. J. Copley/Visuals Unlimited; **18.10 (Thaumaptilon):** © Simon Conway Morris, University of Cambridge; **18.10 (Wiwaxia):** © Albert Copley/Visuals Unlimited; **18.10 (Vauxia):** © Alan Siruinikoff/Photo Researchers, Inc.; **18.11a:** © The Field Museum, CSGEO 75400c; **18.11b:** © John Cancalosi/Peter Arnold, Inc.; **18.11c:** © John Gerlach/Animals Animals/Earth Scenes; **18.12:** © Chase Studio/Photo Researchers, Inc.; **18.13:** © Chase Studio/Photo Researchers, Inc.; **18.14:** © Gianni Dagli Orti/Corbis.

## Chapter 19

**Openers (4 orchids–top left, top right, both bottom):** © Richard L. Stone; **(Top center):** © Brand X Pictures/PunchStock; **19.1 (All):** © Sylvia S. Mader; **9.2a:** Courtesy Uppsala University Library, Sweden; **19.2b:** © Arthur Gurmankin/Visuals Unlimited; **19.2c:** © Dick Poe/Visuals Unlimited; **19Aa:** © Brent Opell; **19Ab:** © Kjell B. Sandved/Visuals Unlimited; **19.8:** © David Dilcher and Ge Sun; **19.12 (Bacteria):** © David M. Phillips/Visuals Unlimited; **19.12 (Archaea):** © Ralph Robinson/Visuals Unlimited; **19.12 (Flower):** © Ed Reschke/Peter Arnold, Inc.; **19.12 (Paramecium):** © M. Abbey/Visuals Unlimited; **19.12 (Mushroom):** © S. Gerig/Tom Stack & Associates; **19.12 (Wolf):** © Art Wolf/Stone/Getty Images.

## Chapter 20

**Opener:** © Eye of Science/Photo Researchers, Inc.; **(Leaf, p. 356):** © B. Runk/S. Schoenberger/Grant Heilman Photography; **20.1a:** © Dr. Hans Gelderblom/Visuals Unlimited; **20.1b:** © Eye of Science/Photo Researchers, Inc.; **20.1c:** © Dr. O. Bradfute/Peter Arnold, Inc.; **20.1d:** © K.G. Murti/Visuals Unlimited; **20.2:** © Ed Degginger/Color Pic Inc.; **20.6:** © RDF/Visuals Unlimited; **20.7a:** Courtesy USDA, photo by Harley W. Moon; **20.7b:** Courtesy C. Brinton, Jr.; **20.8:** © CNRI/SPL/Photo Researchers, Inc.; **20.9a:** © Dr. Richard Kessel & Dr. Gene Shih/Visuals Unlimited; **20.9b:** © Gary Gaugler/Visuals Unlimited; **20.9c:** © SciMAT/Photo Researchers, Inc.; **20.10:** Courtesy Nitragin Company, Inc.; **20.11:** © Alfred Pasieka/SPL/Photo Researchers, Inc.; **20.12a:** © Michael Abbey/Photo Researchers, Inc.; **20.12b:** © Tom Adams/Visuals Unlimited; **20.13a (Main):** © John Sohlden/Visuals Unlimited; **20.13a (Inset):** From J.T. Staley, et al., *Bergey's Manual of Systematic Bacteriology,* Vol. 13 © 1989 Williams & Wilkins Co., Baltimore. Prepared by A. L. Usted. Photography by Dept. of Biophysics, Norwegian Institute of Technology; **20.13b (Main):** © Jeff Lepore/Photo Researchers, Inc.; **20.13b (Inset):** Courtesy Dennis W. Grogan, Univ. of Cincinnati; **20.13c (Main):** © Susan Rosenthal/Corbis RM; **20.13c (Inset):** © Ralph Robinson/Visuals Unlimited.

## Chapter 21

**Opener:** © Rob & Ann Simpson/Visuals Unlimited; **21.2 (Diatoms):** © M.I. Walker/Photo Researchers, Inc.; **21.2 (Nonionina):** © Astrid & Hanns-Frieder Michler/Photo Researchers, Inc.; **21.2 (Synura):** Courtesy Dr. Ronald W. Hoham; **21.2 (Plasmodium):** © Patrick W. Grace/Photo Researchers, Inc.; **21.2 (Blepharisma):** © Eric Grave/Photo Researchers, Inc.; **21.2 (Onychodromus):** Courtesy Dr. Barry Wicklow; **21.2 (Ceratium):** © D.P. Wilson/Photo Researchers, Inc.; **21.2 (Licmorpha):** © Biophoto Associates/Photo Researchers, Inc.; **21.2 (Acetabularia):** © Linda L. Sims/Visuals Unlimited; **21.2 (Amoeba):** © Michael Abbey/Visuals Unlimited; **21.2 (Bossiella):** © Daniel V. Gotschall/Visuals Unlimited; **21.4:** © W.L. Dentler/Biological Photo Service; **21.6 (Top):** © John D. Cunningham/Visuals Unlimited; **21.6 (Right):** © Cabisco/Visuals Unlimited; **21.7a:** © William E. Ferguson; **21.8b:** © M.I. Walker/Science Source/Photo Researchers, Inc.; **21.9a:** © Dr. John D. Cunningham/Visuals Unlimited; **21.9b:** © Kingsley Stern; **21.10:** © Steven P. Lynch; **21.11:** © D.P Wilson/Eric & David Hosking/Photo Researchers, Inc.; **21.12a:** © Dr. Ann Smith/Photo Researchers, Inc.; **21.12b:** © Biophoto Associates/Photo Researchers, Inc.; **21.14:** © James Richardson/Visuals Unlimited; **21.15a:** © C.C. Lockwood/Cactus Clyde Productions; **21.15b:** © Sanford Berry/Visuals Unlimited; **21.16a:** © CABISCO/Phototake; **21.16b:** © Manfred Kage/Peter Arnold, Inc.; **21.16c:** © Eric Grave/Photo Researchers, Inc.; **21.18b:** © Michael Abbey/Visuals Unlimited; **21.19:** © Stanley Erlandsen; **21.20a:** © Eye of Science/Photo Researchers, Inc.; **21.22 (Top left):** © CABISCO/Visuals Unlimited; **21.22 (Top right):** © V. Duran/Visuals Unlimited; **21.24a (Cliffs):** Stockbyte/Getty Images; **21.24a (Inset):** © Manfred Kage/Peter Arnold, Inc.; **21.24b (Tests):** © Dr. Richard Kessel & Dr. Gene Shih/Visuals Unlimited; **p. 389 (Nucleariid)**: Courtesy Dr. Sc. Yuuji Tsukii.

## Chapter 22

**Opener:** © INTERFOTO Pressebildagentur; **22.2a:** © Gary R. Robinson/Visuals Unlimited; **22.2b:** © Dennis Kunkel/Visuals Unlimited; **22.3:** © Dr. Hilda Canter-Lund; **22.4 (Top):** © James W. Richardson/Visuals Unlimited; **22.4 (Bottom), 22.5 (Both):** © David M. Phillips/Visuals Unlimited; **22.6a (Top):** © Walter H. Hodge/Peter Arnold, Inc.; **22.6b (Top left):** © Corbis Royalty-Free; **22.6b (Top right):** © James Richardson/Visuals Unlimited; **22.6c (Top):** © Kingsley Stern; **22.7:** © SciMAT/Photo Researchers, Inc.; 22.8a: © Dr. P. Marazzi/SPL/Photo Researchers, Inc.; **22.8b:** © John Hadfield/SPL/Photo Researchers, Inc.; **22A:** © Patrick Endres/Visuals Unlimited; **22B:** © R. Calentine/Visuals Unlimited; **22.9:** © Biophoto Associates/Photo Researchers, Inc.; **22.10a:** © Glenn Oliver/Visuals Unlimited; **22.10b:** © Larry Lefever/Jane Grushow/Grant Heilman Photography; **22.10c:** © M. Eichelberger/Visuals Unlimited; 22.**10d:** © L. West/Photo Researchers, Inc.; **22.11a:** © Steven P. Lynch; 22.11b: © Arthur M. Siegelman/Visuals Unlimited; **22.12a:** © Digital Vision/Getty Images; **22.12b:** © Steven P. Lynch; **22.12c:** © Kerry T. Givens; **22.13:** © R. Roncadori/Visuals Unlimited.

## Chapter 23

**Opener:** © J. J. Alcalay/Peter Arnold, Inc.; **23.2 (Coleochaete):** © T. Mellichamp/Visuals Unlimited; **23.2 (Chara):** © Heather Angel/Natural Visions; **23.5 (Left):** © Kingsley Stern; **23.5 (Right):** © Andrew Syred/SPL /Photo Researchers, Inc.;

**23.6a:** © Ed Reschke/Peter Arnold, Inc.; **23.6b:** © J.M. Conrarder/National Audubon Society/Photo Researchers, Inc.; **23.6c:** © R. Calentine/Visuals Unlimited; **23.7:** B. Runk/S. Schoenberger/Grant Heilman Photography; **23.8 (Top):** © Heather Angel/Natural Visions ; **23.8 (Bottom):** © Bruce Iverson; **23.9:** Courtesy Hans Steur, the Netherlands; **23.12:** © Robert P. Carr/Bruce Coleman, Inc.; **23.13:** © CABISCO/Phototake; **23.14 (Cinnamon):** © James Strawser/Grant Heilman Photography; **23.14 (Hart's):** © Walter H. Hodge/Peter Arnold, Inc.; **23.14 (Maidenhair):** © Larry Lefever/Jane Grushow/Grant Heilman Photography; **23.15:** © Matt Meadows/Peter Arnold, Inc.; **23.16a:** © Corbis Royalty Free; **23.16b:** © Walt Anderson/Visuals Unlimited; **23.16c:** © The McGraw Hill Companies, Inc./Evelyn Jo Johnson, photographer; 23.17: © Phototake; **23.18a:** © Walter H. Hodge/Peter Arnold, Inc.; **23.18b (Main):** © Runk/Schoenburger/Grant Heilman Photography; **23.18b (Inset):** Courtesy Ken Robertson, University of Illinois/INHS; **23.18c (Main):** © Daniel L. Nickrent; **23.18c (Inset):** Courtesy K.J. Niklas; **23.18d:** © NHPA/Steve Robinson; **23A:** © Sinclair Stammers/SPL/Photo Researchers, Inc.; **23.19:** Courtesy Stephen McCabe/Arboretum at University of California Santa Cruz; **23.20 (Cactus):** © Christi Carter/Grant Heilman Photography; **23.20 (Butterfly weed):** © The McGraw Hill Companies, Inc./Evelyn Jo Johnson, photographer; **23.20 (Water lily):** © Pat Pendarvis; **23.20 (Iris):** © David Cavagnaro/Peter Arnold, Inc.; **23.20 (Trillium):** © Adam Jones/Photo Researchers, Inc.; **23.20 (Apple blossoms):** © Inga Spence/Photo Researchers, Inc.; 23B(Wheat): © Pixtal/age fotostock; **23B (Rice):** © Corbis Royalty Free; **23B(Rice grain):** © Dex Image/Getty RF; **23B (Cornfield):** © Corbis; **23B (Corn ear):** © Nigel Cattlin/Photo Researchers, Inc.; **23Ca:** Heather Angel/Natural Visions; **23Cb:** © Steven King/Peter Arnold, Inc.; **23Cc:** © Dale Jackson/Visuals Unlimited; **23Cd:** © Brand X Pictures/PunchStock.

## Chapter 24

Openers: (Top left): © David Newman/Visuals Unlimited; **(Top right):** © Dwight Kuhn; **(Bottom left):** © Runk-Schoenberger/Grant Heilman Photography; **(Bottom center):** © Ardea London Ltd.; **(Bottom right):** Courtesy George Ellmore, Tufts University; **24.2 (All):** © Dwight Kuhn; **24.4a:** © B. Runk/S. Schoenberger/Grant Heilman Photography; **24.4b:** © J.R. Waaland/Biological Photo Service; **24.4c:** © Kingsley Stern; **24.5 (All):** © Biophoto Associates/Photo Researchers, Inc.; **24.6a:** © J. Robert Waaland/Biological Photo Service; **24.7a:** © George Wilder/Visuals Unlimited; **24.8a (Root tip):** Courtesy Ray F. Evert/University of Wisconsin Madison; **24.8b:** © CABISCO/Phototake; **24.9:** © Dwight Kuhn; **24.10a:** © John D. Cunningham/Visuals Unlimited; **24.10b:** Courtesy George Ellmore, Tufts University; **24.11a:** © Dr. Robert Calentine/Visuals Unlimited; **24.11b:** © Ed Degginger/Color Pic; **24.11c:** © David Newman/Visuals Unlimited; **24.11d:** © Terry Whittaker/Photo Researchers, Inc.; **24.11e (Left):** © Alan and Linda Detrick/Photo Researchers, Inc.; **24.11e (Right):** © David Sieren/Visuals Unlimited; **24A:** © James Schnepf Photography, Inc.; **24.14 (Top):** © Ed Reschke; **24.14 (Bottom):** Courtesy Ray F. Evert/University of Wisconsin Madison; **24.15 (Top):** © CABISCO/Phototake; **24.15 (Bottom):** © Kingsley Stern; **24.17:** © Ed Reschke/Peter Arnold, Inc.; **24.18:** © Ardea London Limited; **24.19a:** © Stanley Schoenberger/Grant Heilman Photography; **24.19b:** © William E. Ferguson; **24.19c, d:** © The McGraw Hill Companies, Inc./Carlyn Iverson, photographer; **24B (Oak):** © Jim Strawser/Grant Heilman Photography, Inc.; **24B (Willow):** © Stephen Owens/SPL/Photo Researchers, Inc.; **24B (Cross Section):** Courtesy Edward F. Gilman/University of Florida; **24.20:** © Jeremy Burgess/SPL/Photo Researchers, Inc.; **24.22a:** © Patti Murray Animals Animals/Earth Scenes; **24.22b:** © Gerald & Buff Corsi/Visuals Unlimited; **24.22c:** © P. Goetgheluck/Peter Arnold, Inc.

## Chapter 25

**Opener:** © Tim Davis/Photo Researchers, Inc.; **25.2(All):** Courtesy Mary E. Doohan; **25.5a:** © CABISCO/Phototake; **25.6 (Top):** © Dwight Kuhn; **25.6 (Circle):** © E.H. Newcomb & S.R. Tardon/Biological Photo Service; **25.7 (Top):** © B. Runk/S. Schoenberger/Grant Heilman Photography; **25.7 (Circle):** © Dana Richter/Visuals Unlimited; **25.8a:** © Kevin Schafer/Corbis; **25.8b (Plant):** © Barry Rice/Visuals Unlimited; **25.8b (Leaf):** © Dr. Jeremy Burgess/Photo Researchers, Inc.; **25Aa (Both):** © Dwight Kuhn; **25.10a, b:** Courtesy Wilfred A. Cote, from H.A. Core, W.A. Cote, and A.C. Day, *Wood: Structure and Identification* 2/e; **25.10c:** Courtesy W. A. Cote, Jr., N. C. Brown Center for Ultrastructure Studies, SUNY-ESF; **25.11:** © Ed Reschke/Peter Arnold, Inc.; **25.13a, b:** © Jeremy Burgess/SPL/Photo Researchers, Inc.; **25B:** Courtesy Gary Banuelos/Agriculture Research Service/USDA; **25.14a:** © M.H. Zimmermann, Courtesy Dr. P.B. Tomlinson, Harvard University; **25.14b:** © Steven P. Lynch.

## Chapter 26

**Opener:** © Kim Taylor/Bruce Coleman, Inc.; **26.4a:** © Robert E. Lyons/Visuals Unlimited; **26.4b:** © Sylvan Whittwer/Visuals Unlimited; **26.5 (All):** Courtesy Alan Darvill and Stefan Eberhard, Complex Carbohydrate Research Center, University of Georgia; **26Aa, 26B (Normal, Mutated center, Right):** Courtesy Elliot Meyerowitz/California Institute of Technology; **26B(Flat):** © Inga Spence/Visuals Unlimited; **26B(Lab):** © Vo Trung Dung/Corbis Sygma; **26.6:** © John Solden/Visuals Unlimited; **26.7:** *Plant Cell* by Donald R. McCarthy. Copyright 1989 by American Society of Plant Biologists. Reproduced with permission of American Society of Plant Biologists in the format Textbook via Copyright Clearance Center; **26.9 (Both), 26.11a:** © Kingsley Stern; 26.11b: Courtesy Malcolm Wilkins, Glascow University; **26.11c:** © BioPhot; 26.13: © John D. Cunningham/Visuals Unlimited; **26.14 (Both):** © John Kaprielian/Photo Researchers, Inc.; **(Venus flytrap, p. 484):** © Steven P. Lynch; **26.15 (Top, both):** © Tom McHugh/Photo Researchers, Inc.; **26.15 (Bottom left):** © BIOS A. Thais/Peter Arnold, Inc.; **26.15 (Bottom right):** © BIOS Pierre Huguet/Peter Arnold, Inc.; **26.17a, b:** © Grant Heilman Photography; **26.19 (Bug):** USDA/Agricultural Research Service, photo by Scott Bauer; **26.19 (Fungus):** © Kingsley Stern; **26.19 (Caterpillar):** © Frans Lanting/Minden Pictures; **26.19 (Butterfly):** © Dwight Kuhn; **(Foxglove, p. 488):** © Steven P. Lynch; **(Leaf spot, p. 489):** © Nigel Cattlin/Photo Researchers, Inc.; **(Acacia ants, p. 489):** © David Thompson/OSF/Animals Animals/Earth Scenes.

## Chapter 27

**Opener:** © Royalty-Free/Corbis; **27.3a:** © Farley Bridges; **27.3b:** © Pat Pendarvis; **27.4a:** © Arthur C. Smith III/Grant Heilman Photography, Inc.; **27.4b:** © Larry Lefever/Grant Heilman Photography, Inc.; **27.5 (Top):** Courtesy Graham Kent; **27.5 (Bottom):** © Ed Reschke; **27.6a:** © George Bernard/Animals Animals/Earth Scenes; **27.6b:** © Simko/Visuals Unlimited; **27.6c:** © Dwight Kuhn; **27Aa:** © Steven P. Lynch; **27Ab:** © Robert Maier/Animals/Animals/Earth Scenes; **27Ba:** © Anthony Mercieca/Photo Researchers, Inc.; **27Bb:** © Merlin D. Tuttle/Bat Conservation International; **27.7 (Proembryo, Globular, Heart):** Courtesy Dr. Chun-Ming Liu; **27.7 (Torpedo):** © Biology Media/Photo Researchers, Inc.; **27.7 (Mature embryo):** © Jack Bostrack/Visuals Unlimited; **27.8a:** © Dwight Kuhn; **27.8b:** Courtesy Ray F. Evert/University of Wisconsin Madison; **27.9a, b:** © Kingsley Stern; **27.9c:** © Dr. James Richardson/Visuals Unlimited; **27.9d:** © James Mauseth; **27.9e:** Courtesy Robert A. Schlising; **27.9f:** © Ingram Publishing/Alamy; **27.10a:** © Marie Read/Animals Animals/Earth Scenes; **27.10b:** © Scott Camazine/Photo Researchers, Inc.; **27.11a:** © Ed Reschke; **27.11b:** © James Mauseth; **27.12:** © G.I. Bernard/Animals Animals/Earth Scenes; **27.13 (All):** Courtesy Prof. Dr. Hans-Ulrich Koop, from *Plant Cell Reports*, 17:601-604; **27.14:** © Kingsley Stern.

## Chapter 28

**Opener:** Courtesy of University of Tennessee Parasitology Laboratory; **28.1 (Top row, zygote to embryo, all):** © Cabisco/Phototake; **28.1 (Adult frog; Bottom row, all):** © Dwight Kuhn; 28.6a: © Andrew J. Martinez/Photo Researchers, Inc.; **28.7a:** © Jeff Rotman; **28.7b:** © J. McCollugh/Visuals Unlimited; **28.8a:** © Azure Computer & Photo Services/Animals Animals/Earth Scenes; **28.8b:** © Ron Taylor/Bruce Coleman; **28.8c:** © Runk/Schoenberger/Grant Heilman Photography; **28.8d:** © Under Watercolours; **28.9:** © CABISCO/Visuals Unlimited; **(Lophophore, p. 520):** © Robert Brons/Biological Photo

Service; **28.10e:** © Tom E. Adams/Peter Arnold, Inc.; **28.11:** © SPL/Photo Researchers, Inc.; **28.14:** © Kjell Sandved/Butterfly Alphabet; **28.15a:** Courtesy Larry S. Roberts; **28.15b:** © Fred Whitehead/Animals/Animals/Earth Scenes; **28.16a:** © M. Gibbs/OSF/Animals Animals/Earth Scenes; 28.16b: © Kenneth W. Fink/Bruce Coleman, Inc.; **28.16c:** © Farley Bridges; **28.16d:** © Ken Lucas/Visuals Unlimited; **28.16e:** © Douglas Faulkner/Photo Researchers, Inc.; **28.16f:** © Georgette Douwma/Photo Researchers, Inc.; **28.17c:** © Roger K. Burnard/Biological Photo Service; **28.18b:** © James H. Carmichael; **28.18c:** © St. Bartholomew's Hospital/SPL/Photo Researchers, Inc.; 28.19a: © Lauritz Jensen/Visuals Unlimited; **28.19b:** © James Solliday/Biological Photo Service; **28.19c:** © Vanessa Vick/The New York Times/Redux; **28.20c:** © OSF/London Scientific Films/Animals Animals/Earth Scenes; **28.21a:** © Michael Lustbader/Photo Researchers, Inc.; **28.21b:** © Bruce Robinson/Corbis; **28.21c:** © Kim Taylor/Bruce Coleman, Inc.; **28.21d:** © Kjell Sandved/Butterfly Alphabet; **28.23a:** © Larry Miller/Photo Researchers, Inc.; **28.23b:** © David Aubrey/Corbis; **28.24 (Mealybug, leafhopper, dragonfly):** © Farley Bridges; **28.24 (Beetle):** © Wolfgang Kaehler/Corbis; **28.24 (Louse):** © Darlyne A. Murawski/Peter Arnold, Inc.; **28.24 (Wasp):** © Johnathan Smith; Cordaiy Photo Library/Corbis; **28.26a:** © Jana R. Jirak/Visuals Unlimited; 28.26b: © Tom McHugh/Photo Researchers, Inc.; **28.26c:** © Ken Lucas; **28.27b:** © Randy Morse, GoldenStateImages.com; **28.27c:** © Alex Kerstitch/Visuals Unlimited; **28.27d:** © Randy Morse/Animals Animals/Earth Scenes.

## Chapter 29

**Opener:** © 2003 Monty Sloan/Wolf Photographer; **29.1 (Bottom):** © Heather Angel/Natural Visions; **29.2:** © Rick Harbo; **29.5:** © Heather Angel; **29.6a:** © James Watt/Animals Animals/Earth Scenes; **29.6b:** © Fred Bavendam/Minden Pictures; **29.7a:** © Ron & Valerie Taylor/Bruce Coleman, Inc.; **29.7b:** © Hal Beral/Visuals Unlimited; **29.7c:** © Jane Burton/Bruce Coleman, Inc.; **29.7d:** © Claus Qvist Jessen; **29.7e:** © Franco Banfi/SeaPics.com; **29.8:** © Peter Scoones/SPL/Photo Researchers, Inc.; **29.11a:** © Suzanne L. Collins & Joseph T. Collins/Photo Researchers, Inc.; **29.11b:** © Joe McDonald/Visuals Unlimited; **29.11c:** © Juan Manuel Renjifo/Animals Animals/Earth Scenes; **29.12b:** © OS21/PhotoDisc; **29.14a:** © H. Hall/OSF/Animals Animals/Earth Scenes; **29.14b:** © Joe McDonald/Visuals Unlimited; **29.14c:** © Joel Sartorie/National Geographic/Getty Images; **29.14d:** © Nathan W. Cohen/Visuals Unlimited; **29.14e:** © Martin Harvey; Gallo Images/Corbis; **29Aa:** © MedioImages/SuperStock; **29Ab:** © Allan Friedlander/SuperStock; **29Ac:** © Account Phototake/Phototake; **29.15b (Both):** © Daniel J. Cox; **29.16a:** © Thomas Kitchin/Tom Stack & Associates; **29.16b:** © Joel McDonald/Corbis; **29.16c:** © Brian Parker/Tom Stack & Associates; **29.16d:** © IT Stock/PunchStock; **29.16e:** © Kirtley Perkins/Visuals Unlimited; **29.17a:** © D. Parer & E. Parer-Cook/Ardea; **29.17b:** © Fritz Prenzel/Animals Animals/Earth Scenes; **29.17c:** © Leonard Lee Rue/Photo Researchers, Inc.; **29.18a:** © Stephen J. Krasemann/Photo Researchers, Inc.; **29.18b:** © Stephen J. Krasemann/DRK Photo; **29.18c:** © Gerald Lacz/Animals Animals/Earth Scenes; **29.18d:** © Mike Bacon; **(Pangolin, p. 555):** © Nigel G. Dennis/Photo Researchers, Inc.

## Chapter 30

**Opener:** © Kazuhiko Sano; **30.1a (Lemur):** © Frans Lanting/Minden Pictures; **30.1a (Tarsier):** © Doug Wechsler; **30.1b (Monkey):** © C.C. Lockwood/DRK Photo; **30.1b (Baboon):** © St. Meyers/Okapia/Photo Researchers, Inc.; **30.1c (Orangutan):** © Tim Davis/Photo Researchers, Inc.; **30.1c (Gibbon):** © Hans & Judy Beste/Animals Animals/Earth Scenes; **30.1c (Chimpanzee, gorilla):** © Martin Harvey/Peter Arnold, Inc.; **30.1d (Humans):** © Comstock Images/JupiterImages; **30.5:** © National Museums of Kenya; **30.8a:** © Dan Dreyfus and Associates; **30.8b:** © John Reader/Photo Researchers, Inc.; **30A:** © Ryan McVay/Getty Images; **30.9:** © National Museums of Kenya; **30.11:** © The Field Museum #A102513c; **30.12:** Transparency #608 Courtesy Dept. of Library Services, American Museum of Natural History; **30.13a:** © PhotoDisc/Getty Images; **30.13b:** © Sylvia S. Mader; **30.13c:** © B & C Alexander/Photo Researchers, Inc.

## Chapter 31

**Opener:** © SPL/Photo Researchers, Inc.; **31.1 (All), 31.3a, b:** © Ed Reschke; **31.3c:** © McGraw-Hill Higher Education, Dennis Strete, photographer; **31.3d, e, 31.5a, c:** © Ed Reschke; **31.5b:** © McGraw-Hill Higher Education, Dennis Strete, photographer; **31.6b:** © Ed Reschke; **31A (Woman):** © Vol. 154/Corbis; **31A (Man):** © Vol. 12/Corbis; **31Ba:** © AP Photo/Elliot D. Novak; **31Bb:** © Diana De Rosa/Press Link; **31.8a:** © John D. Cunningham/Visuals Unlimited; **31.8b:** © Ken Greer/Visuals Unlimited; **31.8c:** © James Stevenson/SPL/Photo Researchers, Inc.

## Chapter 32

**Openers: (Normal):** © Yorgos Nikas/Getty Images; **(Leukemia):** © SPL/Photo Researchers, Inc.; **32.1a:** © CABISCO/Visuals Unlimited; **32.1b:** © B. Runk/S. Schoenberger/Grant Heilman Photography; **32.1c:** © Randy Morse, GoldenStateImages.com; **32.6b:** © SIU/Visuals Unlimited; **32.7b:** © Dr. Don W. Fawcett/Visuals Unlimited; **32.8d:** © Biophoto Associates/Photo Researchers, Inc.; **32.9d:** © David Joel/MacNeal Hospital/Getty Images; **32A:** © Biophoto Associates/Photo Researchers, Inc.; **32.14:** © Eye of Science/Photo Researchers, Inc.

## Chapter 33

**Openers:** © Nicholas Nixon; **33.2 (Marrow):** © R. Calentine/Visuals Unlimited; **33.2 (Thymus, spleen):** © Ed Reschke/Peter Arnold, Inc.; **33.2 (Lymph):** © Fred E. Hossler/Visuals Unlimited; **33.5:** © Dennis Kunkel/Phototake; **33.9:** © Michael Newman/PhotoEdit; **33.10:** © Digital Vision/Getty Images; **33Aa:** © AP/Wide World Photo; **33.14b:** © Steve Gschmeissner/Photo Researchers, Inc.; **(AIDS victim, p. 626):** © A. Ramey/PhotoEdit; **(Chickenpox virus, p. 626):** © George Musil/Visuals Unlimited; **(Pneumonia, p. 626):** © Dr. Dennis Kunkel/Visuals Unlimited; **(Candidiasis, p. 626):** © Everett S. Beneke/Visuals Unlimited; **33.16:** © Richard Anderson; **33.17:** © Dr. Ken Greer/Visuals Unlimited; **33.18:** © Damien Lovegrove/SPL/Photo Researchers, Inc.

## Chapter 34

**Opener:** © Arthur Morris/Visuals Unlimited; **34.8b:** © Ed Reschke/Peter Arnold, Inc.; **34.9(Villi):** © Manfred Kage/Peter Arnold, Inc.; **34.9(Microvilli):** Photo by Susumu Ito, from Charles Flickinger, *Medical Cellular Biology*, W.B. Saunders, 1979; **34.13:** © Amiard/Photocuisine/Corbis; **34.14:** © Benjamin F. Fink, Jr./Brand X/Corbis; **34.15:** Ryan McVay/Getty Images.

## Chapter 35

**Opener:** © Bruce Watkins/Animals Animals/Earth Scenes; **35.4a:** © B. Runk/S. Schoenberger/Grant Heilman Photography; **35.4b(Gills):** © David M. Phillips/Photo Researchers, Inc.; **35.5, 35.6b:** © Ed Reschke; **35.13:** © Andrew Syred/Photo Researchers, Inc.; **35.15a:** © Matt Meadows/Peter Arnold, Inc.; **35.15b:** © SIU/Visuals Unlimited; **35.15c:** © Biophoto Associates/Photo Researchers, Inc.

## Chapter 36

**Opener:** © Georgette Douwma/Photo Researchers, Inc.; **36.3:** © Digital Vision Ltd.; **36.5:** © Bob Calhoun/Bruce Coleman, Inc.; **36.6:** © Eric Hosking/Photo Researchers, Inc.; **36.7:** © James Cavallini/Photo Researchers, Inc.; **36.9b:** © R.G. Kessel and R H. Kardon, *Tissues and Organs: A Text-Atlas of Scanning Electron Microscopy*. W. H. Freeman & Co., San Francisco 1979; *Journal of Ultrastructure Research* by Maunsbach, Arvid B. Copyright 1966 by Elsevier Science & Technology Journals. Reproduced with permission of Elsevier Science & Technology Journals in the format Textbook via Copyright Clearance Center; **36.9c, d:** © 1966 Academic Press, from A.B. Maunsbach, *Journal of Ultrastructural Research* Vol. 15: 242-282; **36.10a:** © Joseph F. Gennaro, Jr./Photo Researchers, Inc.

## Chapter 37

**Opener:** © Ulrich Baumgarten/Vario Images; **37.4a:** © M.B. Bunge/Biological Photo Service; **37.4c:** © Manfred Kage/Peter Arnold, Inc.; **37.7a:** Courtesy Dr. E.R. Lewis, University of California, Berkeley; **37.12c:** © Karl E. Deckart/Phototake; **37A:** © Science VU/Visuals Unlimited.

## Chapter 38

**Opener (Both):** Courtesy of The Virginia Bloodhound Search and Rescue Association; **38.1b (All):** © Omikron/SPL/Photo Researchers, Inc.; **38.3:** © Farley Bridges; **38.4 (Both):** © Heather Angel/Natural Visions; **38.8:** © Lennart Nilsson, from *The Incredible Machine*; **38.9b:** © Biophoto Associates/Photo Researchers, Inc.; **38A (Both):** Courtesy Dr. Yeohash Raphael, University of Michigan, Ann Arbor; **38.11:** © P. Motta/SPL/Photo Researchers, Inc.

## Chapter 39

**Opener:** © 2008 The Associated Press, all rights reserved; **39.2:** © Michael Fogden/OSF/Animals Animals/Earth Scenes; **39.3:** © E. R. Degginger/Photo Researchers, Inc.; **39.4 (Osteocyte):** © Biophoto Associates/Photo Researchers, Inc.; **39.4 (Hyaline cartilage, compact bone):** © Ed Reschke; **39A (Aerobics):** © Jose Luis Pelaez, Inc./Corbis; **39Aa, b:** © Michael Klein/Peter Arnold, Inc.; **39Ac:** © Bill Aaron/PhotoEdit; 39.14(Gymnast): © Royalty-Free/Corbis; **39.14 (Myofibril):** © Biology Media/Photo Researchers, Inc.; **39.15:** © Victor B. Eichler, Ph.D.

## Chapter 40

**Openers (Caterpillar):** © Doug Wechsler/Animals Animals/Earth Scenes; **(Moth):** © Richard Kolar/Animals Animals/Earth Scenes; **40.7a:** © AP/Wide World Photos; **40.7b:** © Ewing Galloway, Inc.; **40.8:** From Clinical Pathological Conference, "Acromegaly, Diabetes, Hypermetabolism, Proteinura and Heart Failure," *American Journal of Medicine*, 20 (1956) 133. Reprinted with permission from Excerpta Medica Inc.; **40.9a:** © Bruce Coleman, Inc./Alamy; **40.9b:** © Medical-on-Line/Alamy; 40.9c: © Dr. P. Marazzi/Photo Researchers, Inc.; 40.13a: © Custom Medical Stock Photos; 40.13b: © NMSB/Custom Medical Stock Photos; **40.14 (Both):** *Atlas of Pediatric Physical Diagnosis*, Second Edition by Zitelli & Davis, 1992. Mosby-Wolfe Europe Limited, London, UK; **40.15:** © Peter Arnold, Inc./Alamy; **40A:** © Bettmann/Corbis; **40.17:** © The McGraw-Hill Companies, Inc./ Evelyn Jo Johnson, photographer.

## Chapter 41

**Opener:** © Rudy Kuiter/OSF/Animals Animals/Earth Scenes; **41.1:** © Dr. Dennis Kunkel/Visuals Unlimited; **41.2:** © Kelvin Aitken/Peter Arnold, Inc.; **41.3:** © Herbert Kehrer/zefa/Corbis; **41.4:** © Anthony Mercieca/Photo Researchers, Inc.; **41.7b:** © Ed Reschke; **41.10:** © Ed Reschke/Peter Arnold, Inc.; **41.14a:** © Saturn Stills/Photo Researchers, Inc.; **41.14b:** © Michael Keller/Corbis; **41.14c:** © LADA/Photo Researchers, Inc.; **41.14d:** © SIU/Visuals Unlimited; **41.14e:** © Keith Brofsky/Getty Images; **41.14f:** © The McGraw-Hill Companies, Inc./Lars A. Niki, photographer; **41.14g:** © Phanie/Photo Researchers, Inc.; **41.14h:** © Getty Images; **41.15:** © CC Studio/SPL/Photo Researchers, Inc.; **41Aa,** b: © Brand X/SuperStock RF; 41.16: © Scott Camazine/Photo Researchers, Inc. **41.17:** © CDC/Peter Arnold, Inc.; **41.18:** © G. W. Willis/Visuals Unlimited; **41.19:** © G.W. Willis/Visuals Unlimited; 41.20: © CNR/SPL/Photo Researchers, Inc.; **41.21a:** © Carroll Weiss/Camera M.D.; **41.21b:** © Centers for Disease Control and Prevention; **41.21c:** © Science VU/Visuals Unlimited; **41B (Left):** Corbis CD Vol. 161; **41B (Right):** Corbis Vol 178; **41C:** © Hartmut Schwarzbach/Peter Arnold, Inc.

## Chapter 42

**Openers (2 days):** Courtesy of the film *Building Babies* © ICAM/Mona Lisa; **(3 weeks):** © Lennart Nilsson, *A Child is Born*, 1990 Delacorte Press, pg. 81; **(6 weeks):** © Claude Edelmann/Photo Researchers, Inc.; **(5 months):** © Derek Bromhall/OSF/Animals Animals/Earth Scenes; **42.1:** © David M. Phillips/Visuals Unlimited; **(Chick, p. 779):** © Photodisc/Getty Images; **42.2a:** © William Jorgensen/Visuals Unlimited; **(Frog, p. 781):** © Photodisc/Getty Images; **42.4b:** Courtesy Kathryn Tosney; **42.9 (All):** Courtesy Steve Paddock, Howard Hughes Medical Research Institute; **42.10a:** Courtesy E.B. Lewis; **42.12 (Fertilization):** © Don W. Fawcett/Photo Researchers, Inc.; **42.12 (2-cell):** © Rawlins-CMSP/Getty Images; **42.12 (Morula):** © RBM Online/epa/Corbis; **42.12 (Implantation):** © Bettmann/Corbis; **42.14a:** © Lennart Nilsson, *A Child is Born*, Dell Publishing; **42.16c:** © Karen Kasmauski/Corbis.

## Chapter 43

**Opener:** © Jennifer Jarvis/Visuals Unlimited; **43.1a:** © Joe McDonald; **43.1b:** Courtesy Refuge for Saving the Wildlife, Inc.; **43.2 (Coastal):** © John Sullivan/Monica Rua/Ribbitt Photography; **43.2 (Inland):** © R. Andrew Odum/Peter Arnold, Inc.; **43.3b, c:** From J.R. Brown et al, "A defect in nurturing mice lacking . . . Gene for fosB" *Cell* v. 86, 1996 pp. 297-308, © Cell Press; **(Imprinting, p. 803):** © Nina Leen/Time Life Pictures/Getty Images; **43.3:** Courtesy Dr. Bernd Heinrich; **43A (Left):** © Alan Carey/Photo Researchers, Inc.; **43A (Right):** © Tom McHugh/Photo Researchers, Inc.; **43.7:** © Gregory G. Dimijian/Photo Researchers, Inc.; 43.8a(Main): © Arco Images/GmbH/Alamy; **43.8a (Inset):** © Fritz Polking/Visuals Unlimited; **43.9:** © Image Source/PunchStock; **43.10 (Firefly):** © Phil Degginger/Alamy; **43.10 (Trees):** © PhotoLink; 43.11: © OSF/Animals Animals/Earth Scenes; **43.12:** © Nicole Duplaix/Peter Arnold, Inc.; **43.14:** © Thomas Dobner 2006/Alamy; **43.15:** © D. Robert & Lorri Franz/Corbis; 43B: © T & P Gardner/Bruce Coleman, Inc.; 43C: Courtesy Gail Patricelli/University of Maryland; **43.16:** © Mark Moffett/Minden Pictures; **(Bat, p. 815):** © Michael Fogden/Animals Animals/Earth Scenes; **43.17:** © J & B Photo/Animals Animals/Earth Scenes.

## Chapter 44

**Opener:** © Vol. 44 PhotoDisc/Getty Images; **44.1:** © David Hall/Photo Researchers, Inc.; **44.2d:** © The McGraw Hill Companies, Inc./Evelyn Jo Johnson, photographer; **44.3a:** © age fotostock/SuperStock; **44.3b:** © Royalty-Free/Corbis; **44.4b:** © Holt Studios/Photo Researchers, Inc.; **44.4c:** © Bruce M. Johnson; **44.4d:** © Digital Vison/Getty RF Images; **44.6a:** © Breck P. Kent/Animals Animals/Earth Scenes; **44.6b:** © Doug Sokell/Visuals Unlimited; **44.9:** © Paul Janosi/Valan Photos; **44.13a, b:** © Michael Fogden/Animals Animals/Earth Scenes; **44.13c:** © Tom McHugh/Photo Researchers, Inc.; **44.13d:** © Matt Meadows/Peter Arnold, Inc.; **44.13e:** © Mike Linley/OSF/Animals Animals/Earth Scenes; **44.14 (Dandelions):** © Ted Levin/Animals Animals/Earth Scenes; **44.14(Bears):** © Michio Hoshino/Minden Pictures; **44Aa:** © Dominique Braud/Animals Animals/Earth Scenes; **44Ab:** © Stephen J. Krasemann/Photo Researchers, Inc.; 44Ac: © John Cancalosi/Peter Arnold, Inc.; **44.15b (Top):** © The McGraw-Hill Companies, Inc./Jill Braaten, photographer; **44.15b (Bottom):** © Robert Harding/Robert Harding World Imagery/Corbis; **44.16c:** © Still Pictures/Peter Arnold, Inc.; **44.17 (LDC):** © Earl & Nazima Kowall/Corbis; **44.17 (MDC):** © Comstock Images/Getty RF.

## Chapter 45

**Openers (Both):** © B. Runk/S. Schoenberger/Grant Heilman Photography; **45.1a (Forest):** © Charlie Ott/Photo Researchers, Inc.; **45.1a (Squirrel):** © Stephen Dalton/Photo Researchers, Inc.; **45.1a (Wolf):** © Renee Lynn/Photo Researchers, Inc.; **45.1b (Rain forest):** © Michael Graybill and Jan Hodder/Biological Photo Service; **45.1b (Kinkajou):** © Alan & Sandy Carey/Photo Researchers, Inc.; **45.1b (Sloth):** © Studio Carlo Dani/Animals Animals Earth Scenes; **45.7:** © Alan Carey/Photo Researchers, Inc.; **45.8a:** © Gustav Verderber/Visuals Unlimited; **45.8b:** © Zig Leszczynski/Animals Animals/Earth Scenes; **45.8c:** © National Audubon Society/A. Cosmos Blank/Photo Researchers, Inc.; **45.9a:** © Edward S. Ross; **45.9b:** © Edward S. Ross; **45.9c:** © James H. Robinson/Photo Researchers, Inc.; **45.9d:** © Edward S. Ross; **45.10:** Courtesy the University of Tennessee Parasitology Laboratory; **45.11:** © Dave B. Fleetham/Visuals Unlimited; **45Aa:** Courtesy Dr. Ian Wyllie; **45.12:** © C. C. Lockwood/Animals Animals; **45.13:** © Bill Wood/Bruce Coleman, Inc.; **45.14 (All):** © Breck P. Kent/Animals Animals/Earth Scenes; **45.16a(Left):** © Ed Reschke/Peter Arnold, Inc.; **45.16a(Right):** © Herman Eisenbeiss/Photo Researchers, Inc.; **45.16b (Left):** © Royalty-free/Corbis; **45.16b (Right):** © Gerald C. Kelley/Photo Researchers, Inc.; **45.16c (Left):** © Bill Beatty/Visuals Unlimited; **45.16c (Right):** © Joe McDonald/Visuals Unlimited; **45.16d (Left):** © SciMAT/Photo Researchers, Inc.; **45.16d (Right):** © Michael Beug; **45.18:** © George D. Lepp/Photo Researchers, Inc.; **45Ca:** Courtesy NASA; **45Cb:** © PhotoDisc Vol. 9/Getty Images.

## Chapter 46

**Openers (Earth):** Courtesy NASA; **(Lake):** © Stephen J. Krasemann/Photo Researchers, Inc.; **(Aspens):** © Dwight Kuhn; **(Sunflowers):** © Jim Brandenburg/Minden Pictures; **(Acacia tree):** © Konrad Wothe/Minden Pictures; **(Rain forest):** © age fotostock/SuperStock; **(Lynx):** © Imagebroker/Alamy; **(Salamander):** © David M. Dennis/Animals Animals/Earth Scenes; **(Buffalo):** © Eastcott Momatiuk/Getty Images; **(Lion):** © age fotostock/SuperStock; **(Toucan):** © age fotostock/SuperStock; **46.6a:** © John Shaw/Tom Stack & Associates; **46.6b:** © John Eastcott/Animals Animals/Earth Scenes; **46.6c:** © John Shaw/Bruce Coleman, Inc.; **46.7 (Taiga):** © Mack Henly/Visuals Unlimited; **46.7 (Moose):** © MaryEllen Silliker/Animals Animals/Earth Scenes; **46.8 (Forest):** © E. R. Degginger/Animals Animals/Earth Scenes; **46.8 (Chipmunk):** © Carmela Lesczynski/Animals Animals/Earth Scenes; **46.8 (Millipede):** © OSF/Animals Animals/Earth Scenes; **46.8 (Bobcat):** © Tom McHugh/Photo Researchers, Inc.; **46.8 (Marigolds):** © Virginia Neefus/Animals Animals/Earth Scenes; **46Aa:** © Porterfield/Chickering/Photo Researchers, Inc.; **46Ab:** © Michio Hoshino/Minden Pictures; **46.10 (Lizard):** © Kjell Sandved/Butterfly Alphabet; **46.10 (Katydid):** © M. Fogden/OSF/Animals Animals/Earth Scenes; **46.10 (Butterfly):** © Kjell Sandved/Butterfly Alphabet; **46.10 (Ocelot):** © Martin Wendler/Peter Arnold, Inc.; **46.10 (Lemur):** © Erwin & Peggy Bauer/Bruce Coleman, Inc.; **46.10 (Macaw):** © Tony Craddock/SPL/Photo Researchers, Inc.; **46.10 (Frog):** © National Geographic/Getty Images; **46.11 (Chaparall):** © Bruce Iverson; **46.11 (Jay):** © H.P. Smith, Jr./VIREO; **46.11 (Chemise):** © Kathy Merrifield/Photo Researchers, Inc.; **46.12 (Prairie):** © Jim Steinberg/Photo Researchers, Inc.; **46.12 (Bison):** © Steven Fuller/Animals Animals/Earth Scenes; **46.13 (Zebra, wildebeest):** © Darla G. Cox; **46.13 (Giraffe):** © George W. Cox; **46.13 (Cheetah):** © Digital Vision/Getty Images; **46.14 (Rat):** © Bob Calhoun/Bruce Coleman, Inc.; **46.14 (Roadrunner):** © Jack Wilburn/Animals Animals/Earth Scenes; **46.14 (Desert):** © John Shaw/Bruce Coleman; **46.14 (Kit fox):** © Jeri Gleiter/Peter Arnold, Inc.; **46.15 (Stonefly):** © Kim Taylor/Bruce Coleman, Inc.; **46.15 (Trout):** © William H. Mullins/Photo Researchers, Inc.; **46.15 (Carp):** © Robert Maier/Animals Animals/Earth Scenes; **46.15 (Crab):** © Gerlach Nature Photography/Animals Animals/Earth Scenes; 46.16a: © Roger Evans/Photo Researchers, Inc.; 46.16b: © Michael Gadomski/Animals Animals/Earth Scenes; **46.18 (Pike):** © Robert Maier/Animals Animals/Earth Scenes; **46.18 (Water strider):** © G.I. Bernard/Animals Animals/Earth Scenes; **46.19a:** © John Eastcott/Yva Momatiuk/Animals Animals/Earth Scenes; **46.19b:** © Theo Allofs/Visuals Unlimited; **46.19c:** © Brandon Cole/Visuals Unlimited.

## Chapter 47

**Opener:** © Allen Blake Sheldon/Animals Animals/Earth Scenes; **47.3 (Periwinkle):** © Kevin Schaefer/Peter Arnold, Inc.; **47.3 (Armadillo):** © John Cancalosi/Peter Arnold, Inc.; **47.3 (Fishermen):** © Herve Donnezan/Photo Researchers, Inc.; **47.3 (Rubber harvest):** © Bryn Campbell/Stone/Getty; **47.3 (Bat):** © Merlin D. Tuttle/Bat Conservation International; **47.3 (Ladybug):** © Anthony Mercieca/Photo Researchers, Inc.; **47.4a:** © William Smithy, Jr.; **47.4b:** © Don and Pat Valenti/DRK; **47.5b:** © Gunter Ziesler/Peter Arnold, Inc.; **47.5c:** Courtesy Woods Hole Research Center; **47.5d:** Courtesy R.O. Bierregaard; **47.5e:** Courtesy Thomas Stone, Woods Hole Research Center; **47.6a:** © Chuck Pratt/Bruce Coleman, Inc.; **47.6b:** © Chris Johns/National Geographic Image Collection; **47.7b:** Courtesy Walter C. Jaap/Florida Fish & Wildlife Conservation Commission; **47A, 47B:** Courtesy Peter Paul van Dijk; **47.8a:** © Shane Moore/Animals Animals/Earth Scenes; 47.8b: © Peter Auster/University of Connecticut; **47.9a:** © Gerard Lacz/Peter Arnold, Inc.; **47.9b (Forest):** © Art Wolfe/Artwolfe.com; **47.9b (Owl):** © Pat & Tom Leeson/Photo Researchers, Inc.; **47.10b:** © Jeff Foott Productions; **47.11 (Panther):** © Tom & Pat Leeson/Photo Researchers, Inc.; **47.11 (Alligator):** © Fritz Polking/Visuals Unlimited; **47.11 (Ibis):** © Stephen G. Maka; **47.11 (Spoonbill):** © Kim Heacox/Peter Arnold, Inc.; **47.11 (Stork):** © Millard H. Sharp/Photo Researchers, Inc.

# LINE ART & TEXT

## Chapter 2

**Ecology Focus, p. 33:** Data from G. Tyler Miller, *Living in the Environment,* 1983, Wadsworth Publishing Company, Belmont, CA; and Lester R. Brown, *State of the World,* 1992, W.W. Norton & Company, Inc., New York, NY. p. 33.

## Chapter 24

**Ecology Focus, p. 443:** Courtesy of Charles Horn.

## Chapter 26

**Science Focus, p. 478:** From Joe Bower, *National Wildlife Magazine,* June/July 2000, Vol. 38, No. 4, Reprinted with permission of the author.

## Chapter 32

**Health Focus, p. 614:** Courtesy of Steven Stanley, The John Hopkins University.

## Chapter 41

**41.17:** Data from Division of STD Prevention, Sexually Transmitted Disease Surveillance, 2006. U.S. Department of Health and Human Services, Public Health Service, Atlanta: Centers for Disease Control and Prevention; **41.18:** Data from Division of STD Prevention, Sexually Transmitted Disease Surveillance, 2006. U.S. Department of Health and Human Services, Public Health Service, Atlanta: Centers for Disease Control of Prevention; **41.19:** Data from Division of STD Prevention, Sexually Transmitted Disease Surveillance, 2006. U.S. Department of Health and Human Services, Public Health Service, Atlanta: Centers for Disease Control and Prevention; **41.20:** Data from Division of STD Prevention, Sexually Transmitted Disease Surveillance, 2006. U.S. Department of Health and Human Services, Public Health Service, Atlanta: Centers for Disease Control and Prevention.

## Chapter 43

**43.2b:** Data from S.J. Arnold, "The Microevolution of Feeding Behavior" in *Foraging Behavior: Ecology, Ethological, and Psychology Approaches,* edited A. Kamil and T. Sargent, 1980, Garland Publishing Company, New York, NY; **Science Focus, p. 812:** Courtesy of Gail Patricelli, University of Maryland.

## Chapter 44

**44.4b:** Data from W.K. Purves, et al., Life: *The Science of Biology,* 4/e, Sinaeur & Associates; **44.4d:** Data from A.K. Hegazy, 1990, "Population Ecology & Implications for Conservation of Cleome Droserifolia: A Threatened Xerophyte, "*Journal of Arid Environments,* 19:269-82; **44.8:** From Raymond Pearl, *The Biology of Population Growth.* Copyright 1925 The McGraw-Hill Companies. All Rights Reserved; **44.9b:** Data from Charles J. Krebs, *Ecology,* 3/e, 1984, Harper & Row; after Scheffer, 1951.

## Chapter 45

**45.1c:** Data from G.G.Simpson, "Species Density of North America Recent Mammals" in *Systemic Zoology,* Vol. 13:57-73, 1964; **45.3:** Data from G.F. Gause, *The Struggle for Existance,* 1934, Williams & Wilkins Company, Baltimore, MD.

## Chapter 47

**47.2:** Redrawn from "Shrimp Stocking, Salmon Collapse, and Eagle Displacement" by C.N. Spencer, B.R. McClelland and J.A. Stanford, Bioscience, 41(1):14-21. Copyright © 1991 American Institute of Biological Sciences; Ecology Focus, p. 873: Courtesy of Stephanie Songer, North Georgia College and State University; **47.7:** Data from David M. Gates, Climate Change and Its Biological Consequences, 1993, Sinauer & Associates, Inc., Sunderland, MA.

# INDEX

Note: Page numbers followed by f and t indicate figures and tables, respectively. Footnotes are indicated by *n*.

## A

# B

# D

# H

# I

# J



# K

# M

# N

# O

# P

# T

# U



# V

# Y



# Z

# ory of Biology

| Year | Name | Country | Contribution |
|---|---|---|---|
| 1628 | William Harvey | Britain | Demonstrates that the blood circulates and the heart is a pump. |
| 1665 | Robert Hooke | Britain | Uses the word *cell* to describe compartments he sees in cork under the microscope. |
| 1668 | Francesco Redi | Italy | Shows that decaying meat protected from flies does not spontaneously produce maggots. |
| 1673 | Antonie van Leeuwenhoek | Holland | Uses microscope to view living microorganisms. |
| 1735 | Carolus Linnaeus | Sweden | Initiates the binomial system of naming organisms. |
| 1809 | Jean B. Lamarck | France | Supports the idea of evolution but thinks there is inheritance of acquired characteristics. |
| 1825 | Georges Cuvier | France | Founds the science of paleontology and shows that fossils are related to living forms. |
| 1828 | Karl E. von Baer | Germany | Establishes the germ layer theory of development. |
| 1838 | Matthias Schleiden | Germany | States that plants are multicellular organisms. |
| 1839 | Theodor Schwann | Germany | States that animals are multicellular organisms. |
| 1851 | Claude Bernard | France | Concludes that a relatively constant internal environment allows organisms to survive under varying conditions. |
| 1858 | Rudolf Virchow | Germany | States that cells come only from preexisting cells. |
| 1858 | Charles Darwin | Britain | Presents evidence that natural selection guides the evolutionary process. |
| 1858 | Alfred R. Wallace | Britain | Independently comes to same conclusions as Darwin. |
| 1865 | Louis Pasteur | France | Disproves the theory of spontaneous generation for bacteria; shows that infections are caused by bacteria, and develops vaccines against rabies and anthrax. |
| 1866 | Gregor Mendel | Austria | Proposes basic laws of genetics based on his experiments with garden peas. |
| 1882 | Robert Koch | Germany | Establishes the germ theory of disease and develops many techniques used in bacteriology. |
| 1900 | Walter Reed | United States | Discovers that the yellow fever virus is transmitted by a mosquito. |
| 1902 | Walter S. Sutton<br>Theodor Boveri | United States<br>Germany | Suggest that genes are on the chromosomes, after noting the similar behavior of genes and chromosomes. |
| 1903 | Karl Landsteiner | Austria | Discovers ABO blood types. |
| 1904 | Ivan Pavlov | Russia | Shows that conditioned reflexes affect behavior, based on experiments with dogs. |
| 1910 | Thomas H. Morgan | United States | States that each gene has a locus on a particular chromosome, based on experiments with *Drosophila.* |
| 1922 | Sir Frederick Banting<br>Charles Best | Canada | Isolate insulin from the pancreas. |
| 1924 | Hans Spemann<br>Hilde Mangold | Germany | Show that induction occurs during development, based on experiments with frog embryos. |
| 1927 | Hermann J. Muller | United States | Proves that X rays cause mutations. |
| 1929 | Sir Alexander Fleming | Britain | Discovers the toxic effect of a mold product he called penicillin on certain bacteria. |

*Antonie van Leeuwenhoek*

*Charles Darwin*

*Louis Pasteur*

*Robert Koch*

*Ivan Pavlov*